INFINITE SERIES

1. $\dfrac{1}{1-x} = 1 + x + x^2 + \cdots + x^n + \ldots, \quad -1 < x < 1$

2. $e^x = 1 + x + \dfrac{x^2}{2!} + \cdots + \dfrac{x^n}{n!} + \ldots, \quad -\infty < x < \infty$

3. $\sin x = x - \dfrac{x^3}{3!} + \dfrac{x^5}{5!} - \cdots + \dfrac{(-1)^n x^{2n+1}}{(2n+1)!} + \ldots, \quad -\infty < x < \infty$

4. $\cos x = 1 - \dfrac{x^2}{2!} + \dfrac{x^4}{4!} - \cdots + \dfrac{(-1)^n x^{2n}}{(2n)!} + \ldots, \quad -\infty < x < \infty$

5. $\ln(1 + x) = x - \dfrac{x^2}{2} + \dfrac{x^3}{3} - \cdots + \dfrac{(-1)^{n+1} x^n}{n} + \ldots, \quad -1 < x \le 1$

6. $\arctan x = x - \dfrac{x^3}{3} + \dfrac{x^5}{5} - \cdots + \dfrac{(-1)^n x^{2n+1}}{2n+1} + \ldots, \quad -1 \le x \le 1$

7. $(1 + x)^r = 1 + rx + \dfrac{r(r-1)}{2!}x^2 + \cdots + \dfrac{r(r-1)\cdots(r-n+1)}{n!}x^n + \ldots, \quad -1 < x < 1$

8. $\cosh x = 1 + \dfrac{x^2}{2!} + \dfrac{x^4}{4!} + \cdots + \dfrac{x^{2n}}{(2n)!} + \ldots, \quad -\infty < x < \infty$

9. $\sinh x = x + \dfrac{x^3}{3!} + \dfrac{x^5}{5!} + \cdots + \dfrac{x^{2n+1}}{(2n+1)!} + \ldots, \quad -\infty < x < \infty$

10. $\arcsin x = x + \dfrac{1}{2}\dfrac{x^3}{3} + \dfrac{1 \cdot 3}{2 \cdot 4}\dfrac{x^5}{5} + \cdots + \dfrac{1 \cdot 3 \cdots (2n-1)}{2 \cdot 4 \cdots (2n)}\dfrac{x^{2n+1}}{2n+1} + \ldots, \quad -1 < x < 1$

HYPERBOLIC FUNCTIONS

1. $\cosh^2 x - \sinh^2 x = 1$
2. $\cosh(-x) = \cosh x$
3. $\sinh(-x) = -\sinh x$
4. $\sinh(x \pm y) = \sinh x \cosh y \pm \cosh x \sinh y$
5. $\cosh(x \pm y) = \cosh x \cosh y \pm \sinh x \sinh y$

Calculus

Calculus

William E. Boyce
Richard C. DiPrima

Rensselaer Polytechnic Institute

JOHN WILEY & SONS
New York Chichester Brisbane
Toronto Singapore

Library of Congress Cataloging in Publication Data:

Boyce, William E.
Calculus.

Bibliography: p.
Includes index.

1. Calculus. I. DiPrima, Richard C. II. Title.
QA303.B8818 1988 515 87-27943
ISBN 0-471-09333-5

Printed in the United States of America

10 9 8 7 6 5 4 3 2 1

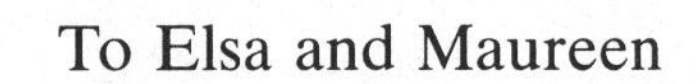

To Elsa and Maureen

preface

Our main goal is to provide a clear and coherent presentation of the topics normally taught in introductory calculus courses. This means an emphasis on the results and methods that make calculus an important problem-solving tool in other fields of inquiry, with a generous sprinkling of applications to illustrate the power and versatility of these methods, and with some attention paid to the logical structure of the subject.

Throughout the book we have concentrated on results and methods that are useful, or at least potentially useful, to those who employ calculus to solve problems in other fields. Accordingly, we have pursued each major topic far enough to reach such results. We have attempted to frame definitions and theorems carefully in what we think is their most useful, although perhaps not their most general, form. We have included numerous examples so that students can see the methods in action.

To become an effective user of calculus we believe that a student must acquire not merely a collection of formulas and rules, but some understanding and appreciation of the way that they fit together, and some capacity to distinguish fairly routine problems from those that involve delicate questions of analysis. Although this is not a particularly theoretical book (there are relatively few deltas and epsilons), we have tried to be honest in pointing out where subtle arguments are needed, and why apparently plausible simple ones may fail.

We have also been quite selective in the proofs presented in the text, including only those that seem instructive. Consequently, we have omitted some proofs because they are too difficult for most students at this level, and others because they are too tedious and repetitive. A good many other proofs are outlined in the problem sets, where instructors can make use of them if they wish. Along with proofs go counterexamples. Quite a few simple ones are included in the text at appropriate points to indicate that the hypotheses listed in certain theorems are indeed necessary to obtain the conclusion. A few more complicated counterexamples also appear in the problem sets.

The view is sometimes advanced that the widespread availability of cheap and efficient computing power reduces the need for present-day students to study calculus. We believe that, if anything, the opposite is the case. In the natural world, wherever variable quantities change in relation to one another, or wherever something can be considered as the result of accumulating a large number of small constituents, there one finds calculus. The fact that most students have enormous computational resources within reach, and perhaps actually on their desks, expands tremendously the range of problems to which they can apply successfully the ideas and concepts of calculus, among others. This makes more urgent the need for students to master these ideas and concepts, and to study their translation into numerical algorithms.

Since the use of calculus for problem-solving often leads eventually to a need for numerical results, it follows that calculus should be taught with the possibility of numerical computation in mind. We recognize this possibility explicitly by the presentation of certain numerical algorithms such as Newton's method, the trapezoidal and Simpson's rules for integration, and the Runge-Kutta method for solving first order initial value problems. Perhaps more important is the fact that the possible need for numerical computation is a thread running through the entire book. It manifests itself in various ways, for instance, in the emphasis given to the estimation of errors or remainders, in the discussion of the speed of convergence of certain infinite series, and simply by asking for numerical answers in some problems and giving them in some of the examples.

A textbook is not a syllabus or lesson plan that must be followed in all details; rather, it is a framework on which an instructor can hang the fruits of his or her own mathematical experience. In this book there are some chapters that we believe should be covered in their entirety, such as Chapter 3 (The Derivative) and Chapter 6 (The Integral). Many other chapters are organized so that the most important topics appear early in the chapter, while less central ones, sometimes including applications, are placed near the end. Thus an instructor has considerable freedom of choice in deciding how far to pursue various topics. There is also some flexibility possible in rearranging the order of some of the chapters.

With respect to applications, this book almost certainly contains more material than can be used in any one course. On the other hand, we have made no effort to be comprehensive in this regard (a hopeless objective, in any event). We have tried to provide a reasonable breadth of applications, so as to make credible the position that calculus is widely applicable and useful in many fields. Applications to mechanics and physics are well-represented, as a result both of our own interests and the long history of interconnected development of these subjects and calculus. Nevertheless, we have also included some discussion of applications outside of the physical sciences. Instructors who find their favorite applications missing are invited to interpolate them at appropriate times in their courses.

Following each section is a set of problems, including a number of fairly straightforward exercises arranged roughly in order of increasing difficulty, followed in most cases by an assortment of other problems. Answers to odd-numbered problems are given at the end of the book, and full solutions to these problems may be found in a separate *Solutions Manual* for students. There is also an *Instructor's Manual* that contains answers and solutions for all problems.

Some problems are labeled with an asterisk, signifying that they are considered to be more difficult, often because they are somewhat more theoretical than most. Other problems, indicated by the symbol ©, are of a computational nature. For most of these a calculator is required, and for some a microcomputer is desirable.

Since we intend this to be an essentially self-contained textbook, we have not prepared a bibliography in the usual sense; however, there are a few specific citations in footnotes. There is also a brief list of references that is included primarily as a guide to students who may wish to read further on subjects related to calculus.

We have inserted historical footnotes and remarks at various points in the book, and have listed some historical works in the references. Our objective is not only to give due credit to the originators of the subject but also to make clear to students that calculus, as we now know it, developed over a period of many years, and resulted only from protracted intellectual struggle by the world's foremost mathematicians with profound and elusive ideas.

Components of the book are numbered in decimal fashion. Thus Section 6.3 is the third section in Chapter 6. Figure 6.3.2 is the second figure in Section 6.3, and similarly for theorems and definitions. The symbols □ and ∎ are used to signify the end of a proof and of an example, respectively.

* * * * *

The manuscript of this book had been nearly completed at the time of Dick DiPrima's death in 1984. Sadly, he was not able to participate in the process of final editing and publication. Nevertheless, during this period, his concern for clarity and thoroughness was constantly before me. As it now appears in print, I trust that he would be pleased with the book on which we worked together for so long.

William E. Boyce
Troy, New York

acknowledgments

As we wrote this book we received a great deal of valuable assistance from many different individuals. In a general sense, we owe much to our colleagues, our students, our own teachers, and other authors, all of whom have influenced our views on calculus and how it may be presented. However, here we would like to identify and to thank those people who have been of specific assistance in the preparation of the book.

Among those who read the manuscript, or parts of it, prior to publication are

RODNEY CARR *University of Michigan*
PETER COLWELL *Iowa State University*
PAUL DAVIS *Worcester Polytechnic Institute*
CHARLES G. DENLINGER *Millersville State College*
BRUCE EDWARDS *University of Florida*
GARRET J. ETGEN *University of Houston*
STUART GOLDENBERG *California Polytechnic State University*
RONALD GRIMMER *Southern Illinois University*
DOUGLAS W. HALL *Michigan State University*
HAL G. MOORE *Brigham Young University*
IVAN NIVEN *University of Oregon*
ALFRED SCHMIDT *Rose Hulman Institute of Technology*
DONALD SHERBERT *University of Illinois*

We express our deep appreciation to each of these individuals for their time and effort. Their suggestions were always carefully considered, and often adopted.

To David Moskovitz, professor emeritus at Carnegie Mellon University, we owe a multiple debt. For one thing, as students in his classes, each of us learned a good deal of mathematics. We also learned something about teaching from the example he provided of a master teacher at work in the classroom. With respect to this book, he read the entire manuscript, a large part of it twice, with meticulous care; his detailed comments were extremely valuable to us in creating the finished product.

The *Solutions Manuals,* one for students and one for instructors, that accompany the book were prepared by Charles W. Haines and Thomas C. Upson of Rochester Institute of Technology. Our special thanks go to them not only for the huge task of compiling, writing, and editing the *Manuals* themselves, but also for reading a near final draft of the manuscript, checking the answer to each individual problem, and bringing many corrections to our attention.

We also thank Keith Howell and Julie Hummon for helping to prepare the sets of Review Problems at the end of each chapter, and Caroline Haddad not only for helping in that task, but also for assistance in correcting the galley proofs.

By far the greater part of the manuscript was typed, with her usual skill and efficiency, by Helen D. Hayes. Others who typed smaller portions were Ava M. Biffer and Victoria R. Lee. To each of them we express our appreciation for a job well and cheerfully done.

Finally, and most important of all, I am grateful to my wife Elsa K. Boyce for checking the galley and page proofs in their entirety, for acting as a general consultant on matters both of analysis and exposition, and above all else, for her unfailing support and encouragement during the years that this project was in progress.

W. E. Boyce

to the student

This book is intended, normally in conjunction with classroom instruction, to guide you in the study of the differential and integral calculus. Although calculus has many ramifications and generalizations that one can pursue far into the realms of advanced mathematics, experience shows that a reasonably well-prepared student can aspire to attain a good working knowledge of calculus within one or two years of study. We hope that this book helps you to do that.

Learning mathematics from a textbook involves reading it in a very active way. Although we have usually tried to indicate all major steps, you should keep paper and pencil at hand, and verify the calculations as you go along if necessary. Definitions and theorems should be examined with special care. Some of them are the result of years of refinement by mathematical scholars of the past, and we have given considerable thought to the wording of each one. Try to make certain that you understand exactly what is being said. Be careful also to think about what is not said; sometimes it is very easy to become careless and to read too much into the statement of a definition or theorem.

Although this book is primarily concerned with the methods and applications of calculus, rather than the theory, we have included the proofs of some theorems in the text and outlined others in the problem sets. We believe that proofs are worth studying, not only to convince yourself of the truth of statements that may not be intuitively obvious, but (perhaps more important) because they provide examples of mathematical thinking, and often contain methods that may be transferable to the solution of other kinds of problems.

Most important of all, if you wish to become proficient with calculus, you should work problems, not just a few problems now and then, but many problems, and on a regular basis. To make it your own, you must use a mathematical idea or method enough so that you become entirely comfortable with it and confident that you know how to use it correctly. It often takes a certain amount of time to attain this level of confidence, but it is usually achieved only by working out problems of various sorts that exploit the idea or method in different ways. Although answers to odd-numbered problems are given at the end of the book, you should refer to them sparingly, and never until you have at least made a serious attempt to solve a problem. Full solutions to these problems are also available in a separate *Solutions Manual* for students.

While working on a problem, you should be thinking about the results you are getting. Ask yourself often, "Does this result seem reasonable?" If so, then it may be right; if not, it is almost certainly wrong. Check your answers whenever you can, possibly by working them out a second time in a different way, or by inserting numerical values to verify a special case.

Those of us who love mathematics enough to devote our time to teaching or writing about it derive part of our pleasure from the artistic side of the subject. A clean and incisive demonstration of a possibly unexpected theorem, or an elegant solution to a difficult problem, is a thing of beauty. We hope that this aspect of mathematics is visible in the pages of this book, and that as a result your own aesthetic appreciation of mathematics will be strengthened.

Finally, although different individuals may have different levels of talent for mathematics (as for other things, such as music or athletics), you can raise your level of performance by hard work and diligent practice. In most cases, you can raise it a good deal, possibly much higher than you might think at first. So, set your goals high, and then give your best efforts to reaching them.

W. E. B.

contents

ONE

functions

Before embarking on the study of calculus proper it is essential to have certain basic concepts firmly in mind. Chief among these is the idea of a function. By now you have encountered numerous examples of elementary functions; for instance, functions such as

$$y = x^2 - 1,$$

or

$$x = \sqrt{2t + 1},$$

as well as many others. It was with expressions such as these in mind that mathematicians first began to use the term "function" in a technical sense. However, the concept of a function is much deeper and more widely inclusive than these simple examples suggest, and the mathematical world came gradually to this realization during the eighteenth and nineteenth centuries. In one way or another functions underlie all aspects of calculus, so this chapter focusses on providing a satisfactory understanding of this crucial idea.

Even though you may already be familiar with much of the material in this chapter, you should look through it with some care, paying particular attention to the definition and discussion of functions in Section 1.4.

1.1 NUMBERS, SETS, INTERVALS, AND ABSOLUTE VALUES

A natural place to begin the study of calculus is with a review of some of the properties of the real number system. Since we assume that you have had long experience with the real numbers, our main purpose here is to review some of the terminology associated with numbers and sets of numbers that will appear throughout the book.

Real numbers

In the elementary grades it is customary to start with the **positive integers** 1, 2, 3, . . . and the operations of addition and multiplication. No further numbers are obtained by applying these operations to the positive integers. The introduction of subtraction, the inverse of addition, leads to **zero** and the **negative integers** -1, -2, -3, ... Further, the introduction of division (except by zero), the inverse of multiplication, leads to the positive and negative fractions, such as $\frac{1}{2}$, $-\frac{4}{5}$, $\frac{7}{3}$, ... Taken all together these numbers form the **rational numbers,** or numbers that can be written as ratios of two integers.

However, the rational numbers alone do not allow us to find the length of the diagonal of a square of unit side, or to solve simple polynomial equations such as $x^2 - 3 = 0$, or to find the circumference of a circle of unit diameter. For such tasks we need the **irrational numbers,** such as $\sqrt{2}$, $\sqrt{3}$, π, . . . , which cannot be expressed as the ratio of two integers. The rational numbers together with the irrational numbers comprise the **real numbers.** The set of all real numbers will be denoted by R^1.

Geometric representation of real numbers

A fundamental concept in mathematics, one that relates algebra to geometry, is the geometric representation of real numbers as points on a straight line. This is done by choosing a point on the line to represent 0 and another point to represent 1. This determines a scale in terms of which the remaining real numbers can be marked off as points on the line. The positive real numbers are conventionally put into correspondence with points to the right of 0 and the negative real numbers with the points to the left of 0, as shown in Figure 1.1.1. Each real number is uniquely associated with a single point on the line, and vice versa. Because of this one-to-one correspondence between the real numbers and points on a straight line we often use the words "real number" and "point" interchangeably, and the line is referred to as the real number line or the real axis.

Sets

It will be convenient from time to time to use some of the terminology associated with **sets** of mathematical objects, especially sets of real numbers having some

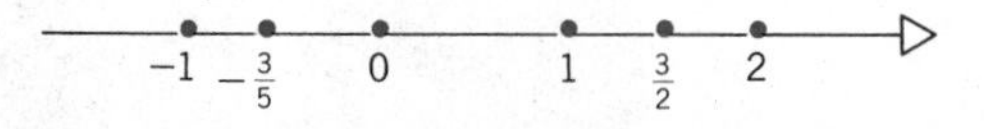

Figure 1.1.1

specified property. For example, we might refer to the set of even positive integers, or to the set of rational numbers. The most important thing to remember about a set in mathematics is that there must be a rule that tells us unambiguously whether any given object is a member of the set or not. For instance, we cannot talk about the set of large positive real numbers. This is not a set in the mathematical sense because we have specified no rule for determining whether any particular positive real number is a member of the set or not. On the other hand, the set of real numbers greater than ten million is a well-defined set.

We will usually use capital letters $A, B, C, \ldots$ to denote sets and lowercase letters $a, b, c, \ldots$ to denote the members or elements of a set. The notation $x \in S$ indicates that x is a member of S; if x is not in S, then we write $x \notin S$. A standard notation for describing a set is to give a generic element of the set, followed by a bar, and then by the rule that determines membership in the set, with the entire expression enclosed in braces. For example, we can identify the set A of all negative real numbers by writing

$$A = \{x | x < 0\}.$$

Alternatively, if there are only a few members in a set, we may simply list them. For example,

$$B = \{2, 4, 6, 8, 10\}.$$

Occasionally, we may use a similar notation even though the set has many elements, as in the case

$$C = \{1, \tfrac{1}{2}, \tfrac{1}{3}, \tfrac{1}{4}, \ldots\}.$$

Here we assume that the rule governing membership in the set is made obvious by listing a few members; the three dots (ellipsis) indicate that the process is to continue indefinitely.

We recall the following important relations and concepts associated with sets.

1. Two sets A and B are **equal,** written $A = B$, if and only if A and B have exactly the same elements.
2. A is contained in B, or A is a **subset** of B, denoted by $A \subset B$, if each element of A is also in B. If $A \subset B$ and $B \subset A$, then $A = B$.
3. The **empty** set, the set containing no elements, is denoted by $\emptyset$.
4. The **union** of A and B, denoted by $A \cup B$, is the set of elements that are either in A or in B, or in both. Thus

$$A \cup B = \{x | x \in A \quad \text{or} \quad x \in B \quad \text{(or both)}\}. \tag{1}$$

5. The **intersection** of A and B, denoted by $A \cap B$, is the set of elements that belong both to A and to B. Thus

$$A \cap B = \{x | x \in A \quad \text{and} \quad x \in B\}. \tag{2}$$

6. The **complement** of A relative to S, denoted by $S - A$, is the set of elements that are in S but not in A. Thus

$$S - A = \{x | x \in S \quad \text{and} \quad x \notin A\}. \tag{3}$$

If S is a nonempty set of real numbers, and if there is a number M such that $x \leq M$ for every $x \in S$, then S is said to be **bounded above.** Similarly, if there is a number m such that $m \leq x$ for every $x \in S$, then S is **bounded below.** The numbers M and m are called **upper** and **lower bounds,** respectively, for S. A set S that is bounded both above and below is said simply to be **bounded;** otherwise, it is called **unbounded.** For example, the set A of positive even integers is bounded below (by zero) but not above; thus A is an unbounded set. On the other hand, the set

$$B = \left\{ x | x = \frac{1}{n} \quad \text{for} \quad n = 1, 2, 3, \ldots \right\}$$

is bounded both above and below (by 1 and 0, respectively) and so is a bounded set.

Intervals

We shall often encounter the special sets of real numbers known as **intervals.** For a given pair of real numbers a and b with $a < b$, the **open interval** (a, b) is the set of real numbers x such that $a < x < b$; see Figure 1.1.2. In symbols

$$(a, b) = \{x | a < x < b\}. \tag{4}$$

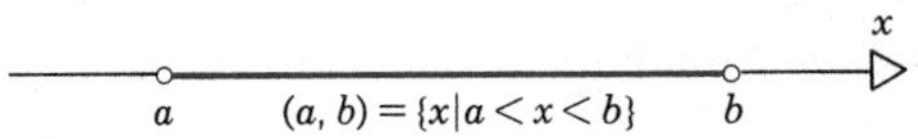

a $(a, b) = \{x | a < x < b\}$ b x

Figure 1.1.2 An open interval.

Similarly, the **closed interval** $[a, b]$ is given by

$$[a, b] = \{x | a \leq x \leq b\} \tag{5}$$

and is shown in Figure 1.1.3. Observe that the closed interval $[a, b]$ contains both of its endpoints, while the open interval (a, b) contains neither endpoint. Note

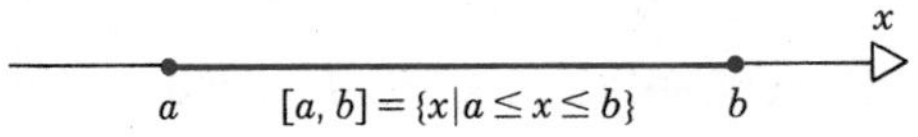

a $[a, b] = \{x | a \leq x \leq b\}$ b x

Figure 1.1.3 A closed interval.

also that we have used brackets or parentheses, respectively, to indicate whether the endpoints are, or are not, contained in the interval. Of course, an interval may include one endpoint, but not the other; such an interval is denoted by $[a, b)$, or by $(a, b]$, depending on which endpoint is included. All of the intervals mentioned so far are **bounded** intervals.

Often we will want to talk about an interval such as $\{x | x > a\}$. In such a case we shall use the "infinity" symbol ∞ and write

$$(a, \infty) = \{x | x > a\} \quad \text{or} \quad [a, \infty) = \{x | x \geq a\}. \tag{6}$$

In a similar way we write

$$(-\infty, b) = \{x | x < b\} \quad \text{or} \quad (-\infty, b] = \{x | x \leq b\}. \tag{7}$$

The set of all real numbers is the interval $(-\infty, \infty)$. We emphasize that the symbols ∞ and $-\infty$ do not stand for real numbers and they do not share the arithmetic properties of the real number system. Intervals such as (a, ∞), $(-\infty, b]$, and $(-\infty, \infty)$ are examples of **unbounded** intervals.

Absolute value

Two important properties of a real number are its sign and its magnitude. Geometrically, the sign tells us the direction, right or left from the origin on the real number line, and the magnitude is the distance from the origin. To isolate the property of magnitude we introduce the **absolute value** of a real number a, denoted by $|a|$, and defined by

$$|a| = \begin{cases} a, & a \geq 0; \\ -a, & a < 0. \end{cases} \tag{8}$$

Thus $|2| = 2$, $|0| = 0$, and $|-\frac{1}{3}| = \frac{1}{3}$. Clearly the absolute value of any real number a is nonnegative, and represents, geometrically, the distance from the origin to the point a (see Figure 1.1.4). Moreover, $|a - b|$ is the distance between the points a

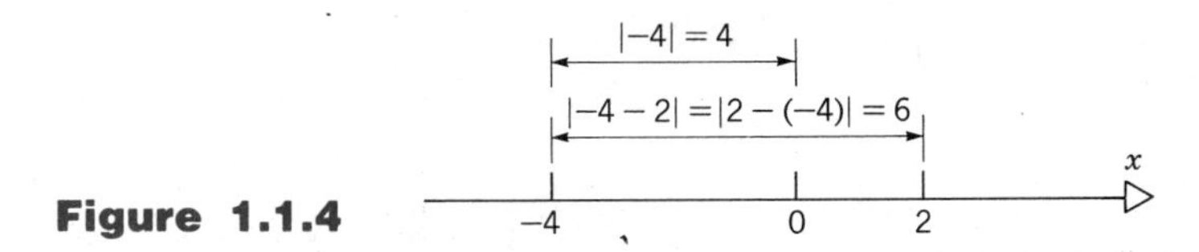

Figure 1.1.4

and b. This is clear if $a > b$, in which case the distance between the points is $a - b = |a - b|$; if $a < b$, then the distance between the points is $b - a = -(a - b) = |a - b|$. In this sense we can think of $|a| = |a - 0|$.

EXAMPLE 1

Solve the equation

$$|2x - 4| = |3 - x|. \tag{9}$$

In the interval $[2, 3]$ both $2x - 4$ and $3 - x$ are nonnegative, so Eq. 9 reduces to

$$2x - 4 = 3 - x,$$

with the solution $x = \frac{7}{3}$. On the other hand, in the intervals $(-\infty, 2)$ and $(3, \infty)$ either $2x - 4$ or $3 - x$ (but not both) is negative, so Eq. 9 becomes

$$2x - 4 = -(3 - x),$$

with the solution $x = 1$. Thus the numbers $x = 1$ and $x = \frac{7}{3}$ are the solutions of Eq. 9. ∎

The absolute value provides an alternative way of defining a bounded set. The set S is bounded if there is a (positive) number K such that $|x| \le K$ for each $x \in S$. In this case K is an upper bound and $-K$ is a lower bound for S.

The absolute value can be expressed in terms of the square root symbol as follows. The radical sign always indicates the nonnegative square root; thus $\sqrt{a}$ stands for the *nonnegative* number whose square is a. For example, $\sqrt{25} = 5$, $\sqrt{(-5)^2} = 5$, and in general

$$\sqrt{x^2} = \begin{cases} x, & x \ge 0; \\ -x, & x < 0. \end{cases} \tag{10}$$

Consequently,

$$|x| = \sqrt{x^2}. \tag{11}$$

A frequent source of confusion and error is the failure to be sufficiently careful of the sign when taking square roots, especially when the sign may change as a variable takes on different values. For example, $\sqrt{(x + 2)^2} = x + 2$ for $x \ge -2$, but $\sqrt{(x + 2)^2} = -(x + 2)$ when $x < -2$.

Under multiplication and division the absolute value has the properties that if x and y are real numbers, then

$$|xy| = |x|\,|y|, \tag{12}$$

and

$$\left|\frac{x}{y}\right| = \frac{|x|}{|y|}, \quad y \ne 0. \tag{13}$$

Equation 12 can be easily established by considering separately the four cases $x \ge 0$, $y \ge 0$; $x < 0$, $y \ge 0$; $x \ge 0$, $y < 0$; and $x < 0$, $y < 0$. The same procedure also suffices to prove Eq. 13, except that $y = 0$ must be excluded.

PROBLEMS

In each of Problems 1 through 4, determine $A \cup B$ and $A \cap B$.

1. $A = (-1, 2)$, $B = [0, 4)$

2. $A = [-3, 3]$, $B = (-5, 1)$

3. $A = (-\infty, 0)$, $B = (0, \infty)$

4. $A = \{x|x^2 \le 0\}$, $B = \{x|(x - 1)^2 = 4\}$

5. If $A = \{0, 2\}$ and $B = \{0, 2, 4\}$, determine whether each of the following statements is true and explain your reasoning.

(a) $2 \in A$ (c) $A \in B$

(b) $2 \subset B$ (d) $A \subset B$.

6. If $A = \{0, 2\}$ and $B = \{0, 2, \{0, 2\}\}$, determine whether each of the following statements is true and explain your reasoning.

(a) $2 \in A$ (c) $A \in B$

(b) $2 \subset B$ (d) $A \subset B$.

In each of Problems 7 through 10, determine $A - B$.

7. $A = [2, 3]$, $B = (2, 3)$

8. $A = R^1$, $B = \{x|x^2 < 4\}$

9. $A = [0, 1]$, $B = [0, \frac{1}{4}) \cup [\frac{3}{4}, 1)$

10. $A = R^1$, $B = [-3, 3] \cap (0, 4)$

In each of Problems 11 through 16, determine whether the given set is bounded above; bounded below.

11. $S = \{0, \frac{1}{2}, 1\}$

12. S is the set of all positive integers.

13. $S = \{x | x \le 1\}$ 14. $S = \left\{\frac{(-1)^n}{n} \middle| n = 1, 2, \ldots\right\}$

15. $S = \left\{1 - \frac{1}{n} \middle| n = 2, 3, \ldots\right\}$

16. $S = \{x | x^2 > 3\}$

17. Determine the set $S = \{s | x^2 + x - s > 0$ for all $x \in R^1\}$.

18. Determine the set $S = \{s | x^2 - 3x + s < 0$ for some $x \in R^1\}$.

In each of Problems 19 through 24, find all values of x that satisfy the given equation.

19. $|2x + 3| = 1$

20. $|3x - 5| = 6$

21. $|1 - x| = x - 1$

22. $|6 - x| = |2x - 7|$

23. $\left|\frac{3 - x}{3 + x}\right| = 4$

24. $|3x - 1| = 5 + 2x$

25. Prove that

$$\tfrac{1}{2}(|x| + x) = \begin{cases} x, & x \ge 0; \\ 0, & x < 0. \end{cases}$$

26. Prove that

$$\tfrac{1}{2}(x + y + |x - y|) = \begin{cases} x, & x \ge y; \\ y, & x < y. \end{cases}$$

In each of Problems 27 and 28, prove the given statement by considering the four cases mentioned in the text.

27. $|xy| = |x|\,|y|$

28. $\left|\frac{x}{y}\right| = \frac{|x|}{|y|}, \quad y \ne 0$

* Completeness of the Real Numbers

A bounded set of real numbers has many upper bounds and many lower bounds. For example, once an upper bound has been found, any larger number is also an upper bound. The smallest of the upper bounds is called the **least upper bound** and the largest of the lower bounds is called the **greatest lower bound.** A vital property of the real number system is that *every nonempty bounded set of real numbers has a least upper bound and a greatest lower bound.* This statement is referred to as the **completeness property** of the real number system. A set that is bounded on one side but not the other has either a greatest lower bound or a least upper bound, as the case may be, but not both.

The completeness property is disarmingly simple to state, but its truth is by no means obvious, and it has numerous profound consequences that can be discussed fully only in more advanced courses on mathematical analysis.

In each of Problems 29 through 32 find the least upper bound and the greatest lower bound (if they exist) of the given set of real numbers.

29. $S = \left\{x \middle| x = \frac{n - 1}{n + 1} \text{ for } n = 1, 2, \ldots\right\}$

30. $S = \left\{x \middle| x = \frac{(-1)^n}{n} \text{ for } n = 1, 2, \ldots\right\}$

31. $S = \{y | y = x^2 + 2x \text{ for } -2 \le x \le 1\}$

32. $S = \left\{y \middle| y = \frac{1}{1 + x^2} \text{ for } -\infty < x < \infty\right\}$

1.2 INEQUALITIES

We will often need to determine the set of real numbers satisfying a given inequality, such as $4 - 2x < 5$ or $-4 < 2 - 3x \le 14$; we will refer to this as solving the inequality. The algebraic treatment of inequalities is based on the following two properties of real numbers.

1. If $a < b$, then $a + c < b + c$ for any real number c.
2. If $a < b$ and if $c > 0$, then $ac < bc$; if $a < b$ and if $c < 0$, then $ac > bc$.

The following examples illustrate the solution of inequalities.

EXAMPLE 1

Find all real numbers satisfying the inequality

$$4 - 2x < 5. \qquad (1)$$

To isolate x we add -4 to both sides of the inequality, obtaining $-2x < 1$. Next, we divide by -2 and reverse the direction of the inequality, since -2 is a negative number; this gives

$$x > -\tfrac{1}{2}. \tag{2}$$

Thus, if x satisfies the inequality $4 - 2x < 5$, then necessarily $x > -\frac{1}{2}$. We have not yet shown that, for every $x > -\frac{1}{2}$, the inequality $4 - 2x < 5$ is satisfied. We can do this by simply reversing our arguments: start with $x > -\frac{1}{2}$, multiply by -2 to obtain $-2x < 1$, and then add 4 to both sides of the inequality to obtain $4 - 2x < 5$. Thus the solution of the inequality (1) is the set $S = \{x|x > -\frac{1}{2}\}$, or the interval $(-\frac{1}{2}, \infty)$. ∎

Example 1 illustrates that there are two steps in "solving" an inequality. First, working from the inequality, we deduce a necessary condition that x is in a certain set S. The second step is to show that the condition is also sufficient, that is, if $x \in S$, then the original inequality is satisfied. Thus the solution of the inequality is the set S. Usually the set S is an interval or a union of intervals; however, it is possible that S consists only of isolated points or even is the empty set. The second part of the argument is carried out by reversing the steps in the first part. Provided that only operations such as addition and multiplication have been used, there will be no difficulty in reversing the steps and this part of the argument is usually omitted.

EXAMPLE 2

Find all real numbers satisfying the inequality

$$-4 < 2 - 3x \leq 14. \tag{3}$$

Observe first that the given inequality is equivalent to the pair of inequalities

$$-4 < 2 - 3x \qquad \text{and} \qquad 2 - 3x \leq 14, \tag{4}$$

each of which can be handled in a manner similar to Example 1. From the first inequality in Eq. 4 we have $3x < 6$ or $x < 2$. The second inequality in Eq. 4 yields $3x \geq -12$ or $x \geq -4$. Thus the solution to both inequalities is the set $S = \{x| -4 \leq x < 2\}$ or the interval $[-4, 2)$.

One can also deal with both sides of Eq. 3 at the same time. First we add -2 to each member, obtaining $-6 < -3x \leq 12$. Then we divide by -3 and reverse the direction of the inequalities: thus $2 > x \geq -4$ or $-4 \leq x < 2$, which is the same as the previous result. ∎

EXAMPLE 3

Find all real numbers satisfying the inequality

$$x^2 - 2x - 1 > 2. \tag{5}$$

First we add -2 to both sides of the inequality, which gives $x^2 - 2x - 3 > 0$; then by factoring the left side, we obtain

$$(x - 3)(x + 1) > 0. \tag{6}$$

Inequality (6) is true when both factors have the same sign; that is, when $x - 3 > 0$ and $x + 1 > 0$, or when $x - 3 < 0$ and $x + 1 < 0$. It is helpful to depict the situation geometrically as shown in Figure 1.2.1. Clearly $x - 3$ and $x + 1$ are both

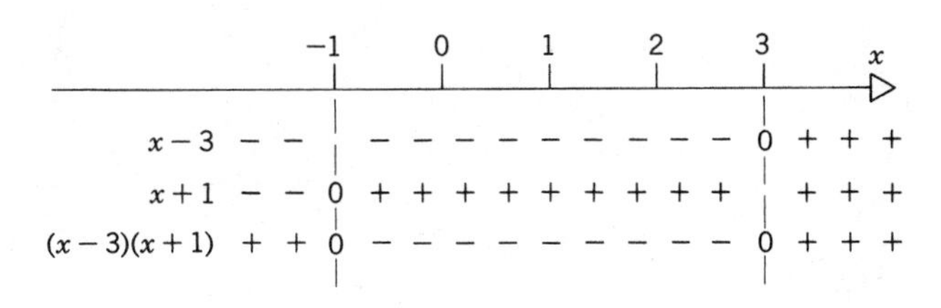

Figure 1.2.1

positive if $x > 3$ and they are both negative if $x < -1$. Therefore the intervals for which the inequality (5) holds are $(-\infty, -1)$ and $(3, \infty)$, and the set of points for which the inequality is satisfied is $(-\infty, -1) \cup (3, \infty)$. ∎

EXAMPLE 4

Find all real numbers satisfying the inequality

$$\frac{(x - 3)(x + 1)}{(x + 2)} > 0, \qquad x \neq -2. \tag{7}$$

This is an extension of Example 3, and can be solved using the diagram shown in Figure 1.2.2. The given expression is positive when all three of the factors are

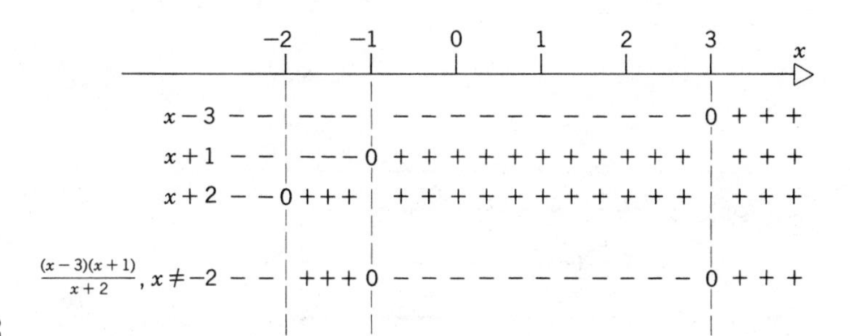

Figure 1.2.2

positive, or when one is positive and two are negative. Thus the inequality (7) is satisfied when $-2 < x < -1$ or when $x > 3$, that is, by the set $S = (-2, -1) \cup (3, \infty)$. ∎

EXAMPLE 5

Find all real numbers satisfying the inequality

$$-\frac{3}{2} < \frac{1}{x} - 2 \leq 2, \qquad x \neq 0. \tag{8}$$

It is convenient to consider this as two separate inequalities. First, if

$$\frac{1}{x} - 2 \leq 2,$$

then

$$\frac{1}{x} \leq 4.$$

Hence

$$\text{either } x < 0 \quad \text{or} \quad x \geq \frac{1}{4}. \tag{9}$$

Next, if

$$-\frac{3}{2} < \frac{1}{x} - 2,$$

then

$$\frac{1}{x} > \frac{1}{2},$$

so

$$0 < x < 2. \tag{10}$$

These results are indicated in Figure 1.2.3. Finally, to satisfy both of Eqs. 9 and 10 we must have

$$1/4 \leq x < 2. \blacksquare \tag{11}$$

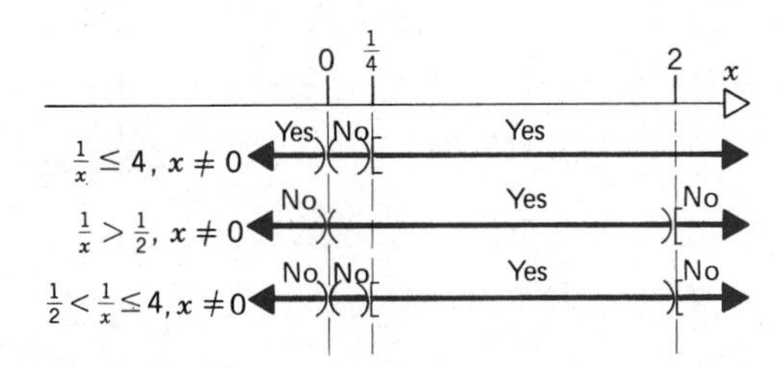

Figure 1.2.3

One error that is often made in solving inequalities is to multiply both sides of an inequality by a quantity that can change sign, such as $x - 2$, without taking account of the fact that if the quantity is negative, then the sign of the inequality must be reversed. Another error that is sometimes made is to assert that if $b > a$, then $b^2 > a^2$; that is, to square an inequality. This is not true unless $b + a > 0$; notice that $4 > -6$, but $16 < 36$.

Inequalities that involve absolute values occur frequently. The following result is basic in such cases.

Theorem 1.2.1

If $r > 0$, then $|x| \le r$ if and only if $-r \le x \le r$; similarly, $|x| \ge r$ if and only if $x \le -r$ or $x \ge r$.

Analogous results can be expressed in terms of $<$ and $>$; for example, $|x| < r$ if and only if $-r < x < r$. A convincing argument that Theorem 1.2.1 is true can be given by an appeal to geometric reasoning. First, if the distance from 0 to x is less than or equal to r, so that $|x| \le r$, then x must lie in the interval $[-r, r]$. Alternatively, if $x \in [-r, r]$, then the distance from 0 to x is less than or equal to r. This is shown in Figure 1.2.4. A similar argument applies to the second part of the theorem. □

Figure 1.2.4

A slight extension of the theorem is illustrated by the result that $|x - a| \le r$ if and only if $-r \le x - a \le r$, or $a - r \le x \le a + r$. Thus

$$|x - a| \le r \quad \text{if and only if} \quad x \in [a - r, a + r]. \tag{12}$$

Geometrically, Eq. 12 states that if the distance from a to x is less than or equal to r, then x must lie in an interval centered at a and extending r units in each direction; and, conversely, if x is in this interval, then it is within r units of a. An open interval $(a - r, a + r)$ is sometimes called a **neighborhood** of a.

EXAMPLE 6

Find all real numbers satisfying the inequality

$$|x + 3| < 2. \tag{13}$$

This inequality says that the distance from -3 to x must be less than 2. Consequently, x is in the interval $(-5, -1)$. Analytically, Eq. 13 is equivalent to $-2 < x + 3 < 2$, so $-5 < x < -1$, or $x \in (-5, -1)$. ∎

EXAMPLE 7

Find all real numbers that satisfy

$$|2x + 3| \leq 5. \tag{14}$$

The inequality can be written in the form

$$|2x + 3| = |2(x + \tfrac{3}{2})| = 2|x + \tfrac{3}{2}| \leq 5$$

from which it follows that

$$|x + \tfrac{3}{2}| \leq \tfrac{5}{2}.$$

Thus the distance from $-\frac{3}{2}$ to x cannot exceed $\frac{5}{2}$; consequently, x must lie in the interval $[-4, 1]$. ∎

EXAMPLE 8

Find all real numbers satisfying the inequality

$$\left|\frac{1}{x} - 2\right| > 3, \qquad x \neq 0. \tag{15}$$

Let $u = 1/x$; then Eq. 15 takes the form $|u - 2| > 3$. Thus u must be more than 3 units from 2; hence, $u > 5$ or $u < -1$. Consequently,

$$\text{if } u > 5, \quad \text{then } \frac{1}{x} > 5,$$

so

$$0 < x < \tfrac{1}{5},$$

and

$$\text{if } u < -1, \quad \text{then } \frac{1}{x} < -1,$$

so

$$-1 < x < 0.$$

Therefore the solution of the inequality (15) is $-1 < x < \frac{1}{5}$, $x \neq 0$; or $x \in (-1, 0) \cup (0, \frac{1}{5})$. ∎

We close this section with a fundamental result that enables us to estimate the absolute value of a sum.

Theorem 1.2.2

(Triangle Inequality)

If x and y are real numbers, then

$$|x + y| \le |x| + |y|. \tag{16}$$

Inequality (16) is a very important result with a simple proof. First, observe that

$$-|x| \le x \le |x| \tag{17}$$

for all values of x; the left side is an equality if $x \le 0$, and the right side is an equality if $x \ge 0$. In a similar way,

$$-|y| \le y \le |y|. \tag{18}$$

By adding Eqs. 17 and 18 we obtain

$$-(|x| + |y|) \le x + y \le |x| + |y|,$$

and Eq. 16 then follows from Theorem 1.2.1. □

Problem 21 below asks for another proof of Theorem 1.2.2, and Problems 22 through 25 indicate some elementary uses of the triangle inequality. It can be generalized considerably and has applications far beyond the present context.

PROBLEMS

In each of Problems 1 through 6, express the given set as an interval or union of intervals.

1. $S = \{x \mid |x + 3| < 3\}$
2. $S = \{x \mid |3 - 2x| < 1\}$
3. $S = \{t \mid 0 < |t - 1| < 1\}$
4. $S = \{w \mid |2 - 3w| > 3\}$
5. $S = \{x \mid |x - a| < \epsilon\}$, where a and ϵ are given numbers, $\epsilon > 0$.
6. $S = \{x \mid 0 < |x - a| < \epsilon\}$, where a and ϵ are given numbers, $\epsilon > 0$.

In each of Problems 7 through 20, determine the set of numbers satisfying the given inequality. Express your answer as an interval or as a union of intervals.

7. $3x - 7 \ge 2 - x$
8. $2x < 5 - 3x$
9. $3 < 4x - 1 \le 6$
10. $-4 \le 2 - 3x \le -2$
11. $\dfrac{x - 2}{x - 3} > 0, \quad x \ne 3$
12. $x^2 + x - 3 > 3$
13. $7 - \dfrac{2}{x} > 3, \quad x \ne 0$
14. $6x^2 + 3x - 8 \le 1$
15. $-1 < 2 - \dfrac{1}{x} < 1, \quad x \ne 0$
16. $\dfrac{x + 1}{(x + 3)(x - 3)} > 0, \quad x \ne \pm 3$
17. $\dfrac{x^2 - 2x + 1}{x^2 - 2x - 3} \le 0, \quad x \ne -1, \quad x \ne 3$
18. $|x - 2| \le |x + 2|$
19. $\left|4 - \dfrac{1}{x}\right| < 1, \quad x \ne 0$
20. $\left|5 + \dfrac{4}{x}\right| \ge 2, \quad x \ne 0$

21. Prove the triangle inequality, Theorem 1.2.2, by considering the four cases $x \geq 0$, $y \geq 0$; $x < 0$, $y < 0$; $x > 0$, $y < 0$; and $x < 0$, $y > 0$.

In each of Problems 22 through 25, the desired result can be obtained by a judicious use of the triangle inequality.

22. If x_1, x_2, and x_3 are real numbers, prove that

$$|x_1 + x_2 + x_3| \leq |x_1| + |x_2| + |x_3|.$$

Can you generalize this result for $x_1, x_2, \ldots, x_n$?

23. If x and y are real numbers, prove that

$$|x - y| \leq |x| + |y|.$$

Under what conditions does the equality hold?

24. If $|x - 1| + |x - 3| < \alpha$ holds for some x, show that $\alpha > 2$. *Hint:* $2 = |3 - 1| = |(x - 1) - (x - 3)|$. Generalize this result to show that if $|x - a| + |x - b| < \alpha$ holds for some x, then $\alpha > |b - a|$.

25. If x and y are any real numbers, prove that

$$|x| - |y| \leq |x - y|.$$

Hint: Consider the identity $|x| = |(x - y) + y|$. Under what conditions does the equality hold?

26. Show that the sum of a positive real number and its reciprocal is greater than or equal to two; that is, $x + (1/x) \geq 2$, for $x > 0$.

27. If $b > a$ and $b + a > 0$, show that $b^2 > a^2$.

28. Show that $a^2 + b^2 \geq 2ab$, and then that $a^2 + b^2 \geq 2|ab|$.

29. Show that

$$(a_1b_1 + a_2b_2)^2 \leq (a_1^2 + a_2^2)(b_1^2 + b_2^2).$$

When does equality occur? This result is known as the Cauchy-Schwarz inequality. Like the triangle inequality, it can be generalized considerably.

* In each of Problems 30 through 34, find the least upper bound and the greatest lower bound (if they exist) of the given set of real numbers (see Problems 29 through 32 of Section 1.1).

30. $S = \{x \mid |x - 2| < 1\}$

31. $S = \{x \mid x^2 - x - 6 < 0\}$

32. $S = \{x \mid x^2 - x - 6 \leq 0\}$

33. $S = \{x \mid x^2 - x - 6 > 0\}$

34. $S = \{x \mid x\,|x + 1| \leq 2\}$

1.3 RELATIONS AND GRAPHS

Cartesian coordinate system

The correspondence between real numbers and geometrical points can be extended so as to deal with points in a plane. We select any two mutually perpendicular lines, assign a positive direction and a unit of measurement on each, and designate their point of intersection O as the **origin.** It is conventional to call these lines the x-axis and y-axis, respectively, and to orient them as shown in Figure 1.3.1. Each

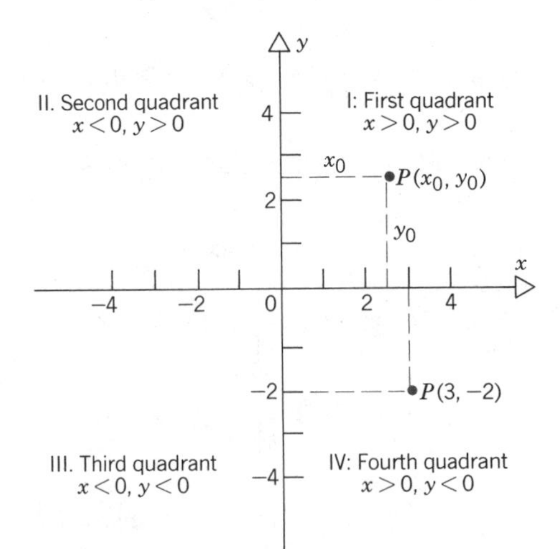

Figure 1.3.1

point P in the plane is assigned a pair of numbers called its **coordinates.** This pair of numbers locates the point in the plane as illustrated in Figure 1.3.1. The first number is called the **x-coordinate,** or abscissa, of the point, and the second number is called the **y-coordinate,** or ordinate, of the point. When we give a pair of numbers for the coordinates of a point, say (x_0, y_0), it is understood that the first number is the x-coordinate and the second number is the y-coordinate. Thus the pair of numbers is called an **ordered pair.** Any point P in the plane can be associated* with a unique ordered pair of real numbers, and conversely any ordered pair of real numbers can be associated with a unique point in the plane. The x- and y-axes divide the xy-plane into four **quadrants** as shown in Figure 1.3.1.

The set of all ordered pairs of real numbers will be denoted by R^2:

$$R^2 = \{(x, y) | x \in R^1 \quad \text{and} \quad y \in R^1\}. \tag{1}$$

Just as in one dimension we will often consider geometric points and their coordinates to be interchangeable; for example, we may often refer to the "point (x_0, y_0)" instead of to the "point whose coordinates are (x_0, y_0)."

Distance

The distance between two points in the plane can be easily computed from the Pythagorean theorem. From Figure 1.3.2 the distance from the origin O to the

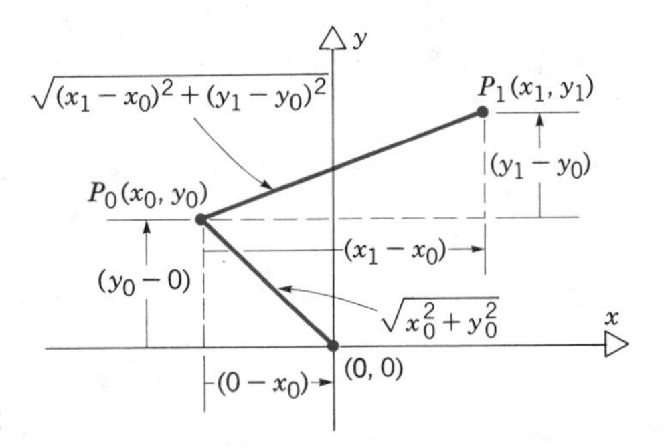

Figure 1.3.2

point P_0 with the coordinates (x_0, y_0) is $\sqrt{x_0^2 + y_0^2}$. As usual, the radical sign denotes the *nonnegative square root* of the quantity under it. Similarly, if the distance from P_0 to P_1 is denoted by $d(P_0, P_1)$, then

*The idea of using pairs of numbers to identify points in a plane and equations to describe curves originated with René Descartes (1596–1650) and Pierre de Fermat (1601–1665). Descartes was a philosopher and scientist who also devoted some time to mathematics. His results and methods in what later came to be known as analytic geometry are contained in his influential book *La Géométrie* published in 1637. Descartes recognized the power of algebra for the solution of geometrical problems, and his thinking represented a significant break with the Greek tradition and viewpoint in geometry. The rectangular, or Cartesian, coordinate system is named for him. Fermat was a jurist and amateur mathematician who lived in Toulouse in southern France. His ideas on coordinate geometry were formulated beginning in 1629, and were circulated through correspondence, but were not published until 1679. He is most famous for the as yet unproved assertion known as Fermat's last theorem, namely, that for integer values of $n > 2$ it is impossible to find positive integers x, y, and z such that $x^n + y^n = z^n$.

$$d(P_0, P_1) = \sqrt{(x_1 - x_0)^2 + (y_1 - y_0)^2}. \tag{2}$$

Note that if $y_0 = y_1$, so that P_0 and P_1 are on a line parallel to the x-axis, then

$$d(P_0, P_1) = \sqrt{(x_1 - x_0)^2} = |x_1 - x_0|. \tag{3}$$

This agrees with the result given in Section 1.1 for the distance between two points on the x-axis.

Variables

We will often want to consider two or more variable quantities that are related in some way. By a **variable** we mean a symbol that may be equal to any one of a particular set of numbers. Alternatively, it is often helpful to think of the variable as a typical or generic member of the allowable set. Usually the set is R^1 or some subset of it, for example, the positive numbers, the numbers between zero and one, and so forth. Letters such as x, y, or z are often used to designate variables that take on values in R^1. Such variables are called real variables.

Relations and graphs

Relations between variables may be of very diverse types and are sometimes extremely complicated. For the present we will be concerned only with relations in R^2 that are expressible in a fairly simple algebraic form. The following are typical examples; each will be considered in more detail later.

$$\begin{aligned} y + 2x &= 3, \\ (x + 2)^2 + (y - 1)^2 &= 9, \\ |y| &= 2|x|, \\ y + 2x &\geq 3, \\ (x + 2)^2 + (y - 1)^2 &< 9\,. \end{aligned}$$

In each case there are some pairs of values of x and y that satisfy the relation, while others do not. In geometric terminology some points in the xy-plane satisfy the relation, but not others. The set of all points in the xy-plane whose coordinates satisfy a given relation constitute the **graph** of the relation.

Questions concerning graphs and relations can be divided into two main categories.

1. Given a relation between two variables x and y, sketch its graph in the xy-plane.
2. Given a geometrical description of a set of points in the xy-plane, find an analytical expression for the relation whose graph is the given set of points.

The first question generally arises when one wishes to visualize, and perhaps to use geometrical intuition to study, a complicated analytical relation. One picture may be worth a thousand words! The second question typically occurs when one wants to take advantage of the powerful computational processes of algebra to study a situation phrased in geometrical terms.

In the remainder of this section we present examples of both kinds of questions. We first consider equations whose graphs are straight lines and circles, respectively.

Straight lines

If (x_1, y_1) and (x_2, y_2) are two points with $x_1 \neq x_2$ (see Figure 1.3.3), then the **slope** m of the line segment joining the two points is defined to be

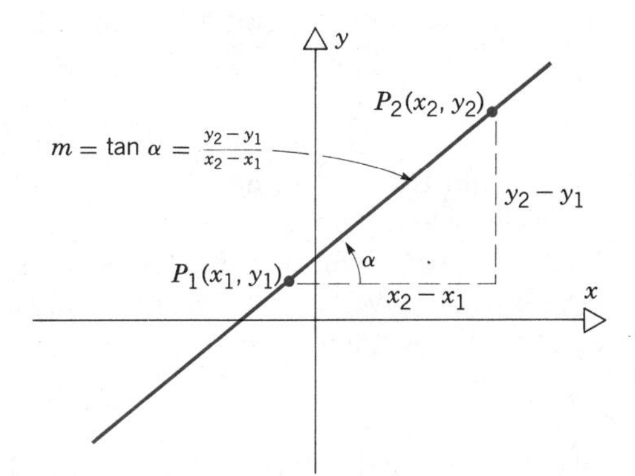

Figure 1.3.3

$$m = \frac{y_2 - y_1}{x_2 - x_1}. \tag{4}$$

A straight line has the property that the slopes of all line segments connecting points on the line are the same. This number is called the slope of the straight line. Clearly the slope of a line is the ratio of the changes in the coordinates x and y in moving from one point to another on the line. The slope is also the tangent of the angle between the line and the positive x-axis. From Figure 1.3.3 we see that

$$m = \tan \alpha. \tag{5}$$

The slope m of a line is positive if the values of y increase as x increases, and is negative if y decreases as x increases. A line parallel to the y-axis is said to have an infinite slope.

Geometrically, a straight line is determined if its slope and one point lying on the line are known. Suppose m is the slope and (x_1, y_1) is the given point. Then, if (x, y) denotes any other point on the line, Eq. 4 yields

$$m = \frac{y - y_1}{x - x_1}, \tag{6}$$

or

$$y - y_1 = m(x - x_1) \tag{7}$$

as the equation of the given line. Equation 7 can also be written in the form

$$y - mx = y_1 - mx_1 \tag{8}$$

or

$$y = mx + b, \tag{9}$$

where b is the y-intercept, or the ordinate of the point at which the line intersects the y-axis.

For a line parallel to the x-axis, the slope m is equal to zero, and Eq. 7 simplifies to

$$y = y_1. \tag{10}$$

Similarly, the equation of a line parallel to the y-axis is of the form

$$x = x_1, \tag{11}$$

although this equation cannot be obtained by assigning a value to m in Eq. 7.

From Eqs. 7, 10, and 11 it is clear that in the equation of a straight line the variables x and y appear to the first power and no products of x and y are present. Such equations, whether they arise geometrically or not, are therefore called **linear equations.** Conversely, the graph of any linear equation in two variables is a straight line. The most general such equation is

$$Ax + By = C, \tag{12}$$

where A, B, and C are constants, and A and B are not both zero. If $B \neq 0$, then upon division by B, Eq. 12 takes the form $y = -(A/B)x + (C/B)$; the slope m is equal to $-A/B$. If $B = 0$, then $A \neq 0$, and Eq. 12 reduces to $x = C/A$, which is a line parallel to the y-axis.

EXAMPLE 1

Find the graph of the relation

$$y + 2x = 3. \tag{13}$$

This is a linear equation and therefore its graph is a straight line. Writing the equation as $y = -2x + 3$, we find that the slope of the line is -2 and that it passes through the point (0, 3). This information completely determines the straight line corresponding to Eq. 13. In practice the easiest way to draw the straight line is to locate two points on it and then draw a line through these points. A second point can be found by setting $y = 0$ in Eq. 13, whence $x = \frac{3}{2}$. The straight line passing through the points (0, 3) and ($\frac{3}{2}$, 0) is shown in Figure 1.3.4. ∎

EXAMPLE 2

Find the equation of the straight line shown in Figure 1.3.5 passing through the points $(-1, 2)$ and $(3, 1)$.

From the coordinates of the given points the slope m of the line is

$$m = \frac{1 - 2}{3 - (-1)} = -\frac{1}{4}. \tag{14}$$

Then from Eq. 7

$$y - 2 = -\tfrac{1}{4}(x + 1), \tag{15}$$

or

$$y = -\tfrac{1}{4}x + \tfrac{7}{4}. \ \blacksquare$$

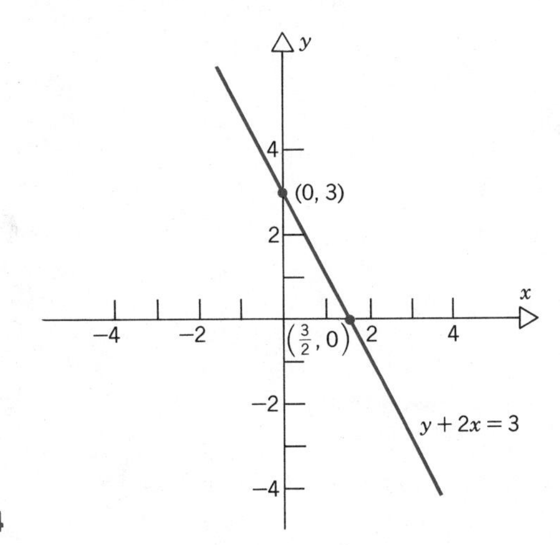

Figure 1.3.4

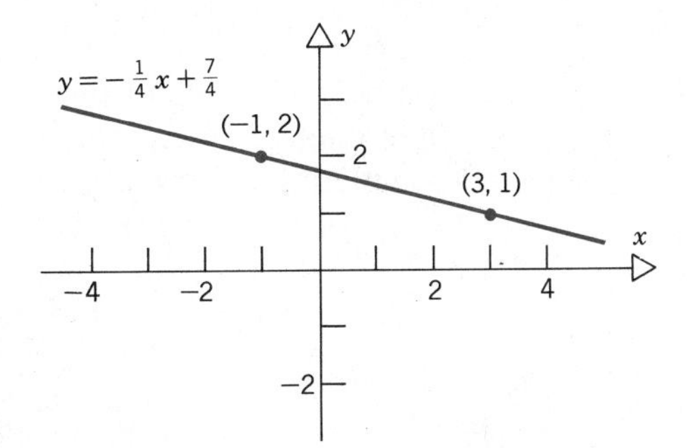

Figure 1.3.5

Circles

A circle is a plane figure consisting of all points that are at some fixed distance (the radius) from a given point called the center. Suppose that (x_0, y_0) is the center of the circle, (x, y) is an arbitrary point on the circle, and r is the radius; then by the distance formula (2),

$$(x - x_0)^2 + (y - y_0)^2 = r^2 \tag{16}$$

is a relation satisfied by points on the circle and by no other points in the xy-plane. For instance, Figure 1.3.6 shows the circle for which $x_0 = -2$, $y_0 = 1$, and $r = 3$.

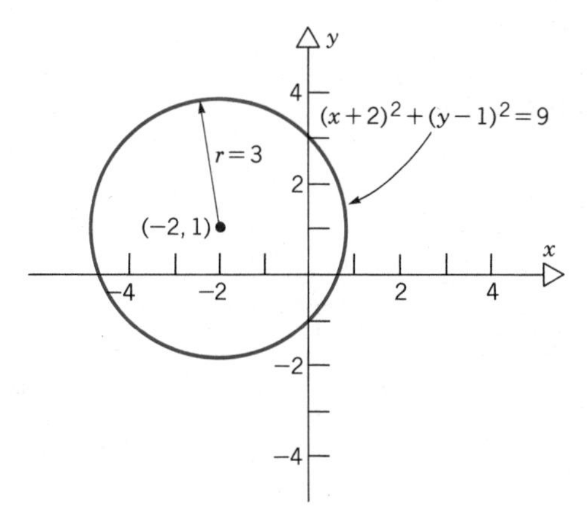

Figure 1.3.6

EXAMPLE 3

Find the graph of the equation

$$x^2 + y^2 + 4x - 2y - 4 = 0. \tag{17}$$

In order to write a general equation that is quadratic in x and y in the form (16) it is necessary that (a) the coefficients of x^2 and y^2 are the same when these terms appear on the same side of the equation, and (b) the product term xy does not appear. Then one can obtain the form (16) by collecting terms involving each variable and completing the square. In the present case we have

$$x^2 + 4x + y^2 - 2y = 4,$$

$$(x^2 + 4x + 4) + (y^2 - 2y + 1) = 4 + 4 + 1,$$

or

$$(x + 2)^2 + (y - 1)^2 = 9. \tag{18}$$

Equation 18 is of the form (16) with $x_0 = -2$, $y_0 = 1$, and $r = 3$. The graph of Eq. 18, and hence of Eq. 17 as well, is therefore the circle of radius three with center at $(-2, 1)$ (see Figure 1.3.6). ∎

Examples of other relations and graphs

Relations may take the form of an equation or an inequality involving two variables, or even several equations and/or inequalities. The graphs of relations have many different forms. The graph of a relation need not be a single curve in the xy-plane or, indeed, may not consist of curves at all.

EXAMPLE 4

Find the graph of the relation

$$|y| = 2|x|. \tag{19}$$

In order to use the definition of the absolute value of a number given in Section 1.1, we must consider separately each of the four quadrants of the xy-plane.

I. If $x \geq 0$, $y \geq 0$, then Eq. 19 is $y = 2x$.

II. If $x < 0$, $y \geq 0$, then Eq. 19 is $y = -2x$.

III. If $x < 0$, $y < 0$, then Eq. 19 is $-y = -2x$ or $y = 2x$.

IV. If $x \geq 0$, $y < 0$, then Eq. 19 is $-y = 2x$.

Combining the information just given, we find that when x and y have the same signs (first and third quadrants), then Eq. 19 reduces to $y = 2x$; when x and y have opposite signs (second and fourth quadrants), then Eq. 19 reduces to $y = -2x$. Thus the graph of Eq. 19 consists of the two intersecting straight lines shown in Figure 1.3.7. ∎

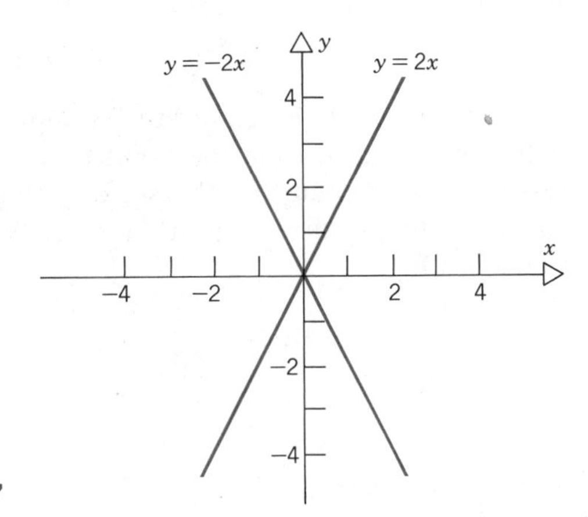

Figure 1.3.7

EXAMPLE 5

Find the graph of the relation

$$y + 2x \geq 3. \tag{20}$$

This example is related to Example 1, and the points on the straight line in Example 1 also satisfy relation (20). Further, if we start from a point on the line $y + 2x = 3$ shown in Figure 1.3.8, and increase either x or y, the relation (20) is still satisfied. However, if we decrease either x or y, the relation (20) is not satisfied. Thus the graph of relation (20) is the half-plane lying above and to the right of the line $y + 2x = 3$ and including the line itself. This half-plane is shown shaded in Figure 1.3.8. ∎

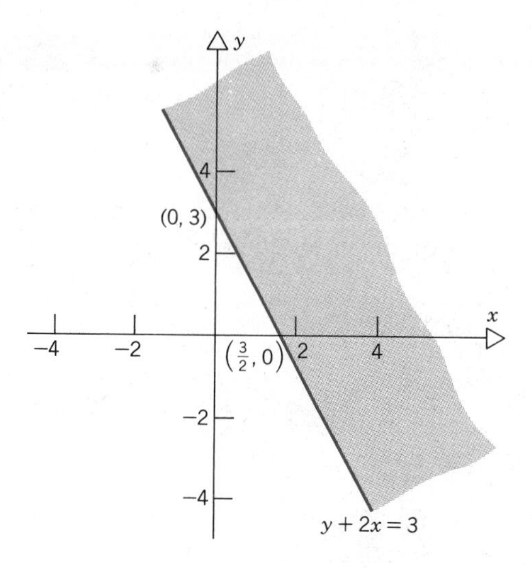

Figure 1.3.8

EXAMPLE 6

Find the graph of the relation

$$(x + 2)^2 + (y - 1)^2 < 9. \tag{21}$$

This example is related to Example 3. A point satisfies relation (21) if and only if its distance from the point $(-2, 1)$ is less than three units. The graph of relation (21) is therefore the set of points that lie inside, but not on, the circle corresponding to the equation $(x + 2)^2 + (y - 1)^2 = 9$. These points are shown shaded in Figure 1.3.9. ∎

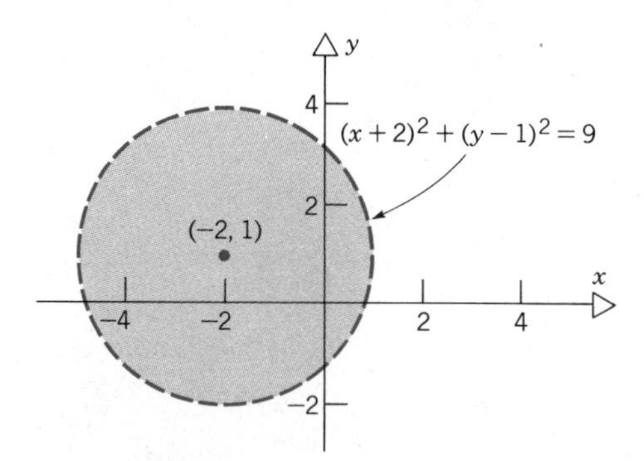

Figure 1.3.9

EXAMPLE 7

Find the graph of the relation

$$y - |y| = 2(x - |x|). \tag{22}$$

As in Example 4, it is advisable to consider each quadrant separately. We indicate below the form of Eq. 22 in each quadrant.

I. If $x \geq 0$, $y \geq 0$, then Eq. 22 is $0 = 0$.

II. If $x \leq 0$, $y \geq 0$, then Eq. 22 is $0 = 4x$.

III. If $x \leq 0$, $y \leq 0$, then Eq. 22 is $2y = 4x$.

IV. If $x \geq 0$, $y \leq 0$, then Eq. 22 is $2y = 0$.

The first statement is an identity, which imposes no restriction upon x and y, so long as both are nonnegative. Thus every point in the first quadrant, including the positive y- and x-axes and the origin, is contained in the graph of Eq. 22. The second and fourth statements again show that the positive y- and x-axes are contained in the graph of Eq. 22. The third statement says that the part of the straight line $y = 2x$ that lies in the third quadrant is also part of the graph. The graph of Eq. 22 is shown in Figure 1.3.10. ∎

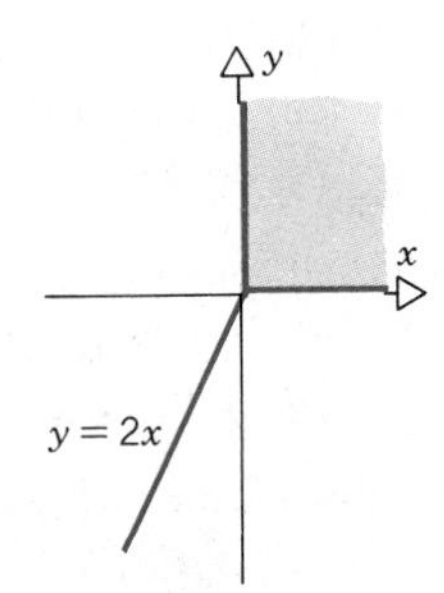

Figure 1.3.10

Finally, note that what may superficially appear to be a relation between two real variables may, in fact, not be one, since there may be no pair of values of x and y satisfying it.

EXAMPLE 8

Find the graph of

$$(x + 2)^2 + (y - 1)^2 + 9 = 0. \tag{23}$$

Since the left side of Eq. 23 is always strictly positive, regardless of the values of x and y, there are no points whose coordinates satisfy this equation. While this fact is obvious from the form of Eq. 23, it is not so obvious if the terms in that equation are expanded by the binomial theorem. If that is done, the equation takes the form

$$x^2 + 4x + y^2 - 2y + 14 = 0, \tag{24}$$

and it is not so clear that no pair of values of x and y satisfies it. ∎

PROBLEMS

In each of Problems 1 through 4, find the distance between the given pair of points.

1. $(2, -3)$ and $(-1, 4)$ 2. $(0, -\pi)$ and $(\pi/2, 0)$
3. $(t, |t|)$ and $(|t|, -t)$ 4. $(t, -1)$ and $(1, t)$
5. Show that the midpoint of the line segment joining the points $P_1(x_1, y_1)$ and $P_2(x_2, y_2)$ is the point
$$P_3\left(\frac{x_1 + x_2}{2}, \frac{y_1 + y_2}{2}\right).$$

In each of Problems 6 through 9, determine whether the triangle with the given points as vertices is a right triangle; an equilateral triangle; an isosceles triangle.

6. $(-6, 4)$, $(-3, -2)$, $(5, 2)$
7. $(-3, -2)$, $(-6, 9)$, $(3, 2)$
8. $(-2, 1)$, $(1, 3)$, $(0, -4)$
9. $(-2, 1)$, $(2, 1)$, $(0, 1 + 2\sqrt{3})$

In each of Problems 10 through 13, determine an equation for the straight line with the given properties.

10. Passing through $(1, 2)$ with slope -2.
11. Passing through the origin with slope 1.
12. Passing through $(-1, 3)$ and $(-1, -1)$.
13. Passing through $(-1, -1)$ and $(2, 3)$.

In each of Problems 14 through 17, sketch the graph of the given equation.

14. $3x + 6y - 6 = 0$ 15. $2y = 4x - 7$
16. $x - y + 2 = 0$ 17. $x + y = 2$
18. **Parallel lines.** Two straight lines in a plane are said to be **parallel** if either they coincide or do not intersect.
   (a) Show that the straight lines $y = 4x - 3$ and $2y - 8x = 1$ are parallel.
   (b) Show that the two straight lines $y = m_1x + b_1$ and $y = m_2x + b_2$ are parallel if and only if $m_1 = m_2$ and that they coincide if and only if $m_1 = m_2$, $b_1 = b_2$.
   (c) Show that the two straight lines $A_1x + B_1y + C_1 = 0$ and $A_2x + B_2y + C_2 = 0$ are parallel if and only if $A_1B_2 = A_2B_1$ (for simplicity assume A_1, B_1, A_2, B_2 are nonzero).
19. **Perpendicular lines.** Two lines are said to be **perpendicular** (or **orthogonal**) if they intersect at right angles; see Figure 1.3.11.

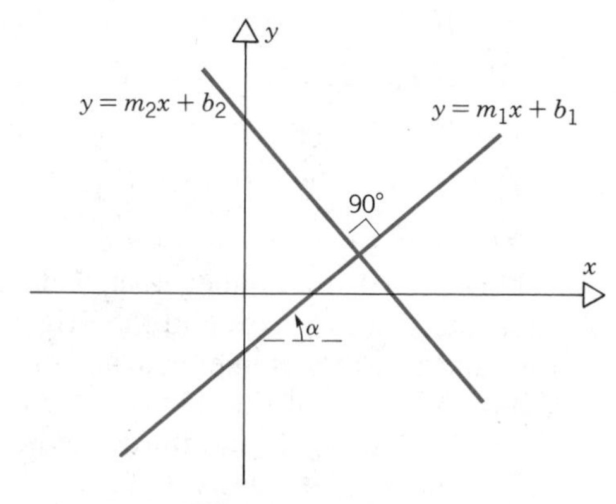

Figure 1.3.11

   (a) If the slope of one line is $m_1 \neq 0$, show that the slope m_2 of the other line is $m_2 = -1/m_1$. *Hint:* $m_1 = \tan \alpha$, $m_2 = ?$
   (b) Show that the two straight lines $A_1x + B_1y + C_1 = 0$ and $A_2x + B_2y + C_2 = 0$ are perpendicular if and only if $A_1A_2 + B_1B_2 = 0$.

In each of Problems 20 through 23, determine an equation for the straight line with the given properties.

20. Parallel to the line $2x - 3y = 4$ and passing through $(2, -5)$.
21. Parallel to the line $x + 2y = -5$ and passing through $(-2, 2)$.
22. Perpendicular to the line $2x - 3y = 4$ and passing through $(2, -5)$.
23. Perpendicular to the line $x + 2y = -5$ and passing through $(-2, 2)$.

In each of Problems 24 through 27, find the center and radius of the given circle.

24. $x^2 + y^2 - 2x + 2y - 1 = 0$
25. $x^2 + y^2 - 6x - 4y + 8 = 0$
26. $4x^2 + 4y^2 - 12x + 8y + 6 = 0$
27. $4y - 4x - x^2 - y^2 = 0$

In each of Problems 28 through 31, find the points of intersection of the graphs of the given pair of equations.

28. $y = 3x - 4$ and $3x + y = 4$
29. $y = 2x$ and $x^2 + y^2 = 9$
30. $y = |x|$ and $2y - x = 2$

31. $x + 2 = y^2$ and $2y - x - 3 = 0$

In each of Problems 32 through 39, sketch the graph of the given relation.

32. $2y - x < 5$

33. $x + 3y \geq 8$

34. $|y - x| = 2$

35. $|y - x| = |y + x|$

36. $(x - 2y)(2x - y) \geq 0$

37. $(x + 3)^2 + (y - 5)^2 > 16$

38. $|x| + |y| = 0$

39. $y - |y| = x + |x|$

In each of Problems 40 through 43, find an equation or inequality satisfied by the coordinates of the given set of points in R^2.

40. All points in the right half-plane not including the y-axis.

41. All points less than two units from the point $(-1, -3)$.

42. All points that are more than three units, but less than four units, from the origin.

43. All points equidistant from $(-1, 3)$ and $(4, 2)$.

44. Sketch the graph of the equations

$$y = |2 + 3x| \quad \text{and} \quad y = |3 - 2x|.$$

Determine the set of real numbers for which the inequality $|2 + 3x| \geq |3 - 2x|$ is satisfied.

45. Determine the set of real numbers for which the inequality $|2x| < |4 - x|$ is satisfied (see Problem 44).

1.4 FUNCTIONS

In this section we begin to develop one of the central concepts of calculus—that of a **function**. This is an idea* that has evolved gradually over a period of centuries, and that can be described in a number of slightly different, although complementary, ways. In ordinary conversation the word "function" has several meanings, one of which signifies some kind of dependence of one quantity upon another. It is this meaning of the word that we wish to clarify sufficiently to use mathematically.

Functions as relations or mappings

At first the word "function" was used to describe simple relations between variables. For example, the volume V of a sphere is a function of the radius r:

$$V = \tfrac{4}{3}\pi r^3. \tag{1}$$

The surface area S of a cube is a function of the side s:

$$S = 6s^2. \tag{2}$$

*The use of the word "function" in a mathematical sense is due to Leibniz, who adopted it beginning in 1673 to describe any quantity varying along a curve. During the following century and a half the concept was gradually expanded and refined, and was occasionally the source of sharp dispute among the leading mathematicians of the eighteenth century, including Euler and Lagrange. The notation $f(x)$ for function values is due to Euler (1734). In 1837 Peter Gustav Lejeune Dirichlet formulated what is essentially the modern definition as given in this section. Dirichlet (1805–1859) was professor at Berlin and later at Göttingen; he is also famous for his work in number theory, Fourier series, and partial differential equations.

The temperature C in degrees Centigrade is a function of the temperature T in degrees Fahrenheit:

$$C = \tfrac{5}{9}(T - 32). \tag{3}$$

Or we might simply have y given as a function of x by some formula, such as

$$y = \sqrt{4 - x^2}. \tag{4}$$

A great advance occurred (during the nineteenth century) when it was realized that the concept of a function did not require that there be an explicit formula such as those given in Eqs. 1 through 4. For example, for any positive real number x there is a certain number p of primes that do not exceed x. If $x = 20$, the prime numbers less than 20 are 2, 3, 5, 7, 11, 13, 17, and 19; hence, corresponding to $x = 20$ the value of p is 8. Simply by listing the prime numbers in order and then counting, it is possible in principle to find the value of p associated with any finite value of x. However, there is no formula that gives the value of p for an arbitrary value of x. Nevertheless, just as we thought of V as a function of r in Eq. 1, it seems reasonable also to consider p as a function of x.

We will first explore the idea of a function informally, and postpone a formal definition until later in the section. In each of the preceding examples there are three features that must be emphasized: two sets X and Y, and a rule that assigns to each element in the set X one and only one element in the set Y. The sets X for each of Eqs. 1 through 4 respectively, are $\{r|r \geq 0\}$, $\{s|s \geq 0\}$, $\{T|T \geq T_0$, where $T_0 \cong -459.7$ corresponds to absolute zero$\}$, and $\{x| -2 \leq x \leq 2\}$.The set Y specifies the type of values that the function assigns. In each of Eqs. 1 through 4 these values are real numbers, so Y can be taken to be R^1. We could also use subsets of R^1 that include all of the values of the function; thus, for Eq. 1 we could take Y to be the set of all nonnegative numbers, or perhaps the set of all numbers greater than -2. In the first fourteen chapters of this book we are concerned only with functions for which the sets X and Y are subsets of R^1. Such functions are called **real-valued functions of a real variable**. However, in the general concept of a function as a relation or dependence there is no requirement that the sets X and Y be sets of real numbers.

Functions are usually denoted by letters such as f, g, h, ϕ, and ψ. If f is a function and $x \in X$, then the element $y \in Y$ assigned to x by the function f is denoted by $f(x)$. The symbol $f(x)$ is read "f of x"; we stress that it does not mean "f times x," but rather is the **value of the function** f at x, or the **image** of the point x under the function f. Thus, if g is the function associated with Eq. 1, then

$$g(r) = \tfrac{4}{3}\pi r^3, \qquad r \geq 0; \tag{5}$$

and, for example,

$$g(2) = \tfrac{4}{3}\pi 2^3 = \tfrac{32}{3}\pi, \qquad g(3) = \tfrac{4}{3}\pi 3^3 = 36\pi.$$

Similarly, if h is the function associated with Eq. 3, then

$$h(T) = \tfrac{5}{9}(T - 32), \qquad T \geq T_0, \tag{6}$$

and, for example,

$$h(77) = \tfrac{5}{9}(45) = 25.$$

The set X is called the **domain** of the function and the set Y is called the **codomain**. Note that the function acts on *each element* in X, but that it is not necessary that every element of Y be a value of $f(x)$. Further, although a function must assign a single value in Y to each element x in X, the same value may be assigned to two or more values of x. For instance, if ϕ is the function associated with Eq. 4, then

$$\phi(x) = \sqrt{4 - x^2}, \qquad -2 \leq x \leq 2, \tag{7}$$

and

$$\phi(-1) = \sqrt{3}, \qquad \phi(1) = \sqrt{3}.$$

Thus the value $\sqrt{3}$ is assigned both to -1 and to 1.

The elements of Y that are actually assigned by f to elements of X form the **range** of f. We use the notation

$$f(X) = \{y | y = f(x), \qquad x \in X\} \tag{8}$$

to denote the range of the function f with domain X. The range $f(X)$ is a subset of the codomain Y. For Eq. 5, $g(X) = [0, \infty)$, and for Eq. 7, $\phi(X) = [0, 2]$; note that in each of these cases we chose Y to be R^1.

A real-valued function f is said to be **bounded** if its range is a bounded set; otherwise, the function is **unbounded**. Thus, f is bounded if there exists a number K such that $|f(x)| \leq K$ for all $x \in X$. For example, the function g of Eq. 5 is unbounded, but the function ϕ of Eq. 7 is bounded.

We often speak of functions in terms of variables. The **independent variable** assumes values in the domain and the **dependent variable** assumes values in the range—a value of the independent variable is given and then the function assigns a value of the dependent variable. Implicit here is the concept that a function is a **transformation** or **mapping** that operates upon elements in its domain X to produce elements in its codomain Y. This viewpoint emphasizes the idea that a function is an active thing. It does something. For example, the function g of Eq. 5 takes a real number in $[0, \infty)$ and transforms it into another real number. This transformation or mapping aspect of a function is illustrated schematically in Figure 1.4.1.

In much the same way it is often helpful to think of a function in terms of a computer or calculator; it accepts *input* (a number in the domain) and, when the

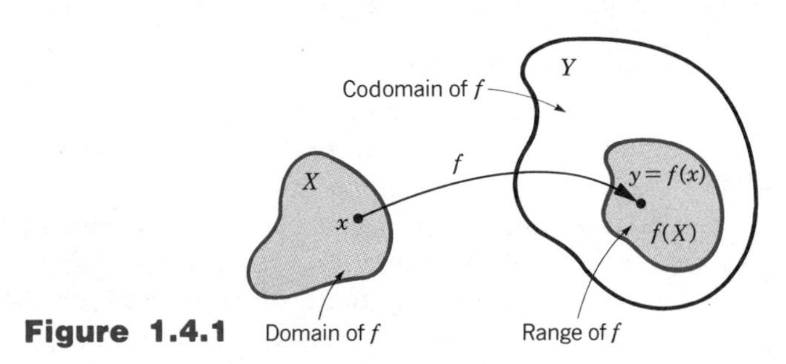

Figure 1.4.1

right buttons are pushed, produces a unique *output* (the corresponding number in the range).

We now formalize our discussion with a definition of a function.

DEFINITION 1.4.1 A function f is a rule (transformation, mapping) that assigns to each element x of a set X one and only one element y of a set Y. The sets X and Y are called the domain and codomain of the function f, respectively. The value of the function f for the element x is denoted by $f(x)$ and the range $f(X) \subset Y$ is given by $f(X) = \{y | y = f(x), x \in X\}$.

If x and y are real numbers, then the ordered pair (x, y) or $(x, f(x))$ for each $x \in X$ can be identified with a point in the xy plane. The set of all such points is the **graph** of the function f and provides a geometrical representation of the function. The graphs of the functions associated with Eqs. 1 and 4 are shown in Figures 1.4.2 and 1.4.3, respectively. The requirement that a single value of y be assigned to each x in the domain X of the function corresponds to the following geometrical statement: If x_0 is any point in the domain of the function f, then the line $x = x_0$, which is parallel to the y-axis, intersects the graph of the function once and only once. Thus the graph of the equation

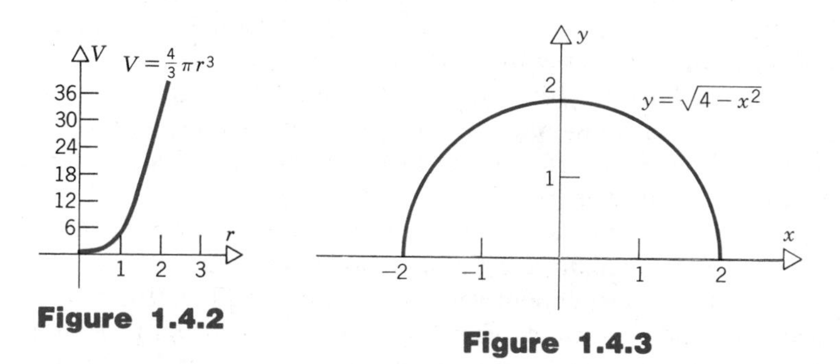

Figure 1.4.2

Figure 1.4.3

$$x^2 + y^2 = 4 \tag{9}$$

shown in Figure 1.4.4 does not represent a function. Of course, it is true that the function ϕ defined by Eq. 7,

$$\phi(x) = \sqrt{4 - x^2}, \qquad x \in [-2, 2]$$

and shown in Figure 1.4.3, satisfies Eq. 9, but there are other functions that also do this. One such function is given by

$$\psi(x) = -\sqrt{4 - x^2}, \qquad x \in [-2, 2] \tag{10}$$

whose graph is the lower semicircle of Figure 1.4.4. The graph of another relation that cannot be the graph of a function is shown in Figure 1.4.5. Observe that the line $x = x_0$ intersects the graph more than once.

We have tried to distinguish carefully between a function f and its value $f(x)$ for the element $x \in X$. These are two different things, as should be clear from the definition given earlier, and the distinction between them must always be kept in mind. Nevertheless, in mathematics, as in other areas of life, somewhat imprecise

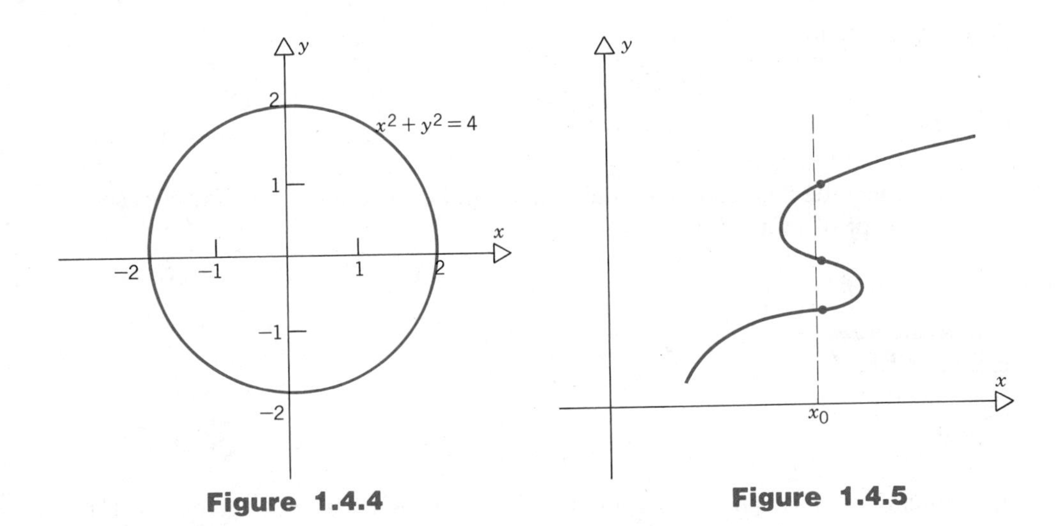

Figure 1.4.4

Figure 1.4.5

expressions are often used, when no confusion is likely to result, in order to save time and to avoid otherwise involved sentences. Thus, for example, it is common to speak of "the function $f(x) = x^2 + 1$" or "the function $x^2 + 1$" instead of "the function f whose values are given by $f(x) = x^2 + 1$." In accord with long-standing mathematical usage we will use such phraseology when no confusion seems likely.

Moreover, the domain X and the codomain Y of a function may not be given explicitly. For example, we might talk about the function

$$\phi(x) = \sqrt{4 - x^2},$$

without specifying the domain X. For this function $\phi(x)$ can be computed only if $4 - x^2 \geq 0$; hence the domain X of ϕ must be a subset of $[-2, 2]$. When the domain is not given it will be understood that *the domain of the function is the set of all values of the independent variable for which functional values can be computed.* Thus, for the function ϕ the domain is understood to be $[-2, 2]$. Again we emphasize that a function must act on each element in its domain and assign to it one and only one value in its codomain.

Algebra of real-valued functions

If f and g are real-valued functions, then algebraic operations on them are defined in terms of the corresponding operations on real numbers. The functions f and g are said to be **equal** if they have the same domain X and if $f(x) = g(x)$ for each $x \in X$. Further, if f and g have domain X, then their sum $f + g$, difference $f - g$, product fg, and quotient f/g are defined in the natural way, namely

$$\begin{aligned}
(f + g)(x) &= f(x) + g(x), && x \in X,\\
(f - g)(x) &= f(x) - g(x), && x \in X,\\
(fg)(x) &= f(x)g(x), && x \in X,\\
(f/g)(x) &= f(x)/g(x), && x \in X, \quad g(x) \neq 0.
\end{aligned}$$

Also, for any constant c,

$$(cf)(x) = cf(x), \qquad x \in X.$$

Since $f(x)$ and $g(x)$ are real numbers, these operations obey the usual commutative, associative, and distributive laws.

We conclude this section with several examples that illustrate different aspects of the concept of a function.

EXAMPLE 1

Consider the function

$$f(x) = \frac{2}{1 + x^2}.$$

The function values can be computed for any $x \in R^1$; hence $X = R^1$. Since the function assigns real numbers to elements in the domain, we can also take $Y = R^1$. It is easy to evaluate $f(0) = 2$, and to observe that $0 < f(x) < 2$ for $x \neq 0$. Further, $f(x)$ takes on smaller and smaller positive values as x moves away from the origin in either direction. Thus the graph of this function must resemble the sketch in Figure 1.4.6. The range of f is $f(X) = (0, 2]$. ∎

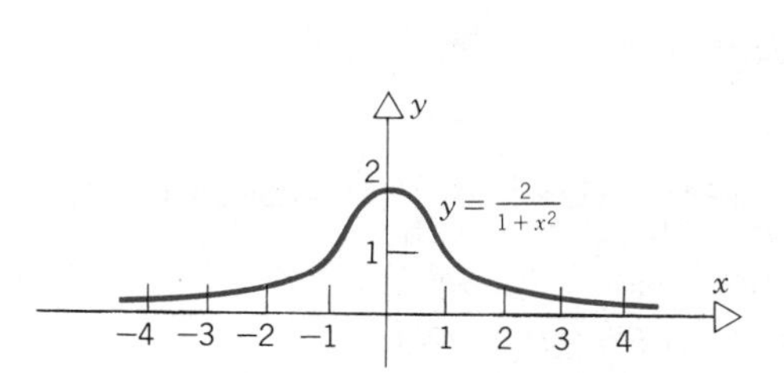

Figure 1.4.6

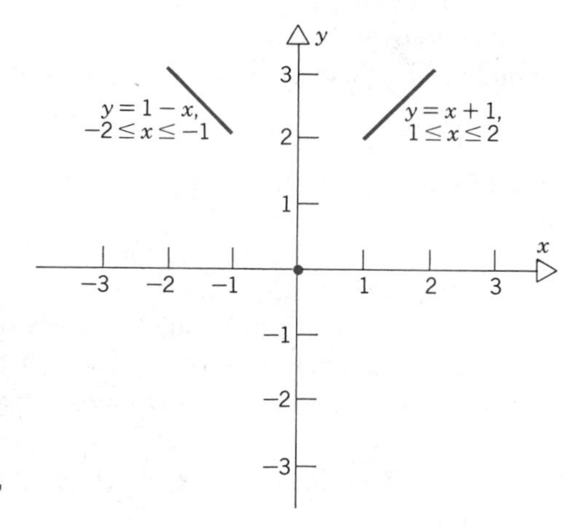

Figure 1.4.7

EXAMPLE 2

Consider the relation

$$y = \begin{cases} x + 1, & 1 \leq x \leq 2 \\ 0, & x = 0 \\ 1 - x, & -2 \leq x \leq -1. \end{cases}$$

This relation defines a function f whose graph is shown in Figure 1.4.7. The domain of f is $X = [-2, 1] \cup \{0\} \cup [1, 2]$ and its range is $f(X) = [2, 3] \cup \{0\}$. This example illustrates that a function need not be given by a single formula; also the domain need not be an interval. ∎

EXAMPLE 3

Given the function

$$f(x) = \sqrt{x + 3},$$

determine $f(s + 7)$, $f(1/t)$, and $f[f(u)]$; state the values of s, t, and u for which each expression is valid.

First, the domain of f consists of those points for which $x + 3 \geq 0$, or the interval $[-3, \infty)$, so this is the domain of f. Next, replacing x by $s + 7$, we obtain

$$f(s + 7) = \sqrt{s + 7 + 3} = \sqrt{s + 10},$$

which is valid for $s \geq -10$. Also, if we replace x by $1/t$, then

$$f\left(\frac{1}{t}\right) = \sqrt{\frac{1}{t} + 3} = \sqrt{\frac{1 + 3t}{t}},$$

for those values of t such that $(1 + 3t)/t \geq 0$. This requires that either $t > 0$ or $t \leq -\frac{1}{3}$. Finally, replacing x by $f(u) = \sqrt{u + 3}$, we have

$$f[f(u)] = \sqrt{\sqrt{u + 3} + 3},$$

provided that $u \geq -3$. ∎

In the next section we continue the discussion of functions and their graphs.

PROBLEMS

1. Let $f(x) = 2 - 3x$. Determine each of the following quantities.

   (a) $f(3)$ (c) $-f(3)$

   (b) $f(-3)$ (d) $-f(-3)$

2. Let $f(x) = x^2 + 4$. Determine each of the following quantities.

   (a) $f(2)$ (c) $2f(\frac{1}{2})$

   (b) $f(\frac{1}{2})$ (d) $\dfrac{f(2)}{2}$

In each of Problems 3 through 8, determine the domain X and the range $f(X)$ of the given function.

3. $f(x) = \dfrac{1}{x - 3}$
4. $f(x) = \dfrac{1}{(x - 3)^2}$
5. $f(x) = \dfrac{1}{x^2 - 9}$
6. $f(x) = \sqrt{x^2 - 9}$
7. $f(x) = \sqrt{9 - x^2}$
8. $f(x) = \dfrac{x}{1 - x}$
9. If $f(x) = \sqrt{1 - x^2}$, $-1 \leq x \leq 1$, find an expression for each of the following, and state for what values of the independent variable the formula is valid.

   (a) $f(s - 1)$ (c) $f\left(\dfrac{1}{u}\right)$

   (b) $f(-t)$ (d) $f[f(w)]$

10. Let $g(x) = 1/(4 - x^2)$, $x \neq \pm 2$. Find an expression for each of the following, and state for what values of the independent variable the formula is valid.

    (a) $g(2 - s)$ (c) $g\left(\dfrac{u}{2}\right)$

    (b) $g(2t)$ (d) $g[g(w)]$

In each of Problems 11 through 16, determine $[f(x + h) - f(x)]/h$, and express the result as simply as possible.

11. $f(x) = 2x - 3$
12. $f(x) = x^2$
13. $f(x) = 2x^2 + 3x - 4$
14. $f(x) = x^3$
15. $f(x) = \frac{1}{x}, \quad x \neq 0$
16. $f(x) = x + \frac{1}{x}, \quad x \neq 0$
17. Let $\phi(x) = |x + 2| + |x - 2|$. Evaluate $\phi(-3)$, $\phi(-1)$, $\phi(1)$, $\phi(3)$. Give a formula for ϕ in which absolute values do not appear. Sketch the graph of the function ϕ.

In each of Problems 18 through 23, determine whether the given relation defines y as a function of x.

18. $y = x^2, \quad x \in R^1$
19. $y^2 = x, \quad x \in [0, \infty)$
20. $|y| = |x|, \quad x \in R^1$
21. $y = |2x|, \quad x \in R^1$
22. $|y| = 2x, \quad x \in [0, \infty)$
23. $y^2 = 1 - x^2, \quad x \in [-1, 1]$
24. The prime number function π is defined as follows: $\pi(x)$ is the number of primes less than or equal to x. The domain of this function is $X = \{x | x > 0\}$. What is $\pi(4)$, $\pi(5.5)$, $\pi(8.9)$? Sketch the graph of the function π for $0 < x \leq 10$.
25. Determine a simple function f that has the following properties. Its domain is R^1, $f(-2) = 0$, $f(3) = 0$, and $f(0) = -36$.
26. Determine a simple function f with the following properties. Its domain is $[-2, 2]$ and $f(-2) = f(0) = f(2) = 0$.

1.5 EXAMPLES OF FUNCTIONS

In this section we discuss several real-valued functions of a real variable that occur often in a study of the calculus. A sketch of the graph of each of the functions considered here can be obtained by using previous experience in graphing, by exercising reasonable judgment, and by evaluating the function at a few points. We will find later that certain techniques of the calculus are very useful in sketching the graphs of functions.

Polynomials

A large and useful class of functions are polynomials. In general, a **polynomial** is a function P that is defined by an equation of the form

$$P(x) = a_0 + a_1x + a_2x^2 + \cdots + a_nx^n, \tag{1}$$

where $a_0, a_1, \ldots, a_n$ are real constants and $a_n \neq 0$. The number n is a nonnegative integer and is called the **degree** of the polynomial. Since Eq. 1 yields a value of $P(x)$ for each value of x, the domain of a polynomial is all of R^1. The range, however, may be R^1 or some subset thereof. In the following examples we consider three simple polynomials: a constant function, a linear function, and a quadratic function.

EXAMPLE 1

Let the function f be defined by

$$f(x) = k \tag{2}$$

where k is a constant. The domain of the constant function is R^1, and every point in the domain is mapped into the point k. Thus the range is the set consisting of the single element k. The graph of the constant function is a straight line parallel to the x-axis. The line is $|k|$ units above the x-axis if $k > 0$ and is $|k|$ units below if $k < 0$ (see Figure 1.5.1). ∎

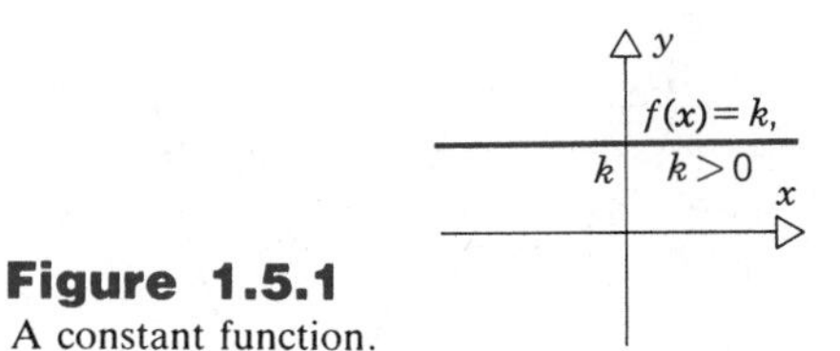

Figure 1.5.1
A constant function.

EXAMPLE 2

Let the function f be defined by

$$f(x) = mx + b, \tag{3}$$

where m and b are constants; thus f is a polynomial of degree one. As shown in Figure 1.5.2, the graph of f is the straight line with slope m that passes through $(0, b)$. The domain of f is R^1. If $m \neq 0$, its range is also R^1; if $m = 0$, then f is a constant function and its range is the single number b. If $m = 1$ and $b = 0$, then Eq. 3 reduces to $f(x) = x$. In this case f is referred to as the identity function; it maps each value of x into itself. Because its graph is a straight line the function f in Eq. 3 is sometimes called a **linear** function. ∎

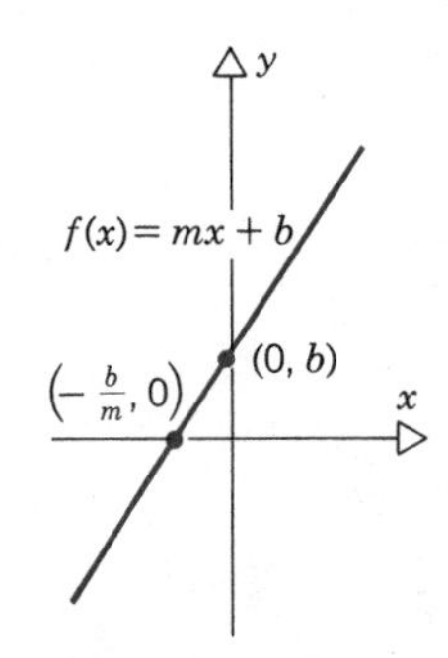

Figure 1.5.2
A linear function.

EXAMPLE 3

Let the function f be defined by

$$f(x) = x^2 - 2x - 3. \tag{4}$$

We can also write $f(x)$ in the factored form

$$f(x) = (x + 1)(x - 3),$$

from which it follows that $f(-1) = f(3) = 0$, $f(x) > 0$ for x in $(-\infty, -1) \cup (3, \infty)$, and $f(x) < 0$ for $-1 < x < 3$. Further, by completing the square in Eq. 4 we obtain

$$f(x) = x^2 - 2x + 1 - 4 = (x - 1)^2 - 4.$$

This shows that $f(1) = -4$ and that $f(x) > -4$ for all other values of x. Since $f(x)$ behaves like x^2 when $|x|$ is large, the range of f is $[-4, \infty)$. The graph of $y = f(x)$ is shown in Figure 1.5.3; it is a parabola with vertex at $(1, -4)$. A function of the form $f(x) = a_0 + a_1x + a_2x^2$, where a_0, a_1, and a_2 are constants with $a_2 \neq 0$ is called a **quadratic** function. The graph of a quadratic function is always a parabola. ∎

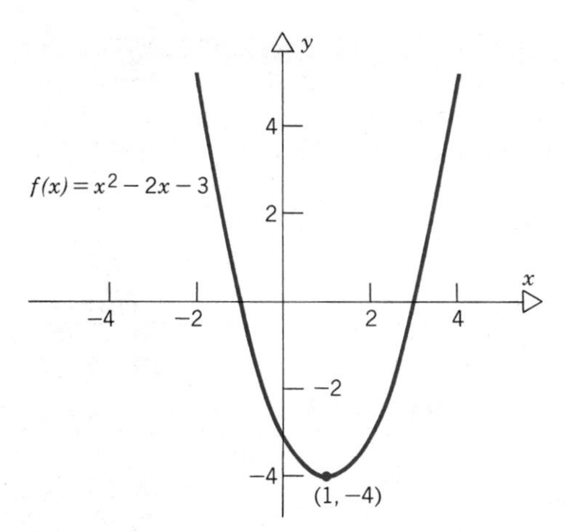

Figure 1.5.3 A quadratic function.

Rational functions

In some ways polynomials are analogous to the integers; for instance, the sum or product of any two polynomials is again a polynomial. However, just as division of integers leads to numbers that are not integers (the rational numbers), so does the division of one polynomial by another lead to functions that are not polynomials. Functions that arise as the ratio of two polynomials are known as **rational functions;** thus a rational function R has the form

$$R(x) = \frac{P(x)}{Q(x)}, \tag{5}$$

where P and Q are polynomials. Since division by zero is not permitted, a rational function is not defined at those points (if any) where the denominator is zero. Thus the domain of a rational function may or may not be all of R^1. The next two examples are simple rational functions.

EXAMPLE 4

Let the function f be defined by

$$f(x) = \frac{x}{x^2 + 1}. \tag{6}$$

Since the denominator is never zero, the domain of this function is all of R^1. The use of calculus is very helpful in sketching the graph of this function; however, a reasonably good sketch can be made by more elementary methods. It is clear that $f(0) = 0$, that $f(x) > 0$ when $x > 0$, and that $f(x) < 0$ when $x < 0$. Hence the graph passes through the origin, and otherwise lies entirely in the first and third quadrants. Further, when x is large and positive, $f(x)$ is small and positive because $x^2 + 1 \cong x^2$ and $f(x) \cong x/x^2 = 1/x$. Hence for large positive x the graph of f lies slightly above the x-axis. The situation is similar for large negative values of x, except that the graph lies slightly below the x-axis. This information along with the fact that $f(1) = \frac{1}{2}$, $f(-1) = -\frac{1}{2}$, $f(2) = \frac{2}{5}$, and $f(-2) = -\frac{2}{5}$ allows us to sketch the graph of the function (6); see Figure 1.5.4. The points P and Q at which $f(x)$ is greatest and least are called maximum and minimum points, respectively; calculus is especially useful in locating such points. For this particular example it turns out that the coordinates of P and Q are $P(1, \frac{1}{2})$ and $Q(-1, -\frac{1}{2})$. Thus the range of the function (6) is $[-\frac{1}{2}, \frac{1}{2}]$. ∎

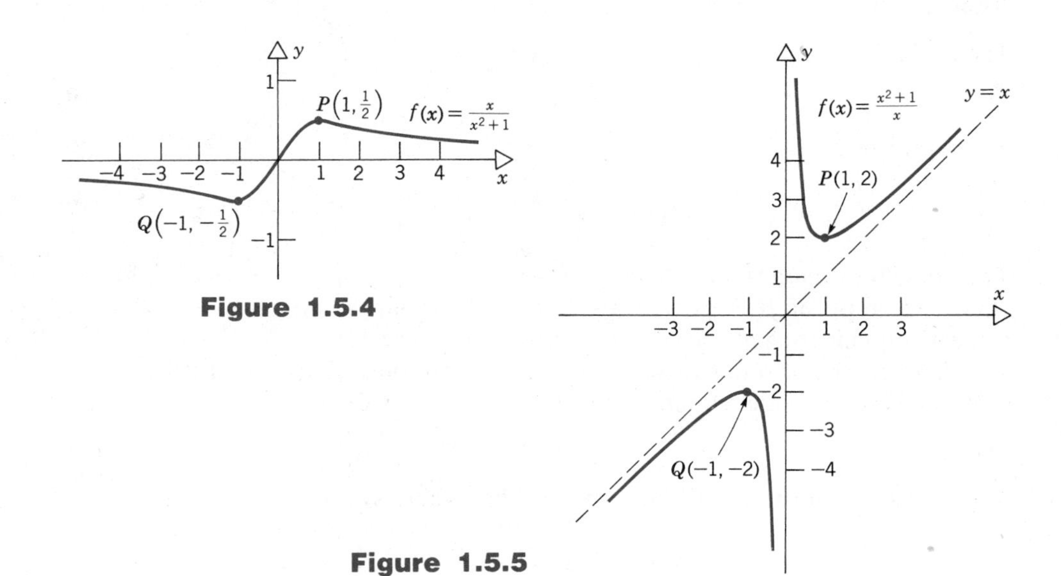

Figure 1.5.4

Figure 1.5.5

EXAMPLE 5

Let the function f be defined by

$$f(x) = \frac{x^2 + 1}{x}. \tag{7}$$

This function is the reciprocal of the function in Example 4. The denominator is zero when $x = 0$ and therefore the domain of f consists of R^1 with the point zero deleted. When x is near zero, $f(x) \cong 1/x$, so $f(x)$ assumes values that are very large in magnitude and are of the same sign as x. If we rewrite Eq. 7 in the form

$$f(x) = x + \frac{1}{x},$$

then we see that, when x is large in magnitude and positive, $f(x)$ is just slightly greater than x. Similarly, when x is large in magnitude and negative, $f(x)$ is just slightly less than x. Finally $f(1) = 2$ and $f(-1) = -2$. The graph of this function is sketched in Figure 1.5.5. Again, calculus would be helpful in sketching the graph of this function, particularly in locating the relative low and high points $P(1, 2)$ and $Q(-1, -2)$. The range of this function is $(-\infty, -2] \cup [2, \infty)$. ∎

Square root function

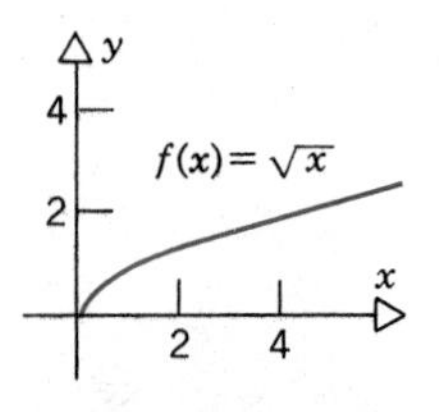

Figure 1.5.6
The square root function.

Consider the function defined by

$$f(x) = \sqrt{x}, \qquad x \geq 0. \tag{8}$$

The square root function given by Eq. 8 associates with each nonnegative real number x the nonnegative real number whose square is x. The square root function is steadily increasing as x increases, and its graph is sketched in Figure 1.5.6.

Absolute value function

The absolute value function is defined by

$$f(x) = |x|, \tag{9}$$

or alternatively

$$f(x) = \begin{cases} x, & x \geq 0, \\ -x, & x < 0. \end{cases} \tag{10}$$

The domain of the absolute value function is $(-\infty, \infty)$, or R^1. Its range is $[0, \infty)$.

The graph of $f(x) = |x|$ consists of two half-lines, with slopes 1 and -1, respectively, meeting at the origin as shown in Figure 1.5.7.

The graphs of other functions involving absolute values can also be readily sketched. For example, suppose that the function g is defined by

$$g(x) = |2x - 3|. \tag{11}$$

To express the function g without the absolute value symbols we note that

$$g(x) = \begin{cases} 2x - 3, & x \geq \frac{3}{2}, \\ -(2x - 3), & x < \frac{3}{2}. \end{cases} \tag{12}$$

The graph of g is also composed of two half-lines, with slopes 2 and -2, respectively, which meet at the point $(\frac{3}{2}, 0)$ on the x-axis. Note that this graph can be obtained from the graph of $y = 2x - 3$ by reflecting about the x-axis the part of the latter graph for which y is negative (see Figure 1.5.8).

Figure 1.5.7
The absolute value function.

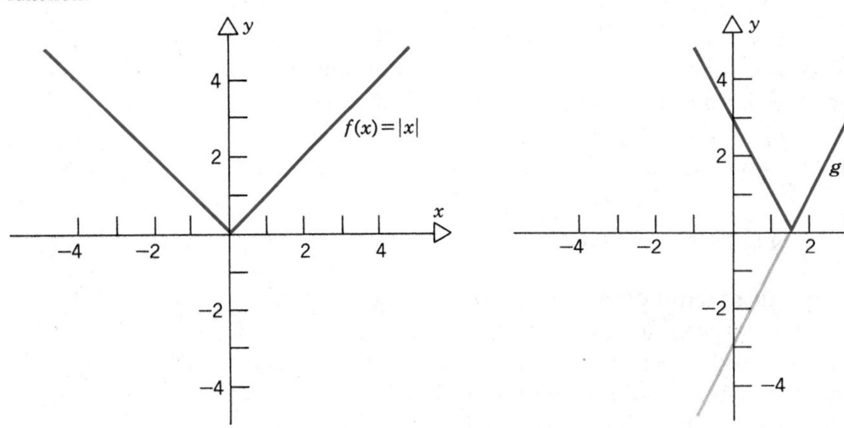

Figure 1.5.8

As a further example, consider the function h given by

$$h(x) = |x^2 - 2x - 3|. \qquad (13)$$

The graph of this function is obtained from that in Figure 1.5.3 by once again reflecting that part of the graph of $f(x) = x^2 - 2x - 3$ for which $f(x) < 0$ about the x-axis. The result is shown in Figure 1.5.9 and is the graph of the function h.

Figure 1.5.9

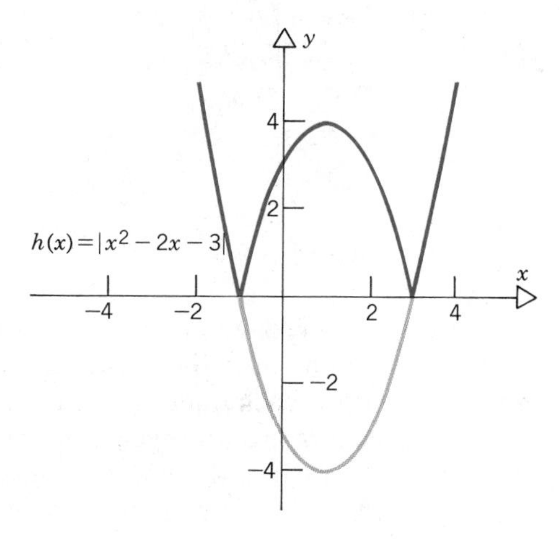

Signum function

This function is defined by

$$\operatorname{sgn}(x) = \begin{cases} \dfrac{x}{|x|}, & x \neq 0, \\ 0, & x = 0. \end{cases} \qquad (14)$$

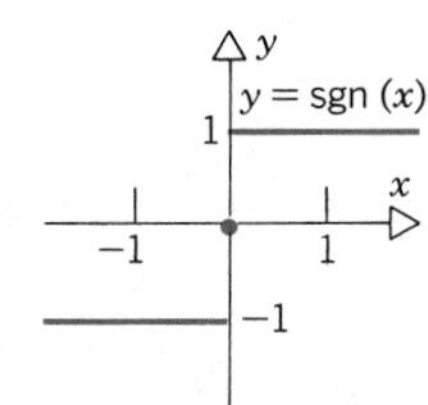

Figure 1.5.10 The signum function.

The graph of the signum function is drawn in Figure 1.5.10. This function has the value 1 if $x > 0$, -1 if $x < 0$, and 0 if $x = 0$. Therefore its range is $\{-1, 0, 1\}$. Note that the value of the signum function can be used to give the sign of the independent variable.

Greatest integer function

The greatest integer function f is defined by

$$f(x) = [\![x]\!] = \text{largest integer } n \text{ such that } n \leq x. \qquad (15)$$

Thus, $[\![-3]\!] = -3$, $[\![-1.5]\!] = -2$, $[\![0]\!] = 0$, $[\![1.1]\!] = 1$. The graph of the greatest integer function is shown in Figure 1.5.11. Its domain is R^1; its range is the set of all integers. The greatest integer function is constant in each interval of the form $[n, n+1)$, where n is an integer. At each integer point $x = n$ the value of the function $[\![x]\!]$ changes from $n-1$ to n; the function is said to have a jump or step of unit magnitude at each integer point. The greatest integer function is an example of a class of functions known as **step functions.**

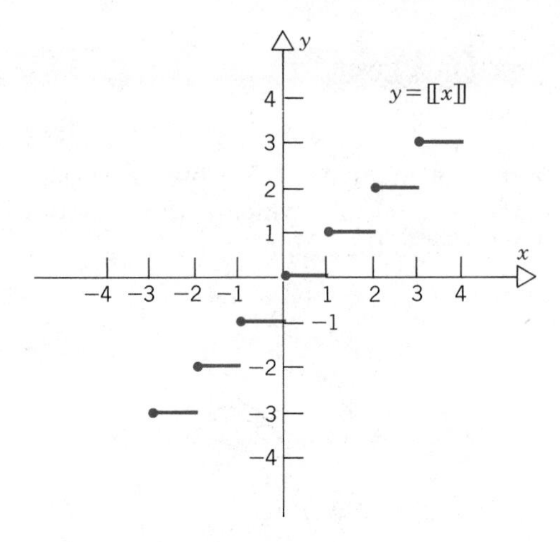

Figure 1.5.11 The greatest integer function.

Several everyday phenomena can be described by the greatest integer function, or variations of it. For instance, the postage for first class mail jumps by a given amount for each ounce of additional weight, but is constant between jumps. Many taxi meters are arranged similarly, with a given increment in fee for each mile or stated fraction thereof. Of course, in these applications the domain of the function is $[0, \infty)$, since negative weights or distances are of no significance.

EXAMPLE 6

Sketch the graph of the function f given by

$$f(x) = [\![2x - 3]\!]. \tag{16}$$

The graph of this function is somewhat similar to that of the function $[\![x]\!]$ itself. The value of $[\![2x - 3]\!]$ increases by one each time the value of $2x - 3$ passes through an integer. It is clear that $2x - 3$ assumes integral values not only when x itself is an integer, but also when x is equal to one half of an odd integer. Thus jumps in $[\![2x - 3]\!]$ occur at the points $x = 0, \pm\frac{1}{2}, \pm1, \pm\frac{3}{2}, \pm2, \ldots$. The graph of f is shown in Figure 1.5.12. ∎

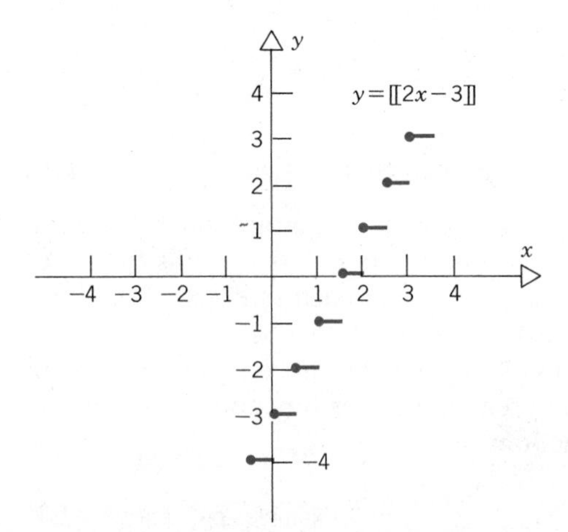

Figure 1.5.12

Factorial function

The domain of the factorial function is the set of positive integers and it is defined by

$$f(n) = n! = 1 \cdot 2 \cdot 3 \cdots n. \tag{17}$$

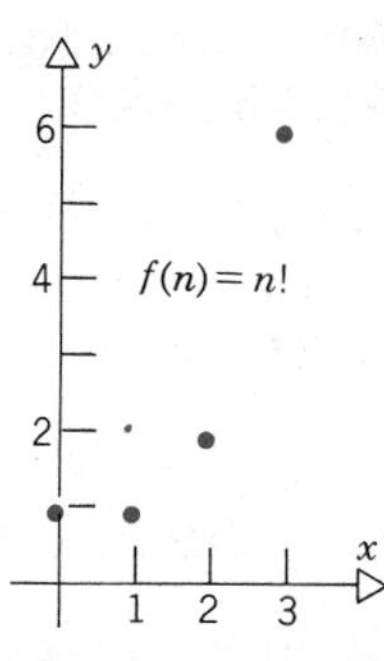

Figure 1.5.13 The factorial function.

The factorial function increases very, very rapidly as n increases. For example, $f(5) = 120$, $f(10) \cong 3.63 \times 10^6$, and $f(20) \cong 2.43 \times 10^{18}$. For convenience in writing formulas involving the factorial function it is often desirable to allow 0 to be in the domain of the function and to define $0! = 1$. The graph of the factorial function consists of the isolated points (0, 1), (1, 1), (2, 2), (3, 6), (4, 24), . . . (see Figure 1.5.13).

A pathological function

All of the functions considered up to now have graphs that are reasonably smooth curves or are composed of broken line segments in the xy-plane, except for the factorial function, which was defined only for the nonnegative integers. However, more complicated situations are possible, as the following example shows.

EXAMPLE 7

Let the function f be given by

$$f(x) = \begin{cases} 1, & \text{if } x \text{ is rational,} \\ 0, & \text{if } x \text{ is irrational.} \end{cases} \tag{18}$$

Equation (18) does indeed define a function on R^1, since for each value of x in R^1 there is associated a corresponding value of $f(x)$, either 0 or 1 as the case may be. Hence the range of f is the set consisting of the two numbers 0 and 1. The graph of this function consists of a set of points on the x-axis together with another set of points on the line $y = 1$; however, it is not a curve and it is impossible to draw a very satisfactory sketch. ∎

Even and odd functions

A function f with domain X is said to be **even** if $-x \in X$ whenever $x \in X$, and if

$$f(-x) = f(x), \qquad x \in X. \tag{19}$$

Similarly, f is called **odd** if $-x \in X$ whenever $x \in X$, and if

$$f(-x) = -f(x), \qquad x \in X. \tag{20}$$

The terms "even" and "odd" arise from the fact that $x^2, x^4, \ldots$ are even functions, while $x, x^3, \ldots$ are odd functions. From Eq. 19 it follows at once that the graph of an even function is **symmetric about the** y**-axis.** Similarly, from Eq. 20 the graph of an odd function is **antisymmetric about the** y**-axis,** or **symmetric about the origin.** Examples of graphs of even and odd functions are shown in Figure 1.5.14.

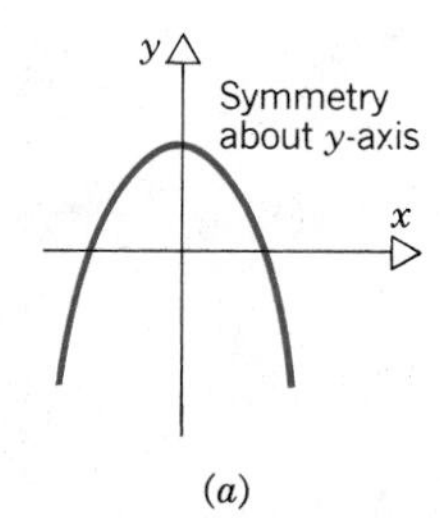

(a)

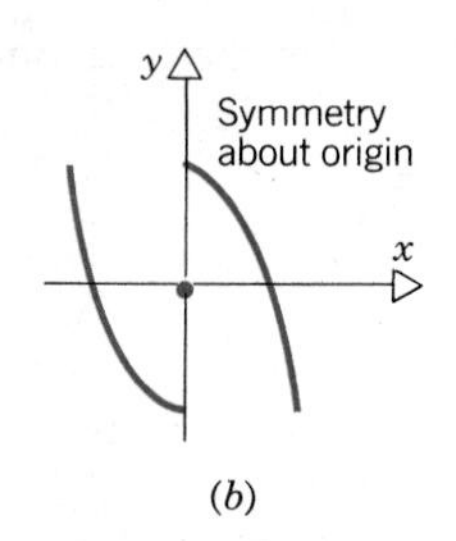

(b)

Figure 1.5.14 (a) An even function. (b) An odd function.

PROBLEMS

In each of Problems 1 through 20, sketch the graph of the given function. Also determine the range of each function.

1. $f(x) = |3 + 2x|$

2. $f(x) = |x + 2| + |x - 1|$

3. $f(x) = |x + 2| - |x - 1|$

4. $f(x) = |x + 1| + |x| + |x - 1|$

5. $g(x) = 2[\![x]\!]$

6. $g(x) = [\![2x]\!]$

7. $g(x) = -[\![x]\!]$

8. $g(x) = [\![-x]\!]$

9. $f(x) = x^2 - 1$

10. $g(x) = |x^2 - 1|$

11. $f(x) = 2x^2 + x - 6$

12. $g(x) = |2x^2 + x - 6|$

13. $g(x) = x(x - 1)(x + 2)$

14. $h(x) = (x + 3)(2x + 1)(4 - x)$

15. $\phi(x) = \sqrt{9 - x^2}, \quad |x| \le 3$

16. $\phi(x) = \sqrt{x^2 - 9}, \quad |x| \ge 3$

17. $\psi(x) = \dfrac{x}{1 - x^2}, \quad x \ne \pm 1$

18. $\psi(x) = \dfrac{1 - x^2}{x}, \quad x \ne 0$

19. $\psi(x) = \dfrac{1 - x}{1 + x}, \quad x \ne -1$

20. $\psi(x) = x|x|$

21. Let the function f be defined by
$$f(x) = \begin{cases} 0, & x < 0 \\ \sqrt{x}, & x \ge 0. \end{cases}$$
Sketch the graphs of $f(x + 2)$ and $f(x - 2)$.

22. The Heaviside unit step function is defined by
$$H(t) = \begin{cases} 1, & t \ge 0 \\ 0, & t < 0. \end{cases}$$
Sketch the graph of $H(t)$. Also sketch the graph of $f(t) = H(t) - H(t - \pi)$.

23. An interval on the x-axis at time t is defined by the set of points at which the function $H(x)H(2t - x)$ is nonzero. What is this interval for $t = \frac{1}{4}, \frac{1}{2}, 1, 2$? How fast is this interval expanding? See Problem 22 for the definition of the function H.

24. The cost of a taxi is 50 cents for the first fifth of a mile and 10 cents for each additional fifth of a mile. Give a formula for the cost C in dollars as a function of the miles traveled x for $0 \le x \le 2$.

In each of Problems 25 through 30, determine whether the given function is even, odd, or neither.

25. $f(x) = -3x$

26. $f(x) = 3x^2 + 4$

27. $f(x) = 2x^3 + 3x^2 - 7x + 4$

28. $f(x) = \dfrac{2x}{4 + 3x^2}$

29. $f(x) = \phi(|x|)$, for any function ϕ

30. $f(x) = \dfrac{x^2}{1 - x}$

31. Show that the sum of two even functions is an even function and that the sum of two odd functions is an odd function. What can you say about the sum of an even function and an odd function?

32. Show that the product of two even functions is an even function, the product of two odd functions is an even function, and the product of an even function and an odd function is an odd function.

33. Let f be a function with the property that whenever x is in its domain, then $-x$ is in its domain. If f is neither even nor odd, show that f is the sum of an even function g and an odd function h.

1.6 TRIGONOMETRIC FUNCTIONS

We assume that you are familiar with the trigonometric functions from the study of geometry and trigonometry. In calculus, and in theoretical science in general, the trigonometric functions have an importance much greater than simply their use in relating sides and angles of triangles. In this section we review the definitions of the trigonometric functions and some of their properties.

Radians

In trigonometry one often uses the degree as a unit in measuring the size of angles; 90° constitute a right angle, 360° form a full circle, and so on. In calculus we invariably use a different unit of measurement called the radian. Consider the circle $x^2 + y^2 = R^2$ shown in Figure 1.6.1. A point P on the circumference of this circle

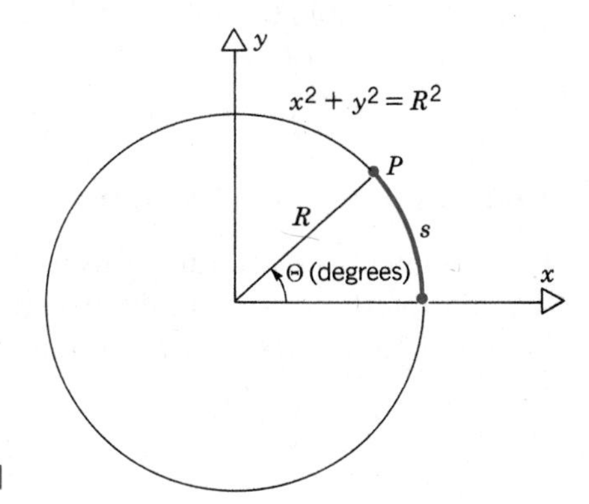

Figure 1.6.1

can be located by giving either an angle Θ in degrees, measured counterclockwise from the positive x-axis if $\Theta > 0$ and clockwise if $\Theta < 0$; or by the distance $|s|$ along the circumference,* measured counterclockwise if $s > 0$ and clockwise if $s < 0$. Figure 1.6.1 shows that the angle Θ in degrees has the same ratio to 360° as the number s has to $2\pi R$, the circumference of the circle:

$$\frac{\Theta}{360°} = \frac{s}{2\pi R}. \tag{1}$$

Note that in Eq. 1 there is no restriction on Θ and s; they can be positive or negative and of any magnitude. Also observe that both sides of Eq. 1 are dimensionless quantities, or pure numbers; that is, they have no physical dimensions. In particular, the right side of Eq. 1 is the ratio of two lengths. The number

$$\theta = \frac{s}{R} = \frac{2\pi}{360°}\Theta \tag{2}$$

can be used as another measure of the size of the given angle. This unit of measurement is called the **radian.** The left side of Eq. 2 is equivalent to

$$s = R\theta, \tag{3}$$

---

*We assume that we know how to measure distance along a circle. In Section 7.3 we take up the question of distance along a curve, including as a special case distance along a circle.

which is the fundamental relation among the radius R, the central angle θ in radians, and the corresponding arc length s. If $s = 2\pi R$, the circumference of the circle, then from Eq. 3 it follows that $\theta = 2\pi$. Consequently, since there are also 360° in a full circle,

$$2\pi \text{ radians are equivalent to } 360°. \tag{4}$$

Similarly, π radians are equivalent to 180°, $\pi/2$ radians to 90°, and so forth. One radian is equivalent to $360°/2\pi$, or about 57°18′. Other equivalencies between radians and degrees can easily be worked out. Hereafter when we mention angles we will always assume that they are measured in radians.

Students who are accustomed to measuring angles in degrees may wonder why it is desirable to introduce the radian. The reason is that certain basic formulas in the calculus of trigonometric functions take their simplest forms only when radians are used.

Definition of the trigonometric functions

In trigonometry these functions were defined for values of an angle; here, they are defined for values of a real variable. Consider the unit circle $x^2 + y^2 = 1$ shown in Figure 1.6.2. Given any real number θ, we locate the point $P(\theta)$ on the circum-

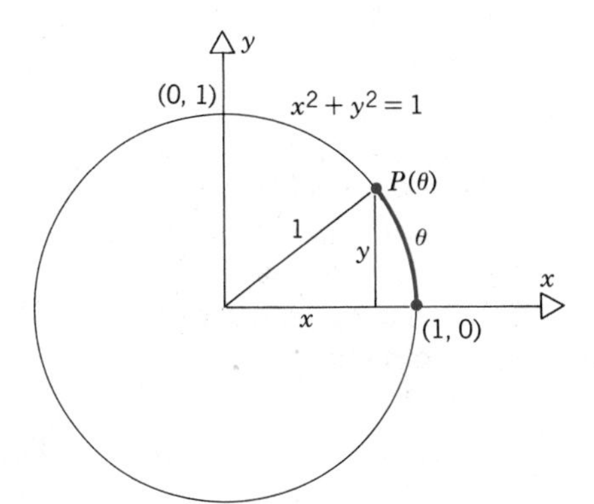

Figure 1.6.2

ference of the unit circle as follows. From the point (1, 0) we measure the distance $|\theta|$ counterclockwise if $\theta > 0$ and clockwise if θ is negative. Since now the radius $R = 1$, it follows that the length of an arc is numerically equal to the magnitude of the subtended angle measured in radians. Hence the angle subtended by the arc corresponding to θ is precisely the angle having the value θ in radians. Let (x, y) be the coordinates of the point $P(\theta)$. The six trigonometric functions are defined as follows.

$$\sin\theta = y, \tag{5a}$$

$$\cos\theta = x, \tag{5b}$$

$$\tan\theta = \frac{y}{x} = \frac{\sin\theta}{\cos\theta}, \qquad x \neq 0 \tag{5c}$$

$$\cot \theta = \frac{x}{y} = \frac{1}{\tan \theta} = \frac{\cos \theta}{\sin \theta}, \qquad y \neq 0 \tag{6a}$$

$$\sec \theta = \frac{1}{x} = \frac{1}{\cos \theta}, \qquad x \neq 0 \tag{6b}$$

$$\csc \theta = \frac{1}{y} = \frac{1}{\sin \theta}, \qquad y \neq 0. \tag{6c}$$

The domains of the sine and cosine functions are all of R^1. The domains of the tangent and secant functions are all of R^1, except those values of θ for which $x = 0$; namely, $\theta = \pm\pi/2, \pm3\pi/2, \ldots$. The domains of the cotangent and cosecant are all of R^1, except those values of θ for which $y = 0$; namely $\theta = 0, \pm\pi, \pm2\pi, \ldots$.

Since the other trigonometric functions can be expressed in terms of $\sin \theta$ and $\cos \theta$, it is natural to concentrate our attention primarily on these latter two functions, and to some extent on the tangent function. Values of the trigonometric functions for $\theta = \pi/6, \pi/4$, and $\pi/3$ can easily be read off from the triangles in Figure 1.6.3. Table 1.1 contains the values of the sine, cosine, and tangent functions for several values of θ. Of course, values of all six functions for any θ can be readily obtained with a pocket calculator.

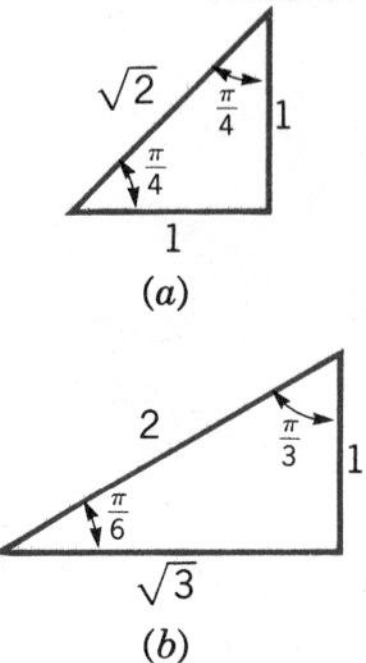

Figure 1.6.3

Table 1.1 Values of $\sin \theta$, $\cos \theta$, and $\tan \theta$ for several values of θ.

θ radians	0	$\frac{\pi}{6}$	$\frac{\pi}{4}$	$\frac{\pi}{3}$	$\frac{\pi}{2}$	$\frac{2\pi}{3}$	$\frac{3\pi}{4}$	$\frac{5\pi}{6}$	π	$\frac{3\pi}{2}$	2π
θ degrees	0°	30°	45°	60°	90°	120°	135°	150°	180°	270°	360°
$\sin \theta$	0	$\frac{1}{2}$	$\frac{\sqrt{2}}{2}$	$\frac{\sqrt{3}}{2}$	1	$\frac{\sqrt{3}}{2}$	$\frac{\sqrt{2}}{2}$	$\frac{1}{2}$	0	-1	0
$\cos \theta$	1	$\frac{\sqrt{3}}{2}$	$\frac{\sqrt{2}}{2}$	$\frac{1}{2}$	0	$-\frac{1}{2}$	$-\frac{\sqrt{2}}{2}$	$-\frac{\sqrt{3}}{2}$	-1	0	1
$\tan \theta$	0	$\frac{\sqrt{3}}{3}$	1	$\sqrt{3}$	not defined	$-\sqrt{3}$	-1	$-\frac{\sqrt{3}}{3}$	0	not defined	0

Periodicity

A function f is said to be **periodic** with period p if for each x in the domain of f the point $x + p$ is also in the domain of f, and $f(x + p) = f(x)$. The smallest positive number p with this property is called the **fundamental period.** Consider the trigonometric functions. If the number θ is increased by 2π the point $P(\theta + 2\pi)$ on the circumference of the unit circle is at the same position as the point $P(\theta)$ and hence has the same coordinates (see Figure 1.6.4). Thus it follows

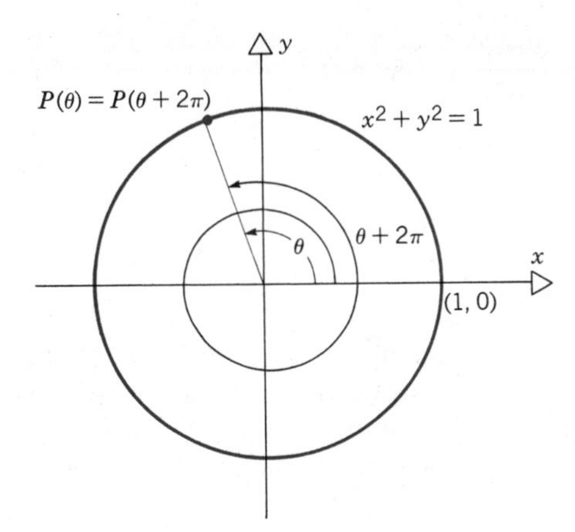

Figure 1.6.4

that all of the trigonometric functions are *periodic with period* 2π for all values of θ for which they are defined. Thus

$$\sin(\theta + 2\pi) = \sin\theta, \tag{7}$$

$$\cos(\theta + 2\pi) = \cos\theta, \tag{8}$$

and so forth. As a consequence, a knowledge of the trigonometric functions for θ in the interval $[0, 2\pi]$ is sufficient to determine these functions at all other points.

A closer examination of the tangent and cotangent functions shows that these two functions are actually periodic with a smaller period π; see Problem 17. Thus π is the fundamental period of the tangent and cotangent; 2π is the fundamental period of the other four functions.

Even and odd functions

If the point $P(\theta)$ has coordinates (x, y), then it is clear from Figure 1.6.5 that the point $P(-\theta)$ has coordinates $(x, -y)$. Hence

$$\cos(-\theta) = x = \cos\theta, \tag{9}$$

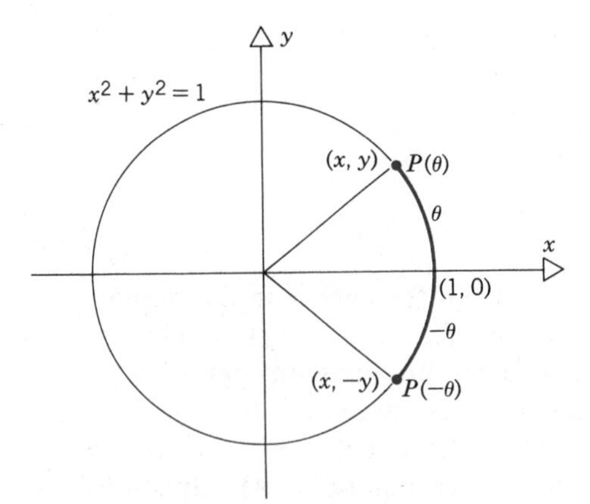

Figure 1.6.5

so the cosine function is an even function. Also

$$\sin(-\theta) = -y = -\sin\theta, \tag{10}$$

$$\tan(-\theta) = \frac{-y}{x} = -\tan\theta, \tag{11}$$

so the sine and tangent functions are odd functions. Similar relations can be derived for the other trigonometric functions. Notice, for example, that if the graph of the sine function is known for θ in $[0, \pi]$, then the fact that it is odd can be used to obtain the graph in $[-\pi, 0]$, and its periodicity with period 2π can be used to obtain the graph for all other values of θ.

Graphs of the trigonometric functions

The graphs of the six trigonometric functions are shown in Figures 1.6.6 through 1.6.11. The sine and cosine functions are bounded functions with range $[-1, 1]$. Each of the other trigonometric functions is unbounded; the range of the tangent and cotangent functions is $(-\infty, \infty)$, and of the secant and cosecant functions is $(-\infty, -1] \cup [1, \infty)$.

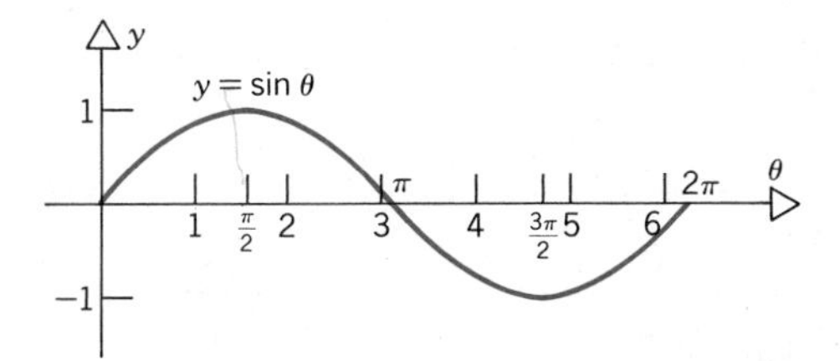

Figure 1.6.6 The sine function.

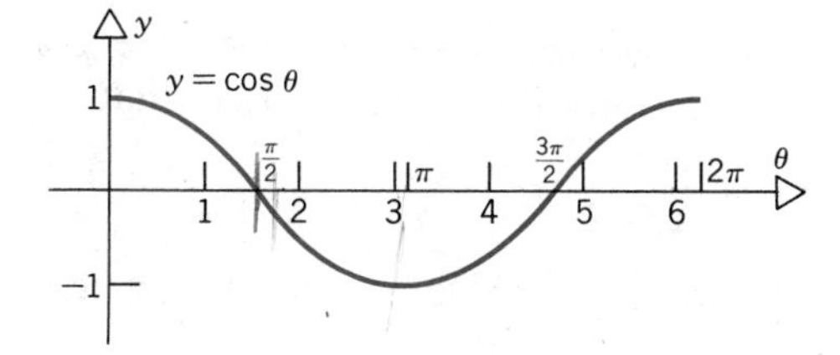

Figure 1.6.7 The cosine function.

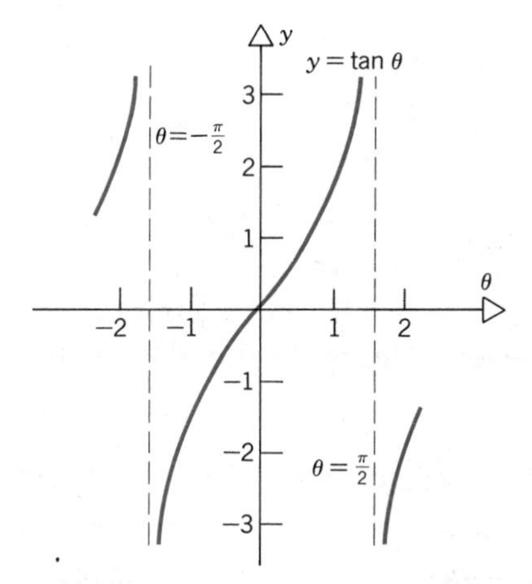

Figure 1.6.8 The tangent function.

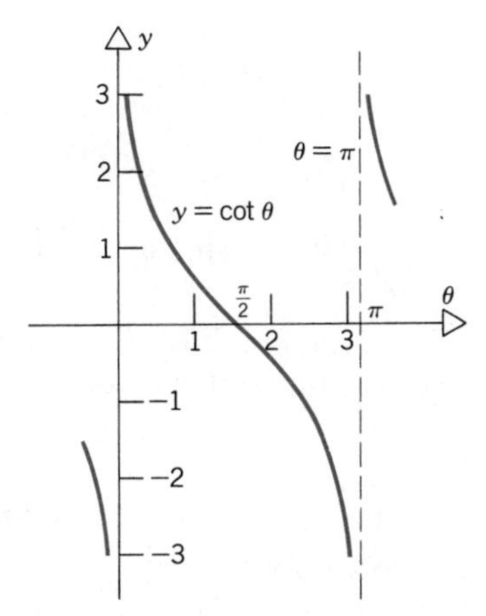

Figure 1.6.9 The cotangent function.

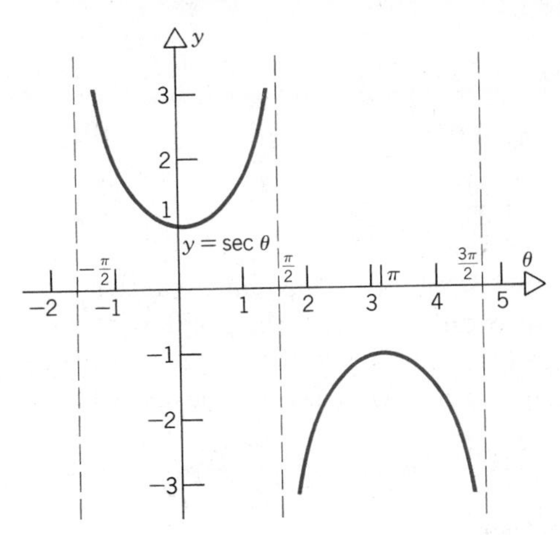

Figure 1.6.10 The secant function.

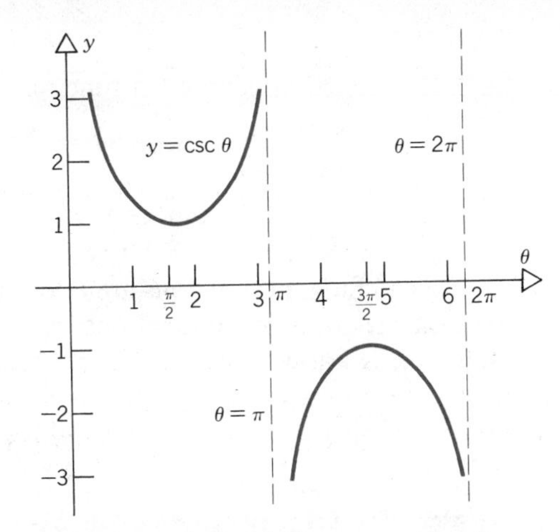

Figure 1.6.11 The cosecant function.

Identities

There are many relations among the six trigonometric functions. We assume that you are reasonably familiar with the common trigonometric identities. Some of the more important identities are listed here for reference.

Pythagorean identities. Since the point $P(\theta)$ is always on the unit circle $x^2 + y^2 = 1$, it follows immediately that

$$\sin^2 \theta + \cos^2 \theta = 1. \tag{12}$$

Dividing by $\cos^2 \theta$ and $\sin^2 \theta$, respectively, gives

$$1 + \tan^2 \theta = \sec^2 \theta, \qquad 1 + \cot^2 \theta = \csc^2 \theta. \tag{13}$$

Translation formulas. If the coordinates of the point $P(\theta)$ are (x, y), then the coordinates of the point $P(\theta + \pi/2)$ are $(-y, x)$ so

$$\cos\left(\theta + \frac{\pi}{2}\right) = -y = -\sin \theta \tag{14}$$

and

$$\sin\left(\theta + \frac{\pi}{2}\right) = x = \cos \theta. \tag{15}$$

Thus, for example, the graph of the cosine function can be obtained by translating the graph of the sine function a distance $\pi/2$ to the left. This is illustrated in Figure 1.6.12.

Law of cosines. Let a, b, and c be the sides of a triangle and let θ be the angle opposite side c; see Figure 1.6.13. Then

$$c^2 = a^2 + b^2 - 2ab \cos \theta. \tag{16}$$

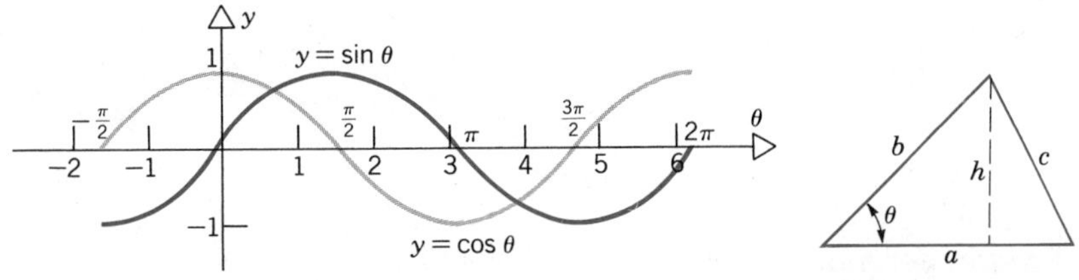

Figure 1.6.12

Figure 1.6.13

This formula can be derived by straightforward trigonometry (see Problem 33).

Sum and difference formulas. Let θ and ϕ be any real numbers; then

$$\cos(\theta + \phi) = \cos\theta\cos\phi - \sin\theta\sin\phi, \tag{17}$$

$$\cos(\theta - \phi) = \cos\theta\cos\phi + \sin\theta\sin\phi, \tag{18}$$

$$\sin(\theta + \phi) = \sin\theta\cos\phi + \cos\theta\sin\phi, \tag{19}$$

$$\sin(\theta - \phi) = \sin\theta\cos\phi - \cos\theta\sin\phi. \tag{20}$$

These identities, which should be very familiar and are fundamental, are derived in Problems 34 and 35. Note that the translation identities (14) and (15) can be derived by setting $\phi = \pi/2$ in Eqs. 17 and 19, respectively.

Double-angle formulas. Setting $\phi = \theta$ in Eqs. 17 and 19, we obtain

$$\cos 2\theta = \cos^2\theta - \sin^2\theta, \tag{21}$$

$$\sin 2\theta = 2\sin\theta\cos\theta. \tag{22}$$

Using the Pythagorean identity, we can also express $\cos 2\theta$ in the form

$$\cos 2\theta = 1 - 2\sin^2\theta \tag{23}$$

or

$$\cos 2\theta = 2\cos^2\theta - 1. \tag{24}$$

Half-angle formulas. If we set $\theta = \phi/2$ in Eqs. 23 and 24 and then solve for $\sin\phi/2$ and $\cos\phi/2$, respectively, we obtain

$$\sin\frac{\phi}{2} = \pm\sqrt{\frac{1 - \cos\phi}{2}}, \tag{25}$$

$$\cos\frac{\phi}{2} = \pm\sqrt{\frac{1 + \cos\phi}{2}}. \tag{26}$$

The choice of the + or − sign depends upon the quadrant in which $\phi/2$ lies. For example,

$$\sin\frac{\pi}{4} = \sin\frac{1}{2}\frac{\pi}{2} = +\sqrt{\frac{1 - \cos\pi/2}{2}} = \frac{1}{\sqrt{2}},$$

and

$$\cos\frac{3\pi}{4} = \cos\frac{1}{2}\frac{3\pi}{2} = -\sqrt{\frac{1+\cos\frac{3\pi}{2}}{2}} = -\frac{1}{\sqrt{2}}.$$

Product formulas. Adding Eqs. 17 and 18, we obtain

$$\cos\theta\cos\phi = \tfrac{1}{2}[\cos(\theta+\phi) + \cos(\theta-\phi)]. \tag{27}$$

Subtracting Eq. 17 from Eq. 18, we obtain

$$\sin\theta\sin\phi = \tfrac{1}{2}[\cos(\theta-\phi) - \cos(\theta+\phi)]. \tag{28}$$

Finally adding Eqs. 19 and 20, we obtain

$$\sin\theta\cos\phi = \tfrac{1}{2}[\sin(\theta+\phi) + \sin(\theta-\phi)]. \tag{29}$$

This list of identities is by no means exhaustive; however, it is sufficient for most purposes as far as this book is concerned. Note that with a knowledge of identities (9), (10), (12), (17), and (19)

$$\begin{aligned}
\cos(-\theta) &= \cos\theta, \\
\sin(-\theta) &= -\sin\theta, \\
\sin^2\theta + \cos^2\theta &= 1, \\
\cos(\theta+\phi) &= \cos\theta\cos\phi - \sin\theta\sin\phi, \\
\sin(\theta+\phi) &= \sin\theta\cos\phi + \cos\theta\sin\phi,
\end{aligned}$$

the other identities, with the exception of the law of cosines, can be readily derived. It will also be helpful for you to be generally familiar with the graphs of at least the sine, cosine, and tangent functions.

We conclude this section with a remark about the classification of functions. A real-valued function f of a real variable x is called **algebraic** if $f(x)$ can be computed by carrying out a finite number of additions, subtractions, multiplications, divisions, and root operations on x. All polynomials, rational functions, and radical functions are algebraic functions. Functions that are not algebraic are called **transcendental.** The trigonometric functions are transcendental functions, although we do not prove this here.

PROBLEMS

In each of Problems 1 through 8, use the identities in the text to evaluate the given quantity. Check your answer by using a calculator to evaluate the given function directly.

1. $\sin\left(\frac{\pi}{2} + \frac{\pi}{4}\right)$

2. $\cos\left(\frac{2\pi}{3} - \frac{3\pi}{2}\right)$

3. $\sin\left(\frac{\pi}{3} - \frac{5\pi}{4}\right)$

4. $\cos\left(\frac{7\pi}{6} + \frac{7\pi}{4}\right)$

5. $\sin\frac{\pi}{8}$

6. $\cos\frac{\pi}{8}$

7. $\cos\frac{\pi}{12}$

8. $\sin\frac{\pi}{12}$

In each of Problems 9 through 14, sketch the graph of the given function.

9. $f(x) = \sin 2x, \qquad 0 \le x \le 2\pi$

10. $f(x) = 2 \sin \frac{1}{2}x, \qquad 0 \le x \le 4\pi$

11. $f(x) = -2 \cos\left(x - \frac{\pi}{4}\right), \qquad 0 \le x \le 2\pi$

12. $f(x) = 3 \cos\left(2x - \frac{\pi}{4}\right), \qquad -\pi \le x \le \pi$

13. $f(x) = 2 \sin 3x, \qquad 0 \le x \le 2\pi$

14. $f(x) = |\sin x|, \qquad -\pi \le x \le \pi$

15. Using the definitions, show that the secant function is an even function, and that the cosecant and cotangent functions are odd functions.

16. Using the definitions, show that if n is a positive integer, then each of the trigonometric functions is periodic with period $2n\pi$; that is, for example, show that $\sin(\theta + 2n\pi) = \sin\theta$.

17. On the unit circle locate the points $P(\theta)$ and $P(\theta + \pi)$. Using the coordinates of these two points show that $\tan(\theta + \pi) = \tan\theta$ and $\cot(\theta + \pi) = \cot\theta$. Consequently the tangent and cotangent functions are periodic with period π.

In each of Problems 18 through 21, find the fundamental period of the given function.

18. $f(x) = \sin 2x$

19. $f(x) = \sin \frac{1}{2}x$

20. $f(x) = 3 \cos 3x$

21. $f(x) = 2 \cos(\frac{1}{3}x + 4)$

22. Show that the functions $f(x) = ax + b$ and $f(x) = ax^2 + bx + c$, $a \neq 0$, are not periodic with any period. More generally it can be shown that the only rational function that is periodic is the constant function $f(x) = c$, which can have any period.

Derive each of the identities in Problems 23 through 26 by letting $\theta = (x + y)/2$ and $\phi = (x - y)/2$ in formulas (27) through (29).

23. $\cos x + \cos y = 2 \cos\left(\frac{x + y}{2}\right) \cos\left(\frac{x - y}{2}\right)$

24. $\cos x - \cos y = -2 \sin\left(\frac{x + y}{2}\right) \sin\left(\frac{x - y}{2}\right)$

25. $\sin x + \sin y = 2 \sin\left(\frac{x + y}{2}\right) \cos\left(\frac{x - y}{2}\right)$

26. $\sin x - \sin y = 2 \sin\left(\frac{x - y}{2}\right) \cos\left(\frac{x + y}{2}\right)$

27. If $a \sin\theta + b \cos\theta = R \sin(\theta + \delta)$ for all values of θ, where a, b, and δ are constants, show that $R = \sqrt{a^2 + b^2}$ and that δ is the angle such that $\cos\delta = a/R$ and $\sin\delta = b/R$. The function $f(\theta) = R \sin(\theta + \delta)$ is periodic with period 2π, and has range $[-R, R]$. Its graph is a sine curve lying between $y = -R$ and $y = R$ and shifted a distance δ to the left from the origin (see Figure 1.6.14). The numbers R and δ are referred to

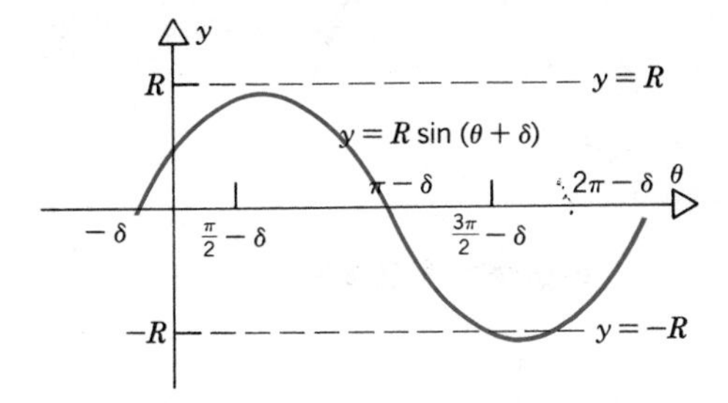

Figure 1.6.14

the **amplitude** and **phase angle,** respectively, of the periodic function.

28. Consider $f(\theta) = \sin a\theta$, where a is a positive constant called the **frequency.** Show that the period of $f(\theta)$ is $2\pi/a$; thus the period decreases as the frequency increases. Similar results hold for the other trigonometric functions.

In each of Problems 29 through 32, use the result of Problem 27 to express each of the given sums or differences in the form $R \sin(\theta + \delta)$. Then sketch the graph of the given function.

29. $\sin\theta + \sqrt{3} \cos\theta$

30. $\frac{3}{\sqrt{2}}(\sin\theta - \cos\theta)$

31. $3 \sin\theta - 2 \cos\theta$

32. $-\sin\theta + \cos\theta$

33. In order to derive the law of cosines, drop a perpendicular from the vertex opposite side a in Figure 1.6.13 to side a. Let the length of this perpendicular be h. Use trigonometry and the Pythagorean theorem to obtain the desired result.

34. In this problem we sketch the derivation of the identity $\cos(\theta - \phi) = \cos\theta \cos\phi + \sin\theta \sin\phi$. The identity for $\cos(\theta + \phi)$ follows by replacing ϕ by $-\phi$. Consider Figure 1.6.15.

   (a) Derive an expression for c^2 using the law of cosines.

   (b) Derive an expression for c^2 using the distance formula between two points. Note that the coordinates of $P(\theta)$ are $(\cos\theta, \sin\theta)$ and the coordinates of $P(\phi)$ are $(\cos\phi, \sin\phi)$.

(c) Equate the two expressions for c^2 to obtain the desired result.

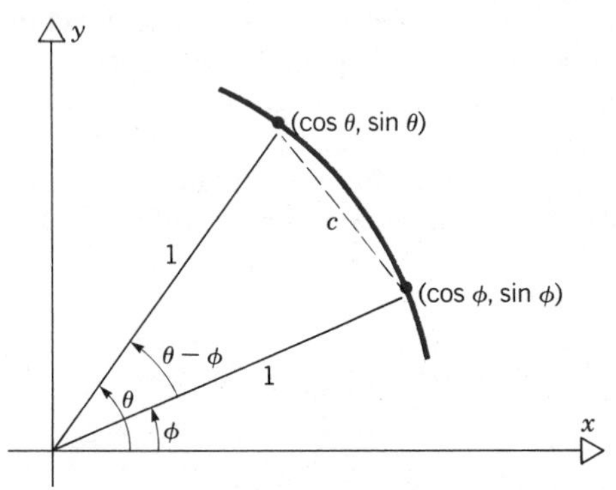

Figure 1.6.15

35. Derive the identity

$$\sin(\theta - \phi) = \sin\theta\cos\phi - \cos\theta\sin\phi$$

by using the result of Problem 34 and relations (14) and (15). The identity for $\sin(\theta + \phi)$ follows by replacing ϕ by $-\phi$.

In each of Problems 36 and 37 prove the given identity; assume that $h \neq 0$.

36. $\dfrac{\sin(x + h) - \sin x}{h} = \dfrac{\sin(h/2)}{h/2}\cos\left(x + \dfrac{h}{2}\right)$

37. $\dfrac{\cos(x + h) - \cos x}{h} = -\dfrac{\sin(h/2)}{h/2}\sin\left(x + \dfrac{h}{2}\right)$

38. If $f(x) = \sqrt{1 - \cos^2 x}$, compute $f(\pi/4)$, $f(3\pi/4)$, $f(5\pi/4)$, and $f(7\pi/4)$. For which of these values of x is $f(x) = \sin x$?

REVIEW PROBLEMS

In each of Problems 1 through 8, sketch the graph of the given relation.

1. $3x - 2y = 12$ **2.** $x + 3y \leq 0$

3. $2x^2 + 2y^2 - 4x + 8y + 1 = 0$

4. $x^2 + y^2 - 6x + 2y > 6$

5. $x + |x| = y + |y|$ **6.** $|y + 2| = |x - 3|$

7. $y = |1 + 2\cos x|$ **8.** $|y| = 1 + 2\cos x$

In each of Problems 9 through 16, find the domain X and the range $f(X)$ of the given function. Also sketch the graph of $y = f(x)$.

9. $f(x) = \dfrac{x + 3}{x + 1}$ **10.** $f(x) = |3x - 4|$

11. $f(x) = x\sin 2x$ **12.** $f(x) = [\![\sin x]\!]$

13. $f(x) = |x| - |x - 1|$

14. $f(x) = |3 - x| + |5 + 2x|$

15. $f(x) = 2 - x - x^2$ **16.** $f(x) = |2 - x - x^2|$

In each of Problems 17 through 24, determine the set of numbers satisfying the given inequality. Express your answer as an interval or as a union of intervals.

17. $x^2 \geq 16$ **18.** $3 - \dfrac{2}{x} \leq 1 + \dfrac{1}{x}$, $x \neq 0$

19. $-7 < \dfrac{6}{x} + x < 5$; $x \neq 0$

20. $|2x - 3| < |x + 1|$ **21.** $|3x + 1| \geq 2$

22. $2 < x^2 + x < 6$ **23.** $\cos 2x \geq \frac{1}{2}$

24. $\sin(x - \pi/6) \leq \frac{1}{2}$

25. Find the equation of the straight line passing through the point (2, 5) and parallel to the line $x + 2y = 4$.

26. Find the equation of the straight line passing through the point $(-2, 1)$ and perpendicular to the line $2x - 3y = 6$.

27. Find the equation of the circle with center $(-3, 1)$ and passing through the point (2, 3).

28. Find the equation of the circle with center (3, 2) that is tangent to the y-axis.

29. Find an equation satisfied by all points twice as far from $(-1, 3)$ as from (4, 2).

30. Find an equation satisfied by all points equidistant from the origin and the line $x = -2$.

In each of Problems 31 through 38, sketch the given function and state its domain X and range $f(X)$ as an interval or a union of intervals.

31. $f(x) = [\![x]\!] + x$ **32.** $f(x) = [\![x]\!] - 2x$

33. $f(x) = x(x - 4)(x + 4)/4$
Is this function even, odd, or neither?

34. $f(x) = |x^3 - 16x|/4$
Is this function even, odd or neither?

35. $f(x) = \dfrac{1}{[\![x]\!]}, \qquad x \geq 1$

36. $g(x) = \left[\!\left[\dfrac{1}{x}\right]\!\right], \qquad x \geq 1$

37. $f(x) = \dfrac{4 - x^2}{x}$

38. $f(x) = \dfrac{x^2 - 4}{x^2}$

In each of Problems 39 through 42, sketch the graph of the given function.

39. $f(x) = \left|\cos\left(2x - \dfrac{\pi}{2}\right)\right|$

40. $f(x) = \dfrac{1 - \cos 2x}{2}$

41. $f(x) = 2 \tan 2x$

42. $f(x) = \dfrac{1}{2} \tan\left(2x + \dfrac{\pi}{4}\right)$

In each of Problems 43 and 44, derive an equivalent expression for the given function.

43. $\sin^2 x$ in terms of $\tan x$.

44. $\csc^2 x$ in terms of $\sec x$.

45. If $b > a > 0$, show that

$$b^2 > ab > a^2,$$

and that

$$\frac{1}{a} > \frac{1}{b}.$$

Is either of these results always true if the condition $b > a > 0$ is replaced by the condition $b > a$?

46. If a and b are nonnegative real numbers, show that if $a^2 < b^2$ then $a < b$.
Hint: $b^2 - a^2 = (b - a)(b + a)$.

47. Show if $a > 0$, $b > 0$, then $\sqrt{a} + \sqrt{b} \geq \sqrt{a + b}$.
Hint: Show that $(\sqrt{a} + \sqrt{b})^2 \geq a + b$ and then use the result of Problem 46.

TWO

limits and continuous functions

Calculus was developed as a means of solving two types of problems, which had appeared in various forms since antiquity. These problems can be illustrated as follows.

1. If an object moves on a straight line in such a way that its position s and time t are related by the equation $s = f(t)$ for a given function f, find the velocity of the object at any time.
2. If f is a given function, find the area bounded by the graph of $y = f(x)$, the x-axis, and the vertical lines $x = a$ and $x = b$.

The first of these problems is solved by the process of *differentiation*, and the second by the process of *integration*. A first course in calculus is mainly concerned with clarifying the nature of these two processes, developing techniques by which they can be efficiently used, and pointing out some of their applications.

Both differentiation and integration are *limiting processes*, that is, they depend on a mathematical concept known as a *limit*. In this chapter we discuss this concept, first setting forth some of the mathematical ideas in an intuitive way, and later refining these ideas and treating them in more detail. In Chapters 3 and 6, respectively, we return to a discussion of differentiation and integration.

2.1 THE NATURAL OCCURRENCE OF LIMITING OPERATIONS

The mathematical concept of a limit is a refinement of an intuitive notion that occurs frequently in our everyday lives. We begin by considering one such occurrence of a limit.

Velocity

Consider a particle (proton, car, rocket) moving along a straight line, the s-axis. Suppose that we know the position s of the particle relative to a fixed origin at each moment of time t, that is, we know $s = f(t)$. Our experience with moving objects provides us with an intuitive idea of velocity, but what precisely does it mean to speak of the velocity v of the particle at time t, and how can we calculate this quantity?

EXAMPLE 1

Galileo (1564–1642) discovered that the position of a freely falling body is given approximately by

$$s = 16t^2, \qquad t \geq 0, \tag{1}$$

where s is measured in feet and t in seconds. Assuming that Eq. 1 correctly gives the position of a certain falling body, find its average velocity for the time intervals from $t = 1$ to $t = 2$ and from $t = 2$ to $t = 3$, respectively. Also determine the velocity at $t = 2$.

The average velocity during the time interval [1, 2] is the change in position during this interval, divided by the elapsed time. Thus, using an obvious notation,

$$v_{\text{av}}(1, 2) = \frac{64 - 16}{2 - 1} = 48 \text{ ft/sec}.$$

Similarly, for the interval [2, 3],

$$v_{\text{av}}(2, 3) = \frac{144 - 64}{3 - 2} = 80 \text{ ft/sec}.$$

To find the velocity at $t = 2$ is a more difficult task, because it clearly makes no sense to form the ratio of the change in position to elapsed time at a single instant. What we can do is to find the average velocity during a very short time interval. For instance, if $t > 2$, then the average velocity during the interval $[2, t]$ is

$$v_{\text{av}}(2, t) = \frac{16t^2 - 64}{t - 2}, \qquad t > 2. \tag{2}$$

Similarly, if $t < 2$, then the average velocity for the interval $[t, 2]$ is

$$v_{\text{av}}(t, 2) = \frac{64 - 16t^2}{2 - t}, \qquad 0 \leq t < 2. \tag{3}$$

Since the right sides of Eqs. 2 and 3 are the same, the average velocity during any interval beginning or ending at 2 is given by

$$v_{av} = h(t) = \frac{16t^2 - 64}{t - 2}, \qquad t \geq 0, \quad t \neq 2. \tag{4}$$

Observe that we have used functional notation to express v_{av} in terms of the endpoint t of the time interval. The domain of the function h is $[0, 2) \cup (2, \infty)$. We must exclude $t = 2$ because, as previously noted, the ratio is meaningless then. However, what happens if we assign values to t that are close to 2? Some results are shown in Table 2.1. Even a casual examination of the data in this table suggests that as t is given values closer and closer to 2, the average velocity takes on values closer and closer to 64. This conclusion is even more apparent if we factor the numerator on the right side of Eq. 4, and write

Table 2.1 **The average velocity, from Eq. 4, for various time intervals.**

t	Time Interval	v_{av}
1	[1, 2]	48
1.5	[1.5, 2]	56
1.9	[1.9, 2]	62.4
1.99	[1.99, 2]	63.84
1.999	[1.999, 2]	63.984
2.001	[2, 2.001]	64.016
2.01	[2, 2.01]	64.16
2.1	[2, 2.1]	65.6
2.5	[2, 2.5]	72
3	[2, 3]	80

$$v_{av} = \frac{16(t - 2)(t + 2)}{t - 2} = 16(t + 2), \qquad t \geq 0, \quad t \neq 2. \tag{5}$$

Observe that $t = 2$ must still be excluded from consideration, since Eq. 5 is derived from Eq. 4. However, there is now no doubt that v_{av} takes on values closer and closer to 64 as t is assigned values closer and closer to 2. Does it now make sense to say that the velocity at 2 is 64 (ft/sec)? Tentatively, at least, we shall answer this question in the affirmative. ∎

Now let us return to the general case where we assume that $s = f(t)$. If the s-axis is horizontal and directed to the right, then a positive or negative value of s corresponds to a position to the right or left of the origin, respectively. For example, the motion might be as shown in Figure 2.1.1*a* with the corresponding graph of f shown in Figure 2.1.1*b*. The average velocity during the time interval from t_1 to t_2 is given by

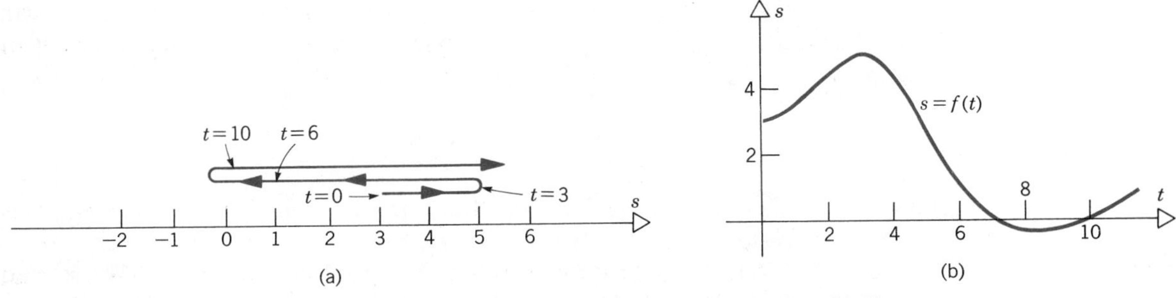

Figure 2.1.1 (a) A particle moving along a straight line. (b) The position of the particle as a function of time.

$$v_{av}(t_1, t_2) = \frac{s_2 - s_1}{t_2 - t_1} = \frac{f(t_2) - f(t_1)}{t_2 - t_1}. \qquad (6)$$

Note that the average velocity is *the net change in position divided by the elapsed time*, which is not necessarily the distance traveled divided by the elapsed time. Of course, these will be the same if the particle always travels in the positive direction. For the situation shown in Figure 2.1.1 the average velocity may be either positive, negative, or zero, depending upon the time interval under consideration. A quotient of the form (6) is called a **difference quotient** because it is a quotient of differences, in this case of position and time.

Although it is of some interest, the average velocity during a time interval does not give the velocity of the particle at a specific instant of time in the interval. Indeed, what do we mean by the velocity at a specific instant t_0, *the instantaneous velocity* at t_0? We have an intuitive conception of instantaneous velocity as contrasted with average velocity during a time interval. However, there is a difficulty in measuring instantaneous velocity because what we measure (or compute) directly is average velocity during a time interval from t_0 to t,

$$v_{av}(t_0, t) = \frac{s - s_0}{t - t_0} = \frac{f(t) - f(t_0)}{t - t_0}, \qquad t \neq t_0. \qquad (7)$$

Observe that, as in Example 1, t may be either greater or less than t_0. The shorter the time interval the closer the average velocity is to our conception of the instantaneous velocity at t_0. However, in trying to take shorter and shorter time intervals we will find that (regardless of the accuracy of our rulers and clocks) the accuracy of our measurements will eventually be open to serious question. Here the mathematician has an advantage, since there is no particular difficulty in *thinking* of arbitrarily short intervals of time and distance. Thus what we must do in Eq. 7 is to assign values to t that are closer and closer to t_0. We describe this process by saying that we let t approach t_0, and we write $t \to t_0$. As $t \to t_0$, both the numerator and denominator in Eq. 7 approach zero, and at the point t_0 the quotient is of the meaningless, or indeterminate, form 0/0. Nevertheless, the quotient may approach a definite number as $t \to t_0$. Indeed, it must do so if our conception of instantaneous velocity is realistic.

The process that we have described here, and illustrated in Example 1, can be summarized in the following way, making use of the customary notation of calculus.

$$v(t_0) = \lim_{t \to t_0} v_{av}(t_0, t) = \lim_{t \to t_0} \frac{f(t) - f(t_0)}{t - t_0} \tag{8}$$

Thus, if the difference quotient in Eq. 8 has a definite limit as $t \to t_0$, then this number is the velocity at t_0.

It is important to understand how Eq. 8 should be interpreted. We started with a conceptual idea of velocity at an instant, drawn from our everyday experience. It might seem reasonable next to try to formulate a precise definition of velocity and then to devise some means of calculating it. Instead, we calculate a related quantity, the average velocity during a short time interval, and then pass to the limit as the interval is made shorter and shorter. If this process yields a definite limiting value, we then *define* this value to be the velocity at the given instant. Thus Eq. 8 is the *definition* of velocity at the instant t_0. Although this manner of proceeding may seem roundabout at first, note that it does lead to a formula for the velocity that one can seek to evaluate in particular cases, as we did in Example 1. An analogous procedure will be followed in numerous other situations that we will encounter later. Of course, a definition such as Eq. 8 should be regarded tentatively at first, until it can be determined whether its use leads to any inconsistencies. In the case of velocity centuries of experience have validated definition (8).

Slope

Let us now examine from a geometrical point of view the procedure we have used to calculate the velocity of a particle moving on a straight line. The geometrical interpretation of Eq. 7 is shown in Figure 2.1.2. The points (t_0, s_0) and (t, s) on

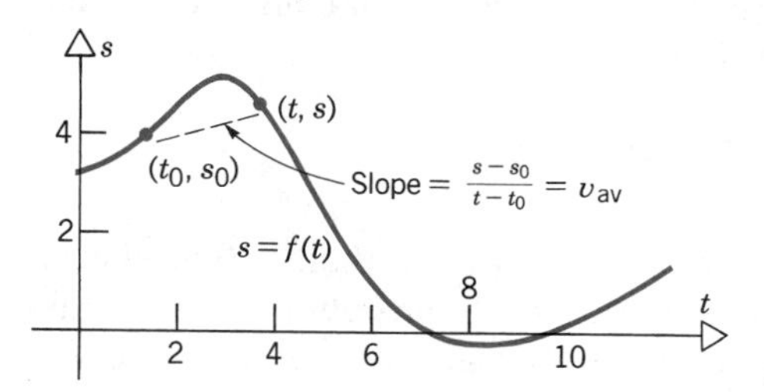

Figure 2.1.2

the graph of $s = f(t)$ are indicated, and the difference quotient $(s - s_0)/(t - t_0)$ in Eq. 7 is precisely the slope of the straight line (the *secant line*) passing through these two points. In other words, given the position s of the particle as a function of time t, *the average velocity during any time interval is the slope of the secant line joining the corresponding pair of points on the graph of* s = f(t).

Now consider the geometrical interpretation of Eq. 8 as $t \to t_0$. In Figure 2.1.3 we have drawn the secant lines with slopes

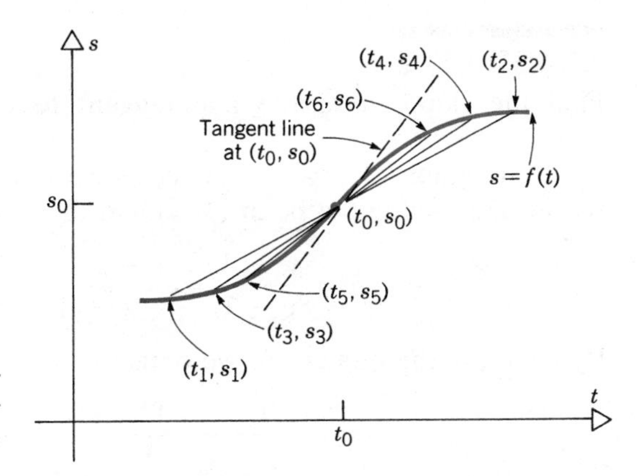

Figure 2.1.3 Secant lines approach the tangent line as $t \to t_0$.

$$\frac{s_2 - s_0}{t_2 - t_0}, \quad \frac{s_4 - s_0}{t_4 - t_0}, \quad \frac{s_6 - s_0}{t_6 - t_0}, \ldots \tag{9}$$

for $t_2, t_4, t_6, \ldots$ greater than but successively closer to t_0, and the secant lines with slopes

$$\frac{s_0 - s_1}{t_0 - t_1}, \quad \frac{s_0 - s_3}{t_0 - t_3}, \quad \frac{s_0 - s_5}{t_0 - t_5}, \ldots \tag{10}$$

for $t_1, t_3, t_5, \ldots$ less than but successively closer to t_0. It seems intuitively reasonable that these slopes approach the slope of the line that is "tangent" to the graph of $s = f(t)$ at (t_0, s_0). In fact, that is how we *define* the slope m of the tangent* line to a curve at a point, namely

$$m = \lim_{t \to t_0} \frac{s - s_0}{t - t_0} = \lim_{t \to t_0} \frac{f(t) - f(t_0)}{t - t_0}. \tag{11}$$

By comparing Eq. 11 with Eq. 8, we conclude that the problems of finding the velocity of a moving particle and the slope of the tangent line to a curve are mathematically identical. This is a fact of central importance. For one thing, it provides a link between mechanics and geometry that may be helpful in the study of each. Of greater significance is that it illustrates the unifying role of mathematics. When fundamental quantities in both mechanics (velocity) and geometry (slope) are given by the same mathematical process, surely it is advisable to study that process and its properties, and then to apply the results in either field. Indeed, as we go along, we will find many other areas where the same mathematical results can be used.

---

*The problem of finding a line tangent to a given curve was considered by the Greeks, but with success only in a few special cases. Apollonius of Perga (c. 262–190 B.C.), for example, knew how to construct tangents to conic sections.

A general method for finding tangent lines was found more or less simultaneously about 1630 by Descartes and Fermat. Descartes' solution was contained in his general treatise on analytical geometry (1637), but Fermat's solution was not published until 1679. Nevertheless, Fermat's work was known to other mathematicians through correspondence, and is somewhat closer to the modern viewpoint. However, neither Descartes nor Fermat dealt satisfactorily with the need for a limiting operation.

EXAMPLE 2

Find the slope m of the line tangent to the graph of $y = 4 - x^2$ at the point (1, 3).

The graph of $y = 4 - x^2$ is shown in Figure 2.1.4. The slope $m_s(x)$ of the secant line joining the point (1, 3) and the arbitrary point (x, y) is

$$m_s(x) = \frac{y - 3}{x - 1} = \frac{4 - x^2 - 3}{x - 1} = \frac{1 - x^2}{x - 1}, \qquad x \neq 1.$$

By factoring the numerator we obtain

$$m_s(x) = \frac{(1 - x)(1 + x)}{x - 1} = -(1 + x), \qquad x \neq 1.$$

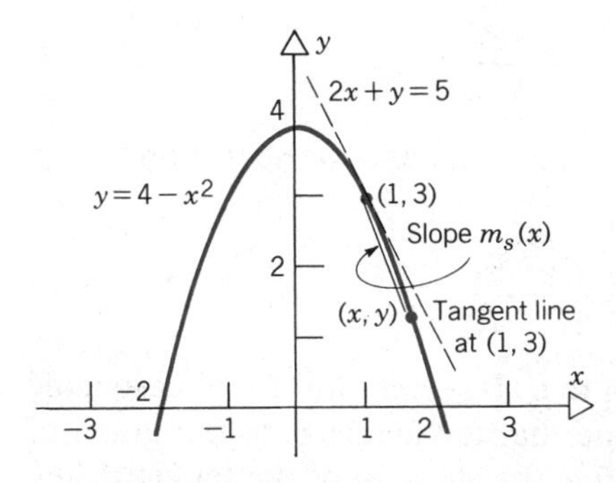

Figure 2.1.4

It is now easy to see that if x is assigned values closer and closer to 1, then $m_s(x)$ takes on values closer and closer to -2. Consequently,

$$m = \lim_{x \to 1} m_s(x) = -2$$

is the slope of the tangent line at (1, 3). The tangent line has the equation

$$y - 3 = -2(x - 1)$$

or

$$2x + y = 5. \blacksquare$$

EXAMPLE 3

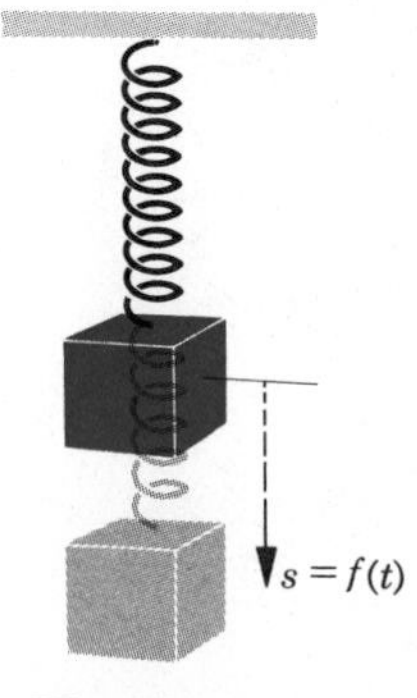

Figure 2.1.5
A vibrating spring–mass system.

Under certain conditions the mass shown in Figure 2.1.5 oscillates in such a way that at any time t its position is

$$s = f(t) = 2 \cos t.$$

Find the velocity v of the mass when $t = t_0 = 2\pi/3$.

The graph of $s = 2 \cos t$ is shown in Figure 2.1.6. The average velocity of the mass during a short time interval from $2\pi/3$ to t is given by the difference quotient

$$\begin{aligned} v_{av}(t) &= \frac{f(t) - f(2\pi/3)}{t - 2\pi/3} \\ &= \frac{2 \cos t - 2 \cos(2\pi/3)}{t - 2\pi/3} \\ &= \frac{2 \cos t + 1}{t - 2\pi/3}, \qquad t \neq \frac{2\pi}{3}, \end{aligned} \tag{12}$$

where t may be either greater or less than $2\pi/3$. Here we cannot simplify the last expression in Eq. 12 by factoring, as we did in the previous examples. However, if we evaluate $v_{av}(t)$ for values of t close to $2\pi/3$, we obtain the data in Table 2.2. From these results it appears that $v_{av}(t)$ approaches a value close to -1.73205 as $t \to 2\pi/3$. Thus we conclude that this value is the velocity of the mass when $t = 2\pi/3$. In fact, it is possible to show that $v = -\sqrt{3}$ when $t = 2\pi/3$. ∎

Table 2.2 The average velocity $v_{av}(t)$, from Eq. 12, for various values of t.

t	$v_{av}(t)$
$\frac{\pi}{2} \cong 1.57080$	−1.909859
2.0	−1.776642
2.05	−1.753676
2.09	−1.734243
2.094	−1.732248
2.0943	−1.732098
2.0944	−1.732048
2.0945	−1.731998
2.095	−1.731748
2.1	−1.729239
2.15	−1.703363
$\frac{5\pi}{6} \cong 2.61799$	−1.398114

Note: $2\pi/3 \cong 2.094395$; $\sqrt{3} \cong 1.732051$.

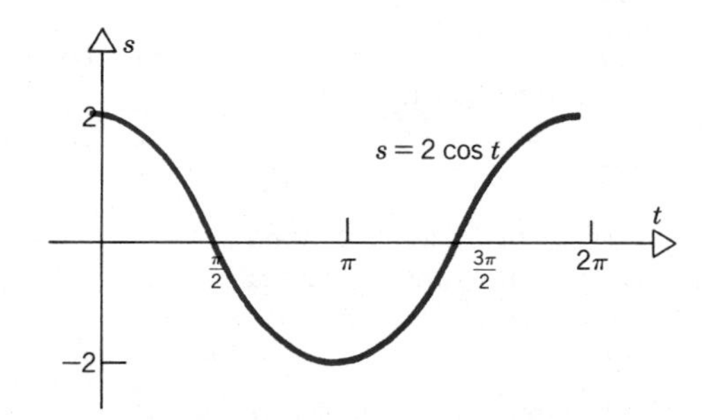

Figure 2.1.6 The position of the mass as a function of time.

In concluding this section, we emphasize the tentative nature of the conclusion we drew in Example 3. While the data in Table 2.2 appear convincing, one should be cautious in drawing firm conclusions about limits solely from numerical calculations. To deal with limits with greater assurance we need to establish a better understanding of the basic properties of limits. This is the purpose of the following sections.

PROBLEMS

In each of Problems 1 through 4, the position s of a moving particle is given as a function of time t. For the given instant t_0 find, in each case, the average velocity v_{av} during the interval $[t_0, t]$ and the velocity v at t_0.

1. $s = 5t^2 - 3, \quad t_0 = 1$
2. $s = t^2 + 3t, \quad t_0 = 2$
3. $s = 2t^3 - t^2, \quad t_0 = 2$
4. $s = t^3 - 3t + 4, \quad t_0 = 3$

In each of Problems 5 through 10, find the slope m of the line tangent to the graph of each of the following equations at the given point. Also find the equation of the tangent line.

5. $y = 3x^2 + 1; \quad (1, 4)$
6. $y = x^3 + 1; \quad (-1, 0)$
7. $y = 2x^2 - 3x + 4; \quad (2, 6)$
8. $y = \dfrac{1}{x}; \quad (2, \tfrac{1}{2})$
9. $y = \dfrac{5}{x + 3}; \quad (2, 1)$
10. $y = \dfrac{x - 3}{x + 2}; \quad (2, -\tfrac{1}{4})$
11. (a) Show that if $s = 16t^2$, as in Example 1, then the average velocity over any time interval is equal to the instantaneous velocity at the mid-point of the interval.

    (b) If $s = t^3$, find the instantaneous velocity at the point $t = t_0$. Also find the average velocity over the interval $[t_0 - \tau, t_0 + \tau]$. Note that in contrast to Part (a) these two quantities are not the same, although the average velocity approaches the instantaneous velocity as $\tau \to 0$.
12. The surface area S of a sphere is related to the radius r by $S = 4\pi r^2$.

    (a) Find the average rate at which S changes as r changes from r_0 to $r_0 + h$.

    (b) Find the instantaneous rate at which S changes with r when $r = r_0$.

    (c) Find the rate at which S changes with r when $r = 2$.
13. If an object is projected vertically upward from the surface of the earth with initial velocity v_0, if the force of gravity is considered constant, and if all other forces are neglected, then the position s of the object at time t is given by

    $$s = -\tfrac{1}{2}gt^2 + v_0 t$$

    where g is the gravitational acceleration.

    (a) By setting up and examining a suitable difference quotient, find the velocity of the object at any time t.

    (b) Find the time at which the object reaches its greatest height.

In each of Problems 14 through 21 form the difference quotient $[f(x) - f(x_0)]/(x - x_0)$. Calculate the value of the difference quotient for several values of x near x_0, and estimate the limiting value as $x \to x_0$.

14. $f(x) = \sqrt{4 - x^2}, \quad x_0 = 1$
15. $f(x) = \sqrt{x^2 + 5}, \quad x_0 = 2$
16. $f(x) = 1/\sqrt{x^2 + 1}, \quad x_0 = 1$
17. $f(x) = (x^2 - 5)^{3/2}, \quad x_0 = 3$
18. $f(x) = \sin x, \quad x_0 = \dfrac{\pi}{6}$
19. $f(x) = \tan x, \quad x_0 = \dfrac{3\pi}{4}$
20. $f(x) = \cos 2x, \quad x_0 = \dfrac{\pi}{6}$
21. $f(x) = \sin 2x, \quad x_0 = \dfrac{\pi}{3}$
22. The gravitational attraction of the earth upon a particle in space a distance r from the center of the earth is given by $f(r) = GMm/r^2$, where M is the mass of the earth, m is the mass of the particle, and $G \cong 6.67 \times 10^{-11}\ \mathrm{N \cdot m^2/kg^2}$ is the universal gravitational constant. Find the rate at which the gravitational attraction is changing with r when the particle is a distance r_0 from the center of the earth. Note that the formula given for $f(r)$ does not apply unless the particle is outside the earth's surface, that is, unless r is greater than approximately 3960 miles, or 6370 kilometers.

2.2 THE LIMIT OF A FUNCTION

In the examples in Section 2.1 we sought to calculate the velocity of a moving particle at a given instant, or the slope of the tangent line to a curve at a given point. In each case it was eventually necessary to consider the limiting behavior of a certain ratio, or difference quotient, as the independent variable was given values closer and closer to a particular value. Here are some additional examples of this kind of limiting operation.

EXAMPLE 1

Find the slope m of the line tangent to the graph of

$$y = g(x) = 0.5x^3 + 1 \tag{1}$$

at the point $(-1, 0.5)$. The graph of this function is shown in Figure 2.2.1.

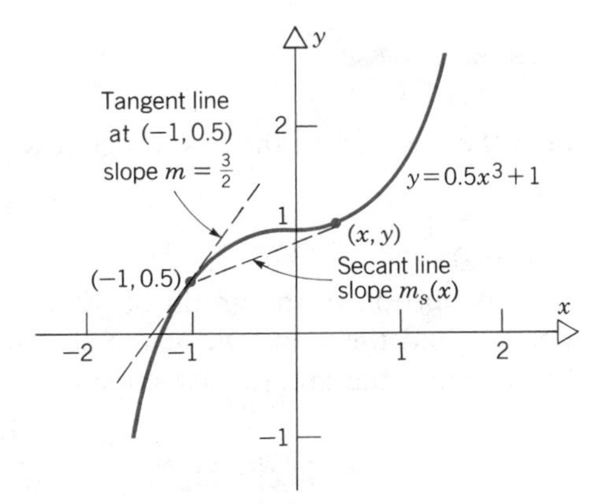

Figure 2.2.1

Proceeding as in Section 2.1, we first find the slope $m_s(x)$ of the secant line from the point $(-1, 0.5)$ to some other point (x, y) on the graph. This slope is given by the difference quotient

$$\begin{aligned} m_s(x) &= \frac{g(x) - g(-1)}{x - (-1)} \\ &= \frac{0.5x^3 + 1 - (-0.5 + 1)}{x + 1} \\ &= \frac{x^3 + 1}{2(x + 1)}, \qquad x \neq -1. \end{aligned} \tag{2}$$

To find the slope of the tangent line we must let $x \to -1$ and determine the limit of the difference quotient (2). In its present form it is not clear how the difference quotient behaves as $x \to -1$; both its numerator and denominator approach zero as $x \to -1$, and the quotient has the indeterminate form 0/0 at $x = -1$. However, the difficulty can be overcome if we notice that the numerator can be written in the factored form

$$x^3 + 1 = (x + 1)(x^2 - x + 1).$$

Then

$$m_s(x) = \frac{(x + 1)(x^2 - x + 1)}{2(x + 1)}$$

$$= \frac{x^2 - x + 1}{2}, \qquad x \neq -1. \tag{3}$$

From Eq. 3 it is not hard to see that $m_s(x)$ takes on values close to $\frac{3}{2}$ when x is assigned values close to -1. Hence we conclude that

$$m = \lim_{x \to -1} m_s(x) = \lim_{x \to -1} \frac{x^3 + 1}{2(x + 1)} = \frac{3}{2} \tag{4}$$

is the slope of the desired tangent line. ∎

EXAMPLE 2

Find the slope m of the line tangent to the graph of

$$y = \sin x$$

at $x = 0$.

A portion of the graph of the sine function is shown in Figure 2.2.2. Again we first find the slope m_s of the secant line from the origin to some other point $(x, \sin x)$ on the graph; this slope is

$$m_s(x) = \frac{\sin x - 0}{x - 0} = \frac{\sin x}{x}, \qquad x \neq 0. \tag{5}$$

To find the slope of the tangent line we must let $x \to 0$, and determine the limit of $m_s(x)$. Since there is no obvious way to simplify the expression for $m_s(x)$ in Eq. 5, we can only estimate the limit by numerical calculation. You should evaluate

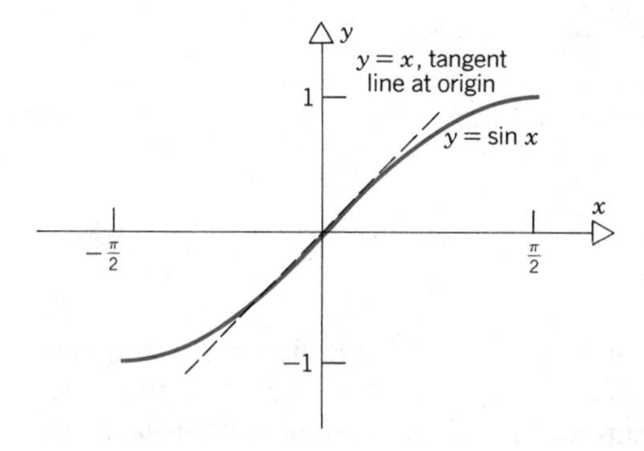

Figure 2.2.2

$m_s(x)$ for a few values of x near 0; remember that x is in radians. After this is done, it should seem plausible that

$$m = \lim_{x \to 0} m_s(x) = \lim_{x \to 0} \frac{\sin x}{x} = 1. \tag{6}$$

We shall tentatively accept this conclusion now, and prove it in the next section. ∎

In each of the previous examples we wanted to calculate the limit of a difference quotient associated with a given function. Since these difference quotients are themselves functions, the mathematical problem that we faced in each instance is to determine the limit of a function as the independent variable approaches a certain value. Up to now we have proceeded in a rather intuitive way, without trying to say precisely what we mean by a limit. To go much further, however, we need to formulate a careful statement of the meaning of this concept. To lead up to this statement, let us recall some common features in the examples that we have considered, both in this section and in the preceding one. We will express these in terms of a function f, depending on the variable x, as x approaches a particular value c.

1. We consider values of x *close to*, and *on both sides of*, the point c where the limit is sought. However, the point c itself is *specifically excluded* from consideration.
2. Indeed, the function f *may not even have a value* at the point c being approached. In fact, in all of the examples so far $f(c)$ has the meaningless form $0/0$.
3. Nevertheless, the function f takes on values *close to some single number* L when x is given values close to c.

Under these circumstances it seems reasonable to say that $f(x)$ approaches the limit L as x approaches c. The notation

$$\lim_{x \to c} f(x) = L \tag{7}$$

is a compact way of making this statement.

We emphasize that the limit L is a *number* associated with a *function* at a *point* (the point c). There are also other numbers associated with this function at the same point, for instance $f(c)$, the value of the function at c. While L and $f(c)$ are frequently equal to each other, they need not be. In fact, either one or both may fail to exist; even if both exist, they may be unequal. The following example should help to clarify the distinction between the *limit* of a function at a point and the *value* of the function there.

EXAMPLE 3

Let

$$f(x) = |x|; \tag{8}$$

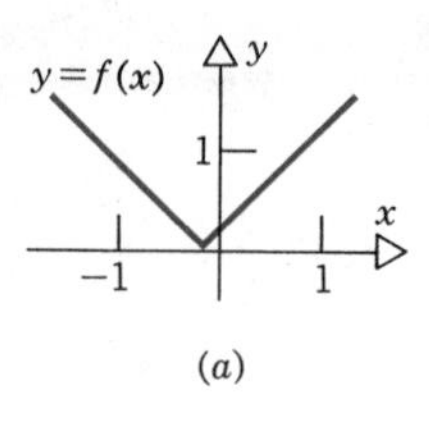

(a)

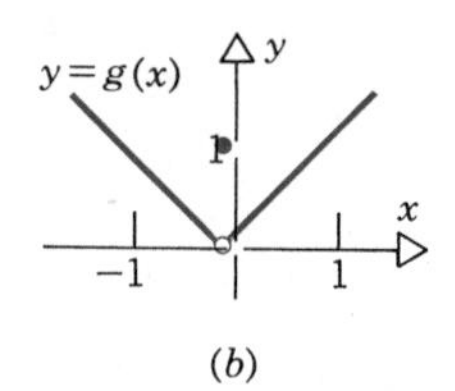

(b)

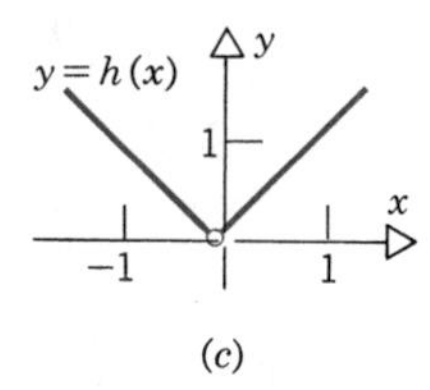

(c)

Figure 2.2.3

see Figure 2.2.3*a*. From the graph of $y = |x|$ we see that if x is close to zero, then so is $f(x)$; indeed, x and $f(x)$ are always the same distance from zero. Thus $f(x) \to 0$ as $x \to 0$, or

$$\lim_{x \to 0} f(x) = 0.$$

Now let

$$g(x) = \begin{cases} |x|, & \text{if } x \neq 0; \\ 1, & \text{if } x = 0; \end{cases} \tag{9}$$

see Figure 2.2.3*b*. Note that $g(x) = f(x)$ for $x \neq 0$. Also, by statement 1, only such points are considered in the limit as x approaches 0. Thus $g(x)$ must have the same limit as $f(x)$ as $x \to 0$, so

$$\lim_{x \to 0} g(x) = 0.$$

This does not alter the fact that $g(0) = 1$, which seems somewhat "unnatural." Nevertheless, $\lim_{x \to 0} g(x) \neq g(0)$.

Finally, consider the function

$$h(x) = |x|, \qquad x \neq 0, \tag{10}$$

which is undefined for $x = 0$ (see Figure 2.2.3*c*). The same argument that we used for $g(x)$ also applies to $h(x)$, so

$$\lim_{x \to 0} h(x) = 0,$$

despite the fact that no value is given for $h(0)$. ∎

Since in all three cases in Example 3 the limit is zero as x approaches zero, it is clear that *the limit of a function at a point has nothing to do with the value that the function has at the point, or even with whether it has any value at all at the point.*

The next two examples illustrate that there are simple functions that do not have limits at certain points.

EXAMPLE 4

Consider the function

$$f(x) = [\![x]\!] \tag{11}$$

defined in Section 1.5 (see Figure 2.2.4). Determine whether f has a limit as $x \to 0$.

From the graph we see that if x is negative and close to zero, then the value of the function at each point is -1. Similarly, if x is positive and close to zero, then the function has the value 0 at each point. The apparent limiting value of f thus depends on whether x is positive or negative. Since there is no *single number* that $f(x)$ approaches as $x \to 0$, we say that $f(x)$ has no limit as $x \to 0$, or that $\lim_{x\to 0} f(x)$ does not exist. Note that $f(0) = 0$, but this has nothing to do with the existence of the limit as $x \to 0$. ∎

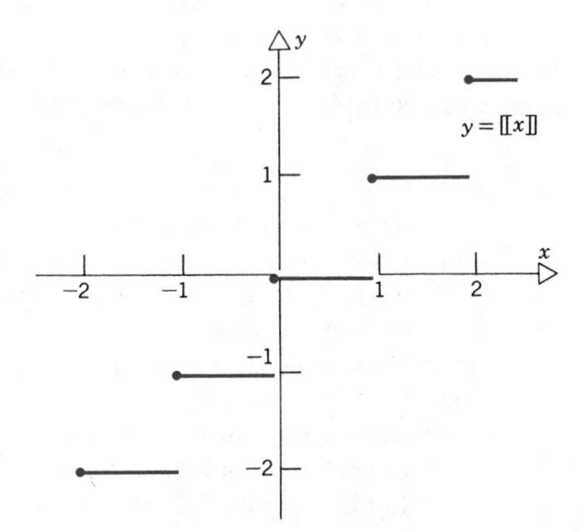

Figure 2.2.4

EXAMPLE 5

Consider the function

$$f(x) = 1 + \frac{1}{x^2}, \qquad x \neq 0. \tag{12}$$

In this case $f(x)$ grows without bound as $x \to 0$; the graph of f is shown in Figure 2.2.5. Since the values of $f(x)$ are not close to any particular number when

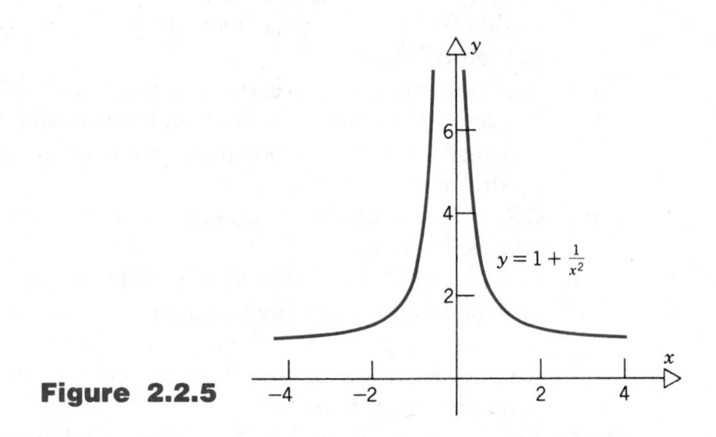

Figure 2.2.5

x is small, we again conclude that $\lim_{x\to 0} f(x)$ does not exist. Observe that in this case $f(0)$ also does not exist. ∎

We now turn to the formal definition of the **limit** of a function at a point, that is, we will give a precise meaning to Eq. 7. The basic idea is that $f(x)$ must take on values close to L when x is close (but not equal) to c. The principal difficulty is to say exactly what is meant by the word "close". How close is "close"?

To show how this difficulty is overcome we introduce the following dialogue between a believer, Dick, and a doubter, Bill. Recall that $|x - c|$ is the distance from x to c and $|f(x) - L|$ is the distance from $f(x)$ to L.

DICK: Consider the function f shown in Figure 2.2.6*a*. Even though it is not defined at $x = c$, I assert that $\lim_{x \to c} f(x) = L$.

BILL: I don't believe you. Can you show me that $f(x)$ is near L when x is near c? In other words, show me that $|f(x) - L|$ is small when $|x - c|$ is small.

DICK: How small do you want $|f(x) - L|$ to be?

BILL: Say, less than $\frac{1}{10}$.

DICK: Let me do a little calculation . . . . For this particular function, I will restrict x so that $0 < |x - c| < \frac{1}{30}$. If I do that, then $|f(x) - L| < \frac{1}{10}$. Look at Figure 2.2.6*b*. Of course, $|x - c| > 0$ is put in because I don't have to worry about the point $x = c$ itself.

BILL: Yes, but notice that there are also points outside of $0 < |x - c| < \frac{1}{30}$ where $|f(x) - L| < \frac{1}{10}$. Look at point P, for example, in Figure 2.2.6*b*.

DICK: So what! I didn't say I had found *all* points where $|f(x) - L| < \frac{1}{10}$. I only said that, for this function, if $0 < |x - c| < \frac{1}{30}$, then $|f(x) - L| < \frac{1}{10}$, which was what you wanted.

BILL: You are right about that, but I am still not convinced that $\lim_{x \to c} f(x) = L$. I will take back $\frac{1}{10}$ and give you $\frac{1}{100}$ instead. Now some of the points in $0 < |x - c| < \frac{1}{30}$ don't work any more; $|f(x) - L| > \frac{1}{100}$ for some of these points.

DICK: That's OK. I can handle that. If you change your number, I will change mine too. Let me do a little more figuring . . . . Now, I will restrict x so that $0 < |x - c| < \frac{1}{70}$; then $|f(x) - L| < \frac{1}{100}$. Do you agree? Look at Figure 2.2.6*c*.

BILL: Yes, but suppose I give you $\frac{1}{1000}$ instead of $\frac{1}{100}$?

DICK: We could keep this up all day and that would be pointless. Why don't you just say ϵ, where ϵ can be any positive number, a tenth, a hundredth, a thousandth, . . . ?

BILL: OK. But don't forget that ϵ can be very, very small; indeed ϵ can be arbitrarily small.

DICK: Fair enough. What I have to do is to determine a number δ that depends on ϵ, and that will usually get smaller as ϵ gets smaller, with the property that for every x in $0 < |x - c| < \delta$, it follows that $|f(x) - L| < \epsilon$. Here is my δ (see Figure 2.2.6*d*) and note how it depends on ϵ. Do you agree that it works?

BILL: Yes, I am afraid so. You have convinced me that $\lim_{x \to c} f(x) = L$.

The definition of limit is essentially given in Dick's last statement. We set it forth here as a formal definition.

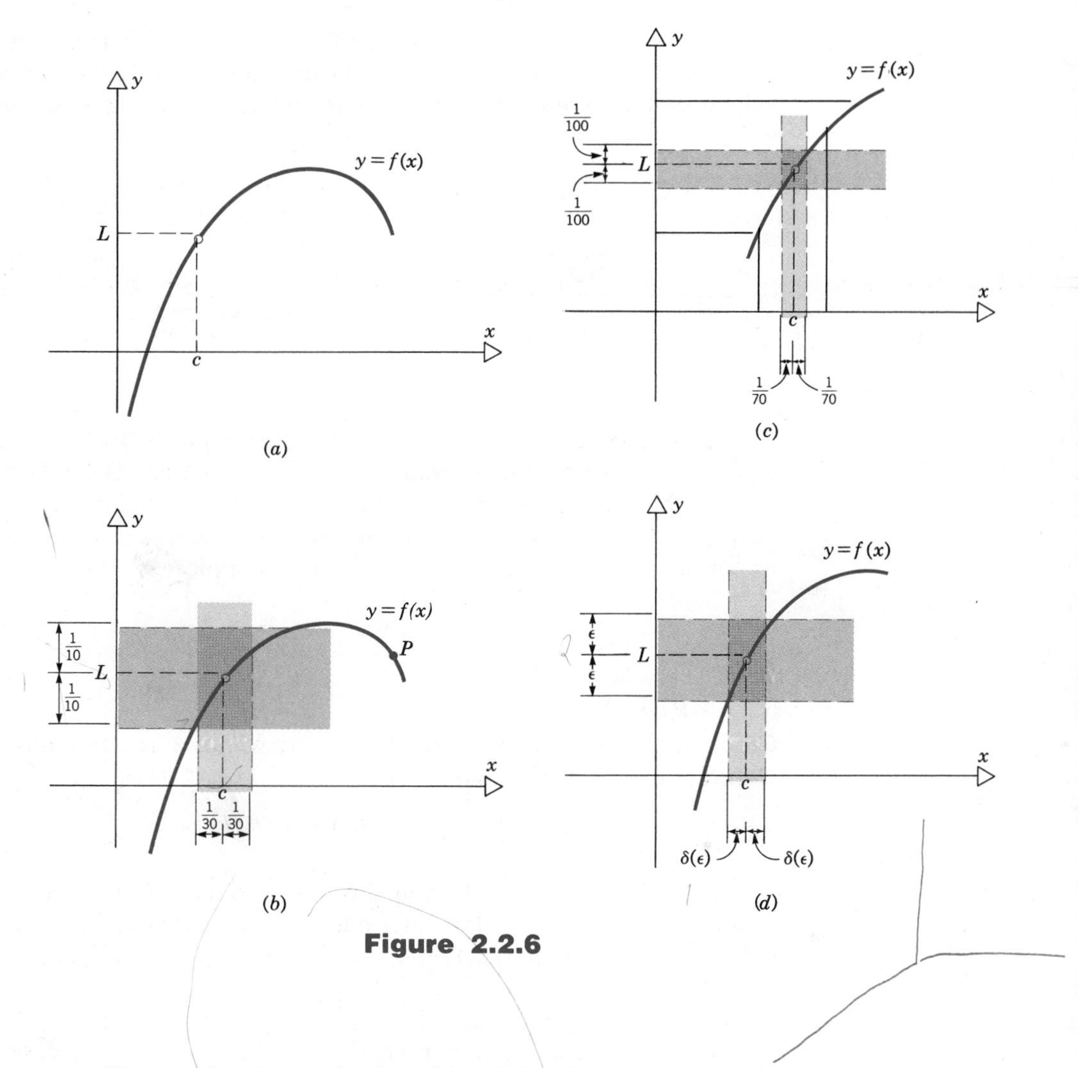

Figure 2.2.6

DEFINITION 2.2.1 (*Limit of a function*). Let f be defined at each point in some open interval containing the point c, except possibly at c itself. Then the limit of $f(x)$ as x approaches c is said to be L, written

$$\lim_{x \to c} f(x) = L,$$

if for each number $\epsilon > 0$, there exists a corresponding number $\delta > 0$ such that if $0 < |x - c| < \delta$, then $|f(x) - L| < \epsilon$.

We stress that the statement

$$\text{“if} \quad 0 < |x - c| < \delta, \qquad \text{then} \quad |f(x) - L| < \epsilon\text{”}$$

means that for each x that satisfies the first inequality, it must follow that $f(x)$ satisfies the second inequality. An equivalent statement is

$$\text{“}|f(x) - L| < \epsilon \qquad \text{for every } x \quad \text{in } 0 < |x - c| < \delta.\text{”}$$

Note also that the set of points $0 < |x - c| < \delta$ is not an interval since the point $x = c$ is not included. We will use the terms **deleted interval** or **deleted neighborhood** to refer to a set of points such as $0 < |x - c| < \delta$, that is, an interval whose midpoint has been deleted (see Figure 2.2.7).

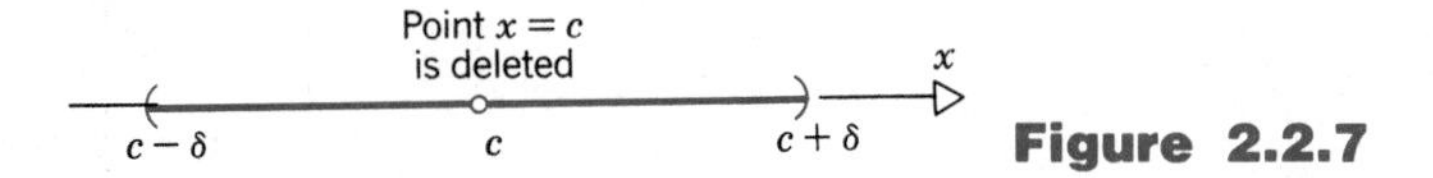

Figure 2.2.7

Definition 2.2.1 states precisely what is meant by the limit* of a function at a point—one of the fundamental ideas of mathematics. Using Definition 2.2.1 we could go back to verify that all of the limits that we found previously in an intuitive geometrical manner do in fact satisfy the statement of the definition. Instead, let us consider the following two very simple examples.

EXAMPLE 6

Consider the constant function $f(x) = k$, where k is any number in R^1. Using Definition 2.2.1, verify that

$$\lim_{x \to c} k = k. \tag{13}$$

Equation 13 asserts that in this case the limit L is equal to k. To confirm this we must show that for each number $\epsilon > 0$, there is a $\delta > 0$ such that if $0 < |x - c| < \delta$, then $|f(x) - k| < \epsilon$ (see Figure 2.2.8). It is clear that for all x

$$|f(x) - k| = |k - k| = 0 \tag{14}$$

and this is always less than ϵ, no matter what ϵ is given. Hence, in this case, δ can be chosen to be any positive number at all. Then, if x satisfies $0 < |x - c| < \delta$, it follows that $|f(x) - k| < \epsilon$ and Eq. 13 is verified. ∎

*Limits and limiting operations were used freely in the seventeenth century by Fermat, Newton, Leibniz, and others. However, the concept was not clearly understood at that time, nor for many years thereafter. During the eighteenth century the methods and applications of calculus were enormously advanced, but the fundamental idea involved in a limit remained imprecise. It was Augustin Louis Cauchy (1789–1857) who formulated a definition of limit substantially equivalent to Definition 2.2.1, who thereby took a giant step toward placing calculus on a firm logical foundation, and who published his work in three influential textbooks in 1821, 1822, and 1829. Cauchy, who lived in Paris for most of his life, was the preeminent French mathematician of his time. His name is associated with many significant theorems in analysis and complex variables.

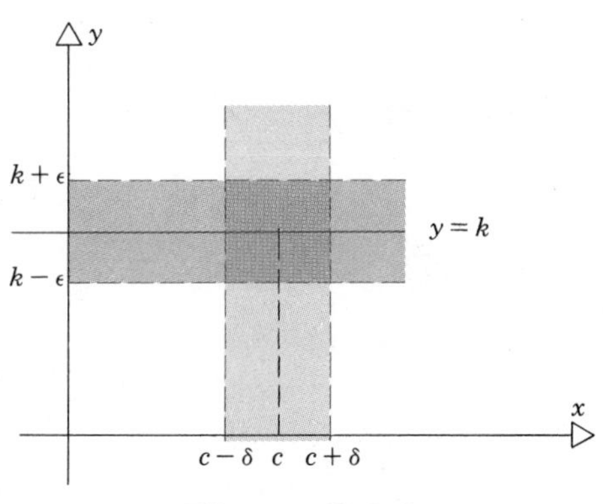

Figure 2.2.8

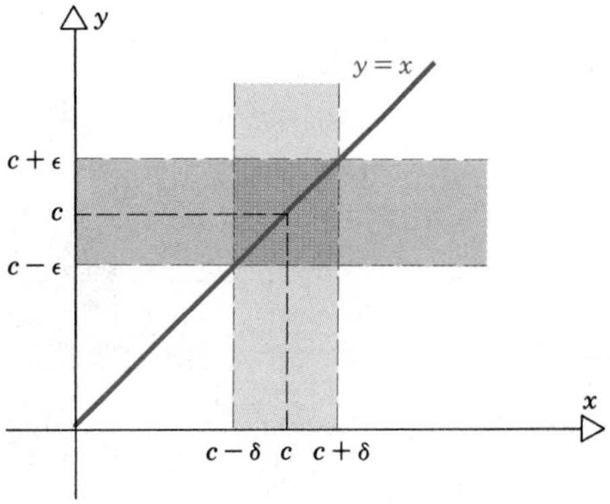

Figure 2.2.9

EXAMPLE 7

Let $f(x) = x$ for each x in R^1. Using Definition 2.2.1, verify that

$$\lim_{x \to c} x = c. \tag{15}$$

Again we must show that for each $\epsilon > 0$ there is a $\delta > 0$ such that if $0 < |x - c| < \delta$, then $|f(x) - c| < \epsilon$ (see Figure 2.2.9). In this case $|f(x) - c| = |x - c|$, and clearly $|x - c| < \epsilon$ when x is near enough to c, nearer than the distance ϵ, in fact. Thus, we can choose $\delta = \epsilon$. Then, if $0 < |x - c| < \delta$, it follows that $|f(x) - c| < \epsilon$, and Eq. 15 is verified. Note that again the choice of δ is not unique; any positive number $\delta < \epsilon$ could have been used just as well.

The definition can also be used to establish the nonexistence of limits, for example for the case considered in Example 4.

Let $f(x) = [\![x]\!]$. Using Definition 2.2.1, show that $\lim_{x \to 0} [\![x]\!]$ does not exist.

All that is needed is to refine slightly the argument given in Example 4. Let us first show that 0 is not the limit. Consider a given $\epsilon > 0$, say $\epsilon = \frac{1}{2}$. Then we must examine whether we can make the quantity

$$|f(x) - L| = |[\![x]\!] - 0| = |[\![x]\!]|$$

less than ϵ simply by restricting x to be close enough to zero. In particular, suppose that x lies in $0 < |x| < \delta$, where δ is some positive number (see Figure 2.2.10). For some of these values of x (those for which $x > 0$ and $f(x) = 0$) we find that $|[\![x]\!]| < \epsilon$, at least when $\delta < 1$. However, for other values of x (those for which

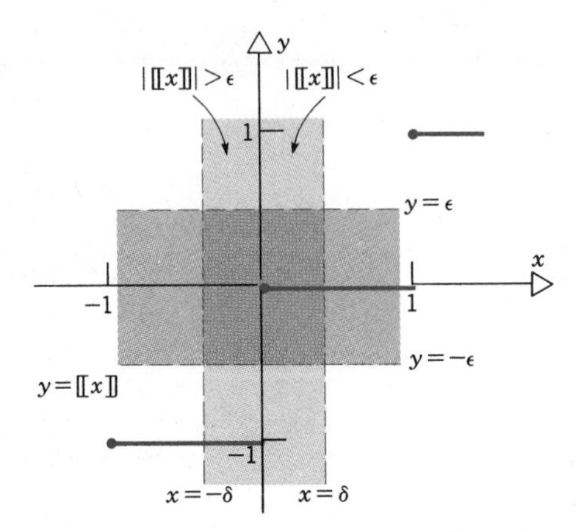

Figure 2.2.10

$x < 0$ and $f(x) = -1$), and *no matter how small δ is chosen*, the inequality $|[\![x]\!]| < \epsilon$ is not true. Since every set $0 < |x| < \delta$ contains points where $x < 0$, we conclude that there is no δ that will satisfy Definition 2.2.1 in this case. Thus we have shown that $L = 0$ is not the limit of $[\![x]\!]$ as $x \to 0$. In a similar manner we can show that no other value of L will work either. Thus we are forced to conclude that $[\![x]\!]$ does not have a limit as $x \to 0$. ∎

The principal use of Definition 2.2.1 does not, however, lie in verifying the existence or nonexistence of limits of specific functions such as those in the examples of this section. Rather, the definition is mainly used to develop general properties of limits that apply to large classes of functions. We discuss some of these properties in the next section. Among other things, we will show how to calculate limits of a very large class of functions (the so-called algebraic functions) using only the results of Examples 6 and 7 and the properties to be discussed in Section 2.3.

Throughout this section we have emphasized that $f(x)$ has a limit at a point c only if $f(x)$ approaches some single number L as x approaches c. Thus a function cannot have two or more different limits at the same point. This fact is stated in the following theorem.

Theorem 2.2.1

(Uniqueness of the Limit)

If

$$\lim_{x\to c} f(x) = L \quad \text{and} \quad \lim_{x\to c} f(x) = M,$$

then $L = M$.

The proof is by contradiction and illustrates how one can base an argument on Definition 2.2.1. Suppose first that $L < M$. Then let $\epsilon = (M - L)/3$ and

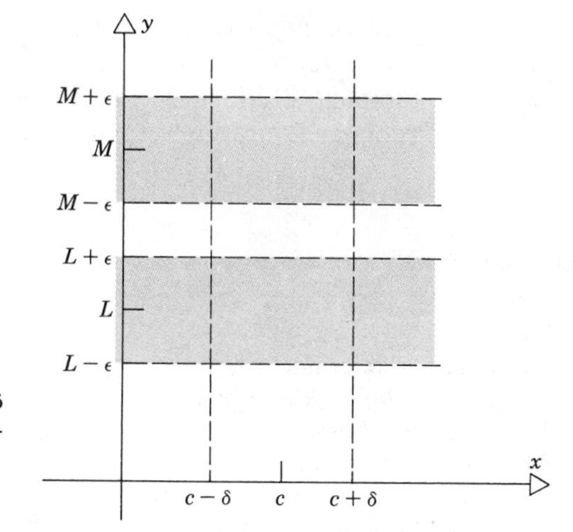

Figure 2.2.11 For $0 < |x - c| < \delta$ the graph of f cannot lie in both shaded strips.

observe that $\epsilon > 0$; see Figure 2.2.11. Then, from Definition 2.2.1, there is a $\delta_1 > 0$ such that

$$\text{if} \quad 0 < |x - c| < \delta_1, \qquad \text{then} \quad L - \epsilon < f(x) < l + \epsilon. \tag{16}$$

Similarly, there is also a $\delta_2 > 0$ such that

$$\text{if} \quad 0 < |x - c| < \delta_2, \qquad \text{then} \quad M - \epsilon < f(x) < M + \epsilon. \tag{17}$$

Let δ be the smaller of δ_1 and δ_2. Then, for $0 < |x - c| < \delta$, the conclusions in both of Eqs. 16 and 17 are true. In particular,

$$\text{if} \quad 0 < |x - c| < \delta, \qquad \text{then} \quad f(x) < L + \epsilon \qquad \text{and} \qquad f(x) > M - \epsilon. \tag{18}$$

However, this last statement is impossible because $L + \epsilon < M - \epsilon$. Geometrically, for $0 < |x - c| < \delta$, the graph of f must simultaneously lie in both of the horizontal strips shown in Figure 2.2.11. This cannot happen because the strips do not overlap and $f(x)$ has only a single value for each x. Consequently, the assumption that $L < M$ must be false. Similarly, one can show that $L > M$ is also impossible. Therefore the only remaining alternative must be true, namely $L = M$, which proves the theorem. □

Finally, we close this section with a word about our philosophy concerning limits. We believe that all students should attempt to achieve a reasonably clear idea of what a limit is, and should see how a precise definition of this concept is formulated. However, it is a historical fact that many important applications of calculus (for example, in mechanics) were developed before the formulation of a good definition of limit. Hence it is not unreasonable for present-day students to aspire to a working knowledge of calculus before attaining skill in using the definition of limit, and before understanding in detail some of its ramifications. Indeed, an intuitive geometrical grasp of a limit is sufficient for most of the material in this book. We will therefore proceed as rapidly as possible to *use* limits in discussing calculus and its applications, rather than to linger over a more detailed consideration of the concept of limit itself. This means that many somewhat related concepts are introduced in this chapter, but are not discussed in great analytical detail. While none of these concepts is particularly difficult by itself, you should proceed slowly enough so as to develop a "feel" for each new idea, and so as not to be overwhelmed by their cumulative effect. More details may be found in some of the problems such as Problems 26 through 33 in the following problem section.

PROBLEMS

In each of Problems 1 through 16, determine whether the given function has a limit at the indicated point. If so, find the limit. Use intuitive or geometrical methods rather than appealing to Definition 2.2.1.

1. $f(x) = \dfrac{x^2 - 4}{x + 2}, \quad x \neq -2; \quad$ limit as $x \to -2$

2. $f(x) = \begin{cases} \dfrac{x^3 - 27}{x - 3}, & x \neq 3; \\ 20, & x = 3; \end{cases}$ limit as $x \to 3$;

3. $f(x) = \dfrac{x^2 - x - 2}{x - 2}, \quad x \neq 2; \quad$ limit as $x \to 2$

4. $f(x) = \begin{cases} \dfrac{x^3 + 3x^2 + 5x + 3}{x + 1}, & x \neq -1; \\ -2, & x = -1; \end{cases}$
limit as $x \to -1$

5. $f(x) = \dfrac{x - 1}{x^2 - 4x + 3}, \quad x \neq 1, 3;$
limit as $x \to 1$; as $x \to 3$

6. $f(x) = \begin{cases} \dfrac{1}{x}, & x \neq 0; \\ 1, & x = 0; \end{cases}$ limit as $x \to 0$; as $x \to 1$

7. $f(x) = \dfrac{x^2 + x - 6}{x^2 - x - 2}, \quad x \neq -1, 2;$
limit as $x \to -1$; as $x \to 2$

8. $f(x) = \dfrac{x}{x + 1}, \quad x \neq -1;$
limit as $x \to -1$; as $x \to 0$; as $x \to 1$

9. $f(x) = \dfrac{\sqrt{x} - 1}{x - 1}, \quad x \geq 0$ and $x \neq 1;$
limit as $x \to 1$
Hint: Multiply by $(\sqrt{x} + 1)/(\sqrt{x} + 1)$.

10. $f(x) = \dfrac{\sqrt{x} - 2}{x - 4}, \quad x \geq 0$ and $x \neq 4;$
limit as $x \to 4$

11. $f(x) = \dfrac{1}{x^2 + 1}; \quad$ limit as $x \to 0$

12. $f(x) = [\![2x]\!]; \quad$ limit as $x \to \frac{1}{2}$

13. $f(x) = [\![2x - 3]\!]; \quad$ limit as $x \to \frac{1}{2}$; as $x \to \frac{1}{4}$

14. $f(x) = \sqrt{x}, \quad x \geq 0; \quad$ limit as $x \to 1$

15. $f(x) = \sqrt{|x|}; \quad$ limit as $x \to 0$

16. $f(x) = \begin{cases} \dfrac{x}{|x|}, & x \neq 0; \\ 0, & x = 0; \end{cases}$ limit as $x \to 0$;

In each of Problems 17 through 22, calculate the values of $f(x)$ for a few values of x near the given point x_0. Use this information to estimate the limit of $f(x)$ as $x \to x_0$.

17. $f(x) = \dfrac{1 - \cos x}{x^2}, \quad x_0 = 0$

18. $f(x) = \dfrac{\sin(x^2)}{x^2}, \quad x_0 = 0$

19. $f(x) = \dfrac{\tan x - 1}{x - \pi/4}, \quad x_0 = \dfrac{\pi}{4}$

20. $f(x) = \dfrac{1}{x} - \csc x, \quad x_0 = 0$

21. $f(x) = \dfrac{x - \sin x}{x^3}, \quad x_0 = 0$

22. $f(x) = \dfrac{\sqrt{1 + 2x} - \sqrt{1 - x}}{x}, \quad x_0 = 0$

23. Consider the function $f(x) = |x|$. Try to find the slope of a line tangent to the graph of f at the origin. Set up the difference quotient and investigate its limiting behavior.

24. Consider the function f given by

$$f(x) = \sin\frac{1}{x}, \qquad x \neq 0.$$

Show that, in any deleted interval $0 < |x| < \delta$ about the origin, $f(x)$ takes on all values in $[-1, 1]$. Then show that $\lim_{x\to 0} \sin(1/x)$ does not exist.

25. Use the substitution $x = c + h$ to show that $\lim_{x\to c} f(x) = \lim_{h\to 0} f(c + h)$. *Hint:* Write out the definition of the two limits that appear in the given equation.

Problems 26 to 33 deal with the use of the definition of limit for specific functions.

26. Let $f(x) = 3x$ and consider points near $x = 2$.

(a) Find δ_1 such that if $0 < |x - 2| < \delta_1$, then $|f(x) - 6| < \frac{1}{10}$.

(b) Find δ_2 such that if $0 < |x - 2| < \delta_2$, then $|f(x) - 6| < \frac{1}{100}$.

(c) Find $\delta(\epsilon)$ such that if $0 < |x - 2| < \delta$, then $|f(x) - 6| < \epsilon$.

27. Let $f(x) = 2x + 1$ with $c = 1$ and $L = 3$.

(a) Find δ_1 such that if $0 < |x - c| < \delta_1$, then $|f(x) - L| < \frac{1}{10}$.

(b) Find δ_2 such that if $0 < |x - c| < \delta_2$, then $|f(x) - L| < \frac{1}{100}$.

(c) Find $\delta(\epsilon)$ such that if $0 < |x - c| < \delta$, then $|f(x) - L| < \epsilon$.

In each of Problems 28 through 33, follow the instructions in Problem 27.

28. $f(x) = 1 - 2x, \quad c = -2, \quad L = 5$

29. $f(x) = 3x - 8, \quad c = 2, \quad L = -2$

30. $f(x) = ax + b$ with $a \neq 0$, $c = 2$, $L = 2a + b$

31. $f(x) = ax + b$ with $a \neq 0$, $c = x_0$, $L = ax_0 + b$

* 32. $f(x) = x^2, \quad c = 2, \quad L = 4$.
Hint: First restrict δ so that $\delta < 1$.

* 33. $f(x) = 1/x, \quad c = 2, \quad L = \frac{1}{2}$.
Hint: First restrict δ so that $\delta < 1$.

OPERATIONS WITH LIMITS

In Section 2.2 we found the limits of several functions by inspecting their graphs. It is frequently necessary to carry out analytical calculations with limits, and this requires a knowledge of some of the properties of limits. These properties are also very useful in finding limits of complicated functions.

The properties of limits stated in Theorems 2.3.1 to 2.3.5 in this section are geometrically plausible and can be proved using the definition of limit given in Section 2.2. The proofs are not particularly difficult, but we wish to concentrate primarily upon the *uses* of these properties rather than their formal proofs. Therefore we have placed some of the proofs in Problems 33 through 35 and have omitted the others.

The first two theorems are concerned with algebraic operations upon limits.

Theorem 2.3.1

Let f and g be functions such that $\lim_{x\to c} f(x) = A$ and $\lim_{x\to c} g(x) = B$. Then

(a) $\lim_{x\to c}[f(x) + g(x)] = \lim_{x\to c} f(x) + \lim_{x\to c} g(x) = A + B;$ (1)

(b) $\lim_{x\to c} \alpha f(x) = \alpha \lim_{x\to c} f(x) = \alpha A,$ for any constant α; (2)

(c) $\lim_{x\to c} f(x)g(x) = [\lim_{x\to c} f(x)][\lim_{x\to c} g(x)] = AB;$ (3)

(d) $\lim_{x\to c} \dfrac{f(x)}{g(x)} = \dfrac{\lim_{x\to c} f(x)}{\lim_{x\to c} g(x)} = \dfrac{A}{B},$ provided $B \neq 0$. (4)

It is worth noting that by combining Parts (a) and (b) of Theorem 2.3.1 we obtain

$$\lim_{x\to c}[\alpha f(x) + \beta g(x)] = \alpha \lim_{x\to c} f(x) + \beta \lim_{x\to c} g(x) = \alpha A + \beta B, \tag{5}$$

for any constants α and β. If $\alpha = 1$ and $\beta = -1$, then Eq. 5 reduces to

$$\lim_{x\to c}[f(x) - g(x)] = \lim_{x\to c} f(x) - \lim_{x\to c} g(x) = A - B. \tag{6}$$

Further, by repeated application of Eqs. 5 and 3, we can easily extend them to any finite number of terms; thus

$$\lim_{x\to c}[\alpha_1 f_1(x) + \cdots + \alpha_n f_n(x)] = \alpha_1 \lim_{x\to c} f_1(x) + \cdots + \alpha_n \lim_{x\to c} f_n(x), \tag{7}$$

and

$$\lim_{x\to c}[f_1(x) \cdots f_n(x)] = [\lim_{x\to c} f_1(x)] \cdots [\lim_{x\to c} f_n(x)]. \tag{8}$$

Theorem 2.3.2

If f is a function such that $\lim_{x\to c} f(x) = L$, if n is a positive integer, and if $L \geq 0$ whenever n is even, then

$$\lim_{x\to c}[f(x)]^{1/n} = [\lim_{x\to c} f(x)]^{1/n} = L^{1/n}. \tag{9}$$

These two theorems state that if the functions f and g have limits at a point c, and if algebraic operations are performed on f and g, then the same operations should be performed on their limits. On the basis of these properties it is easy to find limits of *algebraic functions,* that is, those functions generated by a *finite* number of additions, subtractions, multiplications, divisions, and root extractions. The following examples illustrate this fact.

Recall that, in Examples 6 and 7 of Section 2.2, we showed that

$$\lim_{x\to c} k = k \quad \text{and} \quad \lim_{x\to c} x = c. \tag{10}$$

EXAMPLE 1

Find $\lim_{x\to 2} (3x - 4)$.

If we write $3x - 4 = (3)(x) + (-4)(1)$, then we can identify $3x - 4$ with $\alpha f(x) + \beta g(x)$, where $\alpha = 3$, $\beta = -4$, $f(x) = x$, and $g(x) = 1$. Using Eqs. 5 and 10, we obtain

$$\lim_{x\to 2} (3x - 4) = 3 \lim_{x\to 2} x - 4 \lim_{x\to 2} 1 = 3 \cdot 2 - 4 \cdot 1 = 2. \blacksquare$$

EXAMPLE 2

For each positive integer n, find $\lim_{x\to c} x^n$.

For $n = 1$ the result is given by the second of Eqs. 10. For $n \geq 2$ we let $f_1(x) = x, \ldots, f_n(x) = x$ in Eq. 8 and use the second of Eqs. 10. Then

$$\lim_{x\to c} x^n = (\lim_{x\to c} x) \cdots (\lim_{x\to c} x) = c \cdots c = c^n. \blacksquare$$

EXAMPLE 3

Find $\lim_{x\to 2} (5x^2 - 2x^3)$.

Using Eq. 5 and the result of Example 2, we have

$$\begin{aligned}\lim_{x\to 2} (5x^2 - 2x^3) &= 5 \lim_{x\to 2} x^2 - 2 \lim_{x\to 2} x^3 \\ &= 5 \cdot 4 - 2 \cdot 8 = 4. \blacksquare\end{aligned}$$

EXAMPLE 4

Find $\lim_{x\to -1} (2 - 3x + 4x^2 - x^3)$.

By using Eq. 7 together with the result of Example 2, we find that

$$\begin{aligned}\lim_{x\to -1} (2 - 3x + 4x^2 - x^3) &= \lim_{x\to -1} 2 - 3 \lim_{x\to -1} x + 4 \lim_{x\to -1} x^2 - \lim_{x\to -1} x^3 \\ &= 2 - 3(-1) + 4(1) - (-1) = 10. \blacksquare\end{aligned}$$

The method of Examples 1 through 4 can be used to find the limit of any polynomial at any point, that is,

$$\lim_{x\to c} (k_0 + k_1x + \cdots + k_nx^n) = k_0 + k_1c + \cdots + k_nc^n. \tag{11}$$

In other words, if P is any polynomial, then

$$\lim_{x\to c} P(x) = P(c). \tag{12}$$

Much the same procedure enables us to find limits of rational functions, that is, functions that are the quotients of two polynomials.

EXAMPLE 5

Find

$$\lim_{x\to c} \frac{5x^2 - 2x^3}{x^3 + 1}.$$

According to Theorem 2.3.1(d), we have

$$\lim_{x\to c} \frac{5x^2 - 2x^3}{x^3 + 1} = \frac{\lim_{x\to c}(5x^2 - 2x^3)}{\lim_{x\to c}(x^3 + 1)}, \tag{13}$$

provided that both the limit in the numerator and the limit in the denominator exist, and provided that the latter is not zero. Using Eq. (12) to evaluate the right side of Eq. 13, we find that

$$\lim_{x \to c} \frac{5x^2 - 2x^3}{x^3 + 1} = \frac{5c^2 - 2c^3}{c^3 + 1}, \tag{14}$$

so long as $c^3 + 1 \neq 0$. For instance, if $c = 2$, then

$$\lim_{x \to 2} \frac{5x^2 - 2x^3}{x^3 + 1} = \frac{4}{9}.$$

On the other hand, if $c = -1$, then $\lim_{x \to -1}(x^3 + 1) = 0$, and Theorem 2.3.1(d) does not apply. In fact, it can be shown that $(5x^2 - 2x^3)/(x^3 + 1)$ increases without bound as $x \to -1$, and does not have a limit at this point. ∎

In dealing with rational functions, we usually assume that any common factors appearing in the numerator and denominator are canceled. Thus, if

$$R(x) = \frac{P(x)}{Q(x)}, \tag{15}$$

where P and Q are polynomials, we suppose that P and Q have no common factors. Then, proceeding as in Example 5, we conclude that

$$\lim_{x \to c} R(x) = \frac{P(c)}{Q(c)} = R(c), \tag{16}$$

provided that $Q(c) \neq 0$. If $Q(c) = 0$ (but $P(c) \neq 0$), then $R(x)$ becomes unbounded as $x \to c$ and $\lim_{x \to c} R(x)$ does not exist.

EXAMPLE 6

Find $\lim_{x \to 2} \sqrt{x^2 + 5}$.

From Eq. 12 we have $\lim_{x \to 2}(x^2 + 5) = 9$, and from Theorem 2.3.2 it then follows that

$$\lim_{x \to 2} \sqrt{x^2 + 5} = \sqrt{9} = 3. \ ∎$$

EXAMPLE 7

Find

$$\lim_{x \to 1} \left(\frac{2x^2 + 5x + 1}{x^3 + 2x^2 - 4}\right)^{-1/3}.$$

Note first that

$$\left(\frac{2x^2 + 5x + 1}{x^3 + 2x^2 - 4}\right)^{-1/3} = \left(\frac{x^3 + 2x^2 - 4}{2x^2 + 5x + 1}\right)^{1/3}.$$

Using Eq. 16, we have

$$\lim_{x\to 1} \frac{x^3 + 2x^2 - 4}{2x^2 + 5x + 1} = \frac{\lim_{x\to 1} (x^3 + 2x^2 - 4)}{\lim_{x\to 1} (2x^2 + 5x + 1)} = -\frac{1}{8}.$$

Theorem 2.3.2 then applies with the result that

$$\lim_{x\to 1} \left(\frac{2x^2 + 5x + 1}{x^3 + 2x^2 - 4}\right)^{-1/3} = \left(-\frac{1}{8}\right)^{1/3} = -\frac{1}{2}. \quad \blacksquare$$

Limits such as the one in Example 7 are extremely difficult to verify by using the definition of limit directly; that is, by actually finding a deleted interval $0 < |x - c| < \delta$ in which the requirement $|f(x) - L| < \epsilon$ is satisfied. Consequently, such examples demonstrate the power and usefulness of general theorems for the computation of limits of specific functions.

We emphasize that Theorems 2.3.1 and 2.3.2 say that *if* f and g have limits at a point c, then so do $f + g$, fg, and so forth. On the other hand, the reverse is not necessarily true. For instance, if

$$f(x) = 1 + \frac{1}{x^2}, \qquad g(x) = x^2 - \frac{1}{x^2}, \qquad x \neq 0,$$

then neither f nor g has a limit as $x \to 0$. However,

$$f(x) + g(x) = \left(1 + \frac{1}{x^2}\right) + \left(x^2 - \frac{1}{x^2}\right) = 1 + x^2, \qquad x \neq 0,$$

and therefore the function $f + g$ does have a limit (namely, 1) as $x \to 0$.

Also, the requirement that $B \neq 0$ in Part (d) of Theorem 2.3.1 is essential. Recall that we encountered the limit

$$\lim_{x\to 0} \frac{\sin x}{x} \tag{17}$$

in Example 2 of Section 2.2. Unfortunately, Theorem 2.3.1(d) is not helpful in evaluating this limit, because the denominator approaches zero and we obtain the meaningless expression 0/0. Thus, we must use other methods in such cases. We will return to the limit (17) later in this section, after we discuss some further properties of limits.

Theorem 2.3.3

(a) If $\lim_{x\to c} f(x) = L$, then $\lim_{x\to c} |f(x)| = |L|$.

(b) If $\lim_{x\to c} |f(x)| = 0$, then $\lim_{x\to c} f(x) = 0$.

These properties are plausible on geometrical grounds, and Figures 2.3.1 and 2.3.2 illustrate Parts (a) and (b) of the theorem, respectively. Note that Part (b) may no longer be true if $|f(x)| \to L \neq 0$ as $x \to c$, for in that case $\lim_{x\to c} f(x)$ may not exist. For example, consider the function

$$\operatorname{sgn}(x) = \begin{cases} 1, & x > 0 \\ 0, & x = 0 \\ -1, & x < 0 \end{cases}$$

and its absolute value

$$|\operatorname{sgn}(x)| = \begin{cases} 1, & x \neq 0 \\ 0, & x = 0 \end{cases}$$

whose graphs are shown in Figure 2.3.3. Then

$$\lim_{x\to 0} |\operatorname{sgn}(x)| = 1,$$

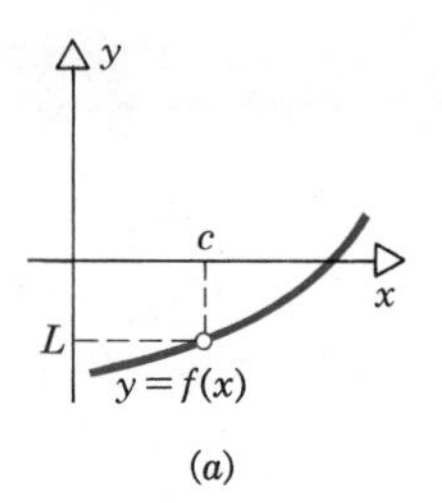

(a)

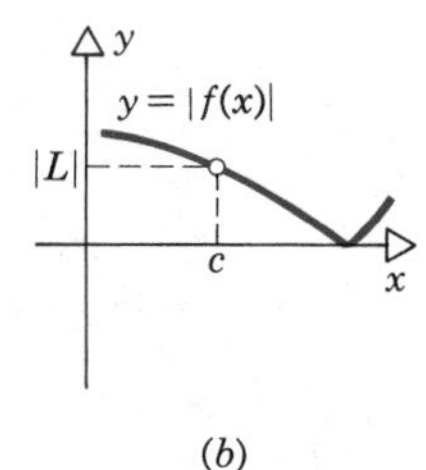

(b)

Figure 2.3.1

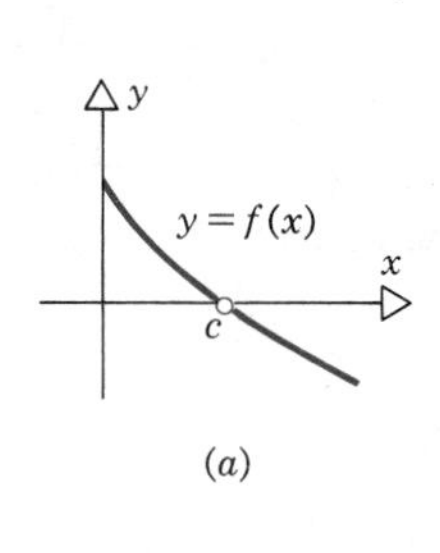

(a)

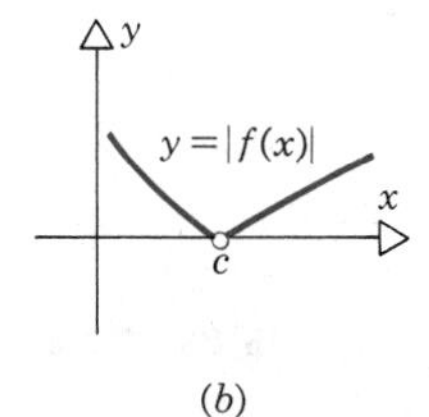

(b)

Figure 2.3.2

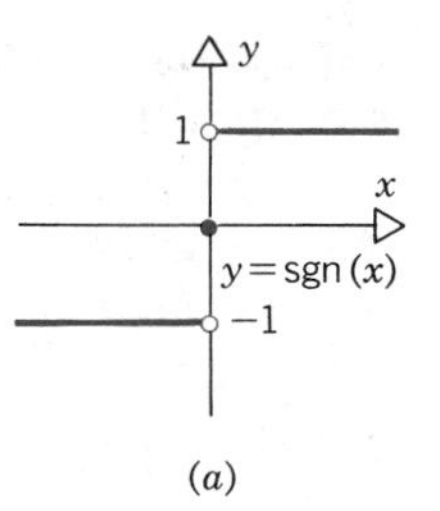

(a)

(b)

Figure 2.3.3

but $\lim_{x\to 0} \operatorname{sgn}(x)$ does not exist.

The following property of limits is sometimes called the sandwich principle.

Theorem 2.3.4

(Sandwich Principle)

Suppose that

(a) $\lim_{x\to c} g(x) = L$ and $\lim_{x\to c} h(x) = L$;

(b) $g(x) \leq f(x) \leq h(x)$ in some deleted interval $0 < |x - c| < r$ about $x = c$;

then $\lim_{x\to c} f(x) = L$.

In other words, if the function f is bounded above and below by other functions that approach the limit L as x approaches c, then f has no alternative but to approach L also (see Figure 2.3.4). The sandwich principle is very useful in finding the limits of certain functions that do not fall into any of the categories already considered. To apply this principle to a particular function f it is necessary to find two other functions g and h, which lie on either side of f, and which are known to approach the same limit L at the point in question. The following examples illustrate this idea.

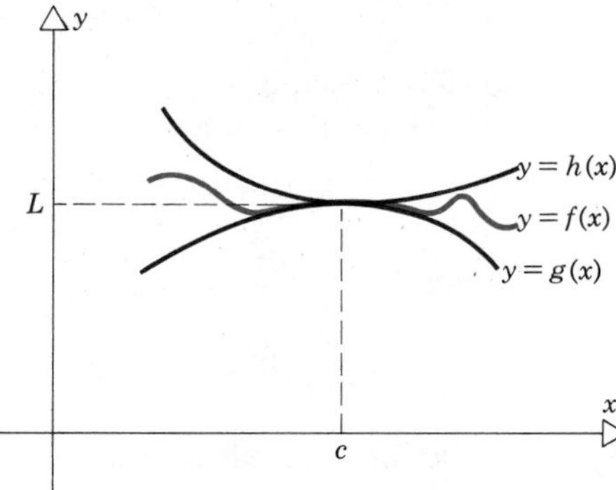

Figure 2.3.4
The sandwich principle.

EXAMPLE 8

Show that

$$\lim_{x \to 0} \sin x = 0, \qquad \lim_{x \to 0} \cos x = 1. \tag{18}$$

Consider triangle PQR in Figure 2.3.5. Clearly

$$\overline{PR}^2 + \overline{RQ}^2 = \overline{PQ}^2. \tag{19}$$

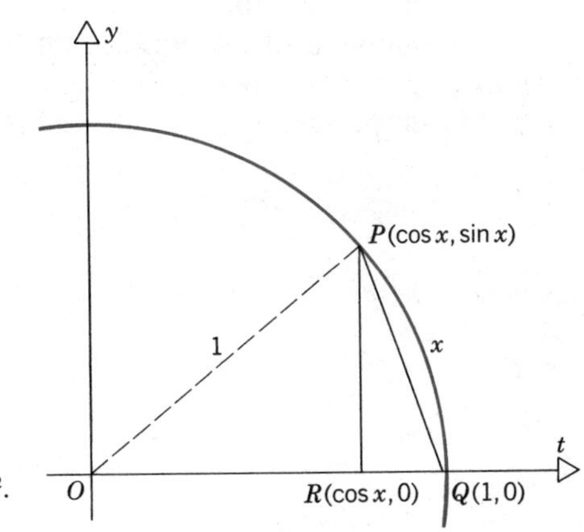

Figure 2.3.5
$\sin^2 x + (1 - \cos x)^2 \leq x^2$.

Since $\overline{PQ} \leq \text{arc } PQ = x$, Eq. 19 becomes

$$\sin^2 x + (1 - \cos x)^2 = \overline{PQ}^2 \leq x^2. \tag{20}$$

Since all terms in Eq. 20 are nonnegative, it follows that

$$\sin^2 x \le x^2 \qquad \text{and} \qquad (1 - \cos x)^2 \le x^2.$$

Consequently, by taking square roots we obtain

$$0 \le |\sin x| \le |x|, \tag{21}$$

and

$$0 \le |1 - \cos x| \le |x|. \tag{22}$$

The sandwich principle applies to Eqs. 21 and 22 with $g(x) = 0$ for each x, and $h(x) = |x|$. Since $|x| \to 0$ as $x \to 0$, we conclude that

$$\lim_{x\to 0} |\sin x| = 0 \qquad \text{and} \qquad \lim_{x\to 0} |1 - \cos x| = 0.$$

Theorem 2.3.3(b) then implies that

$$\lim_{x\to 0} \sin x = 0 \qquad \text{and} \qquad \lim_{x\to 0}(1 - \cos x) = 0.$$

By Eq. 6 the latter result is equivalent to

$$\lim_{x\to 0} \cos x = 1. \blacksquare$$

The results of Example 8 can be restated in the following way:

$$\lim_{x\to 0} \sin x = \sin 0, \qquad \lim_{x\to 0} \cos x = \cos 0. \tag{23}$$

Using these results at $x = 0$, it is possible to obtain corresponding results at any point $x = c$, namely

$$\lim_{x\to c} \sin x = \sin c, \qquad \lim_{x\to c} \cos x = \cos c; \tag{24}$$

see Problems 15 and 16.

In Example 2 of Section 2.2 we found that to compute the slope of the line tangent to $y = \sin x$ at $x = 0$ it was necessary to evaluate $\lim_{x\to 0}(\sin x)/x$. Using the sandwich principle we are now able to do this.

EXAMPLE 9

Find

$$\lim_{x\to 0} \frac{\sin x}{x}.$$

Consider first values of x in $(0, \pi/2)$. In Figure 2.3.6 we compare the areas of the triangle OPR, the sector OPQ of the circle, and the triangle OSQ. Clearly

$$\text{area triangle } OPR < \text{area sector } OPQ < \text{area triangle } OSQ. \tag{25}$$

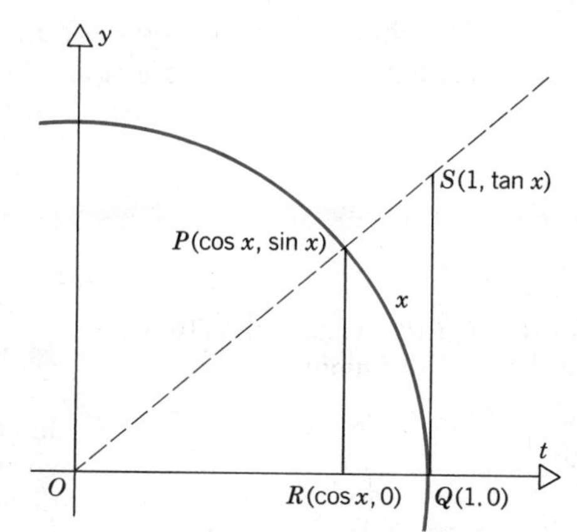

Figure 2.3.6 Area $\Delta OPR <$ area sector $OPQ <$ area ΔOSQ.

Computing these areas in terms of x, we have

$$\tfrac{1}{2}\cos x \sin x < \tfrac{1}{2}x < \tfrac{1}{2}\tan x.$$

Multiplying by $2/\sin x$, a positive number, we obtain

$$\cos x < \frac{x}{\sin x} < \frac{1}{\cos x}.$$

Finally, by taking reciprocals of each term (and reversing the inequalities) we find that

$$\cos x < \frac{\sin x}{x} < \frac{1}{\cos x}. \tag{26}$$

By replacing x by $-x$ we obtain the same inequality for x in $(-\pi/2, 0)$. Since we know that $\lim_{x\to 0} \cos x = 1$ and $\lim_{x\to 0} 1/\cos x = 1$, we can use Eq. 26 in conjunction with the sandwich principle to obtain the result

$$\lim_{x\to 0} \frac{\sin x}{x} = 1. \tag{27}$$

This confirms the tentative conclusion we had reached earlier. ∎

Finally, we mention a weaker result that is related to the sandwich principle.

Theorem 2.3.5

If $\lim_{x\to c} g(x) = L$, and if $f(x) \geq g(x)$ in a deleted interval about c, and if $\lim_{x\to c} f(x)$ exists, then $\lim_{x\to c} f(x) \geq L$.

This theorem gives a lower bound for $\lim_{x\to c} f(x)$, provided it is already known that this limit exists. There is a similar statement for upper bounds.

PROBLEMS

In each of Problems 1 through 14, evaluate the given limit and state which properties of limits are used in your calculation.

1. $\lim\limits_{x\to 3}(2x^2 + x - 5)$

2. $\lim\limits_{x\to -1}(3x^4 - 5x^3 - 2x^2 + 6x + 4)$

3. $\lim\limits_{x\to 1}(x - 2)^5$

4. $\lim\limits_{x\to 1}(x - 2)^{50}$

5. $\lim\limits_{x\to 2}\dfrac{x^2 - 3x - 4}{x + 2}$

6. $\lim\limits_{x\to -1}\dfrac{x^2 + 3x + 4}{2x^2 - x + 5}$

7. $\lim\limits_{x\to -2}\left(x^2 + \dfrac{1}{x^2}\right)$

8. $\lim\limits_{x\to 2}\dfrac{(x - 1)^{64}}{(x - 3)^{87}}$

9. $\lim\limits_{x\to 2}\sqrt{1 + 2x^2}$

10. $\lim\limits_{x\to 3}\sqrt{\dfrac{5x + 1}{x^2 + 16}}$

11. $\lim\limits_{x\to 2}\left(\sqrt{x} + \dfrac{1}{\sqrt{x}}\right)$

12. $\lim\limits_{x\to\sqrt{3}}(x^3 - 2x^2 + 3x + 4)$

13. $\lim\limits_{x\to 1}\left(\dfrac{2x^3 - 4x^2 + 3}{6x^2 + 4x - 1}\right)^{1/2}$

14. $\lim\limits_{x\to -1}\left(\dfrac{x^2 - 3x - 3}{2x^3 + 6x^2 - 3x + 1}\right)^{4/3}$

15. If $x = c + h$, note that

$$\sin x = \sin c \cos h + \cos c \sin h.$$

Also note that if $x \to c$, then $h \to 0$. Use Eqs. 18 to show that

$$\lim_{x\to c}\sin x = \sin c.$$

16. Follow a procedure similar to that in Problem 15 to show that

$$\lim_{x\to c}\cos x = \cos c.$$

In each of Problems 17 through 20, evaluate the given limit. You may use Eqs. 24.

17. $\lim\limits_{x\to\pi/4}(1 - \sin^3 x)$

18. $\lim\limits_{x\to 0}[4\cos^2 x - \sin^2 x]^{1/2}$

19. $\lim\limits_{x\to\pi/2}(x\sin x - \cos x)$

20. $\lim\limits_{x\to\pi}\left(\cos\dfrac{x}{2} - \sin\dfrac{x}{4}\right)$

In each of Problems 21 through 26, first rewrite the given expression so that Eq. 27 can be applied, and then use it to evaluate the required limit.

21. $\lim\limits_{x\to 0}\dfrac{\sin 2x}{x}$

22. $\lim\limits_{x\to 0}\dfrac{\sin x^2}{2x^2}$

23. $\lim\limits_{x\to 0}\dfrac{\sin x^2}{x}$

24. $\lim\limits_{x\to 0}\dfrac{1 - \cos x}{x^2}$

25. $\lim\limits_{x\to 0}\dfrac{\tan x}{x}$

26. $\lim\limits_{x\to 0}\dfrac{\sin 3x}{x\cos 4x}$

27. Let $f(x) = x\sin(1/x)$, $x \neq 0$. Use the sandwich principle to find $\lim\limits_{x\to 0} f(x)$.

28. Use the sandwich principle to show that $2x/(x^2 + 1) \to 0$ as $x \to \infty$.

29. Show by an example that the hypothesis "$\lim\limits_{x\to c} f(x)$ exists" in Theorem 2.3.5 is necessary; that is, find functions f and g such that $f(x) \geq g(x)$ for x near c, g has a limit at c, but f does not.

30. (a) Find functions f and g such that $f + g$ has a limit as $x \to 0$, but neither f nor g has a limit as $x \to 0$.

(b) Is it possible for f and $f + g$ to have limits as $x \to 0$, but not g? If it is possible, give an example; if it is not possible, explain why not.

31. (a) Find functions f and g such that fg has a limit as $x \to 0$, but neither f nor g has a limit as $x \to 0$.

(b) Is it possible for f and fg to have limits as $x \to 0$, but not g? If it is possible, give an example; if it is not possible, explain why not.

32. If $\lim\limits_{x\to c} f(x) = 0$ and $\lim\limits_{x\to c}[f(x)g(x)] = 1$, prove that $\lim\limits_{x\to c} g(x)$ does not exist.

33. **Proof of Theorem 2.3.1(a).** Given $\epsilon > 0$, let $\epsilon' = \epsilon/2$. There is a δ_1 such that if $0 < |x - c| < \delta_1$, then $|f(x) - A| < \epsilon'$. Similarly, there is a δ_2 such that if $0 < |x - c| < \delta_2$, then $|g(x) - B| < \epsilon'$. Let δ be the minimum of δ_1 and δ_2.

    (a) Use the triangle inequality to show that

    $$|f(x) + g(x) - (A + B)| \le |f(x) - A| + |g(x) - B|.$$

    (b) Show that if $0 < |x - c| < \delta$, then $|f(x) + g(x) - (A + B)| < \epsilon$.

34. **Proof of Theorem 2.3.1(b).** Observe that the theorem is true if $\alpha = 0$. If $\alpha \neq 0$, for any given $\epsilon > 0$, let $\epsilon' = \epsilon/|\alpha|$. There is a δ such that if $0 < |x - c| < \delta$, then $|f(x) - A| < \epsilon'$. Show that if $0 < |x - c| < \delta$, then $|\alpha f(x) - \alpha A| < \epsilon$.

35. **Proof of Theorem 2.3.4.** Let $\epsilon > 0$ be given. There is a δ_1 such that if $0 < |x - c| < \delta_1$, then $|g(x) - L| < \epsilon$. Also, there is a δ_2 such that if $0 < |x - c| < \delta_2$, then $|h(x) - L| < \epsilon$. Let δ be the minimum of δ_1, δ_2, and r. Show that if $0 < |x - c| < \delta$, then $|f(x) - L| < \epsilon$.

2.4 ONE-SIDED LIMITS, LIMITS AT INFINITY, AND INFINITE LIMITS

In Section 2.2 we defined the limit of a function at a point, a concept expressed by the equation

$$\lim_{x \to c} f(x) = L. \tag{1}$$

We now explore several generalizations of this idea.

Limit from the right and limit from the left

In defining the meaning of Eq. 1 we were careful to consider points on both sides of c. However, it is also possible to discuss limits in which points on only one side of c are taken into account. Such limits are known as **one-sided limits.**

The number L is called the **right-hand limit** of f at c if $f(x)$ approaches L as x approaches c, subject to the requirement that $x > c$; the **left-hand limit** is similar except that only points $x < c$ are considered.

More precisely, let f be defined on an open interval (c, b). Then f is said to have the limit L from the right, which we write as

$$\lim_{x \to c^+} f(x) = L, \tag{2}$$

if for each $\epsilon > 0$ there exists a $\delta > 0$ such that

$$\text{if} \quad c < x < c + \delta, \quad \text{then} \quad |f(x) - L| < \epsilon. \tag{3}$$

Similarly, if f is defined on an open interval (a, c), then f is said to have the limit L from the left, written

$$\lim_{x \to c^-} f(x) = L, \tag{4}$$

if for each $\epsilon > 0$ there exists a $\delta > 0$ such that

$$\text{if} \quad c - \delta < x < c, \quad \text{then} \quad |f(x) - L| < \epsilon. \tag{5}$$

The following examples illustrate the occurrence of one-sided limits for some simple functions.

EXAMPLE 1

Consider the function

$$f(x) = \sqrt{x}, \qquad x \geq 0.$$

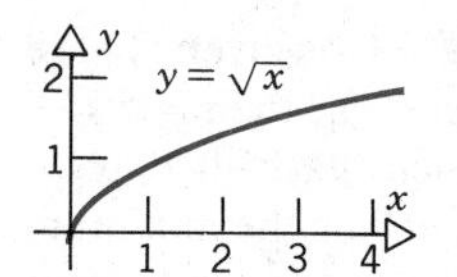

Figure 2.4.1

From the graph of $y = \sqrt{x}$ in Figure 2.4.1, or by calculating $\sqrt{x}$ for a few small positive values of x, we conclude that

$$\lim_{x \to 0^+} \sqrt{x} = 0.$$

See also Problem 33. On the other hand, it is meaningless to talk about the existence (or nonexistence) of $\lim_{x \to 0^-} f(x)$, since f is not defined for points to the left of $x = 0$. ∎

EXAMPLE 2

Consider the function

$$f(x) = \begin{cases} 1 - x^2, & x < 0 \\ \frac{1}{2}, & x = 0 \\ 1 - x, & x > 0 \end{cases}$$

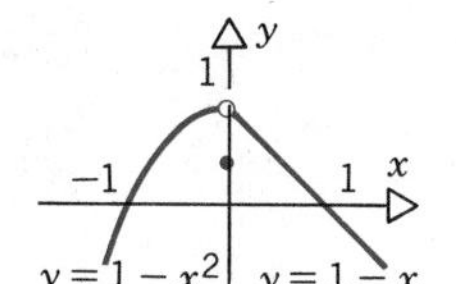

Figure 2.4.2

whose graph is shown in Figure 2.4.2.

In this case

$$\lim_{x \to 0^+} f(x) = 1, \qquad \lim_{x \to 0^-} f(x) = 1,$$

and further

$$\lim_{x \to 0} f(x) = 1.$$

As usual, these results do not depend on the value of $f(0)$, or even on whether or not f is defined at $x = 0$. ∎

EXAMPLE 3

Consider the function

$$f(x) = [\![x]\!].$$

From the graph in Figure 2.4.3 we see that

$$\lim_{x \to 1^+} f(x) = 1, \qquad \lim_{x \to 1^-} f(x) = 0$$

and that $\lim_{x \to 1} f(x)$ does not exist. (Recall Examples 4 and 8 in Section 2.2.) ∎

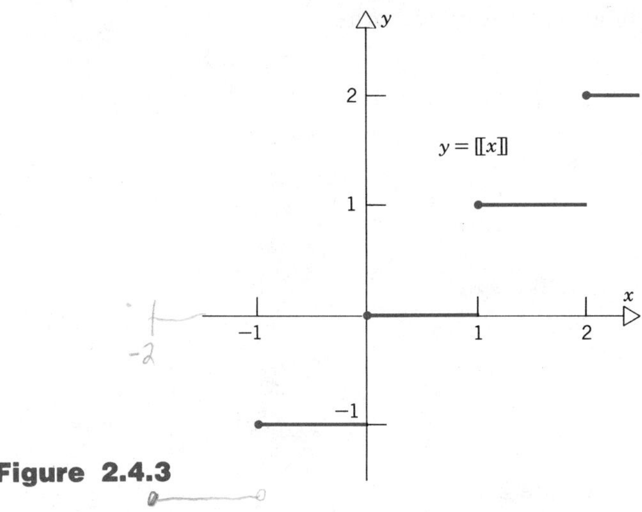

Figure 2.4.3

Examples 2 and 3 suggest the following relation between one-sided limits and the limit of a function as previously defined.

Theorem 2.4.1

The function f has the limit L at the point c if and only if the left-hand limit and the right-hand limit at c exist and are both equal to L. That is,

$$\lim_{x \to c} f(x) = L \quad \text{if and only if} \quad \lim_{x \to c^+} f(x) = \lim_{x \to c^-} f(x) = L. \tag{6}$$

A proof of Theorem 2.4.1 is outlined in Problem 37.

Limits as $x \to \infty$ and as $x \to -\infty$

Until now we have discussed limits only at finite points. However, it is also possible to consider limits of functions at infinity, that is, as the independent variable becomes unbounded. If a variable x increases without bound, we say that x is approaching (positive) infinity and we write $x \to \infty$. Similarly, if x decreases without bound, we say that x is approaching negative infinity, and we write $x \to -\infty$.

If $f(x)$ approaches a number L as $x \to \infty$, then we write

$$\lim_{x \to \infty} f(x) = L. \tag{7}$$

More precisely, Eq. 7 means that for each $\epsilon > 0$, there exists a number $N > 0$, which in general depends on ϵ, such that

$$\text{if} \quad x > N, \qquad \text{then} \quad |f(x) - L| < \epsilon. \tag{8}$$

In a similar way, we write

$$\lim_{x \to -\infty} f(x) = L \tag{9}$$

if, for each $\epsilon > 0$, there exists a number $N > 0$, depending on ϵ, such that

$$\text{if} \quad x < -N, \qquad \text{then} \quad |f(x) - L| < \epsilon. \tag{10}$$

EXAMPLE 4

Let

$$f(x) = \frac{1}{x}, \qquad x \neq 0;$$

see Figure 2.4.4.

As $x \to \infty$, or as $x \to -\infty$, $f(x)$ takes on values closer and closer to zero. Thus

$$\lim_{x \to \infty} \frac{1}{x} = \lim_{x \to -\infty} \frac{1}{x} = 0. \; \blacksquare \tag{11}$$

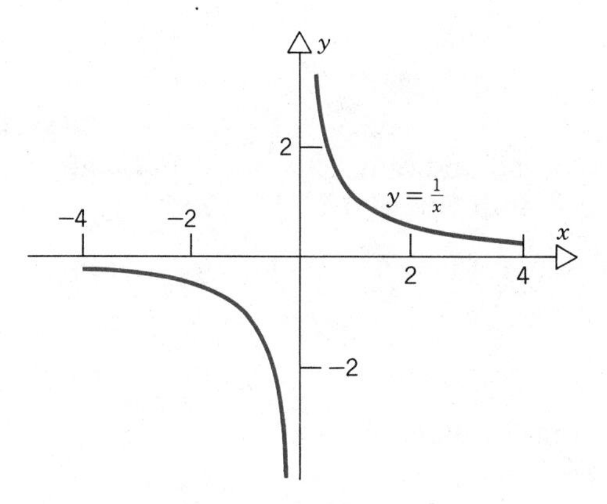

Figure 2.4.4

Algebraic properties

As it is plausible to expect, Theorems 2.3.1 to 2.3.5 remain valid if $\lim_{x \to c}$ is replaced by $\lim_{x \to c^+}$, $\lim_{x \to c^-}$, $\lim_{x \to \infty}$, or $\lim_{x \to -\infty}$. The following examples illustrate how these properties can be used to evaluate limits at infinity.

EXAMPLE 5

If n is a fixed positive integer, then

$$\lim_{x \to \infty} \frac{1}{x^n} = 0. \tag{12}$$

This result follows from Eq. 11 by using Eq. 8 of Section 2.3.

$$\lim_{x\to\infty} \frac{1}{x^n} = \lim_{x\to\infty} \left(\frac{1}{x} \cdot \frac{1}{x} \cdot \cdots \cdot \frac{1}{x}\right)$$

$$= \left(\lim_{x\to\infty} \frac{1}{x}\right)\left(\lim_{x\to\infty} \frac{1}{x}\right) \cdots \left(\lim_{x\to\infty} \frac{1}{x}\right)$$

$$= 0 \cdot 0 \cdot \cdots \cdot 0 = 0.$$

In a similar way

$$\lim_{x\to-\infty} \frac{1}{x^n} = 0. \tag{13}$$

One can also verify Eqs. 12 and 13 by means of the definitions (8) and (10) (see Problem 34). ∎

EXAMPLE 6

Evaluate

$$\lim_{x\to-\infty} \frac{3x^2 + 7x - 6}{4x^2 - 3x + 6}. \tag{14}$$

The quotient rule [Theorem 2.3.1(d)] does not immediately apply, since neither the numerator nor the denominator of the expression in Eq. 13 has a limit; both increase without bound as $x \to -\infty$. In order to use Theorem 2.3.1(d) we must first divide both numerator and denominator in Eq. 14 by x^2. This gives

$$\lim_{x\to-\infty} \frac{3x^2 + 7x - 6}{4x^2 - 3x + 6} = \lim_{x\to-\infty} \frac{3 + (7/x) - (6/x^2)}{4 - (3/x) + (6/x^2)}.$$

Now, assuming the existence of all of the limits that occur (and this can be verified by retracing our steps in the opposite direction), we have

$$\lim_{x\to-\infty} \frac{3x^2 + 7x - 6}{4x^2 - 3x + 6} = \frac{\lim_{x\to-\infty} [3 + (7/x) - (6/x^2)]}{\lim_{x\to-\infty} [4 - (3/x) + (6/x^2)]}$$

$$= \frac{\lim_{x\to-\infty} 3 + 7 \lim_{x\to-\infty} (1/x) - 6 \lim_{x\to-\infty} (1/x^2)}{\lim_{x\to-\infty} 4 - 3 \lim_{x\to-\infty} (1/x) + 6 \lim_{x\to-\infty} (1/x^2)}$$

$$= \frac{3 + 7 \cdot 0 - 6 \cdot 0}{4 - 3 \cdot 0 + 6 \cdot 0} = \frac{3}{4}.$$ ∎

In general, the limit as $x \to \infty$ or as $x \to -\infty$ of any rational function for which the degree of the numerator is less than or equal to the degree of the denominator can be dealt with as in Example 6. Normally, we do not show the calculation of the limit in this much detail; however, we should always keep in mind that it is the knowledge that the individual limits exist at each step that allows us to proceed.

Bounded and unbounded functions

In Section 1.4 we said that a function f is **bounded** on an interval if there exists a number $K > 0$ such that

$$|f(x)| \leq K, \tag{15a}$$

or equivalently,

$$-K \leq f(x) \leq K \tag{15b}$$

for every x in the interval. The interval may be open or closed, bounded or unbounded. If f is not bounded on an interval, then it is said to be **unbounded** there. Geometrically, the graph of a bounded function $y = f(x)$ lies between the two horizontal lines $y = K$ and $y = -K$, as shown in Figure 2.4.5.

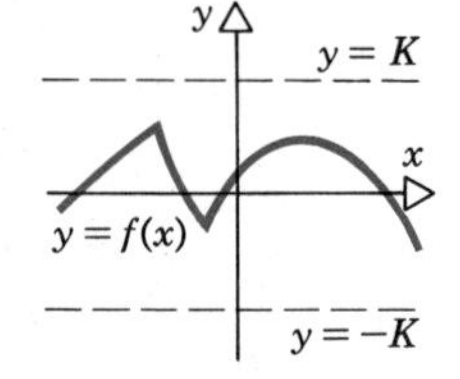

Figure 2.4.5 A bounded function.

For example, the function $f(x) = \sin x$ is bounded on every interval because $|\sin x| \leq 1$ for all x in R^1. On the other hand, the function

$$f(x) = \frac{1}{x^2}, \qquad x \neq 0, \tag{16}$$

whose graph is shown in Figure 2.4.6, is unbounded on the interval $(0, c)$ for any $c > 0$ and is also unbounded on $(d, 0)$ for any $d < 0$. However, on an interval away from the origin, such as $(1, 5)$, the function f is bounded.

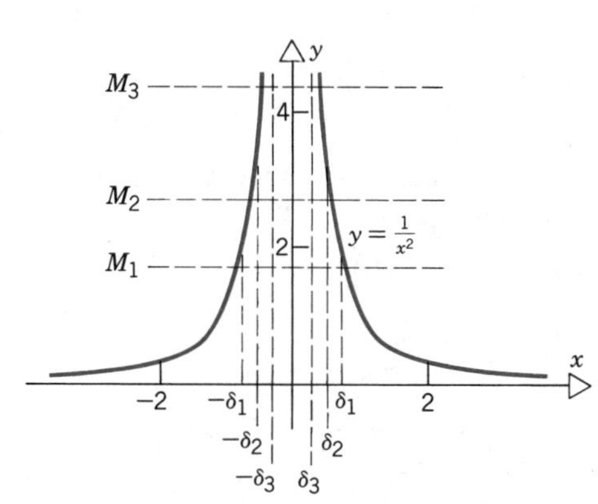

Figure 2.4.6

The function $f(x) = 1/x^2$ has no limit as $x \to 0$ because $f(x)$ increases without bound as $x \to 0$. The fact that the limit does not exist for this particular reason is usually expressed by writing

$$\lim_{x \to 0} \frac{1}{x^2} = \infty.$$

More generally, we say that a function f approaches (positive) ∞ as x approaches c, written

$$\lim_{x \to c} f(x) = \infty, \tag{17}$$

if for each number $M > 0$, there exists a number $\delta > 0$ such that

$$\text{if} \quad 0 < |x - c| < \delta, \qquad \text{then} \quad f(x) > M. \tag{18}$$

This is illustrated geometrically in Figure 2.4.6 for the function $f(x) = 1/x^2$. Note that the larger M is, the smaller δ must be chosen.

Similarly, we say that f approaches $-\infty$ as x approaches c, written

$$\lim_{x \to c} f(x) = -\infty, \tag{19}$$

if for each $M > 0$, there is a $\delta > 0$, such that

$$\text{if} \quad 0 < |x - c| < \delta, \qquad \text{then} \quad f(x) < -M. \tag{20}$$

Obvious modifications can be made in these definitions to cover the cases $x \to c^+$, $x \to c^-$, $x \to \infty$, and $x \to -\infty$.

EXAMPLE 7

Determine the left-hand limit and the right-hand limit as $x \to 0$ for the function

$$f(x) = \frac{2 - x}{x}, \qquad x \neq 0.$$

The graph of $y = f(x)$ is shown in Figure 2.4.7.

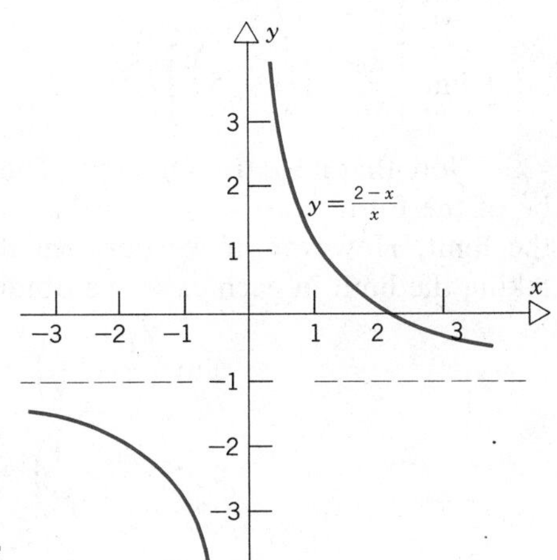

Figure 2.4.7

We have

$$\lim_{x \to 0^+} \frac{2 - x}{x} = \lim_{x \to 0^+} \left(\frac{2}{x} - 1\right) = \infty,$$

since $2/x$ is positive for $x > 0$ and becomes unbounded as $x \to 0^+$. Similarly,

$$\lim_{x \to 0^-} \frac{2 - x}{x} = \lim_{x \to 0^-} \left(\frac{2}{x} - 1\right) = -\infty.$$

Since f becomes unbounded, but in opposite directions, as $x \to 0$ from the right and left, respectively, it is *not correct* to say that $\lim_{x \to 0} f(x) = \infty$. Note the difference between this function and $f(x) = 1/x^2$, whose graph is shown in Figure 2.4.6. ∎

In conclusion we emphasize again that ∞ *and* $-\infty$ *are not numbers*. Therefore Theorem 2.3.1 concerning the limits of sums, differences, products, and quotients must be used with special care in those cases where f, or g, or both approach ∞ or $-\infty$. In particular, Theorem 2.3.1 can be used in the cases $\infty + \infty$, $\infty \pm L$, L/∞, ∞/L, $\infty \cdot \infty$, and $\infty \cdot L$ for $L \neq 0$. In these cases the limits are ∞, ∞, 0, ∞, ∞, and ∞, respectively. It can also be used in the corresponding cases where ∞ is replaced by $-\infty$. However, Theorem 2.3.1 does not apply to the indeterminate cases of the form $\infty - \infty$, ∞/∞, and $0 \cdot \infty$. In these cases, nothing can be said in general about the limits, and each problem must be handled individually, as the following example shows.

EXAMPLE 8

Determine whether each of the following limits exists; if so, find it.

(a) $\lim_{x \to 0^+} \left(\frac{1}{x} - \frac{2}{x}\right)$;

(b) $\lim_{x \to 0^+} \left(\frac{1}{x^2} - \frac{1}{x}\right)$;

(c) $\lim_{x \to 0^+} \left[\frac{2}{x} - \left(\frac{2}{x} - 5\right)\right]$.

Note that if we tried to apply Theorem 2.3.1, each of these expressions would be of the form $\infty - \infty$ as $x \to 0^+$, and we could not draw any conclusions about the limit. However, if we perform the obvious algebraic simplifications before taking the limit in each case, we obtain

$$\lim_{x \to 0^+} \left(\frac{1}{x} - \frac{2}{x}\right) = \lim_{x \to 0^+} \left(-\frac{1}{x}\right) = -\infty; \tag{21}$$

$$\lim_{x \to 0^+} \left(\frac{1}{x^2} - \frac{1}{x}\right) = \lim_{x \to 0^+} \left(\frac{1 - x}{x^2}\right) = \infty; \tag{22}$$

$$\lim_{x \to 0^+} \left[\frac{2}{x} - \left(\frac{2}{x} - 5\right)\right] = \lim_{x \to 0^+} (5) = 5. \tag{23}$$

Hence we conclude that an expression of the form $\infty - \infty$ can approach either ∞, or $-\infty$, or any finite number. A similar situation prevails for the indeterminate forms ∞/∞ and $0 \cdot \infty$. ∎

PROBLEMS

In each of Problems 1 through 12, determine whether $\lim_{x\to c^+} f(x)$, $\lim_{x\to c^-} f(x)$, and $\lim_{x\to c} f(x)$ exist. Eval uate each limit that does exist. Proceed intuitively and geometrically; it is not required that you use the formal definitions.

1. $f(x) = \sqrt{4 - x^2}$; $\quad c = 2, -2, 1$

2. $f(x) = \sqrt{x^2 - 1}$; $\quad c = 1, -1$

3. $f(x) = (4 - x^2)^{1/3}$; $\quad c = 2$

4. $f(x) = \sqrt{|4 - x^2|}$; $\quad c = 2$

5. $f(x) = \sqrt{|x^2 - 1|}$; $\quad c = -1$

6. $f(x) = [\![x]\!] - x$; $\quad c = 1$

7. $f(x) = \begin{cases} 1 + x^2, & x < 0; \\ 0, & x = 0; \\ 1 - x^2, & x > 0; \end{cases} \quad c = 0$

8. $f(x) = \begin{cases} \sqrt{1 - x^2}, & -1 < x < 1; \\ x, & x > 1; \end{cases} \quad c = 1$

9. $f(x) = \begin{cases} \dfrac{1}{x^2}, & x < 0; \\ x^2, & x > 0; \end{cases} \quad c = 0$

10. $f(x) = \begin{cases} \sec x, & -\dfrac{\pi}{2} < x < 0; \\ \csc x, & 0 < x < \dfrac{\pi}{2}; \end{cases} \quad c = 0$

11. $f(x) = \begin{cases} \sin x, & x < 0; \\ 1, & x = 0; \\ \sin 3x, & x > 0; \end{cases} \quad c = 0$

12. $f(x) = \begin{cases} \sin x, & x < \dfrac{\pi}{2}; \\ 0, & x = \dfrac{\pi}{2}; \\ \sin 3x, & x > \dfrac{\pi}{2}; \end{cases} \quad c = \pi/2$

In each of Problems 13 through 30, find the indicated limit or else show that it does not exist. Proceed intuitively and geometrically; it is not required that you use the formal definitions.

13. $\lim_{x\to\infty} \dfrac{1 + x}{1 - x}$

14) $\lim_{x\to-\infty} \dfrac{x^3}{x^3 + 1}$

15. $\lim_{x\to\infty} \dfrac{3x^2 + 4x - 5}{2x^3 - 6x^2 + x + 1}$

16. $\lim_{x\to\infty} \dfrac{-x^3 + 4x^2 + 8}{2x^3 - 5x + 4}$

17. $\lim_{x\to\infty} \dfrac{\sqrt{1 + x^2}}{x}$

18. $\lim_{x\to-\infty} \dfrac{\sqrt{1 + x^2}}{x}$

19. $\lim_{x\to-\infty} \left(\dfrac{2x^2 - x + 1}{x^2 + 4}\right)^{1/2}$

20. $\lim_{x\to\infty} \left(\dfrac{-x^2 + 3x + 2}{8x^2 - 2x + 5}\right)^{1/3}$

21. $\lim_{x\to\infty} \dfrac{\sin x}{x}$

22. $\lim_{x\to\infty} (\sin x - \cos x)$

23. $\lim_{x\to-\infty} \dfrac{x^3 + 2x - 4}{x^2 + 1}$

24. $\lim_{x\to\infty} \dfrac{x^4 + 3x^2 + 1}{x^2 - 9}$

25. $\lim_{x\to0^+} x\sqrt{1 + \dfrac{1}{x}}$

26. $\lim_{x\to\infty} x \sin x$

27. $\lim_{x\to1^-} \dfrac{1 - x^2}{\sqrt{1 - x^4}}$

28. $\lim_{x\to0} (\csc x - \cot x)$

29. $\lim_{x\to0} \dfrac{\csc x - \cot x}{x}$

30. $\lim_{x\to0} \dfrac{1 - \sqrt{1 - x^2}}{x^2}$

31. Draw the graph of a function satisfying all of the following conditions.
   (a) $\lim_{x\to-\infty} f(x) = 2$;
   (b) $\lim_{x\to0^-} f(x) = 0$;
   (c) $f(0) = 1$;
   (d) $\lim_{x\to0^+} f(x) = 2$;
   (e) $\lim_{x\to\infty} f(x) = 0$.

32. Draw the graph of a function satisfying all of the following conditions.
   (a) $\lim_{x\to-\infty} f(x) = -1$;
   (b) $\lim_{x\to0^-} f(x) = -2$;
   (c) $\lim_{x\to0^+} f(x) = 1$;
   (d) $\lim_{x\to1^-} f(x) = \infty$;
   (e) $\lim_{x\to1^+} f(x) = -\infty$;
   (f) $\lim_{x\to\infty} f(x) = 2$.

33. Use definition (3) to show that $\lim_{x\to0^+} \sqrt{x} = 0$; that is, given $\epsilon > 0$, find $\delta > 0$ such that if $0 < x < \delta$, then $0 < \sqrt{x} < \epsilon$.

34. Use definition (8) to show that $\lim_{x\to\infty}(1/x^n) = 0$; that is, given $\epsilon > 0$, find N such that if $x > N$, then $1/x^n < \epsilon$.

35. Use definition (18) to show that $\lim_{x\to 0}(1/x^2) = \infty$; that is, given $M > 0$, find $\delta > 0$ such that if $0 < |x| < \delta$, then $1/x^2 > M$.

36. Use the definition to show that $\lim_{x\to 0+}(1/x^3) = \infty$; that is, given $M > 0$, find $\delta > 0$ such that if $0 < x < \delta$, then $1/x^3 > M$.

37. **Proof of Theorem 2.4.1.** (a) Assume that $\lim_{x\to c+} f(x) = L$ and that $\lim_{x\to c-} f(x) = L$. Given $\epsilon > 0$, there is a δ_1 such that if $c < x < c + \delta_1$, then $|f(x) - L| < \epsilon$. Similarly, there is a δ_2 such that if $c - \delta_2 < x < c$, then $|f(x) - L| < \epsilon$. Let δ be the minimum of δ_1 and δ_2. Show that if $0 < |x - c| < \delta$, then $|f(x) - L| < \epsilon$. Thus $\lim_{x\to c} f(x) = L$.

(b) Assume that $\lim_{x\to c} f(x) = L$. Given $\epsilon > 0$, there is a δ such that if $0 < |x - c| < \delta$, then $|f(x) - L| < \epsilon$. Show that if $c < x < c + \delta$, then $|f(x) - L| < \epsilon$. Thus $\lim_{x\to c+} f(x) = L$. In a similar way show that $\lim_{x\to c-} f(x) = L$.

2.5 CONTINUOUS FUNCTIONS

The set of all functions is extremely large and varied. Indeed, it is so all-inclusive that not very much can be said about the properties of all functions in general. Moreover, in the study of calculus we are mainly interested in functions whose graphs are reasonably smooth curves. Therefore, in order to obtain subclasses of functions with useful and interesting properties, we impose various restrictions that screen out the more irregular and pathological functions. In this section we introduce the concept of a **continuous function,** and show that many functions belong to this class of functions. In Section 2.6 we discuss several important properties of continuous functions.

In Section 2.3 we showed that if P is a polynomial, then

$$\lim_{x\to c} P(x) = P(c) \tag{1}$$

at every point c. Similarly, if $R(x) = P(x)/Q(x)$ is a rational function, the ratio of two polynomials, then

$$\lim_{x\to c} R(x) = R(c), \tag{2}$$

provided that the denominator $Q(c) \neq 0$. In other words, in each of these cases the value of the function at $x = c$ is the same as the limit as x approaches c. The property expressed by Eqs. 1 and 2 is known as *continuity*, and is also possessed by many functions other than polynomials and rational functions. We will see later that the concept of a continuous function* is crucial in the development of calculus,

*The term "continuous function" was used by Euler and other eighteenth-century mathematicians, but in a sense different from the usage today. Bernhard Bolzano (1781–1848), a Bohemian priest and scholar, gave essentially the modern definition of continuity and developed some of its consequences in a paper written in 1817. Unfortunately, his paper, published privately in Prague and not widely circulated, received almost no attention for many years. Meanwhile, Cauchy used the same definition a few years later in the books mentioned previously, and it was through him that the notion of continuity became well known. Complete understanding of the concept proved elusive, however, even to so great a mathematician as Cauchy. Several more decades elapsed before some remaining questions associated with this profound idea were resolved.

but in this section our discussion will be somewhat informal. We start by giving a definition.

DEFINITION 2.5.1 Suppose that the domain of the function f includes an open interval containing the point c. Then f is said to be continuous at c if $\lim_{x \to c} f(x)$ exists, and if

$$\lim_{x \to c} f(x) = f(c). \tag{3}$$

Further, f is said to be continuous on an open interval (a, b) if it is continuous at each point in the interval.

Although Definition 2.5.1 seems simple enough, it turns out that the concept of a continuous function is a subtle one. To understand more clearly what continuity involves, it is helpful to examine some functions that fail to have this property at certain points, that is, functions that are *discontinuous*. There are four commonly occurring ways in which a function may be discontinuous at a point; the three most important of these involve the nonexistence of a limit.

1. ***Jump discontinuity***. A function f has a jump discontinuity at the point c if it has both right- and left-hand limits at c, but these limits are unequal, that is,

$$\lim_{x \to c^+} f(x) \neq \lim_{x \to c^-} f(x). \tag{4}$$

The jump of f at c is the difference between the right- and left-hand limits there. For example, the greatest integer function

$$f(x) = [\![x]\!], \tag{5}$$

whose graph is shown in Figure 2.5.1, has a jump of one at each integer point. Another function with a jump discontinuity is the signum function,

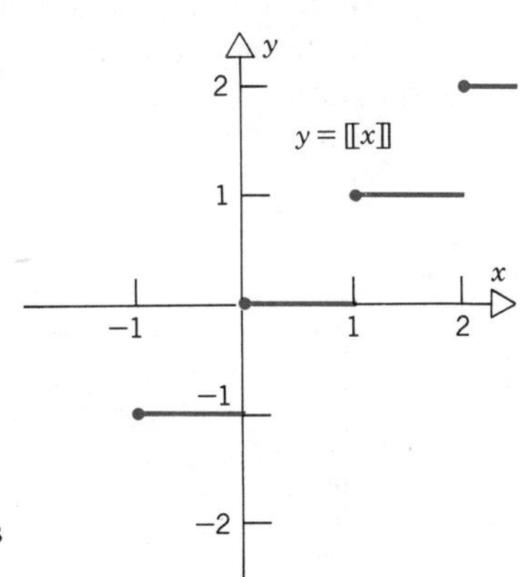

Figure 2.5.1
Jump discontinuities at $x = 0, \pm 1, \ldots$.

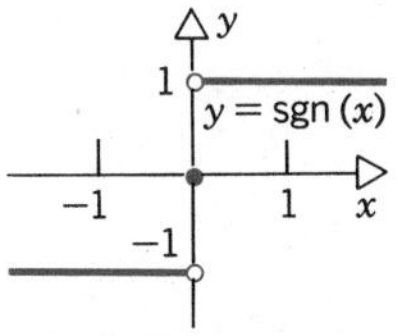

$$\operatorname{sgn}(x) = \begin{cases} -1, & x < 0 \\ 0, & x = 0 \\ 1, & x > 0 \end{cases}. \tag{6}$$

Figure 2.5.2
A jump discontinuity at $x = 0$.

This function has a jump of two at $x = 0$ (see Figure 2.5.2).

2. ***Infinite discontinuity.*** A function f has an infinite discontinuity at c if it is unbounded in every interval about c. For example,

$$f(x) = \frac{1}{x^2}, \qquad x \neq 0 \tag{7}$$

has an infinite discontinuity at $x = 0$ (see Figure 2.5.3). Similarly, the function

$$g(x) = \tan x, \qquad x \neq \pm\frac{\pi}{2},\ \pm\frac{3\pi}{2}, \ldots \tag{8}$$

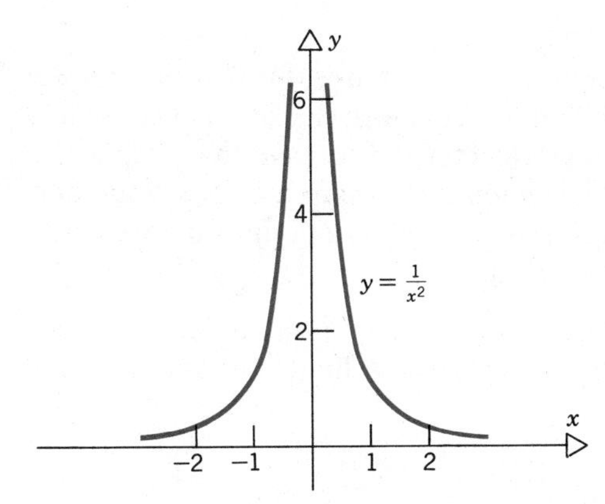

Figure 2.5.3 An infinite discontinuity at $x = 0$.

whose graph is shown in Figure 2.5.4, has an infinite discontinuity at each odd multiple of $\pi/2$. In each of these examples the function is not defined at the point of discontinuity, and hence is also discontinuous for that reason. However, this is a relatively minor matter; regardless of how the function might be defined at such a point, the infinite discontinuity remains. It is also possible for a function to be unbounded only on one side of the point of discontinuity, and to approach a finite limit on the other. For instance, the function

$$f(x) = \frac{1}{x} + \frac{1}{|x|}, \qquad x \neq 0, \tag{9}$$

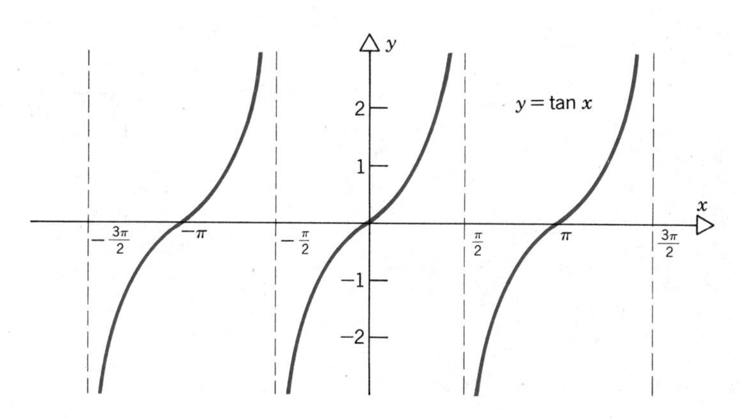

Figure 2.5.4 Infinite discontinuities at $x = \pm\pi/2, \pm3\pi/2, \ldots$.

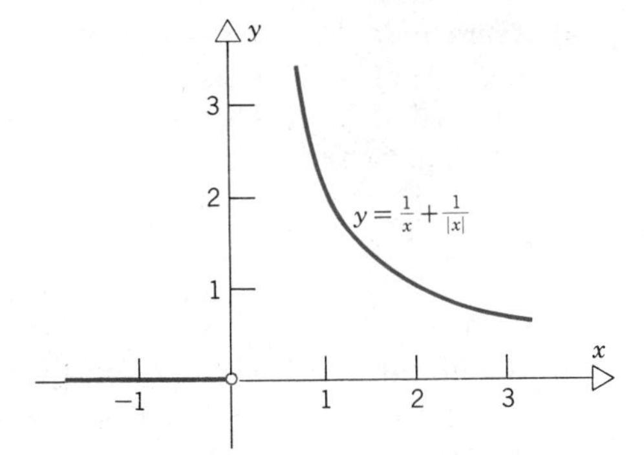

Figure 2.5.5 A one-sided infinite discontinuity at $x = 0$.

has this property, as shown in Figure 2.5.5. Such behavior is also considered to be an infinite discontinuity.

3. ***Oscillatory discontinuity.*** A somewhat more complicated type of discontinuity, called an oscillatory discontinuity, is exhibited by the function

$$f(x) = \sin \frac{1}{x}, \qquad x \neq 0, \tag{10}$$

whose graph is shown in Figure 2.5.6. This function does not have a limit as $x \to 0$ because of its rapidly oscillating behavior in every (deleted) interval about $x = 0$. To establish this fact, let us consider the possibility that f does have a limit as x approaches zero. Suppose first that the limit is zero. We observe, however, that f has the value one at each of the points $x = 2/\pi$, $2/5\pi$, $2/9\pi$, $2/13\pi, \ldots$. Since some of these points are to be found in each deleted interval about $x = 0$, it follows that there is no deleted interval about $x = 0$ within which $f(x)$ always has values close to zero. Hence f does not have the limit zero as x approaches zero. A similar argument can be used to refute the belief that f has any other number for a limit. Hence the function $\sin(1/x)$ has no limit at $x = 0$, and consequently is discontinuous there.

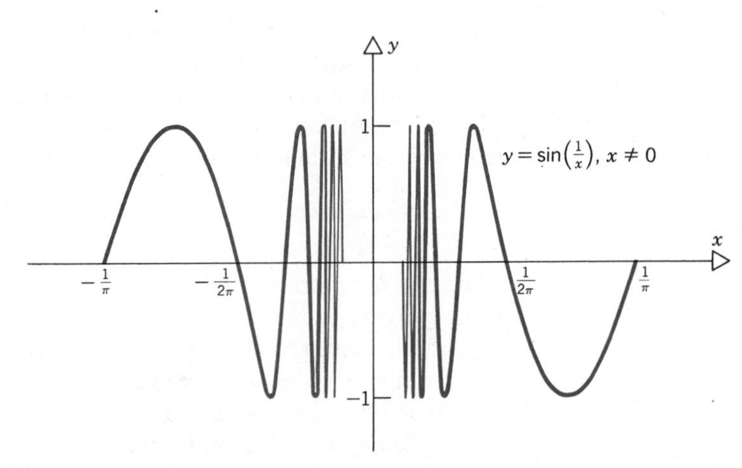

Figure 2.5.6 An oscillatory discontinuity at $x = 0$.

4. ***Removable discontinuity.*** A function f has a removable discontinuity at c if $\lim_{x\to c} f(x) = L$ exists, and if $f(c)$ either does not exist, or has the "wrong" value, that is, if $f(c) \neq L$. The discontinuity can be removed simply by defining, or redefining if necessary, $f(c)$ to have the value L. For example, let

$$f(x) = \frac{\sin x}{x}, \qquad x \neq 0. \tag{11}$$

The graph of f is sketched in Figure 2.5.7. Although $\lim_{x\to 0} f(x) = 1$, the function is discontinuous at $x = 0$ because $f(0)$ is not defined. However, we can define a new function g that is essentially the same as f, and is, moreover, continuous at $x = 0$. Let

$$g(x) = \begin{cases} f(x), & x \neq 0; \\ 1, & x = 0. \end{cases} \tag{12}$$

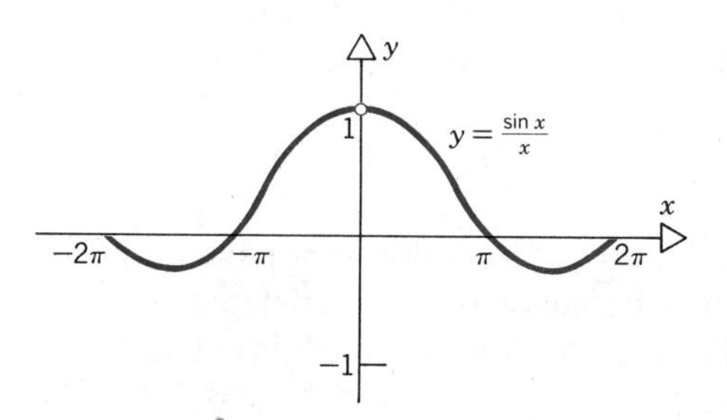

Figure 2.5.7 A removable discontinuity at $x = 0$.

The $\lim_{x\to 0} g(x) = g(0) = 1$, and thus g is continuous at $x = 0$. We can regard g as an "extension" of f, and often it is not necessary to distinguish between them. In geometrical language the function f is discontinuous at $x = 0$ because its graph has a "hole" or "gap" at that point. By properly extending the definition of f, however, we can fill the hole and remove the discontinuity.

This list of types of discontinuities is not altogether comprehensive. For one thing, combinations can occur; for instance, a function can be unbounded on one side of a point and oscillatory on the other. Further, there are some discontinuities that do not fit any of these patterns. For instance, consider again the function mentioned in Example 7 of Section 1.5:

$$f(x) = \begin{cases} 1, & x \text{ rational}, \\ 0, & x \text{ irrational}. \end{cases} \tag{13}$$

While this function is discontinuous everywhere, it does not seem to belong to any of the previously mentioned categories.

In elementary calculus we are almost always concerned with functions that are continuous either everywhere, or at all but a relatively few points. The most important types of discontinuities are jump discontinuities and infinite discontinuities. Removable discontinuities are of no great significance since they can be eliminated if necessary, and the other types occur quite infrequently.

From an analytical point of view, the question of deciding whether a given function f is continuous at a point c is simply a matter of determining $\lim_{x\to c} f(x)$ and checking whether $\lim_{x\to c} f(x) = f(c)$. However, in virtually all cases, one's geometrical intuition can be trusted provided that the graph of the function is available. If the graph shows a jump or becomes unbounded, then the function involved is discontinuous. The idea of continuity is sometimes expressed by stating that a continuous function is one whose graph consists of a single unbroken curve. While this is not very precise mathematically, it is often helpful to think in these terms. If it is not clear what the graph of the function looks like, a more careful analysis based on the definition of limit may be required.

It is an easy matter to adapt the properties of limits in Section 2.3 to continuous functions. This is not surprising in view of the close relationship between limits and continuity.

Theorem 2.5.1

If the functions f and g are continuous at the point c, then the following functions are also continuous there:

(a) $f + g$, $f - g$, fg, $|f|$, and αf, for any constant α;
(b) f/g, provided $g(c) \neq 0$;
(c) $(f)^{1/n}$, where n is a positive integer, provided that $f(c) > 0$ if n is even.

Theorem 2.5.1 is an immediate consequence of the corresponding limit properties in Section 2.3. We will prove one part of this theorem as an illustration of this fact, and leave the remaining proofs as exercises for the reader. Since f and g are continuous at c, the definition of continuity requires that

$$\lim_{x\to c} f(x) = f(c), \qquad \lim_{x\to c} g(x) = g(c). \tag{14}$$

Further, Theorem 2.3.1(a) states that

$$\lim_{x\to c}(f + g)(x) = \lim_{x\to c}[f(x) + g(x)] = \lim_{x\to c} f(x) + \lim_{x\to c} g(x);$$

hence, by Eq. 14, we have

$$\begin{aligned}\lim_{x\to c}(f + g)(x) &= f(c) + g(c)\\ &= (f + g)(c).\end{aligned}$$

Thus the function $f + g$ is also continuous according to the definition of continuity. The other parts of Theorem 2.5.1 are proved in a similar manner. □

Continuity of rational and algebraic functions

At the beginning of this section we noted that polynomials are continuous at every point, and rational functions are continuous at all points where the denominator is nonzero. These facts are expressed in Eqs. 1 and 2, respectively. More general algebraic functions involving root extractions are also continuous at all points where the conditions of Parts (b) and (c) of Theorem 2.5.1 are met.

Continuity of trigonometric functions

It is not difficult to show that the sine and cosine functions are continuous at every point. First recall that in Example 8 of Section 2.3 we showed that

$$\lim_{x\to 0} \sin x = 0 = \sin 0, \tag{15}$$

and

$$\lim_{x\to 0} \cos x = 1 = \cos 0; \tag{16}$$

therefore the sine and cosine functions are both continuous at zero.

To show that $\sin x$ is continuous at an arbitrary point c, note that

$$\begin{aligned} \sin x &= \sin[c + (x - c)] \\ &= \sin c \cos(x - c) + \cos c \sin(x - c) \\ &= \sin c \cos h + \cos c \sin h, \end{aligned}$$

where $h = x - c$. Since the limit as x approaches c is the same as the limit as h approaches zero, we have (by Theorem 2.3.1)

$$\lim_{x\to c} \sin x = \sin c \lim_{h\to 0} \cos h + \cos c \lim_{h\to 0} \sin h.$$

From Eqs. 15 and 16 we obtain

$$\lim_{x\to c} \sin x = \sin c, \tag{17}$$

and hence $\sin x$ is continuous at every point. In a very similar way it is possible to show that $\cos x$ is also continuous at every point.

By writing the other trigonometric functions in terms of the sine and cosine, and using Theorem 2.5.1, we find that the functions $\tan x$, $\cot x$, $\sec x$, and $\csc x$ are also continuous at all points where their respective denominators are nonzero.

The class of continuous functions is much larger than this discussion indicates, and contains many functions other than algebraic and trigonometric functions. We will encounter some of these other functions later, but meanwhile in the next section we will turn to a consideration of some fundamental properties shared by all continuous functions.

PROBLEMS

In each of Problems 1 through 14, find all points where the given function is discontinuous, and identify the type of discontinuity. In each case the domain of the function consists of all points for which the given formula can be evaluated.

1. $f(x) = \dfrac{x}{x^2 - 1}$

2. $f(x) = \dfrac{(x + 2)(x - 1)}{(x + 1)(x - 2)}$

3. $f(x) = \dfrac{x^2}{x^2 - 3x + 2}$

4. $f(x) = \begin{cases} x - 2, & x < 4 \\ 2x - 6, & x > 4 \end{cases}$

5. $f(x) = \begin{cases} x^2 - 1, & x \le 0 \\ x + 1, & x > 0 \end{cases}$

6. $f(x) = \begin{cases} \sin x + \cos x, & x < \dfrac{\pi}{2} \\ \sin x - \cos x, & x \ge \dfrac{\pi}{2} \end{cases}$

7. $f(x) = \left[\!\left[\dfrac{x}{3}\right]\!\right]$

8. $f(x) = [\![x]\!] - x$

9. $f(x) = \operatorname{sgn}(1 - x^2)$

10. $f(x) = \operatorname{sgn}(\sin x)$

11. $f(x) = \operatorname{sgn}(1 + \sin x)$

12. $f(x) = \cos\dfrac{1}{x}, \quad x \neq 0$

13. $f(x) = \begin{cases} \dfrac{1}{x}, & x < 0 \\ x, & x \ge 0 \end{cases}$

14. $f(x) = \dfrac{\sin x}{x}, \quad x \neq 0$

In each of Problems 15 through 20, the given function is not defined at $x = 0$ and hence is discontinuous there. In each case determine whether it is possible to assign a value to $f(0)$ in such a way as to remove the discontinuity.

15. $f(x) = \dfrac{x}{|x|}$

16. $f(x) = \dfrac{x^2 + 2x}{x}$

17. $f(x) = x\left(1 + \dfrac{1}{x}\right)$

18. $f(x) = x\left(1 + \dfrac{1}{x^2}\right)$

19. $f(x) = x\left(1 + \dfrac{1}{\sqrt{|x|}}\right)$

20. $f(x) = \dfrac{|\sin x|}{x}$

21. Suppose that

$$f(x) = \begin{cases} ax + b, & x \le 1 \\ cx^2 + dx + e, & x > 1 \end{cases}$$

where $a, \ldots, e$ are constants. Under what condition on the coefficients is f continuous at $x = 1$?

22. Suppose that

$$f(x) = \begin{cases} \dfrac{a}{x}, & 0 < x < 1 \\ bx + 1, & 1 \le x \le 2 \\ cx^2, & 2 < x \end{cases}$$

Find conditions on a, b and c so that f will be continuous at both $x = 1$ and $x = 2$.

23. Let $f(\epsilon)$ be the degree of the polynomial

$$P(x) = \epsilon x^2 + x - 1.$$

Is f continuous for all ϵ?

24. Find two functions f and g, both of which are discontinuous at a point c, such that $f + g$ is continuous at c.

25. Find two functions f and g, both of which are discontinuous at a point c, such that fg is continuous at c.

26. Find a function $f(x)$ such that f is discontinuous at c but $|f|$ is continuous at c.

27. Show that if $|f(x) - f(t)| < |x - t|$ for each pair of points x and t in the interval (a, b), then f is continuous on (a, b).

2.6 PROPERTIES OF CONTINUOUS FUNCTIONS

Continuous functions have certain properties that are essential in the development of calculus. In this section we state these properties and mention a few of their ramifications. Although the properties themselves are geometrically very plausible, some of their proofs are surprisingly difficult. We include two of the proofs in Problems 31 and 32.

Since some of the properties we wish to consider involve functions that are continuous on closed intervals, we first introduce the idea of one-sided continuity at a point. A function f is said to be **left continuous** at a point c provided that $f(c)$ is defined, $\lim_{x \to c^-} f(x)$ exists, and

$$\lim_{x \to c^-} f(x) = f(c). \tag{1}$$

Similarly, f is **right continuous** at c provided that $f(c)$ is defined, $\lim_{x \to c^+} f(x)$ exists, and

$$\lim_{x \to c^+} f(x) = f(c). \tag{2}$$

For example, the greatest integer function $f(x) = [\![x]\!]$ is right continuous at each point where it has a jump discontinuity, but it is not left continuous at these points; see Figure 2.5.1. From Theorem 2.4.1 it follows that if f is both left continuous and right continuous at c, then it is continuous there, and conversely.

A function f is said to be continuous on the closed bounded interval $[a, b]$ provided that f is continuous at each point in the open interval (a, b), and provided that it is also right continuous at a and left continuous at b.

Theorem 2.6.1

Let f be continuous at c, and suppose that $f(c) \neq 0$. Then there is an open interval $(c - \delta, c + \delta)$ about c in which $f(x)$ has the same sign as $f(c)$.

Another way of stating the idea contained in this theorem is that for a continuous function a small change in the independent variable gives rise to a small change in the value of the function. Figure 2.6.1 shows the situation described in Theorem 2.6.1, and the following example also serves to illustrate this theorem.

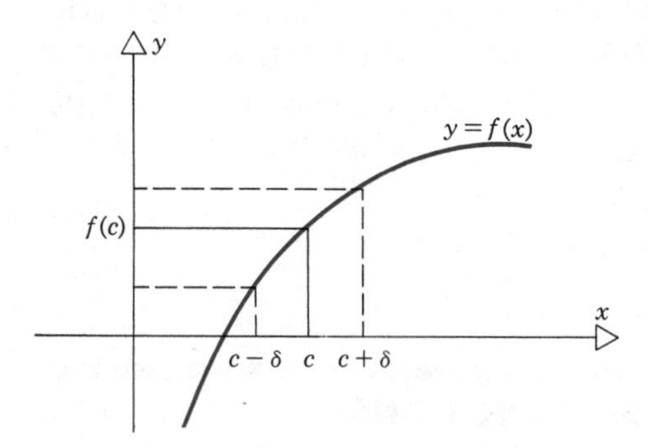

Figure 2.6.1

EXAMPLE 1

The function

$$f(x) = x^2 - 1 \tag{3}$$

has the value $f(1.01) = 0.0201 > 0$ when $x = 1.01$. Find an interval $(1.01 - \delta, 1.01 + \delta)$ in which $f(x) > 0$.

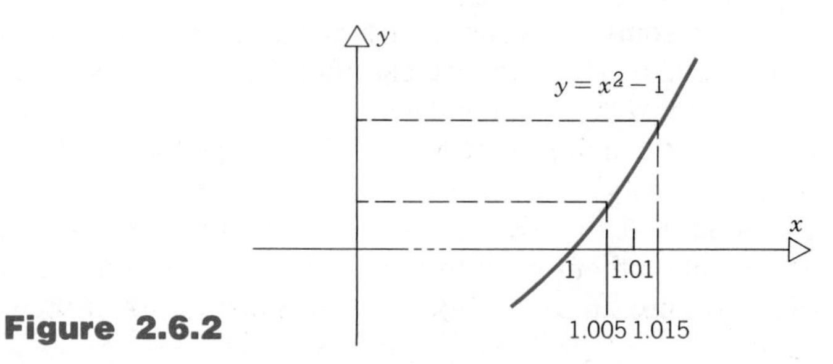

Figure 2.6.2

The graph of $y = f(x)$ is shown in Figure 2.6.2. There are many intervals having the required property. For instance, let $\delta = 0.005$. You can verify that, if x is in the interval $(1.005, 1.015)$, then $f(x)$ takes on only positive values. From the graph it is clear that any other $\delta < 0.01$ would also suffice. ∎

In Example 1 we could easily find the required interval from the graph of f, by noting the points where it crosses the x-axis. This is no longer possible if we lack a formula for $f(x)$, but know only that f is continuous, and that at some point, say $x = 1.01$, it has a nonzero value. Fortunately, it is usually not important actually to find the interval indicated in Theorem 2.6.1. Rather, what is often needed is merely the knowledge that such an interval exists.

The next two theorems shed further light upon the nature of a continuous function.

Theorem 2.6.2

If the function f is continuous on the closed bounded interval $[a, b]$, then it also has the following properties.

(a) **Boundedness Property:** There is a number K such that $|f(x)| \leq K$ for each x in $[a, b]$; that is, f is bounded on $[a, b]$.

(b) **Maximum-Minimum Property:** There are numbers M and m, and also points α and β in $[a, b]$, such that $f(\alpha) = M$, $f(\beta) = m$, and $m \leq f(x) \leq M$ for each x in $[a, b]$; that is, f has a maximum value M and a minimum value m on $[a, b]$, and takes on these values at the points α and β, respectively.

Theorem 2.6.3

(Intermediate Value Property)

Suppose that f is continuous on an interval I (not necessarily closed or bounded). If $x_1 < x_2$ are any two points in I, and if k is any number between $f(x_1)$ and $f(x_2)$, then there is at least one point c in (x_1, x_2) where $f(c) = k$.

If a continuous function f has a maximum at α and a minimum at β, as in Theorem 2.6.2(b), then we can identify α and β with x_1 and x_2 if $\alpha < \beta$, or with x_2 and x_1 if $\alpha > \beta$. Then Theorem 2.6.3 states that f must take on each value k between its maximum M and minimum m at least once.

We will give some applications of these theorems in this section, and in the problems that follow, but their full significance will only become apparent later, as we call on them frequently. The following examples show that if the hypotheses of Theorems 2.6.2 and 2.6.3 are not satisfied, then the conclusions may not hold.

EXAMPLE 2

Let

$$f(x) = \begin{cases} \dfrac{1}{x}, & x \neq 0 \\ 1, & x = 0. \end{cases} \tag{4}$$

The graph of f is shown in Figure 2.6.3.

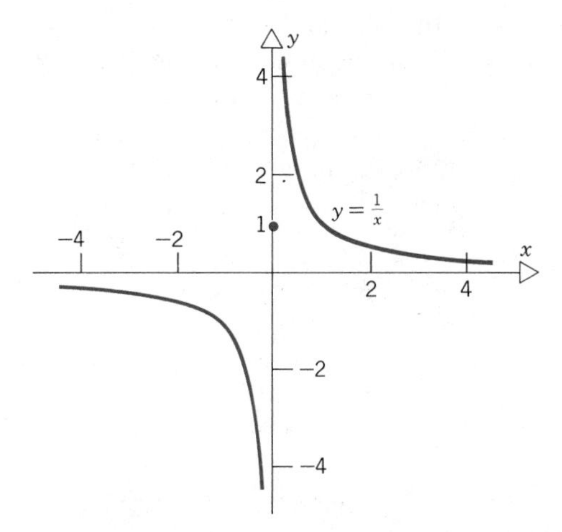

Figure 2.6.3

The hypothesis of Theorem 2.6.2 is violated on the closed bounded interval [0, 1] because the function f is not continuous there. Note that on this interval f is not bounded, nor does $f(x)$ have a maximum value.

The function f is continuous on the interval $[1, \infty)$, but this is an unbounded interval, again violating the conditions of Theorem 2.6.2. Note that on this interval $f(x)$ has no minimum value.

Finally, f is continuous on the unbounded open interval $(0, \infty)$. On this interval $f(x)$ assumes neither a maximum nor a minimum value, nor is f bounded. ∎

EXAMPLE 3

Let

$$f(x) = \begin{cases} x, & 0 < x < 1 \\ \frac{1}{2}, & x = 0 \quad \text{and} \quad x = 1. \end{cases} \tag{5}$$

The graph of f is shown in Figure 2.6.4.

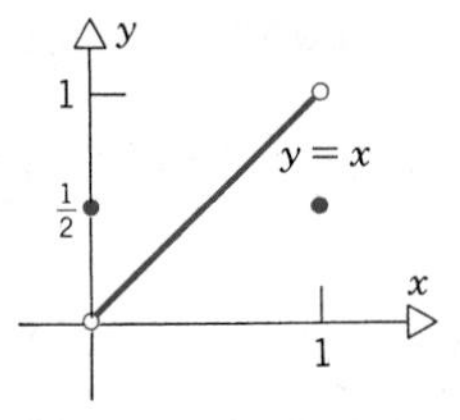

Figure 2.6.4

Again f is not continuous on the closed bounded interval $[0, 1]$. Although f is bounded on $[0, 1]$, $f(x)$ has neither a maximum nor a minimum value.

On the open interval $(0, 1)$, the function f is continuous, but again $f(x)$ has neither a maximum nor a minimum value. ∎

EXAMPLE 4

Let

$$f(x) = \begin{cases} x, & -2 \le x < 0; \\ 1 - x, & 0 \le x \le 2; \end{cases} \qquad (6)$$

see Figure 2.6.5.

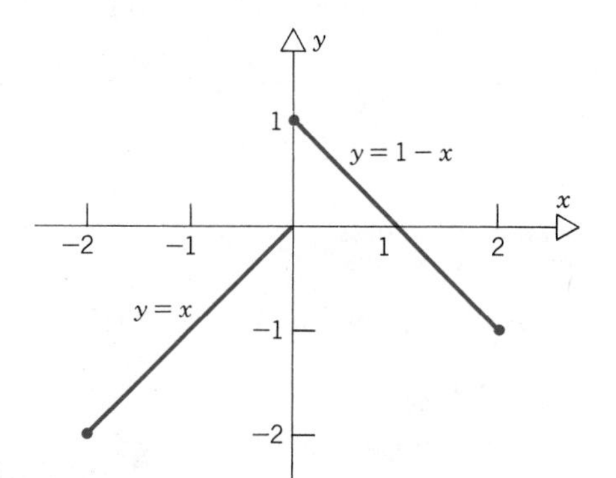

Figure 2.6.5

Observe that f is discontinuous at $x = 0$. As a result, the intermediate value property fails on some subintervals. For instance, consider $-\frac{1}{2} \le x \le \frac{1}{2}$. Then $f(-\frac{1}{2}) = -\frac{1}{2}$ and $f(\frac{1}{2}) = \frac{1}{2}$, but f does not take on all intermediate values. To be specific, for x in $[-\frac{1}{2}, \frac{1}{2}]$, $f(x)$ never has values in $[0, \frac{1}{2})$. Note also that even though f is discontinuous on $[-2, 2]$, it does have both maximum and minimum values and does take on all intermediate values. ∎

The hypotheses of Theorems 2.6.2 and 2.6.3 are *sufficient* to guarantee the stated conclusions, but are not *necessary*. In other words, there are cases in which some (or even all) of the conclusions are true, even though the hypotheses are not satisfied.

Theorem 2.6.3, the intermediate value property of continuous functions, has a number of interesting consequences. The following corollary is sometimes known as Bolzano's theorem.

Corollary

(Bolzano's Theorem)

If f is continuous on a closed bounded interval $[a, b]$, and if $f(a)f(b) < 0$, then there is at least one point c in (a, b) such that $f(c) = 0$. In other words, the equation $f(x) = 0$ has at least one root between $x = a$ and $x = b$.

The hypothesis $f(a)f(b) < 0$ is simply another way of saying that $f(a)$ and $f(b)$ are of opposite signs, and neither is zero. If the given conditions are satisfied, there may be more than one point where $f(x) = 0$. Also $f(x)$ may be zero even if the hypotheses are violated; in other words, the conditions are sufficient but not necessary.

One of the applications of Bolzano's theorem is in isolating roots of equations. It is also the basis for a numerical procedure by which roots can be estimated with steadily increasing accuracy. The following example illustrates these ideas.

EXAMPLE 5

Consider the polynomial equation

$$f(x) = x^3 - 2x^2 + 6x - 7 = 0. \tag{7}$$

By substituting $x = 1$ and $x = 2$ we find that $f(1) = -2$ and $f(2) = 5$. Since these values are of opposite sign, Bolzano's theorem guarantees that Eq. 7 has at least one root in the interval $1 < x < 2$.

To locate this root more accurately we can evaluate $f(x)$ at $x = 1.5$, the midpoint of the interval $(1, 2)$. The result is $f(1.5) = 0.875$, so by Bolzano's theorem we now know that there is at least one root in the interval $(1, 1.5)$. There is no information about the number of roots, if any, in $(1.5, 2)$. Next, we bisect the interval $(1, 1.5)$ and evaluate $f(1.25) = -0.671875$; hence there is at least one root in the interval $(1.25, 1.5)$. The next three steps are

$$f(1.375) \cong 0.068359 \quad \text{so there is a root in} \quad (1.25, 1.375);$$

$$f(1.3125) \cong -0.309326 \quad \text{so there is a root in} \quad (1.3125, 1.375);$$

$$f(1.34375) \cong -0.122467 \quad \text{so there is a root in} \quad (1.34375, 1.375).$$

If we estimate the root by using the midpoint of this last interval, namely 1.359375, then we can be sure that our approximation is in error by no more than 0.015625. If we wish to achieve greater accuracy, we can continue this process indefinitely. At each step we bisect the current interval and choose the subinterval whose endpoints yield opposite signs for $f(x)$. After n steps the root will be confined to an interval of length $1/2^n$. ∎

The procedure in Example 5 is not restricted to polynomial equations, but can be applied to any equation of the form $f(x) = 0$, where f is a continuous function, provided that one can find two points $x = a$ and $x = b$ such that $f(a)f(b) < 0$. Then the bisection method leads systematically to as close a numerical estimate of the root as may be required.

Theorems 2.6.2 and 2.6.3 can be restated more succinctly if we adopt the point of view that a function is a mapping of its domain onto its range.

Theorem 2.6.4

The image of a closed bounded interval $[a, b]$ under a continuous function f is also a closed bounded interval $[m, M]$.

In the usual notation for sets, Theorem 2.6.4 states that the image set S given by

$$S = \{y \mid y = f(x) \quad \text{for} \quad x \text{ in } [a, b]\} \tag{8}$$

is the closed bounded interval $[m, M]$. Alternatively,

$$S = f([a, b]) = [m, M]. \tag{9}$$

To show the relation between Theorems 2.6.2 and 2.6.3 and Theorem 2.6.4, note that if the image set S is a closed bounded interval, then f is bounded, has both maximum and minimum values, and takes on all intermediate values. Conversely, if f has the maximum-minimum and intermediate value properties, then the image set S is a closed bounded interval.

The following example illustrates that corresponding statements cannot be made about the image of an open interval under a continuous function.

EXAMPLE 6

If

$$f(x) = \sin x \tag{10}$$

find the image of $(0, \pi/2)$; of $(0, \pi)$; and of $(-\pi, \pi)$.

From the graph in Figure 2.6.6, it is clear that

Figure 2.6.6

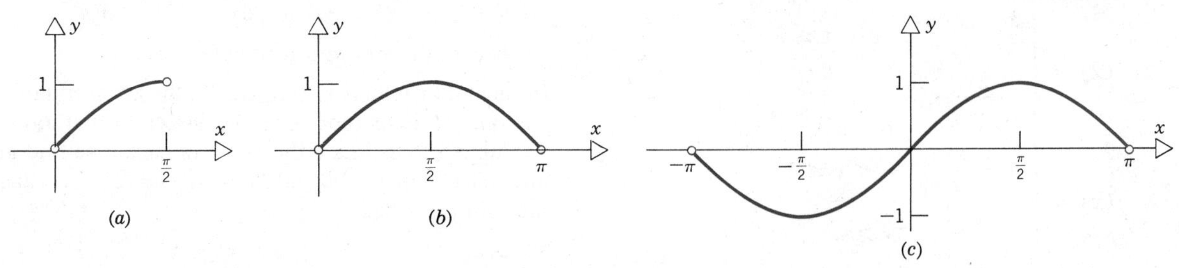

(a) The image of $(0, \pi/2)$ is $(0, 1)$.
(b) The image of $(0, \pi)$ is $(0, 1]$.
(c) The image of $(-\pi, \pi)$ is $[-1, 1]$. ∎

Since the function f in Example 6 is continuous on each of the given intervals, we conclude that the image set of a continuous function on an open interval may be either open, closed, or half open–half closed. It may also be unbounded, as we can see from the function $f(x) = 1/x$ on the interval $0 < x < 1$. In a similar way, nothing definite can be said about the image set of a discontinuous function on a closed bounded interval. The reader should construct examples to illustrate this, or perhaps reinterpret some of those previously given.

PROBLEMS

In each of Problems 1 through 10, state whether the given function is continuous and whether the given interval is closed and/or bounded. Determine which (if any) of the properties given in Theorems 2.6.2 and 2.6.3 are not possessed by the function.

1. $f(x) = \tan x, \quad 0 < x < \frac{\pi}{2}$

2. $f(x) = \begin{cases} \sin x, & 0 \le x \le \frac{\pi}{2} \\ \cos x, & \frac{\pi}{2} < x \le \pi \end{cases}$

3. $f(x) = \begin{cases} 1 - x^2, & -1 < x < 0 \\ 0, & x = 0 \\ 1 - x, & 0 < x < 1 \end{cases}$

4. $f(x) = \begin{cases} x - 2, & -1 < x < 0 \\ 0, & x = 0 \\ x + 2, & 0 < x < 1 \end{cases}$

5. $f(x) = \begin{cases} 4 - x^2, & 0 \le x < 2 \\ x, & 2 \le x \le 4 \end{cases}$

6. $f(x) = \dfrac{x}{3 + x}, \quad x > 0$

7. $f(x) = \begin{cases} \dfrac{x}{3}, & 0 \le x \le 3 \\ \dfrac{2x}{(x - 3)}, & x > 3 \end{cases}$

8. $f(x) = \begin{cases} 1 - \dfrac{x^2}{4}, & 0 < x \le 2 \\ \dfrac{x}{x - 2}, & x > 2 \end{cases}$

9. $f(x) = \begin{cases} \cos x, & 0 < x < \frac{\pi}{2} \\ \sin x, & \frac{\pi}{2} \le x < \pi \end{cases}$

10. $f(x) = \begin{cases} \sin \dfrac{1}{x}, & -1 < x < 0 \text{ or } 0 < x < 1 \\ 0, & x = 0 \end{cases}$

In each of Problems 11 through 14, the given cubic equation has three real roots. For each equation use Bolzano's theorem to find successive integers that bracket each root.

© 11. $2x^3 + x^2 - 18x - 20 = 0$

© 12. $4x^3 - 17x^2 - 12x + 51 = 0$

© 13. $8x^3 - 52x^2 - 0.66x + 45 = 0$

© 14. $8x^3 - 12x^2 - 2x + 3 = 0$

15. Use Bolzano's theorem to show that every polynomial equation of odd degree has at least one real root.

© 16. Find the smallest positive integer n such that the equation

$$x^3 + 3x^2 - 30x - 171 = 0$$

is certain to have a root in $0 < x < n$.

In each of Problems 17 through 20, the given equation has one positive root. Use the bisection method of Example 5 to estimate this root. In each case find an interval of length $0.0078125 = \frac{1}{128} = 2^{-7}$ in which the root must lie.

© 17. $2 \sin x - x = 0$ © 18. $x^3 + x^2 - 7x - 3 = 0$

© **19.** $\frac{2}{x} - \sqrt{x} + 3 = 0$

© **20.** $3 \cos x - x^2 = 0$

21. Show that of all rectangles inscribed in a given circle there is one of maximum area.

22. Show that of all rectangles of given perimeter there is one of maximum area.

23. Suppose that the two functions f and g are both continuous on $[0, 1]$; further, suppose that $f(0) = 2$, $f(1) = -1$, $g(0) = 0$, and $g(1) = 3$. Show that there is at least one point x in $[0, 1]$ where $f(x) = g(x)$.

24. (a) Show that at any instant in time there is at least one pair of points diametrically opposite each other on the equator where the temperature is the same. *Hint:* Assume that the temperature varies continuously, and form an auxiliary function relating the temperature at two diametrically opposite points.

(b) Show that the result of Part (a) also holds for points lying on any circle on the earth's surface.

(c) Do the results of Parts (a) and (b) also apply to barometric pressure? Altitude above sea level?

25. A hiker leaves the rim of the Grand Canyon at 8 A.M. and hikes to the bottom, arriving there at 4 P.M. After camping overnight, the hiker climbs out of the canyon the next day by the same route, again starting at 8 A.M. and reaching the rim at 4 P.M. Show that there is some point on the trail that the hiker passes at the same time on both days.

26. If $f(0) = a$ and $f(1) = b$, where $0 \le a, b \le 1$, and if f is continuous on $[0, 1]$, show that the equation $x = f(x)$ has at least one root in $[0, 1]$. *Hint:* Consider $g(x) = f(x) - x$.

27. Sketch the graph of a discontinuous function on an open bounded interval whose range is (a) an open bounded interval; (b) a closed bounded interval.

28. Sketch the graph of a discontinuous function on a closed bounded interval whose range is (a) an open bounded interval; (b) a closed bounded interval.

29. Sketch the graph of a discontinuous function on a closed bounded interval whose range is an unbounded interval.

30. Sketch the graph of a continuous function on an unbounded interval whose range is a closed bounded interval.

31. **Proof of Theorem 2.6.1.** Suppose that $f(c) > 0$, and let $\epsilon = f(c)/2$. Then there is a δ such that if $|x - c| < \delta$, then $|f(x) - f(c)| < \epsilon$. Show that if $|x - c| < \delta$, then $f(x) > f(c) - \epsilon = f(c)/2 > 0$. How must the argument be modified if $f(c) < 0$?

* **32.** **Proof of Theorem 2.6.3.** For definiteness, suppose that $x_1 < x_2$, and that $f(x_1) > f(x_2)$; other cases can be handled in a similar way. Let k be any number such that $f(x_2) < k < f(x_1)$. Define the set S as follows:

$$S = \{t | f(x) > k \quad \text{for} \quad x_1 \le x \le t\}.$$

(a) Use Theorem 2.6.1 to show that the set S is not empty.

(b) Show that the set S is bounded above, for example, by x_2.

(c) Observe that, by the completeness principle (see discussion preceding Problem 29 in Section 1.1), the set S has a least upper bound; denote it by c.

(d) Show that if $f(c) > k$, then c is not an upper bound for S. Hence $f(c) \le k$.

(e) Show that if $f(c) < k$, then S has smaller upper bounds than c. Hence $f(c) \ge k$.

(f) Show that $f(c) = k$, completing the proof.

REVIEW PROBLEMS

In each of Problems 1 through 4, find the equation of the tangent line for the given equation at the given point. Graph the function and the tangent line.

1. $f(x) = \frac{x^2}{2} + 2x$ at $(2, 6)$

2. $f(x) = \frac{x^2 + 5x + 6}{x - 1}$ at $(-1, -1)$

3. $f(x) = \frac{\tan x}{2}$ at $(\pi, 0)$

4. $f(x) = \cos x$ at $(\pi, -1)$

In each of Problems 5 through 8, find $\lim_{x\to c} [f(x) - f(c)]/(x - c)$ for the given value of c.

5. $f(x) = x^2 + 1, \quad c = 2$
6. $f(x) = 2x^2 - 3x, \quad c = 1$
7. $f(x) = \dfrac{x - 2}{x + 1}, \quad c = 1$
8. $f(x) = \dfrac{x^2 - 3}{x - 2}, \quad c = 1$

In each of Problems 9 through 32, either find the given limit or else determine that it does not exist.

9. $\lim_{x\to 0} \dfrac{x^2 + 2}{x - 1}$
10. $\lim_{x\to 0} \dfrac{x^2 - 2}{x}$
11. $\lim_{x\to 3} \dfrac{(x + 3)(x^2 - 5x + 6)}{x^2 - 9}$
12. $\lim_{x\to 3} \dfrac{(x^2 - 9)(x^2 - 5x + 6)}{(x - 3)^2}$
13. $\lim_{x\to 0} \dfrac{1}{[\![x]\!]}$
14. $\lim_{x\to 0} \left[\!\!\left[\dfrac{1}{x}\right]\!\!\right]$
15. $\lim_{x\to -1} f(x)$, where

$$f(x) = \begin{cases} 0, & x < -3 \\ x^2, & -3 \le x \le 1 \\ \dfrac{x^3 + 2}{x - 1}, & x > 1 \end{cases}$$

16. $\lim_{x\to 2} f(x)$, where $f(x) = \begin{cases} \sqrt{|x - 2|}, & x \ne 2 \\ 2, & x = 2 \end{cases}$
17. $\lim_{x\to 1} \dfrac{x^{3/2} - 3x + 3\sqrt{x} - 1}{(\sqrt{x} - 1)^3(x + 2)}$
18. $\lim_{x\to 4} \dfrac{(x - 4)(x - 2\sqrt{x} + 1)}{x - 4\sqrt{x} + 4}$
19. $\lim_{x\to 4} \dfrac{x^2 - 16}{x - \sqrt{x} - 2}$
20. $\lim_{x\to -3} \dfrac{x^2 + 9}{x^2 - 9}$
21. $\lim_{x\to 0} \dfrac{\sin^2 2x}{2x^2}$
22. $\lim_{x\to \pi} \dfrac{\sin x \sin(x - \pi)}{x - \pi}$
23. $\lim_{x\to 0} \dfrac{x}{\sin 3x}$
24. $\lim_{x\to 0} \dfrac{2}{x \csc 2x^2}$
25. $\lim_{x\to \pi} \dfrac{\sin 3x}{3(x - \pi)}$
26. $\lim_{x\to 1} \dfrac{-3 \sin \pi x}{x - 1}$
27. $\lim_{x\to 1^-} \dfrac{x^2 - 3}{x - 1}$
28. $\lim_{x\to 2} \dfrac{x + 1}{x^2 - 4}$
29. $\lim_{x\to \infty} \dfrac{x^3 + 2x + 1}{3x^3 - 4}$
30. $\lim_{x\to \infty} \dfrac{5x^3 - 5}{x^3 + 1}$
31. $\lim_{x\to 0} \dfrac{x}{x^2 - 4}$
32. $\lim_{x\to \infty} \dfrac{\sin x^2}{x^2}$

In each of Problems 33 through 36, determine whether the given function is continuous at the given point. If not, state the type of discontinuity involved, and if the discontinuity is removable, define $f(c)$ so as to make $f(x)$ continuous at c.

33. $f(x) = \dfrac{\sin \pi x}{x - 1}$ at $c = 1$
34. $f(x) = \dfrac{\tan 2x}{x + (\pi/2)}$ at $c = -\dfrac{\pi}{2}$
35. $f(x) = x \sin \dfrac{1}{x}$ at $c = 0$
36. $f(x) = \sin \dfrac{1}{1 + x} + x \sin \dfrac{1}{1 + x}$ at $c = -1$

In each of Problems 37 and 38, suppose that the function f is continuous on an interval $[a, b]$ except for one discontinuity at a point c, $a < c < b$.

37. (a) Can the boundedness property of Theorem 2.6.2 still hold for the function f? Name one discontinuity that would not allow this.
    (b) If the boundedness property does not hold for the function f, can there exist a "largest" closed interval contained in $[a, c]$ where Theorem 2.6.2 applies? (Try to find such a function.)
38. (a) Can the intermediate value property of Theorem 2.6.3 still hold for the function f? If so, name one discontinuity that might allow this.
    (b) Find a function f for which the discontinuity at $x = c$ is not removable, but Theorem 2.6.3 is applicable on $[c, b]$.

In each of Problems 39 and 40, use Bolzano's theorem to find the two nearest integers surrounding a root r of the equation $f(x) = 0$; then determine the root to two decimal places.

© 39. $f(x) = x^3 - 6x - 16$

© 40. $f(x) = x^4 - 14x^2 + x - 51$ (positive root)

41. (a) Find a function f such that $\lim_{x\to 0} xf(x) = 1$.

(b) Find a function g such that $\lim_{x\to 0} xg(x) = 2$.

(c) Find a function h such that $\lim_{x\to 0} xh(x)$ does not exist.

42. A function w has the properties that $w(x) \neq 0$ for all x in R^1 and $\lim_{x\to 0} w(x) = 0$.

(a) Find a function f such that $\lim_{x\to 0}(wf)(x) = 1$.

(b) Find a function g such that $\lim_{x\to 0}(wg)(x) = -\frac{1}{3}$.

(c) Find a function h such that $\lim_{x\to 0}(wh)(x)$ does not exist.

In each of Problems 43 through 50, find a function f that has exactly the given properties and graph it. *Hint*: This can be done using only rational functions.

43. Removable discontinuity at $x = 2$; unbounded at $x = 0$; $\lim_{x\to 0^-} = -\infty$; $\lim_{x\to -\infty} f(x) = \infty$; $\lim_{x\to\infty} f(x) = \infty$; $f(-1) = 0$.

44. $\lim_{x\to\infty} f(x) = \infty$; $\lim_{x\to -\infty} f(x) = -\infty$; $\lim_{x\to 0} f(x) = -1$; $f(4) = 0$.

45. $\lim_{x\to 2^-} f(x) = -\infty$; $\lim_{x\to 2^+} f(x) = -\infty$; $\lim_{x\to\infty} f(x) = 1$; $\lim_{x\to -\infty} f(x) = 1$; $f(-5) = 0$.

46. $\lim_{x\to -\infty} f(x) = -\infty$; $\lim_{x\to 0^-} f(x) = -\infty$; $\lim_{x\to 0^+} f(x) = \infty$; $\lim_{x\to\infty} f(x) = 0$; $f(3) = 0$.

47. Infinite discontinuity at $x = 2$; jump discontinuity at $x = 0$; $\lim_{x\to -\infty} f(x) = \infty$; $\lim_{x\to\infty} f(x) = 1$.

48. Infinite discontinuity at $x = 1$; infinite discontinuity at $x = -4$; $\lim_{x\to\infty} f(x) = -\infty$; $\lim_{x\to -\infty} f(x) = \infty$.

49. Removable discontinuity at $x = 1$; infinite discontinuity at $x = 0$; $\lim_{x\to\infty} f(x) = 2$; $\lim_{x\to -\infty} f(x) = 2$; $f(-2) = 1$.

50. Removable discontinuity at $x = 0$; jump discontinuity at $x = -4$; $\lim_{x\to -\infty} f(x) = -1$; $\lim_{x\to\infty} f(x) = 0$.

THREE

the derivative

The problem of finding the line tangent to a given curve at a given point originated with the Greek mathematicians more than two thousand years ago. By the middle of the seventeenth century tangent lines had been found in a number of special cases, but a general method was still lacking.

The differential calculus was born sometime during 1665 or 1666, when Isaac Newton* first conceived the process we now know as differentiation, which (among many other things) provided a powerful method for finding the tangent to an essentially arbitrary curve. Since Newton did not publish his findings, they did not become widely known, and

---

*Isaac Newton (1642–1727) grew up at Woolsthorpe, Lincolnshire, England. After graduation from Trinity College of Cambridge University in 1665, he returned to his home because the plague then sweeping England forced Cambridge to close.

In comparative isolation during the next two years, Newton "was in the prime of my age of invention," as he later put it. The results were epochal: the fundamental laws of motion and of gravitation, the beginning of his experiments in optics, and many of the basic results in calculus.

Newton returned to Cambridge when it reopened in 1667 and became professor of mathematics in 1669. He circulated some of his results privately among a few of his friends, but it was not until 1687, at the urging of Edmund Halley, that Newton published *Philosophiae Naturalis Principia Mathematica,* a monumental work in which he set forth the laws of mechanics and gravitation, and laid the foundation for much of the investigation of the physical sciences that has occurred since. (*Continued next pg.*)

calculus was rediscovered independently by Gottfried Wilhelm Leibniz** about eight or ten years later. Among the discoveries of Newton and Leibniz are the rules for differentiating powers (Section 3.1), sums, products, and quotients (Section 3.2), and composite functions (Section 3.5), together with many other results that appear later in this book.

The discovery of calculus by Newton and Leibniz changed the nature and direction of mathematics. A vast array of problems that previously were unsolvable, even by the most outstanding scholars, soon came within the range of people of only ordinary mathematical ability. Coupled with Newton's formulation of the laws of motion and of gravity, the calculus of Newton and Leibniz also provided just the tools needed to launch inquiries in the physical sciences that have revolutionized the modern world.

3.1 DEFINITION OF THE DERIVATIVE; THE POWER RULE

In Section 2.1 we showed that if the position s of a particle moving on a straight line is related to the time t by the function f, that is, if

$$s = f(t), \tag{1}$$

then the velocity v is given by

$$v = \lim_{t \to t_0} \frac{f(t) - f(t_0)}{t - t_0}. \tag{2a}$$

However, Newton's work on calculus remained in manuscript form for many years; for example, his most extensive description of his methods, *De Methodis Serierum et Fluxionum,* was written in 1671 but not published until 1736, nine years after Newton's death. Two shorter summaries of his mathematical work were published in 1704 and 1711, also long after they had been written.

Meanwhile, he had left Cambridge in 1696 to become Warden, later Master, of the British Mint; thereafter he devoted himself largely to nonmathematical pursuits. He died, richly honored by his countrymen and government, in his eighty-fifth year, and is buried in Westminster Abbey.

**Gottfried Wilhelm Leibniz (1646–1716) was a scholar with interests in many fields of study, including philosophy, logic, and law, as well as mathematics and science. Born in Leipzig, he spent most of his life as advisor, librarian, and diplomat for the royal house of Hannover. He traveled widely and engaged in prolific correspondence on many subjects with other European scholars.

During a sojourn in Paris beginning in 1672, Leibniz's interest in mathematics was stimulated by Christian Huygens. During the next four or five years Leibniz rediscovered, independently of Newton, most of the rules and formulas of elementary calculus. The first published work on calculus was Leibniz's paper in 1684 in the journal *Acta Eruditorum*.

Whereas Newton was strongly motivated by the desire to understand the physical world, Leibniz was more of an abstract and philosophical thinker. Throughout his life he sought to formulate a system of reasoning that would enable educated people to engage in rational discourse with the same degree of agreement and certainty that marked numerical calculations.

In the development of calculus in the late seventeenth and early eighteenth centuries Leibniz was more influential than Newton. This was due in part to his frequent correspondence with other scholars, whereas Newton was much more secretive. For example, in their correspondence with each other, Leibniz and the brothers Jakob and Johann Bernoulli discovered many of the elementary methods of solving first order differential equations, a natural outgrowth of calculus itself.

If we let $t = t_0 + h$, then $h \to 0$ as $t \to t_0$, and we can write

$$v = \lim_{h \to 0} \frac{f(t_0 + h) - f(t_0)}{h}. \tag{2b}$$

We also observed that from a geometrical point of view the velocity v is the slope of the line tangent to the graph of Eq. 1 at the point $(t, f(t))$.

However, the limiting process indicated by Eq. 2 can be applied to a function f purely as a mathematical operation, without regard to its possible interpretation in terms of physics or geometry. This mathematical process is known as **differentiation** and it yields a result called a **derivative.** We give the following formal definition.

DEFINITION 3.1.1 Let f be a function whose domain D includes an open interval containing the point x. Then the number $f'(x)$ given by

$$f'(x) = \lim_{h \to 0} \frac{f(x + h) - f(x)}{h}, \tag{3}$$

provided this limit exists, is called the derivative of f at x.

In other words, to each x for which the limit exists, Eq. 3 assigns a unique number $f'(x)$. Consequently, Eq. 3 defines a *function* f', called the *derivative* of f, whose value at x is $f'(x)$. The domain of f' consists of those points of D for which the limit (3) exists. If $f'(x)$ exists at a point x, then f is said to be *differentiable* at that point. Similarly, if f is differentiable at each point of an open interval, then f is said to be differentiable on that interval.

For a differentiable function $y = f(x)$ we will sometimes use y' to denote the value of its derivative at x; that is, $y' = f'(x)$. It is important to remember that f' is a function, whereas $f'(x)$ and y' denote the value of this function at the point x. Sometimes, in discussing differentiation, it is helpful to emphasize the independent variable; thus if x is the independent variable, we may say "derivative with respect to x" instead of merely "derivative." We have already interpreted the derivative as a velocity or as the slope of a tangent line. More generally, if $y = f(x)$, then the derivative f' is the *rate of change* of f with respect to the independent variable x, and $f'(x_0)$ is the rate of change of f at the point $x = x_0$.

The requirement in Definition 3.1.1 that D include an open interval containing x is needed in order to be sure that $x + h$ is also in D when h is sufficiently small, and so that points $x + h$ on both sides of x can be considered during the limiting process. Finally, since the limit in Eq. 3 may fail to exist for some points in D, the domain of f' need not be all of D.

Equation 3 gives us a means of understanding the derivative conceptually, for example, as the limit approached by the slopes of secant lines drawn over shorter and shorter intervals. However, in most cases it is not well suited for the calculation of derivatives of specific functions. Therefore, our primary goal in this chapter is to develop a set of rules and formulas that will enable us to calculate, with little effort, the derivative of any one of a large class of functions. These rules are of two kinds. Some are formulas giving the derivative of a specific function, such as $\sqrt{x}$ or $\cos x$. Others are more general in nature, telling us how to differentiate certain combinations of functions, such as sums or products. We will find that only a relatively few rules and formulas are needed to differentiate most of the functions usually encountered. First, however, we will consider a few examples in which we find the derivatives of certain specific functions by using Definition 3.1.1 directly.

EXAMPLE 1

Consider a constant function

$$f(x) = c, \qquad -\infty < x < \infty \tag{4}$$

where c is any real number; see Figure 3.1.1. Then, for each x,

$$\frac{f(x + h) - f(x)}{h} = \frac{c - c}{h} = 0, \qquad h \neq 0.$$

Consequently

$$f'(x) = \lim_{h \to 0} \frac{f(x + h) - f(x)}{h} = \lim_{h \to 0} 0 = 0. \tag{5}$$

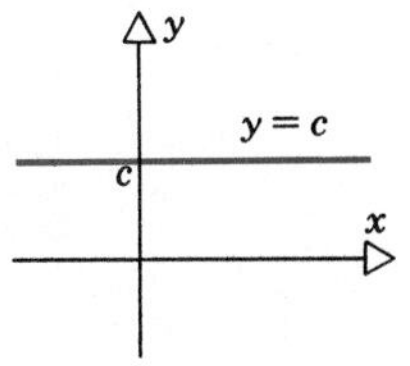

Figure 3.1.1

Thus the derivative of a constant function is everywhere zero. Geometrically, the graph of a constant function is a straight line parallel to the x-axis. The vanishing of the derivative at each point means that the slope of the tangent line at each point is zero, and hence the tangent line is the straight line itself. ∎

EXAMPLE 2

Consider the identity function

$$f(x) = x, \qquad -\infty < x < \infty; \tag{6}$$

see Figure 3.1.2. Then, for each x,

$$\frac{f(x + h) - f(x)}{h} = \frac{(x + h) - x}{h} = 1, \qquad h \neq 0.$$

Therefore

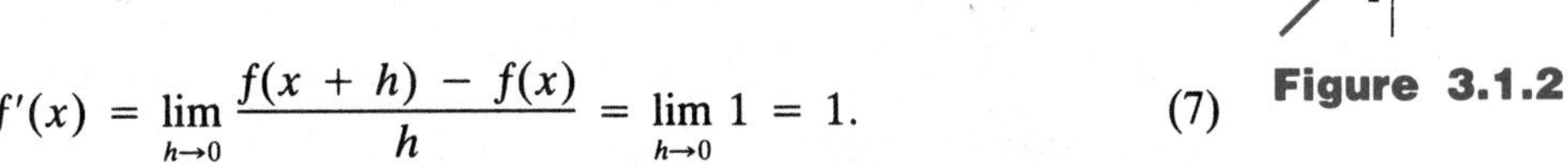

$$f'(x) = \lim_{h \to 0} \frac{f(x + h) - f(x)}{h} = \lim_{h \to 0} 1 = 1. \tag{7}$$

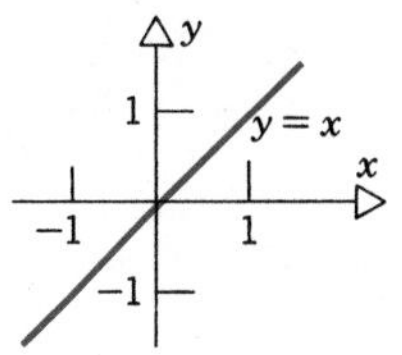

Figure 3.1.2

Hence the derivative of the identity function is everywhere equal to one; that is, the slope of the straight line $y = x$ is one. Again, the tangent line is simply the original line itself. ∎

EXAMPLE 3

Let

$$f(x) = x^3, \qquad -\infty < x < \infty; \tag{8}$$

see Figure 3.1.3. Then

$$\frac{f(x + h) - f(x)}{h} = \frac{(x + h)^3 - x^3}{h}.$$

In order to calculate the limit of this difference quotient it is necessary first to recast it into a more suitable form. Using the binomial theorem we have

$$\begin{aligned}\frac{(x + h)^3 - x^3}{h} &= \frac{(x^3 + 3x^2h + 3xh^2 + h^3) - x^3}{h} \\ &= \frac{3x^2h + 3xh^2 + h^3}{h} \\ &= 3x^2 + 3xh + h^2, \qquad h \neq 0.\end{aligned}$$

Finally, for each x we use the algebraic properties of limits (Theorem 2.3.1) and obtain

$$\begin{aligned}f'(x) &= \lim_{h\to 0} \frac{f(x + h) - f(x)}{h} \\ &= \lim_{h\to 0} (3x^2 + 3xh + h^2) \\ &= 3x^2. \end{aligned} \tag{9}$$

Equation 9 gives us the slope of the line tangent to the graph of $y = x^3$ at each point. For instance, at the point $(-1, -1)$ the tangent line has slope 3, as shown in Figure 3.1.3. ∎

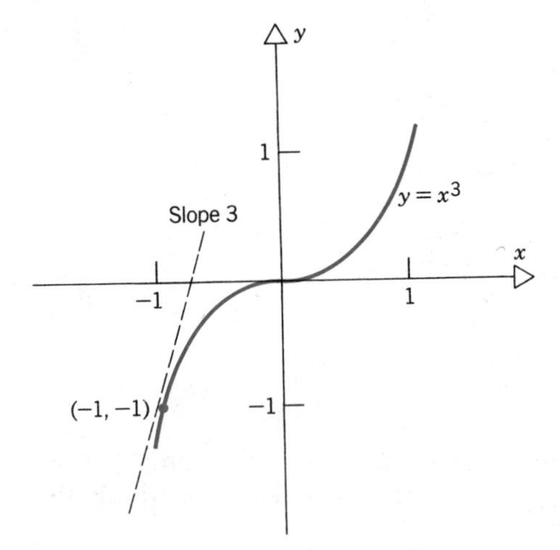

Figure 3.1.3

EXAMPLE 4

Let

$$f(x) = \frac{1}{x}, \qquad x \neq 0; \tag{10}$$

see Figure 3.1.4. Then, for $x \neq 0$,

$$\begin{aligned}
\frac{f(x + h) - f(x)}{h} &= \frac{\dfrac{1}{x + h} - \dfrac{1}{x}}{h} \\
&= \frac{x - (x + h)}{x(x + h)h} \\
&= -\frac{h}{x(x + h)h} \\
&= -\frac{1}{x(x + h)}, \qquad h \neq 0.
\end{aligned}$$

Again making use of the algebraic properties of limits, we have

$$f'(x) = \lim_{h \to 0} \frac{f(x + h) - f(x)}{h} = \lim_{h \to 0} -\frac{1}{x(x + h)} = -\frac{1}{x^2}, \qquad x \neq 0. \tag{11}$$

Note that, in the calculations leading to Eq. 11, we have divided by $x + h$; hence we must make sure that $x + h \neq 0$. Since $x \neq 0$, and since in the limit process we are concerned only with small values of h, it is sufficient to require that $|h| < |x|$; then it is surely true that $x + h \neq 0$. ∎

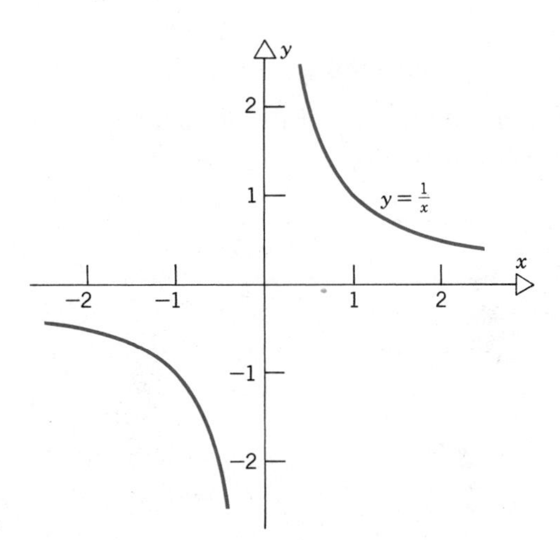

Figure 3.1.4

EXAMPLE 5
Let

$$f(x) = \sqrt{x}, \qquad x \geq 0; \tag{12}$$

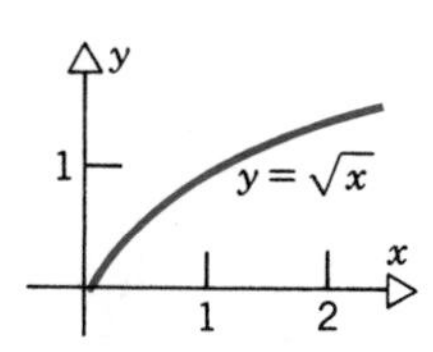

Figure 3.1.5

see Figure 3.1.5. For $x > 0$ and for h small enough so that $x + h \geq 0$, we can form the difference quotient

$$\frac{f(x+h) - f(x)}{h} = \frac{\sqrt{x+h} - \sqrt{x}}{h}.$$

In order to find the limit of this difference quotient as $h \to 0$, it is convenient to use the well-known algebraic device of rationalizing the numerator:

$$\begin{aligned}\frac{f(x+h) - f(x)}{h} &= \frac{\sqrt{x+h} - \sqrt{x}}{h} \cdot \frac{\sqrt{x+h} + \sqrt{x}}{\sqrt{x+h} + \sqrt{x}} \\ &= \frac{x + h - x}{h(\sqrt{x+h} + \sqrt{x})} \\ &= \frac{h}{h(\sqrt{x+h} + \sqrt{x})} \\ &= \frac{1}{\sqrt{x+h} + \sqrt{x}}, \qquad h \neq 0.\end{aligned}$$

Thus, letting $h \to 0$, we obtain

$$f'(x) = \lim_{h \to 0} \frac{f(x+h) - f(x)}{h} = \lim_{h \to 0} \frac{1}{\sqrt{x+h} + \sqrt{x}} = \frac{1}{2\sqrt{x}}, \qquad x > 0. \tag{13}$$

Note that for $x = 0$ we cannot form the difference quotient, since if $h < 0$, $f(0 + h) = \sqrt{h}$ is not a real number. Thus in this case the domain of the function f is $[0, \infty)$, but the domain of the derivative f' is only $(0, \infty)$. ∎

The power rule

An examination of the preceding examples reveals a pattern that suggests a more general result.

Theorem 3.1.1
(Power rule)

$$\text{If} \quad f(x) = x^r, \qquad \text{then} \quad f'(x) = rx^{r-1}; \tag{14}$$

the result holds for any real number r and for all x for which the expression for $f'(x)$ yields a real number.

We will refer to Eq. 14 as the power rule, since it gives the derivative of x^r for a constant exponent r. Observe that the results of Examples 2 through 5 are given by Eq. 14 with $r = 1, 3, -1$, and $\frac{1}{2}$, respectively. In a somewhat degenerate sense the result of Example 1 ($r = 0$) is also included in Eq. 14.

For the present, we will consider only rational values of r in the power rule, since the interpretation of x^r when r is irrational requires further discussion. We deal with irrational exponents in Section 8.3.

The power rule is important because it enables us to avoid calculations such as those in Examples 1 through 5 each time we want to differentiate a function of the form x^r. However, as is usually the case, we must pay attention to the results that we get in order to avoid careless errors. For example, if $f(x) = 1/x$, then $r = -1$, and the power rule gives $f'(x) = -x^{-2}$. As we saw in Example 4, the point $x = 0$ must be excluded, since f is not differentiable there; indeed, $f(0)$ is not even defined. Similarly, if $f(x) = \sqrt{x} = x^{1/2}$, then $r = \frac{1}{2}$, and $f'(x) = \frac{1}{2}x^{-1/2} = 1/2\sqrt{x}$ from the power rule. Of course, this result is valid only for $x > 0$, as we saw in Example 5. Thus, in using the power rule, we must take the normal precautions against dividing by zero or taking an even root of a negative number.

EXAMPLE 6

Determine y' if $y = f(x) = x^{2/3}$.

We use the power rule with $r = \frac{2}{3}$ and obtain

$$y' = \frac{2}{3}x^{-1/3}, \qquad x \neq 0.$$

Although the domain of f is $(-\infty, \infty)$, the point 0 is excluded from the domain of its derivative. The graph of $y = x^{2/3}$ is shown in Figure 3.1.6; it suggests that the tangent line at $x = 0$ is the y-axis, which has an infinite slope. ∎

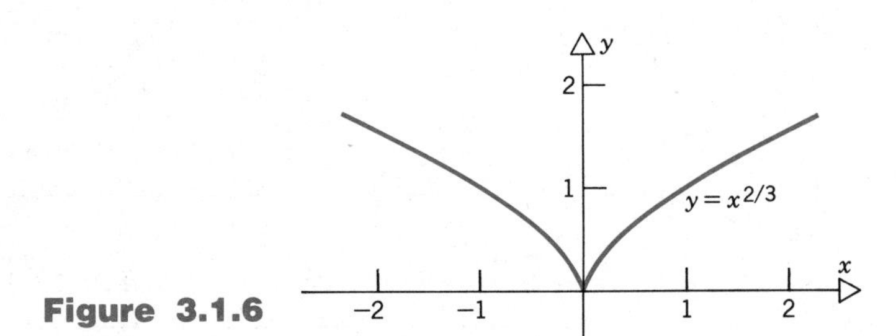

Figure 3.1.6

EXAMPLE 7

If $f(x) = x^{-2}$, $x > 0$, determine an equation of the line tangent to the graph of $y = x^{-2}$ at $(3, \frac{1}{9})$ (see Figure 3.1.7).

By the power rule we have

$$f'(x) = -2x^{-3}, \qquad x > 0.$$

Thus the slope of the tangent line at $(3, \frac{1}{9})$ is $f'(3) = -\frac{2}{27}$. The equation of the tangent line at this point is

$$y - \tfrac{1}{9} = -\tfrac{2}{27}(x - 3),$$

or

$$2x + 27y = 9. \blacksquare$$

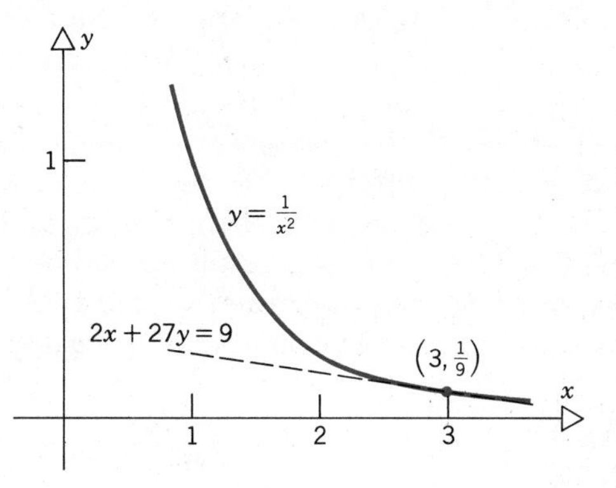

Figure 3.1.7

Proof of Theorem 3.1.1 for rational exponents. We will establish the validity of the power rule in several stages. First, suppose that r is a positive integer n. Since we have already considered the case $n = 1$ in Example 2, let us suppose that $n \geq 2$. Then

$$\frac{f(x + h) - f(x)}{h} = \frac{(x + h)^n - x^n}{h}.$$

Using the binomial theorem as in Example 3 we have

$$\frac{f(x + h) - f(x)}{h}$$

$$= \frac{\left[x^n + nx^{n-1}h + \frac{n(n-1)}{2!}x^{n-2}h^2 + \cdots + nxh^{n-1} + h^n\right] - x^n}{h}$$

$$= \frac{nx^{n-1}h + \frac{n(n-1)}{2!}x^{n-2}h^2 + \cdots + nxh^{n-1} + h^n}{h}$$

$$= nx^{n-1} + \frac{n(n-1)}{2!}x^{n-2}h + \cdots + nxh^{n-2} + h^{n-1}, \qquad h \neq 0. \tag{15}$$

Since every term, except the first, on the right side of Eq. 15 contains a positive power of h, we obtain

$$f'(x) = \lim_{h \to 0} \frac{f(x + h) - f(x)}{h} = nx^{n-1}. \tag{16}$$

Thus the power rule holds for positive integer powers.

Next suppose that r is a negative integer, say $r = -m$, where m is a positive integer; that is, $f(x) = x^{-m}$. Then

$$\frac{f(x+h) - f(x)}{h} = \frac{\dfrac{1}{(x+h)^m} - \dfrac{1}{x^m}}{h}$$

$$= \frac{x^m - (x+h)^m}{hx^m(x+h)^m}$$

$$= -\frac{(x+h)^m - x^m}{h} \cdot \frac{1}{x^m(x+h)^m}.$$

We have seen that $[(x+h)^m - x^m]/h \to mx^{m-1}$ as $h \to 0$; since $(x+h)^m \to x^m$ as $h \to 0$, we obtain

$$f'(x) = \lim_{h\to 0} \frac{f(x+h) - f(x)}{h} = -mx^{m-1}\frac{1}{x^m \cdot x^m}$$

$$= -mx^{-m-1}. \tag{17}$$

Thus, the power rule is also valid for negative integer exponents.

To prove the power rule for fractional exponents we let $r = p/q$, where p and q are nonzero integers. Then

$$\frac{f(x+h) - f(x)}{h} = \frac{(x+h)^{p/q} - x^{p/q}}{h}. \tag{18}$$

The next step is to transform this difference quotient into one we have already seen by making the substitution $u = x^{1/q}$ or $x = u^q$. Further, let $u + k = (x+h)^{1/q}$, so that $(u+k)^q = x + h$ and $h = (u+k)^q - u^q$. Then we can rewrite Eq. 18 as

$$\frac{(x+h)^{p/q} - x^{p/q}}{h} = \frac{(u+k)^p - u^p}{(u+k)^q - u^q} = \frac{[(u+k)^p - u^p]/k}{[(u+k)^q - u^q]/k}. \tag{19}$$

Next we note that if $h \to 0$, then $k \to 0$ also, so that a limit as $h \to 0$ can be replaced by a limit as $k \to 0$. Thus, by combining Eqs. 18 and 19, we obtain

$$f'(x) = \lim_{h\to 0} \frac{f(x+h) - f(x)}{h}$$

$$= \lim_{k\to 0} \frac{[(u+k)^p - u^p]/k}{[(u+k)^q - u^q]/k}$$

$$= \frac{\lim_{k\to 0} [(u+k)^p - u^p]/k}{\lim_{k\to 0} [(u+k)^q - u^q]/k}. \tag{20}$$

The last step leading to Eq. 20 involves the use of Theorem 2.3.1(d) to calculate the limit of a quotient, and requires that the limit in the denominator be nonzero. Whether p and q are positive or negative integers, the limits in the numerator and denominator of Eq. 20 have been determined earlier in the proof (see Eqs. 16 and 17). Consequently, by using these results in Eq. 20, we have

$$f'(x) = \frac{pu^{p-1}}{qu^{q-1}} = \frac{p}{q}u^{p-q}. \tag{21}$$

Finally, we replace u by $x^{1/q}$ and obtain

$$f'(x) = \frac{p}{q}x^{(p-q)/q} = \frac{p}{q}x^{(p/q)-1}, \tag{22}$$

which completes the proof of the power rule for rational exponents. □

PROBLEMS

In each of Problems 1 through 12, use the definition of the derivative as the limit of a difference quotient to determine the derivative of the given function.

1. $f(x) = x^2$
2. $f(s) = s^{-1/2}, \quad s > 0$
3. $g(x) = x^{-2}, \quad x \neq 0$
4. $f(x) = 3x^2 - 2x$
5. $g(s) = s^4 + s^2$
6. $h(t) = t^2 - t - 2$
7. $f(x) = x - \dfrac{1}{x}, \quad x \neq 0$
8. $f(t) = \dfrac{t}{t+1}, \quad t \neq -1$
9. $h(u) = \dfrac{1}{1+\sqrt{u}}, \quad u > 0$
10. $f(x) = x^2 + \dfrac{1}{x}, \quad x \neq 0$
11. $f(x) = \sqrt{x} - \dfrac{1}{x^2}, \quad x > 0$
12. $f(x) = 2x^2 - \dfrac{3}{x}, \quad x \neq 0$

In each of Problems 13 through 22, use the power rule to compute $f'(x)$ for the given function. Also state the domain of f' in each case.

13. $f(x) = x^6$
14. $f(x) = x^{7/3}$
15. $f(x) = x^{3/7}$
16. $f(x) = \sqrt{x^3}, \quad x \geq 0$
17. $f(x) = x^{-4}, \quad x \neq 0$
18. $f(x) = x^{-3/7}, \quad x \neq 0$
19. $f(x) = x^{3/5}$
20. $f(x) = \sqrt[3]{x^4}$
21. $f(x) = x^2\sqrt{x}, \quad x \geq 0$
22. $f(x) = \sqrt[4]{\dfrac{x^{1/4}}{x^2}}, \quad x > 0$

In each of Problems 23 through 26, find an equation of the line tangent to the given curve at the given point.

23. $y = x^3, \quad (2, 8)$
24. $y = x^{1/3}, \quad (-8, -2)$
25. $y = x^{-3}, \quad (-2, -\frac{1}{8})$
26. $y^2 = x^3, \quad (1, 1)$
27. The graphs of $y = x^2$ and $y = 1/x$ intersect at the point (1, 1). Find an equation of the line tangent to each of these curves at this point. Do these two tangent lines intersect at a right angle?
28. Follow the instructions of Problem 27 for $y = \sqrt{x}$ and $y = 1/x^2$, whose graphs also intersect at (1, 1).
29. Find an equation of the line tangent to the graph of $y = 1/x$ at the point $(\frac{1}{2}, 2)$. Then determine where this tangent line intersects the x-axis; the y-axis.
30. Follow the instructions of Problem 29 for the graph of $y = x^{3/2}$ at the point (4, 8).
31. Suppose that f is an even function, and that $f'(1) = 2$. Determine $f'(-1)$. If $a > 0$ and $f'(a) = b$, determine $f'(-a)$.
32. Suppose that f is an odd function, and that $f'(1) = 2$. Determine $f'(-1)$. If $a > 0$ and $f'(a) = b$, determine $f'(-a)$.
33. The line $x - 6y + 9 = 0$ is tangent to the graph of $y = \sqrt{x}$. Find the point of tangency.
34. The line $8x - 3y = 16$ is tangent to the graph of $y = x^{4/3}$. Find the point of tangency.
35. Show that no two different tangent lines to the parabola $y = x^2$ are parallel.

36. The line $4x + y = -4$ is tangent to the parabola $y = x^2$ at $(-2, 4)$. Find an equation of another tangent line to this curve that is perpendicular to the given line, and determine its point of tangency to the parabola.

In each of Problems 37 through 40,

(a) Find $f'(x)$ for each $x \neq 0$;

(b) Form the difference quotient for $x = 0$, and use it to determine whether $f'(0)$ exists;

(c) Sketch the graphs of $y = f(x)$ and $y = f'(x)$.

37. $f(x) = \begin{cases} x, & x < 0 \\ x^2, & x \geq 0 \end{cases}$

38. $f(x) = \begin{cases} x^{2/3}, & x < 0 \\ x^{4/3}, & x \geq 0 \end{cases}$

39. $f(x) = \begin{cases} -x^2, & x < 0 \\ x^2, & x \geq 0 \end{cases}$

40. $f(x) = |x|^3$

3.2 THE ALGEBRA OF DERIVATIVES

The main purpose of this section is to develop rules for differentiating sums, differences, constant multiples, products, and quotients of functions. In some respects the results are similar to those of Theorem 2.3.1 concerning limits of these combinations of functions.

Theorem 3.2.1

Suppose that the functions f and g are differentiable at the point x, and that α is a constant. Then the functions $f + g$, $f - g$, αf, fg, and (provided $g(x) \neq 0$) f/g are also differentiable at x. Further

(a) $[f(x) \pm g(x)]' = f'(x) \pm g'(x)$. (1)

(b) $[\alpha f(x)]' = \alpha f'(x)$, for any constant α. (2)

(c) $[f(x)g(x)]' = f'(x)g(x) + f(x)g'(x)$. (Product Rule) (3)

(d) $[f(x)/g(x)]' = \dfrac{g(x)f'(x) - f(x)g'(x)}{[g(x)]^2}$, if $g(x) \neq 0$. (Quotient Rule) (4)

By combining Parts (a) and (b) of Theorem 3.2.1 we obtain the useful corollary

$$\begin{aligned}[\alpha f(x) + \beta g(x)]' &= [\alpha f(x)]' + [\beta g(x)]' \\ &= \alpha f'(x) + \beta g'(x), \end{aligned} \tag{5}$$

where α and β are any constants. By repeatedly applying Eq. 5 we can extend this result to any finite number of terms. In a similar way, repeated application of Eq. 3 yields a formula for the derivative of the product of any finite number of factors (see Problems 41 and 42).

Equations 1, 2, and 5 have the same form as the corresponding equations for limits, and are therefore what we would naturally expect. However, Eqs. 3 and 4 do not follow any pattern established heretofore, and in particular do not resemble the corresponding results for limits of products and quotients. Precisely because they may be unexpected, it is worthwhile to consider carefully the proofs of these results. We prove all parts of Theorem 3.2.1 later in this section, but meanwhile in the following examples we show how it assists us in the calculation of derivatives.

EXAMPLE 1

If $f(x) = 3x^2 + 2x^{1/3}$, find $f'(x)$.

Showing all of our steps, we have

$$\begin{aligned}(3x^2 + 2x^{1/3})' &= (3x^2)' + (2x^{1/3})' \\ &= 3(x^2)' + 2(x^{1/3})' \\ &= 3(2x) + 2(\tfrac{1}{3}x^{-2/3}) \\ &= 6x + \tfrac{2}{3}x^{-2/3}, \qquad x \neq 0. \end{aligned} \tag{6}$$

Note that in reaching this result we used Eq. 1, Eq. 2, and the power rule in succession. With a little practice we will normally skip most of the intervening steps, and simply write the result almost at once. ∎

EXAMPLE 2

If $y = 2x^3 - 4x^2 + 3x + 5$, find y'.

Proceeding as in Example 1, we have

$$\begin{aligned} y' &= (2x^3 - 4x^2 + 3x + 5)' \\ &= (2x^3 - 4x^2)' + (3x + 5)' \\ &= 2(x^3)' - 4(x^2)' + 3(x)' + (5)' \\ &= 2(3x^2) - 4(2x) + 3(1) + 0 \\ &= 6x^2 - 8x + 3. \end{aligned} \tag{7}$$

Note that by repeated use of Eqs. 1 and 2 we can differentiate the given function by differentiating each term separately, and then combining the results. The same method can be used in differentiating any polynomial. ∎

EXAMPLE 3

If $f(x) = (2x^3 - x)(x^4 + 3x)$, find $f'(x)$.

There are two ways we can proceed here. First, we can use the product rule to differentiate f:

$$\begin{aligned} f'(x) &= (2x^3 - x)'(x^4 + 3x) + (2x^3 - x)(x^4 + 3x)' \\ &= (6x^2 - 1)(x^4 + 3x) + (2x^3 - x)(4x^3 + 3) \\ &= (6x^6 - x^4 + 18x^3 - 3x) + (8x^6 - 4x^4 + 6x^3 - 3x) \\ &= 14x^6 - 5x^4 + 24x^3 - 6x. \end{aligned} \tag{8}$$

Alternatively, we can first carry out the indicated multiplication and then differentiate as in Example 2.

$$\begin{aligned} f'(x) &= (2x^7 - x^5 + 6x^4 - 3x^2)' \\ &= 14x^6 - 5x^4 + 24x^3 - 6x. \quad \blacksquare \end{aligned}$$

EXAMPLE 4

Find $f'(x)$ if

$$f(x) = \frac{x^3 - 4x + 1}{x^2 - 1}, \qquad x \neq -1, 1. \tag{9}$$

From Eq. 4 we have

$$\begin{aligned} f'(x) &= \frac{(x^2 - 1)(x^3 - 4x + 1)' - (x^3 - 4x + 1)(x^2 - 1)'}{(x^2 - 1)^2} \\ &= \frac{(x^2 - 1)(3x^2 - 4) - (x^3 - 4x + 1)(2x)}{(x^2 - 1)^2} \\ &= \frac{(3x^4 - 7x^2 + 4) - (2x^4 - 8x^2 + 2x)}{(x^2 - 1)^2} \\ &= \frac{x^4 + x^2 - 2x + 4}{(x^2 - 1)^2}, \qquad x \neq -1, 1. \quad \blacksquare \end{aligned} \tag{10}$$

EXAMPLE 5

Find y' if

$$y = \frac{3x^2 + 7x - 6}{x}, \qquad x \neq 0. \tag{11}$$

This can be done by using the quotient rule directly on Eq. 11. However, it is simpler to rewrite y in the form

$$y = 3x + 7 - 6x^{-1} \tag{12}$$

before differentiating. From Eq. 12 we immediately obtain

$$y' = 3 + 6x^{-2}, \qquad x \neq 0. \quad \blacksquare \tag{13}$$

EXAMPLE 6

Find y' if

$$y = x\left(1 + \frac{1}{x^2}\right) + \frac{3x^{1/3}}{2 - 7x}, \qquad x \neq 0, \tfrac{2}{7}. \tag{14}$$

From Eq. 1 we first obtain

$$y' = [x(1 + x^{-2})]' + \left[\frac{3x^{1/3}}{2 - 7x}\right]'. \tag{15}$$

Each term on the right side of Eq. 15 can now be computed separately. Using the product rule, we obtain

$$\begin{aligned}[x(1 + x^{-2})]' &= (x + x^{-1})' \\ &= 1 - x^{-2}.\end{aligned} \tag{16}$$

Then, from the quotient rule, we find that

$$\begin{aligned}\left[\frac{3x^{1/3}}{2 - 7x}\right]' &= \frac{(2 - 7x)(3x^{1/3})' - (3x^{1/3})(2 - 7x)'}{(2 - 7x)^2} \\ &= \frac{(2 - 7x)x^{-2/3} - (3x^{1/3})(-7)}{(2 - 7x)^2} \\ &= \frac{2x^{-2/3} - 7x^{1/3} + 21x^{1/3}}{(2 - 7x)^2} \\ &= \frac{2x^{-2/3} + 14x^{1/3}}{(2 - 7x)^2}.\end{aligned} \tag{17}$$

Combining the results of Eqs. 15, 16, and 17, we finally have

$$y' = 1 - \frac{1}{x^2} + \frac{2x^{-2/3} + 14x^{1/3}}{(2 - 7x)^2}, \qquad x \neq 0, \tfrac{2}{7}. \blacksquare \tag{18}$$

Before taking up the proof of Theorem 3.2.1, we need a preliminary result (important in itself) relating the two concepts of continuity and differentiability.

Figure 3.2.1

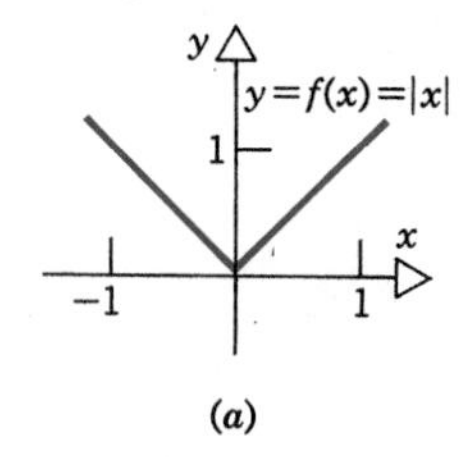

(a)

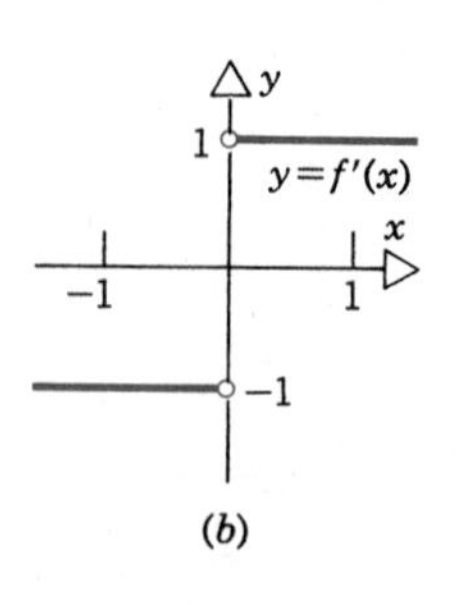

(b)

EXAMPLE 7

Consider the absolute value function

$$f(x) = |x| = \begin{cases} -x, & x < 0, \\ x, & x \geq 0, \end{cases} \tag{19}$$

whose graph is shown in Figure 3.2.1*a*. This function is continuous everywhere, in particular, at $x = 0$.

For all nonzero values of x it is easy to show that f is also differentiable, and that

$$f'(x) = \begin{cases} -1, & x < 0; \\ 1, & x > 0. \end{cases} \tag{20}$$

For instance, one can obtain Eq. 20 by looking at the graph of $y = f(x)$ in Figure 3.2.1*a* and interpreting $f'(x)$ as the slope of the tangent line at the point x. The graph of f' is shown in Figure 3.2.1*b*. To determine whether f is differentiable at $x = 0$ we form the difference quotient

$$\frac{f(0+h)-f(0)}{h} = \frac{|0+h|-|0|}{h} = \frac{|h|}{h} = \begin{cases} -1, & h < 0; \\ 1, & h > 0. \end{cases} \tag{21}$$

As $h \to 0$ the left-hand and right-hand limits of this difference quotient are -1 and 1, respectively; since they are unequal, it follows that the absolute value function is *not* differentiable at $x = 0$. This reflects the geometrical fact that the graph of $y = |x|$ has a "corner" at the origin, where the slope changes abruptly from -1 to 1. ∎

The preceding example shows that a function that is continuous at a point *need not* have a derivative there. The converse, however, is true, as we shall prove. Consequently, the set of functions continuous at a point or on an interval contains the set of functions differentiable at that point or on that interval.

Theorem 3.2.2

If f is differentiable at the point x, then f is also continuous at x.

Proof. To prove Theorem 3.2.2 we must show that $\lim_{t\to x} f(t) = f(x)$, or equivalently, that $\lim_{h\to 0} f(x+h) = f(x)$. To make use of the hypothesis that $f'(x)$ exists, we write

$$\begin{aligned} f(x+h) &= f(x) + [f(x+h) - f(x)] \\ &= f(x) + h\left[\frac{f(x+h)-f(x)}{h}\right], \qquad h \neq 0. \end{aligned} \tag{22}$$

Now taking the limit as $h \to 0$ and using Parts (a) and (c) of Theorem 2.3.1 to find the limit of a sum and then a product, we obtain

$$\begin{aligned} \lim_{h\to 0} f(x+h) &= \lim_{h\to 0} f(x) + \lim_{h\to 0} h \lim_{h\to 0} \frac{f(x+h)-f(x)}{h} \\ &= f(x) + 0 \cdot f'(x) \\ &= f(x). \end{aligned}$$

Thus f is continuous at x and the proof is complete. □

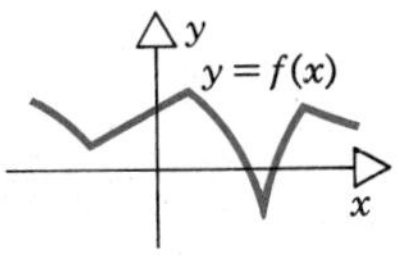

While it is not difficult to imagine a continuous function that has no derivative at a single point, or even at a finite number of points (see Figure 3.2.2), it is a remarkable fact that there exist functions that are continuous at *every* point in an open interval, but which have a derivative at *no* point in the interval. The first example of a continuous but nowhere differentiable function to become generally

Figure 3.2.2

known was due to Weierstrass* in 1872, although one had been constructed earlier by Bolzano. Examples of such functions are too difficult to give here, but may be found in some books on advanced calculus. A little reflection shows that drawing the graph of a continuous nondifferentiable function is a formidable problem indeed. The graph is a single connected "curve," but nowhere does it have a slope or direction (that is, a tangent line). The graph of $y = |x|$ has a "corner" at the origin, but the graph of the Weierstrass function has a corner at every point. The discovery of this function was convincing evidence that the set of continuous functions contains some very peculiar members, and that one should not rely too much on geometrical intuition when studying continuity and its consequences.

Proof of Theorem 3.2.1. We now turn to a proof of the algebraic properties of derivatives.

Proof of Part (a): First consider the case of addition in Eq. 1. From Definition 3.1.1 we have

$$[f(x) + g(x)]' = \lim_{h \to 0} \frac{[f(x + h) + g(x + h)] - [f(x) + g(x)]}{h}$$

$$= \lim_{h \to 0} \left[\frac{f(x + h) - f(x)}{h} + \frac{g(x + h) - g(x)}{h}\right], \tag{23}$$

provided this limit exists. From the hypothesis that f and g are differentiable at x, we know that each of the preceding difference quotients has a limit. Hence, by Theorem 2.3.1(a), we can write the limit of the sum as the sum of the limits taken separately, and we obtain

$$[f(x) + g(x)]' = \lim_{h \to 0} \frac{f(x + h) - f(x)}{h} + \lim_{h \to 0} \frac{g(x + h) - g(x)}{h}$$

$$= f'(x) + g'(x).$$

The proof for subtraction in Part (a) is similar and will be omitted.

Proof of Part (b): Starting from Definition 3.1.1 we have

$$[\alpha f(x)]' = \lim_{h \to 0} \frac{\alpha f(x + h) - \alpha f(x)}{h}$$

$$= \alpha \lim_{h \to 0} \frac{f(x + h) - f(x)}{h}$$

$$= \alpha f'(x),$$

*Karl Weierstrass (1815–1897) was instrumental in clarifying the underlying concepts of calculus; for example, it was he who formulated the definitions of limit and continuity in terms of inequalities and small quantities such as ϵ and δ. He was a secondary school teacher for several years, before becoming professor at the University of Berlin at the age of thirty-nine. He himself published little, and much of his work is preserved only through the notes taken by students at his lectures. His construction of a continuous but nowhere differentiable function shocked the mathematical world, and in 1905 the well-known mathematician E. Picard commented that "If Newton and Leibniz had known that a continuous function need not necessarily have a derivative, the differential calculus would never have been created."

where we have used Theorem 2.3.1(b) and the hypothesis that f is differentiable at x.

Proof of Part (*c*): From Definition 3.1.1 we have

$$[f(x)g(x)]' = \lim_{h\to 0} \frac{f(x+h)g(x+h) - f(x)g(x)}{h}, \tag{24}$$

provided this limit exists. The only information we have to help us evaluate the limit in Eq. 24 is that f and g are differentiable functions. To use this hypothesis we must rewrite the right side of Eq. 24 so that the difference quotients for f and g appear. This can be done by using the device, a common one in mathematics, of adding and then subtracting the same quantity in an equation.* Once we start to think along these lines, we observe that if we add and subtract the quantity $f(x)g(x+h)$ and then regroup the terms in Eq. 24, we obtain

$$\begin{aligned}[f(x)g(x)]' \\ &= \lim_{h\to 0} \frac{f(x+h)g(x+h) - f(x)g(x+h) + f(x)g(x+h) - f(x)g(x)}{h} \\ &= \lim_{h\to 0} \left[\frac{f(x+h) - f(x)}{h} g(x+h) + f(x)\frac{g(x+h) - g(x)}{h}\right].\end{aligned}$$

As $h \to 0$, the difference quotients in the last expression approach $f'(x)$ and $g'(x)$, respectively. Further, since g is differentiable at x, it is also continuous there by Theorem 3.2.2, and therefore $g(x+h) \to g(x)$ as $h \to 0$. Then, since all of the separate limits exist, we obtain

$$\begin{aligned}[f(x)g(x)]' &= \lim_{h\to 0} \frac{f(x+h) - f(x)}{h} \lim_{h\to 0} g(x+h) + f(x) \lim_{h\to 0} \frac{g(x+h) - g(x)}{h} \\ &= f'(x)g(x) + f(x)g'(x),\end{aligned}$$

which is the desired result.

In establishing Eq. 3 we could just as well have added and subtracted the quantity $f(x+h)g(x)$ instead. As we have mentioned, the artifice of adding and then subtracting the same quantity is frequently useful in simplifying a problem. In fact, one of the reasons for studying mathematical proofs is to acquire a mastery of techniques such as this. A knowledge of a number of tricks of this kind can be very helpful when an unfamiliar situation is encountered.

Proof of Part (*d*): We prove the quotient rule in two stages. First we show that if g is differentiable at x, then

$$\left[\frac{1}{g(x)}\right]' = -\frac{g'(x)}{[g(x)]^2}, \qquad g(x) \neq 0. \tag{25}$$

*Note that we have used this device earlier in the proof of Theorem 3.2.2.

From Definition 3.1.1 we have

$$\begin{aligned}\left[\frac{1}{g(x)}\right]' &= \lim_{h\to 0}\frac{1}{h}\left[\frac{1}{g(x+h)}-\frac{1}{g(x)}\right]\\ &= \lim_{h\to 0}\frac{g(x)-g(x+h)}{hg(x)g(x+h)}\\ &= \lim_{h\to 0}\left[-\frac{g(x+h)-g(x)}{h}\cdot\frac{1}{g(x)g(x+h)}\right],\end{aligned}$$

provided this limit exists. Note that we have manipulated the difference quotient for $1/g$ so that the difference quotient for g appears explicitly. The hypothesis that g is differentiable at x assures us that the latter difference quotient has the limit $g'(x)$ as $h \to 0$. Further, $g(x + h) \to g(x)$ as $h \to 0$, since g is differentiable and hence continuous at x. Thus we obtain

$$\begin{aligned}\left[\frac{1}{g(x)}\right]' &= -\frac{1}{g(x)}\left[\lim_{h\to 0}\frac{g(x+h)-g(x)}{h}\right]\left[\lim_{h\to 0}\frac{1}{g(x+h)}\right]\\ &= -\frac{g'(x)}{[g(x)]^2}.\end{aligned}$$

Finally, we remark that in the previous argument we had to know that $g(x + h) \neq 0$ in order to write the difference quotient for $1/g$. We do know this, however, because $g(x) \neq 0$ and g is continuous at x. From Theorem 2.6.1 it follows that, at least for h sufficiently small, $g(x + h) \neq 0$ also. Recall that the same situation occurred in Example 4 of Section 3.1.

To complete the proof of Eq. 4, we write $f(x)/g(x)$ as a product, and then use Eq. 3 and Eq. 25. We obtain

$$\begin{aligned}\left[\frac{f(x)}{g(x)}\right]' &= \left[f(x)\cdot\frac{1}{g(x)}\right]'\\ &= f'(x)\frac{1}{g(x)} + f(x)\left[\frac{1}{g(x)}\right]'\\ &= \frac{f'(x)}{g(x)} - \frac{f(x)g'(x)}{[g(x)]^2}\\ &= \frac{g(x)f'(x) - f(x)g'(x)}{[g(x)]^2}, \qquad g(x) \neq 0.\end{aligned}$$

This completes the proof of Theorem 3.2.1. □

PROBLEMS

Find $f'(x)$ in each of Problems 1 through 20.

1. $f(x) = 3x^2 - 4x + 1$
2. $f(x) = x^4 - 5x^3 + 2x^2 + 6x - 7$
3. $f(x) = 4x^5 + 7x^3 - 6x$

4. $f(x) = x^8 - 4x^4 + 1$

5. $f(x) = (2x^2 + 1)^2$

6. $f(x) = (x^3 + 6x + 1)(x^2 - 1)$

7. $f(x) = (x + 2)(x - 1)(x + 3)$

8. $f(x) = (x - 1)^2(x^2 + 1)^2$

9. $f(x) = (x^2 + 1)(x^2 + 4)$

10. $f(x) = (x - 2)(2x^2 + 3)$

11. $f(x) = \dfrac{x^3 - 6x^2 + 8x - 5}{x^2}, \quad x \neq 0$

12. $f(x) = \dfrac{2x^2 - 4}{x + 3}, \quad x \neq -3$

13. $f(x) = \dfrac{x^4 - 3x^2 + 2}{x^2 - 5x + 6}, \quad x \neq 2, 3$

14. $f(x) = \dfrac{x^3 + 2x}{x^2 + 1}$

15. $f(x) = 3x^2 + x^{4/3} + \dfrac{1}{x} - \dfrac{2}{x^3}, \quad x \neq 0$

16. $f(x) = x^{7/2} + 3x^{1/3} + x^{-2/3}, \quad x > 0$

17. $f(x) = \dfrac{x^{3/2}}{x + 1}, \quad x \geq 0$

18. $f(x) = \dfrac{x^{2/5} - x^{5/2}}{x^2 + 1}, \quad x \geq 0$

19. $f(x) = 3x + (x^{1/3} + 1)(x^{2/3} - 1) + \dfrac{7x}{1 + x^2}$

20. $f(x) = 2x^3 + \dfrac{x^2 - 4}{x^2 + 4} + x^2(1 + x^{4/3})$

21. Let $y = f(x) = x^2 - 6x + 2$.
   (a) Find y'.
   (b) Find an equation of the line tangent to the graph of $y = f(x)$ at the point $(1, -3)$.
   (c) Find the point on the graph of $y = f(x)$ where the tangent line has the slope zero. What relation does this point have to other points on the graph? *Hint:* Sketch the curve and the tangent line in question.

22. Let $y = f(x) = -x^2 + 4x + 7$.
   (a) Find y'.
   (b) Find an equation of the line tangent to the graph of $y = f(x)$ at the point $(1, 10)$.
   (c) Find the point on the graph of $y = f(x)$ where the tangent line has the slope zero. What relation does this point have to other points on the graph? *Hint:* Sketch the curve and the tangent line in question.

23. Let $y = f(x) = \dfrac{3x}{x^2 + 1}$.
   (a) Find y'.
   (b) Find the points where the tangent line to the graph of $y = f(x)$ is horizontal.
   (c) Find the points on the graph of $y = f(x)$ where y achieves its greatest and least values, respectively. *Hint:* Make a rough sketch of the graph of $y = f(x)$.

24. Let $y = f(x) = x^2 - 4x - 7$.
   (a) Find an equation of the line tangent to the graph of $y = f(x)$ at the point $(-2, 5)$.
   (b) Find an equation of the line perpendicular (normal) to the tangent line at $(-2, 5)$.

25. The graphs of $y = x^2 - 3x - 1$ and $y = \sqrt{x} + 1$ intersect at the point $(4, 3)$.
   (a) Find an equation of the line tangent to each curve at the point of intersection.
   (b) Find the tangent of the acute angle formed at $(4, 3)$ by the two lines found in Part (a).

26. Let $y = 1 + x^2$. Find all points on this curve where the tangent line passes through the origin.

27. Let $y = 5 + x - 2x^2$. Find all points on this curve where the tangent line passes through the point $(-1, 10)$.

28. Find all points on the curve $y = x^2 - 3x + 7$ where the tangent line is parallel to the line $y - 5x = 4$.

29. Find all points on the curve $y = x^3 + 2x^2 - 6x + 4$ where the tangent line is parallel to the line $2x + y = 3$.

30. Find all points on the curve $y = x^3 - 2x^2 - 2x - 6$ where the tangent line is perpendicular to the line $x + 2y = 4$.

31. There are two lines through the point $(4, 7)$ that are tangent to the parabola $y = x^2$. Find an equation of each of these lines and determine its point of tangency.

32. Let the point $P(a, b)$ lie below the parabola $y = x^2$; thus $b < a^2$. Show that there are two lines through P tangent to the parabola and find their slopes.

33. The position of a particle moving on a straight line is given by

$$s = t^3 - \tfrac{9}{2}t^2 + 15t + 4,$$

where s is measured in feet and t in seconds.

(a) Find all values of t at which the velocity v has the value 9 ft/sec.

(b) Find the time at which the particle has the smallest velocity.

34. An object dropped from rest in a vacuum falls in time t a distance $s = gt^2/2$, where g is the acceleration due to gravity. Assume $g = 32$ ft/sec^2.

(a) Find the time at which the object reaches a velocity of 128 ft/sec.

(b) If the object is dropped from a height of 128 ft, find the velocity with which it strikes the ground.

35. Let

$$f(x) = \begin{cases} x^2, & x \le 2; \\ ax + b, & x > 2. \end{cases}$$

Find a and b so that f is differentiable at $x = 2$.

36. Let

$$f(x) = \begin{cases} a\sqrt{x}, & x \le 1; \\ x + b, & x > 1. \end{cases}$$

Find a and b so that f is differentiable at $x = 1$.

37. Find the rate of change of the area A of a circle with respect to its radius r.

38. Find the rate of change of the volume V of a sphere with respect to its radius r.

39. Express the volume V of a cube in terms of its surface area S and find the rate of change of V with respect to S.

40. Express the volume V of a sphere in terms of its surface area S and find the rate of change of V with respect to S.

41. (a) Derive a formula for the derivative of the product of three functions. That is, if $f(x) = u(x)v(x)w(x)$, where u, v, and w are differentiable functions, express $f'(x)$ in terms of $u(x)$, $v(x)$, $w(x)$, and their respective derivatives $u'(x)$, $v'(x)$, and $w'(x)$.

(b) Show that

$$\frac{f'(x)}{f(x)} = \frac{u'(x)}{u(x)} + \frac{v'(x)}{v(x)} + \frac{w'(x)}{w(x)}.$$

* 42. Generalize the result of Problem 41 to n functions. In particular, show that if

$$f(x) = u_1(x)u_2(x) \cdots u_n(x),$$

where $u_1, u_2, \ldots, u_n$ are differentiable functions, then

$$\frac{f'(x)}{f(x)} = \frac{u_1'(x)}{u_1(x)} + \frac{u_2'(x)}{u_2(x)} + \cdots + \frac{u_n'(x)}{u_n(x)}.$$

Hint: A formal proof makes use of the principle of mathematical induction (see Section 6.1). If you are not familiar with induction proofs, give a more informal argument.

43. Use the result of Problem 42 to find $f'(x)$ if $f(x) = [u(x)]^n$.

44. Derive the product rule, Eq. 3, by adding and subtracting the quantity $f(x + h)g(x)$ to the difference quotient in Eq. 24.

3.3 HIGHER DERIVATIVES AND ANTIDERIVATIVES

It is, of course, possible to apply the process of differentiation to a function that is itself the derivative of another function. The derivative of f' is $(f')'$, usually written f''; its value at the point x is

$$f''(x) = \lim_{h \to 0} \frac{f'(x + h) - f'(x)}{h}. \tag{1}$$

The function f' is called the (first) derivative of f, and the function f'' the *second* derivative, since it was obtained by differentiating a second time. Higher derivatives can also be calculated. For instance, the derivative of f'' is f''', the *third* derivative of f. In general, the nth derivative of f, obtained by differentiating n times in succession, is denoted by $f^{(n)}$. It is important to remember that the n in parentheses is not an exponent, but an indicator of the number of differentiations to be performed.

EXAMPLE 1

Find the first three derivatives of f, if

$$f(x) = x^4 - 2x^2 + \frac{1}{x} - x^{2/3}, \qquad x \neq 0.$$

Using the power rule and the algebraic properties of derivatives, we find that

$$f'(x) = 4x^3 - 4x - \frac{1}{x^2} - \frac{2}{3}x^{-1/3}, \qquad x \neq 0$$

$$f''(x) = 12x^2 - 4 + \frac{2}{x^3} + \frac{2}{9}x^{-4/3}, \qquad x \neq 0$$

and finally

$$f'''(x) = 24x - \frac{6}{x^4} - \frac{8}{27}x^{-7/3}, \qquad x \neq 0. \blacksquare$$

EXAMPLE 2

If

$$y = \frac{x}{x+1}, \qquad x \neq -1,$$

find y' and y''.

Using the quotient rule, Theorem 3.2.1(d), we obtain

$$\begin{aligned} y' &= \frac{(x+1)(x)' - x(x+1)'}{(x+1)^2} \\ &= \frac{x+1-x}{(x+1)^2} = \frac{1}{(x+1)^2}, \qquad x \neq -1. \end{aligned}$$

To compute y'' we can use either the quotient rule again, or Eq. 25 of Section 3.2 with $g(x) = (x+1)^2 = x^2 + 2x + 1$. Then

$$y'' = -\frac{g'(x)}{[g(x)]^2} = -\frac{2x+2}{(x+1)^4} = -\frac{2}{(x+1)^3}, \qquad x \neq -1. \blacksquare$$

Derivatives of higher than first order also have important physical and/or geometrical interpretations. We will discuss this more fully later, but will mention one example at this time. If

$$s = f(t) \tag{2}$$

gives the position of a moving object as a function of time, then its velocity is given by

$$v = f'(t). \tag{3}$$

The derivative v' of the velocity is called the **acceleration** and is denoted by a. Thus

$$a = v' = s'' = f''(t), \tag{4}$$

so the second derivative with respect to time of the position of the moving object gives the acceleration of the object.

The operator notation for derivatives

At least three different notations are commonly used to designate derivatives. Each has its advantages in different situations; in order to use calculus most effectively, a student should become familiar with all three. We have already used the "prime" notation, introduced by Lagrange, whereby f' is the derivative of f, g' is the derivative of g, and so forth.

A second notation uses the letter D to indicate differentiation. Then Df is the derivative of the function f, and $Df(x)$ is the value of Df at the point x. Thus Df and $Df(x)$ are the same as f' and $f'(x)$, respectively. If several variables appear in a discussion, we sometimes use a subscript to indicate the variable involved in a particular differentiation, for example, $D_x f$ or $D_x f(x)$. The third notation, due to Leibniz, is described in Section 3.4.

In terms of the D-notation, the power rule takes the form

$$D(x^r) = rx^{r-1}, \tag{5}$$

and the algebraic properties of derivatives can be expressed as follows.

$$D(f \pm g) = Df \pm Dg, \tag{6}$$

$$D(\alpha f) = \alpha Df, \tag{7}$$

$$D(fg) = fDg + gDf, \tag{8}$$

$$D\left(\frac{f}{g}\right) = \frac{gDf - fDg}{g^2}, \qquad g \neq 0. \tag{9}$$

The second derivative of f is $D(Df)$ or $D^2 f$. Similarly, the third and higher derivatives are denoted by $D^3 f, D^4 f, \ldots, D^n f$, and so forth.

The D-notation emphasizes that differentiation is a process or operation performed on a function. Thus we sometimes speak of D as the **differentiation operator:** it takes a function f and operates upon it, or transforms it, into a new function Df. In particular, if f and g are differentiable functions and if α and β are any constants, then we have seen (Eq. 5 of Section 3.2) that

$$D[\alpha f(x) + \beta g(x)] = \alpha Df(x) + \beta Dg(x). \tag{10}$$

Any operator or transformation having the property expressed by Eq. 10 is said to be *linear*; thus the differentiation operator D is a *linear operator*. Linear operators or transformations play a very important role in mathematics.

EXAMPLE 3

Find $Df(x)$ and $D^2f(x)$ if

$$f(x) = 2x^3 + x^{-1/2}, \qquad x > 0.$$

Using Eqs. 6 and 7 together with the power rule, we have

$$Df(x) = 6x^2 - \tfrac{1}{2}x^{-3/2}, \qquad x > 0,$$

and

$$D^2f(x) = 12x + \tfrac{3}{4}x^{-5/2}, \qquad x > 0. \blacksquare$$

EXAMPLE 4

Find $D^n(x^n)$, where n is a positive integer.

Using the power rule, we obtain

$$D(x^n) = nx^{n-1},$$

$$D^2(x^n) = n(n-1)x^{n-2},$$

$$D^3(x^n) = n(n-1)(n-2)x^{n-3},$$

and in general

$$\begin{aligned} D^n(x^n) &= n(n-1)(n-2)\cdots[n-(n-1)]x^{n-n} \\ &= n!. \end{aligned} \tag{11}$$

Since $n!$ is a constant, it follows that

$$D^{n+1}(x^n) = 0 \tag{12}$$

for each x. Moreover, it follows from Eq. 12 that if P is a polynomial of degree n, then its $(n + 1)$st and all higher derivatives are everywhere zero. $\blacksquare$

Antiderivatives

Up to now we have been concerned with the problem of finding the derivative f' of some given function f. In practice, however, we are often faced with precisely the reverse problem: given the derivative f', find the function f. In slightly different terminology, given a function g defined on an open interval, find a function f such that $f' = g$, or $Df = g$ on this interval. Such a function f, whose derivative is g, is said to be an **antiderivative** of g. We use the notation D^{-1}, read "D-inverse," to denote antiderivatives. Thus,

$$\text{if} \quad Df = g, \qquad \text{then} \quad f = D^{-1}g. \tag{13}$$

For example,

$$D^{-1}(x^2) = \frac{x^3}{3}, \tag{14}$$

since $D(x^3/3) = x^2$. Note, however, that $x^3/3$ is not the only antiderivative of x^2; for example, the functions $(x^3/3) + 5$, $(x^3/3) - 3$, and indeed

$$\frac{x^3}{3} + c,$$

where c is an arbitrary constant, are also antiderivatives of x^2. That is, each of these functions has x^2 as its derivative. In general, if f is an antiderivative of g, then $f + c$ is also an antiderivative for any constant c. Thus, whereas a differentiable function has precisely one derivative, a given function has more than one antiderivative, assuming that it has any at all.

It is possible to show that *all* antiderivatives of g are of the form $f + c$, where $Df = g$; however, this result requires additional theoretical background and is established in Theorem 4.1.5. Accepting this fact for the present, we see that to find an antiderivative of a function g, we need only find a function f such that $Df = g$; then all possible antiderivatives of g are obtained by adding constants to f.

Further examples of antiderivatives are

$$D^{-1}\left(\frac{1}{x^2}\right) = -\frac{1}{x} + c, \tag{15}$$

$$D^{-1}(x - x^{1/3}) = \frac{x^2}{2} - \frac{3}{4}x^{4/3} + c, \tag{16}$$

and more generally

$$D^{-1}(x^r) = \frac{x^{r+1}}{r + 1} + c, \qquad r \neq -1. \tag{17}$$

The restriction $r \neq -1$ in Eq. 17 means that this equation does not provide an antiderivative for the function $1/x$. This function does have an antiderivative, but not one given by any function so far considered. We fill this gap in Section 8.2.

In general, each differentiation formula can, so to speak, be read backwards as an antidifferentiation formula. The antidifferentiation formulas corresponding to Eqs. 6 and 7, namely,

$$D^{-1}(f \pm g) = D^{-1}f \pm D^{-1}g, \tag{18}$$

and

$$D^{-1}(\alpha f) = \alpha D^{-1}f, \tag{19}$$

respectively, are particularly useful. Equation 18, for example, states that an antiderivative of $f \pm g$ can be found by finding antiderivatives of f and g separately, and then forming their sum or difference, as the case may be. The derivation of Eqs. 18 and 19 is indicated in Problem 36.

For a given function g, the graphs of its antiderivatives

$$y = f(x) + c$$

constitute a family of "parallel" curves in the xy-plane (see Figure 3.3.1). To identify a particular antiderivative we often specify a point (x_0, y_0) through which it passes.

EXAMPLE 5

If $g(x) = x^2$, find the antiderivative of g whose graph contains the point (1, 2).

We have previously remarked that all antiderivatives of x^2 are of the form $(x^3/3) + c$. Thus, among all curves of the family

$$y = \frac{x^3}{3} + c, \tag{20}$$

we must find the one passing through the point (1, 2) (see Figure 3.3.2). Substituting $x = 1$ and $y = 2$ in Eq. 20 we find that

$$2 = \tfrac{1}{3} + c, \qquad \text{so} \quad c = \tfrac{5}{3}.$$

Hence

$$y = \frac{x^3}{3} + \frac{5}{3} = \frac{x^3 + 5}{3} \tag{21}$$

is the desired antiderivative; it is the only one passing through (1, 2). ∎

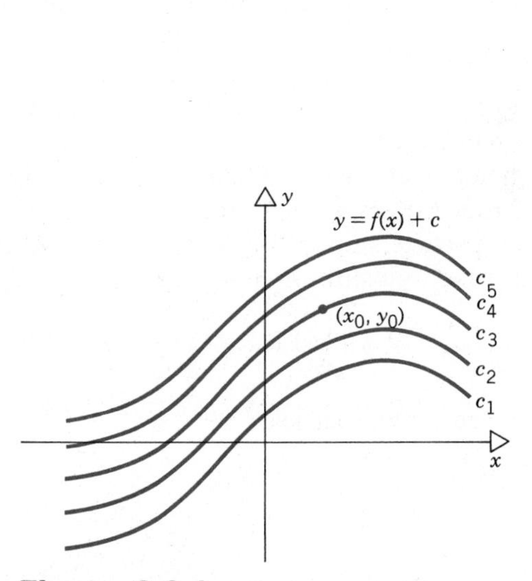

Figure 3.3.1 A family of antiderivatives.

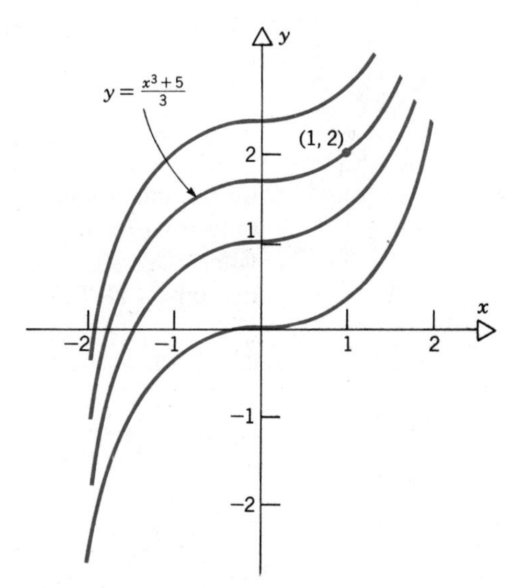

Figure 3.3.2 Graphs of $y = (x^3/3) + c$.

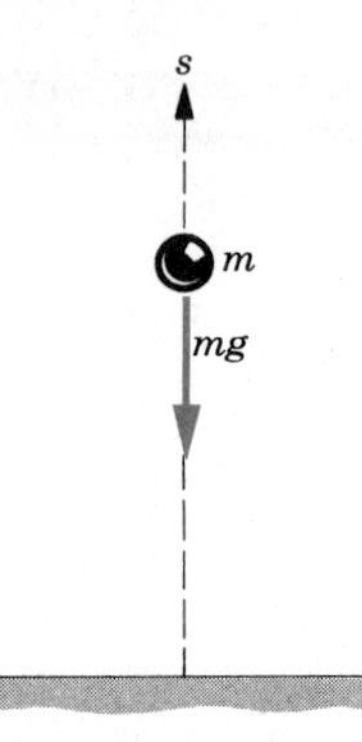

Figure 3.3.3 A point mass acted on by gravity.

EXAMPLE 6

Suppose that a particle of mass m is projected vertically upward from the surface of the earth with initial velocity v_0. Assume that the gravitational attraction of the earth is constant, and neglect air resistance and all other forces acting on the particle. Find the position s of the particle at time t, and determine the maximum height that the particle reaches.

Let s be measured positive upward with the origin of the s-axis at the initial position of the particle as shown in Figure 3.3.3. Then the velocity $v = s'$ is positive when the particle is moving upward and negative when it is falling. According to Newton's law of motion the external force acting on the particle is equal to the product of the mass and acceleration a of the particle; thus

$$F = ma. \tag{22}$$

As we have noted above in Eq. 4, $a = v'$. Further, in the present case, the only external force is the gravitational attraction of the earth, or the weight of the particle. Thus $F = -mg$, where g is the acceleration* due to gravity, and the minus sign is chosen because this force acts in the downward (negative) direction. Substituting for F and a in Eq. 22 we find that the velocity v must satisfy

$$v' = -g. \tag{23}$$

Hence

$$v = -gt + c_1, \tag{24}$$

where c_1 is a constant. Since $v = v_0$ when $t = 0$, we have $c_1 = v_0$, and therefore

$$v = -gt + v_0. \tag{25}$$

Replacing v by s' and solving Eq. 25 for s, we have

$$s = -\tfrac{1}{2}gt^2 + v_0t + c_2, \tag{26}$$

where c_2 is another constant. However, $s = 0$ when $t = 0$, and thus $c_2 = 0$. Hence we obtain the result

$$s = -\tfrac{1}{2}gt^2 + v_0t. \tag{27}$$

If left undisturbed on its trajectory, the particle will return to its starting point on the earth's surface when $s = 0$, that is, when $t = 2v_0/g$. Thus, Eq. 27 is valid only for t in $[0, 2v_0/g]$. From Eq. 25 we see that $v > 0$ when $t < v_0/g$ and $v < 0$ when $t > v_0/g$. Hence the particle stops rising at $t = v_0/g$, and therefore reaches its highest altitude at this time. Substituting $t = v_0/g$ in Eq. 27, we find that

$$s_{\max} = -\frac{1}{2}g\left(\frac{v_0}{g}\right)^2 + v_0\frac{v_0}{g} = \frac{v_0^2}{2g} \tag{28}$$

is the greatest height reached by the particle. ∎

*For the purposes of this text, it is sufficient to assume that $g = 32\ \text{ft/sec}^2$ or $9.8\ \text{m/sec}^2$.

PROBLEMS

In each of Problems 1 through 10, find the indicated derivative of the given function.

1. $f''(x)$ if $f(x) = (x^2 + 1)(x^2 - 2x + 4)$
2. D^3y if $y = x^5 - 3x^4 + 5x^3 - 6x^2 - 2x + 8$
3. $D^2f(x)$ if $f(x) = x^{1/3} - \dfrac{1}{x^2}, \quad x \neq 0$
4. $f'''(x)$ if $f(x) = \dfrac{x^2 + 1}{x^2}, \quad x \neq 0$
5. y''' if $y = \dfrac{1 + x^{-2/3}}{x}, \quad x \neq 0$
6. D^8y if $y = (x^3 + 2x^2 - 6x + 1) \cdot (2x^3 + 5x - 7)$
7. y'' if $y = x^4 - 3x^2 + 6$
8. $f''(x)$ if $f(x) = \dfrac{x - 1}{x + 2}, \quad x \neq -2$
9. y''' if $y = x^2 + 3\sqrt{x} - \dfrac{1}{x} + \dfrac{4}{\sqrt{x}}, \quad x > 0$
10. $D^2f(x)$ if $f(x) = \dfrac{x + 2}{x - 3}, \quad x \neq 3$

In each of Problems 11 through 20, find the indicated antiderivative.

11. $D^{-1}(x - 4)$
12. $D^{-1}(3x^2 - x + 6)$
13. $D^{-1}\left(x^2 + \dfrac{1}{x^2}\right)$
14. $D^{-1}(x^6 - 3x^3 + 6)$
15. $D^{-1}(3x^{2/3} - 4x^{1/3})$
16. $D^{-1}[(x + 1)(x - 2)]$
17. $D^{-1}\left(\dfrac{3}{\sqrt{x}}\right)$
18. $D^{-1}\left[\dfrac{2x}{(1 + x^2)^2}\right]$
19. $D^{-1}\left[\left(\dfrac{1}{2 + x}\right)^2\right]$
20. $D^{-1}\left[-\dfrac{1}{2\sqrt{x}(1 + 2\sqrt{x} + x)}\right]$

In each of Problems 21 through 26, find the function f that satisfies the given conditions.

21. $f'(x) = 2x - 3; \quad f(0) = 4$
22. $f'(x) = 3x^2 + x - 7; \quad f(1) = -2$
23. $f'(x) = 4x^3 - 2x; \quad f(2) = 4$
24. $f'(x) = x^2 + \dfrac{1}{x^2}, \quad x < 0; \quad f(-1) = 1$
25. $f'(x) = \sqrt{x}, \quad x > 0; \quad f(1) = 1$
26. $f'(x) = \dfrac{1}{x^2}, \quad x > 0; \quad f(2) = 0$
27. Find a polynomial P of degree two such that $P(2) = 1$, $P'(0) = -3$, and $P''(4) = 6$.
28. Find a polynomial P of degree three such that $P(0) = 4$, $P'(-1) = 2$, $P'(2) = 5$, and $P''(1) = -5$.

In Problems 29 through 32, assume that the conditions of Example 6 prevail; that is, gravitational attraction is constant, and other forces (in particular, air resistance) are neglected.

29. A stone is dropped from the edge of a cliff and reaches the bottom 3 sec later. How high is the cliff?
30. From what height must an object fall in order to hit the ground with a velocity of 60 mi/hr?
31. A ball with mass 0.25 kg is thrown upward with initial velocity 20 m/sec from the roof of a building 30 m high.
    (a) Find the maximum height above the ground that the ball reaches.
    (b) Assuming that the ball misses the building on the way down, find the time that elapses before it hits the ground.
32. A particle is projected vertically upward from the top of an elevator with an initial velocity of 50 ft/sec. At the same instant the elevator starts to descend at a uniform downward velocity of 10 ft/sec. Find the time at which the particle strikes the top of the elevator again.
33. Show that $(f'g - fg')' = f''g - fg''$.
34. Assuming that f and g are differentiable the required number of times, show that
    (a) $(fg)'' = f''g + 2f'g' + fg''$.
    (b) $(fg)''' = f'''g + 3f''g' + 3f'g'' + fg'''$.

* 35. The results of Problem 34 suggest that

$$(fg)^{(n)} = f^{(n)}g + nf^{(n-1)}g' + \cdots + \binom{n}{k} f^{(n-k)}g^{(k)} + \cdots + fg^{(n)}, \qquad (i)$$

where

$$\binom{n}{k} = \frac{n(n-1)\cdots(n-k+1)}{k!}$$

$$= \frac{n!}{k!(n-k)!} \qquad (ii)$$

is the coefficient of $a^{n-k}b^k$ in $(a+b)^n$. Equation (i) is known as Leibniz's rule. Use mathematical induction to prove this result. *Hint:* If you are not familiar with inductive proofs, see Section 6.1.

* **36.** (a) To derive Eq. 18 from Eq. 6, observe that Eq. 6 is equivalent to

$$f + g = D^{-1}(Df + Dg).$$

Now let $Df = u$ and $Dg = v$. Then show that

$$D^{-1}u + D^{-1}v = D^{-1}(u + v),$$

which is the same as Eq. 18, except for the names of the functions.

(b) Use an argument similar to that in Part (a) to derive Eq. 19 from Eq. 7.

(c) Show that

$$D^{-1}(\alpha u + \beta v) = \alpha D^{-1}u + \beta D^{-1}v$$

for any constants α and β. Thus D^{-1} is also a linear operator.

3.4 LINEAR APPROXIMATIONS

The main purpose of this section is to show how the concept of differentiation leads naturally to the idea of approximating a given function $y = f(x)$ by a function $y = ax + b$ in the neighborhood of some given point. In other words, near the given point we approximate the graph of the function f by a straight line. This type of approximation is called a linear approximation and has many important uses, both in the theory and applications of mathematics.

Recall that for a given function $y = f(x)$, differentiable at $x = x_0$, we can construct the tangent line to the graph of f at the point (x_0, y_0), as shown in Figure 3.4.1. This tangent line has the equation

$$y = y_0 + f'(x_0)(x - x_0). \qquad (1)$$

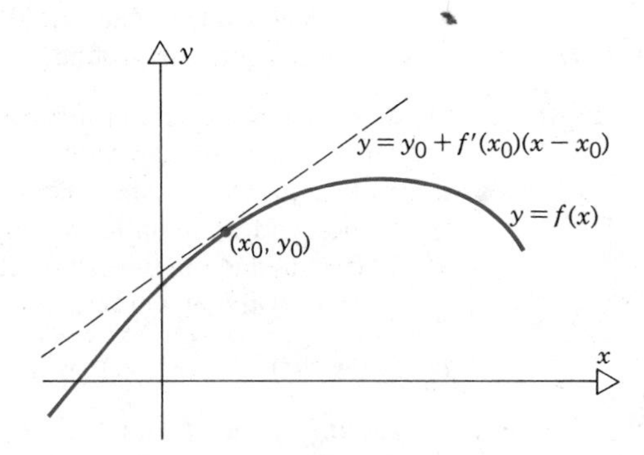

Figure 3.4.1

Thus we have defined a new function, the tangent line function, given by

$$T(x; x_0) = y_0 + f'(x_0)(x - x_0)$$

$$= f(x_0) + f'(x_0)(x - x_0). \qquad (2)$$

The notation $T(x; x_0)$ indicates that T depends both on the independent variable x and also on the point of tangency.

To investigate how well T approximates f we consider the difference $f(x) - T(x; x_0)$. This difference is called the **remainder,** and it is convenient to write it as $r(x; x_0)(x - x_0)$. Thus we have

$$f(x) - T(x; x_0) = r(x; x_0)(x - x_0). \tag{3}$$

Equation 3 serves to define the function r, which we will refer to as the **remainder function.**

EXAMPLE 1

Consider the function

$$f(x) = \tfrac{1}{2}(x^2 - x + 2) \tag{4}$$

whose graph is shown in Figure 3.4.2. For $x_0 = 1$ find the tangent line function T and the remainder function r; investigate the behavior of $r(x; 1)$ as $x \to 1$.

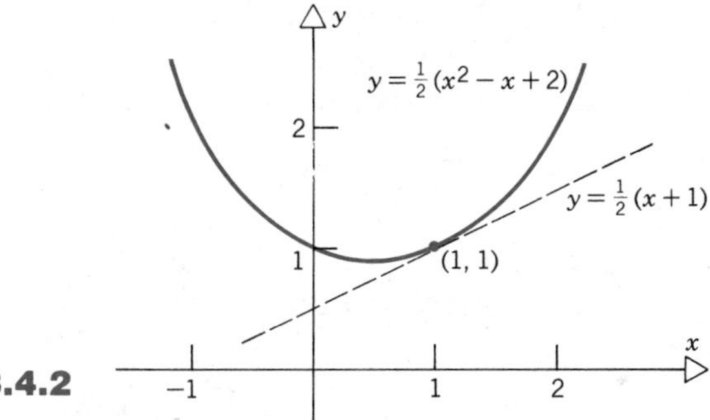

Figure 3.4.2

Since $f'(x) = x - \frac{1}{2}$, we have

$$\begin{aligned} T(x; 1) &= f(1) + f'(1)(x - 1) \\ &= 1 + \tfrac{1}{2}(x - 1) = \tfrac{1}{2}(x + 1). \end{aligned} \tag{5}$$

This function can be used to approximate $f(x)$ in the vicinity of $x = 1$; for example,

$$T(\tfrac{9}{8}; 1) = \tfrac{1}{2}(\tfrac{9}{8} + 1) = \tfrac{17}{16}$$

is an approximation to the actual value

$$f(\tfrac{9}{8}) = \tfrac{1}{2}[(\tfrac{9}{8})^2 - \tfrac{9}{8} + 2] = \tfrac{137}{128}.$$

The remainder in this case is

$$f(\tfrac{9}{8}) - T(\tfrac{9}{8}; 1) = \tfrac{137}{128} - \tfrac{17}{16} = \tfrac{1}{128}.$$

For an arbitrary value of x the remainder is

$$\begin{aligned} f(x) - T(x; 1) &= \tfrac{1}{2}(x^2 - x + 2) - \tfrac{1}{2}(x + 1) \\ &= \tfrac{1}{2}(x^2 - 2x + 1) \\ &= \tfrac{1}{2}(x - 1)^2. \end{aligned} \tag{6}$$

Hence, by Eq. 3, the remainder function $r(x; 1)$ is

$$r(x; 1) = \tfrac{1}{2}(x - 1). \tag{7}$$

Clearly $r(x; 1) \to 0$ as $x \to 1$; as a result, the remainder $r(x; 1)(x - 1) \to 0$ faster than $x - 1$ itself as $x \to 1$. ∎

The results described in Example 1 for the function $f(x) = (x^2 - x + 2)/2$ are typical of the situation for any differentiable function.

Theorem 3.4.1

If f is differentiable at x_0, then

$$f(x) - f(x_0) = f'(x_0)(x - x_0) + r(x; x_0)(x - x_0), \tag{8}$$

where

$$r(x; x_0) \to 0 \quad \text{as} \quad x \to x_0. \tag{9}$$

Proof. Observe that Eq. 8 is obtained by substituting Eq. 2 into Eq. 3. Then, for $x \neq x_0$, we can rewrite Eq. 8 in the form

$$\frac{f(x) - f(x_0)}{x - x_0} = f'(x_0) + r(x; x_0). \tag{10}$$

Since f is differentiable at x_0, the left side of Eq. 10 approaches $f'(x_0)$ as $x \to x_0$, and it follows at once that $r(x; x_0) \to 0$ as $x \to x_0$, as was to be proved. □

It is sometimes useful to state the result of Theorem 3.4.1 in a slightly different form. If we let $x - x_0 = h$, then Eq. 8 becomes

$$f(x_0 + h) - f(x_0) = [f'(x_0) + \bar{r}(h; x_0)]h, \tag{11}$$

where $\bar{r}(h; x_0) = r(x_0 + h; x_0)$ and $\bar{r}(h; x_0) \to 0$ as $h \to 0$.

EXAMPLE 2

For the function $f(x) = (x^2 - x + 2)/2$ in Example 1, find the tangent line function $T(x; x_0)$ and the remainder function $r(x; x_0)$ for an arbitrary value of x_0.

We have

$$f(x_0) = \frac{x_0^2 - x_0 + 2}{2}, \qquad f'(x_0) = x_0 - \frac{1}{2}.$$

Therefore, from Eq. 2,

$$T(x; x_0) = \frac{1}{2}(x_0^2 - x_0 + 2) + \left(x_0 - \frac{1}{2}\right)(x - x_0)$$

$$= 1 - \frac{x_0^2}{2} + \left(x_0 - \frac{1}{2}\right)x. \tag{12}$$

Then the remainder is

$$f(x) - T(x; x_0) = \frac{x^2 - x + 2}{2} - \left(1 - \frac{x_0^2}{2}\right) - \left(x_0 - \frac{1}{2}\right)x$$

$$= \frac{x^2}{2} - x_0x + \frac{x_0^2}{2}$$

$$= \frac{(x - x_0)^2}{2},$$

and it follows from Eq. 3 that

$$r(x; x_0) = \frac{x - x_0}{2}. \tag{13}$$

Again, as Theorem 3.4.1 requires, $r(x; x_0) \to 0$ as $x \to x_0$. ∎

We will use the relation in Eqs. 8 and 9 to define what is meant by a **linear approximation.** Consider a given continuous function f and a given point x_0. Then the function

$$A(x_0) + B(x_0)(x - x_0) \tag{14}$$

is said to approximate $f(x)$ near x_0 if

$$f(x) - A(x_0) = B(x_0)(x - x_0) + r(x; x_0)(x - x_0), \tag{15}$$

where

$$r(x; x_0) \to 0 \quad \text{as} \quad x \to x_0. \tag{16}$$

Theorem 3.4.1 states that if f is differentiable at x_0, then f has such an approximation (the tangent line approximation) with $A(x_0) = f(x_0)$ and $B(x_0) = f'(x_0)$. The converse of Theorem 3.4.1 is also true, that is, if f is continuous and Eqs. 15 and 16 hold, then it must follow that f is also differentiable at x_0, and of course, $A(x_0) = f(x_0)$ and $B(x_0) = f'(x_0)$. The proof of this assertion is indicated in Problem 29. Thus f is differentiable at x_0 if and only if there is a tangent line function that approximates f near x_0 in the sense described here. To express this fact we sometimes say that a differentiable function is **locally linear,** and vice versa.

The following example illustrates how a linear approximation can be used in evaluating (approximately) a more complicated function. We use the terminology "the approximation is correct to n decimal places" to mean that the difference between the exact and approximate values is less (in absolute value) than $5/10^{n+1}$; that is, the error is less than 5 in the $(n + 1)$st place.

EXAMPLE 3

Determine an approximate value of $\sqrt{81.34}$.

If we let $f(x) = x^{1/2}$, then we wish to determine an approximate value of $f(81.34)$. Since 81.34 is near 81 and since $\sqrt{81} = 9$, we construct the tangent line function to $y = f(x) = x^{1/2}$ at $x = 81$. We have

$$f'(x) = \tfrac{1}{2}x^{-1/2}, \qquad f'(81) = \tfrac{1}{18},$$

so the tangent line function at the point (81, 9) is

$$T(x; 81) = 9 + \tfrac{1}{18}(x - 81). \tag{17}$$

Then

$$\begin{aligned} T(81.34; 81) &= 9 + \tfrac{1}{18}(81.34 - 81) \\ &= 9 + \frac{0.34}{18} \cong 9.018889 \end{aligned}$$

is the desired approximate value of $\sqrt{81.34}$. The correct value of $\sqrt{81.34}$ (to six decimal places) is 9.018869, so the linear approximation is accurate to four decimal places. Note also that the tangent line function (17) can be used to obtain an approximate value of $\sqrt{x}$ for *any* x near 81, not just for 81.34. ∎

Leibniz's notation

The idea of a linear approximation to a given function f is helpful in understanding the notation for the derivative that was introduced by Leibniz; this is the last of the three important notations that we mentioned in the preceding section. Leibniz's notation is intuitively appealing as well as useful; for instance, as we shall see in later sections, certain differentiation formulas are most easily remembered when expressed in this notation. Therefore we briefly summarize Leibniz's ideas.

Consider the function

$$y = f(x). \tag{18}$$

Suppose that the independent variable x is changed by an amount Δx, and let Δy (which depends on both x and Δx) be the corresponding change or increment in the dependent variable y (see Figure 3.4.3). Then

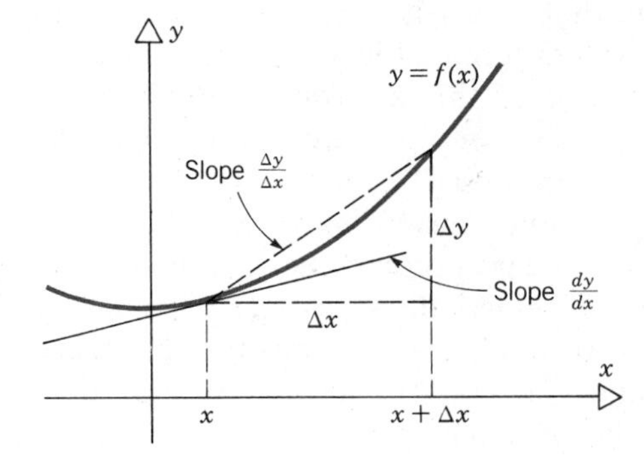

Figure 3.4.3

$$\Delta y = f(x + \Delta x) - f(x) \tag{19}$$

and

$$\frac{\Delta y}{\Delta x} = \frac{f(x + \Delta x) - f(x)}{\Delta x}. \tag{20}$$

Leibniz denoted the limit of this difference quotient by dy/dx (read dy by dx or just $dy\ dx$):

$$\frac{dy}{dx} = \lim_{\Delta x \to 0} \frac{f(x + \Delta x) - f(x)}{\Delta x}. \tag{21}$$

Observe that dy/dx is *not* a quotient of two separate quantities dy and dx; rather it refers to the *limit* of the difference quotient of Eq. 21. Thus dy/dx is exactly the same quantity that we have earlier denoted by y' and Dy.

In the early development of calculus Leibniz and others thought of dx and dy as "infinitesimals" and their quotient as the differential quotient, or derivative. While this concept has been superseded by that of the derivative as the limit of a difference quotient, the notation of Leibniz* is still commonly used. The notation dy/dx now suggests the formation of the difference quotient $\Delta y/\Delta x$, the change in y divided by the change in x, and the passage to the limit as $\Delta x \to 0$.

To emphasize that dy/dx is not a quotient, we may write

$$\frac{d}{dx}y \quad \text{instead of} \quad \frac{dy}{dx}.$$

Thus d/dx is the operator D that we introduced in Section 3.3. The Leibniz notation for the second and higher derivatives is

$$\frac{d^2}{dx^2}y, \quad \frac{d^3}{dx^3}y, \quad \ldots, \quad \frac{d^n}{dx^n}y, \quad \ldots.$$

Once the point has been made that dy/dx is not a true quotient, it is customary to write

$$\frac{dy}{dx}, \quad \frac{d^2y}{dx^2}, \quad \frac{d^3y}{dx^3}, \quad \ldots, \quad \frac{d^ny}{dx^n}, \quad \ldots.$$

For example, if $y = x^r$, then

$$\frac{dy}{dx} = rx^{r-1}, \quad \frac{d^2y}{dx^2} = r(r - 1)x^{r-2},$$

$$\frac{d^3y}{dx^3} = r(r - 1)(r - 2)x^{r-3},$$

and so on.

---

*Leibniz had a keen appreciation of the importance of good notation, and settled upon the use of dy/dx to represent the derivative of y with respect to x only after considerable deliberation. It is a tribute to his insight that after three hundred years his notation is still probably the most useful and frequently used.

We can now summarize the different notations for the derivative. If $y = f(x)$ is a differentiable function, then its derivative can be indicated by any of the following notations:

$$y', f'(x) \qquad \text{(Prime or Lagrange Notation)}$$

$$Dy, Df(x) \qquad \text{(Operator Notation)}$$

$$\frac{dy}{dx}, \frac{df}{dx}(x) \qquad \text{(Leibniz's Notation)}$$

Each is particularly well suited for certain purposes, so it is advisable to become proficient in using all three systems of notation.

Differentials

Let us now reexamine the idea of a tangent line approximation to a given function $y = f(x)$ in the light of the Leibniz notation. This viewpoint is frequently useful in applications. We will sometimes designate a change in x by dx (rather than by Δx). Corresponding to dx (or to Δx) the change in y is Δy and is given by Eq. 19,

$$\begin{aligned}\Delta y &= f(x + \Delta x) - f(x) \\ &= f(x + dx) - f(x).\end{aligned} \tag{22}$$

Geometrically speaking, Δy is the change in y along the curve $y = f(x)$ from x to $x + \Delta x$. On the other hand, the change in y, *measured along the tangent line at the point* x, will be called the **differential** of y and denoted by dy. See Figure 3.4.4

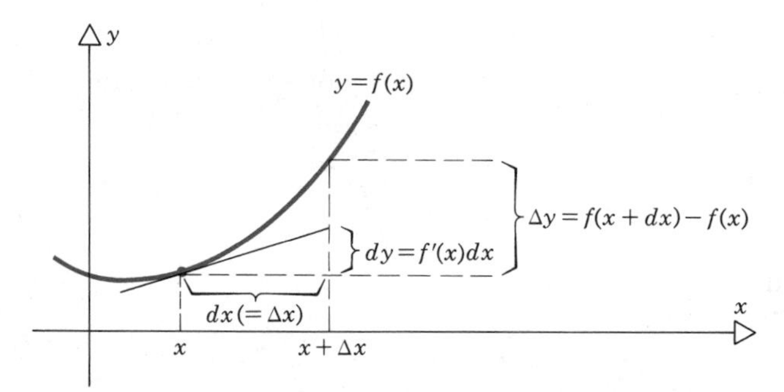

Figure 3.4.4 The increment Δy and the differential dy.

and note carefully the difference between Δy and dy. While dx and Δx are merely two different symbols for the same quantity, dy and Δy are two quite different things: Δy is the actual change in y along the curve $y = f(x)$, but dy is the approximate change in y measured along the tangent line. To emphasize that the differential dy depends both on the point x and the increment dx, we will sometimes write $dy\ (x, dx)$. Thus $dy\ (x, dx)$ is the change in y along the tangent line at x, whose slope is $f'(x)$, corresponding to the increment dx. Consequently,

$$dy\ (x, dx) = f'(x)\, dx,$$

or, more simply,

$$dy = f'(x)\, dx. \tag{23}$$

We have thus defined two quantities dx and dy whose quotient dy/dx is equal to $f'(x)$. We stress, however, that the dx and dy in Eq. 23 are *not* the symbols appearing in Eq. 21. Indeed, as we have said before, the left side of Eq. 21 is *not* the quotient of two distinct quantities dy and dx; rather dy/dx in Eq. 21 must be regarded as the single entity $f'(x)$.

Finally, let us reinterpret Theorem 3.4.1 in terms of differentials. Let $y = f(x)$, $y_0 = f(x_0)$, and $dx = x - x_0$. Then, substituting into Eq. 8, we have

$$\begin{aligned} y - y_0 &= f'(x_0)\, dx + r(x; x_0)\, dx \\ &= dy\ (x_0, dx) + r(x; x_0)\, dx. \end{aligned}$$

Thus the differential dy is the linear approximation to $y - y_0$ that we discussed earlier in this section. For small dx the quantity $y_0 + dy\ (x_0, dx)$ provides a good approximation to the value of y at $x = x_0 + dx$.

EXAMPLE 3

Let

$$y = f(x) = \tfrac{1}{2}(x^2 - x + 2).$$

Find the differential dy as a function of x and dx. Also find dy if $x = 1$ and $dx = \frac{1}{8}$.

Since

$$f'(x) = x - \tfrac{1}{2},$$

according to Eq. 23 we have

$$dy\ (x, dx) = (x - \tfrac{1}{2})\, dx.$$

Thus, for $x = 1$ and $dx = \frac{1}{8}$,

$$dy\ (1, \tfrac{1}{8}) = \tfrac{1}{2} \cdot \tfrac{1}{8} = \tfrac{1}{16}. \tag{24}$$

Noting that $y_0 = f(1) = 1$, we can now find an approximate value for y at $x = \frac{9}{8}$, namely,

$$y = 1 + dy\ (1, \tfrac{1}{8}) = 1 + \tfrac{1}{16} = \tfrac{17}{16}.$$

Of course, this is the same result that we obtained in Example 1. ∎

PROBLEMS

In each of Problems 1 through 6, find an equation of the line tangent to the given curve at the given point.

1. $y = 4 - x^2$; $(1, 3)$
2. $y = \dfrac{x-1}{x+1}$; $(-2, 3)$
3. $y = \dfrac{6x}{2x^2 + 3x + 4}$; $(-1, -2)$
4. $y = \dfrac{2}{x} - \sqrt{x}$; $(1, 1)$
5. $y = x^2 - 3x + 4$; (x_0, y_0)
6. $y = \sqrt{x} + x$; (x_0, y_0) with $x_0 > 0$

In each of Problems 7 through 12, find the tangent line function $T(x; x_0)$ and the remainder function $r(x; x_0)$ for the given value of x_0. Refer to Eqs. 2 and 3 of the text.

7. $f(x) = 4x - 3$; $x_0 = 2$
8. $f(x) = 2x^2 - 3x + 1$; $x_0 = 1$
9. $f(x) = \frac{1}{3}x^2 + 2x - 4$; $x_0 = 0$
10. $f(x) = \dfrac{1}{x}$; $x_0 = 1$
11. $f(x) = \dfrac{x}{x+1}$; $x_0 = 2$
12. $f(x) = \sqrt{x}$; $x_0 = 4$

In each of Problems 13 through 16, use a suitable linear approximation to evaluate (approximately) the given number. Also determine the number of correct decimal places.

13. $(27.16)^{1/3}$
14. $(3.75)^{-1/2}$
15. $(15.82)^{1/4} + (16.12)^{-1/2}$
16. $(4.02)^3 - (3.97)^{3/2}$
17. Use a linear approximation to find approximately the volume of a sphere of radius 0.96 ft.
18. Use a linear approximation to find approximately the area of an equilateral triangle of side 2.06 in.
19. Use a linear approximation to find approximately the value of the function $f(x) = 3x^3 + 6x^2 + 7x - 8$ at $x = 1.01$.
20. Use a linear approximation to find approximately the value of the function $f(x) = \sqrt{x} + (1/\sqrt{x})$ at $x = 3.97$.

In each of Problems 21 through 28, determine dy/dx and d^2y/dx^2.

21. $y = x^4 + 2x^2 - 5$
22. $y = x^3 - 7x^2 + 3x - 8$
23. $y = \dfrac{x^2 - 1}{x^2 + 1}$
24. $y = \sqrt{x} + \dfrac{1}{\sqrt{x}}$, $x > 0$
25. $y = 3x^2 - x^{4/3} + \dfrac{1}{x+1}$, $x \neq -1$
26. $y = x(x - 1)(x + 2)$
27. $y = \dfrac{x+1}{x+2}$, $x \neq -2$
28. $y = (x - 1)^2$

* 29. Let f be continuous at x_0 and let f have a linear approximation near x_0 given by

$$f(x) = A(x_0) + B(x_0)(x - x_0) + r(x; x_0)(x - x_0), \quad (i)$$

where A and B are constants to be determined, and

$$\lim_{x \to x_0} r(x; x_0) = 0. \quad (ii)$$

(a) Let $x \to x_0$ in Eq. (i) and show that $A = f(x_0)$.

(b) Using the result of Part (a), show that

$$\frac{f(x) - f(x_0)}{x - x_0} = B + r(x; x_0). \quad (iii)$$

(c) By letting $x \to x_0$ in Eq. (iii), show that f is differentiable at x_0, and further that $B = f'(x_0)$.

Errors and Relative Errors

Often we want to evaluate a function f at a point x_0. Sometimes, however, due possibly to an error in measurement, we use an incorrect value x instead of x_0; that is, we evaluate $f(x)$ instead of $f(x_0)$, where $x = x_0 + \Delta x$. In this event the difference or increment

$$f(x) - f(x_0) = f(x_0 + \Delta x) - f(x_0) = \Delta f(x_0; \Delta x) \quad (i)$$

is the **error** in f due to the error Δx. The ratio $\Delta f(x_0; \Delta x)/f(x_0)$ is called the **relative error.**

To simplify the calculation it is common practice to replace $\Delta f(x_0; \Delta x)$ by $df(x_0, \Delta x)$, at least if Δx is small. Thus we obtain an approximate relative error

$$\text{Relative error} \cong \frac{df(x_0, \Delta x)}{f(x_0)} = \frac{f'(x_0)\,\Delta x}{f(x_0)}. \quad (ii)$$

Problems 30 through 32 are concerned with errors and relative errors.

30. A carpenter constructs a cubical box with edges of 3 ft with a maximum error of 0.02 ft in each edge.

(a) Find the maximum error in the surface area of the box.

(b) Find the maximum relative error in the surface area.

(c) Use a differential to find the approximate maximum relative error in the surface area.

31. Follow the instructions of Problem 30 for the volume of the box.

32. Suppose that it is desired to build a cubical box of 8 ft^3 capacity. An error of 0.25 ft^3 in the volume of the box can be tolerated.

(a) Determine the maximum error in an edge that can be made without exceeding the allowable error in the volume. Assume that all edges are the same length.

(b) Use a linear approximation to estimate the allowable error that can be tolerated in an edge. Assume that all edges are the same length.

3.5 COMPOSITE FUNCTIONS AND THE CHAIN RULE

For two given functions f and g we discussed in Section 3.2 how to express the derivative of their sum, difference, product, or quotient in terms of f, g, and their derivatives. We now consider a different way of combining two functions f and g, that is, we *substitute* one into the other, and thereby form what is called a **composite function.** We use the notation $f \circ g$ to denote the result of substituting g into f; thus

$$(f \circ g)(x) = f[g(x)]. \tag{1}$$

In order to be able to evaluate $f[g(x)]$ it is necessary that the value $g(x)$ be contained in the domain of f. Thus, to form $f \circ g$ it is necessary that the range of g be contained, at least in part, in the domain of f. The domain of $f \circ g$ consists of all points in the domain of g for which $g(x)$ is in the domain of f. Our main goal in this section is to show how to express the derivative of $f \circ g$ in terms of the derivatives of f and g.

EXAMPLE 1

If $f(x) = \sin x$ and $g(x) = x^2$, then

$$(f \circ g)(x) = \sin(x^2) \tag{2}$$

and the domain is $(-\infty, \infty)$. Also

$$(g \circ f)(x) = (\sin x)^2 = \sin^2 x \tag{3}$$

with domain $(-\infty, \infty)$. Observe that the functions $f \circ g$ and $g \circ f$ are not the same. ∎

EXAMPLE 2

If $f(x) = \sqrt{3x + 5}$ and $g(x) = 1 - 4x$, determine $f \circ g$.

First notice that the domain of g is $(-\infty, \infty)$ and the domain of f is $[-\frac{5}{3}, \infty)$. The composite function $f \circ g$ is given by

$$(f \circ g) = \sqrt{3(1 - 4x) + 5}$$
$$= \sqrt{8 - 12x}. \tag{4}$$

The domain of $f \circ g$ is determined from Eq. 4 by the requirement that $8 - 12x \geq 0$, which means that the domain is $(-\infty, \frac{2}{3}]$. You may verify that this is precisely the set of values of x for which $g(x)$ lies in the domain of f, that is, $1 - 4x \geq -\frac{5}{3}$. ∎

In general, $f \circ g$ and $g \circ f$ are different, as Example 1 shows. Indeed, the fact that it is possible to form $f \circ g$ implies nothing about the possibility of forming $g \circ f$. For example, suppose that

$$f(x) = 4 + \sin x, \qquad -\infty < x < \infty,$$
$$g(x) = \sqrt{1 - x^2}, \qquad -1 \leq x \leq 1.$$

Then

$$(f \circ g)(x) = 4 + \sin \sqrt{1 - x^2} \qquad \text{for} \quad -1 \leq x \leq 1.$$

On the other hand, the range of f is $[3, 5]$, and therefore it has no points in common with the domain of g. Hence $g \circ f$ does not exist.

In certain simple cases we can relate composite functions to a change in the scale of measurement. For instance, suppose that the position s of a particle moving on a straight line is given by

$$s = f(t), \tag{5}$$

where the time t is measured in minutes. If a new time coordinate τ (measured in seconds) is introduced, then t and τ are related by

$$t = \frac{\tau}{60} = g(\tau). \tag{6}$$

If we wish to express s in terms of τ we must write

$$s = f[g(\tau)] = f(\tau/60), \tag{7}$$

a composite function.

In cases such as this, it is reasonable to believe that the change in scale does not affect certain properties of the original function, such as continuity. Thus if the particle just considered moves continuously, we would not expect it to matter whether the time variable is measured in minutes or seconds. In fact, this preservation of continuity is a general property of composite functions, as the following theorem states.

Theorem 3.5.1

Let the function g be continuous at x_0 and let the function f be continuous at z_0, where $z_0 = g(x_0)$. Then the composite function $f \circ g$ is continuous at x_0. That is, if $F(x) = f[g(x)]$, then F is continuous at x_0.

In other words, *a continuous function of a continuous function is continuous.* The proof of Theorem 3.5.1 follows directly from the definition of continuous function, and is not particularly difficult. However, we will omit the proof of this theorem.

Next we take up the problem of differentiating a composite function, the main subject of this section. If $y = f(z)$ and $z = g(x)$, then a change Δx in x will produce a change Δz in z, and the latter will in turn produce a change Δy in y. The question is how to relate the change in y directly to the change in x.

As an example, consider again the moving particle described by Eq. 5. If s is measured in feet, then the velocity

$$\frac{ds}{dt} = f'(t) \tag{8}$$

is measured in feet per minute. To compute the velocity in feet per second we must multiply by a conversion factor corresponding to the change in the time scale; in dimensional terms

$$\frac{\text{feet}}{\text{second}} = \frac{\text{feet}}{\text{minute}} \cdot \frac{\text{minutes}}{\text{second}}. \tag{9}$$

Of course, the conversion factor is $\frac{1}{60}$, the number of minutes per second. Hence, if $ds/d\tau$ is the velocity in feet/second, then

$$\frac{ds}{d\tau} = \frac{ds}{dt} \cdot \frac{1}{60}, \tag{10}$$

or since $dt/d\tau = 1/60$ from Eq. 6,

$$\frac{ds}{d\tau} = \frac{ds}{dt} \cdot \frac{dt}{d\tau}. \tag{11}$$

Thus the derivative of the composite function of Eq. 7 is the product of the derivatives of its two constituents. The relation expressed by Eq. 11 is an example of what is known as the **chain rule** for the differentiation of composite functions.

The chain rule is one of the most important of all differentiation rules. It is a powerful and versatile tool that enables us to differentiate a great variety of functions obtained as compositions of simpler functions. Since it expresses the derivative of a composite function in terms of the derivatives of its constituents, the chain rule allows us to differentiate a composite function without even finding an expression for the composite function itself. This is particularly helpful when it is either impossible or inconvenient to find such an expression. We state the chain rule in the following theorem.

Theorem 3.5.2

(Chain Rule)

If the function $z = g(x)$ has the derivative $g'(x_0)$ at x_0, and the function $y = f(z)$ has the derivative $f'(z_0)$ at $z_0 = g(x_0)$, then the composite function

$$y = F(x) = (f \circ g)(x) = f[g(x)]$$

is differentiable at $x = x_0$, and

$$F'(x_0) = f'(z_0)g'(x_0) = f'[g(x_0)]g'(x_0). \tag{12}$$

In the D-notation Eq. 12 takes the form

$$D_x y = D_z y D_x z, \tag{13}$$

where the subscripts are used to emphasize the variable with respect to which each differentiation is carried out. In the Leibniz notation we have

$$\frac{dy}{dx} = \frac{dy}{dz}\frac{dz}{dx}, \tag{14}$$

which is probably easier to remember than either Eq. 12 or Eq. 13. In words, the chain rule says that if y is a function of z and z in turn is a function of x, then the rate of change of y with respect to x is the rate of change of y with respect to z times the rate of change of z with respect to x. It should be understood in Eq. 14 that dy/dx and dz/dx are evaluated at x_0, and that dy/dz is evaluated at the z_0 corresponding to x_0. A similar statement applies to Eq. 13. The statement (14) of the chain rule using the notation of Leibniz is very suggestive because of the apparent analogy with fractions. Remember, however, that actually there is no cancellation involved; the dz's are merely part of the expressions for the derivatives dy/dz and dz/dx, and have no separate existence of their own.

Before proving the chain rule, let us consider several additional examples in order to illustrate its usefulness in finding derivatives.

EXAMPLE 3

If $F(x) = (3x^2 + 5x - 7)^6$, determine $F'(x)$.

Let $f(z) = z^6$ and $z = g(x) = 3x^2 + 5x - 7$. Then $F(x) = f[g(x)]$. According to Eq. 12 we have

$$F'(x) = f'(z)g'(x)$$

where $z = 3x^2 + 5x - 7$. So

$$\begin{aligned} F'(x) &= 6z^5(6x + 5) \\ &= 6(3x^2 + 5x - 7)^5(6x + 5). \quad \blacksquare \end{aligned}$$

EXAMPLE 4

If $y = [(2x + 1)/(1 - x)]^{1/3}$, find dy/dx.

Let $z = (2x + 1)/(1 - x)$; then $y = z^{1/3}$, and according to Eq. 14 we have

$$\begin{aligned}\frac{dy}{dx} &= \frac{dy}{dz}\frac{dz}{dx} \\ &= \frac{d}{dz} z^{1/3} \frac{d}{dx}\left(\frac{2x+1}{1-x}\right) \\ &= \frac{1}{3} z^{-2/3} \frac{(1-x)2 - (2x+1)(-1)}{(1-x)^2} \\ &= \frac{1}{3}\left(\frac{2x+1}{1-x}\right)^{-2/3} \frac{3}{(1-x)^2} = \frac{1}{(2x+1)^{2/3}(1-x)^{4/3}}\end{aligned}$$

provided $x \neq -\frac{1}{2}$ and $x \neq 1$. ∎

The next result is a combination of the chain rule and the power rule; it includes Examples 3 and 4 as special cases.

Corollary

(Generalized power rule)

If $y = [u(x)]^r$, where u is a differentiable function and r is any real number, then

$$\frac{dy}{dx} = \frac{d}{dx}[u(x)]^r = r[u(x)]^{r-1}u'(x). \tag{15}$$

Equation 15 follows immediately from the chain rule. If we let $z = u(x)$, then $y = z^r$ and

$$\frac{dy}{dx} = rz^{r-1}\frac{dz}{dx},$$

which is the same as Eq. 15.

The preceding corollary illustrates one of the most important ways in which the chain rule is used; hereafter, whenever we derive a new differentiation rule, we will immediately extend it by combining it with the chain rule.

At the same time we also obtain the generalized antidifferentiation rule that corresponds to Eq. 15, namely

$$D^{-1}\{[u(x)]^{r-1}u'(x)\} = \frac{[u(x)]^r}{r} + c, \qquad r \neq 0$$

or as it is usually written,

$$D^{-1}\{[u(x)]^r u'(x)\} = \frac{[u(x)]^{r+1}}{r+1} + c, \qquad r \neq -1. \tag{16}$$

The next example illustrates the use of Eq. 16.

EXAMPLE 5

Find the antiderivative of

$$f(x) = 5x(x^2 + 3)^4.$$

If we let $u(x) = x^2 + 3$, then $u'(x) = 2x$, and

$$f(x) = \tfrac{5}{2}[u(x)]^4 u'(x).$$

Hence, according to Eq. 16,

$$D^{-1}f(x) = \frac{5}{2}\frac{[u(x)]^5}{5} + c = \frac{1}{2}(x^2 + 3)^5 + c. \blacksquare$$

The chain rule may be needed more than once in the calculation of higher derivatives.

EXAMPLE 6

Let $y = z - (1/z)$, where $z = 2x/(x + 1)$. Find dy/dx and d^2y/dx^2.

We have $dy/dz = 1 + z^{-2}$ and (from the quotient rule) $dz/dx = 2(x + 1)^{-2}$; thus, using the chain rule,

$$\frac{dy}{dx} = \frac{dy}{dz}\frac{dz}{dx} = 2(1 + z^{-2})(x + 1)^{-2}. \tag{17}$$

To find d^2y/dx^2 we start by using the product rule; thus

$$\frac{d^2y}{dx^2} = 2\left[\frac{d}{dx}(1 + z^{-2})\right](x + 1)^{-2} + 2(1 + z^{-2})\frac{d}{dx}(x + 1)^{-2}. \tag{18}$$

The derivative in the first term on the right side of Eq. 18 can be found by another application of the chain rule:

$$\begin{aligned}\frac{d}{dx}(1 + z^{-2}) &= \frac{d}{dz}(1 + z^{-2})\frac{dz}{dx} \\ &= -2z^{-3} \cdot 2(x + 1)^{-2} = -4z^{-3}(x + 1)^{-2},\end{aligned}$$

where we have used the expression found earlier for dz/dx. Substituting this result into Eq. 18 and using the generalized power rule to evaluate the derivative in the last term of Eq. 18, we finally obtain

$$\frac{d^2y}{dx^2} = -8z^{-3}(x + 1)^{-4} - 4(1 + z^{-2})(x + 1)^{-3}. \tag{19}$$

Of course, one can substitute for z in terms of x in Eqs. 17 and 19 so as to express dy/dx and d^2y/dx^2 entirely as functions of x if this is required. ∎

EXAMPLE 7

Suppose that $y = (x^2 + 3x + 4)^{4/3}$. Find dy/dx and d^2y/dx^2.

From the generalized power rule we obtain

$$\frac{dy}{dx} = \frac{4}{3}(x^2 + 3x + 4)^{1/3}(2x + 3). \tag{20}$$

To calculate the second derivative we first use the product rule

$$\frac{d^2y}{dx^2} = \frac{4}{3}(2x + 3)\frac{d}{dx}(x^2 + 3x + 4)^{1/3} + \frac{4}{3}(x^2 + 3x + 4)^{1/3}\frac{d}{dx}(2x + 3). \tag{21}$$

To find the derivative in the first term on the right side of Eq. 21 we use the generalized power rule a second time. Thus

$$\frac{d}{dx}(x^2 + 3x + 4)^{1/3} = \frac{1}{3}(x^2 + 3x + 4)^{-2/3}(2x + 3)$$

and, on substituting into Eq. 21, we have

$$\frac{d^2y}{dx^2} = \frac{4}{9}(x^2 + 3x + 4)^{-2/3}(2x + 3)^2 + \frac{8}{3}(x^2 + 3x + 4)^{1/3}. \; ∎ \tag{22}$$

We may need to use the chain rule repeatedly, even in the calculation of a first derivative. This is illustrated in the next example.

EXAMPLE 8

If $y = \sqrt{\sqrt{3x^2 + 4} + \sqrt{x}}$ for $x \geq 0$, determine dy/dx.

First let $z = \sqrt{3x^2 + 4} + \sqrt{x}$, then $y = \sqrt{z}$. Using the chain rule or the generalized power rule, we have

$$\begin{aligned}\frac{dy}{dx} &= \frac{1}{2}z^{-1/2}\frac{d}{dx}[\sqrt{3x^2 + 4} + \sqrt{x}] \\ &= \frac{1}{2\sqrt{\sqrt{3x^2 + 4} + \sqrt{x}}}\left[\frac{d}{dx}\sqrt{3x^2 + 4} + \frac{1}{2\sqrt{x}}\right].\end{aligned}$$

Again using the chain rule or generalized power rule to compute $d\sqrt{3x^2 + 4}/dx$ we obtain

$$\frac{dy}{dx} = \frac{1}{2\sqrt{\sqrt{3x^2 + 4} + \sqrt{x}}}\left[\frac{1}{2}\frac{6x}{\sqrt{3x^2 + 4}} + \frac{1}{2\sqrt{x}}\right]. \tag{23}$$

With a little practice and experience you will find that you can dispense with the intermediate step "let $z = \ldots$" and simply write the result. ∎

We now turn to a proof of the chain rule, using an argument based directly on Theorem 3.4.1.

Proof of Theorem 3.5.2. Referring to Eq. 11 of Section 3.4, recall that if f is differentiable at x, then

$$f(x + h) - f(x) = [f'(x) + r(h; x)]h, \tag{24}$$

where $r(h; x) \to 0$ as $h \to 0$.

Now suppose that $y = f(z)$ and $z = g(x)$. Then

$$y = (f \circ g)(x) = f[g(x)] = F(x).$$

To show that F is differentiable at x we need to consider the difference quotient

$$\frac{F(x + h) - F(x)}{h} = \frac{f[g(x + h)] - f[g(x)]}{h}. \tag{25}$$

Our strategy is to apply Theorem 3.4.1, in the form (24), to both f and g. In doing this, we must make the necessary changes in the names of the variables. Thus

$$g(x + h) - g(x) = [g'(x) + r_1(h; x)]h \tag{26}$$

and

$$f(z + k) - f(z) = [f'(z) + r_2(k; z)]k, \tag{27}$$

where $r_1(h; x) \to 0$ as $h \to 0$ and $r_2(k; z) \to 0$ as $k \to 0$. Further, we require that

$$k = g(x + h) - g(x). \tag{28}$$

so that

$$g(x + h) = g(x) + k = z + k. \tag{29}$$

Observe that $k \to 0$ as $h \to 0$ since g is differentiable and hence continuous at x. By substituting for k from Eq. 28 on the right side of Eq. 27 and then using Eq. 26 we obtain

$$f(z + k) - f(z) = [f'(z) + r_2(k; z)][g'(x) + r_1(h; x)]h. \tag{30}$$

Looking now at the left side of Eq. 30 we see that

$$f(z + k) - f(z) = f[g(x + h)] - f[g(x)] = F(x + h) - F(x). \tag{31}$$

Consequently, by combining Eqs. 30 and 31 we have

$$\frac{F(x+h)-F(x)}{h} = [f'(z) + r_2(k; z)][g'(x) + r_1(h; x)]. \qquad (32)$$

As $h \to 0$, we know that $r_1(h; x) \to 0$ and also that $k \to 0$. Thus $r_2(k; z) \to 0$ as well, and the right side of Eq. 32 has the limit $f'(z)g'(x)$. Hence the left side of Eq. 32 has this same limit. Therefore F is differentiable at the point x, and its derivative $F'(x)$ is given by

$$F'(x) = f'(z)g'(x), \qquad (33)$$

where $z = g(x)$. This completes the proof of the chain rule. □

PROBLEMS

In each of Problems 1 through 10, find $f \circ g$ and $g \circ f$, if they exist, and state their domains.

1. $f(x) = 3x - 2, \quad g(x) = 2x + 7$
2. $f(x) = 2x - 4, \quad g(x) = 5x + 3$
3. $f(x) = 1 - x^2, \quad g(x) = \sqrt{x+1}$
4. $f(x) = \sqrt{1 - x^2}, \quad g(x) = x + 1$
5. $f(x) = \sqrt{x}, \quad g(x) = x^2 - 3x + 2$
6. $f(x) = \sqrt{x}, \quad g(x) = x^2 - 3x + 6$
7. $f(x) = \sqrt{x}, \quad g(x) = -(x^2 - 3x + 6)$
8. $f(x) = \sqrt{x}, \quad g(x) = \dfrac{x+1}{x-1}$
9. $f(x) = \cos 2x, \quad g(x) = \sqrt{x^2 - 4}$
10. $f(x) = \sqrt{x^2 - 4},$
    $g(x) = \sqrt{-(x+3)(x+1)}$

In each of Problems 11 through 16, use the chain rule to find dy/dx. Verify that your answer is correct by first carrying out the indicated substitution and then differentiating.

11. $y = z^2 + 2z - 5, \quad z = x^2 - 7$
12. $y = 3z^2 - 2z + 4, \quad z = 2x^2 + 5x - 3$
13. $y = \dfrac{z^2 - 1}{z^2 + 1}, \quad z = 3x - 2$
14. $y = \dfrac{3z - 5}{z^2 + 4}, \quad z = x^2 - 3$
15. $y = z^2 - 3z + 4, \quad z = \dfrac{x}{x+1}$
16. $y = \dfrac{1}{z - 2}, \quad z = \dfrac{x+1}{x-1}$

In each of Problems 17 through 22, use the chain rule to find dy/dx, and state the values of x for which your answer is valid.

17. $y = u^3 - 6, \quad u = \sqrt{x+1}$
18. $y = s^{3/2}, \quad s = 1 - x$
19. $y = z^{7/3}, \quad z = x^2 + 5$
20. $y = \sqrt{1 - u^2}, \quad u = x + 1$
21. $y = 1 - z^2, \quad z = \sqrt{x+4}$
22. $y = \sqrt{w}, \quad w = \dfrac{x+1}{x-1}$

In each of Problems 23 through 26, find the value of $D(f \circ g)(x)$ at the given value of x.

23. $f(z) = z^2 + 2z - 5, \quad g(x) = x^2 - 7;$
    $x = 2$
24. $f(z) = \sqrt{z}, \quad g(x) = x^2 - 4; \quad x = 3$
25. $f(z) = \sqrt{z^2 - 16}, \quad g(x) = x^2 + 2x + 2;$
    $x = 1$
26. $f(z) = z^2 - 3z + 6, \quad g(x) = \sqrt{9 - x^2};$
    $x = -2$

In each of Problems 27 through 34, find the indicated derivative.

27. $\dfrac{d}{dx}(2x - 1)^3$
28. $\dfrac{d}{dx}\sqrt{x^3 + 1}$
29. $\dfrac{d}{dx}\left(\dfrac{x+1}{2}\right)^{100}$
30. $\dfrac{d}{dx}(x^2 + 1)^3(2x - 3)^4$
31. $\dfrac{d}{dx}(x - 1)^4\sqrt{2x - 3}$
32. $\dfrac{d}{dx}\left(\dfrac{x+2}{x-3}\right)^4$
33. $\dfrac{d}{dx}\left(\dfrac{2x - 1}{x}\right)^{3/2}$
34. $\dfrac{d}{dx}\sqrt{x + \sqrt{x + \sqrt{x}}}$

In each of Problems 35 through 40, find d^2y/dx^2.

35. $y = z^2 + 2z - 5, \quad z = x^2 - 7$

36. $y = z^3 - 5z, \quad z = x^2 - 3$

37. $y = \sqrt{x^3 + 1}$ **38.** $y = \left(\frac{x+2}{x-3}\right)^4$

39. $y = \frac{x^2 - 1}{x^2 + 1}$ **40.** $y = \frac{1}{z - 2}, \quad z = \frac{x+1}{x-1}$

In each of Problems 41 through 46, find the indicated derivative. Assume that the function f has as many derivatives as required.

41. $\frac{d}{dx} f(-x)$ **42.** $\frac{d}{dx} f(x^2)$

43. $\frac{d}{dx} f(ax), \quad a = \text{constant}$

44. $\frac{d}{dx} f\left(\frac{1}{x}\right), \quad x \neq 0$

45. $\frac{d^n}{dx^n} f(ax), \quad a = \text{constant}$ **46.** $\frac{d^2}{dx^2} f(x^2)$

In each of Problems 47 through 50, find the indicated derivative. Assume that u has as many derivatives as required.

47. $\frac{d}{dx} [u'(x)]^2$ **48.** $\frac{d^2}{dx^2} [u'(x)]^2$

49. $\frac{d}{dx} ([u'(x)]^2 u''(x))$ **50.** $\frac{d}{dx} [u(x)]^2(1 + [u'(x)]^2)$

In each of Problems 51 through 58, use Eq. 16 to obtain the indicated antiderivative. Check your answer by differentiating it.

51. $D^{-1} (x - 2)^4$ **52.** $D^{-1} (x + 3)^{-6}$

53. $D^{-1} 2(2x - 1)^5$ **54.** $D^{-1} (3x + 6)^4$

55. $D^{-1} (2x + 1)^{4/3}$ **56.** $D^{-1} x(x^2 + 1)^{4/5}$

57. $D^{-1} 3x\sqrt{x^2 + 4}$ **58.** $D^{-1} x^2(x^3 + 1)^{3/2}$

59. Suppose that f, g, and h are differentiable functions, and that $F(x) = f\{g[h(x)]\}$ exists in some domain D. Find $F'(x)$ in terms of derivatives of f, g, and h.

60. Suppose that f, g, and h are differentiable functions such that the range of g and the range of h are both contained in the domain of f, so that the function $f \circ g + f \circ h$ exists. Find an example to show that $f \circ (g + h)$ may not exist.

61. If $y = f(x)/g(x)$, where $g(x) \neq 0$, find y''. Assume that the functions f and g are twice differentiable.

62. Prove that the derivative of an even function is odd and that the derivative of an odd function is even.

63. The radius of a spherical snowball is given by $r = 7 - 2t$ for $0 \leq t \leq \frac{7}{2}$. Find an expression for the rate of change of the volume of the snowball with respect to t.

64. If the length of the side of an equilateral triangle is given by $s = t^2/3$, find an expression for the rate of change of the area of the triangle with respect to time.

65. The edge of a cube is increasing at a rate of 2 in./sec. Find the rate at which the volume of the cube is increasing when its edge is 6 in. long.

66. Two bicyclists start from the same point at noon. One travels due east at 10 mi/hr, and the other travels due north at 15 mi/hr. Find the rate at which the distance between them is increasing at 1 P.M.

67. In Einstein's special theory of relativity the relation $F = ma = m(dv/dt)$ expressing Newton's law of motion is replaced by

$$F = m_0 \frac{d}{dt} \frac{v}{\sqrt{1 - (v^2/c^2)}}, \qquad (i)$$

where m_0 is the rest mass of the moving particle, v is its velocity, and $c = 3 \times 10^{10}$ cm/sec is the velocity of light. By carrying out the differentiation indicated in Eq. (i), show that

$$F = \frac{m_0 a}{[1 - (v^2/c^2)]^{3/2}}.$$

Thus, the larger the velocity v, the greater is the force F required to produce a given acceleration a. It has been verified experimentally that Eq. (i) accurately describes the motion of particles moving at speeds comparable to the speed of light, whereas Newton's law does not. Newton's law is much simpler, however, and is almost as accurate provided that v/c is small compared to one.

68. Consider a particle moving on a straight line in accord with Einstein's law of motion (see Problem 67)

$$F = m_0 \frac{d}{dt} \frac{v}{\sqrt{1 - (v^2/c^2)}}. \qquad (i)$$

(a) If $F = F_0$, where F_0 is constant, and if $v(0) = 0$, show that

$$v = \frac{cF_0 t}{\sqrt{m_0^2 c^2 + F_0^2 t^2}}. \qquad (ii)$$

(b) If v is given by Eq. (ii), find the limiting velocity that the particle approaches as $t \to \infty$.

(c) Now suppose that, instead of Eq. (i), the particle is governed by Newton's law of motion

$$F = m_0 \frac{dv}{dt}$$

Under the same conditions as in Part (a), find v as a function of t. Compare $\lim_{t\to\infty} v$ with the result found in Part (b).

3.6 DERIVATIVES OF TRIGONOMETRIC FUNCTIONS

We will now derive differentiation formulas for the six trigonometric functions, using some of the properties of derivatives established in Sections 3.2 and 3.5, as well as the definition of the derivative. Remember that in all discussions involving the trigonometric functions radian measure is used (unless otherwise specifically indicated).

Let us first calculate the derivative of the sine function. If $f(x) = \sin x$, then the corresponding difference quotient is

$$\frac{f(x+h) - f(x)}{h} = \frac{\sin(x+h) - \sin x}{h}. \qquad (1)$$

Using the identity

$$\sin(x+h) = \sin x \cos h + \cos x \sin h \qquad (2)$$

we can rewrite Eq. 1 in the form

$$\frac{f(x+h) - f(x)}{h} = \cos x \frac{\sin h}{h} - \sin x \frac{1 - \cos h}{h}.$$

Thus

$$f'(x) = \lim_{h\to 0} \frac{f(x+h) - f(x)}{h}$$

$$= \cos x \lim_{h\to 0} \frac{\sin h}{h} - \sin x \lim_{h\to 0} \frac{1 - \cos h}{h}, \qquad (3)$$

so to determine $f'(x)$ we must find the limits of $(\sin h)/h$ and $(1 - \cos h)/h$, respectively, as $h \to 0$. From Example 9 of Section 2.3 we know that

$$\lim_{h\to 0} \frac{\sin h}{h} = 1. \qquad (4)$$

Further, we can write

$$\frac{1 - \cos h}{h} = \frac{1 - \cos h}{h} \cdot \frac{1 + \cos h}{1 + \cos h}$$

$$= \frac{1 - \cos^2 h}{h(1 + \cos h)}$$

$$= \frac{\sin^2 h}{h(1 + \cos h)}$$

$$= \frac{\sin h}{h} \frac{\sin h}{1 + \cos h}.$$

Thus, using the algebraic properties of limits (Theorem 2.3.1), we conclude that

$$\lim_{h \to 0} \frac{1 - \cos h}{h} = \lim_{h \to 0} \frac{\sin h}{h} \lim_{h \to 0} \frac{\sin h}{1 + \cos h}$$

$$= 1 \cdot 0 = 0. \tag{5}$$

Substituting the results (4) and (5) into Eq. 3, we obtain

$$f'(x) = 1 \cdot \cos x - 0 \cdot \sin x = \cos x. \tag{6}$$

In different notation we have shown that

$$\frac{d}{dx} \sin x = \cos x \qquad \text{or} \qquad D \sin x = \cos x. \tag{7}$$

The differentiation formulas (7) can be generalized immediately by means of the chain rule. Suppose that $y = \sin u$, where u is a differentiable function of x. Then by the chain rule

$$\frac{d}{dx}(\sin u) = \frac{d}{du}(\sin u) \frac{du}{dx}$$

$$= (\cos u) \frac{du}{dx}. \tag{8}$$

EXAMPLE 1

Find dy/dx if $y = \sin(\omega x + \delta)$, where ω and δ are constants.

We write $y = \sin u$, where $u = \omega x + \delta$. Then by the chain rule

$$\frac{dy}{dx} = \frac{dy}{du}\frac{du}{dx} = \frac{d}{du}(\sin u) \frac{d}{dx}(\omega x + \delta) = (\cos u)(\omega)$$

$$= \omega \cos(\omega x + \delta). \tag{9a}$$

In different notation

$$D \sin(\omega x + \delta) = \omega \cos(\omega x + \delta). \; \blacksquare \tag{9b}$$

EXAMPLE 2

Find the rate of change of $y = \sin(x^2 - \pi)$ at the point $x = \sqrt{\pi}$. Also find an equation of the line tangent to the graph of this function at this point.

The required rate of change is given by the derivative dy/dx. According to the chain rule, with $u = x^2 - \pi$, we have

$$\frac{dy}{dx} = \frac{d}{du}(\sin u)\frac{d}{dx}(x^2 - \pi) = (\cos u)(2x)$$

$$= 2x\cos(x^2 - \pi).$$

Evaluating dy/dx at $x = \sqrt{\pi}$ yields $2\sqrt{\pi}\cos(0) = 2\sqrt{\pi}$.

The desired tangent line then has the slope $2\sqrt{\pi}$ and passes through the point with coordinates $x = \sqrt{\pi}$, $y = 0$. Thus an equation of the tangent line is

$$y = 2\sqrt{\pi}(x - \sqrt{\pi}). \blacksquare \tag{10}$$

Next we wish to derive a formula for the derivative of the cosine function. The formula in question can be found in much the same way that Eq. 6 was derived, but we will present an alternative derivation here. Starting from

$$\cos x = \sin\left(\frac{\pi}{2} - x\right),$$

we have, by the chain rule,

$$\frac{d}{dx}\cos x = \frac{d}{dx}\sin\left(\frac{\pi}{2} - x\right)$$

$$= \cos\left(\frac{\pi}{2} - x\right)\cdot(-1)$$

$$= -\sin x,$$

since $\cos(\pi/2 - x) = \sin x$. In other notation,

$$\frac{d}{dx}\cos x = -\sin x \qquad \text{or} \qquad D\cos x = -\sin x. \tag{11}$$

Equation 11 can also be generalized at once by means of the chain rule. If $y = \cos u$, where u is a differentiable function of x, then

$$\frac{d}{dx}(\cos u) = \frac{d}{du}(\cos u)\frac{du}{dx}$$

$$= -\sin u\,\frac{du}{dx}. \tag{12}$$

The derivatives of the other four trigonometric functions can be quickly obtained by combining Eqs. 7 and 11 with the quotient rule. For instance,

$$
\begin{aligned}
\frac{d}{dx}\tan x &= \frac{d}{dx}\frac{\sin x}{\cos x} \\
&= \frac{\cos x(d/dx)(\sin x) - \sin x(d/dx)(\cos x)}{\cos^2 x} \\
&= \frac{(\cos x)(\cos x) - (\sin x)(-\sin x)}{\cos^2 x} \\
&= \frac{\cos^2 x + \sin^2 x}{\cos^2 x} \\
&= \frac{1}{\cos^2 x} \\
&= \sec^2 x. \qquad (13)
\end{aligned}
$$

Similarly,

$$
\begin{aligned}
\frac{d}{dx}\sec x &= \frac{d}{dx}\frac{1}{\cos x} \\
&= \frac{\cos x(d/dx)(1) - 1(d/dx)(\cos x)}{\cos^2 x} \\
&= \frac{(\cos x)(0) - (1)(-\sin x)}{\cos^2 x} \\
&= \frac{\sin x}{\cos^2 x} \\
&= \sec x \tan x. \qquad (14)
\end{aligned}
$$

In the same way we obtain the following differentiation formulas for the cotangent and cosecant.

$$\frac{d}{dx}\cot x = -\csc^2 x \qquad (15)$$

$$\frac{d}{dx}\csc x = -\csc x \cot x. \qquad (16)$$

The chain rule can be used to extend the differentiation formulas (13) to (16) in the same way that it was used earlier to generalize Eqs. 7 and 11. If u is a differentiable function of x, then the basic differentiation formulas for the trigonometric functions can be summarized as follows.

$$\frac{d}{dx}\sin u = \cos u \frac{du}{dx} \qquad (17a)$$

$$\frac{d}{dx}\cos u = -\sin u \frac{du}{dx} \qquad (17b)$$

$$\frac{d}{dx} \tan u = \sec^2 u \frac{du}{dx} \tag{17c}$$

$$\frac{d}{dx} \sec u = \sec u \tan u \frac{du}{dx} \tag{17d}$$

$$\frac{d}{dx} \cot u = -\csc^2 u \frac{du}{dx} \tag{17e}$$

$$\frac{d}{dx} \csc u = -\csc u \cot u \frac{du}{dx} \tag{17f}$$

Of course, it is understood that Eqs. 17(c) to 17(f) are valid only at points where all of the functions that appear are defined. For instance, we cannot use Eq. 17(c) to differentiate $\tan(x^2)$ at $x = \sqrt{\pi/2}$, because the tangent function is not defined at $\pi/2$.

The following examples illustrate the use of some of the preceding differentiation formulas.

EXAMPLE 3

If $y = x^2 + 3 \tan x$, find d^2y/dx^2.

From Eq. 13 we find that

$$\frac{dy}{dx} = 2x + 3 \sec^2 x.$$

Then, differentiating a second time, we have

$$\frac{d^2y}{dx^2} = 2 + 3 \frac{d}{dx} (\sec^2 x)$$

Now we use the generalized power rule from Section 3.5 to obtain

$$\frac{d^2y}{dx^2} = 2 + 3(2 \sec x) \frac{d}{dx} \sec x$$

$$= 2 + 6 \sec^2 x \tan x. \tag{18}$$ ∎

EXAMPLE 4

If $y = x^2 \sec 2x$, find dy/dx.

Using the product rule, we obtain

$$\frac{dy}{dx} = 2x \sec 2x + x^2 \frac{d}{dx} \sec 2x.$$

If we let $u = 2x$, then $\sec 2x = \sec u$, and Eq. 17(d) yields

$$\begin{aligned} \frac{d}{dx} \sec 2x &= \frac{d}{du} (\sec u) \frac{du}{dx} \\ &= (\sec u \tan u)(2) \\ &= 2 \sec 2x \tan 2x. \end{aligned}$$

Hence

$$\frac{dy}{dx} = 2x \sec 2x + 2x^2 \sec 2x \tan 2x. \tag{19}$$

Of course, Eq. 19 is not valid when $2x = \pm\pi/2, \pm 3\pi/2, \ldots$. ∎

EXAMPLE 5

If $y = \cos^2 \sqrt{1 + x^2}$, find dy/dx.

Several applications of the chain rule are required in this case. First we write $y = u^2$, where $u = \cos \sqrt{1 + x^2}$. Then

$$\begin{aligned} \frac{dy}{dx} = \frac{d}{du} (u^2) \frac{du}{dx} &= 2u \frac{d}{dx} \cos \sqrt{1 + x^2} \\ &= 2 \cos \sqrt{1 + x^2} \frac{d}{dx} \cos \sqrt{1 + x^2}. \end{aligned} \tag{20}$$

In order to compute $d(\cos \sqrt{1 + x^2})/dx$, we let $v = \sqrt{1 + x^2}$. Then

$$\begin{aligned} \frac{d}{dx} \cos \sqrt{1 + x^2} &= \frac{d}{dv} (\cos v) \frac{dv}{dx} \\ &= (-\sin v) \frac{d}{dx} \sqrt{1 + x^2} \\ &= -\sin \sqrt{1 + x^2} \frac{d}{dx} \sqrt{1 + x^2}. \end{aligned} \tag{21}$$

Next, to determine $d\sqrt{1 + x^2}/dx$, we let $w = 1 + x^2$. Then the generalized power rule yields

$$\begin{aligned} \frac{d}{dx} \sqrt{1 + x^2} &= \frac{d}{dw} \sqrt{w} \frac{dw}{dx} \\ &= \frac{1}{2\sqrt{w}} \cdot 2x = \frac{x}{\sqrt{1 + x^2}}. \end{aligned} \tag{22}$$

Finally, upon substituting from Eqs. 21 and 22 into (20), we find that

$$\frac{dy}{dx} = -\frac{2x}{\sqrt{1 + x^2}} \sin \sqrt{1 + x^2} \cos \sqrt{1 + x^2}. \ ∎ \tag{23}$$

From the foregoing examples it should be clear that derivatives of very complicated combinations of algebraic and trigonometric functions can be computed by using the formulas derived in this section in conjunction with the chain rule (Section 3.5) and the algebraic properties of derivatives (Section 3.2). We have shown these calculations in considerable detail, particularly by writing out explicitly the substitutions involved in the use of the chain rule. With practice it is usually possible to carry out these substitutions mentally, and thereby obtain such results more quickly. However, whenever there is any danger of confusion, it is advisable to work out the calculations in detail, as we have done here.

In the next example the calculations are shown more compactly.

EXAMPLE 6

If $y = (\sin x^2)\cot[3x - (\pi/4)]$, find dy/dx.

Using the product rule first and then the chain rule, we obtain

$$\begin{aligned}\frac{dy}{dx} &= \left[\frac{d}{dx}(\sin x^2)\right]\cot\left(3x - \frac{\pi}{4}\right) + (\sin x^2)\frac{d}{dx}\cot\left(3x - \frac{\pi}{4}\right) \\ &= (\cos x^2)(2x)\cot\left(3x - \frac{\pi}{4}\right) + (\sin x^2)\left[-\csc^2\left(3x - \frac{\pi}{4}\right)\right](3) \\ &= 2x(\cos x^2)\cot\left(3x - \frac{\pi}{4}\right) - 3(\sin x^2)\csc^2\left(3x - \frac{\pi}{4}\right). \quad ■\end{aligned}$$

Because of the particularly simple form of the differentiation formulas for $\sin x$ and $\cos x$ it follows that it is easy to find higher derivatives of these functions. For instance, the first four derivatives of $\sin x$ are given by

$$\begin{aligned} &D \sin x = \cos x, &&D^2 \sin x = -\sin x, \\ &D^3 \sin x = -\cos x, &&D^4 \sin x = \sin x. \end{aligned} \tag{24}$$

Since $D^4 \sin x = \sin x$, the next four derivatives follow the same pattern, and so on. The derivatives of the cosine function exhibit essentially the same behavior:

$$\begin{aligned} &D \cos x = -\sin x, &&D^2 \cos x = -\cos x, \\ &D^3 \cos x = \sin x, &&D^4 \cos x = \cos x, \ldots . \end{aligned} \tag{25}$$

Thus the sine and cosine functions have the interesting property that they are *closed* under differentiation: repeated differentiation of $\sin x$ and $\cos x$ yields only sines and cosines.

Antiderivatives of trigonometric functions

As we pointed out in Section 3.3, each differentiation formula can also be used in reverse: at the same time it gives the derivative of one function and the antiderivative of another. Upon rewriting the differentiation formulas for the six trigonometric functions as antidifferentiation formulas, we obtain

$$D^{-1}(\cos x) = \sin x + c, \tag{26a}$$

$$D^{-1}(\sin x) = -\cos x + c, \tag{26b}$$

$$D^{-1}(\sec^2 x) = \tan x + c, \tag{26c}$$

$$D^{-1}(\sec x \tan x) = \sec x + c, \tag{26d}$$

$$D^{-1}(\csc^2 x) = -\cot x + c, \tag{26e}$$

$$D^{-1}(\csc x \cot x) = -\csc x + c. \tag{26f}$$

Equations 26 can be generalized by combining them with the chain rule. For instance, from Eq. 17(a) we have

$$D^{-1}[\cos u(x)]u'(x) = \sin u(x) + c, \tag{27}$$

from Eq. 17(b),

$$D^{-1}[\sin u(x)]u'(x) = -\cos u(x) + c, \tag{28}$$

and similarly for the other parts of Eqs. 17. An important special case occurs if $u(x) = ax$, where a is a nonzero constant; then Eq. 27 reduces to

$$D^{-1}a \cos ax = \sin ax + c$$

or

$$D^{-1} \cos ax = \frac{\sin ax}{a} + c. \tag{29}$$

In the same way

$$D^{-1} \sin ax = -\frac{\cos ax}{a} + c. \tag{30}$$

EXAMPLE 7

Find the function whose derivative is $\cos 2x$, and whose graph contains the point $(\pi/4, 1)$.

According to Eq. 29

$$y = D^{-1} \cos 2x = \tfrac{1}{2} \sin 2x + c, \tag{31}$$

where c is a constant to be determined. To find c we substitute $x = \pi/4$ and $y = 1$ into Eq. 31:

$$1 = \tfrac{1}{2} + c.$$

Hence $c = \frac{1}{2}$ and

$$y = \tfrac{1}{2} \sin 2x + \tfrac{1}{2} \tag{32}$$

gives the required function. ∎

EXAMPLE 8

Find the antiderivative of $x \sin(3x^2)$.

If we let $u(x) = 3x^2$, then $u'(x) = 6x$, and

$$x \sin(3x^2) = \tfrac{1}{6}u'(x)\sin u(x).$$

From Eq. 28 we then have

$$D^{-1}x \sin(3x^2) = -\tfrac{1}{6}\cos u(x) + c = -\tfrac{1}{6}\cos(3x^2) + c.$$

Note that the important feature of this example is that the quantity multiplying $\sin(3x^2)$ is (apart from a constant factor) the derivative of $3x^2$. Consequently, once $u(x)$ is identified, the problem reduces immediately to Eq. 28. ∎

PROBLEMS

In each of Problems 1 through 18, find the derivative of the given function.

1. $\cos 4x$

2. $\sin \dfrac{x}{3}$

3. $\sin 4x - 3\cos 2x$

4. $\sin x \cos 2x$

5. $x^2 \sin 2x$

6. $\cot x \csc x$

7. $\sin^2 2x$

8. $\dfrac{x}{1 + \cos 2x}$

9. $x(\sec x - \tan x)$

10. $(1 + \cos^2 2x)^{1/2}$

11. $\dfrac{1 + \sin x}{x - \cos x}$

12. $\left(\dfrac{\sin x}{1 + \cos x}\right)^2$

13. $\sqrt{\dfrac{1 - \cos x}{1 + \cos x}}$

14. $\sin\sqrt{1 - x^2}$

15. $\sin(\cos x)$

16. $\cos^2(\sin^2 x^2)$

17. $\tan(\sec^2 x)$

18. $\cot(\sin 2x)$

In each of Problems 19 through 34, find an antiderivative of the given function. In some cases the problem may be simplified by the use of suitable trigonometric identities.

19. $\sin 2x$

20. $2 \sin x$

21. $2\cos 3x$

22. $3\cos 2x$

23. $\sin\left(x + \dfrac{\pi}{2}\right)$

24. $\cos\left(\omega x - \dfrac{\pi}{4}\right)$, $\omega \neq 0$

25. $\sec \dfrac{x}{2} \tan \dfrac{x}{2}$

26. $\csc^2 \pi x$

27. $\sin 2x \cos 2x$

28. $\cos^2 x - \sin^2 x$

29. $\cos^2 x$

30. $\sin^2 x$

31. $x \sin x^2$

32. $\dfrac{\cos\sqrt{x}}{\sqrt{x}}$

33. $(x - \pi)\cos(x - \pi)^2$

34. $x^2 \sin x^3$

In each of Problems 35 through 44, find the indicated derivative.

35. $D^9 \sin x$

36. $D^{17} \cos 2x$

37. $D^{11} \sin \dfrac{x}{2}$

38. $D^{326} \cos x$

39. $\dfrac{d^2}{dx^2} \tan ax$, $a \neq 0$

40. $\dfrac{d^3}{dx^3} \tan ax$, $a \neq 0$

41. $D^2 \sec x$

42. $D^3 \sec x$

43. $D^2 \cot 2x$

44. $D^2 \csc 3x$

45. Find an equation of the line tangent to the graph of $y = 3 + 4\cos 2x$ at the point $(\pi/6, 5)$.

46. Find an equation of the line tangent to the graph of $y = \tan 2x$ at the point $(\pi/8, 1)$.

47. The position of a moving particle is given by

$$s = 2 \sin 3t - 3 \cos 2t.$$

Find the velocity and acceleration of the particle at any time t.

48. The position of a moving particle is given by

$$s = t \sin 2t.$$

Find the velocity and acceleration of the particle at any time t.

49. Find the function whose derivative is $\sin 2x - 3\cos x$ and whose graph contains the point $(\pi/4, -1)$.

50. Find the function whose graph contains the point $(\pi/4, 4)$ and whose slope is always given by $\cos 2x + \sec^2 x$.

51. Let f be defined by

$$f(x) = \begin{cases} ax + b, & x \leq \dfrac{\pi}{4}; \\ \cos x, & x > \dfrac{\pi}{4}. \end{cases}$$

Determine a and b so that f is differentiable at $x = \pi/4$, and find $f'(\pi/4)$.

52. Let f be defined by

$$f(x) = \begin{cases} 1 + a\cos x, & x \leq \dfrac{\pi}{3}; \\ b + \sin\left(\dfrac{x}{2}\right), & x > \dfrac{\pi}{3}. \end{cases}$$

Determine a and b so that f is differentiable at $x = \pi/3$, and find $f'(\pi/3)$.

53. By considering the appropriate difference quotient, find $D \cos x$.

54. Derive the formulas for $D \cot x$ and $D \csc x$.

In each of Problems 55 through 58, use a linear approximation as in Section 3.4 to estimate the given quantity. Also determine the number of decimal places that are correct in the approximation.

55. $\sin(1.6)$

56. $\cos(0.8)$

57. $\tan(1.0)$

58. $\sin(-0.5)$

59. A hiker starts from a point P and walks around a circular lake of radius 2 mi at a constant rate of 3 mi/hr (see Figure 3.6.1). If the current position of the hiker is R, how fast is the length of the chord PR changing when $\theta = \pi/3$?

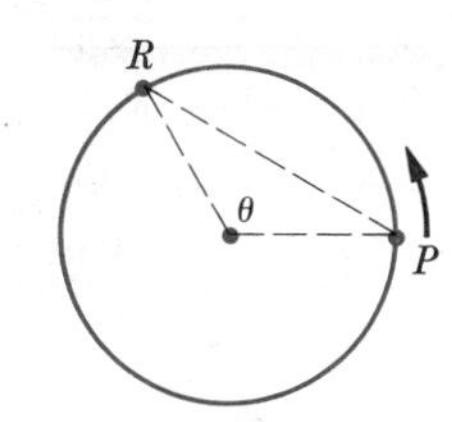

Figure 3.6.1

60. One end of a straight rod of length l moves on a circle of radius r while the other is constrained to move on a straight line through the center of the circle (see Figure 3.6.2). Find a relation between the angular velocity $(d\theta/dt)$ of the point P and the rectilinear velocity (dx/dt) of the point Q.

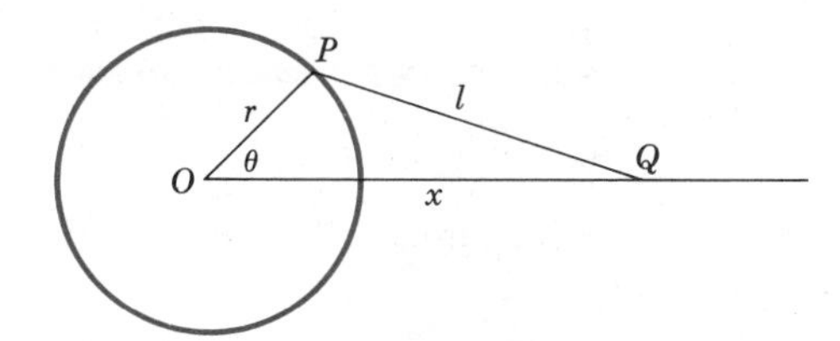

Figure 3.6.2

* 61. Proceeding directly from the difference quotient, show that the function

$$f(x) = \begin{cases} x \sin \dfrac{1}{x}, & x \neq 0 \\ 0, & x = 0 \end{cases}$$

is not differentiable at $x = 0$.

* 62. Using the difference quotient, show that the function

$$g(x) = \begin{cases} x^2 \sin \dfrac{1}{x}, & x \neq 0 \\ 0, & x = 0 \end{cases}$$

is differentiable at $x = 0$, and evaluate $g'(0)$.

3.7 IMPLICIT DIFFERENTIATION

In this chapter we have developed several results that enable us to compute derivatives of many important functions. In each case the function f whose derivative we sought was defined by an explicit formula, such as $f(x) = x^2 + 4$, $f(x) = x \sin x$, and so on. In this section we consider a method, known as implicit differentiation, by which we can sometimes compute the derivative of a function without having an explicit formula for the function.

Suppose that we know, or are willing to assume, that there is a differentiable function f such that if $y = f(x)$, then x and y satisfy an equation of the form

$$F(x, y) = 0. \tag{1}$$

However, suppose also that it is impractical, or impossible, to find a formula for $f(x)$. Can we nevertheless obtain an expression for $f'(x)$? Implicit differentiation is a means for dealing with problems of this type. We first illustrate it by two examples.

EXAMPLE 1

Suppose that there is a differentiable function $y = f(x)$ such that x and y satisfy the equation

$$y^3 + 2y - \sin \pi x - 3 = 0, \tag{2}$$

and such that $f(2) = 1$. Find the value of $f'(2)$.

Note first that the point (2, 1) does indeed satisfy Eq. 2.

$$(1)^3 + 2(1) - \sin 2\pi - 3 = 1 + 2 - 0 - 3 = 0.$$

Without first determining a formula for $f(x)$ we can find the derivative $f'(2)$ by differentiating both sides of Eq. 2. Since y is assumed to be a differentiable function of x, the derivative of y^3 with respect to x (by the chain rule) is $3y^2y'$. Thus, upon differentiating each term in Eq. 2, we obtain

$$3y^2y' + 2y' - \pi \cos \pi x = 0, \tag{3}$$

so that

$$y' = \frac{\pi \cos \pi x}{2 + 3y^2}. \tag{4}$$

Alternatively, we can write

$$f'(x) = \frac{\pi \cos \pi x}{2 + 3[f(x)]^2}. \tag{5}$$

Substituting $x = 2$ into Eq. 5 and noting that $f(2) = 1$, we find that

$$f'(2) = \frac{\pi \cos 2\pi}{2 + 3(1)^2} = \frac{\pi}{5}. \quad \blacksquare$$

EXAMPLE 2

Find the slope of the line tangent to the graph of

$$2x^2y^2 + y^3 \cos \pi x - 1 = 0 \tag{6}$$

at the point (1, 1).

First we observe that the point (1, 1) does lie on the graph. Then we *assume* that there is a differentiable function f such that $y = f(x)$ satisfies Eq. 6 and such that $f(1) = 1$. Then $f'(1)$ is the required slope. To find $y' = f'(x)$ we can proceed as in the preceding example, and differentiate each term in Eq. 6 with respect to x. In this case the product rule as well as the chain rule is needed, since each term on the left side of Eq. 6 is a product of two functions of x. Upon differentiating Eq. 6 we obtain

$$y^2 \frac{d}{dx}(2x^2) + 2x^2 \frac{d}{dx}(y^2) + \cos \pi x \frac{d}{dx}(y^3) + y^3 \frac{d}{dx}(\cos \pi x) - \frac{d}{dx}(1) = 0,$$

or

$$4xy^2 + 2x^2\, 2y \frac{dy}{dx} + \cos \pi x\, 3y^2 \frac{dy}{dx} + y^3(-\sin \pi x)\pi = 0,$$

or

$$(4x^2y + 3y^2 \cos \pi x)\frac{dy}{dx} + 4xy^2 - \pi y^3 \sin \pi x = 0. \tag{7}$$

If $y \neq 0$, we can cancel the factor y from each term in Eq. 7. Then, if $4x^2 + 3y \cos \pi x \neq 0$, it follows that

$$\frac{dy}{dx} = \frac{\pi y^2 \sin \pi x - 4xy}{4x^2 + 3y \cos \pi x}. \tag{8}$$

At the point (1, 1) we obtain

$$\left.\frac{dy}{dx}\right|_{\substack{x=1\\y=1}} = \frac{\pi(1)^2 \sin \pi - 4(1)(1)}{4(1)^2 + 3(1) \cos \pi} = \frac{0 - 4}{4 - 3} = -4. \tag{9}$$

This is the slope of the required tangent line. ∎

The preceding examples illustrate the power of implicit differentiation as a means for calculating derivatives that are difficult or impossible to obtain in other ways. On the other hand, there are also certain drawbacks associated with implicit differentiation that we should not overlook. Let us suppose that there is a differentiable function $y = f(x)$ that satisfies an implicit relation of the form (1)

$$F(x, y) = 0.$$

For example, Eqs. 2 and 6 are of this form. Upon computing y' by implicit differentiation, it is usually the case (as in Examples 1 and 2) that the resulting expression for y' involves both x and y. That is, y' is of the form

$$y' = g(x, y). \tag{10}$$

In order to obtain y' as a function of the independent variable x only, we must solve Eq. 1 for y in terms of x [that is, we must determine $f(x)$] and substitute the result into the right side of Eq. 10. This is usually impossible, as in the preceding examples. Nevertheless, for a particular value of x, say $x = x_0$, we can (if necessary) calculate the corresponding value y_0 numerically from Eq. 1, and then substitute these values of x_0 and y_0 into Eq. 10. This process can be repeated for each value of x_0 of interest. Thus although we cannot (usually) obtain an explicit analytical formula for y' in terms of x only, we can obtain y' for each particular value of x that may be required. Both Examples 1 and 2 illustrate this statement. For instance, in Example 1 we found $f'(2)$, but we did not obtain a general formula for $f'(x)$. Similarly, in Example 2 we found dy/dx only at $x = 1$.

EXAMPLE 3

Suppose that there is a differentiable function $y = f(x)$ such that

$$y^3 + y = x^2 + 1, \tag{11}$$

and such that $f(3) = 2$. Find y' and y'' at the point $x = 3$, $y = 2$.

Again, we verify first that the point (3, 2) satisfies Eq. 11. Then, differentiating Eq. 11 implicitly, we obtain

$$3y^2y' + y' = 2x,$$

or

$$y' = \frac{2x}{3y^2 + 1}. \tag{12}$$

Hence

$$y'\big|_{\substack{x=3\\y=2}} = \frac{2(3)}{3(2^2) + 1} = \frac{6}{13}. \tag{13}$$

To calculate y'' we must differentiate Eq. 12 with respect to x. In differentiating the right side of this equation we must remember that we have assumed that y is a differentiable function of x. Using the quotient rule and the chain rule to differentiate Eq. 12, we find that

$$\begin{aligned} y'' &= \left(\frac{2x}{3y^2 + 1}\right)' \\ &= \frac{(3y^2 + 1)(2) - (2x)(6yy')}{(3y^2 + 1)^2}. \end{aligned} \tag{14}$$

Substituting $x = 3$, $y = 2$, and $y' = \frac{6}{13}$ into Eq. 14, we obtain

$$y''\big|_{\substack{x=3\\y=2}} = -\frac{94}{2197}.$$

Note that this process could be continued to yield higher derivatives, y''', y'''', . . . , although the calculations become more and more complicated. ∎

A sometimes troublesome problem in using implicit differentiation is that there need not be any differentiable function $y = f(x)$ satisfying a given equation of the form (1). For example, consider the equation

$$x^2 + y^2 + 9 = 0. \tag{15}$$

Proceeding formally to differentiate Eq. 15, we obtain

$$2x + 2yy' = 0$$

and hence apparently

$$y' = -\frac{x}{y}. \tag{16}$$

However, since the left side of Eq. 15 is positive for all real values of x and y, it is clear that there is no real-valued function $y = f(x)$ satisfying Eq. 15. Thus, implicit differentiation, used uncritically, has led us to an absurd conclusion: We have a formula that appears to be a derivative, but the function we were presumably differentiating does not exist. Actually, the result (16) is valid if we consider x and y as complex variables. Nevertheless, the danger remains of applying implicit differentiation to an equation that actually is not satisfied by any real-valued differentiable function, and this danger should not be ignored. The problem is that when Eq. 1 is complicated, it may not be easy to tell whether or not there is a real-valued differentiable function $y = f(x)$ satisfying it. This is why we have verified in each example that there do exist real values of the variables that satisfy the given equations.

Another fact to be considered is that there may be many differentiable functions satisfying an equation of the form (1). For example, the equation

$$x^2 + y^2 = 9 \tag{17}$$

is satisfied not only by

$$y = f_1(x) = \sqrt{9 - x^2}, \qquad -3 \le x \le 3 \tag{18}$$

but also by

$$y = f_2(x) = -\sqrt{9 - x^2}, \qquad -3 \le x \le 3 \tag{19}$$

Indeed, Eq. 17 is also satisfied by infinitely many other functions, such as

$$y = f_3(x) = \begin{cases} -\sqrt{9 - x^2}, & -3 \le x \le 0 \\ \sqrt{9 - x^2}, & 0 < x \le 3. \end{cases} \tag{20}$$

The graphs of f_1, f_2, and f_3 are shown in Figures 3.7.1, 3.7.2, and 3.7.3, respectively. Note that f_3 is not continuous at the origin and hence is not differentiable there. However, the expression $y' = -x/y$ obtained by differentiating Eq. 17 correctly gives the derivative of each of these functions at each point where the function is differentiable. That is, if $y = f(x)$ satisfies Eq. 17 and is differentiable at the point x, then

$$f'(x) = -\frac{x}{f(x)}. \tag{21}$$

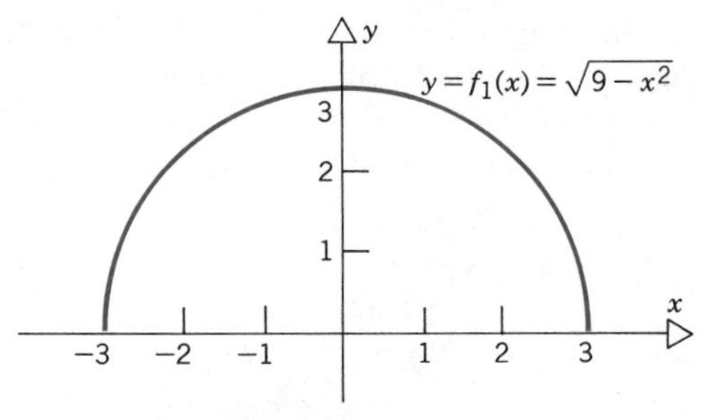

Figure 3.7.1

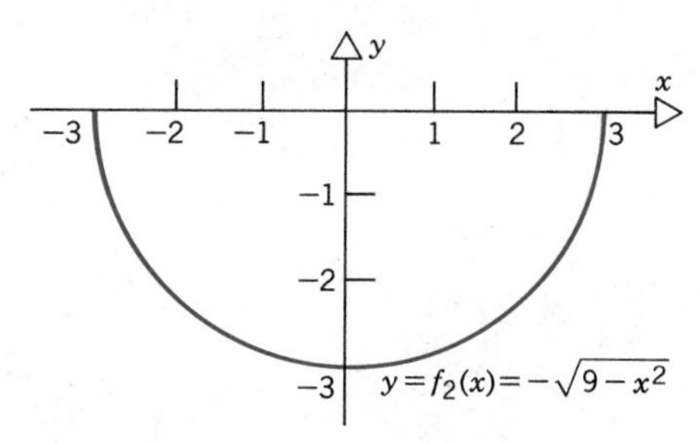

Figure 3.7.2

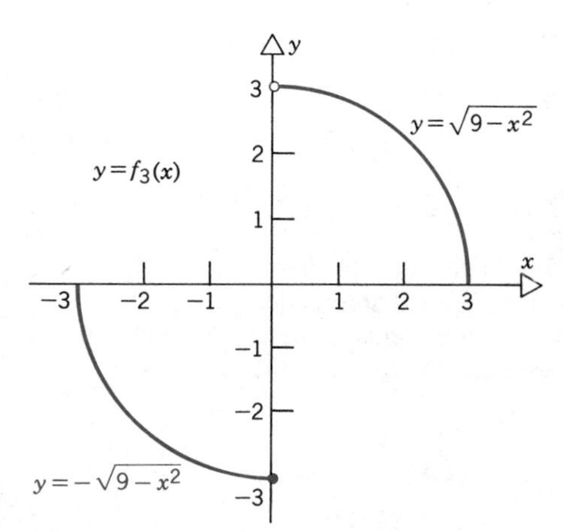

Figure 3.7.3

To summarize, implicit differentiation is an extremely useful procedure. In many cases it permits us to evaluate derivatives that would be inaccessible if we first had to obtain an explicit formula for the function we want to differentiate. Nevertheless, it should be used with an awareness that there may be some delicate questions involved, such as those we have touched on just above. More advanced books contain theorems that answer some of these questions. For the student first learning about implicit differentiation our advice is to make sure that there is a point whose coordinates satisfy the given equation, to compute the derivative only at such a point, and to use the results only when they seem to make sense (for example, when the denominator is nonzero).

PROBLEMS

In each of Problems 1 through 10 there is a differentiable function $y = f(x)$ that satisfies the given equation, and whose graph contains the given point.

(a) Verify that the given point satisfies the given equation.

(b) Use implicit differentiation to find an expression for dy/dx; this expression may depend on both x and y.

(c) Find the value of dy/dx at the given point.

1. $x^2 - xy + y^2 = 1$; $\quad (-1, -1)$

2. $x^2 + 4xy - y^2 = 19$; $\quad (2, 3)$

3. $x^2 + xy + y^2 = 12 - x^2 - y^2$; $(1, 2)$

4. $\dfrac{1 - xy}{1 + x^2y^2} = 1$; $(-1, 1)$

5. $y^3 + y - x = 1$; $(1, 1)$

6. $(x^2 + y^2)^2 = x^2 - y^2$; $\left(\dfrac{\sqrt{6}}{4}, \dfrac{\sqrt{2}}{4}\right)$

7. $(x^2 + y^2)^{3/2} = 2xy$; $\left(\dfrac{3}{4}, \dfrac{\sqrt{3}}{4}\right)$

8. $y \sin 2x - x \sin y = \dfrac{\pi}{4}$; $\left(\dfrac{\pi}{4}, \dfrac{\pi}{2}\right)$

9. $\sin^2 x + \cos^2 y = 1$; $\left(\dfrac{\pi}{4}, \dfrac{\pi}{4}\right)$

10. $3x \sin \pi y + y \sin \pi x = -2$; $(\frac{1}{2}, -\frac{1}{2})$

In each of Problems 11 through 14 there is a differentiable function $y = f(x)$ that satisfies the given equation, and whose graph contains the given point. Use implicit differentiation to find the value of d^2y/dx^2 at the given point.

11. $x^2 - xy + y^2 = 1$; $(-1, -1)$

12. $x^2 + 4xy - y^2 = 19$; $(2, 3)$

13. $y^3 + y - x = 1$; $(1, 1)$

14. $\sin^2 x + \cos^2 y = 1$; $\left(\dfrac{\pi}{4}, \dfrac{\pi}{4}\right)$

In each of Problems 15 through 18, find an equation of the line tangent to the given curve at the given point.

15. $2x^2 - 5xy + 3y^2 = 1$; $(2, 1)$

16. $x^3 - 2xy + y^3 = 11$; $(-1, 2)$

17. $x \sin y + y \cos x = \dfrac{\pi}{2\sqrt{2}}$; $\left(\dfrac{\pi}{4}, \dfrac{\pi}{4}\right)$

18. $x^3 - y^3 - 3xy = 1$; $(2, 1)$

In each of Problems 19 through 22, find an equation of the line normal (perpendicular) to the given curve at the given point.

19. $3x^2 - 4xy + 2y^2 = 9$; $(1, -1)$

20. $x^2y - xy + xy^2 - 2x + 3y + 4 = 0$; $(1, -2)$

21. $2\pi \sin x \sin y = \sqrt{3}(x + y)$; $\left(\dfrac{\pi}{6}, \dfrac{\pi}{3}\right)$

22. $x^3 + y^3 + 6xy = -5$; $(-1, 2)$

23. In a certain vibrating spring–mass system the frequency ω and amplitude A are related by the equation

$$\omega^2 = 1 + \frac{1}{8}A^2 - \frac{1}{A}.$$

Find the rate of change of A with respect to ω.

24. In a certain mechanical system the frequency λ of vibration is related to a support parameter k by the equation

$$\lambda \cos \lambda + k \sin \lambda = 0.$$

Find an expression for $d\lambda/dk$.

25. Assume that the equation $x = f(y)$ defines y implicitly as a function of x. That is, there is a function $y = g(x)$ such that $x = f(y) = f[g(x)]$. The function g is called the *inverse function* of f; inverse functions are discussed further in Section 8.1. Differentiate $x = f(y)$ implicitly with respect to x and show that $dy/dx = 1/f'(y)$; hence $g'(x) = 1/f'(y)$.

In each of Problems 26 through 29, proceed as indicated in Problem 25 to determine dy/dx; express your answer in terms of x only.

26. $x = y^2$; $y \leq 0$, $0 \leq x < \infty$

27. $x = \sqrt{y + 2}$; $y \geq -2$, $0 \leq x < \infty$

28. $x = \sin y$; $-\dfrac{\pi}{2} \leq y \leq \dfrac{\pi}{2}$, $-1 \leq x \leq 1$

29. $x = \tan y$; $-\dfrac{\pi}{2} < y < \dfrac{\pi}{2}$, $-\infty < x < \infty$

REVIEW PROBLEMS

In each of Problems 1 through 6, use Definition 3.1.1 to calculate the derivative of the given function.

1. $f(x) = x + \sqrt{x}$

2. $g(x) = x^{3/2} + \dfrac{1}{x}$

3. $h(x) = \dfrac{1}{x + 1}$

4. $g(s) = \dfrac{s}{s^2 + 1}$

5. $f(t) = \sqrt{t} + \dfrac{1}{t}$

6. $f(x) = x^3 + \sqrt{x} + 2$

In each of Problems 7 through 14, (a) find $f'(x)$; (b) find the equation of the line tangent to the graph of f at the point $(4, f(4))$.

7. $f(x) = 3x^2 - x^{5/2}$

8. $f(x) = -x^{3/2} + x^2$

9. $f(x) = (x^2 + 1)(x - 1)$

10. $f(x) = x(x^3 - 1)$

11. $f(x) = \dfrac{x}{1 + x^2}$

12. $f(x) = \dfrac{x^{1/2}}{1 + x}$

13. $f(x) = \dfrac{x^{3/2}}{x^2 + 1}$

14. $f(x) = -\dfrac{x^2}{\sqrt{x} - 1}$

In each of Problems 15 through 18, find $f'(x)$ and all points where $f'(x) = 0$.

15. $f(x) = x^4 + 2x^2 - 3$

16. $f(x) = x^3 - 3x^2 - 9x$

17. $f(x) = \dfrac{(x - 3)(x - 4)}{x^2 - 4}$

18. $f(x) = \dfrac{x - 1}{x^2 + 9}$

In each of Problems 19 through 22, (a) find the antiderivative of the given function that passes through $(1, -2)$; (b) find the slope of the line tangent to $y = f(x)$ at the point $(4, f(4))$.

19. $f(x) = x^3 + 3$

20. $f(x) = \sqrt{x} - 2$

21. $f(x) = x^{2/3} - \dfrac{1}{x^2}$

22. $f(x) = x^{3/4} + \dfrac{1}{x^3}$

In each of Problems 23 and 24, (a) find the cubic polynomial $f(x)$ satisfying the given conditions; (b) find a quadratic polynomial $g(x) = x^2 + px + q$ such that the function

$$H(x) = \begin{cases} f(x), & x \geq -2 \\ g(x), & x < -2 \end{cases}$$

is differentiable.

23. $f(0) = 2$; $f'(0) = -4$; $f''(1) = 2$; $f'''(1) = 12$.

24. $f(6) = 8$; $f'(2) = 0$; $f''(3) = -1$; $f'''(-1) = 1$.

In each of Problems 25 through 38, find the indicated derivative.

25. $f'(x)$ if $f(x) = \dfrac{(x^2 - 1)^2}{x^{1/2} + 3}$, $x > 0$

26. Dy if $y = \dfrac{(x^3 + 1)^2 + 1}{(x - 1)^2}$, $x \neq 1$

27. $\dfrac{dg}{dx}$ if $g(y) = \cos(3y^2 - 1)$; $y(x) = x^{1/2} - 1$, $x > 0$

28. $\dfrac{dy}{dt}$ if $y(x) = \sin(2x^{1/2})$; $x(t) = 3t^4 + 2$

29. $f''(x)$ if $f(x) = \cos(x^{3/2} + 1)$, $x > 0$

30. $\dfrac{d^2g}{dz^2}$ if $g(z) = \sin\left(1 - \dfrac{1}{z}\right)$, $z \neq 0$

31. $\dfrac{dh}{dt}$ if $h(x) = \dfrac{x^2 - 1}{x}$; $x(t) = t^2 + 1$

32. $\dfrac{dy}{dt}$ if $y(x) = \dfrac{x^{3/2}}{1 + x}$; $x(t) = t^3 - 2$, $t > 1$

33. $f'(x)$ if $f(x) = \dfrac{\cos(ax^2 + b)}{x^2 - 1}$, $x > 1$

34. $Dw(x)$ if $w(x) = \dfrac{\tan(a\sqrt{x})}{x^2}$, $0 < |x| < \dfrac{1}{a^2}$, $a \neq 0$

35. $\dfrac{dh}{ds}$ if $h(y) = \cos(y^3 - 1)$; $y(s) = 4 + \sin^2(s)$

36. $\dfrac{dg}{dt}$ if $g(x) = \sin^2(x^2 + 1)$; $x(t) = 2 - \tan(t^{1/2})$, $0 < t < 1$

37. $Dg(x)$ if $g(x) = \cos(3x^{3/2})\sec[(x - \pi)^2]$

38. $f'(x)$ if $f(x) = \sin^2[(x^{1/2} - 4)^2 + \cos(2x^2)]$

In each of Problems 39 through 46, find an antiderivative of the given function.

39. $2 \sin 3x$

40. $4 \cos(-x)$

41. $2 \sin 2x \cos 2x$

42. $5 \cos 5x \sin 5x$

43. $3x^2 \sin(x^3 + 1)$

44. $(x^3 + 1)\cos\left[\left(\dfrac{x^4}{4}\right) + x\right]$

45. $-(\sin x)^{-2} \cos x$

46. $x(\sin x^2)^{-2} (\cos x^2)$

In each of Problems 47 through 50, (a) determine the tangent line function and the remainder function for any value of x_0; (b) estimate $f(x)$ at the given value of x using an appropriate linear approximation; (c) find the error involved in this estimate and compute $\left|\frac{f(x) - T(x; x_0)}{f(x_0)}\right|$ = relative error.

© **47.** $f(x) = x^3 + \sqrt{x}$; $x = \frac{15}{16}$

© **48.** $f(x) = 2x^{1/3} - x^{-1}$; $x = -\frac{17}{2}$

© **49.** $f(x) = \frac{x + 1}{\sqrt{x}}$; $x = \frac{80}{9}$

© **50.** $f(x) = \frac{x^2 - 1}{x^{1/4}}$; $x = \frac{257}{16}$

In each of Problems 51 through 54, assume that there is a differentiable function $y = f(x)$ that satisfies the given relation. Then (a) check the given pairs of coordinates and ascertain those that satisfy the given equation; (b) find the specified derivatives at all points in Part (a) that satisfy the equation.

51. $x^2 + xy^2 = 3$

(a) $(1, \sqrt{2})$; $(3, -\sqrt{2})$; $\left(4, \frac{\sqrt{13}}{2}\right)$

(b) y'; y''

52. $x^3 + 2xy^3 - 4y^2 = 1$

(a) $(1, -2)$; $(1, 2)$; $(1, 3)$

(b) y'

53. $x^2y + xy^2 = 2$

(a) $(1, -1)$; $(1, 1)$; $(-1, -1)$

(b) y'; y''

54. $1 - x^2y = x^2 + y^2$

(a) $(2, -1)$; $(2, 1)$; $(-1, -1)$

(b) y'

applications of the derivative

In many applications it is eventually necessary to find the largest or smallest values of some particular function; that is, for a given function f with domain D, find those points in D where f assumes its greatest or least values, and then evaluate f at these points. For instance, it may be required to determine how to choose a route between two points that can be traversed in minimum time, or how to select the price of a manufactured product so as to maximize profits, or how to design a container of given shape and volume but of minimum surface area. A few problems of this general type were solved in ancient times, but most could not be handled effectively by methods available then. The discovery of calculus in the seventeenth century, however, provided a simple means of solving many such problems, as we show in this chapter.

4.1 THE MEAN VALUE THEOREM AND SOME OF ITS CONSEQUENCES

Before taking up a detailed treatment of optimization problems in the next section, we need to develop some further properties of differentiable functions that will be indispensable to us later. We begin with a discussion of local maxima and minima. A function f whose domain is an interval I has a **local,** or **relative, maximum** at the interior point c if

$$f(x) \leq f(c) \tag{1}$$

for all x in the interval $(c - \delta, c + \delta)$ for some $\delta > 0$. Of course, δ must also be small enough so that all points in $(c - \delta, c + \delta)$ are in I. The point c is called a *local,* or *relative, maximum point,* and $f(c)$ is a *local,* or *relative, maximum value.* Figure 4.1.1*a* shows the graph of a function with a local maximum at c. Similarly, f has a **local, or relative, minimum** at c if

$$f(x) \geq f(c) \tag{2}$$

for all x in some interval $(c - \delta, c + \delta)$. In this case, c is a *local,* or *relative, minimum point,* and $f(c)$ is the *local,* or *relative, minimum value* (see Figure 4.1.1*b*). A point c that is either a local maximum point or a local minimum point is called a **local extreme point.** The most important thing to remember about local extreme points is that comparisons are made between $f(c)$ and $f(x)$ at *nearby points only.*

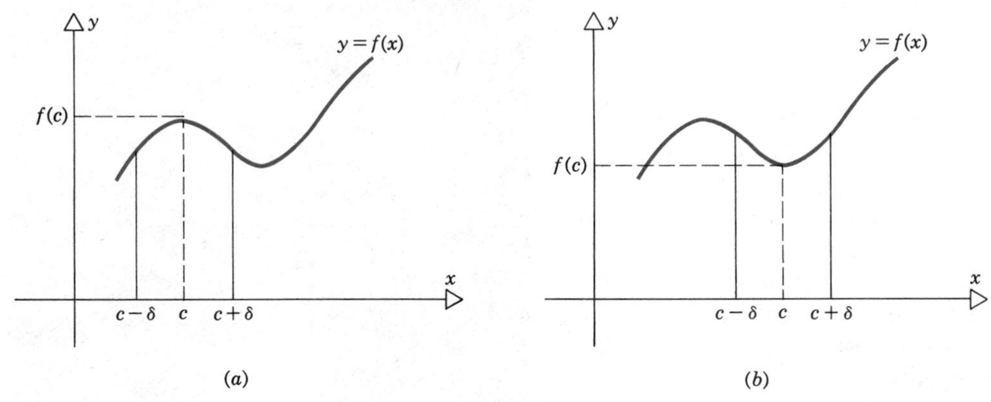

Figure 4.1.1 (*a*) A relative maximum point (*b*) A relative minimum point.

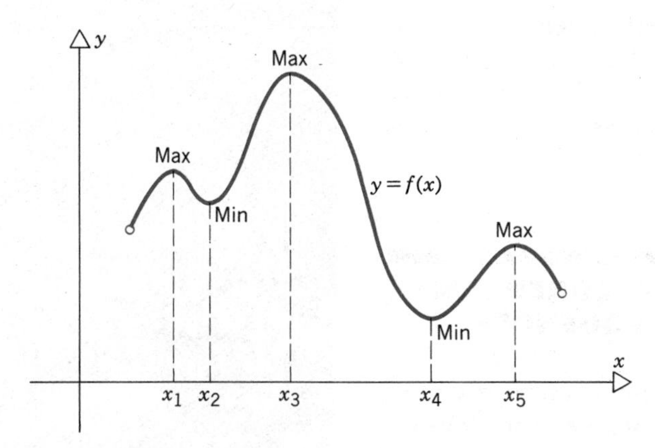

Figure 4.1.2

To illustrate these definitions further, consider the graph of the function f in Figure 4.1.2. The points x_1, x_3, and x_5 are all local maximum points, and the points x_2 and x_4 are local minimum points. Note that x_5, for example, is a local maximum point even though there are many other points in the domain of f where $f(x)$ is larger than $f(x_5)$. There is even a local minimum point, namely x_2, where $f(x)$ is larger than at the local maximum point x_5. Nevertheless, a hilltop is a local maximum even though there may be higher hills, or mountains, or even valleys, elsewhere.

There is a close connection between the local extreme points of a function and its first derivative. Indeed, for a function such as the one shown in Figure 4.1.3, it is geometrically plausible that at a point $[c, f(c)]$ where f has a local

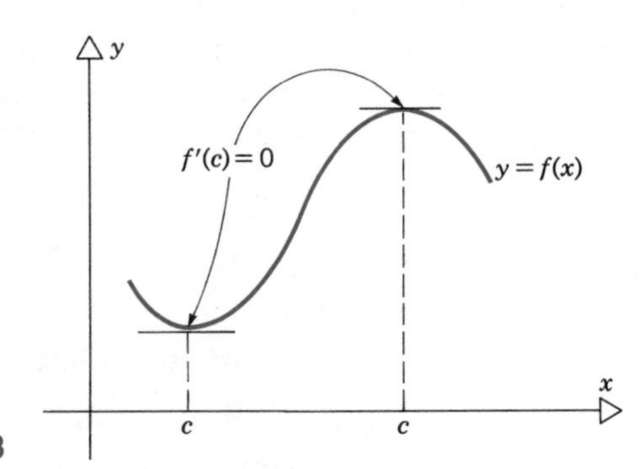

Figure 4.1.3

maximum or minimum the tangent line is horizontal, and hence the derivative $f'(c)$ is zero. The same conclusion is suggested by physical intuition in the case of a particle projected vertically upward from the surface of the earth. It is plausible that the particle reaches its greatest height when its velocity changes from positive to negative, that is, when the velocity is zero. Recall that we used this argument in Example 6 of Section 3.3. The relation between the first derivative and a local maximum or minimum is stated precisely in the following theorem.

Theorem 4.1.1

Suppose that the function f has a local maximum or a local minimum at the point c. If c is an interior point of the domain of f, and if f is differentiable at c, then $f'(c) = 0$.

Proof. Since f is differentiable at c, the difference quotient

$$\frac{f(x) - f(c)}{x - c} \tag{3}$$

must have the limit $f'(c)$ as $x \to c$. This means that the right- and left-hand limits of the difference quotient must both be equal to $f'(c)$. Consider first the case in which c is a local maximum point of f. Then $f(x) \leq f(c)$ for x near c, and the numerator in Eq. 3 is not positive. If $x > c$, then the denominator in Eq. 3 is positive, and the difference quotient is nonpositive. Hence, letting $x \to c$ from the right, we conclude (by Theorem 2.3.5) that $f'(c) \leq 0$. On the other hand, if $x < c$, then the difference quotient is nonnegative, and on taking the limit as $x \to c$ from the left we conclude that $f'(c) \geq 0$. Since f is differentiable, the two limits must agree, and this is possible only if $f'(c) = 0$. If c is a local minimum point, the proof is similar. □

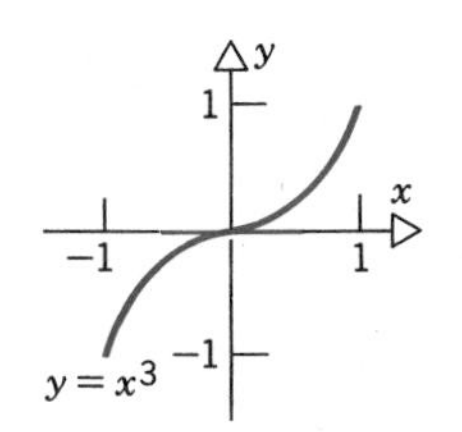

Figure 4.1.4

Note that it is important that c be an interior point, since in the proof it was necessary to take the limit of the difference quotient (3) from both sides of c. Also, note that the converse of Theorem 4.1.1 is not true. That is, if $f'(c) = 0$, the point c need not be an extreme point of f. For example, consider $f(x) = x^3$ on $(-1, 1)$ (see Figure 4.1.4). Clearly $f'(x) = 3x^2$ is zero when $x = 0$, but this point is neither a maximum point nor a minimum point for f on $(-1, 1)$.

The next result is known as Rolle's theorem; it gives some simple conditions that make certain there is a point where the slope of the tangent line is zero.

Theorem 4.1.2

(Rolle's theorem)

Let the function f satisfy the following conditions:

1. f is continuous on the closed bounded interval $[a, b]$;
2. f is differentiable on the open interval (a, b);
3. $f(a) = f(b)$.

Then there is at least one point c in (a, b) such that

$$f'(c) = 0. \tag{4}$$

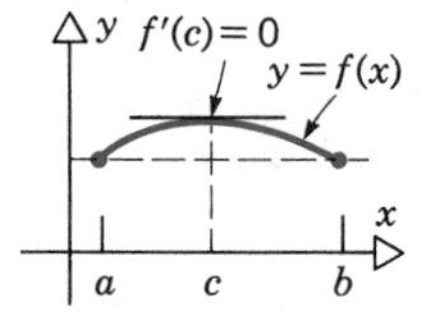

Figure 4.1.5

Figure 4.1.5 suggests that Rolle's theorem is geometrically plausible. The graph of a function f, such as the one shown in Figure 4.1.5, that has a given value at a, and later assumes the same value at b, changes direction at some intermediate point c. At this point the tangent line must be horizontal, and hence $f'(c) = 0$. However, each of the three conditions in Theorem 4.1.2 is necessary in order for the conclusion to be assured. In Figure 4.1.6 we show graphs of three functions, each of which violates exactly one of the hypotheses of Rolle's theorem. Note that

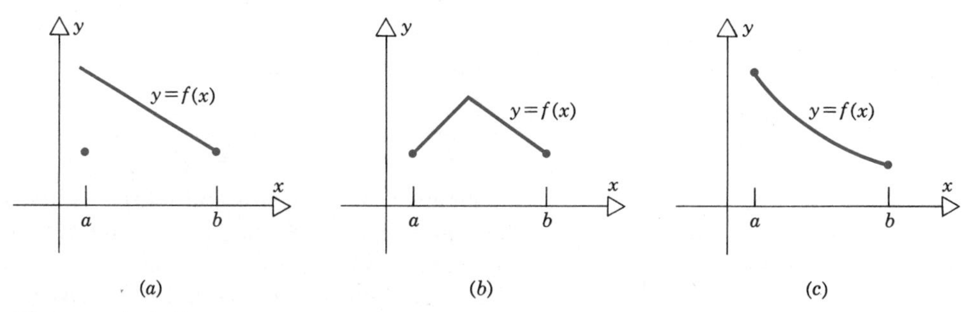

Figure 4.1.6 (*a*) f not continuous on $[a, b]$. (*b*) f not differentiable on (a, b). (*c*) $f(a) \neq f(b)$.

in none of these cases is there a point c where $f'(c) = 0$. Note also that there may be more than one point where the conclusion of Rolle's theorem is true (see Figure 4.1.7).

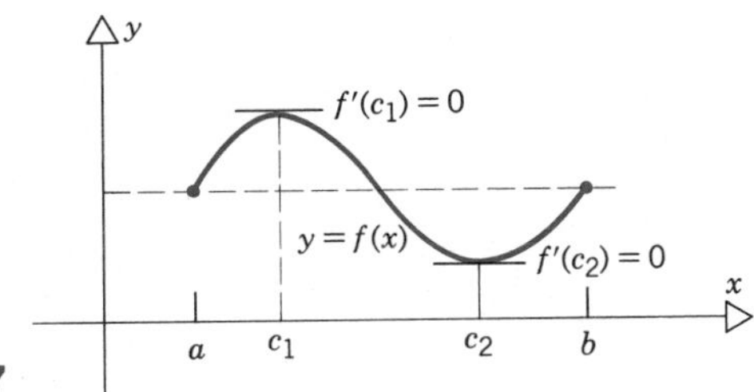

Figure 4.1.7

Proof of Rolle's Theorem. Since f is continuous on the closed bounded interval $[a, b]$, it has (by Theorem 2.6.2) both a maximum and a minimum value on this interval. If both the maximum and minimum occur at the endpoints, then the maximum and minimum values of f are equal, and f is a constant function. Then $f'(x) = 0$ for all x in (a, b), and we can choose c to be any point we wish in (a, b).

On the other hand, if f is not constant, then it must have either a maximum point, or a minimum point, or both, in the open interval (a, b). Let c be such a maximum or minimum point. Then at the point c the conditions of Theorem 4.1.1 are satisfied and hence $f'(c) = 0$, completing the proof. □

If we look again at Figures 4.1.5 and 4.1.7, we may observe the following geometric interpretation of Rolle's theorem: under the given conditions there is at least one point where the tangent line to the graph is parallel to the line segment joining the endpoints of the graph. This observation leads to a generalization of Rolle's theorem, as shown in Figure 4.1.8. Although $f(b) \neq f(a)$, it still appears that there is a point c in (a, b), and a corresponding point $[c, f(c)]$ on the graph, where the tangent line is parallel to the line segment joining the endpoints $[a, f(a)]$ and $[b, f(b)]$. Since the slope of this line segment is $[f(b) - f(a)]/(b - a)$, we are thus led to the following result, known as the mean value theorem.*

---

*Rolle's theorem and the mean value theorem, along with the intermediate value property of continuous functions (Theorem 2.6.3), are examples of *existence theorems* in mathematics. They assert the existence of a number having a certain property, but as a rule, provide no way to calculate the number.

Rolle's theorem was proved in 1691 (for polynomials only) by Michel Rolle (1652–1719), a French mathematician known in his own time mainly for his work in algebra, but remembered today almost solely for the result in Theorem 4.1.2.

The mean value theorem was published in 1797 by Joseph Louis Lagrange (1736–1813), second only to Euler among mathematicians of the eighteenth century. Lagrange was born in Turin of mixed French-Italian ancestry, and was recognized before the age of twenty for his original approach to the calculus of variations. He became professor at the Royal Artillery School in Turin in 1755, was selected to succeed Euler at the Berlin Academy of Sciences in 1766, and was attracted to the Paris Académie of Sciences in 1787. His work in mechanics, differential equations, and the calculus of variations was wide-ranging and of fundamental importance. His most celebrated book, published in 1788, is *Traité de Mécanique Analytique,* an elegant mathematical presentation of a century's progress in Newtonian mechanics.

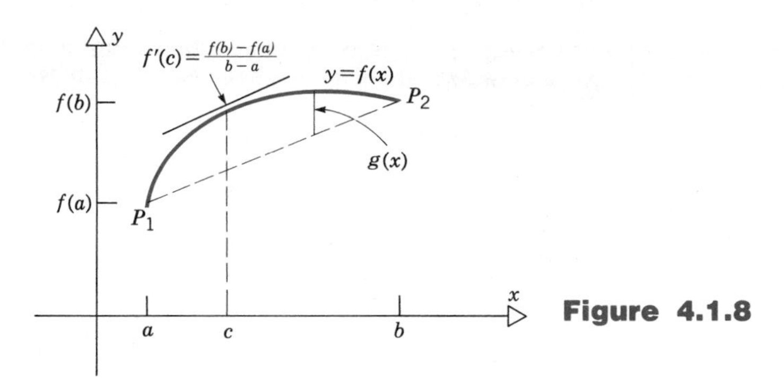

Figure 4.1.8

Theorem 4.1.3

(Mean value theorem)

Let the function f satisfy the following conditions:

1. f is continuous on the closed bounded interval $[a, b]$;
2. f is differentiable on the open interval (a, b).

Then there is at least one point c in (a, b) such that

$$f'(c) = \frac{f(b) - f(a)}{b - a}. \tag{5}$$

Once again, examples such as the ones in Figure 4.1.6*a* and 4.1.6*b* show that the given hypotheses are necessary to ensure the conclusion. Also, as in Rolle's theorem, there may be more than one point c in (a, b) for which Eq. 5 is true.

Proof of Mean Value Theorem. We can reduce the problem to one already solved by constructing a function related to f that satisfies the hypotheses of Rolle's theorem. We must pay particular attention to the third hypothesis. That is, we look for a suitably smooth function that assumes the same values at a and b. One possibility is the function g that describes the vertical difference between the graph of $y = f(x)$ and the graph of the line segment P_1P_2 in Figure 4.1.8. The equation of the straight line passing through P_1 and P_2 is

$$y = f(a) + \frac{f(b) - f(a)}{b - a}(x - a). \tag{6}$$

Hence the function g is defined by

$$g(x) = f(x) - f(a) - \frac{f(b) - f(a)}{b - a}(x - a). \tag{7}$$

Since f satisfies the hypotheses of the mean value theorem, it follows that g satisfies the first two hypotheses of Rolle's theorem. Further, a simple calculation shows that $g(a) = g(b) = 0$. Hence Rolle's theorem applies to g, and states that there is a point c in (a, b) such that $g'(c) = 0$. On differentiating Eq. 7 we find that

$$g'(x) = f'(x) - \frac{f(b) - f(a)}{b - a}. \tag{8}$$

Setting $x = c$ and $g'(c) = 0$, we have

$$0 = f'(c) - \frac{f(b) - f(a)}{b - a}.$$

from which Eq. 5 follows at once, thus proving the mean value theorem. □

Note that the mean value theorem does not explicitly locate the point c. In fact, it is almost never important actually to determine this point in specific cases. What is important in almost all applications of the mean value theorem (or of Rolle's theorem) is the existence of such a point, not its precise location. The following examples point out two ways in which these theorems are sometimes useful.

EXAMPLE 1

Show that the function

$$f(x) = x^3 + 3x^2 + 6x + k \tag{9}$$

has at most one real zero, regardless of the value of the constant k.

We can establish this result by contradiction. Suppose that there are two different values x_1 and x_2 such that $f(x_1) = f(x_2) = 0$. To be definite suppose further that $x_1 < x_2$. Then by Rolle's theorem there must be a point c in the interval (x_1, x_2) where $f'(c) = 0$. However, upon computing $f'(x)$ we have

$$\begin{aligned} f'(x) &= 3x^2 + 6x + 6 \\ &= 3(x^2 + 2x + 2) \\ &= 3[(x + 1)^2 + 1]. \end{aligned} \tag{10}$$

It is clear from Eq. 10 that $f'(x)$ is never zero. Thus there is no such point c, and the hypothesis that f has two different zeros is untenable. Hence f can have at most one zero in $(-\infty, \infty)$.

In fact, $f(x) \to -\infty$ as $x \to -\infty$ and $f(x) \to \infty$ as $x \to \infty$. Since f is continuous on $(-\infty, \infty)$, it follows from the intermediate value property (Theorem 2.6.3) that f must assume the value zero at least once. Hence we conclude that f has exactly one real zero in $(-\infty, \infty)$. ∎

EXAMPLE 2

Show that

$$|\sin x_2 - \sin x_1| \leq |x_2 - x_1| \tag{11}$$

for any real numbers x_1 and x_2.

Suppose that $x_2 > x_1$. Then the hypotheses of the mean value theorem are satisfied for the function $f(x) = \sin x$ on the interval $[x_1, x_2]$. It then follows that

$$f(x_2) - f(x_1) = f'(c)(x_2 - x_1),$$

or

$$\sin x_2 - \sin x_1 = (\cos c)(x_2 - x_1), \tag{12}$$

for some c in (x_1, x_2). Taking the absolute value of both sides of Eq. 12, we obtain

$$|\sin x_2 - \sin x_1| = |\cos c||x_2 - x_1|.$$

Equation 11 follows because $|\cos c| \leq 1$ regardless of the value of c. A similar argument leads to the same conclusion if $x_2 < x_1$. Finally, equality holds in Eq. 11 if $x_2 = x_1$. Note that if $x_1 = 0$ and $x_2 = x$, then Eq. 11 reduces to

$$|\sin x| \leq |x|. \quad \blacksquare \tag{13}$$

Antiderivatives

Recall that in Section 3.3 we stated that the function f is an antiderivative of the function g on the open interval I if

$$f'(x) = g(x) \tag{14}$$

for each x in I. Further, if the function $y = f(x)$ is an antiderivative of some given function g, then so is the function $y = f(x) + c$, where c is any constant. Indeed, as we indicated in Section 3.3, all antiderivatives of g are of this form. We are now able to prove this important fact by making use of the mean value theorem. First, however, we establish a related result that is useful in its own right.

Theorem 4.1.4

Let the function f be continuous on $[a, b]$ and suppose that $f'(x) = 0$ for each x in (a, b). Then f is a constant function on $[a, b]$.

Note that we showed in Section 3.1 that the derivative of any constant function is identically zero. According to Theorem 4.1.4, constant functions are the *only* functions whose derivatives are identically zero. The proof is a simple consequence of the mean value theorem.

Proof of Theorem 4.1.4. Let x_1 and x_2 be any pair of points in $[a, b]$, and suppose that $x_2 > x_1$. Then the hypotheses of the mean value theorem (Theorem 4.1.3) are satisfied by the function f on $[x_1, x_2]$. Hence

$$f(x_2) - f(x_1) = f'(c)(x_2 - x_1) \tag{15}$$

where c is some point in (x_1, x_2). Regardless of where c is located, however, $f'(c) = 0$ and hence $f(x_2) = f(x_1)$. Since x_1 and x_2 are arbitrary, this means that f has the same value at each point, that is, f is a constant function on $[a, b]$. □

Theorem 4.1.5

Let the functions F and G be antiderivatives of the same function g on the interval (a, b). That is,

$$F'(x) = g(x), \qquad G'(x) = g(x) \tag{16}$$

for each x in (a, b). Then there is a constant c such that

$$F(x) = G(x) + c \tag{17}$$

for each x in (a, b).

Proof. Let x_1 and x_2 be an arbitrary pair of points in (a, b) with $x_1 < x_2$, and let $H(x) = F(x) - G(x)$. Then it follows from Eq. 16 that $H'(x) = 0$ for all x in (a, b) and thus for all x in (x_1, x_2). Further, F and G are differentiable and hence continuous for all x in (a, b). Consequently, H is continuous on (a, b) and hence on $[x_1, x_2]$ as well. Theorem 4.1.4 then applies to the function H on the interval $[x_1, x_2]$, and we conclude that H is a constant function on $[x_1, x_2]$; that is, there exists a constant c such that $H(x) = c$ for all x in $[x_1, x_2]$. Since x_1 and x_2 are arbitrary points in (a, b), this conclusion is actually valid for all x in (a, b); that is,

$$F(x) - G(x) = c$$

for all x in (a, b), as was to be shown. □

EXAMPLE 3

We observe that

$$\frac{d}{dx} \sin^2 x = 2 \sin x \cos x$$

and that

$$\begin{aligned} \frac{d}{dx}[-\cos^2 x] &= -2 \cos x(-\sin x) \\ &= 2 \cos x \sin x. \end{aligned}$$

Hence the functions

$$F(x) = \sin^2 x, \qquad G(x) = -\cos^2 x$$

are both antiderivatives of the function $g(x) = 2 \sin x \cos x$. By Theorem 4.1.5 there is a constant c such that

$$\sin^2 x = -\cos^2 x + c.$$

Of course, in this case, $c = 1$. ∎

PROBLEMS

In each of Problems 1 through 6, determine whether the given function satisfies the hypotheses of the mean value theorem. In each case determine whether there is a point c such that Eq. 5 is satisfied. Find all such points in those cases for which they exist. It may be helpful to sketch the graph of the function.

1. $f(x) = x^2, \quad -1 \le x \le 2$
2. $f(x) = |x - 1|, \quad -2 \le x \le 2$
3. $f(x) = |x^2 - 1|, \quad -2 \le x \le 2$
4. $f(x) = \begin{cases} 4 - x^2, & -2 \le x < 0 \\ 2 - x, & 0 \le x \le 1 \end{cases}$
5. $f(x) = \begin{cases} x - 1, & -1 \le x < 0 \\ 0, & x = 0 \\ x + 1, & 0 < x \le 1 \end{cases}$
6. $f(x) = \sin \pi x, \quad -\frac{1}{2} \le x \le 2$
7. Let $f(x) = x^3 - 3x + b$, where b is a constant.
   (a) Show that $f(x)$ is zero at most once for x in $[-1, 1]$.
   (b) Determine those values of b for which $f(x) = 0$ for some x in $[-1, 1]$.
8. Let $f(x) = x^3 - 3a^2x + b$, where a and b are constants with $a > 0$. Show that $f(x)$ is zero at most once for x in $[-a, a]$.
9. Let $f(x) = x^3 + ax + b$, where a and b are constants with $a > 0$. Show that $f(x)$ is zero at most once for x in $(-\infty, \infty)$.
10. Suppose that the function f is continuous on $[a, b]$ and differentiable on (a, b). If $f'(x) > 0$ for each x in (a, b), show that $f(b) > f(a)$.
11. (a) Let the function f be such that f is continuous on $[a, b]$ and $f'(x) > 0$ for each x in (a, b). Show that $f(x)$ is zero at most once in (a, b).
    (b) Let f and f' be continuous on $[a, b]$, and let $f''(x) > 0$ for each x in (a, b). Show that $f(x)$ is zero at most twice in (a, b).
    (c) State a generalization of Parts (a) and (b).

In each of Problems 12 through 15, use the mean value theorem to establish the given inequality.

12. $|\cos x_1 - \cos x_2| \le |x_1 - x_2|$
13. $|1 - \cos x| \le |x|$
14. $2 + \dfrac{x}{2\sqrt{4 + x}} \le \sqrt{4 + x} \le 2 + \dfrac{x}{4}, \qquad x \ge 0$
15. $|\tan x| \ge |x|, \qquad |x| < \dfrac{\pi}{2}$
16. Suppose that f is continuous on $[a, b]$, differentiable on (a, b), and that
$$m < f'(x) < M$$
for each x in (a, b). Show that
$$m(x_2 - x_1) < f(x_2) - f(x_1) < M(x_2 - x_1)$$
where x_1 and x_2 are any two points in $[a, b]$ with $x_2 > x_1$.
17. Suppose that the function f is such that $f'(x) = 1/x$ for each x in $(0, \infty)$, and that $f(1) = 0$. This function is discussed in detail in Section 8.2.
    (a) Show that $\frac{2}{3} < f(3) < 2$.
    (b) If $a > 1$, show that $(a - 1)/a < f(a) < a - 1$.
    (c) If $0 < a < 1$, show again that $(a - 1)/a < f(a) < a - 1$.
18. Suppose that the function f is such that $f'(x) = (1 + x^2)^{-1}$ for all x, and that $f(0) = 0$. This function is discussed further in Section 8.6.
    (a) If $x > 0$, show that $x/(1 + x^2) < f(x) < x$.
    (b) If $x < 0$, show that $x < f(x) < x/(1 + x^2)$.
19. Suppose that the function f is such that $f'(x) = (1 - x^2)^{-1/2}$ for $-1 < x < 1$ and that $f(0) = 0$. This function is discussed further in Section 8.6.
    (a) If $x > 0$, show that $f(x) > x$.
    (b) If $x < 0$, show that $f(x) < x$.

* **20.** This problem deals with an extension of the mean value theorem. Let f and f' be continuous on $[a, b]$ and let f'' exist at each point in (a, b); then there is a point c in (a, b) such that

$$f(b) = f(a) + f'(a)(b - a) + \tfrac{1}{2}f''(c)(b - a)^2. \qquad (i)$$

This result may be proved in the following way.

(a) Define the function g so that

$$g(x) = f(x) - f(a) - f'(a)(x - a) - A(x - a)^2$$

where A is a constant. Show that $g(a) = 0$. Require also that $g(b) = 0$; that is, require that A satisfy

$$f(b) - f(a) - f'(a)(b - a) - A(b - a)^2 = 0. \qquad (ii)$$

(b) Using Rolle's theorem, show that there is a point γ in (a, b) such that $g'(\gamma) = 0$. Show also that $g'(a) = 0$.

(c) Applying Rolle's theorem to the function g', show that there is a point c in (a, γ) such that

$$g''(c) = 0. \qquad (iii)$$

(d) Eliminate A between Eqs. (ii) and (iii), thereby obtaining Eq. (i).

* **21.** This problem deals with another generalization of the mean value theorem. Let the functions f and g both satisfy the conditions of Theorem 4.1.3. Assume further that g' is never zero in (a, b). Then show that there is a point c in (a, b) such that

$$\frac{f(b) - f(a)}{g(b) - g(a)} = \frac{f'(c)}{g'(c)}.$$

Note that this result reduces to Theorem 4.1.3 if $g(x) = x$.

Hint: Consider the function

$$F(x) = [f(b) - f(a)]g(x) - [g(b) - g(a)]f(x).$$

4.2 MAXIMA AND MINIMA OF FUNCTIONS

In Section 4.1 we considered a function f defined on an interval I, and gave the definition of a local maximum or local minimum of f at a point c in the interior of I. These definitions can be extended to the endpoints of I by replacing the interval $(c - \delta, c + \delta)$ by a suitable one-sided interval. For example, if f is defined on $[a, b]$, then f has a local maximum at a if

$$f(x) \le f(a) \qquad (1)$$

for all x in some interval $[a, a + \delta)$. The other possible cases are handled in the same fashion. In Figure 4.2.1 we show the graph of the same function as in Figure

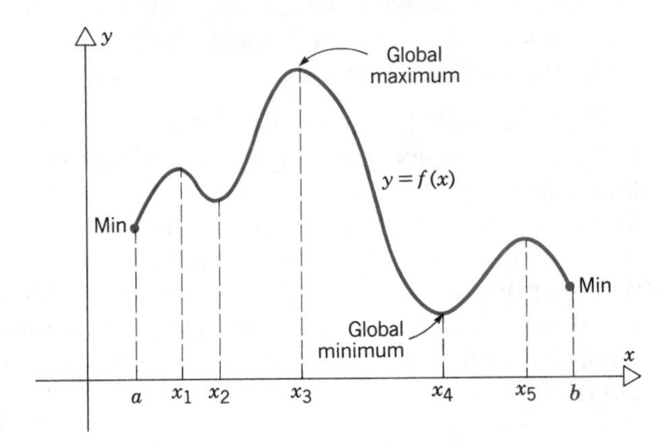

Figure 4.2.1

4.1.2, except that now the endpoints are included in the domain of the function. Both endpoints are local minimum points for this particular function. It may seem plausible that an endpoint is always a local extreme point; however, this is not the case, as the function in Problem 38 shows.

A study of local extreme points, while important, is often not sufficiently far-reaching. Indeed, it is frequently important to consider a function f defined on an interval I, and to compare its value at a local extreme point c with its values at *all other points* in I. We say that f has a **global,** or **absolute, maximum** at c if

$$f(c) \geq f(x) \tag{2}$$

for all x in I. Similarly, f has a **global,** or **absolute, minimum** at c if

$$f(c) \leq f(x) \tag{3}$$

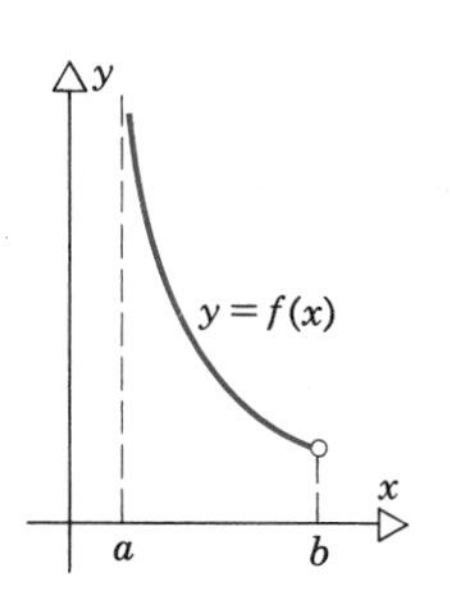

Figure 4.2.2

for all x in I. Thus the global maximum is the largest of the local maxima, and the global minimum is the smallest of the local minima. Referring to Figure 4.2.1 again, the global maximum is at x_3 and the global minimum is at x_4.

Some care must be used in discussing maxima and minima since a function need not have a global maximum or minimum, or even a local maximum or minimum, on a given interval. For example, the function shown in Figure 4.2.2 has no extreme points, either local or global, on the open interval (a, b). As we approach the right endpoint b, the function takes on smaller and smaller values, but it has no smallest (or minimum) value. In the same way it also has no maximum on this interval. However, for a function f that is *continuous on a closed bounded interval* $[a, b]$, we know by Theorem 2.6.2(b) that f has both a global maximum and a global minimum on $[a, b]$.

Critical points

We now consider some theorems that will help us to locate local maximum and minimum points of a given function f defined on an interval I. In the first place, recall Theorem 4.1.1, which states that if c is an extreme point in the interior of I, and if $f'(c)$ exists, then $f'(c) = 0$. Hence to locate local extreme points of f we must consider the following as candidates:

1. Any interior point of I where $f'(x)$ is zero;

and also those points to which Theorem 4.1.1 does not apply, namely:

2. Any interior point of I where $f'(x)$ does not exist;
3. Any endpoint that is in I.

A point satisfying any one of these three conditions is called a **critical point.** For a given function f on a given interval I, the set of critical points includes all local extreme points, but may contain other points as well. If it is known (say from Theorem 2.6.2) that f has a global maximum and a global minimum on I, then these can be found simply by comparing the values of f at the critical points. The points in the set of critical points where f has its largest and smallest values are the global maximum point and the global minimum point, respectively. Note that f may take on its global maximum at more than one point, and similarly for its global minimum.

The following three examples are introduced now to illustrate the ways in which critical points may occur. We will return to each of these examples later to demonstrate how the results of this section can be used to investigate the nature of a critical point.

EXAMPLE 1

Find all critical points on $(-\infty, \infty)$ of the function

$$f(x) = 2x^3 - 3x^2 - 12x + 19. \tag{4}$$

The function f is a polynomial, so there are no points where $f'(x)$ fails to exist; in this example, there are also no endpoints. Thus the only critical points are points where $f'(x) = 0$. Since

$$\begin{aligned} f'(x) &= 6x^2 - 6x - 12 \\ &= 6(x^2 - x - 2) \\ &= 6(x - 2)(x + 1), \end{aligned} \tag{5}$$

it follows that $x = 2$ and $x = -1$ are the only critical points.

If we evaluate f at the critical points we find that $f(2) = -1$ and $f(-1) = 26$. However, Theorem 2.6.2 does not apply because the interval is unbounded, so further investigation is required to determine whether there are any global extreme points. The graph of f is shown here in Figure 4.2.3 for reference; most of the information on which the sketch is based is derived later in this section. ∎

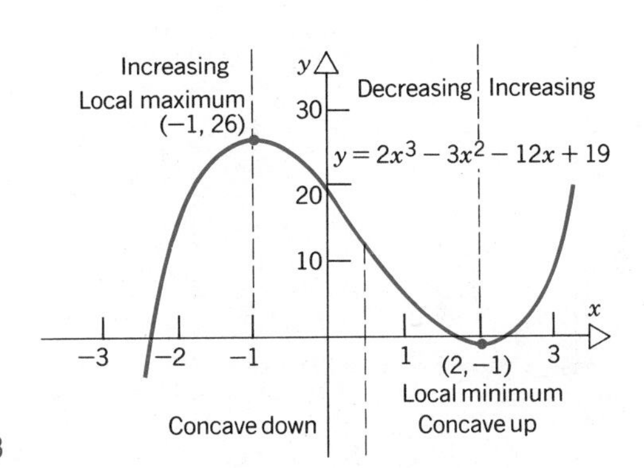

Figure 4.2.3

EXAMPLE 2

Consider the function f defined by

$$f(x) = x - 2\sin x, \qquad 0 \le x \le 2\pi. \tag{6}$$

Find all critical points of this function, and if possible, determine the global extreme points.

If we differentiate f we obtain

$$f'(x) = 1 - 2\cos x. \tag{7}$$

Thus $f'(x) = 0$ when $\cos x = \frac{1}{2}$ and $0 \leq x \leq 2\pi$, that is, when $x = \pi/3$ and $x = 5\pi/3$. There are no interior points where $f'(x)$ does not exist, so the only other critical points are the endpoints $x = 0$ and $x = 2\pi$. Hence the set of critical points is

$$S = \left\{0, \frac{\pi}{3}, \frac{5\pi}{3}, 2\pi\right\}.$$

The function f is continuous on $[0, 2\pi]$, so Theorem 2.6.2(b) applies, guaranteeing that there is a global maximum and a global minimum on that interval. If we evaluate f at each critical point, we find that

$$f(0) = 0, \qquad f\left(\frac{\pi}{3}\right) = \frac{\pi}{3} - \sqrt{3} \cong -0.68485,$$

$$f\left(\frac{5\pi}{3}\right) = \frac{5\pi}{3} + \sqrt{3} \cong 6.9680, \qquad f(2\pi) = 2\pi \cong 6.2832.$$

Therefore $x = 5\pi/3$ is the global maximum point and $x = \pi/3$ is the global minimum point of f on $[0, 2\pi]$. The points $x = 0$ and $x = 2\pi$ are not global extreme points. The graph of f is shown in Figure 4.2.4. Again some of the information in the graph is derived later in the section. ∎

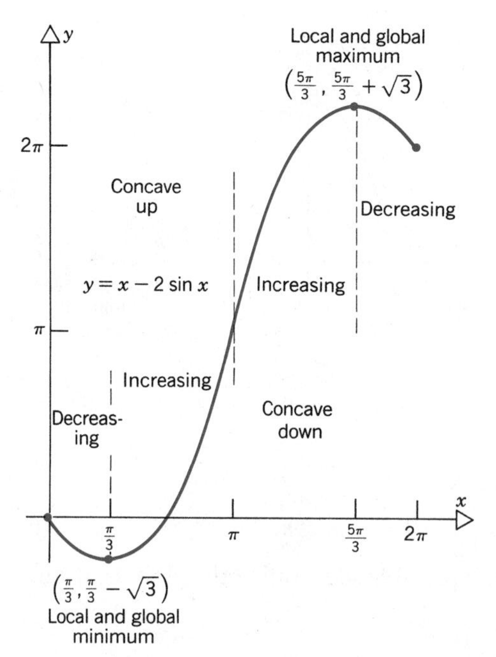

Figure 4.2.4

EXAMPLE 3

Find all critical points of the function

$$f(x) = |x^2 - 1| \tag{8}$$

on the interval $[-2, 2]$. If possible, also determine the global extreme points. The graph of f is shown in Figure 4.2.5; it can be sketched by drawing the parabola $y = x^2 - 1$, and then reflecting the part of the curve for which $y < 0$ about the x-axis.

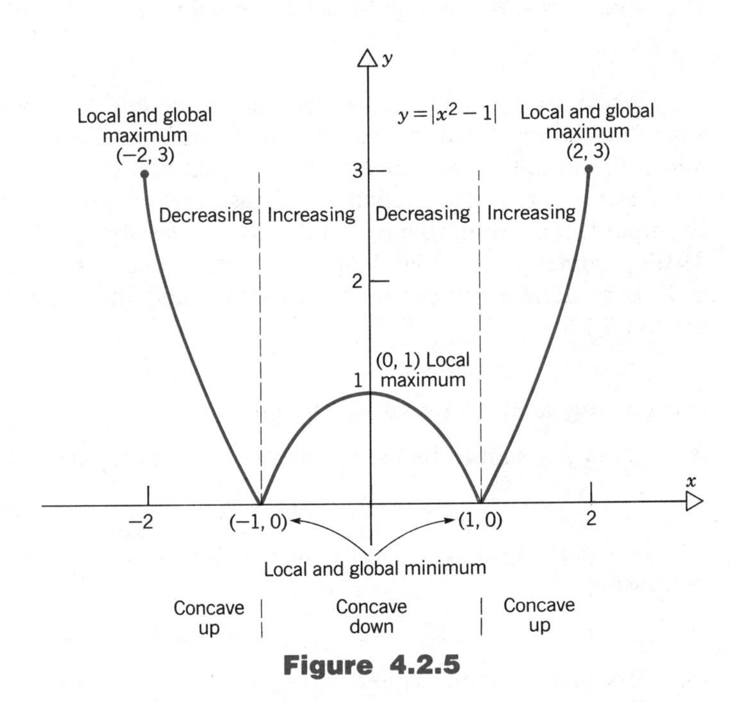

Figure 4.2.5

If we write $f(x)$ without absolute value bars, we have

$$f(x) = \begin{cases} x^2 - 1, & x < -1 \quad \text{or} \quad x > 1; \\ 1 - x^2, & -1 \le x \le 1. \end{cases} \tag{9}$$

Then

$$f'(x) = \begin{cases} 2x, & x < -1 \quad \text{or} \quad x > 1; \\ -2x, & -1 < x < 1. \end{cases} \tag{10}$$

The derivative of f does not exist at the points $x = \pm 1$. This can be shown in much the same way as in Example 7 of Section 3.2 by examining the left- and right-hand limits of the appropriate difference quotient. The graph of f has a corner at each of these points, as indicated in Figure 4.2.5.

Therefore the critical points of the function f on the interval $[-2, 2]$ are the following:

$x = 0$, because $f'(x)$ is zero there;

$x = -1$ and $x = 1$, because $f'(x)$ does not exist there;

$x = -2$ and $x = 2$, because these are the endpoints of the given interval.

As in Example 2, the function (8) is continuous on the interval $[-2, 2]$, so by Theorem 2.6.2(b) there are certain to be global maximum and global minimum points. If we evaluate f at each critical point we obtain

$$f(-2) = f(2) = 3, \qquad f(-1) = f(1) = 0, \qquad f(0) = 1.$$

Hence, $x = \pm 2$ are global maximum points and $x = \pm 1$ are global minimum points. The point $x = 0$ is not a global extreme point. ∎

While the procedure previously suggested and illustrated in Examples 2 and 3 is often the most economical one to follow in seeking global extreme points, it is insufficient unless we know in advance that the function in question actually has a global maximum and minimum. In case this information is not available, as in Example 1, it is useful to have other ways of classifying the critical points. We now develop some results that help to do this. These results will also be needed in a more detailed discussion of functions and their graphs that we present in Section 5.1.

Increasing and decreasing functions

A function f is said to be **increasing** on an interval I if

$$f(x_2) > f(x_1) \tag{11}$$

for every pair of points x_1 and x_2 in I such that $x_2 > x_1$. Similarly, f is said to be **decreasing** on I if

$$f(x_2) < f(x_1) \tag{12}$$

for every pair of points x_1 and x_2 in I with $x_2 > x_1$. If equalities are permitted in Eqs. 11 and 12, then f is said to be **nondecreasing** and **nonincreasing,** respectively. There is a close connection between increasing and decreasing functions and the sign of their first derivatives, as the following theorem states.

Theorem 4.2.1

Let the function f be continuous on $[a, b]$ and differentiable on (a, b). If $f'(x) > 0$ for each x in (a, b), then f is increasing on $[a, b]$, and if $f'(x) < 0$ for each x in (a, b), then f is decreasing on $[a, b]$.

Proof. The proof involves a straightforward application of the mean value theorem (Theorem 4.1.3). Choose *any* pair of points x_1 and x_2 in $[a, b]$ with $x_2 > x_1$. The mean value theorem applies to the function f on the interval $[x_1, x_2]$. Thus

$$f(x_2) - f(x_1) = f'(c)(x_2 - x_1), \tag{13}$$

where c is some point in the open interval (x_1, x_2). If $f'(x) > 0$ for all x in (a, b), then $f'(c) > 0$, and hence $f(x_2) > f(x_1)$, that is, f is increasing. Similarly, if $f'(x) < 0$ for all x in (a, b), then $f(x_2) < f(x_1)$, and f is decreasing. □

The converse of Theorem 4.2.1 is not true. For instance, even if f is increasing, its derivative need not be positive at all points. To see this, consider $f(x) = x^3$, which is increasing on every interval. However, $f'(x) = 3x^2$ is not positive at $x = 0$. See also Problem 37.

The next theorem provides a means of classifying critical points on the basis of information about the first derivative near a critical point.

Theorem 4.2.2

(First derivative test)

Let the function f be continuous on some interval $(c - \delta, c + \delta)$ containing the critical point c.

(a) If $f'(x) > 0$ for x in $(c - \delta, c)$ and $f'(x) < 0$ for x in $(c, c + \delta)$, then f has a local maximum at c.

(b) If $f'(x) < 0$ for x in $(c - \delta, c)$ and $f'(x) > 0$ for x in $(c, c + \delta)$, then f has a local minimum at c.

Proof. We prove Part (a) only; the proof of Part (b) is similar. The given conditions imply that f is increasing to the left of c and decreasing to the right of c, so it is clear geometrically that f has a local maximum at c (see Figure 4.2.6). More

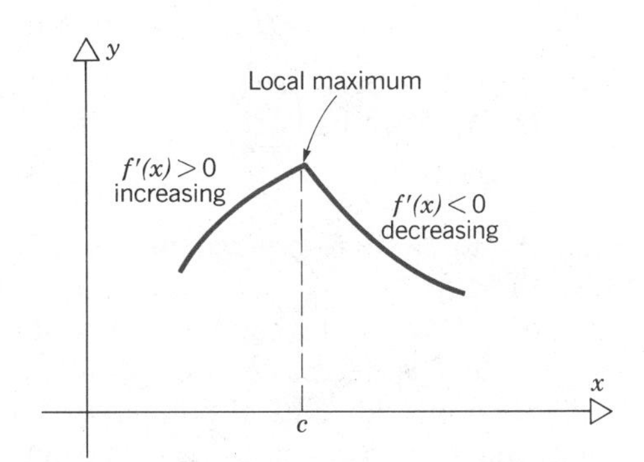

Figure 4.2.6
The first derivative test.

precisely, if $f'(x) > 0$ on $(c - \delta, c)$ and if f is continuous on $(c - \delta, c]$, then, according to Theorem 4.2.1, f is increasing on $(c - \delta, c]$. Hence $f(c) \geq f(x)$ for each x in $(c - \delta, c]$. Similarly, f is decreasing on $[c, c + \delta)$, so $f(c) \geq f(x)$ for each x in $[c, c + \delta)$. Consequently, $f(c) \geq f(x)$ for all x in $(c - \delta, c + \delta)$, and therefore f has a local maximum at c. □

EXAMPLE 1 (*Continued*)
Find intervals in which the function

$$f(x) = 2x^3 - 3x^2 - 12x + 19, \qquad -\infty < x < \infty$$

is increasing or decreasing. Determine the nature of the critical points of this function.

We previously showed (Eq. 5) that

$$f'(x) = 6(x - 2)(x + 1),$$

and that the only critical points are $x = 2$ and $x = -1$. It is easy to see that $f'(x) > 0$ for $x < -1$ and for $x > 2$, and that $f'(x) < 0$ for $-1 < x < 2$. Hence, by Theorem 4.2.1, f is increasing on $(-\infty, -1]$ and $[2, \infty)$, and f is decreasing on $[-1, 2]$. Consequently, by Theorem 4.2.2, $x = -1$ is a local maximum point and $x = 2$ is a local minimum point. Since f takes on larger and larger values as x increases without bound, this function has no global maximum; for a similar reason it also has no global minimum. The information that we have found here is incorporated in the graph of f in Figure 4.2.3. ∎

EXAMPLE 2 (*Continued*)
Find intervals in which the function

$$f(x) = x - 2 \sin x, \qquad 0 \le x \le 2\pi$$

is increasing or decreasing. Use Theorem 4.2.2 to classify the critical points of f.

In Eq. 7 we had

$$f'(x) = 1 - 2 \cos x.$$

Thus $f'(x) > 0$ is equivalent to $\cos x < \frac{1}{2}$. This inequality is satisfied for $\pi/3 < x < 5\pi/3$, so by Theorem 4.2.1 f is increasing on the interval $[\pi/3, 5\pi/3]$. Similarly, $f'(x) < 0$ when $\cos x > \frac{1}{2}$, that is, for $0 \le x < \pi/3$ or for $5\pi/3 < x \le 2\pi$. Hence f is decreasing on $[0, \pi/3]$ and $[5\pi/3, 2\pi]$. Applying Theorem 4.2.2 to the two interior critical points, we conclude that $x = \pi/3$ is a local minimum point and $x = 5\pi/3$ is a local maximum point. We can use Theorem 4.2.1 to determine the nature of the endpoints. Since $f'(x) < 0$ to the right of the left endpoint $x = 0$, it follows that f is decreasing there, and consequently $x = 0$ is a local maximum point. A similar argument shows that the right endpoint $x = 2\pi$ is a local minimum point. The information that we have found here is shown in Figure 4.2.4. Note also that $f'(x)$ is a periodic function with period 2π. Thus the pattern of intervals in which f increases or decreases repeats itself if the domain of f is extended beyond $[0, 2\pi]$. ∎

EXAMPLE 3 (*Continued*)
For the function

$$f(x) = |x^2 - 1|, \qquad -2 \le x \le 2$$

find intervals in which f increases or decreases. Also classify the critical points of f.

From Eq. 10 we have

$$f'(x) = \begin{cases} 2x, & x < -1 \text{ or } x > 1; \\ -2x, & -1 < x < 1. \end{cases}$$

Thus, by Theorem 4.2.1, f is increasing on $[-1, 0]$ and $[1, 2]$, while f is decreasing on $[-2, -1]$ and $[0, 1]$. Consequently, by Theorem 4.2.2, the points $x = \pm 1$ are local minimum points, and $x = 0$ is a local maximum point. The endpoints $x = \pm 2$ are both local maximum points since, by Theorem 4.2.1, f is decreasing to the right of $x = -2$ and increasing to the left of $x = 2$. Refer to the graph of f in Figure 4.2.5. Among other things, this example illustrates that the first derivative test can be used to classify critical points ($x = \pm 1$ in this case) where f is not differentiable, so long as it is continuous at such points. ∎

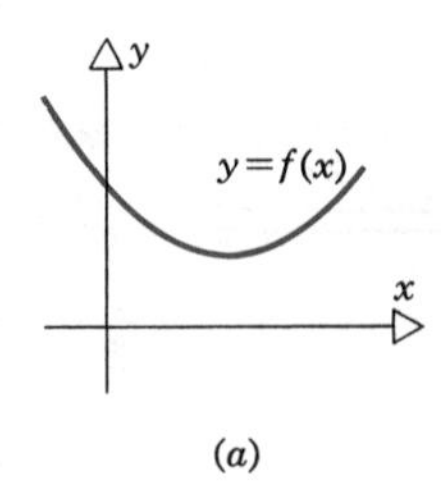

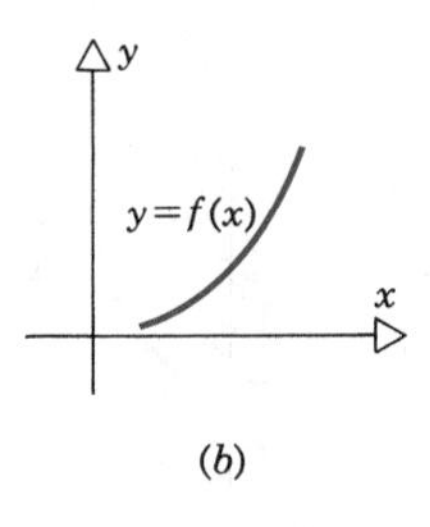

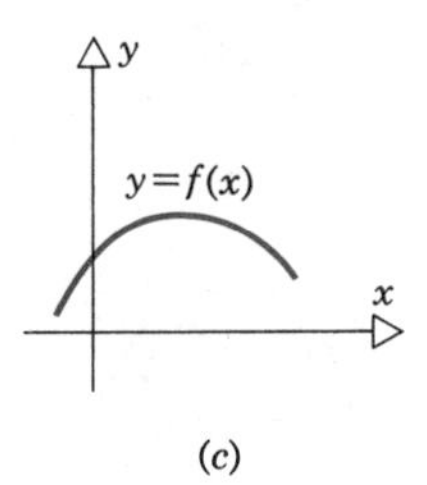

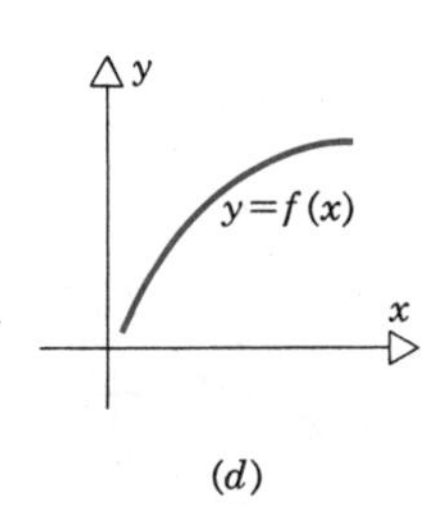

Figure 4.2.7 (*a*) f is concave up. (*b*) f is concave up. (*c*) f is concave down. (*d*) f is concave down.

Concavity

An inspection of Figure 4.2.7 suggests that arcs of curves can be divided into two classes. The curves in Figure 4.2.7*a* and 4.2.7*b* are said to be concave up, while those in Figure 4.2.7*c* and 4.2.7*d* are concave down. More precisely, suppose that f is continuous on $[a, b]$ and differentiable on (a, b). If, moreover, f' is increasing on (a, b), then f is **concave up** on $[a, b]$, while is f' is decreasing on (a, b), then f is **concave down** on $[a, b]$.

Now let us apply Theorem 4.2.1 to f'. It follows that if f is twice differentiable, and if $f''(x) > 0$ on (a, b), then f' is increasing and f is concave up. On the other hand, if $f''(x) < 0$ on (a, b), then f' is decreasing and f is concave down. We state this result as a theorem.

Theorem 4.2.3

Let f be continuous on $[a, b]$ and at least twice differentiable on (a, b). If $f''(x) > 0$ on (a, b), then f is concave up on $[a, b]$, while if $f''(x) < 0$ on (a, b), then f is concave down on $[a, b]$.

By referring to Figures 4.2.3 and 4.2.4 we can observe that if c is a critical point where $f'(c) = 0$, then c is a local minimum point if f is concave up in an interval about c, and a local maximum point if f is concave down in such an interval. This is the substance of the next theorem.

Theorem 4.2.4

(Second derivative test)
Let the function f be defined on an open interval containing the critical point c where $f'(c) = 0$, and let f'' be continuous on this interval.

(a) If $f''(c) < 0$, then c is a local maximum point.
(b) If $f''(c) > 0$, then c is a local minimum point.
(c) If $f''(c) = 0$, then no conclusion is possible without further investigation.

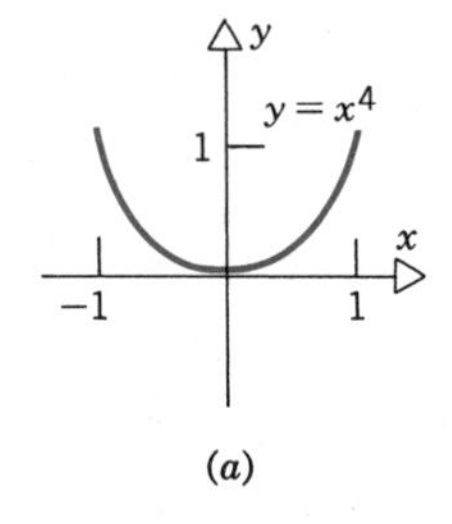

(a)

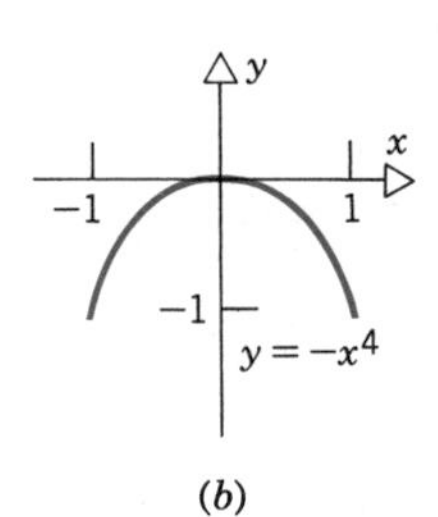

(b)

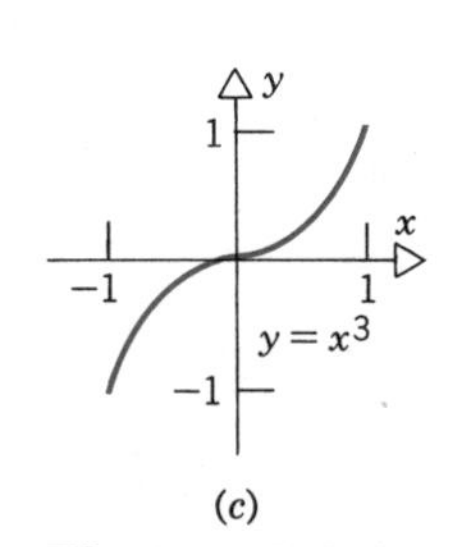

(c)

Figure 4.2.8
In each case $f''(0) = 0$. (*a*) Relative minimum at (0, 0). (*b*) Relative maximum at (0, 0). (*c*) No extremum at (0, 0).

Proof. Consider Part (a). If $f''(c) < 0$ and f'' is continuous on some interval containing c, then according to Theorem 2.6.1 there is an interval $(c - \delta, c + \delta)$ in which $f''(x)$ is always negative. Hence, by Theorem 4.2.1 applied to f', it follows that f' is decreasing on this interval. The combination that f' is decreasing and that $f'(c) = 0$ implies that $f'(x) > 0$ on $(c - \delta, c)$ and $f'(x) < 0$ on $(c, c + \delta)$. These last two results guarantee, by Part (a) of Theorem 4.2.2, that c is a local maximum point. The proof of Part (b) is similar.

We can prove Part (c) by means of examples. Let $f_1(x) = x^4$, $f_2(x) = -x^4$, and $f_3(x) = x^3$, and consider the interval $(-\infty, \infty)$ (See Figure 4.2.8). In each case both the first and second derivatives are zero at $x = 0$. However, f_1 has a local minimum and f_2 a local maximum at $x = 0$; further f_3 has neither a maximum nor a minimum there. Hence, if $f''(x)$ is zero at a critical point, no conclusion can be drawn without further investigation. □

Note that Theorem 4.2.4 applies only to critical points where $f'(x)$ is zero, whereas Theorem 4.2.2 also applies to critical points where f' does not exist. Thus the first derivative test is somewhat more general.

EXAMPLE 1 (*Continued*)

Determine where the function

$$f(x) = 2x^3 - 3x^2 - 12x + 19, \qquad -\infty < x < \infty$$

is concave up and where it is concave down. Apply the second derivative test to the critical points of f.

By referring to Eq. 5 and differentiating a second time, we obtain

$$f''(x) = 6(2x - 1). \tag{14}$$

Thus, $f''(x) < 0$ for $x < \frac{1}{2}$, so by Theorem 4.2.3 f is concave down on $(-\infty, \frac{1}{2}]$. Similarly, f is concave up on $[\frac{1}{2}, \infty)$. The concavity of f is indicated in Figure 4.2.3. If we evaluate $f''(x)$ at the critical points $x = -1$ and $x = 2$, we find that

$$f''(-1) = -12, \qquad f''(2) = 18.$$

Therefore, by Theorem 4.2.4, $x = -1$ is a local maximum point and $x = 2$ is a local minimum point. ∎

EXAMPLE 2 *(Continued)*

Determine where the function

$$f(x) = x - 2 \sin x, \qquad 0 \leq x \leq 2\pi$$

is concave up and concave down, respectively, and classify its critical points by means of Theorem 4.2.4.

From Eq. 7 we obtain

$$f''(x) = 2 \sin x. \tag{15}$$

Therefore, by Theorem 4.2.3, f is concave up on $[0, \pi]$ and concave down on $[\pi, 2\pi]$, as shown in Figure 4.2.4. At the critical points $x = \pi/3$ and $x = 5\pi/3$ we have

$$f''\left(\frac{\pi}{3}\right) = \sqrt{3}, \qquad f''\left(\frac{5\pi}{3}\right) = -\sqrt{3}.$$

Hence, by Theorem 4.2.4, the former is a local minimum point and the latter is a local maximum point. Again, the periodicity of f'' means that the pattern of concavity is repeated if the domain of f is extended. ∎

EXAMPLE 3 *(Continued)*

Consider again the function

$$f(x) = |x^2 - 1|, \qquad -2 \leq x \leq 2.$$

Determine where f is concave up and concave down, respectively, and investigate its critical points by using Theorem 4.2.4.

From Eq. 10 we obtain

$$f''(x) = \begin{cases} 2, & x < -1 \quad \text{or} \quad x > 1; \\ -2, & -1 < x < 1. \end{cases} \tag{16}$$

Thus f is concave up on $[-2, -1]$ and $[1, 2]$, while it is concave down in $[-1, 1]$ (see Figure 4.2.5). The second derivative test is applicable only to the critical point at $x = 0$. Since $f''(0) = -2$, this point is a local maximum point. ∎

We close this section with a few words of summary and advice.

1. If f is a continuous function on a closed bounded interval, then its global extrema can be found by locating all critical points, evaluating f at each one, and identifying the largest and smallest of the resulting set of numbers.
2. If f is continuous, but the interval is not closed and bounded, then a global maximum, or a global minimum, or both, may fail to exist. To investigate these possibilities, examine the behavior of f as the independent variable approaches the endpoints, or $\pm\infty$, as the case may be, as well as at the critical points.
3. Theorem 4.2.2 (first derivative test) and Theorem 4.2.4 (second derivative test) can be used to investigate the local behavior of f near a critical point. The first derivative test is more general, but often both are applicable. In that event, use whichever one is the more convenient.

PROBLEMS

In each of Problems 1 through 14

(a) Find all critical points of the given function.

(b) Find intervals where the function is increasing; decreasing.

(c) Find intervals where the function is concave up; concave down.

(d) Determine whether each critical point is a local maximum point, a local minimum point, or neither.

1. $f(x) = x^2 + 4x + 2$

2. $f(x) = -2x^2 + 3x - 4$

3. $f(x) = x^3 - 3x^2 - 9x + 4$

4. $f(x) = 3x^5 - 25x^3 + 60x - 36$

5. $f(x) = \begin{cases} \dfrac{1}{x+1}, & x \neq -1 \\ 0, & x = -1 \end{cases}$

6. $f(x) = x + \dfrac{4}{x}, \quad x \neq 0$

7. $f(x) = 1 - x^{2/3}$

8. $f(x) = \begin{cases} \dfrac{x+2}{x-2}, & x \neq 2 \\ 2, & x = 2 \end{cases}$

9. $f(x) = x - 2\cos x, \quad -\pi \leq x \leq \pi$

10. $f(x) = 2\cos^2 x - \sin^2 x, \quad 0 \leq x \leq 2\pi$

11. $f(x) = |x^3 - 12x|, \quad -4 \leq x \leq 4$

12. $f(x) = \left|\dfrac{x}{x+1}\right|, \quad x \neq -1$

13. $f(x) = \sqrt{4 - x^2}, \quad -2 \leq x \leq 2$

14. $f(x) = \sin x + \cos x, \quad 0 \leq x \leq 2\pi$

In each of Problems 15 through 28, find the points (if there are any) in the given interval where the given function takes on its global maximum and global minimum values, respectively. These are the same functions as in Problems 1 through 14.

15. $f(x) = x^2 + 4x + 2$ (a) on $[-4, 0)$ (b) on $(-\infty, \infty)$

16. $f(x) = -2x^2 + 3x - 4$ (a) on $(-\infty, 2]$ (b) on $[-2, 2]$

17. $f(x) = x^3 - 3x^2 - 9x + 4$ (a) on $[-1, 5]$ (b) on $(-\infty, \infty)$

18. $f(x) = 3x^5 - 25x^3 + 60x - 36$ (a) on $[-1, 2]$ (b) on $[0, 3]$

19. $f(x) = \begin{cases} \dfrac{1}{x+1}, & x \neq -1 \\ 0, & x = -1 \end{cases}$ (a) on $[-1, 1]$ (b) on $[0, 2]$

20. $f(x) = x + \dfrac{4}{x}$ (a) on $(0, 4)$ (b) on $[1, 3]$

21. $f(x) = 1 - x^{2/3}$ (a) on $(-\infty, \infty)$ (b) on $[-1, 1]$

22. $f(x) = \begin{cases} \dfrac{x+2}{x-2}, & x \neq 2 \\ 2, & x = 2 \end{cases}$ (a) on $[2, \infty)$ (b) on $[0, 1]$

23. $f(x) = x - 2\cos x$ (a) on $[-\pi, \pi]$
(b) on $\left[-\dfrac{\pi}{2}, \dfrac{\pi}{2}\right]$

24. $f(x) = 2\cos^2 x - \sin^2 x$ (a) on $[0, 2\pi]$
(b) on $(0, 2\pi)$

25. $f(x) = |x^3 - 12x|$ (a) on $[-4, 4]$
(b) on $(-3, 3)$

26. $f(x) = \left|\dfrac{x}{x+1}\right|$ (a) on $(-1, \infty)$
(b) on $[-\frac{1}{2}, \frac{1}{2}]$

27. $f(x) = \sqrt{4 - x^2}$ (a) on $[-2, 2]$
(b) on $(0, 1)$

28. $f(x) = \sin x + \cos x$ (a) on $[0, 2\pi]$
(b) on $\left[-\dfrac{\pi}{2}, \dfrac{\pi}{2}\right]$

In each of Problems 29 through 34, draw the graph of a differentiable function f having the given properties.

29. Domain is $(-\infty, \infty)$; local maximum at -2; global maximum at 3; local minimum at 0; no global minimum.

30. Domain is $(0, \infty)$; global minimum at 4; local maximum at 6; no global maximum; $f(x) \to 3$ as $x \to \infty$.

31. Domain is $[-2, 2]$; global minimum at -1; global maximum at -2; local maximum at 1; concave up on $[-2, 0]$; concave down on $[0, 2]$.

32. Domain is $(-\infty, \infty)$; global maximum at 0; global minima at -2 and 2.

33. Domain is $(0, 2]$; global maximum at 1; local minimum at 2; no global minimum.

34. Domain is $(-2, \infty)$; concave down for $(-2, 1]$; concave up for $[1, \infty)$; local maximum at 0; local minimum at 3; no global maximum or global minimum.

35. Determine a, b, and c so that the graph of

$$y = ax^2 + bx + c$$

passes through the points $(0, 1)$ and $(3, 0)$, and has a maximum at $x = 1$.

36. Find a and b so that $f(x) = ax/(x^2 + b^2)$ has a local maximum at $x = 3$ and so that $f'(0) = \frac{1}{3}$.

37. Prove the following partial converse of Theorem 4.2.1: If the function f is differentiable and increasing on the interval (a, b), then $f'(x) \geq 0$ at each point x in (a, b).
Hint: Assume the contrary, namely, that there is a point c in (a, b) such that $f'(c) < 0$. Then use the definition of $f'(c)$ to obtain a contradiction.

38. Let

$$f(x) = \begin{cases} x \sin \dfrac{1}{x}, & x > 0; \\ 0, & x = 0. \end{cases}$$

Show that the endpoint $x = 0$ is neither a local minimum point nor a local maximum point.
Hint: Show that, in every interval $[0, \delta)$, there are points where $f(x) > 0$ and other points where $f(x) < 0$.

4.3 APPLICATIONS OF MAXIMA AND MINIMA

The observation that at a maximum or minimum point the derivative, if it exists, must be zero (Theorem 4.1.1) is the key to solving a large class of optimization problems. Indeed, it provides the starting point for methods of dealing with classes of problems considerably more complicated than those to be mentioned here. In this section we present several examples of simple optimization problems. While the illustrative problems to follow are quite varied in their origin, all are similar in the following respects.

In the first place, each is initially stated in words, including a statement as to which quantity is to be optimized. The first step is to formulate the problem mathematically; this involves identifying the variables and parameters in the problem, assigning a letter to each, and writing one or more equations relating them.

Frequently it is helpful to draw a sketch or a diagram to help in visualizing the situation. The goal is to obtain a function of a single variable to be maximized or minimized on some specified domain. In some cases all of the necessary information is contained in the statement of the problem, while in other cases if may be necessary to use other information, for example, facts drawn from geometry or elementary physics. The formulation of the problem is sometimes complicated by the need to carry out suitable algebraic manipulations in order to express the quantity to be optimized as a function of a single variable. This first step of setting up the problem in mathematical terms is often the most difficult precisely because there are no formal rules for doing it—the procedure depends on the problem, and in some cases a good deal of imagination and insight may be required. The term *mathematical modeling* is often used to describe this process whereby one in effect translates a problem from words to mathematical symbols and equations.

Having formulated the problem as previously indicated, we must next solve it mathematically. Here one calls upon the discussion and theorems about maxima and minima presented in Section 4.2. The degree of difficulty involved in this step depends entirely on how complicated the function is that must be optimized. In any case, however, there is a definite procedure to be followed.

Finally, having solved the problem in a mathematical sense, it is necessary to interpret the solution in the context of the problem as originally posed. Particular care should be taken to note whether the solution seems "reasonable." This serves as a very valuable check on the mathematical work. If the answer obtained is clearly unreasonable in terms of the original statement of the problem, then this indicates that a mistake has been made either in the mathematical formulation or solution. Of course, even if the answer is reasonable, this is no guarantee that it is right, but it does provide a degree of corroboration.

EXAMPLE 1

(constant sum—maximum product problem)

Find two nonnegative numbers whose sum is 10 and whose product is as large as possible.

We begin to formulate this problem by letting x and y be real numbers, as yet undetermined. Then $P = xy$ is the quantity to be maximized, where $x + y = 10$. By expressing y in terms of x we can write

$$\begin{aligned} P(x) &= x(10 - x) \\ &= 10x - x^2. \end{aligned} \tag{1}$$

Since both x and $y = 10 - x$ are to be nonnegative, we must restrict x to the interval $0 \leq x \leq 10$. This completes the formulation of the problem in mathematical terms: we want to maximize the function P given by Eq. 1 on the domain $[0, 10]$.

To solve the problem we first seek the critical points. The derivative of P is

$$P'(x) = 10 - 2x; \tag{2}$$

hence $P'(x) = 0$ when $x = 5$. This point, together with the endpoints $x = 0$ and $x = 10$, constitute the set of critical points for this problem.

Since P is continuous on the closed bounded interval $[0, 10]$, we can determine the maximum of P by evaluating $P(x)$ at each critical point. From Eq. 1 we have

$$P(0) = 0, \qquad P(5) = 25, \qquad P(10) = 0.$$

Thus the maximum of P is 25 when $x = 5$, and the two numbers that solve the problem are $x = 5$ and $y = 5$. ∎

EXAMPLE 2

(constant volume—minimum area problem)

A manufacturer wishes to construct a right circular cylindrical can so as to hold a given volume V. How should the radius and height be chosen so as to minimize the surface area S of the can, including the top and bottom?

Let r be the radius of the base and h the height of the can (see Figure 4.3.1).

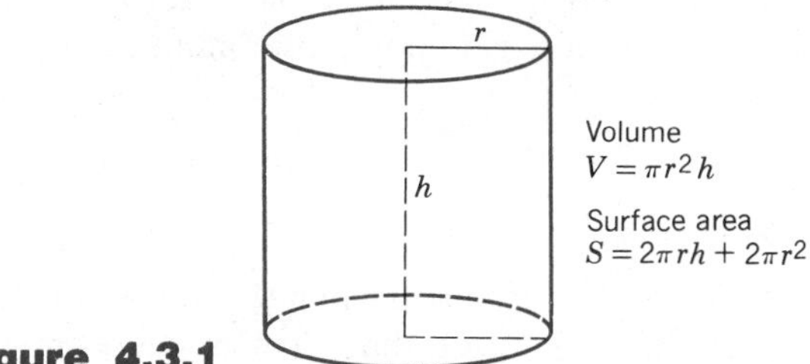

Figure 4.3.1

Then the surface area S is given by

$$S = 2\pi rh + 2\pi r^2, \tag{3}$$

where the first term is the area of the lateral surface, while the second is the area of the top and bottom of the can. Observe that S depends on the two variables r and h. To write S in terms of a single variable we need another relation between r and h. This comes from the expression

$$V = \pi r^2 h \tag{4}$$

for the volume of the can. If we solve Eq. 4 for h and substitute into Eq. 3 we find that

$$S = \frac{2V}{r} + 2\pi r^2. \tag{5}$$

Since V is a constant, Eq. 5 does give S as a function of the single variable r. Further, r must be positive, so we will assume that the domain of S is $(0, \infty)$. Thus the formulation is now complete: to minimize S, given by Eq. 5, on the domain $0 < r < \infty$.

Let us think about the problem qualitatively for a moment before proceeding with the mathematical solution. Clearly if r is very small the can will be tall and thin; from Eq. 5 the area S will be very large, since in this case the lateral surface is great. Similarly, if r is large, the can is short and fat; again it follows from Eq. 5 that S is large, since in this case the top and bottom are large. We suspect that for some intermediate value of r, the area S will be minimum.

To solve the problem we differentiate Eq. 5, obtaining

$$\frac{dS}{dr} = -\frac{2V}{r^2} + 4\pi r. \tag{6}$$

Hence $dS/dr = 0$ when $4\pi r^3 = 2V$, or when

$$r = r_c = (V/2\pi)^{1/3}. \tag{7}$$

This is the only critical point for this problem, since there are no endpoints and the derivative exists at each point in $(0, \infty)$. On computing the second derivative, we have

$$\frac{d^2S}{dr^2} = \frac{4V}{r^3} + 4\pi > 0, \tag{8}$$

and the point $r_c = (V/2\pi)^{1/3}$ is therefore a local minimum point. Actually, we can conclude that r_c is a global minimum point. To show this, we rewrite Eq. 6 in the form

$$\frac{dS}{dr} = \frac{4\pi}{r^2}\left(r^3 - \frac{V}{2\pi}\right) = \frac{4\pi}{r^2}(r^3 - r_c^3); \tag{9}$$

consequently $dS/dr < 0$ when $0 < r < r_c$ and $dS/dr > 0$ when $r > r_c$. It then follows from Theorem 4.2.1 that $S(r)$ is decreasing for $0 < r \leq r_c$ and increasing for $r \geq r_c$. Hence $r_c = (V/2\pi)^{1/3}$ is a global minimum point.

From Eq. 4 the corresponding value of h is

$$h = 2\,(V/2\pi)^{1/3}. \tag{10}$$

Thus for any volume V the can of minimum surface area is attained by making the height equal to the diameter. The minimum value of the surface area is $S = 3(2\pi V^2)^{1/3}$. For instance, for a can whose volume is one quart (57.75 in.^{3}) we have $r_c \cong 2.095$ in. The corresponding values of h and S are $h \cong 4.189$ in. and $S \cong 82.71$ in.^{2}

Experience shows that most cans are not made with the ratio of height/diameter equal to the theoretical optimum value of 1.00. For instance, measurements on a 1-qt paint can, a 1-qt oil can, and a 12-oz soft-drink can gave values of approximately 1.12, 1.38, and 1.81, respectively.

Now suppose that it is decided that a 1-qt can is to have a maximum radius of 1.75 in., so as to permit most people to grasp the can easily in one hand. Find the dimensions of the can so as to minimize the surface area under these conditions.

Proceeding as above we find that $dS/dr = 0$ when $r = (V/2\pi)^{1/3} \cong 2.095$ in. However, r is restricted to the interval $[0, 1.75]$, so this value of r is no longer an acceptable solution. Since $dS/dr < 0$ throughout $(0, 1.75]$, the minimum value of S for r in this interval is attained at the endpoint $r = 1.75$. The corresponding values of h and S are $h \cong 6.002$ in. and $S \cong 85.24$ in^2. The latter value is slightly more than 3 percent higher than the theoretical minimum of $S = 82.71$ in.^{2} obtained before. ∎

EXAMPLE 3

(traffic light problem)

Suppose that a traffic light of weight W lb is to be suspended from the center of a cable stretched across a street of width $2a$ ft (see Figure 4.3.2). Suppose further that the cost of the cable in dollars per foot is equal to kT, where T is the tension (in pounds) in the cable, and k is a given constant. Determine the angle of inclination θ of the cable so as to minimize the cost.

If we look at the problem qualitatively first, we observe that if we make the cable nearly as short as possible, then we must stretch it tightly, the cost per foot will be large, and the total cost will also be large. On the other hand, if we allow it to sag deeply, then the tension (and hence the cost per foot) will be relatively small, but a very long cable will be required, and again the total cost will be large. We may suspect that we can achieve the minimum cost by properly balancing these two effects.

Let x be the sag at the center of the cable. Then from Figure 4.3.2 the length of the cable is $2\sqrt{a^2 + x^2}$ and the total cost is

$$C = 2kT\sqrt{a^2 + x^2}. \tag{11}$$

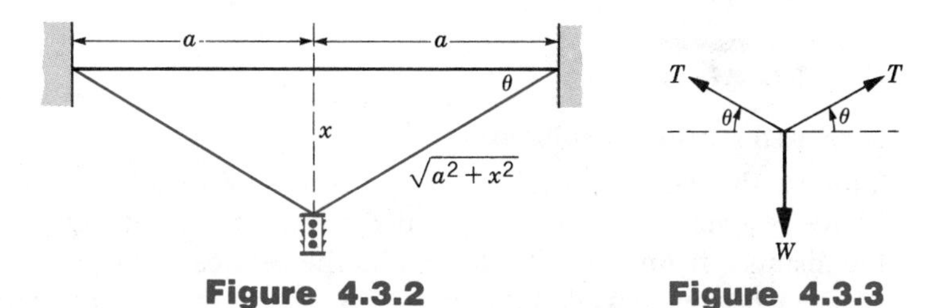

Figure 4.3.2

Figure 4.3.3

However, T depends upon x. To determine this relation a principle from elementary physics is required. The forces acting on the point of support of the traffic light are shown in Figure 4.3.3. If the light is to hang motionless (in equilibrium), then the vertical forces must be balanced; that is, the weight of the light must equal the vertical component of the tension in the cable. Therefore we have

$$\begin{aligned} W &= 2T \sin \theta \\ &= 2T \frac{x}{\sqrt{a^2 + x^2}}. \end{aligned} \tag{12}$$

Solving Eq. 12 for T and substituting for T in Eq. 11 yields

$$C = kW \frac{a^2 + x^2}{x} = kW \left(\frac{a^2}{x} + x \right), \tag{13}$$

where x is in the interval $(0, \infty)$. This completes the formulation of the problem: To minimize the function C in Eq. 13 on the domain $(0, \infty)$. The situation is as our qualitative analysis indicated; from Eq. 13 C will be very large if x is either very large (because of the term x) or very small (because of the term a^2/x).

Differentiating Eq. 13 we obtain

$$\frac{dC}{dx} = kW\left(-\frac{a^2}{x^2} + 1\right); \tag{14}$$

hence $dC/dx = 0$ when $x = a$. Since

$$\frac{d^2C}{dx^2} = \frac{2kWa^2}{x^3} > 0, \tag{15}$$

this critical point is indeed a local minimum point. An argument similar to that in Example 2 shows that $x = a$ is actually a global minimum point. We conclude that the most economical configuration will occur when the sag is equal to one-half the span, that is, for $\theta = \pi/4$.

Of course, other factors may need to be considered; for example, supporting posts may be required on each side of the street. If the traffic light is required to be, say, 15 ft above the ground, and if the street is 40 ft wide, then the preceding analysis would lead to supports 35 ft high. It may prove to be better to stretch the cable more tightly so as to be able to use shorter supports at each side. Problem 10 explores this question. ∎

EXAMPLE 4

(stranded motorist problem)

Suppose that your car runs out of gas at point P on a lonely road (see Figure 4.3.4). There is a service station at point Q. Further, the distance from P to O is 3 mi, the distance from O to Q is 6 mi, and the service station closes in 2.5 hr. You must decide whether to walk along the road, or whether to head straight across country, or whether to walk to an intermediate point R between O and Q and then go along the road to the service station. If you can walk 4 mi/hr on the road, and 2 mi/hr across country, what route should you choose so as to reach the service station as soon as possible? Most important, can you get there at all before it closes?

If x is the distance from O to R, then the time required to walk straight from P to R is $\sqrt{9 + x^2}/2$; the time required to walk from R to Q is $(6 - x)/4$. Hence the total time T is given by

$$T(x) = \frac{\sqrt{9 + x^2}}{2} + \frac{6 - x}{4}. \tag{16}$$

Differentiating Eq. 16 yields

$$T'(x) = \frac{1}{2}\frac{x}{\sqrt{9 + x^2}} - \frac{1}{4}, \tag{17}$$

and it follows that $T'(x) = 0$ when $x = \sqrt{3}$. Further

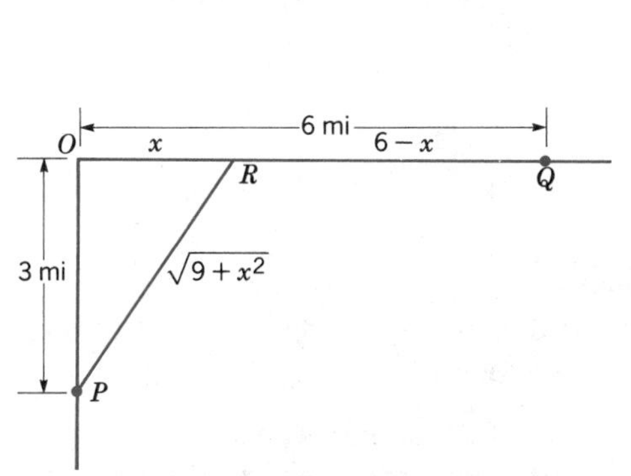

Figure 4.3.4

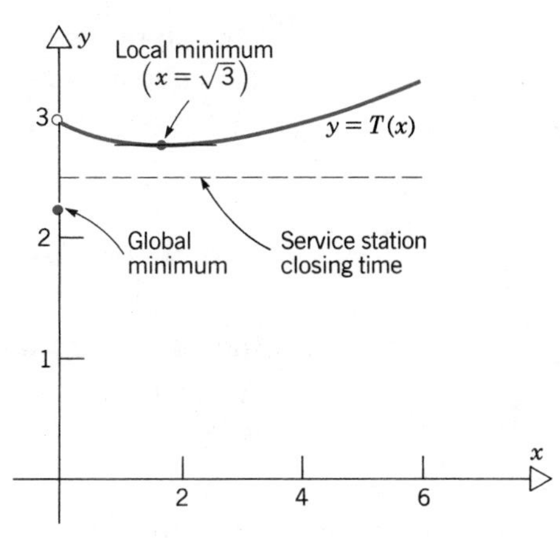

Figure 4.3.5

$$T''(x) = \frac{9}{2(9 + x^2)^{3/2}} > 0, \tag{18}$$

so that the critical point $x = \sqrt{3}$ is a local minimum point. Evaluating Eq. 16 when $x = \sqrt{3}$, we have

$$T(\sqrt{3}) = \tfrac{1}{4}(3\sqrt{3} + 6) \cong 2 \text{ hr } 48 \text{ min}. \tag{19}$$

There are no other points where $T'(x)$ is zero, so we might be tempted to argue as in previous examples that $x = \sqrt{3}$ is actually a global minimum point. However, such an argument would be erroneous in the present example.

To see that this is so, let us consider the endpoints. If $x = 6$, then from Eq. 16 the time of travel is

$$T(6) = \frac{\sqrt{45}}{2} \cong 3 \text{ hr } 21 \text{ min}. \tag{20}$$

On the other hand, if $x = 0$, then T cannot be computed from Eq. 16; substituting $x = 0$ into Eq. 16 corresponds to walking along the road from P to O at only 2 mi/hr rather than 4 mi/hr. For $x = 0$, the proper value of T is found by dividing the total distance along the road (9 mi) by 4 mi/hr; thus the correct definition of the function T is

$$T(x) = \begin{cases} \dfrac{\sqrt{9 + x^2}}{2} + \dfrac{6 - x}{4}, & 0 < x \le 6 \\[2ex] \dfrac{9}{4}, & x = 0. \end{cases} \tag{21}$$

Hence, if you stay on the road, you *can* reach the service station with time to spare. The graph of Eq. 21 is shown in Figure 4.3.5. It is the discontinuity in T at the left endpoint that causes $x = \sqrt{3}$ not to be a global minimum point.

In this example we deliberately failed to define the function T completely in the beginning. Our purpose was to emphasize to you as strongly as possible that care must always be taken to identify the domain of the function involved, and to see that the function is correctly defined throughout its entire domain. In particular, the endpoints must sometimes receive special treatment. ∎

EXAMPLE 5

(maximum profit problem)

A certain wholesale paint dealer, in buying and distributing x cases of paint per week, incurs the following expenses.

(a) Fixed costs (rent, etc.) of \$1200 per week.
(b) An expense of \$$60x$ per week representing the cost of the paint itself to the dealer.
(c) A cost of \$$x^2/24$ per week for storing the inventory, handling accounts, etc.

Sales can be maintained at a rate of x cases per week at a price of p dollars per case, where

$$x = 2160 - 24p. \tag{22}$$

Finally, due to space and other limitations, the dealer's maximum level of operation is the distribution of 1000 cases per week. Determine the price p at which the dealer should sell each case in order to maximize the weekly profit of the business.

The dealer's total cost per week $C(x)$ is given by the sum of the individual costs listed above, namely

$$C(x) = 1200 + 60x + \frac{x^2}{24}. \tag{23}$$

The gross weekly income $R(x)$ from selling x cases at p dollars per case is px. To express R in terms of x we first solve Eq. 22 for p,

$$p = \frac{2160 - x}{24} = 90 - \frac{x}{24}. \tag{24}$$

Then

$$R(x) = x\left(90 - \frac{x}{24}\right) = 90x - \frac{x^2}{24}. \tag{25}$$

The dealer's weekly profit $P(x)$ is given by

$$\begin{aligned} P(x) &= R(x) - C(x) \\ &= 90x - \frac{x^2}{24} - \left(1200 + 60x + \frac{x^2}{24}\right) \\ &= -\frac{x^2}{12} + 30x - 1200. \end{aligned} \tag{26}$$

The domain of the function P is determined by the facts that x is intrinsically nonnegative and that also $x \leq 1000$. Thus we seek the maximum of P on the

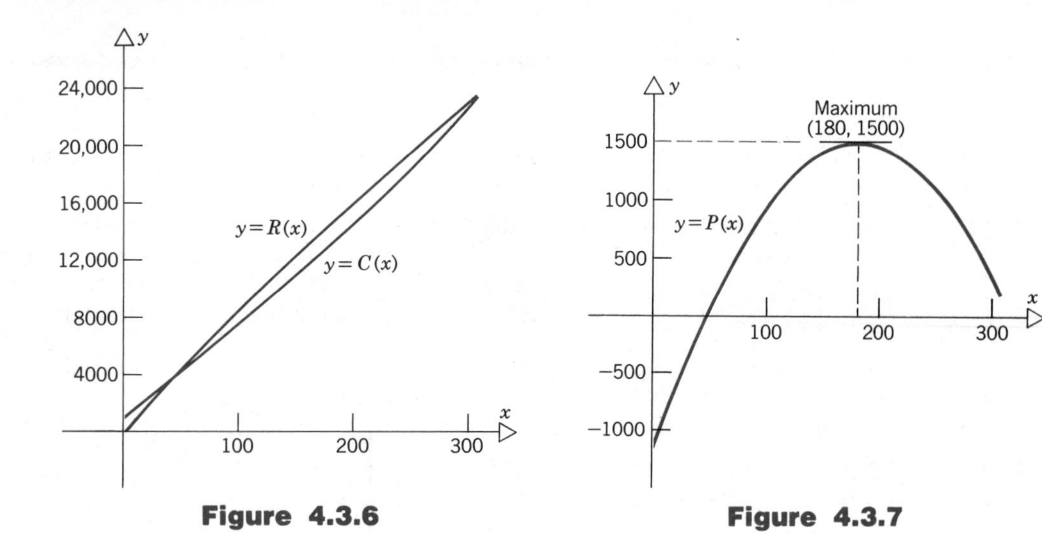

Figure 4.3.6

Figure 4.3.7

interval [0, 1000]. Graphs of the cost and income functions C and R, respectively, are shown in Figure 4.3.6; the profit function P is shown in Figure 4.3.7.

By differentiating Eq. 26 we obtain

$$P'(x) = -\frac{x}{6} + 30; \tag{27}$$

hence $P'(x) = 0$ when $x = 180$. Since $P'(x)$ exists at all points in (0, 1000), the only critical points are $x = 180$ and the endpoints $x = 0$ and $x = 1000$. From Eq. 27 it is clear that $P'(x) > 0$ for $0 < x < 180$ and $P'(x) < 0$ for $180 < x < 1000$. Thus $x = 180$ is the global maximum point for P on the interval [0, 1000]. The corresponding value of P is $P(180) = \$1500$. The optimum price per case p is obtained by setting $x = 180$ in Eq. 24; thus

$$P = 90 - \frac{180}{24} = \$82.50 \text{ per case} \tag{28}$$

is the price that the dealer should charge in order to maximize profits under the given conditions.

In general, the business world is a great deal more complex than this example may suggest. For one thing, in the real world, the relation between the unit price p and the level of sales x is unlikely to be a simple linear equation. Nevertheless, the example illustrates an important economic principle, namely, profits are maximized when

$$C'(x) = R'(x).$$

In words, the optimum level of operation occurs when the **marginal cost** $C'(x)$ is exactly equal to the **marginal income** $R'(x)$. Otherwise, it will pay either to expand or to contract the level of business activity, depending on the sign of $R'(x) - C'(x)$. ∎

PROBLEMS

1. Find two positive numbers x and y whose product is k and whose sum is as small as possible.
2. Of all rectangles of given diagonal, show that the square has the greatest area.
3. Of all rectangles of given perimeter, show that the square has the largest area.
4. Of all isosceles triangles that can be inscribed in a given circle, show that the equilateral triangle has the largest area.
5. Of all rectangles that can be inscribed in a given circle, show that the square has the largest area.
6. Find the radius r and height h of a right circular cylindrical can whose surface area S (including top and bottom) is fixed, and whose volume V is as large as possible.
7. Consider a right circular cylindrical can of radius r and height h. Suppose that the lateral surface is formed from a single rectangular piece of metal, and that the top and bottom are cut from square pieces of side $2r$. Find the dimensions of the can of given volume V constructed from a minimum amount of material (taking into account the wasted corners of the square pieces). Find the height/diameter ratio h/d for the minimizing can.
8. Suppose that a right circular cylindrical can, including top and bottom, is constructed from material costing α dollars per square inch. Suppose further that there is an additional cost of fabrication given by β dollars per inch of the circumference of the top and bottom of the can. Find an algebraic equation whose solution is the radius r of the can of given volume V and of minimum cost.
9. A box with a square base of side x and with height h is to be constructed so as to hold volume V. Find the dimensions of the box of minimum surface area if the box is (a) closed; (b) open on top.
10. Referring to Example 3, suppose that the traffic light is to be hung at a height of h ft above the ground, and that supports at each side of the street cost m dollars per foot of height. Show that in this case the most economical configuration is for the cable to sag a distance x given by

$$x = a\left(\frac{kW}{2m + kW}\right)^{1/2}.$$

Note that $x < a$, and that x decreases as m increases.

11. A hallway of width a feet intersects a corridor of width b feet at right angles (see Figure 4.3.8). Find the length l of the longest straight piece of pipe that can be carried horizontally from the hallway into the corridor.

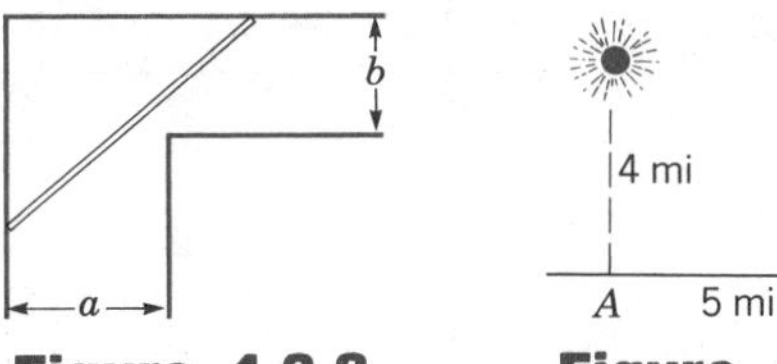

Figure 4.3.8 **Figure 4.3.9**

12. A lighthouse is located 4 mi away from the nearest point A on a straight shoreline. The lighthouse keeper obtains supplies at a store at point B, located 5 mi along the shore from A (see Figure 4.3.9). The keeper can row the lighthouse boat at a rate of 2 mi/hr and can walk along the beach at a rate of 3 mi/hr. Assuming that the boat can be beached at any point on the shore between A and B, where should the keeper land in order to reach the store in the shortest possible time?
13. A square piece of tin of side a is formed into a box by cutting square pieces of side x from each corner and folding up the sides (see Figure 4.3.10). Find the volume of the largest box that can be formed in this way.

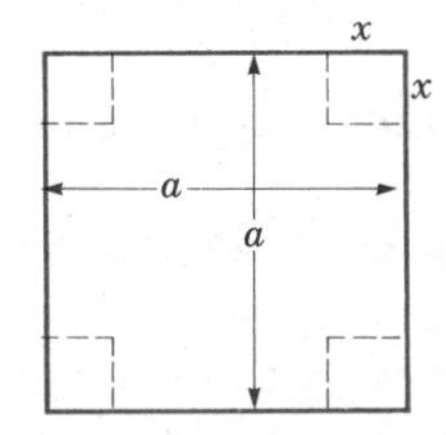

Figure 4.3.10

14. Suppose that a rectangular piece of tin of length a and width b is to be formed into a box by cutting squares of side x from each corner and folding up the sides. Find the value of x that yields the box of maximum volume.
15. A piece of wire of length l is cut into two pieces. One piece is bent into a square and the other into

an equilateral triangle. Determine how to cut the wire so that the total area enclosed by the square and triangle is (a) minimum; (b) maximum.

16. A long sheet of metal of width $3l$ is to be made into a channel of trapezoidal cross section by bending a portion of width l along each side at an angle θ; see Figure 4.3.11 for a view of the cross section. Find θ so that the area of the cross section is as great as possible.

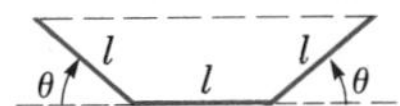

17. Ship A is 30 mi east of point O at midnight and is steaming west at 15 mi/hr. Ship B is 20 mi north of O at midnight and is steaming south at 15 mi/hr. Determine when the two ships are closest to each other and find this minimum distance.

18. A contractor wishes to build a house at some point x on a straight road between factories located a distance l apart (see Figure 4.3.12). The factories emit smoke with strengths α and β, respectively. In each case the intensity of the smoke is inversely proportional to distance from the factory. Find the point x at which the house should be built in order to minimize the smoke intensity at the house.

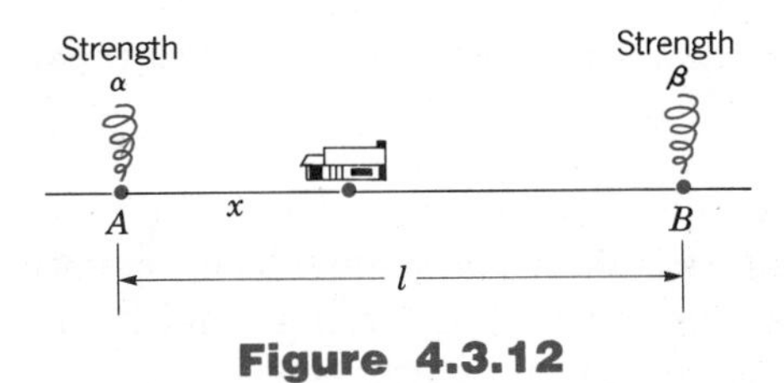

Figure 4.3.12

19. Two campers have pitched their tents at points A and B on the shore of a circular lake of radius 1 mi (see Figure 4.3.13). Where on the lake shore should a third camper locate a tent to maximize the sum of the straight line distances from this tent to those of the first two campers?

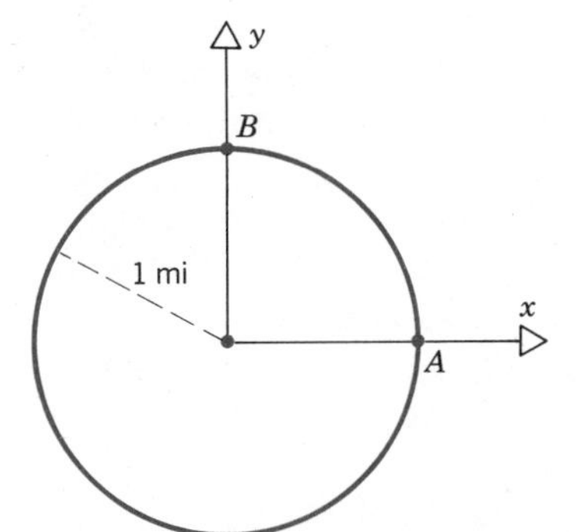

Figure 4.3.13

20. (a) Find the point, or points, on the graph of $x = y^2$ that is closest to the point $(1, 0)$.

    (b) Find the point, or points, on the graph of $x = y^2$ that are closest to the point $(a, 0)$, where a is any real number.

21. Let $y = f(x)$ be differentiable for $-\infty < x < \infty$, and let (a, b) be any point not on the graph of f.

    (a) If (x_0, y_0) is the point on the curve nearest to (a, b), show that x_0 satisfies the equation

    $$x_0 - a + f'(x_0)[f(x_0) - b] = 0. \qquad (i)$$

    (However, there may be other solutions of Eq. (i) that do not correspond to such minimum points.)

    (b) Show that (a, b) lies on the straight line that is normal (perpendicular) to the curve at the point (x_0, y_0).

22. (a) Find the points on the curve $x^2 - xy + y^2 = 1$ that are nearest the origin; farthest from the origin. *Hint:* Let $x = r\cos\theta$, $y = r\sin\theta$. What range of θ must be considered?

    (b) Show that the line from the origin to each of the points found in (a) is normal to the given curve.

23. A building contractor builds x houses per year, and sells them at a price of 80 kilodollars each. The contractor has a fixed cost of 25 kilodollars per year for office expenses and a cost of 65 kilodollars per house per year for supplies and wages. There are also additional costs that increase with the number of houses built of $x^2/3$ kilodollars per year.

    (a) Find the number of houses the contractor should build per year in order to obtain the maximum profit. Note that an integral number of houses must be built.

    (b) Assume that the builder's fixed costs are α kilodollars per year with other costs as previously given. Show that the optimum level of construction is independent of α. If the builder wishes to show an annual profit of at least 125 kilodollars, determine the maximum acceptable value of α.

(c) Assume now that the builder can sell x houses per year at price p kilodollars where

$$x = 4(100 - p).$$

Costs are as given originally. Find the number of houses the contractor should build per year in order to obtain the maximum profit.

24. (a) A trucking company wishes to operate over a distance of 500 mi at an average speed of x mi/hr. Speed laws require that $40 \leq x \leq 55$. There is a fixed cost of \$3 per hour of operation of the truck to cover insurance, maintenance, depreciation, etc.; in addition the driver's wages are \$10 per hour. Fuel costs one dollar per gallon and is consumed at a rate of $x^2/200$ gal/hr. Find the average speed at which the truck should be driven in order to minimize the cost of the trip.

(b) Suppose that the conditions of Part (a) are unchanged except that the driver's wages are w dollars/hour. Find the value of w above which it is most economical to drive the truck at the maximum legal speed.

(c) Suppose that the conditions of Part (a) are unchanged except that the cost of fuel is α cents/gallon. Find the value of α above which it is most economical to drive the truck at the minimum legal speed.

25. The manager of an engineering research office has discovered by experimentation that if x engineers are assigned desk space in a single large office, then the average number of hours w that each engineer works usefully per day is given by

$$w = 8\left[1 - \frac{x^2}{x^2 + 130}\right],$$

where $x \geq 1$. How many engineers should be assigned to the office if the total number of hours of useful work by all of the engineers in the office is to be as large as possible?

26. Consider an experiment in which the probability of success is p, and of failure is $1 - p$. For example, in a roll of a fair die the probability of rolling a one (success) is $\frac{1}{6}$; the probability of not rolling a one (failure) is $\frac{5}{6}$.

It can be shown that, in n repetitions of this experiment, the probability P_{nk} of achieving exactly k successes is

$$P_{nk} = \frac{n!}{k!(n-k)!}\, p^k(1 - p)^{n-k}, \qquad 0 \leq k \leq n.$$

Determine p so that P_{nk} is as large as possible.

4.4 RELATED RATE PROBLEMS

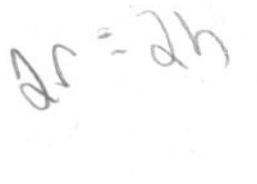

A fairly common type of problem is one involving two or more dependent variables that are functions of a single independent variable, usually time. The time rate of change of one variable is given and it is required to find the time rate of change of another. Such problems are often referred to as related rate problems. The following are typical examples.

EXAMPLE 1

Sand is poured on a conical pile at a rate of 10 ft³/min. The diameter of the base of the pile is always 50 percent greater than its height. Find how fast the height of the pile is rising when the pile is 5 ft high.

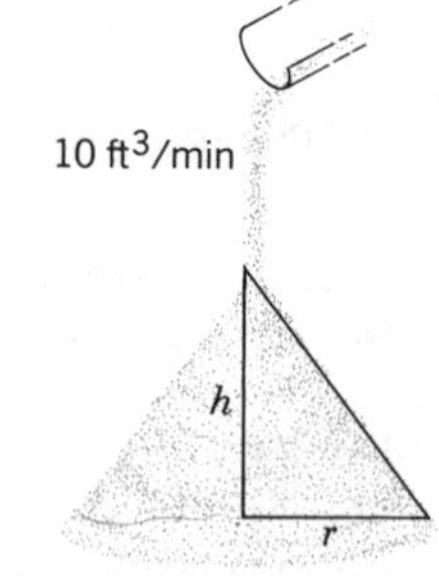

Figure 4.4.1

Referring to Figure 4.4.1, let r be the radius of the base and h the height of the pile. Both r and h depend on time t. We know the rate at which the volume V of the pile is changing, namely, $dV/dt = 10$ ft³/min, and we want to find dh/dt. To obtain a relation between dV/dt and dh/dt, we first need a relation between V and h. The volume V of the conical pile of sand is given by

$$V = \frac{1}{3}\pi r^2 h, \tag{1}$$

and depends on both r and h. However, we are given that $2r = \frac{3}{2}h$, so $r = \frac{3}{4}h$. Consequently, substituting for r in Eq. 1, we obtain

$$V = \frac{1}{3}\pi\left(\frac{9}{16}h^2\right)h = \frac{3\pi}{16}h^3, \tag{2}$$

which is the desired relation between V and h. We look on Eq. 2 as defining V as a composite function of t, that is, V is a function of h, where h is a function of t. Therefore, by the chain rule,

$$\begin{aligned}\frac{dV}{dt} &= \frac{dV}{dh}\frac{dh}{dt} = \frac{3\pi}{16}3h^2\frac{dh}{dt} \\ &= \frac{9\pi}{16}h^2\frac{dh}{dt}.\end{aligned} \tag{3}$$

Finally, we substitute the values given in the statement of the problem for h and dV/dt, and find that

$$10 = \frac{9\pi}{16}\cdot 25 \cdot \left.\frac{dh}{dt}\right|_{h=5},$$

or, solving for dh/dt,

$$\left.\frac{dh}{dt}\right|_{h=5} = \frac{2}{5}\cdot\frac{16}{9\pi} = \frac{32}{45\pi} \cong 0.2264\,\frac{\text{ft}}{\text{min}} \cong 2.7\,\frac{\text{in}}{\text{min}}. \;\blacksquare \tag{4}$$

EXAMPLE 2

A 32-ft-long ladder leans against the side of a building. The lower end of the ladder begins to slip horizontally at a uniform speed of 1 ft/sec. Find the velocity (in the downward direction) of the top of the ladder when the bottom of the ladder is 24 ft from the building.

Further, suppose that a painter is standing 20 ft up the ladder (point P in Figure 4.4.2). Find the vertical component of the painter's velocity under the conditions stated in the previous paragraph.

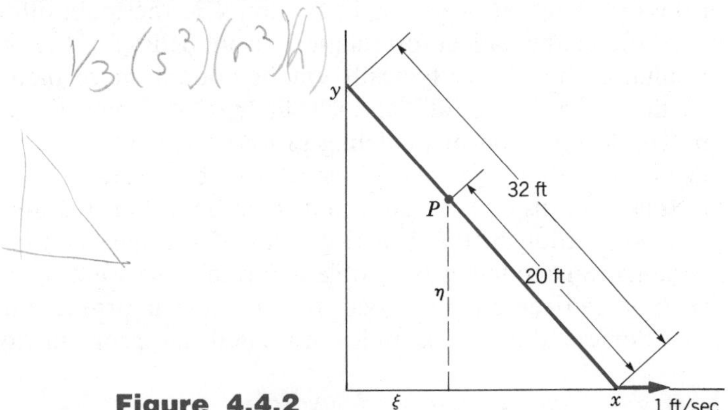

Figure 4.4.2

Referring to Figure 4.4.2, let x and y be the distances from the base of the building wall to the bottom and top of the ladder, respectively. Then

$$x^2 + y^2 = 1024. \tag{5}$$

Thinking of both x and y as functions of t, and using the chain rule to differentiate Eq. 5 with respect to t, we obtain

$$2x\frac{dx}{dt} + 2y\frac{dy}{dt} = 0,$$

or

$$\frac{dy}{dt} = -\frac{x}{y}\frac{dx}{dt}. \tag{6}$$

We are given that $dx/dt = 1$. Further, from Eq. 5, it follows that $y = \sqrt{448}$ when $x = 24$. Substituting these values in Eq. 6, we obtain

$$\left.\frac{dy}{dt}\right|_{x=24} = -\frac{24}{\sqrt{448}} = -1.134 \text{ ft/sec}. \tag{7}$$

Now let ξ and η be the horizontal and vertical coordinates of the painter; next we wish to determine $d\eta/dt$. By noting that the larger and smaller triangles in Figure 4.4.2 are similar, we obtain

$$\frac{\eta}{y} = \frac{20}{32}, \qquad \text{or} \quad \eta = \frac{5}{8}y. \tag{8}$$

Thus, $d\eta/dy = \frac{5}{8}$, and by the chain rule

$$\left.\frac{d\eta}{dt}\right|_{x=24} = \left.\frac{d\eta}{dy}\frac{dy}{dt}\right|_{x=24} = \frac{5}{8}(-1.134) = -0.709 \text{ ft/sec}. \tag{9}$$

This is the painter's (downward) velocity at the instant specified in the statement of the problem. ∎

In related rate problems, such as those just given, perhaps even more than in the optimization problems in Section 4.3, the main difficulty (if there is one) lies in the mathematical formulation, or modeling, of the problem, rather than in its solution. The latter typically entails nothing more than a straightforward application of the chain rule, and is usually routine once the problem is properly set up. The formulation or modeling process, however, can require ingenuity, often mixed with some trial and error. It can be frustrating, especially if, for some unexplained reason, you fail to notice an important relation among the variables. Remember, though, that practice helps; if you have not had much experience in mathematical modeling or problem formulation (and many students have not), then you can reasonably expect to see some improvement in your ability fairly soon. Some additional examples of related rate problems follow.

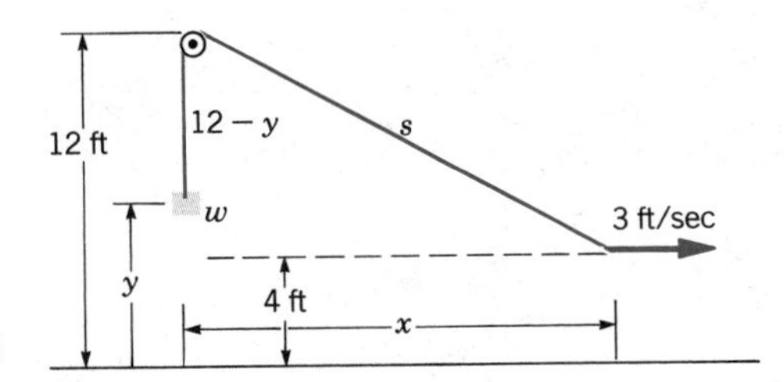

Figure 4.4.3

EXAMPLE 3

A weight w is suspended from a rope that is passed over a pulley, as shown in Figure 4.4.3. The free end of the rope is being pulled horizontally at a rate of 3 ft/sec at a height of 4 ft above ground level. If the pulley is at a height of 12 ft and the rope is 24 ft long, find how fast the weight is rising when it is 6 ft above the ground.

Let y be the height of the weight above the ground, and let x be the horizontal distance from the weight to the free end of the rope (see Figure 4.4.3). It is given that $dx/dt = 3$ ft/sec, and it is required to find dy/dt. We need first to obtain a relation between y and x. We can do this in two stages. Note that the total length of the rope consists of a part of length $12 - y$ between the weight and the pulley, and a part of length s between the pulley and the free end. Thus

$$24 = 12 - y + s,$$

or

$$y = s - 12. \tag{10}$$

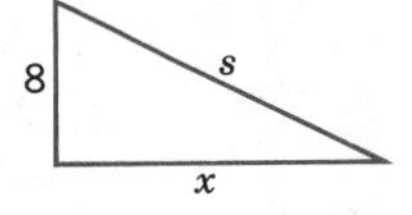

Figure 4.4.4

Further, s is related to x as shown by the diagram in Figure 4.4.4:

$$s^2 = x^2 + 64. \tag{11}$$

Now let us think of y as a function of s, Eq. 10, where s is a function of x, Eq. 11. Then dy/dt can be calculated by the chain rule,

$$\frac{dy}{dt} = \frac{dy}{ds}\frac{ds}{dx}\frac{dx}{dt} = \frac{x}{s}\frac{dx}{dt}, \tag{12}$$

since $dy/ds = 1$ from Eq. 10 and $ds/dx = x/s$ from Eq. 11. Equation 12 is the desired relation between dx/dt and dy/dt. We are given that $dx/dt = 3$ ft/sec; further, when $y = 6$ ft we have $s = 18$ ft from Eq. 10 and then $x = \sqrt{260}$ ft from Eq. 11. Substituting these values in Eq. 12, we obtain

$$\left.\frac{dy}{dt}\right|_{y=6} = \frac{\sqrt{260}}{18} \cong 2.6874 \text{ ft/sec.} \;\blacksquare \tag{13}$$

EXAMPLE 4

A searchlight is mounted 800 ft offshore from a straight shoreline, and rotates at a constant angular speed of four revolutions per minute. Determine how fast the

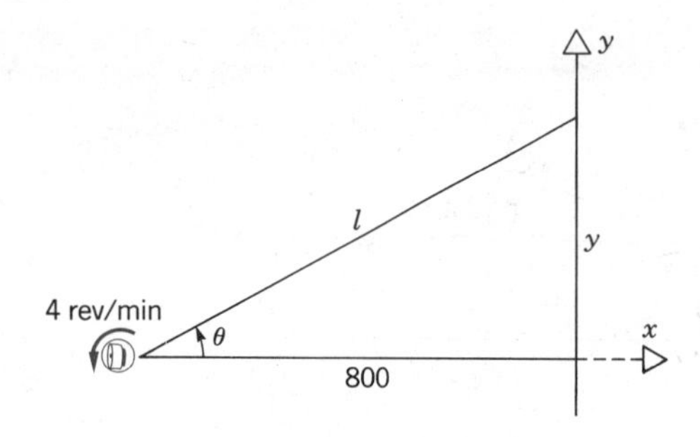

Figure 4.4.5

spot of light is moving along the shoreline when it reaches a point 1000 ft from the light.

Choose a coordinate system as shown in Figure 4.4.5. The origin is at the searchlight and the x-axis is directed toward the shoreline and perpendicular to it. Then let y represent the current position of the spot of light on the shoreline, and let θ be the angle between the positive x-axis and the direction of the searchlight. We know that $d\theta/dt = 8\pi$ radians/min, and we wish to find dy/dt. From Figure 4.4.5

$$y = 800 \tan \theta, \tag{14}$$

so

$$\frac{dy}{d\theta} = 800 \sec^2 \theta. \tag{15}$$

Thus, by the chain rule,

$$\frac{dy}{dt} = \frac{dy}{d\theta}\frac{d\theta}{dt} = 6400\pi \sec^2 \theta. \tag{16}$$

While $l = 1000$, it follows that $\cos \theta = \frac{4}{5}$, so $\sec \theta = \frac{5}{4}$. Substituting this value into Eq. 16, we obtain

$$\left.\frac{dy}{dt}\right|_{l=1000} = 6400\pi \left(\frac{5}{4}\right)^2 \cong 31416 \text{ ft/min}, \tag{17}$$

or

$$\left.\frac{dy}{dt}\right|_{l=1000} \cong 523.6 \text{ ft/sec.} \blacksquare \tag{18}$$

PROBLEMS

1. A spherical balloon is expanding at a rate of $2 \text{ ft}^3/\text{min}$, while always maintaining its shape. How fast is the radius r of the balloon increasing when $r = 3$ ft?

2. How fast is the surface area of the balloon increasing under the conditions of Problem 1?

3. If the length of a side of an equilateral triangle is increasing at a rate of 2 in./min, how fast is the area of the triangle increasing when the side is 7 in. long?

4. A circle is expanding in such a way that its area is increasing at a rate of $2 \text{ m}^2/\text{min}$. How fast is

the circumference changing when the radius is 5 m?

5. A particle is traversing the curve $x^2 + 4y^2 = 8$. When the particle passes through the point $(-2, 1)$, it is known that $dx/dt = 3$. Find the value of dy/dt at this point.

6. A particle is moving on the curve $x^2 + xy + y^2 = 3$. When the particle passes through the point $(-1, -1)$, it is known that $dy/dt = \sqrt{3}$. Find the value of dx/dt at this point.

7. The pressure p and volume V of an ideal diatomic gas undergoing an adiabatic process satisfy the equation $pV^{7/5} = c$, where c is a constant. Suppose that $V = 32$ in.^{3}, that $p = 20$ lb/in.^{2}, and that the gas is being compressed at a rate of 0.1 in.^{3}/sec. Find the rate at which the pressure is changing.

8. A particle moves around the circular orbit $x^2 + y^2 = 4$ with a constant angular speed of one revolution per minute. Find dx/dt and dy/dt when the particle passes through the point $(\sqrt{3}, 1)$.

9. A rectangle is expanding in such a way that its length is always twice its width. If the perimeter of the rectangle is increasing at a rate of 3 in./min, find the rate of change of the area of the rectangle when the area is 24 in.^{2}

10. An isosceles triangle has a base of 6 in. Its altitude is increasing at a rate of 1 in./min. Find the rate of change of the vertex angle when the altitude is 6 in.

11. A 6-ft man walks away from a street light, which is 20 ft above the ground, at a rate of 4 ft/sec. Find the rate at which the length of his shadow is increasing when he is 30 ft away from the base of the lamppost.

12. A girl 5 ft tall casts a shadow 6 ft long when she stands 12 ft from a lamppost. She then walks directly away from the lamppost at a rate of 3 ft/sec. How fast is the length of her shadow changing when she is 24 ft from the lamppost?

13. The radius r of a sphere (in centimeters) is changing at the rate of $(1 + r^2)^{-1}$ cm/min. Find the rate at which the volume is changing when $r = 3$ cm.

14. The side s of a square (in inches) changes at the rate $s^{-1/2}$ in./min. Find the rate at which the area changes when $s = 4$ in.

15. A square is inscribed in a circle. If the radius of the circle is increasing at a rate of 3 cm/min, find how fast the area of the square is increasing when the radius is 10 cm.

16. A cube is inscribed in a sphere whose radius is increasing at a rate of 3 cm/min. How fast is the surface area of the cube changing when the radius of the sphere is 8 cm?

17. A conical water tank (vertex down) is initially full of water. The tank is 8 ft deep and the radius of the top is 2 ft. Water is being pumped into the tank at a rate of 0.5 ft^3/min. Find how fast the water level is rising when the water is 5 ft deep at the center of the tank.

18. A water tank has the shape of a square pyramid with the vertex downward. The base (top) of the tank has a side of 3 ft, and the depth of the tank at the center is 9 ft. When the water is 6 ft deep at the center of the tank, it is observed that the water level is dropping at the rate of 1 in./min due to a leak. How fast is water leaking from the tank?

19. A boat is pulled to a dock by a rope wound on a winch, which is 6 ft higher than the point where the rope is fastened to the boat. If the rope is wound on the winch at the rate of 4 ft/min, how fast is the horizontal distance between the boat and dock changing when this distance is 9 ft?

20. An airplane at an altitude of 3 mi flies directly over an observer on the ground. The airplane maintains a constant speed of 360 mi/hr in level flight. When the airplane is 2 mi past the point directly over the observer, what is the rate of change of the angle between the observer's line of sight and the horizontal?

21. A water trough is 8 ft long and has a cross section in the shape of an isosceles triangle whose base (top) is 2 ft wide and whose depth is also 2 ft. Water is pumped into the trough at a rate of 3 ft^3/min. Determine how fast the water level is rising when the water is 1.5 ft deep at the center of the trough.

22. A water trough is 12 ft long and has a semicircular cross section (flat side up) of radius 3 ft. Water is being drained out of the trough at a rate of 10 ft^3/min. What is the rate of change of the water level when the water is 1 ft deep in the center of the trough?

4.5 NEWTON'S METHOD

Frequently in applications of mathematics we encounter the problem of finding real roots of an equation $f(x) = 0$. Except in special cases (for instance, if f is a quadratic polynomial) such equations are impossible to solve exactly. One method of approximating a real root of an equation is the method of interval bisection introduced in Section 2.6. Another method, which originated with Newton, uses linear approximations to generate successive approximations to a root; often this method is an efficient and accurate means of finding a root with as much accuracy as may be required.

Suppose that f is differentiable and that r is a root of $f(x) = 0$ (see Figure 4.5.1). Suppose further that we do not know r, but do have an approximate value

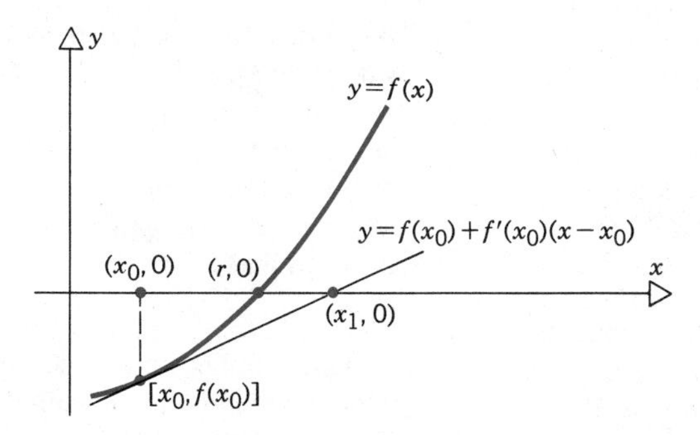

Figure 4.5.1

x_0 for r; for example, the approximate value x_0 may have been determined from a rough graph of $y = f(x)$; the first step in developing Newton's method is to construct the tangent line at $[x_0, f(x_0)]$:

$$y - f(x_0) = f'(x_0)(x - x_0). \tag{1}$$

This line approximates the graph of $y = f(x)$ in the vicinity of $x = x_0$. Therefore we obtain a new approximation x_1 to r by finding the point at which the tangent line crosses the x-axis. To do this we set $y = 0$ and solve Eq. 1 for x. If we denote the resulting value of x by x_1, we have

$$-f(x_0) = f'(x_0)(x_1 - x_0),$$

or

$$x_1 = x_0 - \frac{f(x_0)}{f'(x_0)}. \tag{2}$$

Next we construct the tangent line at $[x_1, f(x_1)]$,

$$y - f(x_1) = f'(x_1)(x - x_1)$$

(see Figure 4.5.2). A new approximation x_2 to r is obtained as the point where this tangent line crosses the x-axis. Thus

$$-f(x_1) = f'(x_1)(x_2 - x_1)$$

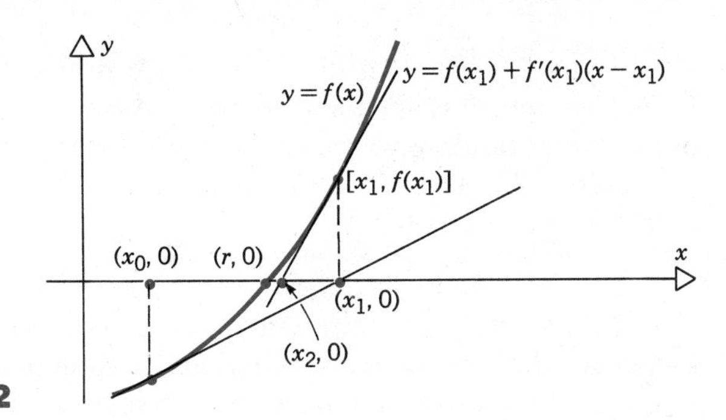

Figure 4.5.2

and

$$x_2 = x_1 - \frac{f(x_1)}{f'(x_1)}. \tag{3}$$

Continuing this process, we obtain the iteration formula

(4)

The technique of using Eq. 4 repeatedly to compute successive approximations to a real root of the equation $f(x) = 0$ is known as Newton's method.* The following examples illustrate the use of this method.

EXAMPLE 1

Use Newton's method to find a real root of the equation

$$f(x) = x^3 - 2x^2 + 6x - 7 = 0. \tag{5}$$

Observe that this is the same equation that we discussed in Example 5 of Section 2.6. It is easy to see that $f(1) = -2$ and $f(2) = 5$, so there must be at least one root of Eq. 5 in the interval $1 < x < 2$. Let us choose our initial estimate of the root to be $x_0 = 1.5$. The resulting computations are summarized in Table 4.1. For example, in the first row of the table it is shown that $f(x_0) = 0.875$ and

Table 4.1 Newton's Method for $f(x) = x^3 - 2x^2 + 6x - 7 = 0$.

n	x_n	$f(x_n)$	$f'(x_n)$
0	1.5	0.875	6.75
1	1.3703704	0.0398313	6.1522634
2	1.3638961	0.0000882	6.1250534
3	1.3638817	7×10^{-10}	6.1249931
4	1.3638817		

*Newton's method is contained in his books *De analysi* and *De Methodis Serierum et Fluxionum*. These books were written about 1670, but were not published until 1711 and 1736, respectively.

$f'(x_0) = 6.75$. Then, using Newton's formula (4) with $n = 0$, we obtain $x_1 = 1.3703704$, which is shown in the second row of Table 4.1. Three more repetitions of the computation give the successive values $x_2 = 1.3638961$, $x_3 = 1.3638817$, and $x_4 = 1.3638817$. The process is terminated at this stage since $x_4 = x_3$, and no further corrections can be obtained (unless more decimal places are retained). ∎

In executing a computation using Newton's method, you should note the successive differences $|x_{n+1} - x_n|$, and terminate the procedure when $|x_{n+1} - x_n|$ is less than the error you are willing to tolerate. In most cases, this will mean that x_n is accurate to a corresponding number of decimal places. In Example 1, the third iterate x_3 correctly approximates the actual root of Eq. 5 to seven decimal places.

Comparing Newton's method for this problem with the method of interval bisection (Section 2.6), we see that Newton's method gives seven decimal place accuracy after only three steps, while interval bisection yielded two decimal place accuracy after five steps. Thus, Newton's method approaches the root considerably faster. On the other hand, at each step Newton's method requires somewhat more computation, since $f'(x_n)$ as well as $f(x_n)$ must be evaluated. On balance, for this problem Newton's method is considerably more efficient.

EXAMPLE 2

Find a positive root of

$$f(x) = x - 2 \sin x = 0. \tag{6}$$

This function was discussed in some detail in Example 2 of Section 4.2. From that discussion, or from the graph in Figure 4.2.4, it is clear that there is only one positive root, and that it lies between the local minimum point at $x = \pi/3$ and the local maximum point at $x = 5\pi/3$. Further, since $f(\pi/2) = \pi/2 - 2 < 0$ and $f(\pi) = \pi > 0$, the root actually lies in the interval $(\pi/2, \pi)$. Let us choose $x_0 = 2.0$ as our initial approximation. Then Table 4.2 shows the ensuing calculations. As in Example 1, only three iterations are required to obtain the root correct to seven decimal places.

In this case our initial estimate x_0 was fairly good; it turned out to be within 6 percent of the actual root. To see the effect of using a much poorer initial approximation, suppose that we start with $x_0 = 3.0$. The resulting calculations are

Table 4.2 **Newton's Method for $f(x) = x - 2 \sin x = 0$.**

n	x_n	$f(x_n)$	$f'(x_n)$
0	2.0	0.1814051	1.8322937
1	1.9009956	0.0090401	1.6484631
2	1.8955116	0.0000285	1.638078
3	1.8954943	-1×10^{-10}	1.6380451
4	1.8954943		

Table 4.3 Newton's Method for $f(x) = x - 2 \sin x = 0$.

n	x_n	$f(x_n)$	$f'(x_n)$
0	3.0	2.71776	2.979985
1	2.0879954	0.3495804	1.988895
2	1.9122293	0.0276776	1.6696753
3	1.8956526	0.0002594	1.6383452
4	1.8954943		

shown in Table 4.3. Despite the fact that x_0 is in error by about 58%, the fourth iterate x_4 gives the root correctly to seven decimal places. The effect of using a poor initial approximation is to increase by only one the number of iterations that are needed. ∎

Examples 1 and 2 illustrate that Newton's method often works very well. By this we mean that the differences $|x_{n+1} - x_n|$ between successive approximations quickly become small, and that the procedure does not depend strongly on the choice of the initial approximation x_0. However, we should keep in mind that Newton's method is not always this effective, and that some conditions must be imposed on the function f to assure the validity of the procedure. Certainly, f must be differentiable. Moreover, $f'(x_n)$ must be nonzero at each point x_n generated during the process. If, in fact, $f'(x_n) = 0$ for some n, the tangent line at $[x_n, f(x_n)]$ is parallel to the x-axis and the next approximation x_{n+1} cannot be calculated (see Figure 4.5.3). Even if $f'(x_n)$ is never zero, Newton's method will usually fail if $f'(x)$ is small in the interval of interest. A more subtle difficulty is revealed by the following example.

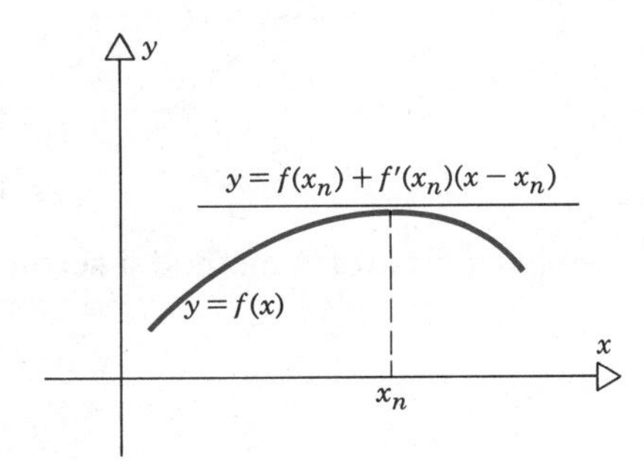

Figure 4.5.3

EXAMPLE 3

Apply Newton's method to the function

$$f(x) = \begin{cases} \sqrt{x - 2}, & x \geq 2, \\ -\sqrt{2 - x}, & x < 2, \end{cases} \tag{7}$$

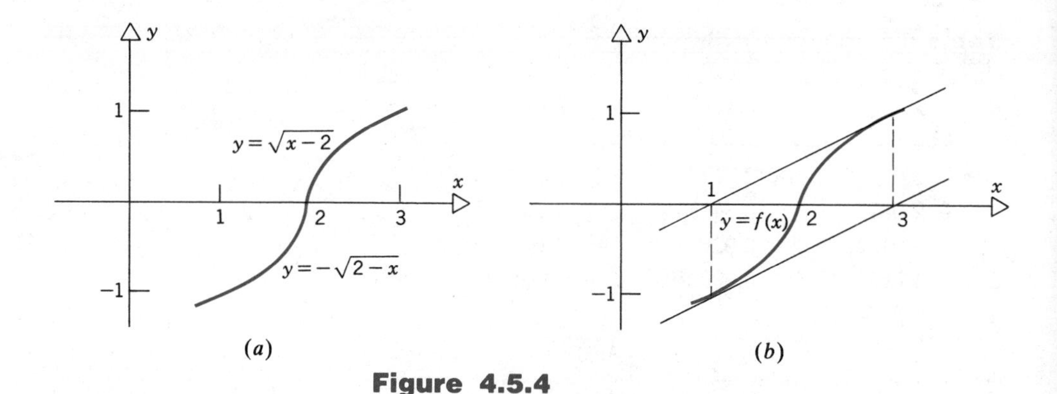

Figure 4.5.4

whose graph is shown in Figure 4.5.4*a*.

From Eq. 7 or from Figure 4.5.4*a* we see that $r = 2$ is the only root of $f(x) = 0$. The function f is differentiable except at $x = 2$, and

$$f'(x) = \begin{cases} \dfrac{1}{2\sqrt{x-2}}, & x > 2, \\ \dfrac{1}{2\sqrt{2-x}}, & x < 2. \end{cases} \tag{8}$$

The fact that $f'(2)$ does not exist casts doubt upon the applicability of Newton's method. However, we will proceed anyway and see what happens. If the initial approximation is $x_0 = 3$, then the next approximation x_1 is given by

$$\begin{aligned} x_1 &= x_0 - \frac{f(x_0)}{f'(x_0)} \\ &= 3 - \frac{\sqrt{3-2}}{1/2\sqrt{3-2}} \\ &= 1. \end{aligned}$$

Applying Newton's method a second time to obtain x_2, we find that

$$\begin{aligned} x_2 &= x_1 - \frac{f(x_1)}{f'(x_1)} \\ &= 1 - \frac{-\sqrt{2-1}}{1/2\sqrt{2-1}} \\ &= 3. \end{aligned}$$

Thus the successive approximations will cycle back and forth between the two values 1 and 3 without ever approaching the root $r = 2$ (see Figure 4.5.4*b*). You may wish to verify that the same situation occurs if a different choice of x_0 is made (see Problem 22). ∎

There are theorems that give conditions for the validity of Newton's method, as well as estimates of the rapidity with which the successive approximations x_0, $x_1, \ldots, x_n, \ldots$ approach the root r. We will not give such a theorem, however, since in many applications we do not use it anyway. Rather, we check that f is differentiable and that $f'(x) \neq 0$, and then proceed somewhat tentatively to use Newton's method. We keep a close watch on the successive differences $|x_{n+1} - x_n|$, and if these numbers decrease rapidly in magnitude, then we conclude that the procedure is working satisfactorily. Thus to a considerable extent we rely on common sense rather than on a formal theorem.

PROBLEMS

Most of the following problems ask for calculations using Newton's method. While the calculations can be done on a pocket calculator, the computations will be less laborious on a programmable calculator or a personal computer.

In each of Problems 1 through 12, the equation $f(x) = 0$ has a root in the given interval. Choose an initial approximation x_0, and use Newton's method to calculate an approximate value of this root. Continue the procedure at least until $|x_{n+1} - x_n| < 10^{-5}$.

© **1.** $f(x) = x^2 + 2x - 4, \quad 0 \le x \le 3$

© **2.** $f(x) = x^2 + 3x - 1, \quad -5 \le x \le -1$

© **3.** $f(x) = x^4 - 2, \quad 1 \le x \le 2$

© **4.** $f(x) = x^3 - 3x^2 + 5x - 4, \quad 0 \le x \le 2$

© **5.** $f(x) = x^3 + 5x^2 - 4, \quad 0 \le x \le 2$

© **6.** $f(x) = \sqrt{1 + x^2} - 3x^2 + 2, \quad 0 \le x \le 2$

© **7.** $f(x) = x \sin x - \cos x, \quad 0 \le x \le \dfrac{\pi}{2}$

© **8.** $f(x) = x \cos x - \sin x, \quad \pi \le x \le \dfrac{3\pi}{2}$

© **9.** $f(x) = x \cos x + 2 \sin x, \quad \dfrac{\pi}{2} \le x \le \pi$

© **10.** $f(x) = 3 \cos x - x \sin x, \quad 0 \le x \le \dfrac{\pi}{2}$

© **11.** $f(x) = \sqrt{2x} - \dfrac{1}{x} - 1, \quad 0 < x \le 2$

© **12.** $f(x) = \dfrac{4x}{1 + x^2} - \sin x, \quad 2\pi \le x \le \dfrac{5\pi}{2}$

13. Using Newton's method we can derive an efficient procedure for computing $\sqrt{a}$, where a is an arbitrary positive number. Consider the function $f(x) = x^2 - a$ whose zeros are $x = \pm\sqrt{a}$. If $x_0 > 0$ is any initial approximation to $\sqrt{a}$, show that Newton's method yields the iteration formula

$$x_{n+1} = \frac{1}{2}\left(x_n + \frac{a}{x_n}\right), \qquad n \ge 0$$

for computing further approximations. This procedure for computing approximate values of $\sqrt{a}$ is equivalent to the Old Babylonian square root algorithm (see C. B. Boyer, *A History of Mathematics*, Wiley, New York, 1968, pp. 31, 449).

14. Use Newton's method to derive the iteration formula

$$x_{n+1} = \frac{1}{p}\left[(p - 1)x_n + \frac{a}{x_n^{p-1}}\right]$$

for the calculation of $a^{1/p}$, where a is an arbitrary positive number, and $p \ge 2$ is an integer. Observe that this formula reduces to the one in Problem 13 when $p = 2$.

In each of Problems 15 through 20, use the result of Problem 13 or Problem 14 to calculate an approximation to the given number. Continue the process at least until $|x_{n+1} - x_n| < 10^{-5}$.

© **15.** $\sqrt{53}$ © **16.** $\sqrt{172}$ © **17.** $\sqrt[3]{5}$

© **18.** $\sqrt[4]{7}$ © **19.** $\sqrt[5]{50}$ © **20.** $\sqrt[3]{100}$

21. Suppose that we wish to compute $1/a$, the reciprocal of the number a. On a computer the process of division is much more time-consuming than those of multiplication and addition. Therefore it is important to try to recast a division problem into one involving only multiplication and addition. By using Newton's method on the function $f(x) = a - (1/x)$, derive an iteration formula for the calculation of $1/a$ that involves only multiplication and addition.

22. Consider the equation $f(x) = 0$, where f is given by Eq. 7. Using Newton's method, show that if $x_0 = 2 + \alpha$, then $x_1 = 2 - \alpha$ and $x_2 = 2 + \alpha$. Thus the successive approximations cycle back and forth regardless of how small α is, that is, no matter how close the initial approximation x_0 is to the actual root $r = 2$.

© 23. Use Newton's method to solve the equation

$$f(x) = x^3 + 5x^2 - 4 = 0$$

for each of the following choices of the initial point x_0.

(a) 0.5 (b) 0.1 (c) −0.1 (d) −0.5

Draw a graph of $y = f(x)$ and use it to explain why the results are as they are.

© 24. Use Newton's method to solve the equation

$$f(x) = x \cos x - \sin x = 0$$

for each of the following choices of the initial point x_0.

(a) 3.6 (c) 3.5 (e) 3.2

(b) 3.51 (d) 3.25 (f) 3.0

Draw a graph of $y = f(x)$ and use it to explain why the results are as they are.

REVIEW PROBLEMS

1. Find conditions on a and b implying that

$$f(x) = ax^3 + bx^2 + x + k$$

has at most one real zero.

2. Find conditions on b and c implying that

$$f(x) = x^3 + bx^2 + cx + k$$

has at most one real zero.

In each of Problems 3 through 6, either establish or refute the given inequality.

3. $|\sin^2 x_2 - \sin^2 x_1| \le |x_2 - x_1|$

4. $|\cos^2 x - \cos^2 y| \le |x - y|$

5. $|\tan x_2 - \tan x_1| \le |x_2 - x_1|$ on $\left(-\frac{\pi}{2}, \frac{\pi}{2}\right)$

6. $\sqrt{y} - \sqrt{x} \le \dfrac{y - x}{(3/2)\sqrt{y}}$, $\quad 0 < x < y$

In each of Problems 7 through 10, determine whether the given pairs of functions are antiderivatives of the same function.

7. $f(x) = 2\cos^2 x - 1, \quad g(x) = \sec^2 x$

8. $f(x) = \tan^2 x, \quad g(x) = \sec^2 x$

9. $f(x) = \cos^2 x, \quad g(x) = \tan^2 x - 1$

10. $f(x) = (x + 1)(x - 1), \quad g(x) = x^2 + 3x + 1$

In each of Problems 11 through 16, determine whether the functions satisfy the hypotheses of the mean value theorem. Determine whether there is a point c such that

$$f'(c) = \frac{f(b) - f(a)}{b - a}.$$

Find all c in those cases for which they exist.

11. $f(x) = \sqrt{x + 1}, \quad 0 \le x \le 1$

12. $f(x) = \sin x - \cos x, \quad -\frac{\pi}{2} \le x \le \frac{\pi}{2}$

13. $f(x) = \tan x + x, \quad -\frac{\pi}{4} \le x \le \frac{\pi}{4}$

14. $f(x) = |\sin x|, \quad 0 \le x \le 2\pi$

15. $f(x) = \begin{cases} |x|, & -1 \le x < 0 \\ \sqrt{x}, & 0 \le x \le 1 \end{cases}$

$f(x) = \begin{cases} x^3, & -1 \le x \le 0 \\ \sqrt{x}, & 0 < x \le 1 \end{cases}$

For any real a, b, d, and k prove that there exists a $c \in (u, v)$ such that

$$f'(c) = \frac{f(v) - f(u)}{v - u},$$

where $f(x) = ax^3 + bx^2 + dx + k$, $u \le x \le v$, and

(a) $[u, v] = [0, 1]$.

(b) $[u, v] = [0, \epsilon], \quad \epsilon > 0$.

(c) u, v arbitrary.

For each of Problems 18 and 19, show that the hypothesis of Problem 20 of Section 4.1 is satisfied and find all points c such that

$$f(b) = f(a) + f'(a)(b - a) + (\tfrac{1}{2})f''(c)(b - a)^2.$$

18. $f(x) = x^2, \quad -1 \le x \le 1$

19. $f(x) = \sin x, \quad -\pi \le x \le \pi$

For each of Problems 20 and 21, show that the given pair of functions satisfies the hypothesis of Problem 21 of Section 4.1 and find all c such that

$$\frac{f(b) - f(a)}{g(b) - g(a)} = \frac{f'(c)}{g'(c)}.$$

20. $f(x) = \sin x, \quad g(x) = x, \quad 0 \le x \le \pi$

21. $f(x) = 1 + x + \tfrac{1}{2}x^2,$
$g(x) = 1 + x + \tfrac{1}{2}x^2 + \tfrac{1}{3}x^3, \quad -1 \le x \le 1$

In each of Problems 22 through 27 find the following, for the given function.

(a) All critical points.

(b) Intervals where the function increases; decreases.

(c) Intervals where the function is concave up; concave down.

(d) All local and global maximum and minimum points.

22. $f(x) = \sin x + x, \quad 0 \le x \le 2\pi$

23. $f(x) = |\sin x|, \quad -\dfrac{\pi}{2} \le x \le \dfrac{\pi}{2}$

24. $f(x) = ax^3 + 3ax^2 + 3ax + d, \quad a > 0$

25. $f(x) = \begin{cases} x^3, & -1 \le x \le 0 \\ \sqrt{x}, & 0 < x \le 1 \end{cases}$

26. $f(x) = \dfrac{1}{x^2 + a^2}, \quad a > 0, \quad -a \le x \le a$

27. $f(x) = \begin{cases} \dfrac{x + 2}{x + 1}, & x < -1 \\ \dfrac{1}{\sqrt{1 - x^2}}, & -1 < x < 1 \\ \dfrac{x - 2}{x - 1}, & x > 1 \end{cases}$

28. In each of the following cases, find the area of the largest rectangle with vertices on the x-axis and on the given curve.

(a) The semicircle $y = \sqrt{4 - x^2}$.

(b) The parabola $y = 9 - x^2$.

(c) The triangle with vertices $(\pm a, 0)$, $(0, b)$.

29. In each of the following cases, find the point on the given curve closest to the given point.

(a) $y = \dfrac{1}{x}, \quad (0, 0), \quad x > 0$

(b) $2y = 2x^2 + 1, \quad (1, 1)$

(c) $y = \sin x, \quad \left(\dfrac{\pi}{2}, 2\right)$

(d) $y = |x|, \quad (0, b)$

30. Divide 10 into two parts such that the product of the two parts plus their difference is a maximum.

31. Of all isosceles triangles of a given area A find the one with least perimeter.

32. If the altitude x of a moving particle is

$$x = -\tfrac{1}{2}gt^2 + v_0t + x_0,$$

where g is acceleration due to gravity, v_0 is the particle's initial velocity, and x_0 its initial position, find the maximum height of the particle.

33. If the particle in Problem 32 is initially on the ground and is propelled upward with velocity of 20 m/sec, how long will it take for the particle to hit the ground? What will be the total distance traveled? (Assume $g = 10$ m/sec.)

34. The equation

$$T = m\left(\frac{v^2}{r} + g \cos \theta\right)$$

describes the tension T in a cord of length r attached to a small mass m being spun in a vertical circle with velocity v, where g is the acceleration due to gravity and θ is the angle of the cord with the downward vertical.

(a) If v is constant, find the angle of minimum tension.

(b) If v changes with θ so that T remains constant, find $dv/d\theta$.

35. A satellite moves about the earth with period

$$P = \left(\frac{2\pi}{R\sqrt{g}}\right) r^{3/2},$$

velocity $v = R\sqrt{g/r}$, and radial acceleration $a_r = (R/r)^2g$, where R is the earth's radius, g the acceleration due to gravity, and r the distance from the satellite to the earth's center.

(a) How does the period change with respect to r?

(b) How does the period change with respect to velocity?

(c) How does the radial acceleration change with respect to velocity?

36. Coulomb's law states that the force between two charged particles x, y is directly proportional to their product and inversely proportional to the square of the distance between them. Write an equation for Coulomb's law. How does the force change with respect to distance? If distance changes in time by the rule

$$\frac{dr}{dt} = 1 + 2r,$$

how does the force change in time?

37. A rectangular door of given height, h, and perimeter, p, is to be constructed so that there is a circular window centered on the vertical with the center $h/4$ down from the top. If the radius of the window is $h/8$, find the base of the door so that the area of the door minus the area of the window is a maximum.

38. A rectangular window surmounted by a semicircular window is to be built with perimeter p. Find the base of the window so that area is a maximum.

39. A satellite travels in a circular orbit of radius R. If its x coordinate decreases at a rate of 2 units/sec at the point (a, b), how fast is the y coordinate changing? How fast is y changing with respect to x?

40. Find the rate of change of area with respect to perimeter for

(a) An equilateral triangle. (c) A square

(b) A circle.

41. Find the rate of change of volume with respect to surface area for

(a) A sphere. (b) A cube.

42. Find the rectangular box with square base and given surface area whose volume is a maximum.

43. Find the height of the box of greatest volume that can be constructed by 108 square feet of wood if the base is twice as long as the width.

44. The Pacific Lumber Company has determined that it can selectively cut an area of land at a cost of

$$h = 500x + 15{,}000 \text{ dollars},$$

where x is the number (in hundreds) of acres cut per year. If they can sell the lumber for

$$R = 3000 - 10x \text{ dollars per acre},$$

what harvest level maximizes profits?

45. If the Pacific Lumber Company finds that it costs

$$h = 500x + 15{,}000 + 25x^2 \text{ dollars}$$

to harvest and reforest the land, and other conditions are as in Problem 44, find the harvest level that maximizes profits.

46. The California Seafood Company has found that it costs

$$c = a + bx + dx^2 \text{ dollars}$$

to fish a certain species of rockfish, where a is the cost of labor, b is the cost per fish of locating a harvest area, d is the change in cost per fish for the harvest as the population is depleted, and x is the number of fish caught. If the fish can be sold for

$$R = q - rx \text{ dollars per fish},$$

How many fish must be caught to maximize profits?

47. A biologist has determined that the interaction of two species is governed by the system

$$\frac{dx}{dt} = ay, \qquad \frac{dy}{dt} = bx,$$

where x and y are the populations of the two species and a and b are constants.

(a) Find dy/dx.

(b) Find all critical points of the equation found in Part (a) by considering dy/dx as a function of x and differentiating with respect to x.

(c) Show that the solution of Part (b) is also a solution of the original equations.

FIVE

some geometrical topics

In applications of mathematics it is often desirable to draw a graph that represents geometrically a function or equation. One way to proceed is by computing the coordinates of a large number of points lying on the graph, plotting these points, and then drawing a curve containing them. If appropriate computing equipment is available, this can be done automatically even for very complicated functions or equations. In many cases, however, what is required is not an extremely accurate graph, but rather one that is qualitatively correct, and adequately shows at a glance the main features of the function or equation involved. Then an analytical approach to curve sketching may be very helpful, particularly if the function or equation is relatively simple. Even for complicated graphs, a brief preliminary analytical investigation may make it possible to carry out required computations more efficiently and economically.

In this chapter we first discuss curve sketching in general, emphasizing properties of curves that are often easy to identify. Later, we review the conic sections, and then consider curves described by sets of parametric equations.

5.1 CURVE SKETCHING

To sketch a graph quickly it is important to be aware of and to seek out the important characteristics of the graph that can be readily identified. It turns out that calculus is a very useful tool for this purpose. We now describe several of these characteristics, not all of which are pertinent to every graph.

Excluded intervals and intercepts

It is sometimes possible to tell at a glance that the graph of an equation lies only in certain regions of the xy-plane. Alternatively, it may be impossible for one or both variables to have values in certain intervals, called **excluded intervals.** If the graph is the graph of a function $y = f(x)$, then this is the same as determining the domain and range of f.

Sometimes it is possible to proceed a bit further by easily identifying a few specific points on a graph. Often the easiest points to locate are the points where the graph intersects one of the coordinate axes; these points are known as **intercepts.** The intercepts on the x-axis are found by setting $y = 0$ and solving the given equation for x, while the intercepts on the y-axis are found by setting $x = 0$ and solving for y.

EXAMPLE 1

Find excluded intervals and intercepts for the graph of

$$y = \frac{x^2}{x^2 + 1}. \tag{1}$$

Since x can have any real value, there are no excluded intervals for x. However, since $x^2 \geq 0$ and $x^2 < x^2 + 1$, it follows that y takes on values only in $0 \leq y < 1$, so $(-\infty, 0)$ and $[1, \infty)$ are excluded intervals for y. This is shown in Figure 5.1.1; the graph must lie in the unshaded region. If $x = 0$, then $y = 0$ also, and vice versa. Hence the origin is the only intercept of the graph of Eq. 1. This point is also indicated in Figure 5.1.1. ∎

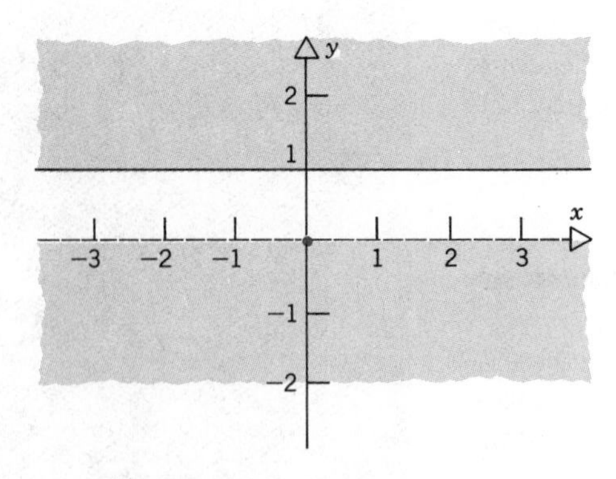

Figure 5.1.1

EXAMPLE 2

Find excluded intervals and intercepts for the graph of

$$y = -\sqrt{x^2 - 1}. \qquad (2)$$

Since $x^2 - 1$ must be nonnegative, it follows that $(-1, 1)$ is an excluded interval for x. Further, y cannot be positive, so $(0, \infty)$ is an excluded interval for y. In Figure 5.1.2 the shaded region is excluded for the graph of Eq. 2. There is no intercept on the y-axis since $x = 0$ lies in the excluded x-interval. However, if $y = 0$, then $x = \pm 1$, so there are intercepts on the x-axis at $(-1, 0)$ and $(1, 0)$. These points are plotted in Figure 5.1.2 as well. ∎

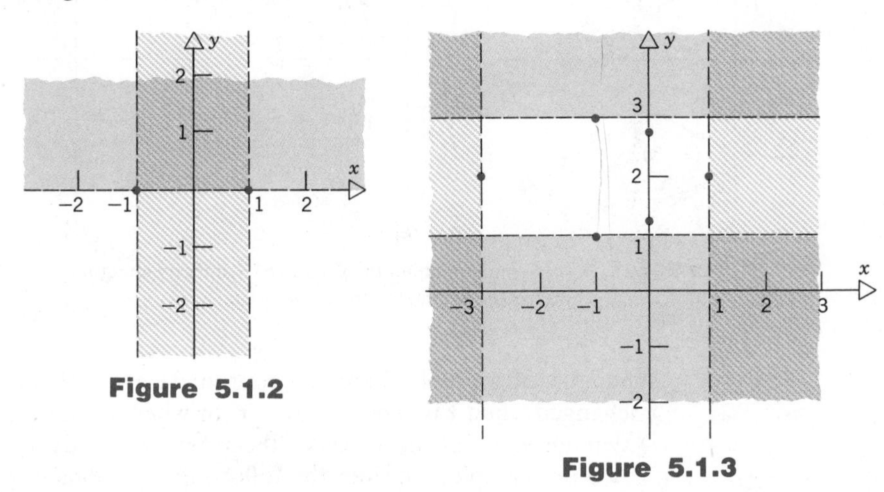

Figure 5.1.2

Figure 5.1.3

EXAMPLE 3

Find excluded intervals and intercepts for the graph of

$$(x + 1)^2 + 4(y - 2)^2 = 4. \qquad (3)$$

Each term on the left side of Eq. 3 is nonnegative, so neither can have a value greater than 4. Hence $(x + 1)^2 \leq 4$ and $4(y - 2)^2 \leq 4$; alternatively $|x + 1| \leq 2$ and $|y - 2| \leq 1$. Thus $-2 \leq x + 1 \leq 2$, so x must be in the interval $[-3, 1]$; similarly y must be in the interval $[1, 3]$. Thus the graph must lie in the unshaded region in Figure 5.1.3. There is no intercept on the x-axis, since $y = 0$ lies in an excluded interval. However, if $x = 0$, then Eq. 3 becomes

$$1 + 4(y - 2)^2 = 4$$

from which we obtain

$$y = 2 \pm \frac{\sqrt{3}}{2}.$$

Thus there are intercepts on the y-axis at the points $(0, 2 - \sqrt{3}/2)$ and $(0, 2 + \sqrt{3}/2)$. It is also easy to see that when $y = 2$, then $x = 1$ or $x = -3$, so that the graph contains $(1, 2)$ and $(-3, 2)$. Similarly, when $x = -1$ we obtain the points $(-1, 1)$ and $(-1, 3)$. ∎

Symmetry

The sketching of a graph is often simplified greatly by the knowledge that it possesses some form of symmetry. A graph is said to be **symmetric about the x-axis** if it contains the point $(x, -y)$ whenever it contains the point (x, y). Similarly, a graph is **symmetric about the y-axis** if it contains the point $(-x, y)$ whenever it contains the point (x, y) (See Figure 5.1.4*a* and 5.1.4*b*). It is easy to test for

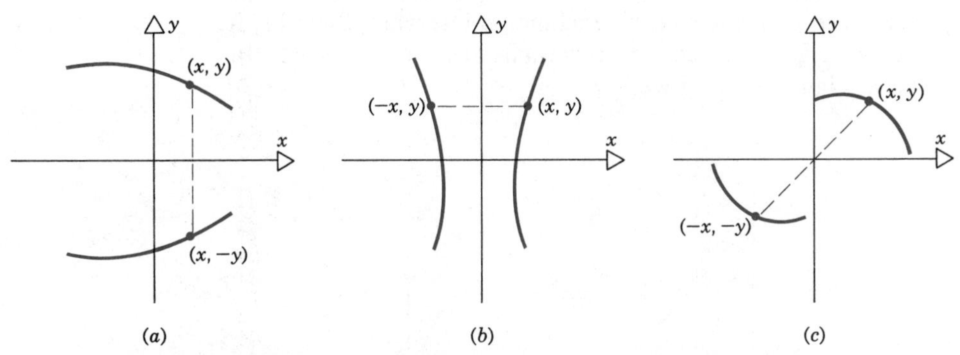

Figure 5.1.4 (*a*) Symmetry about the x-axis. (*b*) Symmetry about the y-axis. (*c*) Symmetry about the origin.

symmetry of a graph about the coordinate axes. If the equation of the graph is unchanged when x is replaced by $-x$, or when y is replaced by $-y$, then the graph is symmetric about the y-axis or the x-axis, respectively.

For example, consider the following equations:

$$x^2 - 3y^2 + y^3 = 4, \tag{4}$$

$$2x^2 + y^2 = 6, \tag{5}$$

$$x - 3x^3 + 2y - y^3 = 4. \tag{6}$$

If in Eq. 4 we replace x by $-x$, then we obtain

$$(-x)^2 - 3y^2 + y^3 = 4$$

or

$$x^2 - 3y^2 + y^3 = 4,$$

which is the same as the original equation. Hence the graph of Eq. 4 is symmetric about the y-axis. However, replacing y by $-y$ in Eq. 4, we find that

$$x^2 - 3(-y)^2 + (-y)^3 = 4$$

or

$$x^2 - 3y^2 - y^3 = 4,$$

which is not the same as Eq. 4. Hence the graph of Eq. 4 is not symmetric about the x-axis. In a similar way we can show that the graph of Eq. 5 is symmetric about both axes, and that the graph of Eq. 6 is symmetric about neither.

Another kind of symmetry that is easy to recognize is symmetry about the origin. A graph is said to be **symmetric about the origin** if it contains the point $(-x, -y)$ whenever it contains the point (x, y) (see Figure 5.1.4*c*). This type of symmetry can also be recognized easily from the equation of the graph; it is present whenever both x can be replaced by $-x$ and y can be replaced by $-y$ without changing the equation. For example, the graph of Eq. 5 is symmetric about the origin, but those of Eqs. 4 and 6 are not.

Graphs may also be symmetric about other lines or points, but such symmetry properties are more difficult to identify from an equation.

From the point of view of sketching graphs the importance of symmetry is that it greatly reduces the labor involved. For instance, if a graph is known to be symmetric about both axes, it is necessary only to sketch the portion in the first quadrant in order to determine the graph completely.

Asymptotes

Let $P(x, y)$ be a point on the graph of a given equation and let L be a given nonvertical line. The distance d from P to L is measured along the line through P perpendicular to L (see Figure 5.1.5). The line L is said to be an **asymptote** of the

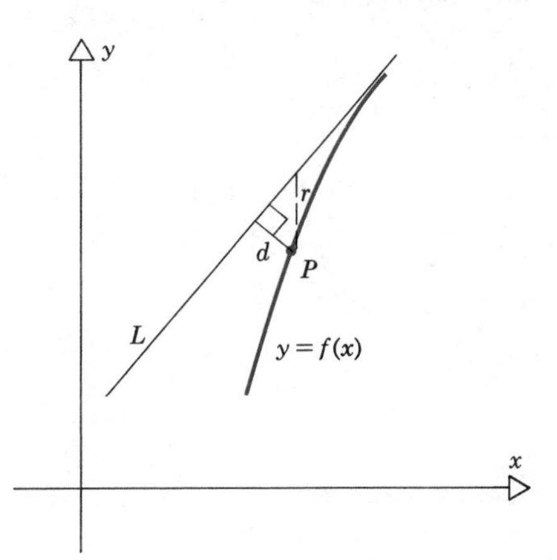

Figure 5.1.5 The line L is an asymptote of the graph of f.

graph if $d \to 0$ as $x \to \infty$ or as $x \to -\infty$. A vertical line $x = a$ is an asymptote of a graph if the y-coordinate of a point on the graph approaches $+\infty$ or $-\infty$ as x approaches a from at least one side.

Vertical and horizontal asymptotes are usually easy to identify. Vertical asymptotes of the graph of a function f are found by locating points where f becomes unbounded. On the other hand, if $\lim_{x\to\infty} f(x) = L$, or if $\lim_{x\to-\infty} f(x) = L$, then $y = L$ is a horizontal asymptote.

Asymptotes that are not vertical or horizontal may be more difficult to determine. Referring again to Figure 5.1.5, observe that the distance d from the point P to the line L is certainly no greater than the distance r from P to L measured along the vertical line through P. Thus, if $r \to 0$ as $x \to \infty$ or as $x \to -\infty$, then d will also do so; this suffices to show that L is an asymptote of the graph of f. If the equation of the line L is written in the form $y = mx + b$, then r is given by

$$r = \pm[mx + b - f(x)], \tag{7}$$

where the sign is chosen according to whether the asymptote is above or below the graph. Note that in determining asymptotes that are neither vertical nor horizontal one must somehow discover the asymptote and then verify that it is correct by taking the limit as $x \to \infty$ or as $x \to -\infty$ of the quantity given in Eq. 7. In contrast, one finds vertical and horizontal asymptotes by examining limits involving only $f(x)$ itself.

EXAMPLE 4

Find the vertical and horizontal asymptotes of the graph of

$$y = \frac{1}{x^2}. \tag{8}$$

The line $x = 0$ is a vertical asymptote since y becomes unbounded as $x \to 0$. The line $y = 0$ is a horizontal asymptote since $y \to 0$ as $x \to -\infty$ and as $x \to \infty$. The graph of Eq. 8 is shown in Figure 5.1.6. Note that the graph is symmetric about the y-axis, that $y > 0$ always, and that there are no intercepts on either axis.

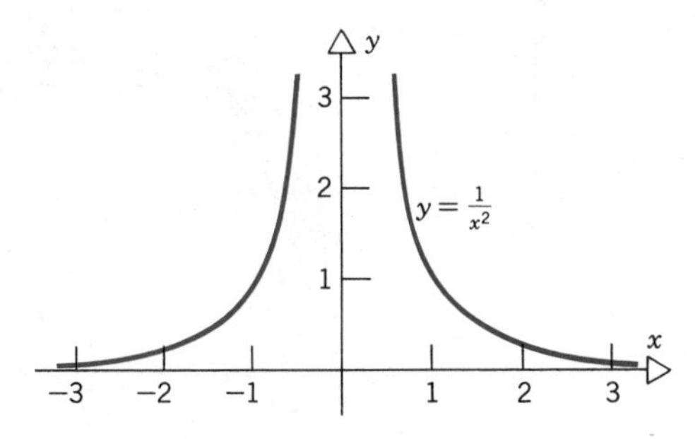

Figure 5.1.6

EXAMPLE 5

Find vertical and horizontal asymptotes of the graph of

$$y = \frac{x^2}{x^2 - 1}. \tag{9}$$

The lines $x = 1$ and $x = -1$ are vertical asymptotes since y becomes unbounded as $x \to \pm 1$. The line $y = 1$ is a horizontal asymptote because

$$\lim_{x \to \pm\infty} \frac{x^2}{x^2 - 1} = \lim_{x \to \pm\infty} \frac{1}{1 - \dfrac{1}{x^2}} = 1.$$

The graph of Eq. 9 is shown in Figure 5.1.7. Note that the graph is symmetric about the y-axis, and that the origin is the only intercept on either axis. Other features of the graph can be determined by methods explained later in this section. ∎

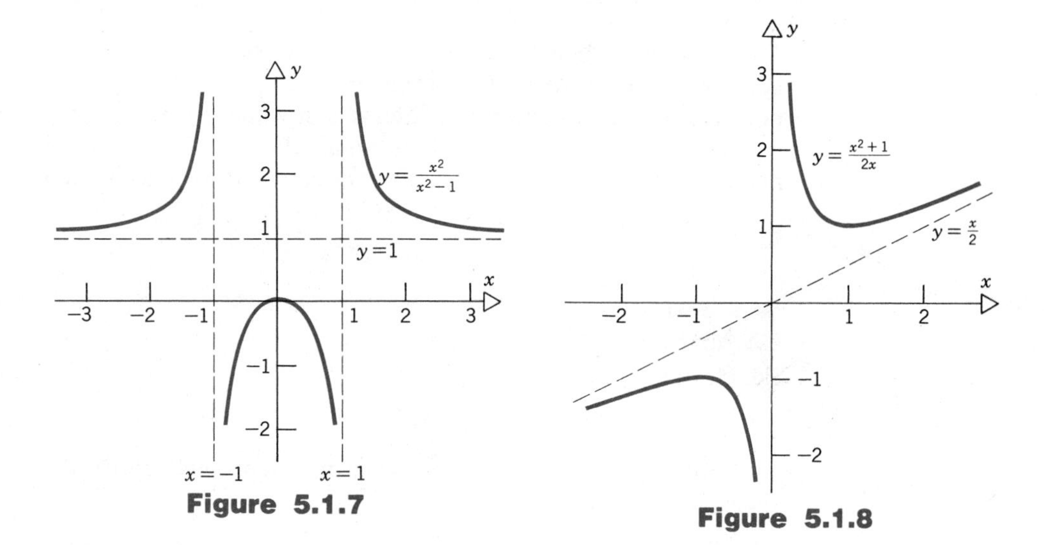

Figure 5.1.7

Figure 5.1.8

EXAMPLE 6

Find all asymptotes of the graph of

$$f(x) = \frac{x^2 + 1}{2x}. \tag{10}$$

First, note that the line $x = 0$ is a vertical asymptote. To search for other asymptotes it is helpful to write Eq. 10 in the form

$$f(x) = \frac{x}{2} + \frac{1}{2x}. \tag{11}$$

Then we see that there are no horizontal asymptotes since $f(x) \to \infty$ as $x \to \infty$, and $f(x) \to -\infty$ as $x \to -\infty$. Nevertheless, it is also apparent from Eq. 11 that $f(x)$ is close to $x/2$ for large x. Hence we conjecture that $y = x/2$ is an asymptote. To verify this, note that

$$\begin{aligned} \lim_{x\to\infty}\left[\frac{x}{2} - f(x)\right] &= \lim_{x\to\infty}\left[\frac{x}{2} - \left(\frac{x}{2} + \frac{1}{2x}\right)\right] \\ &= \lim_{x\to\infty}\left(-\frac{1}{2x}\right) = 0. \end{aligned} \tag{12}$$

Thus $y = x/2$ is indeed an asymptote as $x \to \infty$. A similar calculation shows that $y = x/2$ is also an asymptote as $x \to -\infty$. The graph of $y = f(x)$ is shown in Figure 5.1.8. It can be seen from Eq. 10 that the graph is symmetric about the origin. Other features of the graph can be found by methods developed later in this section. ∎

Intervals of increase or decrease

Consider a function f whose domain includes some interval I. Recall that in Section 4.2 we established that if $f'(x) > 0$ on I, then f is increasing on I, and similarly if

$f'(x) < 0$ on I, then f is decreasing on I. In sketching the graph of a function f it is often useful to identify intervals in which f' is either always positive or always negative, that is, to determine intervals in which f is increasing or decreasing. For a differentiable function this determines the location of all local extreme points, and the general shape of the graph can then be drawn, as in the following example.

EXAMPLE 7

Sketch the graph of $y = f(x)$, where

$$f(x) = 2x^3 + 3x^2 - 12x - 5. \tag{13}$$

On checking each of the properties mentioned up to now, we find the following.

(a) There are no excluded intervals for x.
(b) Since $y \to -\infty$ as $x \to -\infty$ and $y \to \infty$ as $x \to \infty$, and since y is a continuous function of x, there are also no excluded intervals for y.
(c) If $x = 0$, then $y = -5$, so the point $(0, -5)$ is the only intercept on the y-axis.
(d) If $y = 0$, then x satisfies $2x^3 + 3x^2 - 12x - 5 = 0$. The roots of this equation are not obvious, so we will not find the intercepts on the x-axis.
(e) The graph is not symmetric with respect to either axis, or with respect to the origin.
(f) The graph has no vertical or horizontal asymptotes.

To obtain further information about the graph of Eq. 13 we differentiate with respect to x and find that

$$\begin{aligned}\frac{dy}{dx} &= f'(x) = 6x^2 + 6x - 12 \\ &= 6(x^2 + x - 2) \\ &= 6(x + 2)(x - 1).\end{aligned} \tag{14}$$

Hence $f'(x) > 0$ on the intervals $(-\infty, -2)$ and $(1, \infty)$, so f is increasing on those intervals. Similarly, f is decreasing on $(-2, 1)$ because $f'(x) < 0$ there. Since f is increasing to the left of $x = -2$ and decreasing to the right of $x = -2$, it follows from the first derivative test (Theorem 4.2.2) that $x = -2$ is a local maximum point. Similarly, $x = 1$ is a local minimum point. The corresponding points on the graph of f are $(-2, 15)$ and $(1, -12)$; these are the only local extrema of f.

Using this information, as well as the facts noted under items (a) to (f), we obtain the sketch shown in Figure 5.1.9. We refine this sketch a little further later in the section. If yet more detailed information is required about some portion of the graph, then one should plot several points lying on that part of the graph. For the present we emphasize that an examination of $f'(x)$ enabled us to find the two points (the two local extrema) that are most significant in determining the general shape of the graph. This is much more efficient than merely plotting points chosen at random. ∎

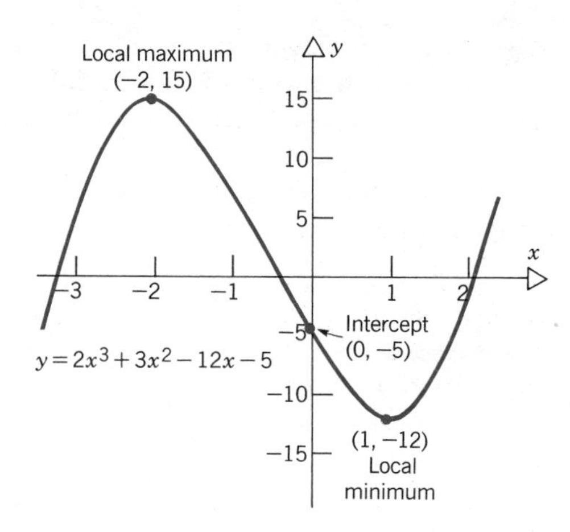

Figure 5.1.9

Concavity

In Section 4.2 we showed that if $f''(x) > 0$ on an interval I, then f is concave up on I, while if $f''(x) < 0$ on I, then f is concave down there. In sketching the graph of a function it is useful to identify points where the direction of concavity changes. A point c is called an **inflection point** of f if

1. f is continuous at c;
2. the graph of f has a tangent line (possibly vertical) at $[c, f(c)]$;
3. f is concave up on one side of c and concave down on the other side.

Figure 5.1.10 shows several possibilities. The points c_1 and c_2 are inflection points because the function is continuous, has a tangent line (which is vertical at c_2), and changes concavity at each of these points. On the other hand, the points c_3 and c_4 are not inflection points, even though there is a change in concavity; there is no tangent line at c_3 and f is not even continuous at c_4.

We will be concerned mainly with functions that have at least two continuous derivatives except possibly at certain isolated points. If concavity changes at c, then $f''(x) > 0$ on one side of c and $f''(x) < 0$ on the other side. If $f''(c)$ exists, then an argument similar to that used in proving Theorem 4.1.1 can be used to show that $f''(c) = 0$. Consequently, inflection points of a function f are found among those points where $f''(x) = 0$ or where $f''(x)$ does not exist. However, not all such points are inflection points.

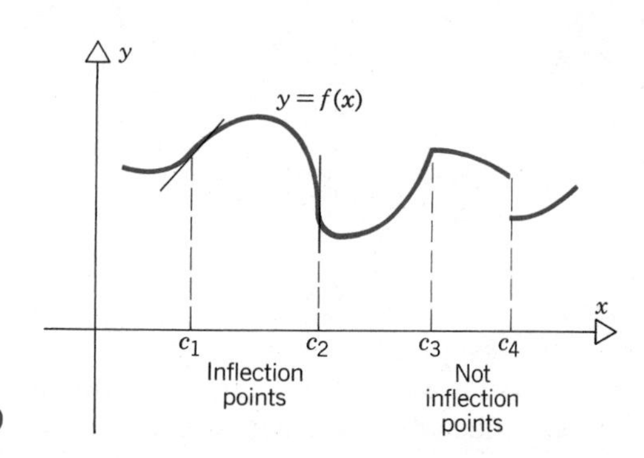

Figure 5.1.10

For example, if $f(x) = x^4$, then $f''(x) = 12x^2$ and $f''(x) = 0$ when $x = 0$. Nevertheless, the origin is not an inflection point because the graph of $y = x^4$ is always concave up. The relation between inflection points and zeros of f'' is similar to the relation between local extrema and zeros of f'.

In the next example we refine the graph in Figure 5.1.9 by determining its inflection points.

EXAMPLE 7 (*Continued*)
Investigate the concavity of the graph of the function f given in Eq. 13,

$$f(x) = 2x^3 + 3x^2 - 12x - 5,$$

and locate its inflection points.

Differentiating f twice, we obtain

$$f''(x) = 12x + 6. \tag{15}$$

Setting $f''(x)$ equal to zero and solving for x, we find that $x = -\frac{1}{2}$ is the only possible inflection point. The corresponding value of f is $f(-\frac{1}{2}) = \frac{3}{2}$. From Eq. 15 it follows that $f''(x) < 0$ for $x < -\frac{1}{2}$ and $f''(x) > 0$ for $x > -\frac{1}{2}$; hence f is concave down on $(-\infty, -\frac{1}{2}]$ and concave up on $[-\frac{1}{2}, \infty)$. Since the graph changes its concavity at $x = -\frac{1}{2}$, this point is indeed an inflection point.

We have used this information in drawing Figure 5.1.11, which is a refined version of Figure 5.1.9. If still more accuracy is required in some portion of the graph, it can be attained by plotting a few additional points. ∎

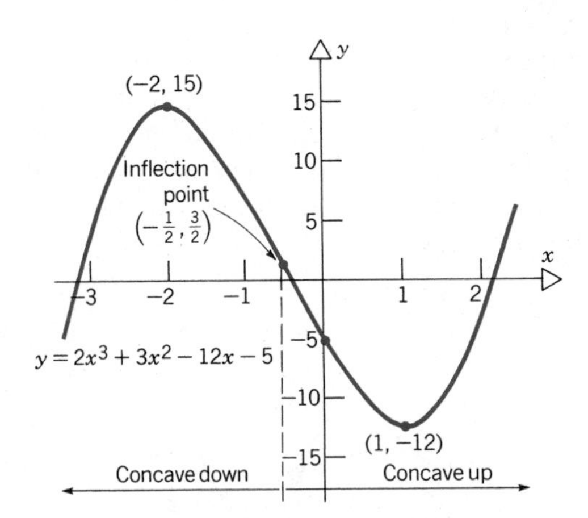

Figure 5.1.11

We now consider some additional examples of curve sketching, making use of calculus whenever appropriate.

EXAMPLE 8

Sketch the graph of the function f given by

$$y = f(x) = x^4 - 2x^2. \tag{16}$$

We will first check each of the properties mentioned in this section and record the information we obtain.

(a) Equation 16 implies no restriction on x, so there are no excluded intervals on the x-axis. However, it is not obvious at a glance what the range of f is.

(b) If $x = 0$, then $y = 0$, so the origin is the only intercept on the y-axis. If $y = 0$, then $x = 0$ or $x = \pm\sqrt{2}$. Thus $(-\sqrt{2}, 0)$, $(0, 0)$, and $(\sqrt{2}, 0)$ are the x-intercepts. If we write Eq. 16 in the form

$$f(x) = x^2(x + \sqrt{2})(x - \sqrt{2}) \tag{17}$$

then it is clear that $f(x) > 0$ if $x < -\sqrt{2}$ or if $x > \sqrt{2}$, and that $f(x) < 0$ if $-\sqrt{2} < x < 0$ or if $0 < x < \sqrt{2}$.

(c) Equation 16 is unchanged if x is replaced by $-x$, so the graph is symmetric about the y-axis. The graph is not symmetric about the x-axis or about the origin.

(d) There are no vertical asymptotes since $f(x)$ remains bounded for all finite values of x. There are also no horizontal asymptotes since $f(x) \to \infty$ as $x \to -\infty$ and as $x \to \infty$.

(e) Upon differentiating Eq. 16 we obtain

$$\begin{aligned} f'(x) &= 4x^3 - 4x \\ &= 4x(x^2 - 1) \\ &= 4x(x + 1)(x - 1). \end{aligned} \tag{18}$$

Hence the points $x = -1, 0, 1$ are critical points. The corresponding values of $f(x)$ are $f(-1) = -1$, $f(0) = 0$, and $f(1) = -1$, respectively. From Eq. 18 we see that $f'(x) < 0$ if $x < -1$ or if $0 < x < 1$; thus f is decreasing in these intervals. Similarly $f'(x) > 0$ if $-1 < x < 0$ or if $x > 1$, so f is increasing in these intervals. Consequently the point $(0, 0)$ is a local maximum and the points $(-1, -1)$ and $(1, -1)$ are local minima.

(f) By differentiating Eq. 16 a second time we find that

$$\begin{aligned} f''(x) &= 12x^2 - 4 \\ &= 12\left(x^2 - \frac{1}{3}\right) \\ &= 12\left(x + \frac{1}{\sqrt{3}}\right)\left(x - \frac{1}{\sqrt{3}}\right). \end{aligned} \tag{19}$$

Since $f''(x) > 0$ if $x < -1/\sqrt{3}$ or if $x > 1/\sqrt{3}$, the graph is concave up in these intervals. Similarly, $f''(x) < 0$ if $-1/\sqrt{3} < x < 1/\sqrt{3}$, so the graph is concave down there. Inflection points are found by setting $f''(x) = 0$; this yields $x = \pm 1/\sqrt{3}$ and the corresponding points on the graph are

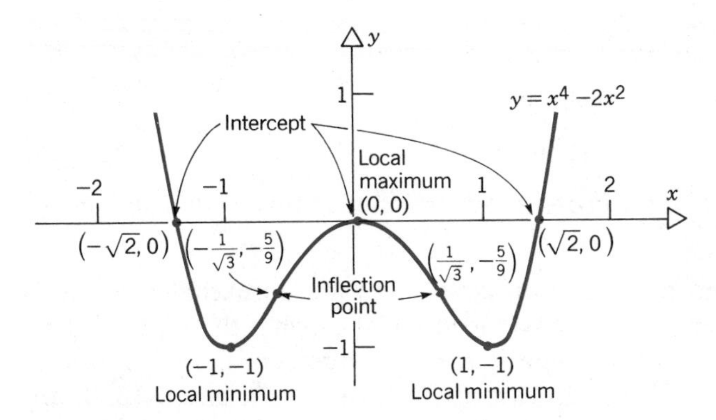

Figure 5.1.12

$(-1/\sqrt{3}, -5/9)$ and $(1/\sqrt{3}, -5/9)$. We have already noted that the concavity changes at these points, so they are indeed inflection points.

The graph of Eq. 16 is shown in Figure 5.1.12. You should verify that each of the bits of information we have found has been used in sketching this graph.

If a more accurate graph is required, then additional points can be plotted, or greater attention paid to the *values* of the slopes of the tangent lines at various points, rather than merely their signs.

The same methods can be applied to sketch the graph of any polynomial, although it will usually not be so easy to find the extreme points and inflection points, especially for polynomials of high degree. ∎

EXAMPLE 9

Sketch the graph of

$$y = \frac{x}{x^2 + 1}. \tag{20}$$

Again we check each property and record the corresponding information, omitting the explanation in some cases.

(a) There are no excluded intervals on the x-axis, but excluded intervals on the y-axis are not obvious.

(b) The origin is the only intercept on either axis. Note that $y < 0$ when $x < 0$ and $y > 0$ when $x > 0$.

(c) The graph is symmetric about the origin, but not about either coordinate axis. Hence it is sufficient to consider the part of the graph for which $x \geq 0$.

(d) There are no vertical asymptotes since the denominator of Eq. 20 is never zero. However, $y \to 0$ as $x \to -\infty$ and as $x \to \infty$, so the x-axis is a horizontal asymptote on both sides of the graph.

(e) On differentiating Eq. 20 we have

$$y' = \frac{(x^2 + 1) \cdot 1 - x(2x)}{(x^2 + 1)^2}$$

$$= \frac{1 - x^2}{(x^2 + 1)^2}. \tag{21}$$

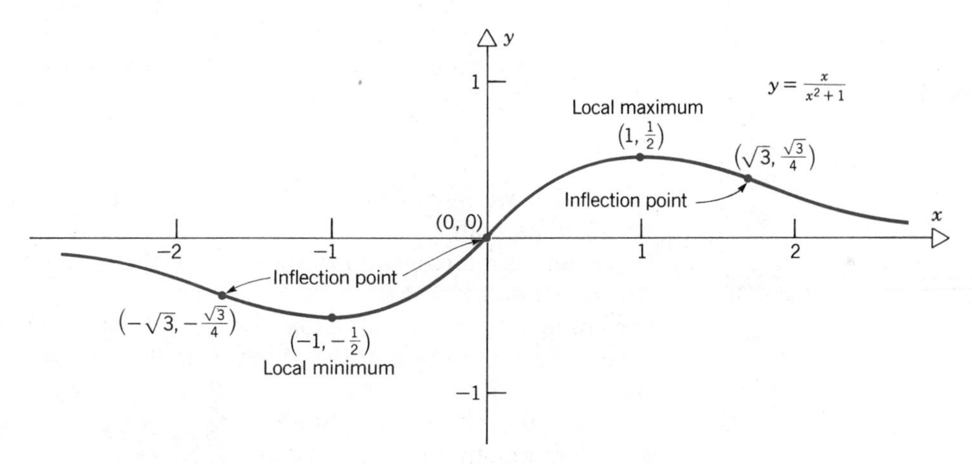

Figure 5.1.13

Considering $x \geq 0$, we see that $y' = 0$ at $x = 1$, that $y' > 0$ for $0 \leq x < 1$, and that $y' < 0$ for $x > 1$. Hence y is increasing on $[0, 1]$ and decreasing on $[1, \infty)$, so the point $(1, \frac{1}{2})$ is a local maximum.

(f) Differentiating a second time, we obtain

$$y'' = \frac{(x^2 + 1)^2(-2x) - (1 - x^2)2(x^2 + 1)2x}{(x^2 + 1)^4}$$

$$= \frac{2x(x^2 - 3)}{(x^2 + 1)^3}. \tag{22}$$

Again considering only $x \geq 0$, we note that $y'' < 0$ for $0 < x < \sqrt{3}$, so the graph is concave down there. Similarly, $y'' > 0$ for $x > \sqrt{3}$, so the graph is concave up in that interval. The point $x = \sqrt{3}$ is an inflection point. Since $y'' > 0$ for $-\sqrt{3} < x < 0$, the graph is also concave up there, and the origin is also an inflection point.

The graph in Figure 5.1.13 has been drawn on the basis of the information just listed. Note that the portion of the graph for $x < 0$ has been drawn by using the symmetry of the graph about the origin. ∎

EXAMPLE 10

Sketch the graph of the equation

$$x^2(4 - y^2) = 4. \tag{23}$$

(a) Rewrite Eq. 23 in the form

$$y^2 = 4\left(1 - \frac{1}{x^2}\right). \tag{24}$$

Since $y^2 \geq 0$, it follows that we must restrict x so that $x^2 \geq 1$. Hence $-1 < x < 1$ is an excluded interval. Similarly, if we rewrite Eq. 23 as

$$x^2 = \frac{4}{4 - y^2} \tag{25}$$

we see that we must have $4 - y^2 > 0$ so that $y \leq -2$ and $y \geq 2$ are also excluded intervals.

(b) There are no intercepts on the y-axis, but there are intercepts on the x-axis at $x = -1$ and $x = 1$.

(c) The graph is symmetric about both axes and the origin. Hence it is sufficient to consider the portion of the graph lying in the first quadrant.

(d) From Eq. 25 it follows that $y = 2$ and $y = -2$ are horizontal asymptotes. There are no vertical asymptotes.

(e) By differentiating Eq. 24 we obtain

$$2yy' = \frac{8}{x^3}$$

or

$$y' = \frac{4}{x^3y}. \tag{26}$$

Hence, in the first quadrant, $y' > 0$ and therefore y is an increasing function of x there. Further, as $x \to 1$, it follows from Eq. 24 that $y \to 0^+$, and then from Eq. 26 that $y' \to \infty$; thus the tangent line is vertical at the intercept $(1, 0)$.

(f) Differentiating Eq. 26 with respect to x, we find that

$$\begin{aligned} y'' &= \frac{x^3y(0) - 4(3x^2y + x^3y')}{x^6y^2} \\ &= \frac{-4(3x^2y^2 + 4)}{x^6y^3}. \end{aligned} \tag{27}$$

Hence, in the first quadrant, $y'' < 0$ and the graph is concave down.

Using the preceding information, we can sketch the graph shown in Figure

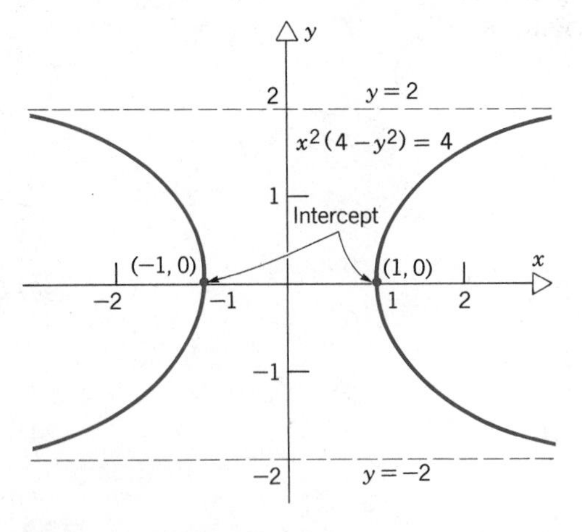

Figure 5.1.14

5.1.14. Note that the portion of the graph lying in the first quadrant was drawn first, the remainder being sketched afterward using the symmetry of the graph. ∎

PROBLEMS

In each of Problems 1 through 10, determine whether the graph of the given equation is symmetric about the x-axis; the y-axis; the origin.

1. $xy = 4$
2. $\dfrac{x^2}{4} + \dfrac{y^2}{9} = 1$
3. $(x - 1)^2 + (y + 2)^2 = 1$
4. $x^3 + y^3 - x - y = 0$
5. $y = x \sin x$
6. $\cos y - \sin x = 1$
7. $\cos y \sin x = 1$
8. $y = |x|$
9. $|y| = |x|$
10. $|y + 2| = x$

In each of Problems 11 through 16, determine all vertical and horizontal asymptotes for the given function or equation.

11. $f(x) = \dfrac{x - 1}{x + 2}$
12. $y = \dfrac{2x + 3}{3x - 2}$
13. $f(x) = \dfrac{x^2 + 2x - 3}{x^2 + x + 2}$
14. $y = \dfrac{x + 1}{x^2 + 1}$
15. $y^2 = \dfrac{x - 1}{x + 1}$
16. $1 + y^2 = \dfrac{3x - 2}{2x + 4}$

In each of Problems 17 through 22, find all asymptotes (including those that are neither vertical nor horizontal) of the graph of the given function or equation.

17. $f(x) = \dfrac{x^2 - 1}{2x}$
18. $y = \dfrac{x^2 + 1}{x + 1}$
19. $y = x^2 + \dfrac{1}{x^2}$
20. $y^2 - x^2 = 1$
21. $x^2 - 2y^2 = 4$
22. $y^2 = x^2 + \dfrac{1}{x^2}$

In each of Problems 23 through 32, determine intervals in which the given function is concave down or concave up. Also determine all inflection points.

23. $f(x) = x^3 + 3x^2 - 4x + 5$
24. $f(x) = 4x^3 - 6x^2 + 6x - 8$
25. $f(x) = (x - 2)^4$
26. $f(x) = \sqrt{x^2 + 1}$
27. $f(x) = \dfrac{1}{x + 2}, \quad x \neq -2$
28. $f(x) = \dfrac{1}{x^2 - 1}, \quad x \neq \pm 1$
29. $f(x) = \sqrt{x}\,(2 + x), \quad x \geq 0$
30. $f(x) = x^{4/3} - 2x^{2/3}$
31. $f(x) = \sin^2 x, \quad 0 \leq x \leq 2\pi$
32. $f(x) = \sec x + \tan x, \quad 0 \leq x \leq 2\pi,$ $x \neq \dfrac{\pi}{2}, \dfrac{3\pi}{2}$

In each of Problems 33 through 50, sketch the graph of the given equation.

33. $y = x^3 - 3x^2 - 9x + 7$
34. $y = x^3 - 6x^2 + 9x - 4$
35. $y = (x - 1)^2(x + 1)^2$
36. $y^2 = (x - 1)^2(x + 1)^2$
37. $x^2 + 4y^2 = 4$
38. $xy = 4$
39. $y = \dfrac{1}{x + 2}$
40. $y = \dfrac{1}{x^2 - 1}$
41. $y = \dfrac{x}{(x + 1)(x - 2)}$
42. $x^3 + y^3 - x - y = 0$
43. $x^{2/3} + y^{2/3} = a^{2/3}, \quad a > 0$
44. $y^3 = x^2$
45. $(y - 2)(x + 2) = 1$
46. $y(y - 2)(x - 1) = 1$
47. $y = \sin x + \cos x$
48. $y = x - \sin x$
49. $y = \sin^2 x$
50. $y = \sec x + \tan x$

In each of Problems 51 through 54, sketch the graph of an equation that has the given set of properties.

51. Symmetric about y-axis; vertical asymptote at $x = 2$; no horizontal asymptote; local minima at $(1, 0)$ and $(4, 2)$; local maximum at $(0, 3)$; inflection point at $(\frac{1}{2}, 2)$.
52. Symmetric about x-axis; vertical asymptote at $x = -1$; horizontal asymptote at $y = 2$; no points on graph for $x \leq -1$.
53. Local maximum at $(-2, 3)$; local minimum at $(4, -1)$; inflection points at $(-4, 0)$, $(0, 1)$, and $(6, 1)$; horizontal asymptote at $y = 3$ as $x \to \infty$;

no horizontal asymptote as $x \to -\infty$; no vertical asymptotes.

54. Symmetric about the origin; local maximum at $(3, 2)$; inflection point at $(5, 1)$; horizontal asymptote at $y = 0$; no vertical asymptotes.

55. Let $f(x) = ax^3 + bx^2 + cx + d$, where a, b, c, and d are real constants with $a \neq 0$.

(a) Show that f has exactly one inflection point.

* (b) Show that the inflection point is the arithmetic mean of the three zeros of f. This is true even if the zeros are not all real. *Hint:* Write $f(x) = a(x - r_1)(x - r_2)(x - r_3)$.

56. Determine the coefficients a, b, c, and d so that

$$f(x) = ax^3 + bx^2 + cx + d$$

has an inflection point at $x = 2$ and a local minimum at $x = -1$.

5.2 QUADRATIC EQUATIONS, PARABOLAS, TRANSLATIONS OF AXES

While the methods of the previous section are useful in sketching the graph of almost any equation, it is desirable to be familiar with the graphs of certain simple equations that occur often in practice. We have already discussed the graphs of linear equations, and in this and the following section we consider the graphs of quadratic equations in two variables. The most general such equation is of the form

$$Ax^2 + Bxy + Cy^2 + Dx + Ey = F, \qquad (1)$$

where $A, \ldots, F$ are constants. It is shown in Problems 25 to 28 that it is always possible to choose a Cartesian coordinate system in which the term Bxy does not appear. We therefore assume that this has been done already; that is, we assume that $B = 0$. Further, for Eq. 1 actually to be quadratic, we must assume that at least one of the coefficients A, C is different from zero. There are then three possibilities to consider.

(a) Only one of A and C is nonzero.
(b) A and C are of the same sign.
(c) A and C are of opposite sign.

Apart from a few degenerate cases, each possibility corresponds to a class of curves of a distinct kind, known as parabolas, ellipses (including circles as special cases), and hyperbolas, respectively. Circles were considered in Section 1.3. We discuss parabolas in this section and take up ellipses and hyperbolas in Sections 5.3 and 5.4, respectively. It turns out that these curves, which are known collectively as **conic sections** (because they also occur if a cone is intersected by a plane), have many interesting and important geometric properties. In each case we describe briefly how the curve may be obtained by a simple geometric construction. Further geometric properties are developed in some of the problems at the end of this and the following two sections.

Parabolas

We first consider the case in which only one of A and C is nonzero; to be specific, suppose that $A \neq 0$ and $C = 0$, leaving the other possibility until later. Then Eq. 1 has the form

$$Ax^2 + Dx + Ey = F. \qquad (2)$$

We also assume* that $E \neq 0$, so that both of the variables x and y actually occur in Eq. 2. To reduce Eq. 2 to a more convenient form, we can complete the square in x:

$$A\left(x^2 + \frac{D}{A}x\right) + Ey = F,$$

$$A\left(x^2 + \frac{D}{A}x + \frac{D^2}{4A^2}\right) + Ey = F + \frac{D^2}{4A},$$

or

$$A\left(x + \frac{D}{2A}\right)^2 + E\left(y - \frac{F}{E} - \frac{D^2}{4AE}\right) = 0. \tag{3}$$

Hence Eq. 3 can be put in the form

$$(x - h)^2 = \alpha(y - k), \tag{4}$$

where

$$h = -\frac{D}{2A}, \qquad k = \frac{F}{E} + \frac{D^2}{4AE}, \qquad \alpha = -\frac{E}{A}. \tag{5}$$

Note that the quantities h and k may be of either sign or zero; on the other hand, α may be either positive or negative, but cannot be zero.

Equation 4 is one of the standard forms of the equation of a **parabola.** Its graph is especially simple to sketch when $h = k = 0$, in which case Eq. 4 reduces to

$$x^2 = \alpha y. \tag{6}$$

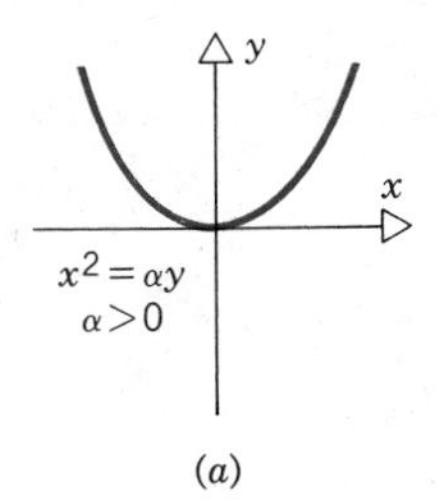

(a)

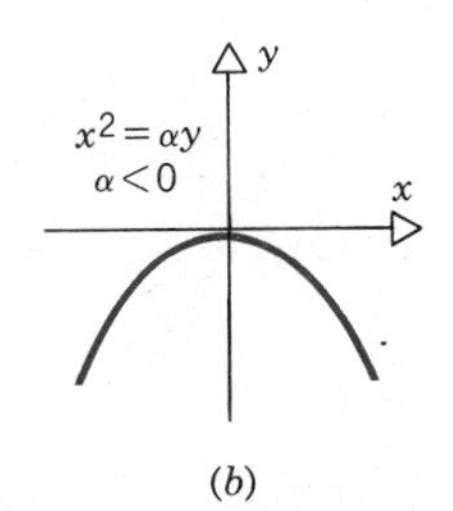

(b)

Figure 5.2.1

The graph of Eq. 6 is symmetric about the y-axis, so we need only consider $x \geq 0$. Let us suppose first that $\alpha > 0$; then it follows that y is nonnegative. Further, $y' = 2x/\alpha$, so the slope of the graph is zero at the origin, and increases as x increases. A second differentiation yields $y'' = 2/\alpha$, and it follows that the graph is concave up. The graph of Eq. 6 for $\alpha > 0$ is shown in Figure 5.2.1*a*; symmetry has been used to sketch the portion of the graph corresponding to $x < 0$. In this case, the origin is known as the **vertex.** Also, note that if α is large, then the parabola is wide and flat, while if α is small, then the parabola is narrow and steep. The situation is much the same if $\alpha < 0$, except that in this case the parabola opens downward instead of upward. The graph of Eq. 6 when $\alpha < 0$ is shown in Figure 5.2.1*b*.

Now let us return to Eq. 1 and consider the case in which $A = 0$ but $C \neq 0$; then Eq. 1 is

$$Cy^2 + Dx + Ey = F. \tag{7}$$

---

*If $E = 0$, then Eq. 2 is $Ax^2 + Dx - F = 0$, which has either no, one, or two real roots. These roots correspond to straight lines parallel to the y-axis. This is one of the degenerate cases previously mentioned.

Assuming in this case that $D \neq 0$, we complete the square in y and obtain

$$(y - k)^2 = \alpha(x - h), \tag{8}$$

where now

$$h = \frac{F}{D} + \frac{E^2}{4CD}, \qquad k = -\frac{E}{2C}, \qquad \alpha = -\frac{D}{C}. \tag{9}$$

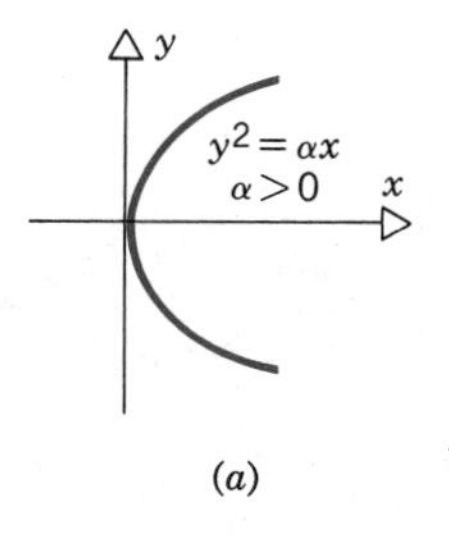

(a)

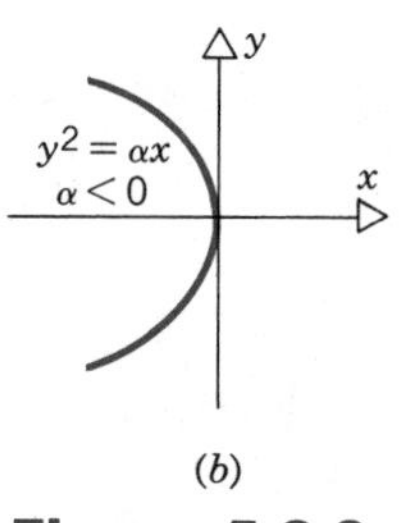

(b)

Figure 5.2.2

Equation 8 is similar in structure to Eq. 4 and is the other standard form of the equation of a parabola.

Again, if $h = k = 0$, Eq. 8 reduces to the simpler form

$$y^2 = \alpha x. \tag{10}$$

The graph of Eq. 10 can be sketched in much the same way as that of Eq. 6; essentially all that is needed is to reverse the roles of the two variables and hence of the two axes. The two possibilities, corresponding to $\alpha > 0$ and $\alpha < 0$, are shown in Figure 5.2.2. If $\alpha > 0$ the graph of Eq. 10 opens to the right while if $\alpha < 0$ the graph opens to the left. Again the origin is called the vertex of the parabola, and the parabola is narrow or wide according to whether $|\alpha|$ is small or large.

In order to deal with the somewhat more general equations (4) and (8), respectively, it is helpful to introduce the idea of a change of coordinates, in particular, a type of coordinate transformation known as a translation. We now discuss this topic, applying it to Eqs. 4 and 8.

Translation of axes

Let new coordinates u and v be defined by the equations

$$u = x - h, \qquad v = y - k. \tag{11}$$

Then the v-axis (on which $u = 0$) is the line $x = h$, and is thus parallel to the y-axis and a distance $|h|$ away from it—to the right if $h > 0$ and to the left if $h < 0$. Similarly, the u-axis (on which $v = 0$) is the line $y = k$, parallel to the x-axis and a distance $|k|$ from it—above if $k > 0$ and below if $k < 0$. The origin in the uv-coordinate system is at the point $x = h$, $y = k$ in the xy-coordinate system. The uv-axes can be regarded as obtained by moving the xy-axes parallel to themselves. Such a transformation of coordinate axes is known as a **translation.** Figure 5.2.3 shows the particular case in which $h = 2$ and $k = -1$, so that $u = x - 2$ and $v = y + 1$. Any point can be identified by giving either its xy- or its uv-coordinates. For instance, the point P shown in Figure 5.2.3 with coordinates $(x = 1, y = 1)$ also has the coordinates $(u = -1, v = 2)$.

Now let us return to Eq. 4. Introducing the new uv-coordinates defined by Eq. 11, we have

$$u^2 = \alpha v. \tag{12}$$

Thus, Eq. 12 has the same form with respect to the uv-coordinate system as Eq. 6 with respect to the xy-coordinate system. Hence we can sketch the graph of the parabola as before, using the uv-coordinates. Once the graph is drawn, we can

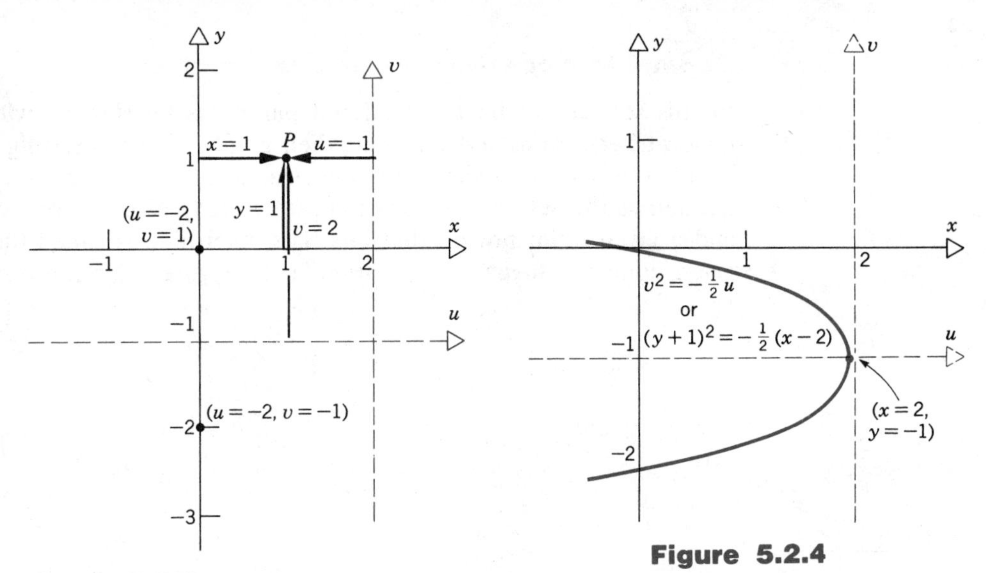

Figure 5.2.4

Figure 5.2.3
A translated coordinate system.

erase (if we wish) the *uv*-axes, leaving only the original *xy*-coordinate system. The following example illustrates this procedure.

EXAMPLE 1

Sketch the graph of

$$2y^2 + x + 4y = 0. \tag{13}$$

By completing the square we obtain

$$2(y^2 + 2y) + x = 0,$$

$$2(y^2 + 2y + 1) + x = 2,$$

or

$$(y + 1)^2 = -\tfrac{1}{2}(x - 2). \tag{14}$$

The form of Eq. 14 suggests that we introduce a *uv*-coordinate system defined by

$$u = x - 2, \qquad v = y + 1 \tag{15}$$

(see Figure 5.2.3). Then Eq. 14 becomes

$$v^2 = -\tfrac{1}{2}u. \tag{16}$$

The graph of Eq. 16 is symmetric about the *u*-axis, and opens to the left with vertex at the origin in the *uv*-coordinate system. The width of the parabola is adequately determined by plotting two further points in addition to the vertex. For example, if $u = -2$, then $v = -1$ or $v = 1$. The graph of Eq. 16, or of Eq. 13, is shown in Figure 5.2.4. ∎

Geometrical construction of a parabola

In this section we have considered parabolas by starting with certain types of quadratic equations and sketching their graphs. It is interesting that parabolas also result from a certain simple geometrical construction. In fact, a parabola can be defined as the set of all points that are equally distant from a given straight line and a given point not on the line. The given line is called the **directrix** and the given point the **focus** of the parabola. In Figure 5.2.5 the distances FP and QP

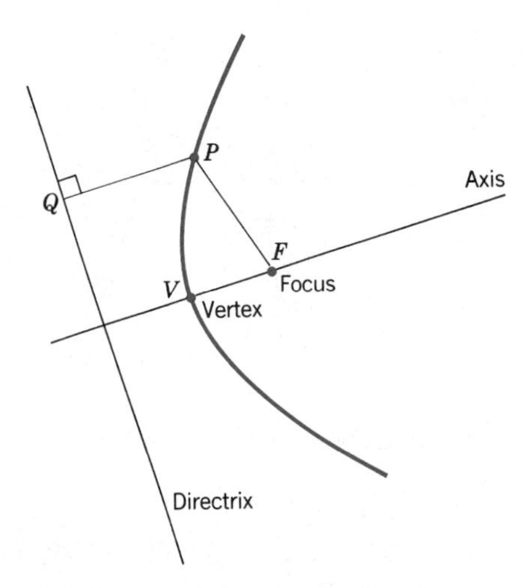

Figure 5.2.5
The geometrical construction of a parabola: $|PQ| = |PF|$.

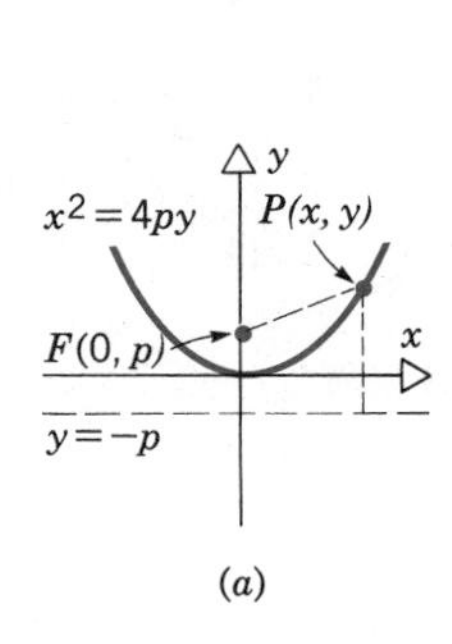

(a)

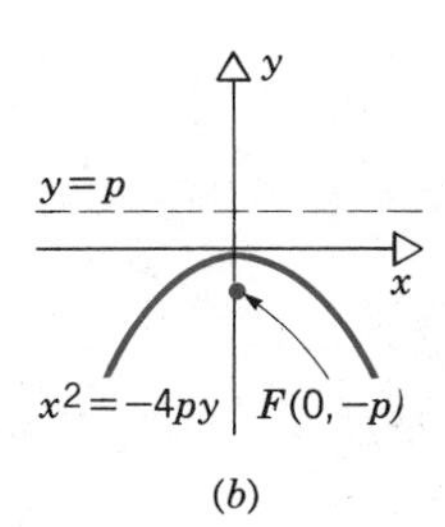

(b)

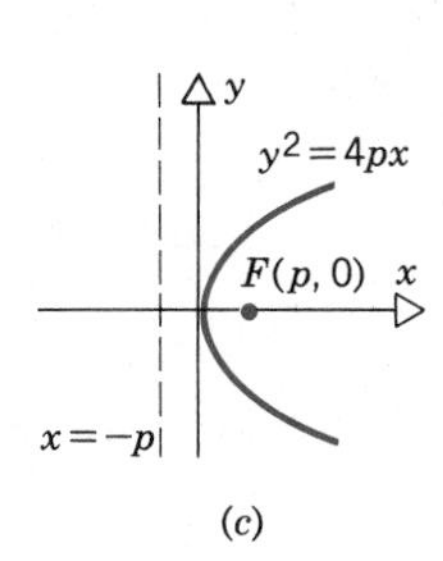

(c)

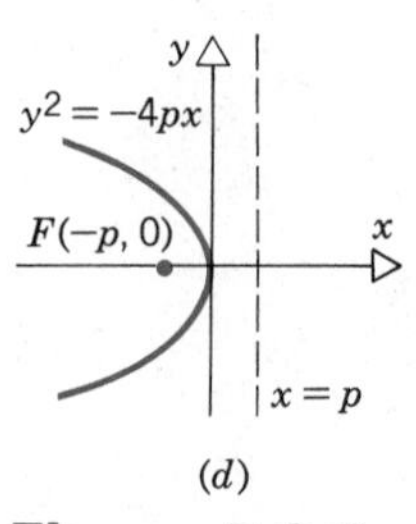

(d)

Figure 5.2.6

must be equal for each point P on the parabola. The line of symmetry, or **axis,** of the parabola must be perpendicular to the directrix and must contain the focus; further, the vertex must be on the axis halfway between the focus and directrix.

To show the equivalence of the two definitions we need to determine the equation satisfied by the points equidistant from the given point and the given line. It is convenient to let $2p$ be the distance from focus to directrix and to introduce a coordinate system as shown in Figure 5.2.6a. The directrix has the equation $y = -p$ and the focus F has the coordinates $(0, p)$. Equating the distances from an arbitrary point $P(x, y)$ on the parabola to the focus and to the directrix, we have

$$y + p = \sqrt{(x - 0)^2 + (y - p)^2}. \tag{17}$$

Squaring both sides, we obtain

$$(y + p)^2 = x^2 + (y - p)^2,$$

or

$$y^2 + 2py + p^2 = x^2 + y^2 - 2py + p^2,$$

and finally

$$x^2 = 4py. \tag{18}$$

Other convenient choices of the coordinate system are shown in Figures 5.2.6b–5.2.6d; in each case the distance from focus to directrix is $2p$. The equations of the resulting parabolas are

$$x^2 = -4py, \qquad y^2 = 4px, \qquad y^2 = -4px \tag{19}$$

respectively. Equations 18 and 19 are the same as Eqs. 6 and 10, provided we identify α with $4p$ or with $-4p$ as the case may be. Hence, in all cases,

$$\alpha = \pm 4p \qquad \text{or} \qquad p = \frac{|\alpha|}{4}.$$

Thus, in terms of α, the distance from focus to directrix is $|\alpha|/2$ and the distance from focus to vertex is $|\alpha|/4$. When the directrix is near the focus, p is small, $|\alpha|$ is small, and the resulting parabola is narrow. On the other hand, if the directrix and focus are far apart, the p is large, $|\alpha|$ is large, and the parabola is broad. Given appropriate information about the vertex, focus, directrix, and axis, it is possible to write the equation of the corresponding parabola, as in the following examples.

EXAMPLE 2

Find the equation of the parabola whose vertex is $(-1, 2)$ and whose focus is $(-1, 1)$.

Since the focus is below the vertex, the parabola must open downward. Thus its equation is of the form

$$(x - h)^2 = \alpha(y - k),$$

where $\alpha < 0$ and (h, k) is the vertex. Substituting $h = -1$, $k = 2$, we obtain

$$(x + 1)^2 = \alpha(y - 2).$$

To determine α we recall that the distance from focus to vertex is always $|\alpha|/4$. In this case this distance is one unit, so $|\alpha| = 4$, and $\alpha = -4$. Thus the equation of the parabola is

$$(x + 1)^2 = -4(y - 2).$$

Its graph is shown in Figure 5.2.7. ◀

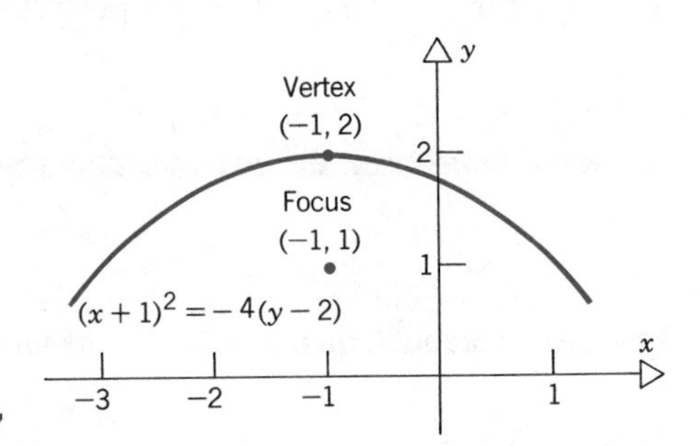

Figure 5.2.7

EXAMPLE 3

Find the equation of the parabola whose axis is horizontal, whose focus is $(2, 1)$, and that passes through $(8, 9)$.

Let us first find the directrix of the parabola. Since the axis is horizontal, the directrix must be vertical. Further, since the parabola contains the point (8, 9) lying to the right of the focus, it follows that the directrix must lie to the left of the focus. Finally, the point (8, 9) must be equidistant from focus and directrix. The distance from (2, 1) to (8, 9) is

$$\sqrt{(8 - 2)^2 + (9 - 1)^2} = \sqrt{36 + 64} = 10.$$

Therefore the directrix is the vertical line 10 units to the left of the point (8, 9), that is, the line $x = -2$. The point (0, 1) halfway between focus and directrix is the vertex. Hence the equation of the parabola is of the form

$$(y - 1)^2 = \alpha x, \tag{20}$$

where $\alpha > 0$. We determine α by substituting $x = 8$ and $y = 9$ in Eq. 20. Thus $\alpha = 8$, and the equation of the parabola is

$$(y - 1)^2 = 8x.$$

This parabola is shown in Figure 5.2.8. ∎

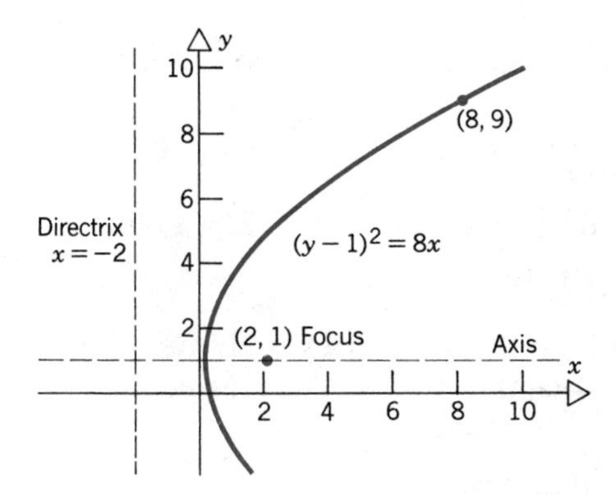

Figure 5.2.8

Several important properties of parabolas are indicated in the problems; perhaps the most important property for applications is discussed in Problem 21.

PROBLEMS

In each of Problems 1 through 8, find the vertex, focus, and directrix of the given parabola; then sketch its graph.

1. $y = -4x^2$

2. $x = \frac{1}{4}y^2$

3. $y = 1 + 2x^2$

4. $y^2 = x + 2$

5. $2y^2 + 3x - 4y + 8 = 0$

6. $3x^2 = 12x - y + 6$

7. $3y = 4x^2 - 6x + 9$

8. $y^2 = 4x + 4y - 8$

In each of Problems 9 through 16, find the equation of the parabola having the given properties.

9. Axis parallel to y-axis; vertex is (2, 1); passes through (−1, −1).

10. Vertex is (−1, 2); focus is (1, 2).

11. Vertex is (1, 1); directrix is $y = -2$.

12. Directrix is $x = 2$; focus is (3, 1).

13. Vertex is $(2, -1)$; passes through $(0, 2)$ and $(4, 2)$.

14. Vertex is $(2, -3)$; focus is $(2, 1)$.

15. Axis parallel to x-axis; vertex is $(-2, 3)$; distance from focus to directrix is 4; opens to right.

16. Axis is $x = 2$; tangent to $y = 8x - 20$ at $(4, 3)$.

17. Find the equation of the parabola whose points are equidistant from the line $x = -3$ and the point $(2, 0)$.

18. Find the equation of the parabola whose points are equidistant from the line $y = 2$ and the point $(3, -1)$.

19. Find the equation satisfied by points whose distance from the line $x = -3$ is two units greater than their distance from the point $(3, 2)$.

20. Find the equation satisfied by points that are twice as far from the line $y = 2$ as from the point $(1, -2)$.

Reflection Property of the Parabola

It was discovered by the Greeks that all parabolas have the following geometrical property. A ray (for example, of light) that emanates from the focus is reflected by the parabola along a path parallel to the axis of the parabola, regardless of the point of reflection. Alternatively, a ray parallel to the axis and reflected by the parabola always passes through the focus. This fact is useful in the construction of flashlights, automobile headlights, and searchlights, where the reflector has a parabolic cross section and the light source is placed at the focus. Similarly, in telescopes and radar receivers, signals from a remote source enter parallel to the axis and are reflected through the focus by a parabolic reflector. The powerful concentration resulting from a large parabolic reflector, as in a radio telescope, makes it possible to detect and to analyze very weak signals.

21. To establish this property of the parabola let us suppose that the vertex is at the origin, that the x-axis is the axis of the parabola, and that the parabola opens to the right (see Figure 5.2.9). Then the equation of the parabola is $y^2 = 4px$, where $p > 0$, and the focus F has the coordinates $(p, 0)$. Let $P(x_0, y_0)$ be an arbitrary point on the parabola (other than the vertex) and let T be the line tangent to the parabola at P. Let ϕ be the angle between T and the x-axis, and let θ be the angle between T and the line PF. The reflection property previously stated can then be proved by showing that $\theta = \phi$. We accomplish this by showing that $\tan\theta = \tan\phi$.

   (a) Let ψ be the angle between FP and the positive x-axis. Show that $\theta = \psi - \phi$.

   (b) Differentiate the equation of the parabola to show that
$$\tan\phi = \frac{2p}{y_0}.$$

   (c) Use the definition of slope to show that
$$\tan\psi = \frac{y_0}{x_0 - p}.$$

   (d) Determine $\tan\theta$ from the identity
$$\tan\theta = \frac{\tan\psi - \tan\phi}{1 + \tan\psi\tan\phi},$$
and show that $\tan\theta = \tan\phi$.

22. Show that the equation
$$2x^2 + 3y^2 + 4x - 6y = 12$$
can be transformed into
$$\alpha u^2 + \gamma v^2 = \kappa$$
by a suitable translation of axes $u = x - h$, $v = y - k$. Find h, k, α, γ, and κ.

23. Show that the equation
$$Ax^2 + Cy^2 + Dx + Ey = F,$$

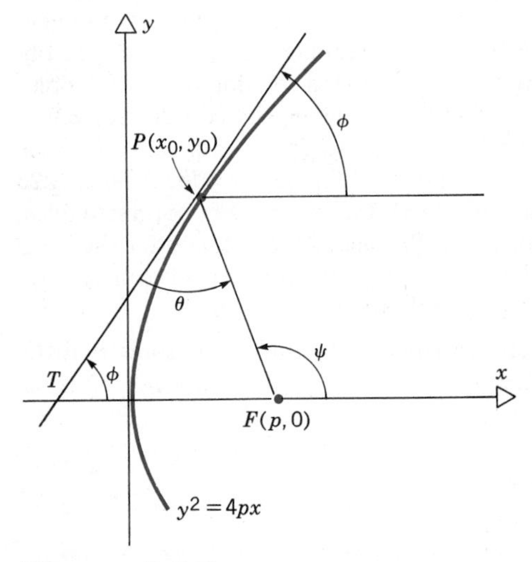

Figure 5.2.9
Reflection property of the parabola; $\phi = \theta$.

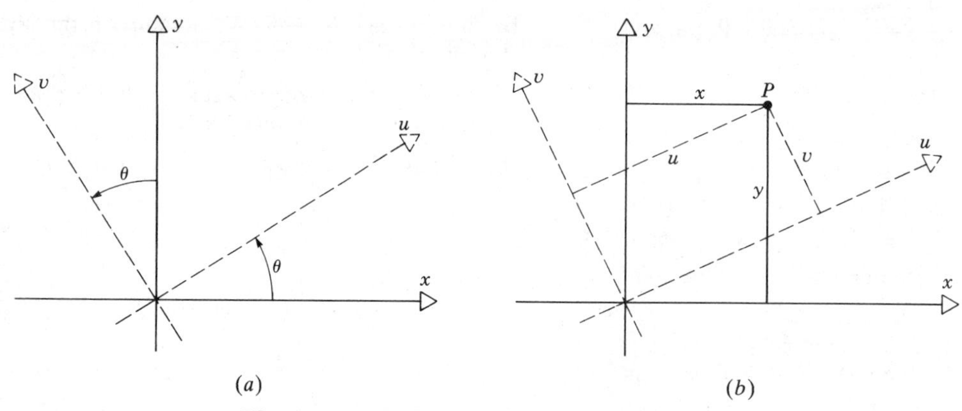

Figure 5.2.10 A rotated coordinate system.

with $A \neq 0$ and $C \neq 0$, can be transformed into

$$\alpha u^2 + \gamma v^2 = \kappa$$

by a suitable translation of axes $u = x - h$, $v = y - k$. Find h, k, α, γ, and κ in terms of A, C, D, E, and F.

24. Find a translation of axes $u = x - h$, $v = y - k$ such that the straight line $Ax + By = C$ passes through the origin in the uv-plane.

Rotation of Axes

Consider two Cartesian coordinate systems, the xy-system and the uv-system, which have the common origin O. Let the positive u- and v-axes be obtained by rotating the positive x- and y-axes, respectively, through the angle θ (see Figure 5.2.10a). In this event we say that the xy-system and the uv-system are obtained from each other by a rotation of axes. A point P can be identified by giving its coordinates either with respect to the xy-system or with respect to the uv-system, as shown in Figure 5.2.10b. Problems 25 through 28 deal with the rotation of coordinate axes. In particular, in Problem 28 we show that the term Bxy in the general quadratic equation (1) can be eliminated by a suitable rotation of axes.

25. (a) By referring to Figure 5.2.11, show that the xy-coordinates of a point P are related to its uv-coordinates by the equations

$$x = OQ = OR - QR$$
$$= u \cos \theta - v \sin \theta,$$
$$y = QP = QS + SP$$
$$= u \sin \theta + v \cos \theta.$$

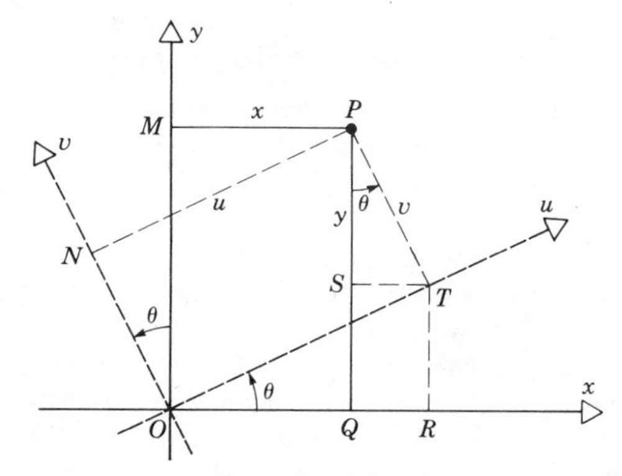

Figure 5.2.11

(b) By drawing a sketch similar to Figure 5.2.11, or by solving the equations in Part (a), show that

$$u = x \cos \theta + y \sin \theta,$$
$$v = -x \sin \theta + y \cos \theta.$$

26. Consider the equation

$$x^2 + 2\sqrt{3}\, xy + 3y^2 - \sqrt{3}\, x + y = 6. \qquad (i)$$

Let the uv-coordinate system be obtained by a rotation through the angle $\theta = \pi/3$. Then, referring to Problem 25(a),

$$x = \tfrac{1}{2}(u - \sqrt{3}\, v),$$
$$y = \tfrac{1}{2}(\sqrt{3}\, u + v).$$

(a) Express Eq. (i) in terms of u and v.

(b) Sketch the graph of the equation obtained in Part (a) with respect to the uv-coordinate system. Note that by superimposing the two coordinate systems, this also gives the graph of Eq. (i) with respect to the x- and y-axes.

27. Consider the equation

$$x^2 - 2\sqrt{3}\,xy + 3y^2 - 2\sqrt{3}\,x - 2y + 8 = 0.$$

Let the uv-system be obtained from the xy-system by a rotation through the angle θ (see Problem 25).

(a) Transform the given equation into one involving u and v, rather than x and y.

(b) Determine θ so that the equation obtained in Part (a) has no term of the form βuv.

(c) For the smallest positive value of θ found in (b), find the equation of the parabola in terms of u and v. Then sketch its graph, showing both the uv-axes and the xy-axes.

28. Consider the equation

$$Ax^2 + Bxy + Cy^2 + Dx + Ey = F, \qquad (i)$$

where A, B, C, D, E, and F are constants. Let the uv-coordinate system be obtained by rotating the xy-system through the angle θ.

(a) Show that in terms of u and v, Eq. (i) has the form

$$\alpha u^2 + \beta uv + \gamma v^2 + \delta u + \epsilon v = \kappa, \qquad (ii)$$

where α, β, γ, δ, ϵ, and κ are constants. Show that α, β, γ, δ, ϵ, and κ are given in terms of A, B, C, D, E, and F by the equations

$$\alpha = A\cos^2\theta + B\sin\theta\cos\theta + C\sin^2\theta,$$

$$\beta = -2A\sin\theta\cos\theta + B(\cos^2\theta - \sin^2\theta) + 2C\sin\theta\cos\theta,$$

$$\gamma = A\sin^2\theta - B\sin\theta\cos\theta + C\cos^2\theta,$$

$$\delta = D\cos\theta + E\sin\theta,$$

$$\epsilon = -D\sin\theta + E\cos\theta,$$

$$\kappa = F.$$

(b) Show that it is always possible to determine θ so that $\beta = 0$.

29. The chord perpendicular to the axis of a parabola and passing through its focus is called the **latus rectum.** Show that the length of the latus rectum is twice the distance from the focus to the directrix. This property can be helpful in sketching the parabola.

Hint: Consider the case $x^2 = \alpha y$ with $\alpha > 0$. By a suitable choice of axes the equation of any parabola can be put into this form.

30. Show that the lines tangent to a parabola at the ends of the latus rectum (see Problem 29) intersect on the directrix and are perpendicular to each other.

Hint: See Hint in Problem 29.

5.3 ELLIPSES

In this section we consider further the equation

$$Ax^2 + Cy^2 + Dx + Ey = F, \qquad (1)$$

where we now assume that both A and C are nonzero and of the same sign. Let us suppose for simplicity that both A and C are positive (if both are negative, multiply Eq. 1 by -1). To reduce Eq. 1 to a more convenient form we complete the square in both x and y. Thus we obtain

$$A\left(x^2 + \frac{D}{A}x + \frac{D^2}{4A^2}\right) + C\left(y^2 + \frac{E}{C}y + \frac{E^2}{4C^2}\right) = F + \frac{D^2}{4A} + \frac{E^2}{4C}, \qquad (2)$$

or

$$A(x-h)^2 + C(y-k)^2 = M, \qquad (3)$$

where

$$h = -\frac{D}{2A}, \qquad k = -\frac{E}{2C}, \qquad M = F + \frac{D^2}{4A} + \frac{E^2}{4C}. \qquad (4)$$

If it turns out that $M < 0$, then Eq. 3 has no graph because the left side is certainly nonnegative. Similarly, if $M = 0$ the graph consists of only the single point $x = h$, $y = k$. These degenerate cases are of little importance. The interesting situation occurs when $M > 0$; in this event, we can divide both sides of Eq. 3 by M to obtain

$$\frac{(x-h)^2}{a^2} + \frac{(y-k)^2}{b^2} = 1, \tag{5}$$

where $a^2 = M/A$, $b^2 = M/C$, $a = \sqrt{M/A} > 0$, and $b = \sqrt{M/C} > 0$.

Equation 5 is the standard form of the equation of an **ellipse.** We can simplify Eq. 5 further by introducing the translated coordinates u, v defined by

$$u = x - h, \qquad v = y - k. \tag{6}$$

As we showed in Section 5.2 the u- and v-axes are parallel to the x- and y-axes, respectively, and the origin of the uv-coordinate system is the point with coordinates $x = h$, $y = k$ in the xy-system. In terms of u and v, Eq. 5 becomes

$$\frac{u^2}{a^2} + \frac{v^2}{b^2} = 1. \tag{7}$$

In the special case when $b = a$, Eq. 5 and Eq. 7 reduce to

$$(x-h)^2 + (y-k)^2 = a^2 \qquad \text{or} \qquad u^2 + v^2 = a^2,$$

which correspond to the circle with center at the point $(x = h, y = k)$ and with radius a.

A good sketch of the graph of Eq. 7 can be made by taking account of the following facts. First note that each term on the left side of Eq. 7 is nonnegative; hence neither can have a value exceeding one. Consequently all points on the graph of Eq. 7 satisfy $|u| \le a$ and $|v| \le b$. In other words, the graph of Eq. 7 lies entirely within the auxiliary rectangle R bounded by the lines $u = \pm a$ and $v = \pm b$. When $v = 0$, we have $u = \pm a$, so that there are intercepts on the u-axis at $(-a, 0)$ and $(a, 0)$. Similarly there are intercepts on the v-axis at $(-b, 0)$ and $(b, 0)$.

The graph of Eq. 7 is symmetric about both the u- and v-axes; therefore it is sufficient to consider in detail only the portion of the graph lying in the first quadrant in the uv-plane. Differentiating Eq. 7 implicitly with respect to u, we find that

$$\frac{2u}{a^2} + \frac{2vv'}{b^2} = 0, \tag{8}$$

or

$$v' = -\frac{b^2}{a^2}\frac{u}{v}, \tag{9}$$

where $v' = dv/du$. From Eq. 9 we see that in the first quadrant $v' < 0$, so v decreases as u increases there. It also follows from Eq. 9 that the tangent line to the graph is horizontal when $u = 0$ and is vertical when $v = 0$. Thus the graph of Eq. 7 is tangent to the side of the rectangle R at each intercept.

A second differentiation with respect to u yields

$$v'' = -\frac{b^2}{a^2}\left(\frac{1}{v} - \frac{uv'}{v^2}\right). \tag{10}$$

In the first quadrant $u > 0$, $v > 0$, and $v' < 0$; therefore $v'' < 0$ there, and this portion of the graph is concave down. Using all of this information, it is not difficult to sketch the graph of Eq. 5, or of Eq. 7. Figures 5.3.1*a* and 5.3.1*b* show the cases $a^2 > b^2$ and $a^2 < b^2$, respectively.

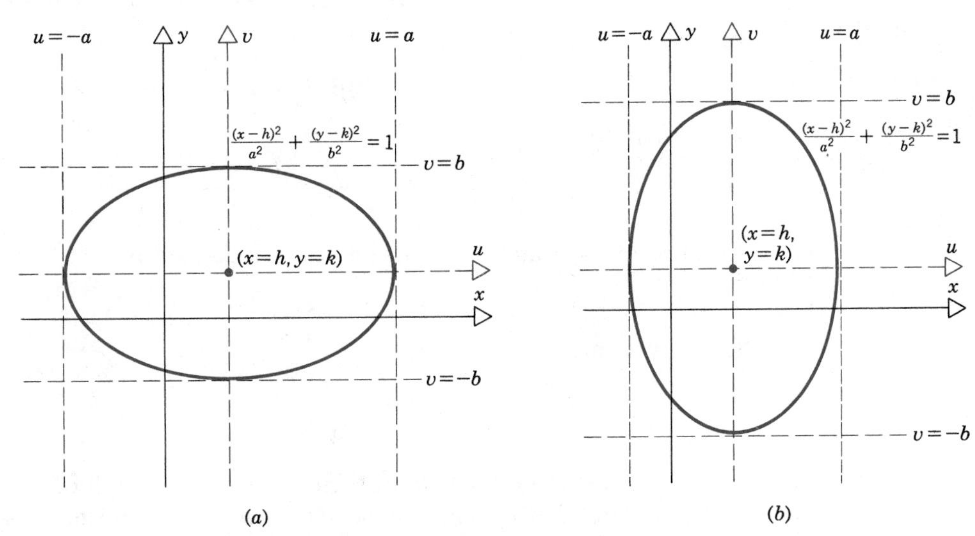

Figure 5.3.1 (*a*) An ellipse, $a^2 > b^2$. (*b*) An ellipse, $a^2 < b^2$.

If $a^2 > b^2$, the oval graph of the ellipse is elongated along the u-axis, as shown in Figure 5.3.1*a*, while if $a^2 < b^2$, the graph is elongated along the v-axis, as shown in Figure 5.3.1*b*. The axis along which the figure is elongated is called the **major axis,** and the other axis is the **minor axis.** Thus, if $a^2 > b^2$, the u-axis is the major axis and the v-axis is the minor axis. The numbers a and b are referred to as the lengths of the semimajor axis and semiminor axis, respectively, in this case. The situation is reversed if $a^2 < b^2$. The origin in the uv coordinate system is known as the **center** of the ellipse, and the intercepts on the major axis are called **vertices.** Because of their symmetry with respect to the center, ellipses and hyperbolas (see Section 5.4) are known as **central conics.**

Perhaps the easiest way to sketch the graph of the ellipse given by Eq. 5 is as follows.

1. Draw x- and y-axes.
2. Determine h and k.
3. Draw u- and v-axes.
4. Draw the auxiliary rectangle whose sides are $u = \pm a$, $v = \pm b$.
5. Inscribe the ellipse in the auxiliary rectangle, as shown in Figure 5.3.1.

Once the sketch is completed the rectangle as well as the u- and v-axes can be erased, if desired, leaving the sketch related to the x- and y-axes only.

EXAMPLE 1

Sketch the graph of

$$9x^2 + 4y^2 - 36x + 8y = -4. \tag{11}$$

Determine the major axis, the minor axis, the center, and the vertices.

Completing the square in x and y leads to

$$9(x^2 - 4x + 4) + 4(y^2 + 2y + 1) = -4 + 36 + 4$$

or

$$9(x - 2)^2 + 4(y + 1)^2 = 36,$$

and finally to

$$\frac{(x - 2)^2}{4} + \frac{(y + 1)^2}{9} = 1. \tag{12}$$

In terms of the translated coordinates u, v defined by

$$u = x - 2, \qquad v = y + 1,$$

Eq. 12 becomes

$$\frac{u^2}{4} + \frac{v^2}{9} = 1. \tag{13}$$

The u- and v-axes are shown in Figure 5.3.2. The graph of Eq. 13 lies within the rectangle bounded by the lines $u = \pm 2$, $v = \pm 3$, and has intercepts at $(\pm 2, 0)$

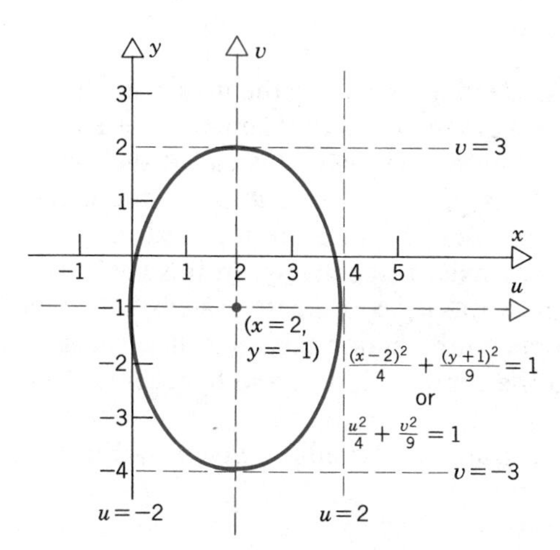

Figure 5.3.2

and $(0, \pm 3)$ in the uv coordinate system. Once these points are located, the rectangle can be drawn and the ellipse can be inscribed within it without difficulty.

In the xy coordinate system the major and minor axes are the lines $x = 2$ and $y = -1$, respectively. The center is $(2, -1)$ and the vertices are $(2, 2)$ and $(2, -4)$. ∎

Geometrical construction of the ellipse

An ellipse can be defined in an alternative way as the set of all points for which the sum of distances from two fixed points is a constant. To determine the equation satisfied by this set of points it is convenient to choose the coordinate system shown in Figure 5.3.3. Then the two fixed points are $(c, 0)$ and $(-c, 0)$, where $c > 0$.

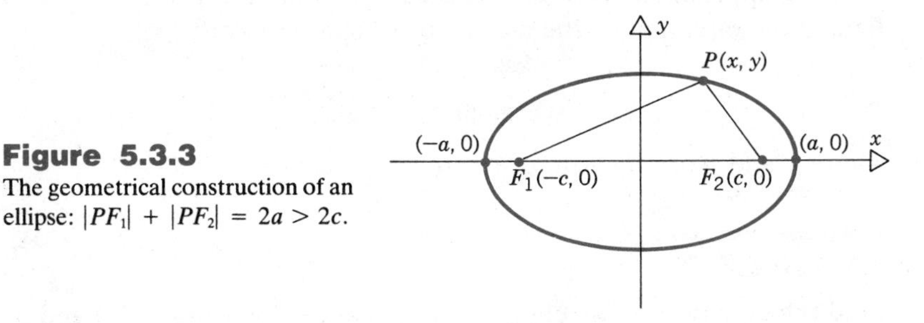

Figure 5.3.3
The geometrical construction of an ellipse: $|PF_1| + |PF_2| = 2a > 2c$.

Let the fixed distance be $2a$, where $2a > 2c$. Then $|PF_1| + |PF_2| = 2a$, or

$$\sqrt{(x + c)^2 + y^2} + \sqrt{(x - c)^2 + y^2} = 2a.$$

Transposing the second radical to the right side, and squaring the resulting equation, we obtain

$$(x + c)^2 + y^2 = 4a^2 - 4a\sqrt{(x - c)^2 + y^2} + (x - c)^2 + y^2;$$

a little algebraic manipulation reduces this to

$$cx - a^2 = -a\sqrt{(x - c)^2 + y^2}.$$

By squaring again we find that

$$c^2x^2 - 2a^2cx + a^4 = a^2(x^2 - 2cx + c^2 + y^2).$$

Finally, we let $b^2 = a^2 - c^2$; then it follows that

$$b^2x^2 + a^2y^2 = a^2b^2,$$

or

$$\frac{x^2}{a^2} + \frac{y^2}{b^2} = 1. \tag{14}$$

On the other hand, if we choose coordinates so that the fixed points are $(0, c)$ and $(0, -c)$, and the fixed distance is $2b > 2c > 0$, then a similar calculation leads again to Eq. 14 with $a^2 = b^2 - c^2$. In any case the two fixed points are called the **foci** of the ellipse; they always lie on the major axis, and the distance c from the center of the ellipse to either focus is called the **focal length.** In general, c is given by

$$c^2 = |a^2 - b^2|. \tag{15}$$

Note that the foci are always closer to the center than the vertices. For the ellipse in Example 1 the focal length is $c = \sqrt{5}$ and the foci are the points $(2, -1 \pm \sqrt{5})$.

Ellipses have a number of interesting applications. For instance, archways are sometimes constructed with an elliptical cross section. However, probably the best-known occurrence of ellipses in nature is in the solar system itself. Each planet travels around the sun on an elliptical path with the sun at one focus. The orbits differ considerably in shape, from nearly circular to fairly elongated. In the case of the earth the orbit is almost circular; the lengths of the semimajor and semiminor axes are approximately 93.004 and 92.991 million miles, respectively. Using these figures the equation of the earth's orbit can be written as

$$\frac{x^2}{8649.74} + \frac{y^2}{8647.33} = 1 \tag{16}$$

where x and y are measured in millions of miles.

EXAMPLE 2

Find the equation of the ellipse with foci at $(-3, 3)$ and $(1, 3)$ and passing through $(-1, 5)$.

The center of the ellipse is the midpoint between the two foci, namely $(-1, 3)$. Thus the ellipse has an equation of the form

$$\frac{(x + 1)^2}{a^2} + \frac{(y - 3)^2}{b^2} = 1. \tag{17}$$

To determine the constants a and b we need two conditions. One comes from the fact that the focal length of the given ellipse is half the distance between the two foci, so $c = 2$; consequently

$$a^2 - b^2 = 4. \tag{18}$$

Note that we have written Eq. 18 so that $a > b$; the major axis contains the foci, so in this case it is parallel to the x-axis. The second condition is obtained by substituting $x = -1$ and $y = 5$ in Eq. 17 with the result that

$$0 + \frac{4}{b^2} = 1.$$

Hence $b^2 = 4$ and, from Eq. 18, $a^2 = 8$. Therefore the equation of the ellipse is

$$\frac{(x + 1)^2}{8} + \frac{(y - 3)^2}{4} = 1. \tag{19}$$

The ellipse is shown in Figure 5.3.4. ∎

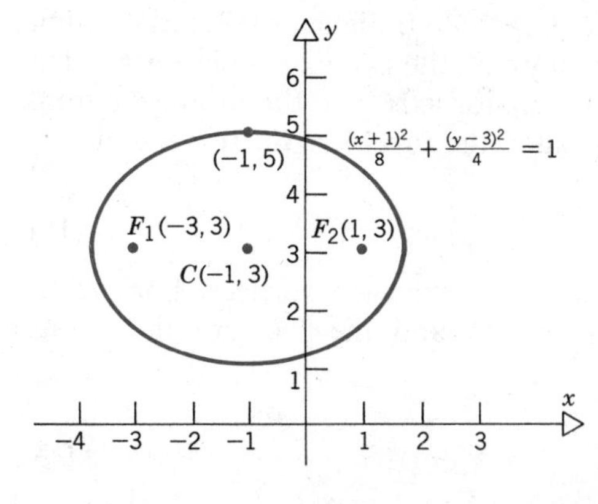

Figure 5.3.4

PROBLEMS

In each of Problems 1 through 8, find the center, the foci, and the lengths of the semimajor and semiminor axes for the given ellipse. Then sketch the graph.

1. $\dfrac{x^2}{16} + \dfrac{y^2}{25} = 1$ **2.** $\dfrac{x^2}{16} + \dfrac{y^2}{9} = 1$

3. $\dfrac{(x-1)^2}{4} + \dfrac{(y+2)^2}{9} = 1$

4. $x^2 + 2y^2 - 2x + 4y = 6$

5. $2x^2 + 2y^2 + 4x - 12y = 5$

6. $4x^2 + 3y^2 + 8x + 18y = -6$

7. $x^2 + 4y^2 - 4x + 16y = -4$

8. $4x^2 + 64x + 9y^2 - 18y = 23$

In each of Problems 9 through 14, find the equation of the ellipse having the given properties.

9. Foci at $(-1, 2)$ and $(3, 2)$; one vertex at $(5, 2)$.

10. Vertices at $(-1, -2)$ and $(-1, 6)$; focal length $\sqrt{7}$.

11. Center at $(1, 2)$; one focus at $(1, 5)$; one vertex at $(1, -3)$.

12. Foci at $(2, 3)$ and $(2, -2)$; tangent to line $y = 5$.

13. Semiminor axis is 3 units long; one focus at $(-2, 3)$; one vertex at $(-3, 3)$.

14. Vertices at $(1, -3)$ and $(1, 7)$; focal length is 3.

15. Find the equation satisfied by points $P(x, y)$ the sum of whose distances from $(-2, 0)$ and $(2, 0)$ is 10.

16. Find the equation satisfied by points $P(x, y)$ the sum of whose distances from $(-2, -2)$ and $(-2, 4)$ is 12.

17. Find the equation satisfied by points $P(x, y)$ the sum of whose distances from $(-1, 2)$ and $(3, -1)$ is 8.

18. Find the equation satisfied by points $P(x, y)$ the sum of whose distances from $(-2, -1)$ and $(2, 1)$ is 6.

19. The **eccentricity** e of an ellipse is defined as the ratio of its focal length to the length of the semimajor axis. For the ellipse

$$\frac{(x-h)^2}{a^2} + \frac{(y-k)^2}{b^2} = 1 \qquad (i)$$

with $a > b$, we have

$$e = \frac{c}{a} = \frac{\sqrt{a^2 - b^2}}{a}. \qquad (ii)$$

Similarly, for the ellipse (*i*) with $b > a$,

$$e = \frac{c}{b} = \frac{\sqrt{b^2 - a^2}}{b}. \qquad (iii)$$

(a) Show that, in either case,

$$0 < e < 1.$$

(b) Show that as $e \to 0$ the ellipse approaches a circle. What happens to the two foci?

(c) Determine the limiting shape of the ellipse as $e \to 1$.

In each of Problems 20 through 23, find the eccentricity of the given ellipse.

20. $\dfrac{x^2}{16} + \dfrac{y^2}{9} = 1$ (see Problem 2)

21. $\dfrac{x^2}{16} + \dfrac{y^2}{25} = 1$ (see Problem 1)

22. $x^2 + 2y^2 - 2x + 4y = 6$ (see Problem 4)

23. $x^2 + 4y^2 - 4x + 16y = -4$ (see Problem 7)

In each of Problems 24 through 27, find the equation of the ellipse having the given properties.

24. Center at $(2, 1)$; one focus at $(-1, 1)$; eccentricity $3/5$.

25. Foci at $(-1, 2)$ and $(3, 2)$; eccentricity $2/3$.

26. One vertex at $(1, 4)$; one focus at $(-3, 2)$; eccentricity $2/\sqrt{5}$; axes parallel to coordinate axes.

27. Center at $(-1, -3)$; one vertex at $(3, -3)$; eccentricity $1/2$.

28. Referring to Eq. 16 of the text, determine the eccentricity of the earth's orbit.

29. **Reflection property of the ellipse.** Show that a ray emanating from one focus of an ellipse is reflected by the ellipse along a line passing through the other focus.

Hint: Choose a coordinate system so that the foci F_1 and F_2 are at $(\pm c, 0)$. Choose an arbitrary point $P(x_0, y_0)$ on the ellipse and find the slope of the normal line N at this point. Show that the angle between N and PF_1 is equal to the angle between N and PF_2.

30. A rod of length l is divided by a point P into two parts of length rl and $(1 - r)l$, respectively, where $0 < r < 1$. If the rod moves so that one end is always on the x-axis, and the other end is always on the y-axis, find an equation of the path traced by the point P.

5.4 HYPERBOLAS

We continue to study the equation

$$Ax^2 + Cy^2 + Dx + Ey = F, \tag{1}$$

where now we assume that A and C have opposite signs. Proceeding exactly as in Section 5.3, we complete the square in both x and y and obtain

$$A(x - h)^2 + C(y - k)^2 = M, \tag{2}$$

where again

$$h = -\frac{D}{2A}, \qquad k = -\frac{E}{2C}, \qquad M = F + \frac{D^2}{4A} + \frac{E^2}{4C}. \tag{3}$$

If $M \neq 0$, then we can divide Eq. 2 by M. Since A and C are of opposite signs, the quotients A/M and C/M also have opposite signs. Let $a^2 = |M/A|$ and $b^2 = |M/C|$. Then Eq. 2 has the form

$$\frac{(x - h)^2}{a^2} - \frac{(y - k)^2}{b^2} = 1 \tag{4}$$

if A and M have the same sign, and the form

$$-\frac{(x - h)^2}{a^2} + \frac{(y - k)^2}{b^2} = 1 \tag{5}$$

if A and M have opposite signs. Note that $a = \sqrt{|M/A|} > 0$ and $b = \sqrt{|M/C|} > 0$.

Equations 4 and 5 are standard forms for the equation of a **hyperbola.** We first consider the graph of Eq. 4. To simplify this equation still further let us again introduce a translated coordinate system defined by

$$u = x - h, \qquad v = y - k; \tag{6}$$

then Eq. 4 becomes

$$\frac{u^2}{a^2} - \frac{v^2}{b^2} = 1. \tag{7}$$

Since $u^2/a^2 \geq 1$ from Eq. 7, it follows that no part of the graph is in the region where $|u| < a$. Further there are intercepts on the u-axis at $(-a, 0)$ and $(a, 0)$, but there are no intercepts on the v-axis. It also follows immediately from Eq. 7 that the graph is symmetric about both the u- and v-axes, so that it is sufficient to consider only the portion of the graph in the first quadrant of the uv-plane.

Differentiating Eq. 7 implicitly with respect to u, we obtain

$$\frac{2u}{a^2} - \frac{2vv'}{b^2} = 0$$

or

$$v' = \frac{b^2}{a^2}\frac{u}{v}. \tag{8}$$

Thus, in the first quadrant, $v' > 0$ and consequently v increases as u does. Since $v' \to \infty$ when $v \to 0+$, it follows that the graph has a vertical tangent at the intercept $(a, 0)$.

Differentiating a second time with respect to u, we have

$$v'' = \frac{b^2}{a^2}\left(\frac{1}{v} - \frac{uv'}{v^2}\right).$$

Substituting for v' from Eq. 8 and also using Eq. 7, it follows that

$$v'' = -\frac{b^4}{a^2v^3}. \tag{9}$$

Therefore in the first quadrant $v'' < 0$ and the graph of Eq. 8 is concave down there.

To determine the shape of the graph in the first quadrant when u and v are large, we solve Eq. 7 for v; thus

$$v = b\sqrt{\frac{u^2}{a^2} - 1}, \tag{10}$$

where the positive square root is required. If u is very large, then u^2/a^2 is much greater than 1, and thus $v \cong bu/a$. Indeed, we now show that in the first quadrant the graph is asymptotic to the line $v = bu/a$ as $u \to \infty$, that is, we show that the distance d between the hyperbola and the straight line $v = bu/a$ tends to zero as $u \to \infty$. This distance is certainly no greater than the vertical distance r between two points, one on the hyperbola and one on the line $v = bu/a$, having the same u-coordinate (see Figure 5.4.1). Therefore it is sufficient to show that $r \to 0$ as $u \to \infty$. This distance r is given by

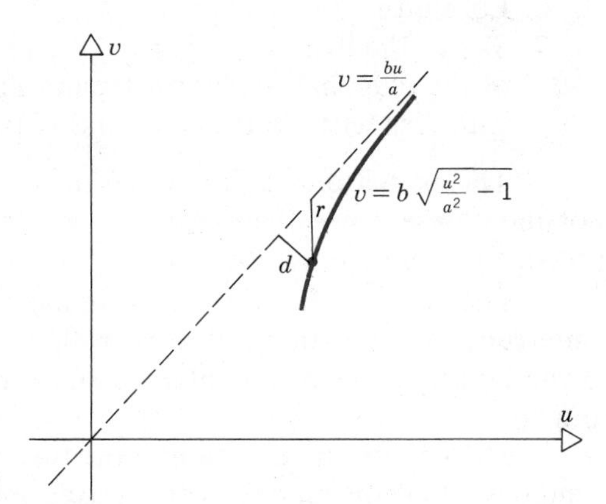

Figure 5.4.1

$$r = \frac{b}{a}u - b\sqrt{\frac{u^2}{a^2} - 1};$$

it is the difference between two quantities each of which becomes infinite as $u \to \infty$. To examine this difference more carefully it is useful to rationalize the expression for r. Thus

$$r = \left[\frac{b}{a}u - b\sqrt{\frac{u^2}{a^2} - 1}\right]\frac{\left[\frac{b}{a}u + b\sqrt{\frac{u^2}{a^2} - 1}\right]}{\left[\frac{b}{a}u + b\sqrt{\frac{u^2}{a^2} - 1}\right]}$$

$$= \frac{b^2}{\frac{b}{a}u + b\sqrt{\frac{u^2}{a^2} - 1}}. \tag{11}$$

As $u \to \infty$ it is now clear from Eq. 11 that $r \to 0$; therefore the line $v = bu/a$ is indeed an asymptote of the hyperbola.

As in the case of the ellipse, the auxiliary rectangle bounded by $u = \pm a$, $v = \pm b$ is very helpful in sketching the hyperbola given by Eq. 7. The hyperbola lies outside of this rectangle, but is tangent to it at the intercepts. Further, the asymptotes of the hyperbola are the extensions of the diagonals of the rectangle. Note that the asymptotes separate the uv-plane into four sectors, and that the hyperbola lies entirely in two of these. Finding the intercepts on the u- or v-axis is a convenient way to determine in which sectors the hyperbola is located.

The graph of Eq. 4, or of Eq. 7, is easily sketched on the basis of the information obtained above by using the following procedure.

1. Draw x- and y-axes.
2. Determine h and k.
3. Draw u- and v-axes.
4. Draw the auxiliary rectangle whose sides are $u = \pm a$, $v = \pm b$.
5. Draw the diagonals of the auxiliary rectangle and extend them beyond the rectangle.
6. Locate the intercepts at $u = \pm a$, $v = 0$.
7. Sketch the hyperbola through the intercepts and asymptotic to the extensions of the diagonals of the rectangle, making use of the symmetry of the graph with respect to both the u- and v-axes.

The sketch of the graph of Eq. 4 is shown in Figure 5.4.2*a*. As in the case of the ellipse, once the sketch is completed the u- and v-axes and the auxiliary rectangle can be erased if desired.

The graph of Eq. 5 is quite similar. The only difference is that in Step 6 the intercepts are now located at $u = 0$, $v = \pm b$. Hence the two branches of the hyperbola open vertically rather than horizontally. The graph of Eq. 5 is shown in Figure 5.4.2*b*.

The point of intersection of the two axes of symmetry is known as the **center,** and the intercepts on one of these axes are called **vertices.** The line containing the

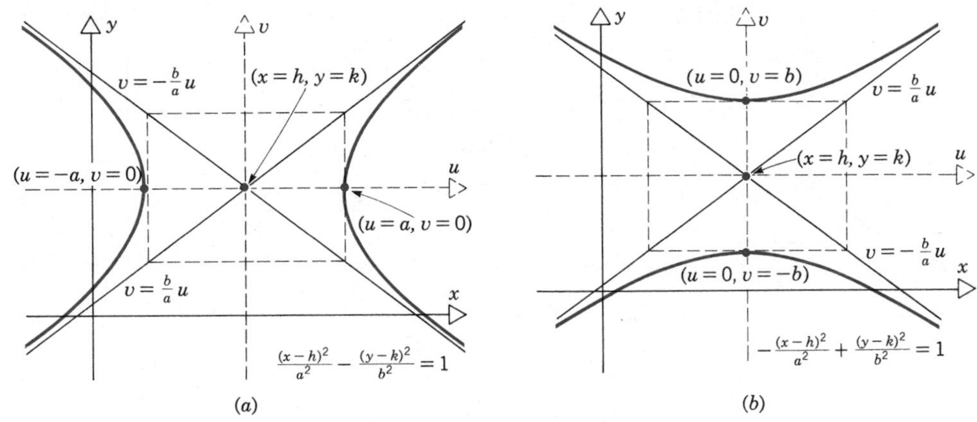

Figure 5.4.2

vertices is called the **transverse axis,** and the distance between the vertices is known as the **transverse diameter.** The line through the center perpendicular to the transverse axis is the **conjugate axis.**

A special degenerate case occurs if, in Eq. 2, A and C are of opposite signs and $M = 0$. In this case, Eq. 2 can be written as

$$A(x - h)^2 = -C(y - k)^2$$

or, since A and C have opposite signs and hence $-(A/C) > 0$,

$$y - k = \pm\sqrt{-\left(\frac{A}{C}\right)}\,(x - h). \tag{12}$$

Thus the graph consists of two intersecting straight lines of slope $\pm\sqrt{-(A/C)}$ and passing through the point $(x = h, y = k)$.

EXAMPLE 1

Sketch the graph of

$$-x^2 + 4y^2 - 4x - 8y = 16. \tag{13}$$

By completing the square in x and y we obtain

$$-(x^2 + 4x + 4) + 4\,(y^2 - 2y + 1) = 16 - 4 + 4$$

or

$$-(x + 2)^2 + 4(y - 1)^2 = 16$$

and finally

$$-\frac{(x + 2)^2}{16} + \frac{(y - 1)^2}{4} = 1. \tag{14}$$

Thus $h = -2$, $k = 1$, and the translated coordinate system is defined by

$$u = x + 2, \qquad v = y - 1. \tag{15}$$

The u- and v-axes are shown by dashed lines in Figure 5.4.3. In terms of these coordinates, Eq. 14 becomes

$$-\frac{u^2}{16} + \frac{v^2}{4} = 1. \tag{16}$$

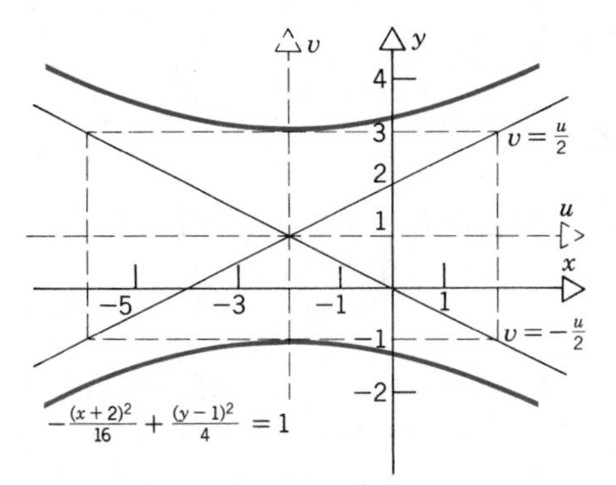

Figure 5.4.3

The auxiliary rectangle is bounded by the lines $u = \pm 4$, $v = \pm 2$, and the asymptotes are given by $v = \pm u/2$. Finally, the vertices are on the v-axis at the points $(0, -2)$ and $(0, 2)$. Using this information, as well as symmetry properties, we can sketch the graph shown in Figure 5.4.3. ∎

Geometrical construction of the hyperbola

Geometrically, a hyperbola can be defined as the set of all points $P(x, y)$ for which the difference of distances from two fixed points is a constant. To obtain an equation satisfied by this set of points it is convenient to choose a coordinate system so that the two fixed points are $(-c, 0)$ and $(c, 0)$ with $c > 0$, as shown in Figure 5.4.4. If the fixed distance is $2a$, with $a < c$, then a derivation almost exactly like the one in Section 5.3 leads to the equation

$$\frac{x^2}{a^2} - \frac{y^2}{b^2} = 1, \tag{17}$$

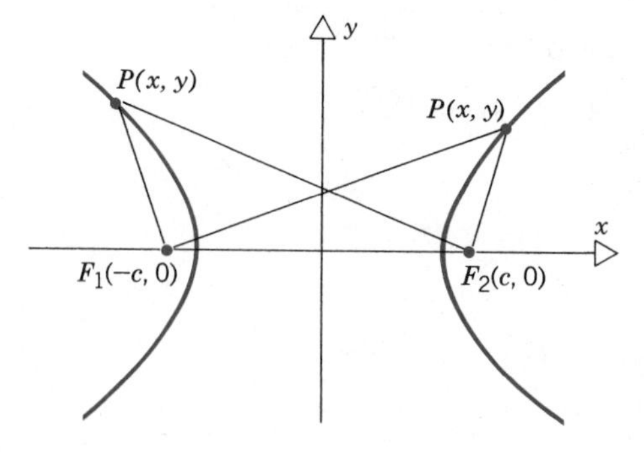

Figure 5.4.4
The geometrical construction of a hyperbola: $|PF_1| - |PF_2| = 2a < 2c$ or $|PF_2| - |PF_1| = 2a < 2c$.

where $b^2 = c^2 - a^2$. On the other hand, if the fixed points are located at $(0, -c)$ and $(0, c)$, and if the fixed distance is $2b < 2c$, then one obtains

$$-\frac{x^2}{a^2} + \frac{y^2}{b^2} = 1, \tag{18}$$

where $a^2 = c^2 - b^2$. In any case the two fixed points are called **foci,** and are always located on the transverse axis equally distant from the center. The distance c from the center to either focus is the **focal length,** and is always given by

$$c^2 = a^2 + b^2. \tag{19}$$

Note that for a hyperbola the foci are always farther from the center than the vertices. For the hyperbola in Example 1 the focal length is $c = \sqrt{20} = 2\sqrt{5}$. The foci are the points $(-2, 1 + 2\sqrt{5})$ and $(-2, 1 - 2\sqrt{5})$.

EXAMPLE 2

Find an equation for the hyperbola whose vertices are $(3, 1)$ and $(-5, 1)$ and with one asymptote having slope $-3/4$.

The center of the hyperbola is the midpoint between the vertices, namely $(-1, 1)$. Also, the transverse axis is the line $y = 1$, the transverse diameter is 8, and consequently $a = 4$. Finally, since the asymptotes have slopes $\pm b/a$ in general, it follows that in this case $b = 3$. Therefore the equation of the hyperbola is

$$\frac{(x + 1)^2}{16} - \frac{(y - 1)^2}{9} = 1. \tag{20}$$

Its graph is shown in Figure 5.4.5. ∎

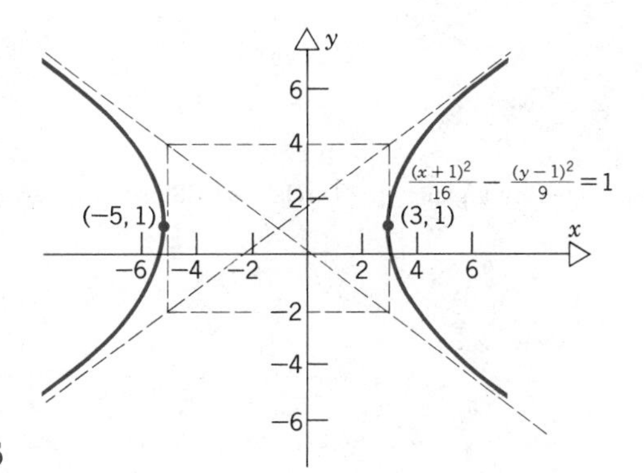

Figure 5.4.5

Conic sections

In addition to their appearance as the graphs of quadratic equations in two variables, and as the result of certain geometric constructions, parabolas, ellipses, and hyperbolas also occur in at least one other important way in mathematics. They (along with circles and straight lines in certain special cases) are the curves of intersection of a cone with a plane, that is, they are sections of a cone, or conic sections (see Figure 5.4.6). Greek mathematicians carried out extensive and sophisticated investigations of these curves, and discovered most of their properties that are familiar today. Of special note is Apollonius of Perga, who wrote a lengthy treatise on them. These ancient scholars worked from a purely geometric point of view, since they did not have the benefit of coordinate algebra, which was developed many centuries later. This makes the extent of their research all the more remarkable to a modern student. To show the equivalence of the curves defined geometrically as sections of a cone with those occurring as the graphs of quadratic equations requires some fairly difficult analysis, which we will not pursue here.

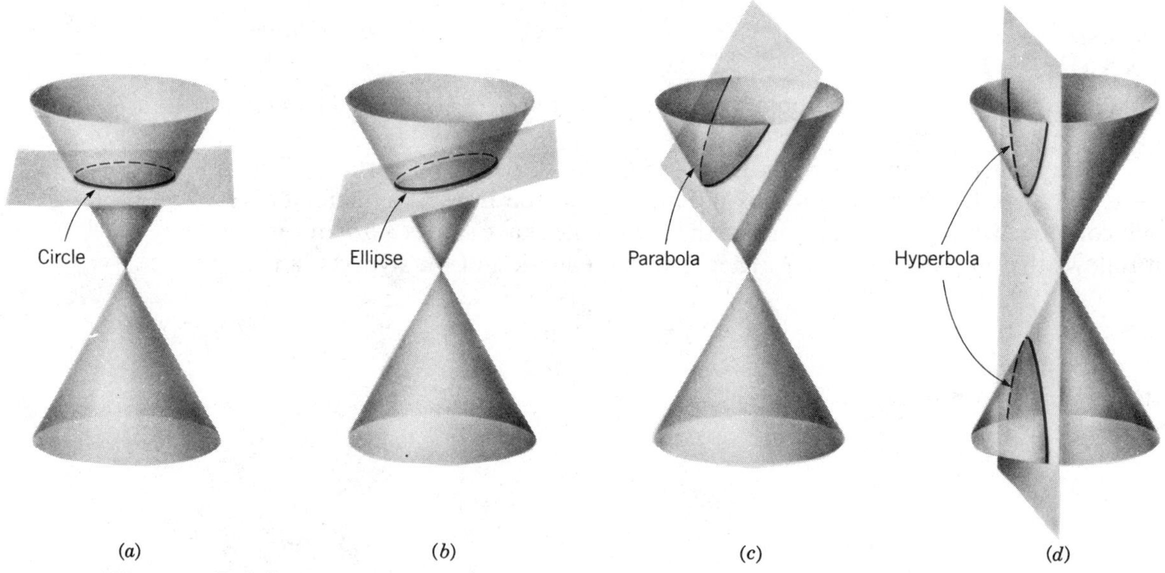

Figure 5.4.6
A cone intersected by a plane. (*a*) Circle. (*b*) Ellipse. (*c*) Parabola. (*d*) Hyperbola.

PROBLEMS

In each of Problems 1 through 8, find the center, the vertices, and the slopes of the asymptotes for the given hyperbola. Then sketch the graph.

1. $\dfrac{x^2}{16} - \dfrac{y^2}{25} = 1$

2. $-\dfrac{x^2}{16} + \dfrac{y^2}{25} = 1$

3. $\frac{(x-2)^2}{9} - \frac{(y+2)^2}{16} = 1$

4. $-(x+1)^2 + 4(y-2)^2 = 4$

5. $-4x^2 + y^2 + 8x + 6y = 11$

6. $x^2 - 4y^2 - 4x + 16y = 28$

7. $4x^2 - 3y^2 - 8x + 18y = 59$

8. $3x^2 + 12x - y^2 + 6y = -39$

In each of Problems 9 through 14, find the equation of the hyperbola having the given properties.

9. One vertex at (2, 0); foci at (3, 0) and (−3, 0).

10. Center at (2, −1); one asymptote is $2(y+1) = 3(x-2)$; one vertex at (2, −4).

11. Foci at (−1, 3) and (−1, −5); one asymptote is $x - 2y = 1$.

12. One vertex at (2, 3); one focus at (2, −1); transverse diameter is 6 units long.

13. Asymptotes are $3x - 4y = -11$ and $3x + 4y = 5$; one focus is (4, 2).

14. Center at (−2, 1); focal length 3; one vertex at (0, 1).

15. Find an equation satisfied by points $P(x, y)$ the difference of whose distances from (0, −2) and (0, 2) is 2.

16. Find an equation satisfied by points $P(x, y)$ the difference of whose distances from (−3, 3) and (3, 3) is 4.

17. Find an equation satisfied by points $P(x, y)$ the difference of whose distances from (−1, 2) and (3, −1) is 3.

18. Find an equation satisfied by the points $P(x, y)$ the difference of whose distances from (−2, 1) and (2, −1) is 4.

19. The **eccentricity** e of a hyperbola is defined as the ratio of the focal length to one half of the transverse diameter. For the hyperbola

$$\frac{(x-h)^2}{a^2} - \frac{(y-k)^2}{b^2} = 1 \qquad (i)$$

the eccentricity is given by

$$e = \frac{c}{a} = \frac{\sqrt{a^2 + b^2}}{a}, \qquad (ii)$$

while for the hyperbola

$$-\frac{(x-h)^2}{a^2} + \frac{(y-k)^2}{b^2} = 1 \qquad (iii)$$

we have

$$e = \frac{c}{b} = \frac{\sqrt{a^2 + b^2}}{b}. \qquad (iv)$$

(a) Show that $e > 1$ in either case.

(b) Determine the limiting shape of the hyperbola as $e \to 1$ and as $e \to \infty$.

Hint: Consider c fixed.

In each of Problems 20 through 23, find the eccentricity of the given hyperbola.

20. $-(x+1)^2 + 4(y-2)^2 = 4$ (Problem 4)

21. $\frac{x^2}{16} - \frac{y^2}{25} = 1$ (Problem 1)

22. $x^2 - 4y^2 - 4x + 16y = 28$ (Problem 6)

23. $4x^2 - 3y^2 - 8x + 18y = 59$ (Problem 7)

In each of Problems 24 through 27, find an equation of the hyperbola having the given properties.

24. Vertices at (1, −2) and (1, 4); eccentricity 2.

25. Foci at (−1, 2) and (3, 2); eccentricity 3/2.

26. Center at (−2, 3); one focus at (4, 3); eccentricity 6/5.

27. One asymptote is $y + 1 = 2(x - 3)$; one focus is (6, −1); eccentricity 3/2.

28. Find the eccentricity of a hyperbola whose asymptotes are $2(y-2) = \pm 3(x+1)$.

* 29. In Problem 28 of Section 5.2 it was shown that it is always possible to transform the equation

$$Ax^2 + Bxy + Cy^2 + Dx + Ey = F \qquad (i)$$

into

$$\alpha u^2 + \beta uv + \gamma v^2 + \delta u + \epsilon v = \kappa \qquad (ii)$$

by the rotation of axes

$$x = u\cos\theta - v\sin\theta,$$
$$y = u\sin\theta + v\cos\theta. \qquad (iii)$$

(a) Using the results of the problem just mentioned, show that

$$B^2 - 4AC = \beta^2 - 4\alpha\gamma$$

for all values of the rotation angle θ. If θ is chosen so as to make $\beta = 0$, then $B^2 - 4AC = -4\alpha\gamma$.

(b) Show that Eq. (*ii*) with $\beta = 0$ represents an ellipse, parabola, or hyperbola, according to whether $\alpha\gamma$ is positive, zero, or negative, respectively.

(c) Show that Eq. (i) represents an ellipse, parabola, or hyperbola according to whether $B^2 - 4AC$ is negative, zero, or positive, respectively.

5.5 PARAMETRIC EQUATIONS

In earlier sections we have discussed how to sketch a plane curve arising as the graph of a function

$$y = f(x), \tag{1}$$

or as the graph of an equation

$$\phi(x, y) = 0. \tag{2}$$

We will call Eqs. 1 and 2 *Cartesian* equations. The advantage of Eq. 1 is that it is relatively easy to calculate the value of y corresponding to a given value of x. However, curves having an equation of the form (1) are restricted to those that are intersected at most once by a straight line parallel to the y-axis. Equations of the form (2) are associated with a larger variety of curves, for example, ellipses and hyperbolas. However, they have the disadvantage that it may be difficult to compute the value or values of y corresponding to a given x, since this involves solving Eq. 2 for y. In short, the *explicit* Eq. 1 is simpler, but the *implicit* Eq. 2 is more general.

In many cases we can obtain both simplicity and generality by using a third way of giving an analytical description of a curve. This type of description is known as a parametric representation, and the equations involved are known as **parametric equations.** We introduce the idea of a parametric representation in a physical context.

Consider the motion of a particle in the xy-plane. The instantaneous position of the particle is given by its x- and y-coordinates. As the particle moves about in the plane, its coordinates change with time. In describing the motion of the particle it is natural to express its coordinates as functions of time. Thus we write

$$x = f(t), \qquad y = g(t), \qquad \text{for } t \text{ in } I, \tag{3}$$

where t denotes time and where I is some interval on which the functions f and g are defined. Instead of relating x and y directly to each other, Eqs. 3 express both x and y as functions of a third variable, the *parameter* t. For each t, Eqs. 3 determine a *point;* the collection of these points is the *path* of the moving particle. Thus Eqs. 3 are said to constitute a set of parametric equations for the trajectory of the particle.

In using parametric representations one must be careful not only to give equations of the form (3), but also to state the interval I over which the parameter is permitted to vary; in other words, it is essential to indicate the domain of the functions f and g. If f and g are continuous on an interval I, then a set of parametric equations (3) is said to define an **arc,** or **curve,** in the xy-plane. The *positive direction* on the arc is the direction corresponding to increasing values of the parameter.

Let us now consider some examples of parametric representations, beginning with some familiar curves.

EXAMPLE 1

Let

$$x = 2t - 1, \qquad y = -t + 2. \tag{4}$$

Describe the curve in the xy-plane corresponding to this set of parametric equations for t in the interval $(-\infty, \infty)$; for t in the interval $[0, 3]$.

Let us start by plotting a few points on the curve. Table 5.1 shows the x and y values for several values of t, and the corresponding points are shown in Figure 5.5.1a. It appears from this figure that the six plotted points lie on a straight line.

Table 5.1 Values from $x = 2t - 1$, $y = -t + 2$.

t	x	y
-2	-5	4
-1	-3	3
0	-1	2
1	1	1
2	3	0
3	5	-1

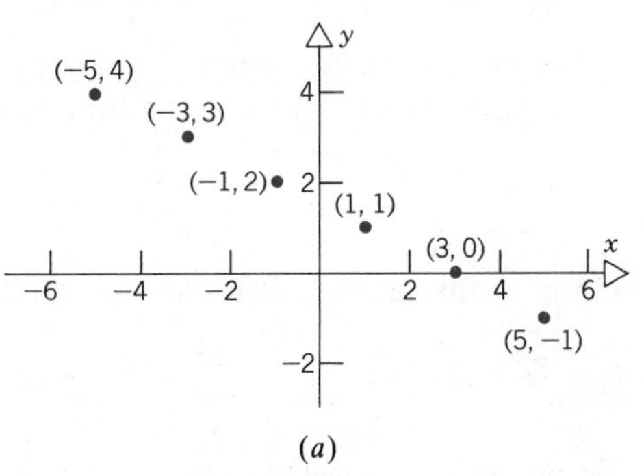

Figure 5.5.1

To show that this is true, we can proceed in the following way. Solve the second of Eqs. 4 for t, obtaining $t = 2 - y$; then substitute this expression for t in the first of Eqs. 4. The result is the Cartesian equation

$$x = 2(2 - y) - 1,$$

or

$$x + 2y = 3, \tag{5}$$

which is the equation of a straight line. The line is shown in Figure 5.5.1b. As t ranges from $-\infty$ to ∞, the entire line is traversed in the direction indicated by the arrow.

Next, consider the interval $0 \leq t \leq 3$. The points corresponding to $t = 0$ and $t = 3$ are $(-1, 2)$ and $(5, -1)$, respectively. For t in $[0, 3]$ we obtain the line segment shown in Figure 5.5.1c that joins $(-1, 2)$ and $(5, -1)$ inclusive of the endpoints. ∎

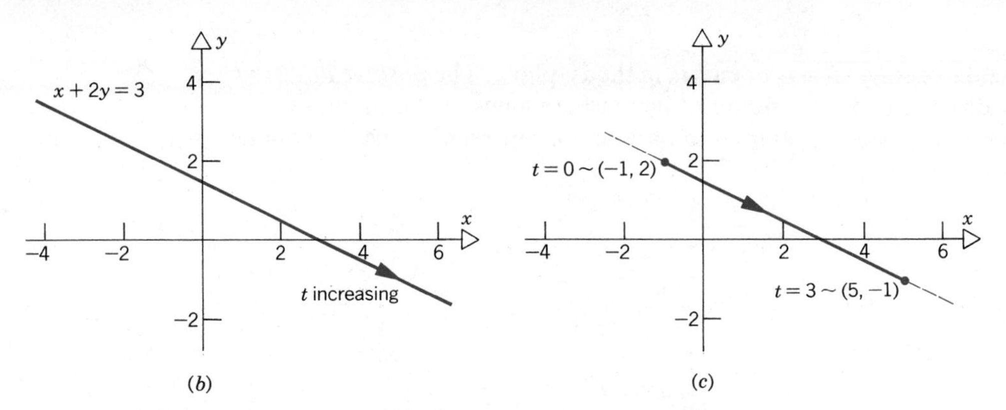

Figure 5.5.1

EXAMPLE 2

Describe the curve corresponding to the parametric equations

$$x = 3 \sin t, \qquad y = 4 \cos t \tag{6}$$

for $-\pi \le t \le 0$; for $-\pi \le t \le 2\pi$.

In order to eliminate the parameter t we first write Eqs. 6 in the form

$$\frac{x}{3} = \sin t, \qquad \frac{y}{4} = \cos t.$$

Upon squaring and adding these equations, we obtain

$$\frac{x^2}{9} + \frac{y^2}{16} = \sin^2 t + \cos^2 t = 1. \tag{7}$$

Equation 7 has for its graph the ellipse shown in Figure 5.5.2*a*. For t in the interval $[-\pi, 0]$ it follows that x is always negative or zero, and that y increases from -4

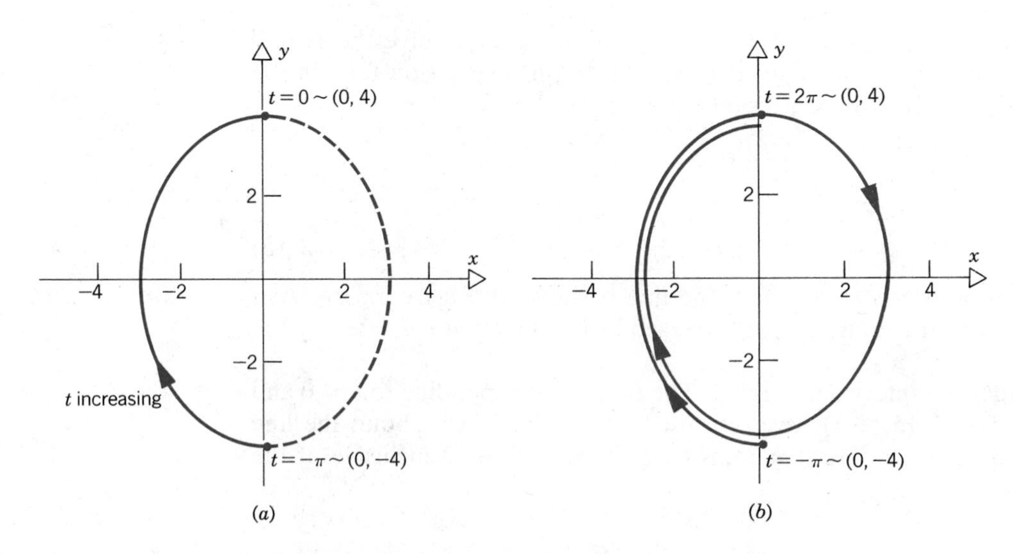

Figure 5.5.2

to 4. Hence Eqs. 6 correspond to the left half of the ellipse, traversed in the clockwise direction as t increases.

On the other hand, for t in $[-\pi, 2\pi]$, Eqs. 6 correspond to one and a half circuits of the ellipse (7), in the clockwise direction, starting at $(0, -4)$ and terminating at $(0, 4)$. This arc is indicated in Figure 5.5.2*b*. ∎

EXAMPLE 3

Consider the parametric equations

$$x = -\frac{3(1-\tau^2)}{1+\tau^2}, \qquad y = \frac{8\tau}{1+\tau^2} \tag{8}$$

for $-1 \le \tau \le 1$. Identify the arc described by these equations.

In this case it may not be immediately obvious how to proceed in order to eliminate the parameter τ from Eqs. 8. However, if we write Eqs. 8 in the form

$$\frac{x}{3} = -\frac{1-\tau^2}{1+\tau^2}, \qquad \frac{y}{4} = \frac{2\tau}{1+\tau^2},$$

then, upon squaring and adding, we obtain

$$\begin{aligned}\frac{x^2}{9} + \frac{y^2}{16} &= \frac{(1-\tau^2)^2}{(1+\tau^2)^2} + \frac{4\tau^2}{(1+\tau^2)^2} \\ &= \frac{1 - 2\tau^2 + \tau^4 + 4\tau^2}{(1+\tau^2)^2} \\ &= \frac{1 + 2\tau^2 + \tau^4}{(1+\tau^2)^2} = 1.\end{aligned} \tag{9}$$

Therefore points satisfying Eqs. 8 lie on the ellipse given by Eq. 9. The values $\tau = -1$ and $\tau = 1$ correspond to the points $(0, -4)$ and $(0, 4)$, respectively. Since $x \le 0$ for $-1 \le \tau \le 1$, it follows that this τ-interval corresponds to the left half of the ellipse only, traversed in a clockwise direction as τ increases. Observe that this is the same as the first arc in Example 2; however, Eqs. 8 are quite different from Eqs. 6. This illustrates that the same curve may be given by completely different sets of parametric equations. ∎

When faced with a curve described by a set of parametric equations, one way to proceed is to eliminate the parameter, thereby obtaining a Cartesian equation of the form (2). The graph can then be sketched by methods developed earlier in this chapter. This approach was convenient, for instance, in each of the preceding examples. However, in more complicated situations it is often difficult or impossible to eliminate the parameter and to obtain a corresponding Cartesian equation. Then it is desirable to sketch the graph directly from the parametric equations themselves. This can be done by plotting points that correspond to several values of t, and then joining them by a smooth curve in the order of increasing t. As in the case of

Cartesian equations, it may also be helpful to seek other information that will assist in drawing a good sketch. The determination of the slope of the curve at each point is of particular importance.

Suppose, then, that we want to find the slope of the arc described by Eqs. 3

$$x = f(t), \qquad y = g(t), \qquad \text{for } t \text{ in } I$$

at a point (x, y) corresponding to a value t in I. For an increment Δt in t there are respective increments Δx in x and Δy in y given by

$$\Delta x = f(t + \Delta t) - f(t), \qquad \Delta y = g(t + \Delta t) - g(t). \tag{10}$$

Thus $(x + \Delta x, y + \Delta y)$ is the point on the curve associated with the parameter value $t + \Delta t$ (see Figure 5.5.3). The ratio $\Delta y/\Delta x$ is the slope of the line segment

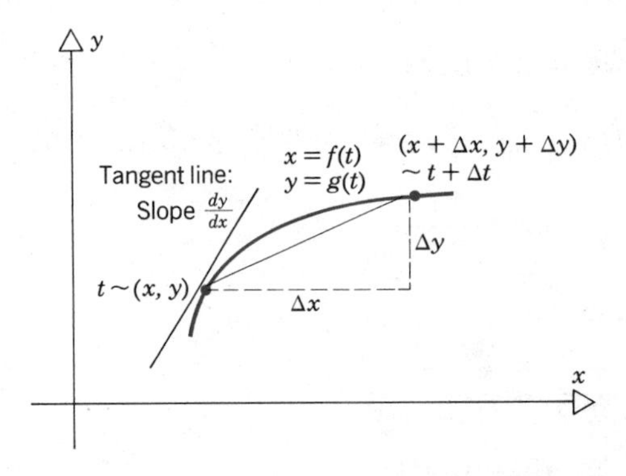

Figure 5.5.3

joining the points (x, y) and $(x + \Delta x, y + \Delta y)$. The limit of $\Delta y/\Delta x$ as $\Delta x \to 0$ yields the required slope. In order to calculate this limit we proceed as follows. If f and g are differentiable at t, then their increments Δx and Δy can be expressed conveniently using the theory of linear approximation discussed in Section 3.4. Referring to Theorem 3.4.1 we have

$$\Delta x = f'(t)\Delta t + r(t + \Delta t, t)\Delta t, \tag{11}$$

$$\Delta y = g'(t)\Delta t + \bar{r}(t + \Delta t, t)\Delta t, \tag{12}$$

where both $r(t + \Delta t, t) \to 0$ and $\bar{r}(t + \Delta t, t) \to 0$ as $\Delta t \to 0$. Finally, we observe that Δx and Δt approach zero together. Hence

$$\begin{aligned} \lim_{\Delta x \to 0} \frac{\Delta y}{\Delta x} &= \lim_{\Delta t \to 0} \frac{g'(t)\Delta t + \bar{r}(t + \Delta t, t)\Delta t}{f'(t)\Delta t + r(t + \Delta t, t)\Delta t} \\ &= \lim_{\Delta t \to 0} \frac{g'(t) + \bar{r}(t + \Delta t, t)}{f'(t) + r(t + \Delta t, t)} \\ &= \frac{g'(t)}{f'(t)}, \end{aligned} \tag{13}$$

provided only that $f'(t) \neq 0$. Equation 13 is the desired expression for the slope of the curve given by Eqs. 3; it involves only the derivatives of the functions f and g appearing in the parametric representation (3).

If the equations $x = f(t)$, $y = g(t)$ define a function $y = F(x)$, then the slope of the graph of F is given by $F'(x) = dy/dx$. From Eq. 13 we conclude that

$$F'(x) = \frac{g'(t)}{f'(t)}, \tag{14}$$

where it is understood that the left side of Eq. 14 is evaluated at the value of x given by $x = f(t)$. If for some value of t it happens that $f'(t) = 0$ but $g'(t) \neq 0$, then the slope of the curve is infinite and its tangent line is vertical. If both $f'(t)$ and $g'(t)$ are zero at the same point, then Eq. 14 is meaningless and cannot be used to determine the slope there; the behavior of the curve at such points must be found in some other way. To eliminate such points from consideration we sometimes require that $f'^2(t) + g'^2(t) > 0$.

If Eq. 14 is written in the Leibniz notation, it takes the form

$$\frac{dy}{dx} = \frac{dy/dt}{dx/dt}. \tag{15}$$

Thus the rate of change of y with respect to x is the rate of change of y with respect to t divided by the rate of change of x with respect to t. The analogy between Eq. 15 and the arithmetic of fractions is one of the beauties of the Leibniz notation. Higher derivatives can be calculated by differentiating Eq. 15 with respect to x. For example, using Eq. 15 with y replaced by dy/dx, we obtain

$$\begin{aligned}\frac{d^2y}{dx^2} = \frac{d}{dx}\left(\frac{dy}{dx}\right) &= \frac{d(dy/dx)/dt}{dx/dt} \\ &= \frac{(d/dt)[g'(t)/f'(t)]}{(d/dt)[f(t)]} \\ &= \frac{f'(t)g''(t) - g'(t)f''(t)}{[f'(t)]^3}.\end{aligned} \tag{16}$$

The arc defined by Eqs. 3 is said to be **smooth** on any interval in which the arc has a continuously varying tangent line. To be sure that this is so, it is sufficient to require that f' and g' are continuous functions and that $f'^2(t) + g'^2(t) > 0$ throughout the interval. If the latter condition is not satisfied, then the arc may or may not be smooth. Of course, once dy/dx and possibly d^2y/dx^2 have been found, conclusions can be drawn about whether y is increasing or decreasing, and whether the curve is concave up or down, by referring to Section 5.1.

We will now consider some further examples of parametric equations.

EXAMPLE 4

For the arc given by Eqs. 6 in Example 2

$$x = 3 \sin t, \qquad y = 4 \cos t; \qquad -\pi \leq t \leq 0$$

find the slope at the point where $t = -\pi/3$.

This can be done, of course, by eliminating the parameter, as in Example 2, and then differentiating the resulting Eq. 7. However, we wish to illustrate the procedure based on the parametric equations themselves. From Eq. 15 we have

$$\frac{dy}{dx} = \frac{dy/dt}{dx/dt} = \frac{-4 \sin t}{3 \cos t}.$$

When $t = -\pi/3$ we obtain

$$\left.\frac{dy}{dx}\right|_{t=-\pi/3} = \frac{-4(-\sqrt{3}/2)}{3(1/2)} = \frac{4}{\sqrt{3}},$$

which is the desired slope. The reader may verify that this result can also be obtained from Eq. 7. ∎

EXAMPLE 5

Find the slope of the curve given by

$$x = 3(\cos t + t \sin t), \qquad y = 3(\sin t - t \cos t) \tag{17}$$

at the point where $t = \pi/6$.

Proceeding directly from the given parametric equations, we have

$$\frac{dx}{dt} = 3(-\sin t + \sin t + t \cos t) = 3t \cos t,$$

$$\frac{dy}{dt} = 3(\cos t - \cos t + t \sin t) = 3t \sin t.$$

Hence, from Eq. 15,

$$\frac{dy}{dx} = \frac{3t \sin t}{3t \cos t} = \tan t.$$

At $t = \pi/6$ we obtain

$$\left.\frac{dy}{dx}\right|_{t=\pi/6} = \tan \frac{\pi}{6} = \frac{1}{\sqrt{3}},$$

which is the slope of the tangent line at the given point. ∎

EXAMPLE 6

(cycloid)

Suppose that a circle of radius a is originally located with its center at $(0, a)$; it then is tangent to the x-axis at the origin. Let the circle roll without slipping in the positive x direction. Find parametric equations of the path followed by the point P on the circle originally located at the origin (see Figure 5.5.4).

The curve in question is known as a **cycloid,** and its parametric equations can be found as follows. Let x and y be the coordinates of the point P, and let θ be

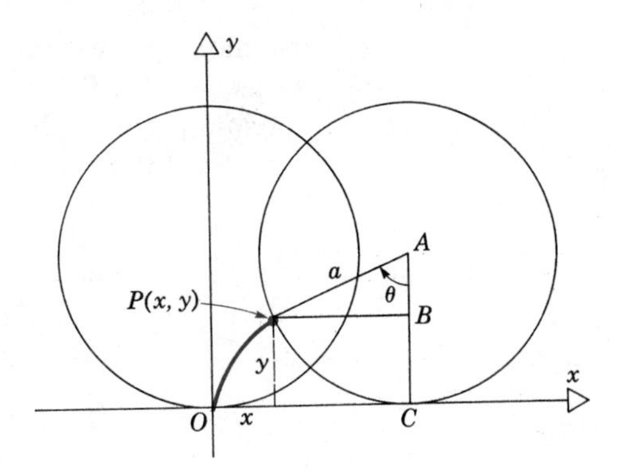

Figure 5.5.4
The geometrical construction of a cycloid.

the angle between the vertical line AC and the radius AP. Since the circle rolls without slipping, we have

$$OC = \text{arc } PC = a\theta.$$

Then

$$x = OC - PB = a\theta - a \sin \theta$$

and

$$y = CA - BA = a - a \cos \theta.$$

Thus the equations

$$x = a(\theta - \sin \theta), \qquad y = a(1 - \cos \theta); \qquad 0 \le \theta < \infty \tag{18}$$

are a set of parametric equations for the cycloid with the central angle θ as parameter.

Although it is possible to eliminate θ from Eqs. 18, thereby obtaining an equation in x and y only, this Cartesian equation is considerably more complicated and is rarely used. The cycloid is an example of a curve studied most easily through its parametric representation.

Let us now discuss the graph of the cycloid. Note first that an increase of 2π in θ results in no change in $\sin \theta$ and $\cos \theta$ because of their periodic character. Thus an increase in 2π in θ leads to no change in y, but to an increase of $2\pi a$ in x. In other words, as a function of x, y is periodic with period $2\pi a$. Hence it is sufficient to sketch the graph over an x-interval of length $2\pi a$, say from $x = 0$ to $x = 2\pi a$. We now restrict ourselves to this interval; that is, we restrict θ to the interval $0 \le \theta \le 2\pi$.

Next, observe that it follows at once from the second of Eqs. 18 that $y \ge 0$ always. Further, within the interval $0 \le \theta \le 2\pi$ we have $y = 0$ only for $\theta = 0$ and for $0 = 2\pi$; thus $y = 0$ for $x = 0$ and $x = 2\pi a$.

Now let us find the slope dy/dx by differentiating Eqs. 18. We have

$$\frac{dy}{dx} = \frac{dy/d\theta}{dx/d\theta} = \frac{a \sin \theta}{a(1 - \cos \theta)} = \frac{\sin \theta}{1 - \cos \theta}. \tag{19}$$

Equation 19 gives dy/dx except for $\theta = 0$ and $\theta = 2\pi$, at which points both numerator and denominator are zero. Disregarding these exceptional points for the moment, we see from Eq. 19 that dy/dx has the same sign as $\sin\theta$. Thus $dy/dx > 0$ for $0 < \theta < \pi$ and $dy/dx < 0$ for $\pi < \theta < 2\pi$. Hence the graph has a maximum point when $\theta = \pi$, that is, at the point $x = \pi a$, $y = 2a$.

By differentiating Eq. 19 we obtain

$$\begin{aligned}\frac{d^2y}{dx^2} &= \frac{d(dy/dx)/d\theta}{dx/d\theta} \\ &= \frac{(1 - \cos\theta)\cos\theta - \sin\theta(\sin\theta)}{(1 - \cos\theta)^2} \cdot \frac{1}{a(1 - \cos\theta)} \\ &= \frac{1}{a}\frac{\cos\theta - 1}{(1 - \cos\theta)^3} = -\frac{1}{a(1 - \cos\theta)^2}.\end{aligned} \tag{20}$$

This quantity is negative for $0 < \theta < 2\pi$, so dy/dx is decreasing on this interval. Hence the graph is concave down for $0 \le x \le 2\pi a$.

Finally, let us determine the shape of the graph near the exceptional points $\theta = 0$ and $\theta = 2\pi$. Equation 19 is not helpful as it stands, because both numerator and denominator approach zero as θ approaches 0 or 2π. However, we can rewrite Eq. 19 in the following way:

$$\begin{aligned}\frac{dy}{dx} &= \frac{\sin\theta}{1 - \cos\theta}\,\frac{1 + \cos\theta}{1 + \cos\theta} \\ &= \frac{\sin\theta(1 + \cos\theta)}{1 - \cos^2\theta} \\ &= \frac{\sin\theta(1 + \cos\theta)}{\sin^2\theta} \\ &= \frac{1 + \cos\theta}{\sin\theta}.\end{aligned} \tag{21}$$

Since $\sin\theta$ approaches zero through positive values as θ approaches zero from the right, it follows that $dy/dx \to \infty$ as $\theta \to 0$ from the right. Similarly, $dy/dx \to -\infty$ as $\theta \to 2\pi$ from the left. Thus the graph has a vertical tangent at the points $(0, 0)$ and $(2\pi a, 0)$ corresponding to $\theta = 0$ and $\theta = 2\pi$, respectively.

Observe that $dx/d\theta = a(1 - \cos\theta)$ and $dy/d\theta = a\sin\theta$ are continuous everywhere, but that both of them are zero at $\theta = 0$ and $\theta = 2\pi$. Thus the curve may not be smooth at these points, and in fact, the graph exhibits a sharp point, or cusp, there.

Putting together all the information we have found we can draw a good sketch of the cycloid. On the interval $0 \le x \le 2\pi a$ it has the general shape of an arch, as shown in Figure 5.5.5. Outside of the interval $0 \le x \le 2\pi a$ the graph is quickly drawn by making use of its periodic character. ∎

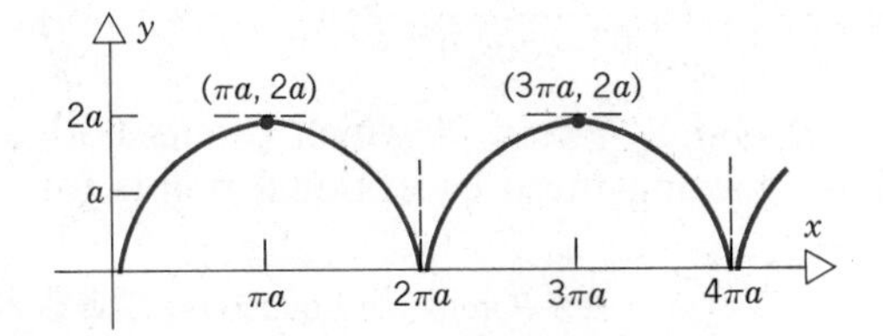

Figure 5.5.5 A cycloid.

PROBLEMS

In each of Problems 1 through 20, sketch the arc represented by the given set of parametric equations and indicate the positive direction. Also find a corresponding Cartesian equation.

1. $x = 3t - 6, \quad y = t - 2; \quad -\infty < t < 2$
2. $x = -2t + 1, \quad y = 3t + 4; \quad -3 < t \le 1$
3. $x = 2 \sin t, \quad y = 2 \cos t; \quad -\frac{\pi}{2} \le t \le \frac{\pi}{2}$
4. $x = 2 \cos t, \quad y = -3 \sin t; \quad -\frac{\pi}{2} < t < \frac{\pi}{2}$
5. $x = 2 \sin 3t, \quad y = -2 \cos 3t; \quad 0 \le t \le \pi$
6. $x = 3 \sin 2t, \quad y = 4 \cos 2t; \quad 0 < t < 2\pi$
7. $x = 2 \sin t, \quad y = 4 \sin t; \quad 0 \le t \le 2\pi$
8. $x = t^2 + 3, \quad y = 2t - 1; \quad 0 \le t < \infty$
9. $x = \sec t, \quad y = 2 \tan t; \quad 0 < t < \frac{\pi}{2}$
10. $x = \tan 2t, \quad y = \sec 2t; \quad \frac{\pi}{4} < t < \frac{3\pi}{4}$
11. $x = \dfrac{1}{\sqrt{1 + t^2}}, \quad y = \dfrac{t}{\sqrt{1 + t^2}}; \quad 0 \le t < \infty$
12. $x = \dfrac{a}{1 + t}, \quad y = \dfrac{bt}{1 + t}; \quad 0 \le t \le 4,$
    where a, b are nonzero constants.
13. $x = 1 + 3 \sin t, \quad y = -2 + 4 \cos t;$
    $0 \le t \le 2\pi$
14. $x = -2 + \sec 2t, \quad y = 1 + 2 \tan 2t;$
    $-\frac{\pi}{4} < t < \frac{\pi}{4}$
15. $x = \cos t, \quad y = \cos 2t; \quad 0 \le t \le \pi$
16. $x = 3t, \quad y = t^3; \quad -\infty < t < \infty$
17. $x = t^2, \quad y = t^3; \quad -\infty < t \le 0$
18. $x = \sqrt{t} + 2, \quad y = t + \sqrt{2}; \quad 0 \le t < \infty$
19. $x = \dfrac{1 + t}{1 + t^2}, \quad y = \dfrac{1 - t}{1 + t^2}; \quad -\infty < t < \infty$
20. $x = \sin t, \quad y = \sin 2t; \quad \frac{\pi}{2} \le t \le \frac{3\pi}{2}$

In each of Problems 21 through 26, find an equation of the line tangent to the given curve at the given point.

21. $x = 2 \cos t, \quad y = -3 \sin t; \quad t = \frac{\pi}{4}$
22. $x = t^2 + 3, \quad y = 2t - 1; \quad t = 3$
23. $x = \tan 2t, \quad y = \sec 2t; \quad t = \frac{\pi}{6}$
24. $x = t^2 + 3t, \quad y = t^3; \quad t = 2$
25. $x = \sqrt{t} + 2, \quad y = t + \sqrt{2}; \quad t = 1$
26. $x = \cos t, \quad y = \cos 2t; \quad t = \frac{\pi}{3}$

In each of Problems 27 through 32, find dy/dx and d^2y/dx^2 as functions of the parameter t.

27. $x = 3 \sin 2t, \quad y = 4 \cos 2t$
28. $x = \sec t, \quad y = -2 \tan t$
29. $x = 1 + 3 \cos t, \quad y = -2 + 4 \sin t$
30. $x = 2t^2 - 3t + 1, \quad y = -t^2 + 2t + 2$
31. $x = t^3 + 1, \quad y = t^2 - 2t$
32. $x = \sin t, \quad y = \cos 2t$
33. Find a Cartesian equation that corresponds to the set of parametric equations
    $$x = 2t^2 - 3t + 1, \quad y = -t^2 + 2t + 2,$$
    and identify the curve that they represent. Do not sketch the graph.

    Hint: Eliminate t^2 by finding the quantity $x + 2y$ in terms of t. Then refer to Problem 29 of Section 5.4.
34. Show that the parametric equations
    $$x = a_1t^2 + a_2t + a_3,$$
    $$y = b_1t^2 + b_2t + b_3,$$
    correspond to a parabola for any choice of the constants $a_1, \ldots, b_3$ such that $a_1b_2 - a_2b_1 \neq 0$.

    Hint: Combine the given equations so as to obtain an equation not containing t^2. Then refer to Problem 29 of Section 5.4.
35. A circle of radius a rolls without slipping on the x-axis. The point P is originally located a distance $b \neq a$ directly below the center of the circle. If $b < a$, P may be visualized as a point on a spoke of a bicycle wheel; if $b > a$, P may be visualized as a point on the flange of a railway wheel. As

the circle moves, the point P traces a curve known as a *trochoid.*

(a) Using a sketch similar to Figure 5.5.4, find a parametric representation for the trochoid.

(b) Sketch the graph of the trochoid when $b < a$; when $b > a$.

* **36.** The path of a point P on a circle of radius b as the circle rolls without slipping on the inside of a larger circle of radius $a > b$ is called a *hypocycloid.* Using Figure 5.5.6, show that a parametric representation for the hypocycloid is

$$x = (a - b)\cos\theta + b\cos\left(\frac{a-b}{b}\right)\theta,$$

$$y = (a - b)\sin\theta - b\sin\left(\frac{a-b}{b}\right)\theta.$$

Hint: Note that $a\theta = b\phi$, where θ and ϕ are the angles shown in Figure 5.5.6.

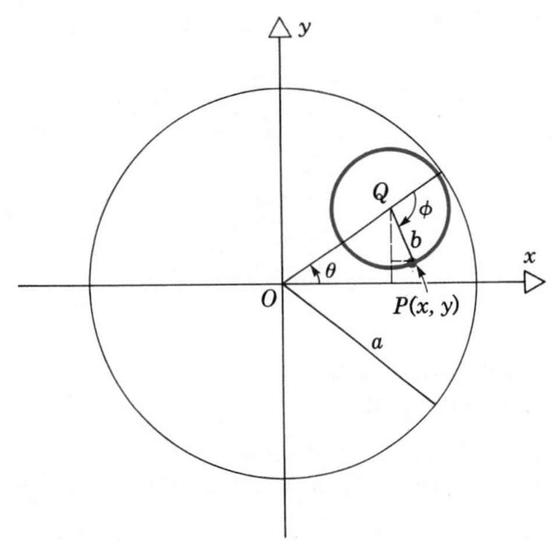

Figure 5.5.6
Construction of a hypocycloid.

* **37.** Suppose that $a = 4b$ in Problem 36.

(a) Using the result stated there, find parametric equations for the hypocycloid in this case.

(b) Sketch the graph of this hypocycloid. It is known as the hypocycloid of four cusps.

(c) Find a Cartesian equation for the hypocycloid of four cusps.

* **38.** Suppose that $a = nb$ in Problem 36, where n is a positive integer. Then there is a cusp on the larger circle each time the smaller circle completes a full revolution. Then as the smaller circle makes a complete circuit around the larger circle, a hypocycloid of n cusps is produced. Sketch the hypocycloid for $n = 2$; for $n = 6$.

* **39.** Consider the hypocycloid (Problem 36) for the case when a/b is rational, but not an integer, that is, let $qa = pb$, where p and q are integers. Then the smaller circle makes p full revolutions as it makes q circuits around the circumference of the larger circle. Sketch the hypocycloid corresponding to $p = 5$ and $q = 2$.

* **40.** Suppose that a circle of radius b rolls without slipping on the outside of the circumference of a circle of radius a, where $a > b$. The curve traced by a point on the smaller circle is called an *epicycloid.* Find a parametric representation for the epicycloid (see Figure 5.5.7).

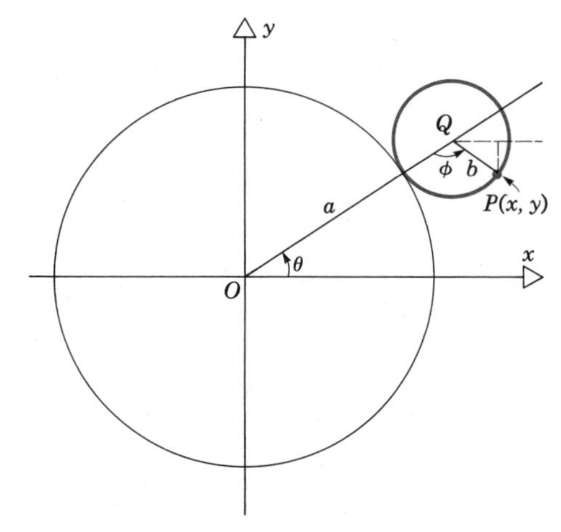

Figure 5.5.7
Construction of an epicycloid.

REVIEW PROBLEMS

In each of Problems 1 through 4, determine whether the given relation is symmetric about the x-axis; the y-axis; the origin.

1. $y^3 = \dfrac{x^2 - 1}{x^3}$

2. $(xy)^3 = x\sin(xy) + x^2$

3. $\cos x = \sin|y| + \pi$

4. $y = x\sin x + 1$

In each of Problems 5 through 8, find all asymptotes (horizontal, vertical, or skew).

5. $x^2y^2 + 2 = (x + y)(x - y)$

6. $f(x) = \dfrac{3x^2 + x - 10}{x^2 - 4} + 1, \qquad x \neq \pm 2$

7. $4x(4x + 3y) = 4 + 12xy + 9y^2$

8. $y^2 - 1 = x^2y^2$

In each of Problems 9 through 12, graph the given relation, showing (where possible) local minima and maxima, inflection points, asymptotes, and zeros.

9. $y^2 - x^2 = 2y - x - 1$

10. $f(x) = \dfrac{x^3 - 3x + 2}{x^3 - 5x^2 + 3x + 9}$

11. $f(x) = \dfrac{\sin x}{x - \pi/2}$

12. $f(x) = \dfrac{x - 1}{(x - 3)(x + 1)}$

In each of Problems 13 through 24, sketch the given conic section. Include and label

on a parabola:	focus, vertex, directrix
on an ellipse:	foci, center, vertices, lengths of semimajor and semiminor axes
on a hyperbola:	foci, center, vertices, asymptotes

13. $x^2 - 18x - 3y + 78 = 0$

14. $9x^2 + y^2 - 18x = 0$

15. $8x^2 - y^2 - 4y - 12 = 0$

16. $y^2 - 4x - 4y - 6 = 0$

17. $4x^2 - 36y^2 - 32x + 36y + 91 = 0$

18. $4x^2 + 81y^2 + 24x - 324y + 36 = 0$

19. $9x^2 - 8y^2 + 36x + 48y - 72 = 0$

20. $4y^2 - x + 4y + 2 = 0$

21. $4x^2 + 5y^2 + 40x - 50y + 205 = 0$

22. $y^2 + 10x - 2y - 99 = 0$

23. $x^2 - 36y^2 - 12x + 144y - 72 = 0$

24. $13x^2 + 12y^2 - 26x - 143 = 0$

In each of Problems 25 through 39, find an equation of the conic section with the given properties.

25. (Parabola) directrix is $y = 2$; focus is $(1, -4)$.

26. (Ellipse) focal length is $\sqrt{2}$; length of semimajor axis is 3 (parallel to x-axis); center is $(2, 0)$.

27. (Hyperbola) transverse axis is $y = 1$; focal length is $\sqrt{29}$; one asymptote is $y = (5/2)x - 4$.

28. (Ellipse) foci are $(3, -2)$ and $(3, -5)$; one vertex is $(3, -\frac{1}{2})$.

29. (Hyperbola) center is $(-1, 0)$; one focus is $(5, 0)$; eccentricity is 6.

30. (Parabola) directrix is $x = -4$; vertex is $(-10, -2)$.

31. (Hyperbola) transverse diameter is 6; foci are $(-3, 8)$ and $(-3, -2)$.

32. (Ellipse) vertices are $(3, 0)$ and $(3, -8)$; length of semiminor axis is 4.

33. (Two parabolas) vertex is $(1, -3)$; axis of symmetry is $x = 1$; focal length is $3/4$.

34. (Two parabolas) vertex is $(5, -2)$; curve goes through $(1, -4)$.

35. (Two ellipses) length of semimajor axis is 3; length of semiminor axis is 1; one vertex is $(3, -2)$; minor and major axes parallel to x- and y-axes, respectively.

36. (Two hyperbolas) asymptotes are $y = (3/4)x - 6$ and $y = (-3/4)x$; focal length is 5; transverse and conjugate axes parallel to x- or y-axes.

37. The set of all points equidistant from the point $(-1, 2)$ and the line $x = -4$.

38. The set of all points, the sum of whose distances from $(-3, 3)$ and $(5, 3)$ is 16.

39. The set of all points for which the absolute value of the difference of the distances to $(2, 1)$ and $(6, 1)$ is 1.

In each of Problems 40 through 42, sketch the given conic section. Refer to Problems 25 through 28 of Section 5.2.

40. $2\sqrt{2}\,x^2 + 2\sqrt{2}\,y^2 - 4\sqrt{2}\,xy + 7x - 9y + 4\sqrt{2} = 0$

41. $13x^2 + 7y^2 - 6\sqrt{3}\,xy - 16 = 0$

42. $8x^2 + 8y^2 - 20xy - 18 = 0$

43. Find bounds on the value of the x-coordinate, x_0, of the center of a nondegenerate hyperbola with the following properties:

    transverse axis is $y = -2$;

    transverse diameter is 12;

    focal length is $2|x_0|$.

In each of Problems 44 through 51 find the Cartesian equation from the given parametric equations, and sketch the graph for the given interval in t.

44. $x = 2t^2, \quad y = 2t^4 + 1; \quad -1 \le t \le 1$

45. $x = 3 + 2t, \quad y = 5 - t; \quad -3 \le t \le 3$

46. $x = \sin t \cos t, \quad y = 4 \sin 2t;$
$0 \le t \le \pi/2$

47. $x = \dfrac{6 + \sec t}{3}, \quad y = \tan^2 t - 1;$
$-\dfrac{\pi}{2} < t < \dfrac{\pi}{2}$

48. $x = \cos t + \sin t, \quad y = \sin 2t + 1;$
$0 \le t \le \pi$

49. $x = \dfrac{1}{\sqrt{1 + t^2}}, \quad y = t^2; \quad 0 \le t < \infty$

50. $x = 1 + \sqrt{\dfrac{\sin^2 t}{9} + 4}, \quad y = \dfrac{\cos t - 4}{2};$
$0 \le t \le \pi$

51. $x = t^3 - 1; \quad y = t^2 + 2; \quad 0 \le t \le 2$

In each of Problems 52 through 55, graph the parametric equations without finding the Cartesian equation.

52. $x = \sin t, \quad y = 1 + \dfrac{t^2}{\pi^2}; \quad -3\pi \le t \le 3\pi$

53. $x = \dfrac{1}{\pi - t}, \quad y = \sin t; \quad 0 < t < \pi$

54. $x = |t| \sin t, \quad y = t^2; \quad -\dfrac{3\pi}{2} \le t \le \dfrac{3\pi}{2}$

55. $x = t \sin^2 t, \quad y = \cos t; \quad -2\pi \le t \le 2\pi$

In each of Problems 56 through 60, find the values of dy/dx and d^2y/dx^2 as functions of t. Find the equation of the line tangent to the curve defined by the parametric equations at $t = \frac{1}{2}$.

56. $x = \sqrt{1 + t}, \quad y = \dfrac{1}{t}$

57. $x = \sin^2(\pi t), \quad y = \cos(\pi t)$

58. $x = \dfrac{1}{1 - t}, \quad y = t^{1/2}$

59. $x = t^3, \quad y = 1 + t^2$

60. $x = t^3, \quad y = \cos^2(\pi t)$

SIX

the integral

The area of a region bounded entirely by straight line segments can be found with relative ease. For instance, there is a simple formula for the area of a triangle, while the area of a more complicated polygonal region can be calculated by subdividing it into triangles, as suggested in Figure 6.1.1. However, the problem is much more difficult if the region is bounded, at least in part, by a curve rather than by straight lines. Some problems of this type were solved by Greek mathematicians, but only with the discovery of calculus did a powerful and general method become available.

The method for finding the area of an essentially arbitrary region utilizes the second basic operation of calculus, known as integration. While each of the two fundamental operations of differentiation (for finding tangents) and integration (for finding areas) has important properties and applications, what is of crucial importance is that they are, in fact, intimately related. This relationship, first clearly perceived by Newton and articulated by Leibniz, is the basis for much of the conceptual beauty and computational power of calculus. The relation between differentiation and integration is known as the fundamental theorem of calculus, and can be expressed in more than one way. We give two complementary statements in Theorem 6.4.2 and 6.4.3 later in this chapter. As its name suggests, the fundamental theorem of calculus is the cornerstone of the subject; much of what has come before is preparation for it, and much of what follows is an exploration of its consequences.

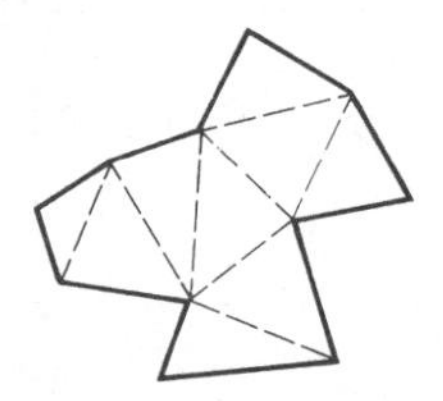

Figure 6.1.1 Area of a polygon by reduction to triangles.

6.1 SUMMATION NOTATION: AREA

Before taking up integration, the second major operation of calculus, it is necessary to discuss some preliminary notation and ideas.

Summation notation

We often need to deal with sums, such as

$$1 + 4 + 9 + \cdots + n^2, \tag{1}$$

or

$$1 + x + x^2 + x^3 + \cdots + x^n, \tag{2}$$

or

$$f(x_1) + f(x_2) + \cdots + f(x_n), \tag{3}$$

where in each case n is some positive integer. A shorthand notation is commonly used to make it easier to handle such expressions. The symbol Σ denotes a sum; thus

$$\sum_{i=1}^{n} u_i = u_1 + u_2 + u_3 + \cdots + u_n. \tag{4}$$

In words, the integers from 1 through n are substituted for i in u_i and the resulting expressions are added together. For the symbol $\Sigma_{i=1}^{n} u_i$ we may say "the sum of u_i from $i = 1$ to $i = n$." The quantity u_i is called the **summand** and the letter i the **index of summation.** Observe that instead of Eq. 4 we could just as well write

$$\sum_{j=1}^{n} u_j = u_1 + u_2 + \cdots + u_n. \tag{5}$$

Equations 4 and 5 illustrate that the letter assigned to the index of summation is immaterial in the sense that it does not appear in the final result; thus the index of summation is sometimes called a "dummy index." The set of values taken on by the index of summation is called the **range of summation.** The range is from 1 to n in Eq. 4 but in general may extend from one arbitrary integer to another. Thus, for

$$\sum_{j=\alpha}^{\beta} u_j = u_\alpha + u_{\alpha+1} + \cdots + u_\beta, \tag{6}$$

the range of summation is from α to β. In Eq. 6 the numbers α and β are called the **lower limit** and the **upper limit** of summation, respectively.

In terms of the Σ notation Eqs. 1, 2, and 3 can be written

$$1 + 4 + 9 + \cdots + n^2 = \sum_{k=1}^{n} k^2, \tag{7}$$

$$1 + x + x^2 + \cdots + x^n = \sum_{j=0}^{n} x^j, \tag{8}$$

and

$$f(x_1) + f(x_2) + \cdots + f(x_n) = \sum_{i=1}^{n} f(x_i), \tag{9}$$

respectively. Note that we have used different letters for the index of summation in each case and that in Eq. 8 the lower limit is 0 rather than 1.

Some familiar properties of sums, expressed in Σ notation, are

$$\sum_{i=1}^{n} cu_i = c \sum_{i=1}^{n} u_i \tag{10}$$

and

$$\sum_{i=1}^{n} (u_i + v_i) = \sum_{i=1}^{n} u_i + \sum_{i=1}^{n} v_i. \tag{11}$$

Equation 10 says, in words, that a common factor can be taken out of each term in a sum while Eq. 11 is a distributive property of addition. A less obvious result is

$$\sum_{i=1}^{n} (u_{i+1} - u_i) = u_{n+1} - u_1. \tag{12}$$

To see this, note that the left side of Eq. 12 in expanded form is just

$$\sum_{i=1}^{n} (u_{i+1} - u_i) = (u_2 - u_1) + (u_3 - u_2) + (u_4 - u_3) + \cdots + (u_{n+1} - u_n).$$

The first term in each set of parentheses cancels the second term in the following parentheses, leading to the result (12). Because of this cancellation property, such a sum is referred to as a *telescoping sum*. Finally, note that

$$\sum_{i=1}^{n} c = \underbrace{c + c + c + \cdots + c}_{n \text{ terms}} = nc. \tag{13}$$

If the summand does not depend on the index of summation, it is simply repeated the proper number of times, which is equivalent to a multiplication.

Sometimes it is useful to shift the index of summation. For example, if we want the sum (8) to start with an index value of 1 rather than 0, we can let $k = j + 1$ or $j = k - 1$. Then $j = 0$ corresponds to $k = 1$, $j = n$ corresponds to $k = n + 1$, and of course $x^j = x^{k-1}$. Then, instead of Eq. 8, we can write

$$1 + x + x^2 + \cdots + x^n = \sum_{k=1}^{n+1} x^{k-1}. \tag{14}$$

If we wish, we can now replace the dummy index k by some other letter, such as j, with the result

$$1 + x + x^2 + \cdots + x^n = \sum_{j=1}^{n+1} x^{j-1}. \tag{15}$$

With a little practice one can go directly from Eq. 8 to Eq. 15 in one step, rather than using two steps as shown here.

When manipulating sums using the Σ notation, especially when altering the index of summation, it is good practice to pause from time to time to check your work. Do this by writing out a few terms of the sum to make sure that it starts and stops where you want it to, and that the terms have the proper form.

EXAMPLE 1

Evaluate

$$\sum_{k=1}^{12} (2^k - 2^{k-1}).$$

This is a telescoping sum of the form (12). Thus

$$\sum_{k=1}^{12} (2^k - 2^{k-1}) = (2 - 2^0) + (2^2 - 2) + \cdots + (2^{12} - 2^{11})$$

$$= 2^{12} - 2^0 = 4096 - 1 = 4095. \blacksquare$$

EXAMPLE 2

Rewrite the sum

$$\sum_{j=1}^{n} a_j x^{j-1} \tag{16}$$

so that the summand involves the factor x^j, rather than x^{j-1}.

First let $j - 1 = k$ or $j = k + 1$. Then

$$\sum_{j=1}^{n} a_j x^{j-1} = \sum_{k=0}^{n-1} a_{k+1} x^k.$$

Next we simply replace the dummy index k by j, obtaining

$$\sum_{k=0}^{n-1} a_{k+1} x^k = \sum_{j=0}^{n-1} a_{j+1} x^j, \tag{17}$$

the required expression. You should check that the right side of Eq. 17 is equivalent to the original expression (16) by writing out a few terms of each. $\blacksquare$

EXAMPLE 3

Write the expression

$$1^2 + 3^2 + 5^2 + \cdots + 15^2 \tag{18}$$

using Σ notation.

The expression (18) is the sum of the squares of the first eight *odd* positive integers. To rewrite it in terms of Σ, we need a way of generating only odd numbers as an index j runs through a range of integers. Observe that $2j - 1$ is always an odd number when j is an integer; further $2j - 1 = 1, 3, 5, \ldots, 15$ when $j = 1, 2, 3, \ldots, 8$. Thus

$$1^2 + 3^2 + 5^2 + \cdots + 15^2 = \sum_{j=1}^{8} (2j - 1)^2.$$

More generally,

$$\sum_{j=1}^{n} (2j - 1)^2 = 1^2 + 3^2 + 5^2 + \cdots + (2n - 1)^2. \tag{19}$$

Similarly, if only even numbers appear in a sum, then one should use $2j$ in writing it in terms of Σ. For instance,

$$2^3 + 4^3 + \cdots + 18^3 = \sum_{j=1}^{9} (2j)^3. \quad \blacksquare$$

Area

We are accustomed to thinking and talking about areas from the early grades in school, but it turns out that it is more difficult than one might expect even to define the area of a more or less arbitrary two-dimensional region. Nevertheless, we do have an intuitive conception of area, which includes the following three properties:

1. *Rectangle property*. The area of a rectangle is the product of its base and its altitude.
2. *Addition property*. The area of a region consisting of a finite number of nonoverlapping rectangles (which may have common boundaries) is the sum of the areas of the separate rectangles.
3. *Comparison property*. If the region R_1 is contained in the region R_2, then the area of R_1 is not greater than the area of R_2.

Now suppose that we are given a function f on an interval $[a, b]$, and suppose further that f is continuous and nonnegative. Intuitively, we believe that the region R bounded by the lines $x = a$, $x = b$, $y = 0$, and the curve $y = f(x)$ shown in Figure 6.1.2*a* has an area. How do we define this area, and how can we compute it?

The basic idea is to approximate the region by the union of a large number of nonoverlapping thin rectangles (see Figure 6.1.2*b*). The area of the desired

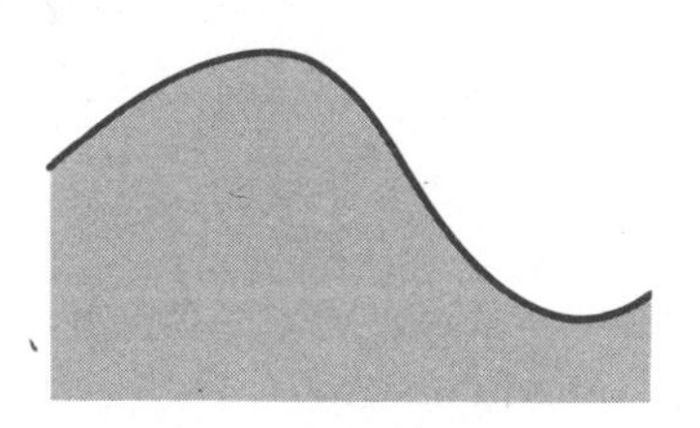

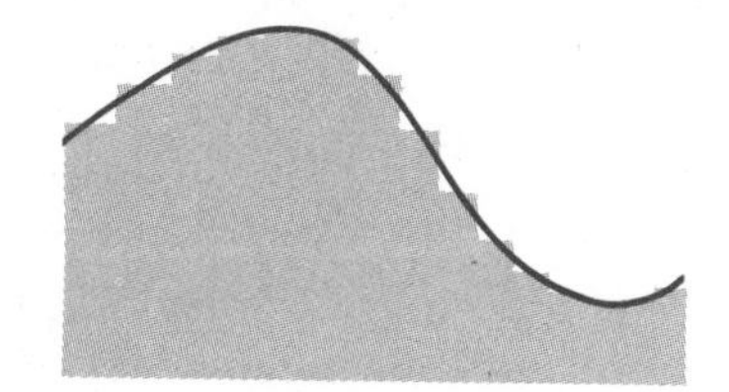

Figure 6.1.2

region is then approximated by the sum of the areas of the rectangles. If more and more rectangles with shorter and shorter bases are used, it is reasonable to expect that the sum of their areas will approximate more and more closely the area of the given region.* The following example illustrates the procedure.

EXAMPLE 4

Find the area of the region in the xy-plane bounded by the x-axis, the line $x = b > 0$, and the parabola $y = x^2$ (see Figure 6.1.3).

The first step is to approximate this region by a number of rectangles. This can be done in various ways, but it is probably simplest to use vertical rectangles of equal width. Thus we begin by subdividing the interval $[0, b]$ on the x-axis into n equal parts, each of length $h = b/n$. We label the points of subdivision $x_0 = 0$, $x_1 = h, x_2 = 2h, \ldots, x_n = nh = b$. Four subintervals are shown in Figure 6.1.3.

Consider first the subinterval $[0, h]$. Let us construct a rectangle whose base is the interval $[0, h]$ and whose height is the distance up to the graph of $y = x^2$ for some value of x in $[0, h]$. For example, we can use the midpoint $x = h/2$ of the base to calculate the height of the rectangle, as shown in Figure 6.1.4. In the first subinterval the height of the rectangle is $(h/2)^2$ and its area is $h(h/2)^2 = h^3/4$. If we proceed in the same way in the second subinterval $[h, 2h]$ we find that

*This is essentially the method of exhaustion, which was originated by Eudoxus (408?–355? B.C.). It was used with particular skill by Archimedes (287?–212 B.C.), the foremost mathematician of antiquity, who is frequently ranked, with Newton and Gauss, among the three greatest mathematicians of all time. Archimedes lived in Syracuse in what is now Sicily. His use of the method of exhaustion to compute the areas and volumes of many plane and solid figures foreshadowed the development of the integral calculus many centuries later. He also was the first person in Western history to compute π in a systematic manner; in particular, he found that $223/71 < \pi < 220/70$, the upper and lower bounds differing by 1/497. He developed many properties of a certain spiral curve known as the spiral of Archimedes. Among other things, he showed that this curve can be used to solve two ancient problems—the trisection of an angle, and the squaring of a circle—although not by using a compass and straightedge alone. However, he is probably best known for his work with levers and for the development of the principles of hydrostatics. He lost his life when Syracuse fell to the Romans during the Second Punic War, after having invented a number of ingenious weapons that helped to prolong the siege for more than two years.

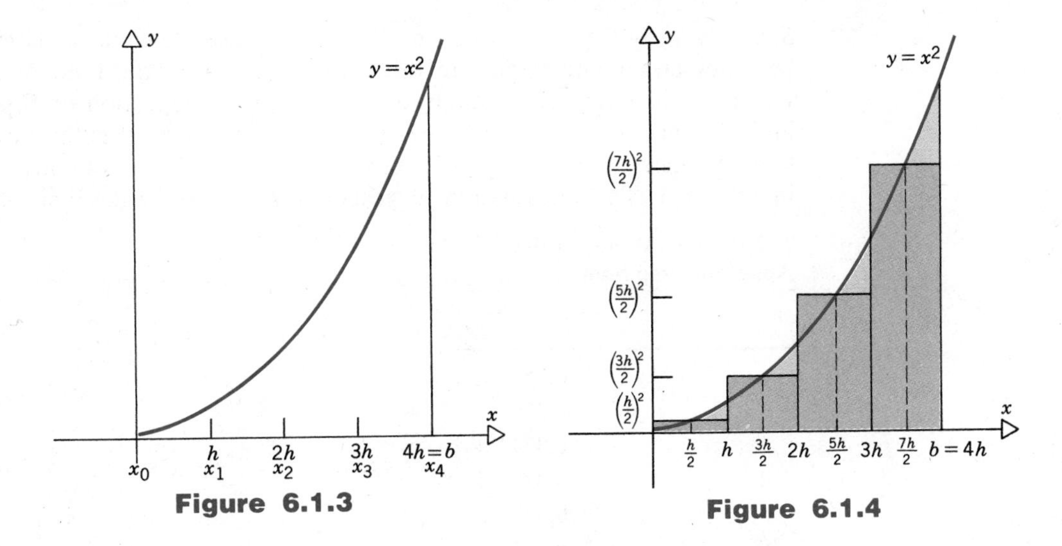

Figure 6.1.3

Figure 6.1.4

the height of the rectangle is $(3h/2)^2$ and that its area is $h(3h/2)^2 = 9(h^3/4)$. By constructing a rectangle in this manner on each of the n subintervals, adding their areas, and then using Eq. 19, we obtain the sum

$$S_n = \frac{h^3}{4}[1 + 9 + 25 + \cdots + (2n - 1)^2] = \frac{h^3}{4}\sum_{k=1}^{n}(2k - 1)^2, \qquad (20)$$

which is an approximation to the area of the given region. Recalling that $h = b/n$, we can also write S_n as

$$S_n = \frac{b^3}{4}\frac{[1 + 3^2 + 5^2 + \cdots + (2n - 1)^2]}{n^3} = \frac{b^3}{4}\cdot\frac{1}{n^3}\sum_{k=1}^{n}(2k - 1)^2. \qquad (21)$$

As we increase the number of rectangles, we expect to obtain a better approximation to the area under the parabola; for instance, compare Figure 6.1.5 with twelve rectangles and Figure 6.1.4 with only four. It is reasonable to believe that

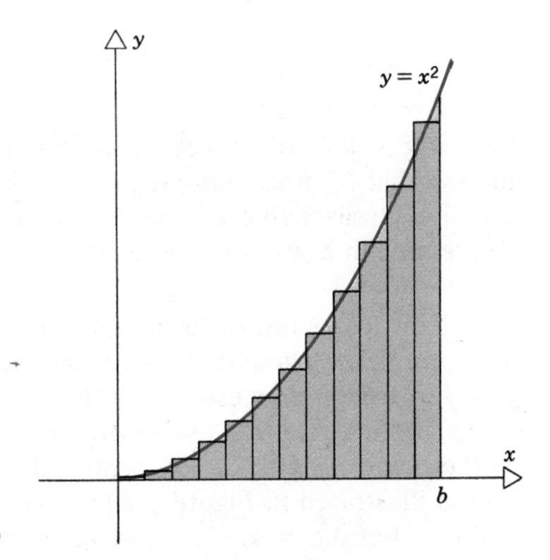

Figure 6.1.5

S_n as given by Eq. 21 approaches closer and closer to some specific number as n becomes larger and larger. In other words, we seek the limit of S_n as $n \to \infty$. However, it is not immediately obvious from an inspection of Eq. 21 what this limit is, since for large values of n the right side of Eq. 21 is the ratio of two very large numbers. If we compute S_n for several values of n, we obtain the results listed in Table 6.1. This information suggests that $b^3/3$ may be the limit of S_n as $n \to \infty$.

Table 6.1 Values of the Approximating Sum S_n.

n	S_n/b^3
1	0.25
2	0.3125
3	0.32407407
4	0.328125
5	0.330
10	0.3325
100	0.333325

To establish this conjecture as fact it is necessary to find a more useful expression for the sum $1 + 3^2 + 5^2 + \cdots + (2n - 1)^2$. This can be done by means of the principle of mathematical induction. We do not discuss this principle here since it would take us too far afield. See, however, Problems 23 through 30. Using the principle of induction it is possible to show that for each positive integer n,

$$1 + 3^2 + 5^2 + \cdots + (2n - 1)^2 = \tfrac{1}{3}n(2n - 1)(2n + 1). \tag{22}$$

Then we can rewrite S_n as

$$S_n = \frac{b^3}{4} \cdot \frac{1}{3} \frac{n(2n - 1)(2n + 1)}{n^3} \tag{23}$$

$$= \frac{b^3}{4} \cdot \frac{1}{3} \cdot \frac{n}{n} \cdot \frac{2n - 1}{n} \cdot \frac{2n + 1}{n}$$

$$= \frac{b^3}{4} \cdot \frac{1}{3} \cdot 1 \cdot \left(2 - \frac{1}{n}\right)\left(2 + \frac{1}{n}\right). \tag{24}$$

For large values of n each of the last two factors approaches two, and hence it follows that S_n does approach $b^3/3$ as n increases without bound. Therefore it is intuitively reasonable to accept $b^3/3$ as the area under the parabola. Later, we *define* area in a way that yields this result for this region. ∎

Now let us return to the question of finding the area $A(R)$ of the region R in Figure 6.1.2a, bounded by the lines $x = a$ and $x = b$, the x-axis, and the curve $y = f(x)$. We follow essentially the same procedure as in Example 4.

The first step is to subdivide the interval $[a, b]$ into a set of n subintervals by the points $x_0, x_1, \ldots, x_n$ arranged so that $a = x_0 < x_1 < x_2 < \cdots < x_n = b$. This is illustrated in Figure 6.1.6a. The spacing between these points may not be uniform. Let $\Delta x_i = x_i - x_{i-1}$ be the length of the ith subinterval. To approximate

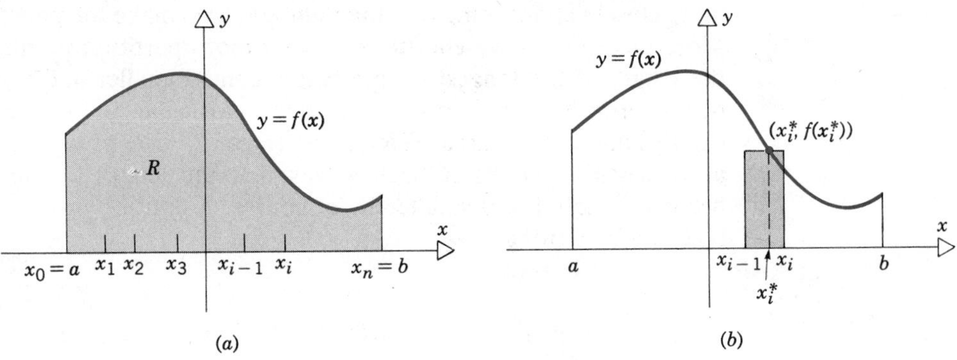

Figure 6.1.6

the area bounded by $x = x_{i-1}$, $x = x_i$, $y = 0$, and $y = f(x)$ shown in Figure 6.1.6*b* we choose any point in $[x_{i-1}, x_i]$, label it x_i^*, call it a star point, and form the rectangle with height $f(x_i^*)$. The area, $f(x_i^*)\,\Delta x_i$, of this rectangle is an approximation to the area bounded by $x = x_{i-1}$, $x = x_i$, $y = 0$, and $y = f(x)$. Proceeding in the same way in the other subintervals and then adding the results together, we conclude that

$$S = f(x_1^*)\,\Delta x_1 + f(x_2^*)\,\Delta x_2 + \cdots + f(x_n^*)\,\Delta x_n$$

$$= \sum_{i=1}^{n} f(x_i^*)\,\Delta x_i \qquad (25)$$

is an approximation to the area of R (see Figure 6.1.7). The extra areas that have been included are shaded more heavily, while the areas that have been missed are shaded lightly.

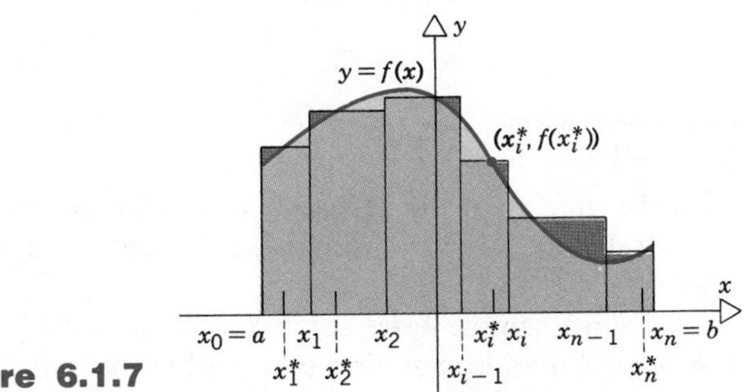

Figure 6.1.7

In general the value of S that we obtain in Eq. 25 depends upon the following.

1. The number of partition points that we choose.
2. How the partition points are spaced.
3. How the star point x_i^* is chosen in each subinterval.

The fundamental question that we must answer is whether the values of S corresponding to different partitions and different star points tend to cluster more and more closely about some definite number as we make the partitions finer and finer. More precisely, if we choose more and more partition points in such a way that the length of the longest subinterval becomes smaller and smaller, do the values of S approach closer and closer to some particular number? If so, it is natural to call this number the area $A(R)$ of the region R. The process of choosing more and more partition points in such a way that the length of the longest subinterval becomes smaller and smaller is, of course, a limiting process; we denote it symbolically by writing $n \to \infty$, $\Delta x_i \to 0$.

Thus we have

$$A(R) = \lim_{\substack{n\to\infty \\ \Delta x_i \to 0}} \sum_{i=1}^{n} f(x_i^*)\, \Delta x_i. \tag{26}$$

Although we have some prior notion of what area is, Eq. 26 is actually the definition of area for the region R.

The limiting process in Eq. 26 is considerably more complicated than the limit of a function at a point. Consequently, Eq. 26 is not usually an effective way actually to compute the area of a given region.

Note that in Example 4 we did not consider the limit (26) in any general way. In that example, we used only one type of subdivision (uniform subintervals) and one choice of star points (the midpoint of each subinterval). We did not consider whether a different type of subdivision or a different choice of star points might lead to a different result.

A further possible disadvantage of the preceding procedure is that, although for each n the value of S_n is an approximation to the desired area, it may not be easy to tell how good an approximation it is, or even whether it is too high or too low unless the limit in Eq. 26 can be evaluated. This shortcoming can be overcome to some extent by using inscribed and circumscribed rectangles, whose tops always lie, respectively, below and above the given curve.

EXAMPLE 5

Find the area A of the region bounded by the x=axis, the line $x = b > 0$, and the parabola $y = x^2$ by using (a) inscribed rectangles and (b) circumscribed rectangles.

As in Example 4, we subdivide the interval $[0, b]$ into n equal parts of length $h = b/n$. An approximation by inscribed rectangles, which lie beneath the curve, is shown in Figure 6.1.8*a*. The area of the jth rectangle, Figure 6.1.8*b*, is the product of its base h and its height $(j - 1)^2h^2$. The sum of the areas of all of the inscribed rectangles is

$$\begin{aligned} s_n &= h^3 \sum_{j=1}^{n} (j-1)^2 = h^3[1^2 + 2^2 + 3^2 + \cdots + (n-1)^2] \\ &= \frac{b^3}{n^3}[1^2 + 2^2 + 3^2 + \cdots + (n-1)^2]. \end{aligned} \tag{27}$$

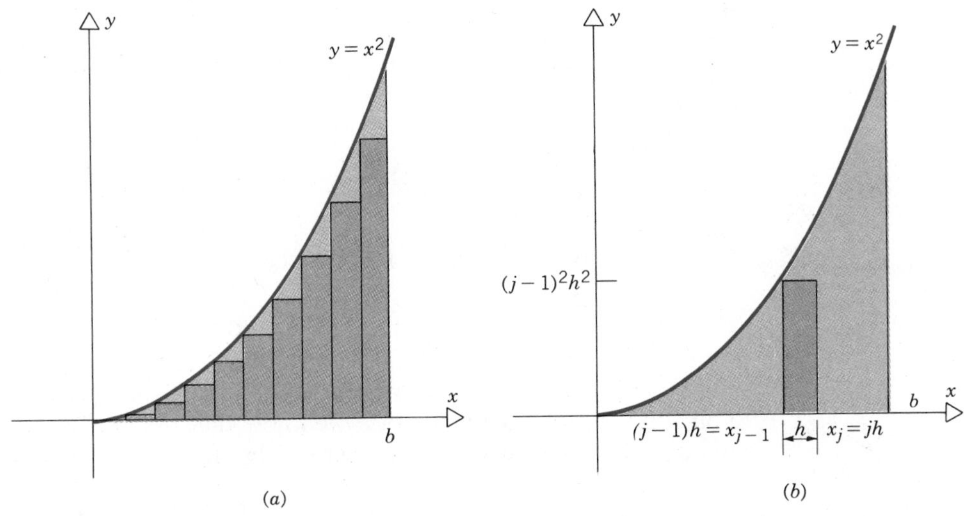

Figure 6.1.8 Approximation of a region by inscribed rectangles.

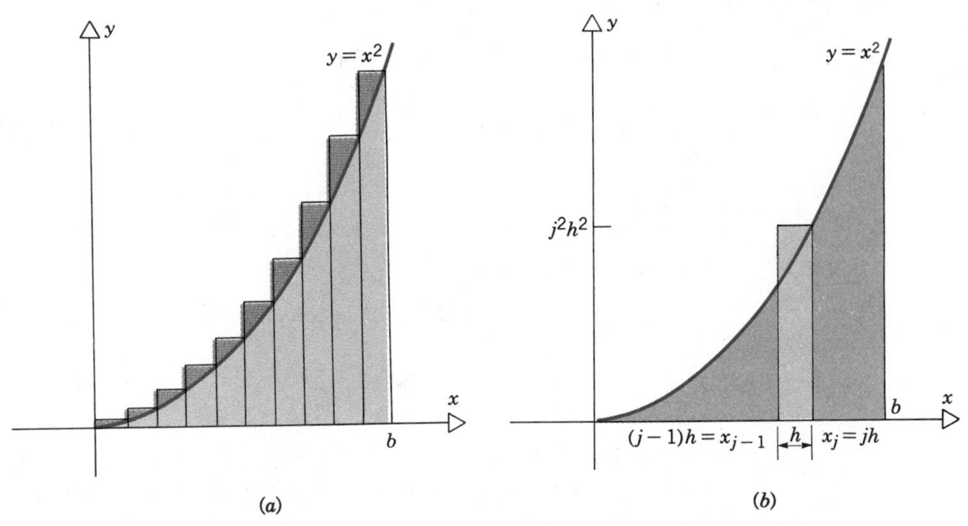

Figure 6.1.9 Approximation of a region by circumscribed rectangles.

Similarly, the area of the jth circumscribed rectangle (see Figure 6.1.9b) is the product of its base h and height j^2h^2, and the area of all of the circumscribed rectangles is (Figure 6.1.9a)

$$\sigma_n = h^3 \sum_{j=1}^{n} j^2 = h^3(1^2 + 2^2 + 3^2 + \cdots + n^2)$$

$$= \frac{b^3}{n^3}(1^2 + 2^2 + 3^2 + \cdots + n^2). \tag{28}$$

By making use of the identity (see Problem 25)

$$1^2 + 2^2 + 3^2 + \cdots + n^2 = n(n + 1)(2n + 1)/6, \tag{29}$$

we can rewrite s_n and σ_n as

$$s_n = \frac{b^3}{n^3} \cdot \frac{1}{6}(n - 1)n(2n - 1) = \frac{b^3}{6}\left(1 - \frac{1}{n}\right) \cdot 1 \cdot \left(2 - \frac{1}{n}\right), \tag{30}$$

$$\sigma_n = \frac{b^3}{n^3} \cdot \frac{1}{6} n(n + 1)(2n + 1) = \frac{b^3}{6} \cdot 1 \cdot \left(1 + \frac{1}{n}\right)\left(2 + \frac{1}{n}\right). \tag{31}$$

Then, as $n \to \infty$, it follows that both s_n and σ_n approach $b^3/3$, which agrees with the result of Example 4.

However, the use of inscribed and circumscribed rectangles enables us to go a bit further. For each value of n we know that the area A always lies between s_n and σ_n. Moreover, if we approximate A by the average of s_n and σ_n,

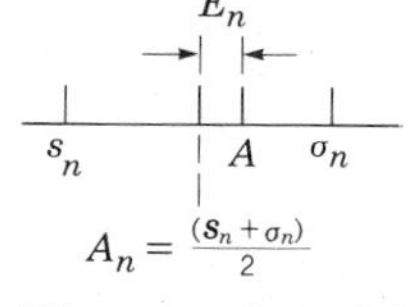

Figure 6.1.10

$$A \cong A_n = \frac{s_n + \sigma_n}{2}, \tag{32}$$

then there is a simple bound (see Figure 6.1.10) for the error $E_n = A - A_n$, namely,

$$|E_n| \le \frac{\sigma_n - s_n}{2}. \tag{33}$$

In the present example, we have $\sigma_n - s_n = b^3/n$ from Eqs. 27 and 28, so

$$|E_n| \le \frac{b^3}{2n}. \tag{34}$$

For instance, if $b = 2$ and $n = 50$, then

$$s_n = 2.5872, \qquad \sigma_n = 2.7472, \qquad A_n = 2.6672, \tag{35}$$

and

$$|E_n| \le 0.08. \tag{36}$$

Actually the approximation $A \cong 2.6672$ is much better than the error bound (36) suggests. Since the exact value of A is $\frac{8}{3}$, the approximate value of 2.6672 is in error by only 0.0005333. ∎

PROBLEMS

In each of Problems 1 through 12, evaluate the given sum.

1. $\displaystyle\sum_{k=1}^{4} \frac{1}{k}$

2. $\displaystyle\sum_{r=0}^{3} 3^r$

3. $\displaystyle\sum_{k=2}^{3} k^k$

4. $\displaystyle\sum_{i=1}^{4} (-1)^{i+1} \sin \frac{i\pi}{2}$

5. $\displaystyle\sum_{k=1}^{n} \left(\frac{1}{k + 1} - \frac{1}{k}\right)$

6. $\displaystyle\sum_{i=1}^{4} \left(\frac{2}{i} + 2i\right)$

7. $\displaystyle\sum_{i=1}^{1000} 1$

8. $\displaystyle\sum_{i=1}^{n} 1000$

9. $\displaystyle\sum_{k=1}^{50} \frac{1}{k(k + 1)}$

Hint: $\displaystyle\frac{1}{k(k + 1)} = \frac{?}{k} - \frac{?}{k + 1}$

10. $\sum_{r=0}^{3} 2r(1 + r)$ **11.** $\sum_{n=1}^{87} [\sqrt{3n+1} - \sqrt{3n-2}]$

12. $\sum_{i=1}^{n} \frac{1}{4i^2 - 1}$

Hint: $\frac{1}{4i^2 - 1} = \frac{?}{2i - 1} - \frac{?}{2i + 1}$

13. Show that

$$\sum_{k=1}^{n} u_k = \sum_{r=0}^{n-1} u_{r+1} = \sum_{j=2}^{n+1} u_{j-1} = \sum_{i=0}^{n-1} u_{n-i}$$

14. Show that

$$\sum_{k=m}^{n} a_k x^k = \sum_{r=0}^{n-m} a_{m+r} x^{m+r}.$$

In each of Problems 15 through 18, express the given sum in expanded form.

15. $\sum_{k=1}^{n} (-1)^{k+1} x^k$ **16.** $\sum_{k=0}^{n} \frac{(-1)^k x^{2k+1}}{(2k + 1)!}$

17. $\sum_{i=0}^{n-1} f\left[a + \frac{i(b - a)}{n}\right] \frac{b - a}{n}$ **18.** $\sum_{i=1}^{n} \frac{x^{n-i}}{n + i}$

In each of Problems 19 through 22, express the given sum in Σ notation.

19. $\frac{1}{2} - \frac{1}{3} + \frac{1}{4} - \frac{1}{5} + \cdots + \frac{1}{26}$

20. $\frac{1}{1 \cdot 3} + \frac{1}{2 \cdot 4} + \frac{1}{3 \cdot 5} + \cdots + \frac{1}{n(n + 2)}$

21. $\frac{x^3}{3} - \frac{x^5}{5} + \frac{x^7}{7} - \cdots - \frac{x^{29}}{29}$

22. $x - \frac{x^3}{9} + \frac{x^5}{25} - \cdots + \frac{x^{21}}{441}$

The principle of induction provides a means of establishing the validity of certain statements for all positive integers. Let $P(n)$ denote a statement, or proposition, that depends on the positive integer n. If

1. $P(1)$ is true;
2. $P(k + 1)$ is true whenever $P(k)$ is true;

then $P(n)$ is true for each positive integer n.

Problems 23 to 30 illustrate the use of the principle of induction.

23. Consider the proposition $P(n)$:

$$1 + 2 + 3 + \cdots + n = \tfrac{1}{2}n(n + 1).$$

(a) Show that $P(1)$ is true.
(b) Assume that $P(k)$ is true, that is,

$$1 + 2 + 3 + \cdots + k = \tfrac{1}{2}k(k + 1).$$

By adding $k + 1$ to each side of the last equation, show that

$$1 + 2 + 3 + \cdots + k + (k + 1) = \tfrac{1}{2}(k + 1)(k + 2),$$

and hence that $P(k + 1)$ is true. By the principle of induction $P(n)$ is therefore established for each positive integer.

24. In Example 4 in the text it was helpful to use the formula

$$1^2 + 3^2 + 5^2 + \cdots + (2n - 1)^2 = \tfrac{1}{3}n(2n - 1)(2n + 1).$$

Use the principle of induction to establish this result for each positive integer n.

In each of Problems 25 through 28, use the principle of induction to prove the given statement, where n is an arbitrary positive integer.

25. $1^2 + 2^2 + 3^2 + \cdots + n^2 = \tfrac{1}{6}n(n + 1)(2n + 1)$

26. $1^3 + 2^3 + 3^3 + \cdots + n^3 = \tfrac{1}{4}n^2(n + 1)^2$

27. $1 \cdot 3 + 3 \cdot 5 + 5 \cdot 7 + \cdots + (2n - 1)(2n + 1) = \tfrac{1}{3}n(4n^2 + 6n - 1)$

28. $\left(1 - \frac{1}{4}\right)\left(1 - \frac{1}{9}\right)\left(1 - \frac{1}{16}\right) \cdots \left(1 - \frac{1}{n^2}\right) = \frac{n + 1}{2n}, \quad n \geq 2$

In each of Problems 29 and 30, compute the given expression for a few small values of n, then guess a general formula, and finally prove your conjecture by using the principle of induction.

29. $1 + 3 + 5 + \cdots + (2n - 1) = ?$

30. $\left(1 - \frac{1}{2}\right)\left(1 - \frac{1}{3}\right)\left(1 - \frac{1}{4}\right) \cdots \left(1 - \frac{1}{n}\right) = ?$

In each of Problems 31 through 34, evaluate the given sum. You may use the results of Problems 23 through 28.

31. $\sum_{i=1}^{n} (3 - 2i)$ **32.** $\sum_{i=1}^{n} (1 + i)^2$

33. $\sum_{k=1}^{n} 2k(1 + k)^2$ **34.** $\sum_{r=0}^{n} (1 - 2r + 6r^2)$

35. Consider the triangular region R bounded by the lines $y = 0$, $x = b > 0$, and $y = x$. Subdivide $[0, b]$ into equal subintervals.

(a) Use inscribed and circumscribed rectangles to form approximating sums s_n and σ_n, respectively.

(b) Find a bound for the error E_n if the area is approximated by $A_n = (s_n + \sigma_n)/2$.

(c) Find the area of R.

36. Follow the instructions of Problem 35 for the region bounded by the lines $y = 0$, $x = b > 0$, and the curve $y = x^3$.

37. For the triangular region of Problem 35 form an approximating sum S_n by choosing as star points the midpoint of each subinterval. Then find the area of R.

38. Follow the instructions of Problem 37 for the region described in Problem 36. *Hint:* You will need an expression for the sum of the cubes of the first n odd integers. Use the result of Problem 26 to derive the required expression.

39. Show that

$$2 \sin \tfrac{1}{2}x \cos kx = \sin(k + \tfrac{1}{2})x - \sin(k - \tfrac{1}{2})x,$$

and hence establish the identity

$$2 \sin \tfrac{1}{2}x \sum_{k=1}^{n} \cos kx = \sin(n + \tfrac{1}{2})x - \sin \tfrac{1}{2}x.$$

In particular if $\sin \frac{1}{2}x \neq 0$, then

$$\sum_{k=1}^{n} \cos kx = \frac{\sin(n + \frac{1}{2})x - \sin \frac{1}{2}x}{2 \sin \frac{1}{2}x}.$$

6.2 THE RIEMANN INTEGRAL

Our discussion up to this point has been motivated by the desire to find the area of a region bounded in part by a more or less arbitrary continuous curve. Ultimately we obtained the expression

$$A(R) = \lim_{\substack{n\to\infty \\ \Delta x_i \to 0}} \sum_{i=1}^{n} f(x_i^*)\, \Delta x_i \tag{1}$$

for the area of such a region. It turns out that many other quantities of interest can also be expressed by limits of sums similar to Eq. 1; for example, see Problems 23 through 26. Therefore it is important to consider this mathematical process independent of any particular application. We now reexamine the procedure leading to Eq. 1 from a purely analytical point of view, that is, without using considerations of geometry.

Suppose that f is a bounded function on the interval $[a, b]$. Observe that we do not require f to be continuous or nonnegative. A **partition** of the interval $[a, b]$ is a finite set of points that divides $[a, b]$ into subintervals, with $a = x_0 < x_1 < x_2 < \cdots < x_n = b$. The length of the ith subinterval is $\Delta x_i = x_i - x_{i-1}$. The symbol Δ is customarily used to denote a partition:

$$\Delta = \{x_0, x_1, \ldots, x_n\}.$$

Next, we need a convenient number that measures the *fineness* of a partition. The best choice for this purpose is the length of the longest subinterval. This number is called the **norm** of the partition and is denoted by $\|\Delta\|$; thus

$$\|\Delta\| = \max \Delta x_i, \qquad i = 1, 2, \ldots, n. \tag{2}$$

For example, if the interval $[-1, 1]$ is partitioned by the points $-1 < -\frac{3}{4} < 0 < \frac{1}{4} < \frac{1}{2} < 1$, then $\Delta = \{-1, -\frac{3}{4}, 0, \frac{1}{4}, \frac{1}{2}, 1\}$. The longest subinterval is the one from $-\frac{3}{4}$ to 0; hence $\|\Delta\| = \frac{3}{4}$. Observe that there are many other partitions of

$[-1, 1]$ for which $\|\Delta\| = \frac{3}{4}$, namely, any partition whose longest subinterval has length $\frac{3}{4}$.

The next step is to choose a star point x_i^* in each subinterval, and to form the product $f(x_i^*)\,\Delta x_i$. Then we sum over all of the subintervals, obtaining

$$S(f, \Delta) = \sum_{i=1}^{n} f(x_i^*)\,\Delta x_i. \tag{3}$$

The notation $S(f, \Delta)$ emphasizes that the sum depends not only on the function f but also on the partition Δ, and of course on the choice of the star points, as well. A sum of the form (3) is called a **Riemann* sum.** Note that it is the same as the sum in Eq. 1, although now we are not assuming that $f(x)$ is nonnegative, so we do not interpret the individual terms as areas of rectangles.

Finally, we need to consider the limit of the Riemann sum (3) as $n \to \infty$ and $\Delta x_i \to 0$, or equivalently, as $\|\Delta\| \to 0$. If, for a given function f, the Riemann sum (3) approaches a single number I as $\|\Delta\| \to 0$, then I is called the **integral** (or Riemann integral) of f from a to b, and f is said to be **integrable** (or Riemann integrable) on $[a, b]$. Instead of I, the notation

$$\int_a^b f(x)\,dx$$

is usually used to denote the integral of f from a to b; the symbol $\int$ is called an **integral** sign,** f is the **integrand,** and a and b are the **lower** and **upper limits** of integration, respectively. The (x) and dx play no essential role in the notation at this stage and they are sometimes omitted; later they will be convenient in certain situations. In any event, the x is a dummy variable of integration, similar to the index of summation, and may be replaced by any other letter, should this be convenient. Summarizing the preceding statements, we have the following definition of the Riemann integral.

DEFINITION 6.2.1 If f is a bounded function on the interval $[a, b]$, then

$$\int_a^b f(x)\,dx = \lim_{\|\Delta\| \to 0} \sum_{i=1}^{n} f(x_i^*)\,\Delta x_i, \tag{4}$$

---

*Georg Friedrich Bernhard Riemann (1826–1866), a student and, after 1859, a professor at Göttingen, was a remarkably gifted and versatile mathematician. His development of integration, based on what are now known as Riemann sums, is contained in a paper on Fourier series written in 1854, but not published until 1867.

His doctoral dissertation in 1851 contained fundamental results in the theory of functions of a complex variable, including the celebrated Riemann mapping theorem.

Shortly afterward (1854) Riemann presented a landmark paper on non-Euclidean geometry in which he explored the properties of curved (or Riemannian) spaces, later used by Einstein in the theory of general relativity.

In 1859 Riemann formulated a statement concerning the zeros of a certain function important in number theory. Riemann's hypothesis has since been supported by much numerical evidence, but remains unproved despite the best efforts of generations of mathematicians. It is one of the most famous unsolved problems in mathematics.

Unfortunately, Riemann was not physically robust, and died of tuberculosis in his fortieth year.

**The integral sign is another bit of notation introduced by Leibniz; it was first used by him in 1675.

provided this limit exists, is the Riemann integral of f on $[a, b]$. In this case f is said to be integrable on $[a, b]$. On the other hand, if the limit does not exist, then f is not integrable on $[a, b]$.

In terms of the definition of a limit, Eq. 4 states that $\int_a^b f(x)\,dx$ is the number I with the following property: for each $\epsilon > 0$ there is a $\delta > 0$ such that for all partitions with $\|\Delta\| < \delta$ and for all choices of the star points x_i^* for each such partition, it follows that

$$\left| \sum_{i=1}^{n} f(x_i^*)\,\Delta x_i - I \right| < \epsilon. \tag{5}$$

There are three main questions that we wish to consider concerning integrals.

1. Are all functions integrable? If not, can we specify a large and useful class of functions that are integrable?
2. Are there some useful properties of integrals?
3. Given a particular function f, is there an effective way to evaluate $\int_a^b f(x)\,dx$?

We give a partial answer to the first question in this section, and take up the second question in Section 6.3. With respect to the evaluation of integrals, one possibility is to use Definition 6.2.1. However, the limiting process involved in this definition is a complicated one. For a given function f, a given interval $[a, b]$, and a given value of $\|\Delta\|$, there are infinitely many partitions of $[a, b]$, and for each such partition there are infinitely many ways of choosing the star points x_i^*. Consequently, the definition is rarely an efficient way to calculate the value of an integral. Fortunately, other ways are available. Beginning in Section 6.4 we present a powerful analytical method for evaluating many integrals. Numerical approximations can also be calculated, as in Example 5 of Section 6.1 and in Example 3 following, or by using algorithms such as those in Section 6.6.

In some particularly simple cases, however, the definition can be used to evaluate the integral of a function.

EXAMPLE 1

Suppose that f is the constant function $f(x) = c$ on $[a, b]$. Then for any partition of $[a, b]$ and any choice of the star points we have $f(x_i^*) = c$, so

$$\sum_{i=1}^{n} f(x_i^*)\,\Delta x_i = \sum_{i=1}^{n} c\,\Delta x_i = c \sum_{i=1}^{n} \Delta x_i = c(b - a).$$

For this function every Riemann sum is equal to $c(b - a)$; hence the constant function is integrable and

$$\int_a^b c\,dx = c(b - a).$$

Not surprisingly, if $c > 0$, this calculation gives the area of a rectangle of height c and base $b - a$. ∎

EXAMPLE 2

Suppose that the function f is defined as follows on $[a, b]$:

$$f(x) = \begin{cases} 1, & \text{if } x \text{ is rational;} \\ 0, & \text{if } x \text{ is irrational.} \end{cases}$$

Consider any partition of $[a, b]$. In each subinterval there are always both rational and irrational numbers. If we form a Riemann sum choosing each star point to be a rational number, we obtain

$$\sum_{i=1}^{n} f(x_i^*)\,\Delta x_i = \sum_{i=1}^{n} 1\,\Delta x_i = b - a.$$

On the other hand, if we form a Riemann sum choosing each star point to be an irrational number, we obtain

$$\sum_{i=1}^{n} f(x_i^*)\,\Delta x_i = \sum_{i=1}^{n} 0\,\Delta x_i = 0.$$

Thus for every partition of $[a, b]$, no matter how small $\|\Delta\|$ is, we can always find Riemann sums whose values are $b - a$ and 0. The Riemann sums do not approach any particular number; therefore this function is not integrable. ∎

Example 2 furnishes the answer to the first part of question one: not all functions are integrable. It is difficult to state exactly which functions *are* integrable, except by repeating Definition 6.2.1. However, widely applicable sufficient conditions for integrability are contained in the following two theorems. We omit the proofs of these theorems, although Problem 27 indicates part of the proof of Theorem 6.2.2.

Theorem 6.2.1

Suppose that f is bounded on the interval $[a, b]$, and is also continuous on this interval except for at most a finite number of points. Then f is integrable on $[a, b]$.

Theorem 6.2.2

Suppose that f is bounded on $[a, b]$ and that $[a, b]$ can be partitioned into a finite number of subintervals in each of which f is either nondecreasing or nonincreasing. Then f is integrable on $[a, b]$.

It follows from Theorems 6.2.1 or 6.2.2 that a polynomial of any degree is integrable on any finite interval; a rational function is integrable on any finite interval that does not contain a zero of the denominator; the sine and cosine functions are integrable on any finite interval; and the tangent, cotangent, secant, and cosecant functions are integrable on any finite interval that does not contain an infinite discontinuity of the function. Also functions such as $f(x) = [\![x]\!]$ that are bounded and continuous except at a finite number of points on any interval $[a, b]$ are integrable.

We emphasize that Theorems 6.2.1 and 6.2.2 give sufficient conditions only. There are functions that are integrable on an interval $[a, b]$ that do not satisfy the hypotheses of either of these theorems.

Notice also that the definitions and theorems of this section apply to functions that are bounded on the interval $[a, b]$. Later on, in Chapter 11, we consider the questions of the integrability of unbounded functions and the integrability of bounded functions on unbounded intervals. However, for the present we will restrict our discussion to the integration of bounded functions on bounded intervals.

Our development of the Riemann integral originated in the problem of determining the area of certain regions. However, the definition of the integral that we have given is free of geometrical language and does not depend on geometrical concepts. Therefore we can now rephrase the definition of area contained in Eq. 26 of Section 6.1.

DEFINITION 6.2.2 If f is continuous and nonnegative on $[a, b]$, then the area $A(R)$ of the region R bounded by the lines $x = a$ and $x = b$, the x-axis, and the graph of $y = f(x)$ is

$$A(R) = \int_a^b f(x)\, dx. \tag{6}$$

Estimation of integrals; upper and lower sums

Frequently it is desirable to estimate the value of an integral, especially when its exact value is difficult or impossible to calculate. A crude, but sometimes useful result can be obtained if we know upper and lower bounds for the integrand, that is, if we know that, for all x in $[a, b]$,

$$m \le f(x) \le M. \tag{7}$$

Then, for any partition Δ and any choice of the star points, the corresponding Riemann sum $S(f, \Delta)$ satisfies

$$S(f, \Delta) = \sum_{i=1}^{n} f(x_i^*)\, \Delta x_i \le \sum_{i=1}^{n} M\, \Delta x_i = M \sum_{i=1}^{n} \Delta x_i = M(b - a).$$

Combining this result with the similar lower bound, we have

$$m(b - a) \le S(f, \Delta) \le M(b - a). \tag{8}$$

If we assume that f is integrable, and take the limit as $\|\Delta\| \to 0$, we obtain

$$m(b - a) \le \int_a^b f(x)\, dx \le M(b - a). \tag{9}$$

Much more accurate bounds can also be obtained, though at the cost of more extensive calculation. Recall that in Example 5 of Section 6.1 we used inscribed and circumscribed rectangles to obtain lower and upper bounds, respectively, for the area under a parabola. The same approach can be used to estimate the values of integrals in general.

Let the integrable function f be given on $[a, b]$; for convenience we assume that f is continuous. Let Δ be an arbitrary partition of $[a, b]$. Denote by M_i and m_i, respectively, the largest and smallest value of f in the ith subinterval $[x_{i-1}, x_i]$. Then the largest possible Riemann sum for this partition is obtained by choosing x_i^* so that $f(x_i^*) = M_i$ in each subinterval. This sum is called the **upper sum** for the partition Δ and is given by

$$\sigma(f, \Delta) = \sum_{i=1}^{n} M_i \, \Delta x_i. \tag{10}$$

Similarly, the **lower sum** for Δ is

$$s(f, \Delta) = \sum_{i=1}^{n} m_i \, \Delta x_i. \tag{11}$$

Clearly, any other Riemann sum

$$S(f, \Delta) = \sum_{i=1}^{n} f(x_i^*) \, \Delta x_i \tag{12}$$

for this partition must lie between the upper and lower sums:

$$s(f, \Delta) \le S(f, \Delta) \le \sigma(f, \Delta). \tag{13}$$

Furthermore, it is possible to show that

$$s(f, \Delta) \le \int_a^b f(x)\, dx \le \sigma(f, \Delta). \tag{14}$$

Inequality (14) is similar to the corresponding relation for area in Section 6.1, and provides a means for estimating the value of the integral. Usually it is simplest to use a uniform partition, and to take the partition points closer and closer together until the difference $\sigma(f, \Delta) - s(f, \Delta)$ is as small as desired.

EXAMPLE 3

Use upper and lower sums to estimate the value of

$$\int_0^{\pi/2} \sin x \, dx. \tag{15}$$

The integrand is continuous and increasing on the interval $[0, \pi/2]$; therefore, by either of Theorems 6.2.1 or 6.2.2, the integral (15) is known to exist. To estimate its value we can divide the interval $[0, \pi/2]$ into n equal subintervals, each one of length $h = \pi/2n$. The partition points are $x_i = ih = i\pi/2n$ for $i = 0, 1, 2, \ldots, n$. In the ith subinterval $\sin x$ has its maximum and minimum values at the right and left endpoints, respectively; that is, $M_i = \sin x_i$ and $m_i = \sin x_{i-1}$. Thus the upper sum for this partition is

$$\sigma_n = \sum_{i=1}^{n} \sin x_i \, \Delta x_i = h \sum_{i=1}^{n} \sin(ih). \tag{16}$$

Similarly, the lower sum is

$$\begin{aligned} s_n &= \sum_{i=1}^{n} \sin x_{i-1} \, \Delta x_i = h \sum_{i=1}^{n} \sin[(i - 1)h] \\ &= h \sum_{i=1}^{n-1} \sin(ih). \end{aligned} \tag{17}$$

Their difference is

$$\sigma_n - s_n = h \sin nh = \frac{\pi}{2n} \sin \frac{\pi}{2} = \frac{\pi}{2n}, \tag{18}$$

which, by Eq. 14, provides an upper bound for the error in using either σ_n or s_n to approximate the given integral.

If we evaluate the sums σ_n and s_n for several values of n, we obtain the information in the second and third columns of Table 6.2. From the data in the table it appears that the upper and lower sums (and consequently all of the Riemann sums) approach a number close to 1. In fact, it is easy to show by the method to be introduced in Section 6.4 that the value of the integral (15) is exactly 1. From the tabulated values we see that s_n and σ_n are in error by less than one percent for $n = 80$, which is consistent with the inequalities (14). However, if we use the mean value of s_n and σ_n to approximate the integral (fourth column of Table 6.2), we obtain much greater accuracy: with only ten subintervals the result is accurate to within one quarter of one percent.

Another way to proceed is to calculate the Riemann sum corresponding to some particular choice of the star points. For instance, if we choose each x_i^* to be the midpoint of the subinterval, we obtain the results in Table 6.3. These approximations are extremely good; for instance, for $n = 10$ the error is barely more than one tenth of one percent.

While the calculations leading to Tables 6.2 and 6.3 were done on a microcomputer, they would be quite feasible on a pocket calculator at least up to $n = 10$. ∎

Table 6.2 **Upper and Lower Sums for $\int_0^{\pi/2} \sin x \, dx$.**

n	Lower Sum s_n	Upper Sum σ_n	Mean $(s_n + \sigma_n)/2$
10	0.9194032	1.076483	0.997943
20	0.9602161	1.038756	0.999486
40	0.9802366	1.019506	0.999871
80	0.9901504	1.009785	0.999968

Table 6.3 **Riemann Sums for $\int_0^{\pi/2} \sin x \, dx$.**

n	Riemann Sum S_n
10	1.001029
20	1.000257
40	1.000064
80	1.000016

PROBLEMS

In each of Problems 1 through 4, determine the norm of the partition of the given interval.

1. $[0, 2]$, $\quad 0 < \frac{1}{4} < \frac{1}{2} < \frac{7}{8} < 1 < \frac{3}{2} < 2$

2. $[1, 4]$, $\quad 1 < 1.1 < 1.5 < 2 < 2.01 < 2.02 < 2.1 < 2.5 < 3 < 4$

3. $[a, b]$, $\quad x_i = a + \frac{i}{n}(b - a),$

   $i = 0, 1, 2, \ldots, n$

4. $[a, b]$, $\quad x_i = a + \frac{i}{2^n}(b - a),$

   $i = 0, 1, 2, \ldots, 2^n$

In each of Problems 5 through 8, obtain an approximate value for the given integral using the given partition and with the star points $x_i^* = x_{i-1} + \frac{1}{2}\Delta x_i$.

© 5. $\displaystyle\int_1^2 \frac{x}{1 + x}\,dx, \quad \Delta = \{1, \tfrac{5}{4}, \tfrac{3}{2}, 2\}$

© 6. $\displaystyle\int_0^1 \cos\frac{x}{2}\,dx, \quad \Delta = \{0, 0.2, 0.4, 0.6, 0.8, 1\}$

© 7. $\displaystyle\int_{-1/2}^1 |x|\,dx, \quad \Delta = \{-\tfrac{1}{2}, 0, 1\}$

© 8. $\displaystyle\int_{-1}^1 f(x)\,dx,$

$$f(x) = \begin{cases} x^2, & -1 \le x < 0 \\ \frac{1}{2}, & x = 0 \\ 1 + x^2, & 0 < x \le 1 \end{cases}$$

$\Delta = \{-1, -\tfrac{1}{2}, \tfrac{1}{2}, 1\}$

In each of Problems 9 through 14, compute the upper and lower sums, $\sigma(f, \Delta)$ and $s(f, \Delta)$, respectively, for the given integrand and partition.

© 9. $\displaystyle\int_1^2 \frac{1}{x}\,dx, \quad \Delta = \{1, \tfrac{5}{4}, \tfrac{3}{2}, \tfrac{7}{4}, 2\}$

© 10. $\displaystyle\int_0^{\pi/2} \sin x\,dx, \quad \Delta = \left\{0, \frac{\pi}{4}, \frac{\pi}{2}\right\}$

© 11. $\displaystyle\int_{-2}^0 \frac{1}{1 + x^2}\,dx, \quad \Delta = \{-2, -\tfrac{3}{2}, -1, -\tfrac{1}{2}, 0\}$

© 12. $\displaystyle\int_2^4 \sqrt{1 + x}\,dx, \quad \Delta = \{2, 2.5, 3, 3.5, 4\}$

© 13. $\displaystyle\int_0^2 f(x)\,dx, \quad f(x) = \begin{cases} x^2, & 0 \le x \le 1 \\ \frac{1}{2}, & 1 < x \le 2 \end{cases}$

$\Delta = \{0, \tfrac{1}{2}, 1, \tfrac{3}{2}, 2\}$

© 14. $\displaystyle\int_1^3 (4 - x^2)\,dx, \quad \Delta = \{1, \tfrac{3}{2}, 2, \tfrac{5}{2}, 3\}$

In each of Problems 15 through 20, carry out the following procedure. Introduce a partition dividing the interval of integration into n equal subintervals.

(a) Evaluate the upper sum σ_n and the lower sum s_n for $n = 10$.

(b) Repeat Part (a) for $n = 20, 40,$ and 80.

(c) Find $(\sigma_n + s_n)/2$ for $n = 10, 20, 40,$ and 80.

(d) Choosing the star points as the midpoint of each subinterval, evaluate the corresponding Riemann sum S_n for $n = 10, 20, 40,$ and 80. Compare each Riemann sum with the corresponding value found in Part (c).

© 15. $\displaystyle\int_0^1 \sqrt{x}\,dx$

© 16. $\displaystyle\int_0^2 \frac{dx}{1 + x^2}$

© 17. $\displaystyle\int_0^3 \sqrt{1 + x^2}\,dx$

© 18. $\displaystyle\int_0^{\pi/3} \tan x\,dx$

© 19. $\displaystyle\int_1^3 (4 - x^2)\,dx$

© 20. $\displaystyle\int_{0.5}^2 \left(x - \frac{1}{x}\right)dx$

In Problems 21 and 22 the integrand is neither increasing nor decreasing throughout the interval of integration. This makes the upper and lower sums a little harder to find. Using a uniform partition with n equal subintervals, and choosing the star points as the midpoint of each subinterval, find the corresponding Riemann sum s_n for each of the given integrals for $n = 10, 20, 40,$ and 80.

© 21. $\displaystyle\int_0^{\pi} x\cos x\,dx$

© 22. $\displaystyle\int_0^{\pi/2} \sqrt{x}\sin 2x\,dx$

23. **Distance traveled.** A particle moving along a straight line with a constant velocity $v > 0$ travels a distance vT in the time interval $[0, T]$. Suppose now that the velocity of the particle is a function of time, $v = v(t)$, and that $v(t) \ge 0$, so that the particle always moves in the same direction. Form a partition Δ of $[0, T]$ and show that the distance d traveled in $[0, T]$ is given approximately by

$$d \cong S(v, \Delta) = \sum_{i=1}^{n} v(t_i^*)\,\Delta t_i.$$

If v is integrable, then

$$d = \lim_{\|\Delta\| \to 0} S(v, \Delta) = \int_0^T v(t)\,dt.$$

24. **Volume.** The volume of a rod of length l and uniform cross-sectional area A is Al. Suppose now that the cross-sectional area varies with position, so that $A = A(x)$ for $0 \leq x \leq l$. Form a partition Δ of $[0, l]$ and show that

$$V \cong S(A, \Delta) = \sum_{i=1}^{n} A(x_i^*) \, \Delta x_i.$$

If A is integrable, then

$$V = \lim_{\|\Delta\| \to 0} S(A, \Delta) = \int_0^l A(x) \, dx.$$

25. **Work.** The work done by a constant force in moving a particle a distance s along a straight line is Fs, where F is the component of the force in the direction of the motion. Suppose now that the force varies with position on the line, so that $F = F(x)$, with $0 \leq x \leq s$. Form a partition Δ of $[0, s]$, and show that the work W done by the force in moving a particle from $x = 0$ to $x = s$ is given approximately by

$$W \cong S(F, \Delta) = \sum_{i=1}^{n} F(x_i^*) \, \Delta x_i.$$

If F is integrable, then

$$W = \lim_{\|\Delta\| \to 0} S(F, \Delta) = \int_0^s F(x) \, dx.$$

26. **Mass.** The mass of a wire of length l and constant density (mass per unit length) ρ is ρl. Suppose now that the density varies with position, so that $\rho = \rho(x)$ for $0 \leq x \leq l$. Form a partition Δ of $[0, l]$ and show that the mass m of the wire is given approximately by

$$m \cong S(\rho, \Delta) = \sum_{i=1}^{n} \rho(x_i^*) \, \Delta x_i.$$

If ρ is integrable, then

$$m = \lim_{\|\Delta\| \to 0} S(\rho, \Delta) = \int_0^l \rho(x) \, dx.$$

***27.** In this problem we prove a result related to Theorem 6.2.2. Suppose that f is bounded and nondecreasing on $[a, b]$, and that Δ is an arbitrary partition of $[a, b]$. Let $\sigma(f, \Delta)$ and $s(f, \Delta)$ be the associated upper and lower sums, respectively.

(a) Show that

$$\sigma(f, \Delta) = \sum_{i=1}^{n} f(x_i) \, \Delta x_i,$$

$$s(f, \Delta) = \sum_{i=1}^{n} f(x_{i-1}) \, \Delta x_i.$$

(b) Recall that $\|\Delta\|$ is the length of the longest subinterval in the partition Δ, and show that

$$\sigma(f, \Delta) - s(f, \Delta) = \sum_{i=1}^{n} [f(x_i) - f(x_{i-1})] \, \Delta x_i$$

$$\leq \|\Delta\| \sum_{i=1}^{n} [f(x_i) - f(x_{i-1})].$$

(c) Use the result of Part (b) to show that

$$\sigma(f, \Delta) - s(f, \Delta) \leq \|\Delta\| [f(b) - f(a)].$$

Therefore, if $\|\Delta\| \to 0$, it follows that $\sigma(f, \Delta) - s(f, \Delta) \to 0$ also. Which theorem justifies this conclusion?

(d) Modify the argument to reach the same conclusion if f is nonincreasing on $[a, b]$.

6.3 PROPERTIES OF THE RIEMANN INTEGRAL

In motivating the concept of integration in Sections 6.1 and 6.2 we relied heavily on an intuitive idea of the area under a curve. However, as we have emphasized, Definition 6.2.1 of the Riemann integral is a purely analytical one and does not depend on geometrical properties or terminology. This made it possible, in Definition 6.2.2, to define area in terms of an integral. The validity of this definition rests on the fact that it can be verified (and we will do so later in many cases) that Definition 6.2.2, when applied to elementary regions bounded by straight lines and circles, gives the same results as the formulas of plane geometry. For the present we note that the interpretation of certain integrals as areas of simple figures provides a way to evaluate these integrals.

EXAMPLE 1

If $c > 0$, then $\int_a^b c\,dx$ is the area of the rectangle of base $b - a$ and height c shown in Figure 6.3.1. Thus

$$\int_a^b c\,dx = c(b - a). \tag{1}$$

Of course, this agrees with the result of Example 1 of Section 6.2. ∎

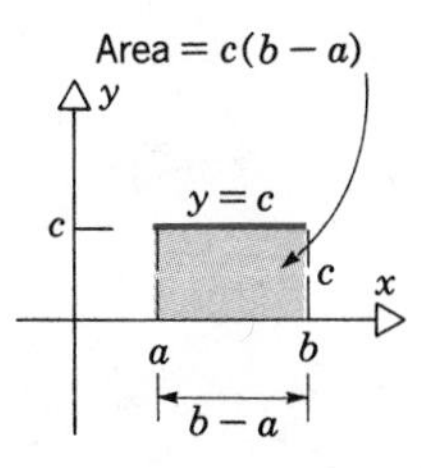

Figure 6.3.1

EXAMPLE 2

The integral $\int_a^b x\,dx$ is the area of the trapezoid shown in Figure 6.3.2. The average height of the trapezoid is $(b + a)/2$ and its base is $b - a$, so

$$\int_a^b x\,dx = \tfrac{1}{2}(b + a)(b - a) = \tfrac{1}{2}(b^2 - a^2). \text{ ∎} \tag{2}$$

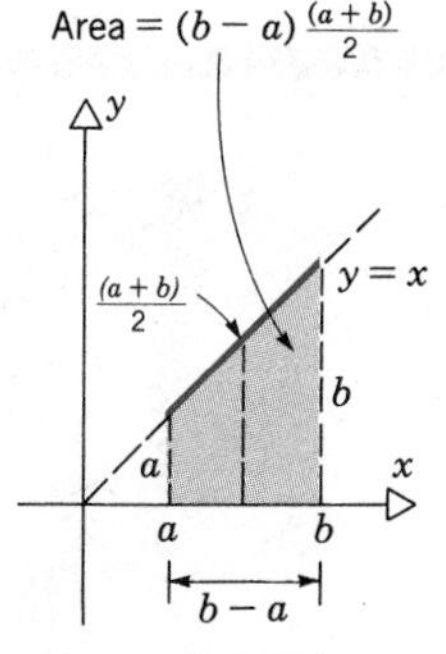

Figure 6.3.2

EXAMPLE 3

For $a > 0$ the integral $\int_0^a \sqrt{a^2 - x^2}\,dx$ can be identified as the area of the quarter circle of radius a in Figure 6.3.3. Thus

$$\int_0^a \sqrt{a^2 - x^2}\,dx = \frac{\pi a^2}{4}, \qquad a > 0. \tag{3}$$

Later, in Problem 31 of Section 6.5, we evaluate this integral by analytical methods and obtain the same result. On the other hand, if we evaluate the integral by the numerical procedure indicated in Problem 23, we can obtain accurate numerical approximations to π. ∎

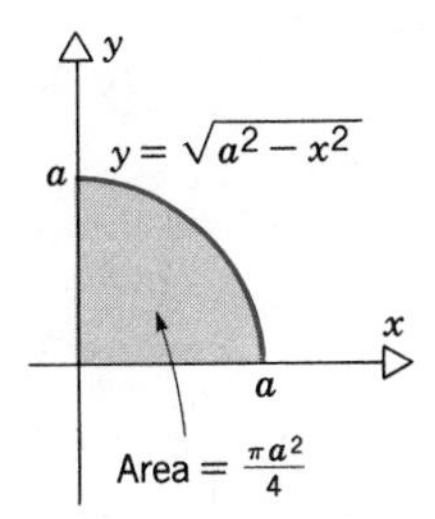

Figure 6.3.3

We now turn to the development of several properties of the integral. Although these properties are true for more general integrands, it may help in visualizing them to think of the integrand as continuous and nonnegative and to interpret the integrals as areas.

First, if f is integrable on the interval $[a, b]$, then it can be shown that f is also integrable on any subinterval of $[a, b]$. Further, up to now we have always assumed that $b > a$. We will also need to consider cases in which $b = a$ and $b < a$.

DEFINITION 6.3.1 If $f(a)$ exists, then

$$\int_a^a f(x)\,dx = 0. \tag{4}$$

Geometrically, Eq. 4 says that there is no area below a point.

If $b > a$ and $\int_a^b f(x)\,dx$ exists, then

$$\int_b^a f(x)\,dx = -\int_a^b f(x)\,dx. \tag{5}$$

In words, reversing the limits of integration changes the sign of the integral.

The first theorem states that the value of an integral is not affected by the values of the integrand on any finite set of points.

Theorem 6.3.1

If f and g are bounded functions, integrable on $[a, b]$, and if $f(x) = g(x)$ except for a finite set of points in $[a, b]$, then

$$\int_a^b f(x)\,dx = \int_a^b g(x)\,dx. \tag{6}$$

Proof. Suppose first that there is only one point c in $[a, b]$ where $f(c) \neq g(c)$. Then, for a particular partition and choice of star points, the Riemann sums for f and g, $S(f, \Delta)$ and $S(g, \Delta)$, respectively, can differ only by those terms that involve c. There can be only one such term unless c is a partition point, in which case there may be two. Thus

$$|S(f, \Delta) - S(g, \Delta)| \leq 2\|\Delta\||f(c) - g(c)|. \tag{7}$$

The right side of Eq. 7 approaches zero as $\|\Delta\| \to 0$; consequently, $S(f, \Delta)$ and $S(g, \Delta)$ must approach the same number, which means that Eq. 6 is true. The argument can be directly extended to cover a finite number of points where $f(x)$ and $g(x)$ are unequal. □

The next two theorems are of great benefit in the evaluation of integrals, as we will see later.

Theorem 6.3.2

(Linearity)

If the bounded functions f and g are integrable on $[a, b]$, then so is the function $\alpha f + \beta g$ for any real constants α and β; further

$$\int_a^b [\alpha f(x) + \beta g(x)]\,dx = \alpha \int_a^b f(x)\,dx + \beta \int_a^b g(x)\,dx. \tag{8}$$

By repeated application of Theorem 6.3.2, the conclusion (8) can be extended to any finite number of terms. The result of Theorem 6.3.2 is similar to the corresponding property of limits at a point, expressed in Eq. 5 of Section 2.3. Since integration involves a limiting process, one might think that the proof of Theorem 6.3.2 can be reduced immediately to that earlier result. However, this is not the case because the limiting process in integration, which involves different partitions and different choices of star points for a fixed value of $\|\Delta\|$, is more complicated than the limit process for functions. The proof of Theorem 6.3.2 is fairly lengthy and is therefore omitted. The next example illustrates a typical application of Theorem 6.3.2.

EXAMPLE 4

Evaluate

$$\int_0^3 (2x + \sqrt{9 - x^2})\, dx.$$

Since the function $f(x) = 2x + \sqrt{9 - x^2}$ is continuous on $[0, 3]$, the integral exists. According to Theorem 6.3.2, we have

$$\int_0^3 (2x + \sqrt{9 - x^2})\, dx = 2\int_0^3 x\, dx + \int_0^3 \sqrt{9 - x^2}\, dx. \tag{9}$$

The individual integrals can now be evaluated by referring to the results of Examples 2 and 3. From Eq. 2 with $a = 0$ and $b = 3$ we have

$$\int_0^3 x\, dx = \tfrac{9}{2}. \tag{10}$$

Next, by Eq. 3, we have

$$\int_0^3 \sqrt{9 - x^2}\, dx = \frac{9\pi}{4}. \tag{11}$$

Finally, substituting from Eqs. 10 and 11 in Eq. 9, we obtain

$$\int_0^3 (2x + \sqrt{9 - x^2})\, dx = 2\left(\frac{9}{2}\right) + \frac{9\pi}{4} = 9\left(1 + \frac{\pi}{4}\right) \cong 16.07. \quad \blacksquare \tag{12}$$

Theorem 6.3.3

If a, b, and c are any real numbers, and if any two of the following integrals exist, then

$$\int_a^c f(x)\, dx + \int_c^b f(x)\, dx = \int_a^b f(x)\, dx. \tag{13}$$

If $a < c < b$ and if f is continuous and nonnegative, then Theorem 6.3.3 says that the area below the curve $y = f(x)$ from $x = a$ to $x = c$ plus the area from $x = c$ to $x = b$ is equal to the area from $x = a$ to $x = b$. This is illustrated in Figure 6.3.4. However, Theorem 6.3.3 is true regardless of the order of the numbers a, b, and c; in particular, c need not be between a and b. Also, f need not be

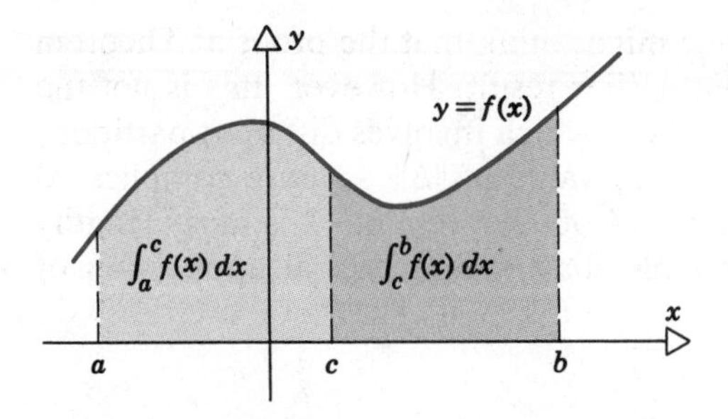

Figure 6.3.4

continuous or nonnegative. Thus Theorem 6.3.3 is a generalization of the additive property of area cited in Section 6.1. While Theorem 6.3.3 is made plausible by Figure 6.3.4, a detailed proof is omitted.

Notice that the statement of Theorem 6.3.3 is consistent with Definition 6.3.1. For example, if $a = b = c$, then Eq. 13 reduces to

$$2\int_a^a f(x)\,dx = \int_a^a f(x)\,dx,$$

and necessarily $\int_a^a f(x)\,dx = 0$. Further, if $b = a$, then Eq. 13 reduces to

$$\int_a^c f(x)\,dx + \int_c^a f(x)\,dx = 0,$$

so

$$\int_a^c f(x)\,dx = -\int_c^a f(x)\,dx.$$

The following examples illustrate how Theorem 6.3.3 can be used.

EXAMPLE 5

Evaluate the integral from -2 to 3 of

$$f(x) = \begin{cases} x + 2, & -2 \le x \le 0; \\ \sqrt{9 - x^2}, & 0 < x \le 3. \end{cases} \tag{14}$$

Since the function f is bounded on $[-2, 3]$ and is continuous except at one point the integral exists. The graph of f is shown in Figure 6.3.5.

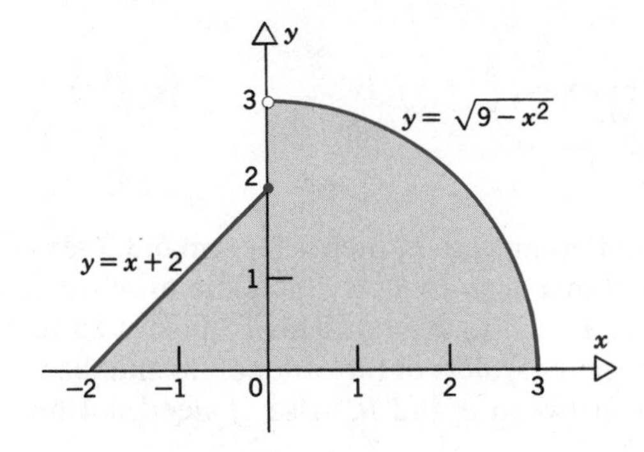

Figure 6.3.5

According to Theorem 6.3.3,

$$\int_{-2}^{3} f(x)\,dx = \int_{-2}^{0} f(x)\,dx + \int_{0}^{3} f(x)\,dx. \tag{15}$$

From Figure 6.3.5 the first integral is the area of the triangle with vertices $(-2, 0)$, $(0, 0)$, and $(0, 2)$; thus

$$\int_{-2}^{0} f(x)\,dx = \int_{-2}^{0} (x + 2)\,dx = \tfrac{1}{2} \cdot 2 \cdot 2 = 2. \tag{16}$$

By Eq. 11 the second integral is

$$\int_{0}^{3} f(x)\,dx = \int_{0}^{3} \sqrt{9 - x^2}\,dx = \frac{9\pi}{4}. \tag{17}$$

In evaluating the integral (17) it is permissible to write the integrand as $\sqrt{9 - x^2}$ even though $f(0) = 2 \neq \sqrt{9 - 0}$; remember that by Theorem 6.3.1 the value of f at one point does not affect the value of the integral. Hence, substituting from Eqs. 16 and 17 in Eq. 15,

$$\int_{-2}^{3} f(x)\,dx = 2 + \tfrac{9}{4}\pi \cong 9.07. \ ∎ \tag{18}$$

EXAMPLE 6

Find the value of

$$\int_{2}^{5} x^2\,dx. \tag{19}$$

In Examples 4 and 5 of Section 6.1 we found that the area of the region between the x-axis, the line $x = b$, and the parabola $y = x^2$ is $b^3/3$. Expressing this result as an integral, we have

$$\int_{0}^{b} x^2\,dx = \frac{b^3}{3}. \tag{20}$$

Now we use Theorem 6.3.3 to write the integral (19) as

$$\int_{2}^{5} x^2\,dx = \int_{2}^{0} x^2\,dx + \int_{0}^{5} x^2\,dx,$$

or, using Definition 6.3.1,

$$\int_{2}^{5} x^2\,dx = \int_{0}^{5} x^2\,dx - \int_{0}^{2} x^2\,dx.$$

This relation is shown in Figure 6.3.6. Then, making use of Eq. 20 with $b = 5$ and $b = 2$, respectively, we have

$$\int_{2}^{5} x^2\,dx = \frac{125}{3} - \frac{8}{3} = \frac{117}{3} = 39. \ ∎$$

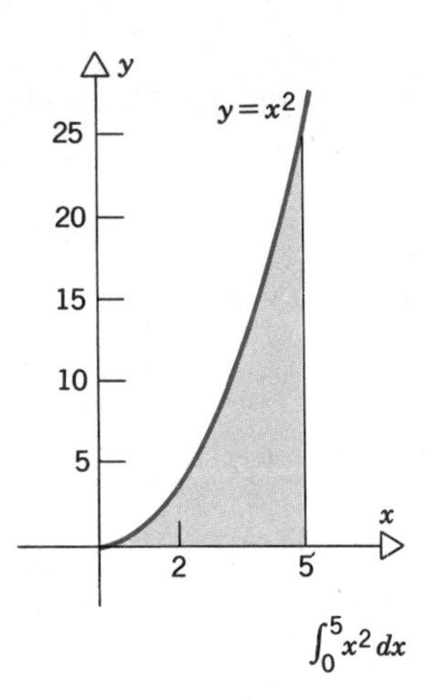

−

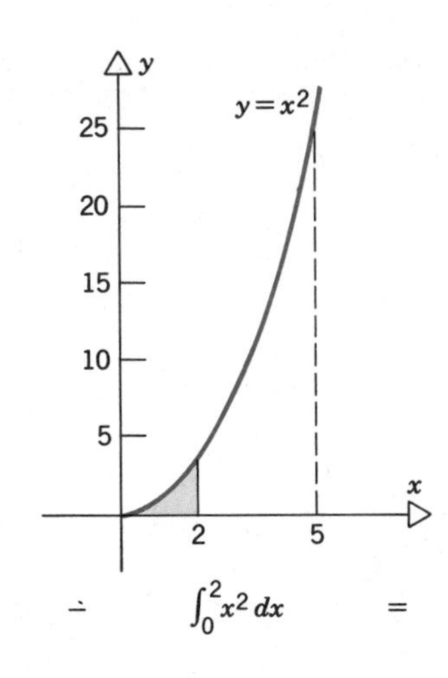

=

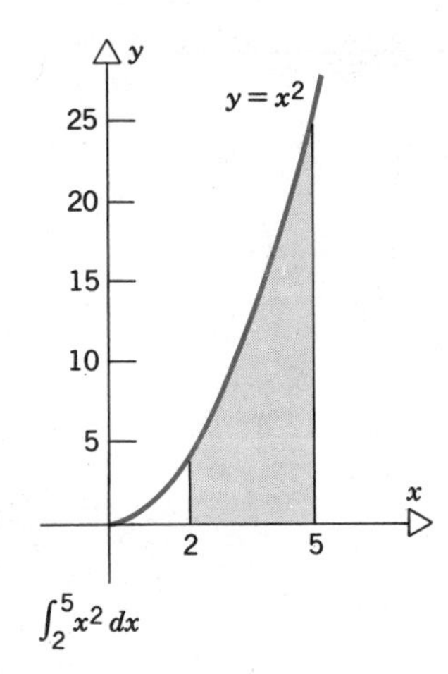

Figure 6.3.6

Observe that the procedure in Example 6 can also be used to show that

$$\int_a^b x^2\,dx = \frac{b^3 - a^3}{3}. \tag{21}$$

Theorem 6.3.4

(Comparison)

If the bounded function f is integrable on $[a, b]$, and if

$$f(x) \geq 0 \tag{22}$$

for each x in $[a, b]$, then

$$\int_a^b f(x)\,dx \geq 0. \tag{23}$$

The proof of this theorem is quite simple. Since f is integrable, the Riemann sums approach a definite limiting value as $\|\Delta\| \to 0$. Further, each Riemann sum is nonnegative because $f(x) \geq 0$, and therefore the limiting value must also be nonnegative. □

The obvious geometric interpretation of Theorem 6.3.4, if f is continuous, is that area is nonnegative. The following corollaries are immediate consequences of this theorem.

Corollary 6.3.1

If the bounded functions f and g are integrable on $[a, b]$ and if $f(x) \geq g(x)$ for each x in $[a, b]$, then

$$\int_a^b f(x)\,dx \geq \int_a^b g(x)\,dx. \tag{24}$$

Proof. We have $f(x) - g(x) \geq 0$, so by Theorem 6.3.4

$$\int_a^b [f(x) - g(x)]\, dx \geq 0.$$

Hence, by Theorem 6.3.2,

$$\int_a^b f(x)\, dx - \int_a^b g(x)\, dx \geq 0,$$

and Eq. 24 follows immediately. □

Observe that if f and g are continuous and nonnegative so that the integrals in Eq. 24 can be interpreted as areas, then this corollary expresses the comparison property of area mentioned in Section 6.1 (see Figure 6.3.7).

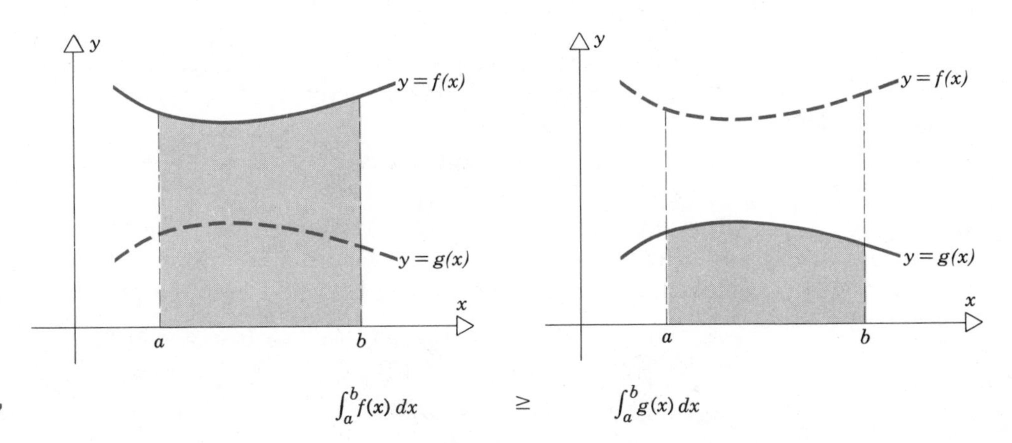

Figure 6.3.7

Corollary 6.3.2

If the bounded function f is integrable on $[a, b]$, then

$$\left| \int_a^b f(x)\, dx \right| \leq \int_a^b |f(x)|\, dx. \tag{25}$$

Proof. If f is bounded and integrable on $[a, b]$, then it can be shown (although we will not do so) that $|f|$ is also integrable on $[a, b]$. Starting from the inequalities

$$-|f(x)| \leq f(x) \leq |f(x)|,$$

it follows from Corollary 6.3.1 that

$$-\int_a^b |f(x)|\, dx \leq \int_a^b f(x)\, dx \leq \int_a^b |f(x)|\, dx.$$

Since the statements $-a \leq u \leq a$ and $|u| \leq a$ are equivalent, the desired result follows immediately. □

Corollaries 6.3.1 and 6.3.2 are often useful in practice and in theoretical discussions for establishing bounds on an integral. This is illustrated in the next example.

EXAMPLE 7

Obtain an upper bound for

$$\int_0^{\pi/2} \sin x \, dx.$$

Since $\sin x \leq 1$ for all x, we readily have

$$\int_0^{\pi/2} \sin x \, dx \leq \int_0^{\pi/2} 1 \, dx = \frac{\pi}{2} \cong 1.571.$$

A more accurate upper bound can be obtained if we use the sharper inequality $0 \leq \sin x \leq x$ on $[0, \pi/2]$; see Example 2 of Section 4.1. Then we have

$$\int_0^{\pi/2} \sin x \, dx \leq \int_0^{\pi/2} x \, dx = \frac{(\pi/2)^2}{2} \cong 1.234,$$

where we have used Eq. 2 to evaluate the integral. The bounding functions $y = x$ and $y = 1$ are shown in Figure 6.3.8. From this figure it can be seen that an even better upper bound can be found from the function

$$f(x) = \begin{cases} x, & 0 \leq x \leq 1; \\ 1, & 1 \leq x \leq \dfrac{\pi}{2}. \end{cases}$$

Using this function, we obtain

$$\int_0^{\pi/2} \sin x \, dx \leq \int_0^1 x \, dx + \int_1^{\pi/2} 1 \, dx = 0.5 + \frac{\pi}{2} - 1 \cong 1.071.$$

As we noted in Example 3 of Section 6.2, the exact value of $\int_0^{\pi/2} \sin x \, dx$ is one. ∎

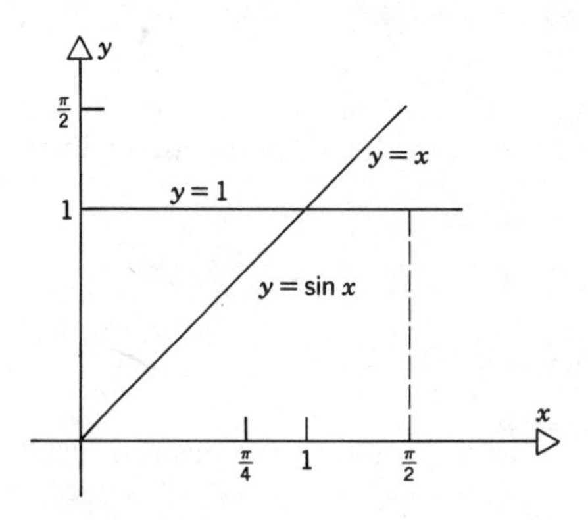

Figure 6.3.8 Estimation of $\int_0^{\pi/2} \sin x \, dx$.

Theorem 6.3.5

(Mean Value Theorem)
If the function f is continuous on $[a, b]$, then there is at least one point c in (a, b) such that

$$f(c) = \frac{1}{b-a}\int_a^b f(x)\,dx. \tag{26}$$

Proof. Since f is continuous on $[a, b]$, it has a minimum m and a maximum M on $[a, b]$,

$$m \le f(x) \le M.$$

By Eq. 9 of Section 6.2 we have

$$m(b-a) \le \int_a^b f(x)\,dx \le M(b-a),$$

or

$$m \le \frac{1}{b-a}\int_a^b f(x)\,dx \le M.$$

Since f is continuous, we know from Theorem 2.6.3 that f takes on all values between its minimum and maximum; hence there is a point c in (a, b) such that

$$f(c) = \frac{1}{b-a}\int_a^b f(x)\,dx. \quad \square$$

To interpret this result geometrically consider the case $f(x) > 0$ as depicted in Figure 6.3.9. The area under the curve is equal to the area of the rectangle of base $b - a$ and height $f(c)$. As a consequence the lightly shaded area below the line $y = f(c)$ and above the curve $y = f(x)$ must be equal to the more heavily shaded area above the line $y = f(c)$ and below the curve $y = f(x)$. Also notice that for the sketch shown in Figure 6.3.9 there are two points that could be used

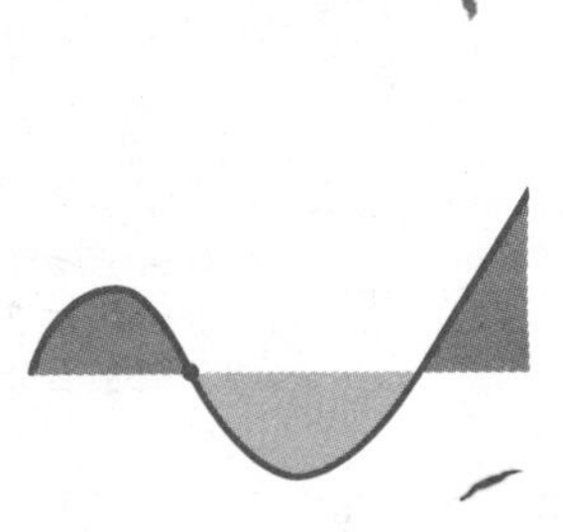

Figure 6.3.9 $\int_a^b f(x)\,dx = (b-a)f(c)$.

for c—the one shown and another point to the right. Theorem 6.3.5 does not tell us how to find the point c; what is usually important in applications of the mean value theorem, such as those given in the next section, is simply the existence of the point c, not its location.

Regardless of the sign of f, the quantity

$$\frac{1}{b-a}\int_a^b f(x)\,dx$$

is called the **mean value** (or average value) of the function f over the interval $[a, b]$. This definition is a generalization of the arithmetic mean $\bar{x}$ of a set of numbers $x_1, x_2, \ldots, x_n$; that is, $\bar{x} = (x_1 + x_2 + \cdots + x_n)/n$. If f is continuous, then Theorem 6.3.5 says that f assumes its average value at least once in the interval. For example, if v is the velocity of a particle moving on a straight line, then

$$\frac{1}{T}\int_0^T v(t)\,dt$$

is the average velocity over the time interval $[0, T]$. If v is continuous, then at some instant of time the particle must have a velocity equal to its average velocity.

PROBLEMS

In each of Problems 1 through 10, use the properties of the integral and the geometric interpretation of the integral of a nonnegative continuous function as an area to evaluate the given integral.

1. $\int_1^3 (3x - 2)\,dx$

2. $\int_{-1}^1 (6 - 3x)\,dx$

3. $\int_{-2}^3 |x|\,dx$

4. $\int_{-2}^2 (2 + \sqrt{4 - x^2})\,dx$

5. $\int_{-1}^2 f(x)\,dx,$

$$f(x) = \begin{cases} \sqrt{1 - x^2}, & -1 \le x \le 0 \\ 1, & 0 < x \le 1 \\ 2x - 1, & 1 < x \le 2 \end{cases}$$

6. $\int_0^4 g(x)\,dx,$

$$g(x) = \begin{cases} x, & 0 \le x < 1 \\ 1 + \sqrt{1 - (x - 2)^2}, & 1 \le x \le 3 \\ -x + 4, & 3 < x \le 4 \end{cases}$$

7. $\int_1^{-1} (1 - \sqrt{1 - x^2})\,dx$

8. $\int_{-1}^1 (2 + \sqrt{1 - x^2})\,dx$

9. $\int_0^3 (2 - |x - 1|)\,dx$

10. $\int_0^2 \sqrt{2x - x^2}\,dx$

In each of Problems 11 through 16, use Eqs. 1, 2, 8, and 21 to evaluate the given integral.

11. $\int_{-3}^6 (2 - 3x)\,dx$

12. $\int_{-1}^2 (2 - 3x + 4x^2)\,dx$

13. $\int_0^1 (2x + 1)^2\,dx$

14. $\int_{-1}^1 (1 - x)^2\,dx$

15. $\int_3^1 (2x^2 + 3x)\,dx$

16. $\int_{-2}^3 f(x)\,dx,$

$$f(x) = \begin{cases} 0, & x = -2 \\ 2x^2, & -2 < x \le 1 \\ 1 - 2x, & 1 < x \le 3 \end{cases}$$

17. If the function f is integrable on $[a, b]$ and if $-M \le f(x) \le M$, show that

$$\left|\int_a^b f(x)\,dx\right| \le \int_a^b |f(x)|\,dx \le M(b - a).$$

18. If the function f is integrable on $[a, b]$ and if $|f(x)| \leq M$, show that

$$\left| \int_c^{c+h} f(x)\, dx \right| \leq M|h|,$$

where c and $c + h$ are in $[a, b]$. Note that h may be negative. This result shows that the absolute value of an integral is always less than or equal to a bound on the absolute value of the integrand times the length of the interval of integration.

In each of Problems 19 through 22, derive the given inequality.

19. $\dfrac{2}{5} \leq \displaystyle\int_0^2 \frac{1}{1 + x^2}\, dx \leq 2$

20. $\displaystyle\int_\pi^{3\pi/2} 2 \left| \sin \frac{x}{10} \right| dx \leq \pi$

21. $\displaystyle\int_0^1 x^4\, dx \geq \int_0^1 x^6\, dx$

22. $2\sqrt{2} \leq \displaystyle\int_1^3 \sqrt{1 + x}\, dx \leq 4$

© 23. In this problem we use the fact that

$$\int_0^1 \sqrt{1 - x^2}\, dx = \frac{\pi}{4}$$

to approximate the value of π.

(a) Partition the interval $[0, 1]$ into n equal subintervals. Find the lower sum s_n and the upper sum σ_n for $n = 10, 20, 40$, and 80.

(b) Determine $(s_n + \sigma_n)/2$ for $n = 10, 20, 40$, and 80.

(c) Choosing the star points as the midpoint of each subinterval, find the corresponding Riemann sum for $n = 10, 20, 40$, and 80.

24. What is the average value of the linear function $f(x) = mx + c$ over the interval $[a, b]$? At what point or points does f take on its average value?

25. The velocity of a particle is given by $v(t) = 2 + 3t^2$ in ft/sec for $0 \leq t \leq 10$. What is the average velocity for the time interval $[2, 5]$?

26. A car travels 25 mi/hr for one hour and then 40 mi/hr for one hour. Draw simple graphs of the instantaneous velocity as a function of time and as a function of distance traveled.

(a) What is the average velocity with respect to time?

(b) What is the average velocity with respect to distance?

(c) Why is the average velocity with respect to distance greater than the average velocity with respect to time?

This problem illustrates that in computing the average of a quantity the independent variable must be specified.

27. If the function f is continuous on $[a, b]$ and if $\int_a^b f(x)\, dx = 0$, show that there is a point c in $[a, b]$ such that $f(c) = 0$.

28. **A generalized mean value theorem.** Suppose that f and g are continuous on $[a, b]$, $g(x) \geq 0$ on $[a, b]$, and

$$\int_a^b g(x)dx > 0.$$

Show that there is a point c in (a, b) such that

$$\int_a^b f(x)g(x)\, dx = f(c) \int_a^b g(x)\, dx.$$

This is a more general form of Theorem 6.3.5.

6.4 THE FUNDAMENTAL THEOREM OF CALCULUS

So far we have discussed the processes of differentiation and integration independently of each other. Now we turn to the close relation between them, a result known as the fundamental theorem of calculus. Among other things, this relation enables us to evaluate many integrals quickly and exactly by means of antiderivatives, rather than resorting to one or another process of numerical approximation.

Suppose that the function f is integrable on $[a, b]$ and hence on the subinterval $[a, x]$ for any x in $[a, b]$. Let

$$F(x) = \int_a^x f(s)\, ds, \qquad a \leq x \leq b. \tag{1}$$

Thus the function F is the integral of f from a to the variable upper limit x. Observe that there are two different variables in Eq. 1, denoted by x and s. The independent variable is x, since for each value of x, Eq. 1 defines the corresponding value of $F(x)$. The other variable is the dummy integration variable. To avoid confusion it should be denoted by some letter other than x; here we have used s. If f is continuous and nonnegative, then $F(x)$ is the area of the region in the sy-plane bounded by the lines $s = a$ and $s = x$, the s-axis, and the graph of $y = f(s)$. This region is shown in Figure 6.4.1.

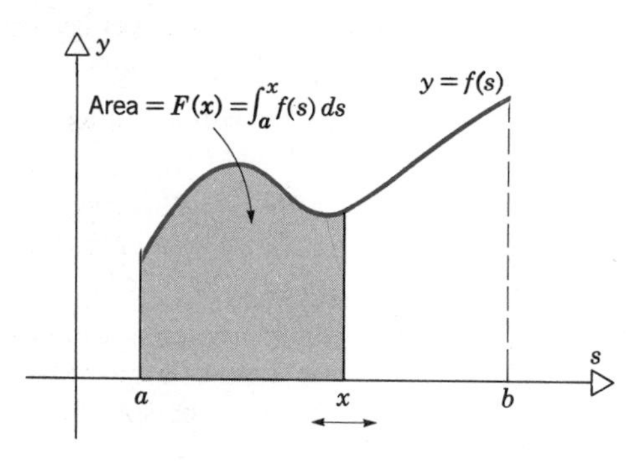

Figure 6.4.1

EXAMPLE 1

The function $f(x) = x^2$ is integrable on any bounded interval. By Eq. 21 of Section 6.3 with b replaced by x, we have

$$F(x) = \int_a^x s^2\, ds = \frac{x^3 - a^3}{3}.$$

For instance, if $a = -1$, then

$$F(x) = \frac{x^3 + 1}{3}. \quad ■$$

We now explore some important properties of the function F.

Theorem 6.4.1

Let the function f be bounded and integrable on $[a, b]$. Then the function

$$F(x) = \int_a^x f(s)\, ds, \qquad a \le x \le b,$$

is continuous on $[a, b]$.

Proof. We must show that if x is any point in $[a, b]$, then $\lim_{h \to 0} F(x + h) = F(x)$. We have

$$F(x + h) - F(x) = \int_a^{x+h} f(s)\, ds - \int_a^x f(s)\, ds.$$

Hence, according to Theorem 6.3.3,

$$F(x + h) - F(x) = \int_x^{x+h} f(s)\, ds. \tag{2}$$

This relation is illustrated in Figure 6.4.2. The conclusion now seems obvious geometrically. As $h \to 0$, the width of the shaded region in Figure 6.4.2 shrinks

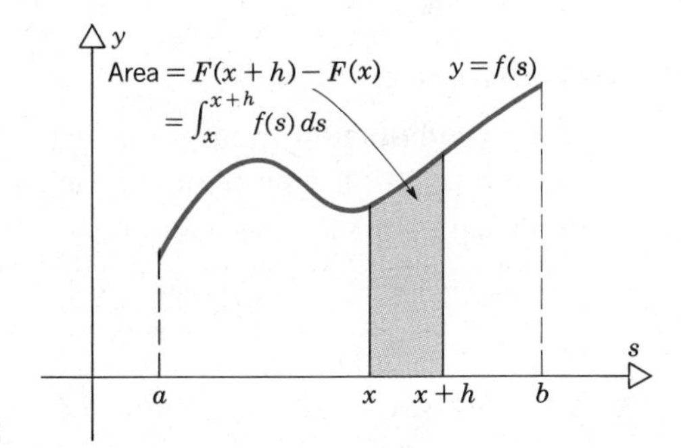

Figure 6.4.2

and its area approaches zero, which is the desired conclusion. To show this analytically, we can proceed as follows. Since f is bounded on $[a, b]$, there exists a constant M such than $|f(x)| \leq M$ for all x in $[a, b]$. For $h > 0$ it follows from Corollary 6.3.2 that

$$\begin{aligned} |F(x + h) - F(x)| &= \left| \int_x^{x+h} f(s)\, ds \right| \\ &\leq \int_x^{x+h} |f(s)|\, ds \leq \int_x^{x+h} M\, ds = Mh; \end{aligned} \tag{3a}$$

similarly, for $h < 0$,

$$\begin{aligned} |F(x + h) - F(x)| &= \left| - \int_{x+h}^{x} f(s)\, ds \right| \\ &= \left| \int_{x+h}^{x} f(s)\, ds \right| \\ &\leq \int_{x+h}^{x} |f(s)|\, ds \leq \int_{x+h}^{x} M\, ds = M(-h). \end{aligned} \tag{3b}$$

Combining these results we obtain

$$|F(x + h) - F(x)| \leq M|h|. \tag{4}$$

Consequently, as $h \to 0$, it follows from Eq. 4 that $F(x + h) - F(x) \to 0$, so the function F is continuous at x.

It is understood that if x is the point a, then necessarily $h > 0$, so that the limit is from the right; similarly if x is the point b, then necessarily $h < 0$, so that the limit is from the left. □

Note that Theorem 6.4.1 states that integration is a *smoothing* process: even though the integrand f has discontinuities, the integral F does not. This is illustrated by Example 2 later in this section.

The next theorem is the first form of the fundamental theorem of calculus.

Theorem 6.4.2

(First fundamental theorem of calculus)
Let the function f be bounded on $[a, b]$ and continuous on (a, b). Then the function

$$F(x) = \int_a^x f(s)\,ds, \qquad a \leq x \leq b,$$

has a derivative at each point in (a, b), and moreover

$$F'(x) = f(x), \qquad a < x < b. \tag{5}$$

This theorem states that integration of a continuous function f to a variable upper limit produces a function F that is an antiderivative of the integrand. We can also rewrite Eq. 5 in the form

$$\frac{d}{dx}\int_a^x f(x)\,ds = f(x), \tag{6}$$

which makes it clear that *differentiation is the inverse of integration*. That is, if we start with a continuous function f, integrate, and then differentiate, we return to the original function.

Proof. To prove this theorem we use the definition of the derivative as the limit of a difference quotient. Let x be any point in (a, b) and form

$$\begin{aligned} F(x + h) - F(x) &= \int_a^{x+h} f(s)\,ds - \int_a^x f(s)\,ds \\ &= \int_x^{x+h} f(s)\,ds. \end{aligned} \tag{7}$$

We now need a sharper estimate of this integral than in the proof of Theorem 6.4.1. It is understood that h is restricted to be sufficiently small so that $x + h$ is also in (a, b). Since f is continuous on (a, b), it is continuous on the interval x to $x + h$, so we can use the mean value theorem for integrals (Theorem 6.3.5) to write Eq. (7) in the form

$$F(x + h) - F(x) = f(c)\,(x + h - x) = hf(c),$$

where c is a point between x and $x + h$ (see Figure 6.4.3). Then

$$\lim_{h \to 0} \frac{F(x + h) - F(x)}{h} = \lim_{h \to 0} \frac{hf(c)}{h} = \lim_{h \to 0} f(c).$$

Clearly $c \to x$ as $h \to 0$. Since f is continuous, the limit of $f(c)$ as h approaches zero exists and is equal to $f(x)$. Thus $F'(x) = f(x)$, for each x in (a, b). □

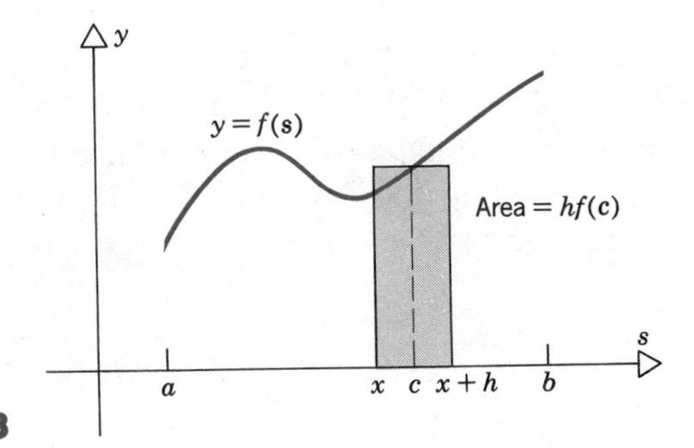

Figure 6.4.3

EXAMPLE 2

Consider the function

$$f(s) = \begin{cases} 1, & 0 \le s \le 1, \\ 2, & 1 < s \le 2, \end{cases}$$

whose graph is shown in Figure 6.4.4*a*.

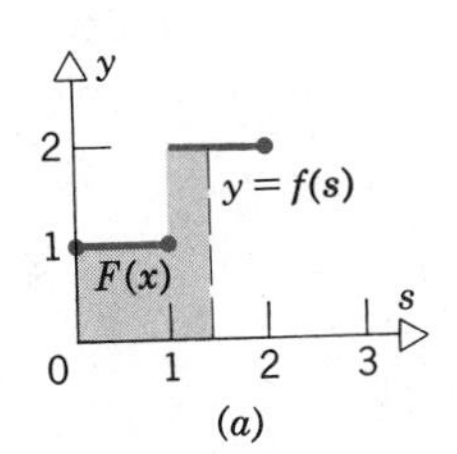

Figure 6.4.4

Let us evaluate

$$F(x) = \int_0^x f(s)\,ds, \qquad 0 \le x \le 2.$$

Using Eq. 1 of Section 6.3, we have for $0 \le x \le 1$

$$\int_0^x f(s)\,ds = \int_0^x 1\,ds = x;$$

for $1 < x \le 2$

$$\int_0^x f(s)\,ds = \int_0^1 1\,ds + \int_1^x 2\,ds = 1 + 2(x - 1)$$

$$= 2x - 1.$$

Hence

$$F(x) = \begin{cases} x, & 0 \le x \le 1 \\ 2x - 1, & 1 < x \le 2. \end{cases}$$

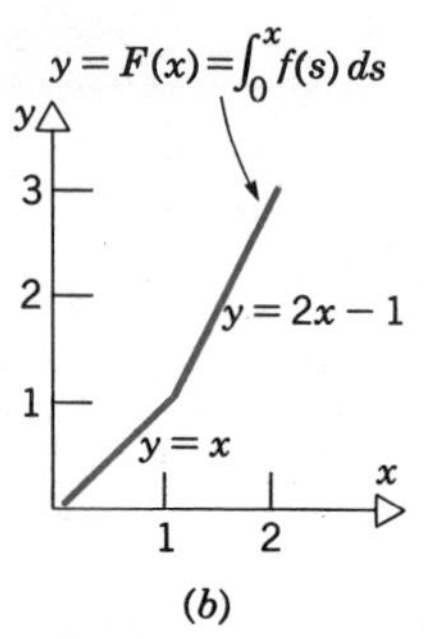

(b)

Figure 6.4.4

Notice that even though f is discontinuous at 1 the function F is continuous there as Theorem 6.4.1 states. Moreover, in $0 < x < 1$ we have $F'(x) = 1 = f(x)$, and in $1 < x < 2$ we have $F'(x) = 2 = f(x)$. It is clear from Figure 6.4.4*b* that F does not have a derivative at 1 and, of course, this is a point of discontinuity of f. ∎

Now we turn to the second form of the fundamental relation between differentiation and integration.

Theorem 6.4.3

(Second fundamental theorem of calculus)
Let the function f be bounded on $[a, b]$ and continuous on (a, b). Let G be any function that is continuous on $[a, b]$ and such that $G'(x) = f(x)$ on (a, b). Then

$$\int_a^b f(x)\,dx = G(b) - G(a). \tag{8}$$

Proof. Recall that according to Theorems 6.4.1 and 6.4.2 the function F, defined by

$$F(x) = \int_a^x f(s)\,ds, \tag{9}$$

is continuous on $[a, b]$, differentiable on (a, b), and $F'(x) = f(x)$ on (a, b). Thus both F and G are antiderivatives of f on (a, b). Consequently, by Theorem 4.1.5, they can differ at most by a constant, or

$$F(x) = G(x) + c. \tag{10}$$

If we set $x = a$ and note that $F(a) = 0$, it follows that $c = -G(a)$, and hence that

$$F(x) = G(x) - G(a). \tag{11}$$

Substituting this result for $F(x)$ in Eq. 9, we obtain

$$G(x) - G(a) = \int_a^x f(s)\,ds. \tag{12}$$

Finally, setting $x = b$ in Eq. 12, we obtain Eq. 8. □

Since $f(s) = G'(s)$, Eq. 12 can also be written as

$$\int_a^x G'(s)\,ds = G(x) - G(a), \tag{13}$$

provided that G' is bounded on $[a, b]$ and continuous on (a, b). Equation 13 states that, under the given hypotheses and up to an additive constant, *integration is the inverse of differentiation*. That is, if we start with the function G, differentiate, and then integrate the result, we obtain again the original function G, augmented at most by a constant.

Under the conditions of Theorem 6.4.3 the integral $\int_a^b f(x)\,dx$ can be evaluated if we can find a function G that is continuous on $[a, b]$ and such that $G'(x) = f(x)$ on (a, b). Modifying slightly the definition in Section 3.3, we will call such a function G an antiderivative of f.

The term **indefinite integral** is also frequently used to refer to G, and we will henceforth use the words "indefinite integral" and "antiderivative" interchangeably. The notation

$$G(x) = \int f(x)\,dx + c, \tag{14}$$

where the integral sign appears without limits of integration, is commonly used to denote the indefinite integrals or antiderivatives of f. The expression $\int f(x)\,dx$ stands for some particular representative of the class of indefinite integrals or antiderivatives of f; by adding the arbitrary constant c, we include all members of this class of functions. Thus the symbol $\int$, used without limits of integration, has the same meaning as the operator D^{-1} introduced in Section 3.3.

Where it is necessary for clarity, the integral $\int_a^b f(x)\,dx$, with numerical limits of integration, is called a **definite integral.** However, the modifiers "indefinite" and "definite" are often omitted when it is clear from the context which one is intended, so the word "integral" alone may mean either a definite or indefinite integral. Remember that the former is a number, while the latter is a function.

In Sections 3.2 and 3.3 we found that

$$D(x^r) = rx^{r-1}, \tag{15}$$

and that

$$D^{-1}x^r = \frac{x^{r+1}}{r+1} + c, \qquad r \neq -1, \tag{16}$$

where c is an arbitrary constant. Using the notation of Eq. 14 for indefinite integrals, we can express Eq. 16 as

$$\int x^r\,dx = \frac{x^{r+1}}{r+1} + c, \qquad r \neq -1. \tag{17}$$

In the same way, in Section 3.6 we found that

$$D \sin x = \cos x, \qquad D \cos x = -\sin x, \tag{18}$$

and that

$$D^{-1} \cos x = \sin x + c, \qquad D^{-1} \sin x = -\cos x + c. \tag{19}$$

From now on we will usually write Eq. 19 as

$$\int \cos x\,dx = \sin x + c, \qquad \int \sin x\,dx = -\cos x + c. \tag{20}$$

Another bit of notation that is commonly used is

$$G(x)\Big|_a^b \quad \text{instead of} \quad G(b) - G(a).$$

Thus Eq. 13 can be written

$$\int_a^b f(x)\,dx = G(x)\Big|_a^b, \tag{21}$$

where G is an antiderivative of f.

Notice that, in using Eq. 13 or Eq. 21 to evaluate a definite integral, it is not necessary to include the arbitrary constant c in $G(x)$. If it is included, it will be canceled in the difference $G(b) - G(a)$.

We now consider a few examples of the use of Theorem 6.4.3.

EXAMPLE 3

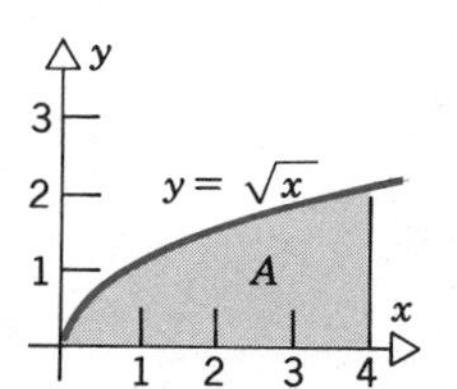

Figure 6.4.5

Determine the area of the region bounded by the x-axis, the line $x = 4$, and the curve $y = \sqrt{x}$.

The desired area is given by

$$A = \int_0^4 \sqrt{x}\,dx$$

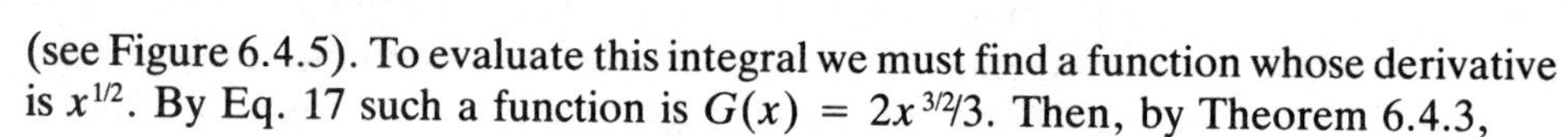

(see Figure 6.4.5). To evaluate this integral we must find a function whose derivative is $x^{1/2}$. By Eq. 17 such a function is $G(x) = 2x^{3/2}/3$. Then, by Theorem 6.4.3,

$$A = \tfrac{2}{3}(4)^{3/2} - \tfrac{2}{3}(0)^{3/2} = \tfrac{16}{3}. \blacksquare$$

EXAMPLE 4

Evaluate

$$\int_0^{\pi/2} (2x - 3\sin x)\,dx.$$

Using Theorem 6.3.2, we can write the given integral as

$$\int_0^{\pi/2} (2x - 3\sin x)\,dx = 2\int_0^{\pi/2} x\,dx - 3\int_0^{\pi/2} \sin x\,dx. \tag{22}$$

The integrals on the right side of Eq. 22 can be evaluated by using Eqs. 17 and 20, respectively, with the result

$$\int_0^{\pi/2} (2x - 3\sin x\,dx = 2\frac{x^2}{2}\Big|_0^{\pi/2} - 3(-\cos x)\Big|_0^{\pi/2}$$

$$= 2\left[\frac{\pi^2}{8} - 0\right] - 3[0 - (-1)] = \frac{\pi^2}{4} - 3 \cong -0.5326. \blacksquare$$

Discontinuous integrands

Theorem 6.4.3 applies to integrands that are continuous on the interval of integration except possibly at the endpoints. However, it is equally easy to evaluate integrals whose integrands have a finite number of discontinuities in the interior of the interval of integration. For instance, suppose that f is discontinuous at c, where $a < c < b$. Then, by Theorem 6.3.3,

$$\int_a^b f(x)\,dx = \int_a^c f(x)\,dx + \int_c^b f(x)\,dx, \tag{23}$$

and each of the integrals on the right side of Eq. 23 can be evaluated by Theorem 6.4.3, provided that the appropriate antiderivatives can be found. Similarly, if f is discontinuous at $c_1, c_2, \ldots, c_n$, where $a < c_1 < c_2 < \cdots < c_n < b$, then

$$\int_a^b f(x)\,dx = \int_a^{c_1} f(x)\,dx + \int_{c_1}^{c_2} f(x)\,dx + \cdots + \int_{c_n}^b f(x)\,dx, \tag{24}$$

and Theorem 6.4.3 can be applied to each term on the right. We have already illustrated this procedure in Example 2. Here is another example.

EXAMPLE 5

Evaluate $\int_{-1}^{1} f(x)\,dx$, where

$$f(x) = \begin{cases} x+1, & -1 \le x < 0; \\ \frac{1}{2}, & x = 0; \\ x^2, & 0 < x \le 1. \end{cases} \tag{25}$$

The graph of f is shown in Figure 6.4.6.

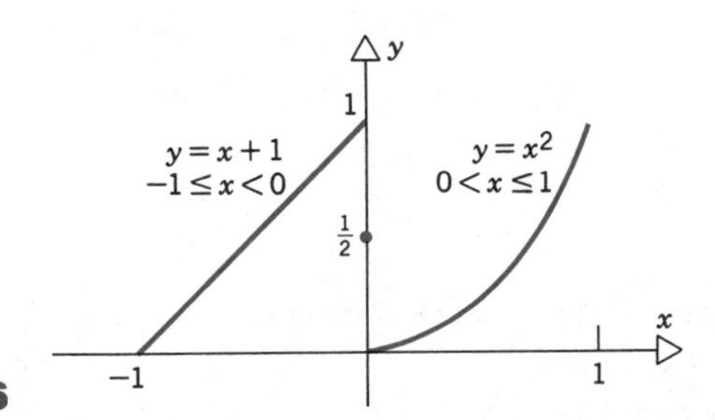

Figure 6.4.6

The function f is bounded on $[-1, 1]$ and continuous except at $x = 0$, so the integral exists. To evaluate it we write

$$\begin{aligned} \int_{-1}^{1} f(x)\,dx &= \int_{-1}^{0} f(x)\,dx + \int_0^1 f(x)\,dx \\ &= \int_{-1}^{0} (x+1)\,dx + \int_0^1 x^2\,dx. \end{aligned} \tag{26}$$

Theorem 6.4.3 applies to each of the latter integrals, and

$$\int_{-1}^{1} f(x)\,dx = \left[\frac{x^2}{2} + x\right]\Bigg|_{-1}^{0} + \frac{x^3}{3}\Bigg|_{0}^{1}$$

$$= 0 - (-\tfrac{1}{2}) + \tfrac{1}{3} - 0 = \tfrac{5}{6}. \blacksquare$$

PROBLEMS

In each of Problems 1 through 18, evaluate the given integral.

1. $\int_0^2 3x^4\,dx$

2. $\int_{-4}^{-1} \frac{3}{x^2}\,dx$

3. $\int_0^{\pi/2} (2 - \cos x)\,dx$

4. $\int_0^4 (x^2 + x^{1/2})\,dx$

5. $\int_{-1}^{1} s(2 + s^2)\,ds$

6. $\int_0^4 (2t - 1)(t + 1)\,dt$

7. $\int_0^{\pi/4} (3 + 6s + 2\sin s)\,ds$

8. $\int_{-2}^{-1} \frac{2 - 3x^2 + 6x^4}{2x^2}\,dx$

9. $\int_{-\pi/2}^{\pi} (3\cos x - \tfrac{1}{3}\sin x)\,dx$

10. $\int_1^4 (2x^{-1/2} - \tfrac{1}{2}x^{3/2})\,dx$

11. $\int_0^x (2t + t^{1/2})\,dt, \qquad x > 0$

12. $\int_{-1}^{x} (2t^2 + 4t^{1/3})\,dt$

13. $\int_0^x (3 - \cos t)\,dt$

14. $\int_0^x (2 + 5\sin t)\,dt$

15. $\int_0^{x^2} (1 + 2t)\,dt$

16. $\int_1^{3x} (2 - 3t + t^2)\,dt$

17. $\int_{2x}^{x^2} (3 - 4t)\,dt$

18. $\int_{x^2}^{x^3} \cos t\,dt$

In each of Problems 19 through 26, evaluate the given integral.

19. $\int_0^2 f(x)\,dx,$

$$f(x) = \begin{cases} 2 + x, & 0 \le x \le 1 \\ 2 - x, & 1 < x \le 2 \end{cases}$$

20. $\int_{-\pi/2}^{\pi/2} g(t)\,dt,$

$$g(t) = \begin{cases} \cos t, & -\dfrac{\pi}{2} \le t < 0 \\ t, & 0 \le t \le \dfrac{\pi}{2} \end{cases}$$

21. $\int_{-10}^{10} f(x)\,dx,$

$$f(x) = \begin{cases} 0, & x = \pm n, \\ x, & x \ne \pm n \end{cases} \qquad n = 0, 1, 2, \ldots, 10$$

22. $\int_{\pi/2}^{3\pi/2} h(s)\,ds,$

$$h(s) = \begin{cases} \tfrac{1}{2}\cos s, & \dfrac{\pi}{2} \le s \le \pi \\ 2\sin s, & \pi < s \le \dfrac{3\pi}{2} \end{cases}$$

23. $\int_{-1}^{1} |2x + 1|\,dx$

24. $\int_{-1}^{2} |x(x - 1)|\,dx$

25. $\int_{-2}^{2} |1 - x^2|\,dx$

26. $\int_{-1}^{2} (|x| + 3|x - 1|)\,dx$

27. Show that if p and q are positive integers then

$$\int_0^1 (x^{p/q} + x^{q/p})\,dx = 1.$$

28. Determine the average value of the function $f(x) = 2x^{1/3} + 3x$ over the interval [1, 3].

29. (a) The velocity of a particle moving on a straight line is given by $v = A\sin\omega t$ where A and ω are constants. Observe that the velocity is periodic with period $2\pi/\omega$ and determine the average velocity over one period.

(b) If the velocity is $v = A\sin^2\omega t$, determine the average velocity over one period. *Hint:* Use a trigonometric identity to write $\sin^2\omega t$ as the sum of two functions that you know how to integrate.

30. The gravitational force acting on a particle is $-k/r^2$, where k is a positive constant and r is the distance from the center of the earth. The minus sign indicates that the force is directed toward the earth (see Figure 6.4.7). What is the work w done by the attractive force in moving the particle from $r = r_A$ to $r = r_B$?

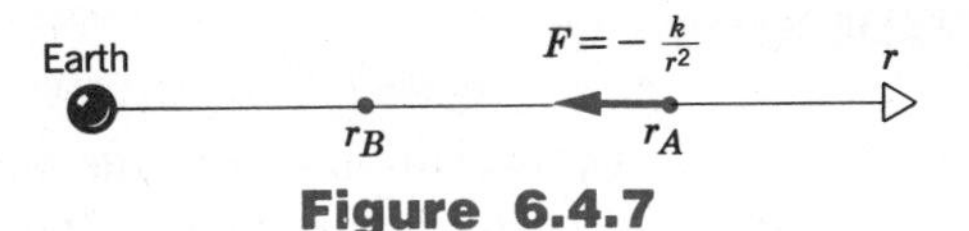

Figure 6.4.7

31. (a) Sketch the function

$$f(x) = \begin{cases} 3 - x, & 0 \le x < 1; \\ 2 + x, & 1 \le x \le 2. \end{cases}$$

(b) Find the function

$$F(x) = \int_0^x f(t)\,dt, \qquad 0 \le x \le 2,$$

and sketch its graph.

(c) Describe the behavior of f and F in the neighborhood of $x = 1$.

32. Show that

$$\int_0^x |t|\,dt = \tfrac{1}{2}x|x|, \qquad \text{for all } x.$$

Differentiation of Integrals

Problems 33 through 44 provide applications and, in some cases, extensions of Theorem 6.4.2.

33. If

$$F(x) = \int_0^x \frac{\sin 2t}{1 + t^2}\,dt,$$

determine $F'(x)$.

34. If

$$F(x) = \int_x^0 \frac{\sin 2t}{1 + t^2}\,dt,$$

determine $F'(x)$.

35. If

$$F(x) = 2x + \int_0^x \frac{\sin 2t}{1 + t^2}\,dt,$$

determine (a) $F(0)$, (b) $F'(0)$, (c) $F''(0)$.

36. If

$$F(x) = 2 + x + \int_0^{x^2} \frac{\sin 2t}{1 + t^2}\,dt,$$

determine $F'(x)$.

Hint: Let $u = x^2$ and use the chain rule in computing the derivative of the given integral.

37. If

$$F(x) = \int_{1-x^2}^{1+x^2} \frac{\sin 2t}{1 + t^2}\,dt, \qquad 0 \le x < 10,$$

determine $F'(x)$.

Hint: Write the integral as the sum of two integrals, from $1 - x^2$ to 1 and from 1 to $1 + x^2$ respectively.

38. If

$$G(x) = \int_x^{x^2} \frac{2}{(1 + t^2)^2}\,dt$$

for $x > 1$, find $G'(x)$.

39. If

$$F(x) = \int_0^{1+x^2} t\,f(t)\,dt$$

for $x > 0$, find $F'(x)$.

40. Assume that the function f is continuous and that the functions u and v are differentiable. Show that

(a) $$\frac{d}{dx}\int_a^{u(x)} f(t)\,dt = f[u(x)]u'(x).$$

(b) $$\frac{d}{dx}\int_{v(x)}^{u(x)} f(t)\,dt = f[u(x)]u'(x) - f[v(x)]v'(x).$$

41. If

$$F(x) = \int_0^x \left(t \int_1^t f(s)\,ds\right) dt$$

where the function f is continuous, determine (a) $F'(x)$, (b) $F'(1)$, (c) $F''(x)$, (d) $F''(1)$.

In each of Problems 42 through 44, assume that the function f is continuous and that the equation is true for all $x \ge 0$.

42. If

$$\int_0^x f(t)\,dt = \sqrt{1 + x^2} - 1,$$

determine $f(1)$.

43. If

$$\int_0^{x^2} f(t)\,dt = \sqrt{1 + x^2} - 1,$$

determine $f(\pi/2)$.

44. If

$$\int_0^x tf(t)\,dt = x^2 + x\sin x + \cos x - 1,$$

determine $f(\pi/2)$.

6.5 INTEGRATION BY SUBSTITUTIONS

We have seen that the chain rule is perhaps the most powerful and versatile single tool for the calculation of derivatives. The counterpart of the chain rule for integrals is the use of substitutions, or changes of the integration variable. This method is similarly powerful and general, and is adaptable to a wide variety of definite and indefinite integrals. In Sections 3.5 and 3.6 we have already given some antidifferentiation formulas derived from the chain rule. The three most important are Eqs. 16 of Section 3.5 and Eqs. 27 and 28 of Section 3.6. We rewrite them here in integral notation.

$$\int [u(x)]^r u'(x)\,dx = \frac{[u(x)]^{r+1}}{r+1} + c, \qquad r \neq -1, \tag{1}$$

$$\int [\cos u(x)]u'(x)\,dx = \sin u(x) + c, \tag{2}$$

$$\int [\sin u(x)]u'(x)\,dx = -\cos u(x) + c. \tag{3}$$

In this section we approach the problem of finding antiderivatives from a somewhat different point of view, and also extend the process to definite integrals.

EXAMPLE 1

Evaluate

$$\int 3x\cos x^2\,dx. \tag{4}$$

Suppose that we introduce the new variable $u = x^2$; then $du = 2x\,dx$, or $x\,dx = du/2$. Substituting for x^2 and for $x\,dx$ in the integral (4), we have

$$\int 3x\cos x^2\,dx = \int 3\cos u\,\frac{du}{2} = \frac{3}{2}\int \cos u\,du. \tag{5}$$

An indefinite integral, or antiderivative, of $\cos u$ is $\sin u$, so we obtain

$$\int 3x\cos x^2\,dx = \tfrac{3}{2}\sin u + c, \tag{6}$$

where c is an arbitrary constant. Finally, we substitute x^2 for u in Eq. 6, which gives

$$\int 3x\cos x^2\,dx = \tfrac{3}{2}\sin x^2 + c. \blacksquare \tag{7}$$

EXAMPLE 2

Evaluate

$$\int 2x\sqrt{x^2 + 1}\, dx. \tag{8}$$

Here we introduce the new variable $u = x^2 + 1$, so $du = 2x\, dx$, and the integral (8) becomes

$$\int 2x\sqrt{x^2 + 1}\, dx = \int u^{1/2}\, du. \tag{9}$$

Then we have

$$\begin{aligned} \int 2x\sqrt{x^2 + 1}\, dx &= \tfrac{2}{3}u^{3/2} + c \\ &= \tfrac{2}{3}(x^2 + 1)^{3/2} + c. \blacksquare \end{aligned} \tag{10}$$

EXAMPLE 3

Evaluate

$$\int \sin^8 x \cos x\, dx. \tag{11}$$

In this case we can let $u = \sin x$. Then $du = \cos x\, dx$, and by substituting into the integral (11) we obtain

$$\begin{aligned} \int \sin^8 x \cos x\, dx &= \int u^8\, du \\ &= \tfrac{1}{9}u^9 + c \\ &= \tfrac{1}{9}\sin^9 x + c. \blacksquare \end{aligned} \tag{12}$$

In each of Examples 1, 2, and 3 one can verify the correctness of the result by differentiating it (via the chain rule) and showing that the original integrand is obtained. In each of these examples it is also possible, perhaps after a little trial and error, to discover the required antiderivative more or less by inspection. However, in all three cases, the substitution immediately reduces the problem to an elementary integral. This illustrates the usual goal of a substitution, namely, to simplify the integral, and if possible, to reduce it to one solvable by an elementary integration formula.

Now let us look at substitutions and their relation to the chain rule from a more general point of view than in Examples 1, 2, and 3. Assume that we are faced with the problem of determining an indefinite integral $\int g(x)\, dx$, and that the required function is not immediately obvious. Further, assume that we can recognize a function $u = u(x)$ such that

$$g(x) = f[u(x)]u'(x) \tag{13}$$

for some function f. Then, if F is an antiderivative of f, it follows from the chain rule that

$$\frac{d}{dx} F[u(x)] = f[u(x)]u'(x). \tag{14}$$

Hence $F[u(x)]$ is the desired antiderivative of g, and

$$\begin{aligned}\int g(x)\,dx &= \int f[u(x)]u'(x)\,dx = \int \frac{d}{dx} F[u(x)]\,dx \\ &= F[u(x)] + c.\end{aligned} \tag{15}$$

Observe that (as in Examples 1, 2, and 3) the use of the Leibniz notation for the differential makes the procedure expressed by Eq. 15 appear even simpler. Recall that if $u = u(x)$, then $du = u'(x)\,dx$, and instead of Eq. 15 we can write

$$\begin{aligned}\int g(x)\,dx &= \int f(u)\,du = \int F'(u)\,du \\ &= F(u) + c = F[u(x)] + c.\end{aligned} \tag{16}$$

In using substitutions to find indefinite integrals, some questions naturally arise, such as the following.

1. For a given integral, how can we be sure that a substitution will be effective?
2. Assuming that a substitution might be helpful, how can we find the right substitution in each particular case?

As we proceed with the study of integration we will find that certain types of integrals lend themselves to substitutions of particular kinds. However, for the present we answer the first question by saying that one cannot always foresee whether a substitution will be useful or not. Nevertheless, this is a widely applicable approach and should certainly be kept in mind for any integral whose antiderivative is not immediately evident. As for what substitution to use, in general it must be indicated by the way in which the integration variable appears in the integrand. Sometimes the best substitution is not obvious. In such a case, it may be a good idea simply to try some substitution that appears plausible. If it works, fine; even if it does not, perhaps by its failure it will suggest something else that will be better.

Thus the procedure is to choose a u that looks reasonable, then make the substitution to determine the function f consistent with Eq. 13. If a function F such that $F'(u) = f(u)$ is known, then we can complete the integration. If no antiderivative for f is evident, then we consider whether we can make a better choice for u. This is the essence of the method of change of variable.

A natural reaction is to feel that if it is difficult (or impossible) to recognize the desired indefinite integral by inspection, then it must be equally difficult (or impossible) to find the function u of Eq. 13. Fortunately, experience shows that this is not the case. The choice for the function u must be suggested by the form of $g(x)$, our knowledge of functions f for which we can recognize antiderivatives, and our analytical skill. Practice is a great help.

Note that in the preceding discussion we have always said "introduce a change of variable $u = u(x)$." It is equally correct to say "introduce a change of variable

$x = x(u)$." Implicit in the procedure is that we must be able to go from the variable x to the variable u and from the variable u to the variable x. However the new variable u is introduced, the object is to transform $g(x)\,dx$ into $f(u)\,du$ in such a way that we know an antiderivative for f. Sometimes this can be accomplished in more than one way, as the following example shows.

EXAMPLE 4

Evaluate

$$\int x^2\sqrt{x+1}\,dx. \tag{17}$$

The difficulty in evaluating this integral is the presence of the square root of a sum, $\sqrt{x+1}$. However, if the integrand were of the form $\sqrt{x}$ times a polynomial, there would be no difficulty. Thus, we try

$$u = x + 1 \quad \text{so} \quad du = dx, \quad \text{and} \quad x = u - 1. \tag{18}$$

Then we have

$$\begin{aligned}
\int x^2\sqrt{x+1}\,dx &= \int (u-1)^2\sqrt{u}\,du \\
&= \int (u^2 - 2u + 1)u^{1/2}\,du \\
&= \int (u^{5/2} - 2u^{3/2} + u^{1/2})\,du \\
&= \frac{u^{7/2}}{7/2} - 2\frac{u^{5/2}}{5/2} + \frac{u^{3/2}}{3/2} + c \\
&= \tfrac{2}{7}(x+1)^{7/2} - \tfrac{4}{5}(x+1)^{5/2} + \tfrac{2}{3}(x+1)^{3/2} + c \\
&= 2(x+1)^{3/2}[\tfrac{1}{7}(x+1)^2 - \tfrac{2}{5}(x+1) + \tfrac{1}{3}] + c.
\end{aligned} \tag{19}$$

An alternative substitution is

$$u = \sqrt{x+1}; \quad \text{then} \quad x + 1 = u^2 \quad \text{and} \quad dx = 2u\,du. \tag{20}$$

In this case we have

$$\begin{aligned}
\int x^2\sqrt{x+1}\,dx &= \int (u^2-1)^2\,u\,2u\,du \\
&= 2\int u^2(u^4 - 2u^2 + 1)\,du \\
&= 2\int (u^6 - 2u^4 + u^2)\,du \\
&= 2\left(\frac{u^7}{7} - \frac{2u^5}{5} + \frac{u^3}{3}\right) + c \\
&= 2u^3\left(\frac{u^4}{7} - \frac{2u^2}{5} + \frac{1}{3}\right) + c \\
&= 2(x+1)^{3/2}[\tfrac{1}{7}(x+1)^2 - \tfrac{2}{5}(x+1) + \tfrac{1}{3}] + c. \blacksquare
\end{aligned} \tag{21}$$

EXAMPLE 5

Evaluate

$$\int_0^{\sqrt{\pi}/2} 3x \cos x^2 \, dx. \tag{22}$$

In Example 1 we found that $\frac{3}{2} \sin x^2 + c$ is an antiderivative for the integrand. Recall that we do not need the arbitrary constant to evaluate the definite integral (22). Hence

$$\begin{aligned} \int_0^{\sqrt{\pi}/2} 3x \cos x^2 \, dx &= \frac{3}{2} \sin x^2 \Big|_0^{\sqrt{\pi}/2} = \frac{3}{2}\left(\sin \frac{\pi}{4} - \sin 0\right) \\ &= \frac{3}{2}\frac{\sqrt{2}}{2} = \frac{3\sqrt{2}}{4} \cong 1.061. \blacksquare \end{aligned} \tag{23}$$

In Example 5 we evaluated the integral (22) by using an antiderivative found earlier through the substitution $u = x^2$. We could also have used the same substitution to transform the limits of integration so as to obtain a new definite integral expressed in terms of the variable u. We will illustrate this for Example 5 and then state a general theorem.

If $u = x^2$, then corresponding to the x limits of integration 0 and $\sqrt{\pi}/2$ we obtain the u values 0 and $\pi/4$, respectively. Thus

$$\begin{aligned} \int_0^{\sqrt{\pi}/2} 3x \cos x^2 \, dx &= \frac{3}{2} \int_0^{\pi/4} \cos u \, du \\ &= \frac{3}{2} \sin u \Big|_0^{\pi/4} = \frac{3\sqrt{2}}{4}, \end{aligned} \tag{24}$$

which agrees with Eq. 23.

The general theorem for substitutions in a definite integral is as follows.

Theorem 6.5.1

Assume that the following conditions are satisfied:

1. $u = u(x)$ is continuous and has a continuous derivative on $[a, b]$, with $u(a) = c$ and $u(b) = d$.
2. f is continuous on the set of values taken on by $u = u(x)$.

Then

$$\int_a^b f[u(x)]u'(x) \, dx = \int_c^d f(u) \, du. \tag{25}$$

Proof. Observe that as x ranges from a to b, the continuity of $u(x)$ ensures that u takes on all values in the interval $[c, d]$, but in addition u may also take on values outside of $[c, d]$, as illustrated in Figure 6.5.1. This is why f is required to be continuous on the set of values taken on by u, rather than just on $[c, d]$.

Now, since f is continuous on $[c, d]$, it has an antiderivative F and

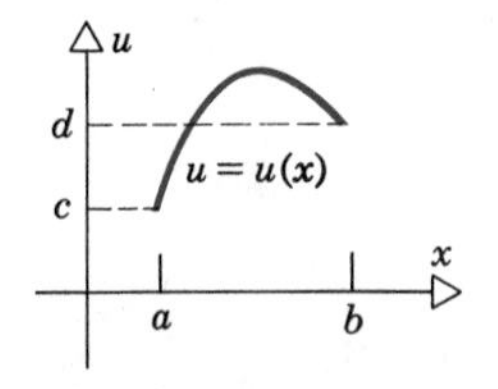

Figure 6.5.1

$$\int_c^d f(u)\, du = F(d) - F(c). \tag{26}$$

On the other hand, by the chain rule,

$$\frac{d}{dx} F[u(x)] = F'[u(x)]u'(x) = f[u(x)]u'(x). \tag{27}$$

Therefore

$$\int_a^b f[u(x)]u'(x)\, dx = F[u(x)]\Big|_a^b = F[u(b)] - F[u(a)] = F(d) - F(c). \tag{28}$$

Combining Eqs. 26 and 28, we obtain the result (25). □

EXAMPLE 6

Evaluate

$$\int_1^4 \left(1 + \frac{2}{x}\right)^{-1/2} \frac{1}{x^2}\, dx. \tag{29}$$

If we note that $1/x^2$ is, except for sign, the derivative of $1/x$, this may suggest the substitution $u = x^{-1}$. Then $du = -x^{-2}\, dx$, and the limits of integration become $u = 1$ and $u = \frac{1}{4}$, corresponding to $x = 1$ and $x = 4$, respectively. Then, substituting into the integral (29), we obtain

$$\begin{aligned} \int_1^4 \left(1 + \frac{2}{x}\right)^{-1/2} \frac{1}{x^2}\, dx &= -\int_1^{1/4} (1 + 2u)^{-1/2}\, du \\ &= \int_{1/4}^1 (1 + 2u)^{-1/2}\, du. \end{aligned} \tag{30}$$

Observe that in the last step we have changed the sign of the integral by reversing the limits of integration. To evaluate the integral (30) a second substitution is helpful. Let $v = 1 + 2u$; then $dv = 2du$ or $du = dv/2$. The limits of integration now become $v = \frac{3}{2}$ and $v = 3$, corresponding to $u = \frac{1}{4}$ and $u = 1$, respectively. Thus

$$\begin{aligned} \int_{1/4}^1 (1 + 2u)^{-1/2}\, du &= \tfrac{1}{2} \int_{3/2}^3 v^{-1/2}\, dv = \tfrac{1}{2} \cdot 2v^{1/2}\Big|_{3/2}^3 \\ &= \sqrt{3} - \sqrt{\tfrac{3}{2}} \cong 0.5073. \end{aligned} \tag{31}$$

By Eq. 30 this is also the value of the original integral (29).

We evaluated the integral in this example in two stages, using a different substitution at each step. Of course, if we had introduced the substitution $v = 1 + (2/x)$ at the start, then we would have obtained the result more promptly. However, this example shows that we do not necessarily have to find the optimum substitution in the beginning. If we can find a substitution that simplifies the problem to some extent, then whatever further steps are needed may become clearer. In any event, it is sometimes desirable to break a complicated procedure down into a sequence of simple steps, rather than trying to do everything at once. ∎

PROBLEMS

In each of Problems 1 through 30, use an appropriate substitution to evaluate the given indefinite or definite integral.

1. $\int \sqrt{3 + 2x}\, dx$
2. $\int \frac{dx}{(3 + 2x)^2}$
3. $\int x(9 + x^2)^3\, dx$
4. $\int \frac{x}{\sqrt{4 + x^2}}\, dx$
5. $\int \frac{t}{(1 + 4t^2)^3}\, dt$
6. $\int \frac{t}{(9 - t^2)^2}\, dt$
7. $\int x^3(4 - x^4)^3\, dx$
8. $\int \frac{x^2}{(1 + x^3)^2}\, dx$
9. $\int \cos 3\left(x - \frac{\pi}{4}\right) dx$
10. $\int 3x \sin 5x^2\, dx$
11. $\int \tan \pi\theta \sec \pi\theta\, d\theta$
12. $\int \frac{x \sin \sqrt{9 - x^2}}{\sqrt{9 - x^2}}\, dx$
13. $\int \cos^3 2x \sin 2x\, dx$
14. $\int \tan 3\theta \sec^2 3\theta\, d\theta$
15. $\int \frac{dx}{\sqrt{x}(1 + \sqrt{x})^2}$
16. $\int s(1 + s)^{1/3}\, ds$
17. $\int x^5 \sqrt{2 + x^2}\, dx$
18. $\int (x - 2)^2(x + 3)^{11}\, dx$
19. $\int x(a^2 + x^2)^r\, dx, \quad r \neq -1$
20. $\int x^2(a^3 + x^3)^r\, dx, \quad r \neq -1$
21. $\int_0^4 \sqrt{4 + 3x}\, dx$
22. $\int_1^6 \frac{dx}{(2x + 3)^2}$
23. $\int_0^4 x\sqrt{16 - x^2}\, dx$
24. $\int_0^3 \frac{3x^2}{\sqrt{1 + x}}\, dx$
25. $\int_{\pi/4}^{\pi/2} \cot \theta \csc^2 \theta\, d\theta$
26. $\int_0^{\sqrt{\pi}} t \sin t^2\, dt$
27. $\int_1^4 \frac{ds}{\sqrt{s}(4 + \sqrt{s})^3}\, ds$
28. $\int_0^a x\sqrt{a^2 - x^2}\, dx$
29. $\int_0^a \frac{x}{\sqrt{a^2 + x^2}}\, dx$
30. $\int_0^{a/2} \frac{x}{\sqrt{a^2 - x^2}}\, dx$

31. In Example 3 of Section 6.3 we interpreted $\int_0^a \sqrt{a^2 - x^2}\, dx$ as the area of a quarter circle of radius a, and therefore concluded that the value of this integral is $\pi a^2/4$. Here we indicate a way of evaluating this integral analytically.

(a) Let $x = a \sin \theta$ and show that

$$\int_0^a \sqrt{a^2 - x^2}\, dx = a^2 \int_0^{\pi/2} \cos^2 \theta\, d\theta.$$

(b) Use the half-angle formula $\cos^2 \theta = (1 + \cos 2\theta)/2$ to simplify the integral obtained in Part (a).

(c) Let $u = 2\theta$, and thereby show that

$$\int_0^a \sqrt{a^2 - x^2}\, dx = \frac{\pi a^2}{4}.$$

In each of Problems 32 through 34, use the substitution $x = a \sin \theta$ to evaluate the given integral.

32. $\int_0^{a/2} \frac{dx}{\sqrt{a^2 - x^2}}$
33. $\int_0^{a/2} \frac{dx}{(a^2 - x^2)^{3/2}}$
34. $\int_0^{a/2} \frac{x^2}{\sqrt{a^2 - x^2}}\, dx$

35. (a) If f is an even integrable function, show that

$$\int_{-a}^a f(x)\, dx = 2 \int_0^a f(x)\, dx.$$

(b) If f is an odd integrable function, show that

$$\int_{-a}^{a} f(x)\, dx = 0.$$

36. If m and n are positive integers, show that

$$\int_0^1 x^n(1-x)^m\, dx = \int_0^1 x^m(1-x)^n\, dx.$$

6.6 NUMERICAL INTEGRATION

The fundamental theorem of calculus, Theorem 6.4.3, gives us a means of evaluating any definite integral for which we can identify an antiderivative of the integrand. The class of such integrals was considerably enlarged in Section 6.5 through the use of substitutions in the integrand. It is enlarged still further by other methods that we discuss later, principally in Chapter 9. Nevertheless, there remain many integrals that do not yield to this approach because their integrands do not have antiderivatives that are expressible in any simple form. If the value of such an integral is required, it is often necessary to use a numerical approximation method. Recall that we have already proceeded from this viewpoint in Sections 6.1 and 6.2, where we estimated the values of a few integrals by evaluating some particular Riemann sum.

We now discuss two other methods, known as the trapezoidal rule and Simpson's rule, for the numerical evaluation of integrals. The trapezoidal rule is easy to derive and simple to use. Unfortunately, its accuracy is often no better than a Riemann sum with a similar number of terms. The derivation of Simpson's rule follows the same lines as for the trapezoidal rule, but is considerably more complicated. However, Simpson's rule is no more difficult to use than the trapezoidal rule or Riemann sums, and it usually gives a much more accurate approximation. Thus, among the alternatives mentioned here, Simpson's rule is normally the method of choice. Both the trapezoidal rule and Simpson's rule have associated *a priori* error estimates that sometimes can be used to estimate in advance how to use the procedure in order to achieve a desired accuracy.

Suppose that we wish to approximate the value of

$$\int_a^b f(x)\, dx, \tag{1}$$

where f is a given bounded integrable function. We introduce a uniform partition

$$a = x_0 < x_1 < \cdots < x_{n-1} < x_n = b$$

that subdivides the interval $[a, b]$ into n equal subintervals, each of length

$$h = \frac{b-a}{n}. \tag{2}$$

The trapezoidal rule and Simpson's rule result from two different ways of approximating the integrand in each subinterval.

Trapezoidal rule

To help in visualizing the meaning of the formulas that follow we illustrate them for the case in which the integrand f is continuous and nonnegative. Then the integral (1) can be interpreted as the area under the graph of f from a to b. However, the formulas apply also to integrands that may be negative or discontinuous.

Recall that in forming a Riemann sum we approximate the area under the graph of f on the ith subinterval $[x_{i-1}, x_i]$ by the area $f(x_i^*)\,h$ of a rectangle of base h and altitude $f(x_i^*)$ (see Figure 6.6.1). Alternatively, we can instead approximate the area under the graph by the area of the trapezoid shown in Figure 6.6.2. Put another way, we approximate the graph of $y = f(x)$ by the straight line

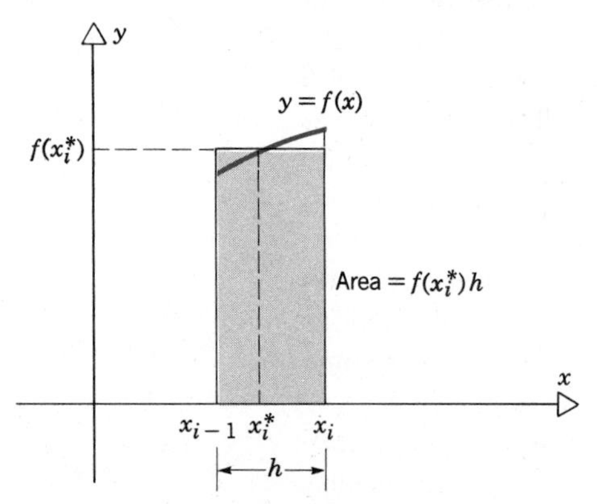

Figure 6.6.1 Approximation by a rectangle.

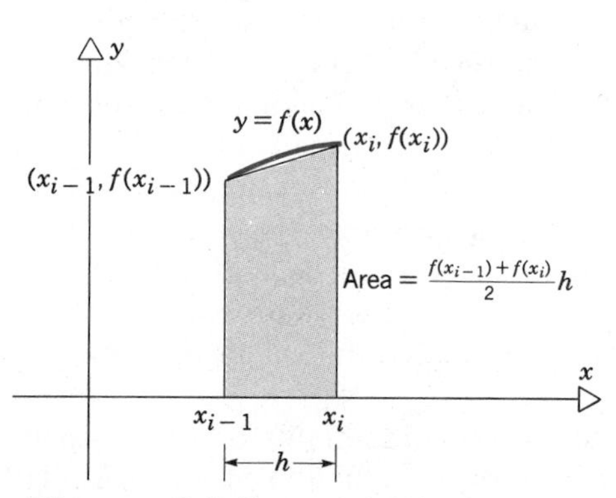

Figure 6.6.2 Approximation by a trapezoid.

segment passing through the points $[x_{i-1}, f(x_{i-1})]$ and $[x_i, f(x_i)]$. The area of the trapezoid on the interval $[x_{i-1}, x_i]$ is

$$[f(x_{i-1}) + f(x_i)]\,\frac{h}{2}. \tag{3}$$

To approximate the integral (1) we apply the formula (3) in each subinterval, and add the results. As indicated in Figure 6.6.3, this is equivalent to approximating the graph of f from $x = a$ to $x = b$ by the polygonal line passing through the

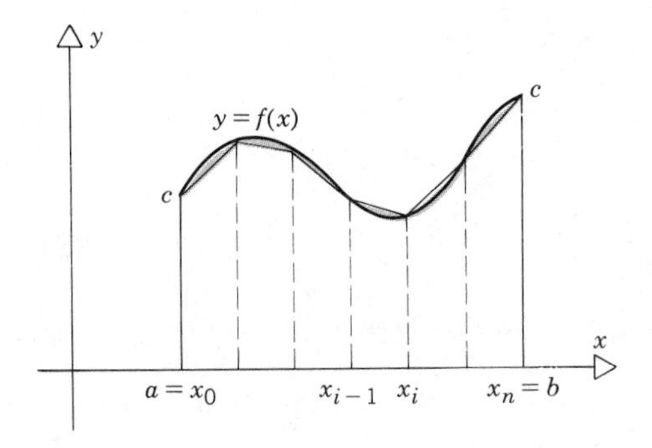

Figure 6.6.3 A polygonal approximation.

points on the graph corresponding to the partition points $x_1, \ldots, x_n$ on the x-axis. This result is

$$\int_a^b f(x)\,dx \cong \sum_{i=1}^{n} [f(x_{i-1}) + f(x_i)]\,\frac{h}{2}$$

$$= \frac{h}{2}\{[f(x_0) + f(x_1)] + [f(x_1) + f(x_2)] + \cdots + [f(x_{n-1}) + f(x_n)]\}$$

$$= \frac{h}{2}[f(x_0) + 2f(x_1) + 2f(x_2) + \cdots + 2f(x_{n-1}) + f(x_n)]. \tag{4}$$

If we let $y_i = f(x_i)$ and denote the trapezoidal approximation to the integral (1) by T_n, then Eq. 4 can be written as

$$\int_a^b f(x)\,dx \cong T_n = \frac{h}{2}(y_0 + 2y_1 + 2y_2 + \cdots + 2y_{n-1} + y_n). \tag{5}$$

Equation 5 is known as the **trapezoidal rule.** Note that $n + 1$ function evaluations are required—one more than in using a Riemann sum on the same partition. We mention the number of times that the function f must be evaluated, since this is the most important consideration in developing a general procedure for evaluating a definite integral numerically. Most of the computational time involved in evaluating T_n from Eq. 5, or in using any similar numerical integration formula, is consumed in evaluating f on a set of points.

EXAMPLE 1

Use the trapezoidal rule to approximate the value of

$$\int_0^{\pi/2} \sin x\,dx. \tag{6}$$

Recall that in Example 3 of Section 6.2 we used various Riemann sums to approximate the given integral. Of course, the exact value of the integral, by Theorem 6.4.3, is

$$\int_0^{\pi/2} \sin x\,dx = -\cos x \Big|_0^{\pi/2} = 1. \tag{7}$$

The results of using the trapezoidal rule for several values of n to evaluate this integral are shown in Table 6.4. We can compare these results with those in Table 6.3 in Section 6.2 obtained from Riemann sums with the star points chosen as the midpoint of each subinterval. For this integral the Riemann sums give a more accurate approximation than the trapezoidal rule. If we compare Table 6.4 with

Table 6.4 Approximations to $\int_0^{\pi/2} \sin x \, dx$ Using the Trapezoidal Rule.

n	T_n
10	0.9979431
20	0.9994860
40	0.9998716
80	0.9999679

the last column in Table 6.2, we note that the results agree except for possible differences in the last decimal place, presumably caused by different rounding procedures. That is, the trapezoidal rule gives the same results as the mean value of the upper and lower sums. This is always the case for integrands that are either nondecreasing or nonincreasing on the interval of integration. ∎

It is natural, and important, to ask whether a general statement can be made as to how accurately the trapezoidal formula approximates the corresponding integral. The following result can be established by methods that are somewhat more advanced than are appropriate in this book.

Theorem 6.6.1

If f is continuous on $[a, b]$, if f'' exists on (a, b), if T_n is given by the trapezoidal formula (5), and if E_n is the error, then

$$\int_a^b f(x)\, dx = T_n + E_n, \qquad E_n = -\frac{f''(\xi)(b-a)h^2}{12}, \tag{8}$$

where $h = (b-a)/n$, n is the number of equal subintervals in the partition of $[a, b]$, and ξ is some number in the interval (a, b).

Since the theorem gives us no information on how to determine the number ξ, the error formula is usually used to provide a bound on the error. If there is a number M_2 such that $|f''(x)| \le M_2$ for $a < x < b$, then

$$\left| \int_a^b f(x)\, dx - T_n \right| = |E_n| = \frac{|f''(\xi)|(b-a)h^2}{12}$$

$$\le \frac{M_2(b-a)h^2}{12} = \frac{M_2(b-a)^3}{12n^2}. \tag{9}$$

Notice that M_2 does not depend on n. Hence, for fixed a and b, the error E_n approaches zero as $n \to \infty$, which we intuitively anticipated. (Also see Problem 23). Moreover, E_n is proportional to the square of the step size h. Reducing the step size by a factor of $\frac{1}{2}$ (doubling the number of steps) reduces the error by a factor of $\frac{1}{4}$; reducing the step size by a factor of $\frac{1}{10}$ reduces the error by a factor of $\frac{1}{100}$. The price for this reduction in the error is, of course, the additional number of times that $f(x)$ must be calculated as n is increased.

In using formula (9) to estimate the accuracy of the approximation T_n there are two things we should note.

1. In general, in finding a bound M_2 for $|f''(x)|$ we will usually not find the best possible bound.
2. Even if we do find the best possible value for M_2, it is unlikely that $M_2 = f''(\xi)$, where ξ is the point at which the equality is attained in Eq. 8.

Thus, in general, our bound on the error is just that—a bound. A very useful and simple formula for estimating, but not bounding, the error in using the trapezoidal rule is given in Problem 25.

EXAMPLE 2

Estimate the error in using the trapezoidal rule to evaluate the integral $\int_a^b \sin x\, dx$ in Example 1.

In this case, $f(x) = \sin x$, so $f''(\xi) = -\sin \xi$. On the interval $[0, \pi/2]$ we are sure only that $0 \leq \sin \xi \leq 1$, so $M_2 = 1$ is the best possible value. Then, from Eq. 9, we obtain

$$|E_n| \leq \frac{(\pi/2)^3}{12n^2} \cong \frac{0.322982}{n^2}. \tag{10}$$

The actual errors in using the trapezoidal rule for this integral can be read off from Table 6.4, and are shown in Table 6.5, along with the error bounds calculated from

Table 6.5 Errors in Using the Trapezoidal Rule to Approximate $\int_0^{\pi/2} \sin x\, dx$.

n	Actual Error	Error Bound from Eq. 10
10	0.0020569	0.0032298
20	0.0005140	0.0008075
40	0.0001284	0.0002019
80	0.0000321	0.0000505

Eq. 10. Observe that, even though we used the best possible value of M_2, the error bound (10) overestimates the actual error by more than 50 percent. Such a result is by no means unusual. ∎

EXAMPLE 3
If the integral

$$I = \int_0^2 \frac{dx}{2 + x^2}$$

is to be calculated using the trapezoidal rule with an error less than 5×10^{-4}, what value of n should be used?

From Eq. 9 we know that

$$|E_n| \leq \frac{M_2(2 - 0)^3}{12n^2} = \frac{2M_2}{3n^2}, \tag{11}$$

where M_2 is a bound for $|f''(x)|$ on $(0, 2)$ with $f(x) = (2 + x^2)^{-1}$. We must find M_2 and then choose n so that $2M_2/3n^2 < 5 \times 10^{-4}$. We have

$$f'(x) = \frac{-2x}{(2 + x^2)^2},$$

$$f''(x) = \frac{-2}{(2 + x^2)^2} + \frac{(-2x)(-2)(2x)}{(2 + x^2)^3} = \frac{-2(2 - 3x^2)}{(2 + x^2)^3}.$$

To find a bound for

$$|f''(x)| = \frac{2|2 - 3x^2|}{(2 + x^2)^3} \quad \text{on} \quad 0 < x < 2$$

we make $|2 - 3x^2|$ as big as possible, namely 10 at $x = 2$, and $(2 + x^2)^3$ as small as possible, namely 8 at $x = 0$. Hence, on $0 < x < 2$,

$$|f''(x)| \leq \frac{2(10)}{8} = 2.5 = M_2. \tag{12}$$

Next, we choose n so that the inequality

$$\frac{2(2.5)}{3n^2} < 5 \times 10^{-4}$$

is satisfied. This requires that

$$n^2 > \frac{10^4}{3},$$

or

$$n > 57.73.$$

Thus, if we choose $n \geq 58$, we can be sure that the error in evaluating I is less than 5×10^{-4}. We *cannot* assert that if $n < 58$, then the error will be greater than 5×10^{-4}. It is quite likely that values of n much smaller than 58 can be used, as we have not obtained the best possible value for M_2. With a little more work, we can show that the maximum of $|f''(x)|$ on $0 \leq x \leq 2$ occurs at $x = 0$ and is $\frac{1}{2}$.

If we use $\frac{1}{2}$ for M_2 instead of 2.5, then we find that $n \geq 26$ is required. Still, this value of n is probably too large, since M_2 is probably greater than the value of $f''(\xi)$ that provides equality in Eq. 8. ∎

Simpson's rule

This is an approximation scheme that is considerably more accurate than the trapezoidal rule, but requires no more computational effort. Rather than approximating the graph of $y = f(x)$ by a straight line segment on each subinterval $[x_{i-1}, x_i]$, we approximate $y = f(x)$ by an arc of a parabola on each of the subintervals $[x_{i-1}, x_{i+1}]$. Since we now use two partition intervals for each subinterval, we must take n to be an *even* integer.

On each subinterval $[x_{i-1}, x_{i+1}]$, the three coefficients in the approximating parabola are chosen so that the parabola passes through the points (x_{i-1}, y_{i-1}), (x_i, y_i), and (x_{i+1}, y_{i+1}). This is illustrated in Figure 6.6.4. Thus, if the approximating parabola on $[x_{i-1}, x_{i+1}]$ is $Ax^2 + Bx + C$, then we choose A, B, and C, so that

$$\begin{aligned} A(x_i - h)^2 + B(x_i - h) + C &= f(x_{i-1}) = y_{i-1}, \\ Ax_i^2 + Bx_i + C &= f(x_i) = y_i, \\ A(x_i + h)^2 + B(x_i + h) + C &= f(x_{i+1}) = y_{i+1}. \end{aligned} \tag{13}$$

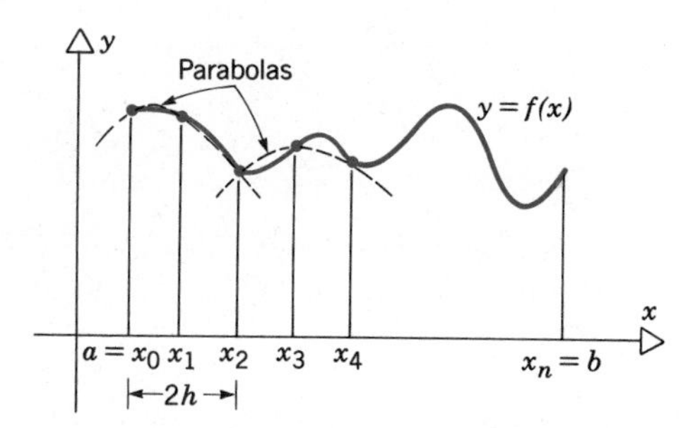

Figure 6.6.4 A parabolic approximation.

In Eqs. 13 we have written x_{i-1} as $x_i - h$ and x_{i+1} as $x_i + h$, where h is the uniform width of the partition. We leave aside for a moment the algebra of solving Eqs. 13 for A, B, and C.

Next we approximate the integral of f on $[x_{i-1}, x_{i+1}]$ by

$$\int_{x_{i-1}}^{x_{i+1}} f(x)\,dx \simeq \int_{x_{i-1}}^{x_{i+1}} (Ax^2 + Bx + C)\,dx = \int_{x_i-h}^{x_i+h} (Ax^2 + Bx + C)\,dx. \tag{14}$$

It turns out that when one solves Eqs. 13 for A, B, and C, substitutes in Eq. 14, and carries out the integration the result is

$$\int_{x_{i-1}}^{x_{i+1}} f(x)\,dx \simeq \frac{h}{3}(y_{i-1} + 4y_i + y_{i+1}). \tag{15}$$

The details of the somewhat lengthy algebra are outlined in Problem 26.

We can now use formula (15) to derive Simpson's* rule. We have

$$\int_a^b f(x)\,dx = \int_{x_0}^{x_2} f(x)\,dx + \int_{x_2}^{x_4} f(x)\,dx + \cdots + \int_{x_{n-2}}^{x_n} f(x)\,dx$$

$$\cong \frac{h}{3}(y_0 + 4y_1 + y_2) + \frac{h}{3}(y_2 + 4y_3 + y_4)$$

$$+ \cdots + \frac{h}{3}(y_{n-2} + 4y_{n-1} + y_n)$$

and therefore

$$\int_a^b f(x)\,dx \cong S_n = \frac{h}{3}(y_0 + 4y_1 + 2y_2 + 4y_3 + 2y_4 + \cdots + 2y_{n-2} + 4y_{n-1} + y_n). \quad (16)$$

Equation 16 is called Simpson's rule. Observe that it involves the same number of function evaluations as the trapezoidal rule. A formula for the error in Simpson's approximation can also be given.

Theorem 6.6.2

If f is continuous on $[a, b]$, if f^{iv} exists on (a, b), if S_n is given by Eq. 16, and if E_n is the error, then

$$\int_a^b f(x)\,dx = S_n + E_n, \qquad E_n = -\frac{f^{iv}(\xi)(b-a)h^4}{180}, \quad (17)$$

where $h = (b - a)/n$, n is the (even) number of equal subintervals in the partition of $[a, b]$, and ξ is some number in the interval (a, b).

As in the case of the trapezoidal rule, we can use Eq. 17 to obtain a bound for the error. If there is a number M_4 such that $|f^{iv}(x)| \le M_4$ for $a < x < b$, then

$$\left| \int_a^b f(x)\,dx - S_n \right| = |E_n| = \frac{|f^{iv}(\xi)|(b-a)h^4}{180}$$

$$\le \frac{M_4(b-a)h^4}{180} = \frac{M_4(b-a)^5}{180n^4}. \quad (18)$$

*The idea of evaluating an integral by replacing the integrand by a suitable interpolating polynomial appeared in 1687 in Newton's *Principia*. The method was further developed by some of Newton's followers, including Roger Cotes (1682–1716) and James Stirling (1692–1770), who were familiar with what is now called Simpson's rule. Some years later Thomas Simpson (1710–1761) rediscovered this approximate integration formula, published it (1743) in a book of mathematical essays, and it is now associated with his name. In his own time Simpson was best known as the author of successful textbooks on algebra, geometry, and trigonometry, which continued to be published for many years after his death.

From Eq. 18 the error bound in using Simpson's rule is proportional to h^4, whereas for the trapezoidal rule it is proportional to h^2. Thus, although Simpson's rule is no more complicated to use than the trapezoidal rule, it is much more accurate. Indeed, reducing the step size by a factor of $\frac{1}{2}$ reduces the error bound by a factor of $\frac{1}{16}$; reducing the step size h by a factor of $\frac{1}{10}$ reduces the error bound by a factor of $\frac{1}{10,000}$. It is probably safe to say that Simpson's rule, or a variant of it, is the most commonly used formula for numerical integration. Unfortunately, it is not always easy to use the error estimate (18), since it is necessary to obtain a bound M_4 for $|f^{iv}(x)|$, and f^{iv} may be a complicated function. Finally, note that if f is a polynomial of degree less than four, then $f^{iv}(x) = 0$ for all x, and the error must be zero, so Simpson's rule gives the exact value of the integral.

EXAMPLE 4

Use Simpson's rule to evaluate the integral $\int_0^{\pi/2} \sin x \, dx$ from Example 1, and use the error bound (18) to estimate the error.

The results of using Simpson's rule are shown in Table 6.6. Note that one obtains a much more accurate approximation by using Simpson's rule with $n = 10$

Table 6.6 Approximations to $\int_0^{\pi/2} \sin x \, dx$ Using Simpson's Rule.

n	S_n	Error Bound from Eq. 19
10	1.000003	0.000005
20	1.000000	0.000000322
40	1.000000	0.0000000207
80	1.000000	0.0000000012

than by using the trapezoidal rule with $n = 80$. Indeed, Simpson's rule is accurate to six decimal places with $n = 20$. The error bound (18) is easy to use in this case since $f^{iv}(x) = \sin x$, and therefore we can use $M_4 = 1$. Hence

$$|E_n| \leq \frac{(\pi/2)^5}{180n^4} \cong \frac{0.053128}{n^4}. \tag{19}$$

The bounds from Eq. 19 are shown in the last column of Table 6.6. A comparison of the results in Table 6.6 with those in Tables 6.4 and 6.5 show clearly the gain in accuracy that is obtained by using Simpson's rule rather than the trapezoidal rule. ∎

If it is difficult to use the error bound (18) because $f^{iv}(x)$ is a complicated function, we can proceed in a more *ad hoc* manner. For example, if we double the number of partition points and find that the result does not change (to a certain number of decimal places), then we can conclude with reasonable assurance that those decimal places are correctly calculated.

EXAMPLE 5

Use the trapezoidal rule and Simpson's rule to evaluate

$$\int_0^{\pi/3} \tan x \, dx. \tag{20}$$

We do not yet have a way to determine an antiderivative of $\tan x$, so we cannot evaluate this integral by Theorem 6.4.3 at this time. If we apply the trapezoidal rule and Simpson's rule, we obtain the results in Table 6.7. Since the values

Table 6.7 **Approximations to $\int_0^{\pi/3} \tan x \, dx$ Using the Trapezoidal Rule and Simpson's Rule.**

n	T_n	S_n
10	0.6958760	0.6931951
20	0.6938318	0.6931503
40	0.6933186	0.6931474
80	0.6931900	0.6931472

obtained from Simpson's rule for $n = 40$ and for $n = 80$ agree to six decimal places, we can be fairly confident that these are correct and that

$$\int_0^{\pi/3} \tan x \, dx \cong 0.693147. \tag{21}$$

Observe also from Table 6.7 that Simpson's rule with only 10 subintervals gives very nearly the same accuracy as the trapezoidal rule with 80 subintervals, again illustrating the much greater accuracy that Simpson's rule provides. ∎

PROBLEMS

Problems 1 through 16 ask for the numerical evaluation of various definite integrals. The required calculations can be carried out with a pocket calculator for $n = 10$, but for larger values of n a programmable calculator, or preferably a computer, is certainly desirable.

In each of Problems 1 through 16, use the trapezoidal rule and Simpson's rule to find approximate values of the given integral. Let $n = 10, 20, 40$, and 80. Where possible, compare your results with the exact value of the integral, obtained by using Theorem 6.4.3. Also, whenever appropriate, compare them with the results of Problems 15 through 22 in Section 6.2.

© **1.** $\int_0^1 \sqrt{x} \, dx$

© **2.** $\int_0^3 \sqrt{1 + x^2} \, dx$

© **3.** $\int_0^2 \frac{dx}{1 + x^2}$

© **4.** $\int_1^3 (4 - x^2) \, dx$

© 5. $\int_{1/2}^{2} \left(x - \frac{1}{x}\right) dx$

© 6. $\int_{0}^{\pi} x \cos x \, dx$

© 7. $\int_{0}^{\pi/2} \sqrt{x} \sin 2x \, dx$

© 8. $\int_{0}^{2} \frac{dx}{(1 + x)^2}$

© 9. $\int_{0}^{1} x^2 \sin x \, dx$

© 10. $\int_{0}^{1} \frac{dx}{\sqrt{1 + x^4}}$

© 11. $\int_{0}^{\pi} \sqrt{\sin \frac{x}{2}} \, dx$

© 12. $\int_{0}^{\pi} \sin^2 x \, dx$

© 13. $\int_{0}^{2} \sqrt{1 + 2x} \, dx$

© 14. $\int_{0}^{\pi/2} \sqrt{1 + \cos x} \, dx$

© 15. $\int_{0}^{\pi} \cos(\theta - 0.5 \sin \theta) \, d\theta$

© 16. $\int_{0}^{\pi/2} \frac{dx}{\sin x + \cos x} dx$

In each of Problems 17 through 20, use the appropriate error bound to determine how large n must be so that the error in evaluating the given integral is less than 10^{-4}. Do this for each of the following.

(a) The trapezoidal rule.

(b) Simpson's rule.

Compare your answers with the results of the indicated problems.

17. $\int_{1/2}^{2} \left(x - \frac{1}{x}\right) dx$; see Problem 5

18. $\int_{0}^{2} \frac{dx}{(1 + x)^2}$; see Problem 8

19. $\int_{0}^{2} \sqrt{1 + 2x} \, dx$; see Problem 13

20. $\int_{0}^{\pi} \sin^2 x \, dx$; see Problem 12

© 21. Sometimes it is necessary to evaluate an integral whose integrand is not given by an analytical formula. For example, calculations show* that for certain operating conditions the ratio p of the pressure in a one-dimensional gas slider bearing to the ambient pressure is given as a function of the position x by the data in Table 6.8. The load carrying capacity W of the bearing is given by $W/BLp_a = \int_0^1 [p(x) - 1] \, dx$, where B and L are the breadth and length of the bearing, respectively, and p_a is the ambient pressure. Determine an approximate value of W using Simpson's rule.

Table 6.8 Data for Problem 21.

x	p	x	p
0	1.000	$\frac{7}{12}$	1.303
$\frac{1}{12}$	1.035	$\frac{8}{12}$	1.352
$\frac{2}{12}$	1.072	$\frac{9}{12}$	1.394
$\frac{3}{12}$	1.113	$\frac{10}{12}$	1.411
$\frac{4}{12}$	1.157	$\frac{11}{12}$	1.355
$\frac{5}{12}$	1.204	1	1.000
$\frac{6}{12}$	1.253		

*W. A. Gross, *Gas Film Lubrication*, Wiley, New York, 1962, p. 71.

© 22. Verify that using Simpson's rule with $n = 4$ and $n = 8$ to evaluate $\int_{-1}^{3} x^3 \, dx$ gives the exact answer. The result is, of course, true for any even n.

23. Show that the trapezoidal formula (5) can be written in the form

$$T_n = \sum_{i=1}^{n} f(x_i) \, \Delta x_i + \tfrac{1}{2}[f(a) - f(b)]h,$$

$$\Delta x_i = h.$$

Since the sum on the right side of this equation is a Riemann sum for $\int_a^b f(x) \, dx$, and since f is integrable on $[a, b]$, it follows that

$$\lim_{n \to \infty} T_n = \int_a^b f(x) \, dx.$$

24. Show that, if f is increasing on $[a, b]$, or decreasing on $[a, b]$, then for each n the trapezoidal rule yields the same approximation to $\int_a^b f(x) \, dx$ as the mean value of the upper and lower sums.

25. We have noted in the text that, while Eq. 9 provides a bound for the error in using the trapezoidal formula, it may not provide an accurate estimate of the error. A useful estimate of E_n is given* by

$$E_n \cong \frac{-(b - a)^2}{12n^2} [f'(b) - f'(a)]. \qquad (i)$$

Note that this is an estimate of the error, not a bound on the error. However, it is very simple to use since it is not necessary to calculate and bound $f''(x)$. Moreover, the formula (i) does not involve

*E. Rozema, *American Mathematical Monthly,* **87,** 1980, pp. 124–128.

an absolute value and hence gives the direction of the error.

(a) Use Eq. (*i*) to estimate the error in the calculated values of $\int_0^{\pi/2} \sin x\, dx$ in Example 1. Compare these estimates with the data in the second column of Table 6.5.

(b) Use Eq. (*i*) to estimate the even value of n required to evaluate $\int_0^2 (2 + x^2)^{-1}\, dx$ with an error "less" than 5×10^{-4}, and compare your result with that obtained in Example 3.

26. In this problem we derive formula (15). That is, if $y = g(x) = Ax^2 + Bx + C$, then

$$\int_{x_i-h}^{x_i+h} (Ax^2 + Bx + C)\, dx = \frac{h}{3}(y_{i-1} + 4y_i + y_{i+1}), \quad (i)$$

where $y_i = g(x_i)$. The derivation is especially simple for the case $x_i = 0$ and we consider that case first.

(a) Show that

$$\int_{-h}^{h} (Ax^2 + Bx + C)\, dx = \tfrac{2}{3}Ah^3 + 2Ch. \quad (ii)$$

Thus for $x_i = 0$, B does not appear in the result, so it is only necessary to solve the system of Eqs. 13 for A and C. Using $x_i = 0$, we have immediately that $C = y_i$. Show that $A = (y_{i-1} + y_{i+1} - 2y_i)/2h^2$. Substituting for A and C in Eq. (*ii*) gives the formula (*i*).

(b) For the case $x_i \neq 0$ show that

$$\int_{x_i-h}^{x_i+h} (Ax^2 + Bx + C)\, dx = \frac{h}{3}[A(6x_i^2 + 2h^2) + B(6x_i) + 6C]. \quad (iii)$$

Solving the second of Eqs. 13 for C and substituting in Eq. (*iii*) gives

$$\int_{x_i-h}^{x_i+h} (Ax^2 + Bx + C)\, dx = \frac{h}{3}(2h^2A + 6y_i). \quad (iv)$$

Thus we only need solve Eqs. 13 for A. Show that $2h^2A = y_{i-1} + y_{i+1} - 2y_i$. Substituting for $2h^2A$ in Eq. (*iv*) gives the formula (*i*).

REVIEW PROBLEMS

In each of Problems 1 through 8, evaluate the given sum.

1. $\displaystyle\sum_{j=1}^{5} \frac{1}{j^2 + 1}$

2. $\displaystyle\sum_{j=12}^{14} \sin^2 \frac{j\pi}{4}$

3. $\displaystyle\sum_{k=1}^{10} \frac{1}{(k + 2)(k + 3)}$

Hint: $\displaystyle\frac{1}{(k + 2)(k + 3)} = \frac{?}{k + 2} - \frac{?}{k + 3}$

4. $\displaystyle\sum_{k=2}^{16} \frac{1}{k^2(k + 1)(k - 1)}$

Hint: $\displaystyle\frac{1}{k^2(k + 1)(k - 1)} = \frac{?}{k^2 - 1} - \frac{?}{k^2}$

5. $\displaystyle\sum_{k=0}^{52} (\sqrt{2k + 1} - \sqrt{2k + 3})$

6. $\displaystyle\sum_{k=1}^{29} (k^2 + 1)$

7. $\displaystyle\sum_{n=1}^{32} (2n + 1)(n - 3)$

8. $\displaystyle\sum_{k=3}^{20} k^2$

In each of Problems 9 through 12, express the given sum in Σ-notation.

9. $\sin(3x + 1) + \sin(4x + 3) + \sin(5x + 5) + \cdots + \sin(11x + 17)$

10. $(-1) + (4x^2 - 2) + (8x^2 - 3) + (12x^2 - 4) + \cdots + (56x^2 - 15)$

11. $\displaystyle\frac{1}{x^2} - \frac{2}{2x^2} + \frac{4}{3x^2} - \frac{8}{4x^2} + \cdots - \frac{128}{8x^2}$

12. $\displaystyle\frac{1}{x^2} + \frac{3}{2x^2} - \frac{5}{3x^2} - \frac{7}{4x^2} + \frac{9}{5x^2} + \frac{11}{6x^2} - \cdots + \frac{25}{13x^2} + \frac{27}{14x^2} - \frac{29}{15x^2} - \frac{31}{16x^2}$

In each of Problems 13 through 17, compare the upper sum σ, the lower sum s, and the Riemann sum S using $x_i^* = x_i + (\Delta x_i/2)$. Which is the best estimate?

13. $\int_1^6 (x^2 + 3)\, dx$;
$\Delta = \{1, 1.5, 2, 3, 4.5, 4.75, 5.5, 6\}$

14. $\int_{-1}^3 \frac{1}{2 + x}\, dx$;
$\Delta = \{-1, -0.5, 0, 2, 2.25, 3\}$

15. $\int_{-3}^3 x^3\, dx$; $\Delta = \{-3, -2, -1, 0, 1, 2, 3\}$

16. $\int_1^2 \left(1 + \frac{1}{x}\right) dx$;
$\Delta = \{1, 1.1, 1.2, 1.3, 1.4, 1.5, 1.75, 2\}$

17. $\int_{-2}^5 (x - 1)(x + 2)(x - 4)\, dx$;
$\Delta = \{-2, -1, 0, 1, 2, 3, 4, 5\}$

In each of Problems 18 through 22, evaluate the given integral.

18. $\int_0^4 (\sqrt{16 - x^2} + 3x^2)\, dx$

19. $\int_0^2 [(1 + 2x)^2 - \sqrt{4 - x^2}]\, dx$

20. $\int_{-3}^8 f(x)\, dx$,

$$\text{where } f(x) = \begin{cases} x, & -3 \le x \le 0 \\ -\sqrt{9 - x^2}, & 0 < x < 3 \\ 1 + x^2, & 3 \le x \le 8 \end{cases}$$

21. $\int_0^{10} f(x)\, dx$,

$$\text{where } f(x) = \begin{cases} \sqrt{25 - x^2}, & 0 \le x \le 5 \\ (3x - 2)^2, & 5 < x \le 10 \end{cases}$$

22. $\int_0^3 (x^2 + \sqrt{9 - x^2})\, dx$

In each of Problems 23 and 24, show that the given inequality is true.

23. $\frac{8}{\sqrt{109}} \le \int_2^{10} \frac{dx}{\sqrt{x^2 + 9}} \le \frac{8}{\sqrt{13}}$

24. $10 \le \int_{-2}^3 (x^2 + 7x + 12)\, dx \le 210$

In each of Problems 25 through 27, calculate the average value of the given integral.

25. $\int_{2\pi}^{5\pi/2} \sin 2x\, dx$ 26. $\int_0^{10} (x^2 - 10x + 21)\, dx$

27. $\int_0^4 \sqrt{16 - x^2}\, dx$

In each of Problems 28 through 50, evaluate the given integral, using appropriate substitutions if necessary.

28. $\int_{-2}^6 (3x^4 - 2x + 1)\, dx$

29. $\int_{-5}^{10} f(x)\, dx$,

$$\text{where } f(x) = \begin{cases} x^2 + 2x, & -5 \le x < 0 \\ \sqrt{50 - 2x^2}, & 0 \le x \le 5 \\ x, & 5 < x \le 10 \end{cases}$$

30. $\int_6^{10} |x^2 - 16x + 63|\, dx$ 31. $\int_{\pi/2}^{7\pi/12} \sin^2 3x\, dx$

32. $\int_1^3 \frac{3x^5 + x + 1}{x^3}\, dx$ 33. $\int_a^x \sin^2\left(t + \frac{\pi}{4}\right) dt$

34. $\int_a^x t^2 \sin(t^3 + \pi)\, dt$

35. $\int_{-\sqrt{\pi}/2}^{2\sqrt{\pi}/3} t \sin^2(3t^2)\cos(3t^2)$
dt

36. $\int_1^4 \frac{t^{1/2}}{\sqrt{1 + t^{3/2}}}\, dt$ 37. $\int_0^{\pi^2/4} t^{-1/2} \sin(\pi + t^{1/2})\, dt$

38. $\int_1^4 t^{3/2}(1 + t^{5/2})^2\, dt$

39. $\int_{-1/2}^0 \frac{1}{\sqrt{1 - 4t}}\, dt$

40. $\int_0^{\pi/2} [\sin(\cos x + x)\sin x - \sin(\cos x + x)]\, dx$

41. $\int_0^1 t(t^{1/2} + t^2)^2\, dt$ 42. $\int_1^2 \frac{\sqrt{3t}}{(t^{3/2} + 1)^2}\, dt$

43. $\int_0^1 \frac{1 - t^6}{1 + t^{3/2}}\, dt$ 44. $\int_a^b t \cot(t^2)\csc^3(t^2)\, dt$

45. $\int_{-1}^3 t^2(1 + t)^{1/2}\, dt$ 46. $\int_4^6 \frac{x\, dx}{\sqrt{x^2 - 4}}$

47. $\int_a^b 2x^2 \sqrt{\frac{x^3}{2} + 1}\, dx$ 48. $\int_2^4 t(2 + 3t)^{-1/3}\, dt$

49. $\int_0^{\pi/2} (x + \pi)\cos(x^2 + 2\pi x + \pi^2)\, dx$

50. $\int_0^{\pi/4} \tan x \sec^4 x\, dx$

In each of Problems 51 and 52, use the error estimates for the trapezoidal rule and Simpson's rule to find an n sufficient to ensure that $|E_n| < 0.001$.

51. $\int_0^\pi \sin x^2 \, dx$

52. $\int_1^4 \frac{x^7 + 1}{x} \, dx$

In each of Problems 53 through 56, estimate the value of the integral using the trapezoidal rule and Simpson's rule with (a) $n = 10$; (b) $n = 20$; (c) $n = 40$; (d) $n = 80$.

© 53. $\int_{-\pi}^{\pi} x^2 \sin^2 x \, dx$

© 54. $\int_1^3 \frac{1}{x^2\sqrt{x + 1}} \, dx$

© 55. $\int_0^{10} x^{3/2} \cos^2 x \, dx$

© 56. $\int_0^1 \frac{x}{(x + 1)^{1/2}} \, dx$

SEVEN

applications of the integral

In Chapter 6 the area of an "arbitrary" region was found by approximating the region by many thin rectangles and then passing to the limit as the number of rectangles approaches infinity and the area of each one approaches zero. Integration is the mathematical abstraction of this process, defined by restating the procedure in language free of terminology associated with geometrical or other applications. Once integration is understood as a purely mathematical operation, it can be applied to the calculation of a wide variety of quantities that can be expressed as an appropriate sum of a large number of individually small terms.

We now consider a number of situations in which integration arises naturally in this way. The features common to each problem are that a quantity of interest is expressed as a sum, and then a limit is taken as the number of terms increases without bound and the size of each term shrinks to zero. If, as the approximation is refined, there is a balance between the increasing number of terms and the diminishing size of each one, then the result is an expression for the given quantity as an integral. This is an enormous advantage, because the methods developed in Sections 6.4 through 6.6, as well as those to be discussed later in Chapters 8 and 9, can be used to evaluate the required integrals systematically.

7.1 MORE ON AREAS OF PLANE REGIONS

A plane region R is said to be **bounded** if it can be contained within a rectangle of sufficient size. On the other hand, if no such rectangle exists, then R is **unbounded.** In Section 6.2 we expressed the area of the region (see Figure 7.1.1) bounded by the graph of a nonnegative continuous function f, the x-axis, and the lines $x = a$ and $x = b$ as the integral

$$A = \int_a^b f(x)\, dx. \tag{1}$$

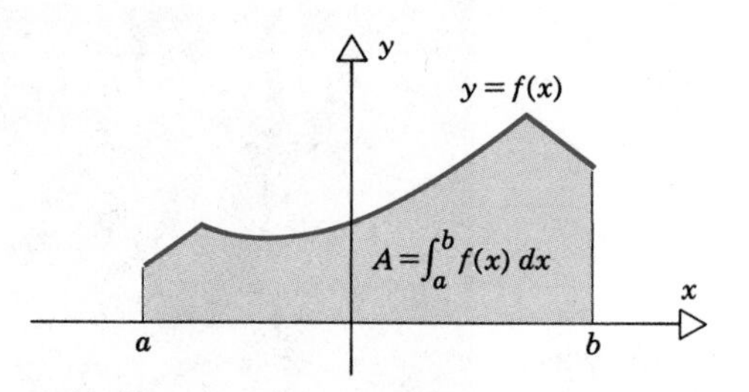

Figure 7.1.1 Area under the graph of a nonnegative continuous function.

Now we wish to show how to calculate the areas of somewhat more general bounded regions.

Suppose that the functions f and g are continuous and satisfy $g(x) \le f(x)$ on $[a, b]$. How do we define the area of the region, depicted in Figure 7.1.2, between the graphs of these functions and the lines $x = a$ and $x = b$?

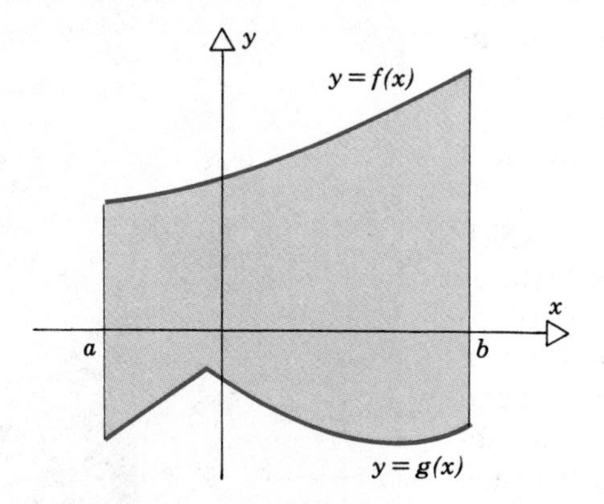

Figure 7.1.2 Area between the graphs of f and g.

It is natural to proceed as we did earlier in defining the area under the graph of a nonnegative continuous function. We partition the interval $[a, b]$, choose star points, and construct approximating rectangles as shown in Figure 7.1.3*a*. The particular rectangle with base Δx_i and height $f(x_i^*) - g(x_i^*)$ is shown in Figure 7.1.3*b*. The corresponding Riemann sum

$$\sum_{i=1}^{n} [f(x_i^*) - g(x_i^*)]\, \Delta x_i$$

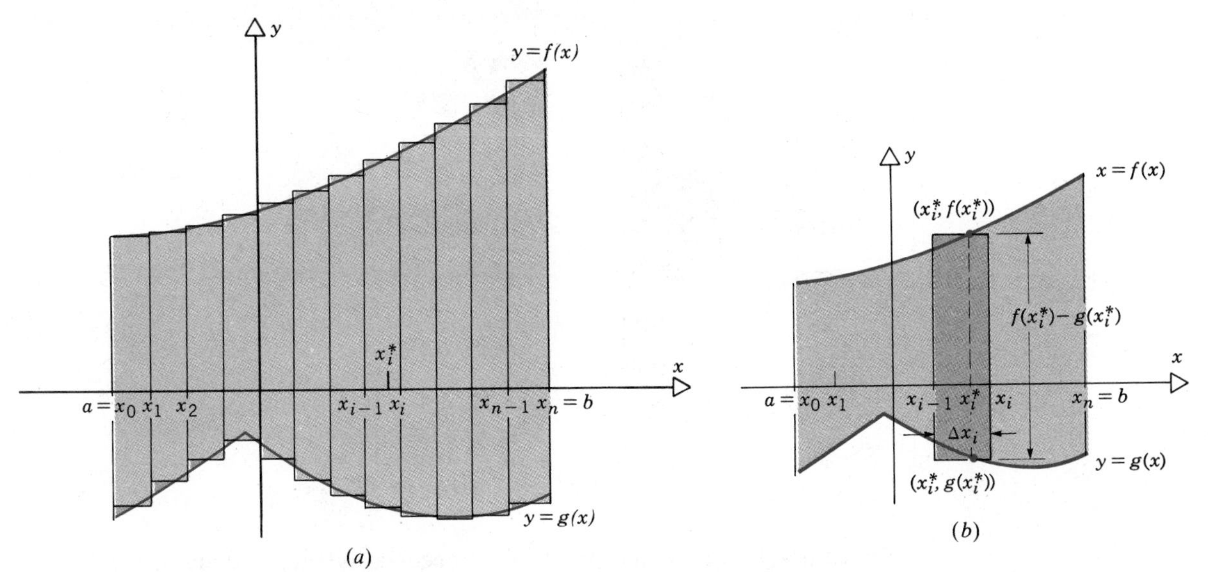

Figure 7.1.3 (a) A typical set of approximating rectangles. (b) The ith rectangle.

is an approximation of what we intuitively consider to be the area of the region in question. Further, this approximation becomes better as the norm of the partition is decreased. Thus *we define the area of the region between the graphs of f and g, and the lines x = a and x = b to be the number*

$$A = \int_a^b [f(x) - g(x)]\, dx. \tag{2}$$

EXAMPLE 1

Determine the area of the region bounded by $y = x^2/2$ and $y = 3 \sin x$ and the lines $x = \pi$ and $x = 3\pi/2$.

The region of interest is shown in Figure 7.1.4. We have

$$\begin{aligned} A &= \int_{\pi}^{3\pi/2} \left(\frac{x^2}{2} - 3 \sin x\right) dx \\ &= \left(\frac{x^3}{6} + 3 \cos x\right)\Bigg|_{\pi}^{3\pi/2} \\ &= \frac{27\pi^3}{48} - \frac{\pi^3}{6} + 3 \\ &= 3 + \frac{19\pi^3}{48} \cong 15.273. \end{aligned} \tag{3}$$

∎

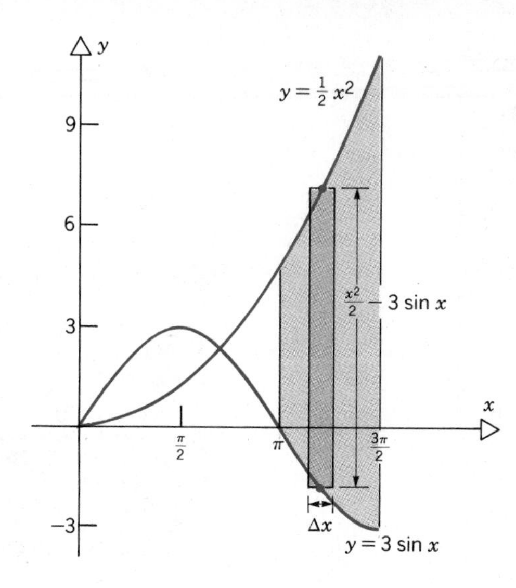

Figure 7.1.4

Notice that if $g(x) = 0$ in Eq. 2, then necessarily $f(x) \geq 0$ and Eq. 2 reduces to Eq. 1 as it should. On the other hand, if $f(x) = 0$ in Eq. 2 then necessarily $g(x) \leq 0$ and Eq. 2 states that *the area bounded by a continuous nonpositive function g, the x-axis, and the lines $x = a$ and $x = b$ is given by*

$$A = \int_a^b [-g(x)]\, dx. \tag{4}$$

EXAMPLE 2

Determine the area of the region between the graph of the function $f(x) = \frac{1}{2}x^2 - 2$ and the x-axis from $x = -2$ to $x = 3$.

The curve $y = \frac{1}{2}x^2 - 2$ and the region of interest are shown in Figure 7.1.5. Notice that $f(x) \leq 0$ on $[-2, 2]$ and $f(x) \geq 0$ on $[2, 3]$. The generic elements in

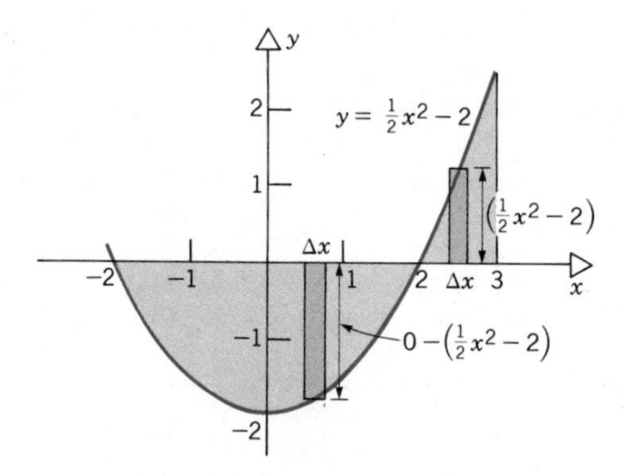

Figure 7.1.5

the Riemann sum that approximates the area on the intervals for which $f(x) \leq 0$ and for which $f(x) \geq 0$ are also shown. We have

$$\begin{aligned}
A &= \int_{-2}^{2} [-f(x)]\,dx + \int_{2}^{3} f(x)\,dx \\
&= \int_{-2}^{2} \left(-\frac{1}{2}x^2 + 2\right) dx + \int_{2}^{3} \left(\frac{1}{2}x^2 - 2\right) dx \\
&= \left(-\frac{x^3}{6} + 2x\right)\Bigg|_{-2}^{2} + \left(\frac{x^3}{6} - 2x\right)\Bigg|_{2}^{3} \\
&= \left[\left(-\frac{8}{6} + 4\right) - \left(\frac{8}{6} - 4\right)\right] + \left[\left(\frac{27}{6} - 6\right) - \left(\frac{8}{6} - 4\right)\right] \\
&= -\frac{8}{3} + 8 + \frac{19}{6} - 2 = \frac{13}{2}. \quad \blacksquare
\end{aligned} \tag{5}$$

The results given in Eqs. 1 and 4 can be combined as follows. *The area A of the region between the graph of a continuous function f and the x-axis from $x = a$ to $x = b$ is*

$$A = \int_a^b |f(x)|\,dx. \tag{6}$$

In intervals where $f(x) \geq 0$ we have $|f(x)| = f(x)$ as in Eq. 1, and in intervals where $f(x) \leq 0$ we have $|f(x)| = -f(x)$ as in Eq. 4. If the intervals for which $f(x) \geq 0$ and for which $f(x) \leq 0$ can be determined, then the integral in Eq. 6 can be written as a sum of integrals over the respective intervals.

We can also restate the result embodied in Eq. 2 in a more general form. Let f and g be two continuous functions on $[a, b]$. *Then the area A of the region lying between the graphs of f and g from $x = a$ to $x = b$ is*

$$A = \int_a^b |f(x) - g(x)|\,dx. \tag{7}$$

Again, in evaluating the integral in Eq. 7, we must determine the intervals on which $f(x) - g(x) \geq 0$ and the intervals on which $f(x) - g(x) \leq 0$, and then write the integral of Eq. 7 as a sum of integrals over the respective intervals.

Before considering several additional examples, we point out a few guidelines that should be followed in determining the area of a region bounded by the graphs of continuous functions.

1. Draw a sketch of the region and its boundaries. Note whether the region is symmetric about one of the coordinate axes, or perhaps about some other line.
2. For problems involving a single function determine the points (if any) at which the curve crosses the coordinate axis.
3. If two curves are involved, determine the points of intersection (if any).
4. Set up a generic element for the area of a rectangle in the Riemann sum that approximates the area of the region. Check your formula for the height (≥ 0) of the generic rectangle as you traverse the interval of integration.

5. Set up an integral or sum of integrals for the area of the region. Take advantage of any symmetry properties the region may possess.
6. Evaluate the integral or integrals.

EXAMPLE 3

Find the area of the bounded region above the graph of $y = 3|x|/2$ and below the parabola $y = 1 + x^2/2$.

The region is shown in Figurc 7.1.6. To determine the points of intersection

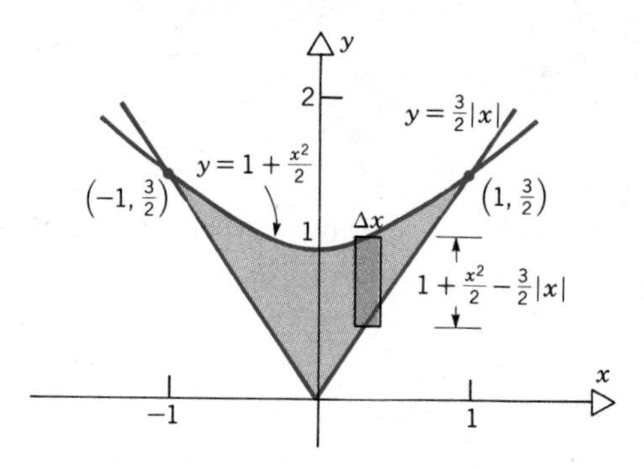

Figure 7.1.6

we must solve the two equations simultaneously. For $x \geq 0$ we have $|x| = x$; consequently

$$1 + \frac{x^2}{2} = \frac{3x}{2},$$

or

$$x^2 - 3x + 2 = (x - 2)(x - 1) = 0.$$

The relevant root is $x = 1$, corresponding to $y = \frac{3}{2}$. Similarly, for $x < 0$ we obtain $x = -1$, $y = \frac{3}{2}$. Observe that the region is symmetric about the y-axis. Therefore we can proceed by finding the area of one half of the region (that part for which $x \geq 0$) and then multiplying by two. Thus we have

$$\begin{aligned} A &= 2\int_0^1 \left(1 + \frac{x^2}{2} - \frac{3}{2}x\right) dx \\ &= 2\left(x + \frac{x^3}{6} - \frac{3}{4}x^2\right)\Bigg|_0^1 \\ &= 2\left(1 + \frac{1}{6} - \frac{3}{4}\right) = 2 \cdot \frac{5}{12} = \frac{5}{6}. \end{aligned} \tag{8}$$

Note that, if we fail to take advantage of the symmetry of the region, then the calculation becomes a little longer. In that case we have, from Eq. 2,

$$A = \int_{-1}^1 \left(1 + \frac{x^2}{2} - \frac{3}{2}|x|\right) dx.$$

We must then split the integral into two parts because $|x|$ has different expressions depending on whether $x \geq 0$ or $x \leq 0$. Hence

$$A = \int_{-1}^{0} \left(1 + \frac{x^2}{2} + \frac{3}{2}x\right) dx + \int_{0}^{1} \left(1 + \frac{x^2}{2} - \frac{3}{2}x\right) dx,$$

and evaluating these integrals yields the same result as before. ∎

EXAMPLE 4

Determine the area of the region bounded by the curves $y = x$ and $y = x^2$ from $x = 0$ to $x = 2$.

First we draw a sketch of the two curves as shown in Figure 7.1.7. Note that they intersect at (1, 1). Next we draw two generic rectangles and observe from Figure 7.1.7 that the height of the rectangle is $x - x^2$ for $0 \leq x \leq 1$ and is $x^2 - x$ for $1 \leq x \leq 2$. Thus

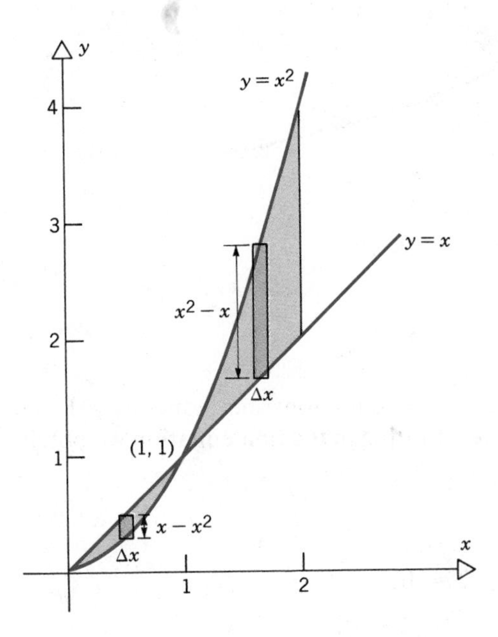

Figure 7.1.7

$$\begin{aligned} A &= \int_{0}^{1} (x - x^2)\, dx + \int_{1}^{2} (x^2 - x)\, dx \\ &= \left(\frac{x^2}{2} - \frac{x^3}{3}\right)\Bigg|_0^1 + \left(\frac{x^3}{3} - \frac{x^2}{2}\right)\Bigg|_1^2 \\ &= \left(\frac{1}{2} - \frac{1}{3}\right) + \left[\left(\frac{8}{3} - 2\right) - \left(\frac{1}{3} - \frac{1}{2}\right)\right] \\ &= 1. \ \blacksquare \end{aligned} \tag{9}$$

EXAMPLE 5

Determine the area of the bounded region between the curves $y^2 = x + 1$ and $y = x - 1$.

First we sketch the curves. The curve $y^2 = x + 1$ is a parabola that opens to the right, and the curve $y = x - 1$ is a straight line (see Figure 7.1.8*a*). The

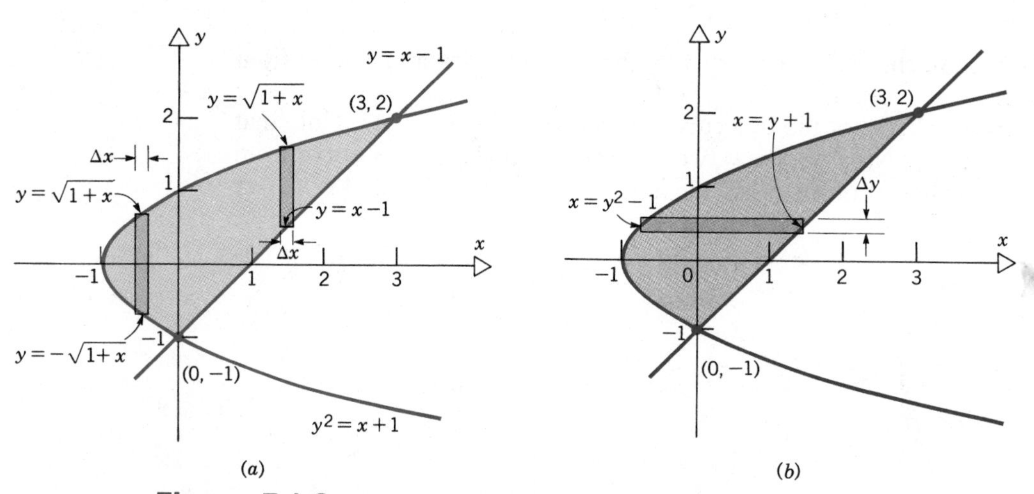

Figure 7.1.8 (*a*) Vertical area elements. (*b*) Horizontal area elements.

points of intersection are found by solving the two equations simultaneously. From the second equation we have $x = 1 + y$; substituting in the first equation we obtain

$$y^2 = (1 + y) + 1,$$

or

$$y^2 - y - 2 = 0,$$

or

$$(y - 2)(y + 1) = 0.$$

Corresponding to $y = 2$ we find $x = 3$, and corresponding to $y = -1$ we find $x = 0$. Thus the points of intersection are $(0, -1)$ and $(3, 2)$.

Next we draw a generic element of the approximating sum. Notice that for $-1 \le x \le 0$ the region is bounded by the upper branch of the parabola, $y = \sqrt{1 + x}$, and the lower branch of the parabola, $y = -\sqrt{1 + x}$. For $0 \le x \le 3$, however, the lower boundary of the region is $y = x - 1$. Thus we express the total area as the sum of the areas of two regions.

$$A = \int_{-1}^{0} [\sqrt{1+x} - (-\sqrt{1+x})]\,dx + \int_{0}^{3} [\sqrt{1+x} - (x-1)]\,dx$$

$$= 2\int_{-1}^{0} \sqrt{1+x}\,dx + \int_{0}^{3} [\sqrt{1+x} - (x-1)]\,dx$$

$$= 2\left[\frac{(1+x)^{3/2}}{3/2}\right]\Bigg|_{-1}^{0} + \left[\frac{(1+x)^{3/2}}{3/2} - \left(\frac{x^2}{2} - x\right)\right]\Bigg|_{0}^{3}$$

$$= 2\left(\frac{2}{3}\right) + \left[\frac{2}{3}(8-1) - \left(\frac{9}{2} - 3\right)\right] = \frac{9}{2}. \tag{10}$$

It is slightly easier to compute the area in question if we form Riemann sums with a generic element as shown in Figure 7.1.8*b*; that is, if we integrate with respect to y rather than with respect to x. Now the interval of integration is $-1 \le y \le 2$. The length of the generic element is $(1 + y) - (-1 + y^2)$ for all y in the interval of integration. Thus

$$A = \int_{-1}^{2} [(1+y) - (-1+y^2)]\,dy$$

$$= \int_{-1}^{2} (2 + y - y^2)\,dy$$

$$= \left[2y + \frac{y^2}{2} - \frac{y^3}{3}\right]\Bigg|_{-1}^{2}$$

$$= \left(4 + 2 - \frac{8}{3}\right) - \left(-2 + \frac{1}{2} + \frac{1}{3}\right) = \frac{9}{2}. \blacksquare \tag{11}$$

Example 5 illustrates that sometimes one can determine an area by integrating either with respect to x or with respect to y. In some cases one way turns out to be significantly easier than the other, so in approaching an integration problem one should consider both possibilities. In Example 5 integration with respect to x required the evaluation of two integrals while integration with respect to y required only one.

In computing the areas of fairly complicated regions it is frequently desirable to subdivide the original region into subregions and then to add the areas of the subregions. Often this can be done in more than one way. It is quite possible that one way of subdividing may lead to integrals that are difficult to evaluate, while another may lead to simpler integrals. Thus, in any moderately complicated problem, one should give some thought to how to minimize the required computational effort.

We conclude this section with a few additional remarks about the concept of area. Recall that in Sections 6.1 and 6.2 we used our intuitive understanding of area to motivate a definition of the Riemann integral. However, this definition was given analytically and without recourse to geometric ideas or language. Then, once the integral was defined, we defined the area of a given region as the value of a corresponding integral. So far we have restricted ourselves to certain regions bounded by the graphs of continuous functions and perhaps by portions of the coordinate axes or other straight line segments. However, once areas have been expressed as integrals, it is natural to extend the concept of area to any region for which the corresponding integral exists, even though the integrand may have numerous discontinuities. Indeed, one can replace the word "continuous" by "integrable" in Definition 6.2.2 and in the similar statements in this section. For example, if f and g are two integrable functions on the interval $[a, b]$, then the area between their graphs from $x = a$ to $x = b$ can be defined as

$$A = \int_a^b |f(x) - g(x)|\, dx, \tag{12}$$

which agrees with the previous definition (7) if f and g are continuous. In this way we can assign an area to some very complicated sets of points in the xy-plane, including some that we might not expect to have an area, as we usually think of the term.

PROBLEMS

In each of Problems 1 through 18, determine the area of the bounded region that satisfies the given conditions. It may be helpful to sketch the region and to show a typical rectangular element of area.

1. The region in the first quadrant bounded by $y = 2x + x^2/2$, $y = 0$, and $x = 2$.
2. The region between $y = \cos 2x$ and $y = 2$ from $x = \pi$ to $x = 2\pi$.
3. The region between the x-axis and $y = \sin x$ from $x = 0$ to $x = \pi$.
4. The region between $y = 2x$ and $y = -x^2$ from $x = 1$ to $x = 3$.
5. The region bounded by $y = 1 - x^2$ and $y = -3$.
6. The region bounded by $x = -y^2$ and $x = -4$.
7. The region below $y = 2 - x^2$ and above
$$y = \begin{cases} |x|, & x \le 0 \\ 0, & x \ge 0. \end{cases}$$
8. The region bounded by $x - 1 = y^2$, $x = 4y - 3$, and the x-axis.
9. The region bounded by $y = \sqrt{x}$, $y = 2 - x$, and the x-axis.
10. The region bounded by $y = (1 + x)^{-1/2}$, $y = 1$, and $x = 3$.
11. The region between $y = \sin x$ and $y = \sin(x - \pi/2)$ from $x = 0$ to $x = \pi$.
12. The region between $y = x^3 + 1$ and $y = x + 1$.
13. The region in the first quadrant bounded by $y = x^{-2}$, $y = x$, and $x = 3$.
14. The region between $x = y^2$ and $x = 4 - y^2$.
15. The region between $x = |y|$ and $x = 2 - y^2$.
16. The region bounded by $y = x^2 - 2x$ and $y = 3x$.
17. The region between $y = \sin x$ and $y = \cos x$ from $x = 0$ to $x = 2\pi$.
18. The region bounded by $x = 2y - y^2$ and $x + y = 0$.

In each of Problems 19 through 24, determine the area of the given region in two different ways; that is, use area elements parallel to the y-axis and then use elements parallel to the x-axis.

19. The region bounded by $x = y^2$ and $x = 1$.
20. The region bounded by $x = y^2$, $x + y = 2$, and $y = -1$ that contains the point $(1, 0)$.
21. The region in the first quadrant bounded by $y = x^4$, $y = 2 - x^4$, and $x = 0$.
22. The region bounded by $y = x^3$, $y = 2 - x$, and $y = 3x + 2$ that contains the point $(0, 1)$.

23. The region bounded by $y = x^3$, $x + y = 2$, and the y-axis.

24. The region bounded by $y = 4 - x^2$, $y = 3x$, and the x-axis.

In each of Problems 25 through 28, set up an integral whose value is the area of the given region. To set up an integral means to determine the integrand and the limits of integration: do not evaluate the integral.

25. The region in the first quadrant bounded by the y-axis, the line $y = \sqrt{3}x$, and the circle $x^2 + y^2 = 4$.

26. The region in the first quadrant bounded by the x-axis, the parabola $y = x^2/3$, and the circle $x^2 + y^2 = 4$.

27. The region common to the circles $x^2 + y^2 = a^2$ and $(x - a)^2 + (y - a)^2 = a^2$.

28. The region inside the circle $x^2 + y^2 = \frac{5}{2}$ and above the branch of the hyperbola $xy = 1$ that is in the first quadrant.

29. The region bounded by the curve $y^2 = x - 1$ and the line $x = 5$ is divided into two parts of equal area by the line $x = a$. Determine the value of a.

7.2 VOLUME

In this section we discuss the calculation of volumes by extending the procedures developed in the previous section for finding areas. We start with solids of revolution. A **solid of revolution** is obtained by revolving a region in the plane about a line in the plane. The line is called the **axis of revolution.** In Figure 7.2.1 we show the solids of revolution obtained by revolving the region bounded by the semicircle $y = \sqrt{4 - x^2}$ and the x-axis about different axes of revolution.

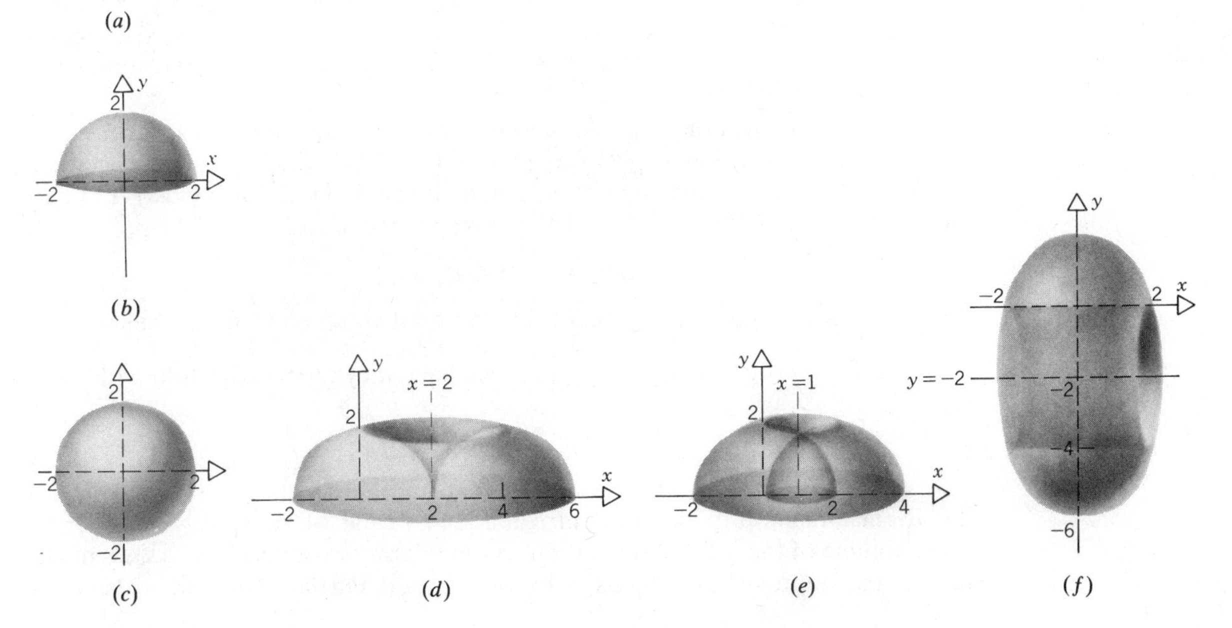

Figure 7.2.1 (*a*) A semicircular region R.
(*b*) R rotated about the y-axis.
(*c*) R rotated about the x-axis.
(*d*) R rotated about $x = 2$.
(*e*) R rotated about $x = 1$.
(*f*) R rotated about $y = -2$.

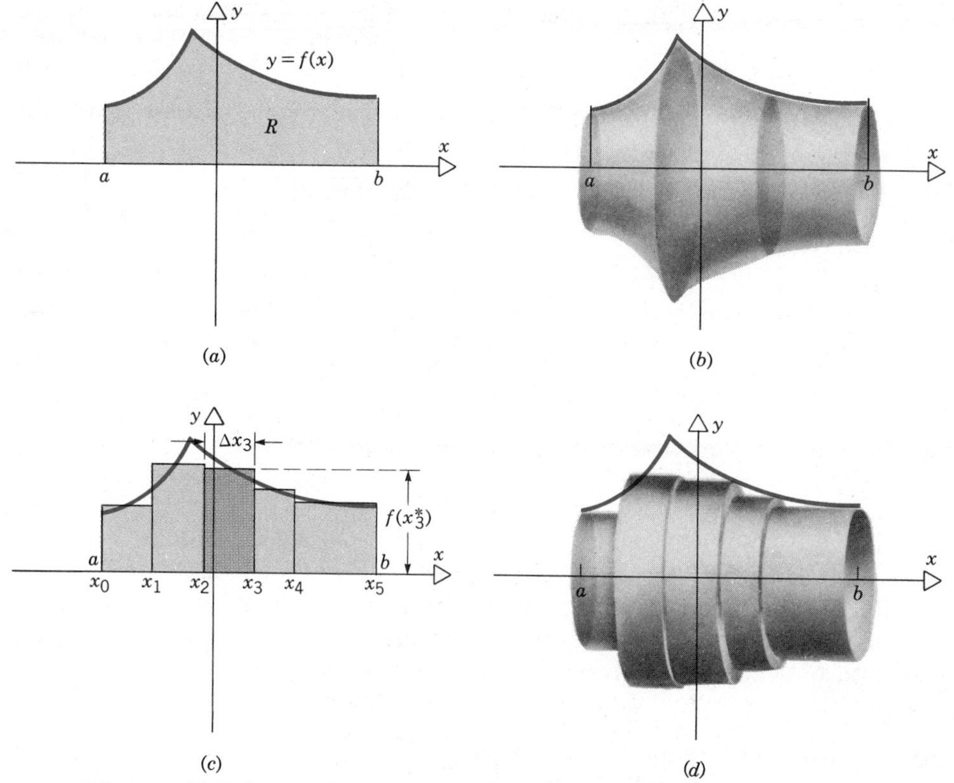

Figure 7.2.2 (*a*) The region R under the graph of f. (*b*) R rotated about the x-axis. (*c*) Rectangular area elements. (*d*) Cylindrical volume elements.

Suppose that the function f is nonnegative and continuous on $[a, b]$. Consider the solid of revolution obtained by revolving the region bounded by the curve $y = f(x)$, the x-axis, and the lines $x = a$ and $x = b$ about the x-axis; see Figures 7.2.2*a* and 7.2.2*b*. What do we mean by the volume of this solid of revolution and how do we compute it?

Let Δ be a partition of $[a, b]$, choose star points x_i^*, and construct the rectangles with bases Δx_i and heights $f(x_i^*)$ (see Figure 7.2.2*c*). When the ith rectangle is revolved about the x-axis we obtain a thin circular cylinder of radius $f(x_i^*)$ and thickness Δx_i (see Figure 7.2.2*d*). The volume of this circular cylinder is

$$\Delta V_i = \pi[f(x_i^*)]^2 \, \Delta x_i. \tag{1}$$

For the partition shown in Figure 7.2.2*c*, the total volume of the five cylinders is $\sum_{i=1}^{5} \Delta V_i$. For a partition $\{x_0, x_1, \ldots, x_n\}$ we have n cylinders with total volume

$$\sum_{i=1}^{n} \Delta V_i = \sum_{i=1}^{n} \pi[f(x_i^*)]^2 \, \Delta x_i. \tag{2}$$

This Riemann sum provides an approximation to what we intuitively consider to be the volume of the solid of revolution. As the norm of the partition, $\|\Delta\|$, is made smaller, this approximation tends to become better. Further, since f^2 is also continuous and hence integrable, we know that the Riemann sums (2) approach the corresponding definite integral as $\|\Delta\| \to 0$. Thus we state the following definition.

DEFINITION 7.2.1 Let the function f be continuous and nonnegative on $[a, b]$. The volume V of the solid of revolution obtained by revolving the region bounded by the curve $y = f(x)$, the x-axis, and the lines $x = a$ and $x = b$ about the x-axis is

$$V = \int_a^b \pi f^2(x)\, dx. \tag{3}$$

Note that $\pi f^2(x)$ is the area of the circular cross section of the solid of revolution at the point x on the axis of revolution. If we write $\pi f^2(x) = A(x)$ for this area, then we can express Eq. 3 in the form

$$V = \int_a^b A(x)\, dx \tag{4}$$

(see Figure 7.2.3). Thus to find the volume of a solid of revolution we determine the formula for the area of the cross section and then integrate along the axis of revolution.

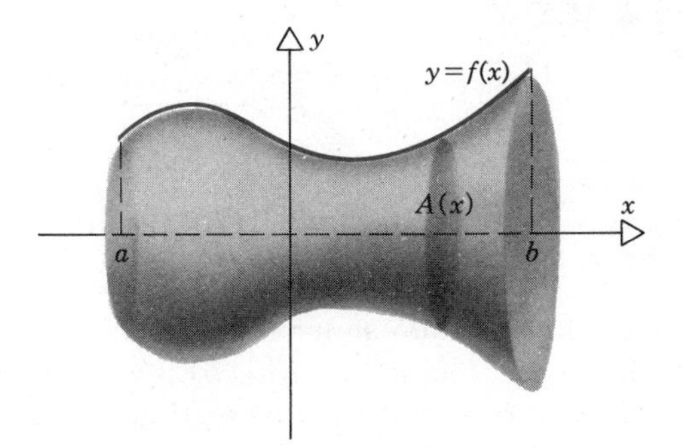

Figure 7.2.3

EXAMPLE 1

Determine the volume of a sphere of radius a by revolving a semicircular region about its diameter.

We rotate the semicircular region bounded by $y = f(x) = \sqrt{a^2 - x^2}$, $-a \le x \le a$, about the x-axis (see Figure 7.2.4). According to Eq. 3, the volume of the sphere is

$$\begin{aligned} V &= \int_{-a}^{a} \pi[\sqrt{a^2 - x^2}]^2\, dx = \pi \int_{-a}^{a} (a^2 - x^2)\, dx \\ &= \pi\left(a^2 x - \frac{x^3}{3}\right)\Bigg|_{-a}^{a} \\ &= \pi\left[\left(a^3 - \frac{a^3}{3}\right) - \left(-a^3 + \frac{a^3}{3}\right)\right] \\ &= \frac{4\pi a^3}{3}. \end{aligned}$$

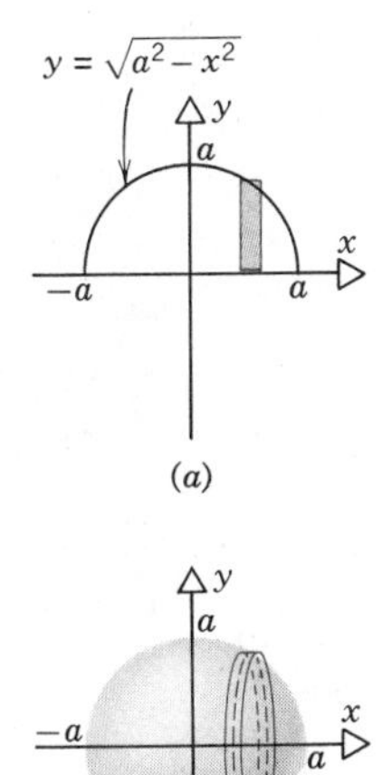

Figure 7.2.4

Because of the obvious symmetry of the semicircular region about the line $x = 0$ we could also have found the volume of a hemisphere and then multiplied by 2; that is,

$$V = 2\pi \int_0^a (a^2 - x^2)\, dx. \blacksquare$$

Method of slices

Let us now examine Eq. 4 more carefully. Suppose that we have a solid, not necessarily a solid of revolution, that is bounded by two parallel planes perpendicular to the x-axis at $x = a$ and $x = b$. Also suppose that we know the area $A(x)$ of the perpendicular cross section at the point x, and that the function A is continuous on $[a, b]$ (See Figure 7.2.5a). The volume of the solid is defined, and

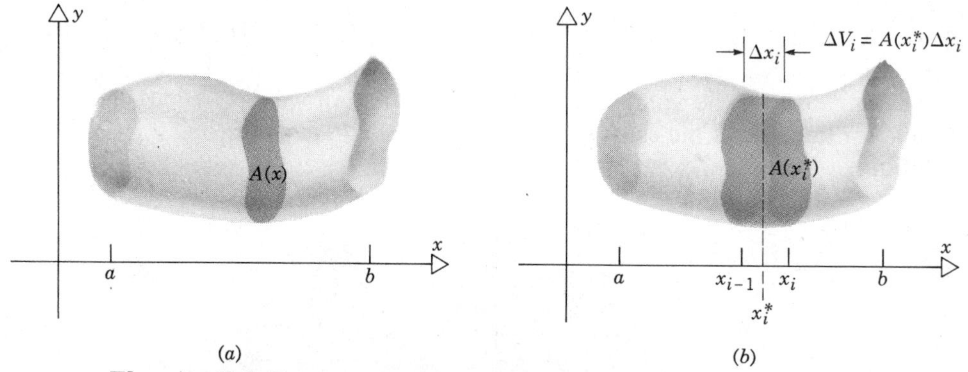

(a) (b)

Figure 7.2.5 (a) A volume with cross section $A(x)$. (b) A typical thin slice.

computed, in a manner exactly similar to the volume of the solid of revolution just discussed. The interval $[a, b]$ is partitioned, star points x_i^* are chosen, the areas of the corresponding cross sections $A(x_i^*)$ are determined, and elements of volume of base area $A(x_i^*)$ and width Δx_i are constructed as in Figure 7.2.5b. The Riemann sum

$$\sum_{i=1}^{n} \Delta V_i = \sum_{i=1}^{n} A(x_i^*)\, \Delta x_i$$

provides an approximation, which becomes better as $\|\Delta\|$ decreases, to what we anticipate to be the volume of the solid. Thus we state the following definition.

DEFINITION 7.2.2 Let a solid be bounded by two planes perpendicular to the x-axis at $x = a$ and $x = b$. Let $A(x)$ be the area of the cross section cut from the solid by a plane perpendicular to the x-axis at the point x. If the function A is continuous, then the volume of the solid is

$$V = \int_a^b A(x)\, dx. \tag{5}$$

This method of computing volumes is called the method of slices.

In the following examples we illustrate the use of the method of slices.

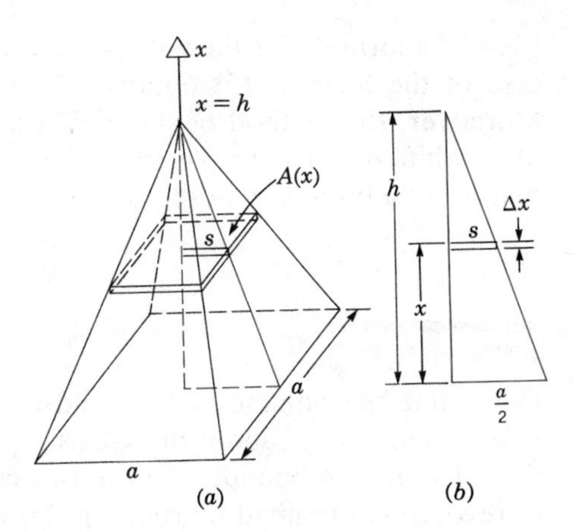

Figure 7.2.6

EXAMPLE 2

Determine the volume of a pyramid of square base of edge a and of height h.

We choose our axis as indicated in Figure 7.2.6a. To determine the area $A(x)$ of the cross section of the pyramid at height x above the base we make use of similar triangles as shown in Figure 7.2.6b. At height x, the area of the perpendicular cross section is $(2s)^2$. From Figure 7.2.6b we have, by similar triangles,

$$\frac{s}{(a/2)} = \frac{h - x}{h}.$$

Hence

$$s = \frac{a(h - x)}{2h} = \frac{a}{2}\left(1 - \frac{x}{h}\right).$$

Thus

$$A(x) = a^2\left(1 - \frac{x}{h}\right)^2,$$

and

$$\begin{aligned} V = \int_0^h A(x)\,dx &= a^2 \int_0^h \left(1 - \frac{x}{h}\right)^2 dx \\ &= a^2 \int_0^h \left(1 - \frac{2x}{h} + \frac{x^2}{h^2}\right) dx \\ &= a^2 \left(x - \frac{x^2}{h} + \frac{x^3}{3h^2}\right)\Bigg|_0^h \\ &= a^2 \left(h - h + \frac{h}{3}\right) \\ &= \frac{1}{3}a^2 h. \ \blacksquare \end{aligned}$$

The formula for the volume of a solid of revolution given in Eq. 3 is a special case of the formula (5) obtained by the method of slices with $A(x) = \pi f^2(x)$. Moreover, the method of slices allows us to compute the volume of other solids of revolution; for example, one obtained by rotating planar regions about a line that is not a boundary of the region. This is illustrated in the next three examples.

EXAMPLE 3

Determine the volume of the solid obtained by rotating the region bounded by $y = x^2$ and $x = y^2$ about the x-axis.

The region bounded by the two curves is shown in Figure 7.2.7*a*. The solid of revolution obtained by rotating this region about the x-axis is shown in Figure 7.2.7*b*. The first step in computing the volume of the solid by the method of slices is to compute the area of the cross section of the solid at position x, $0 \le x \le 1$. It is clear from Figure 7.2.7*b* that the cross section is a circular disk with a hole cut out of the center, like a washer (see Figure 7.2.7*c*). Such a region is called an annulus. Its area is the area of the outer circle minus the area of the inner circle. If r_0 is the outer radius and r_i is the inner radius, then the area is given by

$$\pi r_0^2 - \pi r_i^2 = \pi(r_0^2 - r_i^2).$$

For this particular problem we have $r_0 = \sqrt{x}$ and $r_i = x^2$. Thus

$$A(x) = \pi(x - x^4),$$

and

$$V = \int_0^1 A(x)\,dx = \pi \int_0^1 (x - x^4)\,dx$$

$$= \pi\left(\frac{x^2}{2} - \frac{x^5}{5}\right)\Bigg|_0^1$$

$$= \pi\left(\frac{1}{2} - \frac{1}{5}\right) = \frac{3\pi}{10}. \quad \blacksquare$$

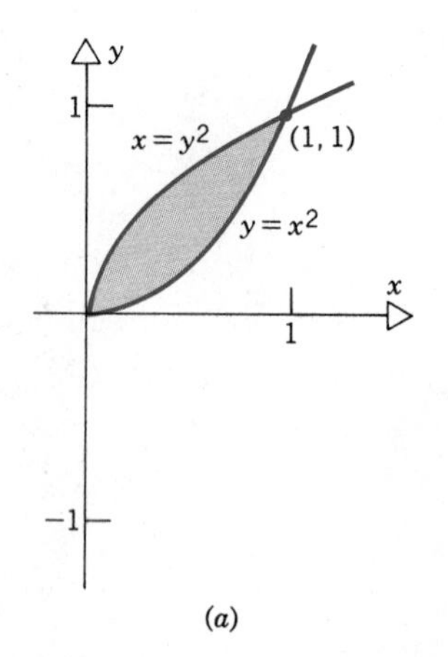

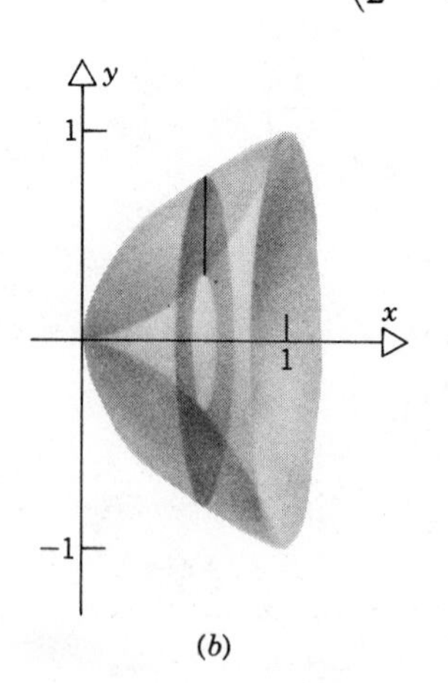

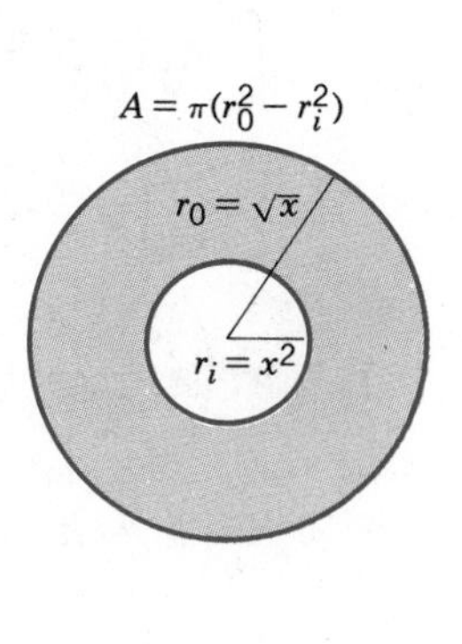

Figure 7.2.7

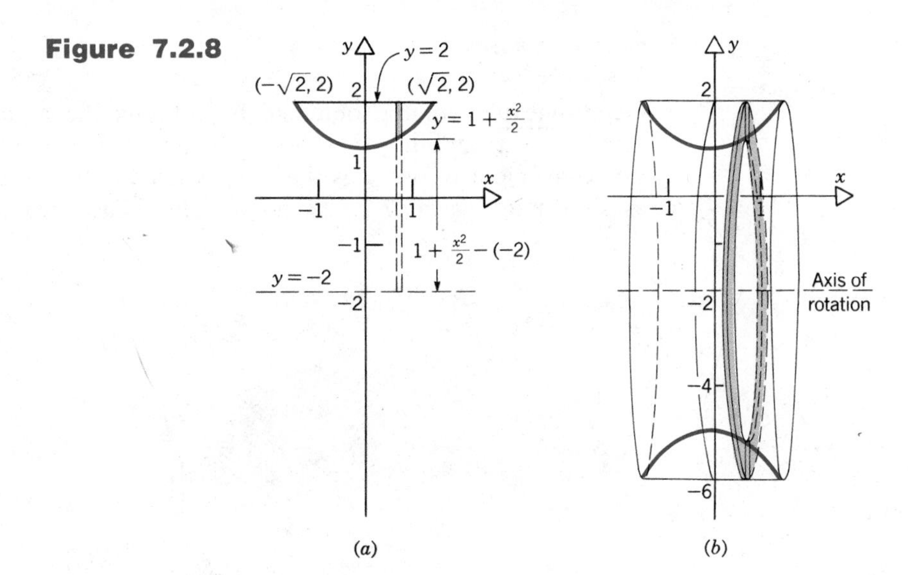

Figure 7.2.8

EXAMPLE 4

Determine the volume of the solid obtained by rotating the region bounded by $y - 1 = x^2/2$ and the line $y = 2$ about the line $y = -2$.

The planar region bounded by $y - 1 = x^2/2$ and the line $y = 2$ is sketched in Figure 7.2.8*a*. The points of intersection of the two curves are obtained by solving the two equations simultaneously. This gives $x = \pm\sqrt{2}$; hence the points of intersection are $(-\sqrt{2}, 2)$ and $(\sqrt{2}, 2)$. The solid of revolution obtained by rotating this region about the line $y = -2$ is shown in Figure 7.2.8*b*.

Again the area of the cross section of the solid at position x is an annulus. For this problem $r_0 = 2 - (-2) = 4$ and $r_i = (1 + \frac{1}{2}x^2) - (-2) = 3 + \frac{1}{2}x^2$. Thus

$$A(x) = \pi[4^2 - (3 + \tfrac{1}{2}x^2)^2],$$

and

$$V = \int_{-\sqrt{2}}^{\sqrt{2}} A(x)\,dx = 2\pi \int_0^{\sqrt{2}} [4^2 - (3 + \tfrac{1}{2}x^2)^2]\,dx,$$

where we have used the symmetry of the region about the y-axis. By evaluating the latter integral we obtain

$$\begin{aligned} V &= 2\pi \int_0^{\sqrt{2}} (7 - 3x^2 - \tfrac{1}{4}x^4)\,dx \\ &= 2\pi \left[7x - x^3 - \frac{x^5}{20}\right]\Bigg|_0^{\sqrt{2}} \\ &= 2\pi \left(7\sqrt{2} - 2\sqrt{2} - \frac{4\sqrt{2}}{20}\right) \\ &= 48\pi \frac{\sqrt{2}}{5}. \ \blacksquare \end{aligned}$$

EXAMPLE 5

Determine the volume obtained by rotating the region bounded by the curve $y = \sqrt{4 - x^2}$ and the x-axis about the line $x = 2$ (see Figure 7.2.1d). In this case it is convenient to use y as the integration variable. The generic element of area as shown in Figure 7.2.9 is an annulus. The area of the generic element is

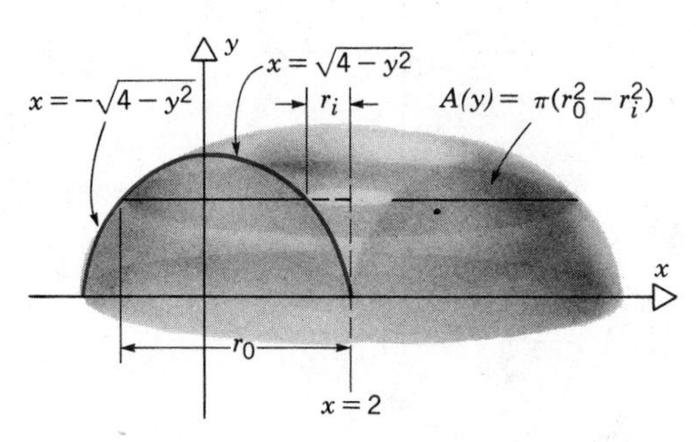

Figure 7.2.9

$\pi(r_0^2 - r_i^2)$ where r_0 is the outer radius and r_i is the inner radius. For a fixed value of y, we must find the corresponding x coordinates of the points on the curve closest to $x = 2$ and farthest away from $x = 2$. Since $y = \sqrt{4 - x^2}$ if follows that $y^2 = 4 - x^2$ and that

$$x = \pm\sqrt{4 - y^2}.$$

It is evident that $r_i = 2 - (\sqrt{4 - y^2})$ and $r_0 = 2 - (-\sqrt{4 - y^2})$. Thus

$$A(y) = \pi[(2 + \sqrt{4 - y^2})^2 - (2 - \sqrt{4 - y^2})^2],$$

and

$$V = \int_0^2 A(y)\, dy = \pi \int_0^2 [(2 + \sqrt{4 - y^2})^2 - (2 - \sqrt{4 - y^2})^2]\, dy$$

$$= \pi \int_0^2 \{[4 + 4\sqrt{4 - y^2} + (4 - y^2)] - [4 - 4\sqrt{4 - y^2} + (4 - y^2)]\}\, dy$$

$$= 8\pi \int_0^2 \sqrt{4 - y^2}\, dy.$$

As in Example 3 of Section 6.3, we can evaluate this integral by recognizing that it is the area of one quarter of a circle of radius 2. Thus

$$V = 8\pi \cdot \tfrac{1}{4}\pi \cdot 2^2 = 8\pi^2. \blacksquare$$

EXAMPLE 6

A hole of radius ρ is drilled through a sphere of radius $a > \rho$. The axis of the hole coincides with a diameter of the sphere. Determine the volume of the portion of the sphere that remains.

The sphere with the hole removed is shown in Figure 7.2.10a. We measure x along the axis of the hole (and along the diameter of the sphere) from the center of the sphere. First we determine the area of a cross section of the solid perpendicular to the x-axis. This area is the area of the section of the sphere minus the area of the section of the hole. Referring to Figure 7.2.10a we have

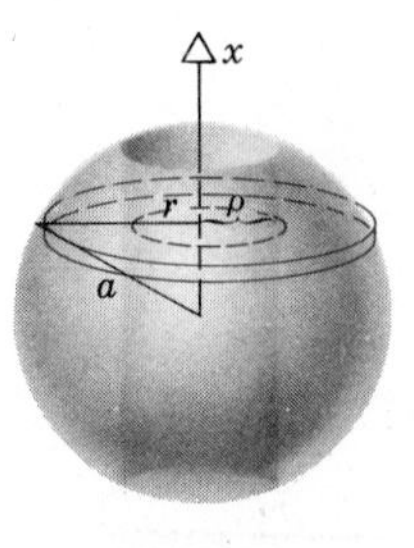

(a)

$$A(x) = \pi r^2 - \pi\rho^2$$
$$= \pi(a^2 - x^2) - \pi\rho^2.$$

Next we determine the interval of integration (the length of the hole). From Figure 7.2.10b it is clear that the upper limit of integration is $x = \sqrt{a^2 - \rho^2}$. Thus

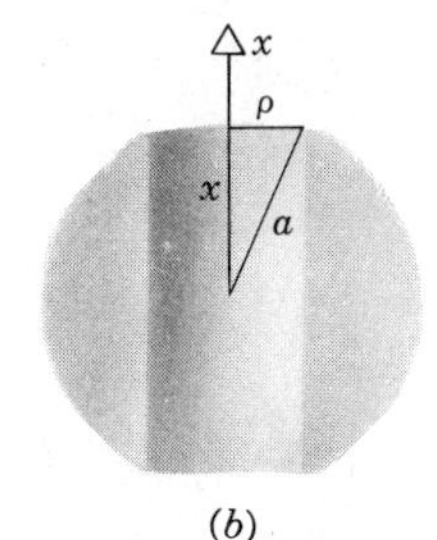

(b)

Figure 7.2.10

$$V = 2\int_0^{\sqrt{a^2-\rho^2}} \pi(a^2 - x^2 - \rho^2)\, dx$$
$$= 2\pi\left[(a^2 - \rho^2)\, x - \frac{x^3}{3}\right]\Bigg|_0^{\sqrt{a^2-\rho^2}}$$
$$= 2\pi\left[(a^2 - \rho^2)^{3/2} - \frac{1}{3}(a^2 - \rho^2)^{3/2}\right]$$
$$= \frac{4\pi}{3}(a^2 - \rho^2)^{3/2}.$$

When $\rho = 0$ this formula reduces to $4\pi a^3/3$, as it should. ∎

Method of Cylindrical Shells

Another method of calculating the volume of a solid of revolution is known as the method of cylindrical shells. In using this method we approximate the solid by a number of thin cylindrical shells rather than by a number of thin solid cylinders, as in the method of slices.

First we calculate the volume of a single cylindrical shell. Suppose that the rectangle shown in Figure 7.2.11a is revolved about the y-axis, thus forming the cylindrical shell in Figure 7.2.11b. The base of the rectangle is $\Delta x_i = x_i - x_{i-1}$

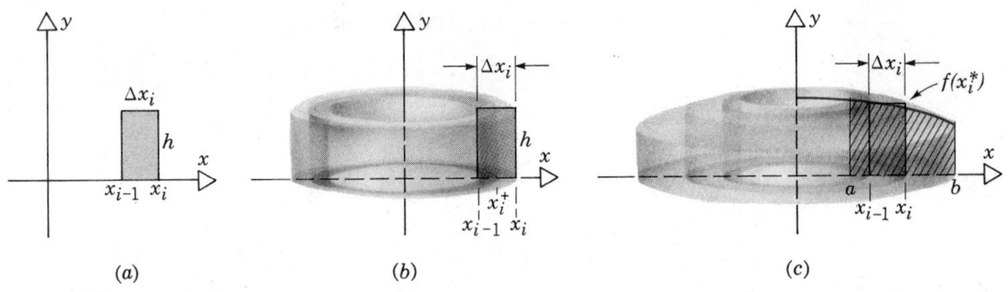

Figure 7.2.11 (a) A rectangular area element. (b) A cylindrical shell volume element. (c) A volume approximated by cylindrical shells.

and its height is h. The volume V_s of the shell is the difference between the volume of the outer cylinder and the volume of the inner cylinder, or

$$V_s = \pi x_i^2 h - \pi x_{i-1}^2 h = \pi(x_i^2 - x_{i-1}^2)h$$
$$= \pi(x_i - x_{i-1})(x_i + x_{i-1})h = 2\pi x_i^+ h \Delta x_i, \tag{6}$$

where $x_i^+ = (x_i + x_{i-1})/2$ is the mean radius of the shell, Δx_i is its thickness, and h is its height.

Next suppose that f is a continuous nonnegative function on $[a, b]$ with $a \geq 0$, and let R be the region bounded by the graph of f, the x-axis, and the lines $x = a$ and $x = b$. We wish to find the volume of the solid of revolution formed by revolving R about the y-axis as shown in Figure 7.2.11c. To approximate the volume of this solid we partition the interval $[a, b]$ and for each subinterval choose $x_i^* = x_i^+ = (x_i + x_{i-1})/2$. Then we form the rectangle with base Δx_i and height $f(x_i^+)$ and revolve it about the y-axis. The resulting cylindrical shell is shown in Figure 7.2.11c and has volume $2\pi x_i^+ f(x_i^+)\,\Delta x_i$. The total volume of all of the n shells formed in this manner is

$$2\pi \sum_{i=1}^{n} x_i^+ f(x_i^+)\,\Delta x_i. \tag{7}$$

The sum (7) approximates the volume V of the solid of revolution, and in general the approximation becomes better as the partition is made finer. Since $xf(x)$ is continuous, the Riemann sum (7) approaches the value of the corresponding integral as $\|\Delta\| \to 0$, so we obtain

$$V = 2\pi \int_a^b xf(x)\,dx. \tag{8}$$

The method of shells provides an alternative way of calculating the volume of certain solids. Sometimes the integral (8) proves to be simpler than the one resulting from the method of slices.

EXAMPLE 7

Consider the region in Figure 7.2.12 between the parabola $y = 4 - x^2$ and the x-axis from $x = 1$ to $x = 2$. Find the volume of the solid formed by revolving this region about the y-axis.

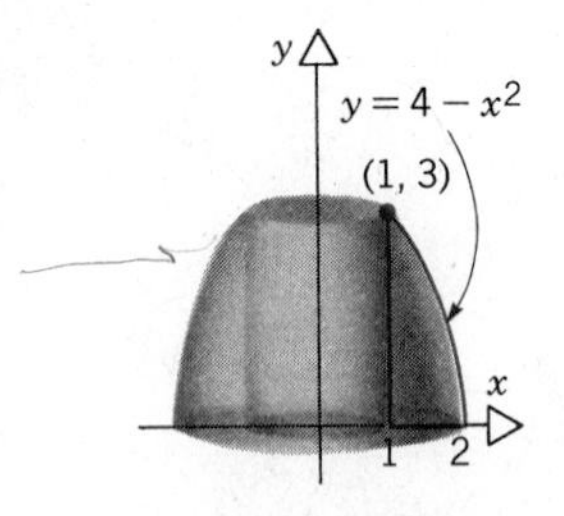

Figure 7.2.12

From Eq. 8 the desired volume is

$$V = 2\pi \int_1^2 x(4 - x^2)\,dx = 2\pi \int_1^2 (4x - x^3)\,dx$$
$$= 2\pi \left(2x^2 - \frac{x^4}{4}\right)\Bigg|_1^2$$
$$= 2\pi \left(4 - \frac{7}{4}\right) = \frac{9\pi}{2}.$$

The same value is also readily obtained by using the method of slices and integrating with respect to y. ∎

PROBLEMS

In each of Problems 1 through 16, determine the volume of the solid generated by rotating the given planar region about the given axis. It may be helpful to sketch the solid of revolution with a generic cross section for each problem.

1. The region bounded by $y = 2x^{1/4}$, the x-axis, and the line $x = 4$ rotated about the x-axis.
2. The region bounded by $y = x^2$, the x-axis, and the line $x = 2$ rotated about the x-axis.
3. The region in Problem 2 rotated about the line $x = 2$.
4. The triangle in the xy-plane with vertices at $(0, 0)$, $(h, 0)$, and (h, a) rotated about the x-axis. (Note that this is a circular cone of height h and radius a.)
5. The region bounded by $y = \cos x$, $0 \le x \le \pi/2$, the x-axis, and the y-axis rotated about the x-axis.
6. The region bounded by $y = \sqrt{x}$, the x-axis, and the line $x = 4$ rotated about the line $x = 4$.
7. The region in Problem 6 rotated about the line $x = 0$.
8. The region in Problem 6 rotated about the line $y = 2$.
9. The region bounded by $y = x^2$ and $y = x$ rotated about the x-axis.
10. The region in Problem 9 rotated about the y-axis.
11. The region in Problem 9 rotated about the line $x = 1$.
12. The region in Problem 9 rotated about the line $y = 1$.
13. The region bounded by $y = 2/(\cos 2x)$, the y-axis, the x-axis, and the line $x = \pi/6$ rotated about the x-axis.
14. The region bounded by $y = \cos x$, $y = \sin x$, and the y-axis, $0 \le x \le \pi/4$, rotated about the x-axis.
15. The region bounded by $x = y^2$, the line $x = 4$, and the x-axis rotated about the line $x = 6$.
16. The region in Problem 15 rotated about the line $x = 2$.

In each of Problems 17 through 20, use the method of cylindrical shells to compute the volume of the solid generated by rotating the given planar region about the given axis.

17. The region bounded by $y = x^2$, $y = 0$, and $x = 2$ rotated about the y-axis.
18. The region bounded by $y = x^2$, $y = 1$, and $x = 0$ rotated about the y-axis.
19. The region bounded by $y = x^2$ and $y = 2x$ rotated about the y-axis.
20. The region bounded by $x - 1 = y^2$ and $y = \frac{1}{2}(x - 1)$ rotated about the x-axis.

In each of Problems 21 and 22, set up an integral (or integrals) using the method of slices and the method of cylindrical shells for the volume of the solid of revolution generated by rotating the given planar region about the given axis. Use the simpler expression to determine the volume.

21. The region bounded by $y = x^2 - 2x$ and $y = 3x$ rotated about the y-axis.
22. The region bounded by $y = x^2 - 2x$ and $y = 3x$ rotated about the line $y = -1$.
23. Find the volume obtained by rotating the region bounded by $y = 3x - x^2$ and $y = -x$ about the y-axis.
24. Determine the volume of an ellipsoid of revolution obtained by rotating the ellipse
$$\frac{x^2}{a^2} + \frac{y^2}{b^2} = 1$$
about the x-axis.
25. The base of a solid is the disk $x^2 + y^2 \le a^2$. Each plane section of the solid cut out by a plane perpendicular to the x-axis is an equilateral triangle with one edge of the triangle in the base of the solid. Find the volume of the solid.
26. The base of a solid is the region $b^2x^2 + a^2y^2 \le a^2b^2$. Each plane section of the solid cut out by a plane perpendicular to the x-axis is a square with one edge of the square in the base of the solid. Find the volume of the solid.
27. A hemispherical bowl of radius a is filled with water to a depth $h < a$. Find the volume of the water.
28. The circle $x^2 + y^2 = a^2$ is rotated about the line $x = b$, where $b > a$, to form a torus (doughnut). Find the volume of the solid.

29. A wedge of cheese is cut from a right circular cylinder of radius a by two planes as shown in Figure 7.2.13. Find the volume of the wedge.

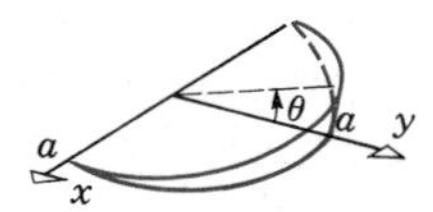

Figure 7.2.13

30. Two right circular cylinders of radius a intersect at right angles; that is, their axes meet at right angles (see Figure 7.2.14). What is the volume of the solid of intersection?

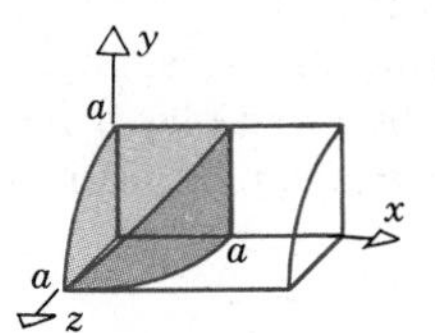

Figure 7.2.14

31. Suppose that a solid is bounded by the planes $x = 0$ and $x = h$, and that the cross section of the solid in a plane perpendicular to the x-axis has area $A(x) = a + bx + cx^2$ for each x. Here a, b, and c are constants. Show that the volume V of the solid is

$$V = \tfrac{1}{6}(B_0 + 4M + B_1)h$$

where B_0, M, and B are the areas of the cross sections at $x = 0$, $x = h/2$, and $x = h$, respectively. Such a solid is called a **prismatoid** and the formula for the volume is called the prismoidal formula. Verify that this formula agrees with that for the volume of a frustum of the cone shown in Figure 7.2.15 that was derived in solid geometry.

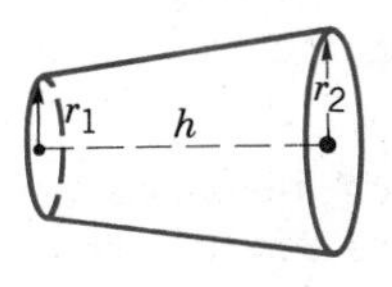

Figure 7.2.15

32. Sketch the curve $y = x^{-2/3}$ for $x > 0$.

(a) Show that the area A_b bounded by the curve $y = x^{-2/3}$, the x-axis, and the lines $x = 1$ and $x = b (b > 1)$ is $A_b = 3(b^{1/3} - 1)$.

(b) Show that the volume V_b obtained by rotating the region of Part (a) about the x-axis is $V_b = 3\pi(1 - b^{-1/3})$.

(c) Show that $A_b \to \infty$ as $b \to \infty$, but that $V_b \to 3\pi$ as $b \to \infty$. Thus we have the remarkable situation that an "infinite area" rotated about an axis gives a finite volume.

(d) Show that the area A_ϵ bounded by the curve $y = x^{-2/3}$, the x-axis, and the lines $x = \epsilon$ $(0 < \epsilon < 1)$ and $x = 1$ is $A_\epsilon = 3(1 - \epsilon^{1/3})$.

(e) Show that the volume V_ϵ obtained by rotating the region of Part (d) about the x-axis is $V_\epsilon = 3\pi(\epsilon^{-1/3} - 1)$.

(f) Show that $A_\epsilon \to 3$ as $\epsilon \to 0+$, but that $V_\epsilon \to \infty$ as $\epsilon \to 0+$. Thus we have the opposite situation that a finite area rotated about an axis gives an "infinite volume."

7.3 ARC LENGTH AND SURFACE AREA

In this section we take up two interesting geometric problems: that of finding the length of the arc between two points on a given curve, and that of finding the area of a surface of revolution.

Arc length

We start by asking how we can derive a formula for the length of the graph of the function $y = f(x)$ on the interval $[a, b]$. For the moment we assume that the function f is continuous; later we will also assume that f has a continuous derivative.

Since we know how to calculate the length of a straight line segment, it is natural to generalize this procedure so as to obtain the length of a curved arc. To do this we partition the interval $[a, b]$, join the successive points on the curve by straight lines, and consider the sum of the lengths of these straight line segments

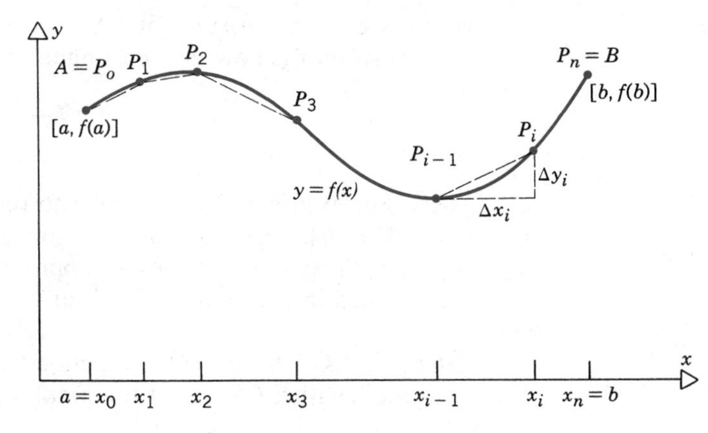

Figure 7.3.1 A polygonal approximation to the graph of f.

as an approximation to the length of the curve (see Figure 7.3.1). The length of the straight line segment connecting the points P_{i-1} and P_i is

$$\sqrt{(\Delta x_i)^2 + (\Delta y_i)^2},$$

where $\Delta x_i = x_i - x_{i-1}$ and $\Delta y_i = f(x_i) - f(x_{i-1})$. The sum

$$\sum_{i=1}^{n} \sqrt{(\Delta x_i)^2 + (\Delta y_i)^2} \tag{1}$$

provides an approximation to our intuitive concept of the length of the curve from P_0 to P_n. The next step is to consider what happens as the norm $\|\Delta\|$ of the partition is decreased. If the sum in Eq. 1 approaches a definite number as $\|\Delta\| \to 0$, then we define this number to be the length of the curve $y = f(x)$ from $A[a, f(a)]$ to $B[b, f(b)]$:

$$L(A, B) = \lim_{\|\Delta\| \to 0} \sum_{i=1}^{n} \sqrt{(\Delta x_i)^2 + (\Delta y_i)^2}. \tag{2}$$

If the limit in Eq. 2 exists, then the curve is said to be **rectifiable,** otherwise it is called **nonrectifiable.**

For computational purposes it is desirable to restate Eq. (2) as an integral. Unfortunately, the sum in Eq. 2 is not in the form of a Riemann sum, so we cannot immediately pass to an integral formula for the length of the curve. However, if we assume that f is differentiable on (a, b), then we can rewrite Eq. 2 in a suitable form. According to the mean value theorem (Theorem 4.1.3), there exists an x_i^* in each subinterval (x_{i-1}, x_i) such that

$$\begin{aligned} \Delta y_i = f(x_i) - f(x_{i-1}) &= f'(x_i^*)(x_i - x_{i-1}) \\ &= f'(x_i^*)\,\Delta x_i. \end{aligned} \tag{3}$$

Hence, the sum (1) can be rewritten in the form

$$\sum_{i=1}^{n} \sqrt{1 + [f'(x_i^*)]^2}\,\Delta x_i, \tag{4}$$

which is a Riemann sum for the function $g(x) = \sqrt{1 + [f'(x)]^2}$. Some further hypothesis on f is now needed to guarantee that the corresponding integral

$$\int_a^b \sqrt{1 + [f'(x)]^2}\, dx \tag{5}$$

exists. For example, it is sufficient to require that f' be continuous on the closed interval $[a, b]$. Then, as $\|\Delta\| \to 0$, any Riemann sum of the form (4), no matter how the star points are chosen, approaches the integral (5). Thus we obtain the result stated in Definition 7.3.1, our working definition of arc length.

DEFINITION 7.3.1 If the function f is such that f' is continuous on $[a, b]$, then the length $L(A, B)$ of the graph of f from $x = a$ to $x = b$ is

$$L(A, B) = \int_a^b \sqrt{1 + [f'(x)]^2}\, dx = \int_a^b \sqrt{1 + \left(\frac{dy}{dx}\right)^2}\, dx. \tag{6}$$

EXAMPLE 1

Determine the length of the curve $y = \frac{1}{3}x^{3/2}$ from (0, 0) to (9, 9).

The curve is sketched in Figure 7.3.2. Since

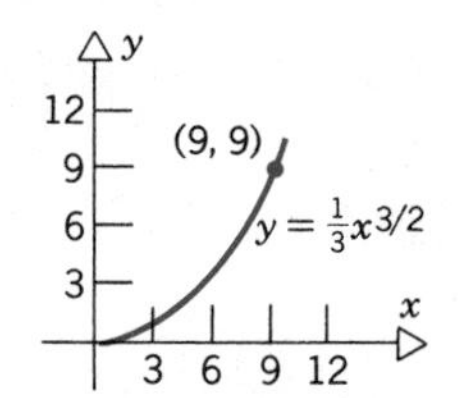

Figure 7.3.2

$$\frac{dy}{dx} = \frac{1}{3} \cdot \frac{3}{2} x^{1/2} = \frac{1}{2} x^{1/2},$$

we have, from Eq. 6,

$$L = \int_0^9 \sqrt{1 + \tfrac{1}{4}x}\, dx.$$

This integral can be evaluated by the substitution $u = 1 + x/4$. Then $du = dx/4$; further, $x = 0$ and $x = 9$ correspond to $u = 1$ and $u = 13/4$, respectively. Therefore

$$L = 4\int_1^{13/4} u^{1/2}\, du = 4 \cdot \frac{2}{3} u^{3/2} \Big|_1^{13/4}$$

$$= \frac{8}{3}\left[\left(\frac{13}{4}\right)^{3/2} - 1\right] \cong 12.96. \blacksquare$$

For curves that can be written as $x = g(y)$, where $g'(y)$ is bounded and continuous, the roles of x and y can be reversed and the formula (6) for the length of a curve takes the form

$$L(A, B) = \int_c^d \sqrt{1 + [g'(y)]^2}\, dy. \tag{7}$$

If the curve can be written either as $y = f(x)$ or as $x = g(y)$, then we can compute its length by using either Eq. 6 or Eq. 7. Sometimes it is much easier to use one rather than the other, as the next example illustrates.

EXAMPLE 2

Determine the length of the curve $y = (x - 1)^{2/3}$ between $x = 1$ and $x = 9$.

The curve is sketched in Figure 7.3.3. On taking the derivative of y, we have

$$\frac{dy}{dx} = \frac{2}{3}(x - 1)^{-1/3}.$$

Note that dy/dx becomes unbounded as $x \to 1+$; hence, we cannot use Eq. 6 on the interval $1 \le x \le 9$ for determining the length of the curve.

On the other hand, if we rewrite the equation of the curve as $x = 1 + y^{3/2}$, then the calculation is quite simple. Corresponding to $x = 1$ we have $y = 0$, and to $x = 9$ we have $y = 4$; thus the interval of integration is $[0, 4]$ if y is considered as the independent variable. Further

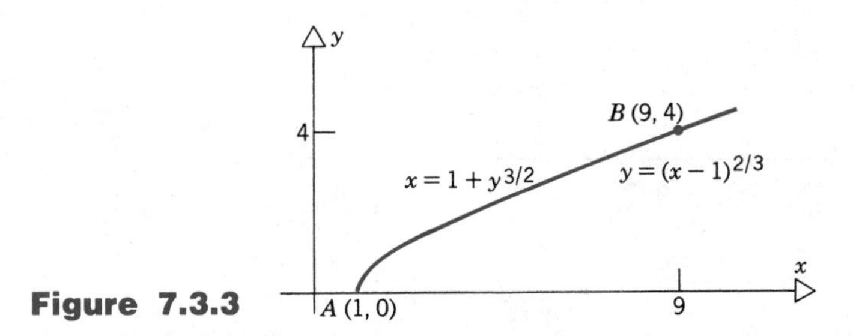

Figure 7.3.3

$$\frac{dx}{dy} = \frac{3}{2}y^{1/2},$$

which is continuous for $y \geq 0$, so we can use Eq. 7. Consequently

$$L = \int_0^4 \sqrt{1 + (\tfrac{3}{2}y^{1/2})^2}\, dy$$

$$= \int_0^4 \sqrt{1 + \tfrac{9}{4}y}\, dy.$$

This integral is similar to the one in Example 1. Using the substitution $u = 1 + 9y/4$, we have

$$L = \frac{8}{27}\left(1 + \frac{9}{4}y\right)^{3/2}\Bigg|_0^4$$

$$= \frac{8}{27}[10\sqrt{10} - 1] \cong 9.07. \blacksquare$$

Arc length using parametric equations

Now consider the case of a curve given by the parametric equations

$$x = f(t), \qquad y = g(t), \qquad \alpha \le t \le \beta, \tag{8}$$

where we assume that f and g have continuous first derivatives. To determine the length of this curve we subdivide $[\alpha, \beta]$ by a partition Δ, calculate the corresponding values $x_i = x(t_i)$ and $y_i = y(t_i)$ and proceed as before (see Figure 7.3.4). Let A and B be the end points $[f(\alpha), g(\alpha)]$ and $[f(\beta), g(\beta)]$, respectively, of the arc.

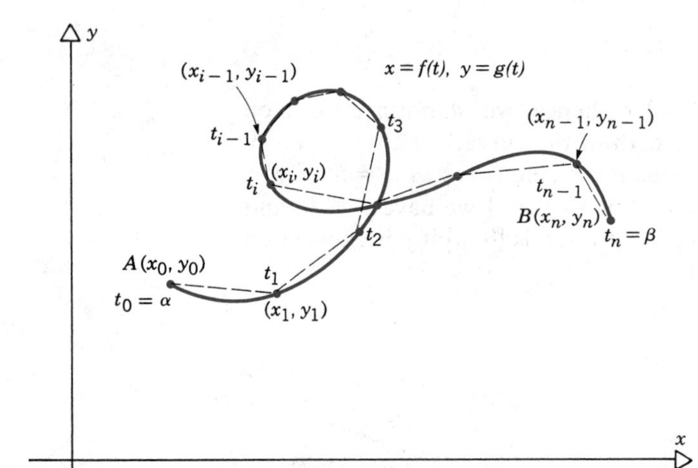

Figure 7.3.4 Polygonal approximation to the graph of $x = f(t), y = g(t)$.

Then

$$L(A, B) = \lim_{\|\Delta\| \to 0} \sum_{i=1}^{n} \sqrt{(\Delta x_i)^2 + (\Delta y_i)^2}$$

$$= \lim_{\|\Delta\| \to 0} \sum_{i=1}^{n} \sqrt{[f(t_i) - f(t_{i-1})]^2 + [g(t_i) - g(t_{i-1})]^2}. \tag{9}$$

Now according to the mean value theorem there exists a point t_i^* in (t_{i-1}, t_i) such that

$$f(t_i) - f(t_{i-1}) = f'(t_i^*)(t_i - t_{i-1})$$

and a point t_i^{**} in (t_{i-1}, t_i) such that

$$g(t_i) - g(t_{i-1}) = g'(t_i^{**})(t_i - t_{i-1}).$$

Thus

$$L(A, B) = \lim_{\|\Delta\| \to 0} \sum_{i=1}^{n} \sqrt{[f'(t_i^*)\, \Delta t_i]^2 + [g'(t_i^{**})\, \Delta t_i]^2}$$

$$= \lim_{\|\Delta\| \to 0} \sum_{i=1}^{n} \sqrt{[f'(t_i^*)]^2 + [g'(t_i^{**})]^2}\, \Delta t_i. \tag{10}$$

The last equation is almost, but not quite, in the proper form for a Riemann sum. The difficulty is that, in general, the points t_i^* and t_i^{**} are not the same. Nevertheless, by using the continuity of the functions f' and g', it can be shown that the limit of the sum of Eq. 10 is the corresponding Riemann integral. Hence

$$L(A, B) = \int_\alpha^\beta \sqrt{[f'(t)]^2 + [g'(t)]^2}\, dt = \int_\alpha^\beta \sqrt{\left(\frac{dx}{dt}\right)^2 + \left(\frac{dy}{dt}\right)^2}\, dt. \qquad (11)$$

EXAMPLE 3

Show that the circumference of the circle

$$x^2 + y^2 = r^2$$

is $2\pi r$.

Although it is possible to carry out the calculation in rectangular coordinates, it is simpler to use the natural parametric representation for the circle indicated in Figure 7.3.5, namely

$$x = r\cos\theta, \qquad y = r\sin\theta, \qquad 0 \le \theta \le 2\pi.$$

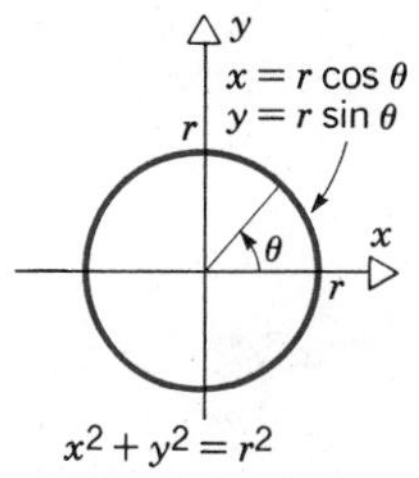

Figure 7.3.5 Parametric representation of the circle $x^2 + y^2 = r^2$.

Then

$$\frac{dx}{d\theta} = -r\sin\theta, \qquad \frac{dy}{d\theta} = r\cos\theta$$

and, by Eq. 11,

$$L = \int_0^{2\pi} \sqrt{(-r\sin\theta)^2 + (r\cos\theta)^2}\, d\theta = \int_0^{2\pi} r\, d\theta$$
$$= 2\pi r. \ \blacksquare$$

EXAMPLE 4

Determine the length L of one arch of the cycloid

$$x = a(\theta - \sin\theta), \qquad y = a(1 - \cos\theta), \qquad a > 0,$$

that was discussed in Example 6 of Section 5.5 (see Figure 7.3.6).

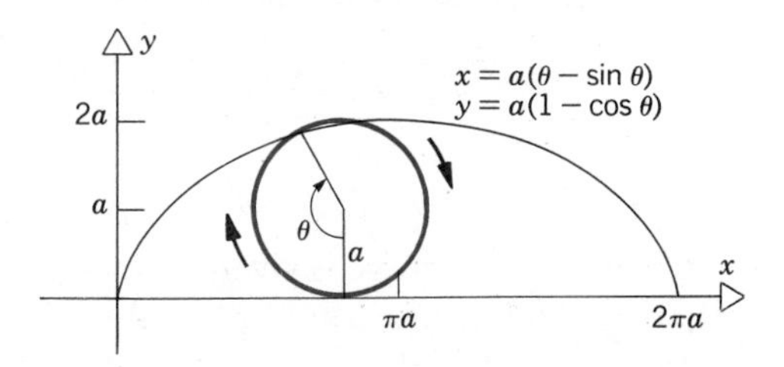

Figure 7.3.6 Parametric representation of a cycloid.

One arch of the cycloid corresponds to $0 \le \theta \le 2\pi$. We have

$$\frac{dx}{d\theta} = a(1 - \cos\theta), \qquad \frac{dy}{d\theta} = a\sin\theta,$$

so

$$\left(\frac{dx}{d\theta}\right)^2 + \left(\frac{dy}{d\theta}\right)^2 = a^2[1 - 2\cos\theta + \cos^2\theta + \sin^2\theta]$$
$$= 2a^2(1 - \cos\theta).$$

Hence

$$L = \int_0^{2\pi} \sqrt{2}\, a \sqrt{1 - \cos\theta}\, d\theta.$$

In order to evaluate this integral we make use of the trigonometric identity $\sin^2(\theta/2) = \frac{1}{2}(1 - \cos\theta)$. Then

$$L = \sqrt{2}\, a \int_0^{2\pi} \sqrt{2\sin^2\frac{\theta}{2}}\, d\theta = 2a \int_0^{2\pi} \left|\sin\frac{\theta}{2}\right| d\theta.$$

Since $\sin(\theta/2) \ge 0$ for $0 \le \theta \le 2\pi$, we can remove the absolute value bars in the integrand to obtain

$$L = 2a \int_0^{2\pi} \sin\frac{\theta}{2}\, d\theta = 2a\, 2\left(-\cos\frac{\theta}{2}\right)\Big|_0^{2\pi}$$
$$= 4a[-(-1) + 1] = 8a.$$

Sir Christopher Wren, the famous architect of St. Paul's Cathedral and many other churches in England, was the first (1658) to obtain this result—the length of the arch of a cycloid is eight times the radius of the generating circle. ∎

Consider the curve $y = f(z)$ for $a \le z \le b$ and suppose that the function f' is continuous. Then the function s defined by

$$s(x) = \int_a^x \sqrt{1 + [f'(z)]^2}\, dz = \int_a^x \sqrt{1 + \left(\frac{dy}{dz}\right)^2}\, dz, \qquad a \le x \le b, \quad (12)$$

gives the length of the curve from $[a, f(a)]$ to $[x, f(x)]$. The function s is called the **arc length function,** and by Theorem 6.4.2,

$$\frac{ds(x)}{dx} = \sqrt{1 + [f'(x)]^2} = \sqrt{1 + \left(\frac{dy}{dx}\right)^2}. \quad (13)$$

Formulas similar to Eqs. 12 and 13 hold if the curve is defined by $x = g(z)$ for $c \le z \le d$. If the curve is given parametrically by $x = f(z)$, $y = g(z)$ for $\alpha \le z \le \beta$, where f' and g' are continuous, then

$$s(t) = \int_\alpha^t \sqrt{[f'(z)]^2 + [g'(z)]^2}\, dz$$

$$= \int_\alpha^t \sqrt{\left(\frac{dx}{dz}\right)^2 + \left(\frac{dy}{dz}\right)^2}\, dz, \qquad \alpha \le t \le \beta, \tag{14}$$

and

$$\frac{ds(t)}{dt} = \sqrt{[f'(t)]^2 + [g'(t)]^2} = \sqrt{\left(\frac{dx}{dt}\right)^2 + \left(\frac{dy}{dt}\right)^2}. \tag{15}$$

A useful way to remember formulas (12) through (15) is by the relation

$$ds = \sqrt{(dx)^2 + (dy)^2}, \tag{16}$$

which can be thought of as a distorted Pythagorean relation (see Figure 7.3.7). In this formula dx and dy are differentials and ds is called the differential of arc length. In particular, if $y = f(x)$, then $dy = f'(x)\, dx$, $ds = s'(x)\, dx$ and, substituting for dy and ds in Eq. 16, we immediately obtain Eq. 13. Equation (15) is obtained in a similar manner. The length L of the curve is obtained by integrating the differential of arc length between the appropriate limits.

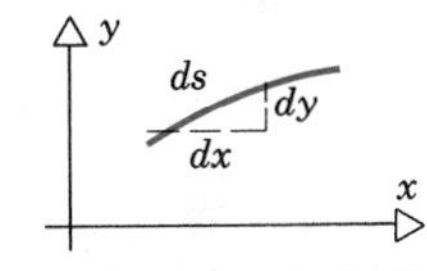

Figure 7.3.7

Now let us summarize what we have done. First, we have introduced the concept of a rectifiable curve. Next, we argued that if the appropriate functions $y = f(x)$, or $x = g(y)$, or $x = f(t)$, $y = g(t)$ have continuous derivatives, then the length of the corresponding curve can be computed from Eqs. 6, 7, and 11, respectively. The condition that the functions have continuous derivatives on the interval of interest is sufficient for the corresponding curve to be rectifiable but not necessary. However, some condition of this type is needed, since a function that is merely continuous on a closed interval may have a nonrectifiable graph. An example appears later in Problem 16 of Section 12.5.

Area of a surface of revolution

We will now derive a formula for the area of a surface of revolution. In our derivation we will need the following formula for the lateral area A of a frustum of a cone (see Figure 7.3.8):

$$A = \pi(r_1 + r_2)l. \tag{17}$$

In Eq. 16, r_1 and r_2 are the radii of the bases and l is the slant height.

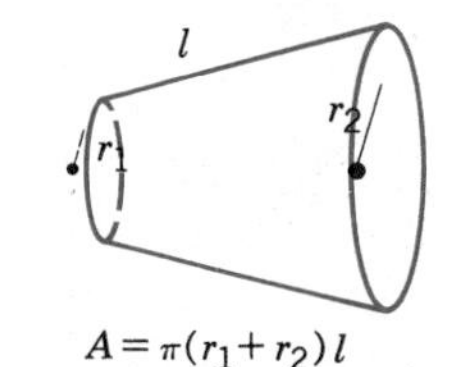

$A = \pi(r_1 + r_2)l$

Figure 7.3.8

Now suppose that the curve $y = f(x)$, $f(x) > 0$, is revolved about the x-axis (see Figure 7.3.9*a*). To compute the area of the surface of revolution we proceed in the now familiar way. We partition the interval $[a, b]$, let $y_i = f(x_i)$, and construct the straight lines l_i connecting the points (x_{i-1}, y_{i-1}) and (x_i, y_i) for $i = 1, 2, \ldots, n$. The area S of the surface is approximated by the sum of the areas of the individual frustums as shown in Figure 7.3.9*b*; thus

$$S \cong \sum_{i=1}^{n} \pi(y_{i-1} + y_i)l_i. \tag{18}$$

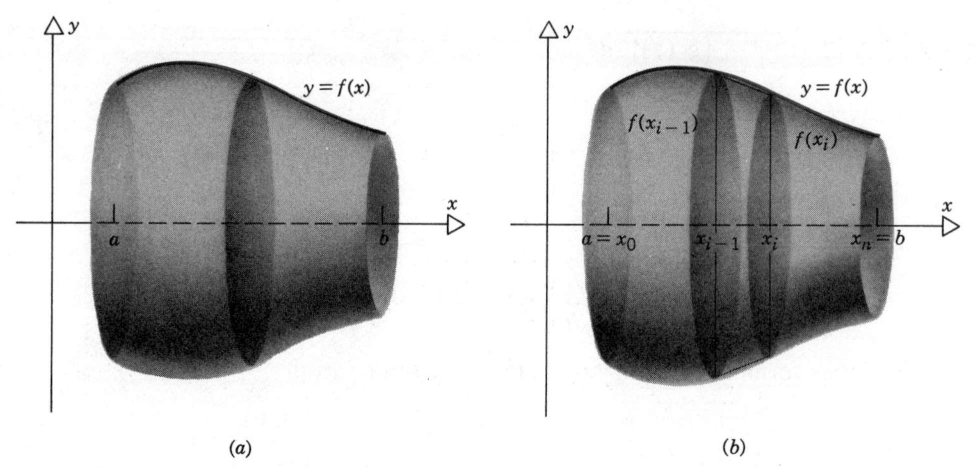

Figure 7.3.9 (*a*) Surface of revolution generated by $y = f(x)$. (*b*) Surface of revolution approximated by conical surface elements.

Now

$$\begin{aligned} l_i &= \sqrt{(y_i - y_{i-1})^2 + (x_i - x_{i-1})^2} \\ &= \sqrt{(\Delta y_i)^2 + (\Delta x_i)^2}, \end{aligned} \tag{19}$$

so Eq. 18 can be rewritten in the form

$$S \cong \pi \sum_{i=1}^{n} [f(x_{i-1}) + f(x_i)] \sqrt{1 + \left(\frac{\Delta y_i}{\Delta x_i}\right)^2}\, \Delta x_i. \tag{20}$$

The next step is to put Eq. 20 in a form so that the limit as $\|\Delta\| \to 0$ can be determined. Assuming that f has a continuous derivative, we can again use the mean value theorem to write

$$\begin{aligned} \Delta y_i = y_i - y_{i-1} &= f(x_i) - f(x_{i-1}) \\ &= f'(x_i^*)(x_i - x_{i-1}) \\ &= f'(x_i^*)\, \Delta x_i, \qquad x_i^* \in (x_{i-1}, x_i). \end{aligned} \tag{21}$$

Next we turn to the term $[f(x_{i-1}) + f(x_i)]$. Since f is differentiable, it is also continuous; therefore, by the intermediate value theorem (Theorem 2.6.3) f takes on all values between its maximum and minimum. In particular there is a point $x_i^+ \in [x_{i-1}, x_i]$ at which f is equal to the average of its values at the endpoints:

$$f(x_i^+) = \tfrac{1}{2}[f(x_i) + f(x_{i-1})]. \tag{22}$$

Substituting from Eqs. 21 and 22 in Eq. 20, we obtain

$$S \cong 2\pi \sum_{i=1}^{n} f(x_i^+) \sqrt{1 + [f'(x_i^*)]^2}\, \Delta x_i. \tag{23}$$

Unfortunately, Eq. 23 is not a Riemann sum, since both x_i^+ and x_i^* appear in the expression multiplying Δx_i. However, as we have mentioned in this section, using the continuity of f and f' it can be shown that as $\|\Delta\| \to 0$ the right side of Eq. 23 approaches the corresponding integral from a to b. Thus we are led to the following definition.

DEFINITION 7.3.2 Let the function f be nonnegative and have a continuous derivative on $[a, b]$. Then the area S of the surface generated by revolving the graph $y = f(x)$ about the x-axis is

$$S = 2\pi \int_a^b f(x) \sqrt{1 + [f'(x)]^2}\, dx. \tag{24}$$

EXAMPLE 5

Determine the surface area of a sphere of radius a.

We can generate a sphere of radius a by revolving a semicircle $y = f(x) = \sqrt{a^2 - x^2}$ about the x-axis (see Figure 7.3.10). Since

$$f'(x) = \frac{1}{2}(a^2 - x^2)^{-1/2}(-2x) = \frac{-x}{(a^2 - x^2)^{1/2}},$$

it follows from Eq. 24 that

$$\begin{aligned} S &= 2\pi \int_{-a}^{a} \sqrt{a^2 - x^2} \sqrt{1 + \frac{x^2}{a^2 - x^2}}\, dx \\ &= 2\pi \int_{-a}^{a} a\, dx \\ &= 2\pi a\, x \Big|_{-a}^{a} = 4\pi a^2. \end{aligned}$$ ∎

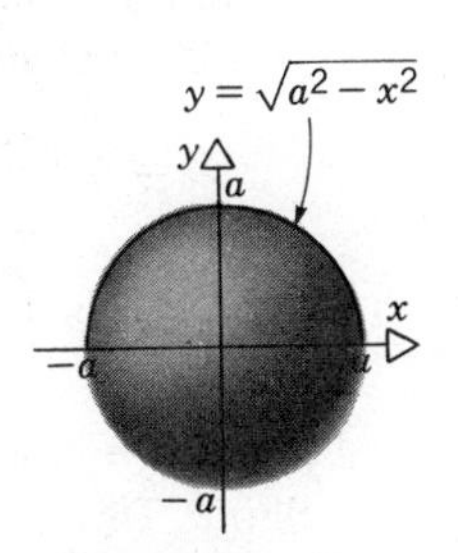

Figure 7.3.10

There are various generalizations of Eq. 24 for the case when a curve is revolved about the y-axis or about a line in the plane or when the curve is given in parametric form. These are discussed in Problems 29 through 34.

PROBLEMS

In each of Problems 1 through 10, determine the arc length of the given curve.

1. $y = 3x - 1$ from $x = 0$ to $x = 3$
2. $x = \sin t$, $y = 1 - \cos t$ from $t = 0$ to $t = \pi$
3. $y = 2x^{3/2} + 3$ from $x = 1$ to $x = 4$
4. $y^2 = 4(x + 1)^3$ from $(-1, 0)$ to $(3, 16)$
5. $4y^3 = 9x^2$ from $(0, 0)$ to $(2, 3^{2/3})$
6. $x = \frac{1}{2}t^2$, $y = \frac{1}{3}(2t + 1)^{3/2}$ from $t = 1$ to $t = 4$
7. $x = a(\cos t + t \sin t)$, $y = a(\sin t - t \cos t)$ from $t = 0$ to $t = 2\pi$, $a > 0$.
8. $x = t^3$, $y = 2t^2$ from $t = -1$ to $t = 0$
9. $y = \dfrac{x^3}{6} + \dfrac{1}{2x}$ from $x = 2$ to $x = 4$
10. $x = a\cos^3\theta$, $y = a\sin^3\theta$ from $\theta = 0$ to $\theta = \dfrac{\pi}{2}$, $a > 0$

11. Find the length (circumference) of the hypocycloid $x^{2/3} + y^{2/3} = a^{2/3}$ where $a > 0$. *Hint:* Introduce an appropriate parametric representation.

12. Find the length of the curve $(x/a)^{2/3} + (y/b)^{2/3} = 1$ in the first quadrant, where a and b are positive constants. *Hint:* Introduce an appropriate parametric representation.

13. Find the length of the arc of the curve $y = \sqrt{x}|x - 3|/3$ connecting $(1, \frac{2}{3})$ and $(3, 0)$.

14. Find the length of the arc of the curve $y = |x|\sqrt{2x + 3}/3$ connecting $(-1, \frac{1}{3})$ and $(3, 3)$.

15. Consider the ellipse $(x/a)^2 + (y/b)^2 = 1$, where $0 < b < a$.

(a) Introduce a suitable parametric representation and show that the circumference L of the ellipse is given by

$$L = 4\int_0^{\pi/2} \sqrt{a^2 \cos^2 t + b^2 \sin^2 t}\, dt$$

$$= 4a \int_0^{\pi/2} \sqrt{1 - e^2 \sin^2 t}\, dt,$$

where $e = \sqrt{a^2 - b^2}/a$ is the eccentricity of the ellipse. This integral belongs to a class of integrals known as **elliptic integrals.** They also arise in problems such as determining the motion of a simple pendulum and the gravitational attraction of an ellipse.

(b) Using the formula $\sqrt{1 - u} \cong 1 - \frac{1}{2}u$ when $|u|$ is small, show that

$$L \cong 2\pi a\left(1 - \frac{1}{4}e^2\right) = 2\pi a\left[1 - \frac{a^2 - b^2}{4a^2}\right]$$

when e is small. Evaluate $L/4a$ for $e = 0.1, 0.25, 0.5$ and 0.8. Note that as $e \to 0$ the circumference of the ellipse approaches the circumference of a circle of radius a.

© **16.** Use Simpson's rule to evaluate the elliptic integral (see Problem 15)

$$\int_0^{\pi/2} \sqrt{1 - e^2 \sin^2 t}\, dt$$

(a) for $e = 0.1$; (b) for $e = 0.25$;
(c) for $e = 0.5$; (d) for $e = 0.8$

Compare your results with those of Problem 15(b).

© **17.** Use Simpson's rule to find the arc length of the parabola $y = x^2$ from $(0, 0)$ to $(2, 4)$.

© **18.** Use Simpson's rule to find the arc length of the hyperbola $y^2 - x^2 = 1$ from $(0, 1)$ to $(2, \sqrt{5})$.

In each of Problems 19 through 25, determine the area of the surface formed by rotating the given curve about the x-axis.

19. $y = 2x, \quad 0 \le x \le 4$

20. $y = x^3, \quad 1 \le x \le 3$

21. $y = \sqrt{r^2 - x^2}, \quad a \le x \le b$, where $-r < a < b < r$

22. $y^2 = x, \quad 1 \le x \le 2, \quad y > 0$

23. $y = \dfrac{x^3}{6} + \dfrac{1}{2x}, \quad 1 \le x \le 4$

24. $y = \frac{2}{3}x^{3/2} - \frac{1}{2}x^{1/2}, \quad 1 \le x \le 3$

25. $y = \sqrt{x + 1}, \quad 0 \le x \le 3$

26. Determine the surface area of a circular cone of height h and radius r by rotating the line $y = rx/h$, $0 \le x \le h$, about the x-axis.

© **27.** Use Simpson's rule to find the area of the surface generated by rotating the parabola $y = 4 - x^2$ from $x = 0$ to $x = 2$ about the x-axis.

© **28.** Use Simpson's rule to find the area of the surface generated by rotating the hyperbola $y = \sqrt{1 + x^2}$ for $0 \le x \le 2$ about the x-axis.

29. Suppose that the graph of the function $y = f(x)$, $a \le x \le b$, lies above the line $y = A$. Show that if f' is continuous on $[a, b]$, then the area of the surface formed by rotating the curve about the line $y = A$ is

$$S = 2\pi \int_a^b [f(x) - A]\sqrt{1 + [f'(x)]^2}\, dx.$$

30. Determine the area of the surface obtained by rotating the curve $y = \frac{2}{3}x^{3/2} - \frac{1}{2}x^{1/2}$, $1 \le x \le 3$, about the line $y = -2$. See Problem 29.

31. Develop an integral formula similar to that of Eq. 24 for the area of a surface formed by rotating a plane curve about the y-axis. Assume that the plane curve lies completely on one side of the y-axis and satisfies the equation $x = g(y)$, where g' is continuous for $c \le y \le d$. Determine the area of the surface formed by rotating the curve $y = x^2/4$ from $(2, 1)$ to $(4, 4)$ about the y-axis.

32. An arc of a curve is given by $x = f(t)$, $y = g(t)$ for $\alpha \le t \le \beta$. Assume that the functions f and g are nonnegative and have continuous first derivatives. Show that the area of the surface obtained by rotating this curve about the x-axis is

$$S = 2\pi \int_\alpha^\beta g(t)\sqrt{[f'(t)]^2 + [g'(t)]^2}\, dt.$$

In each of Problems 33 and 34, use the result of Problem 32 to determine the area of the surface formed by rotating the given curve about the x-axis.

33. $x = \sin t, \quad y = 1 - \cos t, \quad 0 \le t \le \frac{\pi}{2}$

34. $x = a\cos^3\theta, \quad y = a\sin^3\theta, \quad 0 \le \theta \le \frac{\pi}{2},$ $a > 0$

35. Sketch the graph of the curve $y = x^{-2/3}$ for $1 \le x \le b$ where $b > 1$. Let V_b and S_b be the volume of the solid and the area of the surface obtained by rotating this curve about the x-axis.

(a) Show that

$$V_b = 3\pi\left(1 - \frac{1}{b^{1/3}}\right)$$

and that $V_b \to 3\pi$ as $b \to \infty$.

(b) Show that $S_b = 2\pi \int_1^b x^{-2/3}\sqrt{1 + \frac{4}{9}x^{-10/3}}\,dx$.

(c) Use the fact that $1 + \frac{4}{9}x^{-10/3} > 1$ for $x > 1$ to show that $S_b > 6\pi(b^{1/3} - 1)$. Then show that $S_b \to \infty$ as $b \to \infty$.

The results of Parts (a), (b), and (c) show that this solid has a finite volume but an infinite surface area. As a container, it would hold a finite amount of paint, but would require an infinite amount to cover its surface.

7.4 APPLICATIONS IN MECHANICS

In this section we consider several examples in the field of mechanics in which integration arises naturally.

Displacement and distance

Suppose that a particle moves on a straight line with a velocity $v = f(t)$, where t is time (see Figure 7.4.1). If $v > 0$ the particle is moving to the right and if $v < 0$

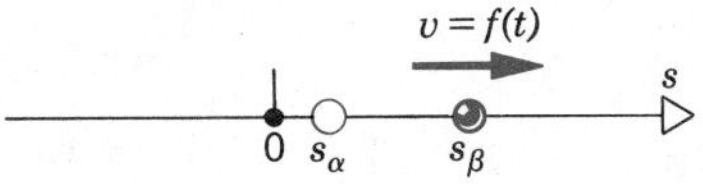

$v = f(t)$

s

$0 \quad s_\alpha \quad s_\beta$

Figure 7.4.1 A particle moving on a straight line.

the particle is moving to the left. If the particle is at position s_α at time $t = \alpha$, relative to a fixed origin, what is the position s_β of the particle at time $t = \beta > \alpha$?

If the velocity v is constant, then we know from elementary physics that

$$\frac{s_\beta - s_\alpha}{\beta - \alpha} = v, \tag{1}$$

and it follows that

$$s_\beta = s_\alpha + v(\beta - \alpha). \tag{2}$$

We wish to generalize Eq. 2 to the case in which v is not constant. We proceed in exactly the same way as we did in defining area and volume. Let Δ be a partition of $[\alpha, \beta]$ and choose a star point t_i^* in each subinterval $[t_{i-1}, t_i]$. An approximation to the change in position or displacement Δs_i in the time interval Δt_i from t_{i-1} to t_i is given by

$$\Delta s_i = s(t_i) - s(t_{i-1}) \cong f(t_i^*)\,\Delta t_i. \tag{3}$$

This leads to the approximation

$$\sum_{i=1}^{n} \Delta s_i \cong \sum_{i=1}^{n} f(t_i^*)\, \Delta t_i \tag{4}$$

for the displacement from $t = \alpha$ to $t = \beta$. The smaller the norm of the partition Δ, the closer the Riemann sums (4) approximate the total displacement. Provided that f is integrable, these Riemann sums approach the integral of f from $t = \alpha$ to $t = \beta$. Thus, if the velocity function f is integrable on $[\alpha, \beta]$, then the displacement $s_\beta - s_\alpha$ is given by

$$s_\beta - s_\alpha = \int_\alpha^\beta f(t)\, dt. \tag{5a}$$

Therefore the position of the particle at time $t = \beta$ is

$$s_\beta = s_\alpha + \int_\alpha^\beta f(t)\, dt. \tag{5b}$$

EXAMPLE 1

The velocity of a particle moving on a straight line is given by

$$v = (2t^2 + \tfrac{1}{2}) \text{ cm/sec}$$

and at $t = 0$ the particle is at $s = -1$ cm. What is the position of the particle at $t = 3$ sec, and what is the displacement in the time interval from $t = 0$ to $t = 3$?

From Eq. 5 we have

$$\begin{aligned} s_3 &= s_0 + \int_0^3 \left(2t^2 + \frac{1}{2}\right) dt \\ &= -1 + \left[\frac{2t^3}{3} + \frac{t}{2}\right]\Bigg|_0^3 \\ &= -1 + \left(18 + \frac{3}{2}\right) = 18.5 \text{ cm.} \end{aligned}$$

The displacement is $s_3 - s_0 = 19.5$ cm. ∎

It is important to emphasize that formula (5a) gives the net change in position, or displacement, and that this is not necessarily the total distance traveled. For example, suppose that a particle is initially at $s = 0$ and moves with a velocity 1 from $t = 0$ to $t = 1$, and then with a velocity -1 from $t = 1$ to $t = 2$. Clearly the particle is back at the origin at time $t = 2$, so there is no change in its position, but it has traveled a distance of two units. In calculating distance traveled it is necessary to use the *speed* of the particle rather than its velocity. Since the speed of the particle is the absolute value of the velocity, we have

$$\text{Distance traveled in the time interval } [\alpha, \beta] = \int_{\alpha}^{\beta} |f(t)|\, dt, \qquad v = f(t). \quad (6)$$

Work

When an object is moved along a straight line under the action of a constant force in the direction of motion, the product of the force and the displacement is called work. If we call the line on which the object moves the x-axis, then the work W done by a constant force F in the positive x direction in moving the object from the point $x = a$ to the point $x = b$ is $W = F(b - a)$ (see Figure 7.4.2). In the

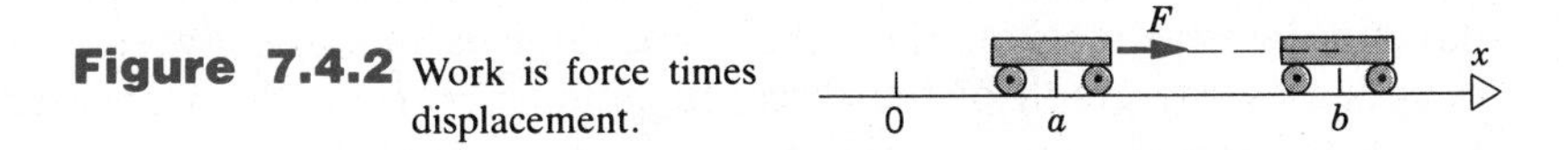

Figure 7.4.2 Work is force times displacement.

English system force is measured in pounds and distance in feet, so work is measured in foot-pounds; in the cgs system, force is measured in dynes and distance is measured in centimeters, so work is measured in dyne-centimeters, called ergs; and in the mks system, force is measured in newtons and distance in meters, so work is measured in newton-meters, called joules. The relationship between the units of measurement of work is 1 joule $= 10^7$ ergs $= 0.7376$ ft-lb. For example, if we take the upward direction as positive, the work done in carrying a desk that weighs 84 lb up a flight of stairs of height 6 ft is

$$W = 84 \text{ lb} \times 6 \text{ ft} = 504 \text{ ft-lb}.$$

Now suppose that the force is not constant, but is a function of position $F(x)$. For example, in space flight the gravitational attraction of the earth on a space ship varies inversely as the square of the distance from the center of the earth.

How do we define the work done by a variable force $F(x)$ in moving an object from $x = a$ to $x = b$? At this stage it undoubtedly would be boring if we were to say more than (1) partition the interval, (2) choose star points, (3) form approximating Riemann sums, (4) let the norm of the partition approach zero, and (5) conclude that if F is integrable, then the work W done in moving the object from $x = a$ to $x = b$ is

$$W = \int_{a}^{b} f(x)\, dx. \quad (7)$$

EXAMPLE 2

It is known from experimental evidence that the force required to stretch a spring a distance x beyond its natural length is proportional to x. If a force of 5 lb extends the spring 1 in., how much work is done in extending the spring 3 in.?

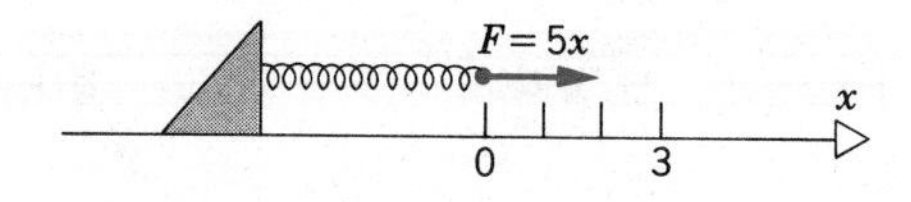

Figure 7.4.3

If we measure x from the natural position of the spring, then the force is $f(x) = cx$, where c is a constant of proportionality. When $x = 1$ in., $F = 5$ lb; hence, it follows that $c = 5$ lb/in. and $F(x) = 5x$ (see Figure 7.4.3). Thus

$$W = \int_0^3 5x\,dx = \left.\frac{5x^2}{2}\right|_0^3 = 22.5 \text{ in.-lb} = 1.875 \text{ ft-lb.} \quad \blacksquare$$

The geometric interpretation of Eq. 7 is that when the force is in the same direction as the motion, then the work done is numerically equal to the area bounded by the graph of the force function, $y = f(x)$, the x-axis, and the lines $x = a$ and $x = b$.

Physically, the work done by a force acting on a particle is equal to the change in the kinetic energy of the particle (see Problem 20). Consider a mechanical system in which the force required to move a mass in the positive direction is different from the force required to move the mass in the negative direction; see curves C_1 and C_2, respectively, of Figure 7.4.4. The shaded area represents the difference

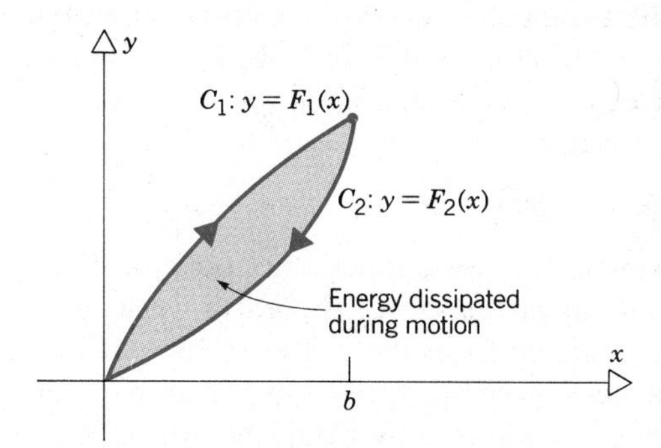

Figure 7.4.4

between the work required to move the mass from $x = 0$ to $x = b$ and back to $x = 0$; hence it is the difference in the kinetic energy from the beginning of the motion to the end of the motion. This energy may be dissipated as heat. It is this principle that allows the dissipation of energy of impact by a shock absorber.

EXAMPLE 3

An irrigation well is 50 ft deep and has a circular cross section with radius 3 ft. If the well is half full, determine the work done by a pump in emptying this well if the water is lifted from the surface level. The weight-density w of water is approximately 62.5 lb/ft^3.

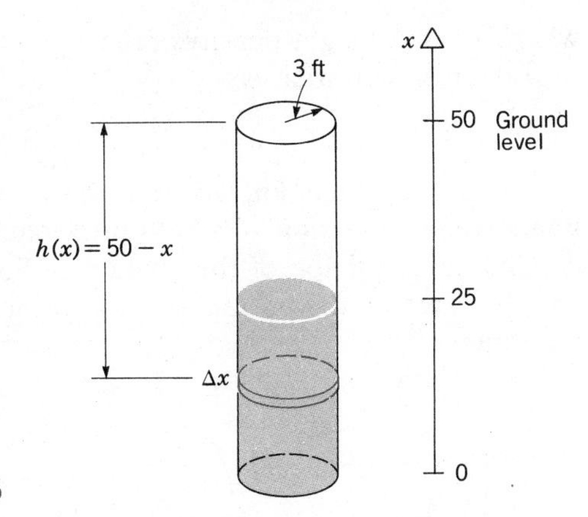

Figure 7.4.5

We must take account of the fact that different elements of volume of water are lifted different heights. Consider the picture shown in Figure 7.4.5. The generic element has volume $9\pi\ \Delta x$ ft^3 and weight $9\pi w\ \Delta x$ lb. Since this element is lifted a height $h = 50 - x$ ft, the work done is $9\pi w\ \Delta x\ (50 - x)$ ft-lb. Thus the total work in emptying the well is

$$\begin{aligned} W &= \int_0^{25} 9\pi w(50 - x)\, dx \\ &= 9\pi w \left(50x - \frac{x^2}{2}\right)\bigg|_0^{25} \\ &= 9\pi w \left[50 \cdot 25 - \frac{25 \cdot 25}{2}\right] \\ &= \frac{3}{4} 9\pi w \times 50 \times 25 = 8437.5\pi w \text{ ft-lb.} \end{aligned} \tag{8}$$

If x is measured from the top of the well rather than the bottom, then the formula for W is

$$W = \int_{25}^{50} 9\pi w x\, dx. \tag{9}$$

Of course, this gives the same result as Eq. 8. ∎

Hydrostatic pressure

Suppose that a plate with area A is placed horizontally a depth h below the surface of a container filled with liquid. Then the downward force exerted on the plate by the liquid above it is

$$F = whA, \tag{10}$$

where w is the weight-density of the liquid ($w \cong 62.5$ lb/ft³ for water). The **pressure,** or force per unit area, is

$$p = wh. \tag{11}$$

Two important facts known from the laws of physics are (a) the pressure is the same in each direction—Pascal's principle,* and (b) the pressure and force do not depend on the shape of the container. For example, the pressure at the bottom and the force on the bottom of each of the containers depicted in Figure 7.4.6 are the same.

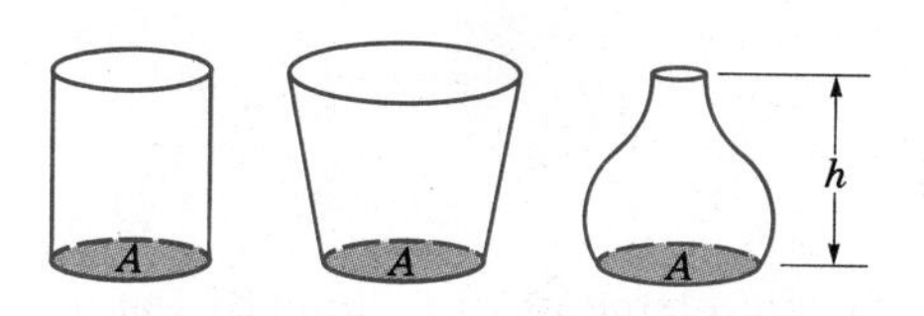

Figure 7.4.6 The pressure and the force on the bottom is the same for all three containers; each has the same base A and the same height h.

Now suppose that the plate is submerged vertically, so that the pressure at different depths is different. How do we compute the force acting on the plate? Consider the plate shown in Figure 7.4.7, where x is measured downward from

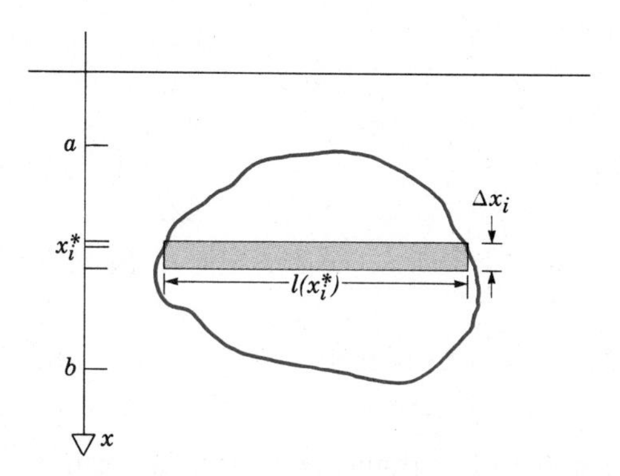

Figure 7.4.7

the surface of the water. Let $l(x)$ be the width of the plate at depth x. Corresponding to a partition Δ of $[a, b]$ and choice of star points $x_i{}^*$, the area of the generic element shown in Figure 7.4.7 is $\Delta A_i = l(x_i{}^*)\,\Delta x_i$. An approximate value of the force $F(x_i{}^*)$ acting on this generic element is

$$F(x_i^*) = p(x_i^*)\,\Delta A_i = wx_i^*\,l(x_i^*)\,\Delta x_i.$$

The sum

$$F \cong \sum_{i=1}^{n} F(x_i^*) = \sum_{i=1}^{n} wx_i{}^*\,l(x_i{}^*)\,\Delta x_i$$

*Although this result is stated as Pascal's principle, it is a consequence of the laws of fluid mechanics and not an independent principle.

provides an approximation to the total force exerted by the liquid on the plate. Carrying out the usual limiting procedure, we are led to define the force exerted on the plate by the fluid as

$$F = \int_a^b wxl(x)\,dx. \tag{12}$$

EXAMPLE 4

The Glen Canyon Dam on the Colorado River near the border between Utah and Arizona was constructed in 1956–1963. It has a height of approximately 600 ft and a crest length of approximately 1200 ft. Neglecting the slope of the dam, the face of the dam on the side of the reservoir has the shape shown in Figure 7.4.8. Assuming that the reservoir is at full level, determine the force on the dam.

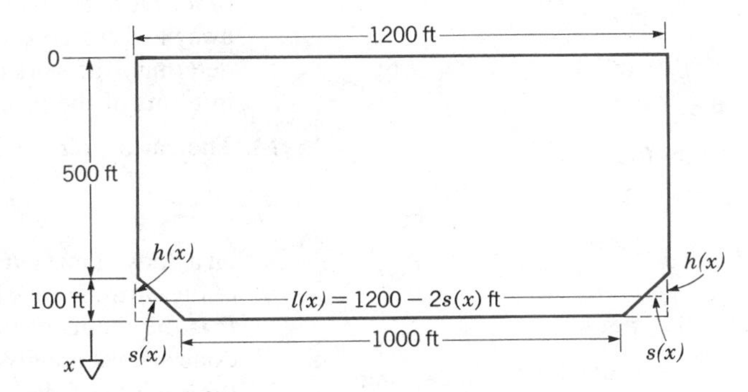

Figure 7.4.8 Diagram of the Glen Canyon Dam.

We measure x downward from the top of the dam (see Figure 7.4.8). First we determine the width $l(x)$ of the dam. We have

$$l(x) = 1200, \qquad 0 \leq x \leq 500.$$

For $500 \leq x \leq 600$ we make use of the fact that the triangular corners that have been removed are isoceles triangles. Then

$$l(x) = 1200 - 2s(x),$$

and $s(x) = h(x)$. Further, $h(x) = x - 500$, so $s(x) = x - 500$, and

$$l(x) = 1200 - 2(x - 500) = 2200 - 2x, \qquad 500 \leq x \leq 600.$$

Using Eq. 12 we can write the force on the dam as

$$F = \int_0^{600} wxl(x)\,dx = \int_0^{500} 1200wx\,dx + \int_{500}^{600} wx(2200 - 2x)\,dx$$

$$= 1200w \left.\frac{x^2}{2}\right|_0^{500} + w\left(1100x^2 - \frac{2}{3}x^3\right)\Bigg|_{500}^{600}$$

$$\cong 210.33w \times 10^6 \cong 1.31 \times 10^{10} \text{ lb}.$$

We could just as well have measured x upward from the bottom of the dam. In this case the formula for F is

$$F = \int_0^{100} w(600 - x)(1000 + 2x)\,dx + \int_{100}^{600} 1200w(600 - x)\,dx. \blacksquare$$

PROBLEMS

In each of Problems 1 through 8, the velocity v of a particle moving on a straight line is given as a function of the time t. Assume that v is measured in ft/sec and that t is measured in sec. Determine (a) the displacement of the particle and (b) the total distance traveled in the given time interval.

1. $v = 1 - \sin t, \quad 0 \le t \le 2\pi$

2. $v = t^2 + 3t + 2, \quad 0 \le t \le 3$

3. $v = t^2 + t - 2, \quad 0 \le t \le 2$

4. $v = 2\cos \frac{1}{2}t, \quad 0 \le t \le 2\pi$

5. $v = t\sqrt{1 + t^2}, \quad 0 \le t \le 3$

6. $v = |t - 2|, \quad 1 \le t \le 3$

7. $v = t - \dfrac{2}{t^2}, \quad 1 \le t \le 2$

8. $v = -2 + \sqrt{t}, \quad 0 \le t \le 4$

9. The force necessary to extend (compress) a spring is proportional to the extension (compression). If the natural length of the spring is 12 in. and a force of 16 lb is required to extend the spring 1 in., how much work is done in extending the spring to a length of 15 in.?

10. A well is cylindrical for a depth of 50 ft with a radius of 3 ft, and then it has a conical shape tapering to a point at a depth of 75 ft. If the level of water is 25 ft below the surface of the ground, determine the work done in emptying the well assuming that no additional water enters the well.

11. A swimming pool is 25 ft by 75 ft. The bottom of the pool is an inclined plane, and the pool has a depth of 4 ft at the shallow end and 10 ft at the deep end, which is 75 ft away. If the pool is full of water to within 1 ft of the top, determine the amount of work required to empty it over the side of the pool.

12. The force necessary to extend (compress) an automobile coil spring is proportional to the extension (compression). If the natural height of the spring is 9 in. and if a weight of 600 lb compresses the spring 1 in., determine the work done in compressing the spring to one-half of its natural length.

13. Two electrons repel each other with a force inversely proportional to the square of the distance between them. If one electron is fixed and the other electron is moved from a distance of 5 cm away to 2 cm along a straight line connecting them, determine the work done against the repelling force in terms of the proportionality constant k.

14. The earth exerts a gravitational force

$$F = \frac{km}{r^2}$$

on a body of mass m at a distance r from the center of the earth. Here k is a constant and $r > R$, where R is the radius of the earth. Show that the work done against gravity in lifting a space capsule from the surface of the earth to a height h above the earth is $W = mgRh/(R + h)$, where g is the acceleration of gravity on the earth's surface. Assuming that $R = 4000$ mi, calculate the work done against gravity in lifting a space capsule that weighs 150 lb on the surface of the earth to a height of 200 mi. Compare this result with that obtained assuming that the gravitational attraction of the earth is constant.

15. If a straight hole could be bored through the center of the earth, then a particle of mass m in this hole would be attracted toward the center with a force $F = mgr/R$, where r is the distance from the center, R is the radius of the earth, and g is the acceleration of gravity at the surface of the earth. How much work is done by this attractive force on a mass as it falls from the surface to the center?

16. A chain that weighs 15 lb/ft is hanging from the top of an 80-ft building to the ground. How much work is done in pulling the chain to the top of the building?

17. A circular cylinder with cross-sectional area A is filled with a gas that is expanded or compressed by the motion of a piston at one end. If p is the pressure in the gas, then the force on the piston is pA. Show that the work done in compressing the gas from a height h_1 to a height h_2 is

$$W = \int_{h_1}^{h_2} pA\,dh = \int_{V_1}^{V_2} p\,dV$$

where V is the volume of the gas. Observe that p must be expressed as a function of h in the first integral, and as a function of V in the second.

18. Assuming that the relation between the pressure and volume in a gas is of the form

$$pV^{\gamma} = C$$

where γ and C are constant, use the result of Problem 17 to show that the work done in changing it from volume V_1 and pressure p_1 to volume V_2 and pressure p_2 is

$$W = \frac{p_2V_2 - p_1V_1}{1 - \gamma}, \qquad \gamma \neq 1.$$

The parameter γ is the ratio of the specific heat at constant pressure to the specific heat at constant volume. For an adiabatic process, one carried out without gain or loss of heat, $\gamma = 1.4$. For an isothermal process, one carried out at constant temperature, $\gamma = 1$.

19. A spherical oil tank of radius 10 ft is half full of oil with a weight density of 60 lb/ft^3. Find the work done in pumping the oil to the top of the tank.

20. In this problem we will show that the work done on a mass m is equal to the change in its kinetic energy. Suppose that at position $x = a$ the mass has velocity v_a and at position $x = b$ the mass has velocity v_b. Use Newton's second law, $F = m\,dv/dt$, and the chain rule,

$$\frac{dv}{dt} = \frac{dv}{dx}\frac{dx}{dt} = v\frac{dv}{dx},$$

to show that

$$W = \int_a^b F\,dx = \tfrac{1}{2}mv_b^2 - \tfrac{1}{2}mv_a^2.$$

21. A gate in an irrigation ditch has the shape of an isosceles triangle. The upper edge of the triangle has length 4 ft and the equal sides of the triangle have length 3 ft. Determine the force on the gate if the irrigation ditch is full of water.

22. A gasoline tank car has the shape of a horizontal circular cylinder. If the radius of the cylinder is 10 ft and if the weight density of gasoline is 45 lb/ft^3, determine the force on the end of the tank when the tank is full. *Hint:* See Example 3 of Section 6.3.

23. Assume that the cross section of the gasoline tank car of Problem 22 is elliptical with major axis (horizontal direction) of 16 ft and minor axis of 6 ft. Set up an integral for the force on the end of the tank when the tank is full.

24. A semicircular plate is submerged in water with its plane vertical and its diameter at the top and 2 ft below the surface of the water. If the diameter of the plate is 2 ft, determine the force on one side of the plate. *Hint:* See Example 3 of Section 6.3.

25. The vertical face of a dam has the form of a trapezoid 1000 ft long at the top, 700 ft long at the bottom, and 80 ft deep. Determine the force on the dam under the following conditions.

(a) The reservoir behind the dam is full.

(b) The reservoir behind the dam has a depth of 60 ft.

7.5 APPLICATIONS IN THE BIOLOGICAL AND SOCIAL SCIENCES

In this section we consider a few examples of how the definite integral arises in the biological and social sciences. In each case the idea of forming a Riemann sum to represent an approximate value of the quantity that we wish to compute, and then refining the partition to obtain a definite integral is the same as in the fundamental geometric examples discussed in the previous sections. For that reason we omit some of the details that should be completely familiar to you by now.

Flow in a capillary

Consider a long thin circular tube of radius a and length l filled with a viscous incompressible fluid such as water. If there is a difference Δp in the pressures at the ends of the tube, then fluid will flow in the direction of the lower pressure. The velocity of the flow is independent of the axial direction and at each cross section has the form

$$u(r) = \frac{\Delta p}{4\mu l}(a^2 - r^2), \tag{1}$$

where r is a radial coordinate and μ is the viscosity of the fluid (see Figure 7.5.1). Notice that $U_0 = (\Delta p)\, a^2/4\mu l$ is the velocity at the center of the tube ($r = 0$). The velocity distribution (1) for flow in a tube is known as *Poiseuille flow.*

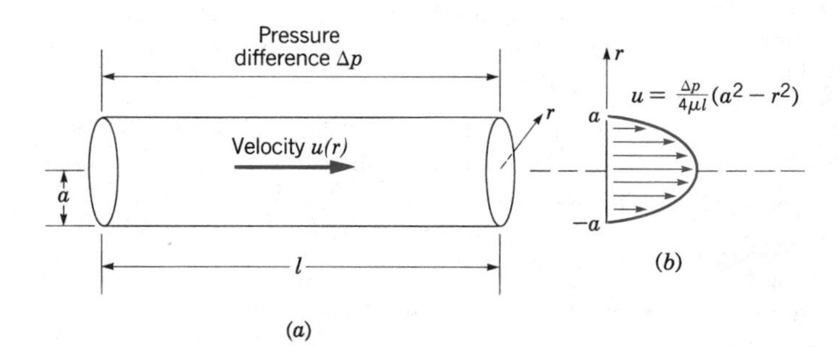

Figure 7.5.1 (*a*) Pressure-driven flow in a tube. (*b*) Velocity profile for Poiseuille flow.

It is of interest to compute the volume of fluid, known as the *flux* Q, that passes through a given cross section per unit of time. If the velocity were constant, say U, then the total volume of fluid passing a given cross section in a unit of time would be

$$\begin{aligned} Q &= U \times \text{area of cross section} \\ &= \pi a^2 U, \end{aligned} \tag{2}$$

with dimensions of volume/time.

When the velocity is not constant, we divide the area of the cross section into appropriate small subareas, compute Q for each subarea by approximating u by some appropriate constant value in each subarea, and then add all of these terms together. From Eq. 1 we note that u is constant on each circle with center at the origin. Therefore we use circular rings as the subareas. If we partition the interval $[0, a]$ by the points $r_0, r_1, \ldots, r_n$, the area of the ring lying between r_{i-1} and r_i (see Figure 7.5.2) is

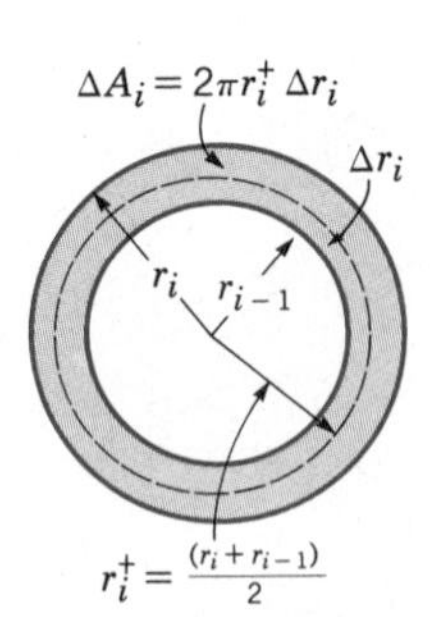

Figure 7.5.2

$$
\begin{aligned}
\Delta A_i &= \pi(r_i^2 - r_{i-1}^2) \\
&= \pi[(r_{i-1} + \Delta r_i)^2 - r_{i-1}^2] = \pi[2r_{i-1}\,\Delta r_i + (\Delta r_i)^2] \\
&= 2\pi\left(r_{i-1} + \frac{\Delta r_i}{2}\right)\Delta r_i \\
&= 2\pi r_i^+\,\Delta r_i. \qquad (3)
\end{aligned}
$$

Here r_i^+ is midway between the inner radius r_{i-1} and the outer radius r_i. Now choose a star point in each subinterval so that $r_i^* = r_i^+$. Then the flux associated with the ith subarea is approximately $u(r_i^+)2\pi r_i^+\,\Delta r_i$ and the total flux is given approximately by

$$
Q \cong \sum_{i=1}^{n} 2\pi u(r_i^+) r_i^+\,\Delta r_i. \qquad (4)
$$

The sum (4) is a Riemann sum for the function $2\pi r u(r)$, so if u is continuous, or even merely integrable, on $[0, a]$, then as $\|\Delta\| \to 0$ the sum on the right side of Eq. 4 approaches the corresponding integral. Thus

$$
Q = \int_0^a 2\pi u(r) r\,dr. \qquad (5)
$$

For the velocity distribution given in Eq. 1 we have

$$
\begin{aligned}
Q &= 2\pi \frac{\Delta p}{4\mu l}\int_0^a (a^2 - r^2) r\,dr \\
&= \frac{\pi\Delta p}{2\mu l}\left(\frac{a^2r^2}{2} - \frac{r^4}{4}\right)\Bigg|_0^a = \frac{\pi(\Delta p)a^4}{8\mu l}. \qquad (6)
\end{aligned}
$$

It is instructive to compare Eqs. 2 and 6. If the velocity were constant, the flux would be proportional to the square of the radius; but the correct velocity distribution yields a flux that is proportional to the fourth power of the radius. Thus if the radius of the tube is reduced by a factor of $\frac{1}{2}$ the amount of fluid passing a cross section per unit time is reduced by a factor of $\frac{1}{16}$ as compared to only a factor of $\frac{1}{4}$ for the simplified model with a constant velocity. This illustrates the inaccuracies that can arise if too simple a model is used. Another interesting observation is the following. While blood is not as simple a fluid as a viscous incompressible fluid and an artery is not a long straight tube, the results using the velocity distribution (1) suggest the serious consequences of even a small constriction in an artery.

Osmosis

One of the fundamental problems in physiology, the study of the functions and vital processes of living organisms, is to understand how fluids pass through a semipermeable membrane, such as the wall of a living cell, into a solution of higher concentration so as to equalize concentrations on both sides of the membrane. This is the process of *osmosis*.

Suppose that we have a long thin-walled tube of length L and radius a filled with salt water and that the concentration of salt $c(x)$ in the tube is a function of the axial coordinate x. Suppose further that the tube is surrounded by salt water in which the concentration of salt is a constant c_0, and that the wall of the tube is permeable to water but not to salt (see Figure 7.5.3*a*). We know from experiments that the rate at which water passes through the wall of the tube is proportional to the difference in the concentrations, $c(x) - c_0$.

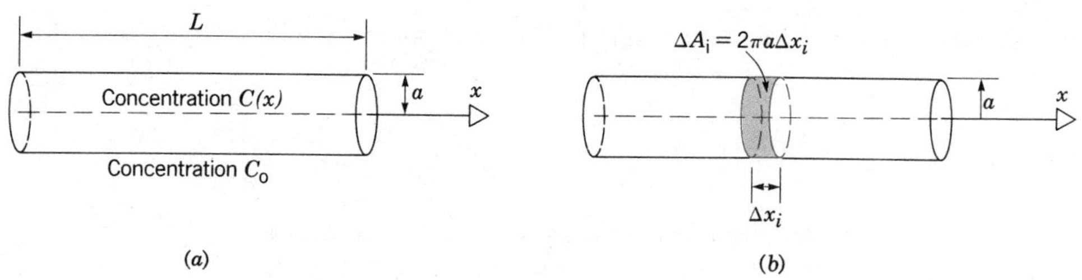

Figure 7.5.3 A tube bounded by a semipermeable membrane.

To find the flux Q through the wall of the tube we divide the tube into cylinders as shown in Figure 7.5.3*b*. The surface area of the ith cylinder is $2\pi a\, \Delta x_i$. Choosing star points in each subinterval we conclude that

$$Q \cong \sum_{i=1}^{n} [c(x_i^*) - c_0] 2\pi a\, \Delta x_i, \tag{7}$$

and upon letting $\|\Delta\| \to 0$, that

$$Q = 2\pi a \int_0^L [c(x) - c_0]\, dx. \tag{8}$$

For an application of this result, see Problem 4.

Reliability theory

Suppose that we consider a large collection of some manufactured product, such as light bulbs. Let $F(t)$ be the proportion of bulbs that fail within the first t hours of use. Alternatively, $F(t)$ represents the probability that a particular bulb will fail within t hours. The function F is defined for $t \geq 0$ and has the following properties.

1. $F(0) = 0$.
2. $F(t) \geq 0$.
3. $F(t)$ is nondecreasing.
4. $F(t) \to 1$ as $t \to \infty$.

In reliability theory the function F is called a **failure distribution;** more generally, such a function is called a **probability distribution.**

The proportion of products that are still serviceable at time t but fail before time $t + \Delta t$, where $\Delta t \geq 0$, that is, the proportion that fail in the interval $[t, t + \Delta t)$, is given by $F(t + \Delta t) - F(t)$. Assuming that F is differentiable and that Δt is small, we can write

$$F(t + \Delta t) - F(t) = \frac{F(t + \Delta t) - F(t)}{\Delta t} \Delta t \cong F'(t)\,\Delta t.$$

Therefore the proportion of failures in the interval $[t, t + \Delta t)$ is approximately $F'(t)\,\Delta t$. Thus $F'(t)$ is a measure of the tendency of the product to fail near time t. The function $f(t) = F'(t)$ is known in reliability theory as the **failure density;** more generally, such a function is called a **probability density.** Since $F(0) = 0$, it follows from Theorem 6.4.3 that

$$F(t) = \int_0^t f(s)\,ds; \tag{9}$$

further, since $F(t) \to 1$ as $t \to \infty$,

$$\lim_{t\to\infty} \int_0^t f(s)\,ds = 1. \tag{10}$$

EXAMPLE 1

Suppose it has been determined by experimental testing that for a certain type of light bulb the failure density is given by

$$f(t) = \begin{cases} At^2(100 - t)^2, & 0 \leq t \leq 100; \\ 0, & t > 100; \end{cases} \tag{11}$$

where A is a constant. Determine the failure distribution, the value of the constant A, and the proportion of light bulbs that fail in the time interval $60 \leq t < 70$.

From Eq. 9 it follows that

$$\begin{aligned} F(t) &= \int_0^t As^2(100 - s)^2\,ds \\ &= A \int_0^t (10^4 s^2 - 2 \times 10^2 s^3 + s^4)\,ds \\ &= A\left[\frac{10^4 t^3}{3} - \frac{10^2 t^4}{2} + \frac{t^5}{5}\right], \qquad 0 \leq t \leq 100. \end{aligned}$$

By Eq. 11 every bulb fails by the time $t = 100$. Therefore $F(100) = 1$, and

$$1 = A\left(\frac{10^{10}}{3} - \frac{10^{10}}{2} + \frac{10^{10}}{5}\right),$$

so

$$A = 3 \times 10^{-9}.$$

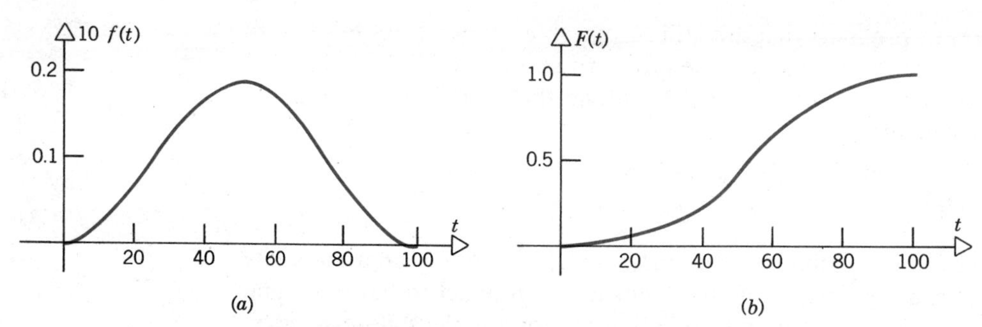

Figure 7.5.4 (*a*) Failure density function. (*b*) Failure distribution function.

Hence

$$F(t) = (3 \times 10^{-9})\left[\frac{10^4t^3}{3} - \frac{10^2t^4}{2} + \frac{t^5}{5}\right], \qquad 0 \le t \le 100. \tag{12}$$

Of course, $F(t) = 1$ for $t > 100$. The failure density and failure distribution are sketched in Figure 7.5.4.

The proportion of light bulbs that fail in $60 \le t < 70$ is $F(70) - F(60)$. Using Eq. 11 and carrying out the necessary arithmetic, we find that $F(70) - F(60) \cong 0.154$. ∎

A question of primary interest is to determine the average, or expected, time of failure for a product having a given failure density. To answer this question we assume that all of the products fail by time T, where T is some large positive number. Then we partition the interval $[0, T]$ with points $0 = t_0 < t_1 < t_2 < \cdots < t_n = T$. The proportion of products that fail in the time interval $[t_{i-1}, t_i)$ is $F(t_i) - F(t_{i-1})$. If the norm of the partition is very small, we may suppose that all of the products that fail in the interval $[t_{i-1}, t_i)$ actually fail at some particular time t_i^+ in $[t_{i-1}, t_i)$. Then the average time of failure τ is given approximately by

$$\tau \cong \sum_{i=1}^{n} t_i^+ \cdot \{\text{proportion of products that fail in } [t_{i-1}, t_i)\}$$

$$= \sum_{i=1}^{n} t_i^+[F(t_i) - F(t_{i-1})].$$

We again use the mean value theorem (Theorem 4.1.3) to write

$$\tau = \sum_{i=1}^{n} t_i^+ F'(t_i^*)\,\Delta t_i = \sum_{i=1}^{n} t_i^+ f(t_i^*)\,\Delta t_i, \qquad t_i^* \in (t_{i-1}, t_i). \tag{13}$$

As in several previous examples the sum (13) is not exactly a Riemann sum; however, it becomes one if in each subinterval we select $t_i^+ = t_i^*$. Then

$$\tau = \lim_{\|\Delta\| \to 0} \sum_{i=1}^{n} t_i^* f(t_i^*)\,\Delta t_i = \int_0^T tf(t)\,dt. \tag{14}$$

EXAMPLE 2

Find the expected time of failure τ of the light bulbs in Example 1, whose failure density is given by Eq. 11.

From Eq. 14 we obtain

$$\begin{aligned}\tau &= (3 \times 10^{-9}) \int_0^{100} t \cdot t^2(100 - t)^2\, dt \\ &= (3 \times 10^{-9}) \int_0^{100} (10^4 t^3 - 200t^4 + t^5)\, dt \\ &= (3 \times 10^{-9}) \left(\frac{10^4 \cdot 10^8}{4} - \frac{2 \cdot 10^2 \cdot 10^{10}}{5} + \frac{10^{12}}{6} \right) \\ &= 50.\end{aligned}$$

In this problem the expected value τ is the midpoint of the interval $(0, 100)$ in which f is nonzero. This is due to the symmetry of f in this case, and is not true in general. ▮

Consumer surplus and producer surplus

Let x denote the number of some item, which we assume varies continuously, and let p be the unit price of the item. A demand curve gives the relation between the price and the number of units that will be bought at that price. A supply curve gives the relation between the price and the number of units that will be produced at that price. Typical demand and supply curves are shown in Figure 7.5.5*a*. In the case of pure competition, market equilibrium is determined as the intersection of the demand and supply curves.

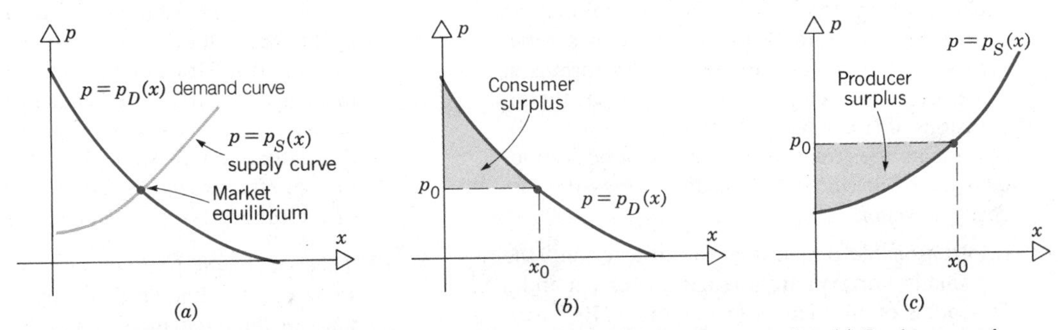

Figure 7.5.5 (*a*) Demand and supply curves. (*b*) Consumer surplus. (*c*) Producer surplus.

Now suppose that a market demand x_0 and a corresponding price p_0 are determined in pure competition or otherwise. Then x_0 items are sold at a price p_0 for a total amount of money equal to p_0x_0. Consumers who would have been willing to pay more than p_0 have gained by having had the price set at p_0 rather than at the maximum price they would have been willing to pay. The total gain by the consumers, known as the **consumer's surplus,** is represented by the area below the demand curve but above the price line $p = p_0$ (see Figure 7.5.5*b*). Thus

$$\text{C.S.} = \text{consumer surplus} = \int_0^{x_0} [p_D(x) - p_0]\,dx$$
$$= \int_0^{x_0} p_D(x)\,dx - p_0 x_0. \tag{15}$$

The consumer surplus is a measure of the well-being of the consuming public. From Figure 7.5.5*b* we see that if the market demand were greater ($x > x_0$), then the product could be produced at a lower unit cost. Consequently, more of the product would be available at a lower unit cost. We assume, of course, that the additional production is possible without harm to natural resources, or at least, less harm than benefit to society.

Similarly, if the quantity supplied x_0 and the price p_0 are set, producers who would have been willing to sell the commodity at a price below p_0 have gained. The total producer's gain, known as the **producer's surplus,** is the area below the price line $p = p_0$ and above the supply curve (see Figure 7.5.5*c*). Hence

$$\text{P.S.} = \text{producer surplus} = \int_0^{x_0} [p_0 - p_S(x)]\,dx$$
$$= p_0 x_0 - \int_0^{x_0} p_S(x)\,dx. \tag{16}$$

It is interesting to observe that it is possible to have both a consumer surplus and a producer surplus; this is an advantage of trading.

PROBLEMS

1. This problem is concerned with several qualitative results about the flow of blood in the human circulatory system that can be obtained by making use of the results for Poiseuille flow in a tube. First we note that pressure has the dimensions of force/area (dyne/cm^2) and viscosity μ has the dimensions of (force/area) $\times$ time (dyne-sec/cm^2). A value for the effective viscosity of blood is about 0.04 dyne-sec/cm^2, which is four times the viscosity of water.

   (a) Reasonable dimensions for an arteriole (the smallest artery) are a length of 0.5 cm and a radius of 30 microns (1 micron $= 10^{-4}$ cm). The width of a red blood cell is about 2 microns. If the pressure drop across an arteriole is 50 mm of mercury (1 mm of mercury $= 1.33 \times 10^3$ dyne/cm^2), determine the volume of blood flow per unit time in the arteriole.

   (b) The resistance R of a tube to flow is defined as $R = \Delta p/Q$, where Δp is the pressure drop in the tube and Q is the fluid flux. For Poiseuille flow show that $R = 8\mu l/\pi a^4$ dyne-sec/cm^5.

   (c) A reasonable value for the rate at which the heart pumps blood is 6 liters/min $= 6000$ cm^3/min. It is also reasonable to take the pressure in the artery leaving the heart as 100 mm of mercury (mmHg) and that in the vein entering the heart as nearly zero. What is the resistance of the circulatory system? Assuming that the circulatory system can be thought of as a long tube of effective length 2 m, what is the radius of the tube?

   (d) This part of the problem is a continuation of Part (c). It is known that most of the resistance of the circulatory system is provided by the flow in the arterioles and capillaries rather than in the arteries. Consider the following crude model: a single artery from the heart that breaks into N arterioles, which then connect into a vein returning to the heart. Suppose further that the blood pressure is 100 mmHg at the artery leaving the heart, 90 mmHg at the arterioles, and 25 mm at the end of the arterioles. This 25 mm of pressure is lost in the capillaries and venous return to the heart. Also assume that the arterioles have length 0.5 cm and radius 30 microns. Determine the number of arterioles. Note that the result is particularly sensitive to the assumption about the radius of the arterioles.

2. Blood flows through a tube of length l and radius a. In the course of this flow, oxygen diffuses through the sides of the tube and combines with the fluid in the outer half of the tube ($a/2 \le r \le a$). Assuming Poiseuille flow, determine the percentage of the blood that is oxygenated (oxygen molecules combined with the hemoglobin).

3. This is an optimization problem that requires the use of the resistance formula derived in Problem 1(b). Suppose that a blood vessel of radius a connects points O and P and that another blood vessel of radius $b(<a)$ branches off to connect with another point Q (see Figure 7.5.6). Determine the branching point by determining the angle θ that will minimize the total resistance of the flow from O to Q.

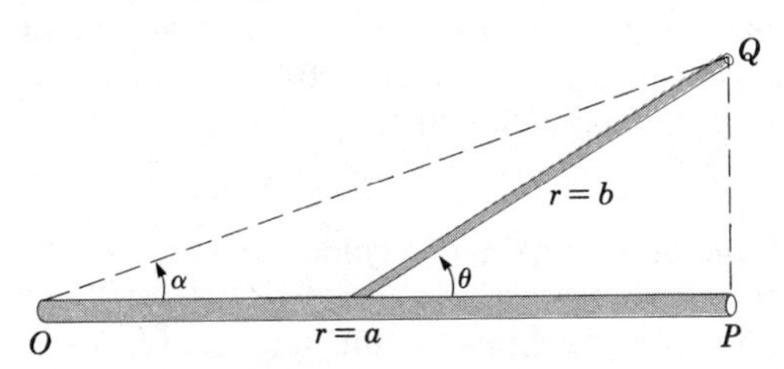

Figure 7.5.6

4. A thin-walled permeable tube of length L stands vertically in a salt-water solution with uniform concentration c_0. The inside of the tube is filled with salt water with a concentration of salt that varies linearly from $1.4c_0$ at the bottom to $0.8c_0$ at the top. If the tube has radius a, determine the flux of water through the sides of the tube.

5. (a) In conducting a poll an interviewing agency knows that the time required for an interview decreases with the number of interviews. The time T required by their best interviewer for the nth interview is $T(n) = 2n^{-1/2}$ hours. Thus the time for 400 interviews is

$$\sum_{n=1}^{400} 2n^{-1/2}.$$

Calculate an approximation to this sum by thinking of n as a continuous variable on the interval [1, 401] and replacing the sum by an appropriate Riemann integral. If this interviewer is paid \$5/hour, what does it cost to complete the 400 interviews?

(b) The time required for a slower interviewer is $T(n) = 2.5n^{-1/3}$. If the slower interviewer is paid \$4.00/hour, what does it cost to carry out the interviews?

(c) Relative to the slower interviewer, how much could one justify paying the faster interviewer?

6. A plumbing corporation has leased a computer that should substantially reduce the cost of their inventory control. They estimate that the savings in employee salaries will be \$21,000/year. The fixed rental cost of the machine is \$500/month and they estimate that maintenance costs will increase with time and will be given by $1.5t^{3/2}$ thousands of dollars/year. In this problem assume that all costs and savings vary continuously with time.

(a) Assuming that the computer is used until it is no longer profitable, how long will the company use the computer and how much money will they save?

(b) If they could reduce the maintenance rate to $1.5t$ thousands of dollars/year, how long could they profitably use the computer and how much money would they save? Compare with the previous case.

(c) A more sophisticated analysis of the longevity of the computer and the savings for the company would take into account salary increases. Suppose that the salaries of the employees to be replaced by the computer can be approximated by \$21,000 $(1 + 0.05\,t)$, and that maintenance costs are $1.5t^{3/2}$ thousands of dollars per year. Show that the computer can be profitably operated for more than 5 years but less than 6 years, and calculate the saving for a 5-year period.

7. (a) Determine the constant k so that the function

$$f(t) = \begin{cases} k, & 0 \le t \le 2 \\ 0, & \text{otherwise} \end{cases}$$

is a possible failure density function.

(b) Find the corresponding failure distribution function.

(c) Find the expected time of failure.

8. (a) Determine the constant k so that the function

$$f(t) = \begin{cases} \dfrac{k}{t^3}, & 1 \le t \le 4 \\ 0, & \text{otherwise} \end{cases}$$

is a possible failure density function.

(b) Find the corresponding failure distribution function.

(c) Find the expected time of failure.

9. Suppose that the failure density for a large group of special filament light bulbs is given by

$$f(t) = \begin{cases} At, & 0 \le t \le 1 \\ A(2 - t), & 1 < t \le 2 \\ 0, & \text{otherwise} \end{cases}$$

where t is measured in years.

(a) Determine the constant A.

(b) Find the failure distribution.

(c) Find the expected life of a light bulb.

10. If air resistance is neglected, the horizontal distance s traveled by a projectile fired from a cannon with an initial velocity v and at an angle θ with the horizontal is $s(\theta) = (v^2 \sin 2\theta)/g$ (see Figure 7.5.7). Suppose that the muzzle velocity v

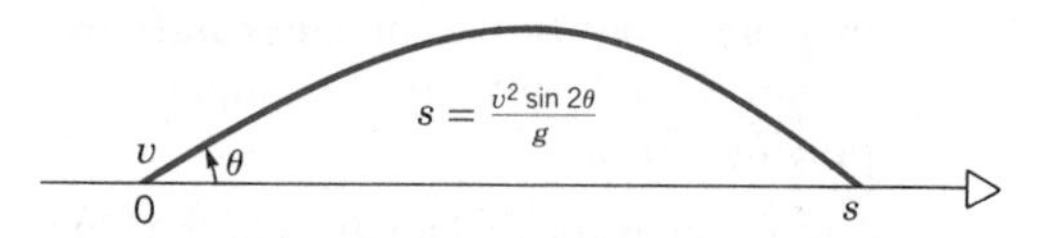

Figure 7.5.7 Trajectory of a projectile.

is known accurately but that the angle of inclination θ is known only to be in the interval $[\theta_1, \theta_2]$. Assuming that any value of θ in this interval is equally likely, then the probability density function for the angle of inclination is

$$f(\theta) = \begin{cases} \dfrac{1}{\theta_2 - \theta_1}, & \theta_1 \le \theta \le \theta_2 \\ 0, & \theta < \theta_1 \text{ or } \theta > \theta_2. \end{cases}$$

(a) Compute $E(\theta)$, the expected value of θ.

(b) Compute $s[E(\theta)]$.

(c) By constructing the appropriate Riemann sum, and generalizing the argument preceding Eq. 14, show that $E(s)$, the expected value of s, is

$$\int_{\theta_1}^{\theta_2} \frac{v^2}{g} (\sin 2\theta) f(\theta)\, d\theta.$$

(d) Evaluate $E(s)$ and observe that $E(s) \neq s[E(\theta)]$.

11. Let $s(t)$, called the *survival function*, denote the fraction of a population that survives to time $t > 0$. Note that $s(t)$ is a decreasing function with $0 \le s(t) \le 1$, that $s(0) = 1$, and that $\lim_{t\to\infty} s(t) = 0$. Let $r(t)$ denote the rate at which new members are added to the population; that is, in a time interval t_{i-1} to t_i approximately $r(t_i^*)\,\Delta t_i$ members are added to the population. The function r is known as the *renewal function*. Let $f(t)$ denote the size of the population at any time t. By partitioning the interval $[0, T]$ and forming appropriate Riemann sums show that

$$f(T) = f(0)s(T) + \int_0^T s(T - t)r(t)\, dt.$$

12. The initial population of a colony of insects is 10^3. If the survival function is $s(t) = 1/(1 + t)^3$ and the renewal function is $r(t) = 100$, determine the size of the population at time T. Refer to Problem 11.

13. The initial population of a colony of insects is 10^4. If the survival function is $s(t) = (1 + t^2)^{-2}$ and the renewal function on the interval $[0, T]$ is $r(t) = 30(T - t)$, determine the size of the population at time T. Refer to Problem 11.

14. If the demand curve is given by $p_D(x) = 81 - x^2$, $0 \le x \le 9$, determine the consumer's surplus if

(a) $x_0 = 6$, (b) $p_0 = 64$.

15. If the supply curve is given by $p_S(x) = 3 + x$, determine the producer's surplus if

(a) $x_0 = 15$, (b) $p_0 = 8$.

16. For a certain product the market is determined under conditions of pure competition. In each case determine the consumer's surplus and the producer's surplus.

(a) $p_D(x) = \dfrac{72}{(1 + x)^2}$, $p_S(x) = \dfrac{2}{5}x$

(b) $p_D(x) = 240 - 2x - x^2$, $p_S(x) = \dfrac{x^2}{2}$

REVIEW PROBLEMS

In each of Problems 1 through 8, determine the area of the bounded region that satisfies the given conditions.

1. Region bounded by $f(x) = -x^2 + 2$, $g(x) = 3$, $x = -1$, and $x = 3$.
2. Region bounded by $f(x) = x^{-2}$, $g(x) = 2 - x^2$, and $h(x) = 4$.
3. Region including $(0, 0)$ bounded by $x^2 + y^2 = 16$ and $16 + 8x = y^2$.
4. Region bounded by $f(x) = \cos 2x$ and $g(x) = (16x^2/\pi^2) - 1$.
5. Region bounded by $f(x) = x^3 - 2x^2 - x + 2$ and $y = 0$.
6. Region between $x = 2\cos(\pi y/2)$ and $x^2 + y^2 = 1$.
7. Region bounded by $f(x) = -x + 2$, $g(x) = x^3$, and $x = -2$.
8. Region bounded above by $f(x) = \sqrt{1 - x^2}$ and below by $g(x) = 1 - x^2$.

In each of Problems 9 through 13, calculate the area of the given region (a) by integrating with respect to x, and (b) by integrating with respect to y.

9. Region bounded by $f(x) = 2 - x$ and $g(x) = x^2$.
10. Region bounded by $f(x) = x + 5$, $g(x) = 2x - 1$, and $h(x) = 1 - x$.
11. Region bounded by $f(x) = x^3$ and $g(x) = 2x$.
12. Region bounded by $f(x) = \sqrt{4 - x^2}$ and $g(x) = -x^2$.
13. Region bounded by $f(x) = |2 + x|$ and $g(x) = x^2$.

In each of Problems 14 through 19, the region R above the x-axis, below the curve $x = y^2$, and with $0 \leq x \leq 4$ is rotated about a given line. Find the volume of the resulting solid using (a) disks, and (b) shells.

14. R is rotated about the y-axis.
15. R is rotated about the x-axis.
16. R is rotated about $x = -2$.
17. R is rotated about $x = 4$.
18. R is rotated about $y = 3$.
19. R is rotated about $y = -1$.

In each of Problems 20 through 22, the region R above the x-axis and below $y = 2 - 2|x|$ is rotated about a given line. Find the volume of the resulting solid using (a) disks, and (b) shells.

20. R is rotated about the x-axis.
21. R is rotated about $y = 3$.
22. R is rotated about $x = 2$.

In each of Problems 23 through 25, a given solid is described. (a) Sketch the solid. Then set up (but do not evaluate) integrals giving the volume of the solid using (b) disks, and (c) shells.

23. The solid formed by rotating about the y-axis the region bounded by lines $x = 0$, $y = 6$, $y = 0$ and
$$y = \begin{cases} 8 - 4x, & \frac{1}{2} \leq x < 1 \\ -(x - 1)^2 + 4, & 1 \leq x \leq 3. \end{cases}$$
24. The solid formed by rotating the region bounded by $x = 0$, $y = 0$, and $y = -(x - 1)^3 + 2$ about the y-axis.
25. The solid formed by rotating about the y-axis the region bounded above by the smaller of $y = 20 - \frac{1}{2}x$ and $y = 16 - (x - 9)^2$, below by $y = 0$, and to the left by $x = 0$.

In each of Problems 26 through 30, determine the arc length of the given curve.

26. $y = 4x^{3/2}$ from $x = 0$ to $x = 2$.
27. $x = -\sin 2t$, $y = \cos 2t + 1$ from $t = 0$ to $t = \pi$.
28. $x = \sin^2 t$, $y = \cos^2 t$ from $t = 0$ to $t = \pi/2$.
29. $2y^3 = 5x^2$ from $(0, 0)$ to $(\sqrt{2/5}, 1)$.
30. $3x^2 = 2(y - 1)^3$ from $(0, 1)$ to $(3\sqrt{2}, 4)$.

In each of Problems 31 through 34, find the area of the surface formed by rotating the given curve about the x-axis.

31. $y = x^3/6$ with $1 \leq x \leq 4$.
32. $y^2 = 9x$ with $0 \leq x \leq 1$ and $y > 0$.
33. $y = 3x + 1$ with $0 \leq x \leq 10$.
34. $y = 2x^{1/2}$ with $1 \leq x \leq 4$.

In each of Problems 35 through 38, the velocity of a particle moving on a straight line is given as a function of time. Assume the velocity v is measured in ft/sec and t is measured in sec. Determine (a) the displacement of the particle, and (b) the total distance traveled.

35. $v = \cos 2t + \frac{1}{2}$, $0 \le t \le \pi$.

36. $v = 2t^2 - 5t + 2$, $0 \le t \le 4$.

37. $v = 2t^{-1/2}$, $1 \le t \le 9$.

38. $v = t^2(3 + t^3)^{1/2}$, $1 \le t \le 2$.

39. The natural length of a spring is 18 in. A force of 12 lb is required to compress the spring one inch. How much work is done in compressing the spring to two-thirds of its original length?

40. The top of a horizontal mine shaft lies 1000 ft below the surface of the earth. A section one-half mile long has 3 ft of water in it. A vertical cross section is a rectangle 4 ft wide and 5 ft high. Find the work done in removing the water (to the surface) assuming no additional water enters the shaft.

41. A fish pond in the shape of a paraboloid ($y = x^2/4$, in feet, rotated about the y-axis) is 5 ft deep, with water 4 ft deep at the center. Find the work required to lower the water level by 2 feet.

42. A 150-lb person is holding onto the bottom of a 100-ft rope. The rope weighs 2 lb/ft. The person climbs up to the top of the rope, still clutching the end of the rope. How much work did the person do in getting to the top of the rope?

43. A car is submerged in a pond, with the top of the rear window 10 ft below the surface of the water. The rear window is vertical, 3 ft wide at the top, 2 ft high, and 4 ft wide at the bottom, in the shape of an isosceles trapezoid. Find the force on the outside of the window.

44. A box 4 in. long, 3 in. wide, and 3 in. high contains mercury 2 in. deep. Mercury has a density of 13.6 times that of water. Find the force on each side of the box.

EIGHT

elementary transcendental functions

Differentiation and integration are operations that are applied to functions, and therefore become more useful as we become acquainted with more and more different functions. Up to now the specific functions that we have considered include polynomials, rational functions, other algebraic functions involving rational exponents, and the six trigonometric functions. Now it is time to enlarge the class of functions* to which we can apply the processes of calculus.

*The recognition of the importance of functions in calculus is due, more than anyone else, to Leonhard Euler (1707–1783), foremost among eighteenth-century mathematicians. Euler grew up near Basel, Switzerland, went to the university there, and was a student of Johann Bernoulli. He followed his friend Daniel Bernoulli to St. Petersburg in 1727, and thereafter was associated with the royal academies of St. Petersburg (1727–1741 and 1766–1783) and of Berlin (1741–1766).

Euler's published work is more extensive (by a substantial margin) than that of any other mathematician in history. He ranged over all areas of mathematics known in his time, including some that he created, but his contributions were particularly numerous in analysis, differential equations, and their applications to mechanics. Although he was blind for the last seventeen years of his life, Euler's mathematical work continued, undiminished in quantity or excellence, until the very day of his death. Indeed, for more than forty years afterward, the St. Petersburg Academy continued to publish manuscripts that he had left behind.
(*continued*)

8.1 INVERSE FUNCTIONS

In this section we consider the possibility of finding a function that is the inverse of some given function. If we think of a function as an operation or transformation from its domain to its range, then in general terms an inverse function must be such as to reverse the effect of this operation or transformation. Before making this idea more precise, let us look at some examples.

EXAMPLE 1

Let the function f be defined by

$$y = f(x) = 2x - 3 \tag{1}$$

with domain $[-2, 4]$. The graph of f is shown in Figure 8.1.1*a*. If possible, find a function that reverses the transformation corresponding to f.

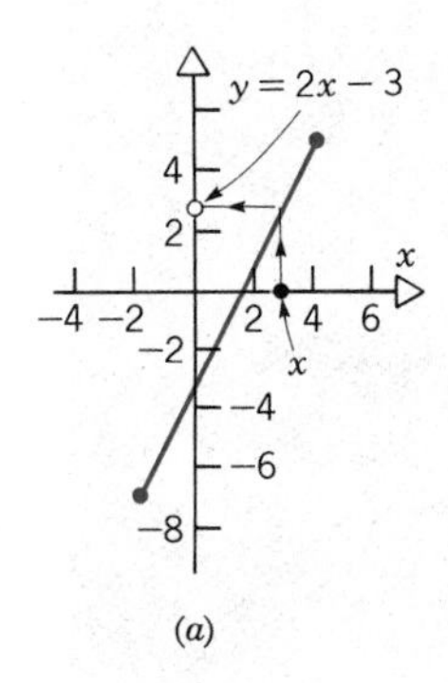

(*a*)

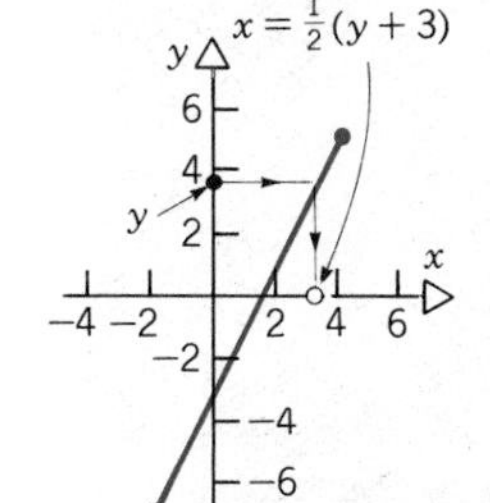

(*b*)

Figure 8.1.1

Observe that the range of f is $[-7, 5]$, and that for each point in the domain of f, Eq. 1 determines precisely one point in the range of f. To reverse the effect of this function, we must find a function or transformation that works in the opposite direction. If we solve Eq. 1 for x, we obtain

$$x = \frac{1}{2}(y + 3) = g(y); \qquad y \text{ in } [-7, 5]. \tag{2}$$

Thus with each point in the range of f we can also associate a unique point in the domain of f (see Figure 8.1.1*b*). This association is given by Eq. 2 and defines a new function g, which is said to be the inverse of f.

EXAMPLE 2

Let the function f be defined by

$$y = f(x) = \frac{x^3}{8} \tag{3}$$

Euler wrote several extremely influential books on calculus; the first, *Introductio in analysin infinitorum,* was published in 1748, and other volumes followed in 1755 and 1768–1770. The *Introductio* was the first calculus text in which functions occupy a central position; in it Euler made it clear that sines, cosines, logarithms, and exponentials, among others, should be viewed as functions, rather than as the result of a geometric or some other construction. Much of our modern notation in calculus was also invented or popularized by Euler; for instance, the letter e for the base of the natural logarithm system, and π for the ratio of the circumference to the diameter of a circle.

Euler was a master of the analytical techniques involving infinite series, infinite products, continued fractions, and integral representations of functions. Formulas and theorems associated with his name abound in many areas of mathematics; possibly the most familiar elementary one is the relation between the exponential and trigonometric functions, namely, $e^{ix} = \cos x + i \sin x$.

Euler's place in mathematics can be summarized by quoting the advice of the eminent French mathematician Pierre Simon de Laplace (1749–1827) to a young contemporary, "Read Euler: he is our master in everything."

with domain $(-\infty, \infty)$, (see Figure 8.1.2a). If possible, find a function that is the inverse of f.

Note that the range of f is also $(-\infty, \infty)$. Proceeding as in Example 1, we solve Eq. 3 for x, thereby obtaining

$$x = 2y^{1/3} = g(y). \tag{4}$$

Again, we can use Eq. 4 to associate with each point in the range of f a unique point in the domain of f, as shown in Figure 8.1.2b. This association defines a new function g, the inverse function of f. ∎

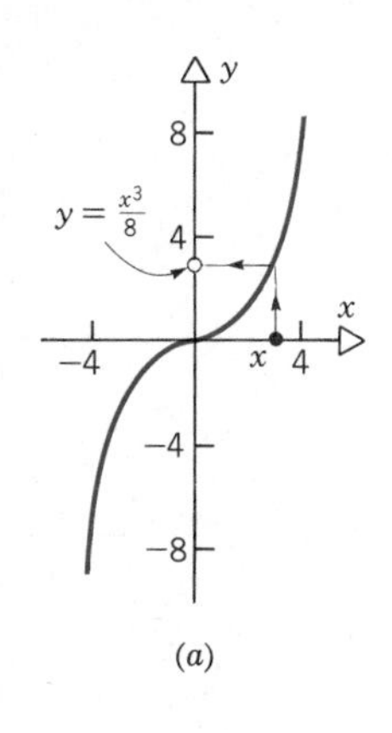

(a)

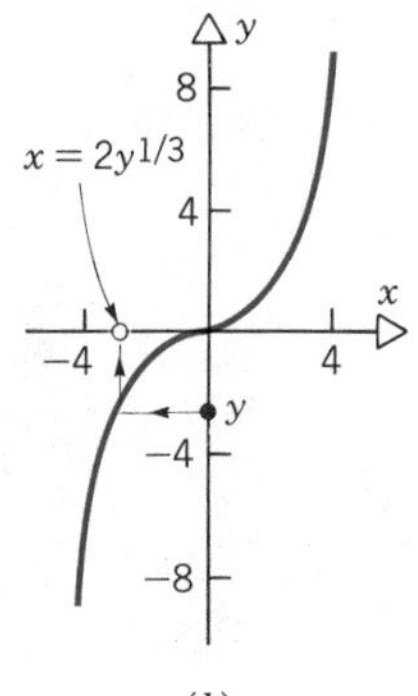

(b)

Figure 8.1.2

Now let us consider a more general situation. Let f be a function with domain D and range R. Then for each x in D the function f associates exactly one y in R. Suppose further that for each y in R there is also exactly one x in D such that $y = f(x)$. Then we say that the function f is a **one-to-one** function. Moreover, the association of a unique x with each y in R enables us to define a new function g given by

$$x = g(y), \qquad y \text{ in } R \tag{5}$$

where x is that point in D such that

$$y = f(x), \qquad x \text{ in } D. \tag{6}$$

The function g defined in this manner is said to be the **inverse** of f. In the same way f is said to be the inverse of g. Clearly the range of f is the domain of g, and vice versa.

On the other hand, if there is some y in R for which there are two or more values of x in D such that $y = f(x)$, then the function f has no inverse. In this case, f is also not one-to-one. Consequently, *f has an inverse if and only if it is a one-to-one function from its domain to its range.*

It is sometimes helpful to observe a simple consequence of Eqs. 5 and 6. Since y in R implies that x is in D, we can substitute for y in Eq. 5 by using Eq. 6. We thereby obtain

$$x = g[f(x)], \qquad x \text{ in } D. \tag{7}$$

Equation 7 expresses the fact that g reverses the transformation performed by f: starting with a given value of x in D and applying the functions f and g in succession, we return to the original value. In exactly the same way we can substitute for x in Eq. 6 by using Eq. 5; this gives

$$y = f[g(y)], \qquad y \text{ in } R. \tag{8}$$

Equation 8 is similar to Eq. 7, but the functions f and g are applied in the opposite order. The reader may verify that Eqs. 7 and 8 are satisfied by the pairs of functions given in Examples 1 and 2.*

*It may also be instructive for you to experiment a bit with a calculator having several function keys and an inverse function key. That is, start with a value of x, press a function key, and then find the inverse of the same function. Do you always obtain the original value of x?

The symbol f^{-1} is commonly used to designate the function that is the inverse of f. Note that the -1 in this notation is a superscript, *not an exponent.* In general, the inverse function f^{-1} and the reciprocal function $1/f$ are not the same.

The previous discussion states the relation that exists between a pair of inverse functions, but it does not include a satisfactory way of determining whether a given function has an inverse. In order to do this, it is helpful to look at the question from a geometrical viewpoint. Geometrically, for a given function $y = f(x)$, the condition that there is a unique value of x in D associated with each value of y in R means that *a line parallel to the x-axis intersects the graph of f at most once.* The graphs of two functions are shown in Figure 8.1.3*a* and 8.1.3*b*; the former satisfies this condition and hence has an inverse, but the latter does not.

Now let us translate this geometrical condition into a more useful form. Observe that it is certainly satisfied if f is increasing, for then $f(x_1) = f(x_2)$ only if $x_1 = x_2$. The same is true if f is decreasing. The word **monotone** is used to describe a function that is either increasing on an interval, or decreasing there. According to Theorem 4.2.1, a simple sufficient condition for f to be monotone is that its derivative be either always positive or always negative. Hence, if f is such that on the entire domain of f, the derivative $f'(x)$ is either always positive or always negative, then f has an inverse. This condition is not *necessary,* as Example 2 shows. Also, it is easy to see that if f is monotone, then its inverse f^{-1} is also monotone, and in the same direction as f. Again, observe that the function whose graph is shown in Figure 8.1.3*a* is monotone, whereas that shown in Figure 8.1.3*b* is not.

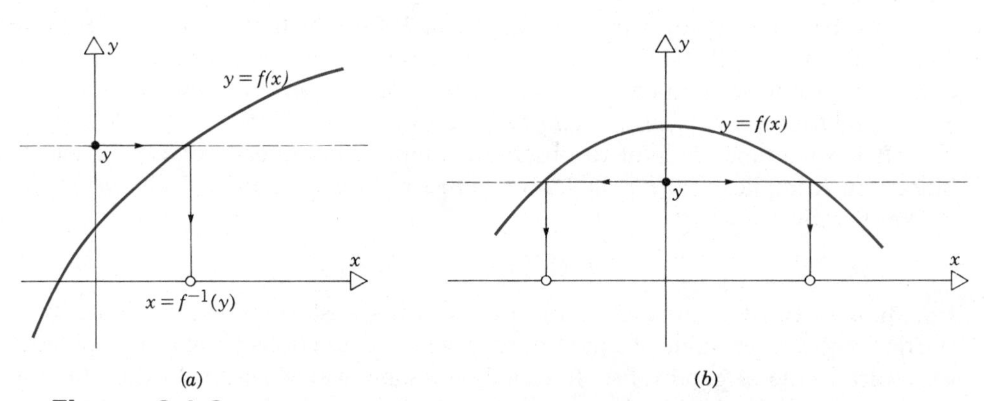

Figure 8.1.3 (*a*) A function that has an inverse. (*b*) A function that does not have an inverse.

Now suppose that a given function $y = f(x)$ with domain D does not have an inverse because for at least some values of y there is more than one corresponding value of x in D. In such a case it is possible to construct a related function simply by restricting the domain; if this is done properly, this new function may have an inverse. This procedure is illustrated by the following example.

EXAMPLE 3

Consider the function defined by

$$y = f(x) = x^2, \qquad x \text{ in } (-\infty, \infty) \tag{9}$$

whose graph is shown in Figure 8.1.4*a*. Clearly the range of f is $[0, \infty)$.

In trying to find an inverse by solving Eq. 9 for x, we immediately encounter the problem that for each y in $(0, \infty)$ there are two possible values of x, namely $+\sqrt{y}$ and $-\sqrt{y}$, in $(-\infty, \infty)$. Thus the function f does not have an inverse.

Now consider the function F given by

$$y = F(x) = x^2, \qquad x \text{ in } (-\infty, 0] \tag{10}$$

whose graph is shown in Figure 8.1.4*b*. This function is obtained from f by a restriction of its domain. The equation

$$y = F(x)$$

has a unique solution for each y in $[0, \infty)$, namely

$$x = -\sqrt{y}.$$

The function F^{-1} defined by

$$F^{-1}(y) = -\sqrt{y}, \qquad y \text{ in } [0, \infty) \tag{11}$$

is thus the inverse function of F. The function F^{-1} is not an inverse of the original function f, however.

Clearly, other restrictions on the domain of f could have been used. For instance, if $H(x) = x^2$ for x in $[0, \infty)$, then $H^{-1}(y) = \sqrt{y}$ for y in $[0, \infty)$ (see Figure 8.1.4*c*).

Note that the functions F and H are monotone on their entire domains and therefore must have inverses. In contrast, the function f is not monotone on its domain. ∎

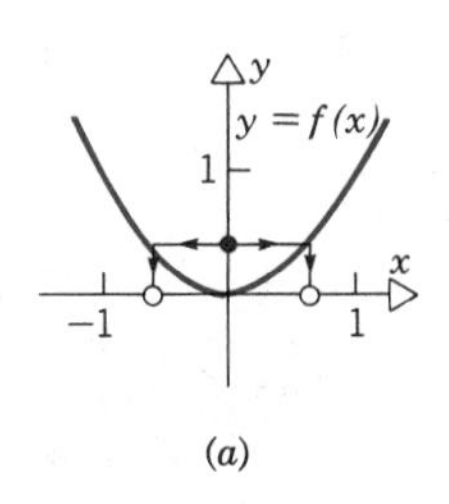

(*a*)

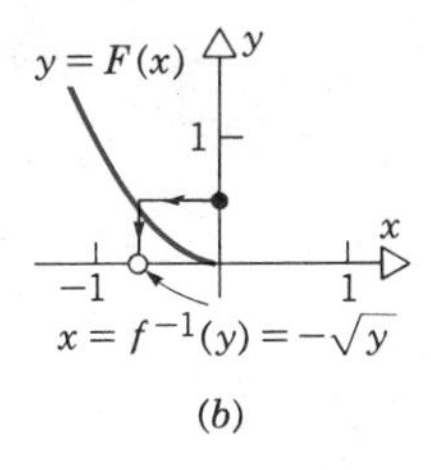

(*b*)

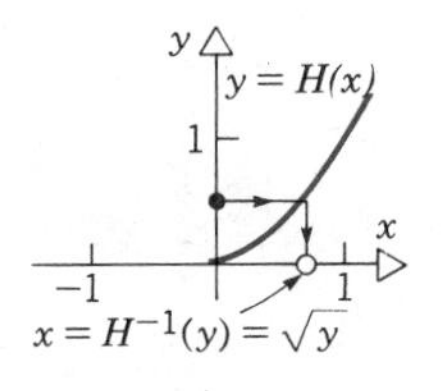

(*c*)

Figure 8.1.4
(*a*) $y = x^2$, $-\infty < x < \infty$.
(*b*) $y = x^2$, $-\infty < x \leq 0$.
(*c*) $y = x^2$, $0 \leq x < \infty$.

Even if a given function f has an inverse, it may be difficult or impossible to find an explicit formula for f^{-1}. Frequently, it is impossible to solve the equation $y = f(x)$ for x in terms of y except by numerical procedures such as Newton's method (Section 4.5). Nevertheless, we can study certain properties of f^{-1} even without having an explicit formula for it. For example, if f is continuous and f^{-1} exists, then f^{-1} is also continuous. Further, if f is differentiable and $f'(x)$ is either always positive or always negative, then f^{-1} is also a differentiable function, and its derivative can be expressed in terms of f'. We will not prove that f^{-1} is differentiable, but we will derive the relationship between the derivatives of f and f^{-1}.

For example, for the function f given by Eq. 3 we have

$$f'(x) = \frac{3x^2}{8}, \tag{12}$$

and for the inverse function g given by Eq. 4,

$$g'(y) = 2 \cdot \frac{1}{3} y^{-2/3} = \frac{2}{3y^{2/3}}. \tag{13}$$

Substituting $y = x^3/8$ in Eq. 13 we find that

$$\begin{aligned} g'[f(x)] = g'[x^3/8] &= \frac{2}{3(x^3/8)^{2/3}} \\ &= \frac{2}{3(x^2/4)} = \frac{8}{3x^2} \\ &= \frac{1}{f'(x)}. \end{aligned} \tag{14}$$

Thus, for this example, the derivative of the inverse function g at a point y is the reciprocal of the derivative of the original function f at the corresponding point x.

Now consider an arbitrary pair of inverse functions f and f^{-1}; in this case we must have, according to Eq. 7,

$$f^{-1}[f(x)] = x. \tag{15}$$

We will assume that both f and f^{-1} are differentiable, and differentiate Eq. 15 with respect to x. Letting $y = f(x)$, and using the chain rule to find the derivative of the left side of Eq. 15, we have

$$\frac{d}{dy} f^{-1}[y] \frac{d}{dx} f(x) = 1; \qquad y = f(x). \tag{16}$$

Hence

$$\frac{d}{dy}[f^{-1}(y)] = \frac{1}{f'(x)}, \qquad y = f(x). \tag{17}$$

Thus, in the general case also, *the derivatives of f and f^{-1} are reciprocals,* provided proper account is taken of the relation between the independent variables in each case. The relation embodied in Eqs. 16 and 17 is probably easier to remember if we observe that $y = f(x)$ and $x = f^{-1}(y)$, and write Eqs. 16 and 17 as

$$\frac{dx}{dy} \cdot \frac{dy}{dx} = 1, \qquad \text{or} \qquad \frac{dx}{dy} = \frac{1}{dy/dx}. \tag{18}$$

We emphasize that dy/dx and dx/dy in Eq. 18 must be evaluated at corresponding points x and y related by $y = f(x)$.

We can summarize the results we have obtained about inverse functions in the following theorem.

Theorem 8.1.1

Let the function f have domain D and range R, and suppose that either $f'(x) > 0$ for each x in D, or that $f'(x) < 0$ for each x in D. Then f has an inverse function f^{-1} with domain R and range D. The function f^{-1} is also differentiable and its derivative satisfies Eq. 17,

$$\frac{d}{dy}[f^{-1}(y)] = \frac{1}{f'(x)}; \qquad y = f(x).$$

As an illustration of this theorem, consider the following example.

EXAMPLE 4

Suppose that the function f is given by

$$y = f(x) = \frac{(x^3 - 3x^2 - 9x + 7)}{4}. \tag{19}$$

Show that f has an inverse on the interval $[-1, 3]$, and determine the domain and range of the inverse function. Also find an expression for the derivative of f^{-1}, and evaluate this derivative at $y = -1$.

Upon differentiating f, we have

$$\begin{aligned} f'(x) &= \frac{(3x^2 - 6x - 9)}{4} \\ &= \frac{3(x + 1)(x - 3)}{4}. \end{aligned} \tag{20}$$

Hence $f'(x) < 0$ for all x in the interval $(-1, 3)$, and therefore (by Theorem 4.2.1) f is decreasing on $[-1, 3]$. Thus f has an inverse there. Since f is decreasing on $[-1, 3]$, its range can be found by inserting the endpoints $x = -1$ and $x = 3$, respectively, into Eq. 19. In this way we obtain the interval $[-5, 3]$ as the range of f. Hence f^{-1} has domain $[-5, 3]$ and range $[-1, 3]$.

If we write $x = f^{-1}(y)$, then by Theorem 8.1.1

$$\begin{aligned} \frac{dx}{dy} = \frac{d}{dy} f^{-1}(y) &= \frac{1}{f'(x)} \\ &= \frac{4}{3(x + 1)(x - 3)}. \end{aligned} \tag{21}$$

In order to express dx/dy in terms of y, it is necessary to solve Eq. 19 for x in terms of y and then substitute for x in Eq. 21. It is impractical to do this analytically, but it is certainly possible (using numerical approximation methods if necessary) to solve Eq. 19 for the value of x associated with a given value of y. For instance, one can show that $y = -1$ implies that $x = 1$. Hence

$$\left.\frac{d}{dy} f^{-1}(y)\right|_{y=-1} = \left.\frac{4}{3(x + 1)(x - 3)}\right|_{x=1} = -\frac{1}{3}. \tag{22}$$

The same sort of calculations can be carried out repeatedly if it is desired to construct an extensive table of values of $f^{-1}(y)$ and its derivative for various values of y.

If we knew in advance that the function f given by Eq. 19 had an inverse $x = f^{-1}(y)$, then we could find dx/dy simply by differentiating Eq. 19 implicitly with respect to y. In this way we obtain

$$\frac{d}{dy} y = \frac{d}{dy} \frac{(x^3 - 3x^2 - 9x + 7)}{4}$$

or

$$1 = \frac{d}{dx} \frac{(x^3 - 3x^2 - 9x + 7)}{4} \frac{dx}{dy}$$

$$= \frac{(3x^2 - 6x - 9)}{4} \frac{dx}{dy},$$

from which Eq. 21 follows at once.

Graphing inverse functions

We close this section with a few words about graphs of inverse functions. Let us start with a function f that is monotone on some interval, and hence has an inverse. Figure 8.1.5 contains the graph of the equation $y = f(x)$ for such a function f. Suppose that the point (2, 1) lies on the graph of $y = f(x)$. The same graph can also be described by the equation $x = f^{-1}(y)$, and so it still contains the point (2, 1). However the independent variable is now plotted on the vertical y-axis, and the dependent variable on the horizontal x-axis. If we wish to plot the independent variable on the horizontal axis, as usual, then instead of the equation $x = f^{-1}(y)$, we would write $y = f^{-1}(x)$. The graph of $y = f^{-1}(x)$ then contains the point (1, 2). In general, if (a, b) is a point on the graph of $y = f(x)$, then (b, a) is a point on the graph of $y = f^{-1}(x)$. The graph of $y = f^{-1}(x)$ is also shown in Figure 8.1.5.

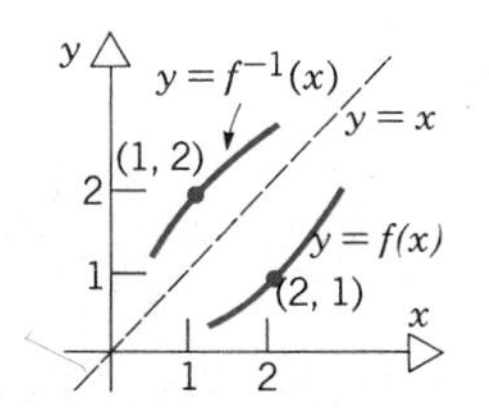

Figure 8.1.5
The graph of $f^{-1}(x)$ is obtained by reflecting the graph of $f(x)$ in the line $y = x$.

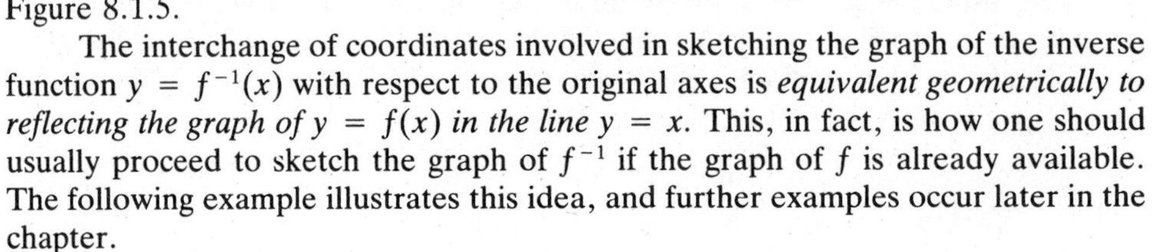

The interchange of coordinates involved in sketching the graph of the inverse function $y = f^{-1}(x)$ with respect to the original axes is *equivalent geometrically to reflecting the graph of* $y = f(x)$ *in the line* $y = x$. This, in fact, is how one should usually proceed to sketch the graph of f^{-1} if the graph of f is already available. The following example illustrates this idea, and further examples occur later in the chapter.

EXAMPLE 5

Sketch the graph of the function

$$y = f(x) = \frac{(x^3 - 3x^2 - 9x + 7)}{4}$$

in Example 4 for $-1 \le x \le 3$, and on the same axes also sketch the graph of its inverse function.

The two graphs are shown in Figure 8.1.6. The graph of f was sketched first, and the graph of its inverse function f^{-1} was then drawn by reflecting the graph

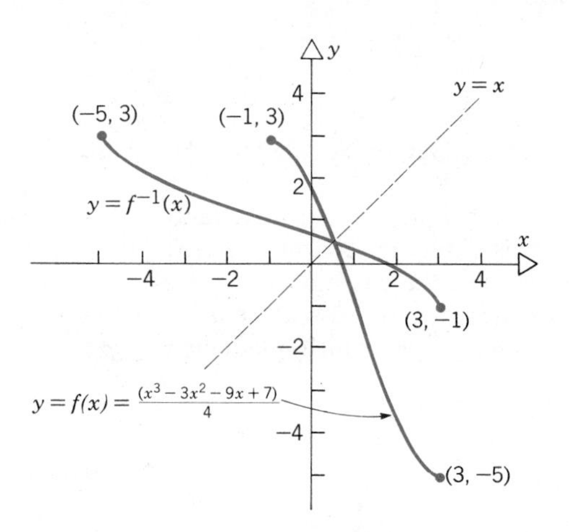

Figure 8.1.6

of f in the line $y = x$. Observe that it is not necessary to find the inverse function explicitly in order to draw its graph. ∎

PROBLEMS

In each of Problems 1 through 10, determine the inverse of the given function. In each case state the domain and range of f^{-1} and sketch the graphs of $y = f(x)$ and of $y = f^{-1}(x)$. Also verify that $f[f^{-1}(y)] = y$ and $f^{-1}[f(x)] = x$.

1. $y = f(x) = x - 3; \quad x$ in $[-2, 4]$.

2. $y = f(x) = 3x + 2; \quad x$ in $[0, 5)$.

3. $y = f(x) = 5 - 2x; \quad x$ in $(1, \infty)$.

4. $y = f(x) = x^2 - 4; \quad x$ in $[0, 4]$.

5. $y = f(x) = x^2 - 2x + 3; \quad x$ in $(-4, 1]$.

6. $y = f(x) = 4x - x^2; \quad x$ in $[2, 3)$.

7. $y = f(x) = \dfrac{x - 1}{x + 2}; \quad x$ in $(0, 3]$.

8. $y = f(x) = \dfrac{x + 3}{x + 1}; \quad x$ in $(-1, \infty)$.

9. $y = f(x) = \sqrt{x + 2}; \quad x$ in $[2, 7]$.

10. $y = f(x) = 1 - \sqrt{x}; \quad x$ in $(1, 4)$.

In each of Problems 11 through 20, use Eqs. 16 through 18 to calculate the derivative of f^{-1}. Check your answer by finding the inverse function and then differentiating it.

11. $y = f(x) = x - 3$; see Problem 1.

12. $y = f(x) = 3x + 2$; see Problem 2.

13. $y = f(x) = 5 - 2x$; see Problem 3.

14. $y = f(x) = x^2 - 4$; x in $(0, \infty)$; see Problem 4.

15. $y = f(x) = x^2 - 2x + 3$; x in $(-\infty, 1)$; see Problem 5.

16. $y = f(x) = 4x - x^2$; x in $(2, \infty)$; see Problem 6.

17. $y = f(x) = (x - 1)/(x + 2)$; $x \neq -2$; see Problem 7.

18. $y = f(x) = (x + 3)/(x + 1)$; x in $(-1, \infty)$; see Problem 8.

19. $y = f(x) = \sqrt{x + 2}$; x in $(-2, \infty)$; see Problem 9.

20. $y = f(x) = 1 - \sqrt{x}$; x in $(0, \infty)$; see Problem 10.

In each of Problems 21 through 26, determine whether the given function has an inverse. If $x = f^{-1}(y)$ exists, find its derivative; express your answer entirely in terms of y whenever this is feasible.

21. $y = f(x) = x^3 - 12x + 4; \quad x$ in $(-2, 2)$.

22. $y = f(x) = 2x^3 + 3x^2 - 12x + 5$; x in $(-2, 2)$.

23. $y = f(x) = \sqrt{4 - x^2}; \quad x$ in $(0, 2)$.

24. $y = f(x) = x^2 + \dfrac{1}{x^2}; \quad x$ in $(1, \infty)$.

25. $y = \cos x; \quad x$ in $[0, 2\pi]$.

26. $y = \tan x; \quad x \text{ in } \left(-\frac{\pi}{2}, \frac{\pi}{2}\right)$.

27. Let $y = f(x) = \sin x$. Determine whether this function has an inverse if x is restricted to each of the intervals given below. If the inverse function $x = f^{-1}(y)$ exists, find its derivative dx/dy in terms of x. Then use the equation $y = \sin x$ and trigonometric identities to express dx/dy in terms of y. Be careful to note which quadrant x is in.

(a) $0 < x < \frac{\pi}{2}$.

(b) $0 < x < \pi$.

(c) $\frac{\pi}{2} < x < 3\frac{\pi}{2}$.

28. If $y = f(x)$ has an inverse function $x = f^{-1}(y)$, and if f^{-1} is twice differentiable, show that

$$\frac{d^2x}{dy^2} = -\frac{d^2y/dx^2}{(dy/dx)^3}.$$

In each of Problems 29 through 36, assume that $x = f^{-1}(y)$ exists and is twice differentiable. Find d^2x/dy^2 in terms of x. *Hint:* See Problem 28, if necessary.

29. $y = 5 - 2x; \quad x \text{ in } (1, \infty)$.

30. $y = 4x - x^2; \quad x \text{ in } (2, 3)$.

31. $y = \frac{x - 1}{x + 2}; \quad x \text{ in } (-2, \infty)$.

32. $y = x + \frac{2}{x}; \quad x \text{ in } (0, \sqrt{2})$.

33. $y = \sqrt{x + 2}; \quad x \text{ in } (0, 5)$.

34. $y = x^3 - 12x + 4; \quad x \text{ in } (-2, 2)$.

35. $y = \sqrt{4 - x^2}; \quad x \text{ in } (0, 2)$.

36. $y = x^2 + \frac{1}{x^2}; \quad x \text{ in } (1, \infty)$.

37. In all parts of this problem assume that the function $y = f(x)$ has an inverse $x = f^{-1}(y)$.

(a) If f has domain $[0, 1]$ and range $[0, 1]$, does $y = [f(x)]^2$ have an inverse? Explain.

(b) If f has domain $[0, 1]$ and range $[-1, 1]$, does $y = [f(x)]^2$ have an inverse? Explain.

(c) If f has domain $[-1, 1]$ and range $[0, 1]$, does $y = [f(x)]^2$ have an inverse? Explain.

38. Draw the graph of a function that has an inverse but is neither increasing nor decreasing on its entire domain. Is it possible for such a function to be continuous at all points?

39. Determine the constants a and b so that the function f given by $f(x) = (x + a)/(x + b)$ is equal to its own inverse.

40. Determine the constants a, b, c, and d so that the function

$$f(x) = \frac{ax + b}{cx + d}$$

is equal to its own inverse.

41. Let the function f be given by

$$y = f(x) = \int_a^x g(t)\, dt,$$

where g is defined on the interval $a \le t \le b$.

(a) State a set of conditions on g that guarantees that f has an inverse.

(b) Assuming that $x = f^{-1}(y)$ exists, what conditions must g satisfy in order for f^{-1} to be differentiable?

(c) Find dx/dy.

42. Let the function f be given by

$$y = f(x) = \int_0^x \frac{dt}{\sqrt{1 + 3t^2}}, \qquad -\infty < x < \infty.$$

(a) Show that $x = f^{-1}(y)$ exists.

(b) Determine $D_y x$ and $D_y^2 x$.

43. Find a sufficient condition on the constants a, b, c, and d so that the function $y = ax^3 + bx^2 + cx + d$ has an inverse.

* 44. Assume that $f(x + y) = f(x)f(y)$, and that f^{-1} exists. Prove that

$$f^{-1}(u) + f^{-1}(v) = f^{-1}(uv).$$

Hint: Let $u = f(x)$ and $v = f(y)$.

8.2 THE NATURAL LOGARITHM FUNCTION

In Section 3.3 we found that (using now the integral notation for the antiderivative)

$$\int x^r\,dx = \frac{x^{r+1}}{r+1} + c,$$

provided that $r \neq -1$. However, so far we have found no differentiation formula that yields x^{-1} as the result; hence we have no antiderivative for $1/x$, and hence no analytical way to evaluate an integral whose integrand is $1/x$.

There is also a conspicuous class of (elementary) functions for which we have not yet constructed differentiation formulas, namely logarithm and exponential functions. Moreover, in Section 3.1 and elsewhere we refrained even from discussing the meaning of x^r when r is irrational. In this and the next section we fill all of these gaps in our knowledge at the same time; more specifically, we show that the antiderivative of x^{-1} is a certain logarithm function, and that we can use it in a simple way to define x^r when r is irrational.

We assume that you have some acquaintance with logarithms,* at least to base 10. If $b > 0$, we denote the logarithm of a number a to base b by $\log_b a$. Logarithms to base 10 will be written without identifying the base, that is, $\log a$ means the logarithm of a to base 10. Regardless of the base used, the characteristic property of logarithms is that they convert products into sums

$$\log_b(a_1 a_2) = \log_b a_1 + \log_b a_2. \tag{1}$$

To begin to relate logarithms with calculus, let us explore the properties of an arbitrary function f that satisfies the logarithm condition (1). That is, let us suppose that

$$f(cx) = f(x) + f(c), \tag{2}$$

where x and c are nonzero and c is a constant. In passing, note that if $x = 1$, then Eq. 2 implies that $f(1) = 0$. Assuming that f is differentiable, we can differentiate both sides of Eq. 2 with respect to x; this gives

$$cf'(cx) = f'(x). \tag{3}$$

By setting $x = 1$ in Eq. 3 we obtain

$$f'(c) = \frac{f'(1)}{c}.$$

*Logarithms were discovered by John Napier (1550–1617), a Scottish baron, who exploited the relation between geometric and arithmetic progressions to develop a means of replacing products by sums in numerical computations. Napier's investigation extended over a twenty year period and was described in books published in 1614 and 1619. The word "logarithm" is also due to Napier.

The computational advantages of logarithms to base ten were pointed out by Henry Briggs (1561?–1631), professor of mathematics at Oxford, who published an extensive 14-place table in 1624. Logarithms rapidly came into widespread use, as their manifest advantages in easing lengthy computations were quickly apparent to astronomers and others. However, it was somewhat longer before logarithms were thought of as values of a function, and more than a century passed before Euler was able to resolve some of the questions concerning this logarithm function.

Since c is an arbitrary number, we replace c by x and write

$$f'(x) = \frac{f'(1)}{x}, \qquad x \neq 0. \tag{4}$$

Consequently, the function $g(x) = f(x)/f'(1)$ is an antiderivative of x^{-1}. Thus we have established the plausibility of a relationship between functions that satisfy the logarithmic relation (2) and functions whose derivatives are x^{-1}. Of course, we have not yet established that there exist any differentiable functions that satisfy Eq. 2.

Table 8.1
A Few Values of the Natural Logarithm Function

x	$\ln x$
1.0	0.00000
1.5	0.40547
2.0	0.69315
3.0	1.09861
4.0	1.38629
5.0	1.60944
10.0	2.30259
50.0	3.91203
100.0	4.60517
1000.0	6.90776

It is possible to extract further information from Eq. 2, but it is more fruitful to shift our point of view and to consider functions that are antiderivatives of x^{-1}. Such antiderivatives exist because x^{-1} is continuous on any interval not containing the origin. According to the remark following Theorem 6.4.2 antiderivatives can be expressed as integrals. In particular, let a be a positive constant, and define the function f so that

$$f(x) = \int_a^x \frac{dt}{t}, \qquad x > 0. \tag{5}$$

Then $f'(x) = 1/x$ by Theorem 6.4.2, and hence f is an antiderivative of x^{-1} for $x > 0$. Note that two functions given by Eq. 5 for different values of a differ from each other only by an additive constant. Hence it is sufficient to examine only one such function in detail; it is convenient to consider the function for which $a = 1$. This function is known as the **natural logarithm function,** and is denoted by the symbol ln. Thus we have the following definition.

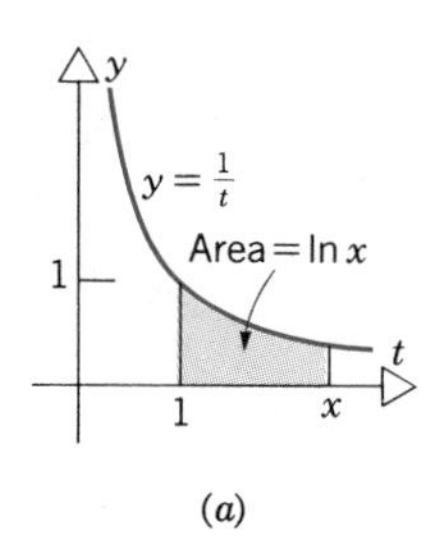

(a)

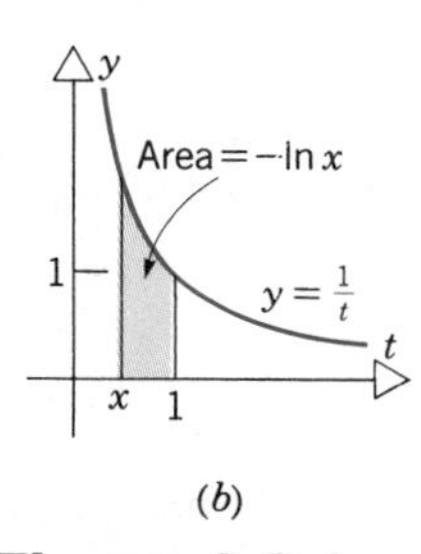

(b)

Figure 8.2.1
(a) The natural logarithm in terms of area for $x > 1$. (b) The natural logarithm in terms of area for $0 < x < 1$.

DEFINITION 8.2.1 The natural logarithm function ln is given by

$$\ln x = \int_1^x \frac{dt}{t}, \qquad x > 0. \tag{6}$$

We will find that we can apply some of the general theorems of calculus to derive many properties of the function ln. Among other things, we discuss the graph of $y = \ln x$, and we show that the function ln has the properties previously associated with logarithms; in particular, it satisfies the logarithmic relation (2).

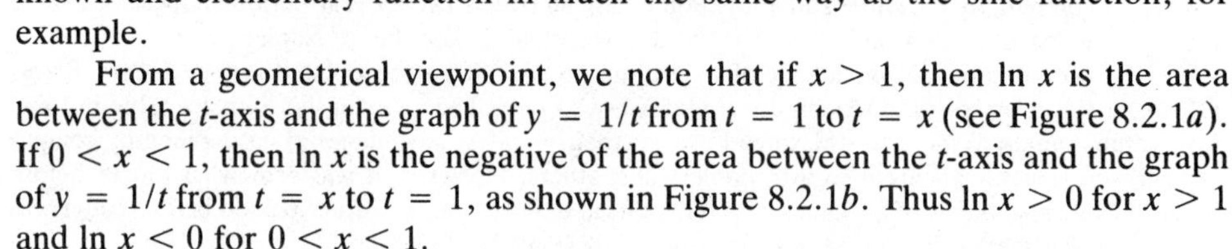

The value of $\ln x$ for a given value of x can be computed in various ways, for instance, by numerical evaluation of the integral in Eq. 6, as explained in Section 6.6. Of course, calculators have built-in routines to generate values of $\ln x$ automatically. A few representative values of $\ln x$ are shown in Table 8.1. Observe that $\ln x$ increases much more slowly than x itself. We will soon regard ln as a well-known and elementary function in much the same way as the sine function, for example.

From a geometrical viewpoint, we note that if $x > 1$, then $\ln x$ is the area between the t-axis and the graph of $y = 1/t$ from $t = 1$ to $t = x$ (see Figure 8.2.1a). If $0 < x < 1$, then $\ln x$ is the negative of the area between the t-axis and the graph of $y = 1/t$ from $t = x$ to $t = 1$, as shown in Figure 8.2.1b. Thus $\ln x > 0$ for $x > 1$ and $\ln x < 0$ for $0 < x < 1$.

Properties of ln

For ready reference we list several properties of the natural logarithm function.

1. ***Domain.*** From Eq. 6 it is clear that the domain of ln is the interval $(0, \infty)$. Other points are excluded because the integral of Eq. 6 does not exist when x is not positive because of the infinite discontinuity in the integrand at the origin.
2. ***Continuity.*** The function ln is continuous on $(0, \infty)$ by Theorem 6.4.1.
3. ***Differentiability.*** Since the integrand in Eq. 6 is continuous, the function ln is differentiable on $(0, \infty)$ by Theorem 6.4.2. Further, for each $x > 0$,

$$\frac{d}{dx} \ln x = \frac{1}{x}. \tag{7}$$

   If u is a differentiable function, and $u(x) > 0$, then by the chain rule

$$\frac{d}{dx} \ln u(x) = \frac{u'(x)}{u(x)}. \tag{8}$$

4. ***Monotonic property.*** Since $(d/dx) \ln x > 0$ for $x > 0$, the function ln is monotone increasing throughout its domain by Theorem 4.2.1. Since $\ln 1 = 0$, it is again clear that $\ln x > 0$ for $x > 1$ and $\ln x < 0$ for $0 < x < 1$.
5. ***Concavity.*** From Eq. 7 we immediately find that

$$\frac{d^2}{dx^2} \ln x = -\frac{1}{x^2}, \qquad x > 0. \tag{9}$$

   Since $(d^2/dx^2) \ln x < 0$ for $x > 0$, then according to Theorem 4.2.3 the graph of the function ln is concave down throughout its domain.
6. ***Range.*** The range of ln is the interval $(-\infty, \infty)$. This fact is not obvious at the moment, since nothing that we have said so far requires that $\ln x$ become unbounded as $x \to 0+$ and as $x \to \infty$. We will return to this question later in the section.

As a result of Properties 1 through 6, the graph of the equation $y = \ln x$ must have the general form sketched in Figure 8.2.2. In particular, Property 6 implies that $y = \ln x$ has no horizontal asymptote as $x \to \infty$, but is asymptotic to the negative y-axis as $x \to 0+$.

We now verify that the function ln has the characteristic property of logarithms, that is, it converts products into sums.

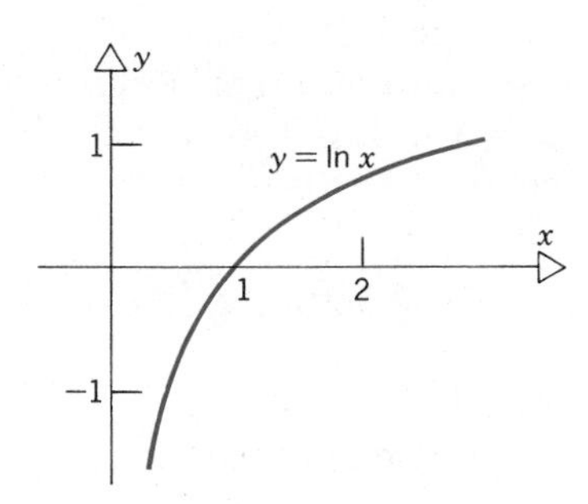

Figure 8.2.2 Graph of $y = \ln x$ for $x > 0$.

Theorem 8.2.1
The natural logarithm function, defined by Eq. 6

$$\ln x = \int_1^x \frac{dt}{t}, \qquad x > 0,$$

satisfies the relation

$$\ln cx = \ln x + \ln c \tag{10}$$

for each $x > 0$ and $c > 0$.

Proof. To establish Eq. 10 let us take the view that x is a variable, and c is a constant. Then Eq. 10 states that two functions of x, namely $\ln cx$ and $\ln x$, differ by the constant $\ln c$. To show that two functions differ by a constant, it is sufficient (by Theorem 4.1.5) to show that their derivatives are equal. Thus let us determine the derivative of $\ln cx$ with respect to x. Using the chain rule, and remembering that c is constant, we obtain

$$\begin{aligned}\frac{d}{dx}\ln cx &= \frac{1}{cx}\frac{d(cx)}{dx} = \frac{1}{cx}\cdot c\\ &= \frac{1}{x}.\end{aligned}$$

Hence

$$\frac{d}{dx}\ln cx = \frac{d}{dx}\ln x,$$

and consequently

$$\ln cx = \ln x + k, \tag{11}$$

where k is a constant. To evaluate k we can set $x = 1$ in Eq. 11, which gives $k = \ln c$. Substituting this value for k in Eq. 11, we obtain Eq. 10, thus proving the theorem. □

Note that we have now answered the question raised after deriving Eq. 4, namely, is there a differentiable function that satisfies Eq. 2? The answer is yes, and one such function is $f(x) = \ln x$.

Several other useful results follow quickly from Eq. 10. For any fixed x, let $c = x$ in Eq. 9; then

$$\ln x^2 = \ln x + \ln x = 2\ln x. \tag{12}$$

Similarly, letting $c = x^2$ in Eq. 10 and using Eq. 12, we find that

$$\ln x^3 = \ln x + \ln x^2 = 3\ln x. \tag{13}$$

Continuing in this way, we obtain

$$\ln x^n = n\ln x \tag{14}$$

for each $x > 0$ and for each positive integer n. Note that Eq. 14 is also valid if $n = 0$.

If we let $c = 1/x$ in Eq. 10, then

$$\ln 1 = \ln x + \ln \frac{1}{x};$$

since $\ln 1 = 0$, it follows that

$$\ln \frac{1}{x} = -\ln x. \tag{15}$$

A more general version of Eq. 14 is also true, namely,

$$\ln x^r = r \ln x \tag{16}$$

for each $x > 0$ and for each rational number r. We can prove Eq. 16 by proceeding as in the proof of Theorem 8.2.1. Using Eq. 8 to differentiate each side of Eq. 16, we obtain

$$\frac{d}{dx} \ln x^r = \frac{rx^{r-1}}{x^r} = \frac{r}{x}$$

and

$$\frac{d}{dx} r \ln x = \frac{r}{x}.$$

Therefore, by Theorem 4.1.5,

$$\ln x^r = r \ln x + c, \tag{17}$$

where c is a constant. By setting $x = 1$ in Eq. 17 we find that $c = 0$, so Eq. 16 is proved.

We are now in a position to establish Property 6: the range of ln is $(-\infty, \infty)$. Since ln is continuous and monotone, we need only show that

$$\lim_{x \to \infty} \ln x = \infty, \qquad \lim_{x \to 0^+} \ln x = -\infty. \tag{18}$$

According to the definition of limit as $x \to \infty$ given in Section 2.4, the first of Eqs. 18 is established by showing that for any number $M > 0$ there is an x_0 such that $\ln x > M$ whenever $x > x_0$. However, the fact that ln is an increasing function assures us that if $x > x_0$, then $\ln x > \ln x_0$. Hence we will know that $\ln x > M$ for all $x > x_0$, provided we determine x_0 so that $\ln x_0 > M$. Given an arbitrary $M > 0$, we can find such an x_0 in the following way. Letting $x = 2$ in Eq. 16 gives

$$\ln 2^r = r \ln 2. \tag{19}$$

Then, since $\ln 2 > 0$, it is possible to choose r large enough so that $r \ln 2 > M$; indeed, r may be chosen to be an integer for simplicity. Using this value of r, let $x_0 = 2^r$; then clearly $\ln x_0 > M$.

The second of Eqs. 18 is verified in a very similar way, making use of Eq. 15.

Differentiation and integration formulas

We have already seen in Eq. 7 that

$$\frac{d}{dx} \ln x = \frac{1}{x}, \qquad x > 0.$$

Thus we have a function whose derivative is x^{-1} for $x > 0$. Can we also find a function whose derivative is x^{-1} for $x < 0$? Clearly $\ln x$ is not such a function, since $\ln x$ is not defined for $x < 0$. We will show that the correct function is $\ln(-x)$, which does exist for $x < 0$. By the chain rule the derivative of $\ln(-x)$ is found as follows:

$$\begin{aligned} \frac{d}{dx} \ln(-x) &= \frac{1}{(-x)} \frac{d(-x)}{dx} \\ &= \frac{1}{(-x)}(-1) = \frac{1}{x}, \qquad x < 0. \end{aligned} \tag{20}$$

Combining Eqs. 7 and 20 we obtain the more compact formula

$$\frac{d}{dx} \ln|x| = \frac{1}{x}, \qquad x \neq 0. \tag{21}$$

Equation 21 is the basic differentiation formula for the natural logarithm function. The graph of $y = \ln|x|$ is symmetric about the y-axis and is shown in Figure 8.2.3.

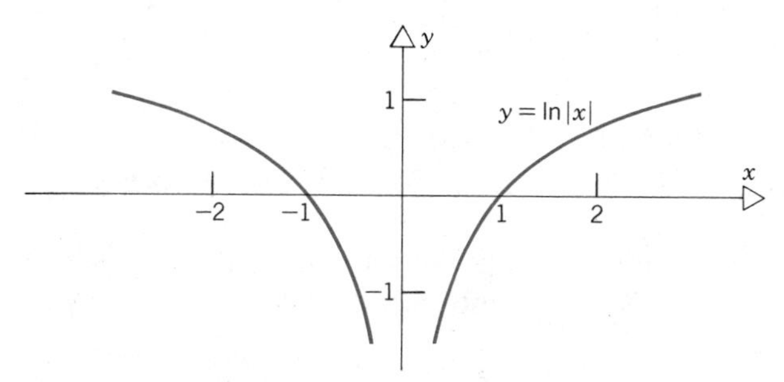

Figure 8.2.3 Graph of $y = \ln|x|$ for $x \neq 0$.

Equation 8 can be generalized in a similar way. If u is a differentiable function and if $u(x)$ is not zero, then

$$\frac{d}{dx} \ln|u(x)| = \frac{u'(x)}{u(x)}, \qquad u(x) \neq 0. \tag{22}$$

If $u(x)$ is zero at isolated points, then Eq. 22 is valid at all points except those where $u(x) = 0$.

EXAMPLE 1

Find the derivative of $f(x) = \ln(x^2 + 9)$.

In this case $u(x) = x^2 + 9$. By Eq. 22 we have

$$\frac{d}{dx}\ln(x^2 + 9) = \frac{1}{x^2 + 9}\frac{d}{dx}(x^2 + 9) = \frac{2x}{x^2 + 9}. \blacksquare \tag{23}$$

EXAMPLE 2

Find the derivative of $f(x) = \ln|\sin x|$.

Now we have $u(x) = \sin x$. Note that the absolute value is needed since $\sin x$ takes on both positive and negative values. Using Eq. 22, we obtain

$$\frac{d}{dx}\ln|\sin x| = \frac{\cos x}{\sin x} = \cot x. \tag{24}$$

The result is valid so long as $\sin x \neq 0$, that is, for $x \neq 0, \pm\pi, \pm 2\pi, \ldots$. ∎

EXAMPLE 3

Find the derivative of $f(x) = \ln(\ln x)$.

The domain of the function f is $(1, \infty)$. On this interval

$$\frac{d}{dx}\ln(\ln x) = \frac{1}{\ln x}\frac{d}{dx}(\ln x) = \frac{1}{x \ln x}. \blacksquare \tag{25}$$

Corresponding to Eq. 21 there is the following antidifferentiation, or integration, formula valid for negative as well as positive values of x.

$$\int \frac{dx}{x} = \ln|x| + c, \qquad x \neq 0. \tag{26}$$

More generally, corresponding to Eq. 22 we have

$$\int \frac{u'(x)}{u(x)}\, dx = \ln|u(x)| + c, \qquad u(x) \neq 0. \tag{27}$$

Equation 27 yields an antiderivative for any expression that can be identified as the ratio $u'(x)/u(x)$ for some differentiable function u. Of course, in using Eq. 26 to evaluate an integral, one must be sure that the interval of integration does not include the origin. Similarly, in using Eq. 27, one must avoid intervals containing a point where $u(x) = 0$.

EXAMPLE 4

Evaluate

$$\int \frac{dx}{2x + 5}.$$

We let $u(x) = 2x + 5$; then $u'(x) = 2$ and $du = 2dx$, or $dx = du/2$. Consequently,

$$\int \frac{dx}{2x + 5} = \int \frac{du/2}{u} = \frac{1}{2} \ln|u| + c = \frac{1}{2} \ln|2x + 5| + c. \tag{28}$$

The result holds on any interval not containing $x = -\frac{5}{2}$. ∎

EXAMPLE 5

Evaluate

$$\int \tan x \, dx.$$

First, it is helpful to write $\tan x = \sin x/\cos x$. Then we let $u(x) = \cos x$, so that $u'(x) = -\sin x$ and $du = -\sin x \, dx$. Thus we obtain

$$\int \tan x \, dx = \int \frac{\sin x}{\cos x} dx = -\int \frac{du}{u}.$$

Therefore

$$\int \tan x \, dx = -\ln|u| + c = -\ln|\cos x| + c.$$

The result is valid on any interval where $\cos x$ is never zero. ∎

In calculating antiderivatives such as these in Examples 4 and 5 the most difficult step is the choice of $u(x)$. Sometimes the appropriate substitution is fairly obvious, sometimes it is more obscure. In general, one may expect to develop proficiency in such manipulations through practice. The following examples require the evaluation of definite integrals, rather than the determination of antiderivatives.

EXAMPLE 6

Find the area A of the region bounded by the x-axis, the lines $x = 1$ and $x = 3$, and the curve $y = x/(x^2 + 7)$. This region is shown in Figure 8.2.4.

The area A is given by the integral

$$A = \int_1^3 \frac{x}{x^2 + 7} dx.$$

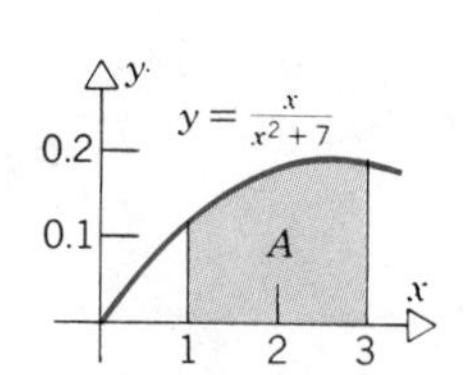

Figure 8.2.4

To evaluate this integral we first find an antiderivative of the integrand. If we let $u = x^2 + 7$, then $u'(x) = 2x$, and

$$\int \frac{x}{x^2 + 7} dx = \frac{1}{2} \int \frac{u'(x)}{u(x)} dx$$

$$= \frac{1}{2} \ln u(x) + c = \frac{1}{2} \ln(x^2 + 7) + c.$$

The value of A is then given by

$$A = \frac{1}{2}\ln(x^2 + 7)\Big|_1^3$$

$$= \frac{\ln 16 - \ln 8}{2} = \frac{\ln 2}{2} \cong 0.34658. \blacksquare$$

EXAMPLE 7

Find the value of

$$I = \int_{-6}^{-3} \frac{dx}{x + 2}.$$

The graph of the integrand is shown in Figure 8.2.5.

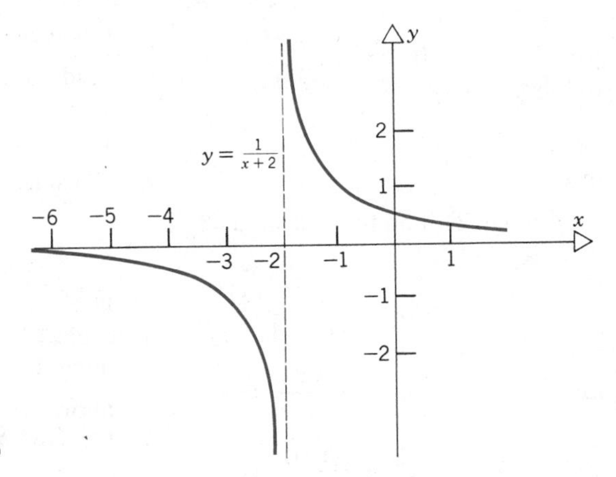

Figure 8.2.5

Note that the integrand becomes unbounded as $x \to -2$, but this causes no trouble since the interval of integration does not contain this point. However, the fact that the integrand is negative on the interval of integration means that we must be careful to retain the absolute value bars in the antiderivative. By letting $u(x) = x + 2$, we obtain

$$I = \ln|x + 2|\Big|_{-6}^{-3} = \ln|-1| - \ln|-4|$$

$$= \ln 1 - \ln 4 = -\ln 4 \cong -1.38629. \blacksquare$$

PROBLEMS

In each of Problems 1 through 16, find the derivative of the given function.

1. $f(x) = \ln|x + 4|; \quad x \neq -4.$
2. $f(x) = \ln|2x - 3|; \quad x \neq \frac{3}{2}.$

3. $f(x) = \ln(x^4 + 1)$.

4. $f(x) = \ln(x^2 + x + 1)$.

5. $f(x) = x \ln(x^2 + 9) + \dfrac{3}{x}$.

6. $f(x) = \ln\left|\dfrac{x - 1}{x + 1}\right|; \quad x \neq \pm 1$.

7. $f(x) = \ln \dfrac{2|x|}{x^2 + 1}; \quad x \neq 0$.

8. $f(x) = \sqrt{x^2 + 4} + \ln \sqrt{x^2 + 4}$.

9. $f(x) = \ln(1 + \sqrt{x}); \quad x > 0$.

10. $f(x) = \ln|\cos x|; \quad x \neq \pm \dfrac{\pi}{2}, \pm \dfrac{3\pi}{2}, \ldots$.

11. $f(x) = x \ln|x| - x; \quad x \neq 0$.

12. $f(x) = \ln [\sqrt{a^2 + x^2} + x]$.

13. $f(x) = \ln|\ln x|; \quad x > 0$ and $x \neq 1$.

14. $f(x) = \cos x \ln|\sin x|; \quad x \neq 0, \pm\pi, \ldots$.

15. $f(x) = \sin(\ln|x|); \quad x \neq 0$.

16. $f(x) = \dfrac{\ln x}{x} + \dfrac{x}{\ln x}; \quad x > 1$.

In each of Problems 17 through 24, find the indicated antiderivative.

17. $\displaystyle\int \frac{dx}{x - 3}$

18. $\displaystyle\int \frac{dx}{2x + 1}$

19. $\displaystyle\int \frac{2x - 3}{x^2 - 3x + 7}\, dx$

20. $\displaystyle\int \frac{x^2}{x^3 + 1}\, dx$

21. $\displaystyle\int \frac{dx}{\sqrt{x}\,(1 + \sqrt{x})}$

22. $\displaystyle\int \frac{(\ln|x|)^3}{x}\, dx$

23. $\displaystyle\int \cot 2x\, dx$

24. $\displaystyle\int \frac{\sec^2 x}{\tan x}\, dx$

In each of Problems 25 through 32, evaluate the given integral.

25. $\displaystyle\int_2^6 \frac{dx}{x - 1}$

26. $\displaystyle\int_0^4 \frac{dx}{3x + 2}$

27. $\displaystyle\int_2^4 \frac{4x - 6}{x^2 - 3x + 4}\, dx$

28. $\displaystyle\int_{-3}^0 \frac{dx}{x - 2}$

29. $\displaystyle\int_{-1}^1 \frac{dx}{2x - 5}$

30. $\displaystyle\int_1^4 \frac{dx}{\sqrt{x}(1 + \sqrt{x})}$

31. $\displaystyle\int_{\pi/6}^{\pi/3} \cot x\, dx$

32. $\displaystyle\int_{-\pi/4}^{-\pi/6} \cot x\, dx$

In each of Problems 33 through 36, find the nth derivative of the given function.

33. $y = \ln|x|, \quad x \neq 0$.

34. $y = \ln|1 + x|, \quad x \neq -1$.

35. $y = \ln|1 - x|, \quad x \neq 1$.

36. $y = \ln|1 - x^2|, \quad x \neq \pm 1$.

In each of Problems 37 through 40, find the maximum and minimum values (if they exist) of the given function on the given interval.

37. $f(x) = x - \ln x$ on $(0, 2]$.

38. $f(x) = x - \ln x$ on $[\frac{1}{2}, 2]$.

39. $f(x) = \ln \sin x$ on $(0, \pi)$.

40. $f(x) = \dfrac{1}{x} + \ln \sqrt{x}$ on $[1, 3]$.

41. Consider the region bounded by the lines $x = 0$, $y = 0$, $x = 8$, and by the curve $y = (1 + x)^{-1/2}$. Find the volume of the solid formed by rotating this region about the x-axis.

42. Find the area of the region bounded by the lines $x = \frac{1}{2}$, $y = 0$, $x = 2$, and by the curve $y = x + x^{-1}$.

43. Consider the integral in Example 4 of the text,

$$I = \int_{-6}^{-3} \frac{dx}{x + 2}.$$

Use the symmetry of the graph in Figure 8.2.5 to replace the integral by an equivalent one whose integrand is positive.

44. Show that $\lim_{x\to 0^+} \ln x = -\infty$. That is, show that for any $M > 0$ there is an x_0 in the interval $0 < x_0 < 1$ such that if $0 < x < x_0$, then $\ln x < -M$.

45. Show that

$$\lim_{x\to 0} \frac{\ln(1 + x)}{x} = 1.$$

Hint: Write

$$\frac{\ln(1 + x)}{x} = \frac{\ln(1 + x) - \ln 1}{x},$$

and identify the required limit in terms of the derivative of $\ln x$ at $x = 1$.

46. (a) Use the mean value theorem on the function $f(x) = \ln(1 + x)$ to show that

$$\frac{x}{1 + x} < \ln(1 + x) < x$$

for all $x > -1$. It may be helpful to consider the cases $x > 0$ and $x < 0$ separately.

(b) Use the result of Part (a) to show that

$$\lim_{x \to 0} \frac{\ln(1 + x)}{x} = 1.$$

47. **Logarithmic differentiation.** The differentiation of products, quotients, and powers can be facilitated by first taking logarithms, a process known as logarithmic differentiation. Let

$$y = f(x) = g_1(x)g_2(x) \cdots g_n(x)$$

where $g_1, \ldots, g_n$ are differentiable functions. Then

$$\ln|y| = \sum_{k=1}^{n} \ln|g_k(x)|,$$

so long as no $g_k(x)$ is zero. Show that

$$\frac{1}{y}\frac{dy}{dx} = \sum_{k=1}^{n} \frac{g_k'(x)}{g_k(x)} = \frac{g_1'(x)}{g_1(x)} + \cdots + \frac{g_n'(x)}{g_n(x)},$$

and that

$$\frac{dy}{dx} = g_1(x) \cdots g_n(x) \sum_{k=1}^{n} \frac{g_k'(x)}{g_k(x)}.$$

In each of Problems 48 through 50, use logarithmic differentiation (see Problem 47) to find the derivative of the given function.

48. $f(x) = (x + 1)(x - 3)(x^2 + 4)$

49. $f(x) = x(x^2 + x + 1)\sin 2x$

50. $f(x) = (x - 2)\sqrt{x^2 + 1}(x^2 + 9)x^3$

51. (a) Use Corollary 1 of Section 6.3 to show that

$$\int_1^x \frac{dt}{t} \le \int_1^x \frac{dt}{\sqrt{t}}, \qquad x > 1.$$

(b) Use the result of Part (a) to show that

$$0 \le \frac{\ln x}{x} \le \frac{2}{\sqrt{x}} - \frac{2}{x}, \qquad x > 1;$$

then show that

$$\lim_{x \to \infty} \frac{\ln x}{x} = 0.$$

(c) Using the result of Part (b), show that

$$\lim_{x \to 0^+} x \ln x = 0.$$

Hint: Let $x = 1/y$.

*(d) Extend the results of Parts (b) and (c) to show that

$$\lim_{x \to \infty} \frac{\ln x}{x^\epsilon} = 0, \qquad \lim_{x \to 0^+} x^\epsilon \ln x = 0$$

for any $\epsilon > 0$. In other words, $|\ln x|$ grows more slowly than any positive power of x as $x \to \infty$ or as $x \to 0^+$.
Hint: Let $x = u^\epsilon$ in the limits established in Parts (b) and (c).

8.3 THE EXPONENTIAL FUNCTION

In the last section we showed that the natural logarithm is a continuous monotone function with domain $(0, \infty)$ and range $(-\infty, \infty)$. Hence it has a continuous monotone inverse function with domain $(-\infty, \infty)$ and range $(0, \infty)$. We will denote this inverse function by exp, and call it the **exponential function**. The appropriateness of this name will be apparent later in this section.

DEFINITION 8.3.1 The exponential function exp is defined for each real number x by

$$y = \exp x \quad \text{if and only if} \quad x = \ln y. \tag{1}$$

Thus to calculate the number $y = \exp x$ for a given x, we must solve the equation

$$x = \ln y = \int_1^y \frac{dt}{t} \tag{2}$$

for y. In practice, values of the exponential function are calculated by other methods that we do not discuss here. Pocket calculators have built in routines that produce values of exp automatically. A few representative values of $\exp x$ are shown in Table 8.2. It is clear that $\exp x$ grows very rapidly as x increases. In fact, it is possible to show (Problems 47 and 48) that $\exp x$ grows faster than any power of x (however great) as $x \to \infty$.

Table 8.2 A Few Values of the Exponential Function

x	$\exp x$
−10	0.00005
− 5	0.00674
− 4	0.01832
− 3	0.04979
− 2	0.13534
− 1	0.36788
− 0.5	0.60653
0	1.0000
0.5	1.6487
1	2.7183
2	7.3891
3	20.086
4	54.598
5	148.41
10	22026.5

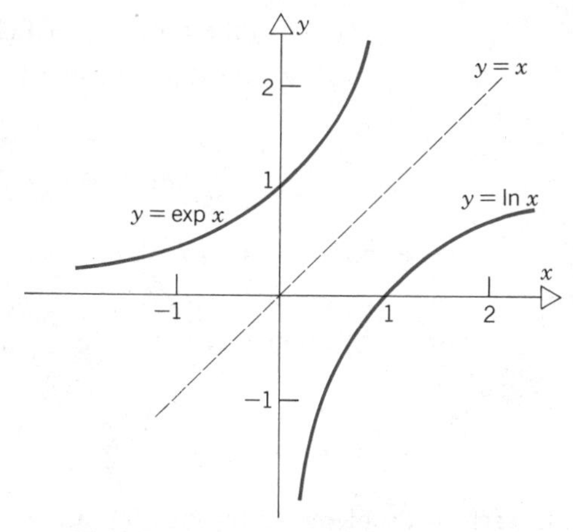

Figure 8.3.1 Graph of $y = \exp x$.

The graph of $y = \exp x$ is sketched in Figure 8.3.1; as described at the end of Section 8.1, it was obtained by reflecting the graph of $y = \ln x$ about the line $y = x$. Note that the graph of $y = \exp x$ is increasing and concave up; further, $\exp x \to 0$ as $x \to -\infty$ and $\exp x \to \infty$ as $x \to \infty$. Also, $\exp 0 = 1$. These properties follow immediately from properties of the natural logarithm function discussed in the previous section.

The graph of $y = \exp(-x)$ is obtained from that of $y = \exp x$ simply by reversing the direction of the positive x-axis, or alternatively, by rotating the graph of $y = \exp x$ about the y-axis. The graph of $y = \exp(-x)$ is shown in Figure 8.3.2.

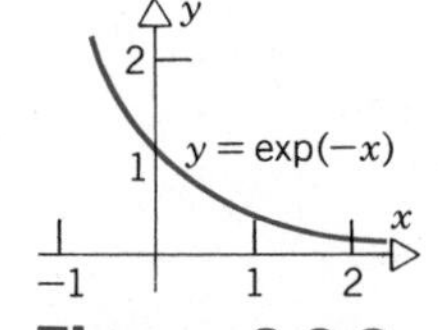

Figure 8.3.2 Graph of $y = \exp(-x)$.

Since the natural logarithm and exponential functions are inverses of each other, it follows immediately from Eq. 1 that

$$x = \ln(\exp x), \qquad \text{for any } \; x \tag{3}$$

and that

$$y = \exp(\ln y), \qquad \text{for } \; y > 0. \tag{4}$$

Algebraic properties of the exponential function

The exponential function has the following properties, sometimes called the laws of exponents:

$$\exp(u + v) = (\exp u)(\exp v), \tag{5}$$

$$\exp(-u) = \frac{1}{\exp u}, \tag{6}$$

$$\exp(u - v) = \frac{\exp u}{\exp v}, \tag{7}$$

$$(\exp u)^r = \exp(ru), \qquad r \text{ rational}, \tag{8}$$

where u and v denote any real numbers.

To prove Eq. 5 we start from the corresponding property of the natural logarithm function, that is,

$$\ln(xy) = \ln x + \ln y. \tag{9}$$

For any real numbers u and v, let

$$x = \exp u, \qquad y = \exp v; \tag{10}$$

note that $x > 0$ and $y > 0$. Then

$$u = \ln x, \qquad v = \ln y. \tag{11}$$

Applying the exponential function to both sides of Eq. 9, we obtain

$$\exp[\ln(xy)] = \exp[\ln x + \ln y],$$

or, since exp and ln are inverse functions,

$$xy = \exp[\ln x + \ln y]. \tag{12}$$

Using Eqs. 10 and 11 to express Eq. 12 in terms of u and v, we have

$$(\exp u)(\exp v) = \exp(u + v),$$

thus proving Eq. 5.

Equations 6 through 8 follow quickly from Eq. 5 (see Problem 58).

Irrational exponents

Recall that in the previous section we showed that

$$\ln(x^r) = r \ln x \tag{13}$$

for each $x > 0$ and each *rational* r. Applying the exponential function to both sides of Eq. 13, we obtain

$$\exp[\ln(x^r)] = \exp(r \ln x)$$

or, from Eq. 4,

$$x^r = \exp(r \ln x). \tag{14}$$

Equation 14 relates the quantity x^r, for rational r, to the natural logarithm and exponential functions. However, the crucial fact about Eq. 14 is that the right side of this equation is well defined for each $x > 0$ and for every real number r, whether rational or not. That is, for a given $x > 0$ we have defined $\ln x$ in Section 8.2, we can then multiply by any real number r to obtain $r \ln x$, and finally in this section we have defined $\exp u$ for every real number u. Thus the right side of Eq. 14 provides a way of *defining* x^r when r is irrational.

DEFINITION 8.3.2 For any positive x and for any real number r (whether rational or irrational)

$$x^r = \exp(r \ln x).$$

In practice, roundoff errors prevent the *exact* calculation of x^r from Definition 8.3.2. From a theoretical point of view the important thing is that with Definition 8.3.2 it is possible to show that all of the familiar laws of exponents and algebraic operations are valid for irrational exponents.

The Number *e*

The number whose natural logarithm is one is a very important number in mathematics, perhaps second only to π in the significance and variety of its appearances. It is sufficiently important so that it is always designated by the special symbol e. Thus e is the number such that

$$\ln e = 1 \qquad \text{or} \qquad e = \exp 1. \tag{15}$$

From Eq. 2 it follows that e satisfies the equation

$$\int_1^e \frac{dt}{t} = 1. \tag{16}$$

Thus if we visualize sweeping out the area between the t-axis and the graph of $y = 1/t$, starting at $t = 1$ and going to the right, then we must stop at $t = e$ in order to have an area of unit size (see Figure 8.3.3).

There are several ways to compute a good approximation to e. For example, one can approximate the integral in Eq. 16 by suitable Riemann sums, as outlined in Problem 56. By using rather crude estimates of this sort it is possible to show that e lies between 2.5 and 3.

In Problem 49 we show that e is also given by the important limiting formula

$$e = \lim_{k\to\infty} \left(1 + \frac{1}{k}\right)^k. \tag{17}$$

Equation 17 can also be used to compute approximate values of e; a few such values are recorded in Table 8.3. This tabulation suggests that e is slightly larger than 2.7.

Other and more efficient ways of computing e emerge later, especially as by-products of a study of infinite series (Example 3 of Section 13.2). Therefore, for the present, we will merely state that e can be accurately calculated, and that its

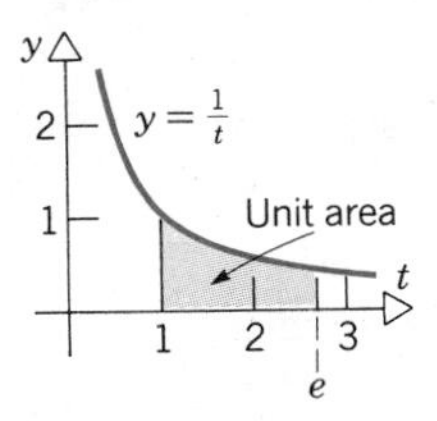

Figure 8.3.3 The number e in terms of an area.

Table 8.3 Some Approximations to e

k	$\left(1 + \frac{1}{k}\right)^k$
1	2
2	2.25
4	2.4414
10	2.5937
25	2.6658
100	2.7048
1000	2.7169
10^4	2.71815
10^5	2.71827
10^6	2.71828

value to five decimal places is

$$e = 2.71828\ldots. \tag{18}$$

Now let us recall Eq. 14

$$x^r = \exp(r \ln x),$$

a relation that is valid for all positive x and for all real r. If we set $x = e$ in Eq. 14 and use Eq. 15, we obtain

$$e^r = \exp r. \tag{19}$$

Since r can be any real number, we have thus identified the value of the exponential function at each point as the number e raised to the corresponding exponent. If, as is customary, we use the letter x to denote the independent variable, then Eq. 19 takes the form

$$\exp x = e^x. \tag{20}$$

Hereafter we will usually use the notation e^x rather than the equivalent $\exp x$; the latter notation, however, is sometimes better when the exponent is a complicated expression.

In view of Eq. 20, we can rewrite Eqs. 1 as

$$y = e^x \qquad \text{if and only if} \qquad x = \ln y. \tag{21}$$

Thus it is apparent that the natural logarithm is the logarithm to base e. In terms of e the expression in Definition 8.3.2 becomes

$$x^r = e^{r \ln x}; \qquad x > 0, \qquad r \text{ real}. \tag{22}$$

Also, the relations (5) to (8) may now be rewritten as

$$e^{u+v} = e^u e^v,$$

$$e^{-u} = \frac{1}{e^u},$$

$$e^{u-v} = \frac{e^u}{e^v},$$

$$(e^u)^r = e^{ru}.$$

Differentiation and integration formulas for the exponential function

Recall from Section 8.2 that $d(\ln y)/dy = 1/y > 0$ for $y > 0$. Therefore, by Theorem 8.1.1, the inverse of the natural logarithm function is also differentiable. In other words, $y = e^x$ is a differentiable function of x. To find the derivative of $y = e^x$ we can differentiate implicitly the equation $x = \ln y$ with respect to x. By using the chain rule we obtain

$$1 = \frac{d}{dx} \ln y = \frac{1}{y}\frac{dy}{dx},$$

so

$$\frac{dy}{dx} = y.$$

Finally by substituting $y = e^x$, we obtain the differentiation formula

$$\frac{d}{dx} e^x = e^x. \tag{23}$$

In other words, the derivative of the exponential function is the exponential function itself. By repeated differentiation of Eq. 23 we find that, for every positive integer n,

$$\frac{d^n}{dx^n} e^x = e^x. \tag{24}$$

Since $e^x > 0$ it follows that the nth derivative of $y = e^x$ is positive for each value of x. For $n = 1$ and $n = 2$ this provides additional confirmation that the graph of $y = e^x$ is increasing and concave up.

The differentiation formula (23) can be generalized immediately by means of the chain rule. Thus, if u is any differentiable function,

$$\frac{d}{dx} e^{u(x)} = e^{u(x)} u'(x). \tag{25}$$

The following examples illustrate the use of Eq. 25.

EXAMPLE 1

Find the derivative of $f(x) = e^{-3x}$.

In this case $u(x) = -3x$ and $u'(x) = -3$, so

$$\frac{d}{dx} e^{-3x} = -3e^{-3x}. \blacksquare \tag{26}$$

EXAMPLE 2

Find the derivative of $f(x) = e^{x^2}$.

Now $u(x) = x^2$ so by Eq. (25)

$$\frac{d}{dx} e^{x^2} = e^{x^2} \frac{d}{dx} x^2 = 2xe^{x^2}. \blacksquare \tag{27}$$

EXAMPLE 3

Find the derivative of $f(x) = e^{\sin 2x}$.

By Eq. 25 we have

$$\frac{d}{dx} e^{\sin 2x} = e^{\sin 2x} \frac{d}{dx} \sin 2x = 2 (\cos 2x)e^{\sin 2x}. \blacksquare \tag{28}$$

Corresponding to Eq. 23 we have the integration formula

$$\int e^x \, dx = e^x + c, \tag{29}$$

and to Eq. 25 the more general integration formula

$$\int e^{u(x)}u'(x) \, dx = e^{u(x)} + c. \tag{30}$$

EXAMPLE 4

Determine the antiderivative

$$\int \frac{e^{\sqrt{x}}}{\sqrt{x}} \, dx.$$

If we let $u(x) = \sqrt{x}$, then $u'(x) = 1/2\sqrt{x}$. Hence $du = dx/2\sqrt{x}$, or $2du = dx/\sqrt{x}$. Then

$$\int \frac{e^{\sqrt{x}}}{\sqrt{x}} \, dx = 2 \int e^u \, du = 2e^u + c = 2e^{\sqrt{x}} + c. \blacksquare \tag{31}$$

EXAMPLE 5

Determine the antiderivative

$$\int e^{\cos(x/3)} \sin\left(\frac{x}{3}\right) dx.$$

In this case we let $u(x) = \cos(x/3)$; then $u'(x) = -\frac{1}{3}\sin(x/3)$ and $du = -\frac{1}{3}\sin(x/3)\, dx$. Consequently,

$$\int e^{\cos(x/3)} \sin\left(\frac{x}{3}\right) dx = -3 \int e^u\, du = -3e^u + c = -3e^{\cos(x/3)} + c. \quad ∎ \tag{32}$$

EXAMPLE 6

Find the value of

$$I = \int_0^2 x^2 e^{x^3/2}\, dx. \tag{33}$$

In this case it is natural to choose $u(x) = x^3/2$. Then $u'(x) = 3x^2/2$ and $x^2\, dx = 2du/3$. Further, $x = 0$ and $x = 2$ correspond to $u = 0$ and $u = 4$, respectively. Thus the integral I can be written as

$$\begin{aligned} I &= \frac{2}{3}\int_0^4 e^u\, du \\ &= \frac{2}{3} e^u \Big|_0^4 = \frac{2}{3}(e^4 - 1) \cong 35.73. \end{aligned}$$

Note that if the integral in Eq. 33 is changed to

$$J = \int_0^2 x e^{x^3/2}\, dx$$

then the problem becomes much harder. The substitution $u(x) = x^3/2$ no longer reduces the integral to an elementary one. No other choice of $u(x)$ suggests itself; indeed, J cannot be found by elementary (analytical) methods. ∎

Extension of the power rule to irrational exponents

In Section 3.1 we derived the differentiation formula $D_x(x^r) = rx^{r-1}$ when r is rational. We can now show that the same formula also holds when r is irrational. Recall that if r is any real number, then x^r is given by Eq. 22,

$$x^r = e^{r \ln x}.$$

Differentiating Eq. 22 we have

$$\frac{d}{dx} x^r = \frac{d}{dx} e^{r \ln x}$$

$$= e^{r \ln x} \frac{d}{dx} r \ln x$$

$$= x^r \cdot \frac{r}{x}$$

$$= rx^{r-1}. \tag{34}$$

Thus the power rule is also valid for irrational exponents. For example,

$$\frac{d}{dx} x^{\pi} = \pi x^{\pi - 1},$$

$$\frac{d}{dx} x^{\sqrt{2}} = \sqrt{2}\, x^{\sqrt{2} - 1},$$

and so on. Equation 34 also establishes the corresponding integration formula

$$\int x^r \, dx = \frac{x^{r+1}}{r+1} + c, \qquad r \neq -1 \tag{35}$$

for irrational values of the exponent r.

PROBLEMS

In each of Problems 1 through 16, find the derivative of the given function.

1. $f(x) = e^{3x}$

2. $f(x) = x^2 + e^{-x^2}$

3. $f(x) = xe^{-x} + \cos 2x$

4. $f(x) = \frac{e^{2x}}{x}, \quad x \neq 0$

5. $f(x) = \exp \sqrt{1 - x^2}, \quad |x| < 1$

6. $f(x) = \frac{1}{2}(e^x + e^{-x})$

7. $f(x) = \frac{1}{2}(e^x - e^{-x})$

8. $f(x) = \frac{1 - e^{2x}}{1 + e^{2x}}$

9. $f(x) = \ln(1 + e^{-x})$

10. $f(x) = e^{e^x}$

11. $f(x) = e^{1/x}, \quad x \neq 0$

12. $f(x) = \int_0^x e^{-t^2} \, dt,$

13. $f(x) = \frac{x}{2} e^{2/x}, \quad x \neq 0$

14. $f(x) = \cos(e^{x^2}) + 2 \ln(1 + x^2)$

15. $f(x) = e^{-x} \sin 2x$

16. $f(x) = \frac{xe^{2x}}{1 + x^2}$

In each of Problems 17 through 26, find the indicated antiderivative.

17. $\int e^{2x} \, dx$

18. $\int xe^{x^2} \, dx$

19. $\int x^2 e^{-x^3} \, dx$

20. $\int (\cos x) e^{\sin x} \, dx$

21. $\int e^x (1 + e^x)^3 \, dx$

22. $\int \frac{e^{-x}}{1 + e^{-x}} \, dx$

23. $\int e^{-x} \cos(e^{-x}) \, dx$

24. $\int \frac{e^x - e^{-x}}{e^x + e^{-x}} \, dx$

25. $\int \frac{e^{2x} - 1}{e^{2x} + 1} \, dx$

Hint: Convert the integrand to that of Problem 24.

26. $\int \frac{1}{1 + e^{-x}} \, dx$

Hint: Multiply numerator and denominator by e^x.

In each of Problems 27 through 34, evaluate the given integral.

27. $\int_0^3 e^{3x}\,dx$

28. $\int_{\ln 3}^{\ln 8} \frac{e^x}{e^x+1}\,dx$

29. $\int_0^1 xe^{-2x^2}\,dx$

30. $\int_{\ln 2}^{\ln 5} e^{-2x}\,dx$

31. $\int_{\ln 2}^{\ln 4} e^{2x}(e^{2x}+4)^2\,dx$

32. $\int_{\ln 2}^{\ln 3} e^{-x}(1+e^{-x})^2\,dx$

33. $\int_0^{\pi/2} (\sin x)e^{\cos x}\,dx$

34. $\int_1^4 \frac{e^{\sqrt{x}}}{\sqrt{x}}\,dx$

In each of Problems 35 through 42, sketch the graph of the given function.

35. $f(x) = e^{-x^2}$

36. $f(x) = \frac{e^x + e^{-x}}{2}$

37. $f(x) = \frac{e^x - e^{-x}}{2}$

38. $f(x) = xe^{-x}$

39. $f(x) = \frac{1}{1+e^{-x}}$

40. $f(x) = \exp\frac{1}{x}$

41. $f(x) = \exp(-1/|x|)$

42. $f(x) = \exp\frac{1}{x^2}$

43. Find the area of the region bounded by the coordinate axes, the line $x = \ln 10$, and the graph of $y = e^{-x}$.

44. Find the volume of the solid formed by rotating the region in Problem 43 about the x-axis.

45. In Problem 11 of Section 7.5 it was shown that if a given population has a survival function $s(t)$ and a renewal function $r(t)$, then the size $f(t)$ of the population at time t is given by

$$f(t) = f(0)\,s(t) + \int_0^t s(t-\tau)r(\tau)\,d\tau.$$

Determine $f(t)$ if $s(t) = e^{-kt}$ and $r(t) = R$, where k and R are positive constants. Also determine the limiting value L of $f(t)$ as $t \to \infty$.

46. Find $\lim_{x\to 0}(e^x - 1)/x$.
Hint: Identify this quantity as a derivative.

47. Use the fact that $(\ln t)/t \to 0$ as $t \to \infty$ (Problem 51(b) of Section 8.2) to show that

$$\lim_{x\to\infty} \frac{x}{e^{\alpha x}} = 0, \qquad \alpha > 0.$$

Hint: Let $t = e^{\alpha x}$.

48. Show that

$$\lim_{x\to\infty} \frac{x^M}{e^x} = 0,$$

where M is an arbitrary positive number. In other words, e^x grows more rapidly than any positive power of x as $x \to \infty$.
Hint: Write $x^M/e^x = [x/e^{x/M}]^M$ and refer to Problem 47.

* 49. In this problem we show how to establish the formula

$$e = \lim_{k\to\infty}\left(1 + \frac{1}{k}\right)^k. \qquad (i)$$

(a) Let $k = 1/h$ and show that Eq. (i) is equivalent to

$$e = \lim_{h\to 0}(1+h)^{1/h}. \qquad (ii)$$

(b) Show that

$$\lim_{h\to 0}(1+h)^{1/h} = \lim_{h\to 0}\exp\left[\frac{\ln(1+h)}{h}\right].$$

(c) Why is it true that

$$\lim_{h\to 0}\exp\left[\frac{\ln(1+h)}{h}\right] = \exp\left[\lim_{h\to 0}\frac{\ln(1+h)}{h}\right]?$$

In other words, why is it valid to take the limit inside the exponential function?

(d) Show that

$$\lim_{h\to 0}\frac{\ln(1+h)}{h} = 1,$$

thereby proving Eq. (ii).
Hint: See Problem 45 of Section 8.2.

In each of Problems 50 through 55, use Eq. 17 or the method of Problem 49 to evaluate the given limit.

50. $\lim_{k\to\infty}\left(1 - \frac{1}{k}\right)^k$

51. $\lim_{k\to\infty}\left(1 + \frac{r}{k}\right)^k$, $\quad r$ a real number

52. $\lim_{h\to 0}(1 - rh)^{1/h}$, $\quad r$ a real number

53. $\lim_{k\to\infty}\left(1 - \frac{1}{k^2}\right)^k$

54. $\lim_{k\to\infty}\left(1 + \frac{1}{k^2}\right)^k$

55. $\lim_{k\to\infty}\left(1 + \frac{r}{k}\right)^{kt}$, $\quad r$ and t real numbers

© **56. Numerical estimation of *e*.** From Eq. 16 of the text we know that e satisfies the equation

$$\int_1^e \frac{dt}{t} = 1.$$

(a) Show that

$$\int_1^{2.5} \frac{dt}{t} \leq 0.996$$

by finding an upper sum for this integral using a step size of $h = 0.25$. Hence show that $e > 2.5$.

(b) By using $h = 0.25$ and finding a lower sum, show that

$$\int_1^3 \frac{dt}{t} \geq 1.019.$$

Hence show that $e < 3$.

© **57.** Estimate e more closely than in Problem 56 by using Simpson's rule to calculate $\int_1^b (dt/t)$ for several values of b between 2.7 and 2.75.

58. Properties of the exponential function. In this problem we indicate the proofs of Eqs. 6, 7, and 8, using only Eq. 5 and the fact that that $\exp(0) = 1$. Throughout this problem u and v denote arbitrary real numbers.

(a) By setting $v = -u$ in Eq. 5, show that

$$\exp(-u) = \frac{1}{\exp u}.$$

(b) Write $u - v = u + (-v)$, and thereby show that

$$\exp(u - v) = \frac{\exp u}{\exp v}.$$

(c) Show that

$$(\exp u)^p = \exp(pu)$$

for every positive integer p.

(d) Show that

$$(\exp u)^{1/q} = \exp\left(\frac{u}{q}\right)$$

for every positive integer q.
Hint: $u = q(u/q)$.

(e) If $r = p/q$ is any rational number, show that

$$(\exp u)^r = \exp ru.$$

Be sure to consider the case $r < 0$.

* **A functional equation for the exponential function.**
The equation

$$f(x + y) = f(x) f(y) \qquad (i)$$

is an example of what are called functional equations; it relates the values of the function f at different points, in this case the points x, y, and $x + y$. Recall that we encountered a corresponding equation in Section 8.2 in connection with the logarithm function. Since $\exp(x + y) = (\exp x)(\exp y)$, it is plausible to expect that solutions of Eq. (i) are of exponential type. Indeed, considerable information about f can be established on the basis of Eq. (i). Problems 59 through 61 indicate how this may be done.

59. Let f be a function with domain $(-\infty, \infty)$ that satisfies the functional equation

$$f(x + y) = f(x) f(y).$$

(a) By setting $x = y = 0$, show that either $f(0) = 0$ or else $f(0) = 1$.

(b) If $f(0) = 0$, show that $f(x) = 0$ for every x.

(c) If $f(0) = 1$, show that $f(x)$ is never zero.

(d) If $f(a) \neq 0$ for some $a \neq 0$, show that $f(0) = 1$, and hence that $f(x)$ is never zero.

(e) If $f(a) = 0$ for some a, show that $f(x) = 0$ for every x.

(f) If f is not a constant function, show that $f(0) = 1$.

60. Let f be a nonconstant function with domain $(-\infty, \infty)$ satisfying

$$f(x + y) = f(x) f(y).$$

Show that

$$f(r) = [f(1)]^r$$

for any rational number r. If f is continuous and $f(1) = e$, then $f(r) = e^r$, and f is the exponential function exp.
Hint: Use the results of Problem 59 to show that $f(1) \neq 0$. Then use an approach similar to that in Problem 58.

61. Let f be a nonconstant function with domain $(-\infty, \infty)$ satisfying

$$f(x + y) = f(x) f(y).$$

(a) Show that

$$\frac{f(x + h) - f(x)}{h} = f(x) \frac{f(h) - f(0)}{h}.$$

(b) If $f'(0)$ exists, show that $f'(x)$ exists for each x, and that

$$f'(x) = kf(x),$$

where $k = f'(0)$.

(c) If $k = 1$, then $f'(x) = f(x)$, and from Problem 59(f) we have $f(0) = 1$.
Observe that these conditions are satisfied if $f(x) = e^x$. In Section 8.4 we show that no other function satisfies them.

8.4 SOME APPLICATIONS OF THE EXPONENTIAL FUNCTION

In many applications it is important to study a quantity Q that varies at a rate proportional to Q itself. For instance, the mass of a body of radioactive material diminishes at a rate proportional to the current mass of the body. Alternatively, in a study of an isolated population (such as a bacteria culture) it is often assumed that the rate of increase of the population is proportional to the population itself.

In such cases $Q'(t)$ is proportional to $Q(t)$, and therefore Q satisfies the equation

$$Q'(t) = rQ(t), \tag{1}$$

where r is a proportionality constant. An equation such as Eq. 1, in which a derivative of the unknown function appears, is called a **differential equation.** Since only the first derivative of Q appears in Eq. 1, this equation is said to be a first order differential equation. It follows from Eq. 1 that r has the dimension of (1/time). If $r > 0$, then $Q'(t) > 0$, so Q is increasing, as in the population of the bacteria culture; similarly, if $r < 0$, then Q is decreasing, as in radioactive decay.

Equation 1 can be solved in the following way. Assuming temporarily that $Q(t)$ is positive, we divide both sides of Eq. 1 by $Q(t)$, so that

$$\frac{Q'(t)}{Q(t)} = r. \tag{2}$$

Then, recognizing that $Q'(t)/Q(t)$ is the derivative of $\ln Q(t)$, we can write Eq. 2 as

$$\frac{d}{dt} \ln Q(t) = r. \tag{3}$$

Therefore, by integrating both sides of Eq. 3, we obtain

$$\ln Q(t) = rt + C,$$

where C is an arbitrary constant of integration. Consequently,

$$\begin{aligned} Q(t) &= \exp(rt + C) = (\exp C)e^{rt} \\ &= ce^{rt}, \end{aligned} \tag{4}$$

where $c = \exp C$.

The expression $Q(t) = ce^{rt}$ actually contains all solutions of Eq. 1, and is therefore called the **general solution.** This can be shown by a slight extension of the preceding argument (to cover the possibility that $Q(t)$ may not be positive), but later in this section we also give a different derivation of this result.

To determine the integration constant c in Eq. 4 we need to know the value of Q at some specific time; for example, suppose that

$$Q(0) = Q_0, \tag{5}$$

where Q_0 is a known value. Equation 5 is called an **initial condition.** Then, substituting 0 for t and Q_0 for Q in Eq. 4, we find that

$$Q_0 = c. \tag{6}$$

Using this value of c in Eq. 4, we finally obtain

$$Q(t) = Q_0 e^{rt}. \tag{7}$$

The expression (7) is referred to as the **particular solution** of the differential equation (1) that also satisfies the initial condition (5).

The differential equation (1) and the initial condition (5) together form an **initial value problem.** The only solution of this initial value problem is given by Eq. 7. The exponential function is important for many reasons, but for the applied scientist one of the most significant is its appearance in the solution (7) of the initial value problem (1), (5). This problem is basic to the study of such diverse fields as population dynamics, chemical reactions, radioactive decay, and finance.

EXAMPLE 1

The radioactive isotope plutonium-241 decays so as to satisfy the differential equation

$$\frac{dQ}{dt} = -0.0525Q, \tag{8}$$

where Q is measured in milligrams and t in years. If 50 milligrams of plutonium-241 are present today, determine how much will remain in ten years.

According to Eq. 4 the solution of Eq. 8 is

$$Q(t) = ce^{-0.0525t}, \tag{9}$$

where c is a constant to be determined. If $t = 0$ refers to the present time, then $Q(t)$ must also satisfy the initial condition

$$Q(0) = 50; \tag{10}$$

therefore $c = 50$ and

$$Q(t) = 50e^{-0.0525t}. \tag{11}$$

Finally, by substituting $t = 10$ into Eq. 11, we obtain

$$Q(10) = 50e^{-0.525}$$

$$\cong 29.6 \text{ milligrams.} \ \blacksquare \tag{12}$$

From Eq. 7 it is clear that the solution of the initial value problem (1), (5) depends on the two parameters Q_0 and r, which are, respectively, the initial value of $Q(t)$ and the constant determining the rate of growth or decay. The latter quantity may be difficult to measure directly, and in decay problems it is often convenient to introduce a parameter τ, known as the **half-life** of the material, which is much easier to measure. The half-life τ is defined as the time interval during which a given amount of the material is reduced by one-half. It is easy to determine the relation between τ and k from Eq. 7. If Q_0 is the amount of the material present at $t = 0$, and if τ is the half-life, then $Q(t)$ must satisfy

$$Q(\tau) = \frac{Q_0}{2}. \tag{13}$$

Thus from Eq. 7 we obtain

$$\frac{Q_0}{2} = Q_0 e^{r\tau},$$

from which it follows that

$$r\tau = -\ln 2. \tag{14}$$

If either r or τ is known, then the other can be found from Eq. 14. For instance, for plutonium-241 (Example 1) $r = -0.0525$ per year, so from Eq. 12 the half-life is

$$\tau = (\ln 2)/0.0525 \cong 13.2 \text{ years}.$$

If neither τ nor r is given, then in order to determine them the value of $Q(t)$ must be observed at some second time instant in addition to the reading taken at the initial time. The following example illustrates this.

EXAMPLE 2

The radioactive isotope thorium-234 disintegrates at a rate proportional to the amount present. If 100 milligrams of this material is reduced to 82.04 milligrams in one week, find the decay constant r and the half-life τ. Also find an expression giving the amount of thorium-234 present at any time.

Let $Q(t)$ be the amount of thorium-234 present at time t, where Q is measured in milligrams and t in days. Then Q satisfies the differential equation

$$\frac{dQ}{dt} = rQ, \tag{15}$$

where r is a negative constant that must be determined. We seek the solution of Eq. 15 that also satisfies the initial condition

$$Q(0) = 100 \tag{16}$$

as well as the condition

$$Q(7) = 82.04. \tag{17}$$

According to Eq. 7 the solution of Eqs. 15 and 16 is

$$Q(t) = 100e^{rt}. \tag{18}$$

In order to satisfy Eq. 17 we set $t = 7$ and $Q = 82.04$ in Eq. 18; this gives

$$82.04 = 100e^{7r},$$

and hence

$$r = \frac{\ln 0.8204}{7} \cong -0.02828 \text{ (days)}^{-1}. \tag{19}$$

Thus the decay constant r has been determined. The half-life τ is found from Eq. 14, namely

$$\tau = -\frac{\ln 2}{r} \cong 24.5 \text{ days}. \tag{20}$$

Substituting the value of r given by Eq. 19 in Eq. 18, we obtain

$$Q(t) \cong 100e^{-0.02828t}, \tag{21}$$

which gives the value of $Q(t)$ at any time. ∎

We now show in a different way that all solutions of Eq. 1 are included in Eq. 4. Let f be any function that satisfies Eq. 1, that is,

$$f'(t) = rf(t), \tag{22}$$

but we do not suppose that $f(t)$ is given by Eq. 4, nor do we assume anything about the sign of f, as we did earlier. In order to determine f it is convenient to introduce the function g by

$$g(t) = f(t)e^{-rt}. \tag{23}$$

We will determine g by first finding $g'(t)$. Using the product rule for derivatives, we have

$$g'(t) = [f'(t) - rf(t)]e^{-rt} = 0$$

because of Eq. 22. Therefore, by Theorem 4.1.4, g is a constant function. Denoting this constant by c, we have $g(t) = c$, or

$$f(t) = ce^{rt}. \tag{24}$$

Thus, if f satisfies Eq. 22, then it must be given by Eq. 24, as was to be shown.

EXAMPLE 3

The population $Q(t)$ of a certain insect species in a given locality grows at a rate proportional to the current population. The population is also affected by a net outward migration at a rate k. If the population at time $t = 0$ is Q_0, find an expression for the population at any later time t.

The population $Q(t)$ satisfies the differential equation

$$Q' = rQ - k \tag{25}$$

and the initial condition

$$Q(0) = Q_0. \tag{26}$$

The first term on the right side of Eq. 25 expresses the rate of growth of the population and the second term describes the effect of the outward migration. We assume that both r and k are given positive constants, determined by prior observation of the species.

We can obtain the solution of the initial value problem (25), (26) in the same way as we solved the initial value problem (1), (5). However, let us proceed in another way—one suggested by the argument starting from Eq. 22. First, we multiply Eq. 25 by e^{-rt} and then rewrite the result in the form

$$e^{-rt}(Q' - rQ) = -ke^{-rt}. \tag{27}$$

The left side of Eq. 27 is the derivative of Qe^{-rt}, so that

$$\frac{d}{dt}(Qe^{-rt}) = -ke^{-rt}.$$

and therefore

$$Qe^{-rt} = \frac{k}{r}e^{-rt} + c.$$

Hence

$$Q = \frac{k}{r} + ce^{rt} \tag{28}$$

is the general solution of Eq. 25. To determine the integration constant c we use the initial condition (26). Thus

$$Q_0 = \frac{k}{r} + c. \tag{29}$$

so

$$Q = \frac{k}{r} + \left(Q_0 - \frac{k}{r}\right)e^{rt} = \frac{1}{r}[k + (Q_0 r - k)e^{rt}] \tag{30}$$

is the solution of the initial value problem (25), (26).

The behavior of the population Q as t increases depends on the sign of $Q_0 r - k$. If $k < Q_0 r$, then the loss of population through emigration is less than the natural population growth, so $Q(t)$ increases with time. On the other hand, if $k > Q_0 r$, then the emigration rate is dominant, and the population decreases. In fact, according to Eq. 30, the population reaches zero at the time $(1/r)\ln[k/(k - Q_0 r)]$. However, this conclusion should be viewed skeptically since the conditions of the problem probably no longer apply when the population Q is small. ∎

EXAMPLE 4

Assume that the population of the United States grows at a rate proportional to the present population.

(a) According to official U.S. census figures the population was 75.99 million in 1900 and 131.67 million in 1940. Find the annual growth rate during this period.

Let $P(t)$ denote the population at time t, with P in millions, t in years, and with 1900 as $t = 0$. Then P satisfies the initial value problem

$$\frac{dP}{dt} = rP, \qquad P(0) = 75.99, \tag{31}$$

and hence

$$P(t) = 75.99e^{rt}. \tag{32}$$

The growth rate r is determined by using the known population in 1940, that is, when $t = 40$:

$$131.67 = 75.99e^{40r}. \tag{33}$$

This gives

$$r = \frac{1}{40} \ln \frac{131.67}{75.99} \cong 0.01374. \tag{34}$$

Hence the average growth rate was about 1.37 percent per year during the period from 1900 to 1940.

(b) Using the growth rate found in Part (a), determine the predicted population in 1980 and in 2000.

This is done by evaluating $P(80)$ and $P(100)$ from Eq. 32 using r as given by Eq. 34. We have

$$P(80) = 75.99e^{(0.01374)(80)} \cong 228.10 \text{ million} \tag{35}$$

and

$$P(100) = 75.99e^{(0.01374)(100)} \cong 300.25 \text{ million}. \tag{36}$$

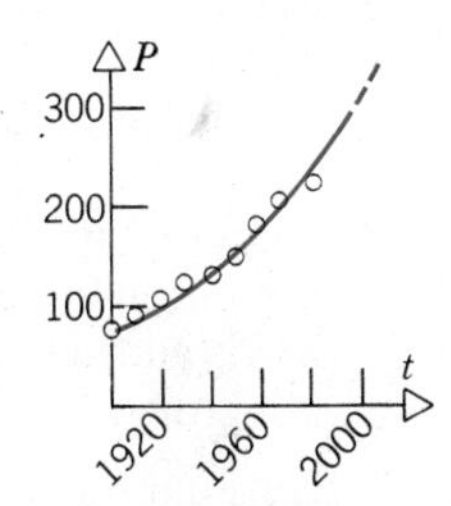

Figure 8.4.1 The dependence of the population of the United States on time. The circles correspond to census data; the curve is an exponential approximation.

According to the official census the actual population in 1980 was 226.50 million, which differs from the result of Eq. 35 by 0.71 percent.

Figure 8.4.1 shows the relation between population and time given by Eq. 32. The small circles show the actual population as given by the U.S. census. Observe that the apparently small annual growth rate of 1.37 percent results in a doubling of the population in about 50 years. If this rate of growth is continued through the twenty-first century, the population of the United States will be about 1.2 billion by the year 2100. Such is the effect of exponential growth. ∎

PROBLEMS

In each of Problems 1 through 10, solve the given initial value problem.

1. $\frac{dQ}{dt} = 2Q, \qquad Q(0) = 4$

2. $\frac{dQ}{dt} = -\frac{Q}{3}, \qquad Q(0) = 5$

3. $\frac{du}{dt} + 3u = 0, \qquad u(2) = -1$

4. $\frac{dx}{dt} - \frac{x}{2} = 0, \qquad x(-1) = 3$

5. $\frac{dQ}{dt} - 2Q = 7, \qquad Q(0) = 3$

6. $\frac{dQ}{dt} + 3Q = 5e^{-t}, \qquad Q(0) = 1$

7. $\frac{du}{dt} - \frac{1}{2}u = 2te^{t/2}, \qquad u(0) = -2$

8. $\frac{du}{dt} + u = e^{-t}\sin 2t, \qquad u(0) = 4$

9. $\frac{dQ}{dt} = rQ, \qquad Q(t_0) = Q_0$

10. $\frac{dQ}{dt} = rQ + k, \qquad Q(t_0) = Q_0$

11. Find the half-life of einsteinium-253 if this material loses one-third of its mass in 11.7 days.

12. Radium-226 has a half-life of 1620 years. Find the time period during which a body of this material is reduced to three quarters of its original size.

13. Suppose that 100 mg of thorium-234 are initially present in a closed container, and that thorium-234 is added to the container at a constant rate of 1 mg/day.
    (a) Find the amount $Q(t)$ of thorium-234 in the container at any time. Recall that the decay rate for thorium-234 was found in Example 2.
    (b) Find the limiting amount Q_l of thorium-234 in the container as $t \to \infty$.
    (c) How long a time period must elapse before the amount of thorium-234 in the container drops to within 0.5 mg of the limiting value Q_l?

14. Suppose that 100 mg of thorium-234 are initially present in a container and that additional thorium-234 is introduced into the container at a rate of k mg/day. Determine k so that a constant level of 100 mg of thorium-234 is maintained in the container. Recall that the decay rate for thorium-234 was found in Example 2.

15. An isolated population of insects is observed to increase at a rate proportional to the current population.
    (a) If there are 10^4 insects initially, and two weeks later there are 2×10^4, find an expression for the number of insects $P(t)$ at any time t.
    (b) How many insects will there be in six months (26 weeks)?
    (c) What is the behavior of $P(t)$ as $t \to \infty$? Does this appear to be realistic?
    (d) Suggest some additional influences that may become important in the study of this insect population for large values of t.

16. In the decade from 1970 to 1980 the population of the United States increased at an average annual rate of 1.081 percent; the official census reported a population of 226.50 million in 1980. Determine the population in the year 2000 if the average growth rate of the 1970s is maintained until that time.

17. According to Newton's law of cooling, the temperature u of a body changes at a rate proportional to the difference between u and the temperature of its surroundings (or ambient temperature) T. Thus

$$\frac{du}{dt} = k(u - T), \qquad (i)$$

where k is the proportionality constant. Note that k is always negative since u tends to decrease if $u > T$.
    (a) Show that the function

$$u(t) = T + ce^{kt}, \qquad (ii)$$

where c is an arbitrary constant, satisfies Eq. (i).
    (b) Determine c if $u(t_0) = u_0$.
    (c) Determine $\lim_{t\to\infty} u(t)$. Does your answer agree with what you would expect intuitively?

18. Assume that a cup of coffee has a temperature of 190°F when freshly poured, and that it cools according to Newton's law of cooling (see Problem 17) in a room whose temperature is 70°F. After 4 minutes the temperature of the coffee is observed to be 160°F. How long will it take for the coffee to reach a temperature of 130°F?

Compound Interest

Suppose that a sum $S(t)$ of money is on deposit in a bank that pays interest at a rate r. Banks normally pay interest on a quarterly, monthly, or perhaps daily, basis. However, the result is nearly the same, over periods up to several years, if we assume that interest is paid and compounded continuously. Then

$$dS/dt = rS.$$

If we also assume that funds are deposited or withdrawn at a constant rate k, then

$$dS/dt = rS + k,$$

where $k > 0$ corresponds to deposits and $k < 0$ to withdrawals.

19. Suppose that an amount S_0 is deposited at an interest rate r compounded continuously.

(a) Find the time T (as a function of r) required for the original sum to double in value.

(b) Find the interest rate that must be obtained if the investment is to double in eight years.

20. A young person with no initial capital invests k dollars per year at an interest rate r. Assume that investments are made and interest is compounded continuously.

(a) Determine the sum $S(t)$ accumulated at time t.

(b) If $r = 7$ percent, determine k so that $100,000 will be available in ten years.

21. (a) At the time of retirement a certain professor has $120,000 in a bank that pays interest at a rate r. The professor wishes to withdraw $1000 per month for living expenses. Assume that the interest payments and withdrawals are made continuously. Determine the time period T during which the professor can withdraw money at this rate before exhausting the bank account.

(b) Under the conditions of Part (a) determine T if $r = 7$ percent; if $r = 9$ percent.

* **22. An alternate approach to the exponential function.** Instead of defining the exponential function as the inverse of the natural logarithm function (the approach used in Section 8.3), it is possible to define it as the solution of a certain initial value problem. In particular, assume that the function E is differentiable and that it satisfies the initial value problem

$$E'(x) = E(x), \qquad E(0) = 1. \qquad (i)$$

It is then possible to show that the function E has the properties previously associated with the exponential function e^x. In this problem we indicate how to begin to do this.

(a) Using the chain rule, show that

$$\frac{d}{dx} E(-x) = -E(-x) \qquad (ii)$$

and that

$$\frac{d}{dx} E(x + c) = E(x + c), \qquad (iii)$$

where c is a constant.

(b) Let

$$u(x) = E(-x)E(x + c). \qquad (iv)$$

Show that, for each x,

$$u'(x) = 0$$

and hence $u(x)$ is a constant.

(c) Show that $u(x) = E(c)$ for each x; then use Eq. (iv) to show that

$$E(c) = E(-x)E(x + c). \qquad (v)$$

(d) Set $c = 0$ in Eq. (v) and thereby show that

$$E(x)E(-x) = 1. \qquad (vi)$$

Thus the function E has one of the important properties of the exponential function, namely, $e^x e^{-x} = 1$.

(e) Using Eqs. (v) and (vi), show that

$$E(x + c) = E(x)E(c). \qquad (vii)$$

Thus E has another fundamental property of the exponential function, that is, $e^{x+c} = e^x e^c$. Once Eq. (vii) has been established, one can proceed to develop other properties of $E(x)$ as in Problems 59 through 61 of Section 8.3.

8.5 LOGARITHMS AND EXPONENTIALS TO OTHER BASES

The discussion in Sections 8.2 and 8.3 concerning logarithms and exponentials to base e can be easily extended to other bases as well. Consider first the expression a^x, where a and x are real numbers and $a > 0$. In Section 8.3 we adopted the definition

$$a^x = e^{x \ln a}, \tag{1}$$

which reduces to the usual algebraic expression involving powers and roots when x is rational. Since the exponential function is always positive, it follows that $a^x > 0$ for all values of x.

From the definition in Eq. 1 and the properties of the exponential function it follows that all of the usual laws of exponents are valid for a^x; in particular

$$a^0 = 1, \tag{2a}$$

$$a^x a^y = a^{x+y}, \tag{2b}$$

$$a^{-x} = \frac{1}{a^x}, \tag{2c}$$

$$(a^x)^r = a^{rx}; \qquad r \text{ any real number.} \tag{2d}$$

The derivative of a^x can be found from Eq. 1:

$$\begin{aligned}\frac{d}{dx} a^x &= \frac{d}{dx} e^{x \ln a} \\ &= e^{x \ln a} \frac{d}{dx}(x \ln a) \\ &= e^{x \ln a} \ln a \\ &= a^x \ln a. \end{aligned} \tag{3}$$

The second derivative is

$$\begin{aligned}\frac{d^2}{dx^2} a^x &= \frac{d}{dx} a^x \ln a \\ &= a^x (\ln a)^2, \end{aligned} \tag{4}$$

and so on for higher derivatives. Equation 3 can be generalized by means of the chain rule; that is, for any differentiable function u,

$$\begin{aligned}\frac{d}{dx} a^{u(x)} &= \frac{d}{dx} e^{u(x) \ln a} \\ &= e^{u(x) \ln a} \frac{d}{dx}[u(x) \ln a] \\ &= a^{u(x)} u'(x) \ln a. \end{aligned} \tag{5}$$

Observe that it is not really necessary to learn the differentiation formulas (3) and (5). It is sufficient to know Eq. 1, provided that one also knows how to differentiate the exponential function.

EXAMPLE 1

Find the derivative of 2^x.

From Eq. 1 we have

$$2^x = e^{x \ln 2}.$$

Consequently

$$\frac{d}{dx} 2^x = e^{x \ln 2} \ln 2 = 2^x \ln 2. \quad \blacksquare \tag{6}$$

EXAMPLE 2

If $y = 5^{x^2/2}$, find dy/dx.

First we write y in the form

$$y = \exp\left[\left(\frac{x^2}{2}\right) \ln 5\right].$$

Then

$$\frac{dy}{dx} = \exp\left[\left(\frac{x^2}{2}\right) \ln 5\right] x \ln 5 = 5^{x^2/2} x \ln 5. \quad \blacksquare \tag{7}$$

EXAMPLE 3

If $f(x) = 10^{\cos x}$, find $f'(x)$.

Expressing $f(x)$ in terms of exp, we have

$$f(x) = 10^{\cos x} = \exp[(\cos x) \ln 10].$$

Hence

$$\begin{aligned} f'(x) &= \exp[(\cos x) \ln 10](-\sin x) \ln 10 \\ &= -\ln 10(\sin x)10^{\cos x}. \quad \blacksquare \end{aligned} \tag{8}$$

A qualitative sketch of the graph of $y = a^x$ can be readily obtained on the basis of Eqs. 1, 3, and 4. It is useful to distinguish the three cases $a = 1$, $a > 1$, and $a < 1$. If $a = 1$, then $a^x = 1$ for all x, and the graph is simply the straight line $y = 1$. If $a > 1$, then $\ln a > 0$ and the graph of $y = a^x$ resembles that of $y = e^x$ in that it is positive, increasing, concave up, and intersects the y-axis at the point

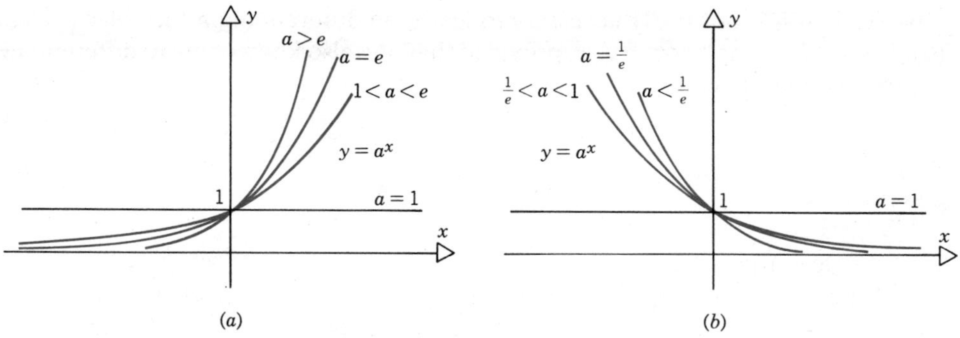

Figure 8.5.1 (*a*) Graphs of $y = a^x$ with $a \geq 1$. (*b*) Graphs of $y = a^x$ with $0 < a \leq 1$.

(0, 1); further $y \to 0$ as $x \to -\infty$ and $y \to \infty$ as $x \to \infty$. Figure 8.5.1*a* shows the graph of $y = a^x$ for several values of a with $a \geq 1$. On the other hand, if $0 < a < 1$, then $\ln a < 0$ and the graph of $y = a^x$ is similar to that of $y = e^{-x}$. In this case the graph is again positive, concave up, and passes through the point (0, 1). However, now $y = a^x$ is a decreasing function; further, $y \to \infty$ as $x \to -\infty$ and $y \to 0$ as $x \to \infty$. The graph of $y = a^x$ for several values of a in $0 < a \leq 1$ is shown in Figure 8.5.1*b*.

Earlier we showed how to differentiate the function a^x by first writing it as an exponential to the base e. We now mention a slightly different procedure, referred to as **logarithmic differentiation.** Starting from the equation

$$y = a^x \tag{9}$$

we take the natural logarithm of each side, and obtain

$$\ln y = \ln a^x = x \ln a. \tag{10}$$

Now let us differentiate Eq. 10 implicitly with respect to x, keeping in mind that y is a function of x and hence the derivative of $\ln y$ must be found by the chain rule. We have

$$\frac{d}{dx}(\ln y) = \frac{d}{dy}(\ln y)\frac{dy}{dx} = \frac{d}{dx}(x \ln a)$$

or

$$\frac{1}{y}\frac{dy}{dx} = \ln a. \tag{11}$$

Hence

$$\frac{dy}{dx} = y \ln a = a^x \ln a. \tag{12}$$

This result, of course, agrees with Eq. 3.

The procedure used for defining a^x and calculating its derivative can be extended to more general functions. For example, consider the function $f(x) = x^x$. We can evaluate this function by expressing it in terms of the natural logarithm and exponential functions, that is,

$$f(x) = x^x = e^{x \ln x}; \tag{13}$$

this latter expression can be evaluated for all $x > 0$, which is the domain of f. To differentiate f we also use Eq. 13:

$$\begin{aligned}\frac{d}{dx}(x^x) &= \frac{d}{dx} e^{x \ln x} \\ &= e^{x \ln x} \frac{d}{dx}(x \ln x) \\ &= x^x (\ln x + 1). \qquad (14)\end{aligned}$$

The same result can also be obtained by first taking the logarithm of each side of $y = x^x$ and then differentiating implicitly.

In a similar way we can define the expression $u(x)^{v(x)}$, provided that $u(x) > 0$. We have

$$u(x)^{v(x)} = e^{v(x) \ln u(x)}. \qquad (15)$$

If u and v are differentiable functions, then the derivative of $u(x)^{v(x)}$ can easily be found from Eq. 15:

$$\begin{aligned}\frac{d}{dx} u(x)^{v(x)} &= \frac{d}{dx} e^{v(x) \ln u(x)} \\ &= e^{v(x) \ln u(x)} \frac{d}{dx}[v(x) \ln u(x)] \\ &= u(x)^{v(x)}\left[v'(x) \ln u(x) + \frac{v(x)u'(x)}{u(x)}\right]. \qquad (16)\end{aligned}$$

Again, one need not learn Eq. 16. The important thing to remember is that the derivative of any exponential-type function can be found by first expressing the function in terms of the exponential function to base e, and then differentiating it. Alternatively, one can use logarithmic differentiation.

EXAMPLE 4

Find the derivative of

$$f(x) = (4 + x^2)^{\cos 2x}.$$

We first rewrite $f(x)$ in the form

$$f(x) = \exp[(\cos 2x)\ln(4 + x^2)].$$

Then

$$\begin{aligned}f'(x) &= \exp[(\cos 2x)\ln(4 + x^2)] \frac{d}{dx}[(\cos 2x)\ln(4 + x^2)] \\ &= \exp[(\cos 2x)\ln(4 + x^2)]\left[-2(\sin 2x)\ln(4 + x^2) + \frac{2x \cos 2x}{4 + x^2}\right] \\ &= (4 + x^2)^{\cos 2x}\left[-2(\sin 2x)\ln(4 + x^2) + \frac{2x \cos 2x}{4 + x^2}\right]. \ \blacksquare\end{aligned}$$

With each differentiation formula there is an associated integration formula. Thus, corresponding to Eqs. 3 and 5, we have

$$\int a^x\,dx = \frac{a^x}{\ln a} + c, \qquad a > 0 \text{ and } a \neq 1 \tag{17}$$

and

$$\int a^{u(x)}u'(x)\,dx = \frac{a^{u(x)}}{\ln a} + c, \qquad a > 0 \text{ and } a \neq 1. \tag{18}$$

However, these results are not particularly useful since they do not enable us to find any antiderivatives that we could not have found before. For instance, Eq. 17 could have been derived by using the integration formula for e^x as follows:

$$\int a^x\,dx = \int e^{x \ln a}\,dx = \frac{e^{x \ln a}}{\ln a} + c = \frac{a^x}{\ln a} + c,$$

and similarly for Eq. 18.

We have seen that for each positive a except $a = 1$ the exponential function to base a is monotone with domain $(-\infty, \infty)$ and range $(0, \infty)$. Consequently, it has an inverse function with domain $(0, \infty)$ and range $(-\infty, \infty)$. It is natural to call this inverse function the logarithm to the base a, and we will denote it by $\log_a x$. Thus, for $x > 0$,

$$y = \log_a x \qquad \text{if and only if} \quad x = a^y. \tag{19}$$

The graph of $y = \log_a x$ can be sketched by reflecting the graph of $y = a^x$ about the line $y = x$; see Figure 8.5.2a and 8.5.2b for the cases $a > 1$ and $0 < a < 1$, respectively. Note that if $a > 1$, then $y \to \infty$ as $x \to \infty$ and $y \to -\infty$ as $x \to 0+$. However, these limits are reversed if $0 < a < 1$.

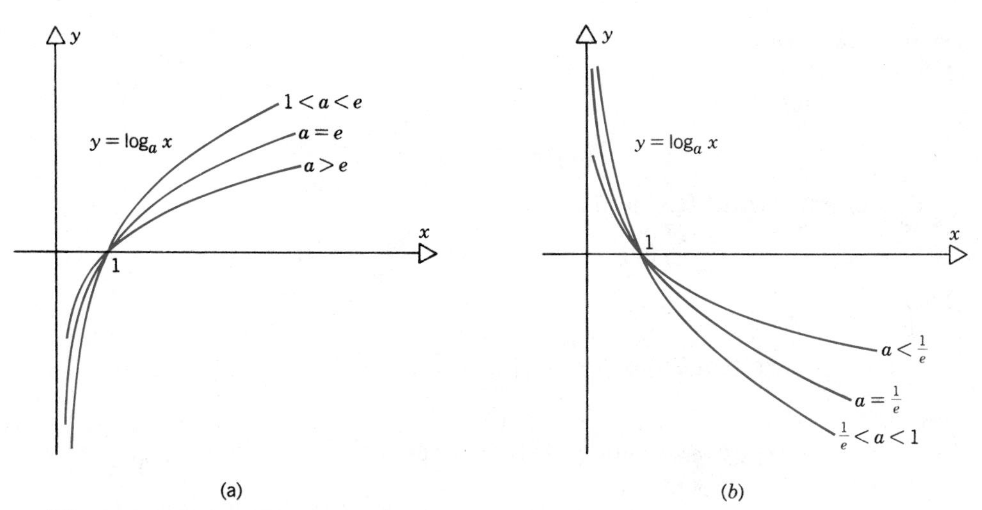

Figure 8.5.2 (a) Graphs of $y = \log_a x$, with $a > 1$. (b) Graphs of $y = \log_a x$, with $0 < a < 1$.

Logarithms to other bases obey the same basic rules as do natural logarithms. For example,

$$\log_a 1 = 0 \tag{20a}$$

$$\log_a(xy) = \log_a x + \log_a y \tag{20b}$$

$$\log_a\left(\frac{1}{x}\right) = -\log_a x \tag{20c}$$

$$\log_a(x^r) = r \log_a x; \qquad r \text{ any real number.} \tag{20d}$$

The proofs of these rules follow readily from the definition of $\log_a x$ and the corresponding rules for the exponential function.

There is also a close relationship between natural logarithms and logarithms to other bases. To derive this relationship let

$$y = \log_a x; \tag{21}$$

then

$$x = a^y. \tag{22}$$

Taking the natural logarithm of each side of Eq. 22, we obtain

$$\ln x = \ln(a^y) = y \ln a.$$

Finally, solving for y and using Eq. 21, we find that

$$\log_a x = \frac{\ln x}{\ln a}. \tag{23}$$

If we set $x = e$ in Eq. 23 we obtain the useful corollary

$$\log_a e = \frac{1}{\ln a}. \tag{24}$$

Now let us consider the calculation of the derivative of the logarithm function to base a. From Eq. 23 it follows at once that

$$\frac{d}{dx} \log_a x = \frac{1}{x \ln a}, \qquad x > 0. \tag{25}$$

As in the case of the natural logarithm, we can also consider the derivative of $\log_a(-x)$ for $x < 0$. Instead of Eq. 23 we then have

$$\log_a(-x) = \frac{\ln(-x)}{\ln a}, \qquad x < 0, \tag{26}$$

so, using the chain rule, we find that

$$\begin{aligned} \frac{d}{dx} \log_a(-x) &= \frac{1}{(-x) \ln a} \frac{d}{dx}(-x) \\ &= \frac{1}{x \ln a}, \qquad x < 0. \end{aligned} \tag{27}$$

If we combine Eqs. 25 and 27 into a single statement, we obtain

$$\frac{d}{dx}\log_a|x| = \frac{1}{x \ln a}, \qquad x \neq 0. \tag{28}$$

As usual, the chain rule can be invoked to generalize Eq. 28:

$$\begin{aligned}\frac{d}{dx}\log_a|u(x)| &= \frac{1}{u(x)\ln a}\frac{d}{dx}u(x)\\ &= \frac{u'(x)}{u(x)\ln a}, \qquad u(x) \neq 0 \end{aligned} \tag{29}$$

for any nonzero differentiable function u.

EXAMPLE 5

If $y = \log_{10}(x^2 + 1)$, find dy/dx.

We make use of Eq. 29 with $u(x) = x^2 + 1$. Then

$$\frac{dy}{dx} = \frac{2x}{(x^2 + 1)\ln 10}. \quad ■ \tag{30}$$

EXAMPLE 6

If $f(x) = \log_{10}|\sin x|$, find $f'(x)$.

In this case $u(x) = \sin x$. Hence, by Eq. 29,

$$f'(x) = \frac{\cos x}{(\sin x)\ln 10}; \qquad x \neq 0, \pm\pi, \pm 2\pi, \ldots . \quad ■ \tag{31}$$

Corresponding to Eqs. 28 and 29 there are the integration formulas

$$\int \frac{dx}{x} = (\log_a|x|)(\ln a) + c \tag{32}$$

and

$$\int \frac{u'(x)}{u(x)}\,dx = (\log_a|u(x)|)(\ln a) + c. \tag{33}$$

However, these results are not useful since they merely give us more complicated antiderivatives than we already had for $1/x$ and $u'(x)/u(x)$, respectively.

PROBLEMS

In each of Problems 1 through 20, find the derivative of the given function.

1. $y = 5^x$

2. $y = 2^{x^2}$

3. $y = 10^{\ln x}, \quad x > 0$

4. $y = x^3 3^x$

5. $y = x^{-2} + 2^{-x}, \quad x \neq 0$

6. $y = \pi^{\sin x}$

7. $y = 4^{\log_2 x}, \quad x > 0$

8. $y = e^{-x} + \log_2|x|, \quad x \neq 0$

9. $y = \log_{10}(x^2 + x + 1)$

10. $y = 2x^2 + x \log_{10} e^x$

11. $y = \log_5(25^{x^2})$

12. $y = \log_2 \left|\frac{1}{x}\right|, \quad x \neq 0$

13. $y = \log_{10}(x^2 + 1)^2$

14. $y = \sin(\log_{10} x^2)$

15. $y = x^{\sqrt{x}}, \quad x > 0$

16. $y = (\sqrt{x})^x, \quad x > 0$

17. $y = x^{\sin x}, \quad x > 0$

18. $y = x^{x^x}, \quad x > 0$

19. $y = x^{1+x^2}, \quad x > 0$

20. $y = \log_x 10, \quad x > 0$

In each of Problems 21 through 26, find the indicated antiderivative.

21. $\int 3^x \, dx$

22. $\int 2^{-x} \, dx$

23. $\int 4^{2x} \, dx$

24. $\int x\pi^{x^2} \, dx$

25. $\int \sin x \cos x \, 10^{\cos 2x} \, dx$

26. $\int 2^x 2^{2^x} \, dx$

In each of Problems 27 through 32, evaluate the given integral.

27. $\int_1^4 2^x \, dx$

28. $\int_0^1 3^{-x} \, dx$

29. $\int_{-1}^1 10^{2x} \, dx$

30. $\int_0^1 x2^{x^2+1} \, dx$

31. $\int_0^5 5^{x/5} \, dx$

32. $\int_1^3 \frac{2^{\ln x}}{x \ln 2} \, dx$

33. By writing $a^x = e^{x \ln a}$ and using properties of the exponential function to base e, show that

(a) $a^x a^y = a^{x+y}$

(b) $a^{-x} = \dfrac{1}{a^x}$

(c) $(a^x)^r = a^{rx}; \quad r$ real

34. If $x > 0$ and $y > 0$, show that the logarithm to base a has the following properties; assume that $a > 0$ and $a \neq 1$.

(a) $\log_a(xy) = \log_a x + \log_a y$

(b) $\log_a \left(\dfrac{1}{x}\right) = -\log_a x$

(c) $\log_a(x^r) = r \log_a x; \quad r$ real

Hint: Make use of the laws of exponents to base a given in Problem 33.

35. Let a and b be positive numbers with $a \neq 1$ and $b \neq 1$; assume also that $x \neq 0$. Show that

$$(\ln a)\log_a|x| = (\ln b)\log_b|x|.$$

36. Find the equation of (a) the tangent line and (b) the normal line to the graph of $y = 2^x$ at the point where $x = 2$.

37. Find the area of the region bounded by the coordinate axes, the line $x = 1$, and the graph of $y = 5^{-x}$.

38. Find the volume of the solid formed by rotating the region in Problem 37 about the x-axis.

In each of Problems 39 through 42, find the inverse of the given function, and determine the domain of the inverse function.

39. $f(y) = 2^{y+3}$

40. $f(y) = \log_2[1 - \sqrt{4 - y^2}]; \quad \sqrt{3} < y \le 2$

41. $f(y) = \log_a\left(\dfrac{y-1}{2}\right); \quad y > 1$

42. $f(y) = \frac{1}{2}(a^y + a^{-y}); \quad a > 1, \quad y \ge 0$

43. Show that for $a > 1$ each of the following equations has a unique solution.

(a) $x + a^x = 0$

(b) $x + \log_a x = 0$

44. Find $\lim_{x \to 0+} x^x$. *Hint:* See Problem 51(c) of Section 8.2.

8.6 THE INVERSE TRIGONOMETRIC FUNCTIONS

In this section we investigate the possibility of defining an inverse for each of the six trigonometric functions. In order to do this some care must be taken and some restrictions must be imposed. Of particular interest is the fact that the derivatives of the inverse trigonometric functions are relatively simple *algebraic* functions. Consequently, integration formulas for these algebraic functions necessarily involve the inverse trigonometric functions; this is one of the principal reasons for introducing these functions in elementary calculus. The inverse sine and inverse tangent are of greatest importance, and will be discussed in some detail. The other four inverse trigonometric functions are of lesser significance, and will be disposed of more quickly. As usual, all angles are measured in radians.

Inverse sine function

On the domain $(-\infty, \infty)$ the sine function is oscillatory, and there are many points in its domain that are associated with each point in its range. In fact, for each y in $[-1, 1]$, there are infinitely many values of x in $(-\infty, \infty)$ for which $y = \sin x$ (see Figure 8.6.1). Thus the sine function on the domain $(-\infty, \infty)$ has no inverse.

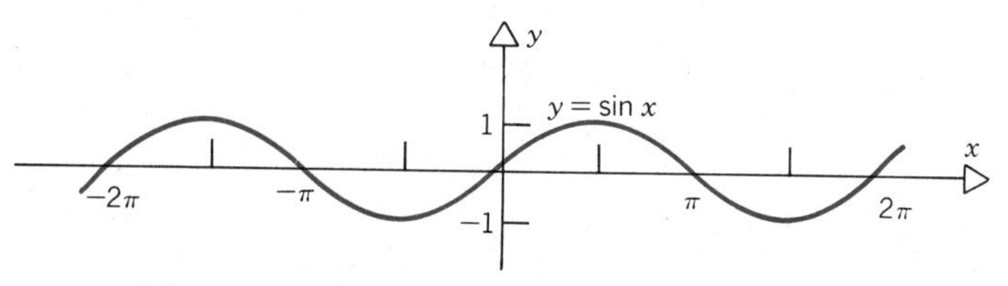

Figure 8.6.1 Graph of $y = \sin x$ on $(-\infty, \infty)$.

In order to obtain a function that does have an inverse, we restrict the domain of the sine function to some interval in which the sine is monotone. At the same time we wish to choose an interval that is associated with the entire range $[-1, 1]$. There are, of course, many possible intervals, such as $[-\pi/2, \pi/2]$, $[\pi/2, 3\pi/2]$, $[-3\pi/2, -\pi/2]$, and so forth. It is customary to choose the interval $[-\pi/2, \pi/2]$; thus we consider the function f given by

$$f(x) = \sin x, \qquad x \text{ in } \left[-\frac{\pi}{2}, \frac{\pi}{2}\right]. \tag{1}$$

The graph of $y = f(x)$ is shown by the solid curve in Figure 8.6.2*a*, while the graph of $y = \sin x$ for x in $(-\infty, \infty)$ is given by the dashed curve. Clearly f is monotone increasing on its domain and thus it has an inverse. The inverse function

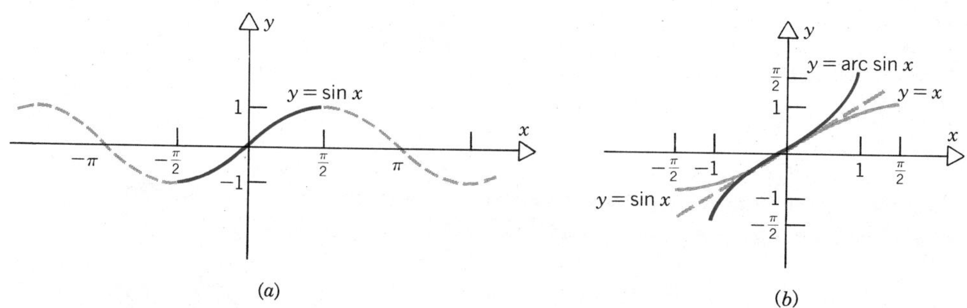

Figure 8.6.2 (*a*) Graph of $y = \sin x$ on $[-\pi/2, \pi/2]$. (*b*) Graph of $y = \arcsin x$.

f^{-1} is known as the **inverse sine function,** and is denoted either by $\sin^{-1}$ or by arcsin. In this book we use arcsin to denote the inverse sine function. The domain of arcsin is $[-1, 1]$ and its range is $[-\pi/2, \pi/2]$. Hence

$$y = \arcsin x, \qquad x \in [-1, 1] \tag{2}$$

means that

$$x = \sin y, \qquad y \in \left[-\frac{\pi}{2}, \frac{\pi}{2}\right]. \tag{3}$$

Often it is helpful to identify $y = \arcsin x$ as the angle whose sine is x. The graph of $y = \arcsin x$ is shown in Figure 8.5.2*b*. It was obtained by reflecting the graph of $y = f(x)$, as given by Eq. 1, about the line $y = x$.

Care must be taken to observe that when a discussion concerns both the sine function and its inverse, then it is understood that the domain of the sine is $[-\pi/2, \pi/2]$. However, when the inverse sine function is not involved, then the domain of the sine is usually $(-\infty, \infty)$.

EXAMPLE 1

Find $\arcsin \frac{1}{2}$ and $\arcsin(-\sqrt{3}/2)$.

If $y = \arcsin \frac{1}{2}$, then $\sin y = \frac{1}{2}$ and $y \in [-\pi/2, \pi/2]$. Hence we have $y = \pi/6$. Similarly, if $y = \arcsin(-\sqrt{3}/2)$, then $\sin y = -\sqrt{3}/2$ and $y \in [-\pi/2, \pi/2]$. Thus we conclude that $y = -\pi/3$. ∎

EXAMPLE 2

Find $\sin(\arcsin x)$.

To determine $\arcsin x$ we must restrict x to $[-1, 1]$. Then, for any x in $[-1, 1]$, we find $\arcsin x = y$. Thus $x = \sin y$ and $y \in [-\pi/2, \pi/2]$. Hence

$$\sin(\arcsin x) = \sin y = x, \qquad x \in [-1, 1]. \tag{4}$$

In other words, Eq. 4 holds for all x in the domain of $\arcsin x$. ∎

EXAMPLE 3

Find $\arcsin(\sin 3\pi/2)$.

Since $\sin 3\pi/2 = -1$, and $\arcsin(-1) = -\pi/2$, we have

$$\arcsin\left(\sin \frac{3\pi}{2}\right) = \arcsin(-1) = -\frac{\pi}{2}.$$

Thus we see that

$$\arcsin(\sin x) \neq x \tag{5}$$

unless x is in the interval $[-\pi/2, \pi/2]$, that is, in the range of arcsin. For values of x not in this interval both sides of Eq. 5 can be calculated, but they are not equal. ∎

We now consider the question of differentiating the inverse sine function. The function $x = \sin y$ is differentiable, so by Theorem 8.1.1, the inverse function $y = \arcsin x$ is also differentiable except at the points where $dx/dy = \cos y$ is zero. To find the derivative of $y = \arcsin x$ we can differentiate $x = \sin y$ implicitly. This gives

$$\frac{d}{dx}(x) = \frac{d}{dx}(\sin y). \tag{6}$$

Making use of the chain rule on the right side, we obtain

$$1 = (\cos y)\frac{dy}{dx},$$

and hence

$$\frac{dy}{dx} = \frac{1}{\cos y}. \tag{7}$$

Note that Eq. 7 can also be obtained by the rule (Eq. 18 of Section 8.1) for differentiating inverse functions:

$$\frac{dy}{dx} = \frac{1}{dx/dy}, \tag{8}$$

where in this case $dx/dy = \cos y$. Equation 7 is valid only for y in $(-\pi/2, \pi/2)$, since the cosine is zero at the endpoints of the interval. To express the right side of Eq. 7 in terms of x, recall that $\sin^2 y + \cos^2 y = 1$, and hence $\cos y = \pm(1 - \sin^2 y)^{1/2}$. The nonnegative square root must be used because $\cos y \geq 0$ for y in $[-\pi/2, \pi/2]$. We then have, from Eq. 7,

$$\frac{dy}{dx} = \frac{1}{\sqrt{1 - \sin^2 y}} = \frac{1}{\sqrt{1 - x^2}}.$$

In other words, we have obtained the differentiation formula

$$\frac{d}{dx}\arcsin x = \frac{1}{\sqrt{1 - x^2}}, \qquad x \in (-1, 1). \tag{9}$$

The derivative does not exist at $x = \pm 1$; this corresponds to the fact that the tangent line to the graph of $y = \arcsin x$ in Figure 8.5.2*b* is parallel to the y-axis at these points.

Equation 9 can be immediately generalized by combining it with the chain rule. Thus, if u is a differentiable function and $|u(x)| < 1$, then

$$\frac{d}{dx}\arcsin u(x) = \frac{1}{\sqrt{1 - u^2(x)}}\frac{d}{dx}u(x) = \frac{u'(x)}{\sqrt{1 - u^2(x)}}, \qquad |u(x)| < 1. \tag{10}$$

EXAMPLE 4

If $y = \arcsin(x^2)$, find dy/dx.

If we let $u(x) = x^2$, it follows from Eq. 10 that

$$\frac{dy}{dx} = \frac{d}{dx}\arcsin(x^2) = \frac{2x}{\sqrt{1 - x^4}}, \qquad -1 < x < 1. \blacksquare \tag{11}$$

EXAMPLE 5

If $y = \arcsin(x/a)$, where $a > 0$, find dy/dx.

In this case $u(x) = x/a$, so by Eq. 10

$$\frac{dy}{dx} = \frac{d}{dx}\arcsin\left(\frac{x}{a}\right) = \frac{1/a}{\sqrt{1 - (x/a)^2}}$$

$$= \frac{1}{\sqrt{a^2 - x^2}}, \qquad -a < x < a. \blacksquare \tag{12}$$

EXAMPLE 6

If $f(x) = \arcsin(1/x)$, find $f'(x)$.

We have $u(x) = 1/x$, so

$$f'(x) = \frac{-1/x^2}{\sqrt{1 - (1/x)^2}} = -\frac{1}{x^2}\frac{1}{\sqrt{x^2 - 1}/\sqrt{x^2}}$$

$$= -\frac{1}{|x|\sqrt{x^2 - 1}}, \qquad |x| > 1. \tag{13}$$

Note that in deriving Eq. 13 we used the fact that $\sqrt{x^2} = |x|$. ∎

Equations 9 and 10 give rise immediately to corresponding integration formulas, namely,

$$\int \frac{dx}{\sqrt{1 - x^2}} = \arcsin x + c, \qquad |x| < 1, \tag{14}$$

$$\int \frac{u'(x)}{\sqrt{1 - u^2(x)}}\,dx = \arcsin u(x) + c, \qquad |u(x)| < 1. \tag{15}$$

In using Eq. 14 to evaluate integrals, we must make sure that the interval of integration is contained in the open interval $(-1, 1)$; similarly, in using Eq. 15 we must make sure that $|u(x)| < 1$ throughout the interval of integration.

The result of Example 5, if read backward, yields an important special case of Eq. 15, namely

$$\int \frac{dx}{\sqrt{a^2 - x^2}} = \arcsin\left(\frac{x}{a}\right) + c; \qquad |x| < a \quad \text{and} \quad a > 0. \tag{16}$$

EXAMPLE 7

Determine the antiderivative

$$\int \frac{dx}{\sqrt{9 - 4x^2}}. \tag{17}$$

We can put this integral in the form (16) by factoring the 4 from the radical, that is,

$$\int \frac{dx}{\sqrt{9 - 4x^2}} = \int \frac{dx}{2\sqrt{(\frac{9}{4}) - x^2}}. \tag{18}$$

Then, choosing $a = \frac{3}{2}$, we can use Eq. 16 to obtain

$$\int \frac{dx}{\sqrt{9 - 4x^2}} = \frac{1}{2} \arcsin\left(\frac{2x}{3}\right) + c. \tag{19}$$

Equation 19 is valid for $2x/3 \in (-1, 1)$ or for $x \in (-\frac{3}{2}, \frac{3}{2})$. ∎

Inverse cosine function

In a very similar way we can define an inverse of the cosine function, at least in a restricted sense. In order for an inverse of the cosine to exist, we restrict the domain to an interval in which the cosine function is monotone and takes on all values in its range $[-1, 1]$. It is usual to take this interval to be $[0, \pi]$. Therefore, we consider the function given by

$$y = \cos x, \qquad x \in [0, \pi] \tag{20}$$

whose range is clearly $[-1, 1]$. The graph of Eq. 20 is shown by the solid curve in Figure 8.6.3*a*, and the graph of $y = \cos x$ for x in $(-\infty, \infty)$ by the dashed curve. The restriction of the cosine function given by Eq. 20 has an inverse that is called the **inverse cosine function** and is denoted by arccos. The domain of arccos is $[-1, 1]$ and its range is $[0, \pi]$. Hence

$$y = \arccos x, \qquad x \in [-1, 1] \tag{21}$$

means that

$$x = \cos y, \qquad y \in [0, \pi]. \tag{22}$$

The graph of $y = \arccos x$ is shown in Figure 8.6.3*b*.

The derivative of the inverse cosine function can be found in much the same way as that of the inverse sine, either by differentiating Eq. 22 implicitly with

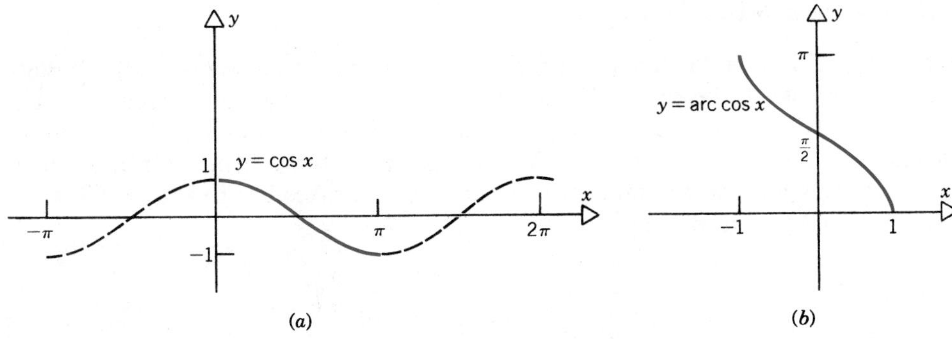

Figure 8.6.3 (*a*) Graph of $y = \cos x$ on $[0, \pi]$. (*b*) Graph of $y = \arccos x$.

respect to x or by using Eq. 8. We find that

$$\frac{d}{dx} \arccos x = -\frac{1}{\sqrt{1 - x^2}}, \qquad x \in (-1, 1). \tag{23}$$

Hence the corresponding integration formula is

$$\int \frac{dx}{\sqrt{1 - x^2}} = -\arccos x + c, \qquad |x| < 1. \tag{24}$$

Equation 24 essentially duplicates the result given by Eq. 14. We will normally use the inverse sine function for purposes of integration; therefore, we will not discuss the inverse cosine function at any length.

We will now derive a simple relation between the arcsin and arccos functions. From Eqs. 9 and 23 we have

$$\frac{d}{dx} \arcsin x + \frac{d}{dx} \arccos x = \frac{d}{dx} (\arcsin x + \arccos x) = 0.$$

Hence (by Theorem 4.1.4)

$$\arcsin x + \arccos x = C, \tag{25}$$

where C is a constant. To evaluate C we can substitute any convenient value of x in $(-1, 1)$ in Eq. 25. Using $x = 0$, we find that

$$\arcsin 0 + \arccos 0 = 0 + \frac{\pi}{2} = C.$$

Hence $C = \pi/2$ and Eq. 25 becomes

$$\arcsin x + \arccos x = \frac{\pi}{2}. \tag{26}$$

The derivation just given establishes Eq. 26 for x in the interval $(-1, 1)$ where arcsin and arccos are differentiable. By setting $x = 1$ and $x = -1$ in turn, however, we see that Eq. 26 is also valid at these points. Hence Eq. 26 holds throughout the interval $[-1, 1]$ in which arcsin and arccos are defined. In different words, Eq. 26 simply states the familiar fact that the angle whose sine is x is the complement of the angle whose cosine is x.

Inverse tangent function

On its entire domain the tangent function takes on each value in $(-\infty, \infty)$ infinitely many times, as can be seen in Figure 8.6.4*a*; consequently, the tangent function does not have an inverse unless its domain is suitably restricted. To obtain an inverse it is convenient to restrict the domain of the tangent function to $(-\pi/2, \pi/2)$. On this interval the tangent is monotone and takes each value in $(-\infty, \infty)$ exactly once. The graph of

$$y = \tan x, \qquad x \in \left(-\frac{\pi}{2}, \frac{\pi}{2}\right) \tag{27}$$

is shown by the solid curve in Figure 8.6.4*a*. The interval is open because $\tan x$ is not defined at $x = \pm\pi/2$. The dashed curve in Figure 8.6.4*b* is the graph of $y = \tan x$ for x in $(-\infty, \infty)$. The function given by Eq. 27 does have an inverse.

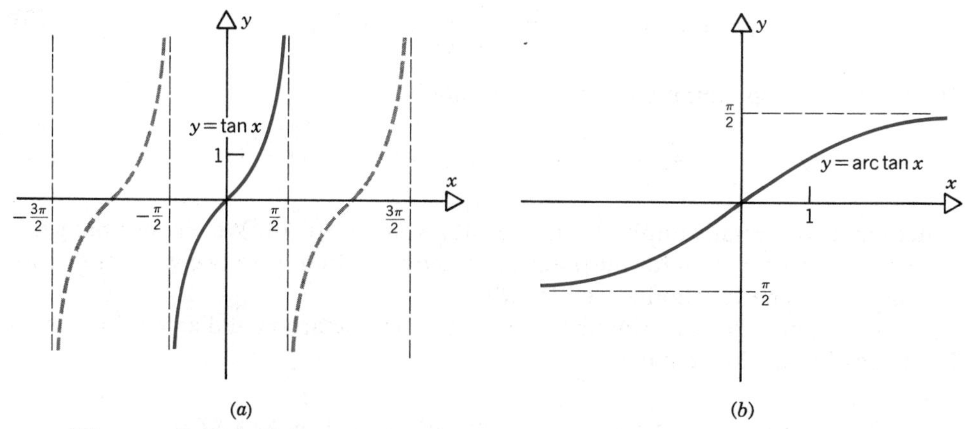

Figure 8.6.4 (*a*) Graph of $y = \tan x$ on $(-\pi/2, \pi/2)$. (*b*) Graph of $y = \arctan x$.

It is called the **inverse tangent function** and is denoted by arctan. The domain of arctan is $(-\infty, \infty)$ and its range is $(-\pi/2, \pi/2)$. Thus

$$y = \arctan x, \qquad x \in (-\infty, \infty) \tag{28}$$

means that

$$x = \tan y, \qquad y \in \left(-\frac{\pi}{2}, \frac{\pi}{2}\right). \tag{29}$$

The graph of $y = \arctan x$ is shown in Figure 8.6.4*b*.

The derivative of $\arctan x$ exists for all x by Theorem 8.1.1. It can be found by differentiating Eq. 29 implicitly with respect to x, or by using Eq. 8. This yields

$$\begin{aligned} 1 &= \frac{d}{dy}(\tan y)\frac{dy}{dx} \\ &= (\sec^2 y)\frac{dy}{dx}. \end{aligned} \tag{30}$$

However,

$$\sec^2 y = 1 + \tan^2 y = 1 + x^2,$$

and therefore from Eq. 30 we have

$$\frac{dy}{dx} = \frac{d}{dx} \arctan x = \frac{1}{1 + x^2}, \qquad x \in (-\infty, \infty). \tag{31}$$

Equation 31 can be generalized by combining it with the chain rule:

$$\frac{d}{dx} \arctan u(x) = \frac{u'(x)}{1 + u^2(x)}, \tag{32}$$

where u is any differentiable function.

EXAMPLE 8

If $y = \arctan(3x^2)$, find dy/dx.

We let $u(x) = 3x^2$; then, by Eq. 32, we obtain

$$\frac{dy}{dx} = \frac{d}{dx} \arctan(3x^2) = \frac{6x}{1 + 9x^4}. \blacksquare \tag{33}$$

EXAMPLE 9

If $y = \arctan(x/a)$, find dy/dx.

In this case $u(x) = x/a$. Thus

$$\frac{dy}{dx} = \frac{d}{dx} \arctan\left(\frac{x}{a}\right) = \frac{1/a}{1 + (x/a)^2} = \frac{a}{a^2 + x^2}, \qquad a \neq 0. \blacksquare \tag{34}$$

Corresponding to Eqs. 31 and 32 we have the integration formulas

$$\int \frac{dx}{1 + x^2} = \arctan x + c, \tag{35}$$

$$\int \frac{u'(x)}{1 + u^2(x)}\, dx = \arctan u(x) + c. \tag{36}$$

These results, together with Eqs. 14 and 15 involving arcsin, are among the more basic integration formulas. From Eq. 34 we obtain the important variation of Eq. 35

$$\int \frac{dx}{a^2 + x^2} = \frac{1}{a} \arctan\left(\frac{x}{a}\right) + c, \qquad a \neq 0. \tag{37}$$

As in the case of the corresponding inverse sine integration formula, some preliminary manipulations are sometimes required before using Eqs. 36 or 37.

Inverse cotangent function

The **inverse cotangent function,** denoted by arccot, is defined so that

$$y = \operatorname{arccot} x, \qquad x \in (-\infty, \infty) \tag{38}$$

means that

$$x = \cot y, \qquad y \in (0, \pi). \tag{39}$$

Thus arccot is the inverse of the cotangent function on the domain $(0, \pi)$. The graphs of these functions are shown in Figure 8.6.5*a* and 8.6.5*b*.

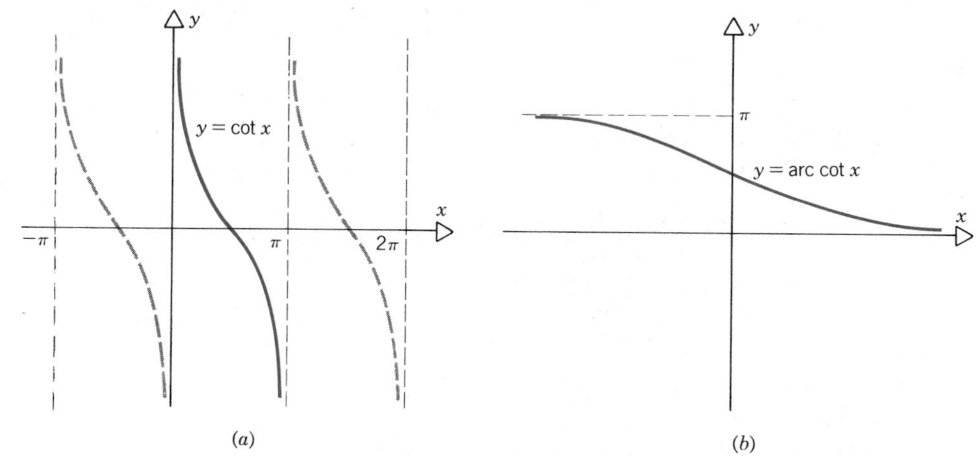

Figure 8.6.5 (*a*) Graph of $y = \cot x$ on $(0, \pi)$. (*b*) Graph of $y = \operatorname{arccot} x$.

The derivative of the inverse cotangent function can be found in essentially the same way as the derivative of arctan. Differentiating Eq. 39 implicitly gives

$$\frac{d}{dx} \operatorname{arccot} x = -\frac{1}{1 + x^2}; \tag{40}$$

the corresponding integration formula duplicates Eq. 35 and will not be used.

Inverse secant function

To define the inverse secant we first consider the function

$$y = \sec x, \qquad x \in \left[0, \frac{\pi}{2}\right) \cup \left(\frac{\pi}{2}, \pi\right] \tag{41}$$

whose graph is shown by the solid portion of the curve in Figure 8.6.6*a*. Note that in restricting the domain of the secant function the point $\pi/2$ is not included. The **inverse secant function,** denoted by arcsec, is the inverse of the function given by

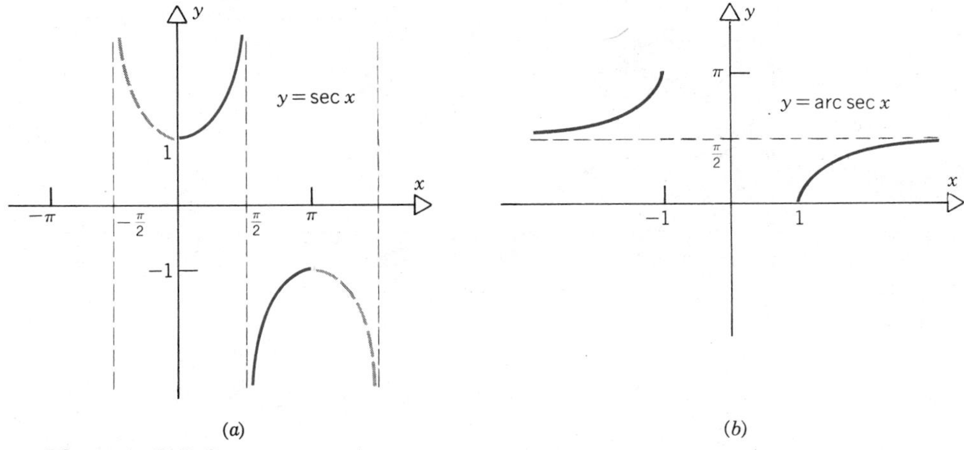

Figure 8.6.6 (*a*) Graph of $y = \sec x$ on $[0, \pi/2) \cup (\pi/2, \pi]$. (*b*) Graph of $y = \operatorname{arcsec} x$.

Eq. 41. Thus

$$y = \operatorname{arcsec} x, \qquad |x| \geq 1 \tag{42}$$

means that

$$x = \sec y, \qquad y \in \left[0, \frac{\pi}{2}\right) \cup \left(\frac{\pi}{2}, \pi\right]. \tag{43}$$

The domain of arcsec consists of the two intervals $(-\infty, -1]$ and $[1, \infty)$ and its range is composed of the two intervals $[0, \pi/2)$ and $(\pi/2, \pi]$. The graph of $y = \operatorname{arcsec} x$ is shown in Figure 8.6.6*b*.

The derivative of arcsec can be found by differentiating Eq. 43 in the now familiar way with the result

$$\frac{d}{dx} \operatorname{arcsec} x = \frac{1}{|x|\sqrt{x^2 - 1}}, \qquad |x| > 1. \tag{44}$$

The corresponding integration formula is

$$\int \frac{dx}{|x|\sqrt{x^2 - 1}} = \operatorname{arcsec} x + c, \qquad |x| > 1. \tag{45}$$

These results are of less importance than those for the arcsin and arctan functions.

Inverse cosecant function

The **inverse cosecant function,** denoted by arccsc, is defined so that

$$y = \operatorname{arccsc} x, \qquad |x| \geq 1 \tag{46}$$

means that

$$x = \csc y, \qquad y \in \left[-\frac{\pi}{2}, 0\right) \cup \left(0, \frac{\pi}{2}\right]. \tag{47}$$

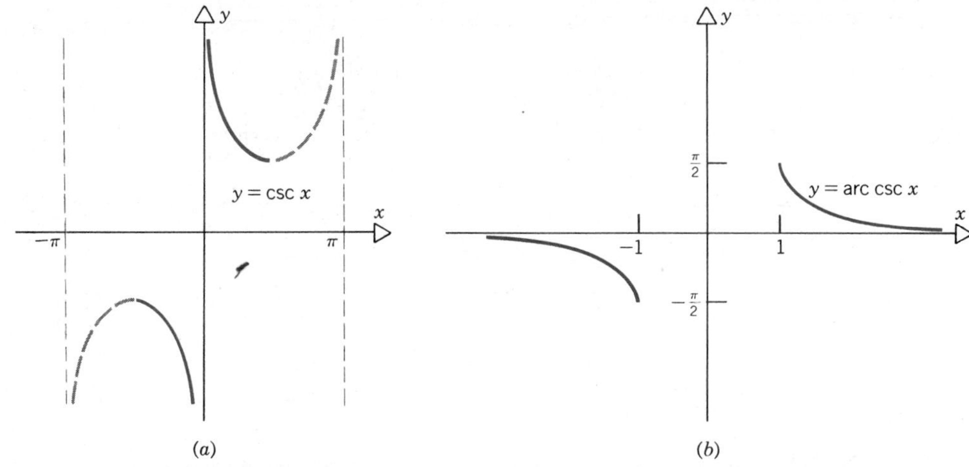

Figure 8.6.7 (*a*) Graph of $y = \csc x$ on $[-\pi/2, 0) \cup (0, \pi/2]$. (*b*) Graph of $y = \operatorname{arccsc} x$.

The arccsc function is thus the inverse of the cosecant function on the domain $[-\pi/2, \pi/2]$ with the origin deleted (see Figure 8.6.7*a* and 8.6.7*b*).

The derivative of arccsc can be found by differentiating Eq. 47; we find that

$$\frac{d}{dx} \operatorname{arccsc} x = -\frac{1}{|x|\sqrt{x^2 - 1}}, \qquad |x| > 1. \tag{48}$$

The corresponding integration formula duplicates that for arcsec and is of little interest to us.

PROBLEMS

In each of Problems 1 through 12, find the value of the given quantity.

1. $\arctan(1)$
2. $\arcsin(-\frac{1}{2})$
3. $\arcsin\left(\frac{1}{\sqrt{2}}\right)$
4. $\arccos\left(-\frac{\sqrt{3}}{2}\right)$
5. $\arcsin(-1)$
6. $\operatorname{arccot}(\sqrt{3})$
7. $\arcsin\left(\cos\frac{\pi}{3}\right)$
8. $\arctan\left(\sin\frac{\pi}{2}\right)$
9. $\cos(\arcsin \frac{1}{2})$
10. $\sin(\arctan \sqrt{3})$
11. $\cos(\arctan x)$
12. $\sin(\arccos x)$

In each of Problems 13 through 18, establish the given identity.

13. $\arcsin x + \arcsin y = \arcsin (x\sqrt{1 - y^2} + y\sqrt{1 - x^2}); |x|, |y| \leq 1$
14. $\arccos x + \arccos y = \arccos(xy - \sqrt{1 - x^2}\sqrt{1 - y^2}); \quad |x|, |y| \leq 1$
15. $\arctan x + \arctan y = \arctan\left(\frac{x + y}{1 - xy}\right)$
16. $\operatorname{arccot} x + \arctan x = \frac{\pi}{2}$
17. $\arcsin \frac{1}{x} = \operatorname{arccsc} x, \quad |x| \geq 1$
18. $\arcsin \frac{x}{\sqrt{1 + x^2}} = \arctan x$

In each of Problems 19 through 34, find the derivative of the given function.

19. $f(x) = \arcsin\left(\frac{x^3}{3}\right)$
20. $f(x) = \arctan(2x)$

21. $f(x) = \left[\arctan\left(\frac{x}{2}\right)\right]^2$

22. $f(x) = e^{x^2} + x \arcsin(x^2)$

23. $f(x) = \arccos(1 - x^2)$

24. $f(x) = \arctan\left(\frac{1 - x}{1 + x}\right)$

25. $f(x) = \operatorname{arcsec}(x^2)$

26. $f(x) = e^x \arctan\left(\frac{1}{x}\right)$

27. $f(x) = \sqrt{1 - x^2} \arcsin x$

28. $f(x) = \sin(\arctan x)$

29. $f(x) = \arctan(2 \tan x)$

30. $f(x) = \arcsin\left(\frac{x}{\sqrt{1 + x^2}}\right)$

31. $f(x) = \arctan\left(\frac{2x}{\sqrt{1 - 4x^2}}\right)$

32. $f(x) = \arcsin \frac{x}{2} + \sqrt{4 - x^2}$

33. $f(x) = \ln[\arctan(2x)]$

34. $f(x) = \sqrt{x} \arctan \sqrt{x}$

In each of Problems 35 through 46, find the indicated antiderivative.

35. $\int \frac{dx}{\sqrt{9 - x^2}}$

36. $\int \frac{dx}{16 + x^2}$

37. $\int \frac{dx}{\sqrt{1 - 4x^2}}$

38. $\int \frac{dx}{1 + 4x^2}$

39. $\int \frac{dx}{4 + 9x^2}$

40. $\int \frac{dx}{\sqrt{16 - 9x^2}}$

41. $\int \frac{dx}{|x| \sqrt{x^2 - 4}}$

42. $\int \frac{dx}{1 + (x - 2)^2}$

43. $\int \frac{dx}{\sqrt{1 - (x + 1)^2}}$

44. $\int \frac{\cos x}{4 + \sin^2 x}\, dx$

45. $\int \frac{[\arcsin(x)]^2}{\sqrt{1 - x^2}}\, dx, \qquad |x| < 1$

46. $\int \frac{\arctan(2x)}{1 + 4x^2}\, dx$

In each of Problems 47 through 50, derive the given differentiation formula.

47. $\frac{d}{dx} \arccos x = -\frac{1}{\sqrt{1 - x^2}}, \qquad |x| < 1$

48. $\frac{d}{dx} \operatorname{arccot} x = -\frac{1}{1 + x^2}$

49. $\frac{d}{dx} \operatorname{arcsec} x = \frac{1}{|x| \sqrt{x^2 - 1}}, \qquad |x| > 1$

50. $\frac{d}{dx} \operatorname{arccsc} x = -\frac{1}{|x| \sqrt{x^2 - 1}}, \qquad |x| > 1$

51. Let $f(x) = \arcsin(\cos x)$ for $-\infty < x < \infty$.

(a) Show that

$$f'(x) = -\frac{\sin x}{|\sin x|}; \qquad x \neq 0, \pm\pi, \ldots .$$

(b) Sketch the graph of $y = f'(x)$.

(c) Sketch the graph of $y = f(x)$. Is f a continuous function?

52. (a) Find the area between the x-axis and the graph of $y = (1 + x^2)^{-1}$ from $x = 0$ to $x = R$ (see Figure 8.6.8).

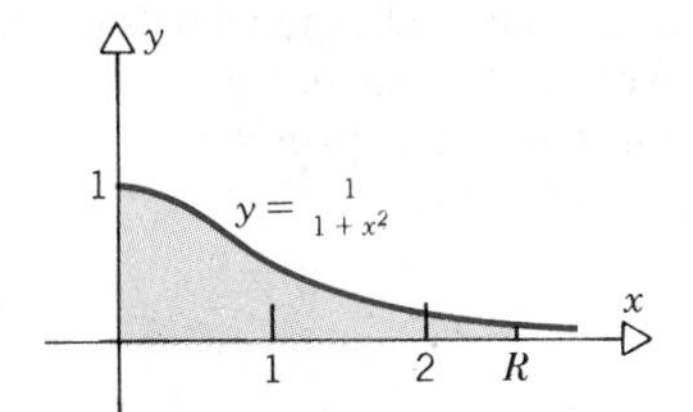

Figure 8.6.8

(b) Use the result of Part (a) and find the limit as $R \to \infty$. This is an example of a region extending to infinity whose area is finite.

53. (a) Find the area between the x-axis and the graph of $y = (1 - x^2)^{-1/2}$ from $x = 0$ to $x = \alpha$, where $0 < \alpha < 1$ (see Figure 8.6.9).

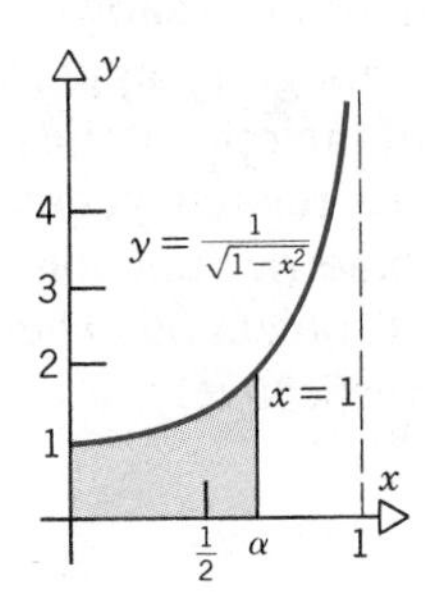

Figure 8.6.9

(b) Use the result of Part (a) and find the limit as $\alpha \to 1$ from the left. Note that even though the integrand becomes unbounded as $x \to 1$, the area under the graph remains finite as $\alpha \to 1$.

54. Find the volume of the solid of revolution formed by rotating the portion of the graph of $y = 1/\sqrt{4 + x^2}$ between $x = 2$ and $x = 6$ about the x-axis.

55. A painting is hung so that the top is b feet above eye level while the bottom is a feet above eye level. At what distance from the painting should a viewer stand so as to maximize the angle subtended by the painting at the viewer's eye?

56. Consider the function f given by

$$f(x) = \int_0^x \frac{dt}{\sqrt{1 - t^2}}, \qquad -1 < x < 1. \qquad (i)$$

We recognize this function as the inverse sine function, of course. Ignoring this fact, however, we can simply start with the function given by Eq. (i) and eventually define the trigonometric functions and derive all of their properties. Such a development avoids all use of geometrical definitions and arguments. This treatment parallels the discussion of logarithms and exponentials given earlier in this chapter. We will not go into this derivation in detail, but in this problem we indicate a few of the initial steps.

(a) Show that

$$f(0) = 0,$$
$$f(-x) = -f(x),$$
$$f'(x) = \frac{1}{\sqrt{1 - x^2}},$$

f is an increasing function on $(-1, 1)$.

(b) Letting $y = f(x)$, show that f has an inverse. We will denote this inverse function by $x = s(y)$ and call it the sine function. Show that s is differentiable and that

$$\frac{dx}{dy} = \frac{1}{dy/dx} = \sqrt{1 - x^2},$$

or that

$$\frac{ds}{dy} = \sqrt{1 - s^2(y)}.$$

(c) Define the function $c(y)$ by the equation

$$c(y) = s'(y)$$

and call this function the cosine function. Show that the function c is differentiable and that

$$c'(y) = -s(y).$$

Also show that

$$c^2(y) + s^2(y) = 1.$$

Further properties of the s and c functions can also be established. We emphasize that these functions are the same as the sine and cosine functions defined geometrically.

8.7 THE HYPERBOLIC AND INVERSE HYPERBOLIC FUNCTIONS

Experience has shown that certain combinations of the exponential function occur often enough in applications to warrant a separate identification and a study of some of their properties. These functions are known as **hyperbolic functions** because they are related geometrically to a hyperbola in much the same way that the trigonometric functions are associated with a circle. This relationship is pointed out in Problem 39. The hyperbolic sine and cosine functions, denoted by sinh and cosh, respectively, are defined for all values of x by the formulas

$$\sinh x = \tfrac{1}{2}(e^x - e^{-x}), \qquad -\infty < x < \infty \qquad (1)$$

$$\cosh x = \tfrac{1}{2}(e^x + e^{-x}), \qquad -\infty < x < \infty. \qquad (2)$$

By recalling properties of the exponential function developed in Section 8.3 and by using Eqs. 1 and 2, we can immediately derive the following properties of sinh and cosh

$$\sinh 0 = 0, \qquad (3)$$

$$\cosh 0 = 1. \tag{4}$$

$$\sinh(-x) = -\sinh x, \qquad \text{(sinh is odd)} \tag{5}$$

$$\cosh(-x) = \cosh x, \qquad \text{(cosh is even)} \tag{6}$$

$$\sinh x \cong \tfrac{1}{2}e^{x} \quad \text{for large positive } x \tag{7}$$

$$\cosh x \cong \tfrac{1}{2}e^{x} \quad \text{for large positive } x \tag{8}$$

$$\sinh x \cong -\tfrac{1}{2}e^{-x} \quad \text{for large negative } x \tag{9}$$

$$\cosh x \cong \tfrac{1}{2}e^{-x} \quad \text{for large negative } x. \tag{10}$$

Further, the range of sinh is $(-\infty, \infty)$ and the range of cosh is $[1, \infty)$. The graphs of $y = \sinh x$ and $y = \cosh x$ are shown in Figures 8.7.1 and 8.7.2, respectively.

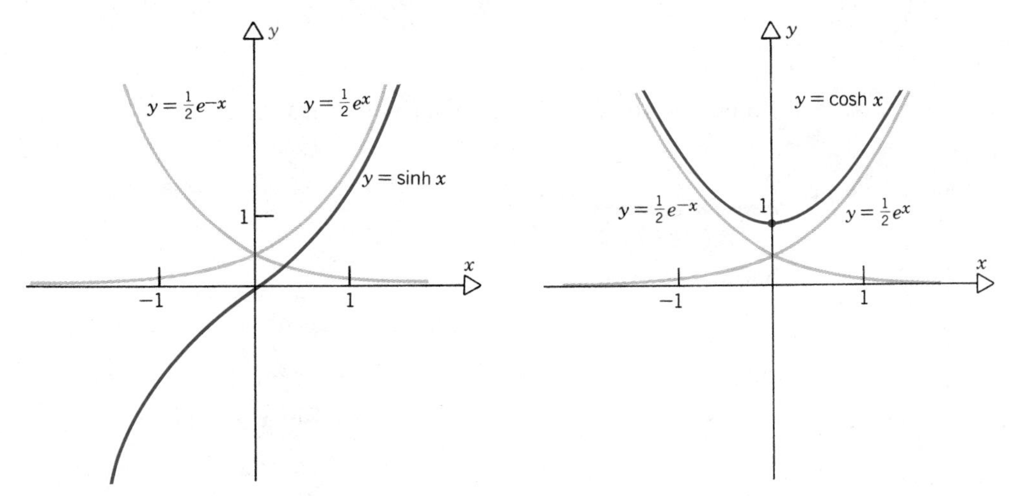

Figure 8.7.1 Graph of $y = \sinh x$.

Figure 8.7.2 Graph of $y = \cosh x$.

The other four hyperbolic functions are defined in terms of sinh and cosh by analogy with the trigonometric functions:

$$\tanh x = \frac{\sinh x}{\cosh x}, \qquad -\infty < x < \infty, \tag{11}$$

$$\operatorname{sech} x = \frac{1}{\cosh x}, \qquad -\infty < x < \infty, \tag{12}$$

$$\coth x = \frac{\cosh x}{\sinh x}, \qquad x \neq 0, \tag{13}$$

$$\operatorname{csch} x = \frac{1}{\sinh x}, \qquad x \neq 0. \tag{14}$$

Figure 8.7.3 Graph of $y = \tanh x$.

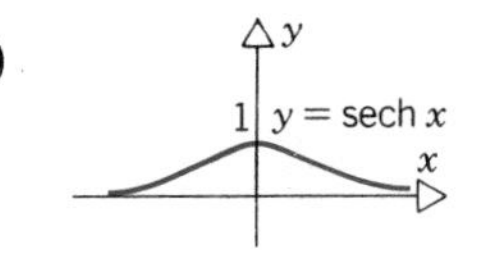

Figure 8.7.4 Graph of $y = \operatorname{sech} x$.

The graphs of these functions are shown in Figures 8.7.3 through 8.7.6.

The values of the hyperbolic functions can be calculated from the exponential function by using Eqs. 1, 2, 11, 12, 13, and 14. Thus the hyperbolic functions are

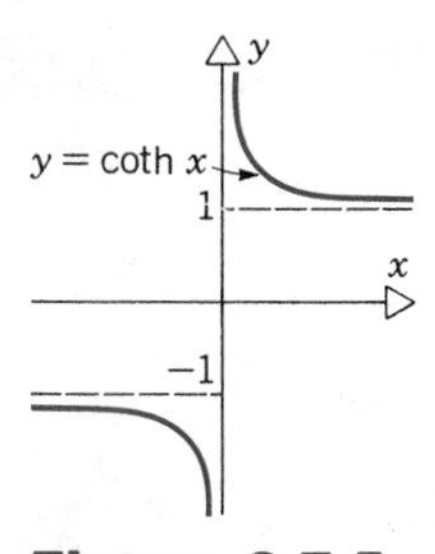

Figure 8.7.5
Graph of $y = \coth x$.

regarded as known in the same sense that the exponential and trigonometric functions are considered to be known functions.

The hyperbolic functions are related by numerous identities that resemble very closely those involving the trigonometric functions. For example, using the definitions of sinh and cosh it is not difficult to show that

$$\cosh^2 x - \sinh^2 x = 1, \tag{15}$$

$$\sinh(x \pm y) = \sinh x \cosh y \pm \cosh x \sinh y, \tag{16}$$

$$\cosh(x \pm y) = \cosh x \cosh y \pm \sinh x \sinh y. \tag{17}$$

For example, to verify Eq. 15 we have

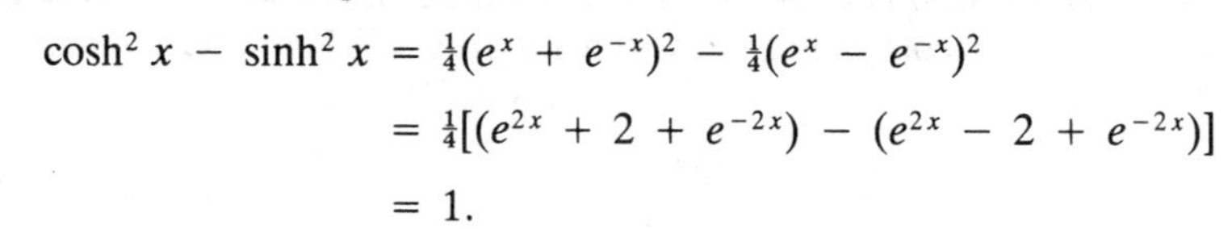

$$\begin{aligned}\cosh^2 x - \sinh^2 x &= \tfrac{1}{4}(e^x + e^{-x})^2 - \tfrac{1}{4}(e^x - e^{-x})^2 \\ &= \tfrac{1}{4}[(e^{2x} + 2 + e^{-2x}) - (e^{2x} - 2 + e^{-2x})] \\ &= 1.\end{aligned}$$

Equations 16 and 17 can be verified in a similar way.

Figure 8.7.6
Graph of $y = \operatorname{csch} x$.

EXAMPLE 1

Show that

$$\sinh 2x = 2 \sinh x \cosh x, \tag{18}$$

and

$$\cosh 2x = \cosh^2 x + \sinh^2 x \tag{19a}$$

$$= 2\cosh^2 x - 1 \tag{19b}$$

$$= 2\sinh^2 x + 1. \tag{19c}$$

Equation 18 follows from Eq. 16 by setting $y = x$ and using the plus sign on both sides of the equation. Equation 19(a) follows from Eq. 17 in the same way. Equations 19(b) and 19(c) are obtained from Eq. 19(a) by using Eq. 15. ∎

A good rule of thumb is that each trigonometric identity has its counterpart for hyperbolic functions, possibly with a different combination of algebraic signs. Lists of such identities are found in standard books of mathematical tables. However, the most useful identities are those given in Eq. 15 through Eq. 19(c).

Since the hyperbolic functions consist of sums, differences, or quotients of the exponential functions e^x and e^{-x}, it follows that each of the hyperbolic functions is differentiable, except that coth x and csch x are not differentiable at $x = 0$. From Eqs. 1 and 2 we have

$$\begin{aligned}\frac{d}{dx}\sinh x &= \frac{1}{2}\frac{d}{dx}(e^x - e^{-x}) \\ &= \tfrac{1}{2}(e^x + e^{-x}) \\ &= \cosh x.\end{aligned} \tag{20}$$

Similarly,

$$\begin{aligned}\frac{d}{dx}\cosh x &= \frac{1}{2}\frac{d}{dx}(e^x + e^{-x})\\ &= \tfrac{1}{2}(e^x - e^{-x})\\ &= \sinh x. \qquad (21)\end{aligned}$$

Thus the derivatives of sinh and cosh follow the same pattern as those of sin and cos, but without the appearance of a minus sign. Further, if u is any differentiable function, then we obtain from the chain rule the following generalizations of Eqs. 20 and 21:

$$\frac{d}{dx}\sinh u(x) = u'(x)\cosh u(x), \qquad (22)$$

$$\frac{d}{dx}\cosh u(x) = u'(x)\sinh u(x). \qquad (23)$$

EXAMPLE 2

Find

$$\frac{d}{dx}\sinh\sqrt{x^2 + 1}.$$

Using the chain rule, we have

$$\begin{aligned}\frac{d}{dx}\sinh\sqrt{x^2+1} &= (\cosh\sqrt{x^2+1})\frac{d}{dx}\sqrt{x^2+1}\\ &= (\cosh\sqrt{x^2+1})\frac{x}{\sqrt{x^2+1}}. \quad \blacksquare\end{aligned}$$

The integration formulas corresponding to Eq. 20 through Eq. 23 are

$$\int \sinh x\, dx = \cosh x + c, \qquad \int \cosh x\, dx = \sinh x + c \qquad (24)$$

and

$$\begin{aligned}\int u'(x)\sinh u(x)\, dx &= \cosh u(x) + c,\\ \int u'(x)\cosh u(x)\, dx &= \sinh u(x) + c. \qquad (25)\end{aligned}$$

The other hyperbolic functions can be differentiated by applying the quotient rule to Eq. 11 through Eq. 14, and making use of Eqs. 20 and 21. The results are

$$\frac{d}{dx}\tanh x = \operatorname{sech}^2 x, \qquad \int \operatorname{sech}^2 x\, dx = \tanh x + c, \qquad (26)$$

$$\frac{d}{dx}\coth x = -\operatorname{csch}^2 x,$$

$$\int \operatorname{csch}^2 x\, dx = -\coth x + c, \qquad x \neq 0, \tag{27}$$

$$\frac{d}{dx}\operatorname{sech} x = -\operatorname{sech} x \tanh x,$$

$$\int \operatorname{sech} x \tanh x\, dx = -\operatorname{sech} x + c, \tag{28}$$

$$\frac{d}{dx}\operatorname{csch} x = -\operatorname{csch} x \coth x,$$

$$\int \operatorname{csch} x \coth x\, dx = -\operatorname{csch} x + c, \qquad x \neq 0. \tag{29}$$

Of course, Eq. 26 through Eq. 29 can also be generalized in a now familiar way by use of the chain rule. We will not record these more general differentiation and integration formulas.

Inverse hyperbolic functions

In much the same way as for the trigonometric functions we can define an inverse of each of the six hyperbolic functions discussed above. As in the case of the trigonometric functions, it is worthwhile to consider the inverse hyperbolic functions mainly because their derivatives are relatively simple algebraic functions. Hence antiderivatives of these algebraic functions can be expressed in terms of the inverse hyperbolic functions.

In some cases it will be necessary to restrict the domain of a hyperbolic function before defining its inverse. However, as shown in Figure 8.7.1 the hyperbolic sine function is monotone on $(-\infty, \infty)$, so no restriction of its domain is required in order to construct its inverse. The inverse hyperbolic sine function is denoted by arcsinh and

$$y = \operatorname{arcsinh} x, \qquad x \in (-\infty, \infty) \tag{30}$$

means that

$$x = \sinh y, \qquad y \in (-\infty, \infty). \tag{31}$$

The graph of $y = \operatorname{arcsinh} x$ is shown in Figure 8.7.7; it is obtained by reflecting the graph of $y = \sinh x$ about the line $y = x$. Both the domain and range of arcsinh are $(-\infty, \infty)$.

The arcsinh function is differentiable everywhere (by Theorem 8.1.1) because sinh is differentiable everywhere and its derivative is never zero. To find the derivative of arcsinh x we proceed as follows. From Eqs. 30 and 31 and the rule for differentiating inverse functions we have

$$\frac{dy}{dx} = \frac{1}{dx/dy} = \frac{1}{(d/dy)(\sinh y)} = \frac{1}{\cosh y}. \tag{32}$$

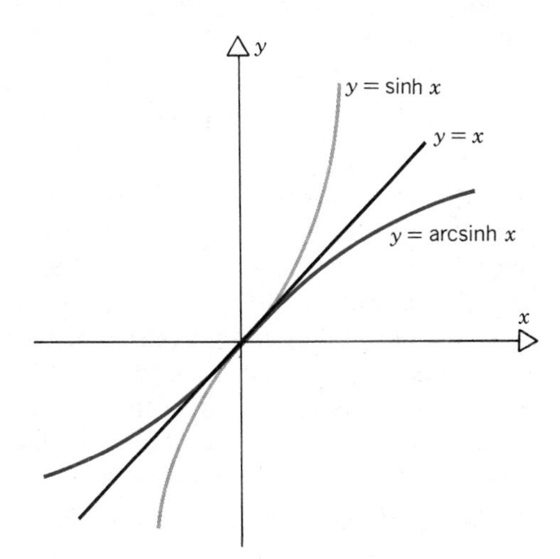

Figure 8.7.7 Graph of $y = \text{arcsinh } x$.

From Eq. 15 it follows that

$$\begin{aligned} \cosh y &= \sqrt{1 + \sinh^2 y} \\ &= \sqrt{1 + x^2}. \end{aligned} \tag{33}$$

Note that we have used the positive square root because $\cosh y > 0$ for all y. Upon combining Eqs. 32 and 33, we obtain the differentiation formula

$$\frac{d}{dx} \text{arcsinh } x = \frac{1}{\sqrt{1 + x^2}}, \qquad x \in (-\infty, \infty). \tag{34}$$

More generally, if u is any differentiable function, then

$$\frac{d}{dx} \text{arcsinh } u(x) = \frac{u'(x)}{\sqrt{1 + u^2(x)}}. \tag{35}$$

The corresponding integration formulas are

$$\int \frac{dx}{\sqrt{1 + x^2}} = \text{arcsinh } x + c, \tag{36}$$

and

$$\int \frac{u'(x)\,dx}{\sqrt{1 + u^2(x)}} = \text{arcsinh } u(x) + c. \tag{37}$$

In particular, if $u(x) = x/a$, where a is a nonzero constant, then Eqs. 35 and 37 yield the useful formulas

$$\begin{aligned} \frac{d}{dx} \text{arcsinh}\left(\frac{x}{a}\right) &= \frac{1/a}{\sqrt{1 + (x/a)^2}} \\ &= \frac{1}{\sqrt{a^2 + x^2}} \end{aligned} \tag{38}$$

and

$$\int \frac{dx}{\sqrt{a^2 + x^2}} = \operatorname{arcsinh}\left(\frac{x}{a}\right) + c. \tag{39}$$

We can proceed in much the same way to discuss inverses of the other hyperbolic functions, but we will restrict ourselves to a brief mention of the inverse hyperbolic cosine only. See also Problem 37 for some information about the inverse hyperbolic tangent.

If the domain is not restricted, then $y = \cosh x$ has no inverse, because for each value of $y > 1$ there are two possible values of x (see Figure 8.7.2). However, if the domain is restricted to the interval $[0, \infty)$, then we obtain the graph shown in Figure 8.7.8*a*. This restriction of the hyperbolic cosine function does have an inverse. The inverse function is denoted by arccosh and has domain $[1, \infty)$ and range $[0, \infty)$. Thus

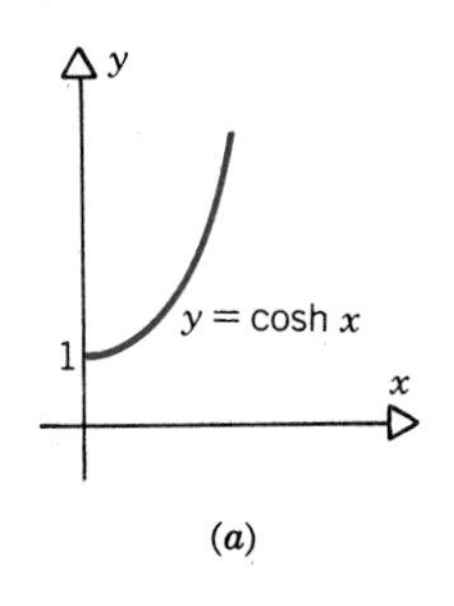

(*a*)

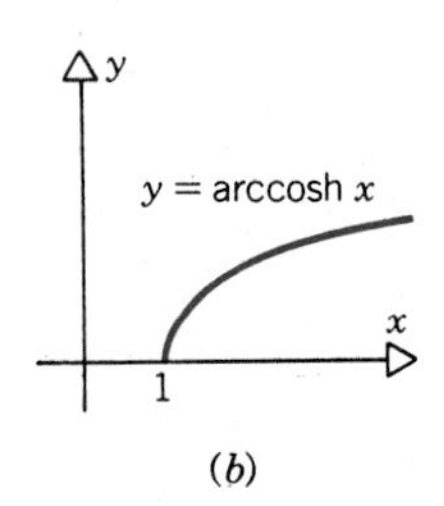

(*b*)

Figure 8.7.8
(*a*) Graph of $y = \cosh x$ on $[0, \infty)$. (*b*) Graph of $y = \operatorname{arccosh} x$.

$$y = \operatorname{arccosh} x, \qquad x \in [1, \infty) \tag{40}$$

means that

$$x = \cosh y, \qquad y \in [0, \infty). \tag{41}$$

The graph of Eq. 40 is shown in Figure 8.7.8*b*. Differentiating Eq. 40 by the inverse function rule, we find that

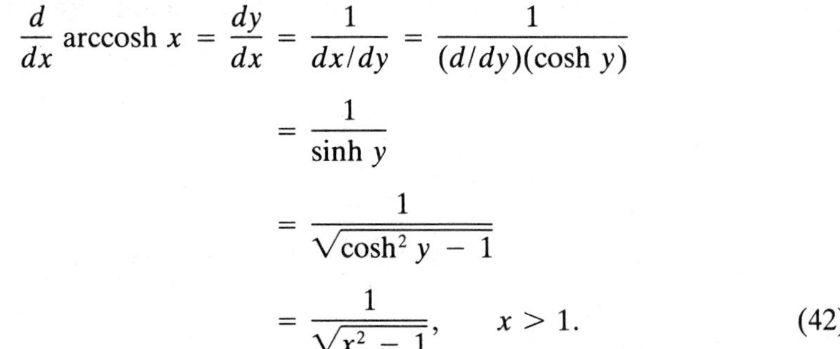
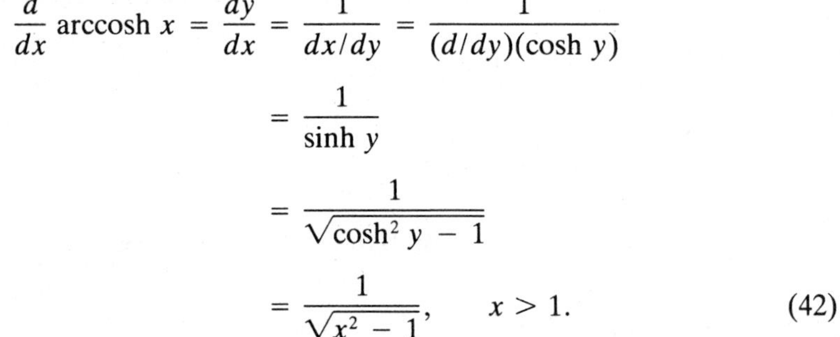

$$\begin{aligned}\frac{d}{dx}\operatorname{arccosh} x = \frac{dy}{dx} = \frac{1}{dx/dy} &= \frac{1}{(d/dy)(\cosh y)}\\ &= \frac{1}{\sinh y}\\ &= \frac{1}{\sqrt{\cosh^2 y - 1}}\\ &= \frac{1}{\sqrt{x^2 - 1}}, \qquad x > 1. \end{aligned}\tag{42}$$

Note that the positive square root is used because $\sinh y > 0$ for $y > 0$. More generally, if u is any differentiable function with $u(x) > 1$, then we have

$$\frac{d}{dx}\operatorname{arccosh} u(x) = \frac{u'(x)}{\sqrt{u^2(x) - 1}}, \qquad u(x) > 1. \tag{43}$$

The corresponding integration formulas are

$$\int \frac{dx}{\sqrt{x^2 - 1}} = \operatorname{arccosh} x + c, \qquad x > 1 \tag{44}$$

and

$$\int \frac{u'(x)}{\sqrt{u^2(x) - 1}}\, dx = \operatorname{arccosh} u(x) + c, \qquad u(x) > 1. \tag{45}$$

An important special case of Eq. 45, similar to Eq. 39, is

$$\int \frac{dx}{\sqrt{x^2 - a^2}} = \operatorname{arccosh}\left(\frac{x}{a}\right) + c, \qquad x > a. \tag{46}$$

EXAMPLE 3

Evaluate the integral

$$\int_6^{12} \frac{dx}{\sqrt{x^2 - 9}}.$$

The integrand is of the form appearing in Eq. 46 with $a = 3$. Thus we have

$$\int_6^{12} \frac{dx}{\sqrt{x^2 - 9}} = \operatorname{arccosh}\left(\frac{x}{3}\right)\Bigg|_6^{12}$$

$$= \operatorname{arccosh} 4 - \operatorname{arccosh} 2 \cong 0.7465. \blacksquare$$

We conclude this section by pointing out that the inverse hyperbolic sine function can be expressed in terms of the natural logarithm function. This is perhaps not surprising since the hyperbolic sine was defined in terms of the exponential function, and the inverse of the exponential function is the logarithm. To show how $\operatorname{arcsinh} x$ and $\ln x$ are related, we start from the equation

$$x = \sinh y = \frac{e^y - e^{-y}}{2} \tag{47}$$

so that $y = \operatorname{arcsinh} x$. We now solve for y in terms of x. From Eq. 47 we have

$$e^y - 2x - e^{-y} = 0,$$

or, multiplying by e^y,

$$(e^y)^2 - 2xe^y - 1 = 0. \tag{48}$$

Equation 48 is quadratic in e^y and hence

$$e^y = \frac{2x \pm \sqrt{4x^2 + 4}}{2} = x \pm \sqrt{x^2 + 1}. \tag{49}$$

The plus sign must be chosen in Eq. 49 because e^y is always positive. Finally, taking the natural logarithm of both sides of Eq. 49, we obtain

$$y = \ln(x + \sqrt{x^2 + 1}). \tag{50}$$

Hence

$$\operatorname{arcsinh} x = \ln(x + \sqrt{x^2 + 1}), \qquad -\infty < x < \infty. \tag{51}$$

In a similar way the other inverse hyperbolic functions can also be expressed in terms of logarithms.

PROBLEMS

In each of Problems 1 through 6, establish the given identity.

1. $\sinh(x \pm y) = \sinh x \cosh y \pm \cosh x \sinh y$
2. $\cosh(x \pm y) = \cosh x \cosh y \pm \sinh x \sinh y$
3. $\sinh\left(\frac{x}{2}\right) = \pm\sqrt{\frac{\cosh x - 1}{2}}$; use + sign for $x > 0$, − sign for $x < 0$.
4. $\cosh\left(\frac{x}{2}\right) = \sqrt{\frac{\cosh x + 1}{2}}$
5. $\operatorname{arccosh} x = \ln[x + \sqrt{x^2 - 1}]$, $x \geq 1$. Why is $\ln[x - \sqrt{x^2 - 1}]$ incorrect?
6. $\sinh(\operatorname{arccosh} x) = \sqrt{x^2 - 1}$, $x \geq 1$

In each of Problems 7 through 16, find the derivative of the given function.

7. $f(x) = \sinh 3x$
8. $f(x) = \cosh 2x - 2\cos x$
9. $f(x) = x \cosh x$
10. $f(x) = \cosh(e^x)$
11. $f(x) = (\sinh x)e^{\cosh x}$
12. $f(x) = \sinh^2 2x$
13. $f(x) = \dfrac{\cosh x}{4 + \sinh^2 x}$
14. $f(x) = \sqrt{\cosh x + \sinh x}$
15. $f(x) = \operatorname{arcsinh}\left(\frac{x}{2}\right)$
16. $f(x) = \operatorname{arccosh} 3x$

In each of Problems 17 through 26, find the given antiderivative.

17. $\int \sinh 2x \, dx$
18. $\int \cosh \frac{x}{3} \, dx$
19. $\int \sinh^2 x \cosh x \, dx$
20. $\int \tanh x \, dx$
21. $\int \coth 2x \, dx$
22. $\int \frac{dx}{\sqrt{x^2 - 4}}$
23. $\int \frac{dx}{\sqrt{4x^2 - 1}}$
24. $\int \frac{dx}{\sqrt{9x^2 - 4}}$
25. $\int \frac{dx}{\sqrt{x^2 + 16}}$
26. $\int \frac{dx}{\sqrt{16x^2 + 25}}$

In each of Problems 27 through 32, evaluate the given integral.

27. $\int_{\ln 2}^{\ln 3} \cosh x \, dx$
28. $\int_{\ln 4}^{\ln 16} \sinh \frac{x}{2} \, dx$
29. $\int_{\ln 8}^{\ln 64} \coth \frac{x}{3} \, dx$
30. $\int_0^{\sqrt{3}} \frac{dx}{\sqrt{x^2 + 1}}$
31. $\int_1^2 \frac{dx}{\sqrt{9x^2 - 1}}$
32. $\int_0^{3/4} \frac{dx}{\sqrt{16x^2 + 9}}$

In each of Problems 33 through 36, establish the given differentiation formula.

33. $\frac{d}{dx} \tanh x = \operatorname{sech}^2 x$
34. $\frac{d}{dx} \coth x = -\operatorname{csch}^2 x$, $x \neq 0$
35. $\frac{d}{dx} \operatorname{sech} x = -\operatorname{sech} x \tanh x$
36. $\frac{d}{dx} \operatorname{csch} x = -\operatorname{csch} x \coth x$, $x \neq 0$

37. (a) Define the function arctanh x. What is its domain and range? Show that

$$\frac{d}{dx} \operatorname{arctanh} x = \frac{1}{1 - x^2}, \qquad |x| < 1.$$

Sketch the graph of $y = \operatorname{arctanh} x$.

(b) Find

$$\frac{d}{dx}\left[\frac{1}{2}\ln\left|\frac{1 + x}{1 - x}\right|\right].$$

(c) Show that

$$\operatorname{arctanh} x = \frac{1}{2}\ln\left(\frac{1 + x}{1 - x}\right), \qquad |x| < 1.$$

38. Evaluate

$$\int_{-10}^{-5} \frac{dx}{\sqrt{x^2 - 1}}.$$

Hint: Let $x = -u$.

* 39. In this problem we explore the relation between hyperbolic functions and hyperbolas. Consider the equation $x^2 - y^2 = 1$ whose graph is the hyperbola shown in Figure 8.7.9*a*.

(a) Show that $x = \cosh t$, $y = \sinh t$ for $-\infty < t < \infty$ is a parametric representation of the right-hand branch of the hyperbola. We wish

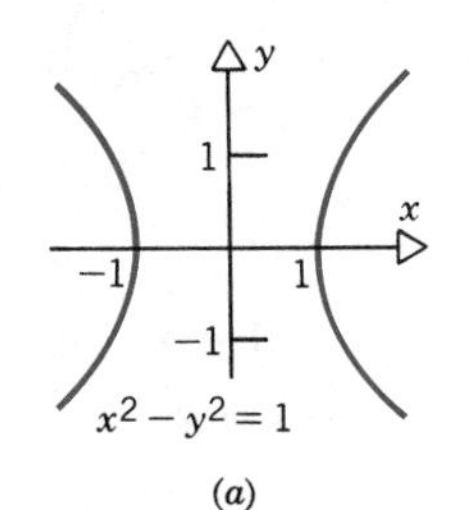

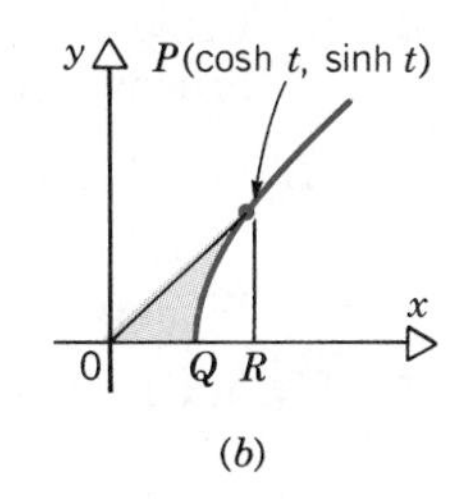

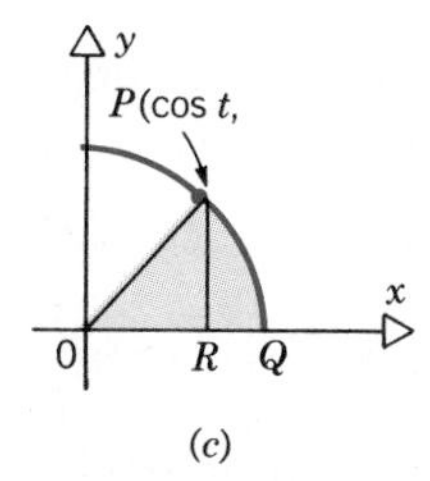

Figure 8.7.9

to give a geometric interpretation to the parameter t.

(b) For a fixed value of $t > 0$ the corresponding point P with coordinates ($\cosh t$, $\sinh t$) lies on the part of the hyperbola in the first quadrant (see Figure 8.7.9*b*). Show that the area $A(t)$ of the shaded region OPQ in Figure 8.7.9*b* is given by

$$A(t) = \tfrac{1}{2}\sinh t \cosh t - \int_1^{\cosh t} \sqrt{x^2 - 1}\, dx.$$

Hint: Observe that $\frac{1}{2}\sinh t \cosh t$ is the area of triangle OPR.

(c) Refer to Problem 40(a) of Section 6.4 and show that, for each t,

$$A'(t) = \tfrac{1}{2}.$$

(d) From the result of Part (c) it follows that $A(t) = (t/2) + c$, where c is a constant. Show that, in fact, $A(t) = t/2$. Thus the parameter t is twice the area bounded by the x-axis, the hyperbola, and the line segment OP.

(e) Start from the equation $x^2 + y^2 = 1$ of the unit circle and its parametric representation $x = \cos t$, $y = \sin t$ (see Figure 8.7.9*c*). Carry out an analysis similar to Parts (b) through (d) and show that t has an analogous interpretation in this case. Thus, in this respect, hyperbolic functions are related to a hyperbola in the same way that trigonometric functions are related to a circle.

(f) Show that for the circle t can also be interpreted as the length of the arc QP. Does this interpretation of t also hold in the case of the hyperbola?

REVIEW PROBLEMS

In each of Problems 1 through 4, find $(f^{-1})'(x)$, assuming it exists. Also find $f(x)$ and $f^{-1}(x)$.

1. $2ff' = 1, \quad f(0) = 0$

2. $f' = f, \quad f(0) = 1$

3. $f' - f^2 = 1, \quad f(0) = 0$

4. $(f')^2 - f^2 = -1, \quad f(0) = 0$

In each of Problems 5 through 8, find the largest interval of the form $\alpha \le x < \infty$ for which the given function has an inverse. Also find $f^{-1}(x)$.

5. $y = x^2$

6. $y = 2x^2 + x + 1$

7. $y = ax^2 + bx + c$

8. $y = x^4 + 1$

In each of Problems 9 through 12, find the inverse of the given function, or else show that it has no inverse.

9. $f(x) = \sqrt{|x|}, \quad -\infty < x < \infty$

10. $f(x) = \sqrt{(x-2)^2}, \quad -\infty < x < \infty$

11. $f(x) = \dfrac{e^x + e^{-x}}{e^x - e^{-x}}, \quad x \neq 0$

12. $f(x) = \ln\sqrt{1+x} - \ln\sqrt{1-x}, \quad -1 < x < 1$

In each of Problems 13 through 22, find the derivative y' of the given function.

13. $y = 2^{\ln(x^x)}$

14. $y = \ln 2^{x^x}$

15. $y = \arcsin(e^{-x})$

16. $y = \ln\left(\dfrac{ax+b}{cx+d}\right)$

17. $y = \log_a e^x$

18. $y = \sinh(\log_{10} \cosh 2^x)$

19. $y = \operatorname{arcsinh}(\ln \sinh x)$

20. $y = \log_3 \arctan x$

21. $y = \arctan(\sinh x)$

22. $y = \exp(\log_{10} \sinh x)$

In each of Problems 23 through 34, evaluate the given integral.

23. $\int (ax + b)e^{(ax+b)^2}\, dx$

24. $\int \frac{e^x}{1 + e^{2x}}\, dx$

25. $\int (1 + \tan^2 x)2^{\tan x}\, dx$

26. $\int (e^x - e^{-x})2^{e^x + e^{-x}}\, dx$

27. $\int (\cos x)\, \pi^{\sin x}\, dx$

28. $\int_0^1 (2^x \ln 2 + 3^x \ln 3)e^{2^x + 3^x}\, dx$

29. $\int_1^2 \left(\frac{1}{x^2}\right) \cos\left(\frac{\pi}{x}\right) 2^{\sin(\pi/x)}\, dx$

30. $\int_0^{\ln 4} 2^{\log_4 e^x}\, dx$

31. $\int \frac{dx}{|x - 4| \sqrt{(x - 4)^2 - 4}}$

32. $\int \frac{x\, dx}{\sqrt{x^4 + 1}}$

33. $\int \frac{dx}{\sqrt{x}\,(1 + x)}$

34. $\int \frac{e^x}{\sqrt{1 - e^{2x}}}\, dx$

In each of Problems 35 through 40, evaluate the given limit.

35. $\lim_{x \to 1} \frac{\sin(\ln x)}{\ln x}$

36. $\lim_{x \to 0^+} x^{(\alpha/\ln x)}, \qquad \alpha > 0$

37. $\lim_{t \to 0} (1/t^M)e^{-(1/t)}, \qquad M > 0$

38. $\lim_{x \to 0} \frac{e^{nx} - 1}{x}, \qquad n > 0$

39. $\lim_{x \to 0} \frac{\arcsin x}{x}$

40. $\lim_{x \to 0} \frac{2 \arctan x}{x}$

41. (a) Using the mean value theorem for integrals (Theorem 6.3.5), show that for $x > 0$

$$\frac{x}{1 + x^2} < \arctan x < x.$$

(b) What is the corresponding result when $x < 0$?

42. Use the result of Problem 41 to show that

$$\frac{1}{2} \ln(1 + x^2) < \int_0^x \arctan t\, dt < \frac{x^2}{2}.$$

43. Find all values of x such that

$$(\ln x)^2 - \ln(x^2) = 0.$$

44. Find all values of x such that

$$(\ln x)^2 + \ln \frac{1}{x^2} + 1 = 0.$$

© 45. Use Newton's method to solve for the root $x = r$ of

$$\sinh^2 x - 2\sqrt{\cosh^2 x - 1} + 1 = 0,$$

using $x_0 = 1$ as an initial guess. Does Newton's method converge to the correct value of r if $x_0 = 0$ is used as an initial guess? Explain your results.

© 46. Use Newton's method to solve

$$e^{(4/3)x} - 2e^{(2/3)x} + 1 = \tan x, \qquad -\frac{\pi}{2} < x < \frac{\pi}{2}.$$

47. Use the result of Problem 35 in Section 8.5 to find the relationship between ln 10 and $\log_{10} e$.

For the given function in each of Problems 48 through 53, find all critical points, intervals where the function is increasing or decreasing, intervals where the function is concave up or concave down, and all local and global maxima and minima. Sketch the graph of the given function.

48. $f(x) = x \ln x - x, \qquad x > 0$

49. $f(x) = e^{(\ln x)^2}, \qquad x > 0$

50. $f(x) = \ln(x + \sqrt{x^2 - 1}), \qquad x \geq 1$

51. $f(x) = e^{\sin x}, \qquad 0 \leq x \leq 2\pi$

52. $f(x) = \arctan(\sinh x)$

53. $f(x) = \sinh x + \cosh x$

NINE

methods of integration

The mathematical formulation of many problems in the physical, natural, and social sciences leads eventually to the problem of evaluating either a definite integral

$$\int_a^b f(x)\,dx$$

or an indefinite integral (antiderivative)

$$\int f(x)\,dx.$$

According to Theorem 6.4.3 (the fundamental theorem of calculus), one way to evaluate a definite integral is to recognize an antiderivative of the integrand. Sometimes it is possible to do this immediately, but often some preliminary manipulation is required. In this chapter we discuss several methods that are frequently useful in finding antiderivatives, or at least in converting integrals to a simpler form.

An important tool in dealing with both definite and indefinite integrals is a reasonably extensive table of integrals. Table 9.1 contains a list of some of the integration formulas developed so far in this book. The first seven are indispensable, and frequent use should make them easy to recall whenever needed. The other entries in Table 9.1 are also important, but do not occur as often. The more comprehensive table of integrals on the inside covers of the book is more than adequate for a calculus course, and provides a reference for possible later use. Much more

Table 9.1 Some Elementary Integration Formulas

1. $\int u^r \, du = \dfrac{u^{r+1}}{r+1} + c, \qquad r \neq -1$

2. $\int \dfrac{du}{u} = \ln |u| + c$

3. $\int \sin u \, du = -\cos u + c$

4. $\int \cos u \, du = \sin u + c$

5. $\int e^u \, du = e^u + c$

6. $\int \dfrac{du}{\sqrt{1-u^2}} = \arcsin u + c,$

7. $\int \dfrac{du}{1+u^2} = \arctan u + c$

8. $\int \sinh u \, du = \cosh u + c$

9. $\int \cosh u \, du = \sinh u + c$

10. $\int \sec^2 u \, du = \tan u + c$

11. $\int \sec u \tan u \, du = \sec u + c$

12. $\int \dfrac{du}{\sqrt{1+u^2}} = \operatorname{arcsinh} u + c$

13. $\int \dfrac{du}{\sqrt{u^2-1}} = \operatorname{arccosh} u + c$

Note: In each formula u is a differentiable function of x.

extensive sets of integration formulas are also available in various mathematical handbooks.

In recent years several software packages for symbolic computation have been developed that include integration routines. The best of these determine the indefinite integral of any function whose integral can be expressed in terms of the elementary functions—the algebraic, trigonometric, exponential, and logarithmic functions—or else state that none exists (among this class of functions). These programs are capable of producing the indefinite integrals of the elementary functions that appear in a standard table of integrals and of evaluating many integrals that are not in such tables. They are of tremendous assistance to individuals who must deal with complicated integrations frequently.

Despite the availability of integral tables and symbolic integration packages, it is still worthwhile to be conversant with some of the more important integration techniques. For relatively simple integrals it may well be quicker and more convenient to evaluate the integral by hand using one of the methods of this chapter than to search through an extensive table, or to log on to a computer and call up the necessary software.

9.1 SUBSTITUTIONS REVISITED

In Section 6.5 we pointed out that the tasks of determining an antiderivative, or of evaluating a definite integral, can sometimes be eased considerably by a judicious substitution, or change of the integration variable. For example, suppose that we wish to determine

$$\int g(x)\,dx.$$

According to the discussion in Section 6.5, we can let $u = u(x)$, find $du = u'(x)\,dx$, and seek to write

$$g(x)\,dx = f[u(x)]u'(x)\,dx = f(u)\,du, \tag{1}$$

where f has a recognizable antiderivative. If F is an antiderivative of f, then

$$\int g(x)\,dx = \int f(u)\,du = F(u) + k = F[u(x)] + k, \tag{2}$$

where k is an arbitrary constant. Moreover, by Theorem 6.5.1,

$$\int_a^b g(x)\,dx = \int_c^d f(u)\,du = F(d) - F(c), \tag{3}$$

provided that $c = u(a)$, $d = u(b)$, and Eq. (1) holds.

In Section 6.5 our use of substitutions was somewhat restricted because we had developed comparatively few differentiation and integration formulas at that time. Now that we have many more such formulas available, it is appropriate to look at some further examples of substitutions as a means of evaluating or simplifying integrals. As we have indicated before, a good substitution is often suggested by some feature of the integrand. It is also helpful to be able to recognize integrals that have elementary solutions. For this reason a knowledge of at least the integration formulas in Table 9.1 is highly desirable.

EXAMPLE 1

Evaluate

$$\int \tan ax\,dx, \tag{4}$$

where a is a positive constant.

Since this integral does not appear in Table 9.1, we must first rewrite the integrand in a more convenient form. A first step is to express the tangent in terms of the sine and cosine, that is,

$$\int \tan ax\,dx = \int \frac{\sin ax}{\cos ax}\,dx. \tag{5}$$

Next, we observe that on the right side of Eq. 5 the numerator is related to the derivative of the denominator. This suggests that we try to make a substitution that would enable us to use formula 2 in Table 9.1. Accordingly, we let $u = \cos ax$, so $du = -a \sin ax\,dx$ and $\sin ax\,dx = -du/a$. Then the integral becomes

$$\int \frac{\sin ax}{\cos ax}\,dx = \int \frac{1}{u}\left(\frac{-du}{a}\right) = -\frac{1}{a}\int \frac{du}{u}$$
$$= -\frac{1}{a}\ln|u| + c.$$

Thus

$$\int \tan ax\,dx = -\frac{1}{a}\ln|\cos ax| + c. \blacksquare \tag{6}$$

EXAMPLE 2

Evaluate

$$\int \frac{dx}{2x^2 - 8x + 26}. \tag{7}$$

An inspection of Table 9.1 shows no integral of this form; the closest is formula 7 for the integral of $(1 + u^2)^{-1}$. We can reduce the integrand in Eq. 7 to that form by completing the square, followed by a substitution. First we observe that

$$2x^2 - 8x + 26 = 2[x^2 - 4x + 13] = 2[x^2 - 4x + 4 + 9]$$
$$= 2[(x-2)^2 + 9] = 18\left[1 + \left(\frac{x-2}{3}\right)^2\right].$$

Thus

$$\int \frac{dx}{2x^2 - 8x + 26} = \frac{1}{18}\int \frac{dx}{[(x-2)/3]^2 + 1}.$$

Now let $u = (x-2)/3$, so $du = dx/3$ and $dx = 3du$. Then

$$\int \frac{dx}{2x^2 - 8x + 26} = \frac{1}{6}\int \frac{du}{u^2 + 1} = \frac{1}{6}\arctan u + c$$
$$= \frac{1}{6}\arctan \frac{x-2}{3} + c. \blacksquare \tag{8}$$

By following the procedure of Example 2 and making use of integrals 6, 7, 12, and 13 in Table 9.1, we can evaluate all integrals of the forms

$$\int \frac{dx}{\alpha x^2 + \beta x + \delta} \quad \text{and} \quad \int \frac{dx}{\sqrt{\alpha x^2 + \beta x + \delta}} \tag{9}$$

except those that reduce to

$$\int \frac{dx}{a^2 - x^2}. \tag{10}$$

Integrals reducible to the form (10) can be handled by the method of partial fractions that we discuss in Section 9.6.

EXAMPLE 3

Evaluate

$$\int_0^1 \frac{5e^{2x}}{(1 + e^{2x})^{1/3}}\, dx. \tag{11}$$

This example illustrates the situation in which an expression is raised to a power, in particular, a fractional one. In such cases it is often helpful to use this expression as the basis for a substitution. Thus we let $u = 1 + e^{2x}$, so that $du = 2e^{2x}\, dx$, and $5e^{2x}\, dx = 5\, du/2$. Further $u = 2$ when $x = 0$, and $u = 1 + e^2$ when $x = 1$. Consequently, by Theorem 6.5.1, the integral (11) is transformed into

$$\begin{aligned} \int_0^1 \frac{5e^{2x}}{(1 + e^{2x})^{1/3}}\, dx &= \frac{5}{2} \int_2^{1+e^2} \frac{du}{u^{1/3}} = \frac{5}{2} \int_2^{1+e^2} u^{-1/3}\, du \\ &= \frac{15}{4} u^{2/3} \Big|_2^{1+e^2} \\ &= \frac{15}{4} [(1 + e^2)^{2/3} - (2)^{2/3}] \cong 9.53. \blacksquare \end{aligned} \tag{12}$$

EXAMPLE 4

Evaluate

$$\int_1^2 \frac{dx}{\sqrt{3 + 2x - x^2}}. \tag{13}$$

This example is somewhat similar to Example 2. The entry in Table 9.1 that is most similar to the integral (13) is formula 6. As in Example 2, a preliminary manipulation followed by a substitution reduces the integral in Eq. 13 to the desired form. First, we observe that

$$
\begin{aligned}
3 + 2x - x^2 &= 3 - (x^2 - 2x) \\
&= 3 - (x^2 - 2x + 1 - 1) \\
&= 4 - (x - 1)^2 = 4\left[1 - \left(\frac{x-1}{2}\right)^2\right].
\end{aligned}
$$

Thus

$$\int_1^2 \frac{dx}{\sqrt{3 + 2x - x^2}} = \frac{1}{2}\int_1^2 \frac{dx}{\sqrt{1 - [(x-1)/2]^2}}.$$

Next, we make the change of variable $u = (x - 1)/2$, so $du = dx/2$ and $dx = 2\,du$. Further, if $x = 1$, then $u = 0$, and if $x = 2$, then $u = \frac{1}{2}$. Hence

$$
\begin{aligned}
\int_1^2 \frac{dx}{\sqrt{3 + 2x - x^2}} &= \int_0^{1/2} \frac{du}{\sqrt{1 - u^2}} \\
&= \arcsin u \Big|_0^{1/2} = \arcsin \frac{1}{2} - \arcsin 0 \\
&= \frac{\pi}{6}. \quad \blacksquare \qquad (14)
\end{aligned}
$$

In each of these examples we have shown a successful substitution. In practice, however, and especially for somewhat complicated integrands, it is not uncommon for a substitution to fail to simplify an integral significantly. Nevertheless, if one is alert, even an unsuccessful substitution can be helpful either in suggesting a better substitution or perhaps another method of attack altogether. Do not hesitate to try a substitution just because you cannot foresee that it will be successful.

Further, keep in mind that you can always check the answer obtained by evaluating an indefinite integral. Simply differentiate the result and make sure that the derivative agrees with the original integrand.

PROBLEMS

In each of Problems 1 through 6, use the given substitution to evaluate the indefinite integral.

1. $\displaystyle\int \frac{3x}{1 + 4x^2}\,dx, \qquad u = 1 + 4x^2$

2. $\displaystyle\int \frac{3^{\sqrt{x}}}{\sqrt{x}}\,dx, \qquad u = \sqrt{x}$

3. $\displaystyle\int \frac{e^x}{\sqrt{1 + e^x}}\,dx, \qquad u = e^x$

4. $\displaystyle\int \frac{s}{\sqrt{3 - 4s^2}}\,ds, \qquad u = 3 - 4s^2$

5. $\displaystyle\int \frac{(\ln x)^3}{x}\,dx, \qquad u = \ln x$

6. $\displaystyle\int \frac{dx}{1 + \sqrt{x}}, \qquad u = 1 + \sqrt{x}$

In each of Problems 7 through 26, choose an appropriate substitution to evaluate the given indefinite or definite integral.

7. $\int \cot x \, dx$

8. $\int \frac{e^{1/x}}{x^2} \, dx$

9. $\int \frac{2 \cos x \, dx}{1 + \sin^2 x}$

10. $\int_0^1 \cos \frac{\pi}{4}(x + 3) \, dx$

11. $\int \frac{e^{2x}}{\sqrt{1 - e^{4x}}} \, dx$

12. $\int_0^2 3xe^{-x^2} \, dx$

13. $\int_1^3 \frac{x \, dx}{1 + x}$

14. $\int \frac{s^3 \, ds}{2 + 3s^4}$

15. $\int \cosh \frac{x}{2} \sinh^4 \frac{x}{2} \, dx$

16. $\int_1^c \frac{\ln x^3}{x} \, dx, \quad c > 1$

17. $\int \frac{\cos(\ln x)}{x} \, dx$

18. $\int \frac{a \sin x}{b + k \cos x} \, dx, \quad a, b, k \neq 0$

19. $\int x^2(7 + 2x^3)^{1/3} \, dx$

20. $\int \frac{\arctan x}{1 + x^2} \, dx$

21. $\int_2^e \frac{dx}{x \ln x}$

22. $\int \frac{x \, dx}{(x^2 + 1)^{3/2}}$

23. $\int \tanh 2x \, dx$

24. $\int e^x e^{e^x} \, dx$

25. $\int_0^2 x^3 \sqrt{x^2 + 4} \, dx$

26. $\int_4^7 \frac{dx}{x + \sqrt{x}}$

In each of Problems 27 through 34, evaluate the given integral.

27. $\int \frac{dx}{\sqrt{4x - x^2}}$

28. $\int \frac{dx}{2x^2 - 12x + 36}$

29. $\int \frac{4x + 18}{x^2 + 4x + 29} \, dx$

30. $\int \frac{3}{\sqrt{8 - 2x - x^2}} \, dx$

31. $\int \frac{dx}{\sqrt{4x^2 + 16x + 17}}$

32. $\int \frac{dx}{x^2 - 8x + 21}$

33. $\int_1^4 \frac{3 \, dx}{2x^2 + 8x + 26}$

34. $\int_2^4 \frac{dx}{\sqrt{x^2 - 4x + 8}}$

* 35. If f is a continuous function, show that

$$\int_0^\pi \phi f(\sin \phi) \, d\phi = \frac{\pi}{2} \int_0^\pi f(\sin \phi) \, d\phi.$$

Hint: Try the change of variable $\phi = \pi - \theta$.

9.2 INTEGRATION BY PARTS

One of the most important and useful integration techniques is **integration by parts.** This technique is based on the formula for the derivative of a product: if u and v are differentiable functions, then

$$\frac{d}{dx}(uv) = u \frac{dv}{dx} + v \frac{du}{dx}. \tag{1}$$

By integrating Eq. 1 and omitting the integration constant, we obtain

$$uv = \int u \frac{dv}{dx} \, dx + \int v \frac{du}{dx} \, dx. \tag{2}$$

Equation 2 can be rewritten in the form

$$\int u \frac{dv}{dx} \, dx = uv - \int v \frac{du}{dx} \, dx \tag{3a}$$

or

$$\int u(x)v'(x) \, dx = u(x)v(x) - \int v(x)u'(x) \, dx. \tag{3b}$$

The formulas for definite integrals corresponding to Eqs. 3 are

$$\int_a^b u \frac{dv}{dx} \, dx = uv \Big|_a^b - \int_a^b v \frac{du}{dx} \, dx \tag{4a}$$

or

$$\int_a^b u(x)v'(x)\,dx = u(b)v(b) - u(a)v(a) - \int_a^b v(x)u'(x)\,dx. \tag{4b}$$

The geometric interpretation of Eq. 4(b) is shown in Figure 9.2.1.

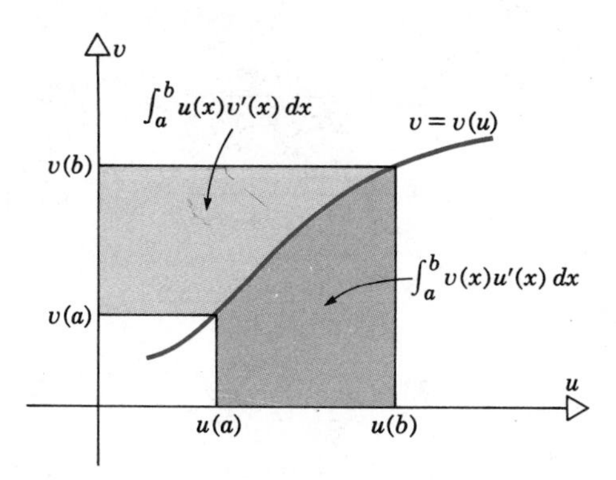

Figure 9.2.1

It is also possible to express the relation in Eqs. 3 in terms of u and v. Since $du = u'(x)\,dx$ and $dv = v'(x)\,dx$, it follows immediately from either Eq. 3(a) or Eq. 3(b) that

$$\int u\,dv = uv - \int v\,du. \tag{5}$$

This is the integration by parts formula that one usually remembers. The importance of Eqs. 3, 4, or 5 is that they enable us to express one integral in terms of another. Surprisingly often, it is possible to arrange for the second integral to be simpler than the first one. Indeed, it is part of the folklore of mathematics to say that if one is in doubt, or if other methods fail, try integration by parts. We now illustrate this technique by several examples.

EXAMPLE 1

Evaluate

$$\int x \sin x\,dx. \tag{6}$$

To use integration by parts we need to split up the integrand by choosing u and dv. In this problem there are two natural choices we could make, namely $u = x$, $dv = \sin x\,dx$ or $u = \sin x$, $dv = x\,dx$. If we let $u = x$ and $dv = \sin x\,dx$, then $du = dx$ and $v = -\cos x$. Hence, from Eq. 5, we obtain

$$\int \underbrace{x}_{u} \underbrace{\sin x\,dx}_{dv} = \underbrace{x}_{u}\,(\underbrace{-\cos x}_{v}) - \int (\underbrace{-\cos x}_{v})\,\underbrace{dx}_{du} = -x\cos x + \int \cos x\,dx,$$

and it readily follows that

$$\int x \sin x\,dx = -x\cos x + \sin x + c. \tag{7}$$

On the other hand, if we initially choose $u = \sin x$ and $dv = x\,dx$, then $du = \cos x\,dx$ and $v = x^2/2$. In that event Eq. 5 yields

$$\int x \sin x\,dx = \frac{x^2}{2}\sin x - \frac{1}{2}\int x^2 \cos x\,dx.$$

This result is correct, but not helpful, since it expresses the original integral in terms of another integral that is more complicated, rather than simpler. ∎

In Example 1 there were two obvious possibilities for u and dv, one of which is effective while the other fails. Can we see in advance which choice of u and dv is preferable? Observe that the use of integration by parts requires us to differentiate the factor we call u and to integrate the factor we call dv. In Example 1 it makes no difference whether we differentiate or integrate $\sin x$; we obtain either $\cos x$ or $-\cos x$ in any case. However, differentiating x yields the result dx, which is simpler than the result, $x^2/2$, obtained by integrating x. Thus we should choose $u = x$ and $dv = \sin x\,dx$ so that we differentiate x and integrate $\sin x$ rather than the other way around. Similar considerations are often helpful in choosing u and dv.

Note also that in finding v from dv we normally omit the constant of integration. This is because we are usually interested in obtaining the simplest possible form for v, not the most general one. Occasionally, however, it may be desirable to assign the integration constant some nonzero value (see Problems 41 through 43).

EXAMPLE 2

Evaluate

$$\int x\,e^{ax}\,dx. \tag{8}$$

As in Example 1, it is better to differentiate x rather than to integrate it. Thus for the integral (8) we choose $u = x$ and $dv = e^{ax}\,dx$; then $du = dx$ and $v = (1/a)\,e^{ax}$. Hence we have

$$\int \underbrace{x}_{u}\,\underbrace{e^{ax}\,dx}_{dv} = \underbrace{x}_{u}\underbrace{\left(\frac{1}{a}\,e^{ax}\right)}_{v} - \int \underbrace{\frac{1}{a}\,e^{ax}}_{v}\,\underbrace{dx}_{du}$$

$$= \frac{1}{a}\,x\,e^{ax} - \frac{1}{a^2}\,e^{ax} + c,$$

or

$$\int x\,e^{ax}\,dx = \frac{1}{a^2}\,e^{ax}(ax - 1) + c. \quad ∎ \tag{9}$$

EXAMPLE 3

Evaluate

$$\int x^n e^{ax}\,dx, \qquad n \text{ a positive integer.} \tag{10}$$

From our experience in Example 2, we try $u = x^n$, $dv = e^{ax}\,dx$; thus $du = nx^{n-1}\,dx$, $v = (1/a)\,e^{ax}$. Hence

$$\begin{aligned}\int x^n e^{ax}\,dx &= x^n \frac{1}{a} e^{ax} - \int \frac{1}{a} e^{ax} n x^{n-1}\,dx \\ &= \frac{1}{a} x^n e^{ax} - \frac{n}{a}\int x^{n-1} e^{ax}\,dx.\end{aligned} \tag{11}$$

The integral on the right side of Eq. 11 can be evaluated by replacing n by $n - 1$ in the formula just derived. Equation 11 is an example of a **reduction formula.** The original integral is evaluated by repeated use of the reduction formula. For example, for $n = 3$ we have

$$\begin{aligned}\int x^3 e^{ax}\,dx &= \frac{1}{a} x^3 e^{ax} - \frac{3}{a}\int x^2 e^{ax}\,dx \\ &= \frac{1}{a} x^3 e^{ax} - \frac{3}{a}\left[\frac{1}{a} x^2 e^{ax} - \frac{2}{a}\int x\, e^{ax}\,dx\right] \\ &= \frac{1}{a} x^3 e^{ax} - \frac{3}{a^2} x^2 e^{ax} + \frac{6}{a^2}\left[\frac{1}{a} x\, e^{ax} - \frac{1}{a^2} e^{ax}\right] + c \\ &= \frac{1}{a} x^3 e^{ax} - \frac{3}{a^2} x^2 e^{ax} + \frac{6}{a^3} x\, e^{ax} - \frac{6}{a^4} e^{ax} + c. \ \blacksquare\end{aligned} \tag{12}$$

EXAMPLE 4

Evaluate

$$\int_0^{1/2} \arcsin x\,dx. \tag{13}$$

This time the integrand is a function for which we do not recognize an antiderivative but for which we do know the derivative. Thus we let $u = \arcsin x$, $dv = dx$; then $du = (1 - x^2)^{-1/2}\,dx$, $v = x$. Hence

$$\begin{aligned}\int_0^{1/2} \arcsin x\,dx &= (\arcsin x)x\Big|_0^{1/2} - \int_0^{1/2} \frac{x}{\sqrt{1 - x^2}}\,dx \\ &= \frac{1}{2}\cdot\frac{\pi}{6} - \int_0^{1/2} \frac{x}{\sqrt{1 - x^2}}\,dx.\end{aligned}$$

The last integral can be evaluated by making the change of variable $w = 1 - x^2$, or by inspection. The result is

$$\int_0^1 \arcsin x\, dx = \frac{\pi}{12} + \sqrt{1 - x^2}\Big|_0^{1/2} = \frac{\pi}{12} + \frac{\sqrt{3}}{2} - 1 \cong 0.1278. \quad \blacksquare \tag{14}$$

EXAMPLE 5

Evaluate

$$\int e^{-x} \sin 2x\, dx. \tag{15}$$

In this case it makes no difference whether we choose u to be e^{-x} or $\sin 2x$, since either choice works equally well. Suppose we take $u = e^{-x}$ and $dv = \sin 2x\, dx$. Then $du = -e^{-x}\, dx$ and $v = -(\cos 2x)/2$. Therefore

$$\int e^{-x} \sin 2x\, dx = -\frac{1}{2} e^{-x} \cos 2x - \frac{1}{2} \int e^{-x} \cos 2x\, dx. \tag{16}$$

Observe that the integral remaining on the right side of Eq. 16 is different from Eq. 15 but of similar difficulty, so it appears that integration by parts has not been helpful in this case. However, let us try the process once more, letting $u = e^{-x}$ and $dv = \cos 2x\, dx$; then $du = -e^{-x}\, dx$ and $v = (\sin 2x)/2$. Using the integration by parts formula (5) on the last integral in Eq. 16, we obtain

$$\begin{aligned} \int e^{-x} \sin 2x\, dx &= -\frac{1}{2} e^{-x} \cos 2x - \frac{1}{2}\left[\frac{1}{2} e^{-x} \sin 2x + \frac{1}{2} \int e^{-x} \sin 2x\, dx\right] \\ &= -\frac{1}{2} e^{-x} \cos 2x - \frac{1}{4} e^{-x} \sin 2x - \frac{1}{4} \int e^{-x} \sin 2x\, dx. \end{aligned} \tag{17}$$

The integral remaining on the right side of Eq. 17 is the same as the one on the left side; hence we can solve Eq. 17 to obtain the desired result, namely,

$$\int e^{-x} \sin 2x\, dx = -\frac{2e^{-x} \cos 2x + e^{-x} \sin 2x}{5} + c; \tag{18}$$

note that we have added an arbitrary integration constant in Eq. 18. ∎

The procedure in Example 5 can be used to evaluate any integral of the form

$$\int e^{ax} \cos bx\, dx \quad \text{or} \quad \int e^{ax} \sin bx\, dx, \tag{19}$$

where a and b are constants. While initially u may be chosen to be either the exponential or trigonometric factor in the integrand, it is essential (as in Example 5) that the same choice also be made when integration by parts is used the second time. Otherwise no useful result is obtained.

EXAMPLE 6

Evaluate

$$\int x^3 e^{-x^2}\, dx. \tag{20}$$

For this example, the choice of u and dv is not immediately obvious. A reasonable first choice is $u = x^3$ and $dv = e^{-x^2}\, dx$ so $du = 3x^2\, dx$, but we cannot calculate v. If we recall that $d(e^{-x^2})/dx = -2xe^{-x^2}$, this suggests the grouping

$$\int x^3 e^{-x^2}\, dx = \int x^2 (xe^{-x^2})\, dx.$$

We can now set $u = x^2$, $dv = xe^{-x^2}\, dx$, so $du = 2x\, dx$, $v = -\frac{1}{2}e^{-x^2}$. Hence

$$\begin{aligned}\int x^3 e^{-x^2}\, dx &= x^2\left(-\frac{1}{2}e^{-x^2}\right) - \int \left(-\frac{1}{2}e^{-x^2}\right) 2x\, dx \\ &= -\frac{1}{2}x^2 e^{-x^2} + \int xe^{-x^2}\, dx \\ &= -\frac{1}{2}x^2 e^{-x^2} - \frac{1}{2}e^{-x^2} + c \\ &= -\frac{1}{2}e^{-x^2}(x^2 + 1) + c. \quad \blacksquare \end{aligned} \tag{21}$$

PROBLEMS

In each of Problems 1 through 24, evaluate the given definite or indefinite integral.

1. $\int x \cos x\, dx$

2. $\int xe^{2x}\, dx$

3. $\int x^2 \sin \pi x\, dx$

4. $\int e^{2x} \cos x\, dx$

5. $\int_0^{1/2} x \sin \pi x\, dx$

6. $\int_0^1 xe^{-x}\, dx$

7. $\int_0^{\pi/2} e^x \sin 2x\, dx$

8. $\int_0^{\pi} x^2 \cos x\, dx$

9. $\int_0^2 x^2 e^{-x}\, dx$

10. $\int \ln x\, dx$

11. $\int \arctan x\, dx$

12. $\int x \ln x\, dx$

13. $\int \sin(\ln x)\, dx$

14. $\int (\ln x)^2\, dx$

15. $\int \dfrac{x^2}{\sqrt{1 - x}}\, dx$

16. $\int x(\ln x)^2\, dx$

17. $\int x2^x\, dx$

18. $\int x^2 \ln x\, dx$

19. $\int_1^3 \dfrac{\ln x}{\sqrt{x}}\, dx$

20. $\int x(x + 10)^{50}\, dx$

21. $\int_0^1 x \ln(9 + x^2)\, dx$

22. $\int \cos \sqrt{x}\, dx$

23. $\int x^5 (x^3 - 1)^{1/2}\, dx$

24. $\int e^{\sqrt{x}}\, dx$

In each of Problems 25 through 34, derive the given integration formula.

25. $\displaystyle\int e^{ax} \sin bx\, dx = \frac{e^{ax}(a \sin bx - b \cos bx)}{a^2 + b^2} + c$

26. $\displaystyle\int e^{ax} \cos bx\, dx = \frac{e^{ax}(a \cos bx + b \sin bx)}{a^2 + b^2} + c$

27. $\displaystyle\int x^p \ln x\, dx = \frac{x^{p+1}}{p + 1} \ln x - \frac{x^{p+1}}{(p + 1)^2} + c, \quad p \neq -1$

28. $\int x^p(\ln x)^2\,dx = \frac{x^{p+1}}{p+1}(\ln x)^2$
$- 2\frac{x^{p+1}}{(p+1)^2}\ln x + 2\frac{x^{p+1}}{(p+1)^3} + c, \qquad p \neq -1$

29. $\int x^m \sin ax\,dx$
$= -\frac{x^m \cos ax}{a} + \frac{m}{a}\int x^{m-1}\cos ax\,dx$

30. $\int x^m \cos ax\,dx$
$= \frac{x^m \sin ax}{a} - \frac{m}{a}\int x^{m-1}\sin ax\,dx$

31. $\int x^n e^{ax}\,dx = \frac{x^n e^{ax}}{a} - \frac{n}{a}\int x^{n-1}e^{ax}\,dx$

32. $\int \sin^n ax\,dx$
$= \frac{-\cos ax \sin^{n-1} ax}{na} + \frac{n-1}{n}\int \sin^{n-2} ax\,dx$

33. $\int (\ln x)^n\,dx = x(\ln x)^n - n\int (\ln x)^{n-1}\,dx$

34. $\int (a^2 - x^2)^n\,dx$
$= x(a^2 - x^2)^n + 2n\int x^2(a^2 - x^2)^{n-1}\,dx$

35. Use the results of Problems 29 and 30 to show that

$$\int x^2 \sin 3x\,dx = -\frac{x^2\cos 3x}{3} + \frac{2x\sin 3x}{9} + \frac{2\cos 3x}{27} + c$$

36. Use the result of Problem 31 to show that

$$\int x^4 e^{2x}\,dx = \left(\frac{x^4}{2} - x^3 + \frac{3x^2}{2} - \frac{3x}{2} + \frac{3}{4}\right)e^{2x} + c.$$

37. Use the results of Problems 29 and 30 to show that

$$\int_0^\pi x^3 \sin x\,dx = \pi^3 - 6\pi.$$

38. Evaluate $\int x^3(1 - x^2)^n\,dx$, where n is an arbitrary positive integer.

39. Use the result of Problem 32 to show that

$$\int_0^{\pi/2} \sin^{2m} x\,dx = \frac{2m-1}{2m}\cdot\frac{2m-3}{2m-2}\cdots\frac{3}{4}\cdot\frac{1}{2}\cdot\frac{\pi}{2}.$$

40. If f'' and g'' are continuous, and if $f(a) = f(b) = g(a) = g(b) = 0$, show that

$$\int_a^b f(x)g''(x)\,dx = \int_a^b f''(x)g(x)\,dx.$$

Problems 41, 42, and 43 illustrate that sometimes by a suitable choice of the integration constant in choosing v the calculations can be simplified or a neater formula can be derived.

41. Consider

$$\int \ln(x+1)\,dx.$$

(a) Evaluate this integral by choosing $u = \ln(x+1)$, $dv = dx$ with $v = x$.

(b) Evaluate this integral by choosing $u = \ln(x+1)$, $dv = dx$ with $v = x + 1$, which differs from the v of Part (a) by a constant. Also observe that the problem can be simplified by first introducing the change of variable $s = x + 1$, and then using integration by parts.

42. Show that

$$\int x \arctan x\,dx = -\frac{x}{2} + \frac{1}{2}(x^2+1)\arctan x + c.$$

If $dv = x\,dx$, then $v = x^2/2 + C$. By a proper choice of C the calculation becomes especially simple. Note that if C is taken to be zero, then the calculation can still be done by observing that

$$\frac{x^2}{1+x^2} = 1 - \frac{1}{1+x^2}.$$

* 43. Starting with the identity

$$f(b) - f(a) = \int_a^b f'(t)\,dt,$$

derive the following generalizations of the mean value theorem:

(a) $f(b) - f(a)$
$= f'(a)(b-a) - \int_a^b f''(t)(t-b)\,dt,$

(b) $f(b) - f(a) = f'(a)(b-a)$
$+ \frac{f''(a)}{2}(b-a)^2 + \int_a^b \frac{f'''(t)}{2}(t-b)^2\,dt.$

* **44.** If $y = f(x)$ has the inverse function given by $x = f^{-1}(y)$, show that

$$\int_a^b f(x)\,dx + \int_{f(a)}^{f(b)} f^{-1}(y)\,dy = bf(b) - af(a).$$

This relationship allows us to choose whichever of the two integrals is the easier to evaluate.

45. Suppose that we want to calculate the indefinite integral of $e^{2x}(2\sin 3x + 6\cos 3x)$. On reflecting on what type of functions have derivatives $e^{2x}\sin 3x$ and $e^{2x}\cos 3x$, it is reasonable to conjecture that

$$\int e^{2x}(2\sin 3x + 6\cos 3x)\,dx = Ae^{2x}\sin 3x + Be^{2x}\cos 3x + c,$$

where the constants A and B are to be determined and c is an arbitrary constant. By differentiating both sides of this equation show that $A = \frac{22}{13}$ and $B = \frac{6}{13}$.

9.3 INTEGRALS INVOLVING SINES AND COSINES

In this section we consider techniques for evaluating the integrals of certain combinations of sines and cosines.

$\int \sin^m x \cos^n x\,dx$, where m or n is a positive odd integer

Suppose that n is a positive odd integer. If $n = 1$, then we have

$$\int \sin^m x \cos x\,dx. \tag{1}$$

This integral can be evaluated by letting $u = \sin x$, so $du = \cos x\,dx$, and

$$\int \sin^m x \cos x\,dx = \int u^m\,du = \frac{u^{m+1}}{m+1} + c = \frac{\sin^{m+1} x}{m+1} + c, \qquad m \neq -1. \tag{2}$$

Note that if $m = -1$, then the integrand is $\cot x$, and the integral must be evaluated in a different way.

For $n = 3$ we first rewrite the integrand as

$$\int \sin^m x \cos^3 x\,dx = \int \sin^m x \cos^2 x \cos x\,dx = \int \sin^m x(1 - \sin^2 x)\cos x\,dx.$$

Then let $u = \sin x$, so $du = \cos x\,dx$ and

$$\begin{aligned}\int \sin^m x \cos^3 x\,dx &= \int u^m(1 - u^2)\,du \\ &= \frac{u^{m+1}}{m+1} - \frac{u^{m+3}}{m+3} + c \\ &= \frac{\sin^{m+1} x}{m+1} - \frac{\sin^{m+3} x}{m+3} + c, \qquad m \neq -1, -3.\end{aligned} \tag{3}$$

Again, if $m = -1$ or $m = -3$, the integral must be handled differently.

In general, when n is an odd positive integer we write $\cos^n x = \cos^{n-1} x \cos x$ where $n - 1$ is an even integer or zero. We can then let $n - 1 = 2k$, where k is a positive integer or zero. Hence

$$\begin{aligned}
\int \sin^m x \cos^n x \, dx &= \int \sin^m x \cos^{n-1} x \cos x \, dx \\
&= \int \sin^m x \cos^{2k} x \cos x \, dx \\
&= \int \sin^m x (\cos^2 x)^k \cos x \, dx \\
&= \int \sin^m x (1 - \sin^2 x)^k \cos x \, dx.
\end{aligned}$$

Next, let $u = \sin x$ so $du = \cos x \, dx$. Then

$$\int \sin^m x \cos^n x \, dx = \int u^m (1 - u^2)^k \, du. \tag{4}$$

The integral on the right side of Eq. 4 can readily be evaluated for specific values of m and $k = (n - 1)/2$ by using the binomial theorem to expand $(1 - u^2)^k$. Observe that m need not be positive or an integer.

If m is an odd positive integer and n is not, then the procedure is the same as we have just described with the role of the sin and cos interchanged. This is illustrated in Example 1. If both m and n are odd positive integers, then we can proceed in either way.

EXAMPLE 1

Evaluate

$$\int \sin^3 x \, dx.$$

For this integral $m = 3$ and $n = 0$. First we write

$$\begin{aligned}
\int \sin^3 x \, dx &= \int \sin^2 x \sin x \, dx \\
&= \int (1 - \cos^2 x) \sin x \, dx.
\end{aligned}$$

Now let $u = \cos x$ so $du = -\sin x \, dx$. Then

$$\begin{aligned}
\int \sin^3 x \, dx &= \int (1 - u^2)(-du) \\
&= -\left(u - \frac{u^3}{3}\right) + c \\
&= -\cos x + \frac{1}{3} \cos^3 x + c. \; \blacksquare
\end{aligned} \tag{5}$$

EXAMPLE 2

Evaluate

$$\int \sin^{1/2} x \cos^3 x \, dx.$$

For this integral $m = \frac{1}{2}$ and $n = 3$. We have

$$\int \sin^{1/2} x \cos^3 x \, dx = \int \sin^{1/2} x \cos^2 x \cos x \, dx$$
$$= \int \sin^{1/2} x(1 - \sin^2 x)\cos x \, dx.$$

Now let $u = \sin x$ so $du = \cos x \, dx$. Then

$$\int \sin^{1/2} x \cos^3 x \, dx = \int u^{1/2}(1 - u^2) \, du$$
$$= \int (u^{1/2} - u^{5/2}) \, du$$
$$= \frac{u^{3/2}}{3/2} - \frac{u^{7/2}}{7/2} + c$$
$$= \frac{2}{3} \sin^{3/2} x - \frac{2}{7} \sin^{7/2} x + c. \quad \blacksquare \tag{6}$$

$\int \sin^m x \cos^n x \, dx$, where m and n are both positive even integers or zero

In this case the integrand can be simplified by using the half-angle trigonometric identities

$$\cos^2 x = \frac{1 + \cos 2x}{2}, \qquad \sin^2 x = \frac{1 - \cos 2x}{2}. \tag{7}$$

For example,

$$\int \cos^2 x \, dx = \int \frac{1 + \cos 2x}{2} \, dx = \frac{x}{2} + \frac{1}{4} \sin 2x + c, \tag{8}$$

$$\int \sin^2 x \, dx = \int \frac{1 - \cos 2x}{2} \, dx = \frac{x}{2} - \frac{1}{4} \sin 2x + c. \tag{9}$$

It may be necessary to use the identities (7) several times. The procedure is illustrated in the following two examples.

EXAMPLE 3

Evaluate

$$\int \cos^4 x \, dx.$$

We have

$$\int \cos^4 x\,dx = \int (\cos^2 x)^2\,dx$$
$$= \int \left(\frac{1 + \cos 2x}{2}\right)^2 dx$$
$$= \frac{1}{4}\int (1 + 2\cos 2x + \cos^2 2x)\,dx.$$

We know how to integrate the first two terms. For the $\cos^2 2x$ term we use the half-angle formula again to obtain

$$\int \cos^4 x\,dx = \frac{1}{4}\int \left(1 + 2\cos 2x + \frac{1 + \cos 4x}{2}\right) dx$$
$$= \frac{1}{4}\int \left(\frac{3}{2} + 2\cos 2x + \frac{1}{2}\cos 4x\right) dx$$
$$= \frac{1}{4}\left[\frac{3}{2}x + \sin 2x + \frac{1}{8}\sin 4x\right] + c$$
$$= \frac{3}{8}x + \frac{1}{4}\sin 2x + \frac{1}{32}\sin 4x + c. \quad \blacksquare \tag{10}$$

EXAMPLE 4

Evaluate

$$\int \sin^4 x \cos^2 x\,dx.$$

We have

$$\int \sin^4 x \cos^2 x\,dx = \int (\sin^2 x)^2 \cos^2 x\,dx$$
$$= \int \left(\frac{1 - \cos 2x}{2}\right)^2 \left(\frac{1 + \cos 2x}{2}\right) dx$$
$$= \frac{1}{8}\int (1 - 2\cos 2x + \cos^2 2x)(1 + \cos 2x)\,dx$$
$$= \frac{1}{8}\int (1 - \cos 2x - \cos^2 2x + \cos^3 2x)\,dx.$$

For the term $\cos^2 2x$ we again use the half-angle formula; for the term $\cos^3 2x$ we proceed as in Example 2, since we have an odd power of the cosine. Thus we have

$$\int \sin^4 x \cos^2 x\,dx = \frac{1}{8}\int \left[1 - \cos 2x - \frac{1 + \cos 4x}{2}\right] dx$$
$$+ \frac{1}{8}\int (\cos^2 2x)\cos 2x\,dx$$

$$= \frac{1}{8}\int \left(\frac{1}{2} - \cos 2x - \frac{1}{2}\cos 4x\right) dx$$

$$+ \frac{1}{8}\int (1 - \sin^2 2x)\cos 2x\, dx.$$

In the last integral we make the substitution $u = \sin 2x$, $du = 2\cos 2x\, dx$. Thus we obtain

$$\int \sin^4 x \cos^2 x\, dx = \frac{1}{8}\left(\frac{x}{2} - \frac{\sin 2x}{2} - \frac{\sin 4x}{8}\right) + \frac{1}{8}\int (1 - u^2)\frac{du}{2}$$

$$= \frac{1}{16}x - \frac{1}{16}\sin 2x - \frac{1}{64}\sin 4x + \frac{1}{16}\left(u - \frac{u^3}{3}\right) + c$$

$$= \frac{1}{16}x - \frac{1}{16}\sin 2x - \frac{1}{64}\sin 4x$$

$$+ \frac{1}{16}\sin 2x - \frac{1}{48}\sin^3 2x + c$$

$$= \frac{1}{16}\left(x - \frac{1}{4}\sin 4x - \frac{1}{3}\sin^3 2x\right) + c. \blacksquare \qquad (11)$$

Table 9.2 summarizes the results obtained so far in this section.

Table 9.2 **Evaluation of $\int \sin^m x \cos^n x\, dx$**

Integrand	Strategy	Useful Identity
m is positive odd integer; n is arbitrary	Let $u = \cos x$, $du = -\sin x\, dx$	$\sin^2 x = 1 - \cos^2 x$
n is positive odd integer; m is arbitrary	Let $u = \sin x$, $du = \cos x\, dx$	$\cos^2 x = 1 - \sin^2 x$
m and n are both nonnegative even integers	Reduce exponents	$\sin^2 x = \dfrac{1 - \cos 2x}{2}$, $\cos^2 x = \dfrac{1 + \cos 2x}{2}$

Integrals of products of sines and/or cosines of different angles

An integral of one of the forms

$$\int \sin ax \cos bx\, dx, \qquad \int \sin ax \sin bx\, dx, \qquad \int \cos ax \cos bx\, dx$$

can be evaluated easily by using one of the following trigonometric identities:

$$\sin ax \cos bx = \tfrac{1}{2}[\sin(a - b)x + \sin(a + b)x], \tag{12a}$$

$$\sin ax \sin bx = \tfrac{1}{2}[\cos(a - b)x - \cos(a + b)x], \tag{12b}$$

$$\cos ax \cos bx = \tfrac{1}{2}[\cos(a - b)x + \cos(a + b)x]. \tag{12c}$$

These identities can be derived easily by appropriate combinations of the sum and difference formulas for the cosine and sine:

$$\cos(a \pm b)x = \cos ax \cos bx \mp \sin ax \sin bx, \tag{13a}$$

$$\sin(a \pm b)x = \sin ax \cos bx \pm \cos ax \sin bx. \tag{13b}$$

For example, adding the formulas for $\cos(a \pm b)x$ gives the formula (12c).

EXAMPLE 5

Evaluate

$$\int \cos x \cos 2x\, dx \quad \text{and} \quad \int \sin 3x \cos 2x\, dx.$$

On using formula (12c) with $a = 1$ and $b = 2$, we have

$$\begin{aligned} \int \cos x \cos 2x\, dx &= \int \frac{1}{2}[\cos(-x) + \cos 3x]\, dx \\ &= \frac{1}{2}\int (\cos x + \cos 3x)\, dx \\ &= \frac{1}{2}\sin x + \frac{1}{6}\sin 3x + c. \end{aligned} \tag{14}$$

Similarly, on using formula (12a) with $a = 3$ and $b = 2$, we obtain

$$\begin{aligned} \int \sin 3x \cos 2x\, dx &= \int \frac{1}{2}[\sin x + \sin 5x]\, dx \\ &= -\frac{1}{2}\cos x - \frac{1}{10}\cos 5x + c. \; ■ \end{aligned} \tag{15}$$

PROBLEMS

In each of Problems 1 through 24, evaluate the given indefinite or definite integral.

1. $\int_0^{\pi/4} \cos^2 \theta\, d\theta$
2. $\int \cos^2 x \sin x\, dx$
3. $\int \sin^3 x \cos^2 x\, dx$
4. $\int \sin^2 \theta \cos^2 \theta\, d\theta$
5. $\int \cos^{1/3} x \sin^3 x\, dx$
6. $\int \cos^5 x\, dx$

7. $\int \cos^4 x \sin^2 x \, dx$

8. $\int \sin^6 x \, dx$

9. $\int_0^{\pi/2} \sqrt{\cos x} \sin^3 x \, dx$

10. $\int_0^{\pi} \sqrt{\sin x} \cos x \, dx$

11. $\int_{\pi/6}^{\pi/2} \frac{\cos x}{\sqrt{\sin x}} \, dx$

12. $\int \cos^{-4} x \sin^3 x \, dx$

13. $\int_0^{\pi} \sin^4 x \, dx$

14. $\int \sin^2 3x \cos^5 3x \, dx$

15. $\int \sin^{-2} x \cos^3 x \, dx$

16. $\int_0^{\pi/2} \sin 4x \cos 2x \, dx$

17. $\int \sin 3x \sin x \, dx$

18. $\int_0^{1/2} \cos \pi x \cos \frac{\pi}{2} x \, dx$

19. $\int (\cos 4x - \cos 2x)^2 \, dx$

20. $\int \sin \frac{1}{2}x \sin 2x \, dx$

21. $\int \cos 3x \sin 2x \, dx$

22. $\int \cos 3x \cos 5x \, dx$

23. $\int \sin 2x \sin 4x \sin 6x \, dx$

24. $\int \sin x \cos 3x \sin 5x \, dx$

25. Evaluate $\int x^3(1 - x^2)^n \, dx$, where n is an arbitrary positive integer, by using the substitution $x = \sin u$. Compare your result with that of Problem 38 in Section 9.2.

26. Find the volume of the solid formed by rotating the area between the curve $y = \sin x$ and the x-axis for $0 \le x \le \pi$ about the x-axis.

27. Find the volume of the solid formed by rotating the area between the curves $y = \cos x$ and $y = \sin x$ for $0 \le x \le \pi/4$ about the x-axis.

Some Properties of a Cycloid

The parametric equations of a cycloid are (see Example 6 of Section 5.5)

$$x = a(\theta - \sin \theta), \qquad y = a(1 - \cos \theta).$$

28. Find the area of the region between the x-axis and one arch of a cycloid.

29. Find the volume of the solid formed by rotating the region in Problem 28 about the x-axis.

30. Find the arc length of one arch of a cycloid.

31. Find the surface area of the solid described in Problem 29.

32. Derive the following integration formulas.

(a) $$\int \sin ax \cos bx \, dx = \frac{\cos(b - a)x}{2(b - a)} - \frac{\cos(b + a)x}{2(b + a)}, \qquad b^2 \neq a^2.$$

(b) $$\int \sin ax \sin bx \, dx = \frac{\sin(a - b)x}{2(a - b)} - \frac{\sin(a + b)x}{2(a + b)}, \qquad b^2 \neq a^2.$$

(c) $$\int \cos ax \cos bx \, dx = \frac{\sin(a - b)x}{2(a - b)} + \frac{\sin(a + b)x}{2(a + b)}, \qquad b^2 \neq a^2.$$

33. Derive the following integration formulas, where m and n are any integers except zero.

(a) $$\int_{-\pi}^{\pi} \cos nx \cos mx \, dx = \begin{cases} 0, & m \neq n \\ \pi, & m = n. \end{cases}$$

(b) $$\int_{-\pi}^{\pi} \cos nx \sin mx \, dx = 0.$$

(c) $$\int_{-\pi}^{\pi} \sin nx \sin mx \, dx = \begin{cases} 0, & m \neq n \\ \pi, & m = n. \end{cases}$$

These relationships are known as the orthogonality properties of the sines and cosines. They are important in the subject known as Fourier analysis where one studies the possibility of expressing an arbitrary periodic function as a sum of sines and cosines.

9.4 INTEGRALS INVOLVING OTHER TRIGONOMETRIC FUNCTIONS

Techniques similar to those used in Section 9.3 can be used to evaluate integrals of other trigonometric functions. In particular, for integrands involving tan x, cot x, sec x, and csc x it is often convenient to use the identities

$$\sec^2 x - \tan^2 x = 1 \quad \text{and} \quad \csc^2 x - \cot^2 x = 1 \tag{1}$$

and to recall the differentiation formulas

$$\begin{aligned} (\tan x)' &= \sec^2 x, & (\sec x)' &= \sec x \tan x, \\ (\cot x)' &= -\csc^2 x, & (\csc x)' &= -\csc x \cot x. \end{aligned} \tag{2}$$

$\int \tan^n x\, dx$ and $\int \cot^n x\, dx$, n a positive integer

We will discuss only $\int \tan^n x\, dx$; evaluation of $\int \cot^n x\, dx$ is similar. For $n = 1$ we know (Example 1 of Section 9.1) that $\int \tan x\, dx = -\ln|\cos x| + c$. Thus we assume that $n \geq 2$. Then

$$\begin{aligned} \int \tan^n x\, dx &= \int \tan^{n-2} x \tan^2 x\, dx \\ &= \int \tan^{n-2} x(\sec^2 x - 1)\, dx \\ &= \int \tan^{n-2} x \sec^2 x\, dx - \int \tan^{n-2} x\, dx. \end{aligned}$$

In the first integral we let $u = \tan x$, $du = \sec^2 x\, dx$; thus we obtain

$$\begin{aligned} \int \tan^n x\, dx &= \int u^{n-2}\, du - \int \tan^{n-2} x\, dx \\ &= \frac{u^{n-1}}{n-1} - \int \tan^{n-2} x\, dx \end{aligned}$$

or

$$\int \tan^n x\, dx = \frac{\tan^{n-1} x}{n-1} - \int \tan^{n-2} x\, dx, \qquad n \geq 2. \tag{3}$$

By repeated application of the reduction formula (3) one eventually obtains the integral $\int \tan x\, dx$ if n is odd or the integral $\int \tan^0 x\, dx = \int dx$ if n is even. Thus the integral $\int \tan^n x\, dx$ can be evaluated for any positive integer n.

The corresponding reduction formula for $\cot^n x$, which is left as an exercise, is

$$\int \cot^n x\, dx = -\frac{\cot^{n-1} x}{n-1} - \int \cot^{n-2} x\, dx, \qquad n \geq 2. \tag{4}$$

Rather than memorizing the reduction formulas (3) and (4), you are encouraged to remember the *method* of integrating integral powers of tan x and cot x.

EXAMPLE 1
Evaluate

$$\int \tan^3 x \, dx.$$

We have

$$\int \tan^3 x \, dx = \int \tan x(\sec^2 x - 1) \, dx$$
$$= \int \tan x \sec^2 x \, dx - \int \tan x \, dx.$$

In the first integral we let $u = \tan x$ so $du = \sec^2 x \, dx$; we already know the answer for the second integral. Thus we obtain

$$\int \tan^3 x \, dx = \int u \, du + \ln|\cos x|$$
$$= \frac{1}{2} u^2 + \ln|\cos x| + c$$
$$= \frac{1}{2} \tan^2 x + \ln|\cos x| + c. \quad \blacksquare \tag{5}$$

$\int \sec^n x \, dx$ and $\int \csc^n x \, dx$, n a positive integer

We discuss only $\int \sec^n x \, dx$; evaluation of $\int \csc^n x \, dx$ is similar. First, we must take care of the case $n = 1$, that is, we must find $\int \sec x \, dx$. There appears to be no obvious method* for evaluating this integral; the various ways of deriving the result all depend on some mathematical trick that at first may well seem obscure. One way to proceed is to multiply and divide the integrand by $\sec x + \tan x$. This gives

$$\int \sec x \, dx = \int \sec x \frac{\sec x + \tan x}{\sec x + \tan x} \, dx$$
$$= \int \frac{\sec x \tan x + \sec^2 x}{\sec x + \tan x} \, dx.$$

Next, we let $u = \sec x + \tan x$, so that $du = (\sec x \tan x + \sec^2 x) \, dx$. Then

$$\int \sec x \, dx = \int \frac{du}{u} = \ln|u| + c$$
$$= \ln|\sec x + \tan x| + c. \tag{6}$$

For $n > 1$, we consider the cases n even and n odd separately. If n is even,

*An account of the history of $\int \sec x \, dx$ and its relation to navigation and to mapmaking has been given by V. Frederick Rickey and Philip Tuchinsky, "An Application of Geography to Mathematics: History of the Integral of the Secant," *Mathematics Magazine,* **53,** 1980, pp. 162–166.

let $n = 2k$ and write $\sec^n x\, dx = \sec^{2k-2} x \sec^2 x\, dx$; this gives

$$\begin{aligned}\int \sec^n x\, dx &= \int \sec^{2k-2} x \sec^2 x\, dx \\ &= \int (1 + \tan^2 x)^{(2k-2)/2} \sec^2 x\, dx \\ &= \int (1 + \tan^2 x)^{k-1} \sec^2 x\, dx.\end{aligned}$$

Note that since n is even and $n > 1$, it follows that $k - 1$ is zero or a positive integer. On letting $u = \tan x$ and $du = \sec^2 x\, dx$, we obtain

$$\int \sec^n x\, dx = \int (1 + u^2)^{k-1}\, du, \tag{7}$$

which can be readily integrated for a given value of k by using the binomial theorem.

For n an odd integer greater than one, we use integration by parts to obtain a reduction formula:

$$\begin{aligned}\int \sec^n x\, dx &= \int \underbrace{\sec^{n-2} x}_{u}\, \underbrace{\sec^2 x\, dx}_{dv} \\ &= \sec^{n-2} x \tan x - \int (n - 2)\sec^{n-3} x \sec x \tan x \tan x\, dx \\ &= \sec^{n-2} x \tan x - (n - 2) \int \sec^{n-2} x \tan^2 x\, dx.\end{aligned}$$

In the last integral we write $\tan^2 x$ in terms of $\sec^2 x$, and obtain

$$\begin{aligned}\int \sec^n x\, dx &= \sec^{n-2} x \tan x - (n - 2) \int \sec^{n-2} x(\sec^2 x - 1)\, dx \\ &= \sec^{n-2} x \tan x - (n - 2) \int \sec^n x\, dx + (n - 2) \int \sec^{n-2} x\, dx.\end{aligned}$$

On solving this equation for $\int \sec^n x\, dx$, we obtain the formula

$$\int \sec^n x\, dx = \frac{\sec^{n-2} x \tan x}{n - 1} + \frac{n - 2}{n - 1} \int \sec^{n-2} x\, dx, \qquad n > 1. \tag{8}$$

The reduction formula (8) is valid for any integer $n > 1$; however, if n is even, it is probably easier to use formula (7).

The corresponding results for $\int \csc^n x\, dx$ are

$$\int \csc x\, dx = \ln|\csc x - \cot x| + c, \tag{9}$$

and

$$\int \csc^n x\, dx = -\frac{\csc^{n-2} x \cot x}{n - 1} + \frac{n - 2}{n - 1} \int \csc^{n-2} x\, dx. \tag{10}$$

Again, it is better to study the method than to memorize formulas (8) and (10).

EXAMPLE 2

Evaluate

$$\int \sec^4 x\, dx.$$

We illustrate the use of Eq. 7 with $n = 4$. The result is

$$\begin{aligned}\int \sec^4 x\, dx &= \int (1 + u^2)\, du, \qquad u = \tan x \\ &= u + \frac{1}{3}u^3 + c \\ &= \tan x + \frac{1}{3}\tan^3 x + c. \blacksquare \qquad (11)\end{aligned}$$

$\int \tan^m x \sec^n x\, dx$ and $\int \cot^m x \csc^n x\, dx$

Again we consider only the first integral; the situation is similar for the second integral. We must distinguish several cases depending on the values of m and n. We state the procedure to be used and illustrate it with an example, but we will not derive the general formulas.

n is an even integer, m is arbitrary and not necessarily an integer. In this case we associate a $\sec^2 x$ with dx, express the remainder of the integrand in terms of $\tan x$ and then let $u = \tan x$. Thus

$$\begin{aligned}\int \tan^m x \sec^n x\, dx &= \int \tan^m x \sec^{n-2} x \sec^2 x\, dx \\ &= \int \tan^m x (1 + \tan^2 x)^{(n-2)/2} \sec^2 x\, dx \\ &= \int u^m (1 + u^2)^{(n-2)/2}\, du, \qquad u = \tan x, \qquad (12)\end{aligned}$$

where $(n - 2)/2$ is zero or a positive integer.

EXAMPLE 3

Evaluate $\int \tan^2 x \sec^4 x\, dx$.

First we write

$$\begin{aligned}\int \tan^2 x \sec^4 x\, dx &= \int \tan^2 x \sec^2 x \sec^2 x\, dx \\ &= \int \tan^2 x (1 + \tan^2 x) \sec^2 x\, dx.\end{aligned}$$

Then we let $u = \tan x$, $du = \sec^2 x\, dx$ and obtain

$$\int \tan^2 x \sec^4 x\, dx = \int u^2(1 + u^2)\, du$$

$$= \frac{1}{3} u^3 + \frac{1}{5} u^5 + c$$

$$= \frac{1}{3} \tan^3 x + \frac{1}{5} \tan^5 x + c. \blacksquare \qquad (13)$$

m is an odd integer, n is arbitrary and not necessarily an integer. In this case we associate $\tan x \sec x$ with dx, express the remainder of the integrand in terms of $\sec x$, and let $u = \sec x$. Thus

$$\int \tan^m x \sec^n x\, dx = \int \tan^{m-1} x \sec^{n-1} x(\tan x \sec x)\, dx$$

$$= \int (\sec^2 x - 1)^{(m-1)/2} \sec^{n-1} x(\tan x \sec x)\, dx$$

$$= \int (u^2 - 1)^{(m-1)/2} u^{n-1}\, du, \qquad u = \sec x, \qquad (14)$$

where $m - 1$ and $n - 1$ are zero or positive integers.

EXAMPLE 4

Evaluate $\int \tan^5 x \sec^3 x\, dx$.

First we rewrite the integral as

$$\int \tan^5 x \sec^3 x\, dx = \int \tan^4 x \sec^2 x(\tan x \sec x)\, dx$$

$$= \int (\sec^2 x - 1)^2 \sec^2 x(\tan x \sec x)\, dx.$$

Then we let $u = \sec x$, $du = \tan x \sec x\, dx$ and find that

$$\int \tan^5 x \sec^3 x\, dx = \int (u^2 - 1)^2 u^2\, du$$

$$= \int (u^6 - 2u^4 + u^2)\, du$$

$$= \frac{1}{7} u^7 - \frac{2}{5} u^5 + \frac{1}{3} u^3 + c$$

$$= \frac{1}{7} \sec^7 x - \frac{2}{5} \sec^5 x + \frac{1}{3} \sec^3 x + c. \blacksquare \qquad (15)$$

n is odd and m is even. We express the integrand entirely in terms of sec x and use the reduction formula (8). We have

$$\int \tan^m x \sec^n x \, dx = \int (\sec^2 x - 1)^{m/2} \sec^n x \, dx. \tag{16}$$

Since $m/2$ is an integer, the integrand is a polynomial in sec x; the integral can be evaluated by methods discussed earlier.

EXAMPLE 5

Evaluate $\int \tan^2 x \sec^3 x \, dx$.

First we write

$$\int \tan^2 x \sec^3 x \, dx = \int (\sec^2 x - 1)\sec^3 x \, dx$$
$$= \int (\sec^5 x - \sec^3 x) \, dx.$$

Now we use the reduction formula (8) to obtain

$$\int \tan^2 x \sec^3 x \, dx = \frac{\sec^3 x \tan x}{4} + \frac{3}{4}\int \sec^3 x \, dx - \int \sec^3 x \, dx$$
$$= \frac{\sec^3 x \tan x}{4} - \frac{1}{4}\left[\frac{\sec x \tan x}{2} + \frac{1}{2}\int \sec x \, dx\right]$$
$$= \frac{\sec^3 x \tan x}{4} - \frac{\sec x \tan x}{8} - \frac{1}{8}\ln|\sec x + \tan x| + c. \quad \blacksquare \tag{17}$$

Some of the results in this section are summarized in Table 9.3.

Table 9.3 Evaluation of $\int \tan^m x \sec^n x \, dx$

Integrand	Strategy	Useful Identity
n is positive even integer; m is arbitrary	Put $\sec^2 x$ with dx; let $u = \tan x$, $du = \sec^2 x \, dx$	$\sec^2 x = 1 + \tan^2 x$
m is positive odd integer; n is arbitrary	Put $\sec x \tan x$ with dx; let $u = \sec x$, $du = \sec x \tan x \, dx$	$\tan^2 x = \sec^2 x - 1$
m is nonnegative even integer; n is positive odd integer	Express integrand entirely in terms of sec x; use reduction formula (8)	$\tan^2 x = \sec^2 x - 1$

PROBLEMS

In each of Problems 1 through 18, evaluate the given integral.

1. $\int \sec ax\, dx$
2. $\int \tan^2 (3x - 2)\, dx$
3. $\int \tan^2 x \sec^2 x\, dx$
4. $\int \sec^3 2x\, dx$
5. $\int \tan^4 x\, dx$
6. $\int \tan^4 x \sec x\, dx$
7. $\int \tan^3 2x \sec^3 2x\, dx$
8. $\int \tan^3 x \sec^4 x\, dx$
9. $\int \tan x \sec^5 x\, dx$
10. $\int \tan^3 \pi x \sec^2 \pi x\, dx$
11. $\int \cot^5 x \csc^3 x\, dx$
12. $\int \cot^2 x \csc^4 x\, dx$
13. $\int \cot^2 x \csc x\, dx$
14. $\int \cot^4 x \csc^4 x\, dx$
15. $\int \tan^{1/2} x \sec^4 x\, dx$
16. $\int \tan^{-2} x \sec^6 x\, dx$
17. $\int \tan^3 x \sec^{1/2} x\, dx$
18. $\int \tan^5 x \sec^{-3} x\, dx$

19. Starting from the relation $1 - \sin^2 x = \cos^2 x$, show that

$$\frac{1 - \sin x}{\cos x} = \frac{\cos x}{1 + \sin x},$$

and hence that

$$\sec x = \tan x + \frac{\cos x}{1 + \sin x}.$$

Use the latter result to evaluate $\int \sec x\, dx$.

20. (a) Show that

$$\int \csc x\, dx = \ln|\csc x - \cot x| + c$$

in a manner similar to the derivation of Eq. 6 in the text.

(b) Evaluate $\int \csc x\, dx$ in a way similar to that in Problem 19.

21. Derive the reduction formula

$$\int \cot^n x\, dx = -\frac{\cot^{n-1} x}{n - 1} - \int \cot^{n-2} x\, dx, \qquad n \geq 2.$$

In each of Problems 22 through 25, use the reduction formula of Problem 21 to evaluate the given integral.

22. $\int \cot^2 x\, dx$
23. $\int \cot^3 x\, dx$
24. $\int \cot^4 x\, dx$
25. $\int \cot^5 x\, dx$

26. Show that

$$\int \csc^n x\, dx = -\frac{\csc^{n-2} x \cot x}{n - 1} + \frac{n - 2}{n - 1}\int \csc^{n-2} x\, dx + c.$$

In each of Problems 27 through 29, use the formulas of Problems 20 and 26 to evaluate the given integral.

27. $\int \csc^3 x\, dx$
28. $\int \csc^4 x\, dx$
29. $\int \csc^5 x\, dx$

In each of Problems 30 and 31, use the half-angle formula to evaluate the given integral.

30. $\int \frac{d\theta}{1 + \cos\theta}$
31. $\int \frac{d\theta}{1 - \cos\theta}$

9.5 TRIGONOMETRIC SUBSTITUTIONS

If the integrand contains $\sqrt{a^2 - x^2}$, $\sqrt{a^2 + x^2}$, or $\sqrt{x^2 - a^2}$, where $a > 0$, a trigonometric substitution is often useful. We consider these three cases in turn.

Integrals involving $\sqrt{a^2 - x^2}$

Examples of such integrals are

$$\int x^3\sqrt{a^2 - x^2}\, dx \quad \text{and} \quad \int \frac{x^2\, dx}{(a^2 - x^2)^{3/2}}. \tag{1}$$

Implicit in writing $\sqrt{a^2 - x^2}$ is that $-a \le x \le a$ or in some cases $-a < x < a$, as in the second integral in Eq. 1. Of course, x may be further restricted to a smaller interval, but certainly $|x|$ cannot exceed a, for then $\sqrt{a^2 - x^2}$ would be imaginary.

The main obstacle to evaluating integrals such as those in Eq. 1 is the presence of the radical $\sqrt{a^2 - x^2}$ in the integrand. Accordingly, we seek a substitution that simplifies the radical, and therefore makes such an integral easier to evaluate. A possible substitution is suggested by the relation $\sin^2\theta + \cos^2\theta = 1$, from which it follows that $a^2\cos^2\theta = a^2 - a^2\sin^2\theta$. Thus let us try the substitution

$$x = a\sin\theta, \qquad -\frac{\pi}{2} \le \theta \le \frac{\pi}{2} \tag{2}$$

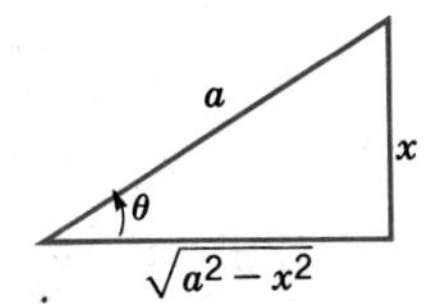

Figure 9.5.1 The substitution $x = a\sin\theta$.

(see Figure 9.5.1). Then

$$dx = a\cos\theta\, d\theta \tag{3}$$

and

$$\begin{aligned}\sqrt{a^2 - x^2} &= \sqrt{a^2 - a^2\sin^2\theta} = \sqrt{a^2\cos^2\theta} \\ &= a|\cos\theta| = a\cos\theta,\end{aligned} \tag{4}$$

where we can drop the absolute value bars on $\cos\theta$ since $\cos\theta \ge 0$ on $[-\pi/2, \pi/2]$. The following examples illustrate the effectiveness of this substitution.

EXAMPLE 1

Evaluate

$$\int \frac{x^2\, dx}{\sqrt{4 - x^2}}. \tag{5}$$

Let $x = 2\sin\theta$, so $dx = 2\cos\theta\, d\theta$ and $\sqrt{4 - x^2} = 2\cos\theta$. Hence

$$\begin{aligned}\int \frac{x^2\, dx}{\sqrt{4 - x^2}} &= \int \frac{(2\sin\theta)^2 2\cos\theta\, d\theta}{2\cos\theta} \\ &= 4\int \sin^2\theta\, d\theta \\ &= 4\int \left(\frac{1}{2} - \frac{1}{2}\cos 2\theta\right) d\theta \\ &= 4\left(\frac{1}{2}\theta - \frac{1}{4}\sin 2\theta\right) + c \\ &= 2\theta - 2\sin\theta\cos\theta + c.\end{aligned}$$

Now observe that $\sin\theta = x/2$ and $\cos\theta = (\sqrt{4 - x^2})/2$. Therefore

$$\int \frac{x^2\, dx}{\sqrt{4 - x^2}} = 2\arcsin\frac{x}{2} - \frac{x\sqrt{4 - x^2}}{2} + c. \;\blacksquare \tag{6}$$

EXAMPLE 2

Evaluate

$$\int_0^{3/2} \sqrt{9 - x^2}\, dx. \tag{7}$$

In this case let $x = 3 \sin \theta$; then $dx = 3 \cos \theta \, d\theta$ and $\sqrt{9 - x^2} = 3 \cos \theta$. Further, $x = 0$ corresponds to $\theta = 0$ and $x = \frac{3}{2}$ to $\theta = \pi/6$. Hence

$$\begin{aligned}
\int_0^{3/2} \sqrt{9 - x^2}\, dx &= \int_0^{\pi/6} (3 \cos \theta)(3 \cos \theta \, d\theta) \\
&= 9 \int_0^{\pi/6} \cos^2 \theta \, d\theta \\
&= \frac{9}{2} \int_0^{\pi/6} (1 + \cos 2\theta)\, d\theta \\
&= \frac{9}{2} \left(\theta + \frac{\sin 2\theta}{2} \right) \Bigg|_0^{\pi/6} \\
&= \frac{3\pi}{4} + \frac{9\sqrt{3}}{8} \cong 4.3048. \quad ■
\end{aligned} \tag{8}$$

Integrals involving $\sqrt{a^2 + x^2}$

Examples of such integrals are

$$\int \sqrt{a^2 + x^2}\, dx \qquad \text{and} \qquad \int \frac{x^2}{(a^2 + x^2)^{3/2}}\, dx. \tag{9}$$

There is no restriction on the value of x. To simplify the radical in this case we let

$$x = a \tan \theta, \qquad -\frac{\pi}{2} < \theta < \frac{\pi}{2} \tag{10}$$

(see Figure 9.5.2). Then

$$dx = a \sec^2 \theta \, d\theta \tag{11}$$

and

$$\begin{aligned}
\sqrt{a^2 + x^2} &= \sqrt{a^2 + a^2 \tan^2 \theta} = \sqrt{a^2(1 + \tan^2 \theta)} \\
&= \sqrt{a^2 \sec^2 \theta} = a|\sec \theta| = a \sec \theta.
\end{aligned} \tag{12}$$

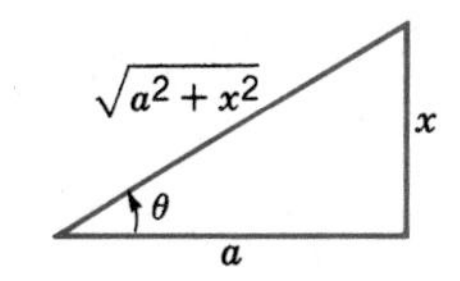

Figure 9.5.2 The substitution $x = a \tan \theta$.

Note that $a \sec \theta > 0$ for $-\pi/2 < \theta < \pi/2$. This is consistent with the fact that $\sqrt{a^2 + x^2} > 0$.

EXAMPLE 3

Evaluate

$$\int \sqrt{x^2 + 16}\, dx. \tag{13}$$

Let $x = 4 \tan \theta$, so $dx = 4 \sec^2 \theta\, d\theta$ and $\sqrt{x^2 + 16} = 4 \sec \theta$. Hence

$$\int \sqrt{x^2 + 16}\, dx = \int (4 \sec \theta) 4 \sec^2 \theta\, d\theta = 16 \int \sec^3 \theta\, d\theta.$$

To evaluate the last integral we use the reduction formula given by Eq. 8 of Section 9.4:

$$\begin{aligned}\int \sec^3 \theta\, d\theta &= \frac{\sec \theta \tan \theta}{2} + \frac{1}{2} \int \sec \theta\, d\theta \\ &= \frac{1}{2} \sec \theta \tan \theta + \frac{1}{2} \ln|\sec \theta + \tan \theta| + c.\end{aligned}$$

Hence

$$\int \sqrt{x^2 + 16}\, dx = 8 \sec \theta \tan \theta + 8 \ln|\sec \theta + \tan \theta| + c. \tag{14}$$

In order to express the result in terms of x, we observe that $\tan \theta = x/4$ and $\sec \theta = (\sqrt{x^2 + 16})/4$. Thus

$$\begin{aligned}\int \sqrt{x^2 + 16}\, dx &= 8 \frac{\sqrt{x^2 + 16}}{4} \frac{x}{4} + 8 \ln\left|\frac{\sqrt{x^2 + 16}}{4} + \frac{x}{4}\right| + c \\ &= \frac{1}{2} x\sqrt{x^2 + 16} + 8 \ln|\sqrt{x^2 + 16} + x| - 8 \ln 4 + c.\end{aligned}$$

Finally, note that the constant $-8 \ln 4$ can be included as part of the arbitrary constant of integration. Thus

$$\int \sqrt{x^2 + 16}\, dx = \frac{1}{2} x\sqrt{x^2 + 16} + 8 \ln|x + \sqrt{x^2 + 16}| + c. \blacksquare \tag{15}$$

Integrals involving $\sqrt{x^2 - a^2}$

Examples of such integrals are

$$\int \frac{dx}{\sqrt{x^2 - a^2}} \quad \text{and} \quad \int \frac{x^2}{(x^2 - a^2)^{3/2}}\, dx. \tag{16}$$

The situation here is slightly more complicated than in the two previous cases, because we must distinguish whether $x \geq a$ or $x \leq -a$. However, if $x \leq -a$ we can make the substitution $u = -x$ and obtain an integral in which $u \geq a$. Thus we need only consider the case $x \geq a$.

EXAMPLE 4

Transform the integral

$$\int_{-8}^{-6} \frac{x^3 + 2}{\sqrt{x^2 - 4}}\, dx$$

into one for which the limits of integration are positive.

If we let $u = -x$, then we have

$$\int_{-8}^{-6} \frac{x^3 + 2}{\sqrt{x^2 - 4}}\, dx = \int_{8}^{6} \frac{-u^3 + 2}{\sqrt{u^2 - 4}}(-du) = \int_{6}^{8} \frac{-u^3 + 2}{\sqrt{u^2 - 4}}\, du. \blacksquare$$

We now return to the general situation. If $x \geq a$, then we let

$$x = a \sec \theta, \qquad 0 \leq \theta < \frac{\pi}{2} \tag{17}$$

(see Figure 9.5.3). Then

$$dx = a \sec \theta \tan \theta \, d\theta \tag{18}$$

and

$$\sqrt{x^2 - a^2} = \sqrt{a^2 \sec^2 \theta - a^2} = a \tan \theta. \tag{19}$$

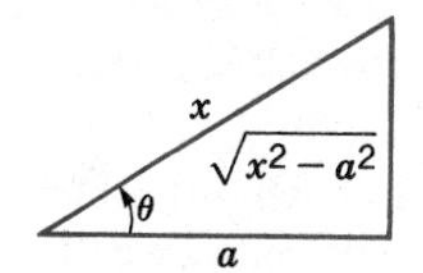

Figure 9.5.3 The substitution $x = a \sec \theta$.

EXAMPLE 5

Evaluate

$$\int \frac{dx}{\sqrt{x^2 - a^2}}. \tag{20}$$

If $x > a$, then we let $x = a \sec \theta$, $0 < \theta < \pi/2$. Thus $dx = a \sec \theta \tan \theta \, d\theta$ and $\sqrt{x^2 - a^2} = a \tan \theta$. Hence

$$\int \frac{dx}{\sqrt{x^2 - a^2}} = \int \frac{a \sec \theta \tan \theta \, d\theta}{a \tan \theta} = \int \sec \theta \, d\theta$$

$$= \ln|\sec \theta + \tan \theta| + c = \ln\left|\frac{x}{a} + \frac{\sqrt{x^2 - a^2}}{a}\right| + c$$

$$= \ln|x + \sqrt{x^2 - a^2}| - \ln a + c.$$

The constant $-\ln a$ can be included in the arbitrary constant of integration, so

$$\int \frac{dx}{\sqrt{x^2 - a^2}} = \ln|x + \sqrt{x^2 - a^2}| + c, \qquad x > a. \blacksquare \tag{21}$$

Table 9.4 summarizes the results in this section.

Table 9.4 Evaluation of integrals containing $\sqrt{a^2 - x^2}$, $\sqrt{a^2 + x^2}$, and $\sqrt{x^2 - a^2}$.

Integrand Contains	Substitution	Useful Identity
$\sqrt{a^2 - x^2}$	$x = a \sin\theta$ $dx = a\cos\theta\, d\theta$	$\sqrt{a^2 - x^2} = a\cos\theta$
$\sqrt{a^2 + x^2}$	$x = a\tan\theta$ $dx = a\sec^2\theta\, d\theta$	$\sqrt{a^2 + x^2} = a\sec\theta$
$\sqrt{x^2 - a^2}$, $x \geq a > 0$	$x = a\sec\theta$ $dx = a\sec\theta\tan\theta\, d\theta$ If $x \leq -a < 0$, first let $x = -u$, then $u = a\sec\theta$	$\sqrt{x^2 - a^2} = a\tan\theta$

PROBLEMS

In each of Problems 1 through 20, evaluate the given indefinite or definite integral.

1. $\int \sqrt{1 - x^2}\, dx$

2. $\int \frac{2x^3\, dx}{\sqrt{1 - x^2}}$

3. $\int \frac{dx}{(2x^2 + 5)^{3/2}}$

4. $\int_0^2 \frac{3x^2}{(9 - x^2)^{3/2}}\, dx$

5. $\int \frac{dx}{(x^2 - 4)^{3/2}}$, $x < -2$

6. $\int \frac{x^2}{\sqrt{9 + x^2}}\, dx$

7. $\int \frac{\sqrt{4 - x^2}}{x}\, dx$

8. $\int \sqrt{x^2 - 4}\, dx$, $x < -2$

9. $\int \frac{3x - 1}{\sqrt{x^2 - 9}}\, dx$, $x > 3$

10. $\int \frac{\sqrt{1 - 4x^2}}{x^2}\, dx$

11. $\int \frac{dx}{x\sqrt{3 + x^2}}$

12. $\int \frac{x^2 + 2x + 5}{\sqrt{4 - x^2}}\, dx$

13. $\int_2^4 \frac{x^2}{\sqrt{x^2 - 1}}\, dx$

14. $\int \frac{x\, dx}{\sqrt{2x - x^2}}$

15. $\int_{-5}^{-3} \frac{dx}{(x^2 - 1)^{3/2}}$

16. $\int \frac{x}{\sqrt{x^2 + 2x + 5}}\, dx$

17. $\int \frac{\sqrt{x^2 - 3}}{x}\, dx$, $x > \sqrt{3}$

18. $\int \frac{dx}{(1 - 4x^2)^{3/2}}$

19. $\int_{-2}^{2} \frac{dx}{\sqrt{4x^2 + 9}}$

20. $\int \frac{dx}{x^2\sqrt{16 - x^2}}$

In each of Problems 21 through 32, derive the given integration formula.

21. $\int \sqrt{a^2 - x^2}\, dx = \frac{x\sqrt{a^2 - x^2}}{2} + \frac{a^2}{2}\arcsin\frac{x}{a} + c$

22. $\int \frac{dx}{(a^2 - x^2)^{3/2}} = \frac{1}{a^2}\frac{x}{\sqrt{a^2 - x^2}} + c$

23. $\int \frac{x^2\, dx}{\sqrt{a^2 - x^2}} = \frac{-x\sqrt{a^2 - x^2}}{2} + \frac{a^2}{2}\arcsin\frac{x}{a} + c$

24. $\int \frac{dx}{x\sqrt{a^2 - x^2}} = -\frac{1}{a}\ln\left|\frac{a + \sqrt{a^2 - x^2}}{x}\right| + c$

25. $\int \sqrt{a^2 + x^2}\, dx = \frac{x\sqrt{a^2 + x^2}}{2} + \frac{a^2}{2}\ln(x + \sqrt{a^2 + x^2}) + c$

26. $\displaystyle\int \frac{dx}{\sqrt{a^2 + x^2}} = \ln|x + \sqrt{a^2 + x^2}| + c$

27. $\displaystyle\int \frac{dx}{(a^2 + x^2)^{3/2}} = \frac{1}{a^2}\frac{x}{\sqrt{a^2 + x^2}} + c$

28. $\displaystyle\int \frac{x^2\,dx}{\sqrt{a^2 + x^2}}$
$\displaystyle = \frac{x\sqrt{a^2 + x^2}}{2} - \frac{a^2}{2}\ln(x + \sqrt{a^2 + x^2}) + c$

29. $\displaystyle\int \frac{dx}{x\sqrt{a^2 + x^2}} = \frac{1}{a}\ln\left|\frac{\sqrt{a^2 + x^2} - a}{x}\right| + c$

30. $\displaystyle\int \frac{dx}{x\sqrt{x^2 - a^2}} = \frac{1}{a}\operatorname{arcsec}\frac{x}{a} + c, \quad x > a$

31. $\displaystyle\int \sqrt{x^2 - a^2}\,dx = \frac{x\sqrt{x^2 - a^2}}{2} - \frac{a^2}{2}\ln|x + \sqrt{x^2 - a^2}| + c, \quad x > a$

32. $\displaystyle\int \frac{dx}{(x^2 - a^2)^{3/2}} = -\frac{1}{a^2}\frac{x}{\sqrt{x^2 - a^2}} + c, \quad x > a$

33. Find the area of the region enclosed by the ellipse $(x^2/a^2) + (y^2/b^2) = 1$.

34. Find the area of the region between the two branches of the hyperbola $(x^2/a^2) - (y^2/b^2) = 1$ and between the two lines $y = \pm b$.

9.6 PARTIAL FRACTIONS

In this section we consider a method for evaluating the integral of a rational function—the quotient of two polynomials. The method makes use of an algebraic procedure known as a partial fractions expansion to rewrite the rational function in a form more amenable to integration. For example, if we are given the function

$$f(x) = \frac{3}{x} + \frac{1}{x - 1} - \frac{2}{x + 2}, \tag{1}$$

it is a simple matter to calculate its indefinite integral

$$\int f(x)\,dx = 3\ln|x| + \ln|x - 1| - 2\ln|x + 2| + c.$$

On the other hand, if we were to put the fractions in Eq. 1 over a common denominator we would have

$$f(x) = \frac{3(x - 1)(x + 2) + x(x + 2) - 2x(x - 1)}{x(x - 1)(x + 2)} = \frac{2x^2 + 7x - 6}{x^3 + x^2 - 2x}. \tag{2}$$

When $f(x)$ has the form (2), its integral is not readily evident. Given a quotient of two polynomials such as that in Eq. 2, we would like to reverse the preceding procedure and rewrite the quotient in a form similar to Eq. 1.

Suppose that

$$f(x) = \frac{P(x)}{Q(x)} \tag{3}$$

is a rational function, where $P(x)$ and $Q(x)$ are polynomials with real coefficients and with no common factors. It is sufficient to consider so-called *proper* rational functions, in which the degree of the denominator $Q(x)$ is greater than the degree of the numerator $P(x)$. If this is not the case, then we divide $Q(x)$ into $P(x)$ so as to obtain a quotient (which is a polynomial and hence easy to integrate) plus a remainder that is a proper rational function. For instance,

$$\frac{6x^4 + 3x^3 - 2x + 1}{x^3 - 3x + 2} = 6x + 3 + \frac{18x^2 - 5x - 5}{x^3 - 3x + 2}.$$

Then we apply the procedure to the remainder.

Also, we assume that the coefficient of the highest power term in $Q(x)$ is one. If this is not the case, then this coefficient can simply be factored out of the entire expression. For instance,

$$\frac{P(x)}{Q(x)} = \frac{2x - 5}{3x^2 + (9/2)x - 3} = \frac{1}{3} \cdot \frac{2x - 5}{x^2 + (3/2)x - 1}.$$

The first step in finding the desired expansion is to factor $Q(x)$. It is known from algebra that *every polynomial with real coefficients can be expressed in a unique way as a product of real linear and irreducible quadratic factors.* A linear factor is of the form $x - a$, where a is real, and corresponds to the real zero $x = a$ of Q. Such a factor may occur more than once; if it occurs r times, then $(x - a)^r$ is a factor of Q. An irreducible quadratic factor is of the form $x^2 + bx + c$, where b and c are real and the discriminant $b^2 - 4c$ is negative. The latter condition assures that the quadratic factor cannot be further decomposed into real linear factors; hence the name irreducible. Irreducible quadratic factors are associated with complex zeros of Q. If $\alpha \pm i\beta$ are a conjugate pair of zeros of Q, then

$$\begin{aligned}[x - (\alpha + i\beta)][x - (\alpha - i\beta)] &= x^2 - 2\alpha x + (\alpha^2 + \beta^2), \\ &= x^2 + bx + c,\end{aligned}$$

provided that $b = -2\alpha$ and $c = \alpha^2 + \beta^2$. In this case

$$b^2 - 4c = 4\alpha^2 - 4(\alpha^2 + \beta^2) = -4\beta^2 < 0.$$

Quadratic factors may also be repeated, corresponding to repeated complex zeros of Q.

Here are a few examples of factored polynomials:

$$\begin{aligned}x^3 + x^2 - 2x &= x(x - 1)(x + 2) \\ &\qquad \text{with zeros } 0, 1, -2; \\ x^3 + x^2 + x + 1 &= (x + 1)(x^2 + 1) \\ &\qquad \text{with zeros } -1, i, -i; \\ x^3 - 3x + 2 &= (x - 1)^2(x + 2) \\ &\qquad \text{with zeros } 1, 1, -2; \\ x^5 - 6x^4 + 22x^3 - 48x^2 + 65x - 50 &= (x - 2)(x^2 - 2x + 5)^2 \\ &\qquad \text{with zeros } 2, 1 \pm 2i, 1 \pm 2i.\end{aligned}$$

Once $Q(x)$ has been factored, it is possible to show that every proper rational function $P(x)/Q(x)$ can be expressed as a sum of fractions of the form

$$\frac{A}{(x-a)^r} \quad \text{and} \quad \frac{Bx+C}{(x^2+bx+c)^s}, \tag{4}$$

where $x - a$ and $x^2 + bx + c$ are real linear and irreducible quadratic factors of Q. This sum of fractions is called the **partial fractions decomposition** of $P(x)/Q(x)$. We will show how to integrate any expression of the form (4). It then follows that every rational function can be integrated by this procedure.

The exact form of the partial fractions representation of P/Q depends on the factorization of Q. There are two rules.

RULE 1. If a factor $x - a$ appears r times in the factorization of $Q(x)$, then assume that the partial fraction representation of $P(x)/Q(x)$ contains the terms

$$\frac{A_1}{(x-a)} + \frac{A_2}{(x-a)^2} + \cdots + \frac{A_r}{(x-a)^r}. \tag{5}$$

We will explain later how to determine the A's.

RULE 2. If an irreducible factor $x^2 + bx + c$ appears s times in the factorization of $Q(x)$, then assume that the partial fraction representation of $P(x)/Q(x)$ contains the terms

$$\frac{B_1x+C_1}{x^2+bx+c} + \frac{B_2x+C_2}{(x^2+bx+c)^2} + \cdots + \frac{B_sx+C_s}{(x^2+bx+c)^s}. \tag{6}$$

Again, we will explain later how to determine the B's and C's.

The following examples illustrate how these rules are to be applied in forming the partial fraction representation for a rational function.

$$\frac{3}{(x-1)(x+2)} = \frac{A}{x-1} + \frac{B}{x+2}; \quad A \text{ and } B \text{ to be determined}$$

$$\frac{2x+4}{(x-1)^2(x+2)} = \frac{A_1}{x-1} + \frac{A_2}{(x-1)^2} + \frac{B_1}{x+2}; \quad A_1, A_2, \text{ and } B_1 \text{ to be determined}$$

$$\frac{3-4x^2}{x(x-1)^2(x^2+1)} = \frac{A}{x} + \frac{B_1}{x-1} + \frac{B_2}{(x-1)^2} + \frac{Cx+D}{x^2+1}; \quad A, B_1, B_2, C, \text{ and } D \text{ to be determined}$$

$$\frac{2-3x+x^2-6x^3}{x(x-1)^3(x^2+1)^2} = \frac{A}{x} + \frac{B_1}{x-1} + \frac{B_2}{(x-1)^2} + \frac{B_3}{(x-1)^3} + \frac{C_1x+D_1}{x^2+1} + \frac{C_2x+D_2}{(x^2+1)^2};$$

$A, B_1, B_2, B_3, C_1, D_1, C_2,$ and D_2 to be determined.

The method of determining the coefficients (the A's, B's, C's, and D's in the previous expressions) is illustrated in the following examples.

EXAMPLE 1

Evaluate

$$\int \frac{2x - 1}{x^2 + x - 2}\, dx.$$

First we observe that $x^2 + x - 2 = (x + 2)(x - 1)$. Next, according to Rule 1, we have

$$\frac{2x - 1}{x^2 + x - 2} = \frac{2x - 1}{(x - 1)(x + 2)} = \frac{A}{x - 1} + \frac{B}{x + 2}. \tag{7}$$

To determine the constants A and B, we multiply both sides of this equation by $(x - 1)(x + 2)$, obtaining

$$2x - 1 = A(x + 2) + B(x - 1). \tag{8}$$

Since Eq. 8 is to be satisfied for all x, it follows that the coefficients of like powers of x must be the same on each side of the equation. That is, the coefficient of x on the left must be the same as the coefficient of x on the right, and the constant term on the left must be the same as the constant term on the right. Thus we conclude that

$$\begin{aligned} &\text{for } x^1: && 2 = A + B, \\ &\text{for } x^0: && -1 = 2A - B. \end{aligned}$$

Solving these equations for A and B we find $A = \frac{1}{3}$ and $B = \frac{5}{3}$. Hence

$$\frac{2x - 1}{x^2 + x - 2} = \frac{1/3}{x - 1} + \frac{5/3}{x + 2},$$

and

$$\begin{aligned} \int \frac{2x - 1}{x^2 + x - 2}\, dx &= \int \left(\frac{1/3}{x - 1} + \frac{5/3}{x + 2}\right) dx \\ &= \frac{1}{3} \ln|x - 1| + \frac{5}{3} \ln|x + 2| + c. \end{aligned} \tag{9}$$

An alternative way of determining A and B from Eq. 8 is to choose values of x that make the separate factors zero. Setting $x = -2$ in Eq. 8 gives $-5 = -3B$ and setting $x = 1$ gives $1 = 3A$. This can always be done when the factors of $Q(x)$ are linear and nonrepeated. ∎

EXAMPLE 2

Evaluate

$$\int \frac{3x^2 + 4x + 2}{x(x + 1)^3}\, dx.$$

According to Rule 1, it is possible to write

$$\frac{3x^2 + 4x + 2}{x(x + 1)^3} = \frac{A}{x} + \frac{B}{x + 1} + \frac{C}{(x + 1)^2} + \frac{D}{(x + 1)^3}. \tag{10}$$

To determine A, B, C, and D we multiply both sides of Eq. 10 by $x(x + 1)^3$. This gives

$$3x^2 + 4x + 2 = A(x + 1)^3 + Bx(x + 1)^2 + Cx(x + 1) + Dx \tag{11a}$$

$$= A(x^3 + 3x^2 + 3x + 1) + B(x^3 + 2x^2 + x) + C(x^2 + x) + Dx \tag{11b}$$

$$= x^3 (A + B) + x^2(3A + 2B + C) + x(3A + B + C + D) + A. \tag{11c}$$

Again, since Eq. 11(c) is to be satisfied for all x, it follows that coefficients of like powers of x on each side of this equation must agree. Hence

$$\begin{aligned} &\text{for } x^3: && A + B = 0, \\ &\text{for } x^2: && 3A + 2B + C = 3, \\ &\text{for } x^1: && 3A + B + C + D = 4, \\ &\text{for } x^0: && A = 2. \end{aligned} \tag{12}$$

Thus we have four linear algebraic equations for A, B, C, and D. From the last equation $A = 2$, then from the first equation $B = -2$, then from the second equation $C = 1$, and finally from the third equation $D = -1$. Therefore

$$\frac{3x^2 + 4x + 2}{x(x + 1)^3} = \frac{2}{x} - \frac{2}{x + 1} + \frac{1}{(x + 1)^2} - \frac{1}{(x + 1)^3},$$

and

$$\int \frac{3x^2 + 4x + 2}{x(x + 1)^3}\, dx = \int \left[\frac{2}{x} - \frac{2}{x + 1} + \frac{1}{(x + 1)^2} - \frac{1}{(x + 1)^3}\right] dx$$

$$= 2 \ln|x| - 2 \ln|x + 1| - \frac{1}{(x + 1)} + \frac{1}{2(x + 1)^2} + c. \tag{13}$$

In this case it was fairly easy to solve Eqs. 12 for A, B, C, and D; in other problems it may be more difficult. However, we can avoid the problem of solving for all four constants simultaneously. The constants A and D can be obtained directly from Eq. 11(a) by setting $x = 0$ and then $x = -1$. Two equations for B and C can be obtained by choosing two convenient values of x, not equal to 0 and -1, and evaluating Eq. 11(a) at these points. For this problem we know $A = 2$ and $D = -1$. If we choose $x = 1$ and $x = -2$, we obtain the following two equations for B and C.

$$\text{for } x = 1: \quad 9 = 2(2)^3 + B(2)^2 + C(2) - 1$$

$$\text{for } x = -2: \quad 6 = 2(-1)^3 + B(-2)(-1)^2 + C(-2)(-1) - 1(-2),$$

which reduce to

$$4B + 2C = -6,$$

$$-2B + 2C = 6.$$

Solving these two equations for B and C we find, as expected, $B = -2$ and $C = 1$. ∎

EXAMPLE 3

Evaluate

$$\int \frac{x^4 + 2x^3 + 9x + 6}{x(x^2 - 2x + 3)}\, dx.$$

First we observe that the quadratic factor $x^2 - 2x + 3$ in the denominator is irreducible, since its discriminant is negative; thus the denominator is in the simplest form. Since the degree of the numerator is larger than the degree of the denominator, we use long division to put the integrand into the proper form for using the technique of partial fractions:

$$\begin{array}{r|l} & x + 4 \\ \hline x^3 - 2x^2 + 3x & x^4 + 2x^3 + 0x^2 + 9x + 6 \\ & \underline{x^4 - 2x^3 + 3x^2} \\ & \quad 4x^3 - 3x^2 + 9x \\ & \quad \underline{4x^3 - 8x^2 + 12x} \\ & \qquad 5x^2 - 3x + 6 \end{array}$$

so

$$\frac{x^4 + 2x^3 + 9x + 6}{x(x^2 - 2x + 3)} = x + 4 + \frac{5x^2 - 3x + 6}{x(x^2 - 2x + 3)}. \tag{14}$$

It follows from Rules 1 and 2 that

$$\frac{5x^2 - 3x + 6}{x(x^2 - 2x + 3)} = \frac{A}{x} + \frac{Bx + C}{x^2 - 2x + 3}. \tag{15}$$

Hence

$$5x^2 - 3x + 6 = A(x^2 - 2x + 3) + (Bx + C)x$$
$$= (A + B)x^2 + (-2A + C)x + 3A,$$

so

$$\begin{aligned} \text{for } x^2: &\quad A + B = 5, \\ \text{for } x^1: &\quad -2A + C = -3, \\ \text{for } x^0: &\quad 3A = 6. \end{aligned} \tag{16}$$

On solving Eqs. 16 we find that $A = 2$, $B = 3$, and $C = 1$. As a consequence, it follows from Eqs. 14 and 15 that

$$\int \frac{x^4 + 2x^3 + 9x + 6}{x(x^2 - 2x + 3)}\,dx = \int \left(x + 4 + \frac{2}{x} + \frac{3x + 1}{x^2 - 2x + 3}\right) dx$$
$$= \frac{1}{2}x^2 + 4x + 2\ln|x| + \int \frac{3x + 1}{x^2 - 2x + 3}\,dx. \tag{17}$$

The last integral is evaluated by completing the square in the denominator as follows:

$$\int \frac{3x + 1}{x^2 - 2x + 3}\,dx = \int \frac{3x + 1}{x^2 - 2x + 1 + 2}\,dx = \int \frac{3x + 1}{(x - 1)^2 + 2}\,dx.$$

Then we let $u = x - 1$, $du = dx$; it follows that

$$\int \frac{3x + 1}{(x - 1)^2 + 2}\,dx = \int \frac{3(u + 1) + 1}{u^2 + 2}\,du$$
$$= 3\int \frac{u}{u^2 + 2}\,du + 4\int \frac{du}{u^2 + 2}$$
$$= \frac{3}{2}\ln|u^2 + 2| + \frac{4}{\sqrt{2}}\arctan\frac{u}{\sqrt{2}} + c$$
$$= \frac{3}{2}\ln[(x - 1)^2 + 2] + \frac{4}{\sqrt{2}}\arctan\frac{x - 1}{\sqrt{2}} + c. \tag{18}$$

We have removed the absolute values in the logarithm, since $(x - 1)^2 + 2$ is always positive.

From Eq. 17 the final result is

$$\int \frac{x^4 + 2x^3 + 9x + 6}{x(x^2 - 2x + 3)}\,dx = \frac{1}{2}x^2 + 4x + 2\ln|x|$$
$$+ \frac{3}{2}\ln(x^2 - 2x + 3) + \frac{4}{\sqrt{2}}\arctan\frac{x - 1}{\sqrt{2}} + c. \blacksquare \tag{19}$$

In Examples 1, 2, and 3 we carried out the integration of rational functions when their partial fraction representations consisted of terms of the form

$$\frac{A}{(x-a)^r} \quad \text{and} \quad \frac{Bx+C}{(x^2+bx+c)}.$$

To complete the general analysis we must consider the case when the partial fraction representation contains a term of the form

$$\frac{Bx+C}{(x^2+bx+c)^n}, \qquad n > 1.$$

If such a fraction occurs, we must compute its indefinite integral. We take up this final part of the analysis in the next section.

PROBLEMS

In each of Problems 1 through 16, evaluate the given indefinite integral.

1. $\int \frac{dx}{x(x+1)}$
2. $\int \frac{5x-13}{(x-2)(x-3)}\,dx$
3. $\int \frac{-4\,dx}{(x^2-4)}$
4. $\int \frac{5x+1}{(x-1)^2(x+2)}\,dx$
5. $\int \frac{x^2+4x+5}{(x+1)(x+2)(x+3)}\,dx$
6. $\int \frac{x^3+2x^2}{x^2-x-2}\,dx$
7. $\int \frac{x\,dx}{(x+1)^2(x-1)^2}$
8. $\int \frac{3x+2}{(x+2)(x^2+4)}\,dx$
9. $\int \frac{x^3+3x^2-x+3}{x(x^2+1)}\,dx$
10. $\int \frac{x^2+5x+6}{(x^2+4)(x^2+9)}\,dx$
11. $\int \frac{2x^2+4}{x(x^2+2x+2)}\,dx$
12. $\int \frac{3\,dx}{x^3-x}$
13. $\int \frac{dx}{x^4-1}$
14. $\int \frac{3x^3+2x^2+2x+6}{x^2+4}\,dx$
15. $\int \frac{dx}{(x-1)^2(x+2)^2}$
16. $\int \frac{(3x-4)\,dx}{(x^2+2x+5)(x^2+2)}$

In each of Problems 17 through 32, derive the given integration formula.

17. $\int \frac{dx}{(x+b)(x+d)} = \frac{1}{d-b}\ln\left|\frac{x+b}{x+d}\right| + c, \qquad d \neq b$
18. $\int \frac{dx}{(ax+b)(cx+d)} = \frac{1}{(ad-bc)} \times \ln\left|\frac{ax+b}{cx+d}\right| + C, \qquad ad-bc \neq 0$
19. $\int \frac{dx}{a^2-x^2} = \frac{1}{2a}\ln\left|\frac{a+x}{a-x}\right| + c$
20. $\int \frac{dx}{(a^2-x^2)^2} = \frac{x}{2a^2(a^2-x^2)} + \frac{1}{4a^3}\ln\left|\frac{a+x}{a-x}\right| + c$
21. $\int \frac{x\,dx}{a^2-x^2} = -\frac{1}{2}\ln|a^2-x^2| + c$
22. $\int \frac{x\,dx}{(a^2-x^2)^2} = \frac{1}{2(a^2-x^2)} + c$
23. $\int \frac{x^2\,dx}{a^2-x^2} = -x + \frac{a}{2}\ln\left|\frac{a+x}{a-x}\right| + c$
24. $\int \frac{x^2\,dx}{(a^2-x^2)^2} = \frac{x}{2(a^2-x^2)} - \frac{1}{4a}\ln\left|\frac{a+x}{a-x}\right| + c$
25. $\int \frac{dx}{x(x^2+a^2)} = \frac{1}{2a^2}\ln\frac{x^2}{x^2+a^2} + c$
26. $\int \frac{dx}{x^2(x^2+a^2)} = -\frac{1}{a^2x} - \frac{1}{a^3}\arctan\frac{x}{a} + c$

27. $\displaystyle\int \frac{dx}{x^3(x^2 + a^2)} = -\frac{1}{2a^2x^2} - \frac{1}{2a^4}\ln\frac{x^2}{x^2 + a^2} + c$

28. $\displaystyle\int \frac{x\,dx}{x^2 + a^2} = \frac{1}{2}\ln(x^2 + a^2) + c$

29. $\displaystyle\int \frac{x^2\,dx}{x^2 + a^2} = x - a\arctan\frac{x}{a} + c$

30. $\displaystyle\int \frac{x^3\,dx}{x^2 + a^2} = \frac{x^2}{2} - \frac{a^2}{2}\ln(x^2 + a^2) + c$

31. $\displaystyle\int \frac{x^4\,dx}{x^2 + a^2} = \frac{x^3}{3} - a^2x + a^3\arctan\frac{x}{a} + c$

* 32. $\displaystyle\int \frac{dx}{x^4 + 1} = \frac{1}{4\sqrt{2}}\ln\frac{x^2 + x\sqrt{2} + 1}{x^2 - x\sqrt{2} + 1} + \frac{1}{2\sqrt{2}}\arctan\frac{x\sqrt{2}}{1 - x^2} + c.$

Hint: The fourth roots of -1 are $(\pm 1 \pm i)/\sqrt{2}$.

9.7 PARTIAL FRACTIONS: REPEATED QUADRATIC FACTORS

In this section we first discuss the problem of evaluating

$$I = \int \frac{Bx + C}{(x^2 + bx + c)^n}\,dx, \qquad n > 1 \quad \text{and} \quad b^2 - 4c < 0. \tag{1}$$

Once we have this result, we can complete the integration of rational functions for which the partial fraction representation involves quadratic polynomials raised to powers greater than one.

To evaluate I we proceed in a standard manner by completing the square in the denominator and then making a simple change of variables

$$I = \int \frac{Bx + C}{\left(x^2 + bx + \frac{b^2}{4} + c - \frac{b^2}{4}\right)^n}\,dx = \int \frac{Bx + C}{\left[\left(x + \frac{b}{2}\right)^2 + \left(c - \frac{b^2}{4}\right)\right]^n}\,dx.$$

Since $b^2 - 4c < 0$ by hypothesis, we know that $c - b^2/4 > 0$. Let $k^2 = c - b^2/4$ and make the change of variables

$$s = x + \frac{b}{2}, \quad \text{so} \quad x = s - \frac{b}{2} \quad \text{and} \quad ds = dx.$$

Then

$$I = \int \frac{B(s - b/2) + C}{(s^2 + k^2)^n}\,ds$$

$$= B\int \frac{s}{(s^2 + k^2)^n}\,ds + \left(C - \frac{Bb}{2}\right)\int \frac{ds}{(s^2 + k^2)^n}. \tag{2}$$

The first integral can be evaluated by making the change of variables $w = s^2 + k^2$ with $dw = 2s\,ds$. Then

$$\int \frac{s\,ds}{(s^2 + k^2)^n} = \frac{1}{2}\int \frac{dw}{w^n} = \frac{w^{-n+1}}{2(-n + 1)} + c$$

$$= \frac{-1}{2(n - 1)(s^2 + k^2)^{n-1}} + c, \qquad n > 1. \tag{3}$$

The second integral in Eq. 2 is more difficult. We will derive a reduction formula using integration by parts. Let

$$u = \frac{1}{(s^2 + k^2)^n}, \qquad dv = ds,$$

so

$$du = \frac{-2ns\,ds}{(s^2 + k^2)^{n+1}}, \qquad v = s.$$

Then

$$\begin{aligned}\int \frac{ds}{(s^2 + k^2)^n} &= \frac{s}{(s^2 + k^2)^n} - \int \frac{-2ns^2\,ds}{(s^2 + k^2)^{n+1}} \\ &= \frac{s}{(s^2 + k^2)^n} + 2n \int \frac{(s^2 + k^2) - k^2}{(s^2 + k^2)^{n+1}}\,ds \\ &= \frac{s}{(s^2 + k^2)^n} + 2n \int \frac{ds}{(s^2 + k^2)^n} \\ &\quad - 2nk^2 \int \frac{ds}{(s^2 + k^2)^{n+1}}.\end{aligned}$$

Finally, we solve this equation for the last integral on the right side to obtain the reduction formula

$$\int \frac{ds}{(s^2 + k^2)^{n+1}} = \frac{s}{2nk^2(s^2 + k^2)^n} + \frac{2n - 1}{2nk^2} \int \frac{ds}{(s^2 + k^2)^n}. \tag{4}$$

By repeated use of the reduction formula (4), we can reduce

$$\int \frac{ds}{(s^2 + k^2)^{n+1}} \quad \text{to} \quad \int \frac{ds}{s^2 + k^2} = \frac{1}{k} \arctan \frac{s}{k} + c.$$

Thus I can be evaluated by expressing it in the form given in Eq. 2 and then using the integration formula (3) and the reduction formula (4).

The use of the reduction formula (4) is illustrated in Example 1; in Example 2 we consider the integral of a fairly complicated rational function.

EXAMPLE 1

Evaluate

$$I = \int \frac{dx}{(x^2 + 2x + 7)^3}. \tag{5}$$

First we complete the square, obtaining

$$I = \int \frac{dx}{[x^2 + 2x + 1 + (7 - 1)]^3} = \int \frac{dx}{[(x + 1)^2 + 6]^3}.$$

Then we let $s = x + 1$, $ds = dx$, so that

$$I = \int \frac{ds}{(s^2 + 6)^3}.$$

Now we use the reduction formula (4) with $k^2 = 6$ and $n = 2$, so that $n + 1 = 3$; this gives

$$I = \frac{1}{4 \cdot 6} \frac{s}{(s^2 + 6)^2} + \frac{3}{4 \cdot 6} \int \frac{ds}{(s^2 + 6)^2}.$$

Next, we use the reduction formula (4) again with $k^2 = 6$ and $n = 1$, obtaining

$$\begin{aligned} I &= \frac{s}{24(s^2 + 6)^2} + \frac{1}{8} \left[\frac{1}{2 \cdot 6} \frac{s}{s^2 + 6} + \frac{1}{2 \cdot 6} \int \frac{ds}{s^2 + 6} \right] \\ &= \frac{s}{24(s^2 + 6)^2} + \frac{s}{96(s^2 + 6)} + \frac{1}{96\sqrt{6}} \arctan \frac{s}{\sqrt{6}} + c. \end{aligned}$$

On replacing s by $x + 1$, we obtain

$$\begin{aligned} \int \frac{dx}{(x^2 + 2x + 7)^3} &= \frac{x + 1}{24(x^2 + 2x + 7)^2} + \frac{x + 1}{96(x^2 + 2x + 7)} \\ &\quad + \frac{1}{96\sqrt{6}} \arctan \frac{x + 1}{\sqrt{6}} + c. \ ∎ \end{aligned} \tag{6}$$

EXAMPLE 2

Evaluate

$$I = \int \frac{5x^4 + 6x^3 + 25x^2 + 22x + 17}{(x - 1)^2(x^2 + 4)^2} \, dx. \tag{7}$$

According to Rules 1 and 2 of Section 9.6 we have

$$\begin{aligned} &\frac{5x^4 + 6x^3 + 25x^2 + 22x + 17}{(x - 1)^2(x^2 + 4)^2} \\ &\qquad = \frac{A}{x - 1} + \frac{B}{(x - 1)^2} + \frac{Cx + D}{x^2 + 4} + \frac{Ex + F}{(x^2 + 4)^2}. \end{aligned} \tag{8}$$

To find the constants A, B, C, D, E, and F we multiply Eq. 8 by $(x - 1)^2 \times (x^2 + 4)^2$. This gives

$$\begin{aligned} 5x^4 + 6x^3 + 25x^2 + 22x + 17 &= A(x - 1)(x^2 + 4)^2 + B(x^2 + 4)^2 \\ &\quad + (Cx + D)(x - 1)^2(x^2 + 4) \\ &\quad + (Ex + F)(x - 1)^2 \\ &= (A + C)x^5 + (-A + B - 2C + D)x^4 \\ &\quad + (8A + 5C - 2D + E)x^3 \end{aligned}$$

$$+ (-8A + 8B - 8C + 5D - 2E + F)x^2$$
$$+ (16A + 4C - 8D + E - 2F)x$$
$$+ (-16A + 16B + 4D + F).$$

Equating coefficients of like powers of x, we obtain the system of equations

$$\begin{aligned}
&\text{for } x^5: & A + C &= 0,\\
&\text{for } x^4: & -A + B - 2C + D &= 5,\\
&\text{for } x^3: & 8A + 5C - 2D + E &= 6,\\
&\text{for } x^2: & -8A + 8B - 8C + 5D - 2E + F &= 25,\\
&\text{for } x^1: & 16A + 4C - 8D + E - 2F &= 22,\\
&\text{for } x^0: & -16A + 16B + 4D + F &= 17.
\end{aligned} \tag{9}$$

These equations have the solution $A = 2$, $B = 3$, $C = -2$, $D = 0$, $E = 0$, and $F = 1$. Thus

$$I = \int \left[\frac{2}{x - 1} + \frac{3}{(x - 1)^2} - \frac{2x}{x^2 + 4} + \frac{1}{(x^2 + 4)^2}\right] dx. \tag{10}$$

The first three integrals on the right side of Eq. 10 can be readily evaluated:

$$\int \frac{2}{x - 1}\, dx = 2 \ln|x - 1| + c,$$

$$\int \frac{3}{(x - 1)^2}\, dx = \frac{-3}{x - 1} + c,$$

$$\int \frac{-2x}{x^2 + 4}\, dx = -\ln(x^2 + 4) + c.$$

To evaluate the last integral in Eq. 10 we use the reduction formula (4) with $k = 2$ and $n = 1$. We have

$$\begin{aligned}
\int \frac{dx}{(x^2 + 4)^2} &= \frac{1}{2 \cdot 4} \frac{x}{(x^2 + 4)} + \frac{1}{2 \cdot 4} \int \frac{dx}{x^2 + 4}\\
&= \frac{1}{8} \frac{x}{x^2 + 4} + \frac{1}{16} \arctan \frac{x}{2} + c.
\end{aligned}$$

Combining these individual results, we find that

$$\begin{aligned}
\int \frac{5x^4 + 6x^3 + 25x^2 + 22x + 17}{(x - 1)^2(x^2 + 4)^2}\, dx &= 2 \ln|x - 1| - \frac{3}{x - 1} - \ln(x^2 + 4)\\
&\quad + \frac{1}{8} \frac{x}{x^2 + 4} + \frac{1}{16} \arctan \frac{x}{2} + c. \blacksquare
\end{aligned} \tag{11}$$

It is clear from Example 2 that if the partial fraction representation of a rational function contains quadratic factors with powers higher than one, then the problem of calculating the integral becomes somewhat tedious. However, the rules and procedures are straightforward—there is no guesswork once the factors of the denominator have been determined. The principal difficulties are not in doing the integration, but rather in factoring the denominator of the rational function and in determining the constants in the partial fraction representation.

PROBLEMS

In each of Problems 1 through 14, evaluate the given indefinite integral.

1. $\displaystyle\int \frac{dx}{(x^2+9)^2}$

2. $\displaystyle\int \frac{3x-2}{(x^2+9)^2}\,dx$

3. $\displaystyle\int \frac{dx}{(x^2+4)^3}$

4. $\displaystyle\int \frac{4-x}{(x^2+4)^3}\,dx$

5. $\displaystyle\int \frac{dx}{(x^2-4x+8)^2}$

6. $\displaystyle\int \frac{2x}{(x^2-4x+8)^2}\,dx$

7. $\displaystyle\int \frac{dx}{(x^2-2x+2)^2}$

8. $\displaystyle\int \frac{6x+5}{(x^2-2x+2)^2}\,dx$

9. $\displaystyle\int \frac{dx}{(2x^2+6)^3}$

10. $\displaystyle\int \frac{dx}{(x^2+2x+5)^3}$

11. $\displaystyle\int \frac{dx}{x(x^2+4)^2}$

12. $\displaystyle\int \frac{(3x^3+1)\,dx}{x^2(x^2+1)^2}$

13. $\displaystyle\int \frac{2x^2+5x+5}{(x-1)(x^3-1)}\,dx$

14. $\displaystyle\int \frac{x^4+2x^3+25}{x(x^2+2x+5)^2}\,dx$

In each of Problems 15 through 17, use the reduction formula derived in this section, Eq. 4, to obtain the given integration formula.

15. $\displaystyle\int \frac{dx}{(x^2+a^2)^2} = \frac{x}{2a^2(x^2+a^2)} + \frac{1}{2a^3}\arctan\frac{x}{a} + c$

16. $\displaystyle\int \frac{dx}{(x^2+a^2)^3} = \frac{x}{4a^2(x^2+a^2)^2} + \frac{3x}{8a^4(x^2+a^2)} + \frac{3}{8a^5}\arctan\frac{x}{a} + c$

17. $\displaystyle\int \frac{dx}{(x^2+a^2)^4} = \frac{x}{6a^2(x^2+a^2)^3} + \frac{5x}{24a^4(x^2+a^2)^2} + \frac{5x}{16a^6(x^2+a^2)} + \frac{5}{16a^7}\arctan\frac{x}{a} + c$

In each of Problems 18 through 22, derive the given integration formula. These formulas extend the results given in Problems 25 through 27 of Section 9.6.

18. $\displaystyle\int \frac{dx}{x(x^2+a^2)^2} = \frac{1}{2a^2(x^2+a^2)} + \frac{1}{2a^4}\ln\frac{x^2}{a^2+x^2} + c$

* 19. $\displaystyle\int \frac{dx}{x(x^2+a^2)^3} = \frac{1}{4a^2(x^2+a^2)^2} + \frac{1}{2a^4(x^2+a^2)} + \frac{1}{2a^6}\ln\frac{x^2}{a^2+x^2} + c$

20. $\displaystyle\int \frac{dx}{x^2(x^2+a^2)^2} = -\frac{1}{a^4x} - \frac{x}{2a^4(x^2+a^2)} - \frac{3}{2a^5}\arctan\frac{x}{a} + c$

* 21. $\displaystyle\int \frac{dx}{x^2(x^2+a^2)^3} = -\frac{1}{a^6x} - \frac{x}{4a^4(x^2+a^2)^2} - \frac{7x}{8a^6(x^2+a^2)} - \frac{15}{8a^7}\arctan\frac{x}{a} + c$

22. $\displaystyle\int \frac{dx}{x^3(x^2+a^2)^2} = -\frac{1}{2a^4x^2} - \frac{1}{2a^4(x^2+a^2)} - \frac{1}{a^6}\ln\frac{x^2}{x^2+a^2} + c$

In each of Problems 23 through 26, derive the given integration formula. These formulas extend the results given in Problems 28 through 31 of Section 9.6.

23. $\displaystyle\int \frac{x\,dx}{(x^2+a^2)^2} = \frac{-1}{2(x^2+a^2)} + c$

24. $\displaystyle\int \frac{x^2\,dx}{(x^2+a^2)^2} = -\frac{x}{2(x^2+a^2)} + \frac{1}{2a}\arctan\frac{x}{a} + c$

25. $\displaystyle\int \frac{x^3\,dx}{(x^2+a^2)^2} = \frac{a^2}{2(x^2+a^2)} + \frac{1}{2}\ln(x^2+a^2) + c$

26. $\displaystyle\int \frac{x^4\,dx}{(x^2+a^2)^2} = x + \frac{a^2x}{2(x^2+a^2)} - \frac{3a}{2}\arctan\frac{x}{2} + c$

27. Show that the substitution $x = a\tan u$ reduces

$$I_n = \int \frac{dx}{(x^2+a^2)^n} \quad \text{to}$$

$$I_n = \frac{1}{a^{2n-1}}\int \cos^{2(n-1)} u\,du.$$

For $n = 2$, show that this new formula for I_n yields

$$I_2 = \frac{x}{2a^2(x^2+a^2)} + \frac{1}{2a^3}\arctan\frac{x}{a} + c.$$

This is the same result obtained in Problem 15 using the reduction formula. For values of $n > 2$, it is more convenient to use the reduction formula to calculate I_n rather than the formula obtained in this problem.

REVIEW PROBLEMS

In each of Problems 1 through 56, evaluate the given definite or indefinite integral.

1. $\displaystyle\int \frac{dx}{x(1+\ln^2 x)}$

2. $\displaystyle\int \left(\frac{\ln x + 1}{x\ln x}\right) dx$

3. $\displaystyle\int \frac{\ln x + 1}{x}\,dx$

4. $\displaystyle\int \operatorname{arcsinh} x\,dx$

5. $\displaystyle\int_{-\pi/4}^{\pi/4} \tan^5 x\,dx$

6. $\displaystyle\int \frac{\sec x}{\tan x + 2\cot x}\,dx$

7. $\displaystyle\int x^4\sqrt{1-x^2}\,dx$

8. $\displaystyle\int x\arcsin x\,dx$

9. $\displaystyle\int \frac{2x - 2x\tan x^2}{1+\tan x^2}\,dx$

10. $\displaystyle\int \tan^4 x\sec^4 x\,dx$

11. $\displaystyle\int x^3\sqrt{1+x^2}\,dx$

12. $\displaystyle\int \frac{dx}{2x+x^3}$

13. $\displaystyle\int \frac{x^3+x^2-7x-2}{x(x+3)}\,dx$

14. $\displaystyle\int_{\pi/6}^{\pi/2} \frac{\sqrt{4-x^2}}{x^2}\,dx$

15. $\displaystyle\int \sin\frac{x}{2}\cos\frac{\pi x}{4}\,dx$

16. $\displaystyle\int \frac{\cos 2x\sin 4x}{\sin 2x}\,dx$

17. $\displaystyle\int \frac{\tan x\sec^3 x}{1+\sec^6 x}\,dx$

18. $\displaystyle\int \frac{x\,dx}{2+2x^2+x^4}$

19. $\displaystyle\int \cos 3x\sin^3 2x\,dx$

20. $\displaystyle\int_0^{\pi/2} \sin^{1/4} x\cos^3 x\,dx$

21. $\displaystyle\int \frac{2x^4-3x^3-5x^2-4}{x^2(x^2-4)}\,dx$

22. $\displaystyle\int \frac{\sinh 2x + \sin 2x}{\sinh^2 x + \sin^2 x}\,dx$

23. $\displaystyle\int_0^{\pi/2} \frac{x^2\,dx}{\sqrt{4+x^2}}$

24. $\displaystyle\int \frac{x(x-2)}{x^4+x^2-2x+1}\,dx$

25. $\displaystyle\int \frac{dx}{x\sqrt{64-x^3}}$

26. $\displaystyle\int \frac{1+e^x}{1+x^2+2xe^x+e^{2x}}\,dx$

27. $\displaystyle\int \frac{dx}{2x^2+2x+1}$

28. $\displaystyle\int_0^2 \sqrt{(1-2x+x^2)(2+2x-x^2)}\,dx$

29. $\displaystyle\int_0^{\pi/4} \frac{1+\tan^2 x}{1+\tan x}\,dx$

30. $\displaystyle\int_0^{\pi/2} \sin^3 x\cos^3 x\,dx$

31. $\displaystyle\int \frac{x+1+\cos x}{x+1}\sec x\,dx$

32. $\displaystyle\int_0^{\pi} \tan^{-3} x\csc^4 x\,dx$

33. $\displaystyle\int \tan^2 x\sin x\,dx$

34. $\displaystyle\int \frac{\tan ax + \tan bx}{1-\tan ax\tan bx}\,dx$

35. $\displaystyle\int_1^2 \frac{dx}{2x^2-6x+5}$

36. $\displaystyle\int \frac{\ln(\arctan x)}{1+x^2}\,dx$

37. $\displaystyle\int x^3\sinh x^2\,dx$

38. $\displaystyle\int \cot^{1/2} x\csc^4 x\,dx$

39. $\displaystyle\int x\sec^2 x\,dx$

40. $\displaystyle\int \frac{\sqrt{1-x^2}-\sqrt{1+x^2}}{\sqrt{1-x^4}}\,dx$

41. $\displaystyle\int \frac{x^2+2x+2}{x^3+3x^2+2x}\,dx$

42. $\displaystyle\int_0^1 \frac{x\,dx}{\sqrt{3+2x^2-x^4}}$

43. $\displaystyle\int \frac{x^2}{\sqrt{x^2-4x+1}}\,dx$

44. $\displaystyle\int \frac{\cos(\arcsin x)}{\sqrt{1-x^2}}\,dx$

45. $\displaystyle\int \frac{\tan(a \arctan x)}{1+x^2}\,dx$

46. $\displaystyle\int \frac{1-\tan x}{1+\tan x}\,dx$

47. $\displaystyle\int \frac{2x^4-x^3+17x^2-2x+2}{(x+1)(x-1)^2(x^2+2)}\,dx$

48. $\displaystyle\int \frac{2x^2-x+3}{\sqrt{x^2-9}}\,dx$

49. $\displaystyle\int (1-e^{2x})^{3/2}\,dx$

50. $\displaystyle\int_0^1 \frac{x^2\,dx}{\sqrt{2-x^2}}$

51. $\displaystyle\int_0^1 \sqrt{1+x^{2/3}}\,dx$

52. $\displaystyle\int (1+\cos\theta)^{3/2}\,d\theta$

53. $\displaystyle\int \frac{d\theta}{(1-\cos\theta)^{3/2}}$

54. $\displaystyle\int \frac{x^2+3}{(x^2-1)(2x^2+1)}\,dx$

55. $\displaystyle\int_{-1}^1 x^3 e^{x^2}\,dx$

56. $\displaystyle\int x \sin x^2\, e^{x^2}\,dx$

In each of Problems 57 through 60, use the substitutions

$$\arctan u = \frac{x}{2}, \quad dx = \frac{2\,du}{1+u^2}$$

and

$$\sin x = \frac{2u}{1+u^2}, \quad \cos x = \frac{1-u^2}{1+u^2}$$

to evaluate the given integrals.

57. $\displaystyle\int \frac{\sin x\,dx}{(1+\sin x)\cos x}$

58. $\displaystyle\int \frac{1+\sin x}{1+\cos x}\,dx$

59. $\displaystyle\int \frac{dx}{a+\cos x}$

60. $\displaystyle\int \frac{dx}{\sin x+\cos x}$

TEN

first order differential equations

One of the most important uses of integration is for solving differential equations. We have already encountered a few such equations, for example, in Section 8.4 in connection with the exponential function as a model of growth or decay phenomena. In the physical sciences differential equations often arise from the application of a physical law governing the rate of change of one or more quantities of interest. For instance, Newton's law $F = ma$ is a differential equation because the acceleration a is the second time derivative of the position of the moving object.

Since differential equations occur naturally in the description of many significant problems, methods for solving them are of corresponding importance. Although a systematic treatment of differential equations belongs in a later course, in this chapter we discuss some types of equations that can be solved by elementary integration methods.*

*These methods are due to Leibniz more than to anyone else. For instance, he discovered how to solve separable equations (Section 10.2) in 1691 and linear equations (Section 10.1) in 1694. The techniques mentioned in Problem 27 of Section 10.1 and in Problem 22 of Section 10.2 are also due to him.

10.1 LINEAR EQUATIONS

In Section 8.4 we discussed a variety of growth and decay phenomena that can be modeled by the initial value problem

$$Q'(t) = rQ(t) + k, \tag{1}$$

$$Q(0) = Q_0, \tag{2}$$

where r, k, and Q_0 are given constants. In this section and in the next we discuss some other types of first order differential equations.

We consider only equations that have the form

$$\frac{dy}{dt} = f(t, y), \tag{3}$$

where f is some given function of the independent variable t and the dependent variable y. Unfortunately, there is no universal method for solving the differential equation (3). Rather, there are several methods, each of which can be used for certain classes of equations. One of these is the class of linear equations. A first order differential equation is said to be **linear** if the function $f(t, y)$ can be written as

$$f(t, y) = -p(t)y + g(t), \tag{4}$$

where p and g are given functions of the independent variable t only. Then the differential equation (3) becomes

$$\frac{dy}{dt} + p(t)y = g(t), \tag{5}$$

which is the most general form of a first order linear differential equation. Often there is also prescribed an initial condition

$$y(t_0) = y_0. \tag{6}$$

As indicated in Section 8.4, a differential equation, such as Eq. 3 or Eq. 5, together with an initial condition form an initial value problem.

To solve the initial value problem (5), (6) we seek a function $y = \phi(t)$ that satisfies Eq. 5 in some interval containing the point t_0, and that also has the value y_0 at t_0. We assume that the functions p and g are continuous on the interval where the solution ϕ is sought.

Observe that Eq. 1 is a linear equation of a special kind, namely, one for which the coefficient functions p and g are constants. We now consider the more general case in which p and g can be arbitrary continuous functions of t.

In Example 3 of Section 8.4 the first step in solving the equation

$$Q' - rQ = -k \tag{7}$$

was to multiply it by the function e^{-rt}, a step that transformed the equation into an easily integrable form. Let us attempt to do something similar to solve Eq. 5; the problem is then to determine what function to use as a multiplier so as to simplify the equation. Before taking up the general argument we consider an example.

EXAMPLE 1

Solve

$$\frac{dy}{dt} + \frac{2}{t}y = 3, \qquad t < 0 \quad \text{or} \quad t > 0. \tag{8}$$

Note that we cannot attempt to solve the differential equation in any interval containing the origin, since the coefficient $p(t) = 2/t$ is not defined at $t = 0$ and becomes unbounded as this point is approached.

To follow the procedure just indicated, let us multiply the equation by a function μ and then try to choose $\mu(t)$ so as to make the left side of the equation easy to integrate. Multiplying Eq. 8 by $\mu(t)$, we obtain

$$\mu(t)y' + \frac{2}{t}\mu(t)y = 3\mu(t). \tag{9}$$

Our object is to choose μ so that the left side of Eq. 9 can be identified as the derivative of some function. The term $\mu(t)y'$ suggests that the desired function might be the product $\mu(t)y$. To obtain the combination $[\mu(t)y]' = \mu'(t)y + \mu(t)y'$, we must add and subtract the term $\mu'(t)y$ on the left side of Eq. 9. By doing this and collecting terms in a suitable way, we obtain

$$[\mu'(t)y + \mu(t)y'] - \left[\mu'(t) - \frac{2}{t}\mu(t)\right]y = 3\mu(t). \tag{10}$$

Now, if the second term on the left side of Eq. 10 were zero, then Eq. 10 would have the form

$$\frac{d}{dt}[\mu(t)y] = 3\mu(t), \tag{11}$$

and the left side (at least) would be readily integrable. To achieve this situation we must choose μ so that

$$\mu'(t) - \frac{2}{t}\mu(t) = 0. \tag{12}$$

If we assume temporarily that μ is not zero, then we can write Eq. 12 as

$$\frac{\mu'(t)}{\mu(t)} = \frac{2}{t},$$

or

$$\frac{d}{dt}[\ln \mu(t)] = \frac{2}{t}. \tag{13}$$

Thus

$$\ln \mu(t) = 2 \ln t + c = \ln t^2 + c.$$

By choosing c to be zero, we obtain the simplest possible function for μ, namely

$$\mu(t) = t^2. \tag{14}$$

Observe that indeed $\mu(t)$ is not zero on any interval not containing the origin, as required.

Having found the function μ, we return to Eq. 8 and multiply it by $\mu(t) = t^2$, thereby obtaining

$$t^2y' + 2ty = 3t^2. \tag{15}$$

The left side of Eq. 15 is indeed the derivative of t^2y, so we have

$$\frac{d}{dt}(t^2y) = 3t^2.$$

Thus

$$t^2y = t^3 + c$$

and

$$y = t + \frac{c}{t^2}. \tag{16}$$

This is the most general solution of Eq. 8, and is valid either for $t > 0$ or for $t < 0$. Figure 10.1.1 shows the graph of y versus t for several values of c. ∎

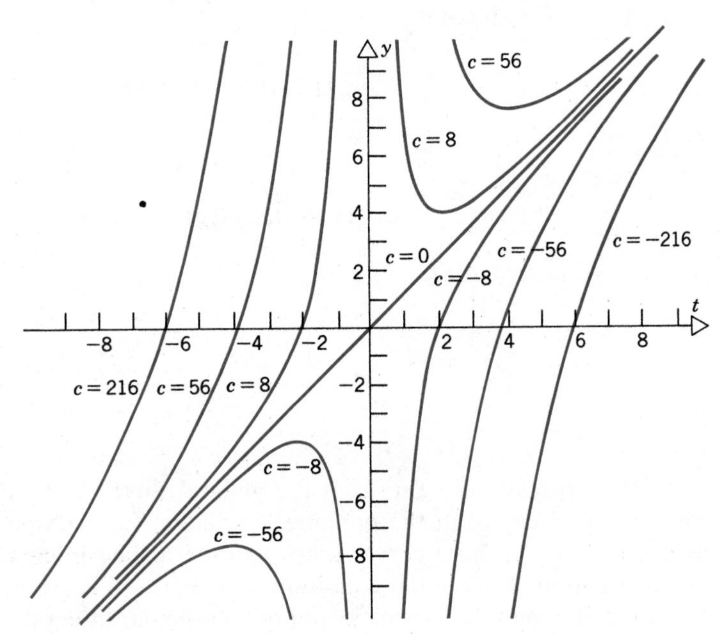

Figure 10.1.1 Some integral curves of $y' + (2/t)y = 3$.

Now let us return to the differential equation (5) and proceed in the same way as in Example 1. First we multiply Eq. 5 by a function $\mu(t)$, obtaining

$$\mu(t)y' + \mu(t)p(t)y = \mu(t)g(t). \tag{17}$$

Next, as in Example 1, we add and subtract the quantity $\mu'(t)y$ and collect terms so that

$$[\mu'(t)y + \mu(t)y'] - [\mu'(t) - \mu(t)p(t)]y = \mu(t)g(t). \tag{18}$$

Then we require that

$$\mu'(t) - \mu(t)p(t) = 0, \tag{19}$$

or

$$\frac{\mu'(t)}{\mu(t)} = \frac{d}{dt}\ln[\mu(t)] = p(t).$$

Hence

$$\ln \mu(t) = \int p(t)\, dt,$$

and

$$\mu(t) = \exp \int p(t)\, dt; \tag{20}$$

note that we have (for convenience) dropped the arbitrary integration constant. The function μ given by Eq. 20 is called an **integrating factor**, since with this choice of $\mu(t)$ Eq. 18 reduces to

$$\frac{d}{dt}[\mu(t)y] = \mu(t)g(t). \tag{21}$$

Hence

$$\mu(t)y = \int \mu(t)g(t)\, dt + c,$$

or

$$y = \frac{\int \mu(t)g(t)\, dt + c}{\mu(t)}. \tag{22}$$

The expression in Eq. 22 is the **general solution** of Eq. 5, and corresponds to a family of curves in the ty-plane, one curve for each possible value of c, just as in Example 1. These curves are sometimes called **integral curves** of the differential equation. If an initial condition such as Eq. 6 is given, then this determines the value of c, and thereby picks out one particular integral curve from the family.

From Eqs. 20 and 22 it should be clear that solving the first order linear differential equation (5) depends on evaluating two integrals: the integral (20) that yields the integrating factor $\mu(t)$, followed by the integral in Eq. 22 that gives the solution y. Of course, the difficulty in evaluating these integrals depends entirely on the nature of the functions p and g in Eq. 5.

Although Eq. 22 gives a formula for the solution of Eq. 5, we do not recommend that you memorize this formula. Rather, you should learn how to find the integrating factor, and then proceed as in the following examples.

EXAMPLE 2

Find the solution of the initial value problem

$$y' - 2ty = t, \qquad y(0) = 0. \tag{23}$$

For this equation the function μ is

$$\mu(t) = \exp\left[\int -2t\, dt\right] = \exp(-t^2).$$

Hence we have

$$e^{-t^2}y' - 2te^{-t^2}y = te^{-t^2},$$

or

$$(e^{-t^2}y)' = te^{-t^2}.$$

Thus

$$e^{-t^2}y = \int te^{-t^2}\, dt = -\frac{1}{2}e^{-t^2} + c,$$

or

$$y = -\frac{1}{2} + ce^{t^2}. \tag{24}$$

To satisfy the initial condition we must choose c so that $0 = -\frac{1}{2} + c$. Hence $c = \frac{1}{2}$, and the solution of the initial value problem is

$$y = -\frac{1}{2} + \frac{1}{2}e^{t^2}. \tag{25}$$

Figure 10.1.2 shows the solution (25) together with the integral curves corresponding to several other values of c. ∎

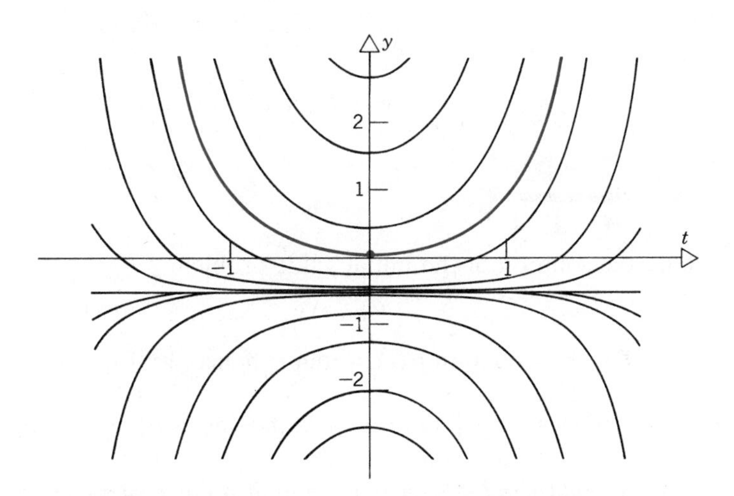

Figure 10.1.2 The solution of the initial value problem $y' - 2ty = t$, $y(0) = 0$, and some other integral curves.

EXAMPLE 3

Find the general solution of

$$ty' - y = t^3 \cos t \tag{26}$$

for $t > 0$.

In this case we must first put the equation in the form (5) by dividing it by t. Thus

$$y' - \frac{1}{t} y = t^2 \cos t. \tag{27}$$

From Eq. 20 we then have, with $p(t) = -1/t$,

$$\mu(t) = \exp \int -\frac{dt}{t} = \exp(-\ln t) = \exp\left[\ln \frac{1}{t}\right] = \frac{1}{t}. \tag{28}$$

Multiplying Eq. 27 by $\mu(t)$, we obtain

$$\frac{1}{t} y' - \frac{1}{t^2} y = t \cos t,$$

or

$$\left(\frac{y}{t}\right)' = t \cos t,$$

so that

$$\frac{y}{t} = \int t \cos t \, dt.$$

Finally, on integrating by parts, we find that

$$\frac{y}{t} = t \sin t + \cos t + c,$$

or

$$y = t^2 \sin t + t \cos t + ct. \blacksquare \tag{29}$$

EXAMPLE 4

Find the solution of the initial value problem

$$y' + (\tan t) y = \sec t, \qquad y(\pi) = -2. \tag{30}$$

For this equation $p(t) = \tan t$; hence, by Eq. 20, we have

$$\mu(t) = \exp \int \tan t \, dt = \exp[-\ln|\cos t|] = \exp[\ln|\sec t|] = |\sec t|. \tag{31}$$

Observe that $\sec t < 0$ near the initial point $t = \pi$. Thus $|\sec t| = -\sec t$ is the integrating factor for this problem. However, when we multiply the differential equation by $-\sec t$, the minus sign can be canceled out, so that we have

$$(\sec t)y' + (\sec t \tan t)y = \sec^2 t.$$

Hence

$$[(\sec t)y]' = \sec^2 t$$

and

$$(\sec t)y = \int \sec^2 t\, dt = \tan t + c.$$

Consequently the general solution of the differential equation is

$$y = \sin t + c \cos t.$$

To satisfy the initial condition we set $t = \pi$ and $y = -2$. Hence $c = 2$, and

$$y = \sin t + 2 \cos t \tag{32}$$

is the solution of the given problem. ∎

Some applications of first order linear equations were discussed in Section 8.4. Further applications are mentioned in Problems 21 through 26.

PROBLEMS

In each of Problems 1 through 12, find the general solution of the given differential equation.

1. $y' - 2y = t^2 e^{2t}$

2. $y' - y = 2e^t$

3. $y' + 3y = t + e^{-t}$

4. $y' + 2y = 3t$

5. $y' + \dfrac{1}{t} y = 3 \cos 2t, \quad t > 0$

6. $y' + \dfrac{2}{t} y = \dfrac{e^{-t}}{t}, \quad t > 0$

7. $ty' + 2y = e^t, \quad t > 0$

8. $y' + (\tan t)y = t \sin 2t, \quad -\dfrac{\pi}{2} < t < \dfrac{\pi}{2}$

9. $t^2 y' + 3ty = \dfrac{\sin t}{t}, \quad t > 0$

10. $ty' + 2y = 2\cos(t^2), \quad t > 0$

11. $ty' + (t - 1)y = te^{-t}, \quad t > 0$

12. $(\sin t)y' + 2(\cos t)y = \dfrac{2 \cos 2t}{\sin t},$
$0 < t < \pi$

In each of Problems 13 through 18, find the solution of the given initial value problem.

13. $y' - y = 2te^t, \quad y(0) = 1$

14. $y' - 2y = t - 3, \quad y(0) = -1$

15. $y' + \dfrac{2}{t} y = \dfrac{\cos t}{t^2}, \quad y(\pi) = 0$

16. $ty' + 2y = t^2 - t + 1, \quad y(1) = \frac{1}{2}$

17. $(t + 1)y' + 2y = t, \quad y(0) = 3$

18. $y' + (\cot t)y = t, \quad y\left(\dfrac{\pi}{2}\right) = 2$

19. Show that if a and λ are positive constants, and b is any real number, then every solution of the equation

$$y' + ay = be^{-\lambda t}$$

has the property that $y \to 0$ as $t \to \infty$. *Hint:* Consider the cases $a = \lambda$ and $a \neq \lambda$ separately.

20. Consider the initial value problem

$$y' - 2ty = 1, \quad y(0) = 1. \tag{i}$$

(a) Show that the general solution of the differential equation is

$$y = e^{t^2} \int_{t_0}^{t} e^{-s^2}\, ds + ce^{t^2}, \qquad (ii)$$

where t_0 is an arbitrary real number.

(b) To determine the constant c so that the initial condition is satisfied, it is convenient to choose $t_0 = 0$ and write

$$y = e^{t^2} \int_{0}^{t} e^{-s^2}\, ds + ce^{t^2}. \qquad (iii)$$

Show that if a different lower limit is used, then formula (*iii*) is changed by a constant $\times$ $\exp(t^2)$, which can be included in the term $c\exp(t^2)$.

(c) Show that the solution of the initial value problem is

$$y = e^{t^2} \int_{0}^{t} e^{-s^2}\, ds + e^{t^2}. \qquad (iv)$$

(d) Note that we have developed no way of finding the indefinite integral in Eq. (*iv*); indeed this integrand has no elementary antiderivative. However, one can resort to numerical integration. Use Simpson's rule (Section 6.6) to find approximately the value of y for $t = 0.2$, $t = 0.4$, $t = 0.6$, $t = 0.8$, and $t = 1.0$. Thus, even though the solution (*iv*) involves an integral, from a computational point of view, it is still a perfectly satisfactory form for the solution of the initial value problem. In fact, the factor e^{t^2} in the solution (*iv*) is also normally calculated numerically. The only difference is that most calculators have a built-in routine for calculating the exponential function automatically.

Determination of the Time of Death *

In the investigation of a homicide or accidental death it is often important to estimate the time of death. Here we describe a simple mathematical model for this problem.

Newton's law of cooling states that the rate of change of temperature of an object is proportional to the difference in temperature between the object and the surrounding medium, say air. If θ is the temperature of the object and T is the temperature of the air, then θ satisfies the differential equation

$$\frac{d\theta}{dt} = -k(\theta - T), \qquad (i)$$

where $k > 0$ is the proportionality constant. The minus sign in this equation is needed because if the object is warmer than its surroundings ($\theta > T$), then θ will tend to decrease, and vice versa. Let t be measured from the time the medical examiner arrives and measures the temperature of the body; suppose that this temperature is θ_0. Then the initial condition corresponding to Eq. (*i*) is

$$\theta(0) = \theta_0. \qquad (ii)$$

21. A corpse is found by the roadside with a body temperature of 28°C. One hour later it has cooled to 22°C. Assuming that $T = 10$°C and remains constant, and that Newton's law applies to this case, determine how long the corpse had been dead before it was discovered. Normal body temperature is 37°C.

22. Suppose that a corpse has temperature θ_0 when it is discovered at time $t = 0$ and has temperature θ_1 at a later time t_1. Assuming that Newton's law of cooling applies, and that the ambient temperature T is constant, find the following:

(a) An expression for the cooling rate k.

(b) The time t_d of death, assuming that the individual had a normal body temperature of 37°C at the moment of death.

Mixing

Many significant applications, including the study of physiological and environmental processes, involve the determination of the concentration of a chemical solution of some kind. As a model of this class of problems we consider a tank containing a salt water solution (see Figure 10.1.3). Let V be the volume of

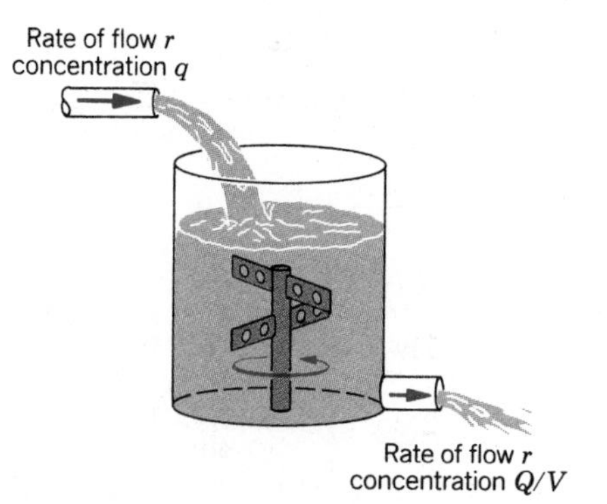

Figure 10.1.3

*See J. F. Hurley, "An Application of Newton's Law of Cooling," *Mathematics Teacher,* **67**, 1974, pp. 141–142; and David A. Smith, "The Homicide Problem Revisited," *The Two Year College Mathematics Journal,* **9**, 1978, pp. 141–145.

water in the tank (which we assume to be constant), and let $Q(t)$ be the amount of salt in the tank at time t. Then $Q(t)/V$ is the concentration; we assume that the tank is kept well-stirred, so that this concentration is maintained uniformly throughout the tank at all times. Suppose that a salt solution of concentration q flows into the tank at the rate r, and that (to maintain constant volume) the solution in the tank is drained at the same rate. Then the rate of change of salt in the tank is given by

$$\frac{dQ}{dt} = \text{rate of flow in} - \text{rate of flow out}$$

$$= rq - \frac{rQ(t)}{V},$$

or

$$Q' + \frac{r}{V}Q = rq,$$

which is a first order linear differential equation. The initial condition is

$$Q(0) = Q_0,$$

where Q_0 is the amount of salt initially in the tank.

23. Consider a tank containing, at time $t = 0$, 20 kg of salt dissolved in 300 liters of water. Assume that water containing 0.1 kg of salt per liter enters the tank at a rate of 5 liters/min, and that the well-stirred mixture leaves at the same rate. Find the amount of salt $Q(t)$ in the tank at any time t.

24. A tank initially contains 120 liters of pure water. A mixture of γ grams per liter of salt enters the tank at a rate of 2 liters per minute, and the well-stirred mixture leaves the tank at the same rate.
    (a) Find the amount of salt $Q(t)$ in the tank at time t.
    (b) Find the limiting value of $Q(t)$ as $t \to \infty$.

25. A tank initially contains 50 gal of pure water. Brine containing γ pounds of salt per gallon is pumped into the tank at a rate of 5 gal/min, and the well-stirred mixture is drained from the tank at the same rate.
    (a) Find the amount of salt $Q(t)$ in the tank at any time.
    (b) Find γ if there are 5 lb of salt in the tank after one hour.

26. A certain water tank initially contains 40 lb of salt dissolved in 100 gal of water. Water containing 0.1 lb of salt per gallon enters the tank at a rate of 5 gal/min, and the well-stirred mixture leaves at the same rate.
    (a) Find the amount of salt $Q(t)$ in the tank at any time.
    (b) Find the time that has elapsed when there are 20 lb of salt in the tank.

27. **Bernoulli equations.** An equation of the form
$$y' + p(t)y = q(t)y^n$$
is called a Bernoulli equation (after Jakob Bernoulli). It is linear if $n = 0$ or if $n = 1$; otherwise, it is nonlinear. Show that if $n \neq 0, 1$, then the substitution $v = y^{1-n}$ reduces Bernoulli's equation to the linear equation
$$v' + (1 - n)p(t)v = (1 - n)\, q(t),$$
which can be solved by the methods of this section.

28. Use the method of Problem 27 to solve the initial value problem
$$y' = (k - ay)y, \qquad y(0) = y_0 > 0,$$
where $k > 0$ and $a > 0$.
This problem arises in the study of population dynamics.

29. Use the method of Problem 27 to solve
$$y' = \epsilon y - \sigma y^3, \qquad \epsilon > 0 \quad \text{and} \quad \sigma > 0.$$
Equations of this form arise in fluid mechanics in studies of transition from laminar to turbulent flow.

10.2 SEPARABLE EQUATIONS

In this section we discuss another class of first order differential equations for which there is a straightforward solution procedure. The general first order equation

$$\frac{dy}{dt} = f(t, y) \tag{1}$$

can always be rewritten in the form

$$M(t, y) + N(t, y)\frac{dy}{dt} = 0. \tag{2}$$

One way to do this is by setting $M(t, y) = -f(t, y)$ and $N(t, y) = 1$, but there may also be other ways. If M is a function of t alone, and N is a function of y alone, then Eq. 2 becomes

$$M(t) + N(y)\frac{dy}{dt} = 0. \tag{3}$$

In this case, the solution of the differential equation can again be reduced to the task of evaluating certain integrals.

EXAMPLE 1

Solve the differential equation

$$\frac{dy}{dt} = \frac{t^2}{1 + 3y^2}. \tag{4}$$

It is easy to see that Eq. 4 can be rewritten as

$$-t^2 + (1 + 3y^2)\frac{dy}{dt} = 0, \tag{5}$$

which is of the form (3). To solve Eq. 5 observe that, by the chain rule,

$$(1 + 3y^2)\frac{dy}{dt} = \frac{d}{dy}(y + y^3)\frac{dy}{dt} = \frac{d}{dt}(y + y^3). \tag{6}$$

Of course, it is also true that

$$-t^2 = \frac{d}{dt}\left(-\frac{t^3}{3}\right). \tag{7}$$

Hence Eq. 5 is the same as

$$\frac{d}{dt}\left(-\frac{t^3}{3}\right) + \frac{d}{dt}(y + y^3) = 0,$$

or

$$\frac{d}{dt}\left(-\frac{t^3}{3} + y + y^3\right) = 0. \tag{8}$$

Therefore

$$-\frac{t^3}{3} + y + y^3 = c, \tag{9}$$

where c is an arbitrary constant.

The procedure that we have used can be expressed more compactly if we write Eq. 5 in the differential form

$$-t^2\,dt + (1 + 3y^2)\,dy = 0. \tag{10}$$

Then, upon integrating each term, we find that

$$-\int t^2\,dt + \int (1 + 3y^2)\,dy = c,$$

or

$$-\frac{t^3}{3} + y + y^3 = c,$$

which is the same as Eq. 9. ∎

Now let us return to the general case. A differential equation that can be put in the form (3) is said to be **separable**. The reason for this name is that if Eq. 3 is written in the differential form

$$M(t)\,dt + N(y)\,dy = 0, \tag{11}$$

then the variables are separated, that is, each appears in only one term. To solve Eq. 3 suppose that we can find functions H_1 and H_2 such that

$$\frac{dH_1(t)}{dt} = M(t) \qquad \text{and} \qquad \frac{dH_2(y)}{dy} = N(y). \tag{12}$$

Then Eq. 3 becomes

$$\frac{dH_1}{dt}(t) + \frac{dH_2}{dy}(y)\frac{dy}{dt} = 0. \tag{13}$$

As in Example 1, the chain rule implies that

$$\frac{dH_2}{dy}(y)\frac{dy}{dt} = \frac{d}{dt}H_2(y). \tag{14}$$

Thus Eq. 13 is

$$\frac{dH_1}{dt}(t) + \frac{d}{dt}H_2(y) = 0, \tag{15}$$

from which it follows that

$$H_1(t) + H_2(y) = c, \tag{16}$$

where c is an arbitrary constant, is the solution of Eq. 13.

Observe that we can also write Eqs. 12 as

$$H_1(t) = \int M(t)\,dt \qquad \text{and} \qquad H_2(y) = \int N(y)\,dy; \tag{17}$$

then the solution (16) becomes

$$\int M(t)\,dt + \int N(y)\,dy = c. \tag{18}$$

Thus, in effect, we solve the separable Eq. 11 simply by integrating each term with respect to the variable appearing in it. The justification for this procedure is the argument that we have given.

Observe that the expressions (16) and (18) give the solution y of Eq. 3 implicitly as a function of t. To obtain the solution in explicit form we must solve Eq. 16 or Eq. 18 for y. In many cases this will be difficult or impossible. Even for a simple equation such as that in Example 1, the solution (9) is a cubic equation and not readily solvable for y. In contrast, the expression (22) in Section 10.1 for the solution of a first order linear differential equation gives the solution explicitly.

We now consider some additional examples of separable equations.

EXAMPLE 2

Find the solution of

$$yy' = 2t\sqrt{y^2 + 4} \tag{19}$$

that passes through the point $(2, -\sqrt{5})$.

Equation 19 can be written in the form

$$y(y^2 + 4)^{-1/2}\, dy = 2t\, dt, \tag{20}$$

and is therefore a separable equation. Upon integrating both sides of Eq. (20), we obtain

$$(y^2 + 4)^{1/2} = t^2 + c, \tag{21}$$

where c is an arbitrary constant. To evaluate c we set $t = 2$ and $y = -\sqrt{5}$ in Eq. 21; then $3 = 4 + c$, so $c = -1$ and

$$(y^2 + 4)^{1/2} = t^2 - 1. \tag{22}$$

In this case we can also obtain the solution in explicit form. By squaring both sides of Eq. 22 we find that

$$y^2 + 4 = (t^2 - 1)^2.$$

Hence

$$y^2 = (t^2 - 1)^2 - 4 = t^4 - 2t^2 - 3,$$

and

$$y = -\sqrt{t^4 - 2t^2 - 3}. \tag{23}$$

Note that the negative square root must be chosen in Eq. 23 so that $y = -\sqrt{5}$ when $t = 2$, as the initial condition requires. ∎

EXAMPLE 3

Find the solution of the initial value problem

$$\frac{dy}{dt} = 2(1 + t)(1 + y^2), \qquad y(0) = 1. \tag{24}$$

First we rewrite the differential equation in the separable form

$$\frac{dy}{1 + y^2} = 2(1 + t)\,dt.$$

Then, by integrating both sides, we find that

$$\arctan y = 2t + t^2 + c. \tag{25}$$

To determine c we set $t = 0$ and $y = 1$ in Eq. 25. Thus $c = \arctan 1 = \pi/4$ and

$$\arctan y = t^2 + 2t + \pi/4,$$

or

$$y = \tan(t^2 + 2t + \pi/4). \tag{26}$$

Again, in this case it was possible to find the solution in explicit form. ∎

The logistic equation

One of the most important separable (but nonlinear) first-order equations is

$$\frac{dy}{dt} = ry\left(1 - \frac{y}{K}\right), \tag{27}$$

where r and K are positive constants. Equation 27 is sometimes called the **logistic equation,** and arises in a number of different applications.

One field in which Eq. 27 occurs is the study of the growth of an isolated population of some species. In this case y is the size of the population, r is the growth rate, and K is the maximum sustainable population. If we write Eq. 27 as

$$\frac{dy}{dt} = \frac{r}{K}y(K - y),$$

then it expresses the hypothesis that the rate of change of the population, dy/dt, is proportional to the product of the current population, y, and the difference, $K - y$, between the current population and the maximum population K.

Another application that involves Eq. 27 is epidemiology, or the study of the spread of infectious diseases. Suppose that y is the proportion of a given population that has a certain disease and can infect others, and x is the proportion of uninfected, but susceptible, persons. Assuming that everyone falls into one of these two groups, it follows that

$$x + y = 1. \tag{28}$$

The simplest model of the spread of the disease is that the rate of change in the proportion of sick people, dy/dt, is proportional to the number of encounters between infectious and susceptible individuals. In turn, if we assume that sick and well people move about freely among each other, then the number of such encounters should be proportional to the product of x and y. In symbols,

$$\frac{dy}{dt} = rxy = ry(1 - y), \tag{29}$$

where we have substituted for x from Eq. 28. Observe that Eq. 29 is the same as Eq. 27 with $K = 1$.

In these applications, and in most others in which Eq. 27 occurs, we are mainly interested in the solution of Eq. 27 for values of y in the interval $0 < y < K$. To solve Eq. 27 we first separate the variables; thus

$$\frac{dy}{y(1 - y/K)} = r\,dt. \tag{30}$$

Next we expand the left side of Eq. 30 in partial fractions, so that

$$\left(\frac{1}{y} + \frac{1/K}{1 - y/K}\right) dy = r\,dt. \tag{31}$$

Then, by integrating both sides of Eq. 31, we obtain

$$\ln y - \ln\left(1 - \frac{y}{K}\right) = rt + c, \tag{32}$$

where c is an arbitrary integration constant. Observe that we have dropped the absolute values in the logarithmic terms, since both y and $1 - y/K$ are positive if $0 < y < K$.

Suppose that an initial condition

$$y(0) = y_0 \tag{33}$$

is also given. Then, by substituting $t = 0$ and $y = y_0$ in Eq. 32, we find that

$$\ln y_0 - \ln\left(1 - \frac{y_0}{K}\right) = c,$$

or

$$c = \ln \frac{y_0}{1 - y_0/K}. \tag{34}$$

Then we can write Eq. 32 as

$$\ln \frac{y}{1 - y/K} = rt + \ln \frac{y_0}{1 - y_0/K}. \tag{35}$$

Taking the exponential of both sides of Eq. 35, we obtain

$$\frac{y}{1 - y/K} = \frac{y_0 e^{rt}}{1 - y_0/K}. \tag{36}$$

Finally, we can solve Eq. 36 for y, with the result that

$$y = \frac{y_0}{y_0/K + (1 - y_0/K)e^{-rt}}. \tag{37}$$

If the initial value y_0 is equal to zero, then y is zero for all t; similarly, if $y_0 = K$, then $y = K$ for all t. If $0 < y_0 < K$, then y increases as t increases, and $y(t) \to K$ as $t \to \infty$. Regardless of the values of r and K, the graphs of the solutions (37) are similar to those in Figure 10.2.1.

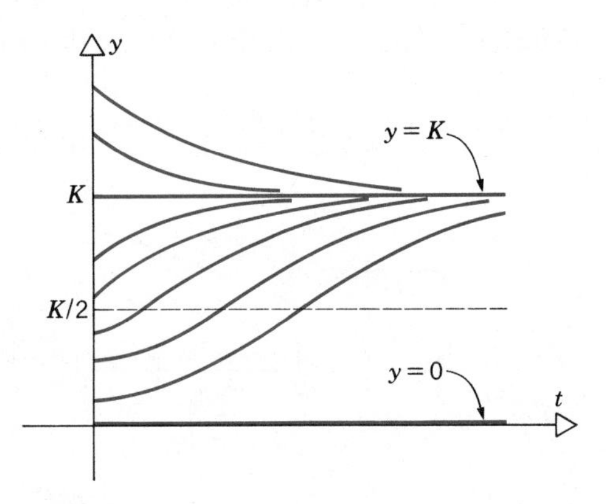

Figure 10.2.1
Solutions of $y' = ry(1 - y/K)$.

If $y < 0$ or if $y > K$, then one can solve Eq. 27 in a similar way. All that is required is the insertion and proper interpretation of absolute values in the logarithmic terms in Eq. 32. It turns out that in the end one again obtains Eq. 37 as the solution of the initial value problem (27), (33). Some solutions for $y_0 > K$ are also shown in Figure 10.2.1.

EXAMPLE 4

A certain population satisfies the logistic equation

$$\frac{dy}{dt} = 0.5y\left(1 - \frac{y}{K}\right), \tag{38}$$

where t is in years. If initially the population is $0.1K$, find how long it takes the population to double in size. Also find the population after five years.

The differential equation is a special case of Eq. 27, so it can be solved in the same way. Rather than repeating this calculation, we can use the solution (37), substituting 0.5 for r and $0.1\ K$ for y_0. Thus

$$y = \frac{0.1K}{0.1 + 0.9e^{-t/2}}. \tag{39}$$

To find the time τ at which the population has doubled we set $y = 0.2K$ and $t = \tau$ in Eq. 39 and solve for τ. We obtain

$$e^{-\tau/2} = \frac{4}{9},$$

so

$$\tau = 2 \ln \frac{9}{4} \cong 1.62 \text{ years}.$$

To find the population after five years set $t = 5$ in Eq. 39. This gives

$$y = \frac{0.1K}{0.1 + 0.9e^{-2.5}} \cong 0.575K.$$

The graph of the solution (39) is shown in Figure 10.2.2. ∎

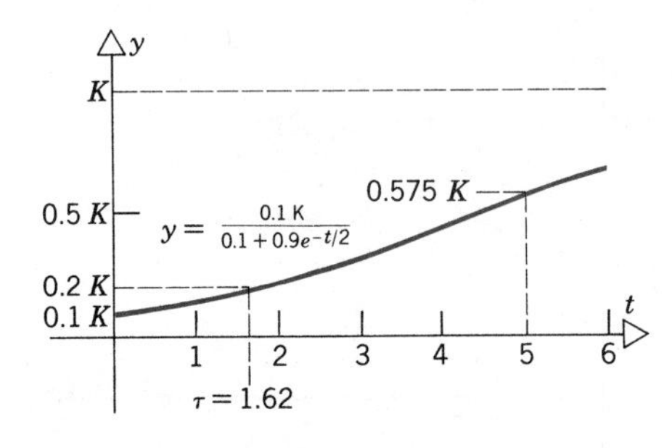

Figure 10.2.2 Solution of the initial value problem $y' = 0.5y(1 - y/K)$, $y(0) = 0.1K$.

PROBLEMS

In each of Problems 1 through 8, solve the given differential equation.

1. $\dfrac{dy}{dt} = \dfrac{t^2}{y}$
2. $\dfrac{dy}{dt} = \dfrac{t^2}{1 + y^2}$
3. $\dfrac{dy}{dt} = \dfrac{2y \cos 2t}{1 + 2y^2}$
4. $\dfrac{dy}{dt} = 1 + t^2 + y^2 + t^2y^2$
5. $\dfrac{dy}{dt} = \dfrac{t^2}{y(1 + t^3)}$
6. $y' + (\sin t)y^2 = 0$
7. $ty' = \sqrt{1 - y^2}$
8. $\dfrac{dy}{dt} = \dfrac{t - e^{-t}}{y + e^y}$

In each of Problems 9 through 18, find the solution $y = \phi(t)$ of the given initial value problem in explicit form.

9. $y' = \dfrac{2t}{y + t^2y}, \quad y(0) = 2$
10. $\dfrac{dy}{dt} = \dfrac{2t}{1 + 2y}, \quad y(2) = 1$
11. $\dfrac{dy}{dt} = ty^3/\sqrt{1 + t^2}, \quad y(0) = 1$
12. $\dfrac{dy}{dt} = -ty^2, \quad y(0) = -2$
13. $\dfrac{dy}{dt} = \dfrac{2t - 1}{2y + 2}, \quad y(3) = -3$
14. $yy' = t(1 + y^2), \quad y(0) = -1$
15. $y' = 2t \exp(3t + y), \quad y(0) = 0$
16. $2(2 + y)y' = t, \quad y(0) = -1$
17. $yy' = 3(2t + 1)^2, \quad y(0) = -1$
18. $y' = \dfrac{t(t^2 + 1)}{4y^3}, \quad y(0) = -\dfrac{1}{\sqrt{2}}$

19. Suppose that a certain population grows according to the logistic equation

$$\frac{dy}{dt} = ry\left(1 - \frac{y}{K}\right).$$

(a) If $y_0 = K/3$, find the time τ at which the initial population has doubled. Also find the value of τ if $r = 0.025$ per year.

(b) If $y_0 = \alpha K$, find the time T at which $y(T) = \beta K$, where $0 < \alpha, \beta < 1$. Find the value of T if $r = 0.025$ per year, $\alpha = 0.1$, and $\beta = 0.9$. Observe that $T \to \infty$ as $\alpha \to 0$ or as $\beta \to 1$.

20. Another equation that has been used to model population growth is

$$\frac{dy}{dt} = ry \ln \frac{K}{y},$$

which is known as the Gompertz equation.

(a) Solve the Gompertz equation subject to the initial condition $y(0) = y_0$.

(b) If $y_0 = K/3$, find the time τ at which the population has doubled. Also find the value of τ for $r = 0.025$ per year. Compare the results with those of Problem 19(a).

21. Solve the equation

$$\frac{dy}{dt} = \frac{at + b}{ct + d}$$

where a, b, c, and d are constants and $c \neq 0$. Also assume $ct + d \neq 0$.

* **22.** The equation

$$\frac{dy}{dt} = \frac{y - 4t}{t - y}$$

is not separable. Show that if the variable y is replaced by a new variable v defined by $v = y/t$, then the equation in t and v is separable. Find the solution of the given equation by this technique.

23. Some diseases (such as typhoid fever) are spread largely by *carriers*, individuals who can transmit the disease but who exhibit no overt symptoms. Let x and y, respectively, denote the proportion of carriers and of well, but susceptible, people in the population at time t. Suppose that carriers are removed from the population at a rate β, so that

$$\frac{dx}{dt} = -\beta x. \qquad (i)$$

Suppose also that the disease spreads at a rate proportional to the product of x and y. Thus

$$\frac{dy}{dt} = -\alpha xy. \qquad (ii)$$

(a) Determine x at any time t by solving Eq. (i) subject to the initial condition $x(0) = x_0$.

(b) Use the result of Part (a) to find y at any time t by solving Eq. (ii) subject to the initial condition $y(0) = y_0$.

(c) Find the proportion of the population that escapes the epidemic; that is, determine the limiting value of y as $t \to \infty$. Also determine how this quantity varies with α and β.

24. Assume that, by evaporation, the volume of a spherical raindrop decreases at a rate proportional to its surface area. If its radius originally is 3 mm, and 20 min later has been reduced to 2 mm, find an expression for the radius of the raindrop at any time t.

Chemical Reactions

A second-order chemical reaction involves the interaction (collision) of one molecule of a substance P with one molecule of a substance Q in order to produce one molecule of a new substance X; $P + Q \to X$ is the usual notation. Suppose that p and q are the initial concentrations of substances P and Q, respectively, and let $x(t)$ be the concentration of X at time t. Then $p - x(t)$ and $q - x(t)$ are the concentrations of P and Q at time t, and the rate at which the reaction occurs is described by the equation

$$\frac{dx}{dt} = \alpha(p - x)(q - x),$$

where α is a positive constant.

25. If X is produced by the chemical reaction previously described, find the concentration $x(t)$ of X at any time t, assuming that $x(0) = 0$.

26. If the substances P and Q are the same, then the reaction equation becomes

$$\frac{dx}{dt} = \alpha(p - x)^2.$$

If $x(0) = 0$, find $x(t)$ at any time t.

Motion According to Newton's Law

Consider a particle of mass m moving on a straight line. Let x, $v = dx/dt$, and $a = dv/dt$ be the position, velocity, and acceleration of the particle. Newton's law states that $F = ma$, where F is the net force acting on the particle. The force F may depend upon the time t, the position x, and the velocity v. Suppose that the particle is moving under the influence of gravity in a medium that offers resistance proportional to the velocity of the particle. Assume that x is positive in the upward direction. Then the forces acting on the particle are shown in Figure 10.2.3. They are as follows:

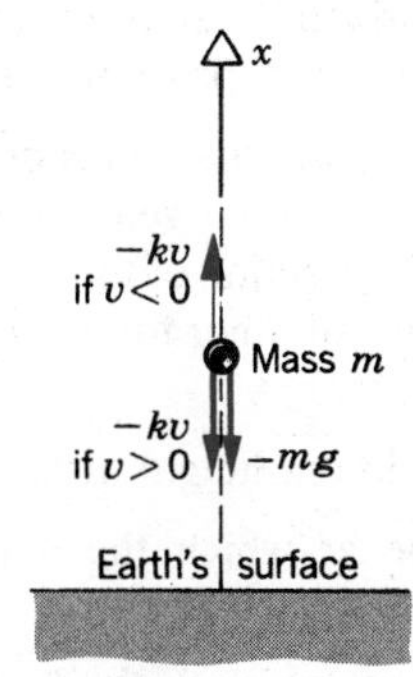

Figure 10.2.3 Forces on a moving particle.

(a) The force of gravity, which always acts downward, and is given by $-mg$, where g is the acceleration due to gravity. In the vicinity of the earth's surface it is sufficient to take g to be a constant with a value of 32 ft/sec^2 or 9.8 m/sec^2.

(b) The resistance force, which always acts in the direction opposite to the direction of motion. This force is given by $-kv$, where k is a constant of proportionality. Note that the resistance is directed downward when v is positive and the particle is therefore moving upward; similarly, the resistance is directed upward when v is negative.

Under these assumptions the equation of motion is

$$m\frac{dv}{dt} + kv = -mg.$$

If the particle is set in motion with an initial velocity v_0, then the initial condition is $v(0) = v_0$.

27. A ball with mass 0.25 kg is thrown upward with initial velocity 20 m/sec from the roof of a building 30 m high. Neglect air resistance.

(a) Find the maximum height above the ground that the ball reaches.

(b) Assuming that the ball misses the building on the way down, find the time that it hits the ground.

28. Assume that the conditions are as in Problem 27 except that there is a force due to air resistance of $|v|/30$, where the velocity v is in meters per second.

(a) Find the maximum height above the ground that the ball reaches.

(b) Find the time that the ball hits the ground.

Hint: Use Newton's method or some other appropriate numerical procedure in Part (b).

29. A body of constant mass m is projected vertically upward with an initial velocity v_0. Assuming the gravitational attraction of the earth to be constant, and neglecting all other forces acting on the body, find the following.

(a) The maximum height attained by the body.

(b) The time at which the maximum height is reached.

(c) The time at which the body returns to its starting point.

* **30.** A body of constant mass m is projected upward from the earth's surface with initial velocity v_0. Assuming that there is no air resistance, but taking account of the inverse square variation of the earth's gravitational field with altitude (see Figure 10.2.4), we have

$$m\frac{dv}{dt} = -\frac{mgR^2}{(x+R)^2}, \qquad v(0) = v_0,$$

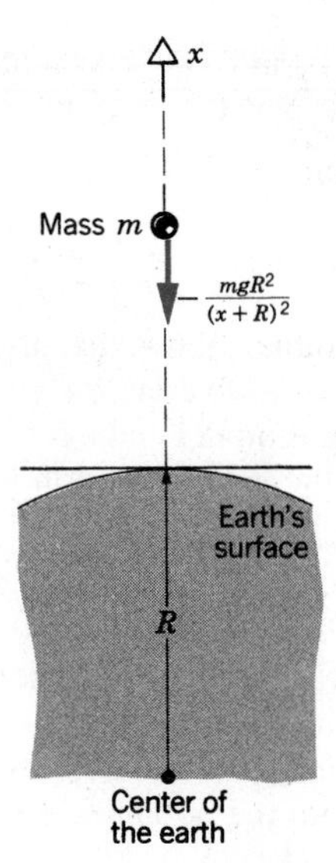

Figure 10.2.4

where R is the radius of the earth and $x(t)$ is measured from the surface of the earth. Since the force is a function of x, it is convenient to think of x, rather than t, as the independent variable. We can do this by noting that

$$\frac{dv}{dt} = \frac{dv}{dx}\frac{dx}{dt} = v\frac{dv}{dx}.$$

Hence

$$v\frac{dv}{dx} = \frac{-gR^2}{(x+R)^2}$$

and $v = v_0$ when $x = 0$.

(a) Show that

$$v^2 = v_0^2 - 2gR + \frac{2gR^2}{x+R}.$$

(b) The **escape velocity** v_e can be found by requiring that the velocity v remain positive for all (positive) x. Also see Problem 36 in Section 11.3. Show that $v_e = (2gR)^{1/2}$, which is approximately 6.9 mi/sec (11.1 km/sec). If the effect of air resistance is included, then the escape velocity is somewhat higher. The effective escape velocity can be significantly reduced if the body is transported a considerable distance above sea level before being launched, since the gravitational force and especially the air resistance decrease with increasing altitude.

10.3 NUMERICAL METHODS

Although many first order differential equations can be solved analytically by the methods discussed in the previous two sections, a great many others are neither linear nor separable, so these methods are not applicable to them. Even for an equation that is linear or separable, it may well be that the integrations required in the analytic solution are too difficult to be performed by elementary methods. For both of these reasons it is desirable to develop numerical methods for solving differential equations.

Consider the initial value problem

$$y' = f(t, y), \qquad y(t_0) = y_0, \tag{1}$$

with solution $y = \phi(t)$. By a numerical method for solving this initial value problem we mean a procedure for calculating approximate values $y_0, y_1, y_2, \ldots, y_n, \ldots$ of the solution ϕ at the points $t_0 < t_1 < t_2 < \cdots < t_n \ldots$ (see Figure 10.3.1). The calculated data may be presented either in tabular or graphical form.

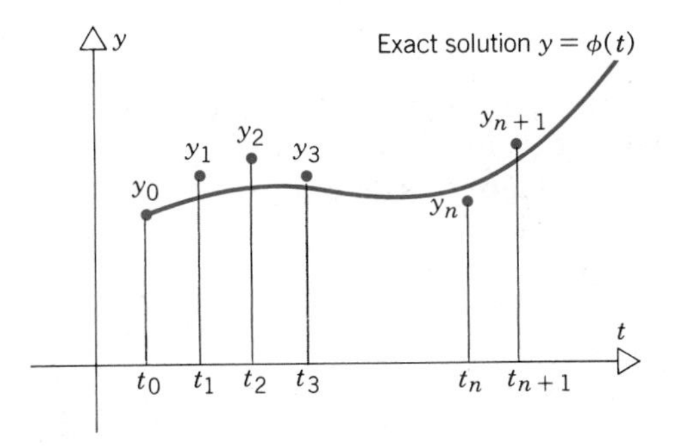

Figure 10.3.1 A numerical approximation to the solution of $y' = f(t, y), y(t_0) = y_0$.

The notation that we use is as follows. The solution of the initial value problem (1) is ϕ [or $y = \phi(t)$], and the value of the solution at t_n is $\phi(t_n)$. The symbols y_n and $y_n' = f(t_n, y_n)$ denote approximate values of the solution and its derivative at t_n. Clearly $y_0 = \phi(t_0)$ and $y_0' = \phi'(t_0)$, but in general $y_n \neq \phi(t_n)$ and $y_n' \neq \phi'(t_n)$ for $n \geq 1$. Further, for convenience, we choose a uniform spacing or step size h on the t-axis. Thus $t_1 = t_0 + h$, $t_2 = t_1 + h = t_0 + 2h$, and in general $t_n = t_0 + nh$.

Euler or tangent line method

The simplest numerical method of solving the initial value problem (1) is the tangent line method, which was first used by Euler.* Since t_0 and y_0 are known, the slope of the tangent line to the solution at t_0, namely $\phi'(t_0) = f(t_0, y_0)$, is also known. Hence we can construct the tangent line to the solution at t_0 and then obtain the

*Among Euler's many important and far-reaching contributions to mathematics was the first systematic use of numerical methods to solve differential equations. He developed and used the tangent line method beginning about 1768.

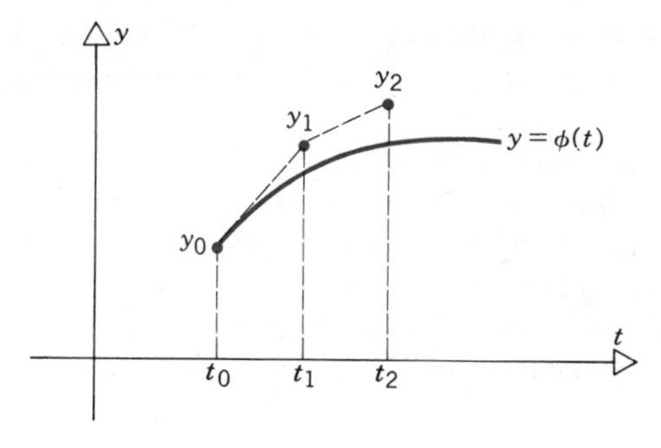

Figure 10.3.2 The Euler or tangent line approximation.

approximate value y_1 of $\phi(t_1)$ by moving along the tangent line to t_1 (see Figure 10.3.2). The tangent line at (t_0, y_0) is

$$y - y_0 = \phi'(t_0)(t - t_0).$$

Hence

$$\begin{aligned} y_1 &= y_0 + \phi'(t_0)(t_1 - t_0) \\ &= y_0 + f(t_0, y_0)h. \end{aligned} \tag{2}$$

Note that this is just the linear approximation to the function ϕ near $t = t_0$ that was discussed in Section 3.4.

Once y_1 is determined, we can compute $y_1' = f(t_1, y_1)$, which is an approximate value of $\phi'(t_1)$. Using this approximate value of the slope, we obtain an approximation to $\phi(t_2)$, namely

$$\begin{aligned} y_2 &= y_1 + (t_2 - t_1)y_1' \\ &= y_1 + hy_1'. \end{aligned}$$

At the nth step we have

$$y_{n+1} = y_n + hy_n', \quad \text{where} \quad y_n' = f(t_n, y_n), \qquad n = 0, 1, 2, \ldots. \tag{3}$$

Equation 3 is called the Euler, or tangent line, formula.

The Euler method consists of nothing more than using the formula (3) over and over again. At each step the calculation of the next approximate value y_{n+1} at t_{n+1} makes use of the result y_n of the preceding step.

EXAMPLE 1

Use the Euler formula (3) and the step size $h = 0.1$ to determine an approximate value of the solution $y = \phi(t)$ at $t = 0.2$ for the initial value problem

$$y' = 1 - t + 4y, \qquad y(0) = 1. \tag{4}$$

First, we observe that the differential equation $y' = 1 - t + 4y$ is linear and can therefore be solved by the method developed in Section 10.1. The solution is

$$y = \phi(t) = \frac{1}{4}t - \frac{3}{16} + \frac{19}{16}e^{4t}. \tag{5}$$

We have chosen for illustrative purposes a simple initial value problem that we can solve easily, so that we can compare the values of the numerical solution with those of the analytical solution.

For the numerical solution, we first compute

$$y_0' = f(0, 1) = 1 - 0 + 4(1) = 5.$$

Then

$$y_1 = y_0 + hy_0' = 1 + (0.1)(5) = 1.5.$$

Next

$$y_1' = f(t_1, y_1) = f(0.1, 1.5)$$
$$= 1 - 0.1 + 4(1.5) = 6.9,$$

and

$$y_2 = y_1 + hy_1' = 1.5 + (0.1)(6.9) = 2.19.$$

The value of $\phi(0.2)$ is 2.5053299 correct through eight digits. Thus the error at $t = 0.2$ is approximately $|2.19 - 2.51| = 0.32$, or a percentage error of about 12.7 percent.

Normally, an error this large is not acceptable. Better approximations can be obtained by using a smaller step size. Some results are shown in Table 10.1. For instance, for $h = 0.01$, which requires 20 steps to reach $t = 0.2$, we obtain the approximate value 2.4645 with a percentage error of about 1.6 percent. The data in Table 10.1 also show that the accuracy deteriorates as t increases. ∎

Table 10.1 A Comparison of Results for the Numerical Solution of $y' = 1 - t + 4y$, $y(0) = 1$ Using the Euler Method for Different Step Sizes h

t	$h = 0.1$	$h = 0.05$	$h = 0.025$	$h = 0.01$	Exact
0	1.0000000	1.0000000	1.0000000	1.0000000	1.0000000
0.1	1.5000000	1.5475000	1.5761188	1.5952901	1.6090418
0.2	2.1900000	2.3249000	2.4080117	2.4644587	2.5053299
0.3	3.1460000	3.4333560	3.6143837	3.7390345	3.8301388
0.4	4.4744000	5.0185326	5.3690304	5.6137120	5.7942260
0.5	6.3241600	7.2901870	7.9264062	8.3766865	8.7120041
0.6	8.9038240	10.550369	11.659058	12.454558	13.052522
0.7	12.505354	15.234032	17.112430	18.478797	19.515518
0.8	17.537495	21.967506	25.085110	27.384136	29.144880
0.9	24.572493	31.652708	36.746308	40.554208	43.497903
1	34.411490	45.588400	53.807866	60.037126	64.897803

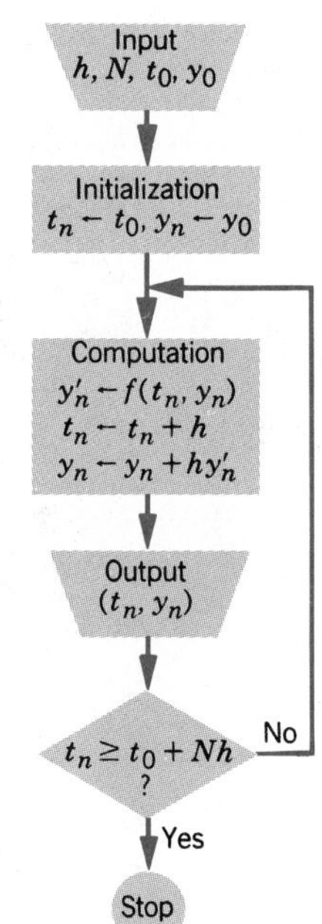

Figure 10.3.3 Flow chart for the Euler method.

The essential steps in solving the initial value problem (1) for $t_0 \leq t \leq t_0 + Nh$ by the Euler method are shown in Figure 10.3.3 in flow chart form. The first box indicates the parameters that must be given values in order to start the computation. In the second box the variables t_n and y_n are set equal to their initial values t_0 and y_0, respectively. The third box represents the calculation of $f(t_n, y_n)$ and the updating of t_n and y_n by the Euler formula. The results may be printed either for each value of n or for some preselected set of values. As long as $t_n < t_0 + Nh$, the computation is repeated using the new values of t_n and y_n obtained in the preceding step. When $t_n = t_0 + Nh$, the process stops.

Although the Euler method is simple in concept and easy to use, the data in Table 10.1 suggest that even for this very simple problem the Euler method is not very accurate. More precisely, it can be shown (using the theory of Taylor polynomials discussed in Chapter 13) that if the value of y_n at t_n is correct, then the error in calculating y_{n+1} at t_{n+1} is proportional to h^2. This error is known as the **local discretization error.** Although the proportionality constant in this error expression usually cannot be determined, it is still useful to know that the local discretization error behaves like a constant times h^2. For example, we can conclude that if the step size h is halved, then the local discretization error is reduced by a factor of 4, and so forth.

The lack of accuracy in the Euler method is due mainly to the fact that the local discretization error is proportional only to h^2. Fortunately, there are other methods that are considerably more accurate than the Euler method, and that are only slightly (if at all) more difficult to use. We will discuss two of these methods, which have local discretization errors proportional to h^3 and h^5, respectively. In each case the outline of the computation is still given by the flow chart in Figure 10.3.3. The only difference is that the new value of y_n is calculated from a different, and more accurate, formula.

Improved Euler method

If $y = \phi(t)$ is the exact solution of the initial value problem (1), and if we integrate the differential equation from t_n to t_{n+1}, we have

$$\int_{t_n}^{t_{n+1}} \phi'(t)\, dt = \int_{t_n}^{t_{n+1}} f[t, \phi(t)]\, dt$$

or

$$\phi(t_{n+1}) = \phi(t_n) + \int_{t_n}^{t_{n+1}} f[t, \phi(t)]\, dt. \tag{6}$$

We can obtain different approximations for $\phi(t_{n+1})$ by making different approximations in calculating the integral on the right side of Eq. 6. For example, to obtain the Euler formula (3) we approximate the integrand by its value at t_n, as shown in Figure 10.3.4. Then

$$\begin{aligned}\phi(t_{n+1}) &\cong \phi(t_n) + \int_{t_n}^{t_{n+1}} f[t_n, \phi(t_n)]\, dt \\ &\cong \phi(t_n) + f[t_n, \phi(t_n)](t_{n+1} - t_n).\end{aligned}$$

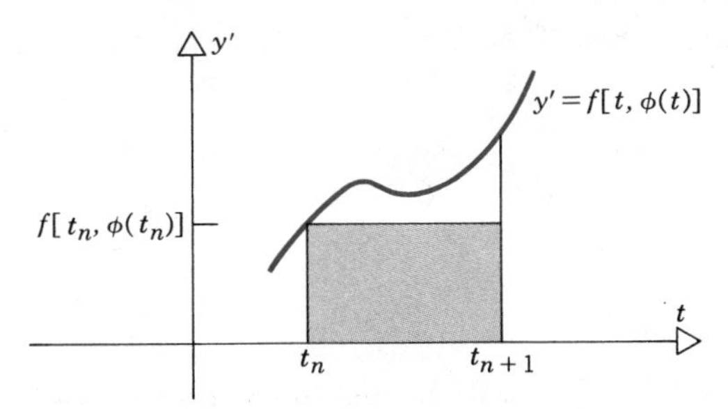

Figure 10.3.4 Integral derivation of the Euler method.

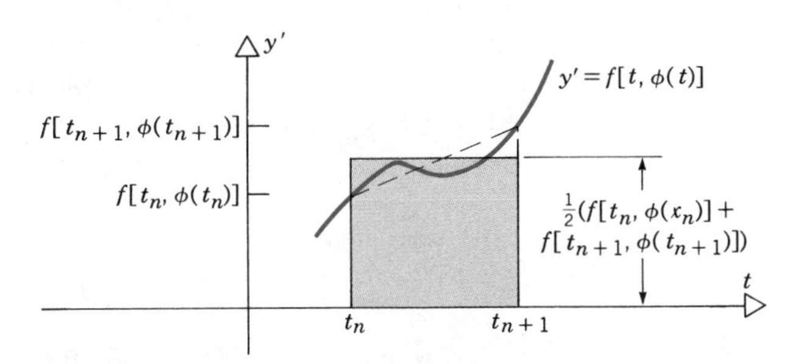

Figure 10.3.5 Derivation of the improved Euler method.

The further approximation $\phi(t_n) \cong y_n$ gives the Euler formula

$$y_{n+1} = y_n + hf(t_n, y_n) = y_n + hy'_n,$$

in agreement with Eq. 3.

We now go back to Eq. 6 and approximate the integral in a different way, by using the average of the values of the integrand at t_n and at t_{n+1}; see Figure 10.3.5. Thus

$$\phi(t_{n+1}) \cong \phi(t_n) + \tfrac{1}{2}\{f[t_n, \phi(t_n)] + f[t_{n+1}, \phi(t_{n+1})]\}(t_{n+1} - t_n).$$

If we again replace $\phi(t_n)$ and $\phi(t_{n+1})$ by the approximate values y_n and y_{n+1}, respectively, then we obtain

$$y_{n+1} = y_n + \tfrac{1}{2}[f(t_n, y_n) + f(t_{n+1}, y_{n+1})]h. \tag{7}$$

Since y_{n+1} appears as one of the arguments of f on the right side of Eq. 7, it will in general be fairly difficult to solve this equation for y_{n+1}. This difficulty can be overcome by approximating y_{n+1} on the right side of Eq. 7 by the value obtained using the Euler formula (3). Thus

$$\begin{aligned} y_{n+1} &= y_n + [f(t_n, y_n) + f(t_{n+1}, y_n + hy'_n)]\frac{h}{2} \\ &= y_n + [y'_n + f(t_{n+1}, Y_{n+1})]\frac{h}{2}, \quad \text{where} \quad Y_{n+1} = y_n + hy'_n. \end{aligned} \tag{8}$$

This formula is known as the improved Euler formula or the Heun formula. It can be shown that the local discretization error is proportional to h^3.

EXAMPLE 2

Use the improved Euler formula (8) to calculate approximate values of the solution of the initial value problem (4)

$$y' = 1 - t + 4y, \qquad y(0) = 1.$$

Let $f(t, y) = 1 - t + 4y$, and suppose that $h = 0.1$. Since $t_0 = 0$ and $y_0 = 1$, we have, as in Example 1, $y_0' = 1 - 0 + 4 = 5$. Then $Y_1 = y_0 + hy_0' = 1 + (0.1)5 = 1.5$ and

$$f(t_1, Y_1) = 1 - 0.1 + 4(1.5) = 6.9.$$

Finally

$$y_1 = 1 + (0.5)(0.1)(5 + 6.9) = 1.595.$$

Further results are shown in Table 10.2. The data in this table show that the results obtained from the improved Euler method with $h = 0.1$ are indeed better than those obtained from the Euler method with $h = 0.1$, and are also better than those obtained from the Euler method with $h = 0.05$. They are even better than the results obtained from the Euler method with $h = 0.025$, and are almost as good as those obtained from the Euler method with $h = 0.01$, as can be seen by comparing Tables 10.1 and 10.2. ∎

Table 10.2 A Comparison of Results Using the Euler and Improved Euler Methods for $h = 0.1$ and $h = 0.05$ for the Initial Value Problem $y' = 1 - t + 4y$, $y(0) = 1$

	Euler		Improved Euler		
t	$h = 0.1$	$h = 0.05$	$h = 0.1$	$h = 0.05$	Exact
0	1.0000000	1.0000000	1.0000000	1.0000000	1.0000000
0.1	1.5000000	1.5475000	1.5950000	1.6049750	1.6090418
0.2	2.1900000	2.3249000	2.4636000	2.4932098	2.5053299
0.3	3.1460000	3.4333560	3.7371280	3.8030484	3.8301388
0.4	4.4744000	5.0185326	5.6099494	5.7404023	5.7942260
0.5	6.3241600	7.2901870	8.3697252	8.6117498	8.7120041
0.6	8.9038240	10.550369	12.442193	12.873253	13.052522
0.7	12.505354	15.234032	18.457446	19.203865	19.515518
0.8	17.537495	21.967506	27.348020	28.614138	29.144880
0.9	24.572493	31.652708	40.494070	42.608178	43.497903
1.0	34.411490	45.588400	59.938223	63.424698	64.897803

For complicated equations the most time-consuming (and therefore most expensive) part of the computation is the repeated evaluation of $f(t, y)$. Consequently, a rough comparison of the time required to use two different methods can be made by counting the number of times that $f(t, y)$ must be evaluated. The improved Euler method requires two function evaluations at each step, whereas the Euler method requires only one. Thus the improved Euler method with step size h should use about the same amount of computing time as the Euler method with step size $h/2$. For the problem (4), and in the great majority of cases, the improved Euler method yields considerably greater accuracy under these conditions.

Runge-Kutta method

One of the most often used numerical methods is the Runge-Kutta* method. It can be derived from Eq. 6 by approximating the integrand by a weighted average of values taken at the points t_n, $t_n + h/2$, and t_{n+1} in a manner similar to approximating the value of an integral using Simpson's rule. This gives rise to a formula involving expressions such as $f[t_n + h/2, \phi(t_n + h/2)]$. It is then necessary to make additional approximations to calculate $\phi(t_n + h/2)$. The resulting formula is

$$y_{n+1} = y_n + \frac{h}{6}(k_{n1} + 2k_{n2} + 2k_{n3} + k_{n4}) \tag{9}$$

where

$$\begin{aligned} k_{n1} &= f(t_n, y_n), \\ k_{n2} &= f(t_n + \tfrac{1}{2}h, y_n + \tfrac{1}{2}hk_{n1}), \\ k_{n3} &= f(t_n + \tfrac{1}{2}h, y_n + \tfrac{1}{2}hk_{n2}), \\ k_{n4} &= f(t_n + h, y_n + hk_{n3}). \end{aligned} \tag{10}$$

The sum $(k_{n1} + 2k_{n2} + 2k_{n3} + k_{n4})/6$, which multiplies h in Eq. 9, can be interpreted as an average slope—that is, as an average value for $f(t, y)$. Note that k_{n1} is the slope at the left end of the interval, k_{n2} is the slope at the midpoint using the Euler formula to go from t_n to $t_n + h/2$, k_{n3} is a second approximation to the slope at $t_n + h/2$ using the slope k_{n2} to go from t_n to $t_n + h/2$, and finally k_{n4} is the slope at the right end of the interval using the Euler formula with the slope k_{n3} to go from t_n to $t_n + h$. It is possible to show that the local discretization error for the Runge-Kutta method is proportional to h^5; hence it is a very accurate method. For example, reducing the step size by a factor of four reduces the error by a factor of 4^5, or 1024.

EXAMPLE 3

Using the Runge-Kutta method with $h = 0.2$, compute an approximate value of the solution $y = \phi(t)$ at $t = 0.2$ for the initial value problem (4)

$$y' = 1 - t + 4y, \qquad y(0) = 1.$$

Since $h = 0.2$, we have

$$k_{01} = f(0, 1) = 5, \qquad hk_{01} = 1$$

$$k_{02} = f(0 + 0.1, 1 + 0.5) = 6.9, \qquad hk_{02} = 1.38$$

*Carl David Runge (1856–1927) was led to consider problems in numerical computation by his work in spectroscopy, and the Runge-Kutta method originated in his paper on the numerical solution of differential equations in 1895. The method was extended in 1901 by M. Wilhelm Kutta (1867–1944), a German mathematician and aerodynamicist, who also made important contributions to the theory of airfoils.

$$k_{03} = f(0 + 0.1, 1 + 0.69) = 7.66, \qquad hk_{03} = 1.532$$

$$k_{04} = f(0 + 0.2, 1 + 1.532) = 10.928.$$

Thus

$$y_1 = 1 + \frac{0.2}{6}[5 + 2(6.9) + 2(7.66) + 10.928]$$

$$= 1 + 1.5016 = 2.5016.$$

Some further results are shown in Table 10.3. These indicate that the Runge-Kutta method is reasonably accurate, even with a rather large step size. For instance, with $h = 0.2$ the Runge-Kutta approximation for this example is in error by less than one percent on the interval $0 \le t \le 1$. Much greater accuracy can be achieved by using a smaller value of h. For this problem, with $h = 0.01$, we obtain results that are correct to at least five decimal places throughout the interval $0 \le t \le 1$.

Table 10.3 Comparison of Results for the Numerical Solution of the Initial Value Problem $y' = 1 - t + 4y$, $y(0) = 1$

t	Euler $h = 0.1$	Improved Euler $h = 0.1$	Runge-Kutta $h = 0.2$	Runge-Kutta $h = 0.1$	Exact
0	1.0000000	1.0000000	1.0000000	1.0000000	1.0000000
0.1	1.5000000	1.5950000		1.6089333	1.6090418
0.2	2.1900000	2.4636000	2.5016000	2.5050062	2.5053299
0.3	3.1460000	3.7371280		3.8294145	3.8301388
0.4	4.4774000	5.6099494	5.7776358	5.7927853	5.7942260
0.5	6.3241600	8.3697252		8.7093175	8.7120041
0.6	8.9038240	12.442193	12.997178	13.047713	13.052522
0.7	12.505354	18.457446		19.507148	19.515518
0.8	17.537495	27.348020	28.980768	29.130609	29.144880
0.9	24.572493	40.494070		43.473954	43.497903
1.0	34.411490	59.938223	64.441579	64.858107	64.897803

Observe that the Runge-Kutta method requires four function evaluations at each step. Thus with $h = 0.1$ we need ten steps (forty function evaluations) to go from $t = 0$ to $t = 1$. For comparison, the improved Euler method with $h = 0.05$ also requires forty function evaluations to traverse the same interval. It is easy to see from Tables 10.2 and 10.3 that the Runge-Kutta method is far superior in terms of accuracy. ∎

As illustrated in this example, the Runge-Kutta method is a practical procedure for the numerical solution of first order differential equations. It is straightforward to program and provides a good combination of simplicity and accuracy.

Until you have advanced a good deal farther in your understanding of the subtleties of differential equations and of numerical computation, the Runge-Kutta method is a good place to start when the numerical solution of a differential equation is called for.

PROBLEMS

The first 16 problems below call for the calculation of approximate values of the solution of a given differential equation. Problems 1 through 8 can be done on a pocket calculator, if necessary, although a programmable calculator is desirable. For Problems 9 through 16 one should have at least a programmable calculator or a microcomputer. In all of these problems the answers are rounded to six figures. In working on these problems you should pay attention to the effects of using different methods and different step sizes.

In each of Problems 1 and 2

(a) Find approximate values of the solution of the given initial value problem at $t = 0.1, 0.2, 0.3$, and 0.4 using the Euler method with $h = 0.1$.

(b) Repeat Part (a) with $h = 0.05$. Compare the results with those in (a).

(c) Repeat Part (a) using the improved Euler method with $h = 0.1$.

(d) Find approximate values of the solution of the given initial value problem at $t = 0.2$ and 0.4 using the Runge-Kutta method with $h = 0.2$.

(e) Find the exact solution $y = \phi(t)$ of the given initial value problem and determine its value at $t = 0.1, 0.2, 0.3$, and 0.4. Compare these values with those of Parts (a) through (d).

© **1.** $y' = 2y - 1, \quad y(0) = 1$

© **2.** $y' = 0.5 - t + 2y, \quad y(0) = 1$

In each of Problems 3 through 8

(a) Find approximate values of the solution of the given initial value problem at $t = t_0 + 0.1$, $t_0 + 0.2$, $t_0 + 0.3$, and $t_0 + 0.4$ using the Euler method with $h = 0.1$.

(b) Repeat Part (a) using the improved Euler method with $h = 0.1$.

(c) Find approximate values of the solution of the given initial value problem at $t = t_0 + 0.2$ and $t_0 + 0.4$ using the Runge-Kutta method with $h = 0.2$.

© **3.** $y' = t^2 + y^2, \quad y(0) = 1$

© **4.** $y' = 5t - 3\sqrt{y}, \quad y(0) = 2$

© **5.** $y' = \sqrt{t + y}, \quad y(1) = 3$

© **6.** $y' = 2t + e^{-ty}, \quad y(0) = 1$

© **7.** $y' = \dfrac{y^2 + 2ty}{3 + t^2}, \quad y(1) = 2$

© **8.** $y' = (t^2 - y^2)\sin y, \quad y(0) = -1$

In each of Problems 9 through 12, find an approximate value of the solution of the given initial value problem at $t = t_0 + 1$.

(a) Using the Euler method with $h = 0.025$.

(b) Using the Euler method with $h = 0.0125$.

(c) Using the improved Euler method with $h = 0.05$.

(d) Using the improved Euler method with $h = 0.025$.

(e) Using the Runge-Kutta method with $h = 0.1$.

(f) Using the Runge-Kutta method with $h = 0.05$.

Note that 40 function evaluations are required in Parts (a), (c), and (e), while 80 are required in Parts (b), (d), and (f). Compare the results obtained by the various methods.

© **9.** $y' = 0.5 - t + 2y, \quad y(0) = 1$

© **10.** $y' = 5t - 3\sqrt{y}, \quad y(0) = 2$

© **11.** $y' = \sqrt{t + y}, \quad y(1) = 3$

© **12.** $y' = 2t + e^{-ty}, \quad y(0) = 1$

In each of Problems 13 through 16, find an approximate value of the solution of the given initial value problem at $t = t_0 + 1$.

(a) Using the Runge-Kutta method with $h = 0.025$.

(b) Using the Runge-Kutta method with $h = 0.01$.

Compare these results with those of Parts (e) and (f) of Problems 9 through 12.

© **13.** $y' = 0.5 - t + 2y, \quad y(0) = 1$

© **14.** $y' = 5t - 3\sqrt{y}, \quad y(0) = 2$

© **15.** $y' = \sqrt{t + y}, \quad y(1) = 3$

© **16.** $y' = 2t + e^{-ty}, \quad y(0) = 1$

* **17.** Consider the differential equation $y' = f(t)$, that is, f does not depend on y. Show that in this case the improved Euler formula reduces to

$$y_{n+1} = y_n + \frac{h}{2}[f(t_n) + f(t_{n+1})].$$

Show that this is equivalent to the trapezoidal rule for evaluating a definite integral.

* **18.** Under the same conditions as in Problem 17, show that the Runge-Kutta method reduces to Simpson's rule for evaluating a definite integral.

REVIEW PROBLEMS

In each of Problems 1 through 10, solve the given initial value problem.

1. $(e^x + e^{-x})\,y' + (e^x - e^{-x})\,y = e^x, \quad y(0) = 1$

2. $xy' + y = \ln x, \quad y(1) = 1$

3. $\cos x\, y' + \sin x\, y = \sec x, \quad y(0) = \frac{1}{2}$

4. $y' - 2ty = e^{t^2}, \quad y(0) = 0$

5. $y' - e^t y = e^{-e^t} y^2, \quad y(0) = 1$

6. $y' = \dfrac{y}{\sqrt{1 - x^2}}, \quad y(\frac{1}{2}) = 1$

7. $y' = \dfrac{y}{\sqrt{1 - x^2}}, \quad y(0) = 0$

8. $yy' + x = 1, \quad y(1) = 1$

9. $\dfrac{dy}{dx} = \dfrac{4 + y^2}{y}, \quad y(0) = -1$

10. $\dfrac{dy}{dx} = \dfrac{e^x(2e^x - 7)}{2y - 4}, \quad y(0) = 4$

In each of Problems 11 through 14, find an approximate value of the solution of the given initial value problem at $t = t_0 + 1$ using

(a) the Euler method with $h = 0.01$;

(b) the improved Euler method with $h = 0.025$ and $h = 0.01$;

(c) the Runge–Kutta method with $h = 0.05$ and $h = 0.025$.

© **11.** $ty' = y + te^{y/t}, \quad y(1) = 0$

© **12.** $2y + 1 + \left(\dfrac{t^2 - y}{t}\right) y' = 0, \quad y(0) = 1$

© **13.** $ty' - y = (ty)^{1/2}, \quad y(1) = 2$

© **14.** $y' = -\dfrac{3t^2y + y^2}{3t^3 + 3yt}, \quad y(1) = -2$

© **15.** Using the Runge–Kutta method with $h = 0.05$, find approximate values of the solution of the initial value problem for the logistic equation

$$\frac{dy}{dt} = 0.5\, y\left(1 - \frac{y}{10}\right), \quad y(0) = 1$$

at $t = 1, 2, 3, 4, 5$, and 6. The exact value of $y(6)$ is 6.905679, to six decimal places.

© **16.** Using the Runge–Kutta method with $h = 0.05$, find approximate values of the solution of the initial value problem for the Gompertz equation

$$\frac{dy}{dt} = 0.5\, y \ln\left(\frac{10}{y}\right), \quad y(0) = 1$$

at $t = 1, 2, 3, 4, 5$, and 6. The exact value of $y(6)$ is 8.916880, to six decimal places. Compare the answers obtained here with those from Problem 15.

17. The radioactive isotope plutonium-241 decays at a rate proportional to the amount present. If the half-life (the time period during which the mass is reduced to one-half of its original value) of plutonium-241 is 13.20 years, find an equation for the decay of the isotope, assuming there are initially P_0 mg present. If 50 mg of plutonium are present today, how much will remain in 10 years?

18. A certain radioactive material has a half-life of 10 days. What time period must elapse before a given mass of this material is reduced to one percent of its original amount?

A second order differential equation is one in which the second derivative of the unknown function is the highest order derivative that appears. In each of Problems 19 through 22, use the change of variable $y' = u$ to reduce the given second order equation to a first order equation. Then solve the initial value problem.

19. $y'' + y' = 0, \quad y(0) = 0, \quad y'(0) = 1$

20. $y'y'' + x = 0, \quad y(0) = 0, \quad y'(0) = 1$

21. $y'y'' = 2, \quad y(0) = 1, \quad y'(0) = 2$

22. $xy'' + y' = 1, \quad x > 0, \quad y(1) = 1,$ $y'(1) = 0$

ELEVEN

indeterminate forms, l'hospital's rule, and improper integrals

In this chapter we extend the calculation of the limit of a function and the evaluation of an integral to some important cases that we have not yet discussed. Theorem 2.3.1 deals with the algebra of limits and, as we have seen repeatedly, is of fundamental importance both for theoretical purposes and for the calculation of limits of specific functions. However, there are many important situations that this theorem does not cover. Among these are limits of what are called indeterminate forms. A typical case arises from the quotient $f(x)/g(x)$ if both $f(x)$ and $g(x)$ approach zero as $x \to a$. Then $f(x)/g(x)$ is said to be an indeterminate form of the type 0/0 as $x \to a$. For example,

$$\frac{\sin x}{x}, \qquad \frac{1 - \cos x}{x^2}, \qquad \text{and} \qquad \frac{\ln(1 + x)}{x^2}$$

are simple expressions of the form 0/0 as $x \to 0$. There is a remarkable theorem, due to Johann Bernoulli, but universally known as L'Hospital's rule, that enables us to find the limit of these and many other indeterminate forms swiftly and elegantly. We explore some of the uses of L'Hospital's rule in the next two sections.

Up to now we have also restricted the operation of integration to integrands that are bounded and to intervals that are finite in length. Sometimes, but not always, it is possible to generalize the concept of integration so that it applies to integrands or to intervals of integration that are unbounded. Such integrals are called improper integrals and are discussed in the last two sections of this chapter.

11.1 INDETERMINATE FORMS AND L'HOSPITAL'S RULE

Theorem 2.3.1(d) states that if $\lim_{x\to a} f(x)$ and $\lim_{x\to a} g(x)$ exist, then

$$\lim_{x\to a} \frac{f(x)}{g(x)} = \frac{\lim\limits_{x\to a} f(x)}{\lim\limits_{x\to a} g(x)}, \tag{1}$$

provided that $\lim_{x\to a} g(x) \neq 0$. If $\lim_{x\to a} g(x) = 0$, and $\lim_{x\to a} f(x) = l \neq 0$, then $f(x)/g(x)$ becomes unbounded and consequently has no limit as $x \to a$. In the terminology of Section 2.4 it may be true in this case that $f(x)/g(x) \to \infty$, or that $f(x)/g(x) \to -\infty$, as $x \to a$, or as $x \to a$ from only one side. We now consider the situation in which both $\lim_{x\to a} f(x) = 0$ and $\lim_{x\to a} g(x) = 0$. In this case the right side of Eq. 1 is the meaningless expression 0/0, and we say that $f(x)/g(x)$ is an **indeterminate form** of the type 0/0 as $x \to a$.

Up to now we have either avoided such situations, or (when it was unavoidable) dealt with each case separately, using whatever means seemed appropriate. For instance, in Example 9 of Section 2.3 we used a geometrical construction as an aid in showing that $(\sin x)/x \to 1$ as $x \to 0$. Now we want to develop more formal and general methods for the determination of the limit of an indeterminate form of the type 0/0, as well as of other types that appear later. We will need to make use of an extension of the mean value theorem for derivatives, so we first turn our attention to that result.

Extended (Cauchy's) mean value theorem

The mean value theorem (Theorem 4.1.3) related the values of a function at the ends of an interval to the value of its derivative at some point in the interior of the interval. The following theorem is similar, except that it involves two functions.

Theorem 11.1.1

Suppose that f and g satisfy the following hypotheses:

(a) f and g are continuous on $[a, b]$.
(b) f and g are differentiable on (a, b).
(c) g' is never zero in (a, b).

Then there is at least one point c in (a, b) such that

$$\frac{f'(c)}{g'(c)} = \frac{f(b) - f(a)}{g(b) - g(a)}. \tag{2}$$

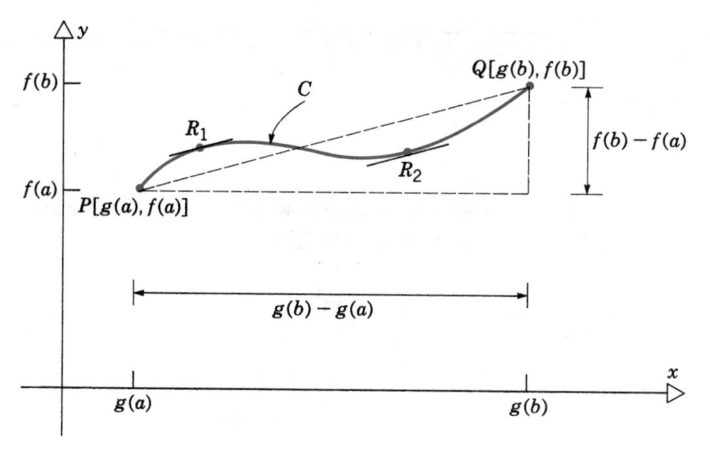

Figure 11.1.1

Clearly Theorem 11.1.1 reduces to Theorem 4.1.3 if $g(x) = x$. Before proving Theorem 11.1.1 we give a geometrical argument to make the result plausible. Let $x = g(t)$, $y = f(t)$ for $a \leq t \leq b$ be parametric equations of an arc C such as that shown in Figure 11.1.1. Then

$$\frac{f(b) - f(a)}{g(b) - g(a)} \tag{3}$$

is the slope of the line segment PQ joining the end points of the arc C. Further, from Section 5.5 we know that $f'(t)/g'(t)$ is the slope of the line tangent to C at the point $[g(t), f(t)]$. The theorem says that there is at least one point on the arc where the tangent line is parallel to the line segment PQ between the endpoints. In the situation shown in Figure 11.1.1 there are two possible values of c, corresponding to the two points R_1 and R_2. The condition that $g'(x)$ is never zero means that there is no point on the arc C where the tangent line is vertical. The result (2) is actually true under somewhat more general conditions. These are indicated in Problem 44.

Proof of Theorem 11.1.1. The theorem can be proved by constructing an auxiliary function to which Rolle's theorem (Theorem 4.1.2) can be applied. Let

$$F(t) = f(t)[g(b) - g(a)] - g(t)[f(b) - f(a)]. \tag{4}$$

Since f and g are continuous on $[a, b]$ and differentiable on (a, b), and since F is a linear combination of f and g, it follows that F also has these properties. By setting $t = a$ and $t = b$, respectively, we find that

$$F(a) = F(b) = f(a)g(b) - f(b)g(a). \tag{5}$$

Hence F satisfies the conditions of Rolle's theorem, and therefore there is at least one point c in (a, b) where $F'(c) = 0$. Differentiating Eq. 5 and setting $t = c$, we obtain

$$F'(c) = f'(c)[g(b) - g(a)] - g'(c)[f(b) - f(a)] = 0. \tag{6}$$

To obtain Eq. 2 from Eq. 6 we must divide by $g'(c)$ and by $[g(b) - g(a)]$; hence we must make sure that neither of these quantities can be zero. Hypothesis

(c) states that $g'(x)$ is never zero; hence $g'(c) \neq 0$ regardless of the location of the point c. Further, if $g(b) = g(a)$, then Rolle's theorem applied to g requires that there be at least one point $\hat{c}$ in (a, b) such that $g'(\hat{c}) = 0$. Since no such point $\hat{c}$ can exist by hypothesis (c), we know that $g(b) \neq g(a)$. Therefore we can divide Eq. 6 by $g'(c)[g(b) - g(a)]$ with the result that

$$\frac{f'(c)}{g'(c)} = \frac{f(b) - f(a)}{g(b) - g(a)}, \tag{7}$$

completing the proof of Theorem 11.1.1. □

L'Hospital's rule

We are now ready to consider the main result of this section, known as L'Hospital's* rule. It provides a means for evaluating limits of many indeterminate forms.

Theorem 11.1.2

(L'Hospital's Rule)

Suppose that f and g are differentiable on an open interval $a < x < b$, and that either

$$\lim_{x \to a^+} f(x) = 0 \quad \text{and} \quad \lim_{x \to a^+} g(x) = 0, \tag{8}$$

or

$$\lim_{x \to a^+} f(x) = \pm\infty \quad \text{and} \quad \lim_{x \to a^+} g(x) = \pm\infty, \tag{9}$$

where all combinations of the plus and minus signs in Eq. 9 are permitted. Suppose further that $g'(x)$ is never zero on (a, b), and that

$$\lim_{x \to a^+} \frac{f'(x)}{g'(x)} = L, \tag{10}$$

where L may be either a finite number, or ∞, or $-\infty$. Then

$$\lim_{x \to a^+} \frac{f(x)}{g(x)} = L. \tag{11}$$

We have stated Theorem 11.1.2 for limits as x approaches a from the right, but if the hypotheses are satisfied on an interval $b < x < a$, then the statement of

*Guillaume F. A. de L'Hospital (1661–1704) was a French marquis and both a student and patron of Johann Bernoulli. In 1696 he published the first textbook on calculus, *L'Analyse des Infiniment Petits pour l'Intelligence des Lignes Courbes,* an influential book that remained a standard reference through much of the eighteenth century. Among other things, it contained a statement of what is now known as L'Hospital's rule, a result that had actually been discovered by Johann Bernoulli in 1694.

the theorem is also true for left-hand limits ($x \to a^-$). If the hypotheses are satisfied on both sides of $x = a$, then the theorem holds for limits as $x \to a$ in an unrestricted manner. For the present we assume that a is finite, although in the next section we consider cases in which $x \to \pm\infty$. The proof of Theorem 11.1.2 is discussed at the end of this section. Meanwhile, we provide some representative examples of its usefulness.

EXAMPLE 1 (0/0)

Find the value of

$$\lim_{x \to 0} \frac{\sin x}{x}. \tag{12}$$

If we let $f(x) = \sin x$ and $g(x) = x$, then the conditions of Theorem 11.1.2 are satisfied in an open interval on each side of $x = 0$. Since $f'(x) = \cos x$ and $g'(x) = 1$, Theorem 11.1.2 tells us that

$$\lim_{x \to 0} \frac{\sin x}{x} = \lim_{x \to 0} \frac{\cos x}{1}, \tag{13}$$

provided that the limit on the right side of Eq. 13 is either a finite number or $\pm\infty$. However, the expression $(\cos x)/1$ is not indeterminate, and has the limit 1 as $x \to 0$. Thus

$$\lim_{x \to 0} \frac{\sin x}{x} = 1. \blacksquare \tag{14}$$

Recall that in Section 3.6 we needed the limiting value of $(\sin x)/x$ in order to find the derivative of $\sin x$. It may be tempting to think that we could have used L'Hospital's rule to avoid the more intricate derivation of this limit that we presented in Example 9 of Section 2.3. However, this would be fallacious thinking. We cannot use L'Hospital's rule to determine $\lim_{x \to 0} (\sin x)/x$ until we have derived the differentiation formula for $\sin x$ in some independent manner.

EXAMPLE 2 (0/0)

Evaluate

$$\lim_{x \to 0} \frac{1 - \cos x}{x^2}. \tag{15}$$

We noted earlier in this section that this is an indeterminate form. Using L'Hospital's rule, we obtain

$$\lim_{x \to 0} \frac{1 - \cos x}{x^2} = \lim_{x \to 0} \frac{\sin x}{2x}, \tag{16}$$

provided that the limit on the right side of Eq. 16 exists, or is $\pm\infty$. The expression $(\sin x)/2x$ is indeterminate as $x \to 0$, but another application of L'Hospital's rule yields

$$\lim_{x\to 0} \frac{\sin x}{2x} = \lim_{x\to 0} \frac{\cos x}{2} = \frac{1}{2}. \tag{17}$$

Thus, combining Eqs. 16 and 17, we have

$$\lim_{x\to 0} \frac{1 - \cos x}{x^2} = \frac{1}{2}. \tag{18}$$

This example illustrates that it may be necessary to use L'Hospital's rule more than once in order to evaluate the limit of an indeterminate form. ∎

EXAMPLE 3 (0/0)

Evaluate

$$\lim_{x\to 0^+} \frac{\tan 2x}{x^2}. \tag{19}$$

This is again an indeterminate form of the type 0/0 as $x \to 0^+$. Using L'Hospital's rule, we obtain

$$\lim_{x\to 0^+} \frac{\tan 2x}{x^2} = \lim_{x\to 0^+} \frac{2 \sec^2 2x}{2x} = \infty, \tag{20}$$

since $\sec 2x \to 1$ as $x \to 0$. This example illustrates the case in which L is infinite.

There is a potential danger in using L'Hospital's rule, namely, the possibility of using it repeatedly without checking that at each stage the expression involved is actually indeterminate. In this example, the expression $(2 \sec^2 2x)/2x$ is not indeterminate. However, if we had failed to notice this, we might have attempted to use L'Hospital's rule a second time; this would have led us to the false conclusion that the required limit is equal to

$$\lim_{x\to 0^+} \frac{8 \sec^2 2x \tan 2x}{2},$$

which is zero. *Thus, before using L'Hospital's rule, one should always check to make sure that the expression is indeed indeterminate.* ∎

EXAMPLE 4 (∞/∞)

Find the value of

$$\lim_{x\to 0^+} \frac{\ln x}{\cot x}. \tag{21}$$

Using L'Hospital's rule, we have

$$\lim_{x\to 0^+} \frac{\ln x}{\cot x} = \lim_{x\to 0^+} \frac{1/x}{-\csc^2 x},$$

provided that we can evaluate the latter limit. However, for $x \neq 0$,

$$\frac{1/x}{-\csc^2 x} = -\frac{\sin^2 x}{x} = -(\sin x)\left(\frac{\sin x}{x}\right) \to -(0)(1) = 0$$

as $x \to 0^+$. Alternatively, we can use L'Hospital's rule a second time to evaluate $\lim_{x\to 0} (\sin^2 x)/x$. In any case, it then follows that

$$\lim_{x\to 0^+} \frac{\ln x}{\cot x} = 0. \blacksquare \tag{22}$$

EXAMPLE 5 (0/0)

Evaluate

$$\lim_{x\to\pi} \frac{\tan x}{(\pi - x)^2}. \tag{23}$$

This example illustrates the use of L'Hospital's rule at a point a other than zero. From L'Hospital's rule we obtain

$$\lim_{x\to\pi} \frac{\tan x}{(\pi - x)^2} = \lim_{x\to\pi} \frac{\sec^2 x}{-2(\pi - x)}. \tag{24}$$

The expression on the right side of Eq. 24 has no limit, since as $x \to \pi$ its numerator approaches one while its denominator approaches zero. Thus $(\tan x)/(\pi - x)^2$ is unbounded as $x \to \pi$. To examine the nature of the unboundedness we consider the corresponding one-sided limits. We find that

$$\lim_{x\to\pi^-} \frac{\tan x}{(\pi - x)^2} = -\infty, \qquad \lim_{x\to\pi^+} \frac{\tan x}{(\pi - x)^2} = +\infty. \blacksquare \tag{25}$$

Sometimes we encounter indeterminate forms that are not of the type 0/0 or ∞/∞. For instance, two other common types of indeterminacy are expressions of the form $\infty - \infty$ or $0 \cdot \infty$. One way to handle such expressions is to rewrite them in the form 0/0 or ∞/∞, and then to apply L'Hospital's rule. The following two examples illustrate this procedure.

EXAMPLE 6 ($\infty - \infty$)

Evaluate

$$\lim_{x\to 0}\left(\frac{1}{x} - \csc x\right). \tag{26}$$

This expression is an example of an indeterminate form of the type $\infty - \infty$ as $x \to 0$. In order to use L'Hospital's rule, we first rewrite the expression (26) so that it has the 0/0 form as $x \to 0$. If $x \neq 0$ and $|x| < \pi$, we have

$$\frac{1}{x} - \csc x = \frac{1}{x} - \frac{1}{\sin x} = \frac{\sin x - x}{x \sin x},$$

and this latter expression is of the form 0/0 as $x \to 0$. On using L'Hospital's rule we obtain

$$\lim_{x\to 0}\left(\frac{1}{x} - \csc x\right) = \lim_{x\to 0}\left(\frac{\sin x - x}{x \sin x}\right)$$
$$= \lim_{x\to 0}\left(\frac{\cos x - 1}{\sin x + x\cos x}\right),$$

provided that the latter limit exists, or is $\pm\infty$. We observe that $\cos x - 1 \to 0$ and $\sin x + x \cos x \to 0$ as $x \to 0$. Therefore we still have an indeterminate form of the type 0/0 as $x \to 0$. We use L'Hospital's rule once more, obtaining

$$\lim_{x\to 0}\left(\frac{\cos x - 1}{\sin x + x\cos x}\right) = \lim_{x\to 0}\left(\frac{-\sin x}{2\cos x - x\sin x}\right).$$

This last limit is zero, since the numerator approaches zero and the denominator approaches two. Therefore we have

$$\lim_{x\to 0}\left(\frac{1}{x} - \csc x\right) = 0. \tag{27}$$

Again, we see that we may need to use L'Hospital's rule more than once in order to find the limit of an indeterminate form. ∎

EXAMPLE 7 ($0 \cdot \infty$)

Evaluate

$$\lim_{x\to 0^+} x \ln x. \tag{28}$$

To use L'Hospital's rule we must first write $x \ln x$ so that it has the form 0/0 or ∞/∞ as $x \to 0^+$. We choose the latter,

$$x \ln x = \frac{\ln x}{1/x},$$

where both numerator and denominator of the right side become infinite as $x \to 0^+$. Using L'Hospital's rule, we obtain

$$\lim_{x\to 0^+} x \ln x = \lim_{x\to 0^+} \frac{\ln x}{1/x} = \lim_{x\to 0^+}\left(\frac{1/x}{-1/x^2}\right)$$
$$= \lim_{x\to 0^+}(-x) = 0. \tag{29}$$

Finally, let us note what happens if we write $x \ln x$ in the 0/0 form. We have

$$x \ln x = \frac{x}{1/\ln x},$$

and attempting to use L'Hospital's rule yields

$$\begin{aligned}\lim_{x\to 0^+} x \ln x &= \lim_{x\to 0^+} \frac{x}{1/\ln x} \\ &= \lim_{x\to 0^+} \frac{1}{-(1/\ln x)^2(1/x)} = -\lim_{x\to 0^+} \frac{x}{(1/\ln x)^2}.\end{aligned}$$

Thus we have obtained another indeterminate form, one that is more complicated than the original problem. Additional efforts to use L'Hospital's rule lead to further complications, but never give the result. This example illustrates that there may be different ways of looking at the same problem, and that one way may be better than another. In dealing with indeterminate forms involving logarithms and polynomials, it is often wise to cast the problem in such a way that differentiation eliminates the logarithm. This was done by the first formulation of this example, but not by the second. ∎

Proof of Theorem 11.1.2. We give a proof for the case in which Eq. 8 holds. Although f and g have zero limits as $x \to a^+$, these functions may not be continuous from the right at a, either because $f(a)$ and $g(a)$ are not defined, or because they are not equal to the limiting value. However, if we define (or redefine if necessary) $f(a) = 0$ and $g(a) = 0$ and choose x in the interval (a, b), then f and g are continuous on the closed interval $[a, x]$. They are also differentiable on the open interval (a, x), and $g'(x)$ is never zero in that interval. Thus Theorem 11.1.1 can be applied to f and g on the interval $[a, x]$, with the result that

$$\frac{f(x) - f(a)}{g(x) - g(a)} = \frac{f'(c)}{g'(c)}, \tag{30}$$

for some c between a and x. Since $f(a) = g(a) = 0$, Eq. 30 becomes

$$\frac{f(x)}{g(x)} = \frac{f'(c)}{g'(c)}.$$

Finally, letting $x \to a^+$ and noting that this requires that $c \to a^+$ as well, we have

$$\lim_{x\to a^+} \frac{f(x)}{g(x)} = \lim_{x\to a^+} \frac{f'(c)}{g'(c)} = \lim_{c\to a^+} \frac{f'(c)}{g'(c)} = L,$$

thus completing the proof for the case where f and g have zero limits as $x \to a^+$. In the cases where f and g approach either $\pm\infty$ the proof is somewhat more difficult, and is omitted; however, see Problem 43. □

PROBLEMS

In each of Problems 1 through 35, evaluate the given limit. In some of these problems L'Hospital's rule is not an appropriate tool, and using it carelessly may lead to an incorrect result.

1. $\lim_{x\to 0} \frac{\sin 2x}{x}$

2. $\lim_{x\to 0} \frac{\sin^2 x}{2x^2}$

3. $\lim_{x\to 0} \frac{1 - \cos^2 x}{x^2}$

4. $\lim_{x\to 0} \frac{1 - \cos(x^2)}{x^4}$

5. $\lim_{x\to 0} \frac{\sin^2 x}{1 + \cos x}$

6. $\lim_{x\to 0} \frac{\arctan x}{x}$

7. $\lim_{x\to(\pi/2)+} \left(x - \frac{\pi}{2}\right)\tan x$

8. $\lim_{x\to(\pi/2)-} \left(\frac{2}{\pi - 2x} - \tan x\right)$

9. $\lim_{x\to 0-} \frac{e^{-x} + 1}{e^x - 1}$

10. $\lim_{x\to 0+} \left(\frac{1}{x} + \ln x\right)$

11. $\lim_{x\to 0+} \frac{\ln(1 + (1/x))}{1/\sqrt{x}}$

12. $\lim_{x\to 0} \frac{e^{2x} - 1}{3x}$

13. $\lim_{x\to 0+} \frac{\ln(1 + x^2)}{x^3}$

14. $\lim_{x\to 0} \frac{1 - e^{-x^2}}{2x^2}$

15. $\lim_{x\to 0+} \frac{\ln x}{x^{-q}}$, $\quad q > 0$

16. $\lim_{x\to \pi} \frac{x - \sin x}{1 - \cos x}$

17. $\lim_{x\to 0+} \frac{e^{1/x}}{1/x}$

18. $\lim_{x\to 0-} \frac{e^{1/x}}{1/x}$

19. $\lim_{x\to 1-} \left(\frac{1}{\ln x} + \frac{1}{\sqrt{1 - x^2}}\right)$

20. $\lim_{x\to 0} \frac{a^x - b^x}{x}$, $\quad a > b > 0$

21. $\lim_{x\to \pi/2} (\sec x - \tan x)$

22. $\lim_{x\to 1} x \ln x$

23. $\lim_{x\to 0} \frac{\sqrt{1 + x^2} - \sqrt{1 - x^2}}{x}$

24. $\lim_{x\to 0} \frac{\sin x - x}{\tan x - x}$

25. $\lim_{x\to 0} \frac{\cot ax}{\cot bx}$, $\quad a > 0, \quad b > 0$

26. $\lim_{x\to 4} \frac{\sqrt{x - 3} - 1}{x^2 - 16}$

27. $\lim_{x\to 2} \frac{x - 2}{(x + 6)^{1/3} - 2}$

28. $\lim_{x\to a} \frac{x^{1/3} - a^{1/3}}{x - a}$, $\quad a \neq 0$

29. $\lim_{x\to 0} \frac{\arctan 2x}{\arcsin x}$

30. $\lim_{x\to 0+} \frac{\ln|\ln x|}{\ln x}$

31. $\lim_{x\to 0} \frac{\sin^2 x - \sin(x^2)}{x^4}$

32. $\lim_{x\to 0} \frac{\arctan x - x}{\arcsin x - x}$

33. $\lim_{x\to 0} \frac{\arcsin 2x - 2\arcsin x}{x \sin^2 x}$

34. $\lim_{x\to \pi/2} \frac{(\sin 2x)(\sin 3x)}{x \sin 4x}$

35. $\lim_{x\to 0+} x(\ln x)^2$

36. (a) Let f be continuous in some neighborhood of the origin. Evaluate

$$\lim_{x\to 0} \frac{\int_0^x f(t)\,dt}{x}.$$

(b) If f is also differentiable in some neighborhood of the origin, and if $f(0) = 0$, evaluate

$$\lim_{x\to 0} \frac{\int_0^x f(t)\,dt}{x^2}.$$

37. Evaluate

$$\lim_{x\to 0} \frac{\left(\int_0^x \sin t\,dt\right)^2}{\int_0^x \sin t^2\,dt}.$$

38. Find the value of a for which

$$\lim_{x\to 0} \frac{\sin ax - \sin x - x}{x^3}$$

is finite, and evaluate the limit in that case.

39. Find the values of a and b for which

$$\lim_{x\to 0} \frac{\cos ax - b}{2x^2} = -1.$$

40. Let P be a point in the first quadrant on the unit circle (see Figure 11.1.2). The line segment AQ is tangent to the circle and the length of AQ is the same as the length of the arc AP. The line through P and Q intersects the x-axis at B. Find the limiting position of B as $P \to A$.

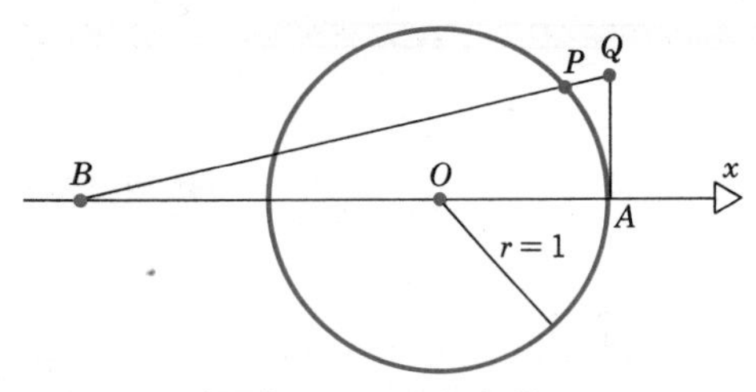

Figure 11.1.2

41. Consider the situation shown in Figure 11.1.3. Let

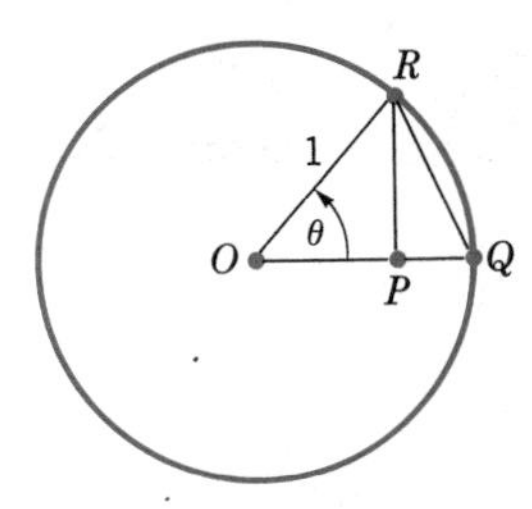

Figure 11.1.3

$$a(\theta) = \text{area of triangle } PQR$$
$$b(\theta) = \text{area of sector } OQR \text{ less area of triangle } OPR$$
$$c(\theta) = \text{area of sector } OQR \text{ less area of triangle } OQR$$

(a) Find $\lim_{\theta\to 0} \dfrac{a(\theta)}{b(\theta)}$. (b) Find $\lim_{\theta\to 0} \dfrac{a(\theta)}{c(\theta)}$.

42. A particle of mass m is dropped from rest under the influence of gravity in a medium offering resistance proportional to the velocity of the particle.

(a) Show that the velocity at time t is the solution of the initial value problem

$$m\frac{dv}{dt} + kv = mg, \qquad v(0) = 0. \qquad (i)$$

See the discussion preceding Problem 27 in Section 10.2.

(b) Find v at any time t and determine the limiting value of v as $k \to 0$, that is, as the resistance approaches zero.

(c) Is the result found in (b) the same as the result found by setting $k = 0$ in Eq. (i) and then solving the initial value problem?

* **43.** Suppose that f and g satisfy the conditions of L'Hospital's rule (Theorem 11.1.2) with

$$\lim_{x\to a+} f(x) = \infty, \qquad \lim_{x\to a+} g(x) = \infty.$$

Assume also that $\lim_{x\to a^+} f'(x)/g'(x) = L$, where $L \neq 0$ or ∞, and that $\lim_{x\to a^+} f(x)/g(x)$ exists. Show that

$$\lim_{x\to a+} \frac{f(x)}{g(x)} = L.$$

Hint: Note that

$$\frac{f(x)}{g(x)} = \frac{1/g(x)}{1/f(x)},$$

and that the latter expression is an indeterminate form of the type 0/0. Note also that in this problem we *assume* the existence of $\lim_{x\to a^+} f(x)/g(x)$. Where is this assumption used in the argument? The assumption was not made in the 0/0 case considered in the text. It is not required here in the ∞/∞ case either, but the proof is considerably more difficult without it.

* **44.** The conclusion of the extended mean value theorem remains true under somewhat more general hypotheses than those given in Theorem 11.1.1 in the text. Suppose that f and g satisfy hypotheses (a) and (b) of the text and also satisfy:

(c) $g(b) \neq g(a)$;

(d) There is no point in (a, b) where both f' and g' are zero.

Show that Eq. 2 of the text is valid under these conditions.

Hint: It is required to show that $g'(c) \neq 0$. Use Eq. 6 of the text and condition (c) above to show that if $g'(c) = 0$, then $f'(c) = 0$ also, violating condition (d).

Note: The significance of this result is that it extends the theorem to arcs that can have vertical tangents. A typical case is shown in Figure 11.1.4, where R_1, R_2, and R_3 indicate points corresponding to possible values of c.

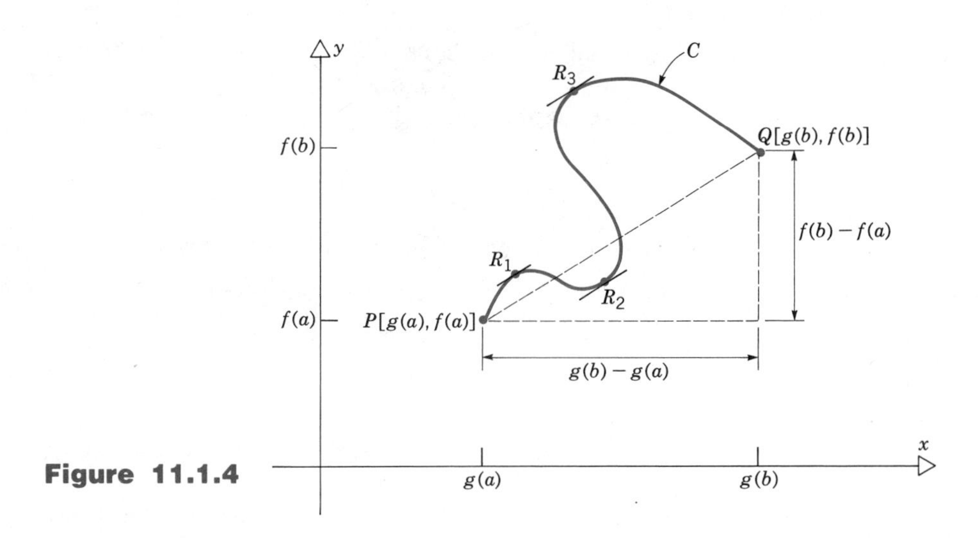

Figure 11.1.4

11.2 OTHER INDETERMINATE FORMS: EXTENSIONS OF L'HOSPITAL'S RULE

In the preceding section we discussed a procedure, known as L'Hospital's rule, for finding the limit of an indeterminate form of the type 0/0 or ∞/∞ at a finite point $x = a$. In Examples 6 and 7 of Section 11.1 we indicated how L'Hospital's rule could also be applied to some indeterminate forms of the type $\infty - \infty$ and $0 \cdot \infty$. In all of these cases the problem is actually one of discovering *whether one of two competing influences is dominant,* or *whether they balance each other.* For example, for a form of the type 0/0, does the numerator approach zero significantly faster than the denominator, or is the reverse the case, or do both numerator and denominator approach zero in a similar way? As we have seen, L'Hospital's rule provides a way of resolving such questions, often quite easily.

In this section we begin by giving an extension of L'Hospital's rule to cases in which the limit is taken at infinity rather than at a finite point.

Theorem 11.2.1

(L'Hospital's Rule)
Suppose that

$$\lim_{x\to\infty} f(x) = 0 \quad \text{and} \quad \lim_{x\to\infty} g(x) = 0, \tag{1}$$

or that

$$\lim_{x\to\infty} f(x) = \pm\infty \quad \text{and} \quad \lim_{x\to\infty} g(x) = \pm\infty, \tag{2}$$

where all combinations of the plus and minus signs in Eq. 2 are permitted. Suppose also that f and g are differentiable for $x > M$ (where M is some positive number), and that $g'(x)$ is never zero for $x > M$. If

$$\lim_{x\to\infty} \frac{f'(x)}{g'(x)} = L,$$

where L may be either a finite number, or ∞, or $-\infty$, then

$$\lim_{x\to\infty} \frac{f(x)}{g(x)} = L.$$

The theorem can be modified in an obvious way if $x \to -\infty$.

Proof of Theorem 11.2.1. For the case of hypothesis (1) a proof can be constructed by transforming the problem into one for which Theorem 11.1.2 can be used. Let $x = 1/t$; then $x \to \infty$ means that $t \to 0^+$. Consequently,

$$\lim_{x\to\infty} \frac{f(x)}{g(x)} = \lim_{t\to 0^+} \frac{f(1/t)}{g(1/t)}. \tag{3}$$

The right side of Eq. 3 involves the limit of an indeterminate form of the type 0/0 at the origin. Thus the version of L'Hospital's rule given in Theorem 11.1.2 can be applied. Using the chain rule to calculate the necessary derivatives, we obtain

$$\begin{aligned} \lim_{t\to 0^+} \frac{f(1/t)}{g(1/t)} &= \lim_{t\to 0^+} \frac{(-1/t^2)f'(1/t)}{(-1/t^2)g'(1/t)} \\ &= \lim_{t\to 0^+} \frac{f'(1/t)}{g'(1/t)}. \end{aligned} \tag{4}$$

Finally, shifting the variable back from t to x, we have

$$\lim_{t\to 0^+} \frac{f'(1/t)}{g'(1/t)} = \lim_{x\to\infty} \frac{f'(x)}{g'(x)} = L. \tag{5}$$

The proof is completed by combining Eqs. 3, 4, and 5.

The proof for the case of hypothesis (2) is considerably more difficult, and is omitted. □

We now consider some examples of Theorem 11.2.1. The first two examples give important qualitative results among some of the elementary functions.

EXAMPLE 1 (∞/∞)

Find the value of

$$\lim_{x\to\infty} \frac{\ln x}{x^p}, \qquad p > 0. \tag{6}$$

Using Theorem 11.2.1, we have

$$\lim_{x\to\infty} \frac{\ln x}{x^p} = \lim_{x\to\infty} \frac{1/x}{px^{p-1}} = \lim_{x\to\infty} \frac{1}{px^p} = 0, \qquad p > 0. \tag{7}$$

Note that the limiting value of zero is independent of p. Thus $\ln x$ grows much more slowly as $x \to \infty$ than does any positive power of x, no matter how small the exponent. In other words, *algebraic growth, no matter how slow, dominates logarithmic growth.* ∎

EXAMPLE 2 (∞/∞)

Find the value of

$$\lim_{x\to\infty} \frac{x^n}{e^x}, \qquad n \text{ a positive integer.} \tag{8}$$

Using Theorem 11.2.1 repeatedly, we have

$$\begin{aligned} \lim_{x\to\infty} \frac{x^n}{e^x} &= \lim_{x\to\infty} \frac{nx^{n-1}}{e^x} = \lim_{x\to\infty} \frac{n(n-1)x^{n-2}}{e^x} \\ &= \cdots = \lim_{x\to\infty} \frac{n(n-1)\cdots(2)x}{e^x} \\ &= \lim_{x\to\infty} \frac{n!}{e^x} = 0. \end{aligned} \tag{9}$$

The result remains true if n is replaced by any positive number ν, that is,

$$\lim_{x\to\infty} \frac{x^\nu}{e^x} = 0, \qquad \nu > 0. \tag{10}$$

To see this we can observe that if n is an integer such that $n \geq \nu$, and if $x \geq 1$, then

$$0 \leq \frac{x^\nu}{e^x} \leq \frac{x^n}{e^x}.$$

Since $x^n/e^x \to 0$ as $x \to \infty$, Eq. 10 follows from the sandwich principle for limits (Theorem 2.3.4). As in Example 1, the result (10) is independent of the exponent ν. Thus, as $x \to \infty$, the exponential function e^x grows much faster than any power of x, no matter how large. In other words, *exponential growth dominates algebraic growth.* ∎

A dramatic example of the significance of logarithmic, algebraic, and exponential growth occurs in connection with numerical algorithms for the solution of problems such as finding the maximum flow in a communication or transportation network. The complexity of an algorithm is expressed in terms of the number of operations required to execute it on a general problem. For instance, for a network problem with n links connecting a set of points a certain algorithm might require a number of steps proportional to n^3, or to 2^n, or to some other function of n. Algorithms that require an exponential number of steps become prohibitively time-consuming for problems of relatively modest size.

Table 11.1 Time Required to Implement Numerical Algorithms of Several Complexities on Problems of Different Sizes

n	10	20	50	100	500	1000
$n \ln n$	0.00002 sec	0.00006 sec	0.0002 sec	0.0005 sec	0.003 sec	0.007 sec
n^2	0.0001 sec	0.0004 sec	0.0025 sec	0.01 sec	0.25 sec	1 sec
n^3	0.001 sec	0.008 sec	0.125 sec	1 sec	2.1 min	17 min
2^n	0.001 sec	1 sec	36 yrs	4×10^{14} centuries		

Assuming that one step is executed every microsecond, Table 11.1 shows the times required to solve problems of various sizes by algorithms whose operation counts are given by $n \ln n$, n^2, n^3, and by 2^n, respectively. A comparison of the entries for $n = 100$ provides striking confirmation of the differences among logarithmic, algebraic, and exponential growth.

EXAMPLE 3 (0/0)

Evaluate

$$\lim_{x\to\infty} \frac{\ln[1 + (1/x)]}{(\pi/2) - \arctan x}. \tag{11}$$

By Theorem 11.2.1 we have

$$\begin{aligned} \lim_{x\to\infty} \frac{\ln[1 + (1/x)]}{(\pi/2) - \arctan x} &= \lim_{x\to\infty} \frac{(-1/x^2)}{[1 + (1/x)]} \frac{(-1)}{[1/(1 + x^2)]} \\ &= \lim_{x\to\infty} \frac{1 + x^2}{x(x + 1)} = 1. \quad \blacksquare \end{aligned} \tag{12}$$

Let us now consider some other types of indeterminate forms. Specifically, we look at expressions that are of the type 0^0, ∞^0, or 1^∞ as the independent variable approaches a finite point or infinity. First we consider an example.

EXAMPLE 4 (0^0)

Evaluate

$$\lim_{x\to 0^+} x^x. \tag{13}$$

The expression x^x, for small positive x, again involves two competing effects. If the x in the exponent is dominant, then one would expect the limit to be one, since $a^0 = 1$ for any positive a. On the other hand, if the x in the base is dominant, then the limit presumably is zero, since $0^b = 0$ for any positive b. Or, if there is some kind of balance, then the limit might be some number between zero and one.

Since x^x is not a ratio, L'Hospital's rule cannot be applied directly to determine the required limit. However, if we take the logarithm of x^x, we obtain

$$\ln x^x = x \ln x. \tag{14}$$

The expression $x \ln x$ is indeterminate of the type $0 \cdot \infty$; it was discussed in Example 7 of Section 11.1, where we showed that

$$\lim_{x \to 0^+} x \ln x = 0. \tag{15}$$

To relate this result to $\lim_{x \to 0^+} x^x$ we take the exponential of Eq. 14; thus

$$x^x = \exp(x \ln x),$$

and

$$\lim_{x \to 0^+} x^x = \lim_{x \to 0^+} \exp(x \ln x). \tag{16}$$

Further, the exponential function is continuous, and therefore the limit can be taken inside the exponential. Consequently,

$$\lim_{x \to 0^+} x^x = \exp\left[\lim_{x \to 0^+} (x \ln x)\right] = \exp(0) = 1. \tag{17}$$

Hence, in this case, the influence of the exponent as it approaches zero is dominant. ∎

More general problems of this type can be handled in the same way. Suppose that

$$f(x) = u(x)^{v(x)},$$

where $u(x) > 0$ for $x \neq a$, and $u(x) \to 0$ and $v(x) \to 0$ as $x \to a$. By taking the natural logarithm of $f(x)$ we obtain

$$\ln f(x) = v(x) \ln u(x),$$

which is indeterminate of the type $0 \cdot \infty$. Assume that, by using L'Hospital's rule or otherwise, we can show that

$$\lim_{x \to a} \ln f(x) = \lim_{x \to a} v(x) \ln u(x) = L. \tag{18}$$

Then, as in Example 4,

$$\lim_{x \to a} u(x)^{v(x)} = \lim_{x \to a} \exp[v(x) \ln u(x)] = \exp\left[\lim_{x \to a} v(x) \ln u(x)\right] = \exp(L). \tag{19}$$

Again the fact that the exponential function is continuous enables us to bring the limit inside the exponential.

The following examples illustrate some other types of indeterminacy that can best be investigated by first taking the logarithm.

EXAMPLE 5 (1^∞)
Evaluate

$$\lim_{x\to\infty}\left(1 + \frac{1}{x}\right)^x. \tag{20}$$

On taking the logarithm we have

$$\ln\left(1 + \frac{1}{x}\right)^x = x\ln\left(1 + \frac{1}{x}\right) = \frac{\ln[1 + (1/x)]}{1/x}. \tag{21}$$

By applying L'Hospital's rule to the last expression in Eq. 21 we find that

$$\lim_{x\to\infty}\frac{\ln[1 + (1/x)]}{1/x} = \lim_{x\to\infty}\frac{-1/x^2}{[1 + (1/x)](-1/x^2)} = \lim_{x\to\infty}\frac{1}{1 + (1/x)} = 1. \tag{22}$$

Finally, we take the exponential of Eq. 21 so as to recover the original expression, with the result that

$$\begin{aligned}\lim_{x\to\infty}\left(1 + \frac{1}{x}\right)^x &= \lim_{x\to\infty}\exp\frac{\ln[1 + (1/x)]}{1/x}\\ &= \exp\left(\lim_{x\to\infty}\frac{\ln[1 + (1/x)]}{1/x}\right)\\ &= \exp(1) = e. \blacksquare\end{aligned} \tag{23}$$

EXAMPLE 6 (1^∞)
Find the value of

$$\lim_{x\to\infty}\left(1 + \frac{2}{x}\right)^{x^2}. \tag{24}$$

By taking the logarithm, we obtain

$$x^2\ln\left(1 + \frac{2}{x}\right),$$

which is of the type $0 \cdot \infty$ as $x \to \infty$. Thus we first write

$$x^2\ln\left(1 + \frac{2}{x}\right) = \frac{\ln[1 + (2/x)]}{x^{-2}}$$

and then apply L'Hospital's rule (Theorem 11.2.1). We find that

$$\begin{aligned}\lim_{x\to\infty} x^2\ln\left(1 + \frac{2}{x}\right) &= \lim_{x\to\infty}\frac{\ln[1 + (2/x)]}{x^{-2}}\\ &= \lim_{x\to\infty}\frac{-2/x^2}{[1 + (2/x)](-2x^{-3})}\\ &= \lim_{x\to\infty}\frac{x}{1 + (2/x)} = \infty.\end{aligned}$$

Thus it is also true that

$$\lim_{x\to\infty}\left(1 + \frac{2}{x}\right)^{x^2} = \lim_{x\to\infty} \exp\left[x^2 \ln\left(1 + \frac{2}{x}\right)\right] = \infty. \blacksquare \tag{25}$$

EXAMPLE 7 (∞^0)

Evaluate

$$\lim_{x\to\infty}(1 + e^{x^2})^{1/x^q} \tag{26}$$

for each positive value of q.

By taking the logarithm we are led to consider

$$\lim_{x\to\infty} \frac{\ln(1 + e^{x^2})}{x^q},$$

which is indeterminate of type ∞/∞. Using Theorem 11.2.1 we obtain

$$\begin{aligned}\lim_{x\to\infty} \frac{\ln(1 + e^{x^2})}{x^q} &= \lim_{x\to\infty} \frac{2xe^{x^2}}{(1 + e^{x^2})qx^{q-1}} \\ &= \frac{2}{q}\left(\lim_{x\to\infty} \frac{e^{x^2}}{1 + e^{x^2}}\right)\left(\lim_{x\to\infty} x^{2-q}\right).\end{aligned}$$

The first limit on the right side of this equation is

$$\lim_{x\to\infty} \frac{e^{x^2}}{1 + e^{x^2}} = \lim_{x\to\infty} \frac{1}{e^{-x^2} + 1} = 1,$$

so the result depends on the second limit, which has a different value, depending on whether $q < 2$, $q = 2$, or $q > 2$. We have

$$\lim_{x\to\infty} x^{2-q} = \begin{cases} \infty, & 0 < q < 2; \\ 1, & q = 2; \\ 0, & q > 2, \end{cases}$$

and hence

$$\lim_{x\to\infty} \frac{\ln(1 + e^{x^2})}{x^q} = \begin{cases} \infty, & 0 < q < 2; \\ 1, & q = 2; \\ 0, & q > 2. \end{cases}$$

Therefore

$$\lim_{x\to\infty}(1 + e^{x^2})^{1/x^q} = \begin{cases} \infty, & 0 < q < 2; \\ e, & q = 2; \\ 1, & q > 2. \end{cases} \quad \blacksquare \tag{27}$$

Summary

We list below the various types of indeterminate forms that we have discussed in these two sections, together with a notation of where an example of that kind can be found.

$0/0$: Examples 1, 2, 3, and 5 of Section 11.1;
Example 3 of Section 11.2.

∞/∞ : Example 4 of Section 11.1;
Examples 1 and 2 of Section 11.2.

$\infty - \infty$: Example 6 of Section 11.1.

$0 \cdot \infty$: Example 7 of Section 11.1.

0^0 : Example 4 of Section 11.2.

1^∞ : Examples 5 and 6 of Section 11.2.

∞^0 : Example 7 of Section 11.2.

PROBLEMS

In each of Problems 1 through 36 evaluate the given limit.

1. $\lim_{x\to\infty} \frac{(\ln x)^n}{x}$, n a positive integer
2. $\lim_{x\to\infty} \frac{x^n}{2^x}$, n a positive integer
3. $\lim_{x\to\infty} \frac{x^n}{(1.01)^x}$, n a positive integer
4. $\lim_{x\to\infty} \frac{x^n}{e^{x/100}}$, n a positive integer
5. $\lim_{x\to\infty} \frac{\ln(1 + e^x)}{x}$
6. $\lim_{x\to\infty} \frac{\ln(\ln x)}{\ln(\ln x^3)}$
7. $\lim_{x\to\infty} \frac{\sin(e^{-x})}{\sin(1/x)}$
8. $\lim_{x\to\infty} \frac{\ln(x^2 + 1)}{\ln(x^2 + 9)}$
9. $\lim_{x\to\infty} \left(\frac{3^x + 5^x}{2}\right)^{1/x}$
10. $\lim_{x\to\infty} x \sin \frac{1}{x}$
11. $\lim_{x\to\infty} \frac{\sqrt{1 + x^2}}{x}$
12. $\lim_{x\to-\infty} \frac{\sqrt{1 + x^2}}{x}$
13. $\lim_{x\to0+} x(\ln x)^n$, n a positive integer
14. $\lim_{x\to\infty} x^n e^{-ax}$, n a positive integer, $a > 0$
15. $\lim_{x\to\infty} \frac{x^a}{b^x}$, $a > 0$, $b > 1$
16. $\lim_{x\to\pi/2} (1 + \cos x)^{\sec x}$
17. $\lim_{x\to\infty} \left(1 + \frac{1}{x^2}\right)^x$
18. $\lim_{x\to\infty} \left(1 - \frac{1}{x}\right)^x$
19. $\lim_{x\to\infty} \left(1 - \frac{1}{x}\right)^{x^2}$
20. $\lim_{x\to\infty} \left(1 - \frac{1}{x^2}\right)^x$
21. $\lim_{x\to\infty} (1 + x^3)^{x^{-2}}$
22. $\lim_{x\to0+} x^{\sin x}$
23. $\lim_{x\to0} (\cos x)^{1/x^2}$
24. $\lim_{x\to0+} x^{x \ln x}$
25. $\lim_{x\to\infty} (1 + e^x)^{x^{-q}}$, $q > 0$
26. $\lim_{x\to\infty} (1 + x^p)^{x^{-q}}$, $p > 0$ and $q > 0$
27. $\lim_{x\to\infty} (1 + x^{-p})^{x^q}$, $p > 0$ and $q > 0$
28. $\lim_{x\to0+} (x^p)^{x^q}$, $p > 0$ and $q > 0$
29. $\lim_{x\to\infty} \left(\sin \frac{1}{x}\right)^{1/x}$
30. $\lim_{x\to\infty} x^{1/x}$
31. $\lim_{x\to1} x^{1/(1-x)}$
32. $\lim_{x\to0} (1 + x^2)^{1/x}$
33. $\lim_{x\to0+} (1 + x)^{1/x^2}$
34. $\lim_{x\to0-} (1 + x)^{1/x^2}$
35. $\lim_{x\to0+} (1 - x)^{1/x}$
36. $\lim_{x\to(\pi/4)-} (\tan x)^{\sec 2x}$

11.3 IMPROPER INTEGRALS

In Chapter 6 we discussed integrals of the form

$$\int_a^b f(x)\,dx,$$

where the interval of integration is finite and the integrand f is bounded on $[a, b]$. Such integrals are sometimes called **proper integrals.** Now we want to extend the concept of integration to include some cases in which the interval of integration is infinite, or the integrand becomes unbounded in the neighborhood of some finite point, or perhaps both. These integrals are called **improper integrals.** For example,

$$\int_0^\infty \frac{dx}{1 + x^4}, \tag{1}$$

$$\int_0^1 \frac{dx}{\sqrt{1 - x^4}}, \tag{2}$$

and

$$\int_0^\infty \frac{dx}{x^4 - 1} \tag{3}$$

are all improper integrals. The first integral is improper because the interval of integration is infinite, the second because the integrand is unbounded near $x = 1$, and the third for both of these reasons.

When an integral is improper for more than one reason, it is convenient to break it up into parts, each of which is improper for a single reason. It is also convenient to arrange for the integrand to become unbounded only in the neighborhood of an endpoint. For example, we can rewrite the integral (3) as

$$\int_0^\infty \frac{dx}{x^4 - 1} = \int_0^1 \frac{dx}{x^4 - 1} + \int_1^2 \frac{dx}{x^4 - 1} + \int_2^\infty \frac{dx}{x^4 - 1}. \tag{4}$$

The first two integrals on the right side of Eq. 4 are improper only because the integrand is unbounded (near an endpoint), while the third is improper only because the interval of integration is infinite. Any point $c > 1$ could be used instead of 2 as the break point between the last two integrals in Eq. 4.

First we consider integrals that are improper only because the interval of integration is infinite. In some (but not all) cases a value can be assigned to such integrals by the following definition.

DEFINITION 11.3.1 Let f be defined for $x \geq a$ and let $I(b) = \int_a^b f(x)\,dx$ exist for each $b > a$. Then the value of the improper integral $\int_a^\infty f(x)\,dx$ is defined by

$$\int_a^\infty f(x)\,dx = \lim_{b\to\infty} I(b) = \lim_{b\to\infty} \int_a^b f(x)\,dx, \tag{5}$$

provided that this limit exists (has a finite value).

If the limit on the right side of Eq. 5 exists, then the improper integral is said to **converge;** otherwise, it **diverges.**

There is a similar definition for integrals whose lower limit of integration is infinite:

$$\int_{-\infty}^{b} f(x)\,dx = \lim_{a\to-\infty} \int_{a}^{b} f(x)\,dx, \tag{6}$$

provided that this limit exists. If both limits of integration are infinite, then we write

$$\int_{-\infty}^{\infty} f(x)\,dx = \int_{-\infty}^{c} f(x)\,dx + \int_{c}^{\infty} f(x)\,dx \tag{7}$$

where c is any convenient finite number. The integral on the left side of Eq. 7 converges if and only if both integrals on the right side of that equation converge.

EXAMPLE 1

Determine whether the improper integral

$$\int_{0}^{\infty} e^{-2x}\,dx \tag{8}$$

converges, and if so, find its value.

For any $b > 0$ we have

$$\int_{0}^{b} e^{-2x}\,dx = -\frac{1}{2}e^{-2x}\Big|_{0}^{b} = \frac{1}{2}(1 - e^{-2b}).$$

Thus, using the definition (5), we find that

$$\begin{aligned}\int_{0}^{\infty} e^{-2x}\,dx &= \lim_{b\to\infty} \int_{0}^{b} e^{-2x}\,dx \\ &= \lim_{b\to\infty} \frac{1}{2}(1 - e^{-2b}) = \frac{1}{2}.\end{aligned} \tag{9}$$

Hence the improper integral (8) converges and has the value $\frac{1}{2}$.

More generally, for any $c > 0$,

$$\begin{aligned}\int_{0}^{\infty} e^{-cx}\,dx &= \lim_{b\to\infty} \int_{0}^{b} e^{-cx}\,dx \\ &= \lim_{b\to\infty} \frac{1}{c}(1 - e^{-cb}) = \frac{1}{c}. \quad \blacksquare\end{aligned} \tag{10}$$

EXAMPLE 2

Determine whether the improper integral

$$\int_0^\infty \frac{2x}{(x^2 + 1)^{3/4}}\, dx \tag{11}$$

converges, and if so, find its value.

For any $b > 0$ we have

$$\int_0^b \frac{2x}{(x^2 + 1)^{3/4}}\, dx = 4(x^2 + 1)^{1/4}\Big|_0^b = 4[(b^2 + 1)^{1/4} - 1].$$

Thus, using the definition (5),

$$\int_0^\infty \frac{2x}{(x^2 + 1)^{3/4}}\, dx = \lim_{b\to\infty} 4[(b^2 + 1)^{1/4} - 1] = \infty,$$

so the given integral diverges. ∎

EXAMPLE 3

Determine whether the improper integral

$$\int_0^\infty \cos 2x\, dx \tag{12}$$

converges, and if so, find its value.

For any $b > 0$ we have

$$\int_0^b \cos 2x\, dx = \frac{1}{2}\sin 2x\Big|_0^b = \frac{1}{2}\sin 2b.$$

However, $\sin 2b$ oscillates back and forth between -1 and 1 as $b \to \infty$ and hence has no limit. Therefore the integral (12) diverges. ∎

When $f(x) \ge 0$, the preceding discussion can be interpreted geometrically in terms of area under a curve. In Figure 11.3.1a, $I(b) = \int_a^b f(x)\, dx$ is the area of the shaded region R. As $b \to \infty$, the right-hand boundary of R recedes to infinity, resulting in the region $\hat{R}$ in Figure 11.3.1b. If the improper integral $\int_a^\infty f(x)\, dx$ converges, then its value is defined to be the area of $\hat{R}$. If $\int_a^\infty f(x)\, dx$ diverges, then $\hat{R}$ is said to have infinite area. The two cases are illustrated by Examples 1 and 2, respectively.

It may seem surprising at first that an unbounded region may have a finite area, but this is often the case. Further, suppose that the region $\hat{R}$ is rotated about one of the coordinate axes so as to create a solid of revolution. Then, even if $\hat{R}$ has finite area, the volume of the corresponding solid of revolution may be infinite. Alternatively, if $\hat{R}$ has infinite area, the solid of revolution may have finite volume. Thus a measure of caution is advisable in applying geometric intuition to improper integrals. See Problems 33 and 34 for a further discussion.

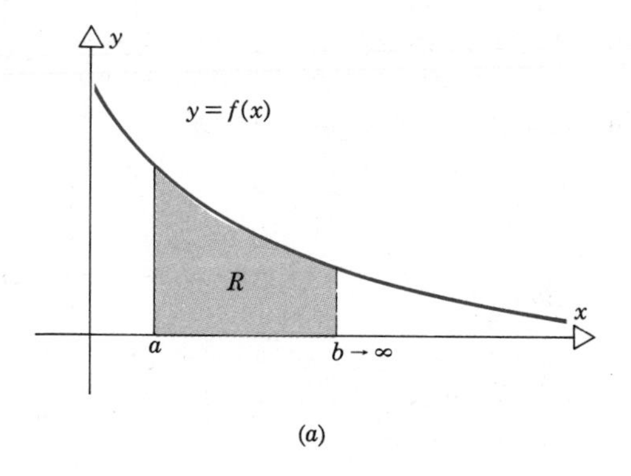

(a)

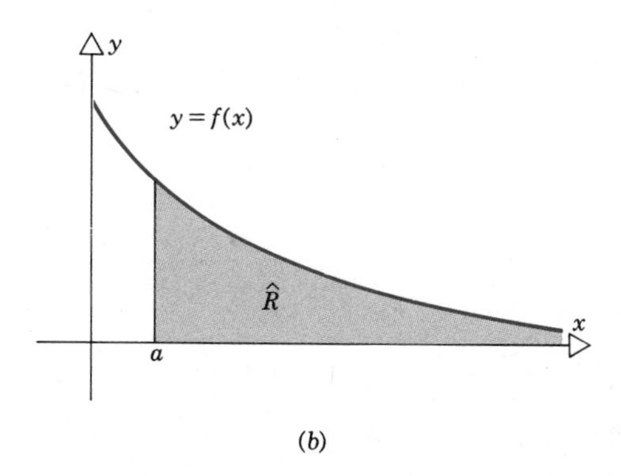

(b)

Figure 11.3.1

Up to now we have been careful to use the expression "improper integral" whenever the integral under discussion was, in fact, improper. Hereafter, in accord with common practice, we will often use the word "integral" alone to refer either to a proper or to an improper integral; normally it is not difficult to tell whether a given integral is improper or not.

EXAMPLE 4

If p is a given real number, determine whether the integral

$$\int_1^\infty \frac{dx}{x^p} \tag{13}$$

converges, and if so, find its value.

If $p \neq 1$, then for any $b > 1$ we have

$$\int_1^b \frac{dx}{x^p} = \left.\frac{x^{1-p}}{1-p}\right|_1^b = \frac{1}{1-p}(b^{1-p} - 1).$$

The limiting behavior of this expression as $b \to \infty$ depends on the sign of $1 - p$. If $1 - p > 0$, that is, if $p < 1$, then $b^{1-p} \to \infty$ as $b \to \infty$. Hence $\lim_{b\to\infty} \int_1^b dx/x^p$ does not exist and the integral (13) diverges. On the other hand, if $1 - p < 0$, that is, if $p > 1$, then $b^{1-p} \to 0$ as $b \to \infty$. Thus, in this case,

$$\int_1^\infty \frac{dx}{x^p} = \lim_{b\to\infty} \int_1^b \frac{dx}{x^p} = \frac{1}{p-1}, \qquad p > 1.$$

If $p = 1$, then we have

$$\int_1^\infty \frac{dx}{x^p} = \int_1^\infty \frac{dx}{x} = \lim_{b\to\infty} \int_1^b \frac{dx}{x}$$

$$= \lim_{b\to\infty} \ln b = \infty,$$

so the integral (13) diverges in this case.

To summarize:

$$\int_1^\infty \frac{dx}{x^p} = \begin{cases} \dfrac{1}{p-1}, & \text{if } p > 1; \\ \text{diverges}, & \text{if } p \leq 1. \end{cases} \quad \blacksquare \tag{14}$$

Now we turn to integrals that are improper only because the integrand becomes unbounded near a finite point. By splitting the original integral into subintervals, if necessary, this point can always be placed at one end of the interval of integration. The discussion follows the same lines as in the case of the unbounded interval.

DEFINITION 11.3.2 Let f be defined on $(a, b]$ and suppose that $f(x)$ becomes unbounded as $x \to a$ from the right. Suppose further that $\int_c^b f(x)\,dx$ exists for each c in (a, b). Then the improper integral $\int_a^b f(x)\,dx$ is defined by

$$\int_a^b f(x)\,dx = \lim_{c\to a^+} \int_c^b f(x)\,dx, \tag{15}$$

provided that this limit exists.

Again, if the limit on the right side of Eq. 15 exists, then the improper integral $\int_a^b f(x)\,dx$ is said to converge; otherwise, it diverges. A similar definition applies if the integral is improper because the integrand becomes unbounded in the neighborhood of the upper limit of integration:

$$\int_a^b f(x)\,dx = \lim_{c\to b^-} \int_a^c f(x)\,dx, \tag{16}$$

provided that this limit exists.

This type of improper integral can also be visualized geometrically in terms of area when $f(x) \geq 0$ (see Figure 11.3.2). The area of the shaded region R is given by $\int_c^b f(x)\,dx$. If the limit in Eq. 15 exists, then the value of the convergent improper integral $\int_a^b f(x)\,dx$ is defined to be the area of the unbounded region obtained from R as $c \to a^+$. A similar interpretation can be given for an integral whose integrand is unbounded as x approaches the upper limit of integration.

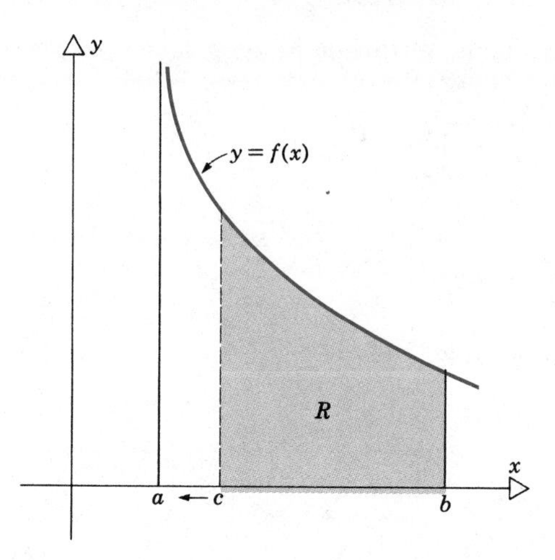

Figure 11.3.2

EXAMPLE 5

Determine whether the integral

$$\int_1^2 \frac{dx}{(x-1)^{1/3}} \tag{17}$$

converges, and if so, find its value.

This integral is improper because the integrand becomes unbounded as $x \to 1^+$. For any c such that $1 < c < 2$ we have

$$\begin{aligned}\int_c^2 \frac{dx}{(x-1)^{1/3}} &= \int_c^2 (x-1)^{-1/3}\,dx \\ &= \frac{3}{2}(x-1)^{2/3}\Big|_c^2 = \frac{3}{2}[1-(c-1)^{2/3}].\end{aligned}$$

As $c \to 1$ from above, $(c-1)^{2/3} \to 0$, and consequently

$$\int_1^2 \frac{dx}{(x-1)^{1/3}} = \lim_{c\to 1^+} \int_c^2 \frac{dx}{(x-1)^{1/3}} = \frac{3}{2}. \blacksquare$$

EXAMPLE 6

If p is a given real number, determine whether the integral

$$\int_0^1 \frac{dx}{x^p} \tag{18}$$

converges, and if so, find its value.

If $p \leq 0$ the integral is not improper, so we will consider only the case $p > 0$ in which the integrand becomes unbounded as $x \to 0^+$. If $p \neq 1$, then for any c in $0 < c < 1$ we have

$$\int_c^1 \frac{dx}{x^p} = \left.\frac{x^{1-p}}{1-p}\right|_c^1 = \frac{1}{1-p}(1 - c^{1-p}).$$

The limiting behavior of this expression as $c \to 0^+$ depends on whether $1 - p$ is positive or negative. If $1 - p > 0$, then $c^{1-p} \to 0$ as $c \to 0^+$. In this case

$$\int_0^1 \frac{dx}{x^p} = \lim_{c \to 0^+} \int_c^1 \frac{dx}{x^p} = \frac{1}{1-p}, \qquad p < 1.$$

However, if $1 - p < 0$, then $c^{1-p} \to \infty$ as $c \to 0^+$, and thus the integral (18) diverges in this case.

If $p = 1$ then

$$\int_0^1 \frac{dx}{x^p} = \int_0^1 \frac{dx}{x} = \lim_{c \to 0^+} \int_c^1 \frac{dx}{x}$$

$$= \lim_{c \to 0^+} (-\ln c) = \infty,$$

and the integral diverges in this case as well.

To summarize:

$$\int_0^1 \frac{dx}{x^p} = \begin{cases} \dfrac{1}{1-p}, & \text{if } p < 1; \\ \text{diverges}, & \text{if } p \geq 1. \end{cases} \tag{19}$$

In exactly the same way it is possible to show that the integrals

$$\int_a^b \frac{dx}{(x-a)^p} \quad \text{and} \quad \int_a^b \frac{dx}{(b-x)^p} \tag{20}$$

converge for $p < 1$ and diverge for $p \geq 1$. ∎

EXAMPLE 7

Determine whether the integral

$$\int_{-1}^1 \frac{dx}{x^2} \tag{21}$$

converges, and if so, find its value.

Since the integral (21) is improper at the origin, we first split it into two parts

$$\int_{-1}^1 \frac{dx}{x^2} = \int_{-1}^0 \frac{dx}{x^2} + \int_0^1 \frac{dx}{x^2}. \tag{22}$$

By the result of Example 6 each integral on the right side of Eq. 22 diverges; therefore the integral (21) also diverges.

The point of this example is to emphasize that

$$\int_{-1}^{1} \frac{dx}{x^2} \neq -\frac{1}{x}\bigg|_{-1}^{1} = -1 - 1 = -2.$$

While this procedure may appear plausible at first glance, it is incorrect because the integrand $1/x^2$ does not satisfy the requirements of the fundamental theorem of calculus (Theorem 6.4.3). The result is also clearly false by Theorem 6.3.4; the integral of a positive function cannot be negative. ∎

We conclude this section with an example of an integral that is improper for both of the reasons previously discussed.

EXAMPLE 8

If p is a given real number, determine whether the integral

$$\int_0^{\infty} \frac{dx}{x^p} \tag{23}$$

converges, and if so, find its value.

Suppose $p > 0$. Then the integral is improper at the origin because the integrand becomes unbounded, and it is also improper because the interval of integration is infinite. Thus we split the integral into two parts

$$\int_0^{\infty} \frac{dx}{x^p} = \int_0^{1} \frac{dx}{x^p} + \int_1^{\infty} \frac{dx}{x^p}. \tag{24}$$

The first integral on the right side of Eq. 24 converges only for $p < 1$ (by Example 6), and the second integral converges only for $p > 1$ (by Example 4). Hence there is no positive p for which both integrals on the right side of Eq. 24 converge, and hence there is no positive p for which the integral (23) converges.

The result is the same if $p \le 0$. In this case, the first integral on the right side of Eq. 24 is no longer improper, but the second integral always diverges.

Thus there is no value of p for which the integral (23) converges. ∎

PROBLEMS

In each of Problems 1 through 24, determine whether the given integral converges or diverges. If it converges, find its value.

1. $\int_0^{\infty} \frac{x}{1 + x^2}\, dx$

2. $\int_0^{\infty} \frac{e^{-x}}{1 + e^{-x}}\, dx$

3. $\int_0^{1} \frac{dx}{(1 - x)^{1/3}}$

4. $\int_0^{\infty} \frac{\arctan x}{1 + x^2}\, dx$

5. $\int_0^{9} \frac{dx}{\sqrt{9 - x}}$

6. $\int_0^{\infty} \frac{x^2}{(1 + x^3)^{1/2}}\, dx$

7. $\int_{-\infty}^{0} xe^{-x^2}\, dx$

8. $\int_0^{\infty} xe^{-x^2/a^2}\, dx, \qquad a \neq 0$

9. $\int_0^{3} \frac{dx}{\sqrt{9 - x^2}}$

10. $\int_{-\infty}^{\infty} \cos^2 x\, dx$

11. $\int_2^\infty \frac{dx}{x \ln x}$

12. $\int_0^9 \frac{dx}{(9 - x)^2}$

13. $\int_0^\infty \frac{dx}{(x - 2)^3}$

14. $\int_0^1 \frac{x^3}{\sqrt{1 - x^4}}\, dx$

15. $\int_0^\infty \frac{x}{(x^2 + 1)^{3/2}}\, dx$

16. $\int_{-\infty}^0 \frac{dx}{1 + x^2}$

17. $\int_0^1 \frac{e^x}{1 - e^x}\, dx$

18. $\int_{-\infty}^\infty \frac{x}{(x^2 + 1)^{5/2}}\, dx$

19. $\int_{-\infty}^\infty \frac{dx}{1 + |x|}$

20. $\int_{-\infty}^0 e^{2x}\, dx$

21. $\int_0^1 \frac{\ln x}{x}\, dx$

22. $\int_{-\infty}^0 \sin \frac{x}{2}\, dx$

23. $\int_{-1}^1 \frac{dx}{x^2 - 1}$

24. $\int_{-\infty}^0 xe^{-x^2}\, dx$

25. **Integration by parts.** Show that if u and v have continuous derivatives for $x \geq a$, then

$$\int_a^\infty u(x)v'(x)\, dx = \lim_{b\to\infty} u(b)v(b) - u(a)v(a) - \int_a^\infty v(x)u'(x)\, dx, \qquad (i)$$

provided that $\lim_{b\to\infty} u(b)v(b)$ exists and either of the two integrals in Eq. (i) converges (in this case, the other integral must also converge).

In each of Problems 26 through 30, use integration by parts (Problem 25) to evaluate the given improper integral.

26. $\int_0^\infty xe^{-x}\, dx$

27. $\int_0^\infty x^n e^{-x}\, dx$, n a positive integer

28. $\int_0^\infty e^{-x} \sin x\, dx$

29. $\int_0^\infty e^{-ax} \sin bx\, dx$, $a > 0$, $b > 0$

30. $\int_0^\infty e^{-ax} \cos bx\, dx$, $a > 0$, $b > 0$

31. Find the value of α for which the integral

$$\int_1^\infty \left(\frac{x}{x^2 + 1} - \frac{\alpha}{2x + 3}\right) dx$$

converges, and evaluate the integral in that case.

32. Find the value of α for which the integral

$$\int_0^\infty \left(\frac{\alpha x}{4x^2 + 1} - \frac{3x^2}{x^3 + 1}\right) dx$$

converges, and evaluate the integral in that case.

33. Let $\hat{R}$ be the region for $x \geq 1$ between the x-axis and the graph of $y = x^{-p}$, where p is a real number.

(a) Find the values of p for which the area of $\hat{R}$ is finite.

(b) Let $\hat{R}$ be rotated about the x-axis. Find the values of p for which the resulting volume is finite.

(c) Let $\hat{R}$ be rotated about the y-axis. Find the values of p for which the resulting volume is finite.

(d) For what values of p can a region with finite area be rotated about one of the coordinate axes so as to give a solid of infinite volume?

(e) For what values of p can a region of infinite area be rotated about one of the coordinate axes so as to give a solid of finite volume?

34. Let R be the region between the x-axis and the graph of $y = x^{-p}$, where p is a real number, for $0 < x \leq 1$.

(a) Find the values of p for which the area of R is finite.

(b) Let R be rotated about the x-axis. Find the values of p for which the resulting volume is finite.

(c) Let R be rotated about the y-axis. Find the values of p for which the resulting volume is finite.

(d) For what values of p can a region of finite area be rotated about one of the coordinate axes so as to give a solid of infinite volume?

(e) For what values of p can a region of infinite area be rotated about one of the coordinate axes so as to give a solid of finite volume?

35. (a) Find the value of $\lim_{R\to\infty} \int_{-R}^R x\, dx$.

(b) Determine whether the integral $\int_{-\infty}^\infty x\, dx$ converges or diverges.

(c) Explain why your answers to Parts (a) and (b) are consistent.

(d) Answer Parts (a), (b), and (c) if the integrand is replaced by x^2.

(e) Answer Parts (a), (b), and (c) if the integrand is replaced by $(1 + x^2)^{-1}$

* **36.** By the inverse square law of gravitational attraction the force exerted on an object of mass m by the earth's gravitational field is

$$F(r) = -mg\left(\frac{R}{r}\right)^2,$$

where g is the acceleration at sea level due to gravity, R is the radius of the earth, and r is the distance of the object from the center of the earth.

(a) Find the work that must be done against F in order to lift the mass m from sea level to an altitude h above sea level.

(b) Find the work that must be done against F to lift the mass m out of the earth's gravitational field, that is, to lift m infinitely far.

(c) Suppose that the mass m is to be lifted from sea level by imparting an initial kinetic energy to the object. Find the initial velocity required to lift the mass infinitely far. This is the escape velocity at sea level; see also Problem 30 in Section 10.2.

11.4 THE COMPARISON TEST FOR IMPROPER INTEGRALS

Consider again the integral

$$\int_a^\infty f(x)\,dx \tag{1}$$

that is improper because the interval of integration is infinite. In Section 11.3 we stated that

$$\int_a^\infty f(x)\,dx = \lim_{b\to\infty}\int_a^b f(x)\,dx, \tag{2}$$

whenever this limit exists. Examples 1 through 4 of that section illustrated the use of definition (2) to determine whether an integral of the form (1) converges, and if so, to what value.

Unfortunately, this procedure is not as widely applicable as we would like, because it requires that $\int_a^b f(x)\,dx$ be evaluated and examined as a function of b. Often this is not possible, and in those cases the definition cannot be used directly as in the examples in the previous section.

The inherent difficulty in using definition (2) can be overcome to a large extent once it is clearly understood that two separate questions are involved. First, *does the improper integral (1) converge or diverge?* Second, *if it converges, what is its value?* These two questions are interconnected in definition (2), but usually they are more amenable to analysis if they are dealt with separately. If an antiderivative of the integrand is available, then both questions can be answered at the same time using the definitions in Section 11.3. If we cannot recognize an antiderivative, then it is necessary to consider the two questions separately.

We will shortly present a result that will enable us to answer the first question, that of convergence or divergence, for many improper integrals. If we conclude that a given improper integral does converge, then the second question, that of evaluation, can be approached by calling on the tools of numerical integration, or perhaps infinite series,* to obtain an adequate approximate value for the integral.

*Infinite series are discussed in Chapters 12 and 13.

This is often a quite difficult problem, and we do not deal with it here, except to say that such approximate methods are usually inappropriate for answering the first and more basic question, the question of convergence or divergence.

We now turn to the question of deciding whether a given improper integral converges, as distinguished from finding its value. It is often possible to do this by comparing a given integral with another one whose convergence or divergence has already been established.

Theorem 11.4.1

(Comparison Test)

Suppose that f and g are integrable on every interval $[a, b]$, where $b > a$, and a is fixed.

(a) If $0 \leq f(x) \leq g(x)$ for all $x \geq a$, and if $\int_a^\infty g(x)\,dx$ converges, then $\int_a^\infty f(x)\,dx$ also converges.

(b) If $0 \leq g(x) \leq f(x)$ for all $x \geq a$, and if $\int_a^\infty g(x)\,dx$ diverges, then $\int_a^\infty f(x)\,dx$ also diverges.

This theorem has a simple geometric interpretation since the integrals can be interpreted as areas. The situation in Part (a) of the theorem is shown in Figure

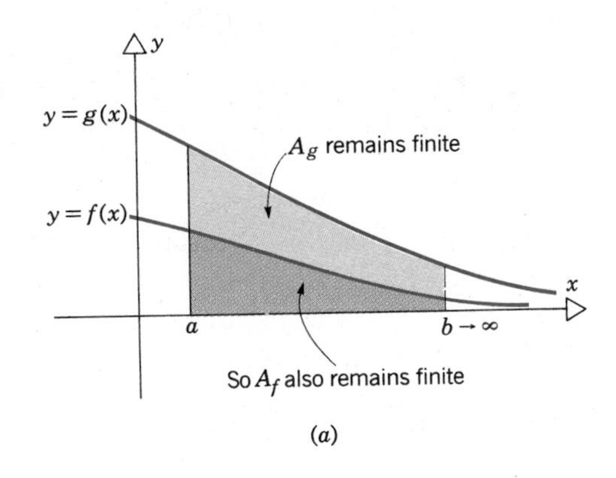

(a)

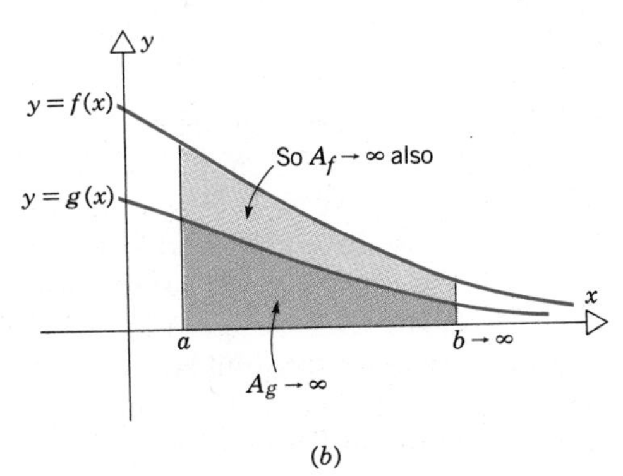

(b)

Figure 11.4.1 Geometrical interpretation of the comparison test for improper integrals.

11.4.1a. The graph of $y = g(x)$ always lies above the graph of $y = f(x)$, so the area A_g under the graph of g from a to b is at least as large as the area A_f under the graph of f on the same interval. Since both f and g are nonnegative, the areas A_f and A_g tend to increase (or at least not to decrease) as b increases. If A_g approaches a finite limit as $b \to \infty$, then the theorem states that the same is true of A_f.

In a similar way Figure 11.4.1b shows the situation in Part (b) of the theorem. In this case $A_f \geq A_g$. Since A_g becomes unbounded as $b \to \infty$, the same must be true for A_f.

A more precise formulation of these geometrical ideas leads to a proof of Theorem 11.4.1; see Problem 39. However, we will omit a further discussion here, since a very similar result, Theorem 12.3.1, is treated in detail in the next chapter.

While Theorem 11.4.1 can be extremely useful, it is sometimes a nuisance to have to deal with the inequalities in its hypotheses. The following variation of the comparison test is frequently easier to use, and is therefore usually preferable.

Theorem 11.4.2

(Limit Comparison Test)

Suppose that $f(x) \geq 0$ and $g(x) > 0$ for $x \geq a$, and that both f and g are integrable on every interval $[a, b]$, where $b > a$. Suppose further that

$$\lim_{x\to\infty} \frac{f(x)}{g(x)} = L, \qquad 0 < L < \infty. \tag{3}$$

Then the improper integrals $\int_a^\infty f(x)\,dx$ and $\int_a^\infty g(x)\,dx$ either both converge or both diverge.

The proof of this theorem is based on the idea that if $f(x)/g(x) \to L$ as $x \to \infty$, then for large x it must follow that

$$f(x) \cong L\,g(x). \tag{4}$$

Consequently, for large R,

$$\int_R^\infty f(x)\,dx \cong L \int_R^\infty g(x)\,dx, \tag{5}$$

and this makes the conclusion of the theorem plausible. Problem 40 indicates how to make this argument more precise. Note that it is important that L lie in the interval $(0, \infty)$, that is, $L \neq 0$ and $L \neq \infty$. The conclusions that can be drawn in these latter cases are indicated in Problem 41.

When we use either Theorem 11.4.1 or Theorem 11.4.2 we draw a conclusion about $\int_a^\infty f(x)\,dx$ from our prior knowledge of the behavior of a comparison integral $\int_a^\infty g(x)\,dx$. The most useful comparison integral is $\int_1^\infty x^{-p}\,dx$, which was discussed in Example 4 of Section 11.3; recall that this integral converges when $p > 1$ and diverges otherwise.

EXAMPLE 1

Determine whether the integral

$$\int_1^\infty \frac{dx}{1 + x^4} \tag{6}$$

converges or diverges.

The convergence or divergence of the integral is determined by the nature of the integrand for large x. Here it seems reasonable to neglect the 1 in the denominator in comparison with x^4 and to use $g(x) = 1/x^4$ as the comparison function. Since

$$\frac{1}{1 + x^4} \leq \frac{1}{x^4} \qquad \text{for} \qquad x \geq 1,$$

and since $\int_1^\infty x^{-4}\, dx$ converges, we conclude from Theorem 11.4.1 that the integral (6) also converges.

Alternatively, we note that if $g(x) = x^{-4}$, then

$$\frac{f(x)}{g(x)} = \frac{1/(1 + x^4)}{1/x^4} = \frac{x^4}{1 + x^4} \to 1 \qquad \text{as} \qquad x \to \infty.$$

Thus Theorem 11.4.2 applies and, in conjunction with the known convergence of $\int_1^\infty x^{-4}\, dx$, shows that the integral (6) converges.

In this example a satisfactory comparison function is fairly obvious, and either Theorem 11.4.1 or Theorem 11.4.2 yields the conclusion about $\int_1^\infty dx/(1 + x^4)$ without difficulty. ∎

EXAMPLE 2

Determine whether the integral

$$\int_2^\infty \frac{x - 2}{x^{3/2} + 1}\, dx \tag{7}$$

converges or diverges.

For large x it is reasonable to believe that

$$\frac{x - 2}{x^{3/2} + 1} \cong \frac{x}{x^{3/2}} = \frac{1}{x^{1/2}}.$$

In fact, using the comparison function $g(x) = 1/x^{1/2}$, we have

$$\frac{f(x)}{g(x)} = \frac{(x - 2)/(x^{3/2} + 1)}{1/x^{1/2}} = \frac{x^{1/2}(x - 2)}{x^{3/2} + 1} \to 1 \qquad \text{as} \qquad x \to \infty.$$

Since $\int_2^\infty dx/x^{1/2}$ diverges, we conclude from Theorem 11.4.2 that the integral (7) also diverges.

In this example it is a bit more difficult to show that the integral (7) diverges by using Theorem 11.4.1, because the obvious comparison function

$g(x) = 1/x^{1/2}$ does not satisfy the right kind of inequality. Indeed, we have

$$\frac{x-2}{x^{3/2}+1} \leq \frac{x}{x^{3/2}} = \frac{1}{x^{1/2}},$$

whereas in order to establish the divergence of the integral (7) we need a comparison function that is *smaller* than the integrand in (7). In Problem 36 we indicate one way to proceed so as to make use of Theorem 11.4.1 in this case instead of Theorem 11.4.2. The main point to remember, however, is that this example illustrates that Theorem 11.4.2 is often more convenient than Theorem 11.4.1. ∎

EXAMPLE 3

Determine whether the integral

$$\int_{-\infty}^{\infty} \frac{dx}{1+e^x} \tag{8}$$

converges or diverges.

First we split the given integral into two parts,

$$\int_{-\infty}^{\infty} \frac{dx}{1+e^x} = \int_{-\infty}^{0} \frac{dx}{1+e^x} + \int_{0}^{\infty} \frac{dx}{1+e^x}, \tag{9}$$

and then we consider each part separately. The second integral on the right side of Eq. 9 can be shown to converge by using the comparison function $g(x) = e^{-x}$ and noting that $\int_0^\infty e^{-x}\,dx$ converges. However, for x large and negative we have $(1+e^x)^{-1} \cong 1$, so for the first integral on the right side of Eq. 9 we choose the comparison function $g(x) = 1$. Then $f(x)/g(x) = 1/(1+e^x) \to 1$ as $x \to -\infty$, and $\int_{-\infty}^0 g(x)\,dx = \int_{-\infty}^0 dx$ diverges, so it follows that $\int_{-\infty}^0 (1+e^x)^{-1}\,dx$ also diverges. Therefore, the integral (8) diverges as well. ∎

There are also comparison tests analogous to Theorems 11.4.1 and 11.4.2 for integrals that are improper because the integrand becomes unbounded near a finite point. We will state these theorems for improper integrals whose integrands become unbounded at the lower endpoint, but it is easy to adapt the theorems to the situation in which the integrands become unbounded at the upper end point. The proofs are analogous to those of Theorems 11.4.1 and 11.4.2 and are omitted.

Theorem 11.4.3

(Comparison Test)

Suppose that f and g are integrable on every interval $[c, b]$, where $a < c < b$.

(a) If $0 \leq f(x) \leq g(x)$ for $a < x \leq b$, and if $\int_a^b g(x)\,dx$ converges, then $\int_a^b f(x)\,dx$ also converges.

(b) If $0 \le g(x) \le f(x)$ for $a < x \le b$, and if $\int_a^b g(x)\,dx$ diverges, then $\int_a^b f(x)\,dx$ also diverges.

Theorem 11.4.4

(Limit Comparison Test)

Suppose that $f(x) \ge 0$ and $g(x) > 0$ for $a < x \le b$, and that both f and g are integrable on every interval $[c, b]$, where $a < c < b$. Suppose further that

$$\lim_{x \to a^+} \frac{f(x)}{g(x)} = L, \qquad 0 < L < \infty. \tag{10}$$

Then the integrals $\int_a^b f(x)\,dx$ and $\int_a^b g(x)\,dx$ either both converge or both diverge.

The following examples illustrate the use of Theorems 11.4.3 and 11.4.4. The most useful comparison integrals are $\int_a^b (x - a)^{-p}\,dx$ and $\int_a^b (b - x)^{-p}\,dx$. From Example 6 of Section 11.3 we know that these integrals converge for $p < 1$ and diverge otherwise.

EXAMPLE 4

Determine whether the integral

$$\int_1^2 \frac{dx}{\sqrt{x^2 - 1}} \tag{11}$$

converges or diverges.

The integral is improper because the integrand becomes unbounded as $x \to 1^+$. To examine the consequences of this it is helpful to factor the integrand

$$\frac{1}{\sqrt{x^2 - 1}} = \frac{1}{\sqrt{x + 1}\,\sqrt{x - 1}} \tag{12}$$

so as to isolate more clearly the part that grows without bound. For $x > 1$ we observe that

$$\frac{1}{\sqrt{x + 1}} < \frac{1}{\sqrt{1 + 1}} = \frac{1}{\sqrt{2}} < 1;$$

hence it follows from Eq. 12 that

$$\frac{1}{\sqrt{x^2 - 1}} \le \frac{1}{\sqrt{x - 1}}, \qquad x > 1.$$

Since we know that $\int_1^2 dx/\sqrt{x - 1}$ converges, we conclude from Theorem 11.4.3 that the given integral (11) also converges. ∎

EXAMPLE 5

Determine whether the integral

$$\int_0^1 \frac{1 - \cos x}{x^{7/2}}\, dx \tag{13}$$

converges or diverges.

In order to draw any conclusions about this integral we must examine the behavior of the integrand near $x = 0$. From Example 2 of Section 11.1 we know that for $|x|$ very small, $1 - \cos x$ is almost proportional to x^2. Hence it is reasonable to conclude that $g(x) = x^{-3/2}$ is a possible comparison function for the integrand in the integral (13). We then have

$$\frac{f(x)}{g(x)} = \frac{(1 - \cos x)/x^{7/2}}{x^{-3/2}} = \frac{1 - \cos x}{x^2}. \tag{14}$$

Hence, using the result of Example 2 of Section 11.1, or applying L'Hospital's rule to the right side of Eq. 14, we obtain

$$\lim_{x \to 0^+} \frac{f(x)}{g(x)} = \frac{1}{2}.$$

Since we know that

$$\int_0^1 g(x)\, dx = \int_0^1 x^{-3/2}\, dx$$

diverges, Theorem 11.4.4 yields the same conclusion about the integral (13). ∎

EXAMPLE 6

Determine the values of p for which the integral

$$\int_0^\infty \frac{dx}{x^p\sqrt{1 + x^p}}, \qquad p > 0 \tag{15}$$

converges.

The integral (15) is improper for two reasons: the interval of integration is infinite, and the integrand also becomes unbounded as $x \to 0^+$. To separate these effects we first write

$$\int_0^\infty \frac{dx}{x^p\sqrt{1 + x^p}} = \int_0^1 \frac{dx}{x^p\sqrt{1 + x^p}} + \int_1^\infty \frac{dx}{x^p\sqrt{1 + x^p}}. \tag{16}$$

Then the integral (15) converges if and only if both integrals on the right side of Eq. 16 converge. We will consider each of them in turn.

Near $x = 0$ we have $\sqrt{1 + x^p} \cong 1$. Hence

$$f(x) = \frac{1}{x^p\sqrt{1 + x^p}} \cong \frac{1}{x^p},$$

so it is reasonable to use $g(x) = x^{-p}$ as a comparison function. Then

$$\frac{f(x)}{g(x)} = \frac{1/(x^p\sqrt{1 + x^p})}{1/x^p} = \frac{1}{\sqrt{1 + x^p}} \to 1 \tag{17}$$

as $x \to 0^+$. Since $\int_0^1 x^{-p}\,dx$ converges if $p < 1$ and diverges otherwise, Theorem 11.4.4 yields the same conclusion about the first integral on the right side of Eq. 16.

Now let us consider the second integral, in which the behavior of the integrand for large x is crucial. We have $\sqrt{1 + x^p} \cong x^{p/2}$ for x very large, so in that case

$$\frac{1}{x^p\sqrt{1 + x^p}} \cong \frac{1}{x^p x^{p/2}} = x^{-3p/2}.$$

Thus we select $g(x) = x^{-3p/2}$ as our comparison function. Then

$$\frac{f(x)}{g(x)} = \frac{1}{x^p\sqrt{1 + x^p}} \Big/ \frac{1}{x^{3p/2}} = \frac{x^{p/2}}{\sqrt{1 + x^p}} \to 1 \tag{18}$$

as $x \to \infty$. Further $\int_1^\infty x^{-3p/2}\,dx$ converges when $3p/2 > 1$, that is, when $p > \frac{2}{3}$, and diverges otherwise. By Theorem 11.4.4 we draw the same conclusion about the second integral on the right side of Eq. 16.

Combining our results we observe that both integrals on the right side of Eq. 16 converge only for those values of p in the interval $\frac{2}{3} < p < 1$. Thus these are the only positive values of p for which the integral (15) converges. We leave to the reader the task (Problem 37) of showing that the integral (15) diverges for all $p \le 0$. ∎

The results in this section apply only to improper integrals with nonnegative integrands, or (if we multiply by -1) with nonpositive integrands. Of course, there are many improper integrals whose integrands fluctuate in sign. Such integrals can sometimes be analyzed by extending the methods presented here, but we leave that subject to more advanced books. The procedure parallels to a considerable extent the discussion for infinite series in Section 12.7.

PROBLEMS

In each of Problems 1 through 34, determine whether the given integral converges or diverges.

1. $\displaystyle\int_1^\infty \frac{x}{4 + x^3}\,dx$

2. $\displaystyle\int_0^3 \frac{dx}{(9 - x^2)^{3/2}}$

3. $\displaystyle\int_1^\infty \frac{x^{3/2}}{4 + x^2}\,dx$

4. $\displaystyle\int_1^\infty \frac{x}{\sqrt{4 + x^3}}\,dx$

5. $\displaystyle\int_0^{\pi/2} \frac{\sin x}{x^{3/2}}\,dx$

6. $\displaystyle\int_1^\infty \frac{\sin^2 x}{4 + x^3}\,dx$

7. $\displaystyle\int_2^5 \frac{x}{x - 2}\,dx$

8. $\displaystyle\int_2^5 \frac{x}{\sqrt{x - 2}}\,dx$

9. $\displaystyle\int_1^\infty \frac{\sin^2 x}{\sqrt{4 + x^3}}\,dx$

10. $\displaystyle\int_2^\infty \frac{x + 1}{x^3 - 1}\,dx$

11. $\displaystyle\int_{-\infty}^\infty \frac{e^x}{e^{2x} + e^{-2x}}\,dx$

12. $\displaystyle\int_0^{\pi/2} \left(\frac{\pi}{2} - x\right)^{1/2} \tan x\,dx$

13. $\displaystyle\int_1^\infty \frac{dx}{(x^2 - 1)^{2/3}}$

14. $\displaystyle\int_{-\infty}^\infty \frac{x^3}{e^x}\,dx$

15. $\displaystyle\int_0^1 \frac{e^{-x}}{\sqrt{1 - x^4}}\,dx$

16. $\displaystyle\int_0^\infty \frac{dx}{x^2 - 1}$

17. $\displaystyle\int_0^\infty \frac{x^n}{e^x}\,dx$, n a positive integer

18. $\displaystyle\int_0^{\pi/2} \sec^2 x\,dx$

19. $\displaystyle\int_1^\infty \frac{\ln x}{x^2}\,dx$

20. $\displaystyle\int_2^\infty \frac{dx}{\sqrt{x}\ln x}$

21. $\displaystyle\int_0^1 \frac{e^x - 1}{x}\,dx$

22. $\displaystyle\int_0^\infty \frac{|\sin(x^2)|}{x^{3/2}}\,dx$

23. $\displaystyle\int_0^\infty \frac{1 - e^{-x}}{x^p}\,dx$, p a real number

24. $\displaystyle\int_0^\infty \frac{\sin^2 x}{x^{3/2}}\,dx$

25. $\displaystyle\int_3^\infty \frac{(x^2 - 9)^p}{1 + x^4}\,dx$

26. $\displaystyle\int_1^\infty \frac{\ln x}{x^{1+p}}\,dx$, $p > 0$

27. $\displaystyle\int_0^1 x \ln x\,dx$

28. $\displaystyle\int_0^1 x^p \ln x\,dx$, p a real number

29. $\displaystyle\int_{-\infty}^\infty \frac{dx}{1 + x^2}$

30. $\displaystyle\int_0^2 \frac{dx}{\sqrt{|x^2 - 3x + 2|}}$

31. $\displaystyle\int_{-\infty}^\infty \frac{e^{-x}}{1 + x^2}\,dx$

32. $\displaystyle\int_{-\infty}^\infty \frac{\sin x}{|x|^{3/2}}\,dx$

33. $\displaystyle\int_1^\infty \frac{x^p}{(4 + x^3)^q}\,dx$, p, q any real numbers

34. $\displaystyle\int_2^\infty \frac{x^p(x - 1)^q}{(1 + x^2)^{3/2}}\,dx$, p, q any real numbers

35. Consider the region R between the x-axis and the graph of $f(x) = x^{-p}$ for $x \geq 1$, where p is a positive real number. Let R be rotated about the x-axis to form a solid D of revolution, and let S be the surface of D. Find the values of p for which the area of S is finite.

Note: In Problem 33(b) of Section 11.3 it was shown that the volume of D is finite when $p > \frac{1}{2}$. Thus for $\frac{1}{2} < p \leq 1$, a finite volume has an infinite surface area. If one were to attempt to paint S, it would require infinitely much paint to cover S completely. However, D can be completely filled with a finite amount of paint. Can you explain this apparent contradiction?

36. Consider again the integral

$$\int_2^\infty \frac{x - 2}{x^{3/2} + 1}\,dx \qquad (i)$$

of Example 2. To show that this integral diverges by using Theorem 11.4.1 we must find a comparison function $g(x)$ such that $g(x) \leq f(x)$ and $\int_a^\infty g(x)\,dx$ diverges.

(a) Show that $x - 2 \geq x/2$ for $x \geq 4$ and that $x^{3/2} + 1 \leq 2x^{3/2}$ for $x \geq 1$. Therefore

$$\frac{x - 2}{x^{3/2} + 1} \geq \frac{x/2}{2x^{3/2}} = \frac{1}{4\sqrt{x}} \qquad \text{for } x \geq 4.$$

(b) Use the result of Part (a) to conclude that the integral (*i*) diverges.

37. Show that the integral of Example 6

$$\int_0^\infty \frac{dx}{x^p\sqrt{1 + x^p}}$$

diverges whenever $p \leq 0$.

* 38. A function of considerable importance, both in mathematics and in its applications, is called the **gamma function** and is defined by the integral

$$\Gamma(p) = \int_0^\infty x^{p-1}e^{-x}\,dx. \qquad (i)$$

The gamma function was first studied extensively by Euler beginning in 1729. The function was named much later by Legendre.

(a) For what real values of p does the integral (*i*) converge? These values of p comprise the domain of Γ.

(b) Show that

$$\Gamma(p + 1) = p\Gamma(p), \qquad p > 0. \qquad (ii)$$

(c) If $p = n$, a positive integer, show that

$$\Gamma(n + 1) = n!. \qquad (iii)$$

Thus the gamma function is an extension of the factorial function to nonintegral values of the independent variable.

* 39. **Proof of Theorem 11.4.1(a).** Define the set S so that

$$S = \{s | s = \int_a^b f(x)\,dx \text{ for some } b > a\}.$$

(a) Show that S is not empty and is bounded above. Therefore S has a least upper bound, which we denote by I (see discussion preceding Problem 29 of Section 1.1).

(b) Show that $\int_a^b f(x)\,dx \leq I$ for each $b > a$.

(c) For each $\epsilon > 0$ show that there is a $B > a$ such that

$$\int_a^B f(x)\,dx \geq I - \epsilon.$$

(d) Show that $\lim_{b\to\infty} \int_a^b f(x)\,dx = I$; in other words, $\int_a^\infty f(x)\,dx$ converges and has the value I.

* **40. Proof of Theorem 11.4.2.**

(a) Suppose that $f(x)/g(x) \to L > 0$ as $x \to \infty$. Then show that there is an R such that

$$\tfrac{1}{2}Lg(x) \leq f(x) \leq \tfrac{3}{2}Lg(x) \qquad (i)$$

for $x \geq R$.

(b) Show that $\int_a^\infty f(x)\,dx$ converges or diverges according to whether $\int_R^\infty f(x)\,dx$ converges or diverges.

(c) Use the right-hand inequality in Eq. (i) and Theorem 11.4.1 to show that if $\int_a^\infty g(x)\,dx$ converges, then so does $\int_a^\infty f(x)\,dx$.

(d) Use the left-hand inequality in Eq. (i) and Theorem 11.4.1 to show that if $\int_a^\infty g(x)\,dx$ diverges, then so does $\int_a^\infty f(x)\,dx$.

* **41.** This problem explores the situation when $L = 0$ or $L = \infty$ in Theorem 11.4.2.

(a) Show that if $L = 0$ and if $\int_a^\infty g(x)\,dx$ converges, then $\int_a^\infty f(x)\,dx$ also converges.

(b) Let $g(x) = x^{-1/2}$, $f(x) = x^{-1}$, and $a > 0$. Show that $L = 0$ and that both $\int_a^\infty f(x)\,dx$ and $\int_a^\infty g(x)\,dx$ diverge.

(c) Let $g(x) = x^{-1/2}$, $f(x) = x^{-2}$, and $a > 0$. Show that $L = 0$, that $\int_a^\infty g(x)\,dx$ diverges, but $\int_a^\infty f(x)\,dx$ converges. Observe that the results of (b) and (c) show that no conclusion about the convergence of $\int_a^\infty f(x)\,dx$ can be drawn when $L = 0$ and $\int_a^\infty g(x)\,dx$ diverges.

(d) Show that if $L = \infty$ and $\int_a^\infty g(x)\,dx$ diverges, then $\int_a^\infty f(x)\,dx$ also diverges.

(e) By means of examples show that if $L = \infty$ and $\int_a^\infty g(x)\,dx$ converges, then no conclusion can be drawn about $\int_a^\infty f(x)\,dx$.

REVIEW PROBLEMS

In each of Problems 1 through 29, evaluate the given limit, or else show that it does not exist.

1. $\displaystyle\lim_{x\to 0} \frac{\cos^2(x - \pi/2)}{x}$

2. $\displaystyle\lim_{x\to 1^-} \frac{\sin \pi x}{|x - 1|}$

3. $\displaystyle\lim_{x\to \pi} \frac{\sin 2x}{|x - \pi|}$

4. $\displaystyle\lim_{x\to \pi/2} \frac{\tan^2 x}{\sec(x - \pi)}$

5. $\displaystyle\lim_{x\to 0} \frac{\ln^2(x + 1)}{x^2 + 2x}$

6. $\displaystyle\lim_{x\to 0} \frac{\sin 2x}{\cos^2 x - 1}$

7. $\displaystyle\lim_{x\to 2} \frac{x^2 - 4}{\ln(2x - 3)}$

8. $\displaystyle\lim_{x\to 0} \frac{e^x - 1}{\sin x}$

9. $\displaystyle\lim_{x\to 0} x^2 \ln(x^{-2} + 1)$

10. $\displaystyle\lim_{x\to 0^+} e^{-1/x} \ln x$

11. $\displaystyle\lim_{x\to \pi^-} \frac{\cos^2 x - 1}{\sqrt{\pi^2 - x^2}}$

12. $\displaystyle\lim_{x\to 1} \frac{e^{x - 1/x} - 1}{\ln x}$

13. $\displaystyle\lim_{x\to 0} \exp\left(\frac{1}{x} - e^{1/x}\right)$

14. $\displaystyle\lim_{x\to 0} \frac{e^{\sin^2 x} - 1}{\sin^2 x \cos x}$

15. $\displaystyle\lim_{x\to 0} \frac{x \arcsin x}{(\pi/2 - \arccos x)^2}$

16. $\displaystyle\lim_{x\to 1^+} (1 - x)^3 e^{1/(x-1)}$

17. $\displaystyle\lim_{x\to 0} \left(\frac{1}{x} - \cot x\right)$

18. $\displaystyle\lim_{x\to 0} \left(\frac{1}{x^p}\right)^{\sin x}$

19. $\displaystyle\lim_{x\to 0^+} \frac{e^{1/\sqrt{x}}}{e^{1/\arcsin x}}$

20. $\displaystyle\lim_{x\to 0^+} (\arcsin x)^x$

21. $\displaystyle\lim_{x\to \infty} \left[\cos\left(\frac{1}{x}\right)\right]^x$

22. $\displaystyle\lim_{x\to 1^+} [\ln(x^2)]^{x-1}$

23. $\displaystyle\lim_{x\to 1^+} \left[\frac{1}{(x^2 - 1)}\right]^{\sin \pi x}$

24. $\displaystyle\lim_{x\to \infty} \frac{\ln(x + \ln x)}{x^p}$

25. $\displaystyle\lim_{x\to 1} (x^{3/2})^{1/(1-x^2)}$

26. $\displaystyle\lim_{x\to \infty} (e^{-1/x})^{1/x}$

27. $\displaystyle\lim_{x\to \infty} \frac{x^p}{\ln(x^2 + 1)}$

28. $\displaystyle\lim_{x\to \infty} (e\sqrt{x-1})^{1/x}$

29. $\displaystyle\lim_{x\to 0} (\cos x)^{1/x^2}$

In each of Problems 30 through 38, determine whether the integral converges or diverges. If it converges, find its value.

30. $\displaystyle\int_1^\infty \frac{x^2}{(x^3 + 1)^{2/3}}\,dx$

31. $\displaystyle\int_0^2 \frac{x}{(x^2 - 4)^{1/3}}\,dx$

32. $\displaystyle\int_0^2 \frac{x^3}{(x^4 - 16)^{1/3}}\,dx$

33. $\displaystyle\int_0^1 x^{-3/2}e^{x^{-1/2}}\,dx$

34. $\displaystyle\int_1^\infty x^{-3/2}e^{x^{-1/2}}\,dx$

35. $\displaystyle\int_0^{2\pi} \frac{\sin 2x}{\sin^{3/2} x}\,dx$

36. $\displaystyle\int_1^\infty \frac{x^2}{(x^3 + 1)^{4/3}}\,dx$

37. $\displaystyle\int_1^\infty \frac{x^{-3/2}}{(x^{-1/2} + 1)^2}\,dx$

38. $\displaystyle\int_1^3 \frac{x - 2}{(x^2 - 4x + 3)^2}\,dx$

In each of Problems 39 and 40, determine the values of p for which the integral converges.

39. $\displaystyle\int_0^4 \frac{x - 2}{(x^2 - 4x)^p}\,dx$

40. $\displaystyle\int_0^\infty x^{p-1}e^{x^p}\,dx$

In each of Problems 41 through 49, determine whether the given integral converges or diverges.

41. $\displaystyle\int_0^\pi \frac{x - \sin x}{x^{3/2}}\,dx$

42. $\displaystyle\int_0^4 \frac{4 - x}{x^{1/2}|x^{3/2} - 8|}\,dx$

43. $\displaystyle\int_2^\infty \frac{x^{3/2}}{(x^{5/2} - 1)^2}\,dx$

44. $\displaystyle\int_2^3 \frac{e^{1/(2-x)}}{x^2 + x - 6}\,dx$

45. $\displaystyle\int_1^2 \frac{x^{3/2}}{(x + \cos \pi x)^2}\,dx$

46. $\displaystyle\int_2^4 \frac{1}{\sqrt{x^2 - 4}}\,dx$

47. $\displaystyle\int_1^\infty e^{1/(x^2-1)}\,dx$

48. $\displaystyle\int_1^\infty \frac{x}{(x^2 + x - 2)^{2/3}}\,dx$

49. $\displaystyle\int_1^\infty \frac{\sin x + x}{(x^3 - 1)^{1/2}}\,dx$

infinite series

In this chapter and the next we discuss some of the properties of infinite sequences and series. In ordinary conversation the words "sequence" and "series" are used almost interchangeably, but in mathematics they have different meanings. Sequences and series are often useful in constructing approximations to numbers or functions that are difficult or impossible to calculate exactly. Usually it is desirable to obtain an approximation that can be systematically improved or refined, if necessary. This often leads us to a consideration of a sequence of successive approximations. A study of sequences and series also provides an opportunity to reexamine some of the ideas associated with limits in a new context.

12.1 SEQUENCES

A **sequence** of real numbers is an ordered set, or list, of numbers

$$\{a_n\} = a_1, a_2, a_3, \ldots, a_n, \ldots. \tag{1}$$

The individual numbers, such as a_1 or a_2, are called the **terms** in the sequence. The characteristic that distinguishes a sequence from other sets of real numbers is that the terms of a sequence *appear in a definite order.* For the sequence (1) a_1 is the first term,

a_2 is the second term, and so forth; a_n is referred to as the general, or nth term. As indicated in Eq. 1, the notation $\{a_n\}$ is used to denote the sequence as a whole. Sometimes it is convenient to index the terms so that the first term corresponds to a value of n different from one. For instance, it is fairly common to start with $n = 0$, and one could equally well start with $n = 2$ or some other value.

We have already encountered sequences on several occasions without making explicit mention of the fact. For example, in Section 4.5 we discussed Newton's method for finding a root of an equation $f(x) = 0$. Starting with an initial approximation x_0 for the root, Newton's method yields further approximations x_1, x_2, ... The successive approximations form a sequence

$$\{x_n\} = x_0, x_1, x_2, \ldots . \tag{2}$$

Again, in Section 6.2, in the course of defining the integral, we formed various Riemann sums. For instance, let f be a continuous function defined on $[a, b]$, and let Δ be a uniform partition of this interval into n subintervals, so that $\Delta x_i = (b - a)/n$. If we choose the star point in each subinterval to be the left endpoint, then $x_i^* = x_{i-1}$. The corresponding Riemann sum is

$$S_n = \sum_{i=1}^{n} f(x_{i-1})\Delta x_i = \frac{b-a}{n} \sum_{i=1}^{n} f(x_{i-1}). \tag{3}$$

Figure 12.1.1

As n takes on the values 1, 2, 3, . . . , Eq. 3 generates a sequence of sums that approximates $\int_a^b f(x)\,dx$.

A third example of a sequence occurred in Section 8.3, where we encountered

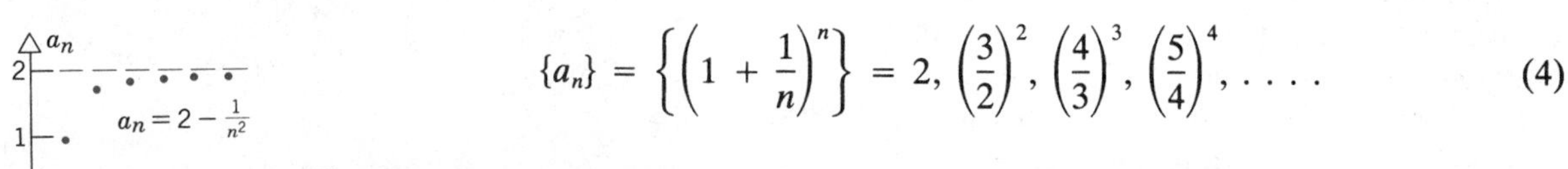

$$\{a_n\} = \left\{\left(1 + \frac{1}{n}\right)^n\right\} = 2, \left(\frac{3}{2}\right)^2, \left(\frac{4}{3}\right)^3, \left(\frac{5}{4}\right)^4, \ldots . \tag{4}$$

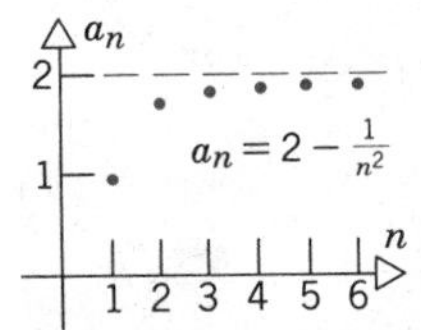

Figure 12.1.2

The first few terms of this sequence are plotted in Figure 12.1.1. We pointed out in Section 8.3 that, as we go farther and farther out in the sequence (4), we obtain increasingly accurate approximations to e.

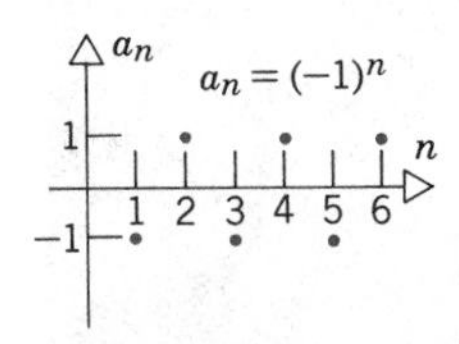

Figure 12.1.3

EXAMPLE 1

Some additional examples of sequences are given below. Figures 12.1.2 through 12.1.7, respectively, show the first few terms in each sequence. In each case we assume that the initial value of n is one.

$$\left\{2 - \left(\frac{1}{n^2}\right)\right\} = 1, \frac{7}{4}, \frac{17}{9}, \frac{31}{16}, \frac{49}{25}, \ldots \tag{5}$$

$$\{(-1)^n\} = -1, 1, -1, 1, -1, \ldots \tag{6}$$

$$\left\{\frac{(-1)^n}{\sqrt{n}}\right\} = -1, \frac{1}{\sqrt{2}}, \frac{-1}{\sqrt{3}}, \frac{1}{2}, \frac{-1}{\sqrt{5}}, \ldots \tag{7}$$

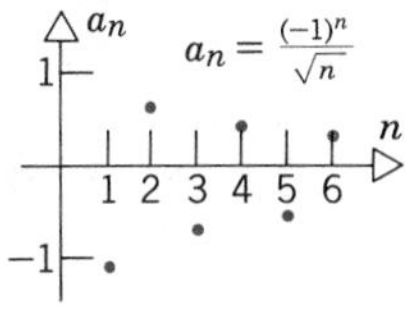

Figure 12.1.4

$$\left\{\sqrt{n} \sin \frac{n\pi}{2}\right\} = 1, 0, -\sqrt{3}, 0, \sqrt{5}, \ldots \tag{8}$$

$$\left\{\frac{n^2 + 1}{n}\right\} = 2, \frac{5}{2}, \frac{10}{3}, \frac{17}{4}, \frac{26}{5}, \ldots \tag{9}$$

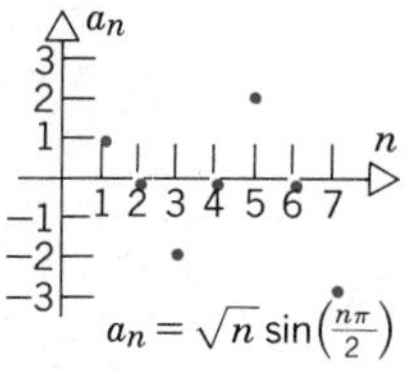

Figure 12.1.5

$$\left\{\frac{4^n}{n!}\right\} = 4, 8, \frac{32}{3}, \frac{32}{3}, \frac{128}{15}, \ldots \;\blacksquare \tag{10}$$

For each of the sequences (4) through (10) there is a formula for the general term a_n. However, this need not be the case. For example, for the sequence (2) obtained from Newton's method, there is instead an iterative formula

$$x_{n+1} = x_n - \frac{f(x_n)}{f'(x_n)} \tag{11}$$

defining each member of the sequence $\{x_n\}$ in terms of the preceding one.

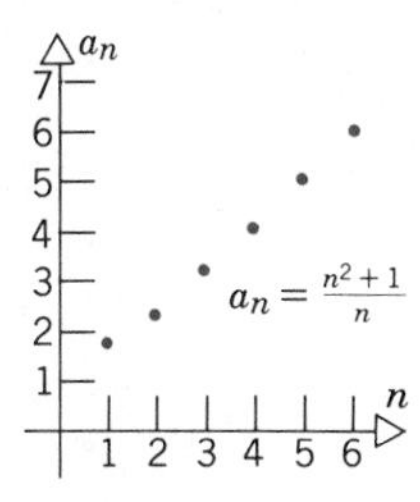

Figure 12.1.6

Convergent and divergent sequences

Usually the most important question to ask about a sequence $\{a_n\}$ is whether or not it has a *limit*, or *converges*, as $n \to \infty$. Intuitively, a sequence $\{a_n\}$ converges to the limit L if the terms in the sequence cluster more and more tightly about the number L as n increases without bound. The formal definition of the limit of a sequence is entirely analogous to the one in Section 2.4 for the limit of a function $f(x)$ as $x \to \infty$. Indeed, we can regard a sequence as a special kind of function, namely, one whose domain is a set of integers. In this event we might write $a_n = f(n)$, and the limit of a sequence is just a special case of the limit of a function.

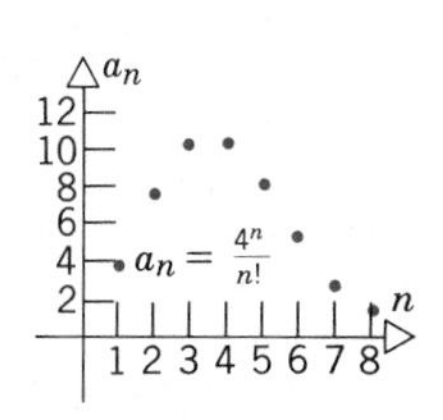

Figure 12.1.7

DEFINITION 12.1.1 The sequence $\{a_n\}$ has the limit L, that is,

$$\lim_{n\to\infty} a_n = L, \tag{12}$$

if for each $\epsilon > 0$ there is an integer N, depending on ϵ, with the property that

$$|a_n - L| < \epsilon \quad \text{for each} \quad n > N. \tag{13}$$

In this case, $\{a_n\}$ is said to converge to L. On the other hand, if there is no number L for which the relation (13) holds, then the sequence $\{a_n\}$ has no limit, and is said to diverge.

The number ϵ in the definition measures the closeness between a_n and L, while N indicates how far out in the sequence one must go in order to achieve the specified degree of closeness (see Figure 12.1.8). Since Eq. 13 involves only the terms in the sequence for which $n > N$, it follows that the possible existence of $\lim_{n\to\infty} a_n$, or its value, does not depend on $a_1, \ldots, a_N$. Put another way, convergence of the sequence $\{a_n\}$ is unaffected by any finite block of terms at the beginning of the sequence; rather, convergence is determined entirely by the remaining terms, those that lie in the "tail" of the sequence.

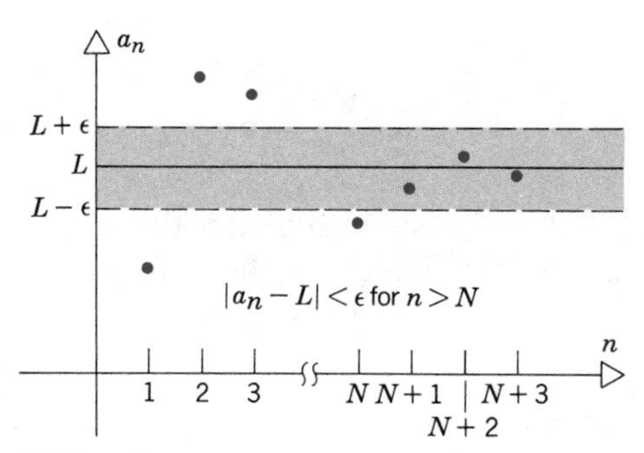

Figure 12.1.8 Convergence of the sequence $\{a_n\}$.

The sequences (5) and (7) have the limits two and zero, respectively. Examples 2 and 3, which follow, illustrate how Definition 12.1.1 can be used to establish these results. The sequence (10) also converges, but this may be less obvious; this sequence is discussed in Example 5.

There are several ways in which a sequence may diverge. One possibility is that its terms become and remain larger than any given positive number as n increases without bound. For instance, this is the case for sequence (9). Stated more formally, for each $M > 0$, there is an integer N, depending on M, with the property that $a_n > M$ for each $n > N$. We then write

$$\lim_{n\to\infty} a_n = \infty. \tag{14}$$

If the sequence behaves in an analogous way but has negative terms, then

$$\lim_{n\to\infty} a_n = -\infty. \tag{15}$$

The sequence (6) does not converge because its terms jump back and forth between the two numbers -1 and 1. Finally, the sequence (8) also experiences repeated sign changes, and the absolute value of its terms also increases without bound. Thus sequence (8) diverges, but neither Eq. 14 nor Eq. 15 can be used to describe its behavior.

The following examples illustrate the use of Definition 12.1.1.

EXAMPLE 2

Use Definition 12.1.1 to show that

$$\lim_{n\to\infty}\left(2-\frac{1}{n^2}\right)=2. \tag{16}$$

Suppose that we are given an $\epsilon > 0$. Then we must consider $|a_n - L|$. We have

$$|a_n - L| = \left|\left(2-\frac{1}{n^2}\right)-2\right| = \left|-\frac{1}{n^2}\right| = \frac{1}{n^2}.$$

If we assume that $n > N$, for some number N, then $1/n < 1/N$, and

$$|a_n - L| < \frac{1}{N^2}.$$

Now, to make $|a_n - L| < \epsilon$ we need only require that $1/N^2 \leq \epsilon$, or that $N^2 \geq 1/\epsilon$, or that

$$N \geq \frac{1}{\sqrt{\epsilon}}. \tag{17}$$

Thus the requirements of Definition 12.1.1 are met provided we choose N to be any integer equal to or larger than $1/\sqrt{\epsilon}$. For example, if $\epsilon = 0.002$, then $1/\sqrt{\epsilon} \cong 22.36$, so we can choose $N = 23$ or any larger integer. ∎

EXAMPLE 3

Use Definition 12.1.1 to show that

$$\lim_{n\to\infty}\frac{(-1)^n}{\sqrt{n}} = 0. \tag{18}$$

Let ϵ be given and suppose that $n > N$. Then

$$|a_n - L| = \left|\frac{(-1)^n}{\sqrt{n}} - 0\right| = \frac{1}{\sqrt{n}} < \frac{1}{\sqrt{N}}.$$

Hence, to make $|a_n - L| < \epsilon$, it is sufficient to require that $1/\sqrt{N} \leq \epsilon$, or

$$N \geq \frac{1}{\epsilon^2}. \tag{19}$$

For instance, if $\epsilon = 0.002$, then $1/\epsilon^2 = 250{,}000$, so we can choose N to be this number, or any larger integer. ∎

Note the difference in the values of N needed in Examples 2 and 3 for the same ϵ. This reflects the fact that the sequence in Example 2 converges much faster than the one in Example 3. The speed of convergence is an important consideration if one wishes to use a sequence for computational purposes.

Some properties of sequences

We now turn to some theorems that enable us to determine the limits of many sequences. The first three theorems correspond to similar theorems about functions, so we will simply state them without proof.

Theorem 12.1.1

If $\{a_n\}$ converges, then the limit L is unique.

Theorem 12.1.2

Suppose that $\lim_{n\to\infty} a_n = A$ and $\lim_{n\to\infty} b_n = B$. Then

(a) $\lim_{n\to\infty}(\alpha a_n + \beta b_n) = \alpha A + \beta B$, for any numbers α and β.
(b) $\lim_{n\to\infty} a_n b_n = AB$.
(c) $\lim_{n\to\infty}(a_n/b_n) = A/B$, provided that $B \neq 0$.

Theorem 12.1.3

(Sandwich principle)

If $a_n \leq b_n \leq c_n$ for all n (or for all $n > N$, for some positive integer N), and if $\lim_{n\to\infty} a_n = \lim_{n\to\infty} c_n = L$, then $\lim_{n\to\infty} b_n = L$.

The following theorem expresses the relation between the limit of a function and the limit of a sequence.

Theorem 12.1.4

If

$$\lim_{x\to\infty} f(x) = L \qquad (\text{or } \infty, \text{ or } -\infty) \tag{20}$$

and if $a_n = f(n)$ for $n = 1, 2, 3, \ldots$, then

$$\lim_{n\to\infty} a_n = L \qquad (\text{or } \infty, \text{ or } -\infty). \tag{21}$$

Proof. The proof follows directly from the definitions of the limits that are involved. From Eq. 20 for a finite limit L, we know that for each $\epsilon > 0$ there is an integer N such that if $x > N$, then

$$|f(x) - L| < \epsilon. \tag{22}$$

Restricting x to integer values n, Eq. 22 becomes

$$|a_n - L| < \epsilon, \tag{23}$$

provided only that $n > N$, and this establishes Eq. 21. The modification of this argument that is required if L is replaced either by ∞ or by $-\infty$ is left to the reader. □

The importance of Theorem 12.1.4 is due to the fact that we may be able to determine $\lim_{x\to\infty} f(x)$, and hence $\lim_{n\to\infty} a_n$, by methods, such as L'Hospital's rule, that cannot be applied directly to sequences.

EXAMPLE 4

Determine

$$\lim_{n\to\infty} \frac{2n^2 - n + 3}{n^2 + 4n + 1}, \tag{24}$$

or else show that this limit does not exist.

We cannot use Theorem 12.1.2 to evaluate the limit (24) as it is given because neither the numerator nor the denominator has a limit. To recast the problem in a better form, we divide both numerator and denominator by n^2. Thus

$$\frac{2n^2 - n + 3}{n^2 + 4n + 1} = \frac{2 - (1/n) + (3/n^2)}{1 + (4/n) + (1/n^2)},$$

so

$$\lim_{n\to\infty} \frac{2n^2 - n + 3}{n^2 + 4n + 1} = \frac{\lim_{n\to\infty}[2 - (1/n) + (3/n^2)]}{\lim_{n\to\infty}[1 + (4/n) + (1/n^2)]}$$

$$= \frac{2 - \lim_{n\to\infty}(1/n) + 3[\lim_{n\to\infty}(1/n)]^2}{1 + 4\lim_{n\to\infty}(1/n) + [\lim_{n\to\infty}(1/n)]^2}.$$

Since each of the limits in this last expression is zero, it follows from Theorem 12.1.2 that

$$\lim_{n\to\infty} \frac{2n^2 - n + 3}{n^2 + 4n + 1} = \frac{2 - 0 + 0}{1 + 0 + 0} = 2. \tag{25}$$

Alternatively, we can evaluate the limit (24) by invoking Theorem 12.1.4 and then using L'Hospital's rule. In this way we obtain

$$\begin{aligned}\lim_{n\to\infty}\frac{2n^2-n+3}{n^2+4n+1} &= \lim_{x\to\infty}\frac{2x^2-x+3}{x^2+4x+1}\\ &= \lim_{x\to\infty}\frac{4x-1}{2x+4}\\ &= \lim_{x\to\infty}\frac{4}{2} = 2. \quad \blacksquare \end{aligned} \tag{26}$$

EXAMPLE 5

Determine

$$\lim_{n\to\infty}\frac{4^n}{n!},$$

or else show that this limit does not exist.

It is convenient to write, for $n \geq 6$,

$$0 < b_n = \frac{4^n}{n!} = \underbrace{\left(\frac{4\cdot 4\cdot 4\cdot 4}{1\cdot 2\cdot 3\cdot 4}\right)}_{=\,c}\underbrace{\left(\frac{4}{5}\cdot\frac{4}{6}\cdots\frac{4}{n-1}\right)}_{<1}\frac{4}{n} < \frac{4c}{n}, \tag{27}$$

where c is a constant; actually $c = \frac{32}{3}$, but this is not important. Let $a_n = 0$ and $c_n = 4c/n$ for each n, and note that $\lim_{n\to\infty} c_n = 0$. Then Theorem 12.1.3 implies that

$$\lim_{n\to\infty} b_n = \lim_{n\to\infty}\frac{4^n}{n!} = 0. \quad \blacksquare \tag{28}$$

EXAMPLE 6

If $-1 < r < 1$, show that

$$\lim_{n\to\infty} r^n = 0. \tag{29}$$

If $r = 0$, the result is obvious, so suppose that $r \neq 0$. Then observe that

$$-|r| \leq r \leq |r|,$$

and hence that

$$-|r|^n \leq r^n \leq |r|^n,$$

where $0 < |r| < 1$. Equation 29 now follows from the sandwich principle (Theorem 12.1.3) provided that

$$\lim_{n\to\infty} |r|^n = 0.$$

To show this, note that

$$\ln |r|^n = n \ln |r|. \tag{30}$$

Since $0 < |r| < 1$ we know that $\ln |r| < 0$; therefore $\ln |r|^n \to -\infty$ as $n \to \infty$. Equation 29 is established by taking the exponential of both sides of Eq. 30. ∎

Bounded monotone sequences

Just as for functions, sequences may also have the properties of boundedness and monotonicity. A sequence $\{a_n\}$ is said to be **bounded** if there is a number K such that $|a_n| \leq K$ for all n, or (equivalently) if there are two numbers A and B such that $A \leq a_n \leq B$ for all n. If $\{a_n\}$ is not bounded, then it is called **unbounded.** For instance, sequences (5), (6), (7), and (10) are bounded, while sequences (8) and (9) are unbounded. For bounded sequences it is often easy to find a bound K for $|a_n|$ merely by inspecting a few terms in the sequence. Thus it should be clear that we can choose $K = 2$ for sequence (5) and $K = 1$ for sequences (6) and (7).

A sequence is said to be **monotone increasing** if $a_{n+1} > a_n$ for each n, and **monotone nondecreasing** if $a_{n+1} \geq a_n$ for each n. Similarly, $\{a_n\}$ is **monotone decreasing** or **monotone nonincreasing** if $a_{n+1} < a_n$, or if $a_{n+1} \leq a_n$, respectively, for each n. A sequence that satisfies any one of these requirements may be called simply **monotone.** It is easy to show that sequences (5) and (9) are monotone increasing; sequence (5) is discussed in Example 7. Sequences (6), (7), and (8) are obviously not monotone because they go back and forth between positive and negative values. Sequence (10) is considered in Example 8.

EXAMPLE 7

Show that the sequence (5) is monotone increasing.

In this case $a_n = 2 - (1/n^2)$. To show that this sequence is monotone we consider the difference $a_{n+1} - a_n$ for an arbitrary positive integer n. We have

$$\begin{aligned} a_{n+1} - a_n &= \left[2 - \frac{1}{(n+1)^2}\right] - \left[2 - \frac{1}{n^2}\right] = \frac{1}{n^2} - \frac{1}{(n+1)^2} \\ &= \frac{(n+1)^2 - n^2}{n^2(n+1)^2} = \frac{2n+1}{n^2(n+1)^2} > 0. \end{aligned} \tag{31}$$

Therefore $a_{n+1} > a_n$ for all n, and the sequence is monotone increasing. ∎

For a sequence with positive terms we can also investigate monotonicity by considering the ratio a_{n+1}/a_n: if $a_{n+1}/a_n < 1$ for all n, then $\{a_n\}$ is monotone decreasing, and if $a_{n+1}/a_n > 1$ for all n, then $\{a_n\}$ is monotone increasing.

EXAMPLE 8

Determine whether the sequence (10), $\{a_n\} = \{4^n/n!\}$, is monotone.

It can be seen from the first few terms displayed in Eq. 10 that the sequence is not monotone; clearly $a_2 > a_1$, but $a_5 < a_4$. However, let us investigate the question further by forming the ratio a_{n+1}/a_n. We find that

$$\frac{a_{n+1}}{a_n} = \frac{4^{n+1}}{(n+1)!}\frac{n!}{4^n} = \frac{4}{n+1}. \tag{32}$$

Therefore $a_{n+1}/a_n < 1$ for $n \geq 4$. Hence the sequence $\{4^n/n!\}$ is monotone decreasing from the fourth term on. ∎

The following two theorems set forth the basic relations among bounded, monotone, and convergent sequences. These relations are also shown in Figure 12.1.9.

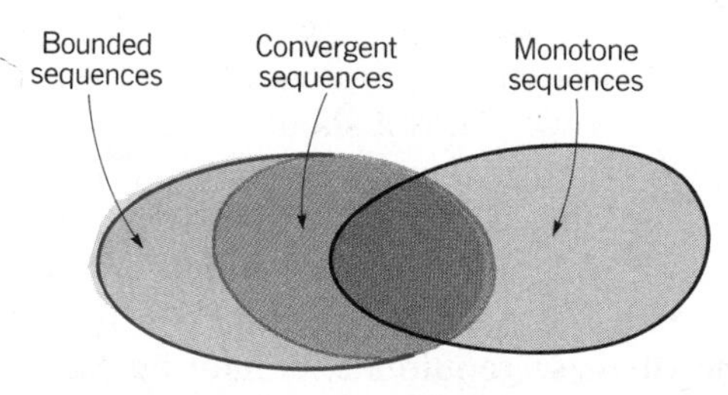

Figure 12.1.9 Relation among bounded, monotone, and convergent sequences.

Theorem 12.1.5
If $\{a_n\}$ converges, then it is bounded.

The proof of this result is not difficult, and is outlined in Problem 47. It follows at once that *if $\{a_n\}$ is unbounded, then it must diverge.* Sequences (8) and (9) illustrate this fact.

Theorem 12.1.6
If $\{a_n\}$ is bounded and monotone, then it converges.

Since convergence of a sequence is independent of the behavior of any finite initial block of terms, the conclusion of Theorem 12.1.6 remains true if the sequence is monotone only for $n > N$ for some N. Theorem 12.1.6 is one of the fundamental theorems of mathematical analysis. It is one of several equivalent statements of what is known as the **completeness property** of the real number system. Another statement of this property appears in the problem set following Section 1.1. A proof of Theorem 12.1.6 requires either a lengthy and sophisticated argument, or else the assumption that one of the other statements of the completeness property is true. Consequently, we will simply accept the truth of this theorem, and leave a discussion of its proof to a more advanced course; however, see Problem 51.

Although it is a profound statement, Theorem 12.1.6 seems plausible on its face. For example, suppose that $\{a_n\}$ is monotone increasing. Then it seems reasonable to conclude that $\{a_n\}$ must either increase without bound and diverge to ∞, or else remain bounded and converge from below to some finite limit. If $\{a_n\}$ is bounded, then the former possibility is excluded, and therefore the sequence must converge.

Theorem 12.1.6 is illustrated by sequence (5), which is bounded and monotone increasing, and by sequence (10), which is bounded and (from the fourth term onward) monotone decreasing, as we showed in Example 8. Observe that sequences (7) and (8) are not monotone; one converges but not the other. Also, sequences (8) and (9) both diverge because they are not bounded, although sequence (9) is monotone. Problems 45 and 46 are further illustrations of the use of this theorem for particular sequences.

In what follows we use Theorem 12.1.6 mainly for theoretical purposes, that is, to help to prove other results that are more immediately useful. However, it does have the following practical significance as well. One potential difficulty in using Definition 12.1.1 to show that a sequence $\{a_n\}$ converges is that one must discern, or guess, the value of the limit L in advance. Often this is easy, but sometimes it is not. In any case, Theorem 12.1.6 provides an alternative; if we can show that $\{a_n\}$ is bounded and monotone, then we know that it converges, although we may have no idea what the limit is.

PROBLEMS

In each of Problems 1 through 10, perform the following tasks.

(a) Write out the first five terms in the given sequence, beginning with $n = 1$.

(b) Determine whether the sequence is bounded.

(c) Determine whether the sequence is monotone.

(d) Determine whether the sequence converges or diverges; if it converges, find its limit L.

1. $a_n = 2 + \dfrac{1}{n}$ **2.** $a_n = \dfrac{(-1)^n}{n^2}$

3. $a_n = \dfrac{n-1}{n+1}$ **4.** $a_n = \dfrac{n+1}{\sqrt{n}}$

5. $a_n = (-1)^{n+1}n^{1/3}$ **6.** $a_n = \cos\dfrac{n\pi}{3}$

7. $a_n = \dfrac{n-4}{n^2+1}$ **8.** $a_n = \sqrt{n+1} - \sqrt{n}$

9. $a_n = \dfrac{2^n}{4n^2}$ **10.** $a_n = \dfrac{1}{n} - \dfrac{1}{n+1}$

In each of Problems 11 through 32, either find the limit of the given sequence, or else show that the sequence diverges.

11. $a_n = \dfrac{n^2+2n-3}{2n^2-3n+1}$ **12.** $a_n = \dfrac{2n^2+(-1)^n\sqrt{n}}{n^2+\sqrt{n}}$

13. $a_n = \dfrac{n^2+1}{n^{3/2}+n}$ **14.** $a_n = (10)^{1/n}$

15. $a_n = \dfrac{\ln n}{n}$ **16.** $a_n = (n)^{1/n}$

17. $a_n = (n)^{1/p}$, where p is a positive integer

18. $a_n = \sqrt{\dfrac{n+1}{3n}}$

19. $a_n = (1.01)^n$

20. $a_n = (1.01)^{\sqrt{n}}$

21. $a_n = \dfrac{n^3}{3^n}$

22. $a_n = \left(1 - \dfrac{1}{n}\right)\left(1 + \dfrac{1}{n}\right)$

23. $a_n = \dfrac{2^{n+1} + n^2}{3^n}$

24. $a_n = (-1)^n \dfrac{2^n}{2^{n-1} + 1}$

25. $a_n = \sqrt{n}(\sqrt{n+1} - \sqrt{n})$

26. $a_n = n \sin \dfrac{1}{n}$

27. $a_n = \ln(n+1) - \ln n$

28. $a_n = n[\ln(n+1) - \ln n]$

29. $a_n = \dfrac{\ln bn}{\ln cn}$, where $b > 0$ and $c > 0$

30. $a_n = \dfrac{n!}{n^n}$

31. $a_n = \dfrac{(2n)!}{n^n}$

32. $a_n = \dfrac{(n)^{2n}}{(2n)^n}$

In each of Problems 33 through 38, find a sequence (other than ones in the text) that has the required properties.

33. Convergent but not monotone
34. Bounded but not convergent
35. Monotone but not bounded
36. Monotone decreasing and unbounded
37. Monotone decreasing and convergent
38. Unbounded but not monotone

In each of Problems 39 through 44, use Definition 12.1.1 to show that the given sequence converges; that is, for an arbitrary ϵ, find L and N such that $|a_n - L| < \epsilon$ for all $n > N$.

39. $a_n = 2 + \dfrac{1}{n}$

40. $a_n = \dfrac{(-1)^n}{n^2}$

41. $a_n = \dfrac{n-1}{n+1}$

42. $a_n = \dfrac{2}{\sqrt{n}}$

43. $a_n = 2^{-n}$

44. $a_n = (-1)^n n^{-1/3}$

* 45. Consider the sequence $\{a_n\}$ for which $a_1 = 1$, and

$$a_n = 1 + \sqrt{a_{n-1}}, \qquad n = 2, 3, 4, \ldots. \qquad (i)$$

This is an example of a sequence defined recursively, for which there is no simple expression for a_n as a function of n.

(a) Show that $\{a_n\}$ is monotone increasing. *Hint:* Use an induction argument to show that $a_{n+1} > a_n$ for each positive integer n.

(b) Show that $\{a_n\}$ is bounded above. *Hint:* Again, an induction argument may be useful. A convenient upper bound is $M = 4$.

(c) The sequence $\{a_n\}$ converges because of (a) and (b); let $L = \lim_{n\to\infty} a_n$. Use Eq. (i) to show that

$$L = 1 + \sqrt{L}, \qquad (ii)$$

and then determine L.

* 46. Consider the sequence $\{a_n\}$ for which $a_1 = 1$, and

$$a_n = \sqrt{3a_{n-1}}, \qquad n = 2, 3, 4, \ldots.$$

(a) Show that $\{a_n\}$ is monotone increasing by showing that $a_{n+1}/a_n > 1$ for each positive integer n.

(b) Show that $\{a_n\}$ is bounded above.

(c) Find $L = \lim_{n\to\infty} a_n$.

* 47. **Proof of Theorem 12.1.5.** Suppose that $a_n \to L$ as $n \to \infty$.

(a) Given $\epsilon > 0$, choose N so that $|a_n - L| < \epsilon$ for $n > N$. Then show that $|a_n| < |L| + \epsilon$ for $n > N$.

(b) Let M_1 be the maximum of the set of numbers $|a_1|, |a_2|, \ldots, |a_N|$. How can you be sure that this maximum exists?

(c) Use the results of (a) and (b) to show that $\{a_n\}$ is bounded.

48. Show that if $\{a_n\}$ converges and $a_n \le M$, then $\lim_{n\to\infty} a_n \le M$.

49. Show that if $\lim_{n\to\infty} a_n = 0$ and $\{b_n\}$ is bounded, then $\{a_n b_n\}$ has the limit zero.

50. (a) Show that if $\{a_n\}$ converges to L, then $\{|a_n|\}$ converges to $|L|$.

(b) By means of examples, show that if $\{|a_n|\}$ converges, then $\{a_n\}$ may or may not converge.

* 51. Theorem 12.1.6 is easily proved if we assume the completeness property in the form preceding Problem 29 in Section 1.1, namely, every

bounded nonempty set S of real numbers has a least upper bound and a greatest lower bound. Let S be defined so that

$$S = \{s \mid s = a_n \quad \text{for some } n\}.$$

Then S is not empty, and if $\{a_n\}$ is bounded, then S is also bounded. Consequently, S has a least upper bound L and a greatest lower bound M. Suppose $\{a_n\}$ is monotone nondecreasing. Then it follows that $\lim_{n\to\infty} a_n = L$.

(a) For a given $\epsilon > 0$, show that there is an N such that $a_N > L - \epsilon$.

(b) For $n \geq N$, show that $|a_n - L| < \epsilon$. In other words, $\lim_{n\to\infty} a_n = L$.
Note: If $\{a_n\}$ is monotone nonincreasing (and bounded), then a similar argument shows that $\{a_n\}$ converges to its greatest lower bound M.

12.2 CONVERGENCE AND DIVERGENCE OF SERIES

An **infinite series** is an expression of the form

$$a_1 + a_2 + \cdots + a_k + \cdots, \tag{1}$$

where the three dots after a_k indicate that the summation never terminates. The quantities $a_1, a_2, \ldots$ are called the **terms** in the series; a_k is referred to as the **general term.** In this chapter we restrict our discussion to series whose terms are real numbers. For brevity we often use the summation notation

$$\sum_{k=1}^{\infty} a_k \tag{2}$$

to denote the sum (1).

Infinite series occur frequently in mathematics and also in its applications. Often the first term represents an initial approximation to some quantity of interest, and further terms are successive corrections of that initial approximation. Computational algorithms to be implemented on a computer are often of this type; the procedure is terminated when sufficient accuracy has been achieved.

It is important to understand the difference between a sequence and a series. As explained in Section 12.1, a sequence is simply a *set* of numbers arranged in a certain order. On the other hand, a series is a *sum.* We can speak of the sequence

$$\{a_k\} = a_1, a_2, \ldots, a_k, \ldots \tag{3}$$

and also of the series

$$\sum_{k=1}^{\infty} a_k = a_1 + a_2 + \cdots + a_k + \cdots \tag{4}$$

whose terms are the same as the terms of the sequence $\{a_k\}$. Nevertheless, the sequence (3) and the series (4) are two quite different things.

Before going any further, we remark that it is not necessary to index the terms in a series so that the first term corresponds to $k = 1$, the second to $k = 2$, and so on. Frequently it is convenient to write the series so that the first term corresponds to $k = 0$, or perhaps to some other value of k. For example, the infinite series

$$\sum_{k=0}^{\infty} \frac{1}{k^2 + 4} = \frac{1}{4} + \frac{1}{5} + \frac{1}{8} + \frac{1}{13} + \cdots,$$

$$\sum_{k=4}^{\infty} \frac{k+1}{k^2} = \frac{5}{16} + \frac{6}{25} + \frac{7}{36} + \frac{8}{49} + \cdots,$$

and

$$\sum_{k=2}^{\infty} \frac{1}{k \ln k} = \frac{1}{2 \ln 2} + \frac{1}{3 \ln 3} + \frac{1}{4 \ln 4} + \cdots$$

all fall within the scope of this chapter.

We cannot assign a sum to the series (4) simply by adding up all of the terms, because this is impossible in any finite time. Recall that we encountered a similar situation in Section 11.3 in connection with the improper integral $\int_a^\infty f(x)\,dx$. In that situation we integrated from a to an arbitrary upper limit b and then considered the limit as $b \to \infty$. Here we do an analogous thing. If we let s_n be the sum of the first n terms of the series (4), then

$$s_1 = a_1,$$

$$s_2 = a_1 + a_2,$$

$$s_3 = a_1 + a_2 + a_3,$$

and so forth. In general,

$$s_n = a_1 + a_2 + a_3 + \cdots + a_n. \tag{5}$$

The numbers $s_1, s_2, \ldots, s_n, \ldots$ form a sequence $\{s_n\}$ called the **sequence of partial sums** of the series (4). As n increases, more and more terms of the series are included in the partial sums. If the sequence $\{s_n\}$ of partial sums has a limit as $n \to \infty$, then we define this limiting value to be the sum of the series (4).

DEFINITION 12.2.1 For the series

$$\sum_{k=1}^{\infty} a_k = a_1 + a_2 + a_3 + \cdots \tag{6}$$

define the sequence $\{s_n\}$ of partial sums so that

$$s_n = \sum_{k=1}^{n} a_k = a_1 + \cdots + a_n; \qquad n = 1, 2, \ldots . \tag{7}$$

The series (6) is said to converge and to have the sum s if and only if the sequence $\{s_n\}$ converges to the limit s. In this case we write

$$\sum_{k=1}^{\infty} a_k = s. \tag{8}$$

On the other hand, the series (6) is said to diverge whenever the sequence (7) diverges.

The following example illustrates the use of Definition 12.2.1.

EXAMPLE 1

Determine whether the series

$$1 + \frac{1}{3} + \frac{1}{9} + \frac{1}{27} + \cdots + \left(\frac{1}{3}\right)^k + \cdots = \sum_{k=0}^{\infty} \left(\frac{1}{3}\right)^k \tag{9}$$

converges, and if so, find its sum.

The sum (9) is an example of a geometric series; each term is a constant multiple ($\frac{1}{3}$) of the preceding term. It is convenient to form the sequence of partial sums as follows:

$$s_0 = 1$$

$$s_1 = 1 + \frac{1}{3}$$

$$s_2 = 1 + \frac{1}{3} + \frac{1}{9}$$

and in general

$$s_n = 1 + \frac{1}{3} + \frac{1}{9} + \cdots + \left(\frac{1}{3}\right)^n, \tag{10}$$

where n is an arbitrary nonnegative integer. We wish to investigate the possible limiting behavior of $\{s_n\}$ as $n \to \infty$. Unfortunately, this is not immediately clear from Eq. 10, which expresses s_n as the sum of many terms. However, let us rewrite s_n in a more useful form. First multiply s_n from Eq. 10 by $\frac{1}{3}$; this gives

$$\frac{1}{3}s_n = \left(\frac{1}{3}\right) + \left(\frac{1}{3}\right)^2 + \left(\frac{1}{3}\right)^3 + \cdots + \left(\frac{1}{3}\right)^n + \left(\frac{1}{3}\right)^{n+1}.$$

Then subtract this equation from Eq. 10 so as to obtain

$$s_n - \frac{1}{3}s_n = 1 - \left(\frac{1}{3}\right)^{n+1},$$

or,

$$s_n = \frac{1 - (\frac{1}{3})^{n+1}}{1 - (\frac{1}{3})}; \qquad n = 0, 1, 2, \ldots. \tag{11}$$

By Example 6 of Section 12.1, we know that $(\frac{1}{3})^{n+1} \to 0$ as $n \to \infty$; hence it follows from Eq. 11 that

$$\lim_{n\to\infty} s_n = \lim_{n\to\infty} \frac{1 - (\frac{1}{3})^{n+1}}{1 - \frac{1}{3}} = \frac{1}{2/3} = \frac{3}{2}. \tag{12}$$

Therefore, by Definition 12.2.1, the series (9) converges and has the sum $\frac{3}{2}$. ∎

Geometric series in general

The series

$$\sum_{k=0}^{\infty} r^k = 1 + r + r^2 + r^3 + \cdots + r^k + \cdots \tag{13}$$

is called the geometric series with common ratio r. It can be handled in the same way as the series (9) in Example 1, for which $r = \frac{1}{3}$. By proceeding as in the derivation of Eq. 11 we can show that, for any positive integer n, the partial sum s_n of the series (13) is given by

$$s_n = \sum_{k=0}^{n} r^k = \frac{1 - r^{n+1}}{1 - r}, \qquad r \neq 1. \tag{14}$$

Several different cases must now be considered. If $|r| < 1$, then $r^{n+1} \to 0$ as $n \to \infty$ by Example 6 of Section 12.1; consequently

$$\lim_{n\to\infty} s_n = \frac{1}{1 - r}, \qquad |r| < 1. \tag{15}$$

If $|r| > 1$, then r^{n+1} becomes unbounded as $n \to \infty$ and as a result $\{s_n\}$ diverges. If $r = -1$, then

$$s_n = \frac{1 - (-1)^{n+1}}{2} = \begin{cases} 0, & n \text{ odd;} \\ 1, & n \text{ even.} \end{cases} \tag{16}$$

Since s_n oscillates between the two values 0 and 1, the sequence $\{s_n\}$ also diverges in this case. Finally, if $r = 1$, the formula (14) does not apply. However, in this case each term in the series (13) has the value 1, and therefore $s_n = n + 1$. Again $\{s_n\}$ diverges as $n \to \infty$.

Summarizing our results, we conclude that

$$\sum_{k=0}^{\infty} r^k \begin{cases} \text{converges to } 1/(1 - r) \text{ for } -1 < r < 1; \\ \text{diverges for } |r| \geq 1. \end{cases} \tag{17}$$

Another illustration of the use of Definition 12.2.1 is given in the following example.

EXAMPLE 2

Determine whether the series

$$\sum_{k=1}^{\infty} \frac{1}{k(k+1)} = \frac{1}{2} + \frac{1}{6} + \frac{1}{12} + \frac{1}{20} + \cdots + \frac{1}{k(k+1)} + \cdots \tag{18}$$

converges, and if so, find its sum.

Let us start by calculating a few partial sums. We have

$$s_1 = \frac{1}{2},$$

$$s_2 = \frac{1}{2} + \frac{1}{6} = \frac{2}{3},$$

$$s_3 = \frac{1}{2} + \frac{1}{6} + \frac{1}{12} = \frac{3}{4},$$

$$s_4 = \frac{1}{2} + \frac{1}{6} + \frac{1}{12} + \frac{1}{20} = \frac{4}{5}.$$

This suggests that

$$s_n = \frac{1}{2} + \frac{1}{6} + \cdots + \frac{1}{n(n+1)} = \frac{n}{n+1}. \tag{19}$$

If Eq. 19 is true, then it follows that $s_n \to 1$ as $n \to \infty$, and therefore the series (18) also converges and has the sum 1.

To establish Eq. 19 we can expand the general term in the series (18) in partial fractions; the result is

$$\frac{1}{k(k+1)} = \frac{1}{k} - \frac{1}{k+1}. \tag{20}$$

Thus we can rewrite the series in the form

$$\sum_{k=1}^{\infty} \frac{1}{k(k+1)} = \left(1 - \frac{1}{2}\right) + \left(\frac{1}{2} - \frac{1}{3}\right) + \left(\frac{1}{3} - \frac{1}{4}\right) + \cdots + \left(\frac{1}{k} - \frac{1}{k+1}\right) + \cdots.$$

In this form it is easy to see that s_n is given by

$$s_n = \left(1 - \frac{1}{2}\right) + \left(\frac{1}{2} - \frac{1}{3}\right) + \left(\frac{1}{3} - \frac{1}{4}\right) + \cdots + \left(\frac{1}{n} - \frac{1}{n+1}\right)$$

$$= 1 - \frac{1}{n+1} = \frac{n}{n+1}.$$

This verifies Eq. 19.

A series whose terms cancel in this manner is often called a *telescoping series.* It is frequently easy to determine the partial sums of such a series by using partial fractions. ∎

Unfortunately, the preceding examples are not typical since Definition 12.2.1 usually is not directly useful as a means of discovering whether a given series converges or diverges, because usually it is not possible to obtain a useful formula for the nth partial sum s_n. Indeed, geometric series and telescoping series are the principal cases in which s_n can be found in an elementary way. As in the case of improper integrals, it is usually advantageous to separate the question of convergence or divergence of an infinite series from the question of calculating its sum (assuming that it does converge). If we are able to determine that a series converges, then we can usually calculate an adequate approximation to its sum by using a computer to add up a sufficient number of terms. What we want to avoid is trying to calculate the sum of a divergent series. The concept of convergence can also be extended to more general series than those having constant terms, as we will see in Chapter 13.

In the remainder of this section we discuss a few general properties of series. In the next five sections we take up several tests that are often useful in determining whether a given series converges or not.

A very important fact is that the convergence or divergence of an infinite series is determined by the terms that appear arbitrarily far along in the series, not by any *finite* block of terms at the beginning of the series. Definition 12.2.1 says that convergence of an infinite series is equivalent to convergence of its sequence of partial sums, that is,

$$\lim_{n\to\infty} s_n = s. \tag{21}$$

For example, suppose that we alter in some way each of the first 1000 terms in the series. Then each partial sum s_n is changed, but for $n \geq 1000$ each s_n is changed by exactly the same amount. Thus the existence or nonexistence of $\lim_{n\to\infty} s_n$ is not affected by the alteration of the first 1000 terms in the series, although the value of the limit is (in general) changed. A consequence of this fact is that the convergence or divergence of a series is not affected if we insert, delete, or alter any finite number of terms in the series. However, if a series converges, then its sum depends on the value of each term in the series, and in general the sum will be altered if even one term is changed.

To make this argument more rigorous recall that Eq. 21 means that, for a given $\epsilon > 0$, there is a positive integer N, depending on ϵ, such that

$$|s_n - s| < \epsilon \tag{22}$$

whenever $n > N$. By substituting for s_n and s from Eqs. 7 and 8, respectively, it follows that Eq. 22 can be rewritten as

$$\left| \sum_{k=n+1}^{\infty} a_k \right| < \epsilon, \tag{23}$$

provided that $n > N$. Thus convergence of the series $\Sigma_{k=1}^{\infty} a_k$ occurs if and only if the inequality (23) is valid. In words, Eq. 23 says that the *remainder* after the first n terms of the series must be arbitrarily small in order for the series to converge. Observe that the first n terms of the series do not appear in Eq. 23, so they have nothing to do with convergence of the series.

Theorem 12.2.1

(kth Term Test)

If the series $\Sigma_{k=1}^{\infty} a_k$ converges, then $a_k \to 0$ as $k \to \infty$.

Proof. To prove this theorem note that, for $k \geq 2$,

$$a_k = s_k - s_{k-1}.$$

Then

$$\lim_{k\to\infty} a_k = \lim_{k\to\infty} (s_k - s_{k-1}).$$

If s is the sum of the series $\Sigma_{k=1}^{\infty} a_k$, then $\lim_{k\to\infty} s_k = s$. Further, $\lim_{k\to\infty} s_{k-1} = s$ also, since $\{s_{k-1}\}$ is the same sequence as $\{s_k\}$; the index has merely been shifted. Hence

$$\lim_{k\to\infty} a_k = \lim_{k\to\infty} s_k - \lim_{k\to\infty} s_{k-1}$$

$$= s - s = 0,$$

as was to be shown. □

It must be clearly understood that Theorem 12.2.1 says that if the series converges, then $a_k \to 0$ as $k \to \infty$. It does *not* say that if $a_k \to 0$ as $k \to \infty$, then the series converges. Indeed, if $a_k \to 0$ as $k \to \infty$, then no conclusion can be drawn from Theorem 12.2.1 about the convergence of $\Sigma_{k=1}^{\infty} a_k$. However, if as $k \to \infty$ the sequence $\{a_k\}$ either diverges or converges to a *nonzero* limit, then the series $\Sigma_{k=1}^{\infty} a_k$ must diverge. The principal use of the theorem is to establish divergence in this way.

EXAMPLE 3

Determine whether the series

$$\sum_{k=0}^{\infty} (-1)^k = 1 - 1 + 1 - 1 + \cdots \tag{24}$$

converges or diverges.

In this case $a_k = (-1)^k$ and the sequence $\{a_k\}$ has no limit as $k \to \infty$ because its terms are alternately $+1$ and -1. Therefore the series (24) diverges.

This conclusion can also be drawn from Definition 12.2.1 by noting that the sequence of partial sums for the series (24) is

$$\{s_n\} = \{1, 0, 1, 0, 1, 0, \cdots\},$$

which diverges by oscillation. ∎

EXAMPLE 4

Determine whether the series

$$\sum_{k=1}^{\infty} \frac{1}{k} = 1 + \frac{1}{2} + \frac{1}{3} + \frac{1}{4} + \cdots + \frac{1}{k} + \cdots \tag{25}$$

converges or diverges.

The series (25) is known as the **harmonic series.** Theorem 12.2.1 is not useful in this case since $a_k = 1/k \to 0$ as $k \to \infty$. However, it is possible to show that the series diverges to infinity despite the fact that its successive terms approach zero monotonically. Note that the terms are all positive; consequently the sequence of partial sums is monotone increasing. If the sequence of partial sums is also bounded, then this sequence must converge by Theorem 12.1.6. Thus to show that the series diverges we need only show that the partial sums become unbounded. We do this by constructing lower bounds for certain of the partial sums. First observe that

$$s_1 = 1,$$

$$s_2 = 1 + \frac{1}{2} = \frac{3}{2},$$

and that

$$s_4 = 1 + \frac{1}{2} + \frac{1}{3} + \frac{1}{4} > 1 + \frac{1}{2} + \frac{1}{4} + \frac{1}{4} = 2.$$

Note that in estimating s_4 we have replaced $a_3(=\frac{1}{3})$ by a smaller number $\frac{1}{4}$, thereby obtaining a lower bound for s_4. Let us now consider s_8. By replacing a_3 by $\frac{1}{4}$ and a_5, a_6, and a_7 by $\frac{1}{8}$, we obtain

$$\begin{aligned} s_8 &= 1 + \frac{1}{2} + \frac{1}{3} + \frac{1}{4} + \frac{1}{5} + \frac{1}{6} + \frac{1}{7} + \frac{1}{8} \\ &> 1 + \frac{1}{2} + \frac{1}{4} + \frac{1}{4} + \frac{1}{8} + \frac{1}{8} + \frac{1}{8} + \frac{1}{8} = 1 + \frac{1}{2} + \frac{1}{2} + \frac{1}{2} = \frac{5}{2}. \end{aligned}$$

In the same way we find that

$$\begin{aligned} s_{16} &> 1 + \frac{1}{2} + \frac{1}{4} + \frac{1}{4} + \frac{1}{8} + \frac{1}{8} + \frac{1}{8} + \frac{1}{8} + \underbrace{\frac{1}{16} + \cdots + \frac{1}{16}}_{\text{eight terms}} \\ &= 1 + \frac{1}{2} + \frac{1}{2} + \frac{1}{2} + \frac{1}{2} = 3. \end{aligned}$$

Proceeding similarly for the partial sum s_{2^m}, where m is an arbitrary positive integer, we have

$$s_{2^m} > 1 + \frac{1}{2} + \frac{1}{4} + \frac{1}{4} + \cdots + \underbrace{\frac{1}{2^m} + \cdots + \frac{1}{2^m}}_{2^{m-1}\text{ terms}} = 1 + \frac{m}{2}. \tag{26}$$

Thus the partial sums s_{2^m} can be made arbitrarily large by choosing m sufficiently large, so the sequence $\{s_n\}$ also increases without bound as $n \to \infty$. Consequently, *the harmonic series diverges.* ∎

The series in Examples 2 and 4 confirm that no conclusion can be drawn about the convergence of $\Sigma_{k=1}^{\infty} a_k$ merely because $a_k \to 0$ as $k \to \infty$. For both of these series it is true that $\lim_{k\to 0} a_k = 0$, but one of them converges while the other diverges.

Theorem 12.2.2

If $\Sigma_{k=1}^{\infty} a_k$ and $\Sigma_{k=1}^{\infty} b_k$ converge to the sums s and S, respectively, and if α and β are real numbers, then $\Sigma_{k=1}^{\infty} (\alpha a_k + \beta b_k)$ also converges and has the sum $\alpha s + \beta S$; that is,

$$\sum_{k=1}^{\infty} (\alpha a_k + \beta b_k) = \alpha \sum_{k=1}^{\infty} a_k + \beta \sum_{k=1}^{\infty} b_k = \alpha s + \beta S. \tag{27}$$

Proof. To prove this theorem let s_n, S_n, and σ_n be the nth partial sums of $\Sigma_{k=1}^{\infty} a_k$, $\Sigma_{k=1}^{\infty} b_k$, and $\Sigma_{k=1}^{\infty} (\alpha a_k + \beta b_k)$, respectively. Then

$$\begin{aligned} \sigma_n &= \sum_{k=1}^{n} (\alpha a_k + \beta b_k) \\ &= \alpha \sum_{k=1}^{n} a_k + \beta \sum_{k=1}^{n} b_k \\ &= \alpha s_n + \beta S_n \\ &\to \alpha s + \beta S, \qquad \text{as } n \to \infty, \end{aligned}$$

by Theorem 12.1.2(a), which completes the proof. □

An important special case of Theorem 12.2.2 occurs if $\beta = 0$, in which case Eq. 27 reduces to

$$\sum_{k=1}^{\infty} \alpha a_k = \alpha \sum_{k=1}^{\infty} a_k = \alpha s. \tag{28}$$

In other words, if each term in a convergent series is multiplied by a constant, then the sum of the series is multiplied by the same constant.

Theorem 12.2.2 is extremely useful in carrying out arithmetic operations on infinite series, and many examples occur later in this chapter.

Finally, we point out that the converse of Theorem 12.2.2 is *not* true: the convergence of $\Sigma_{k=1}^{\infty} (\alpha a_k + \beta b_k)$ does not guarantee convergence of $\Sigma_{k=1}^{\infty} a_k$ and

$\Sigma_{k=1}^{\infty} b_k$. For instance, it was shown in Example 2 that the series

$$\sum_{k=1}^{\infty} \left(\frac{1}{k} - \frac{1}{k+1}\right)$$

converges. However, from Example 4 we know that both of the series $\Sigma_{k=1}^{\infty} 1/k$ and $\Sigma_{k=1}^{\infty} 1/(k+1)$ diverge.

PROBLEMS

In each of Problems 1 through 6, write out a few terms on each side of the equality and thereby verify that the given equation is valid.

1. $\displaystyle\sum_{k=1}^{\infty} a_k = \sum_{i=1}^{\infty} a_i$

2. $\displaystyle\sum_{k=1}^{\infty} a_k = \sum_{k=0}^{\infty} a_{k+1}$

3. $\displaystyle\sum_{k=1}^{\infty} a_k = \sum_{k=2}^{\infty} a_{k-1}$

4. $\displaystyle\sum_{n=0}^{\infty} \frac{1}{(n+4)^2} = \sum_{m=4}^{\infty} \frac{1}{m^2}$

5. $\displaystyle\sum_{k=0}^{\infty} \left(\frac{1}{2}\right)^{k+1} = \sum_{k=2}^{\infty} \left(\frac{1}{2}\right)^{k-1}$

6. $\displaystyle\sum_{k=n}^{\infty} a_{k+r} = \sum_{k=n+r}^{\infty} a_k$

In each of Problems 7 through 20, determine whether the given series converges or diverges. If it converges, find its sum s.

7. $\displaystyle\sum_{k=1}^{\infty} \frac{1}{2^k}$

8. $\displaystyle\sum_{k=0}^{\infty} \frac{(-1)^k}{2^k}$

9. $\displaystyle\sum_{k=1}^{\infty} 2^{k/2}$

10. $\displaystyle\sum_{k=0}^{\infty} \frac{(-1)^k}{3^k}$

11. $\displaystyle\sum_{k=1}^{\infty} \frac{(-1)^{k+1}}{2^{k/2}}$

12. $\displaystyle\sum_{k=2}^{\infty} \frac{1}{k^2-1}$

13. $\displaystyle\sum_{k=0}^{\infty} \frac{1}{(k+1)(k+2)}$

14. $\displaystyle\sum_{k=1}^{\infty} \frac{1}{k^2+3k}$

15. $\displaystyle\sum_{k=1}^{\infty} \frac{1}{k(k+1)(k+2)}$

16. $\displaystyle\sum_{k=1}^{\infty} \ln \frac{k+1}{k}$

17. $\displaystyle\sum_{k=0}^{\infty} a_k,$ where $a_k = \begin{cases} 2^{-k} & k \text{ even} \\ 1, & k \text{ odd} \end{cases}$

18. $\displaystyle\sum_{k=1}^{\infty} a_k,$ where $a_k = \begin{cases} \dfrac{k-5}{k}, & k \text{ odd} \\ 10^{-k}, & k \text{ even} \end{cases}$

19. $\displaystyle\sum_{k=1}^{\infty} \left(\frac{3}{k(k+1)} - \frac{1}{2^k}\right)$

20. $\displaystyle\sum_{k=1}^{\infty} \left(\frac{k}{k+1}\right)^k$

In each of Problems 21 through 24, find all values of α for which the given series converges. If it converges, find its sum s.

21. $\displaystyle\sum_{k=0}^{\infty} (\alpha - 2)^k$

22. $\displaystyle\sum_{k=0}^{\infty} \frac{1}{(\alpha - 2)^k}, \quad \alpha \neq 2$

23. $\displaystyle\sum_{k=0}^{\infty} (2\alpha - 3)^k$

24. $\displaystyle\sum_{k=0}^{\infty} \frac{3}{(3\alpha - 1)^k}, \quad \alpha \neq \tfrac{1}{3}$

Speed of convergence. In each of Problems 25 through 27, find the number of terms that must be retained if the sum s of the given series is to be approximated with an error not exceeding 0.0001. That is, find the smallest N such that $|s - s_n| < 0.0001$ for all $n \geq N$.

© 25. $\displaystyle\sum_{k=0}^{\infty} \frac{1}{3^k}$

© 26. $\displaystyle\sum_{k=0}^{\infty} (0.9)^k$

© 27. $\displaystyle\sum_{k=1}^{\infty} \frac{1}{k(k+1)}$

© 28. For the geometric series $\Sigma_{k=0}^{\infty} r^k$, with $|r| < 1$, find the number of terms that must be retained if the sum s is to be approximated with an error not exceeding ϵ, where ϵ is some given positive number. That is, find the smallest N such that $|s - s_n| < \epsilon$ for all $n \geq N$.

29. Suppose that when a ball is dropped from a height h on a hard, level surface, it always rebounds a distance αh, where $0 < \alpha < 1$. Find the total distance traveled by this ball if it is dropped from a height h.

30. It is well known that repeating decimals represent rational numbers, the ratios of two integers. Here we indicate how to find the rational number corresponding to a repeating decimal.

(a) Let $s = 0.27272727 \ldots$. Show that s can be written in the form

$$s = \frac{27}{10^2} + \frac{27}{10^4} + \frac{27}{10^6} + \frac{27}{10^8} + \cdots$$
$$= \frac{27}{10^2}\left(1 + \frac{1}{10^2} + \frac{1}{10^4} + \frac{1}{10^6} + \cdots\right).$$

(b) Observe that the series obtained in Part (a) is a geometric series with ratio 10^{-2}. Sum this series and show that $s = \frac{3}{11}$.

In each of Problems 31 through 36, use the method outlined in Problem 30 to express the given decimal as a rational number.

31. 0.555555 . . .

32. 0.321321321 . . .

33. 0.39999999 . . .

34. 0.285714285714 . . .

35. 0.8181818181 . . .

36. 0.477477477 . . .

In each of Problems 37 and 38, find the general term in the series corresponding to the given sequence of partial sums. Note that $a_1 = s_1$, $a_2 = s_2 - s_1$, and so forth.

37. $s_n = 1 - \dfrac{1}{(n+1)^2}$

38. $s_n = 1 - \left(\dfrac{1}{3}\right)^n$

39. Derive the expression

$$1 + r + r^2 + \cdots + r^n = \frac{1 - r^{n+1}}{1 - r}, \qquad r \neq 1$$

for the sum of a finite number of terms in a geometric series by the method used in deriving Eq. 11.

12.3 THE COMPARISON TEST

The principal obstacle in using Definition 12.2.1 to determine whether a given series $\Sigma_{k=1}^{\infty} a_k$ converges or diverges is that usually it is impossible to find a useful formula for the partial sum $s_n = \Sigma_{k=1}^{n} a_k$. Because of this, tests for convergence have been developed that do not require a knowledge of s_n. In the remainder of this chapter we discuss the use of some of these tests.

Until further notice we will consider only series having positive terms, that is, $a_k > 0$ for all k. Since convergence is not affected by removal of any finite block of terms, the results also apply to series having a finite number of negative terms. The first test is analogous to the comparison tests for improper integrals that were discussed in Section 11.4.

Theorem 12.3.1

(Comparison test)
If $0 < a_k \le b_k$ for all k, and if $\Sigma_{k=1}^{\infty} b_k$ converges, then $\Sigma_{k=1}^{\infty} a_k$ also converges. On the other hand, if $0 < b_k \le a_k$ for all k, and if $\Sigma_{k=1}^{\infty} b_k$ diverges, then $\Sigma_{k=1}^{\infty} a_k$ also diverges.

Proof. To prove the first part of this theorem let s_n and σ_n be the nth partial sums of the two series:

$$s_n = \sum_{k=1}^{n} a_k, \qquad \sigma_n = \sum_{k=1}^{n} b_k.$$

We are given that σ_n approaches a limit, say B, as $n \to \infty$. We need to show that the sequence $\{s_n\}$ also converges. Since $a_k > 0$, it follows that $\{s_n\}$ is monotone increasing. Further we have

$$0 < s_n \le \sigma_n < B,$$

so that $\{s_n\}$ is bounded. We know (Theorem 12.1.6) that a bounded monotone sequence converges. Hence the sequence $\{s_n\}$ converges, and consequently so does the series $\sum_{k=1}^{\infty} a_k$.

The second part of the theorem is proved by noting that in this case

$$0 < \sigma_n \le s_n$$

and that $\sigma_n \to \infty$ as $n \to \infty$. Consequently $s_n \to \infty$ as $n \to \infty$ also, and hence $\sum_{k=1}^{\infty} a_k$ diverges. □

Theorem 12.3.1 can be extended by noting again that convergence or divergence of $\sum_{k=1}^{\infty} a_k$ is not affected by deleting a finite number of terms from the series. Consequently Theorem 12.3.1 remains true if the hypotheses are satisfied for k sufficiently large, but perhaps not for all k. This result is stated in the following corollary.

Corollary

If $0 < a_k \le b_k$ for k greater than or equal to some integer K, and if $\sum_{k=1}^{\infty} b_k$ converges, then $\sum_{k=1}^{\infty} a_k$ also converges. On the other hand, if $0 < b_k \le a_k$ for $k \ge K$, and if $\sum_{k=1}^{\infty} b_k$ diverges, then $\sum_{k=1}^{\infty} a_k$ also diverges.

As in the case of improper integrals, it is sometimes a nuisance to have to deal with the inequalities in Theorem 12.3.1 or the corollary. This can usually be avoided by using the following limit form of the comparison test. It is similar to the corresponding tests for improper integrals.

Theorem 12.3.2

(Limit comparison test)

Suppose that $a_k > 0$ and $b_k > 0$ for $k \ge K_1$, and that

$$\lim_{k\to\infty} \frac{a_k}{b_k} = L. \tag{1}$$

If $0 < L < \infty$, then the two series $\sum_{k=1}^{\infty} a_k$ and $\sum_{k=1}^{\infty} b_k$ either both converge or both diverge. Further, if $L = 0$ and $\sum_{k=1}^{\infty} b_k$ converges, then $\sum_{k=1}^{\infty} a_k$ also converges. Finally, if $L = \infty$ and $\sum_{k=1}^{\infty} b_k$ diverges, then $\sum_{k=1}^{\infty} a_k$ also diverges. No conclusion can be drawn in the remaining cases.

Proof. We will prove only the first part of this theorem. Suppose that $0 < L < \infty$ and that $\sum_{k=1}^{\infty} b_k$ converges. Since $a_k/b_k \to L$ as $L \to \infty$, we can be sure, for values of k sufficiently large, say $k \geq K_2$, that

$$\frac{a_k}{b_k} \leq 2L$$

or

$$a_k \leq 2Lb_k. \tag{2}$$

By Theorem 12.2.2 the series $\Sigma_{k=1}^{\infty} 2Lb_k = 2L \; \Sigma_{k=1}^{\infty} b_k$ converges whenever $\Sigma_{k=1}^{\infty} b_k$ does. Therefore for $k \geq \max(K_1, K_2)$ the conditions of the corollary to Theorem 12.3.1 are satisfied and hence $\Sigma_{k=1}^{\infty} a_k$ converges.

Now suppose that $0 < L < \infty$ and that $\Sigma_{k=1}^{\infty} b_k$ diverges. Since $a_k/b_k \to L$ as $k \to \infty$, we can now assert that

$$\frac{a_k}{b_k} \geq \frac{L}{2} \tag{3}$$

for k sufficiently large, say for $k \geq K_3$. Since the series $\Sigma_{k=1}^{\infty} (L/2)b_k$ diverges, it now follows from the corollary that $\Sigma_{k=1}^{\infty} a_k$ also diverges. The proofs of the remaining parts of Theorem 12.3.2 are similar and are left as an exercise (Problem 26). □

EXAMPLE 1

Determine whether the series

$$\sum_{k=1}^{\infty} \frac{1}{k!} = 1 + \frac{1}{2!} + \frac{1}{3!} + \frac{1}{4!} + \cdots \tag{4}$$

converges or diverges.

Because $k!$ increases extremely rapidly, it is reasonable to suspect that the series (4) converges. To estimate its behavior more precisely, we note that the successive terms are bounded as follows:

$$\begin{aligned} 1 &\leq 1, \\ \frac{1}{2!} &= \frac{1}{1 \cdot 2} \leq \frac{1}{2}, \\ \frac{1}{3!} &= \frac{1}{1 \cdot 2 \cdot 3} \leq \frac{1}{1 \cdot 2 \cdot 2} = \frac{1}{2^2}, \\ \frac{1}{4!} &= \frac{1}{1 \cdot 2 \cdot 3 \cdot 4} \leq \frac{1}{1 \cdot 2 \cdot 2 \cdot 2} = \frac{1}{2^3}. \end{aligned} \tag{5}$$

In general we have

$$\frac{1}{k!} = \frac{1}{1 \cdot 2 \cdot 3 \cdots k} \leq \frac{1}{1 \cdot 2 \cdot 2 \cdots 2} = \frac{1}{2^{k-1}}, \quad k \geq 1. \tag{6}$$

Thus each term in the series (4) is no greater than the corresponding term in the geometric series

$$1 + \frac{1}{2} + \left(\frac{1}{2}\right)^2 + \left(\frac{1}{2}\right)^3 + \cdots + \left(\frac{1}{2}\right)^{k-1} + \cdots, \tag{7}$$

which is known to converge (Eq. 17 of Section 12.2). Therefore the series (4) also converges by Theorem 12.3.1.

In this case we also know that the sum of the geometric series (7) is 2. Thus it follows from the inequalities (5) and (6) that the sum s of $\Sigma_{k=1}^{\infty} (1/k!)$ is no greater than 2. Further, since all of the terms in the series (4) are positive, the partial sums of this series are monotone increasing. Thus $s_n \leq s \leq 2$ for each n. Once we know that the series (4) converges, we can approximate its sum by adding up terms in the series. Because s_n increases monotonically with n, the more terms we use, the better the approximation. However, even a few terms give a reasonably good approximation to the sum. For example, with only four terms we obtain

$$s_4 = 1 + \frac{1}{2} + \frac{1}{6} + \frac{1}{24} \cong 1.708.$$

In Example 3 of Section 13.2 we show that the series (4) actually has the sum $s = e - 1 \cong 1.718$. Thus the error in using the four-term approximation is about 0.01. ∎

EXAMPLE 2

Determine whether the series

$$\sum_{k=1}^{\infty} \frac{1}{k^2} \tag{8}$$

converges or diverges.

Recall that in Example 2 of Section 12.2 we showed that the rather similar series $\Sigma_{k=1}^{\infty} 1/k(k + 1)$ converges. Hence we may suspect that the series (8) also converges. However,

$$\frac{1}{k^2} \geq \frac{1}{k(k + 1)} \tag{9}$$

so the inequality runs the wrong way for us to use Theorem 12.3.1 without modification. On the other hand,

$$\lim_{k\to\infty} \frac{1/k^2}{1/k(k + 1)} = \lim_{k\to\infty} \frac{k + 1}{k} = 1, \tag{10}$$

so Theorem 12.3.2 does apply and gives the conclusion that $\Sigma_{k=1}^{\infty} (1/k^2)$ converges.

Alternatively, we can use Theorem 12.3.1 provided that the inequality (9) is suitably altered. For instance, we can start from the fact that $k^2 \geq k$ for $k = 1$,

2, ... Then $2k^2 \geq k^2 + k$, that is, $k^2 \geq k(k+1)/2$, and finally

$$\frac{1}{k^2} \leq \frac{2}{k(k+1)}, \qquad k = 1, 2, \ldots . \tag{11}$$

Since

$$\sum_{k=1}^{\infty} \frac{2}{k(k+1)} = 2 \sum_{k=1}^{\infty} \frac{1}{k(k+1)} \tag{12}$$

and the latter series converges, it follows from Theorem 12.3.1 that $\sum_{k=1}^{\infty} (1/k^2)$ also converges. Further, the series in Eq. 12 has the sum 2, so it also follows that

$$\sum_{k=1}^{\infty} \frac{1}{k^2} \leq 2.$$

As in Example 1 the partial sums s_n approach the sum s monotonically from below.

If we calculate an approximate value of s by using only the first four terms, we obtain

$$s_4 = 1 + \frac{1}{4} + \frac{1}{9} + \frac{1}{16} \cong 1.401.$$

For this series it is possible to show by more advanced methods that

$$\sum_{k=1}^{\infty} \frac{1}{k^2} = \frac{\pi^2}{6} \cong 1.645,$$

a result discovered by Euler about 1736. Thus the four-term approximation is not nearly as good here as in Example 1. This reflects the fact that the series (8) converges much more slowly than the series (4). ∎

In order to make effective use of Theorems 12.3.1 and 12.3.2, it is necessary to have a set of potential comparison series whose convergence or divergence has already been established. Geometric series are sometimes useful for comparison purposes, as in Example 1. Perhaps the most useful of all comparison series are those of the form $\sum_{k=1}^{\infty} k^{-p}$, where p is a constant. The following theorem specifies the values of p for which this series converges.

Theorem 12.3.3

The series

$$\sum_{k=1}^{\infty} \frac{1}{k^p}, \tag{13}$$

known as the "p-series", converges for $p > 1$ and diverges for $p \leq 1$.

Proof. The proof of this result is obtained most readily from the integral test, which we discuss in Section 12.5. Meanwhile, we can prove a *part* of Theorem 12.3.3 by a comparison test.

In Example 2 we showed that the series (13) converges for $p = 2$. Further, if $p > 2$, we have

$$\frac{1}{k^p} < \frac{1}{k^2}, \qquad k = 1, 2, \ldots \quad \text{and} \quad p > 2. \tag{14}$$

Therefore, by Theorem 12.3.1, the series $\Sigma_{k=1}^{\infty} k^{-p}$ converges for all $p \geq 2$.

Similarly, in Example 4 of Section 12.2 the harmonic series ($p = 1$) was found to diverge. For $p < 1$ we have

$$\frac{1}{k^p} > \frac{1}{k}, \qquad k = 1, 2, \ldots \quad \text{and} \quad p < 1. \tag{15}$$

Hence, by Theorem 12.3.1, the series (13) diverges for all $p \leq 1$. □

The remaining cases, $1 < p < 2$, are dealt with in Section 12.5. However, we accept the validity of Theorem 12.3.3 and use it in what follows.

A consequence of Theorem 12.3.3 is that the harmonic series $\Sigma_{k=1}^{\infty} (1/k)$ just barely diverges. The harmonic series is obtained by setting $p = 1$ in Eq 13, and using any larger value p produces a convergent series. In other words, the series $\Sigma_{k=1}^{\infty} 1/k^{1+\epsilon}$ converges for any positive ϵ, however small.

We close this section with some further examples of the comparison test.

EXAMPLE 3

Determine whether the series

$$\sum_{k=1}^{\infty} \frac{\sqrt{k+2}}{k^2+1} = \frac{\sqrt{3}}{2} + \frac{2}{5} + \frac{\sqrt{5}}{10} + \cdots \tag{16}$$

converges or diverges.

For large k we have

$$\frac{\sqrt{k+2}}{k^2+1} \cong \frac{\sqrt{k}}{k^2} = k^{-3/2},$$

which suggests that the series converges. To confirm this we let

$$a_k = \frac{\sqrt{k+2}}{k^2+1}, \qquad b_k = \frac{1}{k^{3/2}}$$

and note that $\Sigma_{k=1}^{\infty} b_k$ converges by Theorem 12.3.3. Then

$$\lim_{k\to\infty} \frac{a_k}{b_k} = \lim_{k\to\infty} \frac{\sqrt{k+2}}{k^2+1} \cdot k^{3/2} = \lim_{x\to\infty} \frac{\sqrt{x+2}\, x^{3/2}}{x^2+1}$$

by Theorem 12.1.4. By dividing numerator and denominator by x^2 we find that

$$\frac{\sqrt{x} + 2\,x^{3/2}}{x^2 + 1} = \frac{\sqrt{1 + (2/x)}}{1 + (1/x^2)} \to 1$$

as $x \to \infty$. Consequently,

$$\lim_{k\to\infty} \frac{a_k}{b_k} = 1$$

and it follows from Theorem 12.3.2 that the series (16) also converges. ∎

EXAMPLE 4

Determine whether the series

$$\frac{1}{2} + \frac{1 \cdot 3}{2 \cdot 4} + \frac{1 \cdot 3 \cdot 5}{2 \cdot 4 \cdot 6} + \cdots + \frac{1 \cdot 3 \cdot 5 \cdots (2k - 1)}{2 \cdot 4 \cdot 6 \cdots (2k)} + \cdots \tag{17}$$

converges or diverges.

The crucial idea here is to observe that the general term a_k can be written in the form

$$a_k = \left(\frac{3}{2} \cdot \frac{5}{4} \cdot \frac{7}{6} \cdots \frac{2k - 1}{2k - 2}\right)\left(\frac{1}{2k}\right). \tag{18}$$

Consequently

$$a_k > \frac{1}{2k},$$

and therefore the given series diverges by comparison with the harmonic series. ∎

EXAMPLE 5

Determine whether the series

$$\sum_{k=1}^{\infty} \sin\left(\frac{1}{k^2}\right) = \sin 1 + \sin\frac{1}{4} + \sin\frac{1}{9} + \cdots \tag{19}$$

converges or diverges.

For k large, $1/k^2$ is small and $\sin(1/k^2)$ is close to $1/k^2$. Indeed, if we set $\alpha = 1/k^2$, then

$$\lim_{k\to\infty} \frac{\sin(1/k^2)}{1/k^2} = \lim_{\alpha\to 0} \frac{\sin \alpha}{\alpha} = 1.$$

Since $\Sigma_{k=1}^{\infty} (1/k^2)$ converges, it follows from the limit comparison test that the given series (19) also converges. ∎

Note that in all of these examples our method is to estimate a_k, for large k, in terms of k^{-p}. Then the comparison test tells us that $\Sigma_{k=1}^{\infty} a_k$ converges if $p > 1$ and diverges if $p \leq 1$.

PROBLEMS

In each of Problems 1 through 20, determine whether the given series converges or diverges.

1. $\displaystyle\sum_{k=0}^{\infty} \frac{1}{k^2 + 4}$

2. $\displaystyle\sum_{k=0}^{\infty} \frac{1}{\sqrt{k^2 + 4}}$

3. $\displaystyle\sum_{k=1}^{\infty} \frac{\sqrt{k}}{k + 2}$

4. $\displaystyle\sum_{k=1}^{\infty} \frac{(k^2 + 1)^{1/2}k}{k^4 + 4}$

5. $\displaystyle\sum_{k=0}^{\infty} \frac{(2k - 1)^2}{(k^2 + 4)^{3/2}}$

6. $\displaystyle\sum_{k=1}^{\infty} \frac{\sqrt{k} + (1/\sqrt{k})}{(\sqrt{k} + 1)^{7/2}}$

7. $\displaystyle\sum_{k=2}^{\infty} \frac{(k + 1)^{4/3}}{(2k^4 - 1)^{1/2}}$

8. $\displaystyle\sum_{k=2}^{\infty} \frac{1}{\ln k}$

9. $\displaystyle\sum_{k=1}^{\infty} \sin\left(\frac{1}{k}\right)$

10. $\displaystyle\sum_{k=1}^{\infty} \operatorname{sech} k$

11. $\displaystyle\sum_{k=0}^{\infty} \frac{e^k}{(4 + e^k)^2}$

12. $\displaystyle\sum_{k=1}^{\infty} \frac{k!}{k^k}$

13. $\displaystyle\frac{1}{3} + \frac{1 \cdot 2}{3 \cdot 5} + \frac{1 \cdot 2 \cdot 3}{3 \cdot 5 \cdot 7} + \cdots + \frac{k!}{3 \cdot 5 \cdot 7 \cdots (2k + 1)} + \cdots$

14. $\displaystyle\sum_{k=1}^{\infty} \frac{[\sqrt{k} + (1/k)]^p}{k^2 + 1}$

15. $\displaystyle\sum_{k=1}^{\infty} \frac{(k^2 + 1)^p}{k^q + 4}, \quad p, q > 0$

16. $\displaystyle\sum_{k=2}^{\infty} \frac{(\ln k)^2}{(\ln 2)^k}$

17. $\displaystyle\sum_{k=1}^{\infty} \frac{|\sin k|}{k^2}$

18. $\displaystyle\sum_{k=1}^{\infty} \frac{\sqrt{k + 1} - \sqrt{k}}{k}$

19. $\displaystyle\sum_{k=2}^{\infty} \frac{k \ln k}{4 + k^2 \ln k}$

20. $\displaystyle\sum_{k=1}^{\infty} \frac{1}{\sqrt{k} + \sqrt{k + 1}}$

21. Show that if $\lim_{k\to\infty} ka_k = A \neq 0$, then $\Sigma_{k=1}^{\infty} a_k$ diverges.

22. (a) If $a_k > 0$ and $\Sigma_{k=1}^{\infty} a_k$ converges, show that $\Sigma_{k=1}^{\infty} 1/a_k$ diverges.

    (b) If $a_k > 0$ and $\Sigma_{k=1}^{\infty} a_k$ diverges, show by means of examples that $\Sigma_{k=1}^{\infty} 1/a_k$ may either converge or diverge.

23. (a) If $a_k > 0$ and $\Sigma_{k=1}^{\infty} a_k$ converges, show that $\Sigma_{k=1}^{\infty} a_k^2$ also converges.

    (b) If $a_k > 0$ and $\Sigma_{k=1}^{\infty} a_k^2$ converges, show by means of examples that $\Sigma_{k=1}^{\infty} a_k$ may either converge or diverge.

24. (a) Show that $\Sigma_{k=1} e^{-\alpha k}$ converges for any $\alpha > 0$.

    (b) Show that $\Sigma_{k=1}^{\infty} ke^{-\alpha k}$ converges for any $\alpha > 0$.

    Hint: Refer to Example 2 of Section 11.2 and show that $ke^{-\alpha k/2} \to 0$ as $k \to \infty$. Thus for large k, $ke^{-\alpha k} \leq e^{-\alpha k/2}$.

    (c) Extend the argument in Part (b) to show that $\Sigma_{k=1}^{\infty} k^n e^{-\alpha k}$ converges for any $\alpha > 0$ and for any $n > 0$.

* 25. Show that $\Sigma_{k=2}^{\infty} (\ln k)/k^p$ converges for $p > 1$ and diverges for $p \leq 1$.

Hint: If $p > 1$, let $p = 1 + 2r$, $r > 0$. Then

$$\frac{\ln k}{k^p} = \frac{\ln k}{k^r} \cdot \frac{1}{k^{1+r}}.$$

* 26. Extend the argument given in the text to establish the following parts of Theorem 12.3.2.

    (a) If $L = 0$ in Theorem 12.3.2 and if $\Sigma_{k=1}^{\infty} b_k$ converges, show that $\Sigma_{k=1}^{\infty} a_k$ also converges. If $\Sigma_{k=1}^{\infty} b_k$ diverges, show that no conclusion can be drawn about $\Sigma_{k=1}^{\infty} a_k$.

    (b) If $L = \infty$ in Theorem 12.3.2, and if $\Sigma_{k=1}^{\infty} b_k$ diverges, show that $\Sigma_{k=1}^{\infty} a_k$ also diverges. If $\Sigma_{k=1}^{\infty} b_k$ converges, show that no conclusion can be drawn about $\Sigma_{k=1}^{\infty} a_k$.

* 27. Suppose that $c_k \leq a_k \leq b_k$ for $k = 1, 2, 3, \ldots,$ and that both $\Sigma_{k=1}^{\infty} c_k$ and $\Sigma_{k=1}^{\infty} b_k$ converge. Show that $\Sigma_{k=1}^{\infty} a_k$ also converges. Note that there is no hypothesis about the signs of a_k, b_k, and c_k.

Hint: Consider the series $\Sigma_{k=1}^{\infty} (b_k - c_k)$ and $\Sigma_{k=1}^{\infty} (a_k - c_k)$.

12.4 THE RATIO TEST

In this section we discuss another widely applicable test for convergence or divergence of a given infinite series $\Sigma_{k=1}^{\infty} a_k$ of positive terms. The basic idea is that if the ratio a_{k+1}/a_k of successive terms approaches a limit L as $k \to \infty$, then for large k the given series is very nearly the geometric series with common ratio L, and we might expect it to converge or diverge in a similar manner. This test is known as the ratio test.

Theorem 12.4.1

(Ratio test)
Consider the infinite series $\Sigma_{k=1}^{\infty} a_k$, where $a_k > 0$, and suppose that

$$\lim_{k\to\infty} \frac{a_{k+1}}{a_k} = L. \tag{1}$$

Then

(a) If $L < 1$, the series $\Sigma_{k=1}^{\infty} a_k$ converges.
(b) If $L > 1$, the series $\Sigma_{k=1}^{\infty} a_k$ diverges.
(c) If $L = 1$, the series $\Sigma_{k=1}^{\infty} a_k$ may converge or diverge, so no conclusion is possible from this test.

We prove Theorem 12.4.1 later in this section. First let us look at several examples.

EXAMPLE 1

Determine whether the series

$$\sum_{k=1}^{\infty} \frac{5^k}{k!} \tag{2}$$

converges or diverges.

In this case $a_k = 5^k/k!$, so $a_{k+1} = 5^{k+1}/(k+1)!$. Hence we have

$$\frac{a_{k+1}}{a_k} = \frac{5^{k+1}}{(k+1)!}\frac{k!}{5^k} = \frac{5}{k+1},$$

and therefore

$$\lim_{k\to\infty} \frac{a_{k+1}}{a_k} = 0.$$

Thus, by the ratio test (with $L = 0$) we conclude that the series (2) converges. ∎

EXAMPLE 2

Determine whether the series

$$\sum_{k=1}^{\infty} \frac{3^k}{k^3} \tag{3}$$

converges or diverges.

For this series we have

$$\frac{a_{k+1}}{a_k} = \frac{3^{k+1}/(k+1)^3}{3^k/k^3} = 3\left(\frac{k}{k+1}\right)^3.$$

Thus it follows that

$$\lim_{k\to\infty} \frac{a_{k+1}}{a_k} = 3 \lim_{k\to\infty} \left(\frac{k}{k+1}\right)^3 = 3,$$

so $L = 3$ in this example. Thus the series (3) diverges by the ratio test. ∎

EXAMPLE 3

Determine whether the harmonic series

$$\sum_{k=1}^{\infty} \frac{1}{k} \tag{4}$$

converges or diverges.

Here we have

$$\frac{a_{k+1}}{a_k} = \frac{k}{k+1} \to 1 \quad \text{as} \quad k \to \infty,$$

so the ratio test yields no conclusion. However, from Example 4 in Section 12.2, or from Theorem 12.3.3, we know that the harmonic series (4) diverges. ∎

EXAMPLE 4

Determine whether the series

$$\sum_{k=1}^{\infty} \frac{1}{k^2} \tag{5}$$

converges or diverges.

Again the ratio test is inconclusive, since

$$\frac{a_{k+1}}{a_k} = \left(\frac{k}{k+1}\right)^2 \to 1 \quad \text{as} \quad k \to \infty.$$

In this case, however, the series is already known to converge by Theorem 12.3.3 or by Example 2 in Section 12.3. ∎

Note that the series in Example 3 diverges and the series in Example 4 converges, although in both cases $L = 1$. These examples confirm that no conclusion about convergence can be drawn from the ratio test when $L = 1$.

EXAMPLE 5

Determine whether the series

$$\sum_{k=1}^{\infty} \frac{k!}{k^k} \tag{6}$$

converges or diverges.

Applying the ratio test to the series (6) we obtain

$$\frac{a_{k+1}}{a_k} = \frac{(k+1)!}{(k+1)^{k+1}} \frac{k^k}{k!} = \left(\frac{k}{k+1}\right)^k.$$

To find the limit of this expression, observe that

$$\lim_{k\to\infty} \left(\frac{k+1}{k}\right)^k = \lim_{k\to\infty} \left(1 + \frac{1}{k}\right)^k = e.$$

Consequently,

$$\lim_{k\to\infty} \frac{a_{k+1}}{a_k} = \lim_{k\to\infty} \left(\frac{k}{k+1}\right)^k = \frac{1}{e}.$$

Since $1/e < 1$, it follows that the series (6) converges.

Observe also that the convergence of the series (6) implies, by Theorem 12.2.1, that

$$\lim_{k\to\infty} a_k = \lim_{k\to\infty} \frac{k!}{k^k} = 0.$$

Thus k^k grows much more rapidly than $k!$ as $k \to \infty$. ∎

EXAMPLE 6

Determine whether the series

$$\frac{1}{3} + \frac{1 \cdot 3}{3 \cdot 6} + \frac{1 \cdot 3 \cdot 5}{3 \cdot 6 \cdot 9} + \cdots + \frac{1 \cdot 3 \cdot 5 \cdots (2k-1)}{3 \cdot 6 \cdot 9 \cdots 3k} + \cdots \tag{7}$$

converges or diverges.

We have

$$\begin{aligned}\frac{a_{k+1}}{a_k} &= \frac{1 \cdot 3 \cdot 5 \cdots (2k-1)(2k+1)}{3 \cdot 6 \cdot 9 \cdots 3k(3k+3)} \frac{3 \cdot 6 \cdot 9 \cdots 3k}{1 \cdot 3 \cdot 5 \cdots (2k-1)} \\ &= \frac{2k+1}{3k+3} \to \frac{2}{3} \quad \text{as} \quad k \to \infty.\end{aligned}$$

Thus, by the ratio test, the series (7) converges. ∎

Proof of Theorem 12.4.1. Recall that we are assuming that

$$\lim_{k\to\infty} \frac{a_{k+1}}{a_k} = L; \tag{8}$$

Since $a_k > 0$ for all k, it follows that $L \geq 0$. The convergence part of the ratio test is established by comparing the given series with a suitable geometric series. Suppose then that $L < 1$ in Eq. 8, and choose a number r between L and 1; that is,

$$L < r < 1. \tag{9}$$

Since $a_{k+1}/a_k \to L$ as $k \to \infty$, we know that eventually this ratio must take on values arbitrarily close to L. In particular, the ratio must eventually be no greater than r. Thus for k sufficiently large, say $k \geq M$, where M depends on the choice of r, we have

$$\frac{a_{k+1}}{a_k} \leq r, \qquad k = M, M + 1, \ldots . \tag{10}$$

For $k = M$, Eq. 10 becomes

$$a_{M+1} \leq ra_M.$$

Similarly, for $k = M + 1$,

$$a_{M+2} \leq ra_{M+1} \leq r^2 a_M,$$

and for $k = M + 2$,

$$a_{M+3} \leq ra_{M+2} \leq r^3 a_M.$$

In general, we have

$$a_{M+m} \leq r^m a_M, \qquad m = 0, 1, 2, \ldots . \tag{11}$$

The series

$$\sum_{m=0}^{\infty} a_M r^m = a_M \sum_{m=0}^{\infty} r^m$$

converges because it is a geometric series with $|r| < 1$. Thus the hypotheses of the corollary to Theorem 12.3.1 have been satisfied and the series $\Sigma_{k=1}^{\infty} a_k$ converges.

The divergence part of Theorem 12.4.1 is simpler to prove. If

$$\lim_{k\to\infty} \frac{a_{k+1}}{a_k} = L > 1,$$

then for k sufficiently large, we have

$$a_{k+1} \geq a_k.$$

Thus, beyond a certain point the terms in the series are monotone increasing as well as positive. Hence a_k cannot approach zero as $k \to \infty$. Therefore the series $\Sigma_{k=1}^{\infty} a_k$ diverges by Theorem 12.2.1, the kth term test.

Finally, as we noted earlier in Examples 3 and 4, if $L = 1$, then the series can either converge or diverge. This completes the proof of the theorem. ☐

In contrast to the comparison test, the ratio test has the advantage that it does not require the introduction of an auxiliary comparison series. On the other hand, in some cases the limit L in Eq. 1 turns out to be one, and in such cases the ratio test is useless. Actually, the ratio test and the comparison test complement each other rather well. The ratio test is indicated for series in which a_k involves such quantities as $k!$ or 2^k, for then the ratio a_{k+1}/a_k will be simplified by cancellation of like factors in numerator and denominator, as in Examples 1 and 2. On the other hand, if a_k involves polynomial or other algebraic expressions in k, then the ratio test is likely to fail because $L = 1$. However, such series can usually be handled by the limit comparison test in conjunction with knowledge about the series $\sum_{k=1}^{\infty} k^{-p}$, as in Example 3 in Section 12.3.

PROBLEMS

In each of Problems 1 through 22, determine whether the given series converges or diverges.

1. $\displaystyle\sum_{k=1}^{\infty} \frac{2^k}{k!}$

2. $\displaystyle\sum_{k=1}^{\infty} \frac{k^4}{2^k}$

3. $\displaystyle\sum_{k=1}^{\infty} \frac{k!}{2^{2k}}$

4. $\displaystyle\sum_{k=1}^{\infty} \frac{k!}{2^{k^2}}$

5. $\displaystyle\sum_{k=1}^{\infty} \frac{k!}{1 \cdot 3 \cdot 5 \cdots (2k-1)}$

6. $\displaystyle\sum_{k=1}^{\infty} \frac{k!}{10^k k^3}$

7. $\displaystyle\sum_{k=1}^{\infty} \frac{(k!)^2}{(2k)!}$

8. $\displaystyle\sum_{k=1}^{\infty} \frac{2^k k!}{k^k}$

9. $\displaystyle\sum_{k=1}^{\infty} \frac{3^k k!}{k^k}$

10. $\displaystyle\sum_{k=1}^{\infty} \frac{k! a^k}{(2k)!}, \quad a > 0$

11. $\displaystyle\sum_{k=2}^{\infty} \frac{(\ln k)^3}{(\ln 3)^k}$

12. $\displaystyle\sum_{k=1}^{\infty} k^2 \left(\frac{2}{3}\right)^k$

13. $\displaystyle\sum_{k=1}^{\infty} \frac{2 \cdot 4 \cdot 6 \cdots (2k)}{1 \cdot 4 \cdot 7 \cdots (3k-2)}$

14. $\displaystyle\sum_{k=1}^{\infty} \frac{(2k)!}{(k!)^2 2^k}$

15. $\displaystyle\sum_{k=1}^{\infty} \frac{5^k}{k^2 2^{2k}}$

16. $\displaystyle\sum_{k=1}^{\infty} \frac{k^2 3^k}{2^{2k}}$

17. $\displaystyle\sum_{k=2}^{\infty} \frac{(\ln k)^{10}}{(\ln a)^k}, \quad a > 1$

18. $\displaystyle\sum_{k=1}^{\infty} \frac{(2k)!}{k^{2k}}$

19. $\displaystyle\sum_{k=1}^{\infty} \frac{(2k)!}{(2k)^k}$

20. $\displaystyle\sum_{k=1}^{\infty} \frac{2^k (2k)!}{k^{2k}}$

21. $\displaystyle\sum_{k=1}^{\infty} \frac{(2^k)!}{2^{k!}}$

22. $\displaystyle\sum_{k=1}^{\infty} \frac{1 \cdot 3 \cdot 5 \cdots (2k-1)}{k^k}$

23. Consider the series

$$a + b + a^2 + b^2 + a^3 + b^3 + \cdots$$

where $0 < a < b < 1$. Show that the ratio test fails for this series. Then determine whether the series converges or diverges.

In each of Problems 24 through 27, evaluate the given limit by considering a suitable infinite series.

24. $\displaystyle\lim_{k\to\infty} \frac{10^k}{k!}$

25. $\displaystyle\lim_{k\to\infty} \frac{2^k k!}{k^k}$

26. $\displaystyle\lim_{k\to\infty} \frac{3^k k!}{k^k}$

27. $\displaystyle\lim_{k\to\infty} \frac{(2k)!}{(k!)^2 2^k}$

* 28. (a) Prove the following version of the ratio test: if $a_k > 0$, and if $a_{k+1}/a_k \le r < 1$, for all sufficiently large k, then $\sum_{k=1}^{\infty} a_k$ converges.

(b) Show, by means of an example, that the conclusion of Part (a) may be false if we assume only that $a_k > 0$ and $a_{k+1}/a_k < 1$ for all sufficiently large k.

29. Consider the series

$$1 + \tfrac{1}{2} + \tfrac{1}{6} + \tfrac{1}{12} + \tfrac{1}{36} + \tfrac{1}{72} + \cdots$$

for which

$$\frac{a_{k+1}}{a_k} = \begin{cases} \tfrac{1}{2}, & k \text{ odd}; \\ \tfrac{1}{3}, & k \text{ even}. \end{cases}$$

(a) Show that Theorem 12.4.1 does not apply.

(b) Determine whether the series converges or diverges.

Hint: See Problem 28.

12.5 THE INTEGRAL TEST

There are many similarities between infinite series and improper integrals. The most important relation between them is given in the following theorem, which is most commonly used to deduce convergence or divergence of a series from the known behavior of a certain corresponding integral.

Theorem 12.5.1

(Integral test)

Let f be a continuous, positive, monotone nonincreasing function on the interval $[1, \infty)$. Suppose also that

$$f(k) = a_k \tag{1}$$

for each positive integer $k = 1, 2, \ldots$. Then the series

$$\sum_{k=1}^{\infty} a_k = a_1 + a_2 + \cdots + a_k + \cdots \tag{2}$$

and the integral

$$\int_1^{\infty} f(x)\, dx \tag{3}$$

either both converge or both diverge.

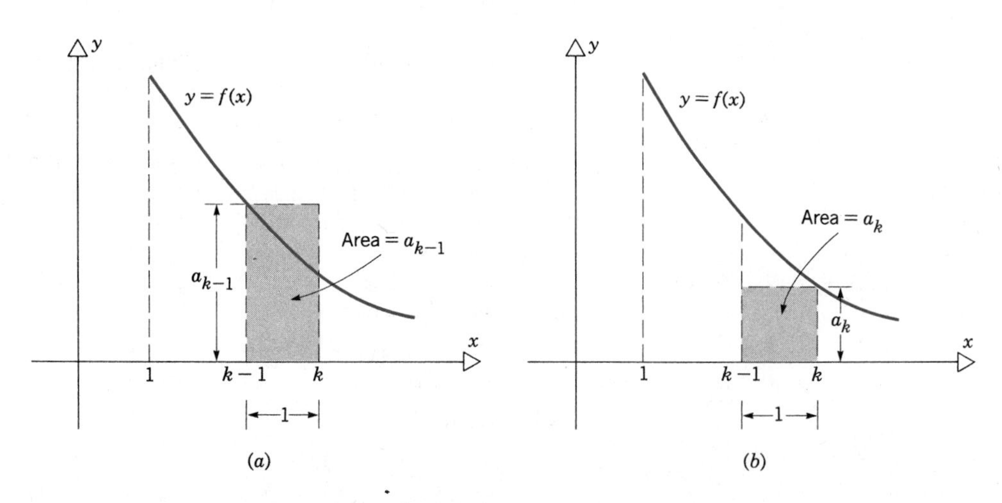

Figure 12.5.1

Proof. To see the connection between the series (2) and the integral (3) it is helpful to refer to Figure 12.5.1. The area between the graph of $y = f(x)$ and the x-axis and between the lines $x = k - 1$ and $x = k$ is given by

$$\int_{k-1}^{k} f(x)\, dx.$$

The area of the inscribed rectangle is a_k and the area of the circumscribed rectangle is a_{k-1}. It is clear from Figure 12.5.1 that these three areas are related by the inequalities

$$a_k \le \int_{k-1}^{k} f(x)\, dx \le a_{k-1}, \tag{4}$$

and this is true for $k = 2, 3, \ldots$. If we write Eq. 4 for $k = 2, 3, \ldots, n$ we have

$$\begin{aligned} a_2 &\le \int_1^2 f(x)\, dx \le a_1, \\ a_3 &\le \int_2^3 f(x)\, dx \le a_2, \\ &\;\;\vdots \\ a_n &\le \int_{n-1}^{n} f(x)\, dx \le a_{n-1}. \end{aligned} \tag{5}$$

By adding these inequalities we obtain

$$a_2 + a_3 + \cdots + a_n \le \int_1^2 f(x)\, dx + \int_2^3 f(x)\, dx + \cdots + \int_{n-1}^{n} f(x)\, dx \le a_1 + a_2 + \cdots + a_{n-1}; \tag{6}$$

this inequality is illustrated in Figure 12.5.2 for the case when $n = 6$. The middle term in Eq. 6 is just $\int_1^n f(x)\, dx$. Since $s_n = a_1 + a_2 + \cdots a_n$, we can rewrite Eq. 6 in the form

$$s_n - a_1 \le \int_1^n f(x)\, dx \le s_n - a_n. \tag{7}$$

Observe that both s_n and $\int_1^n f(x)\, dx$ increase monotonically with n because $f(x) > 0$ for all $x \ge 1$.

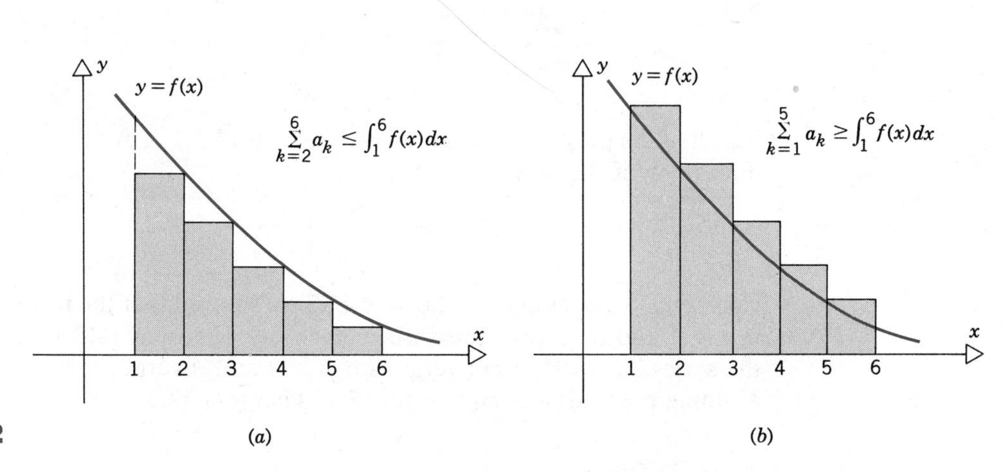

Figure 12.5.2

Now suppose that the integral (3) converges and has the value I. Then by the left inequality in Eq. 7 we have

$$s_n \le a_1 + \int_1^n f(x)\,dx \le a_1 + I. \tag{8}$$

Thus $\{s_n\}$ is a monotone increasing sequence that is bounded above; hence, by Theorem 12.1.6, it must converge. Therefore $\Sigma_{k=1}^{\infty} a_k$ also converges.

On the other hand, if the integral (3) diverges, then we use the right inequality in Eq. 7 to obtain

$$s_n \ge \int_1^n f(x)\,dx + a_n. \tag{9}$$

In this case the right side of Eq. 9 becomes unbounded as $n \to \infty$. Hence $s_n \to \infty$ as $n \to \infty$ and therefore the series (2) also diverges. This completes the proof of Theorem 12.5.1. □

As in other tests for convergence, it is only necessary for the hypotheses of Theorem 12.5.1 to be satisfied for all terms sufficiently far out in the series. Thus the theorem remains valid if the conditions are satisfied for $x \ge M$, where M is some positive number.

EXAMPLE 1

Determine the values of p for which the series

$$\sum_{k=1}^{\infty} \frac{1}{k^p} \tag{10}$$

converges; diverges.

For $p \le 0$ the series diverges because $a_k = k^{-p}$ does not approach zero as $k \to \infty$ (Theorem 12.2.1). Hence it is sufficient to consider only positive values of p. For $p > 0$ the function

$$f(x) = \frac{1}{x^p}, \qquad x \ge 1 \tag{11}$$

has all of the properties specified in Theorem 12.5.1. Thus the series (10) converges if and only if the integral

$$\int_1^{\infty} \frac{dx}{x^p} \tag{12}$$

converges. In Example 4 of Section 11.3 we showed that the integral (12) converges if $p > 1$ and diverges otherwise. Hence, by Theorem 12.5.1, the same is true of the series $\Sigma_{k=1}^{\infty} k^{-p}$: it converges for $p > 1$ and diverges for $p \le 1$. Note that this example provides a complete proof of Theorem 12.3.3. ∎

EXAMPLE 2

Determine whether the series

$$\sum_{k=2}^{\infty} \frac{1}{k \ln k} \tag{13}$$

converges or diverges.

In this case, the ratio test fails and a suitable comparison series is not obvious. However, if we let

$$f(x) = \frac{1}{x \ln x}, \tag{14}$$

then the conditions of the integral test are satisfied for $x \geq 2$. Hence the series (13) converges or diverges according to the behavior of the integral

$$\int_2^{\infty} \frac{dx}{x \ln x}. \tag{15}$$

To evaluate this integral, we write

$$\int_2^{\infty} \frac{dx}{x \ln x} = \lim_{b \to \infty} \int_2^b \frac{dx}{x \ln x},$$

and then use the substitution $u = \ln x$ to obtain

$$\int_2^{\infty} \frac{dx}{x \ln x} = \lim_{b \to \infty} \int_{\ln 2}^{\ln b} \frac{du}{u} = \lim_{b \to \infty} \left(\ln u \Big|_{\ln 2}^{\ln b} \right)$$

$$= \lim_{b \to \infty} [\ln(\ln b) - \ln(\ln 2)] = \infty.$$

Thus the integral (15) diverges, and consequently so does the series (13).

In Example 1 we found that $\sum_{k=1}^{\infty} 1/k^p$ converges whenever $p > 1$, no matter how small the difference $p - 1$ is. Put another way, we must have $k^p > k$ for convergence. In this example we have $k \ln k > k$, at least for $k \geq 3$, but the series $\sum_{k=2}^{\infty} 1/k \ln k$ diverges anyway. This illustrates once again the difference between logarithmic and algebraic growth: $\ln k$ grows more slowly than any positive power of k, and its growth is not sufficiently fast to make the series $\sum_{k=2}^{\infty} 1/k \ln k$ converge. On the other hand, the series $\sum_{k=2}^{\infty} 1/k(\ln k)^{\alpha}$ does converge when α is any number larger than one (Problem 10). ∎

Estimation of the Remainder

In making practical use of an infinite series it is often important to know something about the rate of convergence of the series. This amounts to asking how large we must choose n in order for the partial sum

$$s_n = \sum_{k=1}^{n} a_k \tag{16}$$

to be an adequate approximation of the sum s of the entire series. Put yet another way, we wish to estimate the remainder R_n given by

$$R_n = s - s_n = \sum_{k=1}^{\infty} a_k - \sum_{k=1}^{n} a_k = \sum_{k=n+1}^{\infty} a_k \tag{17}$$

and to choose n so that R_n is negligible for the application at hand. The argument used in proving the integral test can be extended to yield answers to these questions in some cases.

Under the conditions of Theorem 12.5.1, we again have Eq. 4,

$$a_k \leq \int_{k-1}^{k} f(x)\, dx \leq a_{k-1}.$$

If we sum over k in Eq. 4 from $n + 1$ to infinity, we obtain

$$\sum_{k=n+1}^{\infty} a_k \leq \sum_{k=n+1}^{\infty} \int_{k-1}^{k} f(x)\, dx \leq \sum_{k=n+1}^{\infty} a_{k-1}, \tag{18}$$

which is equivalent to

$$R_n \leq \int_{n}^{\infty} f(x)\, dx \leq a_n + R_n. \tag{19}$$

We can rewrite Eq. 19 in the form

$$\int_{n}^{\infty} f(x)\, dx - a_n \leq R_n \leq \int_{n}^{\infty} f(x)\, dx. \tag{20}$$

Note that the left inequality in Eq. 20 is obtained from the right inequality in Eq. 19 and vice versa. Equation 20 provides upper and lower bounds for the remainder R_n. The usefulness of these bounds depends upon whether the integral appearing in Eq. 20 can be conveniently evaluated or estimated. Nevertheless, we have proved the following theorem.

Theorem 12.5.2

Let f be a continuous, positive, monotone nonincreasing function on $[1, \infty)$ with $f(k) = a_k$ for $k = 1, 2, 3, \ldots$. Then the remainder

$$R_n = \sum_{k=n+1}^{\infty} a_k$$

satisfies the inequality (20)

$$\int_{n}^{\infty} f(x)\, dx - a_n \leq R_n \leq \int_{n}^{\infty} f(x)\, dx.$$

By taking the mean of the upper and lower bounds for R_n given by Eq. 20

we obtain the estimate

$$R_n \cong \int_n^\infty f(x)\,dx - \frac{a_n}{2} \tag{21}$$

with a maximum error of $a_n/2$. Since $s = s_n + R_n$, it follows that

$$s \cong s_n + \int_n^\infty f(x)\,dx - \frac{a_n}{2}, \tag{22}$$

also with a maximum error of $a_n/2$. These results are illustrated by the following example.

EXAMPLE 3

Estimate the remainder after ten terms in the series $\Sigma_{k=1}^{\infty} k^{-2}$. Also estimate the sum s of the series.

In this case

$$f(x) = \frac{1}{x^2}, \qquad a_k = f(k) = \frac{1}{k^2}, \tag{23}$$

and all of the conditions of Theorem 12.5.2 are satisfied. Thus Eq. 20 is valid for any positive integer value of n. We have

$$\begin{aligned} \int_n^\infty \frac{dx}{x^2} &= \lim_{b\to\infty} \int_n^b \frac{dx}{x^2} \\ &= \lim_{b\to\infty} \left(-\frac{1}{x}\right)\Bigg|_n^b = \frac{1}{n}. \end{aligned} \tag{24}$$

Also, $a_n = n^{-2}$. Thus, for an arbitrary value of n, Eq. 20 becomes

$$\frac{1}{n} - \frac{1}{n^2} < R_n < \frac{1}{n}, \tag{25}$$

and in particular, for $n = 10$,

$$0.09 < R_{10} < 0.10, \tag{26}$$

which is the required estimate for the remainder.

By adding the first ten terms in the given series we obtain (to five decimal places)

$$s_{10} = 1.54977.$$

To estimate s we note that

$$s = s_{10} + R_{10},$$

so by using the largest and smallest possible values of R_{10} from Eq. 26 we have

$$1.54977 + 0.09 < s < 1.54977 + 0.10$$

or

$$1.63977 < s < 1.64977. \tag{27}$$

Using the mean of the upper and lower bounds for s we have the estimate

$$s \cong 1.64477. \tag{28}$$

Since the difference in the upper and lower bounds for s given by Eq. 27 is 0.01, it follows that the maximum error in using the mean value (28) is 0.005.

In fact the actual value of s is

$$s = \frac{\pi^2}{6} \cong 1.64493, \tag{29}$$

so the estimate (28) is actually in error by only 0.00016, or about one hundredth of 1 percent.

Observe that, by using the estimate of the remainder as we have done here, we obtain a much more accurate value for s than if we had used the partial sum s_{10} alone. To guarantee similar accuracy by using only an appropriate partial sum s_n we would need to use the estimate (25), that is, to choose n so that

$$R_n < \frac{1}{n} < 0.00016.$$

This means using more than 6250 terms in the series rather than 10. ∎

The zeta function

In Example 1 it was shown that the series $\Sigma_{k=1}^{\infty} k^{-p}$ converges for all $p > 1$. We can thus use the series to define a function of p with domain $p > 1$. This function is called the zeta function and is denoted by $\zeta(p)$, that is,

$$\zeta(p) = \sum_{k=1}^{\infty} \frac{1}{k^p}, \qquad p > 1.$$

Some properties of $\zeta(p)$ were discovered by Euler, Gauss, and others, but the zeta function is usually associated with the name of Riemann, who studied it extensively, especially for complex values of p. The domain of $\zeta(p)$ is extended into the complex plane by a process known as analytic continuation. The zeta function turns out to be an important function in mathematics, and among other things, the zeros of $\zeta(p)$ are closely related to the distribution of the prime numbers. A famous conjecture of Riemann is that all of the nonreal zeros of $\zeta(p)$ have a real part of $\frac{1}{2}$. Although more than 120 years have passed since the death of Riemann, intensive efforts by many notable mathematicians have so far failed to produce a proof of this conjecture. Very extensive numerical calculations have also failed to reveal a zero that violates it, so it remains a conjecture, and possibly the most celebrated unsolved problem in mathematics.

PROBLEMS

In each of Problems 1 through 10, use the integral test to determine whether the given series converges or diverges. Note that many of these series can also be handled by the comparison test or the ratio test.

1. $\sum_{k=1}^{\infty} \frac{1}{1 + k^2}$
2. $\sum_{k=1}^{\infty} \frac{1}{(k + 1)^2}$
3. $\sum_{k=1}^{\infty} \frac{k}{1 + k^2}$
4. $\sum_{k=1}^{\infty} \frac{k}{(k^2 + 1)^2}$
5. $\sum_{k=1}^{\infty} \frac{k^3}{k^4 + 1}$
6. $\sum_{k=1}^{\infty} \frac{2k^3}{(k^4 + 1)^2}$
7. $\sum_{k=4}^{\infty} \frac{1}{k \ln k \ln(\ln k)}$
8. $\sum_{k=1}^{\infty} \frac{k}{e^k}$
9. $\sum_{k=1}^{\infty} \frac{\arctan k}{1 + k^2}$
10. $\sum_{k=2}^{\infty} \frac{1}{k(\ln k)^{\alpha}}, \quad \alpha > 0$

© **11.** Consider the series $\Sigma_{k=1}^{\infty} (1/k^4)$.

(a) Find upper and lower bounds for the remainder R_6 after six terms.

(b) Find s_6.

(c) Find an improved estimate for s using the results of (a) and (b). The actual value of s is $\pi^4/90 \cong 1.0823232$.

(d) How large must n be chosen in order to be sure that the estimate given by Eq. 22 has an error less than 10^{-6} in magnitude?

© **12.** Consider the series

$$\sum_{k=2}^{\infty} \frac{1}{k(\ln k)^2}.$$

(a) Use the integral test to show that this series converges.

(b) Find

$$s_{10} = \sum_{k=2}^{10} \frac{1}{k(\ln k)^2}.$$

(c) Find upper and lower bounds for the remainder R_{10}.

(d) Use the results of (b) and (c) to estimate the sum s of the given series. What is the maximum error e in this estimate?

(e) Suppose that it is desired to find a partial sum s_n that approximates s as well as the estimate obtained in (d). How large must n be in order to achieve this?

(f) Assume that your computer performs 10^{12} additions per second. How long will it take to add up the number of terms found in (e)? Observe that this means that it is impossible to obtain more than one decimal place in s merely by adding up terms in the series; it simply converges too slowly for this. To determine s more accurately, one must do something else, such as estimate the remainder.

In Problems 13 and 14 we investigate the growth of the partial sums of certain divergent series. In some cases the partial sums approach infinity *very* slowly.

© **13.** (a) Let $s_n = \Sigma_{k=1}^{n} (1/k)$ be the nth partial sum of the harmonic series. Show that

$$\frac{1}{n} + \ln n < s_n < 1 + \ln n.$$

(b) Calculate upper bounds for the partial sums after 10^4, 10^8, and 10^{12} terms, respectively. Observe that if the partial sums are calculated by adding successive terms on a computer capable of eight decimal place accuracy, then all terms after $k = 10^8$ are negligible to the computer, that is, they are zero in the computer.

© **14.** (a) Consider the series

$$\sum_{k=2}^{\infty} \frac{1}{k \ln k}$$

and let

$$s_n = \sum_{k=2}^{n} \frac{1}{k \ln k}.$$

Show that

$$\ln \ln n - \ln \ln 2 + \frac{1}{n \ln n} < s_n < \ln \ln n - \ln \ln 2 + \frac{1}{2 \ln 2}.$$

(b) Calculate upper bounds for the partial sums for $n = 10^4$, 10^8, and 10^{12}, respectively.

(c) Calculate approximately how many terms are needed to obtain a partial sum as large as ten.

© **15.** In Problem 13(a) it was shown that the nth partial sum of the harmonic series is related to $\ln n$. Here we investigate this relation more carefully. Let

$$\sigma_n = 1 + \frac{1}{2} + \frac{1}{3} + \cdots + \frac{1}{n} - \ln n.$$

Show that the sequence $\{\sigma_n\}$ is monotone decreasing and that $0 < \sigma_n \le 1$ for all positive integral values of n. Hence $\lim_{n\to\infty} \sigma_n$ exists. Calculate σ_1, σ_5, and σ_{10}. The limit of the sequence $\{\sigma_n\}$ is called the Euler-Mascheroni constant and is traditionally denoted by γ; its approximate value is

$$\gamma \cong 0.577215665.$$

A long-standing unsolved problem is to determine whether γ is a rational or irrational number.

* **16. A bounded continuous nonrectifiable curve.**
(a) Consider the function

$$f(x) = \begin{cases} x \sin \dfrac{1}{x}, & x > 0; \\ 0, & x = 0. \end{cases}$$

Show that f is continuous for $x \ge 0$ and differentiable for $x > 0$. Observe that the graph of f is oscillatory with increasing frequency and decreasing amplitude as $x \to 0$, and that $f(x) = 0$ for $x = 1/\pi, 1/2\pi, \ldots, 1/n\pi, \ldots$.

(b) Let s_n be the arc length of the portion of the curve in $1/(n+1)\pi \le x \le 1/n\pi$. Then $s = \sum_{n=2}^{\infty} s_n$ is the arc length of the given curve on the interval $0 < x \le 1/\pi$. By drawing a sketch of the graph of f, or in some other way, show that $s_2 \ge 4/3\pi$. Also show that $s_3 \ge 4/5\pi$, and in general, that

$$s_n \ge \frac{4}{(2n-1)\pi}.$$

(c) Show that

$$s \ge \frac{4}{\pi}\left(\frac{1}{3} + \frac{1}{5} + \cdots + \frac{1}{2n-1} + \cdots\right),$$

and then conclude that the given curve is not rectifiable.

12.6 ALTERNATING SERIES

In this section and the following one we consider series with both positive and negative terms. We examine first a class of series known as **alternating series** in which the positive and negative terms alternate with each other. For example, the series

$$1 - \frac{1}{2} + \frac{1}{3} - \frac{1}{4} + \frac{1}{5} - \frac{1}{6} + \cdots + \frac{(-1)^{k+1}}{k} + \cdots \tag{1}$$

is an alternating series. It is called the *alternating harmonic series*. On the other hand, the series

$$1 + \frac{1}{2} - \frac{1}{3} + \frac{1}{4} + \frac{1}{5} - \frac{1}{6} + \cdots \tag{2}$$

is not an alternating series, because two positive terms occur between each successive pair of negative terms.

We can use the series (1) to illustrate the main properties of alternating series, so we will begin by studying this series in some detail. Observe first that, because the signs alternate with the even-numbered terms always negative, it follows that every odd-numbered partial sum is greater than the immediately following even-numbered sum. Thus we have the inequalities

$$s_1 > s_2,\ s_3 > s_4,\ \ldots,\ s_{2k-1} > s_{2k},\ \ldots, \tag{3}$$

where k is any positive integer.

Further, by grouping the terms in pairs, we obtain the series

$$\left(1 - \frac{1}{2}\right) + \left(\frac{1}{3} - \frac{1}{4}\right) + \left(\frac{1}{5} - \frac{1}{6}\right) + \cdots + \left(\frac{1}{2k - 1} - \frac{1}{2k}\right) + \cdots$$

$$= \frac{1}{2} + \frac{1}{12} + \frac{1}{30} + \cdots + \frac{1}{(2k - 1)(2k)} + \cdots \quad (4)$$

whose partial sums are the even-numbered partial sums of the series (1). Since each term in the series (4) is positive, it follows that the even-numbered partial sums of the series (1) form an increasing sequence

$$s_2 < s_4 < s_6 < \cdots < s_{2k} < \cdots . \quad (5)$$

Now by grouping the terms in a different way we obtain the series

$$1 - \left(\frac{1}{2} - \frac{1}{3}\right) - \left(\frac{1}{4} - \frac{1}{5}\right) - \cdots - \left(\frac{1}{2k - 2} - \frac{1}{2k - 1}\right) - \cdots$$

$$= 1 - \frac{1}{6} - \frac{1}{20} - \cdots - \frac{1}{(2k - 2)(2k - 1)} - \cdots \quad (6)$$

whose partial sums are the odd-numbered partial sums of the series (1). From Eq. 6 it follows that the odd-numbered partial sums of the series (1) form a decreasing sequence

$$s_1 > s_3 > s_5 > \cdots > s_{2k-1} > \cdots . \quad (7)$$

Combining the inequalities (5) and (7) and using the relation $s_{2k-1} > s_{2k}$ from Eq. 3 we obtain, for each positive integer k,

$$s_1 > s_3 > \cdots > s_{2k-3} > s_{2k-1} > s_{2k} > s_{2k-2} > \cdots > s_4 > s_2. \quad (8)$$

The situation is indicated in Figure 12.6.1. Thus the odd-numbered partial sums form a monotone decreasing sequence that is bounded below (by s_2, for example) and hence converges from above to a limit that we will denote by A, where $A \geq s_2 = \frac{1}{2}$. Similarly, the even-numbered partial sums form a monotone increasing sequence that is bounded above (by s_1, for example) and hence converges from below to a limit B, with $B \leq s_1 = 1$.

In order to determine the relation between A and B, we note that

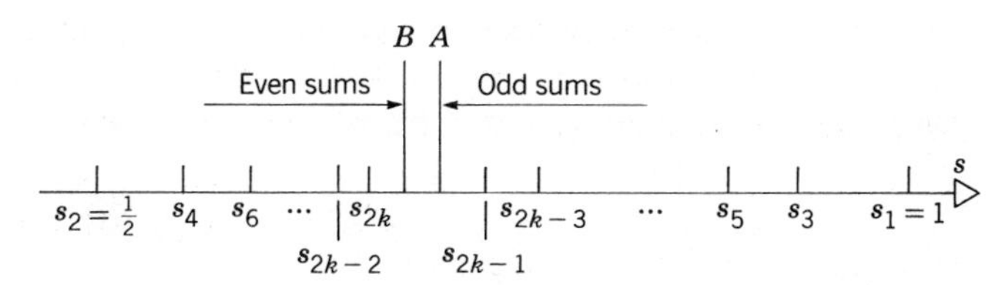

Figure 12.6.1

As $k \to \infty$, the limit of the left side of Eq. 9 is

$$\lim_{k\to\infty}\left(s_{2k-1} - \frac{1}{2k}\right) = \lim_{k\to\infty} s_{2k-1} - \lim_{k\to\infty}\frac{1}{2k} = A - 0 = A, \tag{10}$$

while the limit of the right side of Eq. 9 is B. Hence we must have

$$A = B. \tag{11}$$

Since both the odd-numbered and even-numbered partial sums approach the same limit, it follows that the entire sequence of partial sums must also approach this limit, which we will henceforth denote by s, as usual. In other words, the series (1) converges to the common value s to which the odd- and even-numbered partial sums converge from above and below, respectively.

To estimate s one can resort to the computation of partial sums, remembering that any odd-numbered sum is an upper bound and any even-numbered sum is a lower bound. For example, we have

$$\begin{aligned} s_1 &= 1.0, & s_2 &= 0.5, \\ s_3 &= 0.8333, & s_4 &= 0.5833, \\ s_5 &= 0.7833, & s_6 &= 0.6167, \\ s_7 &= 0.7595, & s_8 &= 0.6345. \end{aligned} \tag{12}$$

This calculation places the sum s in the interval (0.6345, 0.7595). Greater accuracy can be achieved by using more terms in the series.

It is also easy to estimate the error incurred by truncating the summation after any finite number of terms. Suppose that the truncation occurs after an even number of terms, so that the sum obtained is smaller than s. The addition of one more term would result in a value larger than s; hence in stopping after an even number of terms our error is less than the first neglected term. The situation is similar if we stop after an odd number of terms; again the error is less (in absolute value) than the magnitude of the first neglected term.

In the case of the series (1) this error estimate indicates that the series converges rather slowly. For example, after eight terms we have $s_8 = 0.6345$ and know only that the error is less than $a_9 = \frac{1}{9} \cong 0.111$. Moreover, even if we sum the first one thousand terms, we can only conclude that the error is less than $\frac{1}{1001} \cong 0.001$, which suggests that there may yet be an error of one in the third decimal place. Similarly, if one million terms are used, then the error is less than 10^{-6}, so there may still be an error of one in the sixth decimal place. These estimates are conservative since they are based on an upper bound for the error. Nevertheless, they indicate (correctly) that the series (1) is indeed slowly convergent.

In practice, a more accurate estimate for s can be obtained by using the arithmetic average of two successive partial sums, since one of these partial sums (the odd one) is too high and the other (the even one) is too low. For instance, using the partial sums s_7 and s_8 from Eq. 12 we have

$$s \cong \frac{1}{2}(s_7 + s_8) = 0.697.$$

The error in this estimate is less than

$$\frac{1}{2}(s_7 - s_8) = \frac{1}{2}a_8 = 0.0625,$$

or a bit more than one-half of the previous error bound that is associated with the partial sum s_8 alone. In fact, by taking the average of s_7 and s_8 we have done considerably better than that statement suggests. In Example 2 of Section 13.5 we show that actually $s = \ln 2 \cong 0.693147$. Thus the estimate $s \cong 0.697$ is actually in error by only 0.003853, whereas s_8 is in error by 0.058647, or more than fifteen times as much. Nevertheless, the alternating harmonic series does converge slowly, and there are other series that are much better for computing ln 2 (see Problem 17).

Except for the numerical computation of the partial sums, the arguments we have just given apply equally well to *any alternating series whose terms monotonically approach zero in magnitude*. We will usually write such series in the form

$$a_1 - a_2 + a_3 - a_4 + \cdots + (-1)^{k+1}a_k + \cdots = \sum_{k=1}^{\infty} (-1)^{k+1}a_k, \tag{13}$$

with $a_k > 0$ for each k, so that the sign is explicitly indicated. We then have the following results.

Theorem 12.6.1

(Alternating series test)
Suppose that the terms in the alternating series $\Sigma_{k=1}^{\infty} (-1)^{k+1}a_k$ satisfy the following conditions.

(a) $a_k > 0$ for each k.
(b) $a_{k+1} < a_k$ for each k.
(c) $\lim_{k\to\infty} a_k = 0$.

Then the series converges to a sum s, and

$$|s - s_n| < a_{n+1}. \tag{14}$$

Further, if

$$\bar{s} = \frac{s_n + s_{n+1}}{2}, \tag{15}$$

then we have the improved error bound

$$|s - \bar{s}| < \frac{a_{n+1}}{2}. \tag{16}$$

As usual, convergence of a series is not affected by any finite block of terms at the beginning of the series, so the conclusions of Theorem 12.6.1 remain true if the hypotheses are satisfied only for $k > K$, where K is some fixed positive number. The proof of this theorem parallels exactly the discussion of the alternating harmonic series. Note that the theorem gives sufficient conditions for convergence; if these conditions are not satisfied, no conclusion can be drawn from this theorem about convergence or divergence of the series in question.

EXAMPLE 1

Determine whether the series

$$\sum_{k=1}^{\infty} \frac{(-1)^{k+1}}{k^3} \tag{17}$$

converges. If it does, estimate the error made in truncating the summation after one hundred terms.

The series (17) is an alternating series whose terms decrease monotonically to zero in magnitude. Thus Theorem 12.6.1 applies, and the series converges. Using the estimate (14), we have

$$|s - s_{100}| < \frac{1}{(101)^3} < \frac{1}{10^6} = 0.000001. \tag{18}$$

Alternatively, if we let $\bar{s} = (s_{100} + s_{101})/2$ and use Eq. 16, we obtain

$$|s - \bar{s}| < \frac{1}{2} \frac{1}{(101)^3} < 0.0000005. \tag{19}$$

Thus using the estimate $\bar{s}$ gives the sum accurate to at least six decimal places. ∎

EXAMPLE 2

Determine whether the series

$$\sum_{k=1}^{\infty} (-1)^{k+1} \frac{k+1}{2k} \tag{20}$$

converges or diverges.

This is an alternating series, and since

$$a_k = \frac{k+1}{2k} = \frac{1}{2}\left(1 + \frac{1}{k}\right)$$

the terms are monotonically decreasing in magnitude. However,

$$\lim_{k\to\infty} a_k = \lim_{k\to\infty} \frac{k+1}{2k} = \frac{1}{2} \neq 0 \tag{21}$$

so the last hypothesis of Theorem 12.6.1 is not satisfied. Thus no conclusion can be drawn from *that* theorem about the convergence or divergence of the series

(20). However, since a_k does not approach zero as $k \to \infty$, we know from Theorem 12.2.1 that the series must diverge. ∎

EXAMPLE 3

Determine whether the series

$$\sum_{k=1}^{\infty} \frac{(-1)^{k+1}}{\sqrt{|k - 3\pi|}} \tag{22}$$

converges or diverges. Note that the absolute value bars are required to make sure that the quantity under the radical sign is positive.

This is an alternating series whose terms approach zero as $k \to \infty$. However, the absolute values of the terms are not monotone decreasing. In fact, for $k = 1, \ldots, 9$, the terms increase, and only for $k \geq 9$ do they decrease monotonically. Nevertheless, as we emphasized in the remark following Theorem 12.6.1, convergence or divergence is not affected by any finite block of terms. Since the hypotheses of Theorem 12.6.1 are satisfied for $k \geq 9$, we can use that theorem to conclude that the series converges. Observe, however, that this series converges very slowly. For instance, after one million terms, Eq. 14 gives the error estimate

$$|s - s_{10^6}| < (10^6 + 1 - 3\pi)^{-1/2} \cong 10^{-3} = 0.001. \tag{23}$$

Thus, the third decimal place may be inaccurate even after one million terms. ∎

PROBLEMS

In each of Problems 1 through 10, determine whether the given series converges or diverges.

1. $\sum_{k=1}^{\infty} \frac{(-1)^{k+1}}{k^2}$

2. $\sum_{k=0}^{\infty} \frac{(-1)^k}{2^k}$

3. $\sum_{k=0}^{\infty} (-1)^k \left(\frac{3}{2}\right)^k$

4. $\sum_{k=1}^{\infty} \frac{(-1)^{k+1}}{k^{5/4}}$

5. $\sum_{k=1}^{\infty} (-1)^{k+1} \frac{10^k}{k!}$

6. $\sum_{k=2}^{\infty} \frac{(-1)^k}{\ln k}$

7. $\sum_{k=1}^{\infty} \frac{(-1)^{k+1} k!}{k^k}$

8. $\sum_{k=1}^{\infty} \frac{\sin(k\pi/2)}{\sqrt{k}}$

9. $\sum_{k=1}^{\infty} \frac{(-1)^{k+1} \sin[(2k - 1)\pi/2]}{\sqrt{2k - 1}}$

* 10. $\sum_{k=1}^{\infty} (-1)^{k+1} \frac{1 \cdot 3 \cdot 5 \cdots (2k - 1)}{2 \cdot 4 \cdot 6 \cdots (2k)}$

In each of Problems 11 through 16, estimate the number of terms that are needed in order to approximate the sum s with an error less than 10^{-4}. Use both s_n and $\bar{s} = (s_n + s_{n+1})/2$.

© 11. $\sum_{k=1}^{\infty} \frac{(-1)^k}{k^2}$

© 12. $\sum_{k=0}^{\infty} \frac{(-1)^k}{2^k}$

© 13. $\sum_{k=1}^{\infty} \frac{(-1)^{k+1}}{k^{5/4}}$

© 14. $\sum_{k=1}^{\infty} (-1)^{k+1} \frac{10^k}{k!}$

© 15. $\sum_{k=2}^{\infty} \frac{(-1)^k}{\ln k}$

© 16. $\sum_{k=1}^{\infty} \frac{(-1)^{k+1}}{k!}$

© 17. In this problem we give a better series than the alternating harmonic series for calculating the value of ln 2. It can be shown that for $|x| < 1$,

$$\frac{1}{2} \ln \frac{1 + x}{1 - x} = x + \frac{x^3}{3} + \frac{x^5}{5} + \cdots + \frac{x^{2n-1}}{2n - 1} + E_n(x), \qquad (i)$$

where

$$|E_n(x)| \le \frac{|x|^{2n+1}}{(2n+1)(1-x^2)}, \quad |x| < 1. \qquad (ii)$$

(a) Use Eq. (*ii*) to show that $E_n(x) \to 0$ as $n \to \infty$ for each x in $-1 < x < 1$.

(b) What value of x should be used in Eq. (*i*) in order to find an approximate value for ln 2?

(c) Determine n so that Eq. (*i*) yields a value of ln 2 that is in error by less than 10^{-4}. Note that approximately five thousand terms of the alternating harmonic series are required to achieve the same result.

© **18.** Consider the series

$$\sum_{k=1}^{\infty} \frac{(-1)^{k+1}}{k^{1/10}}.$$

(a) Show that the series converges.

(b) Estimate the number n of terms that are needed to be sure that s_n is in error by less than 10^{-4}.

(c) Assuming that 10^{10} terms can be added per second, determine how many years are required to find the partial sum indicated in Part (b). Thus we conclude that some convergent series converge too slowly to be of any practical computational use.

* **19.** Determine whether the series

$$1 + \tfrac{1}{2} - \tfrac{1}{3} + \tfrac{1}{4} + \tfrac{1}{5} - \tfrac{1}{6} + \tfrac{1}{7} + \tfrac{1}{8} - \tfrac{1}{9} + \cdots$$

converges or diverges.

Hint: The alternating series test does not apply (why not?). Try grouping the terms $\frac{1}{2} - \frac{1}{3}$, $\frac{1}{5} - \frac{1}{6}$, $\frac{1}{8} - \frac{1}{9}$, ...

12.7 ABSOLUTE AND CONDITIONAL CONVERGENCE

In studying infinite series it is useful to distinguish two different mechanisms that may cause a given series to converge. In the first place, the terms in the series may diminish in size so rapidly that the series converges even though all the terms are of the same sign. For example, the series

$$\sum_{k=1}^{\infty} \frac{1}{k^2}, \qquad \sum_{k=0}^{\infty} \frac{1}{2^k}, \qquad \sum_{k=1}^{\infty} \frac{2^k}{k!} \qquad (1)$$

are of this type; each of them converges because the terms become small sufficiently rapidly. It may happen that series of this type have variable signs, for instance,

$$\sum_{k=1}^{\infty} \frac{(-1)^{k+1}}{k^2}, \qquad \sum_{k=0}^{\infty} \frac{(-1)^k}{2^k}, \qquad \sum_{k=1}^{\infty} \frac{(-1)^{k+1}\,2^k}{k!}. \qquad (2)$$

If this is true, then the series may converge a bit faster, but the cancellation effect caused by adding positive and negative terms is by no means essential for convergence. The series (2) would converge anyway, even if all the terms were taken to be positive.

On the other hand, there are other series, such as the alternating harmonic series

$$\sum_{k=1}^{\infty} \frac{(-1)^{k+1}}{k} = 1 - \frac{1}{2} + \frac{1}{3} - \frac{1}{4} + \cdots + \frac{(-1)^{k+1}}{k} + \cdots, \qquad (3)$$

that converge only because of the cancellations resulting from the variations in sign of the terms. For such a series the sign changes are essential to convergence. For example, if all of the terms in the series (3) are taken to be positive, then the resulting series is the harmonic series

$$\sum_{k=1}^{\infty} \frac{1}{k} = 1 + \frac{1}{2} + \frac{1}{3} + \frac{1}{4} + \cdots + \frac{1}{k} + \cdots, \tag{4}$$

which diverges.

The first type of convergence, illustrated by Eqs. 1 and 2, is known as **absolute convergence;** the second type, illustrated by Eq. 3, is called **conditional convergence.** To define these terms more precisely it is necessary to associate with a given series $\Sigma_{k=1}^{\infty} a_k$ the series $\Sigma_{k=1}^{\infty} |a_k|$ formed by taking the absolute value of each term. Then we have the following definition.

DEFINITION 12.7.1 If $\Sigma_{k=1}^{\infty} |a_k|$ converges, then $\Sigma_{k=1}^{\infty} a_k$ is said to converge absolutely. If $\Sigma_{k=1}^{\infty} |a_k|$ diverges but $\Sigma_{k=1}^{\infty} a_k$ converges, then $\Sigma_{k=1}^{\infty} a_k$ is said to converge conditionally.

For example, the series in (2) converge absolutely because in each case the corresponding series of absolute values converges. The alternating harmonic series (3) converges conditionally because it converges, and the corresponding series (4) of absolute values diverges.

For series whose terms all have the same sign, convergence and absolute convergence are equivalent because $\Sigma_{k=1}^{\infty} a_k$ and $\Sigma_{k=1}^{\infty} |a_k|$ are the same series, except possibly for multiplication by -1. Thus conditional convergence can only occur for series having infinitely many positive terms, and also infinitely many negative terms.

Since the terms in $\Sigma_{k=1}^{\infty} |a_k|$ are all nonnegative, any of the tests in Sections 12.3 to 12.5 can be used to determine whether this series converges or diverges. The following theorem gives the relation between absolute convergence and ordinary convergence.

Theorem 12.7.1

If $\Sigma_{k=1}^{\infty} |a_k|$ converges, then $\Sigma_{k=1}^{\infty} a_k$ also converges. In other words, if $\Sigma_{k=1}^{\infty} a_k$ converges absolutely, then it converges in the ordinary sense. The converse statement is not true, in general.

Proof. This theorem can be proved by using an auxiliary series $\Sigma_{k=1}^{\infty} b_k$ whose terms are given by

$$b_k = a_k + |a_k| = \begin{cases} 0, & a_k < 0; \\ 2a_k, & a_k \geq 0. \end{cases} \tag{5}$$

Thus it is always true that

$$0 \leq b_k \leq 2|a_k|. \tag{6}$$

Now let us define the partial sums

$$A_n = \sum_{k=1}^{n} |a_k|, \qquad B_n = \sum_{k=1}^{n} b_k. \tag{7}$$

Since $|a_k| \geq 0$ and $b_k \geq 0$, the sequences $\{A_n\}$ and $\{B_n\}$ are monotone nondecreasing. According to the hypothesis of the theorem, the sequence $\{A_n\}$ converges to a limit A. Since $\{A_n\}$ is monotone nondecreasing, it must approach its limit from below, so $A_n \leq A$ for all n. Thus it follows from Eqs. 6 and 7 that

$$0 \leq B_n \leq 2A_n \leq 2A, \qquad n = 1, 2, 3, \ldots. \tag{8}$$

Thus the sequence $\{B_n\}$ is also bounded above as well as monotone nondecreasing; therefore it converges also. Since both of the series $\Sigma_{k=1}^{\infty} |a_k|$ and $\Sigma_{k=1}^{\infty} b_k$ converge, it is a consequence of Theorem 12.2.2 that

$$\sum_{k=1}^{\infty} b_k - \sum_{k=1}^{\infty} |a_k| = \sum_{k=1}^{\infty} [b_k - |a_k|] = \sum_{k=1}^{\infty} a_k \tag{9}$$

also converges, as was to be shown.

To show that convergence does not imply absolute convergence it is sufficient to consider the alternating harmonic series (3), which converges, while the harmonic series (4) does not. This completes the proof of the theorem. □

EXAMPLE 1

Determine whether the series

$$\sum_{k=1}^{\infty} \frac{(-1)^{k+1}}{\sqrt{k}} \tag{10}$$

converges absolutely, converges conditionally, or diverges.

The series (10) converges by the alternating series test. The corresponding series of absolute values

$$\sum_{k=1}^{\infty} \frac{1}{\sqrt{k}} \tag{11}$$

diverges by comparison with the harmonic series, by the integral test, or by Theorem 12.3.3. Therefore the series (10) converges conditionally. ∎

The following version of the ratio test is often useful in determining whether a series converges absolutely.

Theorem 12.7.2
Consider the infinite series $\Sigma_{k=1}^{\infty} a_k$, where $a_k \neq 0$, and let

$$\lim_{k\to\infty} \left|\frac{a_{k+1}}{a_k}\right| = L. \tag{12}$$

Then

(a) If $L < 1$, the series converges absolutely.
(b) If $L > 1$, the series diverges.
(c) If $L = 1$, no conclusion is possible. The series may converge absolutely, converge conditionally, or diverge.

To prove Part (a) let us apply the ratio test to the series $\Sigma_{k=1}^{\infty} |a_k|$. Making use of the hypotheses of the theorem, we obtain

$$\lim_{k\to\infty} \frac{|a_{k+1}|}{|a_k|} = \lim_{k\to\infty} \left|\frac{a_{k+1}}{a_k}\right| = L < 1.$$

Therefore the series $\Sigma_{k=1}^{\infty} |a_k|$ converges by Theorem 12.4.1, and consequently the series $\Sigma_{k=1}^{\infty} a_k$ converges absolutely.

Part (b) is established by observing that if $L > 1$, then the sequence $\{|a_k|\}$ is positive and monotone increasing for k sufficiently large. Hence it is not possible for $|a_k|$ to approach zero as $k \to \infty$. It follows that it is also not possible for a_k to have the limit zero. Therefore, by Theorem 12.2.1, the series $\Sigma_{k=1}^{\infty} a_k$ must diverge.

To prove part (c) note that $L = 1$ for each of the series $\Sigma_{k=1}^{\infty} (-1)^{k+1}/k^2$, $\Sigma_{k=1}^{\infty} (-1)^{k+1}/k$, and $\Sigma_{k=1}^{\infty} 1/k$, and that these series converge absolutely, converge conditionally, and diverge, respectively. □

The usefulness of this theorem is illustrated by the following example.

EXAMPLE 2

Determine the values of α for which the series

$$\sum_{k=1}^{\infty} \frac{\alpha^k}{\sqrt{k}} = \alpha + \frac{\alpha^2}{\sqrt{2}} + \frac{\alpha^3}{\sqrt{3}} + \cdots \tag{13}$$

converges absolutely; converges conditionally; diverges.

To apply Theorem 12.7.2 we must determine L as defined by Eq. 12, namely,

$$L = \lim_{k\to\infty} \left|\frac{\alpha^{k+1}/\sqrt{k+1}}{\alpha^k/\sqrt{k}}\right| = \lim_{k\to\infty} \left|\alpha \sqrt{\frac{k}{k+1}}\right| = |\alpha|. \tag{14}$$

Thus, the series (13) converges absolutely for $|\alpha| < 1$, while it diverges for $|\alpha| > 1$. We must consider separately the cases for which $|\alpha| = 1$. For $\alpha = 1$ the series (13) becomes $\Sigma_{k=1}^{\infty} 1/\sqrt{k}$, which diverges as noted in Example 1. For $\alpha = -1$ the series (13) takes the form $\Sigma_{k=1}^{\infty} (-1)^k/\sqrt{k}$, which converges by the alternating series test or by Example 1. The convergence is conditional since the corresponding series of absolute values $\Sigma_{k=1}^{\infty} 1/\sqrt{k}$ diverges. In summary, the series (13) converges absolutely for $-1 < \alpha < 1$, converges conditionally for $\alpha = -1$, and diverges otherwise. ∎

Rearrangement of series

One reason for distinguishing between absolute and conditional convergence is that absolutely convergent series have several important properties that conditionally convergent series do not share. We will restrict ourselves to a discussion of an example illustrating one such property.

Consider again the alternating harmonic series

$$1 - \frac{1}{2} + \frac{1}{3} - \frac{1}{4} + \frac{1}{5} - \frac{1}{6} + \frac{1}{7} - \frac{1}{8} + \cdots = s. \tag{15}$$

We noted earlier that the value of s is ln 2, but this is immaterial. All we need to know is that $s \neq 0$, which follows from the discussion at the beginning of Section 12.6. Let us first multiply Eq. 15 by $\frac{1}{2}$, thereby obtaining

$$\frac{1}{2} - \frac{1}{4} + \frac{1}{6} - \frac{1}{8} + \cdots = \frac{1}{2}s. \tag{16}$$

This step is justified by Theorem 12.2.2: any convergent series can be multiplied term by term by a constant. Now we insert a zero between each pair of terms in Eq. 16, which gives

$$0 + \frac{1}{2} + 0 - \frac{1}{4} + 0 + \frac{1}{6} + 0 - \frac{1}{8} + \cdots = \frac{1}{2}s. \tag{17}$$

The insertion of zeros does not affect convergence of a series; in this case it just means that each member in the sequence of partial sums is repeated. Finally, we add corresponding terms in Eqs. 15 and 17 to obtain

$$1 + 0 + \frac{1}{3} - \frac{1}{2} + \frac{1}{5} + 0 + \frac{1}{7} - \frac{1}{4} + \cdots = \frac{3}{2}s$$

or

$$1 + \frac{1}{3} - \frac{1}{2} + \frac{1}{5} + \frac{1}{7} - \frac{1}{4} + \cdots = \frac{3}{2}s, \tag{18}$$

where zero terms have been deleted. This addition is justified by Theorem 12.2.2.

An examination of Eq. 18 reveals that the series on the left side contains exactly the same terms as the alternating harmonic series (15). The only difference is that in Eq. 18 the terms are *rearranged* so that two positive terms appear between each pair of negative terms. In other words, the terms in series (18) are just added in a *different order* from those in Eq. 15. Yet the sums of the two series are different; one is half again as large as the other.

The phenomenon illustrated by Eqs. 15 and 18 has no counterpart for sums of a finite number of terms, where the order of addition makes no difference. It is possible to show that absolutely convergent series resemble finite sums in this respect. *The sum of an absolutely convergent series is unchanged no matter how the terms are rearranged.**

The situation is quite different for conditionally convergent series, for which the following remarkable result can be established. *Given any real number σ, and any conditionally convergent series, there is a rearrangement of the series that converges to σ.* That is, by suitably rearranging the terms in a conditionally convergent series, we can make it converge to any number we please. We can even make it diverge to $+\infty$, or to $-\infty$, or by oscillation between two or more numbers. Problem 25 outlines how these results can be obtained.

The question of rearranging terms in a conditionally convergent infinite series illustrates that an infinite limiting process is much more subtle than operations with finite sets of numbers (or of other mathematical objects). One's intuition should be used cautiously when dealing with all infinite processes, since familiar properties may not carry over from the finite to the infinite case. The fact that heuristic reasoning may be misleading emphasizes the need for careful attention to definitions, theorems, and proofs when making use of infinite processes.

PROBLEMS

In each of problems 1 through 18, determine whether the given series converges absolutely, converges conditionally, or diverges.

1. $\sum_{k=0}^{\infty} \frac{(-1)^k}{k^2+4}$
2. $\sum_{k=1}^{\infty} \frac{(-1)^k 2^k}{k!}$
3. $\sum_{k=1}^{\infty} \frac{(-1)^k \sqrt{k}}{k+2}$
4. $\sum_{k=0}^{\infty} (-1)^k \frac{k}{k+1}$
5. $\sum_{k=0}^{\infty} \frac{(-1)^k k^4}{2^k}$
6. $\sum_{k=2}^{\infty} \frac{(-1)^k}{\ln k}$
7. $\sum_{k=1}^{\infty} \frac{(-1)^k (k!)^2}{(2k)!}$
8. $\sum_{k=1}^{\infty} (-1)^k \frac{(3/2)^k}{k^2}$
9. $\sum_{k=1}^{\infty} \frac{(-1)^k k!}{k^k}$
10. $\sum_{k=2}^{\infty} \frac{(-1)^k}{k \ln k}$
11. $\sum_{k=1}^{\infty} \frac{(-1)^{k+1} \arctan k}{k^2}$
12. $\sum_{k=0}^{\infty} \frac{(-1)^k k^3}{k^4+1}$
13. $\sum_{k=1}^{\infty} \frac{(-1)^{k+1} k!}{1 \cdot 3 \cdot 5 \cdots (2k-1)}$
14. $\sum_{k=1}^{\infty} \frac{(-1)^{k+1} k!}{10^k}$
15. $\sum_{k=1}^{\infty} \frac{(-1)^{k+1} (k+1)^2}{k^{1/3}(k^2+4)}$
16. $1 + \frac{1}{4} - \frac{1}{9} + \frac{1}{16} + \frac{1}{25} - \frac{1}{36} + \frac{1}{49} + \frac{1}{64} - \frac{1}{81} + \cdots$
17. $\sum_{k=1}^{\infty} \frac{\sin k}{k^2}$
18. $\sum_{k=0}^{\infty} \frac{(-1)^k e^k}{e^k + e^{-k}}$

In each of Problems 19 through 24, determine for which values of α the given series converges absolutely, converges conditionally, or diverges.

19. $\sum_{k=1}^{\infty} \frac{(-1)^k \alpha^k}{k}$
20. $\sum_{k=1}^{\infty} \frac{(-1)^{k+1} \alpha^k}{k^2}$
21. $\sum_{k=0}^{\infty} \frac{(-1)^k \alpha^k}{2^k}$
22. $\sum_{k=1}^{\infty} \frac{(\alpha - 1)^k}{k}$

*This result, as well as the example in Eqs. 15 and 18, is due to Dirichlet (1837). The results about conditionally convergent series in the following paragraph were discovered by Riemann (1854).

23. $\sum_{k=1}^{\infty} \frac{2^k \alpha^k}{k!}$

24. $\sum_{k=1}^{\infty} \frac{(-1)^{k+1}(\alpha + 2)^k}{\sqrt{k}}$

* 25. Consider the alternating harmonic series

$$\sum_{k=1}^{\infty} \frac{(-1)^{k+1}}{k} = 1 - \tfrac{1}{2} + \tfrac{1}{3} - \tfrac{1}{4} + \cdots \qquad (i)$$

In this problem we indicate how to show that this series can be rearranged so as to converge to an arbitrary sum σ.

(a) Form the series

$$1 + \tfrac{1}{3} + \tfrac{1}{5} + \tfrac{1}{7} + \cdots = \sum_{k=1}^{\infty} p_k, \qquad (ii)$$

$$-\left(\tfrac{1}{2} + \tfrac{1}{4} + \tfrac{1}{6} + \tfrac{1}{8} + \cdots\right) = -\sum_{k=1}^{\infty} q_k, \qquad (iii)$$

by using the positive and negative terms, respectively, of the series (i). Show that the series (ii) diverges to ∞, and that the series (iii) diverges to $-\infty$.

(b) Assuming that $\sigma > 0$, form a rearrangement of the series (i) in the following way. Take just enough positive terms from the series (ii) to form a sum greater than σ, then take just enough negative terms from the series (iii) so that the accumulated sum is less than σ, then take just enough positive terms from the series (ii) so that the accumulated sum exceeds σ, and so on. Show that the series formed in this way converges to σ.

(c) If $\sigma < 0$, describe how to rearrange the series (i) so as to converge to σ.

(d) Describe how to rearrange the series (i) so as to diverge to $+\infty$; to $-\infty$.

(e) Describe how to rearrange the series (i) so that it diverges by oscillation between two values σ_1 and σ_2, where $\sigma_1 \neq \sigma_2$.

12.8 TESTS FOR CONVERGENCE: A SUMMARY

In this chapter we have discussed several tests that have been developed for the purpose of determining whether a given infinite series $\Sigma_{k=1}^{\infty} a_k$ converges or diverges. One difficulty is that sometimes it may not be easy to select a test that is appropriate for a particular series. We indicate below one way of approaching this choice in an orderly and systematic manner. However, a certain amount of trial and error is almost sure to be involved in testing infinite series for convergence or divergence.

1. Does $a_k \to 0$ as $k \to \infty$? If not, then the series diverges, but if so, then no conclusion can be drawn and further tests are needed.
2. Are the terms in the series positive for all (sufficiently large) k? If so, consider the comparison test, the ratio test, or the integral test.
   (a) If a_k contains factorials or quantities raised to a power involving k, try the ratio test.
   (b) If a_k contains expressions involving k raised to various powers, try the limit comparison test with $\Sigma_{k=1}^{\infty} k^{-p}$ as the comparison series.
   (c) If replacing k by x yields a function $f(x)$ that is monotone decreasing to zero and that you can integrate, try the integral test.
3. If the terms in $\Sigma_{k=1}^{\infty} a_k$ are not all positive, replace each term by its absolute value and examine the series $\Sigma_{k=1}^{\infty} |a_k|$ by the methods indicated in paragraph 2. If $\Sigma_{k=1}^{\infty} |a_k|$ converges, then $\Sigma_{k=1}^{\infty} a_k$ converges absolutely. If $\Sigma_{k=1}^{\infty} |a_k|$ diverges, then $\Sigma_{k=1}^{\infty} a_k$ may either diverge or converge conditionally.

4. Do the terms in $\Sigma_{k=1}^{\infty} a_k$ alternate in sign? If so, try the alternating series test. Remember that if the hypotheses of this test are not satisfied, no conclusion can be drawn.
5. If the problem remains unsolved, try to think of some preliminary calculation that will transform a_k into a form more susceptible to analysis. For example, you might group factors in a_k in a certain way, or rationalize the numerator or denominator in a_k so as to eliminate a radical, or split a_k into a sum of two or more terms, or perhaps rewrite a_k in some other way.

The first forty review problems that follow are intended to give you practice in selecting an appropriate test for convergence or divergence for a given series.

REVIEW PROBLEMS

In each of Problems 1 through 40, determine whether the given series converges, converges absolutely, converges conditionally, or diverges.

1. $\sum_{k=1}^{\infty} \frac{1}{k^2 + 9}$

2. $\sum_{k=1}^{\infty} \frac{k!}{10^k}$

3. $\sum_{k=1}^{\infty} \frac{\sqrt{k}}{k^2 + k + 1}$

4. $\sum_{k=0}^{\infty} \frac{1}{(2k + 1)^2}$

5. $\sum_{k=1}^{\infty} \frac{k!}{2^k k^4}$

6. $\sum_{k=1}^{\infty} \frac{(-1)^{k+1} k}{\sqrt{k + 4}}$

7. $\sum_{k=1}^{\infty} \frac{2^{2k}}{(2k)!}$

8. $\sum_{k=1}^{\infty} \frac{\sqrt{k+1} - \sqrt{k}}{\sqrt{k+1} + \sqrt{k}}$

9. $\sum_{k=1}^{\infty} \frac{3^k}{k^2 2^k}$

10. $\sum_{k=1}^{\infty} \frac{(-1)^{k+1} \sqrt{k}}{k + 10}$

11. $\sum_{k=1}^{\infty} \frac{1 \cdot 3 \cdot 5 \cdots (2k - 1)}{1 \cdot 4 \cdot 7 \cdots (3k - 2)}$

12. $\sum_{k=1}^{\infty} \frac{1 - (1/k)}{1 + (1/k^2)}$

13. $\sum_{k=1}^{\infty} \frac{k^k}{k! 2^k}$

14. $\sum_{k=1}^{\infty} \frac{k^k}{(k^2)!}$

15. $\sum_{k=1}^{\infty} \frac{1 \cdot 4 \cdot 7 \cdots (3k - 2)}{2^k k!}$

16. $\sum_{k=1}^{\infty} \left(\sqrt{k} + \frac{1}{\sqrt{k}}\right)^{-3}$

17. $\sum_{k=1}^{\infty} \frac{(-1)^k 2^k}{k^2}$

18. $\sum_{k=0}^{\infty} \frac{(-1)^k k^3}{2^k}$

19. $\sum_{k=1}^{\infty} \frac{[k + (1/k)]^2}{(k^2 + 1)^{3/2}}$

20. $1 - \frac{1}{2^2} - \frac{1}{3^2} + \frac{1}{4^2} - \frac{1}{5^2} - \frac{1}{6^2} + \frac{1}{7^2} - \frac{1}{8^2} - \frac{1}{9^2} + \cdots$

21. $\sum_{k=1}^{\infty} \frac{(-1)^{k^2}}{\sqrt{k}}$

22. $\sum_{k=1}^{\infty} \frac{1}{k^2(\sqrt{k+1} - \sqrt{k})}$

23. $\sum_{k=1}^{\infty} \frac{(-1)^{k+1}}{\sqrt{k+1} - \sqrt{k}}$

24. $\sum_{k=1}^{\infty} \frac{1}{k} \sin \frac{1}{k}$

25. $\sum_{k=1}^{\infty} (-1)^{k+1} [\sqrt{k+1} - \sqrt{k}]$

26. $\sum_{k=1}^{\infty} \frac{(\sqrt{k} + 1)^{3/2}}{k^{5/2} + 4}$

27. $\sum_{k=1}^{\infty} \frac{(-1)^{k+1}(k - 1)!(k + 1)!}{(2k)!}$

28. $\sum_{k=1}^{\infty} \frac{\cos (k\pi/2)}{2k\pi}$

29. $1 + \frac{1}{2} + \frac{1}{4} + \frac{1}{4} + \frac{1}{9} + \frac{1}{8} + \frac{1}{16} + \frac{1}{16} + \frac{1}{25} + \frac{1}{32} + \cdots$

30. $\sum_{k=1}^{\infty} \frac{e^{\sqrt{k}}}{\sqrt{e^k}}$

31. $\sum_{k=1}^{\infty} \frac{(-1)^k 5^{2k}}{2^{2k}(k!)^2}$

32. $\sum_{k=2}^{\infty} (-1)^k \frac{\ln (k^2)}{(\ln k)^2}$

33. $\sum_{k=1}^{\infty} (-1)^{k+1} \frac{3 \cdot 7 \cdot 11 \cdots (4k - 1)}{2^k k!}$

34. $\sum_{k=0}^{\infty} \frac{e^k}{(e^k + 1)^2}$

35. $\sum_{k=1}^{\infty} \frac{k^2}{e^{k^3}}$

36. $\sum_{k=2}^{\infty} \frac{\cos k\pi}{\ln k}$

37. $\sum_{k=1}^{\infty} \frac{\tanh k}{k}$

38. $\sum_{k=1}^{\infty} \frac{k^2}{2^{\sqrt{k}}}$

39. $\sum_{k=2}^{\infty} \frac{1}{2^{\ln k}}$

40. $\sum_{k=2}^{\infty} \frac{(-1)^k \ln k}{\sqrt{k}}$

In each of Problems 41 through 48, either prove the given statement (if it is true), or else provide an example that refutes it (if it is false).

41. If $\{a_n\}$ converges and $\{a_n b_n\}$ diverges, then $\{b_n\}$ diverges.

42. If $\{a_n\}$ and $\{b_n\}$ diverge, then $\{a_n b_n\}$ diverges.

43. If $a_1 + a_2 + a_3 + a_4 + a_5 + a_6 + \cdots$ converges, then $(a_1 + a_2) + (a_3 + a_4) + (a_5 + a_6) + \cdots$ also converges. Is the answer different if $a_k > 0$ for each k?

44. If $(a_1 + a_2) + (a_3 + a_4) + (a_5 + a_6) + \cdots$ converges, then $a_1 + a_2 + a_3 + a_4 + a_5 + a_6 + \cdots$ also converges. Is the answer different if $a_k > 0$ for each k?

45. If $\sum_{k=1}^{\infty} a_k$ converges, then $\sum_{k=1}^{\infty} (-1)^k a_k^3$ converges.

46. If $\sum_{k=1}^{\infty} a_k$ converges absolutely, then $\sum_{k=1}^{\infty} a_k/(1 + a_k)$ also converges absolutely.

47. If $\sum_{k=1}^{\infty} a_k$ converges, then $\sum_{k=1}^{\infty} a_k^2$ also converges. What if $\sum_{k=1}^{\infty} a_k$ converges absolutely?

48. If $\sum_{k=1}^{\infty} a_k$ converges and $\lim_{k\to\infty} c_k = 0$, then $\sum_{k=1}^{\infty} c_k a_k$ converges. What if $\sum_{k=1}^{\infty} a_k$ converges absolutely?

In each of Problems 49 through 52, assume that $f(x)$ is nonnegative and continuous for $x \geq 1$.

49. Define $f(x)$ so that $\int_1^{\infty} f(x)\,dx$ converges but $\sum_{n=1}^{\infty} f(n)$ diverges.

50. Define $f(x)$ so that $\int_1^{\infty} f(x)\,dx$ diverges but $\sum_{n=1}^{\infty} f(n)$ converges.

51. Define $f(x)$ so that $\int_1^{\infty} f(x)\,dx$ converges but $f(x)$ does not approach zero as $x \to \infty$.

52. Define $f(x)$ so that $\int_1^{\infty} f(x)\,dx$ converges but $f(x)$ is not bounded as $x \to \infty$.

THIRTEEN

taylor's approximation and power series

The simplest reasonably large class of functions are the polynomials. Compared to other functions, polynomials are very easy to combine through addition, subtraction, or multiplication, and of course they are also easy to differentiate or to integrate. Moreover, performing any of these operations on polynomials simply produces other polynomials, so that if necessary several of these operations can readily be executed in sequence.

Therefore, if one needs to perform such operations on a more complicated function, it may be advisable to consider replacing it by an approximating polynomial. Then the operations themselves may be greatly simplified. Of course, it is then also necessary to be concerned about how well the polynomial approximates the original function and whether unacceptable errors may have been introduced. Generally speaking, accuracy can be improved by using polynomials of higher degree.

The idea of approximating a function by polynomials of successively higher degrees leads naturally to the exact representation of a function by an infinite series of a type known as power series. As we shall see in this chapter, subject to certain restrictions, power series behave very much like polynomials, which makes them also very valuable analytical tools.

The use of polynomial approximations and power series goes back to the very early days of calculus. They formed a fundamental part of Newton's methodology, although many years passed before a firm theoretical foundation was laid, and questions related to convergence satisfactorily answered. Today polynomial approximations and power series, always useful for analytical investigations, also form the basis for a variety of powerful numerical algorithms.

13.1 TAYLOR POLYNOMIALS

In Section 3.4 we discussed the approximation of a given function f in the neighborhood of a point x_0 by a first-degree polynomial $y = a_0 + a_1(x - x_0)$. Now we want to consider the more general problem of approximating f near x_0 by a polynomial P_n of arbitrary degree n. There are many reasons why it may be useful to do this: for instance, polynomials are easy to evaluate, differentiate, integrate, and manipulate algebraically. The first question that we must consider is how to select the approximating polynomial P_n for a given function f near a given point x_0. Later we will derive estimates for the error incurred by using $P_n(x)$ to approximate the value of $f(x)$.

We determine the approximating polynomial P_n by generalizing the procedure used to find the linear, or tangent line, approximation. The tangent line approximation to f at x_0 is the function $P_1(x) = a_0 + a_1(x - x_0)$ whose value a_0 and slope a_1 at $x = x_0$ agree with the corresponding quantities for f. Thus

$$a_0 = P_1(x_0) = f(x_0), \qquad a_1 = P_1'(x_0) = f'(x_0).$$

To extend this idea to a polynomial P_n of degree n we require that P_n and each of its first n derivatives have the same value at x_0 as f and its corresponding derivative. Thus we must have

$$P_n(x_0) = f(x_0), \quad P_n'(x_0) = f'(x_0), \quad \ldots, \quad P_n^{(n)}(x_0) = f^{(n)}(x_0). \tag{1}$$

Since a polynomial of degree n has $n + 1$ coefficients, it is reasonable to expect the $n + 1$ conditions (1) to determine all of the coefficients in $P_n(x)$.

The determination of the polynomial P_n is greatly simplified by writing it in the form

$$P_n(x) = a_0 + a_1(x - x_0) + a_2(x - x_0)^2 + a_3(x - x_0)^3 + \cdots + a_n(x - x_0)^n, \tag{2}$$

where $a_0, a_1, a_2, \ldots, a_n$ are to be found from the conditions (1). This requires the calculation of the first n derivatives of P_n and their values at x_0. From Eq. 2 we have

$$P_n(x_0) = a_0.$$

Then, by repeatedly differentiating $P_n(x)$ and setting $x = x_0$, we obtain

$$P_n'(x) = a_1 + 2a_2(x - x_0) + 3a_3(x - x_0)^2 + \cdots + na_n(x - x_0)^{n-1},$$

$$P_n'(x_0) = a_1;$$

$$P_n''(x) = 2a_2 + 3 \cdot 2\, a_3(x - x_0) + \cdots + n(n-1)a_n(x - x_0)^{n-2},$$

$$P_n''(x_0) = 2a_2;$$

$$P_n'''(x) = 3 \cdot 2\, a_3 + \cdots + n(n-1)(n-2)a_n(x - x_0)^{n-3},$$

$$P_n'''(x_0) = 3 \cdot 2\, a_3;$$

$$\vdots$$

$$P_n^{(n)}(x) = n!a_n, \qquad P_n^{(n)}(x_0) = n!a_n.$$

The conditions (1) give

$$a_0 = f(x_0), \qquad a_1 = f'(x_0), \qquad a_2 = \frac{f''(x_0)}{2!}, \qquad a_3 = \frac{f'''(x_0)}{3!},$$

and in general

$$a_k = \frac{f^{(k)}(x_0)}{k!}, \qquad k = 0, 1, 2, \ldots, n, \tag{3}$$

where $f^{(0)}(x_0)$ and 0! are to be interpreted as $f(x_0)$ and 1, respectively.

The coefficients (3) are known as the **Taylor coefficients** for the function f at x_0. The corresponding polynomial

$$P_n(x) = f(x_0) + f'(x_0)(x - x_0) + \frac{f''(x_0)}{2!}(x - x_0)^2 + \cdots + \frac{f^{(n)}(x_0)}{n!}(x - x_0)^n \tag{4}$$

is known as the **Taylor polynomial*** of degree n for f about the point x_0. The preceding argument shows that this is the only polynomial of degree n that approximates f near x_0 in the sense of Eqs. 1. Note that the Taylor polynomial P_n depends not only on the function f but also on the point x_0. The dependence on x_0 is sometimes indicated explicitly by writing $P_n(x; x_0)$ instead of $P_n(x)$ as we have done.

Observe that the Taylor polynomial of degree one

$$P_1(x) = f(x_0) + f'(x_0)(x - x_0)$$

is simply the linear approximation to f that we discussed in Section 3.4. As a rule, a Taylor polynomial provides an excellent approximation to f in some neighborhood about x_0. However, the quality of the approximation usually deteriorates as the distance from x_0 to x increases. The following examples show that it is sometimes very easy to determine Taylor polynomials by using Eqs. 3 and 4.

*Taylor polynomials are named for the English mathematician Brook Taylor (1685–1731), who included the general formula for them in his book *Methodus Incrementorum Directa et Inversa,* published in 1715. However, the same (or at least very similar) approximations were known to James Gregory as early as 1670, and to Newton, Leibniz, and Johann Bernoulli before 1700. None of these men, including Taylor, investigated the error involved in using Taylor polynomials.

EXAMPLE 1

Find the Taylor polynomials of degrees one, two, and three for $f(x) = \ln x$ about $x_0 = 2$.

By differentiating f three times we obtain

$$f'(x) = \frac{1}{x}, \qquad f''(x) = -\frac{1}{x^2}, \qquad f'''(x) = \frac{2}{x^3}.$$

Thus, since $x_0 = 2$,

$$a_0 = f(2) = \ln 2,$$

$$a_1 = f'(2) = \frac{1}{2},$$

$$a_2 = \frac{f''(2)}{2!} = -\frac{1/4}{2} = -\frac{1}{8},$$

$$a_3 = \frac{f'''(2)}{3!} = \frac{2/8}{6} = \frac{1}{24}.$$

Hence, from Eq. 4, the desired Taylor polynomials are

$$P_1(x) = \ln 2 + \tfrac{1}{2}(x - 2),$$

$$P_2(x) = \ln 2 + \tfrac{1}{2}(x - 2) - \tfrac{1}{8}(x - 2)^2,$$

$$P_3(x) = \ln 2 + \tfrac{1}{2}(x - 2) - \tfrac{1}{8}(x - 2)^2 + \tfrac{1}{24}(x - 2)^3.$$

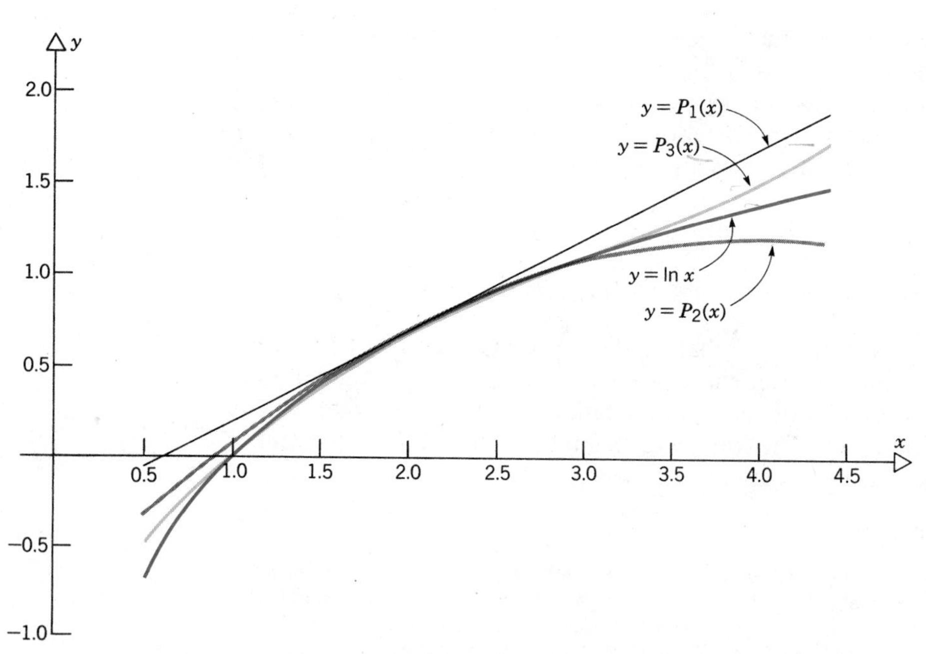

Figure 13.1.1 Polynomial approximations to $y = \ln x$ near $x = 2$.

Table 13.1 gives an idea of how well these polynomials approximate $\ln x$. See Figure 13.1.1 for graphs of $P_1(x)$, $P_2(x)$, $P_3(x)$, and $\ln x$. For example, in $P_3(x)$ the error affects the third decimal place by only one digit over the interval $1.5 \leq x \leq 2.5$. The table also shows that for each polynomial the accuracy decreases as x gets farther away from $x_0 = 2$. ∎

Table 13.1. Values of $\ln x$ and Approximating Taylor Polynomials

x	$\ln x$	$P_1(x)$	$P_2(x)$	$P_3(x)$
1.0	0.0000	0.1931	0.0681	0.0265
1.5	0.4055	0.4431	0.4119	0.4067
2.0	0.6931	0.6931	0.6931	0.6931
2.5	0.9163	0.9431	0.9119	0.9171
3.0	1.0986	1.1931	1.0681	1.1098
4.0	1.3863	1.6931	1.1931	1.5265

EXAMPLE 2

Find the Taylor polynomial of degree n for $f(x) = e^x$ about $x_0 = 0$.

Since $f^{(k)}(x) = e^x$ and $f^{(k)}(0) = 1$ for all k, we have

$$a_0 = 1, \quad a_1 = 1, \quad a_2 = \frac{1}{2!}, \quad \ldots, \quad a_n = \frac{1}{n!}.$$

Thus, from Eq. 4, the desired Taylor polynomial is

$$P_n(x) = 1 + x + \frac{x^2}{2!} + \cdots + \frac{x^n}{n!}. \tag{5}$$

Figure 13.1.2 shows the graphs of e^x and of $P_n(x)$ for $n = 2$, 4, and 8 and for $-4 \leq x \leq 4$. Observe that $P_4(x)$ is a good approximation to e^x for $-1.5 \leq x \leq 1.5$, and that $P_8(x)$ approximates e^x well over almost the whole interval $-4 \leq x \leq 4$. ∎

EXAMPLE 3

Find the Taylor polynomial of degree $2n + 1$ for $f(x) = \sin x$ about $x_0 = 0$.

The successive derivatives of $\sin x$ are

$$\sin x, \cos x, -\sin x, -\cos x, \sin x, \cos x, \ldots.$$

Evaluating them at $x_0 = 0$, we obtain

$$0, 1, 0, -1, 0, 1, 0, -1, \ldots.$$

Hence the Taylor coefficients are

$$a_0 = 0, \quad a_1 = 1, \quad a_2 = 0, \quad a_3 = -\frac{1}{3!}, \quad a_4 = 0, \quad a_5 = \frac{1}{5!}, \ldots.$$

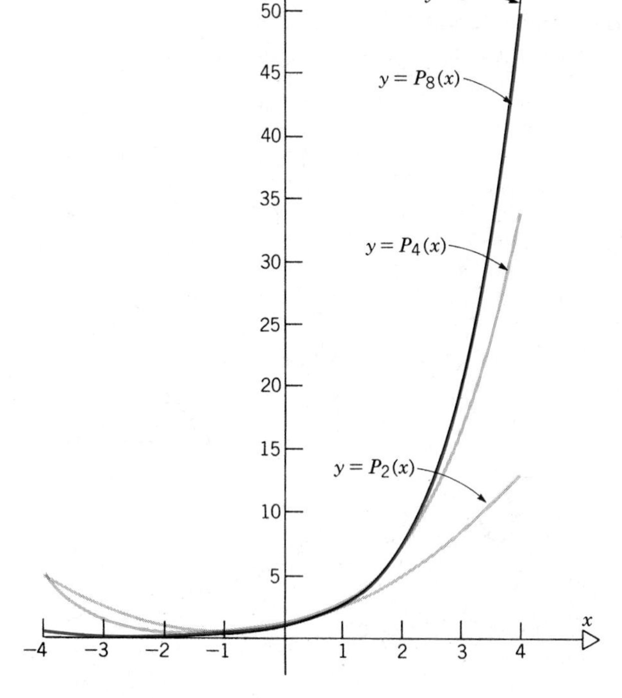

Figure 13.1.2 Polynomial approximations to $y = e^x$ near $x = 0$.

In general, all of the even-numbered coefficients are zero, while the odd-numbered coefficients are given by

$$a_{2k+1} = \frac{(-1)^k}{(2k+1)!}, \qquad k = 0, 1, 2, \ldots .$$

Thus the Taylor polynomial of degree $2n + 1$ is

$$P_{2n+1}(x) = x - \frac{x^3}{3!} + \frac{x^5}{5!} - \frac{x^7}{7!} + \cdots + \frac{(-1)^n x^{2n+1}}{(2n+1)!}.$$

Figure 13.1.3 shows the graphs of $P_1(x)$, $P_3(x)$, $P_5(x)$, and $\sin x$ for $0 \leq x \leq \pi$. It is apparent from the figure that $P_5(x)$ provides a reasonably good approximation to $\sin x$ on the interval $0 \leq x \leq \pi/2$. In fact, $P_5(\pi/2) \cong 1.0045$, and so differs from $\sin(\pi/2) = 1$ by less than one half of one percent.

This simple polynomial approximation can be extended to other values of x by using elementary properties of the sine function. For example, for $\pi/2 \leq x \leq \pi$ we have

$$\sin x = \sin(\pi - x), \tag{6}$$

and then for $\pi \leq x \leq 2\pi$, we can use

$$\sin x = -\sin(x - \pi). \tag{7}$$

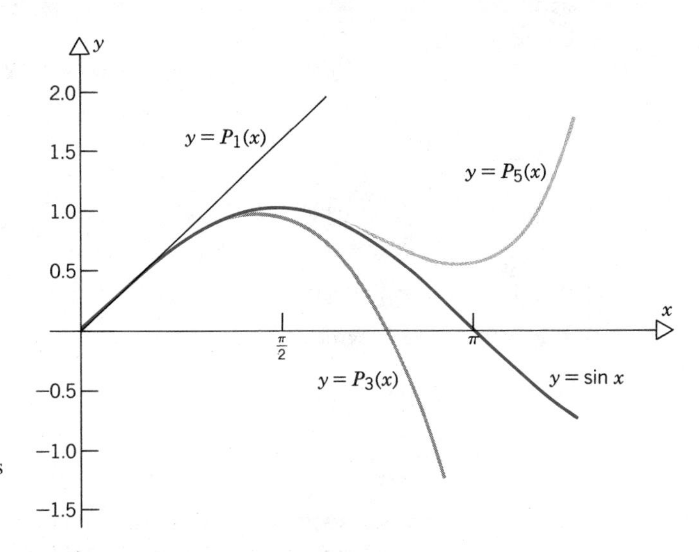

Figure 13.1.3 Polynomial approximations to $y = \sin x$ near $x = 0$.

Finally, we can approximate $\sin x$ for all other values of x from its periodic property,

$$\sin(x + 2\pi) = \sin x. \tag{8}$$

In this manner, the polynomial approximation $P_5(x)$, which is reasonably accurate for $0 \le x \le \pi/2$, can also be used to obtain an approximate value of $\sin x$ for any real x. If greater accuracy is required, one can just use one of the higher degree Taylor polynomials. In fact, this or some similar procedure is how a computer calculates values of the sine function. ∎

In the next section we give a more systematic discussion of the error involved in using a Taylor polynomial to approximate a given function.

PROBLEMS

In each of Problems 1 through 28, find the Taylor polynomial of the indicated degree for the given function about the given point.

1. $f(x) = \sin x;\ x_0 = \dfrac{\pi}{4}$, degree 3
2. $f(x) = \ln(1 - x);\ x_0 = 0$, degree 4
3. $f(x) = \cos 2x;\ x_0 = \dfrac{\pi}{2}$, degree 4
4. $f(x) = \dfrac{1}{x};\ x_0 = 1$, degree 3
5. $f(x) = e^{-x^2};\ x_0 = 0$, degree 4
6. $f(x) = \dfrac{1}{2x + 1};\ x_0 = -1$, degree 3
7. $f(x) = \sqrt{1 + x};\ x_0 = 0$, degree 3
8. $f(x) = x^3 - 2x^2 + 3x - 5;\ x_0 = 1$, degree 3
9. $f(x) = \sqrt{x};\ x_0 = 1$, degree 4
10. $f(x) = (4 - x)^{3/2};\ x_0 = 0$, degree 3
11. $f(x) = \arcsin x;\ x_0 = 0$, degree 3
12. $f(x) = \cosh x;\ x_0 = 0$, degree 4
13. $f(x) = \cos x;\ x_0 = 0$, degree $2n$
14. $f(x) = \sin \pi x;\ x_0 = 0$, degree $2n + 1$

15. $f(x) = \dfrac{1}{x - a}$, where $a \neq 0$; $x_0 = 0$, degree n

16. $f(x) = \dfrac{1}{1 - ax}$, where $a \neq 0$; $x_0 = 0$, degree n

17. $f(x) = \ln(1 - x)$; $x_0 = 0$, degree n

18. $f(x) = e^{x/2}$; $x_0 = 0$, degree n

19. $f(x) = \sqrt{1 + x}$; $x_0 = 0$, degree n

20. $f(x) = e^{-2x}$; $x_0 = -1$, degree n

21. $f(x) = \dfrac{1}{x + 1}$; $x_0 = 1$, degree n

22. $f(x) = \cos \dfrac{x}{2}$; $x_0 = \pi$, degree $2n + 1$

23. $f(x) = \sin 2x$; $x_0 = \dfrac{\pi}{4}$, degree $2n$

24. $f(x) = \sqrt{a - x}$, where $a > 0$; $x_0 = 0$, degree n

25. $f(x) = \sinh x$; $x_0 = 0$, degree $2n + 1$

26. $f(x) = \cosh \dfrac{x}{a}$, where $a > 0$; $x_0 = 0$, degree $2n$

* 27. $f(x) = \dfrac{1}{1 + x^2}$; $x_0 = 0$, degree 4

* 28. $f(x) = \arctan x$; $x_0 = 0$, degree 5

13.2 THE TAYLOR REMAINDER THEOREM AND TAYLOR SERIES

We now turn to the question of estimating the error in approximating a given function f by a Taylor polynomial P_n of degree n. In the last section we showed that $P_n(x)$ has the form

$$P_n(x) = f(x_0) + f'(x_0)(x - x_0) + \cdots + \frac{f^{(n)}(x_0)}{n!}(x - x_0)^n, \tag{1}$$

where x_0 is the point about which the approximation is centered. Our goal is to estimate the difference $|f(x) - P_n(x)|$ between f and P_n for an arbitrary value of x. In general, this difference depends on n, on the point x_0 at which the coefficients in $P_n(x)$ are calculated, and on the point x at which the difference is calculated. We will find an expression for the error by rederiving the approximating polynomial (1) in a different way—one that leads directly to a formula for $f(x) - P_n(x)$. We assume that f has continuous derivatives of all orders that appear in the derivation.

We start from the relation

$$f(x) = f(x_0) + \int_{x_0}^{x} f'(t)\, dt. \tag{2}$$

Our procedure is nothing more than repeated integration by parts of the integral on the right side of Eq. 2. Let $u = f'(t)$ and $dv = dt$. Then $du = f''(t)\, dt$ and $v = t + c$, where c is a constant of integration relative to integration with respect to t. Hence c may depend on x. In fact, it is essential for our purpose to choose $c = -x$, so that $v = t - x$. Then, integrating by parts in Eq. 2, we obtain

$$\begin{aligned} f(x) &= f(x_0) + f'(t)(t - x)\Big|_{t=x_0}^{t=x} - \int_{x_0}^{x} (t - x) f''(t)\, dt \\ &= f(x_0) + f'(x_0)(x - x_0) + \int_{x_0}^{x} (x - t) f''(t)\, dt. \end{aligned} \tag{3}$$

Now we integrate by parts again, letting $u = f''(t)$ and $dv = (x - t)\,dt$. Then $du = f'''(t)\,dt$, $v = -(x - t)^2/2$, and we find that

$$f(x) = f(x_0) + f'(x_0)(x - x_0) - f''(t)\,\frac{(x - t)^2}{2}\bigg|_{t=x_0}^{t=x}$$

$$+ \int_{x_0}^{x} \frac{(x - t)^2}{2}\, f'''(t)\,dt$$

$$= f(x_0) + f'(x_0)(x - x_0) + \frac{f''(x_0)}{2}\,(x - x_0)^2$$

$$+ \int_{x_0}^{x} \frac{(x - t)^2}{2}\, f'''(t)\,dt. \tag{4}$$

Repeating the integration by parts process n times, we finally obtain

$$f(x) = f(x_0) + f'(x_0)(x - x_0) + \cdots + \frac{f^{(n)}(x_0)}{n!}\,(x - x_0)^n$$

$$+ \int_{x_0}^{x} \frac{(x - t)^n}{n!}\, f^{(n+1)}(t)\,dt$$

$$= P_n(x) + \int_{x_0}^{x} \frac{(x - t)^n}{n!}\, f^{(n+1)}(t)\,dt. \tag{5}$$

We can state the result that we have just proved as follows.

Theorem 13.2.1

(Taylor's theorem)

Let f have continuous derivatives up to and including order $n + 1$ on some interval (a, b) containing the point x_0. If x is any other point in (a, b), then

$$f(x) = P_n(x) + R_{n+1}(x), \tag{6}$$

where

$$P_n(x) = f(x_0) + f'(x_0)(x - x_0) + \cdots + \frac{f^{(n)}(x_0)}{n!}\,(x - x_0)^n \tag{7}$$

and

$$R_{n+1}(x) = \frac{1}{n!}\int_{x_0}^{x} (x - t)^n\, f^{(n+1)}(t)\,dt. \tag{8}$$

Observe that Theorem 13.2.1 gives an explicit formula for the difference $R_{n+1}(x)$ between $f(x)$ and $P_n(x)$, that is, for the error that is made if $f(x)$ is approximated by $P_n(x)$. It is customary to refer to $R_{n+1}(x)$ as the **remainder.** Since the evaluation of $R_{n+1}(x)$ from Eq. 8 is often difficult (it involves $n + 1$ differentiations followed by an integration), it is fortunate that there are other formulas for the remainder $R_{n+1}(x)$. The most useful expression for $R_{n+1}(x)$ is given in the following corollary.*

Corollary

Under the conditions of Theorem 13.2.1 there is a point c between x_0 and x such that the remainder $R_{n+1}(x)$ is given by

$$R_{n+1}(x) = \frac{f^{(n+1)}(c)(x - x_0)^{n+1}}{(n+1)!}. \tag{9}$$

The proof of the corollary is outlined in Problem 38. The form (9) of the remainder is easy to remember because of its similarity to the terms in the Taylor polynomial itself: the remainder is just like the next term, except that the derivative is evaluated at c instead of at x_0. The point c arises from the use of a mean value theorem, and its location is not precisely specified. As we indicated before, the remainder depends on n, x_0, and x. Consequently, c also depends on these three quantities.

The following two examples illustrate the calculation of the remainder associated with a Taylor polynomial.

EXAMPLE 1

Find the remainder $R_4(x)$ in using the Taylor polynomial of degree three for the function $f(x) = \ln x$ about $x_0 = 2$. Determine an upper bound for $|R_4(x)|$ on $1 \le x \le 3$.

This is a continuation of Example 1 of Section 13.1. We have $f^{(4)}(x) = -6/x^4$, so from Eq. 9 with $n = 3$

$$R_4(x) = -\frac{6}{c^4}\frac{(x-2)^4}{4!} = -\frac{(x-2)^4}{4c^4}, \tag{10}$$

where c is between 2 and x. An upper bound for

$$|R_4(x)| = \frac{|x-2|^4}{4c^4}$$

*The first appearance of the remainder term in Taylor's formula was in 1797 in Lagrange's book *Théorie des Fonctions Analytiques*. The expression for $R_{n+1}(x)$ given in the corollary is known as Lagrange's formula for the remainder. Taylor expansions are fundamental in all branches of analysis, and are also of vital importance in many applications. For example, the control of errors in many contemporary numerical algorithms is based on estimating the error in a Taylor approximation.

on the interval $1 \le x \le 3$ is obtained by using the worst possible values for x and c. These are the values of x and c that make the numerator in $|R_4(x)|$ as large as possible and the denominator as small as possible. Thus we choose $c = 1$ and $x = 1$, so that, at worst,

$$|R_4(x)| \le \tfrac{1}{4}, \qquad 1 \le x \le 3.$$

This is a rather conservative estimate, since the data given in Example 1 of Section 13.1 indicate that, in fact, $|R_4(x)| \le 0.03$ for $1 \le x \le 3$. ∎

EXAMPLE 2

Find the remainder $R_{2n+2}(x)$ in using the Taylor polynomial of degree $2n + 1$ for $f(x) = \sin x$ about $x_0 = 0$. Also determine the degree of the polynomial that is needed to approximate $\sin x$ with an error less than 0.000001 for all x in $[-\pi/2, \pi/2]$.

This is a continuation of Example 3 in Section 13.1, where $P_{2n+1}(x)$ was found to be

$$P_{2n+1}(x) = x - \frac{x^3}{3!} + \frac{x^5}{5!} - \cdots + \frac{(-1)^n x^{2n+1}}{(2n+1)!}. \tag{11}$$

The remainder $R_{2n+2}(x)$ is easily found from Eq. 9. The $(2n + 2)$nd derivative of $\sin x$ is $(-1)^{n+1} \sin x$, so

$$R_{2n+2}(x) = \frac{(-1)^{n+1}(\sin c)x^{2n+2}}{(2n+2)!}. \tag{12}$$

Since $|\sin c| \le 1$ for all c, it follows from Eq. 12 that

$$|R_{2n+2}(x)| \le \frac{x^{2n+2}}{(2n+2)!}$$

for all x. The absolute value of x is not required since the exponent is always even. For $|x| \le \pi/2$, we have

$$|R_{2n+2}(x)| \le \frac{(\pi/2)^{2n+2}}{(2n+2)!}. \tag{13}$$

By testing a few values of n in Eq. 13 we obtain the results shown in Table 13.2. Thus it is certainly sufficient to choose $n = 5$ in order to make $|R_{2n+2}(x)| \le 0.000001$

Table 13.2. Estimates of the Remainder in Eq. 13

n	$(\pi/2)^{2n+2}/(2n + 2)!$
3	0.0009193
4	0.0000252
5	0.000000471

throughout the interval $[-\pi/2, \pi/2]$. For values of x outside of this interval we can make use of Eqs. 6, 7, and 8 of Section 13.1 in order to find approximate values of $\sin x$. ∎

Taylor series

Let us now suppose that the function f has continuous derivatives of all orders in an interval containing the point x_0. Then it is possible to form the Taylor polynomial

$$P_n(x) = f(x_0) + f'(x_0)(x - x_0) + \cdots + \frac{f^{(n)}(x_0)(x - x_0)^n}{n!}$$

for an arbitrarily large value of n. In addition, suppose that for each x in the given interval it is possible to show that

$$\lim_{n\to\infty} R_{n+1}(x) = 0. \tag{14}$$

Then, on this interval we can write

$$\begin{aligned} f(x) &= f(x_0) + f'(x_0)(x - x_0) + \cdots + \frac{f^{(k)}(x_0)}{k!}(x - x_0)^k + \cdots \\ &= \sum_{k=0}^{\infty} \frac{f^{(k)}(x_0)}{k!}(x - x_0)^k, \end{aligned} \tag{15}$$

where the infinite series on the right side converges to the value of $f(x)$ at each point. This infinite series is known as the **Taylor series** for f about x_0, and often provides a convenient representation for the function. If $x_0 = 0$, then Eq. 15 has the simpler form

$$f(x) = \sum_{k=0}^{\infty} \frac{f^{(k)}(0)}{k!} x^k. \tag{16}$$

In the latter case the series (16) is sometimes called a **Maclaurin series.**

EXAMPLE 3

Find the Taylor series for $f(x) = e^x$ about $x_0 = 0$; also determine where the series converges and represents $f(x)$.

In Example 2 of Section 13.1 we found that the Taylor polynomial of degree n for e^x about $x_0 = 0$ is

$$P_n(x) = 1 + x + \frac{x^2}{2!} + \cdots + \frac{x^n}{n!}. \tag{17}$$

From Eq. 9 it follows that the remainder associated with $P_n(x)$ is

$$R_{n+1}(x) = \frac{e^c\, x^{n+1}}{(n + 1)!}, \tag{18}$$

where c is some point between 0 and x. Now we wish to find the values of x for which $R_{n+1}(x) \to 0$ as $n \to \infty$. Note that c may depend on both n and x. However, since c is always between 0 and x, it is always true that $e^c \le e^{|x|}$, a bound that is independent of n. Thus

$$|R_{n+1}(x)| \le \frac{e^{|x|}\,|x|^{n+1}}{(n+1)!} \tag{19}$$

and we only need to determine where $|x|^{n+1}/(n+1)! \to 0$ as $n \to \infty$. One way to establish the desired result is to consider the infinite series

$$\sum_{k=0}^{\infty} \frac{|x|^{k+1}}{(k+1)!}; \tag{20}$$

if this series converges, then it follows from Theorem 12.2.1 (the kth term test) that

$$\lim_{k\to\infty} \frac{|x|^{k+1}}{(k+1)!} = 0. \tag{21}$$

To show convergence of the series (20) the ratio test is convenient:

$$\lim_{k\to\infty} \frac{|x|^{k+2}}{(k+2)!}\,\frac{(k+1)!}{|x|^{k+1}} = \lim_{k\to\infty} \frac{|x|}{k+2} = 0 \tag{22}$$

for all x. Thus the series (20) converges for all x; hence $R_{n+1}(x) \to 0$ as $n \to \infty$ for all x, and as a result

$$e^x = 1 + x + \frac{x^2}{2} + \cdots + \frac{x^k}{k!} + \cdots = \sum_{k=0}^{\infty} \frac{x^k}{k!} \tag{23}$$

is valid for all x. This is the desired Taylor series for e^x.

In particular, for $x = 1$, Eq. 23 yields

$$e = 1 + 1 + \frac{1}{2!} + \frac{1}{3!} + \cdots + \frac{1}{k!} + \cdots.$$

This series converges rapidly and provides a good way to calculate e; an estimate of the error after any finite number of terms is given by Eq. 19 with $x = 1$.

Observe that we could have shown at once that the series (23) converges by using the ratio test, essentially as in Eq. 22. However, it is possible that even though the series converges, its sum might be different from e^x. An example of this situation is given in Problem 37. To be sure that the sum of the series (23) is actually e^x, one must show that $R_{n+1}(x) \to 0$ as $n \to \infty$. ∎

EXAMPLE 4

Find the Taylor series for $f(x) = \sin x$ about $x_0 = 0$; determine where this series converges and represents $f(x)$.

In Example 3 of Section 13.1 we showed that

$$\sin x = x - \frac{x^3}{3!} + \frac{x^5}{5!} - \cdots + \frac{(-1)^n x^{2n+1}}{(2n+1)!} + R_{2n+2}(x), \tag{24}$$

and in Example 2 of this section we found that

$$|R_{2n+2}(x)| \le \frac{x^{2n+2}}{(2n+2)!}. \tag{25}$$

By an argument similar to that used in Example 3 it follows that $R_{2n+2}(x) \to 0$ as $n \to \infty$ for every fixed x. Thus

$$\begin{aligned} \sin x &= x - \frac{x^3}{3!} + \frac{x^5}{5!} - \cdots + \frac{(-1)^k x^{2k+1}}{(2k+1)!} + \cdots \\ &= \sum_{k=0}^{\infty} \frac{(-1)^k x^{2k+1}}{(2k+1)!}, \end{aligned} \tag{26}$$

and the series converges for all x.

In a similar way it can be shown that

$$\begin{aligned} \cos x &= 1 - \frac{x^2}{2!} + \frac{x^4}{4!} - \cdots + \frac{(-1)^k x^{2k}}{(2k)!} + \cdots \\ &= \sum_{k=0}^{\infty} \frac{(-1)^k x^{2k}}{(2k)!} \end{aligned} \tag{27}$$

and that this series also converges for all x. ∎

EXAMPLE 5

Find the Taylor series for $f(x) = (1 - x)^{-1}$ about $x_0 = 0$. Also determine where the series converges and represents $f(x)$.

To find the Taylor coefficients for $f(x)$ at $x_0 = 0$ we compute the successive derivatives of $f(x)$:

$$f(x) = (1-x)^{-1}, \quad f'(x) = (1-x)^{-2}, \quad f''(x) = 2(1-x)^{-3},$$
$$\ldots, f^{(k)}(x) = k!(1-x)^{-k-1}, \ldots$$

and evaluate them at the origin:

$$f(0) = 1, \quad f'(0) = 1, \quad f''(0) = 2, \quad \ldots, \quad f^{(k)}(0) = k!, \ldots.$$

Then $a_k = f^{(k)}(0)/k! = 1$ for each k, and the required Taylor series for $f(x)$ about $x_0 = 0$ is

$$1 + x + x^2 + \cdots + x^k + \cdots = \sum_{k=0}^{\infty} x^k. \tag{28}$$

Let us now consider the remainder

$$R_{n+1}(x) = \frac{f^{(n+1)}(c)x^{n+1}}{(n+1)!} = \frac{x^{n+1}}{(1-c)^{n+2}}, \tag{29}$$

where c is between 0 and x. If $0 < x < \frac{1}{2}$, then $0 < c < x$ and it follows that

$0 < x < 1 - c$. Consequently, $0 < x/(1 - c) < 1$, and therefore

$$R_{n+1}(x) = \frac{1}{1-c}\left(\frac{x}{1-c}\right)^{n+1} \to 0 \quad \text{as} \quad n \to \infty. \tag{30}$$

Hence the series (28) converges to $f(x) = (1 - x)^{-1}$ at least for $0 < x < \frac{1}{2}$. This investigation of $R_{n+1}(x)$ can be carried further, but it is better to proceed in another way. Observe that the series (28) is just the geometric series with ratio x. From our discussion of the geometric series in Section 12.2 we know that the series (28) converges for $|x| < 1$ and has the sum $(1 - x)^{-1}$; this is the information that was requested. The point is that we obtained this result without examining the remainder R_{n+1} directly. It is often awkward to try to show directly that $R_{n+1}(x) \to 0$ as $n \to \infty$, and it is important to watch for indirect ways, such as the one used here, of establishing the same result. In Sections 13.4 and 13.5 we see many other examples of indirect methods of finding Taylor series for given functions. ∎

PROBLEMS

In each of Problems 1 through 20, use the corollary to Theorem 13.2.1 to find the remainder associated with the Taylor polynomial of the indicated degree for the given function about the given point. The problem number in parentheses indicates the corresponding problem following Section 13.1.

1. $f(x) = \sin x$; $x_0 = \frac{\pi}{4}$, degree 3 (Problem 1)
2. $f(x) = \ln(1 - x)$; $x_0 = 0$, degree 4 (Problem 2)
3. $f(x) = \cos 2x$; $x_0 = \frac{\pi}{2}$, degree 4 (Problem 3)
4. $f(x) = \frac{1}{x}$; $x_0 = 1$, degree 3 (Problem 4)
5. $f(x) = e^{-x^2}$; $x_0 = 0$, degree 4 (Problem 5)
6. $f(x) = \frac{1}{2x + 1}$; $x_0 = -1$, degree 3 (Problem 6)
7. $f(x) = \sqrt{1 + x}$; $x_0 = 0$, degree 3 (Problem 7)
8. $f(x) = x^3 - 2x^2 + 3x - 5$; $x_0 = 1$, degree 3 (Problem 8)
9. $f(x) = (4 - x)^{3/2}$; $x_0 = 0$, degree 3 (Problem 10)
10. $f(x) = \arcsin x$; $x_0 = 0$, degree 3 (Problem 11)
11. $f(x) = \cos x$; $x_0 = 0$, degree $2n$ (Problem 13)
12. $f(x) = \frac{1}{x - a}$, where $a \neq 0$; $x_0 = 0$, degree n (Problem 15)
13. $f(x) = \frac{1}{1 - ax}$, where $a \neq 0$; $x_0 = 0$, degree n (Problem 16)
14. $f(x) = \ln(1 - x)$; $x_0 = 0$, degree n (Problem 17)
15. $f(x) = e^{-2x}$; $x_0 = -1$, degree n (Problem 20)
16. $f(x) = \frac{1}{x + 1}$; $x_0 = 1$, degree n (Problem 21)
17. $f(x) = \sin 2x$; $x_0 = \frac{\pi}{4}$, degree $2n$ (Problem 23)
18. $f(x) = \sqrt{a - x}$, where $a > 0$; $x_0 = 0$, degree n (Problem 24)
19. $f(x) = \sinh x$; $x_0 = 0$, degree $2n + 1$ (Problem 25)
20. $f(x) = \cosh \frac{x}{a}$, where $a > 0$; $x_0 = 0$, degree $2n$ (Problem 26)

21. Let $f(x) = x^{7/3}$.

(a) Find the Taylor polynomials of degrees one and two, respectively, for f about $x_0 = 0$. What is the remainder in each case?

(b) Does f have a Taylor polynomial of degree three about $x_0 = 0$?

(c) Find the Taylor polynomial of degree three for f about $x_0 = 1$.

© 22. Problem 11 required the determination of the Taylor polynomial of degree $2n$ for $\cos x$ about $x_0 = 0$.

(a) Determine how many terms of this polynomial are required in order to approximate $\cos x$ on the interval $0 \le x \le \pi$ with an error not exceeding 10^{-6}.

(b) Is it possible to achieve the same result as in (a) with fewer terms by expanding about $x_0 = \pi/2$?

In each of Problems 23 through 26, use an appropriate Taylor polynomial to calculate the given quantity with an error not exceeding 10^{-4}. State how many terms are needed to achieve this result.

© 23. $\cos\left(\frac{\pi}{4} - 0.1\right)$

© 24. $\sin\left(\frac{\pi}{6} + 0.05\right)$

© 25. $\ln(1.2)$

© 26. $\sqrt{4.07}$

In each of Problems 27 through 36, find the Taylor series for the given function about the given point. Determine where the series converges and represents the given function by showing that the remainder approaches zero as the number of terms increases without bound.

27. $f(x) = \cos x,\ x_0 = 0$

28. $f(x) = \cosh x,\ x_0 = 0$

29. $f(x) = \sinh x,\ x_0 = 0$

30. $f(x) = \sin \frac{x}{2},\ x_0 = 0$

31. $f(x) = \cos 2x,\ x_0 = 0$

32. $f(x) = \sin x,\ x_0 = \frac{\pi}{2}$

33. $f(x) = \cos x,\ x_0 = \frac{\pi}{2}$

34. $f(x) = e^{-2x},\ x_0 = 0$

35. $f(x) = e^{3x},\ x_0 = 0$

36. $f(x) = e^{x},\ x_0 = 2$

* 37. This problem illustrates the situation in which the Taylor series constructed from a function f converges for all x, but converges to $f(x)$ only for $x = 0$. Let

$$f(x) = \begin{cases} e^{-1/x^2}, & x \neq 0 \\ 0, & x = 0 \end{cases}$$

(a) Using the difference quotient

$$\frac{f(h) - f(0)}{h} = \frac{e^{-1/h^2}}{h}$$

and L'Hospital's rule, show that $f'(0) = 0$.

(b) Compute $f'(x)$ for $x \neq 0$.

(c) Using the result of Part (b), the difference quotient $[f'(h) - f'(0)]/h$, and L'Hospital's rule, show that $f''(0) = 0$. In a similar way it can be shown that $f^{(n)}(0) = 0$ for all positive integral values of n.

(d) Write $f(x) = P_n(x) + R_{n+1}(x)$, and determine $P_n(x)$ and $R_{n+1}(x)$. Take the limit as $n \to \infty$, and show that $\lim_{n\to\infty} P_n(x)$ exists for all x, but is not equal to $f(x)$ unless $x = 0$. In other words, $R_{n+1}(x) \to 0$ as $n \to \infty$ only for $x = 0$.

* 38. **Proof of corollary to Theorem 13.2.1.**

(a) If $x > x_0$, show that

$$R_{n+1}(x) = \frac{1}{n!}\int_{x_0}^{x} (x - t)^n f^{(n+1)}(t)\, dt$$

satisfies the conditions of the generalized mean value theorem of Problem 28 of Section 6.3 if $f^{(n+1)}(t)$ is identified with $f(t)$ and $(x - t)^n/n!$ with $g(t)$. Hence show that

$$R_{n+1}(x) = \frac{f^{(n+1)}(c)}{n!}\int_{x_0}^{x} (x - t)^n\, dt,$$

and that Eq. 9 follows.

(b) If $x < x_0$, show that

$$R_{n+1}(x) = \frac{(-1)^{n+1}}{n!}\int_{x}^{x_0} (t - x)^n f^{(n+1)}(t)\, dt;$$

then establish Eq. 9 as in Part (a).

* **39.** In Section 10.3 it was stated that the local formula error in using the Euler method is proportional to h^2, where h is the step size. If $y = \phi(t)$ is the exact solution of

$$y' = f(t, y), \qquad y(t_0) = y_0,$$

show that

$$\phi(t_{n+1}) = \phi(t_n) + \phi'(t_n)h + \phi''(\bar{t}_n)\frac{h^2}{2},$$

where $t_n < \bar{t}_n < t_n + h$. Assuming that $y_n = \phi(t_n)$, show that if $y_{n+1} = y_n + hf(t_n, y_n)$, then $\phi(t_{n+1}) - y_{n+1} = \phi''(\bar{t}_n)h^2/2$.

13.3 POWER SERIES AND THE RADIUS OF CONVERGENCE

Up to now we have always adopted the point of view that a function f is given, and the problem is to find its associated Taylor polynomials or Taylor series about a given point x_0. Our principal tools for this task are the formula

$$a_k = \frac{f^{(k)}(x_0)}{k!} \tag{1}$$

for the Taylor coefficients, and the expression for the remainder $R_{n+1}(x)$ given in the corollary to Theorem 13.2.1.

However, it is also possible to take the viewpoint that the series itself is the basic object of study. Thus, let us suppose that we have a series of the form

$$a_0 + a_1(x - x_0) + \cdots + a_k(x - x_0)^k + \cdots = \sum_{k=0}^{\infty} a_k(x - x_0)^k, \tag{2}$$

where x_0 and the coefficients $a_0, a_1, \ldots, a_k, \ldots$ are given. The series (2) is called a **power series** because it involves successive powers of $x - x_0$. Several questions now arise naturally, including the following.

1. For what values of x does the series (2) converge?
2. Assuming that the series (2) does converge for x in some interval I, its sum is a function f with domain I. Can anything interesting be said about f? For example, is f continuous or differentiable, and does f have a Taylor series about x_0?
3. Assuming that f does have a Taylor series representation about x_0, that series is certainly a power series of the form (2). Is the Taylor series necessarily identical to the series (2), or may a function have two (or more) different power series representations about the same point?

We discuss the first question in the remainder of this section and take up the others in the following sections. However, there are two important comments that we wish to make at the outset. First, it may seem strange to start with the infinite series (2) rather than with a given function f. Nevertheless, it is often important to do this. In many applications, especially those requiring the solution of a differential equation, an extremely useful technique is to seek the solution in the form of a power series, and to try to develop a procedure for finding the coefficients. Once this is done, the next problem is to interpret the results, which basically means answering questions such as (1) and (2) above. Some examples of this type of procedure are given in Problems 26 and 27 in Section 13.5.

The other comment is that an important by-product of this second viewpoint (starting with a series rather than with a function) is that our range of methods for finding Taylor series for particular functions will be considerably enlarged. This is significant because the formula (1) is often unwieldy, since the successive derivatives of f may become more and more complicated.

Now let us turn to the question of convergence. The following example is typical.

EXAMPLE 1

Find the values of x for which the power series

$$\sum_{k=1}^{\infty} \frac{(x-2)^k}{k} = (x-2) + \frac{(x-2)^2}{2} + \frac{(x-2)^3}{3} + \cdots \tag{3}$$

converges.

To study the convergence of a power series it is usually appropriate to start with the ratio test to investigate possible absolute convergence. For the series (3) we obtain

$$\left|\frac{(x-2)^{k+1}}{k+1}\,\frac{k}{(x-2)^k}\right| = \frac{k}{k+1}|x-2| \to |x-2| \quad \text{as} \quad k \to \infty. \tag{4}$$

Therefore the series (3) converges absolutely for $|x - 2| < 1$ and diverges for $|x - 2| > 1$. In other words, the series converges absolutely for $1 < x < 3$ and diverges for $x < 1$ and for $x > 3$. The ratio test is inconclusive for $x = 1$ and $x = 3$, and these points must be examined separately. In fact, for $x = 1$ the series (3) reduces to the alternating harmonic series $\Sigma_{k=1}^{\infty} (-1)^k/k$, and so converges conditionally at this point. At the point $x = 3$ the series reduces to the harmonic series $\Sigma_{k=1}^{\infty} 1/k$, so it diverges there. To summarize, the series (3) converges absolutely for $1 < x < 3$, converges conditionally for $x = 1$, and diverges otherwise. This information is shown in pictorial form in Figure 13.3.1. ∎

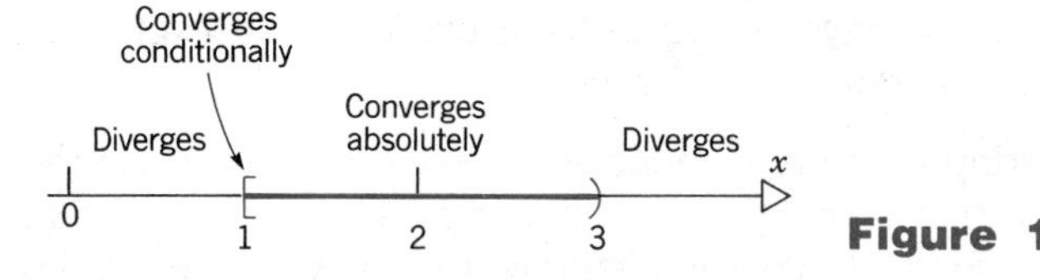

Figure 13.3.1

Now we take up the question of convergence in more generality. For simplicity we will let $x_0 = 0$ and consider

$$\sum_{k=0}^{\infty} a_k x^k = a_0 + a_1 x + a_2 x^2 + \cdots + a_k x^k + \cdots \tag{5}$$

rather than the series (2). Results for the series (5) may be applied also to the series (2) by replacing x by $x - x_0$. The series (5) certainly converges for $x = 0$, since in that case every term after the first is zero. If the series (5) converges for nonzero values of x, it may converge for all x, or (as in Example 1) it may converge for some values of x and diverge for others. We sort out the possibilities in the following three theorems; the first two theorems are preparatory, and the main result is in Theorem 13.3.3.

Theorem 13.3.1

If the series $\sum_{k=0}^{\infty} a_k x^k$ converges for $x = \xi \neq 0$, then it converges absolutely for all x such that $|x| < |\xi|$.

For $\xi > 0$, the theorem is illustrated in Figure 13.3.2. If the series converges for $x = \xi$, then it converges absolutely for $-\xi < x < \xi$, as shown in the figure.

Figure 13.3.2

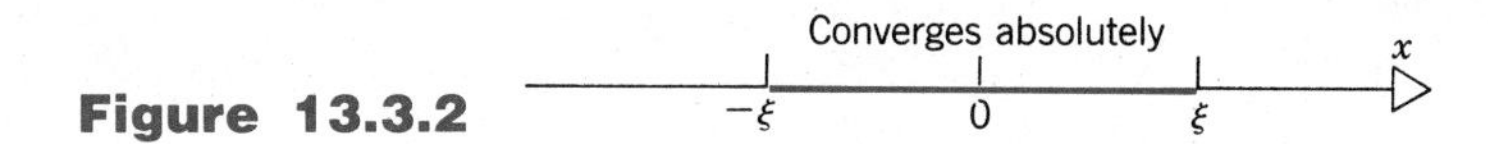

Proof. Since $\sum_{k=0}^{\infty} a_k \xi^k$ converges, we know from Theorem 12.2.1 (the kth term test) that $a_k \xi^k \to 0$ as $k \to \infty$. Hence by Theorem 12.1.5 the terms $a_k \xi^k$ must be bounded; that is, there is a number M such that

$$|a_k \xi^k| \leq M; \qquad k = 1, 2, 3, \ldots. \tag{6}$$

Now let x satisfy $|x| < |\xi|$, and let $r = x/\xi$, so that $|r| < 1$. Then

$$a_k x^k = a_k \xi^k \frac{x^k}{\xi^k} = a_k \xi^k r^k$$

and hence

$$|a_k x^k| = |a_k \xi^k| \, |r|^k \leq M \, |r|^k; \qquad k = 1, 2, 3, \ldots. \tag{7}$$

The series $\sum_{k=0}^{\infty} |r|^k$ is a convergent geometric series, and it follows from the comparison test (Theorem 12.3.1) that $\sum_{k=0}^{\infty} |a_k x^k|$ also converges. Hence the series $\sum_{k=0}^{\infty} a_k x^k$ converges absolutely, and the theorem is proved. □

Theorem 13.3.2

If the series $\sum_{k=0}^{\infty} a_k x^k$ diverges for $x = \eta$, then it diverges for all x such that $|x| > |\eta|$.

For $\eta > 0$ the theorem is illustrated in Figure 13.3.3. If the series diverges for $x = \eta$, then it diverges for $x > \eta$ and for $x < -\eta$.

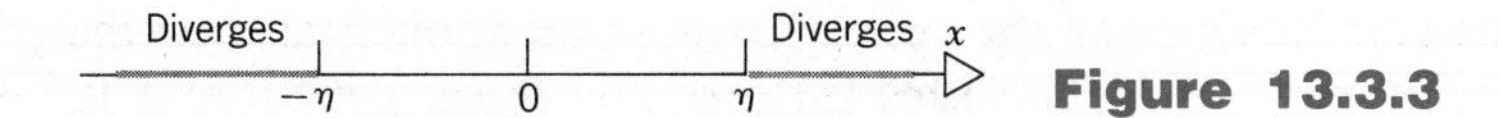

Figure 13.3.3

Proof. The proof is by contradiction. Suppose that the conclusion is false, that is, the series converges for some point $x = \xi$, where $|\xi| > |\eta|$. Then by Theorem 13.3.1 the series converges for all x such that $|x| < |\xi|$, including necessarily the point $x = \eta$. Since this contradicts the hypothesis of the theorem, the proof is complete. □

Theorem 13.3.3

For each power series of the form $\Sigma_{k=0}^{\infty} a_k x^k$ one and only one of the following statements is true.

(a) The series converges only for $x = 0$.
(b) The series converges (absolutely) for all values of x.
(c) There is a number $\rho > 0$ such that the series converges (absolutely) if $|x| < \rho$ and diverges if $|x| > \rho$.

The number ρ mentioned in Theorem 13.3.3 is called the **radius of convergence** of the series $\Sigma_{k=0}^{\infty} a_k x^k$. If the series converges only at $x = 0$, then its radius of convergence is zero. Similarly, if the series converges everywhere, then it has an infinite radius of convergence. With a proper interpretation for the two extreme cases where ρ is either zero or infinite, Theorem 13.3.3 states that each power series $\Sigma_{k=0}^{\infty} a_k x^k$ converges (absolutely) in the interior of an interval of length 2ρ with center at the origin, and diverges outside of this interval. The open interval $-\rho < x < \rho$ is called the **interval of convergence** (see Figure 13.3.4). The theorem says nothing about what happens at the endpoints $x = \pm\rho$ of the interval of convergence, and these must be investigated separately for each series.

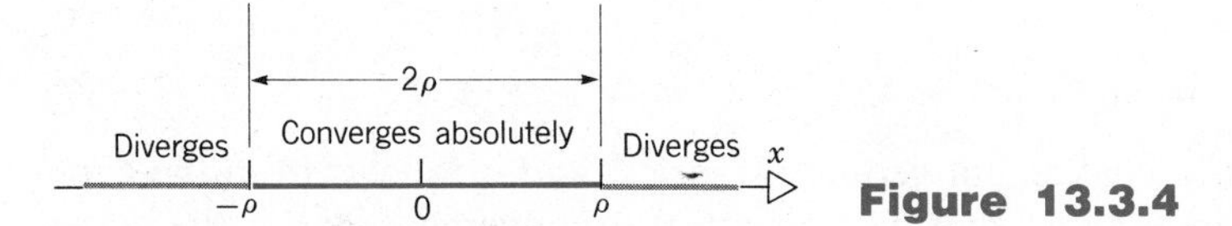

Figure 13.3.4

The same conclusions also hold for series of the form (2), except that the center of the interval of convergence is x_0. Thus the series (2) converges absolutely for $|x - x_0| < \rho$, that is, for $x_0 - \rho < x < x_0 + \rho$. It diverges for $x < x_0 - \rho$ and for $x > x_0 + \rho$, and may either converge or diverge at $x = x_0 \pm \rho$.

Theorems 13.3.1 and 13.3.2 should make the conclusion of Theorem 13.3.3 seem plausible, at the very least. In many cases the radius of convergence ρ can be calculated from the ratio test. If we apply the ratio test to the series $\Sigma_{k=0}^{\infty} |a_k x^k|$, and assume that

$$\lim_{k\to\infty} \left| \frac{a_{k+1}}{a_k} \right| = L, \tag{8}$$

then we obtain

$$\lim_{k\to\infty}\left|\frac{a_{k+1}x^{k+1}}{a_kx^k}\right| = \lim_{k\to\infty}\left|\frac{a_{k+1}}{a_k}\right| |x| = L|x|. \tag{9}$$

Therefore $\sum_{k=0}^{\infty} a_kx^k$ converges absolutely for $|x| < 1/L$ and diverges for $|x| > 1/L$. Hence the radius of convergence is given by

$$\rho = \frac{1}{L} = \lim_{k\to\infty}\left|\frac{a_k}{a_{k+1}}\right|, \tag{10}$$

provided that the limit on the right side exists (as a finite number) or is $+\infty$. Note that this argument does not constitute a proof of Theorem 13.3.3, because there are power series that fail to satisfy the hypothesis about the limit in Eq. 8. A proof is outlined in Problem 22.

Let us now consider some further examples.

EXAMPLE 2

Determine the values of x for which the power series

$$\sum_{k=0}^{\infty} (-1)^k 2^k x^k = 1 - 2x + 4x^2 - 8x^3 + \cdots \tag{11}$$

converges.

Applying the ratio test, we obtain

$$\lim_{k\to\infty}\left|\frac{(-1)^{k+1}2^{k+1}x^{k+1}}{(-1)^k 2^k x^k}\right| = 2|x|. \tag{12}$$

Therefore the series (11) converges for $|x| < \frac{1}{2}$, or for $-\frac{1}{2} < x < \frac{1}{2}$. It diverges for $|x| > \frac{1}{2}$, that is, for $x < -\frac{1}{2}$ and for $x > \frac{1}{2}$. Thus the radius of convergence is $\rho = \frac{1}{2}$, and the center of the interval of convergence is $x_0 = 0$. At the left endpoint $x = -\frac{1}{2}$ the series reduces to $\sum_{k=0}^{\infty} (-1)^{2k}$ and at the right endpoint $x = \frac{1}{2}$ it becomes $\sum_{k=0}^{\infty} (-1)^k$. Both of these series diverge by the kth term test (Theorem 12.2.1). Therefore the series converges absolutely for $-\frac{1}{2} < x < \frac{1}{2}$, and diverges otherwise, as indicated in Figure 13.3.5. ∎

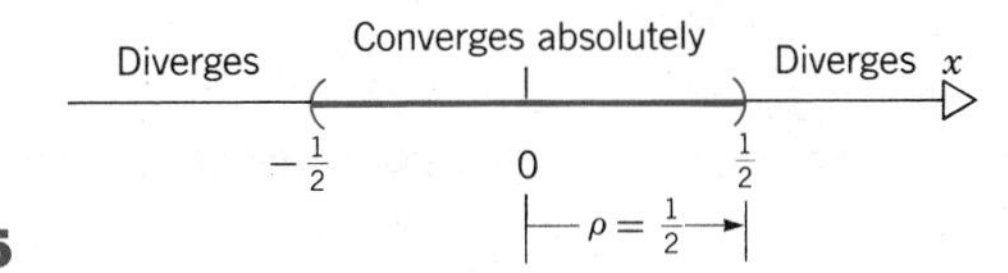

Figure 13.3.5

EXAMPLE 3

Discuss the convergence or divergence of the power series

$$\sum_{k=1}^{\infty} \frac{(x+2)^k}{k^2 2^k}. \tag{13}$$

From the ratio test we obtain

$$\left|\frac{(x+2)^{k+1}}{(k+1)^2\,2^{k+1}}\,\frac{k^2 2^k}{(x+2)^k}\right| = \left(\frac{k}{k+1}\right)^2 \frac{|x+2|}{2} \to \frac{|x+2|}{2} \quad \text{as} \quad k \to \infty. \tag{14}$$

Thus the series converges absolutely for $|x + 2| < 2$, or for $-4 < x < 0$. It diverges if $|x + 2| > 2$, that is, if $x < -4$ or if $x > 0$. The radius of convergence is $\rho = 2$, and the center of the interval of convergence is $x_0 = -2$. At the left endpoint $x = -4$, the series is $\Sigma_{k=1}^{\infty} (-1)^k/k^2$, which converges absolutely. Similarly, at the right endpoint $x = 0$, the series is $\Sigma_{k=1}^{\infty} 1/k^2$, which converges. Thus the given series converges (absolutely) for $-4 \le x \le 0$, and diverges otherwise (see Figure 13.3.6). ∎

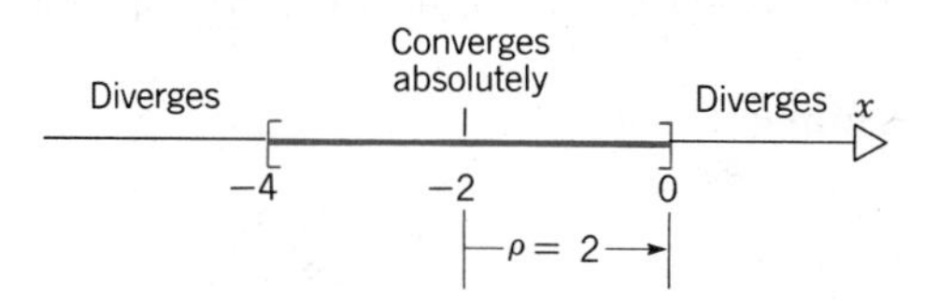

Figure 13.3.6

EXAMPLE 4

Discuss the convergence and divergence of the power series

$$\sum_{k=0}^{\infty} \frac{x^{2k}}{(2k)!}. \tag{15}$$

From the ratio test we have

$$\left|\frac{x^{2k+2}}{(2k+2)!}\,\frac{(2k)!}{x^{2k}}\right| = \frac{x^2}{(2k+1)(2k+2)} \to 0 \tag{16}$$

as $k \to \infty$ for every x. Therefore the given series converges absolutely for all x; its radius of convergence is infinite, and its interval of convergence is $-\infty < x < \infty$. This is indicated in Figure 13.3.7. ∎

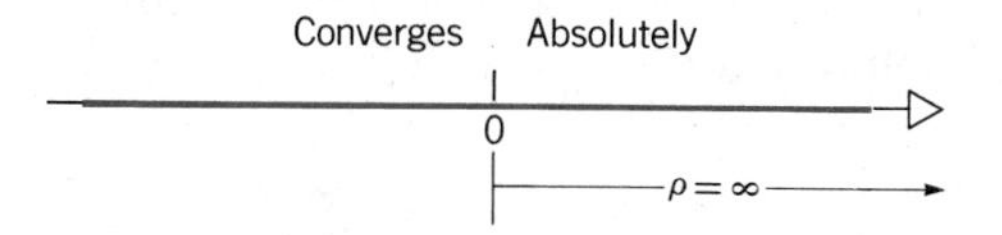

Figure 13.3.7

In each of Problems 1 through 20, find the radius of convergence of the given series, and the set of points for which it converges. Determine whether it converges absolutely or conditionally at each point.

1. $\sum_{k=1}^{\infty} \frac{x^k}{k^2}$

2. $\sum_{k=1}^{\infty} \frac{(x-1)^k}{\sqrt{k}}$

3. $\sum_{k=0}^{\infty} \frac{(x+3)^k}{2^k}$

4. $\sum_{k=1}^{\infty} \frac{(-1)^{k+1} x^k}{k}$

5. $\sum_{k=1}^{\infty} \frac{(-1)^{k+1}(x-2)^k}{\sqrt{k}\,2^{2k}}$

6. $\sum_{k=0}^{\infty} k!x^k$

7. $\sum_{k=0}^{\infty} \frac{x^{2k+1}}{(2k+1)!}$

8. $\sum_{k=1}^{\infty} \frac{(x-\pi)^k}{(\pi/2)^k}$

9. $\sum_{k=0}^{\infty} \frac{x^{2k}}{(k+1)^2}$

10. $\sum_{k=0}^{\infty} \frac{(-1)^k x^{2k+1}}{2k+1}$

11. $\sum_{k=0}^{\infty} \frac{(-1)^k x^{2k}}{k!}$

12. $\sum_{k=1}^{\infty} \frac{k^2(x-5)^k}{k!}$

13. $\sum_{k=1}^{\infty} \frac{(2x-1)^k}{k}$

14. $\sum_{k=0}^{\infty} \frac{(-1)^k x^{2k}}{2^{2k}(k!)^2}$

15. $\sum_{k=0}^{\infty} \frac{2^k\, k!\, x^{2k+1}}{(2k+1)!}$

* 16. $\sum_{k=1}^{\infty} \frac{1 \cdot 3 \cdot 5 \cdots (2k-1)}{2 \cdot 4 \cdot 6 \cdots 2k} x^{2k}$

17. $\sum_{k=0}^{\infty} \frac{(-1)^k x^{2k+1}}{(2k+1)k!}$

* 18. $\sum_{k=1}^{\infty} \frac{(k!)^2 x^{2k}}{(2k)!}$

19. $\sum_{k=1}^{\infty} \frac{k^2(3x+2)^k}{2^k}$

20. $\sum_{k=1}^{\infty} \frac{(-1)^{k+1}(x+1)^{k+1}}{k^{3/2}}$

* 21. (a) Find the radius of convergence of the power series

$$\sum_{k=1}^{\infty} \frac{k!x^k}{k^k}.$$

(b) Determine whether the series converges at the endpoints of the interval of convergence.

Hint: You may wish to use Stirling's approximation to $k!$ for large k: $k! \cong \sqrt{\pi k}\,(k/e)^k$.

* 22. **Proof of Theorem 13.3.3.** It should be clear that no two of statements (a), (b), and (c) in the theorem can be true simultaneously. Hence we need only show that there is no case in which all three are false, that is, there is no fourth possibility to consider. In particular, we assume that statements (a) and (b) are false, and show that statement (c) must then be true.

(a) Define the set S so that

$$S = \{\xi \mid \xi > 0 \text{ and } \sum_{k=0}^{\infty} a_k x^k \text{ converges for } |x| < \xi\}.$$

Show that S is not empty and is bounded above.

Hint: Otherwise either statement (a) or statement (b) in the theorem would be true.

(b) By the completeness principle (see discussion preceding Problem 29 of Section 1.1) S has a least upper bound; let ρ be the least upper bound of S. Then show that $\Sigma_{k=0}^{\infty} a_k x^k$ converges for $|x| < \rho$.

Hint: Otherwise, use Theorem 13.3.2 to show that ρ is not the *least* upper bound of S.

(c) Show that $\Sigma_{k=0}^{\infty} a_k x^k$ diverges for $|x| > \rho$, thus completing the proof.

Hint: Otherwise, use Theorem 13.3.1 to show that ρ is not an upper bound for S.

13.4 ALGEBRAIC PROPERTIES OF POWER SERIES

Power series are useful mathematical tools in large part because they have many properties that make them convenient for algebraic manipulation and numerical calculation. Indeed, within its interval of convergence a power series behaves very much like a polynomial, and may be regarded intuitively as a "polynomial of infinite degree." In this section we discuss some of the more important algebraic properties of power series. Along the way we illustrate how these properties can often be used to determine power series representations for given functions. This provides an alternative that is frequently simpler and more effective than the use of Taylor's formula for the coefficients.

In each example in this section the question can be raised as to whether the power series that we obtain here by algebraic methods is the same as the Taylor series for the same function about the given point. A general (affirmative) answer to this question is given at the end of Section 13.5. Since all of the examples in this section fall within the scope of that result, we omit any detailed discussion here.

Substitutions

One of the most common manipulations is a transformation, or substitution, involving the independent variable. We distinguish three specific kinds, namely, translations, scalings, and power substitutions.

A **translation** moves the center of the interval of convergence from one point to another, but does not change its length. Suppose that we have a power series $\sum_{k=0}^{\infty} a_k t^k$ with radius of convergence ρ. If we let $f(t)$ denote the sum of this series, then

$$f(t) = \sum_{k=0}^{\infty} a_k t^k, \qquad |t| < \rho, \tag{1}$$

where ρ may be infinite. If we let $t = x - x_0$, then

$$f(x - x_0) = \sum_{k=0}^{\infty} a_k (x - x_0)^k, \tag{2}$$

where the latter series converges for $|x - x_0| < \rho$, that is, for $x_0 - \rho < x < x_0 + \rho$.

EXAMPLE 1

Find a power series for $f(x) = \sin x$ about $x_0 = \pi$. In other words, express $\sin x$ as a series of powers of $x - \pi$.

In Example 4 of Section 13.2 we found that (temporarily using t as the independent variable)

$$\sin t = \sum_{k=0}^{\infty} \frac{(-1)^k\, t^{2k+1}}{(2k+1)!}, \qquad -\infty < t < \infty. \tag{3}$$

Letting $t = x - \pi$, we obtain

$$\sin(x - \pi) = \sum_{k=0}^{\infty} \frac{(-1)^k\, (x - \pi)^{2k+1}}{(2k+1)!} \tag{4}$$

and this series converges for $-\infty < x - \pi < \infty$, that is, for all values of x. If we note that

$$\sin(x - \pi) = -\sin x, \tag{5}$$

then it follows from Eq. 4 that

$$\sin x = -\sum_{k=0}^{\infty} \frac{(-1)^k\, (x - \pi)^{2k+1}}{(2k+1)!}, \qquad -\infty < x < \infty. \tag{6}$$

Thus we have obtained a power series representation for $\sin x$ about the point $x_0 = \pi$ rather than about the origin. ∎

A **scaling** multiplies the radius of convergence ρ by a nonzero factor. Suppose again that

$$f(t) = \sum_{k=0}^{\infty} a_k t^k, \qquad |t| < \rho. \tag{7}$$

If we let $t = x/c$, where $c \neq 0$, then

$$f\left(\frac{x}{c}\right) = \sum_{k=0}^{\infty} a_k \left(\frac{x}{c}\right)^k = \sum_{k=0}^{\infty} \frac{a_k}{c^k} x^k, \tag{8}$$

and this series converges for $|x/c| < \rho$, or for $|x| < |c|\rho$. The quantity $|c|$ is known as the scale factor. If $c < 0$, then the corresponding ends of the interval of convergence are reversed, while if $c > 0$ they are unchanged.

EXAMPLE 2

Find a power series for $f(x) = 2/(x + 2)$ about $x_0 = 0$.

First we rewrite $f(x)$ as follows:

$$\frac{2}{x+2} = \frac{1}{(x/2) + 1} = \frac{1}{1 - (-x/2)} = \frac{1}{1 - t}, \tag{9}$$

where $t = -x/2$. From Example 5 in Section 13.2 we know that

$$\frac{1}{1-t} = \sum_{k=0}^{\infty} t^k, \qquad |t| < 1. \tag{10}$$

Hence, making the substitution $t = -x/2$ in Eq. 10, we obtain

$$\frac{2}{x+2} = \sum_{k=0}^{\infty} \left(-\frac{x}{2}\right)^k = \sum_{k=0}^{\infty} \frac{(-1)^k x^k}{2^k}, \qquad |x| < 2. \; \blacksquare \tag{11}$$

EXAMPLE 3

Find a power series for $f(x) = 2/(x + 2)$ about $x_0 = 1$.

Since the series about $x_0 = 1$ must involve powers of $x - 1$, the key step is to rewrite $f(x)$ so that the quantity $x - 1$ appears. This can be done as follows:

$$\frac{2}{x+2} = \frac{2}{x - 1 + 3} = \frac{2}{3}\frac{1}{1 + [(x-1)/3]}$$

$$= \frac{2}{3}\frac{1}{1 - \{-[(x-1)/3]\}} = \frac{2}{3}\frac{1}{1-t}, \tag{12}$$

where $t = -(x - 1)/3$. As in Eq. 10 we have

$$\frac{1}{1-t} = \sum_{k=0}^{\infty} t^k, \qquad |t| < 1.$$

Then, upon substituting for t we obtain

$$\frac{2}{x+2} = \frac{2}{3}\sum_{k=0}^{\infty} \frac{(-1)^k}{3^k}(x-1)^k, \qquad |x-1| < 3. \tag{13}$$

Observe that the substitution used in this example is a combination of a translation and a scaling. ∎

Suppose again that

$$f(t) = \sum_{k=0}^{\infty} a_k t^k, \qquad |t| < \rho. \tag{14}$$

The substitution $t = x^m$, where m is a positive integer, is called a **power substitution.** It leaves the center of the interval of convergence at the origin and changes the radius of convergence to $\rho^{1/m}$. The series (14) is transformed into

$$f(x^m) = \sum_{k=0}^{\infty} a_k x^{mk}, \qquad |x| < \rho^{1/m}. \tag{15}$$

EXAMPLE 4

Find a power series for $\sin(x^2)$ about $x_0 = 0$.

Starting from Eq. 3, we let $t = x^2$ and obtain immediately

$$\sin(x^2) = \sum_{k=0}^{\infty} \frac{(-1)^k x^{4k+2}}{(2k+1)!} = x^2 - \frac{x^6}{3!} + \cdots, \qquad -\infty < x < \infty. \tag{16}$$

In the preceding three examples we could have found the desired series expansion by calculating the Taylor coefficients for the given function at the given point. This procedure would be considerably more difficult here, because the successive derivatives of $\sin(x^2)$ become increasingly complicated. Indirect methods, such as the one used in this example, can be of great benefit in such cases. ∎

EXAMPLE 5

Find a power series for $f(x) = 2/(x^2 + 2)$ about $x_0 = 0$.

To illustrate the use of a power substitution let us make use of the series (11) for $2/(x + 2)$. First we change the variable to t so that we have

$$\frac{2}{t+2} = \sum_{k=0}^{\infty} \frac{(-1)^k t^k}{2^k}, \qquad |t| < 2.$$

Then the substitution $t = x^2$ yields

$$\frac{2}{x^2+2} = \sum_{k=0}^{\infty} \frac{(-1)^k x^{2k}}{2^k} = 1 - \frac{x^2}{2} + \frac{x^4}{4} - \frac{x^6}{8} + \cdots. \tag{17}$$

The same substitution in the inequality $|t| < 2$ gives $x^2 < 2$ or $|x| < \sqrt{2}$. Thus the radius of covergence of the series (17) is $\sqrt{2}$. Observe that, unlike Example 4, the radius of convergence in this example is altered by the power substitution.

The series (17) can also be easily obtained by writing

$$\frac{2}{x^2 + 2} = \frac{1}{1 + (x^2/2)} = \frac{1}{1 - (-x^2/2)},$$

and then letting $t = -x^2/2$ in the geometric series (10). ∎

Sometimes it may be necessary to perform several substitutions in sequence in order to obtain a desired result. With practice it is often possible to combine two or more simpler substitutions in one step.

EXAMPLE 6

Find a power series about $x_0 = 3$ for the function

$$f(x) = e^{-(x-3)^2/2}.$$

From Example 3 of Section 13.2 we have

$$e^t = \sum_{k=0}^{\infty} \frac{t^k}{k!}, \qquad -\infty < t < \infty. \tag{18}$$

We first perform the scaling $t = -u/2$ to obtain

$$e^{-u/2} = \sum_{k=0}^{\infty} \frac{(-1)^k}{2^k} \frac{u^k}{k!}, \qquad -\infty < u < \infty. \tag{19}$$

The power substitution $u = w^2$ gives

$$e^{-w^2/2} = \sum_{k=0}^{\infty} \frac{(-1)^k}{2^k} \frac{w^{2k}}{k!}, \qquad -\infty < w < \infty. \tag{20}$$

Finally, we make the translation $w = x - 3$ with the result that

$$e^{-(x-3)^2/2} = \sum_{k=0}^{\infty} \frac{(-1)^k (x - 3)^{2k}}{2^k\, k!}, \qquad -\infty < x - 3 < \infty. \tag{21}$$

Of course, one can obtain Eq. 21 directly from Eq. 18 by making the single substitution $t = -(x - 3)^2/2$. We have broken the procedure into steps in order to make clear that such a substitution is just a combination of a translation, a scaling, and a power substitution. ∎

Linear combinations

Suppose that

$$f(x) = \sum_{k=0}^{\infty} a_k x^k, \qquad |x| < \rho_1 \tag{22}$$

and

$$g(x) = \sum_{k=0}^{\infty} b_k x^k, \qquad |x| < \rho_2. \tag{23}$$

In other words, the given series have respective radii of convergence ρ_1 and ρ_2, and respective sums $f(x)$ and $g(x)$ within their intervals of convergence. Let $\rho = \min(\rho_1, \rho_2)$ be the smaller of the radii of convergence of the two series. Then for any constants α and β and for each x in the interval $|x| < \rho$ we have

$$\begin{aligned} \alpha f(x) + \beta g(x) &= \alpha \sum_{k=0}^{\infty} a_k x^k + \beta \sum_{k=0}^{\infty} b_k x^k \\ &= \sum_{k=0}^{\infty} (\alpha a_k + \beta b_k) x^k, \qquad |x| < \rho. \end{aligned} \tag{24}$$

This result is established immediately by applying Theorem 12.2.2 to the series (22) and (23) at each fixed x in the interval $|x| < \rho$. In some cases the resulting series may converge in a larger interval (see Problem 27).

EXAMPLE 7

Find a power series for

$$h(x) = \frac{5x + 4}{(x + 2)(x - 1)} \tag{25}$$

that is valid in some interval with center at the origin. Determine the interval of validity.

By expanding $h(x)$ in partial fractions we find that

$$h(x) = \frac{2}{x + 2} + \frac{3}{x - 1}. \tag{26}$$

From Example 2 we know that

$$\frac{2}{x + 2} = \sum_{k=0}^{\infty} \frac{(-1)^k x^k}{2^k}, \qquad |x| < 2, \tag{27}$$

and from Eq. 10 we know that

$$\frac{1}{1 - x} = \sum_{k=0}^{\infty} x^k, \qquad |x| < 1. \tag{28}$$

Thus, for $|x| < 1$ we can combine the series (27) and (28) to obtain

$$\begin{aligned} h(x) &= \sum_{k=0}^{\infty} \frac{(-1)^k x^k}{2^k} - 3 \sum_{k=0}^{\infty} x^k \\ &= \sum_{k=0}^{\infty} \left[\frac{(-1)^k}{2^k} - 3 \right] x^k, \qquad |x| < 1. \ \blacksquare \end{aligned} \tag{29}$$

Multiplication of series

It is also possible to multiply one power series by another within a common interval of convergence. The procedure is similar to that for products of polynomials: form all possible products of pairs of terms with one term from each series, and then collect terms having the same exponents. Thus, if f and g again have the power series (22) and (23), respectively, then

$$\begin{aligned} f(x)g(x) &= \left(\sum_{k=0}^{\infty} a_k x^k\right)\left(\sum_{k=0}^{\infty} b_k x^k\right) \\ &= (a_0 + a_1 x + a_2 x^2 + \cdots)(b_0 + b_1 x + b_2 x^2 + \cdots) \\ &= a_0 b_0 + (a_0 b_1 + a_1 b_0)x + (a_0 b_2 + a_1 b_1 + a_2 b_0)x^2 + \cdots \\ &\qquad + (a_0 b_k + a_1 b_{k-1} + \cdots + a_k b_0)x^k + \cdots \\ &= c_0 + c_1 x + c_2 x^2 + \cdots + c_k x^k + \cdots. \end{aligned} \tag{30}$$

The general coefficient in the product series is

$$c_k = \sum_{j=0}^{k} a_j b_{k-j} = a_0 b_k + a_1 b_{k-1} + \cdots + a_k b_0. \tag{31}$$

Observe that c_k is a sum of $k + 1$ terms; this sum includes all possible products of pairs of coefficients from the original two series whose indices add to k. Since the coefficients in the product series usually tend to become more complicated as k increases, it is often advisable to calculate only as many terms as are actually needed for the intended purpose.

As in the case of linear combinations, the product series converges at least for $|x| < \rho$, where $\rho = \min(\rho_1, \rho_2)$. In other words, the product series certainly converges in the interval of convergence common to both f and g, and may converge in a larger interval (see Problem 28).

EXAMPLE 8

Find a power series for $f(x) = e^x/(1 - x)$ about $x_0 = 0$.

To obtain a series for $f(x)$ we can multiply together the series that have been previously obtained for e^x and $1/(1 - x)$ (see Eqs. 18 and 10). The result is

$$\begin{aligned} \frac{e^x}{1-x} &= \left(1 + x + \frac{x^2}{2!} + \frac{x^3}{3!} + \cdots + \frac{x^k}{k!} + \cdots\right) \\ &\quad \times (1 + x + x^2 + x^3 + \cdots + x^k + \cdots) \\ &= 1 + (1 + 1)x + \left(1 + 1 + \frac{1}{2!}\right)x^2 + \left(1 + 1 + \frac{1}{2!} + \frac{1}{3!}\right)x^3 + \cdots \\ &\quad + \left(1 + 1 + \frac{1}{2!} + \frac{1}{3!} + \cdots + \frac{1}{k!}\right)x^k + \cdots \end{aligned} \tag{32}$$

and the latter series converges at least for $|x| < 1$. ∎

Summary

For easy reference we list here a few of the most important series expansions of elementary functions, ones that are often used in constructing expansions of other functions.

$$e^x = 1 + x + \frac{x^2}{2!} + \cdots + \frac{x^k}{k!} + \cdots = \sum_{k=0}^{\infty} \frac{x^k}{k!}, \qquad -\infty < x < \infty;$$

$$\sin x = x - \frac{x^3}{3!} + \frac{x^5}{5!} - \cdots + (-1)^k \frac{x^{2k+1}}{(2k+1)!} + \cdots$$

$$= \sum_{k=0}^{\infty} (-1)^k \frac{x^{2k+1}}{(2k+1)!}, \qquad -\infty < x < \infty;$$

$$\cos x = 1 - \frac{x^2}{2!} + \frac{x^4}{4!} - \cdots + (-1)^k \frac{x^{2k}}{(2k)!} + \cdots$$

$$= \sum_{k=0}^{\infty} (-1)^k \frac{x^{2k}}{(2k)!}, \qquad -\infty < x < \infty;$$

$$\frac{1}{1-x} = 1 + x + x^2 + \cdots + x^k + \cdots = \sum_{k=0}^{\infty} x^k, \qquad -1 < x < 1.$$

PROBLEMS

In each of Problems 1 through 20, obtain a power series representation for the given function about the given point. Use the methods of this section rather than Taylor's formula for the coefficients. In each case state where the series converges.

1. $f(x) = \cos x, \quad x_0 = \pi$

2. $f(x) = \sin x, \quad x_0 = \frac{\pi}{2}$

3. $f(x) = \cos x, \quad x_0 = \frac{\pi}{2}$

4. $f(x) = \frac{1}{2 - x}, \quad x_0 = 0$

5. $f(x) = \frac{1}{1 + 2x}, \quad x_0 = 0$

6. $f(x) = \frac{1}{1 + 2x}, \quad x_0 = 1$

7. $f(x) = \frac{3x + 4}{(1 + 2x)(2 - x)}, \quad x_0 = 0$

8. $f(x) = \frac{2x + 2}{(x - 1)(x + 3)}, \quad x_0 = 0$

9. $f(x) = \frac{x - 4}{(x - 1)(x - 2)}, \quad x_0 = 0$

10. $f(x) = \sin 3x, \quad x_0 = 0$

11. $f(x) = \cos \frac{x}{2}, \quad x_0 = 0$

12. $f(x) = e^{x/4}, \quad x_0 = 0$

13. $f(x) = \cos^2 x - \sin^2 x, \quad x_0 = 0$

14. $f(x) = e^{-x}, \quad x_0 = 0$

15. $f(x) = e^{-x^2}, \quad x_0 = 0$

16. $f(x) = \cosh x, \quad x_0 = 0$

17. $f(x) = \cosh 3x, \quad x_0 = 0$

18. $f(x) = \sinh \frac{x}{2}, \quad x_0 = 0$

19. $f(x) = \frac{1}{2 - x}, \quad x_0 = 3$

20. $f(x) = e^{x-2}, \quad x_0 = 2$

In each of Problems 21 through 24, use series multiplication to find the first four nonzero terms of a power series about $x_0 = 0$ for the given function.

21. $f(x) = \sin^2 x$

22. $f(x) = e^{-x} \cos x$

23. $f(x) = e^{-2x} \sin 3x$

24. $f(x) = \dfrac{\sin x}{1 - x}$

In each of Problems 25 and 26, assume that $f(x) = \sum_{k=0}^{\infty} a_k x^k$ for $|x| < \rho$, and find the first five terms of a power series about $x_0 = 0$ for the given function.

25. $[f(x)]^2$

26. $[f(x)]^3$

27. (a) From the geometric series (10) for $f(x) = 1/(1 - x)$, derive a power series about $x_0 = 0$ for $g(x) = x^2/(1 - x)$. Observe that the series for both f and g converge only for $|x| < 1$.

(b) Find a power series about $x_0 = 0$ for $h(x) = f(x) - g(x)$. Where does the series for h converge?

28. (a) Recall the series (10) for $f(x) = 1/(1 - x)$, which converges only for $|x| < 1$. Also determine a power series about $x_0 = 0$ for $g(x) = x(1 - x)$.

(b) By multiplication of the series in Part (a) determine a power series about $x_0 = 0$ for $h(x) = f(x)g(x)$. Where does the series for h converge?

13.5 DIFFERENTIATION AND INTEGRATION OF POWER SERIES

It is a remarkable and extremely useful fact that a convergent power series can be differentiated and integrated term by term within its interval of convergence. This means that any function with a convergent power series representation can be differentiated and integrated much as if it were a polynomial. This fact accounts in large part for the important place that power series have both in the theory and in the applications of mathematics.

Theorem 13.5.1

Suppose that

$$f(x) = \sum_{k=0}^{\infty} a_k x^k, \qquad |x| < \rho, \tag{1}$$

where $\rho > 0$. Then

(a) f is a differentiable function for $|x| < \rho$, and

$$f'(x) = \frac{d}{dx} \sum_{k=0}^{\infty} a_k x^k = \frac{d}{dx}\left[a_0 + \sum_{k=1}^{\infty} a_k x^k\right] = \sum_{k=1}^{\infty} \frac{d}{dx} a_k x^k$$

$$= \sum_{k=1}^{\infty} k a_k x^{k-1}, \qquad |x| < \rho; \tag{2}$$

(b) f is integrable on any interval contained in $|x| < \rho$, and

$$\int_0^x f(t)\, dt = \int_0^x \left(\sum_{k=0}^{\infty} a_k t^k\right) dt = \sum_{k=0}^{\infty} \int_0^x a_k t^k\, dt$$

$$= \sum_{k=0}^{\infty} \frac{a_k}{k+1} x^{k+1}, \qquad |x| < \rho. \tag{3}$$

There are several observations about this theorem that deserve emphasis. In the first place, the series obtained through term by term differentiation or integration have exactly the same radius of convergence as the original series, although behavior at the endpoints of the interval of convergence may be different. Furthermore, these series converge to the derivative or integral of the sum of the original series. Since the result of differentiating or integrating Eq. 1 is another convergent power series, the theorem can be applied over and over again. This shows, for example, that f actually has derivatives of all orders, and that they can be calculated by differentiating the series (1) term by term an appropriate number of times. Finally, note that term by term differentiation or integration of a power series is no more difficult than applying the same operations to a polynomial. Thus the functions that have convergent power series expansions in some interval $|x| < \rho$ are particularly nice from an analytical point of view. They have been singled out for a great deal of study and have been given a special name, **analytic** functions.

The proof of Theorem 13.5.1 is usually given in more advanced books as a special case of a more general result that depends on some concepts we have not discussed. Thus we omit a proof of Theorem 13.5.1. The following examples illustrate how it can be used.

EXAMPLE 1

Find a power series for $f(x) = (1 - x)^{-2}$ about $x_0 = 0$, and determine its radius of convergence.

Observe that

$$\frac{1}{(1 - x)^2} = \frac{d}{dx}\frac{1}{1 - x} \tag{4}$$

and recall that (Example 5 of Section 13.2)

$$\frac{1}{1 - x} = \sum_{k=0}^{\infty} x^k = 1 + x + x^2 + \cdots, \qquad |x| < 1. \tag{5}$$

Thus the required power series for $f(x)$ can be obtained by differentiating the series (5) term by term; the result is

$$\frac{1}{(1 - x)^2} = \sum_{k=1}^{\infty} kx^{k-1} = 1 + 2x + 3x^2 + \cdots, \qquad |x| < 1. \tag{6}$$

The term-by-term differentiation is justified by Theorem 13.5.1, which also asserts that the radius of convergence of the resulting series (6) is the same as that of the series (5). ∎

EXAMPLE 2

Find a power series for $f(x) = \ln(1 + x)$ about $x_0 = 0$, and determine its radius of convergence.

We observe that

$$\ln(1 + x) = \int_0^x \frac{dt}{1 + t}, \tag{7}$$

and that the series for $(1 + t)^{-1}$ can be obtained from the series (5) for $(1 - x)^{-1}$ by the scaling substitution $t = -x$. Thus

$$\frac{1}{1 + t} = \sum_{k=0}^{\infty} (-1)^k t^k = 1 - t + t^2 - t^3 + \cdots, \qquad |t| < 1. \tag{8}$$

Substituting the series (8) into Eq. 7 and integrating term by term, we obtain

$$\begin{aligned} \ln(1 + x) &= \int_0^x \left(\sum_{k=0}^{\infty} (-1)^k t^k \right) dt = \sum_{k=0}^{\infty} (-1)^k \int_0^x t^k \, dt \\ &= \sum_{k=0}^{\infty} (-1)^k \frac{x^{k+1}}{k + 1} \\ &= x - \frac{x^2}{2} + \frac{x^3}{3} - \frac{x^4}{4} + \cdots, \qquad |x| < 1. \end{aligned} \tag{9}$$

By Theorem 13.5.1 the radius of convergence of the series (9) is the same as that of the original series (8). This can also be shown by applying the ratio test directly to the series (9).

It is worth noting that at $x = 1$ the series (9) becomes the alternating harmonic series, which we know to converge. If it is legitimate to set $x = 1$ in Eq. 9, then we obtain

$$\ln 2 = 1 - \frac{1}{2} + \frac{1}{3} - \frac{1}{4} + \cdots + \frac{(-1)^k}{k + 1} + \cdots, \tag{10}$$

a result that was mentioned in Section 12.6. However, the validity of setting $x = 1$ in Eq. 9 is by no means obvious, since Eq. 9 was obtained from Eq. 8 by integrating from 0 to x, and $x = 1$ is not within the interval of convergence of the series (8). In fact, Eq. 10 follows from a property of power series known as Abel's theorem, which is discussed in more advanced books. We will simply accept Eq. 10, without pursuing its justification any further. ■

EXAMPLE 3

Find a power series for $f(x) = \arctan x$ about $x_0 = 0$, and determine its radius of convergence.

It is helpful to note that

$$\arctan x = \int_0^x \frac{dt}{1 + t^2}, \tag{11}$$

and that a power series for $(1 + t^2)^{-1}$ can be obtained from Eq. 5 by setting $x = -t^2$. This substitution is a combination of a power substitution and a scaling.

We have

$$\frac{1}{1+t^2} = \sum_{k=0}^{\infty} (-1)^k t^{2k} = 1 - t^2 + t^4 - t^6 + \cdots, \qquad |t| < 1. \tag{12}$$

Finally, we integrate this series term by term, and thereby obtain

$$\begin{aligned} \arctan x &= \sum_{k=0}^{\infty} (-1)^k \int_0^x t^{2k}\, dt \\ &= \sum_{k=0}^{\infty} \frac{(-1)^k x^{2k+1}}{2k+1} \\ &= x - \frac{x^3}{3} + \frac{x^5}{5} - \frac{x^7}{7} + \cdots, \qquad |x| < 1. \end{aligned} \tag{13}$$

The series (13) also converges at $x = \pm 1$ by the alternating series test. As in Example 2, it is possible to justify setting $x = \pm 1$ in Eq. 13 by invoking Abel's theorem. For $x = 1$ the result is

$$\arctan 1 = \frac{\pi}{4} = 1 - \frac{1}{3} + \frac{1}{5} - \frac{1}{7} + \cdots, \tag{14}$$

which is known as Leibniz's series. The series (14) was one of the first infinite series encountered by mathematicians. In principle, it provides a means of calculating π to any degree of accuracy, but in actuality it converges too slowly to be of practical computational use (see Problems 17 and 18). ∎

EXAMPLE 4

Let

$$f(x) = \begin{cases} \dfrac{\sin x}{x}, & x \neq 0; \\ 1, & x = 0. \end{cases} \tag{15}$$

Evaluate

$$I = \int_0^1 f(x)\, dx \tag{16}$$

with an error no greater than $0.000001 = 10^{-6}$.

We cannot use the fundamental theorem of calculus to evaluate the integral I, because no antiderivative of the integrand is available to us. However, if we can find a power series for $f(x)$, we can use Theorem 13.5.1 to calculate I with the required accuracy. From Example 4 of Section 13.2 we have

$$\sin x = \sum_{k=0}^{\infty} \frac{(-1)^k x^{2k+1}}{(2k+1)!} = x - \frac{x^3}{3!} + \frac{x^5}{5!} - \cdots \tag{17}$$

and the series (17) converges for all x. If $x \neq 0$, we can divide Eq. 17 by x to obtain

$$\frac{\sin x}{x} = \sum_{k=0}^{\infty} \frac{(-1)^k x^{2k}}{(2k+1)!} = 1 - \frac{x^2}{3!} + \frac{x^4}{5!} - \cdots, \qquad x \neq 0. \tag{18}$$

However, the series on the right side of Eq. 18 not only converges for all x but also has the value 1 when $x = 0$. Thus it represents the function f given by Eq. 15 for all x. To calculate I we integrate Eq. 18 term by term; this gives

$$\begin{aligned} I &= \int_0^1 \left(\sum_{k=0}^{\infty} \frac{(-1)^k x^{2k}}{(2k+1)!} \right) dx \\ &= \sum_{k=0}^{\infty} \frac{(-1)^k}{(2k+1)!} \int_0^1 x^{2k}\, dx \\ &= \sum_{k=0}^{\infty} \frac{(-1)^k}{(2k+1)!(2k+1)} \\ &= 1 - \frac{1}{3!3} + \frac{1}{5!5} - \frac{1}{7!7} + \cdots. \end{aligned} \tag{19}$$

The alternating series test (Theorem 12.6.1) states that the error incurred by truncating the series (19) is less than the magnitude of the first neglected term. Thus we seek k such that

$$\frac{1}{(2k+1)!(2k+1)} < \frac{1}{10^6}. \tag{20}$$

The smallest value of k that satisfies the inequality (20) is $k = 4$. Thus we need use only the terms corresponding to $k = 0, 1, 2,$ and 3 in Eq. 19 in order to evaluate I with the required accuracy. The result is

$$I \cong 0.946083 \tag{21}$$

with an error of no more than one in the last decimal place. ∎

Relation between power series and Taylor series

We now take up the question of how Taylor series are related to power series. This question was first raised at the beginning of Section 13.4 in connection with the examples in that section. It is clear that every Taylor series

$$\sum_{k=0}^{\infty} \frac{f^{(k)}(x_0)(x - x_0)^k}{k!} \tag{22}$$

has the form of a power series about $x = x_0$. Thus the question is whether a convergent power series must be the Taylor series for the function to which it converges. In other words, if

$$f(x) = \sum_{k=0}^{\infty} a_k (x - x_0)^k, \qquad |x - x_0| < \rho, \tag{23}$$

is it necessarily true that the coefficients a_k are given by Taylor's formula

$$a_k = \frac{f^{(k)}(x_0)}{k!}, \qquad k = 0, 1, 2, \ldots ? \tag{24}$$

To answer this question we will use Theorem 13.5.1 repeatedly to evaluate the coefficients in Eq. 23. First we set $x = x_0$ in Eq. 23, which gives

$$f(x_0) = a_0.$$

Then by differentiating Eq. 23 and setting $x = x_0$, we obtain

$$f'(x_0) = a_1.$$

Applying the process a second time yields

$$\frac{f''(x_0)}{2!} = a_2.$$

In fact, Theorem 13.5.1 justifies repeated differentiation of Eq. 23. By setting $x = x_0$ after each differentiation we obtain Eq. 24. Thus the coefficients in Eq. 23 are precisely the Taylor coefficients for f at the point $x = x_0$, and hence the series (23) is the Taylor series for its sum $f(x)$ about x_0. We conclude that every convergent power series, whether obtained by substitutions, differentiation, integration, or otherwise, is the Taylor series for the function to which it converges. This argument applies to all of the examples in this section and the preceding one.

An important special case of this result occurs if the sum $f(x)$ is zero for all values of x in some open interval containing x_0. Since the coefficients in the series (23) must be given by Eq. 24, it follows that each coefficient is zero. In other words, if

$$\sum_{k=0}^{\infty} a_k(x - x_0)^k = 0$$

for each x in $|x - x_0| < \rho$, then $a_k = 0$ for $k = 0, 1, 2, \ldots$. Alternatively, if

$$\sum_{k=0}^{\infty} a_k(x - x_0)^k = \sum_{k=0}^{\infty} b_k(x - x_0)^k$$

for each x in $|x - x_0| < \rho$, then $a_k = b_k$ for $k = 0, 1, 2, \ldots$.

PROBLEMS

In each of Problems 1 through 16, obtain a power series for the given function about the given point. Use the methods of this and the preceding section rather than Taylor's formula for the coefficients. In each case state where the series converges.

1. $f(x) = \ln(1 - x), \quad x_0 = 0$
2. $f(x) = \ln(1 - 2x), \quad x_0 = 0$
3. $f(x) = \ln\dfrac{1 + x}{1 - x}, \quad x_0 = 0$
4. $f(x) = \ln(1 + x), \quad x_0 = 1$
5. $f(x) = \displaystyle\int_0^x e^{-t^2}\,dt, \quad x_0 = 0$
6. $f(x) = \dfrac{1}{(1 + 2x)^2}, \quad x_0 = 0$

7. $f(x) = \dfrac{1}{(1 + 2x)^3}, \quad x_0 = 0$

8. $f(x) = \dfrac{2x}{(1 + x^2)^2}, \quad x_0 = 0$

9. $f(x) = \arctan \dfrac{x}{2}, \quad x_0 = 0$

10. $f(x) = \displaystyle\int_0^x \arctan t\, dt, \quad x_0 = 0$

11. $f(x) = \displaystyle\int_0^x \ln(1 + t^2)\, dt, \quad x_0 = 0$

12. $f(x) = \begin{cases} \dfrac{e^x - 1}{x}, & x \neq 0; \\ 1, & x = 0; \end{cases} \quad x_0 = 0$

13. $f(x) = \begin{cases} \dfrac{1 - \cos x}{x^2}, & x \neq 0 \\ \dfrac{1}{2}, & x = 0; \end{cases} \quad x_0 = 0$

14. $f(x) = \displaystyle\int_0^x \frac{\sinh t}{t}\, dt, \quad x_0 = 0$

15. $f(x) = \displaystyle\int_0^{x/2} \frac{\ln(1 + t)}{t}\, dt, \quad x_0 = 0$

16. $f(x) = \displaystyle\int_0^x \frac{1 - \cos t}{t^2}\, dt, \quad x_0 = 0$

© 17. (a) Use eight terms in Leibniz's series, Eq. 14, to calculate an approximate value for π.

(b) Use Theorem 12.6.1 to estimate how many terms in Eq. 14 must be retained to obtain a value of π with an error not exceeding 10^{-2}.

© 18. (a) Use four terms of the series (13) to calculate an approximate value for $\arctan \frac{1}{5}$. Use Theorem 12.6.1 to find a bound for the error E.

(b) Let $\theta = \arctan \frac{1}{5}$. Use the identity

$$\tan(A + B) = \frac{\tan A + B}{1 - \tan A \tan B} \qquad (i)$$

to show that $\tan 2\theta = \frac{5}{12}$, $\tan 4\theta = \frac{120}{119}$, and that

$$\tan\left(\frac{\pi}{4} - 4\theta\right) = -\frac{1}{239}. \qquad (ii)$$

(c) From Eq. (*ii*) show that

$$\frac{\pi}{4} = 4\arctan\frac{1}{5} - \arctan\frac{1}{239}. \qquad (iii)$$

(d) Use one term of the series (13) to calculate an approximate value for $\arctan \frac{1}{239}$. Use Theorem 12.6.1 to find a bound for the error $|E|$.

(e) Combine the results of Parts (a), (c), and (d) to obtain an approximate value for π. Find a bound for the error $|E|$.

In each of Problems 19 through 23, find a series representation for the given integral. Determine the number of terms required to approximate the integral with an error E not exceeding the magnitude specified. Calculate an approximate value of the integral using that number of terms.

© 19. $I = \displaystyle\int_0^1 \frac{1 - \cos x}{x^2}\, dx; \quad |E| \leq 10^{-8}$

© 20. $I = \displaystyle\int_0^{\pi/2} \frac{1 - \cos x}{x^2}\, dx; \quad |E| \leq 10^{-6}$

© 21. $I = \displaystyle\int_0^1 e^{-x^2}\, dx; \quad |E| \leq 10^{-5}$

© 22. $I = \displaystyle\int_0^1 \sin(x^2)\, dx; \quad |E| \leq 10^{-8}$

© 23. $I = \displaystyle\int_0^{1/2} \frac{\ln(1 + x)}{x}\, dx; \quad |E| \leq 10^{-4}$

24. Find the value at $x = 0$ of the thirtieth derivative of $\sin(x^2)$.

25. The function

$$J_0(x) = 1 + \sum_{n=1}^{\infty} \frac{(-1)^n x^{2n}}{2^{2n}(n!)^2}$$

is called the Bessel function of the first kind of order zero. Determine the radius of convergence of the series for J_0. Show that $J_0(x)$ satisfies the differential equation

$$xJ_0'' + J_0' + xJ_0 = 0.$$

Bessel functions are important because this equation, and others related to it, occur often in mathematical physics, for example, in problems involving vibrations or heat conduction in a circular cylinder.

* 26. Let us consider the polynomial equation

$$r^3 + r - (2 + \epsilon) = 0 \qquad (i)$$

where $|\epsilon|$ is small. When $\epsilon = 0$, the equation reduces to $r^3 + r - 2 = 0$; in this event the roots are $r = 1$ and $r = (-1 \pm \sqrt{7}\, i)/2$. Suppose that we wish to examine how the real root near $r = 1$ depends on ϵ when ϵ is small but not zero. In this problem we indicate a useful approach to questions of this kind.

(a) Assume that r can be represented by a power series in powers of ϵ, that is,

$$r = a_0 + a_1\epsilon + a_2\epsilon^2 + \cdots, \qquad (ii)$$

where the coefficients $a_0, a_1, a_2, \ldots$ are to be determined so as to satisfy Eq. (*i*). Show that

$$r^3 = a_0^3 + 3a_0^2a_1\epsilon + (3a_0^2a_2 + 3a_0a_1^2)\epsilon^2 + \cdots. \qquad (iii)$$

(b) Substitute from Eqs. (*ii*) and (*iii*) in Eq. (*i*) and collect coefficients of like powers of ϵ. Show that Eq. (*i*) is satisfied if and only if

$$(a_0^3 + a_0 - 2) + (3a_0^2a_1 + a_1 - 1)\epsilon + (3a_0^2a_2 + 3a_0a_1^2 + a_2)\epsilon^2 + \cdots = 0. \qquad (iv)$$

(c) For Eq. (*iv*) to hold for all sufficiently small ϵ it is necessary that the coefficient of each power of ϵ be zero. Why? Thus

$$a_0^3 + a_0 - 2 = 0, \qquad 3a_0^2a_1 + a_1 - 1 = 0,$$
$$3a_0^2a_2 + 3a_0a_1^2 + a_2 = 0, \cdots. \qquad (v)$$

Since we are investigating the root near $x = 1$, choose $a_0 = 1$ and then solve for a_1 and a_2. Note that more coefficients could be found by keeping more terms in Eqs. (*ii*), (*iii*), and (*iv*).

* **27.** In this problem we show how to construct a power series that represents the function satisfying the relation

$$f'(x) - xf(x) - 1 = 0 \qquad (i)$$

with $f(0) = 1$.

(a) Assume that, for $|x| < \rho$,

$$f(x) = \sum_{k=0}^{\infty} a_kx^k = a_0 + a_1x + a_2x^2 + \cdots + a_kx^k + \cdots. \qquad (ii)$$

Show that $a_0 = 1$, that

$$f'(x) = a_1 + 2a_2x + \cdots + (k+1)a_{k+1}x^k + \cdots = \sum_{k=0}^{\infty}(k+1)a_{k+1}x^k, \qquad (iii)$$

and that

$$xf(x) = x + a_1x^2 + \cdots + a_{k-1}x^k + \cdots = x + \sum_{k=2}^{\infty} a_{k-1}x^k. \qquad (iv)$$

(b) Substitute from Eqs. (*iii*) and (*iv*) in Eq. (*i*) and collect coefficients of like powers of x. Show that Eq. (*i*) is satisfied if and only if

$$(a_1 - 1) + (2a_2 - 1)x + (3a_3 - a_1)x^2 + \cdots + [(k+1)a_{k+1} - a_{k-1}]x^k + \cdots = 0. \qquad (v)$$

(c) For Eq. (*v*) to hold for all x in $|x| < \rho$ it is necessary that the coefficient of each power of x be zero. Thus $a_1 - 1 = 0$, $2a_2 - 1 = 0$, $3a_3 - a_1 = 0$, and in general

$$(k+1)a_{k+1} - a_{k-1} = 0, \qquad k = 2, 3, \ldots. \qquad (vi)$$

This set of equations is known as a **recurrence relation** since it expresses each coefficient in terms of a preceding one. Solve the recurrence relation and thereby determine the coefficients in the series (*ii*).

(d) Find the radius of convergence for the series (*ii*).
Hint: Consider the series (*ii*) as the sum of two series consisting of the odd- and even-powered terms, respectively.

13.6 THE BINOMIAL AND SOME OTHER SERIES

In elementary algebra one learns the binomial expansion of $(1 + x)^n$, where n is a positive integer:

$$(1 + x)^n = 1 + nx + \frac{n(n-1)}{2}x^2 + \cdots + \frac{n(n-1)\cdots(n-k+1)}{k!}x^k + \cdots + x^n. \qquad (1)$$

We now want to generalize this expansion to the case in which the exponent n

need not be a positive integer. This will lead us to the very important infinite series known as the binomial series.

Consider the function

$$f(x) = (1 + x)^\alpha, \tag{2}$$

where α is an arbitrary real number. We will construct the Taylor series for $f(x)$ about the origin. To do this we calculate the Taylor coefficients a_k from the formula

$$a_k = \frac{f^{(k)}(0)}{k!}, \qquad k = 0, 1, 2, \ldots. \tag{3}$$

The successive derivatives of f are given by

$$f'(x) = \alpha(1 + x)^{\alpha-1}, \qquad f''(x) = \alpha(\alpha - 1)(1 + x)^{\alpha-2}, \tag{4}$$

and in general by

$$f^{(k)}(x) = \alpha(\alpha - 1) \cdots (\alpha - k + 1)(1 + x)^{\alpha-k}; \qquad k = 1, 2, \ldots. \tag{5}$$

Thus from Eq. 3 we have

$$a_0 = 1, \quad a_1 = \alpha, \quad a_2 = \frac{\alpha(\alpha - 1)}{2!}, \ldots. \tag{6}$$

The general formula for the coefficients is

$$a_k = \frac{\alpha(\alpha - 1) \cdots (\alpha - k + 1)}{k!}, \qquad k = 1, 2, \ldots. \tag{7}$$

Consequently, by using Theorem 13.2.1 (Taylor's theorem) and its corollary, we can write

$$(1 + x)^\alpha = 1 + \alpha x + \frac{\alpha(\alpha - 1)}{2!} x^2 + \cdots + \frac{\alpha(\alpha - 1) \cdots (\alpha - n + 1)}{n!} x^n + R_{n+1}(x), \tag{8}$$

where $R_{n+1}(x)$ is given either by Eq. 8 or by Eq. 9 of Section 13.2. If we can show that

$$\lim_{n\to\infty} R_{n+1}(x) = 0 \tag{9}$$

for some interval $|x| < \rho$, then for that interval it follows from Eq. 8 that

$$\begin{aligned}(1 + x)^\alpha &= 1 + \alpha x + \frac{\alpha(\alpha - 1)}{2!} x^2 + \cdots + \frac{\alpha(\alpha - 1) \cdots (\alpha - k + 1)}{k!} x^k + \cdots \\ &= 1 + \sum_{k=1}^{\infty} \frac{\alpha(\alpha - 1) \cdots (\alpha - k + 1)}{k!} x^k \\ &= \sum_{k=0}^{\infty} \binom{\alpha}{k} x^k, \qquad |x| < \rho.\end{aligned} \tag{10}$$

The series (10) is known as the **binomial series,** and its coefficients are called **binomial coefficients.** In the last line of Eq. 10 we have introduced the conventional notation

$$\binom{\alpha}{0} = 1; \quad \binom{\alpha}{k} = \frac{\alpha(\alpha - 1) \cdots (\alpha - k + 1)}{k!}, \qquad k \geq 1 \tag{11}$$

for the binomial coefficients. The upper number in the expression $\binom{\alpha}{k}$ is the exponent of $(1 + x)$, while the lower number indexes the individual terms in the series.

Observe that if α is a nonnegative integer n, then the series (10) terminates with the term for which $k = n$. Thus the infinite series reduces to a polynomial of degree n, which is identical to the polynomial in Eq. 1. In this case the question of convergence of the series (10) does not arise.

Our main interest is in the case in which α is not a nonnegative integer, in which case the series does not terminate, but is a true infinite series. One way to establish that the series (10) converges to $(1 + x)^{\alpha}$ for some interval $|x| < \rho$ is to verify that Eq. 9 is true in that interval. This turns out to be fairly difficult; some indication of how one might proceed is given in Problem 15. An alternate approach, more in the spirit of Section 13.5, is outlined in Problem 16. We content ourselves here with a determination of the radius of convergence of the binomial series (10). This can be done by applying the ratio test, which gives

$$\begin{aligned}
\left|\frac{\binom{\alpha}{k+1}x^{k+1}}{\binom{\alpha}{k}x^k}\right| &= \left|\frac{\alpha(\alpha - 1) \cdots (\alpha - k + 1)(\alpha - k)x^{k+1}}{(k + 1)!}\right. \\
&\qquad \times \left.\frac{k!}{\alpha(\alpha - 1) \cdots (\alpha - k + 1)x^k}\right| \\
&= \left|\frac{\alpha - k}{k + 1}x\right| \\
&= \frac{k - \alpha}{k + 1}|x|, \qquad \text{for } k > \alpha \\
&\to |x|, \quad \text{as} \quad k \to \infty.
\end{aligned} \tag{12}$$

Thus the series (11) has the radius of convergence $\rho = 1$; the series converges absolutely for $|x| < 1$ and diverges for $|x| > 1$. Its behavior for $x = \pm 1$ is more difficult to investigate and is omitted from our discussion.

EXAMPLE 1

Find the Taylor series for $f(x) = (1 + x)^{-1/2}$ about $x_0 = 0$.

The required series is the binomial series (10) with $\alpha = -\frac{1}{2}$. Thus we have

$$\begin{aligned}
\frac{1}{\sqrt{1 + x}} &= \sum_{k=0}^{\infty} \binom{-\frac{1}{2}}{k} x^k \\
&= 1 + \sum_{k=1}^{\infty} \frac{(-\frac{1}{2})(-\frac{1}{2} - 1) \cdots (-\frac{1}{2} - k + 1)}{k!} x^k, \qquad |x| < 1.
\end{aligned} \tag{13}$$

The coefficients in this series can be written somewhat more simply as follows:

$$\frac{(-\frac{1}{2})(-\frac{1}{2}-1)\cdots(-\frac{1}{2}-k+1)}{k!} = \frac{(-1)^k(\frac{1}{2})(\frac{1}{2}+1)\cdots(\frac{1}{2}+k-1)}{k!}$$

$$= \frac{(-1)^k}{k!}\frac{1}{2}\cdot\frac{3}{2}\cdot\frac{5}{2}\cdots\frac{2k-1}{2}$$

$$= \frac{(-1)^k\, 1\cdot 3\cdot 5\cdots(2k-1)}{2^k k!}. \tag{14}$$

Using Eq. 14 we can rewrite the series (13) in the form

$$\frac{1}{\sqrt{1+x}} = 1 + \sum_{k=1}^{\infty}\frac{(-1)^k\, 1\cdot 3\cdot 5\cdots(2k-1)}{2^k k!}x^k$$

$$= 1 - \frac{1}{2}x + \frac{3}{8}x^2 - \frac{5}{16}x^3 + \cdots, \qquad |x| < 1. \tag{15}$$ ∎

EXAMPLE 2

Find the Taylor series for $f(x) = (1 - x^2)^{-1/2}$ about $x_0 = 0$.

The desired series is obtained by replacing x by $-x^2$ in Eq. 15; such substitutions were discussed in Section 13.4. We obtain

$$\frac{1}{\sqrt{1-x^2}} = 1 + \sum_{k=1}^{\infty}\frac{(-1)^k\, 1\cdot 3\cdot 5\cdots(2k-1)}{2^k k!}(-x^2)^k$$

$$= 1 + \sum_{k=1}^{\infty}\frac{1\cdot 3\cdot 5\cdots(2k-1)}{2^k k!}x^{2k}$$

$$= 1 + \frac{1}{2}x^2 + \frac{3}{8}x^4 + \frac{5}{16}x^6 + \cdots, \qquad |x| < 1. \tag{16}$$ ∎

EXAMPLE 3

Find the Taylor series for $f(x) = \arcsin x$ about $x_0 = 0$.

Recall that

$$\arcsin x = \int_0^x \frac{dt}{\sqrt{1-t^2}}. \tag{17}$$

By using the series (16) and integrating term by term according to Theorem 13.5.1

we find that

$$\arcsin x = \int_0^x \left[1 + \sum_{k=1}^{\infty} \frac{1 \cdot 3 \cdot 5 \cdots (2k-1)}{2^k k!} t^{2k}\right] dt$$

$$= \int_0^x dt + \sum_{k=1}^{\infty} \frac{1 \cdot 3 \cdot 5 \cdots (2k-1)}{2^k k!} \int_0^x t^{2k}\, dt$$

$$= x + \sum_{k=1}^{\infty} \frac{1 \cdot 3 \cdot 5 \cdots (2k-1)}{2^k k!} \frac{x^{2k+1}}{2k+1}$$

$$= x + \frac{1}{6}x^3 + \frac{3}{40}x^5 + \frac{5}{112}x^7 \cdots, \qquad |x| < 1. \blacksquare \tag{18}$$

PROBLEMS

In each of Problems 1 through 14, find the Taylor series for the given function about the given point. Find the radius of convergence of the series.

1. $f(x) = (1 + x)^{1/2}, \quad x_0 = 0$

2. $f(x) = (1 + x^2)^{-1/2}, \quad x_0 = 0$

3. $f(x) = (4 + x)^{-1/2}, \quad x_0 = 0$

4. $f(x) = (1 - 4x)^{-1/2}, \quad x_0 = 0$

5. $f(x) = \ln(\sqrt{1 + x^2} + x), \quad x_0 = 0$

Hint: Find $f'(x)$.

6. $f(x) = \operatorname{arcsinh} x, \quad x_0 = 0$

7. $f(x) = \arcsin \frac{x}{a}, \quad a > 0, \quad x_0 = 0$

8. $f(x) = \sqrt{1 + x} - \sqrt{1 - x}, \quad x_0 = 0$

9. $f(x) = \dfrac{1}{\sqrt{1 + x} + \sqrt{1 - x}}, \quad x_0 = 0$

10. $f(x) = \dfrac{1}{\sqrt{a^2 - x^2}}, \quad a > 0, \quad x_0 = 0$

11. $f(x) = \sqrt{3 + x}, \quad x_0 = 1$

12. $f(x) = (3 - x)^{-1/2}, \quad x_0 = 2$

13. $f(x) = (1 + x)^{2/3}, \quad x_0 = 0$

14. $f(x) = (8 - x)^{1/3}, \quad x_0 = 0$

Convergence of the Binomial Series

In Problems 15 and 16 we indicate two methods for establishing the convergence of the binomial series. Problem 15 involves a direct investigation of the remainder, while Problem 16 makes use of an indirect method.

* **15.** (a) Recall the integral form of the remainder in Taylor's theorem (Eq. 8 of Section 13.2) with $x_0 = 0$:

$$R_{n+1}(x) = \frac{1}{n!} \int_0^x (x - t)^n f^{(n+1)}(t)\, dt.$$

Let $f(t) = (1 + t)^\alpha$ and show that

$$R_{n+1}(x) = \frac{\alpha(\alpha - 1) \cdots (\alpha - n)}{n!} \times \int_0^x \frac{(x - t)^n}{(1 + t)^{n+1-\alpha}}\, dt.$$

(b) If $0 \le t \le x$ or $x \le t \le 0$ and if $|x| < 1$, show that

$$\frac{|x - t|}{1 + t} \le |x|.$$

Hint: Consider separately the cases $x \ge 0$ and $x \le 0$.

(c) Let $0 \le x < 1$. Use the result of Part (b) in the expression for $R_{n+1}(x)$ obtained in Part (a), and show that

$$|R_{n+1}(x)| \le \frac{|\alpha(\alpha - 1) \cdots (\alpha - n)|}{n!} x^n \times \int_0^x (1 + t)^{\alpha - 1}\, dt. \quad (i)$$

(d) By performing the integration indicated in Eq. (*i*), show that

$$|R_{n+1}(x)| \le \frac{|(\alpha - 1) \cdots (\alpha - n)|}{n!} \times x^n |(1 + x)^\alpha - 1|. \quad (ii)$$

(e) From Eq. (*ii*) deduce that $R_{n+1}(x) \to 0$ as $n \to \infty$ for $0 \le x < 1$.

(f) Repeat Parts (c), (d), and (e) for the case $-1 < x < 0$, and show that $R_{n+1}(x) \to 0$ as $n \to \infty$ in this case also. Thus the binomial series converges for $|x| < 1$.

* **16.** Let $f(x)$ be the sum of the binomial series

$$f(x) = \sum_{k=0}^{\infty} \binom{\alpha}{k} x^k, \qquad |x| < 1. \qquad (i)$$

We will outline a method for showing that $f(x) = (1 + x)^\alpha$.

(a) By differentiating Eq. (*i*), and shifting the index of summation, show that

$$f'(x) = \sum_{k=0}^{\infty} \binom{\alpha}{k+1} (k + 1) x^k, \qquad |x| < 1. \quad (ii)$$

(b) Show that

$$(1 + x) f'(x) = \sum_{k=0}^{\infty} \left[\binom{\alpha}{k+1} (k + 1) + \binom{\alpha}{k} k \right] x^k, \qquad |x| < 1. \quad (iii)$$

(c) Show that

$$\binom{\alpha}{k+1} (k + 1) + \binom{\alpha}{k} k = \binom{\alpha}{k} \alpha. \qquad (iv)$$

(d) Substitute from Eq. (*iv*) into Eq. (*iii*) and show that

$$\frac{f'(x)}{f(x)} = \frac{\alpha}{1 + x}. \qquad (v)$$

(e) Solve Eq. (*v*) for $f(x)$. Note that $f(0) = 1$ from Eq. (*i*) and use this condition to evaluate the constant of integration.

* **17. Arc length of an ellipse.** This problem extends the results of Problems 15 and 16 in Section 7.3. Consider the ellipse given parametrically by the equations

$$x = a \sin t, \quad y = b \cos t, \quad 0 \le t \le 2\pi \qquad (i)$$

where $a \ge b > 0$.

(a) Show that the circumference of the ellipse is given by the integral

$$s = 4a \int_0^{\pi/2} (1 - k^2 \sin^2 t)^{1/2}\, dt = 4aE(k) \qquad (ii)$$

where $k^2 = 1 - (b^2/a^2)$ is the square of the eccentricity of the ellipse, and $0 \le k^2 < 1$. The integral in Eq. (*ii*), denoted by $E(k)$, is called a complete **elliptic integral.** Note that when $k = 0$, the ellipse becomes a circle and $E(0) = \pi/2$.

(b) By using the binomial series for the integrand, show that

$$\begin{aligned} E(k) &= \int_0^{\pi/2} \left[1 - \frac{k^2 \sin^2 t}{2} - \sum_{n=2}^{\infty} \frac{1 \cdot 3 \cdot 5 \cdots (2n - 3)}{2^n n!} \times k^{2n} \sin^{2n} t \right] dt \\ &= \frac{\pi}{2} - \frac{k^2}{2} \int_0^{\pi/2} \sin^2 t\, dt - \sum_{n=2}^{\infty} \frac{1 \cdot 3 \cdot 5 \cdots (2n - 3)}{2^n n!} k^{2n} \times \int_0^{\pi/2} \sin^{2n} t\, dt. \end{aligned} \qquad (iii)$$

(c) Approximate $E(k)$ for small k by evaluating the first three terms of the series (*iii*), that is, by including terms up to k^4.

(d) Use the result of Problem 39 of Section 9.2 to show that

$$E(k) = \frac{\pi}{2} \left\{ 1 - \sum_{n=1}^{\infty} \left[\frac{(2n)!}{(2^n n!)^2} \right]^2 \frac{k^{2n}}{2n - 1} \right\}.$$

REVIEW PROBLEMS

In each of Problems 1 through 6, find the interval of convergence of the given power series. Determine whether the series converges absolutely, converges conditionally, or diverges at each endpoint.

1. $\sum_{k=0}^{\infty} \frac{x^k}{(2k+1)^2}$

2. $\sum_{k=0}^{\infty} \frac{x^{2k+1}}{(2k+1)^3}$

3. $\sum_{k=0}^{\infty} \frac{(t+1)^{2k}}{2^k}$

4. $\sum_{k=0}^{\infty} \frac{t^k}{(k+1)(2k+1)}$

5. $\sum_{k=0}^{\infty} \frac{(-1)^{k-1}(k-1)}{k^2+1} x^k$

6. $\sum_{k=1}^{\infty} \frac{(2k)!}{6^k(k!)^2} x^k$

In each of Problems 7 through 12, find the radius of convergence of the given power series.

7. $\sum_{k=0}^{\infty} \frac{k!\, x^k}{2^k}$

8. $\sum_{k=0}^{\infty} \frac{(k+1)!\, x^k}{(2k+1)!}$

9. $\sum_{k=0}^{\infty} \frac{2^k x^k}{\sqrt{k+1}}$

10. $\sum_{k=0}^{\infty} \frac{(-1)^k x^k}{1+\sqrt{k}}$

11. $\sum_{k=2}^{\infty} \frac{2^k x^k}{k(\ln k)^2}$

12. $\sum_{k=1}^{\infty} \sin\left(\frac{\pi}{k^2}\right) x^k$

In each of Problems 13 through 20, find the power series representation for the given function about x_0. State the interval of convergence of the power series.

13. $f(x) = \frac{2}{2+x^3}, \quad x_0 = 0$

14. $f(x) = \cos^2 \frac{x}{2}, \quad x_0 = 0$

15. $f(x) = \frac{1+x^2}{1-x^2}, \quad x_0 = 0$

16. $f(x) = \frac{4x}{(1-x^2)^2}, \quad x_0 = 0$

17. $f(x) = \ln(a+bx), \quad x_0 = 0$

18. $f(x) = \int_0^x \ln(1+t)\,dt, \quad x_0 = 0$

19. $f(x) = (1+x)^{1/3}, \quad x_0 = 0$

20. $f(x) = 1/\sqrt{4-x^2}, \quad x_0 = 0$

In each of Problems 21 through 24, find the first three terms of the power series expansion for the given function, $f(g(x))$, about x_0 by

(a) finding the appropriate Taylor polynomial;

(b) by substituting the series for $g(x)$ into the series for $f(x)$.

21. $f(x) = \sin\left(\frac{1}{1+x}\right), \quad x_0 = 0$

22. $f(x) = \frac{1}{1+\ln(x+1)}, \quad x_0 = 0$

23. $f(x) = e^{\sin x}, \quad x_0 = 0$

24. $f(x) = \sin(\sin x), \quad x_0 = 0$

In each of Problems 25 through 28, find the remainder associated with the Taylor polynomial, $P_n(x)$, of the given function about x_0.

25. $f(x) = (x-1)^{1/3}, \quad x_0 = 2, \quad n = 3$

26. $f(x) = \tanh x, \quad x_0 = 0, \quad n = 2$

27. $f(x) = \ln\left(\frac{1}{1+x}\right), \quad x_0 = 0, \quad n = 3$

28. $f(x) = e^{(\ln x)^2}, \quad x_0 = 1, \quad n = 1$

29. Find the first four nonzero terms in the Taylor expansion for $f(x) = \tanh x$ about $x = 0$. *Hint:* $\tanh x = (e^x - e^{-x})/(e^x + e^{-x})$.

30. Sum the series

$$\sum_{n=1}^{\infty} n^2 x^n, \qquad -1 < x < 1.$$

Hint: Differentiate $f(x) = 1/(1-x)$.

31. Find the first three terms in the power series for

$$f(x) = \frac{e^{x^2}}{1-x} \quad \text{about } x_0 = 0$$

by multiplying the series for $1/(1-x)$ and e^{x^2}.

32. Find the first four terms in the power series for

$$f(x) = \frac{\ln(x+1)}{e^x} \quad \text{about } x_0 = 0$$

by multiplying the series for e^{-x} and $\ln(x+1)$.

33. Prove that

$$\frac{\pi}{6} = \sum_{n=0}^{\infty} \frac{(-1)^n}{3^n\sqrt{3}\,(2n+1)}.$$

Hint: Consider the expansion of arctan x.

34. Is it true that

$$\frac{\pi}{3} = \sqrt{3} \sum_{n=0}^{\infty} \frac{(-1)^n 3^n}{2n+1}?$$

Hint: Consider the expansion of arctan x.

In each of Problems 35 through 38, use the method of Problem 27 in Section 13.5 to construct a solution of the given initial value problem.

35. $y' - 2xy = 0, \quad y(0) = y_0$

36. $y' - y = x, \quad y(0) = -1$

37. $y' - y - e^x = 0, \quad y(0) = 1$

38. $(x + 1)^2 y' + (x + 1) y = 1, \quad y(0) = 0.$

In each of Problems 39 through 42, use the method of Problem 26 of Section 13.5 to approximate the zero of the given polynomial that is near the given value. Keep terms up to order ϵ^2.

39. $x^3 + \epsilon x - 8 = 0; \quad x_0 = 2$

40. $x^2 - (5 + \epsilon)x + 6 + 3\epsilon = 0, \quad x_0 = 2$

41. $x^3 - \epsilon x^2 - 3x + 2 + 2\epsilon = 0, \quad x_0 = -2$

42. $x^2 + \epsilon x - (1 + 2\epsilon) = 0, \quad x_0 = 1$

In each of Problems 43 through 46, show that the given integral exists. Using a power series representation for the integrand, calculate an approximate value of the integral with an error not exceeding 10^{-6} in magnitude.

43. $\displaystyle\int_0^{0.1} \frac{\tan x}{x}\,dx$

44. $\displaystyle\int_0^{0.25} \frac{\arcsin x}{x}\,dx$

45. $\displaystyle\int_0^{0.1} \frac{\ln(\cos x)}{x}\,dx$

46. $\displaystyle\int_0^{0.1} \frac{2^x - 1}{x}\,dx$

FOURTEEN

polar coordinates

The subject of analytic geometry, which relates algebra and geometry, is based on the ability to identify geometrical points by means of numbers, and vice versa. This correspondence in turn depends on the establishment of a coordinate system. Up to now, we have made use only of rectangular (Cartesian) coordinates, which are the most common. However, there are also many other possible coordinate systems, and in some situations one of them may be more convenient than rectangular coordinates. In this chapter we discuss the polar coordinate system.

14.1 THE POLAR COORDINATE SYSTEM

To establish a polar* coordinate system we select a point O in the plane as the **pole,** and choose a ray or half-line emanating from O to be the **polar axis** (see Figure 14.1.1). A point P is identified by the pair of numbers (r, θ), where r is the distance from O to P, and θ is the angle that the line segment OP makes with the polar axis. It is conventional to consider counterclockwise angles as positive and clockwise angles as negative.

*The first use of polar coordinates was by Isaac Newton in *De Methodis Serierum et Fluxionum,* written about 1671 but not published until 1736. The first published account was by Jakob Bernoulli in 1691.

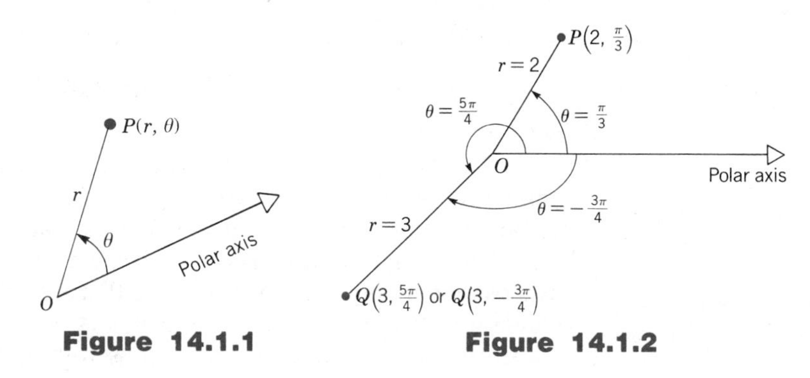

Figure 14.1.1 **Figure 14.1.2**

In Figure 14.1.2 we have plotted several points from their polar coordinates. The point $P(2, \pi/3)$ is two units from the pole along the ray making an angle of $\pi/3$ radians with the polar axis. The point $Q(3, 5\pi/4)$ lies three units from the pole and the line segment OQ makes an angle of $5\pi/4$ radians with the polar axis. The same point Q also has the coordinates $(3, -3\pi/4)$. The angle $-3\pi/4$ is measured in the negative (clockwise) direction from the polar axis to the line segment OQ.

It is also desirable to allow r, as well as θ, to assume a negative value. For example, in Figure 14.1.3 the point $Q(-3, \pi/4)$ is found by locating the ray that makes the angle $\pi/4$ radians with the polar axis, and then proceeding a distance of three units from the pole in the *opposite* direction. It is easy to see that this point is the same as the point Q in Figure 14.1.2.

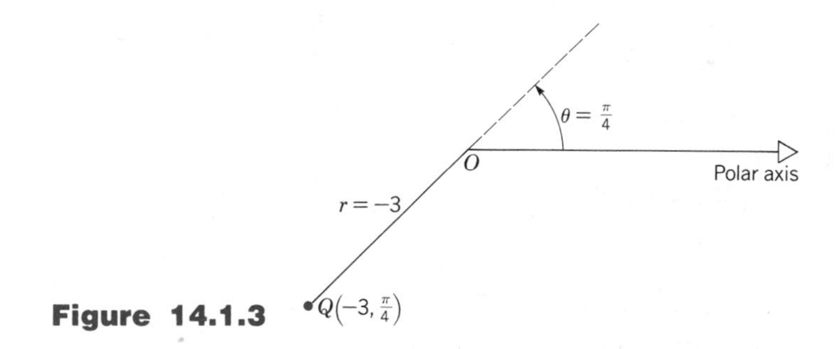

Figure 14.1.3

Normally, in order to locate a point from its polar coordinates (r, θ) it is advisable first to find the ray making the angle θ with the polar axis. Then mark off the distance $|r|$ along the ray if r is positive and in the opposite direction if r is negative; thus along each ray there is a negative as well as a positive direction.

From the foregoing discussion it should be clear that each pair of polar coordinates (r, θ) determines without ambiguity a definite point in the plane. However, each point in the plane has more than one set of polar coordinates (indeed, it has infinitely many). We have already seen that the point Q in Figures 14.1.2 and 14.1.3 has the three sets of polar coordinates $(3, 5\pi/4)$, $(3, -3\pi/4)$, and $(-3, \pi/4)$. Other sets include $(3, 5\pi/4 + 2\pi)$ and $(3, -3\pi/4 - 2\pi)$, where the 2π and -2π terms arise from making a complete circuit of the pole in the positive or negative direction, respectively. Since one can make more than one circuit in either direction, the following sets of polar coordinates all refer to the same geometrical point

$$(r, \theta), \qquad (-r, \theta \pm \pi), \qquad (r, \theta \pm 2n\pi), \qquad (-r, \theta \pm \pi \pm 2n\pi),$$

where n is a positive integer.

The fact that a geometrical point does not correspond to a uniquely determined set of polar coordinates is a significant difference between rectangular and polar coordinate systems. However, by suitably restricting r and θ it is possible to associate a unique set of polar coordinates with each point except the pole. This is most commonly done by requiring that r be positive and that θ lie in a specified interval of length 2π, such as $-\pi < \theta \leq \pi$. However, at the pole $r = 0$ but θ is undetermined, so nothing can be done to remove the ambiguity there.

Polar and rectangular coordinates

In many cases we will want to have both a polar coordinate system and a rectangular coordinate system for the same plane. It is customary to place the pole of the polar system at the origin of the rectangular system, and to choose the positive x-axis as the polar axis, as shown in Figure 14.1.4. Then the equations relating the polar coordinates (r, θ) of a point P and its rectangular coordinates (x, y) can be immediately written down. From Figure 14.1.4 and from the definitions of the sine

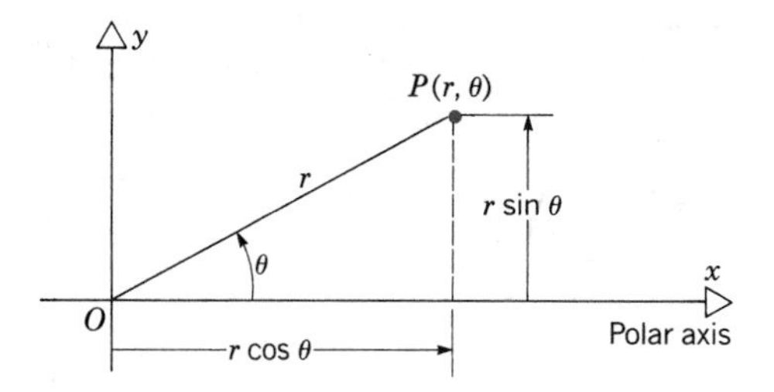

Figure 14.1.4 The relation between rectangular and polar coordinates.

and cosine functions it follows that, for points in the first quadrant,

$$x = r \cos \theta, \qquad y = r \sin \theta. \tag{1}$$

You should redraw Figure 14.1.4, so as to place the point P in each of the other three quadrants, and thereby convince yourself that Eqs. 1 are correct regardless of where P is located. Observe that all pairs of polar coordinates corresponding to the same point P yield the same values of x and y. From Eqs. 1 we readily obtain

$$r = \pm\sqrt{x^2 + y^2}, \qquad \tan \theta = \frac{y}{x}. \tag{2}$$

Equations 2 must be used with some care since they do not uniquely determine r and θ, and not all values of r and θ that satisfy Eqs. 2 are polar coordinates of the same point.

For example, let the point P have rectangular coordinates $x = 3/\sqrt{2}$ and $y = 3/\sqrt{2}$. Then, from Eqs. 2

$$r = \pm 3, \qquad \tan \theta = 1. \tag{3}$$

Since P is in the first quadrant, we can choose $r = 3$ and $\theta = \pi/4$ as its polar coordinates, and these are probably the best ones to use. Of course, $r = 3$ and $\theta = \pi/4 \pm 2n\pi$, where n is a positive integer, are also polar coordinates of the given point. Finally, other coordinates of the same point are $r = -3$ and $\theta = 5\pi/4 \pm 2n\pi$. It is important to observe that the location of P places restrictions on the solutions of Eq. 3. For instance, the values $r = 3$, $\theta = 5\pi/4$ or $r = -3$, $\theta = \pi/4$ satisfy Eq. 3, but are not polar coordinates of P; they correspond to the point three units from the origin along the extension of OP into the third quadrant.

Graphs of polar equations

A polar equation is an equation

$$F(r, \theta) = 0 \tag{4}$$

in which the polar coordinates r and θ appear as variables. Most of the equations that we will encounter will be in the simpler form

$$r = f(\theta). \tag{5}$$

A point is said to lie on the graph of a polar equation if *at least one of its sets of polar coordinates satisfies the equation.* Other sets of polar coordinates for the same point may not satisfy the equation. We now consider several examples of graphs of some simple polar equations.

EXAMPLE 1

Sketch the graph of the equation

$$\theta = \frac{2\pi}{3}. \tag{6}$$

The graph of this equation contains all points whose angular coordinate θ is $2\pi/3$, and whose radial coordinate r may be positive, negative, or zero. Thus the graph is the straight line through the origin shown in Figure 14.1.5.

Note that this straight line has many other polar equations, such as $\theta = 5\pi/3$, $\theta = -\pi/3$, and so forth. Also, for each point on the line, only one of its infinitely many sets of polar coordinates $(r, 2\pi/3 \pm 2n\pi)$, $(-r, -\pi/3 \pm 2n\pi)$ satisfies Eq. 6. ∎

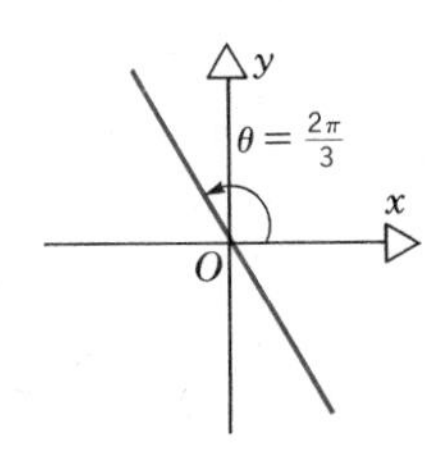

Figure 14.1.5

EXAMPLE 2

Sketch the graph of the equation

$$r = 4. \tag{7}$$

This graph contains all points whose radial coordinate r is 4, and thus is the circle with center at the pole and with radius 4 (see Figure 14.1.6). Observe that the same circle is also the graph of the equation $r = -4$. ∎

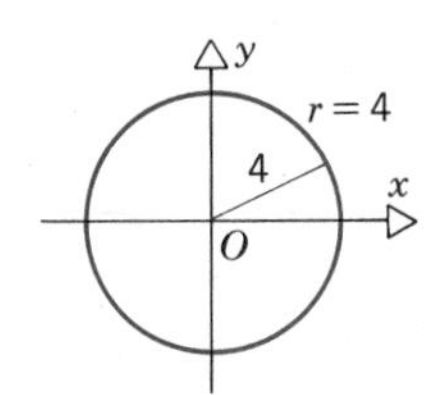

Figure 14.1.6

EXAMPLE 3

Sketch the graph of the equation

$$r = 4 \cos \theta. \tag{8}$$

We start by plotting some points that satisfy Eq. 8. By calculating the values of r from Eq. 8 for several convenient values of θ we obtain Table 14.1.

Table 14.1 Some Values of r and θ That Satisfy $r = 4 \cos \theta$

θ	0	$\frac{\pi}{6}$	$\frac{\pi}{4}$	$\frac{\pi}{3}$	$\frac{\pi}{2}$	$\frac{2\pi}{3}$	$\frac{3\pi}{4}$	$\frac{5\pi}{6}$	π
r	4	$2\sqrt{3}$	$2\sqrt{2}$	2	0	-2	$-2\sqrt{2}$	$-2\sqrt{3}$	-4

For values of θ between π and 2π we do not obtain any new points on the graph. For example, for $\theta = 7\pi/6$ we have $r = -2\sqrt{3}$, which yields the same point that we found for $\theta = \pi/6$. Thus the graph is merely traversed a second time for $\pi \leq \theta \leq 2\pi$.

In Figure 14.1.7 we have plotted the points given in Table 14.1 and joined them by a curve. The curve looks circular, but of course that cannot be established

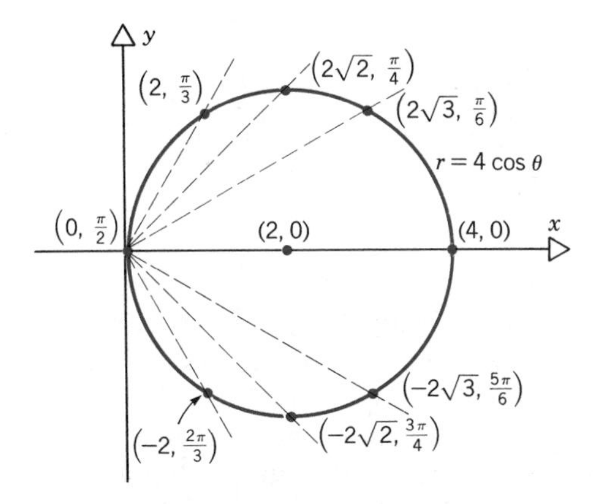

Figure 14.1.7

merely by plotting a few points. To make sure of the identification of the graph of Eq. 8 it is helpful to use Eqs. 1 and 2 to transform Eq. 8 into an equation in the rectangular coordinates x and y. By substituting x/r for $\cos \theta$, multiplying by r, and then replacing r^2 by $x^2 + y^2$ we obtain the equation

$$x^2 + y^2 = 4x.$$

Completing the square in x yields

$$x^2 - 4x + 4 + y^2 = 4$$

or

$$(x - 2)^2 + y^2 = 4. \tag{9}$$

We recognize Eq. 9 as corresponding to the circle with center at the point with rectangular coordinates (2, 0) and with radius 2. ∎

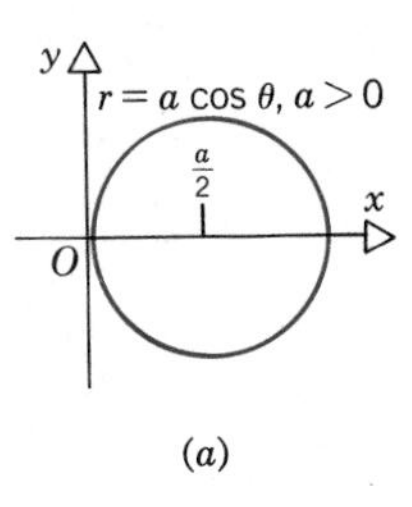

(a)

In the same way as in Example 3 one can show that an equation of the form

$$r = a\cos\theta, \tag{10}$$

where a is a constant, has for its graph the circle with center at the point with rectangular coordinates $(a/2, 0)$, and with radius $|a|/2$. The circle lies to the right of the y-axis if $a > 0$ and to the left if $a < 0$ (see Figure 14.1.8(a, b)).

Similarly, the equation

$$r = b\sin\theta \tag{11}$$

corresponds to the circle with center on the y-axis at the point with rectangular coordinates $(0, b/2)$ and with radius $|b|/2$. It lies above the x-axis if $b > 0$ and below it if $b < 0$ (see Figure 14.1.8(c, d)).

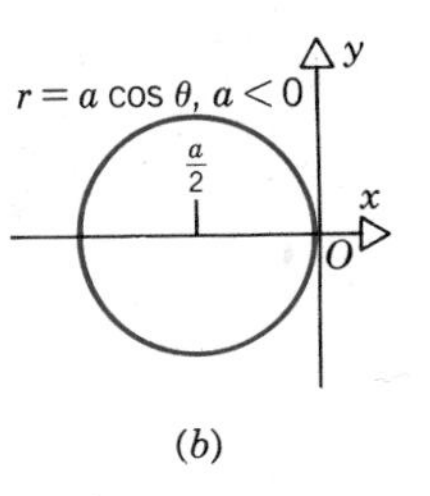

(b)

(c)

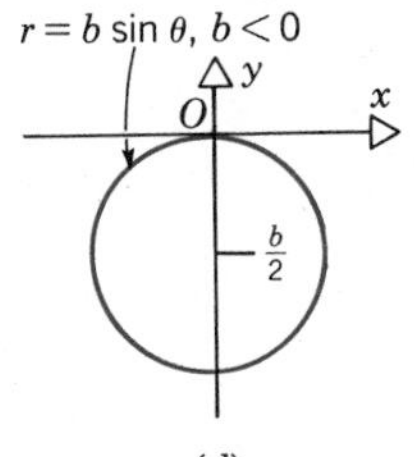

(d)

Figure 14.1.8

EXAMPLE 4

Sketch the graph of the equation

$$r = \theta. \tag{12}$$

Consider first $\theta \geq 0$. If $\theta = 0$, then $r = 0$, so the graph passes through the origin. Then r increases as θ does, so for positive θ, the graph is the spiral shown by the darker curve in Figure 14.1.9. For $\theta < 0$ the angle θ is measured in the

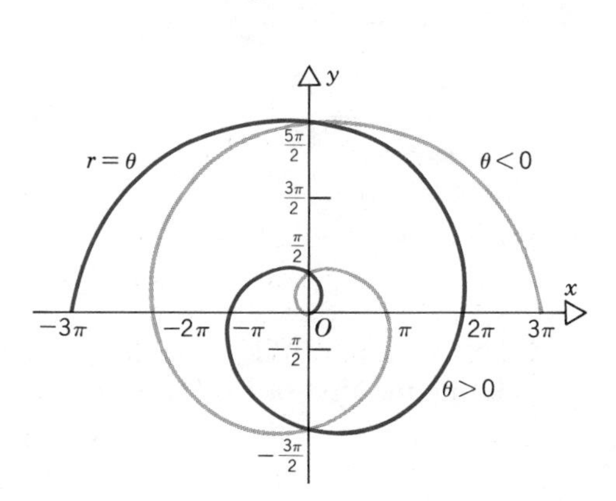

Figure 14.1.9

clockwise direction and r is marked off in the direction opposite to the ray corresponding to θ. Thus we obtain the spiral shown by the lighter curve in Figure 14.1.9. The graph of Eq. 12 consists of both the dark and light curves in Figure 14.1.9. The graph of

$$r = k\theta, \qquad k \neq 0$$

is similar, and is known as the *spiral of Archimedes.* ∎

Transformation from rectangular to polar coordinates, and vice versa

If the equation of a curve is known in rectangular coordinates, Eqs. 1 can be used to transform the equation into polar coordinates. The transformation can also be used in the opposite direction, as we have already seen in Example 3. As a rule, in order to investigate a particular curve, one should choose the coordinate system in which it has the simpler equation.

EXAMPLE 5

Find the polar equation of the curve whose equation in rectangular coordinates is

$$xy = 4. \tag{13}$$

Using Eqs. 1 we have

$$(r\cos\theta)(r\sin\theta) = 4,$$

from which it follows that

$$\tfrac{1}{2}r^2\sin 2\theta = 4,$$

or

$$r^2 = 8\csc 2\theta, \qquad \theta \neq \frac{n\pi}{2} \quad \text{for} \quad n = 0, \pm 1, \pm 2, \ldots \tag{14}$$

In this example the rectangular equation (13) is simpler than the polar equation (14). ∎

EXAMPLE 6

In Section 14.2 we discuss a curve, known as a cardioid, whose equation is

$$r = a(1 + \cos\theta), \qquad a > 0; \tag{15}$$

find the equation of this curve in rectangular coordinates.

Using the first of Eqs. 1 we have

$$r = a\left(1 + \frac{x}{r}\right);$$

then

$$r^2 = a(r + x)$$

or

$$r^2 - ax = ar.$$

Substituting for r from the first of Eqs. 2, we obtain

$$x^2 + y^2 - ax = \pm a\sqrt{x^2 + y^2},$$

and hence

$$(x^2 + y^2 - ax)^2 = a^2(x^2 + y^2). \tag{16}$$

In this case the polar equation (15) is simpler than the rectangular equation (16). ∎

Straight lines in polar coordinates

We have already discussed the equation in polar coordinates of a straight line through the origin. If a line does not contain the origin, its equation in polar coordinates can be found by transforming its rectangular equation.

For example, a line parallel to the x-axis has the equation $y = k$, which becomes $r \sin\theta = k$, or

$$r = k \csc\theta, \qquad \theta \neq n\pi \quad \text{for} \quad n = 0, \pm 1, \pm 2, \ldots .$$

Similarly, the equation $x = c$ of a line parallel to the y-axis is transformed into

$$r = c \sec\theta, \qquad \theta \neq \frac{(2n+1)\pi}{2} \quad \text{for} \quad n = 0, \pm 1, \pm 2, \ldots .$$

Finally, a line inclined to the x- and y-axes has a rectangular equation such as

$$3x - 4y = 7.$$

The corresponding polar equation is easily found by using Eqs. 1; thus

$$r(3\cos\theta - 4\sin\theta) = 7.$$

In all of these cases the rectangular equation is simpler than the polar equation, and thus is usually to be preferred.

PROBLEMS

In each of Problems 1 through 10, plot the point with the given polar coordinates.

1. $\left(4, \frac{\pi}{6}\right)$
2. $(3, \pi)$
3. $(-3, \pi)$
4. $\left(4, \frac{5\pi}{6}\right)$
5. $\left(-4, \frac{7\pi}{6}\right)$
6. $\left(2, -\frac{\pi}{2}\right)$
7. $\left(-2, \frac{\pi}{2}\right)$
8. $\left(-2, -\frac{\pi}{2}\right)$
9. $\left(-3, -\frac{3\pi}{4}\right)$
10. $\left(3, -\frac{3\pi}{4}\right)$

In each of Problems 11 through 18, a set of polar coordinates of a point in the plane is given. Find a set of polar coordinates for the same point in which the sign of the r-coordinate is the opposite to that given. Also find a set of polar coordinates for the same point in which the sign of the θ-coordinate is the opposite to that given.

11. $\left(2, \frac{\pi}{3}\right)$
12. $\left(3, \frac{5\pi}{6}\right)$
13. $\left(-2, \frac{\pi}{4}\right)$
14. $\left(-2, \frac{3\pi}{2}\right)$
15. $\left(4, -\frac{\pi}{4}\right)$
16. $(3, -\pi)$
17. $\left(-1, -\frac{\pi}{4}\right)$
18. $\left(-3, -\frac{2\pi}{3}\right)$

In each of Problems 19 through 28 a set of polar coordinates of a point in the plane is given. Find the rectangular coordinates of the same point.

19. $\left(4, \frac{\pi}{6}\right)$

20. $\left(3, \frac{5\pi}{3}\right)$

21. $\left(2, -\frac{\pi}{4}\right)$

22. $\left(4, \frac{3\pi}{4}\right)$

23. $(5, \pi)$

24. $\left(2, -\frac{3\pi}{2}\right)$

25. $\left(-4, \frac{\pi}{3}\right)$

26. $\left(-3, -\frac{\pi}{6}\right)$

27. $\left(-4, \frac{4\pi}{3}\right)$

28. $(-5, -3\pi)$

In each of Problems 29 through 36, the rectangular coordinates of a point are given. Find the set of polar coordinates of the same point for which $r \geq 0$ and $-\pi < \theta \leq \pi$.

29. $(2\sqrt{3}, 2)$

30. $(0, -3)$

31. $(-2, -2\sqrt{3})$

32. $(3, 0)$

33. $(-2, 2\sqrt{3})$

34. $(1, -\sqrt{3})$

35. $(1, 2)$

36. $(-3, -4)$

In each of Problems 37 through 64, sketch the graph of the given polar equation.

37. $r = 3$

38. $r = -2$

39. $\theta = -\frac{\pi}{3}$

40. $\theta = \frac{5\pi}{6}$

41. $r = 3 \cos \theta$

42. $r = 4 \sin \theta$

43. $r = -6 \sin \theta$

44. $r = -5 \cos \theta$

45. $r = \frac{\theta}{2}$

46. $r = \frac{|\theta|}{2}$

47. $r = e^{|\theta|/2\pi}$

48. $r = e^{-|\theta|/2\pi}$

49. $r = 4 \cos\left(\theta - \frac{\pi}{6}\right)$

50. $r = 6 \sin\left(\theta - \frac{\pi}{4}\right)$

51. $r = 4 \sin\left(\theta + \frac{5\pi}{6}\right)$

52. $r = 3 \cos\left(\theta + \frac{\pi}{3}\right)$

53. $r = 2 \sec \theta$

54. $r = -2 \sec \theta$

55. $r = 2 \csc \theta$

56. $r = -2 \csc \theta$

57. $r \cos\left(\theta - \frac{\pi}{4}\right) = 4$

58. $r \cos\left(\theta + \frac{\pi}{3}\right) = 2$

59. $r = 3 \sec\left(\theta - \frac{\pi}{2}\right)$

60. $r = 2 \sec\left(\theta + \frac{\pi}{6}\right)$

61. $r = 2|\cos \theta|$

62. $r = 3\left|\cos\left(\theta - \frac{\pi}{6}\right)\right|$

63. $r|\cos \theta| = 2$

64. $r\left|\cos\left(\theta - \frac{\pi}{6}\right)\right| = 3$

In each of Problems 65 through 68, find the polar equation of the circle having the given properties. In these problems the given points are specified by their rectangular coordinates.

65. The circle with radius 2 and center at $(0, 2)$.

66. The circle with radius 3 and center at $(-3, 0)$.

67. The circle with radius $\sqrt{2}$ and center at $(1, 1)$.

68. The circle with radius 2 and center at $(-1, \sqrt{3})$.

In each of Problems 69 through 72, find the polar equation of the curve having the given rectangular equation.

69. $2x - y = 5$

70. $9x^2 + 4y^2 = 36$

71. $x^2 - xy + y^2 = 3$

72. $(x - 2)^2 + (y + 2)^2 = 8$

In each of Problems 73 through 76, find the rectangular equation of the curve having the given polar equation.

73. $r = 2(1 - \sin \theta)$

74. $r = 3 + \cos \theta$

75. $r = 3 \sin 2\theta$

76. $r^2 = 8 \cos 2\theta$

77. Find a formula for the distance between the points (r_1, θ_1) and (r_2, θ_2).

14.2 CURVE SKETCHING IN POLAR COORDINATES

In this section we discuss several relatively simple curves whose equations are much more conveniently expressed in polar than in rectangular coordinates. Before looking at these curves individually it is helpful to note how one can often recognize certain symmetry properties in polar coordinates.

The graph of the equation

$$F(r, \theta) = 0 \tag{1}$$

is said to be **symmetric about the pole** if the point $(-r, \theta)$ lies on the graph whenever the point (r, θ) does (see Figure 14.2.1). This means that

$$F(-r, \theta) = 0 \tag{2}$$

for all points (r, θ) satisfying Eq. 1.

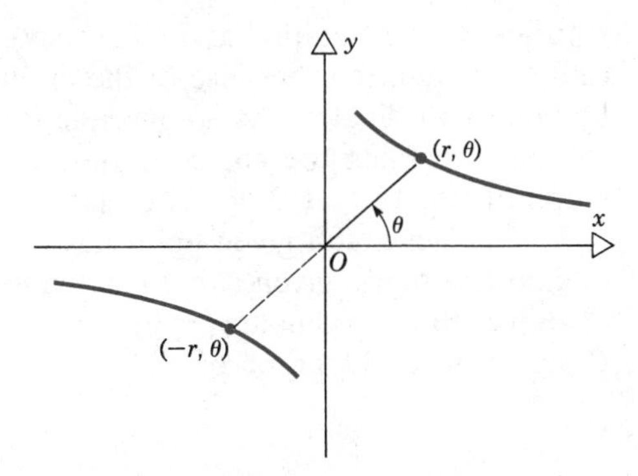

Figure 14.2.1 Symmetry about the pole.

Similarly, the graph of Eq. 1 is said to be **symmetric about the line $\theta = 0$** (the x-axis) if the point $(r, -\theta)$ lies on the graph whenever the point (r, θ) does, as shown in Figure 14.2.2. In this case

$$F(r, -\theta) = 0 \tag{3}$$

for all points (r, θ) that satisfy Eq. 1.

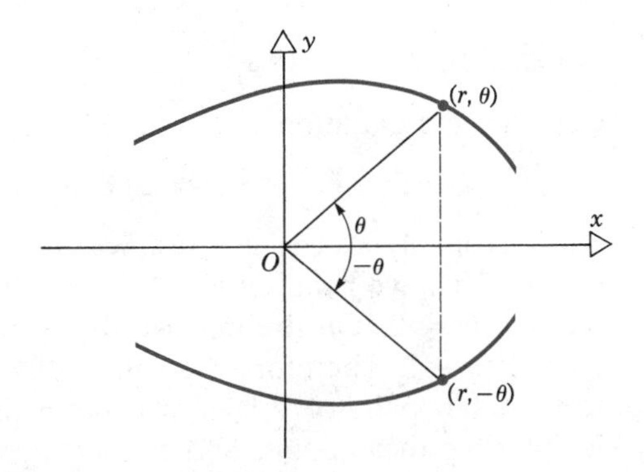

Figure 14.2.2 Symmetry about $\theta = 0$.

Finally, the graph of Eq. 1 is said to be **symmetric about the line $\theta = \pi/2$** (the y-axis) if the point $(r, \pi - \theta)$ lies on the graph whenever the point (r, θ) does (see Figure 14.2.3). Thus

$$F(r, \pi - \theta) = 0 \tag{4}$$

for all points (r, θ) satisfying Eq. 1.

In all three cases one tests for symmetry by replacing r by $-r$, or θ by $-\theta$, or θ by $\pi - \theta$, respectively. If the original equation is left unaltered by the sub-

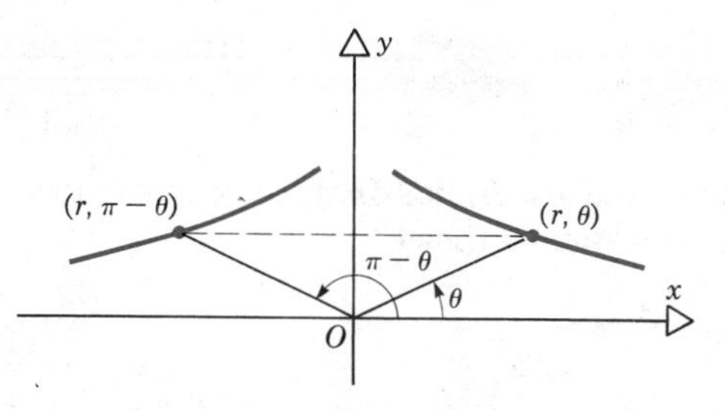

Figure 14.2.3 Symmetry about $\theta = \pi/2$.

stitution, then its graph has the corresponding symmetry property. However, some caution is required, because of the ambiguity in representing a point in the plane by polar coordinates. As a consequence, a curve may have a symmetry property without satisfying the corresponding one of Eqs. 2, 3, or 4. An example is the graph of Eq. 15 that is discussed later in this section.

The fact that a given point has many pairs of polar coordinates leads to the related fact that a given curve may have different (but equivalent) equations. For example, the coordinates (r, θ) and $(-r, \theta + \pi)$ correspond to the same point. Consequently, the graph of

$$r = f(\theta)$$

is the same as the graph of

$$-r = f(\theta + \pi)$$

or of

$$r = -f(\theta + \pi).$$

We will see specific examples a little later.

Cardioids

Consider the equation

$$r = a(1 + \cos\theta), \qquad a > 0. \tag{5}$$

Since $\cos(-\theta) = \cos\theta$ it follows that the graph is symmetric about the line $\theta = 0$. Thus we need only sketch the graph for $0 \le \theta \le \pi$, and then reflect it in the line $\theta = 0$. On the interval $0 \le \theta \le \pi$ the values of $\cos\theta$ steadily decrease from 1 to -1. Therefore the values of r given by Eq. 5 steadily decrease from $2a$ to 0. A few pairs of values of r and θ that satisfy Eq. 5 are given in Table 14.2. By plotting these points and connecting them with a smooth curve we obtain the

Table 14.2 Some Values of r and θ That Satisfy $r = a(1 + \cos\theta)$

θ	0	$\frac{\pi}{6}$	$\frac{\pi}{4}$	$\frac{\pi}{3}$	$\frac{\pi}{2}$	$\frac{3\pi}{4}$	π
r	$2a$	$\left(1 + \frac{\sqrt{3}}{2}\right)a$ $\cong 1.86a$	$\left(1 + \frac{\sqrt{2}}{2}\right)a$ $\cong 1.71a$	$\frac{3}{2}a$ $= 1.5a$	a	$\left(1 - \frac{\sqrt{2}}{2}\right)a$ $\cong 0.29a$	0

graph in Figure 14.2.4. This curve is known as a *cardioid* because of its heartlike shape. The parameter a is a proportionality factor that determines the size of the figure.

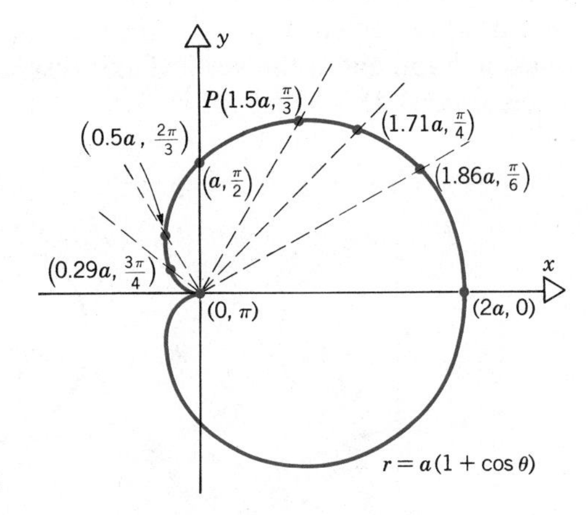

Figure 14.2.4

Several other equations also give rise to cardioids. For example, if we replace (r, θ) by $(-r, \theta + \pi)$, as suggested above, then Eq. 5 becomes

$$-r = a[1 + \cos(\theta + \pi)], \qquad a > 0,$$

or

$$r = a(-1 + \cos\theta), \qquad a > 0. \tag{6}$$

The graph of Eq. 6 is the same as Eq. 5. Of course, for each point on the curve, the polar coordinates that satisfy Eq. 5 are different from the polar coordinates that satisfy Eq. 6.

If we change the sign of the cosine term in Eqs. 5 and 6, we obtain

$$r = a(1 - \cos\theta), \qquad r = a(-1 - \cos\theta), \tag{7}$$

respectively. Each of these equations has the graph shown in Figure 14.2.5.

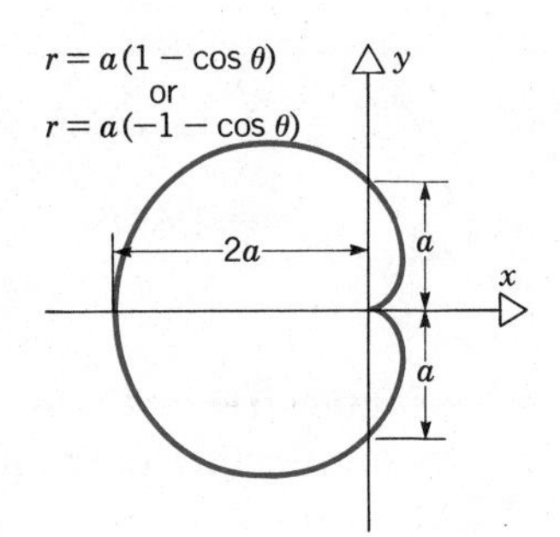

Figure 14.2.5

Finally, if $\cos\theta$ is replaced by $\sin\theta$ in Eqs. 5, 6, and 7, then we obtain the equations

$$r = a(1 + \sin\theta), \qquad r = a(-1 + \sin\theta) \tag{8}$$

and

$$r = a(1 - \sin \theta), \qquad r = a(-1 - \sin \theta) \tag{9}$$

whose graphs are also cardioids. Each of Eqs. 8 has the graph shown in Figure 14.2.6, while each of Eqs. 9 has the the graph shown in Figure 14.2.7. The symmetry of each graph about the vertical axis results from the fact that $\sin(\pi - \theta) = \sin \theta$, so condition (4) is satisfied.

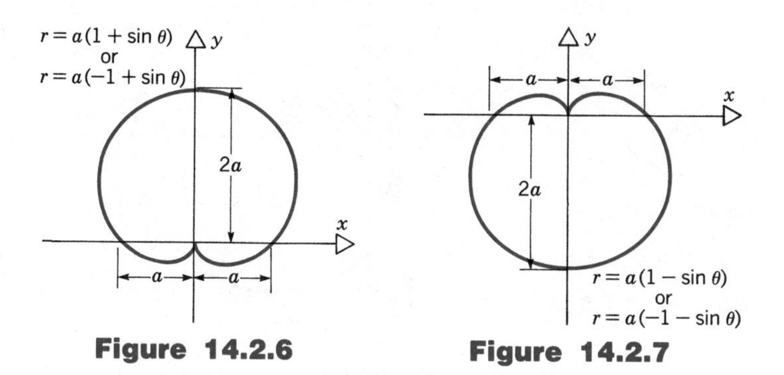

Figure 14.2.6 **Figure 14.2.7**

Limacons

The class of curves known as *limaçons* are generalizations of cardioids. There are two distinct kinds of limaçons, and we will give an illustration of each.

First consider the equation

$$r = a(1 + 0.5 \cos \theta), \qquad a > 0. \tag{10}$$

Again, the graph is symmetric about the line $\theta = 0$, so we need only consider the graph for $0 \le \theta \le \pi$. As θ goes from 0 to π the value of r decreases from $3a/2$ to

Table 14.3 **Some Values of r and θ That Satisfy $r = a(1 + 0.5 \cos \theta)$**

θ	0	$\frac{\pi}{6}$	$\frac{\pi}{4}$	$\frac{\pi}{3}$	$\frac{\pi}{2}$	$\frac{3\pi}{4}$	π
r	$\frac{3a}{2}$	$\left(1 + \frac{\sqrt{3}}{4}\right)a$ $\cong 1.43a$	$\left(1 + \frac{\sqrt{2}}{4}\right)a$ $\cong 1.35a$	$\frac{5a}{4}$	a	$\left(1 - \frac{\sqrt{2}}{4}\right)a$ $\cong 0.65a$	$\frac{a}{2}$

$a/2$, as indicated in Table 14.3. By drawing a smooth curve through these points and then reflecting it about the line $\theta = 0$, we obtain the graph shown in Figure 14.2.8.

Next, consider the equation

$$r = a(1 + 1.5 \cos \theta), \qquad a > 0. \tag{11}$$

Once again, because of symmetry about the line $\theta = 0$, it is sufficient to determine the graph for $0 \le \theta \le \pi$. A few values of r and θ are shown in Table 14.4. The most important difference between this case and the preceding one is that now

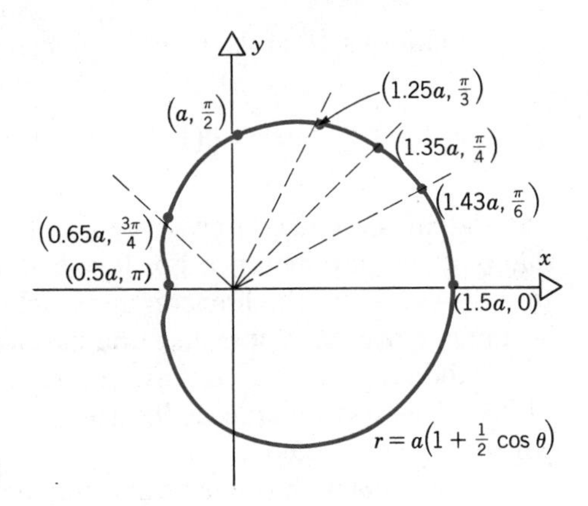

Figure 14.2.8

there is a value of θ in the second quadrant for which $\cos\theta = -\frac{2}{3}$, and for this value of θ it follows that $r = 0$. As θ increases further, r becomes negative and has the value $-a/2$ when $\theta = \pi$. Thus for $0 \leq \theta \leq \pi$ we obtain the graph given

Table 14.4 **Some Values of r and θ That Satisfy $r = a(1 + 1.5\cos\theta)$**

θ	0	$\frac{\pi}{6}$	$\frac{\pi}{4}$	$\frac{\pi}{3}$	$\frac{\pi}{2}$	$\frac{2\pi}{3}$	$\frac{3\pi}{4}$	$\frac{5\pi}{6}$	π
r	$\frac{5a}{2}$	$\left(1 + \frac{3\sqrt{3}}{4}\right)a$ $\cong 2.30a$	$\left(1 + \frac{3\sqrt{2}}{4}\right)a$ $\cong 2.06a$	$\frac{7a}{4}$	a	$\frac{a}{4}$	$\left(1 - \frac{3\sqrt{2}}{4}\right)a$ $\cong -0.06a$	$\left(1 - \frac{3\sqrt{3}}{4}\right)a$ $\cong -0.30a$	$-\frac{a}{2}$

by the darker arc in Figure 14.2.9. The rest of the graph (the lighter portion) is obtained by symmetry about the line $\theta = 0$. Observe that the graph has an interior loop.

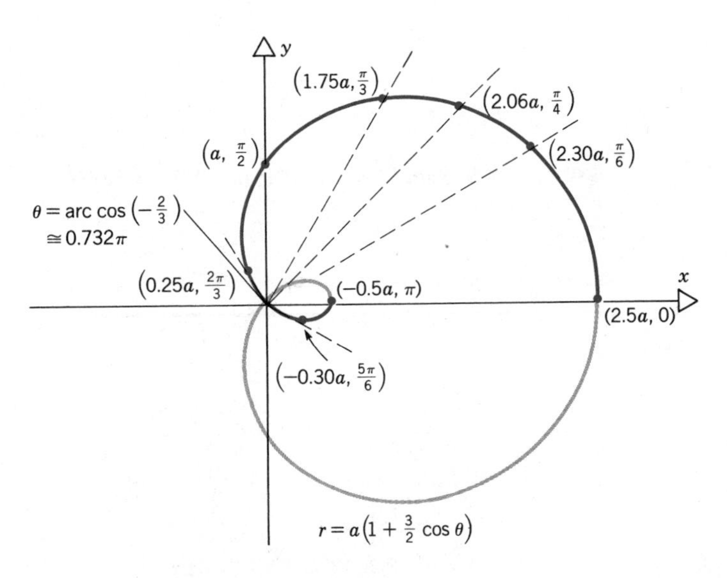

Figure 14.2.9

Equations 10 and 11 are both special cases of the more general equation

$$r = a(1 + b \cos \theta), \qquad a > 0, \qquad b > 0. \tag{12}$$

The size of the limaçon given by Eq. 12 is determined by the parameter a, and its shape by the parameter b. For $0 < b < 1$ the limaçon resembles the one in Figure 14.2.8. As $b \to 0$, the limaçon approaches the circle $r = a$. As $b \to 1$, the limaçon becomes more and more like the cardioid of Eq. 5, shown in Figure 14.2.4. For $b > 1$ the graph of Eq. 12 has an interior loop similar to the one shown in Figure 14.2.9. This loop increases in size as b increases, but never intersects the outer portion of the limaçon.

Limaçons with other orientations can be obtained by modifying Eq. 12 as in the preceding discussion of cardioids.

Leaf curves

Consider the graph of the equation

$$r = a \sin 3\theta, \qquad a > 0. \tag{13}$$

Since $|\sin 3\theta| \le 1$, it follows that $|r| \le a$. As θ goes through the interval from 0 to $\pi/3$, 3θ goes from 0 to π, and $\sin 3\theta$ first increases from 0 to 1 and then decreases to 0 again (see Table 14.5). For this range of θ the graph is the loop in the first quadrant shown in Figure 14.2.10. Because the sine function is symmetric about $\pi/2$, the loop is symmetric about the ray $\theta = \pi/6$; the graph intersects this ray at the point $(a, \pi/6)$.

Table 14.5 Some Values of r and θ That Satisfy $r = a \sin 3\theta$

θ	0	$\frac{\pi}{12}$	$\frac{\pi}{6}$	$\frac{\pi}{4}$	$\frac{\pi}{3}$
r	0	$\frac{a}{\sqrt{2}} \cong 0.71a$	a	$\frac{a}{\sqrt{2}} \cong 0.71a$	0

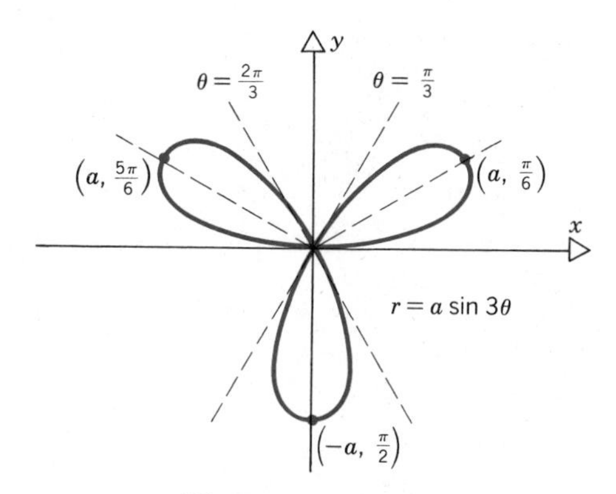

Figure 14.2.10

In a similar way, values of θ in the interval $\pi/3 \le \theta \le 2\pi/3$ yield negative values of r and give rise to the loop in the third and fourth quadrants. Finally, the interval $2\pi/3 \le \theta \le \pi$ produces the loop in the second quadrant. Because of the symmetry and periodic character of the sine function, the three loops are congruent. For values of θ outside of $0 \le \theta \le \pi$ it is straightforward to show that the same curve is retraced.

The curve in Figure 14.2.10 is an example of a *leaf curve.* In general, leaf curves are given by the equations

$$r = a \sin n\theta \quad \text{or} \quad r = a \cos n\theta \tag{14}$$

where $a \neq 0$, and n is an integer greater than one.* The magnitude of a determines the length of the leaves, while n determines their number and their thickness. If n is odd, then there are n leaves, equally spaced, as in the case previously discussed. If n is even, then it can be shown that there are $2n$ leaves, again equally spaced.

To see the effect of an even value of n, consider the equation

$$r = a \sin 2\theta. \tag{15}$$

As θ goes from 0 to $\pi/2$, 2θ goes from 0 to π, and $\sin 2\theta$ goes from 0 to 1 and then back to 0 (see Table 14.6). This gives the leaf in the first quadrant of Figure

Table 14.6 Some Values of r and θ That Satisfy $r = a \sin 2\theta$

θ	0	$\frac{\pi}{8}$	$\frac{\pi}{4}$	$\frac{3\pi}{8}$	$\frac{\pi}{2}$
r	0	$\frac{a}{\sqrt{2}}$ $\cong 0.71a$	a	$\frac{a}{\sqrt{2}}$ $\cong 0.71a$	0

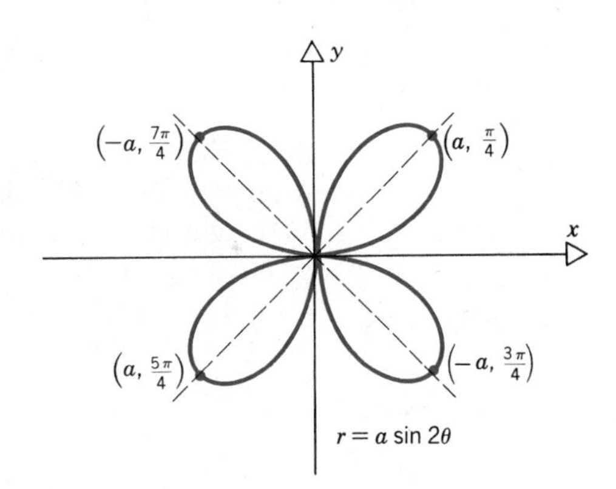

Figure 14.2.11

14.2.11. Similarly, for $\pi/2 \le \theta \le \pi$, we obtain the leaf in the fourth quadrant, since r is negative. The ranges $\pi \le \theta \le 3\pi/2$ and $3\pi/2 \le \theta \le 2\pi$ give the other two leaves.

Observe that the graph in Figure 14.2.11 is symmetric about the origin and about both of the lines $\theta = 0$ and $\theta = \pi/2$, yet Eq. 15 satisfies none of the symmetry conditions (1), (2), or (3). As previously indicated, this situation is a consequence of the multiplicity of polar coordinate pairs for each point. For example, the point $P(a, \pi/4)$ in the first quadrant lies on the curve and its given coordinates satisfy Eq. 15. The point Q in the third quadrant with coordinates $(-a, \pi/4)$ also lies on the curve. However, the coordinates of Q that satisfy Eq. 15 are $(a, 5\pi/4)$ and not $(-a, \pi/4)$.

*If $n = 1$, the graph is a circle as discussed in Section 14.1.

Lemniscates

Let us now consider the equation

$$r^2 = 2a^2 \cos 2\theta, \tag{16}$$

where $a > 0$. Since Eq. 16 satisfies each of the symmetry conditions (2), (3), and (4), it follows that the graph of Eq. 16 is symmetric about the pole and about both the x- and y-axes. Further, $r^2 \geq 0$, so we need only consider values of θ for which $\cos 2\theta \geq 0$. For example, it is sufficient to consider the interval $0 \leq \theta \leq \pi/4$ (see Table 14.7). Thus r goes from $\sqrt{2}a$ to 0 as θ traverses the interval from 0 to $\pi/4$.

Table 14.7 Some Values of r and θ That Satisfy $r^2 = 2a^2 \cos 2\theta$

θ	0	$\frac{\pi}{12}$	$\frac{\pi}{6}$	$\frac{\pi}{4}$
r^2	$2a^2$	$\sqrt{3}\, a^2$	a^2	0
r	$\sqrt{2}\, a$ $\cong 1.41a$	$\sqrt[4]{3}\, a$ $\cong 1.32a$	a	0

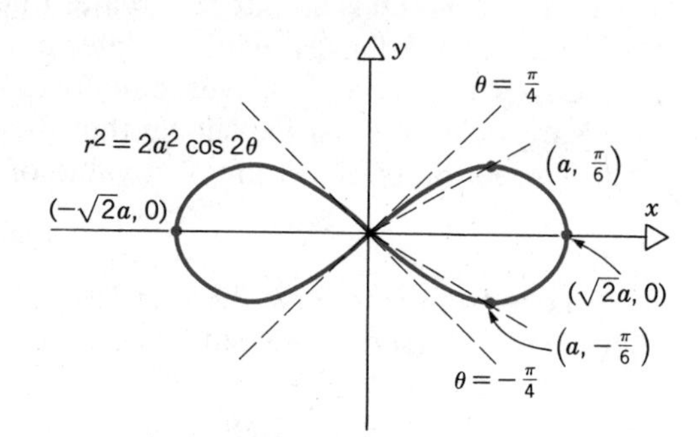

Figure 14.2.12

This produces the upper half of the loop on the right side of Figure 14.2.12. The symmetry about $\theta = 0$ produces the lower half of this loop, and symmetry about the origin yields the loop on the left side of Figure 14.2.12. The graph of Eq. 16 is called a *lemniscate.*

If $\cos 2\theta$ is replaced by $\sin 2\theta$ in Eq. 16, then the graph is a lemniscate congruent to the one in Figure 14.2.12 rotated through an angle of $\pi/4$ radians (see Figure 14.2.13).

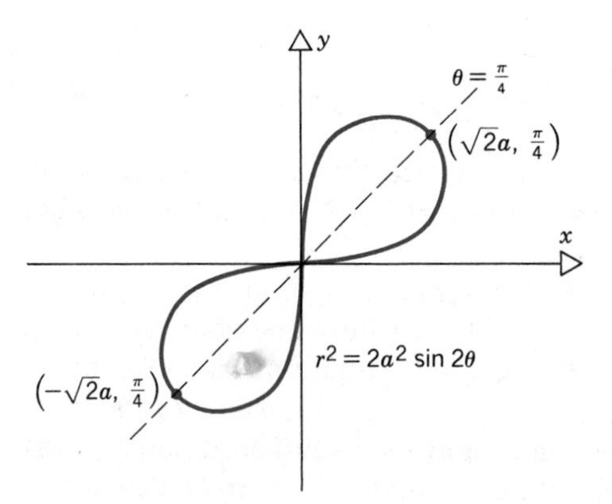

Figure 14.2.13

Intersections of curves in polar coordinates

Geometrically, a point of intersection of two curves is a point that lies on both curves. Analytically, in rectangular coordinates the points of intersection can always be located, at least in principle, by solving simultaneously the equations defining the two curves. In polar coordinates the situation is made more complicated by the fact that a given point of intersection has many sets of polar coordinates. At least one of them must satisfy the equation of one curve, and at least one must satisfy the equation of the other curve, but it may well happen that no single set of coordinates satisfies both equations. If this occurs, then that point of intersection will not be found by simultaneous solution of the two equations.

This difficulty can be illustrated in very simple cases. Consider the two equations

$$r = 4, \qquad \theta = \frac{2\pi}{3} \tag{17}$$

whose graphs are, respectively, the circle and straight line shown in Figure 14.2.14. Clearly there are two points of intersection, denoted by P and Q in the figure. The simultaneous solution of Eqs. 17 gives the coordinates $(4, 2\pi/3)$ of P; Q can have coordinates $(4, -\pi/3)$ or $(-4, 2\pi/3)$, for example, but there are no coordinates of Q that satisfy both of Eqs. 17.

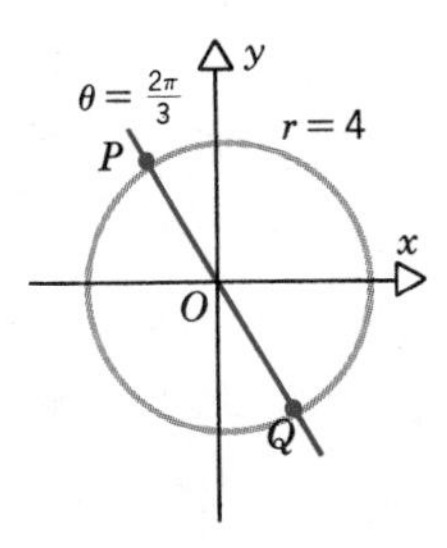

Figure 14.2.14

When searching for points of intersection of curves given by equations in polar coordinates, one should certainly solve the equations simultaneously, if possible. However, one should also bear in mind the possible existence of other points of intersection that cannot be found in this way. If there are such points, a good sketch will usually be very helpful in locating them. To obtain them algebraically it may be necessary to change one or both of the equations into a different form by replacing (r, θ) by an equivalent pair of polar coordinates.

In visualizing the situation, it may be helpful to think of θ as a time variable, and the two curves as the paths traced out by two moving particles. If a point of intersection has the same coordinates for both curves, then the two particles collide there—they reach the same point in the plane at the same time. If a point of intersection has different coordinates for the two curves, then two particles pass through that point at different times, and so do not collide.

EXAMPLE 1

Find the points of intersection of the cardioid

$$r = 1 + \sin\theta \tag{18}$$

and the circle

$$r = 3\sin\theta. \tag{19}$$

The two curves are sketched in Figure 14.2.15. By eliminating r between Eqs. 18 and 19 we obtain

$$1 + \sin\theta = 3\sin\theta$$

from which it follows that

$$\sin\theta = \tfrac{1}{2}. \tag{20}$$

Thus (in the interval $0 \le \theta \le 2\pi$) θ has the values $\theta = \pi/6$ and $\theta = 5\pi/6$. The corresponding value of r is $r = \frac{3}{2}$. These results give us the two points of intersection $(\frac{3}{2}, \pi/6)$ and $(\frac{3}{2}, 5\pi/6)$ labeled P and Q, respectively in Figure 14.2.15. From an examination of the figure it is also clear that the pole is a point of intersection, but it cannot be found by solving Eqs. 18 and 19. For $0 \le \theta \le 2\pi$ the pole has coordinates $(0, 3\pi/2)$ on the cardioid, while on the circle its coordinates are $(0, 0)$ and $(0, \pi)$. ∎

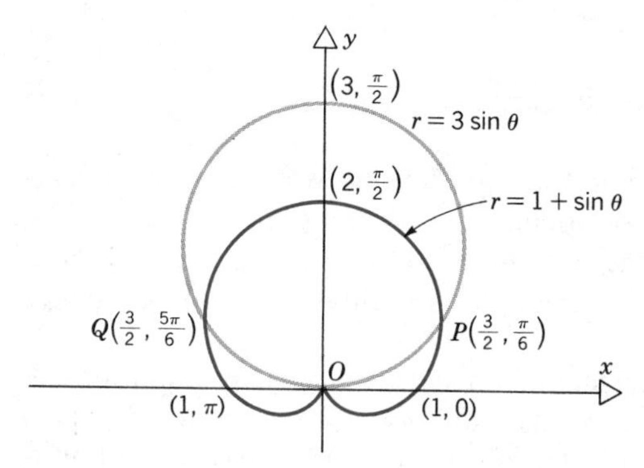

Figure 14.2.15

EXAMPLE 2

Find the points of intersection of the circle

$$r = \cos \theta \tag{21}$$

and the leaf curve

$$r = \sqrt{3} \cos 2\theta. \tag{22}$$

The curves are sketched in Figure 14.2.16. By eliminating r between Eqs. 21 and 22 we find that

$$\sqrt{3} \cos 2\theta - \cos \theta = 0.$$

Using the identity $\cos 2\theta = 2 \cos^2 \theta - 1$, we obtain

$$2\sqrt{3} \cos^2 \theta - \cos \theta - \sqrt{3} = 0, \tag{23}$$

which is a quadratic equation in $\cos \theta$. The quadratic formula gives the solutions

$$\cos \theta = \frac{1 \pm 5}{4\sqrt{3}} = \frac{\sqrt{3}}{2}, \quad -\frac{\sqrt{3}}{3}.$$

For the interval $0 \le \theta < 2\pi$ the first solution $\cos \theta = \sqrt{3}/2$ corresponds to

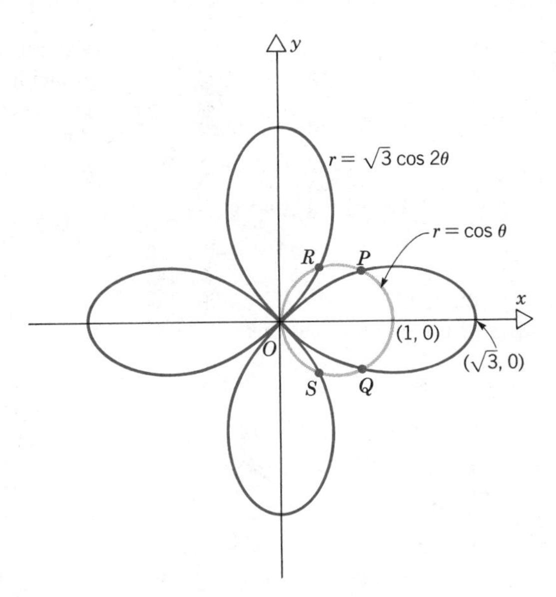

Figure 14.2.16

$\theta = \pi/6$ and $\theta = 11\pi/6$ with the associated value $r = \sqrt{3}/2$. These results give the points P and Q respectively in Figure 14.2.16.

The second solution $\cos\theta = -\sqrt{3}/3 \cong -0.57735$ yields values of θ in the second and third quadrants given by $\theta \cong 2.18628$ radians and $\theta \cong 4.09691$ radians. The corresponding value of r is $-\sqrt{3}/3$. These results give the points S and R, respectively, in Figure 14.2.16.

Note, however, that the interval $0 \le \theta < 2\pi$ corresponds to two passages around the circle $r = \cos\theta$. The points P and S lie on the first circuit, while the points R and Q lie on the second circuit. If we restrict ourselves to one trip around the circle by requiring θ to lie in the interval $0 \le \theta < \pi$, for example, then we must choose the coordinates $(\sqrt{3}/3,\ 0.95532)$ for R and the coordinates $(-\sqrt{3}/2,\ 5\pi/6)$ for Q. In this case the points R and Q have different coordinates on the leaf curve and on the circle.

Finally we observe that the pole is also a point of intersection, although it is not obtained by solving Eqs. 21 and 22. For $0 \le \theta < 2\pi$ the pole has coordinates $(0, \pi/2)$ and $(0, 3\pi/2)$ on the circle, while for the leaf curve its coordinates are $(0, \pi/4)$, $(0, 3\pi/4)$, $(0, 5\pi/4)$, and $(0, 7\pi/4)$. ∎

PROBLEMS

In each of Problems 1 through 16, sketch the graph of the given equation and identify the type of curve.

1) $r = 2 - 2\cos\theta$

2) $r = 2 + \cos\theta$

3) $r = 1 - 2\cos\theta$

4) $r = 3\cos 2\theta$

5. $r = -3 + 3 \sin \theta$

6. $r = 2 \sin 3\theta$

7. $r^2 = 2 \sin 2\theta$

8. $r = 2 + 3 \sin \theta$

9. $r = 3 + 3 \sin \theta$

10. $r = 3 - 3 \sin \theta$

11. $r^2 = 4 \cos 2\theta$

12. $r = 3 - 2 \sin \theta$

13. $r = 4 \sin 4\theta$

14. $r = 3 \cos 5\theta$

15. $r = 2 - 4 \sin \theta$

16. $r^2 = -4 \cos 2\theta$

In each of Problems 17 through 29, find the points of intersection of the graphs of the given equations.

17. $r = 4(1 + \cos \theta), \quad r = 6$

18. $r = 1 - \sin \theta, \quad r = \sin \theta$

19. $r = 3 \sin \theta, \quad r = 3 \cos \theta$

20. $r = 4 \cos 2\theta, \quad \theta = \dfrac{\pi}{6}$

21. $r = 2 - \cos \theta, \quad r = 1 + \cos \theta$

22. $r = 1 + \sin \theta, \quad r = 2 \sin \theta$

23. $r = 2 + 3 \cos \theta, \quad \theta = \dfrac{2\pi}{3}$

24. $r = 2 + 5 \cos \theta, \quad r = \cos \theta$

25. $r = 2 \sin 3\theta, \quad r = \sqrt{2}$, first quadrant

26. $r^2 = 2 \cos 2\theta, \quad r = \cos \theta$

27. $r = \dfrac{1}{\frac{3}{2} - \sin \theta}, \quad r = \frac{3}{2} - \sin \theta$

28. $r = 2 + 2 \cos \theta, \quad \theta = \dfrac{\pi}{4}$

29. $r = 1 + 2 \cos \theta, \quad r = 4 \cos \theta$

Conic Sections

In Sections 5.2 through 5.4 we discussed conic sections in a rectangular coordinate system. In Problems 30 through 32 we consider the representation of conic sections (other than circles and straight lines) in polar coordinates. The discussion is based on the following definition. Let a fixed point F called the focus and a fixed line l (not containing F) called the directrix be given. A conic section consists of the set of all points P for which the ratio of the distance from the focus F to the distance from the directrix l is a constant. The constant ratio is the eccentricity and is denoted by e. If $0 < e < 1$, then the conic section is an ellipse; if $e = 1$, it is a parabola; if $e > 1$, it is a hyperbola. In Problems 30 and 31 we indicate how to derive the standard equations for conic sections in polar coordinates.

30. Establish a polar coordinate system with the pole coinciding with the focus, and with the polar axis perpendicular to the directrix. Suppose that the directrix is to the left of the pole and a distance p from it (see Figure 14.2.17). The definition of

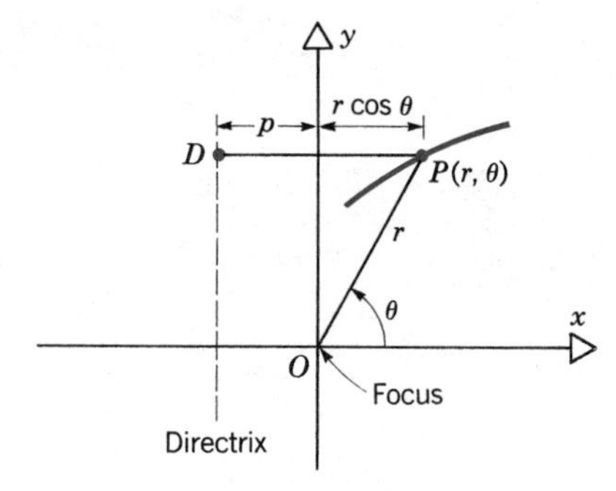

Figure 14.2.17

a conic section says that

$$|OP| = e|DP|. \tag{i}$$

(a) Show that Eq. (i) requires that

$$r = e(r \cos \theta + p)$$

and hence that

$$r = \frac{ep}{1 - e \cos \theta}. \tag{ii}$$

(b) Suppose now that the directrix is a distance p to the right of the focus, but that the situation is otherwise unchanged (see Figure 14.2.18). Show that in this case the equation

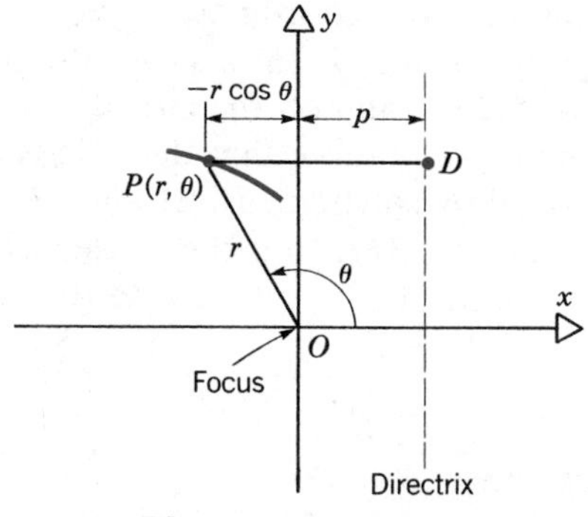

Figure 14.2.18

of a conic section is

$$r = \frac{ep}{1 + e \cos \theta}. \tag{iii}$$

Note: Ellipses and hyperbolas have two foci and two directrices, and satisfy both Eq. (*ii*) and Eq. (*iii*), depending on which focus is placed at the pole. The foci and directrices are on opposite sides of, and equidistant from, the center of the conic. A parabola has only one focus and one directrix, and hence can satisfy only one of Eqs. (*ii*) and (*iii*), depending on whether it opens to the left or to the right.

31. (a) Suppose that the focus is at the pole and the directrix is parallel to the polar axis and a distance p below it. Show that the equation of a conic section is

$$r = \frac{ep}{1 - e\sin\theta}. \qquad (i)$$

(b) Suppose now that the directrix is parallel to the polar axis and a distance p above it. Show that the equation of a conic section is

$$r = \frac{ep}{1 + e\sin\theta}. \qquad (ii)$$

32. Convert the polar equation

$$r = \frac{ep}{1 - e\cos\theta} \qquad (i)$$

of a conic section into an equation in rectangular coordinates.

(a) If $0 < e < 1$, show that the resulting equation can be put into the form of the equation of an ellipse

$$\frac{(x-h)^2}{a^2} + \frac{y^2}{b^2} = 1, \qquad (ii)$$

where

$$h = \frac{pe^2}{1-e^2},$$

$$a^2 = \frac{p^2e^2}{(1-e^2)^2}, \qquad b^2 = \frac{p^2e^2}{1-e^2}. \qquad (iii)$$

(b) If $e > 1$, show that the resulting equation can be put into the form of the equation of a hyperbola

$$\frac{(x-h)^2}{a^2} - \frac{y^2}{b^2} = 1, \qquad (iv)$$

where h and a^2 are given by Eq. (*iii*), and

$$b^2 = \frac{p^2e^2}{e^2-1}. \qquad (v)$$

(c) If $e = 1$, show that the resulting equation can be put into the form of the equation of a parabola

$$x - h = \alpha y^2, \qquad (vi)$$

where

$$h = -\frac{p}{2}, \qquad \alpha = \frac{1}{2p}. \qquad (vii)$$

In each of Problems 33 through 40, sketch the graph of the conic section described by the given equation. Also determine its eccentricity and locate its foci and directrices.

33. $r = \dfrac{4}{2 - \cos\theta}$

34. $r = \dfrac{6}{1 + 2\cos\theta}$

35. $r = \dfrac{4}{3 + 2\cos\theta}$

36. $r = \dfrac{4}{1 - \cos\theta}$

37. $r = \dfrac{15}{2 - 3\cos\theta}$

38. $r = \dfrac{6}{3 - 2\sin\theta}$

39. $r = \dfrac{2}{1 + \sin\theta}$

40. $r = \dfrac{6}{1 + 3\sin\theta}$

41. A conic section with eccentricity $\frac{3}{2}$ has one focus at the pole and $x = 2$ as the corresponding directrix. Write the equation of this conic section and locate the other focus and directrix.

42. A conic section with one focus at the pole has directrices $x = -2$ and $x = 8$. Find the other focus and the eccentricity.

14.3 AREA AND ARC LENGTH IN POLAR COORDINATES

In this section we discuss the use of polar coordinates in two important applications, namely, finding the area of a given region and the arc length of a given curve. A third important question, finding the direction of a curve described by a polar equation, is the subject of Problem 32.

Area

The basic problem that we discuss is that of finding the area of a region R such as the one shown in Figure 14.3.1: a region bounded by the graph of a continuous function $r = f(\theta)$ and by the rays $\theta = \alpha$ and $\theta = \beta$. This corresponds to the

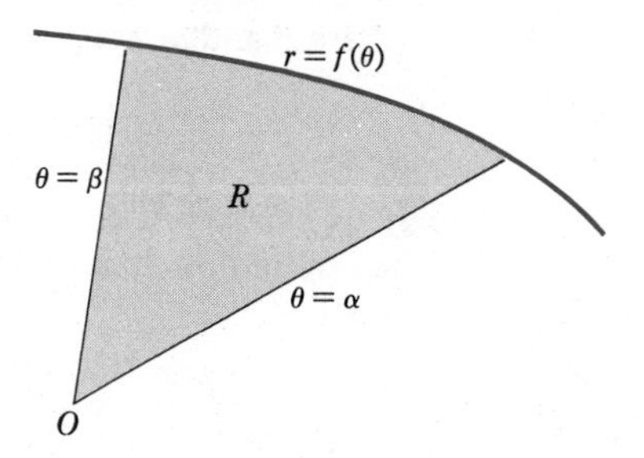

Figure 14.3.1

problem in rectangular coordinates of finding the area of the region between the x-axis and the graph of a continuous function $y = f(x)$, and between the vertical lines $x = a$ and $x = b$. As we have seen in Chapter 6, this latter problem is solved by the definite integral

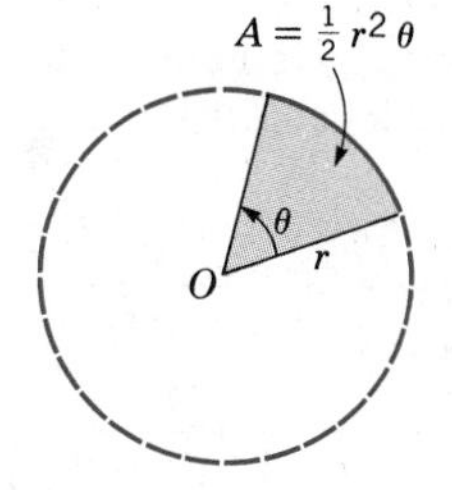

Figure 14.3.2

$$\text{Area} = \int_a^b f(x)\,dx. \tag{1}$$

We wish to derive an analogous integral in polar coordinates that gives the area of the region R shown in Figure 14.3.1.

First we recall the formula for the area A of a circular sector of radius r and central angle θ, as shown in Figure 14.3.2. The area is given by

$$A = (\pi r^2)\left(\frac{\theta}{2\pi}\right) = \frac{r^2\theta}{2}. \tag{2}$$

In the middle term in Eq. 2 the first factor is the area of an entire circle of radius r, while the second factor is the proportion of the circle that is in the given sector.

Next we proceed as in Chapter 6 to form a partition $\Delta = \{\theta_0, \theta_1, \ldots, \theta_n\}$ of the interval $\alpha \leq \theta \leq \beta$, with $\theta_0 = \alpha$ and $\theta_n = \beta$. This corresponds to subdividing

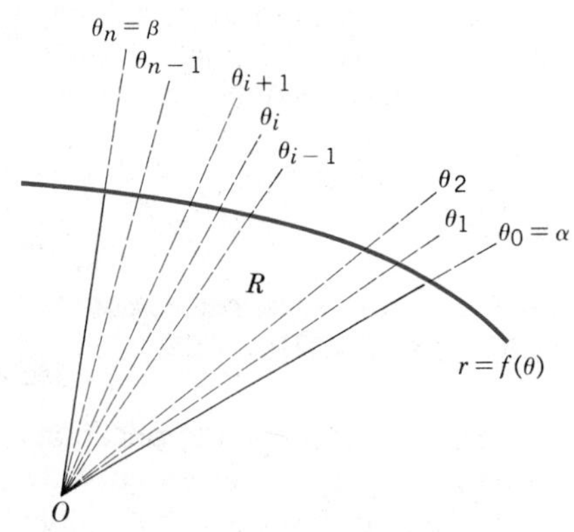

Figure 14.3.3 A partition of the region R.

the region R into n thin slices as shown in Figure 14.3.3. We now focus our attention on estimating the area of a typical one of these slices, for example, the one from θ_{i-1} to θ_i shown in Figure 14.3.4. Choose a value of θ in the interval $[\theta_{i-1}, \theta_i]$, call

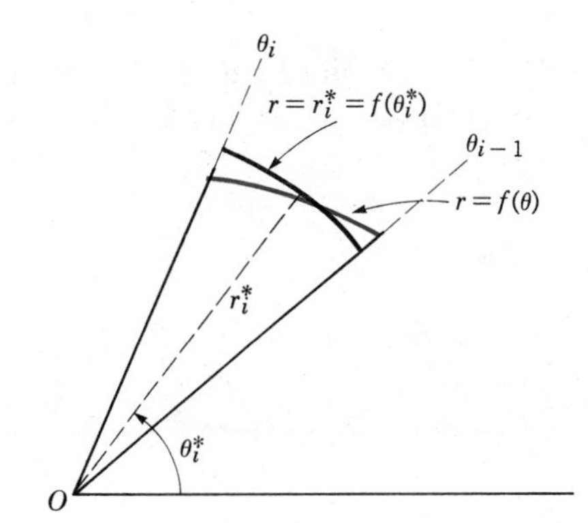

Figure 14.3.4

it θ_i^*, and let

$$r_i^* = f(\theta_i^*). \tag{3}$$

We approximate the area of the portion of R between θ_{i-1} and θ_i by the area of the circular sector of radius r_i^*. By Eq. 2 the area ΔA_i of this circular sector is

$$\begin{aligned} \Delta A_i &= (\pi r_i^{*2})\left(\frac{\theta_i - \theta_{i-1}}{2\pi}\right) \\ &= \tfrac{1}{2} r_i^{*2}\, \Delta\theta_i, \end{aligned} \tag{4}$$

where we have let $\Delta\theta_i = \theta_i - \theta_{i-1}$.

Proceeding in the same way for each of the slices shown in Figure 14.3.3 and adding the terms together, we obtain the following estimate for the area of R:

$$\text{Area of } R \cong \sum_{i=1}^{n} \Delta A_i = \frac{1}{2}\sum_{i=1}^{n} r_i^{*2}\, \Delta\theta_i = \frac{1}{2}\sum_{i=1}^{n} [f(\theta_i^*)]^2\, \Delta\theta_i. \tag{5}$$

The sum on the right side of Eq. 5 is a Riemann sum. Thus, in the limit as $n \to \infty$ and $\max|\Delta\theta_i| \to 0$, the sum approaches the corresponding integral. As in Section 6.2 we then define the area of R to be this integral

$$\text{Area of } R = \frac{1}{2}\int_{\alpha}^{\beta} [f(\theta)]^2\, d\theta. \tag{6}$$

More compactly, we can write

$$\text{Area of } R = \frac{1}{2}\int_{\alpha}^{\beta} r^2\, d\theta \tag{7}$$

where we must remember that $r = f(\theta)$.

EXAMPLE 1

Find the area A enclosed by one loop of the leaf curve

$$r = a \sin 3\theta. \tag{8}$$

The graph of Eq. 8 is shown in Figure 14.3.5. Let us consider the leaf in the first quadrant. That leaf lies between the rays $\theta = 0$ and $\theta = \pi/3$, so from Eq. 6

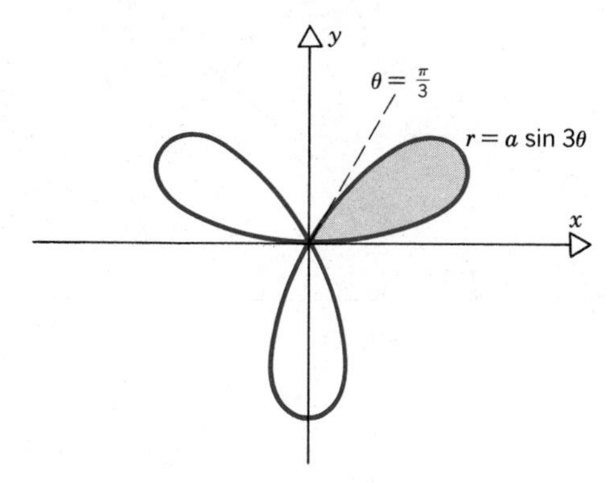

Figure 14.3.5

we obtain the formula

$$A = \frac{1}{2}\int_0^{\pi/3} (a \sin 3\theta)^2 \, d\theta.$$

By using the half angle formula and then integrating, we find that

$$\begin{aligned} A &= \frac{1}{2} a^2 \int_0^{\pi/3} \left(\frac{1}{2} - \frac{1}{2}\cos 6\theta\right) d\theta \\ &= \frac{1}{2} a^2 \left[\frac{1}{2}\theta - \frac{1}{12}\sin 6\theta\right]\Bigg|_0^{\pi/3} \\ &= \frac{\pi a^2}{12}. \quad \blacksquare \end{aligned} \tag{9}$$

EXAMPLE 2

Find the area A of the region that is inside the cardioid

$$r = 2(1 + \cos \theta) \tag{10}$$

and outside the circle

$$r = 3. \tag{11}$$

The region is shaded in Figure 14.3.6. From the symmetry of the region we need only calculate the area in the first quadrant and then multiply by two. To

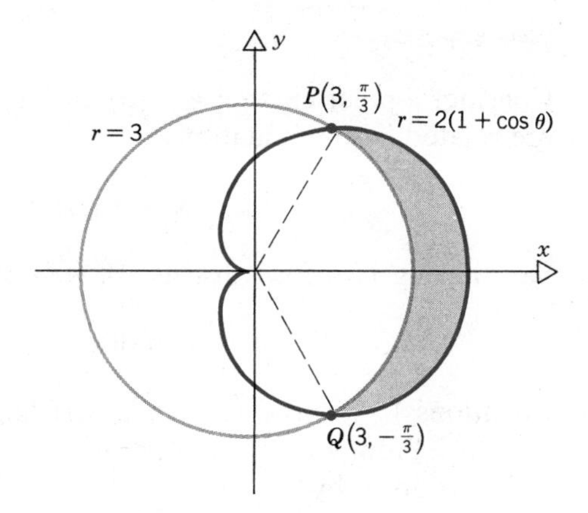

Figure 14.3.6

find the θ-coordinate of the point P where the circle and cardioid intersect we eliminate r between Eqs. 10 and 11. This gives

$$2(1 + \cos \theta) = 3$$

or

$$\cos \theta = \tfrac{1}{2}. \tag{12}$$

The solution in the first quadrant is $\theta = \pi/3$.

We find the desired area by subtracting the area of the appropriate sector of the circle from that of the cardioid. Using Eq. (6), we have

$$\begin{aligned}
A &= 2\int_0^{\pi/3} \frac{1}{2}[4(1 + \cos \theta)^2 - 9]\, d\theta \\
&= \int_0^{\pi/3} (4 + 8 \cos \theta + 4 \cos^2 \theta - 9)\, d\theta \\
&= \int_0^{\pi/3} (-5 + 8 \cos \theta + 2 + 2 \cos 2\theta)\, d\theta \\
&= \int_0^{\pi/3} (-3 + 8 \cos \theta + 2 \cos 2\theta)\, d\theta \\
&= (-3\theta + 8 \sin \theta + \sin 2\theta)\Big|_0^{\pi/3} \\
&= -3\frac{\pi}{3} + 8\frac{\sqrt{3}}{2} + \frac{\sqrt{3}}{2} \\
&= -\pi + \frac{9}{2}\sqrt{3} \cong 4.65. \;\blacksquare
\end{aligned} \tag{13}$$

Arc length

Consider a polar curve $r = f(\theta)$ and recall that rectangular and polar coordinates are related by the equations

$$x = r\cos\theta, \qquad y = r\sin\theta. \tag{14}$$

Substituting $f(\theta)$ for r in Eqs. 14 we obtain

$$x = f(\theta)\cos\theta, \qquad y = f(\theta)\sin\theta. \tag{15}$$

Equations 15 form a set of parametric equations for the curve with θ as the parameter. From Eq. 11 in Section 7.3 the length of the arc between $\theta = \alpha$ and $\theta = \beta$ is given by

$$s = \int_\alpha^\beta \left[\left(\frac{dx}{d\theta}\right)^2 + \left(\frac{dy}{d\theta}\right)^2\right]^{1/2} d\theta. \tag{16}$$

From Eqs. 15 we have

$$\frac{dx}{d\theta} = f'(\theta)\cos\theta - f(\theta)\sin\theta, \tag{17}$$

$$\frac{dy}{d\theta} = f'(\theta)\sin\theta + f(\theta)\cos\theta. \tag{18}$$

Squaring these expressions and adding the results, we obtain

$$\left(\frac{dx}{d\theta}\right)^2 + \left(\frac{dy}{d\theta}\right)^2 = [f'(\theta)]^2 + [f(\theta)]^2. \tag{19}$$

Thus the integral giving the length of arc of the graph of $r = f(\theta)$ is

$$\begin{aligned} s &= \int_\alpha^\beta \{[f(\theta)]^2 + [f'(\theta)]^2\}^{1/2}\, d\theta \\ &= \int_\alpha^\beta \left[r^2 + \left(\frac{dr}{d\theta}\right)^2\right]^{1/2} d\theta. \end{aligned} \tag{20}$$

In differential form, we can write

$$\begin{aligned} ds &= \left[r^2 + \left(\frac{dr}{d\theta}\right)^2\right]^{1/2} d\theta \\ &= [(r\,d\theta)^2 + (dr)^2]^{1/2}, \end{aligned} \tag{21}$$

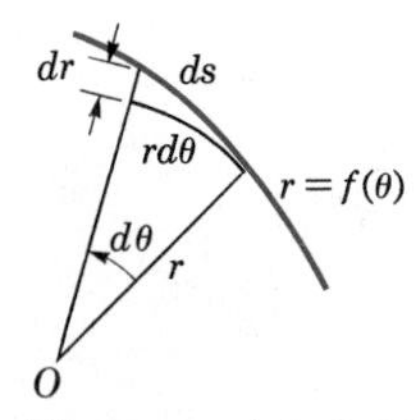

Figure 14.3.7

which is analogous to the expression

$$ds = [(dx)^2 + (dy)^2]^{1/2}$$

in rectangular coordinates. Equation 21 can be remembered by applying Pythagoras' theorem as shown in Figure 14.3.7.

EXAMPLE 3

Find the length of the arc of the cardioid

$$r = a(1 + \cos\theta) \tag{22}$$

that lies in the first two quadrants (see Figure 14.3.8).

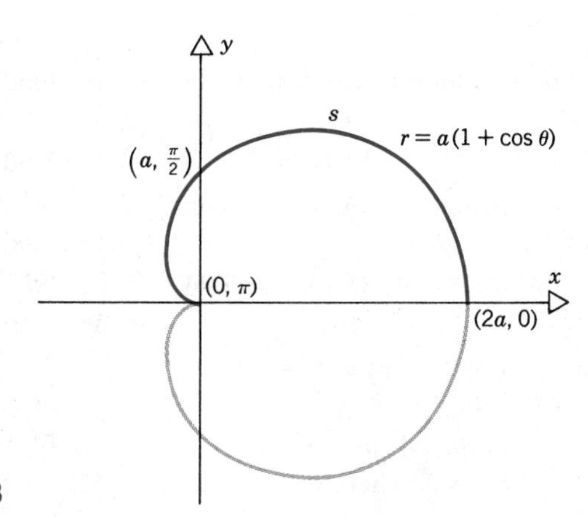

Figure 14.3.8

From Eq. 20 we have

$$\begin{aligned} s &= \int_0^\pi \left[r^2 + \left(\frac{dr}{d\theta}\right)^2\right]^{1/2} d\theta \\ &= \int_0^\pi [a^2(1 + \cos\theta)^2 + a^2(-\sin\theta)^2]^{1/2}\, d\theta \\ &= \sqrt{2}\, a \int_0^\pi [1 + \cos\theta]^{1/2}\, d\theta. \end{aligned}$$

By use of the half-angle identity the last integral is changed into

$$\begin{aligned} s &= \sqrt{2}a \int_0^\pi \left[2\cos^2\left(\frac{\theta}{2}\right)\right]^{1/2} d\theta \\ &= 2a \int_0^\pi \left|\cos\frac{\theta}{2}\right| d\theta. \end{aligned}$$

The absolute value bars are used because it is the nonnegative square root that is required. However, $\cos(\theta/2) \geq 0$ for $0 \leq \theta \leq \pi$, so the absolute value bars can be dropped. Thus we obtain

$$\begin{aligned} s &= 2a \int_0^\pi \cos\frac{\theta}{2}\, d\theta \\ &= 2a \cdot 2\sin\frac{\theta}{2}\Big|_0^\pi = 4a. \end{aligned}$$

By the symmetry of the cardioid about the line $\theta = 0$ it follows that the arc length of the entire cardioid is $8a$. ∎

PROBLEMS

1. Find the area inside one loop of the lemniscate $r^2 = 8 \cos 2\theta$.
2. Find the area of one leaf of $r = 4 \cos 2\theta$.
3. Find the area of the region within the limaçon $r = 1 + 0.5 \sin \theta$.
4. Find the area of the region within the cardioid $r = a(1 + \cos \theta)$.
5. Find the area enclosed by the spiral $r = 3\theta$, for $0 < \theta < \pi$, and by the ray $\theta = \pi$.
6. (a) Find the area of the region within the limaçon $r = 1 + b \cos \theta$, where $0 < b < 1$.

   (b) What are the limiting values of the area as $b \to 0$ and as $b \to 1$, respectively? Note that as $b \to 0$, the limaçon approaches the circle $r = 1$.
7. Consider the limaçon $r = 1 + 2 \cos \theta$.

   (a) Find the total area inside the outer loop.

   (b) Find the area within the smaller loop.

   (c) Find the area between the two loops.
8. Find the area of the region that is outside the circle $r = 2$ and inside the cardioid $r = 2(1 + \cos \theta)$.
9. Find the area of the region that is inside the cardioid $r = a(1 + \sin \theta)$ and outside the circle $r = 3a/2$.
10. Find the area outside the circle $r = 1$ and inside one loop of the leaf curve $r = 2 \sin 3\theta$.
11. Find the area outside the limaçon $r = 2 - \cos \theta$ and inside the cardioid $r = 1 + \cos \theta$.
12. Find the area inside the limaçon $r = 2 - \cos \theta$ and outside the cardioid $r = 1 + \cos \theta$.
13. Find the area within both the limaçon $r = 2 - \cos \theta$ and the cardioid $r = 1 + \cos \theta$.
14. Find the area of the region inside at least one of the limaçon $r = 2 - \cos \theta$ and the cardioid $r = 1 + \cos \theta$.
15. Find the arc length of the cardioid $r = 1 + \cos \theta$ in the first quadrant.
16. Find the arc length of the cardioid $r = a(1 + \sin \theta)$, $a > 0$, for $0 \le \theta \le 2\pi$.
17. Find the arc length of the circle $r = a \cos \theta$, $a > 0$, for $0 \le \theta \le \pi$.
18. Find the arc length of the spiral $r = k\theta$, $k > 0$, for $0 \le \theta \le \pi$.
19. Find the arc length of the graph of $r = a \sec \theta$, $a > 0$, for $0 \le \theta \le \pi/4$.
20. Find the arc length of the graph of $r = \sin^3 (\theta/3)$ for $0 \le \theta \le 3\pi$.
21. Find the arc length of the spiral $r = e^{k\theta}$, $k > 0$, for $0 \le \theta \le \pi$.
22. Find the arc length of the spiral $r = e^{k\theta}$, $k > 0$, for $-\infty < \theta \le 0$.
23. Find the arc length of that part of the cardioid $r = a(1 - \sin \theta)$, $a > 0$, which is in the first and fourth quadrants.
24. Set up (but do not evaluate) an integral giving the arc length of one leaf of the graph of $r = a \sin 3\theta$, $a > 0$.
25. Set up (but do not evaluate) an integral giving the arc length of one loop of the lemniscate $r^2 = 2a^2 \cos 2\theta$, $a > 0$.
26. Set up (but do not evaluate) an integral giving the arc length of the limaçon $r = 1 + b \cos \theta$, $b > 0$, for $0 \le \theta \le 2\pi$.

Surface Area

Suppose that the graph of $r = f(\theta)$ for $a \le \theta \le b$ is rotated about the x-axis (see Figure 14.3.9). The area of the surface thus formed can be found by an integral similar to the one derived in Section 7.3 in rectangular coordinates. The element dS of surface area is again given by $dS = 2\pi\, y\, ds$, where ds is the element of arc length. Rewriting this expression in polar coordinates, we have

$$dS = 2\pi r \sin \theta \sqrt{r^2 + \left(\frac{dr}{d\theta}\right)^2}\, d\theta, \qquad (i)$$

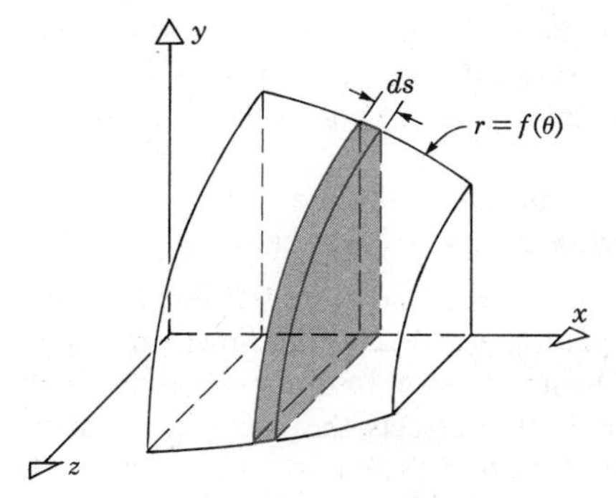

Figure 14.3.9

where $r = f(\theta)$. If we substitute $f(\theta)$ for r and integrate from $\theta = a$ to $\theta = b$, we obtain the required surface area. If the curve is rotated about the y-axis, then $dS = 2\pi\, x\, ds$, so in this case

$$dS = 2\pi r \cos\theta \sqrt{r^2 + \left(\frac{dr}{d\theta}\right)^2}\, d\theta, \qquad (ii)$$

where again $r = f(\theta)$.

In each of Problems 27 through 31, find the area of the surface of revolution formed in the indicated manner.

27. By rotating the circle $r = a \cos\theta$ about the x-axis.

28. By rotating the circle $r = a \cos\theta$ about the y-axis.

29. By rotating the cardioid $r = a(1 + \cos\theta)$ about the x-axis.

30. By rotating the lemniscate $r^2 = 2a^2 \cos 2\theta$ about the x-axis.

31. By rotating the lemniscate $r^2 = 2a^2 \cos 2\theta$ about the y-axis.

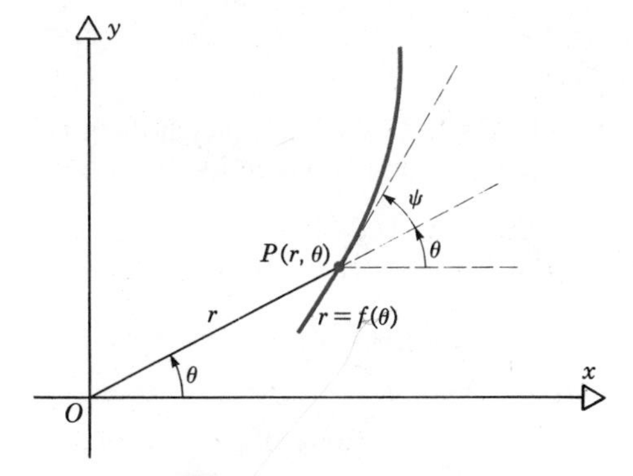

Figure 14.3.10

Tangents and Slopes

For a curve given by the polar equation $r = f(\theta)$, the derivative $dr/d\theta$ is not the slope of the curve, although it does give the rate of change of r with respect to θ. In Problem 32 we outline the derivation of a formula for the slope of a curve in polar coordinates.

Consider the situation shown in Figure 14.3.10. The slope of the graph of $r = f(\theta)$ is the tangent of the angle $(\theta + \psi)$, so we want to find a way to determine ψ.

(a) Refer now to Figure 14.3.11. The point $P(r, \theta)$ is an arbitrary point on the curve. If

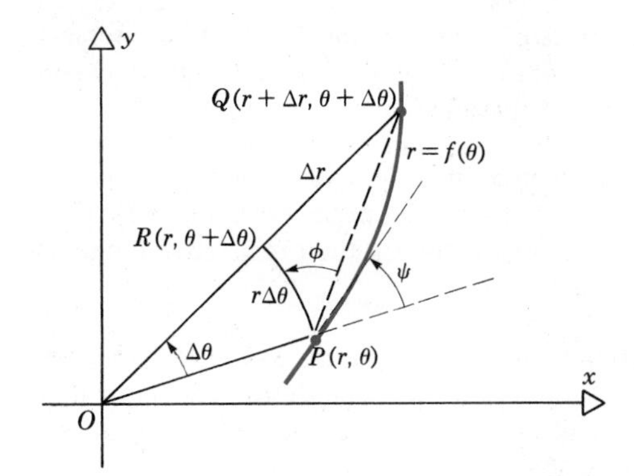

Figure 14.3.11

we change θ by a small amount $\Delta\theta$, then r changes by Δr, and we obtain the point $Q(r + \Delta r, \theta + \Delta\theta)$. The arc PR is an arc of the circle of radius r with center at the pole. Let ϕ be the angle between the line segment PQ and the arc PR. Show that

$$\tan\phi \cong \frac{\Delta r}{r\Delta\theta}.$$

(b) As $\Delta\theta \to 0$, show that

$$\tan\phi \to \frac{1}{r}\frac{dr}{d\theta}.$$

(c) Observe that $\phi + \psi \to \pi/2$ as $\Delta\theta \to 0$, and show that

$$\tan\psi = \frac{r}{dr/d\theta}.$$

It then follows that

$$\text{slope} = \tan(\theta + \psi) = \frac{\tan\theta + \tan\psi}{1 - \tan\theta\tan\psi}.$$

33. Find the slope of the cardioid $r = 1 + \cos\theta$ at the point where $\theta = \pi/4$.

34. Find the slope of the circle $r = 2 \sin \theta$ at the point where $\theta = 2\pi/3$.

35. Find the slope of the limaçon $r = 5 - 2 \cos \theta$ at the point where $\theta = \pi/3$.

36. Find the slope of the limaçon $r = 5 - 2 \cos \theta$ at the point where $\theta = -\pi/3$.

37. Find the slope of the circle $r = 3 \cos \theta$ at the point where $\theta = \pi/3$.

38. Find the slope of the leaf curve $r = a \sin 3\theta$ at the point where $\theta = \pi/12$.

39. The circles $r = 3 \cos \theta$ and $r = 3 \sin \theta$ intersect at the point $(3/\sqrt{2}, \pi/4)$. Find the angle between their respective tangents at this point.

40. The limaçon $r = 2 - \cos \theta$ and the cardioid $r = 1 + \cos \theta$ intersect at the point $(\frac{3}{2}, \pi/3)$. Find the acute angle between their respective tangents at this point.

41. Starting from the parametric equations (15), derive a formula for the slope dy/dx, and show that this result agrees with the one given in Problem 32.

42. Show that the cardioid $r = a(1 + \cos \theta)$ has a vertical tangent for $\theta = 0, 2\pi/3$, and $4\pi/3$, and that it has a horizontal tangent for $\theta = \pi/3, \pi$, and $5\pi/3$. Observe that the points of horizontal and vertical tangency alternate and are uniformly distributed in the angular variable θ.

43. Let $r = ae^{-b\theta}$, where $a > 0$ and $b > 0$. The graph of this equation for $\theta \geq 0$ is shown in Figure 14.3.12. The line tangent to the graph at the point P where $\theta = 0$ intersects the y-axis at Q. Show that the arc length of the curve for $\theta \geq 0$ is equal to the length of the line segment PQ; in other words, if the spiral is unrolled from P, it will exactly cover the segment PQ. Torricelli obtained this result in 1645.

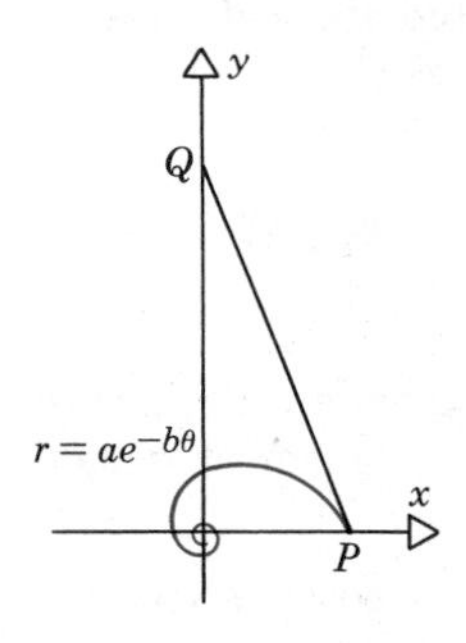

Figure 14.3.12

REVIEW PROBLEMS

In each of Problems 1 through 4, find a set of polar coordinates for the given point in which the sign of the r-coordinate is the opposite to that given, and find a set of polar coordinates for the given point in which the sign of the θ-coordinate is opposite to that given.

1. $\left(-2, \frac{5\pi}{2}\right)$
2. $\left(4, \frac{7\pi}{6}\right)$
3. $\left(3, -\frac{2\pi}{3}\right)$
4. $\left(-4, -\frac{\pi}{6}\right)$

In each of Problems 5 through 8, a set of polar coordinates of a point in the plane is given. Find the rectangular coordinates of the same point.

5. $\left(5, -\frac{\pi}{3}\right)$
6. $\left(4, \frac{7\pi}{6}\right)$
7. $\left(3, -\frac{2\pi}{3}\right)$
8. $\left(-2, \frac{3\pi}{4}\right)$

In each of Problems 9 through 12, the rectangular coordinates of a point are given. Find the set of polar coordinates of the same point for which $r \geq 0$, $-\pi < \theta \leq \pi$.

9. $(\sqrt{12}, 2)$
10. $(3, \sqrt{27})$
11. $(-2, 2\sqrt{3})$
12. $(5\sqrt{2}, -\sqrt{150})$

In each of Problems 13 through 20, sketch the graph of the given polar equation and identify the curve.

13. $r = 3 \sin \left(\theta - \frac{\pi}{6}\right)$
14. $r = 5 \csc \theta$
15. $r = \theta, \quad 0 \leq \theta \leq 2\pi$
16. $r = 2 \sin \theta + 3 \cos \theta$
17. $r = 1 + \cos \theta$
18. $r = 3 \sin 3\theta$
19. $r^2 = -\sin 2\theta$
20. $r = 2 - \cos \theta$

In each of Problems 21 through 24, sketch and identify the graph of the conic section described by the given equation. Also determine its eccentricity e and locate its foci and directrices.

21. $r = \dfrac{9}{1 + 3\sin\theta}$

22. $r = \dfrac{3}{1 + \cos\theta}$

23. $r = \dfrac{6}{4 - \cos\theta}$

24. $r = 2(r\cos\theta + 3)$

In each of Problems 25 through 28, find the polar equation of the curve having the given rectangular equation.

25. $3 = \dfrac{x\sqrt{x^2 + y^2}}{x + \sqrt{x^2 + y^2}}$

26. $1 = \dfrac{x^3}{2y} + \dfrac{y^3}{2x} + xy$

27. $y^2 = \dfrac{x^3}{1 - x}$

28. $x^2 = \dfrac{y^4}{4 - y^2}$

In each of Problems 29 through 32, find the rectangular equation of the curve having the given polar equation.

29. $\theta = \dfrac{11\pi}{6}$

30. $r^2 = \sin 2\theta$

31. $r = 2\cot\theta$

32. $r = \sec\theta - 2\cos\theta$

In each of Problems 33 through 36, find the points of intersection of the given equations.

33. $r = 3\sec\theta + 1, \quad r = 4$

34. $r = 4\sin\theta - 2\cos\theta, \quad r = 2 - 2\cos\theta$

35. $r^2 = -\sin 2\theta, \quad \theta = \dfrac{3\pi}{4}$

36. $r = \sec\theta - \cos\theta, \quad r = \cos\theta$

In each of Problems 37 through 40, find the area of the region bounded by the given equations.

37. $r = 2 + \sin\theta$

38. $r = 1 + \sin\theta$

39. one loop of $r^2 = 4\sin 2\theta$

40. one leaf of $r = 7\cos 5\theta$

In each of Problems 41 through 44, set up an integral for the arc length of the given curve for the given range of θ.

41. $r = 3\csc\theta, \quad \dfrac{\pi}{4} \le \theta \le \dfrac{3\pi}{4}$

42. $r = 5\cos\theta, \quad 0 \le \theta \le 2\pi$

43. $r^2 = 4\cos 2\theta, \quad 0 \le \theta \le \dfrac{\pi}{6}$

44. $r = 7\sin 5\theta, \quad 0 \le \theta \le 2\pi$

In each of Problems 45 through 48, find the slope of the tangent of the given curve at the point where $\theta = \pi/3$.

45. $r = 3 - \cos\theta$

46. $r = 2\cos\theta + 3\sin\theta$

47. $r = \sec\theta - \cos\theta$

48. $r = 7\cos 5\theta$

FIFTEEN

vectors and three-dimensional analytic geometry

There are many important quantities in physics and engineering that have both a magnitude and a direction. For example, force, velocity, acceleration, and momentum have this character. Such quantities are called **vectors**.* In contrast, many other important physical quantities (temperature, pressure, mass, and energy, for instance) have only a magnitude but no direction. These quantities are called **scalars**. A scalar quantity is described mathematically by a real number (referred to some given scale of measurement) and its variation in space or time is

*Vectors were developed in the 1880s by Josiah Willard Gibbs and Oliver Heaviside, working independently on the theory of electricity and magnetism. Gibbs (1839–1903) was professor of mathematical physics at Yale, and is also known for his important contributions to thermodynamics. Heaviside (1850–1925) was English, lacked a university education, and became interested in electricity while working as a telegraph operator as a young man. He is particularly known for the development of the operational calculus and its application to electrical problems. In their treatment of vectors both Gibbs and Heaviside drew heavily on the theory of quaternions that had been devised by William Rowan Hamilton (1805–1865) beginning in 1843. Both were also influenced by James Clerk Maxwell (1831–1879), who used quaternions in his study of electromagnetic phenomena, but who also pointed out their shortcomings as a means of representing physical quantities. After the publication of Gibbs's work on vectors in 1881 and 1884, followed by Heaviside's in 1893, there was a sharp controversy between their followers and the proponents of quaternions. The simplicity and effectiveness of vectors won out, and quaternions receded into history.

described by a function whose values are real numbers. The major purpose of this chapter is to develop the appropriate mathematical framework for dealing with vector quantities. To keep things as simple as possible initially we start with vectors in two dimensions. Beginning in Section 15.3 we extend the discussion to include three-dimensional vectors.

15.1 VECTORS IN TWO DIMENSIONS

In order to distinguish vectors from scalars, we will use bold-faced letters, either capital or lowercase, to denote vectors; for example, **a**, **b**, **u**, **U**, and so forth. Since vectors have both magnitude and direction it is convenient to represent them geometrically as arrows; see Figure 15.1.1, where a vector **a** is shown. The magnitude of the vector **a** is the length of the arrow, and its direction is the direction of the arrow. The initial and terminal points of the vector are labeled P and Q, respectively, in Figure 15.1.1. Sometimes a vector may be required to occupy a fixed location, for example, by requiring a certain point to be the initial point of the vector. On other occasions the location of a vector may not be specified. In either case, and regardless of their location in the plane, two vectors are said to be equal if they have the same magnitude and the same direction.

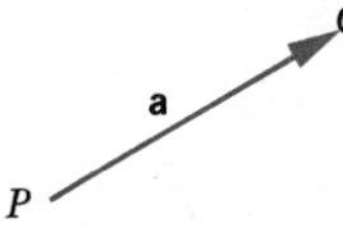

Figure 15.1.1

Let us now place the vector (arrow) **a** so that its initial point lies at the origin O of a rectangular coordinate system, as in Figure 15.1.2. The terminal point of **a**

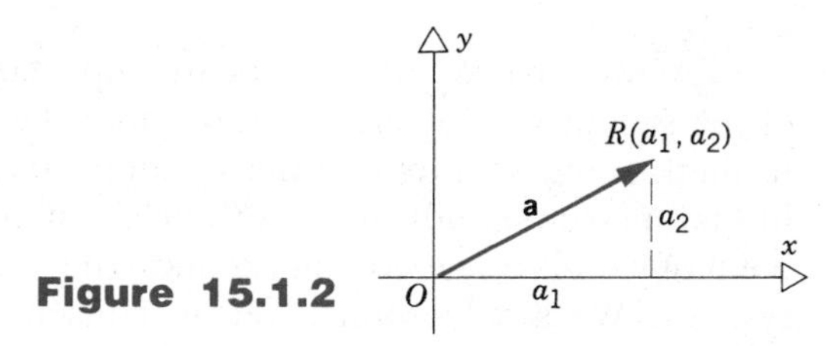

Figure 15.1.2

now coincides with a point R in the xy-plane whose coordinates are (a_1, a_2). When its initial point is placed at the origin, the vector **a** serves to identify the point R, and **a** is then called the **position vector** of R. Conversely, the point R or its coordinates (a_1, a_2) can be used to identify the vector extending from the origin to R. Thus we write

$$\mathbf{a} = (a_1, a_2) \tag{1}$$

to denote the vector from O to R, or any other vector equal to this one but located elsewhere in the plane. The numbers a_1 and a_2 in Eq. 1 are referred to as the ***x*-component** and the ***y*-component,** respectively, of the vector **a**.

In this way we establish a correspondence, expressed by Eq. 1, between vectors and the coordinates of their terminal points (assuming that their initial points are at the origin). This correspondence enables us to represent vectors by pairs of real numbers, and vice versa.

The magnitude or length of the vector **a** is the distance from the origin to the point with coordinates (a_1, a_2). We denote the magnitude of **a** by $\|\mathbf{a}\|$, and it follows that

$$\|\mathbf{a}\| = \sqrt{a_1^2 + a_2^2} \tag{2}$$

(see Figure 15.1.2). For any vector **a** we have $\|\mathbf{a}\| \geq 0$; further, $\|\mathbf{a}\| = 0$ if and only if $a_1 = a_2 = 0$.

For example, the vector $\mathbf{a} = (-2, 3)$ shown in Figure 15.1.3*a* has length

$$\|\mathbf{a}\| = \sqrt{(-2)^2 + 3^2} = \sqrt{13}.$$

On the other hand, the vector **a** in Figure 15.1.3*b* is represented by an arrow 5

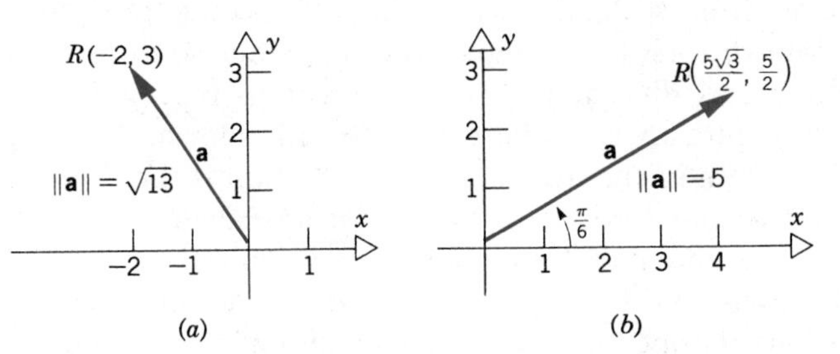

Figure 15.1.3

units long that makes an angle of $\pi/6$ radians with the positive x-axis. The terminal point R has the coordinates $(5\sqrt{3}/2, 5/2)$, so we write

$$\mathbf{a} = \left(\frac{5\sqrt{3}}{2}, \frac{5}{2}\right).$$

As we have noted, vectors are important in physics and engineering because of the significance of the quantities that they represent. It is very often desirable to think about vectors and the relations among them in a geometrical way so as to take maximum advantage of our geometrical intuition. One advantage of this point of view is that it does not require the introduction of any particular coordinate systems. We are therefore free to concentrate on the intrinsic vector relations without the notational or other encumbrances associated with coordinate systems.

On the other hand, vectors are computationally convenient because it is possible to formulate an algebraic system in which it is straightforward to carry out calculations; this does require the presence of a coordinate system. Since we often gain insight by thinking geometrically, and almost always perform calculations algebraically, it is important to be able to shift rapidly from one viewpoint to the other.

We now turn to a description of the algebra of vectors.

(a) ***Equality.*** Two vectors $\mathbf{a} = (a_1, a_2)$ and $\mathbf{b} = (b_1, b_2)$ are equal if and only if they have the same magnitude and the same direction. If their initial points are placed at the origin, then their terminal points must also coincide. Thus

$$\mathbf{a} = \mathbf{b} \quad \text{means} \quad a_1 = b_1 \quad \text{and} \quad a_2 = b_2. \tag{3}$$

In words, **a** and **b** are equal if and only if each component of **a** is equal to the corresponding component of **b**.

(b) ***Zero.*** The zero vector, denoted by **0**, is the vector of length zero. As a result, if its initial point is at the origin, then so is its terminal point, so that

$$\mathbf{0} = (0, 0). \tag{4}$$

Thus the vector **0** is the vector whose components are both zero. The direction of the zero vector is indeterminate.

(c) ***Addition.*** The appropriate definition of the sum of two vectors is suggested by the manner in which forces are added in physics or in mechanics. If the initial point of **b** is placed at the terminal point of **a** as shown in Figure 15.1.4*a*, then **a** + **b** is the vector drawn from the initial point of **a** to the

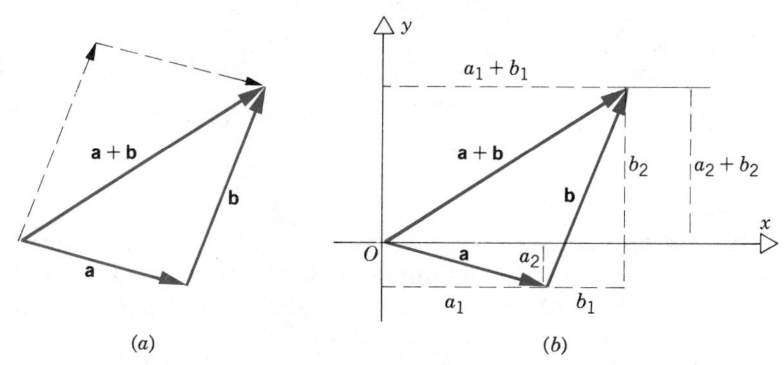

Figure 15.1.4 The parallelogram rule of vector addition.

terminal point of **b**. This is sometimes called the parallelogram rule, since **a** + **b** is the diagonal of the parallelogram with sides **a** and **b**.

Now let us obtain the formula for the components of **a** + **b**. Let $\mathbf{a} = (a_1, a_2)$ and $\mathbf{b} = (b_1, b_2)$. Then from Figure 15.1.4*b* it can be seen that the x-coordinate of the terminal point of **a** + **b** is $a_1 + b_1$. Similarly, the y-coordinate of this point is $a_2 + b_2$, where a_2 is negative in the case shown. By considering other combinations of signs of the components of **a** and **b** it is possible to show in all cases that

$$\mathbf{a} + \mathbf{b} = (a_1 + b_1, a_2 + b_2). \tag{5}$$

In other words, the addition of two vectors is accomplished by adding corresponding components.

Vector addition has the following two important properties of ordinary addition:

$$\mathbf{a} + \mathbf{b} = \mathbf{b} + \mathbf{a} \qquad \text{(Commutative Law)} \tag{6}$$

$$(\mathbf{a} + \mathbf{b}) + \mathbf{c} = \mathbf{a} + (\mathbf{b} + \mathbf{c}) \qquad \text{(Associative Law).} \tag{7}$$

It is easy to establish both of these properties. For example **a** + **b** is given by Eq. 5, while $\mathbf{b} + \mathbf{a} = (b_1 + a_1, b_2 + a_2)$. Since $a_1 + b_1 = b_1 + a_1$ and $a_2 + b_2 = b_2 + a_2$ by the commutative law of addition of real numbers, it follows that Eq. 6 is true. Equation 7 can be established in a similar manner.

EXAMPLE 1

If $\mathbf{a} = (-2, 3)$ and $\mathbf{b} = (4, -1)$, find $\mathbf{a} + \mathbf{b}$.

From Eq. 5 we obtain

$$\mathbf{a} + \mathbf{b} = (-2, 3) + (4, -1) = (-2 + 4, 3 - 1) = (2, 2).$$

The vectors **a**, **b**, and **a** + **b** are shown in Figure 15.1.5. ∎

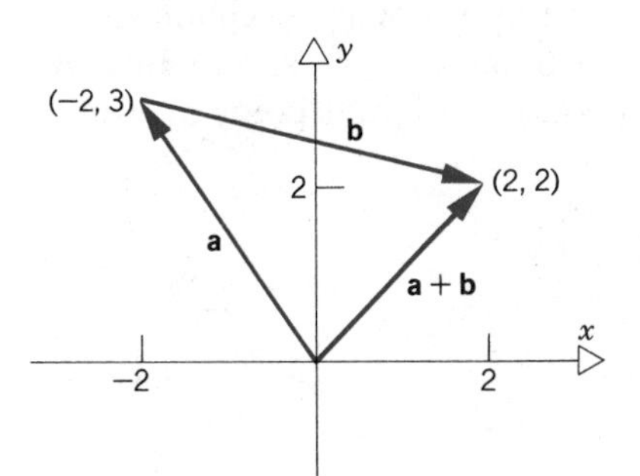

Figure 15.1.5

(d) ***Multiplication of a vector by a scalar.*** Let **a** be a given (nonzero) vector and let λ be a scalar (real number). If $\lambda > 0$, then $\lambda\mathbf{a}$ or $\mathbf{a}\lambda$ is the vector whose direction is the same as that of **a** and whose length is λ times the length of **a** (see Figure 15.1.6*a*). If $\lambda < 0$, then the direction of $\lambda\mathbf{a}$ is opposite to that of **a** and the length of $\lambda\mathbf{a}$ is $|\lambda|$ times the length of **a** (see Figure 15.1.6*b*). If $\lambda = 0$, or if $\mathbf{a} = \mathbf{0}$, then $\lambda\mathbf{a}$ is the zero vector.

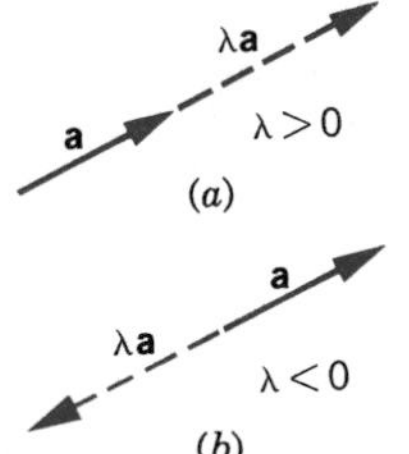

Figure 15.1.6 Multiplication of a vector by a scalar.

To determine the components of $\lambda\mathbf{a}$ let $\mathbf{a} = (a_1, a_2)$ and consider the situation in Figure 15.1.7, where $\lambda > 0$. The two triangles are similar and

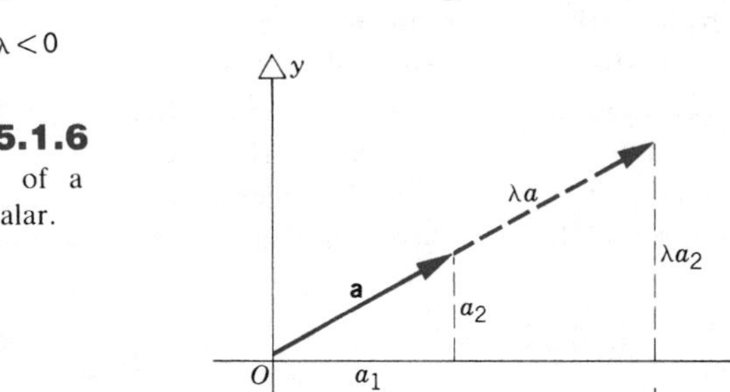

Figure 15.1.7

their hypotenuses have the ratio λ. Thus the other pairs of corresponding sides also have this ratio. This means that the components of $\lambda\mathbf{a}$ are λa_1 and λa_2, and we can write

$$\lambda\mathbf{a} = \lambda(a_1, a_2) = (\lambda a_1, \lambda a_2). \tag{8}$$

A similar argument shows that Eq. 8 is also true when $\lambda < 0$.

The type of multiplication defined by Eq. 8 has the following familiar properties of ordinary multiplication:

$$(\lambda\mu)\mathbf{a} = \lambda(\mu\mathbf{a}), \qquad \text{(Associative Law)} \tag{9}$$

$$(\lambda + \mu)\mathbf{a} = \lambda\mathbf{a} + \mu\mathbf{a}, \qquad \text{(Distributive Law)} \qquad (10)$$

$$\lambda(\mathbf{a} + \mathbf{b}) = \lambda\mathbf{a} + \lambda\mathbf{b}. \qquad \text{(Distributive Law)} \qquad (11)$$

It is easy to prove each of these properties. For example, to establish Eq. 10 note that

$$\begin{aligned}\lambda\mathbf{a} + \mu\mathbf{a} &= (\lambda a_1, \lambda a_2) + (\mu a_1, \mu a_2)\\ &= (\lambda a_1 + \mu a_1, \lambda a_2 + \mu a_2)\\ &= ((\lambda + \mu)a_1, (\lambda + \mu)a_2)\\ &= (\lambda + \mu)\mathbf{a}.\end{aligned}$$

The other properties can be proved in a similar manner.

EXAMPLE 2

If $\mathbf{a} = (2, -1)$ and $\lambda = 3$, find $\lambda\mathbf{a}$.

From Eq. 8 it follows that

$$\lambda\mathbf{a} = 3(2, -1) = (6, -3).$$

The vector $\lambda\mathbf{a}$ has the same direction as $\mathbf{a}$ and is three times as long (see Figure 15.1.8). ∎

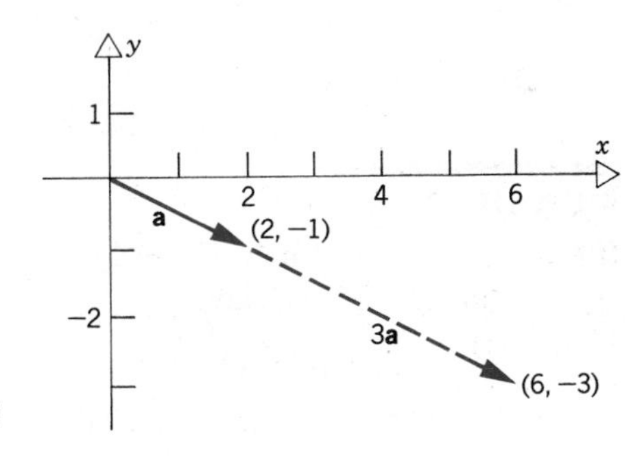

Figure 15.1.8

EXAMPLE 3

If $\mathbf{a} = (2, 3)$, $\mathbf{b} = (-3, 1)$, and $\lambda = -2$, find $\lambda(\mathbf{a} + \mathbf{b})$.

One way to proceed is to find the sum of $\mathbf{a}$ and $\mathbf{b}$

$$\mathbf{a} + \mathbf{b} = (2 - 3, 3 + 1) = (-1, 4).$$

Then from Eq. 8 we have

$$\lambda(\mathbf{a} + \mathbf{b}) = -2(-1, 4) = (2, -8). \qquad (12)$$

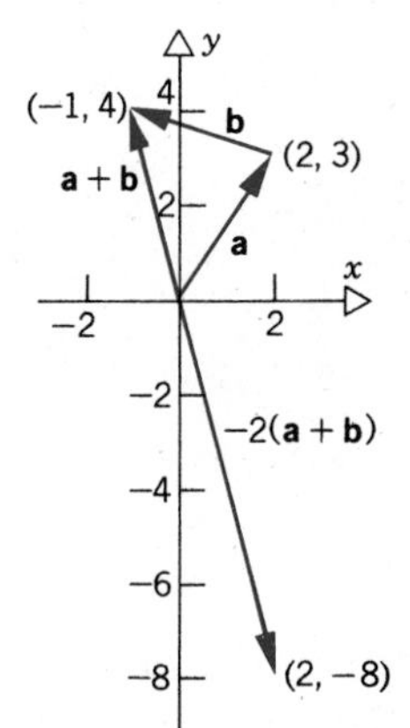

Figure 15.1.9

The vectors $\mathbf{a}$, $\mathbf{b}$, $\mathbf{a} + \mathbf{b}$, and $\lambda(\mathbf{a} + \mathbf{b})$ are shown in Figure 15.1.9.

Alternatively, we can find $\lambda\mathbf{a}$ and $\lambda\mathbf{b}$ separately, and then add them together by the distributive law (11). In this way we obtain

$$\lambda\mathbf{a} = -2(2, 3) = (-4, -6),$$

$$\lambda\mathbf{b} = -2(-3, 1) = (6, -2),$$

and

$$\lambda\mathbf{a} + \lambda\mathbf{b} = (-4, -6) + (6, -2) = (2, -8). \tag{13}$$

Observe that this result agrees with Eq. 12, which confirms the distributive law (11) in this case. ∎

(e) ***Subtraction.*** The difference $\mathbf{c} = \mathbf{a} - \mathbf{b}$ between two vectors $\mathbf{a} = (a_1, a_2)$ and $\mathbf{b} = (b_1, b_2)$ is defined by requiring $\mathbf{c}$ to be the vector that when added to $\mathbf{b}$ gives $\mathbf{a}$ (see Figure 15.1.10). Thus $\mathbf{a} - \mathbf{b}$ is the vector drawn from the terminal point of $\mathbf{b}$ to the terminal point of $\mathbf{a}$. To determine the components of $\mathbf{a} - \mathbf{b}$, let $\mathbf{a} - \mathbf{b} = (c_1, c_2)$, and observe that (from Figure 15.1.10)

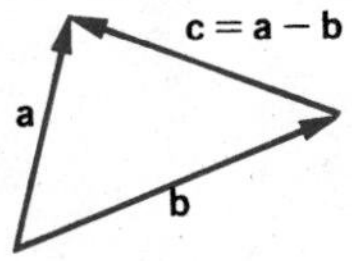

Figure 15.1.10 Vector subtraction.

$$\mathbf{b} + (\mathbf{a} - \mathbf{b}) = \mathbf{a}. \tag{14}$$

Writing Eq. 14 in terms of components, we have

$$b_1 + c_1 = a_1, \qquad b_2 + c_2 = a_2$$

or

$$c_1 = a_1 - b_1, \qquad c_2 = a_2 - b_2.$$

Thus

$$\mathbf{a} - \mathbf{b} = (a_1 - b_1, a_2 - b_2). \tag{15}$$

EXAMPLE 4

If $\mathbf{a} = (1, -2)$ and $\mathbf{b} = (3, -1)$, find $\mathbf{a} - \mathbf{b}$.

From Eq. 15 we have

$$\begin{aligned}\mathbf{a} - \mathbf{b} &= (1, -2) - (3, -1) = (1 - 3, -2 - [-1]) \\ &= (-2, -1).\end{aligned}$$

Figure 15.1.11 shows the vectors $\mathbf{a}$, $\mathbf{b}$, and $\mathbf{a} - \mathbf{b}$.

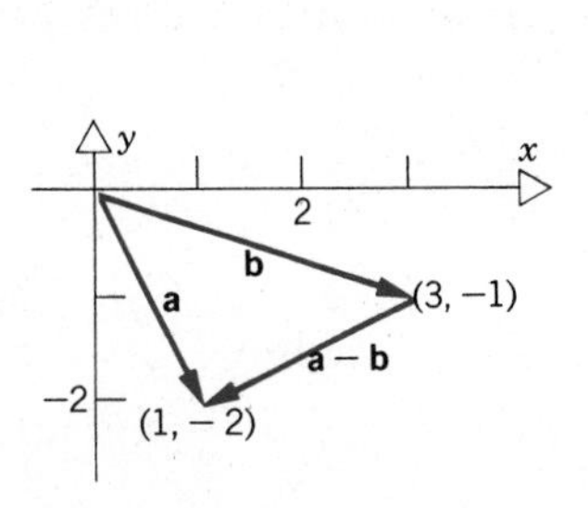

Figure 15.1.11

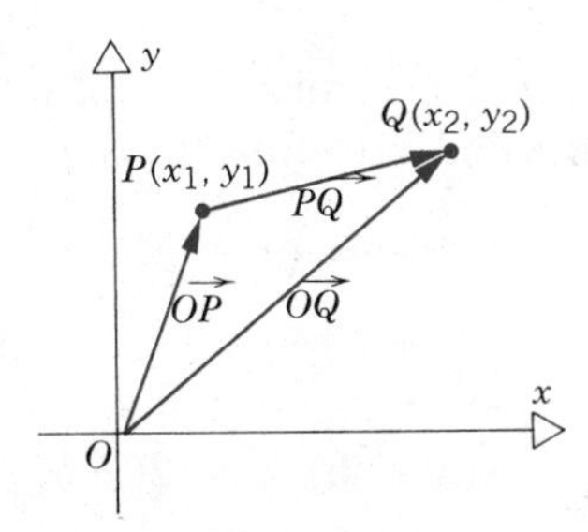

Figure 15.1.12

The vector whose initial point is P and whose terminal point is Q is often referred to as the vector from P to Q and may be denoted by $\overrightarrow{PQ}$. We represent this vector geometrically by drawing an arrow from P to Q, as shown in Figure 15.1.12. To find the components of $\overrightarrow{PQ}$ we introduce the position vectors $\overrightarrow{OP}$ and $\overrightarrow{OQ}$ of P and Q, respectively. Then

$$\overrightarrow{PQ} = \overrightarrow{OQ} - \overrightarrow{OP}. \tag{16}$$

If (x_1, y_1) are the coordinates of P and (x_2, y_2) are the coordinates of Q, then

$$\overrightarrow{OP} = (x_1, y_1), \qquad \overrightarrow{OQ} = (x_2, y_2).$$

Consequently, by Eq. 15,

$$\overrightarrow{PQ} = (x_2, y_2) - (x_1, y_1) = (x_2 - x_1, y_2 - y_1). \tag{17}$$

Thus the components of $\overrightarrow{PQ}$ are found by subtracting the coordinates of its initial point from the corresponding coordinates of its terminal point.

EXAMPLE 5

Find the components of the vector from $P(-2, 1)$ to $Q(4, -3)$.

The points P and Q and the vector $\overrightarrow{PQ}$ are shown in Figure 15.1.13. By Eq. 17

$$\overrightarrow{PQ} = (4, -3) - (-2, 1) = (4 + 2, -3 - 1) = (6, -4). \quad \blacksquare$$

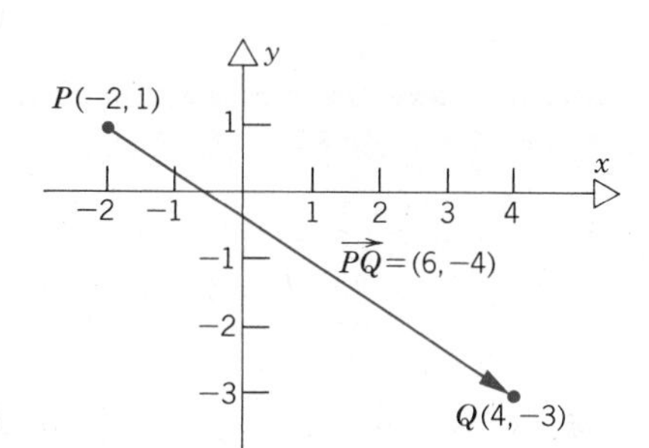

Figure 15.1.13

PROBLEMS

In each of Problems 1 through 4, find $\mathbf{a} + \mathbf{b}$, $\mathbf{a} - \mathbf{b}$, and $2\mathbf{a} - 3\mathbf{b}$.

1. $\mathbf{a} = (-2, 1)$, $\quad \mathbf{b} = (3, -2)$
2. $\mathbf{a} = (1, 0)$, $\quad \mathbf{b} = (-2, 1)$
3. $\mathbf{a} = (1, 1)$, $\quad \mathbf{b} = (1, -1)$
4. $\mathbf{a} = (2, -5)$, $\quad \mathbf{b} = (1, 4)$

In each of Problems 5 through 8, find the vector $\overrightarrow{PQ}$.

5. $P = (2, 1)$, $\quad Q = (4, -2)$
6. $P = (-1, -4)$, $\quad Q = (2, -2)$
7. $P = (2, 4)$, $\quad Q = (-1, 0)$
8. $P = (5, -2)$, $\quad Q = (1, 4)$

9. If $\mathbf{a} = (2, 1)$ is the position vector of the point P and if $\mathbf{b} = (3, -2)$ is the position vector of the point Q, find the components of the vector from P to Q; from Q to P.

10. If $\mathbf{a} = (2, -1)$ is the position vector of the point P and if the vector from P to Q is $\overrightarrow{PQ} = (-3, 2)$, find the position vector $\mathbf{b}$ of the point Q.

11. Find x and y if the vectors $\mathbf{a} = (x + y, x - y)$ and $\mathbf{b} = (2, 3)$ are equal.

12. Find x and y if $\mathbf{a} = 2\mathbf{b}$, where $\mathbf{a} = (x + y, 2)$ and $\mathbf{b} = (3, x - 2y)$.

In each of Problems 13 through 18, find the length $\|\mathbf{a}\|$ of the given vector.

13. $\mathbf{a} = (2, 1)$

14. $\mathbf{a} = (-1, 3)$

15. $\mathbf{a} = (-3, -4)$

16. $\mathbf{a} = (x, x)$, x a real number

17. $\mathbf{a} = (x + y, x - y)$, x and y real numbers

18. $\mathbf{a} = (3x, -4x)$, x a real number

19. Find a vector $\mathbf{b}$ of length 5 that is perpendicular to the vector $\mathbf{a} = (1, 2)$.

20. Find the vector $\mathbf{b}$ of length 8 that has the same direction as the vector from $P(2, 1)$ to $Q(-1, 4)$.

21. Choose λ so that the vectors $(2, -3)$ and $(\lambda, 1)$ have the same length.

22. Prove that $(\mathbf{a} + \mathbf{b}) + \mathbf{c} = \mathbf{a} + (\mathbf{b} + \mathbf{c})$ for any vectors $\mathbf{a}$, $\mathbf{b}$, and $\mathbf{c}$.

23. Prove that $(\lambda\mu)\mathbf{a} = \lambda(\mu\mathbf{a})$ for any vector $\mathbf{a}$ and for any scalars λ and μ.

24. Prove that $\lambda(\mathbf{a} + \mathbf{b}) = \lambda\mathbf{a} + \lambda\mathbf{b}$ for any vectors $\mathbf{a}$ and $\mathbf{b}$, and for any scalar λ.

25. A basic principle of mechanics states that if a body is in equilibrium, then the vector sum of the forces acting on the body must be zero. A weight W is suspended from a cable making an angle θ with the horizontal, as shown in Figure 15.1.14. Find the force T in the cable.

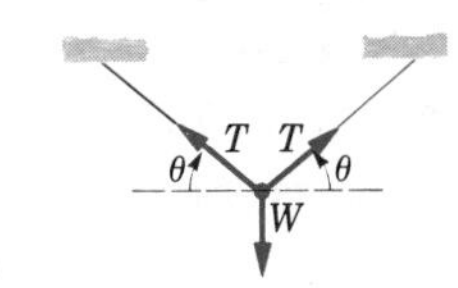

Figure 15.1.14

15.2 THE DOT PRODUCT

We continue the development of properties of vectors in two dimensions. In this section we focus on the multiplication of one vector by another.

Scalar (or dot) product

It turns out to be useful to define two different ways of multiplying two vectors together. The first type of product that we discuss always results in a scalar (or real number), and is called the **scalar product**. Since it is often denoted by a $\cdot$, it is also called the **dot product**. If $\mathbf{a} = (a_1, a_2)$ and $\mathbf{b} = (b_1, b_2)$, then the scalar or dot product of $\mathbf{a}$ and $\mathbf{b}$ is defined to be

$$\mathbf{a} \cdot \mathbf{b} = a_1 b_1 + a_2 b_2. \tag{1}$$

In words, corresponding components of the two vectors are multiplied together, and the results are then added. The scalar product has an important geometrical interpretation that we will discuss a little later.

EXAMPLE 1

If $\mathbf{a} = (3, -2)$ and $\mathbf{b} = (-1, 3)$, find $\mathbf{a} \cdot \mathbf{b}$. From Eq. 1 we obtain

$$\mathbf{a} \cdot \mathbf{b} = (3)(-1) + (-2)(3) = -3 - 6 = -9. \blacksquare$$

The scalar product has the following properties:

$$\mathbf{a} \cdot \mathbf{b} = \mathbf{b} \cdot \mathbf{a}, \qquad \text{(Commutative Law);} \tag{2}$$

$$\mathbf{a} \cdot (\mathbf{b} \pm \mathbf{c}) = \mathbf{a} \cdot \mathbf{b} \pm \mathbf{a} \cdot \mathbf{c}, \qquad \text{(Distributive Law);} \tag{3}$$

and, if λ is a scalar,

$$\lambda(\mathbf{a} \cdot \mathbf{b}) = (\lambda\mathbf{a}) \cdot \mathbf{b} = \mathbf{a} \cdot (\lambda\mathbf{b}), \qquad \text{(Associative Law).} \tag{4}$$

Note that it makes no sense to write an expression such as $(\mathbf{a} \cdot \mathbf{b}) \cdot \mathbf{c}$; the result of the first multiplication is a scalar, and one cannot form the dot product of a scalar $(\mathbf{a} \cdot \mathbf{b})$ and a vector $\mathbf{c}$. The expression $(\mathbf{a} \cdot \mathbf{b})\mathbf{c}$, however, is well-defined; the second multiplication is the product of a scalar and a vector, as discussed in Section 15.1.

Properties (2), (3), and (4) are easily proved. For example, to establish Eq. 3 let us suppose that $\mathbf{a} = (a_1, a_2)$, $\mathbf{b} = (b_1, b_2)$, and $\mathbf{c} = (c_1, c_2)$. Then

$$\begin{aligned}
\mathbf{a} \cdot (\mathbf{b} \pm \mathbf{c}) &= (a_1, a_2) \cdot [(b_1, b_2) \pm (c_1, c_2)] \\
&= (a_1, a_2) \cdot (b_1 \pm c_1, b_2 \pm c_2) \\
&= a_1(b_1 \pm c_1) + a_2(b_2 \pm c_2) \\
&= a_1b_1 \pm a_1c_1 + a_2b_2 \pm a_2c_2 \\
&= (a_1b_1 + a_2b_2) \pm (a_1c_1 + a_2c_2) \\
&= \mathbf{a} \cdot \mathbf{b} \pm \mathbf{a} \cdot \mathbf{c},
\end{aligned}$$

thus proving Eq. 3.

Relation between scalar product and length of a vector

In Section 15.1 we noted that the length of a vector $\mathbf{a} = (a_1, a_2)$ is given by

$$\|\mathbf{a}\| = \sqrt{a_1^2 + a_2^2}. \tag{5}$$

If we form the scalar product of $\mathbf{a}$ with itself, we have

$$\mathbf{a} \cdot \mathbf{a} = a_1^2 + a_2^2 = \|\mathbf{a}\|^2. \tag{6}$$

Also, if λ is a scalar, then

$$\begin{aligned}
\|\lambda\mathbf{a}\| &= \|(\lambda a_1, \lambda a_2)\| = \sqrt{\lambda^2 a_1^2 + \lambda^2 a_2^2} \\
&= |\lambda| \sqrt{a_1^2 + a_2^2} = |\lambda|\,\|\mathbf{a}\|.
\end{aligned} \tag{7}$$

Note that $|\lambda|$ is needed on the right side of Eq. 7, since λ may be negative and $\|\lambda \mathbf{a}\|$ and $\|\mathbf{a}\|$ are not.

Unit vectors

Any vector of length one is called a unit vector. Sometimes it is useful to determine a unit vector $\mathbf{u}$ having the same direction as a given nonzero vector $\mathbf{a}$. To do this we simply divide $\mathbf{a}$ by its own length, so that

$$\mathbf{u} = \frac{\mathbf{a}}{\|\mathbf{a}\|}. \tag{8}$$

Since $\|\mathbf{a}\| > 0$, $\mathbf{u}$ has the same direction as $\mathbf{a}$, and since $1/\|\mathbf{a}\|$ is a scalar, the length of $\mathbf{u}$ is $1/\|\mathbf{a}\|$ times the length of $\mathbf{a}$, or one. Hence $\mathbf{u}$ is the unit vector in the direction of $\mathbf{a}$.

EXAMPLE 2

Find a unit vector $\mathbf{u}$ in the same direction as $\mathbf{a} = (2, -3)$.

From Eq. 5 we have $\|\mathbf{a}\| = \sqrt{13}$, so

$$\mathbf{u} = \frac{1}{\sqrt{13}}(2, -3) = \left(\frac{2}{\sqrt{13}}, -\frac{3}{\sqrt{13}}\right). \blacksquare$$

Geometrical interpretation of the scalar product

Let $\mathbf{a}$ and $\mathbf{b}$ be arbitrary nonzero vectors with the same initial point, and let θ be the angle between them such that $0 \leq \theta \leq \pi$. Consider the triangle shown in Figure 15.2.1 whose sides are $\mathbf{a}$, $\mathbf{b}$, and $\mathbf{a} - \mathbf{b}$, respectively. To understand the geometrical significance of the scalar product we make use of the law of cosines from trigonometry, which (in vector form) states that

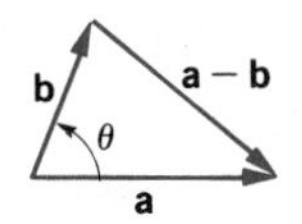

Figure 15.2.1

$$\|\mathbf{a} - \mathbf{b}\|^2 = \|\mathbf{a}\|^2 + \|\mathbf{b}\|^2 - 2\|\mathbf{a}\|\,\|\mathbf{b}\|\cos\theta. \tag{9}$$

Introducing the dot product into the first three terms of Eq. 9 by means of Eq. 6, we obtain

$$(\mathbf{a} - \mathbf{b}) \cdot (\mathbf{a} - \mathbf{b}) = \mathbf{a} \cdot \mathbf{a} + \mathbf{b} \cdot \mathbf{b} - 2\|\mathbf{a}\|\,\|\mathbf{b}\|\cos\theta. \tag{10}$$

Using Eqs. 2 and 3 to expand the left side of Eq. 10, we obtain

$$\mathbf{a} \cdot \mathbf{a} - 2\mathbf{a} \cdot \mathbf{b} + \mathbf{b} \cdot \mathbf{b} = \mathbf{a} \cdot \mathbf{a} + \mathbf{b} \cdot \mathbf{b} - 2\|\mathbf{a}\|\,\|\mathbf{b}\|\cos\theta, \tag{11}$$

so

$$\mathbf{a} \cdot \mathbf{b} = \|\mathbf{a}\|\,\|\mathbf{b}\|\cos\theta. \tag{12}$$

Equation 12 is the fundamental relation between the scalar product and geometrical quantities: the lengths of the vectors $\mathbf{a}$ and $\mathbf{b}$, and the cosine of the angle between them. Observe that Eq. 12 is independent of any particular coordinate system, and expresses the intrinsic geometrical property of the scalar product.

An alternative way of discussing the scalar product is to take the geometrical expression (12) as the definition, and then to reverse the argument we have given here to derive the algebraic expression given by Eq. 1.

By solving Eq. 12 for $\cos\theta$ we obtain

$$\cos\theta = \frac{\mathbf{a}\cdot\mathbf{b}}{\|\mathbf{a}\|\,\|\mathbf{b}\|}, \tag{13}$$

which provides a way to determine the angle between the two vectors **a** and **b.** We can also write Eq. 13 in the form

$$\cos\theta = \left(\frac{\mathbf{a}}{\|\mathbf{a}\|}\right)\cdot\left(\frac{\mathbf{b}}{\|\mathbf{b}\|}\right) = \mathbf{u}\cdot\mathbf{v}, \tag{14}$$

where **u** and **v** are unit vectors. In other words, the dot product of two unit vectors gives the cosine of the angle between them.

EXAMPLE 3

Find the angle θ between the vectors $\mathbf{a} = (2, -1)$ and $\mathbf{b} = (3, 1)$ (see Figure 15.2.2).

From Eq. 13 we find that

$$\cos\theta = \frac{(2)(3) + (-1)(1)}{[(2)^2 + (-1)^2]^{1/2}\,[(3)^2 + (1)^2]^{1/2}}$$

$$= \frac{5}{\sqrt{5}\,\sqrt{10}} = \frac{1}{\sqrt{2}}.$$

Thus $\theta = \pi/4$ radians. ∎

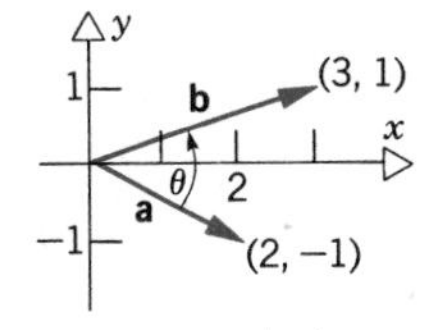

Figure 15.2.2

Suppose that **a** and **b** are nonzero vectors and that $\mathbf{a}\cdot\mathbf{b} = 0$. Then Eq. 13 requires that $\cos\theta = 0$ or that $\theta = \pi/2$. Thus the vectors **a** and **b** are perpendicular, or **orthogonal,** to each other. Conversely, if **a** and **b** are orthogonal, then $\cos\theta = 0$, and $\mathbf{a}\cdot\mathbf{b} = 0$ from Eq. 12. Therefore *nonzero vectors* **a** *and* **b** *are orthogonal if and only if* $\mathbf{a}\cdot\mathbf{b} = 0$. Since the dot product is easy to calculate from the components of **a** and **b**, this is a very convenient way to test the orthogonality of two given vectors.

Equation 12 also gives us a means of calculating the **projection** of a vector **b** on another vector **a**. If **u** is the unit vector in the direction of **a**, then the projection of **b** on **a** is defined as the product of the scalar $\|\mathbf{b}\|\cos\theta$ and the unit vector **u**:

$$\text{proj}_{\mathbf{a}}\,\mathbf{b} = \|\mathbf{b}\|\,(\cos\theta)\mathbf{u} \tag{15}$$

(see Figure 15.2.3). Note that $\text{proj}_{\mathbf{a}}\,\mathbf{b}$ is a vector having the same direction as **a** when $\cos\theta > 0$, and the opposite direction when $\cos\theta < 0$. By using Eq. 12 we can also write $\text{proj}_{\mathbf{a}}\,\mathbf{b}$ as

$$\text{proj}_{\mathbf{a}}\,\mathbf{b} = \frac{\mathbf{b}\cdot\mathbf{a}}{\|\mathbf{a}\|}\,\mathbf{u} = (\mathbf{b}\cdot\mathbf{u})\mathbf{u}. \tag{16}$$

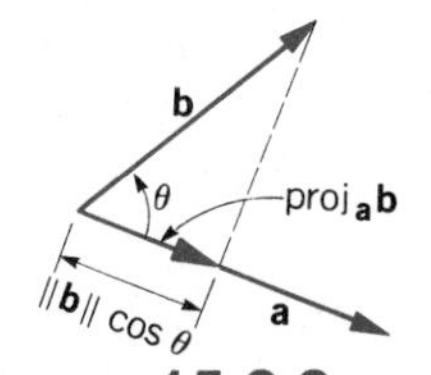

Figure 15.2.3 The projection of **b** on **a**.

The coefficient of $\mathbf{u}$ in Eq. 15 or Eq. 16, namely, $\|\mathbf{b}\| \cos \theta$ or $\mathbf{b} \cdot \mathbf{u}$, is called the **component of b in the direction of a.** Observe that this quantity may be either positive or negative.

EXAMPLE 4

Find the component of $\mathbf{b} = (1, 3)$ in the direction of $\mathbf{a} = (2, 1)$.

The unit vector $\mathbf{u}$ in the direction of $\mathbf{a}$ is

$$\mathbf{u} = \left(\frac{2}{\sqrt{5}}, \frac{1}{\sqrt{5}}\right).$$

Thus from Eq. 16 the component of $\mathbf{b}$ in the direction of $\mathbf{a}$ is

$$\|\mathbf{b}\| \cos \theta = \mathbf{b} \cdot \mathbf{u} = (1, 3) \cdot \left(\frac{2}{\sqrt{5}}, \frac{1}{\sqrt{5}}\right)$$

$$= \frac{(1)(2) + (3)(1)}{\sqrt{5}} = \sqrt{5}. \blacksquare$$

Unit coordinate vectors

Unit vectors in the direction of the x-axis and the y-axis, respectively, are important enough so that the special symbols $\mathbf{i}$ and $\mathbf{j}$ are reserved for them. Their components are

$$\mathbf{i} = (1, 0), \qquad \mathbf{j} = (0, 1). \tag{17}$$

All of the algebraic rules developed in this section apply to $\mathbf{i}$ and $\mathbf{j}$ as special cases; in particular, it is readily verified that

$$\mathbf{i} \cdot \mathbf{i} = 1, \qquad \mathbf{j} \cdot \mathbf{j} = 1, \qquad \mathbf{i} \cdot \mathbf{j} = \mathbf{j} \cdot \mathbf{i} = 0. \tag{18}$$

The first two of Eqs. 18 confirm that $\mathbf{i}$ and $\mathbf{j}$ are of unit length, while the third of Eqs. 18 expresses the fact that $\mathbf{i}$ and $\mathbf{j}$ are orthogonal.

It is often convenient to express a given vector $\mathbf{a} = (a_1, a_2)$ in terms of $\mathbf{i}$ and

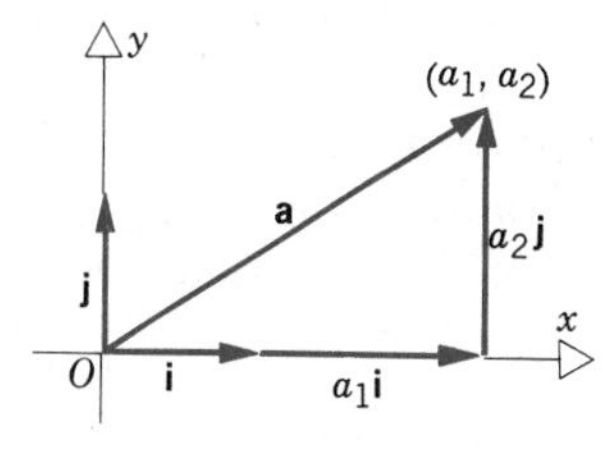

Figure 15.2.4

$\mathbf{j}$ (see Figure 15.2.4). We have

$$\begin{aligned} \mathbf{a} &= (a_1, a_2) = (a_1, 0) + (0, a_2) \\ &= a_1(1, 0) + a_2(0, 1) = a_1\mathbf{i} + a_2\mathbf{j}. \end{aligned} \tag{19}$$

The vectors $a_1\mathbf{i}$ and $a_2\mathbf{j}$ are sometimes called vector components of $\mathbf{a}$ in the x- and y-directions, respectively, to distinguish them from the scalar components a_1 and a_2.

We will frequently write vectors in terms of the coordinate vectors $\mathbf{i}$ and $\mathbf{j}$ as in Eq. 19. All algebraic operations are easy to perform on vectors written in this way. For example, addition and subtraction are handled by collecting terms in $\mathbf{i}$ and $\mathbf{j}$, respectively.

The triangle inequality

Consider the triangle shown in Figure 15.2.5, whose sides are the vectors $\mathbf{a}$, $\mathbf{b}$, and $\mathbf{a} + \mathbf{b}$. A well-known fact from plane geometry is that the length of one side of any triangle is less than or equal to the sum of the lengths of the other two sides, equality holding if and only if the three sides are collinear and the triangle is degenerate. In symbols, this statement takes the form

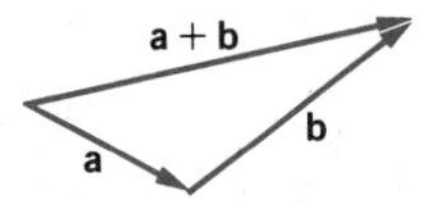

Figure 15.2.5

$$\|\mathbf{a} + \mathbf{b}\| \leq \|\mathbf{a}\| + \|\mathbf{b}\|. \tag{20}$$

Equation 20 is an important relation in vector algebra and is known as the **triangle inequality.**

The result (20) can also be established without an appeal to geometry. To do this it is more convenient to prove the equivalent inequality

$$\|\mathbf{a} + \mathbf{b}\|^2 \leq [\|\mathbf{a}\| + \|\mathbf{b}\|]^2, \tag{21}$$

from which Eq. 20 follows upon taking the nonnegative square root of both sides. To show that Eq. 21 is true we write the left side as

$$\begin{aligned}
\|\mathbf{a} + \mathbf{b}\|^2 &= (\mathbf{a} + \mathbf{b}) \cdot (\mathbf{a} + \mathbf{b}) \\
&= \mathbf{a} \cdot \mathbf{a} + \mathbf{a} \cdot \mathbf{b} + \mathbf{b} \cdot \mathbf{a} + \mathbf{b} \cdot \mathbf{b} \\
&= \|\mathbf{a}\|^2 + 2\mathbf{a} \cdot \mathbf{b} + \|\mathbf{b}\|^2.
\end{aligned} \tag{22}$$

Since $\mathbf{a} \cdot \mathbf{b} = \|\mathbf{a}\|\,\|\mathbf{b}\|\cos\theta$, and $\cos\theta \leq 1$, it follows that $\mathbf{a} \cdot \mathbf{b} \leq \|\mathbf{a}\|\,\|\mathbf{b}\|$. Incorporating this relation in Eq. 22, we have

$$\begin{aligned}
\|\mathbf{a} + \mathbf{b}\|^2 &\leq \|\mathbf{a}\|^2 + 2\|\mathbf{a}\|\,\|\mathbf{b}\| + \|\mathbf{b}\|^2 \\
&= [\|\mathbf{a}\| + \|\mathbf{b}\|]^2,
\end{aligned}$$

which is Eq. 21.

The triangle inequality can be extended to more than two vectors. For example, if $\mathbf{a}$, $\mathbf{b}$, and $\mathbf{c}$ are any three vectors, then

$$\|\mathbf{a} + \mathbf{b} + \mathbf{c}\| \leq \|\mathbf{a}\| + \|\mathbf{b}\| + \|\mathbf{c}\|. \tag{23}$$

PROBLEMS

In each of Problems 1 through 8, find $\mathbf{a} \cdot \mathbf{b}$ and the cosine of the angle between $\mathbf{a}$ and $\mathbf{b}$.

1. $\mathbf{a} = (1, 1)$, $\quad \mathbf{b} = (2, -3)$

2. $\mathbf{a} = (1, 2)$, $\quad \mathbf{b} = (2, 1)$

3. $\mathbf{a} = (1, -2), \quad \mathbf{b} = (2, 1)$

4. $\mathbf{a} = (2, 3), \quad \mathbf{b} = (-1, 2)$

5. $\mathbf{a} = (1, -1), \quad \mathbf{b} = (4, 1)$

6. $\mathbf{a} = 3\mathbf{i} - 2\mathbf{j}, \quad \mathbf{b} = 2\mathbf{i} + 5\mathbf{j}$

7. $\mathbf{a} = 2\mathbf{i} + 2\mathbf{j}, \quad \mathbf{b} = \mathbf{i} + 3\mathbf{j}$

8. $\mathbf{a} = \mathbf{i} - 2\mathbf{j}, \quad \mathbf{b} = 3\mathbf{i} + \mathbf{j}$

In each of Problems 9 through 14, find the unit vector **u** in the direction of the given vector **a**.

9. $\mathbf{a} = (-3, -4)$

10. $\mathbf{a} = (2, -5)$

11. $\mathbf{a} = (3x, -4x), \quad x$ a real number

12. $\mathbf{a} = (x, x), \quad x$ a real number

13. $\mathbf{a} = -5\mathbf{i} + 12\mathbf{j}$

14. $\mathbf{a} = x\mathbf{i} + y\mathbf{j}, \quad x$ and y real numbers

In each of Problems 15 through 18, find the required unit vector **u**.

15. In the direction opposite to $\mathbf{a} = (1, 3)$.

16. In the direction opposite to $\mathbf{a} = 4\mathbf{i} - 3\mathbf{j}$.

17. Perpendicular to $\mathbf{a} = \mathbf{i} - 2\mathbf{j}$.

18. Perpendicular to $\mathbf{a} = (12, 5)$.

In each of Problems 19 through 24, find the component of **b** in the direction of **a**.

19. $\mathbf{a} = (1, 1), \quad \mathbf{b} = (2, 1)$

20. $\mathbf{a} = (2, 1), \quad \mathbf{b} = (1, 1)$

21. $\mathbf{a} = (2, 1), \quad \mathbf{b} = (1, -3)$

22. $\mathbf{a} = (3, 2), \quad \mathbf{b} = (2, -3)$

23. $\mathbf{a} = \mathbf{i} - 2\mathbf{j}, \quad \mathbf{b} = x\mathbf{i} + x\mathbf{j}, \quad x$ a real number

24. $\mathbf{a} = x\mathbf{i} + x\mathbf{j}, \quad \mathbf{b} = \mathbf{i} - 2\mathbf{j}, \quad x$ a real number

In each of Problems 25 through 28, find the projection of **b** on **a**.

25. $\mathbf{a} = (2, -1), \quad \mathbf{b} = (1, -3)$

26. $\mathbf{a} = (-3, 4), \quad \mathbf{b} = (2, -1)$

27. $\mathbf{a} = \mathbf{i} + \mathbf{j}, \quad \mathbf{b} = 2\mathbf{i} + 3\mathbf{j}$

28. $\mathbf{a} = 2\mathbf{i} - 3\mathbf{j}, \quad \mathbf{b} = 2x\mathbf{i} + x\mathbf{j}$, x a real number

29. Prove that $\mathbf{a} \cdot \mathbf{b} = \mathbf{b} \cdot \mathbf{a}$ for any vectors **a** and **b**.

30. Prove that $\lambda(\mathbf{a} \cdot \mathbf{b}) = (\lambda\mathbf{a}) \cdot \mathbf{b}$ for any vectors **a** and **b**, and for any scalar λ.

In each of Problems 31 through 34, verify that the triangle inequality $\|\mathbf{a} + \mathbf{b}\| \leq \|\mathbf{a}\| + \|\mathbf{b}\|$ is satisfied.

31. $\mathbf{a} = (1, 2), \quad \mathbf{b} = (3, -1)$

32. $\mathbf{a} = (1, -1), \quad \mathbf{b} = (2, 3)$

33. $\mathbf{a} = 2\mathbf{i} - \mathbf{j}, \quad \mathbf{b} = 3\mathbf{i} + 4\mathbf{j}$

34. $\mathbf{a} = \mathbf{i} + \mathbf{j}, \quad \mathbf{b} = \mathbf{i} - 3\mathbf{j}$

35. Show that, for any vectors **a** and **b** and for any numbers c_1 and c_2,

$$\|c_1\mathbf{a} + c_2\mathbf{b}\| \leq |c_1|\,\|\mathbf{a}\| + |c_2|\,\|\mathbf{b}\|.$$

36. (a) For any two vectors **a** and **b**, write $\mathbf{a} = \mathbf{b} + (\mathbf{a} - \mathbf{b})$ and draw a sketch to show this relation.

    (b) Use the triangle inequality to show that

$$\|\mathbf{a} - \mathbf{b}\| \geq \|\mathbf{a}\| - \|\mathbf{b}\|. \qquad (i)$$

    (c) By a similar argument, or by interchanging **a** and **b** in Eq. (*i*), show that

$$\|\mathbf{b} - \mathbf{a}\| \geq \|\mathbf{b}\| - \|\mathbf{a}\|. \qquad (ii)$$

    (d) Note that $\|\mathbf{a} - \mathbf{b}\| = \|\mathbf{b} - \mathbf{a}\|$ and show that

$$\|\mathbf{a} - \mathbf{b}\| \geq \big|\,\|\mathbf{a}\| - \|\mathbf{b}\|\,\big|. \qquad (iii)$$

37. Let **a** and **b**, respectively, be the position vectors of two distinct points P and Q.

    (a) Show that the position vector of any point R on the line segment joining P and Q is given by

$$\mathbf{r} = \mathbf{a} + t(\mathbf{b} - \mathbf{a}), \qquad 0 \leq t \leq 1. \qquad (i)$$

    (b) Where are the points whose position vectors are given by Eq. (*i*) with $t < 0$ or $t > 1$?

38. Show that $(\cos\theta)\mathbf{i} + (\sin\theta)\mathbf{j}$ is a unit vector in the direction making an angle θ with the positive x-axis.

15.3 VECTORS IN THREE DIMENSIONS

The discussion in the preceding sections concerning vectors in two dimensions extends readily to three dimensions. We begin by describing a three-dimensional rectangular, or Cartesian, coordinate system.

Select a point O to be the origin of the coordinate system, and choose three mutually perpendicular straight lines intersecting at O to be the x, y, and z coordinate axes. Designate a positive direction and a unit of measurement on each axis. It is usually convenient to orient the axes as shown in Figure 15.3.1*a*. The y- and z-axes are in the plane of the paper, and the x-axis is perpendicular to this plane and points toward the reader. In this orientation the x-, y-, and z-axes form what is called a right-handed coordinate system: if the fingers of the right hand curl from the positive x-axis toward the positive y-axis, then the thumb points in the direction of the positive z-axis. If the coordinate system in Figure 15.3.1*a* is rotated as a rigid body into positions such as those shown in Figures 15.3.1*b* or 15.3.1*c*, it remains a right-handed system. However, if two of the axes are interchanged, as indicated in Figure 15.3.2, the system becomes a left-handed system: if the fingers

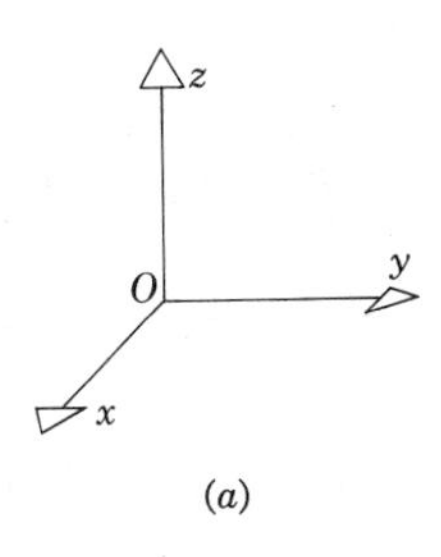

(*a*)

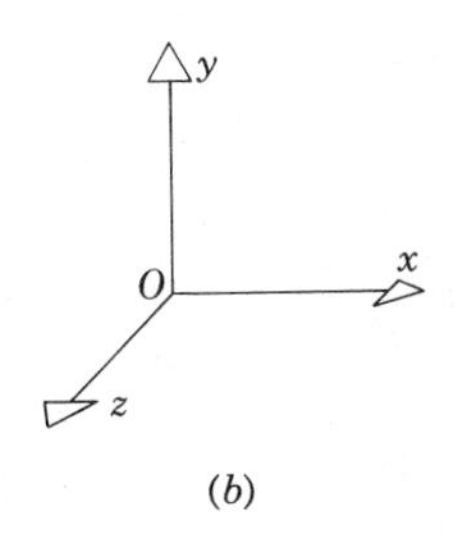

(*b*)

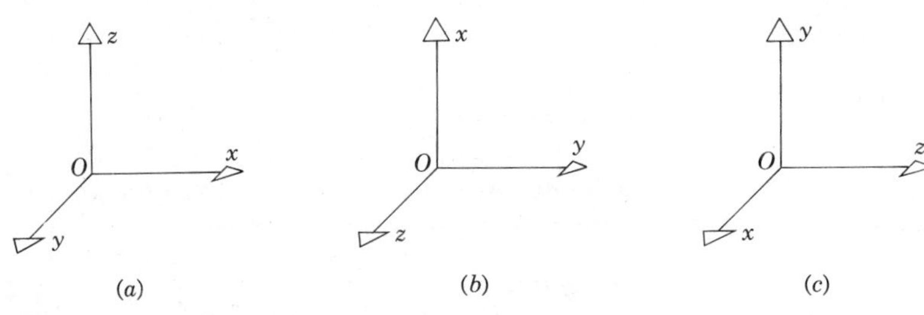

Figure 15.3.2 Left-handed coordinate systems.

of the left hand curl from the positive x-axis to the positive y-axis, then the thumb points in the direction of the positive z-axis. It is customary to use right-handed coordinate systems, and unless otherwise specified, all of our coordinate systems will be right-handed.

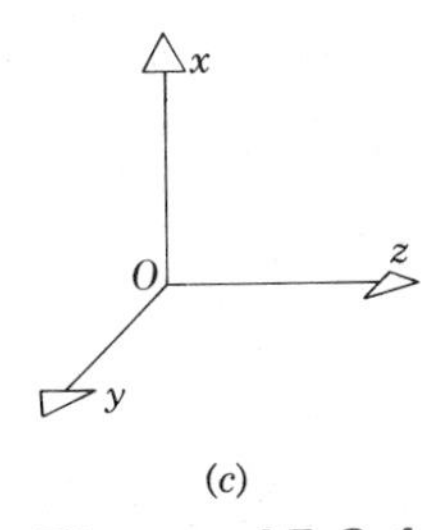

(*c*)

Figure 15.3.1 Right-handed coordinate systems.

A plane containing two of the coordinate axes is called a coordinate plane; the xy-plane contains the x- and y-axes, and similarly for the xz-plane and for the yz-plane. The x, y, and z coordinates of a point P in three-dimensional space are defined in the following way (see Figure 15.3.3). The x coordinate a of P is the

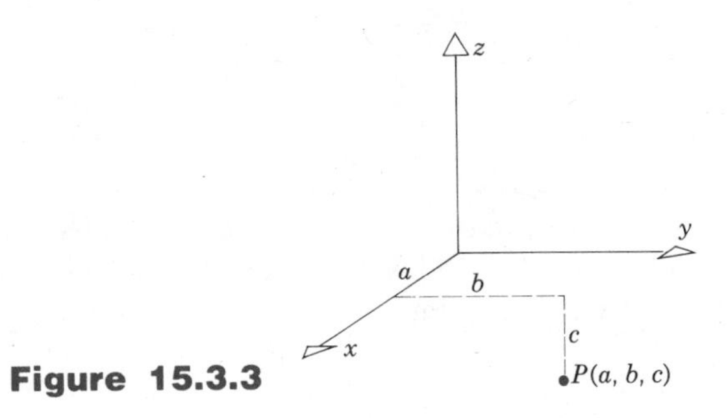

Figure 15.3.3

directed distance from the yz-plane to P. The coordinate is positive if P is on the same side of the yz-plane as the positive x-axis, and negative if P is on the opposite side of the yz-plane. The y coordinate b and the z coordinate c are defined similarly as the directed distances from the xz- and xy-planes, respectively. In the situation shown in Figure 15.3.3 the x and y coordinates are positive and the z coordinate is negative. If one or more of the coordinates is zero, then P lies in the corresponding coordinate plane or planes. Just as in two dimensions, if a rectangular coordinate system is given, then a geometrical point P uniquely determines its coordinates with respect to that coordinate system, and vice versa.

A three-dimensional vector **a** may be thought of as an arrow in three-dimensional xyz-space. Let us place the initial point of **a** at the origin; then the terminal point of **a** coincides with some point P, as shown in Figure 15.3.4, and **a** is said to

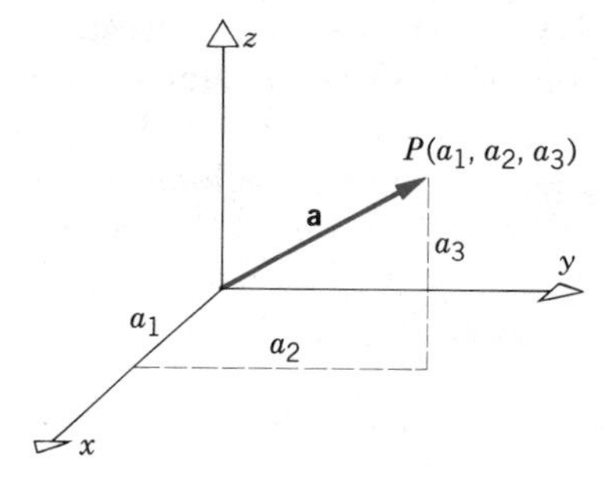

Figure 15.3.4

be the position vector of P. If the coordinates of P are (a_1, a_2, a_3), then these same numbers a_1, a_2, and a_3 are called the components of **a**, and we write

$$\mathbf{a} = (a_1, a_2, a_3). \tag{1}$$

The length of **a** is denoted by $\|\mathbf{a}\|$ and (see Figure 15.3.5) may be calculated from

$$\|\mathbf{a}\| = (a_1^2 + a_2^2 + a_3^2)^{1/2}. \tag{2}$$

The algebraic operations described in the preceding two sections carry over

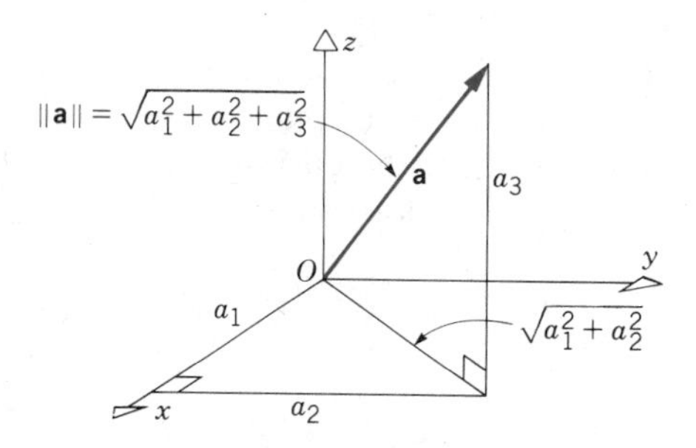

Figure 15.3.5

at once to vectors in three dimensions. Thus we will give only a brief summary of them. In what follows $\mathbf{a} = (a_1, a_2, a_3)$, $\mathbf{b} = (b_1, b_2, b_3)$, and $\mathbf{c} = (c_1, c_2, c_3)$ are arbitrary three-dimensional vectors, and λ and μ are arbitrary scalars.

Equality. $\mathbf{a} = \mathbf{b}$ if and only if $a_1 = b_1$, $a_2 = b_2$, and $a_3 = b_3$.

Zero. $\mathbf{a} = \mathbf{0}$ if and only if $a_1 = 0$, $a_2 = 0$, and $a_3 = 0$.

Addition. $\mathbf{a} + \mathbf{b} = (a_1 + b_1, a_2 + b_2, a_3 + b_3)$. (3)

Subtraction. $\mathbf{a} - \mathbf{b} = (a_1 - b_1, a_2 - b_2, a_3 - b_3)$. (4)

Multiplication by a scalar. $\lambda\mathbf{a} = (\lambda a_1, \lambda a_2, \lambda a_3)$, where λ is a scalar. (5)

Scalar product. $\mathbf{a} \cdot \mathbf{b} = a_1b_1 + a_2b_2 + a_3b_3$. (6)

The following algebraic properties of vector addition, multiplication of a vector by a scalar, and scalar multiplication of two vectors also carry over directly from two to three dimensions:

$$\mathbf{a} + \mathbf{b} = \mathbf{b} + \mathbf{a}, \tag{7}$$

$$(\mathbf{a} + \mathbf{b}) + \mathbf{c} = \mathbf{a} + (\mathbf{b} + \mathbf{c}), \tag{8}$$

$$(\lambda\mu)\mathbf{a} = \lambda(\mu\mathbf{a}), \tag{9}$$

$$(\lambda + \mu)\mathbf{a} = \lambda\mathbf{a} + \mu\mathbf{a}, \tag{10}$$

$$\lambda(\mathbf{a} + \mathbf{b}) = \lambda\mathbf{a} + \lambda\mathbf{b}, \tag{11}$$

$$\mathbf{a} \cdot \mathbf{b} = \mathbf{b} \cdot \mathbf{a}, \tag{12}$$

$$\mathbf{a} \cdot (\mathbf{b} \pm \mathbf{c}) = \mathbf{a} \cdot \mathbf{b} \pm \mathbf{a} \cdot \mathbf{c}, \tag{13}$$

$$\lambda(\mathbf{a} \cdot \mathbf{b}) = (\lambda\mathbf{a}) \cdot \mathbf{b} = \mathbf{a} \cdot (\lambda\mathbf{b}). \tag{14}$$

The geometric properties of the dot product also generalize immediately to three dimensions:

$$\mathbf{a} \cdot \mathbf{b} = \|\mathbf{a}\|\,\|\mathbf{b}\|\cos\theta, \tag{15}$$

$$\mathbf{a} \cdot \mathbf{a} = \|\mathbf{a}\|^2. \tag{16}$$

Finally, the length $\|\mathbf{a}\|$ of the vector **a** has the same properties in three as in two dimensions:

$$\|\mathbf{a}\| \geq 0, \tag{17}$$

$$\|\mathbf{a}\| = 0 \text{ if and only if } \mathbf{a} = \mathbf{0}; \tag{18}$$

$$\|\lambda\mathbf{a}\| = |\lambda|\,\|\mathbf{a}\|. \tag{19}$$

Unit vectors

The unit vector **u** in the same direction as any nonzero vector **a** is given by $\mathbf{u} = \mathbf{a}/\|\mathbf{a}\|$. There are now three unit coordinate vectors, denoted by

$$\mathbf{i} = (1, 0, 0), \qquad \mathbf{j} = (0, 1, 0), \qquad \text{and} \quad \mathbf{k} = (0, 0, 1), \tag{20}$$

and pointing in the positive x, y, and z directions, respectively. In terms of these vectors we can write the vector $\mathbf{a} = (a_1, a_2, a_3)$ as

$$\mathbf{a} = a_1\mathbf{i} + a_2\mathbf{j} + a_3\mathbf{k}. \tag{21}$$

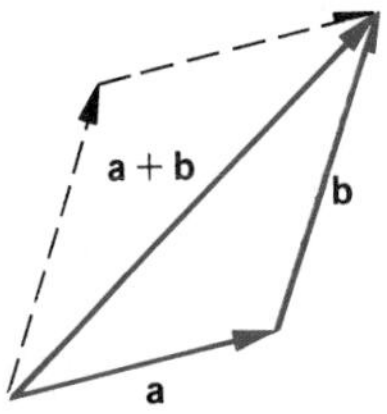

Figure 15.3.6 The parallelogram rule of vector addition.

Geometrical interpretation

The geometrical interpretations of the various vector operations that we have just summarized also generalize from two to three dimensions in a natural way. The sum $\mathbf{a} + \mathbf{b}$ is directed along the diagonal of the parallelogram whose sides are **a** and **b** (see Figure 15.3.6). The difference $\mathbf{a} - \mathbf{b}$ is the vector from the terminal point of **b** to the terminal point of **a** (see Figure 15.3.7). The product $\lambda\mathbf{a}$ is the vector whose length is $|\lambda|$ times $\|\mathbf{a}\|$, and whose direction is the same as that of **a** if $\lambda > 0$, and the opposite if $\lambda < 0$ (see Figure 15.3.8). The associative law of addition, Eq. 8, is illustrated in Figure 15.3.9. Equation 15 for the scalar product provides

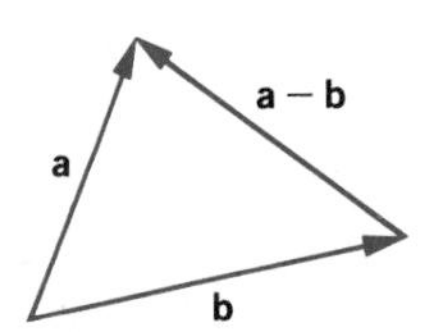

Figure 15.3.7 Vector subtraction.

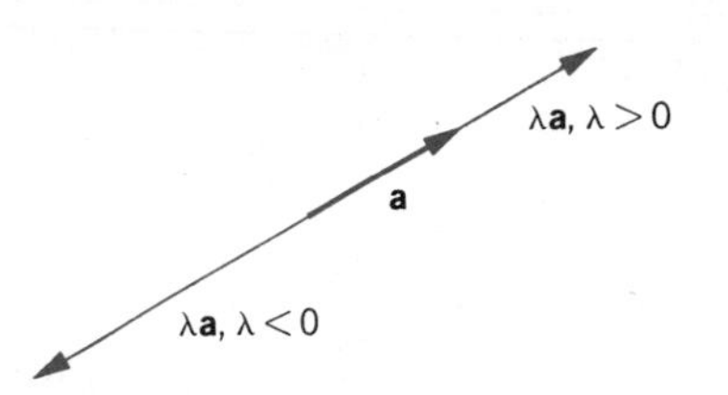

Figure 15.3.8 Multiplication of a vector by a scalar.

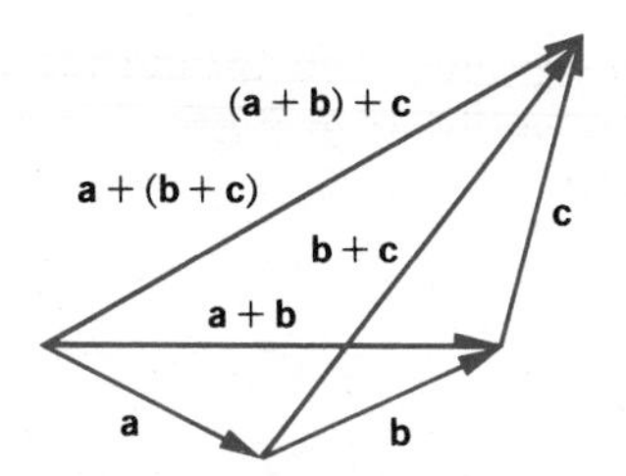

Figure 15.3.9 The associative law of vector addition.

a means of calculating the cosine of the angle between two directions in three dimensions. Of course, if $\cos \theta$ turns out to be zero, then the two directions are perpendicular. The components of $\mathbf{b}$ in the direction of $\mathbf{a}$, or vice versa, given by $\|\mathbf{b}\|\cos \theta$ and $\|\mathbf{a}\|\cos \theta$, respectively, are also readily found from Eq. 15, just as in the two-dimensional case. Finally, the triangle inequality

$$\|\mathbf{a} + \mathbf{b}\| \leq \|\mathbf{a}\| + \|\mathbf{b}\| \tag{22}$$

remains valid in three dimensions. All of these results can be established by the same arguments as in Sections 15.1 and 15.2, taking into account the fact that now the vectors have three components instead of two.

EXAMPLE 1

If

$$\mathbf{a} = 2\mathbf{i} - \mathbf{j} + 3\mathbf{k}, \qquad \mathbf{b} = \mathbf{i} + 3\mathbf{j} - \mathbf{k},$$

find $2\mathbf{a} - 3\mathbf{b}$.

By collecting terms in $\mathbf{i}$, $\mathbf{j}$, and $\mathbf{k}$, we obtain

$$\begin{aligned} 2\mathbf{a} - 3\mathbf{b} &= 2(2\mathbf{i} - \mathbf{j} + 3\mathbf{k}) - 3(\mathbf{i} + 3\mathbf{j} - \mathbf{k}) \\ &= (4\mathbf{i} - 2\mathbf{j} + 6\mathbf{k}) - (3\mathbf{i} + 9\mathbf{j} - 3\mathbf{k}) \\ &= (4 - 3)\mathbf{i} + (-2 - 9)\mathbf{j} + (6 + 3)\mathbf{k} \\ &= \mathbf{i} - 11\mathbf{j} + 9\mathbf{k}. \blacksquare \end{aligned}$$

EXAMPLE 2

If $\mathbf{a}$ and $\mathbf{b}$ are as in Example 1, find $\|\mathbf{a} - \mathbf{b}\|$.

First we find

$$\begin{aligned} \mathbf{a} - \mathbf{b} &= 2\mathbf{i} - \mathbf{j} + 3\mathbf{k} - (\mathbf{i} + 3\mathbf{j} - \mathbf{k}) \\ &= (2 - 1)\mathbf{i} + (-1 - 3)\mathbf{j} + (3 + 1)\mathbf{k} \\ &= \mathbf{i} - 4\mathbf{j} + 4\mathbf{k}. \end{aligned}$$

Now calculating the length of this vector, we obtain

$$\|\mathbf{a} - \mathbf{b}\| = [1^2 + (-4)^2 + 4^2]^{1/2} = \sqrt{33}. \blacksquare$$

EXAMPLE 3

If **a** and **b** are as in Example 1, find $\mathbf{a} \cdot \mathbf{b}$.

From Eq. 6 we have

$$\begin{aligned}\mathbf{a} \cdot \mathbf{b} &= (2\mathbf{i} - \mathbf{j} + 3\mathbf{k}) \cdot (\mathbf{i} + 3\mathbf{j} - \mathbf{k}) \\ &= (2)(1) + (-1)(3) + (3)(-1) = -4. \blacksquare\end{aligned}$$

EXAMPLE 4

If **a** and **b** are as in Example 1, and θ is the angle between them, find $\cos\theta$.

First we calculate the lengths of **a** and **b**:

$$\|\mathbf{a}\| = [2^2 + (-1)^2 + 3^2]^{1/2} = \sqrt{14},$$
$$\|\mathbf{b}\| = [1^2 + 3^2 + (-1)^2]^{1/2} = \sqrt{11}.$$

Then from Eq. 15, and using the result of Example 3, we obtain

$$\begin{aligned}\cos\theta &= \frac{\mathbf{a} \cdot \mathbf{b}}{\|\mathbf{a}\|\,\|\mathbf{b}\|} = \frac{-4}{\sqrt{14}\sqrt{11}} \\ &= -\frac{4}{\sqrt{154}} \cong -0.3223.\end{aligned}$$

Thus $\theta \cong 1.899$ radians $\cong 108.80$ degrees. Note that a negative value of $\cos\theta$ corresponds to an obtuse angle θ. $\blacksquare$

EXAMPLE 5

If $\mathbf{a} = 3\mathbf{i} + 4\mathbf{j} + \mathbf{k}$ and $\mathbf{b} = -\mathbf{i} + 3\mathbf{j} + 2\mathbf{k}$, find the component of **a** in the direction of **b** (see Figure 15.3.10).

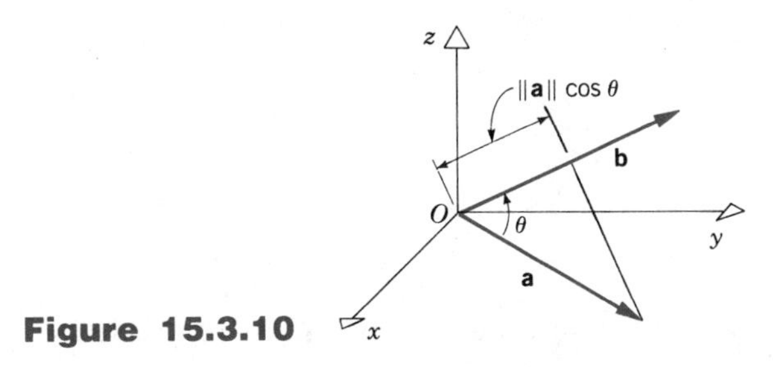

Figure 15.3.10

The desired quantity is given by $\|\mathbf{a}\|\cos\theta$, where θ is the angle between **a** and **b**. From Eq. 15, we have

$$\|\mathbf{a}\|\cos\theta = \frac{\mathbf{a}\cdot\mathbf{b}}{\|\mathbf{b}\|},$$

and consequently

$$\|\mathbf{a}\|\cos\theta = \frac{(3)(-1) + (4)(3) + (1)(2)}{[(-1)^2 + 3^2 + 2^2]^{1/2}}$$

$$= \frac{11}{\sqrt{14}} \cong 2.940.$$

In the same way, the component of **b** in the direction of **a** is given by

$$\|\mathbf{b}\|\cos\theta = \frac{\mathbf{a}\cdot\mathbf{b}}{\|\mathbf{a}\|} = \frac{11}{\sqrt{26}} \cong 2.157. \blacksquare$$

EXAMPLE 6

Show that an angle inscribed in a semicircle must be a right angle.

This example illustrates that many familiar results from geometry can be established efficiently by using vector methods. Consider the vectors **a** and **b** in Figure 15.3.11, drawn from a point P on a circle to the ends of Q and R of a

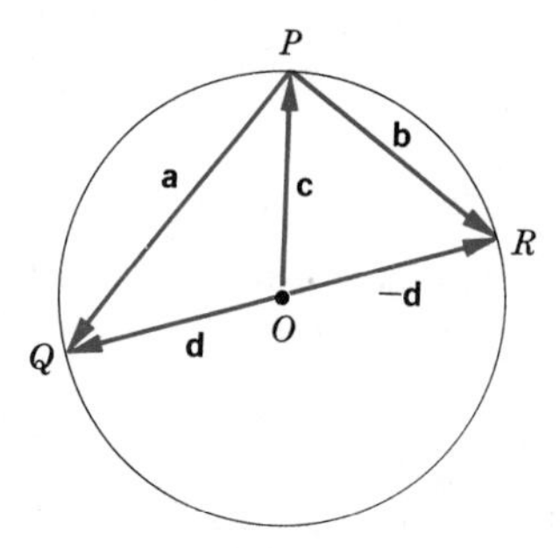

Figure 15.3.11

diameter. Let **c** and **d** be the vectors from the center O of the circle to P and Q, respectively, and note that $\overrightarrow{OR} = -\mathbf{d}$. Then

$$\mathbf{a} = \mathbf{d} - \mathbf{c}, \qquad \mathbf{b} = -\mathbf{d} - \mathbf{c}.$$

To determine the angle between **a** and **b**, we calculate

$$\begin{aligned}\mathbf{a}\cdot\mathbf{b} &= (\mathbf{d} - \mathbf{c})\cdot(-\mathbf{d} - \mathbf{c}) \\ &= (-\mathbf{d}\cdot\mathbf{d} + \mathbf{c}\cdot\mathbf{d} - \mathbf{d}\cdot\mathbf{c} + \mathbf{c}\cdot\mathbf{c}) \\ &= \|\mathbf{c}\|^2 - \|\mathbf{d}\|^2 \\ &= 0,\end{aligned}$$

since the lengths of both **c** and **d** are equal to the radius of the circle. ∎

PROBLEMS

In each of Problems 1 and 2, find the vector $\overrightarrow{PQ}$.

1. $P(2, 1, -1)$, $Q(3, -2, 4)$
2. $P(3, 2, -4)$, $Q(1, -3, 2)$
3. If the position vector of the point P is $2\mathbf{i} - \mathbf{j} + 2\mathbf{k}$, and if $\overrightarrow{PQ} = -\mathbf{i} + 4\mathbf{j} + 2\mathbf{k}$, find the position vector of the point Q.
4. If the position vector of the point Q is $3\mathbf{i} + 4\mathbf{j} - \mathbf{k}$, and if $\overrightarrow{PQ} = 2\mathbf{i} - \mathbf{j} + 3\mathbf{k}$, find the position vector of the point P.

In each of Problems 5 through 8, find $3\mathbf{a} - 2\mathbf{b}$.

5. $\mathbf{a} = (2, -1, 3)$, $\mathbf{b} = (-3, 2, 1)$
6. $\mathbf{a} = (4, 0, 1)$, $\mathbf{b} = (2, -2, 3)$
7. $\mathbf{a} = 2\mathbf{i} + \mathbf{j} - \mathbf{k}$, $\mathbf{b} = -\mathbf{i} + 3\mathbf{k}$
8. $\mathbf{a} = -\mathbf{i} + 2\mathbf{j} - 3\mathbf{k}$, $\mathbf{b} = 2\mathbf{i} + \mathbf{j} - 4\mathbf{k}$

In each of Problems 9 through 14, find the magnitude of the given vector.

9. $\mathbf{a} = (3, -2, 1)$
10. $\mathbf{a} = (2, -2, 3)$
11. $\mathbf{a} = 2\mathbf{i} + \mathbf{j} - 4\mathbf{k}$
12. $\mathbf{a} = 3\mathbf{i} + 5\mathbf{j} - \mathbf{k}$
13. The vector from $P(2, 1, -1)$ to $Q(3, -2, 4)$.
14. The vector from $P(3, 2, -4)$ to $Q(1, -3, 2)$.

In each of Problems 15 through 18, find $\mathbf{a} \cdot \mathbf{b}$ and the cosine of the angle θ between $\mathbf{a}$ and $\mathbf{b}$.

15. $\mathbf{a} = (3, 1, 2)$, $\mathbf{b} = (2, -4, 1)$
16. $\mathbf{a} = (4, 0, 1)$, $\mathbf{b} = (1, -1, 1)$
17. $\mathbf{a} = \mathbf{i} - 2\mathbf{j} - 3\mathbf{k}$, $\mathbf{b} = \mathbf{i} - \mathbf{j} + \mathbf{k}$
18. $\mathbf{a} = 2\mathbf{i} + \mathbf{j} - \mathbf{k}$, $\mathbf{b} = -\mathbf{i} + 3\mathbf{k}$

In each of Problems 19 through 22, find the component of $\mathbf{b}$ in the direction of $\mathbf{a}$.

19. $\mathbf{a} = (2, 1, -3)$, $\mathbf{b} = (1, 1, 2)$
20. $\mathbf{a} = (1, -1, 2)$, $\mathbf{b} = (4, 0, 1)$
21. $\mathbf{a} = 2\mathbf{i} + 3\mathbf{j} + 4\mathbf{k}$, $\mathbf{b} = -\mathbf{i} + \mathbf{j} + \mathbf{k}$
22. $\mathbf{a} = -\mathbf{i} + \mathbf{j} + \mathbf{k}$, $\mathbf{b} = 2\mathbf{i} + 3\mathbf{j} + 4\mathbf{k}$

In each of Problems 23 and 24, choose the number λ so as to make the vectors $\mathbf{a}$ and $\mathbf{b}$ perpendicular (orthogonal).

23. $\mathbf{a} = 3\lambda\mathbf{i} + \mathbf{j} - \mathbf{k}$, $\mathbf{b} = 2\mathbf{i} + \lambda\mathbf{j} + 3\mathbf{k}$
24. $\mathbf{a} = \mathbf{i} + 2\lambda\mathbf{j} - 3\mathbf{k}$, $\mathbf{b} = 2\mathbf{i} - \lambda\mathbf{j} - 4\mathbf{k}$

In each of Problems 25 through 28, find all unit vectors $\mathbf{u}$ that satisfy the given condition.

25. In the same direction as $\mathbf{a} = 3\mathbf{i} + \mathbf{j} - 2\mathbf{k}$.
26. In the direction opposite to $\mathbf{a} = 2\mathbf{i} - 3\mathbf{j} - 4\mathbf{k}$.
27. In the direction from $P(-2, 0, 4)$ to $Q(1, -1, 1)$.
28. Perpendicular (orthogonal) to $\mathbf{a} = 2\mathbf{i} - 3\mathbf{j} + \mathbf{k}$ and $\mathbf{b} = \mathbf{i} + \mathbf{j} + \mathbf{k}$.
29. Use vector methods to show that the diagonals of a parallelogram are perpendicular if and only if the parallelogram is a rhombus, that is, its sides are all the same length.
30. Use vector methods to show that for any parallelogram the sum of the squares of the lengths of the diagonals is equal to the sum of the squares of the lengths of the sides.
31. Find the angle θ between the diagonal of a cube and one of its edges.
32. Find the angle θ between the diagonals of the parallelogram with sides $\mathbf{a} = \mathbf{i} - 2\mathbf{j} + 2\mathbf{k}$ and $\mathbf{b} = -\mathbf{i} + \mathbf{j} + 2\mathbf{k}$.
33. Find the angle between the diagonal of a cube and the diagonal of one of its faces.
34. If A, B, C, D, and E are any given points, prove that $\overrightarrow{AB} + \overrightarrow{BC} + \overrightarrow{CD} + \overrightarrow{DE} = \mathbf{0}$ if and only if E coincides with A. A similar result holds for any number of points.

15.4 THE CROSS PRODUCT

Earlier in this chapter we defined the scalar or dot product of two vectors. This type of multiplication always yields a scalar (number). In three dimensions there is another way of multiplying two vectors in which the result is another vector.

The symbol × is used to designate this kind of multiplication so it is referred to as the **cross product,** or the **vector product.**

The cross product can be defined either geometrically or analytically. The geometrical definition is as follows. First, if either $\mathbf{a} = \mathbf{0}$ or $\mathbf{b} = \mathbf{0}$, then $\mathbf{a} \times \mathbf{b} = \mathbf{0}$. In general, let **a** and **b** be nonzero vectors and let θ be the angle between **a** and **b** such that $0 \leq \theta \leq \pi$ (see Figure 15.4.1). Then $\mathbf{a} \times \mathbf{b}$ is defined to be the vector with the following properties.

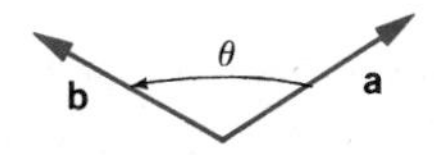

Figure 15.4.1

(a) $\mathbf{a} \times \mathbf{b}$ is perpendicular to both **a** and **b**.
(b) The magnitude of $\mathbf{a} \times \mathbf{b}$ is given by

$$\|\mathbf{a} \times \mathbf{b}\| = \|\mathbf{a}\| \, \|\mathbf{b}\| \sin \theta.$$

(c) The direction of $\mathbf{a} \times \mathbf{b}$ is chosen so that when **a** is rotated into **b** through the angle θ, then **a**, **b**, and $\mathbf{a} \times \mathbf{b}$ form a right-handed system of vectors (see Figure 15.4.2).

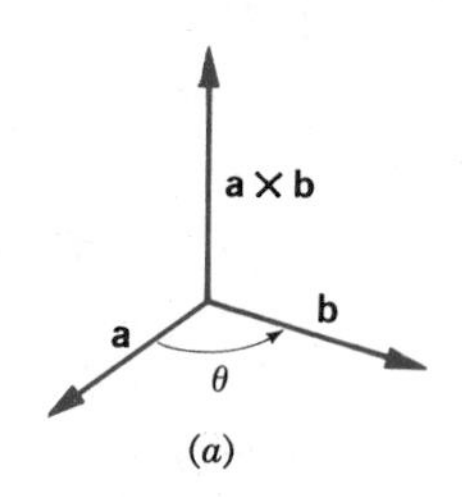

(*a*)

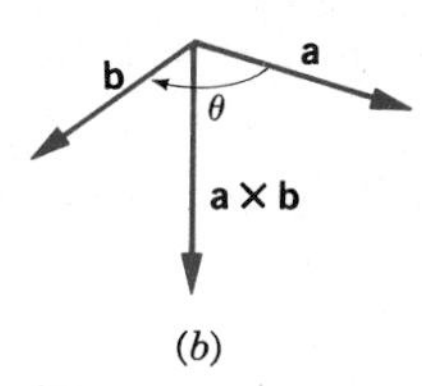

(*b*)

Figure 15.4.2

This geometrical definition of $\mathbf{a} \times \mathbf{b}$ does not depend on the introduction of a coordinate system. Unfortunately, it is also not a particularly convenient way to calculate the cross product of a given pair of vectors **a** and **b**. However, if we do introduce a coordinate system, and if we suppose that

$$\mathbf{a} = a_1\mathbf{i} + a_2\mathbf{j} + a_3\mathbf{k}, \qquad \mathbf{b} = b_1\mathbf{i} + b_2\mathbf{j} + b_3\mathbf{k},$$

then it is possible to show from properties (a), (b), and (c) that $\mathbf{a} \times \mathbf{b}$ is given by

$$\mathbf{a} \times \mathbf{b} = (a_2b_3 - a_3b_2)\mathbf{i} + (a_3b_1 - a_1b_3)\mathbf{j} + (a_1b_2 - a_2b_1)\mathbf{k}. \qquad (1)$$

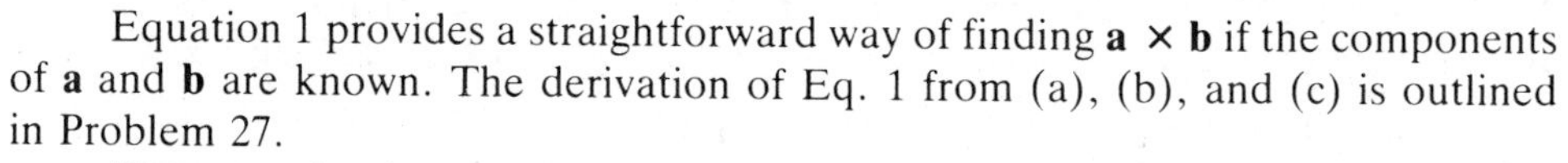

Equation 1 provides a straightforward way of finding $\mathbf{a} \times \mathbf{b}$ if the components of **a** and **b** are known. The derivation of Eq. 1 from (a), (b), and (c) is outlined in Problem 27.

Here we take the opposite point of view. That is, we define the cross product $\mathbf{a} \times \mathbf{b}$ analytically by means of Eq. 1; then we establish some of its properties, including (a), (b), and (c).

First, we derive a formula for $\mathbf{a} \times \mathbf{b}$ that is more easily remembered than Eq. 1. Note that the coefficients of **i**, **j**, and **k** in Eq. 1 can be written as two-by-two determinants, so that

$$\begin{aligned} \mathbf{a} \times \mathbf{b} &= \begin{vmatrix} a_2 & a_3 \\ b_2 & b_3 \end{vmatrix} \mathbf{i} + \begin{vmatrix} a_3 & a_1 \\ b_3 & b_1 \end{vmatrix} \mathbf{j} + \begin{vmatrix} a_1 & a_2 \\ b_1 & b_2 \end{vmatrix} \mathbf{k} \\ &= \begin{vmatrix} a_2 & a_3 \\ b_2 & b_3 \end{vmatrix} \mathbf{i} - \begin{vmatrix} a_1 & a_3 \\ b_1 & b_3 \end{vmatrix} \mathbf{j} + \begin{vmatrix} a_1 & a_2 \\ b_1 & b_2 \end{vmatrix} \mathbf{k}. \qquad (2) \end{aligned}$$

Equation 2 has the same form as the expansion of a three-by-three determinant by cofactors of the elements in its first row. Taking advantage of this observation, we can write $\mathbf{a} \times \mathbf{b}$ in the determinantal form

$$\mathbf{a} \times \mathbf{b} = \begin{vmatrix} \mathbf{i} & \mathbf{j} & \mathbf{k} \\ a_1 & a_2 & a_3 \\ b_1 & b_2 & b_3 \end{vmatrix}. \qquad (3)$$

We have used the notation of determinants in Eq. 3 for convenience. However, we should remember that determinants are numbers, whereas the expression on the right side of Eq. 3 is intended to be the same vector as in Eq. 1, and hence strictly speaking is not a determinant at all. This inconsistency is of no practical significance, and the determinantal notation (3) is extremely useful in dealing with cross products. If there is ever any doubt as to the proper interpretation of Eq. 3, we can always fall back upon Eq. 1.

EXAMPLE 1

If $\mathbf{a} = \mathbf{i} + 2\mathbf{j} - 2\mathbf{k}$ and $\mathbf{b} = 2\mathbf{i} - \mathbf{j} + 3\mathbf{k}$, find $\mathbf{a} \times \mathbf{b}$.

From Eq. 3 we have

$$\begin{aligned}\mathbf{a} \times \mathbf{b} &= \begin{vmatrix} \mathbf{i} & \mathbf{j} & \mathbf{k} \\ 1 & 2 & -2 \\ 2 & -1 & 3 \end{vmatrix} \\ &= \mathbf{i}\begin{vmatrix} 2 & -2 \\ -1 & 3 \end{vmatrix} - \mathbf{j}\begin{vmatrix} 1 & -2 \\ 2 & 3 \end{vmatrix} + \mathbf{k}\begin{vmatrix} 1 & 2 \\ 2 & -1 \end{vmatrix} \\ &= 4\mathbf{i} - 7\mathbf{j} - 5\mathbf{k}. \ \blacksquare\end{aligned}$$

We now turn to a discussion of some of the properties of the cross product. Most of the discussion deals with the case in which $\mathbf{a}$, $\mathbf{b}$, and $\mathbf{a} \times \mathbf{b}$ are nonzero. However, the results remain true (perhaps in a degenerate sense) if any of the vectors are zero.

Property 1. If $\mathbf{a} = \mathbf{0}$ or if $\mathbf{b} = \mathbf{0}$, then $\mathbf{a} \times \mathbf{b} = \mathbf{0}$. For instance, if $\mathbf{a} = \mathbf{0}$, then $a_1 = a_2 = a_3 = 0$, and every term on the right side of Eq. 1 is zero. The argument is similar if $\mathbf{b} = \mathbf{0}$.

Property 2. $\mathbf{a} \times \mathbf{b}$ is perpendicular to both $\mathbf{a}$ and $\mathbf{b}$. To show that $\mathbf{a}$ and $\mathbf{a} \times \mathbf{b}$ are perpendicular we calculate the dot product of the two vectors $\mathbf{a}$ and $\mathbf{a} \times \mathbf{b}$. From Eq. 1 we obtain

$$\begin{aligned}\mathbf{a} \cdot (\mathbf{a} \times \mathbf{b}) &= a_1(a_2b_3 - a_3b_2) + a_2(a_3b_1 - a_1b_3) + a_3(a_1b_2 - a_2b_1) \\ &= 0,\end{aligned}$$

since the terms on the right side cancel in pairs. Alternatively, from the determinantal expression (3) we can write

$$\begin{aligned}\mathbf{a} \cdot (\mathbf{a} \times \mathbf{b}) &= (a_1\mathbf{i} + a_2\mathbf{j} + a_3\mathbf{k}) \cdot \begin{vmatrix} \mathbf{i} & \mathbf{j} & \mathbf{k} \\ a_1 & a_2 & a_3 \\ b_1 & b_2 & b_3 \end{vmatrix} \\ &= a_1\begin{vmatrix} a_2 & a_3 \\ b_2 & b_3 \end{vmatrix} - a_2\begin{vmatrix} a_1 & a_3 \\ b_1 & b_3 \end{vmatrix} + a_3\begin{vmatrix} a_1 & a_2 \\ b_1 & b_2 \end{vmatrix} \\ &= \begin{vmatrix} a_1 & a_2 & a_3 \\ a_1 & a_2 & a_3 \\ b_1 & b_2 & b_3 \end{vmatrix} = 0\end{aligned}$$

because the first and second rows are the same. In a similar way we can show that $\mathbf{b} \cdot (\mathbf{a} \times \mathbf{b}) = 0$, and hence that $\mathbf{a} \times \mathbf{b}$ is also perpendicular to $\mathbf{b}$.

This is one of the most useful properties of the cross product. It is a simple way to produce *a vector that is perpendicular to each of two given vectors.*

EXAMPLE 2

Find a unit vector that is perpendicular both to $\mathbf{a} = \mathbf{i} + 2\mathbf{j} - 2\mathbf{k}$ and to $\mathbf{b} = 2\mathbf{i} - \mathbf{j} + 3\mathbf{k}$.

In Example 1 we found that $\mathbf{a} \times \mathbf{b} = 4\mathbf{i} - 7\mathbf{j} - 5\mathbf{k}$, and from Property 2 we know that $\mathbf{a} \times \mathbf{b}$ is perpendicular to both $\mathbf{a}$ and $\mathbf{b}$. To determine a unit vector in the direction of $\mathbf{a} \times \mathbf{b}$, we must divide it by its own magnitude

$$\|\mathbf{a} \times \mathbf{b}\| = [4^2 + (-7)^2 + (-5)^2]^{1/2} = \sqrt{90}.$$

Thus the desired unit vector is

$$\mathbf{u} = \frac{\mathbf{a} \times \mathbf{b}}{\|\mathbf{a} \times \mathbf{b}\|} = \frac{1}{\sqrt{90}}(4\mathbf{i} - 7\mathbf{j} - 5\mathbf{k}).$$

Note that $-\mathbf{u}$ is also a unit vector that is perpendicular to $\mathbf{a}$ and $\mathbf{b}$. ∎

Property 3. The magnitude of $\mathbf{a} \times \mathbf{b}$ is given by

$$\|\mathbf{a} \times \mathbf{b}\| = \|\mathbf{a}\|\,\|\mathbf{b}\|\sin\theta, \tag{4}$$

where θ is the angle between $\mathbf{a}$ and $\mathbf{b}$ such that $0 \le \theta \le \pi$.

The proof of Eq. 4 involves some moderately complicated algebra. We start by writing

$$\begin{aligned}\|\mathbf{a} \times \mathbf{b}\|^2 &= (\mathbf{a} \times \mathbf{b}) \cdot (\mathbf{a} \times \mathbf{b}) \\ &= (a_2b_3 - a_3b_2)^2 + (a_3b_1 - a_1b_3)^2 + (a_1b_2 - a_2b_1)^2 \\ &= a_2^2b_3^2 - 2a_2a_3b_2b_3 + a_3^2b_2^2 \\ &\quad + a_3^2b_1^2 - 2a_1a_3b_1b_3 + a_1^2b_3^2 \\ &\quad\quad + a_1^2b_2^2 - 2a_1a_2b_1b_2 + a_2^2b_1^2.\end{aligned} \tag{5}$$

The trick now is to add and subtract the quantity

$$a_1^2b_1^2 + a_2^2b_2^2 + a_3^2b_3^2$$

to the right side of Eq. 5 and then to rearrange the terms in a different way. The result is

$$\begin{aligned}\|\mathbf{a} \times \mathbf{b}\|^2 &= (a_1^2 + a_2^2 + a_3^2)(b_1^2 + b_2^2 + b_3^2) - (a_1b_1 + a_2b_2 + a_3b_3)^2 \\ &= \|\mathbf{a}\|^2\|\mathbf{b}\|^2 - (\mathbf{a} \cdot \mathbf{b})^2.\end{aligned} \tag{6}$$

Making use of the fact that

$$\mathbf{a} \cdot \mathbf{b} = \|\mathbf{a}\|\,\|\mathbf{b}\|\cos\theta, \tag{7}$$

we can rewrite Eq. 6 as

$$\begin{aligned}\|\mathbf{a} \times \mathbf{b}\|^2 &= \|\mathbf{a}\|^2\|\mathbf{b}\|^2 - \|\mathbf{a}\|^2\|\mathbf{b}\|^2 \cos^2 \theta \\ &= \|\mathbf{a}\|^2\|\mathbf{b}\|^2(1 - \cos^2 \theta) \\ &= \|\mathbf{a}\|^2\|\mathbf{b}\|^2 \sin^2 \theta,\end{aligned}$$

which is equivalent to Eq. 4.

A simple geometrical interpretation of Eq. 4 is evident from Figure 15.4.3. Two sides of the triangle OAB are represented by the vectors **a** and **b** and the

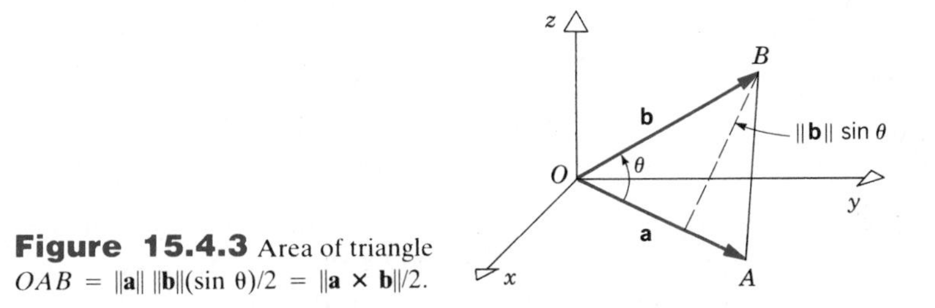

Figure 15.4.3 Area of triangle $OAB = \|\mathbf{a}\|\,\|\mathbf{b}\|(\sin \theta)/2 = \|\mathbf{a} \times \mathbf{b}\|/2$.

lengths of these sides are $\|\mathbf{a}\|$ and $\|\mathbf{b}\|$, respectively. The length of the altitude perpendicular to **a** is $\|\mathbf{b}\|\sin \theta$, and

$$\text{Area of triangle } OAB = \tfrac{1}{2}\|\mathbf{a}\|\,\|\mathbf{b}\|\sin \theta = \tfrac{1}{2}\|\mathbf{a} \times \mathbf{b}\|. \tag{8}$$

In other words, *the magnitude of the cross product* $\|\mathbf{a} \times \mathbf{b}\|$ *is twice the area of the triangle defined by* **a** *and* **b**. Alternatively, $\|\mathbf{a} \times \mathbf{b}\|$ is the area of the parallelogram with adjacent sides **a** and **b**.

EXAMPLE 3

Find the area of the triangle whose vertices are the points $A(-1, 1, 1)$, $B(1, 2, 3)$, and $C(1, -1, -1)$.

First we form the vectors

$$\overrightarrow{AB} = (2, 1, 2), \qquad \overrightarrow{AC} = (2, -2, -2)$$

which lie along two sides of the triangle. Then

$$\overrightarrow{AB} \times \overrightarrow{AC} = \begin{vmatrix} \mathbf{i} & \mathbf{j} & \mathbf{k} \\ 2 & 1 & 2 \\ 2 & -2 & -2 \end{vmatrix} = 2\mathbf{i} + 8\mathbf{j} - 6\mathbf{k},$$

and

$$\|\overrightarrow{AB} \times \overrightarrow{AC}\| = [(2)^2 + (8)^2 + (-6)^2]^{1/2} = \sqrt{104}.$$

Thus the area of the triangle is $\sqrt{104}/2 = \sqrt{26} \cong 5.099$. ■

Property 4. If **a** and **b** are nonzero vectors, then $\mathbf{a} \times \mathbf{b} = \mathbf{0}$ if and only if **a** and **b** are parallel.

Since $\|\mathbf{a}\| \neq 0$ and $\|\mathbf{b}\| \neq 0$, it follows at once from Eq. 4 that $\mathbf{a} \times \mathbf{b} = \mathbf{0}$ if and only if $\sin \theta = 0$.

Thus either $\theta = 0$ or else $\theta = \pi$, so that **a** and **b** have either the same direction or the opposite direction; in either case they are parallel.

This property provides a test for parallelism of two vectors that is analogous to the test for orthogonality based on the dot product.

Property 5. If $\mathbf{a} \times \mathbf{b} \neq \mathbf{0}$, then the vectors **a**, **b**, and $\mathbf{a} \times \mathbf{b}$ form a right-handed system.

Since $\mathbf{a} \times \mathbf{b} \neq \mathbf{0}$, the vectors **a** and **b** are neither zero nor parallel. Let the xy-plane be located so that it contains **a** and **b**, and orient it so that **a** points in the positive x-direction. Then $\mathbf{a} = a_1\mathbf{i}$, where $a_1 > 0$, and $\mathbf{b} = b_1\mathbf{i} + b_2\mathbf{j}$. Consequently

$$\mathbf{a} \times \mathbf{b} = \begin{vmatrix} \mathbf{i} & \mathbf{j} & \mathbf{k} \\ a_1 & 0 & 0 \\ b_1 & b_2 & 0 \end{vmatrix} = a_1 b_2 \mathbf{k}. \tag{9}$$

If $b_2 > 0$, then b lies in the first or second quadrants of the xy-plane; see Figure 15.4.4a. In this case, $\mathbf{a} \times \mathbf{b}$ lies along the positive z-axis, and **a**, **b**, and $\mathbf{a} \times \mathbf{b}$ form a right-handed system.

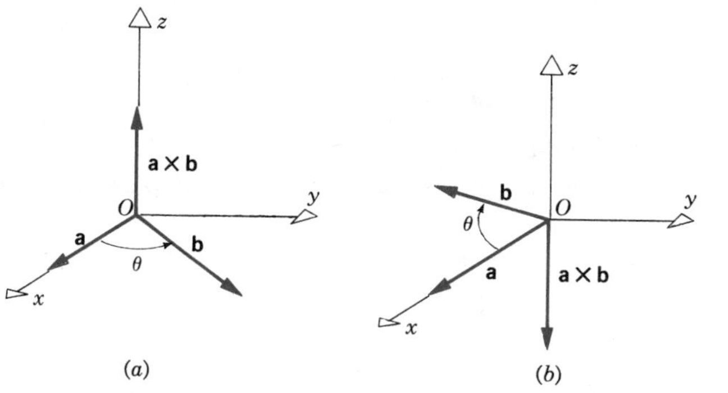

Figure 15.4.4

On the other hand, if $b_2 < 0$, then **b** lies in the third or fourth quadrants of the xy-plane, as shown in Figure 15.4.4b. Now $\mathbf{a} \times \mathbf{b}$ lies along the negative z-axis, but again **a**, **b**, and $\mathbf{a} \times \mathbf{b}$ form a right-handed system.

Note that b_2 cannot be zero since **a** and **b** are not parallel.

Observe that with the derivation of Properties 1 through 5 we have shown that each part of the geometrical definition of the cross product can be established from the analytical definition (1). Also, let **n** be the unit vector perpendicular to **a** and **b** such that **a**, **b**, and **n** form a right-handed system. Then we can write

$$\mathbf{a} \times \mathbf{b} = \|\mathbf{a}\| \, \|\mathbf{b}\| (\sin \theta)\mathbf{n} \tag{10}$$

(see Figure 15.4.5).

Property 6. The cross product is not commutative; in fact,

$$\mathbf{a} \times \mathbf{b} = -(\mathbf{b} \times \mathbf{a}). \tag{11}$$

This result is easily seen from the determinantal expression (3). Reversing the order of the factors in $\mathbf{a} \times \mathbf{b}$ amounts to interchanging the second and third rows of the determinant, which changes its sign. Alternatively, Properties 3 and 5

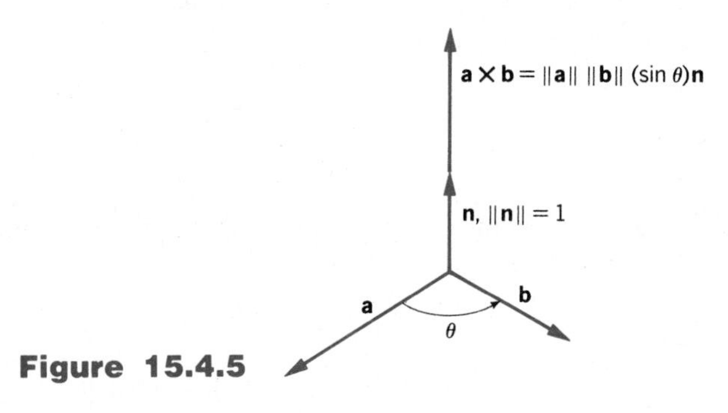

Figure 15.4.5

imply that $\mathbf{a} \times \mathbf{b}$ and $\mathbf{b} \times \mathbf{a}$ have the same magnitude but opposite directions, which is equivalent to Eq. 11.

Property 7. $\mathbf{a} \times \mathbf{a} = \mathbf{0}$.

This is clear from the determinantal expression (3). If $\mathbf{b} = \mathbf{a}$, then the last two rows of the determinant are identical and the determinant is zero. This property also follows at once from Eq. 4, since $\theta = 0$ if $\mathbf{b} = \mathbf{a}$.

Property 8. The unit vectors **i**, **j**, and **k** have the following multiplication table:

$$\mathbf{i} \times \mathbf{j} = \mathbf{k}, \qquad \mathbf{j} \times \mathbf{k} = \mathbf{i}, \qquad \mathbf{k} \times \mathbf{i} = \mathbf{j}, \tag{12}$$

$$\mathbf{j} \times \mathbf{i} = -\mathbf{k}, \qquad \mathbf{k} \times \mathbf{j} = -\mathbf{i}, \qquad \mathbf{i} \times \mathbf{k} = -\mathbf{j}, \tag{13}$$

$$\mathbf{i} \times \mathbf{i} = \mathbf{0}, \qquad \mathbf{j} \times \mathbf{j} = \mathbf{0}, \qquad \mathbf{k} \times \mathbf{k} = \mathbf{0}. \tag{14}$$

Equations 12 can be obtained by substituting $\mathbf{i} = (1, 0, 0)$, $\mathbf{j} = (0, 1, 0)$, and $\mathbf{k} = (0, 0, 1)$ into Eq. 1. Equations 13 and 14 can be obtained in the same way or by using Properties 6 and 7, respectively.

Property 9. The cross product satisfies the distributive laws

$$\mathbf{a} \times (\mathbf{b} + \mathbf{c}) = \mathbf{a} \times \mathbf{b} + \mathbf{a} \times \mathbf{c}, \tag{15a}$$

$$(\mathbf{a} + \mathbf{b}) \times \mathbf{c} = \mathbf{a} \times \mathbf{c} + \mathbf{b} \times \mathbf{c}. \tag{15b}$$

Also, if α is any number, then

$$(\alpha\mathbf{a}) \times \mathbf{b} = \alpha(\mathbf{a} \times \mathbf{b}) \tag{16a}$$

and

$$\mathbf{a} \times (\alpha\mathbf{b}) = \alpha(\mathbf{a} \times \mathbf{b}). \tag{16b}$$

The derivation of these results involves a straightforward use of Eq. 1, and is left to the reader.

Products of three vectors

There are various ways to multiply three vectors through combinations of the dot and cross products. If **a**, **b**, and **c** are arbitrary vectors, then the quantity $\mathbf{a} \cdot (\mathbf{b} \times \mathbf{c})$ is called the **scalar triple product;** it is the dot product of the two vectors **a** and $\mathbf{b} \times \mathbf{c}$ and hence is a scalar. From Eq. 2 we have

$$\mathbf{a} \cdot (\mathbf{b} \times \mathbf{c}) = a_1 \begin{vmatrix} b_2 & b_3 \\ c_2 & c_3 \end{vmatrix} - a_2 \begin{vmatrix} b_1 & b_3 \\ c_1 & c_3 \end{vmatrix} + a_3 \begin{vmatrix} b_1 & b_2 \\ c_1 & c_2 \end{vmatrix}, \tag{17}$$

and therefore we can write

$$\mathbf{a} \cdot (\mathbf{b} \times \mathbf{c}) = \begin{vmatrix} a_1 & a_2 & a_3 \\ b_1 & b_2 & b_3 \\ c_1 & c_2 & c_3 \end{vmatrix}. \tag{18}$$

Equation 17 is just the expansion of the determinant (18) by cofactors of elements in the first row.

In a similar way we obtain

$$\mathbf{b} \cdot (\mathbf{c} \times \mathbf{a}) = \begin{vmatrix} b_1 & b_2 & b_3 \\ c_1 & c_2 & c_3 \\ a_1 & a_2 & a_3 \end{vmatrix} = \begin{vmatrix} a_1 & a_2 & a_3 \\ b_1 & b_2 & b_3 \\ c_1 & c_2 & c_3 \end{vmatrix}; \tag{19}$$

two row interchanges are required to go from the first determinant to the second in Eq. 19, and this leaves the value of the determinant unchanged. Likewise we have

$$\mathbf{c} \cdot (\mathbf{a} \times \mathbf{b}) = \begin{vmatrix} c_1 & c_2 & c_3 \\ a_1 & a_2 & a_3 \\ b_1 & b_2 & b_3 \end{vmatrix} = \begin{vmatrix} a_1 & a_2 & a_3 \\ b_1 & b_2 & b_3 \\ c_1 & c_2 & c_3 \end{vmatrix}. \tag{20}$$

From Eqs. 18, 19, and 20 it follows that

$$\mathbf{a} \cdot (\mathbf{b} \times \mathbf{c}) = \mathbf{b} \cdot (\mathbf{c} \times \mathbf{a}) = \mathbf{c} \cdot (\mathbf{a} \times \mathbf{b}). \tag{21}$$

Further, since the dot product is commutative, the products $(\mathbf{b} \times \mathbf{c}) \cdot \mathbf{a}$, $(\mathbf{c} \times \mathbf{a}) \cdot \mathbf{b}$, and $(\mathbf{a} \times \mathbf{b}) \cdot \mathbf{c}$ are also equal to those in Eq. 21. These six scalar triple products all have the property that their factors appear in the "natural" order **a**, **b**, **c** (or **b**, **c**, **a**, or **c**, **a**, **b**). If the order of any two of the factors is reversed, then the sign of the product is also reversed. For example

$$\mathbf{a} \cdot (\mathbf{c} \times \mathbf{b}) = \begin{vmatrix} a_1 & a_2 & a_3 \\ c_1 & c_2 & c_3 \\ b_1 & b_2 & b_3 \end{vmatrix} = -\begin{vmatrix} a_1 & a_2 & a_3 \\ b_1 & b_2 & b_3 \\ c_1 & c_2 & c_3 \end{vmatrix}. \tag{22}$$

The scalar triple product has a useful geometrical interpretation. Consider the parallelepiped defined by the vectors **a**, **b**, and **c** (see Figure 15.4.6). Let the

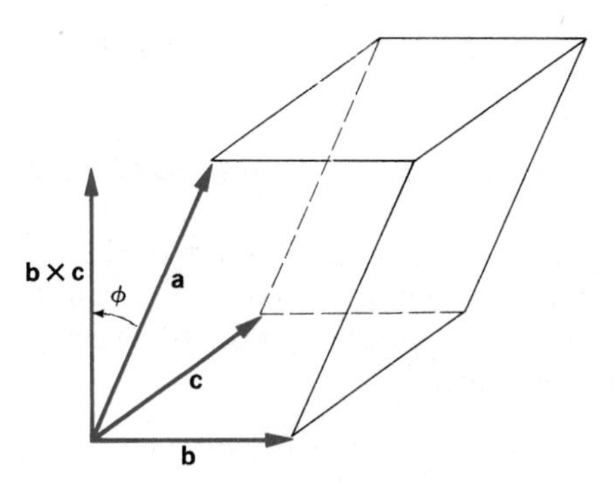

Figure 15.4.6 Volume of parallelepiped $= |\mathbf{a} \cdot (\mathbf{b} \times \mathbf{c})|$.

base of the parallelepiped be the parallelogram defined by **b** and **c**; its area is $\|\mathbf{b} \times \mathbf{c}\|$. The vector $\mathbf{b} \times \mathbf{c}$ is perpendicular to **b** and **c** and the altitude of the parallelepiped is $\|\mathbf{a}\|\,|\cos \phi|$, where ϕ is the angle between **a** and $\mathbf{b} \times \mathbf{c}$. Then the absolute value of the scalar product of **a** and $\mathbf{b} \times \mathbf{c}$ is given by

$$\begin{aligned} |\mathbf{a} \cdot (\mathbf{b} \times \mathbf{c})| &= \|\mathbf{b} \times \mathbf{c}\|\,\|\mathbf{a}\|\,|\cos \phi| \\ &= \text{(area of base)(altitude)} \\ &= \text{volume of parallelepiped.} \end{aligned} \tag{23}$$

If $\mathbf{a} \cdot (\mathbf{b} \times \mathbf{c}) = 0$, then the volume of the parallelepiped is zero and this means that the vectors **a**, **b**, and **c** are coplanar.

EXAMPLE 4

Determine whether the four points $P(-1, 1, 0)$, $Q(2, 0, 3)$, $R(1, 1, -1)$, and $S(1, -1, -3)$ are coplanar. Also find the volume of the parallelepiped with edges PQ, PR, and PS.

First we determine the vectors

$$\begin{aligned} \overrightarrow{PQ} &= 3\mathbf{i} - \mathbf{j} + 3\mathbf{k}, \\ \overrightarrow{PR} &= 2\mathbf{i} - \mathbf{k}, \\ \overrightarrow{PS} &= 2\mathbf{i} - 2\mathbf{j} - 3\mathbf{k}. \end{aligned}$$

Then we calculate the scalar triple product

$$\overrightarrow{PQ} \cdot (\overrightarrow{PR} \times \overrightarrow{PS}) = \begin{vmatrix} 3 & -1 & 3 \\ 2 & 0 & -1 \\ 2 & -2 & -3 \end{vmatrix} = -22.$$

Since $\overrightarrow{PQ} \cdot (\overrightarrow{PR} \times \overrightarrow{PS}) \neq 0$, the points are not coplanar. Further, from Eq. 23 the volume of the parallelepiped is 22. ∎

A second way of multiplying three vectors yields the expression $\mathbf{a} \times (\mathbf{b} \times \mathbf{c})$, which is always a vector, and is called a **vector triple product** of **a**, **b**, and **c**. Some properties of this product are given in Problems 25 and 26.

PROBLEMS

In each of Problems 1 through 6, find $\mathbf{a} \times \mathbf{b}$.

1. $\mathbf{a} = 2\mathbf{i} - \mathbf{j} - \mathbf{k}, \quad \mathbf{b} = \mathbf{i} + 2\mathbf{j} + 4\mathbf{k}$
2. $\mathbf{a} = -\mathbf{i} + 2\mathbf{j} + \mathbf{k}, \quad \mathbf{b} = 3\mathbf{i} + \mathbf{j} - \mathbf{k}$
3. $\mathbf{a} = \mathbf{i} + \mathbf{j} + \mathbf{k}, \quad \mathbf{b} = -2\mathbf{i} - 3\mathbf{j} + \mathbf{k}$
4. $\mathbf{a} = 2\mathbf{i} - 3\mathbf{k}, \quad \mathbf{b} = \mathbf{i} + \mathbf{j} - \mathbf{k}$
5. $\mathbf{a} = a_1\mathbf{i} + a_2\mathbf{j}, \quad \mathbf{b} = b_1\mathbf{i} + b_2\mathbf{j}$
6. $\mathbf{a} = a_1\mathbf{i} + a_2\mathbf{j}, \quad \mathbf{b} = b_3\mathbf{k}$

Let $\mathbf{a} = 2\mathbf{i} - \mathbf{j} - \mathbf{k}$, $\mathbf{b} = \mathbf{i} + 2\mathbf{j} + 3\mathbf{k}$, and $\mathbf{c} = -2\mathbf{i} + 3\mathbf{j} + \mathbf{k}$. Calculate the given expression in each of Problems 7 through 12.

7. $\mathbf{a} \times (\mathbf{b} + \mathbf{c})$

8. $(\mathbf{a} + \mathbf{b}) \times \mathbf{c}$

9. $\mathbf{a} \cdot (\mathbf{b} \times \mathbf{c})$

10. $(\mathbf{a} \times \mathbf{c}) \cdot \mathbf{b}$

11. $\mathbf{a} \times (\mathbf{b} \times \mathbf{c})$

12. $(\mathbf{a} \times \mathbf{b}) \times \mathbf{c}$

In each of Problems 13 and 14, find a unit vector **u** that is perpendicular to **a** and to **b**.

13. $\mathbf{a} = \mathbf{i} - 2\mathbf{j} + \mathbf{k}$, $\mathbf{b} = 3\mathbf{i} + \mathbf{j} - 4\mathbf{k}$

14. $\mathbf{a} = -2\mathbf{i} + 3\mathbf{j} - \mathbf{k}$, $\mathbf{b} = 3\mathbf{i} + 4\mathbf{k}$

In each of Problems 15 and 16, find the area of the triangle whose vertices are the points P, Q, and R.

15. $P(0, 0, 0)$, $Q(0, 3, 0)$, $R(2, 5, 8)$

16. $P(-1, -2, 1)$, $Q(2, 1, 3)$, $R = (1, 4, 0)$

In each of Problems 17 and 18, find the volume of the parallelepiped with edges PQ, PR, and PS; also determine whether the given points are coplanar.

17. $P(0, 0, 0)$, $Q(1, 1, 0)$, $R(-1, 3, 0)$, $S(1, 0, 4)$

18. $P(-1, 2, 2,)$, $Q(2, 1, 4)$, $R(-1, 4, 1)$, $S(3, 0, 5)$

In each of Problems 19 and 20, determine the value of λ for which the points P, Q, R, and S are coplanar.

19. $P(2, 4, -1)$, $Q(0, \lambda, 1)$, $R(-2, 1, 2)$, $S(1, 1, 0)$

20. $P(0, 2, -3)$, $Q(1, 1, 1)$, $R(2, 0, -1)$, $S(\lambda, 2\lambda, -\lambda)$

21. Consider the triangle in the xy-plane whose vertices are the points $P_1(x_1, y_1)$, $P_2(x_2, y_2)$, and $P_3(x_3, y_3)$. Show that the area A of this triangle is given by

$$A = \tfrac{1}{2} \operatorname{abs} \begin{vmatrix} 1 & 1 & 1 \\ x_1 & x_2 & x_3 \\ y_1 & y_2 & y_3 \end{vmatrix},$$

where abs indicates that the absolute value of the determinant must be used.

22. Derive the distributive laws for the cross product

(a) $\mathbf{a} \times (\mathbf{b} + \mathbf{c}) = \mathbf{a} \times \mathbf{b} + \mathbf{a} \times \mathbf{c}$,

(b) $(\mathbf{a} + \mathbf{b}) \times \mathbf{c} = \mathbf{a} \times \mathbf{c} + \mathbf{b} \times \mathbf{c}$.

23. If α is any number, show that

(a) $(\alpha\mathbf{a}) \times \mathbf{b} = \alpha(\mathbf{a} \times \mathbf{b})$

(b) $\mathbf{a} \times (\alpha\mathbf{b}) = \alpha(\mathbf{a} \times \mathbf{b})$

24. Show that $(\mathbf{a} + \mathbf{b}) \times (\mathbf{a} - \mathbf{b}) = 2(\mathbf{b} \times \mathbf{a})$ for any vectors **a** and **b**.

* **25. Vector triple product.** The product $\mathbf{a} \times (\mathbf{b} \times \mathbf{c})$ always yields a vector, and so is called a vector triple product. In this problem we show how to derive a formula for this product.

(a) For any vectors **b** and **c**, show that

$$\mathbf{i} \times (\mathbf{b} \times \mathbf{c}) = c_1\mathbf{b} - b_1\mathbf{c}. \qquad (i)$$

(b) Derive expressions similar to that in Part (a) for $\mathbf{j} \times (\mathbf{b} \times \mathbf{c})$ and $\mathbf{k} \times (\mathbf{b} \times \mathbf{c})$.

(c) Use the results of Parts (a) and (b) to show that

$$\mathbf{a} \times (\mathbf{b} \times \mathbf{c}) = (\mathbf{c} \cdot \mathbf{a})\mathbf{b} - (\mathbf{b} \cdot \mathbf{a})\mathbf{c} \qquad (ii)$$

for any vector **a**.

* **26. Nonassociativity of the cross product.** In this problem we seek to compare $\mathbf{a} \times (\mathbf{b} \times \mathbf{c})$ and $(\mathbf{a} \times \mathbf{b}) \times \mathbf{c}$.

(a) By referring to Problem 25, derive an expression for $(\mathbf{a} \times \mathbf{b}) \times \mathbf{c}$.

(b) By comparing the result of Part (a) with that of Part (c) of Problem 25, determine when $\mathbf{a} \times (\mathbf{b} \times \mathbf{c}) = (\mathbf{a} \times \mathbf{b}) \times \mathbf{c}$.

* **27.** In this problem we start from the geometrical definition of the cross product and indicate how to derive formula (1) of the text. Let $\mathbf{d} = d_1\mathbf{i} + d_2\mathbf{j} + d_3\mathbf{k} = \mathbf{a} \times \mathbf{b}$ and determine d_1, d_2, and d_3 in the following way.

(a) Since **d** is perpendicular to both **a** and **b**, it follows that $\mathbf{d} \cdot \mathbf{a} = 0$ and $\mathbf{d} \cdot \mathbf{b} = 0$. Then show that

$$(a_2b_3 - a_3b_2)d_2 = (a_3b_1 - a_1b_3)d_1, \qquad (i)$$

$$(a_2b_3 - a_3b_2)d_3 = (a_1b_2 - a_2b_1)d_1. \qquad (ii)$$

(b) Since $\|\mathbf{d}\| = \|\mathbf{a}\| \|\mathbf{b}\| |\sin \theta|$, show that

$$\|\mathbf{d}\|^2 = \|\mathbf{a}\|^2 \|\mathbf{b}\|^2 - (\mathbf{a} \cdot \mathbf{b})^2$$
$$= (a_1b_2 - a_2b_1)^2 + (a_3b_1 - a_1b_3)^2 + (a_2b_3 - a_3b_2)^2. \qquad (iii)$$

Observe that this step is essentially the reverse of the derivation of Property 3 in the text.

(c) Assume that $\|\mathbf{d}\| \neq 0$, and use the results of Parts (a) and (b) to show that

$$d_1 = \pm(a_2b_3 - a_3b_2). \qquad (iv)$$

Then, if $d_1 \neq 0$, use the results of Part (a) to show that

$$d_2 = \pm(a_3b_1 - a_1b_3), \qquad (v)$$

$$d_3 = \pm(a_1b_2 - a_2b_1); \qquad (vi)$$

observe that either the plus sign must be used in all three of Eqs. (*iv*), (*v*), and (*vi*), or else the minus sign must be used in all three of these equations.

(d) Let $\mathbf{a} = \mathbf{i}$ and $\mathbf{b} = \mathbf{j}$ and show that the definition requires that $\mathbf{d} = \mathbf{k}$. Show that it follows that the plus sign must be used in Eqs. (*iv*), (*v*), and (*vi*), which establishes Eq. 1.

(e) Modify the preceding argument if $d_1 = 0$ but either d_2 or d_3 is nonzero. Observe that if $d_1 = d_2 = d_3 = 0$, then $\mathbf{a}$ is parallel to $\mathbf{b}$ and $\mathbf{d} = \mathbf{0}$.

15.5 PLANES AND LINES

The algebra of vectors that we have developed so far in this chapter can be used very efficiently to study the properties of planes and straight lines in three-dimensional space.

Planes

A plane can be defined in the following way. Let $P_0(x_0, y_0, z_0)$ be a given point, and let a given vector $\mathbf{n}$ determine a direction at P_0. Then the **plane** that contains P_0 and is perpendicular to $\mathbf{n}$ consists of all those points $P(x, y, z)$ such that the vector from P_0 to P is perpendicular to $\mathbf{n}$. The vector $\mathbf{n}$ is called a **normal vector** to the plane.

The situation is shown in Figure 15.5.1. Let $\mathbf{r}_0$ and $\mathbf{r}$ be the position vectors

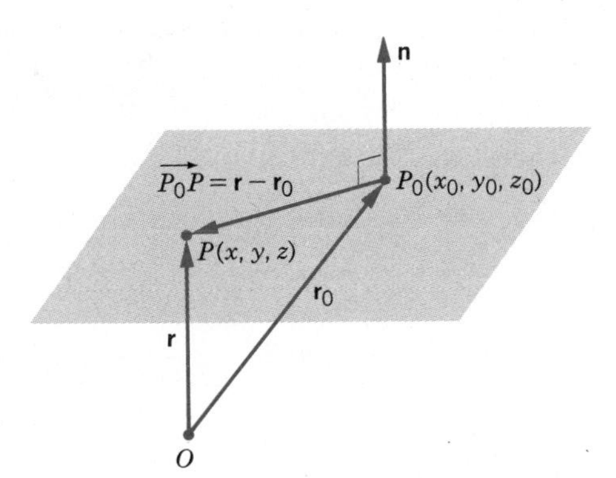

Figure 15.5.1 The plane through P_0 perpendicular to n.

of P_0 and P, respectively; then $\mathbf{r} - \mathbf{r}_0$ is the vector from P_0 to P. The vectors $\mathbf{r} - \mathbf{r}_0$ and $\mathbf{n}$ must be perpendicular, so

$$(\mathbf{r} - \mathbf{r}_0) \cdot \mathbf{n} = 0. \qquad (1)$$

Equation 1 is the basic vector equation of the plane through P_0 that is perpendicular to $\mathbf{n}$. This fundamental relation is also useful in finding equations of planes that satisfy other sets of geometrical conditions. In such cases some preliminary calculations may be required to find a normal vector **n** or the position vector $\mathbf{r}_0$ of a point in the plane.

EXAMPLE 1

Find an equation of the plane perpendicular to the vector $\mathbf{n} = 2\mathbf{i} - \mathbf{j} + 3\mathbf{k}$, and containing the point $P(4, 2, -3)$ (see Figure 15.5.2).

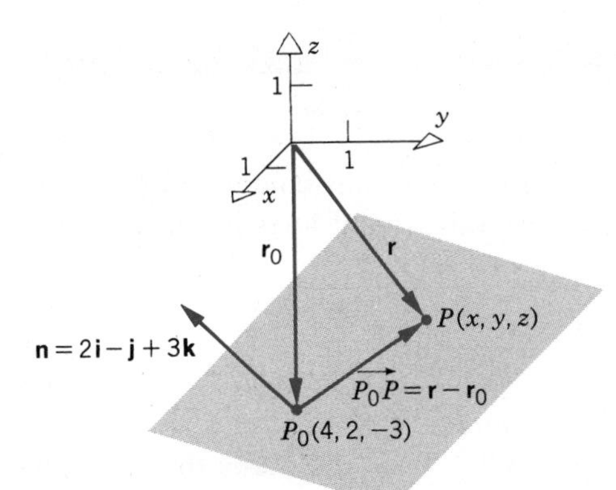

Figure 15.5.2

In this case, the vector $\mathbf{r} - \mathbf{r}_0$ is given by

$$\begin{aligned}\mathbf{r} - \mathbf{r}_0 = \overrightarrow{P_0P} &= (x\mathbf{i} + y\mathbf{j} + z\mathbf{k}) - (4\mathbf{i} + 2\mathbf{j} - 3\mathbf{k}) \\ &= (x - 4)\mathbf{i} + (y - 2)\mathbf{j} + (z + 3)\mathbf{k}. \qquad (2)\end{aligned}$$

Hence Eq. 1 becomes

$$\begin{aligned}(\mathbf{r} - \mathbf{r}_0) \cdot \mathbf{n} &= [(x - 4)\mathbf{i} + (y - 2)\mathbf{j} + (z + 3)\mathbf{k}] \cdot (2\mathbf{i} - \mathbf{j} + 3\mathbf{k}) \\ &= 2(x - 4) - (y - 2) + 3(z + 3) = 0,\end{aligned}$$

or

$$2x - y + 3z = -3. \blacksquare \qquad (3)$$

EXAMPLE 2

Find an equation of the plane that contains the three points $A(1, 3, 3)$, $B(3, 5, 0)$, and $C(2, 5, 2)$ (see Figure 15.5.3).

To find a vector **n** that is normal (perpendicular) to the required plane we first find two vectors in the plane and then form their cross product. For instance, the vectors

$$\overrightarrow{AB} = 2\mathbf{i} + 2\mathbf{j} - 3\mathbf{k}, \qquad \overrightarrow{AC} = \mathbf{i} + 2\mathbf{j} - \mathbf{k}$$

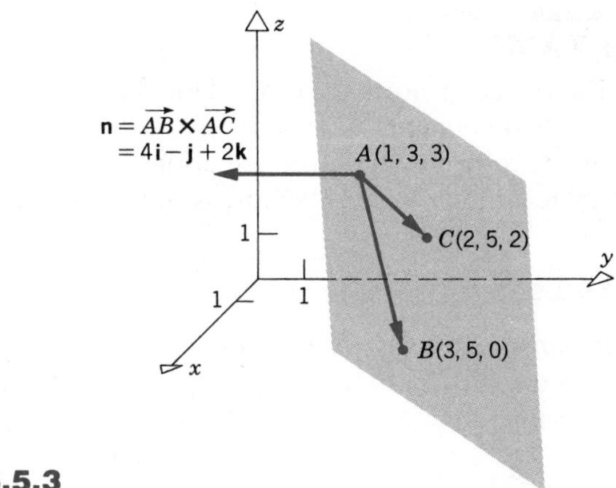

Figure 15.5.3

lie in the plane, so

$$\mathbf{n} = \overrightarrow{AB} \times \overrightarrow{AC} = \begin{vmatrix} \mathbf{i} & \mathbf{j} & \mathbf{k} \\ 2 & 2 & -3 \\ 1 & 2 & -1 \end{vmatrix} = 4\mathbf{i} - \mathbf{j} + 2\mathbf{k} \tag{4}$$

is a normal vector. For $\mathbf{r}_0$ we can choose the position vector of any known point in the plane, for example, the point A; in this case, $\mathbf{r}_0 = \mathbf{i} + 3\mathbf{j} + 3\mathbf{k}$. With these choices of $\mathbf{r}_0$ and $\mathbf{n}$, Eq. 1 becomes

$$\begin{aligned}(\mathbf{r} - \mathbf{r}_0) \cdot \mathbf{n} &= [(x - 1)\mathbf{i} + (y - 3)\mathbf{j} + (z - 3)\mathbf{k}] \cdot (4\mathbf{i} - \mathbf{j} + 2\mathbf{k}) \\ &= 4(x - 1) - (y - 3) + 2(z - 3) = 0,\end{aligned}$$

which simplifies to

$$4x - y + 2z = 7. \blacksquare \tag{5}$$

We now return to a consideration of Eq. 1 in general. Then

$$\mathbf{r} - \mathbf{r}_0 = (x - x_0)\mathbf{i} + (y - y_0)\mathbf{j} + (z - z_0)\mathbf{k} \tag{6}$$

and

$$\mathbf{n} = n_1\mathbf{i} + n_2\mathbf{j} + n_3\mathbf{k}. \tag{7}$$

From Eq. 1 we then obtain

$$n_1(x - x_0) + n_2(y - y_0) + n_3(z - z_0) = 0, \tag{8}$$

or

$$n_1x + n_2y + n_3z = c, \tag{9}$$

where

$$c = n_1x_0 + n_2y_0 + n_3z_0 \tag{10}$$

is a constant. Equation 9 shows that a plane always has an equation that is linear in x, y, and z. Observe that the coefficients of x, y, and z in Eq. 9 are the corresponding components of the normal vector $\mathbf{n}$.

EXAMPLE 3

Find a vector normal to the plane $3x + 2y - 4z = 8$.

Components of such a vector can be read off from the coefficients of x, y, and z, respectively in the equation of the plane. Thus $\mathbf{n} = 3\mathbf{i} + 2\mathbf{j} - 4\mathbf{k}$, or any scalar multiple of this vector, is normal to the given plane. ∎

Lines

A convenient way to describe a straight line is to specify a point on the line and a vector parallel to the line. Let L be the line through the point $P_0(x_0, y_0, z_0)$ having the same direction as the vector $\mathbf{a} = a_1\mathbf{i} + a_2\mathbf{j} + a_3\mathbf{k}$ (see Figure 15.5.4). Let $\mathbf{r}_0 =$

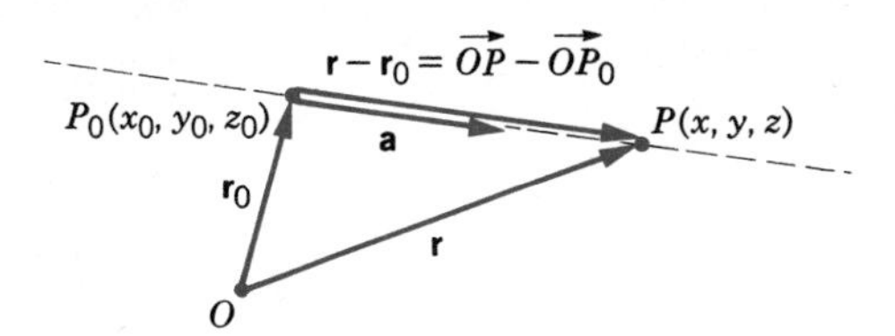

Figure 15.5.4 The line through P_0 in the direction of **a**.

$x_0\mathbf{i} + y_0\mathbf{j} + z_0\mathbf{k}$ be the position vector of P_0, and let $\mathbf{r} = x\mathbf{i} + y\mathbf{j} + z\mathbf{k}$ be the position vector of an arbitrary point P on the line L. Then, as shown in Figure 15.5.4, the difference between $\mathbf{r}$ and $\mathbf{r}_0$ is proportional to $\mathbf{a}$. If we denote the proportionality factor by t, then

$$\mathbf{r} - \mathbf{r}_0 = \mathbf{a}t \quad \text{or} \quad \mathbf{r} = \mathbf{r}_0 + \mathbf{a}t \tag{11}$$

is a **vector parametric equation** of the line L. All points on the line are generated by allowing t to take on all possible real values. Corresponding to Eq. 5 we have the set of **scalar parametric equations**

$$\begin{aligned} x - x_0 &= a_1t & \qquad x &= x_0 + a_1t \\ y - y_0 &= a_2t & \text{or} \qquad y &= y_0 + a_2t \\ z - z_0 &= a_3t & \qquad z &= z_0 + a_3t. \end{aligned} \tag{12}$$

In both Eqs. 11 and 12 the proportionality factor t is the parameter.

If we solve each of Eqs. 12 for t and equate the resulting expressions, we obtain

$$\frac{x - x_0}{a_1} = \frac{y - y_0}{a_2} = \frac{z - z_0}{a_3}. \tag{13}$$

Equations 13 are called the **symmetric,** or **Cartesian, equations** for the line L. Observe that by equating any pair of the terms in Eqs. 13 we obtain the equation of a plane; for example, from the first two terms in Eqs. 13 we have

$$a_2(x - x_0) - a_1(y - y_0) = 0, \tag{14}$$

which corresponds to a plane parallel to the z-axis. We can obtain three such equations from Eqs. 13; however, only two of them are independent. Thus Eqs. 13 describe the line L as the line of intersection of two planes.

In writing Eqs. 13 we have assumed that all of the numbers a_1, a_2, and a_3 are nonzero. If this is not so, then Eqs. 13 must be modified. For example, suppose that $a_1 = 0$. Then Eqs. 12 become

$$x = x_0, \qquad y = y_0 + a_2 t, \qquad z = z_0 + a_3 t, \tag{15}$$

and Eqs. 13 are replaced by

$$x = x_0, \qquad \frac{y - y_0}{a_2} = \frac{z - z_0}{a_3}. \tag{16}$$

The situation is similar if either $a_2 = 0$ or $a_3 = 0$.

Finally, suppose that two of a_1, a_2, and a_3 are zero; for instance, let $a_1 = a_2 = 0$. Then, from Eqs. 12, $x = x_0$, $y = y_0$, and $z = z_0 + a_3 t$, where t is arbitrary. Since x and y are fixed, this corresponds to a line parallel to the z-axis.

EXAMPLE 4

Find vector parametric, scalar parametric, and symmetric equations of the line through the point $(1, -2, 3)$ and parallel to the vector $-2\mathbf{i} + 3\mathbf{j} + \mathbf{k}$.

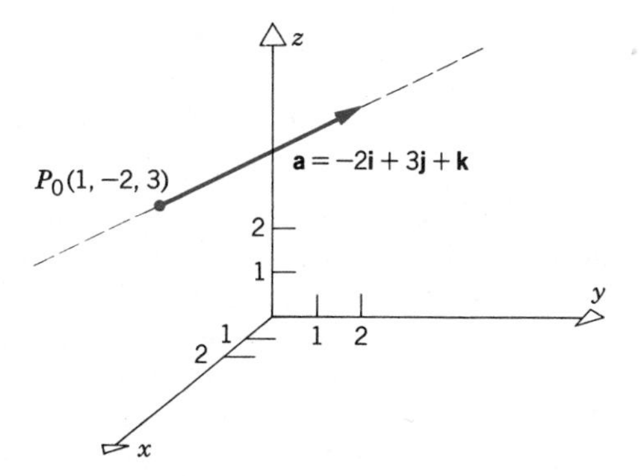

Figure 15.5.5

The line is shown in Figure 15.5.5. From Eq. 11 the vector parametric equation is

$$(x - 1)\mathbf{i} + (y + 2)\mathbf{j} + (z - 3)\mathbf{k} = (-2\mathbf{i} + 3\mathbf{j} + \mathbf{k})t; \tag{17}$$

The corresponding scalar equations are

$$x = 1 - 2t, \qquad y = -2 + 3t, \qquad z = 3 + t. \tag{18}$$

Then, by solving each of these equations for t, we find the symmetric equations

$$\frac{x - 1}{-2} = \frac{y + 2}{3} = \frac{z - 3}{1}. \quad ▮ \tag{19}$$

EXAMPLE 5

Find parametric and symmetric equations of the line containing the two points $P(-2, 1, 5)$ and $Q(2, 3, -2)$ (see Figure 15.5.6).

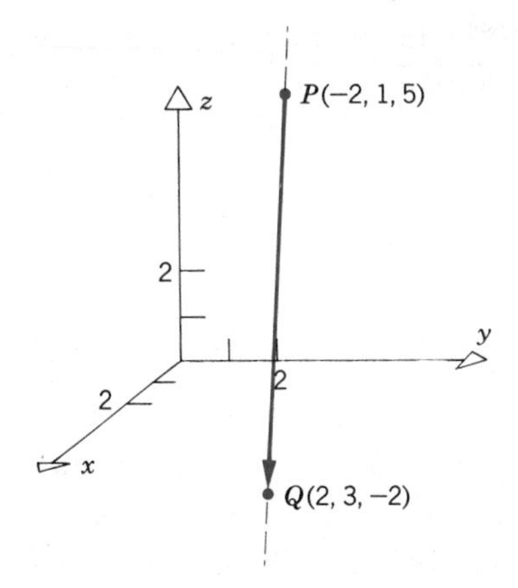

Figure 15.5.6

To determine the equations of the line we need to know a point on the line and a vector parallel to the line. Either the point P or the point Q can serve as the point on the line. A vector parallel to the line is given by $\overrightarrow{PQ} = 4\mathbf{i} + 2\mathbf{j} - 7\mathbf{k}$. Choosing P as the point on the line, we obtain the parametric equations

$$x = -2 + 4t, \qquad y = 1 + 2t, \qquad z = 5 - 7t, \tag{20}$$

and the symmetric equations

$$\frac{x+2}{4} = \frac{y-1}{2} = \frac{z-5}{-7}. \quad \blacksquare \tag{21}$$

Observe that both the vector equation (11) and the parametric equations (12) are linear in the parameter t. Of course, any other letter could be used instead of t to denote the independent variable. Note also that if $\mathbf{a}$ is replaced by any other parallel vector $\lambda\mathbf{a}$, where λ is a constant, then the equations of the line are not changed. This is immediately clear from Eqs. 13 where the effect is to multiply the denominator in each term by λ, which can then be canceled. Similarly, in Eq. 11 we would have $\mathbf{r} = \mathbf{r}_0 + \mathbf{a}\lambda t$; if we define a new parameter $s = \lambda t$, then $\mathbf{r} = \mathbf{r}_0 + \mathbf{a}s$, which is of the same form as Eq. 11.

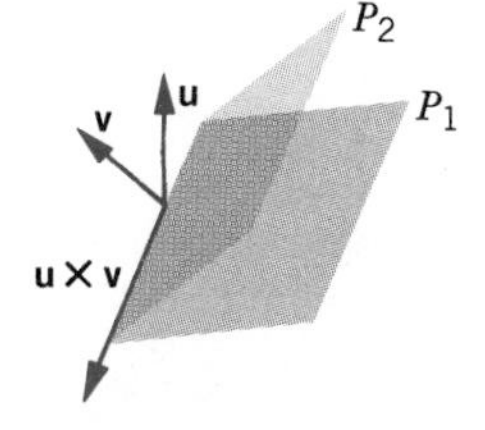

Figure 15.5.7

Sometimes it is required to find parametric or symmetric equations for the line of intersection of two planes P_1 and P_2. In order to do this we need a vector parallel to the line. To find such a vector we start by writing down a vector $\mathbf{u}$ that is normal to the plane P_1 and a vector $\mathbf{v}$ that is normal to the plane P_2. Then the vector $\mathbf{u} \times \mathbf{v}$ is perpendicular to both of the vectors $\mathbf{u}$ and $\mathbf{v}$ and therefore it is parallel to the line of intersection of the planes P_1 and P_2 (see Figure 15.5.7). Using the vector $\mathbf{u} \times \mathbf{v}$ and the coordinates of any point on the line of intersection, we can immediately write down equations for the required line, as illustrated by the following example.

EXAMPLE 6

Find parametric and symmetric equations of the line of intersection of the two planes

$$2x - 3y + z = 8, \qquad x + 2y - z = -3. \tag{22}$$

The vectors $\mathbf{u} = 2\mathbf{i} - 3\mathbf{j} + \mathbf{k}$ and $\mathbf{v} = \mathbf{i} + 2\mathbf{j} - \mathbf{k}$ are respective normals to the given planes. Hence the cross product

$$\mathbf{u} \times \mathbf{v} = \begin{vmatrix} \mathbf{i} & \mathbf{j} & \mathbf{k} \\ 2 & -3 & 1 \\ 1 & 2 & -1 \end{vmatrix} = \mathbf{i} + 3\mathbf{j} + 7\mathbf{k} \tag{23}$$

is parallel to the line of intersection.

We also need a point through which the line passes, that is, a point whose coordinates satisfy both of Eqs. 22. Since this is a set of two equations for three variables, we can usually assign an arbitrary value to one of the variables and then solve for the other two. For example, if $z = 0$, then we have

$$2x - 3y = 8, \qquad x + 2y = -3,$$

which gives $x = 1, y = -2$. Thus the point $(1, -2, 0)$ lies on the line of intersection. Using the coordinates of this point and the vector (23) we obtain the parametric equations

$$x = 1 + t, \qquad y = -2 + 3t, \qquad z = 7t. \tag{24}$$

The corresponding symmetric equations are

$$\frac{x - 1}{1} = \frac{y + 2}{3} = \frac{z}{7}. \quad \blacksquare \tag{25}$$

Another common problem is to find the distance d from a point P to a plane. By this is meant the shortest distance from P to any point in the plane, that is, the distance measured along the line through the given point that is also perpendicular to the given plane (see Figure 15.5.8). This distance is easily found by using vector

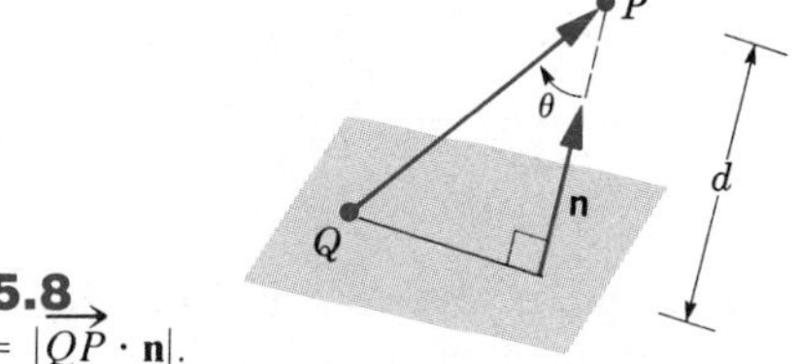

Figure 15.5.8
The distance $d = |\overrightarrow{QP} \cdot \mathbf{n}|$.

methods. Let Q be any point in the plane and let $\overrightarrow{QP}$ be the vector from Q to P. Then the required distance d is the magnitude of the component of $\overrightarrow{QP}$ in the direction normal to the plane. In symbols

$$d = \|\overrightarrow{QP}\|\cos\theta = |\overrightarrow{QP} \cdot \mathbf{n}|, \tag{26}$$

where $\mathbf{n}$ is a unit vector normal to the plane. Observe that $\overrightarrow{QP} \cdot \mathbf{n}$ is positive or negative depending on whether or not $\mathbf{n}$ lies on the same side of the plane as P. The absolute value bars in Eq. 26 take care of both cases at once, so it makes no difference which unit normal vector we use.

EXAMPLE 7

Find the distance from the point $P(2, -1, 4)$ to the plane $3x + 4y - 2z = 6$.

A point Q in the plane can be found (for example) by setting $x = 0$, $y = 0$, and solving the equation of the plane for z. This gives $z = -3$, so the point $Q(0, 0, -3)$ lies in the plane. Then the vector $\overrightarrow{QP}$ is given by

$$\overrightarrow{QP} = 2\mathbf{i} - \mathbf{j} + 7\mathbf{k}.$$

A vector normal to the plane is given by $3\mathbf{i} + 4\mathbf{j} - 2\mathbf{k}$ so the unit vector $\mathbf{n}$ in this direction is

$$\mathbf{n} = \frac{1}{\sqrt{29}}(3\mathbf{i} + 4\mathbf{j} - 2\mathbf{k}).$$

From Eq. 26 we obtain

$$d = \frac{1}{\sqrt{29}}|6 - 4 - 14| = \frac{12}{\sqrt{29}}.$$

Note that $\overrightarrow{QP} \cdot \mathbf{n} < 0$ in this case, so the absolute value is required in order to obtain the distance from the point to the plane. ∎

PROBLEMS

1. Find an equation of the plane that contains the point $(2, -1, 3)$ and is perpendicular to the vector $\mathbf{n} = -\mathbf{i} + 4\mathbf{j} + 5\mathbf{k}$.

2. Find an equation of the plane that contains the point $(3, 0, -4)$ and is perpendicular to the vector $\mathbf{n} = \mathbf{j} + 2\mathbf{k}$.

3. Find an equation of the plane that contains the points $P(1, -2, 3)$, $Q(-2, 0, 1)$, and $R(0, 3, 2)$.

4. Find an equation of the plane that contains the points $P(2, 1, -1)$, $Q(4, 0, 2)$, and $R(-1, -2, 1)$.

5. Find an equation of the plane that contains the point $(3, 1, -2)$ and is parallel to the plane $2x - y + 5z = 7$.

6. Find an equation of the plane that contains the point $(4, -1, 3)$ and is perpendicular to the line
$$\frac{x-2}{3} = \frac{y+1}{-2} = \frac{z}{-1}.$$

7. Find an equation of the plane that contains the two lines
$$\frac{x-2}{3} = \frac{y+1}{-2} = \frac{z}{-1}$$
and
$$\frac{x-2}{-1} = \frac{y+1}{4} = \frac{z}{2}.$$

8. Find an equation of the plane that contains the point $(3, 2, -1)$ and is perpendicular to both of the planes $x - 2y + z = 4$ and $-2x + y + 3z = 7$.

9. Find an equation of the plane that contains the point $(2, 2, -3)$ and the line
$$\frac{x+2}{-1} = \frac{y-3}{2} = \frac{z+1}{3}.$$

10. Find an equation of the plane that contains the point $(1, -3, 2)$ and is perpendicular to the line segment joining that point and $(3, 0, -2)$.

11. Find scalar parametric equations of the line
$$\frac{x-2}{3} = \frac{y+1}{2} = \frac{z-4}{-1}.$$

12. Find symmetric equations of the line $x = -1 + 6t$, $y = 2 - 3t$, $z = 5 + 4t$.

13. Find scalar parametric equations of the line through the point $(2, -1, 3)$ that is parallel to the vector $\mathbf{a} = \mathbf{i} + 2\mathbf{j} - \mathbf{k}$.

14. Find symmetric equations of the line through the point $(-3, 2, 4)$ that is parallel to the vector $\mathbf{a} = 2\mathbf{i} + 3\mathbf{j} - 5\mathbf{k}$.

15. Find symmetric equations of the line through the two points $(4, 2, 3)$ and $(2, -1, 6)$.

16. Find scalar parametric equations of the line through the point $(-1, 0, 3)$ that is parallel to the line

$$\frac{x-4}{2} = \frac{y+2}{-3} = \frac{z}{-1}.$$

17. Find symmetric equations of the line through the point $(-2, 6, 1)$ that is perpendicular to the plane $y + 2z = 7$.

18. Find scalar parametric equations for the line through the point $(3, -1, 4)$ that is perpendicular to the plane $2x + 3y - z = 9$.

19. Find scalar parametric equations of the line of intersection of the planes $x - 2z = 4$ and $3y + z = -9$.

20. Find symmetric equations of the line of intersection of the planes $x - y + z = 7$ and $2x + 3y - 5z = 9$.

21. Find the distance from the point $(2, 1, -5)$ to the plane $x + y + z = 7$.

22. Find the distance from the point $(-1, -3, 2)$ to the plane $2x + 3y - z = 12$.

23. Is the line $x = 2 - 3t$, $y = 2t$, $z = -1 + 4t$ parallel to the plane $2x - y + 2z = 20$?

24. Is the line

$$\frac{x-2}{-2} = \frac{y+1}{3} = \frac{z-2}{-1}$$

parallel to the plane $2x + 3y - z = 11$?

25. Find the point where the line

$$\frac{x-2}{-2} = \frac{y+1}{3} = \frac{z-2}{-1}$$

intersects the plane $2x + 3y - z = 11$.

26. If $\mathbf{p}$, $\mathbf{q}$, and $\mathbf{r}$ are vectors from the origin to three points in a plane, show that the vector $(\mathbf{p} \times \mathbf{q}) + (\mathbf{q} \times \mathbf{r}) + (\mathbf{r} \times \mathbf{p})$ is normal to the plane.

Distance from a Line to a Plane

If a line is parallel to a plane, then the distance d from the line to the plane is the distance from any point P on the line to the nearest point in the plane, that is, the distance measured along the line through P that is perpendicular to the plane; see Figure 15.5.9, where Q is any point in the plane and $\mathbf{n}$ is a unit vector normal to the plane.

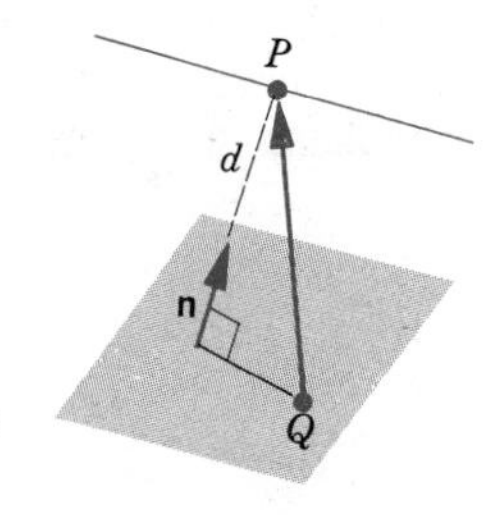

Figure 15.5.9
The distance from a line to a plane.

27. Find the distance from the line

$$\frac{x-2}{-3} = \frac{y}{2} = \frac{z+1}{4}$$

to the plane $2x - y + 2z = 20$.

28. Find the distance from the line $x = 1 + 2t$, $y = -1 + t$, $z = 2 - 3t$ to the plane $5x - 4y + 2z = 12$.

Distance Between Two Lines

In three dimensions two straight lines may be nonintersecting and also nonparallel. The distance d between two such lines is defined to be the minimum distance between two points, where one point is on each line. This distance is the length of the line segment joining the two lines and perpendicular to both of them; see Figure 15.5.10, where P and Q are points on the two lines and $\mathbf{a}$ and $\mathbf{b}$ are vectors parallel to the lines. How can you find a vector perpendicular to both $\mathbf{a}$ and $\mathbf{b}$?

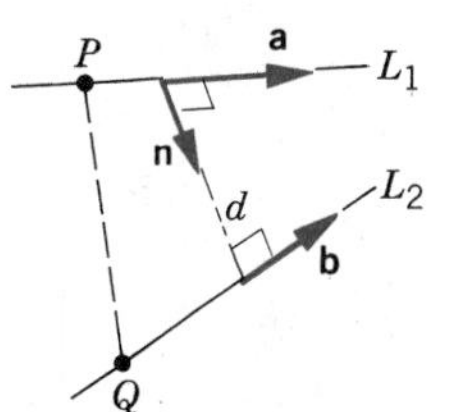

Figure 15.5.10
The distance between two nonintersecting lines.

29. Find the distance between the two lines

$$\frac{x-2}{1} = \frac{y+1}{3} = \frac{z-1}{-2}$$

and

$$\frac{x+1}{4} = \frac{y-2}{-1} = \frac{z+3}{2}.$$

30. Find the distance between the lines $x = 2t$, $y = -1 - t$, $z = 3 + t$ and $x = 2 - t$, $y = -1 + 3t$, $z = t$.

Distance from a Point to a Line

The distance d from a point P to a line L is the distance from P to the nearest point on the line, that is, the distance measured along a line through P that is perpendicular to L. Figure 15.5.11 shows the point P and

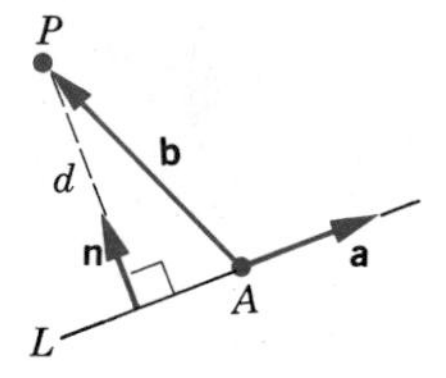

Figure 15.5.11
The distance from a point to a line.

the line L, as well as a vector $\mathbf{n}$ along the required line perpendicular to L. Observe than $\mathbf{n}$ is in the plane determined by P and L.

31. Find the distance from the point (2, 1, 1) to the line

$$\frac{x-1}{2} = \frac{y+1}{3} = \frac{z}{-1}.$$

32. Find the distance from the point $(1, 3, -4)$ to the line $x = -2 + t$, $y = 4 - t$, $z = 1 + 3t$.

33. Let $\mathbf{a}$ be the position vector of a fixed point A other than the origin, and let $\mathbf{r}$ be the position vector of a point P. Describe the surface on which P lies if $\mathbf{r}$ is subject to each of the following conditions.

(a) $\mathbf{a} \cdot \mathbf{r} = 0$.

(b) $|\mathbf{a} \cdot \mathbf{r}| = c\|\mathbf{a}\|, \quad c > 0$.

(c) $(\mathbf{r} - \mathbf{a}) \cdot \mathbf{r} = 0$

34. Show that $\cos\alpha\,\mathbf{i} + \cos\beta\,\mathbf{j} + \cos\gamma\,\mathbf{k}$ is a unit vector making angles α, β, and γ, respectively, with the positive coordinate axes. The coefficients of $\mathbf{i}$, $\mathbf{j}$, and $\mathbf{k}$ are known as the direction cosines of any line parallel to this unit vector.

35. Show that the lines $\mathbf{r} = \mathbf{r}_1 + \lambda\mathbf{a}$ and $\mathbf{r} = \mathbf{r}_2 + \mu\mathbf{b}$ intersect if and only if $(\mathbf{r}_1 - \mathbf{r}_2) \cdot (\mathbf{a} \times \mathbf{b}) = 0$.

15.6 VECTOR FUNCTIONS OF ONE VARIABLE

As in the last section, let $\mathbf{r} = x\mathbf{i} + y\mathbf{j} + z\mathbf{k}$ be the position vector of the point $P(x, y, z)$. We saw that the equation of a straight line in three-dimensional space can be expressed as

$$\mathbf{r} = \mathbf{r}_0 + \mathbf{a}t, \qquad -\infty < t < \infty, \tag{1}$$

where $\mathbf{r}_0 = x_0\mathbf{i} + y_0\mathbf{j} + z_0\mathbf{k}$ is the position vector of a point $P_0(x_0, y_0, z_0)$ on the line, and $\mathbf{a}$ is a vector parallel to the line. The right side of Eq. 1 is a vector-valued function (or simply a vector function) of the single real variable t. It is called a vector function because to each value of the independent variable t, Eq. 1 assigns a unique vector $\mathbf{r}$.

In this section we consider more general vector functions, and begin to examine how the operations of calculus can be applied to them. Suppose that we are given the vector function $\mathbf{F}$ defined by

$$\mathbf{r} = \mathbf{F}(t) = f(t)\,\mathbf{i} + g(t)\,\mathbf{j} + h(t)\,\mathbf{k}, \tag{2}$$

where f, g, and h are scalar functions of the variable t. Since $\mathbf{r} = x\mathbf{i} + y\mathbf{j} + z\mathbf{k}$, it follows by equating the $\mathbf{i}$, $\mathbf{j}$, and $\mathbf{k}$ components, respectively, of Eq. 2 that

$$x = f(t), \qquad y = g(t), \qquad z = h(t). \tag{3}$$

Thus the vector equation (2) is equivalent to the set of three scalar equations (3).

Equations 3 form a set of parametric equations in xyz-space; if f, g, and h are linear functions of t, then we have again the parametric equations of a straight line that appeared in Section 15.5. Equations 3 also generalize the sets of parametric equations in the xy-plane that we discussed in Section 5.5. Much of that earlier discussion can be carried over to the present situation, but we will instead base our treatment here on the vector equation (2). This approach has several advantages. Perhaps the most important is that it provides a geometrical setting in which one's geometrical intuition can often be usefully exploited. Further, vector equations are the same in two or three dimensions, so that one can usually consider both cases at once. Finally, vector notation is compact and leads to an appealing economy of expression.

It is easy to see the geometrical meaning of Eq. 2. Suppose that t lies in some interval $a \leq t \leq b$. Then, for each value of t in the interval, Eq. 2 determines a vector $\mathbf{r}$ that can be regarded as the position vector of a point P. The collection of points obtained in this way form an arc, or curve (see Figure 15.6.1). If we

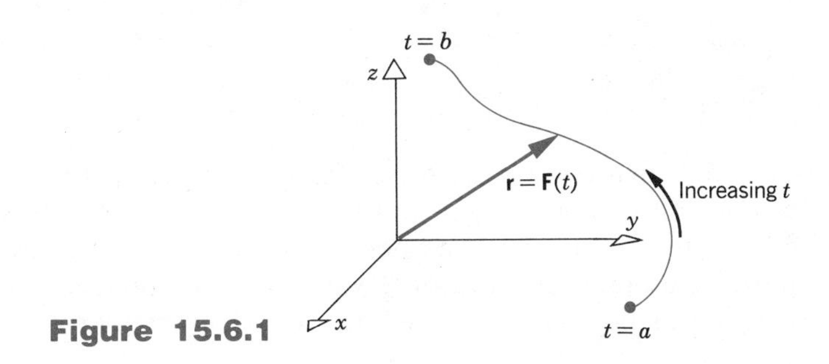

Figure 15.6.1

regard t as representing time, then the curve is the path, or trajectory, traced out by a particle moving in accordance with Eq. 2. The arrow in Figure 15.6.1 indicates the direction of motion. The points $\mathbf{r} = \mathbf{F}(a)$ and $\mathbf{r} = \mathbf{F}(b)$ are called the **initial** and **terminal points,** respectively, of the curve. If they coincide, then the curve is said to be **closed** (see Figure 15.6.2).

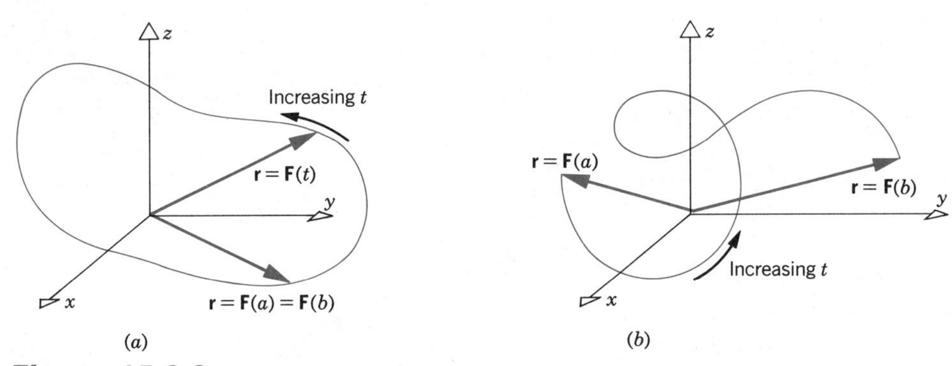

Figure 15.6.2 (a) A closed curve: $\mathbf{F}(b) = \mathbf{F}(a)$.
(b) A curve that is not closed: $\mathbf{F}(b) \neq \mathbf{F}(a)$.

EXAMPLE 1

Sketch the graph of the arc

$$\mathbf{r} = -\frac{t}{2}\mathbf{i} + 2\cos\frac{\pi t}{2}\mathbf{j} + 2t\,\mathbf{k}, \qquad -1 \le t \le 1. \tag{4}$$

The corresponding parametric equations are

$$x = -\frac{t}{2}, \qquad y = 2\cos\frac{\pi t}{2}, \qquad z = 2t.$$

From the first and third of these equations we note that $4x + z = 0$ for all t; thus the arc lies in the plane $4x + z = 0$. In Table 15.1 we record the coordinates of

Table 15.1 Some Values of t and r That Satisfy $\mathbf{r} = -(t/2)\mathbf{i} + 2\cos(\pi t/2)\,\mathbf{j} + 2t\mathbf{k}$ for $-1 \le t \le 1$

t	-1	$-\frac{1}{2}$	0	$\frac{1}{2}$	1
$\mathbf{r}$	$(\frac{1}{2}, 0, -2)$	$(\frac{1}{4}, \sqrt{2}, -1)$	$(0, 2, 0)$	$(-\frac{1}{4}, \sqrt{2}, 1)$	$(-\frac{1}{2}, 0, 2)$

the points that correspond to several values of t. As t ranges from -1 to 1, the x-component of $\mathbf{r}$ decreases, and the z-component increases; the y-component increases for $-1 \le t \le 0$ and decreases for $0 \le t \le 1$. Further, the x- and z-components are odd functions of t, while the y-component is an even function. Using this information, we can draw the sketch shown in Figure 15.6.3. ∎

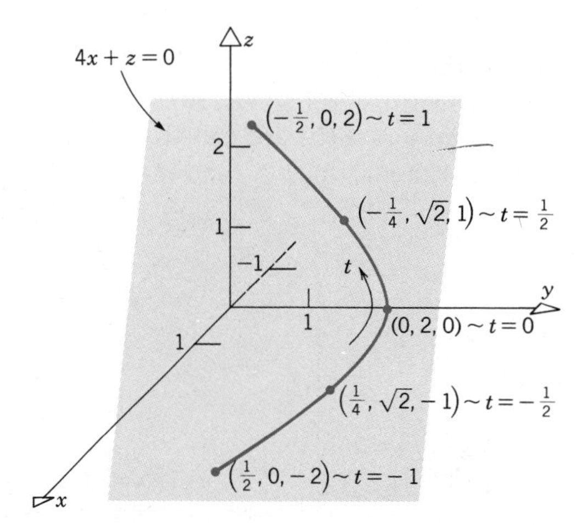

Figure 15.6.3

EXAMPLE 2

Sketch the graph of the curve

$$\mathbf{r} = a\cos\omega t\,\mathbf{i} + a\sin\omega t\,\mathbf{j} + bt\,\mathbf{k}, \qquad -\infty < t < \infty \tag{5}$$

where a, b, and ω are given positive numbers.

Corresponding to Eq. 5 we have the scalar equations

$$x = a\cos\omega t, \qquad y = a\sin\omega t, \qquad z = bt, \tag{6}$$

from which it follows that $x^2 + y^2 = a^2$. Thus all points on the curve are at a distance a from the z-axis; consequently the curve lies on the surface of a right circular cylinder of radius a whose axis is the z-axis. As t increases, a particle following the path (5) moves steadily in the positive z-direction, and at the same time winds around and around the cylindrical surface in the counterclockwise sense (as viewed from the positive z-direction). The graph of Eq. 5 is called a (circular) **helix,** and is shown in Figure 15.6.4. For instance, a particle moving along the

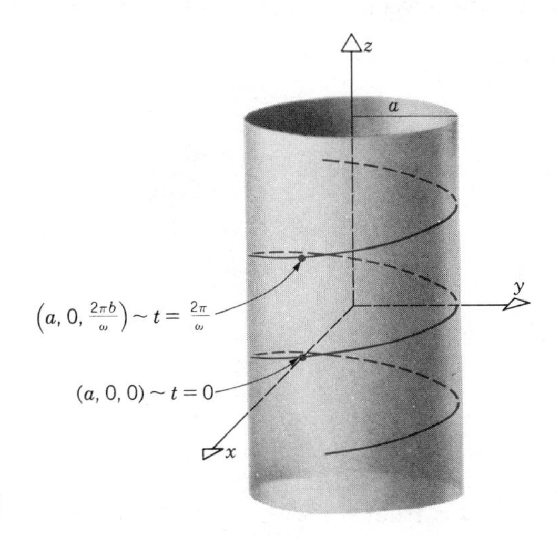

Figure 15.6.4

threads of a bolt follows a helical path. Note that at $t = 0$ the coordinates of the particle are $(a, 0, 0)$, while at $t = 2\pi/\omega$ they are $(a, 0, 2\pi b/\omega)$. Thus, in Eqs. 5 and 6, ω is the frequency of motion around the z-axis and b measures the steepness of the curve in the z-direction. ∎

It is straightforward to extend to vector functions the concept of limit stated in Definition 2.2.1 for scalar functions. The statement

$$\lim_{t \to t_0} \mathbf{F}(t) = \mathbf{L} \tag{7}$$

means that as t approaches t_0, the vector $\mathbf{F}(t)$ approaches the vector $\mathbf{L}$. More precisely, the distance $\|\mathbf{F}(t) - \mathbf{L}\|$ can be made arbitrarily small by requiring $|t - t_0|$ to be sufficiently small, but nonzero. In symbols, for each $\epsilon > 0$, there is a corresponding $\delta > 0$ with the property that if

$$0 < |t - t_0| < \delta \tag{8}$$

then

$$\|\mathbf{F}(t) - \mathbf{L}\| < \epsilon. \tag{9}$$

To see more clearly the meaning of the inequality (9) let us express $\mathbf{F}(t)$ and $\mathbf{L}$ in component form

$$\mathbf{F}(t) = f(t)\mathbf{i} + g(t)\mathbf{j} + h(t)\mathbf{k}, \qquad \mathbf{L} = L_1\mathbf{i} + L_2\mathbf{j} + L_3\mathbf{k}. \tag{10}$$

Then

$$\mathbf{F}(t) - \mathbf{L} = [f(t) - L_1]\mathbf{i} + [g(t) - L_2]\mathbf{j} + [h(t) - L_3]\mathbf{k}$$

and

$$\|\mathbf{F}(t) - \mathbf{L}\| = \{[f(t) - L_1]^2 + [g(t) - L_2]^2 + [h(t) - L_3]^2\}^{1/2}. \tag{11}$$

In geometrical terms the expression in Eq. 9 is the distance from the point P_0 with position vector $\mathbf{L}$ to the point P with position vector $\mathbf{F}(t)$ (see Figure 15.6.5).

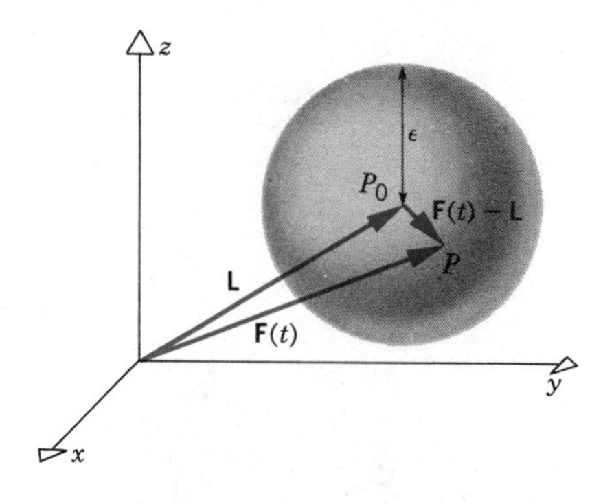

Figure 15.6.5

Equation 9 says that this distance must be less than ϵ; in other words, if Eq. 9 is to be satisfied, then P must lie within the sphere with center at P_0 and radius ϵ. If Eq. 9 is true, then we must also have

$$|f(t) - L_1| < \epsilon, \qquad |g(t) - L_2| < \epsilon, \qquad |h(t) - L_3| < \epsilon. \tag{12}$$

As a consequence, we can assert that if $\lim_{t\to t_0} \mathbf{F}(t) = \mathbf{L}$, then

$$\lim_{t\to t_0} f(t) = L_1, \qquad \lim_{t\to t_0} g(t) = L_2, \qquad \lim_{t\to t_0} h(t) = L_3. \tag{13}$$

It is also possible to modify the argument we have just given to show that Eq. 7 also follows from Eqs. 13. Thus the vector limit (7) is equivalent to the three scalar limits (13). In other words, *limits of vector functions can be taken component by component.* Consequently, all questions related to limits of vector functions can be reduced to corresponding questions about limits of scalar functions. In particular, the familiar results in Theorem 2.3.1(a), (b), (c) about limits of sums, differences, and products of scalar functions can be carried over to limits of vector functions. We omit an explicit listing of these properties here.

EXAMPLE 3

If $\mathbf{F}(t) = (2 + t)\mathbf{i} - t^2\mathbf{j} + \sin(\pi t/2)\mathbf{k}$, find $\lim_{t\to 1} \mathbf{F}(t)$.

From the preceding discussion we have

$$\lim_{t\to 1} \mathbf{F}(t) = \lim_{t\to 1} (2 + t)\mathbf{i} + \lim_{t\to 1} (-t^2)\mathbf{j} + \lim_{t\to 1} \sin\frac{\pi t}{2}\,\mathbf{k}$$

$$= 3\mathbf{i} - \mathbf{j} + \mathbf{k}. \;\blacksquare$$

EXAMPLE 4

If $\mathbf{F}(t)$ is the same as in Example 3, and $\mathbf{G}(t) = (t/2)\mathbf{i} + \cos(\pi t/2)\mathbf{j} - 2t\,\mathbf{k}$, find $\lim_{t\to 1}[\mathbf{F}(t) + \mathbf{G}(t)]$.

We have

$$\lim_{t\to 1} [\mathbf{F}(t) + \mathbf{G}(t)] = \lim_{t\to 1} \mathbf{F}(t) + \lim_{t\to 1} \mathbf{G}(t)$$

where $\lim_{t\to 1} \mathbf{F}(t)$ was found in Example 3, and

$$\lim_{t\to 1} \mathbf{G}(t) = \lim_{t\to 1} \frac{t}{2}\,\mathbf{i} + \lim_{t\to 1} \cos\frac{\pi t}{2}\,\mathbf{j} + \lim_{t\to 1} (-2t)\mathbf{k}$$

$$= \frac{1}{2}\mathbf{i} - 2\mathbf{k}.$$

Then it follows that

$$\lim_{t\to 1} [\mathbf{F}(t) + \mathbf{G}(t)] = (3\mathbf{i} - \mathbf{j} + \mathbf{k}) + \left(\frac{1}{2}\mathbf{i} - 2\mathbf{k}\right) = \frac{7}{2}\mathbf{i} - \mathbf{j} - \mathbf{k}. \;\blacksquare$$

The property of continuity (Definition 2.5.1) also carries over directly from scalar to vector functions. The vector function $\mathbf{F}$ is continuous at t_0 if $\mathbf{F}$ is defined at t_0, and if

$$\lim_{t\to t_0} \mathbf{F}(t) = \mathbf{F}(t_0). \tag{14}$$

This just means that the limit vector $\mathbf{L}$ in the earlier discussion must be the same as $\mathbf{F}(t_0)$. Thus Eq. 14 is equivalent to the three scalar equations

$$\lim_{t\to t_0} f(t) = f(t_0), \qquad \lim_{t\to t_0} g(t) = g(t_0), \qquad \lim_{t\to t_0} h(t) = h(t_0), \tag{15}$$

so $\mathbf{F}$ *is continuous at* t_0 *if and only if each of the three scalar functions* f, g, *and* h *is continuous there.* Questions concerning continuity of vector functions can accordingly be reduced to corresponding questions for scalar functions, and we forego a more elaborate discussion here.

Let us now consider the derivative of a vector function. For a given function $\mathbf{r} = \mathbf{F}(t)$, we first form the difference quotient

$$\frac{\Delta\mathbf{F}}{\Delta t} = \frac{\mathbf{F}(t + \Delta t) - \mathbf{F}(t)}{\Delta t}. \tag{16}$$

Note that $\Delta\mathbf{F}$ is a vector and Δt is a scalar, so that the quotient $\Delta\mathbf{F}/\Delta t$ is a vector. Then we pass to the limit as $\Delta t \to 0$ in order to obtain the derivative, which we denote by $\mathbf{F}'(t)$ or by $d\mathbf{F}/dt$. Thus

$$\frac{d\mathbf{F}}{dt} = \mathbf{F}'(t) = \lim_{\Delta t \to 0} \frac{\mathbf{F}(t + \Delta t) - \mathbf{F}(t)}{\Delta t}, \tag{17}$$

provided the limit exists.

To see the geometrical significance of the derivative of a vector function look at the situation shown in Figure 15.6.6. The vectors $\mathbf{F}(t)$ and $\mathbf{F}(t + \Delta t)$ are the

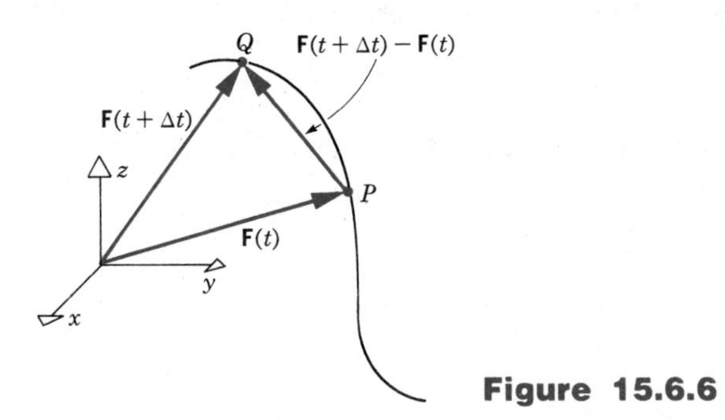

Figure 15.6.6

position vectors of the points P and Q, respectively, on the curve defined by the vector function $\mathbf{F}$. The vector $\mathbf{F}(t + \Delta t) - \mathbf{F}(t)$, or $\overrightarrow{PQ}$, is the secant vector between the two points P and Q on the curve. The vector in the difference quotient (16) is proportional to $\overrightarrow{PQ}$, and is in fact given by $\overrightarrow{PQ}/\Delta t$. If the limit in Eq. 17 exists, that is, if $\overrightarrow{PQ}/\Delta t$ has a limit as $\Delta t \to 0$, and if the limit is not zero, then this limiting vector $\mathbf{F}'(t)$ is tangent to the curve $\mathbf{r} = \mathbf{F}(t)$ at the point P and points in the direction of increasing t (see Figure 15.6.7).

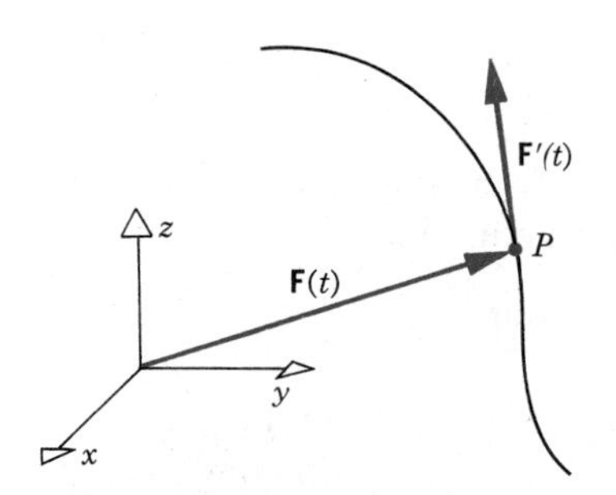

Figure 15.6.7

The line tangent to the curve $\mathbf{r} = \mathbf{F}(t)$ at a point P corresponding to $t = t_0$ is the line through P parallel to the vector $\mathbf{F}'(t_0)$. Similarly, the plane normal to the curve at P is the plane passing through P and perpendicular to $\mathbf{F}'(t_0)$.

Alternatively, if we look upon the equation $\mathbf{r} = \mathbf{F}(t)$ as describing the position of a moving particle at time t, then $\mathbf{F}'(t)$ is the rate of change of the position of the particle with respect to time. In other words, $\mathbf{F}'(t)$ is the **velocity** of the particle, and is tangent to the path on which the particle moves.

Next we derive a formula for $\mathbf{F}'(t)$ in terms of the components of $\mathbf{F}(t)$. Expressing the difference quotient (16) in terms of components, we have

$$\begin{aligned}\frac{\mathbf{F}(t+\Delta t)-\mathbf{F}(t)}{\Delta t} &= \frac{1}{\Delta t}\Big\{f(t+\Delta t)\mathbf{i}+g(t+\Delta t)\mathbf{j}+h(t+\Delta t)\mathbf{k} \\ &\qquad -[f(t)\mathbf{i}+g(t)\mathbf{j}+h(t)\mathbf{k}]\Big\} \\ &= \frac{f(t+\Delta t)-f(t)}{\Delta t}\mathbf{i}+\frac{g(t+\Delta t)-g(t)}{\Delta t}\mathbf{j} \\ &\qquad +\frac{h(t+\Delta t)-h(t)}{\Delta t}\mathbf{k}. \end{aligned} \tag{18}$$

The limit of the right side of Eq. 18 as $\Delta t \to 0$ can be calculated component by component. In this way we obtain

$$\begin{aligned}\mathbf{F}'(t) &= \lim_{\Delta t\to 0}\frac{\mathbf{F}(t+\Delta t)-\mathbf{F}(t)}{\Delta t} \\ &= f'(t)\mathbf{i}+g'(t)\mathbf{j}+h'(t)\mathbf{k}. \end{aligned} \tag{19}$$

Thus $\mathbf{F}'(t)$ *is the vector whose components are the derivatives of the corresponding components of* $\mathbf{F}(t)$. Consequently, $\mathbf{F}'(t)$ can be calculated by applying to each component of $\mathbf{F}(t)$ the methods and formulas already discussed for differentiating scalar functions.

EXAMPLE 5

Find the derivative of

$$\mathbf{F}(t) = 3\cos\frac{\pi t}{2}\,\mathbf{i} + 3\sin\frac{\pi t}{2}\,\mathbf{j} + 2t\,\mathbf{k}. \tag{20}$$

Also find the normal plane and the tangent line at the point (0, 3, 2).

The graph of Eq. 20 is the helix shown in Figure 15.6.8; see Example 2 and

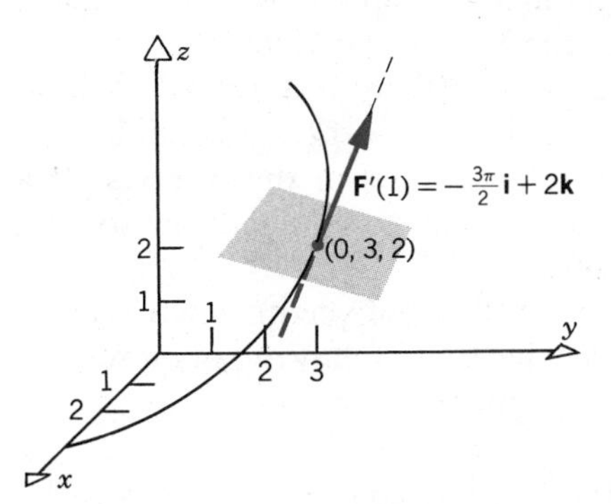

Figure 15.6.8

set $a = 3$, $b = 2$, and $\omega = \pi/2$. By differentiating Eq. 20 component by component, we obtain

$$\mathbf{F}'(t) = -3\frac{\pi}{2}\sin\frac{\pi t}{2}\,\mathbf{i} + 3\frac{\pi}{2}\cos\frac{\pi t}{2}\,\mathbf{j} + 2\mathbf{k}. \tag{21}$$

The point (0, 3, 2) corresponds to $t = 1$, so a tangent vector there is

$$\mathbf{F}'(1) = -3\frac{\pi}{2}\,\mathbf{i} + 2\mathbf{k}. \tag{22}$$

The normal plane has the equation

$$-3\frac{\pi}{2}(x - 0) + 0(y - 3) + 2(z - 2) = 0$$

or

$$3\pi x - 4z = -8. \tag{23}$$

The tangent line has the vector equation

$$\mathbf{r} = 3\mathbf{j} + 2\mathbf{k} + \left(-3\frac{\pi}{2}\,\mathbf{i} + 2\mathbf{k}\right)\tau; \tag{24}$$

in scalar form the corresponding equations are

$$x = -3\frac{\pi}{2}\tau, \qquad y = 3, \qquad z = 2 + 2\tau. \tag{25}$$

In order to avoid possible confusion between the parametric representations of the curve and the tangent line, we have used τ rather than t as the parameter on the tangent line. Some different symbol would have served equally well. Note that $\tau = 0$ corresponds to the point (0, 3, 2). ∎

If $\mathbf{F}'(t)$ is not defined for some value of t because the limit in Eq. 17 does not exist, then the curve $\mathbf{r} = \mathbf{F}(t)$ does not have a tangent there, nor does the moving particle have a velocity at such a point. This corresponds to the situation in which the derivative of a scalar function fails to exist. However, if $\mathbf{F}'(t) = \mathbf{0}$ for some t, then the tangent to the curve $\mathbf{r} = \mathbf{F}(t)$ is also not determined by $\mathbf{F}'(t)$, because the zero vector does not have a unique direction. In such a case the moving particle has zero velocity at the point, even though the direction of the velocity vector is not specified. This case has no counterpart for scalar functions.

As in Section 5.5 for plane curves, the arc $\mathbf{r} = \mathbf{F}(t)$ for $a \le t \le b$ is said to be **smooth** if it has a continuously varying tangent line. A sufficient condition for this is that $\mathbf{F}'(t)$ is continuous and $\mathbf{F}'(t) \neq \mathbf{0}$ for all points in the interval. If a continuous curve consists of a finite number of smooth arcs joined end-to-end, then it is said to be **piecewise smooth** (see Figure 15.6.9).

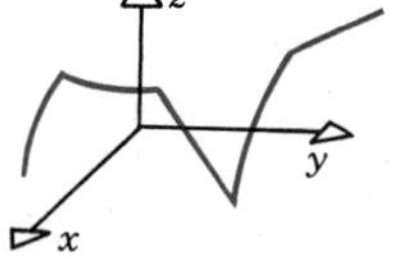

z
y
x

Figure 15.6.9 A piecewise smooth curve.

The differentiation of sums, differences, or products of vector functions proceeds in the same way as for scalar functions. We have the following main results, where $\mathbf{u}(t)$ and $\mathbf{v}(t)$ are arbitrary differentiable vector functions.

$$\frac{d}{dt}[c_1\mathbf{u}(t) + c_2\mathbf{v}(t)] = c_1\frac{d\mathbf{u}}{dt}(t) + c_2\frac{d\mathbf{v}}{dt}(t) \quad \text{for any constants } c_1, c_2; \tag{26}$$

$$\frac{d}{dt}[\mathbf{u}(t)\cdot\mathbf{v}(t)] = \frac{d\mathbf{u}}{dt}(t)\cdot\mathbf{v}(t) + \mathbf{u}(t)\cdot\frac{d\mathbf{v}}{dt}(t); \tag{27}$$

$$\frac{d}{dt}[\mathbf{u}(t)\times\mathbf{v}(t)] = \frac{d\mathbf{u}}{dt}(t)\times\mathbf{v}(t) + \mathbf{u}(t)\times\frac{d\mathbf{v}}{dt}(t). \tag{28}$$

In Eq. 28 the order of the factors is important, while in Eq. 27 it is not. Equations 26 to 28 can be verified by writing out both sides of each equation in terms of the components of $\mathbf{u}(t)$ and $\mathbf{v}(t)$.

EXAMPLE 6

If

$$\mathbf{u}(t) = (2+t)\,\mathbf{i} - t^2\,\mathbf{j} + \sin\frac{\pi t}{2}\,\mathbf{k}$$

and

$$\mathbf{v}(t) = \frac{t}{2}\,\mathbf{i} + \cos\frac{\pi t}{2}\,\mathbf{j} - 2t\,\mathbf{k},$$

find $(d/dt)[\mathbf{u}(t)\cdot\mathbf{v}(t)]$ and evaluate this expression at $t = 1$.

We have

$$\mathbf{u}'(t) = \mathbf{i} - 2t\,\mathbf{j} + \frac{\pi}{2}\cos\frac{\pi t}{2}\,\mathbf{k},$$

$$\mathbf{v}'(t) = \frac{1}{2}\,\mathbf{i} - \frac{\pi}{2}\sin\frac{\pi t}{2}\,\mathbf{j} - 2\mathbf{k}.$$

Then from Eq. 27 we obtain

$$\begin{aligned}\frac{d}{dt}[\mathbf{u}(t)\cdot\mathbf{v}(t)] &= \mathbf{u}'(t)\cdot\mathbf{v}(t) + \mathbf{u}(t)\cdot\mathbf{v}'(t)\\ &= \left(\frac{t}{2} - 2t\cos\frac{\pi t}{2} - 2t\,\frac{\pi}{2}\cos\frac{\pi t}{2}\right)\\ &\quad + \left(\frac{2+t}{2} - \frac{\pi}{2}t^2\sin\frac{\pi t}{2} - 2\sin\frac{\pi t}{2}\right)\\ &= 1 + t - 2t\left(1+\frac{\pi}{2}\right)\cos\frac{\pi t}{2}\\ &\quad - \left(\frac{\pi}{2}t^2 + 2\right)\sin\frac{\pi t}{2}.\end{aligned}$$

At $t = 1$ we have

$$\left.\frac{d}{dt}(\mathbf{u}\cdot\mathbf{v})\right|_{t=1} = 1 + 1 - \left(\frac{\pi}{2} + 2\right) = -\frac{\pi}{2}.$$

PROBLEMS

In each of Problems 1 through 6, find the indicated limit.

1. $\lim_{t\to 0} [e^{-t}\mathbf{i} + 2\cos t\,\mathbf{j} + (t^2 - 1)\mathbf{k}]$

2. $\lim_{t\to \pi} (e^{2t}\,\mathbf{i} - \sin t\,\mathbf{j} + t^3\,\mathbf{k})$

3. $\lim_{t\to 1} \left(\frac{1}{t}\mathbf{i} + \frac{t-2}{t+1}\mathbf{j} + \frac{2t}{t-1}\mathbf{k}\right)$

4. $\lim_{t\to 1} \left(\frac{1}{t}\mathbf{i} + \frac{t-2}{t+1}\mathbf{j} + \frac{t-1}{2t}\mathbf{k}\right)$

5. $\lim_{t\to 0} \frac{1}{t}(\sin 2t\,\mathbf{i} + 3t\,\mathbf{j} + \tan t\,\mathbf{k})$

6. $\lim_{t\to 2} \left(\frac{t^2-4}{t-2}\mathbf{i} + \frac{\sqrt{t}-\sqrt{2}}{t-2}\mathbf{j}\right)$

In each of Problems 7 through 9, assume that both $\lim_{t\to t_0} \mathbf{u}(t)$ and $\lim_{t\to t_0} \mathbf{v}(t)$ exist, and prove the given relation.

7. $\lim_{t\to t_0} [c_1\mathbf{u}(t) + c_2\mathbf{v}(t)] =$

$$c_1 \lim_{t\to t_0} \mathbf{u}(t) + c_2 \lim_{t\to t_0} \mathbf{v}(t) \quad \text{for any} \quad c_1, c_2$$

8. $\lim_{t\to t_0} [\mathbf{u}(t) \cdot \mathbf{v}(t)] = [\lim_{t\to t_0} \mathbf{u}(t)] \cdot [\lim_{t\to t_0} \mathbf{v}(t)]$

9. $\lim_{t\to t_0} [\mathbf{u}(t) \times \mathbf{v}(t)] = [\lim_{t\to t_0} \mathbf{u}(t)] \times [\lim_{t\to t_0} \mathbf{v}(t)]$

In each of Problems 10 through 12, let $\mathbf{u}(t) = e^{-t}\mathbf{i} + 2\cos t\,\mathbf{j} + (t^2 - 1)\mathbf{k}$ and $\mathbf{v}(t) = e^{2t}\,\mathbf{i} - \cos t\,\mathbf{j} + t^3\,\mathbf{k}$. Use the results of Problems 7 through 9 to calculate each of the given limits.

10. $\lim_{t\to 0} [3\mathbf{u}(t) - 2\mathbf{v}(t)]$

11. $\lim_{t\to 0} [\mathbf{u}(t) \cdot \mathbf{v}(t)]$

12. $\lim_{t\to 0} [\mathbf{u}(t) \times \mathbf{v}(t)]$

In each of Problems 13 through 16, find $\mathbf{F}'(t)$.

13. $\mathbf{F}(t) = t^2\,\mathbf{i} + \cos t\,\mathbf{j} + 2\sin t\,\mathbf{k}$

14. $\mathbf{F}(t) = e^{-t}\,\mathbf{i} + te^t\,\mathbf{j} + 4t\,\mathbf{k}$

15. $\mathbf{F}(t) = 2t\sin 2t\,\mathbf{i} - 3\cos t\,\mathbf{j} + 3(2t-1)^2\,\mathbf{k}$

16. $\mathbf{F}(t) = t\cos t\,\mathbf{i} + t\sin t\,\mathbf{j} - 6t\,\mathbf{k}$

In each of Problems 17 through 19, assume that $\mathbf{u}(t) = e^{-t}\,\mathbf{i} + 2\cos t\,\mathbf{j} + (t^2 - 1)\mathbf{k}$, $\mathbf{v}(t) = e^{2t}\,\mathbf{i} - \sin t\,\mathbf{j} + t^3\,\mathbf{k}$, and find the indicated derivative.

17. $\frac{d}{dt}[\mathbf{u}(t) + \mathbf{v}(t)]$

18. $\frac{d}{dt}[\mathbf{u}(t) \cdot \mathbf{v}(t)]$

19. $\frac{d}{dt}[3\mathbf{u}(t) - 2\mathbf{v}(t)]$

In each of Problems 20 through 24, assume that $\mathbf{u}(t) = 2t\,\mathbf{i} - t^2\,\mathbf{j} + t^4\,\mathbf{k}$, $\mathbf{v}(t) = t^2\,\mathbf{i} + 6t\,\mathbf{j} + 5t\,\mathbf{k}$, and find the indicated derivative.

20. $\frac{d}{dt}[4\mathbf{u}(t) - 3\mathbf{v}(t)]$

21. $\frac{d}{dt}[\mathbf{u}(t) \cdot \mathbf{v}(t)]$

22. $\frac{d}{dt}[\mathbf{u}(t) \times \mathbf{v}(t)]$

23. $\frac{d}{dt}[\mathbf{v}(t) \cdot \mathbf{u}(t)]$

24. $\frac{d}{dt}[\mathbf{v}(t) \times \mathbf{u}(t)]$

25. Consider the curve defined by $\mathbf{r} = (2t - 3)\mathbf{i} - 4t^2\,\mathbf{j}$.
    (a) Sketch the curve.
    (b) Find a unit vector tangent to the curve at the point where $t = -1$.
    (c) Find parametric equations of the line tangent to the curve at the point where $t = -1$.
    (d) Find the plane normal to the curve at the point where $t = -1$.

26. Consider the curve defined by $\mathbf{r} = \cosh t\,\mathbf{i} + 2\sinh t\,\mathbf{k}$.
    (a) Sketch the curve.
    (b) Find a unit tangent vector to the curve at the point where $t = 0$.
    (c) Find parametric equations for the line tangent to the curve at the point where $t = 0$.
    (d) Find the plane normal to the curve at the point where $t = 0$.

27. Consider the curve defined by $\mathbf{r} = 2\cos t\,\mathbf{i} + 3\sin t\,\mathbf{j} + 4t\,\mathbf{k}$.
    (a) Find a unit tangent vector to the curve at the point where $t = 0$.
    (b) Find parametric equations for the line tangent to the curve at the point where $t = 0$.
    (c) Find the plane normal to the curve at the point where $t = 0$.

28. Consider the curve defined by $\mathbf{r} = (t^2 + 3)\mathbf{i} - 3t\,\mathbf{j} + 2t^2\,\mathbf{k}$.

(a) Find a unit tangent vector to the curve at the point $(4, -3, 2)$.

(b) Find parametric equations for the line tangent to the curve at $(4, -3, 2)$.

(c) Find the plane normal to the curve at $(4, -3, 2)$.

29. The curves given by $\mathbf{r} = (t^2 + 3)\mathbf{i} - 3t\,\mathbf{j} + 2t^2\,\mathbf{k}$ and $\mathbf{r} = 2\tau\,\mathbf{i} + (\tau - 5)\mathbf{j} + (\tau^2/2)\mathbf{k}$ intersect at the point $(4, -3, 2)$. Find the angle θ between the tangents to the two curves at the point of intersection.

30. The curves given by $\mathbf{r} = \cos \pi t\,\mathbf{i} - 2 \sin \pi t\,\mathbf{j} + (t - 1)\mathbf{k}$ and $\mathbf{r} = (2\tau + 1)\mathbf{i} + 4\tau^2\,\mathbf{j} + \cos \pi\tau\,\mathbf{k}$ intersect at the point $(1, 0, 1)$. Find the angle θ between the tangents to the two curves at the point of intersection.

31. Suppose that the position of a moving particle at time t is given by $\mathbf{r} = a \cos \omega t\,\mathbf{i} + b \sin \omega t\,\mathbf{j}$, where a, b, and ω are positive numbers.

(a) Sketch the path of the particle.

(b) Find the velocity of the particle at any time.

(c) Find the component of the velocity that is directed toward the origin.

32. If $\mathbf{u}(t)$ is a differentiable vector of constant length, show that $d\mathbf{u}/dt$ is perpendicular to $\mathbf{u}$ for every t.

33. Verify Eq. 27 by first calculating $\mathbf{u}(t) \cdot \mathbf{v}(t)$, and then differentiating the resulting expression.

34. Verify Eq. 28 by first calculating $\mathbf{u}(t) \times \mathbf{v}(t)$, and then differentiating the resulting expression.

15.7 ARC LENGTH AND CURVATURE

In this section we discuss some of the geometrical properties of curves in three-dimensional space. Our discussion is based on the description of such a curve by means of a vector function. Let the vector function

$$\mathbf{r} = \mathbf{r}(t) = f(t)\mathbf{i} + g(t)\mathbf{j} + h(t)\mathbf{k} \tag{1}$$

be given, and let C be the corresponding curve. Throughout this section we assume that the functions f, g, and h are differentiable as many times as necessary for the discussion to make sense.

First, let us confine our attention to the portion of the curve for which $a \le t \le b$, and ask how we can define and calculate the length $l(a, b)$ of this arc (see Figure 15.7.1). We have already discussed arc length in two dimensions in Section

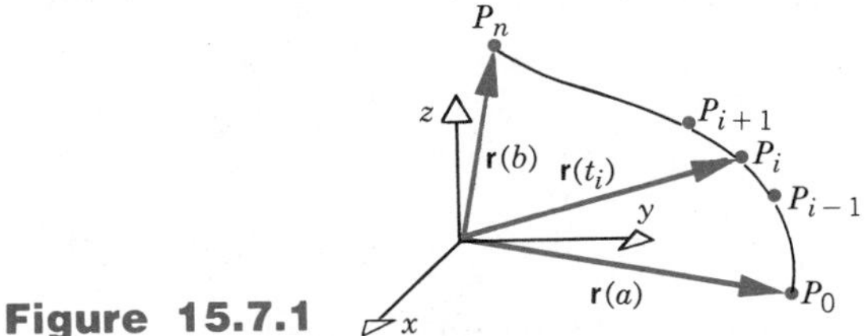

Figure 15.7.1

7.3, and the treatment here is very similar. We begin by partitioning the interval $[a, b]$ into n subintervals so that

$$a = t_0 < t_1 < t_2 < \cdots < t_{n-1} < t_n = b;$$

the norm of the partition is the length of the longest subinterval. For an arbitrary partition point t_i, we can evaluate $\mathbf{r}(t_i)$ and determine the corresponding point P_i

on the curve. As indicated in Figure 15.7.2, we next draw the straight line segments from P_0 to P_1, from P_1 to P_2, and so on. By adding the lengths of all these line

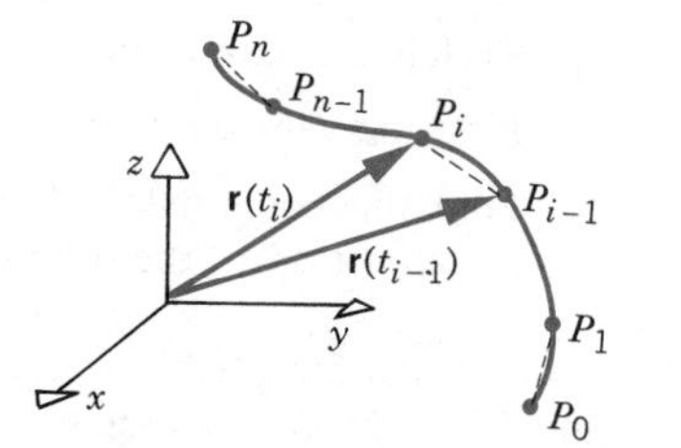

Figure 15.7.2

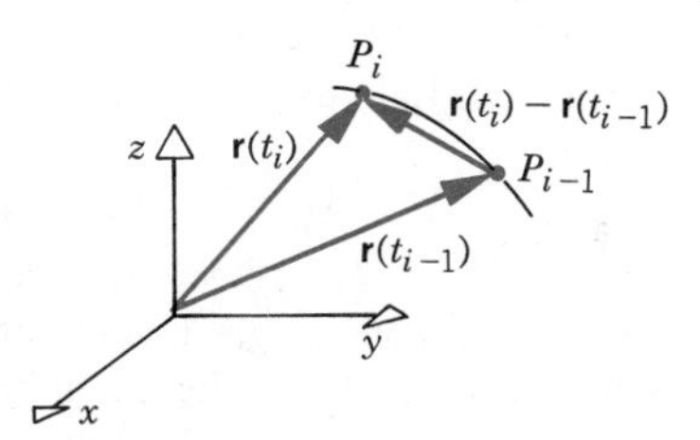

Figure 15.7.3

segments we obtain the length of a polygonal approximation to the given curve. The length of the line segment joining P_{i-1} and P_i is (see Figure 15.7.3)

$$\|\overrightarrow{P_{i-1}P_i}\| = \|\mathbf{r}(t_i) - \mathbf{r}(t_{i-1})\|. \tag{2}$$

Hence the length of the polygonal approximation is

$$\sum_{i=1}^{n} \|\mathbf{r}(t_i) - \mathbf{r}(t_{i-1})\|. \tag{3}$$

If the sum (3) approaches a limit as the norm of the partition approaches zero, then this limiting value is defined to be the length of the curve:

$$l(a, b) = \lim_{\|\Delta\| \to 0} \sum_{i=1}^{n} \|\mathbf{r}(t_i) - \mathbf{r}(t_{i-1})\|. \tag{4}$$

The expression on the right side of Eq. 4 can now be converted into an integral. Multiplying and dividing the ith term in the sum by $\Delta t_i = t_i - t_{i-1} > 0$, we obtain

$$l(a, b) = \lim_{\|\Delta\| \to 0} \sum_{i=1}^{n} \left\| \frac{\mathbf{r}(t_i) - \mathbf{r}(t_{i-1})}{\Delta t_i} \right\| \Delta t_i. \tag{5}$$

As in Section 7.3, it is possible to show that if $\mathbf{r}'(t)$ is continuous, then the limit in Eq. 5 exists, and the arc length l is given by the integral

$$l(a, b) = \int_a^b \|\mathbf{r}'(t)\| \, dt. \tag{6}$$

Since

$$\mathbf{r}'(t) = f'(t)\mathbf{i} + g'(t)\mathbf{j} + h'(t)\mathbf{k}, \tag{7}$$

Eq. 6 can also be put in the form

$$l(a, b) = \int_a^b [f'^2(t) + g'^2(t) + h'^2(t)]^{1/2} \, dt. \tag{8}$$

For a curve in the xy-plane, Eq. 8 reduces to

$$l(a, b) = \int_a^b [f'^2(t) + g'^2(t)]^{1/2} \, dt \tag{9}$$

which is one of the equations derived in Section 7.3. As in the two-dimensional case discussed in Section 7.3, a curve or arc of finite length is said to be **rectifiable.**

EXAMPLE 1

Find the length of the arc given by

$$\mathbf{r}(t) = 2t\,\mathbf{i} + 3t\,\mathbf{j} + \tfrac{2}{3}t^{3/2}\,\mathbf{k} \tag{10}$$

between the origin and the point $Q(2, 3, \frac{2}{3})$ (see Figure 15.7.4).

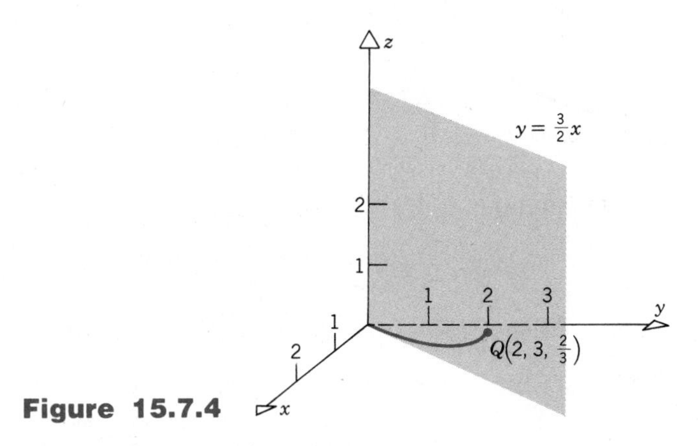

Figure 15.7.4

Note that the curve lies in the plane $y = 3x/2$; this can be seen by eliminating t between the first two components of $\mathbf{r}$. Observe also that $t = 0$ corresponds to the origin and $t = 1$ to the point Q. By differentiating Eq. 10 we find that

$$\mathbf{r}'(t) = 2\mathbf{i} + 3\mathbf{j} + \sqrt{t}\,\mathbf{k}$$

and thus

$$\|\mathbf{r}'(t)\| = (4 + 9 + t)^{1/2} = \sqrt{13 + t}.$$

From Eq. 6 it follows that the required arc length is

$$l(0, 1) = \int_0^1 \sqrt{13 + t}\, dt = \frac{2}{3}(13 + t)^{3/2}\Big|_0^1$$

$$= \frac{2}{3}[(14)^{3/2} - (13)^{3/2}] \cong 3.674. \;\blacksquare \tag{11}$$

EXAMPLE 2

Consider the curve given by

$$\mathbf{r}(t) = \cos t\,\mathbf{i} + \sin t\,\mathbf{j} + \tfrac{1}{2}t^2\,\mathbf{k}. \tag{12}$$

Find the length of the arc joining the point $P_0(1, 0, 0)$ corresponding to $t = 0$ and the point $Q(0, 1, \pi^2/8)$ corresponding to $t = \pi/2$ (see Figure 15.7.5). Also find the length of the arc from $P_0(1, 0, 0)$ to the point P corresponding to an arbitrary value of t.

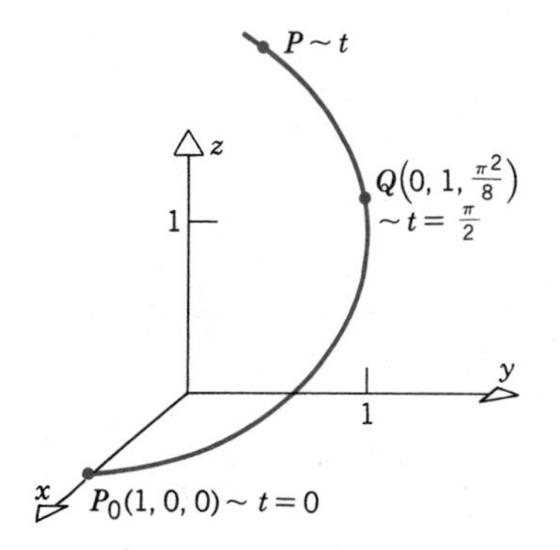

Figure 15.7.5

Differentiating Eq. 12, we obtain

$$\mathbf{r}'(t) = -\sin t\,\mathbf{i} + \cos t\,\mathbf{j} + t\,\mathbf{k}; \tag{13}$$

thus

$$\|\mathbf{r}'(t)\| = [(-\sin t)^2 + (\cos t)^2 + t^2]^{1/2} = \sqrt{1 + t^2}. \tag{14}$$

Then, from Eq. 6, the arc length from P_0 to Q is

$$l\left(0, \frac{\pi}{2}\right) = \int_0^{\pi/2} \sqrt{1 + t^2}\, dt. \tag{15}$$

The substitution $t = \tan\theta$ transforms this integral into

$$l\left(0, \frac{\pi}{2}\right) = \int_0^{\arctan \pi/2} \sec^3\theta\, d\theta,$$

which can be evaluated as described in Section 9.4. The result is

$$\begin{aligned} l\left(0, \frac{\pi}{2}\right) &= \frac{1}{2}[\sec\theta\tan\theta + \ln|\sec\theta + \tan\theta|]\Bigg|_0^{\arctan \pi/2} \\ &= \frac{1}{2}[t\sqrt{1 + t^2} + \ln|\sqrt{1 + t^2} + t|]\Bigg|_0^{\pi/2} \\ &= \frac{1}{2}\left[\frac{\pi}{2}\sqrt{1 + \frac{\pi^2}{4}} + \ln\left(\sqrt{1 + \frac{\pi^2}{4}} + \frac{\pi}{2}\right)\right] \cong 2.079. \end{aligned} \tag{16}$$

Let us denote the arc length from P_0 to P by $s(t)$ to emphasize that it depends on the value of t. Then, using Eq. (6) again, we have

$$\begin{aligned} s(t) = l(0, t) &= \int_0^t \|\mathbf{r}'(\tau)\|\, d\tau \\ &= \int_0^t \sqrt{1 + \tau^2}\, d\tau. \end{aligned} \tag{17}$$

Note that we have changed the integration variable to τ in order to avoid confusion with the independent variable t in the upper limit of integration. The integral in Eq. 17 can be evaluated in the same way as the one in Eq. 15, with the result that

$$s(t) = \tfrac{1}{2}[t\sqrt{1 + t^2} + \ln(\sqrt{1 + t^2} + t)]. \quad ■ \tag{18}$$

For an arbitrary curve C whose equation is

$$\mathbf{r} = \mathbf{r}(t)$$

we can define an arc length function s just as in Example 2. Let the point P_0 corresponding to $t = a$ be the base point from which arc length is measured, and let $s(t)$ be the arc length from P_0 to the point P corresponding to an arbitrary value of t. Then

$$s(t) = l(a, t) = \int_a^t \|\mathbf{r}'(\tau)\| \, d\tau. \tag{19}$$

Differentiating Eq. 19 according to Theorem 6.4.2, we obtain

$$\frac{ds}{dt} = \|\mathbf{r}'(t)\|. \tag{20}$$

If we think of C as the path traced out by a moving particle with position vector $\mathbf{r}(t)$, then $\mathbf{v} = \mathbf{r}'(t)$ is the velocity of the particle, and $\|\mathbf{v}\| = \|\mathbf{r}'(t)\|$ is its speed. Equation 20 states that the speed of the particle is the rate of change of the distance it has moved along C with respect to time: $ds/dt = \|\mathbf{v}\|$.

Recall also that $\mathbf{r}'(t)$ is a vector tangent to the curve C at the point P. It is often convenient to think of $\mathbf{r}$ as a function of arc length s instead of t. If we do this, and if $ds/dt \neq 0$, then by the chain rule we have

$$\frac{d\mathbf{r}}{ds} = \frac{d\mathbf{r}}{dt}\frac{dt}{ds} = \frac{d\mathbf{r}/dt}{ds/dt} = \frac{\mathbf{r}'(t)}{\|\mathbf{r}'(t)\|}. \tag{21}$$

Thus $d\mathbf{r}/ds$ is a unit tangent vector to the curve C, and points in the direction of increasing s (or t). We will use $\mathbf{T}$ to denote the unit tangent vector; thus

$$\mathbf{T} = \frac{d\mathbf{r}}{ds} = \frac{\mathbf{r}'(t)}{\|\mathbf{r}'(t)\|}. \tag{22}$$

EXAMPLE 3

For the curve in Example 2, find the unit tangent vector $\mathbf{T}$ at any point.

We found $\mathbf{r}'(t)$ and $\|\mathbf{r}'(t)\|$ earlier; they are given by Eqs. 13 and 14, respectively. Thus, from Eq. 22, we have

$$\mathbf{T} = \frac{-\sin t\,\mathbf{i} + \cos t\,\mathbf{j} + t\,\mathbf{k}}{\sqrt{1 + t^2}}. \quad ■$$

The direction of the curve C at a point is indicated by the tangent vectors $\mathbf{r}'(t)$ and $\mathbf{T} = d\mathbf{r}/ds$. Now we turn to the definition and calculation of a quantity that measures the rate at which C is changing direction at a given point; more precisely, we measure the rate at which the unit tangent vector $\mathbf{T}$ is changing direction. Observe that $\mathbf{T}$ does not change in magnitude as the curve is traversed; consequently $d\mathbf{T}/ds$ is the rate of change of the direction of $\mathbf{T}$ with respect to s. It is customary to define the **curvature** κ of the curve C as the nonnegative scalar quantity

$$\kappa = \|d\mathbf{T}/ds\|. \tag{23}$$

Unfortunately, the position vector $\mathbf{r}$ of points on C is usually given in terms of time t or some other parameter rather than in terms of arc length s. In such cases, Eq. 23, although simple in appearance, may entail considerable calculation. From Eq. 22 we have, by the chain rule,

$$\begin{aligned}\frac{d\mathbf{T}}{ds} &= \frac{d\mathbf{T}}{dt}\frac{dt}{ds} = \frac{d\mathbf{T}/dt}{ds/dt} \\ &= \frac{1}{\|\mathbf{r}'(t)\|}\frac{d}{dt}\left[\frac{\mathbf{r}'(t)}{\|\mathbf{r}'(t)\|}\right],\end{aligned}$$

and the calculation of the indicated derivative may be tedious.

It is possible to show that the curvature is also given by the expression

$$\kappa = \frac{\|\mathbf{r}'(t) \times \mathbf{r}''(t)\|}{\|\mathbf{r}'(t)\|^3} \tag{24}$$

where $\mathbf{r} = \mathbf{r}(t)$ is the position vector of the curve C. Equation 24 is usually the simplest way to determine κ for a given curve.

We noted earlier that if C is the path of a moving particle with position vector $\mathbf{r}(t)$, then $\mathbf{v} = \mathbf{r}'(t)$ is the velocity of the particle. Similarly, $\mathbf{a} = \mathbf{v}'(t) = \mathbf{r}''(t)$ is its **acceleration.** In terms of $\mathbf{v}$ and $\mathbf{a}$ the expression (24) for the curvature κ takes the form

$$\kappa = \frac{\|\mathbf{v} \times \mathbf{a}\|}{\|\mathbf{v}\|^3} \tag{25}$$

In Problem 31 we outline one way of deriving Eq. 24.

EXAMPLE 4

Find the curvature of a straight line.

A straight line has an equation of the form

$$\mathbf{r}(t) = \mathbf{a} + \mathbf{b}t, \qquad -\infty < t < \infty, \tag{26}$$

where $\mathbf{a}$ and $\mathbf{b}$ are constant vectors with $\mathbf{b} \neq \mathbf{0}$. Then

$$\mathbf{r}'(t) = \mathbf{b}, \qquad \mathbf{r}''(t) = \mathbf{0},$$

and hence $\|\mathbf{r}'(t) \times \mathbf{r}''(t)\| = 0$. Consequently the curvature $\kappa = 0$. Since intuitively a straight line should have zero curvature, this result is partial confirmation that the definition we have given is reasonable. ∎

EXAMPLE 5

Find the curvature of the curve defined by

$$\mathbf{r}(t) = a \cos \omega t\, \mathbf{i} + a \sin \omega t\, \mathbf{j}, \tag{27}$$

where a and ω are positive constants.

Since $\mathbf{r}(t)$ has no $\mathbf{k}$-component, the curve lies in the xy-plane. Further, $x = a \cos \omega t$ and $y = a \sin \omega t$, so $x^2 + y^2 = a^2$. Thus the curve is the circle with center at the origin and with radius a. To find the curvature κ from Eq. 24 we start by calculating

$$\mathbf{r}'(t) = -a\omega \sin \omega t\, \mathbf{i} + a\omega \cos \omega t\, \mathbf{j}, \tag{28a}$$

$$\mathbf{r}''(t) = -a\omega^2 \cos \omega t\, \mathbf{i} - a\omega^2 \sin \omega t\, \mathbf{j}. \tag{28b}$$

Then

$$\mathbf{r}'(t) \times \mathbf{r}''(t) = \begin{vmatrix} \mathbf{i} & \mathbf{j} & \mathbf{k} \\ -a\omega \sin \omega t & a\omega \cos \omega t & 0 \\ -a\omega^2 \cos \omega t & -a\omega^2 \sin \omega t & 0 \end{vmatrix}$$

$$= a^2\omega^3 (\sin^2 \omega t + \cos^2 \omega t)\, \mathbf{k} = a^2\omega^3\, \mathbf{k}. \tag{29}$$

Consequently $\|\mathbf{r}'(t) \times \mathbf{r}''(t)\| = a^2\omega^3$. Also, from Eq. 28(a) we have $\|\mathbf{r}'(t)\| = a\omega$. Substituting these results in Eq. 24 we obtain

$$\kappa = \frac{a^2\omega^3}{(a\omega)^3} = \frac{1}{a}. \tag{30}$$

Thus, as we might anticipate, the curvature of a circle is constant, and is equal to the reciprocal of the radius. Observe that this result does not depend on ω, which can be interpreted as the angular speed of a particle moving in accordance with Eq. 27. ∎

It is sometimes useful to consider a quantity known as the **radius of curvature,** usually denoted by ρ. At an arbitrary point P on a given curve C the radius of curvature is the radius of the circle (called the circle of curvature) that has the same curvature as the given curve at the point P (see Figure 15.7.6). Of course, for a circle the radius of curvature is just the radius of the circle itself. From the result of Example 5 it follows that

$$\rho(s) = \frac{1}{\kappa(s)}. \tag{31}$$

Equation 31 makes it easy to substitute for one of the quantities κ or ρ in terms of the other, so we may use whichever one is more convenient.

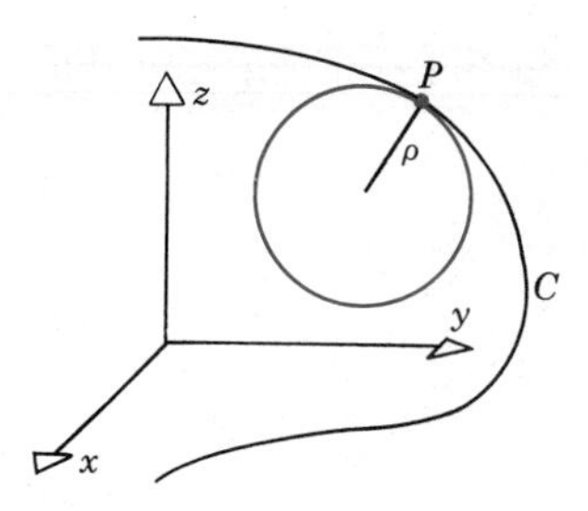

Figure 15.7.6 The radius of curvature ρ.

For a plane curve the formula (24) for κ can be put in a somewhat simpler form. Suppose that the curve is in the xy-plane, in which case Eq. 1 reduces to

$$\mathbf{r} = f(t)\mathbf{i} + g(t)\mathbf{j}, \tag{32}$$

and

$$\mathbf{r}' = f'(t)\mathbf{i} + g'(t)\mathbf{j},$$
$$\mathbf{r}'' = f''(t)\mathbf{i} + g''(t)\mathbf{j}.$$

Then

$$\begin{aligned}\mathbf{r}' \times \mathbf{r}'' &= \begin{vmatrix} \mathbf{i} & \mathbf{j} & \mathbf{k} \\ f'(t) & g'(t) & 0 \\ f''(t) & g''(t) & 0 \end{vmatrix} \\ &= [f'(t)g''(t) - f''(t)g'(t)]\mathbf{k}.\end{aligned}$$

Consequently for the plane curve (32) the curvature is

$$\kappa = \frac{|f'(t)g''(t) - f''(t)g'(t)|}{[f'^2(t) + g'^2(t)]^{3/2}}. \tag{33}$$

The parametric representation (32) allows us to discuss curves that are more general than those that are graphs of functions. However, the formula (33) for κ can be simplified further for a plane curve that is the graph of a function. In this case the curve can be described by an equation of the form

$$y = g(x). \tag{34}$$

Then $x = t$, $y = g(t)$, and Eq. 33 takes the form

$$\kappa = \frac{|g''(x)|}{[1 + g'^2(x)]^{3/2}}. \tag{35}$$

EXAMPLE 6

Find the curvature κ at an arbitrary point on the parabola

$$y = \alpha x^2, \qquad \alpha > 0. \tag{36}$$

Also find the radius of curvature ρ at each point, and determine the point where ρ is smallest.

Equation 36 is of the form (34), so we can find κ from Eq. 35. This gives

$$\kappa = \frac{2\alpha}{(1 + 4\alpha^2x^2)^{3/2}}, \tag{37}$$

where the absolute value bars have been dropped because α is positive. From Eq. 31 we have

$$\rho = \frac{(1 + 4\alpha^2x^2)^{3/2}}{2\alpha}, \tag{38}$$

from which it is clear that ρ is a minimum when $x = 0$, that is, at the vertex of the parabola. At this point ρ has the value $1/2\alpha$.

In Section 5.2 we wrote the equation of the parabola (36) in the form

$$x^2 = 4py, \tag{39}$$

where p is the distance from the vertex to the focus or to the directrix. From Eqs. 36 and 39 we have $\alpha = 1/4p$. Thus the minimum value of ρ is $2p$; at the vertex the radius of curvature is twice the distance to the focus. Figure 15.7.7 shows the parabola (36) for $\alpha = \frac{1}{4}$ and its circle of curvature at the vertex. ∎

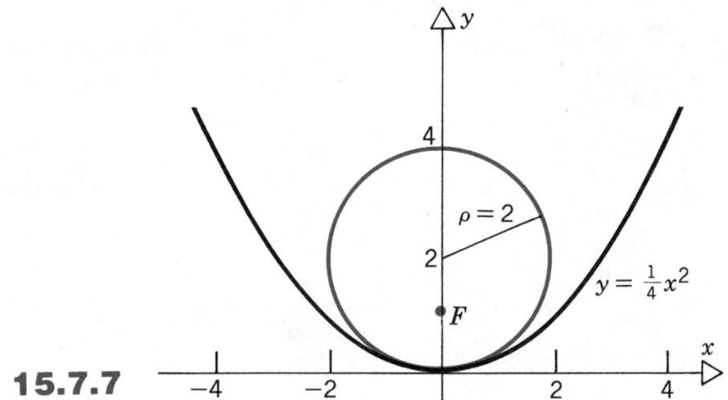

Figure 15.7.7

PROBLEMS

1. Find the length of the curve $\mathbf{r} = \cos t\,\mathbf{i} + \sin t\,\mathbf{j} + 3t\,\mathbf{k}$ between the points where $t = 1$ and $t = 4$, respectively.

2. Find the length of the curve $\mathbf{r} = 2\sin 2t\,\mathbf{i} - 5t\,\mathbf{j} + 2\cos 2t\,\mathbf{k}$ between the points $(0, 0, 2)$ and $(2, -5\pi/4, 0)$.

3. Find the length of the curve $\mathbf{r} = 2\cos t\,\mathbf{i} - 2\sin t\,\mathbf{j} + \frac{2}{3}t^{3/2}\,\mathbf{k}$ between the points where $t = 0$ and $t = 5$, respectively.

4. Find the length of the curve $\mathbf{r} = \frac{4}{3}t^{3/2}\,\mathbf{i} + 3t\,\mathbf{j} - \frac{2}{3}t^{3/2}\,\mathbf{k}$ between the origin and $(\frac{32}{3}, 12, -\frac{16}{3})$.

5. Find the length of the curve $\mathbf{r} = e^t\,\mathbf{i} + e^{-t}\,\mathbf{j} + \sqrt{2}t\,\mathbf{k}$ between the points $(1, 1, 0)$ and $(e^2, e^{-2}, 2\sqrt{2})$.

6. Find the length of the curve $\mathbf{r} = \cos t\,\mathbf{i} - \sin t\,\mathbf{j} + \cosh t\,\mathbf{k}$ between the points where $t = -\pi$ and $t = \pi$, respectively.

In each of Problems 7 and 8, find the length of the indicated portion of the curve $\mathbf{r} = (2\sqrt{2}/3)\,|t|^{3/2}\,\mathbf{i} + t\,\mathbf{j} + \frac{1}{2}t^2\,\mathbf{k}$.

7. From $(0, 0, 0)$ to $(2\sqrt{2}/3, 1, \frac{1}{2})$.
8. From $(16\sqrt{2}/3, -4, 8)$ to $(0, 0, 0)$.

In each of Problems 9 through 12, the position vector $\mathbf{r}$ of a moving particle is given as a function of time t. Find the velocity vector $\mathbf{v}$, the speed $\|\mathbf{v}\|$, and the acceleration $\mathbf{a}$ of the particle at any time.

9. $\mathbf{r} = \cos t\,\mathbf{i} + \sin t\,\mathbf{j} + 3t\,\mathbf{k}$

10. $\mathbf{r} = \cos t\,\mathbf{i} - \sin t\,\mathbf{j} + \cosh t\,\mathbf{k}$

11. $\mathbf{r} = \frac{1}{2}t^2\,\mathbf{i} - 4t\,\mathbf{j} + 3t^3\,\mathbf{k}$

12. $\mathbf{r} = \sqrt{1 + t^2}\,\mathbf{i} + (1 + t^2)^{-1/2}\,\mathbf{j} + (1 + t^2)\mathbf{k}$

In each of Problems 13 through 16, find the unit tangent vector $\mathbf{T}$.

13. $\mathbf{r} = \cos t\,\mathbf{i} + \sin t\,\mathbf{j} + 3t\,\mathbf{k}$

14. $\mathbf{r} = \cos t\,\mathbf{i} - \sin t\,\mathbf{j} + \cosh t\,\mathbf{k}$

15. $\mathbf{r} = \frac{1}{2}t^2\,\mathbf{i} - 4t\,\mathbf{j} + 3t^3\,\mathbf{k}$

16. $\mathbf{r} = \dfrac{2\sqrt{2}}{3}t^{3/2}\,\mathbf{i} + t\,\mathbf{j} + \frac{1}{2}t^2\,\mathbf{k}, \qquad t \geq 0$

In each of Problems 17 through 20, find the curvature κ. Refer to Problems 13 through 16.

17. $\mathbf{r} = \cos t\,\mathbf{i} + \sin t\,\mathbf{j} + 3t\,\mathbf{k}$

18. $\mathbf{r} = \cos t\,\mathbf{i} - \sin t\,\mathbf{j} + \cosh t\,\mathbf{k}$

19. $\mathbf{r} = \frac{1}{2}t^2\,\mathbf{i} - 4t\,\mathbf{j} + 3t^3\,\mathbf{k}$

20. $\mathbf{r} = \dfrac{2\sqrt{2}}{3}t^{3/2}\,\mathbf{i} + t\,\mathbf{j} + \frac{1}{2}t^2\,\mathbf{k}$

In each of Problems 21 through 28, find the curvature κ of the given plane curve.

21. $\mathbf{r} = a\cos t\,\mathbf{i} + b\sin t\,\mathbf{j}$ (Ellipse)

22. $\mathbf{r} = a\cosh t\,\mathbf{i} + b\sinh t\,\mathbf{j}$ (Hyperbola)

23. $\mathbf{r} = a(t - \sin t)\mathbf{i} + a(1 - \cos t)\mathbf{j}$ (Cycloid)

24. $y = \sin x$ **25.** $y = e^{-x}$ **26.** $y = \dfrac{x}{1 + x^2}$

27. $y = 1 - \dfrac{1}{x}$ **28.** $y = a\cosh\dfrac{x}{a}$ (Catenary)

29. Show that, for a particle moving on a circular path, the acceleration vector is directed toward the center of the circle.

30. Suppose that a particle moves so that its velocity vector $\mathbf{v}$ has constant length.

(a) Show that the acceleration vector $\mathbf{a}$ is perpendicular to $\mathbf{v}$.

(b) Find an expression for κ in terms of $\|\mathbf{v}\|$ and $\|\mathbf{a}\|$.

* **31.** In this problem we indicate how to derive the expression (24) for the curvature κ.

(a) From Eqs. 21 and 22 note that

$$\mathbf{T} = \frac{d\mathbf{r}}{ds} = \frac{d\mathbf{r}/dt}{ds/dt} = \frac{\mathbf{r}'(t)}{s'(t)}; \qquad (i)$$

then use the chain rule to show that

$$\frac{d\mathbf{T}}{ds} = \frac{d\mathbf{T}/dt}{ds/dt} = \frac{1}{s'(t)}\left[\frac{\mathbf{r}''(t)}{s'(t)} - \frac{\mathbf{r}'(t)s''(t)}{[s'(t)]^2}\right]. \qquad (ii)$$

(b) Starting from $\|\mathbf{r}'(t)\|^2 = \mathbf{r}'(t)\cdot\mathbf{r}'(t)$, show that

$$\frac{d}{dt}\|\mathbf{r}'(t)\| = \frac{\mathbf{r}'(t)\cdot\mathbf{r}''(t)}{\|\mathbf{r}'(t)\|}. \qquad (iii)$$

(c) Substitute from Eq. 20 and Eq. (*iii*) into Eq. (*ii*), and show that

$$\frac{d\mathbf{T}}{ds} = \frac{1}{\|\mathbf{r}'(t)\|^2}\left[\mathbf{r}''(t) - \frac{\mathbf{r}'(t)\cdot\mathbf{r}''(t)}{\|\mathbf{r}'(t)\|^2}\mathbf{r}'(t)\right]. \qquad (iv)$$

(d) Calculate $(d\mathbf{T}/ds)\cdot(d\mathbf{T}/ds)$ and thereby show that

$$\left\|\frac{d\mathbf{T}}{ds}\right\| = \frac{\{\|\mathbf{r}'(t)\|^2\|\mathbf{r}''(t)\|^2 - [\mathbf{r}'(t)\cdot\mathbf{r}''(t)]^2\}^{1/2}}{\|\mathbf{r}'(t)\|^3} \qquad (v)$$

(e) From Eq. (*v*) show that

$$\kappa = \frac{\|\mathbf{r}'(t) \times \mathbf{r}''(t)\|}{\|\mathbf{r}'(t)\|^3}. \qquad (vi)$$

15.8 DYNAMICS OF PARTICLES

Some of the ideas developed in this chapter can be used very effectively to study the dynamics of a moving particle, or point mass. This is a fundamental branch of mechanics, and its development is closely associated with that of calculus in the seventeenth and eighteenth centuries. In this section and the next we use vector methods to discuss a few aspects of the motion of a point mass on a curved path, with an emphasis on satellite and planetary motion.

Tangential and normal acceleration

We suppose that a particle with mass m moves along a curve C, and that its position at any time t is given by

$$\mathbf{r} = \mathbf{r}(t). \tag{1}$$

As pointed out in Section 15.7, the velocity and acceleration vectors are given by

$$\mathbf{v} = \frac{d\mathbf{r}}{dt}, \qquad \mathbf{a} = \frac{d\mathbf{v}}{dt} = \frac{d^2\mathbf{r}}{dt^2}, \tag{2}$$

respectively, and $\mathbf{v}$ is tangent to the curve C. Further, arc length s satisfies the relation

$$\frac{ds}{dt} = \left\|\frac{d\mathbf{r}}{dt}\right\| = \|\mathbf{v}\|. \tag{3}$$

Thus the unit tangent vector $\mathbf{T}$ is given by any of the expressions

$$\mathbf{T} = \frac{d\mathbf{r}/dt}{\|d\mathbf{r}/dt\|} = \frac{d\mathbf{r}/dt}{ds/dt} = \frac{d\mathbf{r}}{ds} = \frac{\mathbf{v}(t)}{\|\mathbf{v}(t)\|}. \tag{4}$$

Recall also that the curvature κ is defined as

$$\kappa = \left\|\frac{d\mathbf{T}}{ds}\right\|. \tag{5}$$

Instead of resolving the velocity and acceleration vectors into components along the coordinate axes, it is often better to take components in directions tangent and normal to the path of the moving particle. Of course, these directions usually change as the particle moves along the curve.

In order to determine normal components of $\mathbf{v}$ and $\mathbf{a}$, we need to find a vector normal to the curve C. We can do this by starting from the relation

$$\mathbf{T} \cdot \mathbf{T} = 1, \tag{6}$$

which just says that $\mathbf{T}$ is a unit vector. We wish to differentiate Eq. 6, and we choose to take the derivative with respect to s (rather than t), since κ is given in terms of $d\mathbf{T}/ds$ by Eq. 5. Thus we have

$$\frac{d\mathbf{T}}{ds} \cdot \mathbf{T} + \mathbf{T} \cdot \frac{d\mathbf{T}}{ds} = 0,$$

or

$$\mathbf{T} \cdot \frac{d\mathbf{T}}{ds} = 0,$$

since the dot product is commutative. Consequently $d\mathbf{T}/ds$ is a vector perpendicular to $\mathbf{T}$, which means that it is normal to the curve C. The unit vector $\mathbf{N}$ in the direction of $d\mathbf{T}/ds$ is given by

$$\mathbf{N} = \frac{d\mathbf{T}/ds}{\|d\mathbf{T}/ds\|}. \tag{7}$$

The vector **N** is usually referred to as the unit **principal normal vector** (see Figure 15.8.1).Since $\|d\mathbf{T}/ds\| = \kappa$ by Eq. 5, we can write Eq. 7 in the form

$$\frac{d\mathbf{T}}{ds} = \kappa\mathbf{N}. \tag{8}$$

Equation 8 expresses the two facts that $d\mathbf{T}/ds$ is normal to C and has magnitude equal to the curvature.

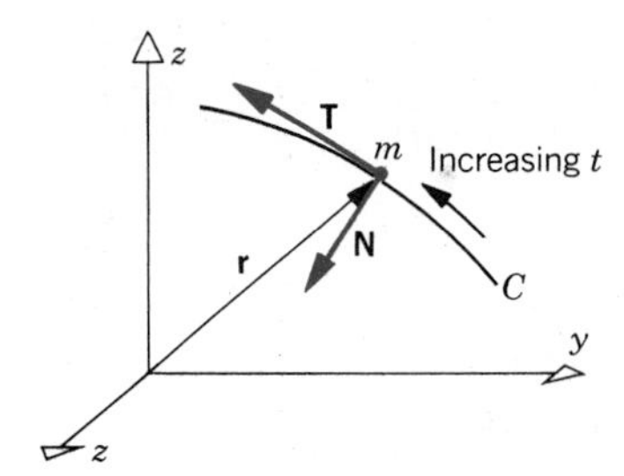

Figure 15.8.1 Unit tangent and principal normal vectors.

From Eq. 4 we can write the velocity vector as

$$\mathbf{v}(t) = \|\mathbf{v}(t)\|\mathbf{T} = \frac{ds}{dt}\mathbf{T}. \tag{9}$$

Then the acceleration vector $\mathbf{a}(t)$ is given by

$$\mathbf{a} = \frac{d\mathbf{v}}{dt} = \frac{d^2s}{dt^2}\mathbf{T} + \frac{ds}{dt}\frac{d\mathbf{T}}{dt}. \tag{10}$$

To complete the calculation of **a** we must find $d\mathbf{T}/dt$. Using the chain rule and Eq. 8, we have

$$\frac{d\mathbf{T}}{dt} = \frac{d\mathbf{T}}{ds}\frac{ds}{dt} = \kappa\frac{ds}{dt}\mathbf{N}. \tag{11}$$

Substituting this expression in Eq. 10, we obtain

$$\mathbf{a} = \frac{d^2s}{dt^2}\mathbf{T} + \kappa\left(\frac{ds}{dt}\right)^2\mathbf{N}. \tag{12}$$

The coefficients of **T** and **N** in Eq. 12 are called the tangential and normal accelerations, and are denoted by a_t and a_n, respectively. Thus

$$a_t = \frac{d^2s}{dt^2}, \qquad a_n = \kappa\left(\frac{ds}{dt}\right)^2. \tag{13}$$

Introducing the radius of curvature ρ (Eq. 31 of Section 15.7) we can also write

$$a_n = \frac{1}{\rho}\left(\frac{ds}{dt}\right)^2 = \frac{\|\mathbf{v}\|^2}{\rho}. \tag{14}$$

EXAMPLE 1

The position of a particle following a helical path is given by

$$\mathbf{r} = a \cos \omega t \, \mathbf{i} + a \sin \omega t \, \mathbf{j} + bt \, \mathbf{k}, \tag{15}$$

where a, b, and ω are positive. Find the velocity $\mathbf{v}$, the acceleration $\mathbf{a}$, the tangential component a_t, and the normal component a_n of acceleration.

Differentiating Eq. 15, we obtain

$$\mathbf{v} = \frac{d\mathbf{r}}{dt} = -a\omega \sin \omega t \, \mathbf{i} + a\omega \cos \omega t \, \mathbf{j} + b \, \mathbf{k} \tag{16}$$

and

$$\mathbf{a} = \frac{d\mathbf{v}}{dt} = -a\omega^2 \cos \omega t \, \mathbf{i} - a\omega^2 \sin \omega t \, \mathbf{j}. \tag{17}$$

Next,

$$\frac{ds}{dt} = \|\mathbf{v}\| = (a^2\omega^2 \sin^2 \omega t + a^2\omega^2 \cos^2 \omega t + b^2)^{1/2} = (a^2\omega^2 + b^2)^{1/2} . \tag{18}$$

Since ds/dt is a constant, it follows that

$$a_t = \frac{d^2s}{dt^2} = 0. \tag{19}$$

Thus $\mathbf{a} = a_n\mathbf{N}$, and since $a_n \geq 0$ from Eq. 13, it follows from Eq. 17 that

$$a_n = \|\mathbf{a}\| = (a^2\omega^4 \cos^2 \omega t + a^2\omega^4 \sin^2 \omega t)^{1/2} = a\omega^2. \; \blacksquare \tag{20}$$

Centripetal acceleration and satellite motion

The normal acceleration a_n is also known as the **centripetal acceleration,** and the force that causes it is the **centripetal force.** For example, it is the centripetal force that keeps objects on the surface of the earth from flying off into space as the earth rotates.

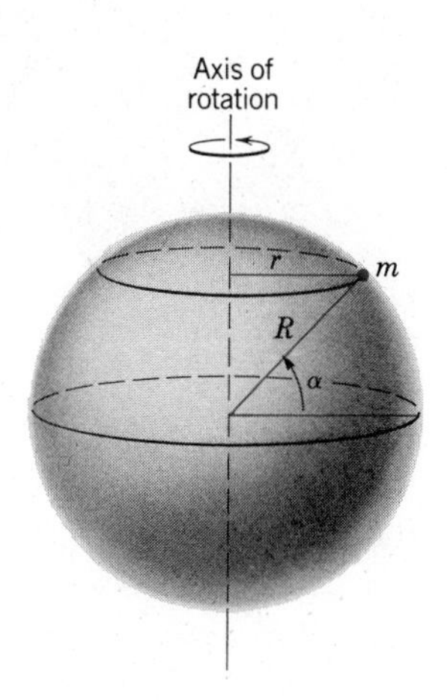

Figure 15.8.2

Let us determine the centripetal acceleration of an object on the surface of the earth. We assume that the earth is a sphere of radius R, and suppose that a point mass m is located at sea level at latitude α (see Figure 15.8.2). As the earth rotates about its axis the object traverses a circular path of radius

$$r = R \cos \alpha. \tag{21}$$

From Eq. 14 the centripetal acceleration of the mass m is

$$a_n = \frac{\|\mathbf{v}\|^2}{r}. \tag{22}$$

Further, the earth rotates at a constant rate, so $\|\mathbf{v}\|$ is a constant. Thus we may calculate it by dividing the distance $2\pi r$ traveled by the mass m in one full revolution by the period T of rotation. Hence

$$\|\mathbf{v}\| = \frac{2\pi r}{T} \tag{23}$$

and

$$a_n = \frac{4\pi^2 r}{T^2} = \frac{4\pi^2 R \cos \alpha}{T^2}. \tag{24}$$

Observe that the centripetal acceleration depends on latitude. It is maximum at the equator ($\alpha = 0$) and diminishes to zero at the poles ($\alpha = \pm\pi/2$).

We now show how the variation of centripetal acceleration with latitude results in a variation of weight as well. The gravitational force $\mathbf{F}$ acting on the point mass m is directed toward the center of the earth, as shown in Figure 15.8.3. If $\mathbf{u}$ is a

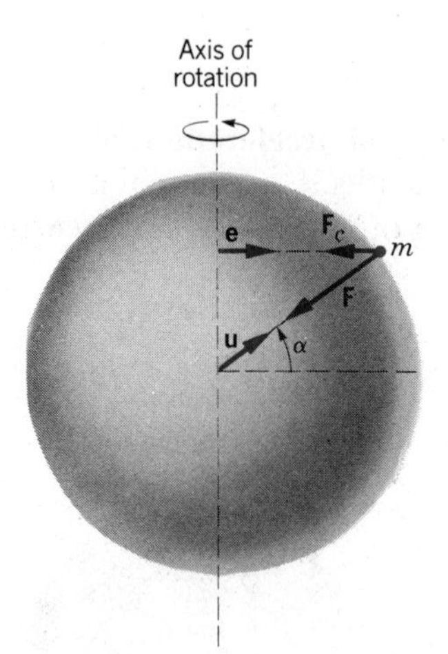

Figure 15.8.3 The force of gravity $\mathbf{F}$ and the centripetal force $\mathbf{F}_c$.

unit vector directed outward from the center of the earth toward the point mass, then

$$\mathbf{F} = -\frac{GMm}{R^2}\,\mathbf{u}, \tag{25}$$

where G is the universal gravitational constant, and M is the mass of the earth. Values of these and other constants occurring in the discussion are given in Table 15.2.

Table 15.2 Values of Certain Physical Constants

$G = 6.67 \times 10^{-11}$ N-m²/kg² (m³/kg-sec²)		
$M = 5.983 \times 10^{24}$ kg		
$R = 6.378 \times 10^{6}$ m	(Equatorial)	
$R = 6.357 \times 10^{6}$ m	(Polar)	
$T = 8.64 \times 10^{4}$ sec	(= 1 day)	

The centripetal force $\mathbf{F}_c$ corresponding to the centripetal acceleration a_n lies in the plane in which the point mass moves, and is directed toward the axis of rotation. If $\mathbf{e}$ is a unit vector in this plane in the direction from the axis of rotation to the point mass, then

$$\mathbf{F}_c = -ma_n\mathbf{e} = -\frac{4\pi^2 mR \cos\alpha}{T^2}\mathbf{e}. \tag{26}$$

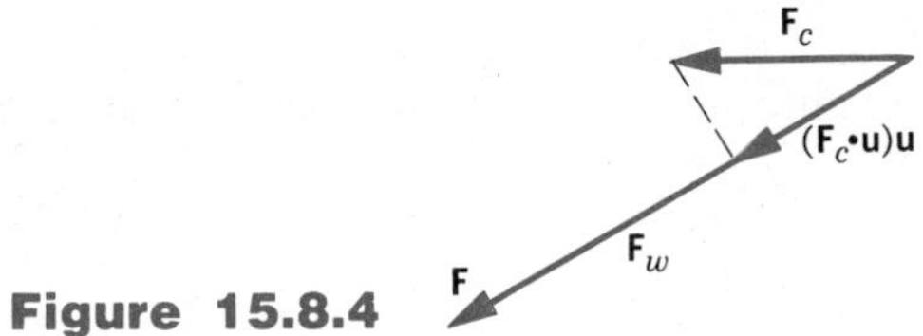

Figure 15.8.4

To see the effect of centripetal acceleration on the weight of the point mass, we express the gravitational attraction $\mathbf{F}$ as the sum of two terms (see Figure 15.8.4),

$$\mathbf{F} = \mathbf{F}_w + (\mathbf{F}_c \cdot \mathbf{u})\mathbf{u}. \tag{27}$$

The second term on the right side of Eq. 27 is the component of $\mathbf{F}_c$ in the direction of $\mathbf{u}$, and thus is the part of $\mathbf{F}$ needed to create the centripetal acceleration a_n. The remaining part of $\mathbf{F}$, denoted by $\mathbf{F}_w$, gives the mass m its weight. By substituting from Eqs. 25 and 26 into 27, we obtain

$$\begin{aligned}\mathbf{F}_w &= \mathbf{F} - (\mathbf{F}_c \cdot \mathbf{u})\mathbf{u} \\ &= -\frac{GMm}{R^2}\mathbf{u} - \left[-\frac{4\pi^2 mR\cos\alpha}{T^2}(\mathbf{e}\cdot\mathbf{u})\right]\mathbf{u} \\ &= -m\left[\frac{GM}{R^2} - \frac{4\pi^2 R\cos^2\alpha}{T^2}\right]\mathbf{u},\end{aligned} \tag{28}$$

since $\mathbf{e} \cdot \mathbf{u} = \cos\alpha$. The magnitude of the vector $\mathbf{F}_w$ is the weight w of the mass m. Thus, as is customary, we can write

$$w = mg, \tag{29}$$

where g is the quantity in brackets in Eq. 28:

$$g = \frac{GM}{R^2} - \frac{4\pi^2 R \cos^2 \alpha}{T^2}. \tag{30}$$

The second term on the right side of Eq. 30 is the correction to the gravitational acceleration due to the rotation of the earth. From Eqs. 29 and 30 it follows that the weight of an object is greatest at the poles and least at the equator. This effect is made slightly more pronounced by the fact that the earth is not a perfect sphere; R is slightly larger in the equatorial plane than along the polar axis, as noted in Table 15.2.

To estimate the effect of centripetal acceleration on g we can calculate the terms on the right side of Eq. 30; we assume that the mass is on the equator ($\alpha = 0$) so that the correction term is maximum. Then

$$\frac{GM}{R^2} = 9.810 \text{ m/sec}^2,$$

$$\frac{4\pi^2 R}{T^2} = 0.03373 \text{ m/sec}^2,$$

so that the ratio of the correction term to the main term is at most about 0.00344, or about a third of one percent.

The same kind of analysis can be applied to the motion of satellites. To do this it is only necessary to replace R by $R + h$, where h is the altitude of the satellite above the earth (see Figure 15.8.5). For example, for a satellite in the

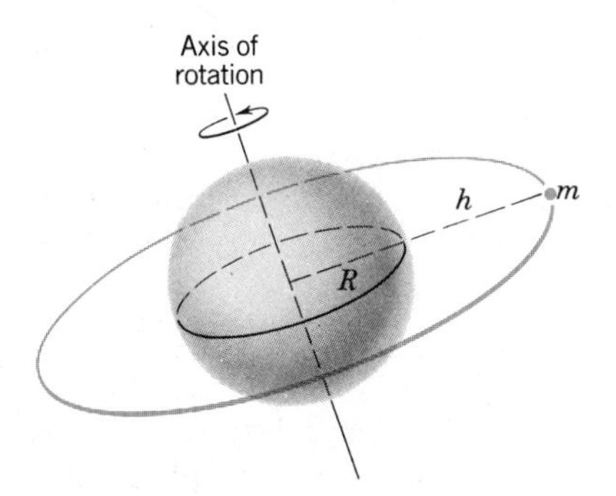

Figure 15.8.5

earth's equatorial plane we replace R by $R + h$ in Eq. 30 and set $\alpha = 0$; then

$$g = \frac{GM}{(R + h)^2} - \frac{4\pi^2(R + h)}{T^2}. \tag{31}$$

For the satellite to remain in a stable circular orbit it must be *weightless*, that is, $\mathbf{F}_w = \mathbf{0}$, and hence $g = 0$. Setting $g = 0$ in Eq. 31 yields

$$GMT^2 = 4\pi^2(R + h)^3, \tag{32}$$

which is a relation between T and h. One may prescribe either the period of the satellite or its altitude, and then use Eq. 32 to determine the other. For example, suppose that the satellite is in an orbit in the equatorial plane at an altitude of

1.609×10^5 m (100 mi) above the earth. Then $R + h = 6.539 \times 10^6$ m, and from Eq. 32 we find that

$$T = 87.65 \text{ min}. \tag{33}$$

A satellite whose period is the same as that of the earth (one day) always remains above the same point on the earth's surface. Such satellites are said to be **geosynchronous**; they have important applications in long distance communications. To find the altitude above the earth of a geosynchronous satellite in the equatorial plane, we set $T = 8.64 \times 10^4$ sec in Eq. 32 and solve for h. The result is

$$h \cong 35.88 \times 10^6 \text{ m} \cong 22{,}300 \text{ mi}. \tag{34}$$

PROBLEMS

1. Show that the normal component of acceleration is given by

$$a_n = \frac{\|\mathbf{v} \times \mathbf{a}\|}{\|\mathbf{v}\|}.$$

2. Show that if a particle moves along a curve at constant speed, then $a_n = \|\mathbf{a}\|$.

3. The position of a particle moving on an ellipse is given by

$$\mathbf{r} = a \cos \omega t\ \mathbf{i} + b \sin \omega t\ \mathbf{j},$$

where $a > b > 0$. Find the velocity, acceleration, and the tangential and normal components of acceleration.

4. The path of a moving particle is given by

$$\mathbf{r} = \cos t\ \mathbf{i} + \sin t\ \mathbf{j} - \cos 2t\ \mathbf{k}.$$

Find the velocity, acceleration, and the tangential and normal components of acceleration.

5. The path of a moving particle is given by

$$\mathbf{r} = (\cos t + \tfrac{1}{4} \cos 2t + \tfrac{1}{4})\mathbf{i} + (\sin t + \tfrac{1}{4} \sin 2t)\mathbf{j}.$$

Find the velocity, acceleration, and the tangential and normal components of acceleration.

6. Suppose that the position of a particle moving along a parabolic path is given by $\mathbf{r} = f(t)\mathbf{i} + [f(t)]^2\,\mathbf{j}$. Find a condition that f must satisfy if the acceleration vector is always normal to the path of motion.

7. Consider a particle moving on a cycloidal path. From Example 6 in Section 5.5 the position of the particle is given by the parametric equations

$$x = a(\theta - \sin \theta), \qquad y = a(1 - \cos \theta)$$

where a is the radius of the rolling circle and θ is its angular displacement. Assume that the circle rolls at a constant speed; then $\theta = \omega t$, where ω is a constant. To consider one full arch of the cycloid, let $0 \le \omega t \le 2\pi$.

(a) Find the velocity vector $\mathbf{v}$ of the particle. Where is the particle moving most rapidly?

(b) Find the acceleration vector $\mathbf{a}$ and its magnitude. What direction does the acceleration vector have?

(c) Show that if $0 < \omega t < \pi$, then $\mathbf{v}$ points toward the highest point on the rolling circle.

(d) Find the tangential and normal components of the acceleration.

8. Find the altitude of a satellite in the earth's equatorial plane whose period T is 120 minutes.

9. Find the altitude of a geosynchronous satellite above a point on the earth's surface whose latitude is (a) 30 degrees; (b) 45 degrees.

10. Generalize Eq. 32 by finding a corresponding relation between the period T and altitude h for a satellite whose position vector from the center of the earth makes an angle α with the equatorial plane.

15.9 KEPLER'S LAWS

The main subject of this section is a discussion of planetary motion and Kepler's laws. As a preliminary step it is desirable to express the velocity and acceleration vectors of a moving particle in terms of polar coordinates.

Velocity and acceleration in polar coordinates

Let $P(r, \theta)$ be an arbitrary point in the plane (other than the origin). We first define unit vectors $\mathbf{u}_r$ and $\mathbf{u}_\theta$ in the radial and circumferential directions, respec-

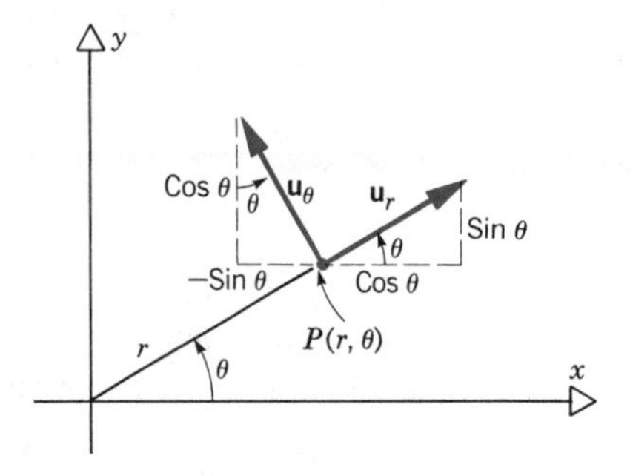

Figure 15.9.1
Unit vectors $\mathbf{u}_r$ and $\mathbf{u}_\theta$.

tively, at P. From Figure 15.9.1 we see that

$$\mathbf{u}_r = \cos\theta\,\mathbf{i} + \sin\theta\,\mathbf{j}, \tag{1}$$

$$\begin{aligned}\mathbf{u}_\theta &= \cos\left(\theta + \frac{\pi}{2}\right)\mathbf{i} + \sin\left(\theta + \frac{\pi}{2}\right)\mathbf{j}\\ &= -\sin\theta\,\mathbf{i} + \cos\theta\,\mathbf{j}.\end{aligned} \tag{2}$$

Further, by differentiating Eqs. 1 and 2, we find that

$$\frac{d\mathbf{u}_r}{d\theta} = -\sin\theta\,\mathbf{i} + \cos\theta\,\mathbf{j} = \mathbf{u}_\theta, \tag{3}$$

$$\frac{d\mathbf{u}_\theta}{d\theta} = -\cos\theta\,\mathbf{i} - \sin\theta\,\mathbf{j} = -\,\mathbf{u}_r. \tag{4}$$

We will soon need to assume that θ depends on t, in which case $\mathbf{u}_r$ and $\mathbf{u}_\theta$ are also functions of t. To calculate derivatives of $\mathbf{u}_r$ and $\mathbf{u}_\theta$, we use the chain rule and obtain

$$\frac{d\mathbf{u}_r}{dt} = \frac{d\mathbf{u}_r}{d\theta}\frac{d\theta}{dt} = \frac{d\theta}{dt}\mathbf{u}_\theta, \tag{5}$$

$$\frac{d\mathbf{u}_\theta}{dt} = \frac{d\mathbf{u}_\theta}{d\theta}\frac{d\theta}{dt} = -\frac{d\theta}{dt}\mathbf{u}_r. \tag{6}$$

The position vector $\mathbf{r}$ of a particle moving in the plane can be written in the form

$$\mathbf{r} = x(t)\mathbf{i} + y(t)\mathbf{j} = r(t)\cos\theta(t)\mathbf{i} + r(t)\sin\theta(t)\mathbf{j}$$
$$= r(t)[\cos\theta(t)\mathbf{i} + \sin\theta(t)\mathbf{j}]$$
$$= r(t)\mathbf{u}_r(t). \tag{7}$$

Then the velocity of the particle is

$$\mathbf{v} = \frac{d\mathbf{r}}{dt} = \frac{dr}{dt}\mathbf{u}_r + r\frac{d\mathbf{u}_r}{dt}$$
$$= \frac{dr}{dt}\mathbf{u}_r + r\frac{d\theta}{dt}\mathbf{u}_\theta, \tag{8}$$

where we have used Eq. 5. Equation 8 gives the resolution of the velocity vector into r and θ components. This is shown in Figure 15.9.2. By differentiating Eq. 8

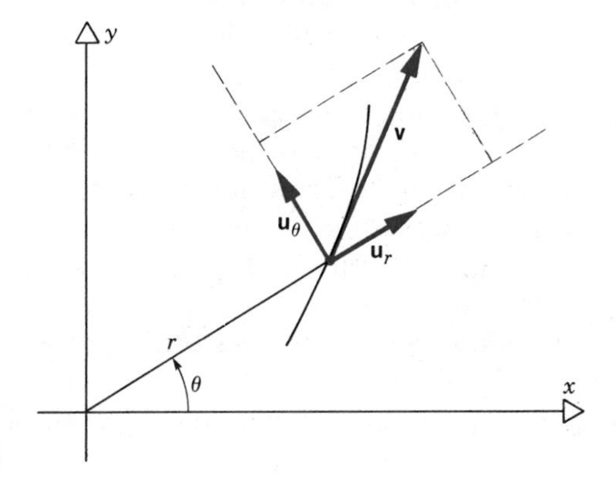

Figure 15.9.2

with respect to t we obtain the acceleration

$$\mathbf{a} = \frac{d\mathbf{v}}{dt} = \frac{d^2r}{dt^2}\mathbf{u}_r + \frac{dr}{dt}\frac{d\mathbf{u}_r}{dt} + \frac{dr}{dt}\frac{d\theta}{dt}\mathbf{u}_\theta + r\frac{d^2\theta}{dt^2}\mathbf{u}_\theta + r\frac{d\theta}{dt}\frac{d\mathbf{u}_\theta}{dt}.$$

Using Eqs. 5 and 6 to substitute for $d\mathbf{u}_r/dt$ and $d\mathbf{u}_\theta/dt$, respectively, we find that

$$\mathbf{a} = \frac{d^2r}{dt^2}\mathbf{u}_r + \frac{dr}{dt}\frac{d\theta}{dt}\mathbf{u}_\theta + \frac{dr}{dt}\frac{d\theta}{dt}\mathbf{u}_\theta + r\frac{d^2\theta}{dt^2}\mathbf{u}_\theta - r\left(\frac{d\theta}{dt}\right)^2\mathbf{u}_r$$
$$= \left[\frac{d^2r}{dt^2} - r\left(\frac{d\theta}{dt}\right)^2\right]\mathbf{u}_r + \left[r\frac{d^2\theta}{dt^2} + 2\frac{dr}{dt}\frac{d\theta}{dt}\right]\mathbf{u}_\theta. \tag{9}$$

The first bracket on the right side of Eq. 9 gives the radial component a_r of the acceleration vector while the second bracket gives the circumferential component a_θ.

Kepler's laws and planetary motion

A significant milestone in the history of science occurred early in the seventeenth century when Johannes Kepler* announced the following three laws of planetary motion.

1. Each planet moves in an elliptical orbit with the sun at one focus (discovered 1605, published 1609).
2. The position vector from the sun to a planet sweeps out area at a constant rate (discovered 1602, published 1609).
3. The square of the period of each planet is proportional to the cube of its mean distance from the sun (discovered 1618, published 1619). For Kepler the "mean distance" was the arithmetic average of the greatest and smallest distances from the sun to the planet, which is just the length of the semimajor axis of the ellipse.

For many centuries prior to Kepler's time astronomers had attempted to explain the behavior of the five known planets (other than the earth) on the basis of orbits consisting of circles or compounded from circular motions. As observations became more exact, simple theories of this kind became untenable, and more and more complex constructions were hypothesized. For example, Copernicus** used five circles to attempt to describe the motion of Mars. Years of painstaking examination of observations made by Tycho Brahe† ultimately led Kepler to the conclusion that the data were better explained by assuming that the planets moved on elliptical orbits. After several more years of laborious calculations, Kepler proposed the three famous laws previously stated as an empirical description of planetary motion that was consistent with all of the data available at that time.

Kepler's laws, elegant and simple, demolished traditions and theories that had evolved over thousands of years. It is hardly surprising that they were not

*Johannes Kepler (1571–1630) was born in Germany and studied astronomy and theology at the University of Tübingen. Fleeing religious persecution, he arrived in Prague in 1600, and was one of Tycho Brahe's assistants during the last year of Brahe's life. Kepler thereby gained access to the wealth of planetary data that Brahe had accumulated, and on which the calculations leading to Kepler's laws were based. Kepler remained in Prague and enjoyed the patronage of Emperor Rudolph II of Bohemia until the latter's death in 1612; Kepler then found himself unwelcome in Prague, and spent most of the rest of his life in Linz. His most influential work was the *Tabulae Rudolphinae*, published in 1627, which contained perpetual tables for the calculation of planetary positions at any past or future date. The availability of these tables led to the widespread acceptance of Kepler's laws of planetary motion.

**Nicolaus Copernicus (1473–1543), a native of Poland, studied law and medicine at the Universities of Bologna, Padua, and Ferrara. He then returned to Poland and spent the last forty years of his life as canon of the cathedral chapter of Frombork. His revolutionary theory that the sun, and not the earth, is the center of the solar system was circulated privately among friends before 1514. Fearing ridicule and rejection, he delayed wider publication until the last year of his life. A detailed explanation of Copernicus' heliocentric theory is contained in his book *De Revolutionibus Orbium Coelestium*, which appeared in 1542, and is one of the classic works in the literature of science.

†Tycho Brahe (1546–1601), a Danish nobleman, was the last great astronomer prior to the discovery of the telescope. He built and operated an observatory on the island of Hven from 1576 to 1597, and made extensive observations of the motions of the moon, planets, and comets. Political differences caused him to leave his native land, and the last two years of his life were spent in and near Prague, where for a short time Johannes Kepler was one of his assistants.

immediately endorsed by other astronomers. In particular, the idea that planets move on ellipses rather than on circles was greeted with considerable skepticism. However, acceptance came as the improved accuracy of the predictions of planetary motions by Kepler's laws became more widely known and appreciated.

Some fifty years later, Newton was strongly motivated by Kepler's laws in formulating his theory of universal gravitation and his laws of motion. Indeed, assuming Newton's second law of motion, one can derive the law of gravity from Kepler's laws, or vice versa. Although historically Kepler's work came first, here we proceed in the reverse order; that is, we assume Newton's laws of motion and of gravitational attraction, and discuss how Kepler's laws may be derived from them.

Accordingly, let us consider the situation shown in Figure 15.9.3. The sun, whose mass is M, is located at the origin. A planet of mass m moves about the

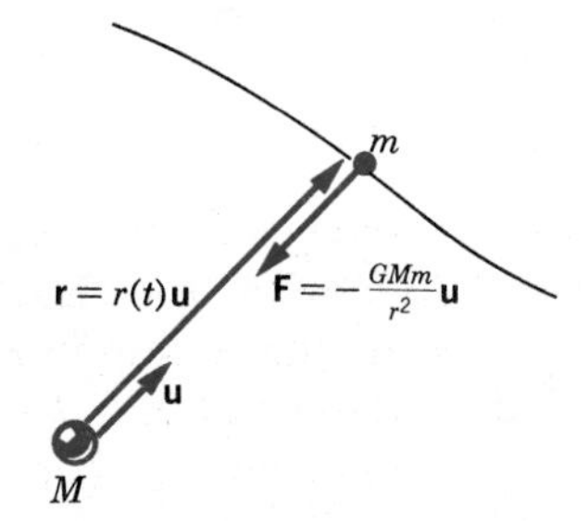

Figure 15.9.3 A planet (mass m) in orbit about the sun (mas M).

sun under the influence of gravity alone. The gravitational force $\mathbf{F}$ is given by

$$\mathbf{F} = -\frac{GMm}{r^2}\mathbf{u}, \tag{10}$$

where $r(t)$ is the current distance of the planet from the sun, $\mathbf{u}(t)$ is a unit vector in the direction from the sun to the planet, and G is the universal gravitational constant. The position vector $\mathbf{r}$ of the planet is given by

$$\mathbf{r} = r(t)\mathbf{u}(t) \tag{11}$$

and its velocity $\mathbf{v}$ and acceleration $\mathbf{a}$ are obtained by differentiating Eq. 11 with respect to t.

Several important properties of planetary motion follow very quickly. In the first place, Newton's second law of motion states that $\mathbf{F} = m\mathbf{a}$; consequently, from Eq. 10,

$$\mathbf{a} = -\frac{GM}{r^2}\mathbf{u}, \tag{12}$$

so *the acceleration vector is always directed toward the sun.*

Next, we consider the vector

$$\mathbf{h} = \mathbf{r} \times m\mathbf{v} = m(\mathbf{r} \times \mathbf{v}); \tag{13}$$

$\mathbf{h}$ is called the **angular momentum** of the planet. We have

$$\begin{aligned}\frac{d\mathbf{h}}{dt} &= m\frac{d}{dt}(\mathbf{r} \times \mathbf{v}) \\ &= m\left(\mathbf{r} \times \frac{d\mathbf{v}}{dt} + \frac{d\mathbf{r}}{dt} \times \mathbf{v}\right) \\ &= m(\mathbf{r} \times \mathbf{a} + \mathbf{v} \times \mathbf{v}) \\ &= m(\mathbf{r} \times \mathbf{a}), \end{aligned} \tag{14}$$

since $\mathbf{v} \times \mathbf{v} = \mathbf{0}$ for any $\mathbf{v}$. However, from Eqs. 11 and 12 we know that $\mathbf{r}$ and $\mathbf{a}$ are parallel, and hence $\mathbf{r} \times \mathbf{a} = \mathbf{0}$ also. Thus

$$\frac{d\mathbf{h}}{dt} = \mathbf{0}, \tag{15}$$

so

$$\mathbf{h}(t) = \mathbf{c}, \tag{16}$$

where $\mathbf{c}$ is a constant vector, which means that *angular momentum is conserved.*

It follows from Eq. 13 that $\mathbf{h}$ is perpendicular to both $\mathbf{r}$ and $\mathbf{v}$. Thus, in particular, the position vector $\mathbf{r}$ is always perpendicular to the constant vector $\mathbf{c}$, which establishes that *the orbit of the planet lies in a plane.*

To proceed further we introduce a polar coordinate system in the plane of motion with the origin at the sun; thus the vector $\mathbf{u}$ is now identified with $\mathbf{u}_r$ as previously defined. Then $\mathbf{r}$, $\mathbf{v}$, and $\mathbf{a}$ are given by Eqs. 7, 8, and 9, respectively, and the law of motion $\mathbf{F} = m\mathbf{a}$ becomes

$$-\frac{GM}{r^2}\mathbf{u}_r = \left[\frac{d^2r}{dt^2} - r\left(\frac{d\theta}{dt}\right)^2\right]\mathbf{u}_r + \left[r\frac{d^2\theta}{dt^2} + 2\frac{dr}{dt}\frac{d\theta}{dt}\right]\mathbf{u}_\theta, \tag{17}$$

where we have canceled m from both sides of Eq. 17.

We can deduce Kepler's second law from the circumferential component of Eq. 17. We have

$$r\frac{d^2\theta}{dt^2} + 2\frac{dr}{dt}\frac{d\theta}{dt} = 0. \tag{18}$$

If we multiply Eq. 18 by r, we obtain

$$r^2\frac{d^2\theta}{dt^2} + 2r\frac{dr}{dt}\frac{d\theta}{dt} = 0, \tag{19}$$

which is the same as

$$\frac{d}{dt}\left(r^2\frac{d\theta}{dt}\right) = 0; \tag{20}$$

consequently

$$r^2\frac{d\theta}{dt} = \text{constant}. \tag{21}$$

In polar coordinates the area element dA is given by

$$dA = \tfrac{1}{2}r^2\,d\theta; \tag{22}$$

see Eq. 7 of Section 14.3. Combining Eqs. 21 and 22, we obtain

$$\frac{dA}{dt} = \frac{1}{2}r^2\frac{d\theta}{dt} = \text{constant}, \tag{23}$$

which establishes Kepler's second law.

The constant in Eq. 23 can be interpreted in terms of the angular momentum $\mathbf{h}$. From Eqs. 13, 7, and 8 we have

$$\begin{aligned}\mathbf{h} &= m(\mathbf{r} \times \mathbf{v}) \\ &= mr\mathbf{u}_r \times \left(\frac{dr}{dt}\mathbf{u}_r + r\frac{d\theta}{dt}\mathbf{u}_\theta\right) \\ &= mr^2\frac{d\theta}{dt}\mathbf{u}_z \end{aligned} \tag{24}$$

where $\mathbf{u}_z = \mathbf{u}_r \times \mathbf{u}_\theta$ is a unit vector perpendicular to the plane of motion. Hence

$$\|\mathbf{h}\| = mr^2\left|\frac{d\theta}{dt}\right|, \tag{25}$$

and consequently

$$\left|\frac{dA}{dt}\right| = \frac{\|\mathbf{h}\|}{2m}. \tag{26}$$

The results derived to this point are valid whenever $\mathbf{F}$ is proportional to $\mathbf{u}$, that is, when $\mathbf{F}$ is a **central force**. The magnitude of $\mathbf{F}$ and its dependence on r have played no part so far. However, to derive Kepler's first and third laws, we need to assume that gravitational attraction has the inverse-square form of Eq. 10.

To establish Kepler's first law we start from the radial component of the equation of motion, Eq. 17, that is

$$\frac{d^2r}{dt^2} - r\left(\frac{d\theta}{dt}\right)^2 = -\frac{GM}{r^2}. \tag{27}$$

The procedure is to obtain from Eq. 27 an expression for r in terms of θ that one can recognize as the equation of an ellipse in polar coordinates. The calculations are too lengthy to present here, but some of the details of this derivation are outlined in Problem 5.

We turn now to Kepler's third law. For the special case of circular orbits we have already obtained Kepler's third law in Eq. 32 of Section 15.8. To consider the more general case of elliptical orbits we assume that the planet moves in the direction of increasing θ, so that $d\theta/dt > 0$ and hence $dA/dt > 0$. Then we can remove the absolute value bars in Eq. 26 and write

$$\frac{dA}{dt} = \frac{\|\mathbf{h}\|}{2m}. \tag{28}$$

Integrating Eq. 28 over the time T required for the planet to make one full revolution, we obtain

$$\text{Area of ellipse} = \frac{\|\mathbf{h}\|}{2m} T. \tag{29}$$

If a and b are the semimajor and semiminor axes, respectively, of the elliptical orbit, then the area of the ellipse is πab, so

$$\pi ab = \frac{\|\mathbf{h}\|}{2m} T. \tag{30}$$

Recall (Problem 19 of Section 5.3) that

$$b = a\sqrt{1 - \epsilon^2}, \tag{31}$$

where ϵ is the eccentricity of the ellipse. Substituting for b in Eq. 30 from Eq. 31, we have

$$\pi a^2\sqrt{1 - \epsilon^2} = \frac{\|\mathbf{h}\|}{2m} T. \tag{32}$$

To complete the derivation of Kepler's third law we must calculate $\|\mathbf{h}\|$. It is possible to show (see Problem 6) that

$$\frac{\|\mathbf{h}\|}{m} = [(1 - \epsilon^2)GMa]^{1/2}. \tag{33}$$

Finally, if we substitute for $\|\mathbf{h}\|$ in Eq. 32, square, and solve for T^2, we obtain

$$T^2 = \frac{4\pi^2 a^3}{GM}. \tag{34}$$

Observe that the proportionality factor $4\pi^2/GM$ in Eq. 34 depends only on the absolute constants G and M; in particular, it does not depend on the eccentricity ϵ. Therefore the relation (34) is the same for all orbits.

It should be borne in mind that Kepler's laws are idealizations of actual planetary motion in that their derivation assumes that the sun is stationary, and that the planet's orbit is determined solely by the gravitational attraction of the sun, neither of which is true. However, the effects of the sun's motion and the gravitational attraction of other celestial bodies are relatively small, and Kepler's laws are therefore highly accurate approximations. It is fortunate that the astronomical data in Kepler's time were accurate enough to discredit the older theories, but not so accurate as to make evident the small departures from Kepler's laws that actually occur. Otherwise, the development of the theories of gravitational attraction and Newtonian mechanics might have been even more difficult to achieve.

PROBLEMS

1. A particle of mass m moves along the curve $r = b \sin \theta$ with constant angular speed ω. Find the radial and circumferential components of the acceleration; also find the angular momentum.

2. A particle of mass m moves along the curve $r = k(1 + \cos \theta)$ with constant angular speed ω. Find the radial and circumferential components of the acceleration vector; also find the angular momentum.

3. A particle of mass m moves along the curve $r = f(\theta)$ with constant angular speed ω. Find the radial and circumferential components of the acceleration vector; also find the angular momentum.

4. A particle of mass m moves on the circle $r = R$ with variable angular speed ω; that is, $\omega = d\theta/dt$ depends on t. Find the radial and circumferential components of the acceleration vector; also find the angular momentum.

5. In this problem we outline one way of solving Eq. 27 of the text,

$$\frac{d^2r}{dt^2} - r\left(\frac{d\theta}{dt}\right)^2 = -\frac{GM}{r^2}, \qquad (i)$$

thereby deriving Kepler's first law.

(a) From Kepler's second law show that

$$\frac{d\theta}{dt} = \frac{H}{r^2}, \qquad (ii)$$

where H is a constant. From Eq. 25 in the text observe that H is the magnitude of the angular momentum per unit mass:

$$H = \frac{\|\mathbf{h}\|}{m}. \qquad (iii)$$

(b) Introduce the new dependent variable $u = 1/r$, and by using the chain rule show that

$$\frac{dr}{dt} = \frac{dr}{du}\frac{du}{d\theta}\frac{d\theta}{dt} = -H\frac{du}{d\theta}. \qquad (iv)$$

In a similar way show that

$$\frac{d^2r}{dt^2} = -H^2u^2\frac{d^2u}{d\theta^2}. \qquad (v)$$

(c) Using the results of Parts (a) and (b), show that Eq. (i) can be rewritten in the form

$$\frac{d^2u}{d\theta^2} + u = \frac{GM}{H^2}. \qquad (vi)$$

(d) Verify that

$$u = C\cos(\theta - \alpha) + \frac{GM}{H^2}, \qquad (vii)$$

where C and α are arbitrary constants, is a solution of Eq. (vi). In fact, it is possible to show that Eq. (vi) has no other solutions.

(e) It is possible to choose the polar axis so that $\alpha = 0$. If this is done, show that Eq. (vii) becomes

$$r = \frac{\epsilon p}{1 + \epsilon \cos \theta}, \qquad (viii)$$

where

$$\epsilon = \frac{H^2C}{GM}, \qquad p = \frac{1}{C}. \qquad (ix)$$

Equation ($viii$) is the polar equation of a conic section (see Problem 30(b) of Section 14.2), where ϵ is the eccentricity and p is the distance from the focus to the directrix. Since planetary orbits are closed, the conic section must be an ellipse. This means that $0 < \epsilon < 1$, and places a restriction on C. The graph of Eq. ($viii$) in this case is shown in Figure 15.9.4.

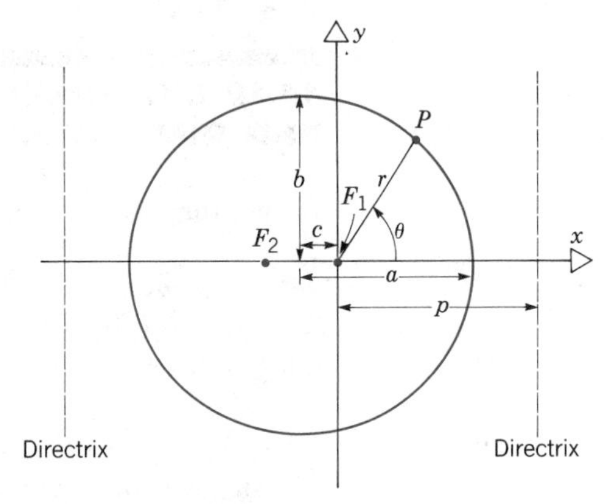

Figure 15.9.4

6. In this problem we indicate how to complete the derivation of Kepler's third law by establishing Eq. 33 in the text.

   (a) Referring to Eq. (*viii*) of Problem 5 and to Figure 15.9.4, observe that $r = a - c$ when $\theta = 0$ and that $r = a + c$ when $\theta = \pi$. Then show that

$$a = \frac{\epsilon p}{1 - \epsilon^2}. \qquad (i)$$

   (b) Use Eqs. (*iii*) and (*ix*) of Problem 5 to show that

$$\epsilon p = \frac{\|\mathbf{h}\|^2}{GMm^2},$$

   and then show that Eq. 33 is valid.

7. In this problem we indicate how to derive the law of gravity from Kepler's laws and Newton's second law of motion.

   (a) Show that Kepler's second law implies that the acceleration **a** lies along the line joining the sun to the planet; that is, the circumferential component of **a** in Eq. 9 is zero.

   (b) From Kepler's first and second laws, respectively, it follows that

$$\frac{1}{r} = \frac{1 + \epsilon \cos \theta}{\epsilon p} \qquad (i)$$

   and

$$\frac{d\theta}{dt} = \frac{H}{r^2} \qquad (ii)$$

   where ϵ, p, and H are constants (see Problem 5). Use Eqs. (*i*) and (*ii*) to show that

$$\frac{dr}{dt} = \frac{H}{p} \sin \theta, \qquad \frac{d^2r}{dt^2} = \frac{H^2}{r^3} - \frac{H^2}{r^2 \epsilon p}. \qquad (iii)$$

   (c) Referring to Eq. 9 of the text, show that the radial component a_r of acceleration is given by

$$a_r = -\frac{(H^2/\epsilon p)}{r^2}. \qquad (iv)$$

   Thus the acceleration vector **a** is directed toward the sun and its magnitude is inversely proportional to r^2. The same is true of the force vector **F** since $\mathbf{F} = m\mathbf{a}$. To complete the argument, we must show that the proportionality factor $H^2/\epsilon p$ is the same for all planets.

   (d) Kepler's third law can be expressed as

$$\frac{T^2}{a^3} = \frac{4\pi^2}{MG}. \qquad (v)$$

   By considering Kepler's second law, show that

$$H = \frac{2\pi ab}{T}. \qquad (vi)$$

   Use the relations

$$b^2 = a^2(1 - \epsilon^2), \qquad \epsilon p = a(1 - \epsilon^2) \qquad (vii)$$

   to show that

$$\frac{H^2}{\epsilon p} = MG; \qquad (viii)$$

   in other words the proportionality factor in Eq. (*iv*) is the same for all planets. This completes the derivation of the inverse square law of gravitational attraction.

15.10 CYLINDERS, SURFACES OF REVOLUTION, AND QUADRIC SURFACES

In two dimensions the graph of an equation

$$F(x, y) = 0 \qquad (1)$$

is a curve (provided degenerate cases are avoided). Similarly, in three dimensions the graph of

$$F(x, y, z) = 0 \qquad (2)$$

is a surface, that is, the set of points whose coordinates satisfy Eq. 2. In Section 5.1 we devoted considerable attention to sketching the graphs of equations of the form (1). In this section we consider the graphs of equations in three variables.

Since the problem of sketching a graph is considerably more complicated in three dimensions than in two, we restrict ourselves to certain types of equations that are of frequent occurrence and for which the corresponding surfaces can be sketched fairly readily.

Cylinders

Let C be a given plane curve and let l be a line that is not in the plane of C. Through each point of C draw a line parallel to l. The set of points that lie on the family of straight lines constructed in this way is called a **cylinder.** The given plane curve is called the **directrix** of the cylinder and the straight lines are its **generators** (see Figure 15.10.1). If the generators are perpendicular to the plane of the directrix

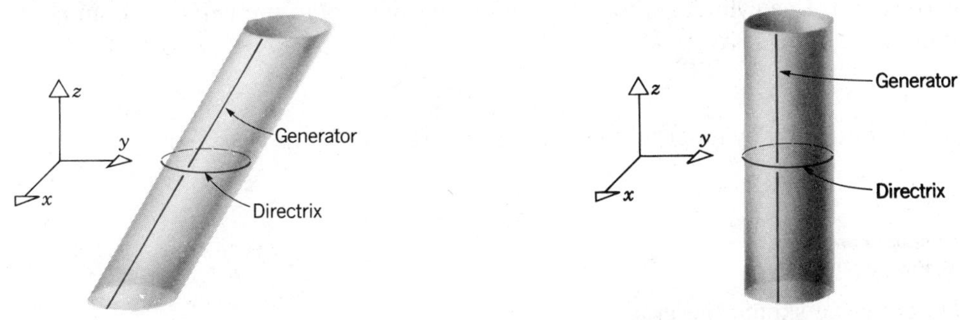

Figure 15.10.1 A cylinder.

Figure 15.10.2 A right circular cylinder.

(Figure 15.10.2), then the cylinder is called a right cylinder. If the directrix is a circle, then the cylinder is called a circular cylinder. In ordinary conversation the word "cylinder" is sometimes used to mean a "right circular cylinder," but we emphasize that the mathematical usage of "cylinder" is as we have just defined it.

We will consider only right cylinders whose generators are parallel to one of the coordinate axes. Thus the directrix of such a cylinder is a curve in a plane parallel to one of the coordinate planes. For example, suppose that the directrix is a curve in the xy-plane having the equation

$$f(x, y) = 0. \tag{3}$$

Consider a point $P_0(x_0, y_0, 0)$ on the directrix (see Figure 15.10.3). Since P_0 is on

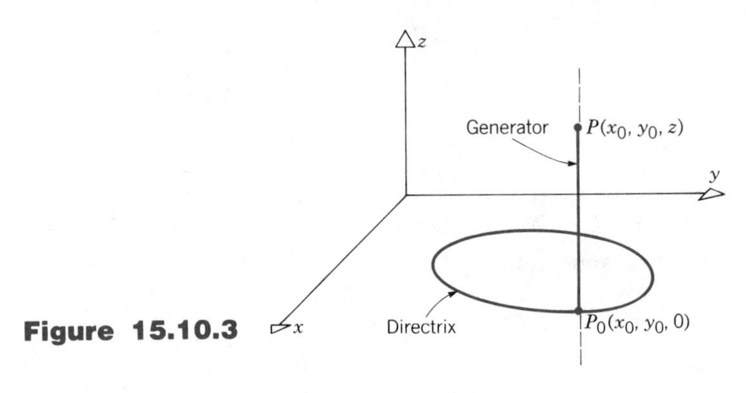

Figure 15.10.3

the directrix, its coordinates satisfy Eq. 3:

$$f(x_0, y_0) = 0. \tag{4}$$

Let P be an arbitrary point on the generator through P_0; then the coordinates of P are (x_0, y_0, z), where z is arbitrary. Since z does not appear in Eq. 3, the coordinates of P also satisfy this equation, just as the coordinates of P_0 do. Thus, viewed as an equation in three variables with one variable (z) missing, Eq. 3 is the equation of the cylinder with generators parallel to the z-axis. The directrix of the cylinder in the xy-plane is given by Eq. 3, which is considered as an equation in the two variables x and y. Similarly, an equation of the form

$$g(x, z) = 0$$

corresponds to a cylinder whose generators are parallel to the y-axis, and an equation of the form

$$h(y, z) = 0$$

corresponds to a cylinder whose generators are parallel to the x-axis.

EXAMPLE 1

Describe the sketch the graph of

$$\frac{x^2}{9} + \frac{z^2}{4} = 1 \tag{5}$$

in three dimensions.

Since the variable y is missing in Eq. 5, the graph is a cylinder with generators parallel to the y-axis. The directrix is the ellipse in the xz-plane having the equation (5). A sketch of a portion of the cylindrical surface is shown in Figure 15.10.4. ∎

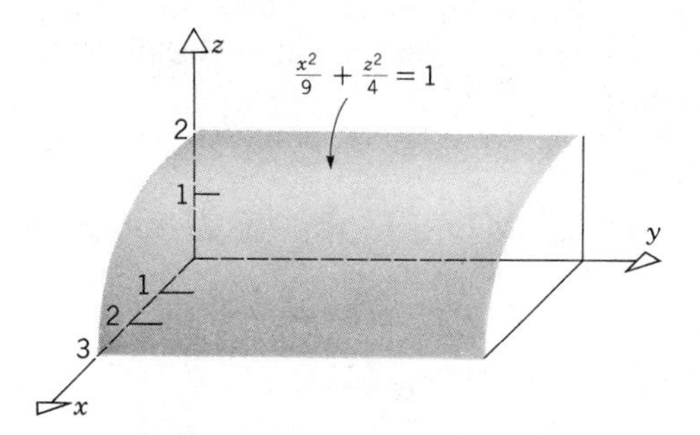

Figure 15.10.4

EXAMPLE 2

Describe and sketch the graph of

$$y = x^3. \tag{6}$$

The variable z is missing so the graph of Eq. 6 is a cylinder with generators parallel to the z-axis. The directrix is the graph of Eq. 6 in the xy-plane. A portion of the surface is shown in Figure 15.10.5. Observe that the directrix of a cylinder need not be a closed curve. ∎

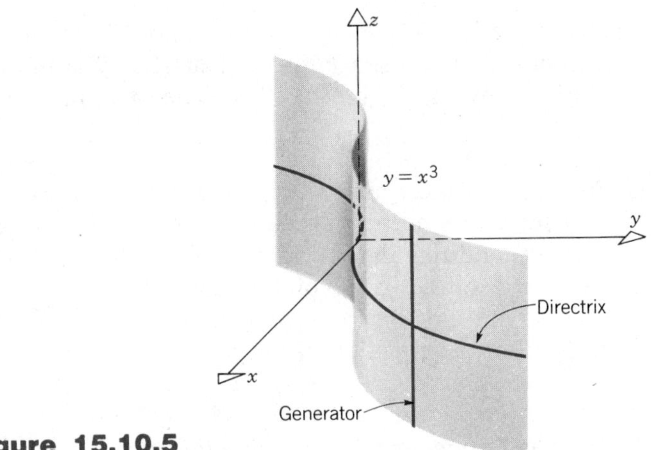

Figure 15.10.5

Surfaces of revolution

A surface of revolution is formed by rotating a plane curve about a line in the plane of the curve. The line is called the axis of rotation or the axis of revolution. If the axis of rotation is one of the coordinate axes, then the equation of the surface has a readily identifiable form.

For example, suppose that the graph of

$$z = y^2, \tag{7}$$

shown in Figure 15.10.6a, is rotated about the z-axis. During the rotation an arbitrary point $P_0(0, y_0, z_0)$ on the curve (7) moves on the circle in the plane $z = z_0$ whose center is $(0, 0, z_0)$ and whose radius is $|y_0|$ (see Figure 15.10.6b).

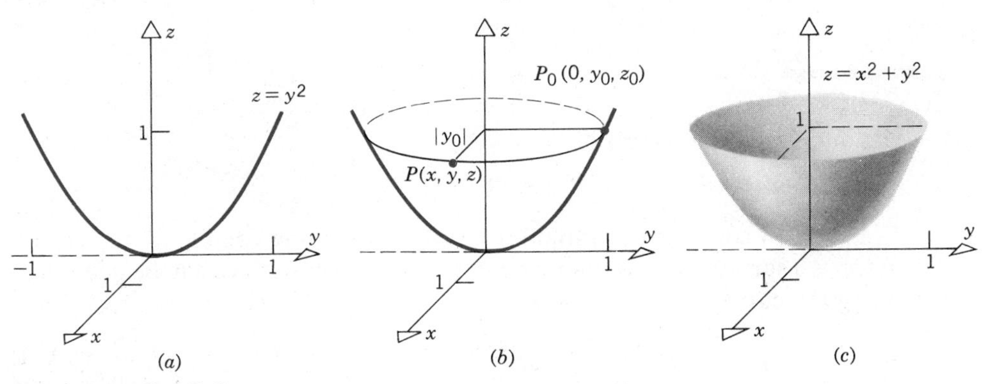

Figure 15.10.6

This circle is described by the equations

$$x^2 + y^2 = y_0^2, \qquad z = z_0. \tag{8}$$

Since $z_0 = y_0^2$, the coordinates of any point $P(x, y, z)$ on the circle must satisfy

$$z = x^2 + y^2. \tag{9}$$

Finally, since Eq. 9 does not contain y_0 or z_0, it remains the same regardless of which point P_0 is chosen initially. Thus Eq. 9 is the equation of the surface of revolution shown in Figure 15.10.6*c*. The same result is obtained if the curve

$$z = x^2 \tag{10}$$

is rotated about the z-axis. Comparing Eq. 9 with Eq. 7 or Eq. 10, we see that in effect y^2 or x^2 has been replaced by $x^2 + y^2$.

By generalizing the previous argument to other curves, we conclude that the equation of a surface of revolution about the z-axis does not depend on x and y separately, but only on the combination $x^2 + y^2$. In other words, such a surface corresponds to an equation of the form

$$F(x^2 + y^2, z) = 0. \tag{11}$$

The surface is obtained by rotating the curve

$$F(x^2, z) = 0$$

in the xz-plane about the z-axis, or the curve

$$F(y^2, z) = 0$$

in the yz-plane about the z-axis.

In a similar way the graphs of the equations

$$G(x^2 + z^2, y) = 0 \tag{12}$$

and

$$H(y^2 + z^2, x) = 0 \tag{13}$$

are surfaces of revolution about the y-axis and the x-axis, respectively.

EXAMPLE 3

Describe and sketch the graph of

$$y + 1 = 2(x^2 + z^2). \tag{14}$$

Equation 14 contains the variables x and z only in the combination $x^2 + z^2$, so its graph is a surface of revolution about the y-axis. The surface can be generated by rotating the curve

$$y + 1 = 2z^2 \tag{15}$$

in the yz-plane about the y-axis. This curve is a parabola whose vertex is at the point $y = -1$, $z = 0$. See Figure 15.10.7 for a sketch of the surface. ∎

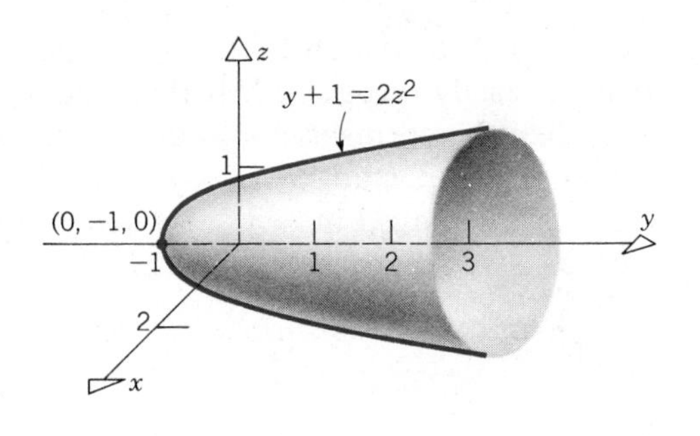

Figure 15.10.7
The graph of $y + 1 = 2(x^2 + z^2)$.

EXAMPLE 4

Describe and sketch the graph of

$$z = (x^2 + y^2)^{1/3}. \tag{16}$$

Since the variables x and y appear only in the combination $x^2 + y^2$, Eq. 16 describes the surface of revolution obtained by rotating the curve

$$z = y^{2/3} \tag{17}$$

in the yz-plane about the z-axis. A sketch of the surface is shown in Figure 15.10.8. ∎

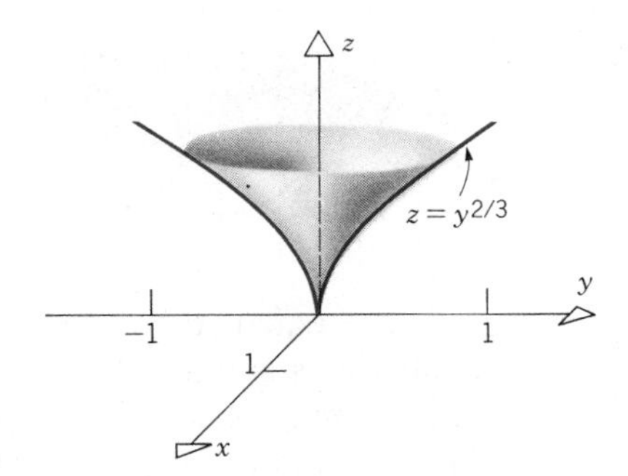

Figure 15.10.8
The graph of $z = (x^2 + y^2)^{1/3}$.

Quadric surfaces

Quadratic equations in the plane correspond to the conic sections, as we have discussed in Sections 5.2 through 5.4. Similarly, in three dimensions the graph of a quadratic equation in three variables is a surface known as a **quadric surface.** Neglecting degenerate cases, there are six distinct types of quadric surfaces. For each case we will give a standard form of the equation and a sketch of the corresponding surface. The exact shape of the surface and its orientation depend on the magnitudes of the coefficients and the arrangement of algebraic signs in the equation. The most important tool in constructing a sketch is to find the cross section, or **trace,** of the surface in each coordinate plane. Once this is done, it is not difficult (with a little practice) to draw a recognizable sketch. You should attempt to develop this skill rather than trying to memorize the various cases that can occur.

In sketching the surfaces it is very helpful to make use of any symmetry properties that the surface may possess. Symmetry with respect to a coordinate plane is easily recognized. If the equation is left unchanged when x is replaced by $-x$, then the corresponding surface is symmetric about the yz-plane. This is true, for example, if x appears only as x^2. Similarly, if the equation is left unchanged when y is replaced by $-y$, or when z is replaced by $-z$, then the corresponding surface is symmetric about the xz-plane or the xy-plane, respectively.

(a) ***Ellipsoid.*** This surface corresponds to an equation of the form

$$\frac{x^2}{a^2} + \frac{y^2}{b^2} + \frac{z^2}{c^2} = 1. \tag{18}$$

The trace in each coordinate plane is one of the ellipses

$$\frac{x^2}{a^2} + \frac{y^2}{b^2} = 1, \qquad \frac{x^2}{a^2} + \frac{z^2}{c^2} = 1, \qquad \frac{y^2}{b^2} + \frac{z^2}{c^2} = 1,$$

and the surface is oval-shaped as shown in Figure 15.10.9. If $a^2 = b^2 = c^2$, then the ellipsoid reduces to a sphere. If any two of the three quantities a^2,

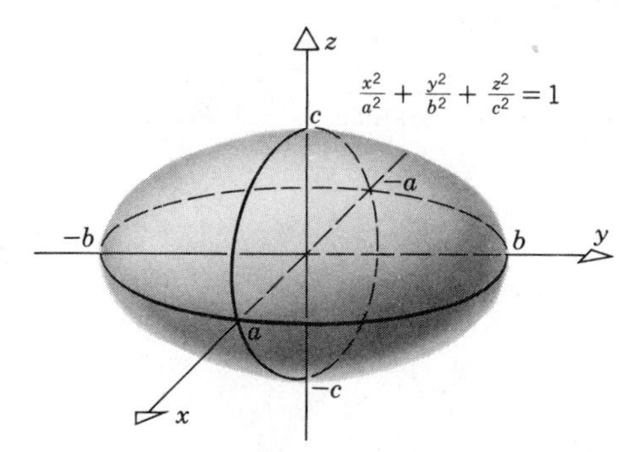

Figure 15.10.9 An ellipsoid.

b^2, c^2 are equal, then the ellipsoid is a surface of revolution about the remaining axis and is called a **spheroid.** If the equal coefficients are less than the remaining one (for example, $a^2 = c^2 < b^2$), then the surface is a **prolate spheroid,** and resembles a football (see Figure 15.10.10a). If the equal coefficients are greater than the remaining one (for example, $a^2 = b^2 > c^2$), then the surface is an **oblate spheroid,** and resembles a pumpkin (see Figure 15.10.10b).

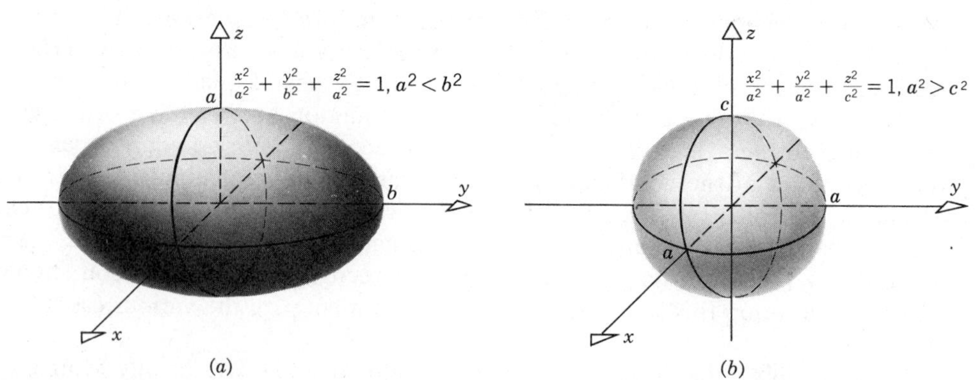

(a) (b)

Figure 15.10.10 (a) A prolate spheroid. (b) An oblate spheroid.

(b) ***Hyperboloid of one sheet.*** The equation of this surface can be illustrated by

$$\frac{x^2}{a^2} + \frac{y^2}{b^2} - \frac{z^2}{c^2} = 1. \tag{19}$$

The traces in the xz- and yz-planes are the hyperbolas

$$\frac{x^2}{a^2} - \frac{z^2}{c^2} = 1, \qquad \frac{y^2}{b^2} - \frac{z^2}{c^2} = 1,$$

respectively, while the trace in the xy-plane is the ellipse

$$\frac{x^2}{a^2} + \frac{y^2}{b^2} = 1.$$

Observe that in the plane $z = z_0$, Eq. 19 becomes

$$\frac{x^2}{a^2} + \frac{y^2}{b^2} = 1 + \frac{z_0^2}{c^2}.$$

Thus in any plane parallel to the xy-plane the cross section of the surface (19) is an ellipse; further, the ellipse expands as $|z_0|$ increases. A sketch of the surface is shown in Figure 15.10.11. If $a^2 = b^2$, then the hyperboloid is a surface of revolution about the z-axis.

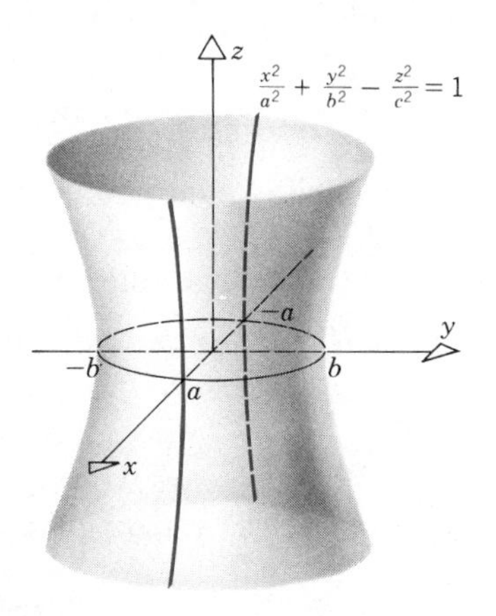

Figure 15.10.11 A hyperboloid of one sheet.

(c) ***Hyperboloid of two sheets.*** The equation of this surface is typified by

$$-\frac{x^2}{a^2} - \frac{y^2}{b^2} + \frac{z^2}{c^2} = 1. \tag{20}$$

In the xz- and yz-planes the traces are the hyperbolas

$$-\frac{x^2}{a^2} + \frac{z^2}{c^2} = 1, \qquad -\frac{y^2}{b^2} + \frac{z^2}{c^2} = 1,$$

respectively. However, in the xy-plane there is no trace since when $z = 0$, Eq. 20 becomes

$$\frac{x^2}{a^2} + \frac{y^2}{b^2} = -1.$$

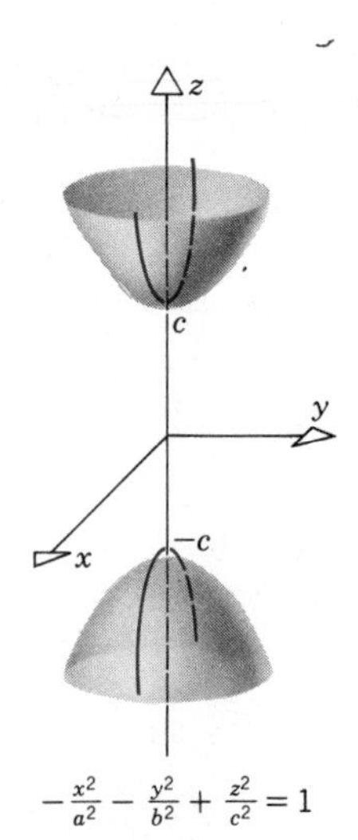

Figure 15.10.12
A hyperboloid of two sheets.

More generally, if $z = z_0$, then Eq. 20 becomes

$$\frac{x^2}{a^2} + \frac{y^2}{b^2} = \frac{z_0^2}{c^2} - 1. \tag{21}$$

If $z_0^2 < c^2$, there are no points that satisfy Eq. 21, so there are no points on the surface for this range of z. However, if $z_0^2 \geq c^2$, then Eq. 21 corresponds to an ellipse, which expands as $|z_0|$ increases. Thus the surface is in two unconnected pieces, as shown in Figure 15.10.12. If $a^2 = b^2$, then the hyperboloid is a surface of revolution about the z-axis.

(d) ***Cone.*** Consider the equation

$$\frac{x^2}{a^2} + \frac{y^2}{b^2} - \frac{z^2}{c^2} = 0. \tag{22}$$

In the yz-plane the trace of the surface is given by

$$\frac{y^2}{b^2} - \frac{z^2}{c^2} = \left(\frac{y}{b} + \frac{z}{c}\right)\left(\frac{y}{b} - \frac{z}{c}\right) = 0$$

and thus consists of the straight lines $y = \pm(b/c)z$. Similarly, the trace in the xz-plane is the pair of lines $x = \pm(a/c)z$. In the xy-plane the trace is just the origin, but the intersection of the surface (22) with any horizontal plane $z = z_0$ is the ellipse

$$\frac{x^2}{a^2} + \frac{y^2}{b^2} = \frac{z_0^2}{c^2},$$

which expands as $|z_0|$ increases. The graph of Eq. 22 is sketched in Figure 15.10.13. If $a^2 = b^2$, then the cone is a surface of revolution and is called a circular cone; otherwise, it is called an elliptic cone.

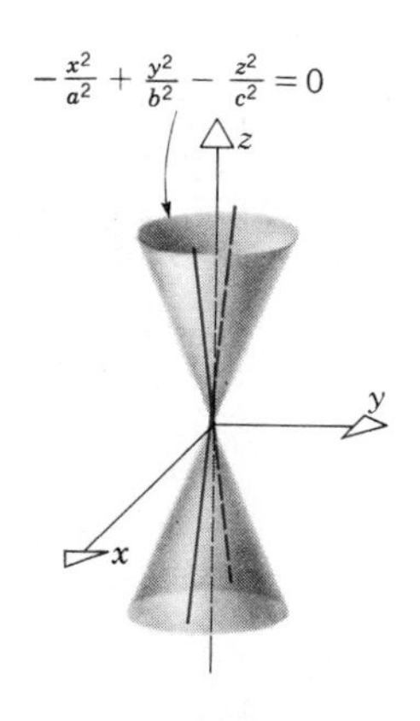

Figure 15.10.13
A cone.

(e) ***Elliptic paraboloid.*** This surface can be illustrated by the equation

$$z = \frac{x^2}{a^2} + \frac{y^2}{b^2}. \tag{23}$$

In the xz- and yz-planes the traces are the parabolas

$$z = \frac{x^2}{a^2}, \qquad z = \frac{y^2}{b^2},$$

respectively. In the xy-plane the trace consists only of the origin, and since z cannot be negative, the graph does not extend below the xy-plane. The intersection of the surface (23) with a horizontal plane $z = z_0 > 0$ is the ellipse

$$\frac{x^2}{a^2} + \frac{y^2}{b^2} = z_0,$$

which expands as z_0 increases. The graph of Eq. 23 is sketched in Figure 15.10.14. If $a^2 = b^2$, then the paraboloid is a surface of revolution and has circular cross sections in horizontal planes; in this case, it is called a circular paraboloid.

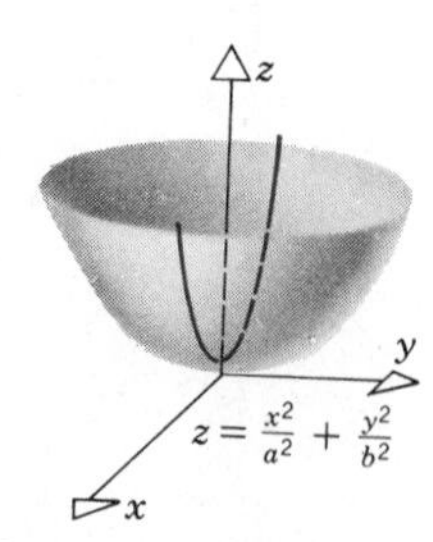

Figure 15.10.14
An elliptic paraboloid.

(f) ***Hyperbolic paraboloid.*** Consider the equation

$$z = -\frac{x^2}{a^2} + \frac{y^2}{b^2}. \tag{24}$$

In the yz-plane the trace is given by

$$z = \frac{y^2}{b^2},$$

which corresponds to a parabola opening up, while in the xz-plane the trace is the downward opening parabola

$$z = -\frac{x^2}{a^2}.$$

In the xy-plane the trace consists of the two straight lines $y = \pm bx/a$, but the intersection of Eq. 24 with any horizontal plane $z = z_0 \neq 0$ is the hyperbola

$$-\frac{x^2}{a^2} + \frac{y^2}{b^2} = z_0. \tag{25}$$

Observe that the orientation of the hyperbola (25) depends on the sign of z_0. If $z_0 > 0$, the intercepts occur when $y = \pm b\sqrt{z_0}$ while if $z_0 < 0$, the intercepts occur when $x = \pm a\sqrt{|z_0|}$. The graph of Eq. 24 is sketched in Figure 15.10.15.

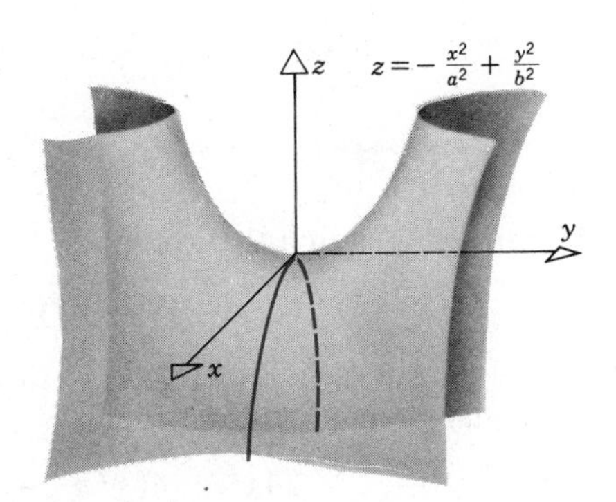

Figure 15.10.15
A hyperbolic paraboloid.

As we have noted, the trace in the xz-plane is a downward opening parabola, so the origin (vertex) is a maximum point. On the other hand, the trace in the yz-plane is an upward opening parabola, so the origin is a minimum point. Near the origin the surface has a shape resembling that of a mountain pass, or a saddle. The surface is often called a saddle surface and the origin a saddle point.

EXAMPLE 5

Identify and sketch the graph of

$$-36x^2 + 9y^2 - 16z^2 = 144. \tag{26}$$

Dividing Eq. 26 by 144, we obtain

$$-\frac{x^2}{4} + \frac{y^2}{16} - \frac{z^2}{9} = 1,$$

which is a quadratic equation in standard form. The traces in the xy- and yz-planes, respectively, are the hyperbolas

$$-\frac{x^2}{4} + \frac{y^2}{16} = 1, \qquad \frac{y^2}{16} - \frac{z^2}{9} = 1.$$

The surface intersects a plane $y = y_0$ only if $|y_0| \geq 4$, in which case the curve of intersection is the ellipse

$$\frac{x^2}{4} + \frac{z^2}{9} = \frac{y_0^2}{16} - 1.$$

The surface is a hyperboloid of two sheets and is sketched in Figure 15.10.16. ∎

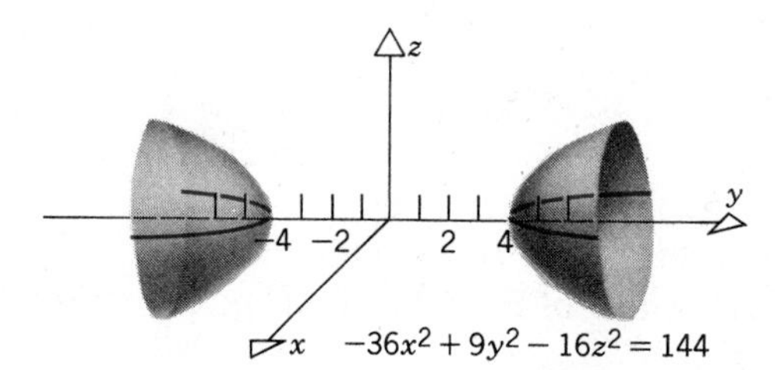

Figure 15.10.16

EXAMPLE 6

Identify and sketch the graph of

$$4x^2 + 4y^2 - 16y - 9z^2 = 20. \tag{27}$$

In order to reduce Eq. 27 to one of the standard forms we must first complete the square in y. Thus we obtain

$$4x^2 + 4(y^2 - 4y + 4) - 9z^2 = 20 + 16,$$

or

$$4x^2 + 4(y - 2)^2 - 9z^2 = 36; \tag{28}$$

dividing both sides of Eq. 28 by 36, we have

$$\frac{x^2}{9} + \frac{(y - 2)^2}{9} - \frac{z^2}{4} = 1. \tag{29}$$

Finally, we let $v = y - 2$, which amounts to translating the xyz-coordinate system two units in the positive y-direction. In terms of x, v, and z, Eq. 29 becomes

$$\frac{x^2}{9} + \frac{v^2}{9} - \frac{z^2}{4} = 1, \tag{30}$$

which is a standard equation for a quadric surface. The trace in the xv-plane is the circle $x^2 + v^2 = 9$, while in the xz- and vz-planes, the traces are hyperbolas. The graph is a hyperboloid of one sheet, and is shown in Figure 15.10.17. Since the coefficients of x^2 and v^2 in Eq. 30 are the same, the surface is a surface of revolution about the line $x = 0$, $y = 2$. ∎

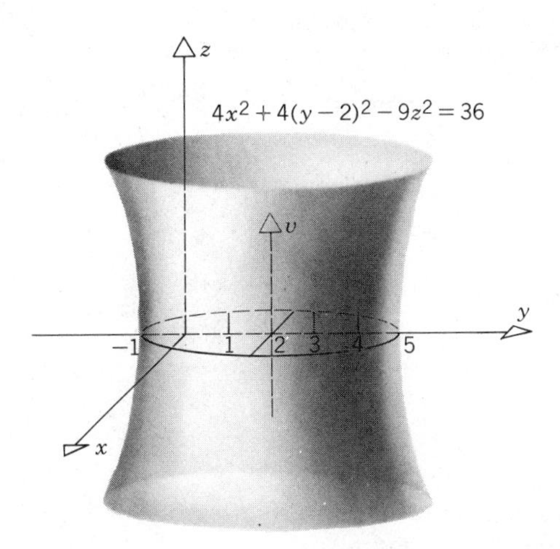

Figure 15.10.17

PROBLEMS

In each of Problems 1 through 24, describe and sketch the graph of the given equation.

1. $y^2 + z^2 = 4$
2. $z = 4x^2 + 4y^2$
3. $x^2 - y^2 = 1$
4. $4x^2 + 9y^2 + 4z^2 = 36$
5. $x - \sin z = 0$
6. $9x^2 - 4y^2 + 36z^2 = 36$
7. $9x^2 - 4y^2 + 36z^2 = 0$
8. $9x^2 - 4y^2 - 36z^2 = 36$
9. $9x^2 - 2y^2 - z = 0$
10. $x^2 + 2z^2 - 4 = 0$
11. $4x^2 + y + 9z^2 = 0$
12. $y^2 - (x^2 + z^2)^{3/2} = 0$
13. $x^2 + 4y^2 + 4z^2 = 16$
14. $3x^2 + 3y^2 - z^2 = 9$
15. $3x^2 + 3y^2 - z^2 = 0$
16. $x^2 - 4y^2 + 9z^2 = -36$
17. $9x^2 + 4y^2 - 18x + 16y = 11$
18. $4x^2 + 8y^2 - z = 3$
19. $6x^2 - 3y^2 + 2z^2 + 24x - 4z = -20$
20. $x^2 + 2y - z^2 = 4$
21. $y^2 - 2z - 6 = 0$
22. $x^2 - 3y + z^2 - 9 = 0$
23. $x^2 + y^2 + z^2 - 6x + 2y - 8z = -10$
24. $-2x^2 + 4y^2 - 9z^2 - 8x - 8y + 54z = 121$

In each of Problems 25 through 32, write an equation for the surface formed by rotating the given curve about the given axis.

25. $y = 1 - x^2$; y-axis
26. $y = 1 - x^2$; x-axis
27. $y = \sqrt{x}$; y-axis
28. $y = \sqrt{x}$; x-axis
29. $y = \dfrac{z^2}{1 + z^2}$; z-axis
30. $z = x^{2/3}$; x-axis
31. $z = x^{2/3}$; z-axis
32. $z = \log|y|$; z-axis

33. Consider the graph of $y = \sqrt{x}$ for $0 \le x \le 4$. We wish to find the equation of the surface formed by rotating this curve about the line $y = 3$ (see Figure 15.10.18).

    (a) Introduce a new variable $u = y - 3$. Express the equation $y = \sqrt{x}$ in terms of u and x.

    (b) Observe that rotating the graph about the line $y = 3$ corresponds to rotating about the line $u = 0$. Determine the equation of the surface of revolution in terms of x, u, and z.

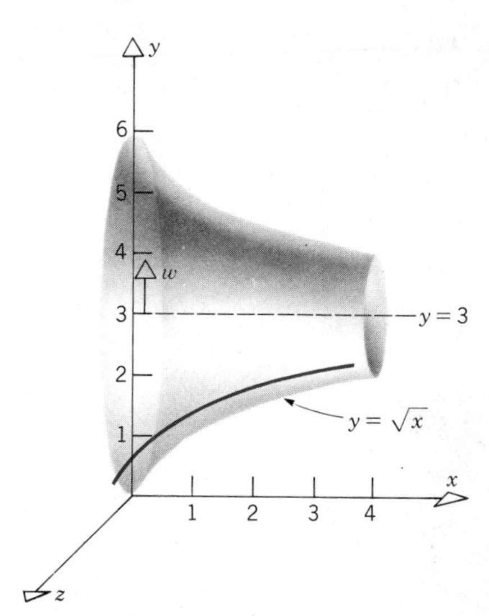

Figure 15.10.18

(c) Express the equation of the surface of revolution in terms of x, y, and z.

In each of Problems 34 through 37, find the equation of surface formed by rotating the given curve about the given line. Refer to Problem 33 if necessary.

34. $y = \sqrt{x}$ about $x = -2$

35. $y = -x^2$ about $y = 1$

36. $z = 1 + x^4$ about $z = -3$

37. $z = \arctan y$ about $z = \pi$

38. The curve given by

$$\mathbf{r}(t) = 2t \cos 3t\, \mathbf{i} + \sqrt{4 - t^2}\, \mathbf{j} - 5t \sin 3t\, \mathbf{k}$$

lies on a quadric surface. Identify the surface and find its equation.

39. The curve given by

$$\mathbf{r}(t) = 2\sqrt{t} \cosh t\, \mathbf{i} + 3\sqrt{t} \sinh t\, \mathbf{j} + 4\sqrt{1 - t}\, \mathbf{k}$$

lies on a quadric surface. Identify the surface and find its equation.

40. (a) Let $\mathbf{a}$ be the position vector of a fixed point A other than the origin, and let $\mathbf{r}$ be the position vector of a point P. Describe the surface on which P lies if $\|\mathbf{a} \times \mathbf{r}\| = \rho\|\mathbf{a}\|$, where $\rho > 0$.

(b) Find an equation for the right circular cylinder whose axis is the line passing through the origin and the point $A(a, b, c)$ and whose radius is ρ.

41. (a) Let $\mathbf{a}$ be the position vector of a fixed point A other than the origin, and let $\mathbf{r}$ be the position vector of a point P. Describe the surface on which P lies if $\|\mathbf{a} \times \mathbf{r}\| = \lambda|\mathbf{a} \cdot \mathbf{r}|$, where $\lambda > 0$.

(b) Find an equation for the cone whose axis is the line passing through the origin and the point $A(a, b, c)$, and whose vertex angle is α.

In Problems 42 through 44, we indicate how to extend the geometric definitions of the conic sections in the plane to surfaces in three dimensions.

42. Let $P(x, y, z)$ be a point the sum of whose distances from two given points is a constant. Find an equation for the surface on which P lies and identify the surface. *Hint:* Choose a coordinate system so that the two given points are $(c, 0, 0)$ and $(-c, 0, 0)$, and let $2a$ be the sum of the distances.

43. Let $P(x, y, z)$ be a point the difference of whose distances from two given points is a constant. Find an equation for the surface on which P lies and identify the surface. *Hint:* Choose a coordinate system so that the two given points are $(c, 0, 0)$ and $(-c, 0, 0)$, and let $2a$ be the difference of the distances.

44. Let $P(x, y, z)$ be a point equally distant from a given point and a given plane not containing the given point. Find an equation for the surface on which P lies and identify it. *Hint:* Choose a coordinate system so that the given point is $(c, 0, 0)$ and the given plane is $x = -c$.

REVIEW PROBLEMS

In each of Problems 1 through 6, perform the indicated operations if $\mathbf{a} = \langle 3, -2, 4\rangle$, $\mathbf{b} = \langle 2, 0, -5\rangle$, and $\mathbf{c} = \langle 1, 3, -2\rangle$.

1. $\mathbf{a} + \mathbf{b} - 2\mathbf{c}$

2. $\mathbf{a} \cdot \mathbf{b}$

3. $\mathbf{b} \times \mathbf{c}$

4. $\|\mathbf{b} \times \mathbf{c}\|$

5. $(\mathbf{a} \times \mathbf{b})(\mathbf{b} \cdot \mathbf{c})$

6. $\|(\mathbf{a} \times \mathbf{b})(\mathbf{b} \cdot \mathbf{c})\|$

In each of Problems 7 through 10, find a scalar λ that satisfies the given condition.

7. $\mathbf{a} = (2\lambda, 2, -3)$ and $\mathbf{b} = (7, 1, \lambda)$ have the same length.

8. $\mathbf{a} = (\lambda, 2, -3)$ and $\mathbf{b} = (7, -1, \lambda)$ are perpendicular.

9. $\mathbf{a} = (1, 0)$ and $\mathbf{b} = (5, \lambda)$ form an angle of $\pi/4$ radians.

10. $P(2, 3, 1)$, $Q(\lambda, 3, 4)$, $R(0, 2, 1)$, and $S(1, 0, 0)$ are coplanar.

In each of Problems 11 through 14, find the indicated limit, where $\mathbf{u} = [(t^2 - 1)/(t - 1), t, \ln(t + 1)]$ and $\mathbf{v} = (\sin \pi t, \cos 2\pi t, e^{t^2})$.

11. $\lim_{t\to 1} (3\mathbf{u} + 4\mathbf{v})(t)$

12. $\lim_{t\to 0} (2\mathbf{u} - 3\mathbf{v})(t)$

13. $\lim_{t\to 0} (\mathbf{u} \cdot \mathbf{v})(t)$

14. $\lim_{t\to 1} (2\mathbf{u} \times \mathbf{v})(t)$

In each of Problems 15 through 18, find the indicated derivatives where $\mathbf{u}(t) = (\cos t, 0, e^{3t^2})$, $\mathbf{v}(t) = (0, 3t^4, \ln t)$

15. $\dfrac{d}{dt}(\mathbf{u} + \mathbf{v})(t)$

16. $\dfrac{d}{dt}(\mathbf{u} \cdot \mathbf{v})(t)$

17. $\dfrac{d}{dt}\|\mathbf{u}(t)\|$

18. $\dfrac{d}{dt}(\mathbf{u} \times \mathbf{v})(t)$

In each of Problems 19 through 22, (a) find the unit vector tangent to the given curve at an arbitrary point; (b) set up an integral giving the length of the curve for the given interval.

19. $\mathbf{r}(t) = 2 \sin t\, \mathbf{i} + \cos^2 t\, \mathbf{j}$,. $\quad 0 \le t \le \dfrac{\pi}{2}$

20. $\mathbf{r}(t) = \tan^2 t\, \mathbf{i} + t^2 e^t\, \mathbf{j} - \sqrt{1 + t^2}\, \mathbf{k}$, $\quad 0 \le t \le 1$

21. $\mathbf{r}(t) = \ln \sin t\, \mathbf{i} + t\, \mathbf{j}$, $\quad \dfrac{\pi}{4} \le t \le \dfrac{3\pi}{4}$

22. $\mathbf{r}(t) = (1 - t^2)\, \mathbf{i} + \ln t\, \mathbf{j} + \sinh t\, \mathbf{k}$, $\quad 3 \le t \le 5$

In each of Problems 23 through 26, identify and sketch the given surface.

23. $x^2 - 8x + 3y - z^2 + 2z = 6$

24. $4x^2 + 8x + 9y^2 + 18y - z = 87$

25. $25x^2 - 50x - 4y^2 + 16y = z^2 + 91$

26. $x^2 - 6x + 2y + y^2 = 10z - z^2 + 9$

In each of Problems 27 through 30, write an equation for the surface found by rotating the given curve about the given line (see Problem 33, Section 15.10).

27. $2y^2 + 5x^2 = 10$ $\quad$ about y-axis

28. $x = \cos y$ $\quad$ about y-axis, $\quad -\pi \le y \le \pi$

29. $z = x^3$ $\quad$ about $z = 1$, $\quad$ or $\le x \le 1$

30. $y = e^z$ $\quad$ about $z = -1$, $\quad 0 \le z \le 5$

31. Show that the line $(x - 1)/6 = y/3 = z$ intersects the plane $x - 2y + 3z = 1$, and find the point of intersection.

32. Find the distance from the point $(1, 1, 1)$ to:
(a) the line $x = 1 + t$, $\; y = 2 - t$, $\; z = 3t$;
(b) the plane $2x + 3y - z = 2$.

33. Find the equation of the plane that contains the points $P(1, 2, 3)$, $Q(1, 0, 1)$, and $R(0, 3, 4)$.

34. Find the equation of the plane that passes through $P(1, 1, 1)$ and has normal $(6, -1, 2)$.

35. Find the equation of the plane that passes through $P(-1, 0, 1)$ and contains the line $(x - 2)/3 = (y - 1)/2 = z/4$.

36. Find the symmetric equation of the line through $P(1, 2, 3)$ and parallel to $2\mathbf{i} + 2\mathbf{j} + \mathbf{k}$.

37. Find the equation of the plane containing the lines $(x - 2)/3 = (y - 1)/2 = z/4$ and $(x - 1)/1 = (y + 2)/3 = (z - 6)/2$.

38. Find the equation of the line of intersection of the two planes $6x + y - 3z = 2$ and $x + z = 3$ in both symmetric and parametric forms.

39. Find the area of the triangle with vertices $(1, 0, 0)$, $(2, 3, 4)$, and $(5, 1, 0)$.

40. A particle moves along the curve $\mathbf{r}(t) = e^t\, \mathbf{i} - t^2\, \mathbf{j} + t^{1/2}\, \mathbf{k}$.
(a) Find the velocity and speed of the particle.
(b) Find the normal and tangential components of the acceleration.
(c) Find the curvature of the path of the particle.

41. A particle of mass m moves along the curve $r = a \cos \theta$ with constant angular speed ω.
(a) Find the radial and circumferential components of the acceleration.
(b) Find the angular momentum.
(c) Find the curvature of the path of the particle.

42. A 575 Newton weight is suspended by two ropes. When the ropes make angles of 24° and 42° with the vertical, respectively, the weight is in equilibrium. Find the magnitude of the tension in each rope.

43. A pilot wants to fly an airplane 6000 km west. In still air the plane has a velocity of 800 km/hr. A wind blows in the direction S45°E with a velocity of 110 km/hr. In what direction should the pilot fly in order to stay on a westerly course? How long will the trip take?

SIXTEEN

differentiation of functions of several variables

Up to now we have examined the fundamental processes of calculus, namely, differentiation and integration, as they relate to problems involving just one independent variable. In order to go much further we need to generalize the concepts of function, derivative, and integral to encompass more variables. While this is desirable and interesting from a mathematical point of view, it is essential if one is interested in the applications of calculus, simply because many physical problems intrinsically require more than one independent variable for their description. For example, the motion of an object in space is described by functions depending on three spatial coordinates and time. If the orientation of the object is important, then three more variables are required.

We begin our discussion by extending the definition of a function to include two or more independent variables, and then focus on the question of what it means to differentiate a function of more than one variable.

16.1 FUNCTIONS, LIMITS, AND CONTINUITY

Functions

The idea of a function introduced in Section 1.4 can be readily generalized. To do this, first let X be some set of points in the xy-plane, and let (x, y) be the coordinates of a typical point in X. Let Z be the set of all real numbers, or possibly some subset of them. Then a **function** f of two real variables is a rule (transformation) that assigns one and only one number z in Z to each point (x, y) in X. In this case we write

$$z = f(x, y). \tag{1}$$

As in the past, the set X is the **domain** of f and Z is the **codomain.** The number z that is associated with a particular point (x, y) in X is the **image** of (x, y), and the set of all such image points is the **range** of f, denoted by $f(X)$. Often, the domain of a function is not explicitly stated; in such cases, we understand it to consist of all points for which the expression $f(x, y)$ makes sense. Also, for the function defined by Eq. 1, we refer to x and y as the **independent variables** and to z as the **dependent variable.**

For example, let X be the entire xy-plane and let

$$f(x, y) = x^2 + y^2. \tag{2}$$

This function assigns a certain nonnegative number to each point in the plane. Indeed, the number is just the square of the distance from the origin to the given point. Thus the value of f corresponding to the point $(3, -2)$ is $f(3, -2) = (3)^2 + (-2)^2 = 13$. The range of f is $f(X) = [0, \infty)$.

As another example, consider the surface area S of a circular cylindrical can of radius r and height h, given by

$$S = 2\pi r^2 + 2\pi rh. \tag{3}$$

We require that $r \geq 0$ and $h \geq 0$, so the domain of this function is the first quadrant of the rh-plane. Again, the range is the set of all nonnegative numbers.

In a similar way we can define a function of three variables

$$w = f(x, y, z), \tag{4}$$

or a function of n variables

$$w = f(x_1, x_2, \ldots, x_n). \tag{5}$$

EXAMPLE 1

Determine the domain of the function

$$f(x, y) = \ln(4 - x^2 - 4y^2). \tag{6}$$

Since the logarithm function is defined only for positive values of its argument, it follows that the domain of f is

$$4 - x^2 - 4y^2 > 0,$$

or

$$x^2 + 4y^2 < 4. \tag{7}$$

Hence the domain of f is the interior of the ellipse shown in Figure 16.1.1. ∎

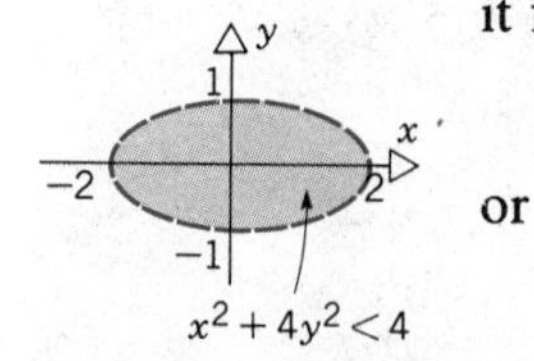

Figure 16.1.1

EXAMPLE 2

Determine the domain of the function

$$g(x, y) = \sqrt{x^2 + y^2 - 1} + \sqrt{4 - x^2 - y^2}. \tag{8}$$

Since each radicand must be nonnegative, we must have

$$1 \le x^2 + y^2 \le 4. \tag{9}$$

Thus the domain of g consists of those points that satisfy both inequalities in Eq. 9, that is, points in the ring shaped region, or annulus, shown in Figure 16.1.2. ∎

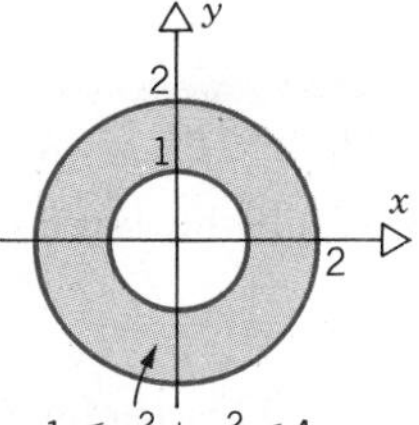

Figure 16.1.2

A function of two variables can be represented geometrically as a surface in three-dimensional space. For example, if

$$z = f(x, y) = \sqrt{1 - (x^2 + y^2)}, \tag{10}$$

then $x^2 + y^2 + z^2 = 1$ with $z \ge 0$. Thus the surface corresponding to this function is the hemisphere shown in Figure 16.1.3. The surface corresponding to the function

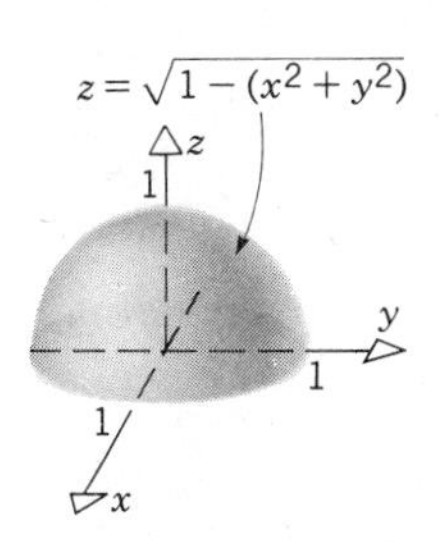

Figure 16.1.3

$$z = f(x, y) = (x^2 + y^2)^{1/3} \tag{11}$$

is shown in Figure 16.1.4. Several of the surfaces corresponding to quadratic polynomials in x, y, and z were sketched in Section 15.10.

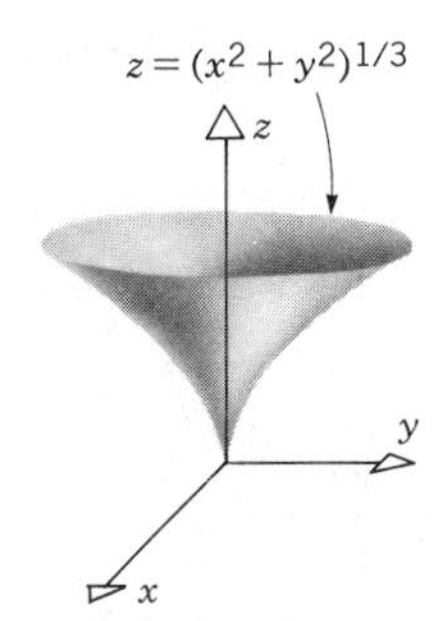

Figure 16.1.4

Unfortunately, for many functions that are only moderately complicated, it can be rather difficult to sketch the corresponding surface. Indeed, even for the quadratic polynomials discussed in Section 15.10, it can sometimes be difficult to get the perspective correct. There are very useful computer graphics packages for plotting surfaces in three dimensions that have found important applications in such problems as designing an automobile body or an airplane. However, for the problems we will discuss only a rough sketch of a surface will be needed.

An alternative procedure that helps us to visualize a surface $z = f(x, y)$ is to draw the **level curves** of the surface. These are the family of curves in the xy-plane corresponding to different constant values of z: $f(x, y) =$ constant. If we think of the xy-plane as representing sea level and z as the altitude above ($z > 0$) or below ($z < 0$) sea level at the point (x, y), then the level curves of the surface give us a topographic map of the terrain. It is customary to construct topographic maps by using a constant altitude difference between adjacent level curves. Then, regions where the level curves are close together are regions where the altitude (value of the function) is changing rapidly. A typical topographic map might look something like Figure 16.1.5, which shows two hills and the pass between them.

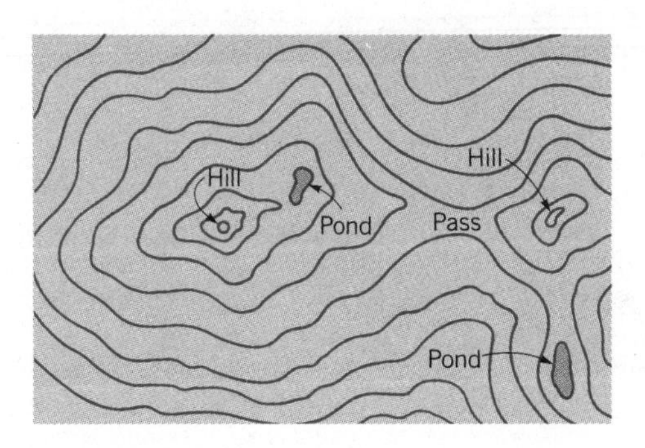

Figure 16.1.5 A typical topographic map.

EXAMPLE 3

Sketch the level curves for the function

$$z = \ln(4 - x^2 - 4y^2).$$

This is the function discussed in Example 1. If we set $z = z_0 = \ln c$, where c is a positive constant, then the level curves are given by

$$c = 4 - x^2 - 4y^2.$$

These are the ellipses

$$\frac{x^2}{4(1 - \frac{1}{4}c)} + \frac{y^2}{1 - \frac{1}{4}c} = 1$$

where $0 < c < 4$. Since $z_0 = \ln c$, we can also write the equation for the level curves in the form

$$\frac{x^2}{4(1 - \frac{1}{4}e^{z_0})} + \frac{y^2}{1 - \frac{1}{4}e^{z_0}} = 1,$$

where $-\infty < z_0 < \ln 4$. The level curves are sketched in Figure 16.1.6*a* for several values of z_0; the surface is sketched in Figure 16.1.6*b*. ∎

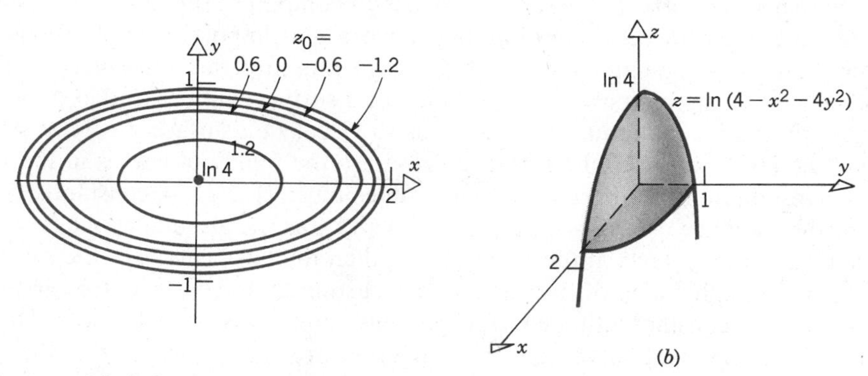

Figure 16.1.6 (*a*) Level curves of $z = \ln(4 - x^2 - 4y^2)$. (*b*) The surface $z = \ln(4 - x^2 - 4y^2)$.

For a function of three or more variables it is no longer possible to draw a realistic sketch of the graph of the function. However, the main conceptual difficulties in extending the ideas of the calculus to higher dimensions occur as one goes from functions of one variable to functions of two variables. In doing this we will make use of geometric intuition as much as possible. Once this transition is made, there is little intrinsic difficulty in thinking in higher dimensions. Indeed, our intuition allows us to think and to visualize relations in higher dimensions even though we cannot draw an accurate geometric representation of them. Most of our discussion will be for functions of two or three independent variables.

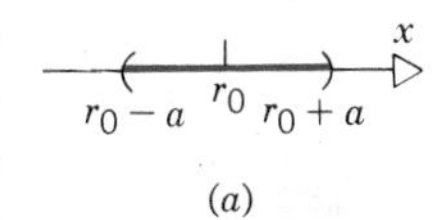

(a)

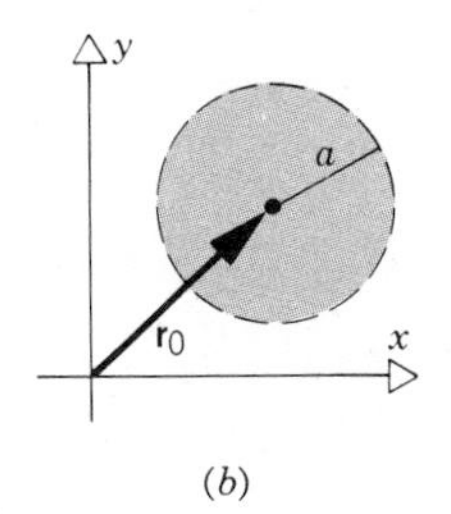

(b)

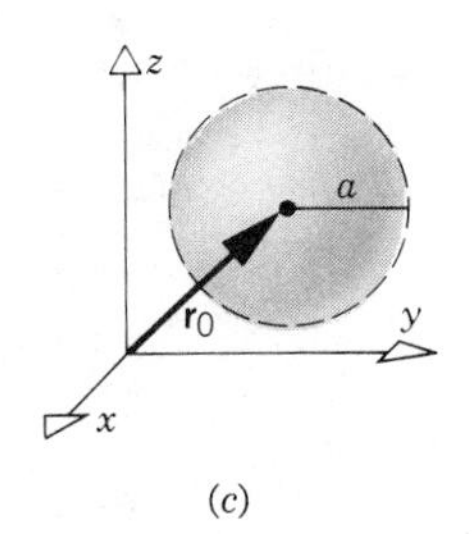

(c)

Figure 16.1.7
(*a*) An open interval.
(*b*) An open disk.
(*c*) An open ball.

Limit of a function

The discussion of limits in Chapter 2 involved the distance $|x - x_0|$ between two points P_0 and P with coordinates x_0 and x, respectively, on the x-axis. In two dimensions the corresponding distance between $P_0(x_0, y_0)$ and $P(x, y)$ is $\sqrt{(x - x_0)^2 + (y - y_0)^2}$. Similarly, in three dimensions the distance between $P_0(x_0, y_0, z_0)$ and $P(x, y, z)$ is $\sqrt{(x - x_0)^2 + (y - y_0)^2 + (z - z_0)^2}$. The notation can be unified by using vectors. Let $\mathbf{r}_0$ and $\mathbf{r}$ be the position vectors of the two points P_0 and P, respectively. Then the distance between the points is always given by $\|\mathbf{r} - \mathbf{r}_0\|$, regardless of the dimension. If $\|\mathbf{r} - \mathbf{r}_0\| \to 0$, then we say that P approaches P_0, or that $\mathbf{r}$ approaches $\mathbf{r}_0$, and write $\mathbf{r} \to \mathbf{r}_0$. If we wish to emphasize the coordinate variables, we may sometimes write $x \to x_0$, or $(x, y) \to (x_0, y_0)$, or $(x, y, z) \to (x_0, y_0, z_0)$, instead of $\mathbf{r} \to \mathbf{r}_0$.

In Section 1.2 we used the word "neighborhood" to refer to an open interval $(x_0 - a, x_0 + a)$ centered at the point x_0. In two dimensions the corresponding geometric entity is the interior of a circle of radius a and center $P_0(x_0, y_0)$, while in three dimensions it is the interior of a sphere of radius a and center $P_0(x_0, y_0, z_0)$. It is sometimes useful to use the notation $N_a(\mathbf{r}_0)$ to denote the neighborhood of the point $P_0(\mathbf{r}_0)$ of radius a. Then, in symbols,

$$N_a(\mathbf{r}_0) = \{\mathbf{r} \mid \|\mathbf{r} - \mathbf{r}_0\| < a\}, \tag{12}$$

where $\mathbf{r}_0$ is the position vector of the point P_0 and a is any positive number.

As mentioned before, a neighborhood in one dimension is an **open interval.** In two dimensions it is an **open disk** and in three dimensions it is an **open sphere** or **ball.** This is illustrated in Figure 16.1.7. If we include equality in Eq. 12, then we have a **closed interval,** or a **closed disk,** or a **closed sphere** or **ball** according to the dimension of the space. The **boundary** of an open or closed interval, disk, or ball is the set of points $\{\mathbf{r} \mid \|\mathbf{r} - \mathbf{r}_0\| = a\}$. Notice that any neighborhood of a boundary point contains both points in the open interval, disk, or ball and also points not in these sets. This is illustrated in Figure 16.1.8 for an open disk and a closed disk. Also observe that an open interval, disk, or ball contains none of its boundary points, while a closed interval, disk, or ball contains all of its boundary points.

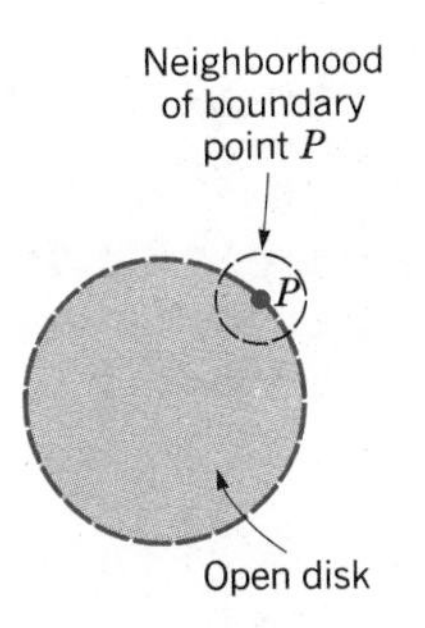

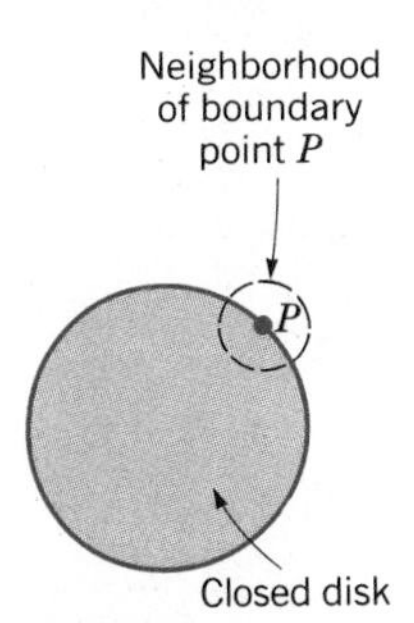

Figure 16.1.8

These observations can be extended to arbitrary sets of points. The point P is called an **interior point** of a set S if P is in S, and if there is some neighborhood of P that contains only points in S. The point P is called a **boundary point** of S if every neighborhood of P contains both points that are in S as well as points that

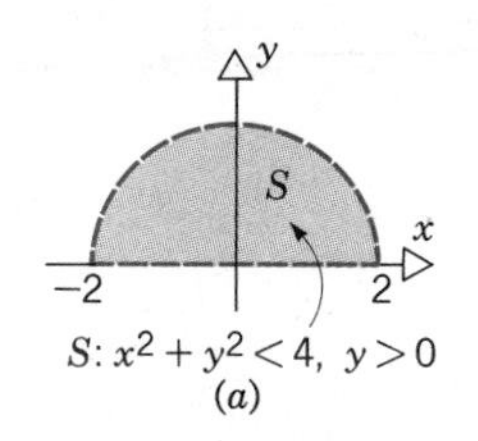

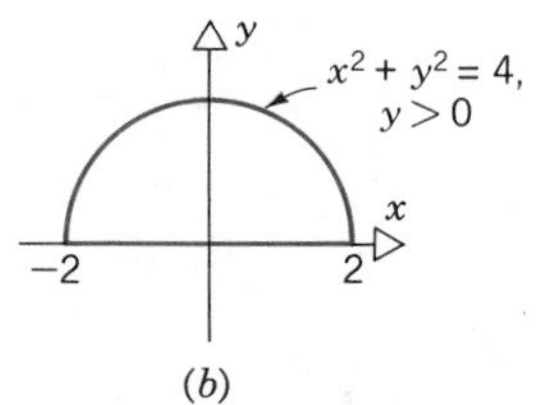

(b)

Figure 16.1.9
(a) An open set S.
(b) The boundary of S.

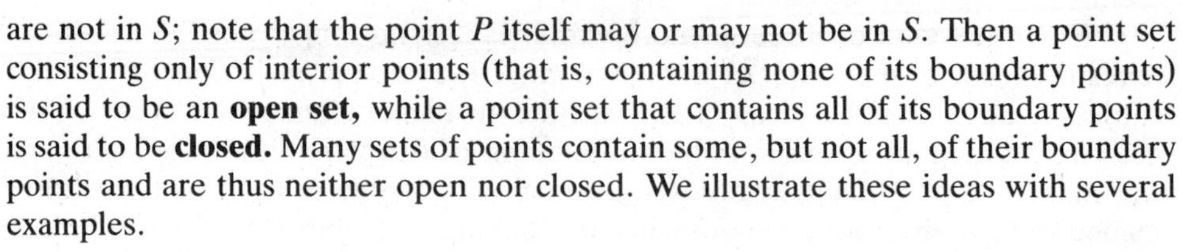

are not in S; note that the point P itself may or may not be in S. Then a point set consisting only of interior points (that is, containing none of its boundary points) is said to be an **open set,** while a point set that contains all of its boundary points is said to be **closed.** Many sets of points contain some, but not all, of their boundary points and are thus neither open nor closed. We illustrate these ideas with several examples.

1. Let $S = \{(x, y)|\ x^2 + y^2 < 4, y > 0$ (see Figure 16.1.9*a*). The boundary of S is the set of points $x^2 + y^2 = 4, y > 0$ and $-2 \le x \le 2, y = 0$ (see Figure 16.1.9*b*). Since S contains none of its boundary points, it is an open set. All points in the set are interior points.
2. Let $S = \{(x, y)|\ 4 < x^2 + y^2 \le 16\}$ (see Figure 16.1.10). The boundary of S is the two circles $x^2 + y^2 = 4$ and $x^2 + y^2 = 16$. Since S contains some, but not all of its boundary points, it is neither an open nor a closed set.
3. Let $S = \{(x, y, z)|\ x^2 + y^2 \le 4,\ 0 \le z \le 2\}$ (see Figure 16.1.11*a*). The boundary of S is the surface of the cylinder $x^2 + y^2 = 4, 0 \le z \le 2$ and the portions of the planes $z = 0, 2$ with $x^2 + y^2 \le 4$ (see Figure 16.1.11*b*). Since S contains all of its boundary points, it is a closed set.

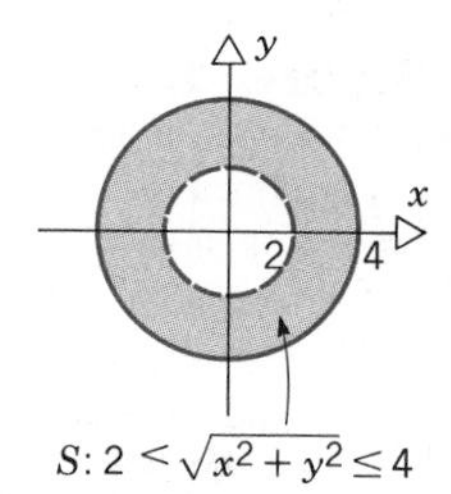

Figure 16.1.10
A set that is neither open nor closed.

Now suppose that f is a function of one or more variables, and that f is defined in some neighborhood of a point $\mathbf{r}_0$, except possibly at the point $\mathbf{r}_0$ itself. Then $f(\mathbf{r})$ has the limit L as $\mathbf{r} \to \mathbf{r}_0$ if $f(\mathbf{r}) \to L$ as $\mathbf{r} \to \mathbf{r}_0$, but is not equal to $\mathbf{r}_0$. The precise definition is as follows.

DEFINITION 16.1.1 (*Limit of a function*). Let f be a function of one or more variables that is defined in some neighborhood of a point $\mathbf{r}_0$, except possibly at $\mathbf{r}_0$ itself. Then the limit of $f(\mathbf{r})$ as $\mathbf{r}$ approaches $\mathbf{r}_0$ is said to be L, written

$$\lim_{\mathbf{r}\to\mathbf{r}_0} f(\mathbf{r}) = L, \tag{13}$$

if for each number $\epsilon > 0$, there exists a corresponding number $\delta > 0$ such that if

$$0 < \|\mathbf{r} - \mathbf{r}_0\| < \delta, \quad \text{then} \quad |f(\mathbf{r}) - L| < \epsilon. \tag{14}$$

It may be useful to compare this definition with Definition 2.2.1 for the limit of a function of one variable and to note how the generalization has been made. We emphasize (again) that the concept of limit does not depend upon the value of $f(\mathbf{r}_0)$ or even upon whether f is defined at $\mathbf{r}_0$.

Continuous functions

Just as for functions of a single variable, the concepts of limit and continuity for a function of two or more variables are closely related.

DEFINITION 16.1.2 (*Continuity of a function*). A function of one or more variables is said to be continuous at $\mathbf{r}_0$ if $f(\mathbf{r}_0)$ is defined, and if

$$\lim_{\mathbf{r}\to\mathbf{r}_0} f(\mathbf{r}) = f(\mathbf{r}_0). \tag{15}$$

Most of the results that were derived for limits and continuity for functions of one variable can be stated and proved for functions of several variables simply by rephrasing the statement and proof. These include such results as the limit and continuity theorems for sums, products, and quotients. Also most of the functions that one encounters are continuous, except possibly where a denominator is zero or where the function is undefined. Some examples of continuous functions are the following.

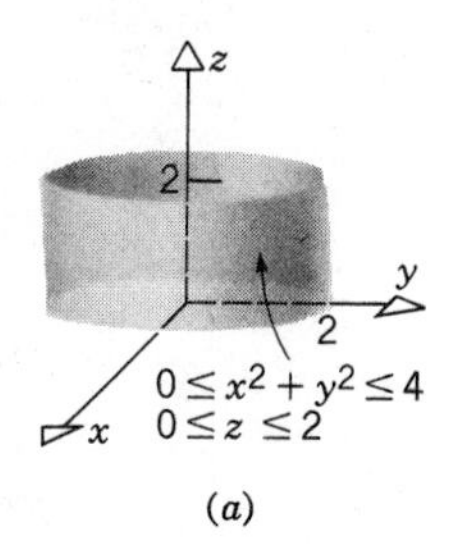

(a)

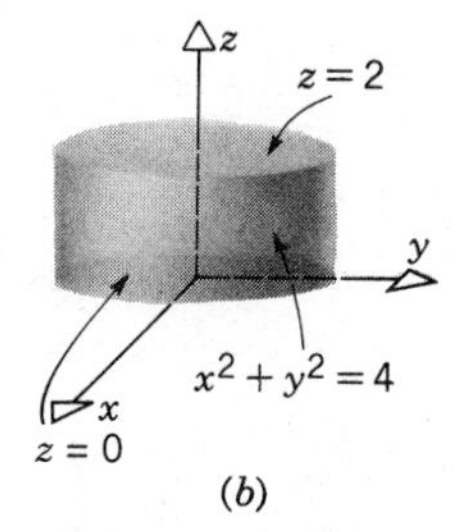

(b)

Figure 16.1.11
(a) A closed set S.
(b) The boundary of S.

1. Polynomial functions, which are functions formed as finite linear combinations of nonnegative integral powers of the independent variables. For example,

$$P(x, y, z) = 3 + 2xy + 6z^2 - xy^2z + 4x^2yz^2$$

is a continuous function for all x, y, and z.

2. Rational functions, which are quotients of polynomial functions, except at zeros of the denominator. For example,

$$R(x, y) = \frac{3 - 2x + xy^2}{4 - x^2 - 2y^2}$$

is a continuous function, except at points on the ellipse $x^2 + 2y^2 = 4$.

3. Functions obtained by taking roots, trigonometric functions, exponential functions, and logarithmic functions, except where the functions are not defined. Some examples of continuous functions are

$$f(x, y, z) = \sqrt{4 - (x^2 + y^2 + z^2)}, \qquad x^2 + y^2 + z^2 \le 4;$$
$$g(x, y) = \sin \frac{xy}{\sqrt{x^2 + y^2}}, \qquad (x, y) \ne (0, 0);$$
$$h(x, y, z) = ze^{x^2+y^2}, \qquad \text{all } (x, y, z);$$
$$\phi(x, y) = \ln|x - y|, \qquad y \ne x.$$

Implicit in the definitions of the limit of a function and the continuity of a function at a point is that there is a neighborhood of $\mathbf{r}_0$ that lies in the domain of f. For a function of one variable this means that $\mathbf{r}_0$ is an interior point, or a point in an open interval (a, b). The concepts of limit and continuity can then be extended to endpoints by considering one-sided limits. For functions of more than one variable we can discuss limits and continuity at a boundary point in a similar way, that is, by considering only points that lie in a neighborhood of the boundary point and also lie in the domain of the function.

In the following example and in several of the problems we illustrate some of the difficulties that can arise in dealing with functions of more than one variable because of the variety of ways in which it is possible to approach a point.

EXAMPLE 4

Show that the function

$$z = f(x, y) = \begin{cases} \dfrac{xy}{x^2 + y^2}, & (x, y) \ne (0, 0) \\ 0, & (x, y) = (0, 0) \end{cases} \tag{16}$$

does not have a limit at the origin, and hence is not continuous there.

Notice that on the x-axis, excluding the origin, $f(x, 0) = 0$; and on the y-axis, excluding the origin, $f(0, y) = 0$. Hence,

$$\lim_{x\to 0} f(x, 0) = \lim_{x\to 0} 0 = 0,$$

$$\lim_{y\to 0} f(0, y) = \lim_{y\to 0} 0 = 0.$$

In these special cases we find 0 to be the limiting value as we approach the origin along the coordinate axes, but this is not sufficient to establish that f has a limit as $(x, y) \to (0, 0)$. If we introduce the polar coordinates $x = r\cos\theta$, $y = r\sin\theta$, then we have, for $r \neq 0$,

$$f(r\cos\theta, r\sin\theta) = \frac{r^2\cos\theta\sin\theta}{r^2\cos^2\theta + r^2\sin^2\theta} = \cos\theta\sin\theta$$
$$= \tfrac{1}{2}\sin 2\theta. \tag{17}$$

Thus f takes on all values between $-\frac{1}{2}$ and $\frac{1}{2}$ as we go around any circle enclosing the origin. Consequently, no matter how small a neighborhood of the origin we choose, there will be a circle (circles) within the neighborhood on which f takes on values between $-\frac{1}{2}$ and $\frac{1}{2}$, and it is impossible to make f close to any number L as $(x, y) \to (0, 0)$.

We could also show that f does not have a limit at $(0, 0)$ by showing that we obtain different limiting values as we approach $(0, 0)$ along two different paths. Consider the set of paths corresponding to the straight lines $\theta =$ constant with $r \to 0$ or alternatively $y = mx$ with $x \to 0$. Then

$$\lim_{x\to 0} f(x, mx) = \lim_{x\to 0} \frac{x(mx)}{x^2 + m^2x^2} = \frac{m}{1 + m^2}. \tag{18}$$

On observing that $m = \tan\theta$, we can show that this is equivalent to the result given in Eq. 17. Since we obtain different limits as $(x, y) \to (0, 0)$ along different straight line paths, we conclude that f does not have a limit at $(0, 0)$. We emphasize that if the result in Eq. 18 had been independent of m, this would not have guaranteed that the limit exists; there are other paths of approach to $(0, 0)$ in addition to straight lines.

The level curves for this function for $(x, y) \neq (0, 0)$ are given by

$$\frac{xy}{x^2 + y^2} = z_0 = \text{a constant}.$$

From Eq. 17 the level curves are given by

$$\sin 2\theta = 2z_0 \quad \text{or} \quad \theta = \tfrac{1}{2}\arcsin 2z_0.$$

Thus they are the straight lines through the origin. Since the level curves correspond to different values of z_0 and since they all come together at the origin, it is clear that the function has no limit at the origin. ∎

PROBLEMS

In each of Problems 1 through 4, evaluate the given function at the given points.

1. $f(x, y) = |x + 1|\sqrt{1 - y}$; (a) $(-1, 0)$ (b) $(0, 1)$ (c) $(-3, -3)$

2. $g(x, y) = \dfrac{3x}{x + y}$; (a) (a^2, b) (b) (a, b^2) (c) (a^{-1}, b^{-1})

3. $f(x, y) = 2y^x + x\sqrt{y^3}$; (a) $(-2, 2)$ (b) (a, a^2) with $a > 0$ (c) $(\frac{1}{3}, 8)$

4. $h(x, y, z) = \dfrac{x + y + z}{|x + y + z|}$; (a) $(1, 1, 1)$ (b) $(-1, 0, -1)$ (c) $(0, 0, -1)$

5. Let $f(x, y) = x \sin y$. Find (a) $f\left(2, -\dfrac{\pi}{2}\right)$, (b) $f(x, x)$, (c) $f(y, x)$, (d) $f(r, s)$

6. Let $f(x, y) = 3x^2\sqrt{y} + 1$. Find (a) $f(2, 4)$, (b) $f(a + b, a - b)$, (c) $f\left(\dfrac{y}{x}, \dfrac{x^2}{y^2}\right)$ for $x \neq 0$, $y \neq 0$

7. Let $F(x, y, z) = xyz/\sqrt{x^2 + y^2 + z^2}$. Find (a) $F(a, a, a)$, $a > 0$, (b) $F(a^{-1}, a^{-1}, a^{-1})$, $a > 0$

8. Let $F(x, y, z) = x^2 + 2y^2 - z^2$. If $x = f(t) = \sin \pi t$, $y = g(t) = \cos \pi t$, and $z = h(t) = 3t$, find (a) $F[f(t), g(t), h(t)]$, (b) $F[f(\frac{1}{2}), g(\frac{1}{2}), h(\frac{1}{2})]$, (c) $F[f(2t), g(2t), h(2t)]$

In each of Problems 9 through 16, determine the domain of the given function.

9. $\phi(x, y) = \sqrt{9 - 2(x^2 + y^2)}$

10. $f(x, y) = \sqrt{x^2 + y^2 - 9}$

11. $\phi(x, y) = 6x^3 - 7y^3 + 4xy$

12. $g(x, y) = \dfrac{1}{\ln(x^2 + y^2)}$

13. $\phi(x, y, z) = \sqrt{z}\, e^{-|xy|}$

14. $g(x, y) = \sqrt{x(1 - |y|)}$

15. $\phi(x, y) = \ln xy$

16. $f(x, y) = \dfrac{(1 - x)^{1/3}}{\sqrt{2 + y} - \sqrt{x}}$

In each of Problems 17 through 22, sketch several of the level curves of the given function.

17. $f(x, y) = x^2 + y^2$

18. $f(x, y) = 1 - 4x^2 - y^2$

19. $f(x, y) = x^2 - y$

20. $f(x, y) = ye^{-|x|}$

21. $f(x, y) = e^{x^2 - y^2}$

22. $f(x, y) = e^{|y^2 - x^2|}$

Problems 23 through 34 deal with limits and continuity for functions of two variables. In answering the questions arguments similar to those given in Example 4 are adequate. To answer questions about limits and continuity at (0, 0) it is often helpful to introduce the polar coordinates $x = r \cos \theta$ and $y = r \sin \theta$ and to study what happens as $r \to 0$; also it may be useful to study what happens on the rays $x = \alpha t$, $y = \beta t$ as $t \to 0$.

In each of Problems 23 through 28, determine where, if anywhere, the function is discontinuous.

23. $\phi(x, y, z) = \sin(3x + 7y + 2z)$

24. $\phi(x, y) = \ln|1 - xy|$

25. $f(x, y) = \dfrac{x^2 + y^2}{x^2 - y^2}$

$f(x, y) = \dfrac{x^2 - y^2}{x^2 + y^2}$

27. $g(x, y) = \begin{cases} x^2 + y^2, & x^2 + y^2 \leq 1 \\ 2 - (x^2 + y^2), & x^2 + y^2 > 1 \end{cases}$

28. $h(x, y, z) = \ln(x^2 + y^2 + z^2)$

29. Is the function

$$g(x, y) = \begin{cases} \sin \dfrac{x}{y}, & x \neq 0 \text{ and } y \neq 0 \\ 1, & x = 0 \text{ or } y = 0 \end{cases}$$

continuous at the origin?

30. Show that the function

$$f(x, y) = \begin{cases} \dfrac{xy}{\sqrt{x^2 + y^2}}, & (x, y) \neq (0, 0) \\ 0, & (x, y) = (0, 0) \end{cases}$$

is continuous at (0, 0).

31. How should the function

$$f(x, y) = \frac{y^3}{x^2 + y^2}, \quad (x, y) \neq (0, 0)$$

be defined at the origin so that f is continuous there?

32. Can the function

$$f(x, y) = \frac{x^3 - y^3}{x - y}, \quad y \neq x$$

be defined on the line $y = x$ so that the resulting function is continuous at all points in the plane?

33. For what values of $\alpha > 0$ can the function

$$f(x, y) = \frac{(x + y)^\alpha}{x^2 + y^2}$$

be defined at the origin so that f is continuous there? What is the value of $f(0, 0)$ in these cases?

34. Consider the function

$$\phi(x, y, z) = \frac{(x + y + z)^2}{x^2 + y^2 + z^2},$$

$$(x, y, z) \neq (0, 0, 0).$$

(a) Show that

$$\lim_{x\to 0} \phi(x, 0, 0) = \lim_{y\to 0} \phi(0, y, 0) = \lim_{z\to 0} \phi(0, 0, z) = 1.$$

(b) No matter how $\phi(0, 0, 0)$ is defined, show that ϕ cannot be continuous at the origin. *Hint:* Consider the limit along the rays $x = \alpha t$, $y = \beta t$, $z = \gamma t$ as $t \to 0$ for arbitrary values of α, β, γ.

16.2 PARTIAL DERIVATIVES

In this section we will extend the computational rules developed so far for finding derivatives to functions of more than one variable. For example, for the surface area S of a circular cylindrical can of radius r and height h,

$$S = 2(\pi r^2) + 2\pi rh = 2\pi r(r + h),$$

we may want to know the rate of change of S with respect to r or with respect to h, while the other remains fixed. Other examples occur frequently in physical applications; for instance, in fluid flow problems the components u, v, and w of fluid velocity are functions of position coordinates x, y, z, and time t. In deriving the equations governing the motion of the fluid it is necessary to introduce the rates of change of each velocity component with respect to each of the position coordinates and time.

We start by considering a function of two variables,

$$z = f(x, y) \tag{1}$$

for (x, y) in some domain Ω of the xy-plane. We ask what is the rate of change of z with respect to x alone, or with respect to y alone, at some point (x_0, y_0). In determining the rate of change of z with respect to y, we hold x constant, that is, we require that $x = x_0$. Similarly, in determining the rate of change of z with respect to x we require that $y = y_0$.

From a geometrical viewpoint the graph of Eq. 1 is a surface in xyz-space and the graph of $x = x_0$ is a plane parallel to the yz-coordinate plane (see Figure 16.2.1). The curve of intersection of this surface and this plane is given by a function of one variable,

$$z = f(x_0, y). \tag{2}$$

The rate of change of z with respect to y at the point (x_0, y_0) is simply the slope of this curve, which we can find by forming the difference quotient

$$\frac{f(x_0, y_0 + k) - f(x_0, y_0)}{k} \tag{3}$$

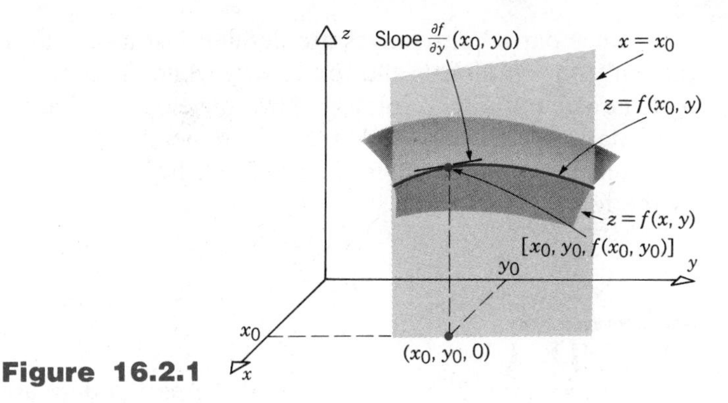

Figure 16.2.1

and passing to the limit as $k \to 0$. If this limit exists, we denote it by

$$\frac{\partial f}{\partial y}(x_0, y_0) = \lim_{k \to 0} \frac{f(x_0, y_0 + k) - f(x_0, y_0)}{k}. \tag{4}$$

The function $\partial f/\partial y$, whose value at (x_0, y_0) is given by Eq. 4, is referred to as the **partial derivative** of f with respect to y. The symbol ∂ is used instead of d to distinguish partial derivatives from ordinary derivatives.

We define the partial derivative of f with respect to x in a similar way. Holding y constant, $y = y_0$, we write

$$\frac{\partial f}{\partial x}(x_0, y_0) = \lim_{h \to 0} \frac{f(x_0 + h, y_0) - f(x_0, y_0)}{h}. \tag{5}$$

If this limit exists, it gives the value at (x_0, y_0) of the function $\partial f/\partial x$, the partial derivative of f with respect to x. The quantity $(\partial f/\partial x)(x_0, y_0)$ is the rate of change of f with respect to x at (x_0, y_0), or the slope of the curve of intersection of the surface $z = f(x, y)$ and the plane $y = y_0$ parallel to the xz-plane (see Figure 16.2.2).

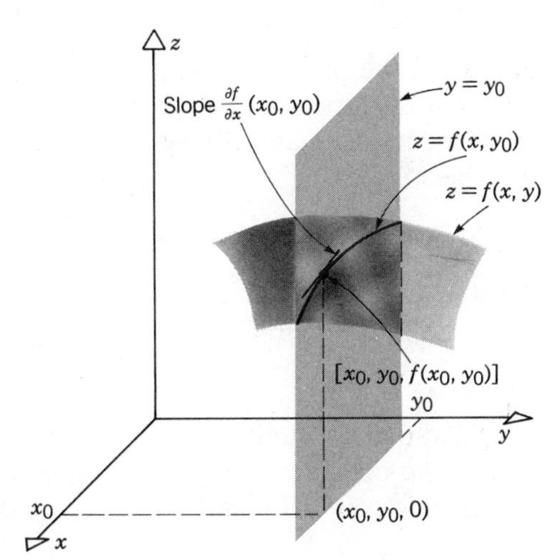

Figure 16.2.2

Since partial derivatives are defined just as are ordinary derivatives, except that only one variable is allowed to vary while the other is held constant, it follows that the computational rules we have developed remain valid for the calculation of partial derivatives, provided that we remember that only one variable is actually changing, and that the other remains fixed during the calculation. This is illustrated by the following example.

EXAMPLE 1

If $f(x, y) = x^4 + 2x^2y^2 + 3xy^3$, find the partial derivatives $\partial f/\partial x$ and $\partial f/\partial y$.

To compute $\partial f/\partial x$ we hold y constant and differentiate f with respect to x, using the power rule and the rule for differentiating sums. We obtain

$$\begin{aligned}\frac{\partial f}{\partial x}(x, y) &= \frac{\partial}{\partial x}(x^4) + \left[\frac{\partial}{\partial x}(2x^2)\right]y^2 + \left[\frac{\partial}{\partial x}(3x)\right]y^3 \\ &= 4x^3 + 4xy^2 + 3y^3. \qquad (6)\end{aligned}$$

Similarly, if we hold x constant and differentiate with respect to y, we find that

$$\frac{\partial f}{\partial y}(x, y) = 4x^2y + 9xy^2. \blacksquare \qquad (7)$$

Often a subscript notation or the D-notation together with a subscript is used to indicate a partial derivative. Thus

$$\frac{\partial f}{\partial x} = f_x = D_x f \quad \text{and} \quad \frac{\partial f}{\partial y} = f_y = D_y f. \qquad (8)$$

Hence the results of Example 1 can also be stated in the following ways.

$$D_x f(x, y) = f_x(x, y) = 4x^3 + 4xy^2 + 3y^3,$$
$$D_y f(x, y) = f_y(x, y) = 4x^2y + 9xy^2.$$

However, the prime symbol, as in $f'(x)$, is not used for partial derivatives.

If f is a function of more than two variables, the concept of a partial derivative is extended in a natural way. Suppose, for example, that f is a function of three variables,

$$w = f(x, y, z). \qquad (9)$$

Then

$$\frac{\partial f}{\partial x}(x_0, y_0, z_0) = \lim_{h \to 0} \frac{f(x_0 + h, y_0, z_0) - f(x_0, y_0, z_0)}{h}, \qquad (10)$$

provided that the limit exists, is the value of the partial derivative of f with respect to x at the point (x_0, y_0, z_0). This quantity is the rate of change of f with respect to x at this point. Similarly, the partial derivatives of f with respect to y and z at the point (x_0, y_0, z_0) are given by

$$\frac{\partial f}{\partial y}(x_0, y_0, z_0) = \lim_{k \to 0} \frac{f(x_0, y_0 + k, z_0) - f(x_0, y_0, z_0)}{k} \tag{11}$$

and

$$\frac{\partial f}{\partial z}(x_0, y_0, z_0) = \lim_{m \to 0} \frac{f(x_0, y_0, z_0 + m) - f(x_0, y_0, z_0)}{m}, \tag{12}$$

provided these limits exist. Observe that in each case only one independent variable is allowed to vary, while the other two are held constant. The definition of partial derivatives for functions of more than three variables is entirely analogous.

EXAMPLE 2

If $f(x, y, z) = (x^2 - 2y)(3x + z^2)$, find the partial derivatives of f with respect to x, y, and z, respectively.

To find $D_x f(x, y, z)$ we hold y and z constant and use the product rule to differentiate f with respect to x. We obtain

$$\begin{aligned} D_x f(x, y, z) &= (x^2 - 2y)D_x(3x + z^2) + (3x + z^2)D_x(x^2 - 2y) \\ &= (x^2 - 2y)(3) + (3x + z^2)(2x) \\ &= 9x^2 - 6y + 2xz^2. \end{aligned} \tag{13}$$

Similarly,

$$\begin{aligned} D_y f(x, y, z) &= (3x + z^2)D_y(x^2 - 2y) \\ &= (3x + z^2)(-2) = -2(3x + z^2), \end{aligned} \tag{14}$$

and

$$\begin{aligned} D_z f(x, y, z) &= (x^2 - 2y)D_z(3x + z^2) \\ &= (x^2 - 2y)(2z) = 2z(x^2 - 2y). \end{aligned} \tag{15}$$

Note that we could also have obtained these results by writing

$$f(x, y, z) = 3x^3 - 6xy + x^2z^2 - 2yz^2$$

and then proceeding much as in Example 1. ∎

EXAMPLE 3

Find $\partial z/\partial x$ and $\partial z/\partial y$ if

$$z = x^2 \sin(xy^2). \tag{16}$$

Holding y constant and using the product rule, we obtain

$$\frac{\partial z}{\partial x} = \left[\frac{\partial}{\partial x}(x^2)\right]\sin(xy^2) + x^2\left[\frac{\partial}{\partial x}\sin(xy^2)\right]. \tag{17}$$

Since y is held constant, the chain rule (Section 3.5) applied to $\sin(xy^2)$ yields

$$\frac{\partial z}{\partial x} = 2x\sin(xy^2) + x^2[\cos(xy^2)]\frac{\partial}{\partial x}(xy^2)$$

so

$$\frac{\partial z}{\partial x} = 2x\sin(xy^2) + x^2y^2\cos(xy^2). \tag{18}$$

Proceeding in much the same way, we have

$$\begin{aligned}\frac{\partial z}{\partial y} &= x^2\frac{\partial}{\partial y}[\sin(xy^2)] \\ &= x^2[\cos(xy^2)]\frac{\partial}{\partial y}(xy^2) \\ &= 2x^3y\cos(xy^2). \blacksquare \end{aligned} \tag{19}$$

It is sometimes convenient to indicate not only the variable with respect to which a partial derivative is taken, but also those that are held constant during the process. This can be done by writing

$$\left(\frac{\partial f}{\partial x}\right)_y$$

to denote the partial derivative with respect to x with y held constant. This notation is frequently seen in thermodynamics, as in the next example.

EXAMPLE 4

In the study of thermodynamics the relation

$$C_p - C_V = T\left(\frac{\partial p}{\partial T}\right)_V\left(\frac{\partial V}{\partial T}\right)_p \tag{20}$$

is derived. In this equation, p, V, and T refer to the pressure, volume, and temperature of a gas, while C_p and C_V are the specific heats of the gas at constant pressure and constant volume, respectively. The quantity $(\partial p/\partial T)_V$ means that p is a function of T and V and the derivative of p with respect to T with V held constant is to be calculated. Similarly $(\partial V/\partial T)_p$ means that V is a function of T and p and the derivative of V with respect to T with p held constant is to be calculated. For a perfect gas the equation of state is

$$pV = RT, \tag{21}$$

where R is a constant. Find an expression for $C_p - C_V$ in this case.

Holding V constant and differentiating Eq. 21 with respect to T, we obtain

$$\left(\frac{\partial p}{\partial T}\right)_V = \frac{R}{V}. \tag{22}$$

Similarly, holding p constant and differentiating Eq. 21 with respect to T, we have

$$\left(\frac{\partial V}{\partial T}\right)_p = \frac{R}{p}. \tag{23}$$

Substituting from Eqs. 22 and 23 into Eq. 20 and using Eq. 21, we find that

$$\begin{aligned} C_p - C_V &= T\frac{R}{V}\frac{R}{p} \\ &= R. \blacksquare \end{aligned}$$

Higher derivatives

Just as we can compute higher derivatives of a function of a single variable, so in general it is possible to calculate partial derivatives of higher than first order for functions of two or more variables. Thus, if $z = f(x, y)$, then the partial derivatives

$$\frac{\partial z}{\partial x} = \frac{\partial f}{\partial x}(x, y), \qquad \frac{\partial z}{\partial y} = \frac{\partial f}{\partial y}(x, y)$$

are also usually functions of both x and y, for instance, see Examples 1 and 3. It then makes sense to talk about derivatives of $\partial z/\partial x$ and $\partial z/\partial y$ with respect to either x or y. We write

$$\frac{\partial}{\partial x}\left(\frac{\partial z}{\partial x}\right) = \frac{\partial^2 z}{\partial x^2} = z_{xx} = D_x^2 z \quad \text{or} \quad \frac{\partial}{\partial x}\left(\frac{\partial f}{\partial x}\right) = \frac{\partial^2 f}{\partial x^2} = f_{xx} = D_x^2 f, \tag{24}$$

$$\frac{\partial}{\partial y}\left(\frac{\partial z}{\partial x}\right) = \frac{\partial^2 z}{\partial y \partial x} = z_{xy} = D_{xy}^2 z \quad \text{or} \quad \frac{\partial}{\partial y}\left(\frac{\partial f}{\partial x}\right) = \frac{\partial^2 f}{\partial y \partial x} = f_{xy} = D_{xy}^2 f, \tag{25}$$

$$\frac{\partial}{\partial x}\left(\frac{\partial z}{\partial y}\right) = \frac{\partial^2 z}{\partial x \partial y} = z_{yx} = D_{yx}^2 z \quad \text{or} \quad \frac{\partial}{\partial x}\left(\frac{\partial f}{\partial y}\right) = \frac{\partial^2 f}{\partial x \partial y} = f_{yx} = D_{yx}^2 f, \tag{26}$$

and

$$\frac{\partial}{\partial y}\left(\frac{\partial z}{\partial y}\right) = \frac{\partial^2 z}{\partial y^2} = z_{yy} = D_y^2 z \quad \text{or} \quad \frac{\partial}{\partial y}\left(\frac{\partial f}{\partial y}\right) = \frac{\partial^2 f}{\partial y^2} = f_{yy} = D_y^2 f. \tag{27}$$

Thus for a function f of two variables there are four second-order partial derivatives. Note that $\partial^2 f/\partial y \partial x$ is obtained by differentiating f first with respect to x, and then with respect to y, whereas $\partial^2 f/\partial x \partial y$ is obtained by performing the same operations in the opposite order. In both cases the order of differentiation is found by working from right to left. In expressions such as z_{xy}, however, the order of differentiation is indicated by the order of the subscripts, reading from left to right. One can think of this operation as though parentheses were present, $(z_x)_y$, in which case it is natural to read z_{xy} as the function z_x being differentiated with respect to y.

In a similar way it is possible to compute derivatives of higher than second order. For example, $f_{xyy} = \partial^3 f/\partial y^2 \partial x$ is obtained by differentiating f once with respect to x and then twice with respect to y. The calculation of partial derivatives of second and higher order can be extended to functions of more than two variables in a natural way.

EXAMPLE 5

If $f(x, y) = x^4 + 2x^2y^2 + 3xy^3$, find the four second partial derivatives of f.

This is the same function that we considered in Example 1; its first partial derivatives are given by Eqs. 6 and 7. Upon differentiating Eq. 6 with respect to x and y in turn, we obtain

$$\frac{\partial^2 f}{\partial x^2}(x, y) = f_{xx}(x, y) = 12x^2 + 4y^2,$$

$$\frac{\partial^2 f}{\partial y \partial x}(x, y) = f_{xy}(x, y) = 8xy + 9y^2.$$

Similarly, by differentiating Eq. 7 with respect to x and y, we find that

$$\frac{\partial^2 f}{\partial x \partial y}(x, y) = f_{yx}(x, y) = 8xy + 9y^2,$$

$$\frac{\partial^2 f}{\partial y^2}(x, y) = f_{yy}(x, y) = 4x^2 + 18xy. \blacksquare$$

EXAMPLE 6

If $z = x^2 \sin(xy^2)$, as in Example 3, find $D^2_{xy}z$ and $D^2_{yx}z$.

We have previously computed $D_x z$ and $D_y z$; they are given by Eqs. 18 and 19, respectively. To find the required second derivatives we must differentiate $D_x z$ with respect to y and $D_y z$ with respect to x. We obtain

$$\begin{aligned} D^2_{xy}z &= 2xD_y[\sin(xy^2)] + x^2[D_y(y^2)]\cos(xy^2) + x^2y^2D_y[\cos(xy^2)] \\ &= (2x)2xy\cos(xy^2) + x^2(2y)\cos(xy^2) + (x^2y^2)(2xy)[-\sin(xy^2)] \\ &= 6x^2y\cos(xy^2) - 2x^3y^3\sin(xy^2). \end{aligned}$$

Similarly, by differentiating Eq. 19 with respect to x we obtain $D^2_{yx}z$:

$$\begin{aligned} D^2_{yx}z &= [D_x(2x^3y)]\cos(xy^2) + 2x^3yD_x[\cos(xy^2)] \\ &= 6x^2y\cos(xy^2) + 2x^3y(y^2)[-\sin(xy^2)] \\ &= 6x^2y\cos(xy^2) - 2x^3y^3\sin(xy^2). \blacksquare \end{aligned}$$

Note that in Example 5

$$\frac{\partial^2 f}{\partial y \partial x}(x, y) = \frac{\partial^2 f}{\partial x \partial y}(x, y)$$

and that in Example 6

$$D_{xy}^2 z = D_{yx}^2 z.$$

In other words, in these examples the mixed second partial derivatives are equal, independent of the order in which the derivatives were taken. In fact, this is almost always the case. However, there are rare situations in which different results are obtained if the order of differentiation is reversed.

It is frequently desirable to interchange the order of differentiation, and to make sure that this is permissible we must impose continuity conditions on the function f and its partial derivatives. Sufficient conditions are given in the following theorem.

Theorem 16.2.1

Consider a function of two variables, $z = f(x, y)$. If f, f_x, f_y, f_{xy}, and f_{yx} are continuous in some neighborhood of a point (x_0, y_0), then

$$f_{xy}(x_0, y_0) = f_{yx}(x_0, y_0). \tag{28}$$

We will omit the proof of this theorem. For a function of three variables the generalization of Theorem 16.2.1 assumes the continuity of f, the three first partial derivatives, and the mixed second partial derivatives. In the rest of this book we assume that the order of differentiation can be changed at will.

In the study of mathematical physics it is frequently important to investigate the solutions of certain partial differential equations, that is, equations containing partial derivatives of the unknown function. A particularly important equation is the *wave equation*,

$$u_{tt} = c^2 u_{xx}, \tag{29}$$

which is fundamental in the study of phenomena involving waves. The positive constant c can be identified as the speed of wave propagation. For example, $u(x, t)$ could denote the transverse displacement of a vibrating string, such as a violin string (see Figure 16.2.3); then c is the velocity at which waves move along the string. While a systematic discussion of partial differential equations is far too difficult to give here, it is often relatively simple to determine whether a given function satisfies a given partial differential equation.

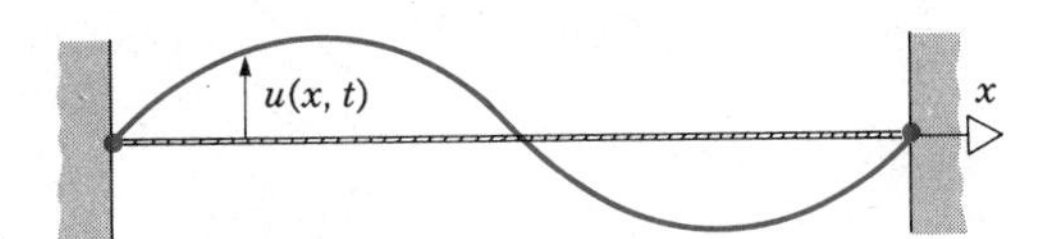

Figure 16.2.3 A vibrating string.

EXAMPLE 7

Show that the function

$$u(x, t) = \sin 2x \cos 2ct \tag{30}$$

is a solution of the wave equation (29).

To verify that $u(x, t)$ satisfies Eq. 29 we need only compute u_{xx} and u_{tt}. We find that

$$u_x(x, t) = 2 \cos 2x \cos 2ct,$$

$$u_{xx}(x, t) = -4 \sin 2x \cos 2ct.$$

Similarly,

$$u_t(x, t) = -2c \sin 2x \sin 2ct,$$

$$u_{tt}(x, t) = -4c^2 \sin 2x \cos 2ct.$$

Hence $u_{tt} = c^2 u_{xx}$, as was to be shown. ∎

PROBLEMS

In each of Problems 1 through 18, find the indicated partial derivatives.

1. $z = 2x^2 - 3xy + y^2$; $\quad z_x, z_y$
2. $z = (x^2 - y^2)(x^2 + 2y^2)$; $\quad \dfrac{\partial z}{\partial x}, \dfrac{\partial z}{\partial y}$
3. $f(x, y) = \dfrac{2xy}{x^2 + y^2 + 1}$; $\quad f_x(x, y)$
4. $w = (2u - v)^2(u + 3v)^3$; $\quad \dfrac{\partial w}{\partial v}$
5. $f(x, y) = x^3 + 4x^2y - 6xy^2 + y^3$; $\quad D_x f(x, y)$, $D_y f(x, y)$
6. $f(x, y) = \sin\sqrt{x^2 + y^2}$; $\quad f_y(x, y)$
7. $u = rs \cos(r - s)$; $\quad D_r u, D_s u$
8. $z = [(2x - y)(x^2 + 2y^2)]^{1/3}$; $\quad \dfrac{\partial z}{\partial y}$
9. $f(x, y) = x^2y^2 \sin(x^2y^2)$; $\quad f_x(x, y)$
10. $w = (xy - 2yz + 3xz)^2$; $\quad \dfrac{\partial w}{\partial x}, \dfrac{\partial w}{\partial z}$
11. $f(x, y, z) = \dfrac{(x - 2y + z)^2}{x^2 + y^2 + z^2}$; $\quad f_y(x, y, z)$
12. $f(x, y, z) = xy \ln(xyz)$; $\quad f_y(x, y, z)$, $f_z(x, y, z)$
13. $f(x, y, z) = (x^2 + y^2 + z^2)^{-1/2}$; $\quad f_x(x, y, z)$
14. $z = e^{x+y^2}$; $\quad z_x, z_y$
15. $z = \dfrac{\ln(x^2 + y^2)}{\sqrt{x^2 + y^2}}$; $\quad z_x, z_y$
16. $z = x^y$, $\quad x > 0$; $\quad z_x, z_y$
17. $z = \arctan \dfrac{y}{x}$; $\quad z_x, z_y$
18. $w = \tan(x^2 + y^3 + z^4)$; $\quad w_x, w_z$

In each of Problems 19 through 22, verify that $z_{xy} = z_{yx}$.

19. $z = 2x^2 - 3xy + y^2$
20. $z = (x^2 + xy + y^2)^{1/2}$
21. $z = x^2y^2 \sin(x^2y^2)$
22. $z = \sin(x + y) + \cos(x - y)$
23. If $z = \ln(x^2 + y^2)^{1/2}$, show that
$$xz_x + yz_y = 1.$$
24. If $w = f(x, y) + g(y, z) + h(x, z)$, show that $\partial^3 w/\partial x \partial y \partial z = 0$.

In each of Problems 25 through 28, verify that the given function satisfies the given partial differential equation.

25. $u(x, t) = \cos 2x \sin 4t$; $\quad 4u_{xx} = u_{tt}$

26. $u(x, y) = \dfrac{x}{x^2 + y^2}$; $\quad u_{xx} + u_{yy} = 0$, $(x, y) \neq (0, 0)$

27. $u(x, y, z) = \dfrac{1}{\sqrt{x^2 + y^2 + z^2}}$; $\quad u_{xx} + u_{yy} + u_{zz} = 0$, $\quad (x, y, z) \neq (0, 0, 0)$

28. $u(r, \theta) = r^n \cos n\theta$; $\quad u_{rr} + \dfrac{1}{r} u_r + \dfrac{1}{r^2} u_{\theta\theta} = 0$, $r \neq 0$

29. Let $w = f(x, y, z)$.

(a) How many second partial derivatives does w have? Assuming that the order of differentiation is immaterial, how many distinct second derivatives are there?

(b) How many partial derivatives of order n does w have? Assuming that the order of differentiation is immaterial, how many distinct derivatives of order n are there?

* 30. For air the equation of state is more accurately given by

$$pV = RT + p(b - cT^n) \qquad (i)$$

than by $pV = RT$. In Eq. (i), known as Callendar's equation, the quantities b, c, and n are constants to be adjusted according to such factors as the humidity of the air. Proceeding as in Example 4, find an expression for $C_p - C_V$.

In each of Problems 31 through 34, show that

$$u_x(x, y) = v_y(x, y) \quad \text{and} \quad u_y(x, y) = -v_x(x, y).$$

These equations are called the Cauchy–Riemann equations and are very important in the study of functions of a complex variable.

31. $u(x, y) = x^2 - y^2$, $\quad v(x, y) = 2xy$

32. $u(x, y) = e^{nx} \cos ny$, $\quad v(x, y) = e^{nx} \sin ny$

33. $u(x, y) = \dfrac{x}{x^2 + y^2}$, $\quad v(x, y) = -\dfrac{y}{x^2 + y^2}$

34. $u(x, y) = \frac{1}{2} \ln(x^2 + y^2)$, $\quad v(x, y) = \arctan \dfrac{y}{x}$

In each of Problems 35 through 40 determine a function $z = f(x, y)$ with the required property.

35. $z_x = \dfrac{1}{y}$

36. $z_y = \dfrac{1}{y} + 2 \sin x$

37. $z_y = e^{xy} + \ln x + yx^2$

38. $z_x = e^{x+y} + x^2y$

39. $z_x = (xy)^2 + x^y$

40. $z_y = x \sin\left(\dfrac{y}{x}\right) + \cos(x - y)$

16.3 DIFFERENTIABLE FUNCTIONS AND THE CHAIN RULE

Our next goal is to develop a statement of what it means for a function of two variables to be differentiable. This turns out to be not quite so simple as we might expect. For instance, we might think that if $f(x, y)$ has partial derivatives f_x and f_y at a point (x_0, y_0), then we should say that f is differentiable there. However, this would be unsatisfactory, for several reasons. One is that to preserve the relation between differentiability and continuity that prevails for functions of one variable, we want a differentiable function of two variables to be continuous. On the other hand, examples such as the one in Problem 39 show that a function $f(x, y)$ may have partial derivatives at a point (x_0, y_0) and yet fail to be continuous there. Consequently, the concept of differentiability for functions of two variables must involve more than the existence of partial derivatives.

The key lies in the concept of a locally linear function, as discussed in Section 3.4 for functions of one variable. There we said that a continuous function f is locally linear at x_0 if there exist constants A and B such that

$$f(x_0 + h) = A + Bh + r(h, x_0)h, \qquad (1)$$

where

$$r(h, x_0) \to 0 \quad \text{as} \quad h \to 0. \tag{2}$$

It follows that f is also differentiable at x_0 and that the constants A and B are given by

$$A = f(x_0), \qquad B = f'(x_0). \tag{3}$$

We also showed in Section 3.4 that if f is differentiable at x_0, that is, if

$$f'(x_0) = \lim_{h \to 0} \frac{f(x_0 + h) - f(x_0)}{h} \tag{4}$$

exists, then f satisfies

$$f(x_0 + h) = f(x_0) + f'(x_0)h + r(h, x_0)h, \tag{5}$$

where

$$\lim_{h \to 0} r(h, x_0) = 0. \tag{6}$$

Consequently, for a function of one variable a differentiable function is locally linear and vice versa. Thus differentiability and local linearity are equivalent properties in the sense that a function having one also has the other. This equivalence provides a means of extending the concept of differentiability to functions of two variables. Instead of trying to generalize Eq. 4, the limit of a difference quotient, let us instead extend the concept of differentiability to functions of two variables by generalizing the statement of local linearity.

Let $z = f(x, y)$ be continuous at (x_0, y_0). Then f is said to be **locally linear** at (x_0, y_0) if

$$f(x_0 + h, y_0 + k) = A + Bh + Ck + r(x_0, y_0, h, k)\sqrt{h^2 + k^2}\,, \tag{7}$$

where A, B, and C are constants, $r \to 0$ as $(h, k) \to (0, 0)$, and, as usual, $h = x - x_0$ and $k = y - y_0$. The constants A, B, and C can be determined as follows. If we let $(h, k) \to (0, 0)$ in Eq. 7 and use the fact that f is continuous at (x_0, y_0), then

$$A = \lim_{(h,k) \to (0,0)} f(x_0 + h, y_0 + k) = f(x_0, y_0). \tag{8}$$

Next we set $k = 0$ in Eq. 7, substitute for A from Eq. 8, solve for B, and let $h \to 0$:

$$\begin{aligned} B &= \lim_{h \to 0} \left[\frac{f(x_0 + h, y_0) - f(x_0, y_0)}{h} - r(x_0, y_0, h, 0)\frac{|h|}{h} \right] \\ &= f_x(x_0, y_0). \end{aligned} \tag{9}$$

In a similar manner we find that

$$C = f_y(x_0, y_0). \tag{10}$$

Thus, if $z = f(x, y)$ is continuous and locally linear at (x_0, y_0), then $f_x(x_0, y_0)$ and

$f_y(x_0, y_0)$ must exist. Thus Eq. 7 becomes

$$f(x_0 + h, y_0 + k) = f(x_0, y_0) + f_x(x_0, y_0)h + f_y(x_0, y_0)k + r(x_0, y_0, h, k)\sqrt{h^2 + k^2}, \quad (11)$$

where

$$r(x_0, y_0, h, k) \to 0 \quad \text{as} \quad (h, k) \to (0, 0). \quad (12)$$

The argument that we have just given makes plausible the following definition of differentiability for a function of two variables. Observe that this definition says nothing about the possible continuity of the function.

DEFINITION 16.3.1 A function f of two variables, defined in some neighborhood of (x_0, y_0), is said to be differentiable at (x_0, y_0) if it satisfies Eqs. 11 and 12.

The relation between differentiability and continuity is an immediate consequence of this definition.

Theorem 16.3.1
If f is differentiable at (x_0, y_0), then f is continuous at (x_0, y_0).

Proof. To prove this result we simply take the limit of Eq. 11 as h and k approach zero; we obtain

$$\lim_{(h,k)\to(0,0)} f(x_0 + h, y_0 + k) = f(x_0, y_0),$$

which means that f is continuous at (x_0, y_0). □

We have already pointed out that the mere existence of the partial derivatives f_x and f_y is not enough to guarantee that a function $f(x, y)$ of two variables is differentiable. However, if the partial derivatives f_x and f_y are continuous, then the function *is* certain to be differentiable. This is a useful result because it may well be simpler to determine that f_x and f_y are continuous than to calculate the function r in Eq. 11, and to ascertain its limit as $(h, k) \to (0, 0)$.

Theorem 16.3.2
If f is a function of two variables, if f_x and f_y exist in some neighborhood of (x_0, y_0), and if f_x and f_y are continuous at (x_0, y_0), then f satisfies Eqs. 11 and 12; hence it is differentiable at (x_0, y_0).

The proof of this theorem is outlined in Problem 40.

The differential

Returning to Eq. 11 and following the pattern established in Section 3.4, we define the **differential** of f, denoted by df, or perhaps by dz, to be the terms that are linear in the increments h and k. Thus, at an arbitary point (x, y),

$$df = dz = f_x(x, y)\,dx + f_y(x, y)\,dy, \tag{13}$$

where we have followed the usual practice of replacing h by dx and k by dy. We regard df as an approximation to the actual increment $\Delta f = f(x + h, y + k) - f(x, y)$. Indeed

$$\Delta f - df = r(x, y, h, k)\sqrt{h^2 + k^2} \to 0 \tag{14}$$

as $(h, k) \to (0, 0)$. We will see in Section 16.5 that df corresponds to approximating the surface $z = f(x, y)$ by the tangent plane at the point $[x, y, f(x, y)]$.

EXAMPLE 1

Let

$$z = f(x, y) = 2x^2 - 3xy + y^2. \tag{15}$$

At an arbitrary point (x, y) find the increment Δz and the differential dz. Also find r and show that f is differentiable by showing that $r \to 0$ as $(h, k) \to (0, 0)$.

We have

$$\begin{aligned} f(x + h, y + k) &= 2(x + h)^2 - 3(x + h)(y + k) + (y + k)^2 \\ &= 2x^2 - 3xy + y^2 + (4x - 3y)h + (-3x + 2y)k \\ &\qquad + 2h^2 - 3hk + k^2, \end{aligned}$$

so

$$\begin{aligned} \Delta z &= f(x + h, y + k) - f(x, y) \\ &= (4x - 3y)h + (-3x + 2y)k + 2h^2 - 3hk + k^2. \end{aligned} \tag{16}$$

Also,

$$f_x(x, y) = 4x - 3y, \qquad f_y(x, y) = -3x + 2y,$$

so

$$dz = (4x - 3y)\,dx + (-3x + 2y)\,dy. \tag{17}$$

Thus

$$\Delta z - dz = 2h^2 - 3hk + k^2 = \frac{2h^2 - 3hk + k^2}{\sqrt{h^2 + k^2}}\sqrt{h^2 + k^2},$$

so

$$r(x, y, h, k) = \frac{2h^2 - 3hk + k^2}{\sqrt{h^2 + k^2}}. \tag{18}$$

Note that in this case r is independent of x and y. To show that $r \to 0$ as $(h, k) \to (0, 0)$ it is helpful to introduce the polar coordinates ρ and θ so that $h = \rho \cos \theta$, $k = \rho \sin \theta$. Then

$$r = \rho(2 \cos^2 \theta - 3 \cos \theta \sin \theta + \sin^2 \theta).$$

As $(h, k) \to (0, 0)$, it follows that $\rho = \sqrt{h^2 + k^2} \to 0$ and consequently $r \to 0$ as well. Thus f is differentiable at each point (x, y). Of course, this also follows from Theorem 16.3.2, since f_x and f_y are continuous everywhere. ∎

The definition of differentiability that we have given for functions of two variables generalizes immediately to functions of three or more variables. Similarly, Theorems 16.3.1 and 16.3.2 and the concept of the differential also have direct extensions to functions of more than two variables.

The chain rule

We turn now to the development of the chain rule for the partial derivatives of composite functions. For example, suppose that $w = f(x, y)$, where

$$x = g(r, s), \qquad y = h(r, s),$$

so that $w = F(r, s) = f[g(r, s), h(r, s)]$. We would like to derive formulas for $\partial w/\partial r$ and $\partial w/\partial s$ involving the original functions f, g, and h without actually calculating the function F. We assume that the functions f, g, and h are differentiable. In this derivation we will use Δ to denote an increment in any of the variables that appear.

To calculate $\partial w/\partial r$ we must calculate Δw corresponding to an increment Δr and then determine the limit of $\Delta w/\Delta r$ as $\Delta r \to 0$. First notice that, corresponding to the increment Δr, there are increments

$$\Delta x = g(r + \Delta r, s) - g(r, s), \qquad \Delta y = h(r + \Delta r, s) - h(r, s).$$

Since the functions g and h are differentiable, we can write

$$\Delta x = g_r(r, s)\, \Delta r + \eta_1\, \Delta r, \qquad \Delta y = h_r(r, s)\, \Delta r + \eta_2\, \Delta r, \tag{19}$$

where $\eta_1 \to 0$ and $\eta_2 \to 0$ as $\Delta r \to 0$. Corresponding to the increment Δr there is also an increment Δw given by

$$\begin{aligned} \Delta w &= F(r + \Delta r, s) - F(r, s) \\ &= f(x + \Delta x, y + \Delta y) - f(x, y). \end{aligned}$$

Since f is also a differentiable function we have

$$\Delta w = f_x(x, y)\, \Delta x + f_y(x, y)\, \Delta y + \eta_3\, (x, y, \Delta x, \Delta y)\sqrt{(\Delta x)^2 + (\Delta y)^2}, \tag{20}$$

where $\eta_3 \to 0$ as $(\Delta x, \Delta y) \to (0, 0)$. Next we substitute from Eqs. 19 into Eq. 20. Leaving out the independent variables in the various functions for brevity, we obtain in this way

$$\begin{aligned} \Delta w = (f_x g_r + f_y h_r)\, \Delta r + f_x \eta_1\, \Delta r + f_y \eta_2\, \Delta r \\ + \eta_3[(g_r + \eta_1)^2 + (h_r + \eta_2)^2]^{1/2}\, |\Delta r|. \end{aligned} \tag{21}$$

Dividing by Δr and taking the limit as $\Delta r \to 0$, we finally obtain

$$\frac{\partial w}{\partial r} = \lim_{\Delta r \to 0} \frac{\Delta w}{\Delta r} = f_x(x, y)g_r(r, s) + f_y(x, y)h_r(r, s). \tag{22}$$

The result is more easily remembered if we write it in the form

$$\frac{\partial w}{\partial r} = \frac{\partial w}{\partial x}\frac{\partial x}{\partial r} + \frac{\partial w}{\partial y}\frac{\partial y}{\partial r}. \tag{23}$$

In a similar way we obtain

$$\frac{\partial w}{\partial s} = f_x(x, y)g_s(r, s) + f_y(x, y)h_s(r, s) = \frac{\partial w}{\partial x}\frac{\partial x}{\partial s} + \frac{\partial w}{\partial y}\frac{\partial y}{\partial s}. \tag{24}$$

The chain rule formulas (22), (23), and (24) can be readily generalized to any combination of variables: w may be a function of one, two, or more variables, each of which is a function of one, two, or more variables. Once one understands the pattern by which the derivatives $\partial w/\partial r$ and $\partial w/\partial s$ are found from Eqs. 22, 23, and 24, it should not be difficult to extend the process to other situations.

EXAMPLE 2

Let $z = 3x^2y^3$, where $x = 2r^2 - s^2$, and $y = 3rs$. Find $\partial z/\partial r$ and $\partial z/\partial s$.

By the chain rule we obtain

$$\frac{\partial z}{\partial r} = \frac{\partial z}{\partial x}\frac{\partial x}{\partial r} + \frac{\partial z}{\partial y}\frac{\partial y}{\partial r} = (6xy^3)(4r) + (9x^2y^2)(3s). \tag{25}$$

In a similar way

$$\frac{\partial z}{\partial s} = \frac{\partial z}{\partial x}\frac{\partial x}{\partial s} + \frac{\partial z}{\partial y}\frac{\partial y}{\partial s} = (6xy^3)(-2s) + (9x^2y^2)(3r). \tag{26}$$

The results can be expressed entirely in terms of r and s by substituting $2r^2 - s^2$ for x and $3rs$ for y in Eqs. 25 and 26. ∎

EXAMPLE 3

Suppose that $z = 2x^2 - xy - y^2$, where $x = r\cos\theta$ and $y = r\sin\theta$. Find $\partial z/\partial\theta$ and evaluate it when $r = 2$ and $\theta = \pi/3$.

From the chain rule

$$\frac{\partial z}{\partial \theta} = \frac{\partial z}{\partial x}\frac{\partial x}{\partial \theta} + \frac{\partial z}{\partial y}\frac{\partial y}{\partial \theta}$$

$$= (4x - y)(-r\sin\theta) - (x + 2y)(r\cos\theta). \tag{27}$$

From $r = 2$ and $\theta = \pi/3$, it follows that $x = 1$ and $y = \sqrt{3}$. Hence

$$\frac{\partial z}{\partial \theta} = (4 - \sqrt{3})(-\sqrt{3}) - (1 + 2\sqrt{3})(1) = 2 - 6\sqrt{3} \cong -8.392. \blacksquare$$

EXAMPLE 4

Suppose that $w = f(x, y, z)$ and

$$x = \rho \sin \phi \cos \theta, \qquad y = \rho \sin \phi \sin \theta, \qquad z = \rho \cos \phi.$$

Determine $\partial w/\partial \phi$ at $\rho = 2$, $\phi = \pi/6$, $\theta = \pi/4$ if $f_x(\sqrt{2}/2, \sqrt{2}/2, \sqrt{3}) = -1$, $f_y(\sqrt{2}/2, \sqrt{2}/2, \sqrt{3}) = 3$, and $f_z(\sqrt{2}/2, \sqrt{2}/2, \sqrt{3}) = 2$.

We have $w = F(\rho, \phi, \theta) = f(\rho \sin \phi \cos \theta, \rho \sin \phi \sin \theta, \rho \cos \phi)$, so

$$\begin{aligned}
\frac{\partial w}{\partial \phi} = \frac{\partial F}{\partial \phi} &= \frac{\partial f}{\partial x}\frac{\partial x}{\partial \phi} + \frac{\partial f}{\partial y}\frac{\partial y}{\partial \phi} + \frac{\partial f}{\partial z}\frac{\partial z}{\partial \phi} \\
&= \frac{\partial f}{\partial x}\frac{\partial}{\partial \phi}(\rho \sin \phi \cos \theta) + \frac{\partial f}{\partial y}\frac{\partial}{\partial \phi}(\rho \sin \phi \sin \theta) + \frac{\partial f}{\partial z}\frac{\partial}{\partial \phi}(\rho \cos \phi) \\
&= \frac{\partial f}{\partial x}(\rho \cos \phi \cos \theta) + \frac{\partial f}{\partial y}(\rho \cos \phi \sin \theta) + \frac{\partial f}{\partial z}(-\rho \sin \phi).
\end{aligned}$$

Since $x = \sqrt{2}/2$, $y = \sqrt{2}/2$, $z = \sqrt{3}$ corresponds to $\rho = 2$, $\phi = \pi/6$, $\theta = \pi/4$, we obtain

$$\begin{aligned}
\frac{\partial w}{\partial \phi} &= (-1)\left(2 \cdot \frac{\sqrt{3}}{2} \cdot \frac{\sqrt{2}}{2}\right) + 3\left(2 \cdot \frac{\sqrt{3}}{2} \cdot \frac{\sqrt{2}}{2}\right) + 2\left(-2 \cdot \frac{1}{2}\right) \\
&= -\frac{\sqrt{6}}{2} + 3\frac{\sqrt{6}}{2} - 2 \\
&= \sqrt{6} - 2 \cong 0.4495 \quad \text{at} \quad \rho = 2, \phi = \frac{\pi}{6}, \theta = \frac{\pi}{4}. \blacksquare
\end{aligned}$$

EXAMPLE 5

Show that if f is a differentiable function, then $u = f(x - ct)$ is a solution of the partial differential equation

$$\frac{\partial u}{\partial t} + c\frac{\partial u}{\partial x} = 0,$$

where c is a constant.

Let $\xi = x - ct$. Then $u = f(\xi)$ is a function of one variable ξ, which in turn is a function of the variables x and t. We have

$$\frac{\partial u}{\partial x} = \frac{du}{d\xi}\frac{\partial \xi}{\partial x} = f'(\xi)\,(1),$$

$$\frac{\partial u}{\partial t} = \frac{du}{d\xi}\frac{\partial \xi}{\partial t} = f'(\xi)(-c).$$

Adding c times the first equation to the second equation gives

$$\frac{\partial u}{\partial t} + c\frac{\partial u}{\partial x} = -cf'(\xi) + cf'(\xi) = 0. \blacksquare$$

Second and higher derivatives of composite functions can be calculated by repeated application of the chain rule. The following examples are typical. The main problem is to organize one's work so that no term is inadvertently omitted.

EXAMPLE 6

If $z = 3x^2y^3$, where $x = 2r^2 - s^2$ and $y = 3rs$, find $\partial^2 z/\partial s^2$.

This is a continuation of Example 2. From Eq. 26 we have

$$\frac{\partial z}{\partial s} = -12xy^3 s + 27x^2y^2r; \tag{28}$$

therefore our task is to differentiate Eq. 28 with respect to s. Observe that in the first term on the right side of Eq. 28 the variable s appears explicitly, and in both terms it is present through the dependence of x and y on s. We start by using the rules for differentiating sums and products; thus

$$\frac{\partial^2 z}{\partial s^2} = -12\frac{\partial x}{\partial s}y^3 s - 12x\left[\frac{\partial}{\partial s}(y^3)\right]s - 12xy^3\frac{\partial s}{\partial s}$$
$$+ 27\left[\frac{\partial}{\partial s}(x^2)\right]y^2r + 27x^2\left[\frac{\partial}{\partial s}(y^2)\right]r. \tag{29}$$

Next we use the chain rule to calculate

$$\frac{\partial}{\partial s}(y^3) = 3y^2\frac{\partial y}{\partial s} = (3y^2)(3r) = 9y^2r.$$

$$\frac{\partial}{\partial s}(x^2) = 2x\frac{\partial x}{\partial s} = (2x)(-2s) = -4xs.$$

$$\frac{\partial}{\partial s}(y^2) = 2y\frac{\partial y}{\partial s} = (2y)(3r) = 6yr.$$

Substituting these expressions in Eq. 29 and using the additional fact that

$\partial s/\partial s = 1$, we finally obtain

$$\frac{\partial^2 z}{\partial s^2} = 24y^3s^2 - 108xy^2rs - 12xy^3 - 108xy^2rs + 162x^2yr^2. \tag{30}$$

Again, we can express the result entirely in terms of r and s (if this is wanted) by substituting the appropriate expressions for x and y in Eq. 30. ∎

EXAMPLE 7

Show that

$$u(x, t) = f(x - ct) + g(x + ct) \tag{31}$$

is a solution of the wave equation

$$u_{tt} = c^2 u_{xx} \tag{32}$$

for arbitrary functions f and g that have continuous second derivatives.

Let $\xi = x - ct$ and $\eta = x + ct$. Then $u = F(\xi, \eta) = f(\xi) + g(\eta)$. We have

$$\frac{\partial u}{\partial t} = \frac{\partial u}{\partial \xi}\frac{\partial \xi}{\partial t} + \frac{\partial u}{\partial \eta}\frac{\partial \eta}{\partial t} = f'(\xi)(-c) + g'(\eta)c.$$

Applying the chain rule again to the function $\partial u/\partial t$, we obtain

$$\begin{aligned}\frac{\partial^2 u}{\partial t^2} &= \left[\frac{\partial}{\partial \xi}\left(\frac{\partial u}{\partial t}\right)\right]\frac{\partial \xi}{\partial t} + \left[\frac{\partial}{\partial \eta}\left(\frac{\partial u}{\partial t}\right)\right]\frac{\partial \eta}{\partial t} = [-cf''(\xi)](-c) + [cg''(\eta)]c \\ &= c^2[f''(\xi) + g''(\eta)].\end{aligned} \tag{33}$$

Similarly

$$\frac{\partial u}{\partial x} = \frac{\partial u}{\partial \xi}\frac{\partial \xi}{\partial x} + \frac{\partial u}{\partial \eta}\frac{\partial \eta}{\partial x} = f'(\xi)(1) + g'(\eta)\,(1),$$

and

$$\frac{\partial^2 u}{\partial x^2} = \left[\frac{\partial}{\partial \xi}\left(\frac{\partial u}{\partial x}\right)\right]\frac{\partial \xi}{\partial x} + \left[\frac{\partial}{\partial \eta}\left(\frac{\partial u}{\partial x}\right)\right]\frac{\partial \eta}{\partial x} = f''(\xi) + g''(\eta). \tag{34}$$

Substituting for u_{tt} and u_{xx} from Eqs. 33 and 34, respectively, in Eq. 32 gives the desired result. ∎

PROBLEMS

In each of Problems 1 through 6, express $f(x + h, y + k)$ in the form of Eq. 11. Determine $r(x, y, h, k)$ and show that $r \to 0$ as $(h, k) \to (0, 0)$. In Problems 5 and 6 a Taylor expansion may be helpful.

1. $f(x, y) = x^2 + 3xy$

2. $f(x, y) = 2x^2 + 3y^2$

3. $f(x, y) = xy^2$

4. $f(x, y) = x^4 + y^4$

5. $f(x, y) = x \sin y$

6. $f(x, y) = y^2 e^x$

In each of Problems 7 through 14, calculate the differential of the given function.

7. $z = \ln(3x^2 - y + 2)$

8. $z = e^x \sin y$

9. $z = x^y, \quad x > 0$

10. $z = e^{1+xy}$

11. $z = \arctan \dfrac{y}{x}$

12. $w = z \tan xy^2$

13. $w = x \operatorname{cox} 2y + 3yz^{1/2}$

14. $w = \sin x \cos 2y \sin 3z$

15. Find the differential dV, where $V = \pi r^2 h/3$ is the volume of a cone of radius r and height h.

16. Find the differential dV, where $V = \pi(R^2 + rR + r^2)h/3$ is the volume of a frustrum of a cone of height h and radii r and R of the two bases.

17. If $r = (x^2 + y^2 + z^2)^{1/2}$, show that

$$xd\left(\frac{x}{r}\right) + yd\left(\frac{y}{r}\right) + zd\left(\frac{z}{r}\right) = 0, \qquad r \neq 0.$$

18. If

$$w = x_1^{p_1} x_2^{p_2} \cdots x_n^{p_n},$$

show that

$$\frac{dw}{w} = p_1 \frac{dx_1}{x_1} + p_2 \frac{dx_2}{x_2} + \cdots + p_n \frac{dx_n}{x_n}.$$

Hence the relative change in w is the (proportionate) sum of the relative changes in the x's.

In each of Problems 19 through 24, use the chain rule to find dz/dt or dw/dt.

19. $z = 3x^2 + 2xy^3, \quad x = \sqrt{t}, \quad y = t^2$

20. $z = \sin(2x - 3y), \quad x = t + t^{-1}, \quad y = t - t^{-1}$

21. $z = e^{1+xy}, \quad x = t^2, \quad y = \sin t$

22. $z = \ln(2y + \sqrt{1 - x^2}), \quad x = t^2, \quad y = t^4$

23. $w = 2x^2 + 3y^3 - 6xyz, \quad x = t^{-1}, \quad y = \sqrt{t}, \quad z = t^{1/3}$

24. $w = xe^{2y} \cos 4z, \quad x = \sqrt{1 - t}, \quad y = t, \quad z = 1 - t^2$

In each of Problems 25 through 28, use the chain rule to find the indicated partial derivatives.

25. $z = 6xy^2, x = r^2 - s^2, y = 2rs; \quad \dfrac{\partial z}{\partial r}, \quad \dfrac{\partial z}{\partial s}$

26. $z = \ln(3x^2 + 4y^2), x = r + s, y = r - s; \quad \dfrac{\partial z}{\partial r}, \quad \dfrac{\partial z}{\partial s}$

27. $z = \arctan(x^2 + y^2), x = e^s \cos t, y = e^s \sin t; \quad \dfrac{\partial z}{\partial s}, \quad \dfrac{\partial z}{\partial t}$

28. $w = (x^2 + y^2 + z^2)^{1/2}, x = e^r \cos s, y = e^r \sin s, z = sr; \quad \dfrac{\partial w}{\partial r}, \quad \dfrac{\partial w}{\partial s}$

29. If $w = f(x - y, y - x)$, show that $w_x + w_y = 0$.

30. If $w = f\left(\dfrac{x}{y}, \dfrac{y}{x}\right)$, show that $xw_x + yw_y = 0$.

31. If $w = f\left(\dfrac{x}{z}, \dfrac{y}{z}\right)$, show that $xw_x + yw_y + zw_z = 0$.

32. If $w = f(x^2 - y^2, y^2 - x^2)$, show that $yw_x + xw_y = 0$.

33. Let $u = f(x, y)$ and $x = r \cos \theta$, $y = r \sin \theta$. Calculate $\partial u/\partial r$ and $\partial u/\partial \theta$ and show that

(a) $\dfrac{\partial u}{\partial x} = \dfrac{\partial u}{\partial r} \cos \theta - \dfrac{\partial u}{\partial \theta} \dfrac{\sin \theta}{r}$,

(b) $\dfrac{\partial u}{\partial y} = \dfrac{\partial u}{\partial r} \sin \theta + \dfrac{\partial u}{\partial \theta} \dfrac{\cos \theta}{r}$.

(c) Use the results of Parts (a) and (b) to show that

$$\left(\frac{\partial u}{\partial x}\right)^2 + \left(\frac{\partial u}{\partial y}\right)^2 = \left(\frac{\partial u}{\partial r}\right)^2 + \left(\frac{1}{r} \frac{\partial u}{\partial \theta}\right)^2.$$

34. If $u = f(x, y)$ and $x = r \cos \theta$, $y = r \sin \theta$, show that if

$$\frac{\partial^2 u}{\partial x^2} + \frac{\partial^2 u}{\partial y^2} = 0, \qquad (i)$$

then

$$\frac{\partial^2 u}{\partial r^2} + \frac{1}{r} \frac{\partial u}{\partial r} + \frac{1}{r^2} \frac{\partial^2 u}{\partial \theta^2} = 0. \qquad (ii)$$

Equations (*i*) and (*ii*) are Laplace's equation, or the potential equation, in rectangular and polar coordinates, respectively.

35. If $z = f(x, y)$ and $x = g(t)$, $y = h(t)$, show that

$$\frac{d^2 z}{dt^2} = \frac{\partial^2 z}{\partial x^2}\left(\frac{dx}{dt}\right)^2 + 2 \frac{\partial^2 z}{\partial x \partial y} \frac{dx}{dt} \frac{dy}{dt} + \frac{\partial^2 z}{\partial y^2}\left(\frac{dy}{dt}\right)^2 + \frac{\partial z}{\partial x} \frac{d^2 x}{dt^2} + \frac{\partial z}{\partial y} \frac{d^2 y}{dt^2}.$$

36. If $w = f(x, y)$ and $x = \xi^2 - \eta^2$, $y = \xi^2 + \eta^2$, calculate

(a) w_ξ (b) w_η (c) $w_{\xi\xi}$ (d) $w_{\xi\eta}$ (e) $w_{\eta\eta}$.

37. Suppose that $w = f(x, y, z)$ and $z = g(x, y)$ so $w = F(x, y) = f[x, y, g(x, y)]$. Determine $\partial F/\partial x$ and $\partial F/\partial y$ in terms of partial derivatives of f and g.

38. Suppose that $w = f(x, y, u, v)$ and that $u = g(x)$, $v = h(x, y)$ so that $w = F(x, y) = f[x, y, g(x), h(x, y)]$. Determine $\partial F/\partial x$ and $\partial F/\partial y$ in terms of partial derivatives of f, g, and h.

39. In this problem we give an example of a function f whose partial derivatives $f_x(0, 0)$ and $f_y(0, 0)$ exist, but which is not differentiable or even continuous at $(0, 0)$. Consider the function

$$f(x, y) = \begin{cases} \dfrac{xy}{x^2 + y^2}, & (x, y) \neq (0, 0); \\ 0, & (x, y) = (0, 0). \end{cases}$$

In Example 4 of Section 16.1, this function was shown to be discontinuous at $(0, 0)$. By Theorem 16.3.1 it is also not differentiable there.

(a) Using the definition of the partial derivative show that $f_x(0, 0) = f_y(0, 0) = 0$.

(b) Show that

$$\Delta f = f(0 + h, 0 + k) - f(0, 0) = \frac{hk}{h^2 + k^2}.$$

(c) Since $f(0, 0) = f_x(0, 0) = f_y(0, 0) = 0$, it follows from Eq. 11 that $r = hk/(h^2 + k^2)^{3/2}$. Show that r does not approach zero as $(h, k) \to (0, 0)$, and thereby confirm that f is not differentiable at the origin.

40. **Proof of Theorem 16.3.2.** Assume that f_x and f_y exist in a neighborhood of (x_0, y_0) and are continuous at (x_0, y_0).

(a) Write Δf in the form

$$\Delta f = [f(x_0 + h, y_0 + k) - f(x_0, y_0 + k)] + [f(x_0, y_0 + k) - f(x_0, y_0)].$$

(b) Apply the mean value theorem (Theorem 4.1.3) to each bracketed expression in Part (a) to show that

$$\Delta f = f_x(\bar{x}, y_0 + k)h + f_y(x_0, \bar{y})k,$$

where $\bar{x}$ and $\bar{y}$ are certain numbers that satisfy $|\bar{x} - x_0| < |h|$ and $|\bar{y} - y_0| < |k|$.

(c) Use the continuity of f_x and f_y at (x_0, y_0) to show that

$$\Delta f = f_x(x_0, y_0)h + f_y(x_0, y_0)k + \zeta_1(x_0, y_0, h, k)h + \zeta_2(x_0, y_0, h, k)k,$$

where $\zeta_1 \to 0$ and $\zeta_2 \to 0$ as $(h, k) \to (0, 0)$.

(d) Show that $r = (\zeta_1 h + \zeta_2 k)/\sqrt{h^2 + k^2} \to 0$ as $(h, k) \to (0, 0)$, and thereby show that f is differentiable at (x_0, y_0).

16.4 THE GRADIENT AND THE DIRECTIONAL DERIVATIVE

The gradient

Definition 16.3.1, stated for a function of three variables, says that $w = f(x, y, z)$ is differentiable if

$$\begin{aligned} \Delta f &= f(x + \Delta x, y + \Delta y, z + \Delta z) - f(x, y, z) \\ &= f_x(x, y, z)\,\Delta x + f_y(x, y, z)\,\Delta y + f_z(x, y, z)\,\Delta z \\ &\quad + \zeta\sqrt{(\Delta x)^2 + (\Delta y)^2 + (\Delta z)^2}, \end{aligned} \tag{1}$$

where ζ depends on x, y, z, Δx, Δy, and Δz, and $\zeta \to 0$ as $(\Delta x, \Delta y, \Delta z) \to (0, 0, 0)$. In this section we are using the Δ notation for increments in the respective variables, rather than separate letters h, k, and so forth. It is useful to rewrite Eq. 1 using vector notation. Let

$$\mathbf{r} = x\mathbf{i} + y\mathbf{j} + z\mathbf{k}, \tag{2}$$

so that

$$\Delta\mathbf{r} = \Delta x\mathbf{i} + \Delta y\mathbf{j} + \Delta z\mathbf{k}, \qquad \|\Delta\mathbf{r}\| = \sqrt{(\Delta x)^2 + (\Delta y)^2 + (\Delta z)^2}. \tag{3}$$

Then Eq. 1 can be written as

$$\begin{aligned} \Delta f &= f(\mathbf{r} + \Delta\mathbf{r}) - f(\mathbf{r}) \\ &= [f_x(\mathbf{r})\mathbf{i} + f_y(\mathbf{r})\mathbf{j} + f_z(\mathbf{r})\mathbf{k}] \cdot \Delta\mathbf{r} + \zeta\|\Delta\mathbf{r}\|. \end{aligned} \tag{4}$$

Equation 4 motivates us to define a vector function, known as the **gradient** of f, by

$$\text{grad } f(\mathbf{r}) = f_x(\mathbf{r})\mathbf{i} + f_y(\mathbf{r})\mathbf{j} + f_z(\mathbf{r})\mathbf{k}. \tag{5}$$

This vector function is called grad f; often it is written as ∇f and is called del f. With this definition, formula (4) for the increment Δf takes the form

$$\Delta f = \text{grad } f(\mathbf{r}) \cdot \Delta\mathbf{r} + \zeta\|\Delta\mathbf{r}\| \tag{6a}$$

or

$$\Delta f = \nabla f(\mathbf{r}) \cdot \Delta\mathbf{r} + \zeta\|\Delta\mathbf{r}\|. \tag{6b}$$

In this form, the formula for Δf has a close resemblance to the corresponding formula for a function of one variable; for example, see Eq. 5 of Section 16.3. It also suggests that, for functions of more than one variable, ∇f plays a role similar to that of f' for a function of one variable.

EXAMPLE 1

If

$$f(\mathbf{r}) = xe^y - ye^z + e^x \sin yz,$$

determine $\nabla f(\mathbf{r})$.

We have

$$\begin{aligned} f_x(x, y, z) &= e^y + e^x \sin yz, \\ f_y(x, y, z) &= xe^y - e^z + ze^x \cos yz, \\ f_z(x, y, z) &= -ye^z + ye^x \cos yz. \end{aligned}$$

Hence

$$\nabla f(\mathbf{r}) = (e^y + e^x \sin yz)\mathbf{i} + (xe^y - e^z + ze^x \cos yz)\mathbf{j} + (-ye^z + ye^x \cos yz)\mathbf{k}. \blacksquare$$

Directional derivatives

We now turn to the question of finding the rate of change of a function in an arbitrary direction. For a differentiable function of two variables, $z = f(x, y) =$

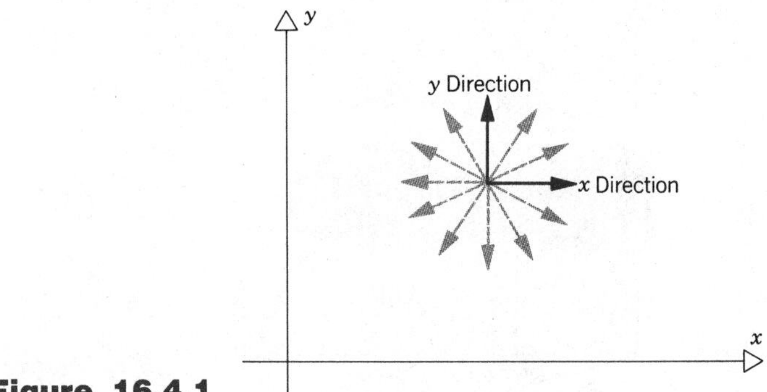

Figure 16.4.1

$f(\mathbf{r})$, the partial derivatives $f_x(x, y)$ and $f_y(x, y)$ give the rate of change of f in the x and y directions, respectively. However, at a point (x, y) there are many other directions, as we indicate in Figure 16.4.1. The rate of change of f in an arbitrary direction is called the **directional derivative** of f in that direction; we will show that it is closely related to the gradient vector ∇f.

How do we calculate the rate of change of $z = f(x, y)$ at a given point in a given direction? We specify the direction by the unit vector

$$\boldsymbol{\lambda} = \lambda_1\mathbf{i} + \lambda_2\mathbf{j} = (\cos\theta)\mathbf{i} + (\sin\theta)\mathbf{j}, \tag{7}$$

as shown in Figure 16.4.2. Further, we require that the increment $\Delta\mathbf{r}$ be proportional to $\boldsymbol{\lambda}$, that is, $\Delta\mathbf{r} = h\boldsymbol{\lambda}$, where h is a scalar multiplier. The rate of change of f in

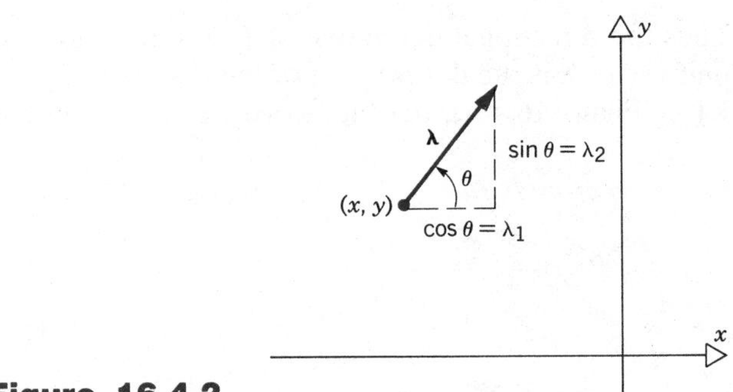

Figure 16.4.2

the direction $\boldsymbol{\lambda}$ is denoted by $D_{\boldsymbol{\lambda}}f(\mathbf{r})$ and is defined to be

$$D_{\boldsymbol{\lambda}}f(\mathbf{r}) = \lim_{h\to 0}\frac{f(\mathbf{r} + h\boldsymbol{\lambda}) - f(\mathbf{r})}{h} = \lim_{h\to 0}\frac{f(x + h\lambda_1, y + h\lambda_2) - f(x, y)}{h}, \tag{8}$$

provided that this limit exists. The geometrical interpretation of $D_{\boldsymbol{\lambda}}f(\mathbf{r})$ is shown in Figure 16.4.3. The surface $z = f(x, y)$ is intersected by the plane parallel to the z-axis that contains the point $\mathbf{r} = (x, y)$ and the vector $\boldsymbol{\lambda}$. If C is the curve of intersection of the surface and the plane, then $D_{\boldsymbol{\lambda}}f(\mathbf{r})$ is the slope of C at the point $(x, y, f(x, y))$. Since f is differentiable, we can calculate $D_{\boldsymbol{\lambda}}f(\mathbf{r})$ by using formula

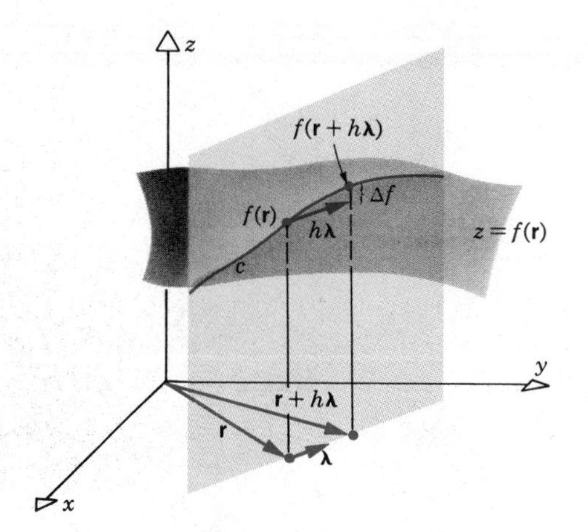

Figure 16.4.3

(6b) for the increment Δf. Then

$$
\begin{aligned}
D_{\boldsymbol{\lambda}} f(\mathbf{r}) &= \lim_{h \to 0} \left[\frac{\nabla f(\mathbf{r}) \cdot h\boldsymbol{\lambda} + \zeta \|h\boldsymbol{\lambda}\|}{h} \right] \\
&= \lim_{h \to 0} \left[\nabla f(\mathbf{r}) \cdot \boldsymbol{\lambda} + \zeta \frac{|h|}{h} \right].
\end{aligned} \tag{9}
$$

The term $\nabla f(\mathbf{r}) \cdot \boldsymbol{\lambda}$ does not depend on h. Further, as $h \to 0$, it follows that $\Delta x = h\lambda_1 \to 0$ and $\Delta y = h\lambda_2 \to 0$; hence $\zeta \to 0$ as well. Therefore we obtain

$$
\begin{aligned}
D_{\boldsymbol{\lambda}} f(\mathbf{r}) &= \nabla f(\mathbf{r}) \cdot \boldsymbol{\lambda} = \operatorname{grad} f(\mathbf{r}) \cdot \boldsymbol{\lambda} \\
&= f_x(x, y)\lambda_1 + f_y(x, y)\lambda_2.
\end{aligned} \tag{10}
$$

Thus the directional derivative of f at a point (x, y) in the direction given by the unit vector $\boldsymbol{\lambda}$ is the dot product of the gradient of f at that point and the unit vector $\boldsymbol{\lambda}$ (see Figure 16.4.4). In other words, *$D_{\boldsymbol{\lambda}} f(\mathbf{r})$ is the component of ∇f in the direction*

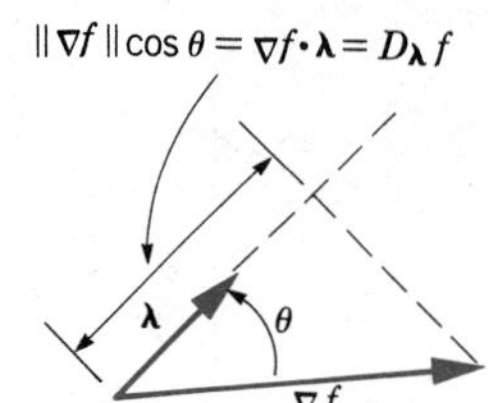

Figure 16.4.4

of $\boldsymbol{\lambda}$. If f is differentiable and if we know ∇f at a given point, then we can quickly calculate the directional derivative of f at this point in any given direction $\boldsymbol{\lambda}$. Note that if $\boldsymbol{\lambda} = \mathbf{i}$, then $D_{\mathbf{i}} f(\mathbf{r}) = f_x(\mathbf{r})$; the rate of change of f in the x direction is, as it should be, $f_x(\mathbf{r})$. Similarly, $D_{\mathbf{j}} f(\mathbf{r}) = f_y(\mathbf{r})$.

EXAMPLE 2

Determine the rate of change of $f(x, y) = 2x^2 + 3xy - 2y^2$ at the point $(1, -2)$ in the direction toward the origin.

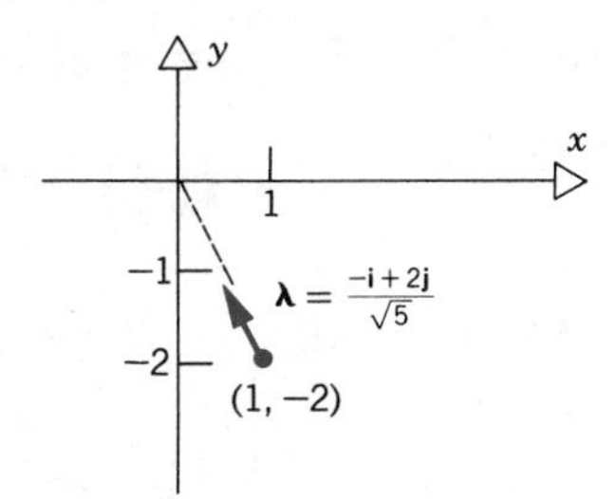

Figure 16.4.5

First, we must calculate the unit vector in the desired direction (see Figure 16.4.5). The vector from $(1, -2)$ to $(0, 0)$ is $-\mathbf{i} + 2\mathbf{j}$; hence

$$\boldsymbol{\lambda} = \frac{-\mathbf{i} + 2\mathbf{j}}{\sqrt{1 + 4}} = \frac{-1}{\sqrt{5}}\mathbf{i} + \frac{2}{\sqrt{5}}\mathbf{j}.$$

Next,

$$f_x(x, y) = 4x + 3y, \quad \text{so} \quad f_x(1, -2) = -2,$$

$$f_y(x, y) = 3x - 4y, \quad \text{so} \quad f_y(1, -2) = 11.$$

Using Eq. 10 we find that

$$D_{\boldsymbol{\lambda}} f(1, -2) = (-2)\left(\frac{-1}{\sqrt{5}}\right) + (11)\left(\frac{2}{\sqrt{5}}\right) = \frac{24}{\sqrt{5}} \cong 10.73. \blacksquare$$

The formula for the directional derivative for a function of three or more variables is given by the vector form of Eq. 10. Thus if $w = f(x, y, z)$ is a differentiable function, then the directional derivative at (x, y, z) in the direction of the unit vector $\boldsymbol{\lambda} = \lambda_1\mathbf{i} + \lambda_2\mathbf{j} + \lambda_3\mathbf{k}$ is

$$D_{\boldsymbol{\lambda}} f(\mathbf{r}) = \nabla f(\mathbf{r}) \cdot \boldsymbol{\lambda} = f_x(x, y, z)\lambda_1 + f_y(x, y, z)\lambda_2 + f_z(x, y, z)\lambda_3. \quad (11)$$

EXAMPLE 3

What is the rate of change of $f(x, y, z) = xy^2e^z$ along the curve $\mathbf{r} = 2 \cos t\, \mathbf{i} + 3 \sin t\, \mathbf{j} + 3t\mathbf{k}$ at the point corresponding to $t = \pi/3$?

By "along the curve" we mean in the direction of the tangent vector to the curve. The point on the curve is

$$\mathbf{r}\left(\frac{\pi}{3}\right) = \mathbf{i} + \frac{3\sqrt{3}}{2}\mathbf{j} + \pi\mathbf{k}.$$

A tangent vector is given by

$$d\mathbf{r}/dt = -2 \sin t\, \mathbf{i} + 3 \cos t\, \mathbf{j} + 3\mathbf{k},$$

so the unit tangent vector is

$$\mathbf{T}(t) = \frac{-2 \sin t\, \mathbf{i} + 3 \cos t\, \mathbf{j} + 3\mathbf{k}}{\sqrt{4 \sin^2 t + 9 \cos^2 t + 9}}.$$

Evaluating $\mathbf{T}$ at $t = \pi/3$, we obtain

$$\boldsymbol{\lambda} = \mathbf{T}\left(\frac{\pi}{3}\right) = \frac{-\sqrt{3}\mathbf{i} + \frac{3}{2}\mathbf{j} + 3\mathbf{k}}{\sqrt{3 + \frac{9}{4} + 9}} = \frac{-2\sqrt{3}\mathbf{i} + 3\mathbf{j} + 6\mathbf{k}}{\sqrt{57}}.$$

Next,

$$f_x(x, y, z) = y^2e^z, \quad \text{so} \quad f_x\left(1, \frac{3\sqrt{3}}{2}, \pi\right) = \frac{27}{4}e^{\pi},$$

$$f_y(x, y, z) = 2xye^z, \quad \text{so} \quad f_y\left(1, \frac{3\sqrt{3}}{2}, \pi\right) = 3\sqrt{3}\,e^{\pi},$$

$$f_z(x, y, z) = xy^2e^z, \quad \text{so} \quad f_z\left(1, \frac{3\sqrt{3}}{2}, \pi\right) = \frac{27}{4}e^{\pi}.$$

Finally, using formula (11) for the directional derivative, we obtain

$$\begin{aligned} D_{\boldsymbol{\lambda}}f\left(1, \frac{3\sqrt{3}}{2}, \pi\right) &= \frac{27}{4}e^{\pi}\left(-\frac{2\sqrt{3}}{\sqrt{57}}\right) + 3\sqrt{3}\,e^{\pi}\left(\frac{3}{\sqrt{57}}\right) + \frac{27}{4}e^{\pi}\left(\frac{6}{\sqrt{57}}\right) \\ &= \frac{e^{\pi}}{\sqrt{57}}\left(-\frac{27}{2}\sqrt{3} + 9\sqrt{3} + \frac{81}{2}\right) \\ &= \frac{e^{\pi}}{2\sqrt{57}}(81 - 9\sqrt{3}) \cong 100.245. \ \blacksquare \end{aligned}$$

Properties of the gradient

Notice that we can write formulas (10) or (11) for the directional derivative as

$$\begin{aligned} D_{\boldsymbol{\lambda}}f(\mathbf{r}) &= \nabla f(\mathbf{r}) \cdot \boldsymbol{\lambda} = \|\nabla f(\mathbf{r})\|\,\|\boldsymbol{\lambda}\| \cos\theta \\ &= \|\nabla f(\mathbf{r})\| \cos\theta, \end{aligned} \tag{12}$$

since $\|\boldsymbol{\lambda}\| = 1$, and where θ is the angle between the vectors $\nabla f(\mathbf{r})$ and $\boldsymbol{\lambda}$ (see Figure 16.4.4). It follows immediately from Eq. 12 that at a given point, $D_{\boldsymbol{\lambda}}f(\mathbf{r})$ is a maximum when $\cos\theta = 1$; that is, when $\theta = 0$ or when $\boldsymbol{\lambda}$ is in the direction of $\nabla f(\mathbf{r})$. Thus

$$[D_{\boldsymbol{\lambda}}f(\mathbf{r})]_{\max} = \|\nabla f(\mathbf{r})\|.$$

Therefore *the magnitude of* ∇f *at a given point* $\mathbf{r}$ *is the maximum rate of change of* f *at that point, and the direction of* ∇f *is the direction in which that maximum rate of change occurs.* Similarly, $-\nabla f(\mathbf{r})$ is a vector in the direction in which f decreases most rapidly and $-\|\nabla f(\mathbf{r})\|$ is the rate of change in this direction.

EXAMPLE 4

If

$$f(x, y) = 3x^2 - 2xy + 4y^2,$$

find the maximum value of the directional derivative of f at the point $(-1, 2)$ and the direction for which it occurs.

To answer these questions we need only compute grad $f(-1, 2)$. We have

$$f_x(x, y) = 6x - 2y, \quad \text{so} \quad f_x(-1, 2) = -10,$$

and

$$f_y(x, y) = -2x + 8y, \quad \text{so} \quad f_y(-1, 2) = 18.$$

Hence

$$\text{grad } f(-1, 2) = -10\mathbf{i} + 18\mathbf{j},$$

and

$$[D_{\boldsymbol{\lambda}}f(-1, 2)]_{\max} = \|\text{grad } f(-1, 2)\| = \sqrt{100 + 324}$$
$$= 2\sqrt{106} \cong 20.59.$$

The unit vector in the direction in which f increases most rapidly is

$$\boldsymbol{\lambda} = \frac{\text{grad } f(-1, 2)}{\|\text{grad } f(-1, 2)\|} = \frac{-5\mathbf{i} + 9\mathbf{j}}{\sqrt{106}}. \quad \blacksquare$$

EXAMPLE 5

In Figure 16.4.6 we show a thin rectangular plate $(0 \le x \le \pi, 0 \le y \le \pi/2)$ with the temperature prescribed on the edges. It can be shown, from the partial dif-

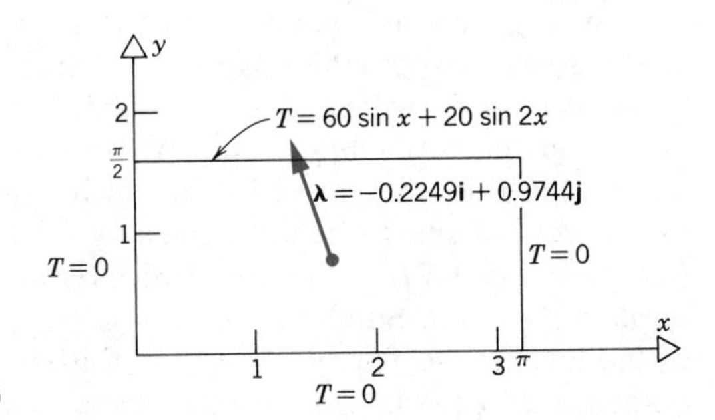

Figure 16.4.6

ferential equation governing the conduction of heat, that the temperature distribution in the plate is

$$T(x, y) = A_1 \sin x \sinh y + A_2 \sin 2x \sinh 2y,$$

where

$$A_1 = 60\left(\sinh \frac{\pi}{2}\right)^{-1} \quad \text{and} \quad A_2 = 20(\sinh \pi)^{-1}.$$

At the middle of the plate, in which direction is the temperature increasing most rapidly and what is this rate of change?

To answer these questions we must calculate grad $T(\pi/2, \pi/4)$. We have

$$\text{grad } T(x, y) = (A_1 \cos x \sinh y + 2A_2 \cos 2x \sinh 2y)\mathbf{i} + (A_1 \sin x \cosh y + 2A_2 \sin 2x \cosh 2y)\mathbf{j}$$

so

$$\begin{aligned}\text{grad } T\left(\frac{\pi}{2}, \frac{\pi}{4}\right) &= \left(-2A_2 \sinh \frac{\pi}{2}\right)\mathbf{i} + \left(A_1 \cosh \frac{\pi}{4}\right)\mathbf{j} \\ &= -\frac{40 \sinh \pi/2}{\sinh \pi}\mathbf{i} + \frac{60 \cosh \pi/4}{\sinh \pi/2}\mathbf{j} \\ &\cong -7.9707\mathbf{i} + 34.5355\mathbf{j}.\end{aligned}$$

The unit vector in the direction in which the temperature is increasing most rapidly is

$$\boldsymbol{\lambda} \cong \frac{-7.9707\mathbf{i} + 34.5355\mathbf{j}}{35.4434} \cong -0.2249\mathbf{i} + 0.9744\mathbf{j},$$

and the rate of change of T in this direction at $(\pi/2, \pi/4)$ is

$$\left\|\text{grad } T\left(\frac{\pi}{2}, \frac{\pi}{4}\right)\right\| \cong 35.44.$$

The direction of most rapid increase is shown in Figure 16.4.6. ∎

In two dimensions the gradient has a simple interpretation in terms of the level curves (topographic map) of the surface $z = f(x, y)$. If we want to climb upwards on this surface along the path that rises most rapidly, then at any point (x, y) on the topographic map we should choose our path in the direction of $\nabla f(x, y)$ as this is the direction in which z increases most rapidly. Conversely, the direction of steepest descent is given by $-\nabla f(x, y)$. We now show that *at any point* (x, y) *the vector* $\nabla f(x, y)$ *is perpendicular to the level curve* $f(x, y) = c_0$ *that passes through the point.* Suppose that $x = \phi(t)$, $y = \psi(t)$ is a parametric representation of the level curve $f(x, y) = c_0$. On this curve $f(x, y) = f[\phi(t), \psi(t)] = F(t) = c_0$; hence $dF/dt \equiv 0$. On the other hand, using the chain rule, we obtain

$$\begin{aligned}\frac{dF}{dt} &= f_x(x, y)\frac{d\phi}{dt} + f_y(x, y)\frac{d\psi}{dt} \\ &= \nabla f(x, y) \cdot \left(\frac{d\phi}{dt}\mathbf{i} + \frac{d\psi}{dt}\mathbf{j}\right) = 0.\end{aligned} \tag{13}$$

The vector $(d\phi/dt)\mathbf{i} + (d\psi/dt)\mathbf{j}$ is tangent to the curve $x = \phi(t)$, $y = \psi(t)$. Assuming that the tangent vector is not the zero vector, it follows immediately that $\nabla f(x, y)$ is perpendicular to the level curve. This is illustrated in Figure 16.4.7.

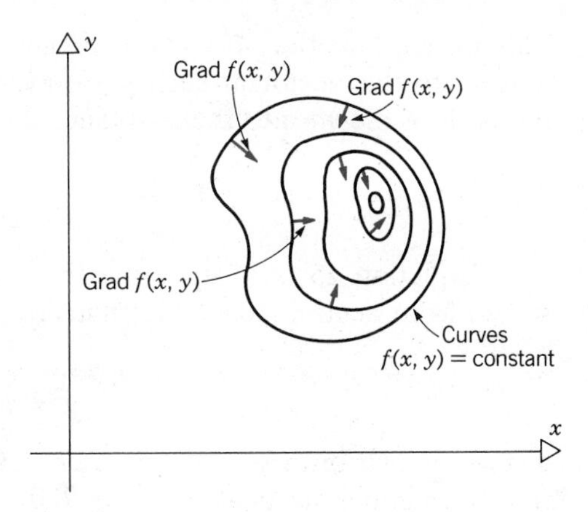

Figure 16.4.7 The gradient is orthogonal to the level curves.

EXAMPLE 6

The height z of a mountain in feet above sea level is given by

$$z = 900 - \frac{x^2}{10{,}000} - \frac{y^2}{8100}. \tag{14}$$

If a climber starts climbing at sea level at $x = y = (27/\sqrt{181}) \times 10^3 \cong 2007$ ft, what is the path in the xy-plane corresponding to a route of steepest ascent up the mountain?

The topographic map for the surface (14) is the set of ellipses shown in Figure 16.4.8a. We know that at each point on the mountain the direction of steepest

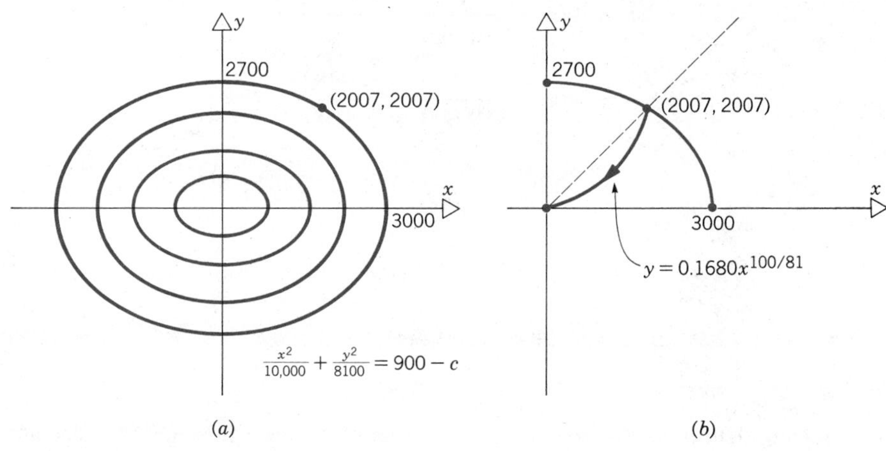

Figure 16.4.8

ascent is given by

$$\text{grad } z = -\frac{x}{5000}\mathbf{i} - \frac{y}{4050}\mathbf{j}.$$

This means that the projection of the path of ascent into the xy-plane must be tangent to this vector at each point (x, y). Thus if $y = g(x)$ is the projection of the path of ascent into the xy-plane, then

$$\frac{dy}{dx} = \frac{-y/4050}{-x/5000} = \frac{100}{81}\frac{y}{x}. \tag{15}$$

Equation 15 is a separable first order differential equation, which can be solved as in Section 10.2. Alternatively, we can simply verify that the solution is given by

$$y = kx^{100/81},$$

where k is an arbitrary constant. We determine the constant k by using the condition that the climber starts at $x = y \cong 2007$ ft. Thus

$$2007 = k(2007)^{100/81}$$

so

$$k = \frac{2007}{11{,}946} = 0.1680.$$

Hence the path in the xy-plane corresponding to the route of steepest ascent up the mountain is given by

$$y = 0.1680x^{100/81}.$$

This path is shown in Figure 16.4.8b. The corresponding height above sea level as the climber ascends the mountain is

$$\begin{aligned} z &= 900 - \frac{x^2}{10{,}000} - \frac{y^2}{8100} \\ &= 900 - \frac{x^2}{10{,}000} - \frac{0.02822}{8100}x^{200/81} \\ &= 900 - \frac{x^2}{10{,}000}(1 + 0.03484x^{38/81}). \quad ■ \end{aligned}$$

PROBLEMS

In each of Problems 1 through 6, determine the gradient of the given function.

1. $f(x, y) = \frac{1}{2}x^2 + 2xy + y^2$
2. $g(x, y) = e^{nx} \cos ny$
3. $f(x, y, z) = xy^2 + 2yz^2 - 3zx^2$
4. $w = \phi(x, y, z) = Ax^{\alpha}y^{\beta}z^{\gamma}$
5. $f(x, y) = \sinh x \tan y$
6. $f(x, y, z) = e^{xy^2/z^3}$

In each of Problems 7 through 12, calculate the directional derivative of the given function at the given point in the given direction.

7. $f(x, y) = \frac{1}{2}x^2 + 2xy + y^2$, $(1, 1)$, $\mathbf{u} = \mathbf{i} + \mathbf{j}$
8. $f(x, y, z) = xy^2 + 2yz + 3zx^2$, $(1, -2, 3)$, $\mathbf{u} = \mathbf{i} - \mathbf{j} + \mathbf{k}$
9. $\phi(x, y) = \ln(x^2 + y^2)^{1/2}$, (a, b), toward the origin

10. $\phi(x, y, z) = (x^2 + y^2 + z^2)^{1/2}$, (a, b, c), toward the origin

11. $f(x, y, z) = ze^{x^2y}$, $(2, \frac{1}{2}, 2)$, toward $(1, -\frac{1}{2}, 3)$

12. $\phi(x, y) = \arctan(x\sqrt{2} - y)$, $(1, 1)$, toward $(-1, 1)$

In each of Problems 13 through 16, determine the unit vector $\boldsymbol{\lambda}$ in the direction in which the given function is increasing most rapidly at the indicated point and determine the rate of change of the function in that direction.

13. $f(x, y) = \frac{1}{2}x^2 + 2xy + y^2$, $\quad (1, 1)$

14. $f(x, y, z) = xy^2 + 2yz^2 + 3zx^2$, $\quad (1, -2, -3)$

15. $\phi(x, y, z) = (x^2 + y^2 + z^2)^{3/2}$, $\quad (3, -4, 12)$

16. $\phi(x, y) = e^x \cos y + e^y \sin x$, $\quad \left(\frac{\pi}{4}, \frac{\pi}{4}\right)$

17. If $\phi(x, y) = \ln(x^2 + y^2)^{1/2}$ and $\mathbf{r} = x\mathbf{i} + y\mathbf{j}$, show that

$$\text{grad}\, \phi(x, y) = \frac{\mathbf{r}}{r^2}, \qquad r \neq 0,$$

where $r = \|\mathbf{r}\|$.

18. (a) If $\phi(x, y, z) = (x^2 + y^2 + z^2)^{1/2}$ and $\mathbf{r} = x\mathbf{i} + y\mathbf{j} + z\mathbf{k}$, show that

$$\text{grad}\, \phi(x, y, z) = \frac{\mathbf{r}}{r}, \qquad r \neq 0,$$

where $r = \|\mathbf{r}\|$.

(b) If $\phi(x, y, z) = \|\mathbf{r}\|^{-n}$ and $\mathbf{r} = x\mathbf{i} + y\mathbf{j} + z\mathbf{k}$, show that

$$\text{grad}\, \phi(x, y, z) = -n\mathbf{r}/r^{n+2}, \qquad r \neq 0,$$

where $r = \|\mathbf{r}\|$.

19. If $f(x, y) = x/y$ and $g(x, y) = \frac{1}{2}x^2 - 4y$, find the rate of change of $f(x, y)$ at the point $(2, -1)$ in the direction in which $g(x, y)$ is increasing most rapidly.

20. If $w = x^2 + y^2$, then at the point (a, b),
(a) find a unit vector in the direction of most rapid increase of w,
(b) find a unit vector in the direction of most rapid decrease of w, and
(c) find a unit vector in the direction(s) of zero rate of change of w. Interpret these vectors relative to the curve $w =$ constant passing through the point (a, b).

21. Determine a unit vector normal to the ellipse

$$\frac{x^2}{a^2} + \frac{y^2}{b^2} = 1$$

at the point $(a/\sqrt{2}, b/\sqrt{2})$.

22. Determine the unit vector normal to the ellipse

$$\frac{x^2}{4} + \frac{y^2}{9} = 1$$

at the point $(1, 3\sqrt{3}/2)$ and directed so that x and y increase along it.

23. Determine a unit vector normal to the hyperbola $xy = 1$ at the point $(a, 1/a)$, $a > 0$.

24. Determine a unit vector normal to the curve $y^2 = 4px$ at the point $(a, 2\sqrt{pa})$.

25. If $w = e^{x^2+y^2}$, what is the rate of change of w along the curve $\mathbf{r} = 2 \cos t\, \mathbf{i} + 2 \sin t\, \mathbf{j} - \sin 2t\, \mathbf{k}$ at the point corresponding to $t = 13\pi/6$?

26. If $w = y^2e^x$, what is the rate of change of w along the curve $\mathbf{r} = (t^2 - 3t)\mathbf{i} + (t + 1)\mathbf{j}$ at the point corresponding to $t = 2$?

27. If $w = y \ln|x| + 2y^2x$, what is the rate of change of w along the curve $\mathbf{r} = (3t + 2)\mathbf{i} - 2t^2\mathbf{j}$ at the point corresponding to $t = -1$?

28. If $w = z/(x + 2y)$, what is the rate of change of w along the curve $\mathbf{r} = 2 \cos t\, \mathbf{i} - 2 \sin t\, \mathbf{j} + \frac{2}{3} t^{3/2}\, \mathbf{k}$ at the point corresponding to $t = \pi/4$?

29. If $w = (y + 3)^2 e^{x^2} \tan(z + \pi/4)$, what is the rate of change of w along the curve $\mathbf{r} = e^{-t}\mathbf{i} + te^t\mathbf{j} + 4t\mathbf{k}$ at the point corresponding to $t = 0$?

30. The temperature distribution in degrees centigrade in a semi-infinite strip $(0 \leq x \leq a, 0 \leq y < \infty)$ for which the edges $x = 0$ and $x = a$ are held at 0 and the edge at $y = 0$ has the prescribed temperature $40 \sin \pi x/a + 20 \sin 2\pi x/a$ is

$$T(x, y) = 40e^{-\pi y/a} \sin \frac{\pi x}{a} + 20e^{-2\pi y/a} \sin \frac{2\pi x}{a}.$$

At the point $(a/2, a)$ determine a unit vector in the direction in which the temperature is increasing most rapidly.

31. In two-dimensional heat conduction an *isotherm* is a curve on which the temperature is constant. The temperature in the semicircular plate $x^2 + y^2 \leq 4, x \geq 0$ is given by $T(x, y) = 3yx^2 - x^3 + 60$. Determine a vector at $(1, -1)$ tangent to the isotherm passing through $(1, -1)$.

32. The second directional derivative of $f(x, y)$ at a given point in the direction of the unit vector $\boldsymbol{\lambda} = \lambda_1\mathbf{i} + \lambda_2\mathbf{j}$ is defined by $D^2_{\boldsymbol{\lambda}}f(x, y) = D_{\boldsymbol{\lambda}}[D_{\boldsymbol{\lambda}}f(x, y)]$. Show that

$$D^2_{\boldsymbol{\lambda}}f(x, y) = f_{xx}(x, y)\lambda_1^2 + 2f_{xy}(x, y)\lambda_1\lambda_2 + f_{yy}(x, y)\lambda_2^2.$$

Verify that $D^2_{\mathbf{i}}f(x, y) = f_{xx}(x, y)$ and $D^2_{\mathbf{j}}f(x, y) = f_{yy}(x, y)$.

In each of Problems 33 through 36, calculate the second directional derivative of the given function at the given point in the given direction.

33. $f(x, y) = \frac{1}{2}x^2 + 2xy + y^2$, $(1, 1)$, $\mathbf{u} = \mathbf{i} + \mathbf{j}$

34. $\phi(x, y) = \ln(x^2 + y^2)^{1/2}$, (a, b), toward the origin

35. $f(x, y) = \sinh x \tan y$, $(0, 0)$, $\mathbf{u} = -\mathbf{i} + 2\mathbf{j}$

36. $f(x, y) = xy$, $(1, 1)$, toward $(2, 3)$

37. The gravitational force $\mathbf{F}$ exerted by a mass M concentrated at the origin on a mass m at the point $\mathbf{r} = x\mathbf{i} + y\mathbf{j} + z\mathbf{k}$ is

$$\mathbf{F} = -GMm\frac{\mathbf{r}}{r^3},$$

where $r = \|\mathbf{r}\|$ and G is a universal constant. Show that $\mathbf{F}$ is the gradient of the scalar field

$$f(\mathbf{r}) = \frac{GMm}{r}.$$

38. When does the directional derivative of $f(\mathbf{r})$ at $\mathbf{r}$ in the direction of the gradient of $g(\mathbf{r})$ equal the directional derivative of $g(\mathbf{r})$ at $\mathbf{r}$ in the direction of the gradient of $f(\mathbf{r})$?

* 39. Determine the projection in the xy-plane of the path of steepest descent on the surface $z = 4x^2 + 12y^2$ from the point $(4, 12, 1792)$ to $(0, 0, 0)$.

* 40. Determine the projection in the xy-plane of the path of steepest descent on the surface $z = (x^2 + y^2)^{1/3}$ from the point $(a, b, (a^2 + b^2)^{1/3})$ to $(0, 0, 0)$. The surface is sketched in Figure 15.10.8 and in Figure 16.1.4.

* 41. Show that the projection in the xy-plane of the paths of steepest descent and steepest ascent on the hyperbolic paraboloid $z = -x^2/16 + y^2/4$ are given by the family of curves $y = Cx^{-4}$, where C is a constant. A hyperbolic paraboloid is sketched in Figure 15.10.15.

* 42. A heat-seeking particle (insect) always moves in the direction of maximum temperature. The temperature distribution is $T(x, y) = T_0e^{-(x^2+3y^2)/5}$, where x and y are measured in feet, and T_0 is a constant. If the insect is released at (a, b), what path does it follow to reach the origin?

43. **The mean value theorem for functions of more than one variable.** In this problem we give the generalization of the mean value theorem for a function of one variable (Section 4.1) to functions of more than one variable. We use vector notation to derive the result for functions of two variables; however, the result is valid for functions of more than two variables. The theorem is as follows. Let f be a differentiable function of two variables on a domain such that if $\mathbf{a} = (a_1, a_2)$ and $\mathbf{b} = (b_1, b_2)$ are in the domain, then all points on the line segment connecting $\mathbf{a}$ and $\mathbf{b}$ are in the domain of f. Then there exists a point $\mathbf{c} = (c_1, c_2)$ on the line segment connecting $\mathbf{a}$ and $\mathbf{b}$ such that

$$f(\mathbf{b}) - f(\mathbf{a}) = \nabla f(\mathbf{c}) \cdot (\mathbf{b} - \mathbf{a}). \quad (i)$$

(a) Verify that the line segment connecting $\mathbf{a}$ and $\mathbf{b}$ can be represented in the parametric form $\mathbf{r} = \mathbf{a} + t(\mathbf{b} - \mathbf{a})$, $0 \le t \le 1$ (see Figure 16.4.9).

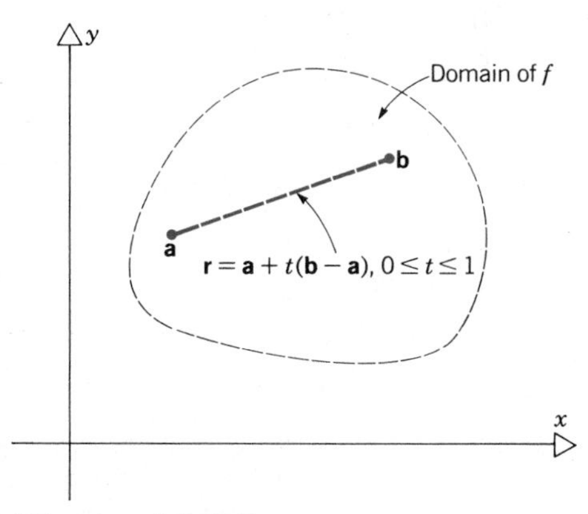

Figure 16.4.9

(b) Let

$$F(t) = f[\mathbf{a} + t(\mathbf{b} - \mathbf{a})], \quad 0 \le t \le 1. \quad (ii)$$

Why is there a number α in the interval $(0, 1)$ such that

$$F(1) - F(0) = F'(\alpha)(1 - 0)? \quad (iii)$$

(c) Use the chain rule to calculate $F'(\alpha)$ in terms of the partial derivatives of f and then deduce the desired result from Eq. (iii). Express **c** in terms of **a**, **b** and α.

44. Use the result of Problem 43 to show that if f is a differentiable function of two variables with a circular domain and if grad $f(x, y) \equiv 0$ in this domain, then $f(x, y)$ is a constant.

16.5 THE TANGENT PLANE AND THE NORMAL LINE

If $z = f(x, y)$ is differentiable, then $f(x, y) - f(x_0, y_0)$, or $z - z_0$, can be approximated near the point (x_0, y_0) by the differential

$$dz = f_x(x_0, y_0)\,dx + f_y(x_0, y_0)\,dy.$$

By replacing dx, dy, and dz by $x - x_0$, $y - y_0$, and $z - z_0$, respectively, we obtain the approximation

$$z - z_0 = f_x(x_0, y_0)(x - x_0) + f_y(x_0, y_0)(y - y_0) \tag{1}$$

as $(x, y) \to (x_0, y_0)$. Alternatively, we can write Eq. 1 in the form

$$f_x(x_0, y_0)(x - x_0) + f_y(x_0, y_0)(y - y_0) - (z - z_0) = 0. \tag{2}$$

Equations 1 and 2 are equations of the plane (see Section 15.5) passing through the point (x_0, y_0, z_0) and with the normal vector

$$\mathbf{N} = f_x(x_0, y_0)\mathbf{i} + f_y(x_0, y_0)\mathbf{j} - \mathbf{k}. \tag{3}$$

This plane is called the **tangent plane** to the surface $z = f(x, y)$ at the point (x_0, y_0, z_0) and the vector **N**, or any scalar multiple of it, is called a **normal vector** to this surface at this point (see Figure 16.5.1).

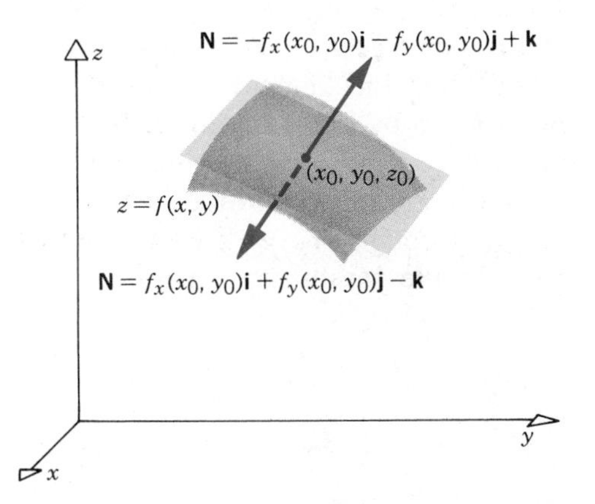

Figure 16.5.1 A normal vector and a tangent plane.

These definitions can be justified as follows. Let C be a curve that lies in the surface $z = f(x, y)$ and passes through the point (x_0, y_0, z_0) as shown in Figure 16.5.2. The curve C can be described by parametric equations of the form

$$x = g(t), \qquad y = h(t), \qquad z = f(x, y) = f[g(t), h(t)],$$

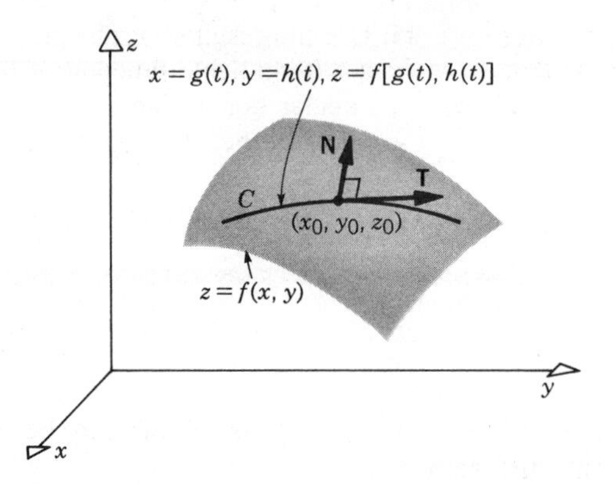

Figure 16.5.2

where t_0 corresponds to the point (x_0, y_0, z_0). If $\mathbf{r} = x\mathbf{i} + y\mathbf{j} + z\mathbf{k}$, then a tangent vector to C is

$$\begin{aligned}\frac{d\mathbf{r}}{dt} &= \frac{dx}{dt}\mathbf{i} + \frac{dy}{dt}\mathbf{j} + \frac{dz}{dt}\mathbf{k} \\ &= g'(t)\mathbf{i} + h'(t)\mathbf{j} + [f_x(x, y)g'(t) + f_y(x, y)h'(t)]\mathbf{k};\end{aligned} \tag{4}$$

note that the chain rule has been used in calculating dz/dt. We assume that not all components of $d\mathbf{r}/dt$ are zero. Then

$$\begin{aligned}\mathbf{N} \cdot \frac{d\mathbf{r}}{dt}(t_0) &= f_x(x_0, y_0)g'(t_0) + f_y(x_0, y_0)h'(t_0) \\ &\quad - [f_x(x_0, y_0)g'(t_0) + f_y(x_0, y_0)h'(t_0)] = 0.\end{aligned}$$

Hence $\mathbf{N}$ is perpendicular to the tangent vector to C at t_0. Since C could be any differentiable curve lying in the surface $z = f(x, y)$ and passing through (x_0, y_0, z_0), it is natural to refer to $\mathbf{N}$ as a normal vector to the surface $z = f(x, y)$ at the point (x_0, y_0, z_0). Also since $d\mathbf{r}(t_0)/dt \cdot \mathbf{N} = 0$ for all such curves C, it follows that the tangent vectors must lie in a plane with normal $\mathbf{N}$, namely the plane given by Eq. 1 or Eq. 2. It is natural to call this plane the tangent plane to the surface $z = f(x, y)$ at the point (x_0, y_0, z_0).

The **normal line** to the surface $z = f(x, y)$ at the point (x_0, y_0, z_0) is the line passing through (x_0, y_0, z_0) and parallel to the vector $\mathbf{N}$. Thus the normal line is given by the parametric equations

$$\begin{aligned}x - x_0 &= f_x(x_0, y_0)t, \\ y - y_0 &= f_y(x_0, y_0)t, \\ z - z_0 &= -t,\end{aligned} \tag{5a}$$

with $-\infty < t < \infty$, or by the symmetric equations

$$\frac{x - x_0}{f_x(x_0, y_0)} = \frac{y - y_0}{f_y(x_0, y_0)} = \frac{z - z_0}{-1}. \tag{5b}$$

EXAMPLE 1

Determine a normal vector, the tangent plane, and the normal line at the point (1, 2, 2) on the hemisphere $z = (9 - x^2 - y^2)^{1/2}$.

First note that

$$z_x = \tfrac{1}{2}(9 - x^2 - y^2)^{-1/2}(-2x), \quad \text{so} \quad z_x(1, 2) = -\tfrac{1}{2};$$

$$z_y = \tfrac{1}{2}(9 - x^2 - y^2)^{-1/2}(-2y), \quad \text{so} \quad z_y(1, 2) = -1.$$

Using Eq. 3, we find that a normal vector at (1, 2, 2) is

$$\mathbf{N} = -\tfrac{1}{2}\mathbf{i} - \mathbf{j} - \mathbf{k}.$$

Then the equation of the tangent plane at (1, 2, 2) is given by Eq. 2:

$$-\tfrac{1}{2}(x - 1) - (y - 2) - (z - 2) = 0,$$

or

$$x + 2y + 2z = 9.$$

The symmetric form of the normal line at (1, 2, 2) follows from Eq. 5(b):

$$\frac{x - 1}{-\frac{1}{2}} = \frac{y - 2}{-1} = \frac{z - 2}{-1}.$$

The normal vector, tangent plane, and normal line are shown in Figure 16.5.3. Note that **N** points inward from the surface of the hemisphere. An equally satis-

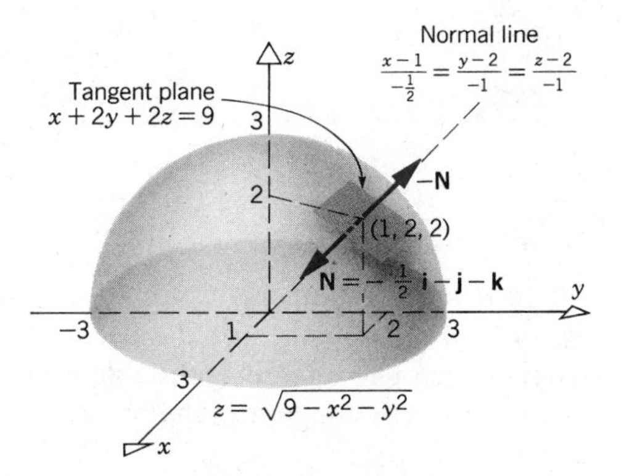

Figure 16.5.3

factory normal vector that points outward is $-\mathbf{N} = \frac{1}{2}\mathbf{i} + \mathbf{j} + \mathbf{k}$. Indeed any non-zero multiple of **N** can serve as a normal vector. Finally notice that **N** is parallel to the radius vector $\overrightarrow{OP}$ from the origin to the point $P(1, 2, 2)$. It seems intuitively correct that a normal vector at a point on a hemisphere or a sphere should be along the line from the center to the point. ∎

Often we are not given an explicit formula $z = f(x, y)$ for a surface, but rather are given an implicit equation $F(x, y, z) = c$, a constant, and are told that

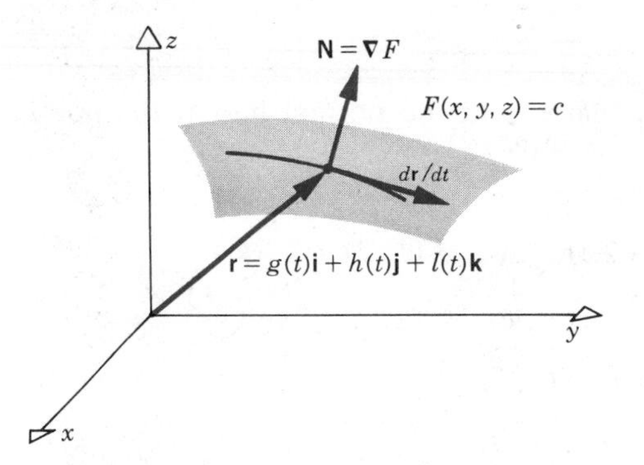

Figure 16.5.4

this equation defines a surface, as shown in Figure 16.5.4. In this case we can show that the vector

$$\mathbf{N} = \nabla F(x, y, z) = F_x(x, y, z)\mathbf{i} + F_y(x, y, z)\mathbf{j} + F_z(x, y, z)\mathbf{k} \tag{6}$$

is perpendicular to the surface $F(x, y, z) = c$ at the point (x, y, z) on the surface. Again let $x = g(t)$, $y = h(t)$, $z = l(t)$ be a parametric representation of any differentiable curve C lying on the surface $F(x, y, z) = c$. On this curve we have

$$\Phi(t) = F[g(t), h(t), l(t)] \equiv c \quad \text{so} \quad \frac{d\Phi}{dt} \equiv 0. \tag{7}$$

On calculating $d\Phi/dt$ by the chain rule we obtain

$$\begin{aligned} \frac{d\Phi}{dt} &= F_x \frac{dx}{dt} + F_y \frac{dy}{dt} + F_z \frac{dz}{dt} \\ &= (\nabla F) \cdot \left(\frac{dx}{dt}\mathbf{i} + \frac{dy}{dt}\mathbf{j} + \frac{dz}{dt}\mathbf{k}\right) \\ &= \mathbf{N} \cdot \frac{d\mathbf{r}}{dt} = 0, \end{aligned} \tag{8}$$

where $d\mathbf{r}/dt$ is a tangent vector to the curve C. Thus it follows that at any point (x_0, y_0, z_0) on the surface, the vector $\mathbf{N} = \nabla F(x_0, y_0, z_0)$ is perpendicular to the tangent vector to any differentiable curve lying on the surface and passing through (x_0, y_0, z_0). The tangent plane at (x_0, y_0, z_0) is given by

$$\nabla F(x_0, y_0, z_0) \cdot [(x - x_0)\mathbf{i} + (y - y_0)\mathbf{j} + (z - z_0)\mathbf{k}] = 0,$$

or

$$F_x(x_0, y_0, z_0)(x - x_0) + F_y(x_0, y_0, z_0)(y - y_0) + F_z(x_0, y_0, z_0)(z - z_0) = 0. \tag{9}$$

The normal line passing through (x_0, y_0, z_0) is

$$\begin{aligned} x - x_0 &= F_x(x_0, y_0, z_0)\, t, \\ y - y_0 &= F_y(x_0, y_0, z_0)\, t, \\ z - z_0 &= F_z(x_0, y_0, z_0)\, t \end{aligned} \tag{10a}$$

or

$$\frac{x - x_0}{F_x(x_0, y_0, z_0)} = \frac{y - y_0}{F_y(x_0, y_0, z_0)} = \frac{z - z_0}{F_z(x_0, y_0, z_0)}. \tag{10b}$$

Notice that for a surface given by $z = f(x, y)$, we can write

$$F(x, y, z) = f(x, y) - z = 0.$$

Then

$$F_x = f_x, \qquad F_y = f_y, \qquad F_z = -1,$$

and Eq. 6 for the normal vector, Eq. 8 for the tangent plane, and Eqs. 10 for the normal line reduce to Eq. 3, Eq. 2, and Eqs. 5, respectively. For this reason it is usually simpler to describe surfaces in the implicit form $F(x, y, z) = c$ with the understanding that in the neighborhood of the point (x_0, y_0, z_0) of interest the implicit equation can be solved for one of the three variables in terms of the other two. We will return to implicit equations in Section 16.8.

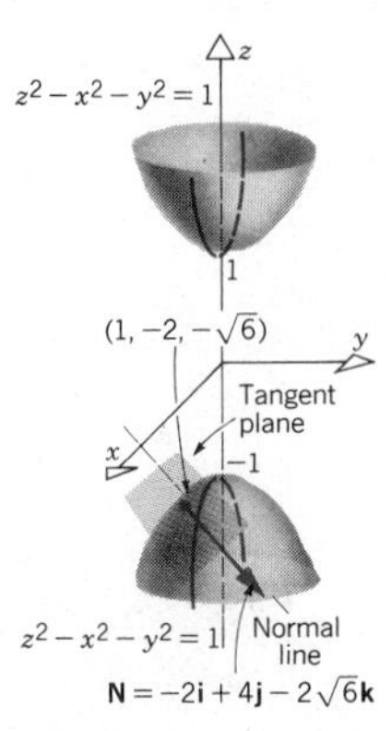

Figure 16.5.5

EXAMPLE 2

Find the tangent plane and normal line at the point $(1, -2, -\sqrt{6})$ on the hyperboloid of two sheets $z^2 - x^2 - y^2 = 1$ (see Figure 16.5.5).

We set $F(x, y, z) = -x^2 - y^2 + z^2$. Then $F_x = -2x$, $F_y = -2y$, $F_z = 2z$, so

$$\mathbf{N} = \text{grad } F(1, -2, -\sqrt{6}) = -2\mathbf{i} + 4\mathbf{j} - 2\sqrt{6}\mathbf{k}.$$

Hence the equation of the tangent plane is

$$-2(x - 1) + 4(y + 2) - 2\sqrt{6}(z + \sqrt{6}) = 0$$

or

$$-x + 2y - \sqrt{6}z = 1.$$

The equation of the normal line is

$$x - 1 = -2t, \qquad y + 2 = 4t, \qquad z + \sqrt{6} = -2\sqrt{6}t, \qquad -\infty < t < \infty,$$

or

$$\frac{x - 1}{-2} = \frac{y + 2}{4} = \frac{z + \sqrt{6}}{-2\sqrt{6}}. \quad \blacksquare$$

EXAMPLE 3

At what points does the normal line through the point $(\sqrt{6}, \sqrt{3}, \sqrt{2})$ on the ellipsoid

$$\frac{x^2}{36} + \frac{y^2}{9} + \frac{z^2}{4} = 1 \tag{11}$$

intersect the sphere

$$x^2 + y^2 + z^2 = 36? \tag{12}$$

The ellipsoid and sphere are shown in Figure 16.5.6. A normal vector at the point (x, y, z) on the ellipsoid is given by $\mathbf{N} = (2x/36)\mathbf{i} + (2y/9)\mathbf{j} + (2z/4)\mathbf{k}$.

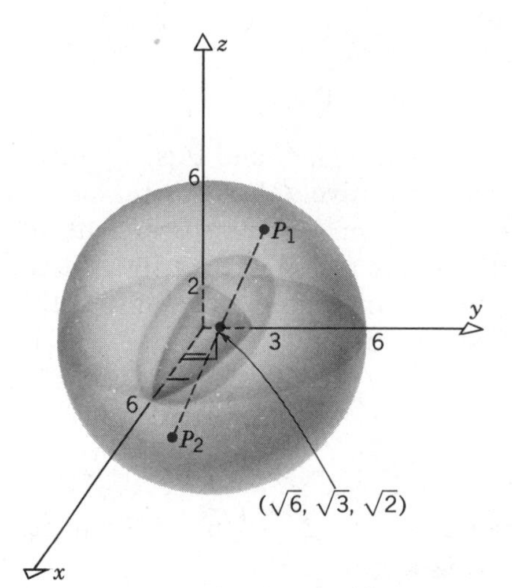

Figure 16.5.6

Thus parametric equations of the normal line passing through the point $(\sqrt{6}, \sqrt{3}, \sqrt{2})$ are

$$\begin{aligned} x - \sqrt{6} &= \frac{2}{36}\sqrt{6}t = \frac{\sqrt{6}}{18}t, \\ y - \sqrt{3} &= \frac{2}{9}\sqrt{3}t, \\ z - \sqrt{2} &= \frac{2}{4}\sqrt{2}t = \frac{\sqrt{2}}{2}t, \end{aligned} \tag{13}$$

$-\infty < t < \infty$. To determine the values of t at which the line given by Eqs. 13 intersects the sphere given by Eq. 12 we substitute Eqs. 13 into Eq. 12. This gives

$$\tfrac{2}{3}t^2 + 4t - 25 = 0,$$

so

$$t = -3 \pm \frac{3}{2}\sqrt{\frac{62}{3}} \cong 3.8191, \ -9.8191.$$

Corresponding to the two values of t we obtain from Eqs. 13 the two points of intersection on the sphere, namely, $P_1(2.9692, 3.2020, 4.1147)$ and $P_2(1.1133, -2.0473, -5.5289)$.

PROBLEMS

In each of Problems 1 through 10, determine the tangent plane and normal line for the given surface at the given point.

1. $z = 3x^2 + 2xy + y^2, \quad (1, 1, 6)$
2. $z = 4xy, \quad (2, -1, -8)$
3. $3x^2 + 2y^2 + z^2 = 6, \quad (0, 1, 2)$
4. $z = e^{2x} \sin \pi y, \quad \left(\frac{1}{2}, \frac{1}{6}, \frac{e}{2}\right)$
5. $x^{1/2} + y^{1/2} + (-z)^{1/2} = 6, \quad (1, 9, -4)$
6. $z^3 + y^3 + x^3 - 6xyz = 0, \quad (2, 1, 3)$
7. $2x^2 - xz + y^2 - zy = -5, \quad (1, 3, 4)$
8. $x^2 + 2y - z^2 = 4, \quad (2, 0, 2)$
9. $z = x^2 \ln(2y^2 - 1) + e^{xy}, \quad (1, 1, e)$
10. $z = xe^{y/x}, \quad (1, -2, e^{-2})$
11. Let (x_0, y_0, z_0) be a point on the ellipsoid

$$\frac{x^2}{a^2} + \frac{y^2}{b^2} + \frac{z^2}{c^2} = 1$$

(see Figure 16.5.7). Show that the equation of the

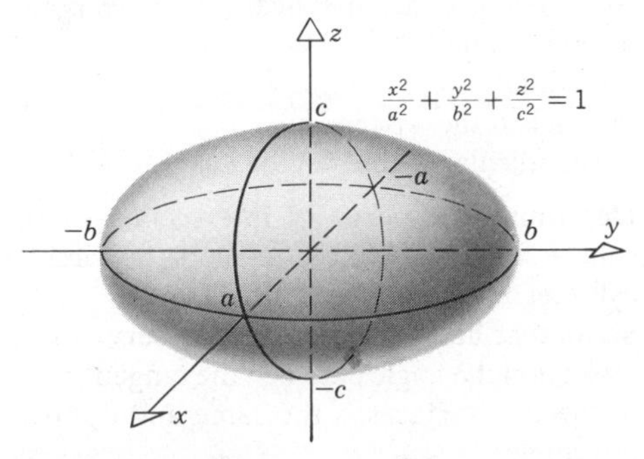

Figure 16.5.7

tangent plane at (x_0, y_0, z_0) is

$$\frac{x_0}{a^2} x + \frac{y_0}{b^2} y + \frac{z_0}{c^2} z = 1.$$

12. Let (x_0, y_0, z_0) be a point on the hyperboloid of one sheet

$$\frac{x^2}{a^2} + \frac{y^2}{b^2} - \frac{z^2}{c^2} = 1$$

(see Figure 16.5.8). Show that the equation of the

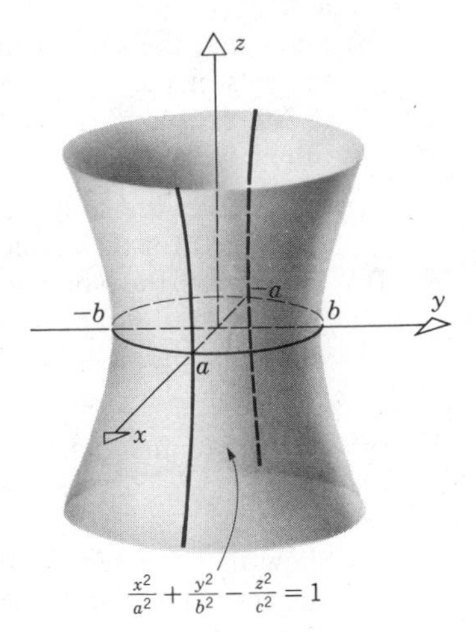

Figure 16.5.8

tangent plane at (x_0, y_0, z_0) is

$$\frac{x_0}{a^2} x + \frac{y_0}{b^2} y - \frac{z_0}{c^2} z = 1.$$

13. Show that the tangent plane at any point (x_0, y_0, z_0) on the surface of the cone

$$\frac{x^2}{a^2} + \frac{y^2}{b^2} - \frac{z^2}{c^2} = 0$$

(Figure 16.5.9) passes through the origin.

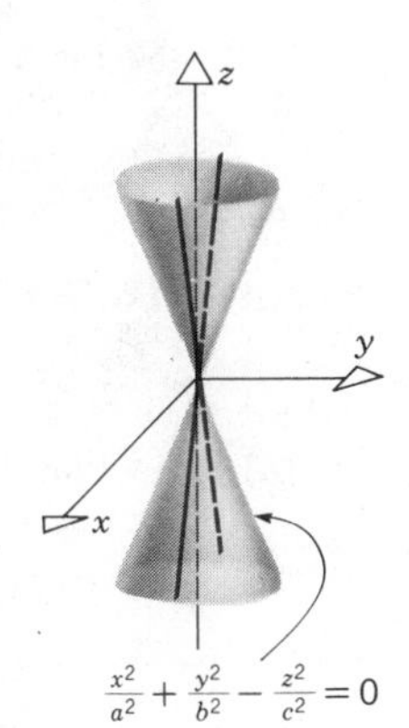

Figure 16.5.9

14. Determine the equation of the tangent plane and normal line for the circular cylinder $x^2 + y^2 = 13$ at the point $(2, 3, \alpha)$.

15. (a) Determine the tangent plane at the point (x_0, y_0, z_0) for the surface $x^{2/3} + y^{2/3} + z^{2/3} = a^{2/3}$.
(b) Show that the sum of the squares of the intercepts of the tangent plane with the coordinate axes is a^2, independent of the choice of (x_0, y_0, z_0).

16. (a) Determine the tangent plant at the point (x_0, y_0, z_0) for the surface $xyz = a^3, a > 0$.
(b) Show that the volume of the tetrahedron formed by the tangent plane and the coordinate planes is $9a^3/2$, independent of the choice of (x_0, y_0, z_0).

17. (a) Determine the tangent plane at the point (x_0, y_0, z_0) for the surface $x^{1/2} + y^{1/2} + z^{1/2} = a^{1/2}$.
(b) Show that the sum of the intercepts of the tangent plane with the coordinate axes is a, independent of the choice of (x_0, y_0, z_0).

18. Suppose that the surfaces $F(x, y, z) = 0$ and $G(x, y, z) = 0$ intersect in a curve; see Figure 16.5.10.

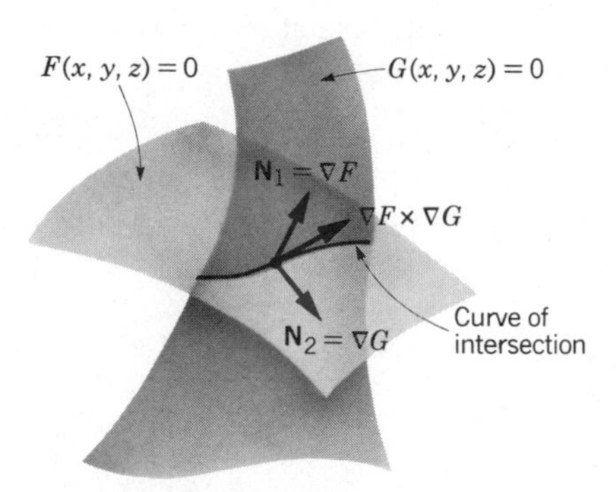

Figure 16.5.10

(a) Show that a tangent vector to this curve at a point (x_0, y_0, z_0) is given by

$$\begin{aligned}\mathbf{u} &= \nabla F(x_0, y_0, z_0) \times \nabla G(x_0, y_0, z_0)\\ &= [(F_yG_z - F_zG_y)\mathbf{i} + (F_zG_x - F_xG_z)\mathbf{j}\\ &\quad + (F_xG_y - F_yG_x)\mathbf{k}]|_{(x_0,y_0,z_0)}.\end{aligned}$$

(b) What is the form of $\mathbf{u}$ if the surfaces are given by $z = f(x, y)$ and $z = g(x, y)$?

In each of Problems 19 through 22, verify that the given point lies on the curve of intersection of the two surfaces and determine the tangent line at this point (see Problem 18).

19. $x + y + z = 0,\ 2x - 3y + 4z = 4,$ $\left(1, -\frac{6}{7}, -\frac{1}{7}\right)$

20. $x^2 + y^2 + z^2 = 14,$ $z = 2x^2 - 4x + y^2 - 4y + 9, \quad (1, 2, 3)$

21. $z = 4xy,\ z = 2xy + 2x - 5y - 13,$ $(2, -1, -8)$

22. $z = -\frac{x^2}{4} + \frac{y^2}{9},\ \frac{x^2}{4} + \frac{y^2}{9} + z^2 = 2,$ $(2, 3, 0)$

23. Consider the surfaces

$$x^2 + y^2 + z^2 = 4$$

$$\text{and} \quad x^2 + y^2 + (z - 4)^2 = 4.$$

Verify that the surfaces intersect at $(0, 0, 2)$. Show that it is not possible to construct a line at this point using the theory developed in Problem 18. Why?

24. At what point on the surface $z = 3x^2 + 2xy + y^2$ is the normal line to the surface parallel to the vector $6\mathbf{i} + 4\mathbf{j} - 2\mathbf{k}$?

25. For the ellipsoid $x^2/a^2 + y^2/b^2 + z^2/c^2 = 1$ (Figure 16.5.7) determine analytically at which points the tangent plane is
(a) horizontal,
(b) parallel to the yz-plane,
(c) perpendicular to the vector $\mathbf{i} + \mathbf{j} + \mathbf{k}$.

26. (a) Determine the curve of intersection of the sphere $x^2 + y^2 + z^2 = a^2$ and the circular cylinder $x^2 + y^2 = a^2/4$ with $z > 0$.
(b) Show that at each point on this curve of intersection the angle between the tangent plane to the two surfaces is the same and evaluate this angle.

27. Show that if f is differentiable, then at each point (x_0, y_0, z_0) on the surface $z = xf(y/x)$ the tangent plane passes through the origin. Note that Problem 10 is a special case of this functional relation.

28. At what points on the surface $z = f(x, y)$ is the tangent plane horizontal?

Angle between a Curve and a Surface

Suppose that the curve $\mathbf{r} = f(t)\mathbf{i} + g(t)\mathbf{j} + h(t)\mathbf{k}$ intersects the surface $F(x, y, z) = 0$ at the point (x_0, y_0, z_0) corresponding to $t = t_0$. The angle between the

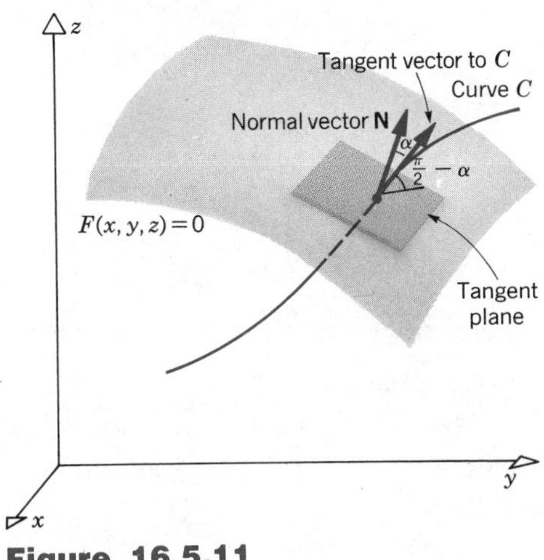

Figure 16.5.11

curve and the surface is defined to be the angle between the tangent vector to the curve and the tangent plane to the surface at (x_0, y_0, z_0) (see Figure 16.5.11). This angle is the complement of the angle between the tangent vector to the curve and the normal vector to the surface.

29. Determine the point (x_0, y_0, z_0) at which the curve $\mathbf{r} = 2\cos \pi t\,\mathbf{i} + 2\sin \pi t\,\mathbf{j} + 6t\,\mathbf{k}$ intersects the paraboloid $z = x^2 + y^2$, and find the angle of intersection.

30. Verify that the curve $\mathbf{r} = \frac{1}{3}t\mathbf{i} + 3\cos\frac{\pi}{3}t\,\mathbf{j} + \frac{1}{2}(t - 1)^2\mathbf{k}$ intersects the ellipsoid $x^2 + y^2/9 + z^2/4 = 3$ at the point $(1, -3, 2)$, and find the angle of intersection.

16.6 MAXIMA AND MINIMA OF FUNCTIONS OF TWO VARIABLES

In this section we extend the discussion of maxima and minima for functions of one variable (Sections 4.1 and 4.2) to functions of two variables.

DEFINITION 16.6.1 A function f of two variables is said to have a relative, or local, maximum at (x_0, y_0) if there is a neighborhood of (x_0, y_0) such that

$$f(x_0, y_0) \geq f(x, y) \tag{1}$$

for all points (x, y) that are in the neighborhood and also in the domain of f. It is said to have a relative, or local, minimum at (x_0, y_0) if there is a neighborhood of (x_0, y_0) such that

$$f(x_0, y_0) \leq f(x, y) \tag{2}$$

for all (x, y) in the neighborhood and in the domain of f. The function is said to have a relative, or local, extremum at (x_0, y_0) if it has either a relative maximum or a relative minimum at (x_0, y_0).

For the surface $z = f(x, y)$ a relative maximum corresponds to the top of a hill and a relative minimum corresponds to the bottom of a pit (see Figure 16.6.1). If a function f of one variable has a local extremum at an interior point x_0, then either $f'(x_0) = 0$ or else $f'(x_0)$ does not exist (Theorem 4.1.1). The corresponding result when f is a function of two variables is stated in terms of ∇f.

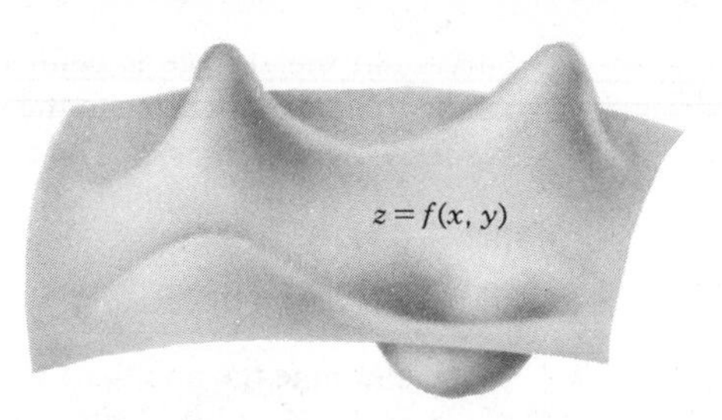

Figure 16.6.1

Theorem 16.6.1
If f is a differentiable function of two variables and if f has a relative extremum at an interior point (x_0, y_0) of its domain, then

$$\nabla f(x_0, y_0) = \mathbf{0}. \tag{3}$$

Proof. Consider the function $g(x) = f(x, y_0)$; see Figure 16.6.2, where $(x_0, y_0) = (0, 0)$. Since f has a relative extremum at (x_0, y_0), it follows that g has a rela-

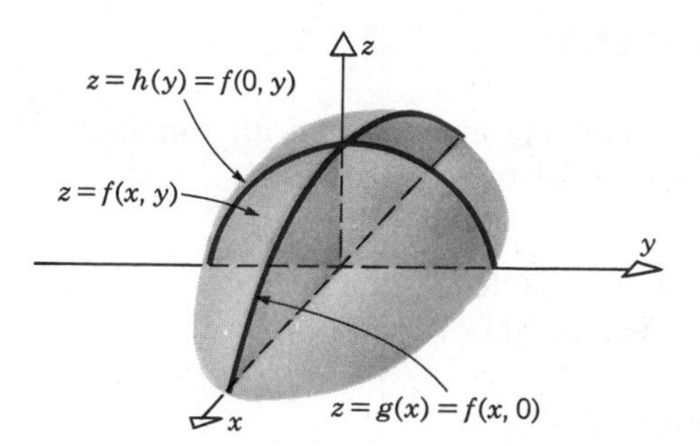

Figure 16.6.2

tive extremum at x_0. Moreover $g'(x) = f_x(x, y_0)$ exists at x_0, and hence $g'(x_0) = f_x(x_0, y_0) = 0$. In a similar manner, using the function $h(y) = f(x_0, y)$, we can show that $f_y(x_0, y_0) = 0$. Hence $\nabla f(x_0, y_0) = \mathbf{0}$.

At such a point we have $D_{\boldsymbol{\lambda}} f(x_0, y_0) = \nabla f(x_0, y_0) \cdot \boldsymbol{\lambda} = 0$, where $\boldsymbol{\lambda}$ is an arbitrary unit vector. That is, at a relative extremum of a differentiable function f the rate of change of f in any direction is zero (see Figure 16.6.3). Also the tangent plane at the point (x_0, y_0, z_0) on the surface $z = f(x, y)$ is

$$f_x(x_0, y_0)\,(x - x_0) + f_y(x_0, y_0)(y - y_0) - (z - z_0) = 0,$$

and this reduces to

$$z = z_0.$$

Thus at a relative extremum the tangent plane to the surface is horizontal. □

Just as for a function of one variable, Theorem 16.6.1 is a necessary but not a sufficient condition for the existence of a relative extremum of a differentiable

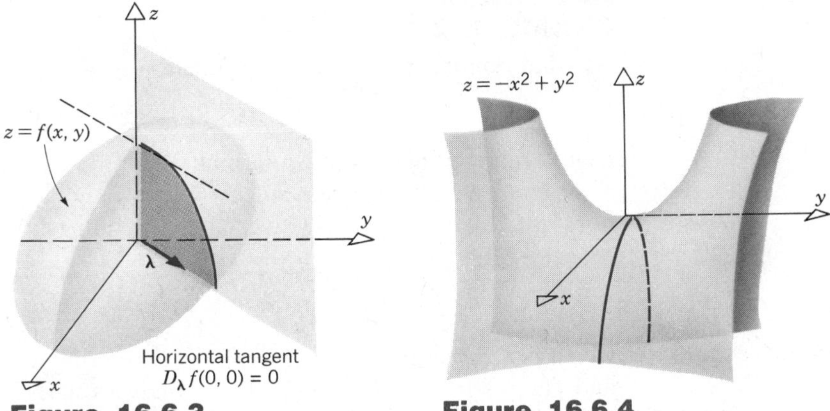

Figure 16.6.3 **Figure 16.6.4**

function. This is illustrated by the hyperbolic paraboloid $z = f(x, y) = -x^2 + y^2$ sketched in Figure 16.6.4. Clearly $f_x(0, 0) = f_y(0, 0) = 0$; however, f has neither a relative maximum nor a relative minimum at $(0, 0)$.

A point at which f is differentiable and $\nabla f(x, y) = \mathbf{0}$ or at which f is not differentiable is called a **critical point*** of f. We shall also refer to a point at which f is differentiable and $\nabla f(x, y) = \mathbf{0}$ as a **stationary point** of f. A stationary point of f that is not an extremum is called a **saddle point**, a terminology that fits naturally with the saddle surface shown in Figure 16.6.4. Given the problem of determining the relative extrema of a function f on an unbounded domain or in the interior of a bounded domain, we must find the critical points of f and then investigate these points as to whether they give a relative maximum, a relative minimum, or neither.

EXAMPLE 1

Determine the critical points of the function

$$f(x, y) = x^2 + y^2.$$

Since f is differentiable for all x and y, we need only find the points at which $f_x(x, y)$ and $f_y(x\ y)$ are equal to zero. We have

$$f_x(x, y) = 2x, \qquad f_y(x, y) = 2y,$$

so on setting $f_x(x, y) = 0$ and $f_y(x, y) = 0$, we find that the only critical point is $(0, 0)$. Since $f(0, 0) = 0$ and $f(x, y) > 0$ for $(x, y) \neq (0, 0)$ it follows that f has a relative minimum at $(0, 0)$. A sketch of $z = x^2 + y^2$ is given in Figure 16.6.5; it confirms that f has a relative minimum at $(0, 0)$. ∎

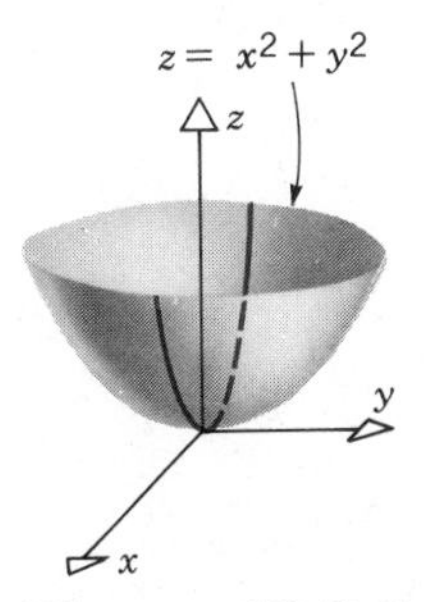

Figure 16.6.5

*The definition of a critical point in this section is not altogether consistent with the definition in Section 4.2, where endpoints were also classified as critical points. Here we wish to deal with boundary points separately (in Section 16.7), so we do not include them among the critical points.

EXAMPLE 2

At what points, if any, does the function

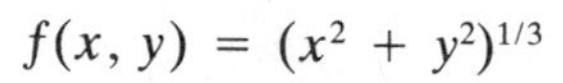

$$f(x, y) = (x^2 + y^2)^{1/3}$$

have a relative maximum or minimum?

First, we observe that since $f(x, 0) = x^{2/3}$ and $f(0, y) = y^{2/3}$ the function f does not have partial derivatives at (0, 0). For $(x, y) \neq (0, 0)$ we have

$$\nabla f(x, y) = \frac{2x}{3(x^2 + y^2)^{2/3}}\mathbf{i} + \frac{2y}{3(x^2 + y^2)^{2/3}}\mathbf{j}, \qquad (x, y) \neq (0, 0),$$

which is *never* equal to $\mathbf{0}$. Thus (0, 0) is the only critical point. Since $f(0, 0) = 0$ and $f(x, y) > 0$ for $(x, y) \neq (0, 0)$, f has a relative minimum at (0, 0). A sketch of $z = (x^2 + y^2)^{1/3}$ is given in Figure 16.6.6. ∎

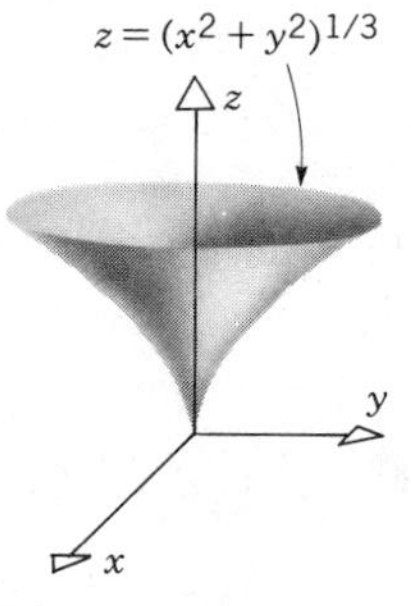

Figure 16.6.6

To determine the stationary points of a differentiable function f we must determine the points (x, y) at which

$$f_x(x, y) = 0 \quad \text{and} \quad f_y(x, y) = 0. \tag{4}$$

In Examples 1 and 2 this was quite easy. However, if Eqs. 4 are not linear in x and y it may be difficult to solve them for the critical points. Indeed, it may be necessary to resort to a numerical procedure to solve Eqs. 4. Here, we wish to focus on the theory of maxima and minima of functions of two variables rather than on numerical procedures for solving two simultaneous equations. Hence, we use only examples and problems for which it is possible to solve Eqs. 4 fairly easily.

To distinguish whether a stationary point (x_0, y_0) of $z = f(x, y)$ is a saddle point, a relative maximum point, or a relative minimum point we need a generalization of the second derivative test for functions of one variable (Theorem 4.2.4). We give a partial derivation, based on the second directional derivative, of this generalization.

Suppose that f has continuous second partial derivatives, and let

$$\boldsymbol{\lambda} = \lambda_1\mathbf{i} + \lambda_2\mathbf{j}$$

be a unit vector in an arbitrary direction. Then the second directional derivative in the direction $\boldsymbol{\lambda}$ is

$$\begin{aligned} D^2_{\boldsymbol{\lambda}}f(x, y) &= D_{\boldsymbol{\lambda}}[D_{\boldsymbol{\lambda}}f(x, y)] \\ &= D_{\boldsymbol{\lambda}}[f_x(x, y)\lambda_1 + f_y(x, y)\lambda_2]. \end{aligned}$$

Calculating the directional derivatives of f_x and f_y separately, we obtain

$$\begin{aligned} D^2_{\boldsymbol{\lambda}}f(x, y) = {} & [f_{xx}(x, y)\lambda_1]\lambda_1 + [f_{xy}(x, y)\lambda_1]\lambda_2 \\ & + [f_{yx}(x, y)\lambda_2]\lambda_1 + [f_{yy}(x, y)\lambda_2]\lambda_2. \end{aligned}$$

The assumption that f has continuous second derivatives means that $f_{xy}(x, y) = f_{yx}(x, y)$. Consequently,

$$D^2_{\boldsymbol{\lambda}}f(x, y) = f_{xx}(x, y)\lambda_1^2 + 2f_{xy}(x, y)\lambda_1\lambda_2 + f_{yy}(x, y)\lambda_2^2. \tag{5}$$

If we now consider the stationary point (x_0, y_0) and let

$$A = f_{xx}(x_0, y_0), \qquad B = f_{xy}(x_0, y_0), \qquad C = f_{yy}(x_0, y_0), \tag{6}$$

then

$$D_\lambda^2 f(x_0, y_0) = A\lambda_1^2 + 2B\lambda_1\lambda_2 + C\lambda_2^2.$$

Assume first that $\lambda_1 \neq 0$. Then we can write

$$D_\lambda^2 f(x_0, y_0) = \lambda_1^2(A + 2Bm + Cm^2), \tag{7}$$

where $m = \lambda_2/\lambda_1$ is the slope of a line through the point (x_0, y_0). Since $\lambda_1^2 > 0$, it follows that the sign of $D_\lambda^2 f(x_0, y_0)$ is the same as the sign of the quadratic expression $Q(m) = A + 2Bm + Cm^2$.

If $B^2 - AC > 0$, then $Q(m)$ has two distinct real zeros, so $Q(m)$ is positive for some values of m and negative for others. Thus, in some directions the trace of f has a minimum at (x_0, y_0), while in others it has a maximum. Consequently, (x_0, y_0) is a saddle point.

On the other hand, if $B^2 - AC < 0$, then $Q(m)$ has no real zeros, so $Q(m)$ is always of one sign, which is the same as the sign of A and C. Hence, if $B^2 - AC < 0$, then (x_0, y_0) is an extreme point. If $A > 0$ and $C > 0$, then $Q(m) > 0$ for all m, and (x_0, y_0) is a relative minimum point, while if $A < 0$ and $C < 0$, then $Q(m) < 0$ for all m, and (x_0, y_0) is a relative maximum point.

If $B^2 - AC = 0$, then $Q(m)$ has a single repeated zero, but does not change sign. In this case no conclusion can be drawn (see Problem 36).

Finally, if λ_1 can be zero, then we assume that $\lambda_2 \neq 0$ and repeat essentially the same argument with the roles of λ_1 and λ_2 reversed. We can summarize these results in the following theorem.

Theorem 16.6.2

(The Second Derivative Test)

Let f be a function of two variables with continuous second partial derivatives in a neighborhood of (x_0, y_0) and suppose that $\nabla f(x_0, y_0) = \mathbf{0}$. Let A, B, and C be given by Eq. 6.

(a) If $B^2 - AC < 0$ with $A < 0$ and $C < 0$, then (x_0, y_0) is a relative maximum point.
(b) If $B^2 - AC < 0$ with $A > 0$ and $C > 0$, then (x_0, y_0) is a relative minimum point.
(c) If $B^2 - AC > 0$, then (x_0, y_0) is a saddle point.
(d) If $B^2 - AC = 0$, then no conclusion can be drawn without further investigation.

The use of this theorem is illustrated by the following examples.

EXAMPLE 3

For the function

$$f(x, y) = x^2 + 2xy - 2y^2 + 3x + 4$$

determine all stationary points and classify them as saddle points, relative maximum points, or relative minimum points.

The stationary points are found from

$$f_x(x, y) = 2x + 2y + 3 = 0, \quad f_y(x, y) = 2x - 4y = 0.$$

The only solution of these equations is $x = -1$, $y = -\frac{1}{2}$, so $(-1, -\frac{1}{2})$ is the only stationary point of f. To classify it we calculate the second derivatives of f; thus $f_{xx}(x, y) = 2$, $f_{xy}(x, y) = 2$, and $f_{yy}(x, y) = -4$. In the notation of Theorem 16.6.2 we have $A = 2$, $B = 2$, and $C = -4$, so $B^2 - AC = 12 > 0$. Therefore $(-1, -\frac{1}{2})$ is a saddle point. ∎

EXAMPLE 4

For the function

$$f(x, y) = \tfrac{1}{3}y^3 + x^2y - 2x^2 - 2y^2 + 6$$

determine all stationary points and classify them as saddle points, relative maximum points, or as relative minimum points.

The stationary points are the solutions of

$$f_x(x, y) = 2xy - 4x = 2x(y - 2) = 0,$$

$$f_y(x, y) = y^2 + x^2 - 4y = 0.$$

From the first equation we must have either $x = 0$ or $y = 2$. If $x = 0$, then the second equation is $y(y - 4) = 0$ with solutions $y = 0$ and $y = 4$. If $y = 2$, then the second equation is $x^2 = 4$ and $x = \pm 2$. Thus there are four stationary points, namely, $(0, 0)$, $(0, 4)$, $(2, 2)$, and $(-2, 2)$.

Next we find the second derivatives:

$$f_{xx}(x, y) = 2y - 4, \qquad f_{xy}(x, y) = 2x, \qquad f_{yy}(x, y) = 2y - 4.$$

Evaluating f_{xx}, f_{xy}, and f_{yy} at each of the critical points and then using Theorem 16.6.2, we obtain the results in Table 16.1.

Table 16.1 Classification of Stationary Points in Example 4

Point	A	B	C	$B^2 - AC$	Conclusion
(0, 0)	−4	0	−4	−16	Relative maximum
(0, 4)	4	0	4	−16	Relative minimum
(2, 2)	0	4	0	16	Saddle point
(−2, 2)	0	−4	0	16	Saddle point

In many applications we are interested in the largest or the smallest possible value of f in a given domain. The function f is said to have an **absolute,** or **global, maximum** at (x_0, y_0) if $f(x_0, y_0) \geq f(x, y)$ for all (x, y) in the domain of f. Similarly, f is said to have an **absolute,** or **global, minimum** at (x_0, y_0) if $f(x_0, y_0) \leq f(x, y)$ for all (x, y) in the domain of f. In general, a function need not have an absolute maximum or an absolute minimum on a given region. However, by the two-dimensional version of Theorem 2.6.2, any function that is *continuous on a closed bounded* region must have an absolute maximum and an absolute minimum in that region. In seeking the absolute maximum, say, of f we must proceed in a manner similar to that for a function of one variable. We study f at its critical points and at the boundary points of the domain of f. Often there are only one or two possibilities and one can determine from the context of the problem that a critical point must give the desired absolute maximum or absolute minimum. ∎

EXAMPLE 5

Find the dimensions of the rectangular parallelepiped of maximum volume with edges parallel to the axes that can be inscribed in the ellipsoid

$$\frac{x^2}{16} + \frac{y^2}{9} + \frac{z^2}{4} = 1. \tag{8}$$

The ellipsoid and rectangular parallelepiped are sketched in Figure 16.6.7. Let V be the volume of the rectangular parallelepiped and let (x, y, z) be the

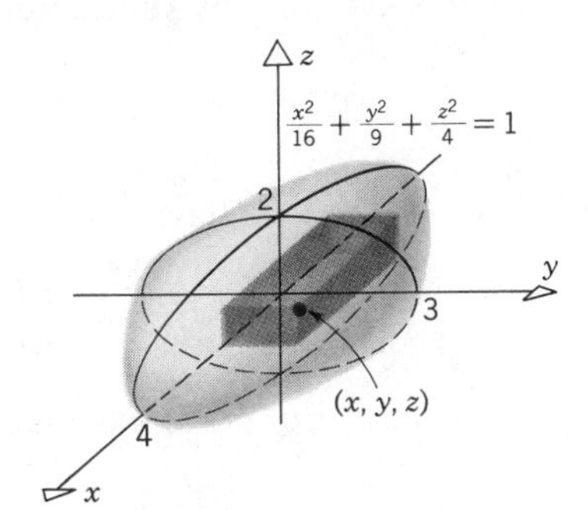

Figure 16.6.7

corner point of the rectangular parallelepiped in the first octant. Then, by symmetry,

$$V = 8xyz. \tag{9}$$

The point (x, y, z) lies on the ellipsoid so $z = 2(1 - x^2/16 - y^2/9)^{1/2}$. Hence

$$V = 16xy\left(1 - \frac{x^2}{16} - \frac{y^2}{9}\right)^{1/2}, \tag{10}$$

where (x, y) lies in the quarter ellipse $x^2/16 + y^2/9 \leq 1$ with $0 \leq x \leq 4$,

$0 \le y \le 3$. It follows from Eq. 10 that $V = 0$ on the boundaries of the quarter ellipse.

To find the stationary points, we calculate V_x and V_y:

$$V_x = 16y\left(1 - \frac{x^2}{16} - \frac{y^2}{9}\right)^{1/2} - \frac{16x^2y}{16}\left(1 - \frac{x^2}{16} - \frac{y^2}{9}\right)^{-1/2},$$

$$V_y = 16x\left(1 - \frac{x^2}{16} - \frac{y^2}{9}\right)^{1/2} - \frac{16xy^2}{9}\left(1 - \frac{x^2}{16} - \frac{y^2}{9}\right)^{-1/2}.$$

Setting $V_x = 0$, $V_y = 0$, and multiplying each equation by $(1 - x^2/16 - y^2/9)^{1/2}$, we obtain

$$16y\left(1 - \frac{x^2}{16} - \frac{y^2}{9}\right) - x^2y = 0,$$

$$16x\left(1 - \frac{x^2}{16} - \frac{y^2}{9}\right) - \frac{16}{9}xy^2 = 0.$$

One solution of these equations is $x = 0$, $y = 0$; however, this corresponds to a parallelepiped of zero volume. Thus, assuming that $x \neq 0$ and $y \neq 0$, we divide the first equation by y, the second equation by x, and combine terms to obtain

$$2x^2 + \frac{16}{9}y^2 = 16,$$

$$x^2 + \frac{32}{9}y^2 = 16.$$

We can solve for x by subtracting twice the first equation from the second equation

$$3x^2 = 16, \quad \text{so} \quad x = \frac{4}{\sqrt{3}}.$$

For $x = 4/\sqrt{3}$ we have

$$y^2 = \frac{9}{16}\left(16 - 2\,\frac{16}{3}\right), \quad \text{so} \quad y = \sqrt{3}.$$

We use Eq. 10 to calculate the corresponding value of V, namely

$$V = 16\,\frac{4}{\sqrt{3}}\,\sqrt{3}\left(1 - \frac{1}{16}\,\frac{16}{3} - \frac{1}{9}\,3\right)^{1/2} = \frac{64}{\sqrt{3}}.$$

Since the boundary points give zero volume, and since, except for $(0, 0)$, there is only one critical point in the quarter ellipse, we conclude that the maximum volume of the rectangular parallelepiped is $64/\sqrt{3} \cong 36.95$ units^3. ∎

In concluding this section we note that while we have restricted our discussion to functions of two variables, the ideas readily generalize to functions of several variables. In particular, if f is differentiable, then the stationary points are given by $\nabla f(\mathbf{r}) = \mathbf{0}$. Of course, the statement of a second derivative test as given in

Theorem 16.6.2 becomes more complicated. However the ideas concerning the second directional derivative $D_{\lambda}^2 f(\mathbf{r})$ are still valid, but in a higher number of dimensions. These generalizations are discussed in some books on advanced calculus and on optimization theory.

PROBLEMS

In each of Problems 1 through 22, locate all the stationary points of the function f and use Theorem 16.6.2 to classify them as saddle points, relative maximum points, or relative minimum points.

1. $f(x, y) = 7 - 2x + 6y + 2x^2 + 3y^2$
2. $f(x, y) = 1 + 3x + 2y - x^2 + xy - 2y^2$
3. $f(x, y) = x^2 - 3xy - 2y^2$
4. $f(x, y) = 5 + 3x + 6y - 3x^2 - 2y^2$
5. $f(x, y) = x + 6y + \frac{1}{2}x^2 + xy + 3y^2$
6. $f(x, y) = 4 + 2x + 6y - x^2 + 2y^2$
7. $f(x, y) = 3 + x - x^2 - 2xy - 2y^2$
8. $f(x, y) = x^2 - y^2 + 3x$
9. $f(x, y) = ye^x - y$
10. $f(x, y) = \frac{1}{2}y^2 + 1 - \cos x$
11. $f(x, y) = y \ln x + 3x^2 - 2y, \quad x > 0$
12. $f(x, y) = 3x^3 + xy - 2x - \frac{1}{6}y^2$
13. $f(x, y) = xy + x^2 + y^3$
14. $f(x, y) = 2x^3 + 2xy^2 + 3x^2 + y^2 - 5$
15. $f(x, y) = e^{x^2} \sin y$
16. $f(x, y) = -\frac{1}{2}x^4 + xy^2 - \frac{1}{4}y^4 + 7$
17. $f(x, y) = e^{-(x^2+y^2)}$
18. $f(x, y) = \dfrac{1}{x} + y^2 + x$
19. $f(x, y) = \ln|y| + x^2 + \frac{1}{2}y^2 + 2xy - 3$
20. $f(x, y) = \dfrac{1}{9}y^3 + 3x^2y + 9x^2 + y^2 + 9$
21. $f(x, y) = x^3 + xy^2 + x^2 + y^2 + 1$
22. $f(x, y) = \sqrt{x^2 + 1} + 3y^2 - 2y + 6$

In several of the following problems it is necessary to minimize or maximize a certain positive quantity, such as a distance. It often turns out that the algebra is somewhat simpler if one studies instead the equivalent problem of minimizing or maximizing the square of the quantity. That the problems are equivalent is demonstrated in Problem 34.

23. If the sum of three positive numbers is a, what are their values if the sum of their squares is a minimum?

24. Show that the rectangular box of maximum volume that can be inscribed in a sphere of radius a is a cube and determine its dimensions and volume.

25. Determine the point on the plane $ax + by + cz = d$ that is closest to the origin and show that the vector from the origin to this point is normal to the plane.

26. Determine the dimensions of a box of length l, breadth b, and height h of fixed volume V that is open on the top and requires the minimum amount of material.

27. Suppose that a rectangular box with an open top is to have a volume of 6 ft^3. If the material for the bottom costs \$1.50/ft^2 and the material for the sides costs \$1.00/ft^2, determine the dimensions of the box that minimizes the cost.

28. Show that among all rectangular parallelepipeds of given surface area the cube has maximum volume.

29. **Method of least squares.** Suppose that a set of N points (X_i, Y_i), with $N > 2$, is given; we wish to draw a straight line $y = mx + b$ so that $y_i = mX_i + b$ provides a reasonable approximation to Y_i for each X_i. How should we choose m and b? In the method of least squares the criterion is to minimize the sum of the squares of the errors

$$S = \sum_{i=1}^{N} (Y_i - y_i)^2.$$

The corresponding line is called the least squares line.

(a) Show that minimizing S leads to the following two equations for m and b:

$$\left(\sum_{i=1}^{N} X_i^2\right)m + \left(\sum_{i=1}^{N} X_i\right)b = \sum_{i=1}^{N} X_iY_i, \qquad (i)$$

$$\left(\sum_{i=1}^{N} X_i\right)m + Nb = \sum_{i=1}^{N} Y_i. \qquad (ii)$$

(b) Determine the least squares line for the set of points (0, 0), (1, 1), and (2, $\frac{3}{2}$), and make a sketch of the points and the line.

(c) For the least squares line we have $y_i = mX_i + b$, where m and b satisfy Eqs. (*i*) and (*ii*). Show that it follows from Eq. (*ii*) that the sum of the deviations, $\Sigma_{i=1}^{N}(Y_i - y_i)$, is zero and verify this result for the specific example in Part (b).

30. Use the procedure described in Problem 29 to determine the least squares line for the set of points (−2, 0), (−1, 2), (0, 3), (1, 3). Make a sketch of the points and the line.

31. In Problem 29 we discussed one procedure for determining a straight line in the xy-plane to approximate a set of data points (X_i, Y_i), $i = 1, 2, \ldots, N$. In this problem we consider how one might approximate a function f by a polynomial $a_0 + a_1x + a_2x^2 + \cdots + a_nx^n$ on an interval $a \leq x \leq b$. One useful procedure is to choose the coefficients $a_0, a_1, \ldots, a_n$ so as to minimize the mean square error of the difference between the function and the polynomial, namely

$$E_n = \int_a^b [f(x) - (a_0 + a_1x + a_2x^2 + \cdots + a_nx^n)]^2dx.$$

How should we choose the a_n if $a = -1$, $b = 1$ and (a) $n = 1$, (b) $n = 2$?

32. A telephone company wants to build a central switching station so as to minimize the sum of the squares of the distances from the station to each of the subscribers it serves. If the locations of the subscribers are (x_i, y_i), $i = 1, 2, \ldots, N$, where should the station be located?

33. Two glove manufacturers in competition with each other have profits p_1 and p_2 that depend upon each company's output and the competitor's output according to the formulas

$$p_1 = 40q_1 - q_1^2 - \tfrac{1}{2}q_2^2,$$
$$p_2 = 60q_2 - 2q_2^2 - \tfrac{1}{2}q_1^2.$$

Here q_1 and q_2 are measured in units of thousands of pairs of gloves, and p_1 and p_2 are measured in tens of dollars.

(a) If the two companies act independently so as to maximize their individual profits, determine the production levels q_1 and q_2, the individual profits p_1 and p_2, and the total profit $p_1 + p_2$.

(b) If the two companies act together so as to maximize the total profit $p_1 + p_2$, determine the corresponding outputs, individual profits, and total profit.

34. Suppose that $f(x, y) > 0$. Let $g(x, y) = f^2(x, y)$. We wish to show that if g has a stationary point at (x_0, y_0), then f has a stationary point at (x_0, y_0), and that the character of the stationary point for f is the same as that for g.

(a) Show that if $g_x(x_0, y_0) = g_y(x_0, y_0) = 0$, then $f_x(x_0, y_0) = f_y(x_0, y_0) = 0$. Hence f has a stationary point at (x_0, y_0).

(b) Let $A_1 = g_{xx}(x_0, y_0)$, $B_1 = g_{xy}(x_0, y_0)$, $C_1 = g_{yy}(x_0, y_0)$, $A_2 = f_{xx}(x_0, y_0)$, $B_2 = f_{xy}(x_0, y_0)$, and $C_2 = f_{yy}(x_0, y_0)$. Show that A_1 and A_2 have the same sign or both are zero, and that

$$B_1^2 - A_1C_1 = 4f^2(x_0, y_0)(B_2^2 - A_2C_2).$$

Since $f(x_0, y_0) > 0$, it follows that $B_1^2 - A_1C_1$ and $B_2^2 - A_2C_2$ have the same sign or both are zero. Hence the character of the stationary point for f is the same as that for g.

*** 35.** Consider the function

$$f(x, y) = (y - x^2)(y - 3x^2).$$

(a) Show that f has a relative minimum at the origin on each straight line $y = mx$.

(b) Show that $f(0, \beta) > f(0, 0) = 0$ for any $\beta \neq 0$ and that $f(\alpha, 2\alpha^2) < 0$ for any $\alpha \neq 0$.

Part (b) shows that even though f has a relative minimum at the origin on each straight line path $y = mx$, it cannot have a relative minimum at (0, 0). However, this does not contradict Theorem 16.6.2 or the discussion leading up to that theorem. Notice that in this case Eq. 5 becomes

$$D_\lambda^2 f(0, 0) = 2\sin^2\theta,$$

and hence is 0 for $\theta = 0$ and π. Also observe that $B^2 - AC = 0$ in Theorem 16.6.2, so the theorem gives no conclusion.

36. In this problem we will show that if (x_0, y_0) is a stationary point of $f(x, y)$ and if $B^2 - AC = 0$, then no conclusion can be drawn about the character of the point (x_0, y_0).

(a) Show that for each of the following functions $(0, 0)$ is a stationary point and $B^2 - AC = 0$:

$$f(x, y) = x^4 + y^4,$$
$$g(x, y) = -(x^4 + y^4),$$
$$h(x, y) = x^4 - y^4.$$

(b) Show that $f(0, 0) \leq f(x, y)$ for all (x, y); hence $(0, 0)$ yields the minimum value of f. Similarly show that $g(0, 0) \geq g(x, y)$ for all (x, y); hence $(0, 0)$ yields the maximum value of g.

(c) Show that in any disc around $(0, 0)$ there are points (x, y) such that $h(0, 0) < h(x, y)$ and points (x, y) such that $h(0, 0) > h(x, y)$; hence $(0, 0)$ is a saddle point for h.

16.7 CONSTRAINED EXTREMUM PROBLEMS

Often we want to find the relative maxima and/or relative minima of $w = f(x, y)$ subject to a constraint (side condition) that the point (x, y) lie on a curve $g(x, y) = 0$. Or we may want to find the relative maxima and/or relative minima of $w = f(x, y, z)$ subject to the constraint that the point (x, y, z) lie on the surface given by $g(x, y, z) = 0$. We refer to such problems as constrained extremum problems.

Such a problem arose in Example 5 of Section 16.6 when we sought the maximum of $V = 8xyz$ subject to the condition that the point (x, y, z) lie on the surface of the ellipsoid

$$\frac{x^2}{16} + \frac{y^2}{9} + \frac{z^2}{4} = 1 \tag{1}$$

in the first octant. In that case we simply solved Eq. 1 for z and substituted in the expression for V to obtain V as a function of two variables. We then used the techniques developed in Section 16.6 for functions of two variables to determine the maximum value of V. For this particular example it was possible to solve for z and to substitute in the expression for V; however, this is not always the case. The function g may be too complicated for us to find an explicit formula for z in terms of x and y, or for any one of the variables in terms of the other two.

We would like to develop an alternative theory for determining the relative maxima and minima of a function of several variables subject to a constraining condition on the variables that does not require us to use the constraining condition to eliminate one of the variables. We will illustrate the procedure for a function of two variables. Thus we consider the problem of determining the relative maxima and minima of $w = f(x, y)$ subject to the constraint that the point (x, y) lie on the curve given by $g(x, y) = 0$. We assume that f and g have continuous first partial derivatives.

Let C denote the curve defined by $g(x, y) = 0$, and recall that the vector

$$\nabla g(x, y) = g_x(x, y)\mathbf{i} + g_y(x, y)\mathbf{j} \tag{2}$$

is normal to C at each point on C. We assume that $\nabla g(x, y) \neq \mathbf{0}$ at points on C. Now suppose that the curve C has the parametric representation $x = x(t)$, $y = y(t)$; then $g[x(t), y(t)] \equiv 0$. Now consider $w = f(x, y)$. On the curve C,

$$w = f[x(t), y(t)] \tag{3}$$

is a function of one variable. Finally, suppose that w has an extremum at t_0 corresponding to the point $x_0 = x(t_0)$, $y_0 = y(t_0)$. Since w is differentiable, we know that $dw(t_0)/dt = 0$. Next we calculate dw/dt using the chain rule:

$$\frac{dw}{dt} = f_x(x, y)\frac{dx}{dt} + f_y(x, y)\frac{dy}{dt}, \tag{4}$$

and, in particular

$$\begin{aligned} \frac{dw(t_0)}{dt} &= f_x(x_0, y_0)\frac{dx(t_0)}{dt} + f_y(x_0, y_0)\frac{dy(t_0)}{dt} \\ &= \nabla f(x_0, y_0) \cdot \left[\frac{dx(t_0)}{dt}\mathbf{i} + \frac{dy(t_0)}{dt}\mathbf{j}\right] = 0. \end{aligned} \tag{5}$$

Since the vector

$$\frac{dx}{dt}\mathbf{i} + \frac{dy}{dt}\mathbf{j}$$

is tangent to the curve C, it follows from Eq. 5 that $\nabla f(x_0, y_0)$ is normal to the curve C at the point (x_0, y_0). Also we know that $\nabla g(x_0, y_0)$ is perpendicular to the curve C. Hence these two gradient vectors at (x_0, y_0) are parallel. Since $\nabla g(x_0, y_0) \neq \mathbf{0}$, there exists a scalar λ such that

$$\nabla f(x_0, y_0) = \lambda \nabla g(x_0, y_0). \tag{6}$$

Thus in seeking the points on $g(x, y) = 0$ at which $f(x, y)$ has a relative extremum we need only consider the points at which ∇f and ∇g are parallel, that is, the points at which ∇f and ∇g satisfy Eq. 6. The constraint $g(x, y) = 0$ and the two equations coming from the vector equation $\nabla f(x, y) = \lambda \nabla g(x, y)$ provide three equations for x, y, and λ.

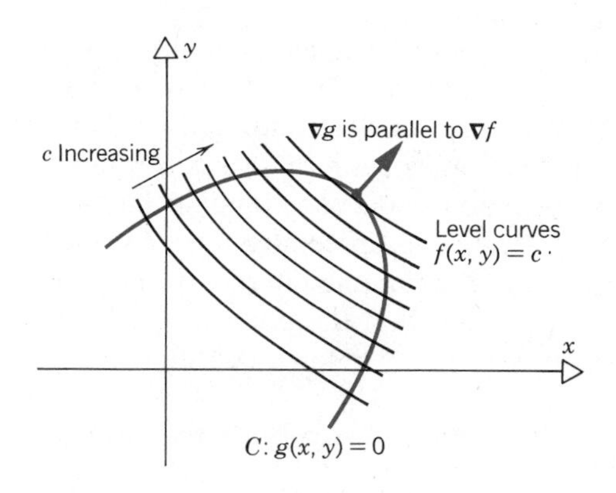

Figure 16.7.1

A geometric interpretation of the situation for the case in which we seek a maximum of $f(x, y)$ subject to the constraint $g(x, y) = 0$ is shown in Figure 16.7.1. At the point at which $f(x, y)$ is a maximum and $g(x, y) = 0$ the level curves of f and the constraint curve are tangent, and hence their normal vectors are parallel.

The parameter λ is called a Lagrange multiplier and the procedure is called the **method** of **Lagrange multipliers**. Though we have discussed the method of Lagrange multipliers only for functions of two variables, the procedure generalizes to the case in which f and g are functions of more than two variables.

Theorem 16.7.1

Let f and g be functions of several variables with continuous first partial derivatives. If $f(\mathbf{r})$ has a relative extremum at $\mathbf{r}_0$ subject to the constraint $g(\mathbf{r}) = 0$, then $\nabla f(\mathbf{r}_0)$ and $\nabla g(\mathbf{r}_0)$ are parallel. Hence, provided $\nabla g(\mathbf{r}_0) \neq \mathbf{0}$, there exists a scalar λ such that

$$\nabla f(\mathbf{r}_0) = \lambda \nabla g(\mathbf{r}_0). \tag{7}$$

For the case that f and g are functions of three variables, the geometric interpretation of Theorem 16.7.1 is as follows. At a point (x_0, y_0, z_0) on the surface $g(x, y, z) = 0$ at which f has a relative extremum the gradient vector $\nabla f(\mathbf{r}_0)$ is parallel to the normal vector $\nabla g(\mathbf{r}_0)$ to the surface $g(x, y, z) = 0$. By analogy with the two-dimensional situation illustrated in Figure 16.7.1, a surface from the family of surfaces $f(x, y, z) = c$ is tangent to the surface $g(x, y, z) = 0$ at the point (x_0, y_0, z_0).

EXAMPLE 1

Find the minimum and maximum of

$$w = f(x, y) = x^2 + y^2 \tag{8}$$

subject to the constraint

$$g(x, y) = x^2 - xy + y^2 - 3 = 0. \tag{9}$$

We illustrate how to solve this constrained maximum-minimum problem using the method of Lagrange multipliers and also by direct elimination. First we use the method of Lagrange multipliers.

We seek values of x, y, and λ such that $\nabla f(x, y) = \lambda \nabla g(x, y)$ and $g(x, y) = 0$. We have

$$\nabla f(x, y) = 2x\mathbf{i} + 2y\mathbf{j}, \qquad \nabla g(x, y) = (2x - y)\mathbf{i} + (-x + 2y)\mathbf{j},$$

so

$$2x = \lambda(2x - y) \tag{10a}$$

and

$$2y = \lambda(-x + 2y). \tag{10b}$$

These two equations together with the constraint (9) give us three equations for x, y, and λ. The simplest way to proceed is to eliminate λ from Eqs. 10, obtaining a relation between x and y that we can then substitute into the constraint (9). First observe that if $y = 2x$, then the right side of Eq. 10a is zero, so we must have $x = 0$, which implies $y = 0$. However, $(0, 0)$ does not satisfy the constraint (9). Thus we can exclude the possibility $y = 2x$. Similarly, working with Eq. 10b, we find that we can exclude the possibility $y = x/2$.

For $y \neq 2x$ and $y \neq x/2$, we can solve each of Eqs. 10 for λ, obtaining

$$\lambda = \frac{2x}{2x - y} = \frac{2y}{-x + 2y}.$$

Hence

$$2x(-x + 2y) = 2y(2x - y)$$

or

$$x^2 = y^2,$$

so

$$y = \pm x.$$

Substituting this result in the constraint (9) we obtain

$$y = x: \quad x^2 - x^2 + x^2 - 3 = 0 \quad \text{so} \quad x = \pm\sqrt{3},\ y = \pm\sqrt{3},$$

$$y = -x: \quad x^2 + x^2 + x^2 - 3 = 0 \quad \text{so} \quad x = \pm 1,\ y = \mp 1.$$

Thus we must consider each of the four points $(\sqrt{3}, \sqrt{3})$, $(-\sqrt{3}, -\sqrt{3})$, $(1, -1)$ and $(-1, 1)$. We evaluate f at each of these points:

$$f(\sqrt{3}, \sqrt{3}) = 3 + 3 = 6, \qquad f(-\sqrt{3}, -\sqrt{3}) = 3 + 3 = 6$$

$$f(1, -1) = 1 + 1 = 2, \qquad f(-1, 1) = 1 + 1 = 2.$$

We conclude that the constrained maximum of f is 6 and it occurs at $(\pm\sqrt{3}, \pm\sqrt{3})$, and the constrained minimum of f is 2 and it occurs at $(\pm 1, \mp 1)$.

We contrast this procedure with the method of elimination. Solving Eq. 9 for y, we obtain

$$y = \frac{x \pm \sqrt{x^2 - 4(x^2 - 3)}}{2} = \frac{x \pm \sqrt{12 - 3x^2}}{2}$$

$$= \frac{x}{2} \pm \frac{\sqrt{3}}{2}\sqrt{4 - x^2}, \qquad -2 \leq x \leq 2.$$

Next we substitute for y in the expression (8) for w:

$$w = x^2 + \left[\frac{x}{2} \pm \frac{\sqrt{3}}{2}\sqrt{4 - x^2}\right]^2$$

$$= x^2 + \frac{x^2}{4} \pm \frac{\sqrt{3}}{2}x\sqrt{4 - x^2} + \frac{3}{4}(4 - x^2)$$

$$= \frac{x^2}{2} \pm \frac{\sqrt{3}}{2}x\sqrt{4 - x^2} + 3, \qquad -2 \le x \le 2. \tag{11}$$

To find the maximum and minimum of w we must study the two functions, corresponding to the $\pm$ sign in Eq. 11, using the techniques for locating maxima and minima of functions of one variable. For this problem the calculation is definitely more cumbersome than using the method of Lagrange multipliers. ∎

EXAMPLE 2

Use the method of Lagrange multipliers to find the dimensions of the rectangular parallelepiped of maximum volume with edges parallel to the axes that can be inscribed in the ellipsoid

$$\frac{x^2}{16} + \frac{y^2}{9} + \frac{z^2}{4} = 1. \tag{12}$$

This is the problem we solved in Example 5 of Section 16.6 by eliminating one of the variables. Recall that we wish to find the maximum of $V = 8xyz$ subject to the constraint (12), and with x, y, z positive. Let $g(x, y, z) = x^2/16 + y^2/9 + z^2/4 - 1$ and $f(x, y, z) = 8xyz$. Then we want to find x, y, z, and λ such that $\nabla f(x, y, z) = \lambda \nabla g(x, y, z)$ and (x, y, z) satisfies the constraint (12). We have

$$\nabla f(x, y, z) = 8(yz\mathbf{i} + xz\mathbf{j} + xy\mathbf{k}), \qquad \nabla g(x, y, z) = \frac{x}{8}\mathbf{i} + \frac{2y}{9}\mathbf{j} + \frac{z}{2}\mathbf{k},$$

so

$$8yz = \lambda\frac{x}{8}, \qquad 8xz = \lambda\frac{2y}{9}, \qquad 8xy = \lambda\frac{z}{2}. \tag{13}$$

Since $x > 0$, $y > 0$, and $z > 0$ we can solve each of Eqs. 13 for λ. The first two of Eqs. 13 give $y^2 = 9x^2/16$ and the first and last of Eqs. 13 give $z^2 = x^2/4$. Substituting for y^2 and z^2 in the constraint (12), we obtain

$$\frac{x^2}{16} + \frac{1}{9}\frac{9x^2}{16} + \frac{1}{4}\frac{x^2}{4} = 1$$

so

$$x^2 = \frac{16}{3}$$

and $x = 4/\sqrt{3}$. The corresponding values of y and z are $\sqrt{3}$ and $2/\sqrt{3}$. This is the only candidate for the maximum point, and it is clear from the context of the problem that the point $(4/\sqrt{3}, \sqrt{3}, 2/\sqrt{3})$ yields the maximum volume of V.

This is the same result as we obtained in Example 5 of Section 16.6; however, the present calculation is somewhat simpler than our earlier calculation. ∎

EXAMPLE 3

Determine the minimum and maximum of

$$f(x, y) = x^2 - xy + y^2 - 3x \tag{14}$$

on the domain $x^2 + y^2 \leq 9$.

We break the problem into two parts. First, we find the relative extrema of f on the open set $x^2 + y^2 < 9$, and then we find the relative extrema of f on the boundary $x^2 + y^2 = 9$. We have

$$f_x(x, y) = 2x - y - 3, \qquad f_y(x, y) = -x + 2y.$$

Setting $f_x(x, y)$ and $f_y(x, y)$ equal to zero gives $x = 2$, $y = 1$. The point $(2, 1)$ lies in the set $x^2 + y^2 < 9$, so it is a candidate for the maximum point or the minimum point.

Next, we consider the constrained maximum–minimum problem for f on the boundary,

$$g(x, y) = x^2 + y^2 - 9 = 0. \tag{15}$$

We have

$$\nabla f(x, y) = (2x - y - 3)\mathbf{i} + (-x + 2y)\mathbf{j}, \qquad \nabla g(x, y) = 2x\mathbf{i} + 2y\mathbf{j},$$

so

$$\begin{aligned} 2x - y - 3 &= 2\lambda x, \\ -x + 2y &= 2\lambda y. \end{aligned} \tag{16}$$

If $x = 0$, then the first of Eqs. 16 is satisfied if we choose $y = -3$, and the second of Eqs. 16 is satisfied if we then take $\lambda = 1$. Since $(0, -3)$ also satisfies the constraint (15), it is a second candidate for the maximum or minimum point. If $y = 0$, then it follows from the second of Eqs. 16 that $x = 0$. However $(0, 0)$ does not satisfy the constraint (15).

For $x \neq 0$ and $y \neq 0$, we can solve each of Eqs. 16 for λ, obtaining

$$\lambda = \frac{2x - y - 3}{2x} = \frac{-x + 2y}{2y}.$$

Hence

$$y(2x - y - 3) = x(-x + 2y),$$

so

$$x^2 = y^2 + 3y. \tag{17}$$

Substituting for x^2 in the constraint (15), we obtain

$$2y^2 + 3y - 9 = 0,$$

so

$$y = -3 \quad \text{and} \quad y = \tfrac{3}{2}.$$

The corresponding values of x are determined from Eq. 17;

$$y = -3: \qquad x^2 = (-3)^2 + 3(-3) \quad \text{so} \quad x = 0;$$

$$y = \frac{3}{2}: \qquad x^2 = \left(\frac{3}{2}\right)^2 + 3\left(\frac{3}{2}\right) \quad \text{so} \quad x = \pm\frac{3\sqrt{3}}{2}.$$

Thus the points $(0, -3)$, $(3\sqrt{3}/2, \frac{3}{2})$, and $(-3\sqrt{3}/2, \frac{3}{2})$ on the boundary are candidates for the maximum point or the minimum point. Notice that we had already recognized $(0, -3)$ as a candidate. To determine the maximum and minimum of f we evaluate f at each of the four points $(2, 1)$, $(0, -3)$, $(3\sqrt{3}/2, \frac{3}{2})$, and $(-3\sqrt{3}/2, \frac{3}{2})$. We have

$$f(2, 1) = 4 - 2 + 1 - 6 = -3,$$

$$f(0, -3) = 0 - 0 + 9 - 0 = 9,$$

$$f\left(\frac{3\sqrt{3}}{2}, \frac{3}{2}\right) = \frac{27}{4} - \frac{9\sqrt{3}}{4} + \frac{9}{4} - \frac{9\sqrt{3}}{2} = 9 - \frac{27\sqrt{3}}{4} \cong -2.69,$$

$$f\left(\frac{-3\sqrt{3}}{2}, \frac{3}{2}\right) = \frac{27}{4} + \frac{9\sqrt{3}}{4} + \frac{9}{4} + \frac{9\sqrt{3}}{2} = 9 + \frac{27\sqrt{3}}{4} \cong 20.69.$$

Thus the maximum of f is $9 + 27\sqrt{3}/4 \cong 20.69$ and it occurs at $(-3\sqrt{3}/2, \frac{3}{2})$; the minimum of f is -3 and it occurs at $(2, 1)$. Notice that the maximum point is on the boundary of the domain $x^2 + y^2 \leq 9$ and the minimum point is in the interior.

The graph of $f(x, y) = x^2 - xy + y^2 - 3x$ on the domain $x^2 + y^2 \leq 9$ is shown in Figure 16.7.2. ∎

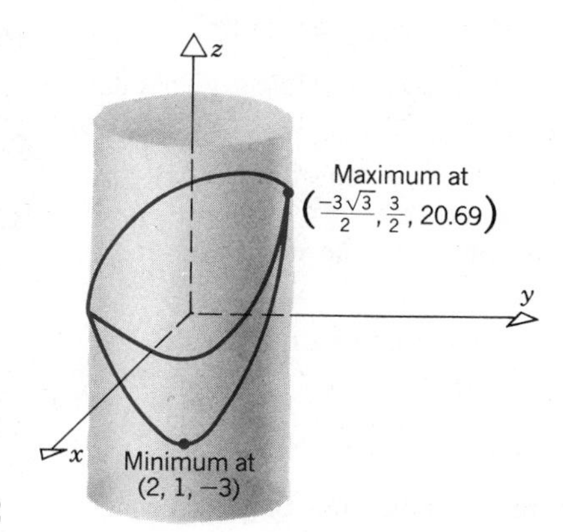

Figure 16.7.2

In summary, there are two general approaches to constrained maximum–minimum problems. The "direct approach" is to solve the constraint equation for one of the variables in terms of the other variables and then substitute for this variable in the function whose extremum is being sought. Since we reduce the number of variables by one, such an approach seems sensible. However, as we have seen in Example 1 here and in Example 5 of Section 16.6 the calculations can become tedious.

The method of Lagrange multipliers is an indirect approach in which we introduce a new variable, the Lagrange multiplier λ, in addition to all of the original variables. One might suspect that this is a step in the wrong direction. However, as illustrated in Examples 1 and 2, the method of Lagrange multipliers is rarely, if ever, more complicated and is often much easier than the direct approach. Of course, this does not mean that it is always easy to solve the resulting equations for the possible maximum and minimum points. Sometimes it is necessary to resort to numerical procedures to solve the equations; however, the examples and problems that appear here do not require this. Nevertheless, in the method of Lagrange we only need to solve the equations $\nabla f(x, y) = \lambda \nabla g(x, y)$ and $g(x, y) = 0$ for *points* (x, y, λ). In the direct method we must solve the equation $g(x, y) = 0$ for y as a *function* of x or for x as a *function* of y.

It is possible to extend the theory of Lagrange multipliers in a straightforward way to problems in which there is more than one constraint. For example, we may wish to determine the maximum of $w = f(x, y, z)$ subject to the constraints $g(x, y, z) = 0$ and $h(x, y, z) = 0$. However, we will not discuss this generalization.

PROBLEMS

In each of Problems 1 through 24, use the method of Lagrange multipliers.

1. Find the maximum and minimum of $f(x, y) = xy$ on the ellipse $x^2/a^2 + y^2/b^2 = 1$.
2. Find the maximum and minimum of $f(x, y) = y^2x$ on the ellipse $x^2/a^2 + y^2/b^2 = 1$.
3. Find the minimum of $f(x, y) = x^2/a^2 + y^2/b^2$ on the hyperbola $xy = 1$.
4. Find the maximum and minimum of $f(x, y, z) = x - y + z$ on the sphere $x^2 + y^2 + z^2 = 9$. At what points on the sphere do the maximum and minimum occur?
5. Find the maximum and minimum of $f(x, y, z) = 2x + 2y - z$ on the sphere $x^2 + y^2 + z^2 = 16$. At what points on the sphere do the maximum and minimum occur?
6. Find the maximum and minimum of $f(x, y) = x^2y^2$ on the ellipse $x^2/a^2 + y^2/b^2 = 1$.
7. Find the maximum and minimum of $f(x, y, z) = x + y - z$ on the ellipsoid $x^2/4 + y^2/4 + z^2 = 1$. At what points on the ellipsoid do the maximum and minimum occur?
8. Find the maximum and minimum of $f(x, y, z) = 3x - y + 2z$ on the ellipsoid $x^2 + y^2/4 + z^2/3 = 1$. At what points on the ellipsoid do the maximum and minimum occur?
9. Find the minimum of $f(x, y, z) = xyz$ on the sphere $x^2 + y^2 + z^2 = a^2$.
10. Find the maximum of $f(x, y, z) = xyz$ on the surface $x^3 + y^3 + z^3 = 1$.
11. Find the minimum of $f(x, y, z) = x^3 + y^3 + z^3$ on the plane $x + y + z = 1$.
12. Find the minimum of $f(x, y, z) = x^4 + y^4 + z^4$ on the plane $x + y + z = 1$.
13. Determine the minimum distance from the origin to the line $ax + by + c = 0$.
14. Determine the minimum distance from the origin to the plane $ax + by + cz = d$, $d \neq 0$.
15. Determine the three dimensional vector $\mathbf{r} = x\mathbf{i} + y\mathbf{j} + z\mathbf{k}$ such that $\mathbf{r} \cdot \mathbf{r} = a^2$ and $x + y + z$ is a maximum.
16. Show that the rectangle of fixed perimeter that has maximum area is a square.

17. Show that among all rectangular parallelepipeds of given surface area the cube has maximum volume. (Problem 28 of Section 16.6.)

18. If the sum of three positive numbers is a, what are their values if the sum of their squares is a minimum? (Problem 23 of Section 16.6.)

19. Show that if $f(x, y)$, subject to the constraint $g(x, y) = 0$, has an extremum at (x_0, y_0), then

$$f_x(x_0, y_0)g_y(x_0, y_0) - f_y(x_0, y_0)g_x(x_0, y_0) = 0.$$

20. Let $a_1, a_2, \ldots, a_n$ be positive numbers. Find the maximum of

$$a_1x_1 + a_2x_2 + \cdots + a_nx_n$$

subject to the constraint

$$x_1^2 + x_2^2 + \cdots + x_n^2 = 1.$$

21. (a) If the sum of two positive numbers is C, what are their values if the square root of their product is a maximum?
(b) Show that if a and b are positive numbers, then $(ab)^{1/2} \leq (a + b)/2$.

22. (a) If the sum of three positive numbers is C, what are their values if the cube root of their product is a maximum?
(b) Show that if a, b, and c are positive, then $(abc)^{1/3} \leq (a + b + c)/3$.

23. If $a_1, a_2, \ldots, a_n$ are positive numbers, show that

$$(a_1 a_2 \cdots a_n)^{1/n} \leq \frac{a_1 + a_2 + \cdots + a_n}{n}.$$

Hint: See Problems 21 and 22.

24. A silo is made in the form of a circular cylinder of radius r and height h with a conical cap of height H. If the radius and volume V of the silo are fixed, how should h and H be chosen so as to minimize the surface area?

25. Determine the maximum and minimum of $f(x, y) = x^2/a^2 + y^2/b^2$ on the domain $x^2 + y^2 \leq r^2$. Assume that $a^2 > b^2$.

26. Determine the maximum and minimum of $f(x, y) = x^2 - xy + y^2 + 4y + 1$ on the domain $x^2 + y^2 \leq 16$.

27. Determine the maximum and minimum of $f(x, y) = 4x^2 - xy + y^2 + y$ on the domain $x^2 + y^2/4 \leq 1$.

28. Determine the maximum and minimum of $f(x,y) = x^4 - y^4$ on the domain $x^2/a^2 + y^2/b^2 \leq 1$.

29. Determine the maximum and minimum of $f(x,y) = \frac{1}{3}y^3 + x^2y - 2x^2 - 2y^2 + 6$ on the domain $x^2 + y^2 \leq 36$. Note that f is the function discussed in Example 4 of Section 16.6.

16.8 IMPLICIT FUNCTIONS

We have often seen that functions we wish to investigate are defined implicitly. For example, we may have y defined implicitly as a function of x by the equation $g(x, y) = 0$, or z defined implicitly as a function of x and y by the equation $h(x, y, z) = 0$. In Section 3.7 we discussed how to calculate dy/dx when y is defined implicitly as a function of x by an equation $g(x, y) = 0$. We now extend those techniques to functions of several variables. For the moment we leave aside the question of when an equation such as $h(x, y, z) = 0$ does indeed define z as a function of x and y, and when this function is differentiable. We start with an example.

EXAMPLE 1

Assume that there is a differentiable function $z = f(x, y)$ defined by the equation

$$h(x, y, z) = x^4z - 2xy^3 + yz^3 - 8 = 0. \tag{1}$$

Verify that $z = 2$ when $x = 1$, $y = 1$ and compute $\partial z/\partial x$ and $\partial z/\partial y$ at the point (1, 1).

We first show by direct substitution that the point (1, 1, 2) does indeed satisfy Eq. 1:

$$h(1, 1, 2) = 1(2) - 2(1)(1) + 1(8) - 8 = 2 - 2 + 8 - 8 = 0.$$

To compute $\partial z/\partial x$, we hold y constant and differentiate Eq. 1 with respect to x. This gives

$$4x^3z + x^4\frac{\partial z}{\partial x} - 2y^3 + 3yz^2\frac{\partial z}{\partial x} = 0.$$

Hence

$$\frac{\partial z}{\partial x} = \frac{2y^3 - 4x^3z}{x^4 + 3yz^2}, \tag{2}$$

provided that the denominator is not zero. To evaluate $\partial z/\partial x$ at $x = 1$, $y = 1$ we must make use of the fact that we know the corresponding value of z, namely $z = 2$. We have

$$\frac{\partial z}{\partial x}(1, 1) = \frac{2(1) - 4(1)(2)}{1 + 3(1)(4)} = -\frac{6}{13}.$$

In a similar manner we calculate $\partial z/\partial y$ by holding x constant and differentiating Eq. 1 with respect to y. This yields

$$x^4\frac{\partial z}{\partial y} - 6xy^2 + z^3 + 3yz^2\frac{\partial z}{\partial y} = 0,$$

and hence

$$\frac{\partial z}{\partial y} = \frac{6xy^2 - z^3}{x^4 + 3yz^2}, \tag{3}$$

provided once again that the denominator is not zero. At $x = 1$, $y = 1$ we have

$$\frac{\partial z}{\partial y}(1, 1) = \frac{6(1)(1) - 8}{1 + 3(1)(4)} = -\frac{2}{13}. \quad ■$$

Now let us consider several general cases.

One equation in two variables

Suppose that the equation $g(x, y) = 0$ defines y as a differentiable function of x, $y = f(x)$. We wish to determine a formula for dy/dx in terms of derivatives of g. Since $y = f(x)$ is a solution of $g(x, y) = 0$, we have

$$g[x, f(x)] \equiv 0, \quad \text{so} \quad \frac{d}{dx}g[x, f(x)] \equiv 0. \tag{4}$$

We now calculate this derivative using the chain rule:

$$\frac{d}{dx} g[x, f(x)] = \frac{\partial g(x, y)}{\partial x}\frac{dx}{dx} + \frac{\partial g(x, y)}{\partial y}\frac{dy}{dx} \equiv 0, \tag{5}$$

with the understanding that $y = f(x)$. Provided that $g_y(x, y) \neq 0$, we can solve Eq. 5 for dy/dx, obtaining

$$\frac{dy}{dx} = -\frac{g_x(x, y)}{g_y(x, y)}. \tag{6}$$

Note that this is a more general statement of the procedure used in Section 3.7 to calculate the derivative of an implicitly defined function. To calculate dy/dx from Eq.(6) for some particular value of x, we must know the corresponding value of y obtained by solving the equation $g(x, y) = 0$. It may be necessary to resort to numerical procedures to do this.

EXAMPLE 2

Assume that the equation

$$g(x, y) = 5ye^y + 2 \sin \frac{\pi}{2}(x + y) + x - 3 = 0 \tag{7}$$

defines y as a differentiable function of x, $y = f(x)$. Show that $x = 1$, $y = 0$ satisfies Eq. 7, and calculate dy/dx at $x = 1$.

First

$$g(1, 0) = 0 + 2 \sin \frac{\pi}{2} + 1 - 3 = 0.$$

Next we can calculate $g_x(x, y)$ and $g_y(x, y)$ and substitute in Eq. 6. However, rather than memorize formula (6) we prefer to carry out the calculation directly. Thus, thinking of y as a function of x, we differentiate Eq. 7 with respect to x:

$$5ye^y \frac{dy}{dx} + 5e^y \frac{dy}{dx} + \left[2 \cos \frac{\pi}{2}(x + y)\right]\left[\frac{\pi}{2}\left(1 + \frac{dy}{dx}\right)\right] + 1 = 0.$$

Solving this equation for dy/dx, we obtain

$$\frac{dy}{dx} = -\frac{\pi \cos \frac{\pi}{2}(x + y) + 1}{5e^y(1 + y) + \pi \cos \frac{\pi}{2}(x + y)}. \tag{8}$$

At $x = 1$, $y = 0$, so we have

$$\frac{dy}{dx}(1) = -\frac{\pi \cos \pi/2 + 1}{5 + \pi \cos \pi/2} = -\frac{1}{5}. \quad ∎$$

One equation in three variables

Suppose that the equation $h(x, y, z) = 0$ defines z as a differentiable function of x and y, $z = f(x, y)$. We wish to determine formulas for z_x and z_y in terms of derivatives of h. Since $z = f(x, y)$ is a solution of $h(x, y, z) = 0$, we have $h[x, y, f(x, y)] \equiv 0$, so

$$\frac{\partial h[x, y, f(x, y)]}{\partial x} \equiv 0 \quad \text{and} \quad \frac{\partial h[x, y, f(x, y)]}{\partial y} \equiv 0. \tag{9}$$

We now calculate the partial derivative with respect to x using the chain rule:

$$\begin{aligned}\frac{\partial h[x, y, f(x, y)]}{\partial x} &= \frac{\partial h(x, y, z)}{\partial x}\frac{\partial x}{\partial x} \\ &\quad + \frac{\partial h(x, y, z)}{\partial y}\frac{\partial y}{\partial x} + \frac{\partial h(x, y, z)}{\partial z}\frac{\partial z}{\partial x} \equiv 0\end{aligned} \tag{10}$$

with the understanding that $z = f(x, y)$. Now $\partial x/\partial x = 1$ and $\partial y/\partial x = 0$, since y is held constant in calculating the partial derivative with respect to x. Solving Eq. 10 for $\partial z/\partial x$ we obtain

$$\frac{\partial z}{\partial x} = -\frac{h_x(x, y, z)}{h_z(x, y, z)}, \tag{11}$$

provided that $h_z(x, y, z) \neq 0$. A similar calculation shows that

$$\frac{\partial z}{\partial y} = -\frac{h_y(x, y, z)}{h_z(x, y, z)}. \tag{12}$$

For Example 1 we have

$$h(x, y, z) = x^4 z - 2xy^3 + yz^3 - 8,$$

so

$$\begin{aligned}h_x(x, y, z) &= 4x^3 z - 2y^3, \\ h_y(x, y, z) &= -6xy^2 + z^3, \\ h_z(x, y, z) &= x^4 + 3yz^2.\end{aligned}$$

If we substitute for h_x, h_y, and h_z in Eqs. 11 and 12, we obtain Eqs. 2 and 3, respectively.

Two equations in four variables

Suppose that the two equations $F(x, y, u, v) = 0$ and $G(x, y, u, v) = 0$ define u and v as differentiable functions of x and y, $u = f(x, y)$ and $v = g(x, y)$. We wish to determine formulas for u_x, u_y, v_x, and v_y in terms of derivatives of F and G. Since $u = f(x, y)$ and $v = g(x, y)$ are solutions of $F(x, y, u, v) = 0$ and $G(x, y, u, v) = 0$, we have

$$F[x, y, f(x, y), g(x, y)] \equiv 0 \quad \text{and} \quad G[x, y, f(x, y), g(x, y)] \equiv 0. \tag{13}$$

Again we observe that the partial derivatives of these functions with respect to x and y are also zero. We calculate the partial derivatives with respect to x using the chain rule:

$$\begin{aligned}
\frac{\partial F[x, y, f(x, y), g(x, y)]}{\partial x} &= \frac{\partial F(x, y, u, v)}{\partial x}\frac{\partial x}{\partial x} + \frac{\partial F(x, y, u, v)}{\partial y}\frac{\partial y}{\partial x} \\
&\quad + \frac{\partial F(x, y, u, v)}{\partial u}\frac{\partial u}{\partial x} + \frac{\partial F(x, y, u, v)}{\partial v}\frac{\partial v}{\partial x} \\
&\equiv 0,
\end{aligned} \tag{14}$$

$$\begin{aligned}
\frac{\partial G[x, y, f(x, y), g(x, y)]}{\partial x} &= \frac{\partial G(x, y, u, v)}{\partial x}\frac{\partial x}{\partial x} + \frac{\partial G(x, y, u, v)}{\partial y}\frac{\partial y}{\partial x} \\
&\quad + \frac{\partial G(x, y, u, v)}{\partial u}\frac{\partial u}{\partial x} + \frac{\partial G(x, y, u, v)}{\partial v}\frac{\partial v}{\partial x} \\
&\equiv 0,
\end{aligned} \tag{15}$$

with the understanding that $u = f(x, y)$ and $v = g(x, y)$. Again $\partial x/\partial x = 1$ and $\partial y/\partial x = 0$. Equations 14 and 15 provide us with two linear nonhomogeneous algebraic equations for u_x and v_x, namely

$$\begin{aligned}
F_u u_x + F_v v_x &= -F_x, \\
G_u u_x + G_v v_x &= -G_x.
\end{aligned} \tag{16}$$

Provided that $F_u G_v - F_v G_u \neq 0$ we can solve these equations for u_x and v_x, obtaining

$$u_x = \frac{-F_x G_v + F_v G_x}{F_u G_v - F_v G_u}, \qquad v_x = \frac{-F_u G_x + F_x G_u}{F_u G_v - F_v G_u}. \tag{17}$$

Thus, to evaluate u_x and v_x at a point (x, y), we must (a) determine the corresponding values of u and v by solving the simultaneous equations $F(x, y, u, v) = 0$ and $G(x, y, u, v) = 0$, (b) calculate the partial derivatives F_x, F_u, F_v, G_x, G_u, and G_v and evaluate them at the point (x, y, u, v), and (c) then evaluate u_x and v_x using Eqs. 17.

We proceed in a similar manner to determine formulas for u_y and v_y, obtaining

$$u_y = \frac{-F_y G_v + F_v G_y}{F_u G_v - F_v G_u}, \qquad v_y = \frac{-F_u G_y + F_y G_u}{F_u G_v - F_v G_u}. \tag{18}$$

In practice we do not need to remember Eqs. 17 and 18, but rather carry out the calculations directly as is illustrated in the following example.

EXAMPLE 3

The variables u and v are defined as differentiable functions of x and y by the equations

$$\begin{aligned}
F(x, y, u, v) &= x^2 + y - u^2 + 2v = 0, \\
G(x, y, u, v) &= xy + 2x - y + 3uv + 10 = 0.
\end{aligned} \tag{19}$$

Verify that $x = 1$, $y = -1$, $u = -2$, $v = 2$ is a solution of these equations and calculate u_y and v_y at $x = 1$, $y = -1$.

We verify that $(1, -1, -2, 2)$ is a solution of Eqs. 19 by direct substitution:

$$F(1, -1, -2, 2) = 1 - 1 - 4 + 4 = 0,$$

$$G(1, -1, -2, 2) = -1 + 2 + 1 - 12 + 10 = 0.$$

Next we differentiate each of Eqs. 19 with respect to y, remembering that x is held fixed and that u and v are functions of x and y. The result is

$$1 - 2uu_y + 2v_y = 0,$$

$$x - 1 + 3vu_y + 3uv_y = 0.$$

Hence

$$\begin{aligned} -2uu_y + 2v_y &= -1, \\ 3vu_y + 3uv_y &= 1 - x. \end{aligned} \tag{20}$$

The solution of these equations is

$$u_y = \frac{3u + 2(1 - x)}{6u^2 + 6v}, \qquad v_y = \frac{2u(1 - x) - 3v}{6u^2 + 6v}, \tag{21}$$

so, using the fact that $u = -2$ and $v = 2$ when $x = 1$ and $y = -1$, we obtain

$$u_y(1, -1) = \frac{-6 + 0}{24 + 12} = -\frac{1}{6},$$

and

$$v_y(1, -1) = \frac{0 - 6}{24 + 12} = -\frac{1}{6}. \quad ■$$

The ideas developed in these three cases can be readily extended to the situation of n variables with $m (m < n)$ equations defining m variables as differentiable functions of the remaining $n - m$ variables. We summarize what one needs to understand in calculating the derivative or partial derivatives of a function or functions defined implicitly.

1. Given m equations with n variables $(m < n)$, decide which variables are independent and which are dependent. In Example 3, x and y are the independent variables and u and v are the dependent variables. The number of dependent variables is equal to the number (m) of equations.
2. Assume that the equations define the m dependent variables as functions of the $n - m$ independent variables for some domain of the independent variables.
3. Differentiate each of the equations with respect to one of the independent variables holding all the other independent variables fixed.
4. The resulting equations are *always linear* in the partial derivatives of the dependent variables with respect to the chosen independent variable, and hence can be readily solved. The results make sense as long as the denominators are not zero.

Once one understands this process, there should be no concern about dealing with any combination of numbers of equations and variables, provided, of course, that the number of equations is less than the number of variables. Moreover, there is no need to memorize formulas. Until one does understand this process, there is the danger of being confused by different cases and complicated formulas. Worse, one may not understand how to deal with some new situation.

In closing this section we state and briefly discuss a theorem that ensures that an equation $g(x, y) = 0$ does define y as a differentiable function of x. Specifically, we ask: Given an equation $g(x, y) = 0$ and a point (x_0, y_0) satisfying this equation, is there a differentiable function $y = f(x)$ such that $f(x_0) = y_0$ and $g[x, f(x)] \equiv 0$ for x in some open interval containing x_0?

Theorem 16.8.1
(Implicit Function Theorem)
Suppose that $g(x, y)$, $g_x(x, y)$, and $g_y(x, y)$ are continuous in a neighborhood of a point (x_0, y_0) such that $g(x_0, y_0) = 0$ and that $g_y(x_0, y_0) \neq 0$. Then there exists a function $y = f(x)$ on some open interval containing x_0 such that $f(x_0) = y_0$ and $g[x, f(x)] \equiv 0$; the function f is differentiable and

$$\frac{dy}{dx} = -\frac{g_x(x, y)}{g_y(x, y)}.$$

Notice that the theorem says that f is defined on some open interval containing x_0; it does not tell us the extent of the interval. Thus the theorem gives us only a "local" result. This is illustrated in Figure 16.8.1. We show by some simple examples

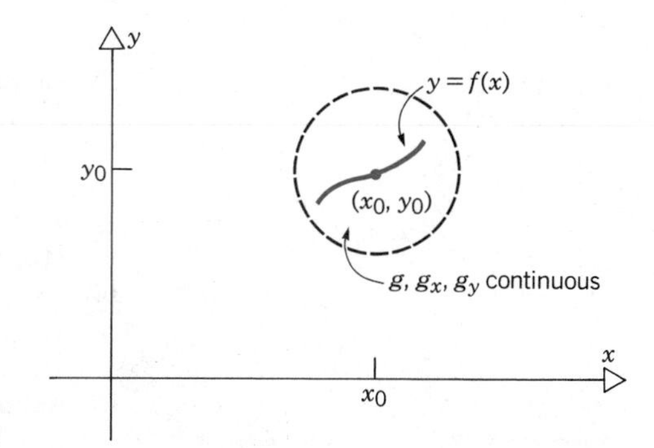

Figure 16.8.1

that conditions such as those in Theorem 16.8.1 are required in order to obtain the result stated there.

Consider

$$g(x, y) = x^2 + y^2 + 1 = 0. \tag{22}$$

Clearly g, g_x, and g_y are continuous everywhere and $g_y(x, y) = 2y$ is zero only on the line $y = 0$. However, there is no differentiable function $y = f(x)$ satisfying

Eq. 22, since there is no point (x_0, y_0) in the plane such that $x_0^2 + y_0^2 + 1 = 0$.

Next, consider

$$g(x, y) = x^2 + y^2 = 0. \tag{23}$$

The only point satisfying this equation is the point (0, 0). However, there is no differentiable function $y = f(x)$ that satisfies Eq. 23 and the condition $f(0) = 0$. Notice that while g, g_x, and g_y are continuous everywhere, $g_y(x, y) = 2y$, and hence is zero at (0, 0).

Finally, consider

$$g(x, y) = x^2 + y^2 - 1 = 0. \tag{24}$$

The point (0, 1) satisfies Eq. 24. Again, g, g_x, and g_y are continuous everywhere and now $g_y(0, 1) = 2$, so there is a differentiable function $y = f(x)$ satisfying $f(0) = 1$ and Eq. 24 for some interval containing 0. This function is $y = f(x) = \sqrt{1 - x^2}$, $-1 < x < 1$. Similarly, the solution of Eq. 24 passing through $(0, -1)$ is $y = -\sqrt{1 - x^2}$, $-1 < x < 1$. However, since $g_y(x, y) = 2y$ is zero on the line $y = 0$, we should not expect that there is a differentiable function $y = f(x)$ satisfying Eq. 24 and the condition $f(1) = 0$. Indeed, there is no such function.

The implicit function theorem generalizes in a straightforward way to n variables and m equations with $m < n$. However, we do not pursue this matter.

PROBLEMS

In each of Problems 1 through 12, find the indicated derivatives. Assume that the equation(s) defines the necessary variable(s) as a differentiable function(s) of the remaining variables.

1. $e^x \sin y + 2y \cos x = 3, \quad \dfrac{dy}{dx}$
2. $e^{xy} \cos 2x + x^2 - 3y = 4, \quad \dfrac{dy}{dx}$
3. $\dfrac{x^2}{y} + \arctan \dfrac{y}{x} = 6, \quad \dfrac{dy}{dx}$
4. $x^2 + 3y^2 + 2z^2 - 3xyz + 2yz + 4xy - 1 = 0, \quad z_x$
5. $3xy - z + e^{x+z} - 2y^2 + 4 = 0, \quad z_x$ and z_y
6. $z = e^{2x} \sin(3y - 2z), \quad z_x$ and z_y
7. $\sin(x + y) + \cos(y + z) + \sin(x + z) = 1, \quad z_y$
8. $x^3y + y^3z + z^3x = 8, \quad x_y$ and x_z
9. $(x^2 + 2y^2)^2 + 2(x^2 + z^2)^2 + (2y^2 + z^2)^2 = 12, \quad z_y$
10. $x^2 - y^2 + 2uv = 0, \quad u^2 + v^2 - 2xy = 0;$ u_x and v_x
11. $x^2 - y^2 + u^2 - v^2 = 0, \quad xyuv = 4;$ u_y and v_y
12. $e^x u + e^y v - uv = 1, \quad yu + xv + \cos 2y = 3; \quad u_x$ and v_x
13. Assume that the equation
$$x^2 - xy + y^3 - 2x + 1 = 0$$
defines $y = f(x)$ with a continuous second derivative in the neighborhood of $x = 2$. Verify that $f(2) = 1$, and calculate (a) $f'(2)$, (b) $f''(2)$.
14. Assume that the equation $g(x, y) = 0$ defines y as a twice differentiable function of x. Determine d^2y/dx^2 in terms of the partial derivatives of g.
15. Assume that the equation
$$x^2 + xy + 4y^2 - 3yz - z^3 = 2$$
defines $z = f(x, y)$ with continuous second partial derivatives in a neighborhood of the point $(1, -1)$. Verify that $x = 1$, $y = -1$, and $z = 2$ satisfies this equation, and calculate f_x, f_{xx}, f_y, and f_{xy} at this point.
16. Assume that the equation $g(x, y, z) = 0$ defines z as a function of x and y with continuous second partial derivatives.

(a) Determine z_{xx} in terms of partial derivatives of g.

(b) Determine z_{xy} in terms of partial derivatives of g.

17. If $z = f(x, y)$, and $g(x, y) = 0$ defines y as a differentiable function of x, determine an expression for dz/dx in terms of the partial derivatives of f and g.

18. The equations $f(x, u, v) = 0$ and $g(x, u, v) = 0$ define u and v as differentiable functions of x. Determine expressions for du/dx and dv/dx.

19. Suppose that the equation $F(x, y, z) = 0$ defines the differentiable functions $z = f(x, y)$, $y = g(x, z)$, and $x = h(y, z)$. Show that

$$\frac{\partial h(y, z)}{\partial y} \cdot \frac{\partial g(x, z)}{\partial z} \cdot \frac{\partial f(x, y)}{\partial x} = -1. \qquad (i)$$

This equation is often written in the simpler form

$$\frac{\partial x}{\partial y} \cdot \frac{\partial y}{\partial z} \cdot \frac{\partial z}{\partial x} = -1, \qquad (ii)$$

where each expression must be interpreted in the sense given by Eq. (i). Also notice, from Eq. (ii) the error that is made if one casually cancels a ∂x "upstairs" with a ∂x "downstairs," and similarly for ∂y and ∂z.

20. Assume that the equation

$$x^2y^2 + xz^3 - x^4 + zy^3 = 6$$

defines differentiable functions $z = f(x, y)$, $y = g(x, z)$, and $x = h(y, z)$. Calculate z_x, y_z, and x_y at the point $(1, -1, 2)$ and verify that at this point $x_y y_z z_x = -1$.

21. (a) Use implicit differentiation to show that if

$$x = r\cos\theta, \qquad y = r\sin\theta, \qquad (i)$$

then

$$\frac{\partial r}{\partial x} = \cos\theta, \qquad \frac{\partial r}{\partial y} = \sin\theta,$$
$$\frac{\partial \theta}{\partial x} = -\frac{\sin\theta}{r}, \qquad \frac{\partial \theta}{\partial y} = \frac{\cos\theta}{r}.$$

(b) Use the results of Part (a) to show that if $w = f(r, \theta)$, then

$$\frac{\partial w}{\partial x} = \frac{\partial w}{\partial r}\cos\theta - \frac{\partial w}{\partial \theta}\frac{\sin\theta}{r},$$
$$\frac{\partial w}{\partial y} = \frac{\partial w}{\partial r}\sin\theta + \frac{\partial w}{\partial \theta}\frac{\cos\theta}{r}.$$

(c) Show that

$$\frac{\partial^2 w}{\partial x^2} + \frac{\partial^2 w}{\partial y^2} = \frac{\partial^2 w}{\partial r^2} + \frac{1}{r}\frac{\partial w}{r} + \frac{1}{r^2}\frac{\partial^2 w}{\partial \theta^2}.$$

This calculation is much easier than solving Eqs. (i) for $r = (x^2 + y^2)^{1/2}$ and $\theta = \arctan(y/x)$ and then calculating $w_{xx} + w_{yy}$.

22. Consider the coordinate transformation

$$x = \cosh u \cos v, \qquad y = \sinh u \sin v.$$

(a) Show that the families of curves u = constant and v = constant are ellipses and hyperbolas, respectively, in the xy-plane.

(b) Use implicit differentiation to show that $u_x = v_y$ and $u_y = -v_x$.

23. Suppose that $x = f(u, v)$ and $y = g(u, v)$ and that it is possible to solve for u and v as differentiable functions of x and y. Determine u_x, u_y, v_x, and v_y in terms of partial derivatives of f and g.

REVIEW PROBLEMS

In each of Problems 1 through 4, evaluate the given function at the given points and determine the domain of the given function.

1. $f(x, y) = \ln(x^2 - y^2)^{3/2}$;
(a) $(1, 0)$,
(b) $(5, 4)$

2. $F(x, y, z) = \exp(\sqrt{x^2 + y^2 + z^2})$;
(a) (a, a, a), $a > 0$
(b) (a^{-1}, a^{-1}, a^{-1}), $a > 0$

3. $\phi(x, y, z) = \arctan\left(\dfrac{3x - 2y}{-2z + x^2}\right)$;
(a) $(1, 1, 1)$,
(b) (a^{-1}, b^{-1}, c^{-1})

4. $g(x, y) = \sin x \sinh y + \cos x \cosh y$;
(a) $\left(\dfrac{\pi}{3}, \ln 2\right)$, (b) $\left(\dfrac{\pi}{6}, \ln 3\right)$

In each of Problems 5 through 8, identify the point(s) of discontinuity for the given function. Define the function at these points to make the function continuous, if possible.

5. $f(x, y) = \dfrac{x^2 - 2y}{2x - y}$

6. $f(x, y) = \dfrac{x^2 + y^2}{\sqrt{x^2 + y^2}}$

7. $f(x, y) = \dfrac{x^3 + y^2}{\ln|x - y|}$

8. $f(x, y) = \cos\left(\dfrac{x^2 - y^2}{|x| + |y|}\right)$

In each of Problems 9 through 12, find the first partial derivatives with respect to each independent variable.

9. $f(x, y) = \ln[(x^2 + y^2)^{1/2}]$

10. $F(u, v) = e^{v \cos u} - uv$

11. $g(r, s, t) = st^2 \sin(rs - t)$

12. $F(u, v, w) = v \arctan(u - w)$

In each of Problems 13 through 16, verify that $z_{xy} = z_{yx}$.

13. $z = \cos(x^2 - y^2)$

14. $z = x^5y^2 - x^2y^5$

15. $z = e^{xy} \sin xy$

16. $z = \ln(4x^2 + 4y^2)$

In each of Problems 17 through 20, determine whether the given functions satisfy the Cauchy–Riemann equations (see Problem 31, Section 16.2).

17. $u(x, y) = -x^3 + 3xy^2$, $v(x, y) = y^3 - 3x^2y$

18. $u(x, y) = 3x^2 + 2y - 3y^2 - 1$,
$v(x, y) = 6xy + 2y$

19. $u(x, y) = \sin(\omega x - y)$, $v(x, y) = x^2e^{\pi y}$

20. $u(x, y) = e^{x^2-y^2} \cos 2xy$, $v(x, y) = e^{x^2-y^2} \sin 2xy$

In each of Problems 21 through 24, find the set of all functions $z = f(x, y)$ that have the given property.

21. $z_y = 2x^3y - 2 \cos 2y$

22. $z_x = \dfrac{y}{x} + (\sin x)e^{\cos x}$

23. $z_y = x/(x^2 + y^2)$

24. $z_x = \dfrac{3y}{\sqrt{1 - (3x - 5y)^2}}$

In each of Problems 25 through 28, express $f(x + h, y + k)$ in the form of Eq. 11, Section 16.3. Determine $r(x, h, k)$ and show that $r \to 0$ as $(h, k) \to (0, 0)$.

25. $f(x, y) = x^3 - 3y^2$

26. $f(x, y) = 2x^2y + 7y^3$

27. $f(x, y) = e^{xy} - x^2y$

28. $f(x, y) = y \cos x$

In each of Problems 29 through 32, calculate the differential of the given function.

29. $z = ye^{\cos^2 x}$

30. $z = \ln[(x^2 + y^2)^{1/2}]$

31. $w = z^2 \arcsin(x - y)$

32. $w = e^x \cos(yz) \ln \dfrac{x^2}{y}$

In each of Problems 33 through 36, use the chain rule to find dz/dt or dw/dt.

33. $z = \sinh(xy - y^2)$, $x = t$, $y = t^{1/2}$

34. $z = \ln(\sqrt{x^2 + y^2} - 1)$, $x = t$, $y = t^2$

35. $w = x^2yz - xy^2z + xyz^3$, $x = t$,
$y = t^2$, $z = t^3$

36. $w = x^2e^y \ln(\cos z)$, $x = t^2$, $y = t^{-1}$,
$z = t$

In each of Problems 37 through 40, use the chain rule to find the first partial derivatives of z or w with respect to r and s.

37. $z = x^3y$, $x = e^r \sin s$, $y = e^s \cos r$

38. $z = \ln(x^2 + y^2)$, $x = r + s$,
$y = r - s$

39. $w = \dfrac{x^2}{a^2} - \dfrac{y^2}{b^2} - \dfrac{z}{c}$, $x = rs$, $y = r^2 + s$,
$z = r - s$; a, b, c are constants

40. $w = \dfrac{x^2}{a^2} + \dfrac{y^2}{b^2} + \dfrac{z^2}{c^2}$, $x = r$, $y = s$,
$z = r + s$; a, b, c are constants

In each of Problems 41 through 44, find the directional derivative of the given function at the given point in the given direction.

41. $f(x, y) = xy^2$, $(1, 1)$, $\mathbf{u} = (2, -3)$

42. $f(x, y) = x^3y - 2x^2y^2 + y$,
$(3, 4)$, toward $(-1, 2)$

43. $g(x, y, z) = x^2 \arctan(zy)$,
$(1, 0, 2)$, $\mathbf{u} = (4, 1, -2)$

44. $g(x, y, z) = z \sinh xy$, $(1, 1, 1)$,
toward $(1, 3, -2)$

In each of Problems 45 through 48, determine the unit vector $\boldsymbol{\lambda}$ in the direction in which the given function is increasing most rapidly, and determine the rate of change of the function in that direction at the given point.

45. $f(x, y) = xy^2$, $(3, -2)$

46. $f(x, y) = x^3y - 2x^2y^2 + y$, $(1, 1)$

47. $g(x, y, z) = x^2 \arcsin(zy)$, $(1, 0, -4)$

48. $g(x, y, z) = z \sinh xy$, $(0, 2, 1)$

In each of Problems 49 through 52, verify that the given point lies on the curve of intersection of the two given surfaces and determine the tangent line to the curve at this point.

49. $2x - y = 0, x^2 + y^2 + z = 9, (1, 2, 4)$

50. $x^2 - y^2 + z = 5,$
$z^2 = (x - 2)^2 + (y - 1)^2, (5, 5, 5)$

51. $\frac{x^2}{4} + \frac{y^2}{9} - \frac{z^2}{4} = 1, x + z = 2, (1, 3, 1)$

52. $x^2 + y^2 + z^2 = 9, z = e^y, (\sqrt{8}, 0, 1)$

In each of Problems 53 through 56, locate and classify the stationary points.

53. $f(x, y) = x^2y - 3xy^2 + 4xy$

54. $f(x, y) = \sin x - xy$

55. $f(x, y) = e^{-x^2+4y^2}$

56. $f(x, y) = \sinh x \cos y, \quad -\pi \leq x \leq \pi,$
$0 \leq y \leq \pi$

In each of Problems 57 through 60, find the indicated derivatives using implicit differentiation.

57. $x^2 - 3xy^2 + 2y^3 = 6; y'$

58. $xy + \sin xy = 2; \frac{dy}{dx}$

59. $z - e^{z \cosh x} + \ln(x^2 - z^2) = 5; z_x, x_z$

60. $u = x \arctan u - y^3; u_x, u_y$

61. Determine the tangent plane and the normal line at the point (3, 0, 1) on the surface $(x^2/4) + (y^2/9) - (z^2/4) = 2$.

62. Determine the tangent plane and the normal line at the point (1, 1, 1) on the surface $z = 2x^2 - 3xy^2 + 2x$.

63. Find the rate of change of $f(x, y) = xy$ at the point (1, 3) in the direction in which $g(x, y) = (x^2/4) + (y^2/9)$ is increasing most rapidly.

64. Use differentials to estimate the error in the volume of a rectangular box that has a 1% error in the height, a 2% error in the length, and a 3% error in the width.

65. Find the dimensions of the rectangle of largest area with sides parallel to the coordinate axes that can be inscribed in the ellipse $(x^2/25) + (y^2/144) = 1$:
(a) by using Lagrange multipliers;
(b) without using Lagrange multipliers.

66. A topless, cylindrical metal tank is to be built to hold 512π cubic liters of water. Determine the dimensions of the tank that will require the least amount of (uniformly thick) metal to build:
(a) by using Lagrange multipliers;
(b) without using Lagrange multipliers.

67. A red car and a blue car start from a point P. The red car drives N60°W and the blue car drives N45°E. At time T, the red car is 100 km from P and moves with a velocity of 60 km/hr; the blue car is 120 km from P and moves with a velocity of 70 km/hr. Determine how fast they are separating at time T.

SEVENTEEN

multiple integrals

The fundamental mathematical operation of integration arises in a variety of circumstances. Up to now, however, all of our integrals have been one-dimensional, or single, integrals. That is, the problems that we considered could always be formulated so that integration was required with respect to only one variable. Now we take up the question of whether and how the process of integration can be generalized to higher dimensions. We will find that the generalization can be accomplished in a natural and straightforward manner, and that multidimensional integrals share the same basic properties as single integrals. This will enable us to deal with many problems that depend intrinsically on two or more variables. Most of this chapter concerns the evaluation of integrals in two and three dimensions, together with various examples of their applications.

17.1 DOUBLE INTEGRALS

In Chapter 6 we discussed the definite integral of a function of one variable. Improper integrals and applications of integration have appeared in subsequent chapters. Now we consider the question of defining and using integrals of functions of several variables. In the first four sections of this chapter we restrict our attention to functions of two variables, but in

Sections 17.5 through 17.7 we discuss integration for functions of three variables. Generalizations to higher dimensions are relatively straightforward, but will not be considered in this book.

Suppose that a bounded function f is given on the rectangle $R : a \leq x \leq b$, $c \leq y \leq d$ shown in Figure 17.1.1*a*. To define the integral of f over R, we follow

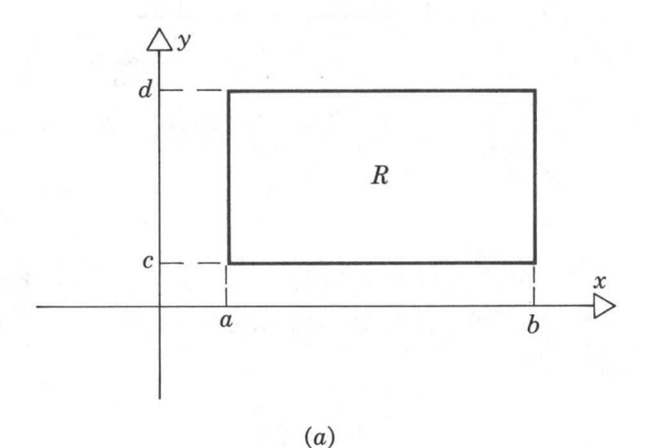

Figure 17.1.1 (*a*)

essentially the same procedure that we described in Chapter 6 for the Riemann integral of a function of one variable. In developing one-dimensional integrals we often made use of the interpretation of such integrals as areas. Similarly, it is frequently helpful to visualize integrals of functions of two variables intuitively in terms of volumes, and we will mention this from time to time. However, the definition of the integral of f over R is purely an analytical one, and does not depend on geometrical ideas.

We begin by partitioning the intervals $[a, b]$ on the x-axis and $[c, d]$ on the y-axis:

$$\begin{aligned} a &= x_0 < x_1 < x_2 < \cdots < x_{m-1} < x_m = b; \\ c &= y_0 < y_1 < y_2 < \cdots < y_{n-1} < y_n = d. \end{aligned} \tag{1}$$

The location of the subdivision points $x_i (i = 1, \ldots, m - 1)$ and $y_j (j = 1, \ldots, n - 1)$ is arbitrary; in particular, they need not be equally spaced on the x- and y-axes. This is illustrated in Figure 17.1.1*b* for a case in which $m = 7$ and $n = 6$.

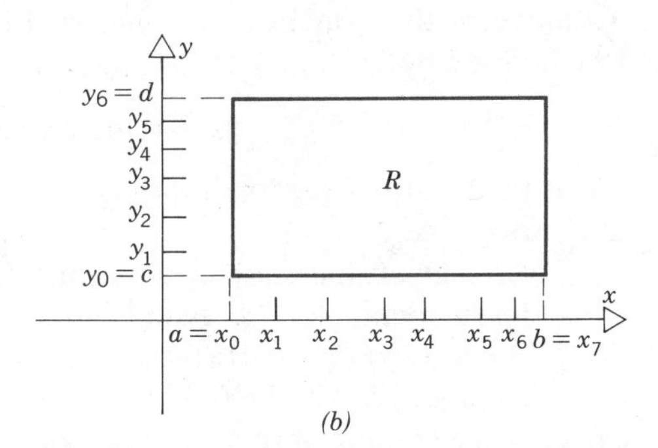

Figure 17.1.1 (*b*)

Then we draw the lines $x = x_i$ and $y = y_j$, parallel to the coordinate axes, and thereby partition the rectangle R into subrectangles, as indicated in Figure 17.1.1*c*.

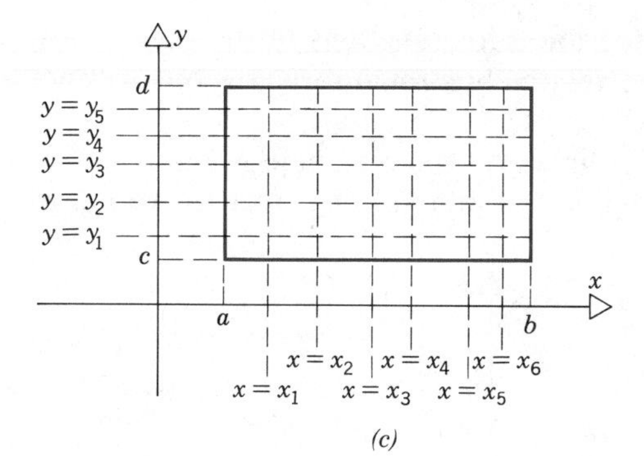

Figure 17.1.1

A typical subrectangle is bounded by the lines $x = x_{i-1}$, $x = x_i$, $y = y_{j-1}$, and $y = y_j$, and will be denoted by R_{ij} (see Figure 17.1.2).

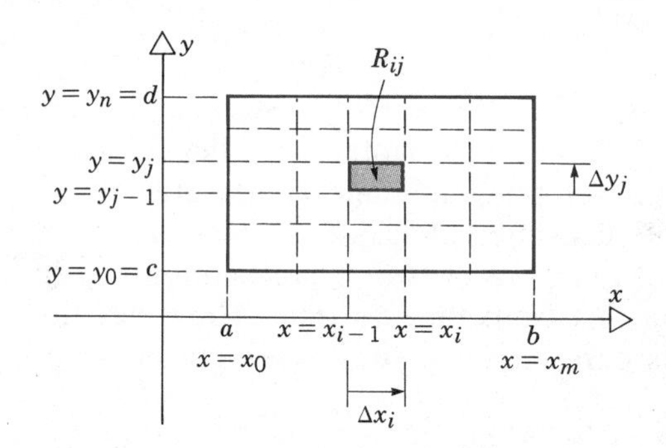

Figure 17.1.2 A typical subrectangle R_{ij}.

The collection of subrectangles

$$\Delta = \{R_{ij}\}, \qquad i = 1, \ldots, m, \qquad j = 1, \ldots, n \tag{2}$$

constitutes a **partition** of the original rectangle R. The area of the subrectangle R_{ij} is

$$\Delta A_{ij} = (x_i - x_{i-1})(y_j - y_{j-1}) = \Delta x_i \, \Delta y_j. \tag{3}$$

To measure the "fineness" of the partition Δ it is convenient to use the number $\|\Delta\|$ defined by

$$\|\Delta\| = \max\sqrt{(\Delta x_i)^2 + (\Delta y_j)^2}; \tag{4}$$

$\|\Delta\|$ is the length of the longest diagonal of all of the subrectangles R_{ij}, and is called the **norm** of Δ.

For a given partition Δ we arbitrarily select a star point $P_{ij}^* = (x_{ij}^*, y_{ij}^*)$ in each closed rectangle R_{ij}, evaluate the function f at P_{ij}^*, and form the product $f(P_{ij}^*) \, \Delta A_{ij}$. If $f(P_{ij}^*) > 0$, then this product is the volume of a rectangular box (parallelepiped) of height $f(P_{ij}^*)$ and with cross-sectional area ΔA_{ij}. Intuitively, we can view the volume of this box as an approximation to the volume of the cylindrical region bounded below by the xy-plane, above by the surface $z = f(x, y)$, and on the sides by the planes $x = x_{i-1}$, $x = x_i$, $y = y_{j-1}$, and $y = y_j$. This is illustrated

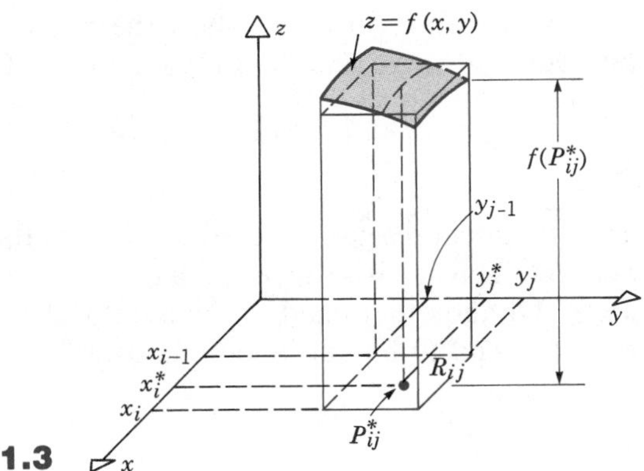

Figure 17.1.3

in Figure 17.1.3. By proceeding in a similar way for each subrectangle and then adding all of these products, we obtain the Riemann sum

$$\sum_{\substack{i=1,\ldots,m \\ j=1,\ldots,n}} f(P_{ij}^*)\,\Delta A_{ij} = f(P_{11}^*)\,\Delta A_{11} + \cdots + f(P_{mn}^*)\,\Delta A_{mn}, \tag{5}$$

or more briefly $\Sigma_{i,j}\, f(P_{ij}^*)\,\Delta A_{ij}$, where it is understood that the sum is extended over all permissible values of i and j. For a nonnegative function f we intuitively expect the sum (5) to be an approximation to the volume of the three-dimensional region bounded below by the rectangle R in the xy-plane, above by the surface $z = f(x, y)$, and on the sides by the planes $x = a$, $x = b$, $y = c$, and $y = d$; see Figure 17.1.4 for a case where $m = n = 2$. While it is often helpful to think

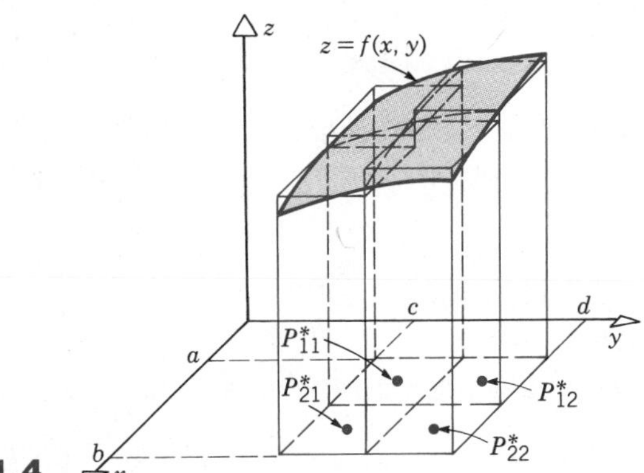

Figure 17.1.4

of Riemann sums in terms of volumes in this way, we emphasize that we can form the sum (5) regardless of the signs of the various terms and without any reference to geometrical ideas. However, we do not interpret it as a volume unless $f(x, y) \geq 0$.

As in the case of one-dimensional integrals, we wish to examine the behavior of the Riemann sum (5) as the partition Δ becomes finer and finer, that is, as $\|\Delta\| \to 0$. If the limit of the sum (5) as $\|\Delta\| \to 0$ exists regardless of how the partition is formed, and regardless of how the star points are selected, then f is said to be **integrable** over R. The limit is called the integral of f over R and is denoted by

$$\iint_R f(x, y)\, dA = \lim_{\|\Delta\|\to 0} \sum_{\substack{i=1,\ldots,m \\ j=1,\ldots,n}} f(P_{ij}^*)\, \Delta A_{ij}. \tag{6}$$

The two integral signs are used to indicate that this is a two-dimensional or **double integral**. This is also suggested by designating the integration element as dA (for area). On the other hand, if the limit of the Riemann sum in Eq. 6 does not exist, then f is said to be not integrable over R.

EXAMPLE 1

Let $f(x, y) = k$ for all points (x, y) in R. Show that f is integrable over R.

For every partition of R and for every choice of the star points we have $f(P_{ij}^*) = k$. Thus

$$\sum_{\substack{i=1,\ldots,m \\ j=1,\ldots,n}} f(P_{ij}^*)\, \Delta A_{ij} = k \sum_{\substack{i=1,\ldots,m \\ j=1,\ldots,n}} \Delta A_{ij} = k(b - a)(d - c).$$

Since in this case all Riemann sums have the same value, it follows that

$$\lim_{\|\Delta\|\to 0} \sum_{\substack{i=1,\ldots,m \\ j=1,\ldots,n}} f(P_{ij}^*)\, \Delta A_{ij} = k(b - a)(d - c),$$

so f is integrable over R, and

$$\iint_R f(x, y)\, dA = \iint_R k\, dA = k(b - a)(d - c).$$

Note that if $k > 0$, then $k(b - a)(d - c)$ is the volume of a rectangular parallelepiped of height k and base $(b - a)$ by $(d - c)$. ∎

EXAMPLE 2

Let f be defined on R so that

$$f(x, y) = \begin{cases} 1, & \text{if } x \text{ and } y \text{ are both rational;} \\ -1, & \text{otherwise.} \end{cases} \tag{7}$$

Show that f is not integrable over R.

For every partition of R it is always possible to choose a star point in each subrectangle such that $f(P_{ij}^*) = 1$. In this case the corresponding Riemann sum is $\Sigma_{i,j}\, 1\, \Delta A_{ij} = (b - a)(d - c)$. On the other hand, for every partition it is also always possible to choose another star point in each subrectangle such that $f(P_{ij}^*) = -1$, in which case the corresponding Riemann sum has the value $-(b - a)(d - c)$. Thus, no matter how small $\|\Delta\|$ is, the Riemann sums do not approach a single number as a limit. Consequently, f is not integrable over R. ∎

Examples 1 and 2 show that some functions are integrable over a rectangle R, while others are not. The following theorem is somewhat analogous to Theorem 6.2.1, and serves to identify a large class of integrable functions.

Theorem 17.1.1

Suppose that f is bounded on the rectangle R and is continuous there except (at most) for the points lying on a finite number of smooth rectifiable arcs. Then f is integrable over R.

Recall that smooth rectifiable arcs are those that have a continuously turning tangent and are of finite length. See Sections 15.6 and 15.7 for fuller discussions of these terms. The graphs of

$$y = \phi(x), \qquad a \le x \le b, \tag{8}$$

or of

$$x = \psi(y), \qquad c \le y \le d, \tag{9}$$

or of the parametric system

$$x = \phi(t), \qquad y = \psi(t), \qquad \alpha \le t \le \beta, \tag{10}$$

are smooth and rectifiable if ϕ' and ψ' are continuous on the given intervals, and in the case of Eqs. 10, $\phi'^2(t) + \psi'^2(t)$ is never zero on $[\alpha, \beta]$.

Theorem 17.1.1 is sufficiently general to guarantee the integrability over rectangles of almost all bounded functions that are encountered in an elementary course and in most applications. The proof is similar to that of the corresponding result in Chapter 6, and is omitted here as well.

EXAMPLE 3

Suppose that f is defined on the rectangle $R : |x| \le 2, |y| \le 1$ in the following way:

$$f(x, y) = \begin{cases} x^2 + y^2, & x^2 + y^2 \le 1; \\ \dfrac{1 - (x^2 + y^2)}{4}, & (x, y) \in R \text{ and } x^2 + y^2 > 1 \end{cases} \tag{11}$$

(see Figure 17.1.5). Determine whether f is integrable over R.

Since f is defined by two polynomial expressions, it is bounded on the rectangle R and continuous except perhaps on the circle $x^2 + y^2 = 1$. From Figure 17.1.5 it is clear that actually f is discontinuous at each point on the circle. However, the circle is a smooth rectifiable arc (of length 2π), so the function f satisfies the conditions of Theorem 17.1.1. Hence f is integrable over R. ∎

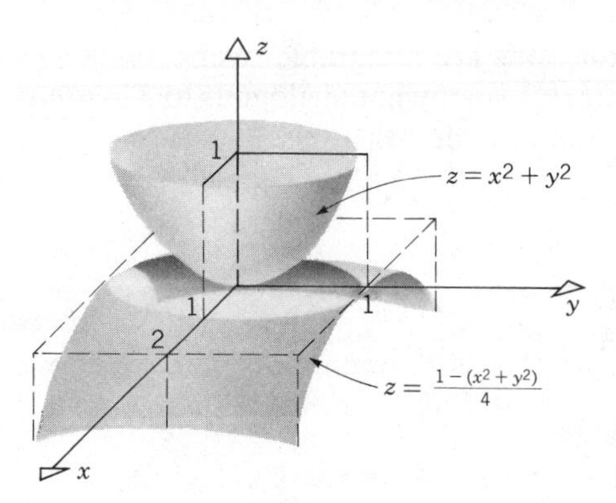

Figure 17.1.5

In one dimension, integration is carried out over intervals, and in two dimensions the natural analog of an interval is a rectangle. However, it is important to be able to integrate over other two-dimensional regions (triangles, circles, . . .) as well, so now we turn to the question of defining a double integral over a more or less arbitrary region in the plane. We denote the region of interest by Ω, and we assume that it is closed and bounded. This means that the boundary of Ω is included in Ω, and that a rectangle R can be found that completely contains Ω (see Figure 17.1.6). Further, we suppose that the boundary of Ω consists of a finite number of smooth rectifiable arcs, each described by equations of the form (8), (9), or (10).

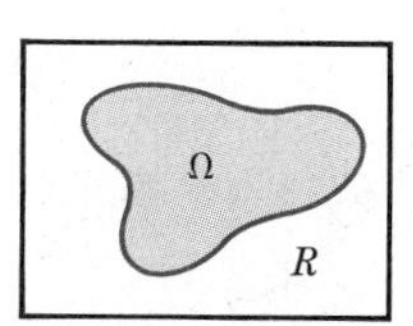

Figure 17.1.6 The bounded region Ω is contained in the rectangle R.

Now suppose that a function f is given that satisfies the hypotheses of Theorem 17.1.1 on Ω. Then we define a new function g in the following way:

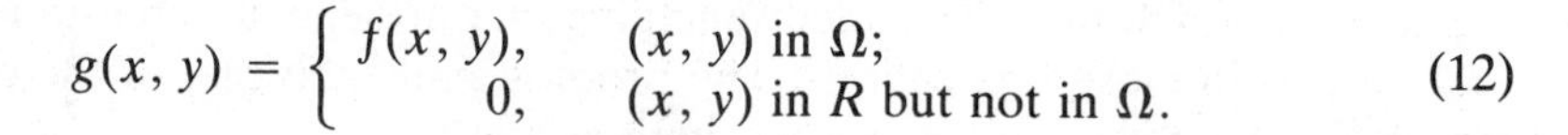

$$g(x, y) = \begin{cases} f(x, y), & (x, y) \text{ in } \Omega; \\ 0, & (x, y) \text{ in } R \text{ but not in } \Omega. \end{cases} \tag{12}$$

Since g is the same as f wherever f is defined (that is, on Ω), and since g is defined on a larger region, we call g an **extension** of f from Ω to R. Further, g is continuous on R except (at most) for two sets of points, namely

1. the points in Ω where f is discontinuous;
2. the boundary of Ω.

Both of these sets of points satisfy the conditions of Theorem 17.1.1, so the integral of g over R exists. Finally, we *define* the integral of f over Ω to have the same value as the integral of g over R; that is

$$\iint_\Omega f(x, y)\, dA = \iint_R g(x, y)\, dA, \tag{13}$$

where g is given by Eq. 12. Further, since the extension of f is zero outside of Ω, it makes no difference which rectangle R we choose, so long as it contains Ω. Equation 13 is valid for any such R. Thus we have obtained the following result.

Theorem 17.1.2

Let Ω be a bounded region in the xy-plane whose boundary consists of a finite number of smooth rectifiable arcs. If f is continuous on Ω except (at most) for a set of points lying on a finite number of smooth rectifiable arcs, then f is integrable over Ω, and the value of the integral is given by Eq. 13.

In the foregoing discussion of double integrals, we have indicated the possible interpretation of certain integrals as volumes. Indeed, volumes are defined in terms of integrals. Consider the cylinder whose cross section in the xy-plane is the region Ω, and whose axis is parallel to the z-axis. The volume V of the portion of this cylinder lying above the xy-plane and below the surface $z = f(x, y)$ is defined to be

$$V = \iint_\Omega f(x, y)\, dA, \tag{14}$$

provided that f is integrable over Ω (see Figure 17.1.7).

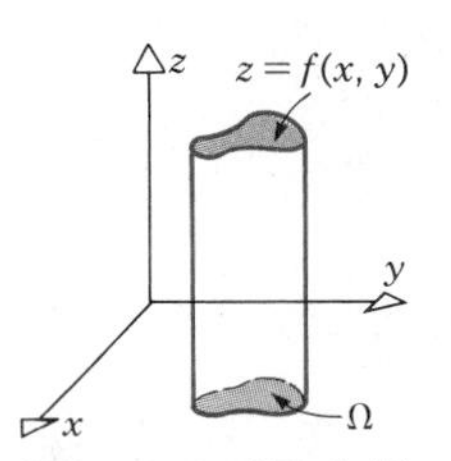

Figure 17.1.7

If $f(x, y) = 1$ for each point in Ω, then Eq. 14 gives the volume of a cylindrical region of unit thickness and cross section Ω. Thus $V = A \cdot 1$, where A is the area of Ω. From Eq. 14 it then follows that

$$A(\Omega) = \iint_\Omega dA. \tag{15}$$

Another important interpretation of a double integral of a nonnegative funtion ρ occurs if we think of $\rho(x, y)$ as the density* (mass per unit area) of a thin sheet of some material that occupies the region Ω. Then $\rho(x_i^*, y_j^*)\, \Delta A_{ij}$ is approximately the mass of the element ΔA_{ij}, and the Riemann sum

$$\sum_{\substack{i=1,\ldots,m \\ j=1,\ldots,n}} \rho(x_i^*, y_j^*)\, \Delta A_{ij} \tag{16}$$

is approximately the total mass M in the region Ω. If $\rho(x, y)$ and Ω satisfy the conditions of Theorem 17.1.2, then the sum (16) has a limit as $\|\Delta\| \to 0$, and the total mass M of the thin sheet is given by

$$M = \iint_\Omega \rho(x, y)\, dA. \tag{17}$$

It is sometimes possible to evaluate double integrals by direct use of the definition. Since such an approach is cumbersome in one dimension, and becomes more so for double integrals, it is hardly ever satisfactory to determine the value of a double integral in this manner. In the next section we develop a method that is far more convenient. There are also numerical methods, similar to those discussed in Section 6.6 for single integrals, but we do not consider them.

*In continuum mechanics the existence and definition of a density function is a fairly subtle question, and requires a careful discussion. Such a treatment would be out of place here, so we will simply assume that the density function is given.

In concluding this section we state several properties of double integrals that are often useful. Each property is directly analogous to a property of single integrals that was discussed in Section 6.3.

Theorem 17.1.3

Suppose that f and g are integrable over Ω.

(a) *Linearity*. If c_1 and c_2 are any constants, then $c_1 f + c_2 g$ is also integrable over Ω, and

$$\iint_\Omega [c_1 f(x, y) + c_2 g(x, y)]\, dA = c_1 \iint_\Omega f(x, y)\, dA + c_2 \iint_\Omega g(x, y)\, dA. \qquad (18)$$

This result generalizes to a sum of any finite number of terms.

(b) If $f(x, y) \geq 0$ on Ω, then

$$\iint_\Omega f(x, y)\, dA \geq 0. \qquad (19)$$

(c) *Comparison*. If $f(x, y) \geq g(x, y)$ on Ω, then

$$\iint_\Omega f(x, y)\, dA \geq \iint_\Omega g(x, y)\, dA. \qquad (20)$$

(d) The function $|f|$ is also integrable over Ω, and

$$\left| \iint_\Omega f(x, y)\, dA \right| \leq \iint_\Omega |f(x, y)|\, dA. \qquad (21)$$

(e) If $|f(x, y)| \leq M$ on Ω for some constant M, then

$$\left| \iint_\Omega f(x, y)\, dA \right| \leq MA, \qquad (22)$$

where A is the area of Ω.

Theorem 17.1.4

(Additivity).

Suppose that Ω is composed of two nonoverlapping parts Ω_1 and Ω_2; suppose further that Ω_1 and Ω_2, as well as Ω, are bounded by a finite number of smooth rectifiable arcs. If any two of the following three integrals exist, then the third also exists, and

$$\iint_\Omega f(x, y)\, dA = \iint_{\Omega_1} f(x, y)\, dA + \iint_{\Omega_2} f(x, y)\, dA. \qquad (23)$$

This result generalizes to the decomposition of Ω into any finite number of nonoverlapping parts $\Omega_1, \ldots, \Omega_n$.

Figures 17.1.8, 17.1.9, and 17.1.10 illustrate the content of Theorem 17.1.4. If one wishes to integrate over the region Ω shown in Figure 17.1.8, it will almost certainly be helpful to separate Ω into nonoverlapping parts and then to integrate over each part separately. Figures 17.1.9 and 17.1.10 show two easy ways of splitting Ω into four nonoverlapping parts that are more convenient for integration.

Figure 17.1.8

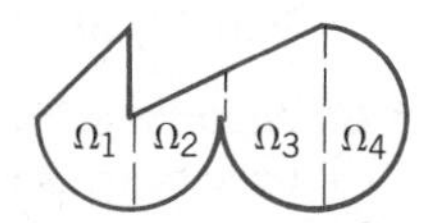

Figure 17.1.9

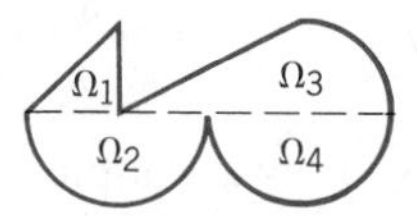

Figure 17.1.10

PROBLEMS

1. Let $f(x, y) = 2xy$ and let R be the square $0 \leq x \leq 2$, $0 \leq y \leq 2$. Subdivide R into four subrectangles by the partition points $x_0 = 0$, $x_1 = 1$, $x_2 = 2$, and $y_0 = 0$, $y_1 = 1$, $y_2 = 2$. Form the corresponding Riemann sum if
   (a) P_{ij}^* is the center of each subrectangle,
   (b) P_{ij}^* is the lower right corner of each subrectangle,
   (c) P_{ij}^* is the upper right corner of each subrectangle.
2. Let $f(x, y) = 2x^2 - y^2$ and let R be the rectangle $0 \leq x \leq 3$, $0 \leq y \leq 2$. Subdivide R into six subrectangles by the partition points $x_0 = 0$, $x_1 = 1$, $x_2 = 2$, $x_3 = 3$, and $y_0 = 0$, $y_1 = 1$, $y_2 = 2$. Form the corresponding Riemann sum if
   (a) P_{ij}^* is the center of each subrectangle,
   (b) P_{ij}^* is chosen in each subrectangle to make $f(P_{ij}^*)$ as large as possible,
   (c) P_{ij}^* is chosen in each subrectangle to make $f(P_{ij}^*)$ as small as possible.
3. Let $f(x, y) = x^2y$ and let R be the rectangle $0 \leq x \leq 2$, $0 \leq y \leq 1$. Find the corresponding Riemann sum if P_{ij}^* is the center of each subrectangle, and R is partitioned into
   (a) two identical squares, each of area 1,
   (b) eight identical squares, each of area $\frac{1}{4}$,
   (c) thirty-two identical squares, each of area $\frac{1}{16}$.
4. Let $f(x, y) = \sin x \cos y$ and let R be the rectangle $0 \leq x \leq \pi$, $0 \leq y \leq \pi/2$. Find the corresponding Riemann sum if P_{ij}^* is the center of each subrectangle, and R is partitioned into
   (a) two identical squares, each of area $(\pi/2)^2$,
   (b) eight identical squares, each of area $(\pi/4)^2$,
   (c) thirty-two identical squares, each of area $(\pi/8)^2$.

In each of Problems 5 through 8, use Part (c), (d), or (e) of Theorem 17.1.3 to find an upper bound for the indicated quantity.

5. $\iint_\Omega \frac{1}{2x^2 + y^2 + 4}\, dA$, where $\Omega : x^2 + y^2 \leq 1$
6. $\left| \iint_\Omega \sin(2x - 3y)\, dA \right|$, where $\Omega : 0 \leq x \leq 2$, $0 \leq y \leq 5$
7. $\iint_\Omega e^{-(x^2+y^2)}\, dA$, where $\Omega : x^2 + y^2 \leq \frac{1}{2}$
8. $\iint_\Omega xy\, dA$, where $\Omega : 1 \leq x \leq 2$, $1 \leq y \leq 3$

Sometimes it is possible to find the value of an integral by identifying it as the volume of an elementary solid figure that is known from geometry. Use this approach to evaluate the integrals in problems 9 through 12.

9. $\iint_\Omega \sqrt{9 - x^2 - y^2}\, dA$, where Ω is the quarter circle $x^2 + y^2 \leq 9$ with $x \geq 0$, $y \geq 0$.
10. $\iint_\Omega (6 - 3x - 2y)\, dA$, where Ω is the triangle bounded by the coordinate axes and the line $3x + 2y = 6$.

11. $\iint_{\Omega} (6 - 3\sqrt{x^2 + y^2})\, dA$, where Ω is the circle $x^2 + y^2 \leq 4$.

12. $\iint_{\Omega} (3 - \frac{3}{4}y)\, dA$, where Ω is the rectangle $0 \leq x \leq 2$, $0 \leq y \leq 4$.

13. Prove Part (b) of Theorem 17.1.3: If f is integrable over Ω, and if $f(x, y) \geq 0$ on Ω, then $\iint_{\Omega} f(x, y)\, dA \geq 0$.

Hint: What can you say about the sign of each of the associated Riemann sums?

14. Assuming that Part (b) of Theorem 17.1.3 has been established, prove Part (c) of that theorem.

Hint: Consider $\iint_{\Omega} [f(x, y) - g(x, y)]\, dA$.

15. If f is integrable over Ω, it is possible to show that $|f|$ is also integrable. Assuming this to be true, and assuming Part (c) of Theorem 17.1.3, prove Part (d) of Theorem 17.1.3.

16. Assuming Parts (c) and (d) of Theorem 17.1.3 have been established, prove Part (e) of that theorem.

17.2 ITERATED INTEGRALS

In the preceding section we defined the double integral $\iint_{\Omega} f(x, y)\, dA$ for a large class of functions f and regions Ω. The discussion followed lines very similar to those in Chapter 6, where the definite integral $\int_a^b f(x)\, dx$ of a function of one variable was treated. In Chapter 6 the next step was to develop a way of evaluating integrals by using an antiderivative $F(x)$ of the integrand $f(x)$, and invoking the fundamental theorem of calculus (Theorem 6.4.3). It is natural to seek to extend this procedure to double integrals and thereby obtain a method for evaluating them.

In calculating partial derivatives of a function f of two variables we hold one variable constant while differentiating with respect to the other. The inverse operation to partial differentiation involves integrating with respect to one variable while holding the other constant. For example, suppose that $f(x, y)$ is defined on the rectangle $R : a \leq x \leq b,\ c \leq y \leq d$. Then

$$g(x) = \int_c^d f(x, y)\, dy \tag{1}$$

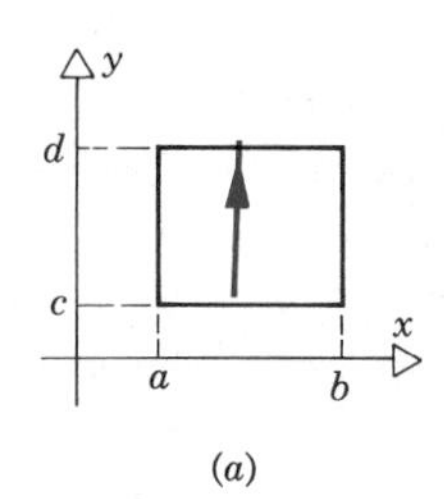

(a)

refers to the result of fixing x and then integrating $f(x, y)$ with respect to y over the interval $c \leq y \leq d$, as suggested in Figure 17.2.1*a*. The notation $g(x)$ indicates that the value of the integral in general depends on x, but it does not depend on y. Similarly,

$$h(y) = \int_a^b f(x, y)\, dx \tag{2}$$

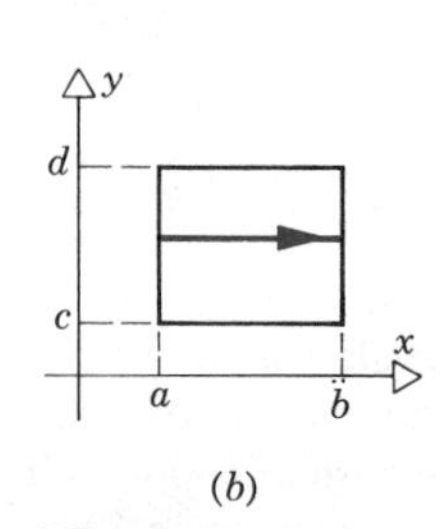

(b)

Figure 17.2.1
(*a*) Integration with respect to y for fixed x. (*b*) Integration with respect to x for fixed y.

is obtained by holding y constant and integrating with respect to x from a to b (see Figure 17.2.1*b*). The fundamental theorem of calculus can be used to evaluate $g(x)$ and $h(y)$ since they involve integration with respect to one variable only.

EXAMPLE 1

Let $f(x, y) = 1 + 2xy + 3x^2y^3$ for $0 \leq x \leq 4$ and $1 \leq y \leq 3$. Find

$$g(x) = \int_1^3 f(x, y)\, dy \qquad \text{and} \qquad h(y) = \int_0^4 f(x, y)\, dx.$$

We have

$$g(x) = \int_1^3 (1 + 2xy + 3x^2y^3)\,dy$$

$$= \left(y + xy^2 + \frac{3}{4}x^2y^4\right)\Bigg|_{y=1}^{y=3}$$

$$= \left(3 + 9x + \frac{243}{4}x^2\right) - \left(1 + x + \frac{3}{4}x^2\right)$$

$$= 2 + 8x + 60x^2, \qquad 0 \le x \le 4. \tag{3}$$

Similarly

$$h(y) = \int_0^4 (1 + 2xy + 3x^2y^3)\,dx$$

$$= (x + x^2y + x^3y^3)\Bigg|_{x=0}^{x=4}$$

$$= 4 + 16y + 64y^3, \qquad 1 \le y \le 3. \blacksquare \tag{4}$$

The functions g and h can now be integrated over the intervals $a \le x \le b$ and $c \le y \le d$, respectively. This gives

$$\int_a^b g(x)\,dx = \int_a^b \left[\int_c^d f(x, y)\,dy\right] dx \tag{5}$$

and

$$\int_c^d h(y)\,dy = \int_c^d \left[\int_a^b f(x, y)\,dx\right] dy. \tag{6}$$

The integrals in Eqs. 5 and 6 are called **iterated integrals.** We have inserted brackets in these equations to emphasize the order of the operations; in Eq. 5 we integrate first with respect to y and then with respect to x, while in Eq. 6 the order is reversed. It is important to understand that each of the integrations on the right side of Eqs. 5 and 6 is with respect to a single variable. Thus, all of the methods presented up to now for the evaluation of integrals of functions of one variable are available for the evaluation of these iterated integrals. Observe that integration works from the inside out, as indicated by the brackets in Eqs. 5 and 6. Usually we will omit the brackets and denote the iterated integrals (5) and (6) by

$$\int_a^b \int_c^d f(x, y)\,dy\,dx \qquad \text{and} \qquad \int_c^d \int_a^b f(x, y)\,dx\,dy, \tag{7}$$

respectively.

EXAMPLE 2

Let $f(x, y)$ be as given in Example 1. Find

$$\int_0^4 \int_1^3 f(x, y)\,dy\,dx \qquad \text{and} \qquad \int_1^3 \int_0^4 f(x, y)\,dx\,dy.$$

In Example 1 we calculated

$$g(x) = \int_1^3 f(x, y)\,dy = 2 + 8x + 60x^2, \qquad 0 \le x \le 4$$

and

$$h(y) = \int_0^4 f(x, y)\,dx = 4 + 16y + 64y^3, \qquad 1 \le y \le 3.$$

Thus

$$\begin{aligned}\int_0^4 \int_1^3 f(x, y)\,dy\,dx &= \int_0^4 g(x)\,dx \\ &= (2x + 4x^2 + 20x^3)\Big|_0^4 \\ &= 8 + 64 + 1280 = 1352.\end{aligned}$$

In a similar way

$$\begin{aligned}\int_1^3 \int_0^4 f(x, y)\,dx\,dy &= \int_1^3 h(y)\,dy \\ &= (4y + 8y^2 + 16y^4)\Big|_1^3 \\ &= (12 + 72 + 1296) - (4 + 8 + 16) = 1352. \blacksquare\end{aligned}$$

For the case considered in Example 2 direct calculation shows that the two sets of iterated integrals give the same result. In other words, the order in which the two integrations are performed does not matter. This raises the question as to whether the two sets of iterated integrals in Eq. 7 are always equal, and if so, whether they also have the same value as the double integral $\iint_R f(x, y)\,dA$ that was discussed in Section 17.1. If the answer to this question is "yes," then iterated integrals provide a means of evaluating double integrals.

We can give a geometrical discussion of the relation between iterated and double integrals when $f(x, y) \ge 0$ on R. Consider the region U, shown in Figure 17.2.2, bounded by the xy-plane, the surface $z = f(x, y)$, and the four planes $x = a$, $x = b$, $y = c$, and $y = d$. In Section 17.1 the volume V of this region was defined as the double integral of f over R:

$$V = \iint_R f(x, y)\,dA. \tag{8}$$

Now let us consider the geometrical interpretation of the iterated integrals (7). First, from Figure 17.2.3,

$$g(x_0) = \int_c^d f(x_0, y)\,dy$$

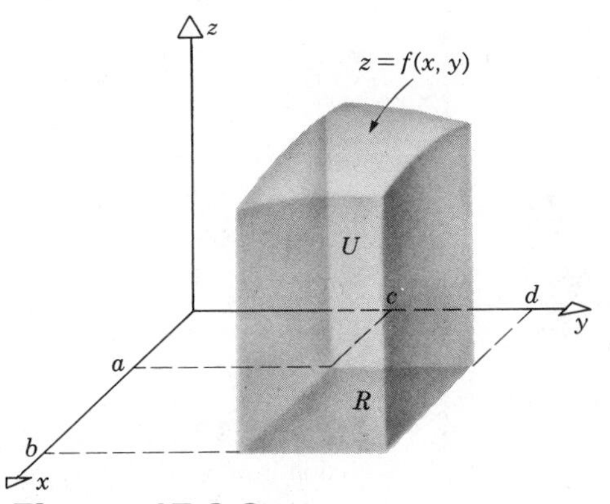

Figure 17.2.2

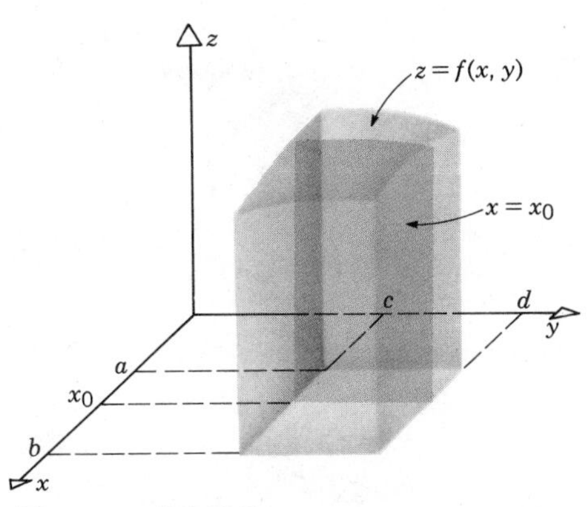

Figure 17.2.3 The area of the shaded cross section is $g(x_0) = \int_c^d f(x_0, y)\, dy$.

is the area of the cross section of U corresponding to $x = x_0$. Then, expressing the integral of $g(x)$ from $x = a$ to $x = b$ in terms of a Riemann sum, we have

$$\int_a^b g(x)\, dx = \lim_{\substack{m\to\infty \\ \|\Delta\|\to 0}} \sum_{i=1}^{m} g(x_i^*)\, \Delta x_i. \tag{9}$$

Each term in the sum can be viewed as the volume of a thin slab whose faces have area $g(x_i^*)$ and whose thickness is Δx_i (see Figure 17.2.4). The Riemann sum in

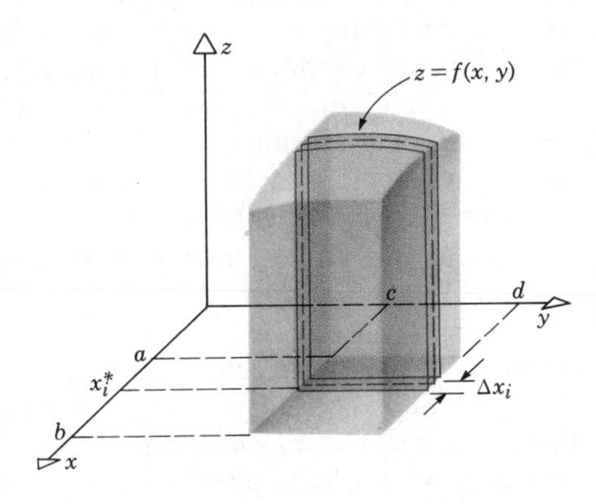

Figure 17.2.4 The volume of the thin slab is $g(x_i^*)\Delta x_i$.

Eq. 9 thus gives an approximation to the volume V of the region U. In the limit the approximation becomes exact, so we have

$$V = \int_a^b g(x)\, dx = \int_a^b \int_c^d f(x, y)\, dy\, dx. \tag{10}$$

In the same way

$$h(y_0) = \int_a^b f(x, y_0)\, dx$$

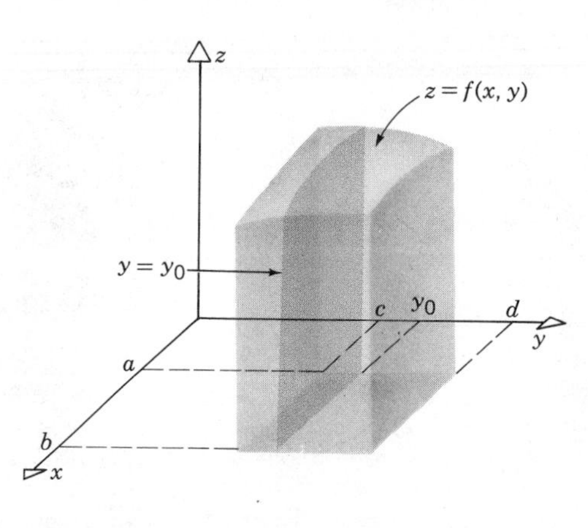

Figure 17.2.5 The area of the shaded cross section is $h(y_0) = \int_a^b f(x, y_0)\, dx$.

is the area of the cross section of U for $y = y_0$ (see Figure 17.2.5). Then V can also be obtained by integrating $h(y)$ from $y = c$ to $y = d$. This corresponds to approximating V by means of thin slabs of thickness Δy_j and area $h(y_j^*)$. Thus we obtain

$$V = \lim_{\substack{n\to\infty \\ \|\Delta\|\to 0}} \sum_{j=1}^{n} h(y_j^*)\, \Delta y_j = \int_c^d h(y)\, dy$$
$$= \int_c^d \int_a^b f(x, y)\, dx\, dy. \tag{11}$$

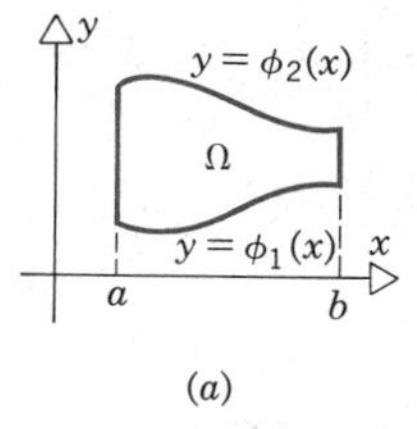

(a)

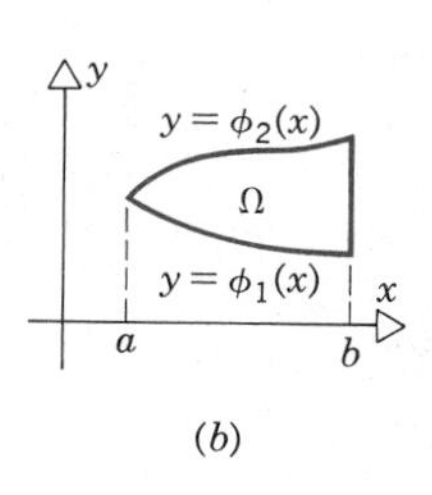

(b)

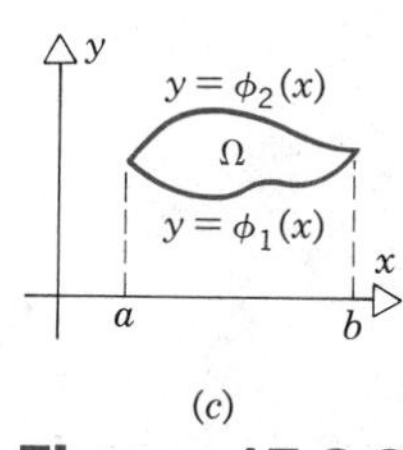

(c)

Figure 17.2.6

Equations 8, 10, and 11 state that the same volume V is given by each set of iterated integrals, as well as by the double integral. Thus we should expect equality to hold among these three integrals, at least when f is nonnegative and satisfies the continuity conditions of Theorem 17.1.1. Indeed this is true; further, this relation is valid for a much wider class of integrals in which the region may not be rectangular and the integrand may not be nonnegative. In order to discuss this more general and useful result, however, we must first extend the use of iterated integrals to nonrectangular regions.

We first consider a region Ω consisting of points (x, y) such that

$$a \le x \le b \quad \text{and} \quad \phi_1(x) \le y \le \phi_2(x), \tag{12}$$

where ϕ_1 and ϕ_2 are given functions. Figure 17.2.6 shows some typical regions of this kind. Suppose that $f(x, y)$ is continuous for each point in Ω. Then for each fixed x in $a \le x \le b$, we can integrate $f(x, y)$ with respect to y from the lower boundary of Ω to the upper boundary, that is, from $y = \phi_1(x)$ to $y = \phi_2(x)$. The result is

$$g(x) = \int_{\phi_1(x)}^{\phi_2(x)} f(x, y)\, dy. \tag{13}$$

Observe that the limits of integration change with the position of x in the interval $a \le x \le b$, but are always given by $\phi_1(x)$ and $\phi_2(x)$, respectively. To complete the integration over the region Ω we integrate $g(x)$ from a to b; thus we obtain

$$\int_a^b g(x)\, dx = \int_a^b \int_{\phi_1(x)}^{\phi_2(x)} f(x, y)\, dy\, dx. \tag{14}$$

It may help to visualize the integration process in the following way. Suppose that the region Ω is to be covered by small thin rectangular tiles of dimensions dx and dy (see Figure 17.2.7*a*). The integration with respect to y corresponds to forming a single strip of tiles parallel to the y-axis, as suggested in Figure 17.2.7*b*. Then the integration with respect to x corresponds to laying many such strips side by side until the region Ω is entirely covered.

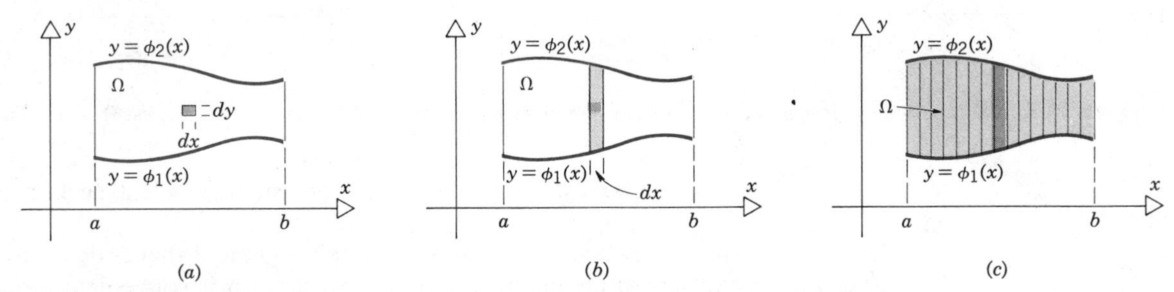

Figure 17.2.7 (*a*) An area element $dx\,dy$. (*b*) A strip of elements parallel to the y-axis. (*c*) The region Ω covered by strips parallel to the y-axis.

Now let us consider a region Ω that is described by the inequalities

$$\psi_1(y) \le x \le \psi_2(y), \qquad c \le y \le d, \tag{15}$$

where ψ_1 and ψ_2 are given functions. Figure 17.2.8 shows some typical regions of

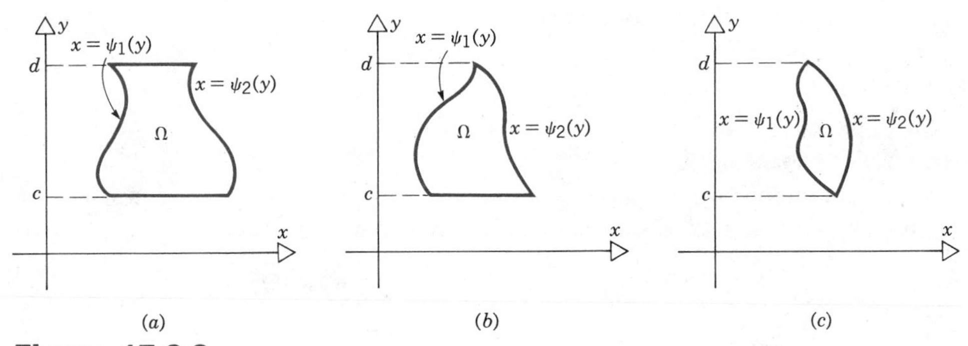

Figure 17.2.8

this kind. Then the derivation leading to Eq. 14 can be repeated, with the roles of x and y reversed. We have

$$h(y) = \int_{\psi_1(y)}^{\psi_2(y)} f(x, y)\, dx \tag{16}$$

and

$$\int_c^d h(y)\, dy = \int_c^d \int_{\psi_1(y)}^{\psi_2(y)} f(x, y)\, dx\, dy, \tag{17}$$

corresponding to Eqs. 13 and 14, respectively. In terms of covering the region Ω by tiles, this corresponds first to forming a single strip of tiles parallel to the x-

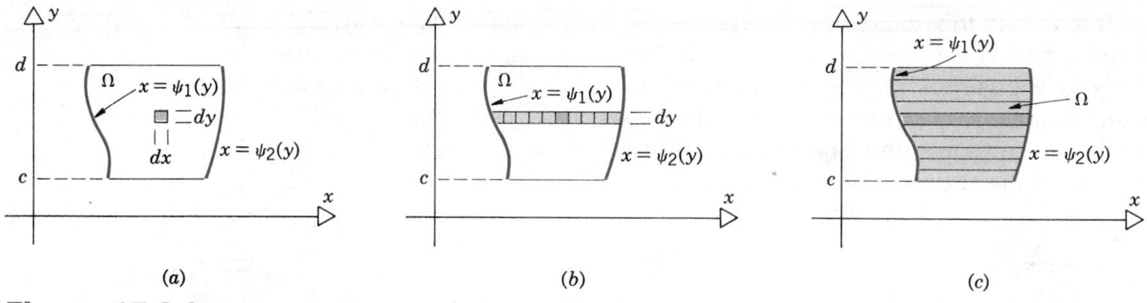

Figure 17.2.9 (*a*) An area element $dx\,dy$. (*b*) A strip of elements parallel to the x-axis. (*c*) The region Ω covered by strips parallel to the x-axis.

axis, and then to covering Ω by many such strips laid side by side (see Figure 17.2.9).

There is no universally accepted term to denote a region Ω that is described by at least one of the sets of inequalities (12) or (15). Since it is convenient to be able to refer to these regions by name, we will call such a region a **standard region.**

Now we turn to the question of whether the iterated integrals on the right sides of Eqs. 14 and 17 are equal to the double integral $\iint_\Omega f(x, y)\,dA$ that was defined in Section 17.1. Based on our previous discussion of this question when the region Ω is a rectangle and $f(x, y) \geq 0$, it seems reasonable to expect that the iterated integrals are indeed equal to the double integral. The following theorem gives a simple set of conditions, adequate for most purposes, that assures the equality of iterated and double integrals.

Theorem 17.2.1

If f is continuous on a region Ω defined by $a \leq x \leq b$, $\phi_1(x) \leq y \leq \phi_2(x)$, where ϕ_1 and ϕ_2 have continuous derivatives on $a \leq x \leq b$, then

$$\iint_\Omega f(x, y)\,dA = \int_a^b \int_{\phi_1(x)}^{\phi_2(x)} f(x, y)\,dy\,dx. \tag{18}$$

If f is continuous on a region Ω defined by $\psi_1(y) \leq x \leq \psi_2(y)$, $c \leq y \leq d$, where ψ_1 and ψ_2 have continuous derivatives on $c \leq y \leq d$, then

$$\iint_\Omega f(x, y)\,dA = \int_c^d \int_{\psi_1(y)}^{\psi_2(y)} f(x, y)\,dx\,dy. \tag{19}$$

The significance of Theorem 17.2.1 is that it provides a means for evaluating double integrals: the corresponding iterated integrals are evaluated *sequentially* by integrating over one variable at a time. The hypotheses in Theorem 17.2.1 are chosen so as to make it easy to conclude from Theorem 17.1.2 that the double integral $\iint_\Omega f(x, y)\,dA$ exists. It is possible to establish more general versions of Theorem 17.2.1; however, even the proof of Theorem 17.2.1 is fairly complicated, and is not discussed here.

EXAMPLE 3

Find the value of

$$\iint_{\Omega} (2y - x^2)\, dA, \tag{20}$$

where Ω is the region $0 \le x \le 2$, $x^2 \le y \le 9 - x$.

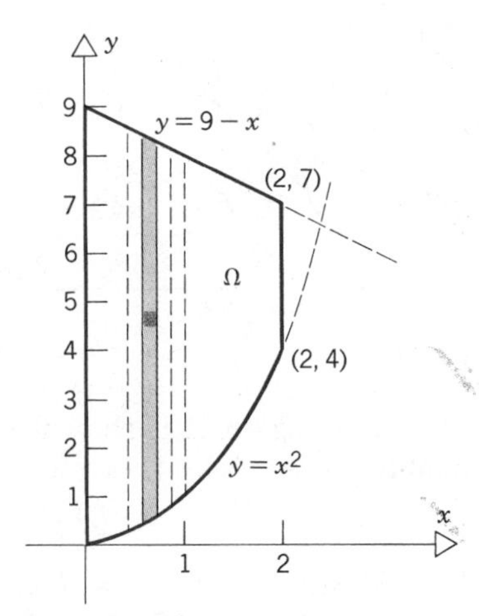

Figure 17.2.10

The region of integration is shown in Figure 17.2.10, and is a standard region of the form (12). Thus we replace the double integral (20) by a set of iterated integrals in which the y-integration is done first:

$$\iint_{\Omega} (2y - x^2)\, dA = \int_0^2 \int_{x^2}^{9-x} (2y - x^2)\, dy\, dx. \tag{21}$$

Evaluating the inside integral in Eq. 21, we obtain

$$\begin{aligned}\int_{x^2}^{9-x} (2y - x^2)\, dy &= (y^2 - x^2y)\Big|_{x^2}^{9-x} \\ &= [(9 - x)^2 - x^2(9 - x)] - [(x^2)^2 - x^2x^2] \\ &= 81 - 18x - 8x^2 + x^3.\end{aligned}$$

Then the second integration in Eq. 21 yields the result

$$\begin{aligned}\iint_{\Omega} (2y - x^2)\, dA &= \int_0^2 (81 - 18x - 8x^2 + x^3)\, dx \\ &= \left(81x - 9x^2 - \frac{8}{3}x^3 + \frac{x^4}{4}\right)\Big|_0^2 \\ &= 162 - 36 - \frac{64}{3} + 4 \\ &= \frac{326}{3}. \quad \blacksquare\end{aligned}$$

EXAMPLE 4

A thin sheet of material whose density (mass per unit area) is proportional to the square of the distance from the origin occupies the triangular region Ω bounded by the x-axis, the line $y = x$, and the line $y = 2 - x$. Find the total mass M of the sheet.

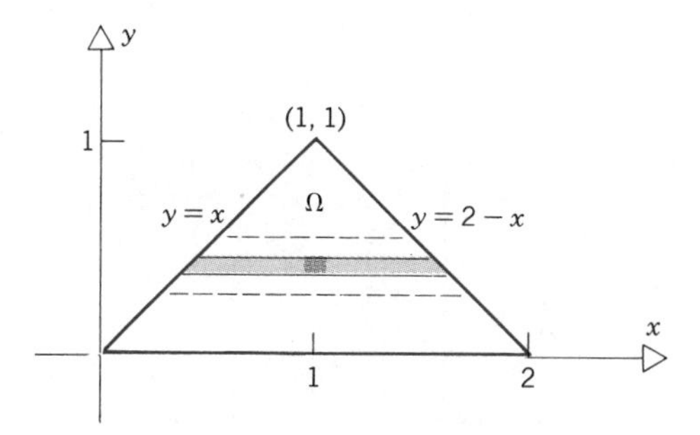

Figure 17.2.11

The region Ω is shown in Figure 17.2.11, and can be described by the inequalities

$$y \leq x \leq 2 - y, \qquad 0 \leq y \leq 1. \tag{22}$$

The density is $\rho(x, y) = k(x^2 + y^2)$, where k is a constant of proportionality. According to Eq. 17 of Section 17.1, the total mass M is found by integrating $\rho(x, y)$ over Ω. We can evaluate this double integral by means of a set of iterated integrals, with the x-integration done first because the inequalities (22) are of the form (15). Thus we obtain

$$M = \iint_{\Omega} \rho(x, y)\, dA = k \int_0^1 \int_y^{2-y} (x^2 + y^2)\, dx\, dy. \tag{23}$$

By evaluating these iterated integrals we find that

$$\begin{aligned} M &= k \int_0^1 \left(\frac{x^3}{3} + xy^2\right)\Bigg|_y^{2-y} dy \\ &= k \int_0^1 \left[\frac{(2 - y)^3}{3} + (2 - y)y^2 - \frac{y^3}{3} - y^3\right] dy \\ &= k \int_0^1 \left(\frac{8}{3} - 4y + 4y^2 - \frac{8}{3}y^3\right) dy \\ &= k \left(\frac{8}{3} - 2 + \frac{4}{3} - \frac{2}{3}\right) = \frac{4}{3}k. \quad ∎ \end{aligned}$$

EXAMPLE 5

Find the volume of the solid in the first octant that is bounded by the xy-plane, the yz-plane, the plane $y = x$, and the cylinder $z = 4 - y^2$.

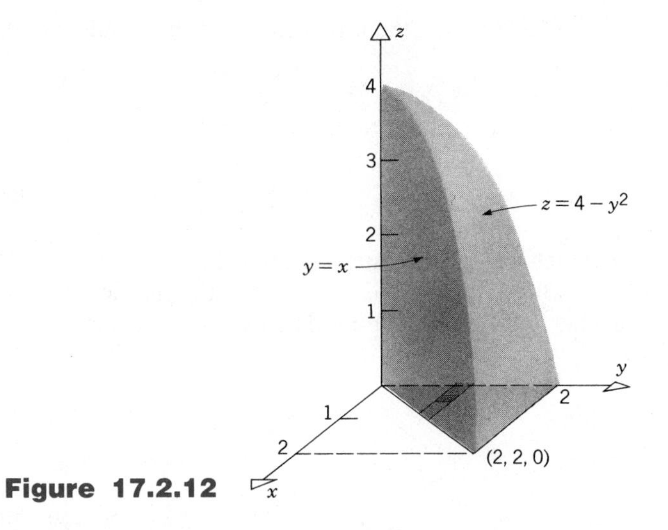

Figure 17.2.12

The three-dimensional solid whose volume is sought is shown in Figure 17.2.12. The integration region Ω in the xy-plane is the triangle bounded by the y-axis, the line $y = x$, and the line $y = 2$. It is shown in Figure 17.2.13, and can

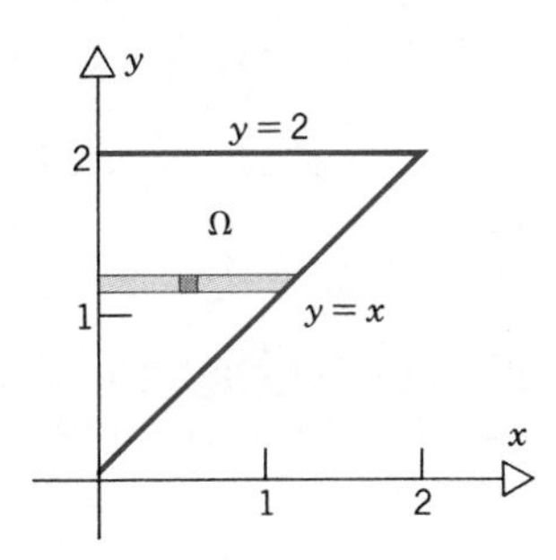

Figure 17.2.13

be described by the inequalities

$$0 \leq x \leq y, \qquad 0 \leq y \leq 2.$$

For each point (x, y) in Ω the altitude of the solid is the distance from the xy-plane to the cylinder $z = 4 - y^2$. Thus the volume V is given by

$$\begin{aligned} V &= \int\int_{\Omega} (4 - y^2)\, dA = \int_0^2 \int_0^y (4 - y^2)\, dx\, dy \\ &= \int_0^2 (4 - y^2)x \Big|_0^y \, dy \\ &= \int_0^2 (4 - y^2)y\, dy \\ &= \left(2y^2 - \frac{y^4}{4}\right)\Big|_0^2 \\ &= 8 - 4 = 4. \end{aligned}$$

Observe that the integration region Ω can also be described by the inequalities

$$0 \le x \le 2, \qquad x \le y \le 2.$$

Thus V can also be found from the integral

$$V = \int_0^2 \int_x^2 (4 - y^2)\, dy\, dx \tag{24}$$

in which the y-integration is done first.

One can also find V by integrating in the yz-plane. In this case Ω is the region in the first quadrant bounded by the coordinate axes and the parabola $z = 4 -$

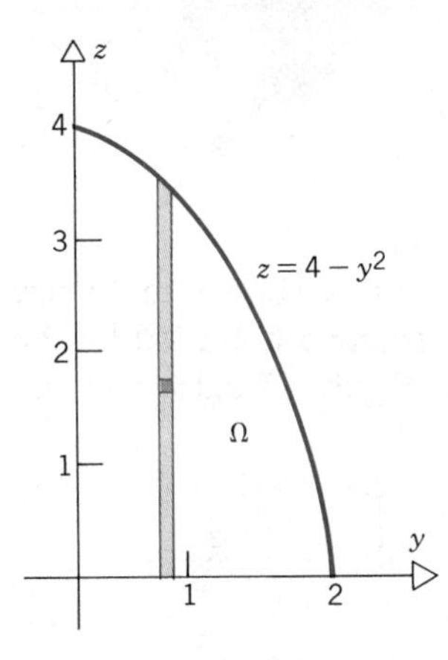

Figure 17.2.14

y^2 shown in Figure 17.2.14. The altitude of the solid now is the distance from the yz-plane to the plane $x = y$. Thus V is given by

$$V = \int_0^2 \int_0^{4-y^2} y\, dz\, dy. \tag{25}$$

You may verify that Eqs. 24 and 25 also yield the value $V = 4$. ∎

As we noted in Section 17.1, the area A of a region Ω is given by

$$A = \iint_\Omega dA.$$

If Ω is described by the inequalities (12), then by Eq. 18 we have

$$\begin{aligned} A &= \int_a^b \int_{\phi_1(x)}^{\phi_2(x)} dy\, dx \\ &= \int_a^b [\phi_2(x) - \phi_1(x)]\, dx. \end{aligned}$$

This is equivalent to the expression for the area of a region that we obtained much earlier (Eq. 2 in Section 7.1). Of course, there is a corresponding result if Ω is described by the inequalities (15).

As in the case of Example 5, it may happen that the integration region Ω can be described by either of the sets of inequalities (12) or (15). In this event, both of Eqs. 18 and 19 are valid, so one can choose either set of iterated integrals as a means of evaluating the double integral. In other words, the integration can be carried out in either order. In Example 5, it made little or no difference which set of iterated integrals we chose to evaluate. However, sometimes one set may be considerably easier than the other, as the following example shows.

EXAMPLE 6

Evaluate the integral

$$\iint_{\Omega} e^{-x^2}\, dA,$$

where Ω is the triangle in the first quadrant bounded by the x-axis, the line $x = 1$, and the line $y = x$ (see Figure 17.2.15).

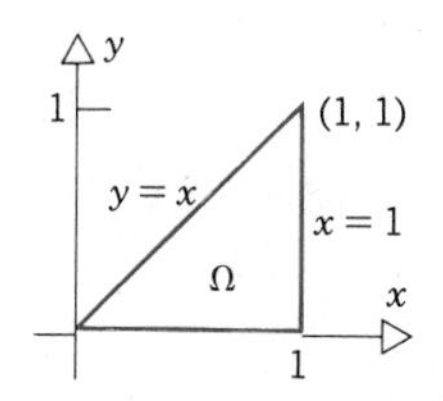

Figure 17.2.15

If we attempt to integrate first with respect to x, we have

$$\iint_{\Omega} e^{-x^2}\, dA = \int_0^1 \int_y^1 e^{-x^2}\, dx\, dy. \tag{26}$$

Since it is impossible to evaluate the inside integral on the right side of Eq. 26 in terms of elementary functions, we would have to resort to numerical procedures, or perhaps a power series expansion of the integrand. On the other hand, if we reverse the order of integration, we have

$$\iint_{\Omega} e^{-x^2}\, dA = \int_0^1 \int_0^x e^{-x^2}\, dy\, dx, \tag{27}$$

and this turns out to be much easier, since the integrand e^{-x^2} is a constant during the first integration. Thus we have

$$\begin{aligned}
\iint_{\Omega} e^{-x^2}\, dA &= \int_0^1 e^{-x^2} y \Big|_{y=0}^{y=x} dx \\
&= \int_0^1 e^{-x^2} x\, dx \\
&= -\frac{1}{2} e^{-x^2} \Big|_0^1 = \frac{1}{2}\left(1 - \frac{1}{e}\right). \quad \blacksquare
\end{aligned}$$

If the region of integration Ω is not a standard region, then one should seek to subdivide it into two or more smaller regions, $\Omega_1, \ldots, \Omega_n$, each of which is standard. Then, by Theorem 17.1.4,

$$\iint_{\Omega} f(x, y)\, dA = \iint_{\Omega_1} f(x, y)\, dA + \cdots + \iint_{\Omega_n} f(x, y)\, dA, \tag{28}$$

where each of the integrals on the right side can be evaluated by means of iterated integrals. Of course, even when Ω is a standard region, one can always subdivide the region of integration and make use of Theorem 17.1.4 if this leads to any simplification in the calculations.

EXAMPLE 7

Set up iterated integrals that are equivalent to the double integral

$$\int\int_{\Omega} (2y + x)\, dA,$$

where Ω is the region bounded by the coordinate axes, the line $y = 7 - x$, and the parabola $y = 1 + x^2$ (see Figure 17.2.16).

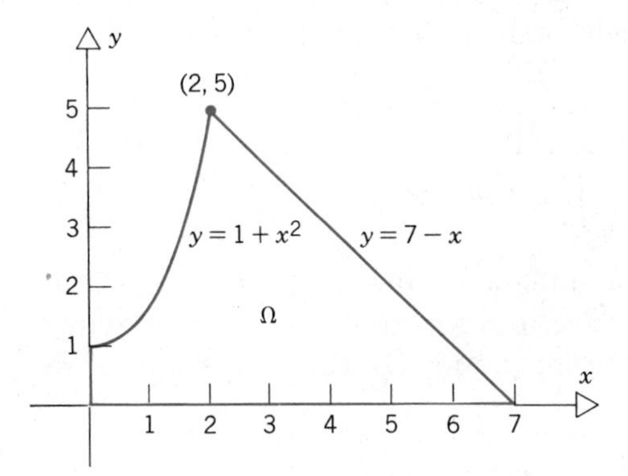

Figure 17.2.16

From an inspection of the figure one can see that Ω is not a standard region. One way of subdividing Ω into standard subregions is to draw the vertical line through the point of intersection of the parabola with the line $y = 7 - x$, as shown in Figure 17.2.17. Then each of the regions Ω_1 and Ω_2 is a standard region of the

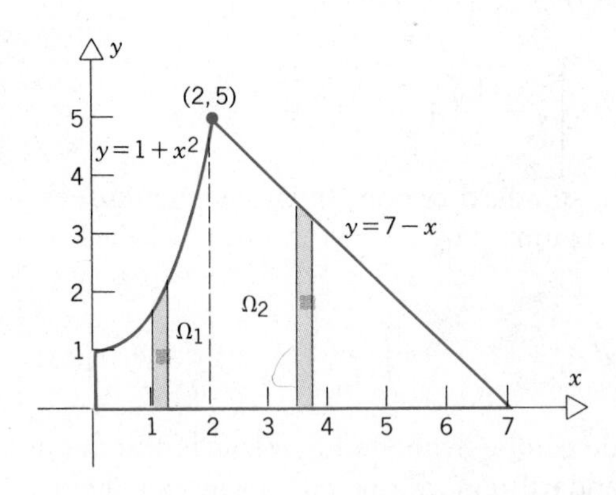

Figure 17.2.17

form (12). To find the point of intersection we solve $y = 1 + x^2$ and $y = 7 - x$ simultaneously, which gives $x = 2$, $y = 5$. Then

$$\begin{aligned}\iint_{\Omega} (2y + x)\, dA &= \iint_{\Omega_1} (2y + x)\, dA + \iint_{\Omega_2} (2y + x)\, dA \\ &= \int_0^2 \int_0^{1+x^2} (2y + x)\, dy\, dx \\ &\quad + \int_2^7 \int_0^{7-x} (2y + x)\, dy\, dx. \end{aligned} \tag{29}$$

Another way to subdivide Ω is by means of the horizontal line $y = 1$, as shown in Figure 17.2.18. Then

$$\begin{aligned}\iint_{\Omega} (2y + x)\, dA &= \iint_{\Omega_1^*} (2y + x)\, dA + \iint_{\Omega_2^*} (2y + x)\, dA \\ &= \int_0^1 \int_0^{7-y} (2y + x)\, dx\, dy \\ &\quad + \int_1^5 \int_{\sqrt{y-1}}^{7-y} (2y + x)\, dx\, dy. \end{aligned} \tag{30}$$

None of these integrals is particularly difficult to evaluate, but it is probably a bit easier to use Eq. 29. ∎

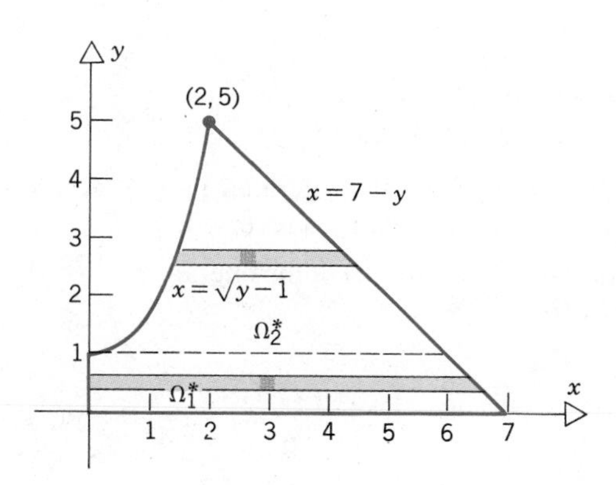

Figure 17.2.18

PROBLEMS

In each of Problems 1 through 10, sketch the region of integration and evaluate the given integral.

1. $\int_0^3 \int_0^2 (3x - 2y)\, dx\, dy$

2. $\int_0^2 \int_1^3 3xy\, dy\, dx$

3. $\int_0^2 \int_{-1}^1 (1 + x^2y^2)\, dx\, dy$

4. $\int_{-2}^3 \int_{-2}^2 x(1 - y)\, dy\, dx$

5. $\int_0^1 \int_{x^2}^x (2x - 5y)\, dy\, dx$

6. $\int_{-1}^1 \int_0^{1-x^2} (y^2 - x^2)\, dy\, dx$

7. $\int_0^{\ln 2} \int_y^1 2xe^{-y}\, dx\, dy$

8. $\int_0^{\pi} \int_0^{\sin x} y\, dy\, dx$

9. $\int_0^{\pi/2} \int_0^y y \sin x \cos y\, dx\, dy$

10. $\displaystyle\int_0^{\pi/2}\int_0^{\cos x} y \sin x \, dy \, dx$

In each of Problems 11 through 14, a double integral over a certain region Ω is given. In each case

(a) set up one or more equivalent iterated integrals with the y-integration done first,

(b) set up one or more equivalent iterated integrals with the x-integration done first,

(c) evaluate the integral or integrals found in either Part (a) or Part (b).

11. $\displaystyle\iint_\Omega 2xy \, dA$, where Ω is bounded by the graphs of $y = x^3$ and $x = y^2$.

12. $\displaystyle\iint_\Omega (2x - y) \, dA$, where Ω is bounded by the line $y = x$ and the parabola $x = 2 - y^2$.

13. $\displaystyle\iint_\Omega (2x - y) \, dA$, where Ω is the triangle bounded by $y = 3x$, $y = x/2$, and $x + 3y = 10$.

14. $\displaystyle\iint_\Omega (1 - y^2) \, dA$, where Ω is bounded by the parabolas $x = y^2$ and $x = 1 + \frac{1}{2} y^2$.

In each of Problems 15 through 20,

(a) sketch the region of integration,

(b) set up one (or more) equivalent iterated integrals with the order of integration reversed,

(c) evaluate either the given integral or the one(s) found in Part (b).

15. $\displaystyle\int_{-1}^{1}\int_0^{1-x^2} dy \, dx$

16. $\displaystyle\int_0^{\sqrt{\pi}}\int_y^{\sqrt{\pi}} \sin x^2 \, dx \, dy$

17. $\displaystyle\int_0^{4}\int_{\sqrt{x}}^{2} \sqrt{1 + y^3} \, dy \, dx$

18. $\displaystyle\int_0^{1}\int_{y^{2/3}}^{1} x^{-3/4} y^{1/6} e^x \, dx \, dy$

19. $\displaystyle\int_0^{1}\int_0^{\sqrt{x}} (x - 2y) \, dy \, dx + \int_1^{2}\int_0^{2-x} (x - 2y) \, dy \, dx$

20. $\displaystyle\int_0^{1}\int_y^{1} \frac{e^x - 1}{x} \, dx \, dy$

21. The region Ω, bounded by $y = x^2$ and $y = \sqrt{x}$, consists of a thin sheet of material of density $\rho(x, y) = k(x^2 + y^2)$, where k is a proportionality constant. Find the total mass in Ω.

22. The region Ω, bounded by $y = 6 - x^2$, $y = x$, and the y-axis, consists of a thin sheet of material of density $\rho(x, y) = kx$, where k is a proportionality constant. Find the total mass in Ω.

23. The region Ω, between the parabolas $y = x^2$ and $y = 8 - 3x^2$, consists of a thin sheet of material of density $\rho(x, y) = ky$, where k is a proportionality constant. Find the total mass in Ω.

24. The region Ω, bounded by $x = y^4$ and $x = y^2 + 2$, consists of a thin sheet of material of density $\rho(x, y) = kx$, where k is a proportionality constant. Find the total mass in Ω.

25. Find the volume of the region bounded by the coordinate planes, the planes $x = 3$ and $y = 2$, and the surface $z = 16 - x^2 - y^2$.

26. Find the volume of the region bounded by the xy-plane, the xz-plane, the cylinder $y = 1 - x^2$, and the plane $z = 2 + x + y$.

27. Find the volume of the region in the first octant bounded by the xy-plane, the yz-plane, the plane $y = 1$, the cylinder $y = x^2$, and the paraboloid $z = 4 - x^2 - y^2$.

28. Find the volume of the region above the xy-plane, below the surface $z = 4 - y^2$, and within the cylinder bounded by the planes $y = 0$, $x = 2$, and $y = x$.

29. Find the volume of the region bounded by the cylinders $y = z^2$ and $y = \sqrt{z}$, and by the planes $x = 0$ and $x = 9 - 2y - z$.

30. Find the volume of the region inside both of the cylinders $x^2 + y^2 = a^2$ and $x^2 + z^2 = a^2$.

31. Find the volume of the region bounded by the coordinate planes, the plane $x + y = a$, and the surface $x^2 + y^2 + z = a^2$.

17.3 APPLICATIONS OF DOUBLE INTEGRALS

In Sections 17.1 and 17.2 we have noted that double integrals can be used for the calculation of masses or volumes. In this section we indicate some further applications of double integrals, and also provide additional examples of setting up and evaluating integrals in two dimensions.

Center of mass

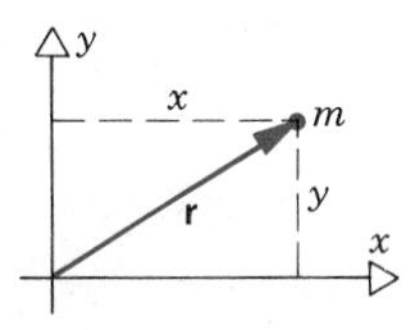

Figure 17.3.1

Consider a particle of mass m whose location in the xy-plane is given by the position vector $\mathbf{r} = x\mathbf{i} + y\mathbf{j}$ (see Figure 17.3.1). Then the **first,** or **linear, moment** of the particle about the origin is the vector $\mathbf{L}$ defined by

$$\begin{aligned}\mathbf{L} &= m\mathbf{r} \\ &= mx\mathbf{i} + my\mathbf{j}.\end{aligned} \tag{1}$$

Observe that the dimensions of the first moment are mass × length.

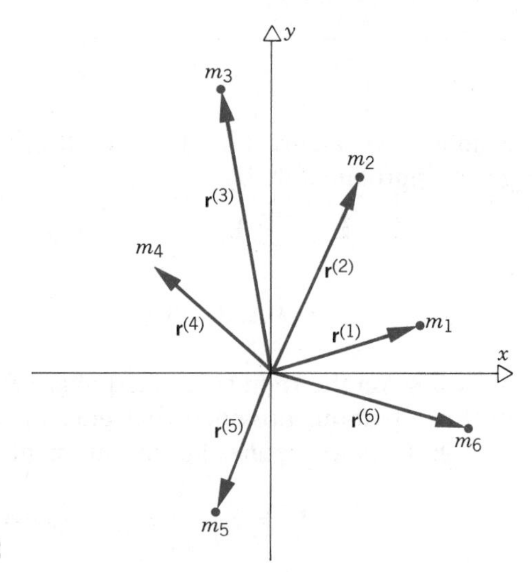

Figure 17.3.2

For a system of n particles with masses m_k and position vectors $\mathbf{r}^{(k)} = x_k\mathbf{i} + y_k\mathbf{j}$ for $k = 1, 2, \ldots, n$, as indicated in Figure 17.3.2, the first moment $\mathbf{L}$ about the origin is the sum of the first moments of the separate particles. Thus

$$\begin{aligned}\mathbf{L} &= \sum_{k=1}^{n} m_k\mathbf{r}^{(k)} \\ &= \sum_{k=1}^{n} m_k(x_k\mathbf{i} + y_k\mathbf{j}) \\ &= \mathbf{i}\sum_{k=1}^{n} m_k x_k + \mathbf{j}\sum_{k=1}^{n} m_k y_k.\end{aligned} \tag{2}$$

How do we extend this concept from a finite set of particles to a thin, continuous sheet of material that occupies a region Ω in the xy-plane, and has mass density $\rho(x, y)$ per unit area? To define the first moment about the origin of the material in Ω, we subdivide Ω into small elements and consider a typical one. The first moment of this element about the origin is given approximately by

$$\mathbf{L}_{ij} \cong \rho(x_i^*, y_j^*)\,\Delta A_{ij}(x_i^*\mathbf{i} + y_j^*\mathbf{j}), \tag{3}$$

where $P_{ij}(x_i^*, y_j^*)$ is an arbitrary point in the element and ΔA_{ij} is the area of the

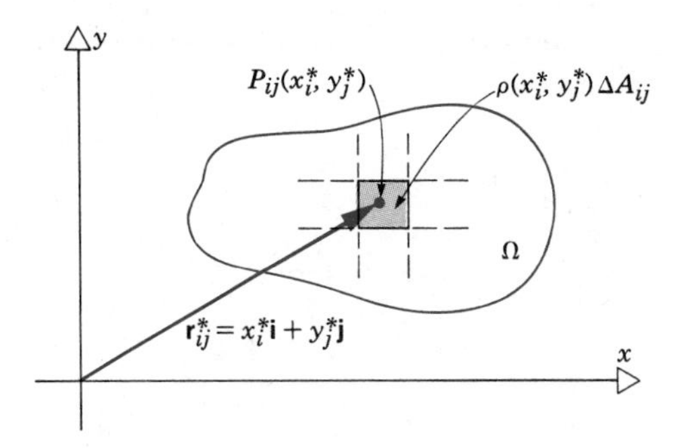

Figure 17.3.3

element (see Figure 17.3.3). Then the first moment of all of the material in Ω is given approximately by

$$\begin{aligned}\mathbf{L} &= \sum_{i,j} \mathbf{L}_{ij} \\ &\cong \mathbf{i} \sum_{i,j} x_i^*\,\rho(x_i^*, y_j^*)\,\Delta A_{ij} + \mathbf{j} \sum_{i,j} y_j^*\,\rho(x_i^*, y_j^*)\,\Delta A_{ij}.\end{aligned} \tag{4}$$

The sums on the right side of Eq. 4 are Riemann sums, so we can pass to the limit in this equation, provided that $\rho(x, y)$ and Ω satisfy the conditions of Theorem 17.1.2. Thus we define the first moment about the origin of the material in Ω as

$$\mathbf{L} = \mathbf{i} \iint_\Omega x\,\rho(x, y)\,dA + \mathbf{j} \iint_\Omega y\,\rho(x, y)\,dA. \tag{5}$$

It is helpful to let

$$L_y = \iint_\Omega x\,\rho(x, y)\,dA. \tag{6}$$

We refer to this quantity as the first moment of the material in Ω about the y-axis. In Eq. 6 the factor x is the position, or moment arm, of the area element dA with respect to the y-axis, and $\rho(x, y)\,dA$ is the mass of the element. Similarly,

$$L_x = \iint_\Omega y\,\rho(x, y)\,dA \tag{7}$$

is the first moment about the x-axis. Then Eq. 5 becomes

$$\mathbf{L} = L_y\mathbf{i} + L_x\mathbf{j}. \tag{8}$$

Note that L_y is the **i**-component of **L** and that L_x is the **j**-component. If this seems peculiar, remember that the moment arm in L_y is x, and that the moment arm in L_x is y.

It is also possible to determine the first moment $\mathbf{L}_0$ of the material in Ω about an arbitrary point $P_0(x_0, y_0)$. One has only to replace the factor x in L_y by $x - x_0$ and the factor y in L_x by $y - y_0$. Thus

$$\mathbf{L}_0 = \mathbf{i} \iint_{\Omega} (x - x_0)\rho(x, y)\, dA + \mathbf{j} \iint_{\Omega} (y - y_0)\rho(x, y)\, dA, \tag{9}$$

which is analogous to Eq. 5.

The **center of mass,** or **mass center,** of the material in Ω is defined to be the point $Q(\bar{x}, \bar{y})$ about which the first moment is zero. Thus the center of mass is the "balance point" of the mass in Ω. To find $\bar{x}$ and $\bar{y}$ we replace P_0 by Q and set $\mathbf{L}_0 = \mathbf{0}$ in Eq. 9. This gives

$$\mathbf{i} \iint_{\Omega} (x - \bar{x})\rho(x, y)\, dA + \mathbf{j} \iint_{\Omega} (y - \bar{y})\rho(x, y)\, dA = \mathbf{0}. \tag{10}$$

The coefficients of **i** and **j** on the left side of Eq. 10 must each be zero. From the coefficient of **i** we find that

$$\iint_{\Omega} (x - \bar{x})\rho(x, y)\, dA = 0 \tag{11}$$

whence

$$\bar{x} = \frac{\displaystyle\iint_{\Omega} x\, \rho(x, y)\, dA}{\displaystyle\iint_{\Omega} \rho(x, y)\, dA} = \frac{L_y}{M}, \tag{12}$$

where

$$M = \iint_{\Omega} \rho(x, y)\, dA \tag{13}$$

is the total mass in Ω. In a similar way

$$\bar{y} = \frac{\displaystyle\iint_{\Omega} y\, \rho(x, y)\, dA}{\displaystyle\iint_{\Omega} \rho(x, y)\, dA} = \frac{L_x}{M}. \tag{14}$$

Further, if we substitute for L_y and L_x from Eqs. 12 and 14 in Eq. 8 we obtain

$$\mathbf{L} = M(\bar{x}\mathbf{i} + \bar{y}\mathbf{j}). \tag{15}$$

The interpretation of Eq. 15 is that *the first moment of the material in Ω is the same as the first moment of a single particle of mass equal to the total mass M in Ω located at the center of mass $(\bar{x}, \bar{y})$.*

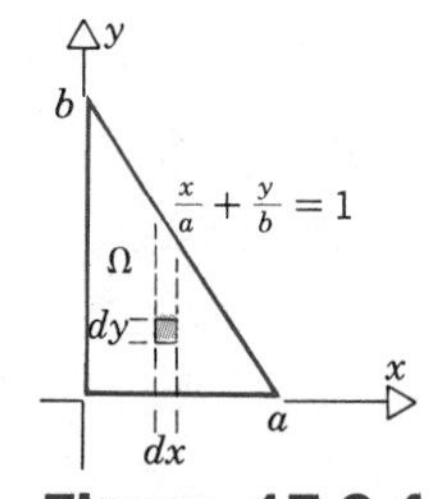

Figure 17.3.4

EXAMPLE 1

Let Ω be the triangle in the first quadrant bounded by the coordinate axes and the line $(x/a) + (y/b) = 1$, where $a > 0$ and $b > 0$ (see Figure 17.3.4). Suppose that the material in Ω has density $\rho(x, y) = k(2b - y)$, where k is a constant. This means that the density diminishes in a linear manner as y increases. Find the center of mass of the material in Ω.

In order to make use of Eqs. 12 and 14 for the coordinates $(\bar{x}, \bar{y})$ of the center of mass, we need to calculate the total mass M and the first moments L_y and L_x about the y- and x-axes, respectively. The mass M is given by

$$M = \iint_\Omega \rho(x, y)\, dA = k \iint_\Omega (2b - y)\, dA. \tag{16}$$

The double integral in Eq. 16 can be evaluated by means of iterated integrals in which the first integration is either with respect to x or y. If we integrate with respect to y first, then (from Figure 17.3.4) for each fixed x we must integrate from the lower boundary, $y = 0$, to the upper boundary, $y = b[1 - (x/a)]$. This is followed by an integration with respect to x from 0 to a. Thus we have

$$\begin{aligned}
M &= k \int_0^a \int_0^{b[1-(x/a)]} (2b - y)\, dy\, dx \\
&= k \int_0^a \left(2by - \frac{y^2}{2}\right)\Bigg|_0^{b[1-(x/a)]} dx \\
&= k \int_0^a \left[2b^2\left(1 - \frac{x}{a}\right) - \frac{1}{2} b^2 \left(1 - \frac{x}{a}\right)^2\right] dx \\
&= kb^2 \int_0^a \left(\frac{3}{2} - \frac{x}{a} - \frac{1}{2}\frac{x^2}{a^2}\right) dx \\
&= kb^2 \left(\frac{3}{2}x - \frac{1}{2}\frac{x^2}{a} - \frac{1}{6}\frac{x^3}{a^2}\right)\Bigg|_0^a \\
&= kb^2 \left(\frac{3}{2}a - \frac{1}{2}a - \frac{1}{6}a\right) = \frac{5}{6}kab^2.
\end{aligned} \tag{17}$$

From Eq. 6 the first moment L_y is given by

$$\begin{aligned}
L_y &= k \int_0^a \int_0^{b[1-(x/a)]} x(2b - y)\, dy\, dx \\
&= k \int_0^a \left(2bxy - \frac{1}{2}xy^2\right)\Bigg|_0^{b[1-(x/a)]} dx \\
&= k \int_0^a \left[2b^2 x\left(1 - \frac{x}{a}\right) - \frac{1}{2} b^2 x \left(1 - \frac{x}{a}\right)^2\right] dx \\
&= kb^2 \int_0^a \left(\frac{3}{2}x - \frac{x^2}{a} - \frac{1}{2}\frac{x^3}{a^2}\right) dx \\
&= kb^2 \left(\frac{3}{4}a^2 - \frac{1}{3}a^2 - \frac{1}{8}a^2\right) \\
&= \frac{7}{24}ka^2b^2.
\end{aligned}$$

Thus the x-coordinate of the center of mass is

$$\bar{x} = \frac{7ka^2b^2/24}{5kab^2/6} = \frac{7}{20}a.$$

In a similar way, from Eq. 7, we have

$$\begin{aligned}
L_x &= k\int_0^a \int_0^{b[1-(x/a)]} y(2b - y)\,dy\,dx \\
&= k\int_0^a \left(by^2 - \frac{1}{3}y^3\right)\Bigg|_0^{b[1-(x/a)]} dx \\
&= k\int_0^a \left[b^3\left(1 - \frac{x}{a}\right)^2 - \frac{1}{3}b^3\left(1 - \frac{x}{a}\right)^3\right] dx \\
&= kb^3\int_0^a \left(\frac{2}{3} - \frac{x}{a} + \frac{1}{3}\frac{x^3}{a^3}\right) dx \\
&= kb^3\left(\frac{2}{3}a - \frac{1}{2}a + \frac{1}{12}a\right) \\
&= \frac{1}{4}kab^3,
\end{aligned}$$

and therefore

$$\bar{y} = \frac{kab^3/4}{5kab^2/6} = \frac{3}{10}b. \quad \blacksquare$$

Center of gravity and centroid

Two other points that are defined in a way similar to the center of mass are the **center of gravity** and the **centroid** of a two-dimensional region Ω. In defining the center of gravity the only difference is that the mass density $\rho(x, y)$ is replaced by the weight density $w(x, y)$, where

$$w(x, y) = \rho(x, y)g(x, y), \tag{18}$$

and g is the acceleration due to gravity at the point (x, y). If (x^*, y^*) is the center of gravity, then corresponding to Eqs. 12 and 14 we have

$$x^* = \frac{\iint_\Omega xw(x, y)\,dA}{\iint_\Omega w(x, y)\,dA}, \qquad y^* = \frac{\iint_\Omega yw(x, y)\,dA}{\iint_\Omega w(x, y)\,dA}. \tag{19}$$

If g is constant, then it can be brought outside each of the integrals in Eqs. 19 and canceled out of the expressions for x^* and y^*. Thus

$$\text{if } g \text{ is constant, then } x^* = \bar{x} \text{ and } y^* = \bar{y}.$$

In other words, *the center of gravity coincides with the center of mass when the gravitational field is uniform.*

The centroid of a two-dimensional region Ω is a purely geometrical property of the region, and has nothing to do with masses or weights. If $(\hat{x}, \hat{y})$ is the centroid of Ω, then $\hat{x}$ and $\hat{y}$ are defined by the equations

$$\hat{x} = \frac{\iint_\Omega x\, dA}{\iint_\Omega dA} = \frac{\hat{L}_y}{A}, \tag{20}$$

$$\hat{y} = \frac{\iint_\Omega y\, dA}{\iint_\Omega dA} = \frac{\hat{L}_x}{A}, \tag{21}$$

which are similar to Eqs. 12 and 14 for the center of mass. Indeed, suppose that Ω is occupied by a thin sheet of material with uniform density, that is, $\rho(x, y)$ is constant. Then the density can be brought outside of the integral signs and canceled out of the expressions (12) and (14) for $\bar{x}$ and $\bar{y}$, leaving Eqs. 20 and 21. Thus

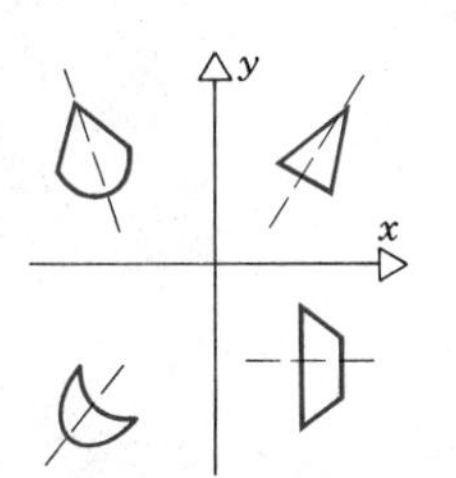

Figure 17.3.5 In each case the centroid lies on the axis of symmetry.

$$\text{if } \rho \text{ is constant, then } \hat{x} = \bar{x} \text{ and } \hat{y} = \bar{y};$$

that is, *if the density is constant, then the center of mass is the same as the centroid.*

The determination of the centroid becomes simpler if Ω has an axis of symmetry, for the centroid must lie on it (see Figure 17.3.5). For example, the centroid of an isosceles triangle lies on the altitude perpendicular to the third side. If Ω has two axes of symmetry, the centroid must lie on both of them, and hence is their point of intersection. Thus the centroid of a circle or of a rectangle is the center.

EXAMPLE 2

Find the centroid of the semicircular region Ω given by $x^2 + y^2 \le a^2$ with $y \ge 0$ (see Figure 17.3.6).

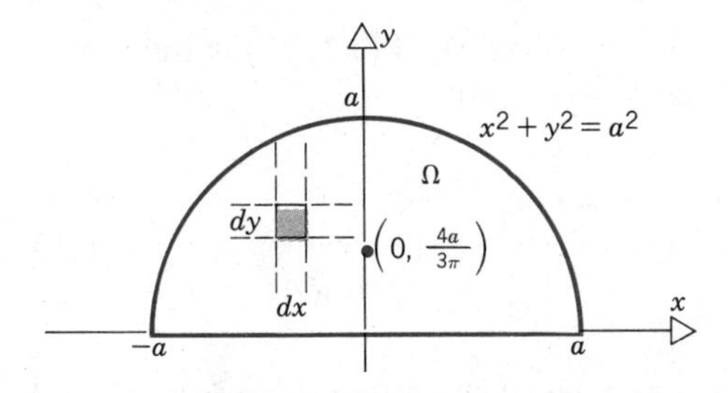

Figure 17.3.6

Because of the symmetry of the region about the y-axis we have

$$\hat{L}_y = \iint_\Omega x\, dA = 0,$$

so $\hat{x} = 0$, and we must only calculate $\hat{y}$. For this we need

$$\hat{L}_x = \iint_\Omega y\, dA = 2\int_0^a \int_0^{\sqrt{a^2-x^2}} y\, dy\, dx$$

$$= \int_0^a (a^2 - x^2)\, dx$$

$$= \left(a^2 x - \frac{1}{3}x^3\right)\Big|_0^a = \frac{2}{3}a^3.$$

Also

$$A = \iint_\Omega dA = \tfrac{1}{2}\pi a^2,$$

since this is just the area of the semicircle. Thus, from Eq. 21

$$\hat{y} = \frac{2a^3/3}{\pi a^2/2} = \frac{4a}{3\pi} \cong 0.4244\, a. \;\blacksquare$$

EXAMPLE 3

Find the centroid of the triangular region Ω in the first quadrant bounded by the coordinate axes and the line $(x/a) + (y/b) = 1$. Note that this is the same region as in Example 1; it is shown in Figure 17.3.4.

From elementary geometry we know that

$$A = \iint_\Omega dA = \frac{ab}{2}.$$

Further,

$$\hat{L}_y = \iint_\Omega x\, dA = \int_0^a \int_0^{b[1-(x/a)]} x\, dy\, dx = b\int_0^a x\left(1 - \frac{x}{a}\right) dx$$

$$= b\left(\frac{x^2}{2} - \frac{x^3}{3a}\right)\Big|_0^a = \frac{a^2 b}{6}.$$

Similarly,

$$\hat{L}_x = \iint_\Omega y\, dA = \int_0^a \int_0^{b[1-(x/a)]} y\, dy\, dx = \frac{b^2}{2}\int_0^a \left(1 - \frac{x}{a}\right)^2 dx$$

$$= \frac{b^2}{2}\left(x - \frac{x^2}{a} + \frac{x^3}{3a^2}\right)\Big|_0^a = \frac{ab^2}{6}.$$

Thus, from Eqs. 20 and 21 we have

$$\hat{x} = \frac{a^2 b/6}{ab/2} = \frac{a}{3}, \qquad \hat{y} = \frac{ab^2/6}{ab/2} = \frac{b}{3}.$$

By comparing these results with those of Example 1 it is clear that the center of mass and the centroid may be different when the mass distribution is nonuniform. Indeed, as in this case, both coordinates of the center of mass may differ from the corresponding coordinates of the centroid even though the density varies only in one direction. ∎

Moment of inertia

Suppose once more that a region Ω in the xy-plane is occupied by a thin sheet of material whose density (mass/area) is $\rho(x, y)$. The **moment of inertia** (or second moment) about the x-axis of the material in Ω is defined to be

$$I_x = \iint_\Omega y^2\, \rho(x, y)\, dA. \tag{22}$$

Similarly the moment of inertia about the y-axis is

$$I_y = \iint_\Omega x^2\, \rho(x, y)\, dA. \tag{23}$$

Finally, the moment of inertia about the origin, usually called the polar moment of inertia, is

$$\begin{aligned} I_0 &= \iint_\Omega (x^2 + y^2)\rho(x, y)\, dA \\ &= I_x + I_y. \end{aligned} \tag{24}$$

Observe that in calculating the moments of inertia I_x and I_y from Eqs. 22 and 23 we use the square of the appropriate position coordinate of the mass element $\rho(x, y)\, dA$, rather than the first power as for the linear moment. From Eqs. 22, 23, and 24 it is also easy to see that the dimensions of moments of inertia are mass $\times$ (length)^2.

The **radius of gyration** about the x-axis, denoted by R_x, is the distance from the x-axis at which the entire mass M in the region Ω should be concentrated in order to produce the same moment of inertia as the actual distributed mass. Thus

$$R_x^2 = \frac{I_x}{M}. \tag{25}$$

In a similar way the radius of gyration about the y-axis is given by

$$R_y^2 = \frac{I_y}{M}. \tag{26}$$

EXAMPLE 4

Consider a thin plate in the shape of a right triangle whose sides have the lengths a, b, and $\sqrt{a^2 + b^2}$. If the plate has uniform density $\rho(x, y) = k$, find the moment of inertia of the plate about the side of length b. Also find the radius of gyration about this axis.

If we locate the plate in the xy-plane as shown in Figure 17.3.7, then we see that the plate occupies the same triangular region Ω that occurred in Examples 1 and 3. Further, the required moment of inertia is just I_y. thus

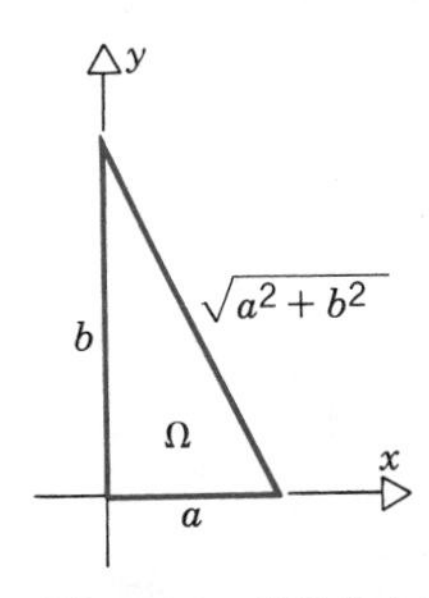

Figure 17.3.7

$$I_y = k \int_0^a \int_0^{b[1-(x/a)]} x^2\, dy\, dx$$

$$= kb \int_0^a x^2 \left(1 - \frac{x}{a}\right) dx$$

$$= kb \left(\frac{1}{3}x^3 - \frac{1}{4}\frac{x^4}{a}\right)\Bigg|_0^a$$

$$= \frac{1}{12} ka^3 b.$$

The total mass of the plate is

$$M = \frac{1}{2}kab,$$

so the radius of gyration R_y is given by Eq. 26:

$$R_y^2 = \frac{ka^3b/12}{kab/2} = \frac{a^2}{6}.$$

Thus

$$R_y = \frac{a}{\sqrt{6}} \cong 0.4082\, a. \blacksquare$$

EXAMPLE 5

A thin plate whose density is proportional to distance from the y-axis occupies the region Ω shown in Figure 17.3.8 that is bounded by the line $y = 4$ and the parabola $y = x^2$. Find the moment of inertia of this plate about the line $y = 4$.

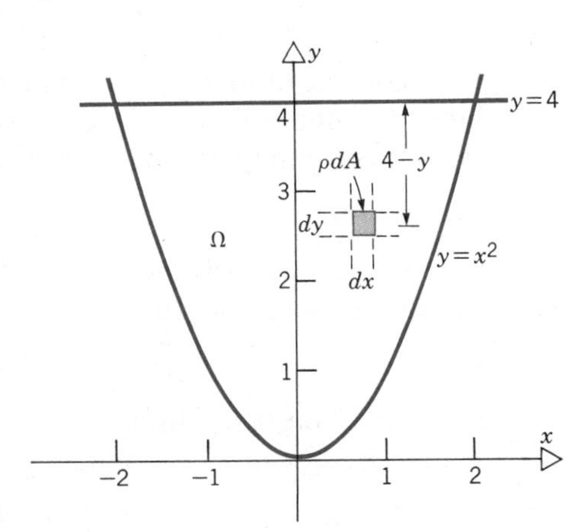

Figure 17.3.8

The density function if $\rho(x, y) = k|x|$, where k is a constant of proportionality, and the absolute value is needed because distance and density are intrinsically nonnegative quantities. The integration element $dA = dx\, dy$ is shown in Figure 17.3.8, and its position relative to the line $y = 4$ is $4 - y$, which is then the moment arm in this problem. Thus the required moment of inertia I is given by

$$\begin{aligned} I &= \iint_{\Omega} (4 - y)^2 \rho(x, y)\, dA \\ &= k \iint_{\Omega} (4 - y)^2 |x|\, dA. \end{aligned}$$

Since Ω is symmetric about the y-axis and the integrand is an even function of x, we can determine I by integrating over the portion of Ω for which $x > 0$ and then multiplying this result by two. Consequently

$$I = 2k \int_0^2 \int_{x^2}^4 (4 - y)^2 |x|\, dy\, dx.$$

Further, we can now replace $|x|$ by x in the integrand because $x > 0$ in the part of Ω over which we are integrating. Thus we finally obtain

$$\begin{aligned} I &= 2k \int_0^2 \int_{x^2}^4 (4 - y)^2 x\, dy\, dx \\ &= 2k \int_0^2 -\frac{(4 - y)^3}{3}\bigg|_{x^2}^4 x\, dx \\ &= \frac{2k}{3} \int_0^2 (4 - x^2)^3 x\, dx \\ &= -\frac{2k}{3} \frac{(4 - x^2)^4}{8}\bigg|_0^2 \\ &= \frac{64k}{3}. \quad \blacksquare \end{aligned}$$

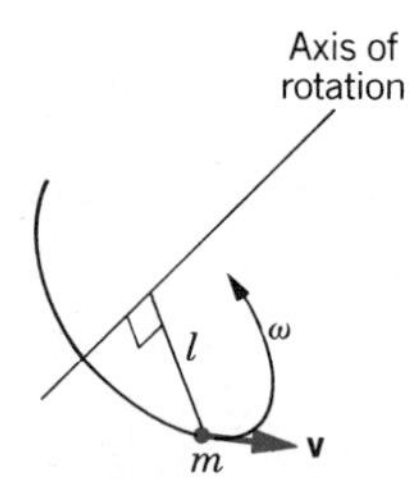

Figure 17.3.9

The moment of inertia is an important quantity in mechanics, especially in situations involving rotation about an axis. Recall that the kinetic energy E of a particle of mass m moving with velocity $\mathbf{v}$ is

$$E = \tfrac{1}{2}m\|\mathbf{v}\|^2. \tag{28}$$

Now suppose that the particle is rotating about a fixed axis (see Figure 17.3.9) with constant angular speed ω. Then $\|\mathbf{v}\|$ and ω are related by

$$\|\mathbf{v}\| = l\omega, \tag{29}$$

where l is the (constant) distance from the axis of rotation to the particle. By combining Eqs. 28 and 29 we obtain

$$E = \tfrac{1}{2}ml^2\omega^2 = \tfrac{1}{2}I\omega^2, \tag{30}$$

where $I = ml^2$ is the moment of inertia of the particle about the axis of rotation.

For planar mass distributions Eqs. 28 and 30 are replaced by more general expressions involving integrals. For a rigid body moving on a straight line with constant speed $\|\mathbf{v}\|$, Eq. 28 is replaced by

$$
\begin{aligned}
E &= \tfrac{1}{2}\|\mathbf{v}\|^2 \int\int_\Omega \rho(x, y)\, dA \\
&= \tfrac{1}{2} M\|\mathbf{v}\|^2. \qquad (31)
\end{aligned}
$$

Similarly, for a rigid body rotating at constant speed ω in the plane of the body, instead of Eq. 30 we have

$$
\begin{aligned}
E &= \tfrac{1}{2}\omega^2 \int\int_\Omega l^2(x, y)\rho(x, y)\, dA \\
&= \tfrac{1}{2}I\omega^2, \qquad (32)
\end{aligned}
$$

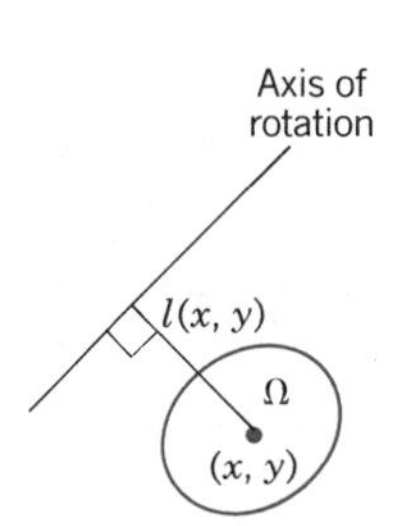

Figure 17.3.10

where $l(x, y)$ is the distance from the axis of rotation to the point (x, y), and the integral in Eq. 32 is the moment of inertia of the material in Ω about the given axis (see Figure 17.3.10). From Eqs. 32 and 31, respectively, we see that in the two expressions for the kinetic energy the moment of inertia (for rotatory motion) and the mass (the linear motion) appear in the same way. Thus *the moment of inertia measures the resistance of a body to rotation in the same sense that the mass measures its resistance to linear motion.*

PROBLEMS

In each of Problems 1 through 5, find the centroid of the given region.

1. The semielliptical region bounded by the x-axis and by the graph of $(x^2/a^2) + (y^2/b^2) = 1$ for $y \geq 0$.
2. The region bounded by the x-axis and by the parabola $y = 1 - (x^2/a^2)$.
3. The region bounded by the x-axis and by the graph of $y = \sin x$ for $0 \leq x \leq \pi$.
4. The region bounded by the x-axis and by one arch of the cycloid $x = a(t - \sin t)$, $y = a(1 - \cos t)$ for $0 \leq t \leq 2\pi$.
5. The triangle with vertices $(0, 0)$, $(a, 0)$, and $(0, a)$.
6. The square $0 \leq x \leq a$, $0 \leq y \leq a$ is occupied by a mass distribution with density $\rho(x, y) = k(2a - x - y)$. Find the center of mass, and observe that it is different from the centroid of the square.
7. The rectangle $0 \leq x \leq a$, $0 \leq y \leq b$ is occupied by a mass distribution with density $\rho(x, y) = k(2b - y)$. Find the center of mass, and observe that it is different from the centroid of the rectangle.
8. Consider a plate in the shape of a triangle with vertices at $(-b/2, 0)$, $(b/2, 0)$, and $(0, h)$, and having density $\rho(x, y) = 1 - (y/h)^2$.
   (a) Find the center of mass of the plate.
   (b) Find the centroid of the triangular region.
9. A rectangular plate of dimensions b and h has uniform density k.
   (a) Find the moment of inertia about a side of length b.
   (b) Find the moment of inertia about a side of length h.
   (c) Find the moment of inertia about a line through the center of the plate parallel to the side of length b.

(d) Find the polar moment of inertia about a corner of the plate.

(e) Find the polar moment of inertia about the center of the plate.

Hint: It may be helpful to choose coordinates differently in different parts of this problem.

10. A right triangular plate of uniform density k has sides of a, a, and $\sqrt{2}\,a$.

(a) Find the moment of inertia about one of the equal sides.

(b) Find the moment of inertia about the hypotenuse.

(c) Find the polar moment of inertia about the vertex opposite the hypotenuse.

(d) Find the moment of inertia about the altitude perpendicular to the hypotenuse.

(e) Find the polar moment of inertia about the center of the hypotenuse.

Hint: It may be helpful to choose coordinates differently in different parts of this problem.

11. Consider an isosceles triangular plate of base b and height h, and with uniform density k.

(a) Find the centroid.

(b) Find the moment of inertia about the base.

(c) Find the moment of inertia about the axis of symmetry of the triangle.

(d) Find the moment of inertia about the line parallel to the base and passing through the centroid.

12. Consider a plate of uniform density k that is bounded by the x-axis and by the parabola $y = 1 - (x^2/a^2)$.

(a) Find the moment of inertia and the radius of gyration about the x-axis.

(b) Find the moment of inertia and the radius of gyration about the y-axis.

13. A plate of uniform density k is bounded by the ellipse $(x^2/a^2) + (y^2/b^2) = 1$.

(a) Find the moment of inertia and the radius of gyration about the x-axis.

(b) Find the moment of inertia and the radius of gyration about the y-axis.

(c) Find the polar moment of inertia.

In each of Problems 14 through 17, a region Ω, a density $\rho(x, y)$, and a line l are given. Set up, but do not evaluate, an integral giving the moment of inertia of the material in Ω about l.

14. Ω: $x^2 + y^2 \le a^2$; $\rho(x, y) = k$; $l{:}y = -a$.

15. Ω: $0 \le x \le \pi$, $0 \le y \le \sin x$; $\rho(x, y) = 1 - y$; $l{:}x = \frac{\pi}{2}$.

16. Ω: $\frac{x^2}{a^2} + \frac{y^2}{b^2} \le 1$; $\rho(x, y) = k$; $l{:}x = -a$.

17. Ω: $\{x^2 + (y - 1)^2 \le 1\} \cap \{(x - 1)^2 + y^2 \le 1\}$; $\rho(x, y) = k(x^2 + y^2)$; $l{:}x = 0$.

18. Center of mass of a composite body. Suppose that a mass distribution occupies the region Ω in the xy-plane, and that Ω can be separated into two nonoverlapping parts Ω_1 and Ω_2 (see Figure 17.3.11). Suppose that the centers of mass of Ω_1

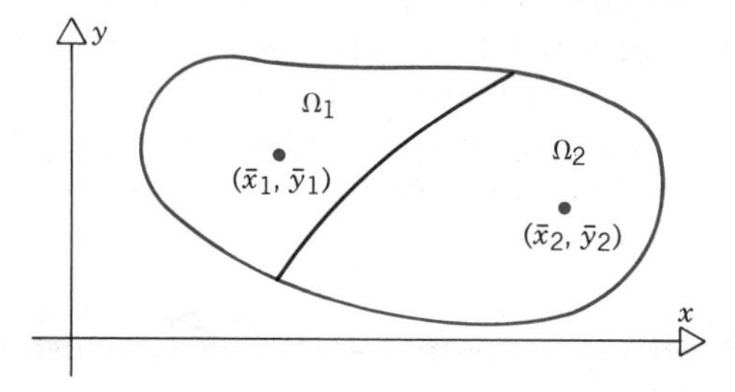

Figure 17.3.11

and Ω_2, respectively, are $(\bar{x}_1, \bar{y}_1)$ and $(\bar{x}_2, \bar{y}_2)$, and that the masses of Ω_1 and Ω_2, respectively, are M_1 and M_2. If $(\bar{x}, \bar{y})$ is the center of mass of the total region Ω, show that

$$\bar{x} = \frac{\bar{x}_1 M_1 + \bar{x}_2 M_2}{M_1 + M_2}, \qquad \bar{y} = \frac{\bar{y}_1 M_1 + \bar{y}_2 M_2}{M_1 + M_2}.$$

Thus the center of mass of a complicated region can sometimes be found by dividing the region into simpler subregions.

In Problems 19 through 22, use the result of Problem 18 to find the center of mass of the material in the given region.

19. The region is shown in Figure 17.3.12 with $\rho(x, y) = k$.

20. The region is shown in Figure 17.3.13 with $\rho(x, y) = k_1$ in the base and $\rho(x, y) = k_2$ in the cross-piece.

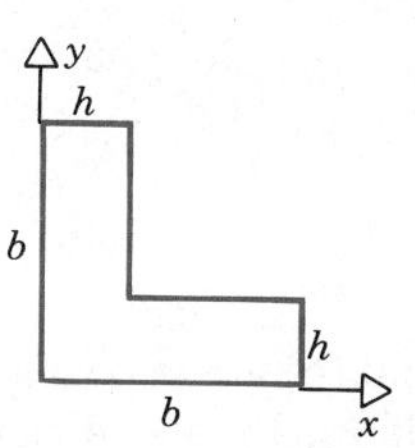

Figure 17.3.12

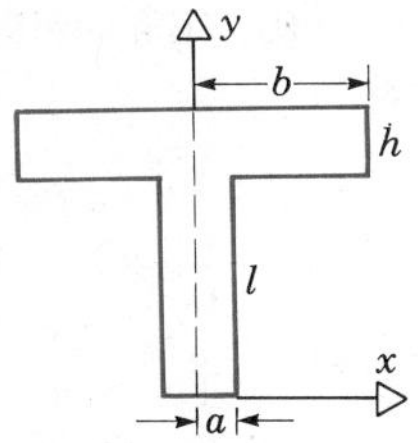

Figure 17.3.13

21. The region is shown in Figure 17.3.14 with $\rho(x, y) = k$.

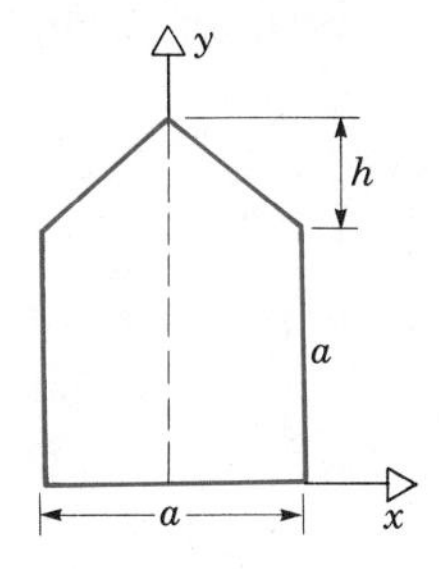

Figure 17.3.14

22. The region is shown in Figure 17.3.15 with $\rho(x, y) = k$.

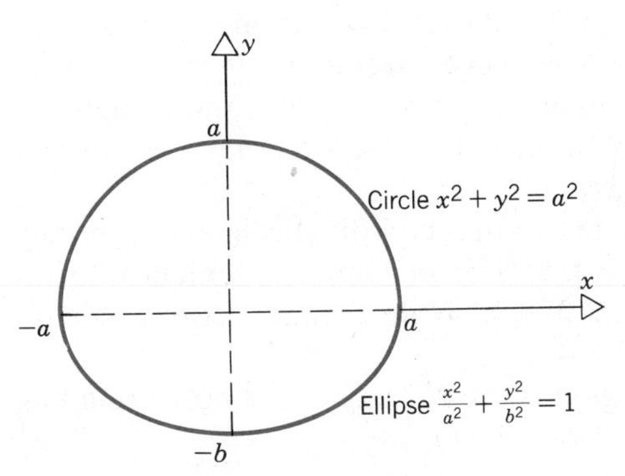

Figure 17.3.15

23. Parallel axes theorem for the moment of inertia. Suppose that the plane region Ω in Figure 17.3.16 is occupied by a mass distribution described by a given density function ρ. Let γ be a given line in the plane and let Γ be the line parallel to γ and passing through the center of mass. Let I_γ and I_Γ be the moments of inertia of Ω about γ and Γ, respectively.

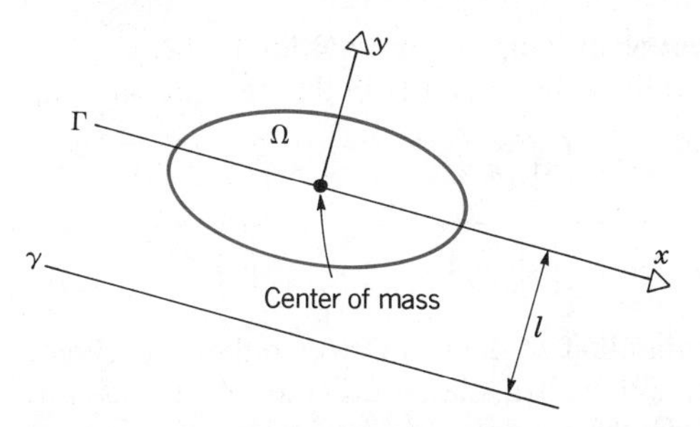

Figure 17.3.16 The parallel axes theorem: $I_\gamma = I_\Gamma + l^2 M$.

(a) Show that

$$I_\gamma = I_\Gamma + l^2 M, \qquad (i)$$

where l is the distance between γ and Γ, and M is the total mass of Ω.

Hint: Choose coordinate axes as shown in Figure 17.3.16.

(b) Use the results of Problem 9(a, c) to verify that (i) is true for the case considered there.

(c) Use the results of Problem 11(b, d) to verify that (i) is true for the case considered there.

24. Use the parallel axes theorem (Problem 23) to find the moment of inertia of a circular plate of radius a and uniform density k about a tangent line. Compare with Problem 14.

25. A right triangular plate of uniform density k has sides of a, a, and $\sqrt{2}\,a$. Use the parallel axes theorem (Problem 23) and the results of Problems 5 and 10 to find the moment of inertia about the line parallel to the hypotenuse and passing through the opposite vertex.

26. A plate of uniform density k has the shape of an ellipse with semimajor axis a and semiminor axis b.

(a) Use the parallel axes theorem (Problem 23) and the results of Problem 13 to find the moment of inertia of the plate about the line tangent to the ellipse at one end of the major axis. Compare with Problem 16.

(b) Find the polar moment of inertia about the point at one end of the major axis.

27. **Theorem of Pappus.** The coordinates $(\hat{x}, \hat{y})$ of the centroid of the region Ω in the xy-plane are

$$\hat{x} = \frac{\iint_\Omega x\,dA}{\iint_\Omega dA}, \qquad \hat{y} = \frac{\iint_\Omega y\,dA}{\iint_\Omega dA}.$$

Assume that Ω does not cross either coordinate axis (although it may touch them); also assume that Ω can be described by inequalities both of the form (12) and of the form (15) in Section 17.2.

(a) Show that

$$\iint_\Omega x\,dA = \frac{V_y}{2\pi},$$

where V_y is the volume of the solid formed by rotating Ω about the y-axis. Then show that

$$\hat{x} = \frac{V_y}{2\pi A},$$

where A is the area of Ω.

(b) In a similar way, show that

$$\hat{y} = \frac{V_x}{2\pi A},$$

where V_x is the volume of the solid formed by rotating Ω about the x-axis.

The results of Parts (a) and (b) are known as the theorem of Pappus. If V_x, V_y, and A can be easily found, then Pappus' theorem is a convenient way to find the centroid of Ω.

In each of Problems 28 through 30, use Pappus' theorem to find the centroid of the given region Ω.

28. Ω is the triangle with vertices at $(0, 0)$, $(a, 0)$, $(0, b)$.

29. Ω is a semicircle of radius a.

30. Ω is the quarter circle $x^2 + y^2 \le a^2$ with $x \ge 0$ and $y \ge 0$.

17.4 INTEGRATION USING POLAR COORDINATES

Up to now we have discussed double integrals and their evaluation using iterated integrals in a rectangular coordinate system. However, iterated integrals in other coordinate systems can also be used to evaluate double integrals, and sometimes it is simpler to do so. In this section we discuss the use of polar coordinates (r, θ) for the evaluation of double integrals.

In order to represent points in the plane by polar coordinates in an essentially unique way, we now require that $r \ge 0$ and that θ lie in some convenient interval of length 2π, usually $0 \le \theta \le 2\pi$ or $-\pi \le \theta \le \pi$. Note that this contrasts with the point of view in Chapter 14.

In Cartesian coordinates the basic region is the rectangle; the corresponding region in polar coordinates is the so-called polar rectangle

$$R\colon a \le r \le b, \qquad c \le \theta \le d \tag{1}$$

shown in Figure 17.4.1.

Now suppose that we wish to evaluate the integral of a function f over the polar rectangle R. First, we form a partition Δ of R consisting of smaller polar subrectangles R_{ij} for $i = 1, \ldots, m$ and $j = 1, \ldots, n$ (see Figure 17.4.2). The norm $\|\Delta\|$ of the partition Δ is the length of the longest diagonal of all the polar subrectangles in Δ. A typical polar subrectangle R_{ij} is bounded by arcs of two circles $r = r_{i-1}$ and $r = r_i$, and by segments of two radial lines, $\theta = \theta_{j-1}$ and $\theta = \theta_j$. Let ΔA_{ij} be the area of R_{ij}, choose a star point $P_{ij}^*(r_i^*, \theta_j^*)$ arbitrarily in each subrec-

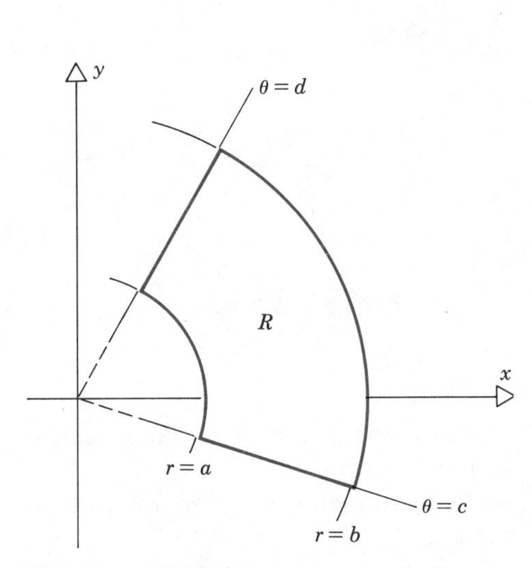

Figure 17.4.1 A polar rectangle R.

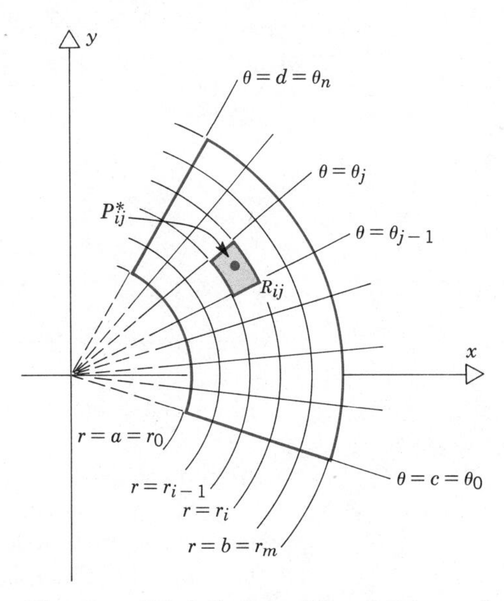

Figure 17.4.2 A partition of R into polar subrectangles.

tangle, and form the Riemann sum

$$\sum_{\substack{i=1,\ldots,m \\ j=1,\ldots,n}} f(r_i^*, \theta_j^*)\,\Delta A_{ij}. \tag{2}$$

If f is continuous on R except (at most) for a set of points lying on a finite number of smooth rectifiable arcs, then by Theorem 17.1.2

$$\iint_R f(r, \theta)\, dA = \lim_{\|\Delta\| \to 0} \sum_{\substack{i=1,\ldots,m \\ j=1,\ldots,n}} f(r_i^*, \theta_j^*)\,\Delta A_{ij}. \tag{3}$$

Further, the limiting value is the same regardless of how the partitioning is done, how the star points are chosen, and how $\|\Delta\| \to 0$. Thus we can carry out these steps in any way that we choose, and by choosing a particularly convenient way, we can replace the double integral (3) by iterated integrals in r and θ.

The first thing that is needed is an expression for ΔA_{ij}. The typical polar subrectangle R_{ij} shown in Figure 17.4.3 is the portion of the circular ring between $r = r_{i-1}$ and $r = r_i$ that subtends the angle $\Delta\theta_j$. Thus its area ΔA_{ij} is given by

$$\begin{aligned} \Delta A_{ij} &= \pi r_i^2 \frac{\Delta\theta_j}{2\pi} - \pi r_{i-1}^2 \frac{\Delta\theta_j}{2\pi} \\ &= \frac{1}{2}(r_i^2 - r_{i-1}^2)\,\Delta\theta_j \\ &= \frac{1}{2}(r_i + r_{i-1})(r_i - r_{i-1})\,\Delta\theta_j \\ &= \frac{1}{2}(r_i + r_{i-1})\,\Delta r_i \Delta\theta_j \\ &= \hat{r}_i \Delta r_i \Delta\theta_j, \end{aligned} \tag{4}$$

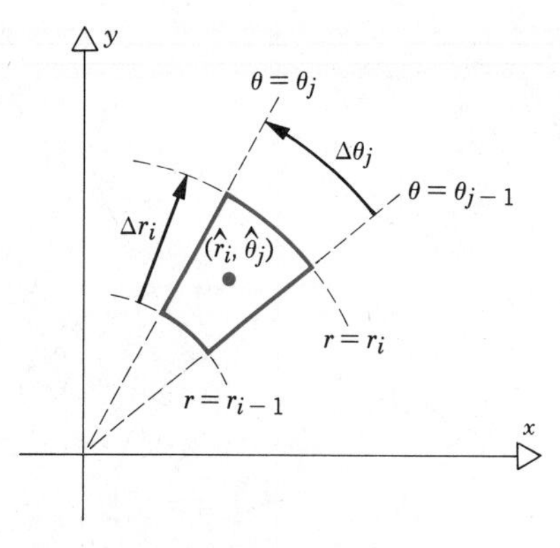

Figure 17.4.3 A typical polar subrectangle.

where $\hat{r}_i = (r_i + r_{i-1})/2$ is the mean radius of the polar subrectangle R_{ij}.

The next step is to select the star points P_{ij}^*. We take P_{ij}^* to be the center of the polar subrectangle R_{ij}, so its coordinates are $r_i^* = \hat{r}_i$ and $\theta_j^* = \hat{\theta}_j$, where $\hat{\theta}_j = (\theta_j + \theta_{j-1})/2$. Then Eq. 3 becomes

$$\iint_R f(r, \theta)\, dA = \lim_{\|\Delta\| \to 0} \sum_{\substack{i=1,\ldots,m \\ j=1,\ldots,n}} f(\hat{r}_i, \hat{\theta}_j)\hat{r}_i \Delta r_i \Delta \theta_j. \tag{5}$$

Finally, we consider the limit on the right side of Eq. 5. We refine the partition first in the r direction by letting $\Delta r_i \to 0$ and $m \to \infty$; this corresponds to integrating with respect to r for fixed θ. Then we refine the partition in the θ direction by letting $\Delta\theta_j \to 0$ and $n \to \infty$, corresponding to an integration with respect to θ. Thus the double integral is evaluated by means of an equivalent set of iterated integrals:

$$\iint_R f(r, \theta)\, dA = \int_c^d \int_a^b f(r, \theta) r\, dr\, d\theta. \tag{6}$$

Note especially that in the iterated integrals on the right side of Eq. 6 the area element dA is replaced by $r\, dr\, d\theta$, not just by $dr\, d\theta$.

Just as for rectangular coordinates we can extend the use of iterated integrals in polar coordinates to a great many regions that are not polar rectangles. For instance, suppose that the region Ω of Figure 17.4.4 is defined by

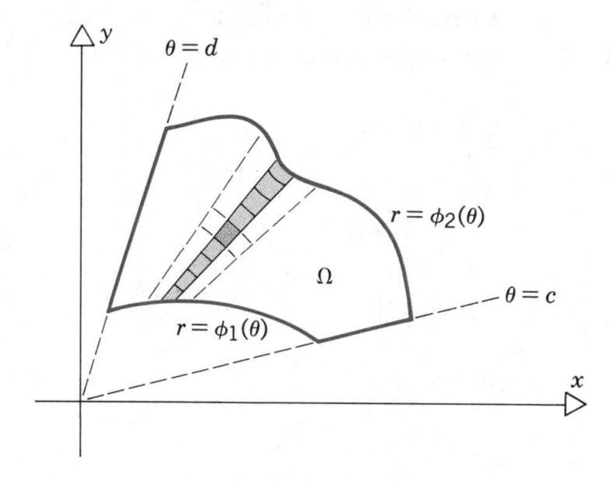

Figure 17.4.4

$$\Omega: \phi_1(\theta) \le r \le \phi_2(\theta), \qquad c \le \theta \le d, \tag{7}$$

where ϕ_1' and ϕ_2' are continuous on $[c, d]$. Then, instead of Eq. 6 we have

$$\iint_\Omega f(r, \theta)\, dA = \int_c^d \int_{\phi_1(\theta)}^{\phi_2(\theta)} f(r, \theta) r\, dr\, d\theta. \tag{8}$$

As in the case of rectangular coordinates, the integration process indicated on the right side of Eq. 8 can be visualized by supposing that the region Ω is to be covered by small tiles having the shape of polar rectangles (see Figure 17.4.4). The integration with respect to r corresponds to forming a thin wedge of such tiles extending from the inner boundary $r = \phi_1(\theta)$ to the outer boundary $r = \phi_2(\theta)$. Then the integration with respect to θ corresponds to placing many such wedges side by side until the region Ω is completely covered.

The interpretation of the integral

$$\iint_\Omega f(r, \theta)\, dA \tag{9}$$

of course depends on what $f(r, \theta)$ represents. For example, the integral (9) could give the value of a volume, a mass, or a moment, as discussed in previous sections. If $f(r, \theta) = 1$ for each point in Ω, then the expression (9) is just the area A of Ω. In this case, Eq. 8 yields

$$A = \int_c^d \int_{\phi_1(\theta)}^{\phi_2(\theta)} r\, dr\, d\theta. \tag{10}$$

The integration with respect to r can be carried out at once with the result that

$$A = \int_c^d \frac{r^2}{2}\bigg|_{\phi_1(\theta)}^{\phi_2(\theta)} d\theta$$

$$= \frac{1}{2}\int_c^d [\phi_2^2(\theta) - \phi_1^2(\theta)]\, d\theta. \tag{11}$$

This is consistent with the result obtained earlier in Section 14.3.

EXAMPLE 1

Find the total mass M in the sector of the ring

$$\Omega: 1 \le r \le 2, \qquad 0 \le \theta \le \frac{\pi}{2}$$

if the density (mass/area) is given by $\rho(r, \theta) = \sin 2\theta$ (see Figure 17.4.5).

The mass M is given by the integral of $\rho(r, \theta)$ over the given region. Thus

$$M = \iint_\Omega \rho(r, \theta)\, dA$$

$$= \int_0^{\pi/2} \int_1^2 (\sin 2\theta) r\, dr\, d\theta.$$

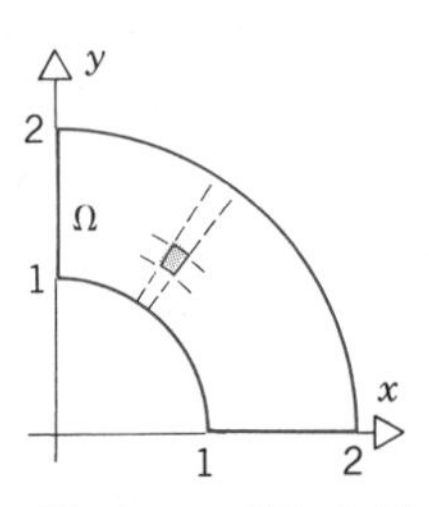

Figure 17.4.5

To determine M we first integrate with respect to r, holding θ constant, and then integrate with respect to θ. We obtain

$$\begin{aligned} M &= \int_0^{\pi/2} (\sin 2\theta) \left.\frac{r^2}{2}\right|_1^2 d\theta \\ &= \int_0^{\pi/2} \frac{3}{2} \sin 2\theta \, d\theta \\ &= \frac{3}{2}\left(-\frac{\cos 2\theta}{2}\right)\Bigg|_0^{\pi/2} \\ &= \frac{3}{4}(1 + 1) = \frac{3}{2}. \quad \blacksquare \end{aligned}$$

EXAMPLE 2

The region Ω is outside the circle $r = 3$ and inside the cardioid $r = 2(1 + \cos \theta)$. Suppose that Ω is occupied by a material whose mass per unit area $\rho(r, \theta)$ is inversely proportional to the distance from the origin. Find the mass M of the material and the coordinates $(\bar{x}, \bar{y})$ of its mass center.

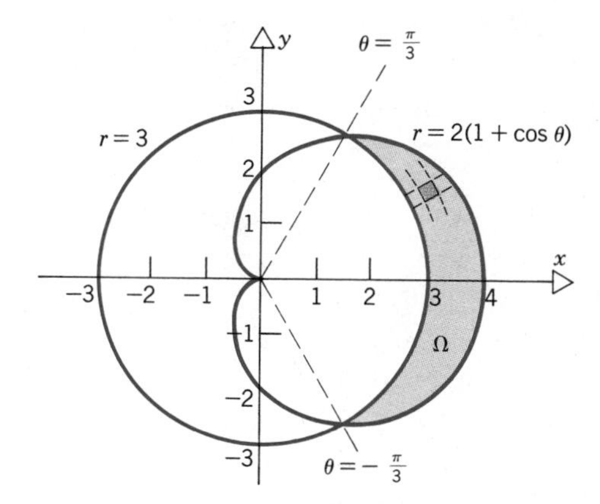

Figure 17.4.6

The region Ω is shown in Figure 17.4.6. The density function is

$$\rho(r, \theta) = \frac{k}{r},$$

where k is a constant of proportionality, so the total mass in Ω is given by

$$M = \int\int_{\Omega} \rho(r, \theta) \, dA.$$

To obtain equivalent iterated integrals we need to describe Ω by inequalities of the form (7), and this requires that we find the points of intersection of the circle and the cardioid. We have

$$r = 3 = 2(1 + \cos\theta),$$

from which it follows that $\cos\theta = \frac{1}{2}$ so $\theta = \pm\pi/3$. Thus Ω is given by

$$\Omega\colon 3 \leq r \leq 2(1 + \cos\theta), \qquad -\frac{\pi}{3} \leq \theta \leq \frac{\pi}{3}, \tag{12}$$

and consequently we have

$$\begin{aligned} M &= \int_{-\pi/3}^{\pi/3} \int_{3}^{2(1+\cos\theta)} \frac{k}{r}\, r\, dr\, d\theta \\ &= 2k \int_{0}^{\pi/3} \int_{3}^{2(1+\cos\theta)} dr\, d\theta. \end{aligned} \tag{13}$$

Observe that in writing Eq. 13 we have integrated over the upper half of Ω and multiplied the result by two; this is permissible because both the region of integration and the integrand are symmetric about the x-axis.

Finally, by evaluating the integrals in Eq. 13 we obtain

$$\begin{aligned} M &= 2k \int_{0}^{\pi/3} (2\cos\theta - 1)\, d\theta \\ &= 2k(2\sin\theta - \theta)\Big|_{0}^{\pi/3} \\ &= 2k\left(\sqrt{3} - \frac{\pi}{3}\right) \cong 1.37\, k. \end{aligned} \tag{14}$$

Now let us find the mass center. First, note that the symmetry of the problem implies that $\bar{y} = 0$, so we only need to determine $\bar{x} = L_y/M$. We have

$$\begin{aligned} L_y &= \iint_{\Omega} x\, \rho(r, \theta)\, dA \\ &= 2 \int_{0}^{\pi/3} \int_{3}^{2(1+\cos\theta)} (r\cos\theta)\left(\frac{k}{r}\right) r\, dr\, d\theta \\ &= 2k \int_{0}^{\pi/3} \int_{3}^{2(1+\cos\theta)} r\cos\theta\, dr\, d\theta. \end{aligned}$$

Evaluating the integral with respect to r, we obtain

$$\begin{aligned} L_y &= k \int_{0}^{\pi/3} \cos\theta\, [4(1 + \cos\theta)^2 - 9]\, d\theta \\ &= k \int_{0}^{\pi/3} \cos\theta\, (-5 + 8\cos\theta + 4\cos^2\theta)\, d\theta \\ &= k \int_{0}^{\pi/3} (-5\cos\theta + 8\cos^2\theta + 4\cos^3\theta)\, d\theta. \end{aligned} \tag{15}$$

The second term in the last integrand can be integrated by using the half-angle formula and the third term by using the fact that $\cos^2\theta = 1 - \sin^2\theta$. Thus we have

$$L_y = k\int_0^{\pi/3} [-5\cos\theta + 4 + 4\cos 2\theta + 4\cos\theta\,(1 - \sin^2\theta)]\,d\theta$$

$$= k\int_0^{\pi/3} (4 - \cos\theta + 4\cos 2\theta - 4\sin^2\theta\cos\theta)\,d\theta$$

$$= k\left(4\theta - \sin\theta + 2\sin 2\theta - \frac{4}{3}\sin^3\theta\right)\Bigg|_0^{\pi/3}$$

$$= k\left(\frac{4}{3}\pi - \frac{\sqrt{3}}{2} + \sqrt{3} - \frac{\sqrt{3}}{2}\right)$$

$$= \frac{4}{3}\pi k \cong 4.19\,k. \tag{16}$$

Finally,

$$\bar{x} = \frac{L_y}{M} = \frac{4k\pi/3}{2k(3\sqrt{3} - \pi)/3} = \frac{2\pi}{3\sqrt{3} - \pi} \cong 3.06. \quad ∎$$

When faced with a problem involving the evaluation of a double integral, one should remember that either rectangular or polar coordinates can be used. The choice of coordinate system should be based on which one leads to the simpler calculations. While this cannot always be foreseen in advance, practice is helpful; further, if one's first choice leads to difficulties, it is worth considering whether the other would result in an improvement. In particular, if the boundaries of the region of integration involve circles with center at the origin, radial lines, cardioids, and so forth, or if the expression $x^2 + y^2$ appears in the integrand, then polar coordinates should be strongly considered.

EXAMPLE 3

Find the volume V of the region that is above the xy-plane, below the paraboloid $z = 4 - x^2 - y^2$, and inside the cylinder $r = 2\sin\theta$.

The three-dimensional region whose volume is sought is shown in Figure 17.4.7*a*. The volume V is given by

$$V = \iint_\Omega (4 - x^2 - y^2)\,dA, \tag{17}$$

where Ω is the region in the xy-plane that is within the circle $r = 2\sin\theta$ (see Figure 17.4.7*b*). The region of integration

$$\Omega\colon 0 \le r \le 2\sin\theta, \qquad 0 \le \theta \le \pi$$

is best described in polar coordinates, and the integrand depends on the combination $x^2 + y^2$, so it is surely best to evaluate the integral (17) by using polar

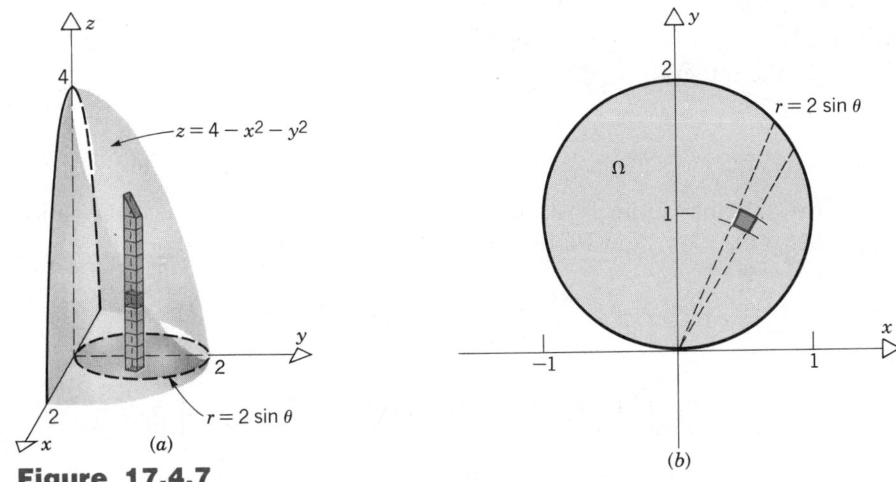

Figure 17.4.7

coordinates. Replacing the integration element dA by $r\ dr\ d\theta$, and writing the integrand in terms of r and θ, we have

$$
\begin{aligned}
V &= \int_0^{\pi}\int_0^{2\sin\theta}(4 - r^2)r\ dr\ d\theta \\
&= -\frac{1}{4}\int_0^{\pi}(4 - r^2)^2\Big|_0^{2\sin\theta}\ d\theta \\
&= -\frac{1}{4}\int_0^{\pi}[(4 - 4\sin^2\theta)^2 - 16]\ d\theta \\
&= 4\int_0^{\pi}(1 - \cos^4\theta)\ d\theta.
\end{aligned}
$$

The integration with respect to θ can be accomplished by repeated use of the half-angle formula:

$$
\begin{aligned}
V &= 4\int_0^{\pi}\left[1 - \frac{1}{4}(1 + \cos 2\theta)^2\right] d\theta \\
&= 4\int_0^{\pi}\left(\frac{3}{4} - \frac{1}{2}\cos 2\theta - \frac{1}{4}\cos^2 2\theta\right) d\theta \\
&= 4\int_0^{\pi}\left[\frac{3}{4} - \frac{1}{2}\cos 2\theta - \frac{1}{8}(1 + \cos 4\theta)\right] d\theta \\
&= 4\int_0^{\pi}\left(\frac{5}{8} - \frac{1}{2}\cos 2\theta - \frac{1}{8}\cos 4\theta\right) d\theta \\
&= 4\left(\frac{5}{8}\theta - \frac{1}{4}\sin 2\theta - \frac{1}{32}\sin 4\theta\right)\Big|_0^{\pi} \\
&= \frac{5\pi}{2}. \ \blacksquare
\end{aligned}
$$

EXAMPLE 4

Evaluate the integral

$$I = \int_0^1 \int_y^{\sqrt{2-y^2}} (x^2 + y^2)\, dx\, dy. \tag{18}$$

If we attempt to deal with this integral as it stands, the first integration (with respect to x) is easy, and we obtain

$$\begin{aligned} I &= \int_0^1 \left(\frac{x^3}{3} + xy^2\right)\Bigg|_y^{\sqrt{2-y^2}} dy \\ &= \int_0^1 \left[\frac{1}{3}(2 - y^2)^{3/2} + y^2\sqrt{2 - y^2} - \frac{4}{3}y^3\right] dy. \end{aligned}$$

The remaining integration is difficult enough so that at this stage, if not before, we should pause and consider possible alternatives. One is to reverse the order of integration, that is, to determine the limits of integration so as to integrate first with respect to y and then with respect to x. However, that is not helpful in this case, as you may confirm.

The presence of $x^2 + y^2$ in the integrand suggests that polar coordinates may be useful. To reformulate the problem in polar coordinates we need to determine the region of integration. This can be read off from the limits in the original integral (18): x goes from $x = y$ to $x = \sqrt{2 - y^2}$ and y from $y = 0$ to $y = 1$. The graph of $x = \sqrt{2 - y^2}$ is part of the circle $x^2 + y^2 = 2$, so the region Ω of integration is as shown in Figure 17.4.8*a*; it is the sector of the disk $x^2 + y^2 \leqq 2$ between the

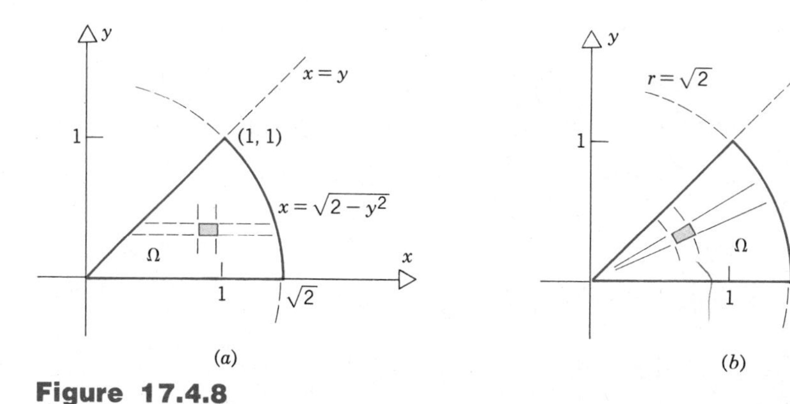

Figure 17.4.8

x-axis and the line $y = x$. In terms of polar coordinates the boundaries of Ω are the circle $r = \sqrt{2}$ and the lines $\theta = 0$ and $\theta = \pi/4$, as shown in Figure 17.4.8*b*. Thus in polar coordinates we have

$$I = \int_0^{\pi/4} \int_0^{\sqrt{2}} r^2\, r\, dr\, d\theta = \int_0^{\pi/4} \int_0^{\sqrt{2}} r^3\, dr\, d\theta. \tag{19}$$

Clearly I is much simpler in polar coordinates and can be immediately evaluated:

$$I = \int_0^{\pi/4} \frac{r^4}{4}\bigg|_0^{\sqrt{2}} d\theta = \int_0^{\pi/4} d\theta = \frac{\pi}{4}. \blacksquare$$

Observe that in setting up the integral (19) in polar coordinates we replaced the original integrand $x^2 + y^2$ by the equivalent expression r^2 in polar coordinates, we replaced the integration element $dx\, dy$ by $r\, dr\, d\theta$, and we set the limits on the integrals that are appropriate for the region of integration. This last step should *never* be done by a mechanical substitution process. Rather, one should *always* sketch the region of integration, and then establish the limits that are needed to integrate over this region.

EXAMPLE 5

Find the moment of inertia of a circular plate of radius a and uniform density $\rho(x, y) = k$ about a diameter.

We choose the origin to be the center of the circle, so that either I_x or I_y gives the required moment of inertia. Further, by symmetry we know that $I_x = I_y$ and consequently $I_x = I_0/2$, where I_0 is the polar moment of inertia. It is simplest to calculate I_0; we have

$$\begin{aligned} I_0 &= \iint_\Omega k(x^2 + y^2)\, dA \\ &= k \int_0^{2\pi} \int_0^a r^2\, r\, dr\, d\theta \\ &= k\frac{a^4}{4} \int_0^{2\pi} d\theta \\ &= \frac{\pi k a^4}{2}. \end{aligned}$$

Thus we obtain

$$I_x = I_y = \frac{\pi k a^4}{4}. \blacksquare$$

PROBLEMS

In each of Problems 1 through 8, sketch the region of integration, and evaluate the given integral.

1. $\displaystyle\int_0^{\pi} \int_0^2 r \sin\theta\, r\, dr\, d\theta$

2. $\displaystyle\int_0^{\pi/4} \int_0^3 \sin\theta \cos\theta\, r\, dr\, d\theta$

3. $\displaystyle\int_0^{\pi/2} \int_0^{1+\sin\theta} \cos\theta\, r\, dr\, d\theta$

4. $\displaystyle\int_0^{\pi/4} \int_0^{2\sin 2\theta} \cos 2\theta \, dr \, d\theta$

5. $\displaystyle\int_{\pi/18}^{5\pi/18} \int_1^{2\sin 3\theta} r \, dr \, d\theta$ 6. $\displaystyle\int_{\pi/2}^{5\pi/6} \int_1^{2\sin\theta} \sin\theta \, dr \, d\theta$

7. $\displaystyle\int_0^{\pi/2} \int_{2\sin\theta}^{2} \sin\theta \, r \, dr \, d\theta$

8. $\displaystyle\int_{-\pi/3}^{\pi/3} \int_{(3/4)\sec\theta}^{1+\cos\theta} \cos\theta \, dr \, d\theta$

In each of Problems 9 through 12, sketch the region of integration and then evaluate the given integral by transforming it into polar coordinates.

9. $\displaystyle\int_0^{\sqrt{2}} \int_y^{\sqrt{4-y^2}} dx \, dy$

10. $\displaystyle\int_0^2 \int_0^{\sqrt{2x-x^2}} \sqrt{x^2+y^2} \, dy \, dx$

11. $\displaystyle\int_0^{3/2} \int_{\sqrt{3}x}^{\sqrt{9-x^2}} e^{-x^2} e^{-y^2} \, dy \, dx$

12. $\displaystyle\int_0^2 \int_0^{\sqrt{4-y^2}} (4 - x^2 - y^2)^{3/2} \, dx \, dy$

13. Find the area of the region inside the circle $r = 2a \sin\theta$ and outside the circle $r = a$.

14. Find the area of the region within the cardioid $r = 1 + \sin\theta$ and above the line $y = \frac{3}{4}$.

15. Find the volume of the region between the xy-plane and the surface $z = 4 - x^2 - y^2$.

16. Find the volume of the region enclosed by the surfaces $z = x^2 + y^2$ and $z = 4 - x^2 - y^2$.

17. Find the volume of the region bounded by the cylinder $r = 1 + \cos\theta$, by the plane $z = 4 - y$, and by the xy-plane.

18. Find the volume of the region that is common to the sphere $x^2 + y^2 + z^2 = a^2$ and the cone $x^2 + y^2 - z^2 = 0$.

19. Find the volume common to the spheres $x^2 + y^2 + z^2 = a^2$ and $x^2 + y^2 + (z - a)^2 = a^2$.

20. Find the centroid of the region enclosed by the cardioid $r = 1 + \cos\theta$.

21. Find the centroid of the region in the first quadrant bounded by $r = a \sin 2\theta$.

22. Find the centroid of the region in the first two quadrants bounded by the x-axis and the graph of $r = \theta$ for $0 \le \theta \le \pi$.

23. A semicircular plate of uniform density k is bounded by the x-axis and by the arc $x^2 + y^2 = a^2$, $y \ge 0$.

(a) Find the moment of inertia about the x-axis.

(b) Find the moment of inertia about the y-axis.

(c) Find the polar moment of inertia.

24. Find the center of mass of the sector of the ring $1 \le r \le 2$, $0 \le \theta \le \pi/2$ if $\rho(r, \theta) = \sin 2\theta$ (see Example 1).

25. A semicircular ring occupies the region $a \le r \le b$, $0 \le \theta \le \pi$, and has density inversely proportional to distance from the origin. Find

(a) the mass,

(b) the center of mass,

(c) the moment of inertia about the x-axis,

(d) the moment of inertia about the y-axis.

26. A circular ring has inner radius a and outer radius b, and its density is inversely proportional to the square of the distance from the center. Find the moment of inertia about

(a) a diameter,

(b) a line tangent to the inner boundary,

(c) a line tangent to the outer boundary.

Hint: In Parts (b) and (c) you may want to refer to Problem 23 of Section 17.3 (the parallel axes theorem).

27. A semicircular plate of constant density k_2 is joined to a square plate of constant density k_1, as shown in Figure 17.4.9. Find the ratio k_2/k_1 for

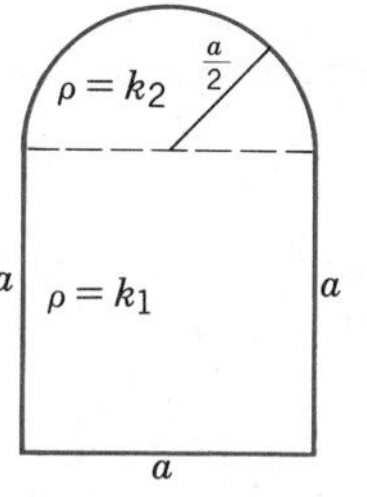

Figure 17.4.9

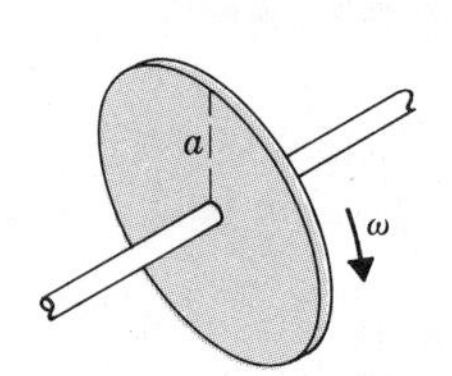

Figure 17.4.10

which the center of mass of the combined structure is on the boundary between the semicircle and the square.

Hint: Refer to problems 18 and 29 in Section 17.3.

28. A flywheel of radius a, uniform thickness, and constant density k is rotating about an axis through its center at a uniform angular velocity ω rad/sec (see Figure 17.4.10). Find the kinetic energy E of the flywheel.

17.5 TRIPLE INTEGRALS

It is important to be able to extend the concept and methods of integration to functions of more than two variables and to regions in spaces of more than two dimensions. Fortunately, the process of integration generalizes in a straightforward manner from two to a higher number of dimensions. In this section we consider integrals of functions of three variables. In order to make the presentation as simple as possible, we will restrict ourselves to functions that are continuous everywhere in the region of integration. We will follow the pattern of Sections 17.1 and 17.2, so our discussion can be brief; there are no new ideas here, just one more dimension.

Let R be the rectangular box, or parallelepiped, $a \le x \le b$, $c \le y \le d$, $p \le z \le q$ shown in Figure 17.5.1, and let f be a function that is continuous at all points

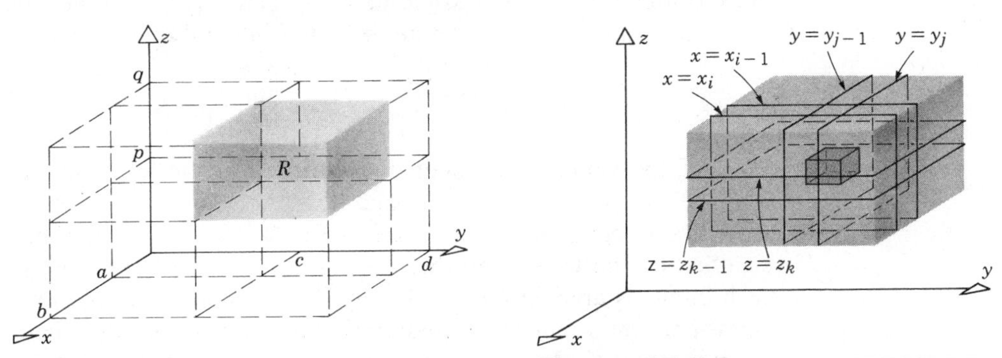

Figure 17.5.1

Figure 17.5.2

in R. Our task is to define, and show how to evaluate, the integral of f over R. First we partition R into a collection Δ of smaller rectangular boxes by means of planes parallel to the coordinate planes: $x = x_i$ for $i = 1, \ldots, l - 1$; $y = y_j$ for $j = 1, \ldots, m - 1$; and $z = z_k$ for $k = 1, \ldots, n - 1$. This partition is suggested in Figure 17.5.2. A typical box in Δ is denoted by R_{ijk} and is bounded by the planes $x = x_{i-1}$, $x = x_i$, $y = y_{j-1}$, $y = y_j$, $z = z_{k-1}$, and $z = z_k$ as shown in Figure 17.5.3. The volume ΔV_{ijk} of R_{ijk} is

$$\begin{aligned} \Delta V_{ijk} &= (x_i - x_{i-1})(y_j - y_{j-1})(z_k - z_{k-1}) \\ &= \Delta x_i \, \Delta y_j \, \Delta z_k. \end{aligned} \tag{1}$$

The norm $\|\Delta\|$ of the partition Δ is the length of the longest diagonal of all the boxes in Δ.

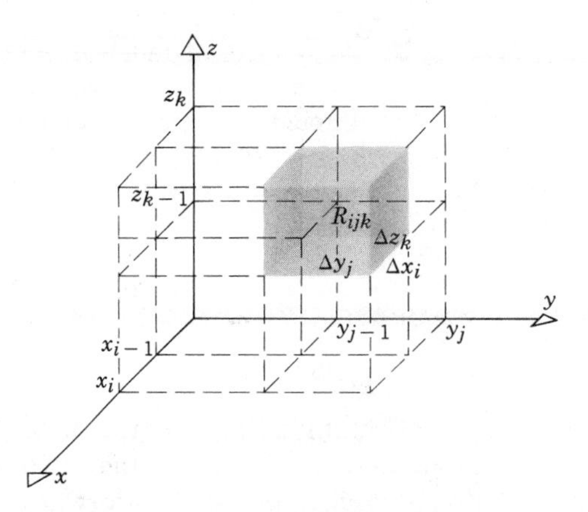

Figure 17.5.3

Next, we choose a star point $P^*_{ijk}(x^*_i, y^*_j, z^*_k)$ arbitrarily in each box R_{ijk} and form the Riemann sum

$$\sum_{i,j,k} f(P^*_{ijk})\, \Delta V_{ijk}. \tag{2}$$

Finally, we consider the limit of the sum (2) as $\|\Delta\| \to 0$. For all continuous functions f (and for many other bounded functions whose discontinuity points are not too numerous) it is possible to show that as $\|\Delta\| \to 0$ the sum (2) approaches a definite limit, which we define to be the integral of f over R:

$$\iiint_R f(x, y, z)\, dV = \lim_{\|\Delta\| \to 0} \sum_{i,j,k} f(P^*_{ijk})\, \Delta V_{ijk}. \tag{3}$$

Further, the value of the limit does not depend on how the partition is formed, how the star points are chosen, and how $\|\Delta\| \to 0$.

The integral on the left side of Eq. 3 is called a **triple integral,** and this is signified by the use of the three integral signs. There are many situations where such integrals arise in a natural way. For example, $f(x, y, z)$ may be the density (mass per unit volume) of a material occupying the box R. Then each term in the Riemann sum (2) is an approximation to the mass of the material in R_{ijk}, and the entire sum is an approximation to the mass in R. The total mass M in R is defined as the limiting value of the Riemann sum as $\|\Delta\| \to 0$; thus

$$M = \iiint_R f(x, y, z)\, dV. \tag{4}$$

Similarly, if $f(x, y, z) = 1$ at each point in R, then the triple integral (3) gives the volume V of the box R.

The value of the triple integral $\iiint_R f(x, y, z)\, dV$ can be found by evaluating a set of iterated single integrals, just as in the two-dimensional case. To visualize what is involved it may be helpful to think along the following lines. Suppose that a room is to be filled with small boxlike tiles. If we let $\|\Delta\| \to 0$ by first refining the mesh in the x direction, which leads to an integral with respect to x, then this corresponds to forming a single strip of tiles parallel to the x-axis (see Figure 17.5.4a). Next let us refine the mesh in the y direction, which leads to an integral with respect to y. This corresponds to combining many strips so as to form a layer

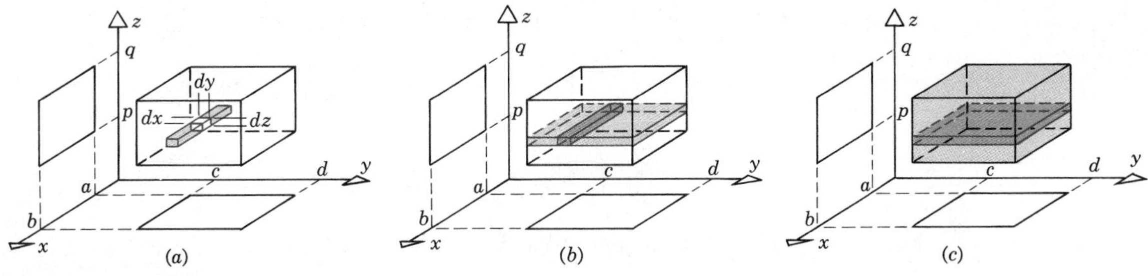

Figure 17.5.4

or slab parallel to the xy-plane (see Figure 17.5.4b). Finally, refining the mesh in the z direction leads to an integral with respect to z, and this corresponds to stacking such layers on top of each other until the room is entirely full of tiles (see Figure 17.5.4c). In this way we obtain

$$\iiint_R f(x, y, z)\, dV = \int_p^q \int_c^d \int_a^b f(x, y, z)\, dx\, dy\, dz. \tag{5}$$

Observe that dV has been replaced by $dx\, dy\, dz$, the volume of a rectangular parallelepiped of sides dx, dy, and dz. The iterated integrals on the right side of Eq. 5 are to be evaluated from the inside out; thus, one first integrates with respect to x from $x = a$ to $x = b$ while holding y and z constant, then one integrates with respect to y from $y = c$ to $y = d$ while holding z constant, and finally one integrates with respect to z from $z = p$ to $z = q$. Of course, the integration can also be done in five other orders; for instance, if we choose to write

$$\iiint_R f(x, y, z)\, dV = \int_a^b \int_c^d \int_p^q f(x, y, z)\, dz\, dy\, dx \tag{6}$$

instead of Eq. 5, then the integration is done first with respect to z, then with respect to y, and finally with respect to x.

EXAMPLE 1

The material in the box

$$R: 0 \le x \le 3, \qquad 0 \le y \le 1, \qquad 0 \le z \le 2$$

has density $\rho(x, y, z) = 6 - x - y - z$. Find the total mass M in the box.

From Eq. 4 we have

$$M = \iiint_R (6 - x - y - z)\, dV.$$

We choose to integrate with respect to z first, then with respect to y, and finally with respect to x. Thus

$$M = \int_0^3 \int_0^1 \int_0^2 (6 - x - y - z)\, dz\, dy\, dx.$$

Integrating with respect to z for fixed x and y, we obtain

$$M = \int_0^3 \int_0^1 \left(6z - xz - yz - \frac{z^2}{2}\right)\Bigg|_{z=0}^{z=2} dy\,dx$$

$$= \int_0^3 \int_0^1 (10 - 2x - 2y)\,dy\,dx.$$

Next, integrating with respect to y for fixed x, we find that

$$M = \int_0^3 (10y - 2xy - y^2)\Bigg|_{y=0}^{y=1} dx$$

$$= \int_0^3 (9 - 2x)\,dx.$$

Finally, the integration with respect to x yields

$$M = (9x - x^2)\Bigg|_0^3 = 18,$$

which is the value of the requested mass.

You may wish to check this result by carrying out the integration in some other order. ∎

The extension of triple integrals from rectangular boxes to other three-dimensional regions Ω is handled in exactly the same way as in Section 17.1 for the two-dimensional case. We enclose Ω in a larger rectangular box R, and define a function g so that $g(x, y, z) = f(x, y, z)$ at points in Ω, and $g(x, y, z) = 0$ elsewhere. Then we define

$$\iiint_\Omega f(x, y, z)\,dV = \iiint_R g(x, y, z)\,dV. \tag{7}$$

With this definition triple integrals have the same properties of linearity, comparison, additivity, and so forth that were stated for double integrals in Theorems 17.1.3 and 17.1.4.

Our main interest is in the evaluation of triple integrals over certain kinds of relatively simple three-dimensional regions by means of iterated integrals. Suppose that Ω is as shown in Figure 17.5.5*a*; this region is described by the inequalities

$$\Omega\colon \chi_1(x, y) \le z \le \chi_2(x, y), \qquad (x, y) \text{ in } \Omega_{xy}. \tag{8}$$

The two-dimensional region Ω_{xy} in the xy-plane is called the **projection** of Ω into the xy-plane. The lower and upper boundaries of Ω are given by the surfaces $z = \chi_1(x, y)$ and $z = \chi_2(x, y)$, respectively. The integral of a given function f over Ω can be expressed as

$$\iiint_\Omega f(x, y, z)\,dV = \iint_{\Omega_{xy}} \int_{\chi_1(x,y)}^{\chi_2(x,y)} f(x, y, z)\,dz\,dA. \tag{9}$$

The first integral on the right side of Eq. 9 represents an integration with respect to z from the lower boundary to the upper boundary of Ω (see Figure 17.5.5b). This is followed by a double integral over the projection Ω_{xy}. For instance, if Ω_{xy} is described by

$$a \le x \le b, \qquad \phi_1(x) \le y \le \phi_2(x), \tag{10}$$

as indicated in Figures 17.5.5c and 17.5.5d, then Eq. 9 becomes

$$\iiint_{\Omega} f(x, y, z)\, dV = \int_a^b \int_{\phi_1(x)}^{\phi_2(x)} \int_{\chi_1(x,y)}^{\chi_2(x,y)} f(x, y, z)\, dz\, dy\, dx. \tag{11}$$

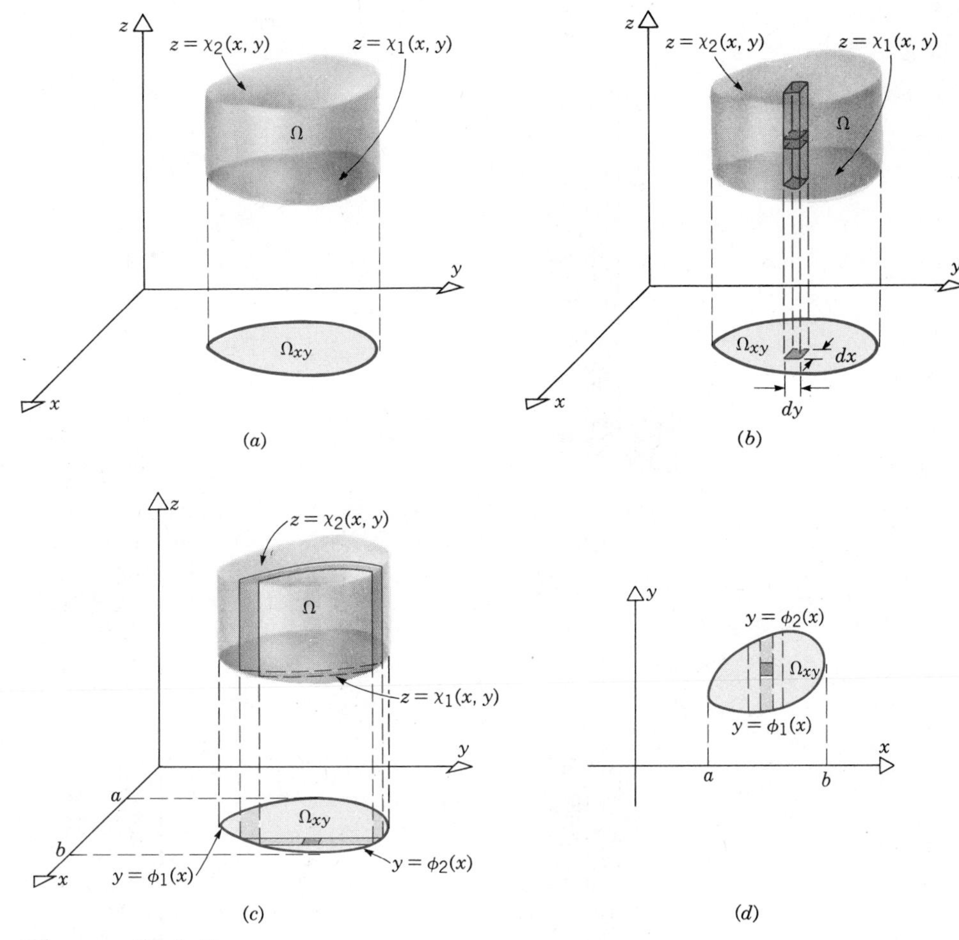

Figure 17.5.5

EXAMPLE 2

Find the volume of the region Ω in the first octant that is bounded by the coordinate planes, the plane $x + 2y = 2$, and the surface $z = 4 - x^2 - y^2$.

The region Ω is shown in Figure 17.5.6*a*. Its volume V can be found by means of a double integral, as in Section 17.2, but for illustrative purposes we will use a triple integral here. It is most convenient to project Ω into the xy-plane, in which case Ω_{xy} is the triangle with vertices at (0, 0), (2, 0), and (0, 1) shown in Figure 17.5.6*b*. Then Ω can be described by the following set of inequalities

$$0 \leq z \leq 4 - x^2 - y^2, \qquad 0 \leq y \leq 1 - \frac{x}{2}, \qquad 0 \leq x \leq 2. \tag{12}$$

The corresponding iterated integrals are

$$V = \int_0^2 \int_0^{1-(x/2)} \int_0^{4-x^2-y^2} dz\, dy\, dx. \tag{13}$$

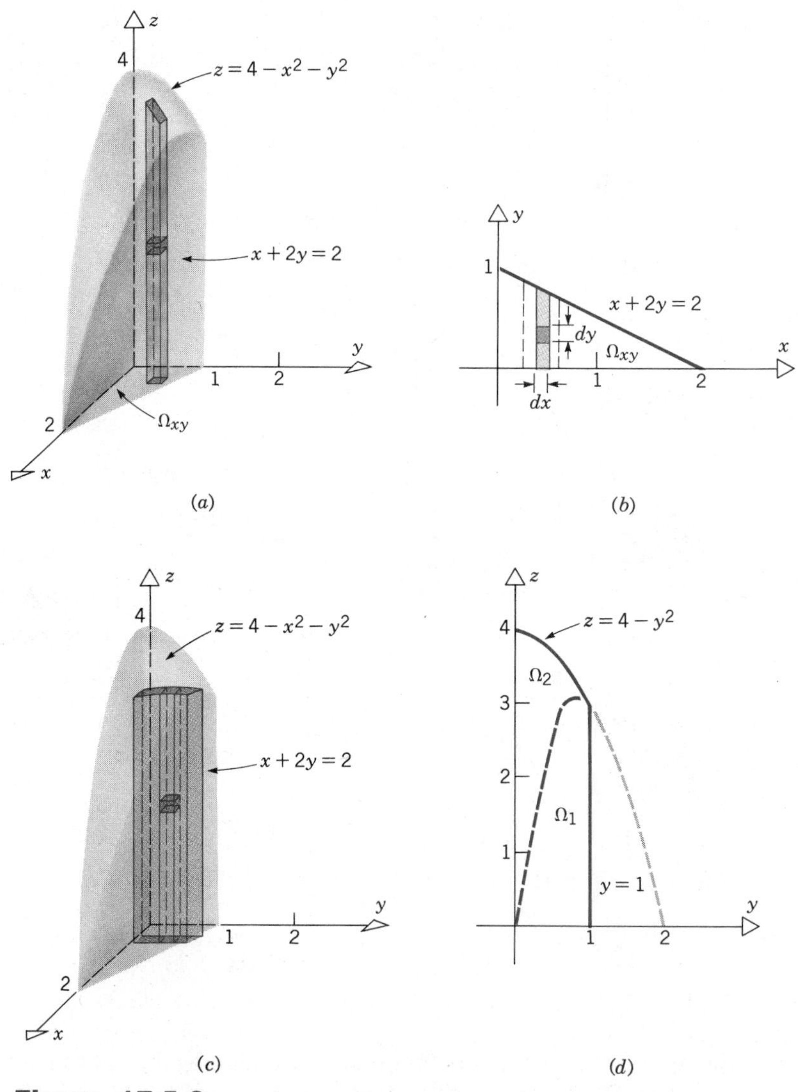

Figure 17.5.6

Observe that the order of integration is z first, y second, and x third, and that this corresponds to the form of the inequalities (12). The integration process is suggested by Figure 17.5.6c: starting with the volume element $dx\,dy\,dz$, the integration with respect to z corresponds to constructing a column parallel to the z-axis, the integration with respect to y corresponds to combining columns into a slab parallel to the yz-plane, and integration with respect to x corresponds to filling Ω with such slabs.

Upon evaluating the first integral in Eq. 13 we have

$$V = \int_0^2 \int_0^{1-(x/2)} (4 - x^2 - y^2)\,dy\,dx,$$

and then the remaining integrals yield

$$\begin{aligned} V &= \int_0^2 \left[(4 - x^2)y - \frac{y^3}{3} \right]\Bigg|_0^{1-(x/2)} dx \\ &= \int_0^2 \left[(4 - x^2)\left(1 - \frac{x}{2}\right) - \frac{1}{3}\left(1 - \frac{x}{2}\right)^3 \right] dx \\ &= \int_0^2 \left(\frac{11}{3} - \frac{3}{2}x - \frac{5}{4}x^2 + \frac{13}{24}x^3 \right) dx \\ &= \frac{22}{3} - 3 - \frac{10}{3} + \frac{13}{6} = \frac{19}{6}. \quad \blacksquare \end{aligned}$$

In many cases there may be more than one way to express the region Ω in terms of x, y, and z, and hence more than one way to carry out the integration. In the preceding discussion and in Example 2, Ω was described by

$$\Omega\colon \chi_1(x, y) \le z \le \chi_2(x, y), \qquad \phi_1(x) \le y \le \phi_2(x), \qquad a \le x \le b. \tag{14}$$

It may also be possible to describe Ω by other sets of inequalities, such as

$$\Omega\colon \psi_1(x, z) \le y \le \psi_2(x, z), \qquad \chi_1(x) \le z \le \chi_2(x), \qquad a \le x \le b, \tag{15}$$

in which case the integral would take the form

$$\iiint_\Omega f(x, y, z)\,dV = \int_a^b \int_{\chi_1(x)}^{\chi_2(x)} \int_{\psi_1(x,z)}^{\psi_2(x,z)} f(x, y, z)\,dy\,dz\,dx. \tag{16}$$

Observe that in Eq. 16 the first integration is with respect to y, corresponding to a projection of Ω into the xz-plane. Similarly, it may also be possible to project into the yz-plane.

In many problems the actual integration is straightforward, although sometimes tedious. Often the most challenging part of a problem is the task of representing a triple integral as a set of iterated integrals in the most convenient way and correctly setting the corresponding limits of integration. It is almost always helpful to draw a sketch, even a primitive one, of all or part of the region of integration, and to describe Ω by means of a set of inequalities such as Eqs. 14 or

15. Frequently, it is useful to sketch both the original region Ω and its projection in one or more of the coordinate planes. Do not merely memorize formulas, and do not proceed simply by substituting values into expressions such as the right sides of Eqs. 11 or 16. Rather, try to understand thoroughly the underlying principles and adapt them to each case that occurs.

In Example 2 one can reverse the order of the integrations with respect to x and y over Ω_{xy} without encountering any particular difficulty. It is also possible, but somewhat more complicated, to integrate first with respect to x or y, rather than with respect to z as we have done. To see where the complication arises, suppose that we wish to integrate first with respect to x, which means that we must project Ω into the yz-plane. The projection Ω_{yz} is shown in Figure 17.5.6*d*. The complication comes from the fact that the upper limit on the first integral, the x-integral, depends on where the point (y, z) is located in Ω_{yz}. If (y, z) is in Ω_1, then the upper limit is obtained from the equation of the plane $x + 2y = 2$, and if (x, y) is in Ω_2, then it is obtained from the equation of the paraboloid $z = 4 - x^2 - y^2$. The point is that one must find the curve that separates Ω_{yz} into Ω_1 and Ω_2, and then set up two integrals in order to cover the entire region of integration. Since this is a complicating factor, it suggests that projecting into the yz-plane will not be helpful and that we should proceed in some other way—as we did.

Moments, centers of mass, centroids, and moments of inertia

The discussion in Section 17.3 also generalizes directly from two dimensions to three. Consider a region Ω occupied by a material with density (mass/volume) $\rho(x, y, z)$. The total mass in Ω is given by

$$M = \iiint_\Omega \rho(x, y, z)\, dV. \tag{17}$$

The first moment about the origin is

$$\begin{aligned}
\mathbf{L} &= \iiint_\Omega \rho(x, y, z)(x\mathbf{i} + y\mathbf{j} + z\mathbf{k})\, dV \\
&= \mathbf{i}\iiint_\Omega x\rho(x, y, z)\, dV + \mathbf{j}\iiint_\Omega y\rho(x, y, z)\, dV \\
&\quad + \mathbf{k}\iiint_\Omega z\rho(x, y, z)\, dV \\
&= L_{yz}\mathbf{i} + L_{xz}\mathbf{j} + L_{xy}\mathbf{k}.
\end{aligned} \tag{18}$$

Here L_{yz} is the first moment with respect to the yz-plane, and similarly for L_{xz} and L_{xy}.

The center of mass of the material in Ω is the point about which the first moment is zero. Expressions for its coordinates $(\bar{x}, \bar{y}, \bar{z})$ can be derived as in Section 17.3:

$$\bar{x} = \frac{L_{yz}}{M} = \frac{1}{M}\iiint_{\Omega} x\rho(x, y, z)\, dV, \tag{19}$$

$$\bar{y} = \frac{L_{xz}}{M} = \frac{1}{M}\iiint_{\Omega} y\rho(x, y, z)\, dV, \tag{20}$$

$$\bar{z} = \frac{L_{xy}}{M} = \frac{1}{M}\iiint_{\Omega} z\rho(x, y, z)\, dV. \tag{21}$$

If $\rho(x, y, z)$ is constant, then it can be taken out of the integrals in Eqs. 19, 20, and 21, and then canceled out of the numerator and denominator. In this way we obtain expressions for the coordinates $(\hat{x}, \hat{y}, \hat{z})$ of the centroid of Ω:

$$\hat{x} = \frac{\hat{L}_{yz}}{V} = \frac{1}{V}\iiint_{\Omega} x\, dV, \qquad \hat{y} = \frac{\hat{L}_{xz}}{V} = \frac{1}{V}\iiint_{\Omega} y\, dV,$$

$$\hat{z} = \frac{\hat{L}_{xy}}{V} = \frac{1}{V}\iiint_{\Omega} z\, dV, \tag{22}$$

where V is the volume of Ω.

The moment of inertia I_x about the x-axis of the material in Ω is

$$I_x = \iiint_{\Omega} (y^2 + z^2)\rho(x, y, z)\, dV. \tag{23}$$

Observe that the integrand is the element of mass $\rho(x, y, z)\, dV$ multiplied by the square of the distance from the x-axis. In the same way, the moments of inertia I_y and I_z about the y- and z-axes, respectively, are

$$I_y = \iiint_{\Omega} (x^2 + z^2)\rho(x, y, z)\, dV, \tag{24}$$

$$I_z = \iiint_{\Omega} (x^2 + y^2)\rho(x, y, z)\, dV. \tag{25}$$

The radii of gyration R_x, R_y, and R_z about each coordinate axis are defined as in the two-dimensional case:

$$R_x^2 = \frac{I_x}{M}, \qquad R_y^2 = \frac{I_y}{M}, \qquad R_z^2 = \frac{I_z}{M}. \tag{26}$$

EXAMPLE 3

The region Ω, bounded by the cylinder $z = y^2$, the plane $x + z = 1$, and the yz-plane, is filled with a material whose density (mass/volume) $\rho(x, y, z)$ is proportional to x. Find the center of mass of the material in Ω.

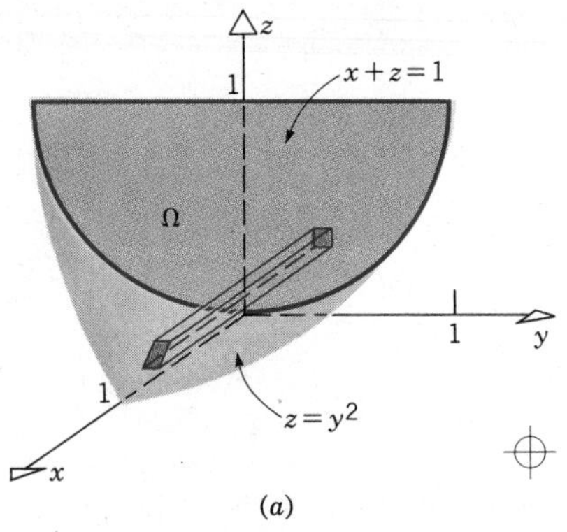

(a)

(b)

Figure 17.5.7

The region Ω is shown in Figure 17.5.7*a*. Observe first that Ω is symmetric with respect to the xz-plane. Further, since the density $\rho(x, y, z)$ does not depend on y, the mass distribution is also symmetric with respect to the xz-plane. Thus the center of mass must lie in the xz-plane, so $\bar{y} = 0$. To determine the remaining coordinates $\bar{x}$ and $\bar{z}$ from Eqs. 19 and 21 we need to calculate the total mass M and the first moments L_{yz} and L_{xy}.

In this problem it is convenient to project into the yz-plane, and the region Ω_{yz} is shown in Figure 17.5.7*b*. The region Ω can be described by the inequalities

$$0 \le x \le 1 - z, \qquad y^2 \le z \le 1, \qquad -1 \le y \le 1. \tag{27}$$

The density of the material in Ω is given by $\rho(x, y, z) = kx$, where k is a constant of proportionality, so the total mass is

$$\begin{aligned} M &= \iiint_\Omega \rho(x, y, z)\, dV \\ &= \int_{-1}^{1} \int_{y^2}^{1} \int_0^{1-z} kx\, dx\, dz\, dy. \end{aligned} \tag{28}$$

Proceeding with the integration, we obtain

$$\begin{aligned} M &= k \int_{-1}^{1} \int_{y^2}^{1} \frac{x^2}{2}\bigg|_0^{1-z} dz\, dy \\ &= \frac{k}{2} \int_{-1}^{1} \int_{y^2}^{1} (1 - z)^2\, dz\, dy \\ &= \frac{k}{2} \int_{-1}^{1} -\frac{(1 - z)^3}{3}\bigg|_{y^2}^{1} dy \\ &= \frac{k}{6} \int_{-1}^{1} (1 - y^2)^3\, dy \\ &= \frac{k}{3} \int_0^1 (1 - y^2)^3\, dy. \end{aligned} \tag{29}$$

In the last step in Eq. 29 we have taken advantage of the facts that the integrand is an even function and the interval of integration is symmetric about the origin. Evaluating the integral in Eq. 29, we find that

$$M = \frac{k}{3}\int_0^1 (1 - 3y^2 + 3y^4 - y^6)\, dy$$

$$= \frac{k}{3}\left(1 - 1 + \frac{3}{5} - \frac{1}{7}\right) = \frac{16}{105}\, k.$$

Next we turn to the calculation of L_{yz}. We have

$$L_{yz} = \iiint_\Omega x\, \rho(x, y, z)\, dV$$

$$= k\int_{-1}^1 \int_{y^2}^1 \int_0^{1-z} x^2\, dx\, dz\, dy.$$

The integration here resembles that involved in calculating M, and we obtain

$$L_{yz} = \frac{k}{3}\int_{-1}^1 \int_{y^2}^1 (1 - z)^3\, dz\, dy$$

$$= \frac{k}{3}\int_{-1}^1 \frac{(1 - y^2)^4}{4}\, dy$$

$$= \frac{k}{6}\int_0^1 (1 - 4y^2 + 6y^4 - 4y^6 + y^8)\, dy$$

$$= \frac{k}{6}\left(1 - \frac{4}{3} + \frac{6}{5} - \frac{4}{7} + \frac{1}{9}\right) = \frac{64}{945}\, k.$$

From Eq. 18 it now follows that

$$\bar{x} = \frac{64k/945}{16k/105} = \frac{4}{9}.$$

In a similar way we have

$$L_{xy} = \iiint_\Omega z\, \rho(x, y, z)\, dV$$

$$= k\int_{-1}^1 \int_{y^2}^1 \int_0^{1-z} z\, x\, dx\, dz\, dy.$$

We omit the evaluation of this integral, since it is not greatly different from the preceding ones in this example. The result is

$$L_{xy} = \frac{16k}{315},$$

and consequently, from Eq. 22, we have

$$\bar{z} = \frac{16k/315}{16k/105} = \frac{1}{3}.$$

Thus the center of mass of the material in Ω is the point $(\frac{4}{9}, 0, \frac{1}{3})$. ∎

EXAMPLE 4

The region Ω is bounded by the xy-plane and the paraboloid $z = 4 - x^2 - y^2$. If this region is filled with a material with constant density k, find the moment of inertia about the z-axis.

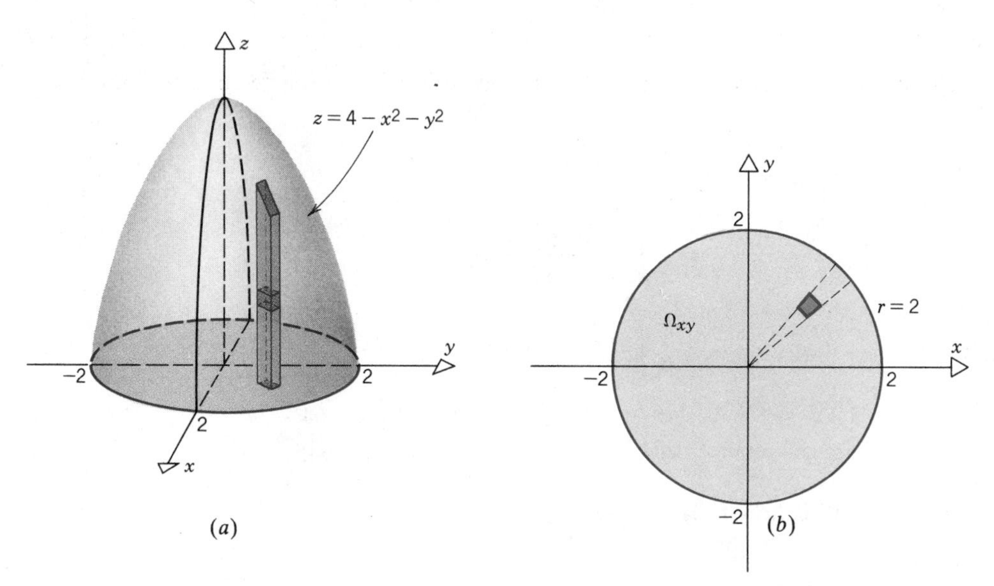

Figure 17.5.8

Figures 17.5.8a and 17.5.8b show the region Ω and its projection Ω_{xy} in the xy-plane, namely the circle with center at the origin and radius two. The moment of inertia I_z is given by Eq. 25:

$$\begin{aligned} I_z &= k \int\int\int_{\Omega} (x^2 + y^2)\, dV \\ &= k \int\int_{\Omega_{xy}} \int_0^{4-x^2-y^2} (x^2 + y^2)\, dz\, dA \\ &= k \int\int_{\Omega_{xy}} (x^2 + y^2)(4 - x^2 - y^2)\, dA. \end{aligned}$$

From this point on the calculation proceeds much more easily if we use polar coordinates (r, θ), so that $x^2 + y^2 = r^2$ and $dA = r\,dr\,d\theta$. Then we obtain

$$\begin{aligned} I_z &= k \int_0^{2\pi} \int_0^2 r^2(4 - r^2) r\,dr\,d\theta \\ &= k \int_0^{2\pi} \int_0^2 (4r^3 - r^5)\,dr\,d\theta \\ &= k \int_0^{2\pi} \left(r^4 - \frac{r^6}{6}\right)\Bigg|_0^2 d\theta \\ &= \frac{16}{3} k \int_0^{2\pi} d\theta = \frac{32\pi k}{3}. \quad \blacksquare \end{aligned}$$

PROBLEMS

In each of Problems 1 through 6, evaluate the given integral.

1. $\displaystyle\int_1^2 \int_0^3 \int_{-1}^2 xyz\,dy\,dz\,dx$

2. $\displaystyle\int_1^3 \int_0^1 \int_0^2 (xy - 2yz)\,dx\,dz\,dy$

3. $\displaystyle\int_0^3 \int_0^2 \int_0^{x+y} 2xy\,dz\,dy\,dx$

4. $\displaystyle\int_0^1 \int_{2y}^2 \int_0^{x-2y} (x - 2z)\,dz\,dx\,dy$

5. $\displaystyle\int_0^{\pi} \int_0^{\sin y} \int_0^{\cos y} (\pi - y)\,dx\,dz\,dy$

6. $\displaystyle\int_0^{\pi/2} \int_0^{\cos z} \int_0^{\sin z} (y - z)\,dx\,dy\,dz$

In each of Problems 7 through 10, sketch the projection of the given region Ω in the indicated plane. Also give a set of inequalities that describes each of the requested projections.

7. Ω is the region above the plane $z = 3$ and inside the sphere $x^2 + y^2 + z^2 = 16$. Find Ω_{xy}.

8. Ω is the region bounded by the coordinate planes, by the plane $x + 2z = 2$, and by the plane $3x + 2y + z = 12$. Find
   (a) Ω_{xy}; (b) Ω_{xz}.

9. Ω is the region bounded by the xz-plane, by the yz-plane, by the plane $4x + 2y + 3z = 12$, and by the cylinder $y = 4 - z^2$. Find
   (a) Ω_{xz}; (b) Ω_{yz}.

10. Ω is above the plane $y + z = 8$ and inside the sphere $x^2 + y^2 + z^2 = 36$. Find Ω_{xy}.

In each of Problems 11 through 16, set up—but do not evaluate—iterated integrals that yield the required quantity.

11. The mass of the region in the first octant bounded by the coordinate planes, the plane $y + 2z = 2$, and the plane $3x + 4y + 6z = 12$; the density is $\rho(x, y, z) = 6 - x - y$.

12. The moment of inertia about the z-axis of the tetrahedron bounded by the coordinate planes and the plane $2x + y + 3z = 6$; the density is $\rho(x, y, z) = k$, a constant.

13. The moment of inertia about a diameter of the base of a right circular cone of height h and base radius r; the density is $\rho(x, y, z) = k$, a constant.

14. The z-coordinate of the center of mass of the region above the xy-plane, below the surface $z = 4 - x^2$, and between the planes $y = 0$ and $y = 4$; the density is $\rho(x, y, z) = 6 - z$.

15. The mass of the region inside the elliptical cylinder $(x^2/a^2) + (y^2/b^2) = 1$ and between the planes $z = 0$ and $z = 2b - y$; the density is proportional to the square of the distance from the origin.

16. The first moment with respect to the xz-plane of the region bounded by one sheet of the hyperboloid $-(x^2/a^2) + (y^2/b^2) - (z^2/c^2) = 1$ and the plane $y = 2b$; the density is $\rho(x, y, z) = k$, a constant.

17. Check the result of Example 1 by integrating first with respect to x, then with respect to z, and finally with respect to y.

18. Find the centroid of the tetrahedron bounded by the coordinate planes and the plane $(x/a) + (y/b) + (z/c) = 1$, where a, b, and c are positive constants.

19. Set up appropriate integrals for solving the problem of Example 2 by projecting into the yz-plane.

20. A rectangular box of length l, width w, and height h has uniform density k.

    (a) Find the moment of inertia about a side of length l.

    (b) Find the moment of inertia about an axis passing through the center and parallel to the sides of length l.

21. Find the center of mass of a pyramid of uniform density k whose height is b and whose base is a square of side a.

22. The region above the xy-plane, below the cylinder $z = 9 - x^2$, and between the planes $y = 0$ and $y = 2$ is filled with a material whose density is $\rho(x, y, z) = kz$, where k is a constant.

    (a) Find the mass.

    (b) Find the center of mass.

23. The region above the xy-plane and below the paraboloid $z = 4 - x^2 - y^2$ is filled with a material of uniform density k. Find the center of mass.

24. The region Ω bounded by the cylinders $x = 4 - z^2$ and $y = 4 - z^2$, and by the planes $x = 0$ and $y = 0$, is filled with a material whose density is $\rho(x, y, z) = k|z|$, where k is a constant.

    (a) Find the total mass in Ω.

    (b) Find the moment of inertia about the z-axis.

    (c) Find the radius of gyration about the z-axis.

25. The region between the paraboloids $z = x^2 + y^2$ and $z = 8 - x^2 - y^2$ is filled with material whose density is $\rho(x, y, z) = k(8 - z)$, where k is a constant.

    (a) Find the total mass.

    (b) Find the center of mass.

26. A pyramid of height a and with a square base of side a is placed on a cube of side a, with the edges aligned as shown in Figure 17.5.9. If the cube has constant density m_1, and the pyramid has constant density m_2, find the center of mass of the combined structure.

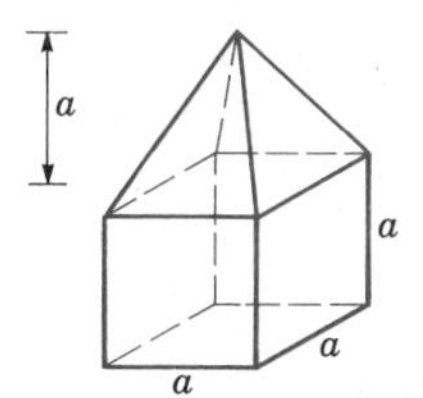

Figure 17.5.9

27. Two square bars of length l and cross-section of side a are joined to form a T, as shown in Figure 17.5.10. If the vertical bar has constant density m_1, and the crossbar has constant density m_2, find the center of mass of the combined structure.

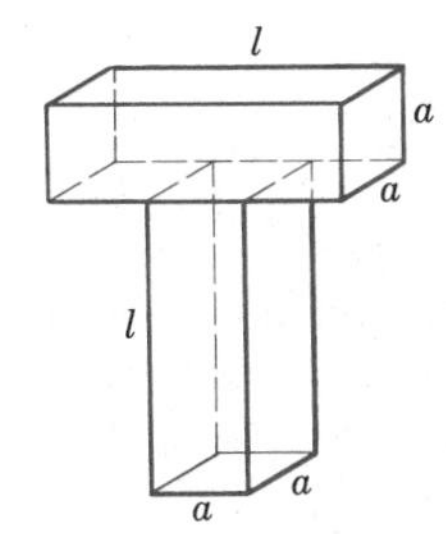

Figure 17.5.10

17.6 INTEGRATION USING CYLINDRICAL COORDINATES

Just as in two dimensions, it is sometimes more convenient to evaluate integrals in three dimensions using coordinate systems other than the rectangular coordinates that were used in Section 17.5. In this section we discuss and illustrate the use of **cylindrical coordinates.**

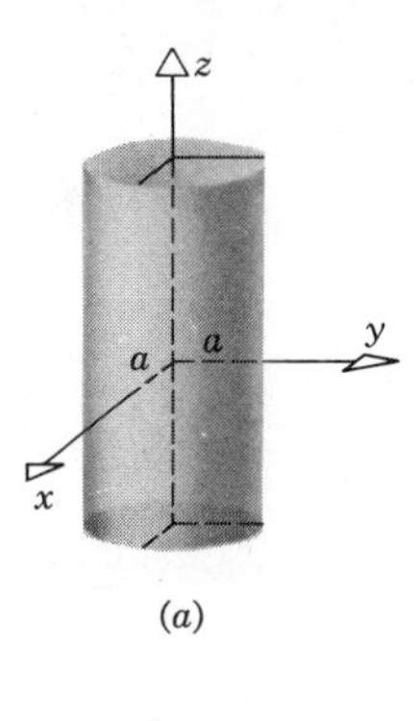

A cylindrical coordinate system consists of polar coordinates (r, θ) in a plane together with a third coordinate (z) measured along an axis perpendicular to the $r\theta$-plane (see Figure 17.6.1). In relating cylindrical and rectangular coordinates it

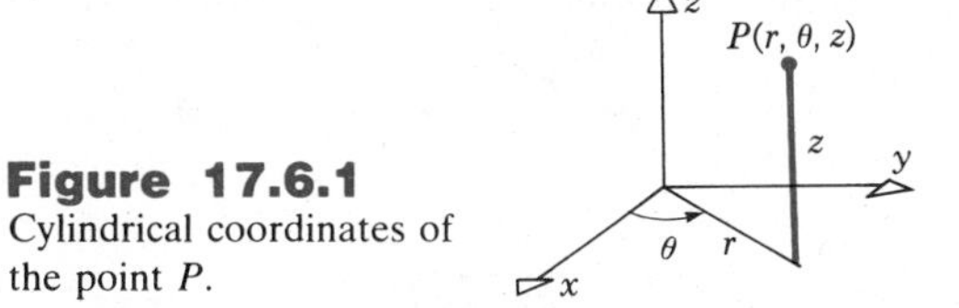

Figure 17.6.1
Cylindrical coordinates of the point P.

is customary to identify the $r\theta$-plane with the xy-plane. This means that the z-coordinate in the cylindrical coordinate system is the same as the z-coordinate in the rectangular system, and that the other coordinates are related by the familiar equations

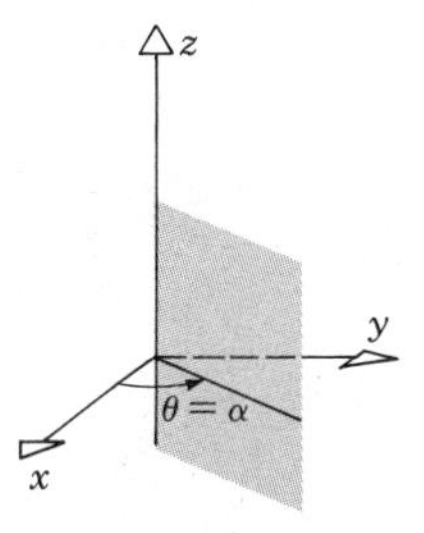

$$x = r \cos \theta, \qquad y = r \sin \theta. \tag{1}$$

As in Section 17.4 we require that $r \geq 0$ and that θ lie in an interval of length 2π.

The level surfaces in cylindrical coordinates corresponding to $r = a$, $\theta = \alpha$, and $z = c$ are shown in Figure 17.6.2a, 17.6.2b, and 17.6.2c. As shown in these figures:

$r = a$ is a right circular cylinder whose axis is the z-axis and every point of which is a distance a from this axis;

$\theta = \alpha$ is the half-plane whose edge is the z-axis, and such that its trace in the xy-plane is the ray making the angle α with the positive x-axis;

$z = c$ is the plane parallel to the xy-plane that contains the point c on the z-axis.

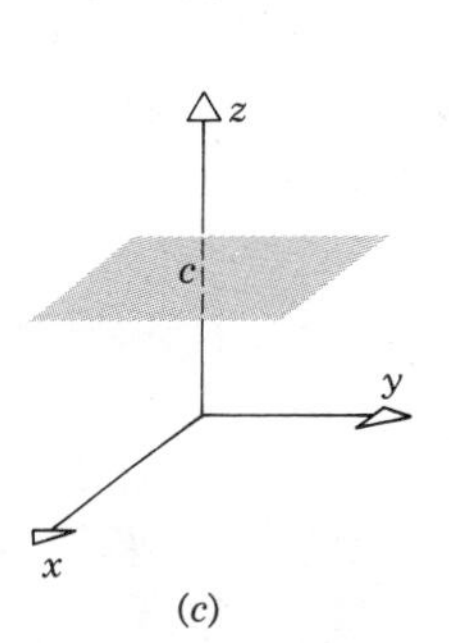

Figure 17.6.2
(a) The cylinder $r = a$.
(b) The half-plane $\theta = \alpha$.
(c) The plane $z = c$.

A point P in three-dimensional space can be identified as the point of intersection of a cylinder $r = a$, a half-plane $\theta = \alpha$, and a plane $z = c$, as indicated in Figure 17.6.3. We will always write the cylindrical coordinates of a point in the

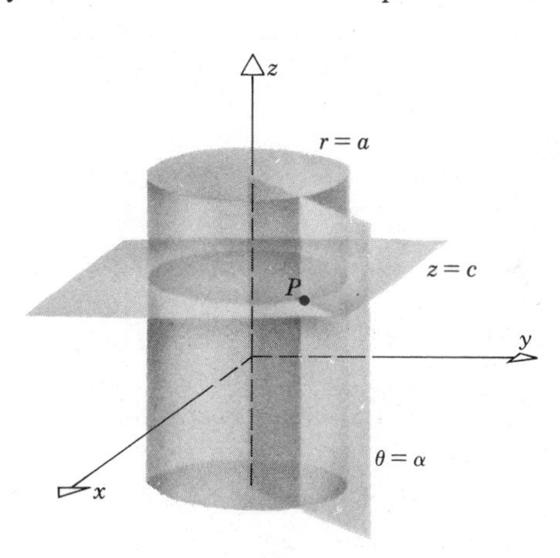

Figure 17.6.3 Coordinate surfaces in cylindrical coordinates.

order r, θ, z. Thus in cylindrical coordinates the point (a, α, c) is the point for which $r = a$, $\theta = \alpha$, and $z = c$.

When using cylindrical coordinates to evaluate triple integrals one thinks of the region of integration as being subdivided into small volume elements by means of a family of concentric cylinders $r = r_i$, a family of half-planes $\theta = \theta_j$ containing the z-axis, and a family of planes $z = z_k$ perpendicular to the z-axis. A typical volume element is shown in Figure 17.6.4; it is bounded by the cylinders $r = r_{i-1}$

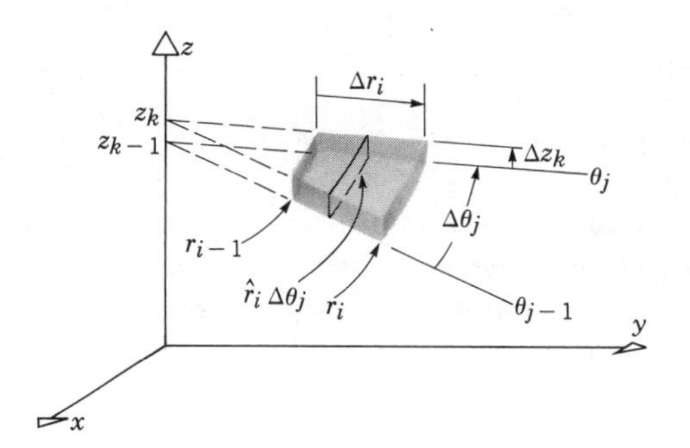

Figure 17.6.4 A volume element in cylindrical coordinates.

and $r = r_i$, the half-planes $\theta = \theta_{j-1}$ and $\theta = \theta_j$, and the planes $z = z_{k-1}$ and $z = z_k$. The volume ΔV_{ijk} of this element is

$$\Delta V_{ijk} = \hat{r}_i \, \Delta r_i \, \Delta\theta_j \, \Delta z_k, \tag{2}$$

where $\hat{r}_i = (r_i + r_{i-1})/2$ is the mean radius of the element. Note that $\hat{r}_i \, \Delta r_i \, \Delta\theta_j$ is the area element in polar coordinates that was found in Eq. 6 of Section 17.4. In the limit as the partition becomes finer and finer the volume element dV is given by

$$dV = r \, dr \, d\theta \, dz. \tag{3}$$

Thus, to evaluate a triple integral

$$\iiint_\Omega f(x, y, z) \, dV \tag{4}$$

by means of a set of iterated integrals in cylindrical coordinates, we must:

1. express the integrand $f(x, y, z)$ in terms of r, θ, and z by using Eqs. 1;
2. replace dV by $r \, dr \, d\theta \, dz$, where dr, $d\theta$, and dz may appear in any order;
3. attach limits to each integral, consistent with the order of integration chosen in (2), so as to integrate over the entire region Ω. Often one integrates first with respect to z, which corresponds to projecting Ω into the $r\theta$-plane.

EXAMPLE 1

Consider a solid right circular cone with altitude h and with a base of radius a. If the density of the cone is proportional to the distance from its axis of symmetry, find the center of mass.

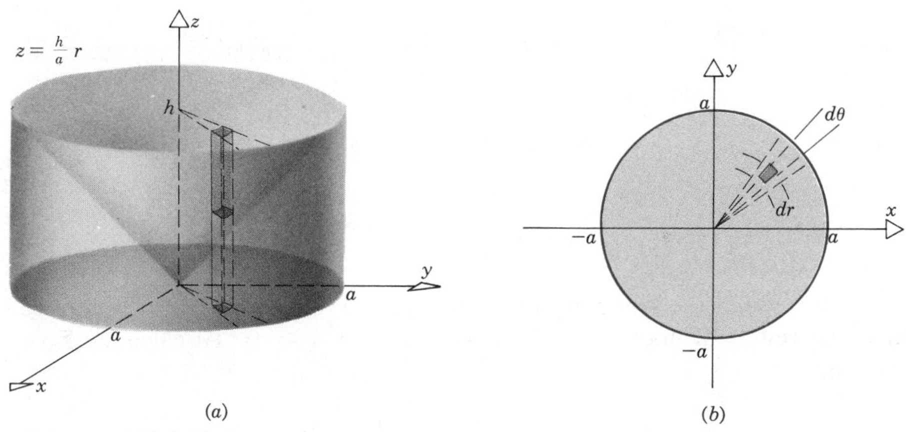

Figure 17.6.5

It is convenient to choose coordinates so that the vertex of the cone is at the origin and the axis of symmetry is the positive z-axis (see Figure 17.6.5a). Then the equation of the lateral surface of the cone is

$$\frac{x^2}{a^2} + \frac{y^2}{a^2} - \frac{z^2}{h^2} = 0,$$

or

$$\frac{r^2}{a^2} - \frac{z^2}{h^2} = 0,$$

or

$$z = \frac{h}{a} r,$$

with $0 \leq r \leq a$ and $0 \leq z \leq h$. The projection of the region of integration into the $r\theta$-plane is the circle with center at the origin and radius a (see Figure 17.6.5b). Thus the cone can be described by the inequalities

$$\frac{hr}{a} \leq z \leq h, \qquad 0 \leq r \leq a, \qquad 0 \leq \theta \leq 2\pi. \tag{5}$$

The density is given by $\rho(r, \theta, z) = kr$, where k is a constant, so the total mass is

$$
\begin{aligned}
M &= \iiint_{\Omega} \rho(r, \theta, z)\, dV \\
&= \int_0^{2\pi} \int_0^a \int_{hr/a}^h (kr)r\, dz\, dr\, d\theta \\
&= k \int_0^{2\pi} \int_0^a h\left(1 - \frac{r}{a}\right) r^2\, dr\, d\theta \\
&= kh \int_0^{2\pi} \left(\frac{r^3}{3} - \frac{r^4}{4a}\right)\Bigg|_0^a d\theta \\
&= \frac{kha^3}{12} \int_0^{2\pi} d\theta = \frac{\pi ka^3h}{6}.
\end{aligned} \tag{6}
$$

Because of the symmetry of the region and of the mass distribution we know that the center of mass is on the z-axis, so $\bar{x} = \bar{y} = 0$. To calculate $\bar{z}$ we must evaluate

$$
\begin{aligned}
L_{xy} &= \int_0^{2\pi} \int_0^a \int_{hr/a}^h (z)(kr)r\, dz\, dr\, d\theta \\
&= k \int_0^{2\pi} \int_0^a \frac{h^2}{2}\left(1 - \frac{r^2}{a^2}\right) r^2\, dr\, d\theta \\
&= \frac{kh^2}{2} \int_0^{2\pi} \left(\frac{r^3}{3} - \frac{r^5}{5a^2}\right)\Bigg|_0^a d\theta \\
&= \frac{kh^2}{2}\frac{2}{15}a^3 \int_0^{2\pi} d\theta = \frac{2}{15}\pi ka^3h^2.
\end{aligned} \tag{7}
$$

Consequently,

$$
\bar{z} = \frac{L_{xy}}{M} = \frac{2\pi ka^3h^2/15}{\pi ka^3h/6} = \frac{4}{5}h. \quad \blacksquare \tag{8}
$$

EXAMPLE 2

Consider the region Ω that is above the xy-plane, inside the sphere $r^2 + z^2 = a^2$, and outside the cylinder $r = a/2$. Find the centroid of Ω.

The region Ω is sketched in Figure 17.6.6a. The centroid must lie on the z-axis because of symmetry; consequently $\hat{x} = 0$, $\hat{y} = 0$, and we need only calculate $\hat{z}$. The projection $\Omega_{r\theta}$ of Ω in the $r\theta$-plane is the ring shown in Figure 17.6.6b. Thus Ω is described by the inequalities

$$
0 \le z \le \sqrt{a^2 - r^2}, \qquad \frac{a}{2} \le r \le a, \qquad 0 \le \theta \le 2\pi. \tag{9}
$$

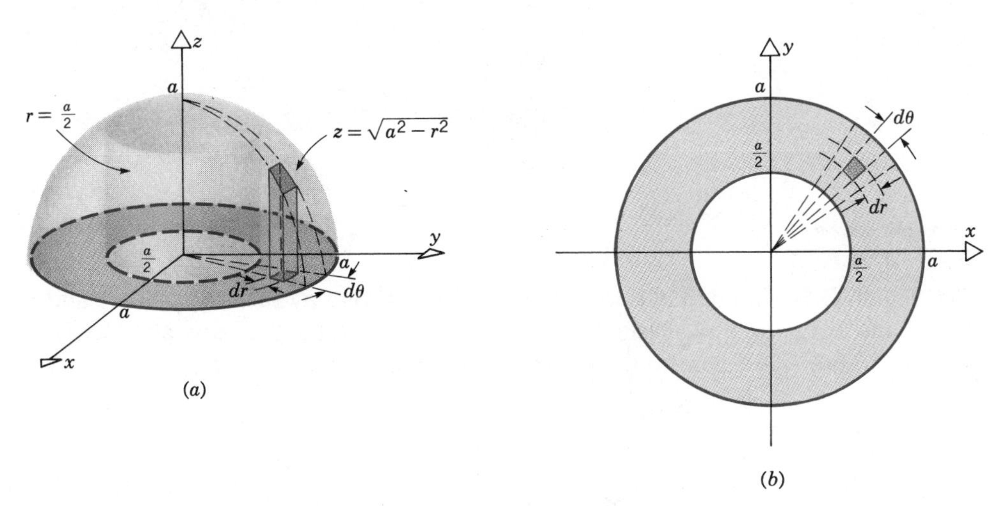

Figure 17.6.6

The volume V of Ω is

$$
\begin{aligned}
V &= \int_{a/2}^{a} \int_{0}^{2\pi} \int_{0}^{\sqrt{a^2-r^2}} r \, dz \, d\theta \, dr \\
&= \int_{a/2}^{a} \int_{0}^{2\pi} r \sqrt{a^2 - r^2} \, d\theta \, dr \\
&= 2\pi \int_{a/2}^{a} r \sqrt{a^2 - r^2} \, dr \\
&= 2\pi \left(-\frac{1}{2}\right) \left(\frac{2}{3}\right) (a^2 - r^2)^{3/2} \Big|_{a/2}^{a} \\
&= \frac{2\pi}{3} \left(\frac{3}{4} a^2\right)^{3/2} = \frac{\sqrt{3}\,\pi}{4} a^3. \qquad (10)
\end{aligned}
$$

Further, we have

$$
\begin{aligned}
\hat{L}_{xy} &= \iiint_{\Omega} z \, dV \\
&= \int_{a/2}^{a} \int_{0}^{2\pi} \int_{0}^{\sqrt{a^2-r^2}} zr \, dz \, d\theta \, dr \\
&= \frac{1}{2} \int_{a/2}^{a} \int_{0}^{2\pi} (a^2 - r^2) r \, d\theta \, dr \\
&= \pi \int_{a/2}^{a} (a^2 - r^2) r \, dr \\
&= \pi \left(\frac{a^2 r^2}{2} - \frac{r^4}{4}\right) \Big|_{a/2}^{a} \\
&= \frac{9\pi}{64} a^4. \qquad (11)
\end{aligned}
$$

Thus

$$\hat{z} = \frac{\hat{L}_{xy}}{V} = \frac{9\pi a^4/64}{\sqrt{3}\pi a^3/4} = \frac{3\sqrt{3}}{16}\, a. \quad \blacksquare \tag{12}$$

EXAMPLE 3

The region Ω is bounded below by the xy-plane, above by the cone $z = r$, and is inside the cylinder $r = 2 \sin \theta$. The material in Ω has density $\rho(r, \theta, z) = k \sin \theta$, where k is a constant. Find the total mass M in Ω.

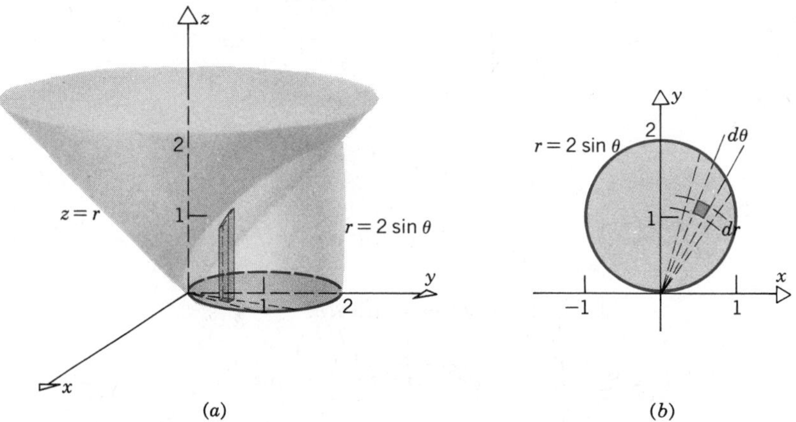

Figure 17.6.7

The region Ω is sketched in Figure 17.6.7*a*, and its projection $\Omega_{r\theta}$ in the $r\theta$-plane is the circle shown in Figure 17.6.7*b*. From the figure we see that Ω is described by the inequalities

$$0 \le z \le r, \qquad 0 \le r \le 2 \sin \theta, \qquad 0 \le \theta \le \pi. \tag{13}$$

Thus the total mass in Ω is

$$\begin{aligned} M &= \iiint_{\Omega} \rho(r, \theta, z)\, dV \\ &= k \int_0^{\pi} \int_0^{2\sin\theta} \int_0^{r} (\sin \theta) r\, dz\, dr\, d\theta. \end{aligned} \tag{14}$$

The integrals with respect to z and r are easy to evaluate, and yield the result

$$\begin{aligned} M &= k \int_0^{\pi} \int_0^{2\sin\theta} (\sin \theta) r^2\, dr\, d\theta \\ &= k \int_0^{\pi} (\sin \theta) \frac{(2 \sin \theta)^3}{3}\, d\theta \\ &= \frac{8k}{3} \int_0^{\pi} \sin^4 \theta\, d\theta. \end{aligned} \tag{15}$$

The remaining integral (15) can be evaluated by repeated use of the half-angle formula. Thus

$$\begin{aligned}
M &= \frac{8k}{3}\int_0^{\pi} (\sin^2\theta)^2\, d\theta \\
&= \frac{8k}{3}\int_0^{\pi} \left(\frac{1}{2} - \frac{1}{2}\cos 2\theta\right)^2 d\theta \\
&= \frac{8k}{3}\int_0^{\pi} \left(\frac{1}{4} - \frac{1}{2}\cos 2\theta + \frac{1}{4}\cos^2 2\theta\right) d\theta \\
&= \frac{8k}{3}\int_0^{\pi} \left[\frac{1}{4} - \frac{1}{2}\cos 2\theta + \frac{1}{4}\left(\frac{1}{2} + \frac{1}{2}\cos 4\theta\right)\right] d\theta \\
&= \frac{8k}{3}\int_0^{\pi} \left(\frac{3}{8} - \frac{1}{2}\cos 2\theta + \frac{1}{8}\cos 4\theta\right) d\theta \\
&= \frac{8k}{3}\left(\frac{3}{8}\theta - \frac{1}{4}\sin 2\theta + \frac{1}{32}\sin 4\theta\right)\Big|_0^{\pi} \\
&= \frac{8k}{3}\,\frac{3}{8}\,\pi = k\pi. \ \blacksquare
\end{aligned} \tag{16}$$

Gravitational attraction

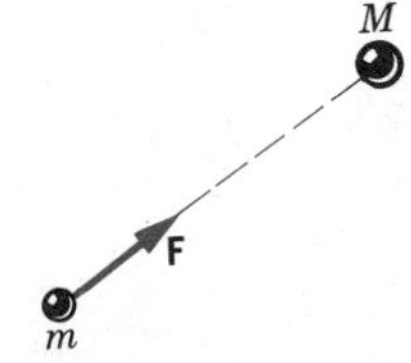

Figure 17.6.8 **F** is the gravitational force exerted by M on m.

According to Newton, the force **F** of gravitational attraction between two particles, of masses m and M, respectively, is directed along the line that joins them and has magnitude

$$\|\mathbf{F}\| = \frac{GmM}{R^2}, \tag{17}$$

where R is the distance between the particles, and $G = 6.67 \times 10^{-11}$ N-m^2/kg^2 is the universal gravitational constant (see Figure 17.6.8). Now we want to calculate the gravitational attraction exerted by the mass distributed over a region upon a particle of mass m.

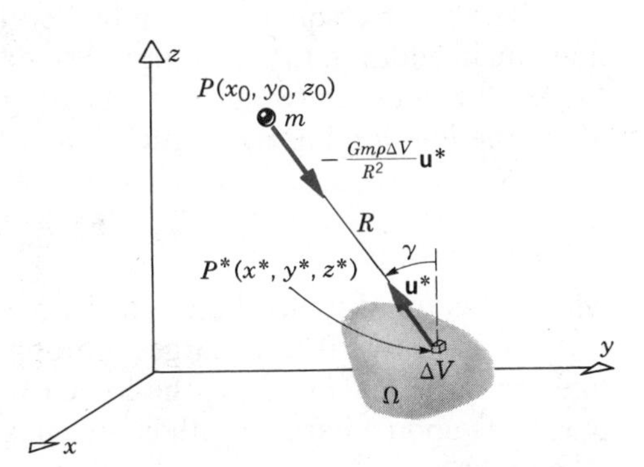

Figure 17.6.9 The gravitational force vector has magnitude $m(\rho\Delta V)/R^2$.

Suppose that the region Ω shown in Figure 17.6.9 contains material of density $\rho(x, y, z)$ and that a particle of mass m is located at the point $P(x_0, y_0, z_0)$. Consider a typical volume element ΔV in Ω. The force exerted by the mass in ΔV upon m is given approximately by

$$-\frac{Gm\rho(x^*, y^*, z^*)\,\Delta V}{R^2(x^*, y^*, z^*)}\,\mathbf{u}^*, \tag{18}$$

where P^* is an arbitrary point in ΔV with coordinates (x^*, y^*, z^*), $R(x^*, y^*, z^*)$ is the distance from P^* to P, $\mathbf{u}^*$ is the unit vector in the direction from P^* to P, and $\rho(x^*, y^*, z^*)\,\Delta V$ is approximately the mass of the element. Since

$$\overrightarrow{P^*P} = (x_0 - x^*)\mathbf{i} + (y_0 - y^*)\mathbf{j} + (z_0 - z^*)\mathbf{k},$$

it follows that

$$R(x^*, y^*, z^*) = \sqrt{(x_0 - x^*)^2 + (y_0 - y^*)^2 + (z_0 - z^*)^2}$$

and

$$\mathbf{u}^* = \frac{(x_0 - x^*)\mathbf{i} + (y_0 - y^*)\mathbf{j} + (z_0 - z^*)\mathbf{k}}{R(x^*, y^*, z^*)}.$$

Summing over all the elements in Ω and then passing to the limit in the usual way, we obtain

$$\mathbf{F} = -Gm\iiint_\Omega \frac{\rho(x, y, z)}{R^3(x, y, z)}\,[(x_0 - x)\mathbf{i} + (y_0 - y)\mathbf{j} + (z_0 - z)\mathbf{k}]\,dV. \tag{19}$$

Sometimes we need to calculate a single component of $\mathbf{F}$. For example, if we denote the z component of $\mathbf{F}$ by F_z, then

$$\begin{aligned} F_z &= -Gm\iiint_\Omega \frac{\rho(x, y, z)(z_0 - z)}{R^3(x, y, z)}\,dV \\ &= -Gm\iiint_\Omega \frac{\rho(x, y, z)\cos\gamma(x, y, z)}{R^2(x, y, z)}\,dV, \end{aligned} \tag{20}$$

where $\gamma(x, y, z)$ is the angle between the positive z-axis and the line from P^* to P.

The inverse square law embodied in Eq. 17 also governs certain other phenomena in addition to gravitation. For example, by Coulomb's law in electrostatics the force $\mathbf{F}$ between two particles bearing charges q and Q, respectively, is directed along the line joining the particles, and has magnitude

$$\|\mathbf{F}\| = K|q|\,\frac{|Q|}{R^2}, \tag{21}$$

where R is the distance between the particles, and $K \cong 9.0 \times 10^9$ N-m^2/C^2. The force is attractive if the charges are opposite in sign, and repulsive if they are of like sign. The magnitude of the force exerted by a body Ω having a charge density $\rho(x, y, z)$ upon a particle with charge q is then given by expressions similar to Eqs. 19 and 20.

EXAMPLE 4

Find the force **F** of gravitational attraction exerted by a right circular cylindrical shell of inner radius a, outer radius b, height h, and constant density k upon a particle of mass m located at the center of one base of the cylinder.

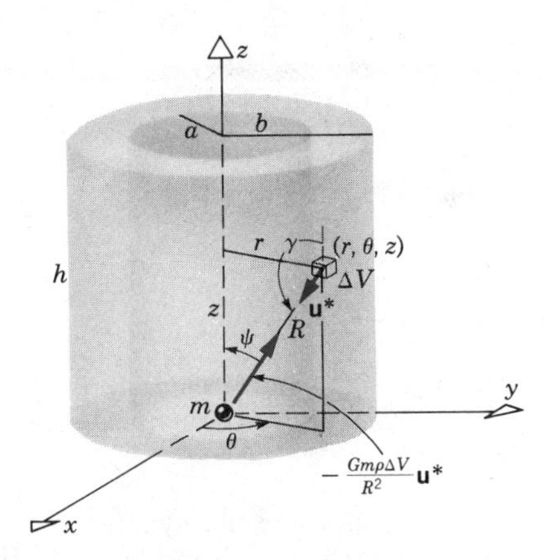

Figure 17.6.10 The gravitational force exerted by a cylindrical shell.

It is convenient to choose the coordinate system as in Figure 17.6.10 so that the positive z-axis is the axis of the cylinder and the mass m is at the origin. It follows from the symmetry of the problem that the x- and y-components of the attraction of the cylinder on the particle are zero, so we need only calculate the z-component, F_z. From Eq. 20 we have

$$F_z = Gmk \int\int\int_\Omega \frac{\cos \psi\,(r, \theta, z)}{R^2(r, \theta, z)}\, dV, \tag{22}$$

where $\psi = \pi - \gamma$. From Figure 17.6.10 we see that $R^2 = r^2 + z^2$ and $\cos \psi = z/R$. By integrating over the cylindrical shell we obtain

$$F_z = Gmk \int_a^b \int_0^h \int_0^{2\pi} \frac{z}{(r^2 + z^2)^{3/2}}\, r\, d\theta\, dz\, dr, \tag{23}$$

where we have chosen to integrate first with respect to θ, next with respect to z, and finally with respect to r. Thus

$$\begin{aligned} F_z &= 2\pi\, Gmk \int_a^b \int_0^h \frac{zr}{(z^2 + r^2)^{3/2}}\, dz\, dr \\ &= 2\pi\, Gmk \int_a^b \left[-\frac{r}{(z^2 + r^2)^{1/2}} \right]\Bigg|_0^h dr \\ &= 2\pi\, Gmk \int_a^b \left[1 - \frac{r}{(h^2 + r^2)^{1/2}} \right] dr \\ &= 2\pi\, Gmk\, [r - (h^2 + r^2)^{1/2}]\Big|_a^b \\ &= 2\pi\, Gmk\, [b - \sqrt{h^2 + b^2} - a + \sqrt{h^2 + a^2}]. \end{aligned} \tag{24}$$

As $a \to 0$, we have

$$F_z \to 2\pi\, Gmk\, [b + h - \sqrt{h^2 + b^2}],$$

the attraction due to a solid cylinder. ∎

PROBLEMS

1. Use cylindrical coordinates to find the volume of the region within the ellipsoid

$$\frac{x^2}{a^2} + \frac{y^2}{a^2} + \frac{z^2}{c^2} = 1.$$

2. Find the volume of the region Ω that is inside the sphere $x^2 + y^2 + z^2 = a^2$ and above the paraboloid $az = \sqrt{12}\,(x^2 + y^2)$.

3. Find the volume of the region inside the hyperboloid $(x^2/a^2) + (y^2/a^2) - (z^2/c^2) = 1$, and between the planes $z = 0$ and $z = h$.

4. A right circular cylinder of radius a and altitude h has a uniform density k. Find the moment of inertia
   (a) about the axis of symmetry;
   (b) about a diameter of one base.

5. The region Ω above the xy-plane, inside the cone $z = r$, and below the paraboloid $z = 6 - r^2$ is filled with a material of constant density k.
   (a) Find the center of mass.
   (b) Find the moment of inertia about the z-axis.

6. The material in the region Ω within both the cylinder $r = a/2$ and the sphere $r^2 + z^2 = a^2$ has density proportional to $|z|$. Find the total mass in Ω.

7. The quarter of the sphere $r^2 + z^2 \le a^2$ for which $0 \le \theta \le \pi/2$ is filled with material whose density is $\rho(r, \theta, z) = k \sin 2\theta$. Find the total mass.

8. The region Ω above the xy-plane, inside the cylinder $r = a$, and outside the cone $z = r$ is filled with a material having uniform density k. Find
   (a) the center of mass;
   (b) the moment of inertia about the z-axis;
   (c) the moment of inertia about the x-axis.

9. A circular cone of uniform density k has altitude h and base of radius a.
   (a) Find the center of mass.
   (b) Find the moment of inertia about the axis of symmetry.
   (c) Find the moment of inertia about a line through the vertex perpendicular to the axis of symmetry.

10. A right circular cone has altitude h, base of radius a, and density proportional to the square of the distance from the base.
    (a) Find the total mass.
    (b) Find the ratio a/h for the cone of greatest mass if the slant height l is fixed.

11. A solid hemisphere of radius a has uniform density k.
    (a) Find the center of mass.
    (b) Find the moment of inertia about the line perpendicular to the plane face and passing through its center.

12. A cylindrical hole of radius $a/2$ is bored through the center of a sphere of radius a and uniform density k. Find the moment of inertia of the remaining material about the axis of symmetry.

13. The region inside the cylinder $r = 2\cos\theta$, above the xy-plane, and below the cone $z = r$ is filled with a material of uniform density k. Find the center of mass.

14. The region Ω above the xy-plane and below the paraboloid $z = 4 - x^2 - y^2$ is filled with material having density $\rho(x, y, z) = kz$, where k is a constant. Find
    (a) the center of mass;
    (b) the moment of inertia about the z-axis.

15. The region Ω above the xy-plane and below the paraboloid $z = 9 - x^2 - y^2$ is filled with material having density $\rho(x, y, z) = k(12 - z)$, where k is a constant. Set up integrals in cylindrical coordinates that give

    (a) the mass;

    (b) the moment of inertia about the z-axis.

16. Consider the cylindrical shell $a \le r \le b$, $0 \le z \le h$, and $0 \le \theta \le 2\pi$, and suppose that its density is $\rho(r, \theta, z) = k/r$, where k is a constant.

    (a) Find the mass M of the shell.

    (b) Find I_z.

    (c) Find I_x.

    (d) Let $c = (a + b)/2$ and $\xi = (b - a)/2$. Note that c is the mean radius of the shell and ξ is one-half of its thickness. Neglect terms of higher than the first power in ξ, and thereby find approximate expressions for I_z and I_x when the shell is thin.

17. A hemispherical water tank of radius a is constructed so that the plane face is the base.

    (a) Find the volume of water in the tank when the depth is h.

    (b) Find the depth at which the tank is half full.

18. A right circular cylinder has constant density k, radius a, and height h.

    (a) Find the gravitational attraction exerted by this cylinder on a particle of mass m located at a distance l from the center of one base along the axial line of the cylinder.

    * (b) In the result of Part (a) let $h \to 0$ and $k \to \infty$ in such a way that hk is always equal to a constant μ, which can be identified as the mass per unit area of a uniform thin disk. In this way find the attraction of a disk on a particle of mass m located a distance l from the disk on the line perpendicular to the disk through its center.

    Hint: Use a Taylor expansion of $[a^2 + (l + h)^2]^{1/2}$ in terms of h.

    * (c) Verify the result of Part (b) by directly calculating the required force.

* 19. In the result of Example 4, Eq. 24, let $a \to b$ and $k \to \infty$ in such a way that $(b - a)k$ is always equal to a constant μ, which can be identified as the mass per unit area of a uniform thin shell. Thus find the gravitational attraction of a thin cylindrical shell on a mass m located at the center of one base of the shell.

    Hint: Use a Taylor expansion of $(a^2 + h^2)^{1/2}$ in terms of $a - b$.

20. (a) Consider the region in the first quadrant of the xz-plane bounded by the z-axis, the parabola $z = x^2$ for $0 \le x \le a$, and the line $z = a^2$. Find $\hat{z}_A$, and z-coordinate of the centroid of this region.

    (b) Rotate the region in Part (a) about the z-axis to form a solid of revolution, and find $\hat{z}_V$, the z-coordinate of the centroid of this solid. Observe that $\hat{z}_V > \hat{z}_A$; thus the centroid of the solid of revolution is higher than the centroid of the plane region.

* 21. (a) Consider the region bounded by the z-axis, the graph of $z = x^p$ for any $p > 0$ and for $0 \le x \le a$, and the line $z = a^p$. Proceed as in Problem 20, and find $\hat{z}_A$ and $\hat{z}_V$ for this case.

    (b) Use the results of Part (a) to show that $\hat{z}_V > \hat{z}_A$ for all $p > 0$; that is, the centroid of the solid of revolution is higher than the centroid of the plane region from which it is obtained. Find the value of p for which $\hat{z}_V/\hat{z}_A$ is maximum.

17.7 INTEGRATION USING SPHERICAL COORDINATES

In addition to rectangular and cylindrical coordinates, there is another coordinate system that is often useful in problems involving a spherical geometry. These coordinates are known as **spherical coordinates,** and are denoted by ρ, θ, and ϕ. The spherical coordinates of a given point P are shown in Figure 17.7.1.

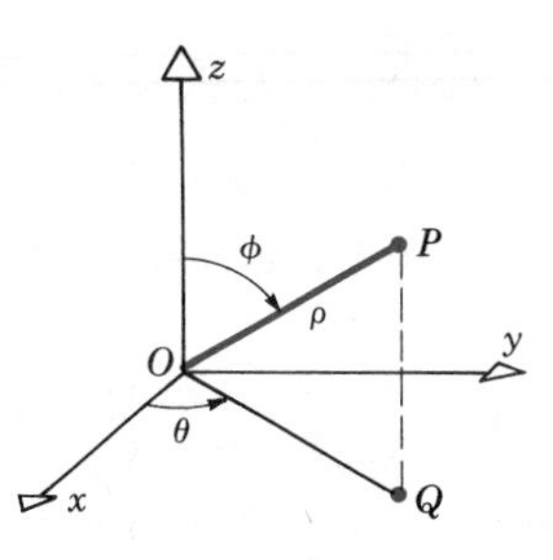

Figure 17.7.1 Spherical coordinates of the point P.

The radial coordinate ρ is the distance from the origin O to P; ρ is always nonnegative and is zero only if P coincides with the origin. The surface $\rho = a$ is the sphere with center at the origin and radius a shown in Figure 17.7.2.

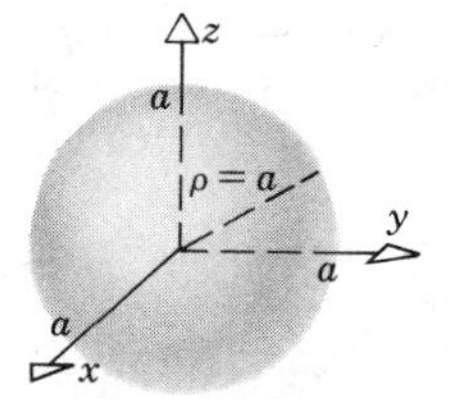

Figure 17.7.2 The sphere $\rho = a$.

The angle θ is the same as in polar and cylindrical coordinates; it is the angle between the positive x-axis and the projection OQ of OP into the xy-plane, as shown in Figure 17.7.1. We will restrict θ to the interval $0 \leq \theta \leq 2\pi$, or possibly to some other interval of length 2π. As in cylindrical coordinates, the surface $\theta = \alpha$ is the half-plane, shown in Figure 17.7.3, containing the z-axis whose trace in the xy-plane is the ray that makes the angle α with the positive x-axis.

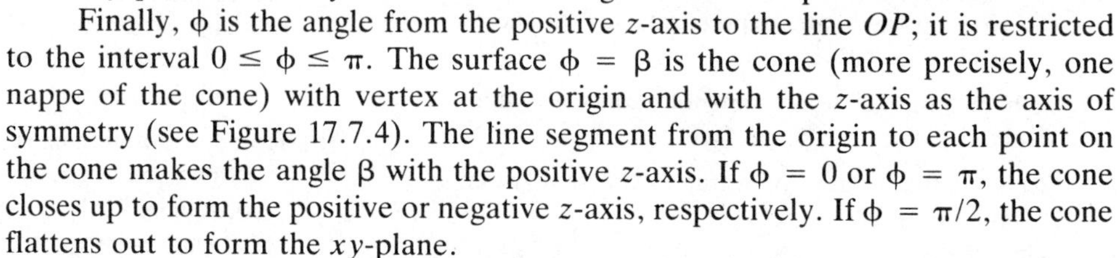

Finally, ϕ is the angle from the positive z-axis to the line OP; it is restricted to the interval $0 \leq \phi \leq \pi$. The surface $\phi = \beta$ is the cone (more precisely, one nappe of the cone) with vertex at the origin and with the z-axis as the axis of symmetry (see Figure 17.7.4). The line segment from the origin to each point on the cone makes the angle β with the positive z-axis. If $\phi = 0$ or $\phi = \pi$, the cone closes up to form the positive or negative z-axis, respectively. If $\phi = \pi/2$, the cone flattens out to form the xy-plane.

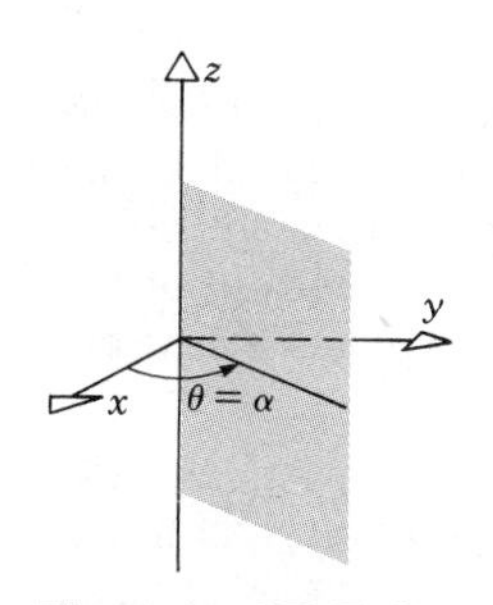

Figure 17.7.3 The half-plane $\theta = \alpha$.

A point P in three-dimensional space can be identified as the point of intersection of a sphere $\rho = a$, a half-plane $\theta = \alpha$, and a cone $\phi = \beta$, as indicated in Figure 17.7.5. We will always give the spherical coordinates of a point in the order

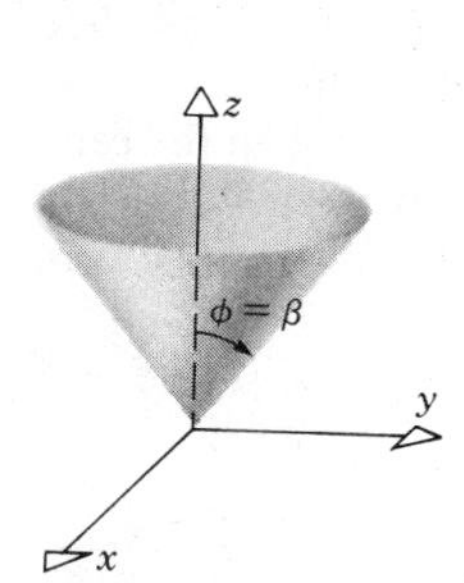

Figure 17.7.4 The cone $\phi = \beta$.

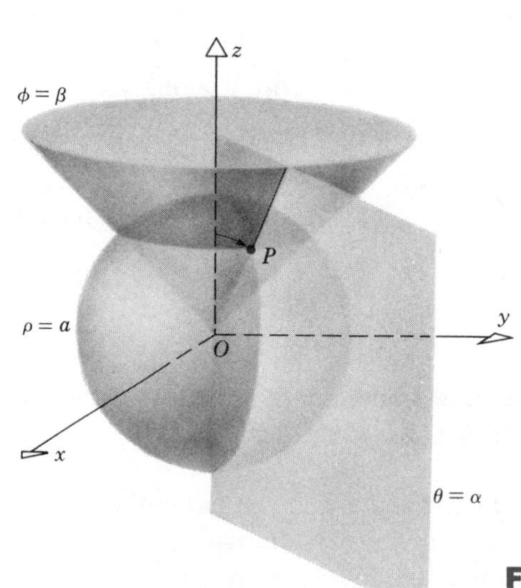

Figure 17.7.5 Coordinate surfaces in spherical coordinates.

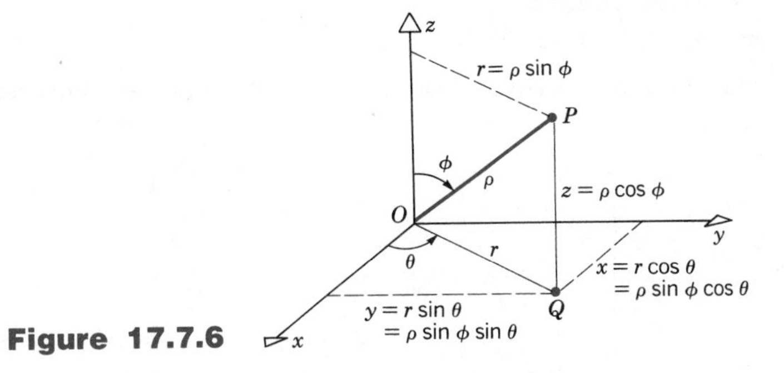

Figure 17.7.6

ρ, θ, ϕ. Thus in spherical coordinates the point (a, α, β) is the point for which $\rho = a$, $\theta = \alpha$, and $\phi = \beta$.

The relation between spherical coordinates (ρ, θ, ϕ) and cylindrical coordinates (r, θ, z) can be seen from Figure 17.7.6. We have

$$\begin{cases} r = \rho \sin\phi, \\ \theta = \theta, \\ z = \rho \cos \phi, \end{cases} \tag{1a}$$

and

$$\begin{cases} \rho = (r^2 + z^2)^{1/2}, \\ \theta = \theta, \\ \tan \phi = \dfrac{r}{z}, \qquad 0 \le \phi \le \pi. \end{cases} \tag{1b}$$

Since

$$x = r \cos \theta, \qquad y = r \sin \theta, \tag{2}$$

the relation between spherical coordinates and rectangular coordinates (x, y, z) is given by

$$\begin{cases} x = \rho \sin \phi \cos \theta, \\ y = \rho \sin \phi \sin \theta, \\ z = \rho \cos \phi, \end{cases} \tag{3a}$$

and

$$\begin{cases} \rho = (x^2 + y^2 + z^2)^{1/2}, \\ \tan \theta = \dfrac{y}{x}, \qquad 0 \le \theta \le 2\pi, \\ \cos \phi = \dfrac{z}{(x^2 + y^2 + z^2)^{1/2}}, \qquad 0 \le \phi \le \pi. \end{cases} \tag{3b}$$

EXAMPLE 1

Find the rectangular coordinates of the point P whose spherical coordinates are $\rho = 4$, $\theta = \pi/3$, and $\phi = \pi/6$ (see Figure 17.7.7).

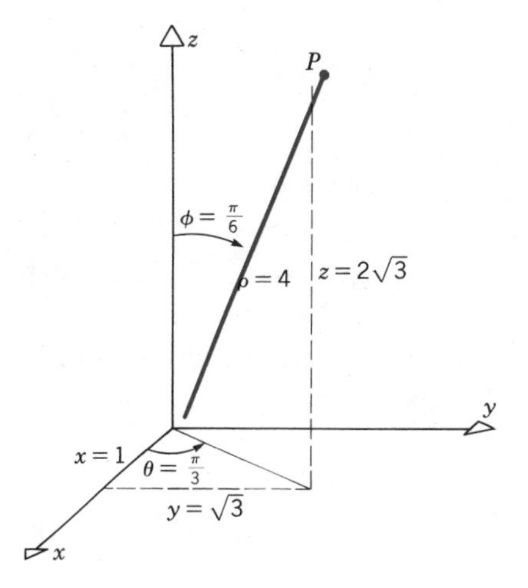

Figure 17.7.7

By substituting the given values of ρ, θ, and ϕ in Eqs. 3(a) we obtain

$$x = (4)\left(\frac{1}{2}\right)\left(\frac{1}{2}\right) = 1,$$

$$y = (4)\left(\frac{1}{2}\right)\left(\frac{\sqrt{3}}{2}\right) = \sqrt{3},$$

$$z = (4)\left(\frac{\sqrt{3}}{2}\right) = 2\sqrt{3}. \blacksquare$$

EXAMPLE 2

Find the spherical coordinates of the point P whose rectangular coordinates are $x = 2$, $y = -2$, and $z = 4$ (see Figure 17.7.8).

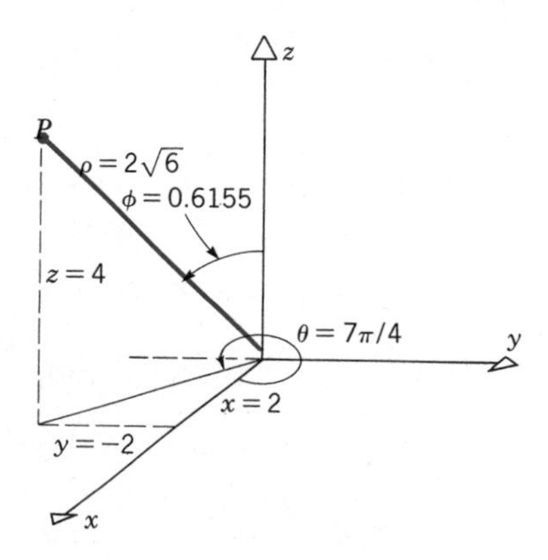

Figure 17.7.8

Here we need to use Eqs. 3(b). From the first of these equations we obtain

$$\rho = \sqrt{(2)^2 + (-2)^2 + (4)^2} = 2\sqrt{6}.$$

From the second we have

$$\tan \theta = \frac{-2}{2} = -1;$$

further, since $x > 0$ and $y < 0$, θ must lie in the fourth quadrant, so $\theta = 7\pi/4$. Finally, from the last of Eqs. 3(b) we find that

$$\cos \phi = \frac{4}{2\sqrt{6}} = \sqrt{\frac{2}{3}} \cong 0.8165,$$

and therefore $\phi \cong 0.6155$ radians. ∎

The angular coordinates θ and ϕ of a point P on a sphere can be related to longitude and latitude, respectively, on the surface of the earth. Referring to Figure 17.7.9, suppose that the positive z-axis is drawn from the center of the earth through

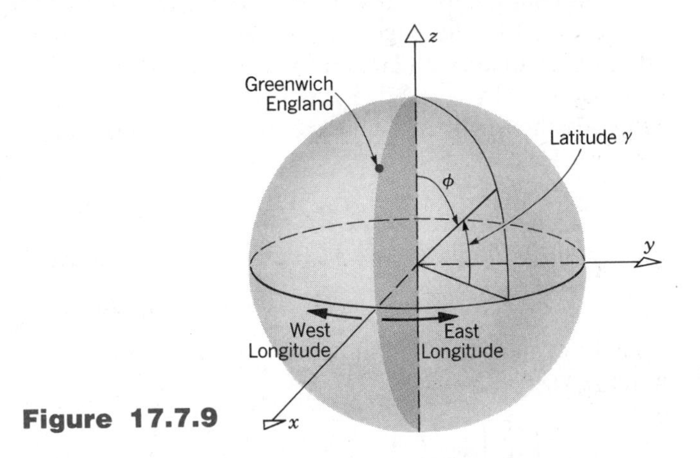

Figure 17.7.9

the north pole, and that the positive x-axis is chosen so that the xz-plane (with $x > 0$) passes through the Royal Observatory in Greenwich, England. Then θ can be identified with longitude, provided that θ is restricted to the interval $[-\pi, \pi]$; east longitude corresponds to positive values of θ and west longitude to negative values. The angle θ is also referred to as the azimuth. On the other hand, the angle ϕ, measured down from the North Pole, is related to latitude, which is measured from the equator. If γ is the latitude, where $-\pi/2 \leq \gamma \leq \pi/2$, then $\phi = (\pi/2) - \gamma$.

Now let us consider the use of spherical coordinates to integrate over a three-dimensional region Ω. We think of Ω as being partitioned into small elements by means of a family of concentric spheres $\rho = \rho_i$, a family of half-planes $\theta = \theta_j$ through the z-axis, and a family of cones $\phi = \phi_k$. A typical volume element is shown in Figure 17.7.10; it is bounded by the spheres $\rho = \rho_{i-1}$ and $\rho = \rho_i$, by the

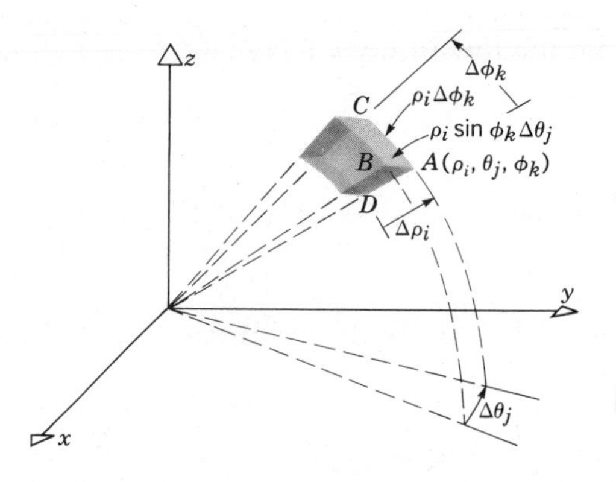

Figure 17.7.10 A volume element in spherical coordinates.

half-planes $\theta = \theta_{j-1}$ and $\theta = \theta_j$, and by the cones $\phi = \phi_{k-1}$ and $\phi = \phi_k$. As usual, we let

$$\Delta\rho_i = \rho_i - \rho_{i-1}, \qquad \Delta\theta_j = \theta_j - \theta_{j-1}, \qquad \Delta\phi_k = \phi_k - \phi_{k-1}. \tag{4}$$

It is then easy to estimate the volume ΔV_{ijk} of the element in the following way. For $\Delta\rho_i$, $\Delta\theta_j$, and $\Delta\phi_k$ sufficiently small the element differs only negligibly from a rectangular box; thus its volume can be determined as the product of the lengths of three contiguous edges, for example, the edges AB, AC, and AD shown in Figure 17.7.10. The edge AD is a straight line segment and has length $\Delta\rho_i$. The edge AC is an arc of a circle of radius ρ_i that subtends the angle $\Delta\phi_k$, so its length is $\rho_i\Delta\phi_k$. Finally, the edge AB is an arc of a circle of radius $r = \rho_i \sin \phi_k$ that subtends the angle $\Delta\theta_j$, and therefore its length is $\rho_i \sin \phi_k \, \Delta\theta_j$. Multiplying the lengths of these sides together, we obtain

$$\Delta V_{ijk} \cong \rho_i^2 \sin \phi_k \, \Delta\rho_i \Delta\theta_j \Delta\phi_k. \tag{5}$$

Finally, in the limit as $\Delta\rho_i$, $\Delta\theta_j$, and $\Delta\phi_k$ all approach zero, we have

$$dV = \rho^2 \sin \phi \, d\rho \, d\theta \, d\phi. \tag{6}$$

Thus to evaluate a triple integral over a region Ω using spherical coordinates we write

$$\iiint_\Omega f(\rho, \theta, \phi)\, dV = \iiint_\Omega f(\rho, \theta, \phi)\rho^2 \sin \phi \, d\rho \, d\theta \, d\phi. \tag{7}$$

Note that not only must dV be expressed in terms of $d\rho$, $d\theta$, and $d\phi$ by Eq. 6, but the integrand f must be written in terms of ρ, θ, and ϕ, and the limits on the integrals must be chosen so as to cover the region Ω.

EXAMPLE 3

The density of a solid hemisphere of radius a is proportional to the distance from the plane face. Find the center of mass of the hemisphere.

Let us choose the coordinate system as shown in Figure 17.7.11. Then the plane face of the hemisphere lies in the xy-plane, its center is at the origin, and

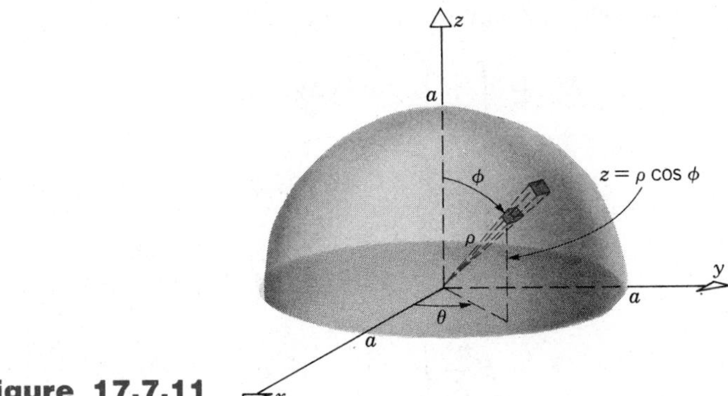

Figure 17.7.11

the hemisphere lies on the side of the xy-plane for which $z \geq 0$. The density* is given by $\mu(x, y, z) = kz = k\,\rho\cos\phi$, where k is a proportionality constant. Because of the symmetry of the problem the center of mass must lie on the z-axis, so it is only necessary to calculate $\bar{z}$. For this purpose we need to determine the total mass M of the hemisphere and its first moment L_{xy} with respect to the xy-plane. The hemisphere is described by the inequalities

$$0 \leq \rho \leq a, \qquad 0 \leq \theta \leq 2\pi, \qquad 0 \leq \phi \leq \frac{\pi}{2}, \tag{8}$$

so the total mass M is given by

$$\begin{aligned} M &= \iiint_\Omega \mu(x, y, z)\,dV \\ &= \iiint_\Omega kz\,dV \\ &= \int_0^{\pi/2}\int_0^{2\pi}\int_0^a (k\rho\cos\phi)\rho^2\sin\phi\,d\rho\,d\theta\,d\phi \\ &= k\int_0^{\pi/2}\int_0^{2\pi}\int_0^a \rho^3\cos\phi\sin\phi\,d\rho\,d\theta\,d\phi. \end{aligned} \tag{9}$$

Upon evaluating these iterated integrals we obtain

$$\begin{aligned} M &= k\frac{a^4}{4}\int_0^{\pi/2}\int_0^{2\pi}\sin\phi\cos\phi\,d\theta\,d\phi \\ &= k\frac{a^4}{4}2\pi\int_0^{\pi/2}\sin\phi\cos\phi\,d\phi \\ &= \frac{\pi}{2}ka^4\frac{\sin^2\phi}{2}\bigg|_0^{\pi/2} = \frac{\pi}{4}ka^4. \end{aligned} \tag{10}$$

*Up to now we have followed common practice in using ρ to denote density. In spherical coordinates, however, ρ is almost universally used for the radial coordinate. Therefore, in this section only, we use μ to denote density.

In a similar way the first moment L_{xy} is given by

$$\begin{aligned}
L_{xy} &= \iiint_{\Omega} z\,\mu(x, y, z)\,dV \\
&= \int_0^{\pi/2}\int_0^{2\pi}\int_0^a (\rho\cos\phi)(k\rho\cos\phi)(\rho^2\sin\phi\,d\rho\,d\theta\,d\phi) \\
&= k\int_0^{\pi/2}\int_0^{2\pi}\int_0^a \rho^4\cos^2\phi\sin\phi\,d\rho\,d\theta\,d\phi \\
&= k\frac{a^5}{5}\int_0^{\pi/2}\int_0^{2\pi}\cos^2\phi\sin\phi\,d\theta\,d\phi \\
&= k\frac{a^5}{5}2\pi\int_0^{\pi/2}\cos^2\phi\sin\phi\,d\phi \\
&= \frac{2}{5}\pi k a^5\left(-\frac{\cos^3\phi}{3}\right)\Bigg|_0^{\pi/2} = \frac{2}{15}\pi k a^5.
\end{aligned} \tag{11}$$

Thus

$$\bar{z} = \frac{L_{xy}}{M} = \frac{2\pi k a^5/15}{\pi k a^4/4} = \frac{8}{15}a. \quad ∎$$

EXAMPLE 4

The region Ω bounded below by the plane $z = b$ and above by the sphere $\rho = a$ is filled with a material whose density is inversely proportional to the distance from the origin. Find the total mass M in Ω.

The region Ω is shown in Figure 17.7.12. In spherical coordinates the plane $z = b$ has the equation $\rho = b/\cos\phi$. Thus Ω is described by the inequalities

$$\frac{b}{\cos\phi} \le \rho \le a, \qquad 0 \le \theta \le 2\pi, \qquad 0 \le \phi \le \arccos\frac{b}{a}. \tag{13}$$

The density is $\mu(\rho, \theta, \phi) = k/\rho$, where k is a constant, so the total mass M is given by

$$\begin{aligned}
M &= \iiint_{\Omega} \mu(\rho, \theta, \phi)\,dV \\
&= \int_0^{\arccos(b/a)}\int_0^{2\pi}\int_{b/\cos\phi}^a \frac{k}{\rho}\rho^2\sin\phi\,d\rho\,d\theta\,d\phi \\
&= k\int_0^{\arccos(b/a)}\int_0^{2\pi}\int_{b/\cos\phi}^a \rho\sin\phi\,d\rho\,d\theta\,d\phi.
\end{aligned} \tag{14}$$

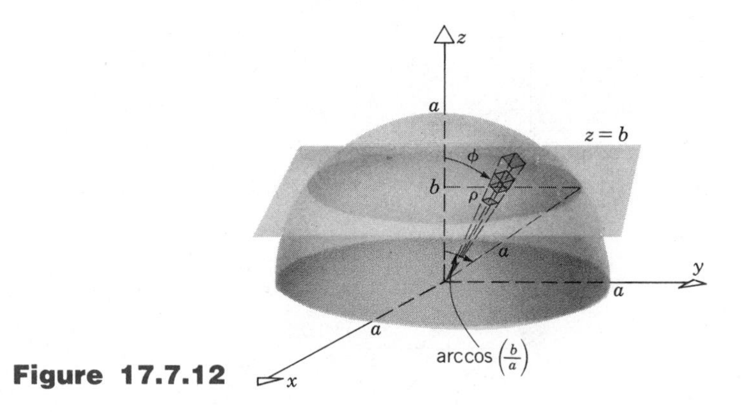

Figure 17.7.12

Evaluating the integrals in Eq. 14, we obtain

$$\begin{aligned}
M &= \frac{k}{2}\int_0^{\operatorname{arc\,cos}(b/a)}\int_0^{2\pi}\left(a^2 - \frac{b^2}{\cos^2\phi}\right)\sin\phi\,d\theta\,d\phi \\
&= \frac{k}{2}2\pi\int_0^{\operatorname{arc\,cos}(b/a)}(a^2 - b^2\cos^{-2}\phi)\sin\phi\,d\phi \\
&= k\pi\left(-a^2\cos\phi - \frac{b^2}{\cos\phi}\right)\Bigg|_0^{\operatorname{arc\,cos}(b/a)} \\
&= k\pi\left(-a^2\frac{b}{a} + a^2 - b^2\frac{a}{b} + b^2\right) \\
&= k\pi(a - b)^2. \qquad (15)
\end{aligned}$$

EXAMPLE 5

Find the force **F** of gravitational attraction exerted by a solid right circular cone of constant density k, height h, and semivertex angle α upon a particle of mass m located at the vertex of the cone.

It is convenient to choose the coordinate system as shown in Figure 17.7.13 so that the vertex of the cone is at the origin and its axis is the positive z-axis. Then the cone is described by the inequalities

$$0 \le \rho \le h\sec\phi, \qquad 0 \le \theta \le 2\pi, \qquad 0 \le \phi \le \alpha. \qquad (16)$$

Because of symmetry we know that the x- and y-components of the attractive force are zero. By Eq. 20 of Section 17.6 and Figure 17.7.13, the z-component is given by

$$F_z = Gmk\iiint_\Omega \frac{\cos\psi}{R^2}\,dV,$$

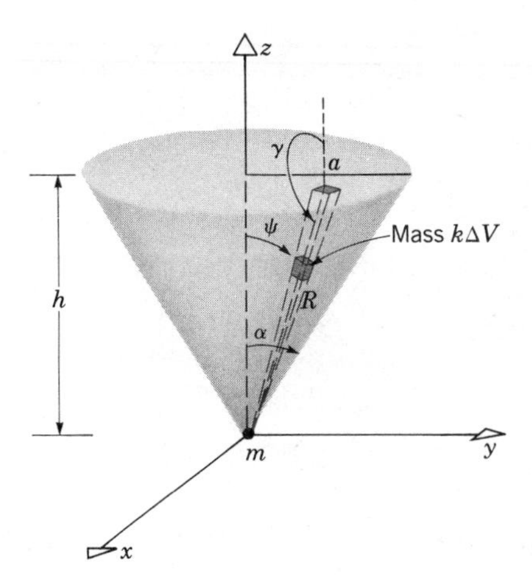

Figure 17.7.13

where $\psi = \pi - \gamma$. Since $\psi = \phi$ and $R = \rho$, this integral reduces to

$$\begin{aligned} F_z &= Gmk \int_0^{\alpha} \int_0^{2\pi} \int_0^{h \sec \phi} \cos \phi \sin \phi \, d\rho \, d\theta \, d\phi \\ &= Gmk \int_0^{\alpha} \int_0^{2\pi} \cos \phi \sin \phi \, [h \sec \phi - 0] \, d\theta \, d\phi \\ &= Gmkh \int_0^{\alpha} \int_0^{2\pi} \sin \phi \, d\theta \, d\phi \\ &= 2\pi Gmkh(-\cos \phi) \Big|_0^{\alpha} \\ &= 2\pi Gmkh(1 - \cos \alpha). \end{aligned} \tag{17}$$

We can also express F_z in terms of the height h and base radius a of the cone. From Figure 17.7.13, we see that $\cos \alpha = h/\sqrt{a^2 + h^2}$, and therefore

$$F_z = 2\pi Gmkh \left(1 - \frac{h}{\sqrt{a^2 + h^2}}\right). \quad \blacksquare \tag{18}$$

PROBLEMS

In each of Problems 1 through 6, find the rectangular coordinates (x, y, z) of the point with the given spherical coordinates (ρ, θ, ϕ).

1. $\left(2, \frac{\pi}{4}, \frac{\pi}{3}\right)$ **2.** $\left(4, \frac{3\pi}{2}, \frac{\pi}{6}\right)$ **3.** $\left(2, \frac{\pi}{6}, \frac{2\pi}{3}\right)$

4. $\left(3, \frac{4\pi}{3}, \frac{\pi}{3}\right)$ **5.** $\left(2, \frac{3\pi}{4}, \frac{3\pi}{4}\right)$ **6.** $\left(4, \frac{\pi}{6}, \frac{\pi}{2}\right)$

In each of Problems 7 through 12, find the spherical coordinates (ρ, θ, ϕ) of the point with the given rectangular coordinates (x, y, z).

7. $(1, -1, \sqrt{2})$ **8.** $(2, 0, -2\sqrt{3})$

9. $\left(\frac{\sqrt{3}}{2}, \frac{3}{2}, -1\right)$ **10.** $(\sqrt{2}, \sqrt{2}, -2\sqrt{3})$

11. $\left(\frac{3\sqrt{3}}{2\sqrt{2}}, \frac{3}{2\sqrt{2}}, \frac{3}{\sqrt{2}}\right)$

12. $\left(-\frac{3}{2}, -\frac{\sqrt{3}}{2}, -1\right)$

In each of Problems 13 through 20, find the equation in spherical coordinates that corresponds to the given equation in rectangular or cylindrical coordinates.

13. $x^2 + y^2 + (z - c)^2 = c^2$

14. $(x - a)^2 + y^2 + z^2 = a^2$

15. $\frac{x^2}{4} + \frac{y^2}{4} - \frac{z^2}{9} = 0, \quad z \geq 0$

16. $x^2 + y^2 = 9$

17. $x = 2y, \quad y \geq 0$

18. $r = 4$

19. $r = 3z$

20. $r = 2 \cos \theta$

In each of Problems 21 through 25, describe the graph of the given equation.

21. $\rho \cos \phi = b$

22. $\rho = 2a \cos \phi$

23. $\rho \sin \phi = b$

24. $\rho \sin \phi = 2a \sin \theta$

25. $(\cos \theta + \sin \theta)\rho \sin \phi = a$

26. Find the centroid of a hemisphere of radius a.

27. Find the moment of inertia of a solid hemisphere of radius a and uniform density k
 (a) about the axis of symmetry;
 (b) about a diameter of the plane face.

28. Find the moment of inertia about a diameter of a spherical shell with inner radius a, outer radius b, and uniform density k.

29. A spherical shell of inner radius a and outer radius b has density $\mu(\rho, \theta, \phi) = k/\rho$, where k is a constant. Find
 (a) the mass of the shell;
 (b) the moment of inertia about a diameter.

30. The region that is inside both the sphere $\rho = a$ and the cone $\phi = \pi/6$ has constant density k. Find
 (a) the mass of the region;
 (b) the center of mass.

31. The region inside both the sphere $\rho = 2a \cos \phi$ and the cone $\phi = \alpha$, where $0 < \alpha < \pi/2$, is filled with a material whose density is proportional to z. Find the total mass in the region.

32. The spherical wedge $0 \leq \rho \leq a$, $\pi/4 \leq \theta \leq \pi/3$, $0 \leq \phi \leq \pi$ is filled with a material whose density is $\mu(\rho, \theta, \phi) = k|\cos \phi|$, where k is a constant. Find
 (a) the total mass;
 (b) the moment of inertia about the z-axis.

33. The hemispherical shell $a/2 \leq \rho \leq a$, $z \geq 0$, has the density $\mu(\rho, \theta, \phi) = k/\rho$, where ρ is a constant. Find
 (a) the total mass;
 (b) the center of mass;
 (c) the moment of inertia about the z-axis.

34. Find the volume of the region within both of the spheres $\rho = 1$ and $\rho = 2 \cos \phi$.

35. Find the mass of a sphere of radius a whose density is proportional to the distance from a given point on its surface.

 Hint: Choose coordinates so that the given point is the origin, and a diameter of the sphere lies along the positive z-axis.

36. (a) A spherical planet of radius a has an atmosphere whose density is $\mu = \mu_0 e^{-\alpha h}$, where h is the altitude above the surface of the planet, μ_0 is the density at sea level, and α is a constant of proportionality. Find the total mass of this planet's atmosphere.
 (b) If we assume that the density is proportional to the pressure, then the density of the earth's atmosphere decays exponentially, as assumed in Part (a), with $\mu_0 \cong 1.20$ kg/m^3 and $\alpha \cong 0.116$ km^{-1}. If the earth is also assumed to be a sphere of radius 6370 km, find the mass of the earth's atmosphere.

* 37. **Gravitational attraction by a sphere.** In this problem we indicate how to calculate the force **F** of gravitational attraction by a uniform sphere of radius a and constant density k upon a particle of mass m located at a distance l from the center of the sphere, where $l > a$ (see Figure 17.7.14).
 (a) With coordinates chosen as in Figure 17.7.14, $F_x = F_y = 0$ and it is necessary to calculate only F_z. Show that

$$\|\mathbf{F}\| = |F_z| = Gmk \int_0^a \int_0^\pi \int_0^{2\pi} \frac{\cos \psi}{R^2} \rho^2 \sin \phi \, d\theta \, d\phi \, d\rho,$$

and, by integrating with respect to θ, that

$$\|\mathbf{F}\| = 2\pi Gmk \int_0^a \int_0^\pi \frac{\cos\psi}{R^2} \rho^2 \sin\phi \, d\phi \, d\rho. \tag{i}$$

(b) To evaluate the integral in Eq. (i) it is necessary to express ψ and R in terms of ρ and ϕ. This can be done by using the law of cosines

$$R^2 = l^2 + \rho^2 - 2l\rho \cos\phi \tag{ii}$$

and the relation

$$l = R \cos\psi + \rho \cos\phi, \tag{iii}$$

which can be seen at once from Figure 17.7.14. However, the resulting integral is difficult, and it is simpler to evaluate $\|\mathbf{F}\|$ by changing the variable of integration from ϕ to R in the inner integral in Eq. (i). Use Eqs. (ii) and (iii) to show that

$$\cos\psi = \frac{l^2 + R^2 - \rho^2}{2lR}. \tag{iv}$$

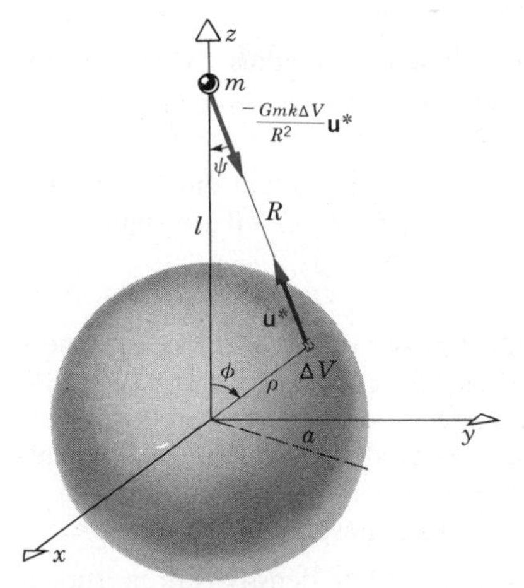

Figure 17.7.14 The gravitational force exerted by a solid sphere.

(c) From Eq. (ii) show that, for fixed ρ,

$$\sin\phi \, d\phi = \frac{R}{l\rho} \, dR, \tag{v}$$

and observe that $R = l - \rho$ when $\phi = 0$ and $R = l + \rho$ when $\phi = \pi$. Then show that

$$\|\mathbf{F}\| = \frac{\pi Gmk}{l^2} \int_0^a \int_{l-\rho}^{l+\rho} \rho \left[1 + \frac{l^2 - \rho^2}{R^2}\right] dR \, d\rho. \tag{vi}$$

(d) Evaluate the integrals in Eq. (vi) and thereby show that

$$\|\mathbf{F}\| = \frac{4}{3}\pi a^3 k \frac{Gm}{l^2} = \frac{GMm}{l^2}, \tag{vii}$$

where $M = 4\pi a^3 k/3$ is the mass of the sphere. Thus, *a uniform solid sphere attracts as though its entire mass were concentrated at the center.*

(e) Suppose now that the sphere is not uniform, but has density $f(\rho)$ depending only on ρ. Show that in this case also $\|\mathbf{F}\| = GMm/l^2$.

* **38. Gravitational attraction by a spherical shell.** Let $\mathbf{F}$ be the attractive force of a uniform spherical shell of inner radius a, outer radius b, and constant density k upon a particle of mass m located at a distance l from the center of the shell. Derive each of the following results by suitably modifying the procedure outlined in Problem 37.

(a) If $l > b$, show that $\|\mathbf{F}\| = GMm/l^2$, where M is the mass of the shell.

(b) If $l < a$, show that $\|\mathbf{F}\| = 0$.

(c) Show that the results of Parts (a) and (b) remain true if the density of the shell is variable, but depends only on ρ.

39. Find the gravitational attraction of a solid hemisphere of radius a and constant density k upon a mass m located at the center of the plane face.

REVIEW PROBLEMS

Let R be the rectangle $0 \le x \le 4$, $2 \le y \le 8$. Subdivide R into six subrectangles by the partition points $x_0 = 0$, $x_1 = 2$, $x_2 = 4$, $y_0 = 2$, $y_1 = 4$, $y_2 = 6$, $y_3 = 8$. In each of Problems 1 through 4, evaluate $\iint_R f(P_{ij}^*) \, dA$ by calculating the corresponding Riemann sum, assuming that P_{ij}^* is the center of each subrectangle.

1. $f(x, y) = \begin{cases} 2, & 0 \le x \le 4, \quad 2 \le y \le 6 \\ 4, & 0 \le x \le 4, \quad 6 < y \le 8 \end{cases}$

2. $f(x, y) = x^2 + y^2$

3. $f(x, y) = xe^y$

4. $f(x, y) = 4x^2 + y$

In each of Problems 5 through 8, evaluate the given integral.

5. $\int_0^2 \int_0^{2-x} e^y \, dy \, dx$

6. $\int_0^2 \int_0^x \tan x^2 \, dy \, dx$

7. $\iint_\Omega (x^2 + 4y^2) \, dA$, where Ω is the square $0 \leq x \leq 3, 0 \leq y \leq 3$

8. $\iint_\Omega (4 - x) \, dA$, where Ω is the elliptical region $2x^2 + y^2 \leq 4$

In each of Problems 9 through 12, sketch the region of integration and evaluate the given integral.

9. $\int_{\pi/6}^{\pi/2} \int_0^{\cot x} \ln(\sin x) \, dy \, dx$

10. $\int_0^4 \int_{-y}^y y^2 e^{xy} \, dx \, dy$

11. $\int_0^1 \int_{x^2}^x \left(\frac{1}{2}x^2 + y\right) dy \, dx$

12. $\int_0^1 \int_{x^2}^{\cosh x^2} x \, dy \, dx$

In each of Problems 13 through 16, for the given double integral over the bounded region Ω:

(a) set up the iterated integral with the y-integration done first;

(b) set up the iterated integral with the x-integration done first;

(c) evaluate either integral.

13. $\iint_\Omega (x - 2y) \, dA$, Ω is bounded by $x = \ln y$, $x = 1, y = 1$.

14. $\iint_\Omega (2 \cos x - 2y) \, dA$, Ω is in the first quadrant and is bounded by $y = \cos x$, $y = \sin x$, $x = 0$.

15. $\iint_\Omega \frac{x + y}{xy} \, dA$, Ω is bounded by $x^2 = y$, $x = y^{1/3}$.

16. $\iint_\Omega 2x\sqrt{x^2 + y^2} \, dA$, Ω is in the first quadrant and is bounded by $y = 1$, $y = x^2$, $x = 0$.

In each of Problems 17 through 20, sketch the region of integration and evaluate the given integral.

17. $\int_0^{2\pi} \int_1^2 r \cos r^2 \, dr \, d\theta$

18. $\int_0^{2\pi} \int_0^1 (4 - r^2)^{-1/2} \, r \, dr \, d\theta$

19. $\int_0^{3\pi/2} \int_0^{\cos\theta} r^2 \, dr \, d\theta$

20. $\int_{\pi/4}^{3\pi/4} \int_0^1 \tan\theta \sec\theta \, dr \, d\theta$

In each of Problems 21 through 24, perform the indicated coordinate changes.

21. Change $(2, \pi/6, 1)$ from cylindrical to rectangular.

22. Change $(3, \pi/4, 3\pi/4)$ from spherical to rectangular.

23. Change $(2, \pi/3, \pi/2)$ from spherical to cylindrical.

24. Change $(0, 2, \sqrt{3})$ from rectangular to spherical.

In each of Problems 25 through 28, sketch and identify the graph of the given equation and make the required coordinate changes.

25. $r^2 + z^2 = 4$ to spherical coordinates.

26. $\rho = \cos\phi \csc^2\phi$ to rectangular coordinates.

27. $\rho = 3 \sec\phi$ to rectangular coordinates.

28. $x^2 - 9y^2 + 4z^2 = 36$ to cylindrical coordinates.

In each of Problems 29 and 30, evaluate the given integral. A change of coordinate systems may be helpful.

29. $\int_{4\pi/3}^{5\pi/4} \int_0^{-2\sec\theta} r^2 \cos\theta \, dr \, d\theta$

30. $\int_{\pi/4}^{\pi/2} \int_0^{\csc\theta} r^5 \sin^3\theta \, dr \, d\theta$

In each of Problems 31 and 32, find the volume of the given solid.

31. The solid bounded above by $x^2 + y^2 + z^2 = 4$ and below by $x^2 + y^2 = z^2$ with $z \geq 0$.

32. The solid bounded above by $x + z = 5$, bounded below by the xy-plane, and inside the cylinder $x^2 + y^2 = 1$.

In each of Problems 33 through 36, find the total mass M in the given region.

33. Ω is the triangle bounded by $y = 2$, $y = 2x$, and $x = 0$, and has density $\rho(x, y) = kxy$.

34. Ω is the region $1 \leq r \leq 2$, $0 \leq \theta \leq \pi/4$ and has density $\rho(r, \theta) = r^2 \sin 2\theta \cos 2\theta$.

35. Ω is the solid bounded above by $r^2 + z^2 = 4$, bounded below by the xy-plane, and inside the cylinder $r = 1$, and the density of Ω is proportional to the square of the distance from the z-axis.

36. Ω is the solid inside both the sphere $\rho = 5$ and the cone $\phi = \pi/3$ with density inversely proportional to the distance from the origin.

In each of Problems 37 through 40, find $\bar{x}$, the x-coordinate of the center of mass of the given region. In each case assume that the density is constant.

37. The region bounded by $y = e^x$, $y = 1$, $x = 0$, and $x = 1$.

38. The region in the first octant that is inside the sphere $\rho = 4$.

39. The wedge in the first octant bounded by $x^2 + 4z^2 = 16$ and $y = x$.

40. The volume of the solid inside both $r^2 + z^2 = 4$ and $z = r$, and above the xy-plane.

41. Find the radius of gyration about the y-axis for the region of Problem 37.

42. Find the radius of gyration about the y-axis for the solid of Problem 38.

43. Find the radius of gyration about the y-axis for the solid of Problem 39.

44. Find the radius of gyration about the y-axis for the solid of Problem 40.

EIGHTEEN

line and surface integrals

In the study of parts of mathematical physics, such as fluid mechanics and electromagnetic theory, it is often desirable to consider integrals on curves or surfaces in space. Such integrals are called line integrals and surface integrals, respectively. In physical applications the integrand usually involves a force, or a velocity, or some other quantity that is best expressed in the language of vectors. To deal effectively with problems in these fields, it is often necessary to combine analytical skill with physical or geometrical insight.

The central mathematical results are two theorems, referred to as the divergence theorem and Stokes' theorem, both of which can be viewed as far-reaching generalizations of the fundamental theorem of calculus. By underscoring once again the crucial importance of the fundamental theorem of Newton and Leibniz, these theorems provide a fitting climax to a book on calculus. They also serve to open the way to the vast and exciting areas of mathematics and its applications that lie beyond.

18.1 LINE INTEGRALS

In this section we develop the concept of an integral along a curve in space. Such an integral is usually referred to as a line integral, although the curve along which the integration takes place may be quite arbitrary, and certainly need not be a straight line. Throughout this chapter we assume, unless otherwise stated, that the curve is smooth, or piecewise smooth, and has finite length (is rectifiable).

Suppose that the curve C is described by the vector equation

$$\mathbf{r}(t) = f(t)\mathbf{i} + g(t)\mathbf{j} + h(t)\mathbf{k}, \qquad a \le t \le b; \tag{1}$$

or in component form

$$x = f(t), \qquad y = g(t), \qquad z = h(t), \qquad a \le t \le b. \tag{2}$$

Recall that if the tangent vector

$$\frac{d\mathbf{r}}{dt} = \frac{dx}{dt}\mathbf{i} + \frac{dy}{dt}\mathbf{j} + \frac{dz}{dt}\mathbf{k} \tag{3}$$

is continuous for $a \le t \le b$, and if $d\mathbf{r}/dt \ne \mathbf{0}$ at every point on C, then C is smooth. Also, we can define an arc length function $s = s(t)$ on C such that

$$\frac{ds}{dt} = \left\|\frac{d\mathbf{r}}{dt}\right\| = \left[\left(\frac{dx}{dt}\right)^2 + \left(\frac{dy}{dt}\right)^2 + \left(\frac{dz}{dt}\right)^2\right]^{1/2}, \tag{4}$$

and $s(a) = 0$. Then the length L of the curve is

$$L = \int_a^b \left\|\frac{d\mathbf{r}}{dt}\right\| dt = \int_a^b \left[\left(\frac{dx}{dt}\right)^2 + \left(\frac{dy}{dt}\right)^2 + \left(\frac{dz}{dt}\right)^2\right]^{1/2} dt. \tag{5}$$

It is often convenient to describe C parametrically in terms of the arc length s, $0 \le s \le L$, as well as in terms of the original parameter t. Then the unit tangent vector $\mathbf{T}$, which points in the direction of increasing s, is given by

$$\mathbf{T} = \frac{d\mathbf{r}}{ds} = \frac{d\mathbf{r}/dt}{\|d\mathbf{r}/dt\|}. \tag{6}$$

Finally, suppose also that a vector function $\mathbf{F}(x, y, z)$, or $\mathbf{F}(\mathbf{r})$, is given in some domain containing the curve C. We will often refer to such a vector function as a **vector field.**

Line integrals occur frequently in physical applications; for instance, consider the following problem. Suppose that a particle is moved along the curve C from the point $\mathbf{r}(a)$ to the point $\mathbf{r}(b)$ by the force $\mathbf{F}(\mathbf{r})$ (see Figure 18.1.1). What is the work W that is done by the force $\mathbf{F}$ on the particle?

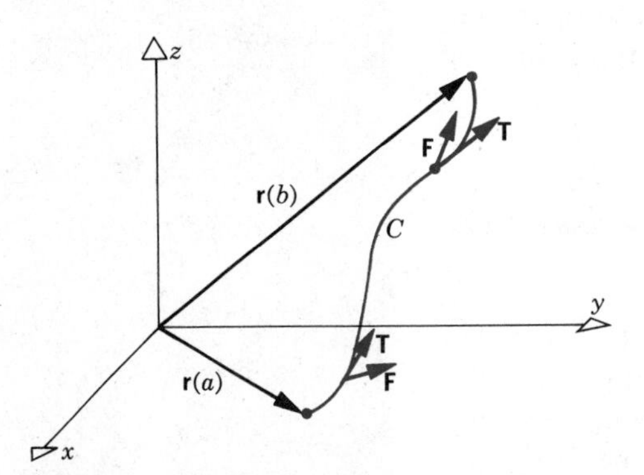

Figure 18.1.1

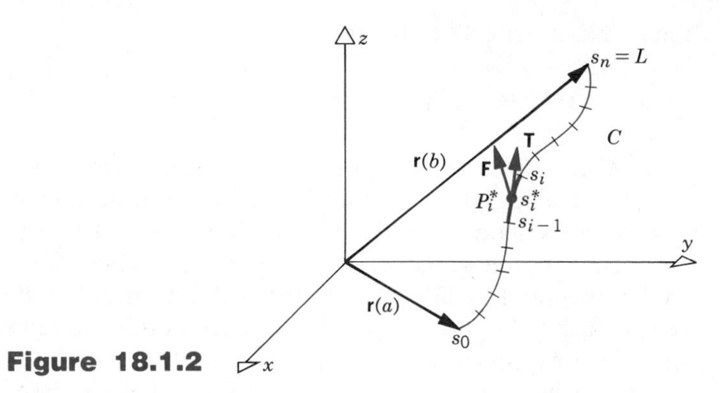

Figure 18.1.2

We can proceed much as we have in the past when faced with similar questions. Let s be arc length measured along the curve C from the initial point $\mathbf{r}(a)$ toward $\mathbf{r}(b)$. Partition C into pieces of length Δs_i for $i = 1, 2, \ldots, n$, and choose a star point P_i^* (corresponding to arc length s_i^*) in each piece, as shown in Figure 18.1.2. At any point s on the curve, the component of force tangent to the curve in the direction of motion is $(\mathbf{F} \cdot \mathbf{T})(s)$. Thus $(\mathbf{F} \cdot \mathbf{T})(s_i^*)\, \Delta s_i$ is an approximation to the work required to move the particle through the portion Δs_i of the curve C, and the Riemann sum

$$\sum_{i=1}^{n} (\mathbf{F} \cdot \mathbf{T})(s_i^*)\, \Delta s_i \tag{7}$$

is an approximation to the total work W. By carrying out the usual limiting procedure on the sum (7), we obtain

$$W = \int_0^L (\mathbf{F} \cdot \mathbf{T})(s)\, ds. \tag{8}$$

We can express W in terms of t rather than s by making use of Eqs. 4 and 6:

$$\begin{aligned} W &= \int_a^b \mathbf{F}[\mathbf{r}(t)] \cdot \frac{d\mathbf{r}(t)/dt}{\|d\mathbf{r}(t)/dt\|} \frac{ds}{dt}\, dt \\ &= \int_a^b \mathbf{F}[\mathbf{r}(t)] \cdot \frac{d\mathbf{r}(t)/dt}{\|d\mathbf{r}(t)/dt\|} \left\| \frac{d\mathbf{r}(t)}{dt} \right\| dt \\ &= \int_a^b \mathbf{F}[\mathbf{r}(t)] \cdot \frac{d\mathbf{r}(t)}{dt}\, dt. \end{aligned} \tag{9}$$

The integrals (8) and (9) are called **line integrals** of **F** along the curve C.

The form of the integral in Eq. 9 also suggests the useful notation

$$W = \int_C \mathbf{F}(\mathbf{r}) \cdot d\mathbf{r} = \int_C \mathbf{F} \cdot d\mathbf{r}, \tag{10}$$

where $\int_C$ emphasizes that the integral depends upon the path of integration C. If

$$\mathbf{F}(x, y, z) = P(x, y, z)\mathbf{i} + Q(x, y, z)\mathbf{j} + R(x, y, z)\mathbf{k},$$

then we can write W in the form

$$W = \int_C P(x, y, z)\, dx + Q(x, y, z)\, dy + R(x, y, z)\, dz. \tag{11}$$

While a line integral involves integration along a curved path in three-dimensional space, the actual evaluation of such an integral does not require any new methods. Notice that the integrals in Eqs. 8 and 9 are simply Riemann integrals of functions of one variable. Thus they can be evaluated by finding an antiderivative of the integrand by any appropriate method, or failing that, by a numerical procedure such as Simpson's rule. The important thing to remember is that everything should be expressed in terms of one variable, usually a parameter t or the arc length s, which reduces the line integral to a definite integral in one variable.

Notice also that in evaluating a line integral there is a definite direction in which the curve C is traversed, that is, from the initial point $\mathbf{r}(a)$ to the terminal point $\mathbf{r}(b)$. *If the direction is reversed,* this causes the limits of integration in Eq. 9 to be interchanged, and consequently *the sign of the result is also reversed.*

Finally, in some cases the path of integration may be described in a way that does not include a set of equations, or parametrization, of the curve. In such a case, the first step is to choose a suitable parametrization to use in evaluating the integral. For any given curve there are many possible parametrizations. It is possible to show (although we do not do it here) that *the value of the resulting line integral is the same for all parametrizations of the path of integration.* Of course, some parametrizations may well lead to integrals that are simpler to evaluate than others, so some attention should be paid to choosing a convenient parametrization when this is possible. The following examples show how to evaluate line integrals, and illustrate some of the preceding remarks.

EXAMPLE 1

Determine the work done on a particle that is moved along the curve

$$C: \quad \mathbf{r}(t) = t^2\mathbf{i} + t\mathbf{j} + t^{-1}\mathbf{k}, \qquad 1 \le t \le 2,$$

from the point $(1, 1, 1)$ to the point $(4, 2, \frac{1}{2})$ by the force

$$\mathbf{F}(x, y, z) = yz\mathbf{i} + xz\mathbf{j} + xy\mathbf{k}.$$

Assume that $\mathbf{F}$ is measured in pounds and lengths are measured in feet.

In scalar form the equations of the curve are $x = t^2$, $y = t$, $z = t^{-1}$. Thus, on the curve we have

$$\begin{aligned} \mathbf{F}[x(t), y(t), z(t)] &= (t)(t^{-1})\mathbf{i} + (t^2)(t^{-1})\mathbf{j} + (t^2)(t)\mathbf{k} \\ &= \mathbf{i} + t\mathbf{j} + t^3\mathbf{k}, \end{aligned}$$

and

$$d\mathbf{r} = \left(\frac{dx}{dt}\mathbf{i} + \frac{dy}{dt}\mathbf{j} + \frac{dz}{dt}\mathbf{k}\right) dt = (2t\mathbf{i} + \mathbf{j} - t^{-2}\mathbf{k})\, dt.$$

Consequently,

$$W = \int_C \mathbf{F} \cdot d\mathbf{r} = \int_1^2 (\mathbf{i} + t\mathbf{j} + t^3\mathbf{k}) \cdot (2t\mathbf{i} + \mathbf{j} - t^{-2}\mathbf{k})\, dt$$

$$= \int_1^2 (2t + t - t)\, dt = t^2 \Big|_1^2 = 3 \text{ ft-lb. } \blacksquare$$

EXAMPLE 2

Evaluate the line integral of $\mathbf{F}(x, y) = y\mathbf{i} - x\mathbf{j}$ along the following paths:

1. the circular arc $x = 2 \cos t$, $y = 2 \sin t$ connecting $(2, 0)$ to $(1, \sqrt{3})$;
2. the straight line connecting $(2, 0)$ to $(1, \sqrt{3})$.

Let C denote the path of integration. Then

$$\int_C \mathbf{F} \cdot d\mathbf{r} = \int_C (y\mathbf{i} - x\mathbf{j}) \cdot (dx\mathbf{i} + dy\mathbf{j}) = \int_C y\, dx - x\, dy. \tag{12}$$

Consider the first path, shown in Figure 18.1.3. On this path we have

$$x = 2 \cos t, \qquad dx = -2 \sin t\, dt,$$

and

$$y = 2 \sin t, \qquad dy = 2 \cos t\, dt.$$

Also, as (x, y) goes from $(2, 0)$ to $(1, \sqrt{3})$, the parameter t goes from 0 to $\pi/3$. Hence, by substituting in Eq. 12, we obtain

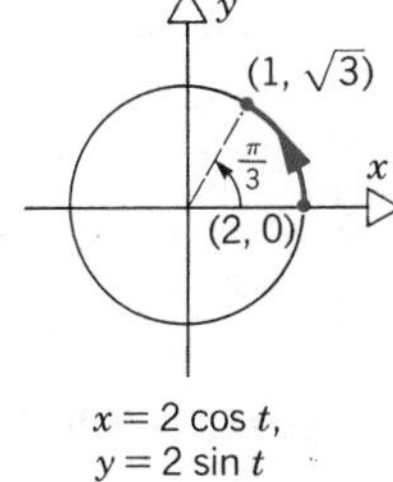

Figure 18.1.3

$$\int_C \mathbf{F} \cdot d\mathbf{r} = \int_0^{\pi/3} [(2 \sin t)(-2 \sin t)\, dt - (2 \cos t)(2 \cos t)\, dt]$$

$$= \int_0^{\pi/3} -4 (\sin^2 t + \cos^2 t)\, dt = -4 \int_0^{\pi/3} dt$$

$$= -\frac{4\pi}{3} \cong -4.189. \tag{13}$$

Now consider the second path. The straight line passing through the two given points is shown in Figure 18.1.4; its Cartesian equation is $y = -\sqrt{3}(x - 2)$. There are many ways in which we can choose a parametric representation of this path of integration. We choose a simple parametric representation such that t increases from 0 to 1 as (x, y) moves from $(2, 0)$ to $(1, \sqrt{3})$, namely

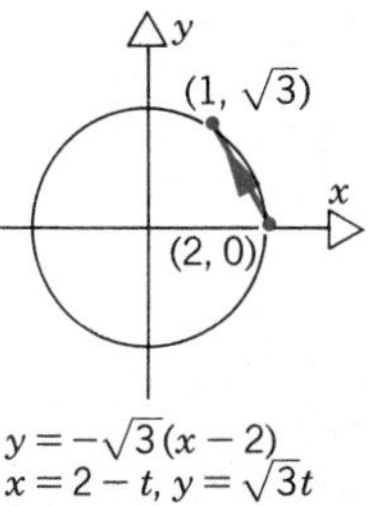

Figure 18.1.4

$$C: \quad x = 2 - t, \qquad y = \sqrt{3}\, t, \qquad 0 \le t \le 1.$$

Hence, substituting in Eq. 12, we have

$$\int_C \mathbf{F} \cdot d\mathbf{r} = \int_0^1 [\sqrt{3}\, t(-dt) - (2 - t)(\sqrt{3}\, dt)]$$

$$= \sqrt{3} \int_0^1 (-t - 2 + t)\, dt$$

$$= -2\sqrt{3} \cong -3.464. \tag{14}$$

The fact that the results given in Eqs. 13 and 14 are different shows clearly that, *in general, the value of a line integral between two given points depends on the path C of integration joining the points.* ∎

EXAMPLE 3

Evaluate the line integral of **F**, where **F** is the same as in Example 2, along the circular arc $x^2 + y^2 = 4$ connecting $(1, \sqrt{3})$ with $(2, 0)$.

Note that this is the same as the problem in part (1) of Example 2, except that the direction of integration is reversed (see Figure 18.1.5). To avoid possible confusion we will refer to the path of integration in the present case as C_-. A possible parametric representation for this path is

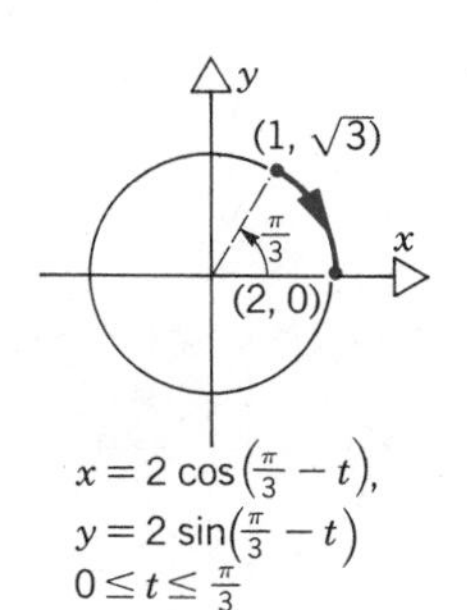

$x = 2\cos\left(\frac{\pi}{3} - t\right)$,
$y = 2\sin\left(\frac{\pi}{3} - t\right)$
$0 \leq t \leq \frac{\pi}{3}$

Figure 18.1.5

$$x = 2\cos\left(\frac{\pi}{3} - t\right), \qquad y = 2\sin\left(\frac{\pi}{3} - t\right),$$

with $0 \leq t \leq \pi/3$. Observe that as t increases from 0 to $\pi/3$, the point (x, y) moves from $(1, \sqrt{3})$ to $(2, 0)$. Hence

$$\int_{C_-} \mathbf{F} \cdot \mathbf{T}\, ds = \int_{C_-} \mathbf{F} \cdot d\mathbf{r} = \int_{C_-} y\, dx - x\, dy$$

$$\int_0^{\pi/3} 4\left[\sin^2\left(\frac{\pi}{3} - t\right) + \cos^2\left(\frac{\pi}{3} - t\right)\right] dt$$

$$= \int_0^{\pi/3} 4\, dt = \frac{4\pi}{3}.$$

Observe that this result is the negative of the one obtained in Eq. 13 of Example 2, as it should be.

An alternative representation for C_- is

$$x = 2\sin t, \qquad y = 2\cos t, \qquad \frac{\pi}{6} \leq t \leq \frac{\pi}{2}.$$

Then

$$\int_{C_-} \mathbf{F} \cdot d\mathbf{r} = \int_{\pi/6}^{\pi/2} 4\,(\cos^2 t + \sin^2 t)\, dt = \frac{4\pi}{3},$$

illustrating that the same result is obtained for two different parametrizations of the same curve.

It is not necessary to use a parametrization for which t increases during the integration. For example, we could describe C_- by the same equations as in Example 2, that is,

$$x = 2\cos t, \qquad y = 2\sin t,$$

where now t decreases from $\pi/3$ to 0 as the path C_- is traversed. Consequently,

$$\int_{C_-} \mathbf{F}\cdot d\mathbf{r} = \int_{\pi/3}^{0} -4(\sin^2 t + \cos^2 t)\,dt$$

$$= -4\left(0 - \frac{\pi}{3}\right) = \frac{4\pi}{3}. \quad \blacksquare$$

In many cases the curve C may be piecewise smooth; that is, C may be made up of a finite number of adjoining smooth arcs $C_1, C_2, \ldots, C_n$. For example, the curve C may be a rectangle, a triangle, or a more general piecewise smooth curve as is shown in Figure 18.1.6. In this case the line integral of $\mathbf{F}$ along C is defined to be

$$\int_C \mathbf{F}\cdot d\mathbf{r} = \int_{C_1} \mathbf{F}\cdot d\mathbf{r} + \int_{C_2} \mathbf{F}\cdot d\mathbf{r} + \cdots + \int_{C_n} \mathbf{F}\cdot d\mathbf{r}. \tag{15}$$

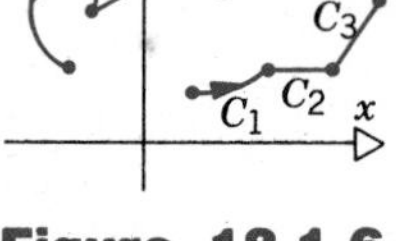

Figure 18.1.6 A piecewise smooth curve.

We also note from Eqs. 10 and 11 that the line integral

$$\int_C P(x, y, z)\,dx$$

is just a special case of $\int_C \mathbf{F}\cdot d\mathbf{r}$ with $\mathbf{F}(x, y, z) = P(x, y, z)\mathbf{i}$. Similarly

$$\int_C Q(x, y, z)\,dy \qquad \text{and} \qquad \int_C R(x, y, z)\,dz$$

correspond to $\mathbf{F}(x, y, z) = Q(x, y, z)\mathbf{j}$ and $\mathbf{F}(x, y, z) = R(x, y, z)\mathbf{k}$, respectively.
We illustrate these ideas in the following examples.

EXAMPLE 4

Evaluate the line integral of

$$\mathbf{F}(x, y) = \sin \pi x\mathbf{i} + xy\mathbf{j}$$

along the path shown in Figure 18.1.7.
The path is composed of three straight line segments, so we have

$$\int_C \mathbf{F}\cdot d\mathbf{r} = \int_{C_1} \mathbf{F}\cdot d\mathbf{r} + \int_{C_2} \mathbf{F}\cdot d\mathbf{r} + \int_{C_3} \mathbf{F}\cdot d\mathbf{r}.$$

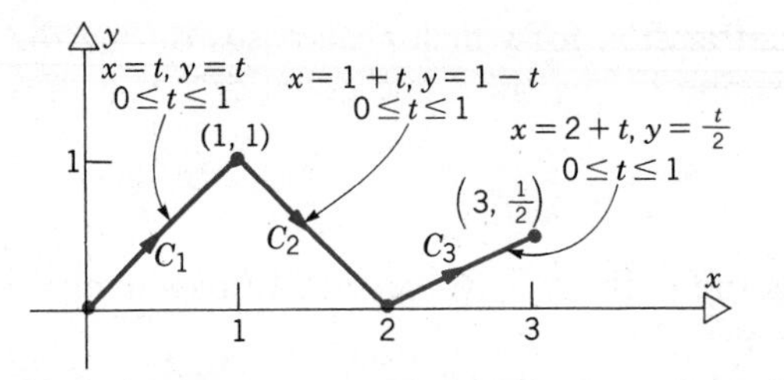

Figure 18.1.7

The path C_1 connecting (0, 0) to (1, 1) is a portion of the straight line $y = x$, which we can describe parametrically by

$$x = t, \qquad y = t, \qquad 0 \leq t \leq 1,$$

with $dx = dy = dt$. Hence

$$\begin{aligned}\int_{C_1} \mathbf{F} \cdot d\mathbf{r} &= \int_{C_1} (\sin \pi x \, dx + xy \, dy) \\ &= \int_0^1 (\sin \pi t + t^2) \, dt = \left[-\frac{\cos \pi t}{\pi} + \frac{t^3}{3} \right] \Bigg|_0^1 \\ &= \frac{2}{\pi} + \frac{1}{3}.\end{aligned}$$

The path C_2 connecting (1, 1) to (2, 0) is a portion of the straight line $y - 1 = -(x - 1)$, which we can describe parametrically by

$$x = 1 + t, \qquad y = 1 - t, \qquad 0 \leq t \leq 1,$$

with $dx = dt$ and $dy = -dt$. Hence

$$\begin{aligned}\int_{C_2} \mathbf{F} \cdot d\mathbf{r} &= \int_0^1 [\sin \pi(1 + t) \, dt + (1 + t)(1 - t)(-dt)] \\ &= \int_0^1 [\sin \pi(1 + t) - (1 - t^2)] \, dt \\ &= \left[\frac{-\cos \pi(1 + t)}{\pi} - t + \frac{t^3}{3} \right] \Bigg|_0^1 \\ &= -\frac{2}{\pi} - \frac{2}{3}.\end{aligned}$$

The path C_3 connecting (2, 0) to $(3, \frac{1}{2})$ is a portion of the straight line $y = (x - 2)/2$, which we can describe parametrically by

$$x = 2 + t, \qquad y = \frac{t}{2}, \qquad 0 \leq t \leq 1,$$

with $dx = dt$ and $dy = dt/2$. Hence

$$\int_{C_3} \mathbf{F} \cdot d\mathbf{r} = \int_0^1 \left[\sin \pi(2 + t)\, dt + (2 + t) \frac{t}{2} \frac{dt}{2} \right]$$

$$= \int_0^1 \left[\sin \pi(2 + t) + \frac{t}{2} + \frac{t^2}{4} \right] dt$$

$$= \left[\frac{-\cos \pi(2 + t)}{\pi} + \frac{t^2}{4} + \frac{t^3}{12} \right] \Bigg|_0^1$$

$$= \frac{2}{\pi} + \frac{1}{3}.$$

Combining these results we obtain

$$\int_C \mathbf{F} \cdot d\mathbf{r} = \left(\frac{2}{\pi} + \frac{1}{3} \right) + \left(-\frac{2}{\pi} - \frac{2}{3} \right) + \left(\frac{2}{\pi} + \frac{1}{3} \right)$$

$$= \frac{2}{\pi} \cong 0.6366. \blacksquare$$

EXAMPLE 5

Evaluate

$$\int_C \left(x + \frac{x^2 y}{1 + y} \right) dx \qquad (16)$$

where the path of integration C is the parabola given by $y = x^2$ from (0, 0) to (2, 4) (see Figure 18.1.8).

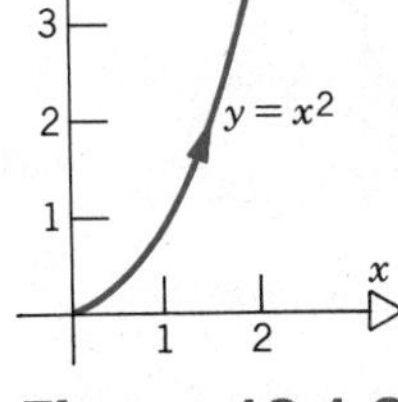

Figure 18.1.8

We set $y = x^2$ in the integral (16) with $0 \leq x \leq 2$, and obtain

$$\int_C \left(x + \frac{x^2 y}{1 + y} \right) dx = \int_0^2 \left(x + \frac{x^4}{1 + x^2} \right) dx$$

$$= \int_0^2 \left(x + x^2 - 1 + \frac{1}{1 + x^2} \right) dx$$

$$= \left[\frac{x^2}{2} + \frac{x^3}{3} - x + \arctan x \right] \Bigg|_0^2$$

$$= 2 + \frac{8}{3} - 2 + \arctan 2 \cong \frac{8}{3} + 1.107 \cong 3.774.$$

Note that x plays the role of the parameter t. ∎

EXAMPLE 6

Evaluate $\int_C \mathbf{F} \cdot d\mathbf{r}$ where

$$\mathbf{F}(x, y) = 3xy^2\mathbf{i} + 2yx^2\mathbf{j},$$

and C is the rectangle shown in Figure 18.1.9 traversed in the counterclockwise direction.

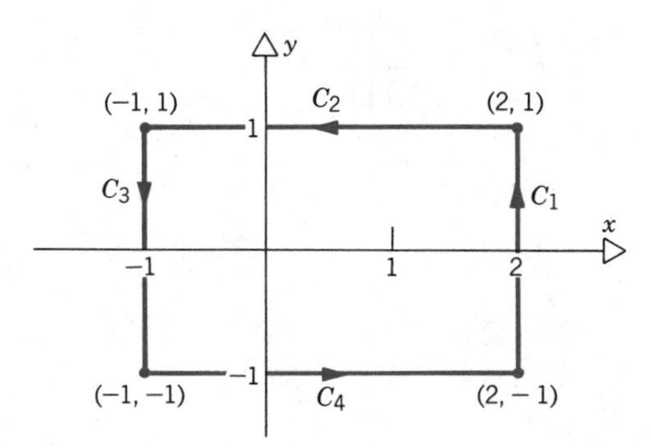

Figure 18.1.9

For this problem it is useful to take advantage of the fact that the path consists of straight line segments parallel either to the x-axis or to the y-axis. We write

$$\int_C \mathbf{F} \cdot d\mathbf{r} = \int_C 3xy^2\,dx + 2yx^2\,dy,$$

and observe that on the sides of the rectangle we have the following relations:

$$\begin{aligned} &C_1\!: \quad x = 2, && dx = 0, && \text{and } y \text{ goes from } -1 \text{ to } 1;\\ &C_2\!: \quad y = 1, && dy = 0, && \text{and } x \text{ goes from } 2 \text{ to } -1;\\ &C_3\!: \quad x = -1, && dx = 0, && \text{and } y \text{ goes from } 1 \text{ to } -1;\\ &C_4\!: \quad y = -1, && dy = 0, && \text{and } x \text{ goes from } -1 \text{ to } 2. \end{aligned}$$

Hence

$$\int_{C_1} \mathbf{F} \cdot d\mathbf{r} = \int_{-1}^{1} 8y\,dy = 0,$$

$$\int_{C_2} \mathbf{F} \cdot d\mathbf{r} = \int_{2}^{-1} 3x(1)^2\,dx = \left.\frac{3x^2}{2}\right|_{2}^{-1} = -\frac{9}{2},$$

$$\int_{C_3} \mathbf{F} \cdot d\mathbf{r} = \int_{1}^{-1} 2y\,dy = 0,$$

$$\int_{C_4} \mathbf{F} \cdot d\mathbf{r} = \int_{-1}^{2} 3x(-1)^2\,dx = \left.\frac{3x^2}{2}\right|_{-1}^{2} = \frac{9}{2},$$

and

$$\int_C \mathbf{F} \cdot d\mathbf{r} = 0 - \frac{9}{2} + 0 + \frac{9}{2} = 0. \quad \blacksquare$$

PROBLEMS

In each of Problems 1 through 12, evaluate the line integral $\int_C \mathbf{F} \cdot d\mathbf{r}$ along the given curve.

1. $\mathbf{F}(x, y) = x\mathbf{i} + y\mathbf{j}$, $\mathbf{r}(t) = t\mathbf{i} + \sin \pi t\,\mathbf{j}$, $0 \le t \le 2$
2. $\mathbf{F}(x, y) = (x - y)\mathbf{i} + (y - x)\mathbf{j}$, $\mathbf{r}(t) = t^2\mathbf{i} + t\mathbf{j}$, $0 \le t \le 1$
3. $\mathbf{F}(x, y, z) = xy\mathbf{i} + y^2\mathbf{j} - zx\mathbf{k}$, $\mathbf{r}(t) = t\mathbf{i} - 2t\mathbf{j} - \ln t\,\mathbf{k}$, $1 \le t \le 3$
4. $\mathbf{F}(x, y) = (x^2 - y^2)\mathbf{i} + 2xy\mathbf{j}$, straight line from $(1, 1)$ to $(3, -1)$
5. $\mathbf{F}(x, y) = e^{x+y}(x\mathbf{i} + y\mathbf{j})$, shorter arc of the circle $x^2 + y^2 = a^2$ from $(a, 0)$ to $(0, a)$
6. $\mathbf{F}(x, y, z) = -x\mathbf{i} + y\mathbf{j} + 2xy\mathbf{k}$, $\mathbf{r}(t) = a\cos t\,\mathbf{i} + b\sin t\,\mathbf{j} + t\mathbf{k}$, $0 \le t \le \pi$
7. $\mathbf{F}(x, y) = 2xy^2\mathbf{i} + 3\sin y\mathbf{j}$, the parabola $y = x^2$ from $(0, 0)$ to $(2, 4)$
8. $\mathbf{F}(x, y, z) = (y + z)\mathbf{i} + (x + z)\mathbf{j} + (x + y)\mathbf{k}$, straight line from $(1, 0, 0)$ to $(2, 1, 3)$
9. $\mathbf{F}(x, y) = y\mathbf{i} - x\mathbf{j}$, the triangle connecting $(1, 0)$, $(0, 1)$, and $(-1, 0)$ traversed counterclockwise
10. $\mathbf{F}(x, y) = y^2x\mathbf{i} + x^2y\mathbf{j}$, the square connecting the points $(1, 0)$, $(1, 2)$, $(-1, 2)$, and $(-1, 0)$ traversed counterclockwise
11. $\mathbf{F}(x, y) = y^2\mathbf{i} + x^2\mathbf{j}$, shorter arc of the ellipse $x^2/a^2 + y^2/b^2 = 1$ from $(a, 0)$ to $(0, b)$
12. $\mathbf{F}(x, y, z) = e^{x-y}\mathbf{i} + e^{y+z}\mathbf{j} - xy\mathbf{k}$, straight line from $(1, -1, 2)$ to $(3, 3, 3)$

In each of Problems 13 through 16, evaluate the given line integral.

13. $\int_C (x + y)^2\,dx$, along the straight line from $(0, 0)$ to $(3, 5)$
14. $\int_C y^2\,dx$, along $x = y^2$ from $(0, 0)$ to $(4, 2)$
15. $\int_C \dfrac{xy}{\sqrt{x^2 + y^2}}\,dy$, counterclockwise along the circle $x^2 + y^2 = a^2$ from $(a, 0)$ to $(0, a)$
16. $\int_C \left(x + \dfrac{x^2y}{1 + xy}\right) dx$, along the curve $x = t^2$, $y = t$ from $(0, 0)$ to $(4, 2)$
17. Evaluate each of the following line integrals along the straight line from $(0, 0)$ to (a, b):

    (a) $\int_C x^2y^3\,dx$ (b) $\int_C x^2y^3\,dy$
18. Evaluate each of the following line integrals along the path $y = x^3$ from $(1, 1)$ to $(2, 8)$:

    (a) $\int_C x\,dy$ (b) $\int_C y\,dx$
19. Evaluate

    $$\int_C (2y - x)\,dx + (3x + y)\,dy$$

    for each of the following paths:
    (a) The straight line from $(0, 0)$ to $(1, 1)$.
    (b) The parabola $y = x^2$ from $(0, 0)$ to $(1, 1)$.
    (c) The straight line segments from $(0, 0)$ to $(1, 0)$ to $(1, 1)$.
    (d) The straight line segments from $(0, 0)$ to $(0, 1)$ to $(1, 1)$.
    (e) The curve $y = \sqrt{x}$ from $(0, 0)$ to $(1, 1)$.
20. Show that

    $$\int_C \frac{x\,dy - y\,dx}{x^2 + y^2} = 2\pi,$$

    where C is the circle $x^2 + y^2 = a^2$ traversed in the counterclockwise direction.
21. Evaluate $\int_C \mathbf{F} \cdot d\mathbf{r}$, where $\mathbf{F}(x, y) = y\mathbf{i} + x\mathbf{j}$ along each of the following paths:
    (a) The shorter arc of the circle $x^2 + y^2 = 4$ from $(-2, 0)$ to $(0, 2)$.
    (b) The complement of the arc of (a) from $(-2, 0)$ to $(0, 2)$.
22. Evaluate

    $$\int_C x^2\,dx + y^2\,dy$$

    along each of the following closed paths:
    (a) The path shown in Figure 18.1.10*a*.
    (b) The path shown in Figure 18.1.10*b*.
23. Evaluate $\int_C \mathbf{F} \cdot d\mathbf{r}$ where $\mathbf{F}(x, y, z) = e^x\mathbf{i} + e^y\mathbf{j} + e^z\mathbf{k}$ along each of the following paths:
    (a) The straight line from $(0, 0, 0)$ to $(1, 1, 1)$.

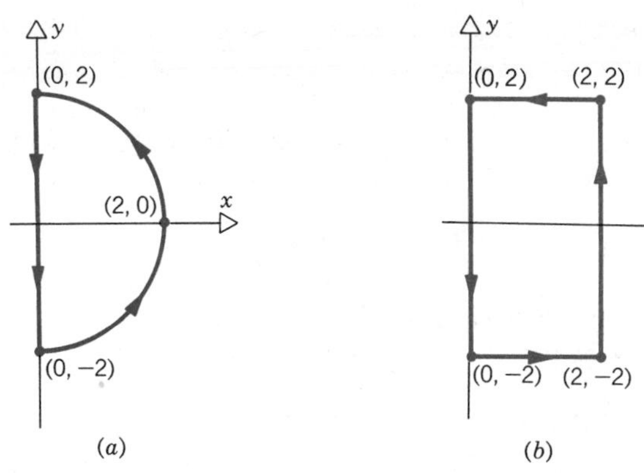

Figure 18.1.10

(b) The straight line segments from (0, 0, 0) to (1, 0, 0) to (1, 1, 0) to (1, 1, 1).

(c) Along the curve $x = t$, $y = t^2$, $z = t^3$ from (0, 0, 0) to (1, 1, 1).

24. An object travels around the ellipse $x^2/a^2 + y^2/b^2 = 1$ in the clockwise direction and is subject to the force

$$\mathbf{F}(x, y) = \frac{1}{2}(-y\mathbf{i} + x\mathbf{j}).$$

Show that the work done in one orbit is numerically equal to the area of the ellipse.

Mass of a Wire.

A wire in the shape of a curve C has density $\rho(s)$ and length L. The total mass of the wire is

$$M = \int_0^L \rho(s)\, ds.$$

In each of Problems 25 through 29 determine M by setting up and evaluating an appropriate integral.

25. A wire has the shape of a semicircle of radius a. If the density of the wire is proportional to the distance from one end of the wire, what is the mass of the wire?

26. For the wire of Problem 25 suppose that the density is proportional to the distance from the diameter connecting the endpoints of the wire. What is the mass of the wire?

27. A wire has the helical shape

$$\mathbf{r}(t) = (2 \cos t)\mathbf{i} + (2 \sin t)\mathbf{j} + 3t\mathbf{k}, \qquad 0 \le t \le 4\pi.$$

If the density of the wire is proportional to the square of the distance from the origin, what is the mass of the wire?

28. A wire has the helical shape

$$\mathbf{r}(t) = (a \cos \omega t)\mathbf{i} + (a \sin \omega t)\mathbf{j} + bt\mathbf{k}.$$

If the density of the wire is kz, determine the mass of the wire from $(a, 0, 0)$ to $(-a, 0, 3b\pi/\omega)$.

29. A wire is shaped into a spiral $r = \theta$, $0 \le \theta \le 2\pi$. If the density of the wire is proportional to θ, determine the mass of the wire. (Arc length in polar coordinates is discussed in Section 14.3.)

Center of Mass and Moments of Inertia.

A thin wire of length L and mass density ρ has the shape of a curve C in the xy-plane. The total mass M of the wire is

$$M = \int_C \rho(s)\, ds.$$

The center of mass $(\bar{x}, \bar{y})$ of the wire is given by

$$\bar{x} = \frac{1}{M}\int_C x\, \rho(s)\, ds, \qquad \bar{y} = \frac{1}{M}\int_C y\, \rho(s)\, ds.$$

If the density of the wire is constant, then the center of mass is the centroid of the wire. The moments of inertia about the x-axis and the y-axis are

$$I_x = \int_C y^2\, \rho(s)\, ds, \qquad I_y = \int_C x^2\, \rho(s)\, ds.$$

30. A wire of constant density is shaped into

(a) a semicircle, $x = a \cos t$, $y = a \sin t$, $0 \le t \le \pi$

(b) the semicircle of Part (a) together with a straight piece connecting the two ends of the semicircle.

Find the centroid of the wire in each case. Compare the results with those of Example 2 of Section 17.3 for a semicircular disk. Explain why the differences are reasonable.

31. For the wire of Problem 30(a) determine I_x and I_y.

32. For the wire of Problem 30(a) determine the center of mass if the density of the wire is proportional to the distance along the wire from the point $(a, 0)$.

33. A uniform wire has the shape of an equilateral triangle of side a (see Figure 18.1.11). Determine the centroid of the wire.

34. A uniform wire has the shape of an isosceles triangle of base b and height h (see Figure 18.1.12).

(a) Determine the centroid of the wire and com-

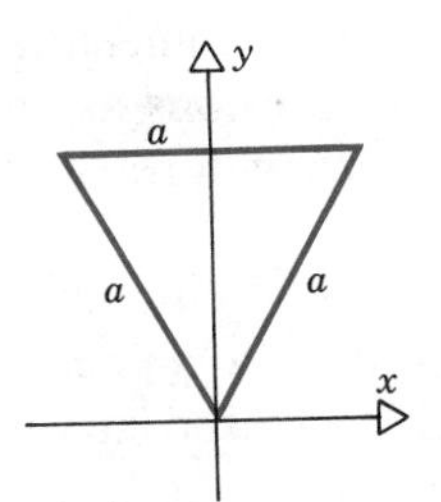

Figure 18.1.11

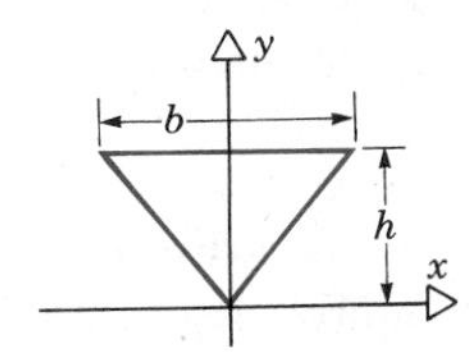

Figure 18.1.12

pare your result with that for the centroid of the corresponding plate (Problem 11 of Section 17.3).

(b) Is there a value of h/b for which the two centroids coincide?

35. A wire has uniform density k and the shape shown in Figure 18.1.13. Determine I_x and I_y.

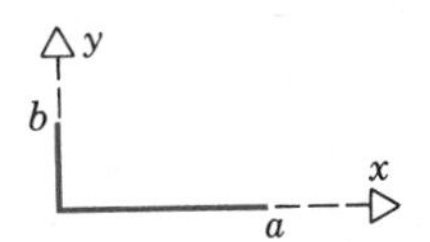

Figure 18.1.13

18.2 INDEPENDENCE OF PATH

In Section 18.1 we emphasized that, in general, the line integral

$$\int_C \mathbf{F}(\mathbf{r}) \cdot d\mathbf{r}$$

between two points in space depends on the path C as well as on the endpoints. However, there is a large class of vector functions $\mathbf{F}$, for which the integral is *independent of the path* connecting the endpoints. These vector functions are of particular importance in many applications.

Before we can develop this and related topics, we must briefly describe some geometrical concepts for point sets in two and three dimensions. Although we do not want to introduce all of the mathematical definitions and notation that are necessary for a rigorous discussion, we do want to provide a correct, but intuitive, understanding of each concept. Recall that in Section 16.1 we introduced the concepts of a neighborhood of a point, open sets, closed sets, and boundary points; in Section 15.6 we explained what was meant by a piecewise smooth curve and a piecewise smooth closed curve. Now we add to these ideas.

An open set of points is said to be **connected** if any two points in the set can be joined by a piecewise smooth curve lying entirely in the set. In Figure 18.2.1

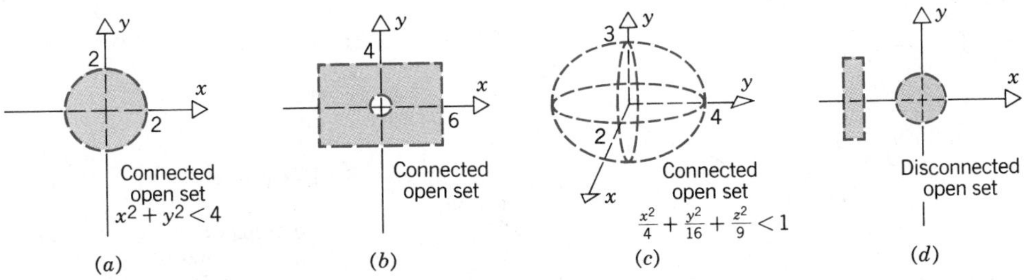

Figure 18.2.1 (*a*), (*b*), and (*c*) show connected open sets; (*d*) shows a disconnected open set.

the open sets (*a*), (*b*), and (*c*) are connected; the open set (*d*) is not connected. For our purposes we can think of a connected open set in two dimensions as an open disk and in three dimensions as an open ball (Section 16.1). We will refer to a connected open set as a **domain.**

A closed curve C is **simple** if, as we trace the curve from an initial point back to the initial point, we do not pass twice through any other point of C. Thus a **simple closed curve** does not touch or cross itself. In Figure 18.2.2 the closed curves

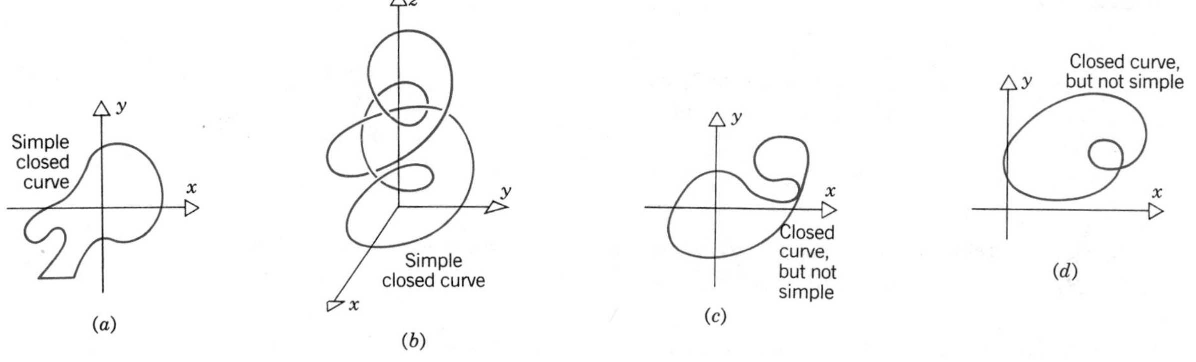

Figure 18.2.2 (*a*) and (*b*) show simple closed curves; (*c*) and (*d*) show closed curves that are not simple.

(*a*) and (*b*) are simple closed curves and (*c*) and (*d*) are not. Note that (*a*) is piecewise smooth. When the path C is a simple closed curve in two dimensions, the line integral is usually denoted by

$$\oint_C \mathbf{F} \cdot d\mathbf{r} \qquad \text{or} \qquad \oint_C \mathbf{F} \cdot d\mathbf{r}$$

according to whether the path C is traversed in the counterclockwise or clockwise direction.

Finally, a domain D is **simply connected** if every simple closed curve in D can be continuously shrunk to a point in D without crossing points not in the domain. Thus, if we think of the simple closed curve as a loop of string, we must be able to contract the loop to a point without leaving the domain. In Figure 18.2.3 the two-dimensional domain (*a*) is simply connected, but the domain (*b*) is not simply connected because of the hole. A string corresponding to, say $x^2 + y^2 = 9$, cannot be contracted to a point without passing through points not in the domain.

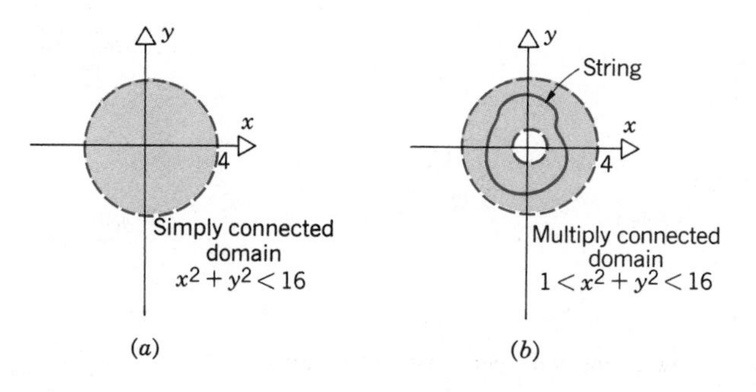

Figure 18.2.3 (*a*) A simply connected domain. (*b*) A multiply connected domain.

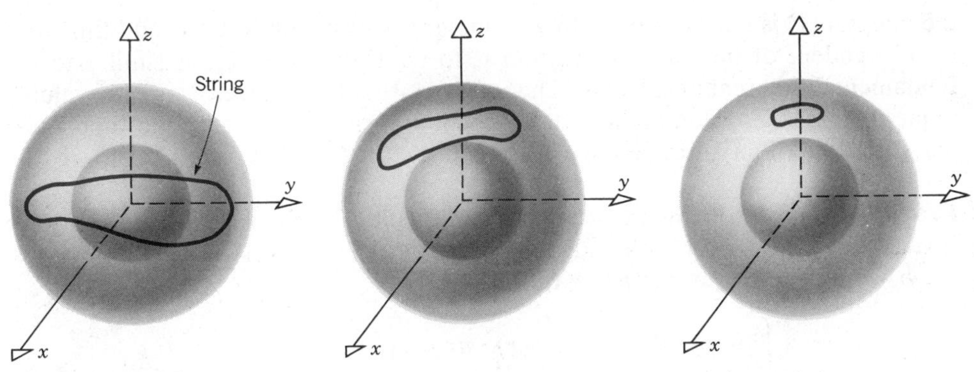

Figure 18.2.4 A simply connected domain in three dimensions may have a hole.

In general, two-dimensional domains with holes are not simply connected. However, note that a domain in three dimensions may be simply connected even though it has a hole; for example, the spherical shell in Figure 18.2.4. An example of a three-dimensional domain that is not simply connected is a torus (doughnut) (see Figure 18.2.5). A domain that is not simply connected is said to be **multiply connected.**

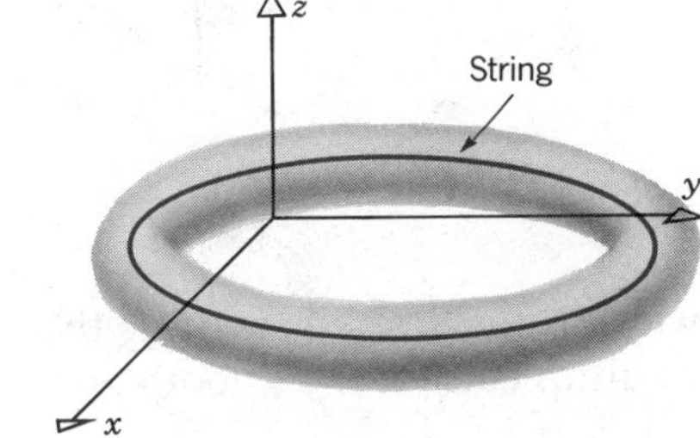

Figure 18.2.5 A multiply connected domain in three dimensions.

With these topological ideas in mind, we can now state and discuss several results about line integrals and independence of path. The following theorem describes a large and important class of integrals that has this property.

Theorem 18.2.1

Let ϕ and its first partial derivatives be continuous on a domain D. Let $\mathbf{r}_a$ and $\mathbf{r}_b$ be two points in D, and let C be any piecewise smooth curve lying in D and connecting $\mathbf{r}_a$ and $\mathbf{r}_b$. If $\mathbf{F} = \nabla\phi$, then $\int_C \mathbf{F} \cdot d\mathbf{r}$ is independent of the path C, and

$$\int_C \mathbf{F}(\mathbf{r}) \cdot d\mathbf{r} = \int_C \nabla\phi(\mathbf{r}) \cdot d\mathbf{r} = \phi(\mathbf{r}_b) - \phi(\mathbf{r}_a). \tag{1}$$

The line integral is independent of the path because its value is determined solely by the values of ϕ at the endpoints. In particular, if $\mathbf{F}$ is a force field, then the integral (1) is the work done by $\mathbf{F}$ in moving a mass from $\mathbf{r}_a$ to $\mathbf{r}_b$, and this work is independent of the path taken from $\mathbf{r}_a$ to $\mathbf{r}_b$. Theorem 18.2.1 is similar to the fundamental theorem of calculus (Theorem 6.4.3) in that it gives a way to evaluate a line integral by evaluating a certain function at the end points. The function ϕ plays the role of an antiderivative.

Proof. The proof of this theorem is not difficult. For the sake of simplicity we first assume that C is smooth and given by $\mathbf{r} = \mathbf{r}(t) = x(t)\mathbf{i} + y(t)\mathbf{j} + z(t)\mathbf{k}$, $a \leq t \leq b$, with $\mathbf{r}(a) = \mathbf{r}_a$ and $\mathbf{r}(b) = \mathbf{r}_b$. Then

$$\begin{aligned}\int_C \mathbf{F}(\mathbf{r}) \cdot d\mathbf{r} &= \int_C \nabla\phi(\mathbf{r}) \cdot d\mathbf{r} \\ &= \int_a^b \nabla\phi[\mathbf{r}(t)] \cdot \frac{d\mathbf{r}}{dt}\, dt \\ &= \int_a^b \left(\frac{\partial\phi}{\partial x}\frac{dx}{dt} + \frac{\partial\phi}{\partial y}\frac{dy}{dt} + \frac{\partial\phi}{\partial z}\frac{dz}{dt}\right) dt, \end{aligned} \qquad (2)$$

where ϕ_x, ϕ_y, and ϕ_z are evaluated at $(x(t), y(t), z(t))$. However, the integrand in the last integral is (by the chain rule) precisely $d\phi\,[\mathbf{r}(t)]/dt$. Hence

$$\begin{aligned}\int_C \mathbf{F}(\mathbf{r}) \cdot d\mathbf{r} &= \int_a^b \frac{d}{dt}\phi[\mathbf{r}(t)]\, dt \\ &= \phi[\mathbf{r}(b) - \phi[\mathbf{r}(a)] \\ &= \phi(\mathbf{r}_b) - \phi(\mathbf{r}_a),\end{aligned}$$

which is the desired result. If the curve C is piecewise smooth, then we apply this result to each of the smooth arcs making up C, and then add the results to obtain Eq. 1. □

If the vector field $\mathbf{F}$ is the gradient of a scalar function, than $\mathbf{F}$ is said to be a **gradient field** or a **conservative field.** The function ϕ, or in mechanics $-\phi$, is referred to as a **potential function.** The reason for this terminology arises from usage in mechanics. A force field for which the work done on a mass is independent of the path of motion is called a conservative force field, and for such fields the function $-\phi$ is the potential energy of the mass. We discuss conservative force fields at the end of this section.

EXAMPLE 1

The gravitational force exerted by a mass M at the origin on a mass m at the point (x, y, z) is

$$\mathbf{f}(x, y, z) = -GMm\,\frac{\mathbf{r}}{r^3} = -GMm\,\frac{x\mathbf{i} + y\mathbf{j} + z\mathbf{k}}{r^3} \qquad (3)$$

where $r = \|\mathbf{r}\| = \sqrt{x^2 + y^2 + z^2}$, and G is the universal gravitational constant. Verify that

$$\mathbf{f} = GMm\, \nabla \left(\frac{1}{r}\right), \tag{4}$$

and calculate the work done by $\mathbf{f}$ in moving the mass m from the point $\mathbf{r}_1$ to the point $\mathbf{r}_2$.

First,

$$\nabla \left(\frac{1}{r}\right) = \frac{\partial}{\partial x}\left(\frac{1}{r}\right)\mathbf{i} + \frac{\partial}{\partial y}\left(\frac{1}{r}\right)\mathbf{j} + \frac{\partial}{\partial z}\left(\frac{1}{r}\right)\mathbf{k}.$$

We have

$$\frac{\partial}{\partial x}\left(\frac{1}{r}\right) = \frac{\partial}{\partial x}(x^2 + y^2 + z^2)^{-1/2} = -\frac{1}{2}(x^2 + y^2 + z^2)^{-3/2}\, 2x = -\frac{x}{r^3},$$

and by symmetry

$$\frac{\partial}{\partial y}\left(\frac{1}{r}\right) = -\frac{y}{r^3}, \qquad \frac{\partial}{\partial z}\left(\frac{1}{r}\right) = -\frac{z}{r^3}.$$

Therefore

$$\nabla \left(\frac{1}{r}\right) = -\frac{x\mathbf{i} + y\mathbf{j} + z\mathbf{k}}{r^3},$$

and Eq. 4 is verified by making use of Eq. 3. Thus the gravitational force field is conservative in any region not including the origin.

The work W done in moving the mass m from $\mathbf{r}_1$ to $\mathbf{r}_2$ is independent of the path, and

$$\begin{aligned} W = \int_C \mathbf{f} \cdot d\mathbf{r} &= \int_C GMm\, \nabla \left(\frac{1}{r}\right) \cdot d\mathbf{r} \\ &= GMm \int_C \nabla \left(\frac{1}{r}\right) \cdot d\mathbf{r} \\ &= GMm \left.\frac{1}{r}\right|_{r_1}^{r_2} \end{aligned}$$

where $r_1 = \|\mathbf{r}_1\|$ and $r_2 = \|\mathbf{r}_2\|$. Hence

$$W = GMm\left(\frac{1}{r_2} - \frac{1}{r_1}\right). \blacksquare \tag{5}$$

The converse of Theorem 18.2.1 is also true. We state this result as a theorem without proof. A proof for the two-dimensional case is developed in Problem 26.

Theorem 18.2.2

Suppose that the vector function $\mathbf{F}$ is continuous on a domain D, and that the line integral of $\mathbf{F}$ is independent of path. That is, for every pair of points $\mathbf{r}_a$ and $\mathbf{r}_b$ in D the line integral

$$\int_C \mathbf{F} \cdot d\mathbf{r}$$

has the same value for every piecewise smooth curve lying in D and connecting $\mathbf{r}_a$ and $\mathbf{r}_b$. Then there exists a scalar function ϕ that is continuous, has continuous first partial derivatives on D, and

$$\mathbf{F}(x, y, z) = \nabla\phi(x, y, z).$$

Theorems 18.2.1 and 18.2.2, taken together, state that $\int_C \mathbf{F} \cdot d\mathbf{r}$ is independent of the path C if and only if $\mathbf{F} = \nabla\phi$ for some potential function ϕ. There is also an equivalence between independence of path and integrals around closed curves. Suppose first that $\int_C \mathbf{F} \cdot d\mathbf{r}$ is independent of the path. Let C be a piecewise smooth simple closed curve, and let P and Q be two points on C (see Figure 18.2.6). Let C_1 and C_2 be two arcs from P to Q, where C consists of C_1 in the positive direction, and then C_2 in the negative direction. Independence of path means that

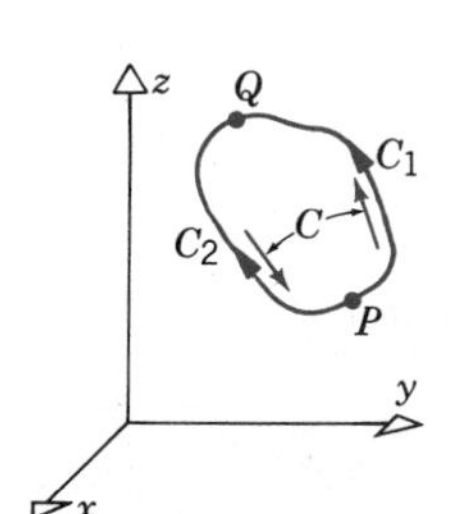

Figure 18.2.6 $\oint_C \mathbf{F} \cdot d\mathbf{r} = \int_{C_1} \mathbf{F} \cdot d\mathbf{r} - \int_{C_2} \mathbf{F} \cdot d\mathbf{r}$.

$$\int_{C_1} \mathbf{F} \cdot d\mathbf{r} = \int_{C_2} \mathbf{F} \cdot d\mathbf{r}, \tag{6}$$

or

$$\left(\int_{C_1} - \int_{C_2}\right) \mathbf{F} \cdot d\mathbf{r} = 0.$$

The latter equation is the same as

$$\oint_C \mathbf{F} \cdot d\mathbf{r} = 0. \tag{7}$$

Thus independence of the path implies that Eq. 7 is valid for every piecewise smooth simple closed curve C. The converse of this statement is also true, since we can reverse the argument just given to derive Eq. 6 from Eq. 7. These results, together with Theorems 18.2.1 and 18.2.2, can be combined in the following statement.

Theorem 18.2.3

Let $\mathbf{F}$ be a continuous vector field on D. Then the following three statements are equivalent.

(a) $\mathbf{F}$ is the gradient of a potential function on D, $\mathbf{F}(x, y, z) = \nabla\phi(x, y, z)$.
(b) The line integral of $\mathbf{F}$ between any two points in D is independent of the path.

(c) The line integral of **F** around any piecewise smooth closed curve in D is zero.

It is clear from Theorem 18.2.1 that it is especially easy to evaluate the line integral $\int_C \mathbf{F} \cdot d\mathbf{r}$ if **F** is a conservative, or gradient, vector field. Hence the following two questions are of practical interest.

1. How can we tell whether a given vector field **F** is conservative? In other words how can we tell whether a potential function ϕ exists such that $\mathbf{F} = \nabla\phi$?
2. If a potential function ϕ does exist, how can we find it?

We will answer the second question first. Suppose that

$$\mathbf{F}(x, y, z) = P(x, y, z)\mathbf{i} + Q(x, y, z)\mathbf{j} + R(x, y, z)\mathbf{k}. \tag{8}$$

If there is a potential ϕ such that $\mathbf{F} = \nabla\phi$, then

$$\frac{\partial\phi}{\partial x} = P, \qquad \frac{\partial\phi}{\partial y} = Q, \qquad \frac{\partial\phi}{\partial z} = R. \tag{9}$$

To find ϕ, we must solve (if possible) these three equations simultaneously. This is done by a process of successive integration that is illustrated in the following two examples.

EXAMPLE 2

Determine whether the vector field

$$\mathbf{F}(x, y, z) = yz\mathbf{i} + (xz + y)\mathbf{j} + (xy + 2)\mathbf{k} \tag{10}$$

is conservative, and if so find a potential function ϕ.

If a potential function ϕ exists, then we must have

$$\phi_x(x, y, z) = yz, \qquad \phi_y(x, y, z) = xz + y, \qquad \phi_z(x, y, z) = xy + 2. \tag{11}$$

A solution of the first of Eqs. 11 is

$$\phi(x, y, z) = xyz + f(y, z), \tag{12}$$

where f is an arbitrary function of y and z, but not x. Substituting this expression into the second of Eqs. 11 gives

$$xz + \frac{\partial f(y, z)}{\partial y} = xz + y.$$

Hence

$$\frac{\partial f(y, z)}{\partial y} = y$$

so

$$f(y, z) = \frac{1}{2}y^2 + g(z).$$

Thus

$$\phi(x, y, z) = xyz + \frac{1}{2}y^2 + g(z), \tag{13}$$

where g is an arbitrary function of z only. Finally, substituting this expression for ϕ in the last of Eqs. 11 gives

$$xy + g'(z) = xy + 2.$$

This equation will be satisfied if we choose

$$g(z) = 2z + k,$$

where k is an arbitrary constant. Substituting for $g(z)$ in Eq. 13, we obtain

$$\phi(x, y, z) = xyz + \frac{1}{2}y^2 + 2z + k. \tag{14}$$

You may verify by direct calculation that in fact $\nabla\phi = \mathbf{F}$. Since we have actually constructed a potential function ϕ for $\mathbf{F}$, we conclude that the vector field (10) is conservative. ∎

EXAMPLE 3

Is the vector field

$$\mathbf{F}(x, y) = (3x^2 + 2xy)\mathbf{i} + (x + y^2)\mathbf{j} \tag{15}$$

conservative?

If so, there must be a function ϕ such that

$$\phi_x(x, y) = 3x^2 + 2xy, \qquad \phi_y(x, y) = x + y^2. \tag{16}$$

From the first of Eqs. 16 we find that

$$\phi(x, y) = x^3 + x^2y + h(y), \tag{17}$$

where h is an arbitrary function of y alone. Substituting this expression for ϕ in the second of Eqs. 16, we have

$$x^2 + h'(y) = x + y^2$$

or

$$h'(y) = -x^2 + x + y^2. \tag{18}$$

Since the right side of Eq. 18 depends on x as well as y, it is impossible to solve Eq. 18 for $h(y)$. Hence the vector field 15 is not conservative. ∎

Now we return to the first question, namely, how can we tell whether a given vector field is conservative? One way is simply to try to construct a potential function. If this can be done, as in Example 2, then the vector field is conservative; otherwise, as in Example 3, it is not. However, it is desirable to have a way to test the components of **F** to determine whether a potential function ϕ exists or not. It is especially desirable to be able to identify when ϕ does not exist, so that we do not waste time trying to solve Eqs. 9 when they are inconsistent.

For the case of a two-dimensional vector field

$$\mathbf{F}(x, y) = P(x, y)\mathbf{i} + Q(x, y)\mathbf{j}$$

that has continuous first partial derivatives in a domain D we can readily derive a necessary condition. Suppose that there does exist a $\phi(x, y)$ such that

$$\phi_x(x, y) = P(x, y), \qquad \phi_y(x, y) = Q(x, y).$$

Since $\phi_{xy} = \phi_{yx}$, it follows from these equations that

$$P_y(x, y) = Q_x(x, y)$$

is a necessary condition for the existence of a potential function ϕ; that is, if **F** has a potential, then $P_y = Q_x$. On the other hand, this condition is not sufficient; even if $P_y = Q_x$ at every point in D, there may still be no potential function ϕ. In order to guarantee the existence of a potential function, we must place an additional restriction on D, namely the domain D must be *simply connected.* We state the appropriate result as a theorem.

Theorem 18.2.4

(Independence of Path)

Let D be a simply connected domain in two- or three-dimensional space, let

$$\mathbf{F}(x, y) = P(x, y)\mathbf{i} + Q(x, y)\mathbf{j} \tag{19}$$

in two dimensions, or

$$\mathbf{F}(x, y, z) = P(x, y, z)\mathbf{i} + Q(x, y, z)\mathbf{j} + R(x, y, z)\mathbf{k} \tag{20}$$

in three dimensions, and let **F** have continuous first partial derivatives in D. A necessary and sufficient condition for the existence of a potential function ϕ such that $\nabla\phi = \mathbf{F}$, and hence for the line integral $\int_C \mathbf{F} \cdot d\mathbf{r}$ to be independent of path, is

$$Q_x(x, y) = P_y(x, y) \tag{21}$$

in two dimensions, and

$$R_y(x, y, z) = Q_z(x, y, z), \qquad P_z(x, y, z) = R_x(x, y, z),$$
$$Q_x(x, y, z) = P_y(x, y, z) \tag{22}$$

in three dimensions.

We have already shown that the condition (21) is necessary. In Section 18.3 we will establish the sufficiency of condition (21) for a simply connected domain in two dimensions. The three-dimensional case will be discussed in Section 18.8; also see Problem 24. We now illustrate that the requirement that D be simply connected is crucial.

EXAMPLE 4

Consider the vector field

$$\mathbf{F}(x, y) = P(x, y)\mathbf{i} + Q(x, y)\mathbf{j} = -\frac{y}{x^2 + y^2}\mathbf{i} + \frac{x}{x^2 + y^2}\mathbf{j} \tag{23}$$

defined on the domain D: $1 < x^2 + y^2 < 9$. Show that $P_y(x, y) = Q_x(x, y)$ at all points in D, but that the line integral $\int_C \mathbf{F} \cdot d\mathbf{r}$ is not independent of the path.

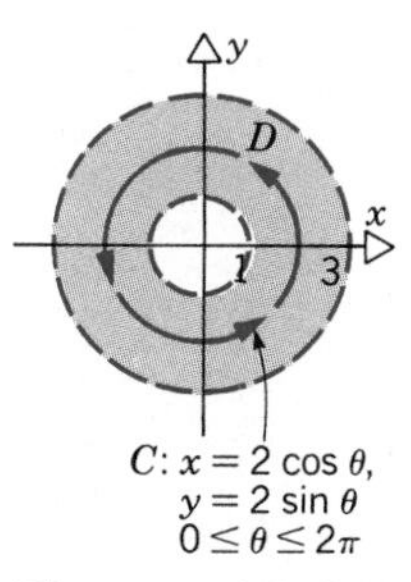

Figure 18.2.7

The domain D is shown in Figure 18.2.7. Notice that the functions P and Q are not bounded near the origin, so it is not possible to define $\mathbf{F}$ as a continuous function on a domain including the origin. As a consequence we have deleted the origin by introducing the hole $x^2 + y^2 \le 1$, and hence the domain D is multiply connected.

It is a simple calculation to show that

$$P_y(x, y) = Q_x(x, y) = \frac{y^2 - x^2}{(x^2 + y^2)^2} \quad \text{in} \quad D. \tag{24}$$

We will show that the line integral $\int_C \mathbf{F} \cdot d\mathbf{r}$ is not independent of path in D by showing that $\oint_C \mathbf{F} \cdot d\mathbf{r} \neq 0$, where C is the circular path $x = 2\cos\theta$, $y = 2\sin\theta$, $0 \le \theta \le 2\pi$ (see Figure 18.2.7). We have

$$\begin{aligned}
\oint_C \mathbf{F} \cdot d\mathbf{r} &= \oint_C P(x, y)\,dx + Q(x, y)\,dy \\
&= \oint_C \frac{-y\,dx + x\,dy}{x^2 + y^2} \\
&= \int_0^{2\pi} \frac{(-2\sin\theta)(-2\sin\theta\,d\theta) + (2\cos\theta)(2\cos\theta\,d\theta)}{4\cos^2\theta + 4\sin^2\theta} \\
&= \int_0^{2\pi} d\theta = 2\pi. \quad \blacksquare
\end{aligned} \tag{25}$$

For evaluating a line integral of a conservative vector field there is an alternative to determining the potential function ϕ. Since the line integral is independent of path it may, in some cases, be easier to choose an especially convenient path on which to carry out the integration. We illustrate this possibility in the following example.

EXAMPLE 5

Evaluate the line integral

$$\int_C P(x, y)\, dx + Q(x, y)\, dy = \int_C \frac{-y\, dx + x\, dy}{x^2 + y^2} \tag{26}$$

along the path C shown in Figure 18.2.8.

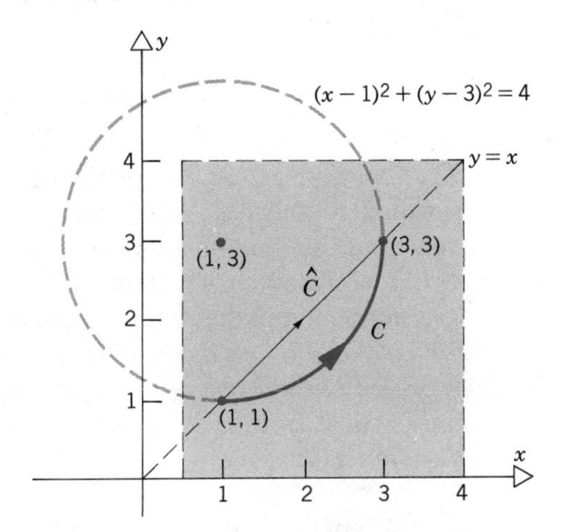

Figure 18.2.8

Notice that P and Q are the same functions as we discussed in Example 4, and that $P_y(x, y) = Q_x(x, y)$, provided $(x, y) \neq (0, 0)$. Thus if we consider a simply connected domain D that includes the path C, say $\frac{1}{2} < x < 4$, $0 < y < 4$, then we know that the line integral (26) is independent of the path in this domain.

We can evaluate the line integral by constructing a potential function ϕ such that

$$\phi_x(x, y) = \frac{-y}{x^2 + y^2}, \qquad \phi_y(x, y) = \frac{x}{x^2 + y^2}. \tag{27}$$

In this case

$$\int_C \frac{-y\, dx + x\, dy}{x^2 + y^2} = \phi(3, 3) - \phi(1, 1). \tag{28}$$

On the other hand, we can simply choose a more convenient path on which to evaluate the line integral. In particular if we choose the path $\hat{C}$: $y = x$, $1 \leq x \leq 3$, then we immediately observe that

$$-y\, dx + x\, dy = -x\, dx + x\, dx = 0.$$

Hence

$$\int_C \frac{-y\, dx + x\, dy}{x^2 + y^2} = 0. \quad \blacksquare$$

Applications to Mechanics. Consider the motion of a particle of mass m in a force field $\mathbf{F}(x, y, z)$. In paragraphs 1, 2, and 3 below assume that the force field is conservative; in paragraph 4 we consider the case when the force field is not conservative.

1. ***Work and potential energy.*** Since the force field is conservative, the work W that is done in moving the particle from any point $\mathbf{r}_a$ to any other point $\mathbf{r}_b$ is independent of the path, and there exists a potential function ϕ such that $\mathbf{F}(x, y, z) = -\nabla\phi(x, y, z)$. Note the introduction of the minus sign. We have

$$W = \int_C \mathbf{F} \cdot d\mathbf{r} = \int_C -\nabla\phi \cdot d\mathbf{r} = \phi(\mathbf{r}_a) - \phi(\mathbf{r}_b). \tag{29}$$

The function ϕ is known as the potential energy. From Eq. 29 the work done is equal to the potential energy at the initial position minus the potential energy at the final position, often referred to as the "loss" in potential energy. If C is a closed curve, then W is zero since $\mathbf{r}_a = \mathbf{r}_b$; hence there is neither a loss nor a gain in the potential energy of the particle during the motion. Thus, *on a closed path in a conservative force field the potential energy is conserved.*

2. ***Work and kinetic energy.*** Another way of calculating the work done in moving the particle from $\mathbf{r}_a$ to $\mathbf{r}_b$ is to make use of Newton's second law of motion: $\mathbf{F} = md^2\mathbf{r}/dt^2$. Then

$$\begin{aligned} W &= \int_C \mathbf{F} \cdot d\mathbf{r} = \int_a^b \mathbf{F}[\mathbf{r}(t)] \cdot \frac{d\mathbf{r}(t)}{dt}\, dt \\ &= \int_a^b m \frac{d^2\mathbf{r}(t)}{dt^2} \cdot \frac{d\mathbf{r}(t)}{dt}\, dt = \frac{1}{2} m \int_a^b \frac{d}{dt}\left[\frac{d\mathbf{r}(t)}{dt} \cdot \frac{d\mathbf{r}(t)}{dt}\right] dt \\ &= \frac{1}{2} m \int_a^b \frac{d}{dt} \|\mathbf{v}(t)\|^2\, dt \\ &= \frac{1}{2} m\|\mathbf{v}(b)\|^2 - \frac{1}{2} m\|\mathbf{v}(a)\|^2, \end{aligned} \tag{30}$$

where $\mathbf{v} = d\mathbf{r}/dt$ is the velocity of the particle. Since $m\|\mathbf{v}\|^2/2$ is the kinetic energy of the particle, the work done in moving the particle from $\mathbf{r}_a$ to $\mathbf{r}_b$ is equal to the kinetic energy at the final time minus the kinetic energy at the initial time, often referred to as the "gain" in kinetic energy. If the particle is moved over a closed curve so that $\mathbf{r}_a = \mathbf{r}_b$, then (since the force field is conservative) the work done is zero. As a consequence, the kinetic energy at the end of the path is the same as at the beginning, so *in a conservative force field the kinetic energy is also conserved on any closed path.*

3. ***Principle of conservation of energy.*** If the force field is conservative, then both Eqs. 29 and 30 are true, and hence

$$\phi(\mathbf{r}_a) - \phi(\mathbf{r}_b) = \frac{1}{2} m\|\mathbf{v}(b)\|^2 - \frac{1}{2} m\|\mathbf{v}(a)\|^2$$

or

$$\phi(\mathbf{r}_a) + \frac{1}{2}m\|\mathbf{v}(a)\|^2 = \phi(\mathbf{r}_b) + \frac{1}{2}m\|\mathbf{v}(b)\|^2. \tag{31}$$

It follows from Eq. 31 that the sum of the potential energy and the kinetic energy, called the mechanical energy, is constant throughout the motion of the particle. A decrease in potential energy must correspond to an increase in kinetic energy, and vice versa. To summarize: *for a conservative force field the potential energy and kinetic energy are conserved separately on any closed path, and their sum is conserved throughout every motion.*

4. ***Nonconservative force field.*** Consider now the situation when the force field is not conservative. An example of such a situation is the decaying oscillation of a pendulum in a viscous fluid (see Figure 18.2.9). Each time the pendulum bob passes through the vertical position, its speed is less than the previous time; thus the bob has different velocities at the same position at different times, and of course the paths are different to reach the same position. If the force field is not conservative, then Eq. 29 no longer holds because there is no potential function ϕ whose gradient is $\mathbf{F}$, but the work done in moving the particle from $\mathbf{r}_a$ to $\mathbf{r}_b$ is still given by Eq. 30. However, we must now specify the path taken by the particle from $\mathbf{r}_a$ and $\mathbf{r}_b$, since the velocity at $\mathbf{r}_b$ may depend on the time taken to reach this position. Even if $\mathbf{r}_b = \mathbf{r}_a$, the work done is not zero, so kinetic energy is not conserved on a closed path. Neither is mechanical energy conserved during the motion; some of the mechanical energy is converted into thermal energy through dissipation.

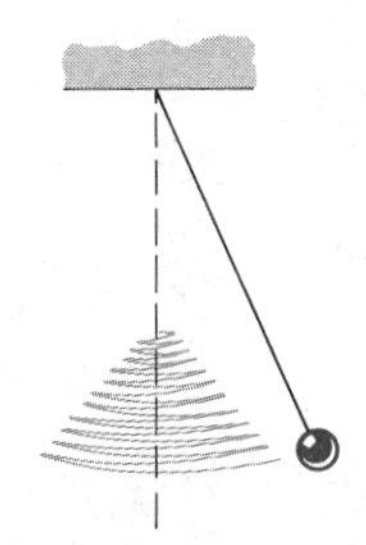

Figure 18.2.9 The decaying oscillations of a simple pendulum in air.

PROBLEMS

In each of Problems 1 through 8, determine whether the vector field is conservative, and, if so, determine the corresponding potential function ϕ.

1. $\mathbf{F}(x, y) = (2xy + 6)\mathbf{i} + (x^2 - 3)\mathbf{j}$
2. $\mathbf{F}(x, y) = (6xy - 4y^2)\mathbf{i} + (3x^2 + 3y^2 - 8xy)\mathbf{j}$
3. $\mathbf{F}(x, y) = (y \cos x + 2xe^y)\mathbf{i} + (\sin x + x^2e^y + 2)\mathbf{j}$
4. $\mathbf{F}(x, y, z) = (2y - 2zx)\mathbf{i} + (z^2 + 2x)\mathbf{j} + (2yz - x^2)\mathbf{k}$
5. $\mathbf{F}(x, y) = (2x + 4y)\mathbf{i} + (2x - 2y)\mathbf{j}$
6. $\mathbf{F}(x, y) = \left(6x + \frac{y}{x}\right)\mathbf{i} + (\ln x - 2)\mathbf{j}$
7. $\mathbf{F}(x, y, z) = \left(\frac{y}{z} - e^z\right)\mathbf{i} + \left(\frac{x}{z} + 3\right)\mathbf{j} - \left(xe^z + \frac{xy}{z^2}\right)\mathbf{k}$
8. $\mathbf{F}(x, y) = (e^x \sin y + 3y)\mathbf{i} - (3x - e^x \sin y)\mathbf{j}$

In each of Problems 9 through 14, show that the line integral is independent of the path, and then evaluate the integral.

9. $\int_C (2 \sin x + y^2)\, dx + (2xy + e^y)\, dy$; from $(0, 1)$ to $(\pi, 2)$
10. $\int_C (2xy^2 + 2y)\, dx + (2x^2y + 2x)\, dy$; from $(0, 0)$ to $(-2, 4)$
11. $\int_C (z^2 - y \sin x)\, dx + (\cos x - 2z)\, dy + (2zx - 2y + z)\, dz$; from $\left(\frac{\pi}{4}, 1, 1\right)$ to $(\pi, 0, 2)$

12. $\int_C \frac{x}{1 + r^2}\,dx + \frac{y}{1 + r^2}\,dy$, $r^2 = x^2 + y^2$; from $(0, 0)$ to (a, b)

13. $\int_C xe^{r^2}\,dx + ye^{r^2}\,dy + ze^{r^2}\,dz$, $r^2 = x^2 + y^2 + z^2$; from $(0, 0, 0)$ to (a, b, c)

14. $\int_C (ye^{xy}\cos z - 2e^{-2x}\sin \pi y)\,dx + (\pi e^{-2x}\cos \pi y + xe^{xy}\cos z)\,dy - (e^{xy}\sin z)\,dz$; from $\left(0, 1, \frac{\pi}{2}\right)$ to $\left(-1, -\frac{1}{2}, \pi\right)$

In each of Problems 15 and 16, determine the value of α for which the given line integral is independent of the path, and then evaluate the integral.

15. $\int_C (xy^2 + \alpha x^2 y)\,dx + (x^3 + x^2 y)\,dy$; from $(0, 0)$ to $(1, -1)$

16. $\int_C (ye^{2xy} + x)\,dx + \alpha xe^{2xy}\,dy$; from $(1, 1)$ to $(2, 0)$

17. Evaluate

$$\int_C \cos x \cos y\,dx - \sin x \sin y\,dy$$

where C is the octagonal path shown in Figure 18.2.10.

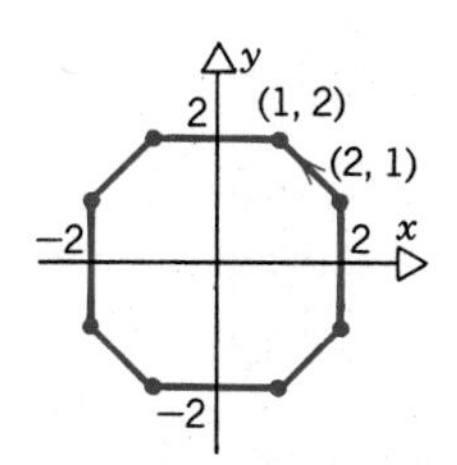

Figure 18.2.10

18. Evaluate

$$\int_C (e^x + y)\,dx + (e^y + x)\,dy$$

where C is the curve from $(0, 0)$ to $(\frac{1}{4}, 1)$ given by $\mathbf{r}(t) = t^2\mathbf{i} + \sin t\mathbf{j}$, $0 \le t \le \frac{1}{2}$.

19. Evaluate

$$\int_C \left(\frac{y}{x} + z - 4x\right)dx + (3 + \ln x)\,dy + x\,dz$$

where C is the curve from $(1, 0, 0)$ to $(e, 2, 1)$ given by $\mathbf{r}(t) = e^{t^2}\mathbf{i} + 2t\mathbf{j} + t\mathbf{k}$, $0 \le t \le 1$.

20. Consider the gravitational force field due to a mass of 200 kg located at the origin. Determine the work done in moving a mass of 5 kg from $(5, 0, 0)$ to $(0, 5, 2)$ along the curve $x = 5\cos t$, $y = 5\sin t$, $z = 4t/\pi$, $0 \le t \le \pi/2$. Here distances are measured in meters, and recall that $G \cong 6.67 \times 10^{-11}$ N-m²/kg².

21. Show that $\int_C \mathbf{r} \cdot d\mathbf{r}$ is independent of path and that the integral from $\mathbf{r}_a$ to $\mathbf{r}_b$ is $(r_b^2 - r_a^2)/2$ where $r = \|\mathbf{r}\|$.

22. A force field of the form

$$\mathbf{F} = f(r^2)(x\mathbf{i} + y\mathbf{j} + z\mathbf{k}), \qquad r^2 = x^2 + y^2 + z^2,$$

where f is a continuous function, is called a **central force field.** Why? Show that a central force field is conservative, and that if $\mathbf{F} = \nabla\phi$, then $\phi(x, y, z) = \frac{1}{2}g(r^2)$ where $g(t) = \int f(t)\,dt$. Problems 12 and 13 are special cases of this result.

23. Evaluate

$$\int_C y \sin(x^2 y^2)\,dx + x\sin(x^2 y^2)\,dy,$$

where C is the straight line from $(0, 2)$ to $(2, 0)$.

24. Let

$$\mathbf{F}(x, y, z) = P(x, y, z)\mathbf{i} + Q(x, y, z)\mathbf{j} + R(x, y, z)\mathbf{k}$$

be a conservative force field with continuous first partial derivatives. Show that

$$R_y(x, y, z) = Q_z(x, y, z),$$
$$P_z(x, y, z) = R_x(x, y, z),$$
$$Q_x(x, y, z) = P_y(x, y, z).$$

This establishes the necessary part of Theorem 18.2.4 for three-dimensional vector fields.

25. This problem provides a generalization of the technique of integration by parts. Suppose that ϕ and ψ have continuous partial derivatives in a three-dimensional domain D. Let C be a piecewise smooth curve, $\mathbf{r} = \mathbf{r}(t)$, $a \le t \le b$, from $\mathbf{r}_a$ to $\mathbf{r}_b$. Show that

$$\int_C \psi\nabla\phi \cdot d\mathbf{r} = \psi(\mathbf{r}_b)\phi(\mathbf{r}_b) - \psi(\mathbf{r}_a)\phi(\mathbf{r}_a) - \int_C \phi\nabla\psi \cdot d\mathbf{r}.$$

26. In this problem we sketch a proof of Theorem 18.2.2. For the sake of simplicity we only consider the case of two dimensions. Suppose that the functions P and Q are continuous in some domain D and that the line integral $\int_C (P(x, y)\,dx + Q(x, y)\,dy)$ is independent of path in D. Then we want to show there exists a function ϕ such that

$$\phi_x(x, y) = P(x, y), \qquad \phi_y(x, y) = Q(x, y).$$

(a) Pick a point (a, b) in D and define ϕ by

$$\phi(x, y) = \int_{(a,b)}^{(x,y)} P(s, t)\,ds + Q(s, t)\,dt.$$

Note that ϕ does not depend on the path connecting (a, b) to (x, y). Show that

$$\phi(x + \Delta x, y) - \phi(x, y) = \int_{(x,y)}^{(x+\Delta x,y)} P(s, y)\,ds,$$

where the integration is along the straight line segment from (x, y) to $(x + \Delta x, y)$, as shown in Figure 18.2.11. Now use the mean value theorem for integrals (Theorem 6.3.5) to show that $\phi_x(x, y) = P(x, y)$.

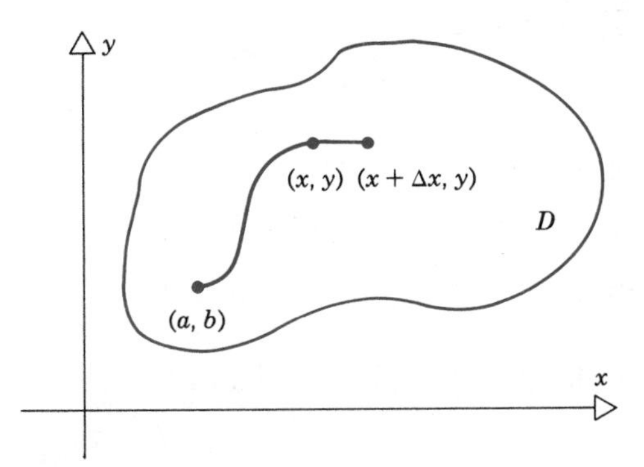

Figure 18.2.11

(b) In a manner analogous to the argument in Part (a), show that

$$\phi(x, y + \Delta y) - \phi(x, y) = \int_{(x,y)}^{(x,y+\Delta y)} Q(x, t)\,dt$$

and use the mean value theorem for integrals to show that $\phi_y(x, y) = Q(x, y)$.

27. A particle of mass m moves along the x-axis. Its position at time t is $x = x(t)$, and it is acted on by a force $F = -kx$ with $k > 0$.

(a) Show that the force field is conservative and determine a potential energy function.

(b) If the particle is at x_a at time $t = a$ and at x_b at time $t = b$ $(b > a)$, determine the work done during the motion.

28. For the particle of Problem 27, suppose that the force field is $F = -kx - cv$ where $v = dx/dt$ is the velocity and c is a positive constant. The term $-cv$ represents a damping force.

(a) Show that the work done in moving the particle from x_a to x_b is

$$W = \int_a^b \left[-kx(t)\frac{dx(t)}{dt} - c\frac{dx(t)}{dt}\frac{dx(t)}{dt} \right] dt$$

$$= \frac{1}{2}k(x_a^2 - x_b^2) - c\int_a^b \left[\frac{dx(t)}{dt}\right]^2 dt.$$

It is not possible to find a function $\phi(x)$ such that $\phi_x[x(t)] = [dx(t)/dt]^2$, and hence this force field is not conservative.

(b) If $x(t) = A \sin t$ and $a = 0$, $b = 2\pi$, determine W.

18.3 GREEN'S THEOREM

In this section we develop an important relation between line integrals along piecewise smooth simple closed curves in a plane and a double integral taken over the region enclosed by the curve.

Theorem 18.3.1

(Green's* Theorem)
Let D be a simply connected domain in the plane, and let C be a piecewise smooth simple closed curve in D that forms the boundary of a region Ω. Suppose that $P(x, y)$ and $Q(x, y)$ are continuous and have continuous first partial derivatives in D. Then

$$\oint_C P\,dx + Q\,dy = \iint_\Omega \left(\frac{\partial Q}{\partial x} - \frac{\partial P}{\partial y}\right) dA. \tag{1}$$

Proof. We will not attempt to prove Green's theorem for the most general case. Rather, we will only consider regions Ω that can be defined both by

$$a \le x \le b, \qquad f_1(x) \le y \le f_2(x) \tag{2a}$$

and

$$c \le y \le d, \qquad g_1(y) \le x \le g_2(y). \tag{2b}$$

Such a region is shown in Figure 18.3.1. Recall that regions of this type were discussed in Section 17.2, where a region that could be described either by Eq. 2(a) or by Eq. 2(b) was called a standard region.

First, consider

$$\iint_\Omega -\frac{\partial P}{\partial y}\,dA.$$

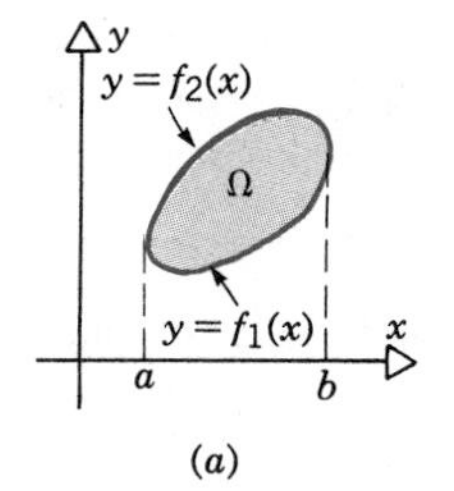

(a)

(b)

Figure 18.3.1

Referring to Figure 18.3.1*a*, we have

$$\begin{aligned}\iint_\Omega -\frac{\partial P}{\partial y}\,dA &= \int_a^b \int_{f_1(x)}^{f_2(x)} -\frac{\partial P(x, y)}{\partial y}\,dy\,dx \\ &= \int_a^b P[x, f_1(x)]\,dx - \int_a^b P[x, f_2(x)]\,dx \\ &= \int_a^b P[x, f_1(x)]\,dx + \int_b^a P[x, f_2(x)]\,dx \\ &= \oint_C P\,dx. \end{aligned} \tag{3}$$

*Green's theorem is named for George Green (1793–1841). Born in Nottingham, England, and poorly educated, Green nevertheless made important contributions to mathematics, especially in an essay on the application of mathematical analysis to electricity and magnetism in 1828. Green's theorem appears in this essay, as well as potential functions, and what are now called Green's functions in differential equations. Green eventually received a bachelor's degree from Caius College of Cambridge University in 1837 and was appointed to its faculty in 1839. Unfortunately, he died shortly afterward. Green's 1828 essay received little circulation and remained virtually unknown during his lifetime. It was republished in the 1850s through the efforts of William Thomson, later known as Lord Kelvin, and Green's important ideas then became well known.

Similarly, referring to Figure 18.3.1*b*, we obtain

$$\int\int_{\Omega} \frac{\partial Q}{\partial x}\, dA = \int_c^d \int_{g_1(y)}^{g_2(y)} \frac{\partial Q(x, y)}{\partial x}\, dx\, dy$$
$$= \int_c^d Q[g_2(y), y]\, dy - \int_c^d Q[g_1(y), y]\, dy$$
$$= \int_c^d Q[g_2(y), y]\, dy + \int_d^c Q[g_1(y), y]\, dy$$
$$= \oint_C Q\, dy. \tag{4}$$

Adding Eqs. 3 and 4 gives Eq. 1. □

We can readily extend this result to regions that can be decomposed into a union of subregions of the form (2); such a region is shown in Figures 18.3.2*a* and

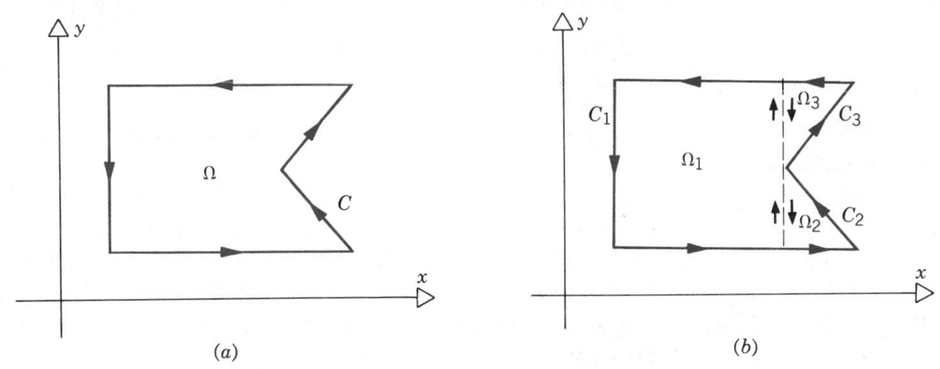

Figure 18.3.2

18.3.2*b*. We apply Green's theorem to each of the subregions Ω_1, Ω_2, and Ω_3 with boundaries C_1, C_2, and C_3, and add the results together to obtain

$$\oint_{C_1} (P\, dx + Q\, dy) + \oint_{C_2} (P\, dx + Q\, dy) + \oint_{C_3} (P\, dx + Q\, dy)$$
$$= \int\int_{\Omega_1} \left(\frac{\partial Q}{\partial x} - \frac{\partial P}{\partial y}\right) dA + \int\int_{\Omega_2} \left(\frac{\partial Q}{\partial x} - \frac{\partial P}{\partial y}\right) dA$$
$$+ \int\int_{\Omega_3} \left(\frac{\partial Q}{\partial x} - \frac{\partial P}{\partial y}\right) dA. \tag{5}$$

Each of the dashed lines in Figure 18.3.2*b* is traversed twice in opposite directions; hence these contributions to the sum of the line integrals on the left side of Eq. 5 cancel. Thus the sum of the line integrals over C_1, C_2, and C_3 reduces to the line integral over C. Also the sum of the double integrals over Ω_1, Ω_2, and Ω_3 is equal to the double integral over Ω. Consequently, for the more general region we have

$$\oint_C P\, dx + Q\, dy = \int\int_{\Omega} \left(\frac{\partial Q}{\partial x} - \frac{\partial P}{\partial y}\right) dA.$$

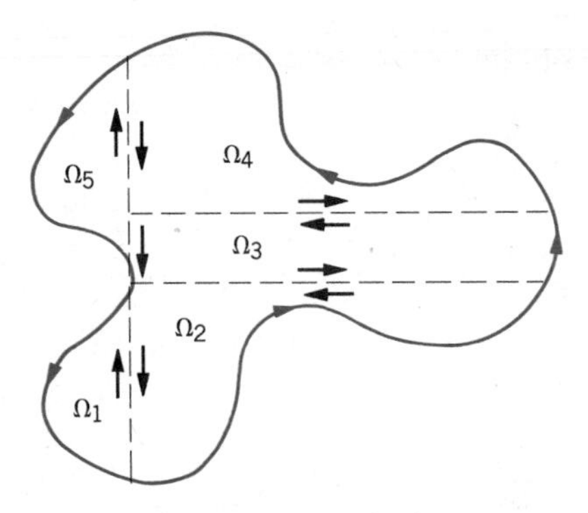

Figure 18.3.3

An even more complicated region and its subdivision into subregions of the form (2) is shown in Figure 18.3.3.

It is clear from Eq. 1 that Green's theorem relates the behavior of the partial derivatives of P and Q in Ω to the values of P and Q themselves on the boundary. In this respect it is similar to the fundamental theorem of calculus in the form

$$\int_a^b \frac{dF(x)}{dx} = F(b) - F(a),$$

in which the integral of the derivative is related to the value of the function at the end (boundary) points.

We can use Green's theorem to prove the sufficient part of Theorem 18.2.4 for two dimensions; namely, if $Q_x - P_y = 0$ in a simply connected domain D, then the line integral $\int_C P\,dx + Q\,dy$ is independent of the path. It follows immediately from Eq. 1 that if $Q_x - P_y = 0$ in D, then $\oint_C P\,dx + Q\,dy = 0$ for every piecewise smooth simple closed curve in D. By Theorem 18.2.3 this is equivalent to independence of path. We also recall that if $\mathbf{F}(x, y) = P(x, y)\mathbf{i} + Q(x, y)\mathbf{j}$ is a conservative vector field ($\mathbf{F} = \nabla\phi$), then the line integral in Eq. 1 is zero. It is easy to confirm that the corresponding double integral is zero, since

$$\int\int_\Omega \left(\frac{\partial Q}{\partial x} - \frac{\partial P}{\partial y}\right) dA = \int\int_\Omega \left[\frac{\partial}{\partial x}\left(\frac{\partial \phi}{\partial y}\right) - \frac{\partial}{\partial y}\left(\frac{\partial \phi}{\partial x}\right)\right] dA = 0. \tag{6}$$

Finally, note that Green's theorem provides a second way of evaluating line integrals and double integrals. Given a line integral along a piecewise smooth simple closed curve, we can either evaluate the line integral directly or else use Green's Theorem and evaluate the corresponding double integral. Alternatively, we can evaluate a double integral of $f(x, y)$ by choosing the functions P and Q so that $\partial Q(x, y)/\partial x - \partial P(x, y)/\partial y = f(x, y)$, and then evaluating the corresponding line integral. We illustrate these ideas in the following examples.

EXAMPLE 1

Use Green's theorem to evaluate

$$\oint_C (2x - y^2)\,dx + (xy - 1)\,dy$$

where C is the triangle shown in Figure 18.3.4.

We have

$$\oint_C (2x - y^2)\,dx + (xy - 1)\,dy = \int\!\!\int_\Omega \left[\frac{\partial}{\partial x}(xy - 1) - \frac{\partial}{\partial y}(2x - y^2)\right] dA$$

$$= \int\!\!\int_\Omega 3y\,dA$$

$$= \int_0^2 \int_0^{-(x-2)/2} 3y\,dy\,dx$$

$$= 3\int_0^2 \frac{1}{2}\cdot\frac{1}{4}(x - 2)^2\,dx$$

$$= \frac{3}{8}\frac{(x-2)^3}{3}\Bigg|_0^2 = 1. \blacksquare$$

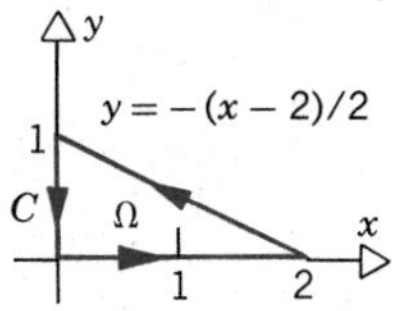

Figure 18.3.4

EXAMPLE 2

Derive a formula for the area of a region Ω bounded by a piecewise smooth simple closed curve C in terms of a line integral around C (see Figure 18.3.5).

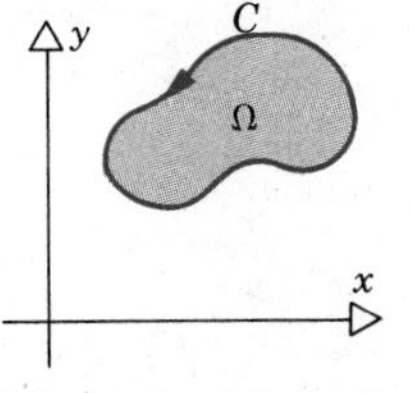

Figure 18.3.5

It follows from Eq. 1 that if we choose P and Q so that

$$\frac{\partial Q}{\partial x} - \frac{\partial P}{\partial y} = k, \tag{7}$$

where k is a constant, then the double integral is equal to $k \times$ (area of Ω). Thus

$$\text{Area of } \Omega = \frac{1}{k}\oint_C P\,dx + Q\,dy. \tag{8}$$

There are many choices for P and Q. Here are three simple ones.

1. $P(x, y) = y$, $Q(x, y) = 0$, so $k = -1$, and

$$\text{Area of } \Omega = \oint_C -y\,dx. \tag{9}$$

2. $P(x, y) = 0$, $Q(x, y) = x$, so $k = 1$, and

$$\text{Area of } \Omega = \oint_C x\,dy. \tag{10}$$

3. $P(x, y) = -y$, $Q(x, y) = x$, so $k = 2$, and

$$\text{Area of } \Omega = \frac{1}{2}\oint_C -y\,dx + x\,dy. \blacksquare \tag{11}$$

EXAMPLE 3

Use Eq. 11 to determine the area of the region within the ellipse $x^2/a^2 + y^2/b^2 = 1$.

A simple parametric representation of this ellipse is

$$x = a \cos t, \qquad y = b \sin t, \qquad 0 \le t \le 2\pi.$$

Using Eq. 11 we have

$$\text{Area} = \frac{1}{2}\int_0^{2\pi} (-b \sin t)(-a \sin t\, dt) + (a \cos t)(b \cos t\, dt)$$

$$= \frac{1}{2}\int_0^{2\pi} ab\, dt = \pi ab. \blacksquare$$

It is not difficult to extend Green's theorem to multiply connected domains—domains with holes. For example, consider the region Ω whose boundary consists of two piecewise smooth simple closed curves C_1 and C_2, as shown in Figure 18.3.6*a*. In order to apply Green's theorem, we introduce a "cut", say from A to B, that joins the outer and inner boundaries C_1 and C_2 (see Figure 18.3.6*b*). The effect of

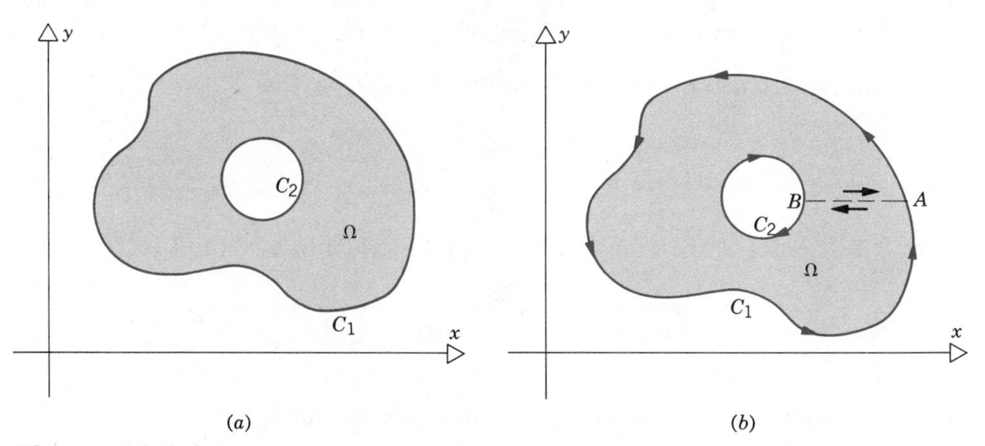

Figure 18.3.6 (*a*) A multiply connected domain. (*b*) Application of Green's theorem to a multiply connected domain.

the cut is to convert Ω into a simply connected region. Applying Green's theorem to the cut region, and starting at the point A, we have

$$\left(\oint_{C_1} + \int_{AB} + \oint_{C_2} + \int_{BA}\right)(P\,dx + Q\,dy) = \iint_\Omega \left(\frac{\partial Q}{\partial x} - \frac{\partial P}{\partial y}\right) dA.$$

Since the cut is traversed twice, once in each direction, the line integrals over the cut cancel, and we obtain

$$\oint_{C_1} (P\,dx + Q\,dy) + \oint_{C_2} (P\,dx + Q\,dy) = \iint_\Omega \left(\frac{\partial Q}{\partial x} - \frac{\partial P}{\partial y}\right) dA. \quad (12)$$

Notice that the boundary curves C_1 and C_2 are traversed so that the region Ω is always to the left. If we use this convention, then we can write Green's theorem in the form

$$\oint_C P\,dx + Q\,dy = \iint_\Omega \left(\frac{\partial Q}{\partial x} - \frac{\partial P}{\partial y}\right) dA, \tag{13}$$

where C is the complete boundary of Ω and is traversed so that Ω always lies to the left.

EXAMPLE 4

Let C be a piecewise smooth simple closed curve. Show that

$$\oint_C \frac{-y}{x^2+y^2}\,dx + \frac{x}{x^2+y^2}\,dy = \begin{cases} 0, & \text{if } C \text{ does not enclose the origin;} \\ 2\pi, & \text{if } C \text{ does enclose the origin.} \end{cases} \tag{14}$$

This line integral was discussed in Example 4 of Section 18.2. There we observed that $P(x, y) = -y/(x^2 + y^2)$ and $Q(x, y) = x/(x^2 + y^2)$ are continuous and have continuous first partial derivatives except at the origin, and that

$$P_y(x, y) = Q_x(x, y) = \frac{y^2 - x^2}{(x^2 + y^2)^2}, \qquad (x, y) \neq (0, 0).$$

If C does not enclose the origin, then Green's theorem for a simply connected domain, Eq. 1, can be used and we have

$$\oint_C P\,dx + Q\,dy = \iint_\Omega \left(\frac{\partial Q}{\partial x} - \frac{\partial P}{\partial y}\right) dA = 0.$$

Now suppose that C encloses the origin. Then draw a circle C_ϵ of radius ϵ about the origin, and choose ϵ sufficiently small so that C_ϵ lies inside C (see Figure 18.3.7). For this annular region, Green's theorem for a multiply connected domain,

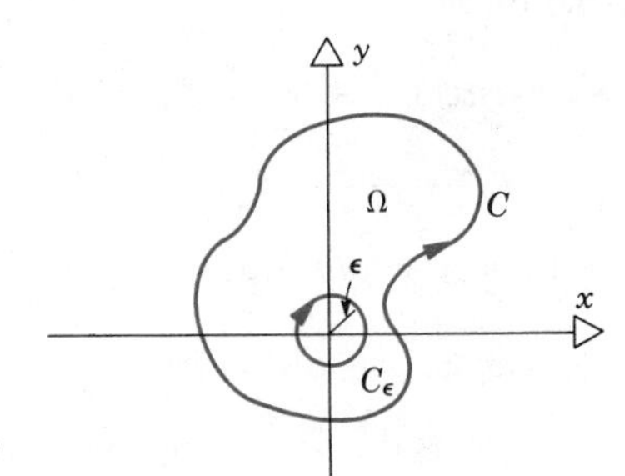

Figure 18.3.7

Eq. 12, can be used and we have

$$\oint_C (P\,dx + Q\,dy) + \oint_{C_\epsilon} (P\,dx + Q\,dy) = \iint_\Omega \left(\frac{\partial Q}{\partial x} - \frac{\partial P}{\partial y}\right) dA = 0.$$

Hence

$$\oint_C P\,dx + Q\,dy = -\oint_{C_\epsilon} P\,dx + Q\,dy = \oint_{C_\epsilon} P\,dx + Q\,dy.$$

We can evaluate the line integral along C_ϵ by taking $x = \epsilon \cos\theta$, $y = \epsilon \sin\theta$, $0 \le \theta \le 2\pi$; see Example 4 of Section 18.2 for a similar calculation. The result is

$$\oint_C P\,dx + Q\,dy = \oint_C \frac{-y\,dx + x\,dy}{x^2 + y^2} = 2\pi. \blacksquare$$

PROBLEMS

In each of Problems 1 through 4, verify that Green's theorem is true by calculating both sides of Eq. 1.

1. $\mathbf{F}(x, y) = y\mathbf{i} + x^2\mathbf{j}$; C is the circle $x^2 + y^2 = a^2$
2. $\mathbf{F}(x, y) = y\mathbf{i} - x\mathbf{j}$; C is the square with vertices at $(\pm a, \pm a)$
3. $\mathbf{F}(x, y) = (2y - x)\mathbf{i} + (3x + y)\mathbf{j}$; Ω is the region bounded by $x^2 \le y \le \sqrt{x}$, $0 \le x \le 1$
4. $\mathbf{F}(x, y) = xy\mathbf{i}$; Ω is the square with vertices $(0, 0)$, $(a, 0)$, (a, a), $(0, a)$

In each of Problems 5 through 10, use Green's theorem to evaluate the given line integral.

5. $\oint_C 2\,dx - 3\,dy$; C is the triangle with vertices at $(0, 0)$, (a, a), and $(0, 2a)$
6. $\oint_C 2y\,dx - 3x\,dy$; C is the triangle with vertices at $(0, 0)$, (a, a), and $(0, 2a)$
7. $\oint_C xy\,dy$; C is the semicircle $x^2 + y^2 = a^2$, $y > 0$ and the line $y = 0$, $-a \le x \le a$
8. $\oint_C (xy + y^2)\,dx + (2xy + x^2)\,dy$; C is the circle $(x - 1)^2 + y^2 = 1$
9. $\oint_C 3xy^2\,dx - 5x^2y\,dy$; C is the square with vertices at $(\pm 1, \pm 1)$
10. $\oint_C e^x \cos y\,dx + e^x \sin y\,dy$; C is the rectangle with vertices $(\pm 1, 0)$ and $(\pm 1, \pi/2)$

In each of Problems 11 through 14, use one of the formulas (9), (10), or (11) to evaluate the area of the given region.

11. The interior of the circle $x^2 + y^2 = a^2$.
12. The interior of the astroid $x^{2/3} + y^{2/3} = a^{2/3}$. *Hint:* Use the parametric representation $x = a\cos^3 t$, $y = a\sin^3 t$.
13. The region bounded by the cycloid $x = a(t - \sin t)$, $y = a(1 - \cos t)$, $0 \le t \le 2\pi$, and the x-axis.
14. The region bounded by $y = x^2$ and $y = x + 2$.
15. If ϕ satisfies Laplace's equation, $\phi_{xx} + \phi_{yy} = 0$, in a simply connected domain D, show that

$$\oint_C \phi_y\,dx - \phi_x\,dy = 0$$

for every piecewise smooth simple closed curve C in D.

16. Suppose that ϕ and ψ are continuous functions of two variables with continuous first partial derivatives in a simply connected domain D. If C is any piecewise smooth simple closed curve in D, show that

$$\oint_C \phi\nabla\psi \cdot d\mathbf{r} + \oint_C \psi\nabla\phi \cdot d\mathbf{r} = 0.$$

17. Let $(\hat{x}, \hat{y})$ be the centroid (Section 17.3) of a plane region Ω with area A that is bounded by a piecewise smooth simple closed curve. Show that

$$\hat{x} = \frac{1}{2A}\oint_C x^2\,dy, \qquad \hat{y} = -\frac{1}{2A}\oint_C y^2\,dx.$$

18. The polar moment of inertia about the z-axis of a region Ω in the xy-plane (Section 17.3) is

$$I_0 = \iint_\Omega (x^2 + y^2)\, dA.$$

If Ω is a region in a simply connected domain with boundary C that is a piecewise smooth simple closed curve, show that

$$I_0 = \frac{1}{3}\oint_C - y^3\, dx + x^3\, dy.$$

19. Consider the line integral

$$\int_C P\, dx + Q\, dy = \int_C \frac{x}{x^2 + y^2}\, dx + \frac{y}{x^2 + y^2}\, dy.$$

(a) Show that $P_y(x, y) = Q_x(x, y)$ for (x, y) in the annulus D: $0 < \epsilon^2 < x^2 + y^2 < R^2$ for any $\epsilon > 0$ and $R > \epsilon$.

(b) Show that $\int_C P\, dx + Q\, dy = 0$, where C is any circle $x^2 + y^2 = a^2$, $a > 0$.

(c) Show that $\int_C P\, dx + Q\, dy = 0$ on any piecewise smooth simple closed curve lying entirely in D, whether C encloses the origin or not. Compare this result with that of Example 4.

(d) Find a potential function ϕ such that $\phi_x(x, y) = P(x, y)$ and $\phi_y(x, y) = Q(x, y)$ for $(x, y) \neq (0, 0)$.

In Problems 20 and 21, let C be a piecewise smooth simple closed curve in a simply connected domain D, and let Ω be the interior of C. Let

$$\mathbf{T} = \frac{dx}{ds}\mathbf{i} + \frac{dy}{ds}\mathbf{j} \quad \text{and} \quad \mathbf{N} = \frac{dy}{ds}\mathbf{i} - \frac{dx}{ds}\mathbf{j}$$

be the unit tangent and outward unit normal vectors, respectively, to the curve C (see Figure 18.3.8). The symbol ∇^2 is defined by $\nabla^2\phi = \phi_{xx} + \phi_{yy}$, and $\partial\phi/\partial n$ denotes the directional derivative of ϕ in the direction of $\mathbf{N}$.

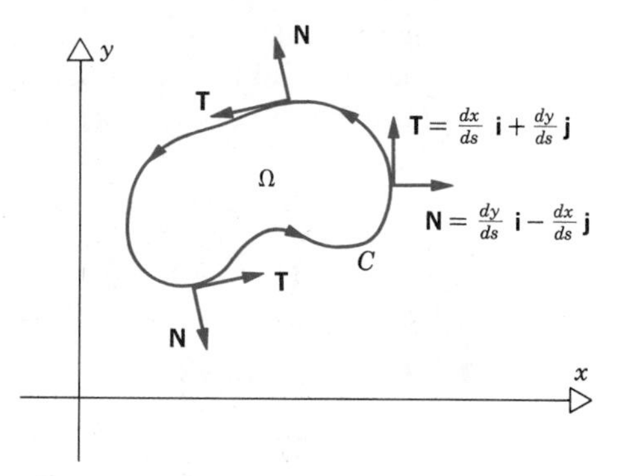

Figure 18.3.8

20. If $\mathbf{F}(x, y) = P(x, y)\mathbf{i} + Q(x, y)\mathbf{j}$, show that

$$\oint_C \mathbf{F} \cdot \mathbf{T}\, ds = \iint_\Omega \left(\frac{\partial Q}{\partial x} - \frac{\partial P}{\partial y}\right) dA$$

and

$$\oint_C \mathbf{F} \cdot \mathbf{N}\, ds = \iint_\Omega \left(\frac{\partial P}{\partial x} + \frac{\partial Q}{\partial y}\right) dA.$$

* **21.** If u and v are functions with continuous second partial derivatives, derive the following identities:

(a) $\displaystyle\oint_C \frac{\partial u}{\partial n}\, ds = \iint_\Omega \nabla^2 u\, dA$

(b) $\displaystyle\oint_C \left(u\frac{\partial v}{\partial n} - v\frac{\partial u}{\partial n}\right) ds = \iint_\Omega (u\nabla^2 v - v\nabla^2 u)\, dA$

(c) $\displaystyle\oint_C u\frac{\partial v}{\partial n}\, ds = \iint_\Omega [u\nabla^2 v + (\nabla u) \cdot (\nabla v)]\, dA$

(d) $\displaystyle\oint_C u\frac{\partial u}{\partial n}\, ds = \iint_\Omega (u\nabla^2 u + \|\nabla u\|^2)\, dA$

Identities (b) and (c) are known as Green's identities.

18.4 SURFACE AREA AND SURFACE INTEGRALS

The main goal of the remaining sections of this chapter is to extend the relation expressed in Green's theorem from two dimensions to three. The first step is to define and interpret a class of integrals over curved surfaces. We begin by showing how to calculate the area of an essentially arbitrary surface, thereby generalizing the result obtained in Section 7.3 for surfaces of revolution. We will proceed in a fairly intuitive way so as to avoid complications that belong only in a more advanced course.

A surface S may be described by an equation of the form

$$F(x, y, z) = 0. \tag{1}$$

Alternatively, we may express one of the variables in terms of the other two, as in

$$z = f(x, y). \tag{2}$$

For a surface S given by Eq. 1 the gradient vector

$$\nabla F = F_x\mathbf{i} + F_y\mathbf{j} + F_z\mathbf{k} \tag{3}$$

is normal to the surface. Thus a unit normal to S is

$$\mathbf{N} = \frac{\nabla F}{\|\nabla F\|} = \frac{F_x\mathbf{i} + F_y\mathbf{j} + F_z\mathbf{k}}{(F_x^2 + F_y^2 + F_z^2)^{1/2}}. \tag{4}$$

Observe that $-\mathbf{N} = -\nabla F/\|\nabla F\|$ is also a possible unit normal vector. For a surface given by Eq. 2 the possible unit normal vectors are

$$\mathbf{N} = \pm\frac{-f_x\mathbf{i} - f_y\mathbf{j} + \mathbf{k}}{(f_x^2 + f_y^2 + 1)^{1/2}}; \tag{5}$$

here the plus sign gives the upward pointing normal, that is, the one with a positive z component.

The surface S is said to be **smooth** if $\mathbf{N}$ varies continuously on S. If S is given by Eq. 1, then it is sufficient for F to have continuous first partial derivatives, and

$$\|\nabla F\| = (F_x^2 + F_y^2 + F_z^2)^{1/2} \neq 0 \tag{6}$$

for every point of S. If S is given by Eq. 2, then it is sufficient for f to have continuous partial derivatives throughout the domain in the xy-plane corresponding to S. Finally, if S consists of a finite number of smooth parts joined together along their edges, then S is called **piecewise smooth.** Thus, for example, an ellipsoid or paraboloid is smooth, while a cube or tetrahedron is piecewise smooth.

We now derive an integral formula for the area of a smooth surface. Consider first the plane element ΔS in the shape of a parallelogram, shown in Figure 18.4.1, that lies above a rectangle ΔR in the xy-plane with sides Δx and Δy and area $\Delta A = \Delta x\, \Delta y$. The sides of the parallelogram are parallel to the xz- and yz-planes, respectively, and are described by the vectors $\mathbf{a}$ and $\mathbf{b}$, where

$$\mathbf{a} = \Delta x\mathbf{i} + a_3\mathbf{k}, \qquad \mathbf{b} = \Delta y\mathbf{j} + b_3\mathbf{k}.$$

Further, the orientation of the parallelogram is given by the unit normal vector $\mathbf{N}$. In terms of $\mathbf{a}$ and $\mathbf{b}$ the upward pointing unit normal vector is

$$\mathbf{N} = \frac{\mathbf{a} \times \mathbf{b}}{\|\mathbf{a} \times \mathbf{b}\|}. \tag{7}$$

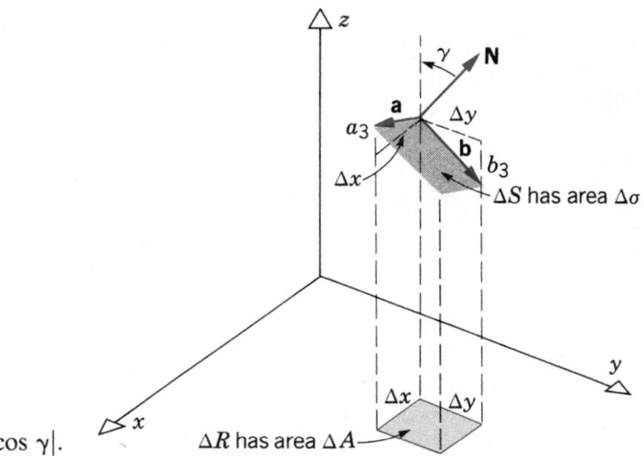

Figure 18.4.1 $\Delta\sigma = \Delta A/|\cos \gamma|$.

Recall from Section 15.4 that $\|\mathbf{a} \times \mathbf{b}\|$ gives the area of the parallelogram, which we will denote by $\Delta\sigma$. Thus Eq. 7 can be written as

$$(\Delta\sigma)\,\mathbf{N} = \mathbf{a} \times \mathbf{b}, \tag{8}$$

and by taking the component of Eq. 8 in the z direction, we obtain

$$\begin{aligned}\Delta\sigma\,(\mathbf{N}\cdot\mathbf{k}) &= \mathbf{a}\times\mathbf{b}\cdot\mathbf{k}\\ &= \begin{vmatrix}\Delta x & 0 & a_3\\ 0 & \Delta y & b_3\\ 0 & 0 & 1\end{vmatrix} = \Delta x\,\Delta y = \Delta A.\end{aligned}$$

If the downward pointing unit normal is used, then $-\Delta\sigma(\mathbf{N}\cdot\mathbf{k}) = \Delta A$. To permit the use of either normal we introduce the absolute value of $\mathbf{N}\cdot\mathbf{k}$, so that in all cases

$$\Delta\sigma = \frac{\Delta A}{|\mathbf{N}\cdot\mathbf{k}|} = \frac{\Delta A}{|\cos\gamma|}, \tag{9}$$

where γ is the angle between $\mathbf{N}$ and the positive z-axis.

Now suppose that R: $a \le x \le b$, $c \le y \le d$ is a rectangle in the xy-plane and that S is the surface defined by

$$z = f(x, y), \qquad (x, y) \in R$$

(see Figure 18.4.2). Assume that f has continuous first partial derivatives throughout R, so that S is smooth. As in the discussion of double integrals in Section 17.1 we form a partition Δ of R, consisting of subrectangles with sides parallel to the x- and y-axes:

$$a = x_0 < x_1 < \cdots < x_m = b, \qquad c = y_0 < y_1 < \cdots < y_n = d.$$

A typical subrectangle is R_{ij}: $x_{i-1} \le x \le x_i$, $y_{j-1} \le y \le y_j$, whose area is $\Delta A_{ij} = \Delta x_i\,\Delta y_j$. The norm $\|\Delta\|$ of the partition is the length of the longest diagonal among all of the subrectangles R_{ij}. Choose an arbitrary star point (x_i^*, y_j^*) in R_{ij} and construct the tangent plane P_{ij} at the corresponding point $[x_i^*, y_j^*, f(x_i^*, y_j^*)]$ on the surface S; this is indicated in Figure 18.4.3. By Eq. 5 the upward pointing unit

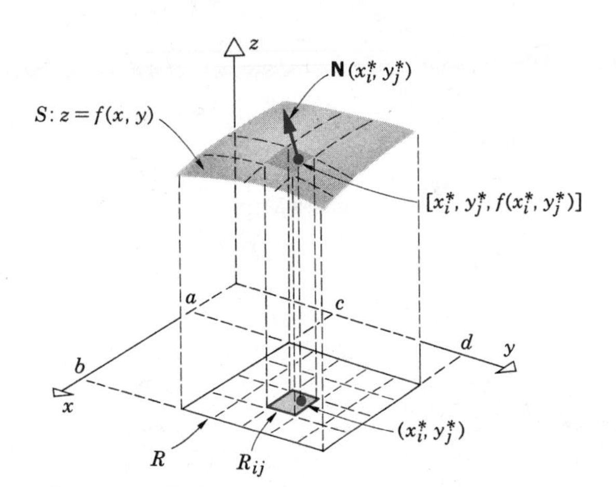

Figure 18.4.2 A partition of the rectangle R also partitions the surface S.

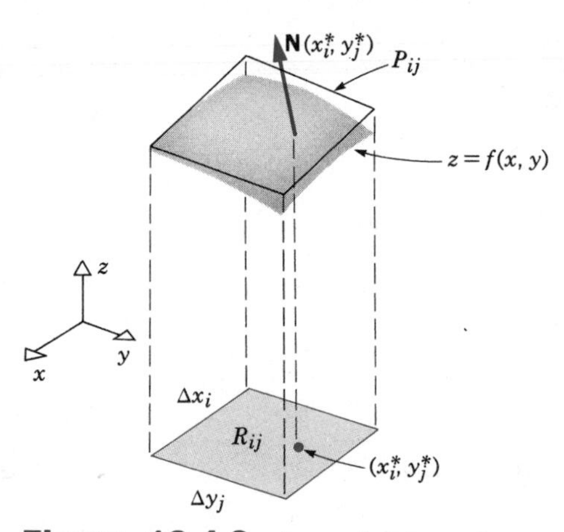

Figure 18.4.3 Approximation of a surface element by a tangent plane element.

normal at this point is

$$\mathbf{N}(x_i^*, y_j^*) = \frac{-f_x(x_i^* y_j^*)\mathbf{i} - f_y(x_i^*, y_j^*)\mathbf{j} + \mathbf{k}}{[f_x^2(x_i^*, y_j^*) + f_y^2(x_i^*, y_j^*) + 1]^{1/2}}. \tag{10}$$

Let $\Delta\sigma_{ij}$ be the area of that portion of the tangent plane P_{ij} that lies directly above R_{ij}. Then from Eq. 9 we have

$$\begin{aligned}\Delta\sigma_{ij} &= \frac{\Delta A_{ij}}{|\mathbf{N}(x_i^*, y_j^*) \cdot \mathbf{k}|}\\ &= [f_x^2(x_i^*, y_j^*) + f_y^2(x_i^*, y_j^*) + 1]^{1/2}\,\Delta A_{ij}.\end{aligned} \tag{11}$$

By proceeding in the same way in each of the subrectangles and then adding the results we obtain the following sum:

$$\sum_{\substack{i=1,\ldots,m\\ j=1,\ldots,n}} [f_x^2(x_i^*, y_j^*) + f_y^2(x_i^*, y_j^*) + 1]^{1/2}\,\Delta A_{ij}. \tag{12}$$

The sum (12) is a two-dimensional Riemann sum, so as $m, n \to \infty$ and $\|\Delta\| \to 0$, it approaches the corresponding integral over the rectangle R. We define the value of this integral to be the area $A(S)$ of the surface S,

$$A(S) = \iint_R [f_x^2(x, y) + f_y^2(x, y) + 1]^{1/2}\,dA. \tag{13}$$

EXAMPLE 1

Find the area of the surface $z = 4 - \frac{2}{3}x^{3/2}$ that lies above the rectangle R: $1 \le x \le 3$, $0 \le y \le 2$ (see Figure 18.4.4).

In this case $f_x = -x^{1/2}$ and $f_y = 0$, so

$$(f_x^2 + f_y^2 + 1)^{1/2} = (1 + x)^{1/2}.$$

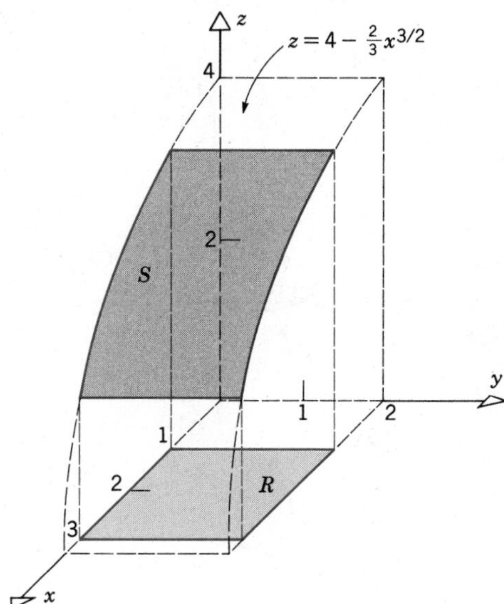

Figure 18.4.4

Then the surface area is given by Eq. 13,

$$A(S) = \iint_R (1 + x)^{1/2}\, dA = \int_1^3 \int_0^2 (1 + x)^{1/2}\, dy\, dx$$

$$= 2\int_1^3 (1 + x)^{1/2}\, dx = \frac{4}{3}(1 + x)^{3/2}\Big|_1^3$$

$$= \frac{4}{3}(8 - 2\sqrt{2}) \cong 6.895. \blacksquare$$

As in Section 17.1, it is possible to extend the concept of surface area so as to integrate over regions other than rectangles. If the surface is given by an equation of the form

$$z = f(x, y), \tag{14}$$

then it is said to be projectible into the xy-plane. Let Ω_{xy} be the region in the xy-plane that is the projection of the surface (see Figure 18.4.5a). Then the area of S is given by

$$A(S) = \iint_{\Omega_{xy}} (1 + f_x^2 + f_y^2)^{1/2}\, dA_{xy}; \tag{15a}$$

the subscripts on the area element dA_{xy} indicate the integration variables. Alternatively, we can write

$$A(S) = \iint_{\Omega_{xy}} |\sec \gamma|\, dA_{xy} = \iint_{\Omega_{xy}} \frac{dA_{xy}}{|\mathbf{N} \cdot \mathbf{k}|}. \tag{15b}$$

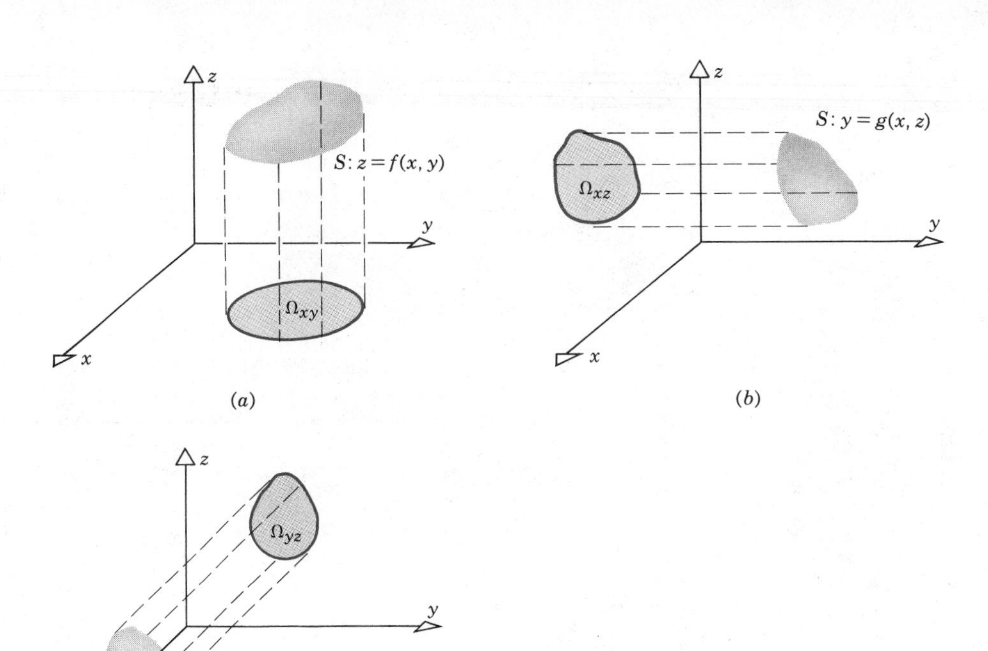

Figure 18.4.5 (*a*) A surface projectible into the xy-plane. (*b*) A surface projectible into the xz-plane. (*c*) A surface projectible into the yz-plane.

Similarly, if the surface is given by

$$y = g(x, z), \tag{16}$$

then it is projectible into the xz-plane (see Figure 18.4.5*b*) and its area is

$$\begin{aligned} A(S) &= \iint_{\Omega_{xz}} |\sec \beta| \, dA_{xz} = \iint_{\Omega_{xz}} \frac{dA_{xz}}{|\mathbf{N} \cdot \mathbf{j}|} \\ &= \iint_{\Omega_{xz}} [1 + g_x^2 + g_z^2]^{1/2} \, dA_{xz}, \end{aligned} \tag{17}$$

where Ω_{xz} is the appropriate region in the xz-plane, and β is the angle between N and the positive y axis. Finally, if the surface has the equation

$$x = h(y, z), \tag{18}$$

then, as shown in Figure 18.4.5*c*, it is projectible into the yz-plane, and has surface area

$$\begin{aligned} A(S) &= \iint_{\Omega_{yz}} |\sec \alpha| \, dA_{yz} = \iint_{\Omega_{yz}} \frac{dA_{yz}}{|\mathbf{N} \cdot \mathbf{i}|} \\ &= \iint_{\Omega_{yz}} [1 + h_y^2 + h_z^2]^{1/2} \, dA_{yz}, \end{aligned} \tag{19}$$

where Ω_{yz} is the corresponding region in the yz plane, and α is the angle between N and the positive x axis. Of course, a given surface may be projectible into two or even all three of the coordinate planes. In this case one should normally choose the projection leading to the simplest integration.

A surface may also be given by an equation of the form

$$F(x, y, z) = 0 \tag{20}$$

and may be projectible into any of the coordinate planes. If it is projectible into the xy-plane, then the element of surface area $d\sigma$ satisfies

$$|\mathbf{N} \cdot \mathbf{k}|\, d\sigma = dA_{xy},$$

where $\mathbf{N}$ is given by Eq. 4, and consequently

$$|\mathbf{N} \cdot \mathbf{k}| = \frac{|F_z|}{(F_x^2 + F_y^2 + F_z^2)^{1/2}}.$$

Thus the surface area is

$$A(S) = \iint_{\Omega_{xy}} \frac{(F_x^2 + F_y^2 + F_z^2)^{1/2}}{|F_z|}\, dA_{xy}, \tag{21}$$

where Ω_{xy} is the projection in the xy-plane of the surface. Observe that in the integrand of Eq. 21 the various partial derivatives of F in general depend on x, y, and z. Before evaluating the integral one must solve Eq. 20 for z in terms of x and y and then substitute for z in the integrand. Recall from Section 16.8 that if $F_z \neq 0$, then Eq. 20 does define z implicitly as a function of x and y.

If the surface (20) is projectible into the xz- or yz-planes, then we obtain other integrals corresponding to Eq. 21 in an obvious way, namely,

$$A(S) = \iint_{\Omega_{xz}} \frac{(F_x^2 + F_y^2 + F_z^2)^{1/2}}{|F_y|}\, dA_{xz} \tag{22}$$

and

$$A(S) = \iint_{\Omega_{yz}} \frac{(F_x^2 + F_y^2 + F_z^2)^{1/2}}{|F_x|}\, dA_{yz}, \tag{23}$$

respectively. In each case one must express the integrand entirely in terms of the two integration variables before evaluating the integral.

EXAMPLE 2

Find the surface area of the part of the hyperbolic paraboloid $z = x^2 - y^2$ that is inside the cylinder $x^2 + y^2 = 4$ (see Figure 18.4.6a).

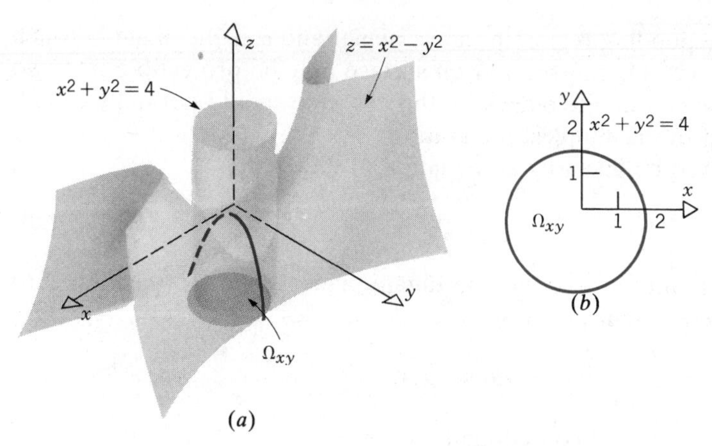

Figure 18.4.6

In this problem $f(x, y) = x^2 - y^2$ and Ω_{xy} is the circular disk with center (0, 0) and radius 2 shown in Figure 18.4.6*b*. Thus, from Eq. 15,

$$A(S) = \iint_{\Omega_{xy}} (1 + f_x^2 + f_y^2)^{1/2}\, dA_{xy}$$

$$= \iint_{\Omega_{xy}} (1 + 4x^2 + 4y^2)^{1/2}\, dA_{xy}.$$

It is convenient to evaluate the integral using polar coordinates. Then

$$A(S) = \int_0^{2\pi} \int_0^2 (1 + 4r^2)^{1/2}\, r\, dr\, d\theta$$

$$= \frac{2\pi}{8}\,\frac{2}{3}(1 + 4r^2)^{3/2}\Big|_0^2$$

$$= \frac{\pi}{6}[(17)^{3/2} - 1] \cong 36.18.$$ ∎

EXAMPLE 3

Find the surface area of the portion of the cylinder $x^2 + z^2 = a^2$ that is inside the cylinder $x^2 + y^2 = a^2$ (see Figure 18.4.7).

Because of the symmetry of the problem, we can find the surface area in the first octant, and then multiply by eight. For the given surface we have

$$F(x, y, z) = x^2 + z^2 - a^2 = 0. \tag{24}$$

Then

$$F_x = 2x, \qquad F_y = 0, \qquad F_z = 2z,$$

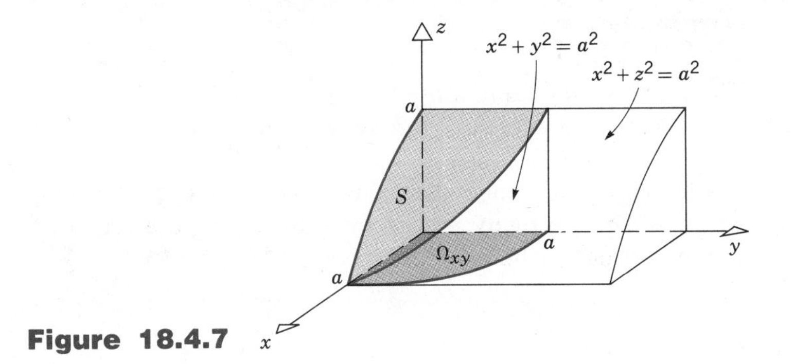

Figure 18.4.7

so

$$\mathbf{N} = \frac{x}{a}\mathbf{i} + \frac{z}{a}\mathbf{k},$$

where we have used Eq. 24 to simplify the expression for $\mathbf{N}$. Consequently,

$$d\sigma = \frac{dA_{xy}}{|\mathbf{N} \cdot \mathbf{k}|} = \frac{a}{z}\,dA_{xy}, \tag{25}$$

and

$$A(S) = \iint_{\Omega_{xy}} \frac{a}{z}\,dA_{xy},$$

where Ω_{xy} is the first quadrant of the disk $x^2 + y^2 \leq a^2$, shown in Figure 18.4.8*a*. Using Eq. 24 we can express the integrand in Eq. 25 in terms of x and y, thereby obtaining

$$A(S) = \int_0^a \int_0^{\sqrt{a^2-x^2}} \frac{a}{\sqrt{a^2 - x^2}}\,dy\,dx. \tag{26}$$

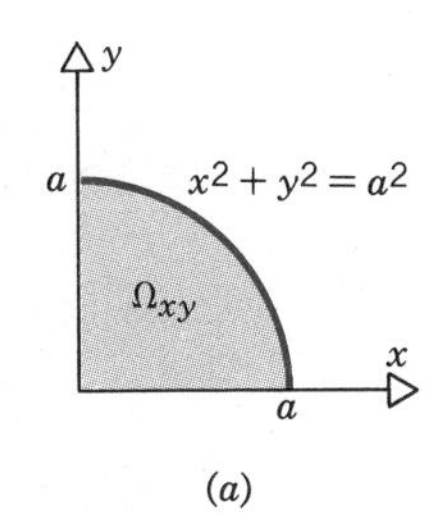

(*a*)

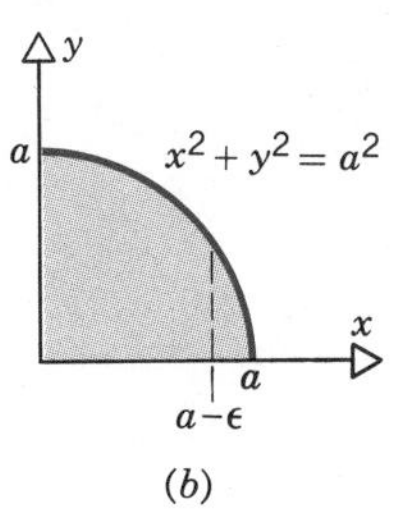

(*b*)

Figure 18.4.8

The same integral is also obtained by direct substitution into Eq. 21. Since the integrand depends only on x, we have chosen to integrate first with respect to y. Note that the integrand also becomes unbounded as $x \to a$, so we evaluate the integral by a limiting process, similar to that used in Section 11.3 for one-dimensional improper integrals. That is, we first restrict x so that $0 \leq x \leq a - \epsilon$, where ϵ is an arbitrary small positive number; this corresponds to integrating over the region shown in Figure 18.4.8*b*. Afterward we let $\epsilon \to 0$. In this way we find that

$$\begin{aligned} A(S) &= \lim_{\epsilon \to 0} \int_0^{a-\epsilon} \int_0^{\sqrt{a^2-x^2}} \frac{a}{\sqrt{a^2 - x^2}}\,dy\,dx \\ &= \lim_{\epsilon \to 0} \int_0^{a-\epsilon} \frac{a}{\sqrt{a^2 - x^2}} \sqrt{a^2 - x^2}\,dx \\ &= \lim_{\epsilon \to 0} a \int_0^{a-\epsilon} dx = \lim_{\epsilon \to 0} a(a - \epsilon) = a^2. \end{aligned}$$

Thus the total area is $8a^2$. ∎

EXAMPLE 4

The graph of the equation

$$-x^2 + y^2 + z^2 = 1$$

is a hyperboloid of one sheet (see Figure 18.4.9). By projecting into the yz- and xz-planes, respectively, set up integrals giving the area of the part of this surface in the first octant between the planes $x = 0$ and $x = 1$.

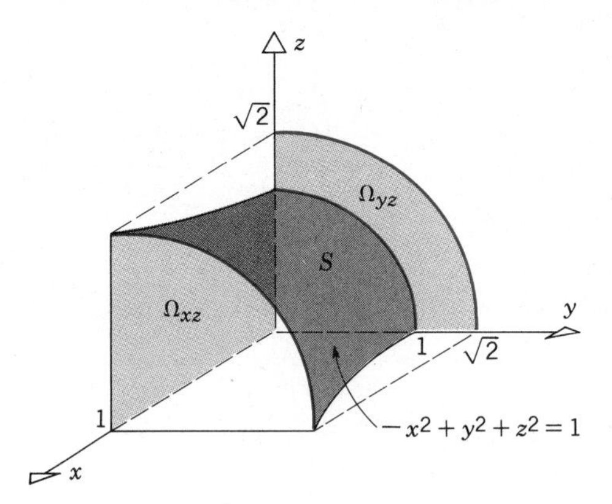

Figure 18.4.9

If we project into the yz-plane, then we can make use of Eq. 23, which yields

$$A(S) = \iint_{\Omega_{yz}} \frac{(x^2 + y^2 + z^2)^{1/2}}{x} \, dA_{yz}$$

$$= \iint_{\Omega_{yz}} \left[\frac{2(y^2 + z^2) - 1}{y^2 + z^2 - 1}\right]^{1/2} dA_{yz},$$

where Ω_{yz} is the quarter annulus shown in Figure 18.4.10. By introducing polar coordinates in the yz-plane we can rewrite the integral as

$$A(S) = \int_1^{\sqrt{2}} \int_0^{\pi/2} \left(\frac{2r^2 - 1}{r^2 - 1}\right)^{1/2} r \, d\theta \, dr$$

$$= \frac{\pi}{2} \int_1^{\sqrt{2}} \left(\frac{2r^2 - 1}{r^2 - 1}\right)^{1/2} r \, dr. \tag{28}$$

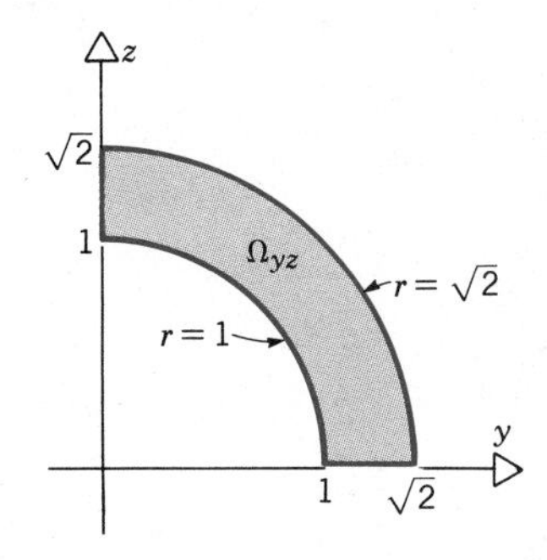

Figure 18.4.10

Observe that the integral (28) is improper because the integrand becomes unbounded as $r \to 1$; however, it is not difficult to show that the integral converges.

If we choose to project into the xz-plane, then the surface area is given by Eq. 22, that is

$$A(S) = \iint_{\Omega_{xz}} \frac{(x^2 + y^2 + z^2)^{1/2}}{y}\, dA_{xz},$$

where Ω_{xz} is shown in Figure 18.4.11. Writing the integrand in terms of x and z,

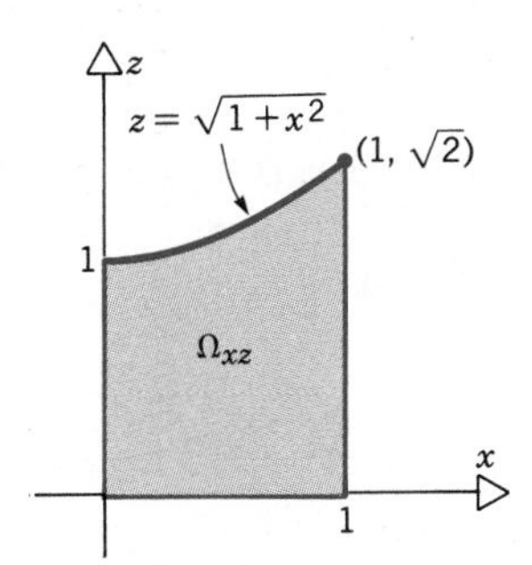

Figure 18.4.11

and inserting the limits of integration, we obtain

$$A(S) = \int_0^1 \int_0^{\sqrt{1+x^2}} \frac{(1 + 2x^2)^{1/2}}{(1 + x^2 - z^2)^{1/2}}\, dz\, dx.$$

The inner integral is again improper, but can be easily evaluated in terms of the arcsin function; thus

$$A(S) = \frac{\pi}{2} \int_0^1 (1 + 2x^2)^{1/2}\, dx.$$

This remaining integral can be evaluated by consulting a table of integrals, or by making the substitution $\sqrt{2}x = \tan u$, with the result that

$$A(S) = \frac{\pi}{2}\left[\frac{\sqrt{3}}{2} + \frac{1}{2\sqrt{2}} \ln(\sqrt{2} + \sqrt{3})\right] \cong 1.997. \quad ■$$

The extension from surface area integrals to integrals of other functions over surfaces is straightforward. Suppose that S is given by

$$z = f(x, y), \qquad (x, y) \in \Omega_{xy}, \tag{27}$$

and that $\phi(x, y, z)$ is continuous in a region containing S. We wish to define $\iint_S \phi(x, y, z)\, d\sigma$, the integral of ϕ on the surface S. Let

$$\Phi(x, y) = \phi[x, y, f(x, y)]$$

and note that $\Phi(x, y)$ is the value of ϕ at the point of S that is directly above (or below) the point (x, y) in Ω_{xy}. Recall that, for a surface described by Eq. 27, the surface area element is

$$d\sigma = (1 + f_x^2 + f_y^2)^{1/2}\, dA_{xy}.$$

Consequently

$$\iint_S \phi(x, y, z)\, d\sigma = \iint_{\Omega_{xy}} \Phi(x, y)\,[1 + f_x^2(x, y) + f_y^2(x, y)]^{1/2}\, dA_{xy}$$

$$= \iint_{\Omega_{xy}} \phi[x, y, f(x, y)]\,[1 + f_x^2(x, y) + f_y^2(x, y)]^{1/2}\, dA_{xy}.$$

In a similar way, we can generalize the other surface area integrals. For instance, if S is the surface

$$F(x, y, z) = 0, \tag{28}$$

and if S is projectible into the region Ω_{xy}, then

$$\iint_S \phi(x, y, z)\, d\sigma$$

$$= \iint_{\Omega_{xy}} \frac{\phi(x, y, z)\,[F_x^2(x, y, z) + F_y^2(x, y, z) + F_z^2(x, y, z)]^{1/2}}{|F_z(x, y, z)|}\, dA_{xy}. \tag{29}$$

Of course, before evaluating the integral (29) we must solve Eq. 28 for z in terms of x and y and substitute into the integrand.

The surface integral $\iint_S \phi(x, y, z)\, d\sigma$ often has a physical interpretation. One possibility is that ϕ is the density, or mass per unit area, of a membrane in the shape of the surface S. Then $\iint_S \phi(x, y, z)\, d\sigma$ is the total mass of the membrane or surface. Some other applications of surface integrals are described in Section 18.7.

EXAMPLE 5

Consider the surface S defined by $x^2 + z^2 = a^2$ and lying in the first octant between the xz-plane and the plane $y = x$ (see Figure 18.4.12a). If the density of this surface is given by $\rho(x, y, z) = a - z$, find the total mass M of the surface.

The total mass is given by

$$M = \iint_S \rho(x, y, z)\, d\sigma.$$

As in Eq. 25 in Example 3,

$$d\sigma = \frac{a}{z}\, dA_{xy},$$

so

$$M = \iint_{\Omega_{xy}} (a - z)\frac{a}{z}\, dA_{xy},$$

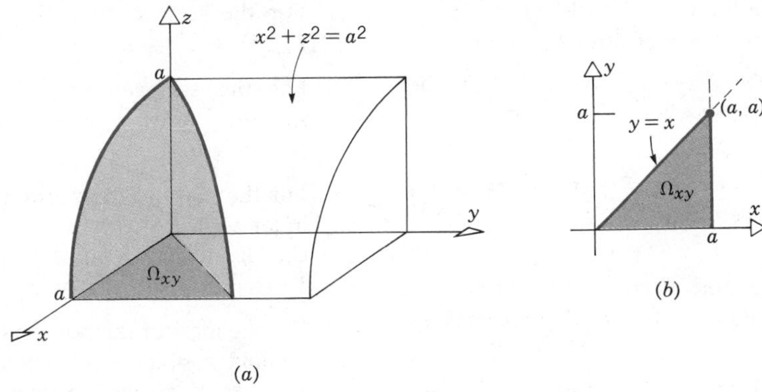

Figure 18.4.12

where Ω_{xy} is the triangle $0 \le y \le x, 0 \le x \le a$ shown in Figure 18.4.12*b*. Expressing z in terms of x and y, and then integrating, we obtain

$$\begin{aligned} M &= a^2 \int_0^a \int_0^x (a^2 - x^2)^{-1/2}\, dy\, dx - a \int_0^a \int_0^x dy\, dx \\ &= a^2 \int_0^a (a^2 - x^2)^{-1/2}\, x\, dx - a \int_0^a x\, dx \\ &= a^3 - \frac{a^3}{2} = \frac{a^3}{2}. \quad \blacksquare \end{aligned}$$

PROBLEMS

In each of Problems 1 through 10, find the area of the given surface.

1. The sphere $x^2 + y^2 + z^2 = a^2$.
2. The portion of the elliptic paraboloid $z = 4 - x^2 - y^2$ for which $z \ge 0$.
3. The portion of the plane
$$\frac{x}{a} + \frac{y}{b} + \frac{z}{c} = 1, \qquad a, b, c > 0$$
that is in the first octant.
4. The portion of the parabolic cylinder $z = 4 - y^2$ that is in the first octant and between the yz-plane and the plane $x = 2y$.
5. The portion of the hyperbolic paraboloid $x = yz$ that is inside the cylinder $y^2 + z^2 = a^2$.
6. The portion of the plane $Ax + By + Cz = D$, with $C \ne 0$, that is inside the cylinder $x^2 + y^2 = a^2$.
7. The portion of the cylinder $x^2 + y^2 = a^2$ that is in the first octant between the xy-plane and the plane $z - 2y = 3$.
8. The portion of the surface $x + \frac{2}{3}z^{3/2} = 1$ that is in the first octant between the xy-plane, the xz-plane, and the plane $y + z = 1$.
9. The portion of the cubic cylinder $y = 8 - x^3$ that is in the first octant between the plane $z = 0$ and the cylinder $z = x^3$.
10. The portion of the cylinder $x^2 + y^2 = a^2$ that is inside the cylinder $(y^2/a^2) + (z^2/b^2) = 1$.
11. Find the area of the lateral surface of a right circular cone of radius R and height H.

In each of Problems 12 through 16, find the mass of the given surface with the given density function.

12. The hemisphere $x^2 + y^2 + z^2 = a^2$, $z \ge 0$, with $\rho(x, y, z) = k(a - z)$, where k is a positive constant.

13. The hemisphere in Problem 12 with $\rho(x, y, z) = k(x^2 + y^2)^{1/2}$, where k is a positive constant.

14. The paraboloid $z = 1 - x^2 - y^2$, $z \geq 0$, with $\rho(x, y, z) = k(1 - z)$, where k is a positive constant.

15. The portion of the cylinder $x^2 + y^2 = a^2$ that is in the first octant between the planes $z = 0$ and $z = y$, with $\rho(x, y, z) = a - x$.

16. The portion of the cone $z^2 = x^2 + y^2$ for which $0 \leq z \leq a$, with $\rho(x, y, z) = kz$, where k is a positive constant.

In each of Problems 17 through 20, set up, but do not evaluate, an appropriate integral.

17. For the surface area of the ellipsoid $(x^2/4) + (y^2/4) + z^2 = 1$.

18. For the surface area of the portion of the sphere $x^2 + y^2 + z^2 = a^2$ cut off by the plane $y + z = a$.

19. For the mass of the portion of the parabolic cylinder $z = 1 - x^2$ with $x \geq 0$ that is between the coordinate planes and the parabolic cylinder $x = 1 - y^2$ and with $\rho(x, y, z) = 1 - z$.

20. For the mass of the portion of the hyperbolic paraboloid $z = x^2 - y^2$ that is inside the cylinder $x^2 + y^2 = a^2$, with $\rho(x, y, z) = k(x^2 + y^2)^{1/2}$, where k is a positive constant.

18.5 PARAMETRIC EQUATIONS OF SURFACES

Surfaces, as well as curves, can be described either by Cartesian or by parametric equations. The advantages of using parametric representations are similar for both curves and surfaces, namely, they provide a more general and unified framework for the discussion. Whereas curves are one-dimensional and therefore require only one parameter, surfaces are intrinsically two-dimensional in nature, so it is necessary to use two parameters to describe them.

EXAMPLE 1

Let

$$x = R \cos\theta \sin\phi, \qquad y = R \sin\theta \sin\phi, \qquad z = R \cos\phi, \tag{1}$$

where $0 \leq \theta \leq 2\pi$, $0 \leq \phi \leq \pi/2$, and R is a positive constant. For each pair of values of θ and ϕ Eqs. 1 give corresponding values for x, y, and z. In this case, it is easy to obtain a Cartesian equation satisfied by these values of x, y, and z. By squaring each of Eqs. 1 and then adding, we obtain

$$\begin{aligned} x^2 + y^2 + z^2 &= R^2 \cos^2\theta \sin^2\phi + R^2 \sin^2\theta \sin^2\phi + R^2 \cos^2\phi \\ &= R^2, \end{aligned} \tag{2}$$

which is an equation for the sphere with center at the origin and radius R. For the given ranges of θ and ϕ, it follows that Eqs. 1 describe the upper hemisphere only (see Figure 18.5.1).

Of course, Eqs. 1 are just the transformation equations from spherical to rectangular coordinates that we used in Section 17.7, with the additional requirement that the radial variable always has the constant value R. As noted in Section 17.7, it is easy to interpret θ and ϕ in terms of longitude and latitude. ■

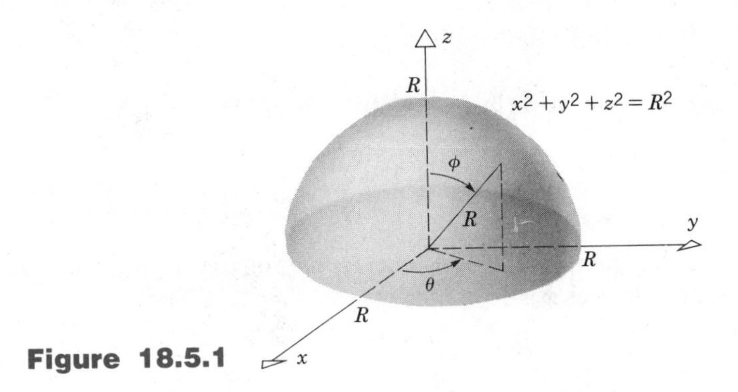

Figure 18.5.1

EXAMPLE 2

Suppose that

$$x = 2 \cosh u \cos v, \qquad y = \cosh u \sin v, \qquad z = 3 \sinh u. \tag{3}$$

Find a Cartesian equation satisfied by x, y, and z. Determine the rectangular coordinates of the point where $u = \ln 2$, $v = \pi/6$.

Note that

$$\left(\frac{x}{2}\right)^2 = \cosh^2 u \cos^2 v, \qquad y^2 = \cosh^2 u \sin^2 v, \qquad \left(\frac{z}{3}\right)^2 = \sinh^2 u.$$

Then it follows at once that

$$\frac{x^2}{4} + y^2 - \frac{z^2}{9} = 1, \tag{4}$$

which is an equation of a hyperboloid of one sheet. The entire surface is covered if $-\infty < u < \infty$ and $0 \leq v \leq 2\pi$ (see Figure 18.5.2).

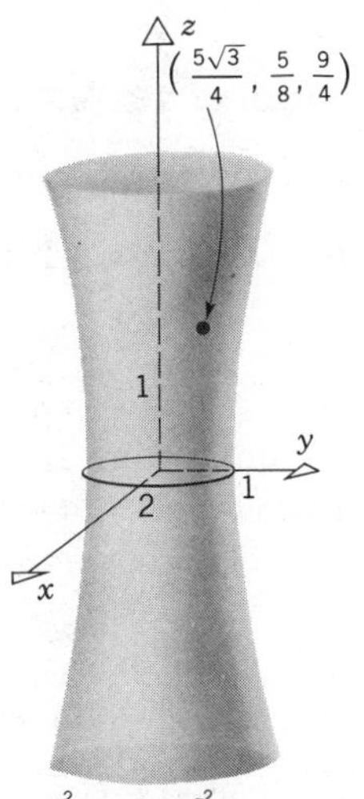

Figure 18.5.2 $\frac{x^2}{4} + y^2 - \frac{z^2}{9} = 1$

If $u = \ln 2$, then

$$\cosh(\ln 2) = \frac{1}{2}(e^{\ln 2} + e^{-\ln 2}) = \frac{1}{2}\left(2 + \frac{1}{2}\right) = \frac{5}{4},$$

$$\sinh(\ln 2) = \frac{1}{2}(e^{\ln 2} - e^{-\ln 2}) = \frac{1}{2}\left(2 - \frac{1}{2}\right) = \frac{3}{4}.$$

Consequently, the point corresponding to the given values of u and v has rectangular coordinates $x = 5\sqrt{3}/4$, $y = \frac{5}{8}$, $z = \frac{9}{4}$. ∎

Examples 1 and 2 illustrate that familiar surfaces can be described by sets of equations of the form

$$x = f(u, v), \qquad y = g(u, v), \qquad z = h(u, v), \qquad (u, v) \in \Omega_{uv}, \tag{5}$$

where f, g, and h are given continuous functions of two variables, and Ω_{uv} is a given region in the uv-plane. For each point (u, v) in Ω_{uv} Eqs. 5 produce a point (x, y, z) in three-dimensional xyz-space. The set of all such points is a surface S. Equations 5 constitute a mapping or transformation from the uv-plane to xyz-space, as suggested by Figure 18.5.3, or a parametric representation of S with parameters u and v.

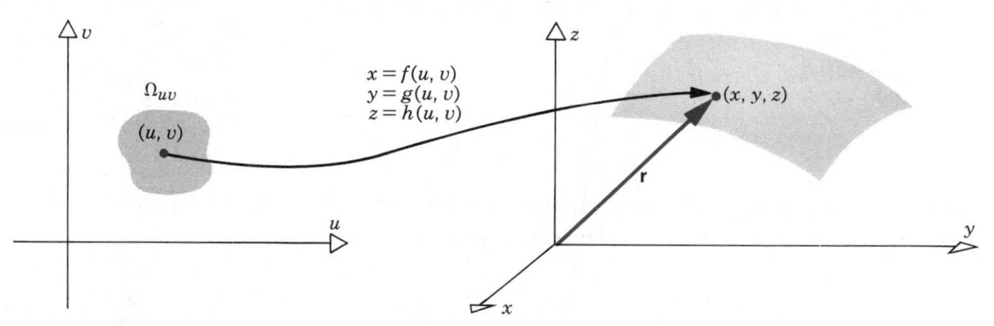

Figure 18.5.3 Parametric representation of a surface S.

Sometimes, as in Examples 1 and 2, it is possible to eliminate the parameters u and v among Eqs. 5, thereby obtaining a Cartesian equation of the form

$$F(x, y, z) = 0. \tag{6}$$

If this is done, we must be careful to note whether all points that satisfy Eq. 6 are obtained from Ω_{uv}, and whether some of them may be obtained more than once. Recall that similar questions occurred in using parametric representations of curves.

The position vector $\mathbf{r}$ of a point on the surface S given by Eqs. 5 is

$$\begin{aligned}\mathbf{r} &= x\mathbf{i} + y\mathbf{j} + z\mathbf{k}\\ &= f(u, v)\mathbf{i} + g(u, v)\mathbf{j} + h(u, v)\mathbf{k}.\end{aligned} \tag{7}$$

If we hold u constant, say $u = u_0$, then the vector equation

$$\mathbf{r} = f(u_0, v)\mathbf{i} + g(u_0, v)\mathbf{j} + h(u_0, v)\mathbf{k}, \tag{8}$$

or the corresponding scalar equations

$$x = f(u_0, v), \qquad y = g(u_0, v), \qquad z = h(u_0, v), \tag{9}$$

depend only on the single parameter v. Consequently, Eqs. 8 and 9 are parametric equations of a curve on the surface S (see Figure 18.5.4). Since such a curve is

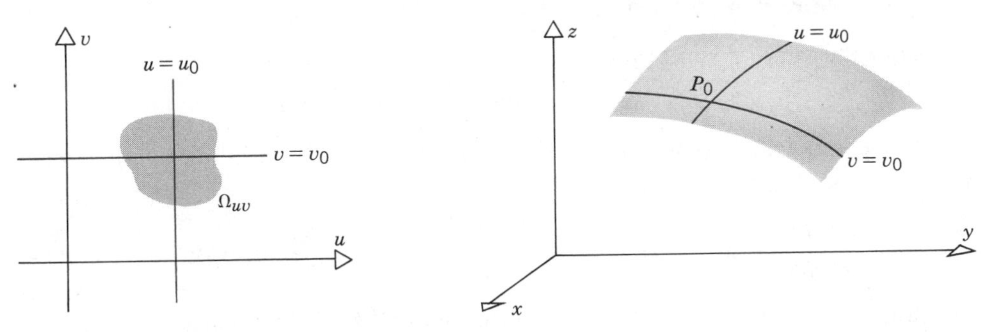

Figure 18.5.4

associated with a particular value of u, namely $u = u_0$, we will refer to it as a u-curve. Similarly, if $v = v_0$, then

$$\mathbf{r} = f(u, v_0)\mathbf{i} + g(u, v_0)\mathbf{j} + h(u, v_0)\mathbf{k}, \tag{10}$$

or

$$x = f(u, v_0), \qquad y = g(u, v_0), \qquad z = h(u, v_0), \tag{11}$$

are parametric representations of a different curve on the surface S. We will call such a curve a v-curve.

Suppose that the curve (8), (9) and the curve (10), (11) intersect at a point P_0 on the surface, as shown in Figure 18.5.4. This point can be identified by its Cartesian coordinates (x, y, z), but it can also be identified by the values u_0 and v_0 associated with the u-curve and the v-curve that intersect at the point. For different constant values of u and v, the u-curves (8), (9) and the v-curves (10), (11) form two families of curves on the surface S, as shown in Figure 18.5.5. These two families of curves can be used as a coordinate system on the surface. Although there may be a few exceptional points, it is typical that a given point on the surface is the point of intersection of one u-curve and one v-curve. Then this point can be identified with the values of u and v, respectively, associated with these curves.

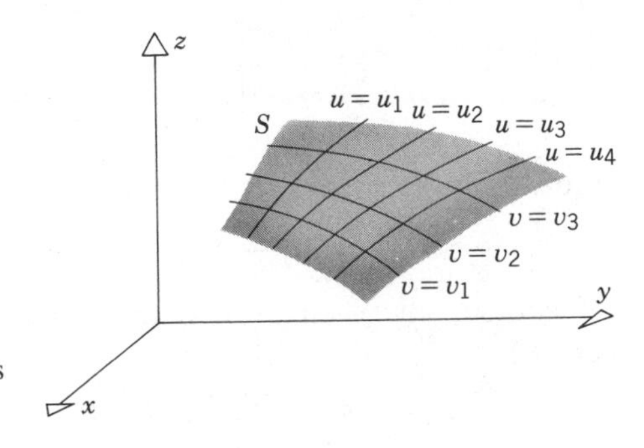

Figure 18.5.5 Parametric coordinates on a surface S.

This is illustrated in a familiar context by the use of meridians of longitude and circles of latitude to identify points on the earth's surface. In this case, the North and South Poles are exceptional points in that every meridian of longitude passes through them. Consequently, longitude is not uniquely determined at the poles.

Tangent Planes and Normal Lines

Consider the curve given by Eq. 10,

$$\mathbf{r} = f(u, v_0)\mathbf{i} + g(u, v_0)\mathbf{j} + h(u, v_0)\mathbf{k},$$

which lies in the surface S defined by Eqs. 5 or 7. By differentiating $\mathbf{r}$ with respect to u while holding v constant, we obtain

$$\frac{\partial \mathbf{r}}{\partial u}(u, v_0) = \frac{\partial f}{\partial u}(u, v_0)\mathbf{i} + \frac{\partial g}{\partial u}(u, v_0)\mathbf{j} + \frac{\partial h}{\partial u}(u, v_0)\mathbf{k}, \tag{12}$$

which is a vector tangent to the curve (see Figure 18.5.6). Similarly,

$$\frac{\partial \mathbf{r}}{\partial v}(u_0, v) = \frac{\partial f}{\partial v}(u_0, v)\mathbf{i} + \frac{\partial g}{\partial v}(u_0, v)\mathbf{j} + \frac{\partial h}{\partial v}(u_0, v)\mathbf{k}, \tag{13}$$

is tangent to the curve given by Eq. 8, which also lies in S. It is now easy to find a vector $\mathbf{n}$ normal to S at the point of intersection of the two curves by finding the

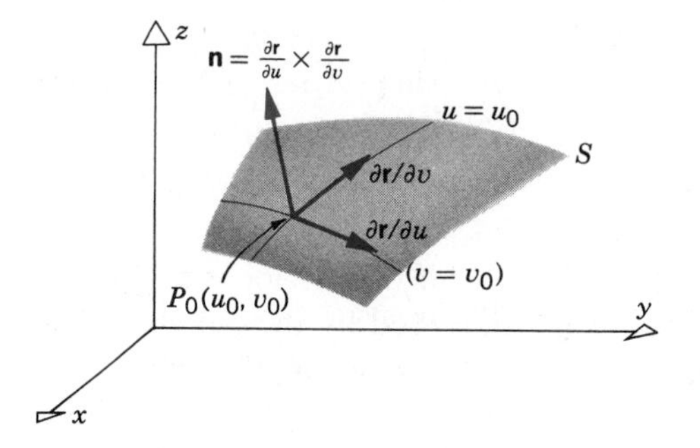

Figure 18.5.6 A normal vector on the surface S.

cross product of $\partial\mathbf{r}/\partial u$ and $\partial\mathbf{r}/\partial v$ at (u_0, v_0). Thus

$$\mathbf{n} = \frac{\partial \mathbf{r}}{\partial u} \times \frac{\partial \mathbf{r}}{\partial v} = \begin{vmatrix} \mathbf{i} & \mathbf{j} & \mathbf{k} \\ f_u & g_u & h_u \\ f_v & g_v & h_v \end{vmatrix}$$

$$= \begin{vmatrix} g_u & h_u \\ g_v & h_v \end{vmatrix}\mathbf{i} + \begin{vmatrix} h_u & f_u \\ h_v & f_v \end{vmatrix}\mathbf{j} + \begin{vmatrix} f_u & g_u \\ f_v & g_v \end{vmatrix}\mathbf{k}. \tag{14}$$

Expanding the determinants in the last expression in Eq. 14, we have

$$\mathbf{n} = (g_u h_v - h_u g_v)\mathbf{i} + (h_u f_v - f_u h_v)\mathbf{j} + (f_u g_v - g_u f_v)\mathbf{k}. \tag{15}$$

Of course, all of the functions in Eqs. 14 and 15 are evaluated at the point (u_0, v_0).

EXAMPLE 3

Find a vector normal to the surface (3) at the point where $u = \ln 2$, $v = \pi/6$. Also find the normal line and the tangent plane to the surface at this point. Observe that the surface is the same as in Example 2.

In this case

$$\mathbf{r} = 2 \cosh u \cos v\mathbf{i} + \cosh u \sin v\mathbf{j} + 3 \sinh u\mathbf{k}. \tag{16}$$

Therefore

$$\frac{\partial \mathbf{r}}{\partial u} = 2 \sinh u \cos v\mathbf{i} + \sinh u \sin v\mathbf{j} + 3 \cosh u\mathbf{k}, \tag{17}$$

$$\frac{\partial \mathbf{r}}{\partial v} = -2 \cosh u \sin v\mathbf{i} + \cosh u \cos v\mathbf{j}. \tag{18}$$

If $u = \ln 2$ and $v = \pi/6$, then

$$\frac{\partial \mathbf{r}}{\partial u}\left(\ln 2, \frac{\pi}{6}\right) = \frac{3\sqrt{3}}{4}\mathbf{i} + \frac{3}{8}\mathbf{j} + \frac{15}{4}\mathbf{k},$$

$$\frac{\partial \mathbf{r}}{\partial v}\left(\ln 2, \frac{\pi}{6}\right) = -\frac{5}{4}\mathbf{i} + \frac{5\sqrt{3}}{8}\mathbf{j},$$

and

$$\left(\frac{\partial \mathbf{r}}{\partial u} \times \frac{\partial \mathbf{r}}{\partial v}\right)\left(\ln 2, \frac{\pi}{6}\right) = \begin{vmatrix} \mathbf{i} & \mathbf{j} & \mathbf{k} \\ \frac{3\sqrt{3}}{4} & \frac{3}{8} & \frac{15}{4} \\ -\frac{5}{4} & \frac{5\sqrt{3}}{8} & 0 \end{vmatrix}$$

$$= \frac{15}{32}(-5\sqrt{3}\mathbf{i} - 10\mathbf{j} + 4\mathbf{k}). \tag{19}$$

Of course, any vector proportional to $\partial \mathbf{r}/\partial u \times \partial \mathbf{r}/\partial v$ can also be used as a normal vector, for example,

$$\mathbf{n} = 5\sqrt{3}\mathbf{i} + 10\mathbf{j} - 4\mathbf{k}. \tag{20}$$

In Example 2 we found that the point on the surface corresponding to $u = \ln 2$, $v = \pi/6$ is $(5\sqrt{3}/4, \frac{5}{8}, \frac{9}{4})$. Therefore, if we use the normal vector (20), then the line normal to the surface at this point is

$$\frac{x - 5\sqrt{3}/4}{5\sqrt{3}} = \frac{y - 5/8}{10} = \frac{z - 9/4}{-4}. \tag{21}$$

The tangent plane is

$$5\sqrt{3}\left(x - \frac{5\sqrt{3}}{4}\right) + 10\left(y - \frac{5}{8}\right) - 4\left(z - \frac{9}{4}\right) = 0,$$

or

$$5\sqrt{3}x + 10y - 4z = 16. \blacksquare \tag{22}$$

Surface Area in Parametric Coordinates

We have seen that the vector equation (7)

$$\mathbf{r} = f(u, v)\mathbf{i} + g(u, v)\mathbf{j} + h(u, v)\mathbf{k},$$

for (u, v) in a region Ω_{uv} of the uv-plane, provides a parametric representation of a surface S. As in Section 18.4, S is said to be smooth if it has a continuously varying normal vector $\mathbf{n}$. To be sure that this is so, it is sufficient to assume that f, g, and h have continuous partial derivatives, and that $\mathbf{n} \neq \mathbf{0}$ throughout Ω_{uv}. We assume that S is at least piecewise smooth, that is, it consists of at most a finite number of smooth pieces, joined along smooth rectifiable arcs, and that the boundary of Ω_{uv} also consists of at most a finite number of smooth rectifiable arcs. Under these conditions we can derive an expression for the area of S in terms of an integral over Ω_{uv}.

The first step is to form a partition Δ of Ω_{uv} by lines parallel to the u and v axes, respectively (see Figure 18.5.7). As usual, the norm of the partition Δ, denoted by $\|\Delta\|$, is the maximum length of the diagonals of the subrectangles formed

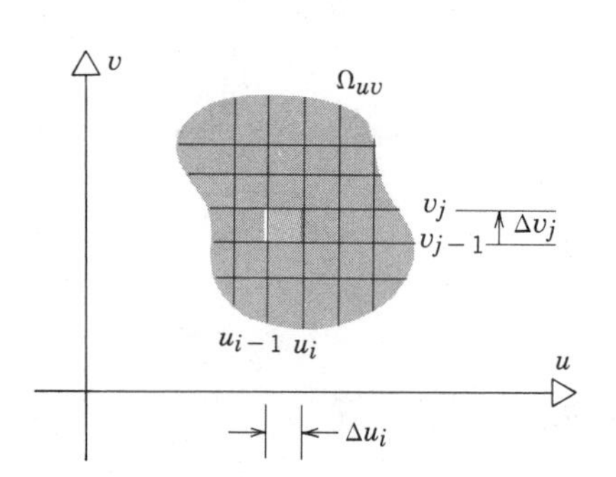

Figure 18.5.7 A partition of the region Ω_{uv}.

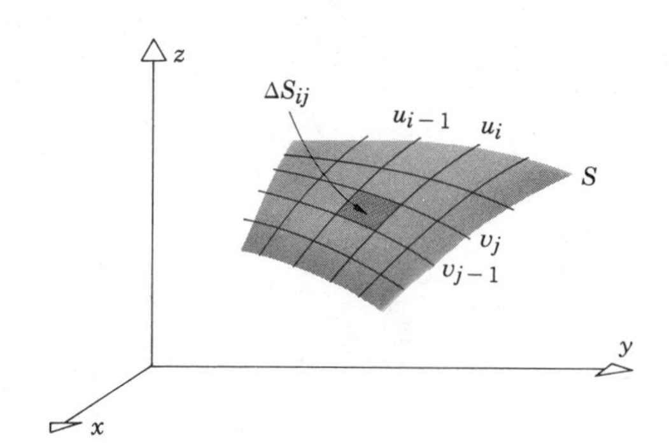

Figure 18.5.8 The corresponding partition of the surface S.

in this way. The partition of Ω_{uv} induces a partition of the surface S by families of u-curves and v-curves, respectively, as indicated in Figure 18.5.8. The shaded region in Figure 18.5.8, denoted by ΔS_{ij}, is the part of the surface bounded by the curves $\mathbf{r}(u_{i-1}, v)$, $\mathbf{r}(u_i, v)$, $\mathbf{r}(u, v_{j-1})$, and $\mathbf{r}(u, v_j)$. It corresponds to the shaded rectangle in Figure 18.5.7 that is bounded by the lines $u = u_{i-1}$, $u = u_i$, $v = v_{j-1}$, and $v = v_j$.

Next we wish to estimate the area $\Delta\sigma_{ij}$ of the surface element ΔS_{ij}. As shown in Figure 18.5.9, ΔS_{ij} is approximately a parallelogram with sides given by the vectors $\Delta\mathbf{r}_u$ and $\Delta\mathbf{r}_v$, where

$$\Delta\mathbf{r}_u = \mathbf{r}(u_i, v_{j-1}) - \mathbf{r}(u_{i-1}, v_{j-1}), \tag{23a}$$

$$\Delta\mathbf{r}_v = \mathbf{r}(u_{i-1}, v_j) - \mathbf{r}(u_{i-1}, v_{j-1}). \tag{23b}$$

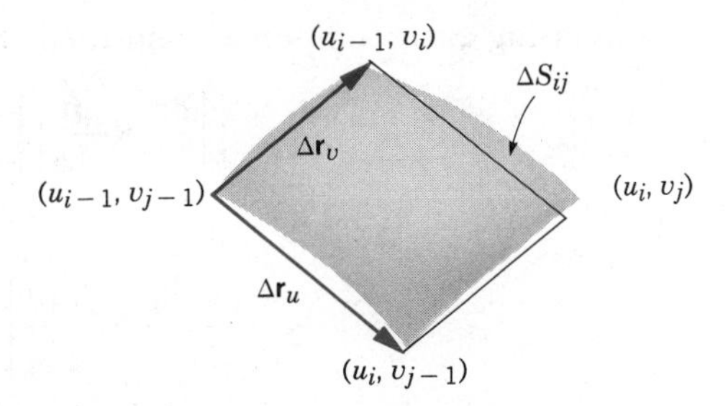

Figure 18.5.9 Approximation of ΔS_{ij} by a parallelogram.

Hence $\Delta\sigma_{ij}$ is given approximately by the area of the parallelogram, that is,

$$\Delta\sigma_{ij} \cong \|\Delta\mathbf{r}_u \times \Delta\mathbf{r}_v\|. \tag{24}$$

A more useful expression can be obtained if we note that

$$\Delta\mathbf{r}_u \cong \frac{\partial\mathbf{r}}{\partial u}(u_{i-1}, v_{j-1})\,\Delta u_i, \qquad \Delta\mathbf{r}_v \cong \frac{\partial\mathbf{r}}{\partial v}(u_{i-1}, v_{j-1})\,\Delta v_j, \tag{25}$$

where $\Delta u_i = u_i - u_{i-1}$ and $\Delta v_j = v_j - v_{j-1}$. Then, by substituting in Eq. 24, we have

$$\Delta\sigma_{ij} \cong \left\|\frac{\partial\mathbf{r}}{\partial u}(u_{i-1}, v_{j-1}) \times \frac{\partial\mathbf{r}}{\partial v}(u_{i-1}, v_{j-1})\right\|\,\Delta u_i\,\Delta v_j. \tag{26}$$

The area of S is then approximated by summing over i and j so as to include the contribution of each surface element. Consequently,

$$A(S) = \sum_{i,j} \Delta\sigma_{ij}, \tag{27}$$

where $\Delta\sigma_{ij}$ is given by Eq. 26. Finally, we pass to the limit as $\|\Delta\| \to 0$ and obtain the integral expression

$$A(S) = \iint_{\Omega_{uv}} d\sigma = \int\int_{\Omega_{uv}} \left\|\frac{\partial\mathbf{r}}{\partial u}(u, v) \times \frac{\partial\mathbf{r}}{\partial v}(u, v)\right\| dA_{uv}. \tag{28}$$

The integral on the right side of Eq. 28 is a double integral over the region Ω_{uv} in the uv-plane. It can be evaluated by replacing it by appropriate iterated integrals, and then integrating with respect to each variable in turn.

For the case of the sphere parametrized by Eq. 1, we have

$$\mathbf{r} = R\cos\theta\sin\phi\mathbf{i} + R\sin\theta\sin\phi\mathbf{j} + R\cos\phi\mathbf{k}.$$

Then

$$\frac{\partial\mathbf{r}}{\partial\theta} = -R\sin\theta\sin\phi\mathbf{i} + R\cos\theta\sin\phi\mathbf{j},$$

$$\frac{\partial\mathbf{r}}{\partial\phi} = R\cos\theta\cos\phi\mathbf{i} + R\sin\theta\cos\phi\mathbf{j} - R\sin\phi\mathbf{k},$$

and

$$\frac{\partial\mathbf{r}}{\partial\theta} \times \frac{\partial\mathbf{r}}{\partial\phi} = -R^2\cos\theta\sin^2\phi\mathbf{i} - R^2\sin\theta\sin^2\phi\mathbf{j} - R^2\sin\phi\cos\phi\mathbf{k}.$$

By squaring each component and then adding, we find that

$$\left\| \frac{\partial \mathbf{r}}{\partial \theta} \times \frac{\partial \mathbf{r}}{\partial \phi} \right\|^2 = R^4 \sin^2 \phi,$$

or

$$\left\| \frac{\partial \mathbf{r}}{\partial \theta} \times \frac{\partial \mathbf{r}}{\partial \phi} \right\| = R^2 \sin \phi.$$

In other words, the element of area on the surface of a sphere of radius R, in terms of the angular coordinates θ and ϕ, is

$$d\sigma = R^2 \sin \phi \, d\theta \, d\phi. \tag{29}$$

EXAMPLE 4

The state of Colorado is bounded by the meridians 102° 3′ and 109° 3′ west longitude, and by the circles 37° and 41° north latitude (see Figure 18.5.10). Assuming that the earth is a sphere of radius 3960 miles, find the area of Colorado.

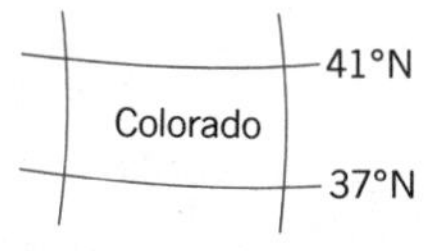

41°N
Colorado
37°N
109°3′W 102°3′W

Figure 18.5.10

Using Eq. 29 for the area of a surface element on a sphere, we have

$$A(S) = \int_{\phi_1}^{\phi_2} \int_{\theta_1}^{\theta_2} R^2 \sin \phi \, d\theta \, d\phi$$

$$= R^2(\theta_2 - \theta_1)(\cos \phi_1 - \cos \phi_2). \tag{30}$$

From the given data it follows that $\theta_2 - \theta_1 = 7° \cong 0.12217305$ radians. The angle ϕ is the complement of the latitude, so $\phi_1 = 49° \cong 0.85521133$ radians, and $\phi_2 = 53° \cong 0.92502450$ radians, respectively. Thus, by evaluating Eq. 30, we obtain

$$A(S) \cong 0.0066271555\, R^2 \cong 103{,}924 \text{ square miles.}$$

The official area of the state of Colorado is 104,247 square miles. ∎

EXAMPLE 5

The surface formed by revolving the circle shown in Figure 18.5.11 about the z-axis is called a **torus;** it resembles a doughnut. By examining Figure 18.5.11 we

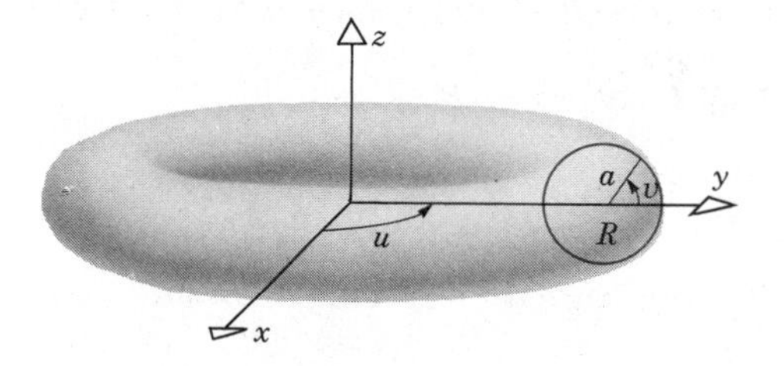

Figure 18.5.11 A torus, formed by revolving a circle about the z-axis.

can write the following parametric representation of the toroidal surface:

$$\mathbf{r} = (R + a \cos v)\cos u\mathbf{i} + (R + a \cos v)\sin u\mathbf{j} + a \sin v\mathbf{k}. \tag{31}$$

The parameter u is the familiar polar angle in the xy-plane (often denoted by θ), a is the radius of the circle that is being revolved, and R is the distance from the axis of revolution to the center of this circle. The quantity $R + a \cos v$ is the radius of the circle that a generic point traverses as it revolves about the z-axis. To cover the entire torus we need to let both u and v range over intervals of length 2π, such as $0 \le u \le 2\pi$ and $0 \le v \le 2\pi$. Find the surface area of the torus.

By differentiating Eq. 31 we obtain

$$\frac{\partial \mathbf{r}}{\partial u} = -(R + a \cos v)\sin u\mathbf{i} + (R + a \cos v)\cos u\mathbf{j},$$

$$\frac{\partial \mathbf{r}}{\partial v} = -a \sin v \cos u\mathbf{i} - a \sin v \sin u\mathbf{j} + a \cos v\mathbf{k}.$$

Then

$$\frac{\partial \mathbf{r}}{\partial u} \times \frac{\partial \mathbf{r}}{\partial v} = a(R + a \cos v)(\cos u \cos v\mathbf{i} + \sin u \cos v\mathbf{j} + \sin v\mathbf{k}),$$

and

$$\left\|\frac{\partial \mathbf{r}}{\partial u} \times \frac{\partial \mathbf{r}}{\partial v}\right\| = a(R + a \cos v). \tag{32}$$

Hence the surface area of the torus is

$$\begin{aligned} A &= a \int_0^{2\pi} \int_0^{2\pi} (R + a \cos v)\, du dv \\ &= 2\pi a \int_0^{2\pi} (R + a \cos v)\, dv \\ &= 2\pi a(Rv + a \sin v)\Big|_0^{2\pi} = 4\pi^2 aR. \end{aligned} \tag{33}$$

Observe that the result is the product $(2\pi a)(2\pi R)$ of the circumferences of two circles: $2\pi a$ is the circumference of the original circle and $2\pi R$ is the circumference of the circle traced by the center of the original circle as it revolves about the z-axis. ■

PROBLEMS

In each of Problems 1 through 6, find a Cartesian equation for the given surface. Name and describe the surface, noting in particular whether the given parametric representation describes all of the surface defined by the Cartesian equation.

1. $x = 2u - v$, $y = u + 2v$, $z = u - v$; $-\infty < u < \infty$, $-\infty < v < \infty$

2. $x = 2 \sin u \cos v$, $y = 3 \sin u \sin v$, $z = 4 \cos u$; $0 \le u \le 2\pi$, $0 \le v \le 2\pi$

3. $x = v\sqrt{1 - u^2}$, $y = \sqrt{u^2 + v^2}$,
$z = u\sqrt{1 + v^2}$; $\quad -1 \le u \le 1, \quad v \ge 0$

4. $x = 2 \sinh u \sin v$, $y = 2 \sinh u \cos v$,
$z = 5 \cosh u$; $\quad 0 \le u < \infty, \quad 0 \le v \le 2\pi$

5. $x = \sin u \sinh v$, $y = 2 \cos u \sinh v$,
$z = \sinh^2 v$; $\quad 0 \le u \le \pi, \quad v \ge 0$

6. $x = u + v$, $y = u - v$, $z = uv$;
$-\infty < u < \infty, \quad -\infty < v < \infty$

In each of Problems 7 through 12, find the normal line and the tangent plane to the given surface at the given point. The surfaces are the same as in Problems 1 through 6, respectively.

7. $x = 2u - v$, $y = u + 2v$, $z = u - v$;
$(1, 8, -1)$

8. $x = 2 \sin u \cos v$, $y = 3 \sin u \sin v$,
$z = 4 \cos u$; $\quad (1, \frac{3}{2}, 2\sqrt{2})$

9. $x = v\sqrt{1 - u^2}$, $y = \sqrt{u^2 + v^2}$,
$z = u\sqrt{1 + v^2}$; $\quad (1, 1, 0)$

10. $x = 2 \sinh u \sin v$, $y = 2 \sinh u \cos v$,
$z = 5 \cosh u$; $\quad \left(-\frac{4}{3}, \frac{4\sqrt{3}}{3}, \frac{25}{3}\right)$

11. $x = \sin u \sinh v$, $y = 2 \cos u \sinh v$,
$z = \sinh^2 v$; $\quad (\frac{3}{4}, 0, \frac{9}{16})$

12. $x = u + v$, $y = u - v$, $z = uv$; $(1, 3, -2)$

13. Find the area of (a) Wyoming (see Figure 18.5.12); and (b) Utah (see Figure 18.5.13).

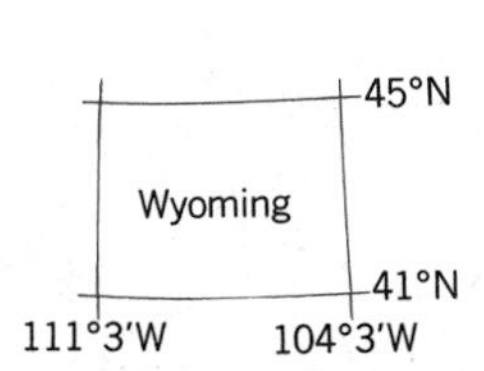

Figure 18.5.12

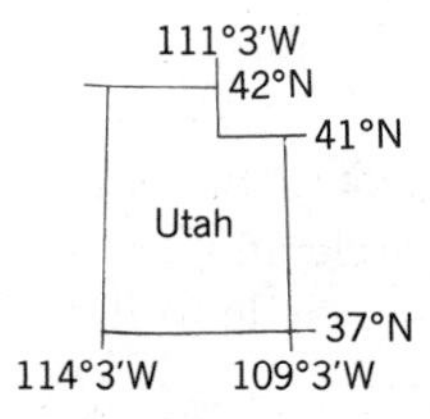

Figure 18.5.13

14. On the surface of the earth the tropics extend from the equator to north and south latitudes 23.5°, the temperate zones from latitude 23.5° to latitude 66.5°, and the polar zones from latitude 66.5° to the poles. Find the percent of the earth's surface that is in the tropics, the temperate zones, and the polar zones, respectively.

15. (a) Show that
$$\mathbf{r} = \rho \sin \phi_0 \cos \theta \mathbf{i} + \rho \sin \phi_0 \sin \theta \mathbf{j} + \rho \cos \phi_0 \mathbf{k},$$
for fixed ϕ_0, is a parametric representation of a right circular cone.
(b) Find the surface area element $d\sigma$ in terms of $d\rho$ and $d\theta$.
(c) Find the area of the lateral surface of a cone of base radius r and altitude h.

16. Consider the torus discussed in Example 5 and a circular cylinder of radius R whose axis is the z-axis.
(a) Find the surface area S_1 of the part of the torus that is outside the cylinder.
(b) Find the surface area S_2 of the part of the torus that is inside the cylinder.
(c) Find the ratio a/R for which $S_1/S_2 = 2$.

17. A certain spiral ramp is described by
$$\mathbf{r} = u \cos v \mathbf{i} + u \sin v \mathbf{j} + cv\mathbf{k}, \quad a \le u \le b, \quad 0 \le v \le 2\pi,$$
where c is a parameter determining the steepness of the ramp. Find the area of the surface of this ramp.

18. Set up an integral for the surface area of the part of the hyperboloid
$$\mathbf{r} = 2 \cosh u \cos v \mathbf{i} + 3 \cosh u \sin v \mathbf{j} + \sinh u \mathbf{k}$$
that is in the first octant between the planes $z = 0$ and $z = \frac{15}{8}$.

19. Set up an integral for the surface area of the ellipsoid
$$\mathbf{r} = a \sin u \cos v \mathbf{i} + b \sin u \sin v \mathbf{j} + c \cos u \mathbf{k}.$$

20. Show that
$$\left\| \frac{\partial \mathbf{r}}{\partial u} \times \frac{\partial \mathbf{r}}{\partial v} \right\|^2 = EG - F^2,$$
where
$$E = \left(\frac{\partial x}{\partial u}\right)^2 + \left(\frac{\partial y}{\partial u}\right)^2 + \left(\frac{\partial z}{\partial u}\right)^2,$$
$$F = \frac{\partial x}{\partial u}\frac{\partial x}{\partial v} + \frac{\partial y}{\partial u}\frac{\partial y}{\partial v} + \frac{\partial z}{\partial u}\frac{\partial z}{\partial v},$$
$$G = \left(\frac{\partial x}{\partial v}\right)^2 + \left(\frac{\partial y}{\partial v}\right)^2 + \left(\frac{\partial z}{\partial v}\right)^2.$$

18.6 THE DIVERGENCE AND THE CURL

The gradient operator ∇, defined by

$$\nabla = \frac{\partial}{\partial x}\mathbf{i} + \frac{\partial}{\partial y}\mathbf{j} + \frac{\partial}{\partial z}\mathbf{k},$$

was introduced in Section 16.4. By applying ∇ to a scalar function f we produce the gradient vector ∇f, whose components are $\partial f/\partial x$, $\partial f/\partial y$, and $\partial f/\partial z$, respectively. In this section we discuss how ∇ can also be used as an operator on vectors. If we treat ∇ formally as a vector, then we can apply it to a vector field $\mathbf{v}(x, y, z)$ in at least two ways.

The first way is to form the "dot product" of ∇ with $\mathbf{v}$ as follows:

$$\begin{aligned} \nabla \cdot \mathbf{v} &= \left(\frac{\partial}{\partial x}\mathbf{i} + \frac{\partial}{\partial y}\mathbf{j} + \frac{\partial}{\partial z}\mathbf{k}\right) \cdot (v_1\mathbf{i} + v_2\mathbf{j} + v_3\mathbf{k}) \\ &= \frac{\partial v_1}{\partial x} + \frac{\partial v_2}{\partial y} + \frac{\partial v_3}{\partial z}. \end{aligned} \tag{1}$$

The quantity $\nabla \cdot \mathbf{v}$ is known as the **divergence** of $\mathbf{v}$, and is sometimes denoted by div $\mathbf{v}$. For a two dimensional vector

$$\mathbf{v}(x, y) = v_1(x, y)\mathbf{i} + v_2(x, y)\mathbf{j}$$

we have

$$\nabla \cdot \mathbf{v} = \frac{\partial v_1}{\partial x} + \frac{\partial v_2}{\partial y}. \tag{2}$$

Observe that $\nabla \cdot \mathbf{v}$ is a scalar quantity, not a vector.

EXAMPLE 1

Find the divergence of

$$\mathbf{v}(x, y, z) = 4x^2y\mathbf{i} + y^2z\mathbf{j} - 3xyz\mathbf{k}.$$

From Eq. 1 we obtain

$$\begin{aligned} \nabla \cdot \mathbf{v} &= \frac{\partial}{\partial x}(4x^2y) + \frac{\partial}{\partial y}(y^2z) + \frac{\partial}{\partial z}(-3xyz) \\ &= 8xy + 2yz - 3xy = 5xy + 2yz. \ \blacksquare \end{aligned}$$

EXAMPLE 2

According to the inverse square law (Example 1 of Section 18.2), the attractive force exerted on a mass m at (x, y, z) by a mass M at the origin is

$$\mathbf{f}(x, y, z) = -GMm\frac{x\mathbf{i} + y\mathbf{j} + z\mathbf{k}}{(x^2 + y^2 + z^2)^{3/2}}. \tag{3}$$

Find the divergence of **f**.

Except for the multiplicative constant $-GMm$, the first term in $\nabla \cdot \mathbf{f}$ is

$$\begin{aligned}\frac{\partial}{\partial x}[x(x^2 + y^2 + z^2)^{-3/2}] &= (x^2 + y^2 + z^2)^{-3/2} + x\left(-\frac{3}{2}\right)(x^2 + y^2 + z^2)^{-5/2}(2x) \\ &= (x^2 + y^2 + z^2)^{-5/2}(x^2 + y^2 + z^2 - 3x^2) \\ &= (x^2 + y^2 + z^2)^{-5/2}(-2x^2 + y^2 + z^2)\end{aligned}$$

for $(x, y, z) \neq (0, 0, 0)$. In a similar way we obtain

$$\frac{\partial}{\partial y}[y(x^2 + y^2 + z^2)^{-3/2}] = (x^2 + y^2 + z^2)^{-5/2}(x^2 - 2y^2 + z^2),$$

$$(x, y, z) \neq (0, 0, 0),$$

$$\frac{\partial}{\partial z}[z(x^2 + y^2 + z^2)^{-3/2}] = (x^2 + y^2 + z^2)^{-5/2}(x^2 + y^2 - 2z^2),$$

$$(x, y, z) \neq (0, 0, 0).$$

Adding these results together we find that

$$\begin{aligned}\nabla \cdot \mathbf{f} &= -GMm(x^2 + y^2 + z^2)^{-5/2} \\ &\quad \times (-2x^2 + y^2 + z^2 + x^2 - 2y^2 + z^2 + x^2 + y^2 - 2z^2) \\ &= 0, \qquad (x, y, z) \neq (0, 0, 0).\end{aligned}$$

In other words, the divergence of **f** is zero at all points except the origin, where it is not defined because the required partial derivatives do not exist. ∎

A second way of applying ∇ to a vector field $\mathbf{v}(x, y, z)$ is to form the "cross product" of ∇ with **v**. Thus

$$\begin{aligned}\nabla \times \mathbf{v} &= \left(\frac{\partial}{\partial x}\mathbf{i} + \frac{\partial}{\partial y}\mathbf{j} + \frac{\partial}{\partial z}\mathbf{k}\right) \times (v_1\mathbf{i} + v_2\mathbf{j} + v_3\mathbf{k}) \\ &= \left(\frac{\partial v_3}{\partial y} - \frac{\partial v_2}{\partial z}\right)\mathbf{i} + \left(\frac{\partial v_1}{\partial z} - \frac{\partial v_3}{\partial x}\right)\mathbf{j} + \left(\frac{\partial v_2}{\partial x} - \frac{\partial v_1}{\partial y}\right)\mathbf{k}. \end{aligned} \tag{4}$$

The vector $\nabla \times \mathbf{v}$ defined in Eq. 4 is known as the **curl** of **v** and is also denoted by curl **v**. To avoid confusing the order or signs of the terms in Eq. 4 we can express $\nabla \times \mathbf{v}$ in the notation of determinants as follows:

$$\nabla \times \mathbf{v} = \begin{vmatrix} \mathbf{i} & \mathbf{j} & \mathbf{k} \\ \dfrac{\partial}{\partial x} & \dfrac{\partial}{\partial y} & \dfrac{\partial}{\partial z} \\ v_1 & v_2 & v_3 \end{vmatrix}. \tag{5}$$

You should verify that we obtain Eq. 4 by formally expanding the expression in Eq. 5 by elements in the first row according to the usual rules for determinants.

EXAMPLE 3

Find the curl of

$$\mathbf{v}(x, y, z) = x^2y\mathbf{i} - 2xyz\mathbf{j} + 2yz^2\mathbf{k}.$$

From Eq. 4 or 5 we obtain

$$\begin{aligned}\nabla \times \mathbf{v} &= \begin{vmatrix} \mathbf{i} & \mathbf{j} & \mathbf{k} \\ \dfrac{\partial}{\partial x} & \dfrac{\partial}{\partial y} & \dfrac{\partial}{\partial z} \\ x^2y & -2xyz & 2yz^2 \end{vmatrix} \\ &= (2z^2 + 2xy)\mathbf{i} + (0 - 0)\mathbf{j} + (-2yz - x^2)\mathbf{k} \\ &= 2(z^2 + xy)\mathbf{i} - (2yz + x^2)\mathbf{k}. \blacksquare\end{aligned}$$

EXAMPLE 4

Find the curl of the force field

$$\mathbf{f}(x, y, z) = -GMm\frac{x\mathbf{i} + y\mathbf{j} + z\mathbf{k}}{(x^2 + y^2 + z^2)^{3/2}}$$

given in Example 2.

The first component of $\nabla \times \mathbf{f}$ is

$$\begin{aligned}\frac{\partial f_3}{\partial y} - \frac{\partial f_2}{\partial z} &= -GMm\left(\frac{\partial}{\partial y}[z(x^2 + y^2 + z^2)^{-3/2}]\right. \\ &\qquad \left. - \frac{\partial}{\partial z}[y(x^2 + y^2 + z^2)^{-3/2}]\right) \\ &= -GMm\left[z\left(-\frac{3}{2}\right)(x^2 + y^2 + z^2)^{-5/2}(2y)\right. \\ &\qquad \left. - y\left(-\frac{3}{2}\right)(x^2 + y^2 + z^2)^{-5/2}(2z)\right] \\ &= 0, \qquad (x, y, z) \neq (0, 0, 0).\end{aligned}$$

The other two components of $\nabla \times \mathbf{f}$ can be calculated similarly. Thus $\nabla \times \mathbf{f} = \mathbf{0}$ at all points except the origin, where it is not defined. ∎

The divergence and curl have important physical interpretations, which we explore to some extent later. For the moment, however, we regard them simply as variations on the theme of differentiation. Thus they should be expected to share certain properties with other differentiation operators, and this is indeed the case. For example, if $\mathbf{u}(x, y, z)$ and $\mathbf{v}(x, y, z)$ are vector fields whose components have continuous partial derivatives, then

$$\nabla \cdot (c_1\mathbf{u} + c_2\mathbf{v}) = c_1\nabla \cdot \mathbf{u} + c_2\nabla \cdot \mathbf{v} \tag{6}$$

and

$$\nabla \times (c_1\mathbf{u} + c_2\mathbf{v}) = c_1\nabla \times \mathbf{u} + c_2\nabla \times \mathbf{v}. \tag{7}$$

The proofs of these formulas are left to the reader; they involve nothing more than writing out the left side of each equation, and collecting terms so as to form the expression on the right side.

If the vector field $\mathbf{u}(x, y, z)$ and the scalar function $\phi(x, y, z)$ have continuous derivatives up to the second order, then the following three relations hold:

$$\nabla \times (\nabla\phi) = \text{curl(grad } \phi) = \mathbf{0}; \tag{8}$$

$$\nabla \cdot (\nabla \times \mathbf{v}) = \text{div(curl } \mathbf{v}) = 0; \tag{9}$$

$$\begin{aligned}\nabla \cdot (\nabla\phi) &= \text{div(grad } \phi) \\ &= \frac{\partial^2\phi}{\partial x^2} + \frac{\partial^2\phi}{\partial y^2} + \frac{\partial^2\phi}{\partial z^2} = \nabla^2\phi.\end{aligned} \tag{10}$$

The proof of each of these results is straightforward. To establish Eq. 8, for example, we can substitute $\nabla\phi$ for $\mathbf{v}$ in Eq. 4. Thus we obtain

$$\nabla \times (\nabla\phi) = (\phi_{yz} - \phi_{zy})\mathbf{i} + (\phi_{zx} - \phi_{xz})\mathbf{j} + (\phi_{xy} - \phi_{yx})\mathbf{k}. \tag{11}$$

If ϕ has continuous partial derivatives of at least second order, then each component on the right side of Eq. 11 is zero, which proves Eq. 8.

In a similar way, we can take the divergence of $\nabla \times \mathbf{v}$ as given by Eq. 4; thus

$$\begin{aligned}\nabla \cdot (\nabla \times \mathbf{v}) &= \frac{\partial^2 v_3}{\partial x\,\partial y} - \frac{\partial^2 v_2}{\partial x\,\partial z} + \frac{\partial^2 v_1}{\partial y\,\partial z} - \frac{\partial^2 v_3}{\partial y\,\partial x} + \frac{\partial^2 v_2}{\partial z\,\partial x} - \frac{\partial^2 v_1}{\partial z\,\partial y} \\ &= 0,\end{aligned}$$

again provided only that the order of differentiation can be reversed. This establishes Eq. 9.

Equation 10 follows immediately from evaluating $\nabla \cdot (\nabla\phi)$. Some further results involving the divergence and curl may be found in Problems 12 and 14 through 18.

Irrotational fields

The position of a particle P moving on a circular orbit of radius a with angular speed ω (radians/second), as shown in Figure 18.6.1, can be described by the equation

$$\mathbf{r} = x\mathbf{i} + y\mathbf{j} = a \cos \omega t\, \mathbf{i} + a \sin \omega t\, \mathbf{j}. \tag{12}$$

The corresponding velocity vector,

$$\begin{aligned}\mathbf{v} = \frac{d\mathbf{r}}{dt} &= -\omega a \sin \omega t\, \mathbf{i} + \omega a \cos \omega t\, \mathbf{j} \\ &= -\omega y\mathbf{i} + \omega x\mathbf{j},\end{aligned} \tag{13}$$

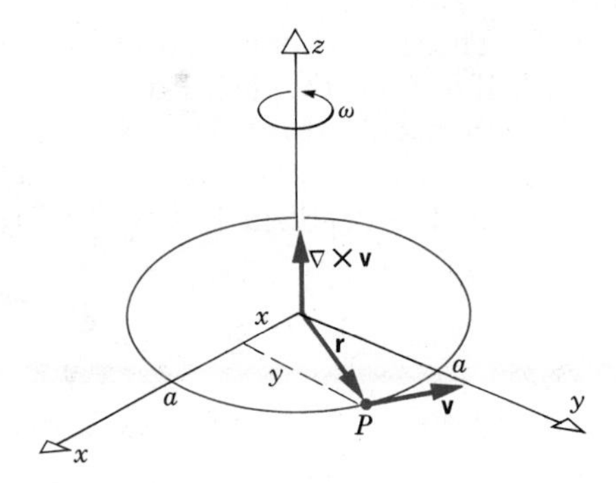

Figure 18.6.1 A particle P moving on a circular orbit.

is tangent to the circular path followed by the particle, and has magnitude $\|\mathbf{v}\| = a\omega$. The curl of $\mathbf{v}$ is given by

$$\nabla \times \mathbf{v} = \begin{vmatrix} \mathbf{i} & \mathbf{j} & \mathbf{k} \\ \dfrac{\partial}{\partial x} & \dfrac{\partial}{\partial y} & \dfrac{\partial}{\partial z} \\ -\omega y & \omega x & 0 \end{vmatrix} = 2\omega\mathbf{k}. \tag{14}$$

Thus the magnitude of $\nabla \times \mathbf{v}$ is twice the angular speed, and its direction is perpendicular to the plane of rotation. In a more general motion the curl of the velocity vector is a measure of the rotational component.

In fluid mechanics it is customary to call

$$\boldsymbol{\omega} = \nabla \times \mathbf{v} \tag{15}$$

the **vorticity** vector. If $\boldsymbol{\omega} = \mathbf{0}$, then the motion of the fluid has no rotational component, and the flow is said to be **irrotational.** This term is also commonly used to describe any vector field whose curl is zero, even if the vector field has nothing to do with fluid mechanics.

Finally, recall that Theorem 18.2.4 states, in part, that if

$$\frac{\partial R}{\partial y} = \frac{\partial Q}{\partial z}, \qquad \frac{\partial P}{\partial z} = \frac{\partial R}{\partial x}, \qquad \frac{\partial Q}{\partial x} = \frac{\partial P}{\partial y} \tag{16}$$

in a simply connected region D, then the vector

$$\mathbf{v} = P\mathbf{i} + Q\mathbf{j} + R\mathbf{k}$$

is the gradient of some scalar potential function ϕ. In vector notation, Eqs. 16 become

$$\nabla \times \mathbf{v} = \mathbf{0}, \tag{17}$$

so, by Theorem 18.2.4, if $\mathbf{v}$ is irrotational in a simply connected region, then $\mathbf{v} = \nabla\phi$ there. This is essentially the converse of Eq. 8, which says that a gradient vector is irrotational (in any region). In other words, in a simply connected region, a vector field is irrotational if and only if it is conservative, that is, if and only if it is the gradient of a scalar potential function.

There is a somewhat analogous result that involves the divergence and the curl. If $\mathbf{u} = \text{curl } \mathbf{v}$, then Eq. 9 states that $\text{div } \mathbf{u} = 0$. Under some mild restrictions, the converse is also true, that is, if $\text{div } \mathbf{u} = 0$, then there is a vector $\mathbf{v}$ whose curl is $\mathbf{u}$. In this case $\mathbf{v}$ is called the vector potential of $\mathbf{u}$. A way of constructing $\mathbf{v}$ is described in Problem 19.

PROBLEMS

In each of Problems 1 through 10, find the divergence and the curl of the given vector field.

1. $\mathbf{v}(x, y, z) = xy \sin z\, \mathbf{i} + x^2 z\mathbf{j} - y \cos z\, \mathbf{k}$

2. $\mathbf{v}(x, y, z) = 2x^2 y\mathbf{i} + 4y^2 z\mathbf{j} + 3z^2 x\mathbf{k}$

3. $\mathbf{v}(x, y, z) = \dfrac{\mathbf{r}}{r} = \dfrac{x\mathbf{i} + y\mathbf{j} + z\mathbf{k}}{r}$,
$r = (x^2 + y^2 + z^2)^{1/2}$

4. $\mathbf{v}(x, y, z) = r\mathbf{r} = r(x\mathbf{i} + y\mathbf{j} + z\mathbf{k})$,
$r = (x^2 + y^2 + z^2)^{1/2}$

5. $\mathbf{v}(x, y) = -ry\mathbf{i} + rx\mathbf{j}, \qquad r = (x^2 + y^2)^{1/2}$

6. $\mathbf{v}(x, y) = \left(\dfrac{x}{x^2 + y^2}\right)\mathbf{i} + \left(\dfrac{y}{x^2 + y^2}\right)\mathbf{j}$

7. $\mathbf{v}(x, y, z) = 2xyz\mathbf{i} - (x^2 + y^2)\mathbf{j} + xyz^2\mathbf{k}$

8. $\mathbf{v}(x, y, z) = xye^z\mathbf{i} + yze^{-x}\mathbf{k}$

9. $\mathbf{v}(x, y, z) = \sin x \sin y\, \mathbf{i} + \cos y \sin z\, \mathbf{j} + \cos z \cos x\mathbf{k}$

10. $\mathbf{v}(x, y, z) = \dfrac{x\mathbf{i} + y\mathbf{j} + z\mathbf{k}}{x^2 + y^2}$

11. If $\mathbf{u}(x, y, z)$ and $\mathbf{v}(x, y, z)$ are differentiable vector fields, and c_1 and c_2 are any constants, show that

(a) $\nabla \cdot (c_1\mathbf{u} + c_2\mathbf{v}) = c_1\nabla \cdot \mathbf{u} + c_2\nabla \cdot \mathbf{v}$.

(b) $\nabla \times (c_1\mathbf{u} + c_2\mathbf{v}) = c_1\nabla \times \mathbf{u} + c_2\nabla \times \mathbf{v}$.

12. If $\mathbf{u}(x, y, z)$ has continuous derivatives up to the second order, show that

$$\nabla \times (\nabla \times \mathbf{u}) = \nabla(\nabla \cdot \mathbf{u}) - \nabla^2\mathbf{u},$$

where

$$\nabla^2\mathbf{u} = \nabla^2 u_1\mathbf{i} + \nabla^2 u_2\mathbf{j} + \nabla^2 u_3\mathbf{k}.$$

13. If $\mathbf{r} = x\mathbf{i} + y\mathbf{j} + z\mathbf{k}$, and $\mathbf{A}$ is a constant vector, show that

$$\nabla \times (\mathbf{A} \times \mathbf{r}) = 2\mathbf{A},$$

and

$$\nabla \cdot (\mathbf{A} \times \mathbf{r}) = 0.$$

In Problems 14 through 18, assume that the vector fields $\mathbf{u}(x, y, z)$, $\mathbf{v}(x, y, z)$ and the scalar functions $\phi(x, y, z)$, $\psi(x, y, z)$ have as many continuous derivatives as necessary. On this basis derive each of the given relations.

14. $\nabla \cdot (\phi\mathbf{u}) = (\nabla\phi) \cdot \mathbf{u} + \phi(\nabla \cdot \mathbf{u})$

15. $\nabla \times (\phi\mathbf{u}) = (\nabla\phi) \times \mathbf{u} + \phi(\nabla \times \mathbf{u})$

16. $\nabla \cdot (\mathbf{u} \times \mathbf{v}) = \mathbf{v} \cdot (\nabla \times \mathbf{u}) - \mathbf{u} \cdot (\nabla \times \mathbf{v})$

17. $\nabla \cdot (\phi\nabla\psi) = \nabla\phi \cdot \nabla\psi + \phi\nabla^2\psi$

18. $\nabla \times (\phi\nabla\psi) = \nabla\phi \times \nabla\psi$

* **19.** In the text we have noted that if $\mathbf{u} = \nabla \times \mathbf{v}$, then $\nabla \cdot \mathbf{u} = 0$. Here we discuss the converse: if $\nabla \cdot \mathbf{u} = 0$ (in a somewhat restricted kind of region), then there exists a vector $\mathbf{v}$ such that $\mathbf{u} = \nabla \times \mathbf{v}$. Suppose that D is an open rectangular box in xyz space, that

$$\mathbf{u}(x, y, z) = L(x, y, z)\mathbf{i} + M(x, y, z)\mathbf{j} + N(x, y, z)\mathbf{k}$$

is defined in D, and that

$$\nabla \cdot \mathbf{u} = \frac{\partial L}{\partial x} + \frac{\partial M}{\partial y} + \frac{\partial N}{\partial z} = 0. \qquad (i)$$

We wish to construct a vector

$$\mathbf{v}(x, y, z) = P(x, y, z)\mathbf{i} + Q(x, y, z)\mathbf{j} + R(x, y, z)\mathbf{k}$$

such that $\mathbf{u} = \nabla \times \mathbf{v}$. In other words, we wish to choose P, Q, and R so that

$$\frac{\partial R}{\partial y} - \frac{\partial Q}{\partial z} = L, \qquad \frac{\partial P}{\partial z} - \frac{\partial R}{\partial x} = M,$$

$$\frac{\partial Q}{\partial x} - \frac{\partial P}{\partial y} = N. \qquad (ii)$$

One way of constructing P, Q, and R is outlined below. The derivation makes use of the fact that if f and $\partial f/\partial x$ are continuous, then

$$\frac{\partial}{\partial x}\int_c^d f(x, y, z)\,dz = \int_c^d \frac{\partial f}{\partial x}(x, y, z)\,dz.$$

(a) Let (x_0, y_0, z_0) and (x, y, z) be points in D. Choose $R(x, y, z) = 0$. Then show that the first two of Eqs. (ii) are satisfied if

$$Q(x, y, z) = -\int_{z_0}^{z} L(x, y, \zeta)\,d\zeta,$$

$$P(x, y, z) = \int_{z_0}^{z} M(x, y, \zeta)\,d\zeta + g(x, y),$$

where g is an arbitrary function of x and y.

(b) Use Eq. (i) to show that the third of Eqs. (ii) is satisfied provided that

$$\frac{\partial g}{\partial y}(x, y) = -N(x, y, z_0),$$

or

$$g(x, y) = -\int_{y_0}^{y} N(x, \eta, z_0)\,d\eta.$$

Observe that $\mathbf{v}$ is determined only up to an additive irrotational vector. Also note that one can start by choosing either P or Q to be zero, rather than R.

In each of Problems 20 through 25, use the method of Problem 19 to find a vector $\mathbf{v}$ such that $\nabla \times \mathbf{v} = \mathbf{u}$, or else show that no such vector exists. Assume that D is all of xyz space.

20. $\mathbf{u}(x, y, z) = yz\mathbf{i} + xz\mathbf{j} + xy\mathbf{k}$

21. $\mathbf{u}(x, y, z) = 2x\mathbf{i} + y\mathbf{j} - 3z\mathbf{k}$

22. $\mathbf{u}(x, y, z) = 2x\mathbf{i} + y\mathbf{j} + 3z\mathbf{k}$

23. $\mathbf{u}(x, y, z) = xy \sin z\,\mathbf{i} + x^2z\mathbf{j} - y \cos z\,\mathbf{k}$

24. $\mathbf{u}(x, y, z) = ye^x\mathbf{i} + \left(yz^2 - \dfrac{y^2e^x}{2}\right)\mathbf{j} - \dfrac{z^3}{3}\mathbf{k}$

25. $\mathbf{u}(x, y, z) = xy \sin z\,\mathbf{i} + x^2z\mathbf{j} + y \cos z\,\mathbf{k}$

18.7 THE DIVERGENCE THEOREM

Many important applications of surface integrals involve vector fields that depend on space variables, and possibly time. For example, the resultant force exerted across a surface, or the rate of flow (mass per unit time) of a fluid through a surface, are given by integrals of this kind.

Suppose that the surface S is described by

$$z = f(x, y), \qquad (x, y) \in \Omega_{xy}. \tag{1}$$

Then, as indicated in Figure 18.7.1, at each point of the surface there are two

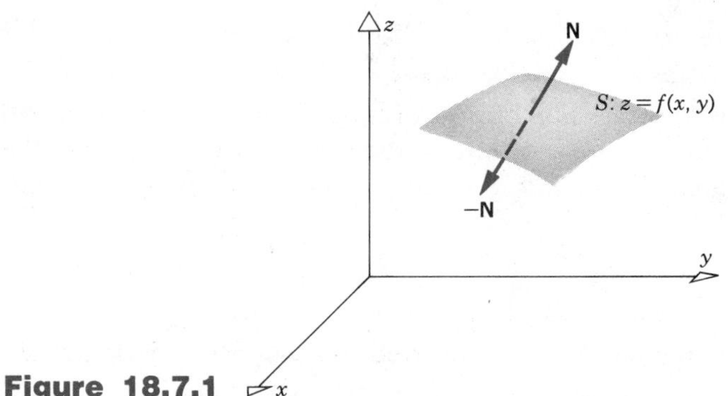

Figure 18.7.1

possible unit normal vectors, namely

$$\mathbf{N} = \pm\frac{-f_x(x, y)\mathbf{i} - f_y(x, y)\mathbf{j} + \mathbf{k}}{[f_x^2(x, y) + f_y^2(x, y) + 1]^{1/2}}. \tag{2}$$

Similarly, if the surface is given by

$$F(x, y, z) = 0, \tag{3}$$

then

$$\mathbf{N} = \pm\frac{F_x(x, y, z)\mathbf{i} + F_y(x, y, z)\mathbf{j} + F_z(x, y, z)\mathbf{k}}{[F_x^2(x, y, z) + F_y^2(x, y, z) + F_z^2(x, y, z)]^{1/2}}. \tag{4}$$

To avoid ambiguity we must choose one of these normals to identify the positive normal direction at each point. For example, we might choose the vector corresponding to the plus sign in Eq. 2 or in Eq. 4 to be the positive unit normal vector $\mathbf{N}$. If S is a closed surface, such as a sphere, we will always choose the outward pointing normal vector as the positive one; otherwise, the choice is arbitrary. However, once the choice of $\mathbf{N}$ has been made, then the side of the surface S on which $\mathbf{N}$ lies is referred to as the positive side and the other side as the negative side.

Throughout this discussion we assume that S actually does have two sides; such surfaces are said to be **orientable.** This is not a trivial remark because some surfaces do not have this property. The most familiar example of a one-sided, or nonorientable, surface is the Möbius band, which can be constructed from a long rectangular strip of paper by twisting it one time and then joining the ends together.

Suppose now that a vector field $\mathbf{v}(x, y, z)$ is defined in some region of xyz space containing S. Then $(\mathbf{v} \cdot \mathbf{N})(x, y, z)$ is the component of $\mathbf{v}$ in the direction of the positive normal vector and

$$\iint_S (\mathbf{v} \cdot \mathbf{N})(x, y, z)\, d\sigma \tag{5}$$

is the total **flux** of $\mathbf{v}$ through S from the negative to the positive side. For example, if $\mathbf{v}(x, y, z)$ is the velocity at the point (x, y, z) in a fluid, then the integral (5) is the rate (volume per unit time) at which fluid flows through the surface S from the negative to the positive side. If $\rho(x, y, z)$ is the density of the fluid, then

$$\iint_S \rho(x, y, z)(\mathbf{v} \cdot \mathbf{N})(x, y, z)\, d\sigma \tag{6}$$

is the mass rate of flow of fluid through S. Similarly, if $\boldsymbol{\tau}(x, y, z)$ represents force per unit area, or **stress,** then $(\boldsymbol{\tau} \cdot \mathbf{N})(x, y, z)\, d\sigma$ is the force exerted across the area element $d\sigma$ in the direction of the positive normal, and

$$F_n = \iint_S (\boldsymbol{\tau} \cdot \mathbf{N})(x, y, z)\, d\sigma \tag{7}$$

is the normal force exerted through S by the material on the positive side upon the material on the negative side. In a similar way,

$$\mathbf{F} = \int\int_S \boldsymbol{\tau}(x, y, z)\, d\sigma \tag{8}$$

is the resultant or total force exerted through S.

EXAMPLE 1

As in Example 2 of Section 18.6, the attractive force exerted on a mass m at (x, y, z) by a mass M at the origin is

$$\mathbf{f}(x, y, z) = -GMm \frac{x\mathbf{i} + y\mathbf{j} + z\mathbf{k}}{(x^2 + y^2 + z^2)^{3/2}}. \tag{9}$$

If the cylindrical surface $x^2 + y^2 = a^2$ has constant density (mass per unit area) ρ, find the resultant force $\mathbf{F}$ exerted by the mass M on the portion of the cylinder for which $0 \le z \le h$.

Figure 18.7.2 indicates the distribution of forces on the cylinder. Because of the symmetry of the forces about the z-axis, the net components F_1 and F_2 in the

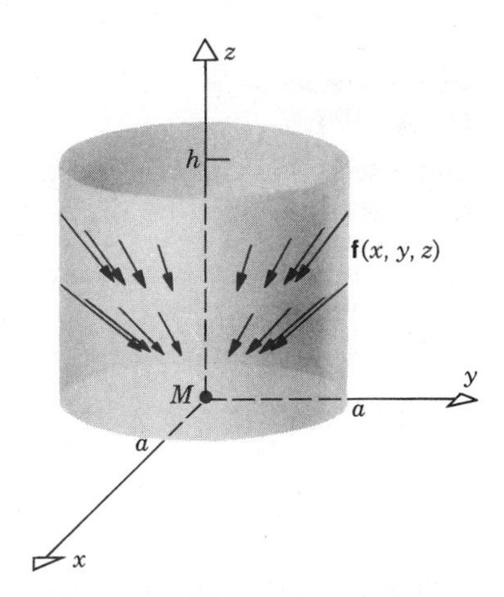

Figure 18.7.2 Gravitational attraction on a cylinder.

x and y directions, respectively, are both zero; thus we need calculate only the z-component F_3 of the resultant force on the cylinder. The mass of an element $d\sigma$ of the cylinder is $\rho d\sigma$. Hence, by using Eq. 9 with m replaced by $\rho d\sigma$, and then integrating over the cylinder, we obtain

$$F_3 = -GM\rho \int\int_S \frac{z}{(x^2 + y^2 + z^2)^{3/2}}\, d\sigma, \tag{10}$$

where S is the surface of the cylinder. Again taking advantage of the symmetry of the configuration, we can determine F_3 by integrating over the part of S in the first octant and then multiplying by four. To evaluate the integral we can project either into the xz- or into the yz-plane. Choosing the latter, we have

$$\begin{aligned}
F_3 &= -4GM\rho \iint_{\Omega_{yz}} \frac{z}{(a^2 + z^2)^{3/2}} \frac{dA_{yz}}{x/a} \\
&= -4GM\rho \int_0^h \int_0^a \frac{z}{(a^2 + z^2)^{3/2}} \frac{a}{(a^2 - y^2)^{1/2}} \, dy \, dz \\
&= -4GM\rho a \left[-(a^2 + z^2)^{-1/2} \Big|_0^h \right] \left[\arcsin \frac{y}{a} \Big|_0^a \right] \\
&= -4GM\rho a \frac{\pi}{2} \left[-(a^2 + h^2)^{-1/2} + a^{-1} \right] \\
&= -2\pi GM\rho \, [1 - a(a^2 + h^2)^{-1/2}]. \qquad (11)
\end{aligned}$$

The result is negative because the force is directed downward (in the negative z direction). Observe that the quantity in brackets approaches 1 as $h \to \infty$. Thus

$$\mathbf{F} = -2\pi GM\rho\mathbf{k}$$

is the resultant force on a semi-infinite cylinder. ∎

EXAMPLE 2

Suppose that a fluid of constant density ρ flows in a circular pipe of radius a. Let the axis of the pipe be the y-axis, as shown in Figure 18.7.3. If the velocity of the

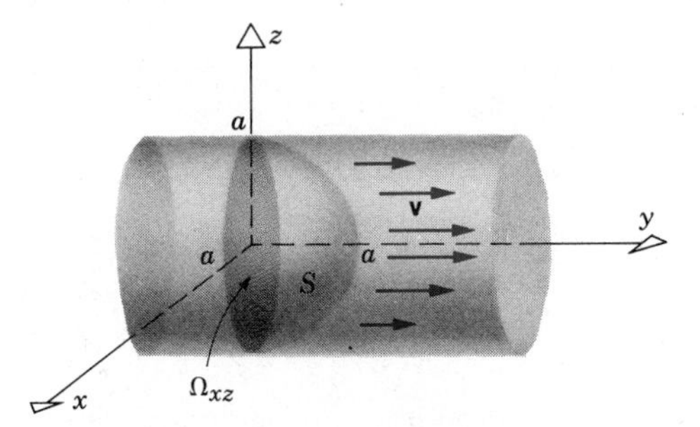

Figure 18.7.3 Flow through a hemispherical surface in a pipe.

fluid is

$$\mathbf{v} = (a^2 - x^2 - z^2)^{1/2}\mathbf{j}, \qquad (12)$$

for $x^2 + z^2 \le a^2$, find the rate at which this fluid flows across the hemispherical surface S,

$$x^2 + y^2 + z^2 = a^2, \qquad y \ge 0.$$

The flux of fluid across S is

$$\rho \iint_S (\mathbf{v} \cdot \mathbf{N})(x, y, z) \, d\sigma.$$

The unit normal vector $\mathbf{N}$, pointing away from the origin, is

$$\mathbf{N} = \frac{x}{a}\mathbf{i} + \frac{y}{a}\mathbf{j} + \frac{z}{a}\mathbf{k}.$$

It is convenient to integrate in the xz-plane, so

$$d\sigma = \frac{dA_{xz}}{|\mathbf{N}\cdot\mathbf{j}|} = \frac{dA_{xz}}{y/a}$$

and the integration region Ω_{xz} is the disk $x^2 + z^2 \leq a^2$. Thus we obtain

$$\begin{aligned}\rho\int\!\!\int_S (\mathbf{v}\cdot\mathbf{N})(x, y, z)\,d\sigma &= \rho\int\!\!\int_{\Omega_{xz}} \frac{(a^2 - x^2 - z^2)^{1/2}\,(y/a)}{y/a}\,dA_{xz} \\ &= \rho\int_0^{2\pi}\int_0^a (a^2 - r^2)^{1/2}\,r\,dr\,d\theta \\ &= 2\pi\rho\left(-\frac{1}{2}\right)\frac{2}{3}(a^2 - r^2)^{3/2}\Big|_0^a \\ &= \frac{2\pi\rho a^3}{3}. \quad \blacksquare\end{aligned}$$

Another look at Green's theorem

As we mentioned at the beginning of Section 18.4, we wish to extend the relation expressed by Green's theorem (Theorem 18.3.1) from two to three dimensions. To accomplish this it is helpful to write down Green's theorem in vector notation, which is essentially independent of dimension. It turns out that there are two ways of doing this, which lead to two important results, both of which can be considered as three-dimensional versions of Green's theorem. The first is considered now, while the second appears in the following section.

Let C be a piecewise smooth simple closed curve in the xy-plane, described parametrically by

$$x = f(s), \qquad y = g(s),$$

where s is arc length measured counterclockwise from some fixed point on the

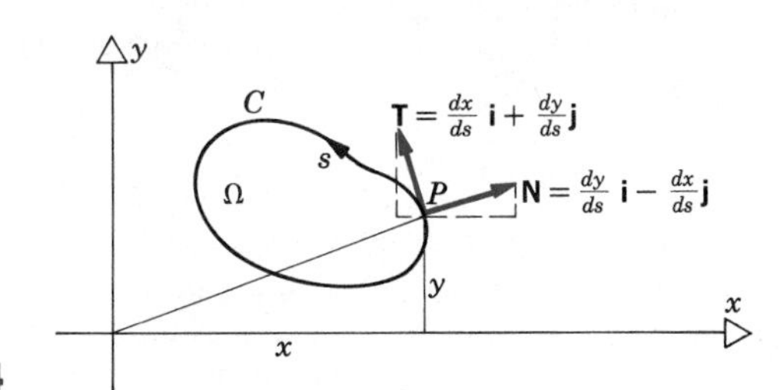

Figure 18.7.4

curve (see Figure 18.7.4). The unit tangent vector $\mathbf{T}$ at a point P is given by

$$\mathbf{T} = \frac{dx}{ds}\mathbf{i} + \frac{dy}{ds}\mathbf{j}$$

and the unit normal vector $\mathbf{N}$ in the outward direction is

$$\mathbf{N} = \frac{dy}{ds}\mathbf{i} - \frac{dx}{ds}\mathbf{j}.$$

If we write $\mathbf{v}(x, y) = v_1(x, y)\mathbf{i} + v_2(x, y)\mathbf{j}$, and then identify $v_1(x, y)$ with $Q(x, y)$ and $v_2(x, y)$ with $-P(x, y)$, we obtain

$$\begin{aligned}\oint_C \mathbf{v} \cdot \mathbf{N}(x, y)\, ds &= \oint_C v_1(x, y)\, dy - v_2(x, y)\, dx \\ &= \oint_C P(x, y)\, dx + Q(x, y)\, dy.\end{aligned}$$

On the other hand, if Ω is the region bounded by C, then

$$\begin{aligned}\iint_\Omega \left(\frac{\partial Q}{\partial x} - \frac{\partial P}{\partial y}\right) dA &= \iint_\Omega \left(\frac{\partial v_1}{\partial x} + \frac{\partial v_2}{\partial y}\right) dA \\ &= \iint_\Omega \nabla \cdot \mathbf{v}\, dA.\end{aligned}$$

Thus the conclusion of Green's theorem can be written in the vector form

$$\oint_C \mathbf{v} \cdot \mathbf{N}\, ds = \iint_\Omega \nabla \cdot \mathbf{v}\, dA. \tag{13}$$

The left side of Eq. 13 is the net outward flux of $\mathbf{v}$ across C. Green's theorem says that this quantity is balanced by the integral of the divergence of $\mathbf{v}$ throughout the interior of C.

The divergence theorem

To extend the result expressed by Eq. 13 to three dimensions it seems natural to consider, instead of C and Ω, a closed surface S containing the region D. At the same time we think of $\mathbf{v}$ as a vector field in three dimensions. Thus we are led to the important result stated in the following theorem.*

Theorem 18.7.1

(Divergence Theorem)

Let S be a piecewise smooth closed orientable surface with unit outer normal vector $\mathbf{N}$, and let D be the region bounded by S. If $\mathbf{v}(x, y, z)$ is a vector field

*The origin of the divergence theorem is somewhat uncertain. As we have shown, it is essentially equivalent to the theorem stated by Green in 1828, but the name of Green is usually reserved for the two-dimensional version of the theorem. The Russian mathematician Mikhail Ostrogradsky (1801–1862) also used the result in a paper given in 1828, and related results appear in earlier work of Gauss and Lagrange. The theorem is sometimes associated with the names of Gauss and Ostrogradsky, but currently it is most commonly identified simply as the divergence theorem.

with continuous partial derivatives in D, then

$$\iint_S \mathbf{v} \cdot \mathbf{N}\, d\sigma = \iiint_D \nabla \cdot \mathbf{v}\, dV. \tag{14}$$

Partial Proof. Observe first that

$$\mathbf{N} = (\cos \alpha)\mathbf{i} + (\cos \beta)\mathbf{j} + (\cos \gamma)\mathbf{k},$$

where α, β, and γ are the angles between $\mathbf{N}$ and the positive x-, y-, and z-axes, respectively. Then

$$\iint_S \mathbf{v} \cdot \mathbf{N}\, d\sigma = \iint_S v_1 \cos \alpha\, d\sigma + \iint_S v_2 \cos \beta\, d\sigma + \iint_S v_3 \cos \gamma\, d\sigma. \tag{15}$$

Also

$$\iiint_D \nabla \cdot \mathbf{v}\, dV = \iiint_D \frac{\partial v_1}{\partial x}\, dV + \iiint_D \frac{\partial v_2}{\partial y}\, dV + \iiint_D \frac{\partial v_3}{\partial z}\, dV. \tag{16}$$

The divergence theorem will now be established provided that we can show that the corresponding terms on the right sides of Eqs. 15 and 16 are equal, that is,

$$\iint_S v_3 \cos \gamma\, d\sigma = \iiint_D \frac{\partial v_3}{\partial z}\, dV, \tag{17}$$

and similarly for the terms involving v_1 and v_2. To prove Eq. 17 we assume that D can be described by the inequalities

$$D : f(x, y) \le z \le g(x, y), \qquad (x, y) \in \Omega_{xy}.$$

Such a region D is said to be projectible into the xy-plane. As indicated in Figure 18.7.5, the boundary S of D consists at least of a lower portion S_1, given by $z = f(x, y)$, and an upper portion S_2 given by $z = g(x, y)$. There may also be a lateral portion S_3 consisting of part of the cylinder with cross section Ω_{xy} and generators parallel to the z-axis. Then, starting with the right side of Eq. 17 and

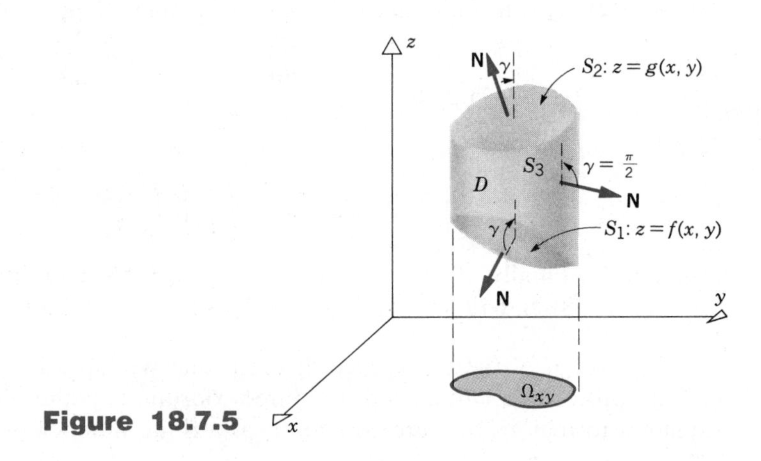

Figure 18.7.5

integrating with respect to z, we obtain

$$\iiint_D \frac{\partial v_3}{\partial z}(x, y, z)\, dV = \iint_{\Omega_{xy}} \int_{f(x,y)}^{g(x,y)} \frac{\partial v_3}{\partial z}(x, y, z)\, dz\, dA_{xy}$$

$$= \iint_{\Omega_{xy}} v_3[x, y, g(x, y)]\, dA_{xy} - \iint_{\Omega_{xy}} v_3[x, y, f(x, y)]\, dA_{xy}.$$

Now looking at the left side of Eq. 17 we have

$$\iint_S v_3 \cos \gamma\, d\sigma = \left(\iint_{S_1} + \iint_{S_2} + \iint_{S_3}\right) v_3 \cos \gamma\, d\sigma.$$

The integral over S_3 is zero because the normal vector on S_3 is parallel to the xy-plane; hence $\cos \gamma = 0$. On S_2, $\cos \gamma \geq 0$, so $(\cos \gamma)\, d\sigma = dA_{xy}$, and

$$\iint_{S_2} v_3 \cos \gamma\, d\sigma = \iint_{\Omega_{xy}} v_3[x, y, g(x, y)]\, dA_{xy}.$$

Finally, on S_1, $\cos \gamma \leq 0$, so $(\cos \gamma)\, d\sigma = -dA_{xy}$, and

$$\iint_{S_1} v_3 \cos \gamma\, d\sigma = -\iint_{\Omega_{xy}} v_3[x, y, f(x, y)]\, dA_{xy}.$$

Hence

$$\iint_S v_3 \cos \gamma\, d\sigma = \iint_{\Omega_{xy}} v_3[x, y, g(x, y)]\, dA_{xy} - \iint_{\Omega_{xy}} v_3[x, y, f(x, y)]\, dA_{xy}.$$

Thus Eq. 17 has been proved for regions that are projectible into the xy-plane. In a similar way, if D can be described by

$$D : h(y, z) \leq x \leq k(y, z), \qquad (y, z) \in \Omega_{yz}$$

or by

$$D : l(x, z) \leq y \leq m(x, z), \qquad (x, z) \in \Omega_{xz},$$

that is, if D is projectible into the yz- or xz-planes, then it follows that

$$\iint_S v_1 \cos \alpha\, d\sigma = \iiint_D \frac{\partial v_1}{\partial x}\, dV, \tag{18}$$

or

$$\iint_S v_2 \cos \beta\, d\sigma = \iiint_D \frac{\partial v_2}{\partial y}\, dV, \tag{19}$$

respectively. Finally, if D is projectible into all three coordinate planes, then all of Eqs. 17, 18, and 19 are valid, and Eq. 14 follows at once. □

The proof of the divergence theorem that we have given applies to regions that are projectible into each of the three coordinate planes. The theorem can be extended to regions that are composed of a finite number of such subregions by an argument similar to that used for Green's theorem in Section 18.3. The divergence theorem can also be extended to regions whose boundary consists of two or more separate surfaces. For example, in Figure 18.7.6*a*, D is the region between the surfaces S_1 and S_2. Then

$$\iiint_D \nabla \cdot \mathbf{v}\, dV = \iint_{S_1} \mathbf{v} \cdot \mathbf{N}\, d\sigma + \iint_{S_2} \mathbf{v} \cdot \mathbf{N}\, d\sigma, \tag{20}$$

where in each integral on the right side of Eq. 20, $\mathbf{N}$ is the unit normal vector directed *outward from D*. Thus $\mathbf{N}$ is directed toward the origin on S_1 and away

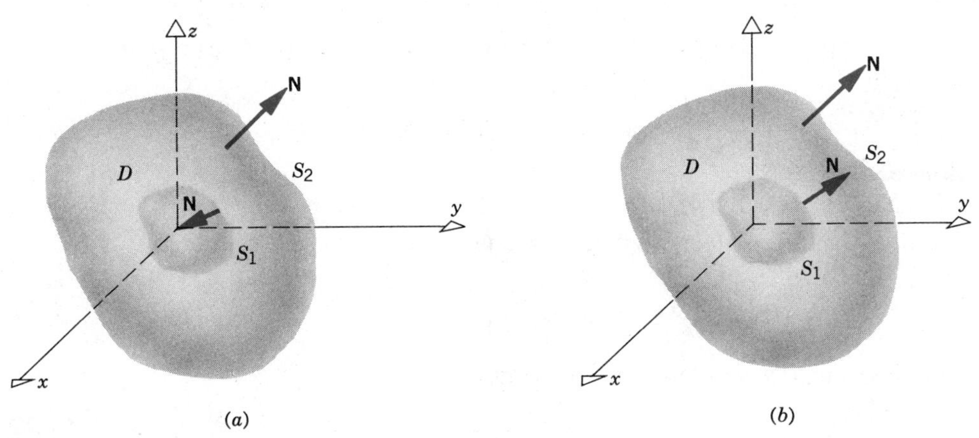

Figure 18.7.6

from it on S_2. Alternatively, we can choose $\mathbf{N}$ to be the normal pointing away from the origin in both cases, as shown in Figure 18.7.6*b*; then, since reversing the direction of $\mathbf{N}$ changes the sign of the integral over S_1,

$$\iiint_D \nabla \cdot \mathbf{v}\, dV = -\iint_{S_1} \mathbf{v} \cdot \mathbf{N}\, d\sigma + \iint_{S_2} \mathbf{v} \cdot \mathbf{N}\, d\sigma. \tag{21}$$

In Figures 18.7.6*a* and 18.7.6*b* both surfaces S_1 and S_2 are shown as enclosing the origin, but Eq. 21 is valid whether or not this is the case.

The divergence theorem is useful in the study of physics and applied mathematics in developing the fundamental equations of fluid mechanics and electromagnetic theory, for example. It is also useful in the evaluation of certain integrals, by providing a different but equivalent expression for them. Most often, this provides a means of evaluating an integral over a closed surface when the vector field has a simple divergence.

EXAMPLE 3

Find the value of

$$\iint_S [(2x + yz)\mathbf{i} + 2xz\mathbf{j} - 3xy\mathbf{k}] \cdot \mathbf{N}\, d\sigma,$$

where S is the boundary of the cylindrical region

$$D : 0 \leq x^2 + y^2 \leq a^2, \quad 0 \leq z \leq h.$$

Note that S includes the ends of the cylinder in the planes $z = 0$ and $z = h$, as well as the lateral surface. If we denote the vector in the integrand by $\mathbf{v}$, then

$$\nabla \cdot \mathbf{v} = \frac{\partial}{\partial x}(2x + yz) + \frac{\partial}{\partial y}(2xz) + \frac{\partial}{\partial z}(-3xy) = 2.$$

Hence, according to the divergence theorem, and noting that the volume of R is $\pi a^2 h$,

$$\iint_S \mathbf{v} \cdot \mathbf{N}\, d\sigma = \iiint_D \nabla \cdot \mathbf{v}\, dV = 2 \iiint_D dV = 2\pi a^2 h. \blacksquare$$

EXAMPLE 4

Consider again the inverse square force given by Eq. 9 of Example 1,

$$\mathbf{f}(x, y, z) = -K \frac{x\mathbf{i} + y\mathbf{j} + z\mathbf{k}}{(x^2 + y^2 + z^2)^{3/2}}, \tag{22}$$

where K is a positive constant. Determine

$$\iint_S \mathbf{f} \cdot \mathbf{N}\, d\sigma, \tag{23}$$

where S is any piecewise smooth closed surface containing the origin (see Figure 18.7.7).

It may seem unlikely that the integral can be evaluated without specifying S more precisely. However, recall that in Example 2 of Section 18.6 we showed that $\nabla \cdot \mathbf{f} = 0$ at all points other than the origin. To take advantage of this fact we can introduce the sphere Σ with center at the origin and radius a, choosing a to be small enough so that Σ is inside of S (see Figure 18.7.7). If D is the region between

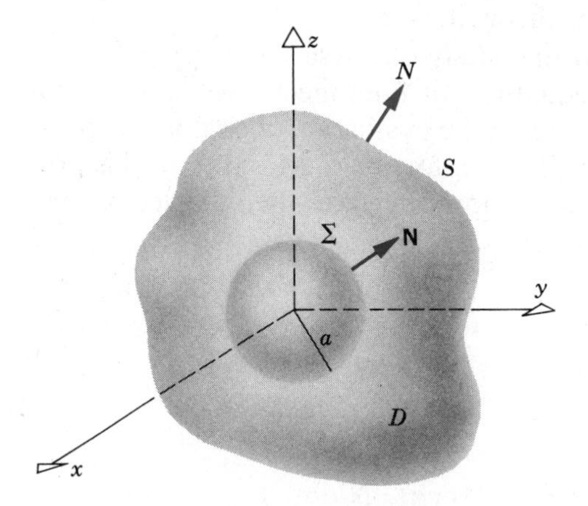

Figure 18.7.7 The normal gravitational force on the surface S is the same as on the sphere Σ.

Σ and S, then $\nabla \cdot \mathbf{f} = 0$ throughout D. Consequently, by applying the divergence theorem to D we obtain

$$\iint_{\Sigma} \mathbf{f} \cdot \mathbf{N}\, d\sigma + \iint_{S} \mathbf{f} \cdot \mathbf{N}\, d\sigma = \iiint_{D} \nabla \cdot \mathbf{f}\, dV = 0,$$

where $\mathbf{N}$ is the unit normal directed outward from D on both Σ and S. By reversing the direction of the normal vector $\mathbf{N}$ on Σ, we have

$$\iint_{S} \mathbf{f} \cdot \mathbf{N}\, d\sigma = \iint_{\Sigma} \mathbf{f} \cdot \mathbf{N}\, d\sigma, \tag{24}$$

where now $\mathbf{N}$ is directed away from the origin on both Σ and S. It happens that the integral over Σ is easy to evaluate; indeed that is why we chose Σ to be a sphere. On Σ we have $x^2 + y^2 + z^2 = a^2$, and

$$\mathbf{N} = \frac{x}{a}\mathbf{i} + \frac{y}{a}\mathbf{j} + \frac{z}{a}\mathbf{k}.$$

Thus

$$\begin{aligned} \iint_{\Sigma} \mathbf{f} \cdot \mathbf{N}\, d\sigma &= -K \iint_{\Sigma} \frac{x^2 + y^2 + z^2}{a(x^2 + y^2 + z^2)^{3/2}}\, d\sigma \\ &= -\frac{K}{a^2} \iint_{\Sigma} d\sigma \\ &= -\frac{K}{a^2}\, 4\pi a^2 = -4\pi K. \end{aligned}$$

From Eq. 24 it follows immediately that

$$\iint_{S} \mathbf{f} \cdot \mathbf{N}\, d\sigma = -4\pi K$$

for an arbitrary surface S surrounding the origin. The integral (23) is therefore independent of the surface in much the same way that certain line integrals are independent of the path. ∎

Finally, we can use the divergence theorem to obtain a clearer understanding of the physical meaning of the divergence. To make our discussion more specific let $\mathbf{v}$ be the velocity of a fluid occupying a certain region in space, let S be a small sphere with center at a point P, and let D be the interior of S. Then, using the divergence theorem and estimating the value of the integral over D by using the value of $\nabla \cdot \mathbf{v}$ at P, we obtain

$$\iint_{S} \mathbf{v} \cdot \mathbf{N}\, d\sigma = \iiint_{D} \nabla \cdot \mathbf{v}\, dV \cong (\nabla \cdot \mathbf{v})(P)V(D),$$

where $V(D)$ is the volume of D. If $\nabla \cdot \mathbf{v}$ is continuous, then the approximation approaches an equality as $V(D) \to 0$. Consequently, we obtain

$$(\nabla \cdot \mathbf{v})(P) = \lim_{V(D)\to 0} \frac{\int\int_S \mathbf{v} \cdot \mathbf{N}\, d\sigma}{V(D)}. \tag{25}$$

Thus $(\nabla \cdot \mathbf{v})(P)$ is the outward flux per unit volume through a small sphere surrounding P, or the flux density at P. Further, if fluid is neither created nor destroyed at P (that is, if there is no source or sink there), then the net flux is zero and so is the divergence. Otherwise, the divergence of $\mathbf{v}$ measures the rate at which fluid is created or destroyed at P.

PROBLEMS

In each of Problems 1 through 4, find the value of the given integral (a) by using the divergence theorem; (b) by direct evaluation of the surface integral.

1. $\int\int_S (2x\mathbf{i} - 3y\mathbf{j} + 4z\mathbf{k}) \cdot \mathbf{N}\, d\sigma$, where S is the surface of the cube $0 \le x \le 1$, $0 \le y \le 1$, $0 \le z \le 1$.
2. $\int\int_S (x\mathbf{i} + 4y\mathbf{j} - 3z\mathbf{k}) \cdot \mathbf{N}\, d\sigma$, where S is the surface of the rectangular parallelepiped $0 \le x \le 2$, $0 \le y \le 3$, $0 \le z \le 1$.
3. $\int\int_S (yz\mathbf{i} + xz\mathbf{j} + xy\mathbf{k}) \cdot \mathbf{N}\, d\sigma$, where S is the surface of the cube $0 \le x \le 1$, $0 \le y \le 1$, $0 \le z \le 1$.
4. $\int\int_S (x^2\mathbf{i} + y^2\mathbf{j} + z^2\mathbf{k}) \cdot \mathbf{N}\, d\sigma$, where S is the surface of the rectangular parallelepiped $0 \le x \le 2$, $0 \le y \le 4$, $0 \le z \le 3$.

In each of Problems 5 through 10, use the divergence theorem to find the value of $\int\int_S \mathbf{v} \cdot \mathbf{N}\, d\sigma$, where $\mathbf{v}$ and S are given.

5. $\mathbf{v}(x, y, z) = 2x\mathbf{i} + y\mathbf{j} + (1 - z)\mathbf{k}$; S is the sphere $x^2 + y^2 + z^2 = a^2$.
6. $\mathbf{v}(x, y, z) = -x\mathbf{i} + (3y + 1)\mathbf{j} + (z - 2)\mathbf{k}$; S consists of the hemisphere $z = \sqrt{a^2 - x^2 - y^2}$ together with the disk $x^2 + y^2 \le a^2$ in the plane $z = 0$.
7. $\mathbf{v}(x, y, z) = x^2\mathbf{i} + y^2\mathbf{j}$; S is the surface of the cylinder $x^2 + y^2 \le 1$, $0 \le z \le 2$, including the top and bottom.
8. $\mathbf{v}(x, y, z) = (x^2 + y^2)\mathbf{i} + 2xy\mathbf{j}$; S is the paraboloid $z = 9 - x^2 - y^2$ for $z \ge 0$ together with the disk $x^2 + y^2 \le 9$ in the plane $z = 0$.
9. $\mathbf{v}(x, y, z) = x^2\mathbf{i} + y^2\mathbf{j} + z^2\mathbf{k}$; S is the surface of the cylinder $x^2 + y^2 \le 4$, $0 \le z \le 3$, including the top and bottom.
10. $\mathbf{v}(x, y, z) = x^3\mathbf{i} + y^3\mathbf{j} + z^3\mathbf{k}$; S is the sphere $x^2 + y^2 + z^2 = a^2$.

In each of Problems 11 through 16, find the value of the given integral.

11. $\int\int_S \mathbf{f} \cdot \mathbf{N}\, d\sigma$, where $\mathbf{f}$ is the inverse square attractive force of Eq. 22 in Example 4, S is the hemispherical surface $x^2 + y^2 + z^2 = a^2$ with $z \ge 0$, and $\mathbf{N}$ is the upper unit normal vector.
12. $\int\int_S \mathbf{v} \cdot \mathbf{N}\, d\sigma$, where $\mathbf{v}(x, y, z) = z(1 - z)\mathbf{j}$, S is the portion of the plane $y = 0$ for which $0 \le x \le 1$ and $0 \le z \le 1$, and $\mathbf{N}$ is the unit normal vector with positive y-component.
13. $\int\int_S \mathbf{v} \cdot \mathbf{N}\, d\sigma$, where $\mathbf{v}(x, y, z) = Uz\mathbf{k}$ for some positive constant U, S is the portion of the paraboloid $z = a^2 - x^2 - y^2$ for which $z \ge 0$, and $\mathbf{N}$ is the upper unit normal vector.
14. $\int\int_S \mathbf{v} \cdot \mathbf{N}\, d\sigma$, where $\mathbf{v}(x, y, z) = Uy\mathbf{j}$ for some positive constant U, S is the hemisphere $x^2 + y^2 + z^2 = a^2$ with $y \ge 0$, and $\mathbf{N}$ is the unit normal vector with positive y-component.
15. $\int\int_S (x\mathbf{i} + y\mathbf{j} + z\mathbf{k}) \cdot \mathbf{N}\, d\sigma$, where S is the portion of the hyperboloid $z^2 - x^2 - y^2 = 1$ for which $1 \le z \le 2$, and $\mathbf{N}$ is the upward pointing unit normal vector.
16. $\int\int_S (yz\mathbf{i} + xz\mathbf{j} + xy\mathbf{k}) \cdot \mathbf{N}\, d\sigma$, where S is the portion of the surface $z = 4 - x^2 - y^2$ that lies in the first quadrant, and $\mathbf{N}$ is the upper unit normal vector.

17. Find the total vertical force on the hemisphere $x^2 + y^2 + z^2 = a^2$, with $z \geq 0$, if the force per unit area is $\mathbf{f}(x, y, z) = \alpha z\mathbf{k}$, where α is a positive constant.

18. Find the total normal force on the portion of the cylinder $x^2 + z^2 = a^2$ with $z \geq 0$ and $0 \leq y \leq b$ if the force per unit area is $\mathbf{f}(x, y, z) = -\alpha(a - z)\mathbf{k}$, where α is a positive constant.

19. Let $\mathbf{v}(x, y, z) = [x/(x^2 + y^2)]\mathbf{i} + [y/(x^2 + y^2)]\mathbf{j}$, and let S be the surface of the cube $-2 \leq x \leq 2$, $-2 \leq y \leq 2$, $-2 \leq z \leq 2$. Find the value of $\int\int_S \mathbf{v} \cdot \mathbf{N}\, d\sigma$.
Hint: Consider replacing S by the surface Σ consisting of the cylinder $x^2 + y^2 = 1$ for $-2 \leq z \leq 2$ together with the two disks $x^2 + y^2 \leq 1$ for $z = 2$ and $z = -2$, respectively.

20. Show that, if $\mathbf{v}$ has continuous second derivatives in a region containing the piecewise smooth closed surface S and its interior, then

$$\int\int_S \nabla \times \mathbf{v} \cdot \mathbf{N}\, d\sigma = 0.$$

21. Show that the volume of a region D is given by

$$V(D) = \frac{1}{3}\int\int_S \mathbf{F} \cdot \mathbf{N}\, d\sigma,$$

where S is the surface of D and $\mathbf{F}(x, y, z) = x\mathbf{i} + y\mathbf{j} + z\mathbf{k}$.

22. If $\mathbf{F}$ is a constant vector, show that

$$\int\int_S \mathbf{F} \cdot \mathbf{N}\, d\sigma = 0$$

for any piecewise smooth closed surface S.

23. In this problem assume that $f(x, y, z)$ and $g(x, y, z)$ have continuous partial derivatives of at least the second order.
(a) Find the divergence of $g\, \nabla f$.
(b) Show that

$$\int\int_S g\frac{\partial f}{\partial n}\, d\sigma = \int\int\int_D (g\nabla^2 f + \nabla f \cdot \nabla g)\, dV,$$

where $\partial f/\partial n$ is the directional derivative of f in the direction of the outer normal vector on S.
(c) Show that

$$\int\int_S \left(g\frac{\partial f}{\partial n} - f\frac{\partial g}{\partial n}\right) d\sigma = \int\int\int_D (g\nabla^2 f - f\nabla^2 g)\, dV.$$

The results stated in Parts (b) and (c) are sometimes called Green's first and second identities, respectively. They are important in the study of partial differential equations. Compare these results with those in Problem 21 of Section 18.3.

24. A function that satisfies the equation $\nabla^2 f = 0$ throughout a region D is said to be **harmonic** in D.
(a) Show that if f is harmonic in D, then $\int\int_S \partial f/\partial n\, d\sigma = 0$.
Hint: Let $g = 1$ in Problem 23(b).
(b) Show that if f is harmonic in D, and if $f = 0$ at each point of S, then $\int\int\int_D \|\nabla f\|^2\, dV = 0$.
Hint: Let $g = f$ in Problem 23(b).
(c) Under the conditions of Part (b) show that $\nabla f = \mathbf{0}$ at each point of D, and then that $f = 0$ at each point of D. Thus the only harmonic function that vanishes on the boundary of a given region is the function that is zero throughout the region. This is a fundamental *uniqueness theorem* in the theory of partial differential equations.

18.8 STOKES' THEOREM

In this section we discuss a second three-dimensional extension of Green's theorem, which is known as Stokes' theorem. It is closely associated with the curl of a vector field, much as the divergence theorem is associated with the divergence. As in

Section 18.7, we begin by expressing Green's theorem in a vector form that facilitates the generalization.

Suppose that

$$\mathbf{v}(x, y) = P(x, y)\mathbf{i} + Q(x, y)\mathbf{j} \tag{1}$$

is a two-dimensional vector field, whose curl is

$$\nabla \times \mathbf{v} = \begin{vmatrix} \mathbf{i} & \mathbf{j} & \mathbf{k} \\ \dfrac{\partial}{\partial x} & \dfrac{\partial}{\partial y} & \dfrac{\partial}{\partial z} \\ P(x, y) & Q(x, y) & 0 \end{vmatrix}$$

$$= \left(\frac{\partial Q}{\partial x} - \frac{\partial P}{\partial y}\right)\mathbf{k}. \tag{2}$$

Suppose further that C is a piecewise smooth simple closed curve in the xy-plane

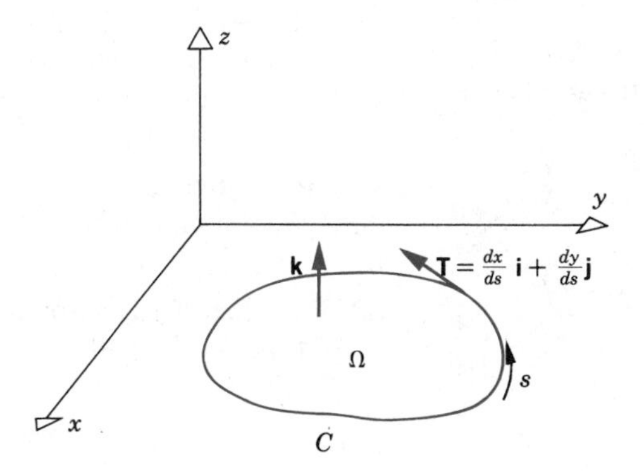

Figure 18.8.1

containing the region Ω (see Figure 18.8.1). The unit tangent vector to C is

$$\mathbf{T} = \frac{dx}{ds}\mathbf{i} + \frac{dy}{ds}\mathbf{j},$$

where s is arc length along C in the counterclockwise direction. Then

$$\oint_C P(x, y)\, dx + Q(x, y)\, dy = \oint_C \mathbf{v} \cdot \mathbf{T}\, ds \tag{3}$$

and

$$\iint_\Omega \left(\frac{\partial Q}{\partial x} - \frac{\partial P}{\partial y}\right) dA = \iint_\Omega \nabla \times \mathbf{v} \cdot \mathbf{k}\, dA. \tag{4}$$

Observe that $\mathbf{k}$ is normal to the plane region of integration Ω. Thus the conclusion of Green's theorem can be expressed in the form

$$\oint_C \mathbf{v} \cdot \mathbf{T}\, ds = \iint_\Omega \nabla \times \mathbf{v} \cdot \mathbf{k}\, dA. \tag{5}$$

To extend this result to three dimensions we can replace Ω by a (piecewise smooth) surface S in space bounded by a (simple closed) curve Γ. The unit normal $\mathbf{N}$ on S

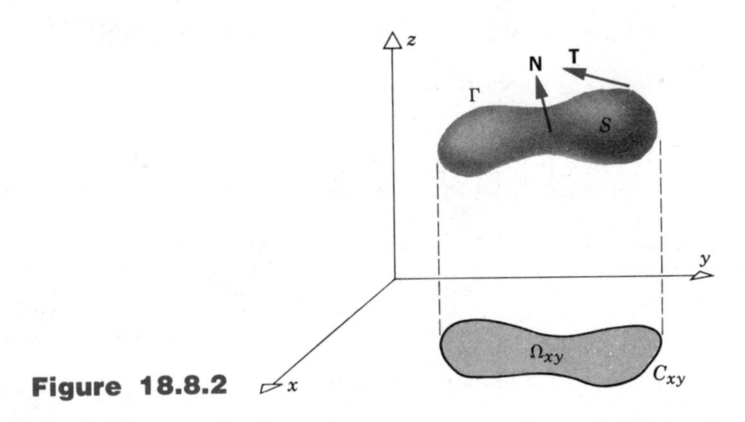

Figure 18.8.2

and the unit tangent **T** on Γ are related as shown in Figure 18.8.2; if one views S from the positive side (the side on which **N** lies), then **T** is directed so that S is on the left as Γ is traversed. In this way we are led to the important result known as Stokes' theorem* in three dimensions.

Theorem 18.8.1

(Stokes' Theorem)

If S is a piecewise smooth orientable surface bounded by a simple closed curve Γ, and if $\mathbf{v}(x, y, z)$ has continuous partial derivatives in a region containing S, then

$$\oint_{\Gamma} \mathbf{v} \cdot \mathbf{T}\, ds = \iint_{S} \nabla \times \mathbf{v} \cdot \mathbf{N}\, d\sigma, \tag{6}$$

where the unit tangent vector **T** on Γ and the unit normal vector **N** on S are oriented as shown in Figure 18.8.2.

Partial Proof of Stokes' Theorem. Let

$$\mathbf{v}(x, y, z) = P(x, y, z)\mathbf{i} + Q(x, y, z)\mathbf{j} + R(x, y, z)\mathbf{k}.$$

*George Gabriel Stokes (1819–1903) was born in Ireland and educated at Cambridge. He became Lucasian professor of mathematics at Cambridge in 1849 and held this position until his death 54 years later. More than any other individual, he was responsible for ending the self-imposed isolation of British mathematicians from developments on the European continent, a situation that had existed for a century or more.

Stokes' theorem was included in a letter from Thomson to Stokes in 1850, but was first published by Stokes as a prize examination question at Cambridge in 1854.

Stokes is also remembered for his pioneering work in fluid mechanics and in the use of asymptotic (divergent) series. The equations of motion of a viscous fluid, the Navier–Stokes equations, are partly named for him.

Then, in terms of P, Q, and R, Eq. 6 takes the form

$$\oint_\Gamma P\,dx + Q\,dy + R\,dz$$

$$= \int\!\!\int_S \left[\left(\frac{\partial R}{\partial y} - \frac{\partial Q}{\partial z}\right)n_1 + \left(\frac{\partial P}{\partial z} - \frac{\partial R}{\partial x}\right)n_2 + \left(\frac{\partial Q}{\partial x} - \frac{\partial P}{\partial y}\right)n_3\right] d\sigma. \tag{7}$$

We can establish Eq. 7 by showing that

$$\oint_\Gamma P\,dx = \int\!\!\int_S \left(\frac{\partial P}{\partial z}n_2 - \frac{\partial P}{\partial y}n_3\right) d\sigma, \tag{8}$$

together with the corresponding equations for Q and R. Suppose that S is given by

$$S : z = f(x, y), \qquad (x, y) \in \Omega_{xy}, \tag{9}$$

where Ω_{xy} is the projection of S in the xy-plane (see Figure 18.8.2). Then

$$\mathbf{N} = \frac{-f_x\mathbf{i} - f_y\mathbf{j} + \mathbf{k}}{(f_x^2 + f_y^2 + 1)^{1/2}}. \tag{10}$$

Consequently,

$$\int\!\!\int_S \left(\frac{\partial P}{\partial z}n_2 - \frac{\partial P}{\partial y}n_3\right) d\sigma = \int\!\!\int_{\Omega_{xy}} \left(-\frac{\partial P}{\partial z}\cdot\frac{\partial f}{\partial y} - \frac{\partial P}{\partial y}\cdot 1\right) dA_{xy}$$

$$= -\int\!\!\int_{\Omega_{xy}} \frac{\partial}{\partial y}P[x, y, f(x, y)]\, dA_{xy},$$

where we have used the chain rule to write the last term. By using Green's theorem in the plane (Theorem 18.3.1) we can convert the last integral over Ω_{xy} into a line integral around its boundary C_{xy}. Thus

$$\begin{aligned}\int\!\!\int_S \left(\frac{\partial P}{\partial z}n_2 - \frac{\partial P}{\partial y}n_3\right) d\sigma &= \oint_{C_{xy}} P[x, y, f(x, y)]\, dx \\ &= \oint_\Gamma P(x, y, z)\, dx.\end{aligned} \tag{11}$$

In a similar way we can show that if

$$S : y = g(x, z), \qquad (x, z) \in \Omega_{xz}, \tag{12}$$

then

$$\int\!\!\int_S \left(-\frac{\partial Q}{\partial z}n_1 + \frac{\partial Q}{\partial x}n_3\right) d\sigma = \oint_\Gamma Q\, dy, \tag{13}$$

and if

$$S : x = h(y, z), \qquad (y, z) \in \Omega_{yz}, \tag{14}$$

then

$$\int\int_S \left(\frac{\partial R}{\partial y}n_1 - \frac{\partial R}{\partial x}n_2\right) d\sigma = \oint_\Gamma R\,dz. \tag{15}$$

If all three of Eqs. 9, 12, and 14 hold, which means that S is projectible into each of the three coordinate planes, then all three of Eqs. 11, 13, and 15 are valid simultaneously. By adding these equations together we obtain Eq. 7, which completes the proof of Stokes' theorem for this class of surfaces. □

For surfaces that are not projectible into all three coordinate planes, one can often establish Stokes' theorem by subdividing the surface into a finite number of parts that do have this property. Stokes' theorem can also be extended to surfaces bounded by two or more curves (that is, surfaces with holes) by using arguments similar to those employed for Green's theorem in analogous circumstances.

An important consequence of Stokes' theorem is the proof of Theorem 18.2.4, as restated in Section 18.6, namely, that if

$$\nabla \times \mathbf{v} = \mathbf{0} \tag{16}$$

in a simply connected region D, then $\mathbf{v} = \nabla\phi$ in D, or equivalently,

$$\oint_\Gamma \mathbf{v} \cdot \mathbf{T}\,ds = 0 \tag{17}$$

for every piecewise continuous simple closed curve Γ in D. If Eq. 16 holds throughout D, then $\nabla \times \mathbf{v} \cdot \mathbf{N} = 0$ on any surface S in D, and Eq. 17 follows by Stokes' theorem. The region must be simply connected, since otherwise a single piecewise continuous simple closed curve Γ might not constitute the entire boundary of a surface in D. Thus Eq. 16 in a simply connected region is sufficient to guarantee that Eq. 17 holds, as well as the other properties equivalent to it.

Even if $\nabla \times \mathbf{v} \neq \mathbf{0}$, Stokes' theorem sometimes provides a means for evaluating certain integrals if one side of Eq. 6 is appreciably simpler to evaluate than the other. The following examples are illustrations of this.

EXAMPLE 1

Find the value of the integral

$$\oint_\Gamma \mathbf{v} \cdot \mathbf{T}\,ds,$$

where

$$\mathbf{v} = y(1 + z)\mathbf{i} + z(2 + x)\mathbf{j} + x(-1 + y)\mathbf{k},$$

and Γ is described by the parametric equations

$$\Gamma : x = \frac{a}{\sqrt{2}}\cos t, \qquad y = \frac{a}{\sqrt{2}}\cos t, \qquad z = a\sin t, \qquad 0 \le t \le 2\pi.$$

Let us begin by visualizing the problem geometrically. Note that the curve Γ lies on the sphere with center at the origin and radius a, since $x^2 + y^2 + z^2 = a^2$. Note further that $y = x$, so it follows that Γ is the circle of intersection of the sphere with this plane (see Figure 18.8.3). For $0 \leq t \leq 2\pi$ we start at the point $(a/\sqrt{2}, a/\sqrt{2}, 0)$ and proceed once around the circle in the direction shown in Figure 18.8.3. In other words, we leave the initial point in the xy-plane, enter the

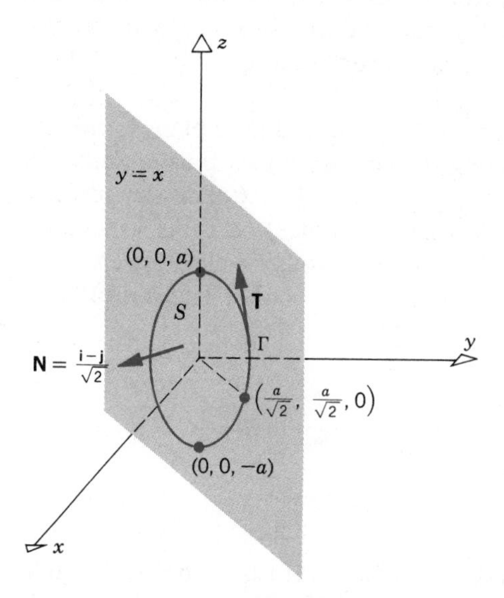

Figure 18.8.3

upper half space $z > 0$, follow the circular arc to the point $(0, 0, a)$ on the positive z-axis, and then continue on around the circle. It is natural to choose for S the disk enclosed by Γ in the plane $y = x$. Then the unit normals to S are given by $\pm(\mathbf{i} - \mathbf{j})/\sqrt{2}$. To preserve the proper relation between $\mathbf{N}$ and $\mathbf{T}$ we must choose

$$\mathbf{N} = \frac{\mathbf{i} - \mathbf{j}}{\sqrt{2}},$$

as indicated in Figure 18.8.3.

Evaluation of the given integral by using the parametric equations for Γ is not particularly difficult, but does require the evaluation of several integrals involving powers of $\sin t$ and $\cos t$. To determine whether we can avoid this task by using Stokes' theorem instead, we calculate the curl of $\mathbf{v}$:

$$\nabla \times \mathbf{v} = \begin{vmatrix} \mathbf{i} & \mathbf{j} & \mathbf{k} \\ \dfrac{\partial}{\partial x} & \dfrac{\partial}{\partial y} & \dfrac{\partial}{\partial z} \\ y(1 + z) & z(2 + x) & x(-1 + y) \end{vmatrix} = -2\mathbf{i} + \mathbf{j} - \mathbf{k}.$$

Thus

$$\nabla \times \mathbf{v} \cdot \mathbf{N} = \frac{-2 - 1}{\sqrt{2}} = \frac{-3}{\sqrt{2}}$$

and

$$\int\!\!\int_S \nabla \times \mathbf{v} \cdot \mathbf{N}\, d\sigma = -\frac{3}{\sqrt{2}}\pi a^2.$$

By Stokes' theorem the given line integral has the same value; this can be verified by evaluating the line integral directly. ∎

EXAMPLE 2

Let

$$\mathbf{v} = z^3\mathbf{i} + x\mathbf{j} + y^2\mathbf{k},$$

and let S be the portion of the paraboloid

$$z = a^2 - x^2 - y^2$$

for which $z \geq 0$. If $\mathbf{N}$ is the unit normal vector on S with positive z-component, find the value of

$$\int\!\!\int_S \nabla \times \mathbf{v} \cdot \mathbf{N}\, d\sigma.$$

It is possible, and relatively straightforward, to evaluate the required integral as a surface integral, as discussed in Section 18.4. However, instead of doing this, we will transform the surface integral into a line integral by means of Stokes' theorem. The geometrical situation is depicted in Figure 18.8.4. The boundary Γ

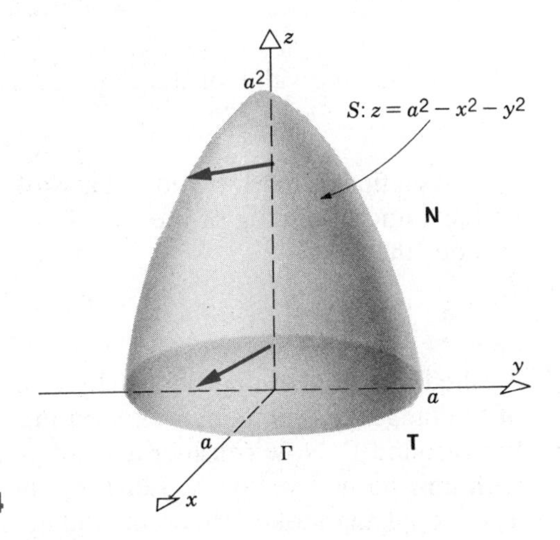

Figure 18.8.4

of S is the circle in the xy-plane with center at the origin and radius a. The curve Γ must be traversed in the direction shown in Figure 18.8.4, that is, counterclockwise as viewed from a point on the positive z-axis. A convenient parametrization of Γ is

$$x = a\cos t, \qquad y = a\sin t, \qquad z = 0, \qquad 0 \leq t \leq 2\pi. \tag{18}$$

By Stokes' theorem we have

$$\iint_S \nabla \times \mathbf{v} \cdot \mathbf{N}\, d\sigma = \oint_\Gamma \mathbf{v} \cdot \mathbf{T}\, ds$$

$$= \oint_\Gamma P\, dx + Q\, dy + R\, dz$$

$$= \oint_\Gamma z^3\, dx + x\, dy + y^2\, dz$$

$$= \oint_\Gamma x\, dy,$$

because on Γ both $z = 0$ and $dz = 0$. If we introduce the parametric equations (18), we obtain

$$\oint_\Gamma x\, dy = \int_0^{2\pi} (a \cos t)(a \cos t)\, dt$$

$$= a^2 \int_0^{2\pi} \cos^2 t\, dt$$

$$= a^2 \int_0^{2\pi} \left(\frac{1}{2} + \frac{1}{2} \cos 2t\right) dt$$

$$= a^2 \left(\frac{1}{2} t + \frac{1}{4} \sin 2t\right) \Bigg|_0^{2\pi}$$

$$= \pi a^2,$$

which is also the value of the required surface integral. ∎

As a final remark we note that Stokes' theorem provides the basis for a better physical understanding of the curl. If $\mathbf{v}(x, y, z)$ is the velocity vector for a fluid in motion, then

$$\gamma(\Gamma) = \oint_\Gamma \mathbf{v} \cdot \mathbf{T}\, ds$$

is called the **circulation** produced by $\mathbf{v}$ on the curve Γ. Observe that γ is the integral of the tangential component of $\mathbf{v}$ on the closed curve Γ, so γ measures the rate of flow around Γ. Now consider a small plane element S containing the point P and with unit normal vector $\mathbf{N}$. Let Γ be the boundary of S and let $A(S)$ be its area. Then, applying Stokes' theorem, and approximating the integrand on S by its value at P, we obtain

$$\gamma(\Gamma) = \oint_\Gamma \mathbf{v} \cdot \mathbf{T}\, ds = \iint_S \nabla \times \mathbf{v} \cdot \mathbf{N}\, d\sigma$$

$$\cong (\nabla \times \mathbf{v} \cdot \mathbf{N})(P) A(S).$$

Solving for $(\nabla \times \mathbf{v} \cdot \mathbf{N})(P)$ and taking the limit as $A(S) \to 0$, we obtain

$$(\nabla \times \mathbf{v} \cdot \mathbf{N})(P) = \lim_{A(S)\to 0} \frac{\gamma(\Gamma)}{A(S)}.$$

Thus at a point P the component of $\nabla \times \mathbf{v}$ in any direction $\mathbf{N}$ is the circulation density (per unit area) in a plane passing through P and perpendicular to $\mathbf{N}$.

PROBLEMS

In each of Problems 1 through 10, use Stokes' theorem to evaluate $\int_\Gamma \mathbf{v} \cdot \mathbf{T}\, ds$ for the given vector $\mathbf{v}(x, y, z)$ and the given curve Γ.

1. $\mathbf{v}(x, y, z) = 2z\mathbf{i} - x\mathbf{j} + 3y\mathbf{k}$; Γ is the triangular path from $(2, 0, 0)$ to $(0, 2, 0)$ to $(0, 0, 3)$ to $(2, 0, 0)$.
2. $\mathbf{v}(x, y, z) = y\mathbf{i} + 2x\mathbf{j} - y\mathbf{k}$; Γ is the rectangular path from $(0, 0, 0)$ to $(1, 2, 0)$ to $(1, 2, 3)$ to $(0, 0, 3)$ to $(0, 0, 0)$.
3. $\mathbf{v}(x, y, z) = 3y\mathbf{i} + 2z\mathbf{j} - x\mathbf{k}$; Γ is the triangular path from $(0, 0, 0)$ to $(3, 1, 1)$ to $(1, 3, 3)$ to $(0, 0, 0)$.
4. $\mathbf{v}(x, y, z) = -z\mathbf{i} + 2x\mathbf{j} + 3y\mathbf{k}$; Γ is the polygonal path from $(2, -1, 0)$ to $(0, 1, 0)$ to $(-1, 1, 2)$ to $(1, -1, 2)$ to $(2, -1, 0)$.
5. $\mathbf{v}(x, y, z) = -z\mathbf{i} + 2x\mathbf{j} + 3y\mathbf{k}$; Γ is the curve of intersection of the cylinder $x^2 + z^2 = 4$ and the plane $y = x$, oriented in the counterclockwise direction as viewed from a point on the positive y-axis.
6. $\mathbf{v}(x, y, z) = (2x - y)\mathbf{i} + (3y + z)\mathbf{j} + (-z + 2x)\mathbf{k}$; Γ is the curve of intersection of the cylinder $x^2 + y^2 = 4$ and the sphere $x^2 + y^2 + z^2 = 8$ for which $z < 0$, oriented in the counterclockwise direction as viewed from the origin.
7. $\mathbf{v}(x, y, z) = 2x\mathbf{i} + z^2\mathbf{j} + 2yz\mathbf{k}$; Γ is the curve of intersection of the ellipsoid $x^2 + y^2 + 4z^2 = 4$ and the plane $y = 2x$, oriented in the counterclockwise direction as viewed from a point on the positive x-axis.
8. $\mathbf{v}(x, y, z) = (2xy + z^2)\mathbf{i} + x^2\mathbf{j} + 2xz\mathbf{k}$; Γ is the curve of intersection of the cylinder $4x^2 + y^2 = 4$ and the plane $z = x + 2$, oriented in the counterclockwise direction as viewed from the origin.
9. $\mathbf{v}(x, y, z) = (2xy + z)\mathbf{i} + x^2\mathbf{j} - 2x\mathbf{k}$; Γ is the same curve as in Problem 8.
10. $\mathbf{v}(x, y, z) = 2y\mathbf{i} + (2x + z)\mathbf{j} + 3y\mathbf{k}$; Γ is the same curve as in Problem 8.

In each of Problems 11 through 16, use Stokes' theorem to find the value of $\int\int_S \nabla \times \mathbf{v} \cdot \mathbf{N}\, d\sigma$ for the given vector field $\mathbf{v}$ and surface S.

11. $\mathbf{v}(x, y, z) = 2yz\mathbf{i} + xz\mathbf{j} + 3xy\mathbf{k}$; S is the surface $z = a^2 - x^2 - y^2$ with $z \geq 0$ and $\mathbf{N}$ is the upward pointing unit normal vector.
12. $\mathbf{v}(x, y, z) = 2yz\mathbf{i} + xz\mathbf{j} + 3xy\mathbf{k}$; S is the portion of the surface $z = a^2 - x^2 - y^2$ in the first octant and $\mathbf{N}$ is the unit normal vector directed away from the origin.
13. $\mathbf{v}(x, y, z) = xyz\mathbf{i} + (x + z)\mathbf{j} + (x^2 - y^2)\mathbf{k}$; S is the surface $z = a^2 - x^2 - y^2$ with $z \geq 0$ and $\mathbf{N}$ is the upward pointing unit normal vector.
14. $\mathbf{v}(x, y, z) = xyz\mathbf{i} + (x + z)\mathbf{j} + (x^2 - y^2)\mathbf{k}$; S is the portion of the surface $z = a^2 - x^2 - y^2$ in the first octant and $\mathbf{N}$ is the unit normal vector directed away from the origin.
15. $\mathbf{v}(x, y, z) = yz^2\mathbf{i} + 2xz^2\mathbf{j} + xyz\mathbf{k}$; S is the surface $x^2 + y^2 + z^2 = a^2$ with $z \geq a/2$ and $\mathbf{N}$ is the upward pointing unit normal vector.
16. $\mathbf{v}(x, y, z) = 2y^2z\mathbf{i} + 3xz\mathbf{j} - xy^2z\mathbf{k}$; S is the portion of the surface $x^2 + y^2 + z^2 = a^2$ with $y \geq \sqrt{3}a/2$ and $\mathbf{N}$ is the unit normal vector with positive y-component.
17. If S is a surface bounded by the curve Γ, and if $\mathbf{v}$ is a constant vector in a region containing S, show that $\oint_\Gamma \mathbf{v} \cdot \mathbf{T}\, ds = 0$.
18. If S_1 and S_2 are two surfaces bounded by the same curve Γ, and having the same orientation with respect to Γ, and if $\mathbf{v}(x, y, z)$ has continuous par-

tial derivatives in a region containing S_1 and S_2, show that

$$\iint_{S_1} \nabla \times \mathbf{v} \cdot \mathbf{N}\, d\sigma = \iint_{S_2} \nabla \times \mathbf{v} \cdot \mathbf{N}\, d\sigma.$$

19. If S is a piecewise smooth closed surface and if $\mathbf{v}(x, y, z)$ has continuous partial derivatives in a region D containing S, show that $\int\int_S \nabla \times \mathbf{v} \cdot \mathbf{N}\, d\sigma = 0$. Compare this result with Problem 20 of Section 18.7.

REVIEW PROBLEMS

In each of Problems 1 through 4, evaluate $\int_C \mathbf{F} \cdot d\mathbf{r}$ along the given curve.

1. $\mathbf{F}(x, y) = \ln x\mathbf{i} + e^y\mathbf{j}$; $\mathbf{r}(t) = t\,\mathbf{i} + \dfrac{t^2}{2}\mathbf{j}$, $1 \le t \le 2$
2. $\mathbf{F}(x, y) = xy\mathbf{i} + (4x^2 - y^2)\mathbf{j}$; the arc of the circle $x^2 + y^2 = 16$ in the first quadrant from $(4, 0)$ to $(0, 4)$
3. $\mathbf{F}(x, y, z) = x^2z\mathbf{i} + yz^2\mathbf{j} + y^2x\mathbf{k}$; $\mathbf{r}(t) = \sin t\,\mathbf{i} - \cos t\,\mathbf{j} + \mathbf{k}$; $0 \le t \le \pi/2$
4. $\mathbf{F}(x, y, z) = (2x - z)\mathbf{i} + (2y - x)\mathbf{j} + (2z - y)\mathbf{k}$; the straight line from $(0, 0, 0)$ to $(1, 1, 1)$.

In each of Problems 5 through 8, evaluate the given line integral.

5. $\displaystyle\int_C (x^2 + y^2)\, dx$, along the curve $x = t$, $y = t^3$; $0 \le t \le 2$
6. $\displaystyle\int_C (x^2 + y^2)^{-1}\, dy$, along the straight line from $(1, 3)$ to $(2, 6)$
7. $\displaystyle\int_C \frac{x}{y}\, dx$, along the parabola $y = x^2 + 1$ from $(0, 0)$ to $(3, 10)$
8. $\displaystyle\int_C (x^3 - y^2x)\, dy$, along the hyperbola $xy = 4$ from $(1, 4)$ to $(4, 1)$

In each of Problems 9 through 12, determine whether the vector field is conservative and, if so, determine the corresponding potential function ϕ.

9. $\mathbf{F}(x, y) = \left(ye^{xy} + 1 - \dfrac{1}{x^2y}\right)\mathbf{i} + \left(xe^{xy} - \dfrac{1}{xy^2}\right)\mathbf{j}$
10. $\mathbf{F}(x, y, z) = ze^x\mathbf{i} + \ln y\mathbf{j} - \cos(xy)\mathbf{k}$
11. $\mathbf{F}(x, y, z) = 2\mathbf{i} + \sin y \sin z\mathbf{j} - \cos y \cos z\mathbf{k}$
12. $\mathbf{F}(x, y, z) = \left(\dfrac{1}{x} - 6xyz\right)\mathbf{i} + \left(\dfrac{1}{y} - 3x^2z\right)\mathbf{j} - 3x^2y\mathbf{k}$

In each of Problems 13 through 16, show that the line integral is independent of path and then evaluate the integral.

13. $\displaystyle\int_C y\left(x + \frac{y}{2}\right) dx + x\left(\frac{x}{2} + y\right) dy$ from $(0, 1)$ to $(1, 0)$
14. $\displaystyle\int_C e^x \sin y\, dx + e^x \cos y\, dy$ from $(0, 0)$ to $\left(1, \dfrac{\pi}{4}\right)$
15. $\displaystyle\int_C \frac{1}{2}(y^2z + z^2y)\, dx + \left(xyz + \frac{xz^2}{2}\right) dy + \left(\frac{xy^2}{2} + xyz\right) dz$ from $(0, 0, 0)$ to $(1, 1, 1)$
16. $\displaystyle\int_C \frac{1}{2}(y^2e^x + z^2)\, dx + \left(ye^x + \frac{z^2}{2}\right) dy + (x + y)z\, dz$ from $(0, 0, 0)$ to $(1, 1, 1)$

In each of Problems 17 through 20, use Green's theorem to evaluate the given line integral.

17. $\displaystyle\oint_C y^2\, dx + x^2\, dy$; C is the circle $x^2 + y^2 = 1$
18. $\displaystyle\oint_C \ln y\, dx - \ln x\, dy$; C is the square with vertices $(1, 1)$, $(1, 2)$, $(2, 1)$, $(2, 2)$
19. $\displaystyle\oint_C -\cos^2 y\, dx + \sin^2 x\, dy$; C is the square with vertices $(0, 0)$, $(0, \pi)$, $(\pi, 0)$, (π, π)
20. $\displaystyle\oint_C -x^3y\, dx + 6yx\, dy$; C is the ellipse $x^2/4 + y^2/9 = 1$

In each of Problems 21 through 24, (a) find the area of the given region Ω, (b) use Problem 17, Section 18.3, to find the centroid of the plane region Ω, (c) use Problem 18, Section 18.3, to evaluate I_0 for Ω.

21. The region bounded by $x^2/4 + y^2/9 = 1$

22. The region bounded by $y = \sqrt{x}$, $y = 0$, $x = y + 2$

23. The region bounded by $y = e^x$, $x = 0$, $x = 1$, $y = 0$

24. The region in the first quadrant bounded by $x = 0$, $y = 0$, and $y = 1 - x^2$

In each of Problems 25 through 28, find the area of the given surface.

25. The portion of the plane $2x + y + 3z = 6$ above the rectangle with vertices $(0, 0)$, $(0, 2)$, $(1, 0)$, $(1, 2)$.

26. The portion of the sphere $x^2 + y^2 + z^2 = 16$ above the circle $x^2 + y^2 = 1$ in the xy-plane.

27. The portion of the cone $4(x^2 + y^2) = z^2$ that is above the triangle with vertices $(0, 0)$, $(1, 0)$, $(1, 1)$.

28. The portion of the paraboloid $x^2 + y^2 = z$ that is cut out by the cylinder $x^2 + (y - 1)^2 = 1$.

In each of Problems 29 through 32, (a) find the divergence and (b) find the curl of the given vector field.

29. $\mathbf{v}(x, y, z) = x^2 y\mathbf{i} + y^2z\mathbf{j} + z^2x\mathbf{k}$

30. $\mathbf{v}(x, y, z) = e^{xyz}\,\mathbf{i} + e^{-xyz}\,\mathbf{j} + \ln(xyz)\mathbf{k}$

31. $\mathbf{v}(x, y, z) = y^2(x^2 + y^2)^{-1/2}\mathbf{i} + z^2(y^2 + z^2)^{-1/2}\mathbf{j} + x^2(x^2 + z^2)^{-1/2}\mathbf{k}$

32. $\mathbf{v}(x, y, z) = \cos x \sin y\mathbf{i} + \cos y \cos z\mathbf{j} + \sin z \cos x\mathbf{k}$

In each of Problems 33 through 36, use the divergence theorem to find the value of $\iint_S \mathbf{v} \cdot \mathbf{N}\, d\sigma$, where $\mathbf{v}$ and S are given.

33. $\mathbf{v}(x, y, z) = xz\mathbf{i} + yz\mathbf{j} + z^2\mathbf{k}$; S is the portion of the sphere $x^2 + y^2 + z^2 = 1$ above the xy-plane; $\mathbf{N}$ is the upward unit normal vector.

34. $\mathbf{v}(x, y, z) = xy\mathbf{i} + y^2\mathbf{j} + zy\mathbf{k}$; S is the surface of the cylinder $x^2 + y^2 = 4$, $0 \le z \le 1$, including the top and bottom; $\mathbf{N}$ is the unit outer normal vector.

35. $\mathbf{v}(x, y, z) = x(2y - z)\mathbf{i} + y(2z - x)\mathbf{j} + z(2x - y)\mathbf{k}$; S is the surface of the parallelipiped $0 \le x \le 3$, $0 \le y \le 4$, $0 \le z \le 2$; $\mathbf{N}$ is the unit outer normal vector.

36. $\mathbf{v}(x, y, z) = yz^2\mathbf{i} + zx^2\mathbf{j} + xy^2\mathbf{k}$; S is the surface of the sphere $x^2 + y^2 + z^2 = 1$ in the first octant; $\mathbf{N}$ is the unit normal vector pointing away from the origin.

In each of Problems 37 through 40, use Stokes' theorem to evaluate $\int_C \mathbf{v} \cdot \mathbf{T}\, ds$ or $\iint_S \nabla \times \mathbf{v} \cdot \mathbf{N}\, d\sigma$ for the given $\mathbf{v}$ and the curve C or the surface S.

37. $\mathbf{v}(x, y, z) = (2y + z)\mathbf{i} + (2z + x)\mathbf{j} + (2x + y)\mathbf{k}$; S is the top surface of the cube with $-2 \le x \le 2$, $-2 \le y \le 2$, $-2 \le z \le 2$; $\mathbf{N}$ is the outer unit normal vector.

38. $\mathbf{v}(x, y, z) = -yz\mathbf{i} + xz\mathbf{j} + xy\mathbf{k}$; S is the surface $z = x^2 + y^2$ below $z = 2$; $\mathbf{N}$ is the upward unit normal vector.

39. $\mathbf{v}(x, y, z) = (yz^{\frac{2}{3}})\mathbf{i} + (xz^{\frac{2}{3}})\mathbf{j} + xyz^2\mathbf{k}$; C is the triangle with vertices $(0, 0, 1)$, $(2, 0, 0)$, and $(0, 2, 0)$ oriented clockwise as viewed from the origin.

40. $\mathbf{v}(x, y, z) = -yz\mathbf{i} + xz\mathbf{j} + xy\mathbf{k}$; C is the intersection of $x^2 + z^2 = 1$ and $y = x$ oriented clockwise as viewed from the negative y-axis.

41. An electron travels along the curve $x = t$, $y = t^2$ subject to a force $\mathbf{F}(x, y) = (x^2 - y)\mathbf{i} + (xy^2 - x^2y)\mathbf{j}$ Newtons. Find the work done by the force $\mathbf{F}$ on the electron in the time period $0 \le t \le 2$.

references and suggestions for further reading

ADVANCED CALCULUS AND ANALYSIS

There are numerous books on calculus and analysis, written from various viewpoints, at levels more advanced than this book. We list a few of these books here. They are good places to look for proofs that we have omitted, or for further discussion and elaboration of the theory. The books by Amazigo and Rubenfeld and by Kaplan are somewhat less theoretically oriented than the other books mentioned.

JOHN C. AMAZIGO and LESTER A. RUBENFELD, *Advanced Calculus and Its Applications to the Engineering and Physical Sciences,* Wiley, New York, 1980.

R. C. BUCK, *Advanced Calculus,* Third Edition, McGraw-Hill, New York, 1978.

WATSON FULKS, *Advanced Calculus: An Introduction to Analysis,* Second Edition, Wiley, New York, 1969.

WILFRED KAPLAN, *Advanced Calculus,* Second Edition, Addison-Wesley, Reading, MA., 1973.

M. H. PROTTER and C. B. MORREY, *A First Course in Real Analysis,* Springer-Verlag, New York, 1977.

WALTER RUDIN, *Principles of Mathematical Analysis,* Third Edition, McGraw-Hill, New York, 1976.

HISTORY

For further reading on the history of calculus one may consult books such as those listed below. The book by Edwards focusses on calculus, as its title suggests; the book by Kline is the most comprehensive.

C. B. BOYER, *A History of Mathematics,* Wiley, New York, 1968.

C. H. EDWARDS, JR., *The Historical Development of the Calculus,* Springer-Verlag, New York, 1979.

M. KLINE, *Mathematical Thought from Ancient to Modern Times,* Oxford, New York, 1972.

The following set of books is a voluminous source of information about the lives and accomplishments of mathematicians of the past.

C. C. GILLESPIE (editor), *Dictionary of Scientific Biography,* fifteen volumes, Scribner's, New York, 1971.

CALCULUS

answers to odd-numbered problems

CHAPTER ONE

1.1 Answers

1. $A \cup B = (-1, 4)$; $A \cap B = [0, 2)$
3. $A \cup B = R^1 - \{0\}$: $A \cap B = \emptyset$
5. (a) True (b) False (c) False (d) True
7. $\{2, 3\}$
9. $[\frac{1}{4}, \frac{3}{4}) \cup \{1\}$
11. Yes, Yes
13. Yes, No
15. Yes, Yes
17. $S = \{s | s < -\frac{1}{4}\}$
19. $x = -1$, $x = -2$
21. $x \geq 1$
23. $x = -\frac{9}{5}$, $x = -5$
29. g. l. b. $= 0$; l. u. b. $= 1$
31. g. l. b. $= -1$; l. u. b. $= 3$

1.2 Answers

1. $(-6, 0)$
3. $(0, 1) \cup (1, 2)$
5. $(a - \epsilon, a + \epsilon)$
7. $[\frac{9}{4}, \infty)$
9. $(1, \frac{7}{4}]$
11. $(-\infty, 2) \cup (3, \infty)$
13. $(-\infty, 0) \cup (\frac{1}{2}, \infty)$
15. $(\frac{1}{3}, 1)$
17. $(-1, 3)$
19. $(\frac{1}{5}, \frac{1}{3})$
23. Equality if $x \geq 0$, $y \leq 0$, or if $x \leq 0$, $y \geq 0$.
25. Equality if $x \geq y \geq 0$ or if $x \leq y \leq 0$.
29. Equality if and only if $a_1b_2 = a_2b_1$.
31. g. l. b. $= -2$; l. u. b. $= 3$
33. No g. l. b. or l. u. b.

1.3 Answers

1. $\sqrt{58}$
3. $2|t|$
7. Isosceles
9. Equilateral
11. $y = x$
13. $y = \frac{4}{3}x + \frac{1}{3}$
15.

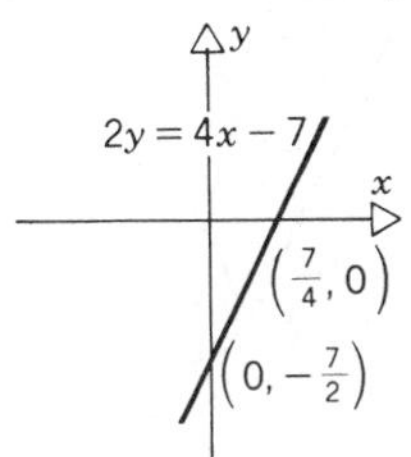

17.

21. $x + 2y = 2$
23. $2x - y = -6$

25. $(3, 2)$; $\sqrt{5}$

27. $(-2, 2)$; $2\sqrt{2}$

29. $\left(\frac{3}{\sqrt{5}}, \frac{6}{\sqrt{5}}\right)$ and $\left(-\frac{3}{\sqrt{5}}, -\frac{6}{\sqrt{5}}\right)$

31. $(-1, 1)$

33.

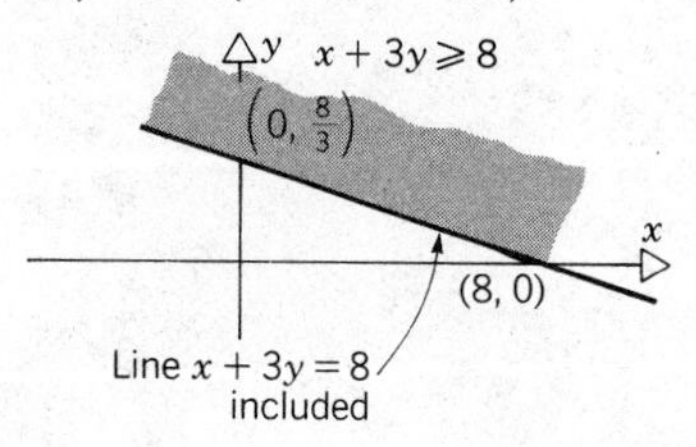

35.

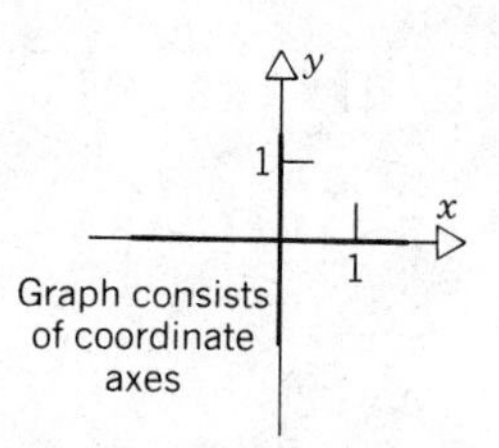

37.

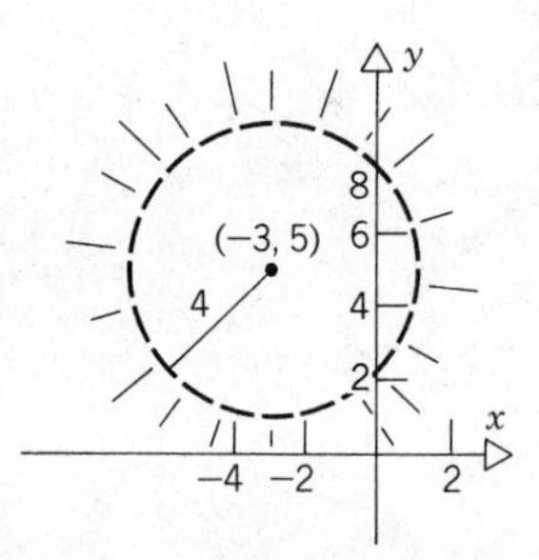

39.

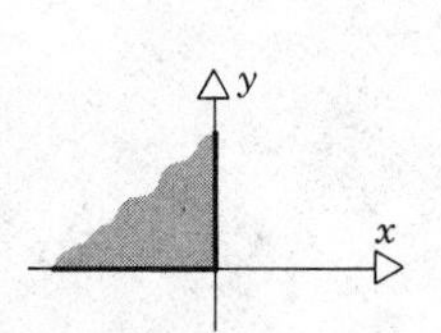

41. $(x + 1)^2 + (y + 3)^2 < 4$

43. $5x - y = 5$

45. $(-4, \frac{4}{3})$

1.4 Answers

1. (a) -7 (b) 11 (c) 7 (d) -11

3. $X = (-\infty, 3) \cup (3, \infty)$; $f(X) = (-\infty, 0) \cup (0, \infty)$

5. $X = \{x \mid x \neq \pm 3\}$, $f(X) = (-\infty, -\frac{1}{9}] \cup (0, \infty)$

7. $X = [-3, 3]$; $f(X) = [0, 3]$

9. (a) $\sqrt{1 - (s - 1)^2}$, $0 \le s \le 2$
   (b) $\sqrt{1 - t^2}$, $-1 \le t \le 1$
   (c) $\sqrt{1 - u^{-2}}$, $|u| \ge 1$
   (d) $|w|$, $-1 \le w \le 1$

11. 2

13. $4x + 3 + 2h$

15. $\dfrac{-1}{(x + h)x}$

17. $\phi(-3) = \phi(3) = 6$, $\phi(-1) = \phi(1) = 4$

$$\phi(x) = \begin{cases} -2x, & -\infty < x < -2 \\ 4, & -2 \le x < 2 \\ 2x, & 2 \le x < \infty \end{cases}$$

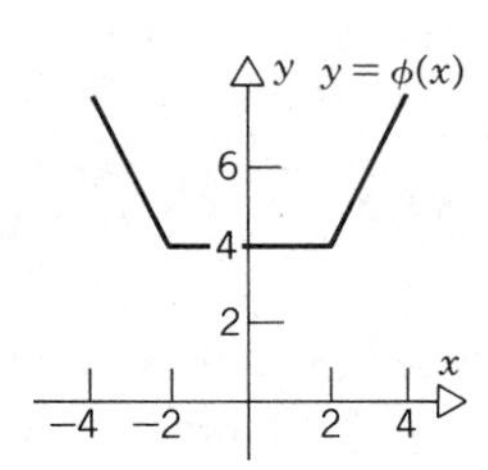

19. No

21. Yes

23. No

25. $f(x) = 6(x + 2)(x - 3)$ is one such function.

1.5 Answers

1. $[0, \infty)$

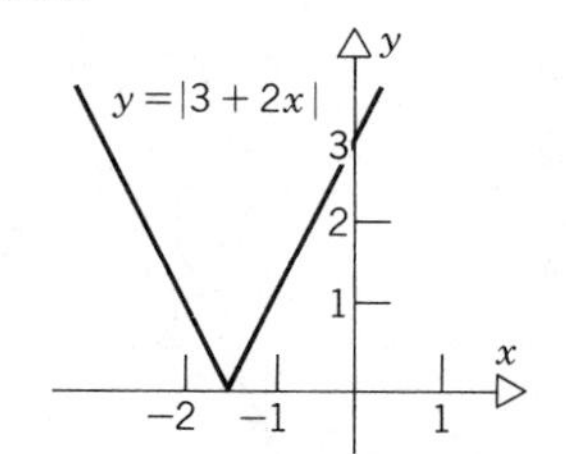

3. $[-3, 3]$

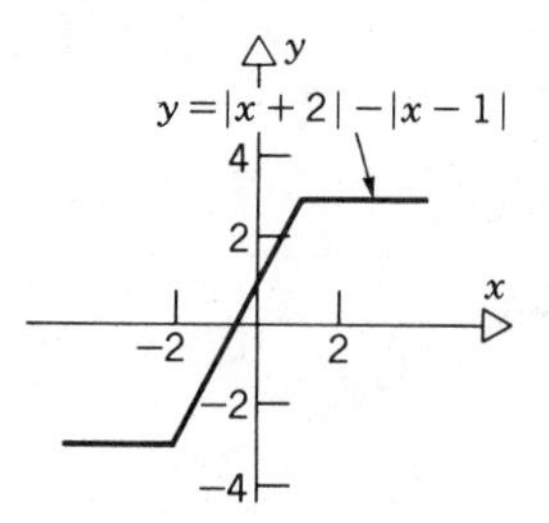

5. $\{\pm 2n \mid n = 0, 1, 2, \ldots\}$

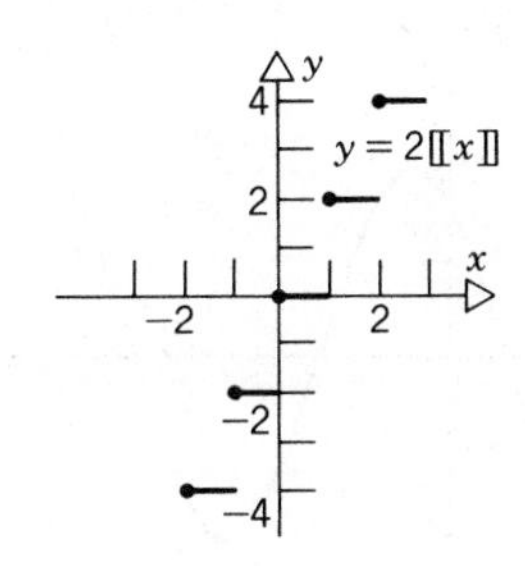

7. $\{\pm n \mid n = 0, 1, 2, \ldots\}$

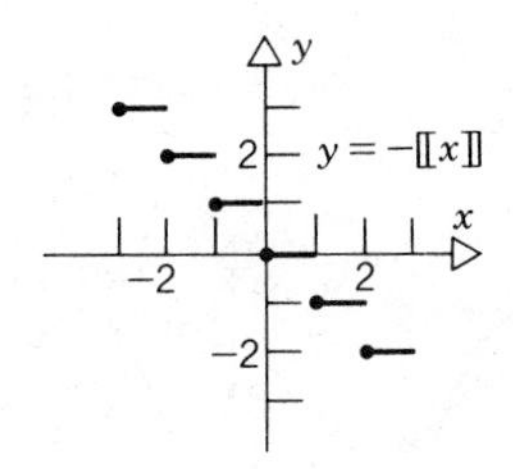

9. $[-1, \infty)$

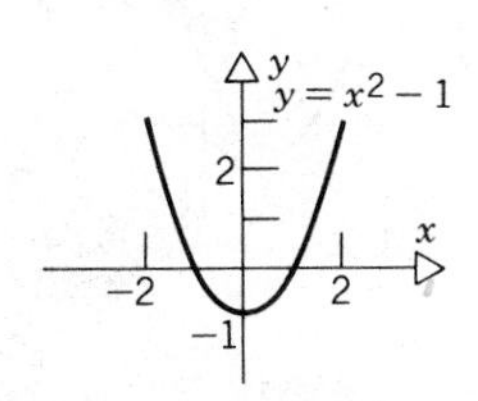

11. $[-\frac{49}{8}, \infty)$

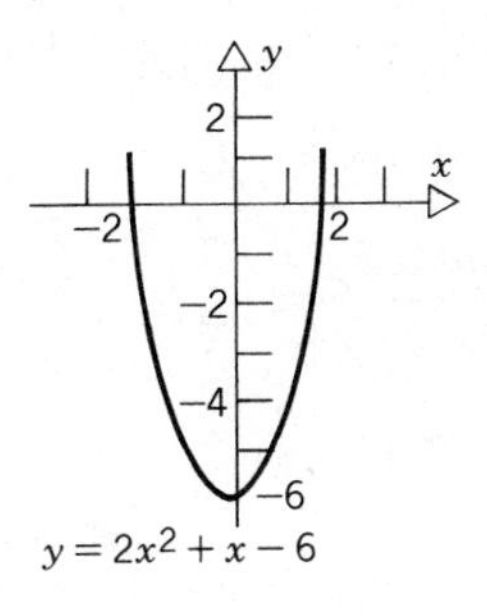

13. $(-\infty, \infty)$

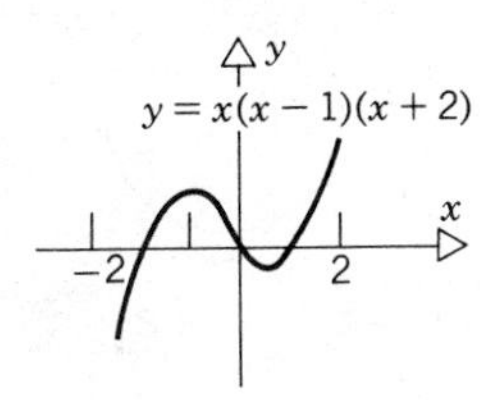

15. $[0, 3]$

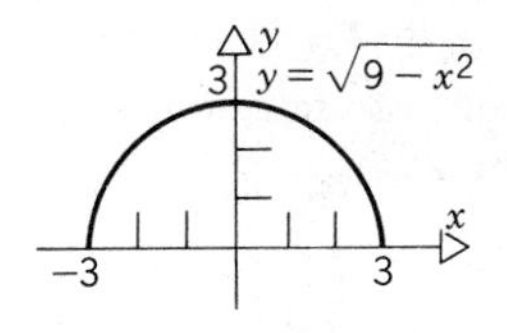

17. $(-\infty, \infty)$

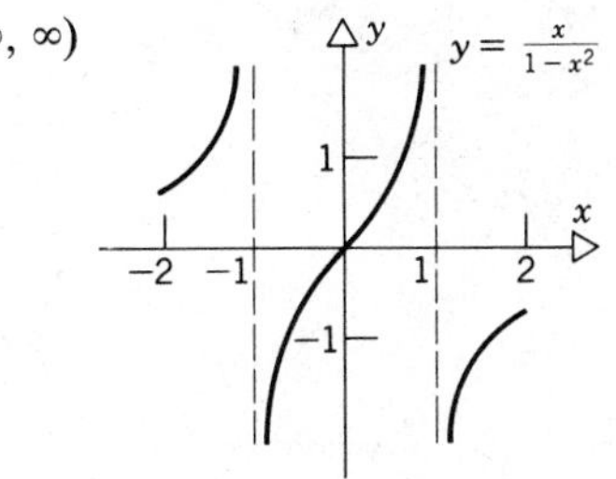

19. $(-\infty, -1) \cup (-1, \infty)$

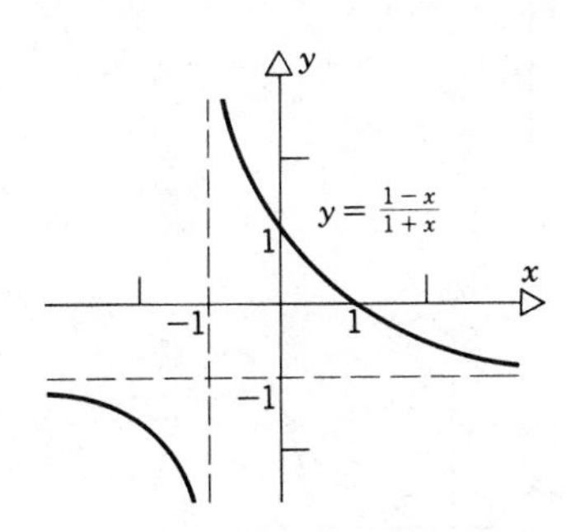

21.

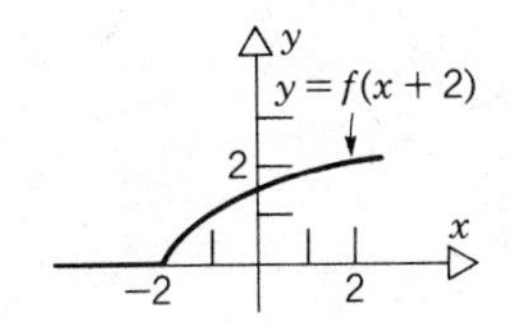

23. $[0, \frac{1}{2}]$, $[0, 1]$, $[0, 2]$, $[0, 4]$, 2 length units per unit time

25. Odd 27. Neither 29. Even

31. It is neither even nor odd, unless one function is everywhere zero.

1.6 Answers

1. $\frac{\sqrt{2}}{2}$

3. $\frac{-\sqrt{6}+\sqrt{2}}{4}$

5. $\frac{\sqrt{2-\sqrt{2}}}{2}$

7. $\frac{\sqrt{2+\sqrt{3}}}{2}$

9.

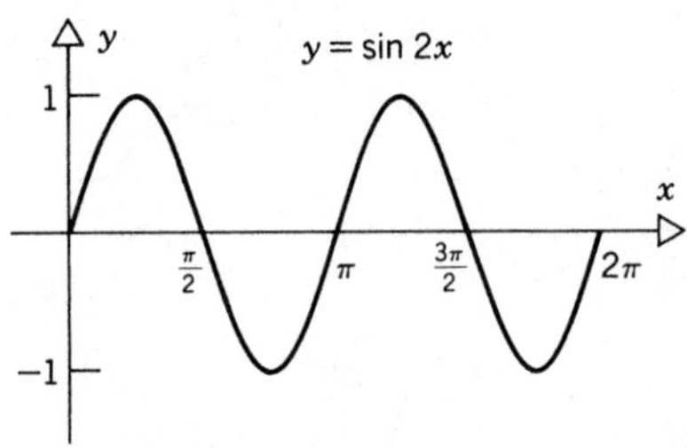

11.

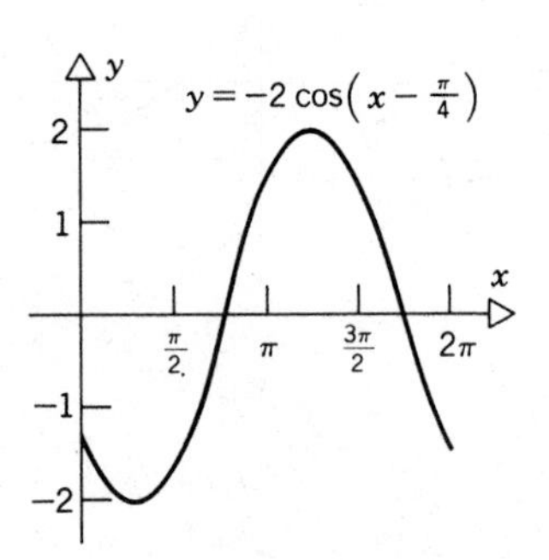

13.

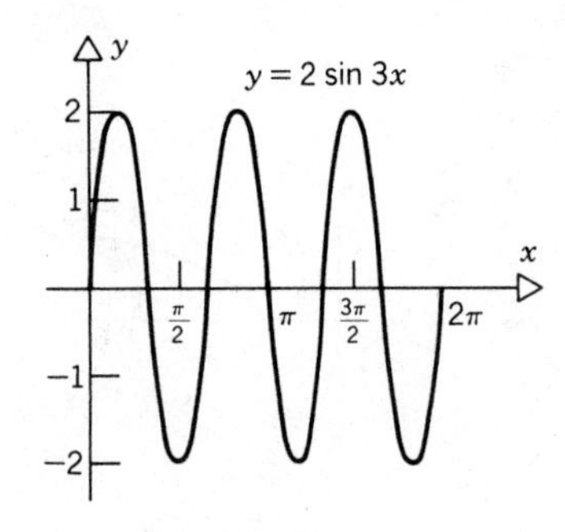

19. 4π

21. 6π

29. 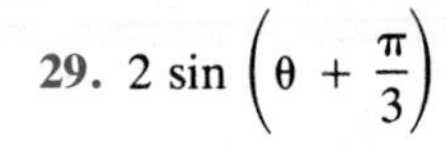 $2\sin\left(\theta+\frac{\pi}{3}\right)$

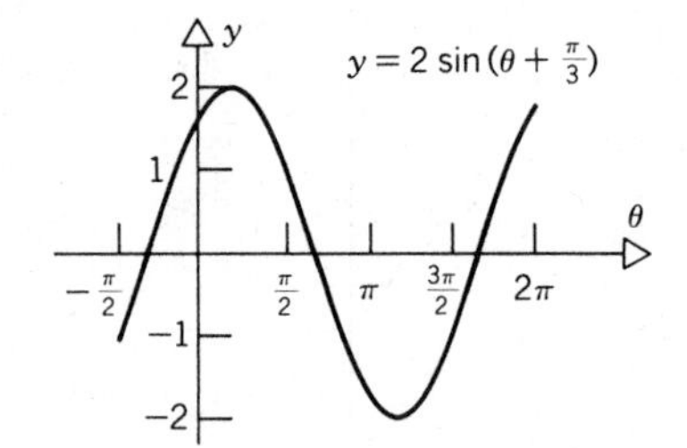

31. $\sqrt{13}\sin(\theta+\delta)$; $\cos\delta=\frac{3}{\sqrt{13}}$ and

$\sin\delta=-\frac{2}{\sqrt{13}}$, so $\delta\cong 5.6952$

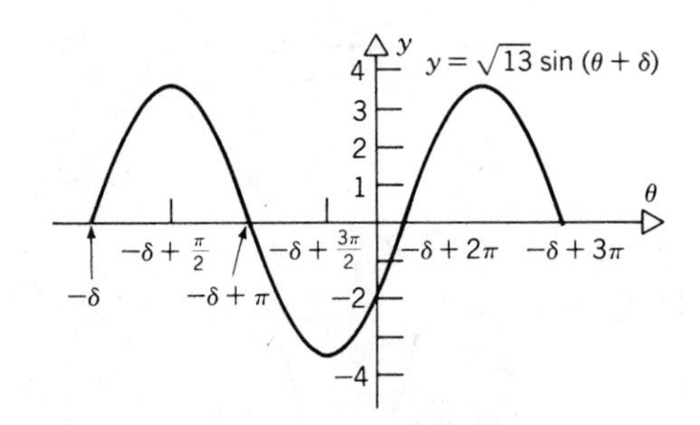

Answers to Review Problems

1.

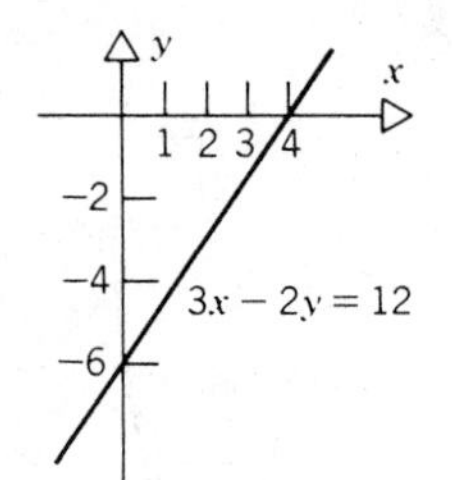

3.

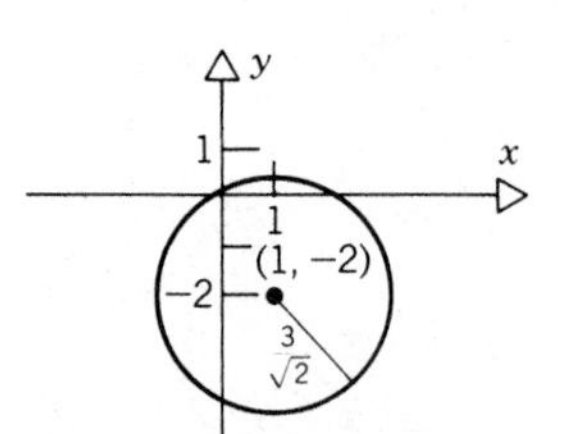

$2x^2+2y^2-4x+8y+1=0$

or

$(x-1)^2+(y+2)^2=\frac{9}{2}$

5.

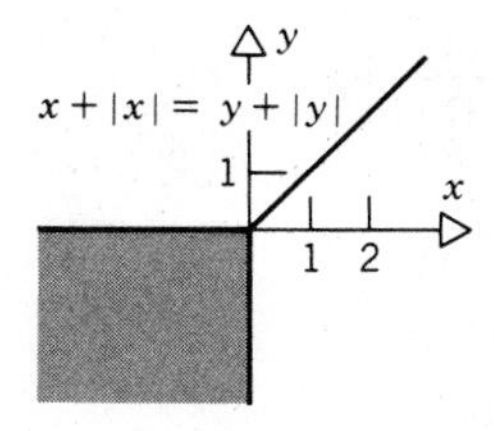

7.

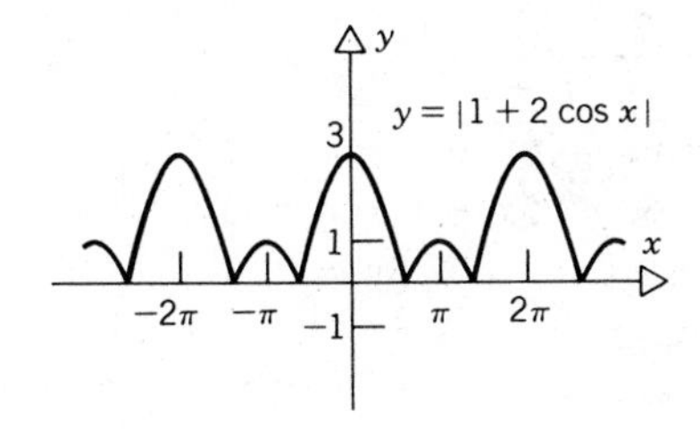

9. $X = (-\infty, -1) \cup (-1, \infty)$,
$f(X) = (-\infty, 1) \cup (1, \infty)$

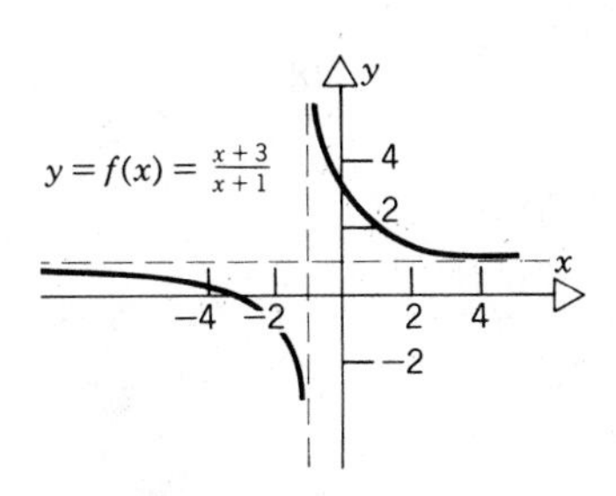

11. $X = (-\infty, \infty)$, $f(X) = (-\infty, \infty)$

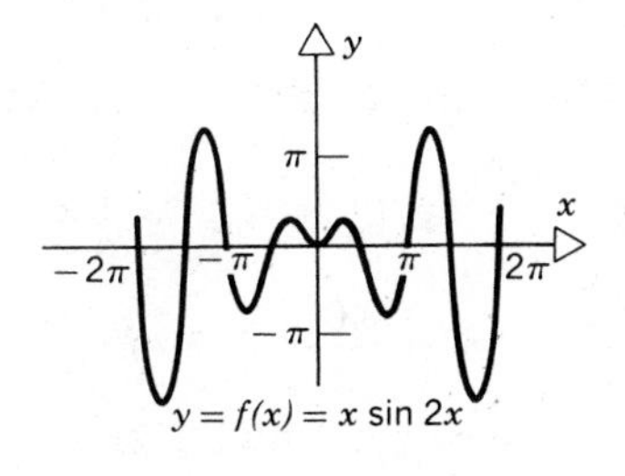

13. $X = (-\infty, \infty)$, $f(X) = [-1, 1]$

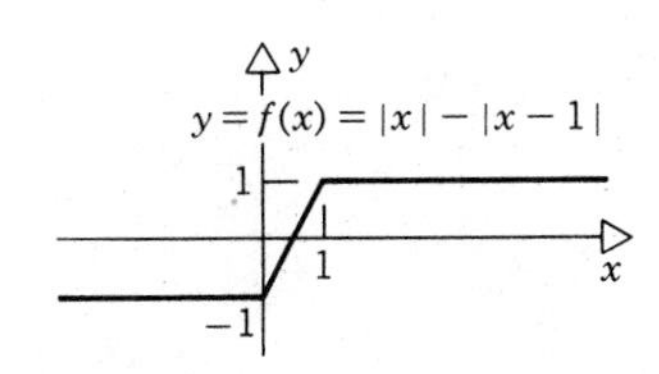

15. $X = (-\infty, \infty)$, $f(X) = (-\infty, \frac{9}{4}]$

$y = f(x) = 2 - x - x^2 = \frac{9}{4} - \left(x + \frac{1}{2}\right)^2$

17. $(-\infty, -4] \cup [4, \infty)$ **19.** $(-6, -1) \cup (2, 3)$

21. $(-\infty, -1] \cup [\frac{1}{3}, \infty)$

23. $\cdots \left[-\pi - \frac{\pi}{6}, -\pi + \frac{\pi}{6}\right] \cup \left[-\frac{\pi}{6}, \frac{\pi}{6}\right] \cup \left[\pi - \frac{\pi}{6}, \pi + \frac{\pi}{6}\right] \cup \left[2\pi - \frac{\pi}{6}, 2\pi + \frac{\pi}{6}\right] \cup \cdots$

25. $x + 2y = 12$

27. $(x + 3)^2 + (y - 1)^2 = 29$

29. $(x + 1)^2 + (y - 3)^2 = 52$

31. $X = (-\infty, \infty)$, $f(X) = \cdots [-4, -3) \cup [-2, -1) \cup [0, 1) \cup [2, 3) \cup \cdots$

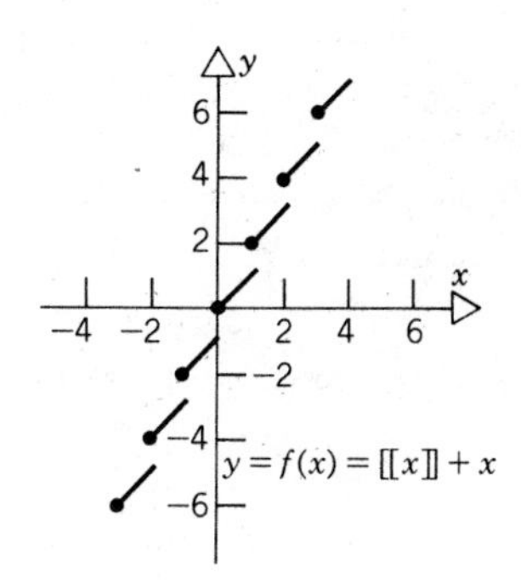

33. $X = (-\infty, \infty)$, $f(X) = (-\infty, \infty)$, odd function

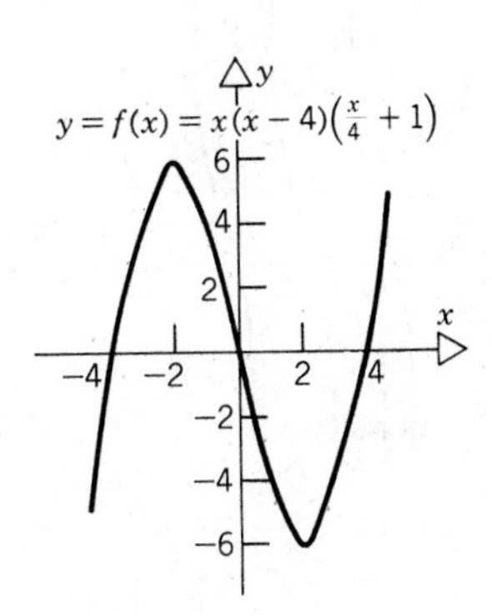

35. $X = [1, \infty)$, $f(X) = \{1, \frac{1}{2}, \frac{1}{3}, \frac{1}{4}, \ldots\}$

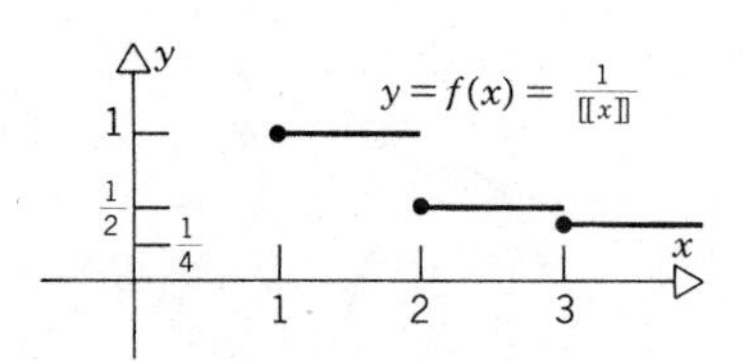

37. $X = (-\infty, 0) \cup (0, \infty)$, $f(X) = (-\infty, \infty)$

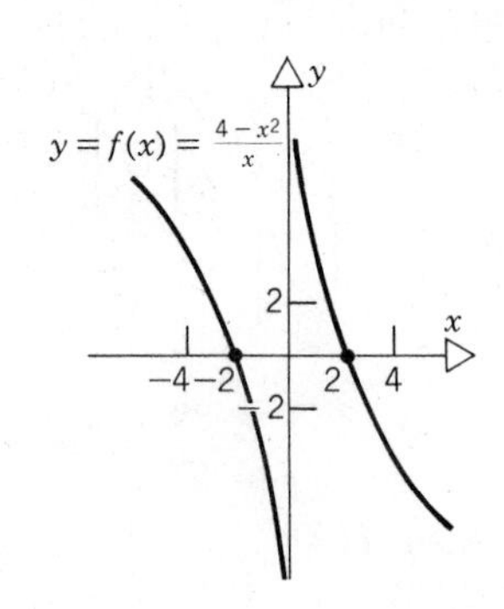

39.

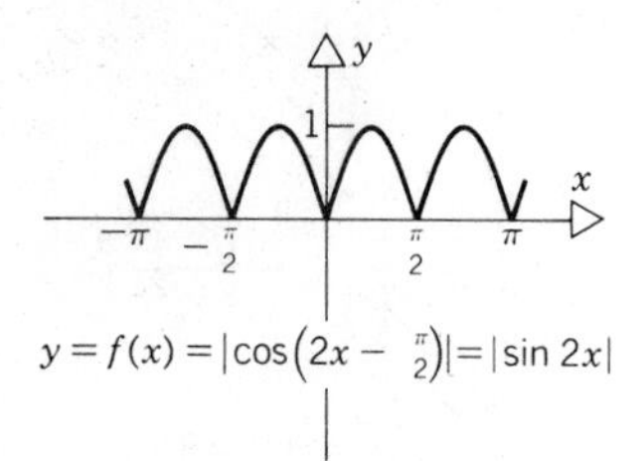

41.

43. $\sin^2 x = \dfrac{\tan^2 x}{1 + \tan^2 x}$

45. No

CHAPTER TWO

2.1 Answers

1. $v_{av} = \dfrac{5t^2 - 5}{t - 1}$, $t \neq 1$; $v = 10$

3. $v_{av} = \dfrac{2t^3 - t^2 - 12}{t - 2}$, $t \neq 2$; $v = 20$

5. $m = 6$, $6x - y = 2$

7. $m = 5$, $5x - y = 4$

9. $m = -\frac{1}{5}$, $x + 5y = 7$

11. (a) $v_{\text{inst}} = v_{av} = 32t_0$
(b) $v_{\text{inst}} = 3t_0^2$, $v_{av} = 3t_0^2 + \tau^2$

13. (a) $v = -gt + v_0$ (b) $t = \dfrac{v_0}{g}$

15. $\dfrac{\sqrt{x^2 + 5} - 3}{x - 2}$; $\dfrac{2}{3}$

17. $\dfrac{(x^2 - 5)^{3/2} - 8}{x - 3}$; 18

19. $\dfrac{\tan x + 1}{x - 3\pi/4}$; 2

21. $\dfrac{\sin 2x - \sqrt{3}/2}{x - \pi/3}$; -1

2.2 Answers

1. -4 3. 3 5. $-\frac{1}{2}$, no limit

7. No limit, $\frac{5}{3}$ 9. $\frac{1}{2}$ 11. 1

13. No limit, -3 15. 0 17. $\frac{1}{2}$

19. 2 21. $\frac{1}{6}$ 23. There is no tangent line.

27. (a) $\delta_1 \leq \frac{1}{20}$ (b) $\delta_2 \leq \frac{1}{200}$ (c) $\delta(\epsilon) \leq \dfrac{\epsilon}{2}$

29. $\delta_1 \leq \frac{1}{30}$, $\delta_2 \leq \frac{1}{300}$, $\delta(\epsilon) \leq \dfrac{\epsilon}{3}$

31. $\delta_1 \leq \dfrac{0.1}{|a|}$, $\delta_2 \leq \dfrac{0.01}{|a|}$, $\delta(\epsilon) \leq \dfrac{\epsilon}{|a|}$

33. $\delta_1 \leq \frac{1}{5}$, $\delta_2 \leq \frac{1}{50}$, $\delta(\epsilon) \leq 2\epsilon$

2.3 Answers

1. 16 3. -1 5. $-\frac{3}{2}$

7. $\frac{17}{4}$ 9. 3 11. $\dfrac{3\sqrt{2}}{2}$

13. $\frac{1}{3}$ 17. $\dfrac{4 - \sqrt{2}}{4}$ 19. $\dfrac{\pi}{2}$

21. 2 23. 0 25. 1 27. 0

29. $g(x) = 0$, $f(x) = 1 + x^{-2}$ as $x \to 0$ is one possibility.

31. (a) One possibility is $f(x) = g(x) = \text{sgn}(x)$.
(b) One possibility is $f(x) = x^2$ and $g(x) = x^{-2}$ as $x \to 0$.

2.4 Answers

In Problems 1 through 11 the answers are given in the following order: right-hand limit, left-hand limit, limit.

1. $c = 2$; meaningless, 0, meaningless
$c = -2$; 0, meaningless, meaningless
$c = 1$; $\sqrt{3}$, $\sqrt{3}$, $\sqrt{3}$

3. 0, 0, 0 5. 0, 0, 0

7. 1, 1, 1

9. 0, does not exist (∞), does not exist

11. 0, 0, 0 13. -1 15. 0

17. 1 19. $\sqrt{2}$ 21. 0

23. Does not exist ($-\infty$) 25. 0 27. 0 29. $\frac{1}{2}$

31.

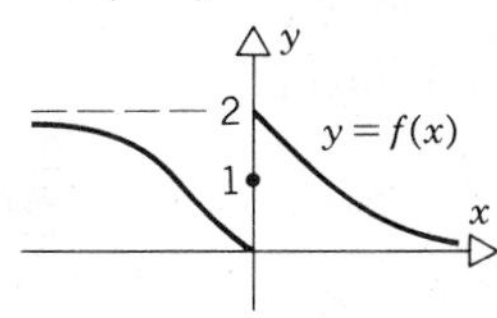

33. $\delta \leq \epsilon^2$ 35. $\delta \leq \sqrt{\dfrac{1}{M}}$

2.5 Answers

1. $x = -1, 1$; infinite

3. $x = 1, 2$; infinite 5. $x = 0$; jump

7. $x = \pm 3n$ for $n = 0, 1, 2, \ldots$; jump

9. $x = -1, 1$; jump

11. $x = \dfrac{3\pi}{2} \pm 2n\pi$ for $n = 0, 1, 2, \ldots$; removable

13. $x = 0$; infinite 15. No

17. Yes: $f(0) = 1$ 19. Yes: $f(0) = 0$

21. $a + b = c + d + e$

23. f is not continuous at $\epsilon = 0$.

25. One such pair of functions is $f(x) = 1$ for $x \geq 0$, $f(x) = -1$ for $x < 0$; and $g(x) = 2$ for $x \geq 0$, and $g(x) = -2$ for $x < 0$; at the point $c = 0$.

2.6 Answers

1. Interval not closed; f is not bounded and has no maximum or minimum.

3. Interval not closed, discontinuous function; f has no maximum, intermediate value property fails on some subintervals.

5. Discontinuous function; f has no minimum, intermediate value property fails on some subintervals.

7. Interval not bounded, discontinuous function; f is not bounded, has no maximum, and intermediate value property fails on some subintervals.

9. Interval not closed, discontinuous function; f has no minimum, intermediate value property fails on some subintervals.

11. $(-3, -2)$, $(-2, -1)$, $(3, 4)$

13. $(-2, -1)$, $(0, 1)$, $(7, 8)$

17. $(1.890625, 1.8984375)$

19. $(10.2109375, 10.21875)$

27.

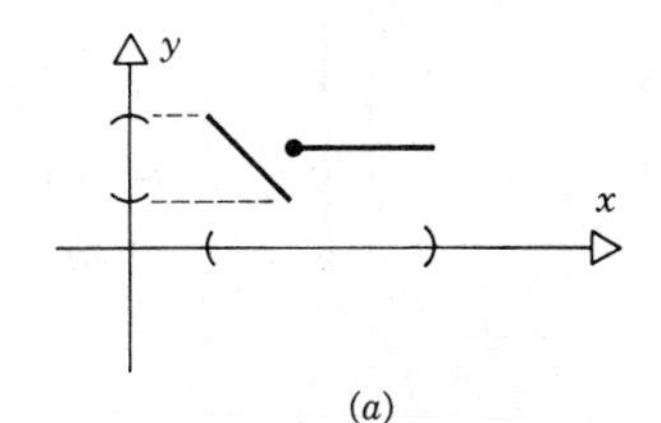

(a)

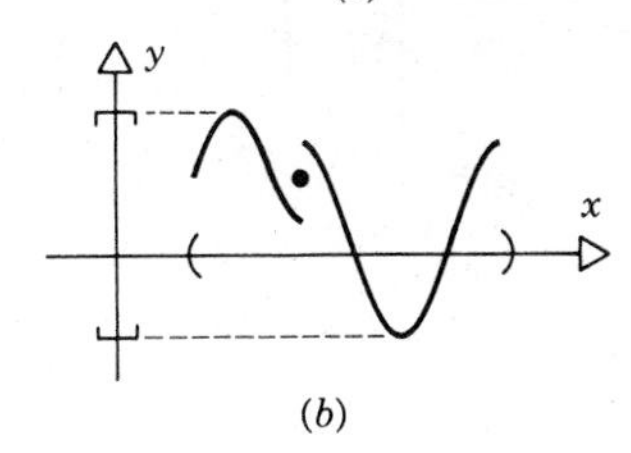

(b)

29.

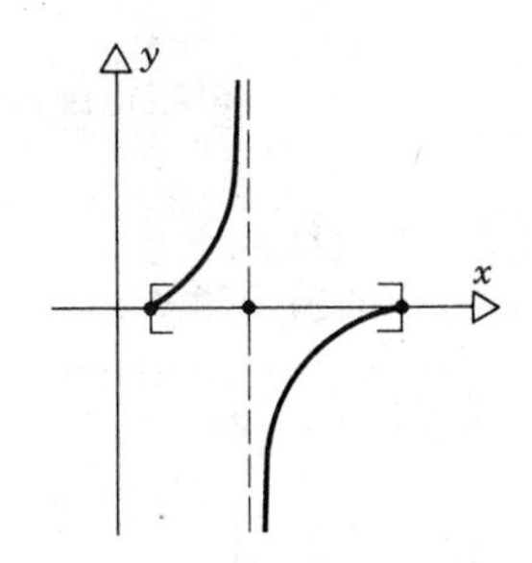

Answers to Review Problems

1. $y = 4x - 2$

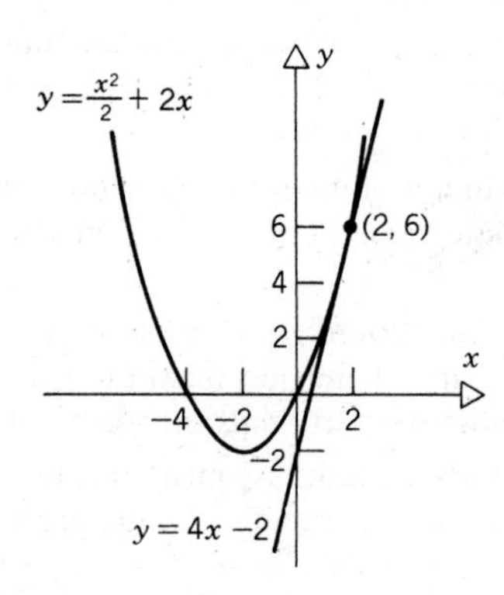

3. $y = \dfrac{x - \pi}{2}$

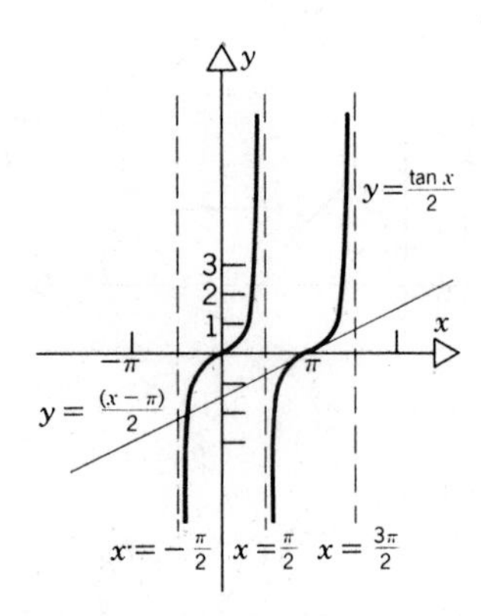

5. 4

7. $\frac{3}{4}$

9. -2

11. 1

13. Does not exist

15. 1

17. $\frac{1}{3}$

19. $\frac{32}{3}$

21. 2

23. $\frac{1}{3}$

25. -1

27. Does not exist $(+\infty)$

29. $\frac{1}{3}$

31. 0

33. Removable; $f(1) = -\pi$

35. Removable; $f(0) = 0$

37. (a) Yes; infinite discontinuity
(b) Yes; for example,

$$f(x) = \begin{cases} 1, & a \le x \le c \\ \dfrac{1}{x - c}, & c < x \le b \end{cases}$$

39. (3, 4), $r \cong 3.29$

41. (a) One such function is $f(x) = \dfrac{1}{x}$.

(b) One such function is $g(x) = \dfrac{2}{x}$.

(c) One such function is $h(x) = \dfrac{1}{x^2}$.

43. One such function is
$f(x) = (x + 1)^3(x - 2)/(x^2 - 2x)$.

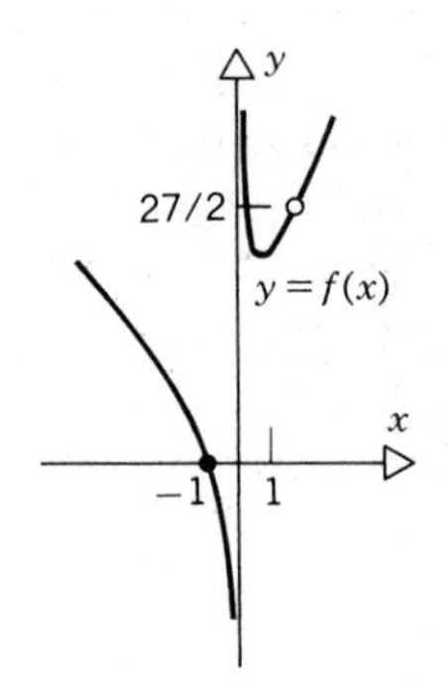

45. One such function is

$$f(x) = \frac{x^2 - 25}{(x - 2)^2}.$$

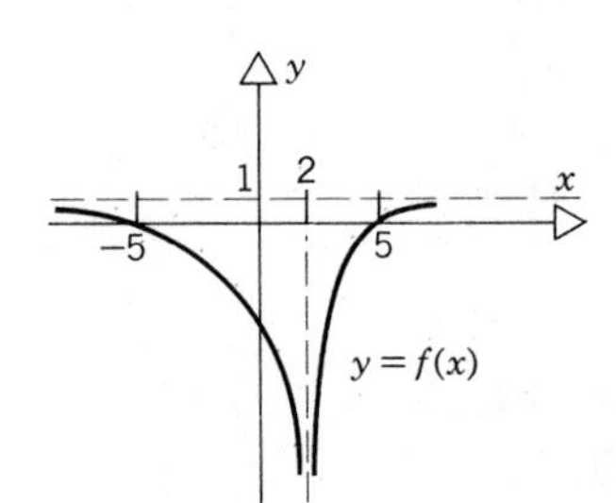

47. One such function is

$$f(x) = \begin{cases} -1 - x, & x < 0; \\ \dfrac{x^2}{(x - 2)^2}, & x \ge 0. \end{cases}$$

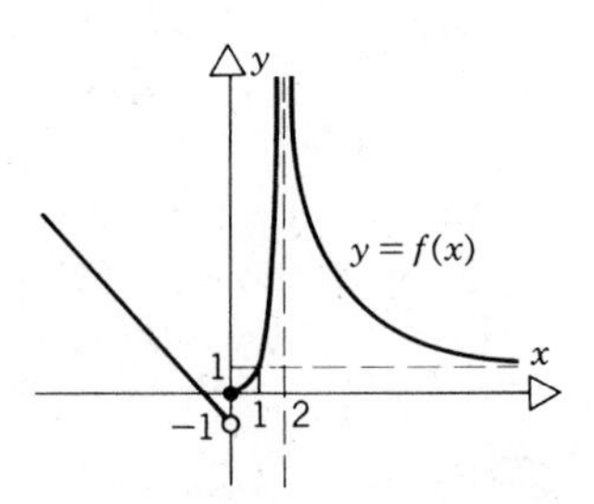

49. One such function is
$f(x) = 2(x^2 - 1)/(x^2 - x)$.

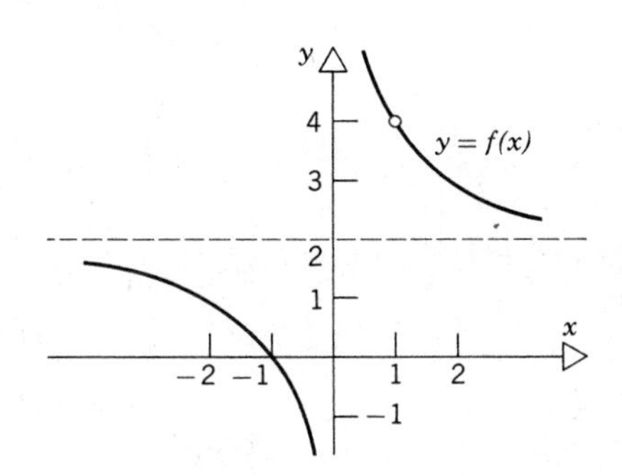

CHAPTER THREE

3.1 Answers

1. $2x$ **3.** $-2x^{-3}$ **5.** $4s^3 + 2s$

7. $1 + x^{-2}$ **9.** $-\dfrac{1}{2\sqrt{u}(1 + \sqrt{u})^2}$

11. $\dfrac{1}{2\sqrt{x}} + 2x^{-3}$ **13.** $6x^5$; all x

15. $\frac{3}{7}x^{-4/7}$; $x \neq 0$ **17.** $-4x^{-5}$; $x \neq 0$

19. $\frac{3}{5}x^{-2/5}$; $x \neq 0$ **21.** $\frac{5}{2}x^{3/2}$; $x > 0$

23. $12x - y = 16$ **25.** $3x + 16y = -8$

27. $2x - y = 1$; $x + y = 2$; no

29. $4x + y = 4$; $(1, 0), (0, 4)$

31. $f'(-1) = -2$; $f'(-a) = -b$

33. $(9, 3)$

37. (a) $f'(x) = \begin{cases} 1, & x < 0 \\ 2x, & x > 0 \end{cases}$ (b) No

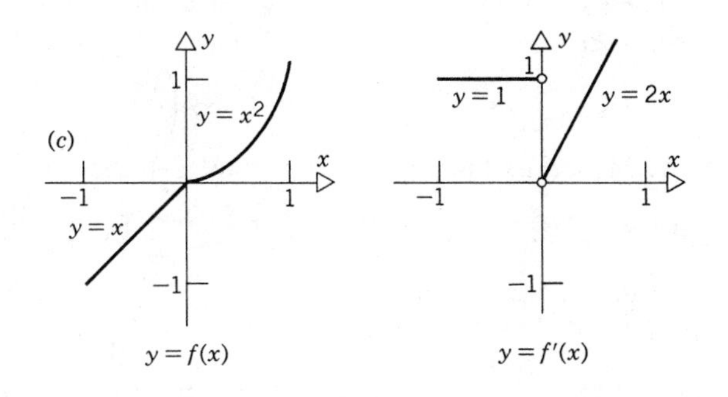

39. (a) $f'(x) = \begin{cases} -2x, & x < 0 \\ 2x, & x > 0 \end{cases}$ (b) $f'(0) = 0$

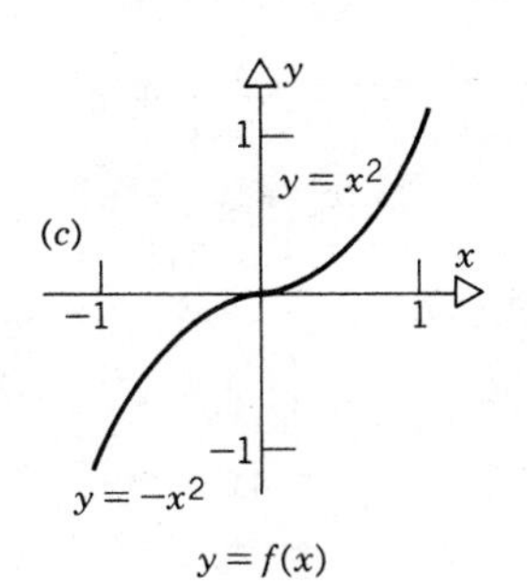

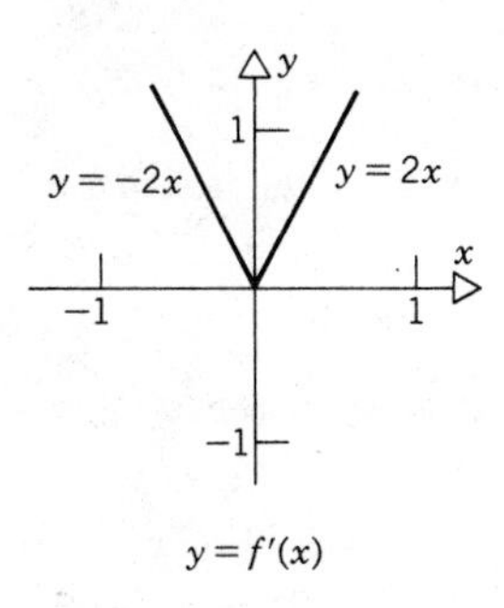

3.2 Answers

1. $6x - 4$ **3.** $20x^4 + 21x^2 - 6$

5. $16x^3 + 8x$ **7.** $3x^2 + 8x + 1$

9. $4x^3 + 10x$ **11.** $1 - 8x^{-2} + 10x^{-3}$

13. $\dfrac{2x^5 - 15x^4 + 24x^3 + 15x^2 - 40x + 10}{(x^2 - 5x + 6)^2}$

15. $6x + \frac{4}{3}x^{1/3} - x^{-2} + 6x^{-4}$ **17.** $\dfrac{x^{1/2}(x + 3)}{2(x + 1)^2}$

19. $4 + \frac{2}{3}x^{-1/3} - \frac{1}{3}x^{-2/3} + \dfrac{7(1 - x^2)}{(1 + x^2)^2}$

21. (a) $2x - 6$ (b) $4x + y = 1$
(c) $(3, -7)$; it is the lowest point on the graph.

23. (a) $\dfrac{3 - 3x^2}{(1 + x^2)^2}$ (b) $(1, \frac{3}{2})$ and $(-1, -\frac{3}{2})$
(c) Greatest value at $(1, \frac{3}{2})$;
least value at $(-1, -\frac{3}{2})$.

25. (a) $5x - y = 17$, $x - 4y = -8$ (b) $19/9$

27. $(-3, -16), (1, 4)$ **29.** $(\frac{2}{3}, \frac{32}{27}), (-2, 16)$

31. $2x - y = 1$ at $(1, 1)$; $14x - y = 49$ at $(7, 49)$

33. (a) $t = 1$, $t = 2$ sec (b) $t = \frac{3}{2}$ sec

35. $a = 4$, $b = -4$ **37.** $A' = 2\pi r$

39. $V = \left(\dfrac{S}{6}\right)^{3/2}$; $V' = \dfrac{\sqrt{S}}{4\sqrt{6}}$

41. (a) $f'(x) = u'(x)v(x)w(x) + u(x)v'(x)w(x) + u(x)v(x)w'(x)$

43. $f'(x) = n[u(x)]^{n-1}u'(x)$

3.3 Answers

1. $12x^2 - 12x + 10$ **3.** $-\frac{2}{9}x^{-5/3} - 6x^{-4}$

5. $-6x^{-4} - \frac{440}{27}x^{-14/3}$ **7.** $12x^2 - 6$

9. $\frac{9}{8}x^{-5/2} + 6x^{-4} - \frac{15}{2}x^{-7/2}$ **11.** $\dfrac{x^2}{2} - 4x + c$

13. $\dfrac{x^3}{3} - x^{-1} + c$ **15.** $\frac{9}{5}x^{5/3} - 3x^{4/3} + c$

17. $6\sqrt{x} + c$

19. $-\dfrac{1}{2 + x} + c$

21. $x^2 - 3x + 4$

23. $x^4 - x^2 - 8$

25. $\frac{2}{3}x^{3/2} + \frac{1}{3}$

27. $3x^2 - 3x - 5$

29. 144 ft

31. (a) 50.4 m (b) 5.25 sec

3.4 Answers

1. $y = -2x + 5$

3. $4x - 3y = 2$

5. $y = y_0 + (2x_0 - 3)(x - x_0)$

7. $T(x; 2) = 4x - 3;\quad r(x; 2) = 0$

9. $T(x; 0) = 2x - 4;\quad r(x; 0) = \dfrac{x}{3}$

11. $T(x; 2) = \dfrac{4 + x}{9};$

$r(x; 2) = -\dfrac{x - 2}{9(x + 1)}, x \neq -1$

13. 3.00593; correct to four decimal places.

15. 2.24344; correct to four decimal places.

17. 3.69 ft^3; correct to one decimal place.

19. 8.280; correct to two decimal places.

21. $4x^3 + 4x;\quad 12x^2 + 4$

23. $\dfrac{4x}{(x^2 + 1)^2};\quad \dfrac{4 - 12x^2}{(x^2 + 1)^3}$

25. $6x - \frac{4}{3}x^{1/3} - (x + 1)^{-2};$
$6 - \frac{4}{9}x^{-2/3} + 2(x + 1)^{-3}$

27. $\dfrac{1}{(x + 2)^2};\quad -\dfrac{2}{(x + 2)^3}$

31. (a) 0.5436 ft^3 (b) $\dfrac{0.5436}{27} \cong 0.0201$

(c) $\dfrac{(27)(0.02)}{27} = 0.02$

3.5 Answers

1. $(f \circ g)(x) = 6x + 19,\quad -\infty < x < \infty;$
$(g \circ f)(x) = 6x + 3,\quad -\infty < x < \infty$

3. $(f \circ g)(x) = -x,\quad -1 \le x < \infty;$
$(g \circ f)(x) = \sqrt{2 - x^2},\quad -\sqrt{2} \le x \le \sqrt{2}$

5. $(f \circ g)(x) = \sqrt{x^2 - 3x + 2},\quad -\infty < x \le 1$ or $2 \le x < \infty$; $(g \circ f)(x) = x - 3\sqrt{x} + 2,$ $0 \le x < \infty$

7. $(f \circ g)(x)$ does not exist;
$(g \circ f)(x) = -(x - 3\sqrt{x} + 6),\quad 0 \le x < \infty$

9. $(f \circ g)(x) = \cos 2\sqrt{x^2 - 4},\quad -\infty < x \le -2$ or $2 \le x < \infty$; $(g \circ f)(x)$ does not exist.

11. $4(z + 1)x$

13. $\dfrac{12z}{(z^2 + 1)^2}$

15. $\dfrac{2z - 3}{(x + 1)^2}$

17. $\frac{3}{2}\sqrt{x + 1},\quad -1 < x < \infty$

19. $\frac{14}{3}x(x^2 + 5)^{4/3},\quad -\infty < x < \infty$

21. $-1,\quad -4 < x < \infty$

23. -16

25. $\frac{20}{3}$

27. $6(2x - 1)^2$

29. $50\left(\dfrac{x + 1}{2}\right)^{99}$

31. $\dfrac{(x - 1)^3(9x - 13)}{\sqrt{2x - 3}}$

33. $\frac{3}{2}x^{-5/2}\sqrt{2x - 1}$

35. $8x^2 + 4(z + 1)$

37. $-\frac{9}{4}x^4(x^3 + 1)^{-3/2} + 3x(x^3 + 1)^{-1/2}$

39. $\dfrac{4(1 - 3x^2)}{(x^2 + 1)^3}$

41. $-f'(-x)$

43. $af'(ax)$

45. $a^n f^{(n)}(ax)$

47. $2u'(x)u''(x)$

49. $2u'(x)[u''(x)]^2 + [u'(x)]^2u'''(x)$

51. $\frac{1}{5}(x - 2)^5 + c$

$\frac{1}{6}(2x - 1)^6 + c$

55. $\frac{3}{14}(2x + 1)^{7/3} + c$

$(x^2 + 4)^{3/2} + c$

59. $f'\{g[h(x)]\}g'[h(x)]h'(x)$

61. $\dfrac{g(gf'' - fg'') - 2g'(gf' - fg')}{g^3}$

63. $-8\pi r^2$

65. 216 in^3/sec

3.6 Answers

1. $-4 \sin 4x$

3. $4 \cos 4x + 6 \sin 2x$

5. $2x \sin 2x + 2x^2 \cos 2x$

7. $4 \sin 2x \cos 2x$

9. $(1 - x \sec x)(\sec x - \tan x)$

11. $\dfrac{x \cos x - 2 \sin x - 2}{(x - \cos x)^2}$

13. $\dfrac{\sin x}{(1 + \cos x)^{3/2}(1 - \cos x)^{1/2}} = \dfrac{\text{sgn}(\sin x)}{1 + \cos x}$

15. $-\sin x \cos(\cos x)$

17. $2 \sec^2(\sec^2 x)\sec^2 x \tan x$

19. $-\frac{1}{2}\cos 2x + c$

21. $\frac{2}{3}\sin 3x + c$

23. $-\cos\left(x + \dfrac{\pi}{2}\right) + c$

25. $2 \sec \dfrac{x}{2} + c$

27. $-\frac{1}{8}\cos 4x + c$

29. $\frac{1}{2}x + \frac{1}{4}\sin 2x + c$

31. $-\frac{1}{2}\cos x^2 + c$

33. $\frac{1}{2}\sin(x - \pi)^2 + c$

35. $\cos x$

37. $-\left(\dfrac{1}{2}\right)^{11}\cos\left(\dfrac{x}{2}\right)$

39. $2a^2 \sec^2 ax \tan ax$

41. $\sec x(\tan^2 x + \sec^2 x)$

43. $8 \csc^2 2x \cot 2x$

45. $y = 5 - 4\sqrt{3}\left(x - \dfrac{\pi}{6}\right)$

47. $v = 6 \cos 3t + 6 \sin 2t;$
$a = -18 \sin 3t + 12 \cos 2t$

49. $f(x) = -\dfrac{1}{2}\cos 2x - 3 \sin x + \dfrac{3}{\sqrt{2}} - 1$

51. $a = -\dfrac{1}{\sqrt{2}},\ b = \dfrac{1 + \pi/4}{\sqrt{2}},\ f'\left(\dfrac{\pi}{4}\right) = -\dfrac{1}{\sqrt{2}}$

55. 1.0000, three decimal places

57. 1.5433, one decimal place

59. $\frac{3\sqrt{3}}{2}$ mi/hr

3.7 Answers

1. (b) $\frac{y - 2x}{2y - x}$ (c) -1

3. (b) $-\frac{4x + y}{x + 4y}$ (c) $-\frac{2}{3}$

5. (b) $\frac{1}{3y^2 + 1}$ (c) $\frac{1}{4}$

7. (b) $-\frac{2y - 3x(x^2 + y^2)^{1/2}}{2x - 3y(x^2 + y^2)^{1/2}}$ (c) $\frac{5\sqrt{3}}{3}$

9. (b) $\frac{\sin x \cos x}{\sin y \cos y}$ (c) 1

11. 6

13. $-\frac{3}{32}$

15. $3x - 4y = 2$

17. $(4 - \pi)x + (4 + \pi)y = 2\pi$

19. $4x + 5y = -1$

21. $(\pi - 2\sqrt{3})\left(x - \frac{\pi}{6}\right) - (3\pi - 2\sqrt{3})\left(y - \frac{\pi}{3}\right) = 0$

23. $\frac{8\omega A^2}{A^3 + 4}$

27. $2x$

29. $(1 + x^2)^{-1}$

Answers to Review Problems

1. $1 + \frac{1}{2}x^{-1/2}$

3. $-(x + 1)^{-2}$

5. $\frac{1}{2}t^{-1/2} - t^{-2}$

7. (a) $6x - \frac{5}{2}x^{3/2}$; (b) $y = 4x$

9. (a) $3x^2 - 2x + 1$; (b) $y = 41x - 113$

11. (a) $\frac{1 - x^2}{(1 + x^2)^2}$; (b) $y = -\frac{15}{289}x + \frac{128}{289}$

13. (a) $\frac{x^{1/2}(3 - x^2)}{2(1 + x^2)^2}$; (b) $y = -\frac{13}{289}x + \frac{188}{289}$

15. $f'(x) = 4x^3 + 4x$; $x = 0$

17. $f'(x) = \frac{7x^2 - 32x + 28}{(x^2 - 4)^2}$; $x \cong 1.179, x \cong 3.392$

19. (a) $\frac{1}{4}x^4 + 3x - \frac{21}{4}$; (b) 48

21. (a) $\frac{3}{5}x^{5/3} + x^{-1} - \frac{18}{5}$; (b) $\frac{3}{5}(4^{-1/3}) + \frac{1}{32} \cong 0.4512$

23. (a) $2x^3 - 5x^2 - 4x + 2$; (b) $x^2 + 44x + 58$

25. $\frac{(x^2 - 1)(7x^2 + 24x^{3/2} + 1)}{2x^{1/2}(x^{1/2} + 3)}$

27. $3(x^{-1/2} - 1)\sin[3(\sqrt{x} - 1)^2 - 1]$

29. $-\frac{9x}{4}\cos(x^{3/2} + 1) - \frac{3}{4\sqrt{x}}\sin(x^{3/2} + 1)$

31. $\frac{2t(t^4 + 2t^2 + 2)}{(t^2 + 1)^2}$

33. $\frac{-2ax(x^2 - 1)\sin(ax^2 + b) - 2x\cos(ax^2 + b)}{(x^2 - 1)^2}$

35. $-6 \sin s \cos s(4 + \sin^2 s)^2 \sin[(4 + \sin^2 s)^3 - 1]$

37. $2(x - \pi)\cos(3x^{3/2}) \sec[(x - \pi)^2] \tan[(x - \pi)^2] - \frac{9}{2}\sqrt{x}\sin(3x^{3/2}) \sec[(x - \pi)^2]$

39. $-\frac{2}{3}\cos 3x + c$

41. $\frac{1}{2}\sin^2 2x + c$

43. $-\cos(x^3 + 1) + c$

45. $[\sin x]^{-1} + c$

47. (a) $T(x; x_0) = x_0^3 + x_0^{1/2} + (3x_0^2 + \frac{1}{2}x_0^{-1/2})(x - x_0)$

$$r(x; x_0) = \frac{(x^3 + x^{1/2}) - (x_0^3 + x_0^{1/2})}{x - x_0} - \left(3x_0^2 + \frac{1}{2}x_0^{-1/2}\right)$$

(b) $x_0 = 1$; $T(\frac{15}{16}; 1) = \frac{57}{32}$
(c) 0.01097; 0.5485% error

49. (a) $T(x; x_0) = \frac{x_0 + 1}{\sqrt{x_0}} + \frac{1}{2}(x_0^{-1/2} - x_0^{-3/2})(x - x_0)$

$$r(x; x_0) = \frac{[(x + 1)/\sqrt{x}] - [(x_0 + 1)/\sqrt{x_0}]}{x - x_0} - \frac{1}{2}(x_0^{-1/2} - x_0^{-3/2})$$

(b) $x_0 = 9$; $T(\frac{80}{9}; 9) = \frac{806}{243}$
(c) -3.83×10^{-5}; 1.15×10^{-3}% error

51. (a) $(1, \sqrt{2})$, (b) $y' = -\sqrt{2}, y'' = \frac{1}{\sqrt{2}}$

53. (a) $(1, 1)$, (b) $y' = -1, y'' = \frac{4}{3}$

CHAPTER FOUR

4.1 Answers

1. Yes; $c = \frac{1}{2}$

3. No; $c = 0$

5. No; no c

7. (a) Proceed as in Example 1. (b) $|b| \leq 2$

4.2 Answers

1. (a) $x = -2$
   (b) Decreasing on $(-\infty, -2]$; increasing on $[-2, \infty)$
   (c) Concave up on $(-\infty, \infty)$
   (d) $x = -2$ is a local minimum point

3. (a) $x = -1$, $x = 3$
   (b) Increasing on $(-\infty, -1]$ and $[3, \infty)$; decreasing on $[-1, 3]$
   (c) Concave down on $(-\infty, 1]$; concave up on $[1, \infty)$
   (d) $x = -1$ is a local maximum point; $x = 3$ is a local minimum point

5. (a) $x = -1$
   (b) Decreasing on $(-\infty, -1)$ and $(-1, \infty)$
   (c) Concave down on $(-\infty, -1)$; concave up on $(-1, \infty)$
   (d) $x = -1$ is neither

7. (a) $x = 0$
   (b) Increasing on $(-\infty, 0]$; decreasing on $[0, \infty)$
   (c) Concave up on $(-\infty, 0]$ and $[0, \infty)$
   (d) $x = 0$ is a local maximum point

9. (a) $x = -\pi, x = -\frac{5\pi}{6}, x = -\frac{\pi}{6}, x = \pi$
   (b) Increasing on $\left[-\pi, -\frac{5\pi}{6}\right]$ and $\left[-\frac{\pi}{6}, \pi\right]$; decreasing on $\left[-\frac{5\pi}{6}, -\frac{\pi}{6}\right]$
   (c) Concave down on $\left[-\pi, -\frac{\pi}{2}\right]$ and $\left[\frac{\pi}{2}, \pi\right]$; concave up on $\left[-\frac{\pi}{2}, \frac{\pi}{2}\right]$
   (d) $x = -\pi$ and $x = -\frac{\pi}{6}$ are local minimum points; $x = -\frac{5\pi}{6}$ and $x = \pi$ are local maximum points.

11. (a) $x = 0$, $x = \pm 2$, $x = \pm 2\sqrt{3}$, $x = \pm 4$
   (b) Decreasing on $[-4, -2\sqrt{3}]$, $[-2, 0]$, and $[2, 2\sqrt{3}]$; increasing on $[-2\sqrt{3}, -2]$, $[0, 2]$, and $[2\sqrt{3}, 4]$
   (c) Concave up on $[-4, -2\sqrt{3}]$ and $[2\sqrt{3}, 4]$; concave down on $[-2\sqrt{3}, 0]$ and $[0, 2\sqrt{3}]$
   (d) $x = \pm 2$ and $x = \pm 4$ are local maximum points; $x = 0$ and $x = \pm 2\sqrt{3}$ are local minimum points.

13. (a) $x = 0$, $x = \pm 2$
   (b) Increasing on $[-2, 0]$; decreasing on $[0, 2]$
   (c) Concave down for $[-2, 2]$
   (d) $x = 0$ is a local maximum point; $x = \pm 2$ are local minimum points.

15. (a) Maximum at $x = -4$; minimum at $x = -2$
   (b) Minimum at $x = -2$; no maximum

17. (a) Maxima at $x = -1$ and $x = 5$; minimum at $x = 3$
   (b) No maximum or minimum

19. (a) Minimum at $x = -1$; no maximum
   (b) Maximum at $x = 0$; minimum at $x = 2$

21. (a) Maximum at $x = 0$; no minimum
   (b) Maximum at $x = 0$; minima at $x = \pm 1$

23. (a) Maximum at $x = \pi$; minimum at $x = -\frac{\pi}{6}$
   (b) Maximum at $x = \frac{\pi}{2}$; minimum at $x = -\frac{\pi}{6}$

25. (a) Maxima at $x = \pm 2$ and $x = \pm 4$; minima at $x = 0$ and $x \pm 2\sqrt{3}$
   (b) Maxima at $x = \pm 2$; minimum at $x = 0$

27. (a) Maximum at $x = 0$; minimum at $x = \pm 2$
   (b) No maximum or minimum

29.

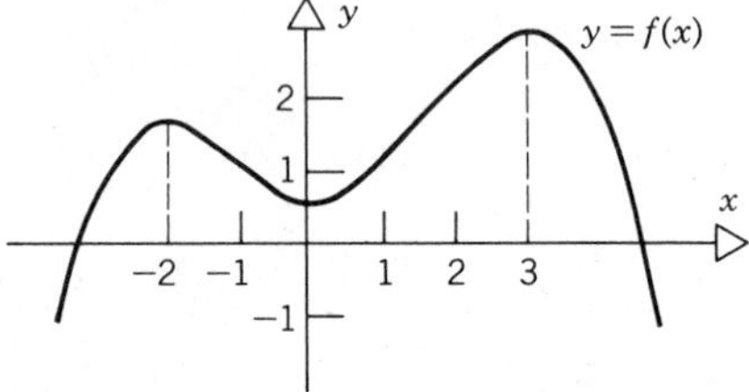

31.

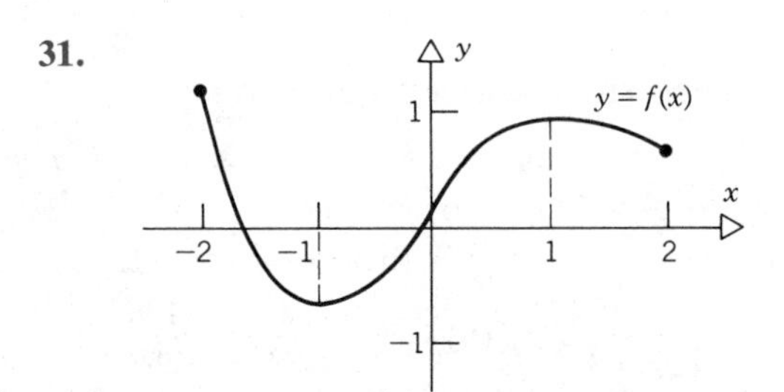

33.

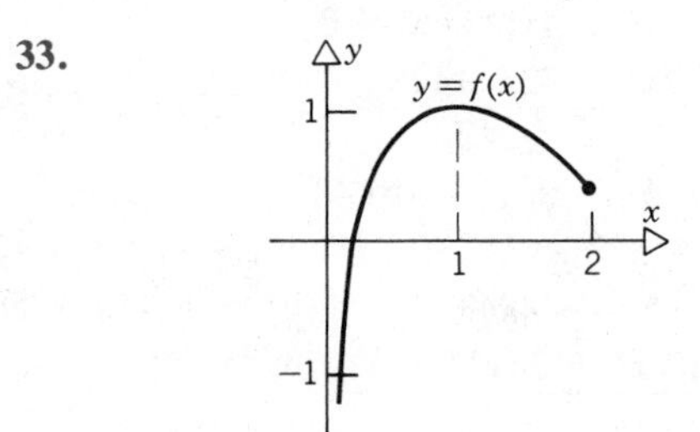

35. $a = -\frac{1}{3}$, $b = \frac{2}{3}$, $c = 1$

4.3 Answers

1. $x = y = \sqrt{k}$

7. $r = \dfrac{V^{1/3}}{2}, \quad h = \dfrac{4V^{1/3}}{\pi}, \quad \dfrac{h}{d} \cong 1.27$

9. (a) $x = h = V^{1/3}$ (b) $x = 2h = (2V)^{1/3}$

11. $l = (a^{2/3} + b^{2/3})^{3/2}$ feet

13. $\dfrac{2a^3}{27}$

15. (a) $\dfrac{9l}{9 + 4\sqrt{3}}$ into triangle
(b) None into triangle

17. 1:40 A.M.; $5\sqrt{2} \cong 7.071$ miles apart

19. $\left(-\dfrac{1}{\sqrt{2}}, -\dfrac{1}{\sqrt{2}}\right)$

23. (a) Either 22 or 23 (b) 43.667 (c) 30

25. 11

4.4 Answers

1. $\dfrac{1}{18\pi} \cong 0.0177$ ft/min

3. $7\sqrt{3} \cong 12.124$ in.^{2}/min

5. $\frac{3}{2}$

7. $\frac{7}{80} \cong 0.0875$ lb/in.^{2} sec

9. $4\sqrt{3} \cong 6.928$ in.^{2}/min

11. $\frac{12}{7}$ ft/sec

13. $\dfrac{18\pi}{5} \cong 11.310$ cm^3/min

15. 120 cm^2/min

17. $\dfrac{8}{25\pi} \cong 0.102$ ft/min

19. -4.81 ft/min

21. $\frac{1}{4}$ ft/min

4.5 Answers

1. 1.2360680
3. 1.1892071
5. 0.8284271
7. 0.8603336
9. 2.2889297
11. 1.4375649
15. 7.2801099
17. 1.7099759
19. 2.1867241

21. $x_{n+1} = 2x_n - ax_n^2$

23. (a) $x_0 = 0.5 \Rightarrow x_n \cong 0.8284271$ for $n \geq 4$
(b) $x_0 = 0.1 \Rightarrow x_n \cong 0.8284271$ for $n \geq 7$
(c) $x_0 = -0.1 \Rightarrow x_n \cong -4.8284271$ for $n \geq 6$
(d) $x_0 = -0.5 \Rightarrow x_n \cong -1.0000000$ for $n \geq 4$

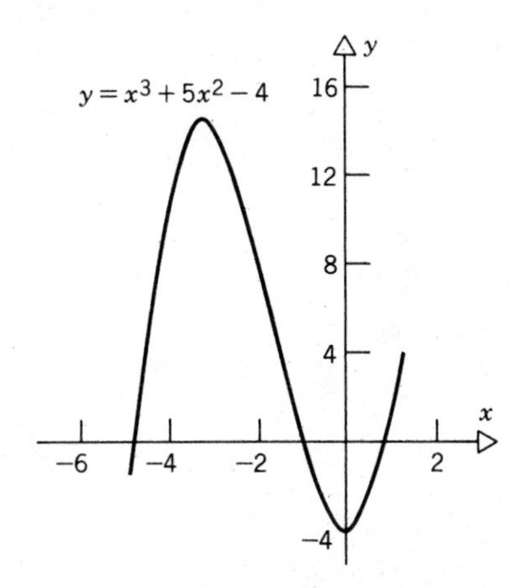

Answers to Review Problems

1. $b^2 < 3a$

3. True

5. False

7. No

9. No

11. Yes, $c = \dfrac{2\sqrt{2} - 1}{4}$

13. Yes, c must satisfy $\tan c = \pm\sqrt{(4/\pi) - 1}$.

15. No

19. c must satisfy $\sin c = -1/\pi$

21. $c = (1 \pm \sqrt{13})/6$

23. Critical points: $(-\pi/2, 1)$, $(0, 0)$, $(\pi/2, 1)$; decreasing on $-\pi/2 \leq x \leq 0$; increasing on $0 \leq x \leq \pi/2$; concave down on $-\pi/2 \leq x \leq 0$ and $0 \leq x \leq \pi/2$; global maxima at $(-\pi/2, 1)$ and $(\pi/2, 1)$; global minimum at $(0, 0)$.

25. Critical points: $(-1, -1)$, $(0, 0)$, $(1, 1)$; increasing on $-1 \leq x \leq 1$; concave down on $-1 \leq x \leq 0$ and $0 \leq x \leq 1$; global minimum at $(-1, -1)$, global maximum at $(1, 1)$.

27. Critical point: $(0, 1)$; decreasing on $-\infty < x < -1$ and $-1 < x \leq 0$, increasing on $0 \leq x < 1$ and $1 < x < \infty$; concave down on $-\infty < x < -1$ and $1 < x < \infty$, concave up on $-1 < x < 1$; local minimum at $(0, 1)$.

29. (a) $(1, 1)$
(b) $(1/2^{1/3}, (1/2) + (1/4^{1/3}))$
(c) $(\pi/2, 1)$
(d) $(\pm b/2, b/2)$ if $b \geq 0$; $(0, 0)$ if $b < 0$

31. Equilateral triangle of side $2\sqrt{A}/(3)^{1/4}$

33. $t = 4$ sec, $x = 40$ m

35. (a) $\dfrac{dP}{dr} = \dfrac{3\pi\sqrt{r}}{R\sqrt{g}}$

(b) $\dfrac{dP}{dv} = -\dfrac{6\pi R^2 g}{v^4}$

(c) $\dfrac{da_r}{dv} = \dfrac{4v^3}{R^2 g}$

37. $b = \dfrac{p[1 + (\pi/32)]}{4 + (\pi/16)}$

39. $\dfrac{2a}{b}, -\dfrac{x}{y}$

41. (a) $\dfrac{r}{2}$; (b) $\dfrac{x}{4}$

43. 4 feet

45. $\dfrac{250}{7}$ acres

47. (a) $\dfrac{dy}{dx} = \dfrac{bx}{ay}$; (b) critical points satisfy $ay^2 - bx^2 = 0$.

CHAPTER FIVE

5.1 Answers

1. Origin

3. None of these

5. y-axis

7. x-axis

9. x-axis, y-axis, origin

11. $x = -2, \quad y = 1$

13. No vertical asymptote, $y = 1$

15. $x = -1, \quad y = -1, \quad y = 1$

17. $x = 0, \quad y = \dfrac{x}{2}$

19. $x = 0$

21. $y = \dfrac{x}{\sqrt{2}}, \quad y = -\dfrac{x}{\sqrt{2}}$

23. Concave down for $(-\infty, -1]$; concave up for $[-1, \infty)$; $x = -1$ is inflection point

25. Concave up for all x; no inflection points

27. Concave down for $(-\infty, -2)$; concave up for $(-2, \infty)$; no inflection points

29. Concave down for $[0, \frac{2}{3}]$; concave up for $[\frac{2}{3}, \infty)$; $x = \frac{2}{3}$ is inflection point

31. Concave up for $\left[0, \dfrac{\pi}{4}\right]$, $\left[\dfrac{3\pi}{4}, \dfrac{5\pi}{4}\right]$, and $\left[\dfrac{7\pi}{4}, 2\pi\right]$; concave down for $\left[\dfrac{\pi}{4}, \dfrac{3\pi}{4}\right]$ and $\left[\dfrac{5\pi}{4}, \dfrac{7\pi}{4}\right]$; $x = \dfrac{\pi}{4}, \dfrac{3\pi}{4}, \dfrac{5\pi}{4}$, and $\dfrac{7\pi}{4}$ are inflection points.

33.

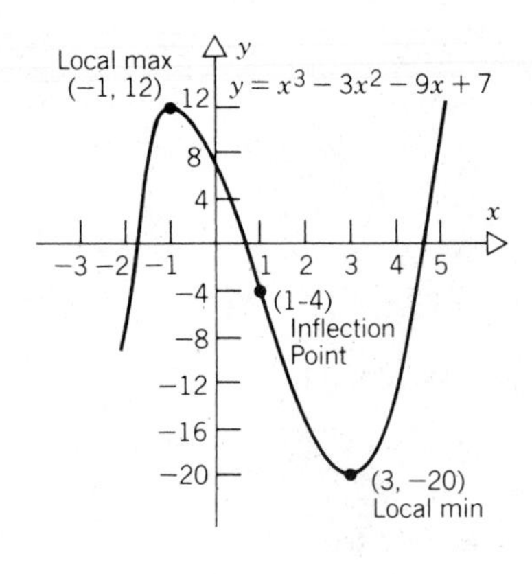

35.

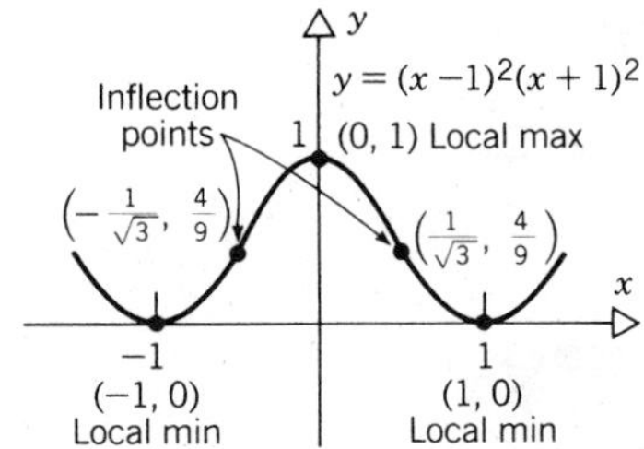

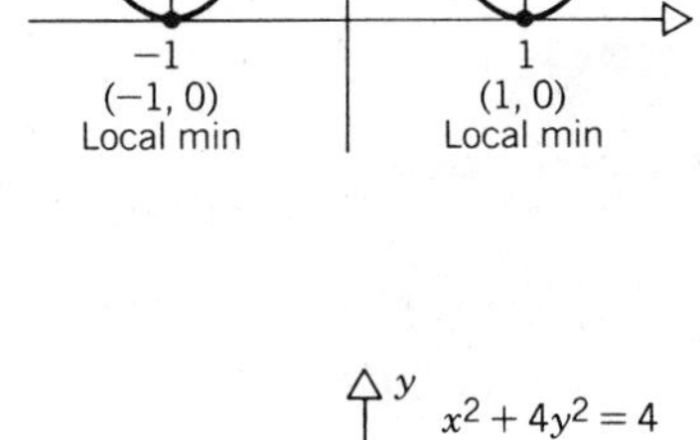

37.

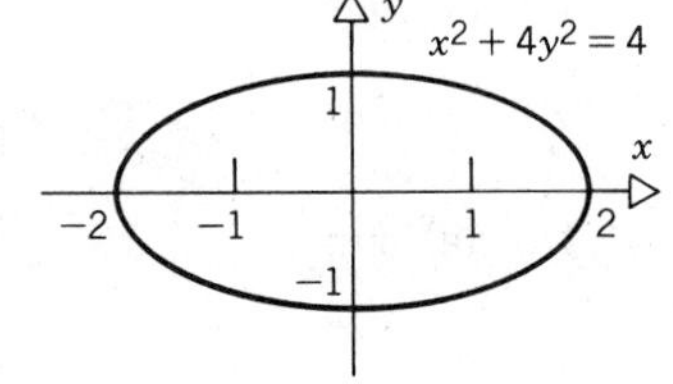

39.

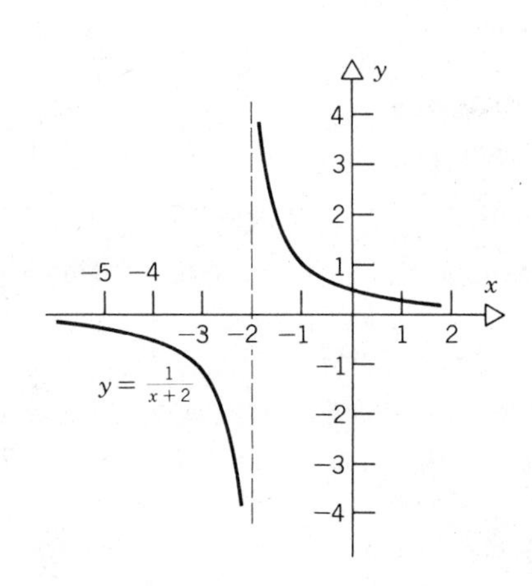

41.

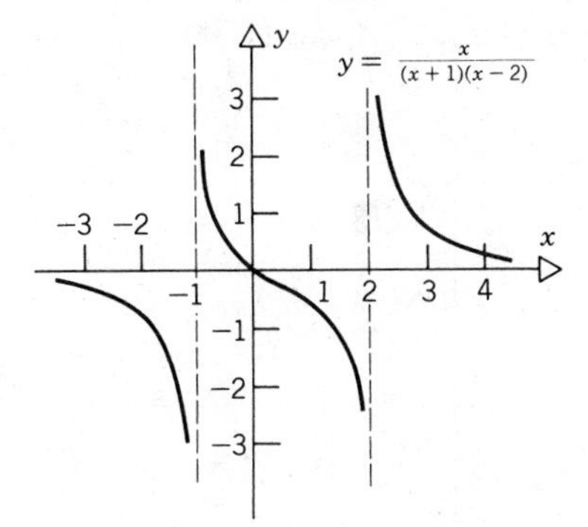

43.

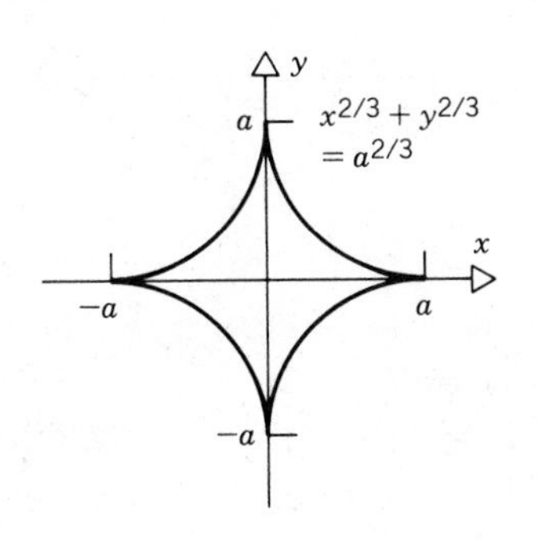

45.

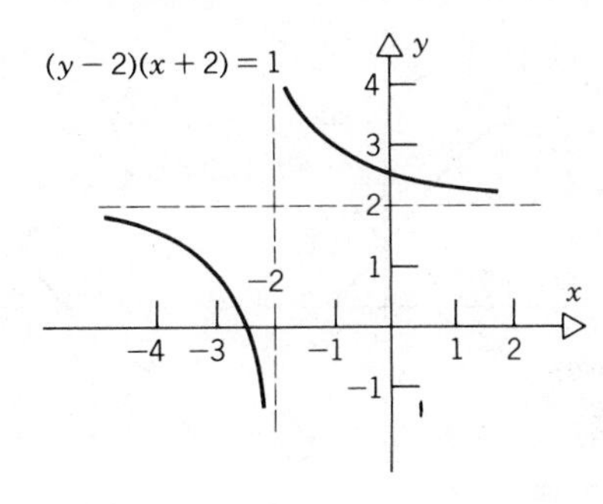

47.

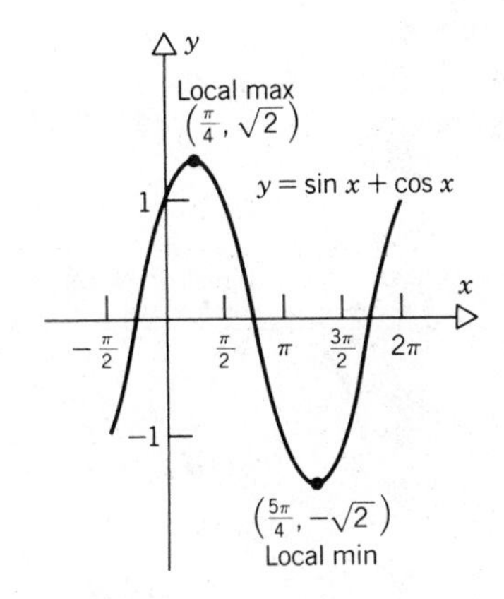

49.

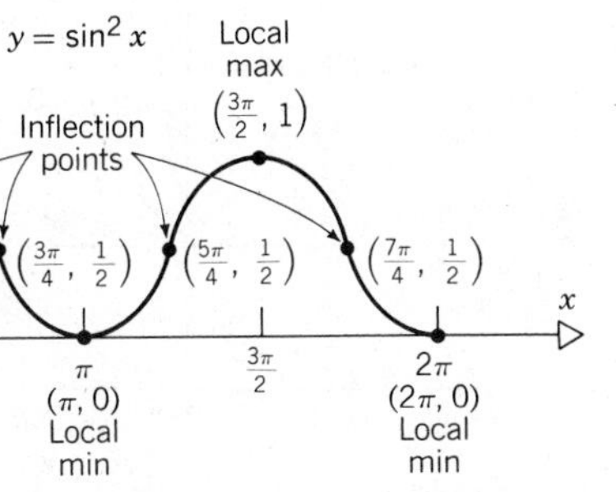

51.

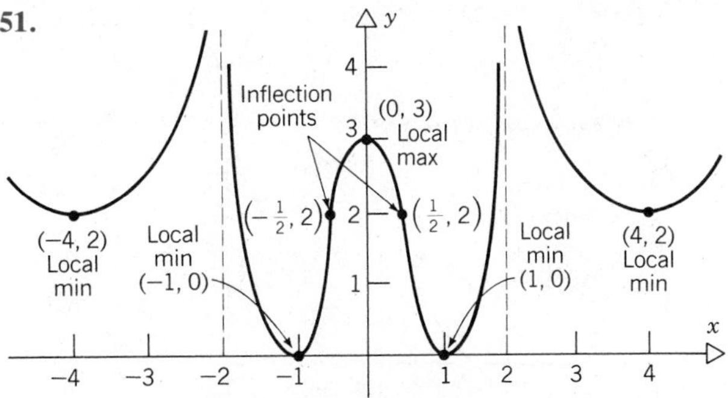

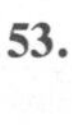
53.

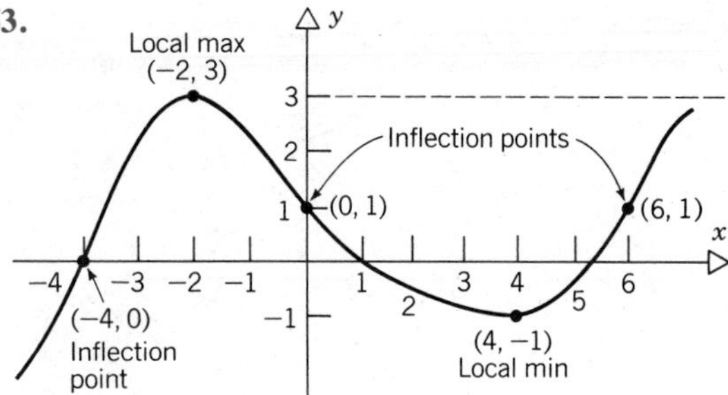

5.2 Answers

1. Vertex (0, 0); focus $(0, -\frac{1}{16})$; directrix $y = \frac{1}{16}$

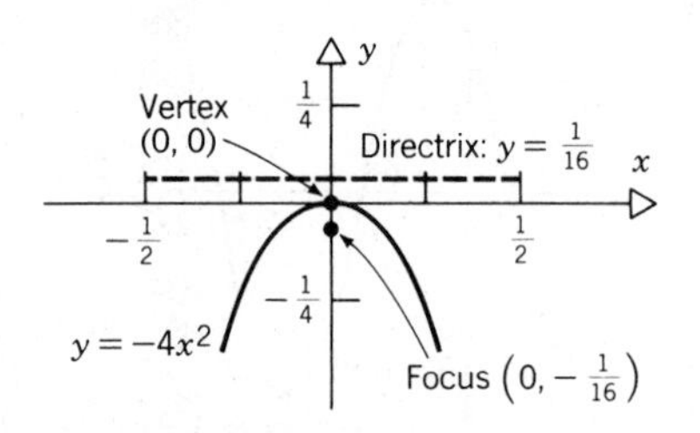

3. Vertex (0, 1); focus $(0, \frac{9}{8})$; directrix $y = \frac{7}{8}$

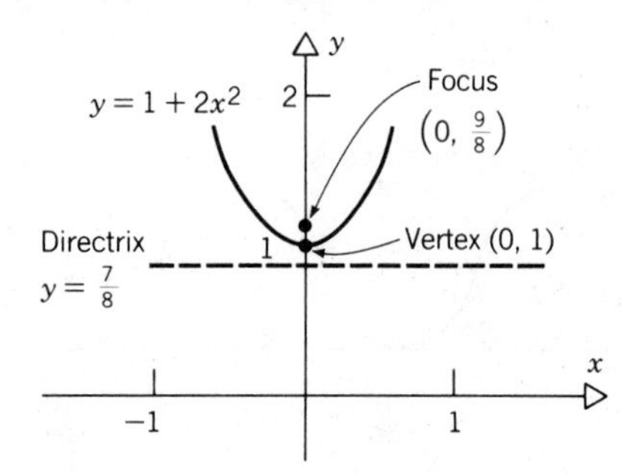

5. Vertex (−2, 1); focus $(-\frac{19}{8}, 1)$; directrix $x = -\frac{13}{8}$

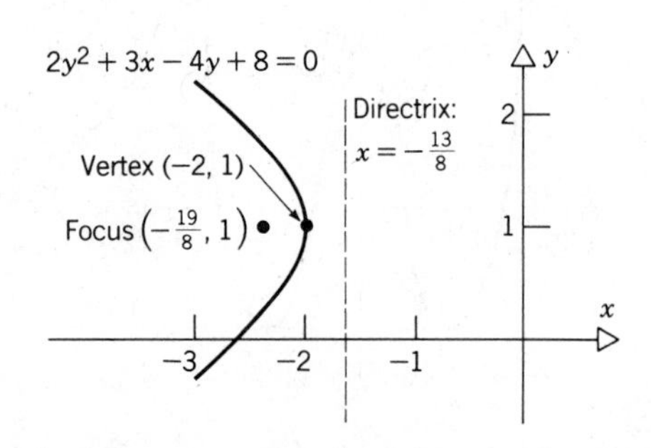

7. Vertex $(\frac{3}{4}, \frac{9}{4})$; focus $(\frac{3}{4}, \frac{39}{16})$; directrix $y = \frac{33}{16}$

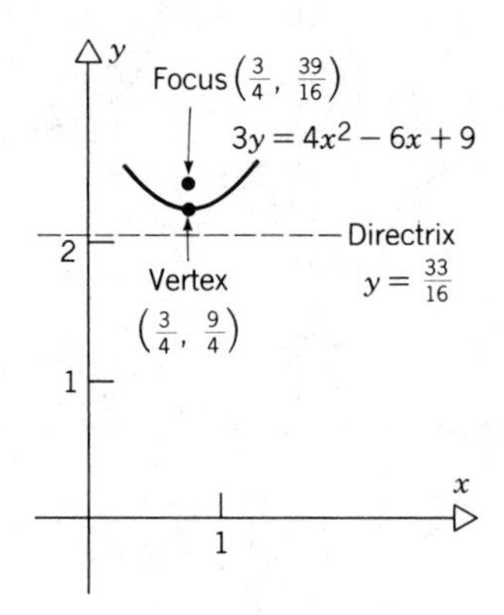

9. $9(y - 1) = -2(x - 2)^2$
11. $12(y - 1) = (x - 1)^2$
13. $4(y + 1) = 3(x - 2)^2$
15. $8(x + 2) = (y - 3)^2$
17. $y^2 = 10x + 5$
19. $8(x - 1) = (y - 2)^2$
23. $h = -\dfrac{D}{2A}$, $k = -\dfrac{E}{2C}$, $\alpha = A$, $\gamma = C$, $\kappa = F + \dfrac{D^2}{4A} + \dfrac{E^2}{4C}$
27. (a) $(\cos^2\theta - 2\sqrt{3}\sin\theta\cos\theta + 3\sin^2\theta)u^2 + (4\sin\theta\cos\theta - 2\sqrt{3}[\cos^2\theta - \sin^2\theta])uv + (\sin^2\theta + 2\sqrt{3}\sin\theta\cos\theta + 3\cos^2\theta)v^2 - 2(\sqrt{3}\cos\theta + \sin\theta)u + 2(\sqrt{3}\sin\theta - \cos\theta)v + 8 = 0$

(b) $\theta = \dfrac{\pi}{6} \pm n\pi$ or $\theta = \dfrac{2\pi}{3} \pm n\pi$

(c) $v^2 = u - 2$ for $\theta = \dfrac{\pi}{6}$

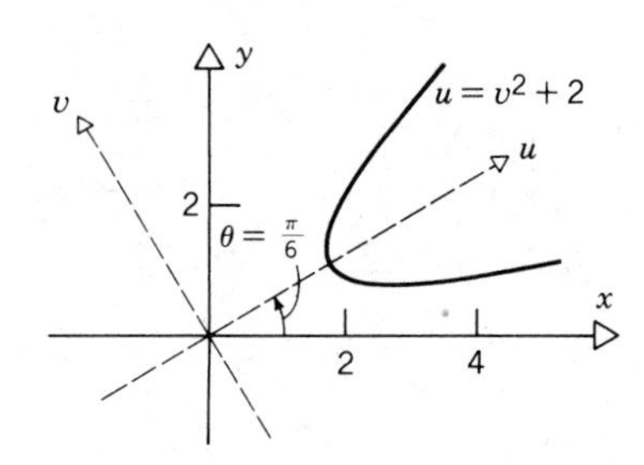

5.3 Answers

1. Center (0, 0); foci (0, ±3); semimajor axis $b = 5$; semiminor axis $a = 4$

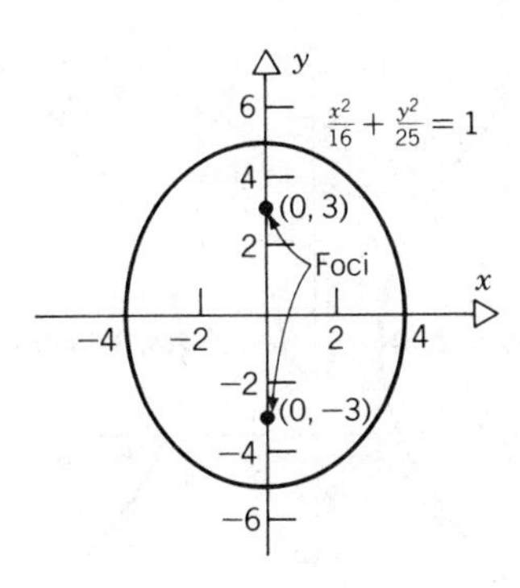

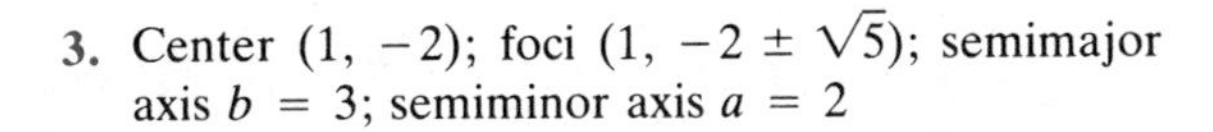
3. Center $(1, -2)$; foci $(1, -2 \pm \sqrt{5})$; semimajor axis $b = 3$; semiminor axis $a = 2$

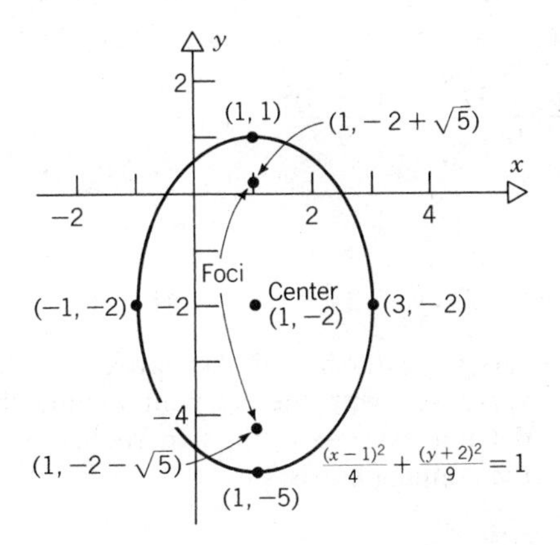

5. Center $(-1, 3)$; $\left(\text{graph is circle of radius } \dfrac{5}{\sqrt{2}}\right)$

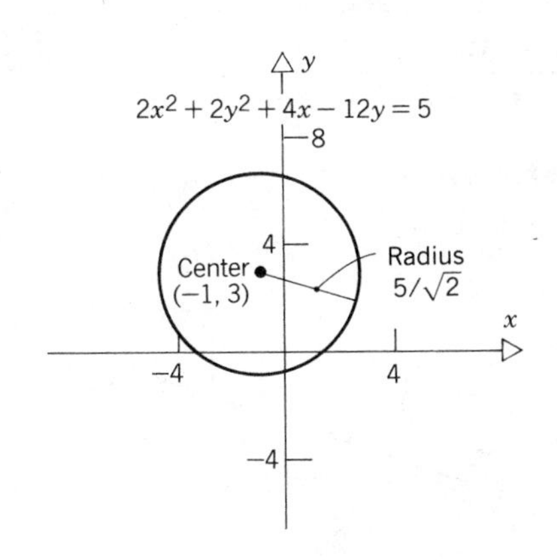

7. Center $(2, -2)$; foci $(2 \pm 2\sqrt{3}, -2)$; semimajor axis $a = 4$; semiminor axis $b = 2$

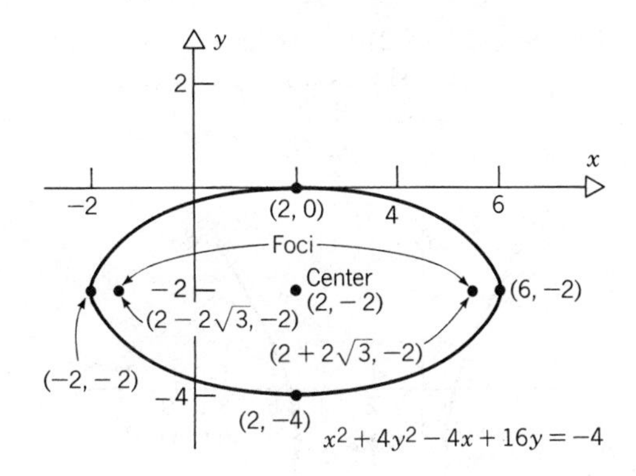

9. $\dfrac{(x-1)^2}{16} + \dfrac{(y-2)^2}{12} = 1$

11. $\dfrac{(x-1)^2}{16} + \dfrac{(y-2)^2}{25} = 1$

13. $\dfrac{(x-2)^2}{25} + \dfrac{(y-3)^2}{9} = 1$ 15. $\dfrac{x^2}{25} + \dfrac{y^2}{21} = 1$

17. $192x^2 + 96xy + 220y^2 - 432x - 316y = 2201$

19. (b) The foci approach the center.
(c) As $e \to 1$ with c fixed, the ellipse more and more nearly coincides with the line segment joining the two foci.

21. $\frac{3}{5}$ 23. $\dfrac{\sqrt{3}}{2}$ 25. $\dfrac{(x-1)^2}{9} + \dfrac{(y-2)^2}{5} = 1$

27. $\dfrac{(x+1)^2}{16} + \dfrac{(y+3)^2}{12} = 1$

5.4 Answers

1. Center $(0, 0)$; vertices $(\pm 4, 0)$; slopes $\pm\frac{5}{4}$

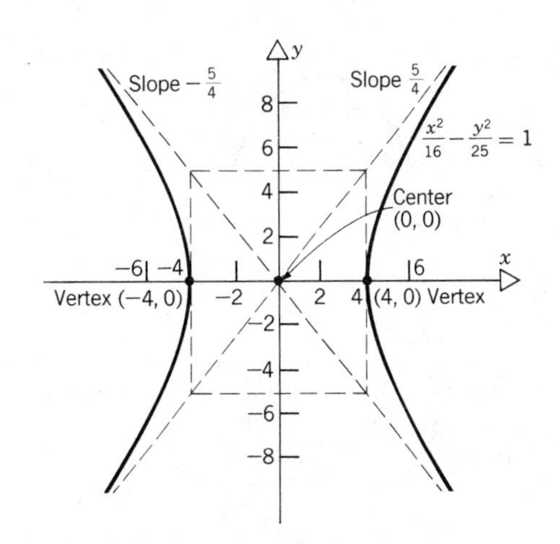

3. Center $(2, -2)$; vertices $(5, -2)$ and $(-1, -2)$; slopes $\pm\frac{4}{3}$

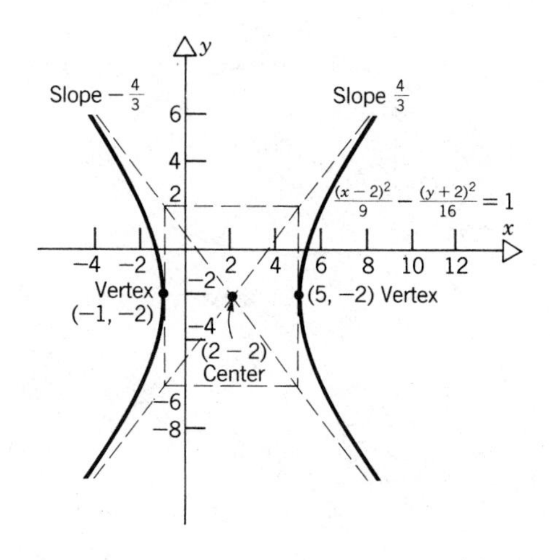

5. Center $(1, -3)$; vertices $(1, -7)$ and $(1, 1)$; slopes ± 2

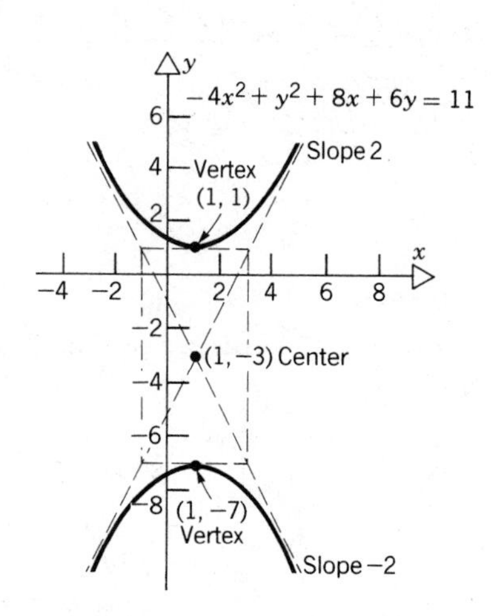

7. Center $(1, 3)$; vertices $(-2, 3)$ and $(4, 3)$; slopes $\pm\dfrac{2\sqrt{3}}{3}$

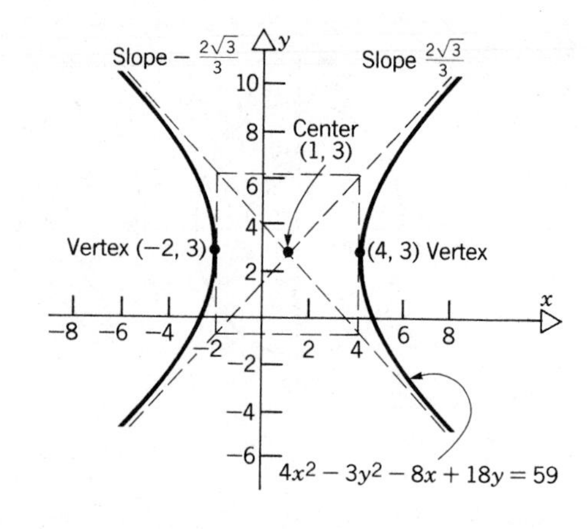

9. $\dfrac{x^2}{4} - \dfrac{y^2}{5} = 1$ 11. $-\dfrac{(x+1)^2}{64/5} + \dfrac{(y+1)^2}{16/5} = 1$

13. $\dfrac{(x+1)^2}{16} - \dfrac{(y-2)^2}{9} = 1$

15. $-\dfrac{x^2}{3} + y^2 = 1$

17. $7x^2 - 24xy - 2x + 24y - 41 = 0$

19. (b) As $e \to 1$, the hyperbola approaches the transverse axis with the segment joining the foci deleted. As $e \to \infty$, the hyperbola approaches the conjugate axis.

21. $\dfrac{\sqrt{41}}{4}$ 23. $\sqrt{\dfrac{7}{3}}$

25. $\dfrac{(x-1)^2}{16/9} - \dfrac{(y-2)^2}{20/9} = 1$

27. $\dfrac{(x-3)^2}{4} - \dfrac{(y+1)^2}{16} = 1$

5.5 Answers

1. $x - 3y = 0$

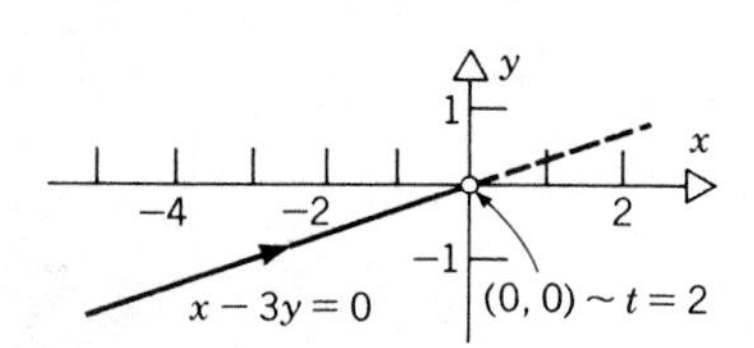

3. $x^2 + y^2 = 4$

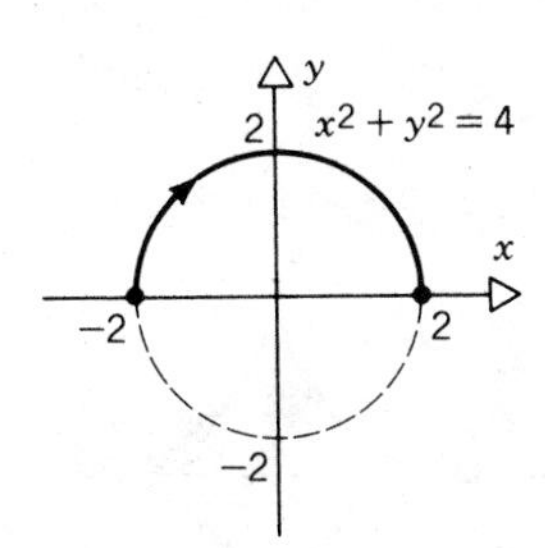

5. $x^2 + y^2 = 4$

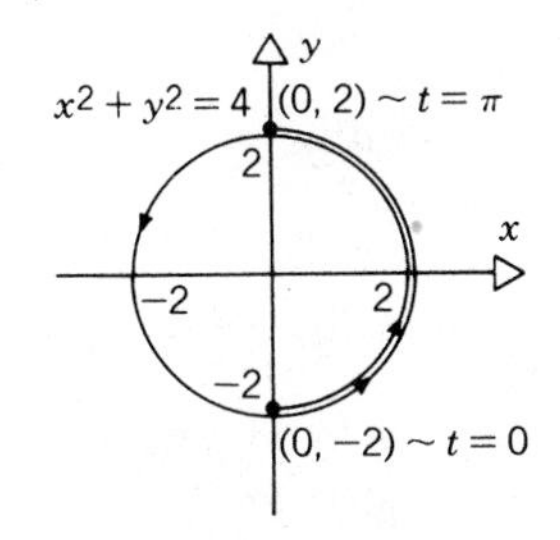

7. $y = 2x$

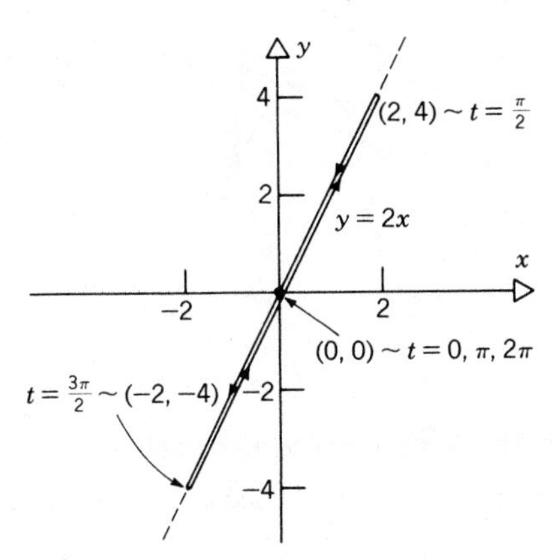

9. $x^2 - \frac{y^2}{4} = 1$

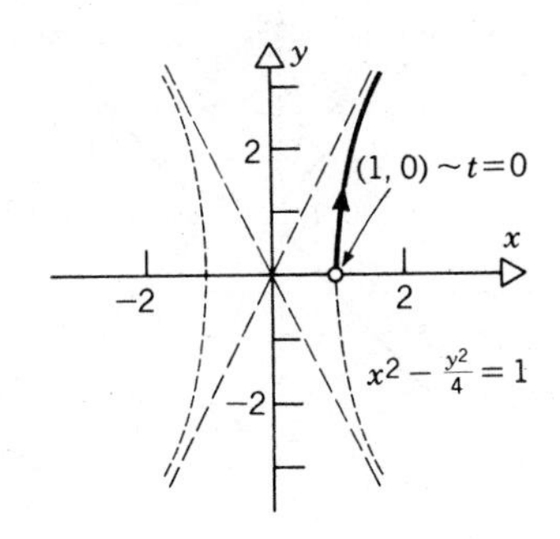

11. $x^2 + y^2 = 1$

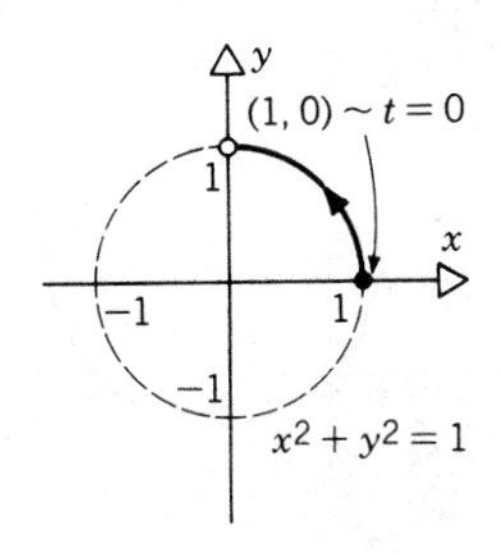

13. $\frac{(x - 1)^2}{9} + \frac{(y + 2)^2}{16} = 1$

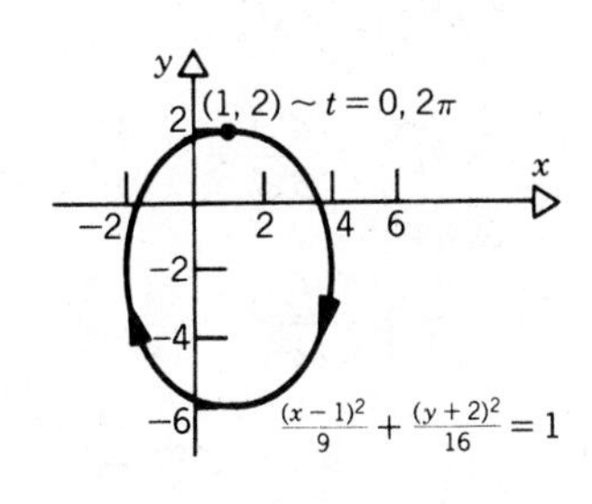

15. $y = 2x^2 - 1$

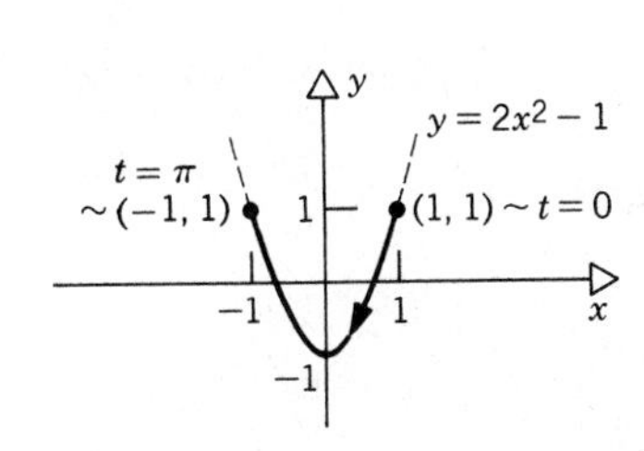

17. $y^2 = x^3$

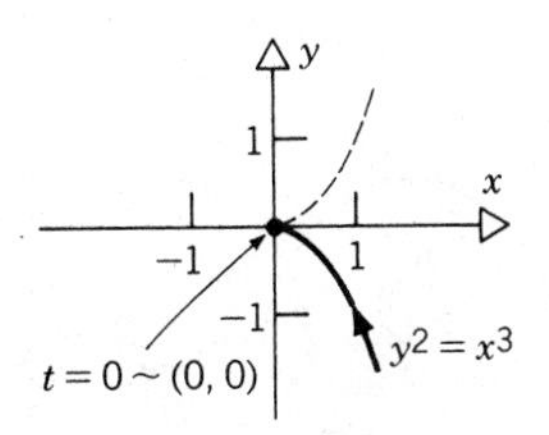

19. $x^2 + y^2 = x + y$

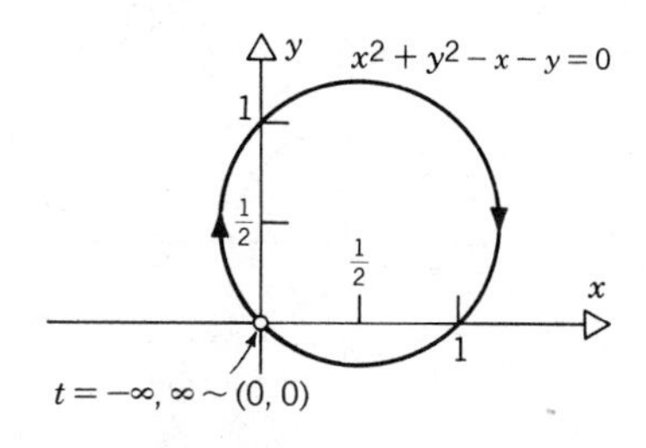

21. $3x - 2y - 6\sqrt{2} = 0$

23. $\sqrt{3}x - 2y + 1 = 0$

25. $2x - y - 5 + \sqrt{2} = 0$

27. $\dfrac{dy}{dx} = -\dfrac{4}{3}\tan 2t;\qquad \dfrac{d^2y}{dx^2} = -\dfrac{4}{9}\sec^3 2t$

29. $\dfrac{dy}{dx} = -\dfrac{4}{3}\cot t;\qquad \dfrac{d^2y}{dx^2} = -\dfrac{4}{9}\csc^3 t$

31. $\dfrac{dy}{dx} = \dfrac{2(t-1)}{3t^2};\qquad \dfrac{d^2y}{dx^2} = \dfrac{2(2-t)}{9t^5}$

33. $x^2 + 4xy + 4y^2 - 12x - 23y + 33 = 0$; parabola

35. (a) $x = a\theta - b\sin\theta,\quad y = a - b\cos\theta,\quad 0 \le \theta \le \infty$

(b)

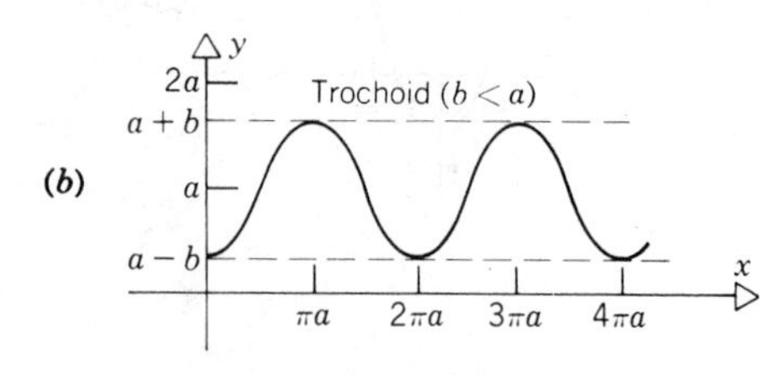

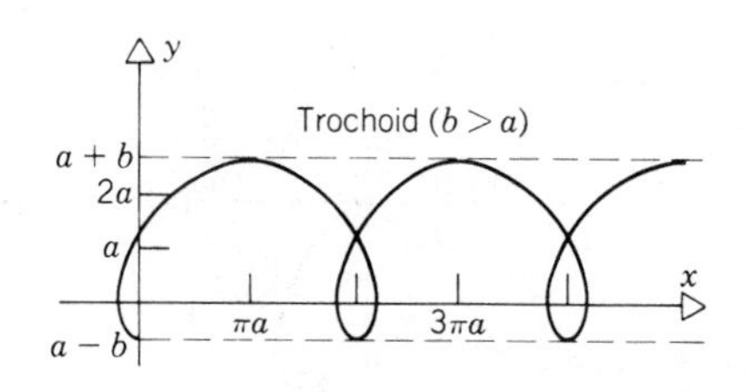

37. (a) $x = a\cos^3\theta,\ y = a\sin^3\theta,\ 0 \le \theta \le 2\pi$

(b)

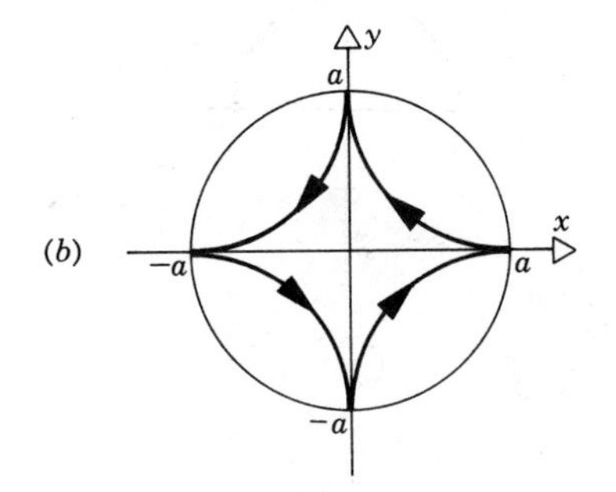

(c) $x^{2/3} + y^{2/3} = a^{2/3}$

39.

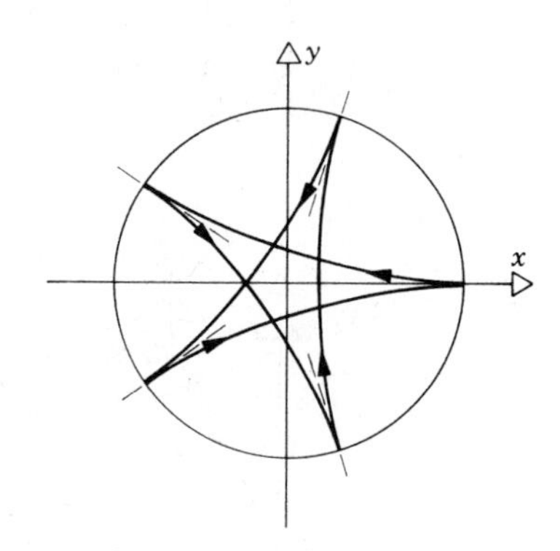

Answers to Review Problems

1. Origin

3. x-axis, y-axis, origin

5. $y = \pm 1$

7. $y = \pm\dfrac{4x}{3}$

9.

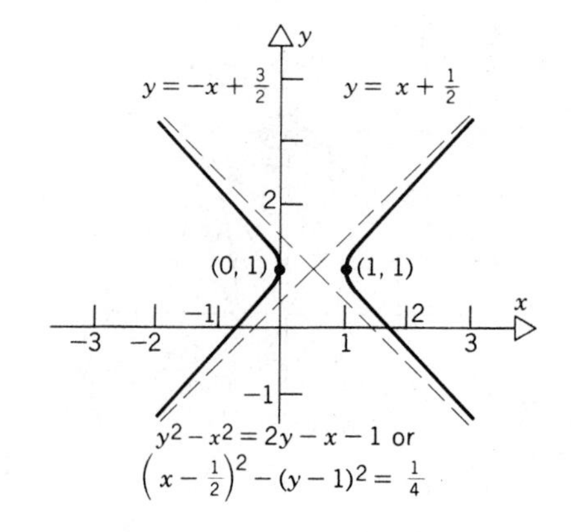

11.

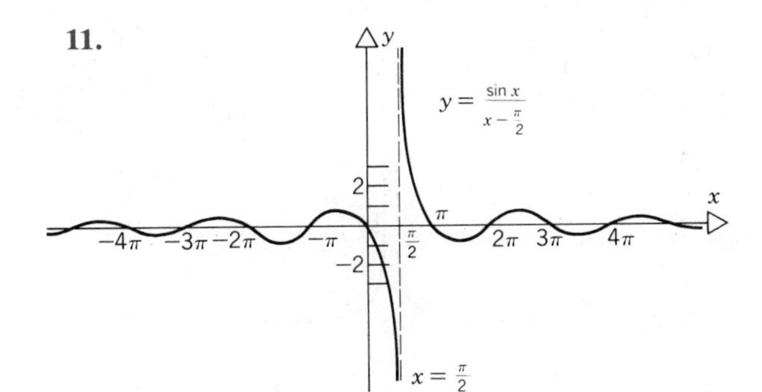

13.

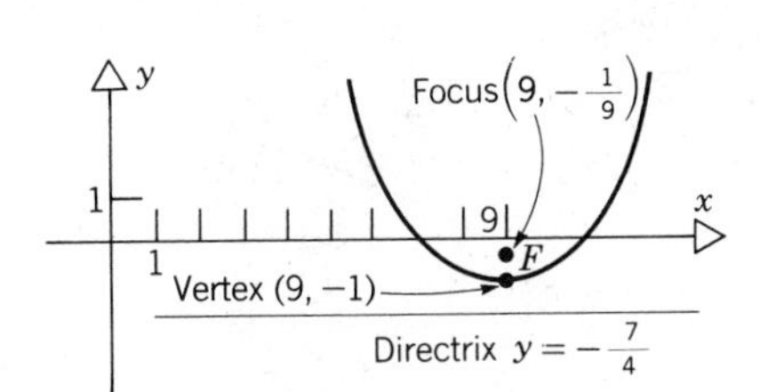

15.

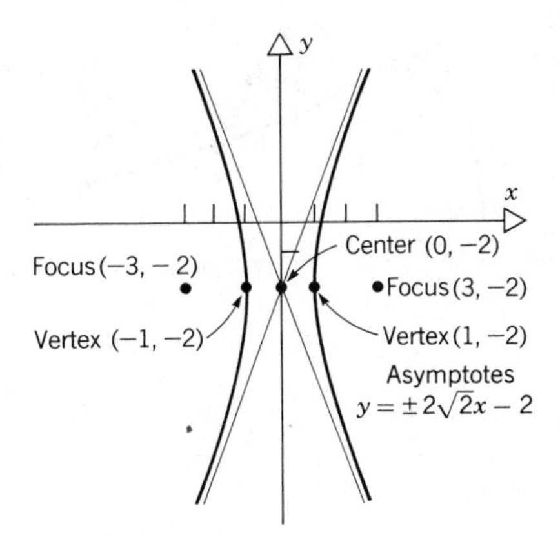

17.

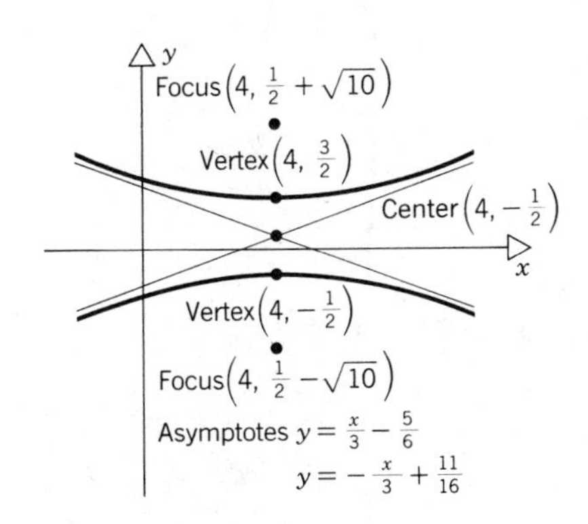

19.

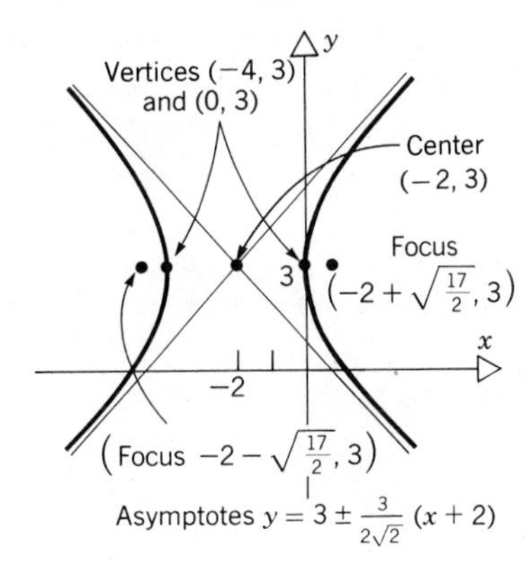

21.

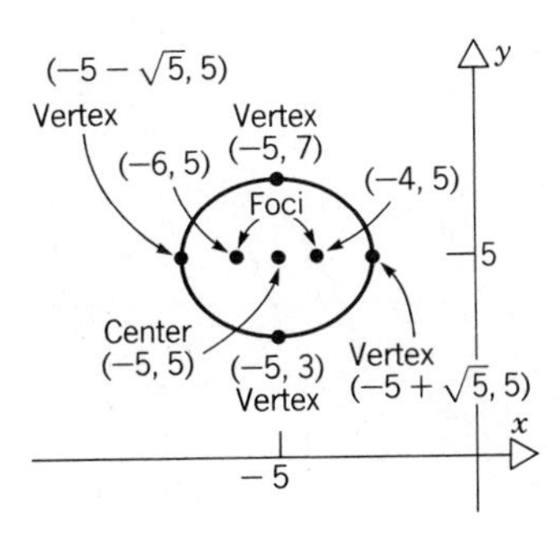

23.

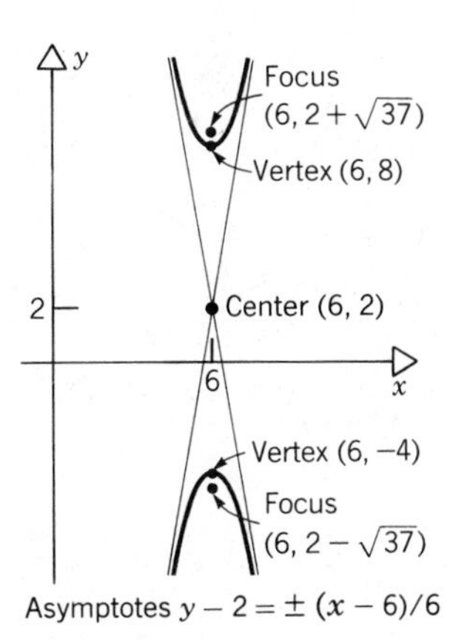

25. $12(y + 1) = -(x - 1)^2$

27. $\frac{(x - 2)^2}{4} - \frac{(y - 1)^2}{25} = 1$

29. $(x + 1)^2 - \frac{y^2}{35} = 1$

31. $\frac{(y - 3)^2}{16} - \frac{(x + 3)^2}{9} = 1$

33. $3(y + 3) = \pm(x - 1)^2$

35. $(x-3)^2 + \dfrac{(y-1)^2}{9} = 1$ or

$(x-3)^2 + \dfrac{(y+5)^2}{9} = 1$

37. $(y-2)^2 = 6\left(x + \dfrac{5}{2}\right)$

39. $\dfrac{(x-4)^2}{1/4} - \dfrac{(y-1)^2}{15/4} = 1$

41. $\theta = -\dfrac{\pi}{6}$; if $x = \dfrac{\sqrt{3}\,u + v}{2}$,

$y = \dfrac{-u + \sqrt{3}\,v}{2}$, then $u^2 + \dfrac{v^2}{4} = 1$

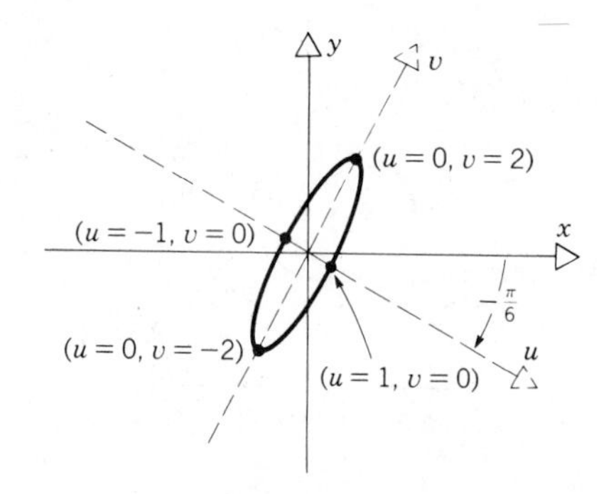

43. $x_0 < -3$ or $x_0 > 3$

45. $x + 2y = 13$

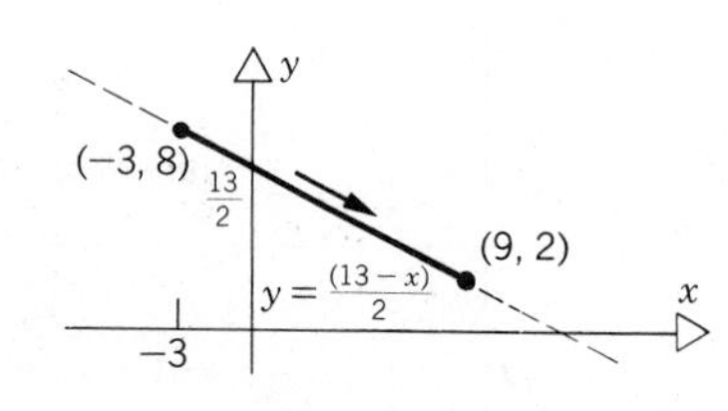

47. $y = 9(x-2)^2$

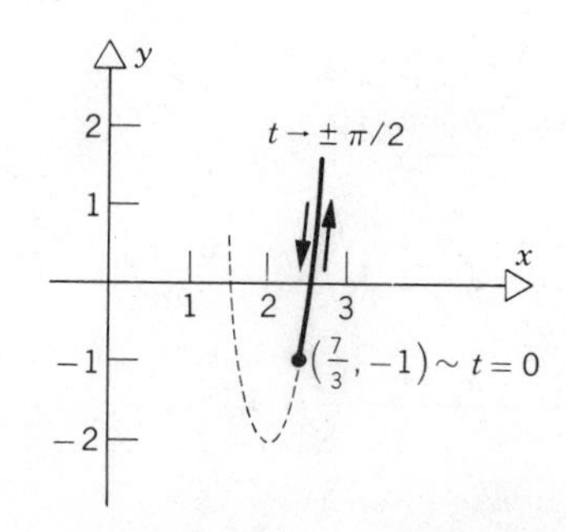

49. $y = \dfrac{1-x^2}{x^2}$

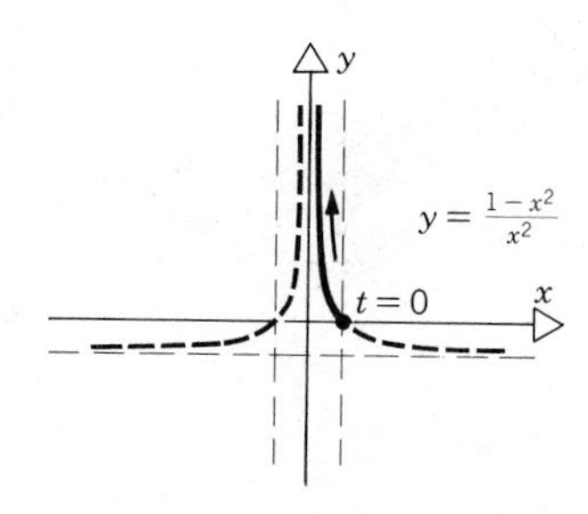

51. $y - 2 = (x+1)^{2/3}$

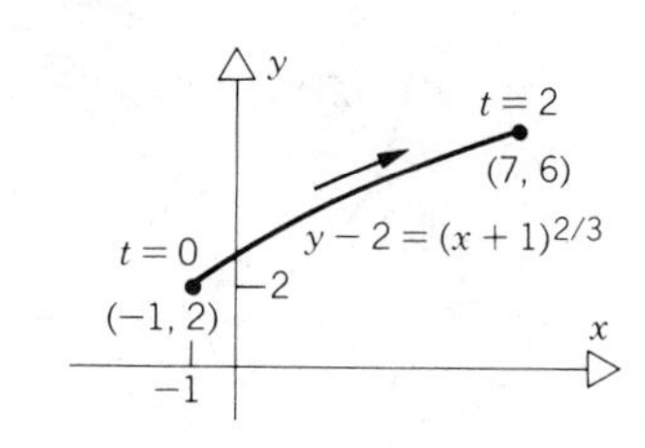

53.

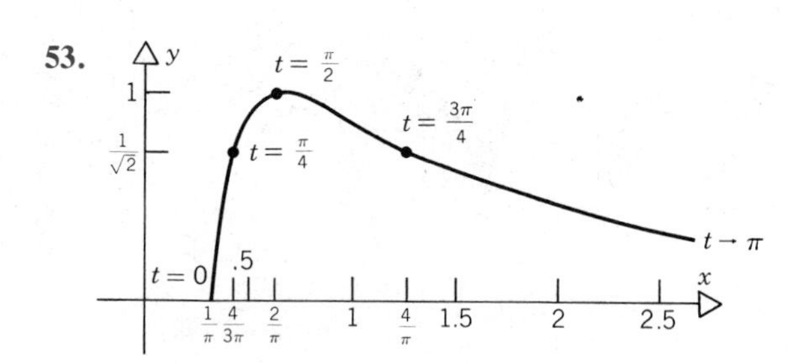

55.

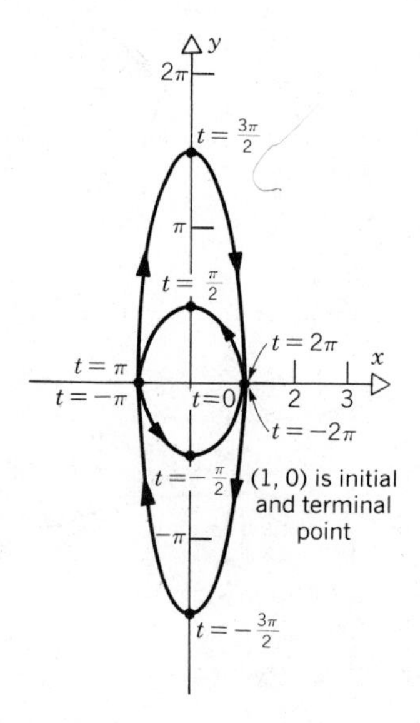

57. $\frac{dy}{dx} = -\frac{1}{2\cos \pi t}$

$\frac{d^2y}{dx^2} = -\frac{1}{4\cos^3 \pi t}$

tangent line is $x = 1$

59. $\frac{dy}{dx} = \frac{2}{3t}$

$\frac{d^2y}{dx^2} = -\frac{2}{9t^4}$

tangent line is $y - \frac{5}{4} = \frac{4}{3}(x - \frac{1}{8})$

CHAPTER SIX

6.1 Answers

1. $\frac{25}{12}$
3. 31
5. $-\frac{n}{n+1}$
7. 1000
9. $\frac{50}{51}$
11. $\sqrt{262} - 1$
15. $x - x^2 + x^3 - x^4 + \cdots + (-1)^{n+1}x^n$
17. $\frac{b-a}{n}\left\{f(a) + f\left[a + \frac{b-a}{n}\right] + f\left[a + \frac{2(b-a)}{n}\right] + \cdots + f\left[a + \frac{(n-1)(b-a)}{n}\right]\right\}$
19. $\sum_{n=2}^{26} \frac{(-1)^n}{n}$
21. $\sum_{k=1}^{14} \frac{(-1)^{k+1}x^{2k+1}}{2k+1}$
29. n^2
31. $n(2 - n)$
33. $\frac{n(n+1)(3n^2 + 11n + 10)}{6}$
35. (a) $s_n = \frac{b^2}{n^2}[1 + 2 + 3 + \cdots + (n-1)]$

$\sigma_n = \frac{b^2}{n^2}(1 + 2 + 3 + \cdots + n)$

(b) $|E_n| \le \frac{b^2}{2n}$

(c) $\frac{b^2}{2}$

37. $S_n = \frac{b^2}{2n^2}[1 + 3 + 5 + \cdots + (2n-1)]$

$= \frac{b^2}{2n^2}(n^2) = \frac{b^2}{2}$. See result of Problem 29.

6.2 Answers

1. $\frac{1}{2}$
3. $\frac{b-a}{n}$
5. $2115/3553 \cong 0.5953$
7. $\frac{5}{8}$
9. $\sigma = \frac{638}{840} \cong 0.7595$, $s = \frac{533}{840} \cong 0.6345$
11. $\sigma = \frac{339}{260} \cong 1.3038$, $s = \frac{47}{52} \cong 0.9038$
13. $\sigma = \frac{11}{8}$, $s = \frac{5}{8}$

15.

n	s_n	σ_n	$(s_n + \sigma_n)/2$	S_n
10	0.610509	0.710509	0.660509	0.668384
20	0.639447	0.689447	0.664447	0.667295
40	0.653371	0.678371	0.665871	0.666894
80	0.660133	0.672633	0.666383	0.666749

17.

n	s_n	σ_n	$(s_n + \sigma_n)/2$	S_n
10	5.335414	5.984097	5.659756	5.649082
20	5.492248	5.816590	5.654419	5.651751
40	5.572000	5.734170	5.653085	5.652417
80	5.612209	5.693293	5.652751	5.652585

19.

n	s_n	σ_n	$(s_n + \sigma_n)/2$	S_n
10	−1.480000	0.120000	−0.680000	−0.660000
20	−1.070000	−0.270000	−0.670000	−0.665000
40	−0.867500	−0.467500	−0.667500	−0.666250
80	−0.766875	−0.566875	−0.666875	−0.666563

21.

n	S_n
10	−1.991704
20	−1.997940
40	−1.999486
80	−1.999872

6.3 Answers

1. 8 3. $\frac{13}{2}$ 5. $3 + \frac{\pi}{4}$ 7. $-2 + \frac{\pi}{2}$

9. $\frac{7}{2}$ 11. $-\frac{45}{2}$ 13. $\frac{13}{3}$ 15. $-\frac{88}{3}$

23.

n	s_n	σ_n	$(s_n + \sigma_n)/2$	S_n
10	0.726130	0.826130	0.776130	0.788103
20	0.757116	0.807116	0.782116	0.786358
40	0.771737	0.796737	0.784237	0.785738
80	0.778737	0.791237	0.784987	0.785518

25. 41 ft/sec

6.4 Answers

1. $\frac{96}{5}$ 3. $\pi - 1$ 5. 0

7. $2 - \sqrt{2} + \frac{3\pi}{4}\left(1 + \frac{\pi}{4}\right)$ 9. $\frac{8}{3}$

11. $x^2 + \frac{2}{3}x^{3/2}$ 13. $3x - \sin x$ 15. $x^2 + x^4$

17. $-6x + 11x^2 - 2x^4$

19. 3 21. 0 23. $\frac{5}{2}$ 25. 4

29. (a) 0 (b) $\frac{A}{2}$

31. (a)

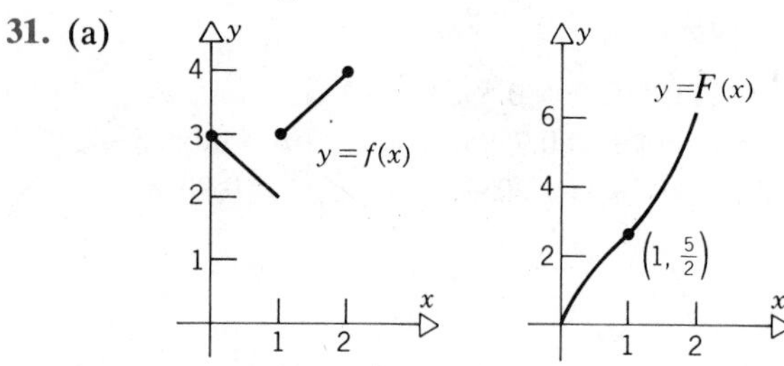

(b)
$$F(x) = \begin{cases} 3x - \dfrac{x^2}{2}, & 0 \le x < 1: \\ 2x + \dfrac{x^2}{2}, & 1 \le x \le 2. \end{cases}$$

(c) f is discontinuous at $x = 1$; F is continuous but not differentiable at $x = 1$.

33. $F'(x) = \dfrac{\sin 2x}{1 + x^2}$

35. $F(0) = 0, \quad F'(0) = 2, \quad F''(0) = 2$

37. $F'(x) = \dfrac{2x \sin 2(1 + x^2)}{1 + (1 + x^2)^2} + \dfrac{2x \sin 2(1 - x^2)}{1 + (1 - x^2)^2}$

39. $F'(x) = 2x(1 + x^2)\, f(1 + x^2)$

41. (a) $F'(x) = x\int_1^x f(s)\, ds$ (b) $F'(1) = 0$
(c) $F''(x) = xf(x) + \int_1^x f(s)\, ds$
(d) $F''(1) = f(1)$

43. $\dfrac{\sqrt{2(\pi + 2)}}{2}$

6.5 Answers

1. $\frac{1}{3}(3 + 2x)^{3/2} + c$ 3. $\frac{1}{8}(9 + x^2)^4 + c$

5. $-\frac{1}{16}(1 + 4t^2)^{-2} + c$ 7. $-\frac{1}{16}(4 - x^4)^4 + c$

9. $\dfrac{1}{3}\sin 3\left(x - \dfrac{\pi}{4}\right) + c$ 11. $\dfrac{1}{\pi}\sec \pi\theta + c$

13. $-\frac{1}{8}\cos^4 2x + c$ 15. $-\dfrac{2}{1 + \sqrt{x}} + c$

17. $\frac{1}{7}(2 + x^2)^{7/2} - \frac{4}{5}(2 + x^2)^{5/2} + \frac{4}{3}(2 + x^2)^{3/2} + c$

19. $\dfrac{(a^2 + x^2)^{r+1}}{2(r + 1)} + c$ $\frac{112}{9}$ $\frac{64}{3}$ $\frac{1}{2}$

27. $\frac{11}{900}$ 29. $(\sqrt{2} - 1)|a|$ 33. $\dfrac{1}{\sqrt{3}a^2}$

6.6 Answers

1.

n	T_n	S_n
10	0.660509	0.664100
20	0.664447	0.665759
40	0.665871	0.666346
80	0.666383	0.666553

3.

n	T_n	S_n
10	1.106616	1.107147
20	1.107016	1.107149
40	1.107115	1.107149
80	1.107140	1.107149

5.

n	T_n	S_n
10	0.481739	0.488480
20	0.486952	0.488690
40	0.488267	0.488705
80	0.488596	0.488706

7.

n	T_n	S_n
10	0.855353	0.861337
20	0.859639	0.861067
40	0.860678	0.861025
80	0.860933	0.861018

9.

n	T_n	S_n
10	0.225098	0.223240
20	0.223708	0.223244
40	0.223360	0.223244
80	0.223273	0.223244

11.

n	T_n	S_n
10	2.370394	2.386178
20	2.387129	2.392707
40	2.393045	2.395018
80	2.395137	2.395834

13.

n	T_n	S_n
10	3.391611	3.393424
20	3.392987	3.393445
40	3.393331	3.393446
80	3.393417	3.393446

15.

n	T_n	S_n
10	0.761109	0.761109
20	0.761109	0.761109
40	0.761109	0.761108
80	0.761109	0.761109

17. (a) $n \geq 213$ (b) $n \geq 24$

19. (a) $n \geq 82$ (b) $n \geq 14$

21. $0.2252BLp_a$

25. (a)

n	*Error estimate*
10	0.0020562
20	0.0005410
40	0.0001285
80	0.0000321

(b) 10

Answers to Review Problems

1. $\frac{1983}{2210}$ **3.** $\frac{10}{39}$ **5.** $1 - \sqrt{107}$ **7.** 20,144

9. $\sum_{n=1}^{9} \sin[(n+2)x + (2n-1)]$

11. $\sum_{n=1}^{8} \frac{(-2)^{n-1}}{nx^2}$

13. $\sigma = 103.828$, $s = 71.234$, $S = 86.234$

15. $\sigma = 27$, $s = -27$, $S = 0$

17. $\sigma = 48.3923$, $s = -20.3923$, $S = 11.375$

19. $\frac{62}{3} - \pi$

21. $\frac{25\pi}{4} + 2195$ **25.** $\frac{2}{\pi}$ **27.** π

29. $\frac{25\sqrt{2}\,\pi}{4} + \frac{325}{6}$ **31.** $\frac{\pi + 2}{24}$

33. $\frac{x-a}{2} + \frac{\cos 2a - \cos 2x}{4}$

35. $-\frac{3\sqrt{3} + 2\sqrt{2}}{144}$ **37.** -2

39. $\frac{\sqrt{3} - 1}{2}$ **41.** $\frac{17}{18}$ **43.** $\frac{147}{220}$ **45.** $\frac{1712}{105}$

47. $\frac{8}{9}\left[\left(\frac{b^3}{2} + 1\right)^{3/2} - \left(\frac{a^3}{2} + 1\right)^{3/2}\right]$

49. $\frac{\sin(9\pi^2/4) - \sin(\pi^2)}{2}$

51. Trapezoidal rule, $n \geq 330$;
Simpson's rule, $n \geq 44$

53.

n	*Trapezoidal*	*Simpson*
10	8.747225	8.856339
20	8.763592	8.769048
40	8.764566	8.764892
80	8.764626	8.764644

55.

n	*Trapezoidal*	*Simpson*
10	68.29303	71.67857
20	70.06278	70.65269
40	70.47419	70.61134
80	70.57512	70.60876

CHAPTER SEVEN

7.1 Answers

1. $\frac{16}{3}$ **3.** 2 **5.** $\frac{32}{3}$

7. $\frac{7 + 8\sqrt{2}}{6}$ **9.** $\frac{7}{6}$ **11.** $2\sqrt{2}$ **13.** $\frac{19}{3}$

15. $\frac{7}{3}$ **17.** $4\sqrt{2}$ **19.** $\frac{4}{3}$ **21.** $\frac{8}{5}$

23. $\frac{5}{4}$ **25.** $A = \int_0^1 (\sqrt{4 - x^2} - \sqrt{3}x)\,dx$

27. $A = \int_0^a [\sqrt{a^2 - x^2} - (a - \sqrt{2ax - x^2})]\,dx$

29. $a = 2^{4/3} + 1 \cong 3.52$

7.2 Answers

1. $\frac{64\pi}{3}$ **3.** $\frac{8\pi}{3}$ **5.** $\frac{\pi^2}{4}$ **7.** $\frac{128\pi}{5}$

9. $\frac{2\pi}{15}$ **11.** $\frac{\pi}{6}$ **13.** $2\pi\sqrt{3}$ **15.** $\frac{192\pi}{5}$

17. 8π **19.** $\frac{8\pi}{3}$ **21.** $\frac{625\pi}{6}$ **23.** $\frac{128\pi}{3}$

25. $\frac{4\sqrt{3}a^3}{3}$ **27.** $\pi h^2\left(a - \frac{h}{3}\right)$ **29.** $\frac{2}{3}a^3 \tan\theta$

7.3 Answers

1. $3\sqrt{10}$ **3.** $\frac{2}{27}[(37)^{3/2} - (10)^{3/2}]$

5. $\frac{2}{3}[(1 + 3^{2/3})^{3/2} - 1]$ **7.** $2a\pi^2$

9. $\frac{227}{24}$ 11. $6a$ 13. $\dfrac{6\sqrt{3} - 4}{3}$

15. (b) $\dfrac{L}{4a} \cong 1.56687, 1.54625, 1.47262, 1.31947$

17. 4.64678 19. $\dfrac{32\pi}{\sqrt{5}}$ 21. $2\pi r(b - a)$

23. $\dfrac{7615\pi}{64}$ 25. $\dfrac{\pi}{6}[(17)^{3/2} - (5)^{3/2}]$ 27. 63.56045

31. $S = 2\pi \int_c^d g(y)\sqrt{1 + [g'(y)]^2}\,dy$,

$\dfrac{8\pi}{3}(5\sqrt{5} - 2\sqrt{2})$

33. $\pi(\pi - 2)$

7.4 Answers

1. (a) and (b) 2π ft
3. (a) $\frac{2}{3}$ ft (b) 3 ft
5. (a) and (b) $\dfrac{10\sqrt{10} - 1}{3}$ ft
7. (a) 0.5 ft (b) $\frac{11}{2} - 6(2)^{-1/3}$ ft
9. 6 ft-lb

11. $47{,}812.5w \cong 2.988 \times 10^6$ ft-lb 13. $\dfrac{3k}{10}$

15. $\dfrac{mgR}{2}$ 19. $55\pi \times 10^4$ ft-lb 21. $\dfrac{10w}{3} \cong 208$ lb

23. $16w \int_{-3}^{3} (3 - y)\sqrt{1 - \dfrac{y^2}{9}}\,dy$

25. (a) $2{,}560{,}000w$ lb (b) $1{,}395{,}000w$ lb

7.5 Answers

1. (a) 1.06×10^{-4} cm^3/sec
 (c) 1330 dyne-sec/cm^5, 0.352 cm
 (d) 7.3×10^5

3. $\cos\theta_{\min} = \left(\dfrac{b}{a}\right)^4$

5. (a) \$380.40 (b) \$800.70 (c) \$10.52 per hour

7. (a) $\frac{1}{2}$ (b) $F(t) = \begin{cases} t/2, & 0 \le t \le 2 \\ 1, & t > 2 \end{cases}$ (c) 1

9. (a) 1

(b) $F(t) = \begin{cases} t^2/2, & 0 \le t \le 1 \\ 2t - (t^2/2) - 1, & 1 \le t \le 2 \\ 1, & t > 2 \end{cases}$ (c) 1

13. $10^4(1 + T^2)^{-2} - 15(1 + T^2)^{-1} + 15$

15. (a) 112.5 (b) 12.5

Answers to Review Problems

1. $\frac{40}{3}$ 3. $8\pi + \frac{32}{3}$ 5. $\frac{37}{12}$ 7. $\frac{45}{4}$

9. $\frac{9}{2}$ 11. 2 13. $\frac{9}{2}$ 15. 8π

17. $\dfrac{256\pi}{15}$ 19. $\dfrac{56\pi}{3}$ 21. $\dfrac{28\pi}{3}$

23. (b) $\pi \int_0^4 (1 + \sqrt{4 - y})^2\,dy + \dfrac{\pi}{16}\int_4^6 (y - 8)^2\,dy$

(c) $\dfrac{3\pi}{2} + 2\pi \int_{1/2}^{1} x(8 - 4x)\,dx$

$+ 2\pi \int_1^3 x[4 - (x - 1)^2]\,dx$

25. (a)

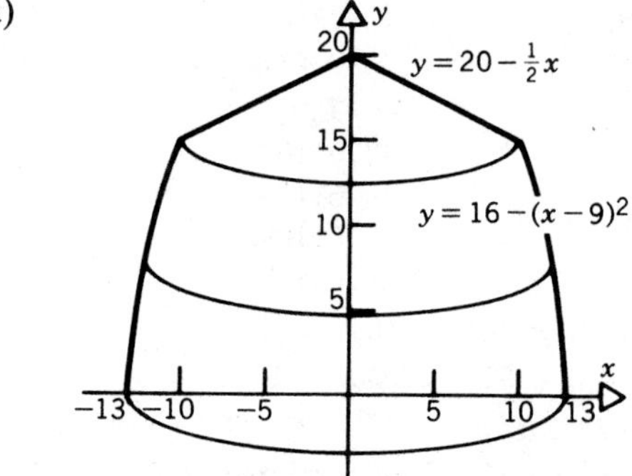

(b) $\pi \int_0^{15} (9 + \sqrt{16 - y})^2\,dy$

$+ 4\pi \int_{15}^{20} (20 - y)^2\,dy$

(c) $2\pi \int_0^{10} x\left(20 - \dfrac{x}{2}\right)dx$

$+ 2\pi \int_{10}^{13} x[16 - (x - 9)^2]\,dx$

27. 2π 29. $\frac{20}{27}[(\frac{19}{10})^{3/2} - 1]$

31. $\dfrac{\pi}{36}[(260)^{3/2} - (5)^{3/2}]$ 33. $320\sqrt{10}\,\pi$

35. (a) $\dfrac{\pi}{2}$ (b) $\dfrac{\pi}{6} + \sqrt{3}$

37. (a) 8 (b) 8 39. 216 in-lb

41. $\dfrac{8500\pi}{3}$ ft-lb 43. $\dfrac{14{,}500}{3}$ lb

CHAPTER EIGHT

8.1 Answers

1. $f^{-1}(y) = y + 3$; domain $[-5, 1]$, range $[-2, 4]$

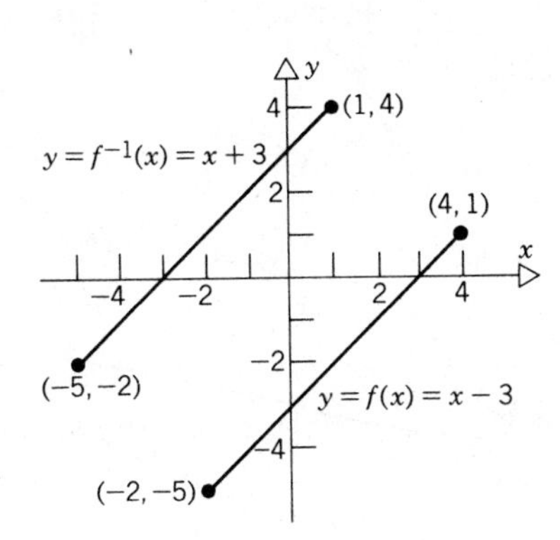

3. $f^{-1}(y) = \dfrac{5 - y}{2}$; domain $(-\infty, 3)$, range $(1, \infty)$

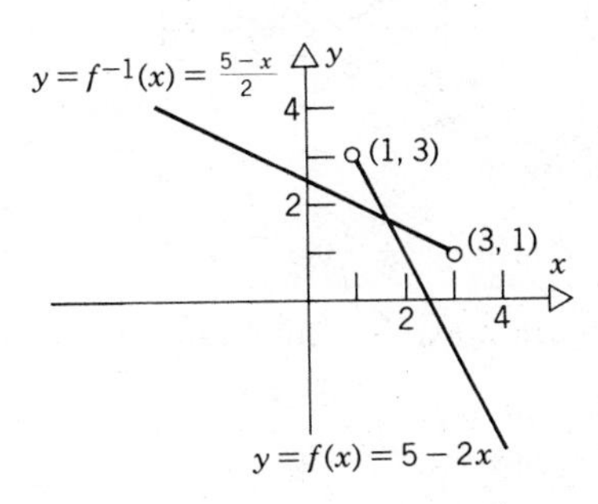

5. $f^{-1}(y) = 1 - \sqrt{y - 2}$; domain $[2, 27)$, range $(-4, 1]$

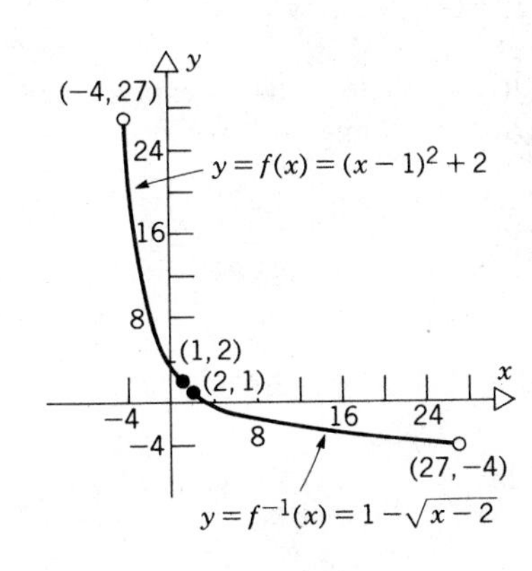

7. $f^{-1}(y) = \dfrac{1 + 2y}{1 - y}$; domain $(-\frac{1}{2}, \frac{2}{5}]$, range $(0, 3]$

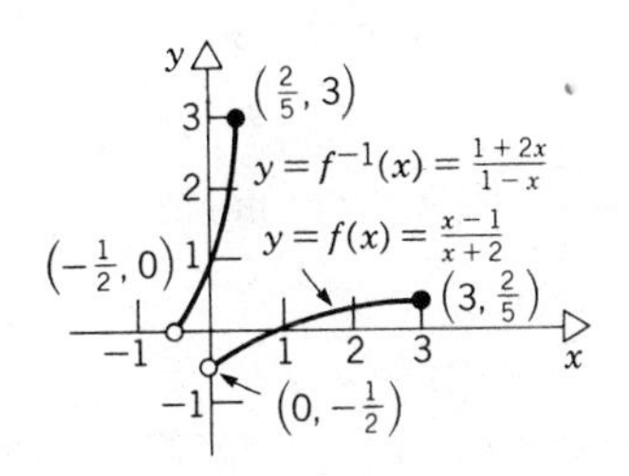

9. $f^{-1}(y) = y^2 - 2$; domain $[2, 3]$, range $[2, 7]$

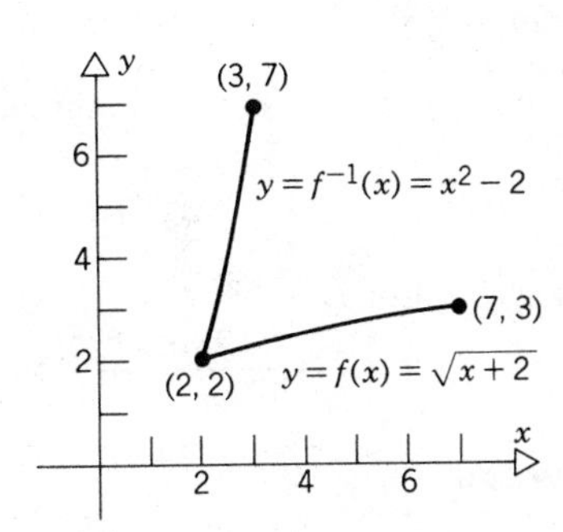

11. $\dfrac{dx}{dy} = 1$ **13.** $\dfrac{dx}{dy} = -\frac{1}{2}$ **15.** $\dfrac{dx}{dy} = \dfrac{1}{2x - 2}$

17. $\dfrac{dx}{dy} = \dfrac{(x + 2)^2}{3}$ **19.** $\dfrac{dx}{dy} = 2\sqrt{x + 2}$

21. $\dfrac{dx}{dy} = \dfrac{1}{3(x^2 - 4)}$ **23.** $\dfrac{dx}{dy} = -\dfrac{y}{\sqrt{4 - y^2}}$

25. No inverse function

27. (a) Yes; $\dfrac{dx}{dy} = \dfrac{1}{\sqrt{1 - y^2}}$ (b) No; (c) Yes; $\dfrac{dx}{dy} = -\dfrac{1}{\sqrt{1 - y^2}}$

29. 0 **31.** $\frac{2}{9}(x + 2)^3$ **33.** 2

35. $-\dfrac{4}{x^3}$ **37.** (a) Yes (b) No (c) Yes

39. $b = -1, \quad a \neq b$

41. (a) g integrable and $g(t)$ always of one sign (b) g continuous and $g(t) \neq 0$ (c) $\dfrac{dx}{dy} = \dfrac{1}{g(x)}$

43. $4b^2 - 12ac < 0$

8.2 Answers

1. $\dfrac{1}{x+4}$ 3. $\dfrac{4x^3}{x^4+1}$

5. $\ln(x^2+9) + \dfrac{2x^2}{x^2+9} - \dfrac{3}{x^2}$ 7. $\dfrac{1-x^2}{x(x^2+1)}$

9. $\dfrac{1}{2(x+\sqrt{x})}$ 11. $\ln|x|$ 13. $\dfrac{1}{x|\ln x|}$

15. $x^{-1}\cos(\ln|x|)$ 17. $\ln|x-3| + c$

19. $\ln(x^2 - 3x + 7) + c$

21. $2\ln(1+\sqrt{x}) + c$ 23. $\frac{1}{2}\ln|\sin 2x| + c$

25. $\ln 5$ 27. $2\ln 4$ 29. $-\frac{1}{2}\ln\frac{7}{3}$

31. $\dfrac{\ln 3}{2}$ 33. $\dfrac{(-1)^{n-1}(n-1)!}{x^n}$ 35. $-\dfrac{(n-1)!}{(1-x)^n}$

37. Minimum is 1 at $x = 1$; no maximum

39. Maximum is 0 at $x = \dfrac{\pi}{2}$; no minimum

41. $\pi \ln 9$ 43. $I = -\displaystyle\int_{-1}^{2}\frac{dx}{x+2}$

49. $x(x^2+x+1)\sin 2x \times \left[\dfrac{1}{x} + \dfrac{2x+1}{x^2+x+1} + 2\cot 2x\right]$

8.3 Answers

1. $3e^{3x}$ 3. $(1-x)e^{-x} - 2\sin 2x$

5. $-x(1-x^2)^{-1/2}\exp\sqrt{1-x^2}$

7. $\dfrac{e^x + e^{-x}}{2}$ 9. $\dfrac{-e^{-x}}{1+e^{-x}}$ 11. $-x^{-2}e^{1/x}$

13. $\left(\dfrac{1}{2} - \dfrac{1}{x}\right)e^{2/x}$ 15. $-e^{-x}\sin 2x + 2e^{-x}\cos 2x$

17. $\frac{1}{2}e^{2x} + c$ 19. $-\frac{1}{3}e^{-x^3} + c$

21. $\dfrac{(1+e^x)^4}{4} + c$

23. $-\sin(e^{-x}) + c$ 25. $\ln(e^x + e^{-x}) + c$

27. $\dfrac{e^9 - 1}{3}$ 29. $\dfrac{1-e^{-2}}{4}$ 31. 1248 33. $e - 1$

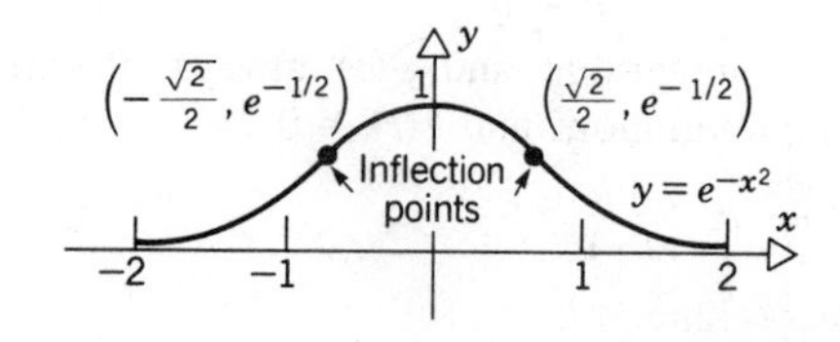

37.

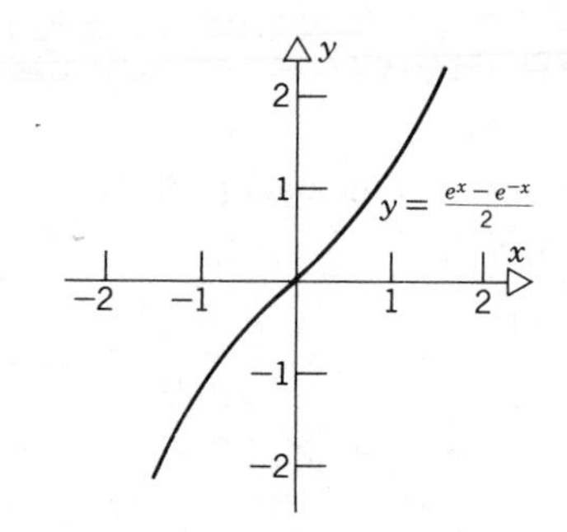

39.

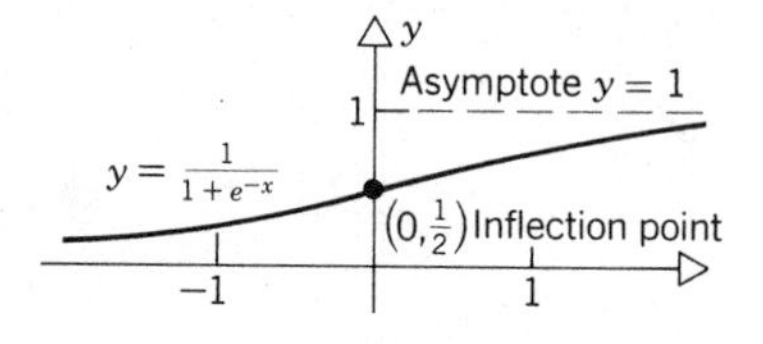

41.

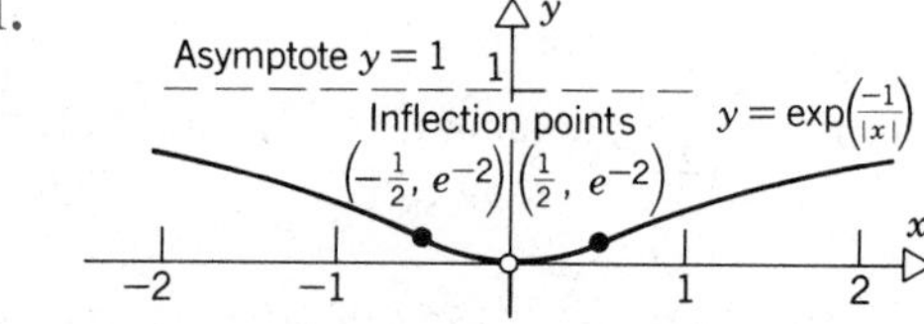

43. $\frac{9}{10}$ square units

45. $f(t) = f(0)e^{-kt} + \dfrac{R}{k}(1 - e^{-kt})$; $L = \dfrac{R}{k}$

49. (c) Because exp is a continuous function.

51. e^r 53. 1 55. e^{rt}

8.4 Answers

1. $Q(t) = 4e^{2t}$ 3. $u(t) = -e^{-3(t-2)}$

5. $Q(t) = -\frac{7}{2} + \frac{13}{2}e^{2t}$ 7. $u(t) = t^2e^{t/2} - 2e^{t/2}$

9. $Q(t) = Q_0e^{r(t-t_0)}$

11. $\dfrac{-11.7\ln 2}{\ln(2/3)} \cong 20.0$ days

13. (a) $Q(t) = 35.36 + 64.64\exp(-0.02828t)$ mg
(b) $Q_l = 35.36$ mg (c) 171.9 days

15. (a) $P(t) = 10^4\exp\left[(\ln 2)\dfrac{t}{2}\right] \cong 10^4\exp(0.3466t)$
(b) 81.92×10^6 (c) $P(t) \to \infty$

17. (b) $c = (u_0 - T)\exp(-kt_0)$ (c) T

19. (a) $T = \dfrac{\ln 2}{r}$ (b) 8.67%

21. (a) $\dfrac{1}{r}\ln\dfrac{1}{1-10r}$ (b) 17.20 years; 25.58 years

8.5 Answers

1. $5^x \ln 5$
3. $x^{-1} 10^{\ln x} \ln 10$
5. $-2x^{-3} - 2^{-x} \ln 2$
7. $2x$
9. $\dfrac{2x + 1}{(x^2 + x + 1)\ln 10}$
11. $4x$
13. $\dfrac{4x}{(x^2 + 1)\ln 10}$
15. $\dfrac{x^{\sqrt{x}}(\ln x + 2)}{2\sqrt{x}}$
17. $x^{\sin x}[(\cos x)(\ln x) + x^{-1} \sin x]$
19. $x^{1+x^2}(2x \ln x + x^{-1} + x)$
21. $\dfrac{3^x}{\ln 3} + c$
23. $\dfrac{4^{2x}}{\ln 16} + c$
25. $-\dfrac{10^{\cos 2x}}{4 \ln 10} + c$
27. $\dfrac{14}{\ln 2}$
29. $\dfrac{9999}{100 \ln 100}$
31. $\dfrac{20}{\ln 5}$
37. $\dfrac{4}{5 \ln 5}$
39. $f^{-1}(x) = \log_2 x - 3; \quad x > 0$
41. $f^{-1}(x) = 2a^x + 1; \quad -\infty < x < \infty$

8.6 Answers

1. $\dfrac{\pi}{4}$
3. $\dfrac{\pi}{4}$
5. $-\dfrac{\pi}{2}$
7. $\dfrac{\pi}{6}$
9. $\dfrac{\sqrt{3}}{2}$
11. $\dfrac{1}{\sqrt{1 + x^2}}$
19. $\dfrac{3x^2}{\sqrt{9 - x^6}}$
21. $\dfrac{4 \arctan (x/2)}{4 + x^2}$
23. $\dfrac{2x}{\sqrt{2x^2 - x^4}}$
25. $\dfrac{2}{x\sqrt{x^4 - 1}}$
27. $1 - \dfrac{x \arcsin x}{\sqrt{1 - x^2}}$
29. $\dfrac{2 \sec^2 x}{1 + 4 \tan^2 x}$
31. $\dfrac{2}{\sqrt{1 - 4x^2}}$
33. $\dfrac{2}{(1 + 4x^2)\arctan 2x}$
35. $\arcsin \dfrac{x}{3} + c$
37. $\frac{1}{2} \arcsin 2x + c$
39. $\dfrac{1}{6} \arctan \dfrac{3x}{2} + c$
41. $\frac{1}{2} \operatorname{arcsec} \dfrac{x}{2} + c$
43. $\arcsin(x + 1) + c$
45. $\frac{1}{3} (\arcsin x)^3 + c$
51. (b)

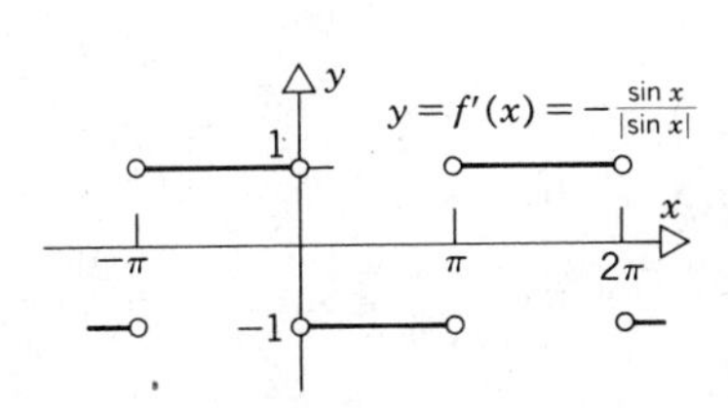

(c) Yes

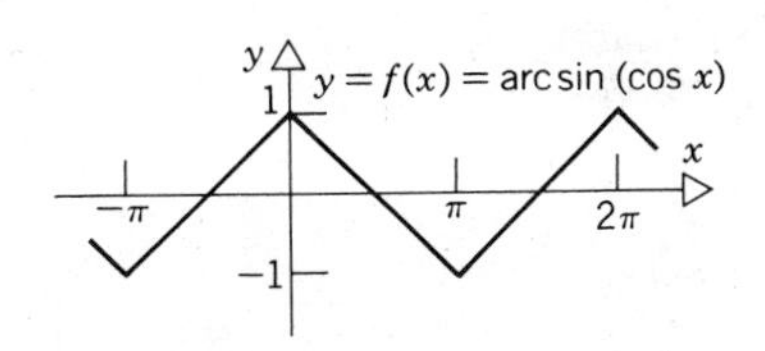

53. (a) $\arcsin \alpha$ (b) $\dfrac{\pi}{2}$
55. $\sqrt{ab}$ feet

8.7 Answers

7. $3 \cosh 3x$
9. $\cosh x + x \sinh x$
11. $(\cosh x + \sinh^2 x)\exp(\cosh x)$
13. $\dfrac{2 \sinh x - \sinh^3 x}{(4 + \sinh^2 x)^2}$
15. $(4 + x^2)^{-1/2}$
17. $\frac{1}{2} \cosh 2x + c$
19. $\frac{1}{3} \sinh^3 x + c$
21. $\frac{1}{2} \ln|\sinh 2x| + c$
23. $\frac{1}{2} \operatorname{arccosh} 2x + c, x \geq \frac{1}{2}$
25. $\operatorname{arcsinh} \dfrac{x}{4} + c$
27. $\sinh(\ln 3) - \sinh(\ln 2) = \frac{7}{12}$
29. $3 \ln \frac{5}{3}$
31. $\dfrac{\operatorname{arccosh} 6 - \operatorname{arccosh} 3}{3}$
37. (a) $y = \operatorname{arctanh} x$ means $x = \tanh y$ with no restriction on y. Domain is $(-1, 1)$; range is $(-\infty, \infty)$.

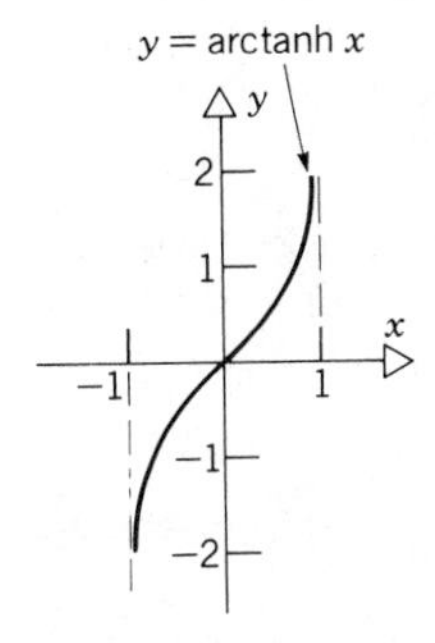

(b) $\dfrac{1}{1 - x^2}$

39. (f) No

Answers to Review Problems

1. $(f^{-1})'(x) = 2x, \quad f(x) = \sqrt{x}, \quad f^{-1}(x) = x^2$

3. $(f^{-1})'(x) = \dfrac{1}{x^2 + 1}$, $\quad f(x) = \tan x$, $f^{-1}(x) = \arctan x$

5. $x \geq 0$; $\quad f^{-1}(x) = \sqrt{x}$

7. $x \geq -\dfrac{b}{2a}$;

$$f^{-1}(x) = \frac{1}{2a}[-b \pm \sqrt{b^2 - 4a(c - x)}]$$

9. No inverse

11. $f^{-1}(x) = \dfrac{1}{2} \ln \dfrac{x + 1}{x - 1}$, $\quad x < -1$ or $x > 1$

13. $2^{x \ln x}(\ln 2)(\ln x + 1)$

15. $-\dfrac{e^{-x}}{\sqrt{1 - e^{-2x}}}$

17. $\dfrac{1}{\ln a}$

19. $\dfrac{\coth x}{\sqrt{(\ln \sinh x)^2 + 1}}$

21. $\dfrac{1}{\cosh x}$

23. $\dfrac{1}{2a} e^{(ax+b)^2} + c$

25. $\dfrac{1}{\ln 2} 2^{\tan x} + c$

27. $\dfrac{\pi^{\sin x}}{\ln \pi} + c$

29. $-\dfrac{1}{\pi \ln 2}$

31. $\frac{1}{2} \operatorname{arcsec}\left(\dfrac{x - 4}{2}\right) + c$

33. $2 \arctan \sqrt{x} + c$

35. 1

37. 0

39. 1

41. (b) $x < \arctan x < \dfrac{x}{1 + x^2}$

43. $x = 1$, $\quad e^2$

45. $r = \operatorname{arcsinh} 1 \cong 0.8814$; $\quad$ no; $\quad r$ is a double root

47. $\log_{10} e = \dfrac{1}{\ln 10}$

49. Critical point at $x = 1$; decreasing on $0 < x \leq 1$, increasing on $1 \leq x < \infty$; concave up on $x > 0$; global min at $x = 1$.

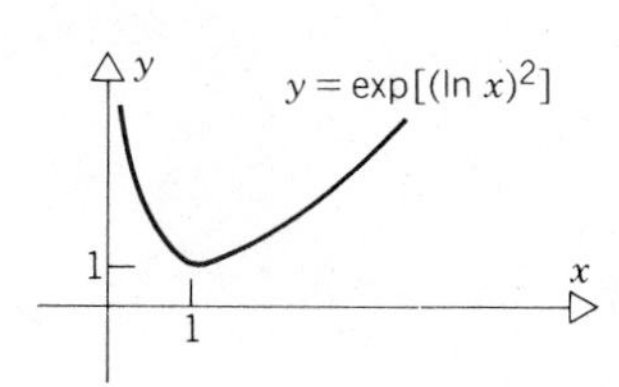

51. Critical points at $x = 0, \pi/2, 3\pi/2, 2\pi$; increasing on $0 \leq x \leq \pi/2$ and $3\pi/2 \leq x \leq 2\pi$, decreasing on $\pi/2 \leq x \leq 3\pi/2$; local min at $x = 0$, local max at $x = 2\pi$, global max at $x = \pi/2$, global min at $x = 3\pi/2$. There are inflection points at $x = \alpha$ and $x = \pi - \alpha$, where $\alpha = \arcsin[(\sqrt{5} - 1)/2] \cong 0.66624$. The function is concave up on $0 \leq x \leq \alpha$ and $\pi - \alpha \leq x \leq 2\pi$; it is concave down on $\alpha \leq x \leq \pi - \alpha$.

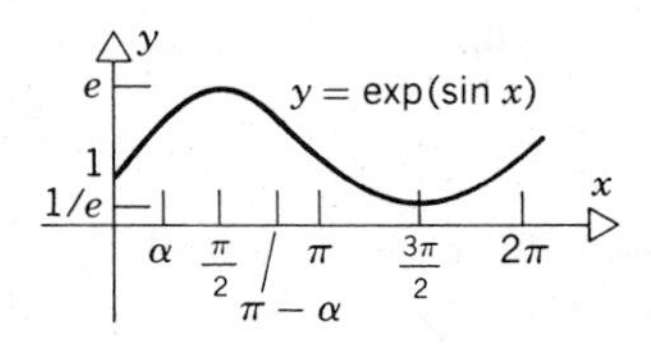

53. Increasing on $-\infty < x < \infty$; $\quad$ concave up on $-\infty < x < \infty$. There are no critical points.

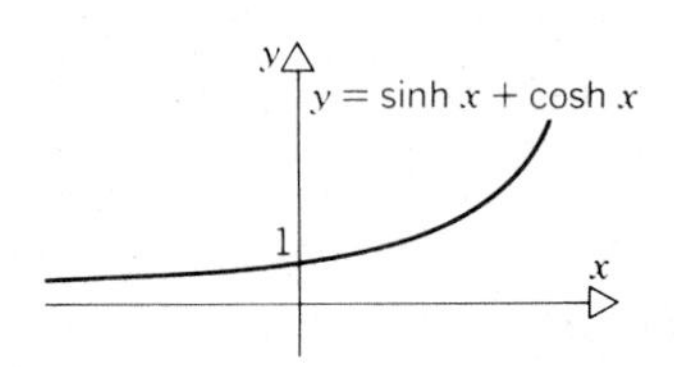

CHAPTER NINE

9.1 Answers

1. $\frac{3}{8} \ln(1 + 4x^2) + c$

3. $2\sqrt{1 + e^x} + c$

5. $\dfrac{(\ln x)^4}{4} + c$

7. $\ln|\sin x| + c$

9. $2 \arctan(\sin x) + c$

11. $\frac{1}{2} \arcsin e^{2x} + c$

13. $2 - \ln 2 \cong 1.3069$

15. $\dfrac{2}{5} \sinh^5\left(\dfrac{x}{2}\right) + c$

17. $\sin(\ln x) + c$

19. $\frac{1}{8}(7 + 2x^3)^{4/3} + c$

21. $-\ln(\ln 2) \cong 0.3665$

23. $\frac{1}{2} \ln(\cosh 2x) + c$

25. $\dfrac{64(\sqrt{2} + 1)}{15} \cong 10.30$

27. $\arcsin\left(\dfrac{x - 2}{2}\right) + c$

29. $2 \ln(x^2 + 4x + 29) + 2 \arctan\left(\dfrac{x + 2}{5}\right) + c$

31. $\frac{1}{2} \operatorname{arcsinh}(2x + 4) + c$

33. $\frac{1}{2}\left(\arctan 2 - \dfrac{\pi}{4}\right) \cong 0.1609$

9.2 Answers

1. $x \sin x + \cos x + c$

3. $-\dfrac{1}{\pi} x^2 \cos \pi x + \dfrac{2}{\pi^2} x \sin \pi x + \dfrac{2}{\pi^3} \cos \pi x + c$

5. $\dfrac{1}{\pi^2} \cong 0.1013$

7. $\frac{2}{5}(e^{\pi/2} + 1) \cong 2.3242$

9. $2 - \frac{10}{e^2} \cong 0.6466$

11. $x \arctan x - \frac{1}{2}\ln(1 + x^2) + c$

13. $\frac{x\sin(\ln x) - x\cos(\ln x)}{2} + c$

15. $-2x^2(1 - x)^{1/2} - \frac{8}{3}x(1 - x)^{3/2} - \frac{16}{15}(1 - x)^{5/2} + c$

17. $\frac{2^x(x\ln 2 - 1)}{(\ln 2)^2} + c$

19. $2\sqrt{3}\ln 3 - 4\sqrt{3} + 4 \cong 0.8775$

21. $\frac{10\ln 10 - 9\ln 9 - 1}{2}$

23. $\frac{2}{9}x^3(x^3 - 1)^{3/2} - \frac{4}{45}(x^3 - 1)^{5/2} + c$

41. $(x + 1)\ln(x + 1) - x + c$

9.3 Answers

1. $\frac{\pi + 2}{8}$

3. $-\frac{1}{3}\cos^3 x + \frac{1}{5}\cos^5 x + c$

5. $-\frac{3}{4}\cos^{4/3} x + \frac{3}{10}\cos^{10/3} x + c$

7. $\frac{1}{16}x - \frac{1}{64}\sin 4x + \frac{1}{48}\sin^3 2x + c$

9. $\frac{8}{21}$

11. $2 - \sqrt{2}$

13. $\frac{3\pi}{8}$

15. $-\sin x - \csc x + c$

17. $\frac{1}{4}\sin 2x - \frac{1}{8}\sin 4x + c$

19. $x - \frac{1}{2}\sin 2x + \frac{1}{8}\sin 4x - \frac{1}{6}\sin 6x + \frac{1}{16}\sin 8x + c$

21. $\frac{1}{2}\cos x - \frac{1}{10}\cos 5x + c$

23. $-\frac{1}{16}\cos 4x - \frac{1}{32}\cos 8x + \frac{1}{48}\cos 12x + c$

25. $-\frac{(1 - x^2)^{n+1}}{2(n + 1)} + \frac{(1 - x^2)^{n+2}}{2(n + 2)} + c$

27. $\frac{\pi}{2}$

29. $5\pi^2 a^3$

31. $\frac{64\pi a^2}{3}$

9.4 Answers

1. $\frac{1}{a}\ln|\sec ax + \tan ax| + c$

3. $\frac{1}{3}\tan^3 x + c$

5. $\frac{1}{3}\tan^3 x - \tan x + x + c$

7. $\frac{1}{10}\sec^5 2x - \frac{1}{6}\sec^3 2x + c$

9. $\frac{1}{5}\sec^5 x + c$

11. $-\frac{1}{7}\csc^7 x + \frac{2}{5}\csc^5 x - \frac{1}{3}\csc x + c$

13. $-\frac{1}{2}\csc x \cot x - \frac{1}{2}\ln|\csc x - \cot x| + c$

15. $\frac{2}{3}\tan^{3/2} x + \frac{2}{7}\tan^{7/2} x + c$

17. $\frac{2}{5}\sec^{5/2} x - 2\sec^{1/2} x + c$

23. $-\frac{1}{2}\cot^2 x - \ln|\sin x| + c$

25. $-\frac{1}{4}\cot^4 x + \frac{1}{2}\cot^2 x + \ln|\sin x| + c$

27. $-\frac{1}{2}\csc x \cot x + \frac{1}{2}\ln|\csc x - \cot x| + c$

29. $-\frac{1}{4}\csc^3 x \cot x - \frac{3}{8}\csc x \cot x + \frac{3}{8}\ln|\csc x - \cot x| + c$

31. $-\cot\frac{\theta}{2} + c$

9.5 Answers

1. $\frac{1}{2}(\arcsin x + x\sqrt{1 - x^2}) + c$

3. $\frac{x}{5\sqrt{2x^2 + 5}} + c$

5. $-\frac{x}{4\sqrt{x^2 - 4}} + c$

7. $2\ln\left|\frac{2 - \sqrt{4 - x^2}}{x}\right| + \sqrt{4 - x^2} + c$

9. $3\sqrt{x^2 - 9} - \ln(x + \sqrt{x^2 - 9}) + c$

11. $\frac{1}{\sqrt{3}}\ln\left|\frac{\sqrt{3 + x^2} - \sqrt{3}}{x}\right| + c$

13. $2\sqrt{15} - \sqrt{3} + \frac{1}{2}\ln\left(\frac{4 + \sqrt{15}}{2 + \sqrt{3}}\right) \cong 6.387$

15. $\frac{3\sqrt{2}}{4} - \frac{5\sqrt{6}}{12} \cong 0.04004$

17. $\sqrt{x^2 - 3} - \sqrt{3}\,\mathrm{arcsec}\frac{x}{\sqrt{3}} + c$

19. $\ln 3 \cong 1.099$

33. πab

9.6 Answers

1. $\ln\left|\frac{x}{x + 1}\right| + c$

3. $\ln\left|\frac{x + 2}{x - 2}\right| + c$

5. $\ln\left|\frac{(x + 1)(x + 3)}{x + 2}\right| + c$

7. $-\frac{1}{2(x^2 - 1)} + c$

9. $x + 3\ln|x| - 2\arctan x + c$

11. $2\ln|x| - 4\arctan(x + 1) + c$

13. $\frac{1}{4}\ln\left|\frac{x - 1}{x + 1}\right| - \frac{1}{2}\arctan x + c$

15. $\frac{2}{27}\ln\left|\frac{x + 2}{x - 1}\right| - \frac{1}{9}(x - 1)^{-1} - \frac{1}{9}(x + 2)^{-1} + c$

9.7 Answers

1. $\frac{x}{18(x^2 + 9)} + \frac{1}{54}\arctan\frac{x}{3} + c$

3. $\frac{x}{16(x^2 + 4)^2} + \frac{3x}{128(x^2 + 4)} + \frac{3}{256}\arctan\frac{x}{2} + c$

5. $\frac{x - 2}{8(x^2 - 4x + 8)} + \frac{1}{16}\arctan\frac{x - 2}{2} + c$

7. $\dfrac{x-1}{2(x^2-2x+2)} + \dfrac{1}{2}\arctan(x-1) + c$

9. $\dfrac{x}{96(x^2+3)^2} + \dfrac{x}{192(x^2+3)} + \dfrac{1}{192\sqrt{3}}\arctan\dfrac{x}{\sqrt{3}} + c$

11. $\dfrac{1}{8(x^2+4)} + \dfrac{1}{16}\ln|x| - \dfrac{1}{32}\ln(x^2+4) + c$

13. $-\ln|x-1| - \dfrac{4}{x-1} + \dfrac{1}{2}\ln(x^2+x+1) - \dfrac{1}{\sqrt{3}}\arctan\left(\dfrac{2x+1}{\sqrt{3}}\right) + c$

Answers to Review Problems

1. $\arctan(\ln x) + c$ 3. $\frac{1}{2}(\ln x + 1)^2 + c$ 5. 0

7. $\frac{1}{16}\arcsin x + \frac{1}{6}x\sqrt{1-x^2}\,(x^2-\frac{3}{4})(x^2+\frac{1}{2}) + c$

9. $\ln|\sin x^2 + \cos x^2| + c$

11. $\dfrac{x^2}{3}(1+x^2)^{3/2} - \dfrac{2}{15}(1+x^2)^{5/2} + c$

13. $\frac{1}{2}(x-2)^2 - \frac{1}{3}\ln[x^2(x+3)] + c$

15. $\dfrac{4}{\pi^2-4}\left(2\cos\dfrac{x}{2}\cos\dfrac{\pi x}{4} + \pi\sin\dfrac{x}{2}\sin\dfrac{\pi x}{4}\right) + c$

17. $\frac{1}{3}\arctan(\sec^3 x) + c$

19. $\frac{3}{8}\cos x + \frac{1}{24}\cos 3x - \frac{3}{40}\cos 5x + \frac{1}{72}\cos 9x + c$

21. $\dfrac{2x^2-1}{x} + \ln\left(\dfrac{|x-2|}{(x^2-4)^2}\right) + c$

23. $\dfrac{\pi}{4}\sqrt{\dfrac{\pi^2}{4}+4} - 2\,\text{arcsinh}\,\dfrac{\pi}{4} \cong 0.5549$

25. $\frac{1}{12}\ln\left|8x^{-3/2} - x^{-3/2}\sqrt{64-x^3}\right| + c$

27. $\arctan(2x+1) + c$ 29. $\ln 2$

31. $\ln|(x+1)(\sec x + \tan x)| + c$

33. $\cos x + \sec x + c$ 35. $\pi/2$

37. $\frac{1}{2}(x^2\cosh x^2 - \sinh x^2) + c$

39. $x\tan x + \ln|\cos x| + c$

41. $\ln\left|\dfrac{x(x+2)}{x+1}\right| + c$

43. $\left(\dfrac{x-2}{2}+4\right)\sqrt{x^2-4x+1} + \dfrac{11}{2}\ln|x + \sqrt{x^2-4x+1} - 2| + c$

45. $\dfrac{1}{a}\ln|\sec(a\arctan x)| + c$

47. $2\ln|x+1| + \dfrac{8}{3}\ln|x-1| - \dfrac{3}{x-1} - \dfrac{4}{3}\ln(x^2+2) - \dfrac{4\sqrt{2}}{3}\arctan\dfrac{x}{\sqrt{2}} + c$

49. $\ln[e^{-x}(1-\sqrt{1-e^{2x}})] + \sqrt{1-e^{2x}} + \dfrac{1}{3}(1-e^{2x})^{3/2} + c$

51. $\dfrac{9}{8}\sqrt{2} - \dfrac{3}{16}\ln(3+2\sqrt{2})$

53. $\dfrac{1}{\sqrt{2}}\left[-\frac{1}{2}\csc\dfrac{\theta}{2}\cot\dfrac{\theta}{2} + \frac{1}{2}\ln\left|\csc\dfrac{\theta}{2} - \cot\dfrac{\theta}{2}\right|\right] + c$

55. 0

57. $\frac{1}{2}\ln\left|\dfrac{\tan(x/2)+1}{\tan(x/2)-1}\right| - \dfrac{\tan(x/2)}{[\tan(x/2)+1]^2} + c$

59. $\dfrac{2}{\sqrt{a^2-1}}\arctan\left(\sqrt{\dfrac{a-1}{a+1}}\tan\dfrac{x}{2}\right) + c, \quad a > 1;$

$\dfrac{1}{\sqrt{1-a^2}}\ln\left|\left(\alpha - \tan\dfrac{x}{2}\right)\Big/\left(\alpha + \tan\dfrac{x}{2}\right)\right| + c,$ $0 < a < 1;$

$\alpha = \sqrt{\dfrac{1+a}{1-a}}$

CHAPTER TEN

10.1 Answers

1. $y = ce^{2t} + \frac{1}{3}t^3e^{2t}$

3. $y = ce^{-3t} + \dfrac{t}{3} - \dfrac{1}{9} + \dfrac{1}{2}e^{-t}$

5. $y = ct^{-1} + \frac{3}{4}t^{-1}\cos 2t + \frac{3}{2}\sin 2t$

7. $y = \dfrac{c + te^t - e^t}{t^2}$ 9. $y = \dfrac{c - \cos t}{t^3}$

11. $y = cte^{-t} + te^{-t}\ln t$ 13. $y = t^2e^t + e^t$

15. $y = \dfrac{\sin t}{t^2}$ 17. $y = \dfrac{18 + 3t^2 + 2t^3}{6(t+1)^2}$

21. One hour 23. $Q(t) = 30 - 10e^{-t/60}$

25. (a) $Q(t) = 50\gamma - 50\gamma e^{-t/10}$

(b) $\gamma = \dfrac{1}{10(1-e^{-6})} \cong 0.10025$

29. $y = \left(\dfrac{\sigma}{\epsilon} + ce^{-2\epsilon t}\right)^{-1/2}$

10.2 Answers

1. $3y^2 - 2t^3 = c$

3. $y^2 + \ln|y| - \sin 2t = c;$ also $y = 0$

5. $3y^2 - 2\ln|1 + t^3| = c$

7. $\arcsin y - \ln|t| = c$

9. $y = \sqrt{2\ln(1 + t^2) + 4}$

11. $y = (3 - 2\sqrt{1 + t^2})^{-1/2}$

13. $y = -1 - \sqrt{t^2 - t - 2}$

15. $y = -\ln\left(\dfrac{-6te^{3t} + 2e^{3t} + 7}{9}\right)$

17. $y = -(2t + 1)^{3/2}$

19. (a) $\tau = \dfrac{1}{r}\ln 4 \cong \dfrac{1.3863}{r}$; 55.452 years

(b) $T = \dfrac{1}{r}\ln\left[\dfrac{\beta(1 - \alpha)}{(1 - \beta)\alpha}\right]$; 175.78 years

21. $y = \dfrac{a}{c}t + \left(\dfrac{bc - ad}{c^2}\right)\ln|ct + d| + k$

23. (a) $x = x_0 e^{-\beta t}$

(b) $y = y_0 \exp\left[-\dfrac{\alpha x_0 (1 - e^{-\beta t})}{\beta}\right]$

(c) $y_0 \exp\left(-\dfrac{\alpha x_0}{\beta}\right)$

25. $x = \dfrac{pq[e^{\alpha(p-q)t} - 1}{pe^{\alpha(p-q)t} - q}$

27. (a) 50.4 m (b) 5.25 sec

29. (a) $x_m = \dfrac{v_0^2}{2g}$ (b) $t_m = \dfrac{v_0}{g}$ (c) $t = \dfrac{2v_0}{g}$

10.3 Answers

1. (a) 1.1, 1.22, 1.364, 1.5368
(b) 1.105, 1.23205, 1.38578, 1.57179
(c) 1.11, 1.2442, 1.40792, 1.60767
(d) 1.24587, 1.61263
(e) $y = \dfrac{1 + e^{2t}}{2}$; 1.11070, 1.24591, 1.41106, 1.61277

3. (a) 1.1, 1.222, 1.37533, 1.57348
(b) 1.111, 1.25153, 1.43606, 1.68801
(c) 1.25299, 1.69592

5. (a) 3.2, 3.40736, 3.62201, 3.84387
(b) 3.20368, 3.41478, 3.63320, 3.85888
(c) 3.41488, 3.85908

7. (a) 2.2, 2.42993, 2.69426, 2.99840
(b) 2.21497, 2.46469, 2.75525, 3.09420
(c) 2.46665, 3.09994

9. (a) 7.53999 (b) 7.70957
(c) 7.86623 (d) 7.88313
(e) 7.88889 (f) 7.88905

11. (a) 5.35218 (b) 5.35712
(c) 5.36194 (d) 5.36203
(e) 5.36206 (f) 5.36206

13. (a) 7.88906 (b) 7.88906

15. (a) 5.36206 (b) 5.36206

Answers to Review Problems

1. $y = (e^x + 1)/(e^x + e^{-x})$

3. $y = \frac{1}{2}[\tan x + \cos x + \cos x \ln|\sec x + \tan x|]$

5. $y = e^{e^t}/(e - t)$

7. $y = 0$

9. $y = -\sqrt{5e^{2x} - 4}$

11. (a) 0.705844
(b) 2.360865, 2.362464
(c) 2.362773, 2.362775

13. (a) 3.636682
(b) 6.200323, 6.200680
(c) 6.200744, 6.200744

15. 1.548281, 2.319693, 3.324278, 4.508531, 5.751209, 6.905679

17. $P(t) = P_0 e^{-kt}$, $k = \dfrac{\ln 2}{13.2} \cong 0.0525$ 29.6 mg

19. $y = 1 - e^{-x}$

21. $y = \dfrac{4}{3}(x + 1)^{3/2} - \dfrac{1}{3}$

CHAPTER ELEVEN

11.1 Answers

1. 2

3. 1

5. 0 (not indeterminate)

7. -1

9. $-\infty$ (not indeterminate)

11. 0

13. ∞

15. 0

17. ∞

19. ∞

21. 0

23. 0

25. $\dfrac{b}{a}$

27. 12

29. 2

31. $-\frac{1}{3}$

33. 1

35. 0

37. 0

39. $a = \pm 2$, $b = 1$

41. (a) $\dfrac{3}{4}$ (b) 3

11.2 Answers

1. 0

3. 0

5. 1

7. 0

9. 5

11. 1

13. 0

15. 0

17. 1

19. 0

21. 1

23. $\dfrac{1}{\sqrt{e}}$

25. ∞ if $0 < q < 1$; e if $q = 1$; 1 if $q > 1$

27. 1 if $q < p$; e if $q = p$; ∞ if $q > p$

29. 1

31. $\dfrac{1}{e}$

33. ∞

35. $\dfrac{1}{e}$

11.3 Answers

1. Diverges
3. Converges; $\frac{3}{2}$
5. Converges; 6
7. Converges; $-\dfrac{1}{2}$
9. Converges; $\dfrac{\pi}{2}$
11. Diverges
13. Diverges
15. Converges; 1
17. Diverges
19. Diverges
21. Diverges
23. Diverges
27. $n!$
29. $\dfrac{b}{a^2 + b^2}$
31. $\alpha = 2$; $\frac{1}{2} \ln \frac{25}{8}$
33. (a) $p > 1$ (b) $p > \frac{1}{2}$ (c) $p > 2$ (d) $1 < p \le 2$ (e) $\frac{1}{2} < p \le 1$
35. (a) 0 (b) Diverges (d) ∞; diverges (e) π; converges

11.4 Answers

1. Converges
3. Diverges
5. Converges
7. Diverges
9. Converges
11. Converges
13. Converges
15. Converges
17. Converges
19. Converges
21. Integral exists (not improper)
23. Converges only for $1 < p < 2$
25. Converges only for $-1 < p < \frac{3}{2}$
27. Integral exists (not improper)
29. Converges
31. Diverges
33. Converges only for $3q - p > 1$
35. $p > 1$

Answers to Review Problems

1. 0
3. Limit does not exist; right-hand limit is 2, left-hand limit is -2
5. 0
7. 2
9. 0
11. 0
13. 0
15. 1
17. 0
19. 0
21. 1
23. 1
25. $e^{-3/4}$
27. 0 if $p \le 0$; ∞ if $p > 0$
29. $e^{-1/2}$
31. Converges; $-\frac{3}{4}(16)^{1/3}$
33. Diverges
35. Converges; 0
37. Converges; 1
39. Converges for $p < 1$, diverges for $p \ge 1$
41. Converges
43. Converges
45. Diverges
47. Diverges
49. Diverges

CHAPTER TWELVE

12.1 Answers

1. (a) $3, \frac{5}{2}, \frac{7}{3}, \frac{9}{4}, \frac{11}{5}$ (b) Bounded (c) Monotone decreasing (d) $L = 2$
3. (a) $0, \frac{1}{3}, \frac{1}{2}, \frac{3}{5}, \frac{2}{3}$ (b) Bounded (c) Monotone increasing (d) $L = 1$
5. (a) $1, -(2)^{1/3}, (3)^{1/3}, -(4)^{1/3}, (5)^{1/3}$ (b) Unbounded (c) Not monotone (d) Diverges
7. (a) $-\frac{3}{2}, -\frac{2}{5}, -\frac{1}{10}, 0, \frac{1}{26}$ (b) Bounded (c) Monotone decreasing for $n \ge 8$ (d) $L = 0$
9. (a) $\frac{1}{2}, \frac{1}{4}, \frac{2}{9}, \frac{1}{4}, \frac{8}{25}$ (b) Unbounded (c) Monotone increasing for $n \ge 3$ (d) Diverges to ∞
11. $\frac{1}{2}$
13. Diverges to ∞
15. 0
17. Diverges to ∞
19. Diverges to ∞
21. 0
23. 0
25. $\frac{1}{2}$
27. 0
29. 1
31. Diverges to ∞
33. $a_n = \dfrac{(-1)^{n+1}}{n}$, for example
35. $a_n = \sqrt{n}$, for example
37. $a_n = \dfrac{1}{n}$, for example
39. $L = 2$; $N \ge \dfrac{1}{\epsilon}$
41. $L = 1$; $N \ge \dfrac{2}{\epsilon} - 1$
43. $L = 0$; $N \ge \dfrac{\ln(1/\epsilon)}{\ln 2}$
45. (c) $\dfrac{3 + \sqrt{5}}{2} \cong 2.618$

12.2 Answers

7. Converges, $s = 1$
9. Diverges
11. Converges, $s = \dfrac{1}{\sqrt{2} + 1}$
13. Converges, $s = 1$
15. Converges, $s = \frac{1}{4}$
17. Diverges
19. Converges, $s = 2$
21. Converges for $1 < \alpha < 3$, $s = \dfrac{1}{3 - \alpha}$
23. Converges for $1 < \alpha < 2$, $s = \dfrac{1}{4 - 2\alpha}$
25. $N = 8$
27. $N = 10{,}000$
29. $\dfrac{(1 + \alpha)h}{1 - \alpha}$
31. $\frac{5}{9}$
33. $\frac{2}{5}$
35. $\frac{9}{11}$
37. $a_k = \dfrac{2k + 1}{k^2(k + 1)^2}$

12.3 Answers

1. Converges
3. Diverges
5. Diverges
7. Diverges
9. Diverges
11. Converges
13. Converges
15. Converges for $q - 2p > 1$; diverges for $q - 2p \leq 1$
17. Converges
19. Diverges
23. (b) Consider $a_k = 1/k$ and $a_k = 1/k^2$, for example.

12.4 Answers

1. Converges
3. Diverges
5. Converges
7. Converges
9. Diverges
11. Converges
13. Converges
15. Diverges
17. Diverges for $1 < a \leq e$, converges for $a > e$
19. Diverges
21. Converges
23. Converges
25. 0
27. ∞
29. Converges

12.5 Answers

1. Converges
3. Diverges
5. Diverges
7. Diverges
9. Converges
11. (a) $0.0007716 < R_6 < 0.0015432$
    (b) $s_6 = 1.0811235$ (c) $s \cong 1.0822809$
    (d) $n \geq 27$
13. (b) $s_{10^4} < 10.22$, $s_{10^8} < 19.43$, $s_{10^{12}} < 28.64$
15. $\sigma_1 = 1$, $\sigma_5 = 0.6738954$, $\sigma_{10} = 0.6263832$

12.6 Answers

1. Converges
3. Diverges (Theorem 12.6.1 does not apply)
5. Converges
7. Converges
9. Diverges (Theorem 12.6.1 does not apply)
11. $n = 100$ for s_n; $n = 70$ for $\bar{s}$
13. $n = 1584$ for s_n; $n = 910$ for $\bar{s}$
15. $n \cong 8.81 \times 10^{4342}$ for s_n;
    $n \cong 2.97 \times 10^{2171}$ for $\bar{s}$
17. (b) $x = \frac{1}{3}$ (c) $n = 3$
19. Diverges

12.7 Answers

1. Converges absolutely
3. Converges conditionally
5. Converges absolutely
7. Converges absolutely
9. Converges absolutely
11. Converges absolutely
13. Converges absolutely
15. Converges conditionally
17. Converges absolutely
19. Converges absolutely for $|\alpha| < 1$; converges conditionally for $\alpha = 1$; diverges otherwise
21. Converges absolutely for $|\alpha| < 2$; diverges otherwise
23. Converges absolutely for all α

Answers to Review Problems

1. Converges
3. Converges
5. Diverges
7. Converges
9. Diverges
11. Converges
13. Diverges
15. Diverges
17. Diverges
19. Diverges
21. Converges conditionally
23. Diverges
25. Converges conditionally
27. Converges absolutely
29. Converges
31. Converges absolutely
33. Diverges
35. Converges
37. Diverges
39. Diverges
41. True
43. True; no

45. False. For example, let $a_k = (-1)^k/k^{1/3}$.

47. False. For example, let $a_k = (-1)^k/\sqrt{k}$. Statement is true if $\sum_{k=1}^{\infty} a_k$ converges absolutely.

CHAPTER THIRTEEN

13.1 Answers

1. $P_3(x) = \frac{1}{\sqrt{2}}\left[1 + \left(x - \frac{\pi}{4}\right) - \left(\frac{1}{2!}\right)\left(x - \frac{\pi}{4}\right)^2 - \left(\frac{1}{3!}\right)\left(x - \frac{\pi}{4}\right)^3\right]$

3. $P_4(x) = -1 + 2\left(x - \frac{\pi}{2}\right)^2 - \frac{2}{3}\left(x - \frac{\pi}{2}\right)^4$

5. $P_4(x) = 1 - x^2 + \frac{1}{2}x^4$

7. $P_3(x) = 1 + \frac{1}{2}x - \frac{1}{8}x^2 + \frac{1}{16}x^3$

9. $P_4(x) = 1 + \frac{1}{2}(x - 1) - \frac{1}{8}(x - 1)^2 + \frac{1}{16}(x - 1)^3 - \frac{5}{128}(x - 1)^4$

11. $P_3(x) = x + \frac{1}{6}x^3$

13. $P_{2n}(x) = 1 - \frac{x^2}{2!} + \frac{x^4}{4!} - \frac{x^6}{6!} + \cdots + \frac{(-1)^n x^{2n}}{(2n)!}$

15. $P_n(x) = -\frac{1}{a}\left[1 + \left(\frac{x}{a}\right) + \left(\frac{x}{a}\right)^2 + \cdots + \left(\frac{x}{a}\right)^n\right]$

17. $P_n(x) = -\left[x + \frac{x^2}{2} + \frac{x^3}{3} + \cdots + \frac{x^n}{n}\right]$

19. $P_n(x) = 1 + \frac{x}{2} - \frac{x^2}{8} + \frac{x^3}{16} - \cdots + \frac{(-1)^{n-1}1 \cdot 3 \cdot 5 \cdots (2n - 3)x^n}{2^n n!}$

21. $P_n(x) = \frac{1}{2} - \frac{1}{4}(x - 1) + \frac{1}{8}(x - 1)^2 - \cdots + \frac{(-1)^n}{2^{n+1}}(x - 1)^n$

23. $P_{2n}(x) = 1 - \frac{2^2}{2!}\left(x - \frac{\pi}{4}\right)^2 + \frac{2^4}{4!}\left(x - \frac{\pi}{4}\right)^4 - \cdots + \frac{(-1)^n 2^{2n}}{(2n)!}\left(x - \frac{\pi}{4}\right)^{2n}$

25. $P_{2n+1}(x) = x + \frac{x^3}{3!} + \frac{x^5}{5!} + \cdots + \frac{x^{2n+1}}{(2n + 1)!}$

27. $P_4(x) = 1 - x^2 + x^4$

13.2 Answers

1. $R_4(x) = \frac{(\sin c)(x - \pi/4)^4}{4!}$

3. $R_5(x) = -2^5(\sin 2c)\,\frac{(x - \pi/2)^5}{5!}$

5. $R_5(x) = (-32c^5 + 160c^3 - 120c)e^{-c^2}\frac{x^5}{5!}$

7. $R_4(x) = -\frac{15x^4}{(16)4!(1 + c)^{7/2}}$

9. $R_4(x) = \frac{9x^4}{(16)4!(4 - c)^{5/2}}$

11. $R_{2n+1}(x) = \frac{(-1)^{n+1}(\sin c)x^{2n+1}}{(2n + 1)!}$

13. $R_{n+1}(x) = \frac{a^{n+1}x^{n+1}}{(1 - ac)^{n+2}}$

15. $R_{n+1}(x) = \frac{(-1)^{n+1}2^{n+1}(x + 1)^{n+1}}{e^{2c}(n + 1)!}$

17. $R_{2n+1}(x) = \frac{(-1)^n 2^{2n+1}(\cos 2c)(x - \pi/4)^{2n+1}}{(2n + 1)!}$

19. $R_{2n+2}(x) = \frac{(\sinh c)x^{2n+2}}{(2n + 2)!}$

21. (a) $P_1(x) = 0, \quad R_2(x) = \frac{14c^{1/3}x^2}{9}$;
$P_2(x) = 0, \quad R_3(x) = \frac{14x^3}{81c^{2/3}}$
(b) No
(c) $P_3(x) = 1 + \frac{7}{3}(x - 1) + \frac{14}{9}(x - 1)^2 + \frac{14}{81}(x - 1)^3$

23. $n = 3$; 0.7742

25. $n = 4$; 0.1823

27. $\cos x = \sum_{k=0}^{\infty}\frac{(-1)^k x^{2k}}{(2k)!}, \quad -\infty < x < \infty$

29. $\sinh x = \sum_{k=0}^{\infty}\frac{x^{2k+1}}{(2k + 1)!}, \quad -\infty < x < \infty$

31. $\cos 2x = \sum_{k=0}^{\infty}\frac{(-1)^k 2^{2k}x^{2k}}{(2k)!}, \quad -\infty < x < \infty$

33. $\cos x = \sum_{k=0}^{\infty}\frac{(-1)^{k+1}(x - \pi/2)^{2k+1}}{(2k + 1)!}, \quad -\infty < x < \infty$

35. $e^{3x} = \sum_{k=0}^{\infty}\frac{3^k x^k}{k!}, \quad -\infty < x < \infty$

13.3 Answers

1. $\rho = 1$; converges absolutely for $-1 \le x \le 1$; diverges otherwise
3. $\rho = 2$; converges absolutely for $-5 < x < -1$; diverges otherwise
5. $\rho = 4$; converges absolutely for $-2 < x < 6$; converges conditionally for $x = 6$; diverges otherwise
7. $\rho = \infty$; converges absolutely for all x
9. $\rho = 1$; converges absolutely for $-1 \le x \le 1$; diverges otherwise
11. $\rho = \infty$; converges absolutely for all x
13. $\rho = \frac{1}{2}$; converges absolutely for $0 < x < 1$; converges conditionally for $x = 0$; diverges otherwise
15. $\rho = \infty$; converges absolutely for all x
17. $\rho = \infty$; converges absolutely for all x
19. $\rho = \frac{2}{3}$; converges absolutely for $-\frac{4}{3} < x < 0$; diverges otherwise
21. (a) $\rho = e$ (b) Diverges at both endpoints

13.4 Answers

1. $f(x) = \sum_{k=0}^{\infty} \frac{(-1)^{k+1}(x-\pi)^{2k}}{(2k)!}, \quad -\infty < x < \infty$
3. $f(x) = \sum_{k=0}^{\infty} \frac{(-1)^{k+1}(x-\pi/2)^{2k+1}}{(2k+1)!}, \quad -\infty < x < \infty$
5. $f(x) = \sum_{k=0}^{\infty} (-1)^k 2^k x^k, \quad -\frac{1}{2} < x < \frac{1}{2}$
7. $f(x) = \sum_{k=0}^{\infty} [(-1)^k 2^k + 2^{-k}]x^k, \quad -\frac{1}{2} < x < \frac{1}{2}$
9. $f(x) = \sum_{k=0}^{\infty} (2^{-k} - 3)x^k, \quad -1 < x < 1$
11. $f(x) = \sum_{k=0}^{\infty} \frac{(-1)^k x^{2k}}{2^{2k}(2k)!}, \quad -\infty < x < \infty$
13. $f(x) = \sum_{k=0}^{\infty} \frac{(-1)^k 2^{2k} x^{2k}}{(2k)!}, \quad -\infty < x < \infty$
15. $f(x) = \sum_{k=0}^{\infty} \frac{(-1)^k x^{2k}}{k!}, \quad -\infty < x < \infty$
17. $f(x) = \sum_{k=0}^{\infty} \frac{3^{2k} x^{2k}}{(2k)!}, \quad -\infty < x < \infty$
19. $f(x) = \sum_{k=0}^{\infty} (-1)^{k+1}(x-3)^k, \quad 2 < x < 4$
21. $f(x) = x^2 - \frac{2}{3!}x^4 + \left[\frac{2}{5!} + \frac{1}{(3!)^2}\right]x^6 - \left[\frac{2}{7!} + \frac{2}{3!5!}\right]x^8 + \cdots$
23. $f(x) = 3x - 6x^2 + \frac{3}{2}x^3 + 5x^4 + \cdots$
25. $[f(x)]^2 = a_0^2 + 2a_0a_1x + (2a_0a_2 + a_1^2)x^2 + (2a_0a_3 + 2a_1a_2)x^3 + (2a_0a_4 + 2a_1a_3 + a_2^2)x^4 + \cdots$
27. (a) $g(x) = \sum_{k=0}^{\infty} x^{k+2}$
    (b) $h(x) = 1 + x$ converges for all x

13.5 Answers

1. $f(x) = -\sum_{k=0}^{\infty} \frac{x^{k+1}}{k+1}, \quad -1 \le x < 1$
3. $f(x) = 2\sum_{k=0}^{\infty} \frac{x^{2k+1}}{2k+1}, \quad -1 < x < 1$
5. $f(x) = \sum_{k=0}^{\infty} \frac{(-1)^k x^{2k+1}}{k!(2k+1)}, \quad -\infty < x < \infty$
7. $f(x) = \sum_{k=0}^{\infty} (-1)^k(k+2)(k+1)2^{k-1}x^k, \quad -\frac{1}{2} < x < \frac{1}{2}$
9. $f(x) = \sum_{k=0}^{\infty} \frac{(-1)^k x^{2k+1}}{2^{2k+1}(2k+1)}, \quad -2 \le x \le 2$
11. $f(x) = \sum_{k=0}^{\infty} \frac{(-1)^k x^{2k+3}}{(k+1)(2k+3)}, \quad -1 \le x \le 1$
13. $f(x) = \sum_{k=0}^{\infty} \frac{(-1)^k x^{2k}}{(2k+2)!}, \quad -\infty < x < \infty$
15. $f(x) = \sum_{k=0}^{\infty} \frac{(-1)^k x^{k+1}}{2^{k+1}(k+1)^2}, \quad -2 \le x \le 2$
17. (a) 3.017
    (b) 200 terms if a partial sum is used; 100 terms if the average of successive partial sums is used
19. $I = \sum_{k=0}^{\infty} \frac{(-1)^k}{(2k+2)!(2k+1)}$; five terms; $I \cong 0.48638538$
21. $I = \sum_{k=0}^{\infty} \frac{(-1)^k}{k!(2k+1)}$; eight terms; $I \cong 0.74682$
23. $I = \sum_{k=0}^{\infty} \frac{(-1)^k}{(k+1)^2 2^{k+1}}$; seven terms; $I \cong 0.4485$
25. (c) $a_0 = 1, \quad a_1 = \frac{1}{4}, \quad a_2 = -\frac{3}{64}$

13.6 Answers

1. $f(x) = 1 + \frac{x}{2} + \sum_{k=2}^{\infty} \frac{(-1)^{k-1} 1 \cdot 3 \cdot 5 \cdots (2k-3)}{2^k k!} x^k, \ \rho = 1$

3. $f(x) = \frac{1}{2}\left[1 + \sum_{k=1}^{\infty} \frac{(-1)^k 1 \cdot 3 \cdot 5 \cdots (2k-1)}{2^{3k} k!} x^k\right],$
$\rho = 4$

5. $f(x) = x + \sum_{k=1}^{\infty} \frac{(-1)^k 1 \cdot 3 \cdot 5 \cdots (2k-1)}{2^k k!(2k+1)} x^{2k+1}, \ \rho = 1$

7. $f(x) = \frac{x}{a} + \sum_{k=1}^{\infty} \frac{1 \cdot 3 \cdot 5 \cdots (2k-1)}{2^k k!(2k+1)} \frac{x^{2k+1}}{a^{2k+1}},$
$\rho = a$

9. $f(x) = \frac{1}{2} + \sum_{k=1}^{\infty} \frac{1 \cdot 3 \cdot 5 \cdots (4k-1)}{2^{2k+1}(2k+1)!} x^{2k},$
$\rho = 1$

11. $f(x) = 2 + \frac{x-1}{4} + \sum_{k=2}^{\infty} \frac{(-1)^{k-1} 1 \cdot 3 \cdot 5 \cdots (2k-3)}{2^{3k-1} k!} (x-1)^k,$
$\rho = 4$

13. $f(x) = 1 + \frac{2x}{3} + 2\sum_{k=2}^{\infty} \frac{(-1)^{k+1} 1 \cdot 4 \cdot 7 \cdots (3k-5)}{3^k k!} x^k, \ \rho = 1$

17. (c) $\frac{\pi}{2}\left(1 - \frac{k^2}{4} - \frac{3k^4}{64}\right)$

(d) $\frac{1 \cdot 3 \cdot 5 \cdots (2n-1)}{2 \cdot 4 \cdot 6 \cdots (2n)} \frac{\pi}{2}$

Answers to Review Problems

1. $-1 < x < 1$; converges absolutely at $x = \pm 1$.
3. $-1 - \sqrt{2} < t < -1 + \sqrt{2}$; diverges at $t = -1 \pm \sqrt{2}$
5. $-1 < x < 1$; diverges at $x = -1$, converges conditionally at $x = 1$.
7. $\rho = 0$ 9. $\rho = 1/2$ 11. $\rho = 1/2$
13. $f(x) = \sum_{k=0}^{\infty} \frac{(-1)^k x^{3k}}{2^k}, \quad -\sqrt[3]{2} < x < \sqrt[3]{2}$
15. $f(x) = 1 + 2\sum_{k=1}^{\infty} x^{2k}, \quad -1 < x < 1$
17. $f(x) = \ln(a) + \sum_{k=1}^{\infty} \frac{(-1)^{k+1}}{k} \left(\frac{b}{a}\right)^k x^k,$
$-\frac{a}{b} < x < \frac{a}{b}$
19. $f(x) = 1 + \frac{x}{3} + \sum_{k=2}^{\infty} (-1)^{k-1} \frac{2 \cdot 5 \cdots (3k-4)}{3^k k!} x^k,$
$-1 < x < 1$

21. $\sin(1) - \cos(1)\, x + \frac{2\cos(1) - \sin(1)}{2} x^2$
23. $1 + x + \frac{1}{2}x^2$ 25. $-\frac{10}{243}(c-1)^{-11/3}(x-2)^4$
27. $\frac{x^4}{4(c+1)^4}$
29. $\tanh x = x - \frac{x^3}{3} + \frac{2x^5}{15} - \frac{17x^7}{315} + \cdots$
31. $f(x) = 1 + x + 2x^2 + \cdots, \quad -1 < x < 1$
35. $y(x) = y_0(1 + x^2 + \frac{1}{2}x^4 + \cdots) = y_0 e^{x^2}$
37. $y(x) = 1 + 2x + (3/2)x^2 + (2/3)x^3 + \cdots = (1 + x)e^x$
39. $x = 2 - (1/6)\epsilon + \cdots$
41. $x = -2 + (2/9)\epsilon - (16/243)\epsilon^2 + \cdots$
43. 0.100111 45. -0.002502

CHAPTER FOURTEEN

14.1 Answers

1.

$(4, \frac{\pi}{6})$, $r = 4$, $\theta = \frac{\pi}{6}$, O, Polar axis

3.

$\theta = \pi$, $(-3, \pi)$, O, $r = -3$, Polar axis

5.

$(-4, \frac{7\pi}{6})$, $r = -4$, $\theta = \frac{7\pi}{6}$, O, Polar axis

7.

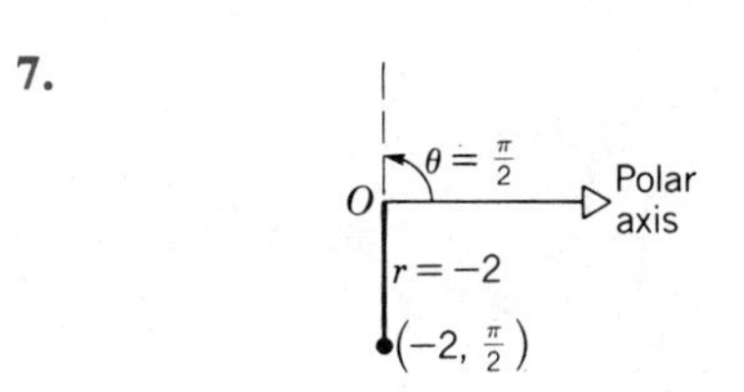

9.

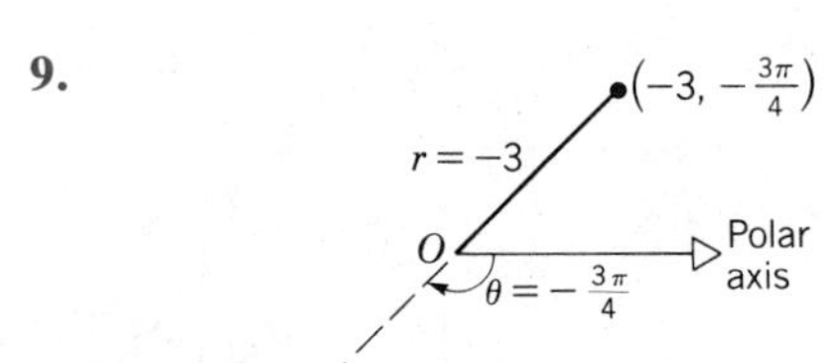

11. $\left(-2, \frac{4\pi}{3}\right)$ and $\left(2, -\frac{5\pi}{3}\right)$

13. $\left(2, \frac{5\pi}{4}\right)$ and $\left(-2, -\frac{7\pi}{4}\right)$

15. $\left(-4, \frac{3\pi}{4}\right)$ and $\left(4, \frac{7\pi}{4}\right)$

17. $\left(1, -\frac{5\pi}{4}\right)$ and $\left(-1, \frac{7\pi}{4}\right)$

19. $(2\sqrt{3}, 2)$ **21.** $(\sqrt{2}, -\sqrt{2})$

23. $(-5, 0)$ **25.** $(-2, -2\sqrt{3})$

27. $(2, 2\sqrt{3})$ **29.** $\left(4, \frac{\pi}{6}\right)$

31. $\left(4, -\frac{2\pi}{3}\right)$ **33.** $\left(4, \frac{2\pi}{3}\right)$

35. $(\sqrt{5}, \arctan 2)$

37.

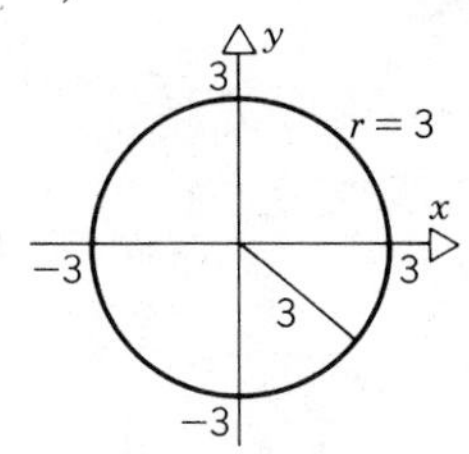

39.

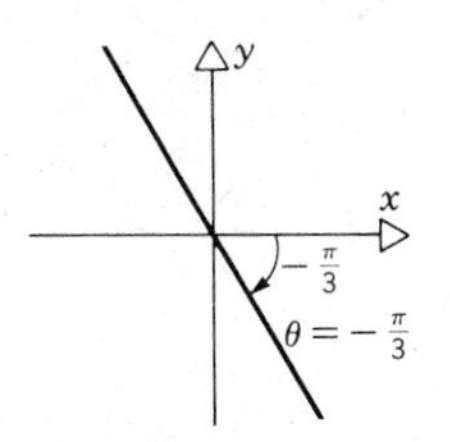

41.

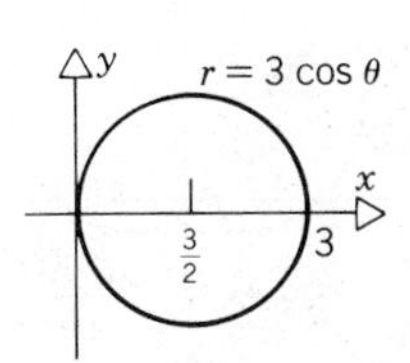

43.

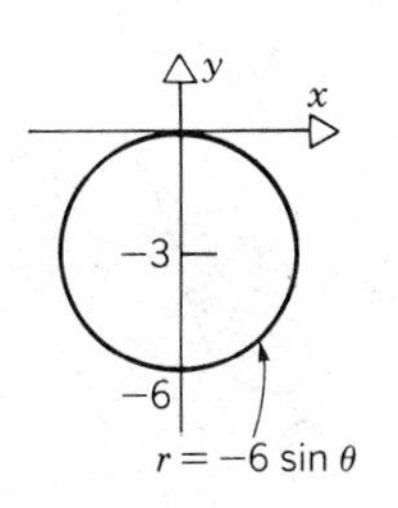

45.

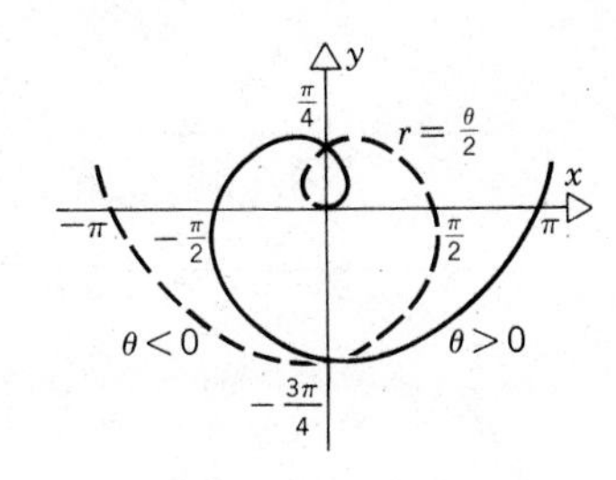

47.

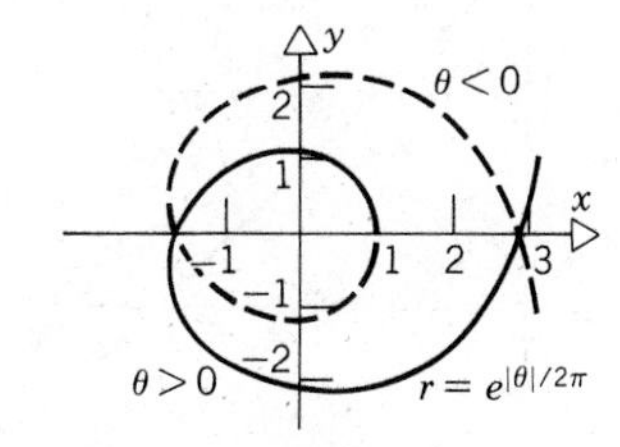

49.

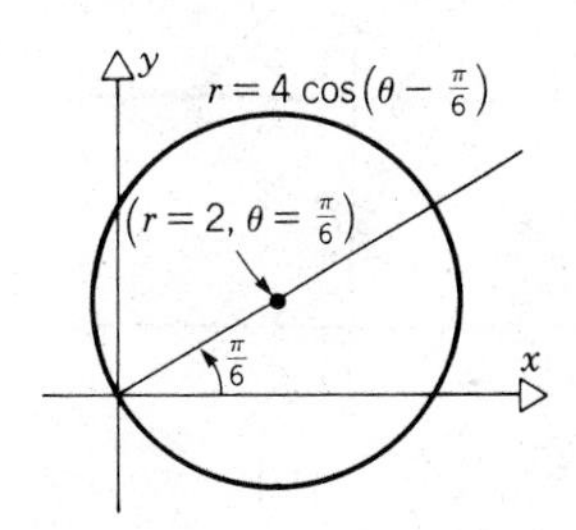

51.

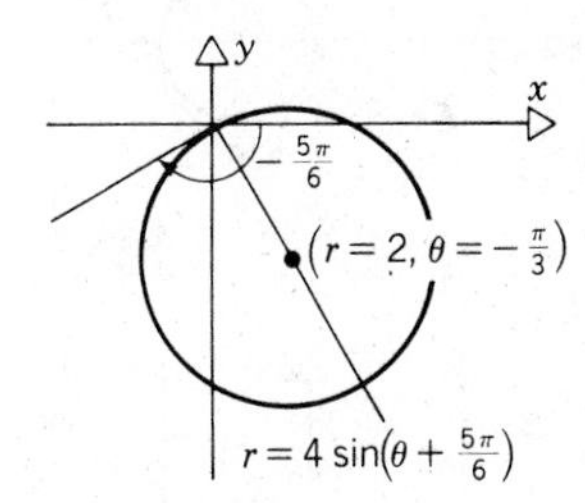

53.

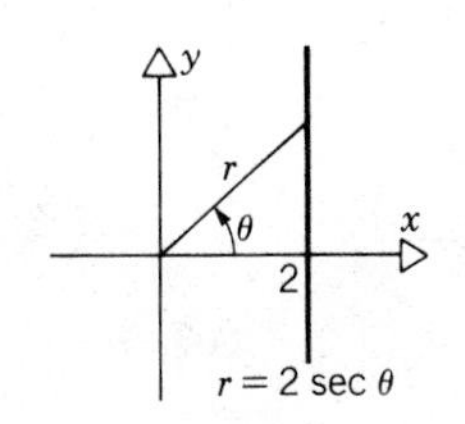

55.

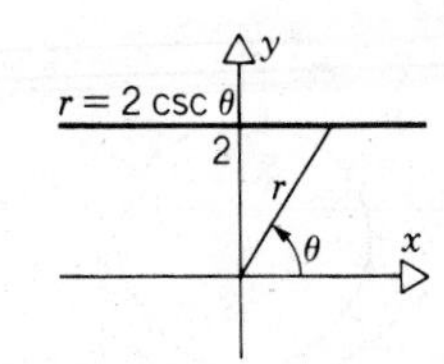

57.

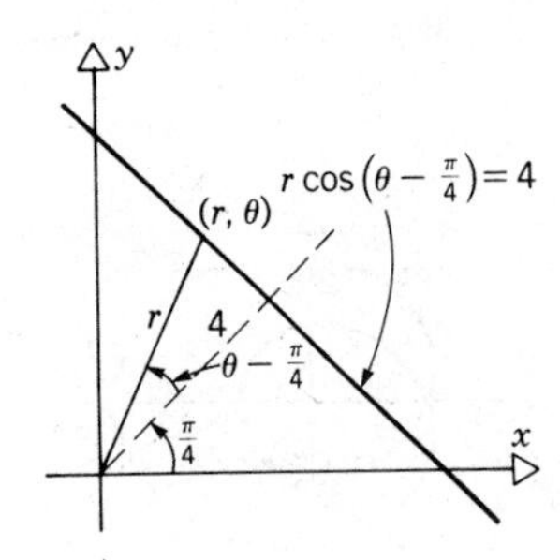

59.

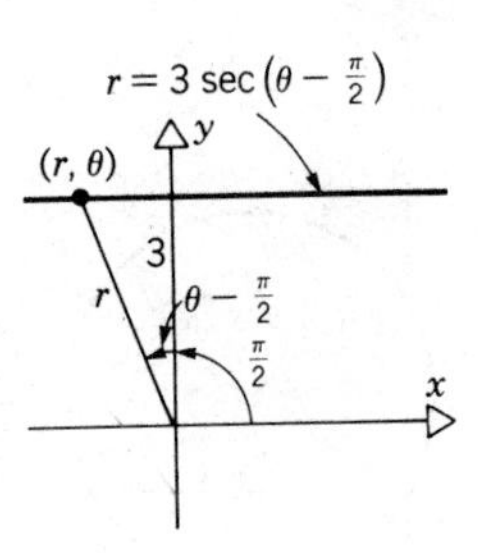

61.

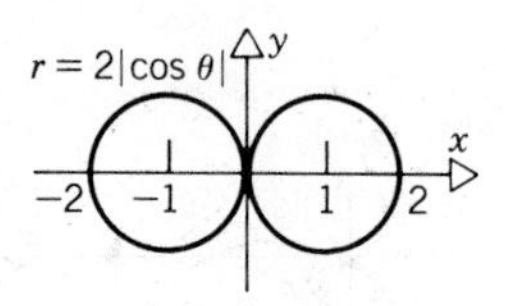

63.

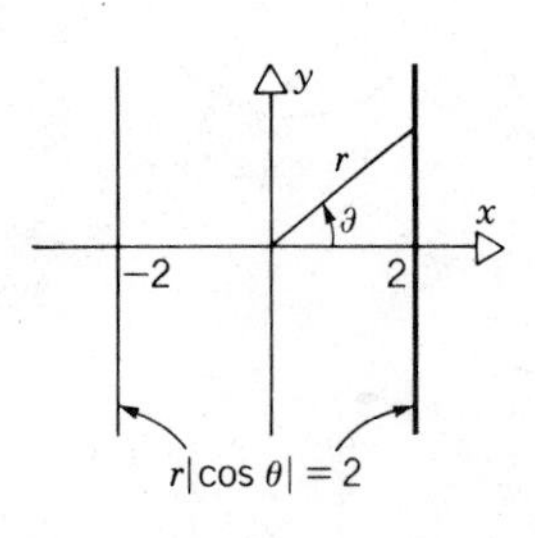

65. $r = 4 \sin \theta$ **67.** $r = 2\sqrt{2} \cos\left(\theta - \frac{\pi}{4}\right)$

69. $r(2 \cos \theta - \sin \theta) = 5$ **71.** $r^2(2 - \sin 2\theta) = 6$

73. $(x^2 + y^2 + 2y)^2 = 4(x^2 + y^2)$

75. $(x^2 + y^2)^3 = 36x^2y^2$

77. $d = [r_1^2 + r_2^2 - 2r_1r_2 \cos(\theta_2 - \theta_1)]^{1/2}$

14.2 Answers

1. Cardioid

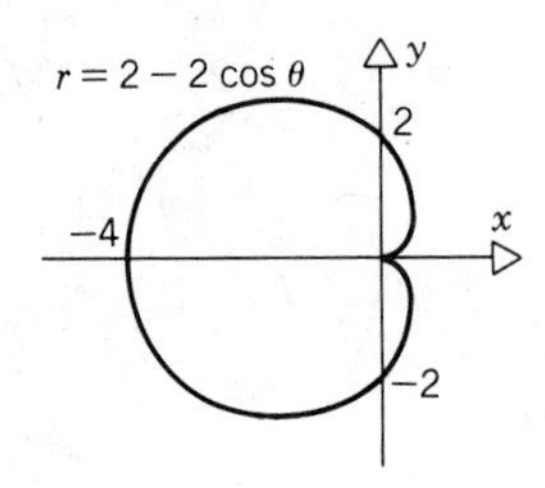

3. Limaçon

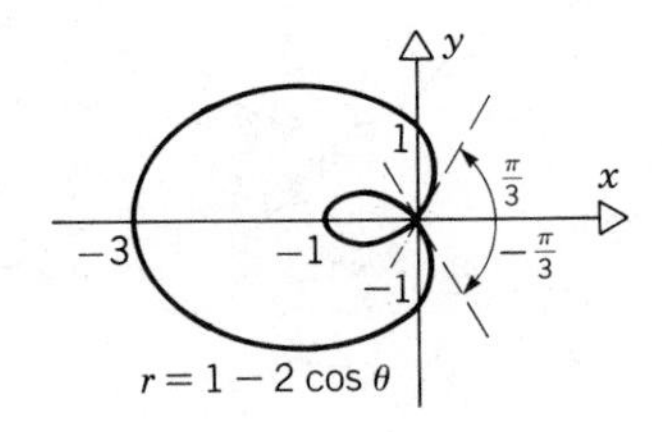

5. Cardioid

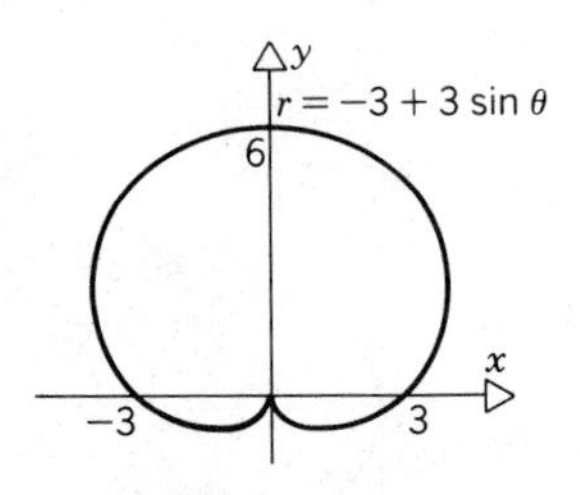

7. Lemniscate

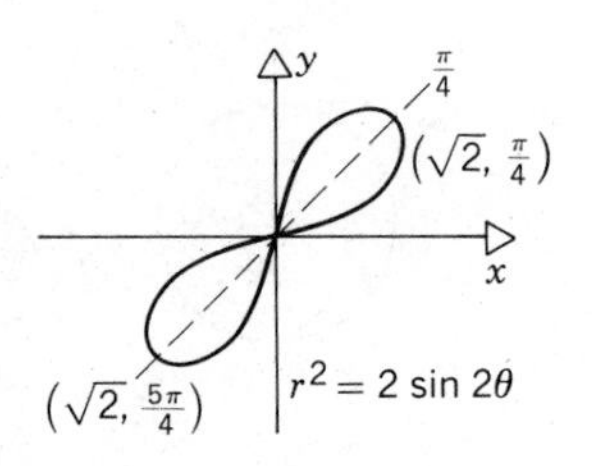

9. Cardioid

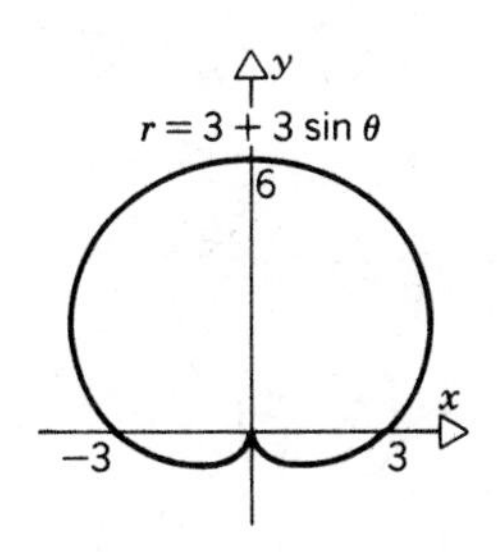

11. Lemniscate

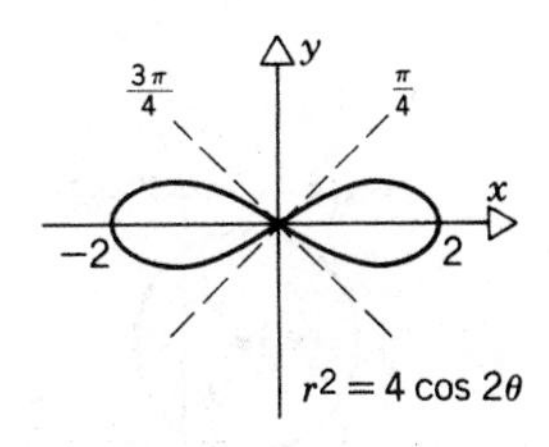

13. Leaf curve

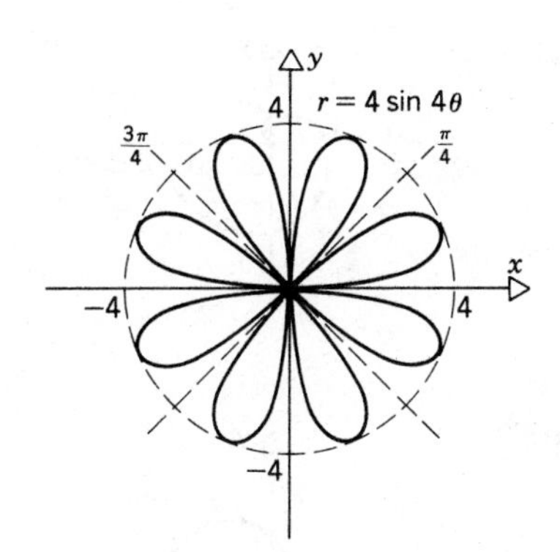

15. Limaçon

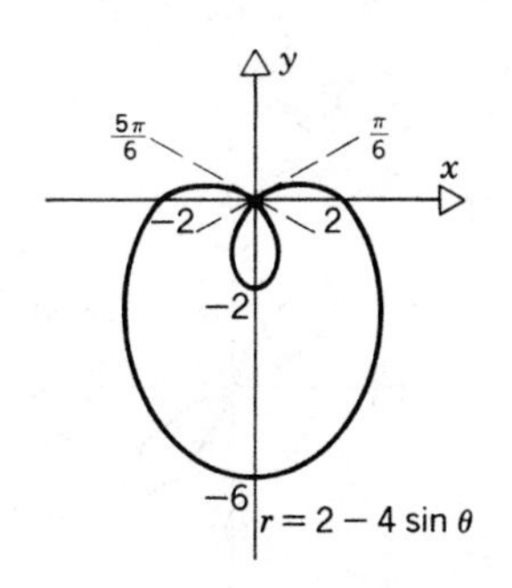

17. $\left(6, \frac{\pi}{3}\right), \quad \left(6, -\frac{\pi}{3}\right)$

19. $\left(\frac{3}{\sqrt{2}}, \frac{\pi}{4}\right)$, pole

21. $\left(\frac{3}{2}, \frac{\pi}{3}\right), \quad \left(\frac{3}{2}, -\frac{\pi}{3}\right)$

23. $\left(\frac{1}{2}, \frac{2\pi}{3}\right), \quad \left(\frac{7}{2}, -\frac{\pi}{3}\right)$, pole

25. $\left(\sqrt{2}, \frac{\pi}{12}\right), \quad \left(\sqrt{2}, \frac{\pi}{4}\right)$

27. $\left(1, \frac{\pi}{6}\right), \quad \left(1, \frac{5\pi}{6}\right)$

29. $\left(2, \frac{\pi}{3}\right), \quad \left(2, \frac{5\pi}{3}\right)$, pole

33. $e = \frac{1}{2}$; foci are $(0, 0)$ and $(\frac{8}{3}, 0)$; directrices are $x = -4$ and $x = \frac{20}{3}$

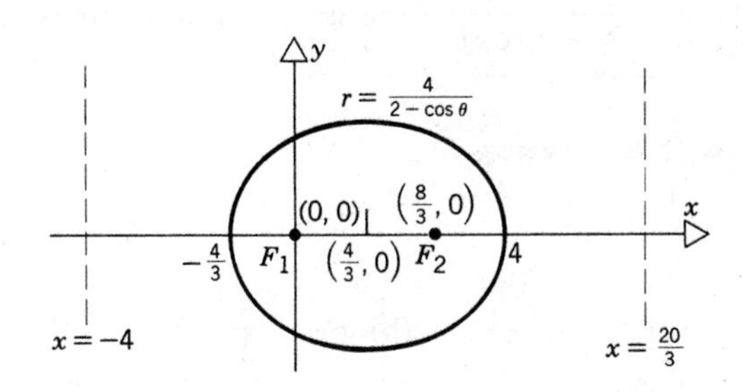

35. $e = \frac{2}{3}$; foci are $(0, 0)$ and $(-\frac{16}{5}, 0)$; directrices are $x = 2$ and $x = -\frac{26}{5}$

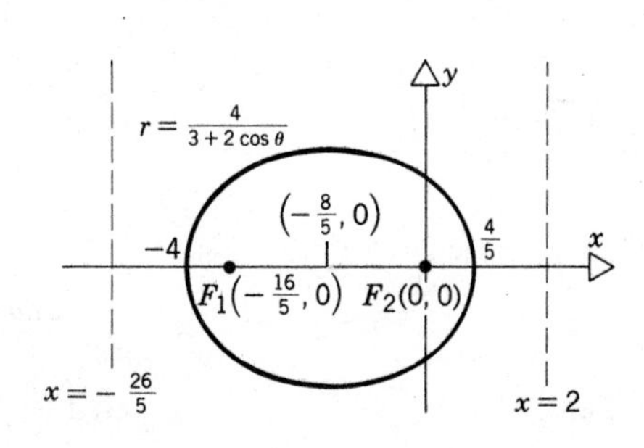

37. $e = \frac{3}{2}$; foci are $(0, 0)$ and $(-18, 0)$; directrices are $x = -5$ and $x = -13$

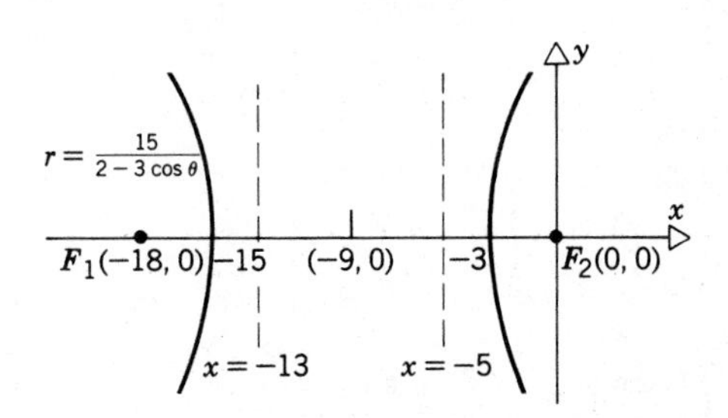

39. $e = 1$; focus is $(0, 0)$; directrix is $y = 2$

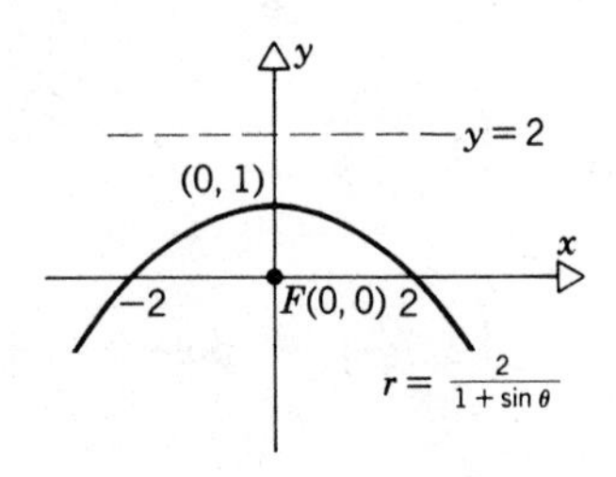

41. $r = \dfrac{6}{2 + 3\cos\theta}$; focus is $(\frac{36}{5}, 0)$; directrix is $x = \frac{26}{5}$

14.3 Answers

1. 4

3. $\dfrac{9\pi}{8}$

5. $\dfrac{3\pi^3}{2}$

7. (a) $2\pi + \dfrac{3\sqrt{3}}{2}$ (b) $\pi - \dfrac{3\sqrt{3}}{2}$ (c) $\pi + 3\sqrt{3}$

9. $a^2\left(-\dfrac{\pi}{4} + \dfrac{9\sqrt{3}}{8}\right)$

11. $3\sqrt{3} - \pi$

13. $\dfrac{5\pi}{2} - 3\sqrt{3}$

15. $2\sqrt{2}$

17. πa

19. a

21. $\dfrac{\sqrt{1 + k^2}(e^{k\pi} - 1)}{k}$

23. $4a$

25. $2\sqrt{2}a \displaystyle\int_0^{\pi/4} (\sec 2\theta)^{1/2}\, d\theta$

27. πa^2

29. $\dfrac{32\pi a^2}{5}$

31. $4\sqrt{2}\pi a^2$

33. $1 - \sqrt{2}$

35. $-7/3\sqrt{3} \cong -1.347$

37. $\dfrac{\sqrt{3}}{3}$

39. $\dfrac{\pi}{2}$

Answers to Review Problems

1. $\left(2, \dfrac{3\pi}{2}\right)$, $\left(-2, -\dfrac{3\pi}{2}\right)$

3. $\left(-3, \dfrac{\pi}{3}\right)$, $\left(3, \dfrac{4\pi}{3}\right)$

5. $\left(\dfrac{5}{2}, -\dfrac{5\sqrt{3}}{2}\right)$

7. $\left(-\dfrac{3}{2}, -\dfrac{3\sqrt{3}}{2}\right)$

9. $\left(4, \dfrac{\pi}{6}\right)$

11. $\left(4, \dfrac{2\pi}{3}\right)$

13. circle

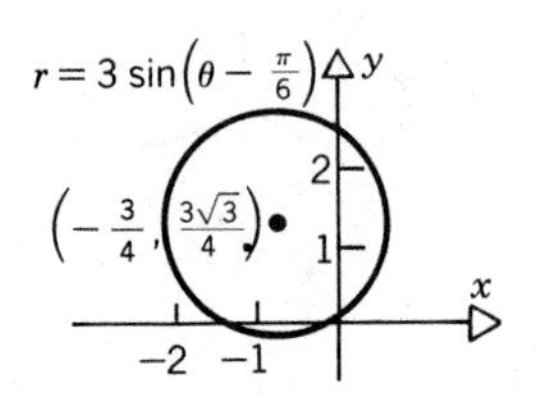

15. spiral

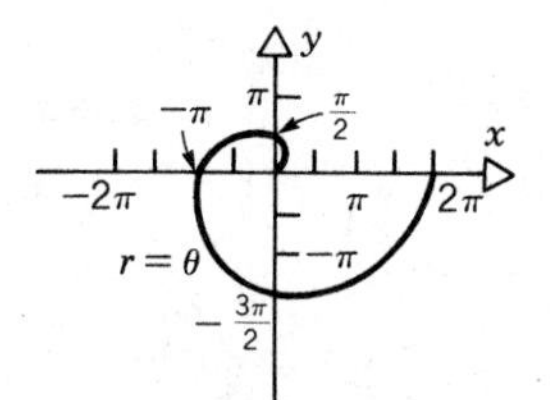

17. cardioid

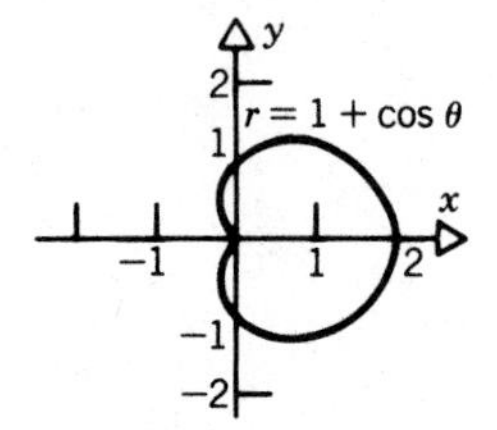

19. lemniscate

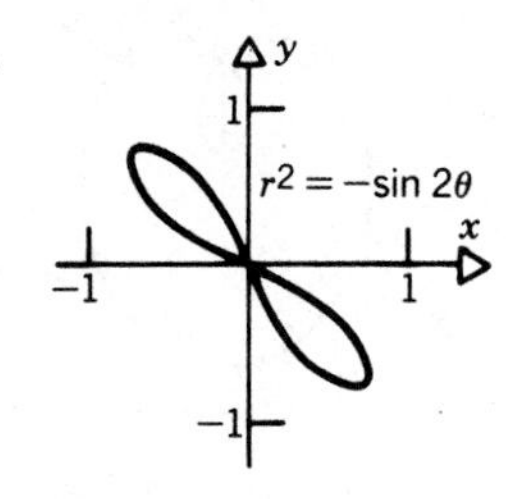

21. Hyperbola; foci are $(0, 0)$ and $(0, \frac{27}{4})$; directrices are $y = 3$ and $y = \frac{15}{4}$; $e = 3$

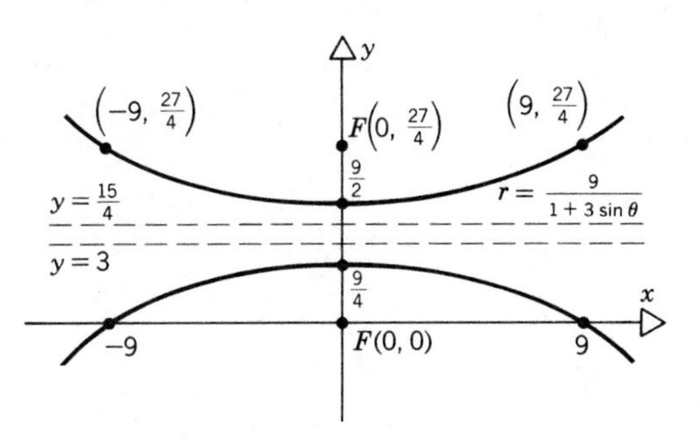

23. Ellipse; foci are $(0, 0)$ and $(4/5, 0)$; directrices are $x = -6$ and $x = 34/5$; $e = \frac{1}{4}$

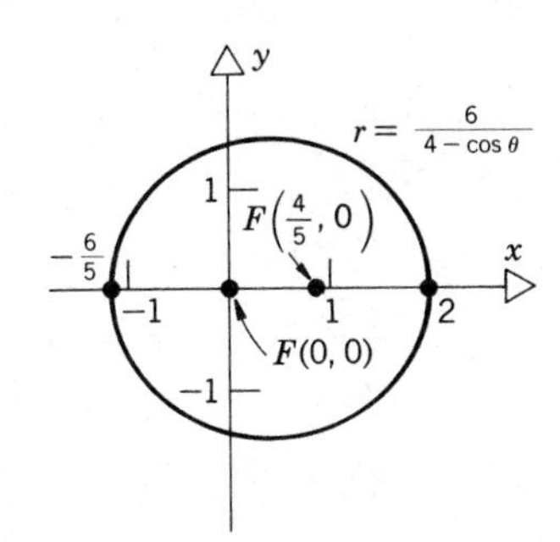

25. $r = 3 \sec \theta + 3$

27. $r = \sec \theta - \cos \theta$

29. $x + \sqrt{3}y = 0$

31. $y^4 + x^2y^2 - 4x^2 = 0$

33. $(4, 0)$

35. $\left(1, \frac{3\pi}{4}\right)$, $\left(-1, \frac{3\pi}{4}\right)$, pole

37. $\frac{9\pi}{2}$

39. 2

41. $L = 3 \int_{\pi/4}^{3\pi/4} \csc^2 \theta \, d\theta$

43. $L = 2 \int_0^{\pi/6} \sqrt{\sec 2\theta} \, d\theta$

45. $-\frac{2\sqrt{3}}{3}$

47. $3\sqrt{3}$

CHAPTER FIFTEEN

15.1 Answers

1. $\mathbf{a} + \mathbf{b} = (1, -1)$, $\mathbf{a} - \mathbf{b} = (-5, 3)$, $2\mathbf{a} - 3\mathbf{b} = (-13, 8)$

3. $\mathbf{a} + \mathbf{b} = (2, 0)$, $\mathbf{a} - \mathbf{b} = (0, 2)$, $2\mathbf{a} - 3\mathbf{b} = (-1, 5)$

5. $\overrightarrow{PQ} = (2, -3)$

7. $\overrightarrow{PQ} = (-3, -4)$

9. $\overrightarrow{PQ} = (1, -3)$, $\overrightarrow{QP} = (-1, 3)$

11. $x = \frac{5}{2}$, $y = -\frac{1}{2}$

13. $\sqrt{5}$

15. 5

17. $\sqrt{2}\sqrt{x^2 + y^2}$

19. $\mathbf{b} = \pm\sqrt{5}(-2, 1)$

21. $\lambda = \pm 2\sqrt{3}$

25. $T = \dfrac{W}{2 \sin \theta}$

15.2 Answers

1. $\mathbf{a} \cdot \mathbf{b} = -1$; $\cos \theta = -\frac{1}{\sqrt{26}} \cong -0.1961$

3. $\mathbf{a} \cdot \mathbf{b} = 0$; $\cos \theta = 0$

5. $\mathbf{a} \cdot \mathbf{b} = 3$; $\cos \theta = \frac{3}{\sqrt{34}} \cong 0.5145$

7. $\mathbf{a} \cdot \mathbf{b} = 8$: $\cos \theta = \frac{2}{\sqrt{5}} \cong 0.8944$

9. $\mathbf{u} = (-\frac{3}{5}, -\frac{4}{5})$

11. $\mathbf{u} = \left(\frac{3x}{5|x|}, -\frac{4x}{5|x|}\right)$

13. $\mathbf{u} = -\frac{5}{13}\mathbf{i} + \frac{12}{13}\mathbf{j}$

15. $\mathbf{u} = \left(-\frac{1}{\sqrt{10}}, -\frac{3}{\sqrt{10}}\right)$

17. $\mathbf{u} = \pm\left[\frac{2}{\sqrt{5}}\mathbf{i} + \frac{1}{\sqrt{5}}\mathbf{j}\right]$

19. $\frac{3}{\sqrt{2}}$

21. $-\frac{1}{\sqrt{5}}$

23. $-\frac{x}{\sqrt{5}}$

25. $(2, -1)$

27. $\frac{5}{2}\mathbf{i} + \frac{5}{2}\mathbf{j}$

31. $\|\mathbf{a} + \mathbf{b}\| = \sqrt{17} \cong 4.123$; $\|\mathbf{a}\| + \|\mathbf{b}\| = \sqrt{5} + \sqrt{10} \cong 5.398$

33. $\|\mathbf{a} + \mathbf{b}\| = \sqrt{34} \cong 5.831$; $\|\mathbf{a}\| + \|\mathbf{b}\| = \sqrt{5} + 5 \cong 7.236$

37. (b) On the straight line through P and Q, but not on the segment from P to Q.

15.3 Answers

1. $(1, -3, 5)$

3. $\mathbf{i} + 3\mathbf{j} + 4\mathbf{k}$

5. $(12, -7, 7)$

7. $8\mathbf{i} + 3\mathbf{j} - 9\mathbf{k}$

9. $\sqrt{14}$

11. $\sqrt{21}$

13. $\sqrt{35}$

15. $\mathbf{a} \cdot \mathbf{b} = 4$; $\cos \theta = \frac{4}{7\sqrt{6}} \cong 0.2333$

17. $\mathbf{a} \cdot \mathbf{b} = 0$; $\cos \theta = 0$

19. $-\frac{3}{\sqrt{14}} \cong -0.8018$

21. $\frac{5}{\sqrt{29}} \cong 0.9285$ 23. $\frac{3}{7}$

25. $\frac{3\mathbf{i} + \mathbf{j} - 2\mathbf{k}}{\sqrt{14}}$ 27. $\frac{3\mathbf{i} - \mathbf{j} - 3\mathbf{k}}{\sqrt{19}}$

31. $\arccos \frac{1}{\sqrt{3}} \cong 0.9553$ radians

33. $\arccos \sqrt{\frac{2}{3}} \cong 0.6155$ radians

15.4 Answers

1. $-2\mathbf{i} - 9\mathbf{j} + 5\mathbf{k}$ 3. $4\mathbf{i} - 3\mathbf{j} - \mathbf{k}$
5. $(a_1b_2 - a_2b_1)\mathbf{k}$ 7. $\mathbf{i} - 7\mathbf{j} + 9\mathbf{k}$
9. -14 11. $-14\mathbf{i} - 7\mathbf{j} - 21\mathbf{k}$
13. $\frac{\pm(\mathbf{i} + \mathbf{j} + \mathbf{k})}{\sqrt{3}}$ 15. $\sqrt{153} \cong 12.369$
17. 16; no 19. -2

15.5 Answers

1. $-x + 4y + 5z = 9$
3. $8x - y - 13z = -29$
5. $2x - y + 5z = -5$
7. $y - 2z = -1$
9. $x - 10y + 7z = -39$
11. $x = 2 + 3t, \quad y = -1 + 2t, \quad z = 4 - t$
13. $x = 2 + t, \quad y = -1 + 2t, \quad z = 3 - t$
15. $\frac{x - 4}{-2} = \frac{y - 2}{-3} = \frac{z - 3}{3}$
17. $x = -2, \quad \frac{y - 6}{1} = \frac{z - 1}{2}$
19. $x = 4 + 6t, \quad y = -3 - t, \quad z = 3t$
21. $3\sqrt{3} \cong 5.196$ 23. Yes
25. $(-2, 5, 0)$ 27. 6
29. $\frac{10}{\sqrt{285}} \cong 0.5923$ 31. $\frac{\sqrt{10}}{2} \cong 1.581$
33. (a) Plane through origin perpendicular to $\mathbf{a}$.
    (b) Plane perpendicular to $\mathbf{a}$ at a distance c from the origin.
    (c) Sphere with center at $\mathbf{a}/2$ and radius $\|\mathbf{a}\|/2$.

15.6 Answers

1. $\mathbf{i} + 2\mathbf{j} - \mathbf{k}$ 3. Limit does not exist.
5. $2\mathbf{i} + 3\mathbf{j} + \mathbf{k}$ 11. -1
13. $2t\mathbf{i} - \sin t\mathbf{j} + 2\cos t\mathbf{k}$
15. $(2\sin 2t + 4t\cos 2t)\mathbf{i} + 3\sin t\mathbf{j} + 12(2t - 1)\mathbf{k}$
17. $(-e^{-t} + 2e^{2t})\mathbf{i} + (-2\sin t - \cos t)\mathbf{j} + (2t + 3t^2)\mathbf{k}$
19. $(-3e^{-t} - 4e^{2t})\mathbf{i} + (-6\sin t + 2\cos t)\mathbf{j} + (6t - 6t^2)\mathbf{k}$
21. $-12t^2 + 25t^4$ 23. $-12t^2 + 25t^4$
25. (a)

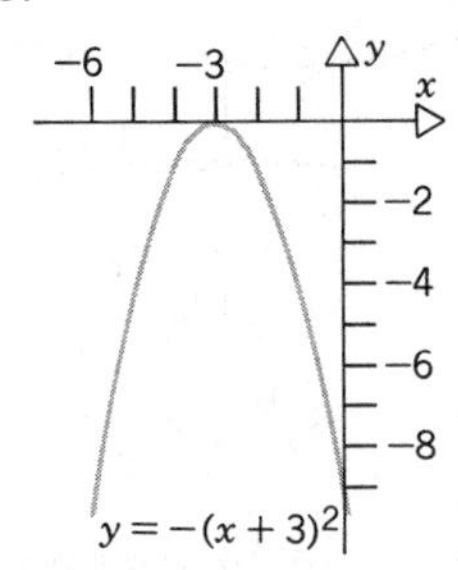

(b) $\frac{\mathbf{i} + 4\mathbf{j}}{\sqrt{17}}$
(c) $x = -5 + \tau, \quad y = -4 + 4\tau, \quad z = 0$
(d) $x + 4y = -21$

27. (a) $\frac{3\mathbf{j} + 4\mathbf{k}}{5}$
(b) $x = 2, \quad y = 3\tau, \quad z = 4\tau$
(c) $3y + 4z = 0$

29. $\arccos \frac{3}{\sqrt{29}} \cong 0.9799$ radians

31. (a)

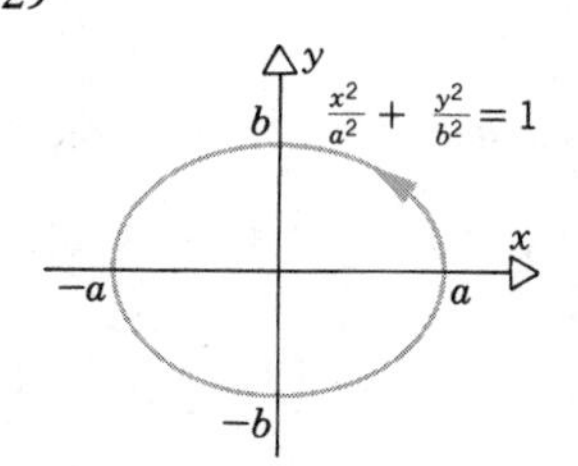

(b) $-a\omega \sin \omega t\,\mathbf{i} + b\omega \cos \omega t\,\mathbf{j}$
(c) $\frac{\omega(a^2 - b^2)\sin \omega t \cos \omega t}{\sqrt{a^2\cos^2 \omega t + b^2 \sin^2 \omega t}}$

15.7 Answers

1. $3\sqrt{10} \cong 9.487$ 3. $\frac{38}{3}$
5. $2\sinh 2 \cong 7.254$ 7. $\frac{3}{2}$
9. $\mathbf{v} = -\sin t\,\mathbf{i} + \cos t\,\mathbf{j} + 3\mathbf{k}; \quad \|\mathbf{v}\| = \sqrt{10};$
   $\mathbf{a} = -\cos t\,\mathbf{i} - \sin t\,\mathbf{j}$
11. $\mathbf{v} = t\mathbf{i} - 4\mathbf{j} + 9t^2\mathbf{k};$
    $\|\mathbf{v}\| = (16 + t^2 + 81t^4)^{1/2};$
    $\mathbf{a} = \mathbf{i} + 18t\mathbf{k}$
13. $\frac{-\sin t\,\mathbf{i} + \cos t\,\mathbf{j} + 3\mathbf{k}}{\sqrt{10}}$

15. $\dfrac{t\mathbf{i} - 4\mathbf{j} + 9t^2\mathbf{k}}{(16 + t^2 + 81t^4)^{1/2}}$

17. $\frac{1}{10}$

19. $\dfrac{(16 + 5184t^2 + 81t^4)^{1/2}}{(16 + t^2 + 81t^4)^{3/2}}$

21. $\dfrac{|ab|}{(a^2 \sin^2 t + b^2 \cos^2 t)^{3/2}}$

23. $\dfrac{1}{2\sqrt{2}a(1 - \cos t)^{1/2}}$

25. $\dfrac{e^{-x}}{(1 + e^{-2x})^{3/2}}$

27. $\dfrac{2x^3}{(x^4 + 1)^{3/2}}$

15.8 Answers

3. $\mathbf{v} = -a\omega \sin \omega t\, \mathbf{i} + b\omega \cos \omega t\, \mathbf{j}$

$\mathbf{a} = -a\omega^2 \cos \omega t\, \mathbf{i} - b\omega^2 \sin \omega t\, \mathbf{j}$

$$a_t = (a^2 - b^2)\frac{\omega^2 \sin \omega t \cos \omega t}{(a^2 \sin^2 \omega t + b^2 \cos^2 \omega t)^{1/2}}$$

$$a_n = \frac{ab\omega^2}{(a^2 \sin^2 \omega t + b^2 \cos^2 \omega t)^{1/2}}$$

5. $\mathbf{v} = -(\sin t + \frac{1}{2} \sin 2t)\mathbf{i} + (\cos t + \frac{1}{2} \cos 2t)\mathbf{j}$

$\mathbf{a} = -(\cos t + \cos 2t)\mathbf{i} - (\sin t + \sin 2t)\mathbf{j}$

$$a_t = -\frac{\sin t}{2(\frac{5}{4} + \cos t)^{1/2}}$$

$$a_n = \frac{\frac{3}{2}|1 + \cos t|}{(\frac{5}{4} + \cos t)^{1/2}}$$

7. (a) $\mathbf{v} = a\omega(1 - \cos \omega t)\mathbf{i} + a\omega \sin \omega t\, \mathbf{j}$;

$\|\mathbf{v}\| = 2a\omega \sin \dfrac{\omega t}{2}$ is maximum when $\omega t = \pi$, that is, at the top of the arch

(b) $\mathbf{a} = a\omega^2(\sin \omega t\, \mathbf{i} + \cos \omega t\, \mathbf{j})$; $\|\mathbf{a}\| = a\omega^2$; $\mathbf{a}$ points toward the center of the rolling circle

(d) $a_t = a\omega^2 \cos \dfrac{\omega t}{2}$; $\quad a_n = a\omega^2 \sin \dfrac{\omega t}{2}$

9. (a) 24,900 miles (b) 29,100 miles

15.9 Answers

1. $a_r = -2b\omega^2 \sin \theta$, $\quad a_\theta = 2b\omega^2 \cos \theta$, $\mathbf{h} = m\omega b^2 \sin^2 \theta\, \mathbf{k}$

3. $a_r = \omega^2[f''(\theta) - f(\theta)]$, $\quad a_\theta = 2\omega^2 f'(\theta)$, $\mathbf{h} = m\omega[f(\theta)]^2\mathbf{k}$

15.10 Answers

1. Cylinder parallel to x-axis; circular cross section, radius 2

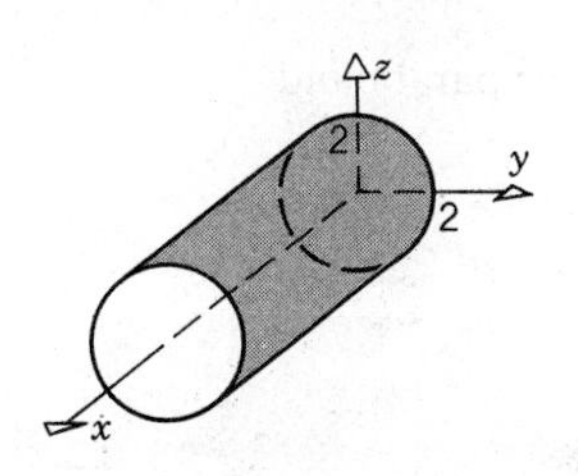

3. Cylinder parallel to z-axis; hyperbolic cross section

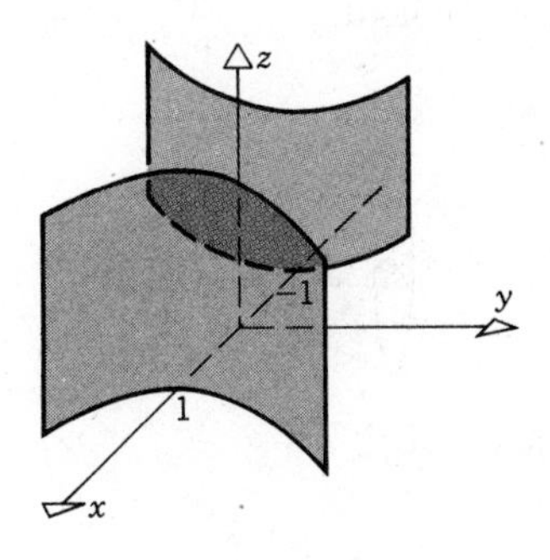

5. Cylinder parallel to y-axis

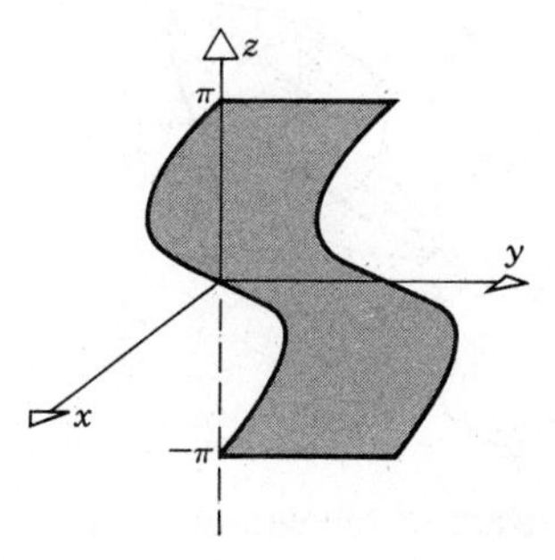

7. Cone

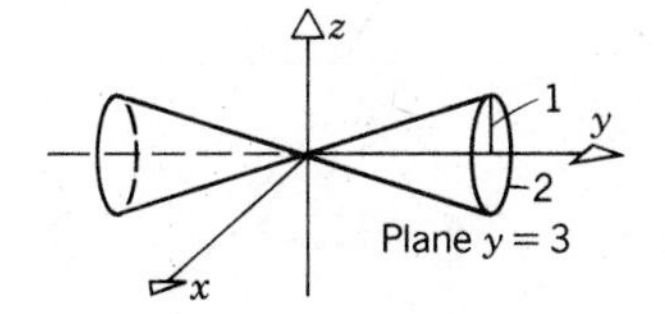

9. Hyperbolic paraboloid

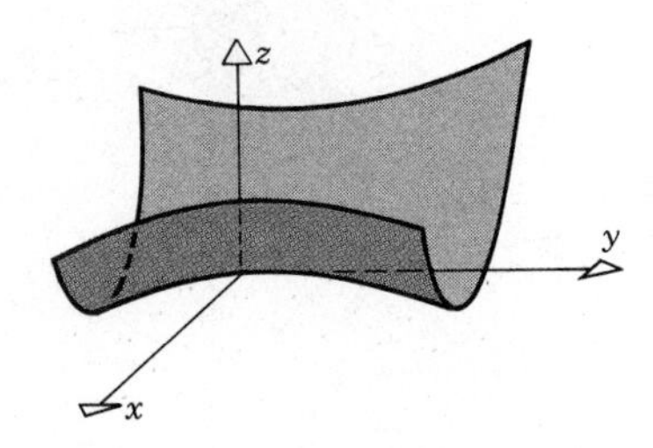

11. Elliptic paraboloid

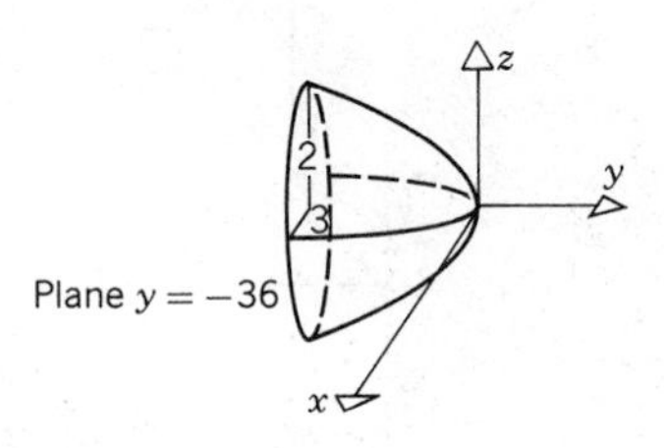

13. Ellipsoid of revolution about x-axis (prolate spheroid)

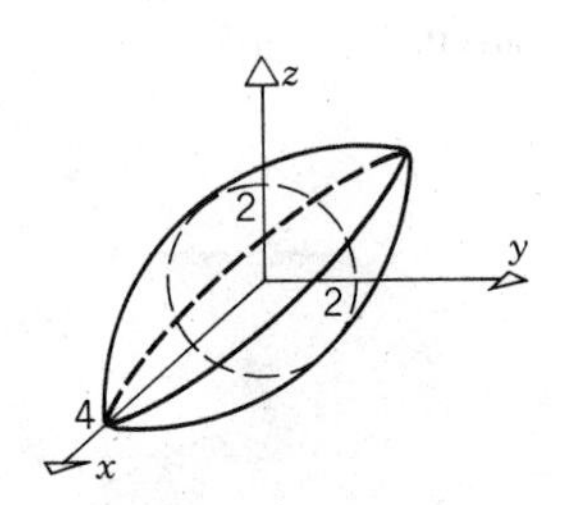

15. Cone; surface of revolution about z-axis

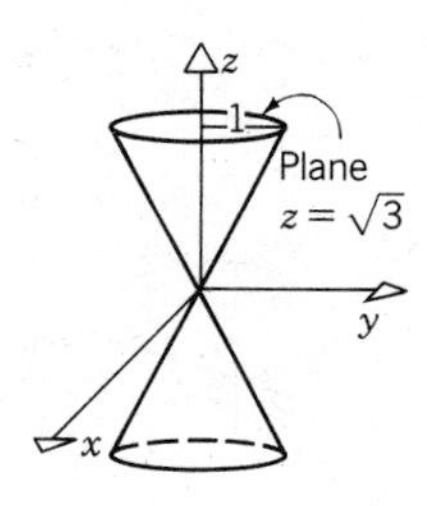

17. Cylinder parallel to z-axis; elliptic cross section

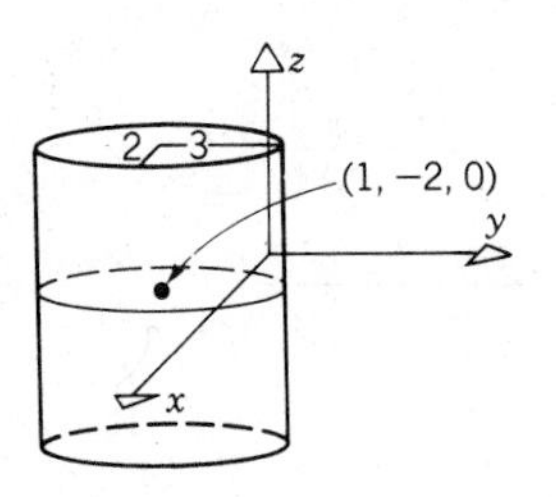

19. Hyperboloid of one sheet

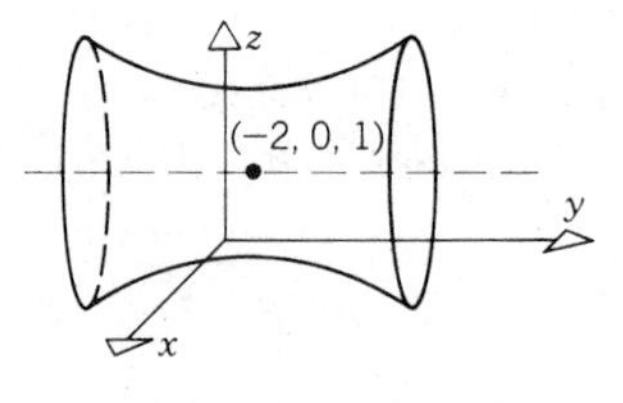

21. Cylinder parallel to x-axis; parabolic cross section

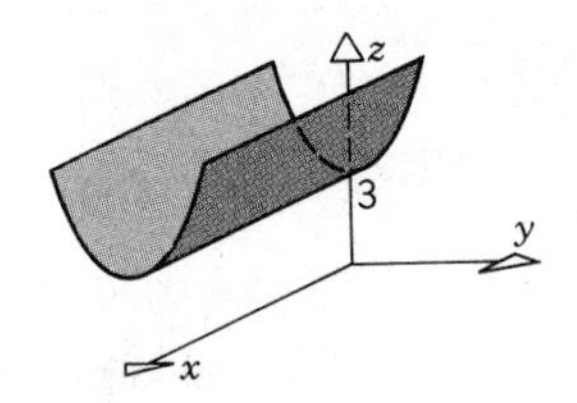

23. Sphere, center $(3, -1, 4)$, radius 4

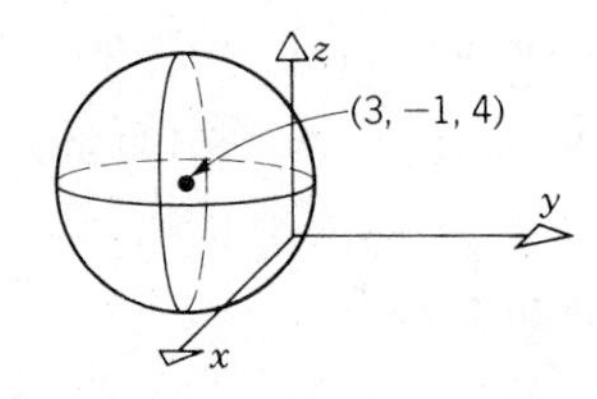

25. $y = 1 - x^2 - z^2$

27. $y^4 = x^2 + z^2$

29. $x^2 + y^2 = \left(\dfrac{z^2}{1 + z^2}\right)^2$

31. $z^3 = x^2 + y^2$

33. (a) $u + 3 = \sqrt{x}$
(b) $u^2 + z^2 = (\sqrt{x} - 3)^2$
(c) $(y - 3)^2 + z^2 = (\sqrt{x} - 3)^2$

35. $(y - 1)^2 + z^2 = (1 + x^2)^2$

37. $x^2 + (z - \pi)^2 = (\arctan y - \pi)^2$

39. $\frac{x^2}{4} - \frac{y^2}{9} + \frac{z^2}{16} = 1$; hyperboloid of one sheet

41. (a) Cone with axis along **a** and vertex angle equal to $\arctan \lambda$
(b) $(-\lambda^2 a^2 + b^2 + c^2)x^2 + (a^2 - \lambda^2 b^2 + c^2)y^2 + (a^2 + b^2 - \lambda^2 c^2)z^2 - 2(1 + \lambda^2)(abxy + acxz + bcyz) = 0$; $\lambda = \tan \alpha$

43. $\frac{x^2}{a^2} - \frac{y^2}{b^2} - \frac{z^2}{b^2} = 1$, where $b^2 = c^2 - a^2$; two sheeted hyperboloid of revolution about x-axis

Answers to Review Problems

1. $(3, -8, 3)$

3. $(15, -1, 6)$

5. $12\ (10, 23, 4)$

7. $\lambda = \pm\sqrt{37/3}$

9. $\lambda = \pm 5$

11. $(6, 7, 3 \ln 2 + 4e)$

13. 0

15. $(-\sin t, 12t^3, 6te^{3t^2} + 1/t)$

17. $(-\sin t \cos t + 6t\, e^{6t^2})\,(\cos^2 t + e^{6t^2})^{-1/2}$

19. (a) $\frac{(1, -\sin t)}{(1 + \sin^2 t)^{1/2}}$
(b) $2\int_0^{\pi/2} \cos t \sqrt{1 + \sin^2 t}\, dt$

21. (a) $(\cos t, \sin t)$
(b) $\int_{\pi/4}^{3\pi/4} \csc t\, dt$

23. hyperbolic paraboloid

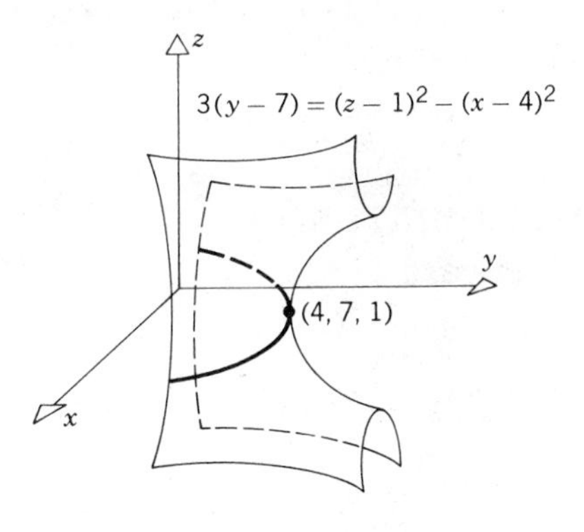

25. hyperboloid of two sheets

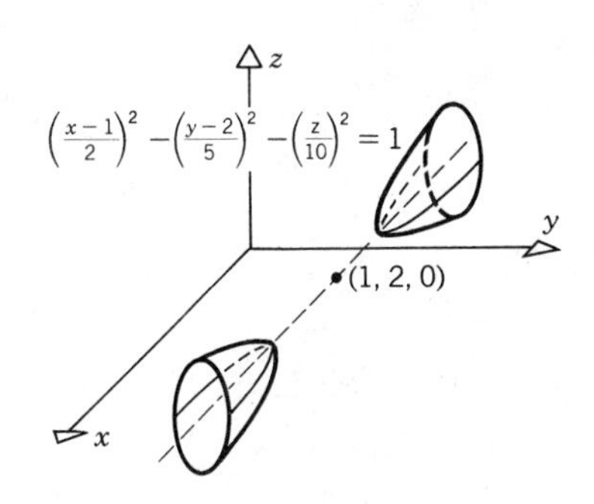

27. $2y^2 + 5x^2 + 5z^2 = 10$

29. $(z - 1)^2 + y^2 = (x^3 - 1)^2$

31. point of intersection: $(1, 0, 0)$

33. $y - z = -1$

35. $2x - 5y + z = -1$

37. $8x + 2y - 7z = 14$

39. $\sqrt{393}/2$

41. (a) $a_r = -2a\omega^2 \cos\theta$
$a_\theta = -2a\omega^2 \sin\theta$
(b) $\mathbf{h} = m\omega r^2 \mathbf{u}_z$
(c) $\kappa = 2/a$

43. N 84.4°W
t = 8 hours 21 minutes

CHAPTER SIXTEEN

16.1 Answers

1. (a) 0 (b) 0 (c) 4

3. (a) $\frac{1}{2} - 4\sqrt{2}$ (b) $2a^{2a} + a^4$
(c) $4 + \frac{16\sqrt{2}}{3}$

5. (a) -2 (b) $x \sin x$ (c) $y \sin x$
(d) $r \sin s$

7. (a) $\frac{a^2}{\sqrt{3}}$ (b) $\frac{1}{a^2\sqrt{3}}$

9. $x^2 + y^2 \leq \frac{9}{2}$

11. $-\infty < x < \infty, \quad -\infty < y < \infty$

13. $z \geq 0$

15. $xy > 0$

17.

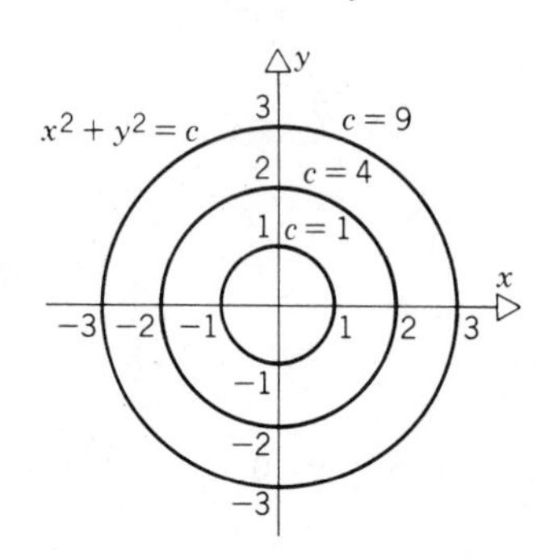

19.

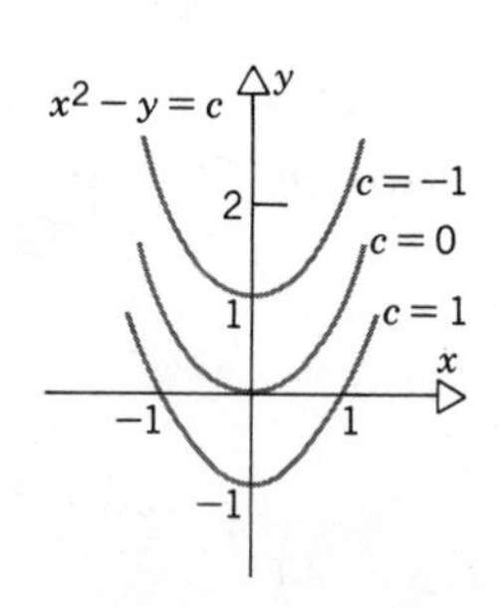

21.

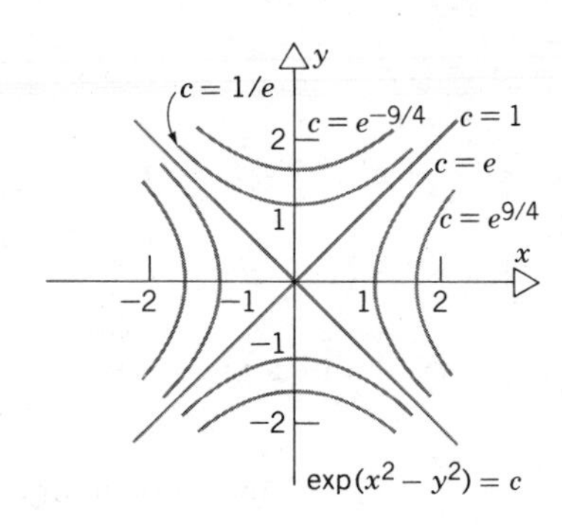

23. Nowhere

25. $y = \pm x$

27. Nowhere

29. No

31. $f(0, 0) = 0$

33. $\alpha > 2, \quad f(0, 0) = 0$

16.2 Answers

1. $z_x = 4x - 3y, \quad z_y = -3x + 2y$

3. $f_x(x, y) = \frac{2y(-x^2 + y^2 + 1)}{(x^2 + y^2 + 1)^2}$

5. $D_x f(x, y) = 3x^2 + 8xy - 6y^2$;
$D_y f(x, y) = 4x^2 - 12xy + 3y^2$

7. $D_r u = s\cos(r - s) - rs\sin(r - s)$,
$D_s u = r\cos(r - s) + rs\sin(r - s)$

9. $f_x(x, y) = 2xy^2 \sin(x^2y^2) + 2x^3y^4 \cos(x^2y^2)$

11. $f_y(x, y, z) = \frac{-2(x - 2y + z)(2x^2 + 2z^2 + xy + yz)}{(x^2 + y^2 + z^2)^2}$

13. $f_x(x, y, z) = -\frac{x}{(x^2 + y^2 + z^2)^{3/2}}$

15. $z_x = \frac{x[2 - \ln(x^2 + y^2)]}{(x^2 + y^2)^{3/2}}$,
$z_y = \frac{y[2 - \ln(x^2 + y^2)]}{(x^2 + y^2)^{3/2}}$

17. $z_x = -\frac{y}{x^2 + y^2}, \quad z_y = \frac{x}{x^2 + y^2}$

19. $z_{xy} = z_{yx} = -3$

21. $z_{xy} = z_{yx} = 4xy(1 - x^4y^4)\sin(x^2y^2) + 12x^3y^3\cos(x^2y^2)$

29. (a) 9, 6 (b) $3^n, \frac{(n + 1)(n + 2)}{2}$

35. $z = \frac{x}{y} + F(y)$

37. $z = \frac{e^{xy}}{x} + y \ln x + \frac{x^2y^2}{2} + F(x)$

39. $z = \dfrac{x^3y^2}{3} + \dfrac{x^{y+1}}{y+1} + F(y)$

16.3 Answers

1. $f(x+h, y+k) = x^2 + 3xy + (2x+3y)h + 3xk + r\sqrt{h^2+k^2}$, $r = (h^2 + 3hk)/\sqrt{h^2+k^2}$

3. $f(x+h, y+k) = xy^2 + y^2h + 2xyk + r\sqrt{h^2+k^2}$, $r = \dfrac{2xyhk + xk^2 + hk^2}{\sqrt{h^2+k^2}}$

5. $f(x+h, y+k) = x\sin y + (\sin y)h + (x\cos y)k + r\sqrt{h^2+k^2}$,
$$r = \frac{[(x+h)(\sin y)(\cos k - 1) + x(\cos y)(\sin k - k) + (\cos y)h\sin k]}{\sqrt{h^2+k^2}}$$

7. $dz = \dfrac{6x\,dx - dy}{3x^2 - y + 2}$

9. $dz = yx^{y-1}\,dx + (\ln x)x^y\,dy$

11. $dz = \dfrac{-y\,dx + x\,dy}{x^2+y^2}$

13. $dw = (\cos 2y)\,dx + (-2\sin 2y + 3\sqrt{z})\,dy + \left(\dfrac{3y}{2\sqrt{z}}\right)dz$

15. $dV = \left(\dfrac{2\pi rh}{3}\right)dr + \left(\dfrac{\pi r^2}{3}\right)dh$

19. $3 + 13t^{11/2}$

21. $(2t\sin t + t^2\cos t)\exp(1 + t^2\sin t)$

23. $-4t^{-3} + \dfrac{9\sqrt{t}}{2} + t^{-7/6}$

25. $\dfrac{\partial z}{\partial r} = 48rs^2(2r^2 - s^2)$, $\dfrac{\partial z}{\partial s} = 48r^2s(r^2 - 2s^2)$

27. $\dfrac{\partial z}{\partial s} = \dfrac{2e^{2s}}{1+e^{4s}}$, $\dfrac{\partial z}{\partial t} = 0$

37. $\dfrac{\partial F}{\partial x} = \dfrac{\partial f}{\partial x} + \dfrac{\partial f}{\partial z}\dfrac{\partial g}{\partial x}$, $\dfrac{\partial F}{\partial y} = \dfrac{\partial f}{\partial y} + \dfrac{\partial f}{\partial z}\dfrac{\partial g}{\partial y}$

16.4 Answers

1. $(x+2y)\mathbf{i} + 2(x+y)\mathbf{j}$

3. $(y^2 - 6xz)\mathbf{i} + 2(xy + z^2)\mathbf{j} + (4yz - 3x^2)\mathbf{k}$

5. $(\cosh x\tan y)\mathbf{i} + (\sinh x\sec^2 y)\mathbf{j}$

7. $\dfrac{7}{\sqrt{2}}$

9. $-(a^2+b^2)^{-1/2}$

11. $-\dfrac{11e^2}{\sqrt{3}} \cong -46.927$

13. $\boldsymbol{\lambda} = \dfrac{3\mathbf{i} + 4\mathbf{j}}{5}$, 5

15. $\boldsymbol{\lambda} = \dfrac{3\mathbf{i} - 4\mathbf{j} + 12\mathbf{k}}{13}$, 507

19. $\dfrac{3}{\sqrt{5}}$

21. $\pm\dfrac{b\mathbf{i} + a\mathbf{j}}{\sqrt{a^2+b^2}}$

23. $\pm\dfrac{(\mathbf{i} + a^2\mathbf{j})}{\sqrt{1+a^4}}$

25. 0

27. $\frac{62}{5}$

29. $\dfrac{20e}{\sqrt{2}}$

31. $\pm(\mathbf{i} + 3\mathbf{j})$

33. $\frac{7}{2}$

35. $-\dfrac{4}{\sqrt{5}}$

39. $y = \dfrac{3x^3}{16}$

43. (c) $\mathbf{c} = \mathbf{a} + \alpha(\mathbf{b} - \mathbf{a})$

16.5 Answers

1. $8x + 4y - z = 6$; $x - 1 = 8t,\ y - 1 = 4t,\ z - 6 = -t$

3. $y + z = 3$; $x = 0,\ y - 1 = z - 2 = t$

5. $6x + 2y - 3z = 36$; $x - 1 = 6t,\ y - 9 = 2t,\ z + 4 = -3t$

7. $y - 2z = -5$; $x = 1,\ y - 3 = t,\ z - 4 = -2t$

9. $ex + (4+e)y - z = 4 + e$; $x - 1 = et,\ y - 1 = (4+e)t,\ z - e = -t$

15. (a) $\dfrac{x}{x_0^{1/3}} + \dfrac{y}{y_0^{1/3}} + \dfrac{z}{z_0^{1/3}} = a^{2/3}$

17. (a) $\dfrac{x}{\sqrt{x_0}} + \dfrac{y}{\sqrt{y_0}} + \dfrac{z}{\sqrt{z_0}} = \sqrt{a}$

19. $x - 1 = 7t,\ y + \frac{6}{7} = -2t,\ z + \frac{1}{7} = -5t$

21. $x - 2 = -9t,\ y + 1 = -4t,\ z + 8 = 4t$

23. The normals are parallel; the surfaces intersect only at a point, not along a curve.

25. (a) $(0, 0, \pm c)$ (b) $(\pm a, 0, 0)$
(c) $\left(\pm\dfrac{a^2}{\sqrt{a^2+b^2+c^2}}, \pm\dfrac{b^2}{\sqrt{a^2+b^2+c^2}}, \pm\dfrac{c^2}{\sqrt{a^2+b^2+c^2}}\right)$

29. $(-1, \sqrt{3}, 4)$, 0.1306 radians

16.6 Answers

1. $(\frac{1}{2}, -1)$, relative minimum

3. $(0, 0)$, saddle

5. $(0, -1)$, relative minimum

7. $(1, -\frac{1}{2})$, relative maximum

9. $(0, 0)$, saddle

11. $(e^2, -6e^4)$, saddle

13. $(0, 0)$, saddle; $(-\frac{1}{12}, \frac{1}{6})$, relative minimum

15. $\left(0, \pm(2n-1)\frac{\pi}{2}\right)$, $n = 0, 1, 2, \ldots$, saddles

17. (0, 0), relative maximum

19. $(-1, 1)$, saddle; $(1, -1)$, saddle

21. (0, 0), relative maximum; $(-\frac{2}{3}, 0)$, saddle

23. Each is $\frac{a}{3}$.

25. $\left(\frac{ad}{r^2}, \frac{bd}{r^2}, \frac{cd}{r^2}\right)$, where $r^2 = a^2 + b^2 + c^2$

27. Bottom is 2 ft by 2 ft; height is 1.5 ft

29. (b) $y = \frac{9x+1}{12}$

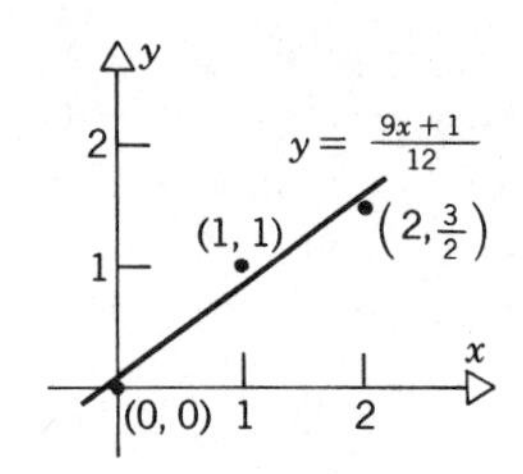

31. (a) $a_0 = \frac{I_0}{2}$, $a_1 = \frac{3I_1}{2}$, where $I_n = \int_{-1}^{1} x^n f(x)\,dx$

(b) $a_0 = \frac{9I_0 - 15I_2}{8}$, $a_1 = \frac{3I_1}{2}$, $a_2 = \frac{45I_2 - 15I_0}{8}$

33. (a) $q_1 = 20$, $q_2 = 15$, $p_1 = 287.5$, $p_2 = 250.00$, $p_1 + p_2 = 537.50$

(b) $q_1 = \frac{40}{3}$, $q_2 = 12$, $p_1 = 283.56$, $p_2 = 343.11$, $p_1 + p_2 = 626.67$

16.7 Answers

1. Maximum is $\frac{ab}{2}$; minimum is $-\frac{ab}{2}$.

3. $\frac{2}{ab}$

5. Maximum is 12 at $(\frac{8}{3}, \frac{8}{3}, -\frac{4}{3})$; minimum is -12 at $(-\frac{8}{3}, -\frac{8}{3}, \frac{4}{3})$.

7. Maximum is 3 at $(\frac{4}{3}, \frac{4}{3}, -\frac{1}{3})$; minimum is -3 at $(-\frac{4}{3}, -\frac{4}{3}, \frac{1}{3})$.

9. $-\frac{a^3}{3\sqrt{3}}$

11. $\frac{1}{9}$

13. $\frac{|c|}{\sqrt{a^2+b^2}}$

15. $x = y = z = \frac{a}{\sqrt{3}}$

21. (a) Each is $\frac{C}{2}$.

25. Maximum is $\frac{r^2}{b^2}$; minimum is 0.

27. Maximum is $(8 + 3\sqrt{3})/2$ at $(-\frac{1}{2}, \sqrt{3})$; minimum is $-4/15$ at $(-1/15, -8/15)$.

29. Maximum is $72\sqrt{2} - 66$ at $(\pm 3\sqrt{2}, 3\sqrt{2})$; minimum is $-72\sqrt{2} - 66$ at $(\pm 3\sqrt{2}, -3\sqrt{2})$.

16.8 Answers

1. $\frac{dy}{dx} = -\frac{e^x \sin y - 2y \sin x}{e^x \cos y + 2 \cos x}$

3. $\frac{dy}{dx} = -\frac{2xy(x^2+y^2) - y^3}{xy^2 - x^2(x^2+y^2)}$

5. $z_x = -\frac{(3y + e^{x+z})}{e^{x+z} - 1}$, $z_y = -\frac{(3x - 4y)}{e^{x+z} - 1}$

7. $z_y = \frac{\cos(x+y) - \sin(y+z)}{\cos(x+z) - \sin(y+z)}$

9. $z_y = -\frac{2y(x^2 + 4y^2 + z^2)}{z(2x^2 + 2y^2 + 3z^2)}$

11. $u_y = \frac{u(y^2 - v^2)}{y(u^2 + v^2)}$, $v_y = -\frac{v(u^2 + y^2)}{y(u^2 + v^2)}$

13. (a) -1 (b) -10

15. $f_x(1, -1) = \frac{1}{9}$, $f_{xx}(1, -1) = \frac{50}{243}$, $f_y(1, -1) = -\frac{13}{9}$, $f_{xy}(1, -1) = \frac{70}{243}$

17. $\frac{dz}{dx} = \frac{f_x g_y - g_x f_y}{g_y}$

23. $u_x = \frac{g_v}{\Delta}$, $v_x = -\frac{g_u}{\Delta}$, $u_y = -\frac{f_v}{\Delta}$, $v_y = \frac{f_u}{\Delta}$, where $\Delta = f_u g_v - f_v g_u$

Answers to Review Problems

1. (a) 0

(b) ln 27
Domain: $|x| > |y|$

3. (a) $-\frac{\pi}{4}$

(b) $\arctan\left(\frac{ac}{b}\frac{3b - 2a}{c - 2a^2}\right)$
Domain: $x, y, z \in R^1$, $z \neq \frac{x^2}{2}$

5. Discontinuous on the line $y = 2x$.

7. Discontinuous on the line $y = x$; define $f(x, x) = 0$.
Discontinuous on the lines $y = x \pm 1$.

9. $f_x = \frac{x}{x^2 + y^2}$, $f_y = \frac{y}{x^2 + y^2}$

11. $g_r = s^2t^2 \cos(rs - t)$
$g_s = rst^2 \cos(rs - t) + t^2 \sin(rs - t)$
$g_t = -st^2 \cos(rs - t) + 2\,st \sin(rs - t)$

13. $z_{xy} = 4xy \cos(x^2 - y^2) = z_{yx}$

15. $z_{xy} = e^{xy}[2xy \cos xy + \sin xy + \cos xy] = z_{yx}$

17. Yes

19. No

21. $z = x^3y^2 - \sin 2y + g(x)$

23. $z = \arctan(y/x) + g(x)$

25. $f(x + h, y + k) = x^3 - 3y^2 + 3x^2h - 6yk + r\sqrt{h^2 + k^2}$,
$r = (3x\,h^2 + h^3 - 3k^2)/\sqrt{h^2 + k^2}$

27. $f(x + h, y + k) = e^{xy} - x^2y + (ye^{xy} - 2xy)h + (xe^{xy} - x^2)k + r\sqrt{h^2 + k^2}$,
$r = \{e^{xy}[e^{hy+hk+kx} - (1 + xk + yh)] - (y + k)h^2 - 2xhk\}/\sqrt{h^2 + k^2}$

29. $dz = -2\cos x(\sin x)y\,e^{\cos^2 x}\,dx + e^{\cos^2 x}\,dy$

31. $dw = z^2[1 - (x - y)^2]^{-1/2}\,dx - z^2[1 - (x - y)^2]^{-1/2}\,dy + 2z\arcsin(x - y)dz$

33. $\dfrac{dz}{dt} = \left(\dfrac{3}{2}t^{1/2} - 1\right)\cosh(t^{3/2} - t)$

35. $\dfrac{dz}{dt} = 7t^6 - 8t^7 + 12t^{11}$

37. $z_r = e^{3r+s}\sin^3 s(3\cos r - \sin r)$
$z_s = e^{3r+s}\sin^2 s\cos r(3\cos s + \sin s)$

39. $w_r = \dfrac{2r\,s^2}{a^2} - \dfrac{4r^3 + 4rs}{b^2} - \dfrac{1}{c}$
$w_s = \dfrac{2r^2s}{a^2} - \dfrac{2r^2 + 2s}{b^2} + \dfrac{1}{c}$

41. $-4/\sqrt{13}$

43. $2/\sqrt{21}$

45. $\lambda = \left(\dfrac{1}{\sqrt{160}}\right)(4, -12)$; Rate: $4\sqrt{10}$

47. $\lambda = (0, -1, 0)$; Rate: 4

49. $\dfrac{x - 1}{1} = \dfrac{y - 2}{2} = \dfrac{z - 4}{-10}$

51. $\dfrac{x - 1}{2} = \dfrac{y - 3}{-3} = \dfrac{z - 1}{-2}$

53. $(0, 0)$, $(0, 4/3)$, and $(-4, 0)$ are saddle points; $\left(-\dfrac{4}{3}, \dfrac{4}{9}\right)$ is a local minimum.

55. $(0, 0)$ is a saddle point.

57. $y' = \dfrac{3y^2 - 2x}{6y^2 - 6xy}$

59. $z_x = \dfrac{(x^2 - z^2)\,z\sinh x\,e^{z\cosh x} - 2x}{(x^2 - z^2) - (x^2 - z^2)\cosh x\,e^{z\cosh x} - 2z}$
$x_z = \dfrac{2z + (x^2 - z^2)\cosh x\,e^{z\cosh x} - (x^2 - z^2)}{2x - (x^2 - z^2)\,z\sinh x\,e^{z\cosh x}}$

61. Tangent Plane: $3x - z = 8$; Normal Line: $2(x - 3)/3 = -2(z - 1)/1$, $y = 0$

63. 13/5

65. $w = 5\sqrt{2}$, $l = 12\sqrt{2}$; $A = 120$

67. c' = rate of separation $\cong$ 103 km/hr

CHAPTER SEVENTEEN

17.1 Answers

1. (a) 8 (b) 6 (c) 18

3. (a) $\frac{5}{4}$ (b) $\frac{21}{16}$ (c) $\frac{85}{64}$

5. $\dfrac{\pi}{4}$

7. $\dfrac{\pi}{2}$

9. $\dfrac{9\pi}{2}$

11. 8π

17.2 Answers

1. 0

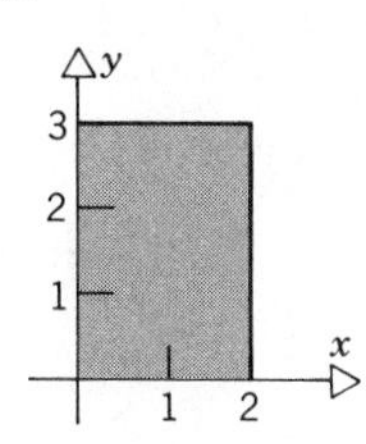

3. $\frac{52}{9}$

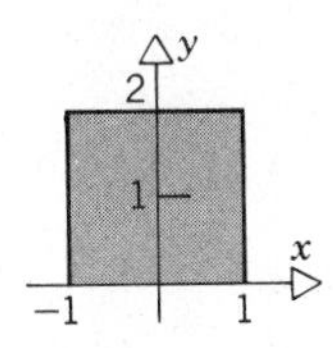

5. $-\frac{1}{6}$

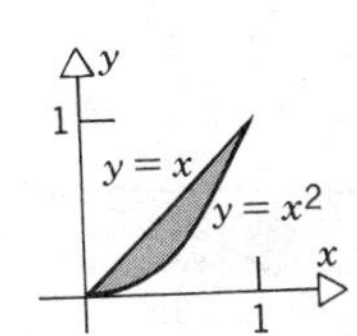

7. $\dfrac{(\ln 2)^2 + 2\ln 2 - 1}{2}$

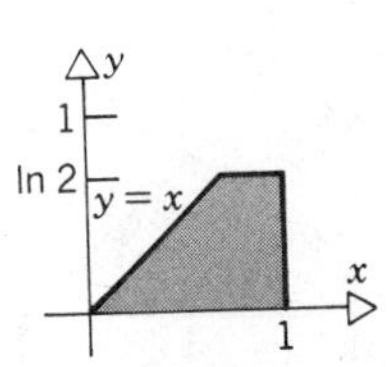

9. $\dfrac{\pi}{2} - \dfrac{\pi^2}{16} - \dfrac{3}{4}$

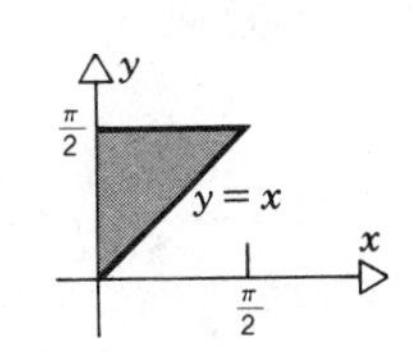

11. (a) $\int_0^1 \int_{x^3}^{\sqrt{x}} 2xy\, dy\, dx$

(b) $\int_0^1 \int_{y^2}^{y^{1/3}} 2xy\, dx\, dy$

(c) 5/24

13. (a) $\int_0^1 \int_{x/2}^{3x} (2x - y)\, dy\, dx + \int_1^4 \int_{x/2}^{(10-x)/3} (2x - y)\, dy\, dx$

(b) $\int_0^2 \int_{y/3}^{2y} (2x - y)\, dx\, dy + \int_2^3 \int_{y/3}^{10-3y} (2x - y)\, dx\, dy$

(c) 25/3

15. (a)

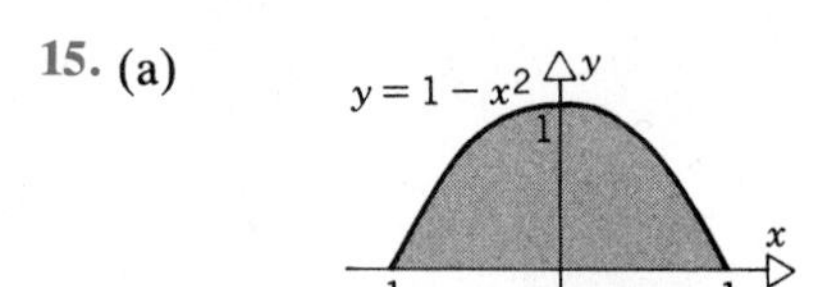

(b) $\int_0^1 \int_{-\sqrt{1-y}}^{\sqrt{1-y}} dx\, dy$; (c) $\frac{4}{3}$

17. (a)

(b) $\int_0^2 \int_0^{y^2} \sqrt{1 + y^3}\, dx\, dy$; (c) $\frac{52}{9}$

19. (a)

(b) $\int_0^1 \int_{y^2}^{2-y} (x - 2y)\, dx\, dy$; (c) $\frac{7}{30}$

21. $\dfrac{6k}{35}$ 23. $\dfrac{192\sqrt{2}k}{5}$ 25. 70

27. $\frac{236}{205}$ 29. $\frac{51}{20}$ 31. $\dfrac{a^4}{3}$

17.3 Answers

1. $\hat{x} = 0, \hat{y} = \dfrac{4b}{3\pi}$ 3. $\hat{x} = \dfrac{\pi}{2}, \hat{y} = \dfrac{\pi}{8}$

5. $\hat{x} = \hat{y} = \dfrac{a}{3}$ 7. $\bar{x} = \dfrac{a}{2}, \bar{y} = \dfrac{4b}{9}$

9. (a) $\dfrac{kbh^3}{3}$ (b) $\dfrac{kb^3h}{3}$ (c) $\dfrac{kbh^3}{12}$ (d) $\dfrac{kbh(b^2 + h^2)}{3}$ (e) $\dfrac{kbh(b^2 + h^2)}{12}$

11. (a) On the center line, at a distance $h/3$ from the base.
(b) $\dfrac{kbh^3}{12}$ (c) $\dfrac{kb^3h}{48}$ (d) $\dfrac{kbh^3}{36}$

13. (a) $I_x = \dfrac{\pi kab^3}{4}$, $R_x = \dfrac{b}{2}$
(b) $I_y = \dfrac{\pi ka^3b}{4}$, $R_y = \dfrac{a}{2}$
(c) $I_0 = \dfrac{\pi kab(a^2 + b^2)}{4}$

15. $\int_0^{\pi} \int_0^{\sin x} \left(x - \dfrac{\pi}{2}\right)^2 (1 - y)\, dy\, dx$

17. $k \int_0^1 \int_{1-\sqrt{1-x^2}}^{\sqrt{1-(x-1)^2}} (x^2 + y^2)x^2\, dy\, dx$

19. $\bar{x} = \bar{y} = \dfrac{b^2 + hb - h^2}{2\,(2b - h)}$

21. $\bar{x} = 0, \quad \bar{y} = \dfrac{a^2 + ah + (h^2/3)}{2a + h}$ 25. $\dfrac{ka^4}{8}$

29. On the line of symmetry, at a distance $4a/3\pi$ from the straight side.

17.4 Answers

1. $\frac{16}{3}$

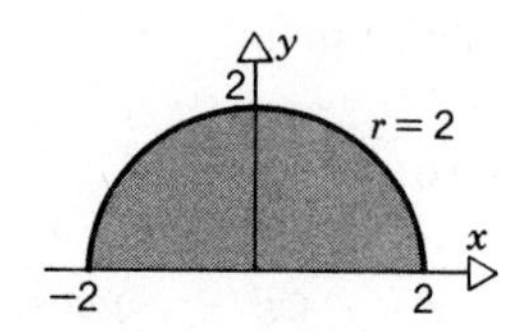

3. $\frac{7}{6}$

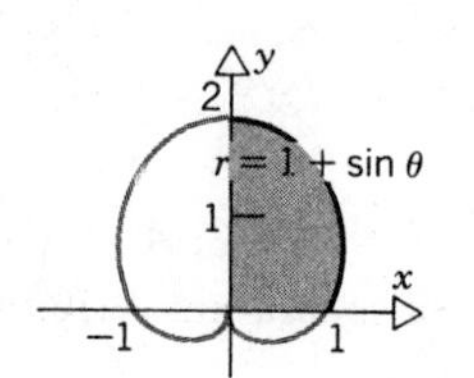

5. $\frac{\pi}{9} + \frac{\sqrt{3}}{6}$

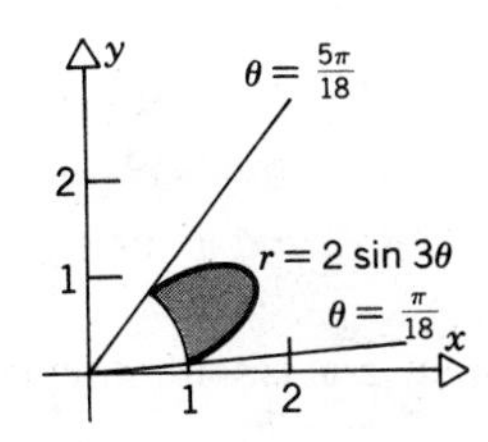

7. $\frac{2}{3}$

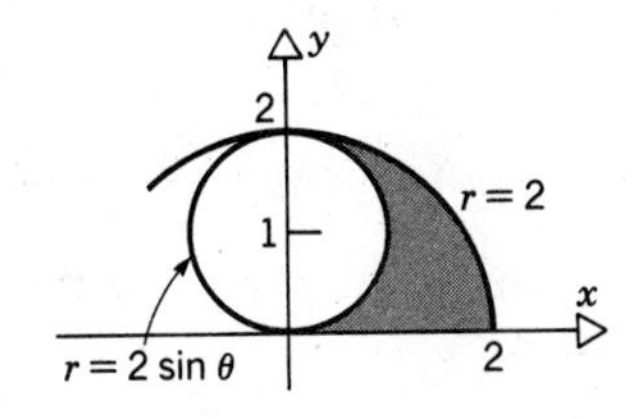

9. $\int_0^{\pi/4} \int_0^2 r\, dr\, d\theta = \frac{\pi}{2}$

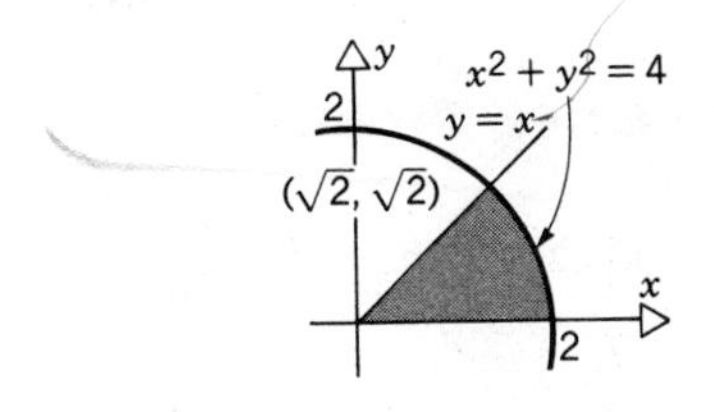

11. $\int_{\pi/3}^{\pi/2} \int_0^3 e^{-r^2} r\, dr\, d\theta = \frac{(1 - e^{-9})\pi}{12}$

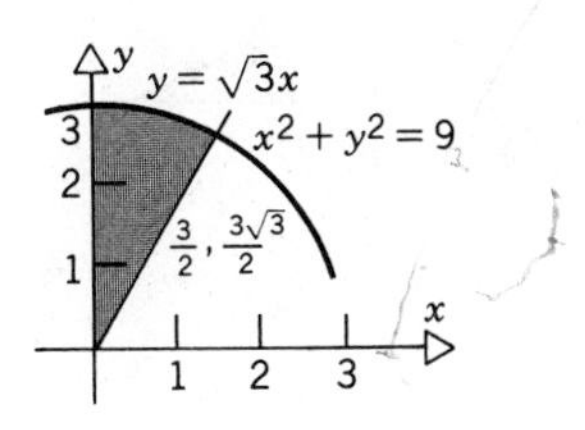

13. $a^2\left(\frac{\pi}{3} + \frac{\sqrt{3}}{2}\right)$　　15. 8π　　17. 6π

19. $\frac{5\pi a^3}{12}$　　21. $\hat{x} = \hat{y} = \frac{128a}{105\pi}$

23. (a) $\frac{\pi k a^4}{8}$　(b) $\frac{\pi k a^4}{8}$　(c) $\frac{\pi k a^4}{4}$

25. (a) $k(b - a)\pi$　(b) $\bar{x} = 0, \quad \bar{y} = \frac{b + a}{\pi}$
(c) $\frac{k\pi(b^3 - a^3)}{6}$　(d) $\frac{k\pi(b^3 - a^3)}{6}$

27. 6

17.5 Answers

1. $\frac{81}{8}$　　3. 60　　5. $\frac{\pi}{4}$

7. $-\sqrt{7 - x^2} \le y \le \sqrt{7 - x^2}; \quad -\sqrt{7} \le x \le \sqrt{7}$

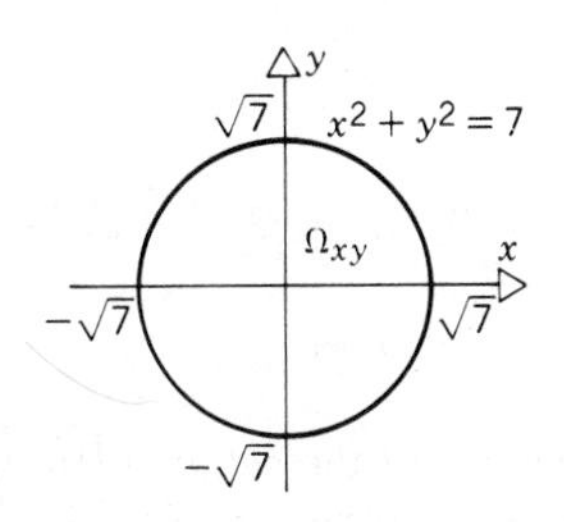

9. (a) $0 \le x \le 3 - \frac{3z}{4}; \quad -2 \le z \le 2$
(b) $0 \le y \le 4 - z^2; \quad -2 \le z \le 2$

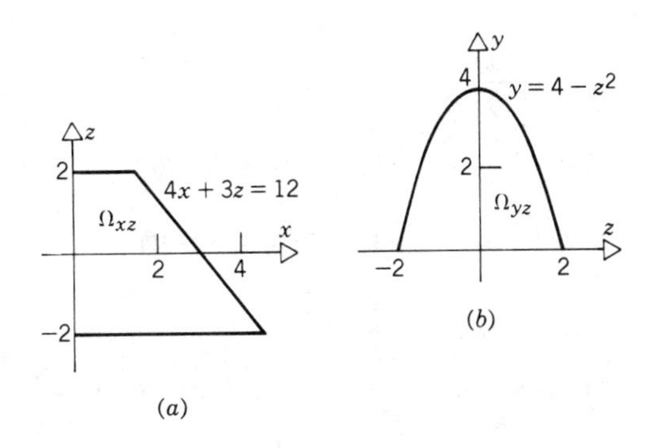

11. $\int_0^1 \int_0^{2-2z} \int_0^{4-(4/3)y-2z} (6 - x - y)\, dx\, dy\, dz$, or other equivalent integrals

13. $4k \int_0^r \int_0^{\sqrt{r^2-x^2}} \int_0^{h-(h/r)\sqrt{x^2+y^2}} (x^2 + z^2)\, dz\, dy\, dx$, or other equivalent integrals

15. $2k \int_0^a \int_{-b\sqrt{1-(x^2/a^2)}}^{b\sqrt{1-(x^2/a^2)}} \int_0^{2b-y} (x^2 + y^2 + z^2)\, dx\, dy\, dz$

17. 18

19. $\int_0^1 \int_0^{8y-5y^2} \int_0^{2-2y} dx\, dz\, dy$

$$+ \int_0^1 \int_{8y-5y^2}^{4-y^2} \int_0^{\sqrt{4-y^2-z}} dx\, dz\, dy$$

21. On the center line, at a height $b/4$ from the base.

23. $\bar{x} = 0, \quad \bar{y} = 0, \quad \bar{z} = \frac{4}{3}$

25. (a) $64k\pi$ (b) $\bar{x} = 0, \quad \bar{y} = 0, \quad \bar{z} = \frac{10}{3}$

27. On the center line, at the height

$$\bar{z} = \frac{m_1 l + 2m_2 l + m_2 a}{2(m_1 + m_2)}$$

17.6 Answers

1. $\dfrac{4\pi a^2 c}{3}$

3. $\pi a^2 h\left(1 + \dfrac{h^2}{3c^2}\right)$

5. (a) $\bar{x} = \bar{y} = 0, \quad \bar{z} = \frac{23}{8}$ (b) $\dfrac{208\pi k}{15}$

7. $\dfrac{2ka^3}{3}$

9. (a) On the axis of the cylinder, three quarters of the distance from the vertex to the base.
   (b) $\dfrac{\pi k a^4 h}{10}$ (c) $\dfrac{\pi k a^2 h(a^2 + 4h^2)}{20}$

11. (a) On the axis of symmetry, $3a/8$ from the plane face.
    (b) $\dfrac{4\pi k a^5}{15}$

13. $\left(\dfrac{6}{5}, 0, \dfrac{27\pi}{128}\right)$

15. (a) $k \int_0^{2\pi} \int_0^3 \int_0^{9-r^2} (12 - z) r\, dz\, dr\, d\theta$
    (b) $k \int_0^{2\pi} \int_0^3 \int_0^{9-r^2} (12 - z) r^3\, dz\, dr\, d\theta$

17. (a) $V = \pi a^2 h - \dfrac{\pi h^3}{3}$
    (b) h satisfies $(h/a)^3 - 3(h/a) + 1 = 0$; $h/a \cong 0.3473$.

19. $2\pi G m \mu(1 - b(b^2 + h^2)^{-1/2}]$

21. (a) $\hat{z}_A = \dfrac{(p + 1)a^p}{2p + 1}, \quad \hat{z}_V = \dfrac{(p + 2)a^p}{2(p + 1)}$
    (b) $\dfrac{\hat{z}_V}{\hat{z}_A} = 1 + \dfrac{p}{2(p + 1)^2}$; $\dfrac{\hat{z}_V}{\hat{z}_A}$ is maximum when $p = 1$

17.7 Answers

1. $\left(\sqrt{\dfrac{3}{2}}, \sqrt{\dfrac{3}{2}}, 1\right)$

3. $\left(\dfrac{3}{2}, \dfrac{\sqrt{3}}{2}, -1\right)$

5. $(-1, 1, -\sqrt{2})$

7. $\left(2, \dfrac{7\pi}{4}, \dfrac{\pi}{4}\right)$

9. $\left(2, \dfrac{\pi}{3}, \dfrac{2\pi}{3}\right)$

11. $\left(3, \dfrac{\pi}{6}, \dfrac{\pi}{4}\right)$

13. $\rho = 2c \cos \phi$

15. $\phi = \arctan \frac{2}{3}$

17. $\theta = \arctan \frac{1}{2}$

19. $\phi = \arctan 3$

21. Plane parallel to, and b units from, xy-plane; $z = b$.

23. Cylinder about z-axis with circular cross section of radius b; $r = b$.

25. Plane parallel to z-axis; $x + y = a$.

27. (a) $\dfrac{4\pi k a^5}{15}$ (b) $\dfrac{4\pi k a^5}{15}$

29. (a) $2\pi k(b^2 - a^2)$ (b) $\dfrac{2\pi k(b^4 - a^4)}{3}$

31. $4k\pi a^4(1 - \cos^6 \alpha)/3$, where k is the proportionality constant.

33. (a) $\dfrac{3\pi k a^2}{4}$ (b) $\bar{x} = \bar{y} = 0, \quad \bar{z} = \dfrac{7a}{18}$
    (c) $\dfrac{5\pi k a^4}{16}$

35. $\dfrac{8\pi k a^4}{5}$

37. $\pi G m k a$

Answers to Review Problems

1. 64
3. $16(e^3 + e^5 + e^7) \cong 20{,}242$
5. $e^2 - 3$
7. 135
9. $-(\ln 2)^2/2$

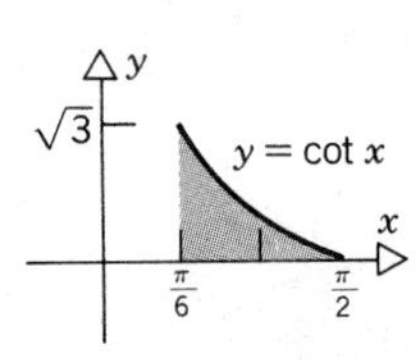

11. 11/120

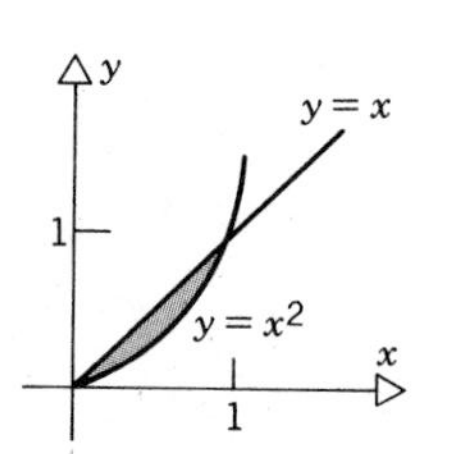

13. (a) $\int_0^1 \int_1^{e^x} (x - 2y)\,dy\,dx$

(b) $\int_1^e \int_{\ln y}^1 (x - 2y)\,dx\,dy$

(c) $2 - (e^2/2)$

15. (a) $\int_0^1 \int_{x^3}^{x^2} \left(\frac{x + y}{xy}\right) dy\,dx$

(b) $\int_0^1 \int_{y^{1/2}}^{y^{1/3}} \left(\frac{x + y}{xy}\right) dx\,dy$

(c) $\frac{7}{6}$

17. $\pi(\sin 4 - \sin 1)$

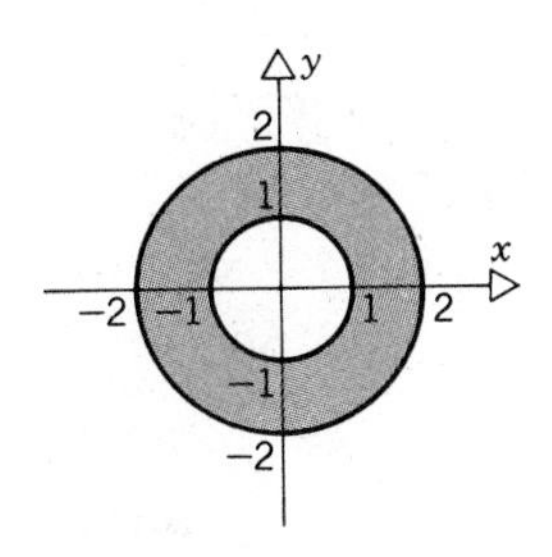

19. $-\frac{2}{9}$

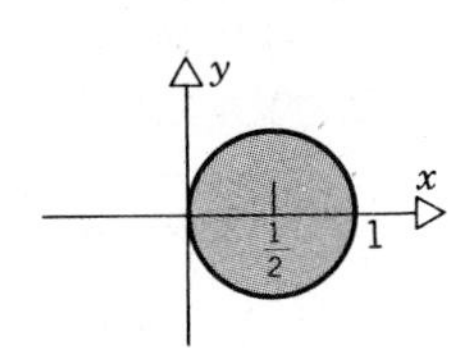

21. $(\sqrt{3}, 1, 1)$

23. $\left(2, \frac{\pi}{3}, 0\right)$

25. $\rho = 2$; sphere

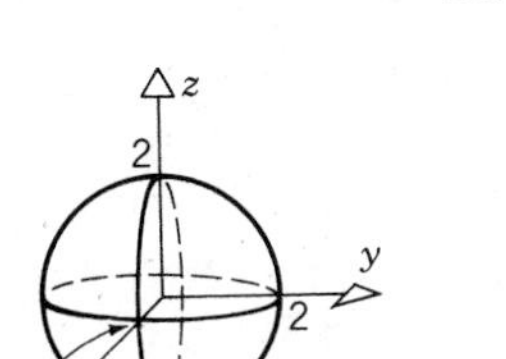

27. $z = 3$; plane

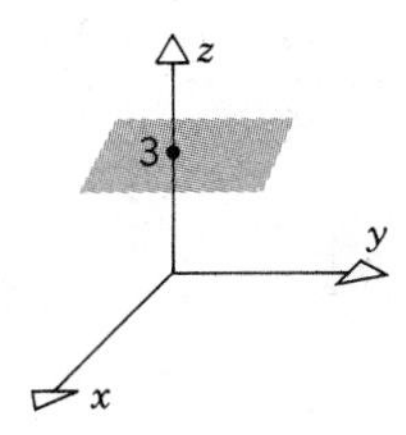

29. $8\frac{(\sqrt{3} - 1)}{3}$

31. $8\frac{(2 - \sqrt{2})\pi}{3}$

33. $\frac{k}{2}$

35. $k\pi(128 - 66\sqrt{3})/15$

37. $\frac{1}{2(e - 2)}$

39. $\frac{3\pi}{4}$

41. $R_y^2 = \frac{e - (7/3)}{e - 2}$

43. $R_y^2 = \frac{36}{5}$

CHAPTER EIGHTEEN

18.1 Answers

1. 2

3. $-\frac{254}{3} - 3\ln 3 \cong -87.96$

5. 0

7. $\frac{73}{3} - 3\cos 4 \cong 26.29$

9. -2

11. $\frac{2ab(a - b)}{3}$

13. 64

15. $\frac{a^2}{3}$

17. (a) $\frac{a^3b^3}{6}$ (b) $\frac{a^2b^4}{6}$

19. (a) $\frac{5}{2}$ (b) $\frac{8}{3}$ (c) 3 (d) 2 (e) $\frac{7}{3}$

21. (a) 0 (b) 0

23. (a) $3e - 3$ (b) $3e - 3$ (c) $3e - 3$

25. $\frac{ka^2\pi^2}{2}$

27. $16\sqrt{13}\pi(1 + 12\pi^2)k \cong 21645.8k$

29. $[(1 + 4\pi^2)^{3/2} - 1]\frac{k}{3} \cong 85.51k$

31. $I_x = \frac{\pi k a^3}{2}, \quad I_y = \frac{\pi k a^3}{2}$

33. $\left(0, \frac{a}{\sqrt{3}}\right)$

35. $I_x = \frac{kb^3}{3}, \quad I_y = \frac{ka^3}{3}$

18.2 Answers

1. $x^2y + 6x - 3y + k$

3. $y\sin x + x^2e^y + 2y + k$

5. Not conservative

7. $\frac{xy}{z} - xe^z + 3y + k$

9. $4(1 + \pi) + e^2 - e \cong 21.24$

11. $\frac{15\pi}{4} + \frac{7}{2} - \frac{\sqrt{2}}{2} \cong 14.574$

13. $\frac{1}{2}\exp(a^2 + b^2 + c^2) - \frac{1}{2}$

15. $\alpha = 3, -\frac{1}{2}$

17. 0

19. $10 + e - 2e^2 \cong -2.060$

23. 0

27. (a) $\frac{kx^2}{2}$ (b) $\frac{k(x_a^2 - x_b^2)}{2}$

18.3 Answers

1. $-\pi a^2$ 3. $\frac{1}{3}$ 5. 0 7. $\dfrac{2a^3}{3}$

9. 0 11. πa^2 13. $3\pi a^2$

19. (d) $\frac{1}{2}\ln(x^2 + y^2)$

18.4 Answers

1. $4\pi a^2$ 3. $\dfrac{(b^2c^2 + c^2a^2 + a^2b^2)^{1/2}}{2}$

5. $\dfrac{2\pi}{3}[(1 + a^2)^{3/2} - 1]$ 7. $\dfrac{3\pi a}{2} + 2a^2$

9. $\dfrac{145^{3/2} - 1}{54} \cong 32.315$ 11. $2\pi R(R^2 + H^2)^{1/2}$

13. $\dfrac{\pi^2 ka^3}{2}$ 15. $\dfrac{a^3}{2}$ 17. $2\pi \displaystyle\int_0^2 \frac{\sqrt{16 - 3r^2}}{\sqrt{4 - r^2}}\, r\, dr$

19. $\displaystyle\int_0^1 \int_0^{1-y^2} x^2\sqrt{4x^2 + 1}\, dx\, dy$

18.5 Answers

1. $3x - y - 5z = 0$; plane through the origin (entire surface)
3. $x^2 - y^2 + z^2 = 0$; cone (portion for which $x \geq 0$, $y \geq 0$)
5. $z = x^2 + (y^2/4)$; elliptic paraboloid (portion for which $x \geq 0$)
7. $\dfrac{x - 1}{-3} = \dfrac{y - 8}{1} = \dfrac{z + 1}{5}$; $3x - y - 5z = 0$
9. $x - 1 = -(y - 1)$, $z = 0$; $x - y = 0$
11. $\dfrac{x - 3/4}{3} = \dfrac{z - 9/16}{-2}$, $y = 0$; $3x - 2z = \frac{9}{8}$
13. (a) 97,801 mi^2 (b) 84,965 mi^2
15. (b) $d\sigma = \rho \sin \phi_0\, d\rho\, d\theta$ (c) $\pi r\sqrt{r^2 + h^2}$
17. $\pi[b\sqrt{b^2 + c^2} - a\sqrt{a^2 + c^2}$
 $+ c^2 \ln(b + \sqrt{b^2 + c^2}) - c^2 \ln(a + \sqrt{a^2 + c^2})]$
19. $\displaystyle\int_0^{\pi} \int_0^{2\pi} \sin u(b^2c^2 \sin^2 u \cos^2 v$
 $+ a^2c^2 \sin^2 u \sin^2 v + a^2b^2 \cos^2 u)^{1/2}\, dv\, du$

18.6 Answers

1. $\nabla \cdot \mathbf{v} = 2y \sin z$; $\nabla \times \mathbf{v} = -(\cos z + x^2)\mathbf{i} + xy \cos z\, \mathbf{j} + x(2z - \sin z)\mathbf{k}$
3. $\nabla \cdot \mathbf{v} = \dfrac{2}{r}$, $r \neq 0$; $\nabla \times \mathbf{v} = \mathbf{0}$, $r \neq 0$
5. $\nabla \cdot \mathbf{v} = 0$; $\nabla \times \mathbf{v} = 3r\mathbf{k}$
7. $\nabla \cdot \mathbf{v} = 2yz - 2y + 2xyz$;
 $\nabla \times \mathbf{v} = xz^2\mathbf{i} + (2xy - yz^2)\mathbf{j} - (2x + 2xz)\mathbf{k}$
9. $\nabla \cdot \mathbf{v} = \cos x \sin y - \sin y \sin z - \sin z \cos x$;
 $\nabla \times \mathbf{v} = -\cos y \cos z\, \mathbf{i} + \cos z \sin x\, \mathbf{j} - \sin x \cos y\, \mathbf{k}$
21. $\mathbf{v}(x, y, z) = yz\mathbf{i} - 2xz\mathbf{j}$
23. No $\mathbf{v}$ exists, since $\nabla \cdot \mathbf{u} \neq 0$
25. $\mathbf{v}(x, y, z) = \left(\dfrac{x^2z^2 - y^2}{2}\right)\mathbf{i} + xy(\cos z - 1)\mathbf{j}$

18.7 Answers

1. 3 3. 0 5. $\dfrac{8\pi a^3}{3}$ 7. 0

9. 36π 11. $-2\pi K$ 13. $\dfrac{\pi U a^4}{2}$

15. 2π 17. $\alpha\pi a^3$ 19. 8π

18.8 Answers

1. 13 3. -16 5. -16π 7. 0

9. 0 11. 0 13. πa^2 15. $\dfrac{3\pi a^4}{16}$

Answers to Review Problems

1. $e^2 - e^{1/2} + \ln 4 - 1$
3. $-\frac{1}{6}$ 5. 440/21 7. $\ln \sqrt{10}$
9. conservative; $\phi(x, y) = e^{xy} + x + (1/xy) + c$
11. conservative, $\phi(x, y, z) = 2x - \cos y \sin z + c$
13. 0 15. 1 17. 0 19. 0
21. (a) 6π
 (b) (0, 0)
 (c) $I_0 = 39\pi/2$
23. (a) $e - 1$
 (b) $\left(\dfrac{1}{e - 1}, \dfrac{e + 1}{4}\right)$
 (c) $(e^3 + 9e - 19)/9$
25. $2\sqrt{14}/3$ 27. $\sqrt{5}/2$
29. (a) $2xy + 2yz + 2zx$
 (b) $-(y^2\mathbf{i} + z^2\mathbf{j} + x^2\mathbf{k})$
31. (a) $-\left[\dfrac{xy^2}{(x^2 + y^2)^{3/2}} + \dfrac{yz^2}{(y^2 + z^2)^{3/2}} + \dfrac{zx^2}{(x^2 + z^2)^{3/2}}\right]$
 (b) $-\left[\dfrac{(z^3 + 2zy^2)}{(y^2 + z^2)^{3/2}}\mathbf{i} + \dfrac{(x^3 + 2xz^2)}{(x^2 + z^2)^{3/2}}\mathbf{j} + \dfrac{(y^3 + 2yx^2)}{(x^2 + y^2)^{3/2}}\mathbf{k}\right]$
33. π 35. 108 37. -16 39. 0
41. $320/21 \cong 15.24$

index

INTEGRATION FORMULAS

Algebraic Integrands

1. $\int x^r dx = \frac{x^{r+1}}{r+1} + c, \qquad r \neq -1$

2. $\int \frac{dx}{x} = \ln|x| + c$

3. $\int \frac{dx}{a^2 + x^2} = \frac{1}{a} \arctan\left(\frac{x}{a}\right) + c$

4. $\int \frac{dx}{\sqrt{a^2 - x^2}} = \arcsin\left(\frac{x}{a}\right) + c$

5. $\int \frac{dx}{a^2 - x^2} = \frac{1}{2a} \ln\left|\frac{a+x}{a-x}\right| + c$

6. $\int \sqrt{x^2 \pm a^2}\, dx = \frac{1}{2}[x\sqrt{x^2 \pm a^2} \pm a^2 \ln|x + \sqrt{x^2 \pm a^2}|] + c$

7. $\int \sqrt{a^2 - x^2}\, dx = \frac{1}{2}\left[x\sqrt{a^2 - x^2} + a^2 \arcsin\left(\frac{x}{a}\right)\right] + c$

8. $\int \frac{dx}{\sqrt{x^2 \pm a^2}} = \ln|x + \sqrt{x^2 \pm a^2}| + c$

9. $\int \frac{dx}{|x|\sqrt{x^2 - a^2}} = \frac{1}{a} \operatorname{arcsec}\left(\frac{x}{a}\right) + c$

10. $\int \frac{\sqrt{a^2 \pm x^2}}{x}\, dx = \sqrt{a^2 \pm x^2} - a \ln\left|\frac{a + \sqrt{a^2 \pm x^2}}{x}\right| + c$

11. $\int \frac{\sqrt{x^2 - a^2}}{x}\, dx = \sqrt{x^2 - a^2} - a \operatorname{arcsec}\left(\frac{x}{a}\right) + c$

12. $\int \frac{dx}{x\sqrt{a^2 \pm x^2}} = -\frac{1}{a} \ln\left|\frac{a + \sqrt{a^2 \pm x^2}}{x}\right| + c$

13. $\int \sqrt{2ax - x^2}\, dx = \frac{x-a}{2}\sqrt{2ax - x^2} + \frac{a^2}{2} \arcsin\left(\frac{x-a}{a}\right) + c$

14. $\int \frac{dx}{\sqrt{2ax - x^2}} = \arcsin\left(\frac{x-a}{a}\right) + c$

15. $\int \frac{dx}{(a^2 + x^2)^{m+1}} = \frac{1}{2ma^2} \frac{x}{(a^2 + x^2)^m} + \frac{2m-1}{2ma^2} \int \frac{dx}{(a^2 + x^2)^m}$

16. $\int (a^2 - x^2)^m\, dx = \frac{x(a^2 - x^2)^m}{2m+1} + \frac{2ma^2}{2m+1} \int (a^2 - x^2)^{m-1}\, dx$

Integrands Involving Exponential and Logarithmic Functions

17. $\int e^x\, dx = e^x + c$

18. $\int \ln x\, dx = x \ln x - x + c$

19. $\int a^x\, dx = \frac{a^x}{\ln a} + c, \qquad a > 0$

20. $\int x e^x\, dx = x e^x - e^x + c$

21. $\int x \ln x\, dx = \frac{x^2}{4}(2 \ln x - 1) + c$

22. $\int x^m e^x\, dx = x^m e^x - m \int x^{m-1} e^x\, dx$

23. $\int x^m (\ln x)^n\, dx = \frac{x^{m+1} (\ln x)^n}{m+1} - \frac{n}{m+1} \int x^m (\ln x)^{n-1}\, dx$

24. $\int e^{ax} \sin bx\, dx = \frac{e^{ax}(a \sin bx - b \cos bx)}{a^2 + b^2} + c$

25. $\int e^{ax} \cos bx\, dx = \frac{e^{ax}(a \cos bx + b \sin bx)}{a^2 + b^2} + c$

Integrands Involving Trigonometric Functions

26. $\int \sin x\, dx = -\cos x + c$

27. $\int \cos x\, dx = \sin x + c$

28. $\int \sin^2 x\, dx = \frac{1}{2}x - \frac{1}{4} \sin 2x + c$

29. $\int \cos^2 x\, dx = \frac{1}{2}x + \frac{1}{4} \sin 2x + c$